THE CONCISE OXFORD
TURKISH
DICTIONARY

# THE CONCISE OXFORD TURKISH DICTIONARY

EDITED BY

A. D. ALDERSON

AND

FAHİR İZ

OXFORD
AT THE CLARENDON PRESS

*Oxford University Press, Walton Street, Oxford OX2 6DP*
*Oxford New York Toronto*
*Delhi Bombay Calcutta Madras Karachi*
*Petaling Jaya Singapore Hong Kong Tokyo*
*Nairobi Dar es Salaam Cape Town*
*Melbourne Auckland*
*and associated companies in*
*Beirut Berlin Ibadan Nicosia*

*Oxford is a trade mark of Oxford University Press*

*Published in the United States*
*by Oxford University Press, New York*

*ISBN 0 19 864109 5*



*First published 1959*
*Reprinted 1968, 1971, 1974, 1975, 1980, 1983, 1984, 1987*



*Printed in Hong Kong*

# PREFACE

THE *Oxford Turkish–English Dictionary* and *English–Turkish Dictionary* have both had great success but, because of their size and cost, they could not satisfactorily meet the demand for a first-class dictionary suitable for school and university pupils. So this abridgement has been prepared, reducing the originals to about one-third of their length.

This economy of material has been achieved only by the careful rearrangement of entries and the elimination of close synonyms, such geographical and scientific terms as should be easily recognizable, and those Arabic and Persian forms now commonly replaced by Turkish ones. Further, no new material has been added. It has thus been possible to retain the wealth of idiom which is the essential characteristic of the larger volumes.

The chief innovation is that the pronunciation of English words is now shown in terms of the modern Turkish alphabet rather than by the use of phonetic symbols.

We should like to take this opportunity to thank Mr. H. C. Hony for allowing us to condense the original material and for his co-operation in the preparation of this volume. Our thanks are also due to the Delegates and staff of the Clarendon Press for their encouragement and advice.

F. I.
A. D. A.

*December 1958*

# INTRODUCTION

## TURKISH–ENGLISH SECTION

*Spelling.* 1. In spelling words of Arabic origin, the original voiced consonants **(b, d, c,)** have been retained, though the Turks do sound them as voiceless **(p, t, ç).** This has been done because the voiced consonants appear in any case when the word is declined **(mektebe, derdim, haricî),** and also in other forms of the Arabic root **(idhal – dahil, takib – müteakıben, cemaat – ictimai).**

2. There is still some confusion in spelling as between certain pairs of letters:

**a/e, ı/i, ö/ü, u/ü; b/p, c/ç, d/t, ğ/v.**

If a word does not appear under one form, it will be found under the other.

*Vocabulary.* 1. The less easily recognizable French and English words introduced into modern Turkish have been given **(şambrnu-var**—chambre noire; **nakaut**—knock-out).

2. Only the more established of the recently invented Turkish words have been included **(ğelenek** for **an'ane; yakıt** for **mahrukat).**

*Arrangement.* 1. As far as possible all compound words are placed under the heading of the main word.

2. The passive (*-ılmak*), causative (*-tırmak*), and reciprocal (*-laşmak*) forms of the verbs are only given where they have special meanings or rather unusual spellings.

3. Idioms are given under one of the main words, preferably the most uncommon.

*Pronunciation.* 1. This is shown in brackets after the word, with a dash (–) for a long syllable and a dot (·) for a short one. Stress is shown by an accent (ˋ) over the stressed syllable.

2. The final long syllable of many Arabic words is short in Turkish but, when declined and followed by a vowel, it becomes long again; such a syllable is indicated by an asterisk (*). Thus **hayat**(·ˋ*) is pronounced (·ˋ), but the accusative **hayatı** is pronounced (·–ˋ).

3. The vowels are *approximately*:

**a** (·) as *u* in *sun*.
**a** (–) and **â** (long Arabic **a**) as *a* in *far*.
**â** (after **ğ, k, l**) as *a* in *bad*.

**e** as *e* in *bed*.
**ı** between the *i* in *big* and the *u* in *bug*.
**i** (·) as *i* in *hit*.
**i** (–) as *ee* in *feet*.
**o** as *o* in *doll* or *au* in *author*.
**ö** as French *eu* in *peu*.
**u** (·) as *u* in *bull*.
**u** (–) as *u* in *rule*.
**û** (after **ǵ, k, l**) as *u* in *cute*.
**ü** as French *u* in *tu*.

4. The consonants are as in English, except:

**c** as *j* in *jam*.
**ç** as *ch* in *choke*.
**ǵ, k** as in English; followed by **â** or **û** are palatalised (**ǵâ, kâ** are almost *gya, kya*).
**ğ** with hard vowels is a very guttural but faint *g* (**dağ** = *dagh*); with soft vowels it is a consonantal *y* (**eğer** = *eyer*).
**j** as *s* in *treasure*.
**ş** as *sh* in *shut*.
**v** is between the English *v* and *w*; in some words it is interchangeable with *ğ* (**dövmek** = **döğmek**).

## ENGLISH–TURKISH SECTION

*Vocabulary*. 1. To save as much space as possible compound words with endings such as: **-logist, -logical, -fication, -ization,** have been omitted.

2. Similarly only the adjectival meanings of past participles are normally given.

3. The past tense and past participle of irregular verbs are given in brackets: **take (took, taken)**; if these have the same form it is only given once: **strike (struck).**

4. One of the richest sources of idiom in English is the compound verb, where the following particle is not a real preposition but affects the meaning of the verb. Compare: **to go** = *gitmek*, and **to go under the bridge** = *bir köprü altından geçmek*, but **to go under** = *batmak*. These varying forms are given separately for each verb; they deserve careful attention.

*Brackets*. 1. In the Turkish translations brackets generally indicate that the English word may be used both as a noun and a verb: **climb** = *tırmanma(k), çıkma(k)*.

2. In English expressions round brackets indicate that the use

of the word(s) enclosed is optional: **The Cape (of Good Hope)** = *Ümid Burnu, Kap.*

3. Square brackets, however, indicate alternative forms of an expression: **to get [stand] in someone's light** = *birisine karanlık etmek.*

*Pronunciation.* 1. The pronunciation of English words has been given as far as possible in terms of the modern Turkish alphabet; however, a few phonetic symbols have been introduced where there is no exact equivalent.

2. The fact that the vowel-sounds in many English words are longer than the Turkish ones is indicated by a circumflex (^).

3. The letters **g** and **r** are enclosed in round brackets when, as final consonants, they are not fully pronounced.

4. Diphthongs are indicated by a (‿) joining the two letters.

5. Table of Pronunciation Symbols:

(i) Vowels/Sesliler.

| | | |
|---|---|---|
| a | hat [hat] | çok açık, yayık bir *e*'ye yakın. |
| â | car [kâ(r)] | *mazi*'deki *a* gibi. |
| e | set [set] | *ev*'deki *e*'den daha kapalı bir *e*. |
| i | hit [hit] | *git*'teki *i*'den daha kapalı bir *i*. |
| î | bee [bî] | *biçare*'deki *i* gibi. |
| o | pot [pot] | *ot*'taki *o*'dan daha açık. |
| ô | form [fôm] | *portakal*'daki *o*'ya yakın. |
| ö(r) | fur [fö(r)] | uzun bir *e* ile uzun bir *ö* arası bir ses. |
| u | book [buk] | *çabuk*'taki *u*'dan daha açık bir *u*. |
| û | too [tû] | *tufan*'daki *u*'ya yakın. |
| ʌ | hut [hʌt] | *kat*'taki *a*'ya yakın. |
| ə | ajar [əcâr] | yarım bir *e* sesi. |

(ii) Diphthongs/Çiftsesliler.

| | | | |
|---|---|---|---|
| a͜u | how [ha͜u]. | o͜u | tow [to͜u]. |
| â͜y | buy [bâ͜y]. | ô͜y | toy [tô͜y]. |
| ê͜y | pay [pê͜y]. | o͜ə | for [fo͜ə]. |
| e͜ə | fair [fe͜ə]. | u͜ə | sure [şu͜ə]. |
| i͜ə | near [ni͜ə]. | | |

(iii) Consonants/Sessizler.

As in Turkish, but for:

| | | |
|---|---|---|
| n(g) | sing [sin(g)] | genizden bir *n*; *k* gibi telaffuz edilmez. |
| θ | thin [θin] | peltek *s*. |
| ð | them [ðem] | peltek *z*. |
| w | one [wʌn] | *duvar*'daki *uv*'a yakın. |

# ABBREVIATIONS

*a.* adjective—sıfat.
*abb.* abbreviation—kısaltma.
*abl.* ablative—'den hali.
*acc.* accusative—'i hali.
*adv.* adverb—zarf.
*Amer.* American—Amerikan.
*anat.* anatomy—anatomi.
**a.o.** **anyone**—bir kimse.
*approx.* approximately—takriben.
*Ar.* Arabic—Arabca.
*arith.* arithmetic—aritmetik.
*baz.* bazan—sometimes.
*biol.* biology—biyoloji.
*bot.* botany—botanik.
*caus.* causative mood—müteaddi, geçişli.
*cf.* compare—karşılaştırınız.
*coll.* colloquial—konuşurken.
*comm.* commercial—ticarî.
*comp.* comparative—kıyasî.
*cont.* contemptuously—istihfaf olarak.
*corr.* corruption—bozulma.
*dat.* dative—'e hali.
*dial.* dialect—lehçe.
*ech.* echoic—yankı gibi.
*elect.* electricity—elektrikî.
*Eng.* English—İngiliz.
*err.* erroneously—yanlış olarak.
*esp.* especially—bilhassa.
*esk.* eski—old.
*etc.* et cetera—ve saire.
**etm.** **etmek.**
*fig.* figuratively—mecaz olarak.
*fin.* financial—maliyeye aid.
*Fr.* French—Fransızca.
*gen.* generally—umumiyetle. genitive—'in hali.
*gram.* grammar—gramer.
*impers.* impersonal—gayrişahsi.
*ind.* indicative—ihbar tasrifi.
*inf.* infinitive—mastar.
*inter.* interjection—nida.
*iron.* ironically—istihzalı.
*iz.* izafet.
*jur.* juridical—hukukî.
*köt.* kötüleyici—pejorative.
*küç.* küçültme—diminutive.
*Lât.* Lâtince—Latin.
*lit.* literally—harfî.
*math.* mathematics—matematik.
*mech.* mechanics—mekanik.
*med.* medical—tıbbî.
*mes.* meselâ—for example.
*mim.* mimarî—architectural.
*mil.* military—askerî.
*mit.* mitoloji—mythology.
*mus.* music—musikî.
*mür.* mürekkeb kelimelerde—in compound words.
*n.* noun—isim.
*naut.* nautical—denizciliğe aid.
*neg.* negative—menfi.
*num.* numeral (adjective)—adedî.
**olm.** **olmak.**
*opp.* opposed—aksi.
*orig.* originally—esasî.
*otom.* otomobil—motoring.
*part.* participle—ismifail (ismimeful).
*pass.* passive—mechul fiil.
*p.p.* past participle—ismimeful.
*Pers.* Persian—Farisî.
*pl.* plural—cemi.
*pop.* popular—halka aid.
*pref.* prefix—ön ek.
*prep.* preposition—harficer.
*pres.* present—hal sıygası.
*pron.* pronoun—zamir.
*prop. n.* proper noun—has isim.
*q.v.* which see—ona bakınız.
*rel. pron.* nisbet zamiri.
*sev.* sevgi ifade eden tabir—affectionately.
*sing.* singular—müfred.
*sl.* slang—argo.
**s.o.**/*s.o.* **someone**—bir kimse.
**stg.**/*stg.* **something**—bir şey.

*suff.* suffix—son ek.
*superl.* superlative—tafdil sıygası.
*şair.* şairane—poetically.
*şak.* şaka yollu—jokingly.
*şim.* şimdi—now.
*v.* verb—fiil.
see—bakınız.
*va.* verb active—nesne alan fiil.
*verb.* verbal—fiile aid.
*vn.* verb neuter—nesne almıyan fiil.
*vulg.* vulgarly—kaba söz.
*yal.* yalnız—only.
⌜ ⌝ proverb(ial expression)—atalar sözü (değerinde tabir).
* the word so marked must have the appropriate personal pronominal suffix: **beli* kırılmak** = to be exhausted; I was exhausted = **belim kırıldı.**

# TURKISH-ENGLISH DICTIONARY

**aba** Coarse woollen stuff; Arab cloak. ···**e ~yı yakmak,** to be 'gone on' *s.o.*

**abajur** (*Fr. abat-jour*) Lampshade.

**abalı** Wearing an **aba**; poor, wretched. ⌜**vur ~ya!**⌝, attack the weak!; put the blame on *s.o.* who can't defend himself!

**abanî** (——́) A mixed tissue of cotton and silk, *generally with a yellow design.*

**abanmak** *vn.* (*with dat.*) Lean over *or* against *stg.*; push against.

**abanoz** Ebony. ~ **kesilmek,** to become as hard *or* as black as ebony.

**abazan** Hungry (*esp.* for love). Hunger; craving for love.

**Abbas** *A name of men.* ⌜**~yolcu**⌝, *said in allusion to one who is going away or being dismissed his post.*

**abdal** Silly fool. (*also* **~sı, ~ımsı**) imbecile; feckless.

**abdest, -ti** Ritual ablution. ~ **almak,** to perform such ablution: ~ **bozmak,** to relieve nature: **~i* gelmek,** to want to relieve nature. **~hane** (··—ˋ), latrine, water-closet. **~siz,** not having performed ritual ablution; impure: ~ **yere basmamak,** to be exceedingly religious.

**abdülleziz** Earth-nut.

**Abdürrahman** ⌜**Koyun bulunmadığı yerde keçiye ~ Çelebi denir**⌝, (i) make the best of what you have got; (ii) ⌜a Triton among minnows⌝.

**abdüsselâm** (···—́) Mandrake.

**abe** (—́·) *Used to call attention; v. also* **ağabey.**

**abes** Useless. Useless thing; nonsense. **~le uğraşmak,** to pass one's time with trifles.

**abıhayat** (—··ˋ), **-tı** The water of life; water of a legendary spring; any excellent water. ~ **içmiş,** *said of one who never grows old.*

**abıru** (—·—́) Honour; self-respect. ~ **dökmek,** to abase oneself by toadying, to truckle.

**abi** Elder brother. *V. also* **ağabey.**

**âbid** (—ˋ), **-bdi** Worshipper; devotee.

**âbide** (—·ˋ) Monument.

**abla** (ˋ·) Elder sister; elderly domestic.

**ablak** *Only in* ~ **yüzlü,** round-faced.

**abluka** (·ˋ·) Blockade. ~ **kaçağı,** blockade-runner: **~yı yarmak,** to run the blockade.

**abone** (·ˋ·) (*Fr. abonné*) Subscriber; subscription.

**aborda** (·ˋ·) *In* ···**e ~ etm.,** (of a ship) to go alongside *another ship etc.*

**abraş** Speckled; piebald; (leaf) spotted by chlorosis.

**abuhava** (—··—́) Climate.

**abuksabuk** (·ˋ·ˋ) Incoherent nonsense.

**abullabut** Stupid; loutish.

**abur cubur** (·ˋ·ˋ) Incongruous mixtures (foods *etc.*); trash. ~ **kimseler,** riffraff.

**abus** (·—́) Sour-faced; grim; frowning.

**acaba** (ˋ··) I wonder!; is it so?

**acaib** *v.* **acayib.**

**acar** Bright (child); self-reliant; plucky.

**acayib** (·—ˋ) Wonderful; strange. How strange!; you don't say so! (*often iron.*).

**aceb** *v.* **acaba.**

**acele** Haste, hurry. Hasty; urgent. ~ **etm.,** to hasten, to be in a hurry: **~ye gelmek,** to be done in a hurry and carelessly: ···**ye getirmek,** to profit by another's haste (to cheat him): ⌜~ **işe şeytan karışır**⌝, haste is of the Devil; ⌜more haste less speed⌝. **~ci,** in a hurry; impatient, restless (person).

**acem** Persian. ~ **kılıcı,** a double-edged sword. **Acemistan** (···—́), Persia.

**acemi** Tyro, novice; raw recruit. Inexperienced, raw. ~ **oğlan,** a recruit in the Janissaries. **~lik,** inexperience; a being a tyro.

**acenta** (·ˋ·) Agent (commercial).

**aceze** *pl. of* **âciz.** The destitute; waifs and strays.

**acı**[1] Bitter; sharp, pungent; rancid (butter, oil); glaring (colour); dismal, pitiable. ~ **bakla,** lupin: ~ **kabuk (kök),** quassia: ~ **su,** hard water; brackish water. **~ağac,** quassia tree. **~hiyar,** colocynth.

**acı**[2] Pain, ache; grief, sorrow. **~sını çekmek,** to pay the penalty *of an action*: **~sını çıkarmak,** to indemnify oneself for a loss, to take revenge for *stg.*; **ben bunun ~sını senden**

çıkarırım, I'll make you pay for this. **~ görmüş,** who has suffered much: **~sı tepesinden* çıkmak,** to suffer great pain: **can ~sı,** physical pain: **evlâd ~sı,** grief for the loss of a child: **içler ~sı,** heartrending.

**acık** Grief. **~lı,** tragic; touching.

**acık·mak** *vn.* Feel hungry. **karnım acıktı,** I am hungry. **~tırmak,** to make hungry.

**acı·lanmak** *vn.* Become bitter *or* rancid. **~lık,** bitterness. **~msı, ~mtrak,** rather bitter.

**acı·mak** *vn.* Feel pain; ache; hurt; feel pity, be sorry. *va.* Grudge *stg.*; regret its loss or waste; pity. **~nmak,** to be pitied: be regretted; grieve. **~nacak,** pitiable; deplorable, regrettable.

**acırga** Horse-radish.

**acıtmak** *va.* Cause pain *or* suffering; hurt.

**acib** (·⌐) Wonderful; strange. **~e** (·–⌐) strange or wonderful thing.

**âcil** (–⌐) Hurried; hasty; urgent; prompt; pressing. **~en** (⌐··), promptly; hastily, urgently.

**aciz, -czi** Incapacity; impotence; poverty.

**âciz** (–⌐) Incapable; weak; impotent; poor, humble. **~leri,** your humble servant: **···den ~ kalmak,** to be incapable of. **~ane** (–·–⌐), **~zî** (–·⌐), humble; modest; *used as polite or modest way of expressing* 'my', *e.g.* **tarafı âcizanemden,** for my part, from me.

**acul** (·⌐) Impatient; precipitate.

**acur** Kind of cucumber.

**acuze** (·⌐·) Old woman; hag.

**acyo** (⌐·) Agio; premium. **~cu** speculator.

**aç**[1] *imperative of* **açmak.**

**aç**[2] Hungry, destitute; covetous, insatiable. **~ açına,** without food, without having a meal: ⌜**~ ayı oynamaz**⌝, a hungry bear won't dance; a discontented man won't work well: **~ bırakmak,** to starve (*va.*): **~biilâc,** starving, utterly destitute. **~ durmak,** to be able to stand hunger; to fast: **~ karnına,** on an empty stomach: **~ından ölmek,** to die of starvation; to be fearfully hungry: ⌜**~ tavuk kendini arpa ambarında sanır**⌝, pleasant illusions, wishful thinking.

**açalya** (·⌐·) Azalea.

**açar** Implement for opening; key; condiment, appetizer.

**açevele** (··⌐·) (*naut.*) Span; stretcher of a hammock.

**açgözlü** (⌐··) Covetous, avaricious; insatiable.

**açı** Angle.

**açık** Open; uncovered; clear; cloudless; vacant, unoccupied; light (colour); free in manner; impudent; obscene (book); audible; distinct, plain, clear; blank (bill, endorsement). Open air; open country; open sea, the high seas; vacant space; deficit; outside (football). **~tan açığa,** openly, frankly: **açığı çıkmak,** to show a deficit: **açığa çıkmak,** to lose one's post, to be without a job: **~ durmak,** to stand aside, not to interfere: **~ itibar,** overdraft. **~ta kalmak,** to be out in the cold; to be uncared for; to be without a job: **~lar livası,** the unemployed (*iron.*): **~tan para kazanmak,** to get an unexpected and unearned addition to one's income: **açığa satış,** short sale: **açığa varmak,** to go out into the open sea: **açığa vurmak,** to proclaim to the world; to declare openly: **~ yer,** open space; vacant post: **başı ~,** bareheaded: **eli ~,** open-handed: **gözü ~,** wide awake: **gözü ~ gitti,** he died without seeing the fulfilment of his hopes: **yolun ~ olsun!,** I wish you a pleasant journey; good luck to you!

**açık·ağız** (·⌐··), babbler; imbecile; astonishment; rocket (flower). **~ça** (·⌐·), openly; clearly: **~sı,** in plain words. **~göz, ~gözlü,** wide awake; sharp, cunning; 'bright fellow', 'cunning chap'. **~gözlülük,** a being wide awake: **~ etm.,** to be wide awake and practical; to 'have an eye on the main chance'. **~kapı** (·⌐··), way of escape; open door (*commercially*). **~lamak,** to make public, divulge. **~lı,** *in* **~ koyulu,** with light and dark colours; variegated; dappled. **~lık,** open space; interval; freedom of manner; indecency; clearness (of weather *or* expression); lightness (of colour); open; spacious. **~meşreb,** libertine; immoral. **~saçık** (·⌐··), immodestly dressed; indecent (words *etc.*); disorderly, untidy.

**açılış** Opening; inauguration; clearance.

**açılmak** *vn.* Be opened; be discovered; be widened; become more spacious; be amused; be refreshed; escape from boredom; recover *from* *faintness etc.*; (of weather) to clear; (of a post) to become vacant; be cleaned; open out; develop; throw off restraint, become more easy in manner; be morally corrupted; draw away from, shun; put to sea; (of a swimmer) to go far out. **açılıp saçılmak,** to be immodestly dressed: **araları açıldı,** they have quarrelled: **başı*** ~, to begin to get thin on top: **birine** ~, to open one's heart to *s.o.*: **içi*** ~, to be cheered up.

**açkarnına** ('...) *adv.* Fasting; on an empty stomach.

**açlık** Hunger; greed. **~tan nefesi* kokmak,** to be destitute.

**açma** Act of opening; field cleared of bushes *etc. ready for cultivation.*

**açmak** *va.* Open; begin (a discussion, a war *etc.*); reveal (secret); solve (difficulty); undo, unravel; whet (the appetite); unfold; set (sails); clear away (obstruction); sink (well); rub up, polish; uncover; bring into cultivation; explain *a subject* more fully; suit, become. *vn.* (Of a flower) to bloom; (of the sky *or* weather) to clear. ⌜**aç gözünü açarlar gözünü**⌝, keep your eyes open or you will have them opened for you: **başına* is** (*or* **oyun**) ~, to create a difficulty for *s.o.*: **bayrak** ~, to unfurl the flag of revolt: **el** ~, to beg: **elektriği (radyoyu)** ~, to turn on the light (wireless): **telefon** ~, to make a telephone call; **yanlış açmışsınız,** you've got the wrong number: **yer** ~, to make room for (*dat.*).

**açmaz** Difficult situation (in chess *etc.*). **~a düşmek,** to fall into a trap: **~a getirmek,** to play a trick on, to lay a trap for: ~ **oynamak,** to lay a trap.

**açtırmak** *va. caus. of* **açmak.** Cause to open *etc.*; clear (land). **göz açtırmamak,** to give no peace; not to give *anyone* a chance to recover himself.

**ad**[1] Name; reputation. **~ına,** in the name of, on behalf of, for: **~ını ağza almamak,** to refuse to mention *s.o.'s* name: **~ı* batmak,** to pass into oblivion: **~ı bile okunmamak,** to be of no importance: **~ı* çıkmak,** to get a good (bad) name: **~ı* ···e çıkmak,** to be known as ...; to get the reputation of ...: **~ı sanı,** one's name and reputation: **~ı sanı yok,** of no account, of no repute: **~ıyle sanıyle,** by his (its) well-known name: ~ **takmak,** to give an unpleasant name: **~ı var,** it exists only in name, *not in reality*: **çocuğun ~ını koymak,** to give a child a name; to come to a decision; to define precisely, to make clear.

**ad**[2], **-ddi** A counting *or* esteeming; number; computation.

**ada** Island; ward *of a town*. **Adalar,** the Aegean Archipelago.

**âdab** (–'–) *pl. of* **edeb.** Customs; proprieties. **~ı muaşeret,** rules of behaviour, 'savoir vivre'.

**adaçayı** (.'..), **-ni** Garden sage.

**âdad** (–'–) *pl. of* **aded.** Numbers.

**adak** Vow; votive offering; threat.

**Adalardenizi, -ni** Aegean Sea.

**adalât** (..'–), **-tı** *pl. of* **adale.**

**adale** Muscle. **~li,** muscular.

**adalet** (.–'), **-ti** Justice. **~li** just. **~siz,** unjust, **~sizlik,** injustice.

**adalı** Islander, *esp.* Aegean islander.

**adalî** (..'–) Muscular.

**adam** Man; human being; person; good, fine man; serving-man; agent; partisan. **~a dönmek,** to become presentable *in dress etc.*: ~ **etm.,** to make a man of, to bring up well: ~ **evlâdı (oğlu),** a well-bred man: ~ **içine çıkmak,** to go out in public: **~ına göre,** according to the individual (*as regards status, worth etc.*): ~ **olmaz,** hopeless, incorrigible: ~ **sen de!,** come along!, you can do it!; don't worry!: ~ **sırasına geçmek,** to become an important person (*iron.*): ~ **yerine koymak** *or* **~dan saymak,** to hold in esteem; to count as a person of consequence.

**adamak** *va.* Vow; offer *stg.* in fulfilment of a vow. ⌜**~la mal tükenmez**⌝, promises cost nothing.

**adamakıllı** (.'...) Proper; reasonable. Thoroughly, fully.

**adam·ca** (.'.), **adamcasına** (.'...) In a human manner; in the proper way; as a man should. **~cağız,** the good fellow; the poor chap. **~cıl,** *lit.* 'not afraid of man', *and thus* 'liable to attack man', man-hater; misanthrope; vicious (horse); *but also* '*not afraid of man*' *and thus* 'tame'. **~lık,** decent and honest behaviour. **~otu** (.'..), **-nu,** mandrake. **~sız,** without servants; without help.

**adasoğanı** (.'...), **-nı** Squill.

**adaş** Of the same name; namesake.

**adatavşanı, -nı** Rabbit.

**aday** Candidate.

**add·etmek** ('..), **-eder** *va.* Count; esteem. **~olunmak** ('...) to be counted.

**aded** Number; numeral. **bir ~,** one piece, one: **yüz ~,** a hundred head of.
**adem** Non-existence; lack of; absence; (*followed by izafet and noun it can generally be translated by* in-). **~i kabul,** non-acceptance (of a bill).
**âdem** (–‿) Adam. **~zade** (–·–‿), mankind.
**adese** Lens.
**âdet** (–‿), **–ti** Custom, habit; menstruation. **~ etm.,** *or* **edinmek,** to contract a habit: **~ görmek,** to menstruate: **~ üzere,** according to custom; as usual: **~ yerini bulsun diye,** as a mere formality. **~a** (‿·–), as usual; simply; merely; sort of; nearly; as good as: walk! (*riding command*).
**adım** Step; pace; pitch *of a screw.* **~ ~,** step by step, gradually: **~ açmak,** to lengthen the step, to go faster: **~ını* alamamak,** to be unable to refrain *from doing stg.*: **~ atmak,** to step, to walk: **~ başında,** at every step, frequently: **~ını tek almak,** to proceed with caution. **cimnastik ~,** at the double. **~lamak,** to measure by pacing. **~lık,** a distance of *so many* steps: **bir ~ yer,** only a step away, quite near.
**âdi** (–‿) Customary, usual; commonplace, common; vulgar, mean. **~ gün,** an ordinary day.
**adil** (‿·), **–dli** Justice; equity. **kâtibi ~,** Notary Public.
**âdil** (–‿) Just; legally competent *witness.*
**adi·leşmek** *vn.* Become common *or* inferior. **~lik,** commonness, vulgarity.
**ad·lamak** *va.* Give a name to. **~lanmak,** to be named. **~lı,** named; famous: **~ adıyle,** by its (his) well-known name: **~ sanlı,** celebrated.
**adlî** (·‿) Pertaining to justice; judicial. **~ye** (·–‿), Ministry of Justice.
**adres** Address. **~ rehberi,** address book; directory.
**adsız** Nameless; unknown; without reputation. **~ parmak,** the ring-finger.
**af, –ffı** Pardon; exemption; excusal; dismissal. **~ olunmaz,** unpardonable: **~fı umumî,** general amnesty.
**afacan** Unruly, undisciplined; 'gamin', whipper-snapper, urchin.
**afallamak, afallaşmak** *vn.* Be astonished, taken aback, stupefied.
**aferim, aferin** (–·‿) Bravo!, well done! (*often used ironically to express disappointment*). **~ almak,** to obtain an honourable mention (*in a school etc.*).
**âfet** (–‿), **–ti** Disaster, calamity; bane, blight; person of bewitching beauty.
**affetmek** (‿··), **–eder** *va.* Pardon, forgive; excuse; give leave to go; dismiss. **affedersiniz!,** I beg your pardon!
**affı** *v.* **af.**
**afi** Showing off, swagger. **~ kesmek** *or* **yapmak,** to cut a dash, to swagger. **~li,** swaggering.
**afif** (·‿) Chaste; virtuous; innocent.
**afiş** (·‿) (*Fr. affiche*) Advertisement, poster.
**afiyet** (–·‿), **–ti** Health. **~ ola!,** 'bon appetit!': **~ olsun!,** may it do you good! (*said to one about to eat or drink, have a bath or a shave etc.*): **~le yemek,** to eat with a good appetite.
**aforoz** Excommunication.
**afsun** (·‿), **efsun** Spell, charm, incantation; crafty tale. **~cu,** sorcerer. **~lamak,** to bewitch.
**aftos** (*sl.*) Sweetheart. **~piyos** (·‿·‿), useless, worthless.
**afur tafur** (·‿··) With a superior air.
**afyon** Opium. **~keş,** opium addict. **~ruhu, –nu,** laudanum.
**ağ** Net; web (of spider); fly (of trousers). **~ gözü,** mesh: **~ gemisi,** mine-sweeper.
**ağa** *Obsolete title given to an illiterate man*; lord, master, gentleman; *obs. title of many officials.* **ak ~,** white eunuch: **harem ~sı,** black eunuch *who used to supervise the harem*: **köy ~sı,** village headman: ⌜**gidene ~m gelene paşam,**⌝ 'you say "Sir" to your outgoing superior, but "My Lord" to his successor' (*said of a toady*): ⌜**sen ~ ben ~ koyunları kim sağa?**⌝, 'you are a gentleman, I am a gentleman, who is to milk the sheep?'
**ağababa** (·‿··) Important person; senior; old-fashioned man.
**ağabey** (·‿·), **abe, abey** Elder brother; senior (colleague *etc.*).
**ağac** Tree; wood, timber. Wooden. **~çileği, –ni,** raspberry. **~kakan,** woodpecker. **~kavunu, –nu,** citron. **~kurdu, –nu,** wood-boring maggot. **~lamak, ~landırmak,** to plant with trees, afforest. **~lık,** wooded place; copse: well-wooded.
**ağa·lanmak** *vn.* Become proud like an Agha; give oneself airs. **~lık,**

state of being an Agha; title of Agha, pride; generosity, magnanimity.

**ağar·mak** *vn.* Become white *or* pale; become light (dawn). **ortalık ağardı,** the dawn broke. **~tı,** a growing grey; whiteness; curd; milk residues. **~tmak,** to whiten; clean; clear *s.o.'s* honour: **···in yüzünü ~,** to do honour to ....

**ağbanı** Cotton stuff embroidered with yellow silk.

**ağda** A semi-solid sweet *made of sugar, honey etc.* **~ koymak,** to apply this sticky substance to the skin *to remove hairs.* **~lanmak, ~laşmak,** to become of the consistency of thick syrup. **~lı,** of the consistency of thick syrup; heavy, involved *speech or writing.*

**ağı, ağu** Poison. **~ otu,** hemlock. **~lamak,** to poison.

**ağıl** Pen, fold *for sheep*; halo. **~lanmak,** to be folded *or* penned; to be surrounded by a halo.

**ağır** Heavy; weighty; serious, dignified; slow; lazy; unwholesome; offensive (words); severe (illness, punishment); valuable; painful; fatiguing; pregnant. **~ almak,** to be slow *or* lazy in doing *stg.*: **~dan almak,** to avoid excess; not to be too keen: **~ ayak,** heavy with child: **~ basmak,** to have influence: **~ ceza mahkemesi,** Criminal Court: **~ davranmak,** to act in a slow or reluctant manner: **~ esvab (mal),** expensive clothes (goods): **~ gelmek,** to be hard to bear, to be trying *or* humiliating: **ağrına* gitmek,** to hurt the feelings of: **~ gövde,** fat: **~ hapis cezası,** imprisonment with hard labour: **~ hava,** unhealthy weather *or* climate: **~ işitmek,** to be hard of hearing: **~ satmak,** to put on an air of importance: **eli (eline) ~,** slow at work; lazy; heavy-handed; severe: **kulağı ~,** hard of hearing.

**ağır·başlı,** (of a man) grave, serious, dignified. **~canlı,** lazy, indifferent. **~ezgi,** slowly. **~kanlı,** lazy, torpid. **~lamak,** *va.* to treat with respect, honour; entertain, show hospitality to. **~laşmak,** to become heavier, more difficult, slower, more serious; (of food) to begin to go bad; (of weather) to become overcast: **dili ~,** (of a drunkard *etc.*) to be incoherent in speech: **kulakları* ~,** to become hard of hearing. **~lık,** weight, heaviness; slowness; stupidity; seriousness; hardness of hearing; nightmare; oppression *caused by heat or boredom*; trousseau; baggage; heavy transport of an army: **~ basmak,** for sleepiness to come upon one; for a nightmare to oppress one; to feel uneasy.

**ağıt** Lament; mourning. **~çı** professional mourner.

**ağız, -ğzı** Mouth; speech, manner of speaking, dialect; opening; edge (of a knife); biestings (colostrum); each successive baking in an oven, *hence* **ilk ~,** the first attempt *at anything, e.g.* the first child. **~ ağza,** brimful; **~ ağza vermek,** (of two people) to speak close to each other *so that no one else may hear*: **ağzından,** verbally, by hearing, unauthentic: **ağzı* açık kalmak,** to gape with astonishment: ⌜**ağzını açacağına gözünü aç!**⌝, instead of gaping with astonishment open your eyes (*and understand what is happening*): **ağza almak,** to mention *stg.*: **ağza alınmaz,** uneatable; unmentionable, obscene: **ağzını* aramak** *or* **yoklamak,** to sound *s.o.*: **ağzına* bakmak,** to hang on *s.o.'s* lips; to act in accordance with *s.o.'s* words: **ağzına beraber,** brimful: **~ birliği etm.,** to tell the same tale: **~ bozmak,** to vituperate, to swear: ⌜**ağzından çıkanı kulağı işitmiyor**⌝, he does not realize what he says: **ağzından* dökülmek,** to be clear from a man's speech *that he is lying etc.*: **ağza düşmek,** to become a matter of common talk: **~ etm.,** to speak one's mind; to try to talk *s.o.* over; to talk in a theatrical manner: **ağzına* geleni söylemek,** to say disagreeable things; to speak without due reflection: **ağzını* havaya açmak,** to be disappointed; to get nothing: **···in ağzının içine bakmak,** to hang on the lips of ...: **~dan işitme,** hearsay, as a rumour only: **ağzı kara,** who delights in giving bad news: **~ kokusu,** foul breath; moodiness, caprice; **ben herkesin ağzının kokusunu çekemem,** I can't put up with everyone's whims *or* criticism: **ağzı kulaklarına varmak,** to grin from ear to ear, to be exceedingly pleased: **~ kullanmak,** to explain away a matter: ⌜**ağzile kuş tutsa faydası yok**⌝, even if he does the impossible it will be of no use: **ağzından lâf (söz) almak,** *by dint of talking of this and that* to get a person to say what one wants: **ağzına lâyık,** fit for your mouth, *i.e.* something really good to eat: **ağzının* suyu akmak,** for one's mouth to water, to long for something: ~ **tadı,** *v.* **ağıztadı: ağzını topla!, behave** yourself!, don't be impudent! ···**den ağzı* yanmak,** to suffer loss *or* trouble from ...: ~ **yapmak,** to try to make out that *stg.* is other than what it really is: **dört yol ağzı,** cross-roads: **düşman ağzı,** calumny: **halk ağzı,** rumour, 'canard': **ilk ~da,** at the first shot, at the first attempt: **kaymak ağzı,** the crust of clotted cream.

**ağız·ağza,** full to the brim; completely, utterly. **~bağı,** mousing (of a hook). **~dolusu,** without restraint, unreservedly; at the top of one's voice. **~kalabalığı, -nı,** a spate of words. **~na getirmek,** to disconcert by a flow of words. **~lık,** mouthpiece (of a pipe *etc.*); cigarette-holder; muzzle; stone surround *of a well*; cover of leaves *over a basket of fruit*. **~otu, -nu,** priming *of a gun*. **~persengi, -ni,** constant refrain; anything constantly harped upon. **~sız,** submissive, docile, meek. **~suyu, -nu,** spittle. **~tadı, -nı,** enjoyment of a meal; general feeling of satisfaction; *often written in two words and* **ağız** *with possessive suffix, e.g.* **ağzımın tadı.**

**ağla·mak** *vn.* Weep: grieve. ⌜**ağlarsa anan ağlar, gerisi yalan ağlar**⌝, don't expect others to worry about you: **anası ~,** to suffer great pain *or* misfortune: ⌜**hem ağlarım hem giderim**⌝, *used of s.o. who pretends to be unwilling when he is really keen.* **~malı,** on the verge of tears. **~şma,** a weeping together: **bir ~ faslı başladı,** they all started to weep. **~şmak,** to weep together; to complain continually. **~tmak,** to cause to weep: **anasını* ~,** to give a good hiding to; to ill-treat. **~yış,** weeping, complaint.

**Agop** (*Armenian*) Jacob. ⌜**Agobun kazı gibi ne bakıyorsun?**⌝, why do you stand and stare in that stupid way and do nothing?

**ağrı** Pain; ache. **~sı tutmak,** for the pains of childbirth to come on: **baş ~sı,** headache; nuisance, pest: **ilk göz ~sı,** one's first love; one's first child. **~lı, ~klı,** aching, painful. **~mak,** to ache, hurt: **başım ağrıyor,** my head aches. **~sız,** without pain: ~ **baş,** a man without cares: ~ **başına derd açmak,** to bring unnecessary trouble on *s.o.* **~tmak,** to cause pain to, hurt.

**Ağrıdağı** Mount Ararat.

**ağu** *v.* **ağı.**

**ağucuk** (*iron.*) Baby; dolly; overgrown boy. ~ **bebecik,** naughty boy!, shame!

**Ağustos** August. ⌜**~ta suya girsem balta kesmez buz olur**⌝, I am always unlucky. **~böceği, -ni,** cicada; chatterbox.

**ağyar** (·–́) *pl. of* **gayr.** Others; strangers; rivals. **~a karşı,** before others, publicly: **yar ve ~,** (lover and rivals) friend and foe, all the world.

**ağzı** *v.* **ağız. ~açık,** gaping idiot. **~bir,** talking the same language; in agreement. **~bozuk,** foul-mouthed, blasphemous. **~gevşek,** talkative, indiscreet. **~kara,** calumnious; bearing bad tidings. **~pek,** who can keep a secret.

**ah**[1] Sigh; curse. ~ **çekmek** *or* **etm.,** to sigh: **~a gelmek** *or* **~ına uğramak,** to suffer one's due retribution: **~ı* kalmamak,** for *s.o.'s* curse to take effect sooner or later: **~ı* tutmak,** for one's curse to take effect.

**ah**[2] Ah!; oh! ~ **deyip ~ işitmek,** to be alone and helpless: ~ **minel'aşk!,** alas! what a tyrant is love!: ~ **minelmevt!,** ah! cruel Death!

**ahali** (·–́) *pl. of* **ehil.** People; the public.

**ahbab** (·́) Friends; acquaintance. **~lık,** acquaintanceship, friendship.

**ahçı, aşçı** Cook. ~ **başı,** head-cook: ~ **dükkânı,** eating-house. **~lık,** the profession of cook.

**ahd·en** (́·) By treaty; as a pact. **~etmek** (́··), **-eder,** *vn.* to take an oath; solemnly promise. **~i,** *v.* **ahid. ~î** (·–́), pertaining to a pact *or* contract. **~üpeyman** (···–́) **etm.,** to take a solemn oath.

**ahenk** (–́) Purpose, intention; accord, harmony; music. **~li,** harmonious, in accord.

**aheste** (–·́) Slow; gentle; calm; 'piano' (*mus.*). ~ **beste,** slowly (*iron.*).

**ahetmek** (́··), **-eder** *vn.* Utter a curse; sigh.

**ahfad** (·́) Grandsons; descendants.

**ahfeş** *Only in* ⌜**~in keçisi gibi başını sallamak**⌝, to agree with everything that *s.o.* says.

**ahım şahım** (–ˋ–·) *In* ~ **değil,** (*iron.*) anything but beautiful *or* excellent.
**ahır** Stable.
**âhır** *v.* **âhir.**
**ahi** (*Formerly*) member of a trade guild.
**ahibba** (··ˊ) *pl. of* **habib** Friends.
**ahid, -hdi** Oath; promise; pact; undertaking; solemn injunction; period, epoch. ~**leşmek,** to take an oath together; to enter into a solemn agreement with one another. ~**li,** bound by promise *or* contract.
**âhir** (–ˋ) The end. Last; latter. ~ **zaman,** the latest time, the present age.
**ahize** (–·ˋ) Receiver (telephone *etc.*).
**ahkâm** (·ˋ) *pl. of* **hüküm.** Judgements; dispositions; laws; inferences. ~ **çıkarmak,** to draw arbitrary conclusions; to put forward ridiculous suppositions.
**ahlâk** (·ˋ), **-kı** *pl. of* **hulk.** Moral qualities; character; morals; ethics. ~**çı** (·–ˋ), moralist. ~**î** (·–ˊ), pertaining to morals, moral. ~**sız,** immoral; amoral.
**ahlamak** *vn. In* **ahlayıp vahlamak,** to sigh and moan.
**ahlat, -tı** Wild pear; boor. ⌜~**ın iyisini dağda ayılar yer**⌝, 'the bears in the mountains eat the best of the pears'; the best things go to those who do not deserve them.
**ahmak** Silly fool, idiot. ~**lık,** stupidity, foolishness. ~**ıslatan,** a fine drizzle.
**ahrar** *pl. of* **hür.** Free men; liberals.
**ahret, -ti** The next world, the future life. ~**te on parmağım yakanda** (**olacak**), you'll pay for this in the next world; it's most unfair. ~ **suali,** endless questions, a 'regular' cross-examination. ~**lik,** anything pertaining to the next world; an orphan brought up by well-to-do people as a servant.
**ahşab** Wood. Wooden.
**ahtapot, -tu** Octopus; polypus; cancerous ulcer.
**ahu** (–ˊ) Gazelle. ~ **gözlü,** with beautiful eyes. ~**baba** (–ˊ··), a nice old fellow; spook. ~**dudu** (–ˊ··), **-nu,** raspberry.
**ahval** (·ˋ), **-li** *pl. of* **hal.** Conditions; circumstances; the state of a person's health. ~**i âlem,** the state of the world: ~**i hazıra,** the present state of affairs: ~ **ne merkezde?,** how are things going?

**ahyan** (·ˊ) *pl. of* **hin.** Times; moments. ~**en,** at times, occasionally.
**aid** (–ˋ) Relative to; concerning; belonging; competent (authority *etc.*). ···**e** ~ **olm.,** to belong to, to concern.
**aidat** (–·ˊ), **-tı** *pl.* Revenues; income; contribution; allowance; remuneration.
**aidiyet** (–··ˋ), **-ti** A belonging to; a concerning; competence (of a court *etc.*). **bize bir** ~**i yoktur,** it has nothing to do with us.
**aile** (–·ˋ) Family; wife. ~**vî** (–··ˊ) *a.,* family, domestic.
**ajan** Agent. ~**s,** news agency.
**ajur** (*Fr. à jour*) Open-work embroidery; hemstitch.
**ak, -kı** *a.* White; clean. *n.* White; white *of an egg or an eye.* ~ **akçe,** silver money: ⌜~ **akçe kara gün içindir**⌝, 'save up for a rainy day!': ~**ı** ~ **karası kara,** with a white complexion but black eyes and hair; ~**la karayı seçmek,** to have the greatest difficulty to do *stg.*: ~ **pak,** bright and clean: **saçına*** ~ **düşüyor,** his hair is beginning to turn white: **yüz** ~**ı,** honour; **yüzü*** ~, his conscience is clear: **yüzünü** ~ **çıkarmak,** to be a source of pride: **bu işten yüzümün** ~**ile çıktım,** I have come out of this affair without a stain on my honour.
**akabinde** (ˋ···) *adv.* Immediately afterwards; subsequently.
**akağa** (ˋ··) White eunuch.
**akağac** (ˋ··) A kind of birch; (?) = **akçaağac.**
**akaid** (·–ˋ) *pl. of* **akide.** Articles of faith, religious precept; catechism.
**akamet** (·–ˋ), **-ti** Sterility.
**akar**[1] Flowing; running (water); leaky. ~ **yakıt,** liquid fuel.
**akar**[2] (·ˊ) Landed property; real estate. ~**at** (·–ˋ), **-tı** *or* ~**et** (·–ˋ), **-ti,** houses *or* flats built by a public institution to let.
**akarsu** Running water; a single necklace of diamonds *or* pearls.
**ak·asma** (ˋ··) White climbing rose; clematis. ~**baba** (ˋ··), vulture. ~**balık** (ˋ··), dace (?). ~**benek** (ˋ··), leucoma. ~**ciğer** (ˋ··), lungs.
**akasya** (·ˋ·) Acacia.
**akça**[1] Whitish; *v.* **akçe.** ~ **kavak,** white poplar: ~ **ağac,** maple: ~ **pakça,** rather pretty.
**akçe, akça**[2] Coin; third of a para; money; sum of money. ~ **kesmek,** to

have a mint of money: **geçer ~,** good currency, genuine.
**akçıl** Whitish; faded.
**Akdeniz** (ʼ··) Mediterranean.
**akdetmek** (ʼ··), **-eder** *va.* Bind, tie; conclude (bargain, treaty); contract (marriage); set up, establish (council); organize (meeting).
**ak·diken** (ʼ··) Hawthorn. **~doğan** (ʼ··), gerfalcon (?). **~günlük** (ʼ··), oil of savin (juniper). **~haşhaş** (ʼ··), white poppy *from which a syrup (diacodium) is made.*
**âkıbet** (–·ʼ), **-ti** End; consequence, result; the near future; destiny. *adv.* (–̀··) Finally.
**akıcı** *a. & n.* Fluid; flowing; fleeting. **~lık,** fluidity; fluency.
**âkıl** (–ʼ) Wise; intelligent.
**akıl, -klı** Reason, intelligence, sense, wisdom; comprehension; prudence; memory; opinion. **~ almaz,** incomprehensible; incredible: **aklı* başına* gelmek,** to come to one's senses: **aklı başında,** he knows what he's about, he's 'all there': **aklı başında olmamak,** to be incapable of clear thought: **aklı* başından* gitmek,** to lose one's head: **aklını* bozmak,** to lose one's reason: **aklı* durmak,** to be dumbfounded: **~ etm.,** to think *of a plan etc.*: **aklı evvel,** very clever: **aklına* gelmek,** to come to one's mind, to occur to one: **aklına* getirmek,** to remind *s.o.*; to recollect: **aklı* kesmek,** to decide, judge: **aklına* koymak,** to have quite made up one's mind: **~ öğretmek,** to give advice: **bu aklı sana kim öğretti?,** who put this idea into your head?: **aklı sıra,** according to him (*iron.*); if he is to be believed. **~da tutmak,** to bear in mind: ⌜**~ ~dan üstündür**⌝, there is always a better brain to be found: ⌜**~ var izan var**⌝ *or* ⌜**~ var yakın var**⌝, it's quite obvious, it's only common sense: **akla yakın,** reasonable, clear: **bir işe aklı* yatmak,** to be satisfied about a matter.
**akıl·dişi, -ni,** wisdom tooth. **~hocalığı, -nı,** a being *s.o.'s* adviser; the giving of pretentious advice: **~ istemem,** I want nobody's interference in this matter. **~hocası, -nı,** mentor, pretentious adviser. **~kutusu, -nu,** trusty adviser; one fertile in expedients. **~lı,** clever, intelligent; reasonable; prudent. **~lılık,** wisdom, intelligence. **~sız,** stupid: ⌜**~ başın zahmetini ayak çeker**⌝, 'little wit in the head makes much work for the feet'.
**akın** Raid; foray; rush; stream *of people etc.* **~ ~,** in crowds, in streams: **hava ~ı,** air-raid. **~cı,** raider; (*formerly*) a corps of light cavalry in the Turkish army.
**akıntı** Current, stream, flow. **~ya kürek çekmek,** to row against the current; to waste one's efforts: **~ zaviyesi,** leeway (due to current).
**akış** Course; inclination; trip.
**akıtma** Blaze (on a horse).
**akıtmak** *va.* Make *or* let flow; shed (tears). **kanını içine ~,** to hide one's resentment.
**akib** (·ʼ) **···in ~inde,** immediately after *or* behind . . .
**akid, -kdi** A tying; tie, knot; compact, treaty; bargain; marriage.
**akide** (·–ʼ) Religious faith, creed; sugar candy. **~yi bozmak,** to go counter to one's faith *or* convictions.
**akik** (·–̀) Agate.
**akim** (·–̀) Sterile, barren; without result.
**akis, -ksi** Opposite; contrary; reflection (in water *etc.*); reverse side; reverberation. **aksine anlamak,** to understand exactly the opposite of what was said: **aksi gibi,** 'just to spite one', *e.g.* **o gün aksi gibi yağmur yağdı,** of course it *would* rain that day.
**aklen** (ʼ·) Reasonably, intelligently.
**akletmek** (ʼ··), **-eder** *va.* Think of.
**aklı**[1] *v.* **akıl.**
**aklı**[2] Spotted *or* mixed with white. **~ karalı,** with white and black spots; piebald.
**aklık** Whiteness; white face-paint; innocence.
**aklısıra** (·ʼ··) According to him; if he is to be believed: **~ şıklaştı,** she dressed in what she thought was a chic way.
**akl·î** (·–̀) Pertaining to the mind *or* reason; reasonable; mental **~iselim,** common sense. **~iye,** mental diseases; mental clinic. **~iyeci,** mental specialist.
**akma** Flow; current. Flowing.
**akmak** *vn.* Flow; ooze; drip; (of textiles) to become frayed *or* unravelled; (of a stocking) to ladder. **···e ~,** to go as far as ...: ⌜**akmazsa da damlar**⌝, it brings in something, if not a lot: ⌜**akan sular durur**⌝, there is nothing more to be said; that clinches the

argument: **gönlü*** ~, to feel an attraction towards *stg.*: **gözü*** ~, to be blinded (by accident); ⌜**iki gözüm önüme aksın**⌝, may my eyes drop out (if what I say is not true); **gözümden uyku su gibi akıyor,** I am terribly sleepy: **··· yüzünden akıyor,** it is obvious that he . . .; you can see by his looks that . . .

**akord** (*Fr. accord*) A being in tune. ~ **etm.,** to tune *a musical instrument.* ~**cu,** tuner.

**akraba** (··⊥) *pl. of* **karib.** Relatives; *also as sing.* relative. ~**lık,** relationship.

**akran** (·⊥) *pl. of* **kırn** *and* (*wrongly*) *as pl. of* **karîn.** Equals, peers; people of the same age, contemporaries; companions. ⌜~ ~**dan azar**⌝, bad habits are infectious: ~**ından kalmak,** to be behind one's fellows *in success etc.*: **ben** ~, people of my own age.

**akreb** Scorpion; hour-hand *of a clock.*

**aksak** Lame; limping; lop-sided. ~**lık,** lameness; limping behind others; hitch, defect. ~**sız,** without limping; without a hitch.

**aksakal** (⸌··) Village elder.

**aksam** (·⊥) *pl. of* **kısım.** Parts.

**aksamak** *vn.* Limp; falter; have a hitch.

**aksan** Accent.

**aksata** Buying and selling.

**aksatmak** *va.* Cause to limp; hinder, delay.

**akse** (*Fr. accès*) Fit; attack; paroxysm.

**akset·mek** (⸌··), **-eder** *va. & vn.* Reflect; be reflected; reverberate; come to the hearing of *s.o.* ~**tirmek,** *va.* reflect; echo.

**aksır·ık** Sneeze. ~**ıklı tıksırıklı,** sneezing and coughing; old and in bad health. ~**mak,** to sneeze. ~**tıcı,** sternutatory, sneeze-provoking.

**aksi**[1] *v.* **akis.**

**aksi**[2], **aksî** (·⊥) Contrary; perverse, contrary; unlucky. ~ **çıkmak,** to happen contrary to expectation: ~ **takdirde,** in the opposite case; on the other hand. ~**lenmek,** to show resentment, to bridle. ~**lik,** perversity, contrariness; obstinacy; difficulty, hitch; misfortune, mishap. ~**ne,** on the contrary; in a contrary manner. ~**seda** (···⊥), echo. ~**tesir** (··–⸌), reaction, counterstroke.

**aksoy** (⸌·), ~**lu** Noble, well-born.

**aksöğüt** (⸌··) White willow.

**aksu** (⸌·) Amaurosis.

**aksülâmel** (··–⸌) Reaction.

**aksülümen** Corrosive sublimate.

**akşam** Evening. In the evening. ~**ı etm.,** to stay till the evening: ~ **sabah,** constantly, all the time: ~ **sabah demez gelir,** he comes at all sorts of times (without any consideration for one): ~**a sabaha,** very shortly, at any moment: ~**üstü** *or* ~ **üzeri,** towards evening. ~**cı,** one who spends his evenings drinking; worker on a night-shift. ~**lamak,** to spend the whole day doing *stg. or* going somewhere; to pass the night at a place. ~**layın** (·⸌··), in the evening. ~**lık,** evening *clothes etc.*

**aktar, attar** Druggist, herbalist; dealer in small wares, haberdasher, mercer.

**aktarma** Change (of train); transhipment; plagiarism. ~**k** *va.,* to move from one receptacle to another; to tranship; to move from one train to another; to turn topsy-turvy: **damı** ~, to re-tile a roof.

**aktavşan** (⸌··) Jerboa.

**akvaryum** Aquarium.

**al**[1] (·) Red, vermilion, crimson; chestnut (horse). Red colour; rouge; erisipelas, puerperal fever. ~ **basmak,** to blush; be stricken with puerperal fever: ~**ı al moru mor,** red in the face *from exertion or confusion*: ~ **sancak,** the Turkish flag.

**al**[2] (·) Fraud, trick.

**al**[3] (·) *In* ~ **aşağı etm.,** to knock down.

**âl** (–) Dynasty, line; family of an important man. ~**i resul,** the posterity of the Prophet.

**ala-** *prefix* (*Fr. à la*). ~**turka,** in the Turkish style: ~**franga,** in the European style.

**ala** Of various colours; speckled; pied. ~**bacak,** with white socks (horse). ~**balık,** trout.

**âlâ** (–⊥) *a. comp. of* **ali.** Higher; highest; good, excellent. **ne** ~!, what a good thing!; *also iron.* what a shame!: **pek** ~, very good!, all right!, excellent!

**alabanda** (··⸌·) The inside of a ship's side; broadside; order to 'about ship!'. ~ **etm.,** to put the helm hard over: ~ **iskele (sancak),** hard to port (starboard).

**alabildiğine** (⸌·····) *adv.* To the utmost. ~ **koşmak,** to run at full speed: ~ **şişman bir adam,** 'the fattest man you ever saw': **göz ~,** as far as the eye can see.

**alabros** (*Fr. à la brosse*) *a.* (Hair) cut very short; 'crew-cut'.

**alabura** (··᾿·), **albura.** ~ **etm.,** to hoist a flag (*naut.*); to give a nautical salute by raising oars on end; to furl sails. ~ **olm.,** to capsize.

**alaca**[1] Of various colours; motley; piebald. A kind of striped stuff. ~ **basma,** chintz: ~ **dostluk,** a fickle friendship: ~ **karanlık,** twilight; dawn: ~**sı içinde,** sly, shifty: **deli ~sı,** crude and clashing colours. ~**bulaca** (··᾿··᾿), daubed with incongruous colours.

**alaca**[2] *In* ~ **verece,** the completion of a purchase.

**alacak** *3rd pers. sing. fut. of* **almak.** *n.* Money owing; credit. **ondan alacağım var,** he owes me money. **alacağı olsun!,** 'I'll make him pay for it!': ⌈~**la verecek ödenmez**⌉, a debt you owe cannot be paid by a debt owed to you. ~**lı,** creditor.

**alafranga** (··᾿·) In the European fashion.

**alageyik** (·᾿··) Fallow deer.

**alâim, alâyim** (·—᾿) *pl. of* **alâmet.** Signs. ~**i sema,** rainbow.

**alâka** (·—᾿) Connexion; relationship; attachment; love; interest. ···**le ~sı* olm.,** to be interested in: **kat'i ~ etm.,** to break off relations, to sever connexion. ~**dar** (·—·᾿), connected, concerned, interested: ~ **etm.,** to interest. ~**landırmak,** to interest, to affect. ~**lı,** in love; connected, interested.

**alakarga** (·᾿··) Jay.

**alâküllihal** (··᾿··) *Only in* ~ **geçinmek** *or* **yaşamak,** to have just enough to live on: ~ **geçiniyoruz,** (*politely*) 'we're not too badly off'.

**alamana** (··᾿·) Small lugger, fishing-smack; large trawl-net.

**alâmeleinnas** (·—᾿··᾿) In public.

**alâmet** (·—᾿), **-ti** Sign; mark; symbol; symptom; trace. Enormous; monstrous. ~**i farika,** distinguishing mark; trademark.

**alan**[1] Who takes *or* receives; purchaser.

**alan**[2] Clearing *in a forest*; open space; square *in a town*; sphere *of work etc.* **spor ~ı,** sports ground, playing-field.

**alantalan, alantaran** (·᾿·᾿) In utter confusion.

**alarga** (·᾿·) Open sea. Distant, apart. ~!, push off!; keep clear! ~**da,** in the offing; clear of: ~ **durmak,** to remain aloof: ~ **etm.,** to put out to sea.

**alaşağı** (᾿···) **etm.** *va.* Beat down, knock down; depose, dethrone.

**alaturka** (··᾿·) In the Turkish fashion.

**alavere** (··᾿·) Utter confusion; jumble; passing *or* throwing a thing from hand to hand; a kind of gangway *for loading coal on to a ship.* ~ **tulumbası,** force-pump.

**alavire** (··᾿·) Speculation (financial). ~**ci,** speculator.

**alay** Troop; crowd; procession; regiment; great quantity; joke, derision. ~ ~, in great numbers: ~**a almak,** to make fun of: ···**le ~ etm.,** to make fun of ...: ~ **etme!,** 'joking apart', 'tell me seriously': ~ **malay,** the whole outfit: **bir işin ~ında olm.,** not to take a matter seriously: ~ **sancağı,** regimental flag. ~**cı,** joker; mocker; mocking, derisive. ~**lı,** ceremonious; pompous; mocking; officer risen from the ranks, as *opp. to* **mektebli,** *hence as a.,* amateur (*iron.*).

**alâyım** *v.* **alâim.**

**alâyiş** (——᾿) Pomp, display; showiness.

**alaz** Flame. ~**lama,** erythema. ~**lamak,** to scorch, singe; brand; speak a few friendly words to *s.o.* ~**lanmak,** to be singed *or* branded; (of the skin) to be inflamed *or* to come out in spots.

**albastı** Puerperal fever.

**albatr** Alabaster.

**albay** Colonel; captain (naval); group captain (air).

**albeni** (᾿··) Attractiveness, charm.

**albura** *v.* **alabura.**

**alçacık** (᾿··) Rather low.

**alçak** Low; short; vile, base, abject; cowardly. ~**tan almak,** to adopt a friendly, modest attitude; to change one's tone: ~ **gönüllü,** humble, modest; *but* ~ **tabiatli,** of a base nature: ~ **ses,** low voice; (*mus.*) low-pitched voice. ~**lamak** *va.,* to despise, to treat with contempt: *vn.* to become low; to stoop. ~**latmak,** to lower, lessen, humiliate. ~**lık,** lowness, shortness; baseness, meanness, cowardice.

**alçalmak** *vn.* Bend down, stoop; abase oneself; condescend.

**alçarak** Lowish.

**alçı** Plaster of Paris. ~**lamak,** to cover with plaster of Paris. ~**taşı, -nı,** gypsum: **sert ~,** anhydrite.

**aldanmak** *vn.* Be deceived; be enticed; be taken in; be mistaken.

**aldat·ıcı** Deceptive; deceiving; cheating. **~maca,** trick, 'sell', catch. **~mak,** to deceive, dupe, cheat.

**aldır·ış ~ etm.,** to pay attention, to take notice (*usually in the negative*). **~mak,** *va. caus. of* **almak** *q.v.*; to make *stg.* fit into *stg. else*; *vn.* to take notice; to pay attention (to = *dat.*): **göze ~,** *as* **göze almak,** *v.* **göz.** **~mamak** (·ˋ··), *va. & vn. neg. of above*: to take no notice; to pretend not to see *or* hear: **aldırma!,** pay no attention!, take no notice!; don't worry!: **burnundan kıl ~,** to be very stand-offish and superior; not to yield an inch. **~mamazlık** (·ˋ···), indifference: **~tan gelmek,** to be indifferent, not to care.

**alel·âcayib** (ˋ··–·) Strange; surprising. **~âcele** (ˋ····), in haste. **~âde** (ˋ·–·), ordinary, normal, usual. **~ûmum** (ˋ··–), generally, in general. **~ûsul** (ˋ··–), in the customary manner; as a mere formality.

**alem** Sign; mark; flag; proper name; peak of a minaret; the crescent and star on top of a mosque. **~ olm.,** to become a distinctive name or epithet for ...; to be known as .... **~dar** (··ˊ), standard-bearer.

**âlem** (–ˋ) World; universe; all the world, everyone; class *of beings*; state of health; amusement, entertainment, party; a surprising state of affairs. **~de,** *expresses a fervent wish, e.g.* **~de o kitablar benim elime geçmeliydi!,** really I *ought* to have got hold of those books: **böyle yapmanın ~i yok,** one shouldn't do this sort of thing: **fahri ~,** the Glory of the World, Mohammed: **kendi ~de olm.,** to be occupied with one's own affairs; to live quietly and contentedly: **ne ~desiniz?,** how are you getting on?: **o bir ~!,** he's quite a character!

**alen·en** (ˋ··) Publicly, openly. **~î** (··ˊ), public; openly said *or* done. **~iyet, -ti,** publicity.

**alesta** (·ˋ·) Ready; prepared. **~!,** stand by!, ready!

**alet** (–ˋ), **-ti** Tool; instrument; member *of the body*.

**alev** Flame; pennant. **~ almak** *or* **kesilmek,** to catch fire; to flare up *in a passion*: **~ makinesi,** flame-thrower. **~li,** flaming; in a passion.

**alevî** (··ˊ) Partisan of Ali; Shiite.

**aleyh** Against. **~inde,** against him *or* you: **~imizde,** against us. **~dar** (··ˊ), opponent: **~lık,** opposition. **~te,** unfavourable.

**aleykümselâm** (··ˋ·ˊ) Peace be to you! (*in reply to the Moslem salutation* **selâmaleyküm.**)

**alfabe** Alphabet.

**algarina** (··ˋ·) Floating crane; sheer-hulk.

**algınlık** *In* **soğuk algınlığı,** a chill.

**algoncar** Sloe; bullace.

**alıc** Neapolitan medlar.

**alıcı** Buyer, customer; receiver (telephone, wireless). Dazzling; attractive. **~ gözü ile bakmak,** to look at *stg.* as a genuine purchaser; to regard earnestly, with serious intention, carefully: **~ kuş,** bird of prey.

**alık** Crazy; imbecile.

**alıkomak, alıkoymak** (·ˋ··) *va.* Keep; detain; stop.

**alım** *n.* Taking; buying, purchase; attraction; range. **~ satım,** business, trade: **göz ~ı,** distance the eye can see. **~lı,** attractive. **~sız,** unattractive, ugly.

**alın, -lnı** Forehead, brow; face; front; boldness, shamelessness. **alnı açık,** honest, honourable: **alnı ak,** innocent, blameless: **alnının akıyle,** without a stain on his character: **alnını* çatmak,** to frown: **~ damarı çatlamak,** to lose all sense of shame: **~ teri,** the sweat of the brow, hard work: **~ yazısı,** destiny: **güneşin alnı,** the heat of the day.

**alıngan** Touchy.

**alınlık** Ornament worn on the forehead; inscription *or* sign on the front of a house.

**alınmak** *vn. pass. of* **almak.** Be taken *etc.*; take offence. **kimse üstüne alınmasın,** let no one think that this refers to him.

**alınteri** *v.* **alın.**

**alış** Action of taking *etc.*, *v.* **almak**; **~ veriş,** buying and selling, trade; connexion; relation.

**alış·ık** Accustomed to; used to; tame; familiar; regular (customer *etc.*); working smoothly (machine). **~kan,** accustomed; familiar: **~lık,** a being accustomed; familiarity. **~mak,** to be accustomed; become familiar; become tame; work smoothly. **~tırmak,** to accustom, familiarize; tame; instruct; train; get into running order. **···i alıştırarak haber**

**vermek,** to break the news gently *to s.o.*
**alışveriş** *v.* **alış.**
**Ali** *prop. n.* Ali. ~ **kıran baş kesen,** bully, despot: ~ **Paşa vergisi,** a present the return of which is demanded: ~ **Veli,** Tom, Dick, and Harry.
**âli** (–⌄) High; exalted. ~**cenab** (––·⌄), magnanimous, generous. ~**lık,** magnanimity, generosity.
**alicengiz** *In* ~ **oyunu oynamak,** to play a dirty trick.
**alikıran** *v.* **Ali.**
**alil** (·⌄) Ill; invalid; blind.
**âlim** (–') Learned; scholar.
**alimallah** (·'·–) By God!; God knows!; I warn you!
**alivre** (*Fr. à livrer*) For future delivery.
**alize** (*Fr. alizé*) Trade-wind.
**alkım** Rainbow.
**alkış** Applause. ~**lamak,** to applaud.
**alkol, -lü** Alcohol. ~**ik,** alcoholic.
**Allah** (·⌄) God. ~ ~!, by God!, Good Lord!: ~**ın** ···**i,** *often used as a euphemism for* 'God-forsaken' *cf. English* 'blessed': ~**tan** *or* ~**tanki,** (i) natural(ly), inborn; (ii) thank Goodness that, it's a good thing that . . .: ~**ü âlem,** probably, God knows (*deprecatingly*): ~ **aşkına,** for the love of God; for God's sake!: ~ **bir,** I swear: ~**indan bulsun!,** may God punish him (*for I cannot*)!: ~ **derim,** words fail me!, it's hopeless!: ⌜~ **dokuzda verdiğini sekizde almaz**⌝, what Fate has decreed cannot be altered: ~**ın evi,** mosque; heart: ~ **göstermesin!,** God forbid!: ~ **için,** *used to strengthen an assertion*: ~**a ısmarladık,** goodbye!: ~**tan kork!,** shame! don't do it!: ~**tan korkmaz,** impious; unjust, cruel: ~ **lâyığını (müstahakkını) versin!,** may God give him his deserts!, curse him!: ~ **ne verdiyse,** (of food *etc.*) whatever is at hand, 'pot-luck': ~**vere,** God grant that, let's hope that ...: ~ **vergisi,** talent: ~ **vermesin!,** God forbid!: ~ **versin,** (i) as ~ **vere**; (ii) *reply to a beggar when refusing alms*; (iii) *an ironical congratulation.*
**allahlık** (·–') Useless but harmless (man); beyond the comprehension of man; beyond human power to remedy; imbecile.
**allâk** Tricky, deceitful; fickle.
**allak bullak** (·'·') Pell-mell; all over the place; in utter confusion, topsy-turvy.
**allâme** (·–') Very learned.
**allem kallem** (·'·') Vague words (*such as 'so-and-so'*); tricks, dodges. ~ **etm.,** to put *s.o.* off with words; to trifle with *s.o.*: **allem etmek kallem etmek,** to try all sorts of tricks to attain one's purpose.
**allı** *In* ~ **pullu** *or* ~ **güllü,** brightly coloured and decorated.
**allık** Redness; red colour; rouge.
**alma** (·') *Word of command to stop rowing,* 'easy all!'
**almak, -lır** *va.* Take; get, obtain; receive; accept; buy; take in, contain; comprehend; conquer, capture. *vn.* Take fire; (of a contagion) to take hold, to affect. **al benden de o kadar,** it's the same with me; 'same here!': **alıp yürümek,** to make swift progress; to become the vogue, to 'catch on': **ateş** ~, to catch fire: **boynuna*** ~, to undertake, to assume responsibility for: ···**in etrafını** ~, to surround: **gözün alabildiği kadar,** as far as the eye can reach; ···**den kendimi alamadım,** I could not help..., I could not refrain from ...: **kız** ~, to marry: **ödünc** ~, to borrow: **söz** ~, to receive a promise: **üstüne*** ~, to take upon oneself: **yol** ~, to take the road, to travel; to advance.
**alman** German. ~**ca** (·'·), German (language). **Almanya** (·'·), Germany.
**alnı** *v.* **alın.**
**alp, -pı** Hero; brave.
**alt, -tı** Lower *or* underpart *of a thing*; underside; bottom. Lower, under. ~ ~**a üst üste,** one on top of the other; ~ ~**a üst üste boğuşma,** tough struggle: ~**ında*,** under him, it *etc.*: ~**tan,** from below: ~**tan** ~**a,** by implication, between the lines; in an underhand way; ~**tan** ~**a anlatmak,** to hint at, to imply: ~**ına almak** *or* ~**ına alıp dövmek,** to give a good thrashing to: ~ **başında,** near, next: ~ **başindan,** from the very bottom: ~**etm.,** to conquer, to overthrow: ~**ına etm.,** to foul one's clothing *or* bed: ~**ından girip üstünden çıkmak,** to put things in disorder; to squander all one's money: ~**ta kalmak,** to get the worst of it; ⌜~**ta kalanın canı çıksın!**⌝, vae victis!, woe to the conquered!: ~**ında kalmamak,** not to be outdone; not to be under an obligation: ~ **olm.,** to be overcome: ~ **tarafı,** the underside; the sequel: '**bunda düşünecek ne var? Alt tarafı**

beş lira', 'Why hesitate? After all, it's only a matter of five pounds': **~ını üstüne getirmek,** to turn topsy-turvy: **ayak ~ı,** the sole of the foot; the ground under the feet.

**altabaşo** (··ˈ·) **~ yakası,** the foot *of a sail.*

**altı** Six. **~lık,** consisting of six (piastres, pounds *etc.*), **~ncı,** sixth.

**altın** Gold; gold coin; wealth. Golden. **~adını bakır etm.,** to bring dishonour upon an honoured name. **~babası, -nı,** immensely wealthy; a Croesus. **~baş,** the name of a choice kind of melon: **~ kefal,** the Golden Mullet (*Mugil auratus*). **~cı,** goldsmith. **~kökü, -nü,** ipecacuanha root. **~lamak,** to gild. **~otu, -nu,** hart's-tongue fern. **~sarısı,** golden blonde. **~top, -pu,** trollius. **~topu, -nu,** (golden ball), *expression used to describe a pretty, chubby baby.*

**altıparmak** Very large **palamut** *q.v.*

**altıpatlar** (··ˈ·) Six-shooter, revolver.

**altışar** Six each; six at a time.

**altlı** *In* **~ üstlü,** topsy-turvy; upside-down.

**altlık** Something put under as a support; pad *put under writing paper.*

**altmış** Sixty. **~altı** (··ˈ·), *a kind of card game*; **~ya bağlamak,** to put *s.o.* off with vague promises; to come to an agreement which satisfies both parties. **~ar,** sixty each; sixty at a time. **~ıncı,** sixtieth. **~lık,** sixty years old; containing sixty parts; worth sixty ... .

**alttaraf** (ˈ··) Continuation; sequel.

**altüst** (ˈ·) Upside-down; topsy-turvy.

**alüfte** (–·ˈ) Tart, cocotte.

**alyans** (*Fr. alliance*) Wedding-ring.

**âm** (–) General; universal; public.

**ama, amma** (ˈ·) But; *sometimes expresses surprise or exaggeration.* **~sı yok** *or* **~sı maması yok,** there's no 'but' about it; 'but me no buts!'

**âmâ** (–ˈ–) Blind. **amâ** (·ˈ–), blindness.

**amac** Target; aim, object.

**âmade** (––ˈ) Ready; prepared. **emre ~,** at (your) disposal; available.

**aman** (·ˈ–) Pardon, mercy. **~!,** help!; mercy!; alas!; for goodness sake: **~ dilemek,** to ask for quarter: **···in önünde ~a gelmek,** to be helpless in the face of ...: **~ vermek,** to grant quarter: **~ zaman dinlemez,** merciless; **~ı zamanı yok!,** there's no getting out of it; you must! **~ın, ~!,** help! **~sız,** pitiless, without mercy.

**Amasya** (·ˈ·) *prop. n.* Amasia. ⌜**~nın bardağı biri olmazsa biri daha**⌝, 'there are as good fish in the sea as ever came out of it', *or* 'it doesn't matter, I can easily get another'.

**ambalaj** (*Fr. emballage*) Packing; wrapping.

**ambar, anbar** Granary; storehouse; magazine; hold *of a ship*; barn; bin *for measuring sand, stones, etc.* **~ ağzı,** hatchway: **~ kapağı,** hatch (on a ship). **~cı,** storekeeper; warehouseman.

**ambargo** (·ˈ·) Embargo.

**amber** Ambergris. **~ balığı,** spermaceti whale: **~ çiçeği,** musk-mallow; musk-seed.

**ambreyaj** (*Fr. embrayage*)* Clutch. **~ yapmak,** to let in the clutch.

**amca, amuca** (ˈ·) Paternal uncle. **~zade** (··–ˈ), cousin.

**amel** Action, deed; work; diarrhœa: **~ olm.,** to have diarrhœa.

**amele** *pl. of* **âmil.** Workers; *also used as sing.* workman, labourer.

**amelî** (··ˈ–) Practical (*as opp. to* **nazarî,** theoretical). **ameliyat, -tı,** *pl. of* **ameliye,** practical deeds (*as opp. to theories*); *as sing.,* surgical operation. **ameliye,** surgical operation; work; chemical process.

**amelmanda, amelimanda** Incapable of work; past work, retired; invalid.

**amenna** (–ˈ–) *interj.* Admitted!, agreed!

**amerikalı** (·ˈ···) *n.* American.

**amerikan** (···ˈ–) *a.* American: **~ bezi,** unbleached calico.

**âmil** (–ˈ) Doing; working; active. Workman; factor; manufacturer.

**amin** (–ˈ) Amen!, so be it! **~ alayı,** *ceremony observed by a school for a new pupil.*

**âmir** (–ˈ) Commanding. Commander; superior; chief. **~lik,** authority; command; a being a chief *or* superior.

**amiral, -li** Admiral. **~gemisi,** flagship.

**amirane** (–·–ˈ) In an imperious *or* commanding manner.

**âmiyane** (–·–ˈ) Vulgar, common.

**amma** (ˈ·) But. *v.* **ama.**

**amme** (–ˈ) The public; everyone.

**âmme** (–ˈ) *fem. of* **âm,** *a.* general; public.

**amortisör** (*Fr. amortisseur*) Shock-absorber.

**ampirik** (*Fr. empirique*) Empirical.

**ampul** (·ˊ) (*Fr. ampoule*) Electric light bulb.
**amuca** (ˋ··) *v.* **amca.**
**amud** (·ˊ) Perpendicular. **~dî** (·–ˊ), perpendicular, vertical.
**amyant** (*Fr. amiante*) Asbestos.
**an** (·) Moment, instant. **bir ~ evvel,** as soon as possible: **her ~,** at any moment.
**ana** Mother; the principal part *or* main body *of a thing.* Main, principal; basic. **~ akçe** *or* **para,** the principal (financial): **~ baba,** parents; **~m babam,** my father and mother; *sometimes* 'my dear fellow!'; **~ baba gün,** great confusion: **~ baba evlâdı,** a darling child: **~dan doğma,** stark naked; **~dan doğma kör,** born blind: **~sını satarım,** (*sl.*) I don't care two hoots!: **~ yarısı,** one who takes the place of a mother.
**anac**[1] Capable of looking after itself (animal *or* child); grown up, big; brought up to a business; experienced; shrewd.
**anac**[2] Stock *for grafting.*
**Ana·dolu** (·ˋ··) Anatolia. **~dolulu** (·ˋ···), **~dollu** (ˋ···), Anatolian.
**anafor** Eddy; something got for nothing; illicit gain; bribe. **~cu,** one who lives at another's expense, parasite; profiteer; one who makes illicit gains. **~lamak** (*sl.*), to make illicit gains, to profiteer.
**anahtar** Key; spanner; electric switch. **~cı,** keeper of keys, warden.
**analık** Motherhood, maternity; a motherly woman; adoptive mother.
**ananas** Pineapple.
**an'ane** Tradition. **~sile,** with full details. **~vî** (···ˊ), traditional.
**anarşi** Anarchy.
**anasıl** (ˋ··) By origin, originally.
**anasır** (·–ˋ) *pl. of* **unsur.** Elements.
**anason** Aniseed.
**anat** (–ˊ), **-tı** *pl.* Shades of meaning.
**anayasa** (·ˋ··) Constitution.
**anayol** (·ˋ·) Main road.
**anayurd** (·ˋ·) Mother country; original home.
**anca** (ˋ·) Hardly, barely; only; just: ⌜**~ beraber kanca beraber**⌝, inseparable *friends etc.*; 'we must stick together'.
**ancak** (ˋ·) Hardly, barely, only, just. But; on the other hand, however.
**and** Oath; vow. **~ bozmak,** to violate an oath: **~ etm.** *or* **içmek,** to take an oath: **~ vermek,** to conjure, to adjure. **~laşma,** treaty, covenant. **~laşmak,** to take an oath with another. **~lı,** who has taken an oath.
**andavallı** (*sl.*) Simpleton.
**andetmek** (ˋ··), **-eder** *vn.* Take an oath; pledge.
**andır·an** Reminiscent of. **~mak,** to bring to mind; to bear a striking resemblance to.
**anfi** (*abb. Fr. amphithéâtre*) Lecture-room.
**angarya** (·ˋ·) Forced labour; unpaid job; hard task; fatigue (*mil.*); burden, infliction. Carelessly, perfunctorily; unwillingly.
**angıt, angut** The Ruddy Sheldduck; (*sl.*) fool.
**angudi** (·–ˊ) Orange-brown.
**anha minha** (ˋ·ˋ·) Approximately.
**anılmak** *vn. pass. of* **anmak.** Be mentioned; be called; be remembered.
**anırmak** *vn.* Bray; boast.
**anıt** Monument; memorial.
**anî** (–ˊ) Instantaneous; momentary; sudden, unexpected. **~de** (ˊ–·), on the instant, instantly.
**anjin** Angina.
**anka** (·ˊ) A legendary bird; phoenix; imaginary thing; will-o'-the-wisp.
**ankesman** (*Fr. encaissement*) Paying-in.
**anket, -ti** (*Fr. enquête*) A series of interviews by a journalist on some particular subject.
**anlamak** *va.* Understand. **···den ~,** to understand, to know about; to appreciate; to find pleasure in: **bu ilâcdan hiç bir şey anlamıyorum,** I don't think much of this medicine: **senin anlıyacağın,** (i) you know what I mean; (ii) in short; to cut a long story short: **söz (lâf) ~,** to be reasonable, sensible: **söz anlamaz,** unreasonable; wilful.
**anlaşamamazlık** (··ˋ···) Misunderstanding.
**anlaşıl·an** So it seems; probably; apparently. **~ır** *a.*, intelligible, clear. **~mak,** to be understood, be evident. **anlaşılır,** it is clear that; it can be inferred that: **şimdi anlaşıldı,** now the matter is clear.
**anlaşma** Agreement, understanding. **~k,** to understand one another; to come to an agreement.
**anlat·ılmak** *vn.* Be recounted; be explained. ⌜**anlatılışa göre fetva verilir**⌝, judgement is given in accordance with the way the case is presented.

~ış, mode of explaining; explanation; description: ⌜~tan ~a fark var⌝, it makes a difference according to how a case is presented. **~mak,** to explain, expound; narrate; make known. **söz ~,** to persuade.

**anlayış** Understanding; intelligence; sagacity. **~lı,** intelligible; intelligent.

**anmak** *va.* Call to mind; mention. **adını hayır ile ~,** to speak well of *s.o.* after his death.

**anne** (ʹ·) Mother. **~ ~,** maternal grandmother. **~ciğim** (ʹ···), dearest mother; mummy!

**anonim** Anonymous. **~ şirket,** joint-stock company.

**anormal** Abnormal.

**ansız** (ʹ·), **~ın** (ʹ··) Sudden; suddenly; without warning.

**antika** (·ʹ·) Antique, rare; comic; eccentric. Figure of fun; hemistitch. **~cı,** dealer in antiques. **~lık,** eccentricity.

**antrasit** Anthracite.

**antre** (*Fr. entrée*) Entry; entrance-fee.

**antrepo** (*Fr. entrepôt*) Bonded warehouse.

**antrenör** (*Fr. entraîneur*) Trainer.

**apaçık** (ʹ··) Wide open; very evident *or* clear.

**apansız** (ʹ··), **~ın** All of a sudden.

**apar topar** (·ʹ·ʹ) With surprising suddenness; headlong; helter-skelter.

**aparmak** *va.* Carry off; make off with.

**apartıman, apartman** Block of flats; flat.

**apartma** Something taken; plagiarism.

**apaşikâr** (ʹ–·–) Very evident.

**apayrı** (ʹ··) Quite different; quite distinct.

**apış** Fork of the body. **~ açmak,** to stand with the legs apart. **~ arası,** *polite for* 'the private parts'. **~ık,** with legs a-straddle; (of a horse) too tired to move; weary; dejected; dazed: **apışıp kalmak,** to be completely nonplussed.

**apiko** (·ʹ·) A ship's having the anchor taut ready to haul up. Ready; on the qui vive, alert, smart.

**apolet, -ti** Epaulet.

**apse** Abscess.

**aptal, aptes** *v.* **abdal, abdest.**

**apukurya** (··ʹ··) Carnival; shrovetide.

**apul apul** (·ʹ·ʹ) In a swaying manner; waddling.

**âr** (–) Shameful deed; shame, bashfulness; modesty. **~ etm.,** to be ashamed: ⌜**~ yılı değil, kâr yılı**⌝, (i) of course one wants to make a profit; (ii) (*sarcastically*) he won't miss a chance of making money!

**ar**[1] (·) (*Fr. art*) Art.

**ar**[2] (·) (*Fr. are*) Are (100 sq. metres = 119·6 sq. yds.).

**ara** Interval *or* space *of place or time*; relation *or* understanding *between two persons*; (*with possessive pronoun and* **-da**) among, amongst; (*with poss. pron. and after two nouns connected by* **ile**) between. **~dan,** in between (of time), meanwhile: **~ları* açık,** their relations are strained, they are not on good terms: **~ya almak,** to surround, to hem in: **~da bir,** here and there: **~larını* bozmak,** to create a rift between ...: **~ bulmak,** to find a way *or* opportunity: **~sını bulmak,** to find a way of arranging a matter; to reconcile: **~da çıkarmak,** *whilst otherwise occupied* to find an opportunity of doing *some necessary thing*: **~dan çıkarmak,** to remove *stg.*; to get *one job* out of the way *in order to be able to do others*: **~dan çıkmak,** to go away from the midst of others, to have nothing more to do with them; not to interfere; to take no further part in *stg.*: **~ya girmek,** to act as mediator; to interfere; to prevent by interference: **~da gitmek,** to pass unnoticed *amongst many others*: **~ya gitmek,** to be lost; to be sacrificed: **~sı* iyi olmamak,** not to like *s.o. or stg.*: **~da kalmak,** to be mixed up in an affair: **~ kapı,** connecting door *between two buildings or rooms*: **birini ~ya koymak,** to put in *s.o.* as intermediary: **~da sırada,** now and again: **~ vermek,** to pause; to make a break; **~ vermeksizin,** without interruption: **bir ~da,** at the same time, simultaneously; in the same place together: **bir ~ya gelmek,** to come together, to meet in one place: **bu ~da,** at this moment; at this place; including, amongst these: **bunlar ~sında,** amongst these: ⌜**işin ~sı soğudu**⌝, the matter has been neglected too long *to do much about it now*: **söz ~sında,** in the course of conversation *or* of a speech.

**arab** Arab; Ethiopian; nigger. **~ aklı,** stupidity: **ak ~,** the true Arab (not black): ⌜**anladımsa ~ olayım**⌝, 'hanged if I understand!': ⌜ne Şamın şekeri ne Arabın yüzü⌝, 'better forgo something pleasant if it involves something unpleasant'. **~sabunu,** soft soap. **~saçı, -nı,** woolly hair; any tangled affair; mess: **~ gibi,** tangled, confused, intricate.

**araba** Carriage; cart; wagon; slide-rest (lathe). **~sını düze çıkarmak,** to put matters straight, to get over difficulties: ⌜**~ kırılınca yol gösteren çok olur**⌝, there is plenty of help to be had when it is too late: **bir ~ lâf,** a lot of talk: **çek ~yı!,** be off with you!, clear out! **küçük ~,** top-slide (lathe). **~cı,** driver, coachman; maker of carts, *etc.* **~lık,** coach-house, cart-shed; cart-load.

**arabağı** (·ˋ··), **-nı** Septum.

**arabî** (··ˍˋ) Arab; Arabian; Arabic.

**Arabistan** (···ˋ) Arabia.

**arabozan** (·ˋ··) Mischief-maker.

**arabulan** (·ˋ··) Peacemaker; mediator.

**Araf** (–ˋ) Purgatory.

**arak** Sweat; raki. **~iye** (·–·ˋ), soft felt cap *worn under a turban*; a kind of fife.

**arak·çı** Pilferer, thief. **~lamak,** to remove *or* steal adroitly; filch.

**aralamak** *va.* Make a space between *two things*; half-open; leave ajar.

**aralık** Space; interval *in place or time*; chink, crevice; opening; passage; interruption, break. Half-open, ajar. **~ (ayı),** (*formerly*) the month of **Zilkade,** (*now*) December: **~ etm.,** to leave half-open, to leave ajar: **aralığa gitmek,** to fall between the cracks, *i.e.* to be lost *or* put away: **~ vermeden konuşmak,** to talk continuously: **bir ~,** for a while; at a suitable moment; at some time or other: **bu ~,** at this time; then; meanwhile.

**aramak** *va.* Seek, look for; look *s.o.* up; search; examine; hope for, long for; miss (feel the absence of). **ağız ~,** to endeavour to get *s.o.* to tell *stg.*: ⌜**bizde para pul arama!**⌝, 'don't think we've got any money!': **birini arayıp sormak,** to make inquiries about *s.o.'s* health, circumstances *etc.*

**aranmak** *vn.* Be sought; be searched; search oneself, one's pockets *etc.*; search one's mind; be sought after; be desired *or* appreciated too late; be missed.

**ararot, -tu** Arrowroot.

**arasıra** (·ˋ··) At intervals; now and then.

**arasız** Without a break, uninterruptedly.

**araşid** Peanut, earth-nut, ground-nut.

**araştırma** Research; investigation. **~k,** to search; investigate; make researches into.

**aratmak** *va.* Make *anyone* regret *or* long for *stg.*

**arattırmak** Cause *s.o.* to search *etc.*; try to get *s.o.* to take *stg. which is not so good as that first offered.* ⌜**gelen gideni arattırır**⌝, 'the newcomer makes one long for his predecessor'.

**arayıcı** Seeker; beachcomber. **~ fişeği,** jumping cracker.

**arayış** *n.* Searching.

**âraz** (–ˋ) Symptoms.

**arazi** (·–ˋ) *pl. of* **arz.** Lands, countries; estates, real property.

**arbede** Quarrel; noise; row.

**ard** Back; behind; hinder part; sequel, end. **~ımca, ~ıma, ~ımda,** behind me, after me: **~ı arası kesilmeden,** continuously, uninterruptedly: **~ı arkası gelmiyen,** endless: **~ına bak!,** look behind you!: **~a düşmek,** to lag behind: **~ında gezmek,** to run after *s.o.*, to try to obtain *stg.*: **~ına kadar açık,** wide open: **~da kalmak,** to remain behind; to survive, to outlive: **~ı kesilmek,** to cease, to come to an end: **···in ~ından koşmak,** to run after ...: **~ı sıra,** one behind the other, in a series; **···in ~ı sıra,** immediately following ...: **~ı sıra yürütmek,** to send as escort *or* companion: ⌜**elinden geleni ~ına koyma!**⌝, 'do your worst!'

**arda** A turner's chisel; a peg *for marking.*

**ardcı** Rearguard.

**ardıc** Juniper. **~ kuşu,** thrush: **~ katranı,** oil of savin.

**ardısıra** *v.* **ard.**

**ardiye** Rent for storage in a warehouse; demurrage; warehouse.

**ardsız** In **~ arasiz,** uninterruptedly.

**arduvaz** (*Fr. ardoise*) Slate.

**argac** Woof *of a tissue.*

**argın** Thin; exhausted. **~lık,** emaciation; exhaustion.

**arı**[1] Bee .**~beyi, -ni,** queen-bee. **~cı,** bee-keeper. **~kovanı** (·ˋ···), **-nı,** beehive. **~kuşu, -nu,** bee-eater.

**arı**[2] Clean; innocent. **~kan** (·ˋ·), pure-blooded; blood (horse). **~lamak,** to cleanse. **~lık,** cleanliness.

**arık** Lean, emaciated; clean, pure, honourable.

**arış** Warp *of a tissue*; forearm, cubit; pole *of an ox-wagon*.
**arıtmak** *va.* Cleanse; purify.
**ârız** (–ʹ) That which occurs *or* befalls; accidental; intercepting, impeding, obstructing. ~ **olm.**, to happen, to befall. **~a**, accident; incident; defect; obstruction. **~alı**, uneven, broken (country); full of difficulties and obstacles; defective. **~asız**, unobstructed; level; free from difficulties; without hitch; without defects, undamaged. **~î** (–·ʹ), accidental; adventitious; casual; temporary.
**arî** (–ʹ) Aryan.
**ârif** (–ʹ) Wise; intelligent; skilled, expert. ⌜**~e tarif ne hacet?**⌝, there is no need to explain to a sensible man. **~ane** (–·–ʹ), skilful, clever (*of an inanimate object or an act*).
**arifane** (–·–ʹ) A picnic *or* feast to which each one supplies his share.
**arife, arefe** Eve *of a festival etc.*; threshold *of an event*.
**arina** Sand for scouring.
**ariyet** (–·ʹ), **-ti**, anything borrowed *or* lent. ~ *or* **ariyeten kullanmak**, to make temporary use of *stg*.
**ariz** (·ʹ) Broad, wide. **~u amik**, fully and in great detail.
**ariza** (·–ʹ) Petition; letter from an inferior to a superior.
**Arjantin** (·ʹ·) Argentina.
**ark, -kı** Irrigation canal.
**arka** Back; back part; reverse side; sequel, end; protection, support; supporter, backer. ~ **~ya**, one after the other: ~ **~ya vermek**, to stand back to back *for mutual protection*: **~dan ~ya**, behind the back, secretly: **bir işin ~sını almak**, to put an end to a matter: **~sından dolaşmak**, to follow up *a matter etc.*: **bir işin ~sına düşmek**, to follow a matter up to its conclusion: **birinin ~sına düşmek**, to go in pursuit of, to run after *s.o.*: **~sını getirememek**, to be unable to carry out an enterprise *which has been started*: **~sından gezmek**, *v.* **~sından dolaşmak**: **~da kalanlar** (**~dakiler**), those left behind: **~sı kesilmek**, to be cut off, terminated; to come to an end; to become extinct: **~sı* olm.**, to have *s.o.* behind one, to have a 'backer', to have 'influence' in one's career: **birine ~ olm.**, *or* **~sını vermek**, to support *or* protect *s.o.*: **~dan söylemek**, to speak of *s.o.* behind his back, to backbite: ~ **tarafı**, continuation, sequel: ~ **üstü yatmak**, to lie on one's back: **~si var**, to be continued; *also* to have 'backing': **~sı yere gelmez**, he can't be 'downed'; he has powerful backers: **ardı ~sı kesilmeden**, incessantly: ⌜**elinden geleni ~na koyma!**⌝, 'do your worst!': **gözü* ~da kalmak**, not to be at rest about a matter; to feel unhappy about *stg.* not done.
**arka·çantası, -nı**, haversack, knapsack. **~daş**, companion, friend; colleague. **~daşlık**, comradeship, friendship. **~lamak** *va.*, to support, to back, to protect. **~lı**, who has supporters *or* 'friends at court'. **~lıç**, porter's saddle. **~lık**, *as* **~lıç**; a kind of jacket; carrier *of a bicycle*; back *of a chair etc.*: **~sıpek**, strong; warmly clad; having protectors.
**arlanmak** *vn.* Be ashamed. **arlanmaz**, shameless, brazen-faced.
**arma** (ʹ·) Armorial bearings; rigging of a ship. ~ **soymak**, to dismantle the rigging: ~ **uçurmak** *or* **budatmak**, (of a storm) to carry away the rigging.
**armador** Rigger.
**armağan** Present, gift.
**armatör** (*Fr. armateur*) Ship-owner; belaying-pin; cleat.
**armoz, armuz** Seam between the planks of a boat. **~lu**, carvel-built.
**armud** Pear; (*sl.*) stupid. ⌜~ **piş ağzıma düş!**⌝, 'pear, ripen and fall into my mouth!' (*purporting to be the prayer of one who hopes to obtain stg. without effort*): ⌜**~un sapı var üzümün çöpü var demek**⌝, to be very hard to please; always to find fault with everything: ⌜**benim elim de ~ devşirmiyor da**⌝, 'my hand is not engaged in picking pears either', *i.e.* I can hit back, I can look after myself.
**arnavud** Albanian. ~ **kaldırımı**, cobbled street. **Arnavudluk,** Albania.
**arpa** Barley. ⌜~ **ektim darı çıktı**⌝, it was a bitter disappointment; I was disillusioned: **bir ~ boyu**, a very short distance: ⌜**atın ölümü ~dan olsun**⌝, you can't have too much of a good thing: **frenk ~sı**, pearl-barley. **~cı**, seller of barley: ~ **kumrusu**, a kind of domestic dove; ~ **kumrusu gibi**, pensive and sad.
**arpacık** Stye; foresight of a gun. ~ **soğanı**, (?) Spanish leek.
**arpalama** (i) Founder (disease of horses' feet); (ii) the curing of a stye

by a charm. **~k,** to be overfed with barley; to be sick.
**arpalık** Field with good barley; barley bin; a kind of fief; mark on a horse's teeth *by which his age can be told.*
**arpasuyu** (·ˈ··), **-nu** Beer.
**arsa** (ˈ·) Building plot.
**arsız** Shameless; impudent; importunate.
**arslan** *v.* **aslan.**[2]
**arş**[1] Forearm; cubit; warp *of a tissue.*
**arş**[2] *mil. command.* March!
**arş**[3] The Throne of God.
**arşın** Turkish yard (= *about 27 in.*). ⌜**(bezi) herkesin ~ına göre vermezler**⌝, 'you won't always be taken at your own evaluation'. **~lamak** *va.*, pace out; measure by yards: *vn.* step out, walk quickly: **sokakları ~,** to loaf, to stroll about. **~lık,** *anything* sold by the yard; *anything* a yard in length.
**arşiv** (·ˈ) Archives.
**art** *v.* **ard.**
**artakalmak** (·ˈ··) *vn.* Remain over. **bin belâdan ~,** to have suffered much.
**artık** Left, remaining; superfluous. What is left, residue, remnant. More. (ˈ·) Well now!, come now!, enough! **~ eksik,** deficiencies; more or less; under or over.
**artırma, arttırma** Act of increasing; economizing; overbidding; auction.
**artırmak, arttırmak** *va.* Increase, augment; economize, save, put by; raise a bid *at an auction*; collect; go too far, exceed the bounds. **boğazdan ~,** to economize by cutting down one's food.
**artma** *verb n. of* **artmak.** Left over, remaining. Residue.
**artmak** *vn.* Increase, multiply; remain over; rise (of prices, the tide *etc.*). **yeter de artar,** enough and more than enough.
**aruz** Prosody.
**arya** (ˈ·) *In* **~ etm.,** to dip flag.
**arz**[1] The earth; land; country.
**arz**[2] Width, breadth; latitude. **~ dairesi,** a parallel of latitude: **~ı şimalî (cenubî),** northern (southern) latitude. **~an** (ˈ·), in width; in latitude. **~anî** (·–ˈ), transverse, cross.
**arz**[3] Representation; petition; expression of opinion; submission *of stg. for consideration.* **~ı hürmet ederim,** I present my respects: **~ ve taleb,** supply and demand. **~etmek** (ˈ··), **-eder,** *va.*, to present *a petition*; submit *a proposal*; offer *an opinion*; express *a sentiment*; present *one's respects.*
**arziye** *v.* **ardiye.**
**arzu** (·ˈ) Wish, desire. **~ etm.,** to wish, to desire: **bir şey yapmak ~sunda olm. (bulunmak),** to have a wish to do *stg.* **~keş, -lu,** wishful, desirous. **~lamak,** desire, long for.
**arzuhal, -li** Written petition *or* representation. **~ci,** writer of a petition; street letter-writer.
**arzullah** *In* **~i vâsia,** a spacious place; vast.
**as**[1] Myrtle.
**as**[2] Ermine (animal).
**as**[3] Ace.
**asâ** (·ˈ) Stick; baton; sceptre.
**asab** Nerve; *pl.* **âsab** (–ˈ). **~î** (··ˈ), nervous; neurotic; irritated; on edge. **~iyeci,** neurologist. **~iyet, -ti,** nervousness; nervous irritation. **~ileşmek,** to have one's nerves on edge, become irritable.
**asalet** (·–ˈ), **-ti** Nobility of birth; performing a duty in person; personal appearance in court. **~en** (·ˈ··), in person; acting as principal (*not as representative*).
**âsan** (–ˈ) Easy.
**asansör** (*Fr. ascenseur*) Lift, elevator.
**asar** (–ˈ) *pl. of* **eser.** Works; monuments; signs, traces, remains; legends. **~ı atika,** ancient monuments, antiquities.
**asayiş** Quiet; repose; security.
**asbaşkan** (ˈ··) Vice-president.
**aselbend** Benzoin. **~ mahlûlü,** friar's balsam.
**ases** Night-watchman; night patrol; (*formerly*) Chief of Police.
**asgarî** Minimum.
**ası**[1] Profit, benefit. **~lı** profitable, advantageous.
**ası**[2] Suspension.
**asığ, asık**[1] Profit; advantage. **~lı,** useful.
**asık**[2] Cross; gruff. **~ yüzlü,** sulky.
**asıl, -slı** Foundation; base; origin; source; root; essential part *or* substance. Real; true; essential; main; original. **aslına bakarsan,** the truth of the matter is: **aslı faslı yok,** entirely unfounded, devoid of all truth; **aslile mukabele etm.,** to compare with the original: **aslı nesli,** his origin and family: **aslı yok,** it is without foundation; untrue.
**asılacak** About to be hanged; gallows-bird, criminal.

**asılı** Hanging; suspended; (arm) in a sling; hanged (executed).
**asılmak** *vn.* Hang; be hung; be suspended; stretch out; lean over; insist; be obstinate; (*with dat.*) cling to; pull.
**asılsız** Without foundation; untrue.
**asılzade** (··–ˈ) Nobleman; aristocrat.
**asım takım** (·ˈ·ˈ) Ornaments.
**asınti** A putting-off, a delaying; unpaid debt.
**asır, -srı** Century; age. **~lık,** (so many) centuries old.
**asi** (–ˈ) Rebel, insurgent. Rebellious; refractory.
**asid** Acid.
**aside** (·–ˈ) A dish made of rice *or* barley meal, meat and okra (**bamya**).
**asîl** (·–ˈ) Of noble birth; noble. **~ memur,** real *as opp. to acting* official. **~lik,** nobility.
**asker** Soldier. **~î,** military. **~lik,** military service.
**askı** Anything hanging *esp.* a bowl for flowers; pendant; bunch of fruit *hung up to ripen*; braces, suspenders; sling (*med.*); king-post; delay, postponement; suspense, doubt. **~ya almak,** to prop up, support: **~da bırakmak,** to leave in doubt *or* suspense: **~da kalmak,** to remain in suspense, to await conclusion: **kâğıdları ~ya çikmış,** their banns have been put up (*before marriage*).
**aslâ** (ˈ–) Never; in no way.
**aslan**[1], **aslen** (ˈ·) Originally; fundamentally; essentially.
**aslan**[2], **arslan** Lion; brave man. **~ payı,** the lion's share: **~ sütü,** raki: **gemi ~ı,** a ship's figurehead: ⌜**her yiğidin kalbinde bir ~ yatar**⌝, everyone has an ambition. **~ağzı, -nı,** snapdragon; stone tap *in the form of a lion's mouth.* **~lık,** great bravery. **~pençesi, -ni,** lady's mantle, alchemilla.
**aslı** *v.* **asıl.**
**aslî** (·–ˈ) Fundamental; radical; essential; substantive (*rank*). **~ye,** *fem. of* **aslî**; Court of First Instance.
**asma** Suspended; pendent. Vine *esp. on a trellis.* **~ kat,** entresol: **~ kilid,** padlock: **~ köprü,** suspension bridge: **~ merdiven,** gangway. **~kabağı, -nı,** a kind of gourd grown on a trellis. **~lık,** place planted with vines. **~yaprağı, -nı,** vine leaves *used in cooking* **dolma.**
**asmak,** *va.* Hang; suspend; put off paying *a debt*; cease *work*; play truant *from school,* 'cut' *a lecture etc.* **asıp kesmek,** to behave in an arbitrary and despotic manner: **astığı astık kestiği kestik olm.,** to wreak one's will, to be answerable to no one for one's actions: **kulak ~,** to pay attention; **kulak asmamak,** to take no notice: **surat ~,** to scowl.
**aso** Ace; star performer.
**asrı** *v.* **asır.**
**asrî** (·ˈ) Modern; up-to-date. **~leşmek,** to modernize. **~lik,** modernization; a being up-to-date.
**assubay** Junior officer.
**astar** Lining; priming, undercoat (*paint*). ⌜**~ı yüzünden pahalı**⌝, 'its lining costs more than the outside material', the game is not worth the candle; **bir işin aslı ~ı,** the real truth of a matter: **aslı ~ı yok,** there is no foundation whatever for it: ⌜**yüz verirsen ~ını da ister**⌝, if you give him an inch he will take an ell. **~lık,** material for lining; suitable for lining.
**asteğmen** (ˈ··) Second-lieutenant. **deniz ~i,** midshipman.
**astragan** Astrakhan.
**asude** (––ˈ) Tranquil; at rest.
**asurî** (–––ˈ), **asurlu** Assyrian.
**asyaî** (··–ˈ) Asiatic.
**aş** Cooked food. **~ (y)ermek,** (*of a pregnant woman*) to long for unusual food: **~ta tuzu bulunmak,** to make a contribution, however small, to *stg.*: **ağzının ~ı mı?,** it is something he could not do: ⌜**azıcık ~ım, kavgasız başım**⌝, not rich but free from care: ⌜**pişmiş ~a soğuk su katmak**⌝, to add cold water to cooked food, *i.e.* to spoil everything at the last moment.
**aşağı** Down; below. Lower; low, common, inferior. The lower part. **~da,** below: **~ya,** downwards, down: **~dan almak,** to begin to sing small; to return a soft answer: **~ görmek (tutmak),** to underestimate; to despise: **~ yukarı,** up and down; to and fro; roughly speaking, more or less; **Ahmed ~ Ahmed yukarı,** it's nothing but Ahmed Ahmed all the time; **kedi ~ kedi yukarı,** *she talks about* nothing but her cat: **baş ~,** headlong; upside-down, topsy-turvy: **beş ~ beş yukarı,** close bargaining: **bir ~ bir yukarı,** up and down, to and fro.
**aşağı·lamak** *va.,* lower, degrade; treat as inferior: *vn.* come down; come lower; be discredited. **~lı,** *in*

~ yukarılı, both upstairs and downstairs. ~lık, lowness; inferior, base, common.
**âşar** (–⌄) *pl. of* **üşür,** Tenths; tithes.
**aşçı** *v.* **ahçı.**
**aşermek** *v.* **aş.**
**aşevi** (⌄..), **–ni, aşhane** (·–⌄) Cookhouse; kitchen; restaurant; soup-kitchen.
**aşı** Graft; vaccination, inoculation.
**aşıboyası** (·⌄...), **–nı** Red ochre paint.
**aşık** Knuckle-bone; helmet. **~ atmak,** *lit.* to play knuckle-bones; to compete: **aşığı cuk oturmak,** to be lucky, to be successful. **~kemiği, –ni,** ankle-bone (astragalus).
**âşık** (–⌄) In love. Lover; wandering minstrel.
**aşı·kalemi** (·⌄...), **–ni** Cutting used for grafting; scion. **~lamak,** graft, bud; inoculate, vaccinate; cool with ice. **~lı,** grafted; inoculated.
**aşındırmak** *va.* Wear out *by friction.* **birinin kapısını ~,** to be always visiting *s.o.*; to pester *s.o.* by frequent visits.
**aşınma** Depreciation, wear and tear (*accountancy*); erosion. **~k,** to be worn away *by friction*; to be effaced; to depreciate.
**aşırı** The space beyond a thing; overseas. Excessive, beyond bounds. **~ gitmek,** to go beyond the bounds, to exceed the limit: **gün ~,** every other day.
**aşırma** Conveyed over; smuggled; stolen. Plagiarism; theft.
**aşırmak** *va.* Convey over a height; pass over; escape (danger); go beyond (limit); smuggle; steal, pilfer; get rid of *s.o.*
**aşırtma** *verb n. of* **aşırtmak.** Passed over; stolen; saddlegirth. **~ topçu ateşi,** indirect fire (*mil.*).
**aşifte** (–·⌄) 'Fast' woman.
**aşikâr** (–·⌄) Clear, evident. **~e** (–·–⌄), clearly; openly, publicly.
**aşina** (–·⌄) Acquaintance. **~lık,** acquaintance; intimacy; expert knowledge: **···le aşinalığı olm.,** to know *s.o.* to speak to.
**aşiret** (·–⌄), **–ti** Tribe.
**aşk, –kı** Love; passion. **~a gelmek,** to get excited: **~ olsun!,** bravo!, well done!; *also expresses disappointment,* that's too bad of you!: **Allah ~ına!,** for God's sake!
**aşketmek** (⌄..), **–eder** *va. In such phrases as*: **bir tokat ~,** to give *s.o.* a box on the ears.
**aşkın** Passing over *or* beyond; excessive; furious, impetuous. **iş başımdan ~,** I am up to the ears in work.
**aşlamak** *v.* **aşılamak.**
**aşmak** *va.* Pass over *or* beyond; (*of a stallion*) to cover. *vn.* Pass the limit; be extravagant; overflow. **iş başından* ~,** to be overwhelmed with work.
**aşna fişne** (···⌄) Coquetry, flirting.
**aşnalık** *v.* **aşinalık. ~ etm.**; to greet, to salute.
**aşure** (·–⌄) The tenth of Muharrem; sweet dish *prepared for that day.*
**aşvades** (·⌄·) Cockle (shell-fish).
**aşyermek** *v.* **aş.**
**at, –tı** Horse. **~lar anası,** large masculine woman: **~ başı beraber gitmek,** to be on a level with, to be neck and neck: **~a binmek,** to ride a horse: ⌜**~ bulunur meydan bulunmaz, meydan bulunur ~ bulunmaz**⌝, *stg.* one needs is sure to be lacking: ⌜**~a deveye değil ya!**⌝, it won't be a great expense: **~ hırsızı gibi,** his appearance is against him: ⌜**~tan inip eşeğe binmek**⌝, to lose position; to come down in the world: **~ kafalı,** stupid: **~ oynatmak,** (i) to have complete mastery of a subject; (ii) to have authority in a place.
**ata** Father; ancestor. **~lar sözü,** proverb.
**atak** Testy; irritable, quick to take offence.
**atalet** (·–⌄), **–ti** Idleness; inertia.
**atcambazı** (⌄...), **–nı** Circus-rider; rough-rider; horse-dealer.
**aterina** (··⌄·) Sand smelt.
**ateş** Fire; heat; ardour, vehemence; fever, temperature. **~ almamak,** to misfire: ⌜**~ almağa mı geldiniz?**⌝, *said to one who comes to a house etc. and only stays a minute*: **~ bacayı (saçağı) sarmak,** *lit.* for a fire to catch the chimney (eaves), *i.e.* to have got beyond control: **~ basmak,** to feel hot *from shame or anger*: **~ten gömlek,** ordeal: **~ kesmek,** to cease fire: **~ pahasına,** at an exorbitant price: **~e vermek,** to set on fire; to cause a panic: **kendini ~e atmak,** to sacrifice oneself.
**ateş·balığı, –nı,** sardine. **~böceği, –ni,** fire-fly; glow-worm. **~çi,** stoker; fireman. **~feşan,** erupting fire. **~gede,** fire-temple. **~in** (–·⌄), hot;

fiery. ~**kayığı, –nı,** a kind of large rowing-boat. ~**kes,** cease-fire. ~**leme,** firing; spark *of a petrol engine.* ~**lenmek,** to catch fire; to fly into a passion. ~**li,** burning, fiery; passionate; having a temperature. ~**pare** (··–ˈ), spark; man full of fire. ~**perest, –ti,** fire-worshipper.

**atfen** (ˈ·) With reference to; considering.

**atfetmek** (ˈ··), **–eder** *va.* Direct, incline; attribute; impute.

**atıcı** Marksman, good shot; braggart, swaggerer. ~**lık,** marksmanship.

**atıf, –tfı** Inclination; an attributing *or* ascribing; the joining of two words *or* two sentences to each other.

**atıfet** (–·ˈ), **–ti** Benevolence; affection; sympathy, pity, mercy. ~**en** (–ˈ··), out of kindness.

**atık** Small churn.

**âtıl** (–ˈ) Idle, lazy; inactive. ~ **kalmak,** to be inactive; to be out of use *or* abandoned.

**atılgan** Dashing, reckless, bold. ~**lık,** dash, pluck.

**atılmak** *vn. pass. of* **atmak.** Be thrown, discharged, thrown away; be discredited; be dismissed *from an office*; burst in *upon a conversation*; rush; hurt oneself on *stg.*

**atım** Discharge (gun); range *of a gun*: charge *of powder and shot*; round. **bir** ~**lık barutu* kalmak (olm.,),** to be down to one's last round, to have nothing in reserve.

**atış** Act of firing *a gun.* ~ **meydanı (sahası),** rifle *or* artillery range.

**atışmak** *vn.* Quarrel; have a tiff.

**atıştırmak** *va.* Bolt (food). *vn.* Begin to snow *or* rain.

**ati** (–ˉ) Future; subsequent. The future. ~**de,** in the future; below (what follows).

**atik** Quick, alert, agile.

**atîk** (·ˉ) Ancient. **asarı** ~**a** ancient monuments,

**Atina** (·ˈ·) Athens.

**atiyye** (·–ˈ) Gift; largesse.

**atkestanesi** (ˈ····), **–ni** Horse-chestnut.

**atkı** Woof, weft; hay-prong; shawl; shoe-buckle.

**atlama** Jump, spring; the omission *or* skipping *of a passage.* ~ **ipi** skipping-rope: ~ **taşı,** stepping-stone.

**atlamak** *va.* Jump over; skip, omit; narrowly escape *a danger etc. vn.* Leap, spring.

**atlanmak**[1] *vn. pass. of* **atlamak.**

**atlanmak**[2] Get a horse; mount a horse.

**atlas** Atlas; satin. **Atlas okyanusu,** the Atlantic Ocean. ~**çiçeği, –ni,** cactus.

**atlatmak** *va.* Cause to jump; pass over; overcome *a danger or an illness*; put off; deceive; put off *or* get rid of *a person* by empty promises.

**atlayış** Jump.

**atlet** Athlete.

**atlı** Rider, horseman. Mounted on horseback; having *so many* horses. ~**karınca,** roundabout, merry-go-round.

**atma** Strap of a shoe.

**atmaca** Sparrow-hawk.

**atmak** *va.* Throw; throw away; drop; put into; postpone; fire (gun); shoot (an arrow *etc.*); blow up (a bridge *etc.*); aim at; give *a kick*; step *a pace*; cast *an imputation*; tell *lies*; drink off *a glass of beer etc. vn.* Splinter, crack, split; (of a gun) to go off; (of a magazine) to blow up; (of the heart) to beat; boast; invent stories; tell lies; (of the dawn) to break; (of a colour) to fade. ⌜**atma Receb!**⌝, 'don't tell such tales to me!': **atıp tutmak,** to rant, declaim, inveigh: ⌜**attığını vuruyor**⌝, he never fails to hit the mark, he always attains his object: **başından*** ~, to reject, to refuse; to get out of a *task*: **can** ~, to desire ardently; to save one's life *by taking refuge*: **söz** ~, to make remarks and innuendos to a girl: **···e taş** ~, to make insinuations against ... .

**atmasyon** (*sl.*) Bragging, boast; bluff.

**atmeydanı** (ˈ·–·), **–nı** Hippodrome; *name of a famous square in Istanbul.*

**atölye** (·ˈ··) (*Fr. atelier*) Workshop; studio.

**atsız** Horseless; on foot.

**atsineği** (ˈ···), **–ni** Horsefly, gadfly.

**attar** *v.* **aktar.**

**atuf** Affectionate; kindly. ~**etlû** (···ˉ), *former official form of address to certain ranks.*

**av** The chase; hunting, shooting, fishing; game; prey; booty. ~ **aramak,** to hunt for game; to search for *stg.* to lay hands upon: ~**a çıkmak,** to go out hunting; to form a line of skirmishers: ~ **havası,** good hunting weather; thieves' opportunity: ~ **uçağı,** fighter plane.

**avadanlık** (Artificer's) set of tools.

**aval**[1] Half-witted.
**aval**[2] *In* ~ **etm.**, to back *a bill.*
**avam** (·⁴) *pl. of* **amme.** The public, the common people. **Avam Kamarası,** the House of Commons. **~firib** (·—·⁄), demagogue.
**avanak** Gullible. Simpleton.
**avans** Advance (money).
**avanta** (·⁴·) Advantage; profit, *esp.* illicit profit. **~cı,** one who makes illicit gains.
**avara** (·⁴·) Cast off!; shove off! (*naut.*); clear out! Free, not in gear. **~ya almak,** to throw out of gear, to disengage: ~ **kasnak,** loose pulley: **~lı kavrama,** loose coupling: ~ **kolu,** lever for striking gear *used with fast-and-loose pulleys.*
**avare** (——⁴) Vagabond; good-for-nothing; out-of-work. ~ **etm.**, to interfere with *s.o.*'s work by *talking etc.* **~lik,** idleness, vagrancy.
**avarya** (·—⁴) Damage to goods in transit; average.
**avaz** Loud voice. **~çıktığı kadar,** at the top of one's voice.
**avcı** Hunter; skirmisher (*mil.*); fighter *plane or pilot.* ~ **kuş,** bird of prey. **~lık,** hunting; shooting.
**avdet, -ti** Return. ~ **etm.**, to return.
**avene, avane** Helpers; accomplices.
**avize** (——⁴) Chandelier; hanging-lamp. ~ **ağacı,** yucca.
**avizo** (·⁴·) Dispatch-boat.
**avköpeği** (⁴···), **-ni** Sporting dog.
**avlak** Good hunting-ground.
**av·lamak** *va.* Hunt; shoot; fish for; entice, allure. *vn.* Go out hunting *etc.* **gafil** ~, to catch unawares: **gönül** ~, to attract: **sinek** ~, to idle about. **~lanmak,** to be hunted; to be caught; to go out hunting.
**avlu** Courtyard.
**avniye** *A kind of* coat *worn in the nineteenth century.*
**avrat, -tı, -dı** Wife; woman.
**avret, -ti** The private parts.
**Avrupa** (·⁴·) Europe. **~lı,** European. **~laşmak,** to become europeanized.
**avuc** Palm of the hand; handful. ~ **açmak,** to beg: ~ **içi,** the hollow of the hand: ~ **içi kadar yer,** a tiny place: **~unun* içine almak,** to have *s.o.* completely in one's power: **~unu* yalamak,** to go away empty-handed: **~unu yala!**, you'll get nothing!: **ele ~a sığmaz,** uncontrollable, intractable: **elinde* ~unda* nesi varsa,** everything he (you) possesses. **~dolusu,** handfuls (of), heaps (of). **~lamak,** to grasp in the hand; to take a handful of.
**avukat, -tı** Advocate; lawyer. Loquacious.
**avunmak** *vn.* Be put off *or* deceived; have the attention distracted; have the mind taken off *stg.*; (of a cow) to be in calf.
**avurd** Hollow inside the cheeks. **~u ~una çökmüş,** with sunken cheeks: ~ **etm.** *or* **satmak,** to give oneself airs: ~ **şişirmek,** to puff out the cheeks with conceit: ~ **zavurd,** bluster, self-assertion: ~ **zavurd etm.**, to assume a blustering and superior air. **~lu,** puffed up; conceited; swaggering.
**Avustralya** (··⁴··) Australia.
**Avusturya** (··⁴··) Austria.
**avutmak** *va.* Keep *a child* quiet by amusing it; delay *or* distract *s.o. by some pretence*; delude; console.
**ay** Moon; month; crescent; beautiful face. ~ **başı,** the first of the month: ~ **dede,** *personification of the moon*: **~ın ondördü,** the full moon: **~ın ondördü gibi,** very beautiful (girl): ~ **parçası,** a beauty: ~ **tutulmak,** for the moon to be eclipsed: **~da yılda bir,** very rarely.
**aya**[1] Palm of the hand; sole of the foot.
**aya**[2] (⁴·) Holy. **Aya Sofya,** St. Sophia.
**ayağınaçabuk** (·⁴···⁴) Agile.
**ayak** Foot; leg; step; foot (measure); stair; round *of a ladder*; outlet *of a lake*; tributary *of a river.* **~la,** afoot, by walking: **~ta,** on foot, standing: **~tan,** 'on the hoof': **···den ayağını* alamamak,** to be unable to give up ...: **···e ayağı* alışmak,** to frequent *a place*: ~ **altı,** just under one's feet; in one's way: ~ **altında kalmak,** to be trodden underfoot: ~ **basmak,** to set foot *in a place*; to put the foot down, to insist: **ayağını* çabuk tutmak,** to hurry: **···den ayağını çekmek,** to give up going *somewhere*: **ayağına* gelmek,** (of luck, success *etc.*) to come to one without any efforts on one's own part: **ayağına* kadar gelmek,** to condescend to visit *s.o.*: **'ayağıma kadar geldi'**, 'he was good enough to come and see me': **'ben senin ayağına gelemem!'**, 'you don't expect me to come and see you, do you?' (*sarcastically*): **kendi ayağıyle gelmek,** to come of one's own accord: **ayağını* giymek,**

to put one's shoes on: ~ **işi,** light work; errand: **(ortalığı) ayağa kaldırmak,** to instigate *or* arouse *or* alarm people: **ayağa kalkmak,** to rise to one's feet, to stand up: **···in ayağına kapanmak (düşmek, sarılmak),** to implore *s.o.* for mercy: **···den ayağını* kesmek,** to cease going *somewhere or other*: **birinin ayağını kesmek,** to prevent *s.o.* from going: **ayağı* kesilmek,** to be prevented from going: **ayağı* suya ermek (girmek),** to understand the position, to realize the truth: ~ **takımı,** the common herd, the rabble: ~ ~ **üstüne almak,** to cross the legs: **ayağı* yere basmamak,** to jump for joy: ~ **yerden kesilmek,** to find some means of transport other than walking: ⌜**bin bir ~ bir ~ üzerinde**⌝, all treading on each other's toes, very crowded: **bir ~ evvel,** as soon as possible: **söz ayağa düşmek,** for affairs of importance to get into the hands of the rabble.

**ayak·altı** (·ˈ··), **-nı,** the sole of the foot; ground under the feet; much frequented place: **~nda,** on this very spot. **~bastı, ~parası,** toll *or* tribute levied on travellers *or* passers-by. **~kabı,** footwear, boots, shoes *etc.*: ⌜**sağlam ~ değil**⌝, he is not to be trusted; a bad lot; (woman) 'no better than she should be'. **~lanmak,** to rise to the feet; to rise in rebellion *or* remonstrance. **~lı,** having feet; on foot; long-legged (animal); movable, ambulatory: ~ **kütübhane,** (a walking library) a very learned man. **~lık,** anything serving as a foot; anything a foot in length; stilts. **~takımı, -nı,** the lower classes, the rabble. **~taş, ~daş,** companion, comrade; accomplice. **~teri, -ni,** doctor's fee; fee *paid to anyone called away from his home to attend to a job.* **~üstü,** standing; in haste; frequented: ~ **anlatılmaz,** it can't be explained all in a moment. **~yolu, -nu,** latrine, water-closet.

**ayal, -li** *pl. of* **aile.** Families; dependants; *as sing.*, wife; family. **evlâdü~ sahibi,** a family man.

**ayan** Manifest, evident. ~ **beyan,** very evident; clearly; in every detail.

**âyan** (–ˈ) Notables; chiefs, senators. ~ **meclisi,** Senate.

**ayar** Standard of fineness *of gold or silver*; carat; gauge; accuracy *of a weighing-machine or watch*; degree; grade; disposition, temper; *in conjunction with another noun can usually be translated by* regulating ..., *e.g.* ~ **musluğu,** regulating cock: ~ **valfı,** regulating valve: ~ **bobini,** tuning-coil (wireless): ~ **etm.,** to regulate *a watch etc.* **~lamak,** to assay; to test; to adjust; to regulate; to verify correctness *of weights* and stamp them. **~lı,** of standard fineness; regulated (watch); adjustable: ~ **bomba,** time-bomb.

**ayar·mak** *vn.* Go astray. **~tmak,** to lead astray, pervert; entice; seduce.

**ayaz** Frost *on a clear winter night or* dry cold *on a winter day.* **~a kalmak,** to miss *stg.* by being too late: **~da kalmak,** to be exposed to frost; to wait in vain. ⌜**~ Paşa kol geziyor**⌝, it's frightfully cold: **işler ~ gitti** (*sl.*), things have gone 'to pot'. **~lamak,** to pass a frosty night in the open; to be cold; to wait in vain. **~lanmak,** (of the night) to become clear and frosty; to catch a chill.

**ayazma** (·ˈ·) Sacred spring.

**aybaşı** (ˈ··), **-nı** The first day of a month; menstruation.

**ayçiçeği** (ˈ···), **-ni** Sunflower.

**aydın** Light; moonlight. Bright; luminous; clear; brilliant; enlightened. **gözünüz ~!,** I wish you joy!, *said on hearing good news concerning the person addressed*: **göz ~a gitmek,** to pay a visit of congratulation. **~latmak,** to illuminate, enlighten. **~lık,** light; daylight; skylight.

**ayet, -ti** Verse of the Koran. **~ülkürsü,** *the name of a verse of the Koran usually recited in times of danger.*

**aygın** *In* ~ **baygın,** languid; half asleep.

**aygır** Stallion; violent man. **deniz ~ı,** sea-horse, hippocampus: **su ~ı,** hippopotamus.

**ayı** Bear; boor; clumsy lout. ~ **gibi,** huge: ⌜**köprüyü geçinceye kadar ~ya dayı demek**⌝, 'to call the bear "uncle" until you have crossed the bridge', to treat anyone with respect as long as he can be useful *or* harmful to you. **~balığı, -nı,** seal. **~cı,** bear-leader; coarse rough man (*esp. in games*).

**ayıb** Shame; disgrace; something 'not done'; fault, defect. Shameful, disgraceful; unmannerly. **~dır!,** for shame!; you ought to be ashamed of yourself: ~ **aramak,** to seek an excuse for blaming: ~ **değil,** there's no harm in it; it's quite permissible: **sormak ~ olmasın,** I hope you don't mind my asking. **~lamak,** to find fault with, censure. **~sız,** without defect; irreproachable; innocent.

**ayık** In full possession of the senses; wide awake; sober *after being drunk.* **~lık,** soberness; recovery *from fainting etc.*; consciousness.

**ayıklamak** *va.* Clear of refuse; clean; clean off; pick and choose, select. ⌜**gel, ayıkla pirincin taşını**⌝, 'come and pick the stones out of the rice!', *i.e.* 'here's a pretty mess!'

**ayılmak** *vn.* Recover *from drunkenness etc.*; come round *after fainting.*

**ayırd** Difference. **~etm.,** to distinguish; make a difference between. **~lamak,** to separate; distinguish; select.

**ayır·mak** *va.* Separate; sever; disconnect; set apart, reserve; choose; distinguish; discriminate. **···den göz ayırmamak,** not to take the eyes off *stg.*: **yerinden ~,** to move from its place. **~tmak,** to put aside; reserve.

**ayışığı** (ˈ···), **-ni** Moonlight.

**âyin** (–ˈ–) Rite; ceremony.

**ayine** (––ˈ) Mirror; *v. also* **ayna.** ⌜**~si iştir kişinin, lâfa bakılmaz**⌝, a man's worth is known by his deeds, not his words: ⌜**bakalım, ~yi devran ne gösterir**⌝, let's see what the future will show.

**aykırı** Crosswise, transverse, athwart; incongruous; perverse; eccentric. **···e ~,** not in accordance with ...: **~ gitmek,** to swerve from the straight path; to go on in an extravagant manner.

**aylak** Unemployed; involuntarily idle. Loafer, tramp. **~çı,** casual labourer.

**aylık** Monthly; a month old; lasting a month; monthly salary. **~çı,** one who receives a monthly salary.

**ayn** Exact copy, counterpart. **~ını çıkarmak,** make a copy of *stg.*

**ayna** Mirror; lathe chuck; *v. also* **ayine. üç ayaklı ~,** 3-jaw chuck. **~lı,** having a mirror; (*sl.*) fine, excellent. **~lık,** transom *of a boat*: **~ tahtası,** the back of a stern thwart. **~sız** (*sl.*), unpleasant; policeman.

**Aynaroz** Mount Athos.

**aynen** (ˈ·) In exactly the same way; textually; in kind (payment).

**aynı** (ˈ·) *v.* **ayn.** Identical; the same; veritable. **~ zamanda,** at the same time.

**aynî** (·ˈ–) Ocular; (*legal*) real *as opp. to personal.*

**aynile** (ˈ··) *v.* **aynen.**

**ayniyat** (··ˈ), **-tı** Goods in hand; inventory; payment in kind.

**ayol** (ˈ·) *Friendly but slightly reproachful acclamation.*

**ayran** *Drink made with* **yoğurt**; buttermilk. **~ı kabarmak,** to lose control of oneself from *temper or drink*; to be infuriated: ⌜**sütten (çorbadan) ağzı yanan ~ı üfler de içer**⌝, ⌜a scalded cat fears cold water⌝.

**ayrı** Apart; separated; alone; isolated; other. Separately; one by one; by oneself. **~ basım,** reprint (of an article *etc.*): **~ gayrı yok,** no need for ceremony; indiscriminately, without any exception, all alike: **~mız gayrımız yok,** we have everything in common; we don't stand upon ceremony: **~ seçi,** difference, disparity: **~ seçi yapmak,** to discriminate.

**ayrı·ca** (ˈ··), separately; apart; otherwise; in addition. **~k,** separated; wide apart (eyes); parted (hair): **~ otu,** couch grass. **~lık,** separation; isolation; difference; absence: **~ çeşmesi,** a well outside a village, *where it is customary to bid farewell to those departing.* **~sız,** indiscriminately.

**ayrılmak** *vn.* Be separated, parted; differ, be distinguished; disagree; depart; crack, split open.

**aysberg** Iceberg.

**Ayşe** *A name of women, esp. of Mohammed's second wife Ayesha.* **~ kadın fasulyesi,** French beans.

**ayva** Quince. **ekmek ~sı,** a juicy kind of quince. **~lık,** quince orchard.

**ayvaz** Steward, major-domo. ⌜**~ kasab hep bir hesab**⌝, it all comes to the same thing.

**ayyar** (·ˈ–) Crafty, deceitful.

**ayyaş** Given to drink; toper.

**Ayyuk** (·ˈ–), **-ku** The star Capella; the highest point of the heavens. **sesi ~a çıkıyordu,** he was shouting at the top of his voice.

**az** Little; few; insignificant. Seldom. **~ buçuk,** somewhat: **~ buz şey değil,** it is no small matter: **~ çok,** in some degree, more or less; a certain number; **~a çoğa bakmamak,** to make the best of it: **~ daha (kaldı),** all but: **~ görmek,** to deem too little: **~ günün adamı değil,** he is a man of experience: **bir ~,** a little, rather: **bir ~dan, in a moment, soon.**

**âza** (–⁻́) *pl. of* **uzuv.** Limbs; members; *as sing.* member. **···e ~ kaydolunmak,** to be enrolled as a member of ... .

**azab** (·⁎) Pain; torture. **~ vermek,** to cause pain *or* annoyance: ⌜**şeytan ~da gerek**⌝, it serves him (me) right!; (*hum.*) 'no peace for the wicked!'

**azacık** (ˋ··), *v.* **azıcık.**

**azad** Free; freeborn. Liberation; dismissal *of children from school*; the giving of a holiday. **~e** (——ˋ), free; released; absolved: **~etmek** (·ˋ··), **-eder, ~lamak,** to set free, emancipate; to let *children* out of school. **~lı,** set at liberty. **~sız,** who cannot be freed.

**âzalık** (——ˋ) Membership.

**azal·mak** *vn.* Diminish; be reduced; be lowered. **~tmak,** *va.* to diminish; lessen; lower.

**azamet, -ti** Greatness; grandeur; pride, conceit. **~li,** great, magnificent, imposing; proud, ostentatious.

**âzamî** Maximum: in the greatest degree.

**azar**[1] Reprimand; reproach. **~lamak,** to scold, reprimand, reproach.

**azar**[2] **azar** (·ˋ·ˋ) Little by little, gradually.

**azdırmak** Lead astray; spoil (a child); rouse, excite; allow a small evil to become serious.

**Azerî** (—·⁻́) Native of *or* belonging to Azerbaijan.

**azgın** Furious; mad; unbridled; extreme, excessive; rebellious; astray; (river) in flood.

**azı, azıdişi** (·ˋ··) Molar tooth; tusk. **gemi ~ya almak,** to take the bit between the teeth, to get out of hand.

**azıcık** (ˋ··) *Dim. of* **az.** A very little.

**azık** Food; provisions; fodder.

**azılı** Furnished with tusks; ferocious, savage; dangerous; violent; unbridled.

**azımsamak, azınsamak** Think *stg.* insufficient.

**azınlık** Minority.

**azışmak** *vn.* Become aggravated.

**azıtmak** *vn.* Grow excessively; overflow; rebel; go astray; go too far; become unreasonable; become insolvent. *Sometimes va. as in* **işi azıttı,** he went too far. **kedi (köpek) ~,** to take an unwanted cat (dog) right away and leave it in a distant place.

**azil, -zli** Dismissal; removal *from office.*

**azim, -zmi** Resolution, determination; firm intention. **~kâr** (··⁻́), resolute, determined.

**azîm** (·⁻́) Great, vast, immense.

**azimet** (·—ˋ), **-ti** Departure; setting out on a journey.

**aziz** (·⁻́) Dear; precious; rare; saintly. Saint. **-e,** *fem. of* **aziz. ~lik,** practical joke; annoyance caused for amusement.

**azletmek** (ˋ··), **-eder** *va.* Dismiss *from an office.*

**azlık** Scarcity; paucity; minority.

**azlolunmak** (ˋ···) *vn.* Be dismissed; be placed in retirement.

**azma** Hybrid; monstrosity.

**azmak** *vn.* Go astray; become furious, mad, unmanageable; become depraved; be on heat, rut; (river) be in flood. *n.* River in spate; overflow of a river dam. **düş ~,** to have nocturnal emissions.

**azman** Monstrous, enormous; hybrid.

**azmetmek** (ˋ··), **-eder** *va.* Resolve upon; intend.

**azmış** Enraged; excited; furious; on heat.

**azot, -tu** Nitrogen.

**Azrail** (·—⁻́) The Angel of Death.

**azvay** (·⁻́·) Aloes.

# B

**bab** (–) Chapter; subdivision; class, category; connexion; subject. **bir ~ ev,** one house: **bu ~da,** on this head, in this connexion.

**baba** Father; forefather; venerable man; head of a religious order; a kind of fit *peculiar to negroes*; knob; post of a staircase; bollard. **~larımı ayağa kaldırma!,** do not infuriate me!: **~ dostu,** an old friend of the family: ⌜**~sının hayrına değil ya**⌝, not 'just for love', not in a disinterested manner: **~dan kalma,** inheritance; inherited: **~ ocağı,** family house: **~m da olur** *or* **bunu ~m bilir** (**yapar** *etc.*), that's easy; anyone knows (can do *etc.*) that: **~na rahmet!,** *a form of thanks*: **~sı tutmak** *or* **~ları üstünde olm.,** (*of negroes only*) to have a kind of fit; to run amok: **~ yurdu,** paternal home: **ha ~m ha!,** *an expression of encouragement or emphasis*: **iskele (Şam, trabzan) ~sı,** *terms used of a father whose authority is not respected*: **konuşur ~m konuşur,** he talks and talks.

**baba·anne,** paternal grandmother. **~can,** kindly, good-natured, easy-going (man). **~ç,** big cock; turkey-cock; swaggering. **~hindi** (·ˋ··), turkey-cock. **~lı,** having as father ...; running amok; irascible: **dokuz ~,** bastard. **~lık,** paternity; fatherly affection; adoptive father; guardian; simple old man; good and sincere; simple, ingenuous. **~yani** (··–ˋ), fatherly, unpretentious, free-and-easy; shabby. **~yiğit,** full-grown strong young man, 'stout fellow'; brave; virile: **···lerin en ~i,** the best of....

**Babıâli** (–·–ˊ) The Sublime Porte.

**Babil** (ˊ·) Babylon; Babel.

**baca** Chimney; skylight; funnel *of a steamship*. **~ başı,** stone mantelpiece: **~ kulağı,** small stone shelf by the side of a fireplace: **~ tomruğu,** the part of the chimney above the roof: **~sı tütmez,** (of a family) in an evil plight, destitute: **kapı ~ açık,** open to all, unguarded: ⌜**kapıdan atsalar ~dan düşer**⌝, a bore from whom one cannot escape: **kapısı ~sı yok,** a tumbledown or jerry-built building.

**bacak** Leg; thigh; knave (cards). **~ kadar,** knee high (of a child): **~ kalemi,** fibula: ⌜**her koyun kendi bacağından asılır**⌝, everyone must answer for his own misdeeds: **~lı,** long-legged; having *such-and-such* legs. **~sız,** short-legged; squat; naughty (child).

**bacanak** Brother-in-law (*only of one of two men who have married sisters*).

**bacı** Elder sister; wife; *title of respect for elderly women*; negro nurse; wife of the chief of a religious order.

**baç, –cı** Toll; tax; customs duty.

**badana** (·ˋ·) Whitewash; distemper. **~lamak,** to whitewash *or* distemper. **~li,** whitewashed, distempered.

**badaş** Grain mixed with earth and stones *left on the ground after hreshing.*

**badaşmak** *vn.* Come to an agreement; make up a quarrel, be reconciled; get on well with one another.

**bade**[1] (–ˋ) Wine-glass; wine; raki.

**bade-**[2] (ˊ·) *prefix.* After ... . **~hu** (ˊ·–), after that, then: **~ma** (ˊ··), henceforth.

**badem** (–ˋ) Almond. Almond-shaped. **~cik,** tonsil. **~ezmesi, –ni,** almond paste, marzipan. **~içi, ~ni,** kernel of the almond. **~lik,** almond orchard. **~şekeri, –ni,** sugar almonds. **~yağı, –nı,** almond oil.

**badi** A kind of duck. **~ ~,** waddling.

**bâdi** (–ˊ) Causing; originating. *n.* Beginning.

**badik** Tiny. **~lemek,** to waddle.

**badire** (–·ˋ) Unexpected calamity; difficult situation.

**badya** (–ˋ) Large wooden *or* earthen bowl; tub.

**bafon** *Corruption of* **vakfon** *q.v.*

**bağ**[1] Bond, tie; bandage; impediment; restraint; bundle. **ayak ~ı,** tether; moral restraint: **ayağının ~ını çözmek,** to divorce one's wife.

**bağ**[2] Vineyard; orchard; garden. **~ bahçe sahibi,** a man of property: **~ bozumu,** the vintage: **~ budamak,** to prune a vineyard: ⌜**bakarsan ~ olur bakmazsan dağ olur**⌝, if you look after your property it will be a garden, if not, a wilderness; success depends upon effort. **~cılık,** viniculture.

**bağa** *Generic name for batrachians and chelonians*; tortoise; tortoiseshell.

**Bağdad** (ˋ·) Baghdad. ⌜**~ harab oldu**⌝, I am very hungry: ⌜**sora sora ~ bulunur**⌝, you can get anywhere by asking: ⌜**yanlış hesab ~dan döner**⌝, an error will be found out sooner or later. **~i,** made of lath and plaster.

**bağdalamak** *va.* Trip (in wrestling).

**bağdaş** Sitting cross-legged *in oriental fashion*. **~ kurmak,** to sit thus.

**bağdaşmak** *v.* **badaşmak.**

**bağı** Spell, incantation. **~cı,** sorcerer.

**bağır, –ğrı** The middle part of the body; bosom; internal organs; middle *or* front *of anything*. **bağrına taş basmak,** to restrain *or* hide one's grief: **bağrı yanık,** afflicted, distressed: **göğüs ~ açık,** with the shirt open; carelessly dressed; disorderly.

**bağırış** *n.* Shouting, yelling. **~mak,** to shout together; to clamour, bawl, raise an outcry.

**bağır·mak** *vn.* Shout, yell; cry out. **~tı,** a shouting.

**bağırsak** Intestine; (*comm.*) casing.

**bağırtlak** The Desert Grouse (*Pterocles*), *also used for the* Sand Grouse (*Syraptes paradoxus*).

**bağış** Gift; donation. **~lamak,** to give gratis; to give in charity, donate; pardon; spare the life of: **adını* ~,** to tell one's name: **gencliğine ~,** to put *stg.* down to one's youth *and forgive on that account.*

**bağlama** *verb. n. of* **bağlamak**; a kind of lute. Tied, bound; *in conjunction with another noun,* connecting ..., coupling ... .
**bağlamak** *va.* Tie, bind, fasten, attach; bandage; form (skin, ice, seed *etc.*); assign (an income *etc.*); do up (parcel); dam (a stream). **...e ~,** to end up by ... : **başını ~,** to give in betrothal or marriage: **başını* bir yere ~,** to find a job for *s.o.*; to save him from destitution: **cerahat ~,** to suppurate: **...e gönül ~,** to set one's heart on; to fall in love with: **göz ~,** *err. for* **göz bağmak,** *q.v.*: **kabuk** *or* **yüz ~,** (of meat *or* milk) to form rind *or* skin; (of a wound) to heal up: **bir işi mukaveleye ~,** to make a contract about *stg.*: **şarta ~,** to make *stg.* depend upon a condition, *i.e.* to attach a condition to *stg.*: **tane ~,** (of a plant) to form seed: **yağ ~,** to put on fat.
**bağlanmak** *vn. pass. of* **bağlamak.** Be tied *etc.*; be obliged; be engaged *to do stg.*: **basireti* ~,** to be blind *or* to lack perception as to one's own interests: **kızının kısmeti bağlandı,** somehow the girl has never found a husband.
**bağlı**[1] Bound; tied; impotent; settled, concluded (agreement *etc.*); assigned (income *etc.*): **...e ~,** dependent on, connected with: **adama ~,** depending on the man concerned: **başı ~,** married; settled; connected *with some office etc.*: **eli kolu ~,** tied hand and foot; who has cut the ground from under his own feet by some injudicious admission: **gözü ~,** blindfolded; bewitched; stupid; inexperienced.
**bağlı**[2] Having vineyards. **~ bahçeli,** rich, well-to-do.
**bağlılık** A being bound; attachment, affection.
**bağmak** *va.* Bewitch. **göz ~,** to hoodwink.
**bağrı** *v.* **bağır.**
**bağrışmak** *vn.* Shout together.
**baha** *v.* **paha.**
**bahadır** (·–') Brave. Hero.
**bahane** (·–') Pretext; excuse. **~sile,** on the pretext that: **~ etm.,** to make an idle excuse.
**bahar**[1] (·'–) Spring (season), *more usually* **ilk~,** (*as opp. to* **son~,** autumn).
**bahar**[2] (·'–) Spice. *pl.* **~at, -tı,** spices. **~lı,** spiced, aromatic.
**bahçe** Garden. **nebatat ~si,** botanical garden: **hayvanat ~si,** zoological garden. **~lik,** full of gardens; pertaining to a garden; garden-plot.
**bahçivan** (··') Gardener.
**bahir, -hri** Sea; a poetical metre.
**bahis, -hsi** Discussion; inquiry; subject of discussion *or* inquiry; bet, wager. **bahsi geçen,** above-mentioned: **ne bahsine istersen girerim,** I'll bet anything you like: **~ tutmak,** to bet.
**bâhname** (––') Pornographic *or* obscene writing.
**bahr·i** *v.* **bahir. ~î,** maritime, naval, nautical.
**bahriye** *fem. of* **bahrî.** Navy. **~ Nezareti (Vekâleti),** the Ministry of Marine: **~ zabiti,** naval officer. **~li,** sailor belonging to the navy.
**bahs·etmek** ('··), **-eder** *va.* (*with abl.*) Discuss; mention; treat of; bet. **~i,** *v.* **bahis: ~ müşterek,** totalizator, 'pari mutuel'.
**bahş** Gift. **~etmek** ('··), **-eder,** to give, grant; forgive, remit. **~iş,** tip, gratuity; gift.
**baht, -tı** Luck, fortune; good luck; destiny. **...den ~ı* açık olm.,** to be lucky in ... : **~ı kara,** unlucky: **ne çıkarsa ~ına,** *v.* **peşrev. ~iyar** (··'–), lucky; happy. **~iyarlık,** good fortune; happiness.
**bahusus** ('–··) Especially; particularly; above all.
**bakakalmak** *v.* **bakmak.**
**bakan** Minister. **~lık,** Ministry.
**bakaya** (·–'–) *pl. of* **bakiye.** Residues, arrears.
**bakıcı** Soothsayer; attendant; nurse.
**bakılmak** *vn. pass. of* **bakmak.** Be attended to; be looked after; be considered; be followed as a rule; (of a lawsuit) to be heard.
**bakım** Look, glance; attention; upkeep. **bir ~dan** *or* **~a,** in one respect; from one point of view. **~evi, -ni,** crèche; hospital for children; home for old people. **~lı,** well-cared for; plump. **~sız,** neglected.
**bakındı** ('··) Just look!
**bakınmak** *vn.* Look around; be bewildered.
**bakır** Copper; copper utensil. **~ çalmak,** to taste of copper *from being cooked in an untinned copper vessel*: **~ pası,** verdigris: ⌜**yer demir gök ~**⌝, a hopeless outlook. **~cı,** coppersmith.

**bakış** Look; care. **kuş ~ı,** bird's-eye view. **~lı,** having *such and such* a way of looking, *e.g.* **cin ~,** having a look of malicious cunning: **şahin ~,** hawk-eyed.
**baki** (–⌐) Permanent, enduring, lasting; remaining. Remainder. Finally; as to the rest.
**bakir** (–⌐) *a.* Virgin; untouched. **~e,** *n.* virgin.
**bakiye** Remainder, residue; arrears; balance; sequel.
**bakkal** Grocer. **~iye,** groceries.
**bakla** Bean, *esp.* broad-bean; link (of a chain). **~yı ağzından* çıkarmak,** to let out something that one has hitherto kept back: **~ dökmek,** to tell fortunes by beans: **ağzında ~ ıslanmaz,** indiscreet talker, chatterbox: **senin ağzında bir ~ var,** you are keeping something back. **~çiçeği,** a dirty yellowish-white colour. **~giller,** leguminous plants. **~kırı,** dapple-grey (horse).
**baklava** A kind of pastry. Lozenge-shaped.
**bakliye** Leguminous. *pl.* **bakliyat,** legumes, peas and beans.
**bakmak** *vn.* & (*with dat.*) *va.* Look; look at; examine; look after, tend; attend to, see to; look to for guidance; face towards, look out on. **bakalım,** we'll see! (*according to context implies doubt, hesitation, threat, or encouragement*): **bakarsın,** in case; it may be: **bakar mısın?,** *a polite way of calling attention*: **baka kalmak,** to stand in astonishment *or* bewilderment: **bakar kör,** a blind man whose eyes appear normal; one who fails to perceive what is going on around him: **bana bak!,** look here!; hi!: **bana bakma!,** never mind what I do!: **işe ~,** to attend to one's job: **kusura bakma!,** excuse me!: **yüzüne* bakmamak,** to have no further consideration for *s.o.*, to have nothing more to do with *s.o.*: **bu iş yüz liraya bakar,** this is a matter of say, £100.
**bakrac** Copper bucket.
**baksana** (·⌐·) Look here!; listen to me!
**bakteri** Bacterium, microbe.
**baktırmak** *va. caus. of* **bakmak.** Make *or* let look; show; cause to look after.
**bal, -lı** Honey; syrup. **~ alacak çiçek,** one from whom something is likely to be had; an act likely to be profitable: **~ başı,** the purest honey: ⌜**~ dök yala**⌝, very clean (*cf.* a floor 'off which one could eat one's meal'): **~ gibi,** pure as honey, unadulterated; easily, properly, certainly: **~ gibi yalan,** a lie pure and simple: ⌜**~ tutan parmak yalar**⌝, 'who touches honey licks his fingers', he is getting something out of it: **ağzından ~ akıyor,** he is quick-witted and intelligent: ⌜**ağzına bir parmak ~ çalmak**⌝, to try to silence a man with a grievance by offering him something quite inadequate in return *or* by some empty compliment: ⌜**bir eli yağda, bir eli ~da**⌝, very comfortably off, well-to-do: **bir parmak ~ olm.,** to be the subject of gossip: **sözünü ~la kestim,** excuse my interrupting!
**bala** Child, baby.
**balaban** Tame bear; enormous man *or* animal; large drum. **~ kuşu,** bittern. **~laşmak,** to become very large *or* fat.
**baladör** Sliding-gear.
**balan** Porch.
**balar** Thin board.
**bal·arısı** (⌐···), **-nı** Honey-bee. **~ayı** (⌐··), **-nı,** honeymoon. **~cı,** dealer in honey.
**balçak** Guard of a sword-hilt.
**balçık** Wet clay; potter's clay; sticky mud. **~ hurması,** crushed dates: ⌜**güneş ~la sıvanmaz**⌝, 'the sun cannot be coated with clay'; ⌜truth will out!⌝
**baldır** Calf of the leg. **~ı çıplak,** a rough, rowdy: **~ kemiği,** tibia, shin-bone. **~ak,** the lower part of the trouser leg. **~ıkara,** maidenhair fern.
**baldıran** Hemlock.
**baldız** Wife's sister, sister-in-law.
**balgam** Mucus; phlegm. **~ taşı,** jasper: **bir ~ atmak,** to drop a malicious hint.
**balık** Fish. ⌜**~ baştan kokar**⌝, 'the fish begins to stink at the head', *i.e.* corruptions starts at the top: **balığa çıkmak,** to go out fishing. **~ağı, -nı,** fishing-net. **~ane, ~hane,** custom-house for dues on fish brought into Istanbul market. **~çı,** fisherman; fishmonger. **~çıl,** piscivorous; heron. **~eti,** well fleshed, neither fat nor thin. **~lama,** dive (head foremost); headlong. **~lava,** fishing-ground. **~nefsi, -ni,** spermaceti. **~pulu, -nu,** fish-scale. **~sırtı, -nı,** ridge; road with a steep camber; hog-backed (road). **~yağı, -nı,** fish-oil;

cod-liver oil. ~**yumurtası, -nı,** fish-roe.
**baliğ** (–ˈ) Reaching; amounting (to); adult. Amount, sum. ~ **olm.,** to reach the age of puberty: ···**e** ~ **olm.,** to amount to ... .
**balina** (·ˈ·) Whale; whalebone. ~ **çubuğu,** whalebone.
**balkabağı** (ˈ···), **-nı** Sweet gourd; idiot.
**balkan** Chain of wooded mountains. **Balkan yarımadası,** the Balkan Peninsula.
**balkımak** *vn.* Shimmer, glitter; throb slightly.
**ballan·dırmak** Smear with honey; extol extravagantly. ~**mak,** to become thick like honey; become sweet and attractive.
**ballı** Honeyed. ~**baba,** dead-nettle.
**balmumu** (ˈ··), **-nu** Wax. ~ **yapıştırmak,** to stick wax *on stg.*; to note down *a thing to be remembered*; to mark *s.o.* down *for revenge or punishment*: **kırmızı dipli** ~**ile davet,** (an invitation with red sealing-wax), a formal invitation (*only used ironically*); ⌜**kırmızı dipli** ~**ile çağırmadım ya**⌝, 'nobody asked you to come!'
**balo** (ˈ·) Ball, dance.
**balon** Balloon; ship's fender; balloon flask.
**baloz** Low-class 'café chantant'.
**balpeteği** (ˈ···), **-ni** Honeycomb.
**balsıra** Manna; a disease of tree leaves.
**balta** Axe. ~**yı asmak,** to worry, to have a down on, to keep on at *s.o.*: ~ **burun,** a curving nose: ~ **görmemiş (girmemiş) orman,** virgin forest: ~ **olm.,** *or* ~**yı kapıya asmak,** to be tiresome, to bore, to pester: **bir** ~**ya sap olm.,** to find a career: ~**yı taşa vurmak,** to do *or* say something unseemly; to make a 'faux pas': ~**sı varmış,** (of a girl) she has a young man. **balta·baş,** the bow of a straight-stemmed ship. ~**cı,** maker *or* seller of axes; wood-cutter; sapper *or* pioneer (*mil.*); halberdier in the old Palace Guard of the Sultans; fireman equipped with axe. ~**lamak,** to cut down with an axe; extirpate, demolish; cut away (hopes *etc.*); sabotage. ~**layıcı,** saboteur. ~**lık,** district within which the inhabitants of a village have the right to cut wood.
**balya** (ˈ·) Bale ~**lı,** baled.
**balyemez** An ancient kind of gun.
**balyos** *Old title of Venetian Ambassador, applied vulgarly also to foreign consuls.*
**balyoz** Sledge-hammer.
**bambaşka** (ˈ··) Utterly different.
**bambu** Bamboo.
**bamteli** (ˈ··), **-ni** Bass note; whiskers, imperial; vital point; sore spot. ~**ne basmak,** to tread on *s.o.'s* corns, to touch him on a sore spot.
**bamya** (ˈ·) Okra, gumbo.
**bana** *dat. of* **ben.** To me. ~ **bakma,** *v.* **bakmak**: ~ **mısın dememek,** to be very thick-skinned; to have no effect.
**bandıra** (ˈ··) Flag (*not used of Turkish flag*).
**bandırma** Sweetmeat *made of grape-juice.*
**bandırmak** *va.* Dip into a fluid, soak.
**bandırol, bandrol** Stamped paper *attached to certain monopolies.*
**bando** (ˈ·) Musical band.
**bangır bangır** Yelling, screaming.
**bâni** (–ˈ) Builder, constructor; founder.
**banka** (ˈ·) Bank (*comm.*). ~**cı,** banker.
**bankiz** (*Fr. banquise*) Ice-floe, ice-pack.
**banliyö** (*Fr. banlieu*) Suburb.
**banmak** *v.* **bandırmak.**
**banotu** (ˈ··), **-nu** Henbane.
**banyo** (ˈ·) Bath, developing dish; development *of a negative.*
**bâr** (–) Burden. ···**e** ~ **olm.,** to be a burden to ... .
**bar**[1] (·) Drinking bar; low cabaret.
**bar**[2] (·) *Name of a popular dance in E. Anatolia.*
**bar bar** (ˈ·) *In* ~ ~ **bağırmak,** to shout loudly, to keep on shouting,
**baraj** Barrage (dam); barrage (*mil.*).
**barak** Long-haired (animal); thick-piled (stuff). Plush; a kind of climbing vine.
**baraka** (·ˈ·) Hut, shed.
**barbar** Barbarian.
**barbunya** (·ˈ·) Red Mullet; a kind of bean.
**barçak** Guard of a sword hilt.
**barda** Cooper's adze.
**bardak** Drinking mug; glass for water. ~ **eriği,** egg-plum: **bir bardak suda fırtına,** a storm in a tea-cup: ⌜**bunun üstüne bir** ~ **soğuk su iç!**⌝, you can write that off!, you'll never see it again!
**barem** (*Fr. barême*) Ready reckoner; scale of official salaries. ~ **kanunu,**

law regulating official salaries: **hayat ~i,** cost of living.
**barı·mak** *va.* Shelter; shield; assist. **~nak,** shelter; hiding-place. **~ndırmak,** to give shelter to, to lodge (*va.*). **~nmak,** to take refuge *or* shelter; to lodge (*vn.*).
**barış** Reconciliation; peace. **~ (görüş) olm.,** to be reconciled. **~ık,** at peace; reconciled; in agreement. **···le yıldızı ~ olm.,** to be on good terms with ...: **⌜yedi kıralla ~⌝,** on good terms with everyone, *but more frequently used of* a woman with many lovers. **~mak,** to be reconciled; to make peace.
**bari** (-́·) For once; at least.
**barikad** Barricade.
**bâriz** (–́·) Manifest; prominent.
**bark** *Only in* **ev ~ sahibi** *or* **evli ~lı bir adam,** a family man. **~lanmak,** to set up house, start family life.
**barka** (́··) Large rowing-boat.
**barko** (́··) Barque.
**barlam** Hake.
**baro** (́··) The (legal) Bar.
**barsak** Intestine. **~ askısı,** mesentery: **ince ~,** small intestine: **kalın ~,** large intestine: **kör ~,** appendix.
**barsam** Sting-fish, weever.
**barudî** (·–́–) Of the colour of gunpowder; slate-coloured.
**barut, -tu** Gunpowder; very pungent drink; hot-tempered man. **~ hakkı,** charge of powder: **~ kesilmek,** to fly into a rage: **⌜~ yoktu kaptanım⌝,** 'no further explanation is needed'. **~hane** (··–́·), powder factory; powder magazine. **~luk,** powder-flask.
**baryum** (́··) Barium.
**basamak** Step, stair; tread, round *of a ladder*; column (of figures).
**basan** Pressing; treading. Nightmare; printer.
**basarık** Treadle.
**basbayağı** (́···) Very common *or* ordinary. Altogether, entirely.
**basık** Compressed; low, dwarf; mumbling.
**basılı** Stamped; printed.
**basılmak** *vn. pass. of* **basmak.** Be pressed *etc.*; be overcome; be extinguished.
**basım** *n.* Printing. **~evi, -ni,** printing-house, press.
**basın** The Press.
**basil** (*Fr. bacille*) Bacillus.
**basir** (·–́) Intelligent, discerning. **~et,** (·–́·), **-ti,** perception, insight; circumspection; foresight. **~i* bağlanmak,** to lose one's common sense; to be blind to the consequences of one's actions.
**basit** Simple; plain; elementary. **~leştirmek,** to simplify.
**baskı** Press; stamp; edition; oppression; restraint. **~ altında,** under discipline; under pressure.
**baskın** Overpowering; superior, surpassing. Sudden attack; night attack; raid *by the police etc.* **~ çıkmak,** to come out best; **···dan ~ çıkmak,** to get the better of: **~ etm.,** to make a surprise attack *or* raid: **su ~ı,** flood.
**baskül** (*Fr. bascule*) Weighing-machine.
**basma** *verb. n. of* **basmak.** Print; printed goods, *esp.* printed cotton hangings. Pressed; printed. **bakır ~sı,** copperplate printing: **taş ~sı,** lithograph.
**basmak** *va.* (*with acc.*) Stamp, print; attack suddenly, surprise; raid; overpower; give (a blow); utter (a cry); (*with dat.*) press; tread on; oppress. *vn.* Tread; be oppressive; settle, sag; (cold *or* darkness) set in; (water) flood, overflow; (of a number of people) to arrive suddenly, to crowd in. **bas!,** 'go to hell!': **basıp gitti,** he went off *without caring a straw*: **ağır ~,** to press heavily; to impose one's will; to reprimand: **ayak ~,** to put down the foot; to set foot *in a place*: to insist: **çürük tahtaya ~,** to run into danger; to compromise oneself: **el** *or* **kitaba el ~,** to take an oath: **kahkahayı ~,** to burst out laughing: **sikke ~,** to coin money: **tam üzerine ~,** to 'hit the nail on the head': **on sekiz yaşına bastı,** he has reached the age of 18: **yaygarayı ~,** to cry out loudly.
**basmakalıb** (·́··) Stereotyped (remark *etc.*); cliché.
**baso** (́··) Bass (*mus.*).
**Basra** (́··) Basra. **~ Körfezi,** the Persian Gulf.
**bastı** Vegetable stew.
**bastıbacak** Short-legged; squat; *used ironically of a child behaving like an adult.*
**bastırmak** *va. caus. of* **basmak.** Push down, press; overwhelm; extinguish; repress; suppress; hide, hush up (scandal *etc.*); catch unawares; tack (sew); relieve (stomach); cry out; give a decisive *reply etc.*; (cold *etc.*) set in. **ateş ~,** for a fever to set in.

**bastika** (·꞉·) Snatch-block.
**baston** Stick; jib-boom. ~ **francalası,** long French bread: ~ **yutmuş gibi,** 'as stiff as a poker'.
**basur** Piles, haemorrhoids. ~ **otu,** lesser celandine: **kanlı** ~, dysentery.
**bâsübâdelmevt** (–·–·꞉) The resurrection of the dead.
**baş** Head; top; knob; bow *of a ship*; beginning, source; head *of cattle etc.*; leader; intelligence, understanding; money-changer's charge. *a.* Head, chief (*generally written with noun as one word, e.g.* **başvekil,** Prime Minister); *often used as a preposition meaning* near, by, *e.g.* **ocak başı,** by the fire; *often used where English would use a personal pronoun, e.g.* **~ıma gelenler,** the things that happened to me; **~ına belâ getirmek,** to bring trouble upon him. ~ **~a,** together, tête-à-tête, face to face; confidentially; ~ **~a kalmak,** to be in consultation with; ~ **~a vermek,** to have a tête-à-tête, to collaborate, to put their heads together: **~a** ~, entirely, completely; just enough; *v. also* **başabaş**: **~ta,** at the top; at the head; first; before everything: **~tan,** again, from the beginning: **~tan ~a,** from end to end; entirely: **~ı açık,** bare-headed: ~ **açmak,** to uncover the head: **···den** ~ **alamamak,** to be too busy with *stg. to be able to do anything else*: **~ını* alıp gitmek,** to go off without notice: **~ının* altından çıkmak,** (of a plan *etc.*) to be hatched out in *s.o.*'s head; ~ **aşağı,** head first, headlong; upside-down; ~ **aşağı gitmek,** (of a business *etc.*) to go steadily down: **~tan aşağı,** from head to foot; entirely: **~ından* atmak,** to put off *or* get rid of *s.o.*: **~ı boş,** vagabond; not tied up (animal); free, unfettered: **~ımla beraber,** 'with the greatest pleasure!': **~ına* bitmek,** to pester: ~ **çekmek,** to take the lead, to guide: **~tan çıkarmak,** to seduce *or* pervert: **···i ~a çıkartmak,** to make a lot of *s.o.*: **···le ~a çıkmak,** to cope with, to master, to succeed with: **~ına* çıkmak,** to presume on *s.o.*'s kindness: **~tan çıkmak,** to throw off restraint, to get out of control; to be led astray: **~döndürücü,** vertiginous: ~ **dönmesi,** giddiness: **~ı dönük,** giddy; bewildered: **···le** ~ **edebilmek,** to cope with ...; **···le** ~ **edememek,** to be unable to cope with ...; ~ **etm.,** to succeed, to overcome: **~tan geçen,** event, adventure: **~ından* geçmek,** (of an event *etc.*) to happen to one: **~a gelmek,** to happen; ⌜**~a gelen çekilir**⌝, ⌜what can't be cured must be endured⌝: ~ **göstermek,** to appear, to arise; (of a revolt *etc.*) to break out: **~ına* hal gelmek,** to be hard put to it: **~ı* için,** for the sake of: ~ **kaldırmak,** to rise in revolt: ~ **kaldırmamak,** to work unceasingly: ~ **tan kara etm.,** (of a ship) to run ashore; to make a last desperate effort: **~tan kara gitmek,** to head for disaster: ~ **kesmek,** to bow: **~ından* korkmak,** to be afraid of being involved in *stg., for fear of the consequences*: **~ında* olm.,** (of a job) to be on one's hands: **~i önünde,** harmless and submissive: **~ınız sağ olsun!,** 'may your head be safe', *the recognized formula of sympathy to one who has lost a relative by death*: **~ı taşa geldi,** he has had a bitter experience. **···e** ~ **tutmak,** to head *a ship* towards ...: ~ **tutmamak,** (of a ship) not to hold her course: ~ **ucunda** at his side, *or, more usually,* at his bedside: ~ **üstünde tutmak,** to honour highly; ~ **üstünde yeri olm.,** to be highly venerated *or* loved: ~ **üstüne!,** 'with pleasure!', 'very good!': ~ **vermek,** (i) to give one's life; (ii) to show its head, to begin to appear: ~ **vurmak,** (i) to have recourse to; (ii) (of a fish) to bite: **~ına* vurmak,** (of drink, wealth *etc.*) to go to one's head: ~ **yastığı,** pillow: **~ı* yerine gelmek,** to collect oneself: ~ **yukarıda (havada),** proud, ambitious: **adam ~ına,** per man, per head: **alt** ~, the lower end of a thing; **alt ~tan başlamak,** to begin at the end: **bir ~tan,** all at once, all together: **büyük** ~, cattle: **küçük** ~, sheep and goats: **iş ~ı,** place of business: **kendi ~ına,** of his own accord; on his own, independently: **saat ~ında,** at the end of every hour, for every hour: **yeni ~tan,** afresh, all over again.
**baş·abaş** (꞉··), at par: **~tan yukarı (aşağı),** above (below) par: *v. also* **başa baş.** **~ağrısı, -nı,** headache; nuisance, worry. **~bakan,** Prime Minister. **~buğ,** Commander-in-Chief; head of irregular troops; leader. **~çavuş,** sergeant-major. **~çı,** seller of cooked sheeps'-heads. **~gedikli,** sergeant-major. **~göz,** ~ **etm.,** to marry: ~ **yarmak,** to bungle, to make a mess of things; to do something un-

seemly. ~ıbozuk, irregular soldier; not in uniform; civilian. ~kan, president; chairman. ~kâtib, first secretary; chief clerk. ~kent, -tı, capital (of a country). ~kumandan, Commander-in-Chief. ~lı, having a head; principal, important; rounded; with a knob; (ship) down by the head: ~ başına, independent, on his own; independently: belli ~, eminent; principal: koyun ~, stupid. ~lık, headship; presidency; headgear; helmet; cowl; bridle; title of a book; heading of an article; top of a mast; (*in parts of Anatolia*) sum paid to a bride's father by the bridegroom: harb başlığı, war-head *of a torpedo*. ~muharrir, editor-in-chief. ~murakıb, Controller-in-Chief (financial). ~örtü, ~örtüsü, -nü, headgear; veil. ~sağlığı, -nı, a saying 'başınız sağ olsun!', (*form of condolence on a bereavement*): ~ dilemek, to offer one's condolences. ~sız, without a head; without a leader. ~vekil, Prime Minister. ~yazar, principal leader-writer.

**başak** Ear of corn. ~ **bağlamak** *or* **tutmak,** to come into ear: ~ **samanı,** chaff. ~**çı,** gleaner. ~**lı,** in ear, bearing spikes.

**başarı** Success. ~**cı,** ~**lı,** successful, enterprising.

**başarmak** *va.* Bring to a successful conclusion; succeed in; accomplish.

**başgöstermek** *v.* **baş.**

**başka** Other; another; different; apart. ···**dan** ~, other than; besides: ~ ~, separately, one by one: **gelmedi, o** ~, the fact that he did not come is beside the point: ~**sı,** another; *s.o.* else. ~**ca** (·ˈ·), separately, independently; rather different; otherwise; further. ~**laşmak,** to alter, to grow different. ~**lık,** diversity; alteration; change of appearance.

**başla·mak** *vn. va.* (*with dat.*). Begin. ~**ngıc,** beginning; preface. ~**nmak**[1], to be begun. ~**tmak,** to cause *or* let begin; to put *a child* to school. ~**yış,** beginning.

**başlanmak**[2] *vn.* Form a head *or* a root (as of an onion *etc.*).

**başlı** *etc. v. after* **baş.**

**başmak** Shoe; slipper. ~**çı,** shoemaker.

**başparmak** (ˈ··) Thumb; great toe.

**başvurmak** (ˈ··) *vn.* Have recourse (to); (of a ship) to pitch.

**bataet, -ti** Slowness, tardiness.

**batak** Quagmire; bog, marsh. Boggy; fraudulent; about to sink *or* perish; desperate (condition, affair). ~**çı,** fraudulent borrower; swindler, cheat: ⌜~**ya mal kaptırmış gibi**⌝, 'as if a swindler had got hold of your property' (*said to a person anxious to get back what he had just lent to a friend*). ~**hane** (··–ˈ), gambling den; den of thieves; dangerous spot. ~**lı,** marshy. ~**lık,** marshy place, bog; quicksand.

**bâtapu** (ˈ··) Holding an authentic title-deed. ~ **sahib,** owner by virtue of a legal title-deed.

**batarya** (·ˈ·) Battery of artillery.

**batası** *A form of the 3rd pers. sing. of the subjunctive of* **batmak,** *only used in curses*; **yere** ~ **âdet,** an accursed custom; **adı** ~, 'that damned fellow'; **adı** ~ **hastalık,** that confounded illness.

**batı** The West; west wind.

**batık** Submerged.

**batıl** (–ˈ) False; erroneous; vain; useless. ~ **itikad,** superstition.

**batın, -tnı** Abdomen, stomach; generation. **bir** ~**da üç çocuk,** three children at a birth.

**batır** Brave.

**batırmak** *va.* Submerge; sink; thrust into; ruin; reduce to bankruptcy; decry, speak ill of.

**batış** A sinking; decline; ruin; setting *of the sun.*

**bati** (·ˈ) Slow, tardy; slothful.

**batmak** *vn.* Sink; go to the bottom; (of the sun *or* a star) set; (of money) be lost; be destroyed; pass out of existence; go bankrupt. *va.* (*with dat.*) Penetrate; enter deeply; prick, sting; get on the nerves of, irk. **bata çıka,** 'sinking and rising', *i.e.* with the greatest difficulty: **boğazına* kadar borca** ~, to be up to one's ears in debt: **göze** ~, to attract unfavourable attention, to offend the eye: **kan tere** ~, to sweat profusely: **para** ~, (of money) (i) to be lost; (ii) to burn a hole in one's pocket.

**batman** Measure of weight, varying from 2 to 8 okkas = $5\frac{1}{2}$ to 22 lb.

**battal** Abrogated, cancelled, void; useless, worthless; unemployed, idle; large and clumsy. Large-sized paper; documents no longer valid. ~ **etm.,** to render null and void; to cancel.

**battaniye** (·–·ˈ) Blanket.

**batur** Brave.

**bavul** Traveller's trunk.
**bay** (*Originally*) a rich man; (*now*) a gentleman. ~ **A.**, Mr. A.
**bayağı** (··ˈ) Common, ordinary; rough, coarse. (ˈ··) Simply, merely, just.
**bayan** Lady. ~ **A.**, Mrs. *or* Miss A.
**bayat** Stale; not fresh; insipid; out-of-date. **hayatı* gidip ~ı* kalmak,** to be overcome with enthusiasm about *stg.* **~î,** *in such expressions as*: **bizimki yine ~ faslından başladı,** our friend started to harp on the same old string.
**baygın** Fainting; unconscious; languid; faint; heavy (smell). **~lık,** swoon, fainting condition; languor; drooping (of a flower). **mide baygınlığı,** sinking sensation at the stomach.
**bayıl·mak** *vn.* Faint, swoon; droop. *va.* Pay; spend. **···e ~,** to be passionately fond of *or* addicted to *stg.*: **içi ~,** to feel faint from hunger: **imam bayıldı,** a dish of stuffed aubergines: **parayı ~,** to pay up, to 'hand over the cash'. **~tıcı,** sickly (taste, smell). **~tmak,** *caus. of* **bayılmak**: **içini ~,** (of food) to be sickly, to cause slight nausea.
**bayındır** Prosperous; developed. **~lık,** prosperity; development. **~ Bakanı,** Minister of Public Works.
**bayır** Slope; ascent, slight rise.
**bayi** (–ˈ), **-yii** Vendor.
**baykuş** Owl. Ill-omened.
**bayrak** Flag, standard. **~ açmak,** to unfurl a flag; to collect volunteers; to revolt; **~ları açmak,** to become insolent and abusive: **~ koşusu,** relay race. **~lı, eli ~,** insolent and abusive. **~tar,** standard-bearer.
**bayram** Religious feast day (*esp.* the festival following the fast of Ramadan). **~dan ~a,** seldom: **~ etm.,** to be very delighted: ⌜**deliye her gün ~**⌝, to a fool every day is a holiday: **millî ~,** national holiday. **~laşmak,** to exchange appropriate greetings on a feast day. **~lık,** pertaining to a festival; new clothes for a feast day; present given on a feast day.
**baytar** Veterinary surgeon.
**bazan, bazen** (ˈ·) Sometimes; now and then.
**bazı** (ˈ·) Some; a few; certain. Sometimes. **~nız,** some of you: **~ ~,** at times, now and then: **~ adam** *or* **adamlar,** some people, certain people: **~ kere,** sometimes.
**bazirgân** *v.* **bezirgân.**
**bazu** (–ˈ) The upper arm; strength. **~bend,** armlet, brassard.
**be** Hi!, I say!
**bebek** Baby; doll. **göz bebeği,** pupil (eye); apple of the eye: **sakallı ~,** one who behaves in a manner unsuitable to his years: **taş ~,** (of a woman) a mere doll; pretty but cold.
**beberuhi** (··ˈ·) Pygmy; dwarf.
**becayiş** (·–ˈ) Exchange of posts between two officials. **~ etm.,** to exchange posts.
**becelleşmek** Struggle *or* fight with one another.
**becerik** Resourcefulness; tact; ability. **~li,** resourceful; capable, efficient; clever. **~siz,** maladroit, incapable.
**becermek** *va.* Carry out *stg.* skilfully; do *stg.* with success; (*iron.*) spoil, ruin.
**becit** Necessary.
**beç** *In* **~ tavuğu,** guinea-fowl.
**bed, bet** Bad; ugly; unseemly. **~ine* gitmek,** to vex, annoy.
**bedava** (·–ˈ) Gratis, for nothing; very cheap, for next to nothing. **~cı,** one who expects to get *stg.* for nothing.
**bedbaht** Unfortunate; unhappy.
**bedbin** Pessimistic; cynical.
**beddua** (··ˈ) Curse, malediction. **···e ~ etm.,** to curse *s.o.*
**bedel** Substitute; equivalent; price; sum paid for exemption from military service; military substitute who serves for somebody else. **···e ~,** in place of ...; in exchange for ...; as a substitute for ...: **dünyalara ~,** worth almost anything: **icar ~i,** rental: **bir ömre ~,** worth a life. **~ci, ~li,** one who has paid military exemption tax.
**beden** Trunk; body; wall of a castle; bight of a rope. **~ bağı,** rolling hitch. **~en** (·ˈ·), in person; personally; physically. **~î** (··ˈ), bodily; corporeal.
**bedesten** Covered market *for the sale of valuable goods*; drapery market.
**bedevî** (··ˈ) Bedouin; nomad.
**bed·hah** (·ˈ) Malevolent.
**bediî** (·–ˈ) Aesthetic. **~yat, -tı,** aesthetics.
**bedir, -dri** Full moon.
**beğendi** A dish of aubergines.
**beğenmek** *va.* Like; approve; admire; select. **beğenmemek,** not to like; to be particular about *stg.*:

beğen beğendiğini!, choose whichever you like, take your choice!: **beğenmiyen beğenmesin!**, if you don't like it you may lump it!: **kendini ~**, to think a lot of oneself; to be presumptuous: '**kime rastgelsem beğenirsiniz**', 'guess whom I met': **yer ~**, to select a place for oneself.
**behemehal** (·ˈ··) In any case; whatever happens.
**beher** To each; for each. **~ine**, for each one, per head: **~ gün**, daily.
**behey** *Interj. expressing irritation.* Hi, you there!
**behim·e** (·—ˈ) Animal; brute. **~î**, pertaining to brutes; animal *i.e. sexual feelings etc.* **~iyet, -ti**, bestiality *esp. sexual.*
**beis** Harm. **~ yok!**, no matter!
**beka** (·ˈ) Permanence; stability; what remains; sequel. **darı~, darül~**, the next world.
**bekâr** Celibate. Bachelor; provincial living alone in a large town. **~et** (·—ˈ), **-ti**, virginity. **~lık**, celibacy; a being unmarried.
**bekaya** (·—ˈ) *pl. of* **bakiye.** Arrears of state revenue; recruits due for military service, but not called up.
**bekçi** Watchman, *esp.* night-watchman; sentry; forest guard.
**bekle·mek** *va.* Await; watch; guard; lie in wait for; expect, hope for. *vn.* Wait; remain in expectation. **beklemiş et**, meat that has been kept *or* hung: **···in yolunu ~**, to lie in wait for *s.o.* **~nmek, ~nilmek**, *pass. of* **beklemek. ~şmek**, to wait *or* keep watch together. **~tmek**, to cause to wait; to have *s.o. or stg.* watched *or* guarded.
**bekri** Habitual drunkard; toper.
**Bektaşi** (·—ˈ) Belonging to the Bektashi sect of dervishes; freethinker; dissolute. **~ sırrı**, a great mystery: **~ üzümü**, gooseberry.
**bel**[1] Waist; loins; middle of the back; midship; mountain pass. **~i* açılmak**, to feel the need to urinate: **~i anahtarlı kadın**, a careful and efficient housewife: **···e ~ bağlamak**, to rely upon, to trust to: **~ini* bükmek**, to break the back of, to defeat; (of an illness) to cripple; (*sl.*) to make short work *of a piece of meat etc.*: **~ vermek**, (of a building) to sag, to show signs of collapse: **yarı ~e kadar**, to the waist.
**bel**[2] Spade; gardener's fork.
**belâ** (·ˈ) Trouble; misfortune; calamity; grief; difficulty; punishment; curse. ... **~sı**, for the sake of .... *e.g.* **namus ~sı**, for honour's sake: **~sını* bulmak**, to get the punishment one deserves: **~ çıkarmak**, to start a quarrel: ⌜**~ geliyorum demez**⌝, misfortunes give no warning: **~lar mübareği**, a blessing in disguise: **~yı satın almak**, to ask for trouble: **~ya uğramak**, to get into trouble, to meet with misfortune: **Allahın ~sı**, a 'perfect pest': **Allah ~sını versin!**, may God punish him!; curse him!: **baş ~sı** *or* **başa ~**, a source of trouble and annoyance: **bin ~ ile**, only with the greatest difficulty: **dilinin* ~sını çekmek**, to get into trouble from inability to hold one's tongue.
**belâgat** (·—ˈ), **-ti** Eloquence.
**belağrısı** (ˈ···), **-nı** Lumbago.
**belâlı** (·—ˈ) Troublesome; difficult; calamitous; quarrelsome. Man kept by a prostitute; (*jokingly*) man who believes himself to be the favourite of a girl.
**belbel** (ˈ·) *In* **~ bakmak**, to look about one stupidly.
**Belçika** (·ˈ·) Belgium. **~lı**, Belgian.
**belde** Town; country.
**beledî** (··ˈ) Belonging to a town; local; municipal; civic. **~ye**, municipality: **~ reisi**, mayor.
**beleş** (*sl.*) Gratis. **~çi**, one who gets *stg.* for nothing, sponger.
**belge** Document; certificate.
**beli** (ˈ·) Yes.
**beliğ** (·ˈ) Eloquent.
**belir·mek** *vn.* Become conspicuous; appear; gaze with eyes wide open *from fear or anger.* **~siz**, indistinct, not clear; uncertain: **adı ~**, an unimportant person: **belli ~**, hardly perceptible: **ne idügü ~**, a doubtful character of unknown antecedents. **~ti**, mark; sign; proof. **~tmek**, to make conspicuous *or* clear; to make known; to expound; to point out; to open *the* eyes in astonishment *etc.*
**belkemiği** (ˈ···), **-ni** Backbone.
**belkemikli** (ˈ···) Vertebrate.
**belki** (ˈ·) Perhaps; may be. Even, but. **~de**, as likely as not.
**belleme** Numnah; horse-blanket.
**bellemek**[1] *va.* Commit to memory, learn by heart.
**bellemek**[2] *va.* Dig with a spade.
**bellenmiş**[1] Committed to memory; well-known.
**bellenmiş**[2] Dug with a spade.

**belli** Known; evident, clear. ~ **başlı,** eminent, notable; chief. **~siz,** unknown; imperceptible: **belli ~,** almost invisible.

**belsoğukluğu** (ˋ....), **-nu** Gonorrhoea.

**bembeyaz** (ˋ..) Extremely white.

**bemol** Flat (*mus.*).

**ben**[1] I. **~im diyen** *in such phrases as*: **~im diyen adam bu işi yapamaz,** this is not just anybody's job. '~ **ettim sen eyleme (etme)!**' *a humble entreaty or apology.*

**ben**[2] Mole; beauty-spot.

**bence** (ˋ.) In my opinion; as for me.

**bencileyin** (ˋ...) Like me (*deprecatingly*).

**bend** Bond, tie, fastening; dyke, dam, reservoir; paragraph; article *in a paper.* **~etmek, -eder,** to bind, fasten; attach to oneself.

**bende** Slave; servant. **~niz,** 'your humble servant' (*a polite form of the first pers. sing.*). **~gân** (..ˊ) *pl. of* **bende,** slaves, servants; household servants of a princely house. **~hane** (..—ˋ), *deferential way of saying* 'my house'.

**benek** Spot on the skin *or* coat. **~li,** spotted.

**bengi** Eternal.

**beni·âdem** (..—ˋ) Sons of Adam, mankind. **~beşer,** human beings.

**benim** *gen. of* **ben.** Mine, my. **~ki,** the one belonging to me, mine: **o sokak senin, bu sokak ~ dolaştı,** he walked about from street to street. **~semek,** to appropriate *or* lay claim to *stg. that does not belong to one*; to adopt as one's own; to assume as a personal obligation *or* interest.

**beniz, -nzi** Colour of the face. **benzi* atmak,** to turn pale: **benzinde* kan kalmamak,** to look very pale.

**benli**[1] Having moles on the face *or* body.

**benli**[2] *v.* **senli.**

**benlik** Egoism; personality; self-respect. ~ **davası,** self-conceit; a feeling of personal superiority: **milli ~,** national pride; national character.

**bensiz** Without me; without my assistance. Without a mole.

**benzemek** *vn.* (*with dat.*) Be like; resemble; *with a participle in the dat.*, to look as if ..., *e.g.* **ümid kesmişe benziyor,** it looks as if he had given up hope. **bir şeye benzemiyor,** it's useless (*cf. Eng.* 'it's like nothing on earth'); **işte şimdi bir şeye benzedi,** now it's beginning to look like something, *i.e.* it's all right, *but* **işte şimdi tam benzedi,** look at the mess it's in now!

**benzer** Resembling. *n.* Likeness; copy. **~i yok** *or* **~siz,** unique: **tam ~i,** an exact similar. **~lik,** resemblance, similarity.

**benzetmek** *va.* Liken; compare; make an imitation *or* copy of; mistake *a person for another*; see a resemblance. ⌜~ **gibi olmasın!**⌝, I don't wish to make a comparison, *said when making a comparison with a dead person or one who is ill or when the comparison might seem to be uncomplimentary to one of the parties*: **...i bir şeye ~,** to make *stg.* look 'something like', to put in order: **mektebden kaçan çocuğu babası iyice benzetmiş,** his father gave the truant a good hiding.

**benzeyiş** Resemblance.

**benzi** *v.* **beniz.**

**benzin** Petrol; benzine.

**beraber** (.—ˋ) Together; in company with. Abreast, equal; in the same direction *or* on the same level. **~e bitmek,** to finish in a dead heat: **~e kalmak,** to be equal in a game: **bununla ~,** at the same time, nevertheless: **yerle ~ olm.,** to be razed to the ground. **~lik,** a being equal *or* abreast; solidarity, co-operation.

**beraet, -ti** Innocence; acquittal. ~ **etm.,** to be acquitted.

**berakis** On the contrary.

**berat** (.ˋ), **-tı** Patent; warrant; order conferring a dignity *or* decoration. ~ **gecesi,** *the Moslem Feast (15th of Shaban), celebrating the revelation of his mission to Mohammed.* **~lı,** patented.

**berbad** Destroyed; scattered; ruined; filthy; soiled. ~ **etm.,** to spoil, to ruin.

**berber** Barber. ~ **aynası,** hand-mirror.

**Berber, berberî** Berber.

**berdevam** (..ˊ) Lasting; continuous. Continuously. **...de ~dır,** he continues to ...: ~ **olm.,** to continue to exist.

**bere**[1] Bruise; dent. **~lemek,** to bruise.

**bere**[2] Beret.

**bereket, -ti** Abundance, increase; blessing. ~ **versin,** may God bless you (*a form of thanks*); fortunately, as good luck would have it. **~li,** fruitful, fertile; abundant.

**bergüzar** Memento.

**berhane** (·—˙) Large rambling house.

**berhava** (··˙) Destroyed. **~ etm.,** to blow up *stg.*: **~ olm.,** to be blown up; to be annihilated.

**berhayat** (··˙) Alive.

**berhurdar, berhudar** (··˙) Enjoying the fruits of one's labour; successful, prosperous. **~ ol!,** I wish you prosperity (*meaning* 'I thank you').

**beri** The near side (*as opp. to* **öte,** the far side); this side; here, hither; the time since; until now. Since. **~deki,** that which is on this side: **~ taraf,** this side: **geldiğimden ~,** since I came: **o günden ~,** since that day. **~ki,** nearest; last mentioned.

**berk** Hard; firm; solid; rugged.

**berkarar** (··˙) Stable; constant; durable. **~olm.,** to remain unchanged.

**berkemal** (··˙) In perfection; complete.

**berkitmek** *va.* Render solid *or* firm; strengthen; back, support; confirm, affirm.

**bermucibi** (˙—··) In conformity with; as required by.

**bermurad** (··˙) Satisfied; happy. **~ olm.,** to have one's wish.

**bermutad** (˙——) As usual.

**berrak** Clear; limpid; sparkling.

**bertaraf** Aside; out of the way; putting aside; apart from. **~ etm.,** to put aside; to get rid of: **~ olm.,** to be put aside; to disappear: **şaka (lâtife) ~,** joking apart.

**berzah** Isthmus; precipice. **belâyı ~,** a great difficulty; a cursed nuisance.

**besa** (˙·) Truce (*esp. in a blood-feud*); compact.

**besbedava** (˙···) Absolutely gratis.

**besbelli** (˙··) Very clear *or* evident.

**besbeter** (˙··) Very much worse.

**besi** Place in which animals are fattened; the fattening of animals. **~ye komak,** to put an animal out to fatten. **~li,** fattened; well-nourished.

**besle·me** Act of feeding *or* fattening; domestic servant *brought up in a house from childhood.* **~ kılıklı,** slatternly. **~mek,** to nourish; to fatten; to rear *an animal*; to support, prop; to take into one's service: **can ~,** to be a gourmet. **~yici,** feeder, fattener; nutritious, wholesome.

**besmele** *The formula* '**bismillah!**' (in the name of God). **~ okumak** *or* **çekmek,** to pronounce this formula *before doing stg.* **~siz,** of evil omen; rogue.

**beste** Bound; fastened; obliged; congealed. Tune; musical composition. **~kâr** (··˙), musical composer. **~lemek,** to compose music, to set to music. **~li,** with music.

**beş** Five. **~ aşağı ~ yukarı,** close bargaining. **~ aşağı ~ yukarı uyuşmak,** to come to some sort of an understanding: **~ para etmez,** worthless.

**beşbıyık** Medlar.

**beşer**[1] Five each.

**beşer**[2] Mankind. **~î** (··˙), human. **~iyet, -ti,** humanity; human nature.

**beşibirlik, beşibiryerde** (···˙·) Turkish gold 5-lira piece, *used as ornament.*

**beşik** Cradle. **~ kertme nişanlı,** engaged to one another while yet in the cradle.

**beş·inci** Fifth. **~li,** having five parts; the five (in cards). **~lik,** five-piastre piece; worth five piastres; five *yards long etc.* **~me,** a kind of cloth with stripes of five colours. **~pençe,** starfish. **~yüzlük,** Ltq. 500 note.

**beşuş** Smiling; happy.

**bet**[1], **-ti** *Only with* **beniz. ~i benzi kalmamış,** he has no colour left.

**bet**[2] *Only in* **~ bereket,** abundance, prosperity.

**bet**[3] *v.* **bed.**

**beter** *comp. of* **bed.** Worse. **~in ~i var,** there is always something worse.

**beton** (*Fr.*) Concrete. **~arme,** reinforced concrete.

**bev·il, -vli** Urine.

**bey** *Formerly a title inferior to Pasha and superior to Agha*; gentleman; prince; chief; ace (cards). **küçük ~,** the son of the house, the 'young master'.

**beyan** (·˙) Declaration; explanation; expression; style. **~ etm.,** to announce, declare, explain. **~at** (·—˙), **-tı,** *pl. of* **beyan**; declarations; *as sing.* discourse, speech: **~ta bulunmak,** to hold forth, make a speech, give an explanation, make an announcement. **~name** (·——˙), manifesto; affidavit.

**beyarı** (˙··) Queen bee.

**beyaz** White. White colour; fair copy. **~a çekmek,** to make a fair copy: **~a çıkarmak,** to stand up for *s.o.*, to clear his character. **~ımsı, ~ımtrak,** whitish. **~lı,** with white in it.

**beyefendi** (ˋ···) *Term of respect in addressing a person*; sir!
**beygir** Horse; horse-power. **~ci, ~ sürücüsü**, a man who lets horses out on hire *or* attends to hired horses: **bostan ~i gibi dönmek (dolaşmak)**, to walk aimlessly round and round; to stick to routine, to make no progress.
**beyhude** (·—ˋ) Vain, useless. In vain.
**beyin, -yni** Brain; intelligence. **beyni* atmak**, to fly into a passion: **beyni* sulanmak**, to become senile *or* muddle-headed: **~ tavası**, brain fritters. **~siz**, brainless, stupid.
**beyit, -yti** Couplet, verse.
**beylerbeyi** (ˋ···) (*Formerly*) governor of a province.
**beylik** Title or status of Bey; gentle birth; district governed by a Bey; 'the Government'. Belonging to the State; conventional, stereotyped. Small blanket *issued to soldiers*. **~ gemi**, a government ship: **~ satmak**, to behave like a little lord, to give oneself airs.
**beyn** *As Turkish* **ara. ~imizde**, amongst us, between us; **~lerinde**, amongst them *etc.* **beyn–**, *Ar. prefix.* inter-: **~elmilel**, international.
**beynamaz** (—·ˊ) (*err. for* **binamaz**). One precluded from prayer *on account of canonical uncleanness*; unbeliever. **~ özürü**, a lame excuse: *v.* **cami.**
**Beyoğlu** (ˋ··), **-nu** Pera.
**beyt, -ti** House; dwelling. **~ullah**, the Cubical House (Kaaba) at Mecca.
**beyzade** (·—ˋ) Son of a Bey; noble.
**beyzî** (·ˊ) Oval.
**bez**[1] Linen *or* cotton material; cloth; duster; canvas. **Amerikan ~i**, calico: **yelken ~i**, sail-cloth: ⌜**kenarına bak ~ini al, anasına bak kızını al**⌝, 'before buying stuff look at a sample, before marrying a girl look at her mother'.
**bez**[2] Gland.
**bezdirmek** Annoy, plague.
**bezek** Ornament; decoration. **~çi**, woman who dresses brides; decorator.
**bezelye** (·ˋ··) Pea(s).
**beze·mek** *va.* Deck out; adorn. **~n**, ornament; embellishment. **~nmek**, to be adorned.
**bezgin** Tired of life; disgusted; discouraged.
**bezir** Linseed; linseed oil. **~yağı, -nı** Linseed oil.
**bezirgân** (··ˊ) Merchant *or* pedlar (*esp.* Jewish). **din (politika) ~ı**, one who exploits religion (politics) for his own ends. **~lık**, cunning dealing.
**bezmek** Become weary and discouraged. **···den ~**, to become sick of ..., disgusted with ... .
**bıcılgan** (··ˋ) Cracked heels (in a horse).
**bıçak** Knife. ⌜**~ kemiğe dayandı**⌝, (the knife has come up against the bone), it can no longer be borne; it has reached the limit: **~ sırtı kadar**, a very small difference; **ölmesine ~ sırtı kalmış**, he all but died: **~ silmek**, to finish off a job: **ağzını ~ açmıyor**, he is very depressed. **~çı**, cutler; a bit too ready with his knife; quarrelsome. **~lamak**, to stab, to knife. **~lı**, armed with a knife: **kanlı ~ olm.**, to be at daggers drawn. **~yeri, -ni**, scar of a knife wound.
**bıçılgan**[1] *v.* **bıcılgan.**
**bıçılgan**[2], **bıçırgan** Burnisher.
**bıçkı** Cross-cut saw. **~ talaşı**, sawdust: **~ tezgâhı**, saw-mill.
**bıçkın** A quarrelsome rough; vagabond, ne'er-do-well.
**bıdık** Short and tubby.
**bıkkın** Disgusted; satiated; bored.
**bıkmak** *vn.* Be disgusted *or* satiated. **···den ~**, to be sick of *stg.*: **bıktım!**, I'm sick of it!; I've had enough of it!
**bıldır** (ˋ·) Last year; a year ago.
**bıldırcın** Quail; plump and attractive little woman.
**bıngıl** *In* **~ ~**, quivering like a jelly, well-nourished, fat. **~dak**, fontanelle. **~damak**, to quiver like a jelly.
**bırakmak** *va.* Leave; leave go, release; quit, abandon; put off; put down; deposit; leave off, cease from; entrust *stg. to s.o.*; allow; grow (moustache *etc.*); divorce; bequeath; keep *a boy* in a class, not to promote him. *vn.* (Of *stg.* stuck on) to come away, to come unstuck. **akıl bırakmamak**, to upset *s.o.'s* thoughts: **···e etmediğini* bırakmamak**, to leave nothing undone to harm *or* annoy *s.o.*: **kendini ~**, to cease to take an interest in oneself (one's dress, one's business *etc.*): **kâr ~**, to show a profit: **mektubu postaya ~**, to post a letter.
**bıyık** Moustache; whisker *of a cat*; tendril. **~ altından gülmek**, to laugh in one's sleeve: **bıyığını balta kesmez olm.**, to be a 'tough' fellow, to fear no one: **~ burmak**, to curl the moustache; (of a young man) to swagger

slightly, to show himself off to the girls: **~larını ele almak,** to become a man. **~lı,** having a moustache: **~ balık,** barbel.

**biat** (–ˋ) Oath of allegiance; homage; (*only correctly used with reference to the Caliph*).

**biber** Pepper; pimento. **~lik,** pepper-pot.

**biblo** (*Fr. bibelot*) Trinket; curio; knick-knack.

**bicili** *In* **cicili ~,** gaudy; all decked out.

**biçare** (––ˋ) Without help; poor, wretched.

**biçilmek** *vn. pass. of* **biçmek.** Be cut *etc.* **biçilmiş kaftan,** appropriate, well-adapted, 'cut out' for the job; 'just made' for *s.o.*

**biçim** Cut *of a coat*; form, shape; manner, sort. **~e sokmak,** to put in order: **ne ~ ...?,** what sort of ...; **bu ne ~ şey?,** what's this?; this is odd! **~li,** well-cut, well-shaped. **~siz,** ill-shaped, unsymmetrical, ugly.

**biçki** Cutting-out *of clothes etc.*

**biçmek** *va.* Cut out; cut up; divide; reap, mow. **ekin ~,** to harvest crops: **kesip ~,** to consider and decide: **ölçüp ~,** to measure and cut out; to take in *stg.* (*a room etc.*) at a glance: **paha ~,** to estimate a price: **pay ~,** to apportion a share; to deduce a conclusion.

**bid'at, -tı** Innovation *which is evil from a religious point of view*; heresy.

**bidayet** (·–ˋ), **-ti** Beginning. **~ mahkemesi,** Court of First Instance.

**bidon** (*Fr.*) Can, drum.

**biftek** Beefsteak.

**bigâne** (––ˋ) Foreign; strange; aloof; uninterested, indifferent.

**bihakkin** (·ˋ·) Rightly, deservedly; as it should be; perfectly.

**bikes** (–ˋ) Friendless; destitute; orphan.

**bikir, -kri** Virginity; virgin. **izalei ~ etm.,** to deflower.

**bilâfasıla** (ˋ––··) Uninterruptedly.

**bilâhara** (ˋ–··) Later on.

**bilâistisna** (ˋ–··–) Without exception.

**bilâkaydüşart** (ˋ–···) Unconditionally.

**bilâkis** (ˋ··) On the contrary.

**bilânço** Balance sheet.

**bilâperva** (ˋ–··) Without fear; boldly; freely.

**bilârdo** Billiards.

**bilâ·sebeb** (ˋ–··) Without cause *or* reason. **~ücret,** without payment *or* reward. **~vasıta** (ˋ–––··), directly, without intermediary. **~veled,** without issue.

**bilcümle** (ˋ··) All. In all; totally.

**bildik** *Past part. of* **bilmek.** Known; not a stranger. Acquaintance. **~ çıkmak,** to turn out to be an old acquaintance: **bildiğini okumak,** to go one's own way, to follow one's own judgement without asking the opinion of others: **bildiğinden şaşmamak,** not to be deterred by others from going one's own way.

**bildirmek** *va.* Make known; communicate; notify.

**bile**[1] Even.

**bile**[2] *In* **~ ~,** knowingly; with one's eyes open.

**bile**[3] *v.* **bilye.**

**bileği** Instrument for sharpening. **~ çarkı,** grindstone: **~ kayışı,** strop: **~ taşı,** whetstone.

**bilek** Wrist; pastern. **~ damarı,** pulse: **~ gibi akmak,** to flow abundantly: **ayak bileği,** ankle; tarsus: **bu işte kimse onun bileğini bükemez,** nobody can equal him in this. **~li,** strong-wristed.

**bilemek** Sharpen, whet, grind. **···e diş ~,** to cherish a grudge against *s.o.*, to await a chance of revenge against *s.o.*

**bilet, -ti** (*Fr.*) Ticket. **~ kesmek,** (of tram-conductor *etc.*) to give a ticket. **~çi,** ticket-collector, tram-conductor.

**bileyici** Knife-grinder.

**bilezik** Bracelet, bangle; collet; stone forming the mouth of a well; ring of metal round a column *or* a gun. **altın ~,** gold bracelet; skill *or* trade *enabling a man to earn his livelihood.*

**bilfarz** (ˋ·) Supposing that.

**bilfiil** (ˋ··) In fact, actually.

**bil·ge** Learned and wise. **~gi,** knowledge. **~giç,** pedant; pretending to be learned. **~giçlik,** pedantry: **~ satmak,** to make a parade of one's knowledge. **~gili,** learned. **~gin,** learned man, scientist.

**bilhassa** (ˋ··) Especially, in particular. **~ ve ~,** above all.

**bililtizam** (ˋ···) On purpose.

**bilin·mek** *vn. pass. of* **bilmek.** Be known, recognized *etc.*; be acceptable, gratefully received. **~mez,** unknown; incomprehensible.

**bilirkişi** Expert.

**bilistifade** (ˋ··—·) Profiting by, taking advantage of.

**biliş** A knowing; knowledge; acquaintance. ~ **tanış,** friends and acquaintances. ~**mek,** to know one another, be acquainted; strike up an acquaintance.

**billahi** (ˋ—·) *In* **vallahi** ~!, I swear by God.

**billur** Crystal; cut-glass. Crystal-clear.

**bilmece** Riddle, enigma.

**bilmedik** (ˋ··) Who does not know; unknown. ~ **kimse kalmadı,** everyone knows it.

**bilmek, -ir** *va.* Know; learn; recognize; guess; consider; believe; (*with the gerund in -a, -e,* to be able to ..., *e.g.* **yapabilmek,** to be able to do; **gidebilmek,** to be able to go). *v. also* **bildik** *and* **bile**[2]. **bilmemezlikten gelmek,** to pretend not to know: **çektiklerimi ben bilirim!,** you can't think what I suffered!: ⌜**kendi bilir!**⌝, (i) just as he wishes!; it's up to him!; (ii) on his head be it!: **kendini bilmez,** insensible; confused; who does not know his place, arrogant, insolent: **kendimi bildim bileli,** from the time I was capable of thinking: ⌜**siz bilirsiniz!**⌝, 'as you will!': ⌜**ya Rabbi, sen bilirsin!**⌝, 'O my Lord, you know!', *said under one's breath and generally to oneself to express annoyance*: (**bunu yapmayın!**) **yaparsanız orası artık sizin bileceğiniz iştir,** if you do it, you will do it at your own risk.

**bilmez** Who does not know; ignorant; ungrateful. ~**lik,** ignorance; feigned ignorance: ~**den** *or* **bilmemezlikten gelmek,** to feign ignorance.

**bil·misil** (ˋ··) In a like manner. **mukabele** ~, tit-for-tat, retaliation. ~**mukabele** (ˋ·—··), in return, in retaliation. ~**ûmum** (ˋ·—), in general; on the whole; all. ~**vasıta** (ˋ—··), indirectly. ~**vesile** (ˋ·—·), on the pretext of; profiting by the occasion.

**bilye** (ˋ·) Marble (child's plaything); billiard ball; ball (*mech.*). ~ **yatağı,** ball-bearing.

**bin**[1] Thousand; a great many. ~**lerce,** in thousands: ~**i bir para,** abundant; very cheap: ~**de bir,** very rarely: ~ **can ile,** with the utmost eagerness: ⌜~**in yarısı beş yüz (o da sende yok)**⌝, *said to one deep in thought, stg. like* 'a penny for your thoughts!', *but meaning rather more* 'what have you got to worry about that you are so pensive?': ~ **yaşasın,** long live ...!

**bin**[2] Son; *chiefly in Arab names, e.g.* Ali bin Hassan, Ali the son of Hassan.

**-bin** (—) *Pers. suffix.* -seeing. **durbin,** far-sighted; telescope.

**bina** Building, edifice; act of building, construction; a basing *a claim etc. on some fact*: chapter on indeclinable words in Arabic grammar; ⌜**benim oğlum** ~ **okur, döner döner yine okur**⌝, *used about one who keeps harping on the same subject, or about stg. which makes no progress*: ~ **etm.,** to build.

**binaen** (·ˊ·) On account of; in consequence of; based on. ~**aleyh,** consequently, therefore.

**binamaz** (—·ˊ) *v.* **beynamaz.**

**binbaşı** (ˋ··) Major; commander (navy); squadron-leader.

**bindallı** Purple velvet embroidered with leaves and flowers.

**bindirme** *verb. n. of* **bindirmek.** Joint; overlap. Mounted on, resting on; clinker-built.

**bindirmek** *va. caus. of* **binmek.** Cause to mount *etc.*; cause to rest on *or* overlap; (of a ship) to collide with, to ram. **üzerine** ~, to fall on; to increase violence; to blame (*wrongly*).

**binek** Connected with riding. A mount (horse *etc.*). ~ **arabası,** carriage: ~ **otomobili,** passenger car. ~**atı, -nı,** saddle-horse.

**biner** A thousand each.

**bingözotu** (·ˋ··), **-nu** Scammony.

**binici** A good horseman; jockey. ~**lik,** horsemanship.

**bininci** Thousandth.

**biniş** Act *or* method of riding; ancient ceremonial riding dress; long full cloak *worn formerly by Ulema.*

**binlerce** In thousands; thousands of.

**binlik** Large bottle holding 1,000 dirhems; a 1,000-lira note.

**binmek** *va.* (*with dat.*). Mount; ride; go on (a train or ship); overlap; assume an attitude *or* condition. **iş fenaya bindi,** the affair has taken a bad turn.

**biperva** (—·ˊ) Fearless; unscrupulous.

**bir** One; a; the same; equal; in such a way; once; alone, only; *sometimes* = just, *e.g.* ~ **gidip bakalım,** let's just have a look; *sometimes adds emphasis, e.g.* ~ **yağmur** ~ **yağmur!,** you never saw such rain!; ~ **döğdü ki,** he gave him such a beating. **~den,** suddenly; together: **~de,** in addition, furthermore: **~ başına,** alone; on his own: **~e bin kazanmak,** to make an enormous profit: **~e ~ gelmek,** to be the one and only cure, to be 'just the thing': **~ dediği iki olmaz,** highly esteemed and beloved; **karısının ~ dediğini iki etmiyordu,** he could refuse nothing to his wife: **~ iki demeden,** (i) without hesitation; (ii) suddenly: **~ olm.,** to be at one, to agree: **⌜~ olur, iki olur⌝,** once or twice *but not always*: **~ şey,** something: **~ şey değil,** don't mention it!; it's nothing, it does not matter: **~ şeyler,** something or other; lots of things: **⌜~ varmış ~ yokmuş⌝,** 'once upon a time' (*beginning a fairy story*); 'that's all a happy memory now': **bu evlerin ikisi de ~,** these houses are identical; it makes no difference which house: **gelmesi ile gitmesi ~ oldu,** he no sooner came than he went: **neler çektiğimi ~ ben bilirim (~ de Cenabı Hak),** what I went through is only known to me and to God: **seninle ~ daha konuşursam iki olsun,** I'll never speak to you again: **yerle ~ olm.,** to be razed to the ground.

**bira** (ʹ·) Beer. **~hane** (··–ʹ), beer-house; public house, brewery.

**birader** (·–ʹ) Brother.

**bir·az** (ʹ·) A little, rather. **~dan,** in a short while, soon; shortly afterwards. **~biri, -ni,** one another. **~çok** (ʹ·), a good few; a lot. **~den,** together; at the same time: **hepsi ~,** all together; the whole of. **~ denbire** (·ʹ··), all of a sudden; all at once.

**birdirbir** Leap-frog.

**birdüziye** (ʹ···) Continuously.

**bire** *v.* **bre.**

**birebir** (ʹ··) Most efficacious (remedy); equal to the occasion; 'just the thing'.

**bir·er** One apiece, one each. **~ ~,** one by one: **~ ikişer,** one or two each: **~ kuruşa,** at one piastre each. **~i,** one of them; an individual: **~ ~,** one another: **~miz,** one of us: **~ gelir ~ gider,** one goes, another comes: **herifin ~,** some fellow or other. **~i biri, -ni,** one another. **~icik** (ʹ··), unique; one only; a small one; a pet one.

**birikinti** Accumulation; heap; garbage. **su ~si,** puddle.

**birik·mek** *vn.* Come together, assemble; accumulate. **~tirmek,** *va.* collect; assemble; amass; save up (money); let *work etc.* accumulate. **içine ~,** to bury *insults or sorrows* in one's heart.

**birim** Unit.

**bir·inci** First; chief. **~incilik,** first prize; first place. **~isi** *v.* **biri.** **~kaç,** a few; some: **~ defa,** several times. **~leşik,** united. **~leşmek** *vn.* unite; reunite; assemble; agree; be reconciled. **~leşmiş,** united: **~ milletler,** United Nations. **~li,** ace *of cards*; the one *in dominoes etc.* **~lik,** unity; union; agreement; association; equality; similarity; identity: **~te,** together, in company: **ağız birliği,** unanimity of expression: **el birliği,** unity in action, co-operation. **~örnek** (ʹ··), of the same pattern.

**birsam** Hallucination.

**bir·şey** *v.* **bir.** **~takım** (ʹ··), a quantity; some. **~teviye** (ʹ···), continuously; regularly. **~türlü** (ʹ··) (*with neg. verb*), by no means; in no way.

**bisiklet, -ti** (*Fr.*) Bicycle.

**bisküvit, -ti** Biscuit.

**bismillâh** (··ʹ) In the name of God! (*said by Moslems before starting on any undertaking*).

**bit, -ti** Louse. **~i kanlanmak (canlanmak),** to recover one's spirits, one's money *etc.*; to become uppish. **yaprak ~i,** aphis.

**bitab** (–ʹ) Exhausted; without strength. **~ u takat,** utterly exhausted. **~i** (––ʹ), exhaustion; weakness.

**bitaraf** (–·ʹ) Impartial; neutral. **~lık,** impartiality; neutrality.

**biteviye** (ʹ···) All of a piece; complete. Uninterruptedly.

**bitik** Worn out; exhausted.

**bitirim** End; conclusion. (*sl.*) 'topping'.

**bitirmek** *va. caus. of* **bitmek.** Finish; complete; terminate, bring to an end; destroy; cause to grow *or* sprout: **işini ~,** to finish *s.o.*'s business; to kill him.

**bitiş·ik** Touching; contiguous; neighbouring. Next-door house; neighbour. **~mek,** to be contiguous; to join; adhere.

**bitki** Plant.

**bitkin** Exhausted, worn out.

**bit·lemek** *va.* Delouse. **~lenmek,** to be infested with lice; to be deloused. **~li,** lousy.

**bitme** *verb. n. of* **bitmek. yerden ~,** short, squat.
**bitmek** *vn.* Come to an end, finish; be exhausted; be ruined *or* destroyed; sprout, grow: **···e ~,** to be very fond of ... : **olup ~,** to happen; **olup bitenler,** events; **⌜oldu da bitti, maşallâh!⌝,** 'well that's done and finished with!' (*said when s.o. has been rushed into giving their unwilling consent*): **ben oldum bittim,** ever since I can remember.
**bitmez** Intermediate, endless.
**bit·otu** (ˈ··), **–nu** Lousewort. **~pazarı, –nı,** old clothes market. **~sirkesi, ~ni,** nit.
**bittabi** (ˈ··) Naturally, of course.
**bitüm** Bitumen.
**bityeniği** (ˈ···), **–ni** *Lit.* louse-bite; tender spot; secret anxiety *due to some wrong done.* **bu işin içinde bir ~ var,** there is *stg.* fishy about this.
**biyel** (*Fr. bielle*) **~ kolu,** connecting-rod.
**biz**[1] We. **···le ~bize kalmak,** to have a tête-à-tête with ... .
**biz**[2] Awl.
**bizaa, bizaat** (·–ˈ) Ability.
**bizar** (–ˈ) Tired, wearied; sick of a thing. **~ etm.,** to distress, annoy.
**bizatihi** (·ˈ··) In itself; intrinsically.
**bizim** Our. **~ için,** for us. **~ki,** ours; my wife *or* my husband.
**bizzarur** (ˈ·–), **bizzarure** Forcibly, necessarily.
**bizzat** (ˈ–) In person, personally.
**blöf** Bluff.
**blûz** Blouse.
**bobin** Bobbin; spool; coil (*elect.*).
**boca** (ˈ·) Act of pouring *or* decanting. **~ etm.,** to turn over; transport; pour out, empty; tilt; veer (ship). **~ alabanda tiramola etm.,** to wear ship. **~lamak** *vn.,* veer, bear away; turn from side to side; lurch, stumble, falter, vacillate; fail; get confused *when speaking.*
**bocurgat** Capstan; winch. **bu adam ~sız iş görmez,** this man will only work under pressure.
**bodoslama** Stem *or* stern post of a ship.
**bodrum** Subterranean vault; dungeon; cellar. **~ katı,** basement,
**bodur** Short, dwarf, squat.
**bofa ~ balığı,** lamprey.
**boğa** Bull. **~ güreşi (dövüşü),** bull-fight.
**boğaça** (·ˈ·) Flaky pastry.
**boğası** Thin twill *used for linings.*
**boğaz** Throat; mountain pass *or* defile; strait; mouth of a river; mouth *requiring food*; food *in general*; board *of a servant or animal.* **~dan bahsetmek,** to talk about food: **~ını çıkarmak,** to earn just enough for one's food: **~ derdine düşmek,** to be mainly concerned with the question of food: **~ından* geçmemek,** to be unable to enjoy a meal *because s.o. was absent*: **~ ~a gelmek,** to fly at one another's throats, to quarrel violently: **~da kalmak,** to stick in the throat: **~ kavgası,** a quarrel about food: **~ olm.,** to have swollen glands: **~ına* sarılmak** *or* **~ını* sıkmak,** to take *s.o.* by the throat, to throttle; to put pressure on, insist: **~ tokluğuna hizmet etm.,** to give service in return for food: **~dan yatmak,** to be laid up with a sore throat: **~ını* yırtmak,** to shout oneself hoarse: **⌜can ~dan gelir⌝,** one must eat to live: **canı* ~ına* gelmek,** to come to the end of one's tether, to lose one's patience: **sık ~ etm.,** to 'keep on at' *s.o. to get stg. done.*
**Boğaziçi** (·ˈ··), **–ni** The Bosphorus.
**boğaz·kesen** (·ˈ··) Fortress commanding a strait; fortress surrounded by water; narrow street; mountain gorge. **~lamak,** to cut the throat of; strangle. **~ lanmak,** *pass. of* **boğazlamak**: to acquire a good appetite; (*sl.*) to be cheated. **~lı,** gluttonous. **~sız,** who eats little.
**boğmaca** (ˈ··) *or* **~ öksürüğü** Whooping-cough.
**boğmak**[1] Node; joint, articulation; a kind of lark.
**boğ·mak**[2] *va.* Constrict; choke; strangle; suffocate; drown. **şakaya ~,** to ridicule, to make a jest of *stg.* **~ucu,** suffocating; stuffy; (*sl.*) cheating. **~uk,** suffocated; hoarse (voice); **beli ~ bardak,** a narrow-waisted glass. **~ulmak** *vn. pass. of* **boğmak,** be choked, drowned *etc.*; gasp for breath; choke with laughter *or* anger; be fleeced.
**boğum** Node; articulation; choke (gun). **~lu,** articulated; knotty; wrinkled.
**boğun** *As* **boğum**; hole in the roof for smoke *or* ventilation.
**boğuntu** Suffocation; oppression; swindling, cheating *at a game.* **~ya**

**getirmek,** to squeeze money out of *s.o. by putting him in an embarrassing position*: **lâfı ~ya getirmek,** to twist words, to quibble.

**boğuşmak** *vn.* Fly at one another's throats; quarrel; fight.

**bohça** Square piece of stuff for wrapping; bundle; parcel; square shawl; selected and finely cut tobacco. **~ etm.,** to wrap up in a bundle: **~sını koltuğuna vermek,** to 'sack' *s.o.*: **parça ~sı,** rag-bag: **parça ~sı gibi,** made of ill-assorted pieces. **~cı,** woman pedlar of small draperies, *who used to visit harems.* **~lamak,** to wrap up, make a parcel of.

**bok, -ku** Excrement; ordure. Worthless. **~tan,** made of rubbish, useless. **~böceği, -ni,** dung-beetle. **~lamak,** to soil, befoul; besmirch, bring into disrepute; mismanage. **~laşmak,** become bad *or* difficult; be annoyed *or* bored; meddle. **~lu,** fouled with dung, filthy. **~luca,** thicket: **~ bülbülü,** wren; pert and talkative person. **~luk,** dungheap; filthy place; state of disorder *or* misery: ⌜**nerede çokluk orada ~**⌝, too many people spoil a party *etc.*; *sometimes* 'too many cooks spoil the broth'.

**boksit, -ti** Bauxite.

**bol**[1] Wide; loose; ample, copious, abundant. **~ ahenk,** fond of gaiety: **~ ~** *or* **~ bolama,** abundantly; generously: **~ bulamak,** to lay on thickly; to give abundantly: **~ keseden,** generously; **~ keseden atmak,** to make extravagant promises; to talk wildly about spending money: **~ paça,** wide trouser legs. **~lanmak,** to become wide *or* loose; become copious *or* abundant. **~luk,** wideness; looseness; easiness of fit; abundance, plenty: **~ bir memleket,** a land of plenty: '**nerede bu ~?**', *a reply to an excessive demand or statement,* 'that's a bit stiff, isn't it?'

**bol**[2], **-lü** A drink of mixed liqueurs and fruit-juice.

**bol**[3], **-lü** The game of bowls.

**bomba** (ˈ·) Bomb. **~lamak,** to bomb.

**bombarda** (·ˈ·) A kind of sailing boat *used in the Aegean.*

**bombardıman** Bombardment; bombing.

**bombok** (ˈ·) Utterly spoilt; quite useless.

**bombort, bonbort** Bombardon.

**bomboş** (ˈ·) Completely empty; utterly nonsensical.

**boncuk** Bead (*esp. a blue bead in a child's hair or a horse's mane to avert the evil eye*). **~ illeti,** infantile convulsions: ⌜**mavi ~ kimde ise benim gönlüm ondadır**⌝, *phrase from a Nasreddin Hodja story meaning* 'to please both parties'.

**bonfile** Best undercut of beef.

**bonjur** Morning-coat.

**bono** (ˈ·) Bond; bill. **açık ~,** blank cheque; 'carte blanche'.

**bonservis** (*Fr. bon service*). Certificate of good character.

**Bor**[1] *A town in Anatolia.* ⌜**geçti ~'un pazarı, sür eşeği Niğde'ye**⌝, (i) it's too late now to do anything; (ii) things have changed now.

**bor**[2] Waste *or* uncultivated land. Wine.

**bor**[3] Boron.

**bora** (ˈ·) Squall, tempest; violent reproach *or* abuse. **~ yemek,** to be exposed to a storm of wind *or* words. **~ğan** (··ˈ), whirlwind; tempest.

**borazan** *Originally 'trumpeter', now only* trumpet. ⌜**sesine (nefesine) güvenen ~cı başı olur**⌝, 'by all means do it, if you think you can!'

**borc** Debt; loan; duty, obligation. **~a alışveriş etm.,** to buy on credit: **~ almak,** to borrow: **~ bini aşmak,** to have a pile of debts: **~ etm.,** to incur a debt: **~ gırtlağa çıkmak,** to be up to the ears in debt: **~ harc,** getting money by hook or by crook: **~ vermek,** to lend: **boynunun* ~u,** one's first duty. **~lanmak,** to become indebted. **~lu,** indebted; debtor; under obligation: **uçan kuşa ~ olm.,** to be in debt all round.

**borda** (ˈ·) Ship's side; broadside. **~ ~ya,** alongside (*naut.*).

**bordro** (ˈ·) (*Fr. bordereau*). Memorandum; account; docket.

**borsa** (ˈ·) Bourse; stock-exchange.

**boru** Tube; pipe; trumpet; idle tale, nonsense. **~ değil,** it's no small matter: **~su ötmek,** to be the 'big noise', to be the most important person in a place: **kalk ~su,** reveille: ⌜**yem ~su çalmak**⌝, to put off with empty promises. **~çiçeği, -ni,** datura.

**bos** *Only in conj. with* **boy,** Figure; statute.

**Bosna** (ˈ·) Bosnia.

**bostan** Vegetable garden (*esp.* melon

garden); cucumber; melon. ~ **dolabı,** irrigation water-wheel: ~ **korkuluğu,** scarecrow; lazy *or* incapable person. ~**cı,** gardener; (*formerly*) one of the Sultan's body-guards: ~ **başı,** officer commanding the Sultan's body-guards.

**boş** Empty; empty-handed; vain, useless; unoccupied; unemployed; divorced; loose, untethered. ~**ta,** unemployed: ~**una,** in vain: ~**unu almak,** to take up the slack; to trim an unlikely story so as to make it plausible: ⌜~ **atıp dolu tutmak**⌝, to 'draw a bow at a venture' and unexpectedly to hit the mark; to make a lucky shot; to learn something one wants to know by subtle questions: ~ **bulunmak,** to be taken unawares; to be surprised: ~**a çıkmak,** to come to nothing, to fail: ~ **düşmek,** (of a woman) to be divorced: ~ **gezmek,** to idle about, to be without work: ⌜~ **gezenin** ~ **kalfası**⌝, ne'er-do-well, vagabond. ~ **kâğıdı,** a written declaration of divorce: ~ **olm.,** to be empty; to be useless; to be unoccupied: ~ **vermek,** to pay no attention: ~ **yere,** in vain: **başı** ~, without bridle or halter; vagabond; not in any occupation, independent: **eli** ~ **durmak,** to remain idle: **vakti** ~**una geçirmek,** to waste time.

**boş·almak** *vn.* Be emptied; run out; (animal) get loose. ~**altmak** *va.* empty; pour out; discharge (firearm). ~**ama,** divorce. ~**amak,** to divorce. ~**anmak,** to be loosed, break loose, escape; be poured; pour with rain; (firearm) be discharged; be emptied; be divorced; empty one's heart; burst into tears: **bardaktan boşanırcasına yağmur yağıyor,** it's raining in torrents: **sinirler** ~, to be unable to refrain from laughing. ~**atmak,** to make *or* let *a wife* be divorced. ~**boğaz,** garrulous, indiscreet. ~**boğazlık,** idle talk, indiscreet babbling. ~**lamak,** to loose, let go; abandon, neglect. ~**luk,** emptiness; vacant space; vacuum; vanity; uselessness; leisure.

**Boşnak** Bosnian.

**bot, -tu** Dinghy.

**boy**[1] Length; height; depth; stature; size. ~ ~, of different sizes *or* qualities: ~**dan** ~**a,** from end to end; completely: ~**una,** lengthwise; continually; along: ~**unca,** lengthwise; according to its length; along; **sahil** ~, along the shore; **tarih** ~, in the course of history; *v. also* **boylu**: ~ **almak (atmak, çekmek, sürmek),** to increase in stature: ~**u bosu,** *v.* **boybos**: ~**a çekmek,** (child) to keep growing taller without growing proportionately broader: ~**u* devrilsin!,** (*a curse*) 'may he die!': ~ **göstermek,** to put in an appearance *and do nothing*; *but* '**gelsin** ~**unu göstersin!**', let him come and we will see what he is made of: ~**unun* ölçüsünü almak,** to get one's deserts; to learn by painful experience: ~ **ölçüşmek,** to compete *with s.o.*: ~**undan utan!,** you should be ashamed to do this at your age: **dere burada** ~ **vermez,** the river is out of one's depth here.

**boy**[2] Branch of a race (people).

**boya** Dye; paint; colour. ~ **tutmak,** to take a dye; (of a dye) to become fast: ~ **vurmak,** to paint *or* dye: **atmaz** ~, fast dye: **göz** ~**sı,** eyewash: **her** ~**dan boyamak,** to be a jack-of-all-trades: **kundura** ~**sı,** boot polish: ~**cı,** dyer; colourman; shoeblack: ~ **küpü,** dyer's vat; ⌜~ **küpü değil**⌝, 'it's not so easy as all that!' ~**hane** (··—‿), dye-house; colourman's shop. ~**lı,** painted; dyed; coloured. ~**ma,** action of painting *etc.*; coloured, painted, dyed; false, imitation. ~**mak,** to paint, dye; swear at *s.o.*: **göz** ~, to hoodwink: **birbirinin gözünü** ~, (of two parties) to pretend to believe *stg.* when each knows the other does not believe it.

**boyan** Liquorice plant. ~**balı, -nı,** liquorice. ~**kökü, -nü,** the root of the liquorice plant.

**boy·bos** (‿·), *acc.* **boyu bosu.** Stature. **boyu bosu yerinde,** tall and well-built. ~**ca** (‿·), as regards height *or* length: ~ **evlâd,** a practically grown-up child: ~ **günaha girmek,** to commit a great sin. ~**lamak,** to measure the length *or* height of; to traverse the length of; (*sl.*) to betake oneself to: **ahreti** ~, to die: **elliyi boyluyorum,** I am getting on for fifty: **mahkemeyi** ~, to have to appear in court. ~**lanmak,** to grow, to become long *or* tall. ~**latmak,** to make growth in height *or* length; to make *s.o.* go to a distant place. ~**lu,** possessing length *or* height: ~ **boslu,** well-built, well-developed: ~ **boyuna,** at full length: ~ **boyunca,** in its entire length: **uzun ~**, tall. **~suz,** short in stature.

**boyna** Single oar used over the stern to steer or scull a boat. **~ ile yürütmek,** to scull *a boat.*

**boynu** *v.* **boyun.**

**boynuz** Horn; cupping-glass. **~ çekmek,** to cup: **···e ~ taktırmak,** to make a cuckold of *s.o.*: **~ vurmak,** to gore. **~lu,** horned; cuckold; pimp; procuress. **~otu, -nu,** hellebore.

**boyun**[1] *gen. of* **boy.**

**boyun**[2], **-ynu** Neck. **boynuna* almak,** to take upon oneself: **boynu altında kalsın!,** (*a curse*) 'may he perish!': **boynuna* atmak,** to impute to: **boynuna* binmek,** to dun, to persecute: **~ borcu,** a binding duty: **~ eğmek,** to bow the neck, to submit; to humiliate oneself: **boynu* eğri,** hanging the head in shame: **~ kesmek,** to bow the head *in respect or humility*: '**boynum kıldan ince**', I am at your mercy; upon my head be it!: **~ noktası,** the point where two hills join: **boynu* tutulmak,** to have a stiff neck: **~ vermek,** to obey: **boynunu* vurmak,** to decapitate: **günahı boynuna*,** on his head be it!

**boyuna** 1. *Dat. of* **boyun.** 2. *Dat. with possessive suffix of* **boy,** *q.v.*

**boyun·atkısı, -nı,** Scarf. **~bağı, -nu,** necktie; dog-collar. **~buran,** wryneck. **~ca,** *v.* **boy.** **~duruk,** yoke: **boyunduruğa vurmak** *or* **~ altına almak,** to put under the yoke, to reduce to servitude *or* submission. **~luk,** scarf; collar.

**boz** Grey. Uncultivated land.

**boza** (ˈ·) Boza, *a fermented drink made of millet.* **ensesinde ~ pişirmek,** to torment; to inflict punishment on.

**bozarmak** *vn.* Become grey; become brown *or* sunburnt. **kızarıp ~,** to turn red and pale alternately *from fear, rage, or shame.*

**bozca** Greyish. Uncultivated land. **Bozca Ada,** Tenedos.

**bozgun** Routed, defeated. Rout, defeat. **~ vermek,** *or* **~a uğramak,** to be routed. **~cu,** defeatist. **~culuk,** defeatism.

**bozkır** Pale grey. Steppe.

**bozkurt** Grey wolf; *legendary wolf said to have led the Turks across a mountain barrier into the open world.*

**bozma** Act of spoiling *etc.* Spoilt; demolished; made out of old materials, reconstructed. **~cı,** one who buys old things to use the materials again.

**bozmak** *va. & vn.* Derange; spoil; ruin, destroy; deprave, corrupt; deflower; disconcert; cancel; break (oath, treaty); change (money); defeat, rout; gather final crop of *grapes etc.*; break up and reconstruct into *stg.* else; go mad; (of weather) to deteriorate. **abdest ~,** to relieve nature: **ağız ~,** to abuse, vituperate: **aklını* ~,** to go off one's head: **and ~,** to violate an oath: **bağ ~,** to gather in the vintage: **birisile ~,** to 'have a down on' *s.o.*: **bir şeyle ~,** to have a bee in one's bonnet about *stg.*: **oruc ~,** to break a fast: **şakaya ~,** to turn into a jest, to make a mockery of: ⌜**Allah yazdıysa bozsun!**⌝, God forbid!: **yapıp ~,** to make and remake, to constantly reconstruct.

**bozuk** Destroyed; spoilt; out of order; ruined; broken; depraved, corrupt. **~ para,** small change: **ahlâkı ~,** depraved: **başı ~** (**başıbozuk**), irregular soldier; civilian. **~düzen,** disordered, irregular. **~luk,** ruin; a being out of order *or* broken down; defeat; small change.

**bozulmak** *vn. pass. of* **bozmak.** Be spoilt *etc.*; break down; deteriorate; (meat *etc.*) go bad; be disconcerted; look vexed; grow thin; become pale and ill.

**bozum** *verb. n. of* **bozmak.** **~ olm.,** to be disconcerted: **~ havası,** the embarrassment caused in company by anyone's discomfiture: **bağ ~u,** the vintage.

**bozuntu** Discomfiture, embarrassment; old materials, scrap; *also as* **bozma** *q.v.* **~ya uğramak,** to be embarrassed *or* discomfited: **~ya vermemek,** not to let oneself be disconcerted: **şair ~su,** a mere parody of a poet.

**bozuş·mak** *vn.* Quarrel with one another; become estranged. **~uk,** who has quarrelled; on bad terms.

**böbrek** Kidney. **~ yağı,** suet.

**böbür** Tiger; *sometimes* leopard. **~lenmek,** to assume an arrogant air, to boast.

**böcek** Insect; crayfish (spiny lobster). **~kabuğu, -nu,** bright greenish-blue colour; iridescent. **~lenmek,** to be infested with insects.

**böğ** A kind of poisonous spider.

**böğrülce** (·ˈ·) Kidney-bean.

**böğür, -ğru** Side; flank. **boş ~,** the small of the back: **eli böğründe,** disheartened; unable to do anything.
**böğürmek** *vn.* Bellow; low.
**böğürtlen** Blackberry.
**bölge** District; zone.
**bölme** Bulkhead; partition; dividing wall. **su ~leri,** watertight compartments.
**bölmek** *va.* Separate; divide; cut up. **ikiye ~,** to divide into two.
**bölük** Part; subdivision; fragment; compartment; company, squadron (*mil.*).
**bölüm** Act of dividing; portion; slice; chapter; article.
**bölünmez** Indivisible.
**bölüşmek** *va.* Divide up; share out.
**bön** Silly; naïve; vacant; imbecile.
**börek** *Name given to various kinds of* pastry *or* pie. **Nemse böreği,** fritter.
**börtmek** *vn.* Be slightly cooked, be half cooked. **sıcaktan ~,** to be half broiled in the heat.
**börülce** Kidney-bean.
**böyle** (ˈ·) Such; similar to this. So, thus, in this way. **~ iken,** while it is *or* was thus; notwithstanding: **~ ise,** if so, in that case: **bundan ~,** henceforth: **şöyle ~,** so so, not too well. **~ce** (ˈ··), somewhat in this fashion. **~likle** (··ˈ·), in this manner.
**branda** Sailor's hammock; awning *or* hood *of a lorry etc.* **~ bezi,** canvas.
**bre** Now then!, hi, you!; *sometimes in admiration,* wonderful! **bekle ~ bekle,** he (we *etc.*) waited and waited.
**Brehmen** Brahmin.
**brik** Brig.
**brom** Bromium. **~ür,** bromide.
**bronş** (*Fr. bronche*) Bronchus, windpipe. **~it, -ti,** bronchitis.
**broş** Brooch.
**bu** This; *pl.* **bunlar,** these.
**bucak** Corner; angle. **~ ~ aramak,** to hunt in every hole and corner: **birine dünyanın kaç ~ olduğunu göstermek,** to teach *s.o.* a lesson, to give him 'what for': **ucu bucağı yok,** vast.
**buçuk** Half (*only after numerals*). **az ~,** just a little: **bir ~,** one and a half: **saat iki ~,** the time is half-past two.
**Buda** (ˈ·) Buddha. Budapest.
**budak** Twig; shoot; knot *in timber.* **gözünü ~tan sakınmamak,** to be fearless. **~lanmak,** to send forth shoots; become knotty *or* snaggy; become complicated *or* difficult. **~lamak,** to bud-graft.
**budala** Silly fool; imbecile; greedy. **futbol ~sı,** mad on football.
**buda·mak** *va.* Prune; lop, trim. **~nmak,** to be pruned *or* trimmed; to apply oneself assiduously to *stg.*
**buğday** Wheat. **~ benizli,** light-complexioned: **mısır ~ı,** maize.
**buğu** Steam; vapour; mist. **~hane** (··–ˈ), vapour bath; sterilizer; oveu *for killing silkworms.* **~lanmak,** to be enveloped in vapour *or* steam; (glass) to be misted over.
**bugün** (ˈ·) Today. **~e ~,** *adds emphasis to the succeeding statement,* 'don't forget that ...': **~ü* yarınına* uymamak,** for one's circumstances (one's mood *etc.*) to be liable to change. **~edek** (ˈ···), up till today. **~kü**(ˈ··), of today, today's. **~lük** *a.*, just for today: **~ yarınlık,** that may happen any moment.
**buhar** (·ˈ–) Steam; vapour; exhalation.
**buhran** (·ˈ–) Crisis. **~lı** (·–ˈ–), critical.
**buhur** (·ˈ–) Incense; fumigation. **~dan** (·–ˈ–), **~luk,** censer. **~umeryem,** cyclamen.
**bukadar** *v.* **kadar.**
**bukalemun** (···ˈ–) Chameleon; fickle *or* changeable person; shot silk.
**buket, -ti** Bouquet.
**bukle** (*Fr. boucle*) Lock of hair; curl.
**bulama** Grape-juice *boiled down to the consistency of honey.*
**bulamac** Thick soup *made with flour, butter, and sugar.*
**bulamak** *va.* Smear; bedaub; dirty, soil; mix; coat *or* cover (*with* = *dat.*).
**bulan·dırmak** *va.* Render turbid *or* muddy; cloud (the eye *or* mind); turn (the stomach). **~ık,** turbid; cloudy; overcast. **~mak,** to become cloudy *or* turbid; become nauseated; (of the eye) to become bloodshot *or* opaque; (of the eye *or* mind) to be clouded *or* dimmed. **~tı.** nausea.
**bulaş·ıcı** Contagious, infectious. **~ık,** smeared over; soiled; infected; contagious; suspect *after being in contact with infection*; (ship) not having a clean bill of health. Dirt; contagion; dirty kitchen utensils: **~ bezi,** dishcloth: **~ suyu,** dirty water; dishwater. **~ıkçı,** dishwasher.
**bulaş·mak** *vn.* Be smeared *or* stuck; become dirty. **···e ~,** to come into contact with *contagion*; be involved

in *an affair*; take in hand; have to do with. **~tırmak,** to smear; stick on; dirty; infect; involve in *stg.* unpleasant: **yüzüne gözüne ~,** to make a mess of *stg.*, to fail.

**buldok** Bulldog.

**bulgur** Boiled and pounded wheat.

**Bulgurlu** *In* ⌜**~ya gelin mi gidecek?**⌝, 'why all this unnecessary fuss?'

**bullak** *v.* **allak.**

**bulmaca** Cross-word puzzle.

**bulmak, -ur** *va.* Find; invent; obtain; reach. **buldukça bunamak,** to be never content but always to ask for more: **Allahtan bulsun!,** may God punish him!: **aradığını buldu,** he got what he was asking for, *i.e.* his deserts: **belâsını* ~,** to meet with well-merited punishment: **(bilmem neden) buldu buldu da beni buldu,** I don't know why he picked on me (of all people): **ettiğini* ~,** to pay for one's misdeeds: **fena ~,** to die; to come to an end: **seksenini* ~,** to reach the age of eighty: **vuku ~,** to happen: **yerini ~,** to reach its proper place: **yüz ~,** to have the presumption, to dare.

**bulundurmak** *va.* Make *or* let be present; have ready; have in stock.

**bulun·mak** *vn. pass. of* **bulmak.** Be found *etc.*; be present; be (*cf. Fr.* '*se trouver*'); **···e ~,** to assist. **~maz,** not to be found; rare, choice: **~ Hind Kumaşı,** *used ironically of stg. erroneously thought to be rare or precious.*

**buluş** Act of finding; invention, discovery.

**buluş·mak** *vn.* Be together with others; meet. **~turmak,** to bring together; arrange a meeting of: **bulup ~,** to find at any cost.

**bulut, -tu** Cloud. **~ gibi sarhoş,** dead drunk. **~lanmak,** to become cloudy *or* opaque: **gözleri ~,** to be on the verge of tears. **-lu,** cloudy, overcast.

**bulûz** Blouse.

**bumba** (ˋ·) Boom (*naut.*).

**bumbar** Intestine; sausage; draught excluder.

**buna** *dat. of* **bu.**

**bunak** Imbecile; in one's second childhood, senile.

**bunalmak** *vn.* Be stupified; be suffocated *with smoke etc.*: be utterly bored *or* wearied.

**bunamak** *vn.* Reach one's dotage. **buldukça ~,** to be always grumbling, never to be satisfied.

**bunca** (–́·) In this fashion; this much. **~ kere,** so many times: **~ zaman,** so long a time. **~layın** (ˋ···), so much, so many; in this way.

**bundan** *abl. of* **bu. ~ böyle,** from now on, henceforth: **~ dolayı,** on account of this.

**bunu** *acc. of* **bu.**

**bunun** *gen. of* **bu. ~la beraber,** however, in spite of this: **~ üzerine,** then, thereupon.

**bura** (–́·) This place; this condition; this point. **~m,** this part of me: **~ya,** to this spot, here: **~dan,** from here: **~ ahalisi,** the people of this place.

**buralı** (ˋ··) Belonging to this place.

**buram ~ ~,** whirling, eddying; like a whirlpool; excessively: **~ ~ terlemek,** to sweat profusely.

**buranda** (·ˋ·) Canvas hammock.

**burası** (ˋ··) This place, here. **~ neresi?,** what place is this?

**burc**[1] Tower; sign of the Zodiac.

**burc**[2] Mistletoe.

**burçak**[1] Vetch.

**burçak**[2] Screwed; wound round itself.

**burçun** Female roedeer.

**burgacık** (ˋ··) *In* **kargacık ~,** bad scrawling (writing); crooked, twisted.

**burgu** Auger; gimlet; corkscrew.

**burjuvazi** Bourgeoisie.

**burkulmak** *vn.* Be sprained. **içi* ~,** to feel a pang *of remorse, pity, or envy.*

**burkutmak** *va.* Sprain.

**burma** Act of twisting; castration; screw; convolution; griping of the stomach; sweet *made of dough, oil, and sugar.* Screwed; twisted; spiral; castrated.

**burmak** *va. & vn.* Twist; wring; castrate; dislocate; (of the bowels) to gripe; bore a hole. **dudak ~,** to screw up the lips.

**burnu** *v.* **burun.**

**burnuz** Burnous, woollen Arab cloak; bath robe.

**burs** (*Fr. bourse*) Scholarship.

**buruk** Twisted; sprained; acrid, sour.

**burulmak** *vn. pass. of* **burmak.** Be twisted; writhe with pain; be offended.

**burum** Torsion; contortion. **~ ~,** contorted; griped.

**burun, -rnu** Nose; beak; promontory; pride, arrogance. **burnu* büyümek,** to become conceited: **burnunu* çekmek,** to sniff; to go away emptyhanded: **burnuna* çıkmak,** to become intolerable: **burnunun* doğrusundan**

**ayrılmamak,** to be above taking any one's advice, to be too conceited to listen to others: **burnundan düşen bin parça,** very conceited; unapproachable; in the worst of tempers: ⌜**(hık demış) burnundan düşmüş**⌝, the very spit of ... (*usually in a derogatory sense*): **burnumdan* geldi,** it completely spoilt my pleasure: **burnundan* getiririm,** I'll make you sorry for it: **burnuna* girmek,** to come too close to *s.o.*: **burnunun* kemiği sızlamak,** to feel very sad: **~ kırmak,** to screw up the nose *in dislike or contempt*: **burnunu* kırmak,** to take the conceit out of *s.o.*: **burnunun* ucunda,** right under one's nose: **burnunun ucunu görmemek,** to be blind with pride; to be very drunk: ⌜**ağzından* girip burnundan* çıkmak**⌝, to try all sorts of tricks on *s.o.* to gain one's end: **canım* burnumdan* geldi,** I was dead tired.

**burun·deliği, -ni** Nostril. **~duruk,** twitch *for holding an unruly horse.* **~kanadı, -nı,** outside of the nostril, ala. **~luk,** nose-ring *of a bull*; iron toe-cap of a boot. **~salık,** muzzle.

**buruntu** Colic.

**buruş·mak** *vn.* Be wrinkled, creased, *or* puckered; have the teeth set on edge. **~turmak,** to crease, wrinkle; corrugate; frown; set the teeth on edge. **~uk,** puckered; wrinkled; ruffled.

**buse, puse** (–̀) Kiss.

**but, bud** Thigh; leg of meat. ⌜**eti ne, budu ne?**⌝, what can you expect from that poor, weak creature ?

**butlan** Error; a being null and void.

**buud, -u'du** Distance; dimension.

**buymak** *vn.* Freeze; freeze to death.

**buyruk, buyuruk** Order, command. **başına ~,** who goes his own way without asking others' permission *or* advice; who is his own master.

**buyrultu** Order; decree; nomination paper *for lower officials.*

**buyurmak** *va.* Order; deign *or* condescend to do; *used as an auxiliary instead of* **etmek** *when addressing a person of rank or when wishing to be polite*; **buyurunuz!,** please *sit down, lead the way etc.*; **ne buyurdunuz?,** I beg your pardon, what did you say?; **ne buyururlar** (*politely*) what is your opinion? (*sarcastically*) what have you got to say about that?: ⌜**buyurun cenaze namazına**⌝, (come to the funeral!), we are done for!: ⌜**tembele iş buyur, sana akıl ögretsin**⌝, 'offer work to a lazy one, let him give you good advice', *said of or to a man who tries to get out of doing stg. by suggesting a different course of action*: **teşrif ~,** to do the honour of visiting.

**buz**[1] *v.* **az.**

**buz**[2] Ice. Very cold; frozen. **~ gibi,** very cold; new and clean; fresh and tender (meat); *also* 'of course', 'as a matter of course': **~ bağlamak,** to form ice, to freeze over: **~ kesmek,** to feel very cold: **~ kesilmek,** to be frozen; to be petrified *with fear, etc.*: **~ üstüne yazmak,** to do *stg.* which will not last, to build on sand: **~ yalağı,** a depression where snow lies. **~hane,** ice-factory; ice-house. **~kıran,** ice-breaker. **~lanmak,** to become ice; be covered with ice; lose polish, become dull. **~lu,** iced; dulled; ground (glass). **~lucam** (·̀·), ground glass. **~luk,** ice-factory; ice-house; ice-box.

**buzağı** Sucking-calf.

**buzul** (·̀) Glacier.

**bücür** Short, squat, dwarf; 'cute' (*of a child, or of an insignificant person from whom one would not expect 'cuteness'*).

**büğrü** *v.* **eğri.**

**büğü** Sorcery.

**bühtan** (·–̀) False accusation; calumny. **~cı,** calumniator.

**bükâ** (·–̀) Weeping.

**büklüm** Twist; curl. **iki ~,** bent double with age.

**bükmek** Twist; spin; curl; contort; bend. **belini ~,** *v.* **bel**: **boyun ~,** to bow; **boynunu* ~,** to resign oneself to the inevitable; to be bowed down by grief: **dudak ~,** to curl the lip *in disdain or contempt*: **kol ~,** to twist the arm; to overcome: **···in kulağını ~,** to drop a hint to *s.o.*

**bükük** Curved; twisted. **boynu ~,** wretched: **kulakları ~,** warned beforehand.

**bükül·mek** *vn. pass. of* bükmek. Be twisted *etc.* **beli ~,** to be bent double with age; to be handicapped by *stg.* **~ü,** bent; twisted; spun.

**büküm** Act of twisting *or* spinning; torsion; twist; once-spun wool *etc.*; skein; yarn.

**büküntü** Fold; twist; knot; hemstitch.

**bülbül** Nightingale ⌜**~ün çektiği dili**

belâsıdır⌉, one's tongue is apt to get one into trouble. ~ **gibi,** fluently.
**bülten** Bulletin.
**bülûğ** (·-) Puberty; maturity; age of discretion. ~**a ermek,** to reach the age of puberty.
**bünye** Edifice; structure (of the body *etc*); constitution.
**büro** Bureau; office.
**bürülü** Wrapped up; enfolded.
**bürüm** A wrapping up, folding; fold. ~**cek,** anything wrapped up like a cocoon. ~**cük,** raw silk; crêpe; gauze.
**bürümek** *va.* Wrap; enfold; cover up; invade, infest. **gözünü* kan ~,** to be near murder with anger; to see red.
**bürüncük** *v.* **bürümcük.**
**bürünmek** *vn.* Wrap oneself up.
**büsbütün** (`··) Altogether; quite.
**bütçe** Budget.
**bütün** Whole, entire, complete. Wholly, completely. The whole. ~ **gün,** the whole day. ~**leme,** a second examination for those who failed the first time. ~**lemek,** to complete; to repair.
**büvelek, büye, büyelik** Gad-fly.
**büyü** Spell, incantation; sorcery. ~**yü bozmak,** to break a spell: ~ **otu,** thorn-apple (datura): ~ **yapmak,** to practise sorcery, to cast a spell. ~**cü,** sorcerer, magician. ~**lemek,** to cast a spell on, to bewitch. ~**lü,** bewitched; having magic powers.
**büyük** Great, large, high; important; elderly. ~ **ana (anne),** grandmother: ~ **baba,** grandfather: ~ **kafalı,** highly intelligent: ~ **söylemek,** to dogmatize; to be too cocksure; to talk in a manner for which one will be sorry later: **gözü*** ~**te olm.,** to aim high, to have great ambitions.
**büyültmek** *va.* Make great; exaggerate.
**büyümek** *vn.* Grow large; grow up; increase in importance. **büyülmüş de küçülmüş,** (of children) appearing to be much older than they are, precocious: **gittikçe burnu büyüyor,** he gets more conceited every day: **gözleri büyümüş bir halde,** with eyes starting out of his head *with terror etc.*
**büyütmek** *va.* Make great; enlarge; exaggerate; nourish and cherish; bring up. **gözde ~,** to exaggerate the importance of *stg.*
**büzgü** Pucker; pleat.
**büzmek** Constrict; pucker. ⌈**her kesin ağzı torba değil ki büzesin**⌉, you can't stop people talking: **lâkırdıyı ezip ~,** to hum and haw without being able to express one's meaning.
**büzük** Constricted; puckered. Anus.
**büzülmek** (··`) *vn.* Contract; shrink; shrivel up; cower, crouch. **bozulup ~,** to be disconcerted and 'retire into one's shell': **ezile büzüle,** humbly, apologetically.

# C

**caba** (`·) Thrown into the bargain. ~ **etm.,** to give gratis: ... **da ~sı,** and into the bargain ...: **bu da ~,** and what's more, and into the bargain. ~**cı,** sponger, parasite.
**cacık** A drink *made of* **ayran** *q.v., with pieces of cucumber in it.*
**cadaloz** A spiteful old hag.
**cadde** Main road; thoroughfare; highway. ~**yi tutmak,** to 'clear out'.
**cadı** Witch; hag; vampire.
**cafcaf** Pompous *or* pretentious speech; ostentation. ~**lı,** pompous; showy; elegant.
**cağ** Spoke of a wheel. ~ **torbası,** spare nosebag *for a horse.*
**-cağız, -ceğiz** *Suffix implying smallness or affection.* **evceğiz,** a little house: **adamcağız,** the good fellow; the poor man.
**cahil** (-`) Ignorant; inexperienced; uneducated; young. **kara ~,** utterly ignorant. ~**iyet, -ti,** the Age of Ignorance, *i.e.* prior to Mohammed. ~**lik,** ignorance; inexperience; youth.
**caiz** (-`) Lawful; permitted, admissible; possible. ~**e,** mark, dash, tick; present, reward.
**caka** (`·) Ostentation; swagger. ~ **satmak (yapmak),** to swagger, to show off. ~**cı,** ~**lı,** swaggering.
**câli** (--) False; imitation; insincere.
**cam** Glass; photographic plate. ~ **evi (yuvası),** rebate *or* groove for glass. ~**cı,** glazier. ~**lı,** glass-covered, glazed.
**camadan** Double-breasted waistcoat; chest for clothes *or* linen; valise; reef-point. ~ **bağı,** reef-knot: ~**ı fora etm.,** to shake out a reef: ~ **vurmak,** to take in a reef.
**cambaz** Acrobat; circus-rider; horse-dealer; swindler. ~**hane** (··-`), circus.

**cam·dolab** (·ˈ·) Glass-fronted case *for books etc.* **~ekân,** shop-window; show-case; garden frame; dressing-room of a bath. **~göbeği,** glass-green. **~göz** (·ˈ), tope; avaricious.

**cami** (–ˈ), **-ii** *or* **-yi** Mosque. ⌜**eceli gelen köpek ~ duvarına siyer** (**işer**)⌝, 'the dog who is doomed to die makes water against the wall of a mosque', ⌜quem deus vult perdere, prius dementat⌝, ⌜**iki ~ arasında beynamaz**⌝, 'who hesitates between two mosques, thereby missing service at either', ⌜falling between two stools⌝.

**câmi** (–ˈ) Collecting; bringing together. **~a,** assembly of people; the community.

**cam·lamak** *va.* Cover with glass; glaze. **~lık,** anything covered with glass; shop-window.

**camus** (·ˈ) Water buffalo.

**can** Soul; life; a soul, a living person; darling, beloved friend; a member (*of certain Moslem sects*); force, vigour, zeal. Agreeable, pleasant. **~dan,** very sincerely: **~ım!,** (i) 'my dear fellow!', 'my good man!' (*expresses a mild reproof*); (ii) *as an adj.*, beloved: **~ acısı,** fearful pain: **~ acıtmak,** to cause suffering, to oppress: **~ı* ağzına* gelmek,** to be half-dead *from anxiety or fatigue*: **~ alacak yer,** the most sensitive spot; **~ alacak** *or* **alıcı nokta,** a vital point: **~ını almak,** to take *s.o.'s* life: **~ atarcasına,** ardently: **···e ~ atmak,** to desire passionately: **~ı aziz,** one's own dear self: **~la başla,** with heart and soul: **~ ~a baş başa,** each for himself, 'sauve qui peut': **~ı* çekilmek,** to feel extreme pain: **~ı* çekmek,** to long for: **~ını* çıkarmak,** to exhaust, to ruin, to ill-treat: **~ı* çıkmak,** to die; to be very tired; (of clothes) to be very worn; ⌜**~ çıkmayınca huy çıkmaz**⌝, ⌜what's bred in the bone won't out of the flesh⌝: **~ damarı,** the vital spot, the most important point: **~ dostu,** a very sincere friend: **~ düşmanı,** a mortal enemy: **~ evi,** the heart; the pit of the stomach; any vital spot: **~ü gönülden,** most willingly: **~ı* istemek,** to feel (physically) like doing *stg., but* **~ın isterse!,** 'all right!', 'as you will!': **~ına* işlemek** (**kâr etm.**), to become intolerable: **~ kalmamak,** to have no life left in one, to be utterly exhausted: **~a ~ katmak,** to make one feel more alive, to refresh, to delight: **~ kurtaran yok mu?,** 'help!': **~ını* yakmak,** to hurt (*esp.* the feelings): **~ına yandığım,** the cursed ... (*sometimes used in admiration*): **~ı* yanmak,** to suffer pain *or* loss: **ne ~ı var?,** he's not got much strength left: **vay ~ına!,** by Heaven!, by Jove!: ⌜**çok veren maldan, az veren ~dan**⌝, it is not the gift that matters, but the spirit in which it is given; 'the widow's mite'.

**canan** (–ˈ) Sweetheart. ⌜**evvela can sonra ~**⌝, everyone thinks first of his own interests; ⌜number one comes first⌝.

**canavar** Monster; brute; wild beast; wild boar. **~ düdüğü,** a warning syren. **~lık,** savagery, ferocity, barbarity. **~otu, -nu,** broom-rape.

**canciğer** *In* **~** (**kuzu sarması**), people very dear to each other.

**caneriği** (ˈ···), **~ni** A sweet and juicy kind of green plum.

**canevi** (ˈ··), **-ni** Any vital part of the body; pit of the stomach; seat of the affections.

**canfes** Taffeta. Made of taffeta.

**canhıraş** (–·ˈ) Harrowing; disagreeable.

**cani** (–ˈ) Criminal.

**can·kulağı** (ˈ···), **-nı** Rapt attention. **~kurtaran,** lifebuoy, lifebelt. Life-saving: **~ otomobili,** motor ambulance. **~lanmak,** to come to life, to become active *or* lively. **~lı,** alive; lively; active; vigorous: **iki ~,** pregnant: **yedi ~,** having seven lives; invincible: **para ~sı,** a lover of money. **~pazarı** (ˈ···), **-nı,** a situation where life is at stake. **~sağlığı, -nı,** health; *v. also* **sağlık. ~sıkıntısı, -nı,** boredom. **~sız,** lifeless; weak, slack, dull.

**cant, -tı** (*Fr. jante*) Felloe; rim *of a bicycle*.

**car**[1] Woman's shawl *or* cloak.

**car**[2] **~ ~,** *used to express loud and continuous noise or talk.* **~car,** chatterbox.

**cari** (–ˈ) Running, flowing; current (money *etc.*); valid; occurring, present; usual.

**cariye** (–·ˈ) Female slave; concubine; *politely for* 'my daughter'.

**carlamak** *vn.* Talk loudly and incessantly.

**cart** *Imitates the sound of something being torn.* **~ curt etm.,** to use threatening words. **~adak,** suddenly and noisily.

**cascavlak** (ˋ··) Completely naked *or* bald. **~ kalmak,** to be left destitute *or* helpless.

**-casına, cesine** *Suffix denoting* 'in the manner of', *e.g.* **hayvancasına,** in a bestial way; **delicesine,** like a madman; *also with verbs, e.g.* **pek eskiden tanışırmışçasına el sıkıştık,** we shook hands in the manner of people who had known one another a long long time.

**casus** (–ˋ) Spy. **~luk,** espionage.

**cavalacos** Worthless.

**cavlak** Naked; bald; featherless, hairless. **~lık,** nudity; baldness; utter destitution.

**caydırmak** *va. caus. of* **caymak.** Cause *s.o.* to renounce *or* change his purpose.

**cayır** *In* ~ ~, *denotes continuous and rather violent action*; willy-nilly; 'jolly well': **~ ~ yanmak,** to burn furiously. **~damak,** to crackle *or* creak. **~tı,** noise of tearing *or* creaking *or* the crackling of flames.

**caymak** *vn.* Swerve; deviate *from a purpose,* change one's mind.

**caz** Jazz. **~bant,** jazz band.

**cazib** (–ˋ) Attractive. **~e,** attractiveness; attraction; charm; the force of gravity. **~eli,** attractive.

**-ce, -çe, -ca, -ça** *Suffix denoting* on the part of; in the manner of; according to: **hükûmetçe,** on the part of the government: **adamca,** in the manner of a man: **bence,** according to me; for my part. *With an adjective it forms the diminutive*: **soğukça,** rather cold.

**ceb** Pocket; purse. **~i delik,** penniless: **birini ~inden çıkarmak,** to outdo *s.o.*: ʳ**istemem yan ~ime koy!**ˀ, *said to or of a person who pretends not to want stg. which he really does want.*

**cebel** Mountain. **Cebelitarık** (··–ˋ), Gibraltar.

**cebelleşmek** *v.* **cedelleşmek.**

**cebellezi** (·ˋ··) *In* **~ etm.,** to appropriate what does not belong to one.

**ceberut** (··ˋ) The majesty of God; despotism, tyranny. Tyrant; bully.

**cebhane** (·–ˋ) Powder magazine; ammunition; munitions of war. **~lik,** ammunition store.

**cebharclığı** (ˋ···), **-nı** Pocket-money.

**ceb·he** Front; forehead. **~den taarruz,** frontal attack: **···e karşı ~ almak,** to take the field against ... . **~hî** (·ˋ), frontal.

**cebiboş, cebidelik** Without a *sou,* penniless.

**cebir, -bri** Force; violence; compulsion; algebra. **cebrile,** by force, compulsorily. **~e** (·–ˋ), surgeon's splint.

**Cebrail** (·–ˋ) The Archangel Gabriel.

**cebr·en** (ˋ·) By force. **~etmek** (ˋ··), **-eder,** to force, compel. **~î** (·ˋ), done by force; compulsory; algebraical. **~inefis, ~fsi,** self-control, self-restraint: **···den ~ etm.,** to restrain one's desire to ..., to refrain from ... .

**ced, -ddi** Grandfather; ancestor. **~di âlâ,** a remote ancestor, the founder of one's family: **~ beced,** from one generation to another: **~dine rahmet!,** well done!: **yedi ~dine,** *adds emphasis to the succeeding word, e.g.* **yedi ~dine lânet!,** curse and damn him!; **yedi ~ine tövbe etm.,** to forswear for good and all.

**cedel** Dispute; argument. **~leşmek,** to dispute, debate, argue; struggle.

**cedvel** Canal; ruled paper *or* book; tabulated form, table, list, schedule; ruler. **~e geçirmek,** to tabulate.

**cefa** (·ˋ) Ill-treatment, unkindness, cruelty. **~ etm.,** to ill-treat, torment. **~kâr** (·–ˋ), (*orig.*) cruel; (*now*) who has suffered much. **~keş** (·–ˋ), suffering, tormented; long-suffering.

**ceffelkalem** (ˋ···) Offhand, without reflection.

**-ceğiz** *v.* **-cağız.**

**cehalet** (·–ˋ), **-ti** Ignorance.

**cehd** A striving; endeavour. **~etmek** (ˋ··), **-eder,** to strive.

**cehennem** Hell; inferno. **~ dibi,** a remote and inaccessible place: ʳ**~e kadar yolu var!**ˀ, he can go to hell for all I care!: **~ kütüğü,** a hardened sinner: **~ ol!,** clear out!, go to hell!: **canı ~e!,** damn him! **~î** (···ˋ), hellish; damned. **~lik,** worthy of hell; hardened sinner; stoke-hole *of a Turkish bath.* **~taşı, -nı,** nitrate of silver.

**cehil, -hli** Ignorance.

**cehiz** *v.* **cihaz.**

**cehre** Spindle; reel.

**cehr·en** (ˋ·) Loudly; clearly; publicly. **~î,** loudly, clearly *or* publicly read *or* spoken.

**cehri** Madder root *and* the dye derived therefrom.

**ceket, -ti** Jacket, coat.

**celâdet** (·—ˋ), **–ti** Sturdiness; intrepidity; moral courage.
**celâl** (·–́) Majesty (*esp.* of God); glory; wrath. **~lenmek,** to get into a rage. **~li,** quick-tempered, irascible.
**celâli** (·—ˋ) *Originally* a provincial rebel; *hence* a rebellious person.
**celb** A procuring; attraction; summons. **~etmek** (ˋ··), **–eder,** to attract, bring, procure; supply; import; summon; drive; transport. **~name** (·—ˋ), summons, citation.
**celeb** Drover; cattle-dealer.
**cellâd** (·–́) Executioner. Merciless.
**celse** A sitting; session.
**cemaat** (·—ˋ), **–ti** Community; party; group; congregation.
**cem'an** (ˋ·) In all; as a total. **~ yekûn,** the sum total.
**cemaziyel·âhir** (·–́··—ˋ) The sixth month of the Moslem year. **~evvel** (·–́···ˋ), the fifth month of the Moslem year. **~ini bilmek,** to know all about *s.o.*'s antecedents (not to his credit).
**cem·etmek** (ˋ··), **–eder** *va.* Bring together; collect; add up. **~i, –m'i,** a collecting, bringing together, adding; plural; total. **~î** (ˋ—), all; everyone; the whole. **~'î** (·–́), collective.
**cemil** (·–́) Beautiful; charming, gracious. **~e** (·—ˋ), a kind *or* gracious act; compliment. **~ckâr** (·—·–́), kind, attentive.
**cemiyet, –ti** Assembly; meeting; society; association; wedding. **~li,** full of people, crowded; comprehensive.
**cemre** (*lit.* a burning ember); the gradual increase of warmth in February. **~ düşmek,** (of the weather) to get warmer.
**cenab** (·–́) Majesty; excellency. **~ları,** His Excellency (*of foreigners only*). **Cenabı Hak,** God.
**cenabet** (·—ˋ), **–ti** A state of impurity; a person in that state; (*as a term of abuse*) foul brute.
**cenah** (·–́) Wing; fin; flank (*mil.*).
**cenaze** (·—ˋ) Corpse. **~ alayı,** funeral procession. **canlı ~,** a living skeleton.
**cenbiye** An Arabian dagger.
**cendere** (ˋ··) Press; roller press; narrow gorge; crowded place, squash. **~ baklası,** a very thin kind of bean: **~ye koymak,** to oppress, torture.
**ceneviz** (·ˋ·) Genoese; artful, cunning.
**cengâver** (·—ˋ) Warlike; brave.
**cengel, cangal** Jungle.
**cenin** (·–́) Foetus, embryo. **ıskatı ~,** miscarriage.
**cenk** War; battle; quarrel. **~leşmek,** to make war, fight, struggle.
**cennet, –ti** Paradise. **~li,** belonging to paradise; the late ...; 'of happy memory'. **~lik,** deserving of Heaven. **~mekân** (···–́), 'of happy memory'; in paradise.
**centilmen** Gentleman. **~ce,** in a gentlemanly way.
**cenub** (·–́) The South. **~u garbî,** South-West; **~u şarkî,** South-East. **~en** (·ˋ·), to *or* from the South. **~î** (·—–́), southern.
**cep–** *v.* **ceb–**.
**cepane** *v.* **cebhane.**
**cer, –rri** A pulling *or* drawing; jack; crane; derrick. **~ atölyesi,** railway repair shop: **~re çıkmak,** (of religious students) to go to the provinces and preach *to gain money for their studies.*
**cerahat** (·—ˋ), **–ti** Pus. **~li,** suppurating.
**cerbeze** Quick-wittedness; readiness of speech; push, go; wiliness. **~li,** go-ahead; loquacious, having 'the gift of the gab'.
**cereme** Fine, penalty; amends. **···in ~sini çekmek,** to pay the penalty of ... .
**cereyan** (··–́) A flowing; a movement; current; draught; an occurring; course. **~ etm.,** to happen, occur; flow; conform.
**cerh** A wounding; a refusing to accept evidence; confutation. **~etmek** (ˋ··), **–eder,** to wound; to declare evidence to be invalid; to refute, contradict.
**cerrah** (·–́) Surgeon. **~î** (·—–́), surgical. **~lık,** surgery.
**cerrar** (·–́) Mendicant preacher; importunate beggar.
**cesamet** (·—ˋ), **–ti** Largeness; hugeness; size; importance; grandeur. **~li,** huge, grandiose.
**cesaret** (·—ˋ), **–ti** Boldness, daring; courage. **···e ~ etm.,** to dare *stg.*
**cesed** Corpse.
**–cesine** *v.* **casına.**
**ceste** *In* **~ ~,** little by little; gradually; by instalments.
**cesur** (·–́) Bold, daring; courageous.
**cet** *v.* **ced.**
**cevab** (·–́) Reply, answer. **red ~ı,** a negative reply, refusal: **sudan ~,** an

unsatisfactory reply. **~en** (·–́·), in reply. **~î** (·—́), replying: **~ mektub,** letter in reply. **~lı,** having *or* needing a reply.
**cevahir** (·—́) *pl. of* **cevher.** Jewels; *as sing.* jewel. **~ yumurtlamak,** to say some pricelessly silly things. **~ci,** jeweller.
**cevelân** (··–́) Revolution; circuit; circulation; a going for a stroll.
**cevher** Essence; substance; nature; disposition; mineral ore; jewel; damascening of steel. **~ini* tüketmek,** to be at the end of one's tether. **~li,** talented; naturally capable; set with jewels; damascened; containing ore.
**cevir, -vri** Injustice; oppression; tyranny.
**ceviz** Walnut; (*naut.*) crown *or* wall knot. **(büyük) Hindistan ~i,** coconut: **(küçük) Hindistan ~i,** nutmeg: **⌜kırdığı ~ bini aştı (geçti)⌝,** his errors *or* stupidities are past counting.
**cevval** (·–́) Active, lively, energetic.
**ceylân** (·–́) Gazelle.
**ceza** (·–́) Punishment; fine; retribution; apodosis. **~sını bulmak (çekmek),** to get one's deserts: **~ evi,** prison: **~ kanunu,** criminal code: **~ kesmek,** to fine: **~ zaptı tutmak,** to report a dereliction of duty *on the part of an official*: **Allah ~sını versin!,** damn him! **~en** (·–́·), by way of punishment. **~lanmak,** to be punished.
**Cezayir** (·—́) Algiers; Algeria. **~ menekşesi,** periwinkle (flower).
**cezb** Attraction; allurement; drawing-in of breath. **~e,** ecstasy, rapture; mystical contemplation. **~etmek** (́··), **-eder,** to draw, attract; imbibe; draw in.
**cezir, -zri** Ebb.
**cezrî** (·–́) Radical; thorough.
**cezve** Pot with long handle *for making coffee.*
**Cezvit, -ti** Jesuit.
**cıcık** Stuffing, *only in* **cıcığını çıkarmak,** to damage by use, to wear to pieces; **cıcığı çıkmış,** worn out, in pieces.
**cıgara** (·́·) Cigarette.
**cıkcık** Squeak.
**-cılayın** *v.* **cileyin.**
**cılgar** Extra yoke of oxen *used for ascending a hill.*
**cılız** Puny; thin; delicate; badly formed *writing.*
**cılk** Addled; rotten; inflamed, festered. **~ çıkmak,** to be addled; to come to nought.
**cılkava** Fur *made from the neck of wolves or foxes*; a kind of wolf.
**cıllık** *v.* **cıcık.**
**cımbız** Tweezers.
**cırcır** Chirping *or* creaking sound; cricket; babbler. **~lık,** chattering, loquacity.
**cırlak** Cricket; shrike; shrew; person with a screeching voice.
**cırlamak, cırtlamak** *vn.* Screech; sing; chatter.
**cırtlak** Braggart; *v. also* **cırlak.**
**cıva** (́·) Mercury, quicksilver. **~lı** (́··), containing mercury; loaded (dice).
**cıvadra** (·́·) Bowsprit.
**cıvata** (·́·) Bolt (screw). **~ anahtarı,** spanner.
**cıvık** Wet; sticky; viscid; clammy; tiresome, importunate, silly and facetious. Any viscid *or* sticky substance.
**cıvıl** *In ~ ~, imitates the twittering or gathering of small birds.* **~damak,** to twitter, chirp. **~tı,** twittering, chirping.
**cıvı·mak** *vn.* Become soft *or* sticky; become insipid *or* tiresome; become too familiar *or* impertinent; be 'done for', be hopelessly spoilt. **~tmak,** to make soft *or* sticky; spoil; become soft *or* sticky; become tiresome *or* over-familiar *like a drunkard etc.*
**cıyak** *In ~ ~, imitates the cry of a kite or the high-pitched shouting of a child.*
**cız** *Imitates a sizzling noise.* **~ sineği,** gadfly: **yüreği* ~ etm.,** to be deeply affected *by sudden bad news etc.* **-bız,** grilled meat; a sort of toffee. **~ıktırmak,** to scrawl. **~ıldamak,** to sizzle. **~ıltı,** a sizzling noise. **~ır, ~ ~,** imitates the sound of sizzling meat, of breaking glass, of the scratching of a pen. **~ırdamak,** *v.* **cızıldamak. ~ırtı,** *v.* **cızıltı. ~lamak,** to burn with a sizzling noise: **yüreği* ~,** to be deeply affected *by sudden bad news etc.*
**cızlam** *In* **~ı çekmek** (*sl.*), to go away.
**-ci, -cü, -cı, -cu** *Suffix which, added to a noun, forms the agent, maker or seller connected with that noun, e.g.* **gemi,** ship, **gemici,** sailor; **yol,** road, **yolcu,** traveller; **saat,** watch, **saatçi,** watchmaker, seller or repairer of watches.
**cibillî** (··–́) Natural; innate, inborn.

**~yet, –ti,** nature; innate character. **~yetsiz,** of base character.

**cibinlik** Mosquito-net.

**cibre** Residue of fruit after pressing.

**Cibril** *v.* **Cebrail.**

**cici** (*childish language*) Good; pretty. Toy; trinket; trifle. **~m,** my darling. **~li,** gaudily decked out with trifles: **~ bicili,** glaring, gaudy.

**cicim** Light carpet *for hanging as a curtain or on walls*; *v. also* **cici.**

**cicoz** *Only in* **para ~,** not a penny left!

**cidal** (·–́), **–li** Dispute; combat.

**cidar** (·–́) Wall.

**Cidde** Jeddah.

**cidd·en** (·́·) Seriously; in earnest; greatly, exceedingly. **~î** (·–́), earnest; serious, not joking; strenuous. **~iyet,** seriousness.

**cife** (–·́) Carcass, carrion; anything disgusting.

**ciğer** Liver; lungs; heart; vitals; affections. **ak~,** lungs: **kara~,** liver: **~ acısı,** a bitter grief: **~i* ağzına* gelmek,** to be terribly frightened: **~i beş para etmez,** a worthless fellow: **~ hastalığı,** consumption: **~ köşesi,** darling, well-beloved: **~i* yanmak,** to be greatly grieved *or* upset. **~ci,** seller of livers. **~otu, –nu,** rock lichen; sea-moss. **~pare** (··–·́), darling.

**ciglör** (*Fr. gicleur*) (Carburettor) jet.

**cihad** Holy War *against non-Moslems*; a fight in any good cause (*cf.* English 'crusade').

**cihan** (·–́) World; universe. **~ harbi,** World War: **iki ~,** this world and the next. **~gir** (·––́), world conqueror. **~nüma** (·–·–́), terrace with extensive view. **~şümul,** world-embracing, mondial.

**cihar** (·–́) Four (at dice).

**cihaz** Apparatus; outfit; trousseau.

**cihet, –ti** Side; quarter; direction, bearing; motive, consideration. **o ~ten,** from that direction, from that consideration *or* that point of view: **~ ~,** from that side and this: **~ tayin etm.,** to take one's bearings.

**–cik, –cuk** *etc. Suffix forming a diminutive either of a noun or an adjective.*

**cilâ** (·–́) Brightness, lustre; polish, varnish. **~ etm.,** to polish, burnish: **~ vurmak,** to varnish. **~cı,** polisher, varnisher. **~lı,** polished, shining, varnished.

**cild** Skin; hide; binding *of a book*; volume. **~î** (·–́), pertaining to the skin, cutaneous. **~lemek,** to bind (book). **~li,** bound (book); in volumes: **iki ~ kitab,** a book in two volumes. **~siz,** unbound (book).

**–cileyin** (··́·), **–cılayın** *Suffix.* Like ..., in the manner of ...; **adamcılayın,** in the manner of a man; **buncalayın,** to this degree; **bencileyin,** like me.

**cilve** Grace, charm, coquetry; phenomenon, spectacle. **kaderin ~si,** the irony of Fate; **tabiatın ~si,** the manifestations of Nature. **~kâr** (··–́), graceful; charming; coquettish. **~li,** graceful; coquettish; capricious.

**cim** *The Arabic letter* jim, *now* **c** *and called 'jeh'* (*jay*). **~ karnında bır nokta,** a nonentity; a matter of no importance.

**cimbekuka** (··́··) Ill-shaped, deformed (man).

**cimcime** A kind of small water-melon. Small and dainty.

**cimri** Mean, miserly.

**cin** Genie, djinn, demon; spirit; intelligent man. **~leri başına* toplanmak,** to become furious: **~ çarpmak,** to have a stroke, be paralysed: **~ gibi,** clever and agile: **~ mısırı,** a dwarf kind of maize: **~ taifesi,** the family of demons and sprites: ⌜**~ler top (cirid) oynuyor**⌝, 'the demons play ball', *i.e.* deserted, haunted: **~i** *or* **~leri tutmak,** to go mad.

**cinaî** (·––́) Criminal.

**cinas** Pun.

**cinayet** (·–·́), **–ti** Crime.

**cinfikirli** (·́···) Shrewd; ingenious.

**cingil** Each individual stalk of a bunch of grapes.

**cingöz** Shrewd; crafty, sly.

**cinnet, –ti** Madness; insanity. **~ getirmek,** to become insane.

**cins** Genus; species; class; race; kind, variety; sex; gender. Pure bred, thoroughbred. **~ ~,** of various kinds. **~i lâtif,** the gentler sex. **~î** (·–́), generic; sexual. **~ibir,** of the same genus *or* kind. **~iyet, –ti,** the belonging to a race *or* genus; nationality; characteristic of sex, sexuality.

**ciranta** (··́·) Endorser *of a bill.*

**cirid** Stick *used as a dart in the game of jerid*; the game itself; javelin (in modern athletics). **~ oynamak,** to move about freely.

**cirm** Body (*not living*); volume, size. ⌜**ateş olsa ~i kadar yer yakar**⌝, 'there's nothing to be feared from him!'

**ciro** (ˋ·) Endorsement. **~lu,** negotiable. **~suz,** not negotiable.
**cisim, -smi** Body; substance; material thing. ⌜**ismi var cismi yok**⌝, 'it has a name but no substance', *stg.* known by name but in fact non-existent.
**cism·anî** (·–ˊ) Corporeal; material. **~en** (ˋ·), as regards the body; in size.
**civan** (·ˊ) A youth; young man. Youthful. **~ kaşı,** a kind of embroidery. **~lık,** youth, adolescence. **~merd,** generous, munificent. **~perçemi, -ni,** yarrow, milfoil.
**civar** (·ˊ) Neighbourhood; environs. Neighbouring. **···in ~ında,** in the neighbourhood of, near to ... .
**civciv** Chick; chirrupping of birds; chatter, noise. **~li,** noisy; lively; crowded; numerous.
**civelek** Strong lively lad; youth in the service of the Janissaries. Brisk, lively, playful.
**Cizvit** *v.* **Cezvit.**
**cizye** Tribute; poll-tax.
**coğraf·î** (·–ˊ) Geographical. **~ya** (·ˊ·), geography.
**cokey** (ˋ·) Jockey.
**conta** (ˋ·) Gasket.
**cop, -pu** Truncheon.
**corum** Shoal of fish.
**coşkun** Ebullient, boiling over; overflowing; exuberant; furious; anxious. **~luk,** ebullition; overflowing; exuberance; enthusiasm; exaltation.
**coşmak** *vn.* Boil up; overflow; become exuberant; be enthusiastic.
**cömerd** Generous. **~lik,** generosity.
**cönk** Anthology; miscellany (*lit.*).
**cudam** Useless sort of person. ⌜**adam değil, ~**⌝, 'don't call him a man!'
**cuma** (·ˊ) Friday. **~ alayı,** (*formerly*) *the ceremonial procession of the Sultan to a mosque on Fridays, the* **selâmlık. ~lık,** pertaining to Friday. **~rtesi** (·ˊ··), Saturday.
**cumba** (ˋ·) Any projecting part of a building; bow-window; balcony.
**cumbadak** (ˋ··) Falling suddenly into water *or* an abyss; *imitates the noise of such a fall.*
**cumbul, cumbur ~ ~,** *imitates the noise of stg. falling into water.*
**cumbur·tu** Plop; loud noise. **~lop,** plop!, splash!
**cumhur** (·ˋ) The mass of the people; populace; republic. **~ Reisi, ~ Başkanı,** the President of the Republic. **~a uymak,** to follow the majority. **~iyet** (·–·ˋ), **-ti,** republic. **~iyetçi, ~iyetperver,** upholder of the Republic, republican.
**cunda** (ˋ·) End of a gaff, peak.
**cuppadak** (ˋ··) *v.* **cumbadak.**
**cura** Two- *or* three-stringed lute; a small kind of hawk; (*sl.*) cigarette stump. **~ zurna,** a small shrill horn.
**curcuna** (·ˋ·) Drunken revel; disorderly dance; orgy; brawl; confused medley.
**curnal** *v.* **jurnal.**
**curnata** (·ˊ·) Mass arrival of quails.
**cuş** (–), **cuşiş** (–ˋ) Effervescence; commotion; enthusiasm. **~u huruş,** commotion, excitement, enthusiasm.
**cübbe** *v.* **cüppe.**
**cüce** Dwarf.
**cücük** Sweet, pleasant; tender. The heart of an onion; tuft of beard (imperial); chick.
**cücümek** *vn.* Become sweet.
**cühelâ** (··ˊ) *pl. of* **cahil.** Ignoramuses.
**cülûs** (·ˊ) Accession to the throne.
**cümbür** *In* **~ cemaat,** all in a body; the 'whole caboodle'.
**cümbüş** Amusement, entertainment, jollity; a kind of metal mandolin. **~ etm.,** to make merry, to enjoy oneself. **~lü,** with jollity and amusement.
**cümle** A total, a whole; category; phrase, sentence. **~si,** all of them: **~ kapısı,** the main *or* front door of a biggish house: **~den,** as an instance of this: **kara ~,** simple arithmetic; **kara ~si bozuk,** half illiterate, very ignorant. **~ten** (ˋ··), wholly, entirely.
**cünha** Serious offence, crime.
**cünüb** Canonically unclean.
**cüppe** Robe *with full sleeves and long skirts.*
**cür'a** A draught *or* gulp of a drink; dregs.
**cür'et, -ti** Boldness, daring. **···e ~ etm.,** to dare to ... . **~li, ~kâr,** bold, daring.
**cürmümeşhud** (···ˊ) Flagrant offence. **~ halinde,** 'in flagrante delicto': **~ yapmak,** to lay a trap to catch *s.o.* red-handed.
**cüruf** (·ˊ) Slag; scoriae.
**cürüm, -rmü** Crime; fault; sin; *v.* **cürmümeşhud.**
**cüsse** Body, trunk. **~li,** big-bodied; huge.
**cüz, -z'ü** Part; section; piece, fragment; one of the thirty sections of the

Koran; section of a book; pamphlet. ~ **kesesi,** a case containing extracts from the Koran, *carried by pupils in old Turkish schools.*

**cüzam** Leprosy. **~lı,** leprous.

**cüzdan** Pocket-book; wallet; portfolio. **hesab ~ı,** pass-book.

**cüz'î** (·ˊ) Trifling, insignificant; partial; fragmentary. **~ce** (·ˊ·), somewhat small, petty.

# Ç

**çaba·lamak** Strive; struggle; endeavour. **nafile çabalama!,** don't waste your efforts!; it's no use! **~lanmak,** to move one's limbs about in an agitated manner; struggle.

**çabucak** (ˈ··), **çabucacık** (·ˈ··) Quickly.

**çabuk** Quick; agile. Quickly; in a hurry; soon. ~ **olm.,** to hurry: **eline ~,** quick at his work. **~luk,** speed; agility; haste. **el çabukluğu,** sleight-of-hand.

**çaça** Sprat; very small mackerel.

**çaçaron** Talkative person; charlatan.

**çadır** Tent. ~ **kurmak,** to pitch a tent: **kıl ~,** horsehair *or* felt tent *used by nomads.*

**çağ** Time; age, period; stature. **orta ~,** the Middle Ages.

**çağan** Camel fetter.

**çağanak, çağnak** (ˈ·) *Only in* **çalgı ~,** a noisy musical party.

**çağanoz** Crab.

**çağdaş** Contemporary.

**çağıl** *In* ~ ~, *imitates the noise of running water*; burbling. **~damak,** to burble, to murmur *as running or boiling water.* **~tı,** the murmur of running water.

**çağırmak** *va. & vn.* Call out; call; invite; sing.

**çağırtkan** Decoy bird.

**çağla** (ˈ·) Green almond *or other* unripe fruit; dull greyish-green colour.

**çağla·ma** Murmuring of water. **~mak,** *v.* **çağıldamak.** **~r,** natural cascade. **~yan,** artificial cascade. **~yık,** bubbling spring; hot spring.

**çağlı** Reaching a certain stage of maturity *or* height. **pehlivan ~,** built like a wrestler; strong, robust.

**çağnak** Castanet; *v.* **çağanak.**

**çağrılmak** *vn.* Be called out; be shouted; be called; be invited; be sung. **geriye ~,** to be recalled, to be removed from office.

**çağrışmak** *vn.* Cry out together; call one another; make a row.

**çakal** Jackal. ~ **eriği,** wild plum.

**çakar** *v.* **çakmak.** Phosphorescence caused by moving objects in water.

**çakı** Pocket-knife. ~ **gibi,** keen and active.

**çakıl** Pebble. ~ **döşemek,** to pave with pebbles. **~dak,** ball of dried dung *hanging on the tail or belly of a sheep*; clapper *of a mill*; rattle. **~damak,** to make a clattering sound. **~lık,** pebbly place; court paved with pebbles. **~tı,** clattering rattling noise.

**çakılmak** *vn. pass. of* **çakmak.** Be stuck *in a place.* **çakılmadan** (*sl.*), without anyone noticing.

**çakır**[1] Any intoxicating drink. **~keyf** (·ˈ·), half-tipsy.

**çakır**[2] Grey with blue streaks; greyish-blue.

**çakır**[3] *In* ~ **çukur,** rattling noise; broken ground. Full of pot-holes, uneven.

**çakış·mak** *vn.* Fit into one another. **~tırmak,** *caus. of* **çakışmak**; *vn.* to drink, to booze; to make merry with drink.

**çakma** Flash; stamp *on silver etc.*

**çakmak**[1] *va.* (*with acc.*) Drive in by blows; nail; tether by a peg; strike *a flint or a match*; snap *the teeth*; palm off *false coin*; fit *one thing into* another; guess, perceive: (*with abl.*) understand, know about *stg.*; be 'ploughed' in *an exam. vn.*, (of lightning) to flash; carouse. **çakar almaz,** that which misfires, *hence* that which does not work: '**birer tane çakalım**' (*sl.*), 'let's have a drink'.

**çakmak**[2] Steel for striking on a flint; pocket-lighter; trigger. **gözleri ~ ~,** heavy-eyed *from sleeplessness or fever.* **~lı,** flintlock (gun). **~taşı, -nı,** flint.

**çakşır** Trousers.

**çaktırmak** *va. caus. of* **çakmak** (*sl.*), **çaktırmadan,** without attracting anyone's attention; without being noticed.

**çal** Escarpment.

**çal–, çala–** *Prefix indicating repeated action of the verb* **çalmak,** *e.g.* **çalçene,** 'chin-wagger', talkative; **çalakürek,** (rapidly striking with the oar), rowing hard.

**çala·kalem** (ˈ···) With flowing pen;

scribbling hastily. **~kaşık** (ˈ···), **~ yemek,** to eat greedily. **~kürek** (ˈ···), rowing hard.

**çalarsaat** (·ˈ··), **-ti** Striking clock; repeater (watch); alarm clock.

**çalçene** Chatterbox, babbler.

**çaldırmak** *va. caus. of* **çalmak.** Cause to strike, steal, play *etc.*; let *a thief* steal; lose by theft.

**çalgı** Musical instrument. **~ çalmak,** to play a musical instrument: **~ takımı,** an orchestra *or* band; their instruments. **~cı,** musician. **~cılık,** music (*as a profession or art*).

**çalı** Bush, shrub; thicket. **kara ~,** *v.* **karaçalı**; one who causes bad blood between others: **süpürge ~sı,** heather. **~çırpı,** brushwood; sticks and thorns *used for fencing.* **~fasulyesi, -ni,** climbing kidney-beans; runner beans. **~kavak,** pollarded poplar *used in basket-making.* **~kuşu, -nu,** golden-crested wren. **~lık,** thicket.

**çalık** Slanting, awry; walking sideways; tainted, spoilt; dismissed, expelled; dazed, deranged. Sheep scab. **rengi ~,** discoloured, faded.

**çalım** Stroke, blow; swagger. **~ına getirmek,** to find a suitable opportunity: **~ satmak,** to swagger, to give oneself airs. **~lı,** swaggering, arrogant.

**çalınmak** *vn. pass. of* **çalmak.** Be struck *etc.*; **···e ~,** smell *or* taste of *stg.*; reach *the ear.*

**çalışkan** Industrious, hard-working.

**çalış·mak** *vn.* Work; try, strive; study; (of wood) to warp. **çalışıp çabalamak,** to strive hard, to do one's best. **~tırmak,** to make *or* let *s.o.* work *or* try; set to work, employ.

**çalkalamak** *v.* **çalkamak.**

**çalkama** *verb. n. of* **çalkamak**; omelet.

**çalkamak** *va.* Agitate; shake *or* toss about; rinse; wash out (mouth); gargle; beat (eggs); churn (milk); (of a hen) to turn *its eggs*; (of tainted milk) to turn the stomach of. **göbek ~,** to shake the belly *in Oriental dance.*

**çalkan·mak** *vn. pass. of* **çalkamak.** Be agitated *or* shaken about; (of a ship) to roll; (of the sea) to become rough; sway about while walking; become addled; fluctuate; (of news) to spread like wildfire. **~tı,** agitation; a shaking; a tossing about *of the sea*; rapids *of a river*; rolling *of a ship*; disturbance *of the stomach.*

**çalkar** Anything that upsets the stomach; purgative; gin *for separating seeds from cotton.*

**çalmak** *va.* Give a blow to; knock (on a door); strike (the hour); ring (bell); play (musical instrument); (of a referee) to blow his whistle for a foul *etc.*; smear (butter *etc.*); mix; steal; have a flavour *or* smell *of stg. else*; verge on (a colour); have a *certain* accent; (of the sun *or* wind) to scorch. **çalmadığım kapı kalmadı,** (there is no door that I have not knocked on), I have left no stone unturned, I have made every effort: **çalmadan oynar,** over-cheerful; gushing; officious: **süt ~,** (of milk) to upset *a baby*: **yere ~,** to throw to the ground.

**çalpara** Castanet; swimming crab. **~lı tulumba,** chain-pump.

**çalyaka** (ˈ··) Seizure *of s.o.* by the collar. **~ etm.,** to seize by the collar.

**çam** Fir; pine. **~ devirmek,** to 'drop a brick', to put one's foot in it: ⌜**eski ~lar bardak oldu**⌝, (old fir-trees have been made into mugs), things are not what they were: **~ yarması gibi,** *said of a strongly built man.*

**çamak** Freshwater bream.

**çamaşır** Underclothing; linen; soiled linen; the washing. **~ değiştirmek,** to change one's linen: **bir kat ~,** a change of linen: **birinin kirli (iç) ~larını ortaya çıkarmak,** to reveal *s.o.*'s misdeeds **~cı,** washerwoman. **~cılık,** profession of laundering; white goods. **~hane** (···–ˈ), laundry. **~lık,** linen-cupboard; laundry; washing-tub; stuff for making underclothing.

**çamçak** Wooden vessel for water; (?) roach.

**çam·fıstığı** (ˈ···), **-nı** Pine kernel. **~lık,** pine wood, **~sakızı** (ˈ···), **-nı,** pine resin; an unescapable bore; sticky: ⌜**çoban armağanı ~**⌝, 'a shepherd's present is a bit of resin', *said when giving a small present.*

**çamuka** (·ˈ·) Sand-smelt.

**çamur** Mud; unmannerly person; 'a bad lot'. **~ atmak,** to vituperate: **~a bulamak,** to cover with disgrace: **~dan çıkarmak,** to help *s.o.* out of the mud, to help out of a difficulty: **~a düşmek,** to fall into adversity: **~ sıvamak,** to cast aspersions: **~a taş atmak,** to invite abuse from an impudent *or* aggressive person: ⌜**(yan bastı) ~a yattı**⌝, *v.* **yan basmak**: **çömlekçi ~u,** potter's clay. **~cuk,** tench.

~cun, teal. ~lamak, to cover with mud; to slander. ~lu, muddy. ~luk, muddy place; gaiters; mudguard; boot-scraper.

**çan** Bell. ~ **çalmak**, to ring a bell; to trumpet abroad: ~ ~ **etm.**, to chatter unceasingly: ~**ına*** **ot tıkamak**, to confound an opponent; to muzzle; to render impotent.

**çanak** Earthenware pot *or* pan, beggar's alms-bowl. ~ **çömlek**, pots and pans, crockery: ~ **tutmak**, by one's own words *or* deeds to bring down insult on oneself; to 'ask for it': ~ **yalamak**, to agree with *s.o.* in order to please him: ~ **yalayıcı**, lick-spittle, toady: **kan çanağı**, barber's bleeding basin; an eye red with rage. ~**çı**, potter.

**Çanakkale** (·˙··) Dardanelles (town). ~ **Boğazı**, the Dardanelles Straits.

**çanaklık** Top of a mast; crow's-nest.

**çançan** Loud and continuous chatter.

**çançiçeği** (˙···), **-ni** Campanula.

**çangıl çungul** *Imitates a harsh broken speech or a foreign or provincial accent.*

**çangır** ~ **çungur**, clink-clank; jingle-jangle. ~**damak**, to make a clanking *or* jangling sound. ~**tı**, clanking, jangling sound.

**çankulesi** (˙···), **-ni** Belfry.

**çanta** (˙·) Bag; case; valise; knapsack. ~**da keklik**, a partridge in the bag, *i.e.* something safely secured: **arka** ~**sı**, haversack: **para** ~**sı**, purse. ~**cı**, maker *or* seller of bags *etc.*; porter employed to carry official correspondence to ministers *etc.*

**çap, -pı** Diameter; bore; calibre (*also used fig. as in English*); plan showing size and boundaries of a building *or* piece of land.

**çapa** (˙·) Hoe; mattock; mortar larry; anchor (with flukes). **fırtına** ~**sı**, sea-anchor, drogue. ~**cı**, hoer.

**çapaçul** Slovenly; untidy; tatterdemalion.

**çapal** Crust round the eyes *in 'blear-eye'*; rough skin *on metal castings*; burr. ~**balığı, -nı**, freshwater bream.

**çapalamak** *va.* Hoe.

**çapan, çapar** One who gallops; courier.

**çapanak** Booty; pillage; smuggled goods.

**çapanoğlu** (·˙··) *In* ⌜(**işi kurcalama**), **alt tarafı** ~ **çıkar**⌝, (better not meddle with the matter) there's a 'fly in the ointment', 'there's a snag somewhere', 'it might raise awkward questions'.

**çapar** Albino. *v.* **çapan.**

**çaparı** Multi-pointed fish-hook; lure.

**çaparız** Obstacle; entanglement. Perverse. ~ **etm.**, to create a difficulty, to block the way.

**çapıcı** Raider, marauder.

**çapkın** Vagabond, rascal; scamp; rake; dissolute; ne'er-do-well; swift horse. ~**lık**, profligacy, debauchery, rascality.

**çaplı** Of a *certain* calibre. **büyük** ~, of a large bore *or* big calibre.

**çaprak** Saddle-cloth.

**çapraş·ık** Crosswise; complicated, entangled. ~**mak**, to be interlaced; be involved *or* difficult.

**çapraz** Waistcoat *etc.* fastened by frogs; anything fastened crosswise. Saw-set. Crossing, crosswise; double-breasted (coat). ~**lama**, ~**vari**, crossed obliquely.

**çapul** Raid; booty, plunder. ~ **etm.**, to sack, pillage; to go on a raid. ~**cu**, raider. ~**lamak**, to go raiding; to pillage, sack.

**çaput, -tu** Rag; patch.

**çar** Czar. ~**lık**, reign *or* government of a Czar: ~ **idaresi**, the Czarist régime.

**çarçabuk** (˙··) Very quickly; in haste.

**çarçur** In a wasteful *or* squandering manner. ~ **etm.**, to squander: ~ **olm.**, to be wasted.

**çardak** Hut *built of brushwood on supports*; trellis; pergola; gallows.

**çare** (–˙) Remedy; means; measure; help. ~ **bulmak**, to find a remedy *or* a means: **bundan başka** ~ **yok**, there is no other way out; this is the only thing one can do: **başının** ~**sine bakşın**, let him look out for himself *or* look after himself: **ne** ~?, what can one do?: **ne** ~ **ki** ..., inevitably ..., it can't be helped but ... . ~**siz**, irremediable; inevitable; necessary; helpless, without means. ~**sizlik**, lack of means; urgency; poverty; helplessness.

**çarhetmek** (˙··) *v.* **çark.**

**çarık** Sandal of raw-hide *or* rope *worn by peasants*; drag for a cart-wheel. ~**lı**, wearing sandals; ignorant and illiterate: ~ **diplomat**, *said of a peasant who, though illiterate, is pretty shrewd*; shrewder than one would suspect.

**çariçe** (·˙·) Czarina.

**çark, -kı** Wheel; paddle-wheel; fly-wheel; anything that revolves; machinery; the celestial sphere; fate, destiny. **~ı bozulmak,** to have one's affairs upset, to meet with misfortune: **~ çevirmek,** to wander around: **~ etm.,** to wheel, to turn: **~a etm.,** *or* **çekmek,** to put *a tool* to the grindstone: **~ işi,** machinery; machine-made: **~ işletmek,** to work a machine; to work a scheme for one's own advantage. **~çı,** engineer, mechanic, engine-driver; knife-grinder; **~başı,** chief engineer. **~ıfelek,** the sphere of heaven; destiny; Catherine-wheel; passion-flower. **~lı,** with paddle-wheels.

**çarmıh** Cross *on which malefactors were crucified*; *v.* **çarmık.** **~a germek,** to crucify.

**çarmık** Shroud *of a mast.*

**çarnaçar** (˙–·) Willy-nilly; of necessity.

**çarpan** A fish of the Weever family (*Trachinus radiatus*).

**çarpı** Whitewash.

**çarpık** Crooked; awry; bent; warped; slanting; deviating; struck with madness *or* paralysis; ill-omened. **~ bacaklı,** bandy-legged: **~ çurpuk,** crooked; deformed; deviating in all directions.

**çarpılmak** *vn. pass. of* **çarpmak,** become crooked *etc.*; be struck with madness *or* paralysis; meet with divine punishment *for blasphemy etc.*; have a fit; behave oddly.

**çarpın·mak** *vn.* Struggle; get flustered. **~tı,** shock; palpitation.

**çarpışma** Collision; conflict; clash. **~k,** to strike one another; come into collision; fight.

**çarpıtmak** *va.* Make crooked *or* awry. **yüzünü gözünü ~,** to make a wry face.

**çarpma** Blow; stroke; multiplication; rough-cast. **cin ~sı,** paralytic stroke *or* sudden attack of insanity.

**çarpmak** *va.* Strike; knock; strike mad *or* paralytic; carry off, burgle; hold up *for robbery*; pick *s.o.*'s pocket; multiply. *vn.* Bump; come into collision; (of the heart) to palpitate; (of wine) to go to the head; (of the sun) to give a sunstroke. **cezaya ~,** to punish: **cin ~,** for an evil spirit to possess one and cause madness *or* paralysis: **göze ~,** to strike the eye, to be conspicuous: **kömür ~,** for charcoal fumes to affect one: **kuran (evliya) ~,** for the Koran (or a saint) to punish a profane act: **yere ~,** to knock down: **yerden yere ~,** to discredit, to run down: **yüzüne* ~,** to cast in *s.o.'s* teeth.

**çarptırmak** *va. caus. of* **•çarpmak.** Cause to collide *etc.*; allow to pillage; have one's pocket picked.

**çarşaf** Sheet (of a bed); dress with veil *formerly worn by Turkish women in public.*

**çarşamba** Wednesday. **~dır ~ demek,** to insist that one is right: **~ karısı,** witch, hag; a very untidy woman, a 'regular fright': ⌜**ayın dört ~sı bir araya gelmiş**⌝, difficulties *or* work have come with a rush: **çıkmaz ayın son ~sı,** the Greek Kalends.

**çarşı** Market; bazaar; street with shops. **~ya çıkarmak,** to expose for sale: **~ya çıkmak,** to go shopping.

**çat** *Imitates a sudden noise.* **~ diye,** all of a sudden: **~ kapı,** unexpectedly, suddenly: **~ orada ~ burada,** now here now there (*of a person or thing always in a different place*): **~ pat,** *v.* **çatpat.**

**çatal** Fork, prong; dilemma. Forked; bifurcated; with a double meaning (word); difficult, delicate (matter); hoarse (noise). **~ çekiç,** claw-hammer: **~ iş,** a vexed question, dilemma: **~ tırnak,** cloven hoof: **~ görmek,** to see double: **geyik ~ı,** antler. **~laşmak, ~lanmak,** to bifurcate; to become ambiguous *or* complicated. **~lı,** forked; ambiguous, doubtful.

**çatana** (·˙·) Small steamboat.

**çatı** Framework; skeleton; roof. **~ katı,** attic; top storey: **bir ~ altında,** under the same roof; intimate with one another; 'in the same box'.

**çatık** Touching; contiguous; fitted together; intimate; sour-faced, scowling. **~kaş,** with eyebrows that join; sulky face.

**çatır** **~ ~,** *imitates a crackling or clashing noise*; willy-nilly. **~damak,** to make a chattering *or* clattering noise; **dişleri* ~,** (of the teeth) to chatter. **~tı,** clattering *or* chattering noise.

**çatışmak** *vn.* Bump into one another, collide; (of joints) to fit into one another; (of animals and insects) to mate; to come up against one another *in dispute or competition.*

**çatkı** Bandage round the head; pile *of rifles.* **~lık,** pole connecting the yokes of two oxen.

**çatkın** Protégé, favourite; bound by ties of interest *or* friendship puckered, creased; frowning. **~lık,** grumpiness, sour looks.

**çatlak** Crevice; crack; fissure. Split; cracked; 'cracked' (crazy); chapped (hand); hoarse (voice). **damarı ~,** brazen-faced, shameless.

**çatla·mak** *vn.* Crack; split; (of a wave) to break; (of an animal) to die from exhaustion; burst *with rage or heat etc.*; weep *or* groan bitterly. **~tmak,** to split; crack; make one's head ache; make *s.o.* nearly burst *with rage etc.*: ride *a horse* to death; pronounce in an exaggerated and pedantic manner.

**çatma** *verb. n. of* **çatmak.** Action of putting together *etc.*; anything put together like framework; parts temporarily and roughly put together; a thick silk cloth *used for furnishing.* Assembled, put together; loosely sewn together. **~kaş,** *v.* **çatıkkaş**: **derme ~,** loosely put together; jerry-built; odds and ends, scraps *of news, knowledge, etc.*

**çatmak** *va. & vn.* Fit together; build; set up; pile (arms); load (an animal); bump up against, collide with; come up against (a difficulty); meet; have a lucky encounter with *an influential man or one who can be of use to one*; tack, sew coarsely; (of a season) to come round; become aggressive. **çattık!,** 'we're up against it!': **···e ~,** to seek a quarrel with ...: **adamına ~,** (of a quarrelsome man) to meet his match: **'tam adamına çattık',** 'we've struck just the one man *who can help us* (*or ruin us*)': **başına ~,** to bind on the head: **gelip ~,** (of a time *or* a regular event) to be due, to come round; (of an unpleasant event) to befall, to happen to one: **sağa sola ~,** to seek a pretext for a quarrel.

**çatpat** (ˈ·) Now and then, rarely; somewhat. **~ franzızca konuşmak,** to have a smattering of French: **~ konuşmak,** (of a child) to be able to talk a little.

**çatra** (ˈ·) *In* **~ patra,** with a clatter. **~patra söylemek,** to speak *a language* incorrectly *or* with difficulty.

**çavdar** Rye. **~ mahmuzu,** ergot.

**çavuş** Sergeant; doorkeeper; messenger; uniformed attendant *of an ambassador or consul.* **~kuşu, -nu,** hoopoe. **~üzümü, -nü,** a kind of large and sweet grape.

**çay**[1] Stream; tributary of a river.

**çay**[2] Tea. **~ üzümü,** bilberry. **~danlık,** teapot. **~hane** (·–ˈ), tea-shop; tea-garden.

**çayır** Meadow; pasture; pasture grass. **~a çıkarmak (salmak),** to put out to grass. **~kuşu, -nu,** lark. **~latmak,** to pasture, to put out to grass. **~lık,** pasture, meadowland. **~otu, -nu,** meadow grass; grass cut and fed green. **~peyniri, -ni,** cream cheese.

**çayır çayır** *v.* **cayır cayır.**

**çaylak** Kite; avaricious, grasping person. **acemi ~,** a clumsy beginner.

**çeç** Heap of winnowed grain.

**çeçe** Tsetse.

**Çeçen** The Tchetchen tribe *of N.E. Caucasus*; talkative.

**çedik** *In* **~ pabuc,** yellow morocco slipper.

**çegane** (·–ˈ) Tambourine.

**çehiz** *v.* **cihaz.**

**çehre** Face, countenance; sour face. **~si* atmak** *or* **~ etm.,** to make a wry face, to sulk: **~ eğmek,** to show disapproval: **~ züğürdü,** ugly: **ne bu ~?,** why do you make such a face? **~li,** having *such and such* a face; scowling.

**Çek, -ki, Çekli** Czech.

**çek, -ki** Cheque.

**çekçek** Small four-wheeled handcart.

**çekecek** Shoehorn. **çizme çekeceği,** bootjack.

**çekeme·mek** (·ˈ··) *va.* Be unable to tolerate; envy. **~mezlik** (·ˈ···), intolerance, envy.

**çeki** A measure of weight (about 500 lb.) *for wood etc.*; horse-load (*of firewood etc.*). **~ye gelmez,** very heavy; unbearable, intolerable; unseemly.

**çekic** Hammer. **şeytan ~i,** a very bright child. **~hane** (··–ˈ), steam-hammer shop *of a factory.*

**çekidüzen** Orderliness, tidiness; toilet. **kendine ~ vermek,** to tidy oneself up; to put one's house in order.

**çekik** Elongated; slanting (eyes). **~ çene,** receding chin.

**çekilir** Endurable, tolerable.

**çekil·mek** *vn. pass. of* **çekmek.** Be pulled *etc.*; withdraw, retire; draw back; shrink; contract; (of water) to recede; to dry up. **çekil oradan!,** clear out of there! **~mez,** unendurable, intolerable.

**çekinecek** Which is to be avoided *or* guarded against.

**çekingen** Timid; shy; reserved; retiring.

**çekinmek** *vn.* Beware; take precautions; draw back; recoil *through fear or dislike*; refrain *from*; hesitate.

**çekirdek** Stone *or* pip *of a fruit*; grain (in weight), nucleus. **~ içi,** kernel: **~ kahve,** coffee beans: **~siz,** stoneless, pipless: **~siz kuru üzüm,** sultana: **~ten yetişmek,** to be accustomed to *or* trained to *stg.* from an early age: **iki dirhem bir ~,** over-elegantly dressed; dressed up to the nines.

**çekirge** Grasshopper; locust; cricket.

**çekiş·mek** *vn.* Pull one another about; struggle; scramble for *stg.*; quarrel, dispute, litigate. **çekişe çekişe pazarlık etm.,** to make a bargain after long haggling: **can ~,** to be in the throes of death: **halat ~,** to have a tug-of-war. **~tirmek,** to slander; criticize; reproach; curse.

**çekitaşı**(··ˈ··), **-nı** The counterweight for weighing a **çeki** *q.v.* **~ gibi,** very heavy, ponderous.

**çekme** Act of drawing *etc.*; drawer; till; overalls; shrinkage. Rolled (iron *etc.*); well-formed; *clothing or boots* that draw on. **~ce,** drawbridge; drawer, till, small coffer; desk; small port of refuge.

**çekmek** *va.* Pull, draw, attract; drag; pull on (boots, trousers *etc.*); withdraw; trace ( a line); undergo, suffer, bear, endure, support; cause to support; decline, conjugate; inhale; absorb; (*sl.*) drink; weigh; make a copy of; mate (an animal). *vn.* Contract, shrink; weigh; last, take *so long.* **···e ~,** to resemble; **babasına çekiyor,** he takes after his father: **çek!,** (*to a chauffeur etc.*) 'go on!': **çek arabayı!,** 'clear out!': **çekip çekiştirmek,** to gossip maliciously about *s.o.* (as a servant about her mistress): **çekip çevirmek,** to run (a house *etc.*); to know how to treat ( a person): **çekip çıkarmak,** to pluck out, eradicate: **çekip uzatma,** prolixity: **canı* ~,** to long for: **duvar ~,** to set up a wall *between two places*: **···den el ~,** to relinquish, renounce, give up *stg.*: **enfiye ~,** to take snuff: **bir fiili ~,** to conjugate a verb: **güçlük ~,** to have difficulty, to suffer difficulties: **ispirto ~,** to distil alcohol: **kafayı ~,** to drink heavily: **kâğıda (deftere) ~,** to copy on to paper (into a notebook): **kahve ~,** to grind coffee: **odun ~,** to weigh wood; **yaş odun ağır çeker,** wet wood weighs heavy: **perde ~,** to separate by a curtain; **perdeyi ~,** to draw the curtain: **resim ~,** to take a photograph: **sözümü fenaya çekti,** he put a bad interpretation on my words: **telgraf ~,** to send a telegram: **bu yol iki saat çeker,** this journey takes two hours: **zahmet ~,** to suffer trouble; to take trouble: **ziyafet ~,** to give a feast.

**çekmez** Unable to bear; unshrinkable.

**çektirmek** *va. caus. of* **çekmek. ···i işten el ~,** to remove *s.o.* temporarily from an office.

**çelebi** *Formerly a title of* a royal prince; educated man; gentleman; *title given to men of certain religious orders.* Well-mannered, courteous.

**çelenk** Wreath; plume; bejewelled aigrette.

**çelik**[1] Short piece of tapered wood; tipcat; belaying-pin; marlinspike; cutting *of plant.* Clipped; bevelled; diverted; ousted. **~ çomak,** the game of tipcat.

**çelik**[2] Steel. **~hane** (··–ˈ), steel foundry.

**çelim** Form, shape. **~li,** well-made; strong. **~siz,** misshapen; uncouth; scraggy; infirm, frail.

**çelme** *verb. n. of* **çelmek.** Trip (with the foot). **~ takmak (atmak),** to trip up.

**çelmek** Strike lightly; clip, lop; divert the mind; persuade; dissuade; confute, rebut *a statement.* **aklını* ~,** to captivate the mind, to bias, to persuade, dissuade: **ayağı ~,** to trip up: **gönlünü* ~,** to try to gain *s.o.*'s affection.

**çeltik** Unhusked rice; rice-field; Glossy Ibis.

**çember** Hoop; ring *of wood or metal*; fillet; neckerchief; the vault of heaven; fortune's wheel. Rounded. **~ içine almak,** to encircle (*mil.*). **~lemek,** to fit with hoops; to encircle.

**çemen** Cummin. *v. also* **çimen.**

**çemre·mek** *va.* Tuck up (clothes); roll up (sleeves *etc.*). **~nmek,** to tuck up one's sleeves; to prepare for action.

**çene** Jaw; chin; loquacity; end of a ship's keel. **birinin ~sini açmak,** to give an opening to speak to one *who you would rather did not speak*: **~si atmak,** to drop the jaw at death, to die: **~ çalmak,** to chatter: **~si düşük,** very talkative: **~ kavafı,** chatterbox: **~ye kuvvet,** by force of words, by dint of speaking: **~si* oynamak,** to eat: **~sini* tutmak,** to hold one's tongue; ⌈**~n tutulsun!**⌉, 'may your jaw be stuck!' (*a curse on one who speaks in an ill-omened way*): **~ yarışı,** a prolonged argument: **~ yarıştırmak,** to enter on a prolonged argument: **~ yormak,** to talk in vain, to waste one's breath: ⌈**alt ~den girip üst ~den çıkmak**⌉, to talk *s.o.* over.

**çene·altı, -nı,** the under side of the chin; dewlap. **~baz, ~li, ~sidüşük,** garrulous, talkative. **~kemiği, -ni,** jaw-bone.

**çengel** Hook. Hooked; crooked. **~lemek,** to hang on a hook: **kafamda bir sual çengellendi,** a question-mark arose in my mind. **~li,** hooked: **~ iğne,** safety-pin. **~sakızı, -ni,** chewing-gum *made from a species of cardoon*.

**çengi** Public dancing-girl. **~ kolu,** a troop of dancing-girls: **bir kol ~,** one who talks too loudly and heartily.

**çenk** Harp.

**çentik** Notch; defect. Notched.

**çentmek** *va.* Notch, nick; mince.

**çepçevre** (ˋ··), **çepeçevre** (·ˋ··) All around.

**çepel** Foul (weather); gloomy, dull; muddy; mixed, adulterated. Storm of wind, rain and sleet; muddy season; mud, dirt; extraneous material.

**çepiç** Year-old goat.

**çepin** Trowel.

**çepken** A kind of short overcoat with wide sleeves.

**çer** *In* **~den çopten,** flimsy.

**çerağ** Lamp; candle; torch.

**çerçeve** Frame; window-frame. **kanunun ~sine sığmaz,** it is not in accordance with the law. **~li,** framed.

**çerçi** Pedlar.

**çerez** Appetizer; hors-d'œuvre. **~lenmek,** to partake of an appetizer.

**çergi** Gipsy tent. **~ci,** stallholder in a market.

**çeri** Soldier; military force. **yeni ~,** Janissary. **~başı** *in error for* **çergibaşı,** gipsy leader.

**Çerkes, Çerkez** Circassian. **~tavuğu, -nu,** a dish of chicken with walnuts.

**çermik** Thermal waters.

**çeşid** Sort; variety; sample; set *of cups etc.* **~ ~,** of various sorts: **~ düzmek,** to buy various sorts of a thing. **~li,** various, assorted.

**çeşm** Eye; sight. **~ibülbül,** *glassware or stuff* decorated with small coloured spots.

**çeşme** Fountain. ⌈**iki gözü iki ~**⌉, in floods of tears.

**çeşni** Taste; flavour; small portion eaten to judge the flavour; sample; assay. **~sine bakmak,** to test the flavour of *stg.*: **~ olur,** it would be a pleasant change. **~ci,** person who tests the flavour of foods; assayer. **~lenmek,** to be properly flavoured. **~li,** tasty; properly flavoured; fired by a percussion-cap.

**çete** Band (of brigands, rebels *etc.*). **~ye cıkmak,** to go forth on a marauding raid: **~ muharebesi,** guerilla warfare. **~ci,** member of a **çete**; comitadji.

**çetele** Tally-stick.

**çetik** *v.* **çedik.**

**çetin** Hard; difficult; harsh; perverse; obstinate. **~ ceviz,** an obstinate and difficult person; a 'tough nut'.

**çetrefil** Badly spoken (language); complicated; difficult to understand.

**çevgen** Polo; polo-stick.

**çevik** Quick, agile, adroit; sound.

**çevirme** Piece of meat roasted on a spit *or* skewer; kebab; a kind of thick jam; turning movement (*mil.*); translation. Translated.

**çevirmek** *va.* Turn; turn round; change *one thing* into *another*; translate; recant; surround, enclose. **çekip ~,** to manage *a business etc.*; to know how to deal with *a person*: **dolab (hile, iş** *etc.*) **~,** to be up to some mischief: **şaşkına ~,** to dumbfound: **···den yüz (baş) ~,** to withdraw one's favour from; to abandon (a project, an opinion).

**çevre** Circumference; circuit; surroundings; contour; embroidered handkerchief. **~sini dolaşmak,** to go all round about *a place*: **···in ~sini sarmak,** to surround. **~lemek,** to surround.

**çevrik** Turned round; surrounded; overturned.

**çevrilmek** *vn. pass. of* **çevirmek.**

**çeyiz** Bride's trousseau. **~çemen,** *as* çeyiz.

**çeyrek** Quarter; quarter of an hour; five-piastre piece.

**çıban** (·‧) Boil; abscess. **Bağdad ~ı,** the Baghdad boil *or* Aleppo button. **~başı, -nı,** head *or* apex of a boil: **çıbanın başını koparmak,** to bring matters to a head.

**çıfıt** Jew (*contemptuously*); mean, stingy. **~ çarşısı,** like a jumble-sale: **içi ~ çarşısı,** evil-minded, malevolent.

**çığ** Avalanche.

**çığıltı** Confused noise of animal cries.

**çığır, -ğrı** Track left by an avalanche; path, way. **~ açmak,** to start a new method, to open the way to *stg.* new; **···e ~ açmak,** to open the way to ..., to give an opportunity to ...: **çığrından çıkmak,** to get off the rails, to fall into disorder.

**çığır·mak** *v.* **çağırmak. ~ğan, ~tkan,** decoy bird; tout; noisy fellow. **~tma,** a kind of fife.

**çığlık** Cry, clamour; scream.

**çığrış** Clamour, outcry. **~mak,** to cry out together *or* against one another.

**çıkagelmek** (·‧··), **~ir** *vn.* Appear suddenly.

**çıkar** Profit, advantage; way out. Leading to success; (of a street) leading to another street. **~ yol bulmak,** to find a way out: **~ yolu yok,** there is no way out of it: **bir işte ~ı olm.,** to have an axe to grind: **kendi ~ına bakmak,** to seek one's personal advantage, to 'look after number one'.

**çıkarmak** *va.* Take out, extract; remove; expel; export; take off (clothes *etc.*); bring out; publish; raise; produce; bring out and offer (food *etc.*); certify *as unfit etc.*; make out, get the sense of; derive, deduce; last *or* serve for *a season etc.* **baştan ~,** to seduce, lead astray: **bulup ~,** to invent, find out, discover: **dil ~,** to put out the tongue in derision: **diş ~,** to extract a tooth; (of a child) to cut a tooth: **ekmeğini* ~,** to gain one's livelihood; **ekmeğini taştan ~,** to be capable of earning his livelihood from anything: **elden ~,** to part with, dispose of, lose: **gözden ~,** to reckon on the loss of *stg.*, to be prepared to sacrifice *stg.*: **birini haklı ~,** to prove *s.o.* right; to maintain that *s o.* is right, to take his part: **hizmetten ~,** to dismiss from service: **iş ~,** to raise difficulties: **işten ~,** to dismiss from a post *etc.*: **işten iş ~,** to raise unnecessary objections: **'köyü bugün çıkarırız',** 'we'll make the village today': **bir sözden mâna ~,** to misinterpret a word *or stg.* said: **masrafını ~,** to recover one's expenses: **mesele ~,** to make a fuss, raise a difficulty; to seek a cause for a quarrel: **meydana ~,** to show, to bring to light, discover: **ortaya ~,** to divulge; to produce: **···e taş ~,** to give points to, to be far superior to: **taşın suyunu ~,** to be extremely strong *or* energetic: **yalancı ~,** to prove *s.o.* to be a liar; to make *s.o.* look like a liar: **bu palto kışı çıkarır,** this overcoat will last me the winter: **kışı Mısırda çıkardı,** 'he passed the winter in Egypt'.

**çıkartma** Transfer (of a coloured print to another surface).

**çıkartmak** *va. caus. of* **çıkarmak.** Cause *or* let remove *etc.*

**çıkaryol** (·‧·) Road leading to another road; the right way out *or* of doing *stg.*

**çıkı** *v.* **çıkın.**

**çıkık** Projecting; dislocated (limb). **~çı,** bone-setter.

**çıkın** Something (*esp.* money) wrapped up in a cloth *or* handkerchief; small bundle; hoard of money. **kirli çıkı,** one who has accumulated money by stinginess. **~lamak,** to tie up in a bundle.

**çıkıntı** Projection; salient; projecting balcony; marginal note. **~lı işaret,** caret (‸): **girintili ~lı,** serrated, zigzag.

**çıkış** Method of going out; exit; start (of a race); sortie (*mil.*); scolding.

**çıkış·mak** *vn.* Enter into competition *or* rivalry with another; undertake something beyond one's power; reach; suffice; burst out into anger *or* reproach; scold. **~tırmak,** to make (money) suffice *for stg.*; procure; cause to reach.

**çıkma** *verb. n. of* **çıkmak.** Act of going out *etc.*; exit; promontory; balcony, bow-window; invention; marginal note *or* word; loin-cloth *used in coming out of a Turkish bath.* Newly appeared *or* invented. **~ takımı,** bath linen: **yeni ~ âdet,** newfangled custom.

**çıkmak** *vn.* Come *or* go out; issue; appear; come into existence; come to pass; result; turn out to be; jut out, be prominent; (*month, season etc.*) pass, be over; set forth, start; come *or* go up, ascend, rise; make, suffice for (a coat *etc.*); (of a rumour) to get about; (of a joint) to be dislocated. ···**den** ~, to leave *one's job*, resign from ...; pass out of *a school, university etc.*: ···**e** ~, amount to ...; fall to the lot of; start on *a journey*; compete with *s.o.*; be received in audience by ...; play the part of, *e.g.* **Otelloya çıktı,** he played the part of Othello: **çık bakalım!** (*sl.*), 'fork out!', 'pay up!': ⌜**çıkmadık candan ümid kesilmez**⌝, 'while there's life there's hope': **adı* fena(ya) çıkmak,** to get a bad name: ···**le başa** ~, to cope with ...: **başına** ~, to presume on *s.o.'s* kindness, to take liberties with *s.o.*: **dediğin çıktı,** it has turned out as you said: **dışarı** ~, to go out, to go out to the lavatory: **dörtten iki çıkarsa,** if two be subtracted from four: **evden çıkmak,** to leave a house, *esp.* to leave for good, to give up a house: **kış çıktı,** winter is over: **ne çıkar!,** what does it matter!; **bir şey çıkmaz,** it doesn't matter: **paradan** ~, to incur expense: **vekile (elçiye)** ~, to go and see a Minister (Ambassador): **yalan** ~, to turn out to be untrue: **yola** ~, to set forth on a journey; **yoldan** ~, to get off the road, to lose one's way: **bu yol nereye çıkar?,** where does this road lead to?: **bu işte kimse ona çıkamaz,** nobody can compete with him in this.

**çıkmaz** Blind alley; dead-end; impasse.

**çıkolata** (··⁻́·) Chocolate.

**çıkrık** Winding-wheel (of a well *etc.*); windlass; pulley; reel; lathe; spinning-wheel.

**çılbır** Dish of poached eggs and yoghourt; chain *or* rope attached to a halter; lunge.

**çıldırmak** Go mad; go off one's head.

**çılgın** Mad; insane; raging.

**çıma** (ˋ·) Hawser; cable; fall of a tackle; end (*as opp. to 'bight'*) of a rope. ~ **dikişi,** short splice. ~**çı** (ˋ··), quayside hand.

**çımarıva** (··ˋ·) Ceremonial manning of the yards *or* decks by a ship's crew.

**çın çın** *Imitates the noise of tinkling glass or thin metal.* ~ ~ **ötmek,** to make an empty, ringing sound; to be empty *or* deserted.

**çınakop, çinekop** Young lüfer fish.

**çınar** Plane-tree.

**çıngar** Quarrel; dispute; lawsuit.

**çıngıl** Small bunch of grapes on a side shoot.

**çıngır** ~ ~, tinkle tinkle. ~**damak,** to give out a ringing sound, to tinkle. ~**tı,** ringing *or* clinking sound.

**çıngırak** Small bell.

**çınlamak** *vn.* Give out a ringing *or* clinking sound; have a singing in the ear. **kulağı çınlasın!,** 'may his ears be singing!' *said when making a kind reference to one who is absent.*

**çıplak** Naked; bare; destitute. ~**lık,** nudity, bareness.

**çıra** (ˋ·) Resinous wood; torch; kindling wood. ~**lı,** resinous.

**çırak, çırağ** Apprentice; pupil; pensioner; *formerly a person brought up as a servant in a great house and subsequently set up in life, usually by being married off*; favourite. ~ **çıkmak (olunmak, edilmek, çıkarılmak),** to quit the service of *s.o.* with provision for the future; ~ **etm.,** to pension off *or* set up in life a **çırak.** ~**lık,** apprenticeship; fee paid by an apprentice; allowance from a patron.

**çıramoz** Torch-holder *for attracting fish by night.*

**çırçıplak** (ˋ··), **çırılçıplak** Stark naked; utterly destitute.

**çırçır** Cotton-gin; rivulet; a kind of wrasse.

**çırpı** Twig; chip; clipping; skim coulter *of a plough*; chalk-line *for marking straight lines on timber.* ~**dan çıkmak,** to get out of line: ~**ya getirmek,** to put into line, to make straight: ~ **ipi,** carpenter's chalk-line: ~ **vurmak,** to mark a straight line with the chalk-line: **bir** ~**da,** easily and quickly; **işi bir** ~**da çıkarmak,** to make nothing of a job, to do a thing easily: **çalı** ~, brushwood.

**çırpıcı** Fuller; man who washes printed stuffs in the sea *to make their colours fast.*

**çırpın·mak** *vn.* Flutter; struggle; fuss about; be all in a fluster. ~**tı,** flurry; fretting *of small waves.*

**çırpış·mak** *vn.* Flutter. ~**tırmak,** to strike lightly with a small stick; to scribble hastily.

**çırpmak** *va.* Strike lightly with a series of small blows; tap; pat; beat (carpet); clap (the hands); flutter (the wings); rinse; bleach; trim *the ends of stg.* **çalıp** ~, to pilfer.

**çıt** *Imitates the sound of a small thing breaking.* ~ **yok,** there is not a sound to be heard.
**çıta** Border, moulding; long narrow strip of wood; bar *for high-jump.* ~**lı geçme,** tongue-and-groove joint.
**çıtçıt, -tı** Press-button.
**çıtıpıtı** Graceful; pretty; dainty.
**çıtır** ~ ~, *imitates a cracking sound.* ~**damak,** to crackle. ~**pıtır** *imitates the agreeable speech of a child or young girl.* ~**tı,** slight sound.
**çıtkırıldım** Effeminate; weak, fragile.
**çıt·lamak** *vn.* Make a slight cracking sound. ~**latmak,** *caus. of* **çıtlamak**; to hint at: **parmak** ~, to crack a finger. ~**pıt,** percussion cap *which goes off when trodden on.*
**çıvğar** Additional animal *for ploughing, or for hauling a gun.*
**çıyan** Centipede. ~ **gözlü,** with cold reptilian eyes: **sarı** ~, *used of a fair-complexioned but unpleasant looking person.* ~**cık,** little centipede; bistort.
**çızıktırmak** *va.* Scribble; scrawl.
**çiçek** Flower; small-pox; fickle *or* tricky man. **çiçeği burnunda,** quite fresh, brand new; in the full bloom of youth: **aman ne** ~!, *said of a girl who unexpectedly becomes rather forward and flighty*: **ne** ~ **olduğunu bilirim,** I know the sort of fellow he is: **su çiçeği,** chicken-pox. ~ **aşısı, -nı,** vaccine. ~**bozuğu, -nu,** pock-mark; pock-marked. ~**çi,** florist. ~**li,** in bloom; decorated with flowers; with a flowery pattern (stuff); suffering from small-pox. ~**lik,** flower garden; flower vase; conservatory. ~**suyu, -nu,** orange-flower essence, neroli.
**çift, -ti** Pair; couple; mate; yoke of oxen; pair of pincers. Even (number); double. ~**bir altı,** double one six (*in giving a telephone number*): ~ **çubuk,** *v.* **çiftçubuk**: ~**e gitmek,** to go ploughing: ~ **koşmak,** to harness to the plough: ~**e koşmak,** *to use for ploughing an animal normally used for other purposes*: ~ **sürmek,** to plough: **bir** ~, a pair: **sana bir** ~ **sözüm var,** I want a word or two with you.
**çift·çi,** ploughman; farmer. ~**çilik,** agriculture. ~**çubuk,** general work of the farm *or* vineyard; farm; estate.
**çifte** Paired; doubled. Double-barrelled gun; pair-oar boat; horses' kick with both hind feet. ~ **atmak,** (of a horse) to lash out with both hind feet: ~ **kumrular,** a pair of inseparable friends: ~ **telli,** belly-dance: **iki** ~ **kayık,** boat with two pairs of oars. ~**lemek,** (of a horse) to lash out. ~**li,** kicking (horse); unlucky; treacherous. ~**nağra,** double-drum.
**çifter** Having a pair each.
**çiftleşmek** *vn.* Mate (of animals).
**çiftlik** Farm.
**çiğ**[1] Dew.
**çiğ**[2] Raw, uncooked; unripe; crude, inexperienced. ~ **toprak,** neglected, uncultivated land: ···**i** ~ ~ **yemek,** to tear *s.o.* to pieces (*fig.*); to nourish a bitter enmity against *s.o.*: ⌜~ **yemedim ki karnım ağrısın**⌝, 'I've done nothing that I should be blamed': ⌜**hatır için** ~ **tavuk yenir**⌝, one would do almost anything for friendship's sake: ⌜**insan** ~ **süt emmiştir**⌝, 'to err is human'.
**çigan** Gipsy.
**çiğdem** Crocus; saffron.
**çiğid** Cotton-seed; freckle.
**çiğlik** Rawness; tactlessness.
**çiğnemek** *va.* Trample on, crush; masticate, chew; ignore. **'bizi çiğneyip geçtiniz'**, 'you passed us without even condescending to notice us' (*jokingly*); **müdürü çiğneyip umum müdüre çıkmış,** he ignored (*or* passed by) the director and went straight to the Director-General: **bir kanunu (kaideyi)** ~, to ignore a law (rule): **lâkırdıyı** ~, to mumble.
**çiğnemik** Food masticated *and then fed to a baby.*
**çihar** Four (*used in calling the throws of dice*).
**çiklet, -ti** Chewing-gum.
**çil** Bright; shiny; speckled, freckled. Spot; freckle; grey partridge. ~ **yavrusu gibi dağılmak,** to scatter like a covey of partridges, to be utterly routed. ~**li,** speckled; freckled.
**çile**[1] Period of forty days religious retirement and fasting; period of penitence; trial; suffering. ~ **çekmek (çıkarmak, doldurmak),** to pass through a severe trial: ~**den çıkarmak,** to infuriate: ~**den çikmak,** to be exasperated.
**çile**[2] Hank *or* skein *of silk etc.*; bowstring.
**çilek** Strawberry. **ağac çileği,** raspberry.
**çile·keş** Suffering; (of a dervish) undergoing period of solitude and fasting. ~**siz,** carefree.

**çilingir** Locksmith. ~ **sofrası,** small table with drinks and hors-d'œuvres on it.
**çillenmek** *vn.* Become freckled *or* speckled.
**çim** Turf; short grass; grass plot; moss.
**çimdik** Pinch. ~**lemek,** to pinch.
**çimen** Meadow; lawn; turf. ~**lik,** ~**zar,** grassy spot; expanse of lawn; meadow.
**çimensiz** *In* çehizsiz ~, (girl) without dowry; without means.
**çimento** Cement.
**çimlenmek** *vn.* Sprout; become grassy; have a nibble; get perquisites *or* pickings.
**çimmek** *vn.* Plunge into water; bathe.
**çimrenmek** *vn.* Tuck up one's trousers *before entering water*; gird oneself up *for some enterprise.*
**çin** *v.* çın.
**Çin** China. ~**ce,** Chinese language. ~**li,** Chinese.
**çingene, çingâne** Gipsy; low, mean, cunning fellow; very small palamut. ~ **borcları,** petty debts: ⌜~ **çalar Kürd oynar**⌝, 'the gipsy plays, the Kurd dances', one is as bad as the other; a disorderly party: ~ **düğünü,** a disorderly, riotous assembly: ~ **sarısı,** very bright yellow. ~**lik,** mode of life and habits of a gipsy; vagabondage; meanness: ~ **etm.,** to behave in a mean, miserly way.
**çini** Porcelain; encaustic tile; enamel ware; indian ink. Tiled (stove *etc.*). ~ **mavi,** bright blue. ~**li,** decorated with painted tiles.
**çinko** Zinc.
**çintan, çintiyan** Wide trousers *worn by peasant women in Turkey.*
**çipil** Blear-eyed; dirty (weather).
**çipo** Anchor-stock.
**çiriş** Shoemakers' *or* bookbinders' paste; size. ~ **gibi** *or* ~ **çanağı gibi,** sticky and bitter: **işim** ~ (*sl.*), I am in a mess. ~**lemek,** to smear with paste *or* size. ~**otu, -nu,** asphodel.
**çirkef** Sink; sewer; dirty water; ill-bred, badly-behaved man. ~**e taş atmak,** to do *stg.* to invite the abuse of an insolent person.
**çirkin** Ugly; unseemly; unpleasant. ~**lik,** ugliness; nasty habit *or* behaviour.
**çiroz** Dried mackerel; person who is mere skin and bone.

**çis** Honeydew; manna.
**çis·elemek,** ~**imek** *vn.* Drizzle. ~**inti,** fine drizzle.
**çiş** Urine. ~ **bezi,** baby's diaper: ~**im* geldi,** I want to make water.
**çişik** (·ˈ) Leveret.
**çit**[1], **-ti** Chintz.
**çit**[2], **-ti** Fence of hurdles *or* brushwood. ~ **çekmek,** to make a fence round a place: **yerli** ~, permanent fence; hedge: ~ **kuşu,** wren.
**çiti** Act of putting together; darning.
**çitilemek** *va.* Squeeze *or* rub *stg.* while washing it.
**çitişmek** *vn.* Become tangled *or* matted.
**çitlembik** Terebinth berry. ~ **gibi,** (of a girl) small and dark.
**çitlik** Material suitable for making a hedge; place enclosed by a hedge.
**çitmek** *va.* Put together; darn; *as* **çitilemek.**
**çitmik** One sprig of a bunch of grapes; a pinch of *stg.*
**çitsarmaşığı** (ˈ····), **-nı** Convolvulus.
**çivi** Nail. ~ **gibi,** brisk, alert: ~ **kesmek,** to shiver with cold: ⌜~ ~**yi söker**⌝, ⌜set a thief to catch a thief⌝; (*sl.*) 'another little drink won't do us any harm'. ~**dişi, -ni,** canine tooth. ~**leme,** dive (feet foremost). ~**lemek,** to nail; to stab. ~**lenmek,** to be nailed; to be glued to the spot.
**çividî** Indigo blue.
**çivit, çivid** Indigo; washing-blue. ~**lemek,** to dye with indigo; to treat *laundry* with blue.
**çiy** *v.* **çiğ.**
**çizecek** Scriber.
**çizgi, çizik** Line; mark; scratch; trace; stripe; wrinkle; dash *in Morse code* (*dot* = **nokta**); hopscotch. ~**li,** striped; ruled (paper).
**çizi** *As* **çizgi**; furrow. ~**nti,** small scratch *or* mark.
**çizme** Top-boot; Wellington. ⌜~**den yukarı çıkmak**⌝, to meddle with things you don't understand (*cf. 'shoemaker, stick to your last!'*). ~**li,** wearing top-boots: ⌜**(orduda) sarı** ~ **Mehmed Ağa**⌝, any unknown ordinary person, 'Tom, Dick, or Harry'.
**çizmek** *va.* Draw a line *or* mark; sketch; rule with lines; scratch; erase, cancel. **yan** ~, to slip away; to 'cut' *s.o.*: **işten yan** ~, to shirk one's work: **yan çizilecek iş değil,** it is not a job to be scamped.

**çoban** Shepherd; rustic; boor. ~ **armağanı,** an unpretentious present: ~. **etm.,** to shepherd, (*iron.*) to look after children *etc.*: ~ **köpeği,** sheepdog. ~**aldatan,** goatsucker. ~**çantası, -nı,** shepherd's-purse. ~**kebabı, -nı,** bits of meat threaded on a stick and cooked over a wood fire. ~**lık,** the calling of a shepherd; a shepherd's pay. ~**püskülü, -nü,** holly. ~**süzgeci, -ni,** goose-grass, cleavers. ~**tarağı, -nı,** teazle. ~**yıldızı, -nı,** the planet Venus.

**çocuk** Infant; child; boy. ~ **almak,** (of a doctor) to deliver a child: ~ **aldırmak,** to have a child delivered *or* aborted: ⌜~**tan al haberi**⌝, if you want to know the truth listen to what children say; 'little pitchers have long ears': ~ **düşürme,** abortion: ~ **oyuncağı,** a toy, a matter of no account; child's-play (very easy thing).

**çocuk·cağız,** the poor little child. ~**ça** (·ˈ·), childish(ly). ~**laşmak,** to be *or* become childish. ~**lu,** having children. ~**luk,** childhood; childishness, folly, silly deed. ~**su,** childish.

**çoğalmak** *vn.* Increase, multiply.

**çoğu** The greater part; most; *v.* **çok.** ~ **kimse,** most people.

**çoğunluk** Majority.

**çoğumsamak** *va.* Deem too many *or* too much.

**çok** Much; many; too much, too many; very. **çoğu,** the greater part, most; **çoğumuz,** most of us: ~**tan** *or* ~**tan beri** *or* ~**tandır,** for a long time, since long ago: ~ **bilmiş,** (of a child) precocious, who knows more than he should; (of a man) who 'knows a thing or two': ~ **gelmek,** to seem excessive: ⌜**çoğu gitti azı kaldı**⌝, it won't be long before ...; the end is near: ~ **görmek,** to deem excessive; to grudge: **çoğa kalmaz,** it won't be long before ...: **bu iş** ~ ~ **bir saat sürer,** this business won't take more than an hour at the most: ~ **şey!,** how odd!; 'you don't say so!': ~ **yaşa!,** long live!

**çokbilmiş** *v.* **çok.**

**çokça** A good many; a good deal; somewhat.

**çokluk** Abundance; crowd. Often.

**çolak** With one arm missing *or* paralysed.

**çolpa** With one leg damaged; clumsy; untidy.

**çolukçocuk** Wife and children; all the family; junior and unimportant people. **çoluğa çocuğa karışmak,** to become a family man.

**çomak** Club; cudgel.

**çomar** Large watch-dog; mastiff; (*sl.*) old publican.

**çopur** Pock-mark. Pock-marked.

**çorab** Stocking. ~ **söküğü gibi,** 'like the unravelling of a stocking', *i.e.* proceeding easily and quickly *or* in rapid succession: **birinin başına** ~ **örmek,** to play a dirty trick on *s.o.* ~**cı,** hosier. ~**cılık,** hosiery.

**çorak** Arid; barren; brackish (water). An impervious kind of clay; saltpetre bed. ~**lık,** land impregnated with salt.

**çorba** Soup; medley, mess. ~ **etm.,** *or* ~**ya döndürmek,** to make a mess of, to turn upside-down: ~ **gibi,** muddy; sloppy; in disorder: ~**da tuzu*** (**maydanozu***) **bulunmak,** to participate in a small way. ~**cı,** Christian notable (*now used jokingly about non-Moslems*); official entertainer of guests in a village. ~**lık,** anything suitable for making soup,

**çotra** Flat wooden bottle *or* mug. ~ **balığı,** file-fish.

**çöğür** A species of eryngium. A kind of lute.

**çökelek** Skim-milk cheese.

**çöker·mek,** ~**tmek** *va.* Make *a camel* kneel; cause to collapse. **diz** ~, to make kneel. ~**tme,** a kind of fishing-net; break-through (*mil.*).

**çök·mek** *vn.* Collapse; fall down; give way; (of sediment) to settle, be deposited; be prostrated *by age or fatigue*; (of darkness, fog, sorrow *etc.*) to descend upon one. **acısı yüreğine*** ~, *for a loss etc.* to be deeply felt: **diz** ~, to kneel. ~**türmek,** *v.* **çökermek.**

**çök·ük** Collapsed; fallen down; caved in; sunk, depressed; precipitated. ~**üntü,** sediment, deposit; debris, dilapidation; depression; subsidence. ~**üşmek,** to fall down together; to run in from all sides and fall upon *stg.*

**çöl** Desert; wilderness. ~**lük,** desert tract of country.

**çömelmek** *vn.* Squat down on one's heels.

**çömez** Boy who works in return for board and lodging; hodja's assistant; follower, disciple; (*school sl.*) swat, sap. **papaz** ~**i,** novice *in the priesthood.*

**çömlek** Earthenware pot. ~ **hesabı,**

clumsy work; accounts made by an illiterate person: ~ **kebabı,** meat roasted in a pipkin: **çanak ~,** pots and pans. **~çi,** potter. **~çilik,** earthenware.

**çöp, -pü** A fragment of vegetable matter; chip; straw; sweepings, litter, rubbish. ~ **arabası,** dustcart: ~ **atlamamak,** to let nothing escape one, to pay the utmost attention: ~ **gibi,** very thin: ~ **kebabı,** small bits of meat grilled on a skewer: ~ ~ **üstüne koymamak,** not to do a stroke of work: **ağzıma** ~ **koymadım,** I haven't eaten a morsel: **gözünü ~ten sakınmaz,** dare-devil, intrepid: ⌜**sakınan göze** ~ **batar**⌝, overcaution is often dangerous.

**çöp·atlamaz,** Punctilious, meticulous. **~çatan,** predestination, foreordination (*only in connexion with marriage*); woman who acts as go-between in arranging a marriage. ~ **çatmak,** to arrange a marriage. **~çü,** dustman, scavenger. **~leme,** hellebore. **~lenmek,** to pick up scraps for a meal; to get pickings *from another's business or in virtue of an official position.* **~lük,** rubbish-heap; dustbin. **~süz,** without a stalk: ~ **üzüm,** a person without relatives (*usually of a marriageable girl or boy who has no relatives and is therefore desirable*).

**çör** *Used only in conjunction with* **çöp.** **~çöp,** sticks and straws; brushwood: **~den çöpten,** 'made of matchwood', flimsily built.

**çörek** *A kind of* shortbread *in the shape of a ring*; anything ring-shaped. **~lenmek,** to be coiled up; to coil oneself up. **~otu, çöreotu, -nu,** seeds of *Nigella sativa, used to flavour* **çörek.**

**çöven¹, çoğan** Soapwort (*the roots of which are used by sweetmakers*).

**çöven²** *v.* **çevgen.**

**çözğü** Warp; cotton sheeting.

**çözme** A kind of silk tissue.

**çöz·mek** *va.* Untie; unravel; undo; solve (problem, cipher). **~ük,** untied; disentangled; loose; thawed. **~ülmek,** to be untied *etc.*; (*mil.*) to withdraw from contact with the enemy, to disengage; melt; thaw: **dili*** ~, to recover one's speech. **~üntü,** a going to pieces; downfall; *débâcle.*

**çubuk** Shoot, twig; staff; metal rod; pipe; stripe (in a textile). **çubuğunu tellendirmek,** to take it easy: **çubuğunu tüttürmek,** to smoke one's pipe; to be lazy and indifferent: **tüfek çubuğu,** ramrod. **~lamak,** to beat (a carpet *etc.*), **~lu,** striped (cloth *etc.*). **~luk,** cupboard *or* rack for pipes.

**çuha** Cloth. **~çiçeği, -ni,** polyanthus.

**çuka** Sterlet.

**çukur** Hole; hollow; ditch; cavity; tomb; dimple. Hollowed out; sunk; concave. **~a düşmek,** to meet with misfortune: ···**in ~unu kazmak,** to prepare the ruin of *s.o.*: **bir ayağı ~da olm.,** to have one foot in the grave: **bostan ~u,** manure-pit: **göz ~u,** eye-socket: **gözleri* ~a kaçmak,** to have sunken eyes, to wear a haggard look. **~luk,** place full of hollows *or* ditches.

**çul** Hair-cloth; horse-cloth; badly made clothes. **~u tutmak,** to become rich, to come into property: ~ **tutmaz,** spendthrift, shiftless. **~ha,** weaver: ~ **kuşu,** long-tailed tit. **~lama,** food covered with dough and then baked. **~lamak,** to cover *a horse* with a rug; (of waves) to break right over a ship. **~lanmak,** to be covered with a horse-cloth; to put on new clothes; to become rich; (*with dat.*) to hurl onself upon *s.o.*, to fall upon; to pester.

**çulluk** Woodcock. **kervan çulluğu,** curlew: **su çulluğu** *or* **efendi çulluğu,** snipe: ~ **tersi,** woodcock 'trail'.

**çulpa** *v.* **çolpa.**

**çulsuz** Destitute.

**çultarı** Quilted saddle-cloth.

**çultutmaz** Shiftless, improvident.

**çurçur** A kind of wrasse.

**çutal** *In* **çatal ~,** confused, all mixed up.

**çuval** Sack. **~dız,** packing-needle: ⌜**iğneyi kendine ~ı karşındakine (başkasına)**⌝, try it on yourself first, *or sometimes,* be sure you are not to blame before blaming others.

**çük** Little boy's penis.

**çünki, çünkü** Because.

**çürük** Rotten; unsound; spoilt; worthless; unsound (argument); unfit for military service. Bruise. ~ **buhar,** waste steam: ~ **çarık,** utterly rotten: ~ **çıkmak,** to turn out rotten, to prove to be unsound *or* false; **çürüğe çıkmak,** (of a soldier) to be invalided out of the army: ~ **tahta,** a risky business: ~ **tahtaya basmak,** to suffer loss, to be cheated: **ipi ~,** undependable: **yaprak çürüğü,** leaf-

mould. ~**lük,** rottenness; putrefaction; rot-heap; common grave in a cemetery.
**çürü·mek** *vn.* Rot; decay; spoil; become worthless *or* unsound; be unsound in credit; (of a soldier) to be rejected for unsoundness; show the mark of a bruise. ~**tmek,** to cause a rot *etc.*; to show *an argument etc.* to be unsound, to refute, rebut.
**çüş** *Sound made to stop an ass or to deride a man behaving like one.*

# D

**-da, -de, -ta, -te** *Particle forming the locative case or the equivalent of the prepositions* in; on.
**da, de, ta, te** And; also; but. **sen de, ben de,** both you and I: **her kes gitti, o da kaldı,** everyone went, but he remained; **üşüdüm de giyindim,** I was cold, so I put on my clothes: **mektub gelse de meraktan kurtulsak,** if only a letter would come and relieve our anxiety: **görsem de tanıyamam,** even if I were to see him I should not recognize him: **ne iyi ettin de geldin,** you've done well to come: **gelse de gelmese de faydası yok,** it's useless whether he comes or not: **anlamadım da sordum,** I didn't understand, so I asked: '**Niçin sordun?**' '**Anlamadım da**', 'Why did you ask?' 'I didn't understand, that's why'.
**dadanmak** *va. with dat.* Acquire a taste for; get fond of; visit *a place* frequently; frequent; make too free use of, abuse.
**dadaş** Village 'blood' (*in E. Anatolia*).
**dadı, dada** Nurse.
**dafire** (·–ˋ) Plexus (*med.*).
**dağ¹** Mountain; *as an attributive noun it often merely means* 'wild', *e.g.* ~ **elması,** crab-apple. ~ **adamı,** mountaineer; rough fellow: ~**lar anası,** a huge woman: ~ **ayısı,** mountain bear; uncouth, boorish fellow: ~ **babaları,** brambling: ~ **başı,** top of the mountain; wild, remote spot: ~**da büyümüş,** bucolic; country bumpkin: ~ **deviren,** coarse and clumsy, a 'bull in a china shop': ~**lara düşmek,** to be in a destitute condition: ~ **eteği,** foothill, lower slopes of a mountain: ⌜~**dan gelip bağdakini kovmak**⌝, of a new arrival to treat the old inhabitants with contempt, *or,* of an upstart official to treat the old stagers cavalierly: ~ **gibi,** huge; a great amount of *anything*: ~**a kaldırmak,** to kidnap: ~ **taş,** all around; in great quantities; greatly; ⌜~**lara taşlara**⌝, *an expression used when talking of a calamity, meaning* 'may such a thing be far from us!': **aralarında** ~**lar kadar fark var,** there is a world of difference between them.
**dağ²** Brand; mark; cautery; scar. ~ **basmak,** to brand.
**dağarcık** Leather bag *or* wallet, *used by shepherds etc.* **dağarcığındakini çıkarmak,** to bring out a remark one has prepared beforehand.
**dağdağa** Trouble and turmoil; confusion; noise. ~**lı,** noisy, confused; troublesome.
**dağdar** Afflicted, grieved; branded, stigmatized.
**dağ·ılmak** *vn.* Scatter, disperse; fall to pieces; be spread, be disseminated; be distributed; (of a room *etc.*) to be untidy. ~**ınık,** scattered; dispersed; wide apart; disorganized; unorganized; untidy. ~ **nizam,** open order (*mil.*). ~**ıtmak** *va.*, scatter, disperse; distribute; break to pieces; create a mess *in a room etc.*
**dağlamak** *va.* Brand; cauterize; (of heat *or* cold) to scorch; wound the feelings, grieve.
**dağlı¹** Mountaineer; uncouth, bucolic.
**dağlı²** Branded; scarred; sore at heart.
**dağlıç** A kind of sheep; the mutton therefrom.
**dağlık** Mountainous country. Mountainous.
**dağsıçanı** (ˋ···), **-nı** Marmot.
**daha** More; a greater number. More, again, besides; further; yet; *preceding an adjective it forms the comparative thereof.* ~ **bu sabah geldi,** he only arrived this morning: ~ **dün,** only yesterday: ~ **erken,** earlier; it is still early: ~ **neler!,** what next!, how absurd!: ~**sı var,** that's not all; to be continued: **bir** ~ **gitmem,** I'll go no more: **bir** ~ **sene,** next year: **bir** ~ **yaz!,** write once again! **iki üç** ~ **beş eder,** two plus three makes five.
**dahdah** Gee-gee.
**dahi** And; also; furthermore; too; even. **ve** ~ **duralar,** and that was the end of it!
**dâhi** (–ˋ) A genius. ~**lik,** the state *or* quality of a genius.

**dahil** (–̀) Who *or* which enters; who *or* which is inside; inner; interior; inside; included. ~ **etm.**, to include: ~ **olm.**, to be included: **bir şeyin ~inde,** inside a thing. **~en** (–̀··), internally; in the interior (of a country). **~î** (–·–̀), internal; inner. **~iye** (–··̀), *fem. of* **dahilî**; Ministry of the Interior; clinic for internal diseases. **~iyeci,** official of the Ministry of the Interior; specialist in internal diseases.

**dahiyane** (–·–̀) Of genius (book, action *etc.*, *but not used of a person*).

**dahl, -li** Entry; interference; participation. **···de ~i olm.,** to have a part in, to be implicated in.

**daim** (–̀) Enduring, permanent; continuous. **~a** (–̀·–), always; perpetually. **~î** (–·–̀), constant; permanent; perpetual.

**dair** (–̀) Revolving. (*with dat.*) Concerning; about; relating to. **~e** (–··̀), circle; suite of rooms; department; offices (of an administration); limit (of jurisdiction *etc.*); tambourine. **edeb ~sinde,** in a decent, mannerly way: **makine ~si,** engine-room.

**dakika** (·–̀), **dakka** (·̀) *n.* Minute. **~sı ~sına,** to the very minute, very punctually.

**daktilo** (·̀·) Typist; typewriting. ~ **etm.,** to type.

**dal**[1] Branch, bough; twig; spray *of diamonds etc.* **~dan ~a atlamak,** to jump from one subject to another: ~ **budak salmak,** to send out ramifications; to grow in size *or* importance: ~ **gibi,** graceful, slender: **bindiği* ~ı kesmek,** to cut the ground from under one's own feet: **bir ~da durmaz,** fickle, capricious: **güvendiği ~lar elinde kalmak,** to be 'let down', to be deceived.

**dal**[2] Shoulder; back. **~ına* basmak,** to irritate *s.o.*: **~ına* binmek,** to put pressure on *s.o.*; to pester.

**dal**[3] Bare; naked.

**dalak** Spleen; honeycomb. ~ **olm.,** to have inflammation of the spleen.

**dalâlet** (·–̀), **-ti** A going astray; deviation, error.

**dalamak** *va.* Bite; sting; prick.

**dalaş** Quarrel. **ağız ~ı,** a bickering, wrangle. **~mak,** to bite one another; to wrangle.

**dalavere, dalavera** (··̀·) Trick; deceit; intrigue.

**dalbastı** *In* ~ **kiraz,** lit. 'cherries that weigh down the branch', *seller's cry* 'Fine cherries!'

**daldırmak** *va. caus. of* **dalmak.** Plunge *into a liquid*; put *into the ground*; layer (a shoot).

**dalga** Wave; undulation; watering (of silk *etc.*); (*sl.*) trick, deceit. **~yı başa almak,** (of a boat) to turn to meet a wave; to breast the waves (*also fig.*): ~ **geçmek,** to be lost in reverie, to be wool-gathering; (*sl.*) to pretend to be doing some work: **···le ~sı olm.,** (*sl.*) to have a love affair with ... .

**dalga·cı,** a wool-gatherer; one who pretends to be working; tricky fellow: ~ **Mahmud,** a lazy chap. **~kıran,** breakwater. **~lanmak,** (of the sea) to become rough; (of a flag) to wave; undulate; become corrugated. **~lı,** covered with waves; rough (sea); undulating; corrugated; watered (silk).

**dalgıç** Diver.

**dalgın** Plunged in thought; absent-minded; somnolent. **~lık,** reverie; brown study; absence of mind; distraction.

**dalkavuk** Toady; sycophant; parasite. **~luk,** sycophancy; flattery; fawning.

**dalkılıc** Who has bared the sword; with bared sword; warrior; swashbuckler.

**dallan·dırmak** *va.* Cause to ramify; complicate; render difficult; exaggerate. **~mak,** to become branched, ramify; become complicated. **dallanıp budaklanmak,** to spread out in all directions, to have far-reaching effects.

**dallı** Branched. ~ **budaklı,** complicated, intricate.

**dalmak** *vn.* Plunge, dive; be lost in thought; be absorbed in an *occupation*. **dala çıka,** sinking and rising, *i.e.* with the greatest difficulty: **hasta gene daldı,** the patient has become unconscious again: **uykuya ~,** to drop off to sleep.

**daltaban** Bare-footed; destitute; vagabond.

**dalya**[1] (̀··) Dahlia.

**dalya**[2] (̀··) Tally. ~ **on,** that makes ten.

**dalyan** Fishing nets, *fastened to poles, on one of which stands a look-out to observe the entry of fish*; fishpond; reserved fishery. ~ **gibi,** well set-up: ~

tarlası, the area enclosed by the **dalyan.**

**dam**[1] Roof (*in Anatolia only a flat roof*); roofed shed; small house; outhouse; stable; prison. **~ altı,** any place covered by a roof, *esp.* a loft *or* garret: **~dan düşer gibi,** suddenly; untimely; out-of-place (remark *etc.*): ⌜**~dan düşen halden bilir**⌝, the only ones who can sympathize with a man's misfortunes are those who have suffered similarly.

**dam**[2] (*Fr. dame*) Lady partner *in a dance.*

**dama** (ˈ·) The game of draughts. **~ demek,** to give up, to accept defeat: **~taşı,** draughtsman; **~ taşı gibi,** who is constantly moving from place to place. **~cı,** draughts-player.

**damacana** (··ˈ·) Large wicker-covered bottle; demijohn.

**damad** (–ˈ) Son-in-law; bridegroom; title given to the son-in-law of a Sultan; a kind of duck.

**damak** Palate. **~ eteği,** the roof of the mouth: **damağını* kaldırmak,** to reassure *a frightened person*: **tadı damağında* kalmak,** to be remembered with longing. **~lı,** having a palate.

**damalı** Having a check pattern.

**damar** Vein; seam (of coal *etc.*); disposition, character; bad humour. **~ı atmak,** of a vein to pulsate: **···in ~ına basmak,** to tread on *s.o.'s* corns, to exasperate *s.o.*: **···in ~ını bulmak,** to find out *s.o.'s* weak spot: **···in ~ına girmek,** to ingratiate oneself with *s.o.*: **~ı* tutmak,** to have a fit of bad temper; to be capricious *or* obstinate: **hasislik ~ı tuttu,** he had a fit of avarice, he was overcome by avarice; *if used of a normally good quality this idiom is ironical, e.g.* **vazife ~ı tuttu,** his sense of duty called him (*iron.*); **~ iltihabı,** phlebitis: ⌜**akacak kan ~da durmaz**⌝, what is fated will happen.

**damar·lı,** veined; with swollen veins; obstinate; perverse. **~sız,** without veins; shameless.

**damdazlak** (ˈ··) Completely bald.

**damga** Instrument for stamping; stamp; mark; stigma. **ayar ~sı,** hallmark: **soğuk ~,** embossed stamp. **~lamak,** to mark with a stamp; to stigmatize. **~lanmak,** to be marked *or* stamped; to be disgraced. **~lı,** stamped, marked: **~ eşek,** *said in contempt of a person known to everyone.* **~pulu, –nu,** stamp (paper); revenue stamp.

**damıtmak** *va.* Distil.

**damız** Breeding-shed; breeding establishment. **~lık,** animal kept for breeding; stallion; used for propagating: **~ fidan,** nursery plant.

**damkoruğu** (ˈ···), **–nu** Houseleek.

**damla** Drop; paralytic stroke; gout. (Medicine) taken with a dropper. **~ inmek,** to have a stroke; *also sometimes* to have an attack of gout: **~ yakut,** ruby of the finest water. **~lık,** medicine dropper; space between two houses on which the eaves drip. **~lamak,** to drip; to appear suddenly, to turn up (when not wanted); to frequent habitually. **~latmak,** to pour out drop by drop; to distil.

**damping** (ˈ·) Dumping *of goods.*

**–dan, –den** *Particle forming the ablative case or the equivalent of the prepositions* from, by.

**dana** Calf. **~ eti,** veal: ⌜**~nın kuyruğu kopmak**⌝, for the crucial moment to come, for the worst to happen: **anasıyle ~,** mother and child. **~burnu, –nu,** cockchafer grub; whitlow.

**dandini** (ˈ··) *Word used when dangling a baby.* **~ bebek,** effeminate dandy, childish person: **ortalık ~,** everything is in a mess.

**dan dun** Bang! bang!

**dane** (–ˈ) *v.* **tane.**

**dangalak** Loutish, boorish, stupid.

**dangıl** *In* **~ dungul,** boorish: **~ dungul konuşmak,** to speak with an ugly provincial accent.

**danışık** Consultation. **~lı,** *in* **~ dövüş,** a 'put up job', a plan previously concocted to deceive *s.o.*

**danışmak** *va. & vn.* Consult, ask advice; confer.

**danıştay** *New name of the old* **Şurayı Devlet**; Court which tries cases between private persons and the government.

**Danimarka** (··ˈ·) Denmark.

**daniska** (·ˈ·) The best of anything. **bunun ~sını bilir,** he knows this from A to Z.

**dank** *In* **kafasına* ~ demek (etmek),** suddenly to understand from some incident something that had previously puzzled one; **kafama ~ dedi,** the truth dawned upon me; **kafasına ~ dedi** *also means* 'he has learnt his lesson'.

**dans** Dance. ~ **etm.,** to dance.
**dantelâ** (·˙·) Lace.
**dapdaracık** (˙···) Very narrow, tight *or* scanty.
**dar** Narrow; tight; difficult; scanty; short (of time). Narrow place; straits; difficulty. **~a ~** *or* **~ ~ına,** with difficulty, hardly: **~a boğmak,** to take advantage of *s.o.'s* difficulties: **~da bulunmak,** to be in financial straits: **~a düşmek,** to be in a difficulty: **~a gelmek,** to be pressed for time; to be forced by circumstances: **~a gelemem,** I won't be rushed!: **~a getirmek,** to hurry or 'rush' *s.o. into doing stg.*: **~da kalmak,** to be in difficulties: **~ kurtulmak,** to narrowly escape: **~ yetişmek,** barely to reach: **···e canını* ~ atmak,** to manage to take refuge in ...: **eli ~,** close-fisted; not having enough to live on (*also* **eli ~da**): **içi ~,** impatient, hasty: **kendisini* bir yere ~ atmak,** to make hastily for a place: **sabahı ~ etti,** he waited impatiently for the morning.
**dara** Tare (weight). **~ya atmak** *or* **çıkarmak,** to regard as unimportant, to take no notice of. **~sız,** net (weight), excluding the tare.
**daracık** (˙··) Rather narrow *or* tight; somewhat scanty.
**daradar** (˙··) With difficulty; only just.
**darağacı** (˙···), **~nı** Gallows; sheerlegs.
**daralmak** *vn.* Become narrow *or* tight; shrink; become scanty; become difficult; be restricted. **nefesi* ~,** to be short of breath: **vakit daraldı,** there is not much time.
**darb** Blow; minting *of coins*; multiplication (*arith.*). **~e,** blow, stroke; turn of fortune; chance: **~i hükûmet,** coup d'état: **~hane,** mint.
**darbımesel** Proverb.
**darbuka** (·˙·) Earthenware kettledrum.
**dardağan** Scattered; in utter confusion. **~ darısı,** the seeds of the tamarisk: **~ darısını saçmak,** to sow discord; to sow the seeds of evil.
**dardarına** (˙···) With difficulty; only just.
**dargın** Offended, cross.
**darı** Millet. **~sı başına,** 'may he (you) follow suit!'; may his turn come next *for a piece of good luck*: **dibine ~ ekmek,** to squander, consume, utterly exterminate.
**darılmaca** *In* **~ yok,** 'you must not get angry'; 'I am warning you beforehand that you may be disappointed'.
**darılmak** *va.* Scold. *vn.* Get cross.
**darifülfül** (—··˙) Long pepper.
**darkafalı** (˙···) Narrow-minded.
**dar·lanmak, ~laşmak** *vn.* Become narrow *or* confined; be in straits. **~latmak, ~laştırmak,** to make narrow; restrict; oppress.
**darlık** Narrowness; narrow place; trouble; poverty. **gönül darlığı,** worry, depression: **nefes darlığı,** shortness of breath; asthma.
**darmadağın** (˙···) In utter confusion; all over the place.
**dar·ülâceze** (—···˙) Poorhouse, infirmary. **~ülfünun** (—··–˙), university. **~üşşafaka** (—···˙), school for orphans.
**darvincilik** Darwinism.
**dasitan** (—·–˙) Story; legend; ballad; epic; adventure; spell. **dillere ~ olm.,** to be the general topic of conversation. **~î** (—·—–˙), epic, legendary.
**-daş** *Suffix implying fellowship or participation, e.g.* **yol~,** fellow-traveller; **din~,** co-religionist.
**daüssıla** (—··˙) Nostalgia.
**dâva** (—–˙) Lawsuit; trial (law); claim; petition; problem; thesis; matter, question. **···e ~ açmak** *or* **···i ~ etm.,** to bring a suit at law against *s.o.*: **~ başı,** the main argument: **~sında bulunmak,** to claim that ...: **~ etm.,** to claim, to pretend to; to be arrogant: **···in ~sını tutmak,** to take the side of, to adhere to the cause of: **~ vekili,** advocate, barrister: **şairlik ~sına düştü,** he sets up to be a poet. **dâva·cı,** claimant; plaintiff: **~yım!,** I'll have the law of you, I'll sue you! **~lı,** which is claimed; who *or* which is the subject of a lawsuit; defendant; pretentious; (book *etc.*) written to uphold a theory.
**davar** Flocks of sheep *or* goats; beasts of burden. **kara ~,** cattle.
**davet** (—˙), **-ti** Summons; invitation; feast. **~ etm.,** to invite; to incite, cause. **~iye,** card of invitation. **~li,** one who is invited, guest.
**davlumbaz** Paddle-box *of a steamer*; hood *of a forge*.
**davranış** Behaviour; attitude.
**davranmak** Bestir oneself; prepare for action; resist; behave; take pains. **davranma!,** don't stir!: **silâha ~,** to prepare to use a weapon.

**davudî** (– – ⌐) Bass *or* baritone voice; fine manly voice.
**davul** Drum. **~a dönmek,** to swell. ⌜**uzaktan ~un sesi hoş gelir**⌝, 'oh! the brave music of a *distant* drum!'
**dayak** Prop; support; beating (*esp.* bastinado). **~ atmak,** to give a thrashing: **~ vurmak,** to put up a prop *to a wall etc.*: **~ yemek,** to get a thrashing: **kapı dayağı,** piece of timber put behind a door to keep it shut. **~lamak,** to shore *or* prop up. **~lık,** suitable as a prop; deserving a beating.
**dayalı** Propped up; leaning against; **döşeli ~,** completely furnished.
**dayamak** *va.* Prop up; support: lean *a thing against stg. as support to it*: (*sl.*) to give (*with an idea of cheating or with some other derogatory sense*). **eline ~,** to thrust into *s.o.'s* hand.
**dayanık·lı** Lasting, enduring; strong. **~sız,** not lasting; temporary; weak.
**dayanmak** *vn.* Endure, last, hold out; (*with dat.*) lean on; rely on; confide in; resist; support; push; (of a road) to end in *a place*; succeed in reaching *a place*. **can olsun dayansın!,** it's more than flesh and blood can stand!: **mahkemeye dayanmışlar,** they ended up in court: **polis kapıya dayandı,** the police came on the scene.
**dayatmak** *va. caus. of* **dayamak.** Cause to lean against *or* prop up; fling (an accusation *or* a refusal) in *s.o.'s* face. *vn.* Be obstinate in refusing to do *stg.* '**ben bunu yapmam dedi dayattı**', he insisted that he would not do it.
**dayı** Maternal uncle; captain of a ship; 'a good fellow'. ⌜**dümende ~sı var**⌝, (his uncle is at the helm), he has powerful supporters, 'he's got influence'. **~zade** (· · – ⌐), cousin.
**dâyin** (– ⌐) Creditor.
**daz** Baldness. Bald. **~lak,** bald.
**de**[1] And; also; *v.* **da**[1].
**de**[2] Hi!; now then! **~ bakalım!,** now then!; well, how about it!
**debagat** (· – ⌐), **–tı** The trade of a tanner.
**debbağ** (· ⌐) Tanner.
**debboy** *v.* **deppoy.**
**debdebe** Noise, clamour; pomp, display. **~li,** magnificent, resplendent, showy.
**debe** Ruptured, suffering from hernia. **~lik,** rupture, hernia.
**debelenmek** *vn.* Struggle and kick.
**Debreli** *In* ⌜**at martini, ~ Hasan, dağlar inlesin!**⌝, 'fire your gun, D.H., and let the mountains echo!', *used about one who is drawing the longbow.*
**debreyaj** (· · ⌐) (*Fr. débrayage*) **~ yapmak,** to declutch.
**dede** Grandfather; old man; dervish.
**dedik** *v.* **düdük** *and* **demek.**
**dedikodu** Tittle-tattle, gossip.
**dedirgin** *v.* **tedirgin.**
**dedirmek, dedirtmek** *va. caus. of* **demek.** Make *or* let say; give occasion for *stg.* to be said.
**ded veyt** *In* **~tonluk,** deadweight tonnage.
**def** *v.* **tef.**
**defa** Time; turn. **bir kaç ~,** on several occasions: **çok ~,** often: **iki ~,** twice.
**defetmek** (⌐ · ·), **–eder** *va.* Drive away; expel; eject; abolish.
**defi, –f'i** A repelling; expulsion; removal; refutation. **def'i belâ kabilinden etm.,** to do *stg.* merely as a precaution against some possible evil: **def'i hacet (tabiî) etm.,** to relieve nature.
**defin, –fni** Burial.
**define** (· – ⌐) Buried treasure; unexpected wealth.
**defne** (⌐ ·) Bay-tree. **~ yaprağı,** bay-leaf.
**defnetmek, –eder** *va.* Bury.
**defolmak** (⌐ · ·), **–olur** *vn.* Be removed; go away. **defol!,** clear out!
**defter** Register; account-book; notebook; list; catalogue. **~ açmak,** to open a list *of subscriptions etc.*: **~i kapamak,** to close the account, to finish with a matter. **~dar,** head of the financial department of a vilayet; accountant.
**değdirmek** *va. caus. of* **değmek.** Cause to reach *etc.*; cause to be worth; have *stg.* valued.
**değer** Value, worth, price; talent. Worth; worthy of. **~li,** valuable; estimable (man). **~siz,** worthless.
**değil** *Negative particle.* Not; not only; not so. **~(–dir),** he is not.
**değin, dek** Until; up to.
**değirmek** *v.* **değdirmek.**
**değirmen** Mill. ⌜**bu sakalı ~de ağartmadım**⌝, 'it was not in a mill that I got this white beard', *i.e.* I am old and experienced: **~de sakal ağartmak,** to be inexperienced and immature; not to have learnt from experience.
**değirmi** Round; circle; square piece *of stuff.* Roundish *or* squarish.
**değiş** Exchange. **~ etm.** *or* **~ tokus**

etm., to exchange, to barter. ~ici, changeable; changing. ~ik, changed; changeable; varied; fickle; new, unusual: bu çocuk ~ olmuş, this child must be a changeling. ~iklik, alteration; variation: ağız değişikliği, a change of food; a change *in general*.

**değişmek** *va.* & *vn.* Change, alter, vary. Exchange.

**değme** Ordinary; chance; any; every; (*sometimes*) the very best, 'super'; (*with neg.*) hardly any; not just any; *v. also* **değmek.**

**değmek** *va.* (*with dat.*) & *vn.* Touch; reach; attain; be worth. **değmez!**, it isn't worth while: **değme gitsin,** *in such phrases as*: **öyle bir ziyafet ki değme gitsin!**, you never saw such a feast!: ⌜**değme keyfine!**⌝, now he should be happy!: **canına*** ~, to be enjoyable; ⌜**babanın canına değsin!**⌝, *phrase used when giving thanks for a kind action*: **eli*** ~, to have an opportunity *or* to find time *to do stg.*; **elim değmişken,** while I have the chance; while I am about it: **el değmemiş,** untouched; virgin: **nazar** (**göz**) ~, for the evil eye to strike one; **nazar değmesin!**, absit omen!; 'touch wood!': '**bu kahve değdi doğrusu!**', 'I *did* enjoy that coffee!' *v. also* **değme.**

**değnek** Stick; rod; beating. ~**yemek,** to get a beating: ⌜**dokuz körün bir değneği**⌝, a most precious thing: **koltuk değneği,** crutch: ⌜**kör değneğini bellemiş gibi**⌝, 'as a blind man knows his stick', he does it as a matter of course, *i.e. without troubling to think whether it is the best course.* ~**çi,** official of a guild.

**deha** (·—́) Great ability; genius.

**dehalet** (·—̀), **-ti** Submission. ~**etm.,** to take refuge; (of rebels *etc.*) to give oneself up.

**dehlemek** Urge on *an animal with cries of 'deh!'*; chase away.

**dehliz** Entrance-hall; vestibule; corridor; ear-passage.

**dehşet, -ti** Terror. ~ **salmak,** to spread terror. ~**li,** terrible: ~ **sıcak,** 'frightfully' hot.

**dek** *v.* **değin.**

**dekan** Dean *of a faculty.*

**dekor** Stage scenery.

**delâlet** (·—̀), **-ti** Acting as a guide *or* pilot; guidance; indication. ~**le,** through, by the agency of; care of (c/o): ~ **etm.,** to guide; to indicate, show, signify.

**deli** Mad; insane; crazy; foolish, rash; wild (*also* of plants *etc.*); violently addicted to, 'mad on'. Madman. ~ **bal,** honey made from poisonous plants: ⌜~ **balkabağından olmaz ya!**⌝, he must be off his head!: ~ **divane,** utterly mad: ~**nin eline değnek vermek,** to give an opportunity to one who is capable of doing harm: ~ **ırmak,** a furious, violent river: ... ~**si olm.,** to be 'mad on' *stg.*: ~ **orman,** a dense forest: ~ **saçması,** absolute nonsense, 'utter rot': ⌜~**nin zoruna bak!**⌝, what an extraordinary idea!: **ne oldum** ~**si,** one whose head is turned by unexpected good luck.

**deli·alacası, -nı,** a wild mixture of colours. ~**bozuk,** erratic, unstable, eccentric. ~**ce,** rather mad; poisonous (mushroom *etc.*); ergot; darnel; (·̀·) furiously, madly, like a madman. ~**dolu,** who talks at random and without reflection. ~**duman,** foolhardy, reckless; crazy, good-for-nothing. ~**fişek,** unbalanced, flippant.

**delik** Hole; opening; (*sl.*) prison. Bored; pierced. ~ **deşik,** full of holes: ~ **kapamak,** to make good a deficiency: **deliğe tıkmak,** to 'put into clink': **kulağı** ~, who keeps his ears open; wide awake.

**delikanlı** (·̀··) Youth; young man. Young; sprightly. ~**lık,** youth; youthfulness.

**delik·deşik** (·̀··) Full of holes; honeycombed. ~**li,** having a hole *or* holes; perforated: ~ **taş,** well-curb formed of one stone. ~**siz,** without holes; sound: ~ **bir uyku,** a sound sleep.

**delil** Guide, pilot; proof, evidence; indication, sign.

**delilik** Madness.

**delinmek** *vn. pass. of* **delmek.** Be holed *or* perforated; be worn through.

**delir·mek** *vn.* Go mad; become insane; be furiously angry; be mad *with love or desire.* ~**tmek,** to drive mad.

**delişmen** A bit crazy; wild (youth *etc.*).

**delme** A piercing *or* perforating. Perforated, holed.

**delmek** *va.* Pierce; hole; bore. **delip geçmek,** to make a hole and go right through: **kulak** ~, to make an ear-splitting noise.

**dem** Breath; vapour; alcoholic rink;

moment, time. ~ **çekmek,** to drink wine *etc.*; (of birds) to sing sweetly: ~ **tutmak,** to accompany music: ···**den** ~ **vurmak,** to talk at random and vaguely about *stg.*; to claim.

**demeç** Statement; words; speech.

**demek (der, diyor, diyecek, diyen)** *va.* Say; tell; mean; call *or* give a name to. **dediği dediktir,** he abides by what he says; he is an obstinate fellow; ⌜**dediği dedik düdügü (çaldığı) düdük**⌝ —*said of a spoilt child whose every whim is satisfied, or of one who expects such treatment,* 'a little despot': **deme gitsin,** indescribable, unbounded; *v. also* **değmek**: **der demez,** just at that moment, he had no sooner spoken than ...: **derken,** *lit.* 'even as he spoke', *i.e.* just then, all at once: ... **deyip, geçme,** 'don't think you can dismiss it by saying ...', don't think ... of no account: ~ **ki** *or* ~ **oluyor ki,** that means to say that : ⌜**bana mı(sın) demedi**⌝, it had no effect: **bu ne ~?,** what does that mean?: **ne ~!,** 'how so?' (*to express disapproval*); 'not at all!', 'certainly you may!' (*as terms of politeness*): **ne dedim de oraya gittim?,** why on earth did I go there?: ⌜**kimin ne demeğe hakkı var?**⌝, 'who can object to that?': ⌜**ne oldum dememeli ne olacağım demeli**⌝, 'don't boast about your present unless you are sure of your future': ⌜**yaşı ben diyeyim 17, siz deyiniz 18**⌝, his age is 17, maybe 18: **yaz demez kış demez,** whether it be winter or summer: ... **diyor da (başka) bir şey demiyor,** he thinks of nothing but ...: **diye** *v.* **diye.**

**demet, ~ti** Sheaf; bunch; faggot; bundle. **~lemek,** to tie up in a bunch *or* a faggot.

**demevî** (··́) Pertaining to the blood; full-blooded, sanguine.

**demin** (́·) Just now; not long ago. **~cek** (́··), just a moment ago.

**demir** Iron; anchor; iron *part of anything,* barrel *of a gun,* blade *of a knife etc.* ~ **almak,** to raise anchor: ~ **atmak,** to cast anchor, to anchor: ~ **boku,** iron slag: ~ **kazık,** the Pole Star: ~ **resmi,** anchor dues: ~ **üzerinde,** with the anchor up, ready to sail: ···**i ~e vurmak,** to put in irons.

**demir·baş,** iron head *of an implement*; persistent *or* obstinate man; movable stock *of a farm or* furnishings *of a shop, let to the tenant to be accounted for to the owner*: ~ **erzak,** iron rations: ~ **yemek,** a standing dish. **~ci,** blacksmith. **~hindi,** tamarind. **~kapı,** gorge of a river; fortified defile. **~lemek,** to anchor; to bar (a door). **~li,** at anchor. **~yeri, -ni,** anchorage. **~yolcu,** railwayman. **~yolu, -nu,** railway.

**demle·mek, ~tmek, ~ndirmek** *va.* Steep (tea). **~nmek,** to drink (spirits *etc.*); (of tea) to be steeped.

**demlik** Tea-pot.

**demokrasi** Democracy.

**-den, -ten, -dan, -tan** *Particle forming the ablative case or the equivalent of a preposition such as* from; than; on account of, *etc.*

**deneme** Trial; test. **~k,** to try, test.

**denilmek** *vn. pass. of* **demek.** Be said; be called.

**deniz** Sea; wave; storm. ~ **olm.,** to be rough (sea): ~ **onu tutuyor,** he is seasick: **açık ~,** the open sea. **~altı, -nı,** *n.* submarine; *a.* submarine; open to the sea, exposed. **~anası, -nı,** jelly-fish. **~aşırı,** overseas. **~aygırı, -nı,** sea-horse, hippocampus. **~ayısı, -nı,** walrus. **~ci,** seaman, sailor; seaworthy. **~cilik,** profession of sailor; aquatic sports; sailing; seaworthiness. **~kadayıfı, -nı,** a kind of seaweed. **~kızı, -nı,** mermaid. **~köpüğü, -nü,** meerschaum. **~lik,** sloping board on the rail of a boat *to keep the water out*; window-sill. **~mili, -ni,** nautical mile. **~otu, -nu,** seaweed. **~piresi, -ni,** crayfish. **~yolları, -nı,** sea-ways; shipping lines.

**denk** Bale; half a horseload; balance, equilibrium; trim; counterpoise. In proper balance; equal. **dengi dengine,** to everyone his due: ~ **etm.,** to make into bales; to pack up; to trim (boat *etc.*); **dengine getirmek,** to choose the psychological moment: **ayağını*** ~ **almak,** to be careful where one is going, to 'watch one's step': **bu adam senin dengin değil,** this man is not your equal (in position *or* capacity): **bu kız o gencin dengi değil,** this girl is not a good match for that young man: **kafa dengi bir insan,** a man of one's own sort, a kindred spirit.

**denk·leşmek,** to be in equilibrium, be balanced *or* trimmed. **~leştirmek,** *caus. of* **denkleşmek;** *also* to manage to find (money). **~siz,** unbalanced; awry.

**denmek, -nir** *vn. pass. of* **demek.** Be said; be named. **buna ne denir?**, what is this called?: **hemen hemen yok denecek kadar az,** so few that one might almost say there were none.

**densiz** Peevish; refusing to listen to reason.

**depo** (ˈ·) Depot; warehouse. **benzin ~su,** petrol tank.

**depozito** (··ˈ·) Deposit; security.

**deppoy** Military store; depot.

**deprem** Earthquake.

**depre·mek, ~tmek** *va.* Move; stir. **~nmek, ~şmek** *vn.*, move; be stirred; rise.

**derbeder** Vagabond; bohemian; living an irregular life; careless *in dress, etc.*

**derc** Insertion. **~etmek,** to insert; to inscribe.

**derceb etm.** *va.* Pocket; put in the pocket.

**derd** Pain; suffering; grief; trouble; grievance; obsession; chronic disease; boil. **~e girmek,** to fall into trouble: **~i günü,** one's pet grievance, obsession: **~i* ne imiş?**, what's his trouble?; what does he want?: **~ ortağı,** fellow-sufferer; one to whom one confides one's woes: **(bir şeyin) ~ine düşmek,** to be quite taken up with *stg.*: **başına* ~ açmak,** to bring trouble on *s.o.'s* head: **başı* ~e düşmek,** to get into trouble: **başı* (kendi) ~ine* düşmek,** to be completely preoccupied with one's own troubles: **canının* ~ine düşmek,** for one's vital interests to be at stake: ⌜**elin ~ görmesin!**⌝, 'may you be free from trouble!', *a form of thanks for help given*: **içine* ~ ~ olm.,** to be a thorn in one's side; to cause regret *or* remorse.

**derd·daş,** one to whom one confides one's troubles. **~etmek,** to let *stg.* prey on one's mind. **~lenmek,** to ache; to be pained, be sorrowful. **~leşmek,** to pour out one's grief to another *or* to one another; to sympathize with one another. **~li,** pained; sorrowful; aggrieved. **~siz,** free from pain *or* sorrow; free from cares: ⌜**~ başa derd almak**⌝, to bring unnecessary trouble upon oneself.

**dere** Valley; stream. **~ tepe,** hills and valleys; **~ tepe dolaşmak,** to wander over hill and dale: ⌜**~ tepe demedik yürüdük gittik**⌝, we stopped at no obstacle: **~den tepeden konuşmak,** to have a long chat about all sorts of things: ⌜**bin ~den su getirmek**⌝, to make all kinds of excuses, to raise innumerable difficulties.

**derebey·i** (·ˈ··), **-ni** Feudal chieftain; despot; bully. **~lik,** feudalism: **~ devrinde,** in the feudal age.

**derece** Step, stair; degree; rank; grade; thermometer. **~ ~,** by degrees: **~ koymak,** to take *s.o.'s* temperature: **çekilmez ~ye kadar,** to an intolerable degree: **son ~,** in the highest degree, utterly.

**dereotu, -nu** Fennel; dil.

**dergâh** (·ˈ) Court of a king; dervish convent.

**dergi** Collection (of poems *etc.*); review, magazine.

**derhal** (ˈ·) At once, immediately.

**deri** Skin; hide; leather. **~sine sığmamak,** to be 'too big for one's boots': **bir ~ bir kemik,** nothing but skin and bones. **~lenmek,** (of a wound *etc.*) to heal up.

**derilmek** *vn. pass. of* **dermek.** Be collected; collect oneself.

**derin** Deep; profound. Deep place. **~den ~e,** from far away; minutely, profoundly: **~den sesler geliyor,** there are far-off sounds. **~lik,** depth; profundity; hollow.

**derkâr** (·ˈ) At work, busy; manifest, evident.

**derken** *Lit.* 'while speaking'; even as he spoke; and then; at that moment.

**derkenar** Marginal note; postscript.

**derle·mek** *va.* Gather together, collect. **~ toplamak,** to tidy up, to clear away (dishes *etc.*). **~nmek,** *pass. of* **derlemek**: **~ toplanmak,** to pull oneself together.

**derli toplu** Gathered together and arranged; in order; compact; tidy.

**derman** (·ˈ) Strength; energy; remedy. **~ bulmak,** to find a remedy: **~ım yok,** I am exhausted; I have not the means. **~sız,** feeble, debilitated; exhausted.

**derme** Gathered together; collected. **~ çatma,** hastily collected and put together; jerry-built; (soldiers) collected at random; scraps of, odds and ends of. **fransızcası pek ~ çatma,** he has only a smattering of French.

**dermek** *as* **derlemek.**

**dermeyan ~ etm.,** to bring forward, to put forward (a proposal *etc.*).

**dernek** Gathering, party; society.

**derpiş etm.** *va.* Take into consideration; suggest.
**ders** Lecture; lesson. **~i asmak,** to play truant: **bu ~i sana kim verdi?,** who put that idea into your head? **~hane** (·–˝), class-room.
**dert** *v.* **derd.**
**dertop** *In* **~ olm.** *or* **büzülmek,** to curl oneself up: **~ etm.** (*sl.*), to round up, to catch.
**derviş** Dervish. Simple; contented; humble. ⌜**~in fikri neyse zikri odur**⌝, 'the dervish keeps reciting that formula of prayer which best coincides with his thoughts', *i.e.* he has a bee in his bonnet about that, he can't keep off that subject: **~ meşreb,** unconventional: ⌜**bekliyen ~ muradına ermiş**⌝, 'everything comes to those who wait'.
**derya** (·˝) Sea; ocean; very learned man. **deniz ~ ayak altında,** *phrase used to describe a fine view near the sea.* **~dil,** large-hearted, magnanimous; always looking at the best side of things.
**desene** (·˝·), **desenize** (·˝··) *Optative of* **demek** *with a vocative ending;* 'isn't it so?'; it seems that . . . : you mean ... .
**desise** (·–˝) Trick, ruse; intrigue.
**despot** Greek bishop, Metropolitan; despot. **~hane** (··–˝), office and residence of Greek Metropolitan.
**destan** *v.* **dasitan.**
**deste** Handle; hilt; bunch; packet; quire; pestle. **~ ~,** in packets, by dozens, in heaps; **~ başı,** choice specimen put on the top of a heap of goods.
**destek** Beam; prop; stand; support. **~ vurmak** *or* **~lemek,** to prop or shore up; to support.
**destur** Formula; code; permission. **~!,** by your leave!, make way!; *also said superstitiously when entering an unpleasant or dangerous place.* **~ var mı?,** have I your permission? **~un!,** 'excuse the expression!'
**deşelemek** *va.* Scratch up *the ground.*
**deş·ik** Hole. Pierced: **delik ~,** full of holes. **~ilmek,** to be pierced; (of an abscess) to be lanced; to burst *of its own accord.* **~mek,** to open by incision, lay open; unearth; open up (a sore point, a vexed question).
**dev** Giant. **~ anası,** giantess, huge woman: **~ gibi,** huge.
**deva** (·˝) Medicine; remedy. **her derde ~,** a panacea: **~yı kül,** panacea.
**devair** (·–˝) *pl. of* **daire.** Offices *etc.*
**devam** (·˝) A continuing; continuation; permanence; assiduity; regular attendance at work. **~ etm.,** to continue, last; persevere; (*with dat.*) to continue *stg.*; frequent (a place); attend *or* follow (a course of lectures *etc.*). **~lı,** continuous, assiduous. **~sız,** without continuity; inconstant; unenduring; not persevering; irregular at work. **~sızlık,** lack of continuity *or* perseverance; irregular attendance at work, absenteeism.
**devaynası, ~nı** Magnifying mirror. ⌜**kendini ~nda görmek**⌝, to exaggerate one's own importance.
**deve** Camel. ⌜**~ bir akçe ~ bin akçe**⌝, 'there are cheap camels and dear camels', *i.e.* it all depends by what standard you are reckoning: **(yok) ~nin başı!,** incredible!, impossible!, out of the question!: ⌜**~yi hamudu ile yutmak**⌝, 'to swallow the camel, bridle and all'; *has the same meaning as* **~ yapmak** *q.v., but is used about a large-scale affair*: **~ hamuru,** 'camel dough', *said of some indigestible food*: **~ kini,** rancour; the nourishing of a grudge: **~de kulak,** *stg.* very small in relation to the whole, a mere trifle: **~ yapmak,** to acquire by deceit: **bu adama lâf anlatmak ~ye hendek atlatmaktan zor,** it's impossible to convince this man: ⌜**ya bu ~yi gütmeli, ya bu diyardan gitmeli**⌝, 'you must pasture this camel or leave this country'; one must adapt oneself to circumstances.
**deve·bağırtan,** (that makes the camel cry out), a steep and stony road. **~boynu, -nu,** S- *or* U-shaped tube; saddle between hills. **~ci,** camel-driver; camel owner. **~dikeni, -ni,** thistle. **~dişi,** having large seeds or grains. **~kinli,** who nourishes an old and bitter grudge. **~kuşu, -nu,** ostrich. **~tabanı, -nı,** coltsfoot. **~tımarı, -nı,** hastily and negligently performed job. **~tüyü, -nü,** camel hair; light brown colour.
**deveran** (··˝) Circulation. **~ etm.,** to go round, rotate, circulate.
**devir, -vri** Rotation; revolution; cycle; circuit; tour; period, epoch; transfer *of a thing from one receptacle to another.* **~ etm.,** *or* **devretmek,** to revolve, circulate; turn upside down; turn over to another: **devri saadet,**

the age of Mohammed, the Golden Age: **devri sabık,** the old régime: **devrin valisi,** the Vali of the day, the then Vali.

**devirmek** *va.* Overturn; reverse; knock down; pull down; drink *a glass etc.* to the dregs. **···e gözlerini ~,** to look daggers at ... .

**devlet, -ti** State; government; kingdom; prosperity, success, good luck. **~le,** prosperously; (*as interj.*) 'good luck to you!': **~ü ikballe,** (*as a parting wish*) 'good luck and prosperity!': **~ memuru,** government official: **ne ~,** what good luck! **~çi,** one who favours state control. **~çilik,** the policy of state control. **~hane** (··–ˈ), 'the house of prosperity' (*very polite for* 'your house'). **~kuşu, -nu,** unexpected good luck. **~leşmek,** to become nationalized. **~li,** prosperous, wealthy; *distinguished title formerly given to certain high officials.* **~lû,** *title given only to the highest officials.*

**devralmak** (ˈ··) *va.* Take over.

**devran** (·ˈ) Time; epoch; the wheel of fortune; fate.

**devre** Cycle; generation; period; session *of Parliament*; period *of an election*; electrical circuit. **kısa ~,** short circuit.

**devren** (ˈ·) By cession; while making a round *of inspection.* **~ kiralık,** subletting.

**devretmek** (ˈ··) *v.* **devir.**

**devri** *v.* **devir.**

**devrî** Rotatory. **senei ~ye,** anniversary.

**devrik** Turned back on itself; inverted: **~ yaka,** turned-down collar.

**devrilmek** *vn. pass. of* **devirmek.** Be overturned. ⌜**boyu devrilesi!**⌝, 'curse him!'

**devriye** A certain class of circuit judges (**ulema**); police round; patrol. **~ gezmek,** to patrol.

**devrü·ferağ etm.,** *va.* Cede, transfer. **~teslim,** transfer (of a trust); handing over (of a post *etc.*).

**devşirme** Act of gathering *or* collecting; (*formerly*) the selection of boys to be brought up as Janissaries. Collected together.

**devşirmek** *va.* Gather, pick; roll up (carpet *etc.*); collect oneself.

**devvar** Revolving continuously. **~ tulumba,** rotary pump.

**deyiş** A kind of folk-song.

**deyn** Debt; obligation.

**deyyus** Cuckold; pander.

**dımdızlak** (ˈ··) Stark naked; destitute; quite bald; (*sl.*) stony-broke.

**dıngıldatmak** *va.* Clang.

**dırahoma** (·ˈ··) *v.* **drahoma.**

**dıral** *Only in* ⌜**~ dedenin düdüğü gibi kalmak**⌝, to be at a loss; not to know what to do.

**dır·dır** Continuous tiresome chatter *or* grumbling. Who thus chatters. **~ıltı,** continuous and annoying chatter; slight squabble: **~ çıkarmak,** to have a slight quarrel. **~lamak, ~lanmak,** to babble tiresomely; make annoying noises; nag. **~laşmak,** to squabble in undertones.

**dış** Outer; exterior. Outside, exterior; outward appearance. **içi ~ı bir,** uniform; sincere; consistent: **içi* dışına* çıkmak,** for one's clothes to become dishevelled.

**dışarda, dışardan** *v.* **dışarı.**

**dışarı** *n.* Outside, exterior; the provinces; the country (*as opp. to town*); abroad. *adv.* Out; outside; abroad. **dışarda,** outside; abroad: **dışardan,** from without; from abroad: **~ çıkmak,** to go out; to go to the closet: **~ya gitmek,** to go into the provinces; to go abroad: **gözü ~da olm.,** to be discontented with one's job and on the look-out for another: **şehrin ~sında,** outside the town: **~ vurmak,** to show; to reflect. **~lık,** the provinces. **~lıklı,** provincial.

**dışı** *Used as a suffix*; outside of ...; extra-. **evlilik ~ birleşmeler,** extramarital unions: **kanun ~,** outside the law, illegitimate: **memleket ~ haklar,** extraterritorial rights.

**dışişleri** (ˈ···) *In* **~ Bakanı,** Minister of Foreign Affairs.

**dışkı** Excrement.

**dışlı** Pertaining to the outside; possessing an outside. **içli ~,** familiar, intimate.

**dışyüz** (ˈ·) Outer surface; outside; appearance.

**dızdık** *Only in* **dızdığının dızdığı** (**dışkapının mandalı**), a distant relative.

**dız·dız** Buzzing, whizzing; pickpocketing. **~dızcı,** a pickpocket *who takes advantage of a hubbub to rob s.o.* **~ıltı,** a humming *or* buzzing. **~lamak,** to hum, buzz.

**dızlak** Naked; bald.

**dibek** Large stone *or* wooden mortar. **~ ~ kahvesi,** coffee ground in a mortar.

**Dicle** The Tigris.

**didik** Teased out, pulled to shreds. **~lemek,** to tease out into fibres *or* shreds; pick to pieces; cut to shreds: **birbirini ~,** to tear one another to pieces (*fig.*).

**didilmek** *vn.* Be picked *or* worn to shreds.

**didinmek** *vn.* Wear oneself out; toil; be excessively eager *or* anxious, fret.

**didişmek** *vn.* Push each other about; quarrel; struggle.

**didon** *Popular name for a Frenchman.* **~ sakallı,** wearing an imperial.

**difteri** Diphtheria.

**diğer** (ˊ·) Other; another; altered, different; next (day). **~endiş, ~kâm** (··ˊ), altruistic.

**dik** Perpendicular; upright; straight; steep; uncompromising, opinionated. **~~ bakmak,** to glare, to look daggers: **~ durmak,** to stand upright: **(suyun) ~ine gitmek,** to do just the opposite *of what one is asked etc.*, to be pig-headed: **~ ses,** harsh, loud voice: **~ sözlü,** who does not mince his words: **~ine tıraş,** shaving against the grain; utterly boring *or* exaggerated talk. **~başlı,** pig-headed, obstinate.

**dikel** Long-handled digging-fork. **~eç,** dibber.

**diken** Thorn; sting *of an insect*; obstacle. **tüyleri* ~ ~ olm.,** for one's hair to stand on end: ⌜**gülü seven ~ine katlanır**⌝, 'no rose without a thorn'. **~dudu, –nu,** blackberry. **~li,** thorny; prickly: **~ tel,** barbed wire.

**dikici** Cobbler.

**dikili** Sewn, stitched; planted; set up. **~ ağacı yok,** he has no children *or* no home: **~ taş,** obelisk.

**dikilmek** *vn. pass. of* **dikmek.** Be sewn; be planted; stand stiff and immobile; (of the eye) to be intently fixed on; (of a horse) to jib. **baş ucuna ~,** to pester *s.o.*

**dikim** Act of sewing *or* planting. **~hane** (··—ˋ), sewing workshop (*esp. a government shop*).

**dikiş** Sewing; planting; stitch; seam; splice. **(bir yerde) ~ tutmak,** to take root *or* settle down in a place: **~tutturamamak,** to be incapable of settling down to a job *or* keeping a post: **bir ~te içmek,** to drink off at a draught: **düşmesine ~ kaldı** he only just saved himself from falling. **~çi,** seamstress.

**dikiz** Roguish look; observation. **~ etm.** *or* **geçmek,** to watch intently: **~ aynası,** observation mirror; mirror of a motor-car. **~lemek,** to keep under observation; spy upon.

**dikkafalı** (ˋ···) Obstinate, pig-headed.

**dikkat, –ti** Attention; care; fineness; subtlety. **~le,** with care; attentively: **~ etm.,** to pay attention; to be careful: **~ kesilmek,** to pay great attention: **···i nazarı ~e almak,** to take into consideration: **···e nazarı ~i celbetmek,** to call attention to .... **~li,** attentive; careful; carefully made. **~siz,** inattentive; careless. **~sizlik,** inattention; carelessness.

**diklemesine** Perpendicularly.

**diklik** A being upright *etc. v.* **dik. aklın dikliğine gitmek,** to be obstinately perverse.

**dikme** *verb. n. of* **dikmek.** Act of sewing *etc.*; young plant; prop; derrick. Sewn; stitched; spliced.

**dikmek** *va.* Sew; stitch; splice; set up; plant; drain (a cup); fix (the eyes *upon stg.*); stick *a thing into stg.*; prick up (the ears); (of hair) to stand on end. **göz ~,** to glare; to cast envious glances on *stg.*, to long to possess *stg.*: **şişeyi ~,** to drink straight from the bottle.

**dikmen** Peak; summit.

**dikte** (*Fr. dicté*) Dictation. **~ etm.,** to dictate.

**dil**[1] Tongue; language; spit of land; index *of a balance*; sheave *of a block*; bolt *of a lock*; reed *of an oboe etc.*; a prisoner of war captured in order to obtain information. **~i* açılmak,** for the tongue to be loosed; (of a person hitherto silent) to start talking: ⌜**~inin* altında bir şey var**⌝, there is something he is keeping back *or* hesitates to say: **~i bağlı,** silent, raising no objections: **~ bilmez,** he does not know Turkish: **~ çıkarmak,** to put out the tongue at *s.o.*: **~ dökmek,** to talk *s.o.* round: **~i* döndüğü kadar,** *to explain* as well as one can: **~i* dönmemek,** to be unable to pronounce a word: **~e** *or* **~lere düşmek,** to become the subject of scandal: **~inden* düşürmemek,** to never cease talking about *stg.*: **~e gelmek,** to become the subject of scandal; to begin to speak: **~e getirmek,** to cause to be talked about; to give utterance to *stg.*: **~ini* kesmek,** to become silent: **~ine sağlamdır,** (i) he does not use bad language; (ii) he is discreet: **~ini* tut-**

**mak,** to hold one's tongue: **~i* tutulmak,** to be speechless *from fear etc.*: **~ ini yutmak,** *as* **küçük ~ini yutmak** *but also* 'to appear to have lost the use of one's tongue': ⌜**ağzı var ~i yok**⌝, innocent; silent: ⌜**Allah kimseyi ~ine düşürmesin!**⌝, 'may God preserve anyone from his tongue!': **küçük ~,** uvula; **küçük ~ini* yutmak,** to be overcome by great surprise *or* fear.

**dil²** Heart. **~ü can,** heart and soul.

**dilak** Clitoris.

**dilatı** (ˈ··), **–nı** Pip (in fowls); pustule under the tongue.

**dilbalığı** (ˈ···), **–nı** Sole (fish).

**dil·baz** (·ˈ–) Talkative and coquettish; (a woman) who can always talk a man over. **~ber,** captivating, charming; sweetheart, darling: **~dudağı,** a kind of sweet cake.

**dilcik** Little tongue; clitoris; pointer *or* needle *of a gauge etc.*

**dil·cu** (·ˈ–) Desirable, sought after. **~dade** (·–ˈ), beloved; lover.

**dilebesi** (ˈ···), **–ni** Talkative, glib.

**dilek** Wish, desire; request. **~ kuyusu,** wishing-well. **~çe,** petition, formal request.

**dilemek** *va.* Wish for, desire; ask for. **özür ~,** to ask pardon.

**dilen·ci** Beggar. **~ çanağı gibi,** full of odds-and-ends: **~ vapuru,** steamer that stops at every port of call. **~cilik,** mendicancy, begging. **~mek,** to be a beggar, to beg.

**dilim** Slice; strip. **~ ~,** in slices.

**dillenmek, dilleşmek** *vn.* Find one's tongue; become chatty; grumble; talk indiscreetly *or* rudely; be the object of talk, be criticized.

**dilli** Having a tongue *etc.*, *v.* **dil¹**; who chatters pleasantly. **tatlı ~,** pleasant-spoken: **uzun ~,** indiscreet in speech. **~düdük,** talkative, chatterbox.

**dilmek** *va.* Cut in slices *or* strips.

**dilpeyniri** (ˈ···), **–ni** *Kind of* cream cheese *made in long strips.*

**dilsiz** Dumb; mute. **~lik,** dumbness.

**dimağ** (·ˈ) Brain; intelligence. **~çe,** cerebellum. **~î** (·–ˈ–), cerebral.

**dimdik** (ˈ·) Bolt upright; quite perpendicular; very steep (downwards).

**dimi** A kind of textile, (?) dimity.

**Dimyat** *prop. n.* Damietta. ⌜**~a pirince giderken evdeki bulgurdan olm.**⌝, to lose what one has in the effort to get more *or* better.

**din** Religion, faith (*esp.* the Moslem). **~i bir uğruna,** for the sake of Islam: **~i bütün,** sincerely religious, good and honest: **~den imandan çıkmak,** to lose all patience, to become exasperated: **~ine yandığım** (*sl.*), the cursèd.

**dinar** An ancient gold coin; Serbian franc; Persian and Iraki coins.

**dinc** Vigorous; robust. **başı* ~,** at peace. **~elmek,** to become vigorous, recover one's strength. **~leşmek,** to feel refreshed. **~lik,** robustness; good health.

**din·dar** Religious, pious. **~darlık,** piety. **~daş,** co-religionist.

**dindirmek** *va. caus. of* **dinmek.** Cause to cease; stop (bleeding *etc.*); calm, quieten.

**dingil** Axle. **~demek,** to rattle, to wobble.

**dinî** (–ˈ–) Pertaining to religion; religious.

**dini·bütün** Good; pious. **~ müslüman,** a true Moslem. **~mübin,** Islam.

**diniyri** (·ˈ··) Diamonds (at cards).

**dink** Fulling-mill.

**dinlemek** *va.* Listen to; hear; pay attention to; obey. **başını* (kafasını*) ~,** to rest (mentally): **kendini o kadar dinleme!,** don't fuss so much about yourself: **söz ~,** to listen to advice, to be docile; **söz dinlemez,** disobedient.

**dinlen·dirmek** *va.* Make *or* let rest; calm. **kafasını* ~,** to rest oneself: **kalıbı ~,** to die. **~mek,** to rest; to become quiet; be heard, be listened to.

**dinletmek** *va.* Cause to hear; make listen; bore; recount (a tale); sing *a song etc.* well. ⌜**külahıma dinlet!**⌝, 'tell that to the Marines!': **sözünü* ~,** to make oneself listened to; to be obeyed.

**dinleyici** Listener. **~ler,** the audience.

**dinmek** *vn.* Cease; leave off (rain, bleeding *etc.*); calm down.

**dinsiz** Without religion; impious; cruel. ⌜**~in hakkından imansız gelir**⌝, 'a villain will be worsted by a greater villain'; 'diamond cut diamond'.

**dip, –bi** *n.* Bottom; lowest part; foot (of a tree *etc.*); anus. *a.* Bottom; lowest. **dibe çökmek,** (of a deposit *etc.*) to sink to the bottom: **~ göstermek** *or* **sömürmek,** to drink to the dregs: **dibinden traş,** a clean shave: **burnunun dibinde,** under one's nose, very close: **tencere dibi,** food adhering to the bottom of a saucepan.

**dipçik** Butt of a rifle. ~ **kuvvetile,** by force.

**dipdiri** (ˈ..) Full of life; very robust; safe and sound.

**dipkoçanı** (ˈ...), **–nı** Counterfoil.

**dipsiz** Bottomless; unfounded, false; inconstant. ~ **testi,** spendthrift.

**dirahmi** The Greek coin 'drachma'.

**dirayet** (.–ˈ), **–ti** Comprehension; intelligence; ability. **~li,** capable, intelligent. **~siz,** stupid; incapable.

**direk** Pole; pillar; column; mast; septum; very tall man. ~ ~ **bağırmak,** to shout at the top of one's voice: **ana ~,** lower mast: **burnunun* direği kırılmak,** to be quite overcome by a smell: **burnunun* direği sızlamak,** to be very sad: **can direği,** the sound-post of a violin: **yarış direği,** 'greasy pole'. **~li,** masted: **üç ~,** three-masted.

**direksiyon** Steering-wheel and mechanism of a motor-car.

**diremek** *va. In* **ayak ~,** to 'put one's foot down'.

**diren** Large-pronged fork, *esp.* winnowing-fork.

**direnmek** *vn.* Disagree; dissent emphatically; insist.

**dirhem** 400th part of an **okka**; ancient Arab coin; shot (in a sporting gun). ~ **kadar,** a very small quantity: ⌜**ağzından söz ~le çıkıyor**⌝, it's hard to get anything out of him: **kendini ~ ~ satmak,** to give oneself airs.

**diri** Alive; fresh, not faded (salad, leaf); energetic, lively; sharp (words); not properly cooked. **~lik,** a being alive; vitality; brusqueness.

**diril·mek** *vn.* Come to life; be resuscitated. **~tmek,** to bring to life, resuscitate; re-invigorate; enliven.

**dirim** Life. **ölüm ~ meselesi,** a matter of life and death.

**dirisa, dirise** The training of a gun.

**dirlik** A living in amity. ~ **düzenlik,** harmonious life, good fellowship. **~siz,** cantankerous (man); inharmonious (family life).

**dirsek** Elbow; bend; knee *of timber*; crank; bracket; outrigger *of a rowing-boat.* ···**e ~ çevirmek,** to take a dislike to; to 'drop' *s.o. when he can be of no further use to one*: ~ **çürütmek,** to wear out the elbows with study.

**diş** Tooth; cog; thread *of a screw*; clove; any tooth-shaped thing; (*sl.*) a pinch *of stg.* ···**e ~ bilemek,** to nourish a hatred for ..., to watch for a chance to harm ...: ~ **çıkarmak,** to pull out a tooth; to cut a tooth: **~e değmedi,** (of food) there was very little of it: **~e dokunur,** profitable, worth while; enjoyable: ···**e ~ geçirememek,** to be unable to influence *or* harm ...: **~e gelir,** an easy prey: **~ine göre,** to one's taste: ~ ~ **kar,** snow, which, instead of lying smoothly, assumes a granular appearance, each flake seeming to stand out by itself: ~ **kirası,** *v.* **dişkirası**: **canını* ~ine* takmak,** to take one's life in one's hands; to make desperate efforts: ⌜**işten artmaz ~ten artar**⌝, better to save money by economical living than to try to make more out of one's job.

**diş·bademi, –ni,** soft-shelled almond. **~budak,** ash tree. **~çi,** dentist; **~lik,** dentistry. **~eti, –ni,** gums. **~fırçası, –nı,** toothbrush. **~hekimi, –ni,** dental surgeon.

**dişi** Female; soft, yielding. ~ **demir,** soft iron.

**diş·karıştıracağı, –nı** Toothpick. **~kirası, –nı,** salary for a sinecure. **~lek,** having prominent teeth; having gaps in one's teeth. **~lemek,** to take a bite out of *stg.* **~li,** toothed; cogged; jagged; having sharp teeth, formidable; influential; hustling; who gets things done: ~ *or* ~ **çark,** cogwheel: ~ **olm.,** to 'show one's teeth': ~ **tren,** cog-wheel railway. **~macunu, –nu,** toothpaste. **~otu, –nu,** sage. **~tabibi, –ni,** dental surgeon. **~tacı, –nı,** crown of a tooth. **~tozu, –nu,** toothpowder.

**ditmek** *va.* Pick into fibres; tease; card; (of birds of prey) to tear to pieces.

**dival** Embroidery in gold thread *or* cotton, mounted on cardboard.

**divan** (–ˈ) Council of State; public sitting *of a governor or council*; collection of poems by one author; (.ˈ) divan, sofa. ~ **durmak,** to stand in a respectful position with hands joined in front.

**divane** (––ˈ) Crazy, insane. **~lik,** madness.

**divan·hane** (–.–ˈ) Council-chamber; hall. **~ıâli** (––.–ˈ), High Court *before which only Ministers and other high officials are tried.* **~ıharb,** court-martial. **~ımuhasebat** (––..–ˈ), **–tı,** Council charged with the examination of the expenditure of Ministries.

**divit** Case for reed pens and ink *for Arabic writing.*

**diyabet, -ti** Diabetes.

**diyakos** Deacon.

**diyanet** (·–ˈ), **-ti** Piety; religious affairs.

**diyar** (·ˈ) *pl. of* **dar.** Houses; countries; *used as sing.*, country; district.

**diyare** Diarrhoea.

**diye** *Gerund of* **demek**; *lit.* 'saying'; *used following direct speech, where it can best be rendered by inverted commas; otherwise it is used to express purpose, reason, hope, supposition etc., e.g.* ⌜**yanlışlık olmasın ~ acele etmem**⌝, 'I do not hurry for fear of making a mistake'.

**diyet**[1], **-ti** Blood-money; ransom. **~ istemek,** to demand retaliation by the law of talion.

**diyet**[2], **-ti** Diet.

**diyez** (*Fr. dièse*) Sharp (*mus.*).

**diz** Knee. **~inin* bağı çözülmek,** to give way at the knees *from fear etc.*: **~ boyu,** knee-deep; all-pervading: **~ çökmek,** to kneel: **~ çukuru,** shallow rifle-pit: **~ini* dövmek,** to repent bitterly: **~e gelmek (varmak),** to kneel in entreaty: **~lerine* kapanmak,** to embrace *s.o.'s* knees in supplication: **~leri* kesilmek,** to give way at the knees from fatigue: **~ üstü,** on the knees, kneeling.

**dizanteri** Dysentery.

**dizbağı, -nı** The tendons of the knee; garter.

**dizdize** Knee to knee.

**dizel** Diesel.

**dizgin** Bridle; rein. **~leri ele vermek** *or* **~leri başkasına kaptırmak,** to hand over the reins to another, to let another take control: **~ini kısmak,** to hold in, to curb: **~leri toplamak,** to rein in, to curb: **dolu ~,** at a gallop, with the utmost speed.

**dizi** Line; row; string (of beads *etc.*); progression (*math.*). **~ci,** one who arranges; compositor. **~li,** arranged in a line *or* row; on a string *or* a skewer; set up (type).

**dizkapağı, -nı** Knee-cap.

**dizlik** Trousers *or* drawers reaching to the knee; knee-cap (protective covering).

**dizmek** *va.* Arrange in a row; string (beads *etc.*); set up (type).

**dobra dobra** (ˈ·ˈ·) Bluntly, frankly, without beating about the bush.

**doçent** Lecturer *or* assistant professor in a university.

**doğan** Falcon. **~cı,** falconer.

**doğma** Birth. Born, by birth. **~ büyüme buralıyım,** I was born and bred here: **anadan ~,** stark naked; **anadan ~ kör,** born blind.

**doğmak** *vn.* Be born; (of sun *or* stars) to rise; come to pass. **doğduğuna pişman,** 'who wishes he had never been born'; tired of life; 'born tired': ⌜**gün doğmadan neler doğar**⌝, 'ere the day break many things may happen': **içime* doğdu,** I had a presentiment.

**doğram** Slice. **~a,** act of slicing; parts of a house made by a carpenter (doors, windows *etc.*): **dişi ~,** dovetail. **~acı,** carpenter.

**doğramak** *va.* Cut up into pieces *or* slices. **bol ~,** to make great promises, to brag.

**doğru**[1] *For* **doğu.** East; east wind.

**doğru**[2] *a.* Straight; upright; level; direct; right; true; honest, faithful, straightforward. *n.* The right; the truth. **···e ~,** towards, near: **~dan ~ya,** directly, without intermediary: **~su,** the truth of the matter: **~dan ayrılmak,** to swerve from the path of right: **~ çıkmak,** to come true; to prove to be right: **~ durmak,** to behave oneself: **~ oturmak,** to sit still, to keep quiet: ⌜**~söyliyeni dokuz köyden kovarlar**⌝, 'he who tells the truth is chased out of nine villages'; home truths are not welcomed: **daha ~su,** or, to be more exact: **ellisine ~,** not far off fifty.

**doğru·ca** (ˈ··), directly; straight. **~lamak,** to correct; to confirm. **~lmak,** to become straight, level *or* true; sit up; come true, be realized: **···e ~,** to set out for, go towards. **~ltmak,** to put straight *or* right; correct; (*sl.*) to manage to earn. **~luk,** straightness; rectitude; honesty; truth. **~su,** in truth: **daha ~,** or, to be more exact.

**doğu** The east; the East.

**doğum** Birth; confinement. **~ kâğıdı,** birth certificate. **~evi, -ni,** maternity home. **~lu,** born in *such and such* a year: **1920 ~,** the 1920 class (*mil.*).

**doğur·mak** *va. & vn.* Give birth (to); bring forth; breed. **~tmak,** to cause to give birth; to assist delivery.

**doğuş** Birth; sunrise.
**dok**[1], **-ku** Dock.
**dok**[2], **-ku** Duck (material).
**doka, duka** (ꜝ·) Duke.
**doksan** Ninety. ~ **dokuzluk tesbih,** rosary of 99 beads. **~ıncı,** ninetieth. **~ar,** ninety each; ninety at a time.
**doktor** Doctor. **~luk,** profession of medicine; title of doctor. **~a,** degree of doctor.
**doku** Tissue.
**dokuma** Weaving; woven tissue; cotton fabric. Woven. **~cı,** weaver. **~cılık,** textile industry.
**dokumak** *va.* Weave. ⌜**ince eleyip sık ~**⌝, to examine too minutely; to be too exact and meticulous.
**dokunaklı** Touching, moving; biting, piquant; harmful; strong (tobacco *etc.*).
**dokundurmak** *va. caus. of* **dokunmak**[2]. Make *or* let touch; hint at.
**dokunmak**[1] *vn. pass. of* **dokumak.** Be woven.
**dokunmak**[2] *va.* (*with dat.*) Touch; come in contact with; (of a fish) to bite; affect; have an evil effect on; injure; meddle with; vex; concern. **···in ···e faydası ~,** to be of use to: **···in ···e yardımı ~,** to help, to contribute to: **namusa ~,** to slur the honour of, to insult: **namusuna** *or* **ırzına ~,** to violate *a woman*: **···in ···e zararı ~,** to do harm to.
**dokunulmazlık** Immunity from arrest.
**dokunur** Touching; affecting; harming. ~ **dokunmaz,** scarcely touching.
**dokurcun** Stack of hay *or* corn; *a game played with beads or pebbles,* merels.
**dokuş** *In* **değiş ~,** exchange, barter.
**dokuşmak** *v.* **tokuşmak.**
**dokuz** Nine. ~ **babalı,** whose father is unknown: ~ **doğurmak,** to fret with impatience; to suffer hardship: ~ **doğurtmak,** to hustle: ⌜**~unda ne ise doksanında da odur**⌝, 'what he was at nine he is at ninety', he'll never improve: ~ **yorgan eskitmek,** 'to wear out nine blankets', to live to a great age.
**dokuz·ar** Nine each; nine at a time. **~taş,** *the game of* nine men's morris *or* merels. **~lu,** containing nine *parts etc.*; the nine (at cards). **~uncu,** ninth.
**dolab** Anything that revolves; waterwheel; treadmill; turnstile; cupboard; trick, plot. ~ **çivisi,** medium-sized nail: ~ **kurmak** *or* **çevirmek,** to set a snare, to lay a trap: **~a girmek,** to fall into a trap, to be cheated. **~cı,** plotter, intriguer. **~lı,** furnished with cupboards; tricky, deceitful: ~ **saat,** grandfather clock.
**dolak** Puttee.
**dolam** One turn *of any coiled thing*; fold *of a turban*. **~a,** act of winding; a kind of wrap; whitlow.
**dolamak** *va.* Twist; wind round; encircle; bandage; lay *a burden or duty on s.o.* **başına* ~,** to saddle *s.o.* with *stg.* **···i diline ~,** to repeat constantly, to make *stg.* the main topic of one's conversation; to be always running *s.o.* down: **···i parmağına ~,** to have an 'idée fixe' about *stg.*, to have 'a bee in one's bonnet' about *stg.*
**dolambaç** Winding; sinuous; flexible; twisting (road). Tortuosity. **~lı,** not straightforward, tortuous: ~ **yer,** labyrinth.
**dolamık** Net; snare.
**dolan** Deceit. **yalan ~,** a pack of lies. **~dırıcı,** swindler; swindling.
**dolan·dırmak** *va.* Make to go round; surround; cheat, swindle; acquire by fraud. **~mak,** to revolve; circulate; saunter about. **···e ~,** to surround.
**dolâr** Dollar.
**dolaş** Tangle; obstacle. Tangled; involved. **~ık,** tortuous; intricate; confused. **~ıklık,** tortuosity; intricacy; obscurity of style *etc.*
**dolaşmak** *va. & vn.* Go around; walk about; go a round *of visits etc.*; make a tour; become tangled *or* confused; (of a street or river) to wind about. **···in ayaklarına ~,** to be a hindrance *or* handicap to *s.o.*; **ayakları* birbirine ~,** to be thrown into confusion; **ettiği fenalık ayağına dolaştı,** he paid for his mistake; his misdeeds were visited upon him: **dili* ~,** to be tongue-tied *or* to stammer *from surprise or confusion*; to speak thickly *when drunk*; **dillerde ~,** to be talked about everywhere: **dört ~,** to be in a quandary: **zihni* ~,** to be confused *or* bewildered.
**dolay** Surroundings; outskirts. **···den ~ı,** on account of, due to: **~ısile,** on account of; in connexion with; as regards; indirectly; consequently.
**doldurmak** *va.* Fill; complete; stuff; load (a gun *etc.*); prime (a person); fill up (a hole); charge (an accumula-

tor); (of a child) to foul its clothes. **çile ~**, to undergo suffering: **denizi ~**, to reclaim land from the sea: **diş ~**, to stop a tooth: **gününü* ~**, to complete one's sentence in prison: **kuzu ~**, to stuff a lamb *for roasting*: **yerini* ~**, to fill a post, to perform the duties of it properly.

**dolgu** The filling *or* stopping of a tooth.

**dolgun** Full, filled; stuffed; abundant; high (wages); spiteful, full of wrath. **etine ~** plump: **kulağı ~**, well-informed. **~luk**, fullness, plenitude; a being overfull; anger, spite: **mide dolgunluğu**, oppression of the stomach, indigestion.

**dolma** A dish of meat *or* vegetables stuffed with rice and forcemeat; anything stuffed; reclaimed land; embankment; lie, invention. Stuffed; filled up with earth *or* stones. **~ kalem**, fountain-pen: **~ yutmak**, to swallow a lie, to be taken in: **ağızdan ~ top**, muzzle-loading gun: **kabak ~sı**, stuffed vegetable marrow: **yaprak ~sı**, vine leaves wrapped round forcemeat.

**dolmak** *vn.* Fill, become full; swell; be completed; be full of anger *or* spite. **dolup dolup boşalmak**, (of a place) to be thronged with people: **gözleri* ~**, to have the eyes full of tears.

**dolmuş** Filled; stuffed. Vehicle *or* boat that only starts when all the seats are taken.

**dolu**[1] Full; solid (*not hollow*); loaded (gun). Contents (of a bottle *etc.*); charge, load (of a gun). **~ dizgin gitmek**, to ride at full speed: **ağız ağıza ~**, brimful: **avuc ~su**, handful: ⌜**boşa koydum dolmadı, ~ya koydum almadı**⌝, I couldn't find any solution of the problem.

**dolu**[2] Hail. **~ yağmak**, to hail: ⌜**yağmurdan kaçıp ~ya tutulmak**⌝, 'out of the frying-pan into the fire'.

**doluk** Hide *used as a waterbottle or a float.*

**dolun·ay** (·ˈ·) Full moon. **~mak**, (of the sun) to set; (moon) to be full.

**domalan** Tumour, abscess; truffle.

**domalmak** *vn.* Project as a hump; bend down with the back protruding; (animal) to lie humped up.

**domates** (·ˈ·) Tomato.

**dombay** Female buffalo.

**dombaz** *v.* **tombaz.**

**domuz** Pig, swine; obstinate and disagreeable man. **~una**, out of spite; (*sl.*) thoroughly: **~ gibi**, obstinate; swinish; strong, healthy; **~ gibi** *or* **~una çalışmak**, to work like a nigger: ⌜**~dan kıl çekmek (koparmak)**⌝, to get something out of a stingy *or* unfriendly man: **~ tırnağı**, crowbar, jemmy: **hind ~u**, guinea-pig.

**domuz·ayağı, -nı**, corkscrew; caltrop (*mil.*). **~budu, -nu**, leg of pork, ham. **~luk**, swinish behaviour; pigheadedness; pigsty. **~yağı, -nı**, lard.

**don**[1] Pair of drawers; coat *or* colour *of a horse.* ⌜**ayağında ~u yok başına fesleğen takar**⌝, 'he has no pants on but wears a sprig of sweet basil on his head', *i.e.* he spends money on luxuries when he cannot afford the necessaries of life.

**don**[2] Frost. **~lar çözülmek**, for a thaw to set in.

**donakalmak** (·ˈ··) *vn.* Be petrified *with horror or fear.*

**donanım** Rigging.

**donanma** A being decked out with flags *etc.*; illumination; fleet; navy.

**donanmak** *vn.* Be decked out, illuminated, equipped, rigged.

**donatım** Equipment.

**donatmak** *va.* Deck out, ornament; equip; rig; illuminate; dress (ship); (*sl.*) abuse. **tepeden tırnağa ~**, to curse *s.o.* up and down.

**dondurma** Frozen; set (concrete). Ice-cream; concrete. **~cı**, ice-cream vendor. **~k** *va.*, to freeze.

**donmak** *vn.* Freeze; become frozen; (of cement) to set.

**donsuz** Without pants; destitute person, vagabond.

**donuk** Frozen; frosted (glass); matt; dull; dim; torpid.

**donyağı, -nı** Tallow. **~yle pekmez**, incompatible.

**dopdolu** (ˈ··) Chock-full.

**doru** Bay (horse). **açık ~**, light bay: **kestane ~su**, chestnut bay: **yağız ~**, brown bay.

**doruk** Summit; peak. Piled up into a cone. **~lama**, heap; in a heap, piled up.

**dosa** Gangway plank (*naut.*).

**dosdoğru** (ˈ··) Absolutely straight; straight ahead; perfectly correct.

**dost, -tu** Friend; lover; mistress. Friendly. ⌜**~ ağlatır düşman güldürür**⌝, the friend brings tears, the enemy smiles; friends criticize, enemies flatter: ⌜**~lar alışverişte görsün**⌝, for the sake of appearances: ⌜**~lar başına!**⌝, may the same befall all my friends!, *said when speaking about another's success*: **iyigün ~u**, a fair-weather friend.

**dost·ane** (·–ˡ), **~ça**, friendly, amical. **~luk**, friendship; friendly act; favour. **bir ~ kaldı**, 'that's the last lot' (*said by a shopman*).

**dosya** (ˡ·) Dossier, file.

**doy·mak** *vn.* Be satiated; be sick of *stg*: **~ bilmez**, insatiable. **~maz**, insatiable, greedy. **~ulmak**, to have enough of (of = *dat.*); be satiated.

**doyum** Satiety. **sonbaharın güzelliğine ~ olmaz**, one never gets tired of the beauty of autumn: ⌜**size ~ olmaz!**⌝, 'one can't have enough of you', *said jokingly when taking leave of s.o.*; *the reply is*: ⌜**size de inan olmaz**⌝ (*I don't believe you*). **~luk**, quantity sufficient to satiate; plunder, booty.

**doyurmak** *va.* Satiate; satisfy; nourish; be beneficial; corrupt by generosity.

**doz** Dose. **~unu kaçırmak**, inadvertently to mix a medicine too strong; (*fig.*) to overdo it.

**döğ–** *v.* **döv–**.

**dökme** Act of pouring *or* casting; a metal casting. Cast (metal); poured out in a heap (wheat *etc.*); (liquid) in bulk. **~ci**, founder, moulder.

**dökmek** *va.* Pour; scatter; throw away; cast (metal); shed (tears, leaves); come out in skin eruptions (small-pox, measles); reject, 'plough' *in an exam.*; change, turn into. **dil ~**, to talk *s.o.* round: ⌜**···in eline su dökemez**⌝, he can't hold a candle to ...: **kâğıda ~**, to put on paper, to write down: **para ~**, to pour out money: **su ~**, to make water: **tüyünü ~**, (of a bird) to moult: **yüzsuyu (âbiru) ~**, to demean oneself, to humiliate oneself *by asking for stg.*

**döktürmek** *va. caus. of* **dökmek**; (*sl.*) write *or* speak well and easily.

**dökük saçık** Dishevelled, unkempt.

**dökülmek** *vn. pass. of* **dökmek.** Be poured *etc.*; fall into decay; disintegrate, go to pieces; (of leaves) to drop off, fall; (of teeth, hair) to fall out; be 'ploughed' *in an exam.* **dökülüp saçılmak**, to unburden oneself, to tell everything; to throw off one's clothes *in undressing*; to spend lavishly: **denize ~**, (of a river) to run into the sea: **kırılıp ~**, to coquet: **iş yeni bir kalıba döküldü**, the affair has taken a new turn: ⌜**tepemden (başımdan) aşağı kaynar sular döküldü**⌝, I felt hot all over (from shame *etc.*): **üstünden* ~**, (of clothes) to be unbecoming: **yola ~**, to take *a certain* course: **yollara ~**, for everyone to come out into the streets.

**döküm** A pouring out; a casting; a shedding (of leaves *etc.*). **~hane** (··–ˡ), foundry.

**dökünmek** *vn.* Throw water over oneself. **soyunup ~**, to undress.

**döküntü** Debris; remnants; reef; stones thrown into the sea *to make a breakwater*; dregs (of the population *etc.*); skin eruption.

**döl** Foetus; semen; germ; race, stock. **~ döş**, progeny, descendants: **~ tutmak**, (of animals *only*) to become pregnant. **~yatağı, –nı**, womb.

**döndürmek** *va. caus. of* **dönmek.** Turn round *or* back; turn inside out; reverse; 'plough' *s.o. in an exam.*

**dönek** Untrustworthy; who never keeps his word; (boy) who always fails in his exams.

**dönemec** Corner; bend. **~li yol**, a winding road.

**döngel** Medlar.

**dönme** Act of turning; retreat; conversion; Jewish convert to Islam. Convert; renegade. **~ dolab**, revolving cupboard *in a wall for conveying food from one room to another*; revolving wheel *at a fun fair.*

**dönmek** *vn.* Turn; turn back; return; change, be transformed; change one's religion; swerve *from a course*; fail to keep *an agreement*; fail *in an exam.*; fail to be promoted to a higher class. **döne döne çıkmak**, to ascend in a spiral: **dönüp dolaşıp**, in the long run, after all: **başı* ~**, to turn giddy: **beyninden* vurulmuşa ~**, to be terribly shocked *by bad news etc.*: **dili* dönmemek**, to be unable to pronounce a word: **gözü* ~**, to be blind with anger *or* passion; **gözleri* ~**, (i) *as* **gözü ~**; (ii) for the eyeballs to be turned up *when at the point of death*: **sözünden ~**, to go back on one's word.

**dönük** With the back turned; turned away. **rengi ~**, faded.

**dönüm¹**· *Superficial measure 40 arshins long by 4 arshins wide = about 97 sq. yards.* **yeni ~**, one hectare (*about* $2\frac{1}{2}$ *acres*).

**dönüm²** Turn; revolution; a time. ~ **noktası,** the turning-point: **gün ~ü,** the solstice: **yıl ~ü,** anniversary.
**dönüş** Act of turning *or* returning; return journey.
**dördayak** (ˋ··) On all fours. ⌈**kedi gibi her zaman ~ üzerine düşmek**⌉, 'like a cat always to fall on one's feet', *i.e.* always to emerge successfully from a difficulty. **~lı,** quadruped.
**dörd·er** Four apiece; four at a time. **~üncü,** fourth.
**dört** Four; (*largely used to express totality, e.g.*: ~ **el ile,** with the greatest energy; ~ **gözle beklemek,** to await with the greatest impatience; ~ **nala gitmek,** to go at full speed; ~ **taraftan,** from every quarter; ~ **bir yana,** in every direction; ~ **yanına bakmak,** to look all round *etc., etc.*): **~ayak,** *v.* dördayak: ~ **başı mamur,** prosperous, flourishing; well-appointed: ~ **dönmek,** to turn a place upside down; to search everywhere; to think of every possible means *of doing stg.*: ~ **ucunu koyuvermek (bırakmak),** to lose heart about *stg.*, to give *stg.* up.
**dört·kaşlı,** with bushy eyebrows; with a budding moustache. **~köşe,** square. **~lü,** possessing four ...; the four *of a suit at cards.* **~lük,** that which is worth *four piastres etc., or* weighs four *pounds etc.*; quatrain. **~nal,** gallop: **~a,** at a gallop, at full speed. **~yolağzı, -nı,** four cross-roads.
**döş** Flank; breast; withers. ~ **tarafından et,** scrag end of meat. **döl ~,** a man's progeny and descendants.
**döşek** Mattress; bed.
**döşeli** Spread; laid down (carpet *etc.*); furnished; ornamented. ~ **dayalı,** fully furnished.
**döşeme** Floor covering; pavement; furniture; upholstery. **~ci,** furniture dealer; upholsterer.
**döşe·mek** *va.* Lay down, spread (carpet *etc.*); carpet; pave; furnish with carpets. **döşeyip dayamak,** to furnish fully. **~nmek,** to be spread out; be laid down; take to one's bed; 'let oneself go', give vent to one's feelings. **~tmek,** *va. caus. of* **doşemek.**
**dövdü** Hammer; mallet; back of an axe.
**döven** Flail; threshing-machine.
**döviz** (*Fr. devise*) Foreign bills; foreign exchange.
**dövme, döğme** Wrought (iron *etc.*); tattooing.
**döv·mek** *va.* Beat; hammer, forge; thresh; thrash; bombard; pound. **~ünmek,** to beat oneself; beat the breast *from pain or sorrow,* lament.
**dövüş, döğüş** Fight. **~ken,** warlike, bellicose, combative. **~mek,** to fight against one another; struggle.
**drahoma** (·ˋ·) Dowry.
**dram** Drama; tragedy.
**drednot, -tu** Dreadnought.
**drezin** (*Fr. draisienne*) Platelayer's trolley.
**dua** (·ˊ) Prayer; blessing; desire. ~ **etm.,** to pray; to bless; to ask for a blessing on: **hayır ~,** benediction: ⌈**olmıyacak ~ya amin demem**⌉, I can't agree to an impracticable suggestion. **~cı,** one who prays for another: **~nız,** your humble servant. **~han** (·–ˊ), ~ **olm.,** to pray.
**duba** (ˋ·) Barge; pontoon; floating bridge. ~ **gibi,** very fat.
**dubara** (·ˋ·) Deuce at dice; trick, fraud. **~cı,** trickster, cheat.
**duçar** (–ˊ) Face to face; subject to; afflicted with; exposed to. **···e ~ olm.,** to be subject *or* exposed to ...; to contract *a disease.*
**dudak** Lip.
**dudu** Old Armenian woman. ~ **kuşu,** parrot.
**duhul** (·ˊ), **-lü** Entrance. **~iye,** entrance fee; ticket of admission; import duty.
**duka** Duke; ducat.
**dul** Widowed. Widow; widower. **~luk,** widowhood.
**dulavratotu** (··ˋ··), **-nu** Burdock.
**dum** *v.* **tım.**
**duman** Smoke; mist; condensation *on a glass of cold water etc.* **~ı üstünde,** with the bloom still upon it, quite fresh: **birine ~ attırmak,** to defeat *s.o.* utterly, to have him at one's mercy: ⌈**(bacası eğri olmuş) ~ı doğru çıksın**⌉, (no matter if the chimney be crooked) as long as the smoke gets out, *i.e.* no matter as long as the result is good: **işi ~dır,** he is in a bad way; he is ruined: **kafa ~ olm.,** to be befuddled: **sonu* ~dır,** he'll come to a bad end: **tozu ~a katmak,** to raise a dust, to proceed at great speed.
**duman·lamak** *va. & vn.*, give out smoke; cover with mist; render turbid: **kafayı ~,** to become fuddled. **~lan-**

**mak,** to be filled with smoke *or* mist; be smoked *or* cured; become confused in mind; become fuddled: **gözü* ~,** to be blind with rage. **~lı,** smoky, misty; tipsy: **başı* ~,** fuddled with drink; deranged by passion: ⌜**kurd ~ havayı sever**⌝, the wolf likes foggy weather, evil-doers shun the light.

**dumur** *Only in* **~a uğramak,** to be atrophied; to disappear.

**dur** Halt!; stop! **selâm ~!,** present arms!; *v.* **durmak.**

**durak** Stopping-place; halt; pause; residence. Stationary. **~ su,** stagnant water. **~lama,** standstill. **~lamak, ~samak,** to hesitate; to break off, to come to a stop.

**duralar** *v.* **dahî.**

**durdurmak** *va. caus. of* **durmak.** Stop; make to stand; cause to wait. **akıllar ~,** to dumbfound.

**durgun** Stationary; stagnant; calm; fatigued; perplexed; at a standstill. **~luk,** stagnation; quiet; heaviness; anxiety. **akıllara ~ verecek derecede,** to an astounding degree.

**durmak, -ur** *vn.* Stop; cease; stand; wait; remain; endure; continue; dwell *on a subject*; (*as an auxiliary verb it expresses continuous action, e.g.* **bakadurmak** *or* **bakıp ~,** to keep on looking). **durup dinlenmeksizin,** unceasingly: **durup durma!,** 'don't keep standing there!': ⌜**durup dururken**⌝, without any reason; without provocation: **durmuş et,** meat that has been kept a few days: **durmuş oturmuş bir hal,** a settled mature state; a being too old for one's years and having the manners of an experienced man: ⌜**dur yok otur yok**⌝, no respite, not a moment's peace: **ayakta ~,** to stand: **boş ~,** to remain idle: **gözüne dizine ~,** to be punished for one's ingratitude: **hiç durmadan,** uninterruptedly; without a moment's hesitation: **içi* durmamak,** not to feel comfortable (until *stg.* is done); to feel bound to do *stg.*: **işler duruyor,** business is stagnant: **karşı ~,** to oppose: **mideye ~,** to lie heavily on the stomach: **sözünde* ~,** to abide by one's word: ... **şöyle dursun** ..., ... let alone ...: **···in üzerinde ~,** to dwell upon *a subject etc.*: **yüzüne* duramamak,** to be unable to refuse.

**duru** Clear; limpid. **~lamak,** to rinse in clean cold water.

**durulmak¹** *vn. impers. of* **durmak.** **burada durulmaz,** one can't stop here; 'no stopping here!'

**durulmak²** *vn.* Become quiet, settle down; become well-behaved; (of liquids) to become clear. **dibe ~,** (of particles in a liquid) to settle to the bottom: **ortalık duruldu,** it became quiet.

**durum** Position; attitude.

**duruş** Posture; attitude; aspect; behaviour. **yan ~u,** side view, profile.

**duruşma** Sitting *or* hearing of a court.

**duruşmak** *vn.* Confront one another (in combat *or* argument).

**duş** Douche; shower-bath.

**dut** Mulberry. ⌜**~ yemiş bülbül gibi**⌝, 'like a nightingale that has eaten mulberries', *i.e.* taciturn. **~luk,** mulberry garden; full of mulberry trees.

**duvak** Veil *worn by brides or new-born babies*; stone *or* earthenware lid. **duvağına doymamak,** of a newly married bride to die *or* be separated from her husband: **~ düşkünü,** a newly married bride who becomes a widow.

**duvar** Wall. **~ arpası,** wild barley: **~ ayağı,** the foot *or* foundation of a wall: **~ çekmek,** to build a wall round: **~ gibi,** very solid; as deaf as a post: ⌜**~a yazıyorum**⌝, don't forget that I have told you this; some day you will learn that I am right: **kapı ~,** no one opened the door; there was no answer *to a knock or ring.*

**duygu** Perception; feeling; sense; sensation. **~lu,** sensitive, impressionable; perceptive, intelligent; well-informed. **~suz,** insensitive; apathetic; ignorant.

**duy·mak** *va.* Feel; perceive; hear; get information of. **~maz,** who does not perceive; imperceptive; insensitive: **vurdum ~,** thick-skinned; 'slow in the uptake'. **~ulmadık** (·ˋ··), unheard-of, strange. **~ulmak,** *pass. of* **duymak,** to be heard *or* felt; be heard of; be known; be promulgated, made public. **~urmak,** to make heard, felt *or* perceived; to make known, divulge. **~uş,** impression, feeling.

**duziko** Greek raki.

**dü** Two (at dice). **~ beş,** double fives.

**Dübbüekber** The Great Bear, Ursa Major.

**dübür, -brü** Arse; anus; back.

**düdük** Whistle, pipe, flute; long hollow tube; silly fellow. Very long and slender. ~ **gibi kalmak,** to be left alone and helpless (through one's own fault): ~ **gibi kıyafet,** clothes too small and tight: ⌜**parayı veren düdüğü çalar**⌝, 'who pays the piper calls the tune'.

**düğme** Button; knob; pimple. ~**li,** with buttons, buttoned. ~**lemek,** to button up.

**düğüm** Knot; bow; knotty problem. ~ **noktası,** crucial *or* vital point.

**düğün** Feast *on the occasion of a wedding or circumcision.* ~ **bayram,** feast, merrymaking: ~ **evi gibi,** *place* full of merrymakers. ~**çiçeği, -ni,** buttercup, ranunculus.

**dükkân** Shop. **ikindiden sonra** ~ **açmak,** to open a shop late in the day, *i.e.* to undertake *stg.* when it is too late. ~**cı,** shopkeeper.

**düldül** *Humorous name for a* horse; nag.

**dülger** Carpenter; builder. ~**balığı, -nı,** John Dory.

**dümbelek** Small drum; idiot.

**dümdar** Rearguard.

**dümdüz** Absolutely flat, level *or* straight.

**dümen** Rudder. ~**i eğri,** walking about in a zigzag manner; acting aimlessly and without plan: ···**in** ~**i elinde olm.,** to be in charge of ...: ~ **kırmak,** to change course; **kır** ~**i,** 'get out!', 'clear off!': ~ **neferi,** the last man in a file; anyone who is left behind others, *e.g.* the laziest boy in a class: ~ **suyu,** the wake of a ship: ⌜**dayısı** ~**de**⌝, 'his uncle is at the helm', *said of one who gets on by favour.* ~**ci,** helmsman; *also as* **dümen neferi.**

**dümtek** *In* ~ **vurmak,** to beat time by slapping the knees with the hands.

**dün** Yesterday. ⌜~ **bir bugün iki**⌝, only just arrived; of recent occurrence; *also used of a new arrival getting above himself,* 'what cheek!': ~**den bugüne,** in a short time: ~**den hazır (razı, teşne),** only too glad; just waiting for it. ~**kü,** o fyesterday: ~ **gün,** yesterday: ~ **çocuk,** still only a beginner; young and inexperienced.

**dünür** *The* relationship *between the parents of a husband and those of his wife and vice-versa*; *such a* relation-at-law.

**dünya** World; the Earth; this life; worldly goods. ~**da** (*with neg.*), never in the world, on no account: ~**nın** ···**i,** a vast quantity of ..., *e.g.* ~**nın parası,** a mint of money: ~ **âlem,** all the world, everybody: ~**lar benim oldu,** I felt on top of the world: ~ **durdukça,** as long as the world lasts: ~ **evi,** marriage: ~**yı gözü* görmemek,** to be completely obsessed by one idea: ~ **gözüyle görmek,** to see *stg.* before one dies: ~ **güzeli,** extremely beautiful *woman*: ~ **kadar,** a world of; very large; ~ **kelâmı etm.,** to talk about things, to talk *in general*: ~**lar kendisinin olmuş gibi sevindi,** he was overcome with joy: ~ **varmış!,** *expression of relief or pleasure,* good!, that's better!: ~**sından* vazgeçmek,** to give up all interest in everything, to neglect one's affairs, one's person *etc.*: **şöhreti** ~**yı tutmak,** to become world-famous.

**dünya·lık,** worldly goods, wealth; money.

**dünyevî** Worldly, mundane.

**düpedüz** (\`··) Absolutely flat; quite level; utterly. ~ **delilik,** sheer madness: '**herif** ~ **beni tahkir etti**', 'the fellow flatly insulted me'.

**dür, dürdane** Pearl.

**dürbün** Field-glasses.

**dürmek** *va.* Roll up. **dürup bükmek,** to fold.

**dürt·mek** *va.* Prod; goad; incite. ~**ücü,** one who prods *or* goads; inciter, instigator. ~**üklemek,** to prod. ~**üş,** push, prod; incitement. ~**üşmek,** to prod one another; to incite. ~**üştürmek,** to keep pushing one another; keep on inciting *or* goading *s.o. to do stg.*

**dürü** Roll; anything rolled up.

**dürüm** Fold; pleat. ~ ~, in folds *or* pleats: ~**ü bozulmamış,** brand new.

**dürüst** Straightforward, honest; correct. **doğru** ~ **konuşmak,** to speak correctly; to speak decently; to speak frankly.

**dürüşt** Coarse; severe; brutal.

**Dürzi** Druze.

**dürzü** *A term of abuse implying treachery or cruelty.*

**düstür** Principle; formula; code of laws; register; precedent; permission; authority; book of medical prescriptions.

**düş** Dream. ~**ü azmak,** to have a seminal emission during sleep.

**düşeş** Double six (at dice); an unexpected bit of luck.

**düşkün** Fallen; broken down, decayed; fallen on bad times, 'come down in the world'; addicted; a slave (to). **boğazına ~,** gluttonous; **fırsat ~ü,** opportunist, 'with an eye to the main chance', profiteer: **kıyafetine ~,** particular about his dress, *but* **kıyafet ~ü,** poorly dressed: **kumara ~,** addicted to gambling: **vücudden ~,** emaciated: **yıldızı ~,** whose star has set, unfortunate. **~lük,** decay, poverty; misfortune; excessive addiction.

**düşman** Enemy, foe. **~ ağzı,** calumny: ⌜**eski dost ~ olmaz**⌝, only old friends are reliable: **kaşık ~ı,** wife (*sl. and jokingly*): ⌜**su uyur ~ uyumaz**⌝, always be on your guard against an enemy. **~lık,** enmity, hostility.

**düşme** Act of falling *etc.* **elden ~,** second-hand.

**düşmek** *vn.* Fall; fall down; fall *in price, esteem or position*; fall on evil days; fall away in health; (*naut.*) drift; fall to one's lot, befall, happen; befit, become, be suitable; fall for, take to *a thing* (*dat.*); appear suddenly; give the impression of being *odd, silly etc.*; (of a baby) to be aborted. *va.* Subtract, deduct. **düşe kalka,** with great difficulty, after great efforts: ⌜**düşmez kalkmaz bir Allah**⌝, 'only God is free from the vicissitudes of Fate': **···le düşüp kalmak,** to live with ..., to be intimate with ...: '**düş önüme!**', 'come along with me!' (*usually used by policemen etc.*): **birinin arkasına ~,** to follow *s.o.* about: **bu bana düşmez,** it is not for me to say *or* do this: **çiğ ~,** to seem crude: **dile ~,** to become the object of gossip: **hesabdan ~,** to be deducted from *or* not included in an account: **birisine işi* ~,** to have recourse to *s.o.*: **küçük ~,** to look small, to feel foolish: **sırası düştü,** the moment has come *to do stg.*: **ev yolun soluna düşüyor,** the house is on the left side of the road: **suya ~,** to come to nought: **bir işin üstüne ~,** to persist in a matter; to be very keen on *stg.*: **birinin üzerine ~,** to 'keep on at' a person; to be unduly concerned about *s.o.*: **yollara ~,** to set forth *in quest of s.o. or stg.*: **halk yollara düştü,** the people poured out on to the roads.

**düşük** Fallen; drooping; low; loose, disconnected (writing); aborted. **~ etek,** untidy, slovenly. **~lük,** looseness *of style*; faultiness *of rhythm etc.*; fall *in prices.*

**düşünce** Thought, reflection; anxiety. **~li,** thoughtful; pensive; worried. **~siz,** thoughtless, unreflecting.

**düşünmek** *va.* Think of; remember; ponder over. *vn.* Be pensive; reflect; be worried. **bir şeyi yapmağı ~,** to think of doing *stg.*; **bir şeyi yapmağa ~,** to hesitate about doing *stg.* (to be reluctant): **düşünüp taşınmak,** to ponder *stg.* well: '**düşünün bir kere!**', 'just think for a moment!': '**çok düşünüyor**', 'it's very pensive' (*said of an animal that is looking ill*).

**düşünüş** Mode of thinking; reflection.

**düşürme** Act of dropping *or* causing to fall; miscarriage. **elden ~,** a chance of buying cheaply; a bargain.

**düşürmek** *va. caus. of* **düşmek.** Cause to fall; drop; beat down; bring down (an airplane); cause to meet; pass *from the body* (a worm *etc.*): **ağzından* düşürmemek,** never to cease talking about *stg.*: **birbirine ~,** to set two people at loggerheads: **çocuk ~,** to have a miscarriage: **elden ~,** to get cheap, to get a bargain: **elden düşürmemek,** to use *stg.* continually: **fırsatını (sırasını) ~,** to seize an opportunity: **itibardan ~,** to discredit *s.o.*: **küçük ~,** to make *s.o.* look small: ⌜**bu yüzüğü ellili raya düşürdüm**⌝, 'I had the luck to get this ring for fifty lira'.

**düşürtmek** *as* **düşürmek.**

**düşüş** Fall; manner of falling.

**düt** *Only in* '**herkesin ~ dediği keçi olmaz**', one doesn't get all one wants.

**düttürü** (\\··) Oddly dressed. Odd dress. **D~ Leylâ,** *the personification of* an eccentrically dressed person; one whose clothes are too tight and too short.

**düve** Heifer.

**düvel** *pl. of* **devlet.** Powers; states. **~i muazzama,** the Great Powers.

**düyun** (·⊥) *pl. of* **deyn.** Debts. **~u umumiye,** the Public Debt.

**düz** Flat; level; smooth; straight; uniform. Flat level place. **~ayak,** on a level with the street *or* ground; without stairs, on one floor.

**düz·elmek** *vn.* Be arranged; be put in order; be improved. **~eltmek, ~etmek,** to make smooth *or* level; put in order; arrange; correct; set on the right road.

**düzen** Order; regularity; harmony; tidiness; toilet; invention, lie, trick. **~i bozuk,** out-of-tune: **~ kurmak,** to use cunning: **~ vermek,** to put in order, to tune. **~baz,** tricky, deceitful; cheat, impostor. **~li,** in order; orderly, neat, tidy; harmonious; tricky; fictitious. **~lik,** harmony: **dirlik ~,** harmonious and quiet life: ⌜**Allah dirlik ~ versin!**⌝, *said in wishing happiness to a newly married couple.* **~siz,** in disorder; untidy; out-of-tune; without guile, straightforward. **~sizlik,** disorder, untidiness; discord.
**düzetmek** *v.* **düzeltmek.**
**düzgün**[1] Smooth; level; in order; arranged; correct; in tune; in unison. **eli ~,** skilful: **eli yüzü ~,** rather pretty (woman). **~lük**[1], order, regularity.
**düzgün**[2] Face-paint. **~ sürmek,** to make up *the face.* **~lü,** made-up, painted (face). **~lük**[2], receptacle for cosmetics.
**düzine** Dozen.
**düzlemek** *va.* Smooth; flatten; level.
**düzlük** Smoothness; levelness; flatness; plainness, simplicity; flat level plain.
**düzme, düzmece** Made up; false; counterfeit; sham.
**düzmek** *va.* Arrange; prepare; put in order; invent (tale *etc.*); counterfeit, forge. **düzüp koşmak,** to arrange and put together.
**düztaban** Flat-footed; ill-omened man, a 'Jonah'.
**düzü** *Only in* **bir ~ye,** uninterruptedly, continuously.
**düzülmek** *vn. pass. of* **düzmek; yola ~,** to set out on a journey.

# E

**eb'ad** (·–́) *pl. of* **buud.** Dimensions.
**ebcem** *In* **~ çüş,** stupid and uncouth.
**ebced** *The first mnemonic formula of Arabic letters according to their numerical value.* **~ hesabı,** numeration by letters of the alphabet.
**ebe** Midwife; 'he' *in such games as 'catch', 'touch-last' etc.* **dil ~si,** garrulous person; quick at repartee. **~gümeci, -ni,** marshmallow. **~kuşağı, -nı,** rainbow. **~lik,** midwifery.
**ebed** Eternity. **~î** (··–́), eternal; without end. **~iyen** (··–́·), eternally. **~iyet, -ti,** eternity.
**ebeveyn** Parents.
**ebleh** Imbecile, stupid.
**ebniye** *pl. of* **bina.** Buildings.
**ebru** (·–́) Marbled (paper). Marbling; marbled paper. **~lu, ~lî,** marbled; changing colour.
**ecdad** (·–́) *pl. of* **ced.** Grandfathers; ancestors.
**ecel** The appointed hour of death; death. **~i* gelmek,** for one's hour of death to have come: **~iyle ölmek,** to die a natural death: **~ine susamak,** to run into the jaws of death, knowingly to face death: **~ teri dökmek,** to be in mortal fear.
**echel** *superl. of* **cahil.** Very ignorant.
**ecinni** Genie; evil spirit.
**ecir, -cri** Reward; recompense. ⌜**Allah ~ sabır versin!**⌝, 'may God recompense you and give you patience!' (*said when expressing sympathy for a death*): **~ sabır dilemek,** to offer one's condolences.
**ecîr** (·–́) Hireling; day labourer; salaried person; employee.
**eciş bücüş** Shapeless; crooked; bent double with age. **~ ~ yazı,** scrawl, bad handwriting.
**ecnebi** Foreign. Foreigner; stranger.
**ecza** (·–́) *pl. of* **cüz.** Parts; components; drugs, chemicals; unbound sections of a book. **~cı,** chemist, druggist. **~hane** (·––́), **~ne,** pharmacy, chemist's shop.
**eda** (·́) Payment; execution (of a deed, a duty); air, tone, manner; affectation. **~ etm.,** to pay (debts); perform (duty). **~lı,** gracious, seductive; having *such and such* an air; affected: **ciddî ~,** with a serious air.
**edat, -tı** Particle (*grammar*).
**edeb** Breeding, manners; education; the science of letters; modesty, shame. **~ erkân,** good manners: **~ öğretmek (göstermek, vermek),** to teach *s.o.* manners, to chastise: ⌜**~dir söylemesi**⌝, 'excuse the expression!': **~ini takın** *or* **~inle otur!,** behave yourself!: **~ yeri,** the private parts. **~lenmek,** to be *or* become well-behaved. **~li,** well-behaved, with good manners. **~siz,** ill-mannered; rude; shameless. **~sizlik,** bad manners; impertinence; shameful act *or* behaviour.
**edevat** (··́) Tools; particles.

**edib** (·⅄) Literary man. Polite, gentlemanly.

**edici** Making; doing; *usually used with other adjectives to form a compound adjective, e.g.* **mest,** drunk; **mestedici,** intoxicating.

**edilmek** *vn. pass. of* **etmek.** Be done; be made; *forms the passive of compound verbs made with* **etmek,** *e.g.* **zannetmek,** to think; **zannedilmek,** to be thought.

**edinmek** *va.* Get, procure. **âdet ~,** to acquire a habit: **dost ~,** to get friends, to make a friend of ...: **oğlu ~,** to get a son; to adopt as a son.

**Edirne** (·⅄·) Adrianople.

**ediş** Manner of doing *or* making.

**edyan** (·⅄) *pl. of* **din.** Religions.

**ef'al** (·⅄), **-li** *pl. of* **fiil.** Actions; verbs.

**efe** *Title borne by Zeybek notables;* Zeybek; swashbuckler. **~lik,** braggadoccio, 'side'.

**efendi** (·⅄·) Master; *formerly used after a name as* 'Mister'; gentleman. **~m,** Sir!; I beg your pardon!; what did you say?; *or just put in for politeness like the French 'monsieur'*: ⌜**~m nerede, ben nerede**⌝, *expression used ironically when the person, to whom you are speaking, quite fails to get your meaning*: ⌜**~me söyliyeyim**⌝, *expression used to link two sentences in speech,* ... and then ...; *sometimes denotes hesitation,* 'what was I going to say?' **~lik,** gentlemanly behaviour.

**Efgan** (⅄·), **efganlı** Afghan.

**efkâr** (·⅄) *pl. of* **fikir.** Opinions; ideas; *as sing.* thinking; anxiety. **~ etm.,** to be worried about *stg.*: **~ı umumiye,** public opinion. **~lanmak,** to become thoughtful *or* anxious. **~lı,** anxious, worried.

**Eflâtun**[1] Plato; learned man.

**eflâtun**[2] Lilac-coloured.

**efrad** (·⅄) *pl. of* **ferd.** Individuals; private soldiers.

**efsane** (·–⅄) Fable; idle tale. **~ ve efsun,** idle talk; wild stories.

**efsun** *v.* **afsun.**

**Ege** (⅄·) Aegean.

**eğe**[1] File. **~ kemiği,** rib. **~lemek,** to file.

**eğe**[2] Master; guardian; relation.

**eğer**[1], **eyer** Saddle.

**eğer**[2] (⅄·) If; whether.

**eğic** Wooden hook (*for gathering fruit etc.*).

**eğik** Inclined; slanting.

**eğilmek** *vn.* Bend; incline; bow; lean out *of a window etc.* **eğilip bükülmek,** to bow and scrape.

**eğirmek** *va.* Spin.

**eğit·im** Education. **~mek,** to educate. **~men,** village teacher.

**eğlemek** *va.* Stop; retard; delay.

**eğlen·ce** Diversion; amusement; plaything; joke; useless *or* unimportant thing; very easy matter. **~celi,** diverting, amusing. **~celik,** *such things as sweets, salted almonds etc. eaten merely as tit-bits for amusement rather than nourishment.* **~mek,** to be diverted *or* amused; amuse oneself; joke; stop; wait.

**eğleşmek** *vn.* Rest oneself; reside.

**eğmek, iğmek** *va.* Bend; incline; persuade. **boyun ~,** to bow the neck, submit.

**eğre** Felt saddle-pad.

**eğreltiotu, -nu** Fern.

**eğreti** *v.* **iğreti.**

**eğri** *v.* **iğri.**

**eğrim** Felt saddle-cloth; small whirlpool.

**egsos, egzos** Exhaust (of an engine).

**ehem, -mmi** Most *or* more important. **~miyet, -ti,** importance: **~le,** efficiently; with interest. **~miyetli,** important. **~miyetsiz,** unimportant.

**ehil, -li** Family, household; friends; people; husband, wife. Possessor of, endowed with. **ehli idrak,** a man of intelligence: **işinin ehli adam,** a man who knows his job: **söz ehli,** eloquent. **ehli·beyt,** the Prophet's family. **~dil,** wise; wise men, sages; as **rint** *q.v.* **~hibre,** expert. **~keyf,** one who knows how to enjoy life properly, *bon viveur.* **E~salib,** the Crusaders. **~vukuf,** connoisseur, expert.

**ehlî** Tame; domesticated. **hayvanatı ~ye,** domestic animals.

**ehliyet, -ti** Capacity; competence. **~li,** capable; competent; qualified; skilled. **~name** (···⅄), certificate of competence.

**ehram**[1] (·⅄) The Pyramids; a pyramid.

**ehram**[2] *v.* **ihram.**

**Ehrimen** Ahriman; devil.

**ehven** Easiest; cheapest; lesser (evil). **~i şer,** the lesser of two evils.

**ejder, ejderha** Dragon; monster.

**ek, -ki** Joint; patch; scar; suffix. **~ini belli etmemek,** to dissimulate; not to give oneself away; to hide the

true state of affairs: **~ten pükten,** made of odd pieces put together.
**ekâbir** (·—ˋ) *pl. of* ekber. Great persons; important people.
**ekal** *comp. of* **kalil.** Lesser; least. **en ~,** the very least. **~liyet, -ti,** minority.
**ekalim** (·—ˋ) *pl. of* **iklim.** Regions; countries.
**ekber** *comp. of* **kebir.** Greater; greatest; elder, eldest. **Allahü ~,** Allah is the greatest.
**ekici** Sower; cultivator.
**ekili** Planted; sown.
**ekim** Sowing; October.
**ekin** Sowing; cultivation; crops. **~ biçmek,** to reap the harvest: **~ vakti,** seed-time. **~ci,** sower, cultivator.
**ekip, -pi** (*Fr. équipe*) Team; crew; gang.
**ekle·mek** *va.* Join *a piece* on to *stg.*: increase by adding a piece; join together; deal (a blow). **~nti,** something added on; suffix. **~tmek,** *caus. of* **eklemek.**
**ekler** (*Fr. éclair*) Zip-fastener.
**ekmek**[1] *va.* Sow; scatter; drop (and lose); deal (blows); give *s.o.* the slip.
**ekmek**[2] Bread; food; livelihood, profession. **~ çarpmak,** for ingratitude to be punished: **ekmeğe el basmak,** to swear upon bread (*as stg. sacred*): ⌜**~ elden su gölden**⌝, 'bread from a stranger, water from the lake', to live without working, to get one's living free: **···i ekmeğinden etmek,** to take the bread out of *a person's* mouth; to cause *s.o.* to lose his job: **~ kapısı,** the place where one works for one's living: **ekmeğinden* olm.,** to lose one's job: ⌜**ekmeğini taştan çıkarmak**⌝, to be able to get one's living out of anything: ⌜**ekmeğine yağ sürmek**⌝, *lit.* to butter his bread, *i.e.* to help *s.o.* inadvertently when one would rather not have helped, to 'play into *s.o.'s* hand': **eli* ~ tutmak,** to earn one's own living.
**ekmek·çi,** baker. **~çilik,** baking, the trade of a baker. **~içi, -ni,** the inside of a loaf; the crumb of bread. **~kabuğu, -nu,** the crust of bread. **~kadayıfı, -nı,** sweetmeat *of thin layers of bread soaked in syrup with clotted cream on top.* **~lik,** suitable for making bread; (*sl.*) a job by which one can live. **~ufağı, -nı,** crumb.
**ekran** (*Fr. écran*) Screen (of a cinema *etc.*).
**ekser**[1] Large nail.
**ekser**[2] (·ˋ) *comp. of* **kesir.** More *or* most numerous *or* frequent; most usual. The majority; the most part. **~î** (ˋ·—), for the most part. **~iya** (ˋ···), generally; more frequently; for the most part. **~iyet, -ti,** majority; greatest part; quorum.
**eksik** Deficient; lacking, absent; defective; incomplete. Deficiency; deficit. **~ etek,** woman, 'petticoats': **~ etmemek,** not to deprive *s.o.*; always to have in stock: **~ gedik,** small deficiencies: **'eksiği ne?',** 'what is there to complain of?': ⌜**~ olma!**⌝ *or* ⌜**Allah ~ etmesin!**⌝, 'may you never be wanting!' (*expresses gratitude*): **~ olmamak,** (of a person) always to turn up *at a party, a meeting etc.*: ⌜**~ olsun!**⌝, 'better without!', 'I would rather not have it!': **artık ~,** over or under, more or less: **on para ~,** ten paras short.
**eksik·lemek,** to render deficient *or* defective. **~lik,** deficiency; defectiveness. **~siz,** without defect; complete; perfect.
**eksilmek** *vn.* Grow less, decrease; be absent. **bunu yapmasa nesi eksilir?,** if he doesn't do it what would he miss?
**eksiltme** A putting up to tender; an asking for the most reduced terms for carrying out a contract. **···in inşaatı kapalı ~ye konmuştur,** the construction of ... has been put up to secret tender.
**eksiltmek** *va.* Diminish; reduce.
**eksiz** Without an addition; in one piece; seamless.
**eksper** Expert.
**ekşi** Sour; acid; fermented; sour-faced. Any sour substance; pickle; leaven. **~mek,** to become sour; ferment; become cross *or* disagreeable; (of the stomach) to be upset; to feel foolish *when proved to have been in the wrong*: to become stale *or* hackneyed. **başına* ~,** to be a burden to *s.o.*: **bu iş elinde ekşidi,** he has let this matter drag on, has neglected it. **~mik,** cheese made from skim milk. **~msi, ~mtrak,** sourish. **~suratlı,** sour-faced. **~tmek, ~lemek,** to render sour; cause to ferment; make *s.o.* look foolish *by exposing his error*: **çehreyi (yüzü)* ~,** to frown, to look cross.
**ekti** Tart; morose; miserly. **~ püktü,** hanger-on.

**ekûl, -lü** Gluttonous, voracious. Glutton.

**ekyeri** (ˈ..), **-ni** Line where one thing is joined to another.

**ekzos** Exhaust (of an engine).

**el**[1] Hand; forefoot; handle; handful; one discharge *of a firearm*; deal *at cards*. **~de,** in possession; in hand, in the course of being done: **~den,** by hand (*as of a letter*); cash payment; directly, without an intermediary, *e.g.* **bir muameleyi ~den takibetmek,** to follow up a matter personally instead of in an official manner: **~ açmak,** to beg *for alms*: **~e alınmaz,** bad in quality; unacceptable: **~ almak,** (of a novice in an order of mystics) to receive permission to initiate others: **~e almak,** to show (mercy, severity *etc.*): **~inin* altında,** in one's power: **~ altından,** in an underhand way, secretly: **~ atmak,** to lay hands upon; **bir işe ~ atmak,** to start, to take a matter up: **~ ayak çekildi,** everyone had retired; the streets were deserted: **~den ayaktan düşmüş,** crippled by illness *or* old age: **~i ayağı tutar,** sound of limb, sturdy: **~ bağlamak,** to join hands together *in a respectful attitude*: **~ bakmak,** to read the hand *in palmistry*: **···in ~ine bakmak,** to read *s.o.'s* hand; to depend on *s.o.* for one's living: **~de bir,** a sure thing, a 'dead cert.'; **o ~de bir,** that's all right; there is no doubt we'll get it: **~den bırakmak (çıkarmak),** to put down; to relinquish; to surrender *stg.*; **'insafı ~den bırakma!',** 'don't be unreasonable!': **~i boş,** empty-handed; without work; without means of subsistence: **~i çabuk,** dexterous: **···den ~ini (ayağını) çekmek,** to give up doing *stg.*; to withdraw from ...: **işten ~ çektirmek,** to suspend *s.o.* from a post: **~den çıkarmak,** to get rid of *stg.*: **~den çıkmak,** to be lost, to pass out of one's possession: **~imde* değil,** it is not within my power; I can't help *doing stg.*: **~den düşme,** secondhand, a bargain: **~ ~de baş başta,** in great confusion; very embarrassed: **~ ~e,** hand in hand: **~ ~e vermek,** to co-operate: **~de etm.,** to obtain, to get hold of: **birine ~ etm.,** to beckon to *s.o.*: **~den geçirmek,** to examine and clean *or* repair *stg.*: **~den geçmek,** to be examined; to be overhauled: **~e geçmek,** to come into one's possession: **ele geçmez,** not easily come by, not often met with: **~den gelmek,** (i) to be possible for one; (ii) to tip; to pay: **~inden* gelmek,** to be possible for one, to be within one's power: **⌜~den ne gelir?⌝,** 'what can one do?': **~lerde gezmek,** to be the fashion, to be the rage: **~ kaldırmak,** to show fight: **~de kalmak,** to be left over; to remain unsold: **~inde kalmak,** to remain undone; to be spoilt by delay: **~ koymak,** to commandeer, to requisition: **~inde* olm.,** to be within one's power; **~inde* olmamak,** to be beyond one's power; **~inde olmıyarak güldü,** he could not help laughing: **bir işte ~i* olm.,** to have a finger in a matter: **bir elden satılmak,** to be sold *en bloc or* at one go: **~inden tutmak,** to help: **~le tutulur,** tangible: **~ üstünde gezmek,** to be popular *or* beloved by the people: **~ üstünde tutmak,** to treat with honour: **~de var bir (beş),** ... and carry one (five) (*arith.*): **~e vermek,** to deliver up; betray, 'give away': **~ vurmak,** to clap the hands; to interfere, meddle with: **işe ~ vurmamak,** not to lift a finger to do *stg.*: **~de yapmak,** to do *stg.* easily *or* offhand.

**el**[2] One other than oneself; people outside one's own family; people *in general*; a people, tribe; the country of a people *or* tribe; stranger. **~ evi,** another's house: **~e (güne) karşı,** in the presence of others, before all.

**elâ** Light-brown (eyes). **açık ~,** greenish-grey: **gök ~,** bluish-grey.

**elâlem** All the world, everybody; strangers.

**elaltından** (ˈ...) Secretly; in an underhand way.

**el'aman** Enough! **···den ~ demek,** to have enough of, to be absolutely sick of.

**el'an, elân** Now, at present; still.

**elarabası** (ˈ....), **-nı** Hand-cart.

**elâstikî** Elastic. **~yet, -ti,** elasticity.

**elayak** (ˈ..) *In* **~ çekilmek,** to be deserted (of streets at night): **~ yürümek,** to walk on all fours.

**elbet** (ˈ. *or* .ˈ), **elbette** (.ˈ. *or* ˈ..) Most certainly; decidedly.

**elbezi** (ˈ..), **-ni** Napkin; towel.

**elbirliği** (ˈ...), **-ni** Agreement; co-operation.

**elbise** Clothes, garments.

**Elcezire** Mesopotamia.

**elçabukluğu** (ˈ....), **-nu** Sleight-of-hand; dexterity.

**elçi** Envoy; ambassador. **büyük ~,** Ambassador: **orta ~,** Minister.
**eldeğirmeni** (ˋ····), **-ni** Coffee-grinder.
**eldiven** Glove.
**elebaşı** (·ˋ··) Chief (of a bandit gang); ringleader.
**eleğimsağma** *For* **alâimi sema,** Rainbow.
**elek** Sieve. **~ten geçirmek,** to sieve; to examine minutely.
**elektrik** Electricity. **~li,** electric; live (wire).
**elele** (ˋ··) *In* **~ vermek,** to shake hands; to agree; to work together.
**elem** Pain; suffering; illness; sorrow. **~ çekmek,** to suffer pain *or* grief.
**eleman** Element; personnel.
**eleme** Sieved; sifted; selected. Act of sieving. **~ imtihanları,** preliminary examination.
**elemek** *va.* Sift; sieve; inspect *or* search carefully; select; wind *yarn* into hanks.
**elhak** Truly, really, indeed.
**elhamdülillah** (·ˋ···) Thank God!
**elhasıl** (ˋ–·) In short; in brief.
**elif** The letter 'A' *in the Arabic alphabet; (elif is a vertical straight line, hence it gives the idea of* thin and straight vertically). ⌜**~i görse mertek (direk) sanır**⌝, 'if he sees an elif he thinks it a post', *i.e.* quite illiterate: **işin ~ini dahi bilmiyor,** he doesn't know the rudiments of the business: **saat ~i ~ine dokuz,** it is exactly nine o'clock.
**elifbe, elifba** (··ˊ) Alphabet.
**elişi** (ˋ··), **-ni** Manual labour; handicraft. Hand-made.
**ellem** *In* **~ kömürü,** hand-picked, selected charcoal.
**elle·mek** *va.* Handle; feel with the hand; pick out by hand; take by the hand and turn out.
**elli**[1] Having hands; having a handle.
**elli**[2] Fifty. **~nci,** fiftieth. **~şer,** fifty at a time; fifty each.
**elma** Apple; round thing. **~ şarabı,** cider: ⌜**bir ~nın bir yarısı biri, bir yarısı biri**⌝, as like as two peas: ⌜**yarım ~ gönül alma**⌝, 'a trifling present wins the heart' (*said when offering a present or stg. to eat*).
**elma·baş,** pochard. **~cık,** small apple; hip-bone; high part of the cheek: **~ kemiği,** cheek-bone. **~kürk,** fur made of the cheek pieces of fox-skins. **~lık,** apple orchard.
**elmas** Diamond; precious; beloved. **~çı,** diamond-merchant. **~traş,** cut glass; diamond glass-cutter; diamond-cutter (man).
**elmasiye** (·–·ˋ) Fruit jelly.
**eloğlu** (ˋ··), **-nu** Stranger; outsider; other people.
**elpençe** *In* **~ divan durmak,** to stand in an attitude of respect, *with the hands clasped in front.*
**elti** *The relationship between the wives of two brothers*; sister-in-law.
**elulağı** (ˋ···), **-nı** Anything the hand can reach; a thing at hand; tool; servant; assistant.
**elvan** (·ˋ) *pl. of* **levin.** Colours. Of various colours.
**elveda** (··ˊ) Farewell!, good-bye!
**elveriş** Sufficiency; suitability. **~li,** sufficient; suitable, well-adapted; useful; convenient; profitable. **~siz,** unsuitable; inconvenient.
**elvermek** *vn.* Suffice; be suitable; be useful; be convenient; happen; be current. **elverir!,** that's enough!: **fırsat elvermemek,** not to have the opportunity: **keseye elvermemek,** to be beyond one's means.
**elyaf** (·ˋ) *pl. of* **lif.** Fibres.
**elyazısı** (ˋ···), **-nı** Handwriting; manuscript.
**elzem** *comp. of* **lâzım.** More *or* most necessary. **~iyet, -ti,** extreme urgency *or* necessity.
**emanet** (·–ˋ), **-ti** Anything entrusted to another; deposit; a government office receiving *or* paying out government money. **~e hiyanet,** breach of trust: **Allaha ~!,** farewell!; good-bye to ... . **~çi,** person with whom a thing has been deposited for safe keeping *or* deliverance; left luggage. **~en** (·ˊ··), for safe keeping; on deposit.
**emare** (·–ˋ) Sign; mark; token; circumstantial evidence.
**emaret** (·–ˋ), **-ti** Chieftainship; territory of an emir.
**emaye** (*Fr. émaillé*) Enamel ware.
**emcik** Nipple; teat.
**emece** By common effort; collectively.
**emek** Work; labour; trouble; fatigue. **emeği* geçmek,** to have contributed great efforts towards achieving *stg.*: **~ vermek (çekmek),** to labour, to take great pains: **el emeği,** manual labour; workmanship. **~çi,** worker; one who takes pains. **~dar,** old servant; veteran. **~daş,** fellow workman.

~**lemek,** to walk with difficulty; (of a baby trying to get about) to shuffle along. ~**li,** retired *officer etc.*; who has been long in service: pensioner: ~**ye ayırmak,** to pension off. ~**siz,** free from labour *or* fatigue; easy (life).

**emel** Longing; desire; ambition; thing wished for, ideal. ~ **etm.,** to long for; to aspire to. ~**siz,** unambitious.

**emi** *In such phrases as*: **kimseye söyleme** ~**!,** don't tell anyone, will you!; **unutma** ~**!,** now don't forget!

**emin** Safe, secure; sure, certain; trustworthy. Steward; custodian; superintendent; controller. **bölük** ~**i,** quartermaster: **sandık** ~**i,** chief cashier.

**emir, -mri** Order, command; matter, business; event, case. ~ **atlısı,** mounted orderly: ~ **eri,** orderly; batman: **emre hazır,** at hand for use; at disposal: ~ **kulu,** one who is under orders: ~ **subayı,** adjutant: **evvel** ~**de,** in the first place; first of all.

**emîr** (·–) Emir; chief; commander. ~**ülmu'minin,** Commander of the Faithful (*title of the Caliphs*).

**emirber** Orderly (*mil.*).

**emircik kuşu, -nu** Kingfisher.

**emirname** (··–·) Written command; decree; *polite form for* 'your letter'.

**emlâk, -ki** *pl. of* **mülk.** Lands; possessions. ~**i miriye,** Crown lands.

**emme** Act of sucking. ~ **tulumba,** suction pump.

**emmek** *va.* Suck. ⌜**anasından emdiği (süt) burnundan geldi**⌝, 'he paid for it'; 'he suffered for it': **kanını** ~, to suck the blood of, to exhaust *or* despoil *s.o.*

**emniyet, -ti** Security; safety; confidence. ···**e** ~ **etm.,** to trust; ···**i** ···**e** ~ **etm.,** to entrust *stg.* to *s.o.* ~ **müdürü,** Chief of Police: ~ **somunu,** lock-nut. ~**li,** safe; trustworthy, reliable. ~**siz,** untrustworthy; insecure; unsafe; distrustful. ~**sizlik,** untrustworthiness; lack of confidence.

**emoraji** Haemorrhage.

**emretmek** (·‿··), **-eder** *va.* Command, order.

**emri** *v.* **emir.** ~**hak, -kkı,** God's will, *euphemism for* 'death'. ~**vaki** (··–·), 'fait accompli'; accomplished fact. ~**yevmi,** order of the day.

**emsal** (·–), **-li** *pl. of* **misil** and **mesel.** Similars; likes; equals; proverbs, tales; coefficients; *used as sing.* a like; precedent. ~ **olmamak şartile,** provided it is not regarded as a precedent. ~**siz,** peerless; unequalled; unprecedented.

**emtaa, emtia** *pl. of* **meta.** Goods.

**emzik** Nipple; teat; baby's bottle; spout. ~**lemek,** to fit a spout *to a vessel.* ~**li,** having a teat *or* spout; with a child at breast (woman).

**emzirmek** *va.* Suckle.

**en**[1] Width; breadth. ~**ince,** according to its width; across its width: ~**ine boyuna,** widthwise and lengthwise; fully; tall, well-built (person): ~**i konu** (*usually one word*), at length, fully, thoroughly: ~**inde sonunda,** in the end, at last: ⌜**kimi** ~**ine çeker kimi boyuna**⌝, everyone has his own opinion.

**en**[2] Most; *forms the superlative.* ~ **evvel,** before everything; first of all.

**enayi** Credulous. Fool. ~ **dümbeleği,** a prize idiot.

**encam** (·–) End; result.

**encek, encik** Puppy, kitten, cub; shin.

**encümen** Council; committee.

**endam** (·–) Body; shape; figure; symmetry; stature. **arzı** ~ **etm.,** to make an appearance, to present oneself. ~**aynası, -nı,** full-length mirror. ~**lı,** well-proportioned, graceful.

**endaze** (·–·) Measure; proportion; *measure of about 26 in.*

**ender** *comp. of* **nadir.** More *or* most rare.

**endişe** (·–·) Thought; anxiety; doubt. ~**li,** thoughtful, anxious, troubled.

**endüstri** Industry.

**enfes** *comp. of* **nefis.** Choice, delightful, delicious.

**enfiye** Snuff. ~ **kutusu,** snuff-box.

**enfüsî** (··–) Subjective.

**engebe** Unevenness of ground; broken ground. ~**li,** steep and broken (ground).

**engel** Obstacle; difficulty; handicap; rival.

**engerek** Adder, viper. ~**otu, -nu,** Viper's bugloss.

**engin** Vast, boundless; the open sea; the high seas.

**enginar** Artichoke.

**enik** The young of a carnivorous animal; whelp, cub, puppy *etc.*

**enikonu** Fully; at length.

**enine** *v.* **en**[1].

**enis** (·–) Sociable. Companion.

**enişte** (·ˋ·) Husband of an aunt *or* a sister.
**enkaz** Ruins; debris; wreck, wreckage. **~cı,** ship-breaker; housebreaker.
**enli** Wide, broad.
**ense** Back of the neck, nape; back. **~sine* binmek,** to persecute, to tyrannize: **~sinden* gitmek,** to follow *s.o.* closely: **~si kalın,** well-off; care-free; influential: **~ kökü,** the lower part of the neck: **~ kökünden gelmek,** to be very close behind one, on top of one: ⌜**~sine vur ekmeğini al elinden**⌝, 'hit him on the neck and take the bread out of his hand', *used to describe a very mild person*: **~ yapmak,** to lead a lazy and comfortable life: **top ~,** hair kept full at the back.
**enselemek** *va.* Seize by the neck; collar.
**ensiz** Without width; narrow.
**entari** Loose robe. **gecelik ~,** nightgown.
**entipüften** Insignificant; ridiculous; futile; flimsy.
**entrika** (·ˋ·) Intrigue; trick. **~ çevirmek,** to resort to tricks, to intrigue.
**enva** (·ˉˋ), **-aı** *pl. of* **nevi.** Sorts, kinds.
**envanter** Inventory.
**epey** (ˋ·) Pretty good, pretty well; a good many; a good deal of. **~ce** (ˋ··), fairly; to some extent; pretty well.
**er**[1] Man; male; husband; private *soldier*: a manly man; a capable man. ⌜**~ oyunu birdir**⌝, once is enough; no need to try again: **~e varmak,** (of a woman) to marry: **~e vermek,** to give *a girl* in marriage: **~oğlu ~,** hero: **iş ~i,** a good man for work: **sözünün ~i,** a man of his word.
**er**[2] Early; soon. **~ geç,** sooner or later.
**erat, -tı** *pl. of* **er.** Private soldiers.
**erazil** *pl. of* **rezil.** Low, vile people.
**erbab** (·ˋ) Expert; *with izafet,* possessing ..., gifted with .... **~ı bilir,** (the connoisseur knows it), first-class: **işlerinin ~ı,** masters of their craft.
**erbain** Forty; *only used in Turkish to mean* the forty days of severe winter, *i.e.* 21 Dec. to 30 Jan.: **~e girmek,** to hibernate.
**erbaş** Non-commissioned officer, N.C.O.
**erce**[1] (ˋ·) In a manly way.
**erce**[2] (·ˋ) Somewhat early.
**erdirmek** *va. caus. of* **ermek.** Cause to reach *or* attain. **···e akıl ~,** to come to understand *stg.*
**Erem** *v.* **İrem.**
**eren** One who has arrived at the truth; wise and virtuous man; *the pl. is used in the same sense and as a mode of address among dervishes.*
**erfane** *v.* **arifane.**
**erganun** Organ (*mus.*).
**ergeç**[1], **erkeç** (·ˋ) He-goat.
**ergeç**[2] (ˋ·) Sooner or later.
**ergen** Youth of marriageable age; bachelor, celibate. **~lik,** the pimples of puberty.
**ergin** Mature; ripe; adult.
**erguvan** Judas-tree; purple colour. **~î** (··—ˋ), purple.
**erik** Plum.
**erimek** *vn.* Melt; fuse; pine away; (of a boil *etc.*) to subside, pass away; (of textiles) to wear out.
**erincik** Lazy; bashful, timid.
**erinmek** *vn.* Melt away; flag.
**erişmek** *vn.* Arrive; attain; mature; reach the age of marriage.
**erişte** (·ˋ·) Freshly made vermicelli.
**eritmek** *va.* Melt; dissolve; cause to waste away; squander.
**erkân** (·ˋ) *pl. of* **rükün.** Rules; recognized rules of procedure *or* behaviour; high officials. **usul ~,** the rules of good behaviour, 'savoir vivre'. **~iharbiye,** General Staff (*mil.*).
**erkeç, -çi** He-goat.
**erkek** Man; male. Manly; virile; courageous; honest and true. **~ canlısı,** a woman who is always running after men: **~ demir,** hard iron. **~lik,** masculinity, virility; manliness; courage. **~si,** masculine (woman). **~siz,** (of a woman) husbandless, without support.
**erken** Early. **~ce,** rather early. **~ci,** early-rising; who comes early.
**erlik** Masculinity; virility; bravery.
**ermek** *vn.* Reach; attain; arrive at maturity; reach religious perfection, become a saint. **···e aklı* ~,** to comprehend *or* grasp *stg.*: ⌜**ayağı* suya ~**⌝, for the feet to reach water, *i.e.* to know where one stands, to realize the position: ⌜**el ermez güc yetmez**⌝, an impossibility: **kemale ~,** to reach perfection *or* completion.
**Ermeni** Armenian. ⌜**~ gelini gibi kırıtmak**⌝, 'to be as coy as an Armenian bride', *i.e.* to hang back *or* be slow in doing *stg.* **~ce,** the Armenian language. **~stan,** Armenia.

**erte, ertesi** The following day *or* year *etc.*

**ervah** (·ˋ) *pl. of* **ruh.** Spirits; souls. **ham ~,** coarse-minded.

**erzak** (·ˋ), **-kı** *pl. of* **rızık.** Provisions; food.

**erze, erz** Cedar.

**erzel** *comp. of* **rezil.** More *or* most despicable.

**esafil** (·–ˋ) *pl. of* **sefil.** The rabble, the lower classes.

**esami** *pl. of* **isim.** Names. **~si okunmaz,** he is of no consequence.

**esans** (*Fr. essence*) Perfume.

**esaret** (·–ˋ), **-ti** Captivity; slavery.

**esas** (·ˋ) Foundation; base; principle; essence. Principal, basic. **~ında,** fundamentally: **~ı yok,** there is no foundation for it (of a rumour *etc.*). **~en** (·ˊ·), fundamentally; essentially; in principle; from the beginning; besides; anyhow. **~î,** fundamental; essential; principal: **kanunu ~,** constitution. **~lı,** based; founded; secure; true; sure; principal; fundamentally right. **~sız,** baseless; unfounded.

**esatir** (·–ˊ) *pl.* Legends; mythology.

**esbab** (·ˋ) *pl. of* **sebeb.** Causes; reasons; means; requisites; materials. **~ı mucibe,** motives; justification: **~ı muhaffife,** extenuating circumstances: **~ı mücbire,** 'force majeure'.

**esbak** Former; late; ex-.

**esef** Regret. **~ etm.,** to regret, to be sorry.

**esen** Blowing; hale, hearty, robust. **~lik,** health, soundness.

**eser** Sign; mark; trace; remains; monument; work (of art, literature *etc.*); effect; action. ... **~ olarak,** as the result of ...: **bir talih ~i,** a piece of good fortune: **buralarda sudan ~ yok,** there is no trace of water hereabouts. **~icedid** (····ˊ), **~ kâgıdı,** foolscap paper.

**esham** (·ˋ) Shares, *esp.* shares of companies.

**esinti** Breeze.

**esir** Captive; prisoner of war; slave; a slave *to drink etc.* **~ almaca,** the game of 'prisoners' base', **~ci,** slave-dealer.

**esirge·mek** *va.* Protect; spare; grudge. ⌜**Allah esirgesin!**⌝, 'may God protect us!', 'God forbid!': **bunu benden esirgeme!,** do not grudge me this!: **sözlerini esirgemedi,** he did not mince his words. **~yici,** who protects *or* spares; grudging; stingy.

**eski** Old; ancient; chronic; out of date; worn; secondhand. **~ler,** the ancients: **~den,** of old: ⌜**~ ağza yeni taam**⌝, *said when eating the first dish of fruit etc. of the year*: **~den beri,** from of old, for a long time past: **~si gibi,** as of old: **~ kafalı,** old-fashioned (person).

**eski·ci,** dealer in secondhand goods; old-clothes man; cobbler. **~mek,** to be worn out; to grow old in service. **~püskü,** old clothes; old and tattered things. **~tmek,** to wear out; to cause to grow old.

**eskrim** (*Fr. escrime*) Fencing.

**eslâf** (·ˋ) *pl. of* **selef,** Predecessors.

**esma** *pl. of* **isim.** Names; attributes of God. ⌜**~yı üstüne sıçratmak**⌝, to bring trouble upon oneself.

**esmek** *vn.* Blow (of the wind *etc.*); come by chance; (of unexpected good fortune) to befall; come into the mind. **esip savurmak,** to shout and bluster: **aklına* ~,** to occur to one: **akıllarına esti İstanbula gittiler,** they took it into their heads to go to Istanbul: **aklına eseni söylüyor,** he says whatever comes into his head.

**esmer** Dark complexioned; brunette.

**esna** Course; interval; time. ... **~sında,** in the course of ..., during ...; **o ~da,** at that time, meanwhile.

**esnaf** *pl. of* **sınıf.** Trades; guilds; *as sing.* tradesman; artisan; prostitute. **~ zihniyeti,** the mentality of a shopkeeper; being commercially minded: **ayak ~ı,** pedlar. **~ça,** mercenary; utilitarian.

**esnek** Elastic. **~lik,** elasticity.

**esne·mek** *vn.* Yawn; stretch and recover shape; bend; yield. **~tmek,** to cause to yawn; bore; to stretch; bend. **~yiş,** yawn; elasticity.

**esrar**[1] *pl. of* **sır.** Secrets; mysteries. **~ kutusu (küpü),** a man of mystery; mysterious person. **~engiz, ~lı,** mysterious.

**esrar**[2] Hashish. **~keş,** hashish addict.

**estağfurullah** (·ˋ···) 'I ask pardon of God', *phrase used in reply to a compliment or an expression of politeness such as* '**bendeniz**'; 'don't mention it!'

**estek** *Only in conjunction with* **köstek.** **~ köstek etm.,** *or* **~ etmek köstek etm.,** to make all sorts of excuses to get out of doing *stg.*

**esvab** *vulg.* **espap** Clothes, clothing. **~cı,** ready-made clothes merchant; second-hand clothes dealer: **~ başı,** Keeper of the Wardrobe. **~lık,** suitable for clothes.

**eş** One of a pair; a similar thing, a thing that matches another; mate; fellow; husband; wife. **~ dost,** one's friends and acquaintances: **~i görülmemiş,** unprecedented: **~ etm.,** to match: **~ olm.,** to be a match: **~ tutmak,** to choose a partner: **~i yok,** peerless.

**eşarp** (*Fr. écharpe*) Scarf; sash.

**eşas** (*Fr. échasse*) Stilt.

**eşcar** (·⅄) *pl. of* **şecer.** Trees.

**eşek** Donkey, ass. **~ arısı,** wasp, hornet, bumble-bee: ⌜**~ başı mıyım?**⌝, 'am I not worth listening to?' (*said as a reproach to those paying no attention to the speaker*): ⌜**~ başı mısın?**⌝, 'why don't you use your authority?': ⌜**eşeğin kuyruğu gibi ne uzar ne kısalır**⌝, 'like a donkey's tail he neither grows longer nor shorter', *said of people who make no progress (from incapacity or bad luck)*: ⌜**~ sudan gelinceye kadar dövmek**⌝, to thrash soundly: **~ şakası,** a coarse practical joke: **eşeğe ters bindirmek,** to pillory, to show *s.o.* up: ⌜**dilini ~ arısı soksun!**⌝, 'may a wasp sting his tongue!' (*used when talking of a foul-mouthed person*).
**eşek·çi,** donkey driver. **~lenmek,** to make an ass of oneself. **~lik,** silliness, asininity.

**eşelemek** *va.* Stir up; scratch about; hunt for; rummage. **bir meseleyi ~,** to rake up some matter; to seek out and inquire into it.

**eşhas** (·⅄) *pl. of* **şahıs.** Persons; characters *in a play.*

**eşik** Threshold; entrance to a palace; bridge (of a violin *etc.*). **eşiğini aşındırmak,** to frequent a place constantly.

**eşinmek** (Of animals) to scratch up the ground.

**eşit** Equal, equivalent; the same.

**eşk, -ki** Tears, weeping.

**eşkâl, -li** *pl. of* **şekil.** Forms; figures. '**adamın ~ini tarif et!**', 'describe the man's appearance!'

**eşkıya** *pl. of* **şakı.** Rebels; brigands; *also as sing.* brigand. **~lık,** rebellion; brigandage.

**eşkin** Canter. Cantering. **~ gitmek (yürümek),** to canter.

**eşkina, eşkine** A Mediterranean fish (*Corvina nigra*).

**eşme** Shallowly dug. Water-hole.

**eşmek** *va.* Dig lightly; scratch up the ground. *vn.* Hurry off *to war or other duty.*

**eşraf** (·⅄) *pl. of* **şerif.** Notables *of a town etc.*

**eşref** *comp. of* **şerif.** More *or* most noble *or* eminent. **~ saat,** the propitious moment.

**eşsiz** Matchless, peerless; without a mate.

**eşya** *pl. of* **şey.** Things; objects; furniture; luggage; belongings; goods. **~lı,** furnished. **~sız,** unfurnished.

**et, -ti** Meat; flesh; pulp of a fruit *etc.* **~ bağlamak,** (of a wound) to begin to heal up: **~i budu yerinde** *or* **~ine dolgun,** well-covered but not fat: ⌜**~i senin, kemiği benim**⌝, 'his flesh is yours, his bones mine' (*said by a parent handing his son over to the care of a schoolmaster and meaning stg. like* 'don't spare the rod!'): **~le tırnak gibi,** a very near relation: **~ tırnaktan ayırmak,** to separate *s.o.* from his closest relations: **başının* ~ini yemek,** to worry the life out of *s.o.*, to nag at him.

**etba, -aı** *pl. of* **tâbi.** Followers; attendants; servants.

**etek** Skirt *of a garment*; foot *of a mountain.* **eteği belinde,** industrious, active (woman): **~ dolusu,** in abundance: **eteğine* düşmek,** to fall at *s.o.'s* feet in entreaty; to turn to *s.o.* for help: **eteğine* sarılmak,** to entreat *s.o.*: **eteği temiz,** honest: **~leri* tutuşmak,** to be exceedingly alarmed: **~leri* zil çalmak,** to be frightfully pleased: **dünyadan elini* eteğini* çekmek,** to retire from active life; to go into seclusion: **eksik ~,** woman, 'petticoats': **eline eteğine doğru,** chaste.
**etek·lemek,** to kiss the skirt of *s.o. in respect or congratulation*; flatter; fan with the skirt. **~lik,** skirt; material for a skirt.

**etepetâ** (·⅄·–) Precocious (child); affectedly serious (adult).

**eter** Ether.

**Eti** Hittite.

**etibba** *pl. of* **tabib.** Physicians.

**etiket, -ti** (*Fr. étiquette*) Label, ticket; etiquette.

**etinedolgun** (··⅄··) Plump, well-covered.

**etkafalı** (ˈ...) Thick-headed.
**etkesimi** (ˈ...) Carnival before Lent.
**etki** Effect; influence.
**etli** With meat in it (pilaf *etc.*); fleshy, plump. ~ **budlu,** well-furnished with flesh, strong: ~**ye sütlüye karışmamak,** not to worry about other people's affairs, to mind one's own business.
**etmek** (**eder, ediyor, edecek** *etc.*) *va.* Do; make; be worth; fetch *a certain price.* **Etmek** *is the verb most commonly used to make a composite verb, chiefly with Arabic nouns, e.g.* **zannetmek,** to think; **sarfetmek,** to spend; *when the noun is of two syllables, the noun and the verb are usually written separately, e.g.* **hizmet etmek,** to serve; **telefon etmek,** to telephone. ⌜**eden bulur**⌝, one pays for what one does: **çocuk oynamadan edemez,** a child must play, cannot do without games: **birini bir şeyden** ~, to cause *s.o.* to lose *stg.*, to deprive *s.o.* of *stg.*
**etraf** (·ˈ) *pl. of* **taraf.** Sides; ends; directions; regions; surroundings; the world around; relatives; details; circumstances. ···**in** ~**ında,** around ...: ~**ta,** in the neighbourhood, around: ~**tan,** from all around, from all directions: ···**in** ~**ını almak,** to surround: ~**a haber vermek,** to give general notice. ~**lı,** detailed; long-winded. ~**lıca** (·ˈ··), fully, with all details; (···ˈ), fairly detailed; in rather a detailed manner.
**etsiz** Without meat; thin.
**etsuyu** (ˈ··), **-nu** Gravy; meat broth.
**ettirmek** *va. caus. of* **etmek.** Cause to do *etc.*
**etvar** (·ˈ) *pl. of* **tavır.** Modes; manners.
**etyaran** Deep-seated whitlow.
**ev** House, dwelling; home; household; family; compartment, pigeon-hole. ~ **açmak,** to set up house: ~ **bark,** house and family: ~ **bark sahibi,** a family man: ~ **ekmeği,** home-made bread: ~ **eşyası,** household effects: ~ **kadını,** housewife, a woman who looks after her home properly: ~ **halkı** (**takımı**), household (family and servants): ~ **sahibi,** master *or* mistress of a house; proprietor of a house, landlord: ⌜~**lere şenlik!**⌝, 'joy to houses!', *a phrase used when mentioning a death or disaster and meaning stg. like* 'may Heaven save you from such disaster!': ~ **yapıcı,** bringing domestic happiness: ~ **yıkmak,** to cause domestic infelicity, to break up a home.
**evc** Apogee; summit.
**evce, evcek** (ˈ·) All the family. With *or* by all the family.
**evci** Boarder at a school who spends the week-end at home.
**evcimen** Fond of one's home, domesticated; thrifty.
**evel** *and derivatives v.* **evvel** *etc.*
**evermek** *va.* Cause to marry; give in marriage.
**evet** (ˈ·) Yes. ~**efendimci,** yes-man. ~**lemek,** to say 'yes'; to keep saying 'yes, yes!'
**evham** (·ˈ) *pl. of* **vehim.** Apprehensions; illusions; hypochondria. ~ **getirmek,** to become a hypochondriac, to have a nervous breakdown. ~**lı,** full of false apprehensions; hypochondriac.
**evirmek** *va. Only in* ~ **çevirmek,** to turn *stg.* over and over; look about one in hesitation; explain in the wrong way; invert, turn inside out. **evire çevire dövmek,** to thrash soundly.
**evkaf** (·ˈ) *pl. of* **vakıf.** Pious foundations; estates in mortmain; the government department in control of these estates.
**evlâ** (·ˉ) Better; preferable; most suitable.
**evlâd** (·ˈ) *pl. of* **veled.** Children; descendants; *as sing.* child, son. ~ **acısı,** grief for the loss of a child: ~ **edinmek,** to become a parent; to adopt a child. ~**iyelik,** heirloom; a thing that will last for many years. ~**lık,** quality of a child; adopted child: **evlâdlığa kabul etm.,** to adopt.
**evlek** Furrow; the quarter of a **dönüm**; draining ditch in a field.
**evlenme** Marriage. ~**k,** to marry: ... **üstüne** ~, to marry in addition to ....
**evleviyet, -ti** Preference. ~**le,** all the more; so much the sooner.
**evli** Married; having houses. ~ **barklı,** married and having a family: **elli** ~ **köy,** a village of fifty houses.
**evliya** (··ˉ) Saint. ~ **otu,** sainfoin: ~**yı umur,** the heads of the administration. ~**lık,** saintliness; fitted to be a saint; innocent, ingenuous.
**evrak** (·ˈ), **-kı** *pl. of* **varak.** Leaves; documents, papers, archives. ~ ~, in sheets: ~**ı nakdiye,** paper-money: ~**ı resmiye,** official documents.
**evsaf** (·ˈ) *pl. of* **vasıf.** Qualities; qualifications.

**evvel, evel** (**evel** *and its derivatives can be spelt with one* **v** *or two*) First; former; foremost; initial. The first part; beginning. Before. First; ago; formerly. ~**den,** from former times; beforehand, previously: ~ **ve ahır,** formerly; on several occasions: ~ **zaman,** in olden times, long ago: **aklı** ~, very clever: **beş sene** ~, five years ago: **bir an** ~, as soon as possible: **bundan** ~, before this, previously: **en** ~, first of all, before anything else: **gitmezden (gitmeden)** ~, before going. **evvel·â** (ˈ··), firstly. ~**allah** (··ˈ–), (with the help of God), certainly, surely. ~**ce** (·ˈ·), a little time before; previously; a little way in front. ~**emirde** (·ˈ···), in the first place, first of all. ~**iyat, -tı,** origins; rudiments, first principles, preamble. ~**ki,** ~**si** (·ˈ·), first, former: ~ **gün,** the previous day; the day before yesterday: ~ **sene,** the previous year; the year before last.

**ey** O!; well!; hi!; eh!

**eyalet** (·–ˈ), **-ti** Province.

**eyer** Saddle.

**eyi, eyici** *v.* **iyi, iyici.**

**eylemek** *va.* Do; *used in compound verbs as* **etmek**; **neyleyim?,** what shall I do?; what am I to do!

**eylûl, -lü** September.

**eytam** (·ˈ) *pl. of* **yetim.**

**eyvah** Alas!

**eyvallah** (ˈ··) Yes; so be it!; thanks! ~**ı olmamak,** to be obliged to no one: **buna da** ~, God's will be done!: **kimseye** ~ **etmezdi,** he sought nobody's favour: **her şeye** ~ **demek,** to agree with everything that is said.

**eyyam** *pl. of* **yevm.** Days. ~**ı bahur,** the dog days: ~ **reisi,** fair-weather sailor; time-server: **bir** ~, at one time, formerly.

**ez** *Pers. prep.* From.

**eza** Vexation; torment.

**ezan** The call to prayer by a **muezzin.**

**ezber** By heart. ~**den,** (i) by heart; (ii) without knowing: ~ **etm.,** to learn by heart: ~**e gitmek,** to proceed blindly without knowing where one is going: ~**e konuşmak,** to talk about *stg.* without understanding it: ~ **okumak,** to recite from memory. ~**ci,** one who learns by heart easily *or* in a parrot-like way. ~**lemek,** to learn by heart, to con.

**ezcümle** (ˈ··) For instance; among other things.

**ezel** Eternity in the past. ~**den,** without hesitation. ~**î** (··ˈ), eternal (without beginning).

**ezgi** Tune, song. ⌜**ince** ~ **fıstıkı makam**⌝, lazily and slowly.

**ezgin** Crushed; trampled on; oppressed.

**ezher** *In* ~ **cihet,** in every respect.

**ezici** Crushing.

**ezik** Crushed; squashed. ~ **büzük,** (fruit) spoilt by crushing in transit.

**ezilmek** *vn. pass. of* **ezmek.** Be crushed, oppressed *etc.* **içi (yüreği)** ~, to feel a sinking sensation in the stomach *from hunger etc.*

**ezinti** Sensation of sinking in the stomach; faintness; breakdown.

**eziyet, -ti** Injury; ill-treatment; pain; vexation; torture. ~**li,** fatiguing; vexatious; painful.

**ezkaza** (ˈ··) By chance.

**ezme** Action of crushing *etc.*; something crushed; paste. **badem** ~**si,** almond paste: **patates** ~**si,** mashed potatoes.

**ezmek** *va.* Crush; pound; triturate; bruise; reduce to poverty *or* impotence. ⌜**ez de suyunu iç!**⌝, (i) 'take it back, I don't want it!' (*in returning a gift in anger or rejecting an offer*); (ii) 'you can write that off!', 'it's worthless!'

**Ezrail** (·–ˈ) The angel of death.

# F

**faal** (·ˈ) Active; industrious. ~**iyet** (·–·ˈ), **-ti,** activity; energy; industry.

**fabrika** (·ˈ·) Factory. ~ **işi,** machine-made goods.

**facia** (–·ˈ) Tragedy; drama; disaster. ~**lı,** tragic; terrible.

**faça** (ˈ·) A turning about; volte-face.

**façeta** (·ˈ·) *v.* **faseta.**

**façuna** (·ˈ·) The serving *or* whipping *of a rope.*

**fağfur** *Ancient title of the Emperor of China*: porcelain. *v,* **dokunmak.**

**fahiş** (–ˈ) Immoral; obscene; excessive. ~**e** (–·ˈ), harlot.

**fahrî** (·ˈ) Honorary.

**faide** (–·ˈ) *v.* **fayda.**

**faik** (–ˈ) Superior. ~**iyet** (–··ˈ), **-ti,** superiority.

**fail** (–ˈ) Who does or acts; author; agent; maker; subject *of a verb.* **ismi** ~, present participle active: noun of the agent. ~**iyet** (—··<sup>·</sup>), **-ti,** action; activity; efficiency; influence.

**faiz** (—<sup>·</sup>) Interest *on capital*. ~**i işlemek,** for *stg.* to bear interest: ~**e yatırmak,** to put out at interest: **yüzde on** ~, interest at ten per cent.

**fak** *Only in* ~**a basmak,** to make a false step, to be deceived.

**fakat** (<sup>·</sup>·) But; only; exclusively.

**fakfon** German silver.

**fakır, -krı** Poverty.

**fakîh, fakıh** Moslem jurist; learned.

**fakir** Poor. Pauper; 'your humble servant'. ~**ane** (··—<sup>·</sup>), pertaining to a poor man; poor; humble. ~**hane** (··—<sup>·</sup>) 'the poor man's house', *i.e.* my house. ~**lik,** poverty.

**fakr·üddem** Anaemia. ~**ühal,** ~ **kağıdı,** certificate showing that the holder is without means of subsistence.

**fal** Omen, augury; fortune. ~ **açıcı,** soothsayer, fortune-teller: ~ **açmak** *or* ~**a bakmak,** to tell a fortune: ~ **tutmak,** to take an omen, to draw a lot: **el** ~**ı,** fortune told from the hand: **falı hayır,** good omen; of good omen. ~**cı,** fortune-teller: **el** ~**lığı,** palmistry.

**falaka** (·<sup>·</sup>·) Piece of wood to which the feet of the victim of bastinado are tied; whippletree. ~**ya çekmek,** to subject *s.o.* to bastinado.

**falan** *v.* **filân.**

**falçeta** (·<sup>·</sup>·) Curved shoemaker's knife.

**falname** (·—<sup>·</sup>) Book of omens *or* oracles.

**falso** (<sup>·</sup>·) False note; error. ~ **vermek,** to make a slip; to fall into error.

**faltaşı** (<sup>·</sup>··), **-nı** A pebble *or* bean from which an omen is taken. **gözünü** ~ **gibi açmak,** to stare in amazement: **gözü** ~ **gibi açılmak,** to learn by bitter experience.

**falya** (<sup>·</sup>·) Touch-hole of a muzzle-loading gun. ~ **çivilemek (tıkamak),** to spike a gun.

**familya** (·<sup>·</sup>·) Family; wife.

**fanfan** *Popular expression for* talking French.

**fâni** Transitory; mortal. ~**lik,** a being transitory *or* perishable.

**fanilâ** (·<sup>·</sup>·) Flannel; flannel vest; blanket

**fantaziye** Ostentation, display; fancy goods.

**fantezi** Fancy goods; (*rarely*) fancy, imagination.

**fanus** (—<sup>·</sup>) Lantern; lamp-glass.

**far** (*Fr. phare*) Head-light *of a motor-car*.

**faraş** Dust-pan; dust-bin.

**faraz·a** (<sup>·</sup>··) Supposing that .... ~**î** (··<sup>·</sup>—), hypothetical. ~**iye,** *pl.* ~**iyat,** supposition, hypothesis.

**farbala** (<sup>·</sup>··) Furbelow; flounce; fringe.

**fare** (—<sup>·</sup>) Mouse; rat.

**farfara** Empty-headed braggart; windbag. ~**lık,** frivolity; idle brag.

**Faris** (—<sup>·</sup>) The province of Fars; Persia. ~**î** (—·<sup>·</sup>—), Persian.

**fariza** (·—<sup>·</sup>) Religious duty.

**fark, -kı** Difference, distinction; discrimination. ~ **gözetmek,** to discriminate, treat differently: ~**ında mısın?,** do you notice?; do you realize?: ~**ında olm.,** to be aware: ~**ına varmak,** to become aware, to perceive, realize, understand: ~ **vermek,** to give change (money). ~**etmek** (<sup>·</sup>··), **-eder,** to distinguish, perceive; differ; alter. ~**lı,** different; changed; better: ~ **farksız,** hardly distinguishable: ~ **tutmak,** to discriminate, to treat differently. ~**sız,** indistinguishable; without difference; equal. ~**sızlık,** resemblance; equality.

**farmason** Freemason. ~**luk,** freemasonry.

**Fars** Persia; ~**ça,** Persian language.

**fart, -tı** Excess; abundance; exaggeration. ~**ı nezaket,** over-politeness.

**farta furta** (<sup>·</sup>···) Brag; empty threats.

**farz** Religious precept; binding duty; hypothesis, supposition. Binding, obligatory. ~**ı ayın,** a duty which must be observed by all: ~**ı kifaye,** a religious duty, the observance of which by some will absolve the rest, *hence stg.*, the performance of which is not very important, *or* a duty which is sure to be fulfilled by some and therefore neglected by others: ~**ı muhal,** supposing the impossible; in the improbable event of.

**farzetmek** (<sup>·</sup>··), **-eder** *va.* Suppose. **farzedelim,** let us suppose.

**Fas** Morocco; Fez. ~**lı,** Moroccan, Moorish.

**fasafiso** (··<sup>·</sup>·) Nonsense.

**faseta** (·<sup>·</sup>·) Facet.

**fasıl, -slı** Chapter; section; decision; season; musical performance. ···**in hallü faslı,** the settlement *of a dispute* *etc.*: ~ **heyeti,** the orchestra (in Oriental music).

**fâsıl** (–ˈ) Separating, dividing. **hattı** ~, dividing line, boundary. ~**a,** separation; partition; interval; interruption: ~ **vermek,** to have an interval *or* a temporary stop; to interrupt: **bilâ** ~, uninterruptedly. ~**alı,** with breaks *or* interruptions: with partitions. ~**asız,** continuous, uninterrupted.

**fasır** *In such expressions as*: **çocuğun bütün vücudü** ~ ~ **kabardı,** swellings came out all over the child's body.

**fâsid** (–ˈ) Corrupt; vicious; perverse; false; mischievous. ~ **daire,** vicious circle.

**fasih** (·ˈ–) Correct and distinct (speech); eloquent.

**fasile** (·–ˈ) Botanical order; classification.

**fasletmek** (ˈ··), **–eder** *va.* Separate; divide; decide (a case); solve (a problem); malign, traduce.

**fassal** Backbiter.

**fasulya, fasulye** (·ˈ·) Bean. **ayşe kadın** ~**sı** *or* **taze** ~, French beans: **çalı** ~**sı,** runner-beans: ⌜**kendini** ~ **gibi nimetten saymak**⌝, (to think that one is a heaven-sent gift like beans), *i.e.* to think onself very important.

**faş** (–) ~**etmek, –eder,** to divulge; to betray (a secret).

**faşır faşır** *Imitates the noise of splashing water.* ~**tı,** a splashing noise.

**fatanet** (·–ˈ), **–ti** Intelligence.

**fâtır** (–ˈ) Creator.

**fâtih** Conqueror; *the title of Mahomet II, the conqueror of Constantinople; the quarter of Istanbul round the* **Fatih Camii,** *the Conqueror's Mosque.* ~**a** (–·ˈ), the opening chapter of the Koran; opening *or* beginning of an undertaking; decision: ···**e** ~ **demek,** to 'say good-bye' to *s.o. or stg.*, to give up as lost: ~ **okumak,** to recite the fatiha, to pray for the soul of *s.o.*; to give up as lost.

**fatura** (·ˈ·) Invoice; book of samples.

**fava** Broad beans mashed *and eaten cold with oil and lemon juice.*

**favl** Foul (in a game).

**favori** Whiskers.

**fayda, faide** Use; profit; advantage. **ne** ~**?**, what's the use?: ~ **yok,** it's no use; there's nothing to be done: ···**in** ~**sı dokunur,** it is useful *or* profitable. ~**cılık,** utilitarianism. ~**lanmak,** ···**den** ~, to derive a profit from; to make use of; to profit by. ~**lı,** useful; profitable; advantageous. ~**sız,** useless; in vain; unprofitable.

**fayrap, faryap** (ˈ·) (Fire up!), *order to a ship's engineer to stoke up for maximum speed.* ~ **etm.,** to stoke up; to get to work quickly on a thing.

**faz** Phase (*elect.*).

**fazıl, –zlı** Superiority; merit. **erbabı** ~, men of culture. **fâzıl** (–ˈ), virtuous; munificent.

**faziha** (·–ˈ) Shameful act; crime.

**fazilet** (·–ˈ), **–ti** Merit; excellence; superiority. ~**kâr,** ~**li,** virtuous, excellent.

**fazla** Remainder; balance; excess. Excessive; superfluous. Beside; more (than). Too much; very much. ~**sile,** abundantly: ⌜~ **mal göz çıkarmaz**⌝, you can't have too much of a good thing: ~ **olarak,** moreover, furthermore: ⌜**her şeyin** ~**sı** ~⌝, 'enough is as good as a feast'. ~**laşmak,** *vn.* to increase.

**febiha** (ˈ·–) Well and good!; so much the better!

**fecaat** (·–ˈ), **–ti** Calamity.

**fecayi** (·–ˈ), **–ii** *pl. of* **facia.** Disasters; tragedies.

**feci** (·ˈ–) Painful; tragic. ~**a** (·–ˈ), grievous calamity, tragedy.

**fecir, –cri** Dawn. **Fecri âti,** *name given to a school of writers about the beginning of the 20th century*: **fecri kâzib,** false dawn: **şimal** ~**i,** Aurora Borealis.

**feda** (·–) Ransom; sacrifice. ~ **etm.,** to sacrifice: ~ **olm.,** to be sacrificed: ~ **olsun,** I will gladly make this sacrifice. ~**i** (·–ˈ–), one who sacrifices his life for a cause; patriot; revolutionary. ~**ilik** (·–·ˈ), self-sacrifice; devotion. ~**kâr** (·–ˈ), self-sacrificing; self-denying; devoted. ~**kârlık,** self-sacrifice; abnegation; devotion.

**fehmetmek** (ˈ··) *va.* Understand.

**fehva** (·ˈ–) Tenor, import. ~**sınca,** as the saying goes; with the meaning of.

**felâh** (·ˈ) Prosperity; deliverance; security. ~ **bulmaz,** hopeless (drunkard *etc.*).

**felâket** (·–ˈ), **–ti** Disaster, catastrophe. ~**li,** disastrous. ~**zede,** victim of a disaster; involved in a calamity.

**felc** Paralysis. ~**e uğramak,** to be paralysed (of a public service *etc.*).

**felek** The firmament; the heavens; fortune, destiny. ~**ten bir gün aşırmak (çalmak),** to pass a very enjoyable day: ⌜~ **bunu da çok gördü**⌝, fate has denied me this: **feleğin çemberinden geçmiş,** who has suffered the ups and downs of fortune; who has seen life: **~ten kâm almak,** to have a very good time: ⌜~ **kimine kavun yedirir, kimine kelek**⌝, Fate deals kindly with some, unkindly with others: **feleğin sillesini yemiş,** who has suffered the buffetings of fate.

**Felemenk** (·˙·) Holland. **~li,** Dutch; Dutchman. **~taşı** (··˙··), **-nı,** diamond.

**felfelek** Betel-nut.

**fellâh** Agriculturalist; Egyptian peasant; Egyptian; negro.

**fellek, fellik** ~ **~,** running confusedly in all directions. ~ ~ **aramak,** to search high and low.

**felsef·e** Philosophy. **~î** (··–), philosophical.

**fen, -nni** Science; art; sort, variety; branch of science; ruse.

**fenâ** (·–) Extinction; dissolution; death. ~ **bulmak,** to come to an end, to die.

**fena** (·–) Bad; ill; unpleasant. Bad thing. **işin ~sı bu ki,** the worst of it is that: **~ya çekmek,** to take *stg.* in a bad sense: ~ **etm.,** to do evil; to do *stg.* badly: ~ **olm.,** to feel ill; to feel like fainting: **~ya sarmak,** to take a turn for the worst: **~ya varmak,** to go from bad to worse: **bu iş ~ma gitti,** this business has exasperated me. **~laşmak,** to become worse, deteriorate, be aggravated; go bad. **~lık,** evil; bad action; injury: **birine** ~ **gelmek,** for *s.o.* to feel ill.

**fend** (*corr. of* **fen**) Trick, ruse. ⌜**kadının ~i erkeği yendi**⌝, 'woman's wiles are too much for a man'.

**fener** Lantern; street-lamp; lighthouse; pinion *of a shaft.* ~ **alayı,** torchlight procession: ~ **çekmek,** to light the way with a lantern: ~ **dubası,** lightship: ~ **gövdesi,** headstock *of a lathe*: ~ **kasnak,** cone-pulley: ~ **mili,** spindle *of a lathe,* mandrel: ⌜**~i nerde söndürdün?**⌝, 'where did you put your lantern out?', *said jokingly to one who arrives late.* **gündüz ~i** (*sl.*), negro. **~balığı, -nı,** angler fish. **~li,** having a lantern; member of the **Fenerbahçe** Sports club; Phanariot (member of the old Greek aristocracy).

**Fenike** (·˙·) Phoenicia. **~li,** Phoenician.

**fenn·en** (˙·) Scientifically; technically. **~enmek,** to be experienced (in evil). **~î** (·–), scientific; technical.

**fer** Pomp; display; splendour, radiance; lustre; ornament.

**ferace** (·–˙) *A kind of* overall *formerly worn by Turkish women when they went out*; cloak *worn by ulema.* **~lik,** material suitable for such garments.

**ferağ** (·˙) Renunciation; cession (of property *etc.*); leisure; tranquillity. ~ **etm.,** to cede, transfer, give up. **~at** (·–˙), **-ti,** abandonment (of a project, of work *etc.*); renunciation; self-sacrifice; abnegation; a being free from care *or* work: ~ **etm.,** to abdicate, renounce, give up; be at ease. **~atkâr,** self-sacrificing.

**ferah** Spacious; open; roomy; cheerful. Easily, with room to spare. Cheerfulness, joy, pleasure. ~ **~,** amply: **içi*** **~,** cheerful, in a good humour. **~lanmak,** to become spacious *or* airy; become cheerful; enjoy oneself. **~lik,** spaciousness, airiness; cheerfulness, enjoyment; distraction; relief. **~nâk,** cheerful, gay; an air in Oriental music.

**feraset, -ti** Sagacity, intuition. **~li,** sagacious, perspicacious.

**ferd** Person, individual; odd number. Single; unique, peerless; odd (number). **~en** (˙·), individually, one by one. **~î,** *a.,* individual. **~iyet, -ti,** individuality. **~iyetçi,** individualist. **~iyetçilik,** individualism.

**ferda** (·–) The morrow, the next day; the future. **~sı,** the next day.

**feri, -r'i** Branch, ramification; secondary matter. **fer'î** (·–), derived; secondary.

**feribot** (˙··), **-tu** Ferry-boat; train ferry.

**ferih** (·–) Cheerful, happy. ~ **fahur yaşamak (geçinmek),** to live in great comfort, to be very well off.

**ferik**[1] (·–) Divisional General.

**ferik**[2] Chick of any game bird; toasted grains of wheat; any small thing. ~ **elması,** a kind of small apple.

**ferma** (˙·) *In* ~ **etm.,** *or* **durmak,** (of a sporting dog) to point *or* set.

**ferman** Command; firman; decree; imperial edict. ~ **sizin,** with pleasure; as you will: ⌜**tozdan dumandan** ~ **okunmuyor**⌝, the situation is so confused that one can do nothing.

**fermejüp** (*Fr.*) Snap-fastener.

**fermene** (ˈ··) Short waistcoat ornamented with braid.

**fersah** League; an hour's journey. ~ ~, greatly (*cf. Eng.* 'miles' better): ···i ~ ~ **geçmek,** to be miles ahead of ...: ~**larca uzaktan,** miles away.

**fersiz** Without radiance; dull. ~ **gözler,** lack-lustre eyes.

**fersude** (·–ˈ) Worn out, old.

**fertik** *In* **fertiği çekmek,** to sneak away. ~**lemek,** to make off.

**feryad** (·ˈ) A cry; wail; cry for help. ~ **etm.,** to cry out; to call for help.

**ferz** The queen *at chess.* ~ **çıkarmak,** to give the queen (as handicap): ~ **çıkmak,** (of a pawn) to queen.

**fes** Fez. ~ **tarağı,** fuller's teazle.

**fesad** (·ˈ) Depravity; corruption; duplicity; malice; intrigue; sedition; disorder. ~ı **ahlâk,** bad morals, demoralization of character: ~ **kurmak,** to plot mischief: ~ **kutusu,** mischief-maker, conspirator: **mide** ~ı, disorder of stomach. ~**cı,** mischief-maker; engaged in subversive activity; conspirator. ~**cılık,** subversive activity, intrigue, conspiracy.

**fesh, -shi** Abolition; cancellation; dissolution. ~**etmek** (ˈ··), **-eder,** to abolish, annul; abrogate; dissolve (parliament *etc,*).

**fesleğen** Sweet-basil; *v.* **don.**

**festekiz** *Only in* **filân** ~, this, that and the other (was said).

**fesübhanallah** (ˈ·–··) 'Then praise to God!'; good God alive!

**feşmekân** *As* **festekiz.**

**fetanet** (·–ˈ), **-ti** Intelligence.

**fethetmek** (ˈ··), **-eder** *va.* Conquer.

**fethimeyyit, -ti** Autopsy.

**fetih, -thi** Conquest. ~**name** (··–ˈ), Bulletin announcing a victory; poem celebrating a victory.

**fettan** Seducer. Alluring; cunning.

**fetva** Opinion *or* decision on a matter of Canon Law *given out by a* **mufti.** ~**hane** (··–ˈ), *formerly the official residence of the Sheikh-ul-islam.*

**fevaid** (·–ˈ) *pl. of* **fayda.** Uses; advantages.

**feveran** A boiling; effervescence; eruption. ~ **etm.,** (of a volcano) to erupt; to boil over with anger.

**fevk, -kı** Top; upper part. ~**alâde** (ˈ···), extraordinary: ~**den terfi etmiş,** he got special promotion. ~**alâdelik,** singularity; a being extraordinary.

**fevr** Haste, hurry. ~**î** (·ˈ), sudden; speedy; impulsive.

**fevt, -ti** Irreparable loss; death. ~**etmek** (ˈ··), **-eder,** to lose *an opportunity etc.*

**feyezan** Overflowing; flood; abundance.

**feylesof** Philosopher.

**feyz** Abundance; munificence; progress. ~ **almak (bulmak),** to make progress, to be successful: ···**den** ~ **almak,** to profit by ...: ~**ü bereket,** abundance *of harvest etc.*: **Allah** ~**ini artırsın!**, I wish you all success. ~**li,** abundant; prosperous; bountiful.

**feza** (·ˈ) Vast empty space; space (firmament).

**fezail** (·–ˈ) *pl. of* **fazilet.** Qualities *etc.*

**fezleke** Résumé; précis; substance of a police report.

**fıçı** Cask, barrel; tub. ~ **balığı,** salted fish in barrels: ~ **dibi** (*sl.*), low pub: ~ **gibi,** very fat: ~ **üzümü,** grapes brought to the market in barrels. ~**cı,** cooper.

**fıkara, fukara** *pl. of* **fakir.** Poor; the poor; pauper. ~**lık,** poverty.

**fıkdan** (·ˈ) Absence (of *stg.*); need; privation.

**fıkıh, -khı** Moslem jurisprudence.

**fıkır** ~ ~, *imitates the sound of boiling water*; coquettishly. ~**dak,** coquetry; flirtatious. ~**damak,** to make a bubbling noise; skip about; sparkle; flirt. ~**tı,** a bubbling noise.

**fıkra** Vertebra; sentence; paragraph; article *in a paper*; anecdote.

**fıldır fıldır** Skipping about; rolling the eyes. ~ ~ **aramak,** to hunt around feverishly for *stg.*

**fındık** Hazel-nut. ~ **kurdu,** nut maggot; tiny (woman). ~**çı,** seller of nuts; girl who uses her charms to get *stg.* out of a man. ~**î,** nut-brown. ~**kıran,** nutcrackers. ~**sıçanı, -nı,** common house-mouse.

**fır** A whirr; a circuit. ~ ~ **dolaşmak,** to go whirring round: **birinin etrafında** ~ **dönmek,** to hover round *s.o.*

**Fırat, -tı** (ˈ·) The Euphrates.

**fırça** (ˈ·) Brush.

**fırdolayı** (ˈ···) All around.

**fırdöndü** (ˈ··) Swivel; lathe carrier.

**fırfır** *v.* **firfir.**

**fırıl** *In* ~ ~ **aramak,** to search high and low. ~**dak,** weather-cock, vane; spinning-top; ventilator; whirligig; deception, ruse: ~ **çevirmek,** to be up to some mischief; to intrigue.

~dakçı, trickster, cunning fellow. ~danmak, to spin round; to move around hurriedly and anxiously.

**fırın** Large oven. ⌈**bunu yapman için kırk ~ ekmek yemen lâzım**⌉, it will be a long long time before you are capable of doing that: **yüksek ~**, blast-furnace. **~cı**, one who looks after a furnace; baker. **~kapağı, -nı**, thick-skinned; unruffled.

**fırışka** (·ʹ·) Light breeze; cat's-paw.

**fırka** Group; party; military division; squadron of a fleet.

**fırlak** Protruding.

**fırla·ma** Act of flying off; bastard. **~mak**, to fly off into space; fly out; leap up; rush; (of prices) to soar. **~tmak**, to hurl. **~yış**, a leaping up; rush; upward rush *of prices*.

**fırsat, -tı** Opportunity; chance. **~ düşkünü**, opportunist: **~ı kaçırmak**, to miss an opportunity: **~ yoksulu**, one who would do evil if he had a chance; waiting for an opportunity. **~çı**, who seeks an opportunity; on the look out for a profitable chance.

**fırt** *Imitates the sound of a bird's flight.* **~ ~ girip çıkmak**, to be continuously going in and out.

**fırtına** (·ʹ·) Gale; storm. **~lı**, stormy.

**fıs, fısfıs** Whisper. In a whisper. **~ıldamak, ~lamak**, to whisper. **~ıltı**, whisper.

**fısk, -kı** Sin; immorality. **~u fücur**, debauchery.

**fıskıye** Jet of water; fountain.

**fıslamak** *va. & vn.* Whisper.

**fıstık** Pistachio nut; ground-nut; peanut; pine-kernel; (*more specifically* **şamfıstığı** *or* **antepfıstığı** = pistachio nut; **yerfıstığı** = ground-nut; peanut; **çamfıstığı** = pine-kernel). **~ çamı**, stone pine: **~ gibi**, plump and healthy. **~i**, pistachio green, light green: ⌈**her boyadan boyadı, ~si kaldı**⌉, 'he's tried every paint, there only remains pistachio green', *sarcastic remark when there is a suggestion of doing stg. absurd and out-of-place*: ⌈**ince ezgi ~ makam**⌉, (*musical terms*), in slow time; 'taking his time about it' (*rather sarcastically*).

**fış fış** *Imitates a hissing or rustling sound.*

**fışfış** *In* **hacı ~**, *a nickname for Arabs.*

**fışıl ~ ~**, *imitates a splashing sound.* **~damak**, to make a splashing sound; rustle. **~tı**, a splashing *or* rustling sound.

**fışır ~ ~**, *imitates a gurgling or rustling sound.* **~damak**, to gurgle; to rustle. **~tı**, a gurgling *or* rustling noise.

**fışkı** Horse dung; manure.

**fışkın** Sucker; shoot.

**fışkır·ık** Squirt; syringe. **~mak**, to gush out; to spurt out; (of a plant) to spring up.

**fıta**[1] (ʹ·) Racing skiff.

**fıta**[2] (ʹ·), **futa** Apron; loin-cloth.

**fıtık, -tkı** Hernia, rupture.

**fıtır, -trı** The breaking of a religious fast. **idi ~**, Moslem festival at the end of the Ramazan fast.

**fıtnat, -tı** Intelligence.

**fıtr·at, -tı** Creation; nature; natural character. **~aten** (ʹ··), by nature, naturally. **~î** (·ʹ), natural; innate.

**fiat** (–ʹ), **-ti** Price; value. **maktu ~**, fixed price.

**fidan** Plant; sapling; shoot. **~ gibi**, straight, well set-up (boy *etc.*). **~lık**, nursery-garden; newly planted vineyard.

**fide** (ʹ·) Seedling plant *for planting out.* **~lemek**, to plant out *seedlings*. **~lik**, nursery-bed.

**fidye, ~inecat, -tı** Ransom.

**figan** (·ʹ) Cry of distress, wail.

**fihris, fihrist, -ti** Index; catalogue; list.

**fiil** Act, action, deed; verb. **~e gelmek**, to become a fact; to be done: **~e getirmek**, to execute, to carry into effect: **kuvveden ~e getirmek**, to turn a potentiality into a fact, to carry out a project; realize *an ambition etc.* **~en** (ʹ·), actually, really. **~î** (·ʹ), actual, real; de facto (*as opp. to* **hukukî**, *de jure*). **~iyat, -tı**, deeds.

**fikir, -kri** Thought, idea; mind; memory. **fikrimce**, in my opinion: **fikrine* koymak**, to decide *to do stg.*: **birinin fikrine koymak**, to put into the mind of another, to suggest to him: **fikri sabit**, fixed idea: **fikre varmak**, to ponder, to give oneself up to reflection: **aklı* fikri* birbirine karışmak**, to be bewildered.

**fikir·li**, having ideas; intelligent; thoughtful: **cin ~**, very shrewd. **~siz**, thoughtless; unintelligent.

**fikr·en** (ʹ·) Intellectually. **~etmek, -eder**, to think of; ponder. **~i**, *v.* **fikir**. **~î**, intellectual; mental.

**fil** Elephant; bishop *at chess.* ⌈**deveden büyük ~ var**⌉, there is a still greater one to be found.

**filâma** *v.* **flâma.**
**filân** So and so; such and such. ~ **festekiz (feşmekân),** *he said* this, that and the other: ~ **fıstık,** this and that; and so forth: **15 lira** ~, somewhere about 15 lira. ~**ca,** so and so; *s.o.* or other. ~**ıncı,** numbered so-and-so; the 'so manyeth'.
**filândıra** *v.* **flândıra.**
**filbahar, filbahri** Mock orange, philadelphus.
**filcan** *v.* **fincan.**
**fildekos** (*Fr. fil d'écosse*) Lisle thread; garments made of the same; vest.
**fildişi** (ˋ··), **-ni** Ivory.
**file** Net; netting.
**filen** *v.* **fiilen.**
**fileto** (·ˋ·) Fillet *of beef etc.*
**filhakika** (ˋ···) In truth, truly.
**filî** *v.* **fiilî.**
**Filibe** (·ˋ·) Philippopolis; Plovdiv.
**filibit, -ti** Phlebitis.
**filigran** Watermark in *paper.*
**filika** (·ˋ·) Ship's boat. ~ **demiri,** grapnel.
**filim, -lmi** Film (cinema). **filmini almak,** (i) to film; (ii) to X-ray: **filme çekmek,** to film.
**filinta** (·ˋ·) Carbine.
**filispit** ('Full-speed'); (*sl.*) dead drunk.
**Filistin** (·ˋ·) Palestine.
**filiyat** *v.* **fiiliyat.**
**filiz** Tendril; bud; young shoot; cutting. ~**i,** bright green. ~**kıran,** cold East wind, *which blows in May.* ~**lemek,** to prune. ~**lenmek,** to send forth shoots, to sprout.
**filo** (ˋ·) Fleet; squadron *of ships.* ~**tillâ** (··ˋ·), flotilla.
**filoş**[1] Floss silk. ~**otu,** a kind of reed *used for wickerwork.*
**filoş**[2] Flush (at cards).
**filozof** Philosopher.
**filvaki** (ˋ··) In fact, actually.
**fincan** Cup; porcelain insulator. ~**cı,** seller of cups: ⌜~**cı katırlarını ürkütmek**⌝, to bring unnecessary trouble upon oneself.
**finfon** *v.* **fanfan.**
**fing, fink** *Only in* ~ **atmak,** to saunter about and enjoy oneself; to flirt around.
**fingir** *In* ~ ~, with a swaying motion; coquettishly. ~**dek,** frivolous; coquettish. ~**demek,** to behave frivolously and coquettishly.
**fino** (ˋ·) Pet dog; lap-dog.

**firak** (·ˋ), **-kı** Separation; sorrow at separation. ~**lı,** sad, melancholy.
**firar** (·ˋ) Flight; desertion. ~ **etm.,** to fly, escape, desert. ~**i** (·–ˋ), deserter.
**Firavun** Pharaoh; a proud and obstinate man. ~ **faresi,** ichneumon: ~ **inciri,** Sycamore fig.
**fire** (ˋ·) Loss of weight by evaporation; inevitable loss *or* wastage. ~ **vermek,** to suffer such wastage.
**firfir** Purple; blossom of the Judas tree. ~**î,** purple.
**firik** Wheat grains plucked green and dried.
**firkat, -ti** Separation; absence; nostalgia.
**firkateyn** Frigate.
**firkete** (·ˋ·) Hairpin.
**firma** (ˋ·) Commercial firm.
**firuz, firuze** (·–ˋ) Turquoise.
**Fisagor** Pythagoras.
**fisebilillah** (ˋ····) (In God's way); gratis; for nothing.
**fisk** *v.* **fısk.**
**fiske** Flick; flip with the fingers; bruise made by a flip; small blister. ~ **dokundurmamak,** to protect *s.o.* from the slightest aggression: ~ ~ **kabarmak (olmak),** to be covered with small bruises: ~ **vurmak,** to give a flip.
**fisket, -ti** Boatswain's pipe. ~ **çalmak,** to pipe (on board ship).
**fiskos** A whispering with evil intent; insinuation. ~ **etm.,** to whisper insinuations; to plot under one's breath
**fislemek** *v.* **fıslamak.**
**fistan** Kilt; skirt, petticoat.
**fiş** (*Fr. fiche*) Counter (for games); slips of paper *or* cards *for an index etc.*; peg; plug (electric). ~**lemek,** to make a card index.
**fişek** Cartridge; rocket; roll of coins; (*sl.*) sexual intercourse. **deli** ~, madcap: **kestane fişeği,** squib. ~**hane** (··–ˋ), cartridge factory. ~**lik,** cartridge-belt; bandolier; ammunition pouch.
**fit**[1], **-ti** Instigation, incitement; an equivalent gain *or* loss at a game of chance. ~ **olm.,** to be quits; to come to an agreement about price; to consent: ~ **vermek,** to put a mischievous idea into another's head, to spread discontent: **dünden** ~, only too glad; only too ready. ~**çi,** mischief-maker.
**fit**[2] *In* ~ **tulumbası,** feed-pump.
**fitil** Wick; fuse; seton; piping (of

dress); (*coll.*) incitement. Drunk. **~i* almak,** to get into a rage, to flare up; to become alarmed: **~ gibi,** as drunk as a lord: **~ vermek,** to work *s.o.* up to great excitement; to incite *s.o.* against another: **burnundan* ~ ~ gelmek,** utterly to spoil one's pleasure. **~lemek,** to light the fuse of a mine, a bomb *etc.*; to enrage; to incite one against another.

**fitle·mek** *va.* Denounce; incite, instigate; set one person against another. **~yici,** inciting, instigating.

**fitne** Instigation; mischief-making. Mischief-maker, instigator. **~ fücur,** a dangerous mischief-maker: **~ vermek,** to make a mischievous suggestion. **~ci,** trouble-maker; seditious person. **~lemek,** to instigate trouble; to make a mischievous remark.

**fitre** Alms given at the close of the Ramazan fast.

**fitret, -ti** Interval between two successive events; interregnum; languor.

**fiyaka** (*sl.*) Showing-off; ostentation. **~sı bozulmak,** to look sheepish when one is shown up: **~ yapmak (satmak),** to show off.

**fiyat** *v.* **fiat.**

**fiyonga, fiyongo** Bow-tie; bow knot.

**fizik** Physics; temperament; constitution; physique. Physical (*also* **fisikî**).

**flâma** (ʾ·) Pennant; surveyor's pole.

**flândıra** (ʾ·) Ship's pennant; red band-fish.

**flâvta** (ʾ·) Flute.

**floka** (ʾ·) Felucca.

**floş** Floss silk; flush (at cards).

**flört, -tü** Flirt.

**flûrya** (ʾ·) Greenfinch; a yellow scented flower.

**flusahmer** *v.* **fülusiahmer.**

**fodla** (ʾ·) A kind of flat round loaf *formerly distributed by theological colleges to the poor.* **~cı,** one who takes a job merely to get food *etc.*

**fodra** (ʾ·) Lining *or* padding *of a coat.*

**fodul** Vain; presumptuous; egotistical. ⌜**hem kel hem ~**⌝, bald but vain, *i.e.* proud though with nothing to be proud of; in the wrong but proud of it.

**fokur ~ ~,** *imitates the sound of boiling water.* **~damak,** to bubble noisily, to boil up.

**fol** Nest-egg. ⌜**ortada ~ yok yumurta ~**⌝, nothing on which to base a claim *or* to prove anything. **~luk,** sitting-box (for hen).

**fondo, fonda** (ʾ·) Let go the anchor!

**font** Cast-iron.

**fora** (ʾ·) Out!; open!; unfurl (sails)! Open; opened up. **~ etm.,** to take down, open up (engine *etc.*); draw (a weapon); unfurl (flag); set (sails): **~ kürek!,** ship oars!

**forma** (ʾ·) Compositor's forme; folio; 16 pages of a book; number, part (publication); school uniform; colours *of a sporting club.* **~ ~ çıkmak** (of a publication) to come out in parts.

**fors**[1] Personal flag *flown on a ship*; national *or* personal emblem.

**fors**[2] (*Fr. force*) Power; prestige.

**forsa** (*Fr. forçat*) Galley-slave; convict.

**fos** False; bad.

**fosfor** Phosphorus. **~lu,** phosphorous, phosphoric.

**foslamak** *vn.* (*sl.*) Fail; be disconcerted.

**fosur fosur ~ ~,** *imitates the noise of smoking a narghile etc.* **~datmak,** to smoke noisily and with enjoyment.

**fotin** Boot. **~bağı, -nı** Bootlace.

**fotoğraf** Photograph; camera alaminüt ~, 'while-you-wait' photo.

**foya** (ʾ·) Foil (for setting off a gem); eyewash; fraud. **~sı meydana çıkmak,** to be shown up: **~ vermek,** to give oneself away.

**frak, -kı** Tail-coat.

**francala** (ʾ··) Fine white bread; roll.

**frank, -gı** Franc. **~lık,** a franc's worth of *stg.*

**Fransa** (ʾ·) France. **~lı,** French, Frenchman.

**fransız** French. **~ca,** the French language; in French.

**fren** (*Fr. frein*) Brake. **~lemek,** to brake.

**frengi** Syphilis; a kind of lock. **~ deliği,** scupper-hole. **~li,** syphilitic.

**Frengistan** Europe.

**frenk, -gi** European. **~gömleği, -ni,** shirt. **~inciri, -ni,** prickly pear. **~üzümü, -nü,** red currant.

**freze** (ʾ·) Milling cutter. **~ makinesi,** milling-machine.

**fuar** (*Fr. foire*) Fair.

**fuh·şiyat, -tı** Obscenities; immoralities. **~uş, -hşu,** prostitution.

**fukara** *v.* **fıkara.**

**ful, -lü** A small kind of bean; mock-orange, syringa.

**fulya** (ʾ·) Jonquil. **~balığı, -nı,** eagle ray.

**fund** Pound (weight).
**funya** (ˈ··) Primer (artillery).
**furgon** (*Fr.*) Luggage-van.
**furş** (*Fr. fourche*) Fork *of a bicycle.*
**furya** (ˈ··) Rush; glut.
**futbol, -lü** Football.
**fuzul** Silly meddlesome man. **~en** (·ˈ·), superfluously; without right; unjustly. **~lî** (·–ˈ), meddling; officious; superfluous.
**füc'eten** (ˈ··) Suddenly (*esp. of dying*).
**fücur** (·ˈ) Immorality, debauchery; incest.
**fülusiahmer** (·ˈ···) *In* **~e muhtac olm.,** not to have 'a red cent'.
**fünun** (·ˈ) *pl. of* **fen.** Sciences; arts. **darül~,** university.
**füsun** (·ˈ) Charm; enchantment, sorcery.
**fütuhat** (··–ˈ), **-tı** *pl. of* **fetih.** Victories, conquests. **~çı,** conqueror; imperialist.
**fütur** (·ˈ) Languor; abatement. **~ gelmek,** to be languid *or* lukewarm; to be discouraged. **~suz,** indifferent; regardless of public opinion; undeterred.

# G

**gaasıb** Who seizes by violence; usurper.
**gabavet** (··–ˈ), **-ti** Stupidity; obtuseness.
**gabi** (·ˈ) Stupid; obtuse.
**gabya** (ˈ··) Topsail. **~ çubuğu,** topmast.
**gacur** *In* **~ gucur,** a creaking noise.
**gaddar** Cruel; perfidious. **~lık,** cruelty; perfidy; sale at exorbitant prices.
**gadir, -dri** Cruelty; tyranny; injustice; perfidy; breach of trust.
**gadretmek** (ˈ··), **-eder** *vn.* Do a wrong; act unjustly; commit a breach of trust.
**gaf** (*Fr. gaffe*) Gaffe, blunder.
**gafil** (–ˈ) Careless; inattentive; unwary. **~ avlamak,** to catch unawares: **~ olm. (bulunmak),** to take no heed.
**gaflet, -ti** Heedlessness; inattention; somnolence. **~ basmak, ~e düşmek,** to be heedless *or* unaware: **~ etm.,** to be negligent; to commit a mistake through absentmindedness; to be unaware of what is going on.
**gaga** Beak. Aquiline; hook-nosed. **~ burun,** aquiline nose. **~lamak,** to peck.
**gâh, kâh** (–) A time; moment. **~ ~,** at times, now and then: **~ ..., ~ ...,** at one time ..., at another ... .
**gaib** (–ˈ) Absent; invisible; hidden; lost. The invisible world; the 3rd person (*gram.*). **~den haber almak,** to practise divination, to foretell the future: **~e ihtar,** warning published in the papers that if a certain person does not present himself before the court he will be tried by default.
**gaile** (–·ˈ) Anxiety; trouble; difficulty; war. **hizmetçi ~si var,** there is the servant problem. **~li,** troubled, worried. **~siz,** without worries, carefree.
**galat, -tı** Error; erroneous expression, barbarism. **···den ~,** a corruption of *a word*; an erroneous version of ...: **~ı hilkat,** a freak of nature, monster.
**galebe** Victory; superiority, predominance, uncontrollable ferocity. **~ etm. (çalmak),** to conquer, overcome.
**galeta** (·ˈ·) Hard biscuit.
**galeyan** Effervescence; rage; excitement. **~ etm.,** to effervesce; boil with rage.
**galib** (–ˈ) Victorious; superior; dominant; prevailing; most usual; probable. **~ olm. (gelmek),** to be victorious, win, surpass. **~a** (ˈ··–), probably; presumably. **~iyet** (–··ˈ), **-ti,** victory; superiority.
**galibarda** (··ˈ·) Bright scarlet colour.
**galiz** (·ˈ) Coarse; thick; rude.
**gam**[1], **-mmi** Care, anxiety; grief. **~ yemek,** to be oppressed with anxiety *or* sorrow: **def'i ~ etm.,** to console oneself, distract oneself: ⌜**ölsem de ~ yemem**⌝, (if only I can do this) I don't mind what happens.
**gam**[2] (*Fr. gamme*) Gamut; scale.
**gamalı** (ˈ··) *In* **~ haç,** swastika.
**gamba** (ˈ·) Kink *in a rope*; upper part of a boot *or* stocking.
**gambot, -tu** Gunboat.
**gam·lanmak** *vn.* Be grieved; fret. **~lı,** sorrowful, grieved; anxious.
**gammaz** Telltale, sneak, informer. **~lamak,** to tell tales about; to spy upon; inform against.
**gamsele** (ˈ··) Mackintosh, oilskin.
**gamsız** Free from grief; carefree.

**gamze** Wink; significant look; twinkle; dimple.
**ganaim** (·—ˈ) *pl. of* ganimet. Spoils *etc.*
**gangıran, gangren** Gangrene.
**gani** Wealthy; independent; free from want; abundant; generous. ~ **gönüllü,** generous: ⌈**Allah ~ ~ rahmet etsin**⌉, 'may God rest his soul' (*only used when mentioning a good deed of a deceased person*).
**ganimet, -ti** Spoils, booty; windfall, godsend.
**gar** (*Fr. gare*) Station.
**garabet** (·—ˈ), **-ti** A being a stranger; absence from home; exile; strangeness; curiousness, singularity.
**garaib** (·—ˈ) *pl. of* **garibe.** Strange things *etc.*
**garaz, garez** Selfish aim *or* motive; spite, rancour, grudge. ···**e ~ bağlamak,** to nourish a spite against ...: ~ ..., the thing is that ...: **bir ~a mebni,** for some private end; because of a spite: ~ **tutmak,** to bear a grudge. **~kâr** (··ˈ), **~lı,** selfish, interested; spiteful, malicious. **~kârlık,** malice, spitefulness; evil intent.
**garb** The West; Europe. **~cı,** one who desires the westernization of Turkey. **~en** (ˈ·), westwards. **~î** (·ˈ), western. **~lı,** western; European. **~lılaşmak,** to become westernized.
**gardfren** (*Fr. garde-frein*) Brakesman.
**gardırop** (*Fr. garde-robe*) Wardrobe; cloak-room.
**garez** *v.* **garaz.**
**gargara** Gargling; gargle.
**garib** (·ˈ) Stranger; away from his own country; poor, needy; strange; curious. **~i** ..., the strange part of it is that ...: **~ine* gitmek,** to appear strange, to strike one as odd. **~lik,** being a stranger; a being without friends; poverty. **~semek,** to find a thing strange; to feel lonely *or* a stranger.
**gark, -kı** A being submerged; drowning; a being overwhelmed. ~ **etm.,** to submerge; to overwhelm (with presents *or* favours): ~ **olm.,** to be submerged; be drowned; be covered *or* overwhelmed.
**garnizon** (*Fr.*) Garrison.
**garson** (*Fr. garçon*) Waiter.
**gasb** Wrongful seizure; usurpation. **~etmek** (ˈ··), **-eder,** to seize by force; snatch away.
**gaseyan** Vomiting.
**gâsıb** *v.* **gaasıb.**
**gas·il, -sli** Washing, *esp.* canonical washing and washing of the dead. **~letmek** (ˈ··), **-eder,** to wash. **~sal,** a washer of the dead.
**gaşy, gaşiy** Fainting, swoon; ecstasy. **~etmek** (ˈ··), **-eder,** to cause to faint; enrapture. **~olmak** (ˈ··), to swoon; be in an ecstacy.
**gavamız** (·—ˈ) Niceties; minutiae, fine points.
**gâvur** Non-Moslem; giaour; infidel; atheist. ~ **etm.,** to waste utterly; to ruin: ~ **gibi inad etm.,** to refuse obstinately: ~ **olm.,** to be a renegade. **~luk,** quality of being a non-Moslem; irreligion; Christian fanaticism; cruelty. **~ca** (·ˈ·), in the manner of an infidel; cruelly; in a European language.
**gaybî** (·ˈ) Pertaining to the unknown *or* occult.
**gayda** A kind of bagpipes, *used by Bulgars.*
**gaye** (—ˈ) Aim, object; end.
**gayet** (—ˈ), **-ti** End; limit; object. Extremely. **~le,** extremely.
**gayr, -rı** *or* **-ri** Another thing *or* person; *used with the izafet it forms a negative prefix, e.g.* **~icaiz,** illicit; (*these adjectives are written as one word and the commonest will be found under* gayri). **~ım,** another than I, *s.o.* else: **ayrımız ~ımız yok,** we have everything in common; there is no difference between us: **ondan ~i,** other than that; moreover, besides.
**gayret, -ti** Zeal; energy; perseverance; jealousy of one's rights *or* honour. ⌈**~ dayıya düştü**⌉, 'it is for uncle to make the effort', 'well, if no one else can do it you (I) must!': **~ine dokunmak,** to goad *s.o.* on by reminding him of his duty, his honour *etc.*: **~i elden bırakmamak,** to persist: ~ **etm.,** to display zeal *or* energy, to bestir oneself: ~ **vermek,** to inspirit, encourage. **~keş,** jealous of one's rights *or* honour; partisan; zealous. **~li,** zealous, persevering. **~siz,** without enthusiasm; slack.
**gayrı**[1] (ˈ·) Henceforth; at length, finally.
**gayrı**[2], **gayri** *v.* **gayr.** *As neg. prefix* = un-, in- *etc.* (*for pronunciation see the simple adjective; the main stress falls on the first syllable of* **gayri**). **~caiz,** illicit; improper. **~ihtiyari,**

involuntary, willy-nilly. **~ilmî,** unscientific. **~kabil,** impossible. **~kâfi,** insufficient. **~memnun,** displeased. **~meşru,** illegitimate. **~muntazam,** irregular; disorderly. **~ mübadil,** not subject to change; not subject to exchange (of populations); established. **~mümkün,** impossible. **~samimî,** insincere. **~tabiî,** unnatural, abnormal.

**gayya** A well in Hell; 'the bottomless pit'.

**gayz** Anger; hatred.

**gaz**[1] Gauze.

**gaz**[2] Gas; paraffin.

**gaza** (·-̀), **gazve** War on behalf of Islam, Holy War; victory over infidels.

**gazab** Wrath. **~ etm.,** *or* **~a gelmek,** to get angry.

**gazel**[1] Lyric poem. **~iyat, -tı,** lyrics.

**gazel**[2] Withered leaves. **~ vakti,** autumn, fall. **~lenmek,** (of leaves) to wither and fall.

**gazete** (·̀·) Newspaper. **~ci,** journalist; newsvendor.

**gazi** (–̀) One who fights for Islam; veteran of a war; *title taken by a victorious Moslem general or ruler.* ⌜**Ey ~ler, yol göründü**⌝, it's about time we were going!

**gaz·lemek** *va.* Gas (*mil.*). **~li,** gaseous; worked by *or* mixed with gas *or* paraffin.

**gazoz** Fizzy lemonade *or* gingerbeer.

**gazve** *v.* **gaza.**

**gebe** Pregnant. ⌜**geceler ~dir**⌝, 'wait, things may be better tomorrow!'

**gebermek** *vn.* Die (*of animals or, contemptuously, of men*). **~tmek,** to kill.

**gebeş** Thick-headed; awkward, uncouth.

**gebre**[1] Hair-cloth glove *for grooming horses.*

**gebre**[2] Caper; caper tree.

**gece** Night. Last night; tonight; by night. **~ baskını,** night raid (by a burglar *or* footpad); night attack (*mil.*): **~ gündüz (demeden),** day and night, continuously: **~kuşu,** owl; nightbird (of a man): **~ yarısı,** midnight: **~ yatısı,** who passes the night at a place; hospitality for the night; **~ yatısına buyurun!,** pray come and stay the night!. **~ci,** night worker; night-watchman. **~lemek,** to pass the night; to become night. **~leyin** (·̀··), by night. **~lik,** pertaining to the night; night-dress, night-gown; food for the night. **~sefası** (·̀···), **-nı,** Marvel of Peru.

**gecik·mek** *vn.* Be late. **~tirmek,** to cause to be late; be slow in doing *stg.*

**geç** Late. **~ kalmak,** to be late: ⌜**~ olsun da güç olmasın**⌝, 'better late than never'; better there should be a delay than a difficulty: **Allah gecinden versin!,** (*used when speaking to a person of his own death*), I hope it won't be for many years.

**geçe** Past (*only in telling the time*); **onu çeyrek ~,** a quarter past ten.

**geçelim** *1st pers. pl. optative of* **geçmek.** Let's pass that by; let's not talk about that.

**geçen** Passing; past; last. **~lerde,** recently: **~ gün,** the other day: **~ sene,** last year.

**geçer** Current (coin *etc.*); saleable. Current value.

**geçici** Passing; temporary; infectious.

**geçid** Place of passage, pass, ford; fairway (*naut.*); act of passing. **~ vermek,** (of a river) to be fordable; (of a pass) to be open.

**geçil·mek** *vn. pass. of* **geçmek.** Be passed; be passable; be given up *or* renounced. **buradan geçilmez,** no passage! **~mez,** impassable; not to be given up: impasse.

**geçim** A living together in agreement; a 'getting on' with one another; compatibility; current value, currency; livelihood. **~ derdi,** the struggle to earn one's daily bread: ⌜**~ dünyası bu!**⌝, 'one's got to live!': **~ seviyesi,** the standard of living. **~li,** easy to get on with; affable. **~siz,** unable to get on with others. **~sizlik,** inability to get on with others *or* with each other; incompatibility.

**geçinecek** Means of subsistence; income.

**geçinmek** *vn.* Live; exist; subsist; get on well with others; pass for, have the reputation of. **birisinden ~,** to live on *s.o.* else, to sponge on *s.o.*

**geçirmek** *va. caus. of* **geçmek.** Make *or* let pass; transport over *stg.*; get rid of (pain *etc.*); go through, experience; appoint (to a post); see off (a friend *etc.*); insert, fit into; pass off as good *or* current; cause to renounce. **... başına* ~,** to hit *s.o.* over the head with ... : **diş ~,** to get one's teeth into; injure, annoy: **ele ~,** to get into one's power: **evini başına geçiririm,** I'll bring the house down about his ears (*a threat*): **gözden ~,** to scrutinize: **hesaba ~,** to enter to an account: **vaz ~,** to persuade *or* make *s.o.* give up *stg.*: **zihinden* ~,** to ponder.

**geçirtmek** *va. caus. of* **geçirmek.**

**geçiştirmek** *va.* Get over (an illness); escape *or* survive (an accident). **sükûtla ~,** to pass over *stg.* in silence.

**geçkin** Over-ripe; past the prime; not so young; over-matured (wood); former, previous (holder of an office); past (a certain age).

**geçme** *verb. noun of* **geçmek.** *a.* Fitting into *stg.* else; made in sections. **sandık ~si,** dovetail.

**geçmek** *vn.* Pass; pass along, over, into *etc.*; pass away, come to an end, expire; deteriorate; fade; (fruit) be over-ripe; be transferred; move; pass as current; be in vogue. *va.* (*with acc.*) Pass (overtake); go beyond; skip, leave out; (*sl.*) denounce, tell tales about; talk about; (*sl.*) *used as an auxiliary to replace* **etmek,** *e.g.* **alay ~,** to mock; **işaret ~,** to make signs. **···den ~,** to pass, pass by, through *etc.*; give up, abandon, renounce: **···e ~,** undertake; take over, succeed to *a post etc.*; penetrate: **geç!,** don't take any notice!; ignore that!: **geçelim,** let's pass by that; let's not talk about that: **geçmez,** *v.* **geçmez**: **geçmiş,** *v.* **geçmiş**: **geçtim olsun!,** I'll drop the idea: **adı ~,** to be mentioned: **adi (bahsi) geçen,** the aforementioned: **artık bizden geçti** *or* **biz artık geçtik,** I'm past that sort of thing, I'm too old for that: **başa ~,** to come to the fore, to become chief: **başından* ~,** to happen to one: **birbirine ~,** to fit into one another; to intertwine; to fall into confusion; to quarrel with one another: ⌜**bu da geçer**⌝, 'this too will pass!'; never mind!: **candan ~,** to sacrifice one's life *or* be ready to give one's life: **çok geçmeden,** before long: **ele ~,** to be arrested; to be obtained: **eline* ~,** to come into one's possession: **gün geçmez ki,** not a day passes but ...: **gün geçtikçe,** as the days pass, in course of time: **hatırından* ~,** to pass through the mind: **ismi geçen,** the aforementioned: **iş işten geçti,** it's all over; there's nothing one can do now: **kendinden* ~,** to lose consciousness; to be beside oneself *with joy etc.*: **kendinden geçmiş,** (of a man) no longer of any use: **sözü* geçiyor,** his word carries weight: **yere** *or* **yerin dibine ~,** to sink into the ground from shame.

**geçmez** That does not pass; non-current; incurable; non-infectious. **su ~,** impermeable; waterproof.

**geçmiş** Past; past the prime; passed away; deceased; over-ripe. **~i kınalı (kandilli),** damn the fellow!; that scoundrel: ⌜**~e mazi (yenmişe kuzu)**⌝, that was a long time ago; things aren't the same now: **~ ola,** it's too late now: ⌜**~ olsun!**⌝, *said to congratulate s.o. on recovery from an illness or escape from an accident*: **birisile ~i olm.,** to have common experiences *or* memories with s.o.

**gedik** Breach; notch; gap; warrant; tenure; *a kind of* leasehold; a licence *for certain trades.* With teeth missing. **~ kapamak,** to fill a gap: **eksik ~,** any kind of deficiency: **eksik ~ tamamlamak,** to make everything good: ⌜**taşı gediğine koymak**⌝, to give as good as one gets; to make a clever retort. **~li,** breached; notched; having a gap; *property* held under a **gedik**; possessing the warrant *or* licence known as **gedik**; regular (customer, visitor), habitué. Regular n.c.o. (*mil.*).

**gedilmek** *vn.* Become notched *or* jagged; have a gap.

**geğir·mek** *vn.* Belch. **~ti,** belch.

**geğrek** Lower rib; false rib. **~ ağrısı (batması),** stitch in the side.

**gelberi** Iron rake *for a fire.*

**gelecek** About to come; future. **~ ay,** next month, *but* **~ salı,** Tuesday after next (*next Tuesday* = **bu salı** *or* **önümüzdeki salı**).

**gelen** Coming; comer. **~ geçen,** coming and going. Passers-by.

**gelenek** Tradition.

**gelgeç** Fickle, inconstant. **~ hanı,** a place where people come and go.

**gelgelelim** (ˋ···) All the same; and yet; however.

**Gelibolu** Gallipoli.

**gelici geçici** Transient, passing.

**gelin** Bride; daughter-in-law. **~ gibi sallana sallana yürümek,** to walk slowly and lazily: **~ odası gibi,** very tidy: ⌜**kendi kendine ~ güvey olm.**⌝, (i) to set out to decide a matter without the authority *or* the competence to do so; (ii) to be ridiculously self-important; (iii) to reckon without one's host: ⌜kızım sana söyliyorum, ~im sen anla!⌝, to talk 'at' *s.o.*; to make remarks intended for *s.o.*, but not directly addressed to him.

**gelince** (·ˋ·) Regarding; as for. **bana ~,** as for me, as far as I am concerned.

**gelin·cik** *Dim. of* **gelin**; weasel; poppy; three-bearded rockling (fish). **~ illeti,** *popularly used to describe* any illness causing swelling of the legs, *e.g.* dropsy. **~havası, -nı,** fine calm weather. **~kuşu, -nu,** a species of lark. **~lik,** quality of a bride; marriageable girl; anything suitable for a bride.

**gelinmek** *vn. impers. of* **gelmek. bir saatte gidilir gelinir,** you can go there and back in an hour: **hakkından gelindi,** he (it) was got the better of, was overcome.

**gelir** *present tense of* **gelmek.** Income, revenue. **~li,** having such and such an income; having a fixed income.

**geliş** Act *or* manner of coming; gait; the way a thing comes *or* happens. **söz ~i,** supposing that; for example; apropos: **sözün (lâfın) ~i,** in the course of conversation. **~igüzel** (··ˋ··), by chance, at random; haphazard.

**gelişme** Development.

**gelişmek** *vn.* Grow up; develop; grow healthy *or* fat; make progress.

**gelme** Act of coming. Come, arrived, brought. **uşaklıktan ~,** an ex-servant.

**gelmek, -ir** *vn.* Come; suit, fit, answer *a purpose*; seem, appear; sham, pretend; endure, bear; (*with the gerund in* **-a** *or* **-e**) happen habitually, *e.g.* **edegelmek,** to do habitually; **olagelmek,** to happen constantly. **geldi geleli,** ever since he came: **gel gelelim,** *v.* **gelgelelim: gelip gitmek,** to come and go, to go to and fro: **gelsin yemek, gitsin yemek,** there was heaps of food: **gel de kızma!,** 'how could anyone not be angry?'; ⌜**gel de bu adama bir daha yardım et!**⌝, 'how could one help this man again (after the way he behaved last time)!': **güleceğim geldi,** I wanted to laugh: **İstanbulu göreceğim geldi,** I long to see Istanbul: **görmemezlikten geldi,** he pretended not to see: **işitmemezlikten geldim,** I pretended not to hear: **yağmur altında durmağa gelmez,** it doesn't do to stop out in the rain: **bu kumaş yıkamağa gelmez, boyası çıkar,** this stuff won't wash, its colour runs: **bu adamla munakaşa etmeğe gelmez,** it doesn't do to argue with this man: ⌜**geleceği varsa göreceği de var!**⌝, 'let him come and see what's waiting for him!'; 'let him try it on!': **bana öyle geliyor ki,** it seems to me that ...: **başına* ~,** to happen to one: **çıka ~,** to appear suddenly: **işe ~,** to be suitable for the job: **işine* ~,** to suit one's purpose *or* desires: **yola ~,** to think better of *stg.*; to come round to the right way of thinking; to submit.

**gem** Bit (of a horse). **~ almak,** to submit to the bit: **~ almaz,** uncontrollable: **~i ağzıya almak,** to take the bit between the teeth, to get out of control: **···e ~ vurmak,** to curb ... .

**gemi** Ship. **~de teslim,** free on board (f.o.b.): **~ci,** sailor: **~ nuru,** St. Elmo's fire. **~cilik,** the profession of a sailor; art of managing a ship; navigation.

**genc** Young. Young man. **~elmek, ~leşmek,** to become youthful *or* vigorous. **~lik,** youth; youthful folly: **gencliğine* doyamamak,** to die young.

**gene**[1], **yine** (ˋ·) Again; moreover; still.

**gene**[2] Tick; castor bean. **buğday ~si,** weevil.

**genel** *a.* General. **~ev,** brothel. **~kurmay,** General Staff.

**geniş** Wide; vast; extensive; abundant; generous, magnanimous; at ease, free from care. **eli ~,** generous: **içi ~,** easy-going; phlegmatic: **bir işi ~ tutmak,** to do *stg.* on a broad scale: **gönlünü ~ tut!,** don't worry!, don't take it too seriously! **~gönüllü,** not easily upset, possessing equanimity. **~lemek,** to widen, extend, ease; become spacious *or* wide; be at ease, in easy circumstances. **~lik,** width, spaciousness; abundance; ease of mind; easy circumstances.

**geniz, -nzi** Nasal passages. **~e kaçmak,** (of food) to go down the wrong way: **~den söylemek,** to speak through the nose.

**gerçek** True; actual; genuine; truthful; in the right. In truth; in earnest; really. Truly!, really! The truth. **~ten,** truly, really; true, genuine: **gerçeğini söylemek,** to tell the truth of it. **~lenmek, ~leşmek,** to turn out to be true.

**gerçi** (ˋ·) Although; granted that; it is true that.

**gerdan** Neck, throat, front of the neck. ~ **kırmak,** to put on coquettish airs; to bow. ~**kıran,** (neck-breaking) stumbling (horse); wryneck. ~**lık,** necklace; neck-band.

**gerdek** Nuptial chamber.

**gerdel** Wooden *or* leather bucket.

**gerek**[1] A necessity, a requisite. Necessary, needed; fitting, proper, due; *with the conditional tense* **gerek** *conveys probability, e.g.* **pek de yanlış olmasa** ~, and it is probably not far off the truth. ~**se,** if it be necessary; if it be proper: ~**tir,** it is necessary; it is right: **gereğince,** as far as necessary; as required; if need be; in accordance with: **gereği gibi,** as is due, properly: **nene** ~**?**, why worry?, it doesn't concern you!: **neme**~**?**, what's that to me?, what do I care?: ⌜**yolcu yolunda** ~⌝, a traveller's place is on his road.

**gerek**[2] ~ ... ~ ..., whether ... or ...: ~ **büyük** ~ **küçük,** whether large or small.

**gereklik** Necessity; fitness.

**gerekmek** *vn.* Be needful, necessary; be lacking; be fitting *or* suitable; be worthy of.

**gergedan** Rhinoceros; rhinoceros horn.

**gergef** Embroiderer's frame.

**gergi** Instrument for stretching; weaver's bar; stretcher *of a rowing-boat.*

**gergin** Stretched; taut; strained (relations). ~**lik,** tension.

**geri** Behind; back; again. Hinder part; rear; remainder. Hinder; posterior; back; backward. ~ **almak,** to take back: ~**sini almak,** to complete *stg.*: ~ **basmak,** to reverse (a car): ~**ye bırakmak,** to put off, postpone: ~ **durmak,** to abstain, refrain; not to interfere: ~ **gitmek,** to go back, recede, decline: ~ **kalmak,** to remain behind, be late; ~ **kalan,** the rest, the remainder: ~ **komak,** to put back; leave undone; postpone: ~ **olm.,** (of a watch) to be slow: ~**sin** ~**ye yürümek,** to walk backwards: **ayakları*** ~ ~ **gitmek,** to draw back *in fear etc.*; to go unwillingly: **ilerisini** ~**sini düşünmek,** to weigh all considerations.

**geridon** (*Fr. guéridon*) Round pedestal table.

**gerilemek** *vn.* Recede; be slow; be late; remain behind; make no progress.

**geri·li** Stretched; taut. ~**lme,** tension: ~ **kuvveti,** tensile strength. ~**lmek,** *pass. of* **germek.** ~**nmek,** to stretch oneself.

**germe** Tension; strain. ~ **somun,** turnbuckle.

**germek** *va.* Stretch; tighten. **göğüs** ~, to put on a good face, to feel confident: ···**e göğüs** ~, to stand up to, resist: **göğsünü gere gere,** proudly, confidently: **haça** ~, to crucify.

**getire** *In* **hak** ~, nil; nothing more; finished; that's all!

**getirmek** *va.* Bring; produce; import. **dile** ~, to mention; to make the subject of talk *or* censure: **hatıra** ~, to remember, think of: **imana** ~, to convert to the faith: **meydana** (**vücude**) ~, to bring into existence, create: **yerine** ~, to carry out, fulfil; put in its place; replace: **yola** ~, to bring to reason; put to rights.

**getirtmek** *va.* Cause to be brought, imported *or* transported; order (book *etc.*).

**getr** Gaiter.

**gevelemek** Chew; hum and haw.

**geven** Tragacanth shrub.

**geveze** Talkative; chattering; gossiping; indiscreet, unable to keep a secret. ~**lik,** babbling, gossip; indiscreet talk.

**geviş** Chewing the cud, rumination. ~ **getirmek,** to chew the cud.

**gevrek** Friable; brittle; crackly. Biscuit. ~ ~ **gülmek,** to laugh in an easy, self-satisfied way.

**gevremek** *vn.* Become crisp and dry.

**gevş·ek** Loose; slack; lax; soft; feeble; weak in health; lukewarm, lacking in zeal. ~ ~ **gülmek,** to laugh in a vulgar and rather too free-and-easy a manner: **ağzı** ~, one who cannot hold his tongue: **bir işi** ~ **tutmak,** not to take a matter seriously. ~**emek,** ~**eklemek,** ~**elmek,** to become loose, slack, feeble, lukewarm; become too familiar. ~**etmek,** to loosen, slacken; weaken; be slack about *stg.*, neglect.

**geyik** Deer; stag. **ala** *or* **yağmurca** ~, fallow-deer: **ulu** ~, red deer. ~**dili, -ni,** hart's-tongue fern.

**gez** Notch in an arrow; backsight of a gun (*foresight* = **arpacık**); rope with knots at intervals for measuring ground; plumb-line. ~**e vurmak,** to level.

**gezdirmek** *va. caus. of* **gezmek,** *q.v.*; **göz ~,** to cast the eye over *stg.*
**gezgin** Travelled; who has seen much of the world. **~ci,** *as* **gezgin**; itinerant pedlar.
**gezi**[1] Tissue of mixed silk and cotton.
**gezi**[2] Promenade; excursion.
**gez·ici** Travelling, touring; itinerant. **~ilmek,** to be gone round; to be visited *or* inspected. **~inmek,** to go about aimlessly; stroll. **~inti,** excursion; walk, stroll; place where one strolls; passage, corridor.
**gezlemek** *va.* Notch (an arrow); measure (ground); adjust, set straight.
**gezme** *verb. n. of* **gezmek**; patrol; watchman.
**gezmek** *vn.* Go about; travel; walk about (*esp. with a view to seeing things or for enjoyment*); go about and inspect: **dillerde ~,** to be on everyone's tongue: **ellerde ~,** to pass from hand to hand, to be a common object: **el üstünde ~,** to be highly valued: **pek ileride ~,** to put forward great pretensions: ⌜**ne gezer**⌝, by no means, not at all; 'not likely!'; out of the question: ⌜**nerelerde geziyor!**⌝, 'what on earth is he about!'
**gıbta** Longing; envy (without ill will). **~ etm.,** to envy: **şayanı ~,** enviable.
**gıcık** A tickling sensation (*esp. in the throat*). **gıcığı*** **tutmak,** to have a tickling of the throat, to wish to cough. **~lamak,** to cause an irritation in the throat.
**gıcır** A kind of chewing-gum. **~ ~,** very white and clean, brand new; *imitates the sound of chewing gum or gnashing the teeth or creaking.* **~damak,** to creak; to rustle; to give out the sound of **gıcır. ~datmak,** to make creak; gnash the teeth. **~tı,** a creaking *or* rustling noise.
**gıda** (·⊥) Food; nourishment; amount of food *or* drink usually taken. **~î,** nutritious; alimentary. **~lanmak,** to be fed *or* nourished. **~lı,** nutritious. **~sız,** not nutritious; without food; undernourished.
**gıdak** *In* **gıt gıt ~,** *imitating the noise of a hen that has laid an egg.* **~lamak,** to cackle.
**gıdık** Tickling; the under side of the chin. **~lamak,** to tickle.
**gık** *In* **~ bile demedi,** without a murmur: **bir şeyden ~ demek,** to be sick of *stg.*: **~ dedirtmek,** to cause *s.o.* to be sick of *stg.*: **~ dedirtmemek,** not to give *s.o.* a chance to speak; not to give *s.o.* a breathing-space; to listen to no objections.
**gıllügış** Malice; rancour; treachery. **~tan âri,** utterly sincere. **~lı,** malicious; untrustworthy. **~sız,** free from malice; open, sincere.
**gılzet, -ti** Coarseness; rudeness thickness.
**gına**[1] (·⊥) Wealth; contentment; sufficiency; satiety; disgust. **~ gelmek,** to have had enough; to be surfeited.
**gına**[2] (·⊥) Nasal twang; song through the nose.
**gıpta** *v.* **gıbta.**
**gır** *The sound of a* snarl, a snore *etc.* **~gır,** snarling; snoring; tiresome noise; small motor-boat; zip-fastener; bag-shaped fishing-net: **~ söylemek,** to harp querulously on *stg.*
**gırıl·damak** (Of the stomach) to rumble. **~tı,** a rumbling noise (*esp. of the stomach*).
**gırla** (⊥·) Abundantly; incessantly; too much.
**gırlamak** *v.* **girildamak.**
**gırt** *Imitates the noise of cutting stg. thick with scissors.*
**gırtlak** Windpipe; throat. **gırtlağına* basmak,** to force *s.o.* to do *stg.*: **gırtlağına düşkün,** greedy: **~ gırtlağa gelmek,** to be at one another's throats: **~ kemiği,** Adam's apple. **~lamak,** to strangle.
**gışa** (·⊥) Membrane; covering; veil.
**gıt** *v.* **gıdak.**
**gıyab** (·⊥) Absence; default. **~en** (·⊥·), by default; in the absence of: **~ tanımak,** to know *s.o.* by name. **~i** (·–⊥), defaulting; not present in court: **~ hüküm,** judgement given in default.
**gıybet, -ti** Speaking ill of *s.o.* in his absence; backbiting; absence; alibi. **~ etm.,** to slander, backbite. **~çi,** slanderer, backbiter.
**gıygıy** *Imitates the sounds of a violin.*
**gibi** The similar; the like. Similar, like. As; as soon as; just as; as though. **geldiği ~,** as soon as he came *or* comes: **onun ~,** like him; **benim ~ler,** people like me: **bu ~ler,** the likes of these; people like this: **gereği ~,** as it should be: **ne ~?,** what sort of?, how?: **bir ses duyar ~ oldum,** I thought I heard a voice: **bilmez ~ soruyor,** he asks as though he does not know, pretending not to know: **yağmur yağacak ~ görünüyor,** it looks as though

it were going to rain: **bu işin sonu iyi olmıyacak ~me geliyor,** I have a feeling that this won't turn out well.

**gider** *present tense of* **gitmek.** Expenditure, outlay. **~ayak** (·ˋ··), at the last moment; at the moment of going. **~ek,** gradually.

**gidermek** *va.* Remove; cause to go; satisfy (a desire).

**gidi** Pander; *interj. expressing abuse, often mild, sometimes almost affectionate.* **seni ~!,** scoundrel!; you little rascal!: **hey ~ günler!,** oh! the good old days!

**gidilmek** *Impers. of* **gitmek. oraya gidilmez,** one can't go there.

**gidiş** A going; movement, gait; conduct. **~ o ~,** that was the last that was seen of him: **~ini beğenmiyorum,** I don't like his conduct. **~at, -tı,** *pseudo-Arabic pl. of* **gidiş**; goings-on.

**gidişmek** *vn.* Itch.

**gidon** (*Fr. guidon*) Burgee; handlebar of a bicycle.

**-gil** *Suffix meaning* belonging to the family of ... .

**girdab** Whirlpool; dangerous place.

**girdi** *In such phrases as*: **bir işin ~sini çıktısını bilmek,** to know the ins and outs of a matter: **bu işin daha bir çok ~si çıktısı var,** there are a lot more complications in this matter.

**girgin** Who knows how to ingratiate himself; pushing. **~lik,** pushfulness; ability to worm one's way into favour.

**girift** Interlaced (writing); involved, intricate. Small flute.

**girilmek** *impers. of* **girmek. girilmez,** no entry!

**girinti** Recess; indentation. **~li,** having recesses: **~ çıkıntılı,** wavy, zigzag, indented.

**giriş** Entry; entrance. **~ supapı,** inlet valve. **~ken,** enterprising, pushful.

**girişmek** *vn.* (*with dat.*) Penetrate; mix oneself up with *a matter,* meddle, interfere; set about, undertake.

**Girit** (ˋ·) Crete. **~li,** Cretan.

**girizgâh** Introduction *to a subject*: introductory part of a poem.

**girme** Act of entering; a breaking into the enemy's position.

**girmek** *vn.* (*with dat.*) Enter; go into, be contained by; enter upon, begin; come into, join, participate. **girdi çıktı,** *v.* **girdi**: ⌜**girmiş çıkmış**⌝, (who has entered a lunatic asylum and come out again), a bit queer, with a screw loose: **araya ~,** to mediate: **birbirine ~,** to be intermixed, confused; to become embroiled, to come to blows: **eline* ~,** to fall into *s.o.'s* hands, to be caught: **yola girmek,** to come right somehow.

**girme·li** Having an entrance; having a recess. **~ çıkmalı,** having places for entrance and exit; indented; zigzag.

**giry·an** *a.* Weeping. **~e,** *n.* Weeping, tears: **~ etm.,** to weep.

**gişe** Guichet; ticket-window; grille; pay-desk.

**gitgide** (ˋ··) Gradually; in the course of time.

**gitmek, gider** *vn.* Go; go away; go on *doing*; fade; perish, die; (*with dat.*) suit, fit; be sufficient for; **gitmek** *is sometimes used with another verb to express finality or certainty, e.g.*: **anlıyamadım gitti,** I just *couldn't* understand. **gide gide,** gradually: **gide gele,** by continually going and returning; with great insistence: **gider ayak,** at the last moment; at the moment of going: ⌜**gitti gider dahi gider**⌝, they are gone for ever: **bu elbise iki sene gider,** this suit will last two years: **elden ~,** to be lost: **öyle ... ki deme (sorma) gitsin!,** you never saw such a ...: ⌜**sen giderken ben geliyordum**⌝, while you were going I was coming back; I've forgotten more than you ever knew; you can't take me in: **yola ~,** to set out on a journey.

**gittikçe** (·ˋ·) By degrees, gradually; more and more.

**giy·dirmek** *va.* Put on (clothes); clothe, dress; abuse, reproach. **~ecek,** clothing, dress. **~ilmek,** (of clothes) to be put on, to be worn. **~im,** garment; clothing, dress: **~ kuşam,** clothes, *esp.* one's best clothes; finery: **bir ~ nal,** a set of horseshoes. **~inmek,** to dress oneself, put on one's clothes: **giyinip kuşanmak,** to dress oneself up, to put on one's best clothes.

**giyme** (*Fr. guillemet*) Inverted commas.

**giymek** *va.* Wear, put on (clothes). **mahkemenin hükmünü ~,** to be condemned by a court.

**gizleme** Concealment; camouflage.

**gizlemek** *va.* Hide; conceal; secrete.

**gizli** Hidden, concealed; secret. **~ din taşımak,** to have a hidden religion (*also fig.*): **~ kapaklı,** very secret, clandestine: **~siz kapaklısız,** frankly, openly: **~ sıtma,** pruritis, itching; a slight tickling of the throat; one who acts in a sly and underhand manner. **~ce,** in a secret manner, secretly.

**gocuk** Sheepskin cloak.

**gocunmak** *v.* **kocunmak.**

**gol, -lü** Goal (football).

**gomalaka** (ʾ···) Shellac.

**gomalâstık** (*Fr. gomme*) India-rubber (for erasure).

**gomba** (ʾ·) Mat made of rushes *or* fibres; coir rope.

**gomene** Cable (measure).

**gonca** *v.* **konca.**

**goril** Gorilla.

**goygoycu** Blind beggar, *who was led round to collect provisions for the 10th of Muharrem.*

**göbek** Navel; belly; paunch; centre; heart (of a plant); central ornament; generation. **~ adı,** *the name given to a Turkish child at birth, later inscribed on his identity papers, but not necessarily used*: **~ atmak,** to dance the belly-dance: **~ bağlamak (salıvermek),** to develop a paunch; **~leri beraber kesilmiş,** inseparable friends: **göbeği* çatlamak,** to exert oneself to the utmost: **göbeği* düşmek,** to develop an umbilical hernia; to carry heavy weights: **göbeğini kesmek,** to cut the navel-string: **kendi göbeğini kendi kesmek,** to rely only on oneself for everything: ⌜**onunla göbeğiniz bitişik değil ya!**⌝, 'must you always do everything together?', 'can't you do this by yourself?'

**göbek·lenmek,** to become paunchy; (of a cabbage *etc.*) to develop a heart. **~li,** with a central boss; having a paunch. **~taşı, -nı,** the raised central platform in a Turkish bath.

**göç** Migration; change of abode. **~ etm.,** to migrate; to strike tents; to pass over to the next world. **~ebe,** nomad. **~er,** nomadic; movable. **~kün,** emigrant.

**göçen, göçken** Leveret; stoat.

**göç·mek** *vn.* Strike tents and move off; change one's abode; die; (of a building) to fall down, cave in. **karnı* ~,** to have the belly sink in from starvation. **~men,** emigrant; refugee. **~ük,** the caving-in *of a mine.* **~ünmek,** to pass on, to die. **~ürmek,** *caus. of* **göçmek.**

**göğde, göğermek** *v.* **gövde, gövermek.**

**göğsü** *v.* **göğüs.**

**göğüs, -ğsü** Breast, chest, bosom; flare of a ship's bow. **~ çukuru,** the pit of the stomach: **~ geçirmek,** to sigh, groan: **~ germek,** to face, stand up to: **~ göğse gelmek,** to come to hand-to-hand fighting: **~ göğse muharebe,** hand-to-hand fight: **~ illeti,** consumption; asthma: **göğsü* kabarmak,** to swell with pride: **~ tahtası,** breastbone, sternum: **elini* göğsüne* koymak,** to lay one's hand on one's heart, to search one's conscience. **~lemek,** to breast (waves *etc.*). **~lü,** broad-chested; having a flared bow (ship): **~ bindirme,** scarf. **~lük,** bib, apron; breastplate; breast harness.

**gök, -kü, -ğü** Sky; heavens. Blue, sky-blue; beautiful. ⌜**~te ararken yerde bulmak**⌝, to meet in an unexpected way *s.o.* one has been searching for; to obtain *stg.* in an unexpected manner: **göğe çıkarmak,** to laud to the skies: **~ gözlü,** blue-eyed: **~ gürlemek,** to thunder: **başı* göğe ermek,** to be in the seventh heaven of delight: **başı göğe erdi sanıyor,** he thinks he has done *stg.* wonderful: ⌜**tecrübeyi göğe çekmemişler**⌝, 'no harm in trying': **yerden göğe kadar,** utterly, completely: **yeri göğü birbirine katmak,** to move heaven and earth (to do *stg*): **yerle ~ bir olsa,** even if the heavens should fall; no matter what happens.

**gök·çe,** somewhat blue; pleasant; rock-dove. **~çeağac,** a kind of willow. **~çek, ~çen,** pretty; pleasant. **~çül,** inclining to blue. **~dere,** the Milky Way. **~elâ,** bluish-grey (eyes). **~gürültüsü, -nü,** thunder. **~kır,** blue-grey, ashen. **~kubbe,** the vault of heaven; the universe. **~lük,** blueness, blue colour. **~taşı, -nı,** turquoise. **~yüzü, -nü,** firmament, heavens.

**göl** Lake; pond; puddle. **~ ayağı,** the outlet of a lake. **~başı, -nı,** head of a lake, stream feeding a lake. **~cük,** pond, puddle. **~ek,** pond, puddle; gnat.

**gölge** Shadow; shade; shading (drawing); protection. **~de bırakmak,** to overshadow, surpass: **~ etm.,** to cast a shadow; to trouble, become an obstacle. **~altı, -nı,** shady place, shade. **~lenmek,** to sit in the shade; become shady. **~li,** shaded, shady. **~lik,** shady spot; arbour.

**gömgök** Intensely blue; dark blue.

**gömlek** Shirt; layer; cover; sleeve (*mech.*); gas mantle; skin of a snake; generation. ~ **değiştirmek,** (of a snake) to change its skin; (*fig.*) to change one's opinions *etc.*: ~ **taşı,** facing-stone: **ateşten** ~, a pitiable situation, *esp.* poverty: **bir** ~ **farklı olm.,** to be little, if any, superior: **şimdiki hali eskisinden bir** ~ **iyidir,** his present condition is hardly any better than his old one: **dosya gömleği,** file cover. **~lik,** shirting.

**gömme** Act of burying. Buried; let-in, recessed, inlaid; flush.

**gömmek** *va.* Bury; let in; inlay.

**gömüldürük** Bow of an ox-yoke; breast-band.

**gömül·mek** *vn.* Be buried; sink deeply *into stg.* **~ü,** buried; hidden; underground; flush.

**gön** Leather.

**gönder** Pole; boom; gaff; goad; flagstaff.

**göndermek** *va.* Send.

**gönlü** *v.* **gönül.** **~nce,** after one's heart; as desired.

**gönül, gönlü** Heart; feelings; affection, *esp.* amorous affection; mind; inclination; courage. ~ **açıklığı,** peace of mind, happiness: **gönlü* açılmak,** to feel at ease, to feel serene, to be cheered up: ~ **açmak,** to cheer *s.o.* up: **gönlü* akmak,** to feel attracted by, to fall in love with: ~ **almak,** to please, to content, to make up to *a child etc. after being severe*: **gönlünü avlamak,** to try to get the attentions *of a girl etc.*; to run after *a girl*: ~ **bolluğu,** generosity: **gönlü* bulanmak,** to feel sick, nauseated; to feel suspicious *about stg. or s.o.*: ~ **bulandırmak,** to nauseate; to arouse suspicion: **gönlü* çekmek,** to desire: ~ **darlığı,** foreboding, anxiety: **gönlüne* doğmak,** to have a presentiment: ~ **eğlencesi,** pleasure; a toy of love; solace: **gönlünü* etm.,** to please, conciliate; to induce *s.o.* to do *stg.*: **gönlünden* geçirmek,** to meditate *doing stg.*, to entertain *an idea*: ~ **gözü,** perception, the power of seeing the truth: **gönlünü* hoş tutmak,** not to worry: ~ **hoşluğu ile,** willingly: **gönlü* kalmak,** (i) to feel resentment, to feel hurt; (ii) to hanker after *stg.*: ~ **kırmak,** to hurt the feelings: **gönlünden* kopmak,** (of a present *or* tip) to be given gladly; '**gönlünüzden ne koparsa veriniz!**' 'give what you feel like giving!': **gönlümden koptu ... verdim**', 'I gladly gave ...': **gönlü* olm.,** to agree (*to* = ···**e**); to be in love (*with* = ···**de**): **gönlü var,** he is willing; he is in love: ~ **vermek,** to give one's heart, to fall in love: **gönlünü* yapmak,** to console; to satisfy: **gönlü yok,** he is unwilling; he is not in love: **canü ~den** *or* **candan ve ~den,** with all one's heart and soul: **iki** ~ **bir olm.,** for two hearts to beat as one; to be in full agreement.

**gönül·lü,** willing; self-assertive. Volunteer; lover; beloved. **alçak** ~, meek, modest; affable. **~süz,** without pride, modest; affable; disinclined, unwilling, **~süzlük,** unwillingness; modesty.

**gönye** Square (drawing instrument). **~sinde olm.,** to be at right angles.

**göre** (*with dat.*) According to; respecting; about; considering; suitable for.

**göre göre** *v.* **göz.**

**görenek** A doing *stg.* because others do it; fashion.

**göresi** *In* ~ **gelmek,** to long to see, to miss. **~mek,** to long for.

**görgü** Experience; breeding, good manners. **~lü,** well-bred, having good manners. **~süz,** ill-bred, common; without manners; uncouth.

**görme** Act of seeing, sight. Seeing; seen. **sonradan** ~, upstart, parvenu. **~ce,** subject to the condition of being seen (of a sale); estimated by sight only.

**görmek, -ür** *va.* See; deem; visit; experience; *used as an auxiliary verb with the gerund in* ···**e,** ···**a,** *it signifies continuous action, e.g.* **söyliyegörmek,** to go on speaking. **görerek ateş,** direct fire (*mil.*); **görmiyerek ateş,** indirect fire: ⌈**gören Allah için söylesin!**⌉, *stg. like* 'I swear it's true!': **görmüş geçirmiş,** a man of great experience: **vaktile görmüş geçirmiş,** one who has seen better days: **görmemiş** *v.* **görmemiş**: ⌈**göreyim seni!**⌉, 'let's see what you are made of': **görsün** *v.* **görsün**: **çok** ~, to deem too much; to regard as being beyond *s.o.'s* deserts: **harcını** ~, to defray the expenses of *stg.*: **hizmet** ~, to serve; to render a service: **hoş** ~, to tolerate, condone: **iş** ~, to work: **münasib** ~, to deem fit; to think it a fitting moment *to do stg.*: **rüya** ~, to dream: **tahsil** ~, to study: **terbiye** ~, to be educated: **zarar** ~, to suffer loss:

'**onun azametini görme!**', 'you never saw anything like his conceit!'

**görmemezlik** (ʼ···) A feigning not to see; indifference; connivance. **~ten** *or* **görmemezliğe gelmek,** to pretend not to see.

**görmemiş** Inexperienced; parvenu.

**görsün** *3rd sing. imperative of* **görmek.** Let him see! **gelmiye ~!**, wait till he comes (and then you'll see)!: **bir kere kızmaya ~,** if he *does* get angry (then you'll regret it)!: **yazmıya ~,** he's only got to write about it (and that would settle the matter).

**görücü** Woman sent to find *or* inspect a prospective bride.

**görüm** Sight; look. **~ce,** husband's sister, sister-in-law. **~lük,** *stg.* to be seen: **yüz görümlüğü,** present given by the bridegroom to the bride.

**görün·mek** *vn.* Show oneself; appear; seem; be visible. ⌜**görünen köye kılavuz istemez**⌝, 'no guide is needed to a village that is in sight', *i.e.* it is too obvious to require explanation: **görünmez olm.,** to disappear. **~ür,** apparent; visible: **~lerde yok,** it is not in sight. **~üş,** appearance; show, parade: **~ü böyle,** such are the appearances: **~ etm.,** to make a show: **~te,** apparently.

**görüş** Mode *or* act of seeing; point of view. **ilk ~te,** at first sight: **onun son ~üm oldu,** that was the last I saw of him. **~me,** interview; talk, conversation.

**görüş·mek** *vn.* See one another; meet and converse; become acquainted. *va.* Discuss. **görüşeni karışanı olmamak,** to be free from interference by others, to be independent: **gene görüşürüz inşallah,** I hope we shall meet again! **~türmek,** to introduce to one another.

**gösteriş** Imposing appearance; show, ostentation, display; demonstration; eyewash. **kuru ~,** mere show. **~li,** of striking appearance; stately, imposing. **~siz,** poor looking, unimposing.

**göstermek** *va.* Show; indicate; expose *to the sun etc. vn.* Appear. **baş ~,** to show its head, to appear: **Allah göstermesin!**, God forbid!: **kendini* ~,** to prove one's worth: **sana gösteririm!**, I'll show you! (I'll teach you not to do that again!): **ufak ~,** to look younger than one is.

**göstermelik** Worth exhibiting; for show only (of goods in a shop window).

**götürmek** *va.* Take away; carry off; lead *or* conduct to; hold, contain; bear, endure, support. **bu mesele su götürür,** that is an open question: **bu iş şaka götürmez,** this is not a joking matter: **içi* götürmemek,** to be unable to bear the misfortunes of others.

**götürü** In a lump sum. **~ almak,** to buy in the lump, to contract at a lump price: **~ pazarlık,** a bargain for the whole lot.

**gövde** Body; trunk; whole carcass. **~ye atmak,** to eat, swallow: **kan ~yi götürüyor,** there's a regular massacre going on; 'there's a hell of a mess'. **~li,** bulky; corpulent.

**göver·mek** *vn.* Turn blue *or* green. **~ti,** a blue spot on the skin.

**göya** *v.* **gûya.**

**göz** Eye; the evil eye; hole; mesh; opening; drawer; compartment; pigeon-hole; tray *of a balance*; spring; arch *of a bridge*; bud. **~ ~,** all holes; porous; reticulated: **~ü açık,** wide awake, shrewd: **~ü açık gitmek,** to die disappointed: **~ünü açıp kapamadan** *or* **kapayıncaya kadar,** in the twinkling of an eye: **~ açtırmamak,** to give *a person* no respite, to give no chance to recover himself *or* take action: **~ alıcı,** striking, dazzling: **~e almak,** to envisage, to bring oneself to *or* resign oneself to *stg.*: **···e ~ atmak,** to glance at: **~ünü* bağlamak,** to blindfold; to hoodwink: **~ bağmak,** to cast a spell, to bewitch: **~ boyamak,** *v.* **boyamak**: **~den çıkarmak,** to be prepared to pay *or* sacrifice *stg.*: **bir şeyin ~ünü çıkarmak,** to reject *stg.* good (for *stg.* inferior): **~den çıkmak,** to fall from favour, to fall in consideration: **~ü çıkmak,** to lose an eye: **~ü daldan budaktan esirgememek (sakınmamak),** to disregard dangers: **~den düşmek,** to fall into disesteem: **~ etm.,** to wink: **~üne* girmek,** to ingratiate oneself with *s.o.*, to curry *s.o.'s* favour: **~ göre göre,** what is obvious to all, *e.g.* **~ göre göre çalmak,** to steal openly; **~ göre göre yalan söylemek,** to tell a barefaced lie; **~ göre göre inkâr etm.,** to deny what one's auditors have already heard one admit: **~ ~ü görmez (karanlık),** pitch dark: **~ hapsi,** a being under surveillance *or* detention; ~ **hapsine almak,** to keep under observation *or* under open arrest: ~**ünün* içine bakmak,** (i) to cherish dearly; (ii) to be at the beck and call of *s.o.*; (iii) to look entreatingly at *s.o.*: ~**ü* kalmak,** to hanker after *stg.*, to envy *stg. possessed by another*: ~**ü* yollarda kalmak,** to have been waiting a long time *for s.o. or stg.*: ~ **kapağı,** eyelid: ~ **karası,** the iris of the eye: ~ **kararı,** judgement by the eye; guess; roughly speaking: ~**ü* kesmek,** to think oneself capable *of doing stg.*; to like; to think suitable: ~**üne* kestirmek,** to think oneself capable of ...; to have an eye on *stg.* as suitable, to mark down as a desirable possession: ~ **koymak,** to cast covetous eyes upon *stg.*: ~ **kulak olm.,** to be all eyes and ears, to be on the qui-vive; ⌜**bu çocuğa ~ kulak oluver!**⌝, 'just keep a sharp eye on this child': ~ **nuru,** work that strains the eyes; ~ **nuru dökmek,** to engage on work that strains the eyes: ~**ünün* nuru,** the light of one's eyes, darling: ~**de olm.,** to be in favour, to be much thought of: ···**de ~ü* olm.,** to desire *stg.* strongly: ~**ü* olmamak,** to have no particular desire for: ~ **önünde bulundurmak,** to keep in view, to bear in mind: ⌜~**ünü seveyim!**⌝, please!; *sometimes* 'well done!': ~**üne* sokmak.** to thrust *stg.* under *s.o.'s* eyes by way of reproof *or* accusation: ~**de tutmak,** to hold in favour: ~**ü* tutmak,** to take a fancy to: ~ **ucu,** the corner of the eye: iki ~**üm,** my dear.

**göz·ağrısı, -nı,** eye-ache; ilk ~, one's first love. ~**akı, -nı,** the white of the eye. ~**bağı, -nı,** magic, spell. ~**bağıcı,** magician; conjuror. ~**bebeği, -ni,** pupil of the eye; 'the apple of one's eye'. ~**cü,** watchman, sentinel; spy. ~**cülük,** watch; keeping guard; spying. ~**çukuru, -nu,** eye-socket. ~**dağı, -nı,** intimidation; fright: ~ **vermek,** to intimidate; to act as a deterrent. ~**de,** favourite, pet. ~**demiri, -ni,** bower anchor.

**gözen** Fallow-deer.

**gözetleme** Observation (*mil.*). ~**k,** to observe, spy upon.

**gözetmek** *va.* Mind, look after, take care of; pay regard to; observe (duty); envisage. **hatır ~,** to respect the feelings of.

**göz·evi, -ni,** eye-socket. ~**gü,** mirror. ~**leme,** an eyeing, a watching for; fritter, pancake. ~**lemek,** to watch for, wait for; keep an eye on. ~**lü,** having eyes; having drawers *or* pigeon-holes: **beş ~,** (bridge) having five arches: **para ~,** fond of money. ~**lük,** spectacles: ~ **otu,** honesty (plant). ~**lükçü,** optician. ~**lüklü,** bespectacled. ~**taşı, -nı,** copper sulphate. ~**ükmek,** to appear, show oneself. ~**yaşı, -nı,** tears: ~ **dökmek,** to shed tears. ~**yılgınlığı, -nı,** dread, terror.

**grandi** (ˋ··) Mainmast.

**grev** Strike. ~**ci,** striker.

**griva** Cat-head. **demiri ~ya almak,** to cat the anchor.

**gron** A kind of heavy silk cloth.

**grup, -pu** Group.

**gudd·e** Gland. ~**î** (·ˊ), glandular.

**gudubet** (·–ˋ)**,-ti** Ugly face. ~ **bozuntusu,** hideously ugly.

**gufran** Mercy (*of God only*).

**guguk** Cuckoo; cry of derision.

**gugurik** Odd; ridiculous.

**gul** Ghoul, ogre.

**gulâm** (·ˊ) Boy; youth; male slave. ~**para,** pederast.

**guluklamak** *vn.* (Of a hen) to cluck.

**gulyabani** (··–ˋ) Ogre.

**gumena** (ˋ··) Cable (length).

**gupilya** (·ˋ·) Split pin.

**gurbet, -ti** Absence from home; exile; foreign travel. ~ **çekmek,** to feel homesick. **ihtiyarı ~ etm.,** voluntarily to exile oneself. ~**zede,** exiled; living abroad.

**gureba** (··ˊ) *pl. of* **garib.** Strangers; people living out of their own country; paupers. ~ **hastanesi,** infirmary.

**gurk** Broody (hen); turkey-cock.

**gurlamak, guruldamak** *vn.* Rumble.

**gurub** Sunset; setting of a star. ~ **etm.,** to set.

**gurultu** Rumbling noise.

**gurur** Pride; conceit. ~**lu,** arrogant.

**gussa** Sorrow; anxiety.

**gusül, -slü** Ritual ablution. ~**hane** (··–ˋ), bath for ritual washing.

**gûya** As if, as though; it seemed that; one would think that; supposedly.

**gübre** Dung; manure. ~ **şerbeti,** liquid manure. ~**lemek,** to dung, manure. ~**lik,** dung-hill.

**güc** Strength; force, violence; difficulty. ~**üne* gitmek,** to offend, annoy, hurt *s.o.'s* feelings: ~**ü* yetmek,** to be strong enough, to be able:

**iş ~,** business, occupation (*generally written in the nominative as one word but declined as two words, e.g.* **işine* ~üne bakmak,** to attend to one's work; **işi ~ü yok,** he has no occupation).

**güc·enmek,** to be offended, hurt *or* angry. **~lü,** strong; violent: **~ kuvvetli,** very strong and healthy: ⌜**hem suçlu hem ~**⌝, not merely at fault but offensive about it. **~süz,** *in* **işsiz ~,** out-of-work, without a job: **~ kuvvetsiz,** without strength. **~süzlük,** weakness.

**güç** Difficult, hard. With difficulty. **~ belâ** *or* **~ hal ile,** with great difficulty: **sabahı ~ etm.,** to wait impatiently for the morning.

**güderi** Chamois leather; deerskin.

**güdük** Tailless; docked; stumpy; incomplete. **~ kalmak,** to be incomplete *or* unfinished; to be childless.

**güdümlü** Controlled, *esp.* government controlled. **~ mermi,** guided missile.

**güfte** The words of a song.

**güğül** Cocoon (from which the moth has emerged).

**güğüm** Copper vessel with a long handle.

**güherçile** Saltpetre.

**gül** Rose; dial *of a compass.* **~ gibi,** neat; charming; finely, 'swimmingly': **···in üstüne ~ koklamaz,** he wouldn't dream of making love to anyone but ... .
**gül·ab** (·⌄), rose-water. **~bayramı, -nı,** the Feast of Tabernacles; Feast of Pentecost. **~beşeker** (·⌄··), conserve of roses. **~çehreli,** rosy-cheeked. **~dan,** vase for flowers.

**gülbank, -kı** Prayer *or* song uttered by many in unison; chant; slogan; war-cry.

**güldür güldür** *Imitates fluent reading or steady burning.*

**güldür·mek** *va.* Make laugh; amuse. **kendini âleme (her kese) ~,** to be a laughing-stock: **yüzünü* ~,** to rejoice the heart of. **~ücü,** causing to laugh, amusing.

**gülecek** Laughable; ridiculous.

**güleç** Smiling.

**güler** Smiling; given to laughter. **~ yüz,** a smiling cheerful face. **~yüzlü,** merry, cheerful; affable.

**güleş** *v.* **güreş.**

**gül·gün** Rosy, rose-coloured. **~hatmi,** hollyhock. **~istan,** rose-garden. **~istanlık,** *in* ⌜**dünyayı güllük ~ görmek**⌝, to see everything through rose-coloured glasses. **~kurusu, -nu,** dried rose leaves *for jam*; the colour of dried rose leaves.

**güllâbi, güllâbici** Warden in a lunatic asylum. ⌜**ben deli ~si değilim**⌝, 'I can't deal with idiots'.

**güllac** Sweet *made with starch wafers, filled with cream and flavoured with rose-water.*

**gülle** Cannon-ball; shell; bar-shot; any very heavy thing. **~ atmak,** to put the weight.

**gül·lü** Surrounded by *or* decorated with roses. **~lük,** rose-garden.

**gülmek** *vn.* Laugh; smile; be pleased. ⌜**güle güle!**⌝, 'good-bye and good luck!': **güle güle** (*with a verb*) to do anything happily or with success; **güle güle kullanınız (giyiniz)!,** *said to one who has just acquired stg. new*; **güle güle kirleniniz!,** *a greeting to one just out of a bath*: **bir gözü* ~,** to have mixed feelings: **yüze ~,** to feign friendship, to dissimulate: **yüzü* ~,** to be merry and cheerful, to be prosperous and happy.

**gülmez** Sullen; sour-faced; severe.

**gül·rengi** (⌄··), **-ni** Rose colour. **~suyu** (⌄··), **-nu,** rose-water.

**gülümsemek** *vn.* Smile.

**gülün·ç** Ridiculous. **~ecek,** ridiculous; odd. **~mek,** to laugh to oneself; to be a subject of laughter: **buna gülünmez,** this is not a thing to laugh about.

**gülüstan** Rose-garden.

**gülüşmek** *vn.* Laugh together; laugh at one another.

**gül·yağı** (⌄··), **-nı** Attar of roses. **~yanaklı** (⌄···), rosy-cheeked. **~zar** (·⌄), rose-garden; flower garden.

**güm**[1] A thing buried; ruin. **~e gitmek** (*sl.*), to go to ruin; perish.

**güm**[2] Hollow booming noise.

**güman** Doubt; suspicion.

**gümbür ~ ~,** *imitates* a booming noise. **~demek,** to boom, thunder, reverberate; (*sl.*) to 'pop off', die. **~tü,** a booming noise: '**seyreyle sen ~yü!**', now for the crash!, now there will be a to-do!

**gümec** Honeycomb.

**gümlemek** *As* **gümbürdemek.**

**gümrah** Dense; copious; luxuriant. **~lık,** abundance, luxuriance.

**gümrük** Customs; custom-house. ⌜**~ten mal kaçırır gibi**⌝, unnecessarily hurried and flustered: ~ **resmi**, customs dues: **kara gümrüğü**, octroi. ~**çü**, customs officer.

**gümüş** Silver. ~ **kaplama**, silver-plated: ~**takımı**, silver plate, set of silver. ~**balığı, -nı**, sand smelt. ~**lemek**, to silver-plate. ~**selvi**, reflection of the moon on water. ~**suyu**, crystal-clear water. ~**ü**, silver grey; silvery.

**gün** Day; time; sun; light; feast-day. ~**den** ~**e**, from day to day: ~**lerden bir** ~, once upon a time: ~**ü** ~**üne**, to the very day: ~**ünü** ~ **etm.**, to enjoy oneself properly; to make the best use of one's time (*iron.*): ~ **aşırı**, every other day: ~**ün birinde**, then, one day (unexpectedly): ···**e** ~ **doğmak**, for *s.o.'s* day to come, *e.g.* ⌜**harb çıktı muhtekirlere** ~ **doğdu**⌝, 'war broke out, it was the profiteers' chance': ~ **geçmek**, for a day to pass; for the sun to scorch *or* tan: ~ **görmez**, sunless (place): ~ **görmüş**, who has seen better days; who has held important posts *or* been a man of consequence: ~ **görmemiş**, of no standing, with no career behind him: ~**ünü görürsün!**, (*a threat*) 'you'll pay for this!': **bir** ~ **evvel**, as soon as possible: **ele** ~**e karşı**, before all, publicly: **geçen** ~, the other day, not long ago: ⌜**ne** ~**e duruyor?**⌝, 'why not use it?'; 'why not do this?': **o** ~ **bugündür**, ever since that day: **öbür** ~, the day after tomorrow: **öteki** ~, the day before yesterday; the other day.

**günah** Sin; fault. ⌜~**ı boynuna!**⌝, 'well, *you* must take the consequences!': ~ **çıkarmak**, (of a priest) to hear a confession: **bu** ~**ına (bile) değmez**, the game is not worth the candle: ~**ından* geçmek**, to pass over *s.o.'s* sin, to forgive him: ~**a girmek**, to sin: ~**ına* girmek**, to accuse wrongfully, to wrong: ~ **vebali**, the whole responsibility for an evil deed: ⌜**beni** ~**a sokma!**⌝, 'don't drive me to blasphemy (or other sin)!'; 'don't insist upon this!'

**günah·kâr**, sinner; culpable; prostitute. ~**kârlık**, sinfulness; guiltiness; prostitution. ~**lı**, culpable; sinful. ~**sız**, blameless; without sin.

**gün·aşırı**, every other day. ~**aydın**, good morning! ~**balığı, -nı**, rainbow wrasse. ~**batısı, -nı**, sunset; West; west wind. ~**begün**, day by day. ~**delik**, daily wage; daily (paper *etc.*); everyday (wear); ephemeral. ~**delikçi**, day-labourer. ~**dem**, agenda. ~**doğrusu**, ~**doğuşu, -nu**, East; south-east wind. ~**dönümü, -nü**, solstice. ~**düz**, daytime; by day: ~ **feneri**, a nigger. ~**düzlü**, *in* **geceli** ~, going on night and day, continuous. ~**düzsefası, -nı**, convolvulus. ~**düzün**, by day; in the daytime.

**güneş** Sun; sunshine. ⌜~**i balçıkla sıvamak**⌝, (to plaster over the sun with clay), to try to hide the truth: ~ **gibi**, plain, manifest: ~ **tutulmak**, for the sun to be eclipsed: ~ **vurmak**, for the sun to cause sunstroke: **başına** ~ **geçmiş**, he is feeling the effect of the sun. ~**lenmek**, to bask in the sun; sunbathe; be spread in the sun to dry. ~**lik**, sunny place; sunshade; sun-hat; peak *of a cap*.

**güney** South.

**gün·kü** Of the day. **geçen** ~ **gazete**, the paper of a few days ago: **her** ~, everyday. ~**lemek**, to pass the day *in a place*. ~**lük**[1], sufficient for *so many* days; *so many* days old; **on** ~ **zahire**, provisions for ten days; **iki** ~ **çocuk**, a two-day-old baby: ~ **emir**, order of the day (*mil.*): ~ **güneşlik**, bright sunny weather: ~ **yumurta**, new-laid egg. ~**lük**[2], liquidambar tree; incense, frankincense. ~**übirlik**, day visit (without spending the night). ~**ügününe**, to the very day; right on time.

**günye** *v.* **gönye.**

**güpegündüz** (ˋ···) In broad daylight.

**güpeşte** *v.* **küpeşte.**

**gür**[1] Abundant; dense; rank. ~ **sesli**, with a fine strong voice.

**gür**[2] *In* ~ ~, *imitates a gurgling or a humming noise.*

**gürbüz** Sturdy; robust. ~**lük**, sturdiness; healthiness.

**Gürcü** Georgian. ~**stan**, Georgia.

**güreş** Wrestling. ~**ci**, wrestler. ~**mek**, to wrestle: **başa** ~, to wrestle for the championship; (*fig.*) to concern oneself only with important matters.

**gürgen** Hornbeam; made of hornbeam; beautiful; showy; (*commercially* **gürgen** *is often used for* beechwood).

**gürleme** Loud noise. **gök** ~, thunder. ~**k**, to make a loud noise: **gök** ~, to thunder.

**gürlük** Abundance; exuberance;

luxuriance. ⌜**Allah son gürlüğü versin!**⌝, 'may God give him happiness and prosperity in the end!'

**güruh** Class, group (*always derogatory*).

**gürül** *In* ~ ~, bubbling, gurgling; in a loud and rich voice. **~demek,** to make a loud noise; to thunder; (of cattle) to low.

**gürültü** Loud noise; uproar; confusion. **~ye gitmek,** to be lost in the confusion; to suffer punishment *or* loss through no fault of one's own; for the innocent to suffer with the guilty. **~ye pabuc bırakmamak,** not to be intimidated by mere threats. **~cü,** noisy; boisterous (person). **~lü,** noisy, tumultuous, crowded. **~süz,** noiseless, quiet.

**gürz** Iron club; mace.

**güteperka** (··᾿·) Guttapercha.

**gütmek, -der** *va.* Drive before one; drive *an animal* to pasture; cherish; nurse (a project *or* a grievance). **hedef ~,** to pursue an aim: **kan ~,** to nurse revenge for a murder, to have a blood feud.

**güve** Clothes-moth.

**güvec** Earthenware cooking-pot; casserole. **türlü ~,** meat and vegetables *cooked in such a pot.*

**güven** Confidence; reliance. **~lik,** security. **~mek,** (*with dat.*) trust in; rely on; be confident; dare: ⌜**güvendiğim dağlara kar yağdı**⌝, I have been let down, sadly disappointed.

**güvercin** Pigeon. **~ gerdanı,** shot colour. **~lik,** dove-cot, pigeon-loft; small cupboard *in the stern of a boat*; small fort *in the form of a tower.* **~otu, -nu,** vervain; verbena.

**güverte** (·᾿·) Deck. **~ yolcusu,** deck passenger (without a cabin). **~li,** decked.

**güvey** Bridegroom; son-in-law. **~ feneri,** winter-cherry: **~ girmek,** to marry.

**güveyeniği** (·᾿···), **-nı** Moth-eaten place.

**güz** Autumn.

**güzaf** (·–᾿) Idle talk; bombast. **lâfü ~,** empty words; brag.

**güzel** Beautiful, pretty; good; nice. A beauty. **~ hava,** fine weather: **çok ~!,** very good! **~avratotu** (···᾿··), **-nu,** deadly nightshade; belladonna. **~ce** (··᾿), pretty fair; fairly well; (·᾿·), thoroughly. **~im,** that beautiful (thing) of mine, of his *etc.* **~lik,** beauty; goodness, agreeableness; fragrancy; happiness: **~ ile,** gently, without using force: ⌜**zorla ~ olmaz**⌝, *lit* 'good cannot be imposed by force', *but the meaning is rather* 'this is something which cannot be achieved by force'.

**güzergâh** Place of passage; ford; ferry; route (of a bus *etc.*).

**güzi·de** (·–᾿) Choice; elect; select. **~delik,** choice part, the best part.

**güz·lük** Autumn-sown (crops). **~ün** (᾿·), in the autumn.

# H

**ha** *interj. Calling attention, expressing surprise, asking a rhetorical question; as English* 'ah!', 'hi!', 'ha!'. **~ babam konuşuyor!,** he talks and talks!: **~ bugün ~ yarın,** either today or tomorrow, it doesn't matter which: ⌜**~ bugün ~ yarın yaparım diye beni oyaladı**⌝, 'he put me off by saying he would do it within the next day or two': ⌜**~ deyince bulunmaz**⌝, you can't find it on the spur of the moment: ⌜**böyle söyledi ~?**⌝, 'so he said that, did he?': ⌜**gitti ~?**⌝, 'he's gone, has he?': **herkes harb ~ çıktı ~ çıkacak diye telaş içinde idi,** everyone was perturbed about the probability of war: **manzara da manzara ~!,** what a wonderful view!

**habbe** Grain; seed. **~si kalmadı,** there was nothing left: ⌜**~yi kubbe yapmak**⌝, 'to make a mountain out of a molehill': **su ~si,** bubble.

**haber** Knowledge, information; news; anecdote; predicate (*gram.*). **~ almak,** to receive information, to learn; to make inquiry: **~ etm.,** to inform, give notice: **~im olmadan,** unknown to me, without my knowing: **~im var,** I know, I am aware: **~im yok,** I know nothing about it; I have not heard: **kara ~,** bad news. **~ci,** messenger; telltale. **~dar,** possessed of information: **~ etm.,** to inform, to bring to the knowledge of. **~leşmek,** to keep one another informed; to correspond.

**Habeş, habeşî** Abyssinian; dark-olive coloured. **~istan,** Abyssinian.

**habib** (·᾿) Lover, beloved; friend.

**Habil** (–᾿) Abel. **~ ile Kabil,** Cain and Abel.

**habire** (·–᾿) Continuously.

**habîs** (·᾿) Wicked; abominable;

vicious; malignant (tumour). Wretch, scoundrel.

**hac, -ccı** The pilgrimage to Mecca. **~ca gitmek (varmak)**, to make the pilgrimage.

**hacalet** (·–˘), **-ti** Shame; mortification.

**hacamat, -tı** A bleeding *or* cupping.

**haccetmek** (˘··), **-eder** *vn.* Make the pilgrimage to Mecca.

**hacet** (–˘), **-ti** Need; necessity; requirement, want; prayer *to God or some supernatural power, asking for stg.* **~ dilemek,** to make such a prayer: **~ kapısı (penceresi)**, door *or* window of a saint's tomb, *where people pray for the fulfilment of a wish*: **~ yok,** there is no need for it: **defi ~ etm.,** to relieve nature: **ne ~?**, what need is there?; for what purpose?

**hacı** One who has made the pilgrimage to Mecca, Hadji; pilgrim. ⌜**~sı hocası**⌝, 'all the ecclesiastics': ⌜**seni gören ~ olur**⌝, 'you're quite a stranger'. **~ağa,** a provincial nouveau riche. **~baba,** a venerable old man who has made the pilgrimage. **~laryolu,** the Milky Way. **~otu, -nu,** mandrake plant. **~yağı, -nı,** a kind of cheap perfume. **~yatmaz,** *a toy loaded at the bottom, so that when knocked over it gets up again,* tumbler; restless, mischievous child.

**hacım** *v.* **hacim.**

**hacim, -cmi** Volume, bulk; tonnage.

**hacir, -cri** Prohibition, interdiction, *esp. the legal interdiction of a person's control over property etc.*

**Hacivad** *A character in Karagöz representing the official type.*

**haciz, -czi** Sequestration; distraint.

**hacmen** (˘·) In volume; in size.

**haç, -çı** Cross; crucifix. **~ çıkarmak,** to cross oneself. **~lı,** having a cross: **~lar,** the Crusaders.

**had, -ddi** Boundary; limit; degree, rank. **~dini* bilmek,** to know one's place: **~dini* bilmemek,** to be above oneself, to presume: **···e ~dini bildirmek,** to put *s.o.* in his place, to teach him how to behave: **~ çizmek,** to set a limit to: **~dim değil,** it is not for me (to say *or* do *stg.*): **~dine mi düşmüş!**, he wouldn't dare!: **~dim olmıyarak,** if I may be so presumptuous; although it is not for me to say: **~dim yok,** I dare not, I have no right: **~di zatında,** in itself, essentially: **~den ziyade,** beyond the limit, excessively: **kimin ~dine?**, who would dare?: **ne ~dimize!**, how could I presume to do such a thing?; I wouldn't dare.

**hâdde** Wire-drawer's plate; rolling-mill.

**haddi** *v.* **had.**

**hademe** Servant (*in offices, schools, etc.*).

**hadım** Eunuch. **~ etm.,** to castrate. **~ağası, -nı,** chief eunuch in a palace *or* great house; black eunuch.

**hadi** (˘·) *v.* **haydi.**

**hadis** (·˘) Religious tradition of the Prophet.

**hâdis** (–˘) Of recent occurrence. **~ olm.,** to occur; to come into existence. **~ e** (*pl.* **hâdisat**), event; incident; accident: **~ çıkarmak,** to provoke an incident.

**hadsiz** Unlimited. **~ hesabsız,** innumerable.

**haf** Half-back (football).

**hafakan** Palpitation. **~lar basmak (bozmak)**, to be exasperated.

**hafazanallah** (··˘··) 'May God preserve us!', *used when talking about a disaster.*

**hafız** (–˘) Keeper; protector; one who has committed the Koran to memory; (*sl.*) simpleton: **~ı kütüb,** librarian. **~a** (–·˘), memory.

**hafid** (·˘) Grandson. **~e** (·–˘), granddaughter.

**hafif** Light; easy; flighty. **~ten almak,** to make light of, not to take seriously: **~ tertib,** slightly, just **a** little. **~lemek,** to become lighter *or* easier; to lose weight; diminish. **~lik,** lightness; flightiness; relief, ease of mind. **~meşreb,** flighty, frivolous; dissolute. **~meşreblik,** levity, frivolity; looseness of morals.

**hafi·ye** Detective; spy; secret agent. **~yyen** (·˘·), secretly; stealthily.

**hafr·etmek** (˘··), **-eder** *va.* Excavate. **~iyat, -tı,** excavations.

**hafta** Week. **~larca,** for weeks on end: **~sına,** the same day the following week: **~sına kalmaz,** within a week: **~ya,** a week today. **~başı, -nı,** the first day of the week; payday. **~lık,** weekly; per week; weekly wages. **~lıkçı, ~lıklı,** worker paid by the week.

**haftaym** (˘·) Half-time (football).

**haham** Rabbi. **~başı, -nı,** Chief Rabbi.

**hain** (–˘) Traitor. Treacherous; deceitful; ungrateful; mischievous. **tuz**

**ekmek ~i,** who repays his host with ingratitude. **~leşmek,** to become a traitor, behave treacherously. **~lik,** treachery, perfidy; an act of treachery.

**haiz** (–̀) Possessing; obtaining; furnished with. **bu şeraiti ~ olanlar,** those who fulfil these conditions.

**hâk** (–), **-ki** Earth, soil. **~ ile yeksan etm.,** to raze to the ground.

**hak[1]** (·), **-kkı** Truth; right; justice; due; respect, relation. Right; true; proper; equitable. **~kında,** concerning; **bunun ~kında,** concerning this: **~kını* almak,** to get one's due, to get one's fare share: **···in ~kından gelmek,** to get the better of *s.o. or stg.*; to pay *s.o.* out: **~kı* için,** for the sake of: **~ kazanmak,** to be proved right; to deserve: **~kı sükut,** hush-money: **~ üzere,** according to equity, by rights: **~kı* var,** he is right; **onda bir lira ~kım var,** he owes me a lira: **···e ~ vermek,** to acknowledge *s.o.* to be right: **···in ~kını* yemek,** to cheat *s.o.* of his rights, to wrong *s.o.*: **~ yerini buldu,** justice prevailed: **~kı yerine getirmek,** to make right prevail, to do justice: **ana (baba, hoca) ~kı,** the debt one owes to one's mother (father, teacher): **namusum ~kı için,** upon my honour: **telif ~kı,** copyright: '**vazifeyi ihmal etmeniz ~kınızda iyi olmaz**', 'it will be bad for you if you neglect your duty'.

**Hak[2]**, **-kkı** *or* **Cenabı ~,** God.

**hak[3]** (·), **-kki** Engraving; erasing.

**hakan** (–̀) Oriental potentate; Sultan.

**hakaret** (·–̀), **-ti** Insult; contempt. **~ etm.,** to insult: **~ görmek,** to be insulted. **~amiz** (·–·–̀), insulting.

**hakem** Arbitrator; referee, umpire.

**haketmek** (̀··), **-eder** *va.* Deserve; be entitled to.

**hakî** (–́) Earth-coloured; khaki.

**hakikat** ·–̀), **-ti** Truth; reality, sincerity. Truly, really. **~i hal,** the truth of the matter. **~en** (·́··), in truth; really. **~li,** true; sincere; faithful. **~siz,** false; insincere.

**hakikî** (·–́) True; real; genuine; sincere. **~ mermi,** live cartridge.

**hâkim** (–̀) Ruling; dominating; overlooking. Ruler; governor; judge. **~i mutlak,** absolute ruler: **···e ~ olm.,** to overlook *a place etc.* **~ane** (–·–̀), lordly; as befits a ruler. **~iyet** (–··̀), **-ti,** sovereignty; domination; rule: **~i milliye,** the sovereignty of the people. **~lik,** the office of a judge *or* ruler; domination.

**hakîm** (·́) Very wise *or* learned. Sage; philosopher. **~ane** (·––̀), wise; prudent; philosophic; in a wise, prudent *or* philosophic manner.

**hakîr** (·́) Despicable; of no account; 'your humble servant'. **~ görmek,** to despise. **~ane** (·––̀), humble, modest: in a humble *or* modest way.

**hakkâk** (·́), **-ki** Engraver.

**hakk·alinsaf** (̀···) In a fair and just manner. **~aniyet** (·–·̀), **-ti,** justice; equity. **~aniyetli,** just, equitable.

**hakketmek** (̀··), **-eder** *va.* Engrave; erase.

**hakk·ı** *v.* **hak[1]**. **~ında,** concerning; with regard to. **~ile, ~iyle,** rightfully, properly.

**hakkuran** *In* **~ kafesi,** a tumble-down house.

**hak·lamak** *va.* Destroy; finish (a meal *or* a man). **~laşmak,** to settle mutual accounts; to be quits.

**hak·lı** Right; who is right; having a right. **~sız,** unjust; wrong; having no rightful claim. **~sızlık,** injustice, wrong. **~şinas,** who knows the truth; just.

**Haktaalâ** (··–̀) God most High.

**hal[1]**, **-li** Condition; situation, state, circumstance; strength; quality; attribute; ecstasy; trouble; the present time; present tense (*gram.*); case (*gram.*). **~de** (*following a participle*), although, *e.g.* **geleceğimi bildiği ~de beklememiş,** although he knew I was coming he did not wait: **~ ile,** consequently; as a matter of course: **~den anlamak,** to be capable of understanding and sympathizing with others: ⌈**~e bak!**⌉, (*referring to an impudent claim*) 'what cheek!'; (*referring to a well-deserved failure*) 'ah! look at the result!', *or* 'what a lesson!': **~ine bakmadan,** *to do or say stg.* without regard to one's abilities *or* circumstances: **~ böyle iken,** and yet, and even under these circumstances: **~i hazır,** the present time; the present tense; the 'status quo': **~i* kalmamak,** to have no strength left: **~ olm.,** (of dervishes *etc.*) to be in a trance, to be in convulsions; ⌈**sana bir ~ olmuş!**⌉, 'what's come over you?', 'what on earth has happened to you *that you should behave thus*?'; ⌈**şayed bana bir ~ olursa**⌉, 'if anything should happen to me'; ⌜**çocuğu susturuncaya kadar bir ~ oldum**⌝, 'I had a job to quieten the child': ···**in ~ini sormak,** to inquire after *s.o.*, to ask after his health *etc.*: **~ hatır sormak,** to ask formal questions showing a general interest in a person *or* his family: ⌜**senin ~in neye varacak!**⌝, 'what will become of you I don't know (if you go on like this)': ⌜**ne ~in varsa, gör!**⌝, 'well, go your own way!': ⌜**aşağı inmeğe ~im yok**⌝, 'I don't feel like coming downstairs': **~e yola** (*or* **~ine yoluna**) **koymak,** to put to rights, to put in order: ⌜**bu adamın ~lerini beğenmiyorum**⌝, 'I don't like the way this man carries on': ⌜**onun her ~i sinirime·dokunuyor**⌝, 'everything about him gets on my nerves': **başına* ~ler gelmek,** for misfortunes to befall one: **her ~de,** in any case; at any rate: **kendi halinde,** quiet and inoffensive; insignificant; ⌜**onu kendi ~ine bırak!**⌝, 'let him be!', 'leave him alone!': ⌜**kimse ~in nedir demedi**⌝, 'no one took any interest in me': **o ~de,** in that case: **şimdiki ~de,** under present circumstances.

**hal², -lli** Solution *of a question or problem*; a melting, liquefaction.

**hal³, -li** (*Fr. halles*) Covered market-place.

**hala** (ˈ·) Paternal aunt.

**halâ** (·ˈ) Void, vacuum; latrine, closet.

**hâlâ** (ˈ–) Up till now; still; yet.

**halâs** (·ˈ) Salvation. **~ bulmak,** to be saved: **~ etm.,** to save, deliver. **~kâr** (·–ˈ), saviour, deliverer.

**halat, -tı** Rope; hawser.

**halâvet** (·–ˈ), **-ti** Sweetness; agreeableness. **~li,** sweet; agreeable. **~siz,** disagreeable, ugly.

**halayık** Female servant *or* slave.

**halazade** (··–ˈ) Cousin (child of a paternal aunt).

**halbuki** (·ˈ·) Nevertheless; whereas.

**hale** (–ˈ) Halo *round the moon.*

**Haleb** (ˈ·) Aleppo. **~ çıbanı,** the Aleppo button *or* boil: ⌜**~ orada ise arşın burada**⌝, 'well, prove it!'; 'let's put it to the test!'.

**halecan** Palpitation *of the heart*; excitement, agitation, anxiety. **~lanmak,** to be agitated *or* excited.

**halef** Successor; posterity; substitute. **~ selef olm.,** to succeed *another in office etc.* **~iyet, -ti,** succession; subrogation.

**halel** Defect; injury; prejudice. **~ gelmek,** for injury *or* prejudice to occur: ···**e ~ getirmek (vermek),** to cause harm to, to prejudice. **~dar,** harmful, prejudicial.

**halen** (ˈ·) Now; at present. **~ ve kalen,** in manner and speech.

**halet** (–ˈ), **-ti** Situation; state; condition. **~i ruhiye** *or* **ruh ~i,** state of mind.

**hal'etmek** (ˈ··), **-eder** *va.* Dethrone, depose.

**halezon** *v.* **helezon.**

**halhal, -li** Anklet *or* bangle *worn by women.*

**halı** Carpet. **~cı,** maker *or* seller of carpets.

**hali¹, -l'i** Dethronement, deposition.

**hali², hâli** (–ˈ) Vacant, empty; unoccupied. ···**den ~,** free from, exempt from.

**halic** Strait; estuary; canal. **H~,** the Golden Horn.

**halife** (·–ˈ) The Caliph. **~lik,** the Caliphate.

**hali·harb** State of war. **~hazırda** (ˈ·–··), at the present moment.

**halik** (–ˈ), **-kı** Creator.

**halîm** (·ˈ) Mild; gentle; patient. **~ selim,** quiet and good-tempered.

**halis** (–ˈ) Pure, unadulterated; genuine; sincere. **~ muhlis,** true, authentic. **~ane** (–·–ˈ), sincere (words *etc.*); sincerely; without ulterior motive. **~üddem** (ˈ···), pure-blooded, thoroughbred.

**halita** (·–ˈ) Mixed substance; alloy.

**hali(y)le** Consequently; as a matter of course.

**halk, -kı** People; crowd; the common people. **ev ~ı,** the household.

**halka** Ring; hoop; circle; door-knocker; link *of a chain.* **~yı burnuna takmak,** to bring into submission: **~ olm.,** to form a ring: **ayın ~sı,** halo round the moon: **ders ~sı,** a class of students. **~lamak,** to furnish with a ring; encircle; fasten with a ring on to a hook (door *etc.*). **~lı,** ringed; linked; in coils.

**halkâri** Robe embroidered with gold.

**halkçı** An upholder of the rights of the people; democrat. **~lık,** democracy.

**halketmek** (ˈ··), **-eder** *va.* Create.

**halkıyat, -tı** Folklore.

**hallac** Wool-carder. **~ pamuğu gibi dağıtmak,** to scatter about in all directions.

**hall·enmek** *vn.* Manifest interest *in a thing or a person.* **~eşmek,** to confide troubles to one another, to have a good talk.
**halletmek** (ʾ··), **–eder** *va.* Undo; solve (a question); explain; dissolve; analyse.
**halli**[1] *acc. of* **hal**[2], *q.v.*
**halli**[2] Of *or* in a *certain* condition. **orta ~,** not too well-off. **~ce,** slightly better; pretty well-off.
**hallihamur** (···ʾ) Of the same substance; one with; 'part and parcel' (of = **ile**).
**halsiz** Weak; exhausted, tired out. **~lik,** exhaustion.
**halt, –tı** Impertinence; stupid and improper speech *or* deed. **~ etm.,** *v.* **haltetmek**: **~ karıştırmak (yemek),** to make a great blunder, to 'put one's foot in it' badly. **~etmek** (ʾ··), **–eder,** to do *or* say *stg.* stupid *or* out-of-place: ⌜**ne halteder ağanın beygiri!**⌝, 'what an absurd mess!' *or* 'how are we to get out of this mess?': ⌜**ne haltetmeğe oraya gittin?**⌝, 'what the dickens did you go there for?': **ona ~ düşer!,** he has no right at all to interfere; it's nothing to do with him!: **yanında haltetmiş,** he is nothing in comparison.
**halter** (*Fr. haltère*) Dumb-bell.
**halûk** (·ʾ̲) Of good character; well-disposed, decent.
**halükâr** (–·ʾ̲) *In* **her ~da,** in all circumstances.
**halva** *In* ⌜**helva demesini de bilirim ~ demesini de**⌝, 'I can adjust myself to all circumstances'.
**halvet, –ti** Solitude; retirement; privacy; private room; private room in a bath. **~e dönmek,** (of a room) to become hot and stuffy: **~ etm.,** to make a room private by turning out other persons; to retire into a private room (*also* **~e girmek**).
**ham** Unripe; immature; raw, crude; out of training; vain, useless; unreasonable. **~ halat,** coarse, clumsy (man): **~ toprak,** uncultivated land: **hayali ~,** vain illusion; hopeless task.
**hamail** (·–ʾ) Band *or* cord worn over one shoulder *for carrying a weapon etc.*; amulet *thus worn*; any amulet *or* charm. **~î,** suspended by a cord *or* belt worn over one shoulder and under the other.
**hamak** Hammock.
**hamakat, –ti** Stupidity; folly.
**hamal** Porter; carrier; common coarse fellow. **~ camal,** the lowest class of people. **~başı** (·ʾ··), **–nı,** foreman of a group of porters. **~iye,** porterage. **~lık,** profession of a porter; coarse behaviour; toiling and slaving; unnecessary burden.
**hamam** Bath, *esp.* a Turkish bath. **~ anası,** a huge woman: **~ takımı,** bath towels. **~böceği, –ni,** the so-called 'German' cockroach. **~cı,** proprietor of a public bath; canonically unclean and in need of a ritual bath.
**hamarat** Hardworking, industrious.
**hamas·et** (·–ʾ), **–ti** Heroism. **~î,** epic.
**hamd** A giving thanks and praise to God. **~etmek** (ʾ··), **–eder,** to give thanks and praise to God. **~olsun** (ʾ··), thank God!; thanks be! **~üsena** (···ʾ̲), thanks and glory to God!
**hamel** The Zodiac sign Aries.
**hamhalat** (ʾ··) Loutish, boorish.
**hamhum** Humming and hawing. **~ şarolop,** a lot of nonsense; swindle.
**hami** (–ʾ) Protecting, guarding. Protector; patron.
**hamî** (–ʾ̲) Hamite. Hamitic.
**hamil, –mli** Pregnancy. **vaz'ı ~,** parturition: **~ vakti,** the moment of parturition.
**hâmil** (–ʾ) Bearing; bringing. Bearer (of a document *etc.*). **~e,** pregnant woman: **···den ~ kalmak,** to be with child by ... . **~en** (ʾ̲··) (*with acc.*), having with one; furnished with.
**hâminne** (·ʾ·) 'Grannie'.
**hâmis** (–ʾ) Fifth. **~en,** fifthly.
**hâmiş** (–ʾ) Marginal note; postscript.
**hamiyet, –ti** Zeal; public spirit; patriotism. **~li,** public-spirited; philanthropic.
**hamlac** Blowpipe.
**hamlacı** Chief rower; stroke oarsman; *title of the Palace boatmen in fomer times.*
**hamlamak, hamlaşmak** *vn.* Get out of condition *or* out of practice; become soft from lack of work.
**hamle** Attack, onslaught; effort; dash, 'élan'; turn (at chess, draughts *etc.*). **~ etm.,** to attack; make a great effort: **ilk ~de,** at the first onslaught; 'at the first go'.
**hamletmek** (ʾ··), **–eder** *va.* Load; ascribe, impute (to = *dat.*).
**hamlık** Unripeness; crudeness; inexperience; lack of condition.

**hampa** *v.* **hempa.**
**hamsi** Anchovy (fresh).
**hamsin** Fifty; the last fifty days of winter; the hot summer wind of Egypt.
**hamud** Collar *of a horse's harness.*
**hamule** (·–ˈ) Cargo; load.
**hamur** Dough; leaven; essence, nature; quality (of paper); anything of the consistency of dough. Half-baked (bread). ~ **açmak,** to roll out dough: ~ **gibi,** limp, flabby: ~ **işi,** pastry: ~ **tutmak,** to knead dough *for bread or pastry*: ⌜**elinin ~iyle erkek işine karışmak**⌝, to set out to do *stg.* beyond one's power; (of a woman) to try to do a man's job: **hal ve** ~, *v.* **hallihamur.**
**hamur·kâr,** baker's assistant. **~lamak,** to cover with dough; lute. **~lu,** covered *or* made with dough; leavened; fermented. **~suz,** unleavened; unleavened bread: ~ **bayramı,** the Jewish Feast of Passover.
**han**[1] Sovereign; *oriental title of princes etc.,* Khan.
**han**[2] Inn; caravanserai; large commercial building. ~ **gibi,** vast: ~ **hamam sahibi,** a man of property: ⌜**burası yol geçen ~ı değil**⌝, 'where do you think you are!'; 'you must behave yourself here!' **~cı,** innkeeper: ⌜**ben ~ sen yolcu iken**⌝, you may be glad of my help some day.
**hançer** Dagger.
**hançere** Larynx.
**handiyse** (·ˈ·) Shortly; any moment now.
**hane** (–ˈ) House; dwelling; family *or* household; compartment; subdivision; column (for figures); square *of a chessboard*; sign of the Zodiac. ~ ~, in separate compartments. *Used as a suffix and often abbreviated to* **–ane** *or after* **a** *to* **–ne,** house of ..., place of ..., *e.g.* **hastahane** *or* **hastane,** hospital; **postane,** post office.
**hanedan** (–·ˈ) Great family; dynasty. Of illustrious descent, noble; courteous; hospitable.
**hanefi** (··ˈ) Of the Hanefi sect; orthodox.
**haneharab** (–··ˈ) Whose home is ruined; miserable; good-for-nothing.
**haneli** (–·ˈ) Having *so many* houses; in squares, check (pattern). **dört** ~ **rakam,** a four-figure sum: **kırk** ~ **köy,** a village of forty houses.
**hanende** (–·ˈ) Singer.
**hangi** (ˈ·) Which?; whichever. **~niz,** which of you?; whichever of you: **~si,** which of them?; whichever of them.
**hanım** Lady. ~ **evlâdı,** mollycoddle; a rather too well-behaved child. **~böceği, –ni,** ladybird. **~efendi,** madam. **~eli, –ni,** honeysuckle. **~hanımcık,** a model housewife.
**hani**[1] Sea-perch; comber.
**hani**[2] (ˈ·) Where?; well?; you know!; and then. ~ **bana?,** what about me?, where do I come in?: **~dir,** it's a long time now since ...: ⌜**~ dün bize gelecektin**⌝, 'I thought you were coming to see us yesterday (but you didn't)': ~ **o günler!,** (*expressing regret*) 'ah! those days are over!'; 'I only wish we could!'; (*iron.*) 'what an idea!'; 'out of the question!': ⌜**~ yok mu?**⌝, 'you know what I mean': **güzelliğine de güzel ~!,** there's no question about her beauty!
**haniya** *v.* **hani.**[2]
**hanlık** Title *or* territory of a Khan; khanate.
**hantal** Big; clumsy; badly made; coarse; clownish, boorish.
**hanüman** (··ˈ) Home; family and belongings.
**Hanya** (ˈ·) Canea (in Crete). ⌜**~yı Konya'yı anlamak**⌝, to learn by bitter experience: ⌜**sana ~yı Konya'yı göstereri m!**⌝, I'll teach you a lesson! (*threateningly*)'.
**hap, –pı** Pill. **~ı yutmak,** to be 'done for': ⌜**tam mânasile ~ı yuttuk**⌝, we're properly 'in the soup'; it's all up with us! **~ıcık,** *in* ~ **yapmak,** to catch *a sweet, a cherry etc.* in the mouth.
**hapis, –psi** Confinement, imprisonment; prison. Imprisoned; detained. ~ **giymek,** to be sentenced to prison: **···i göz hapsinde tutmak,** not to let *s.o.* out of one's sight; to keep under observation. **~hane, ~ane** (··–ˈ), prison.
**hapsetmek** (ˈ··), **–eder** *va.* Imprison; confine.
**hapş·u** *Imitates a sneeze,* 'atishoo!' **~ırmak,** to sneeze.
**har** (·) *In* ⌜**~ vurup harman savurmak**⌝, to spend money prodigally, to 'blue' one's money.
**hara** (·ˈ) Stud farm; stock farm.
**harab** (·ˈ) A ruining *or* destroying. Ruined, devastated; decayed; desolate. ~ **etm.,** to destroy, devastate: ~ **olm.,** to be devastated; to fall into

ruin; to be impoverished; to be desperately in love: **hali*** ~, he is ruined; he is in a bad way.
**harab·at** (·–ˈ), **–tı** *pl. of* **harabe.** ruins; *as sing.* tavern. **~ati** (·––ˈ), (*originally*) dissolute drunkard; (*now*) careless about his dress and unconventional in his habits; bohemian. **~e** (·–ˈ), a ruin; tumble-down house *or* town. **~i** (·–ˈ), ruin; poverty, misery. **~iyet** (·–·ˈ), **–ti,** **~lık,** a state of ruin; destruction.

**harac** Tax; tribute; public auction; (*formerly*) tax paid by non-Moslems in lieu of military service. ~ **~!,** going! going! (*at an auction*): **~a çıkarmak,** to put up to auction: **~a kesmek,** to levy a tribute on *a place*; to extort heavy taxes; to oppress: ~ **mezad,** selling by auction. **~cı,** collector of tribute.

**haram** (·ˈ) Forbidden by religion; unlawful; sacred, inviolable. Forbidden deed. ~ **etm.,** to forbid the use *or* enjoyment *of stg.*: ~ **mal,** property unlawfully acquired; ill-gotten gains: ⌜~ **olsun!**⌝, 'may Heaven punish you!' (*said to one who wrongfully seizes another's property*): ~ **yemek,** to enrich oneself unlawfully. **~i** (·–ˈ), brigand, thief. **~lik** (·––ˈ), brigandage. **~zade** (·––ˈ), bastard; scoundrel.

**harar** Large sack made of haircloth.

**hararet** (·–ˈ), **–ti** Heat; fever, temperature; thirst; fervour, exaltation. ~ **basmak,** to feel thirsty: ~ **söndürmek,** to quench one's thirst: ~ **vermek,** to make thirsty. **~li,** irascible; heated (argument); feverish; enthusiastic.

**harb, –bi** War. ~ **hali,** state of war: ~ **ilânı,** declaration of war: ~ **zengini,** war profiteer. **~ci,** warlike; warmonger. **~e,** javelin, pike. **~en** (ˈ·), by war, by force of arms. **~etmek** (ˈ··), **–eder,** to go to war, fight. **~i,** ramrod. **~iye,** War Academy; Ministry of War: **erkânı** ~, General Staff. **~sonrası, –nı, ~sonu, –nu,** post-war period; aftermath of war.

**harc** Expenditure, expenses, outlay; raw material; ingredients; trimmings (braid *etc.* for a dress); anything within one's power *or* means; mortar, plaster; soil mixture, compost. **~ı âlem,** within everybody's means; in common use, ordinary, everyday: **akıllı adamın ~ı,** the way a wise man would act: **benim ~ım değil,** that's not for me; it's beyond my means: **mahkeme ~ı,** legal costs.

**harc·amak,** to expend, spend; use; designedly to put a person in danger, get rid of, kill. **~etmek** (ˈ··), **–eder,** to spend, disburse. **~ıâlem,** *v.* **harcı âlem.** **~ırah,** journey-money, travelling expenses. **~li,** made with mortar *or* compost; trimmed *with braid etc.*; expensive. **~lık,** pocket-money.

**hardal** Mustard. ~ **lâpası,** mustard poultice: ~ **yakası,** mustard plaster. **~iye** (·–·ˈ), grape-juice flavoured with mustard. **~lık,** mustard-pot.

**hare** (–ˈ) Moiré; watering (of silk *etc.*).

**harekât** (··ˈ), **–tı** (i) *pl. of* **hareket.** Movements. (ii) *pl. of* **hareke.** Vowel points.

**hareke** Vowel point (in Arabic writing). **~lemek,** to insert the vowel points. **~li,** with the vowel points inserted.

**hareket, –ti** Movement; act; behaviour; departure; excitement. **~i arz,** earthquake: ~ **etm.,** to move, act; set out: **~e gelmek,** to begin to move; to come into play: **~e getirmek,** to set in motion: ~ **kolu,** starting-handle: ⌜**nerede** ~ **orada bereket**⌝, activity brings prosperity. **~siz,** motionless. **~sizlik,** immobility.

**hareli** (–·ˈ) Watered (silk *etc.*), moiré.

**harem** The women's apartments *in a Moslem house*; a sacred territory *esp. that of Medina and Mecca.* ~ **selâmlık olm.,** for men and women to form separate groups. **~ağası, –nı,** black eunuch. **~eyn,** the two sacred places of Islam, Mecca and Medina. **~işerif** (····ˈ), the Prophet's tomb at Medina.

**harf, –fi** Letter *of the alphabet*; particle (*gram.*); word; speech, language; allusion; witticism. **~i ~ine,** to the very letter; word for word: ~ **inkılâbı,** the reform of the Turkish alphabet, (*change from Arabic to Latin script,* 1928). **~î** (·ˈ), pertaining to letters, literal. **~itarif** (···ˈ), the definite article (*gram.*). **~iyen** (·ˈ·), literally.

**harhara** Death-rattle.

**harıl** *In* ~ ~, assiduously; ~ ~ **yanmak,** to burn furiously and continuously. **~tı,** loud and continuous noise *esp. that of burning.*

**harın** Restive, refractory (horse); obstinate. **~lamak,** (of a horse) to become restive after staying too long in the stable.

**haric** (–ˈ) The outside, exterior; abroad. External, outside; not included. ~ **etm.**, to exclude: ~ **olm.**, to be excluded: ~ **ez kanun**, extra-legal: ~ **ez memleket**, extra-territorial: **harb ~i kalmak**, to remain out of the war: **hasmı muharebe ~i bırakmak**, to put the enemy out of action: **ihtimalden ~**, improbable.

~**en** (ˈ··), externally. ~**î** (–·ˈ), external; foreign; heretic. ~**iye** (–··ˈ), foreign affairs; external diseases. ~**ciyeci**, member of the Foreign Service; specialist in external diseases.

**harika** (–·ˈ) Wonder, prodigy. Marvellous.

**harikulâde** (ˈ··–·) Extraordinary; unusual; prodigious.

**harim** (·ˈ) Most intimate and private place; any place that a man is bound to protect and defend. Private; intimate.

**haris** (·ˈ) Greedy; avaricious; ambitious.

**harita** (·ˈ·) Map; plan. ~**da olmamak**, to be unexpected, unforeseen (of difficulties *etc.*).

**harl·amak** *vn.* Be in flames; burn furiously. ~**ı**, burning furiously.

**harman** Operation of threshing grain; heap of grain for threshing; threshing-floor; harvest time; blend (of tea, tobacco *etc.*); heap of printed leaves *ready to be put into a book.* ~ **dövmek**, to thresh: ~ **döveni**, a kind of sledge driven over the grain to thresh it out: ~ **etm.**, to blend; to sort and arrange: ~ **savurmak**, to winnow: ~ **sonu**, residue of grain *mixed with stones and dust, left after threshing*; remnants of a fortune *or* business: ~ **yeri**, threshing floor.

**harmancı** Thresher; blender *of tobacco etc.*

**harmani** (·–ˈ) Long cloak.

**harman·lamak** *va.* Blend (tea, tobacco *etc.*). ~**lanmak, ay ~**, for a halo to form round the moon.

**harp**[1], **-pı, harpa** (ˈ·) Harp.

**harp**[2] *v.* **harb.**

**harrangürre** (ˈ·ˈ·) In a disorderly and noisy manner.

**harrü** (ˈ·) *In* ⌜**ya ~ ya marrü**⌝, 'well, we'll have a shot at it; perhaps it will come off, perhaps it won't'.

**hars** Culture; education. ~**î** (·ˈ), cultural.

**harta** (ˈ·) Map; plan.

**hartadak** *Imitates the sound of stg. being bitten or suddenly seized.*

**hartuç** Cartridge (for cannon).

**harub** Carob, locust-bean. ~**iye**, a drink made of locust-beans.

**has** (–), **-ssı** Special; peculiar *to*; private; pure, unmixed; fast (dye). The upper class; *formerly* a fief of a yearly value of over 100,000 **akçe.**

**hasad** Reaping; harvest. ~ **etm.**, to reap.

**hasar** (·ˈ) Damage; loss. ~**a uğramak**, to suffer loss *or* damage. ~**at** (·–ˈ), **-tı**, losses.

**hasb·elicab** (ˈ···) According to need; necessarily. ~**eliktıza** (ˈ···–), according to the requirements of the case. ~**ellüzum**, according to need. ~**elvazife** (ˈ··–·), as required by duty.

**hasbetenlillah** (ˈ····) (For the sake of God); disinterestedly; without expecting anything in return; just for the love of it.

**hasbıhal, -li** Private and friendly chat. ~ **etm.**, to have a friendly chat; to exchange confidences.

**hasbi** (·ˈ) Disinterested; gratuitous; without reason. ~ **geçmek** (*sl.*), not to care.

**haseb** Personal qualities; merit. ~ **ve neseb**, distinction by personal qualities as well as by birth.

**hasebi(y)le** By reason of ... .

**hased** Envy; jealousy. ~ **etm.**, to envy. ~**ci**, envious; jealous.

**haseki** (*Formerly*) one of the Sultan's personal bodyguards; favourite wife of the Sultan. ~ **küpesi**, columbine, aquilegia.

**hasenat, -tı**, good works, pious deeds; charitable institutions.

**hâsıl** (–ˈ) Resulting; happening; produced; growing, produce; product; crop; result; profit. ~**ı** *or* ~**ı kelâm**, to sum up, in short. ~ **etm.**, to produce; to acquire: ~ **olm.**, to result, ensue; be produced, be obtained.

**hasıl·a** (–·ˈ), produce. ~**at** (–·ˈ), **-tı**, produce; products; revenue; profit: **safi ~**, net profit. ~**atlı**, productive; profitable. ~**lanmak**, (of children and crops) to grow up. ~**sız**, producing *or* yielding nothing; unprofitable; useless.

**hasım, -smı** Enemy, adversary. ~**lık**, enmity, hostility.

**hasır**[1] (–ˈ), **-srı** A restraining; restriction; a devoting *or* consecrating to one purpose.

**hasır**[2] Rush mat *or* matting. ~ **kaplamak,** to cover with wickerwork: ~ **koltuk,** wicker chair: ~ **şapka,** straw-hat. ~**altı** (·ˋ··), (*lit.* 'under the mat'): ~ **etm.,** to hide; to leave *a request* unanswered; to shelve *a matter*; to hush *stg.* up. ~**lamak,** to cover with matting. ~**lı,** covered with matting; made of wickerwork: large bottle covered with wickerwork. ~**otu, -nu,** rush.

**hasis** (·ˋ) Stingy; vile; mean, petty. ~**lik,** stinginess, vileness.

**hasiyet** (–·ˋ), **-ti** A special quality *or* virtue; good effect on the body. ~**li,** having a special quality; beneficial to the health; savoury.

**haslet, -ti, haslat** Moral quality; character.

**hasm·ane** (·–ˋ) Hostile (attitude *etc.*). ~**ıcan,** a deadly foe, a mortal enemy.

**hasna** (·ˋ) A beauty (woman); an agreeable kind of woman. ~ **müstesna,** an exceptional beauty.

**haspa** Minx; rascal (*used affectionately and generally of a girl*).

**hasret, -ti** Regret *for stg. lost*; longing *for stg. not yet gained*; longing *for a person or place.* Feeling a loss. ···**in** ~**ini çekmek,** to long to see *a person or thing* again: ···**e** ~ **kalmak,** to feel the loss *or* absence of ... . ~**li,** suffering from separation; longing for *s.o. or stg.*

**hasretmek** (ˋ··), **-eder** *va.* Restrict; restrain; appropriate; devote *or* consecrate to one thing.

**hassa**[1] Quality; property; peculiarity.

**hassa**[2] (–ˋ) *fem. of* **has.** A ruler's bodyguard; anything especially belonging to the sovereign. **hazinei** ~, the Privy Purse.

**hassas** (·ˋ) Very sensitive; delicate in feeling; scrupulous. ~**iyet** (·–··), **-ti,** ~**lık,** sensibility; perceptivity; sensitivity; touchiness; quickness of response; delicacy of feeling.

**hassaten** (ˋ··) Specially, particularly.

**hâsse** (–ˋ) Sense; each of the five senses.

**hasta** Sick, ill. ~**bakıcı,** hospital attendant; trained nurse. ~**lık,** illness, disease. ~**lıklı,** ailing; in ill health. ~**ne** (·–ˋ), ~**hane** (··–ˋ), hospital.

**hasud** (·ˋ) Jealous, envious.

**haşa** Saddle-cloth.

**hâşa** (–ˋ) God forbid!. ~ **minhuzur** (**minelhuzur** *or* **huzurdan**), 'with all due respect'; 'excuse the expression!

**haşarat** (··ˋ), **-tı** *pl. of* **haşere.** Insects; reptiles; small creeping things; vermin; mob; blackguards.

**haşarı** Wild; dissolute; out-of-hand; naughty. ~**lık,** unruly behaviour, wild prank.

**haşhaş** Poppy. ~**iye** (·–·ˋ), the poppy family, Papaveraceae.

**haşır** *v.* **hışır** *and* **haşir.**

**haşin** Harsh; rough; bad-tempered.

**haşir, -şri** The assembly of all men for the Day of Judgement; the Day of Judgement. **haşre kadar,** till the Day of Judgement: ···**le** ~ **neşir olm.,** to be in close contact with ..., to be cheek by jowl with ... .

**haşiv, -şvi** Redundant words; padding.

**haşiye** (–·ˋ) Marginal note; comment; postscript.

**haşlama** Boiled. Boiled meat.

**haşlamak** *va.* Boil; scald; (of an insect) to sting; (of frost) to nip; scold severely.

**haşmet, -ti** Majesty; pomp. ~**li,** majestic, grand. ~**lû,** *title given to European sovereigns,* His Majesty. ~**meab** (···ˋ), Your Majesty.

**haşr·etmek, -eder** *va.* Collect people together; to assemble mankind for the Last Judgement. ~**üneşr** *v.* **haşir.**

**haşviyat** (··ˋ), **-tı** Redundant words; verbiage.

**haşyet, -ti** Fear; awe.

**hat**[1], **-ttı** Scratch; line; railway line; mark; writing; decree. ~**tı hareket,** line of conduct, method of proceeding: ~**tı şerif** *or* **humayun,** royal mandate *sent by the Sultan to the Grand Vizier*: **dar** ~, narrow-gauge railway: **geniş** ~, normal gauge railway: **hüsnü** ~, calligraphy.

**hat**[2] *v.* **had.**

**hata** (·ˋ) Mistake; fault; offence. ~ **etm.,** to err; to miss (in shooting). ~**en** (·ˋ·), by mistake, in error. ~**lı,** erroneous.

**hatıl** Beam *or* course of bricks in a stone wall.

**hatır**[1] Thought; idea; memory; mind; feelings; influence; the consideration that one person expects from another. ~**ını* almak,** to content *s.o.*; to have a kindly thought for *s.o.*: ~**dan çıkarmak,** to forget: ~**ından* çıkmak,** to pass out of one's mind, to be forgotten: ~ına* gelmek, to occur to one's mind: ~ına* getirmek, to remind *s.o.*; ~a gönüle bakmamak, to act independently, to be impartial: ~ gönül bilmemek, to take no account of others' feelings: ~ gözetmek, to have regard for the feelings: ~ını* hoş tutmak, to keep one's mind at ease about *stg.*; to keep *another* satisfied: ~ için, as a favour; ···in ~ı için, for the sake of ..., out of regard for ...: ~ı* kalmak, to feel hurt, to be offended; ~nız kalmasın!, don't be offended!: ~ı* olm., to be of account, to be considered, *e.g.* ⌜~ım yok mu?⌝, 'have I no weight, don't I count for anything?': ~larda olm., to be in everyone's mind, to be generally remembered: ~ı* sayılmak, to have one's feelings respected; to have influence; ~ı sayılır, respected; considerable: ~ saymak, to take another's feelings into account; to esteem: ~ sormak, to inquire about *s.o.'s* health: ~da tutmak, to bear in mind: ~ını* yapmak, to placate, to make amends to *s.o.*: sefai ~la, with tranquillity; with pleasure.

**hatır²** *In ~ ~, imitates the noise of eating a raw vegetable etc.*; raw, crude. ~ **hutur,** coarse, unpolished (man).

**hâtıra** (–·˙) Memory; remembrance; souvenir. ~ **defteri,** diary: ···in ~sı **olarak,** in memory of... . ~**at, –tı,** memories; memoirs.

**hatır·lamak** *va.* Remember. ~**latmak,** to remind. ~**lı,** influential; who must be considered; esteemed. ~**nüvaz,** ~**şinas,** considerate, obliging, courteous.

**hatib** Preacher; orator. ~**lik,** oratory.

**hatif** (–˙) An invisible speaker, a mysterious voice; echo.

**hatim, –tmi** A reading *or* reciting the Koran from end to end. ~ **duası,** prayer after reading the whole Koran: ~ **indirmek,** to finish the reading of the Koran: **hatmi kelâm etm.,** to finish speaking.

**hâtime** (–·˙) End, conclusion; epilogue; peroration. ~ **çekmek,** to conclude.

**hatmetmek** (˙··), **–eder** *va.* Conclude; complete; read through *the Kuran or any other book* from end to end; to read *a book* again and again.

**hatmi** Marsh-mallow; hollyhock.

**hats, hatsî** *v.* **hads, hadsî.**

**hattâ** (˙·) Even; so much so that; to the extent that.

**hattat, –tı** Calligrapher; ~**lık,** calligraphy.

**hattı** *v.* **hat.** ~**humayun** (···–˙), royal mandate. ~**istiva,** ~**üstüva** (····˙), the Equator. ~**şerif** (···˙), royal mandate.

**hatun** (–˙) Lady; woman. ~ **kişi,** woman.

**hav** Down (feathers); nap; skin (of a peach).

**hava** Air; weather; wind; atmosphere; climate; tune; liking, desire, whim, fancy; hobby. ~**dan,** for nothing; as a windfall: ~ **almak,** (i) to take the air, to have an airing; (ii) to let in air; (iii) to get nothing: ~ **çalmak,** to play a tune: ~**ya gitmek,** to be in vain, to be wasted: ~ **ve heves,** fancies; pleasures: ~ **oyunu,** speculation on the Stock Exchange: ~ **parası,** premium ('key-money') charged in addition to rent: ~ **payı,** room (space); ~ **payı bırakmak,** to leave a loophole (*fig.*): ~**dan sudan mevzular,** subjects of no importance: ⌜**bana göre** ~ **hoş**⌝, it's all the same to me; it suits me: **başı*** ~**da,** proudly: **kendi** ~**sında* olm.,** to follow one's own fancies; to think only of oneself. **hava·cı,** aviator, airman. ~**cılık,** aviation. ~**cıva,** alkanet; trifles; nought. ~**dar,** airy. ~**fişeği, –ni,** rocket (*mil. signal*). ~**gazı, –nı,** coal-gas. ~**î** (·–˙), aerial; fanciful, flighty; sky-blue: ~ **fişek,** rocket: ~ **hat,** aerial railway: ~ **meşreb,** not serious, frivolous. ~**ilik** (·––˙), flightiness, frivolity. ~**iyat** (·–·˙), **–tı,** futilities; trifles. ~**lanmak,** to take the air; be aired; be ventilated; take to the air, fly, be airborne; (of a girl) to become flighty and frivolous.

**havadis** (·–˙) News.

**havale** (·–˙) Assignment; the referring *or* transfer *of a matter*; bill of exchange; infantile convulsions; eclampsia; lattice fence; overhang. ~ **etm.,** to transfer (a debt, a business *etc.*); to refer *a matter to another person or department*; to point *a weapon* towards *s.o.*: ~ **olm.,** to be referred *to s.o. else*; (of an infant) to have convulsions.

**havale·li** (·–·˙) Docketed; unwieldy, bulky, top-heavy, overhanging; (ship) with a high superstructure; surrounded by a palisade; given to convulsions. **~name** (·—·—̀), order for payment; bill of exchange; money order.

**havalı** Airy; having *such and such* a climate.

**havali** (·—́) Environs, neighbourhood.

**havan** Mortar *for pounding*; mortar (gun); tobacco-cutting machine. **~ eli,** pestle: ⌜**~da su dövmek**⌝, to engage in useless discussions *or* fruitless work: **~ topu,** mortar, howitzer.

**havari** (·—̀) Apostle (*pl.* **~yun,** the Apostles).

**havas** (·́) The upper classes; people of distinction; intimate friends; special qualities; private domains of the Sultan. **~ ve avam,** the upper and the lower classes.

**havasız** Airless, badly ventilated, close.

**havayic** (·—̀) Necessities; wants. **~i zaruriye,** the bare necessities of life.

**havayolları, -nı** Air Lines; Airways.

**haves** *v.* **heves.**

**havhav** *n.* Barking; baying.

**havi** (—́) Containing ... (*with acc.*). **~ olm.,** to contain.

**havil, -vli** *In* **can havliyle,** in terror of one's life, desperately.

**havlamak** *vn,* Bark.

**havlı** Downy; having a nap *or* pile. Towel.

**havli** *v.* **havil.**

**havlican** Galingale (aromatic root).

**havlu** Towel. *v. also* **avlu.**

**havra** Synagogue. **~ya dönmek,** (of a place) to become very noisy and crowded.

**havsal·a** Pelvis; crop *of a bird*; gizzard; comprehension, intelligence. **~sı* almamak,** to be unable to comprehend: **~ya sığmaz,** incomprehensible. **~î** (··́), pelvic.

**havsız** Without down *or* nap; shiny (clothes).

**havşa** Countersink. **~ açmak,** to countersink, counterbore. **~lı,** countersunk.

**havuc** Carrot. **yabani ~,** parsnip.

**havut** Camel pack-saddle.

**havuz** Artificial basin *or* pond; dock. **sabih ~,** floating-dock. **~lamak** *va.,* to dock (a ship).

**Havva** (·́) Eve.

**havya** (̀·) Soldering-iron.

**havyar** Caviare. **~ kesmek,** to idle around, moon about.

**havza** River-basin, catchment area; sphere; domain; territory *of a state or town.*

**hay[1]** Hey!; alas! **~ ~,** certainly! by all means!; right ho!

**hay[2]** *In* ⌜**~dan gelen huya gider**⌝, 'easy come, easy go' (money).

**haya** Testicle.

**hayâ** (·́) Shame; modesty, bashfulness. **~lı,** bashful; modest. **~sız,** shameless, impudent.

**hayal** (·̀), **-li** Spectre, phantom; reflection; image; fancy, imagination; shadow pantomime. **~ etm.,** to imagine *stg.*: **~ kuvveti,** the power of imagination, a lively imagination: **sukutu ~** (**inkisarı ~, ~ kırıklığı**), disappointment.

**hayal·at** (·—̀), **-tı,** vain imaginings: **~a kapılmak,** to be fed on illusions. **~ci,** giver of a puppet-show; visionary, day-dreamer. **~en** (·́·), in imagination. **~et** (·—̀), **-ti,** phantom, ghost, apparition. **~hane** (·——̀), imagination, fancy. **~î** (·—́), fantastic, imaginary; puppet-player. **~ifener,** a 'mere skeleton'. **~meyal,** hardly perceptible: **~ görmek,** to be hardly able to make *stg.* out *from darkness or distance*: **~ hatırlamak,** to remember faintly. **~perest,** visionary; castle-builder.

**hayat[1], -tı** Covered court; vestibule.

**hayat[2]** (·̀), **-tı** (·—̀) Life; living. **kaydı ~ şartile,** for life (appointment *etc.*). **~î** (·—́), vital. **~iyat** (·—·̀), **-tı,** biology. **~iyet** (·—·̀), **-ti,** vitality.

**hayda·lamak, ~mak[1]** *va.* Drive on *animals with loud shouts.* **~mak[2],** cattle-lifter, thief.

**hayderî** (··́), **~ye** Kind of waistcoat *formerly worn in the house.*

**haydi, hadi, haydin** (̀·) Come!; be off! **~ ~,** at the most; easily: **~ gidelim,** come along!, let's be going!: **~ git!,** clear out! **~sene** (···̀), come along!

**haydud** Brigand. **~ yatağı,** brigands' den. **~luk,** brigandage.

**hayhay** Certainly!, by all means!

**hayhuy** *v.* **hayühuy**; *also* **hay[2].**

**hayıf** Alas!; what a pity! **~lanmak** *vn.,* to bemoan, lament.

**hayır[1]** (̀·) No!

**hayır[2], -yrı** Good; prosperity; health; excellence; profit, advantage. Good; advantageous; auspicious. **~lar!,** all well! (*as an answer to* '**ne var ne yok?**'): **hayra alâmet değil,** it bodes

no good: **hayrı dokunmak,** to be of use: ~ **dua,** a blessing: **hayrını görmek,** to enjoy the advantages *or* profit of *stg.*: **çocuklarının hayrını görmedi,** his children turned out badly; **hayrını gör!,** may it bring you good luck! (*said by the vendor to the purchaser*); (*sarcastically*) well, I wish you the luck of it!: ~**hasenat,** pious *or* charitable institutions: ~**dır inşallah!,** I hope all is well!: **hayrı kalmamak,** to be no longer of any use: ~ **ola!,** 'good news, I hope!': ~ **sahibi,** a philanthropist: **hayrı yok,** there is no good to be expected of him (it); **gemiden ~ yoktu,** there was no hope of saving the ship: ⌜**ağzını hayra aç!**⌝, 'say something cheerful for a change!' (*as a rebuke to one who utters gloomy forebodings*).
**hayır·hah,** benevolent. ~**hahlık,** benevolence. ~**laşmak,** to conclude a sale with the formula '**hayrını gör!**'. ~**lı,** good; advantageous; auspicious: ~ **olsun!** *or* **haydi ~sı!,** let's hope for the best; good luck to it! ~**sız,** useless, good-for-nothing; unproductive; ill-omened.
**hayıt** Agnus-castus tree.
**hayız, -yzı** Menstruation.
**hayide** (——̀) Chewed; stale (joke); hackneyed (expression).
**haykırış, haykırma** A shouting, bawling.
**haykırmak** *va. & vn.* Shout, bawl, cry out.
**haylaz** Idle, lazy. Lazy man; vagabond.
**hayli** (̀·) Much; many; very; pretty (fairly). ~**den ~ye,** a lot of; very much. ~**ca** (̀··), a good many; considerably.
**haymana** (̀··) A large open plain where animals are turned out to graze. ~ **beygiri gibi dolaşmak,** to wander about aimlessly.
**hayran** (·—́) Astonished; perplexed; filled with admiration. ⌜**ben sana ~, sen cama tırman**⌝, *a meaningless expression used to describe an impracticable and purely romantic idea.* ~**lık,** amazement; admiration.
**hayrat** (·̀), **-tı** Pious deeds; pious foundations. **eski ~ı da berbad etm.,** in trying to improve a thing to make it worse.
**hayret, -ti** Amazement, stupor; admiration. ~**te bırakmak,** to astound: ~ **etm.,** to be perplexed; to be lost in admiration *or* astonishment (at = *dat.*): ~**te kalmak,** to be lost in amazement.
**hayrı** *v.* **hayır.**
**hayrola** Good news, I hope!; well, how did it go off?
**hayrülhalef** Worthy successor.
**haysiyet, -ti** Personal dignity; amour propre. ~ **divanı,** a court of honour *to decide on questions relating to the conduct of members of a profession or an association.* ~**li,** jealous of his personal honour; self-respecting. ~**şiken,** that hurts one's self-respect.
**hayt, -tı** Fibre; filament.
**hayta** Out-of-hand, mischievous child; young hooligan.
**hayühuy** (—·—́·), **hayıhuy** Worries and troubles; the humdrum of everyday life.
**hayvan** Living creature; animal; beast of burden; stupid fool. ~**at** (·—̀), **-tı,** animals: ~ **bahçesi,** zoological garden. ~**ca** (·̀·), ~**casına,** like an animal; bestially; stupidly. ~**cı,** cattle-dealer *or* breeder. ~**î** (·——́), animal, bestial. ~**iyet, -ti,** ~**lık,** quality of an animal; bestiality; stupidity; stupid *or* brutal action.
**haz, -zzı** Pleasure; contentment; enjoyment. ~ **duymak,** to feel pleasure: ~**zı nefs,** sensual pleasure; luxury.
**hâzâ** (—́—) *Lit.* 'this (is)', *used in such espressions as*: ~ **kibar,** a real gentleman; ~ **ev,** a perfect house; ~ **sersem,** a complete idiot.
**hazan** Autumn. ~ **yaprağı gibi titremek,** to tremble like an aspen leaf.
**hazar**[1] Big saw. **su ~ı,** sawmill worked by water.
**hazar**[2] Peace. ~ **mevcudu,** peace establishment *of an army.* ~**î** (··—́), peaceful: ~ **kuvvet,** peacetime strength of the army.
**hazele** Rogues. ~ **bezele,** hooligans.
**hazer** Precaution. **···den ~ etm.,** to be on one's guard against ... .
**hazerat** (··—́), **-tı** *pl. of* **hazret.** Excellencies.
**Hazerdenizi** (·̀···), **-ni** The Caspian Sea.
**hazf, -fi** Elision; suppression. ~**etmek** (̀··), **-eder,** to elide; suppress.
**hâzım** (—̀) Who digests; digestive; long-suffering. **hazım, -zmı,** digestion; patience under insult. ~**lı,** patient, long-suffering; tolerant. ~**sız,** indigestible; irritable, touchy. ~**sızlık,** indigestion.

**hazır** Present (not absent); present (time); ready, prepared; ready-made. ~ **ekmek yemek,** to live without working: ~ **etm.,** to prepare: ~**a konmak,** to settle down to enjoy what is already prepared; to enjoy the fruits of others' labours: ~ **olm.,** to be ready; to be present; ⌜~ **ol!**⌝, 'attention!' (*mil.*); 'get ready!': ~ **para,** money in hand, the sum of money that a man has at his disposal; ⌜~ **paraya dağlar dayanmaz**⌝, no one can long go on living on his capital: ~ **yemek,** food already prepared (*such as tinned soups etc.*): ~**dan yemek,** to live on one's capital: ~ **yiyici,** one who lives on his capital; as **hazırcı** (ii): ⌜~ **sokağa giderken şu mektubu da postaya atıver!**⌝, 'since you are going out anyhow, just post this letter for me!'

**hazır·cevab,** quick at reply and repartee. ~**cı,** (i) seller of ready-made clothes; (ii) one who likes the good things of this world if he can get them without trouble to himself. ~**lamak,** to prepare. ~**lık,** readiness; preparedness; preparation: ~ **görmek,** to make preparations: ~ **tahkikatı,** preliminary investigations (legal). ~**lop,** hard-boiled egg; a lucky find. ~**ol,** attention! (*mil.*): ~ **vaziyetinde durmak,** to stand at attention. ~**un** (—·⏑́), *pl. of* **hazır**; those present.

**hazin** Sad; melancholy.

**hazine** (·—́) Treasure; treasury; buried treasure; storehouse; reservoir, cistern; chamber of a gun. ~**i humayun,** the Imperial Treasury: ~**li tüfek,** magazine rifle. ~**dar,** treasurer. ~**ievrak, -kı,** state archives.

**haziran** (·—́) June.

**hazm·etmek** (́··), **-eder** *va.* Digest; swallow (an affront). ~**î** (·—́), digestive.

**hazne** *v.* **hazine.** Tank, reservoir.

**hazret, -ti** Excellency (title); *title of the earlier Caliphs.* ~**i Peygamber,** the Prophet: ~**leri,** His Excellency (*after a title*); *used as a jocular address to a friend* 'old fellow!', 'old man!'

**hazzetmek** (́··), **-eder** *vn.* Rejoice; be pleased. ···**den** ~, to be fond of ....

**heba** (·—́) Waste; loss. ~ **etm.,** to waste; spoil: ~ **olm.,** to be wasted; to be sacrificed in vain.

**hebenneka** *The name of an historic fool*; a fool who thinks himself clever.

**hece** Syllable. ~ **vezni,** syllabic metre. ~**lemek,** to spell out by syllables. ~**li,** having *so many* syllables.

**hecin** Dromedary. ~ **süvar,** member of a Camel Corps.

**hedaya** (·——́) *pl. of* **hediye.** Presents.

**hedef** Mark; target; object, aim. ~**e isabet etm.,** to hit the target; to attain one's object.

**heder** Waste. ~ **etm.,** to waste. ~ **olm.,** to be sacrificed uselessly; to be wasted.

**hedik** Snowshoe.

**hediye** Present, gift; prize. ~**lik,** fit for a present; a choice thing. ~**ten** (·́··), as a present.

**hedmetmek** (́··), **-eder** *va.* Demolish.

**hekim** Doctor. ~**lik,** profession of doctor; medical science.

**hektar** Hectare (= 2·471 *acres*).

**helâ** (·—́) Closet, privy.

**helâk** (·—́), **-ki** Destruction; death; exhaustion. ~ **etm.,** to destroy; to wear out with fatigue: ~ **olm.,** to perish; to be utterly done up.

**helâl** (·—́), **-li, -lı** A permitted *or* legitimate act *or* person; lawful spouse. Lawful; legitimate. ~**im,** my lawful spouse: ~**inden,** legitimately; honestly earned; as a free gift: ~ **etm.,** to declare *stg.* lawful; to give up a legitimate claim to another: ~ **olsun!,** may it be your lawful right and property! (*said when concluding a bargain or when giving stg.*): ~ **süt emmiş,** entirely trustworthy: ⌜**ananın sütü gibi** ~ **olsun!**⌝, you are quite entitled to it; you need have no qualms about accepting it: **kanı** ~, outlaw, proscribed person.

**helâl·î** (··—́), a tissue of silk with cotton *or* wool; old-fashioned pinchbeck watch. ~**laşmak,** mutually to forgive one another (*when dying or setting out on a long journey*). ~**lık,** forgiving an unlawful act; person whom one may legitimately marry; lawful spouse: **helâllığa almak,** to take as one's lawful wife: ~ **dilemek,** to ask forgiveness for an unlawful act. ~**zade** (··——́), legitimate offspring; an honest man.

**helâvet** *v.* **halâvet.**

**hele** (́·) Above all, especially; at least; at last. Hi! listen to me!; now just look here! ~ ~, now tell me truly!: ~ **bir göreyim,** just let me see!: ~ **bunu yapma, sana gösteririm!,**

don't you dare do it, or I'll give you what for!

**helecan** *v.* **halecan.**

**helezon** Snail; snailshell; spiral; propellor. **~î** (··–⌐), spiral, helical.

**helile** (·–⌐) Myrobalan (*used as a purgative*).

**helke** Pail; bucket.

**helme** Thick liquid *or* paste *made by boiling starchy substances.* **~ ~ olm.**, to become like paste: **~ dökmek** *or* **~lenmek,** to become a jelly, become like paste.

**helva** Sweetmeat *made of sesame flour, butter, and honey; there are many kinds such as* **koz ~sı** (with walnuts), **badem ~sı** (with almonds), *etc.* **~ sohbeti,** a social gathering *where* **helva** *is offered to the guests*: **kudret ~sı,** manna: **~cı,** maker *or* vendor of **helva**: **~ kabağı,** a kind of white-fleshed pumpkin. **~hane** (··–⌐), large shallow pan in which **helva** is made.

**hem** And also; too. **~ ben ~ sen,** both you and I.

**hem–** *Pers. prefix indicating similarity or company.* **~ahenk,** harmonious. **~cins,** of the same kind *or* race. **~dem,** constant companion; crony; confederate. **~derd,** fellow sufferer.

**hemen** (⌐·) At once; exactly; just; just now; continually; about, nearly; only just; only. **~ ~,** almost, very nearly. **~cecik** (·⌐··), **~cek** (⌐··), at once.

**hem·fikir** Like-minded. **~hal, –li,** in the same condition; fellow sufferer. **~hudud,** having the same boundary *or* frontier; contiguous. **~pa** (·⌐), accomplice, confederate. **~şeri,** fellow townsman; compatriot.

**hemşire** Sister; trained hospital nurse. **~zade** (···–⌐), sister's child, nephew, niece.

**hendek** Ditch; moat; trench.

**hendes·e** Geometry; mathematics. **~eli, ~î** (··⌐), geometrical.

**hengâm** Time; season; period.

**hengâme** Uproar; tumult.

**henüz** (·⌐) Yet; still; (*with neg.*) not yet: (⌐·), just now; a little while ago.

**hep** All; the whole. Wholly; always. **~imiz,** all of us; **~iniz,** all of you: **~ten,** entirely: **~ bitti,** it is entirely finished. **~si,** the whole of it; all of it; all of them: **~nden ziyade,** above all. **~yek,** double-one (in dice).

**her** Every; each. *Many of the compound expressions of* **her** *are variously written as one word or as two or three. Those which are almost always written as one are given separate headings, q.v.* **~ an,** always, any moment: **~ bir,** every; each; each one; whoever; any-one of: **~ gün,** every day; continually: **~ günkü,** everyday: **~ günlük,** everyday clothes: **~ halde,** in any case, under any circumstances; for sure; apparently: **~ kim,** whosoever, whoever: **~ nasıl,** in whatever way: **~ nasılsa,** in whatever way may be; somehow or other: **~ ne,** whatever; **~ ne hal ise,** well anyhow ...: **~ ne pahasına olursa olsun,** at whatever cost: **~ nedense,** for some reason or other; I don't know why: **~ nekadar,** however much: **~ nerede,** wherever: **~ ne türlü,** of whatever kind; in whatever way: **~ ne vakit,** whenever; every time that ...: **~ neyse,** whatever it be; however it be; anyhow: **~ yerde,** everywhere: **~ zaman,** every time; always.

**herc** Confusion; turmoil. **~ümerc,** a confused, disordered mass.

**hercai** (·–⌐) Ubiquitous; roving; inconstant. **~ menekşe,** pansy.

**herdem** (⌐·) At every moment. **~taze** (·⌐–·), evergreen; ageless.

**herek** Prop *for vines etc.*; temporary platform *for drying raisins, tobacco etc.*

**herfane** (·–⌐) *v.* **arifane.**

**hergele** Herd of animals; a lot of roughs; rough coarse fellow. **~ci,** glossy ibis.

**herhangi** (·⌐·) Whoever; whatever.

**herif** Fellow (*always derogatory*). **~in biri,** some fellow or other. **~ci,** *in* **~ oğlu,** (*an insulting expression*) 'the fellow'.

**herkes** (⌐·) Everyone.

**hernedense** (··⌐·) *v.* **her.**

**herrü** *v.* **harrü.**

**Hersek** Herzogovina.

**herze** Nonsense. **~ yemek,** to talk nonsense. **~vekil,** busybody; twaddler.

**hesab** (·⌐) A counting; reckoning; calculation; account; bill; accounts; arithmetic. **~ını bilmek,** to be careful and economical: **~a çekmek,** to call to account, to hold responsible: **~ etm.,** to calculate: **~a** *or* **~ına* gelmek,** to suit: **~a gelmez,** countless; unbounded: **~ görmek,** to have a reckoning, to settle up: **~ işi,** (sewing) not judging by eye where to insert needle, but counting the threads in the stuff sewn: ~ı **kesmek,** to settle an account definitely; to sever relations with *s.o.*: ~ **kitab,** after full consideration: ~**a kitaba uygun,** suitable to one's means; ~ı **kitabı yok (bilmez),** uncontrolled; unlimited: ~ **meydanda,** it's quite obvious: ⌈**evdeki ~ çarşıya uymaz**⌉, things don't turn out as one reckons; 'don't count your chickens before they are hatched'.

**hesab·ca** (·ˋ·), according to the reckoning; normally, properly speaking. ~**cı,** calculating; careful; miserly. ~**î** (·—ˊ), pertaining to accounts; who calculates; economical; stingy. ~**lamak,** to reckon; estimate. ~**lı,** calculated; well considered. ~**sız,** countless; uncertain, problematical; without reflection; ~ **kitabsız,** uncontrolled (expenditure *etc.*); casual, at random, thoughtlessly.

**hethüt** *Imitates a loud and menacing way of speaking.* ~ **etm.,** to behave in a bullying manner, to browbeat.

**hevenk, -gi** Bunches *of grapes etc.* hung up.

**heves** Desire; inclination; mania; zeal. ~**ini almak,** to satisfy a desire. ~**kâr,** ~**li,** desirous; eager; having aspirations; dilettante. ~**kârlık,** a passing desire; longing; hobby. ~**lendirmek,** to awake the desire *in s.o. to do stg.* ~**lenmek,** to long for, desire. ~**siz,** disinclined.

**hey** Hi!; *also expresses regret, reproach or admiration.*

**heyamola** (··ˋ·) *Cry in unison of sailors at work*; with much difficulty and fuss. ~ **ile iş görmek,** to work only when hustled.

**heybe** Saddle-bag; wallet.

**heybet, -ti** Awe; majesty; imposing air. ~**li,** imposing, majestic; awe-inspiring.

**heyecan** Excitement; enthusiasm; emotion. ~**a gelmek** (~ **bulmak**), to get excited, be enthusiastic. ~**lı,** excited; exciting; enthusiastic.

**heyelân** Landslide.

**heyet, -ti** Shape, form; state, aspect; assembly, commission, committee; astronomy. ~**i mecmuası itibarile,** taken as a whole: ~**i umumiyesile,** taken as a whole. ~**i vükela** *or* **vekile,** Council of Ministers, Cabinet.

**heyhat** Alas!

**heyhey**[1] Amusement; orgy; song.

**heyhey**[2] *In* ~**ler geçirmek,** to be very agitated, to have a fit of nerves: **içime** ~**ler geliyor,** I can't stand it any longer; I shall have a nervous breakdown.

**hey'î** (·ˊ) Astronomical.

**heykel** Statue. ~**traş,** sculptor. ~**traşlık,** sculpture.

**heyulâ** (·—ˊ) Spectre, bogey.

**hezar**[1] (·ˊ) Thousand. ~**fen, -fenni,** Jack-of-all-trades; omniscient; versatile.

**hezar**[2] (·ˊ) Nightingale.

**hezaren** Cane; rattan; delphinium.

**hezen arısı** Bumble-bee.

**hezeyan** Talking nonsense; delirium.

**hezimet** (·—ˋ), **-ti** Utter defeat, rout.

**hıçkır·ık** Hiccough; sob. ~ **tutmak,** to have the hiccoughs. ~**mak,** to sob.

**hıdırellez, hıdrellez** The beginning of summer (6th of May).

**hıdiv** (·ˊ) Khedive. ~**î** (·—ˊ), Khedivial. ~**iyet, -ti,** Khediviate.

**hıfız, -fzı** A guarding *or* protecting; protection; a committing to memory; committing the Koran to memory. **hıfza çalışmak,** to be trained in learning the Koran by heart.

**hıfz·etmek** (ˋ··), **-eder** *va.* Protect; preserve; commit to memory. ~**ıssıhha,** hygiene.

**hık** *In* ⌈~ **demiş burnundan düşmüş**⌉, he is the very spit of him; *v. also* **hınk.**

**hık mık etm.** *vn.* Hum and haw. **hık mık yok,** there is no question about it!

**hıl'at, -ti** Robe of honour.

**hımbıl** Silly; slow.

**hımhım** Talking through the nose. ⌈~**la burunsuz birbirinden uğursuz**⌉, 'birds of a feather'.

**hınc** Rancour; hatred. ~ **beslemek,** to nourish a grudge: ···**den** ~**ını çıkarmak (almak),** to take revenge on, to vent one's spleen on ...: ···**in** ~**ını çıkarmak,** to take revenge for ....

**hıncahınç** Chock-a-block.

**hınk** *In* ⌈**kahve döğücünün ~ deyicisi**⌉, one who only helps by making a lot of noise; one who, to curry favour, agrees with all another says.

**hınzır** Swine; foul fellow; *sometimes used in joking admiration.* Swinish. ~**lık,** a dirty trick.

**hır** Row, quarrel. ~ **çıkarmak,** to start a quarrel.

**hırçın** Ill-tempered; (of the sea) angry. ~**lık,** bad temper; obstinacy.

**hırdavat, -tı** *pl.* Small wares; ironmongery. ~**çı,** seller of small wares, pedlar; ironmonger.

**hırgür** Snarling; noisy quarrel.
**hırhıra** *In* ~ **kemiği,** the Adam's apple.
**hırıl·damak** *vn.* Growl; purr; have a rattle (râle) in the throat. ~**daşmak,** to snarl at one another; quarrel without reason. ~**tı,** growling, snarling; squabble; râle, death-rattle.
**hırızma** (·'·) Nose-ring worn by Arab women; nose-ring (of animals). ~**sı çıkmış,** all skin and bones.
**hıristiyan** Christian. ~**lık,** Christianity.
**hırka** Short cloak, *usually quilted*; dervish's coat. **Hırkai Şerif,** the Prophet's Mantle, *preserved as a relic at Istanbul*: ⌜**bir lokma bir** ~⌝, a morsel of food, a rag of clothing, *i.e.* enough to keep body and soul together.
**hırla·mak** *vn.* Growl, snarl. ~**nmak,** to growl *or* snarl (*of a man only*). ~**şmak,** to snarl at one another; to squabble noisily.
**hırlı** *Only in* ~ **hırsız,** honest men and thieves.
**hırpa·lamak** *va.* Ill-treat; misuse. ~**lanmak,** to be ill-treated; be upset.
**hırpani** (·—') Ruffled; untidy, unkempt.
**hırpo** (*sl.*) Big clumsy fellow.
**hırs** Inordinate desire; greed; ambition (bad); anger. ~**ından* çatlamak,** to be ready to burst with anger: ~**ını*** ···**den çıkarmak,** to vent one's spleen on ... .
**hırsız** Thief. Thieving. ~ **feneri,** dark-lantern: ~ **malı,** stolen goods: ~**a yol göstermek,** imprudently *or* inadvertently to help a wrongdoer. ~**lama,** stealing; like a thief, furtively, stealthily, surreptitiously. ~**lamak,** to steal. ~**lık,** theft, thieving. ~**yatağı, –nı,** thieves' den; receiver of stolen goods.
**hırs·lanmak** *vn.* Get angry. ~**lı,** angry; desirous; avaricious.
**hırt** Coarse and vulgar yet conceited.
**hırtıpırtı** Trifles, rubbish; old clothes.
**hırtlamba** (·'·) Poorly and untidily clothed. ~**sı çıkmış,** in rags and tatters; thin and bony: ~ **gibi giyinmek,** to be untidily dressed with an excess of clothing.
**Hırvat** Croat; Croatian; great big man. ~**istan,** Croatia.
**hısım, –smı** A relative. ~**lık,** relationship.
**hışıl·damak** *vn.* Make a wheezing *or* rustling noise. ~**tı,** a wheezing *or* rustling noise.
**hışım, –şmı** Anger; indignation. **hışmına* uğramak,** to be the object of *s.o.'s* anger.
**hışır** Unripe melon; rind of a melon. Stupid and gullible. ~ ~ **etm.,** to make a harsh grating sound. ~**damak,** to make the noise produced by dry leaves, by paper *or* silk, *or* a snake moving in dry grass. ~**tı,** such a noise.
**hıtta** Region; country.
**hıyaban** (·—') Alley; avenue.
**hıyanet** *v.* **hiyanet.**
**hıyar** Cucumber; dolt. Uncouth and stupid. *v. also* **hiyar. yabani** ~, squirting cucumber. ~**cik,** bubo; tumour.
**hız** Speed; impetus. ~ **almak,** to get up speed; *but* ~**ını almak,** to slow down; to take it easy: ~**ını alamamak,** to be unable to slow down, not to be able to stop oneself.
**hızar** *v.* **hazar**[1].
**Hızır** *A legendary person who was reputed to arrive and help in critical moments.* ~ **gibi yetişmek,** to be a 'deus ex machina', a timely help; to come as a godsend: ⌜**kul daralmayınca (sıkışmayınca)** ~ **yetişmez**⌝, *said when unexpected help arrives to one in difficulties.*
**Hızırilyas** *v.* **hıdirellez.**
**hız·lanmak** *vn.* Gain speed *or* impetus. ~**lı,** swift; violent; loud.
**hibe** Gift.
**hibre** *In* **ehli** ~, expert.
**hicab** (·') Modesty, shame. ~**lı,** veiled; modest; bashful.
**Hicaz**[1] Hedjaz.
**hicaz**[2] *or* ~**kâr** A mode in Oriental music.
**hicir, –cri** *v.* **hicran.**
**hiciv, –cvi** Satire; lampoon.
**hicran** (·') Separation; absence from one's family; mental pain; bitterness of heart.
**hicr·et, –ti** Abandoning one's country; emigration; *the emigration of Mohammed from Mekka to Medina, i.e.* the Hegira (A.D. 622). ~**î** (·—'), pertaining to the Hegira: **senei** ~**ye,** any year of the Hegira era.
**hicv·etmek** ('··), **–eder** *va.* Satirize. ~**î** (·—'), satirical. ~**iye,** satire, satirical poem.
**hiç** No. Nothing. Not at all; never; (*without neg.*) ever. ~ (*or* ~ **bir**) **kimse,**

nobody: ~ **bir şey,** nothing at all: ~ **değilse (olmazsa),** at least: ~ **mi** ~, absolutely nothing: ~ **olur mu?,** does it ever happen?, is it possible?: ~**e saymak,** to hold of no account: ~ **yoktan (yüzünden),** for no reason.
**hiç·lik,** nullity. ~**ten,** *abl. of* **hiç;** sprung from nothing, parvenu; useless; got for nothing; trifling, insignificant: ~ **sebeblerle,** under absurd pretexts.
**hidayet** (·–ˈ), **–ti** The right way, *esp.* the way to Islam; a searching for the right way.
**hiddet, –ti** Violence; impetuosity; anger, fury. ~**e gelmek,** to fly into a passion. ~**lenmek,** to become angry. ~**li,** angry; violent.
**hidemat** (··ˈ), **–tı** *pl. of* **hizmet.** Services, *etc.* ~**ı şakka,** hard labour (penal).
**hidiv** *v.* **hıdiv.**
**hikâye** A relating, narration; story; novel. ~ **etm.,** to tell, narrate. ~**ci,** story-teller; short-story writer. ~**cilik,** the art of story-telling.
**hikmet, –ti** The ultimate hidden cause *for existence or occurrence*; the Divine Wisdom; inner meaning *or* object; wisdom; philosophy; wise saying; reason; physics. **Allahın** ~**i** *or* ~**i Huda,** the divine dispensation; 'Heaven knows why!'; strangely enough!'; 'for some mysterious reason': **işin** ~**ini bilmiyorum,** I don't know the real truth of the matter: ~**inden sual olmaz (olunmaz),** 'Heaven only knows why!': **bu çocuk bir** ~**!,** this child is hopeless; there's something wrong about this child: **her gün gelirdi ne** ~**se bugün gelmedi,** he has been coming every day, it's odd he hasn't come today: ⌜**pahalıdır** ~**i var, ucuzdur illeti var**⌝, 'if it is dear, there is a good reason for it; if it is cheap, there is something wrong with it', *i.e.* it always pays to buy the best.
**hikmet·ivücud** (····ˈ), **–dü,** the real reason; the *raison d'être.*
**hilâf** (·ˈ) The contrary, opposite; contravention; opposition; lie. ~**ına,** contrary to ..., against ...: ~**ı hakikat,** contrary to the truth: ~**ım varsa,** if I lie.
**hilâfet** (·–ˈ), **–ti** The Caliphate.
**hilâl** (·ˈ), **–li** Crescent. **Hilâli Ahmer,** the Red Crescent.
**hile** (–ˈ) Trick; wile; stratagem; fraud. ~ **etm.,** to make use of a trick: ···**den** ~ **sezmek,** to suspect ... of laying a trap: ~**baz,** ~**li,** ~**kâr** (–·ˈ), wily, deceitful; fraudulent. ~**kârlık** (–·–ˈ), deceit, trickery, fraud. ~**siz,** genuine; above-board.
**hilk·at, –ti** Creation; natural form *or* disposition. ~**aten** (ˈ··), by nature. ~**î** (·ˈ), natural, inborn, congenital.
**himaye** (·–ˈ) Protection; defence. ~ **etm.,** to protect: ~ **usulü,** protectionist system (*as opposed to free trade*). ~**kâr,** protective; patronizing. ~**li,** escorted (convoy *etc.*).
**himmet, –ti** Effort, zeal, endeavour; influence; moral support; benevolence. ~**inizle,** thanks to your assistance: ~ **etm.,** to exert oneself, to take trouble: ~ **etmemek,** to be unwilling to help, to make no effort.
**hin** A kind of djinn. ~**oğlu hin,** son of the devil, scoundrel; very crafty fellow.
**Hind** India. ~**istan,** India. ~**î** (·ˈ), Indian.
**hindi** Turkey (bird).
**hindiba** (··ˈ) Chicory.
**hindistancevizi, –ni (büyük)** ~, coconut; **(küçük)** nutmeg.
**hinihacette** (ˈ····) When the need arises.
**hinoğlu** (ˈ··), **–nu** *v.* **hin.**
**hint·kumaşı** (ˈ···), **–nı** Very rare and precious thing (*usually used jocularly*). ~**yağı, –nı,** castor-oil.
**his, –ssi** Sense; perception; faculty; feeling, sensation; sentiment. ~**lerine* kapılmak,** to be swayed by one's feelings.
**hisa** (ˈ·), **etm.** *va.* Hoist (flag *etc.*); toss (oars).
**hisar** Castle; fortress.
**hisli** Sensitive.
**hisse** Share; allotted portion; share *in a company.* ···**den** ~ **kapmak,** to learn a lesson from ...: ~ **senedi,** share certificate: ~**i şayia,** co-ownership: **kıssadan** ~, the moral of the story is .... ~**dar,** participator; shareholder. ~**li,** having shares; divided into portions; belonging to various people. ~**leşmek,** to divide into portions; to share together.
**hisset, –ti** Avarice; stinginess; a stingy person.
**hiss·etmek** (ˈ··), **–eder** *va.* Feel; perceive. ~**î** (·ˈ), perceptible; that can be felt; psychological; sentimental. ~**ikablelvuku, –uu** (*err. for* **hiss-**

kablelvuku) premonition; presentiment. ~**iyat** (··⁄), sensations; feelings; perceptions.

**hissiz** Insensitive, unfeeling, callous.

**hiş, hişt** *Interj. used to call attention*; hist! ~ **piş etm.**, to call out 'hist!'; to call out to a girl.

**hitab** (·⁄) An addressing a person; address; allocution. ~ **etm.**, to address; to make a remark to: ⌜**bu adam kabili** ~ **değil** *or* **gayrikabili** ~**dır**⌝, 'one can't talk to this man' (he's too stupid *or* he can't be trusted). ~**e** (·—⁄), address, speech. ~**en** (·⁄·), addressing: **bana** ~, addressing me. ~**et** (·—⁄), **-ti**; office of a preacher; oratory.

**hitam** (·⁄) Conclusion; completion. ~ **bulmak**, to come to an end: ···**e** ~ **vermek**, to bring to an end.

**hitan** (·⁄) Circumcision.

**hiyanet** (·—⁄), **-ti** Treachery, perfidy. Perfidious; basely ungrateful. ~**i vataniye**, high treason. ~**lik**, malicious act.

**hiyar** (·⁄) Option. **hakkı** ~, the right of option: *v. also* **hıyar.**

**hiyerarşi** Hierarchy.

**hiza** (·⁄) The point opposite *or* on the same level; line; level. ···**in** ~**sına**, on a level with ..., in a line with ...: ~**sını almak**, to take the bearing *or* the level of ...: ···**in** ~**sına kadar**, up to the level of ...: ~**ya gelmek**, to get into line: **bir** ~**da**, in one line, on one level.

**hizib, -zbi** Clique; party; portion of the Koran.

**hizmet, -ti** Service; duty; employment; function. ···**in** ~**inde bulunmak**, to be in the service of ...: ···**e** ~ **etm.**, to render service to ...: ~**ini görmek**, to serve for, take the place of. ~**çi**, servant. ~**cilik**, the position and duties of a servant: ~ **etm.**, to be a servant *with s.o.* ~**kâr** (··⁄), servant.

**hoca** (⁄·) Moslem priest; hodja; schoolmaster. **akıl** ~**sı**, one who sets up to give good advice to others: ⌜**ha** ~ **Ali ha Ali** ~⌝, it's much of a muchness. ~**lık**, the quality *or* profession of a hodja *or* schoolmaster: **ahlâk hocalığı etm.**, to set oneself up as a guardian of morals: **akıl hocalığı yapmak**, to give good advice (pretentiously). ~**nım**, *for* **hoca hanım**, woman teacher.

**hodan** Borage.

**hod·bin** Egotistical, selfish; conceited. ~**binlik**, self-conceit; selfishness. ~**gâm**, ~**kâm**, wilful; egotistical. ~**perest**, egotistical. ~**pesend**, self-satisfied, conceited.

**Hodiri** (⁄··), **Hodri, Hodori** *In* ⌜~ **meydan**⌝, 'come and try!' (*a challenge*).

**hohlamak** *va.* Breathe upon.

**hokka** Inkpot. ~ **agızlı**, with a small pretty mouth: **şeker** ~**sı**, sugar-bowl: **tükürük** ~**sı**, spitoon.

**hokkabaz** Conjuror; cheat, knave. ~**lık**, conjuring; trickery.

**hol, -lü** Hall.

**Holanda** (·⁄·) Holland. ~**lı**, Dutchman.

**homur·danmak**, *vn.* Mutter; grumble. ~**tu**, a muttering, grumbling.

**hona** Stag.

**hop** *In* ~ **oturup** ~ **kalkmak**, to keep on jumping up and down *from excitement.* ~**lamak**, to jump about.

**hoparlör** (*Fr. haut parleur*) Loudspeaker.

**hoppa** Volatile, flighty, flippant. ~**lık**, levity, flightiness.

**hoppala** (⁄··) *Used when dangling a child*; 'up she goes!'; *exclamation of surprise.* ~ **bebek**, a childish person.

**hopurdatmak** *v.* **höpürdetmek.**

**hor** Contemptible. ~**a geçmek**, to be appreciated: ~**a gelmez**, that will not stand rough usage: ~ **görmek (bakmak)**, to look down upon, to treat as of no account: ~ **kullanmak**, to misuse; to use for the commonest purposes.

**hora** (⁄·) A kind of round dance; noisy party. ~ **tepmek**, to dance the hora; to dance about noisily and clumsily (*as of a lot of tipsy people*).

**horan** A kind of folk dance *on the Black Sea coast.* ~ **etm.**, *as* **hora tepmek.**

**Horasan** (·⁄·) Khorasan. **horasan**, mortar made of brickdust and lime. ~**î** (··—⁄), native of Khorasan; kind of turban formerly worn by government clerks; santonin; dull red colour.

**horlamak**[1] *vn.* Snore.

**hor·lamak**[2] *va.* Treat with contempt; ill-treat. ~**luk**, contemptibility.

**horon** *v.* **horan.**

**horosbinâ** Butterfly blenny.

**horoz** Cock; hammer *of a gun.* ~**dan kaçmak**, (of a girl) to be coy and bashful: ~ **kuyruğu**, cock's tail;

French bean: ~ mantarı, chanterelle: ~ şekeri, a sweet in the form of a cock *stuck on the end of a stick*: ⌜~ ölmüş, gözü çöplükte kalmış⌝, 'the cock is dead, but his eye remains on the dung-heap', *used of one who looks back regretfully to things that are now lost to him*: ~ yumurtası, a very small egg, pullet's egg: yabani ~, blackcock.
**horoz·ayağı, -nı,** cartridge-extractor. **~cuk,** a little cock; youth; tansy (?). **~gözü, -nü,** a kind of camomile. **~ibiği, -ni,** cock's-comb (*Celosia*); hoopoe; bright red; *when meaning* a cock's comb *it is written as two words.* **~lanmak,** to strut about. **~oğlu,** abnormal, crazy.
**hort·lak** *A corpse which is vulgarly supposed to arise from its grave and frighten people in the dark*; spook; vampire. **~lamak,** to rise from the grave and haunt people.
**hortum** Elephant's trunk; hose-pipe; waterspout.
**hortzort** *v.* zartzurt.
**horul** *Imitates the noise of snoring.* ~ ~ **horlamak,** to snore noisily: ~ ~ **uyumak,** to sleep soundly; to be supine. **~damak,** to snore. **~tu,** snore, snoring.
**hoş** Pleasant, agreeable; quaint. *conj.* Well; even. ~ **geçinmek,** to get on well *with s.o.*: ⌜~ **geldiniz!**⌝, 'you are welcome!', *greeting to one just arrived, the correct reply being* ⌜~ **bulduk!**⌝: ~ **gelmek,** to be agreeable, to be liked, *e.g.* **bu bana ~ geliyor,** I like this: ~ **görmek,** to condone, overlook, tolerate; ~ **görürlük,** tolerance: **~una* gitmek,** to please, to be agreeable to: ···**le başı* ~ olmamak,** to dislike: **bir ~ olm.,** to feel uncomfortable; to be disconcerted *or* offended: ⌜**iyi misiniz ~ musunuz?**⌝, 'I hope you are well!': ⌜~ **ben gitmiyecektim ya!**⌝, 'well, I shouldn't have gone anyhow!': ⌜~ **gelse ne faydası var?**⌝, 'even if he does come what use will it be?'
**hoş·amedi** (·—·ʾ) *n.* Welcome. **~beş,** friendly greeting; friendly chat. **~bu** (·ʾ—), pleasantly scented. **~ça** (ʾ··), pleasantly, nicely: ~ **kalın!,** good-bye! **~etmek** (ʾ··), **-eder,** *in* ···**in hatırını (gönlünü) ~,** to satisfy *or* please *s.o.* **~lanmak,** ···**dan ~,** to like. **~laşmak,** to get to like (*with abl.*); to like one another; to be disconcerted *or,* offended; (of food) to turn sour. **~luk,** pleasantness; happiness; comfort; quaintness: **bir hoşluğum var,** I am feeling rather queer: **gönül hoşluğu ile,** gladly, willingly. **~nud,** contented, pleased. **~nudluk,** contentment; pleasure. **~nudsuz,** discontented, displeased. **~nudsuzluk,** discontent, dissatisfaction. **~sohbet, -ti,** good company; conversationalist.
**hoşaf** Fruit in syrup. ⌜**eşek ~tan ne anlar?**⌝, 'it's throwing pearls before swine': ···**de ~ın yağı kesilmek,** to be dumbfounded.
**hoşmerim** *A dish made with unsalted fresh cheese.*
**hoşt** *Noise made to frighten away a dog.*
**hotoz** (*Corr. of* **kotaz** = yak); yak's tail *as a banner*; crest *of a bird*; a kind of ancient headgear for women; bun, topknot (hair).
**hov** Hawking, falconry. **~ağası,** falconer. **~lamak,** to throw a hawk.
**hovarda** Scapegrace; rake; spendthrift. **~lık,** dissoluteness; rakishness; being free with money.
**hoyrat** Vulgar; rough; coarse and clumsy man.
**höcre** *v.* **hücre.**
**hödük** Boorish; clumsy. Lout; bumpkin.
**hökelek** Peevish; contrary; boastful.
**höpürdetmek** *va.* Drink noisily.
**hörgüç** Hump *of a camel*; any protuberance. **~lük,** camel saddle.
**höykürmek** *vn.* Intone religious formulae.
**höyük** *v.* **öyük.**
**hu** Hi!; I say!
**hububat** (·—ʾ), **-tı** Cereals.
**hud'a** Fraud; deceit.
**hudayinabit** (·—·—ʾ) That which grows by itself; of spontaneous growth; in its natural and wild state; (of children) running wild; untaught; amateur.
**hudud** *pl. of* **had.** Boundaries; limits; *as sing.* frontier.
**huğ** Mud hut.
**hukuk** (·ʾ), **-ku** *pl. of* **hak.** Rights; dues; laws. **~u düvel,** international law: ~ **Fakültesi,** the Law Faculty: ⌜**onunla aramızda ~umuz var**⌝, 'we are old friends'. **~çu,** jurist. **~î** (·—ʾ), legal; juridical; 'de jure'.
**hulâsa** (·—ʾ) The extract *of a substance*; quintessence; abstract, summary. In short, in fine. **~sını almak,** to sum up: ~ **etm.,** to summarize, to make a précis of: **~i kelâm,** the sum

and substance of what has been said, in short. ~**aten** (·–́··), in short; to sum up.

**hulefa** (··–́) *pl. of* **halife.** Successors; Caliphs; *formerly* superior clerks in a government office. ~**yi Raşidin,** the first four Caliphs

**hulk, -ku** Moral quality; character.

**hulûl** (·*), **-lü** An entering *or* penetrating; appearance; beginning *of a season*; reincarnation; osmosis. ~ **etm.,** to penetrate into; to enter (another body); to occur; to gain ascendancy over another: ~**i muslihane,** pacific penetration: ~ **siyaseti,** the policy of penetration.

**hulûs** (·*) Sincerity; devotion; esteem *for a superior*; flattery. **birine** ~ **çakmak,** to toady to, to make up to *s.o. by subtle flattery.* ~**kâr,** sincere; sincere friend; hypocritical; sycophant.

**hulya** (·–́) Day-dream. ⌜~ **bu ya**⌝, 'if by any possible chance ...'; 'let's suppose that ...'. ~**lı,** dreamy, romantic.

**humayun** *v.* **hümayun.**

**humma** (·–́) Fever; typhus; typhoid. **kazıklı** ~, tetanus: **lekeli** ~, purpura; typhus; cerebrospinal meningitis. ~**lı,** feverish.

**hummaz** Sorrel; *also* = **loğusa şekeri,** *q.v.*

**humus**[1], **-msu** A fifth.

**humus**[2] Mashed chickpeas.

**hun** Blood. ~**har,** ~**riz,** bloodthirsty.

**huni** Funnel (*for pouring liquids*).

**hunnab** *v.* **hünnab.**

**hunnak, -kı, -ğı** Quinsy; angina.

**hunsa** (·–́) Hermaphrodite.

**hurada** *v.* **kurada.**

**huraf·e** (·–́) Silly tale; superstition; myth. ~**at, -tı,** superstitions. ~**eperest,** superstitious.

**hurc** Large leather saddle-bag; holdall.

**hurd** *In* ~ **hurdavat,** old rubbish. ~**a,** old iron, scrap metal; small, fine. ~ **fiatine,** at scrap price; for the value of the materials. ~**acı,** secondhand metal-dealer. ~**lanmak,** to be broken into small pieces.

**hurde** Trifle; small bit; fine point, nicety. Small; trifling; minute. ~**haş** (··–́), in fragments.

**huri** (–́) Houri; beautiful girl.

**hurma** Date (fruit). ~**dorusu,** of a brown bay colour (horse). ~**lık,** date-grove.

**hurrem** Happy, gay.

**huruc** (·*) An issuing forth; exit; sortie; rebellion; start of a new movement. ~ **etm.,** to come out; to make a sortie; to rise in rebellion.

**hurya** *v.* **hürya.**

**husran** *v.* **hüsran.**

**husuf** (·*) Eclipse of the moon.

**husul** (·*), **-lü** An occurring, appearing *or* being produced; attainment. ~ **bulmak** *or* ~**e gelmek,** to be accomplished *or* attained; ~**e getirmek,** to accomplish.

**husumet** (·–́), **-ti** Enmity, hostility.

**husus** (·*) Particularity; peculiarity; matter; connexion; particular. ~**iyle,** especially: ~**unda,** with reference to: **bu** ~**ta,** in this matter, in this connexion. ~**î** (·–́), special, particular, private, personal. ~**ilik** (·–·́), a being peculiar. ~**iyet** (·–·́), **-ti,** peculiarity; speciality; intimacy.

**husye** Testicle.

**huş** Birch-tree.

**huşu** (·–́), **-uu** Deep and humble reverence.

**huşunet** (·–́), **-ti** Harshness; roughness; coarseness.

**hutbe** The sermon and prayer delivered by the official preacher in a mosque on Fridays.

**huy** Disposition; temper; habit; bad habit. ~**unu husunu bilmiyorum,** I don't understand his temperament: ~ **edinmek,** to acquire a habit: ~ **kapmak,** to contract a bad habit: ⌜~ **canın altındadır**⌝, you can't eradicate what is innate, 'what's bred in the bone will out in the flesh': ···**in** ~**una suyuna gitmek,** to treat *s.o.* tactfully. ~**etmek** (́··), **-eder,** to contract the bad habit of ... . ~**lanmak,** to get into bad habits; (of an animal) to become restive *or* obstinate; (of a person) to become touchy *or* nervous. ~**lu,** of such-and-such a temper; suspicious, touchy; bad-tempered, fractious: **iyi** ~, good-tempered, well-behaved: ~ **huyundan vazgeçmez,** it is difficult to get out of a bad habit. ~**suz,** bad-tempered, fractious. ~**suzluk,** bad temper; obstinacy.

**huzur** (·*) Presence; repose, quiet, freedom from anxiety. ···**in** ~**una çıkmak,** to enter the presence (of a great man), to have an audience with ...: **hakkı** ~, honorarium *paid to the member of a Commission.* ~**lu,** at ease; comfortable; tranquil. ~**suz,** uneasy, troubled.

**hüccet, -ti** Argument; proof; title-deed. **~im elinde,** I am your slave.
**hücra** (·ˉ) Remote, solitary (place).
**hücre** Small room; cell; cell (biology); niche, alcove. **~vî** (··ˉ), cellular.
**hücum** (·ˊ) Attack, assault; rush *of blood.* **···e ~ etm.,** to attack.
**Hüda** *v.* **Huda.**
**hükm·en** (ˊ·) By the decision of a judge; legally; in accordance with rules *or* regulations. **~etmek** (ˊ··), **-eder** *vn.* (*with dat.*) *va.,* decide on; judge; exert influence; master; rule; command. **~î** (·ˉ), judicial; done in accordance with a rule; nominal (*as opp. to actual*): **~ şahıs,** juridical person. **~ünce** (·ˊ·), in conformity with; according to the requirements of.
**hükûmet, -ti** (*pl.* hükûmat) Government; administration; state; authority. **~ darbesi,** coup d'état: **~ etm.** (**sürmek**), to rule, govern: **~ kapısına düşmek,** to have dealings with the Government.
**hüküm, -kmü** Rule; authority; government; command, edict; judicial sentence *or* decision; judgement; tenor, import; effect, influence; importance. **~den düşmek,** to be no longer valid: **hükmünü geçirmek,** to assert one's authority: **hükmü* geçmek,** (of a person) to have authority, for his word to carry weight; (of an order) to be enforceable, *but also* to have lost its validity: **~ giydirmek,** to pass sentence: **~ giymek,** to lose one's case at law; to be condemned: **hükmü* olm.,** for one's word to go, to be of importance; to be valid *or* effective: **hükmünde olm.,** to be equivalent to, to have the same effect as: **~ sürmek,** to reign; to prevail: **fırtınanın hükmü geçti,** the worst of the storm is over: ⌜**ne hükmü var?**⌝, 'what does it matter?', 'it's of no importance!'
**hüküm·dar,** monarch, ruler. **~darlık,** sovereignty. **~ran,** ruler, sovereign. **~ranlık,** sovereignty, dominion. **~süz,** no longer in force; null.
**hülle** A suit of clothes; a mock marriage *which enables a divorced woman to return to her previous husband.* **~ci,** *the man whom she marries on this occasion and who then divorces her.*
**hüma** (·ˉ) A fabulous bird of good omen; phoenix; good luck.
**hümayun** (·–ˊ) Felicitous; imperial.
**hüner** Skill; dexterity; ability; art; talent. **~li,** skilful; talented; cleverly made. **~siz,** without talent; inartistic; clumsily made.
**hüngür** *In* **~ ~ ağlamak,** to weep bitterly. **~demek,** to sob. **~tü,** sobbing.
**hünkâr** Sultan. **~ imamı gibi,** with an air of great importance. **~beğendi,** *a dish made with aubergines.*
**hünnab** Jujube.
**hünsa** (·ˉ) Hermaphrodite.
**hür, -rrü** Free; well-born.
**hürmet, -ti** Respect; veneration. **···e ~ etm.,** to respect, honour. **~en** (ˊ··), out of respect (*for = dat.*). **~kâr,** respectful. **~li,** worthy of respect, venerable; respectable; rather big (*iron.*); **~ce,** *only in the ironical sense.* **~siz,** irreverent; disrespectful.
**hürriyet, -ti** Freedom, liberty. **~perver,** loving liberty; liberal.
**hürya** (ˊ·) In a rush; in a sudden burst (of a crowd *etc.*).
**hüsnü** *v.* **hüsün.** *As prefix,* good-. **~ahlâk, -kı,** good morals; morality; good character. **~hal, -li,** good conduct; **~ varakası** (**kâğıdı**), certificate of good conduct. **~hat, -ttı,** calligraphy. **~kabul, -lü,** friendly reception. **~kuruntu,** fond imagination; wishful thinking. **~muamele,** good treatment. **~netice,** good result. **~niyet, -ti,** good intention; goodwill. **~tabiat, -tı,** good taste. **~tabir** (··–ˉ), a happy expression (*only used ironically*). **~tefsir,** favourable interpretation. **~tesadüf** (···–ˊ), a happy coincidence. **~tesir** (··–ˊ), good impression. **~yusuf,** Sweet-William. **~zan, -nnı,** good opinion: ⌜**bu sizin ~nınız!**⌝, 'that's what *you* think (*but you're wrong*)'.
**hüsran** (·ˊ) Moral loss; disappointment; frustration.
**hüsün, -snü** Goodness; beauty; agreeableness.
**Hüt** *Only in* **~ dağı gibi,** very swollen.
**hüthüt kuşu, -nu** Hoopoe.
**hüviyet, -ti** Identity. **~i mechul,** identity unknown: **~ cüzdanı** (**varakası**), identity papers.
**hüyük, höyük** Mound; barrow (ancient grave).
**hüzün, -znü** Sadness, melancholy, grief. **~lenmek,** to be grieved or sad. **~lü,** sad.
**hüzzam** *A mode in Oriental music.*

# I

**ığıl ığıl** Gurgling.
**ığrıb** A medium-sized fishing-smack.
**ıh** *Cry used to make a camel kneel.* **~lamak,** to call out 'ıh!' to make a camel kneel; to breathe loudly. **~latmak, ~tırmak,** to make a camel kneel by saying 'ıh!'
**ıhlamur** Lime-tree; made of lime wood; infusion of lime flowers.
**ık** *v.* hık.
**ıkın·mak** *vn.* To hold the breath when making a great physical effort. **ıkına sıkına,** grunting and groaning; with great effort. **~tı,** great effort.
**ıklamak** *vn.* Breathe with difficulty; sigh, groan.
**ıldırgıç** Deceit, swindle.
**ılgar** Gallop; cavalry charge; foray, raid. **~cı,** raider. **~lamak,** to make a raid.
**ılgın** Tamarisk.
**ılgıncar** Wild cherry.
**ılı** Tepid, lukewarm. **~ca,** hot spring. **~cak, ~k,** tepid. **~mak,** to become lukewarm; to cool down. **~ndırmak, ~tmak, ~ştırmak,** to make tepid, to cool.
**ımızganmak** *vn.* Doze; (of a fire) to be almost out.
**ıncalız** A kind of wild bulb *used for pickling.*
**Irak**[1], **-kı** Irak; Mesopotamia. **~lı,** inhabitant of Irak.
**ırak**[2] Distant. ⌜**dostlar başından ~!**⌝, 'may it be far from our friends!' (*formula used on hearing of a catastrophe*): **üstümüzden ~!**, *said on hearing news of a misfortune.*
**ırga·lamak, ~mak** *va.* Move, shake. **~lanmak, ~nmak** *vn.*, to move, be moved; vibrate; be shaken.
**ırgat** Labourer, workman; bricklayer; capstan, windlass. **~ pazarı,** place where labourers collect to be hired. **~başı, -nı,** foreman. **~lık,** profession *or* pay of a labourer.
**ırıp** Large fishing net.
**ırk, -kı** Race; lineage. **~an** (ˈ·), by race, racially. **~çılık,** racialism. **~î** (·ˈ), racial.
**ırmak**[1] Large river.
**ırmak**[2] *va.* Scatter; rout.
**ırz** Honour; chastity; modesty. **~ düşmanı,** a vicious rake: **~ına* dokunmak,** to dishonour: **~ına* geçmek (tecavüz etm.),** to violate.
**ısdar** (·ˈ) **etm.** *va.* Issue; put forth.
**ısgara** (·ˈ·) Gridiron; grill; grate; grating; any framework in the form of a grating. Grilled (meat *etc.*).
**ısı** Heat. Warm. **~lık,** heat; heatspots; rash.
**ısın·dırmak** *va.* Warm, heat; cause to like. **gönül ~,** to warm the affections, to cause to have friendly feelings. **~mak,** to grow warm: **…e ~,** to have a warm affection for; to get accustomed to.
**ısırgan** Nettle.
**ısırık** Mark *or* wound made by a bite.
**ısır·mak** *va.* Bite. **dilimi ısırdım,** I all but said …: **…i gözüm ısırıyor,** I feel I know *that man etc.*: **parmak ~,** to be surprised, to marvel. **~tmak,** *caus. of* ısırmak: **parmak ~,** to astonish.
**ısıtmak** *va.* Heat. **yerini ~,** to be unwilling to give up one's post.
**ıska** Miss. **~ geçmek,** to miss the ball at football *etc.*; to fail; to fail to see *or* hear.
**ıskaça** (·ˈ·) Step of a mast.
**ıskala** (·ˈ·) Musical scale.
**ıskalara** (··ˈ·) Ratlines *of a ship*; port gangway (of a man-of-war).
**ıskara** *v.* **ısgara.**
**ıskarça** (·ˈ·) Stowed tight, packed.
**ıskarmoz** Rowlock; thole pin; rib *or* frame of a ship; barracuda (*sphyræna*).
**ıskarta** (·ˈ·) Discard (in card games). Discarded, scrapped. **~ya çıkarmak,** to discard; to scrap.
**ıskat**[1] (·ˈ), **-tı** A throwing down. **~ı cenin,** miscarriage: **~ etm.,** to throw down, to bring to nought; dispossess; annul; reject.
**ıskat**[2], **-tı** Alms given on behalf of the dead *as compensation for neglected religious duties.* **~çı,** priest *or* beggar receiving such alms.
**ıskorpit, -ti, ıskorbüt** Scurvy.
**ıskota** (·ˈ·) Sheet *of a sail.* **~ bağı,** sheet-bend: **~ yakası,** clew *of a sail.*
**ıslah** (·ˈ) Improvement, reform. **~ etm.,** to improve, put to rights, reform. **~at** (·–ˈ), **-tı,** improvements, reforms. **~atçı,** reformer. **~hane** (··–ˈ), reformatory.
**ıslak** Wet. **~ karga,** a chickenhearted fellow.
**ısla·mak, ~tmak** *va.* Wet; punish; flog. **~nmak,** to become wet; be wetted.
**ıslık** Whistle. **~ çalmak,** to whistle. **~lamak,** to boo (*Fr. siffler*).

**ısmarlama** Ordered; bespoken. A thing ordered *or* made to order; a stereotyped expression *etc.*
**ısmarlamak** *va.* Order; bespeak; commend; recommend. **Allaha ısmarladık** (*pronounced* **Allahsmalladık**), good-bye!
**ıspanak** Spinach.
**ısparçana** (··ˋ·) Serving of a cable; strand of a rope. ~ **etm.**, to serve *a rope.*
**ıspari** (·ˋ·) Kind of sea-bream (*Sparus annularis*).
**ısparmaca** (··ˋ·) Fouled cables.
**ıspatı** Clubs (in cards).
**ıspavlı** (·ˋ·) Twine.
**ıspazmoz** Spasm.
**ıspor** Sport.
**ısrar** (·ˋ) Insistence. ~ **etm.**, to insist.
**ıssız** Lonely, desolate (place); without an owner; without signs of man, deserted. ~**lık**, desolation.
**ıstağfurullah** *v.* **estağfurullah.**
**ıstampa** (·ˋ·) Inking-pad.
**ıstavroz** Cross; sign of the cross. ~ **çıkarmak**, to cross oneself.
**ıstıfa** (··ˊ) A choosing; selection; natural selection (survival of the fittest). ~ **etm.**, to choose, to prefer.
**ıstılah** (··ˋ) Technical term; conventional name. ~ **paralamak**, to use unusual *or* scientific language. ~**at** (··–ˋ), **-tı**, technical terms.
**ıstırab, ıstırar** *v.* **ıztırab, ıztırar.**
**ışık** Light; lamp. Bright, light.
**ışıl** ~ ~, shining brightly; sparkling. ~**amak**, ~**damak**, to shine, sparkle, flash. ~**atmak**, ~**datmak**, to cause to shine, flash. ~**dak**, searchlight; torch. ~**tı**, brightness; shining; flash.
**ışımak** *v.* **ışılamak.**
**ışın** Gleam; flash.
**ışkırlak** The pointed cap worn by **Karagöz**, *q.v.*
**ıtır, -trı** Perfume, aroma; fragrant plant (*esp.* the Rose-scented Geranium). ~**şahı**, sweet-pea. ~**yağı, -nı**, attar of roses.
**ıtrî** (·ˊ) Perfumed; aromatic. ~**yat, -tı**, perfumes; perfumery. ~**yatçı**, perfumer.
**ıttıla** (··ˊ), **-aı** Information; cognizance.
**ıttırad** (··ˊ) Regularity; uniformity.
**ıvır zıvır** Rubbish, nonsense; unimportant things. Nonsensical.
**ızbandut** Bandit; brigand; huge terrifying man.
**ızgara** *v.* **ısgara.**
**ızmar** (·ˋ) Dissimulation. ~ **etm.**, to hide one's feelings.
**ızrar** (·ˋ) A causing harm. ~ **etm.**, to harm, prejudice.
**ıztırab** (··ˋ) Distress; anxiety; pain.
**ıztırar** (··ˋ) Need; compulsion. ~**ında kalmak**, to find oneself compelled *to do stg.* ~**an** (··ˊ·), from sheer necessity. ~**î** (··–ˊ), forced; involuntary.

# İ

**iade** (·–ˋ) Restoration; giving back; repetition. ~ **etm.**, to give back, return; to restore (*peace etc.*): ~**i muhakeme**, retrial. ~**li**, *in* ~ **taahüdlü mektub**, registered letter receipt of which is returned to the sender.
**iane** (·–ˋ) Help; subsidy; donation, subscription *to a charitable fund etc.* ~ **toplamak**, to collect subscriptions. ~**ten** (·ˊ··), as a help *or* donation.
**iare** (·–ˋ) Loan. ~ **etm.**, to lend.
**iaşe** (·–ˋ) Subsistence; victualling. ~ **etm.**, to sustain, feed.
**ibad** *pl. of* **abid.** Servants; servants of God, men. ~**ullah** (·ˊ··), 'servants of God!', *used to express satisfaction*; a lot of *stg.*, *e.g.* ⌜**bu sene uskumru** ~⌝, this year there is an abundance of mackerel.
**ibadet** (·–ˋ), **-ti** Worship, prayer.
**ibare** (·–ˋ) Sentence; clause.
**ibaret** (·–ˋ) Consisting (of); composed (of). ···**den** ~ **olm.**, to consist of ...; to be equivalent to ...; to be nothing but ... .
**ibda** (·ˊ), **-aı** A creating; an inventing.
**ibham** Obscurity, ambiguity.
**ibik** Comb (of a cock *etc.*).
**ibiş** Idiot.
**ibka** (·ˊ) A making permanent; a keeping an official in his post. ~ **etm.**, to maintain; to confirm *a person* in his office; to keep *a pupil* in the same class without promoting him.
**iblağ** (·ˊ) **etm.** *va.* Cause to reach; cause to amount to; communicate (to).
**iblis** (·ˊ) Satan; devil; devilish man.
**ibne** Catamite.
**ibni** Son of ... . ~**vakit, -kti**, opportunist.
**ibra** (·ˊ) Discharge. ~ **etm.**, to discharge *a debt*; to free from claim; to

pass *accounts*. ~ **kâğıdı** *or* ~**name** (·——ˈ), certificate of receipt *or* discharge.
**ibranî** (·—ˈ) Hebrew. ~**ce** (·—ˈ·), Hebrew (language).
**ibraz** (·ˈ) Manifestation; display (of feelings *etc.*); presentation (of documents).
**ibre** Needle, *esp.* magnetic needle; pointer; stamen.
**ibret, -ti** Example; warning; admonition. ~ **almak,** to take warning: ~ **gözüyle bakmak,** to look at *stg.* in order to get a warning *or* a lesson from it: ~ **kudreti,** odd and ugly: ···**e** ~ **olm.,** to be a lesson to ...: **bu adam bir** ~**!,** what an impossible man!
**ibrik** Vessel with a handle and spout; kettle; ewer.
**ibrişim** Silk thread. Made of silk thread.
**ibtal** (·ˈ), **-li** A rendering null and void. ~ **etm.,** to annul; to bring to nought: **hissi** ~ **etm.,** to anaesthetize.
**ibtida** (··ˈ) Beginning. As a beginning. ~ **etm.,** to begin. ~**î** (··—ˈ), primary; primitive; elementary.
**ibtilâ** (··ˈ) Addiction. ~ **etm.,** to be addicted to *stg.*
**ibtizal, -li** A becoming of little account *from being in abundant supply*; depreciation. ~**e düşmek (uğramak),** to become commonplace *or* vulgarized.
**ibzal** (·ˈ) **etm.** *va.* Give without stint.
**icab** (—ˈ) A rendering necessary, a requiring; exigency. ···**in** ~**ı,** through, on account of, as required by, *e.g.* **arkadaşlık** ~**ı,** as friendship requires: ~**ında,** in case of need: ~**ına bakmak,** to do what is necessary; ⌜~**ına bakarım!**⌝, 'I'll settle his hash!'; 'I'll soon settle that!': ~ **etm.,** to render necessary, require; to be necessary: ~ **eden tedbirler almak,** to take the necessary measures: ~**ı hale göre,** as required by circumstances.
**icabet** (·—ˈ), **-ti** Acceptance; favourable answer. ~ **etm.,** to accept (an invitation); to accede (to a request).
**icad** (—ˈ) Invention; fabrication. ~ **etm.,** to invent; fabricate, trump up.
**icar** (—ˈ) A letting *or* leasing. ~ **etm.,** *or* ~**a vermek,** to let out, to let out on lease. ~**e** (·—ˈ), rent, rental.
**icazet** (·—ˈ), **-ti** Authorization; formal permission; certificate, diploma. ~**name** (·—·—ˈ), diploma; authorization to teach.
**icbar** (·ˈ) A compelling, constraining. ~ **etm.,** to force, compel.
**icik cicik** *In* **iciğini ciciğini bilmek,** to know a thing inside out.
**iclâs** (·ˈ) Enthronement.
**icma** (·ˈ), **-aı** Agreement. ~**ı ümmet,** a consensus of opinion *on religious matters.*
**icmal** (·ˈ), **-li** Summary; *résumé*; adding up. ~ **etm.,** to summarize. ~**en** (·ˈ·), in a summarized form; briefly.
**icra** (·ˈ) Execution; performance; accomplishment. ~ **etm.,** to carry out, perform, execute: ~ **heyeti,** committee charged with the execution of judicial decisions: ~ **kuvveti,** the executive power: ~ **memuru,** executive officer (judicial official): ~ **vekilleri heyeti,** Council of Ministers, Cabinet. ~**at, -tı,** operations, performances; judicial acts; affairs. ~**î** (·—ˈ), executive.
**ictihad** (··ˈ) Interpretation *of a legal or religious point*; doctrine; legal ruling; opinion, conviction. ... ~**ında bulunmak,** to be of the opinion that ...: ⌜~ **kapısı kapandı**⌝, the last word has been said upon that point.
**ictima** (··ˈ), **-aı** Assembly, gathering, meeting. ~**î** (··—ˈ), social. ~**iyat, -tı,** social sciences, sociology.
**ictinab** (··ˈ) Avoidance; abstention. ~ **etm.,** to avoid, abstain from: ~**ı gayrikabil,** inevitable, unavoidable.
**iç** Inside, interior; stomach; heart. Inner, interior. [**iç** *is often used where English would use a personal pronoun, e.g.* **içi sıkıldı,** he was bored; **içim bulanıyor,** I feel sick]. ~ ~**e,** one within the other; *one room* opening into another: ~**ten** ~**e,** secretly; to the innermost recesses: ~**imde,** within me: ~**imizde,** amongst us: ···**in** ~**inde,** inside ..., within ...; **üc gün** ~**inde,** within three days: ~**ler acısı,** tragic, heartrending: ~ **açıcı,** cheering (news *etc.*): ~ **açmak,** to cheer up, to set at ease; ~**ini*** **açmak,** (i) to cheer *s.o.* up; (ii) to unburden oneself: ~**ine almak,** to contain; to include: ~**i*** **almamak,** to feel an aversion for *some kind of food*: ~**ine*** **atmak,** to endure in silence: ~**i*** **bayılmak,** to feel faint *from hunger etc.*: ~ **bezelye,** shelled peas: ~**i*** **bulanmak,** to feel nauseated; to wish to vomit: ~ **çamaşırı,** underclothing: ~ *or* ~**ini*** **çekmek,** to sigh: ~**i*** **çekmek,** to long for: ~**inden çıkamamak,** to be

unable to get at the root of *a matter*; to be unable to settle *stg.*; ~**inden çıkılmaz bir mesele,** an insoluble problem, an incomprehensible matter: ~**inden* doğmak,** to have a sudden impulse to do *some good act*: ~**ine* doğmak,** to have a feeling *or* presentiment *that stg. will happen*: ~**ini* dökmek,** to unburden oneself: ~**i* dönmek,** to be nauseated: ~ **etm.,** to 'bag', to appropriate to one's own use: ~**i* geçirmek,** to sigh: ~**inden* geçirmek,** to review in one's mind: ~**i* geçmek,** to doze; (of a fruit) to become over-ripe: ~**inden* geçmek,** to pass through the mind, to occur to one: ~**i geçmiş,** lethargic: ~**inden* gelmek,** to feel (mentally, not physically) like doing *stg.*: **bir şey için** ~**i* gitmek,** to desire a thing strongly: ~**i* götürememek,** not to have the heart to: ~**i* kalmak,** for the gorge to rise, to have a feeling of nausea: ~**i* kan ağlamak** *or* ~**inden* kan gitmek,** to grieve bitterly, to pine away from grief: ~**inden okumak,** to read to oneself: ~**i* rahat olm.,** to feel at ease about *stg.*: ~**i* sıkılmak,** to be bored: **badem** ~**i,** the kernel of an almond: **ekmek** ~**i,** the crumb of bread: **işin** ~**inden iş çıkarmak,** to keep raising difficulties: **işin** ~**inden çıkmak,** to get out of a difficulty; ⌜... **deyip işin** ~**inden çıktı**⌝, 'he avoided further discussion by saying ...': **işin** ~**inde iş var,** there's something behind it all.

**içecek** Drinkable. Drink, beverage.

**içerde, içeride** Within; inside.

**içeri** The inside; interior; stomach. In; to the inside. ~**den,** from the inside: ~**si,** his (its) interior: ~**ye,** to the inside: ~**(ye) buyurun!,** please come in!: ~ **girmek,** to enter; **bin lira** ~ **girdim** (*sl.*), I have lost £1,000.

**içerlek** Standing back (of a house); secluded.

**içerlemek** *vn.* Be grieved; be annoyed, be angry *without showing it.*

**içgüvey, içgüveyisi, -ni** Husband who lives with his wife's parents. ⌜~**sinden hallice**⌝, so so, not too well.

**içiçe** *v.* iç.

**içim**[1] *v.* iç.

**içim**[2] A draught *of water etc.* **bir** ~ **su,** delicious (*generally of a woman*): **bu suyun** ~**i iyidir,** this water has a good taste. ~**li,** pleasant to the taste (of a drink).

**için**[1] For; on account of; in order to. **Allah** ~, for God's sake: **bunun** ~, for this reason; for this purpose: **onun** ~, for him, for his sake; on account of him *or* that: **yaşamak** ~, in order to live: **ne için,** for what reason, why.

**için**[2] *In* ~ ~, internally; secretly; in a hidden and imperceptible manner.

**içinde, içinden, içine** *v.* **iç.**

**içindekiler** The contents.

**içirmek** *va. caus. of* **içmek.** Cause to drink *etc.*; add a liquid so that it is absorbed.

**içkale** (ˋ..) Citadel.

**içkapı** (ˋ..) Inner door.

**içki** Drink, *esp.* alcoholic drink. ~**ye düşkün,** addicted to drink. ~**li,** licensed to sell alcoholic drinks; drunk.

**içlenmek** *vn.* Be affected (emotionally). **büsbütün içleniyorum,** I am quite overcome by my feelings.

**içli** Having an inside, a kernel, a pulp *etc.*; reticent; sensitive; touching; emotional; sad, hurt. ~ **dışlı,** intimate, familiar.

**içlik** Pertaining to the inside.

**içme** *verb. n. of* **içmek**; spring of mineral water (*gen. used in the pl.*).

**içmek** *va. & vn.* Drink. ⌜**içtikleri su ayrı gitmez**⌝, they are very intimate friends: **and** ~, to take an oath: ⌜**bunun üstüne soğuk su iç!**⌝, 'you can write that off!', 'you'll never see that again!': **çok içer,** he drinks a lot, he is a toper: **tutun** ~, to smoke (tobacco).

**içoğlanı** (ˋ...), **-nı** A young page *in training for domestic service.*

**içre** (ˋ.) Interior. In, within, among.

**içsıkıntısı** (ˋ....), **-nı** Boredom.

**içten** *v.* iç; sincere, from the heart.

**içtihad** *etc. v.* **ictihad** *etc.*

**içyağı** (ˋ..), **-nı** Suet.

**içyüz** (ˋ.) The inside of a matter; the inner meaning; the real truth.

**ida** (–ˋ), **-aı** A depositing for safe keeping.

**idadi** (– – ˋ) Preparatory; secondary. ~ *or* **idadiye,** secondary school.

**idam** (–ˋ) Capital punishment; execution. ~ **etm.,** to condemn to death.

**idame** (.–ˋ) Continuance. ~ **etm.,** to preserve; to prolong.

**idare** (.–ˋ) Management, direction, superintendence; administration; economizing; night-light. ~ **etm.,** to administer, manage, take charge of; to economize, make ends meet; suffice; **bir adamı** ~ **etm.,** to handle a man carefully (*because he is easily upset,* *etc.*); to keep an eye on a man (*because he is not to be trusted to do a thing properly*); **otomobil ~ etm.**, to drive a motor-car: **~i beytiye**, domestic economy: **~ ile geçinmek**, to live economically: **~ kandili**, night-light: **~i maslahet etm.**, to be a skilful negotiator, to be clever at dealing with difficult situations; to get along somehow or other: **~ meclisi**, board of management of a business: **kendini ~ etm.**, to manage for oneself; **kendini ~den âciz**, utterly incapable.

**idare·ci**, a good manager; a tactful person; administrator, organizer; specialist in administrative law. **~hane** (·—·—́), office; administration. **~li**, economical; good at managing; efficient. **~siz**, wasteful; who manages badly. **~ten** (·—́-), administratively; departmentally.

**idarî** (·—́) Pertaining to the administration; administrative.

**idbar** (·*) Adversity; a falling into disgrace.

**iddia** (··́) Claim; pretension; assertion; bet. **~ etm.**, to claim, to set up a pretension: **İddia Makamı**, the Public Prosecutor (as a party in a trial): **~ya tutuşmak (girişmek)**, to assert *stg. in contradiction of another's assertion*. **~cı**, obstinate; assertive; dogmatic. **~lı**, on which a bet has been made; about which claims *or* disputes have arisen. **~name** (··——́), the formal charge *against a person in a Court of Law*. **~sız**, unpretentious.

**idhal** (·*), **–li** An introducing, importing, inserting. **~ etm.**, to import, introduce, insert. **~at** (·—́), **–tı**, imports.

**idiş, idiç** Gelding. Castrated; emasculated. **~ etm.**, to geld.

**idman** (·*) Physical exercise; training; sport. **~lı**, in good training; well-trained; accustomed.

**idrâk** (·*), **–ki** Perception, intelligence; collection, getting-in (of harvest). **~ etm.**, to catch up, overtake, reach; be contemporaneous with; perceive, comprehend. **~li**, intelligent, perceptive. **~siz**, dense, unintelligent.

**idrar** (·*) Urine. **~ yolu**, urethra: **~ zorluğu**, retention of urine.

**idrojen** Hydrogen.

**idüğü** *Used in the phrase* ⌜**ne ~ belirsiz bir adam**⌝, a man of doubtful antecedents; a man of no consequence.

**ifa** (—́) Performance; fulfilment; payment. **~ etm.**, to execute, fulfil, pay.

**ifade** (·—́) Explanation; expression; deposition *of evidence*. **~ etm.**, to express, explain, depose, instruct: **~i meram**, the exposition of thought *or* intention: **~i meram etm.**, to express oneself.

**ifakat** (·—́), **–tı** Convalescence. **~ bulmak**, to convalesce, to get well.

**ifate** (·—́) Loss; waste of time.

**iffet, –ti** Chastity; honesty, uprightness. **~li**, chaste; virtuous; loyal. **~siz**, unchaste, dissolute; dishonest.

**iflâh** (·*) Salvation. **~ı kesilmek**, to be exhausted: **Allah ~ etsin!**, may God reform him!: **bu adam ~ olmaz**, this man is incorrigible.

**iflâs** (·*) Bankruptcy, insolvency. **~ etm.**, to go bankrupt.

**ifrat** (·*), **–tı** Excess; doing too much (*as opp. to* **tefrit** = *doing too little*). **~la**, excessively: **~ etm.**, to overdo, to go to excess.

**ifraz** (·́) A separating; secretion. **~ etm.**, to secrete (a liquid *etc.*); to allot, set aside. **~at** (·—*), **–tı**, secretions *of the body*.

**ifrit, –ti** Demon. Malicious, devilish. **~ olm. (kesilmek)**, to be mad with fury.

**ifsad** (·*) A spoiling, corrupting, inciting. **~etm.**, to corrupt, seduce, incite to revolt. **~at** (·—*), **–tı**, *pl. of* **ifsad**. **~cı**, agitator, political inciter.

**ifşa** (·́) Divulgation; disclosure. **~ etm.**, to divulge, reveal. **~at** (·—*), **–tı**, revelations: **~ta bulunmak**, to reveal, divulge.

**iftar** (·*) The breaking of a fast; the meal taken at sundown *during the fast of Ramazan*. **~ etm.**, to break one's fast: **~ topu**, the gun fired at sunset during Ramazan. **~lik**, suitable to the breaking of the fast; hors-d'œuvres *etc.* eaten at the **iftar**.

**iftihar** (··*) Laudable pride. **…le ~ etm.**, to glory in, to be proud of.

**iftira** (··́) Slander; fabrication, forgery. **~ etm.**, to slander.

**iğ** Spindle. **~ ağacı**, spindle-tree; **~ taşı**, millstone.

**iğbirar** (··*) Annoyance; disappointment.

**iğde** Wild olive, oleaster.

**iğdiç, iğdiş** *v.* **idiş**.

**iğfal** (·*), **–li** Deception; seduction. **~ etm.**, to delude; to take advantage of another's negligence; seduce. **~kâr,** who deludes *or* seduces.

**iğmaz** (·⁴) A winking *or* closing of the eye.

**iğmek** *v.* eğmek.

**iğne** Needle; pin; thorn; sting; pintle *of a rudder*; fish-hook; hypodermic injection; (*fig.*) pinprick. ⌜**~ atsan yere düşmez**⌝, a very crowded place: **~ deliği,** the eye of a needle: ⌜**~ deliğine kaçmak**⌝, to hide oneself in confusion: **~ iplik,** mere skin and bone: **~den ipliğe kadar,** everything required: **~den ipliğe (sürmeye) kadar,** with full details, in a circumstantial way: ⌜**~ ile kuyu kazmak**⌝, to use a needle to dig a well, *i.e.* to do *stg.* in an unpractical way *or* to undertake *stg.* requiring great care and much time: **~ yemek,** to have an injection: **~ yutmuş köpek gibi,** exhausted, upset.

**iğne·ardı, -nı,** back-stitch. **~cik,** pintle *or* brace of a rudder. **~dan, ~lik,** needle-case. **~lemek,** to fasten with a pin; to prick; to give a hypodermic injection; to wound with words. **~leyici,** stinging, pricking. **~li,** having pins; having a needle *or* a sting; biting (words).

**iğrelti** Fern.

**iğrenc** Disgust, loathing. Disgusting, loathsome, repulsive.

**iğren·mek** *vn.* Be disgusted; feel aversion. **~ecek,** disgusting.

**iğreti** Borrowed; makeshift; temporary; false, artificial. **~ almak,** to borrow for temporary use; **~ye almak,** to prop up temporarily: ⌜**~ ata binen tez iner**⌝, 'one doesn't enjoy the use of a borrowed thing for long', *i.e.* a temporary job is not as good as a permanent one.

**iğri, eğri** Crooked; bent; awry; perverse. Curved timber *esp. of a ship*. **~ büğrü,** twisted; gnarled: **~ gitmek,** to deviate; to go wrong: **~ kalem,** carving chisel. **~bacak,** bow-legged. **~lik,** crookedness; curvature; perversity; dishonesty. **~lmek,** to become bent, to incline. **~ltmek, ~tmek,** to make crooked; bend; twist: **çehre ~,** to put on a sour face.

**iğva** (·⁻) A leading astray. **~ etm.,** to tempt, to lead astray. **~at, -tı,** temptations: **~a kapılmak,** to yield to temptation, to be led astray.

**ihale** (·—⁾) A referring *or* delegating; adjudication. **~ etm.,** to hand over, transfer, refer, delegate; award (contract). **~ten** (·⁻··), by award, by adjudication.

**ihanet** (·—⁾), **-ti** Treason; infidelity.

**ihata** (·—⁾) A surrounding *or* embracing; comprehension; erudition, comprehensive knowledge. **~ etm.,** to embrace; surround; comprehend. **~lı,** widely read, erudite.

**ihbar** (·⁴) A communicating; notification; denunciation. **~ etm.,** to communicate, convey (information, news); inform; warn; denounce. **~name** (··—⁾), notification; declaration.

**ihda** (·⁻) **etm.** *va.* Give as a present.

**ihdas** (·⁴) **etm.** *va.* Produce; invent; raise (difficulty); introduce (a new thing); create (a new post).

**ihkak** (·⁴) *In* **~ı hak etm.,** to see that the right prevails, to ensure justice: **bizzat ~ı hak etm.,** to take the law into one's own hands.

**ihlâl** (·⁴), **-li** A spoiling; infraction (of the law, of a treaty); non-observance. **~ etm.,** to spoil; to violate (treaty), break (the law).

**ihmal** (·⁴), **-li** Negligence. **~ etm.,** to neglect; to act negligently, be careless. **~ci, ~kâr,** negligent, careless.

**ihrac** (·⁴) Extraction; exportation; expulsion; emission; disembarkation. **~ etm.,** to emit, expel, export, disembark. **~at** (·—⁾), **-tı,** exports. **~atçı,** exporter.

**ihram** Woollen cloak *worn by Arabs and pilgrims at Mecca*; sofa covering.

**ihraz** (·⁻) **etm.** *va.* Obtain; attain.

**ihsan** (·⁴) Kindness; favour; benevolence. **~ etm.,** to do a kindness; bestow *a favour*; give *a present*.

**ihsas** (·⁴) Sensation. **~ etm.,** to cause to feel; inspire; hint, insinuate; feel, perceive.

**ihtar** (·⁴) A reminding; warning **~ etm.,** to remind; to call attention; to warn. **~name** (··—⁾), written warning; official warning *to cease a practice*.

**ihtısas** (··⁴) Specialization. **~ kesbetmek,** to specialize: **~ mahkemesi,** (*formerly*) court for the suppression of contraband: ···**de ~ peyda etm.,** to specialize in ...: **~ sahibi olm.,** to be a specialist in ... . **~siz,** non-specialist: **~ işçi,** unskilled labourer.

**ihtibas** Repression (psychological).

**ihtida** (··⁻) Conversion to Islam.

**ihtifal** (··⁴), **-li** Ceremony *in memory of s.o. or on some anniversary*. **~ci,** the organizer of such a ceremony.

**ihtikan** (··⁎) Administration of a clyster; enema; congestion (*med.*).
**ihtikâr** (··⁎) Cornering for profit; profiteering. **~cı**, profiteer.
**ihtilâc** Agitation; convulsion.
**ihtilâf** (··⁎) Difference; disagreement. **~ı manzara**, parallax. **~lı**, controversial.
**ihtilâl** (··⁎), **-li** Riot; rebellion; revolution. **~ çıkarmak**, to raise a rebellion: **~i şuur**, mental disturbance. **~ci**, revolutionary.
**ihtilâm** (··⁎) Nocturnal emission.
**ihtilâs** (··⁎) Embezzlement; malversation.
**ihtilât** (··⁎), **-tı** Confusion; complication (of diseases *etc.*); social intercourse. **~ etm.**, to mix, to mingle with others.
**ihtimâl, -li** Probability; possibility; hypothesis. Probable; possible. Probably, most likely. **~at** (··–⁎), **-tı**, probabilities. **~î** (··––́), probable.
**ihtimam** (··⁎) Care; carefulness; taking pains; solicitude. **~ etm.**, to take pains, work carefully. **~kâr**, **~lı**, taking pains; solicitous.
**ihtira** (··–́) Invention. **~ beratı**, patent: **~ etm.**, to invent.
**ihtiram** (··⁎) Respect, veneration. **~ etm.**, to respect, venerate, honour.
**ihtiras** (··⁎) Violent longing; passion; greed; ambition.
**ihtiraz** (··⁎) Precaution; avoidance, abstention. **~ etm.**, to guard against, take precautions; avoid: **~ kaydı**, reservation.
**ihtisar** (··⁎) Abbreviation; reduction of a fraction. **~ etm.**, to shorten, abridge. **~en** (··–́·), in brief; in an abridged form.
**ihtisas** (··⁎) Sensation; impression.
**ihtişam** (··⁎) Pomp; magnificence.
**ihtiva** (··–́) **etm.** *va.* Contain; include.
**ihtiyac** (··⁎) Want; necessity; poverty; need. **···e ~ı* olm.** to be in need of ... .
**ihtiyar**[1] Old. Old man. **~ heyeti** council of elders (village council). **~lamak**, to grow old. **~lık**, old age.
**ihtiyar**[2] (··⁎) Choice, selection; option. **~ etm.**, to choose, prefer; incur (expense, trouble): **~ı elden gitmek.**, to be overcome by emotion; to lose one's self-control.
**ihtiyarî** (··––́) Optional; voluntary; not obligatory.
**ihtiyat** (··⁎), **-tı** Precaution; providing oneself with a reserve; reserve (*also mil.*). **~ akçesi**, reserve fund: **~ ordusu**, reserve of troops: **kaydi ~la**, with reserve. **~en** (··–́·), by way of precaution; as a reserve. **~î**, precautionary. **~lı**, **~kâr**, cautious, prudent. **~sız**, imprudent; improvident; incautious. **~sızlık**, imprudence; improvidence.
**ihtizar** (··–́) **etm.** *vn.* Be at the point of death.
**ihvan** (·⁎) *pl.* Brethren; friends.
**ihya** (·–́) A bringing to life, resuscitation. **~ etm.**, to bring to life, reinvigorate, enliven; load with benefits.
**ihzar** (·⁎) Preparation; a causing to be present. **~ etm.**, to cause to be present, summon, cite: **~ emri** *or* **müzekkeresi**, summons to appear before a court. **~î** (·––́), preparatory. **~iye**, fee for a summons.
**ikame** (·–́) A setting up *or* establishing; substitution. **~ etm.**, to set up; to post (sentinel *etc.*); to substitute: **dâva ~ etm.**, to bring an action *against s.o.*
**ikamet** (·––́), **-ti** Residence; dwelling; a staying at a place. **~ etm.**, to dwell, to stay: **~e memur**, ordered to reside *at a certain place*: **~ tezkeresi**, permit to reside. **~gâh** (·–·–́), place of residence.
**ikaz** (–⁎) A rousing; warning. **~ etm.**, to awake, arouse; caution, warn.
**ikbal** (·⁎), **-li** Good fortune; success; prosperity; (*formerly*) a slave girl about to become the Sultan's concubine.
**iken** While being; while; though; (*when joined to a participle it is generally abbreviated to* **-ken**, *e.g.* **giderken**, while going). **yazacak ~**, while on the point of writing *or* instead of writing: **hal böyle ~**, even though this was the situation; in spite of this situation.
**iki** Two. **~miz**, both of us: **~si**, both of them: **~ baştan olm.**, to be possible only if both parties show goodwill: **~de bir (birde)**, at frequent intervals, constantly: **~ ···de bir**, one in two ..., every other ..., *but* (*illogically*) **~ günde bir**, every third day (*but nowadays usually* every other day); *also* frequently; every now and again: **~ canlı**, pregnant: ⌜**~ el bir baş içindir**⌝, 'two hands are for one head', *i.e.* one must work oneself and not depend on others: **bir ~ derken,**

then all of a sudden; ⌜**bir** ~ **derken işler fenaya varır**⌝, if you allow small faults to pass unnoticed they will become bad habits.

**iki·ağızlı**, two-edged (sword). **~çenekliler**, dicotyledons. **~çifte**, boat with two pairs of oars. **~lemek**, to make two; to get another *thing in addition to that which one already has*. **~li**, having two (parts *etc.*); the two *of a suit of cards*. **~lik**, consisting of two; costing two (piastres *etc.*); disunion.

**ikinci** Second.

**ikindi** The time of the afternoon prayer; afternoon. ⌜**~den sonra dükkân açmak**⌝, to leave doing a thing until too late. **~yin** (·˘··), in the afternoon.

**ikişer** Two each; two at a time. **~ ~ koşmak**, to harness a pair of horses *or* oxen *to a cart etc.*

**ikiyüzlü** (·˘··) Having two faces; double-faced; hypocrite.

**ikiz** Twins. Twin. **~leme**, **~li**, having twins; double.

**iklim** Region; country; climate. **yedi ~ dört bucak**, all over the world.

**ikmal** (·*), **-li** Completion. **~ etm.**, to complete: **~ efradı**, (*mil.*) draft, replacements: **~ hatları**, (*mil.*) lines of communication: **~ kolları**, (*mil.*) supply columns: **gününü ~ etm.**, (of a convicted man) to complete his sentence: **noksanı ~ etm.**, to make good a deficiency. **~ci**, pupil who has to undergo a second examination after failure at the first.

**ikna** (·–̀) **etm.** *va.* Satisfy; convince; persuade.

**ikrah** (·*) Disgust, loathing. **~ etm.**, to loathe. **~en** (·–̀·), with loathing; much against one's will.

**ikram** (·*) A showing honour and respect (*esp. to a guest*); kindness; gift; discount, abatement on price. **~ etm.**, to show honour to; to give *a present*; to offer (a cigarette *etc.*); to make a reduction in price. **~iye** (·–·˘), bonus; gratuity; prize *in a lottery*. **~lı**, hospitable.

**ikrar** (·*) Declaration; confession; acknowledgement. **~etm.**, to confess, acknowledge.

**ikraz** (·*) Loan. **~ etm.**, to lend money. **~en** (·–̀·), as a loan.

**iksir** Elixir; philosopher's stone; any magic substance; cordial.

**ikta'** (·–̀) Fief.

**iktibas** (··*) Quotation; adaptation *of a novel or play*. **~ etm.**, to borrow; to use as a quotation.

**iktidar** (··*) Ability, capacity; power. **~ı* olm.**, to be capable of, able to do: **~da olm.**, (of a political party) to be in power: **~ partisi**, the party in power, the government: **ademi ~**, impotence. **~lı**, capable. **~sız**, incapable.

**iktifa** (··–̀) **etm.** *vn.* Content oneself (with = **ile**); be satisfied; suffice.

**iktiham** (··–̀) **etm.** *va.* Surmount; overcome; endure.

**iktiran** (··*) Approach. ...**e ~ etm.**, to meet with (approval *etc.*): **tasdiki âliye ~ etm.**, to receive the approval *or* signature of the head of the state.

**iktisab** (··*) Acquisition, gain. **~ etm.**, to acquire.

**iktisad** (··*) Economy; political economy. **~ etm.**, to economize. **~cı**, economist. **~î** (··–̀), economic; pertaining to economics.

**iktiza** (··–̀) Necessity; requirement. ... **~sı**, on account of ..., as rendered necessary by ...: **~sınca**, according to its requirements: **~sına göre**, as may be necessary, as occasion requires: **~ etm.**, to be requisite: **~ ettirmek**, to render necessary, require: **~sı olm.**, to be necessary.

**il** Province; country; vilayet; (*for* **el**) people.

**ilâahiri** *v.* **ilâhiri**.

**ilâc** (·–̀) Remedy; medicine; device. **~ için yok**, *said of stg. not to be found*: **la ~**, through sheer necessity, there being no way out. **~lı**, medicated. **~sız**, incurable.

**ilâh**[1] *Abb. for* **ilâahiri**. Etcetera.

**ilâh**[2] God. **~e** (·–˘) goddess. **~i** (·–˘), hymns, chants; *interj. expressing admiration or astonishment*. **~î** (·––̀), divine. **~ici**, singer of chants, dervish. **~iyat** (·–·*), **-tı**, theology. **~iyun** (·–·–̀), theologians.

**ilâhiri** Etcetera.

**ilâm** (–*) Sentence; judicial decree; official decision. **~ etm.**, to notify officially.

**ilân** (–*) Declaration, notice, announcement, proclamation; advertisement. **~ı harb**, declaration of war: **~ etm.**, to declare, proclaim, announce: **~ pulu**, revenue stamp on advertisements. **~at**, **-tı**, advertisements: **~ idarehanesi**, adver-

tising agents. ~**cılık,** publicity, advertising.
**ilânihaye** (⸍..–.) To the end.
**ilâve** (.–⸍) Addition; supplement; postscript; exaggeration. ~ **etm.,** to add. ~**li,** having an addition *or* supplement. ~**ten** (.⸍..), in addition; as an addition.
**ilca** (.⸍) Constraint, compulsion. ~ **etm.,** to compel.
**ilçe** Administrative district formerly known as a **kaza.**
**ile, ilen** (**–le, –yle**) With; by means of; and. **benimle refikam,** my wife and I: **kalemle,** with a pen; **kalemile** *or* **kalemiyle,** with his pen: **söylemekle,** by saying: **bunu söylemekle beraber,** while saying this, in spite of saying this: **bunu söylemesile,** just as he had said this, he had hardly said this when ... .
**ilelebed** (⸍...) For ever.
**ilen·c** Curse. ~**mek,** to curse.
**iler** *Only in* ~ **tutar yer bırakmamak,** to tear to tatters, to maltreat, to 'take it out of' one; *and* ~ **tutar yeri kalmamak,** to be torn to tatters, to be in a pitiable condition.
**ilerde, ilerden** *v.* **ileri.**
**ileri** The forward part; the front; the future. Advanced; in advance; forward; in front; future; fast (of a clock). ~ **almak,** to promote (a person); to put forward (a clock): ~ **gelmek,** to come forward, to make progress, to surpass; *v. also* **ilerigelmek**: ~ **gelenler,** notables, important people: ~ **gitmek,** to go too far: ···**in** ~**sine gitmek,** to follow *a matter* up; to go deeply into *stg.*: ~ **geri sözler,** unseemly, inappropriate, fault-finding words: ~ **sürmek,** to drive forward; to advance (a reason, an argument); to advance *a person* in rank. ~ **varmak,** to go ahead; to go too far.
**iler·ide** In front; in future. ~**igelen** (..⸍..), notable; important person. ~**igelmek** (..⸍..), ···**den** ~, to result from, be caused by; *v. also* **ileri.** ~**leme,** advance, progress; feed (of a tool). ~**lemek,** to advance, progress; (of a clock) to be fast. ~**letmek,** to cause to advance; make progress in (an art *etc.*); put forward (of a clock).
**iletmek** *va.* Send *or* give *to another*; send *to a place*; carry off, make off with.
**ilga** (.⸍) Abolition; annulment. ~**etm.,** to abolish, annul, abrogate.
**ilgi** Interest. ~**lenmek,** to be interested (in = **ile**). ~**li,** interested; connected *with.*
**ilh** *Abb. for* **ilâahiri.** Etcetera.
**ilhak** (.⸍), **–kı** Annexation. ~ **etm.,** to annex; join on.
**ilham** (.⸍) Inspiration. ~ **etm.,** to inspire.
**ilhan** Prince; emperor.
**ilik**[1] Loop *for button or hook*; buttonhole. ~**lemek,** to button up; to fasten by hook and eye. ~**li**[1], buttoned; fastened by hook.
**ilik**[2] Marrow. Delicious. ~ **gibi,** tasty, appetizing: **iliğe işlemek,** (of cold *etc.*) to penetrate to the marrow; to make a great impression: **iliğini* kurutmak,** to wear *s.o.* out. ~**li**[2], containing marrow.
**ilim, –lmi** Knowledge; science.
**ilinti** Roughly sewing together, tacking; knot.
**ilişik** Connexion; relation; bond; impediment, hitch; suspense account; liability, obligation. Connected; relative; attached; enclosed herewith. **ilişiği* kalmamak,** (of a matter) to be settled up; (of a person) to sever his connexion, to have no further interest: **ilişiğini* kesmek,** to sever one's connexion with; to sever *s.o. else's* connexion, *i.e.* dismiss, discharge. ~**siz,** unattached, free.
**iliş·mek** *va.* Touch; interfere with. *vn.* Be fastened *to*; hold on slightly *to*; sit uncomfortably *on the edge of stg.*; remain a short time; have a disagreement, quarrel. **ceketim çiviye ilişti,** my coat caught on a nail: **göze** ~, to catch the eye. ~**tirmek,** to fasten, attach.
**ilk** First; initial; primary (school); beginning. ~ **ağızda,** at the first attempt, at the first shot: ~ **önce,** first of all.
**ilka** (.⸍) A throwing *or* dropping into; suggestion. ~ **etm.,** to suggest, put an idea *into s.o.'s head*; sow (discord); throw *into danger*; lower *a boat* into the sea.
**ilkah** (.⸍) Fertilization; insemination.
**ilk·bahar** (⸍..) The spring. ~**çağ** (⸍.), ancient times. ~**in** (⸍.), first; in the first place. ~**mekteb,** ~**okul** (⸍..), primary school. ~**söz** (⸍.), preface.
**illâ** (⸍.) *v. also* **ille.** Except; or else; whatever happens; without fail. ~ **ki**

**oraya gitmiyesin!**, whatever you do, don't go there! : ~ **velâkin**, but on the other hand: **bunu böyle yapmak lâzımdır ve ~ felâ**, you must do it this way or not at all.

**illallah** (ˈ··) *Expresses annoyance or disgust.* ~**i ve resulihi!**, I am sick to death of it (you *etc.*): **senden ~!**, I'm sick of you!

**ille** (ˈ·) *v. also* **illâ.** Absolutely; just; especially. ~ **bu gün mü olmalı?**, *must* it be today ? : ~ **gideceğim diyor,** he says he simply must go.

**illet, -ti** Disease; defect; cause, reason. ~ **edinmek,** to acquire a tiresome habit: **frengi ~i,** syphilis. **~li,** ill, diseased; defective; having some annoying habit.

**ilmek, ilmik** Loop; bow; noose; slip-knot. **~lemek,** to tie loosely. **~li,** lightly tied; in a bow.

**ilmi** *v.* ilim. **~hal,** book for teaching to children the elements of religion. **~ye,** a member of the hierarchy of Moslem jurisprudents and religious teachers.

**ilmî** (·ˈ–) Scientific; pertaining to knowledge.

**ilmik** *v.* ilmek.

**ilmühaber** Identity papers; certificate; receipt.

**iltibas** (··ˈ) Confusion of one word *or* phrase with another. **~a mahal kalmamak için,** to avoid any ambiguity. **~lı,** ambiguous.

**iltica** (··ˈ–) **etm.** *vn.* Take refuge; ask for quarter, surrender.

**iltifat** (··ˈ), **-tı** Courteous *or* kind treatment; favour. ~ **etm.,** to take notice of, greet; treat with kindness. **~çı,** kind, affable. **~kârane,** courteous; in a kind and affable manner.

**iltihab** (··ˈ) Inflammation. **~î** (··–ˈ–), inflammatory. **~lanmak,** to become inflamed.

**iltihak** (··ˈ), **-kı** Action of joining *or* attaching oneself. ···**e ~ etm.,** to join; to connect oneself with; to adhere to.

**iltimas** (··ˈ) Request, prayer; recommendation; favouritism. ~ **etm.,** to request, solicit; ask a favour for a protégé, recommend him for a post. **~çı,** one who asks such favours, patron, protector. **~lı,** one who gets a job through favouritism.

**iltizam** (··ˈ) A looking after the interests of *s.o.*; a farming of a branch of the public revenue. ~ **etm.,** to take the part of, uphold, prefer, favour; to farm a branch of the revenue: **~a vermek,** to lease out the collection of certain revenues. **~cı,** one who farms a branch of the revenue. **~î** (··–ˈ–), done on purpose; showing favour, partial, sympathetic: optative *tense* (**yazayım, gideyim**): ~ **olarak,** on purpose, designedly.

**ilzam** (·ˈ–) **etm.** *va.* Silence by argument.

**ima** (–ˈ–) Allusion, hint. ~ **etm.,** to allude to, hint at.

**imal** (–ˈ), **-li** Manufacture; making. ~ **etm.,** to make, manufacture, produce, prepare: **~i fikir etm.,** to go into a matter deeply, to think it out. **~at** (––ˈ), **-tı,** manufactured goods; production of same. **~athane,** factory, workshop.

**imale** (·–ˈ) A bending *or* inclining; the pronunciation of a short vowel as a long one *for the purposes of prosody.* ~ **etm.,** to incline, persuade, convince: ···**e ~i nazar etm.,** to examine *stg.*

**imalı** (––ˈ) Alluding to; containing a hint *or* implication. **bu ~ sözler,** words alluding to this, this sort of words.

**imam** (·ˈ) Leader in public worship; religious leader; Imam. ⌜**~ evinden aş ölü gözünden yaş**⌝, 'you might as well expect tears from a corpse as food from an Imam' (*allusion to their alleged stinginess*): ⌜**cami** (*or* **cemaat**) **ne kadar büyük olsa ~ bildiğini okur**⌝, 'however large the mosque (congregation) the Imam will read the appointed prayers', *i.e.* rules are rules; *also sometimes* 'he goes his own way without paying attention to others'. **imam·bayıldı,** a dish of aubergines with oil. **~e** (·–ˈ), stem of a rosary; mouthpiece of a pipe; turban. **~et** (·–ˈ), **-ti,** quality and office of an Imam. **~suyu** (·ˈ··), **-nu,** raki. **~zade** (··–ˈ), son of an Imam.

**iman** (–ˈ) Belief; faith; religion. **~ım,** *a term of address among the lower classes*: **~ı* ağlamak** (*sl.*), to suffer, to undergo hardship: ~ **etm.,** to have faith in God (*esp. in accord with Moslem tenets*): **~a gelmek,** to be converted to the true faith; to see reason: **~a getirmek,** to convert to Islam; to bring *s.o.* to see reason: ···**e ~ getirmek,** to believe *in a dogma etc.*: **~ı gevremek** (*sl.*), to undergo hard-

ship; to be exhausted: **~ına kadar,** to the utmost degree: **~ tahtası,** the breast: **~ı yok,** *a term of abuse.*
**imansız** (—·˟) Unbelieving; atheist; cruel, inhuman.
**imar** (—˟) Improvement by cultivating *or* building. **~ etm.,** to improve; to render prosperous. **~et** (·—˟), **-ti,** prosperous condition; kitchen for the distribution of food to the poor.
**imbat**[1], **-tı** A cool sea-breeze *which prevails in the Aegean in the summer.*
**imbat**[2] *v.* **inbat.**
**imbik** Retort, still. **~ etm.** *or* **~ten çekmek,** to distil.
**imdad** (·˟) Help, assistance; reinforcement. **~ etm.,** to come to *s.o.'s* help: **~a yetişmek,** to come to the help of, to reinforce. **~cı,** one who comes to another's help; reinforcement.
**imdi** (˟·) Now; in a short time; thus; and so.
**imece** Work done for the community by a whole village; corvée. By the united efforts of the community.
**imha** (·˟) Destruction; effacement. **~ etm.,** to obliterate, destroy, cancel.
**imik** Soft place on a baby's skull, fontanel.
**imizgenmek** *vn.* Doze; (fire) to be nearly out.
**imkân** (·˟) Possibility; practicability. **~ dahilinde,** within the bounds of possibility, as far as possible: **~ı yok,** it is impossible. **~lı,** possible. **~sız,** impossible. **~sızlık,** impossibility.
**imlâ** (·˟) Action of filling; spelling, orthography. **~ etm.,** to spell; to dictate: **~ makinesi,** dictaphone: **bu çocuk ~ya gelmez,** this child is incorrigible. **~sız,** misspelt.
**imparator** Emperor. **~içe** (···˟·), empress. **~luk,** title and position of emperor; empire.
**imrenmek** *va.* (*with dat.*) Long for; envy, covet.
**imsak** (·˟), **-ki** Fasting; abstention; abstinence, continence; the hour at which the Ramazan fast begins each day. **~ etm.,** to fast, abstain, refrain. **~li davranmak,** to be moderate, to avoid excess. **~iye,** time-table giving the hour when the fast begins, *v.* **imsak.**
**imtihan** (··˟) Trial, test; examination. **~a çekmek,** to examine *a pupil* thoroughly: **~ etm.,** to examine (a pupil *etc*): **~a girmek** *or* **~ olm.,** to sit for an exam.: **~da kalmak,** to fail in an exam.: **~ vermek,** to pass an exam.
**imtisal** (··˟), **-li** A conforming to rule. **···e ~ etm.,** to conform to; comply with; follow *an example.* **nümunei ~,** an example to follow, exemplary. **~en** (··˟·), conformably, in compliance with.
**imtiyaz** (··˟) Distinction; privilege; concession; autonomy; diploma. **~ bulmak,** to acquire distinction: **~ sahibi,** concessionaire; privileged person. **~lı,** privileged; autonomous.
**imtizac** (··˟) A blending; a fitting together; a getting on well together. **···le ~ etm.,** to blend with, fit, get on well together with; get accustomed to. **~lı,** harmonious; well-fitting (door *etc.*). **~sızlık,** incompatibility of temperament.
**imza** (·˟) Signature. **···i ~ etm.,···e ~ atmak** *or* **koymak** to sign: **~ sahibi,** signatory. **~lamak,** to sign.
**in**[1] Human being, *only used in conjunction with* **cin.** ⌜**~ misin cin misin nesin**⌝ whatever you are, a man or a spirit: ⌜**~ cin yok**⌝, there's not a soul there.
**in**[2] Den *or* lair *of a wild beast.*
**inabe** (·—˟) Appointment of a deputy; entrance into a religious fraternity.
**inad** (·˟) Obstinacy. **~ına** (·—·˟), out of obstinacy; out of contrariness; just to spite one; it would just happen that; as luck would have it: **···e ~,** despite ...: **~ etm.,** to be obstinate, to persist: **···in ~ına yapmak,** to do *stg.* just to spite *s.o.*: ⌜**iş ~a bindi**⌝, it is a matter of sheer obstinacy; we'll see who can hold out longest, **~cı,** obstinate, pig-headed.
**inan** Belief; trust. **~ olsun!,** believe me!, rest assured that ... . **~c,** belief; confidence. **~dırmak,** to cause to believe; persuade; deceive: **Allah inandırsın!,** believe me! **~ılacak, ~ılır,** credible. **~ılmaz,** incredible. **~ış,** belief; credulity.
**inanmak** *va.* (*with dat.*) Believe; trust.
**inayet** (·—˟), **-ti** Kindness; favour; grace; care, effort. **~ola!,** *said to a beggar when refusing to give alms.* **~en** (·˟··), as a favour. **~kâr, -li,** kind; gracious; obliging.
**ince** Slender; thin; fine; slight; subtle; delicate. **~den ~ye,** minutely; in a subtle manner: **~ elemek,** to pass

through a fine sieve; ⌜~ **eleyip sık dokumak**⌝, to be too particular, to be very meticulous: ~ **hastalık**, tuberculosis: ⌜~**sini ipe kalınını çöpe dizmek**⌝, to go into details, to be too particular.
**ince·cik** ('··), very slender *or* fine; minutely, finely. ~**lemek**, to examine minutely; go into *a matter* carefully. ~**lik**, fineness; delicacy; refinement; subtlety; ingenuity. ~**lmek**, to become fine *or* thin; be refined *or* delicate; be too subtle, quibble. ~**ltmek**, to make fine *or* slender; refine; subtilize; make nice distinctions. ~**rek**, slender, elegant; rather thin. ~**saz**, Turkish orchestra of stringed instruments.
**inci** Pearl. ~ **dizisi**, a string of pearls: ~ **balığı**, bleak. ~**çiçeği**, **-ni**, lily of the valley.
**incik**[1] Shin.
**incik**[2] Slightly bruised *or* broken. Sprain; bruise.
**incil** Gospel, New Testament.
**incinmek** *vn.* Be hurt; be offended.
**incir** Fig. ~ **çekirdeğini doldurmaz**, trifling, insignificant: ⌜**bir çuval** ~**i berbad etm.**⌝, to undo everything, to 'upset the apple-cart': **frenk** ~**i**, prickly pear: **kavak** ~**i**, purple fig: **lop** ~**i**, green fig: **ocağına** ~ **dikmek**, to break up a home; to ruin, exterminate. ~**ağacı**, **-nı**, fig-tree. ~**kuşu**, **-nu**, beccafico.
**incitici** Harmful; hurting; offending.
**incitmek** *va.* Hurt; touch; offend, vex.
**ind** Side. **benim** ~**imde**, in my opinion; **onun** ~**inde**, in his opinion.
**indî** (·⁻) Personal; subjective; arbitrary.
**indinde** *v.* **ind.**
**indirme** Act of lowering; unloading (from a ship *etc.*); landing of troops from the air.
**indirmek** *va.* Cause to descend; lower; calm; give (a blow). **denize** ~, to launch *a ship*.
**inek** Cow. ~**çi**, cowman, cowherd. ~**lik**, cowshed; idiotic behaviour.
**ineze** Sickly, debilitated.
**infaz** (·*) Execution of an order. ~ **etm.**, to carry out *an order*.
**infial** (··*), **-li** Annoyance; anger, indignation. ~ **etm.**, to be annoyed *or* indignant.
**infilâk** (··*), **-kı** Explosion. ~ **etm.**, to burst, to explode.
**infirad** (··*) Isolation. ~ **etm.**, to be isolated. ~**cı**, isolationist.
**infisah** (··*) Disintegration; dissolution; cancellation. ~ **etm.**, (of an assembly) to be dissolved.
**ingiliz** English. Englishman; pound sterling. ~ **anahtarı**, spanner. ~**ce**, English language; in English; in the English fashion.
**İngiltere** (··'·) England.
**inha** (·⁻) Official memorandum to a superior department recommending the appointment *or* promotion of an official.
**inhilâl** (··*), **-li** A being undone, solved, dissolved. ~ **etm.**, to be dissolved, decomposed; to become void.
**inhimâk** (··*), **-ki** Addiction.
**inhina** (··⁻) A being curved *or* bent; abasement.
**inhiraf** (··*) Deviation. ~ **etm.**, to deviate; to be altered. ~**ı hatır**, displeasure: ~**ı mizac**, indisposition.
**inhisar** (··*) Monopoly; limitation, restriction. ~**a almak** *or* **kendine** ~ **ettirmek**, to monopolize: ···**e** ~ **etm.**, to be restricted to ... .
**inhitat** (··*), **-tı** Decline; degradation; degeneration.
**inhizam** (··*) Defeat, rout.
**ini** Younger brother.
**inik**[1] Cub; pup.
**inik**[2] Lowered (of a tent *etc*).
**inildemek** *vn.* Echo; resound; moan.
**inilmek** *vn. impers. of* **inmek. buradan inilir**, one goes down here.
**inilti** Moan, groan; echo.
**inim inim** With groans and laments.
**iniş** Descent; slope; decline; landing *of an aeroplane.* ~ **çıkışlar**, ups and downs.
**inkâr** (·*) Denial; refusal. ~ **etm.**, to deny, refuse.
**inkıbaz** (··*) Constipation. ~**a uğramak**, to be constipated.
**inkılâb** (··*) Radical change; revolution; transformation. ···**e** ~ **etm.**, to be transformed *or* turned into ... : ~ **geçirmek**, to undergo a transformation: **harf** ~**ı**, the reform of the *Turkish* alphabet.
**inkıraz** (··*) The decline and extinction (of a family *or* dynasty). ~ **bulmak**, to become extinct.
**inkıta** (··⁻), **-aı** Cessation; interruption. ~**a uğramak**, to cease; to be interrupted.
**inkıyad** (··*) Submission; obedience. ~ **etm.**, to submit, obey (*with dat.*).

**inkisar** (··ʹ) A breaking; refraction; vexation; curse. ~**etm.**, to refract; to be vexed; to curse. ~**ı hayal,** disappointment: ~**ı* tutmak,** for one's curse to take effect.
**inkişaf** (··ʹ) Development. ~ **etm.,** to develop.
**inle·mek** *vn.* Moan, groan. ~**tmek,** to cause to moan; oppress; make resound.
**inme** Act of descending; landing of an aeroplane; launching of a ship; fall; apoplexy, stroke. Fallen, dropped. **gökten** ~, heavensent. ···**e** ~ **inmek,** to have a stroke.
**inmek** *vn.* Descend; fall down; alight; (aeroplane) land; subside; (price) fall; (apoplexy) attack, strike. *va.* Plant (a blow); strike: **yüreğine*** ~, to die of a heart attack; to suffer a terrible blow; **az daha yüreğime iniyordu,** I nearly died of shame.
**inmeli** Struck by apoplexy; paralysed.
**insaf** (·ʹ) Justice; moderation; reasonableness, fairness. ~ !, 'be reasonable!', 'be fair!': ~ **etm.,** to act with justice: ~**a gelmek,** to come to reason; to be fair; to show moderation *or* pity: **artık** ~**ına kalmış,** it now all depends on his sense of fairness; it is at his discretion. ~**kâr,** ~**lı,** just, equitable; humane; reasonable, fair. ~**sız,** unjust; unfair; without a conscience; cruel. ~**sızlık,** injustice; inhumanity; unfairness.
**insan** Human being, man; fine type of man. Humane; upright. ~ **içine çıkmak,** to go out in public, to mix with one's fellow-beings. ~**ca** (·ʹ·), humane: as a man should. ~**iyet** (·—·ʹ), **-ti,** ~**lık,** humanity; humankind; kindness. ~**iyetli,** humane. ~**iyetsiz,** inhuman, cruel. ~**oğlu, -nu,** man, mankind.
**insicam** (··ʹ) Coherence *in speech or writing*; harmony; regularity. ~**lı,** coherent. ~**sız,** incoherent.
**insiyak** (··ʹ), **-kı** A being driven *or* guided; instinct. ~**î** (··—ʹ), instinctive.
**inşa** (·ʹ) Construction; creation; literary composition; book giving models for letter writing. ~ **etm.,** to construct, build: **hali** ~**da,** in course of construction. ~**at** (·—ʹ), **-tı,** buildings; works. ~**î** (·—ʹ), pertaining to building *or* shipbuilding.
**inşad** Recitation; repetition (in a school). ~ **etm.,** to recite.
**inşallah** (ʹ··) *Lit.* 'if God pleases'; I hope; it is to be hoped that.
**inşirah** (··ʹ) Cheerfulness; exhilaration; relief.
**intac** (·ʹ) **etm.** *va.* Result in; cause; bring to a conclusion; bring forth *young.*
**intıba** (··ʹ), **-aı** Impression; feeling.
**intıbak** (··ʹ), **-kı** Adaptation; adjustment. ~ **etm.,** to be adapted, to adapt *or* adjust oneself; to conform. ~**sızlık,** lack of adaptability.
**intibah** (··ʹ) An awakening; vigilance; circumspection.
**intihab** (··ʹ) Selection, choice; election; preference. ~ **dairesi,** electoral district, constituency: ~ **etm.,** to choose; elect; prefer. ~**at,** ~**tı,** elections.
**intihal** (··ʹ), **-li** Plagiarism. ~ **etm.,** to plagiarize.
**intihar** (··ʹ) Suicide. ~ **etm.,** to commit suicide.
**intikal** (··ʹ), **-li** Transition; passing from one place to another; transfer of property (by sale *or* inheritance); passing away to another world; perception, understanding. ~ **etm.,** to pass to another place; pass away (die); (of a conversation) to change to another subject; perceive: **süratı** ~, quickness of perception.
**intikam** (··ʹ) Revenge. ···**den** ~ **almak,** to revenge oneself on, take revenge on ... : ~**ını almak,** to avenge ... . ~**cı,** vindictive.
**intisab** (··ʹ) Relation; attachment. ~ **etm.,** to be connected with; to join *a political party etc.*: to enter (a career).
**intişar** (··ʹ) Publication; dissemination; ~ **etm.** *or* **bulmak,** to be spread *or* disseminated.
**intizam** (··ʹ) Regularity; order; arrangement. ~ **bulmak,** to become well arranged, be set in order: ~ **üzere olm.,** to be in good order *or* well disciplined. ~**sız,** irregular; disordered.
**intizar** (··ʹ) Expectation; waiting.
**inzibat** (··ʹ), **-tı** Discipline, ~ **memuru,** military policeman. ~**î** (··—ʹ), disciplinary.
**inzimam** (··ʹ) Act of being added *or* joined to. ~ **etm.,** to be added.
**inziva** (··ʹ) A retirement into seclusion; a leading the life of a hermit. ~**ya çekilmek,** to retire into seclusion.
**ip, -pi** Rope; cord; string. ~ **atla-**

**mak,** to skip: ~ **cambazı,** tight-rope walker: ~**e çekmek,** 'to string up', hang: …**i** ~**le çekmek,** to await *stg.* anxiously: ~**ini çözmek,** to sever *s.o.'s* connexion with *stg.*: ~**e gelmek,** to come to the gallows: ~ **kaçkını,** jailbird: ~**e kazığa vurmak,** to execute by hanging *or* impaling: ~**ten kazıktan kurtulmus,** gallows-bird: ~**i kırık,** vagabond: ~**ini* kırmak,** to slip away; to get out of hand: ~ **koparmak,** to break the link *with stg.*: ~**ini* koparmak,** to make off: ⌜**onun** ~**iyle kuyuya inilmez**⌝, he is not to be relied on: ~**i sapı yok** *or* ~**e sapa gelmez,** without connexions, vagabond; incoherent, irrelevant (*also* **ipsiz sapsız**): ~ **ucu,** *v.* **ipucu**: ~**in ucunu kaçırmak,** to lose the thread of *stg.*; to lose control of *stg.*: ~**e un sermek,** to make vain excuses; to be lazy: ~**ini üstüne atmak** (*sl.*), to give *s.o.* his head, to leave him to his own devices: **ayağına*** ~ **takmak,** to run *s.o.* down, to gossip maliciously about *s.o.*

**ipek** Silk; silken. ~ **kozası,** silk cocoon: **ham** ~, raw silk. ~**böceği, -ni,** silkworm. ~**çi,** silk merchant *or* manufacturer: ~**çilik,** silk industry. ~**çiçeği, -ni,** the portulaca and mesembryanthemum classes of flowering plants. ~**hane** (··–ˊ), silk factory. ~**li,** of silk.

**ipeka** (·ˋ·) Ipecacuanha.

**ipham** *v.* **ibham.**

**ipipullah** (·ˊ··) *In* ⌜~ **sivri külâh**⌝, stonybroke, destitute.

**ipiri** Huge.

**ipka** *v.* **ibka.**

**iplemek** *va.* Bind with a rope; (*sl.*) respect, pay attention to.

**iplik** Thread; sewing-cotton. Made of linen. ~ ~, thread by thread; ~ ~ **olm.,** to become frayed *or* threadbare: **ipliği* pazara çıkmak,** to get a bad name; to be shown up: **anasının ipliğini pazara çıkarmış,** an unprincipled scoundrel: **pamuk ipliğile bağlamak,** to settle *a matter* in an unsound manner.

**iplik·cik,** threadlike worm *attacking the flesh of human beings in hot countries* (? *filaria*). ~**hane** (··–ˋ), spinning-mill. ~**lenmek,** to become unravelled. ~**peyniri, -ni,** *a kind of cheese showing the marks of the cheese-cloth.*

**ipotek** Mortgage.

**ipsiz** Having no connexions, vagabond. ~ **sapsız,** without house or home; irrelevant.

**iptal, iptida** *etc. v.* **ibtal, ibtida** *etc.*

**ipucu** (ˋ··), **-nu** Clue; motive. *v. also* **ip.**

**irab** (–ˋ) The declension of an Arab word; the case-endings. ~**da mahalli yok,** of no account, insignificant.

**irad** (–ˋ) Income, revenue; a citing *or* quoting. ~ **etm.,** to quote; deliver (a speech); adduce (proof): **Hazineye** ~ **kaydedilmek (yazılmak),** (of a deposit *etc.*) to be forfeited to the Treasury: **sabit** ~, fixed income.

**irade** (·–ˋ) Will; will-power; command, decree. ~**si* elden gitmek** *or* ~**sine* hâkim olmamak,** to lose one's self-control: ~**si haricinde,** over which he had no control. ~**li,** strong-willed.

**iradî** (·–ˊ) Voluntary.

**irak** *v.* **ırak.**

**İran** Persia (ˊ·), ~**î** (––ˊ), ~**lı,** Persian.

**irca** (·ˊ), **-aı** A sending back. ~ **etm.,** to cause to return; to reduce; to ascribe.

**irem, irembağı** A fabulous paradise in Arabia; the garden of Eden.

**irfan** Knowledge; culture; refinement; spiritual knowledge.

**iri** Huge; voluminous; coarse. ~ **yarı,** big, powerfully built *man.* ~**baş,** tadpole. ~**ce,** somewhat large *or* coarse. ~**leşmek,** to become large. ~**li,** *in* ~ **ufaklı,** mixed large and small, fine and coarse. ~**lik,** largeness; size.

**irin** Pus; filth. ~**lenmek,** to suppurate.

**iriş** Warp *of a tissue*; pole *of a wagon.*

**irk·ilmek** *vn.* Be startled, draw back *in fear*; (water) collect, become stagnant; become inflamed. ~**inti,** stagnant pool. ~**mek,** (of water) to collect and stagnate.

**İrlânda** Ireland. ~**lı,** Irish.

**irmik** Semolina.

**irs** Inheritance; hereditary quality. ~**en** (ˋ·), by inheritance. ~**î** (·ˊ), hereditary.

**irşad** (·ˊ) Act of guiding, showing the right way; enlightenment. ~ **etm.,** to direct; guide with advice; initiate (a novice). ~**at** (·–ˋ), **-tı,** *in* ~**ta bulunmak,** to give advice.

**irtibat** (··ˋ), **-tı** Connexion; tie; communication (*mil.*); liaison. ~ **hatları,** lines of communication.

**irtica** (··⌐), **-aı** Political reaction. **~î** (··—⌐), reactionary.
**irtical** (··⌐), **-li** Improvisation. **~en** (··⌐·), extempore.
**irtidad** (··⌐) Apostasy.
**irtifa** (··⌐), **-aı** Elevation; height, altitude. **~en** (··⌐·), in altitude.
**irtifak** (··⌐), **-kı** *In* ~ **hakkı**, easement (legal).
**irtihal** (··⌐), **-li** A passing away. **~i dârıbeka** **etm.**, to pass into the next world: ~ **etm.**, to pass away, die.
**irtikâb** (··⌐) Perpetration; corruption, embezzlement. ~ **etm.**, to commit (a crime), perpetrate; to embezzle; take a bribe.
**irtisam** (··⌐) Projection (map). ~ **etm.**, to be delineated *or* portrayed.
**irtişa** (··⌐) Corruption; the taking of bribes.
**is** Soot. ~ **kokmak**, (of food) to be slightly burnt.
**İsa** Jesus.
**isabet** (·—⌐), **-ti** A hitting the mark; a thing said *or* done just right. ~ !, well done!, capital!: ... **de** ~ !, 'and a good thing too!': ~ **etm.**, to hit the mark; to do *or* say just the right thing; to guess rightly: ~ **oldu da**, it was a good thing that ...: ⌜**piyangoda bana yüz lira** ~ **etti**⌝, 'I won a hundred lira in the lottery'.
**is'af** (·⌐) **etm.** *va.* Grant (a request).
**isbat** (·⌐), **-tı** Proof; confirmation; maintenance of *stg.* in its present condition. ~ **etm.**, to prove, demonstrate, confirm: **~ı vücud etm.**, to appear in person, to put in an appearance.
**ise** 1. *Verbal suffix forming the 3rd pers. sing. of a conditional tense, generally abbreviated to* **-se, -sa.** 2. *conj.* However; as for. ~ **de**, although: **ben** ~, as for me: ···**mekten** ~, instead of ...ing: ⌜**okudunsa okudun artık bir az dinlen**⌝, 'well, you've done your reading, now take a rest': ⌜**gittimse gittim, sana ne?**⌝, 'well, if I did go, what's that to you?'
**isevî** (—·⌐) Christian. **~lik**, Christianity.
**isfenks** Sphinx.
**ishak, -kı** Scops' owl.
**ishal** (·⌐), **-li** Purging; diarrhoea. ~ **olm.**, to have diarrhoea.
**isilik** Heat-spots; rash.
**isim, -smi** Name; noun. **ismini cismini bilmem**, I know nothing about him: **ismi fail**, active participle: **ismi geçen**, aforesaid, above-mentioned: **ismi has**, proper noun: **ismi mef'ul**, passive participle. **~lendirmek**, to name, call.
**iska** (·⌐) A watering *or* irrigating. ~ **etm.**, to water *or* irrigate.
**iskambil** Playing card; *name of a card game.* ~ (**kâğıdı**) **gibi dağıtmak**, to scatter in all directions.
**iskân** (·⌐) A causing to settle *or* inhabit; settling in; inhabiting. ~ **etm.**, to settle *emigrants etc.*; to inhabit, dwell.
**iskandil** Sounding-lead. ~ **etm.**, to sound; to probe.
**iskarpelâ** Carpenter's chisel.
**iskarpin** (Woman's) shoe.
**iskeç** Sketch.
**iskefe** (·⌐·) Skull-cap. The stretching and softening of leather.
**iskele** (·⌐·) Landing-place; quay; port of call; ladder; scaffolding; port (*opp. to starboard*). ~ **verilmek**, for a gangway to be attached to a ship.
**iskelet, -ti** Skeleton.
**iskemle** Chair; stool.
**İskender** Alexander the Great. **~iye** (···⌐·), Alexandria. **~un**, Alexandretta.
**iskete** (·⌐·) Finch.
**İskoç** Scotch. **~ya** (·⌐·), Scotland. **~yalı**, Scotch, Scotchman.
**iskonto** (·⌐·) Discount.
**iskorbut, -tü** Scurvy.
**iskorpit, -ti** Scorpion fish (*scorpaena*).
**islâm** Islam. **~a gelmek**, to become a Moslem: **ehli** ~, Moslems; a Moslem. **~iyet, -ti**, the religion of Islam; the Moslem world.
**is·lemek** *va.* Blacken with soot; smoke (fish *etc.*); burn (food). **~lenmek**, to become black with soot; smell of soot. **~li**, sooty: ~ **balık**, smoked fish.
**islim** Steam. ~ **üzerinde**, with steam up: ⌜~ **arkadan gelsin!**⌝, *used of* (i) *people who do not understand what they are talking about*; (ii) *people who enter upon an undertaking without proper preparation.*
**ismen** (⌐·) By name.
**ismi** *v.* isim. **~has**, proper name.
**isnad** (—⌐) Imputation. ~ **etm.**, to impute, ascribe.
**İspanya** (·⌐·) Spain.
**ispanyol** Spanish; Spaniard. **~ca**, Spanish (language).
**ispanyolet, -ti** Window latch.
**ispat** *v.* **isbat.**

**ispenc, ispec** Bantam. ~ **horozu gibi,** 'cocky'.
**ispermeçet, -ti** Spermaceti.
**ispinoz** Chaffinch.
**ispir** Groom.
**ispirto** (·˙·) Alcohol. **~luk,** spirit-lamp.
**isporcu** Sportsman.
**israf** (·˙) Wasteful expenditure; prodigality. ~ **etm.,** to waste, squander.
**İsrafil** (·—˙) *The name of* the Archangel who will sound the Last Trump (surü ~).
**İsrail** (·—˙) Israel. **beni ~,** the Israelites.
**istaka** (·˙·) Billiard cue.
**istakoz** Lobster.
**istalâktit, -ti** Stalactite; pendant (architecture).
**istalya** (·˙·) Demurrage.
**istalyoz** (·˙·) Awning stanchions.
**istampa** (·˙·) Stamp; inking-pad.
**istanbulin** Kind of frock-coat *formerly worn by higher Turkish officials.*
**istasyon** Station.
**istavrit, -ti** Horse mackerel. ~ **azmanı,** tunny.
**istek** Wish; longing; appetite. **~li,** desirous; bidder *at an auction.* **~siz,** unwilling, reluctant; apathetic.
**istemek** *va.* Wish for, desire; ask for; require, need. **istemez !,** it is not necessary; **arsızlık istemez !,** none of your cheek! : **ister ... ister...,** whether ... or ...; **ister gelsin ister gelmesin,** whether he comes or not *or* I don't care whether he comes or not: **ister inan ister inanma,** believe it or not, just as you wish: **ister istemez,** whether he will or not, willingly or unwillingly: ⌜**ister misin şimdi gelsin!**⌝, 'I am afraid he may come now', *but it may also mean* 'what a pleasant surprise it would be if he came now': **istersen !** *to one who is unwilling to do as he is asked,* 'well have it your own way!': ⌜**isteyenin bir yüzü kara, vermeyinin iki yüzü kara**⌝, it is more embarrassing to refuse than to ask: **birinin kızını ~,** to ask *s.o.* for the hand of his daughter: **canım istemiyor !,** I don't like it !: **çok zaman ister,** it will require a long time.
**istenmek, istenilmek** *vn. pass. of* **istemek.**
**ister** *v.* **istemek.**
**isteri** Hysteria.
**isterlin** Pound sterling.
**istetmek** *va. caus. of* **istemek.**
**istiab** (··˙) Capacity. ~ **etm.,** to contain, to hold *so much.*
**istibdad** (··˙) Despotism; absolute rule.
**istical** (·—˙), **-li** Haste. ~ **etm.,** to make haste, hurry.
**isticar** (·—˙) Hiring. ~ **etm.,** to take on hire.
**isticvab** (··˙) Interrogation. ~ **etm.,** to interrogate.
**istida** (··˙) Demand; petition, *esp.* an official request. ~ **etm.,** to request, make a formal demand. **~name** (··——˙), formal written petition.
**istidad** (·—˙) Aptitude; readiness to learn; talent. **~lı,** promising, talented.
**istidlâl** (··˙), **-li** Deduction, inference. ~ **etm.,** to deduce, infer. **~en** (··˙·), by deduction.
**istif** Stowage; arrangement of goods *in a warehouse or ship.* **~ini* bozmak,** to disturb, disconcert; **~ini* bozmadan,** without being upset, quite unperturbed; **~ini* bozmamak,** to be quite unperturbed; to look on with indifference: ~ **etm.,** to pack, stow: **balık ~i,** packed like sardines.
**istifa** (·—˙) Resignation *from an office.* ~ **etm.,** to resign. **~name** (·——˙), letter of resignation.
**istifade** (··—˙) Profit, advantage. **···den ~ etm.,** to benefit by, profit by, take advantage of. **~li,** profitable, advantageous.
**istifçi** Packer; stevedore; hoarder. **~lik,** stowage, packing; hoarding for profit.
**istifham** (··˙) An asking for explanation; interrogation. ~ **işareti (alameti),** interrogation mark.
**istiflemek** *va.* Stow; pack; hoard.
**istifrağ** (··˙) **etm.** *va.* Vomit.
**istifsar** (··˙) An asking for information *or* explanation. **~ı hatır etm.,** to inquire after *s.o.*'s health *etc.*
**istiğfar** (··˙) **etm.** *vn.* Ask God's pardon.
**istiğna** (··˙) A being able to do without; independence; disdain. ~ **göstermek,** to show that one has no need of a thing; to be disinterested.
**istihale** (··—˙) Transformation; metamorphosis.
**istihare** (··—˙) Oneiromancy; divination by dreams. **~ye yatmak,** to pray and go to sleep hoping for a dream in which God will show his will.

**istihbar** (··‿) An asking for information. ~ **etm.**, to get information, to make inquiries. ~**at** (··—‿), **-tı**, news: ~ **hizmeti**, intelligence service.
**istihdaf** (··‿) **etm.** *va.* Aim at, pursue *an object*.
**istihdam** (··‿) **etm.** *va.* Take into service, employ.
**istihfaf** (··‿) Contempt. ~ **etm.**, to despise; consider unimportant.
**istihkak** (··‿), **-kı** Merit; that which is due; fee, remuneration; ration. ~ **etm.**, to deserve, have a right to: ~ **kesbetmek**, to earn.
**istihkâm** (··‿) Fortification; military engineering. ~ **bulmak**, to be consolidated: ~ **subayı**, engineer officer (*mil.*). ~**at**, **-tı**, fortifications. ~**cılık**, military engineering.
**istihkar** (··‿) Scorn, contempt. ~ **etm.**, to despise, treat with contempt: **hayatını** ~ **etm.**, to scorn death, to sacrifice one's life.
**istihlâk** (··‿), **-ki** Consumption; using up. ~ **etm.**, to consume, use up: **dahilî** ~ **için**, for internal consumption.
**istihlâs** (··‿) Rescue. ~ **etm.**, to rescue; to appropriate to oneself.
**istihrac** (··‿) Deduction; divination. ~ **etm.**, to try to get the meaning of. ~**at** (··—‿), **-tı**, auguries: ~**ta bulunmak**, to draw arbitrary conclusions.
**istihsal** (··‿), **-li** Act of producing *or* acquiring. ~ **etm.**, to produce, to obtain. ~**at** (··—‿), **-tı**, products.
**istihza** (··‿) Ridicule, mockery; irony, sarcasm. ···**le** ~ **etm.**, to ridicule, mock: ~ **tarikile**, ironically: **mukadderatın** ~**sı**, the irony of fate.
**istihzar** (··‿) **etm.** *va.* Prepare. ~**at** (··—‿), **-tı**, preparations.
**istikamet** (··—‿), **-ti** Direction; uprightness, integrity.
**istikbal** (··‿), **-li** Future; a going to meet *s.o.* ~**e çıkmak** *or* ~ **etm.**, to go forth to meet *s.o., esp. as a formal ceremony*.
**istiklâl** (··‿), **-li** Independence. ~ **harbi**, the War of Independence (1922): ~ **marşı**, the Turkish National Anthem. ~**iyet**, **-ti**, the state of being independent.
**istikra** (··‿) Inductive reasoning.
**istikrah** (··‿) Aversion. ···**den** ~ **etm.**, to loathe.
**istikrar** (··‿) A becoming established; stabilization; stability. ~ **etm.**, to become fixed, established. ~**lı**, settled, established, stabilized. ~**sız**, unstable, unsettled; inconsistent.
**istikraz** (··‿) Loan. ~ **etm.**, to borrow money.
**istikşaf** (··‿) Reconnaissance. ~ **etm.**, to reconnoitre; to endeavour to discover.
**istilâ** (··‿) Invasion. ~ **etm.**, to invade; to flood. ~**î** (··—‿), invading.
**istilâm** (·—‿) **etm.** *va.* Ask for information about.
**istilzam** (··‿) **etm.** *va.* Render necessary; make inevitable; involve, entail.
**istim** Steam. ~**ini tutmak**, to have steam up; (*sl.*) to boil over with anger *etc.*; to be drunk. *v. also* islim.
**istimal** (·—‿), **-li** A making use of. ~ **etm.**, to employ, make use of.
**istimbot**, **-tu** Small steamboat.
**istimdad** (··‿) An asking for help. ~ **etm.**, to ask for assistance.
**istimlâk** (··‿), **-ki** Legal expropriation (against payment). ~ **etm.**, to expropriate, to acquire property by force of law.
**istimna** (··‿) Self-abuse, masturbation.
**istimzac** (··‿) **etm.** *va.* Make polite inquiries of *s.o.*; ask about *s.o.'s* tastes and feelings; to inquire whether a person is *persona grata* (to another Government).
**istinabe** (··—‿) Appointment of a proxy; the taking, on commission, of the evidence of an absent witness. ~ **suretile**, by proxy.
**istinad** (··‿) A relying upon, a being supported. ···**e** ~ **etm.**, to rely on, lean on, be supported by, be based on. ~**en** (··‿·), ···**e** ~, based on, supported by. ~**gâh**, point of support.
**istinaf** (·—‿) Appeal *at law.* ~ **etm.**, to appeal *against a legal decision*: ~ **mahkemesi**, court of appeal. ~**en** (·—‿·), on appeal. ~**î** (·——‿), pertaining to an appeal.
**istinas** (·—‿) A being friendly *or* familiar; tameness; practice. ~ **etm.** *or* ~ **peyda etm.**, to become familiar *or* tame; to have practice *in doing stg.*
**istinga** (·‿·) Brail; rope for closing a purseseine net.
**istinkâf** (··‿) Rejection, refusal. ~ **etm.**, to draw back, refuse, abstain.
**istinsah** (··‿) **etm.** *va.* Make a copy of.

**istintac** (··–́) **etm.** *va.* Conclude; infer.
**istintak** (··–́), **-kı** Interrogation; cross-examination. ~ **etm.**, to interrogate. ~**name** (···–́), official record of evidence taken by an examining magistrate from an accused person.
**istirahat** (··–́), **-tı** Repose. ~ **etm.**, to rest, to take one's ease.
**istirdad** (··–́) A retaking, recovery *of stg.* ~ **etm.**, to retake, recover.
**istirham** (··–́) A begging for mercy; an asking a favour; petition. ~ **etm.**, to beg, petition. ~**name** (···–́), written petition.
**istiridye** (··–́·) Oyster.
**istirkab** (··–́) **etmek** *va.* Be jealous of; regard as a rival.
**istisgar** (··–́) **etm.** *va.* Regard as insignificant; despise; underestimate.
**istiska** (··–́) Dropsy.
**istiskal** (··–́), **-li** A giving *s.o.* a cold reception. ~ **etm.**, to show *s.o.* that he is not welcome; to be disagreeable.
**istismar** (··–́) **etm.** *va.* Bring to fruition; exploit, profit by. ~**cı**, one who exploits something for his own ends.
**istisna** (··–́) Exception. ~ **etm.**, to except, exclude. ~**î** (··––́), exceptional.
**istişare** (··–́) Consultation. ~ **etm.**, to hold a consultation.
**istitrad** (··–́) Digression. ~**en** (··–́·), by way of digression. ~**î** (··––́), digressive, parenthetical.
**istiva** (··–́) **hattı**. The Equator.
**istizah** (·–́) An asking for an explanation; interpellation. ~ **etm.**, to ask for an explanation, *esp.* to question a Minister.
**istofa** (·–́·) Brocade.
**istok, -ku** Stock *of goods.*
**istop** Stop. ~ **etm.**, to stop.
**istor** Roller blind. ~**lu yazıhane**, roll-top desk.
**istralya** (·–́·) Stay of a mast.
**istromaça** (··–́·) Twisted ropes used as a fender.
**istrongilos** (·–́··) Kind of sea-bream (*Smaris vulgaris*).
**İsveç** Sweden. ~**li**, Swedish.
**İsviçre** (·–́·) Switzerland. ~**li**, Swiss.
**isyan** (·–́) Rebellion. ~ **etm.**, to rebel. ~**kâr**, rebellious, refractory.
**iş** Work; action; business; occupation, profession; affair, matter. ~**imiz** ~, all goes well (*often ironically used to mean the opposite*): ~ **adamı**, a business man; a businesslike man: ~**im Allaha kaldı**, I am done for (*often jokingly*): ~**ten anlamak**, to be an expert: **sen** ~**ine bak!**, mind your own business!: ~ **başa gelmek** (**düşmek**), for *stg.* to have to be done personally: ~ **başında**, at one's work: ~ **başındakiler**, those in authority: ~ **başına geçmek**, to take the lead; to take control: ···**in** ~**ini bitirmek**, to finish *s.o.* off (kill): ~ **çıkarmak**, to raise unnecessary difficulties: ~**i çıkarmak**, to finish a job: ~**ten değil** *or* ~ **deme değil**, it's an easy matter, it's a mere nothing; *also in such phrases as* **çıldırmak** ~**ten değil, it's** enough to drive one mad: ⌜**adama o kadar kızdım ki öldürmek** ~ **deme değil**⌝, 'I was so angry with the man that I could have killed him': ···**e** ~**i* düşmek**, to be obliged to apply to *s.o.* for help, advice *etc.*; for one's business to take one *to a certain place or person*: ~ ~**ten geçti**, it's all over now, it's too late to do anything: ~**ine göre**, according to his work *or* skill; according to its workmanship *or* quality; 'it all depends!': ~ **görmek**, to work; to perform a service: ···**e** ~ **göstermek**, to give *s.o.* a job of work: ~ **güç**, *v.* **işgüç**: ~**in içinde** ~ **var!**, there's *stg.* behind all this!: ~ ~**i olm.**, for things to go well with one: ~ **ola** *or* ~ **olsun diye**, *used sarcastically of one who does stg. only for show or as eyewash; to do stg.* unnecessarily *or* officiously *or* in order to appear busy: ~**inden* olm.**, to lose one's job: ~**im mi yok!**, (*as a refusal*) 'not for me!'; 'I don't feel like it!': ~**in mi yok** *much as* ~ **im mi yok**, *but also* 'what an idea!'; 'what an absurd suggestion!': ~**im var**, I am busy: **bu adamla** ~**im var**, I've a job with this man!; he wants a lot of looking after; he's a damned nuisance!: **bunda bir** ~ **var!**, there's *stg.* funny about this!: **birinin başına** ~ **açmak** (**çıkarmak**), to get *s.o.* into difficulty: **bu** ~ **yürümez** (**sökmez**), this sort of thing can't go on!
**işaa** (·–́) **etm.** *va.* Spread; publish; divulge.
**iş'ar** (·–́) A making known. ~**ı ahire değin**, until further notice: ~ **etm.**, to communicate, notify.
**işar·et** (·–́), **-ti** Sign; signal; mark; punctuation mark. ~ **çekmek**, to hoist a signal: ~ **etm.**, to make a

sign; beckon; make a mark: ~ **kolu** (**değneği**), traffic indicator: ~ **memuru,** traffic policeman: ~ **zamiri,** demonstrative pronoun. ~**leşmek,** to make signs to one another, *esp. in a furtive flirtation.* ~**î** (·–⊥), conveyed by a signal, indicated.

**işba** (·⊥), **–aı** Saturation. ~ **etm.,** to satiate, saturate.

**iş·başı, –nı** Head of a business; foreman; hour at which work begins; place of business. ~**bilir,** who knows his work; businesslike. ~**birliği, –ni,** co-operation. ~**bölümü, –nü,** division of labour.

**işbu** (⊥·) This; the present (*year etc.*).

**işçi** Workman, labourer; workwoman. ~**lik,** occupation *or* pay of a workman.

**işemek** *vn.* Urinate, make water.

**işgal** (·⊥), **–li** A causing to be occupied; occupation. ~ **etm.,** to keep busy; engage the attention of; keep *s.o.* from his work; occupy (*mil.*).

**işgüc** (*Decline both words*) Occupation, employment. **bir adamın işi gücü,** a man's job, his daily work.

**işgüzar** Who does his work; efficient; officious. ~**lık,** officiousness.

**işhad** (·⊥) **etm.** *va.* Call to witness; cause to testify.

**işit·ilmek** *vn. pass. of* **işitmek.** be heard; be heard of by all. ~**ilmiş,** heard; hearsay.

**işit·mek** (**işidiyorum** *or* **işitiyorum**) *va. & vn.* Hear; listen. ⌜**ağzından çıkanı kulağı işitmez**⌝, he speaks without weighing his words: **azar** (**lâf**) ~, to get a scolding. ~**memezlik** (**işitmezlik**), a not hearing: ~**ten gelmek,** to pretend not to hear, to feign deafness. ~**tirmek,** to cause to hear *or* be heard; announce; communicate.

**işkampaviye** (··⊥··) Longboat, pinnace.

**işkembe** (·⊥·) Paunch; tripe. ~**den** (*or* ~**i kübradan**) **atmak** (**söylemek**), to invent *a story etc.*, to exaggerate, to embroider *a tale*: ~**sini* şişirmek,** to eat greedily: ~ **suratlı,** pockmarked.

**işkence** Torture.

**işkil** Doubt, suspicion. ~**lenmek,** to be dubious; be anxious. ~**li,** suspicious, anxiously doubtful. ~**lik,** dubiousness; mistrust.

**işlek** Working well and easily; good flowing *handwriting*; busy *thoroughfare*; experienced.

**işleme** Work, workmanship; handiwork; embroidery; carving; engraving.

**işlemek** *va. & vn.* Work; function; work up; manipulate; take effect; embroider; carve; engrave; be frequented; penetrate (*with dat.*); commit (crime); (of a road) to be open *or* much used; (of a ship *or* vehicle) to ply; (of a boil) to discharge; (of a dye) to penetrate. **canına*** (**içine***) ~, to hurt one's feelings, to affect one painfully: **kurşun işlemez,** bulletproof.

**işle·meli,** embroidered, decorated with needlework. ~**nmek,** *pass. of* **işlemek**; be embroidered, worked up. ~**tme,** the working *of a railway, a mine etc.*; the running *of an hotel, a business etc*,: ~ **Bakanı,** Minister of Industry: ~ **dairesi,** the traffic department *of a railway*: ~ **malzemesi,** rolling-stock. ~**tmek,** *caus. of* **işlemek**; cause to work; work (a machine, railway, mine *etc.*); make fun of *s.o.* ~ **yici,** penetrating.

**işmar** (·⊥) Sign; nod; wink.

**işmizaz** (·–⊥) Grimace; look of disgust.

**işporta** (·⊥·) Large basket. ~ **malı,** shoddy goods. ~**cı,** hawker, pedlar.

**işret,** ~**ti** Drinking, carousal; festival. ~ **etm.,** to drink (wine *etc.*), to make merry.

**işsiz** Without work, unemployed. ~**lik,** unemployment.

**iştah, iştiha** Appetite; desire; greed. ~ **açan,** appetizer, apéritif: ~ **açmak,** to whet the appetite: ~**ım yok,** I have no appetite. ~**lı,** having an appetite, hungry; desirous; too keen, officious. ~**sız,** without appetite; without desire.

**işte** Look!; here!; now; thus.

**iştigal** (··⊥), **–li** An occupying oneself. ···**le** ~ **etm.,** to busy oneself with … .

**iştiha** (··⊥) *v.* **iştah.**

**iştikak** (··⊥), **–kı** Derivation; etymology. ···**den** ~ **etm.,** to be derived from … .

**iştirak** (··⊥), **–ki** Participation. ···**e** ~ **etm.,** to share, participate in: ~ **üzere,** jointly, in participation.

**iştiyak** (··⊥), **–kı** Longing. ···**e** ~ **çekmek,** to long for.

**işünuş** (⊥··) Revel, carousal.

**işve** Coquetry, flirting. **~baz, ~kâr,** coquettish, amorous.

**it, -ti** Dog. ⌜**~e atsan yemez**⌝, a dog wouldn't eat it; very nasty: **~ canlı,** very tough and enduring: ⌜**~ ~e, ~ de kuyruğuna**⌝, 'a dog asked another to do *stg.*, the other dog asked his tail', everyone trying to get another to do the job: **~ nişanı,** an excrescence on a horse's fetlock: **~oğlu ~,** a scoundrel, a cunning fellow: **~ sürüsü,** a pack of scoundrels; a whole crowd (*pej.*).

**ita** (–⸗) A giving *or* paying. **~ âmiri,** the official authorized to sign an **~ emri,** an order for payment of government money: **~ etm.,** to give; to pay: **ahzü ~,** (taking and giving), commerce.

**itaat** (·–⸗), **-ti** Obedience. **~ etm.,** to obey. **~li,** obedient. **~siz,** disobedient.

**itboğan** Meadow saffron, colchicum.

**itburnu** (⸗··), **-nu** Hip of the wild rose.

**itdirseği** (⸗···), **-ni** Stye.

**iteklemek** *va.* Treat roughly, manhandle.

**itfa** (·⸗) An extinguishing. **~iye,** fire-brigade: **~ neferi,** fireman.

**ithaf** (·⸗) Presentation; dedication. **~ etm.,** to present *stg.* *rare,* to dedicate *a poem etc.* **~iye** (·–·⸗), dedication (of a book).

**ithal** *v.* **idhal.**

**itham** (·⸗) Imputation; accusation. **~ etm.,** to suspect; to accuse. **~name** (··–⸗), indictment.

**ithiyarı** (⸗···), **-nı** Colocynth.

**itibar** (–·⸗) Esteem; consideration; regard; credit; nominal value; hypothesis. **··· ~ile,** as regards ...: **~dan düşmek,** to be discredited: **~ etm.,** to esteem, consider, deem: **~ görmek** to be respected; to be in demand: **~ mektubu,** letter of credit: **millî ~,** national credit: **~ı olm.,** to be held in esteem: **~ı var,** his credit is good: **açık ~,** overdraft: **bu ~la,** and therefore, and so: **esas ~ile,** essentially: **nazarı ~a almak,** to take into account.

**itibar·en** (–·⸗·) ,from, dating from. **~î** (–·–⸗), theoretical; conventional; nominal. **~lı,** esteemed, trusted. **~sız,** discredited; not held in esteem.

**itidal** (–·⸗), **-li** Moderation; temperance; equilibrium; equanimity; equinox. **~ kesbetmek,** to moderate (*vn.*). **~siz,** immoderate, extreme.

**itikad** (–·⸗) Belief; creed. **···e ~ etm.,** to believe in. **~lı,** believing. **~sız,** unbelieving, irreligious.

**itikâf** (–·⸗) A going into religious seclusion for a day or two. **~a çekilmek,** to go into seclusion *or* retirement.

**itilâ** (–·⸗) An ascending, a being elevated; progress to a higher standard; exaltation. **~ etm.,** (of a people *etc.*) to progress.

**itilâf** (–·⸗) Agreement; understanding, entente, friendship. **Hürriyet ve ~ Fırkası,** *the name of the party in opposition to the* **İttihad ve Terakki Cemiyeti** *in the years succeeding the Constitution of 1908.* **~cı,** a member of that party.

**itimad** (–·⸗) Confidence; reliance. **···e ~ etm.,** to have confidence in, to rely on: **~ telkin etm.,** to inspire confidence. **~name** (–··–⸗), letter giving credentials.

**itina** (–·⸗) Care; attention. **~ etm.,** to pay great attention, be very careful.

**itiraf** (–·⸗) Confession, admission. **~ etm.,** to confess, acknowledge, admit.

**itiraz** (–·⸗) Objection; (*pl.* **~at**). **~ etm.,** to raise an objection, to demur.

**itişmek** *vn.* Push one another; brawl; skylark. **itişe kakışa,** pushing and shoving one another.

**itiyad** (–·⸗) Habit. **~ etm.,** to accustom oneself to; to make a habit of: **~ üzere,** habitually.

**itizal** (–·⸗), **-li** Schism. **~ etm.,** to secede.

**itizar** (–·⸗) **etm.** *vn.* Make excuses.

**itlik** Quality of a dog; villainy, vileness.

**itmek** *va.* Push. **itip kakmak,** to push and shove, to elbow one's way.

**ittırad** (··⸗) Regularity; continuity; monotonous rhythm. **~sız,** irregular, discontinuous.

**ittiba** (··⸗), **-aı** A following; a copying *or* obeying another. **~ etm.,** to follow, obey, copy. **~en** (··⸗·), in conformity with.

**ittifak** (··⸗), **-kı** Harmony, concord; alliance; agreement; coincidence, chance. **~ı âra,** unanimity: **tedafüî (tecavüzî) ~,** a defensive (offensive) alliance. **~sızlık,** disagreement.

**ittihad** (··⸗) Union. **İttihad ve Terakki Cemiyeti,** *the famous* Party of Union and Progress, *mainly in power after the Constitution of 1908*.

**ittiham** (··ˎ) A being accused. ···**le** ~ **etm.**, to be accused of ...; (*sometimes confused with* **itham** *which means 'an accusing'*).

**ittihaz** (··⌐) **etm.** *va.* Procure; take; adopt (a proposal *etc.*).

**ittisal, -li** A being in contact; contiguity. ~ **bulmak,** to get in contact with; be joined.

**itüzümü** (ˎ···), **-nü** Black nightshade.

**ityatağı** (ˎ···), **-nı** Dog's bed; meeting-place of vagabonds; filthy spot.

**ivaz** An exchanging; thing given in exchange. **bilâ** ~ *or* ~**sız,** without anything being given in exchange; without expecting anything in return, disinterested.

**ivdirmek** *va.* Hasten.

**ividi** In a hurry.

**ivik** A kind of ski *or* snowshoe.

**ivirmek** *va.* Hasten.

**iviz** Gad-fly.

**ivmek** *vn.* Be in a hurry.

**iyi, eyi** Good; well; in good health. The good; the good side of a thing. ~**den** ~**ye,** properly, thoroughly: ~ **etm.,** to make well, to cure; to do well: ~**siniz inşallah!,** I hope you are well!: ~ **kötü,** mediocre; more or less: ~**si mi,** the best thing to do is ...; if I were you ...: **en** ~**si,** the best of it (*or* them): ⌜**kime rastgelsem** ~**?**⌝, 'guess whom I met!': **pek** ~**!,** very good!, all right!: ⌜**sizden** ~ **olmasın!**⌝, *said when praising a person to another, stg. like* 'of course I don't like him as much as I do you', *or* 'present company excepted'.

**iyi·ce, eyi·ce** (··ˎ), pretty well; rather good; (·ˎ·) thoroughly. ~**leşmek,** to improve, to get better; to recover from an illness. ~**lik,** goodness; kindness; good health; good part *or* side: ~ **bilmek,** to be grateful: ~ **görmek,** to experience kindness *or* favour. ~**mser,** optimistic; eulogistic.

**iyod** Iodine. ~**ür,** iodide.

**iz** Footprint; track; trace. ~**i belirsiz olm.,** to disappear without leaving a trace: ~**ini kaybetmek,** to lose track of: ~ **toz,** the tracks of *s.o.*; ⌜**ne** ~**i belli ne tozu**⌝, he disappeared without trace: ~**ine uymak,** to follow in the footsteps of: ⌜**karda yürür de** ~**ini belli etmez**⌝, he is very cunning and cautious: **tekerlek** ~**i,** rut.

**izabe** (·–ˎ) **etm.** *va.* Melt, fuse.

**iz'ac etm.** *va.* Vex, worry, disturb.

**izafe** (·–ˎ) An attributing, attaching. ~ **etm.,** to attribute, attach; to join two words together. ~**ten** (·⌐··), referring, attributing.

**izafî** (·–⌐) Relative; extrinsic; nominal. ~ **siklet,** specific gravity. ~**yet, -ti,** relativity.

**izah** (–ˎ) Explanation; manifestation; elucidation. ~ **etm.,** to manifest, explain, elucidate. ~**en** (–⌐·), in explanation, by way of elucidation. ~**at, -tı,** explanations: ~ **vermek,** to give a full explanation, give full details. ~**name** (–·–ˎ), manifesto.

**izale** (·–ˎ) A removing, a causing to disappear. ~ **etm.,** to remove, to put an end to, to destroy.

**izam, îzam** (–ˎ) An enlarging, exaggerating. ~ **etm.,** to exaggerate.

**izan, iz'an** (·ˎ) Quickness of understanding; consideration for others. ~**lı,** intelligent; considerate. ~**sız,** inconsiderate.

**izaz** (–ˎ) **etm.** *va.* Honour, treat with respect; entertain, regale.

**izbarço** (·ˎ·) ~**bağı,** bowline (knot).

**izbe** Hovel, hut; basement. Dark and dirty.

**izbiro** (·ˎ·) Sling *for lifting bulky goods*.

**izci** Tracker; boy-scout.

**izdiham** (··ˎ) Crowd. **seyrüsefer** ~**ı,** congestion of traffic.

**izdivac** (··ˎ) Mating; matrimony. ~ **etm.,** to marry.

**izhar** (·ˎ) **etm.** *va.* Manifest, display.

**izin, -zni** Permission; leave; discharge. ~ **vermek,** to grant leave; give permission; dismiss. ~**li,** on leave; one who has inherited the gift of curing by charms *or* **üfürük,** *q.v.* ~**name** (··–ˎ), permit; licence; marriage licence. ~**siz,** without permission, unauthorized; 'kept in' (of school-children); school punishment of being 'kept in': ~ **almak,** to receive such punishment: ~ **kalmak,** to be 'kept in'.

**izlemek** *va.* Track, trace.

**izli** Leaving a trace. ~ **mermi,** tracer shell.

**izmarit** (··⌐), **-ti** Sea-bream (*Smaris alcedo*); cigarette stump.

**izmihlâl** (··ˎ), **-li** Disappearance; annihilation.

**izzet, -ti** Might; glory, honour, dignity. **~inefis,** self-respect; 'amour propre'. **~lemek,** to treat with respect, honour. **~li, ~lû,** honourable (*title formerly given to officers and officials of a certain rank*).

# J

**jaketatay** (*Fr. jaquette à taille*) Morning coat.
**jale** (-ˈ) Dew.
**jaluzi** (*Fr. jalousie*) Venetian blind.
**jant** Rim.
**Japon** Japan; Japanese. **~ca,** in Japanese. **~ya** (·ˈ·), Japan.
**jartiye** Garter.
**jest, -ti** (*Fr. geste*) Gesture.
**jilet, -ti** Safety-razor.
**jurnal** Report of a delator; diary. **~cı,** denouncer, delator.
**jüt, -tü** Jute.

# K

**kâad** *v.* **kâğıd.**
**kaadir** (-ˈ) Powerful; Almighty God; capable (of = *dat.*).
**kaatil** (-ˈ) Murderer. **~lik,** a being a murderer; murder.
**kab** Cover; envelope; vessel, receptacle; dish, portion (of food) . ···**e ~ geçirmek,** to put a cover on; to bind (book): **~ kacak,** pots and pans; **~ yok kacak yok,** bare of the simplest necessities: **bir ~a kotaramamak** (*sometimes err.* **kurtaramamak**), to be unable to find a solution *to a problem etc.*: **~ına sığmamak,** to be uncontrollably impatient *or* ambitious: **üç ~ yemek,** three portions of food.
**kâb** (-) Ankle; knuckle-bone; cube; dice. ···**e ~ında olmamak,** not to reach the standard of ... : **~ına varılmaz,** unrivalled, peerless.
**kaba** Large but light; puffed out; spongy; coarse; common, vulgar; rough (calculation, guess *etc.*). **~sını almak,** to tidy up roughly, to get rid of the worst *of the dirt etc.*; to trim roughly, to rough hew: **~ et,** the buttocks: **~ döşek,** a soft downy mattress: **~kumpas,** dead-reckoning (*naut.*): **~ saba,** common, coarse: **~ sakal,** a bushy beard: **~ toprak,** freshly dug soil. **~ca** (··ˈ), rather bigger *or* older; (·ˈ·) roughly, coarsely.
**kabadayı** (·ˈ··) Swashbuckler, bully; 'tough'; having 'guts'; the best of anything. **muflisin ~sı,** one who spends money he can't afford, in order to show off.
**kabahat, -ti** Unseemly act; fault; offence. **~ etm.** (**işlemek**), to commit an offence. **~li,** guilty. **~siz,** innocent.
**kabak** Pumpkin; marrow. Bald, bare; close-shaven (head); tasteless. **~ başına* patlamak,** to suffer a disaster: **~ çekmek,** to smoke hashish: ⌈**~ çiçeği gibi açılmak**⌉, (of a new-comer) to be too forward; suddenly to become too free-and-easy: **~ tadı vermek,** to become a bore: **karpuz ~ çıktı,** the watermelon turned out to be tasteless.
**kabakulak** (·ˈ··) Mumps.
**kabalak** A kind of military hat, *worn by the Turkish Army in the First World War.*
**kabalaşmak** *vn.* Become coarse *or* vulgar.
**kabalık** Sponginess; bushiness; coarseness; vulgarity.
**kabara** (·ˈ·) Hob-nail; ornamental brass-headed nail.
**kabarcık** Bubble; pimple; pustule. **kara ~,** carbuncle; anthrax.
**kabare** Cabaret.
**kabarık** Swollen; blistered; puffy; loose, *i.e.* not compressed *or* compact. Blister; swelling. **~ deniz,** high tide.
**kabar·mak** *vn.* Swell; be puffed out; become fluffy; blister; be raised; be increased; be puffed up, swell with importance; rise when boiling; (sea) become rough. **içi*** (**midesi, safrası**) **~,** to become bilious, to feel sick: **koltuğu* ~,** to be puffed up with pride. **~tı,** swelling; puffiness. **~tma,** embossed (design); raised in relief. **~tmak,** *caus. of* **kabarmak:** **göğsünü ~,** to puff out the chest with pride: **kulak ~,** to prick up one's ears: **toprağı ~,** to break up *or* loosen ground.
**kaba·saba** (·ˈ··) Coarse; rough; common. **~sakal,** having a bushy beard. **~soğan,** a 'rotter', a coward at heart. **~sorta, ~ donanımlı,** square-rigged. **~taslak,** roughly drawn; in outline without details. **~yel,** south wind. **~yonca,** lucerne.

**Kâbe** (–ˈ) The Kaaba, cubical temple at Mecca.
**kabız, -bzı** Constipation. ~ **olm.**, to be constipated. ~**lık,** constipation.
**Kabil** (–ˈ) Cain.
**Kâbil** (ˈ·) Kabul.
**kabil**[1] (–ˈ) Capable; possible. *As prefix with the izafet* = capable of..., admitting ..., *e.g.* ~**i istifade,** profitable. ~**dir,** it is possible, it may be: **temyizi** ~ **olarak,** with right to appeal (law).
**kabil**[2] (·ˈ) Sort; category. ... ~**inden,** on the lines of, something like ... : **bu** ~**den,** of this sort.
**kabile** (–·ˈ) Midwife. ~**lik,** midwifery.
**kabîle** (·–ˈ) Tribe.
**kabiliyet** (–··ˈ), **-ti** Capability; possibility. ~**li,** intelligent, skilful. ~**siz,** incapable, unintelligent.
**kabine** (·ˈ·) Cabinet; small room; office; water-closet.
**kabir, -bri** Grave; tomb. ~ **suali,** endless questioning.
**kablî** (·ˈ) A priori.
**kablo** (ˈ·) Telegraphic cable.
**kabotaj** (*Fr. cabotage*) Coasting trade.
**kabran** Slack; without energy. Trunk (of a tree).
**kabri** *v.* **kabir.** ~**stan,** cemetery.
**kabuk** The outer covering *of anything*; bark, rind, peel, skin, shell, crust. ~ **bağlamak,** to form a skin *or* crust. ~**lanmak,** to form a skin *etc.* ~**lu,** having a shell, skin *etc.*
**kabul, -lü** Acceptance; reception; consent. ~**ümdür,** I accept: ~ **etm.,** to accept, receive, consent, admit: ~ **etmem,** I won't accept that; I don't agree; I will not consent: **ne desen** ~**ümüz,** I'll do whatever you say: **kaydü** ~, admittance *or* entry **to** *a school.* ~**lenmek,** to seize for oneself, appropriate.
**kaburğa** (·ˈ·) The ribs; a rib; frame of a ship.
**kâbus** (–ˈ) Nightmare.
**kabza** Handle; hilt.
**kabzetmek** (ˈ··), **-eder** *va.* Grasp; take.
**kabzı** *v.* **kabız.** ~**mal,** middleman (*esp. in fruit and vegetables*).
**kaç** How many?; how much? ~**a?,** what is the price?: ~**a sınız?,** what's the score?, how does the game stand?: ~**ın kur'ası,** one who 'knows a thing or two', an old hand: **ayın** ~**ında?,** on what day of the month?: **bir** ~, a few, some: **saat** ~ **?,** what is the time?
**kaçaburuk** Bradawl; shoemaker's awl.
**kaçak** Fugitive; deserter; contraband; leakage (*elect.*). Smuggled. ~ **inşaat,** unlicensed building: **muayene kaçağı,** that has escaped inspection; who has escaped medical inspection. ~**çı,** smuggler. ~**çılık,** smuggling. ~**lık,** desertion (*mil.*).
**kaçamak** Flight; evasion; subterfuge; refuge; shelter; a kind of hasty pudding of maize flour. ~ **yolu, an** excuse *or* subterfuge for getting out of doing *stg.* ~**lı,** evasive.
**kaçan**[1] When; at the time that ... .
**kaçan**[2] Running away. **kulağa** ~, earwig.
**kaçar** How many each? ~**a?,** for how much each?: **bu evler** ~ **kattır?,** how many stories has each of these houses?
**kaçarula** (··ˈ·) Casserole.
**kaçgöç** *The Moslem practice of women covering their faces in the presence of men.*
**kaçık** Crazy; receding. Ladder *in a stocking.* **sağa** ~, leaning to the right.
**kaçıncı** Of what number (*in a series*)? ~ **katta?,** on what floor?
**kaçınmak** *vn.* Abstain; be reluctant. ···**den** ~, to avoid, keep away from.
**kaçırmak** *va.* Make *or* let escape; drive away; smuggle; kidnap, elope with; hide *possessions etc. from the tax-gatherer etc.*; miss (train *etc.*); lose (an opportunity); let slip (a remark). *vn.* Go off one's head. **ağzından*** ~, inadvertently to let out *stg.* that one did not want to: **aklını*** ~, to go off one's head: **altına (yatağına)** ~, (of a child) to mess its clothes (bed): **fazla** ~, to have a drop too much; to let (pour) out more than one intends: **kelepir** ~, to miss a golden opportunity: **rahatını*** ~, to disturb the peace of, to be a source of worry to ... : **tadını** ~, to spoil the taste of *stg.*; to be very tiresome about *stg.*; to spoil the pleasure *of a party etc.*
**kaçırmaz** Tight; that cannot slip.
**kaçışmak** *vn.* Disperse; flee in confusion.
**kaçkın** Fugitive; deserter; crazy. **dayak** ~**ı,** one who deserves a beating: **medrese** ~**ı,** fanatic; reactionary: **mekteb** ~**ı,** ignorant and uneducated: **tımarhane** ~**ı,** almost a lunatic.
**kaçlı** Having how many? ~**sınız?,**

what is your year of birth (*i.e.* to which class do you belong for the military call-up ?) : **elindeki kâğıd ~?**, what is the value of the card you hold ?

**kaçlık** Worth how many pounds?; at which price ?

**kaçmak** *vn.* Flee, run away; escape; desert; (of a woman) to veil herself before men; seem (rude, inopportune *etc.*); (of a stocking) to ladder. **···den ~**, to avoid: **···e ~**, to slip into: **ağzından* ~**, (of a word) to slip out inadvertently: **gözüme toz kaçtı**, dust got into my eye: **işten ~**, to avoid work: **keyfi* ~**, to become dispirited *or* gloomy: **geç oldu; ben artık kaçayım!**', 'it's late, I must be going': **gösterişe (lükse) ~**, to be prone to, given to ostentation (luxury): **bu söz tatsız kaçtı**, that remark sounded rather out of taste: **tuz bir az fazla kaçtı**, there was a bit too much salt.

**kaçsız** *Only in* **~ göçsüz**, not respecting the rule about women veiling in the presence of men.

**kad, -ddi** Form, figure. **~dükamet**, stature.

**kadana**[1] (·ˈ·) Fetters *of a prisoner.*

**kadana**[2] (·ˈ·) Heavy horse; artillery horse; huge woman.

**kadar** Quantity, amount; degree. As much as; as big as; about. **···e ~**, as much as; until; up to; as far as. ⌜**al benden de o ~**⌝, I agree; 'same here!': **anladığım ~**, as far as I understand: **bu ~**, so much; to this amount *or* number: **bunun ~ büyük**, as big as this: ⌜**bu ~ olur**⌝ *or* ⌜**bu ~ı da fazla**⌝, 'this is a bit too much of a good thing!': **gelinceye ~**, until he comes: **ne ~**, how much?; however much : **o ~**, so; so much; **sekiz ~**, about eight; **sekize ~**, up to eight: **şu ~ ki**, only, but: **yaşadığım ~**, as long as I live.

**kadarcık** A small amount. **bu ~ şeyi bilmeli idiniz!**, you ought to know better!

**kadastro** (·ˈ·) Land survey.

**kadavra** (·ˈ·) Corpse; carcass.

**kadayıf** *Name of various kinds of sweet pastry.*

**kadeh** Glass; cup; wine-glass.

**kadem** Foot; foot (measure); pace; good luck. **···e ~ basmak**, to set foot in, enter: **~ getirmek** *or* **~i yaramak**, to bring good luck: **sırra ~ basmak**, to disappear. **~e**, step; stair; rung *of a ladder*: **~ ~** *or* **~ nizamı**, en échelon. **~eli**, en échelon. **~hane** (··–ˈ), latrine. **~li**, lucky, auspicious: **uğurlu ~ olsun!**, good luck to you! **~siz**, unlucky, inauspicious.

**kader** Destiny, fate; providence; worth; dignity; power. **~i ilâhî**, Divine providence: **~ine küs!**, it's a bit of bad luck!; you're not to blame!: **~de varmış**, it was inevitable; it was fated thus: **hükmü ~**, the decrees of fate. **~iye**, fatalism. **~iyeci**, fatalist.

**kadı** Cadi; Moslem judge. ⌜**bu kadar kusur ~ kızında da bulunur**⌝, that's a very trifling fault.

**kadın** Lady; woman; matron. **asçı ~**, woman cook. **~budu** (·ˈ··), **-nu**, meat and rice rissoles fried in egg batter. **~cık**, *only in* **kadın ~**, quiet, domestically minded woman. **~göbeği** (·ˈ···), **-ni**, sweet dish made with semolina and eggs. **~lık**, quality of a lady; ladylike behaviour. **~tuzluğu** (·ˈ···), **-nu**, barberry.

**kadırga** (·ˈ·) Galley; ancient warship. **~balığı, -nı**, cachalot.

**kadid** Skeleton; skin and bone. **~i çıkmak**, to be all skin and bones **kuru ~**, a mere skeleton.

**kadife** Velvet.

**kadih** (·ˈ) Reproach; slander.

**kadim** (·ˈ) Old; ancient; eternal (in the past). The olden time; bygone days. **kârı ~**, old style. **~en**, in olden times; from the beginning of time.

**kadinne** (·ˈ·) (*For* **kadın nine**) grandmother; old woman.

**kadir**[1] (–ˈ) *v.* **kaadir.**

**kadir**[2], **-dri** Worth; personal value; rank, dignity. **···in kadrini bilmek**, to know the value of, to appreciate: **~ gecesi**, the Night of Power (the 27th of Ramazan, *when the Koran was revealed*): **anası ~ gecesi doğurmuş**, very lucky. **~şinas**, who appreciates value *or* merit; appreciative.

**kadran** (*Fr. cadran*) Face, dial (of a clock *etc.*).

**kadril** Quadrille.

**kadro** (ˈ·) (*Fr. cadre*) Staff; roll; establishment; framework. **~ya dahil**, on the permanent staff: **~ harici**, not employed, on half-pay 'en disponibilité'.

**kafa** Head; back of the head, nape; intelligence. (*In many idioms* **kafa** *and* **baş** *are interchangeable*). **~sı* almamak**, not to understand; to be unable to take *stg.* in: **~sından* ayrılmamak**, to follow *s.o.* wherever he goes: ~sı* boş, empty-headed, stupid: ~yı çekmek, to drink heavily: ~m durdu, my mind won't work, I am too tired: ~ kâğıdı, identity card: ~sına* koymak, to make up one's mind, to decide *to do stg.*: ~ sallamak, to flatter: ~sına* söz girmez, he is stupid *or* obstinate: ~sı* taşa çarpmak, to suffer for one's mistake: ~sını* taştan taşa çarpmak, to repent bitterly: ~ tutmak, to be obstinate *or* defiant: ···e karşı ~ tutmak, to resist obstinately: ~ yormak, to ponder, to think hard: ~ yorucu, head-splitting: bu ~da adamlar, this kind of people: bu ~yı bırak!, you must give up these sorts of ideas: her ~dan bir ses, everyone expressing a different opinion. **kafa·dar,** intimate friend; intimate; like-minded. **~lı,** having *a certain* head; intelligent: **boş ~,** stupid: **eski ~,** old-fashioned; narrow-minded. **~ sız,** stupid. **~tası** (·ˈ··), **-nı,** skull.

**Kafdağı, -nı** *A fabulous mountain inhabited by djinns (often used to express enormous difference, great obstacle).* **~na kadar,** to the end of the world: **burnu ~nda,** very conceited.

**kafes** Cage; lattice; grating; framework of a wooden house; (*sl.*) prison. **~ gibi,** a mere skeleton: **~e girmek,** to be duped: **~e koymak,** to deceive, to take in: **~ tamiri,** extensive repairs to a building: **tel ~,** wire netting. **~lemek** (*sl.*), to cheat; to get by cheating.

**kâffe** The whole; all; everyone. **~ten** (ˈ··), wholly, entirely.

**kâfi** (–ˈ) Sufficient, enough.

**kafile** (–·ˈ) Caravan; convoy; gang.

**kâfir** (–ˈ) Misbeliever; non-Moslem; who denies. **~i nimet,** ungrateful: **vay ~ !,** *an exclamation of surprised admiration,* 'oh!, well done!' **~lik,** disbelief, irreligion; cruelty.

**kafiye** (–·ˈ) Rhyme. **~li,** rhyming.

**Kafkasya** (·ˈ·) Caucasus; a dark variety of walnut wood. **~lı,** Caucasian.

**kaftan** Robe of honour; robe. **biçilmiş ~,** well-suited, appropriate; 'cut out for'. **~böceği, -ni,** ladybird.

**kâfur** (–ˈ) Camphor; anything very white. **~î** (–·ˈ), **~lu,** pertaining to camphor, camphorated. **~u,** spirits of camphor.

**kâğıd** Paper; letter; playing-card; (*sl.*) pound note. Of paper. **~ balığı,** Deal-fish (*Trachypterus*): **~ hamuru,** cellulose: **~ helvası,** a kind of pastry in thin layers: **~ sepeti,** waste-paper basket: **~ üzerinde,** on paper only, theoretical. **~cı,** stationer.

**kâğıthane** (··–ˈ) Paper-mill; *the valley of the Sweet Waters of Europe, at the top of the Golden Horn.*

**kâgir** (–ˈ) Built of brick *or* stone (house).

**kağnı** Two-wheeled ox-cart.

**kağşa·k** Dry and crackling; about to collapse. **~mak,** to crack with dryness; become old and wizened; be about to collapse.

**kâh, gâh** (–) A time; moment; place. At one time; sometimes. **~ ... ~ ...,** at one time ... and at another ... .

**kahhar** Overpowering; irresistible; all-powerful (God).

**kahır, -hrı** An overpowering; subjugation; anxiety, distress. **···in kahrını çekmek,** to suffer anxiety *or* trouble on account of ... : **⌜~ yüzünden lûtuf⌝,** good fortune arising from misfortune: **kahrından* ölmek,** to die of a broken heart. **~lanmak,** to be grieved *or* distressed.

**kâhin** (–ˈ) Soothsayer; seer; oracle. **~lik,** profession *or* quality of a soothsayer.

**kahir** (–ˈ) Overpowering; dominant; irresistible.

**Kahire** (–·ˈ) Cairo.

**kahkaha** Loud laughter. **~ atmak,** to burst out laughing: **~ çiçeği,** convolvulus: **~ ile gülmek,** to roar with laughter.

**kahpe** Prostitute. Perfidious, deceitful. **~lik,** prostitution; dirty trick; perfidious behaviour. **~oğlu, -nu,** 'son of a bitch'; low-down treacherous fellow.

**kahraman** Hero; gallant fellow; hero (of a book *etc.*). **~lık,** heroism.

**kahren** (ˈ·) By force, violently.

**kahr·etmek** (ˈ··), **-eder** *va.* Overpower, subdue. *vn.* Be distressed; fret. **···i Allah kahretsin!,** may God curse ... . **~ı,** *v.* **kahır.** **~olsun,** damn him!

**kahvaltı** Breakfast. **ikindi ~sı,** light refreshment in the afternoon.

**kahve** Coffee; café. **çekilmiş ~,** ground coffee: **kuru (çekirdek) ~,** coffee beans: **sade ~,** black coffee without sugar: **~ ocağı,** room where coffee is made: **~ parası,** tip, pourboire: **mahalle ~sine dönmek,** to become disorderly and slack. **~ci,** keeper of a coffee-shop *or* café; coffee-maker in a large establishment. **~hane** (··–`), coffee-shop; café. **~rengi, -ni,** coffee-colour; brown.

**kâhya** Steward; major-domo; bailiff; warden of a trade-guild. **~ kadın,** housekeeper: **~ kesilmek,** to set oneself up as the adviser of others; to interfere in other people's affairs: ⌜**~sı yok**⌝, he can do as he likes: ⌜**keyfimin ~sı mısın?**⌝, what right have you to interfere in my affairs? **~lık,** office of a **kâhya**: **···e ~ etm.,** to meddle with the affairs of ... .

**kaide** (–·`) Base; pedestal; rule, principle, custom. **~ten** (–`··), according to rule; in principle.

**kail** (–`) Saying; consenting, agreeing. Speaker. **~ etm.,** to persuade: **···e ~ olm.,** to consent *or* agree to.

**kaim** (–`) Standing; lasting; taking the place of; perpendicular; rectangular. **···le ~ olm.,** to exist thanks to ..., to be dependent on ...: **···in yerine ~ olm.,** to act for, take the place of ... . **~e** (–·`), bank-note; bill, account; right angle. **~en** (–··), standing; perpendicularly.

**kaka** (*Childish language*) 'gaga'; excrement. **kötü ~ olm.,** to be discredited.

**kakao** (·`·) Cocoa.

**kakavan** Tiresome, peevish and old (*gen.* of a woman); stupid.

**kakıç** Fisherman's gaff.

**kakılı** Nailed; driven in. **olduğu yerde ~ kaldı,** he was rooted to the spot.

**kakım** Ermine, stoat.

**kakır** Dry and crackling. Noise of a dry crackling thing.

**kakırca** (·`·) Dormouse (?).

**kakır·dak** Making a crackling noise. Skin of suet and sheep's fat after melting down the fat. **~damak,** to rattle, rustle, crackle, make a harsh sound; become dry and hard; be without life and spirit; die. **~datmak,** *caus. of* **kakırdamak**; (*sl.*) to kill. **~tı,** crackling noise *made by dry things.*

**kakışmak** *vn.* Push one another about.

**kakma** Worked in relief; repoussé. Repoussé work. **~lı,** with a decoration in relief; inlaid; encrusted *with jewels etc.*

**kakmak** *va.* Push; prod; tap; nail; strike metal to raise relief on the other side; inlay; encrust *with jewels etc.*; embroider with a raised pattern of gold *or* silver *or* mother-of-pearl. **başa ~,** to remind *s.o.* of a former kindness, twit with, taunt.

**kaknem** (*sl.*) Very ugly.

**kakule** (·–`) Cardamon.

**kâkül** Lock of hair, *esp.* a side-lock. **~lü,** having pendent locks.

**kal, -i** Speech; talk. **~e almak,** to take into consideration.

**kala** *Gerund of* **kalmak.** *In such phrases as*: **saat dokuza beş ~,** five minutes to nine: **üç gün ~,** three days wanting to ... : **~ ~ ... kaldı,** there only remains ...; all that is left is ... .

**kalabalık** Crowd; throng; confused mass; a mass of furniture *or* belongings. Crowded; bustling; thronged. **~ etm.,** to be in the way: **kalabalığı kaldırmak,** to tidy up: **ağız kalabalığı,** a torrent of words; **ağız kalabalığına getirmek,** to bewilder by many words: **başı ~,** who has to deal with a lot of people: **kuru ~,** unnecessary presence, a being present and doing nothing.

**kalafat, -tı** Caulking; spurious decoration *to hide repairs*; trick. **~çı,** caulker. **~lamak,** to caulk; paint; hide defects *by superficial repair or paint.*

**kalakalmak** (·`··) *vn.* Stand petrified *with fear or amazement.*

**kalamar** Large squid, cuttle-fish.

**kalamış** Reed-bed.

**kalantor** One having the appearance of a well-to-do and important man.

**kalas** Beam; rafter; plank.

**kalavra** (·`·) Rough leather shoe; patched shoe; old leather goods.

**kalay** Tin; tinsel; scolding. **~ atmak (vermek),** to give a good scolding: **~ basmak,** to abuse: ⌜**altı alay üstü ~**⌝, with smart outer clothes (*for ostentation, though underneath may be shabby clothes*). **~cı,** tinsmith; fraud (man), imposter. **~lamak,** to tin; adorn superficially, cover with sham decoration; abuse. **~lı,** tinned; tinsel; sham. **~sız,** untinned; without sham decoration.

**kalb** Heart; core; kernel. **~ ~e karşıdır,** friendship is mutual. **~en** (`·), from the heart, sincerely. **~gâh** (·–), heart (*fig.*); centre. **~î** (·–), cardiac; cordial. **~lı,** -hearted; having a weak heart.

**kalbetmek** (ˋ··), **-eder** *va.* Transform; transpose; convert.
**kalbur** Sieve, riddle, screen. **~a dönmek,** to be riddled: **~dan geçirmek,** to sieve: **~la su taşımak,** to make futile efforts: **~ üstü,** 'the cream', the choicest *or* most distinguished: ⌜**evvel zaman içinde ~ saman içinde, deve tellal iken**⌝ *etc., formula for the beginning of a story* (*cf.* '*once upon a time*'). **~lamak** *va.*, to sieve, sift.
**kalça, kalçak** Hip.
**kalçete** (·ˋ·) Gasket (*naut.*).
**kalçın** Long felt hose *reaching to the hip*; felt boot *worn inside jack boots.*
**kaldırım** Pavement; causeway. **~ mühendisi,** loafer, ne'er-do-well: **~ taşı,** paving-stone: **~ yaması (kargası),** loose woman. **~cı,** paviour; loafer. **~cılık,** picking the pockets of *s.o.* whose attention is distracted by an accomplice. **~lı,** paved. **~sız,** unpaved.
**kaldırmak** *va.* Raise, erect; lift; carry; remove; abolish, abrogate; tolerate; let start *or* sail; cause to get up; cause to recover from an illness. **atı dört nala ~,** to put one's horse into a gallop: **ayağa ~,** to cause a commotion: **baş ~,** to rise in rebellion: **dansa ~,** to invite to dance: **derse ~,** (teacher) to hear a pupil say his lesson: ⌜**beni erken kaldır!**⌝, 'call me early!': **lâkırdı (söz) kaldırmamak,** to be easily offended, to be unable to stand criticism: **masraf ~,** to bear an expense: **ortadan ~,** to do away with: **bir malı piyasadan ~,** to corner goods: **tabanı ~,** to take to one's heels: **vergı ~,** be capable of paying *so much* tax: **birinin vücudünü ~,** to do away with *s.o.*
**kaldırtmak** *va. caus. of* **kaldırmak.**
**kale** Fortress; castle; wall round a fortress; goal (in football *etc.*). **~ gibi,** as firm as a castle: **çift ~ oynamak,** to play a proper game of football (*as opp. to* **tek ~,** using only one goal for practice). **~bent,** confined in a fortress. **~ci,** goalkeeper.
**kalem** Reed; pen; paint-brush; cutting *of a plant*; fine chisel, turning tool; office; style; item *or* entry *in an account*; category; vaccination tube. **~e almak,** to write, to draw up, edit: **~ aşısı,** graft: **(üstüne) ~ çekmek,** to draw the pen through, cancel: **~ efendisi,** (*formerly*) clerk in a government office: **~e gelmez,** indescribable; unreasonable: **~inden kan damlıyor,** his style is brilliant: **~ kömürü,** high quality charcoal: **~ kulaklı,** having small upright ears (horse): **~i mahsus,** the private secretariat of a Minister: **~ oynatmak,** to write; to correct; to spoil by altering: **bir ~de,** at one effort; instantly: **eli ~ tutar,** capable of expressing himself in writing: **taş ~,** slate pencil: **taşçı ~i,** a mason's chisel: **üç ~ eşya,** three items.
**kalembek** A yellow scented wood (aloes-wood); a kind of maize.
**kalem·dan** Pen-case. **~en** (·ˋ·), in writing. **~iye,** office fees. **~kâr** (··ˋ), a painter of designs on muslin *etc.*; an engraver on gold *or* silver. **~kârî** (··—ˋ), painted *or* engraved by hand. **~lik,** lathe tool-post. **~traş,** penknife; pencil-sharpener. **~ucu** (·ˋ··), **-nu,** nib; steel pen.
**kalender** A wandering mendicant; dervish who has renounced the world; unconventional person. **~ane** (···—ˋ), unconventional; unconcerned; free-and-easy. **~lik,** unconventionality; free-and-easiness; a bohemian existence: **işi kalenderliğe vurmak,** to take things philosophically.
**kaleska** (·ˋ·) Small open carriage.
**kaleta** (·ˋ·) A kind of biscuit.
**kalfa** (ˋ·) Usher (of school); master builder; qualified workman; head clerk. **~ kadın,** elderly domestic servant: ⌜**boş gezenin boş ~sı**⌝, lazy and out-of-work.
**kalgımak** *v.* **kalkımak.**
**kalhane** (·—ˋ) Metal refinery.
**kalıb** Mould; form; die; model; shoe-tree. **~ının adamı olmamak,** not to come up to appearances: **~ını basmak,** to be certain about, to guarantee: **~ı dinlendirmek,** (*sl.*) to die: **~dan ~a girmek,** to keep changing one's profession *or* one's ideas: **~ kesilmek,** to be petrified: **~ı kıyafeti yerinde,** well set up and well-dressed: **bir ~ sabun,** a cake of soap: **~ gibi yatmak,** to lie in bed *or* elsewhere from laziness. **~lamak,** to make in a mould; block (a hat *etc.*).
**kalık** Defective; wanting. Omission; defect.
**kalım** *In* **ölüm ~,** life-and-death.
**kalın** Thick; stout; coarse; dense. **~ barsak,** the large intestine: **~ kafa-(lı),** thick-headed: **~ ses,** deep voice. **~lık,** thickness; coarseness; stupidity.

**kalınmak** *vn. impers of* **kalmak**; **burada kalınır mı?**, can one stay here?
**kalıntı** Remnant, remainder; survival.
**kalibre** (*Fr. calibre*) Jig (*mech.*).
**kalite** (*Fr. qualité*) Quality. **~si bozuk**, good-for-nothing, degenerate.
**kalkan** Shield. **~ balığı**, turbot.
**kalkık** Raised; tilted, upturned (nose).
**kalkınma** Recovery *of a nation etc.*; progress.
**kalkınmak** *vn.* Pick up *after an illness*; (of a nation *etc.*) to make a material recovery.
**kalkış** Departure.
**kalkışmak** *vn.* Attempt *stg.* beyond one's powers; pretend to be able to do *stg.*
**kalkıtmak** *va.* Cause to jump *or* start; cause to prance.
**kalkmak** *vn.* Rise; get up; start *to do stg.*; take it into one's head *to do stg.*; set out *on a journey*; (steamer) sail; be annulled *or* cancelled; be removed, be done away with; stand up; rise in rebellion. **kalkıp kalkıp oturmak,** to show one's anger by one's movements: **kalk gidelim etm.** (*sl.*), to pinch *stg.*: **âdet ~,** of a custom to fall into disuse: **bir şeyin altından ~,** to get out of *a difficulty etc.*: **yüreği* ~,** to feel sick.
**kallâvi** Ceremonial turban *formerly worn by Ministers*; large coffee-cup. Huge; weighty.
**kalleş** Untrustworthy; mean, caddish fellow who lets one down. **~lik,** a dirty trick.
**kalma** Act of remaining; passing the night at a place. Remaining (from); dating (from). **babadan ~,** inherited.
**kalmak, -ır** *vn.* Remain; be left; be left over; survive; halt; cease; be abandoned; be postponed; stay the night; (of the wind) to drop. [*Used as an auxiliary verb it implies the continuation of an action, e.g.* **bakakalmak,** to keep on looking]. **···den ~,** to be prevented from *doing stg.*: **kala kala,** there only remains; all that is left: **kala ~,** suddenly to find oneself in a difficult position; to be disconcerted; to find oneself left in the lurch: **kaldı ki,** there remains the fact that ...; there only remains to say ...; and moreover: **kalsın!**, leave it!; it doesn't matter!: **açıkta ~,** to be left without employment; to be destitute: **ağzı* açık ~,** to be left gaping: **altı aya kalmaz sulh olur,** in less than six months there will be peace: **az kaldı,** nearly, all but: **babadan ~,** to be inherited: **bana kalırsa (kalsa)**, if it were left to me, in my opinion: **bildiğinden kalmamak,** to go one's own way: **demesine (sormasına** *etc.*) **kalmadan,** before he could say (ask *etc.*): **işinden* ~,** to be kept from one's work: **bu kadarla kalsa iyi,** it would not have mattered if he had gone no further *or* if that had been all: **nerede kaldı!**, well, what about it!: **nerede kaldı ki,** how much less .. ; let alone the fact that ...: **uykuya ~,** to fall asleep: **üstünde* ~,** (of a lot at an auction) to fall to one's bid: **···den kalır yeri yok,** he (it) is not much better than ... .
**kalmaz** *v.* **kalmak**; ⌜**aman demeye ~**⌝, before I (he *etc.*) could say a word ... .
**kaloma** (·ˈ·) Slack (of a rope or anchor chain). **~ etm.,** to pay out (rope).
**kalorifer** Stove; central heating.
**kaloş** Galosh.
**kalp**[1] *v.* **kalb.**
**kalp**[2] False; spurious; adulterated; blustering; insincere, untrustworthy.
**kalpak** *Headgear shaped like a fez but made of fur or astrakhan.*
**kalpazan** False coiner, counterfeiter; liar; cheat. **~lık,** manufacture of false coins; lying, cheating.
**kalsiyum** Calcium.
**kaltaban** Pander; cuckold; charlatan.
**kaltak** Saddle-tree; saddle; whore. **~kaşı** (·ˈ··), **-nı,** the raised part in front of and behind an eastern saddle. **~lık,** dirty trick, mean behaviour (*only of a woman*).
**kalûbelâ** (–··ˈ) *In* **~dan beri,** from time immemorial.
**kalya**[1] (ˈ··) Potash.
**kalya**[2] (ˈ··) Vegetable stew.
**kalyon** Galleon. **~cu,** sailor; (*sl.*) old sea-dog.
**kâm** (–) Desire, wish. **···den ~ almak,** to enjoy *stg.*
**kama** Wedge; dagger; breech-block *of a gun*; key (*mech.*); mark *made by the winner on the loser's face in certain games.* **~ basmak,** to win a game: **~ ~ya gelmek,** to come to blows with knives. **~cı,** artillery artificer. **~lamak,** to stab.
**kamara** (·ˈ·) Ship's cabin; House *of Commons or Lords.*

**kamarot, -tu** Ship's steward.
**kamaş·mak** Be dazzled; (of teeth) to be set on edge. **~tırmak,** to dazzle; to set *the teeth* on edge.
**Kamber** *v.* Kanber.
**kambiyal, -li** Bill of Exchange.
**kambiyo** (ꜝ··) Foreign exchange.
**kambur** Hunchback(ed); round-backed; crooked, warped. **~u* çıkmak,** to become hunchbacked: **~unu* çıkartmak,** to cause to become hunchbacked; to hunch the shoulders, to stoop: **~ felek,** cruel Fate: **~ üstüne ~,** one trouble after another: **~ zambur,** bumpy, uneven.
**kamçı** Whip. **~ bağı,** rolling-hitch. **~lamak,** to whip; stimulate.
**kamer** Moon. **~î** (··ꜝ), lunar. **~iye,** arbour: **senei ~,** lunar year.
**kamet** (–ꜝ), **-ti** Stature; fathom; prayer before a Moslem service; outcry. **~i artırmak,** to become presumptuous *or* bumptious; become more insistent.
**kamış** Reed; cane; fishing-rod; penis. **~ bayramı,** (the Jewish) Feast of Tabernacles: **~ böceği,** razor-shell. **~çık,** jeweller's blowpipe; nozzle of bellows. **~lı,** furnished with a reed *etc.*; large-stemmed. **~lık,** reed-bed.
**kâmil** (–ꜝ) Perfect; complete; mature; of mature years; well-educated; well-conducted. **~en** (ꜝ··), perfectly; fully, entirely.
**kamineto** (··ꜝ·) Methylated spirit lamp.
**kâmkâr** (–ꜝ) Prosperous; successful; august; despotic.
**kamlı** Having a cam. **~ mil,** cam-shaft.
**kampana** (·ꜝ·) Bell.
**kampanya** (·ꜝ·) Cropping-season (*esp.* of sugar-beet).
**kâmran** (–ꜝ) Successful; fortunate.
**kamu** All; the whole.
**kâmus** (–ꜝ) Lexicon.
**kamyon** Lorry.
**kan** Blood; bloodshed; revenge for murder. **~ ağlamak,** to weep bitterly; to be in deep distress: **~ aktarmak,** to give a blood transfusion: **~ almak,** to let blood, bleed (*va.*): **~ boğmak,** to die of cerebral haemorrhage: **~ı bozuk,** suffering from unhealthy blood *or* indigestion; degenerate, base: **~a boyamak, to cause** bloodshed: **~a bulamak (boyanmak),** to be bloodstained: **~ dâvası,** blood-feud, vendetta: **~ına* dokunmak,** to make one's blood boil: **~ına girmek,** to have *s.o.'s* blood on one's hands: **~ gütmek,** to cherish a vendetta: **~ını* içine* akıtmak,** to hide one's sorrows: **~ı* kurumak,** to be beside oneself with anxiety *or* suffering: **~ olm.,** to be bloodstained; to cause bloodshed: **~ oturmak,** for blood to be effused (bruise *etc.*); **elime ~ oturdu,** my hand was severely bruised: **~ı* pahasına,** at the cost of one's life: **~a susamış,** bloodthirsty; **kendi ~ına susamak,** to seem anxious to sacrifice one's life: **~ tutmak,** (i) to faint at the sight of blood; (ii) for a murder to haunt the mind of the murderer: **~ında* var,** inborn, innate: **içinden* ~ gitmek,** to suffer secretly: **iki eli* ~da,** in straits; over-burdened with work.
**kana**[1] Waterline marks *on the stem and stern of a ship.*
**kana**[2] *In* **~ ~,** to repletion; abundantly.
**kanaat** (·–ꜝ), **-ti** Contentment; conviction; opinion. **~indeyim,** I am of the opinion that ...: **~ etm., to be** satisfied: **~ getirmek,** to be convinced *or* persuaded: **~ sahibi,** contented with what he has. **~kâr** (·–·ꜝ), contented; satisfied with little. **~kârlık,** contentment.
**kanad** Wing; fin; leaf *of a door*; fold *of a screen*; flap *of a tent*; sail *of a windmill*; paddle *of a waterwheel*. **emniyet ~ı,** safety-catch *of a rifle*: **tutar kolu ~ı olmamak,** *or* **kolu ~ı kırılmak,** to be helpless *or* broken. **~lanmak,** to take wing, fly away; be fledged. **~lı,** winged; finned; folded.
**Kanada** (·ꜝ·) Canada. **~lı,** Canadian.
**kanal** Canal.
**kanamak** *vn.* Bleed. ⌜**kimsenin burnu kanamadan**⌝, without hurting anyone; bloodlessly.
**kanape** Sofa.
**kanarya** (·ꜝ·) Canary. **~lık,** stern-galley *on a ship*. **~otu, -nu,** groundsel.
**kanata** (·ꜝ·) Small earthenware *or* tin receptacle *for measuring liquids.*
**kanatmak** *va.* Make bleed. **burun kanatmamak,** to act gently, without bloodshed.
**kanaviçe** (··ꜝ·), **kanava** (·ꜝ·) Coarsely woven linen; fine canvas.
**Kanber** *In* ⌜**~siz düğün olmaz**⌝, *used*

*of s.o. who turns up on every occasion*; 'of course he was (will be) there!'

**kanca** (ˈ·) Large hook; meat-hook; boat-hook. **~yı atmak,** to make a grab at *stg.*: **~yı takmak,** to 'have one's knife' into *s.o.*, to 'have a down on' *s.o.* **~lamak,** to grapple with a hook; to put on a hook. **~lı,** hooked: **~ kurd,** hook-worm.

**kancık** Bitch; any female animal; perverse treacherous person. **~lık,** treachery, deceit.

**kan·çanağı** (ˈ···), **-nı** Bleeding-basin; eye red with rage *or* weeping. **~çıbanı** (ˈ···), **-nı,** boil.

**kançılar** Vice-Consul; consular assistant. **~ya** (··ˈ·), Consular office.

**kandaş** One of the same blood.

**kandır·ıcı** Convincing; satisfying. **~mak,** to satisfy; convince, persuade; satiate; take in, cheat.

**kandil** *Old-fashioned* oil lamp. **kör ~,** very drunk. **~ gecesi,** the night of a Moslem feast *when the minarets are illuminated.*

**kandilisa** (··ˈ·) Halyard.

**kandil·leşmek** *vn.* Greet one another on the feast of **kandilgecesi. ~li,** decorated with lamps, illuminated. **geçmişi (ölüsü) ~,** the cursed fellow: **~ selâm (temenna),** *old-fashioned* very polite salutation, *raising the hand from the ground several times*: **~ sümbül,** grape-hyacinth.

**kangal** Coil; skein. **~ etm.** *or* **~lamak,** to coil, wind in a skein.

**kanık** Content; satisfied. **~samak, ~sımak,** to be satiated; to become inured.

**kanırık** *v.* **kanrık.**

**kanırmak** *va.* Force back; bend; attempt to force open.

**kan·ısıcak** Warm-hearted, friendly. **~ısoğuk,** cold in manner; antipathetic. **~kardeşi** (ˈ···), **-ni,** intimate friend. **~kırmızı** (ˈ···), blood-red. **~kurutan** (ˈ···), mandrake. **~lamak,** to stain with blood. **~lanmak,** (i) to be stained with blood; (ii) to become healthy.

**kani** (–ˈ) Content; satisfied; convinced.

**kanlı** Bloody; underdone (meat); guilty of murder; full-blooded. **~ canlı,** full of health.

**kanmak** *vn.* Be satiated; be content; be persuaded. **···e ~,** to be taken in by ...: **kana kana,** to repletion: **kana ~,** to shed blood unsparingly: **paraya kanmamak,** to have an insatiable desire for money.

**kanotye** (*Fr. canotier*) Straw-hat; boater.

**kanrevan** Flowing with blood; sanguinary.

**kanrık** Perverse; very obstinate.

**kanser** Cancer; canker; gall; (potato) wart disease.

**kansız** Anaemic. **~lık,** anaemia.

**kantar** Weighing-machine; steelyard; weight of 40 okes (about 120 lb.). **~ı belinde,** wide awake, acute: **~a çekmek,** to weigh in one's mind: **~ topu,** the ball of a steelyard; **~ın topunu kaçırmak,** to go too far, to overdo *stg.*: **yeni ~,** modern weight of 100 kilos. **~cı,** public weighing official. **~iye,** fee for weighing. **~lı,** *in* **~ küfür,** violent abuse.

**kantara** Arch; centre (architecture).

**kantarma** (·ˈ·) Heavy curb; curb-rein. **~k,** to pull up *a horse.*

**kantaron** Centaury.

**kantaşı** (ˈ··), **-nı** Agate.

**kanter** *v.* **ter.**

**kantin** Canteen.

**kanto** Song (in cabaret *or* theatre).

**kanun**[1] (–ˈ) (*Formerly*) Military policeman.

**kanun**[2] (–ˈ) Dulcimer. **~î**[1], player of the dulcimer.

**kanun**[3] (–ˈ) Rule; law; code of laws. **~u esasî,** Constitution: **~ hükmünde olm.,** to have the force of law, to be as good as a law: **~ vazetmek,** to lay down a law *or* a code. **~en** (–ˈ·), according to law; legally. **~î**[2] (––ˈ), legal; legislative; *name given to the Sultan Suleiman I (the Lawgiver).* **~iyet** (–––ˈ), **-ti,** legality; force of law.

**kânun** (·ˈ) *In* **~uevvel (birinci ~)** and **~usani (ikinci ~),** *old names of* December and January.

**kanyak** (*Fr. cognac*) Brandy.

**kanyot, -tu** (*Fr. cagnotte*) Pool, pot (in gambling).

**kap** Cape; mantle.

**kapak** Cover; lid. **kapağı bir yere atmak,** to take refuge in a place; to succeed in getting to a place: **ambar kapağı,** hatch (of a ship): **dış ~,** outer part of a leg of mutton: **işe ~ vurmak,** to hush a matter up.

**kapaklanmak** *vn.* Stumble and fall on one's face; capsize; overturn.

**kapaklı** Provided with a lid *or* cover; concealed; clandestine. Early breech-loading gun. **gizli ~,** secret and clandestine.

**kapalı** Shut; covered; secluded, out-of-the-world; reserved; overcast (sky); obscure. **~ kutu,** a closed box; a secret; (of a man) inscrutable: **~ zarf usuliyle,** by sealed tender: **gözü ~,** inexperienced, without knowledge of the world; thoughtless: **bir noktayı ~ geçmek,** to pass over a point without mention: **üstü ~,** sous entendu; indirectly. **~çarşı,** covered market; bazaar.

**kapama** Act of shutting *etc.*; complete suit of ready-made clothes; stew of lamb and onions. **~ca,** *in* **kapı ~,** everyone in the house. **~cı,** dealer in all kinds of ready-made clothing, boots *etc.*

**kapamak** *va.* Shut, close, shut up; confine; cover up; hush up (a matter); turn off (tap *etc.*); fill up (hole *etc.*); close (an account). **göz ~,** to close the eyes; pretend not to see; cut off the view from another's house.

**kapan**[1] Who seizes *or* grabs. **~ ~a,** a general scramble: **~ın elinde kalmak,** to be in great demand: ⌜**~ da kaçan mı**⌝ (*dial. for* **kapar da kaçar mısın**), 'nothing doing!'; 'no you don't!'.

**kapan**[2] Trap; wicker covering for tobacco plants. **~a kısılmak,** to be caught in a trap: **~ kurmak,** to set a trap. **~ca,** small trap for birds.

**kapanık** Shut in, confined (place); cloudy, overcast; dark; unsociable, shy; gloomy.

**kapanmak** *vn. pass. of* **kapamak.** Be shut *or* closed; be shut up *or* confined; be covered; (of the weather) to be dull and cloudy; (of a subject *or* discussion) to be closed; shut oneself up, not go out; (of a woman) to veil herself before men; (of a factory *or* business) to cease work; fall down, stumble; (of a wound) to heal up; (of a ship) to broach to. **ağzı ~,** to be reduced to silence: **ayağa ~,** to supplicate: **içi* ~,** to feel depressed: **kitaba** *etc.* **~,** to be absorbed in a book *etc.*: **yere ~,** to prostrate oneself.

**kaparo** (·ˈ·) Earnest money.

**kaparoz** Illicit gain. **~cu,** one who picks up what he can, who lives on his wits.

**kapatma** Shut up, confined. Kept mistress.

**kapatmak** *va. caus. of* **kapamak,** *but usually used as* **kapamak** *except in the following meanings*: acquire by a trick; get *stg.* cheap; keep (a mistress). **içini* ~,** to depress *s.o.*: ⌜**o faslı kapat!**⌝, 'let's close that subject'.

**kapı, kapu** Door; gate; situation, employment; place of employment; point *at backgammon*; (*popularly*) 'the Government'. **~yı açmak,** to open the door; to start *stg.*: **~yı büyük açmak,** to start an expensive undertaking, to spend money prodigally: **~yı çekmek,** to shut a door: **~ dışarı etm.,** to show *s.o.* the door, eject: **~ dolaşmak,** to go from door to door, to have recourse to all sorts of places: **~ gibi,** large, powerful (person): **~ halkı,** the household *of a great house*: **~sından* olm.,** (of a servant) to lose his job: **~ yapmak,** to prepare the way for *stg.* one is going to say; to capture a space *at backgammon*: **~ yoldaşı,** a fellow servant: **ayni bir ~ya çıkar,** it all leads to the same thing: ⌜**başka ~ya (müracaat)!**⌝, 'you'd better try somewhere else!': **bu iş beş liranın ~sıdır,** this business will cost a fiver: **kırk (seksen) ~nın ipini çekmek,** to knock at many doors.

**kapı ağası, -nı,** *formerly* the Chief White Eunuch *in the Imperia Palace.* **~cı**[1], doorkeeper, porter, concierge. **~cılık,** the occupation and duty of a doorkeeper. **~kulu, -nu,** *formerly* a lifeguardsman of the Janissaries; palace servant. **~lanmak,** to secure a situation; to enter *s.o.'s* service.

**kapıcı**[2] One who seizes; attractive. **~ kuş,** bird of prey.

**kapılmak** *vn. pass. of* **kapmak** *q.v.* Be seized; be deceived. **bu zanna kapılanlar,** those who are carried away by this idea: **söze ~,** to be taken in by fair words.

**kapısız** Without a door; without a job.

**kapış** Manner of seizing; looting; scramble. **~ ~ gitmek,** to sell like hot cakes: **~ ~ yemek,** to eat greedily.

**kapışmak** *va. & vn.* Snatch *stg.* from one another; scramble for *stg.*; quarrel, wrestle. **···le ~,** to get to grips with ...: **bu malı kapışıyorlar,** there is a rush on these goods.

**kapkacak** Pots and pans, household utensils.

**kapkaç** Snatch-thief; stealing by snatching.

**kapkara** (ˈ··) Exceedingly black; pitch dark.

**kaplama** Act of covering *etc.*; covering, coating; plate; crowning (tooth); skin *of a boat.* Covered; lined; faced; plated. **gümüş ~,** silver-plated. **~cı,** silver-plater.
**kaplamak** *va.* Cover over; line; face; plate; bind (book); envelop, surround; include, comprise.
**kaplan** Tiger.
**kaplayıcı** Covering. Silver-plater.
**kaplı** Covered; bound (book). **bakir ~,** copper-sheathed: **kara ~ kitab,** orthodox, formal, traditional way; the law.
**kaplıca**[1] (·ˋ·) Thermal spring.
**kaplıca**[2] Spelt.
**kaplumbağa** (·ˋ··) Tortoise; turtle.
**kapma** Act of seizing. Seized.
**kapmaca** Puss-in-the-corner (children's game).
**kapmak** *va.* Snatch; seize; carry off; acquire; scramble for *stg.*; learn quickly, pick up. **kapan,** *v.* **kapan**: **ağızdan ~,** to learn by listening to others: **ağzından* ~,** to catch *s.o.* off his guard and learn a secret which he does not wish to divulge: **hastalık ~,** to catch a disease: **nem ~,** to get damp; to be easily offended: **su ~,** for a blister to form; to fester.
**kapot, -tu** Hood *of a car.*
**kapsol** Percussion cap.
**kapsül** Capsule.
**kaptan** Captain (of a ship). **K~ Paşa,** *formerly the title of the head of the Navy*: **~ oynamak,** to play marbles: ⌈**gemisini kurtaran ~(dir)**⌉, a clever *or* skilful man will find his way out of a difficulty.
**kaptıkaçtı** A small motorbus; a kind of card game.
**kaptırmak** *caus. of* **kapmak** *q.v.* **···e gönlünü* ~,** to fall in love with: **kendini eğlenceye (içkiye** *etc.*) **~,** to give oneself up to amusement (drink *etc.*).
**kapu** *v.* **kapı.**
**kapurta** (·ˋ·) Hatchway; skylight *of a ship's cabin.*
**kapuska** (·ˋ·) Cabbage stew.
**kaput, -tu** Military cloak; French letter. **~ bezi,** coarse calico.
**kar** Snow. **~ yağmak,** to snow: **~ tutmadı,** the snow did not lie: ⌈**bu sıcağa (buna) ~ mı dayanır?**⌉, such an expenditure, *or* such a consumption, had obviously soon to come to an end: ⌈**güvendiği dağlara ~ yağdı**⌉, what he relied upon has failed him.
**kâr** Work; business; gain, profit; effect. **bu iş akıl ~ı değildir,** this is unreasonable, without sense: **···den ~ çıkarmak,** to profit by ...: **~ etmez,** it's no good; it's useless: **~ı kadim,** old-fashioned; of an old type: **~dan zarar,** not to get all one hoped, but still to get *stg.*: ⌈**zararın neresinden dönülürse ~dır**⌉, any diminution of loss may be counted a profit.
**-kâr** (–) *Pers. suffix.* Who does ...; who makes ..., *e.g.* **hilekâr,** a trickster; **sanatkâr,** an artist.
**kara**[1] Land; mainland; shore. **~ Ataşesi,** Military Attaché: **~ya çıkmak,** to go ashore: **~ya düşmek (oturmak),** (of a ship) to run aground, to be stranded: **~dan gitmek,** to go by land: **~ suları,** territorial waters: **baştan ~ etm.,** (of a ship) to be run ashore intentionally; (of a man) to be in a fix; to take desperate measures.
**kara**[2] Black; gloomy; ill-omened. Black colour; negro. **~lar,** mourning clothes: **~ çalı,** an obstacle to the reconciliation of two parties; *v. also* **karaçalı**: **~sı elinde,** a habitual backbiter: **~ et,** lean and sinewy meat: **~ gün dostu,** a friend in need: **akla ~yı seçmek,** to be in a predicament, to have great difficulty *in doing stg.*: **ayağına* ~ su inmek,** to be kept standing for a long time: **yüzünün ~sı,** a source of shame to one; **yüzünü ~ çıkarmak,** to let *s.o.* down. [*Note: most compounds of* **kara** *are written as one word, see below.*]
**kara·ağac,** elm. **~basan,** nightmare. **~baş,** monk; French lavender; Anatolian sheep-dog. **~batak,** cormorant: **kaz karabatağı,** darter (*anhinga rufa*).
**karabet** (·–ˋ), **-ti** Near relationship.
**karabiber** Black pepper; dark man.
**karabina** (··ˋ·) Carbine; blunderbuss.
**kara·borsa** Black market. **~boya,** sulphate of iron; copper sulphate; sulphuric acid; lamp-black. **~buğday,** buckwheat. **~bulut, -tu,** black rain-cloud. **~ca**[1], somewhat black; swarthy: **~ ot,** black hellebore, Christmas rose. **Karacaahmed,** *name of the largest cemetery in Istanbul.*
**karaca**[2] Roe deer.
**karaca**[3] The upper arm. **~ kemiği,** humerus.
**karacı** Backbiter.
**kara·ciğer** Liver. **~cümle,** *v.* **cümle.** **~çalı,** gorse. **~çam,** Austrian pine.

**~çayır,** rye-grass. **K~dağ,** Montenegro. **~demir,** wrought iron. **K~ deniz,** Black Sea: ⌜**~de gemilerin mi battı?**⌝, 'why so worried?' **~dut,** black mulberry. **~fatma,** cockroach, 'blackbeetle'. **~fik,** a kind of vetch.
**karağı** Fire-rake.
**karagöz** Black-eyed person; gipsy; the Turkish Punch. **~ oynatmak,** to present the shadow-show of **Karagöz**; to play a trick, deceive. **~balığı, -nı,** (?) black sea-bream.
**karagümrüğü, -nü** Custom-house for goods coming by land.
**kara·gürgen** Beech. **~haber,** news of death *or* disaster. **~horasan,** finest Damascus steel. **~humma,** typhus. **~iğne,** small black ant.
**karakaçan** *A community of Roumanian and Bulgarian nomads.*
**kara·kafes** Comfrey. **~kalem,** pencil *or* charcoal drawing; having a black design (stuff, porcelain *etc.*). **~karga,** carrion crow. **~kaş,** black-eyebrowed. **~kavak,** black poplar. **~kış,** severe winter; the depth of winter.
**karakol** Patrol; guard; sentry; guard-room; police-station. **~ gezmek,** to patrol. **~hane** (··–᾿), guard-room.
**karakoncolos** Bogey; vampire; very ugly person.
**karakulak** Lynx. *A noted spring of water near Istanbul.* A kind of dagger.
**kara·kurbağa** Toad. **~kuş,** eagle; a disease of horses' feet. **~kuşî**(··–⁻), arbitrary, high-handed: **hükmü ~,** despotic rule; arbitrary behaviour; might is right.
**kara·lama** Act of blackening *etc.*; writing exercise; copybook writing. **~lamak,** to blacken; dirty; scribble; write exercises in a copybook. **~lık,** blackness.
**karaltı** Blackness; a figure in the dark; an indistinct figure; slight stain.
**Karaman** *A former principality in Asia Minor; the town of that name in the Konya Vilayet;* the fat-tailed sheep *of that district and* the mutton *therefrom*; a swarthy complexioned man; a heavy sledgehammer. ⌜**~ın koyunu, sonra çıkar oyunu**⌝, 'not so innocent as he looks'. **~lamak,** (of a sail) to flap wildly about.
**karambol** (*Fr. carambole*) Cannon *at billiards.* **~yapmak,** to cannon into *s.o.*
**karamelâ** Caramel.
**karamuk** Corn-cockle.
**karamürsel, karamusal** *In* ⌜**~ sepeti sanmak**⌝, to underestimate *s.o.*
**karanfil** Pink (flower); carnation; clove.
**karanlık** Darkness; a dark place. Dark. **~ olm.** (**basmak**), to become dark, for night to fall: **karanlığa çıkmak,** to go out in the dark: **bu nokta bir az ~,** this point is rather obscure: **sabah karanlığı,** before daybreak.
**karantina** (··᾿·) Quarantine. **~ vazetmek,** to impose quarantine.
**kara·oğlan** A dark youth; gipsy. **~pazı,** orach, mountain spinach. **~pelin,** black wormwood.
**karar** Decision; sentence; resolution; agreement; constancy. **~ınca,** as much as is required: **~ında,** moderately: **~larında,** about, approximately (in time): **~ bulmak,** to become settled, to be decided: **bir şeyde ~ kılmak,** to abide by a decision; to settle down to *stg.*: **~ nısabı,** a quorum: **~ vermek,** to decide; to sentence; **bir ~da** *or* **bir ~ üzere,** in an unvarying degree, in a uniform manner: **göz ~iyle,** judging by eye: **tam ~,** just right.
**karar·gâh,** military headquarters. **~lama,** estimated; by rule of thumb. **~lamak,** to estimate by eye. **~ laşmak,** to be agreed upon, be decided. **~laştırmak,** to decide, resolve. **~lı,** fixed, settled, decided.
**kararmak** *vn.* Become black; become dark; be indistinctly perceived; become overclouded *or* misty. **gönlü*** (**yüreği**) **~,** to become pessimistic, to be tired of life: **gözü* ~,** for one's sight to become dim; to feel giddy; to be beside oneself with anger *or* despair: **içi* ~,** to despair: **sular kararıyordu,** night was falling.
**karar·name** (··–᾿) Decree; legal decision. **~sız,** unstable; undecided; changeable; restless; hesitating. **~ sızlık,** instability; indecision; fickleness; restlessness.
**karart·ı** *v.* **karaltı.** **~ma,** black-out.
**kara·saban** Primitive plough. **~sakız,** pitch. **~sevda,** hypochondria, melancholy. **~sinek,** common house-fly.
**kara·su** Glaucoma; a disease of the legs *in animals.* **ayaklarıma ~ indi,** I have been kept waiting for a long

time; I have been kept standing. **~suları, -nı,** territorial waters. **~tavuk,** blackbird.
**karavana** Flat copper pan; mess-tin, dixie; a miss *in target shooting*; a kind of flat diamond.
**karavide** (··́·) Freshwater crayfish.
**kara·yağız** Very dark-complexioned; swarthy and sturdy boy. **~yazı,** evil fate, ill luck. **~yel,** north-west wind; north-west. **~yonca,** (?) black medick, yellow trefoil. **~yüzlü,** shameless, depraved.
**kardeş, kardaş** Brother; sister; fellow; like; comrade. **~ten ileriyiz,** we are bosom friends: **ana baba bir ~,** whole brother *or* sister: **beş ~,** a box-on-the-ear; ⌜**beş ~in tadını tatmamış**⌝, he wants a good smacking: **din ~i,** co-religionist: **süt ~i,** foster-brother. **~çe,** brotherly, fraternal. **~kanı, -nı,** dragon's-blood. **~lik,** brotherhood; friendship.
**kare** (*Fr. carré*) Square; the four *at cards*. **~li,** in squares; chequered.
**kâretmek** (́··), **-eder** *vn.* Win; profit; produce an effect, tell.
**karfiçe** (·́·) French nail.
**karga** Crow; rook. ⌜**~ bokunu yemeden**⌝, very early in the morning: **~ tulumba etm.,** to carry *s.o.* by arms and legs: **alaca ~,** jackdaw (?): ⌜**besle ~yı oysun gözünü**⌝, to nourish a viper in one's bosom: **ekin ~sı,** rook: **korkak ~,** a cowardly fellow: **yeşil ~,** roller. **~burun, -rnu,** Roman nose; curved forceps. **~ büken,** nux vomica. **~cık,** small crow; very bad writing, scrawl; **~ burgacık,** little misshapen thing; scrawl. **~delen,** a kind of soft-shelled almond. **~derneği, -ni,** a crowd of roughs.
**kargaşa, kargaşalık** Confusion; disorder, tumult.
**kargı** Pike; javelin; lance.
**kargılık** Cartridge bag *or* belt.
**kargın** Carpenter's large plane.
**kârgir** *v.* **kâgir.**
**kârhane** (·—́) Brothel; (*originally*) workshop. **~ci,** brothel-keeper, pimp.
**ka'rı** Sea-bottom.
**karı** Woman; wife. **~ almak,** to marry: **~ koca,** wife and husband, married couple; **koca ~,** old woman: **koca ~ lâkırdısı,** old wives' tale, silly nonsense. **~lık,** womanhood, wifehood: **~ etm.,** to play a dirty trick.
**karın, -rnı** Belly, stomach; womb; inside *of anything*; protuberant part; bulge of a ship's hull: **karnım* aç (acıktı),** I am hungry: **karnı* ağrımak,** to have stomach-ache: **karnı karnına geçmiş,** very thin: **karnı geniş,** easy-going, tolerant: **karnım* tok,** I am full, I have had enough; ⌜**böyle yalanlara karnım tok**⌝, 'I won't be taken in by this sort of lie': ⌜**ben senin karnındakini ne bileyim?**⌝, 'how am I to know what is in your mind?'
**karın·ağrısı, -nı,** stomach-ache; pest; tiresome child *or* person: ... **ne ~dır,** ..., or whatever the name of the thing is; or whatever the fellow calls himself.
**karınca** (·́·) Ant; blow-hole *in a moulding*; pit *caused by rust*. **~ duası,** 'the ant's prayer', a kind of written charm; very illegible writing: ⌜**~ kararınca (kaderince)**⌝, 'as a modest contribution'; 'as much as I can afford'. **~lanmak,** to have the feeling of formication, have pins-and-needles; feel benumbed; be full of blow-holes *or* blisters; be pitted by rust. **~lı,** infested by ants; full of blow-holes; pitted by rust.
**karındaş** *v.* **kardeş.**
**karınlı** Having a paunch; pot-bellied.
**karınzarı, -nı** Peritoneum.
**karış**[1] Span (*about 9 in.*); the third of an **arşın.** **~ ~ aramak,** to search every inch of the ground: **~ ~ bilmek,** to know every inch of a place: **~ ~ ölçmek,** to measure very carefully; to calculate very closely: **aklı başından bir ~ yukarı,** doing whatever comes into one's head without reflection: **burnu bir ~ havada,** very conceited.
**karış**[2] Confusion, turmoil. **~ ~ etm.,** to throw into utter confusion.
**karışık** Mixed; adulterated; confused, in disorder. **~ bir adam,** a man about whom one has one's doubts. **~lık,** confusion, disorder; tumult, riot.
**karışılmak** *vn. impers. of* **karışmak.** **buna karışılmaz,** one shouldn't interfere with this.
**karışlamak** *va.* Measure by the span. **alnını* ~,** to defy; to challenge *s.o.* to do *stg.*
**karışmak** *vn.* Mix (with = **ile**); become confused *or* disordered; inter-

fere; …e ~, interfere with, meddle with; exercise control over. **karışanı görüşeni yok,** he is free from interference, he can act independently: **araya lâkırdı karıştı,** another subject cropped up *and the conversation changed*: ⌜**ben karışmam!**⌝, (*after giving advice or warning*) 'well, don't blame me if things go wrong!': **bu işe hangi daire karışır?,** what department deals with this?: **tarihe ~,** to become a matter of history, to be a thing of the past: **toprağa ~,** to be dead and buried.

**karıştır·acak** *In* **diş karıştıracağı,** toothpick. **~ıcı,** causing confusion *or* tumult.

**karıştırmak** *va.* Mix (with = **ile**); stir up; confuse; allow to interfere (with = *dat.*); add (to = *dat.*). **diş ~,** to pick the teeth.

**kari** (–̀), **-ii** Reader.

**-kâri** (**kar** *with possessive suffix used as a suffix*). Worked in ...; work of ...; in the style of ...; **telkâri,** worked in wire, woven of gold *or* silver thread.

**karides** (·̀·) Shrimp.

**kariha** (·–̀) Fertile mind; imaginative power.

**karina** (·̀·) Underwater hull, bottom of a ship. **~ etm.,** *or* **~ya basmak,** to careen *a ship*.

**karine** (·–̀) Context; deduction. **bir ~ ile anlamak,** to infer from an accompanying circumstance.

**kârlı** Profiting; profitable, advantageous. **~ çıkmak,** to come out a gainer; to turn out profitably.

**karlı** Covered with snow; inclined to snow. **~k,** vessel for cooling water with snow; snow-pit.

**karma** Mixed.

**karmak** *va.* Make a mash of; knead; mix (cement *etc.*); shuffle (cards).

**karmakarışık** (̀····), **karmakarış, karmançorman** All mixed up; in utter disorder.

**karmanyola** (··̀·) A nocturnal aggression. **~ etm.,** to attack by night in order to rob.

**karmık, karmuk** Grappling iron; small dam.

**karnabahar** Cauliflower.

**karne** (*Fr. carnet*) Book of tickets *etc.*; schoolboy's report and list of marks.

**karnı** *v.* **karın.** **~yarık,** seeds of flea-wort; a dish of aubergines stuffed with mincemeat; snatch-block.

**karniye** Cornea.

**karo** (*Fr. carreau*) Diamond (cards).

**karola** (·̀·) *In* **~ yakası,** tack of a sail.

**karoseri** (*Fr. carrosserie*) Coachwork; bodywork (of a car *etc.*).

**karpit** Calcium carbide.

**karpuz** Water-melon; anything round; globe *of a lamp*. **… in ayağının altına ~ kabuğu koymak,** to lay a trap for *s.o.*; to cause *s.o.* to lose his job: ⌜**iki ~ bir koltuğa sığmaz**⌝, one can't do two things at the same time.

**kârsız** Unprofitable.

**karşı** Opposite; opposed. Opposite side *or* direction. In any opposite way *or* direction. Against; towards; opposite to. **~ya,** to the opposite side *or* bank: **~ ~ya,** face to face; exactly opposite: **~dan ~ya,** from one side to another: **~nızda*,** opposite you; against you: **ona ~,** against him: **denize ~ oturmak,** to sit facing the sea: **…e ~ çıkmak,** to go to meet *s.o.*; to prepare to oppose *s.o.*: **~sına* çıkmak,** to appear suddenly in front of one: **…e ~ gelmek,** to oppose; to answer back (impertinently): **…e ~ komak,** to make a stand against, to resist: **…e ~ söylemek,** to speak against, oppose: **yüzüne* ~,** to his face.

**karşı·lamak** *va.,* to go out to meet; oppose; reply to; meet (a need *etc.*). **~lanmak,** to be met; to come opposite *or* face to face. **~laşmak,** to confront one another, meet face to face; balance, be equivalent: **…le ~,** to come up against (difficulties *etc.*). **~laştırmak,** to cause to meet *or* balance; to confront *one person* with *another*; to compare. **~lık,** equivalent; reply, retort; recompense; allocation: ⌜**~ istemez!**⌝, 'no backchat!': **karşılığını yapmak,** to give the equivalent, recompense, reciprocate. **~lıklı,** equivalent; reciprocal; corresponding; balanced; done in return; in reply; facing one another.

**kart**[1], **–tı** Card.

**kart**[2] Dry; hard; tough; wizened; old. **~almak, ~lanmak, ~laşmak,** to become dry, tough, shrivelled, old. **~lık,** dryness; toughness; loss of the freshness of youth.

**kartal** Eagle.

**karter** (*Fr. carter*) Crank-case; gear-case.

**karton·pat** (*Fr. carton-pâte*) Mill-

board. ~**piyer** (*Fr. carton-pierre*), thick cardboard.
**kartopu** (ʹ··), **-nu** Snowball; guelder-rose.
**kartuk** Large rake.
**karuçe** (·ʹ·) Cart.
**Karun** (–ʹ) Korah; a very rich person, a Croesus.
**karyağdılı** 'As though it had been snowed upon', speckled; pepper-and-salt.
**karyola** (·ʹ·) Bedstead; bed.
**kasa** (ʹ·) Chest; coffer; packing-case; safe; till; cashier's office; banker *at cards*: spliced eye *or* loop of a rope. ~ **dairesi**, safe-deposit: ~ **etm.**, to pull taut; to frap: **atlama** ~**sı**, spring-board (gymnastics).
**kasab** Butcher. ⌜**yüzü** ~ **süngerile silinmiş**⌝, unblushing, brazen-faced.
**kasaba**[1] Small town. ~**lı**, belonging to a small town.
**kasaba**[2] Reed; pipe; windpipe.
**kasab·iye** *Formerly* a tax on slaughtered sheep; butcher's fee for slaughtering. ~**lık**, trade of a butcher; butcher's fee for slaughtering; butchery, massacre; fit for slaughter. ~ **koyun gibi**, 'like a lamb to the slaughter', mild and uncomplaining.
**kasadar** Cashier; treasurer.
**kasara** Deck-cabin. **baş** ~**sı**, fo'c'sle; **kıç** ~**sı**, quarter-deck. ~**üstü**, **-nü**, poop-deck.
**kasatura** (··ʹ·) Sword-bayonet.
**kasavet** (·–ʹ), **-ti** *Pop. form of* **kasvet.**
**kasd** Intention; endeavour; premeditation; attempt on *s.o.'s* life. ···**den** ~, what is meant by ...: ⌜**bana** ~**ın var mı?**⌝, 'have you evil intentions against me?' ~**en** (ʹ·), intentionally, deliberately. ~**etmek** (ʹ··), **-eder**, to purpose, intend; have a design against *s.o.*; mean; express ~**î** (·–́), premeditated, deliberate.
**kâse** Bowl; basin.
**kasem** Oath.
**kasık** Groin. ~ **bağı**, truss *for hernia*: ~ **biti**, crab-louse: ~ **çatlağı** (**yarığı**), rupture, hernia. ~**otu**, **-nu**, agrimony.
**kasılmak** *vn.* Be stretched; contract; diminish.
**kasım** November.
**Kasımpaşa** *Name of a district of Istanbul.* ~ **ağzı**, 'billingsgate'. ~**lı**, foul-mouthed.
**kasımpatı** (·ʹ··), **-nı** Chrysanthemum.
**kasın·mak** *vn.* Shrink; contract. ~**tı**, tightness; tension; a taking in *of a dress to make it fit.*
**kasır**, **-srı** Castle; palace; summer-house; pavilion. **kasrı hümayun**, imperial palace.
**kasırga** (·ʹ·) Whirlwind; cyclone; waterspout.
**kaside** (·–ʹ) Eulogy; ode.
**kaskatı** (ʹ··) Very hard; rigid; benumbed; petrified.
**kasket**, **-ti** (*Fr. casquette*) Cap.
**kasmak** *va.* Stretch tight; curtail. ~ **kavramak**, to hold in with a tight rein, keep a tight hold on: **kasıp kavurmak**, to turn topsy-turvy; plague, torment, tyrannize: **kendini** ~, to draw oneself up in a superior manner.
**kasnak** Rim *or* hoop (of tambourine, sieve); embroidery-frame; embroidery; any contrivance for stretching tight; pulley; drum *of a cupola.*
**kasnı** A resinous gum (galbanum).
**kaspeannek** (ʹ···), **kaspannek** By force; willy-nilly; on purpose.
**kassam** (·–́) *Moslem functionary, whose duty is to fix the shares of an inheritance.*
**kastanyola** (··ʹ·) Pawl.
**kastarlamak** *va.* Bleach.
**kastor** Beaver.
**kasvet**, **-ti** Depression, melancholy; gloom, oppressiveness. ~ **basmak**, to become dejected: ~ **çekmek**, to be anxious *or* distressed. ~**li**, oppressive, gloomy.
**kaş** Eyebrow; a curved thing; collet (of a ring); accolade (*mus.*). ~ **çatmak**, to frown: ~ **göz etm.**, to wink, to make a sign with the eye and eyebrow: ~**la göz arasında**, in the twinkling of an eye: ⌜~ **yapayım derken göz çıkarmak**⌝, when trying to effect an improvement to spoil the lot: **eyer** ~**ı**, the pommel of a saddle: ⌜**gözünün* üstünde** ~**ın var dememek**⌝, to raise not the slightest objection: **kalem** ~**ı**, slender eyebrows: **kılıc** ~**ı**, the guard of a sword: **samur** ~, thick and regular eyebrows.
**kaşa-ğı** Currycomb. ~**klamak**, to curry, groom.
**kaşamak** *va.* Curry, groom.
**kaşan** Urine (of a horse). ~**ı gelmek**, (of a horse) to wish to stale. ~**mak**, to stale.

**kâşane** (——ʹ) Luxurious dwelling; mansion.

**kaşar, kaşer** Kosher (meat lawful for Jews); a kind of flat cheese *made in the Balkans.*

**kaşar·lanmak** *vn.* Become old *or* worn out; become callous. **~lanmış,** experienced; hardened, callous; cunning. **~lı,** old, worn-out; insensitive.

**kaşe** Cachet; pill; seal; stamp, mark; shade.

**kaşer** *v.* **kaşar.**

**kaşgöz** *v.* **kaş.**

**kaşık** Spoon; spoonful; castanet. **~ atmak,** to eat heartily: **~ düşmanı,** one's wife, 'the missus': ⌜**kaşığı ile yedirip sapı ile göz çıkarmak**⌝, to feed *s.o.* with one's spoon and knock his eye out with the handle thereof, *i.e.* to spoil a good deed by a bad one: **···in ağzının kaşığı olmamak,** to be too deep to be understood by ...; to be above *s.o.'s* head *or* capacity: ⌜**elinden gelse beni bir ~ suda boğar**⌝, he hates me like poison: ⌜**herkes ~ yapar ama sapını ortaya getiremez**⌝, it's not as easy as it looks: ⌜**kısmetinde olan kaşığında çıkar**⌝, you get what Fate brings you: ⌜**suratı ~ kadar kaldı**⌝, emaciated.

**kaşık·çı,** spoonmaker. **~ elması,** *the largest diamond in the Turkish regalia*: **~ kuşu,** pelican. **~çın,** shoveller duck. **~lamak,** to eat spoonfuls *out of a dish etc.*: (*fig.*) to hasten to avail oneself of *stg.* **~otu, -nu,** scurvy-grass.

**kaşımak** *va.* Scratch. **başını kaşımağa vakit bulamamak,** to be too busy *to attend to stg. else.*

**kaşınma** Itching.

**kaşın·mak** *vn.* Scratch oneself; itch. **dayak yemek için kaşınıyor,** he's just asking for a beating: **kavgaya ~,** to be itching for a quarrel. **~tı,** itching.

**kâşif** (—ʹ) A discoverer *or* revealer. **~i esrar,** a revealer of secrets.

**kaşkariko** (··ʹ·) Trick; deceit; buffoonery.

**kaşkaval** A Balkan cheese *in round cakes*; fid.

**kaşkol** (*Fr. cache-col*) Scarf.

**kaşkorse** (*Fr. cache-corset*) Camisole; under-bodice.

**kaşlı** Having eyebrows; having a ... as stone (ring). **~ gözlü,** complete in every way (of a living thing); pretty: **dört ~,** with bushy eyebrows: **elmas ~,** diamond mounted (ring): **kalem ~,** with thin eyebrows.

**kaşmer** Buffoon, clown. **~lik,** buffoonery.

**kaşmir** Kerseymere (soft material resembling cashmere).

**kat, -tı** Fold; layer; coating; story *of a building*; quantity; time *of repetition*; opinion. **~ ~,** in layers; time after time; many times more, much more: **~ çıkmak,** to add a story *to a building*: **~ etm.,** to fold; to stow in layers *or* tiers: **alt ~,** the lower story; first layer; first coat *of paint etc.*: **bin ~,** a thousandfold: **bir ~ elbise,** a suit of clothes: **iki ~,** double; folded double; bent double with age; two stories; two-storied; **iki ~ olm.,** to bow to the ground: **···in yüksek ~ına sunmak,** to offer *thanks etc.* to ... (some great person).

**katakulli, katakofti** (··ʹ·) Act of cheating; hoax. **~ye gelmek,** to be taken in: **Hacı ~!,** old humbug!; swindler!

**katar** String *or* file (of camels *etc.*); railway train. **eşya ~ı,** goods train: **surat ~ı,** express train. **~lamak,** to make a file *or* train *of camels, carts etc.*

**kat'etmek** (ʹ··), **-eder** *va.* Cut; interrupt; terminate; travel over, traverse.

**katı**[1] Hard; violent; dry; strong. Very. Gizzard.

**katı**[2], **katı', -t'ı** A cutting *or* cutting off; interruption; a deciding; a terminating. **kat'ı alâka etm.,** to cease to take interest; to discontinue relations: **~ etm.,** *v.* **kat'etmek.**

**katık** Added. Anything eaten with bread as a relish; condiment; exaggeration. **~ etm.,** to eat *stg.* with bread; not to eat up all one's **katık** before one's bread, to eke out *one's cheese etc.* **~sız,** unmixed, unadulterated: **~ ekmek,** dry bread.

**katı·lanmak, ~laşmak,** to become hard *or* heavy; to coagulate. **~lık,** hardness, dryness; severity.

**katılmak**[1] *vn. pass. of* **katmak.** Be added; be mixed; be driven along; join oneself to others. **su katılmadık bir Türk,** a pure-blooded Turk; a Turk of the Turks.

**katılmak**[2] *vn.* (*from* **katı,** hard). Become hard; get out of breath *from laughing or weeping.* **katıla katıla ağlamak,** to choke with tears: **katıla**

katıla *or* katılırcasına gülmek, to split one's sides with laughter: soğuktan (içi) ~, to be chilled to the marrow.

**katıltmak** *va. caus. of* katılmak². gülmekten ~, to make *s.o.* split his sides with laughing.

**katım** An adding; a joining; a mixing. **koç ~ı**, the season for putting the rams to the ewes *and* the act of doing so.

**katır¹** *In* ~ **kutur**, *v.* **hatır**.

**katır²** Mule; obstinate, ungrateful *or* malicious man; a kind of strong shoe *with iron heel tips.* **~boncuğu, -nu**, blue bead *often hung round animals' necks.* **~cı**, muleteer. **~tırnağı, -nı**, broom, genista.

**katış·ık** Mixed. **~mak**, to join in; to mix with others.

**katı·yağ** Oil usually found in a solid state, *e.g. paraffin wax.* **~yürekli**, hard-hearted.

**kat'î** (·⊥) Definite; decisive; absolute.

**kâtib** (–⌄) Clerk; secretary. **~ane** (–·–⌄), in a correct literary manner (*iron.*). **~iadil**, Notary Public. **~lik**, quality *or* profession of a clerk *or* secretary.

**katil¹, -tli** A killing; assassination; murder.

**katil²** (–⌄), *v.* **kaatil**.

**kat'ileşmek** *vn.* Become definite *or* decisive.

**kat'iyet, -ti** Definiteness; precision. **~yen, ~en** (·⌄·), definitely, absolutely; finally.

**katlamak** *va.* Fold; pleat; put layer upon layer; repeat.

**katlandırmak** *va.* Bend; fold; pleat; cause to consent *or* acquiesce; cause to cringe *or* writhe.

**katlanmak** *vn.* Bend; fold; pucker; become stratified. *va.* (*with dat.*). Undergo; suffer; put up with; acquiesce in. ⌜**göz görmeyince gönül katlanır**⌝, 'what the eye does not see the heart does not rue'.

**katletmek** (⌄··), **-eder** *va.* Kill.

**katliâm** (··⊥) Massacre.

**katma** Joined on; supplementary. Addition; appendage.

**katmak** *va.* Add; join; mix; embroil. **bire bin ~**, to exaggerate greatly: **biribirine ~**, to incite one against the other: **geceyi gündüze ~**, to work day and night, to work incessantly: **···i hesaba ~**, to take ... into account: **önüne* ~**, to drive in front of one: **···i yanına* ~**, to attach ... to *s.o. as escort or companion.*

**katmer** A having folds; double flower; flaky pastry. **~li**, having many folds; manifold; multiplied; multiple; double (flower); flaky (pastry).

**katra** Drop. **~ ~**, drop by drop.

**katran** Tar. **~ ağacı**, cedar tree: **~ ruhu**, creosote. **~köpüğü, -nü**, a kind of fungus. **~lamak**, to tar. **~lı**, tarred.

**katrat, -tı** Quadrat; quad.

**kauçuk** Caoutchouc, unvulcanized rubber.

**kav¹** Tinder; touchwood. **~ çakmak**, tinder and flint: **~ gibi**, soft; inflammable: **mantar ~ı**, amadou.

**kav²**(*Fr. cave*) Pool (in a game).

**kavaf** Dealer in ready-made boots. **~ işi**, coarsely made: **~ malı**, fraudulent rubbish: **ağız (çene) ~ı**, one who tries to deceive by talking a great deal: **ayak ~ı**, a person who is always to be seen about: **kâğıd ~ı**, petty official: **söz (lâkırdı) ~ı**, chatterbox. **~iye** (·–·⌄), ready-made shoe shop; goods at such a shop.

**kavak** Poplar. ⌜**balık kavağa cıktığı vakit**⌝, 'when pigs have wings', 'at the Greek Kalends': ⌜**başında ~ yelleri esmek**⌝, to have childish *or* fantastic ideas *or* projects; not to know what one is doing. **~lık**, poplar wood *or* grove.

**kaval** Shepherd's pipe; any hollow pipe. ⌜**(koyun) ~ dinler gibi**⌝, listening without understanding: **~ kemiği**, fibula: **~ tüfek**, smoothbore.

**kavalye** (·⌄·) (*Fr. cavalier*) Male partner *in a dance.*

**kavança, kavanço** (·⌄·) Transhipment *of goods*; transfer; handing over.

**kavanoz** Glass *or* earthenware jar; pot.

**kavas** Guard *of an embassy or consulate.*

**kavasya** (·⌄·) Quassia.

**kavata** (·⌄·) A bitter kind of tomato *used for pickling.*

**kavelâ** Dowel.

**kavga** Tumult; brawl, quarrel; fight; battle. **~ etm.**, to quarrel, to fight. **~cı**, quarrelsome. **~laşmak**, to quarrel. **~lı**, quarrelling; angry.

**kavi** (·⊥) Strong; robust.

**kavil, -vli** Word; assertion; agreement. **kavlince**, according to the assertion of *so-and-so*: ~ **etm.**, to agree; to promise: **kavli mücerred,** unwarranted assertion; mere words. **~leşmek,** to come to an agreement.

**kavilya** (·ˈ·) Marlinspike.

**kavim, -vmi** Tribe; people; nation.

**kavis, -vsi** Bow; arc; curve; the constellation Sagittarius.

**kavla·k** Having the skin *or* bark peeled off. **~mak,** (of bark, skin *etc.*) to dry and fall off.

**kavmantarı**(ˈ···), **-nı** Fungus from which tinder is obtained, *Fomes fomentarius*.

**kavram** Conception; notion; coupling (*mech.*).

**kavrama** *Verb. n. of* **kavramak.** Coupling (*mech.*).

**kavra·mak** *va.* Seize; grasp; understand. **~yışlı,** quick at understanding.

**kavruk** Scorched; dried up; stunted.

**kavrulmak** *vn. pass. of* **kavurmak.** Be roasted, be fried; be scorched; be withered *or* stunted. **···e** *or* ... **için yanıp ~,** to have an obsession *or* to be extremely keen about ... .

**kavşak**[1] *v.* **kavurak.**

**kavşak**[2], **kavşamak** *v.* **kağrak.**

**kavuk** Large wadded headgear *formerly worn by Turks*; bladder. Hollow; rotten. ⌜**kavuğuma dinlet!**⌝ 'tell that to the Marines!': ~ **giydirmek,** to cheat, deceive: ~ **sallamak,** to acquiesce unhesitatingly; to toady. **~çu,** toady; hypocrite. **~lu,** wearing a **kavuk**; *name of a character in the* **orta oyunu.**

**kavun** Melon. **~içi, -ni,** flesh of the melon; dark yellow with a pinkish tinge.

**kavurma** Act of frying. Broiled *or* fried meat. Fried; roast (coffee). **~ç,** parched wheat.

**kavurmak** *va.* Fry; roast (coffee, corn *etc.*); scorch. **kasıp ~,** to turn topsy-turvy; to plague, tyrannize.

**kavuşak** Junction.

**kavuş·mak** *vn.* Come together: **···e ~,** reach; attain; obtain; join; touch; meet; meet again after a long absence. **Allahına ~,** to go to one's Maker: **gün kavuşuyor,** the sun is setting. **~turmak,** to bring together; unite; join; cause *two people* to meet: ⌜**Allah kavuştursun!**⌝, 'may God bring him back to you!' (*said to one whose relative has left on a long journey*): **el ~.** to fold the hands on the chest *in a deferential attitude*: **birini bir şeye ~,** to enable *s.o.* to get a thing; to grant a thing to *s.o.*: **kolları ~,** to fold the arms: **önünü ~,** to button up one's coat. **~uk,** joining; touching.

**kavut** Roasted wheat ground to flour; a gruel made of such flour.

**kay, -yyı** Vomiting.

**kaya** Rock; rocky cliff *or* hill. **~balığı, -nı,** goby; **dere ~,** gudgeon: **tatlısu ~sı,** tench. **~başı, -nı,** a kind of rustic song. **~cık,** small rock; hop hornbeam. **~ğan,** a kind of laminated rock; slate; *v. also* **kaygan.** **~keleri, -ni,** chameleon. **~koruğu, -nu,** stone-crop. **~lık,** rocky place. **~tuzu, -nu,** rock salt.

**kayak** Ski.

**kayan** Mountain torrent. Swift; violent.

**kayar** Horseshoe with special nails *for walking on ice.*

**kayb·etmek** (ˈ··), **-eder** *va.* Lose. **kendini ~,** to lose consciousness; to lose one's head. **~ı,** *v.* **kayıb.** **~olmak,** to be lost; disappear.

**kayd·etmek** (ˈ··), **-eder** *va.* Enrol; enregister; notice; note down. **onun sözünü ihtiyatla kaydetmeli,** what he says should be accepted with caution. **~ı,** *v.* **kayıd.** **~iye,** registration fee. **~olmak, ~olunmak,** to be enrolled *or* enregistered.

**kaydırak** Flat circular stone *used in a game resembling quoits.*

**kaydırmak** *va. caus. of* **kaymak.** Cause to slip *etc.*; cause *s.o.* to lose his post. **birinin ayağını ~,** to cause a person to lose his job: **gözlerini ~,** to slant the eyes.

**kayetmek** (ˈ··), **-eder** *vn.* Vomit.

**kaygan** Slippery; polished; fickle, mercurial. **~taşı, -nı,** slate.

**kaygana** (ˈ··) Omelette.

**kaygı, kaygu** Care; anxiety; grief. ⌜**kasaba yağ ~sı, keçiye can ~sı**⌝, 'the butcher thinks about his suet, the goat about his life'; everyone thinks of his own interests. **~lı,** anxious, worried; causing anxiety. **~sız,** without care *or* anxiety; carefree. **~sızlık,** freedom from care.

**kayıb** *v.* **gaib.** **~lara karışmak,** to disappear, abscond.

**kayıd, -ydı** Restriction; reservation; enrolment; registration; record; a caring, a paying attention; a brooding over a thing. **kaydile,** with the reservation that; provided that: ~ **altına girememek,** to refuse to be bound by restrictions, to be independent: **kaydı hayat şartıle,** for life, life *interest etc.*: **bir şeyin kaydına düşmek,** to be mainly concerned with *stg.*: **kaydı ihtirazî serdetmek,** to make a reservation: **kaydı ihtiyatla telakki etm.,** to accept *a rumour etc.* with reserve.

**kayıd·lı,** restricted; with reservation; careful; registered. **~sız,** unregistered; careless, indifferent; carefree. **~sızlık,** indifference, carelessness; freedom from care; a not being registered.

**kayık** Boat; caique. **~ tabak,** oval dish. **~çı,** boatman. **~hane** (·ˈ–·), boathouse.

**kayın[1]** Beech.

**kayın[2], ~birader** Brother-in-law. **~peder,** father-in-law. **~valide,** mother-in-law.

**kayıntı** Crack; fissure; ebullition.

**kayır** Sandbank.

**kayır·ıcı** Protector, supporter; who looks after. **~lık,** favouritism. **~ma,** protection; backing: **adam ~,** favouritism.

**kayırmak** *va.* Look after; take care of; protect; back, support; give a job to; employ. **bir vazifeye kayırılmak,** to be given a post by favour.

**kayısı** Apricot.

**kayış[1]** Act of slipping.

**kayış[2]** Strap; strop; belt. (Of meat) like leather. **~ balığı,** Bearded Ophidium: **~a çekmek,** to strop. **~kıran,** rest-harrow.

**kaykılmak** *vn.* Lean; lean back.

**kaylule** Sleep in the forenoon; *also* siesta.

**kaymak[1]** *vn.* Slip; slide; glide; skate; become awry. **gözü* ~,** (i) to squint slightly; (ii) to see by chance.

**kaymak[2]** Cream, *esp.* clotted cream; the cream *of anything*; essence; hard crust *left on the earth after rain.* Very soft and white. **~ altı,** skim milk: **~ tutmak,** to form cream. **~lı,** creamy, made with cream. **~yağı, -nı,** fresh butter.

**kaymakam** The governor of a **kaza** (administrative district); acting representative; (*formerly*) lieutenant-colonel. **~lık,** office *or* district of a kaymakam.

**kaynak[1]** Spring, fountain; source.

**kaynak[2]** Place where two things join; where the buttocks join; welding. **~ yapmak,** to weld. **~çı,** welder.

**kaynama** Act of boiling *or* welding. Boiled.

**kaynamak** *vn.* Boil; be boiled; effervesce; spout up; join; (of mischief) to be brewing; swarm; be perpetually moving; be lost *or* ruined; (of a ship) to founder; (*sl.*) be 'pinched'. **arada ~,** to be lost in the midst of a confusion: ⌜**başından* (tepesinden) aşağı kaynar sular dökülmek**⌝, to be overcome by shame *or* embarrassment: **kanı* ~,** to be active and exuberant; **···e kanı* ~,** to 'take to' *s.o.*, to like him.

**kaynana** (ˈ··) Mother-in-law. **~ zırıltısı,** child's rattle.

**kaynaş·mak** *vn.* Unite; be welded; unite in friendship; (of a crowd) to swarm. **~tırmak,** to cause to unite; weld together.

**kaynata** (ˈ··) Father-in-law.

**kaynatmak** *va.* Cause to boil; boil; weld; cause *a ship* to founder; plot; become friends; (*sl.*) 'pinch', pilfer.

**kaypak** Slippery; unreliable, fickle.

**kaypamak, kaypımak** *vn.* Slip away.

**kayran** Clearing in a forest.

**kayrı** *v.* **gayrı.**

**kayrılmak** *vn. pass. of* **kayırmak.**

**kaytan** Cotton *or* silk cord; braid. **~ bıyıklı,** having a thin curling moustache.

**kaytarmak** *va.* Dodge *payment*; get out of *doing stg.*

**kayyum, kayyım** Caretaker of a mosque.

**kaz** Goose; silly fool. ⌜**~ın ayağı öyle değil**⌝, the fact of the matter is really quite otherwise: ⌜**~ gelen yerden tavuk esirgenmez**⌝, 'don't grudge a penny where you may get a pound': ⌜**çevir ~ yanmasın!**⌝, 'don't try to get out of what you said!'

**kaza** (·ˈ–) Accident, mischance; chance; office and functions of a judge; divine judgement, fate; administrative district governed by a kaymakam; performance of a religious duty omitted at the proper time; performance of an act; payment of a debt. **~ ile,** by chance, by accident: **~ kuvveti,** judicial power: **~ya uğramak,** to meet with an accident: **görünür ~,** an event which can be foreseen and prevented: **görünmez ~,** that cannot be seen and prevented.

**kaza·en** (·ˌ·), by chance, by accident. **~î** (·—ˌ), judicial.
**Kazak** *Name of a Turkish tribe*: Cossack. **kazak,** jersey; a husband who rules his wife (*opp. to* **kılıbık** = henpecked).
**kazalı** (·—ˌ) Causing accidents; dangerous.
**kazan** Cauldron; boiler. ~ **devirmek (kaldırmak),** to overturn (remove) the kettle, *both signs of mutiny among the Janissaries, hence* 'to mutiny': **başı* (kafası*)** ~ **olm.,** to have a buzzing in the head *after a noisy environment*: **bir ~da kaynamak,** to be in complete agreement; to get on very well together.
**kazanc** Gain; profits; earnings.
**kazan·cı** Boilermaker. **~dibi, -ni,** a sweet *made of burnt milk stuck to the bottom of a cooking-pot.*
**kazanmak** *va.* Earn; win; gain.
**kazara** (·ˌ—) By chance.
**kazasker** *Originally the chief military judge; later a high official in the hierarchy of the Moslem Judiciary.*
**kazayağı** (ˌ···), **-nı** Hook with several prongs; branching halyard; three-ended rope; the herb Good King Henry.
**kazazede** (·—·ˌ) Struck down by misfortune; shipwrecked.
**kaz·beyinli** (ˌ···) Stupid; idiot. ~ **boku** (ˌ··), **-nu,** (goose dung) greenish-yellow.
**kazel** Dry leaf remaining on tree. ~ **mevsimi,** autumn.
**kazemat, -tı** Casemate.
**kazevi** Receptacle made of palm-leaves *or* reeds.
**kazı** Excavation. **~cı,** excavator; engraver.
**kazık** Stake, peg; pile; impalement; trick, swindle. **~!,** 'what a fraud!': ~ **atmak** (*sl.*), to cheat: ~ **bağı,** clove-hitch: ~ **kakmak,** to drive in a stake; to establish oneself firmly: ~ **kesilmek,** to be petrified: ~ **kök,** tap-root: **kazığını koparmak,** to make one's escape; (of an animal) to get loose: **kazığa vurmak,** to impale: ~ **yemek,** to be cheated: **demir ~,** iron stake; the Pole Star: ⌜**dünyaya ~ kakacak değil ya!**⌝, he won't live for ever: **sağlam ~,** a secure foundation *for an undertaking.*
**kazık·çı** swindler. **~lamak,** to impale; to cheat, dupe. **~lı,** having stakes; ~ **humma,** tetanus.
**kazılı** Excavated.
**kazı·mak** *va.* Erase by scraping; scratch; shave off completely; eradicate. **~nmak,** scratch oneself; be scratched *etc.* **~ntı,** scrapings; erasure: **tekne ~sı,** the last child of a family. **~ntılı,** scraped; erased. **~tmak,** *caus. of* **kazımak.**
**kaziye** Question, affair; judicial decision.
**kazkanadı** (ˌ···), **-nı** *Name of a wrestling hold.*
**kazma** Act of digging; pickaxe, mattock. Dug; excavated. ~ **kürek,** digging implements.
**kazmak** *va.* Dig, excavate; engrave. **···in kuyusunu ~,** to dig a pit for *s.o., i.e.* to try to ruin him.
**kazmir** Kerseymere.
**kazurat** (·—ˌ), **-tı** Faeces.
**kazzaz** Silk manufacturer.
**kebab** Roast meat. ~ **kestane,** roast chestnuts: **çömlek ~ı,** stewed mutton with vegetables: **döner ~,** meat wrapped round a skewer and roasted: **şiş ~ı,** pieces of meat roasted on a skewer: **tas ~ı,** pieces of meat stewed with onions. **~cı,** proprietor of a small restaurant.
**kebe** Very thick kind of felt; cloak made from such felt; felt carpet. **ter ~,** thick horse-blanket.
**kebed** Liver. **~î** (··ˌ), pertaining to the liver, hepatic.
**kebeş** Ram.
**kebise** (·—ˌ) Leap-year.
**keçe** Felt; carpet, mat. Made of felt. **~külâh,** conical felt hat. **~lenmek, ~leşmek,** to become matted; become numb, be benumbed. **~li**[1], made of felt.
**keçeli**[2] *In* **iki ~,** on both sides.
**keçi** Goat. Obstinate. **yaban** *or* **dağ ~si,** ibex. **~boynuzu, -nu,** goat's horn; carob bean: ~ **gibi,** insipid. **~sakalı, -nı,** salsify. **~yolu, -nu,** narrow path.
**keder** Care; grief; affliction. ~ **etm.,** to be troubled *or* grieved (at = *dat.*). **~lenmek,** to be sorrowful *or* anxious. **~li,** sorrowful, grieved; grievous.
**kedi** Cat. ⌜~ **ciğere bakar gibi**⌝, with intense longing: ⌜~ **ne, budu ne?**⌝, 'well, what can you expect from such a poor creature?; *sometimes* 'you've nothing to fear from him!': ⌜**~ye peynir ısmarlamak**⌝, to entrust *stg.* to an untrustworthy person: ⌜~ **uzanamadığı ciğere pis der**⌝, the fox

said the grapes were sour: ⌜aralarına* bir kara ~ sokmak⌝, to cause bad blood between two persons; 'aranızdan kara ~ mi geçti?', 'have you quarrelled?': ⌜sermayeyi ~ye yükletmek⌝, to go bankrupt: ⌜süt dökmüş ~ gibi⌝, keeping quiet with rather a guilty conscience; subdued. **kedi·balığı, -nı,** lesser spotted dogfish. **~otu, -nu,** valerian.

**keene** (·ˋ·) As if; as though.

**keenlemyekün** As though it had never been.

**kefafınefis** (·—··ˋ) Sufficiency of food to live. **~ etm.,** to be satisfied with very little.

**kefal** Grey mullet. **has ~,** female grey mullet in season, *from which botargo (salted fish roe) is obtained*: **tatlısu ~ı,** chub.

**kefalet** (·—ˋ), **-ti** A being sponsor, bail *or* security; bail, security. **~ etm.,** to make oneself responsible, to stand as surety: **~e rabtetmek,** to take bail *or* security for ...: **~le salıvermek,** to release on bail. **~en** (·ˊ··), as bail *or* security. **~name** (·—·—ˋ), written guarantee *or* agreement to stand as surety.

**kefaret** (·—ˋ), **-ti** Atonement; penance; indemnity.

**kefe**[1] Scale of a balance.

**kefe**[2] Hair glove *used for grooming horses*. **~lemek,** to groom a horse with a hair glove.

**kefeki** Tartar (on teeth). **~ye dönmek,** to be all in holes. **~taşı, -nı,** coarse sandstone.

**kefen** Shroud, winding-sheet. **~i yırtmak,** (to tear the shroud), to cheat death; to recover from a serious illness. **~ci,** maker and seller of shrouds; thief who steals grave-clothes; extortioner. **~lemek,** to wrap in a shroud; cover *a fowl etc.* in batter before roasting. **~li,** wrapped in a shroud; covered with batter.

**kefere** *pl. of* **kâfir.** Unbelievers.

**keferet** *v.* **kefaret.** ⌜**~i budur**⌝, *formula pronounced by quack healers.*

**kefil** (·ˋ) Surety; security; bail. **~ olm.,** to stand as surety.

**kefiye** Light shawl *worn as head-dress by Arabs.*

**kehanet** (·—ˋ), **-ti** Soothsaying; augury. **~ etm.,** to predict the future.

**kehf** Cave. **eshabı ~,** the Seven Sleepers.

**kehkeşan** The Milky Way.

**kehlibar, kehrübar** Amber. **~ balı,** clear yellow honey: **kara ~,** jet.

**kek, -ki** Cake.

**kekâh** *Exclamation expressing a feel- of comfort.*

**keke** Stammering. **~lemek,** to stammer, stutter; be at the last gasp. **~lik,** stammer. **~me,** having a stammer.

**kekik** Thyme. **~yağı, -nı,** oil of thyme.

**keklik** Red-legged partridge.

**kekre** Acrid; pungent; setting the teeth on edge. **~lik,** acridity. **~msi, ~si,** somewhat acrid.

**kel** Ringworm; bald spot. Bald; bare of vegetation; mangy; poor, miserable. ⌜**~ başa şimşir tarak**⌝, 'a boxwood comb for a bald head', an out-of-place luxury: **~i görünmek,** for a defect to be shown up: **~ kâhya,** busybody: **~i kızmak,** to lose one's temper (of one not easily angered): **~den köseye yardım,** the blind leading the blind: **~i körü toplamak,** to fill *an office etc.* with incompetent people: ⌜**~ tavuk ~ horozla**⌝, ⌜birds of a feather⌝.

**kelâm** (·ˊ) Word; speech; sentence; language. **ilmi ~,** the study of the Koran, theology. **~ıkadim** (·—··ˊ), the Koran. **~ullah** (·—·ˋ), the Word of God, the Koran.

**kelebek** Butterfly; gaily dressed girl; disease affecting the liver of sheep, fluke. **~ gözlük,** pince-nez.

**kelek**[1] Raft *made of inflated sheep-skins.*

**kelek**[2] Unripe melon. Partly bald; not properly developed, immature.

**kelem** Cabbage; cabbage stew.

**kelepçe** (·ˋ·) Handcuffs. **···e ~ vurmak,** to handcuff. **~li,** handcuffed.

**kelepir** Something acquired for nothing *or* very cheaply; bargain; bad bargain, useless thing; step-child. **~ci,** bargain hunter; opportunist.

**kelepser** Martingale.

**keler** Lizard; reptile. **~balığı, -nı,** monk fish. **~derisi, -ni,** sharkskin (*used as sandpaper*); shagreen.

**keleş** Ringwormy; bald; dirty.

**kelime** Word (*pl.* **kelimat**).

**kelle** Head; sheep's head; sugar loaf; cake *of cheese.* ⌜**~ götürür gibi**⌝, with unnecessary haste and fuss: **~sini**

**koltuğuna almak,** to take one's life in one's hands: ⌜~ **kulak yerinde**⌝, *phrase used to describe an exceptionally well-built man or a rather ostentatiously wealthy man,* 'ah! there's not much wrong with him!': **~yi vermek,** to lose one's head (be killed): ⌜**bu iş böyle değilse ~mi keserim**⌝, 'if this is not so, I'll eat my hat': **pişmiş ~ gibi sırıtmak,** to grin foolishly.

**kelli**[1] (*dial.*) Since, because; as.

**kelli**[2] Affected with ringworm.

**kellifelli** (·\`··) Well-dressed; serious, dignified; showy.

**kellik** Baldness; bare waste land.

**kellim** *Only in* ⌜~ ~ **lâ yenfa**⌝, I kept on saying it, but it was no use.

**keloğlan** (\`··) *A popular hero of Turkish folk-tales, who starts as an unknown and poor boy, but, thanks to his talents, eventually achieves success.*

**kelpe** Vine prop.

**kelpeten** *v.* **kerpeten.**

**kem** Few; deficient; bad ~ **göz (nazar),** the evil eye: **~nazarla bakmak,** to look at *s.o.* with evil intentions.

**kemal** (·\`), **–li** Perfection; maturity; cultural attainment; moral quality; worth, value, price; the most that can be said of a thing *or* a person. **~i beş lira,** five pounds at the most: **~ini bulmak,** to attain to its perfection: **~e ermek,** to reach perfection, to attain maturity: **~i ihtiram,** all honour and respect: **~i muhabbet,** sincere affection: '**~i ne?**', 'well, after all it's not a great expense!'.

**keman** (·\`) Bow (archery); violin. ~ **kaşlı,** with arched eyebrows. **~cı,** maker of violins; violinist. **~e** (·–\`), bow *for violin.* **~î** (·–\`), violinist (*in Oriental music*). **~keş,** archer.

**kemençe** (·\`·) Small violin with three strings; instrument for sowing artificial manures.

**kemend** Rope with a noose at the end; lasso; halter; snare; (*poet.*) tress of hair.

**kemer** Belt; girdle; arch. **su ~i,** aqueduct. **~altı, –nı,** vaulted bazaar. **~e,** beam under a ship's deck. **~li,** belted, girdled; arched, vaulted. **~lik,** leather belt *used by lemonade-sellers etc. for holding glasses.* **~taşı, –nı,** keystone *of an arch.*

**kemik** Bone. ~ **atmak,** to throw a bone *to a dog*; to appease by a favour: **kemiği* çıkmak,** to have a bone dislocated: ~ **gibi,** as hard as bone: ~ **hastalığı,** rickets: **kemiğe işlemek,** (of cold) to penetrate to the very marrow: ~ **kapmak,** to 'get *stg.* out of it' (*rather contemptuously*): **bir deri bir kemik,** mere skin and bone: ⌜**dilin kemiği yok**⌝, one can never be sure that people won't blab. **~li,** having bones, bony; strongly built.

**kemirici** Rodent; corrosive.

**kemirmek** *va.* Gnaw; nibble; corrode. **içi* içini*** ~, to be consumed by anxiety *etc.*

**kemiyet, –ti** Quantity; number (*gram.*). ~ **ve keyfiyet,** quantity and quality.

**kemküm** Hesitatingly; confusedly (of speech). ~ **etm.,** to hum and haw.

**kem·lik** Evil; malice.

**kemmî** (·–\`) Quantitative.

**kem·niyet** (\`··), **–ti** Evil design. **~söz** (\`·), evil *or* inauspicious language.

**kemter** *comp. of* **kem.** Less; inferior; worse; (*modestly*) I. **~leri,** your humble servant.

**–ken** *v.* **iken.**

**kenar** Edge, border; bank, shore; marginal note; retired place, corner; suburb. Out of the way, remote. **~a atmak,** to put aside: **~a çekilmek,** to get out of the way; to retire apart: ~ **dilberi,** a suburban beauty: **~da kalmak,** to remain in an inferior job: **~da köşede,** in unlikely spots: **~da oturmak,** to sit apart.

**kenar·cı,** shore fisherman. **~lı,** having an edge *or* margin.

**kende** Dug; excavated; engraved. Ditch, moat, trench.

**kendi** Self. **~m,** myself: **~si,** himself: **~niz,** yourselves: **~ler,** themselves: ~ **~me,** to myself; by myself: ~ **~sine,** to himself; by himself; all alone: **~nden** *or* **~sinden,** from itself; automatically; naturally: ~ **evim,** my own house: ~ **kitabı,** his own book: **~ni* beğenmek,** to think a lot of oneself, to be conceited: **~ni* bilmek,** to be conscious; to have self-respect; **~sini bilmez,** presumptuous; one who does not know his place; unconscious: **~ne* etm.,** to harm oneself; ⌜**ne ettiyse ~ne etti**⌝, in doing this he only harmed himself: ~ **sinden* geçmek,** to lose consciousness; to be in ecstasy; to become slack: **~ne gel!,** pull yourself together!: ~ **gelen,** a godsend: ~

halinde, occupied by his own thoughts; quiet, inoffensive.
**kendiişler** Automatic.
**kendilik** One's own personality; initiative. **kendiliğinden yapmak,** to do *stg.* of one's own accord *or* by oneself.
**kendir** Hemp.
**kene** Tick. **~göz** (·⅄·), small-eyed. **~otu, -nu,** castor-oil plant.
**kenet** Metal clamp *for holding together masonry or cracked earthenware.* **~lemek,** to clamp together; bind tightly. **~li,** clamped together; closely united.
**kenevir** Hemp; hempseed.
**kent** Fort; town.
**kental** Quintal.
**kepaze** (·–⅄) Vile; contemptible; scoffed at. Ridiculous *or* contemptible person. **~lik,** vileness; degradation; ignominy.
**kepbastı** A kind of net *used in fishponds.*
**kepçe** Skimmer; ladle; landing-net; butterfly-net. ⌜**İstanbul kazan, ben ~ aradım**⌝, I have thoroughly searched Istanbul. **~kulak,** having large prominent ears. **~kuyruk** (*sl.*), sponger. **~surat,** having a small face.
**kepek** Bran; scurf; dandruff.
**kepenek** Moth; coarse felt cape.
**kepenk** Large pull-down shutter; wooden cover.
**kepir** Barren, bare.
**kerahat** (·–⅄), **-tı** A being abominable; repugnance, aversion; a lawful but blameworthy act. **~ etm.,** to abominate, detest: **vakti ~,** drinking-time.
**kerake** (·–⅄) A kind of light cloak. ⌜**anlaşıldı Vehbinin kerrakesi**⌝, now all is clear!: **zırdeli mor ~,** completely mad.
**keramet** (·–⅄), **-ti** Miracle; a word *or* deed so opportune that it appears to be divinely inspired. **~ buyurdunuz!,** your words are wonderful! (*formerly used in flattery, now only ironically*): **~ göstermek,** to work a miracle: **~ sahibi,** prophet; miracle-worker: ⌜**~im yok ya!**⌝, 'how could I possibly know!': ⌜**şeyhin ~i kendinden menkul**⌝, 'the sheikh's miracles are related by himself', *i.e.* I want further confirmation before I believe it.
**kerata** Shoehorn; cuckold; pander; scoundrel; *also used affectionately.* '**bak ~ya!**', 'look at the little rascal!' **~lık,** quality of a pander *etc.*; villainy.
**keraviye** (··⅄·) Caraway.
**Kerbelâ** The town Kerbela in Irak. **~ sıkıntısı,** lack of water.
**kere** A time; bracket, parenthesis. **~ içinde,** in brackets: **bir ~,** once; just; let it be said that ...; for one thing ...; to begin with ...; **bir ~ daha,** once again, once more: **bu ~,** this time; now; recently: **kaç? ~,** how often?.
**kerem** Nobility, magnanimity; munificence; kindness; favour. **~ buyurun,** be so kind!; I beg of you.
**kerempe** Rocky spit *running out into the sea.*
**keres** Large bowl.
**kereste** (·⅄·) Timber used for building; any kind of material. **~si kavi** *or* **~li,** strongly built (man).
**kerevet**[1], **-ti** Wooden bedstead; couch.
**kerevet**[2], **-ti** (*Fr. crevette*) Prawn.
**kerevid, kerevides** Crayfish.
**kereviz** Celery.
**kerhane** (·–⅄) Brothel. **~ci,** brothel-keeper, pimp.
**kerhen** (⅄·) With repugnance; against one's will.
**kerih** Disgusting; detestable. **~e** (·–⅄), disgusting *or* abominable thing.
**kerim** Noble; generous; honoured, illustrious. ⌜**Allah~!**⌝, 'never mind, it will come all right!' **~e** (·–⅄), daughter (*as a term of respect*).
**keriz**[1] Drain.
**keriz**[2] *In* **~ alayı,** a troop of gipsy musicians.
**kerkenes** Hawk, (?) kestrel.
**kerki** Large axe.
**kerliferli** *v.* **kellifelli.**
**kerpeten** Pincers.
**kerpic** Sun-dried brick. Made of sun-dried bricks; hard; dry. **~ kesilmek,** to be petrified *with fear etc.*
**kerrake** (·–⅄) *v.* **kerake.**
**kerrat** (·⅄), **-tı** *pl. of* **kere.** Times. **~ cedveli,** multiplication table. **~la,** repeatedly.
**kerte** Rhumb; one of the 32 points of the compass; point; degree; mark, sign; best state *or* quality, right moment. **~sini almak,** to take the bearing *of an object*: **~sini geçmek,** to pass the exact degree, to be overdone: **~sine gelmek,** to come to the right degree, to the point of perfection; **~ye gelmek,** to come to such a point

*or* degree that ...: **o** ~ (*sl.*), at that moment.

**kertenkele** (·ˈ··), **kertenkeler** Lizard.

**kerteriz** Bearing *of an object or star.*

**kerti** *v.* **kerte.**

**kertik** Notch; gash; tally; fraction *of a pound etc.* Notched. **~li,** giving fractions, *i.e. the exact sum, not just a round figure.*

**kertmek** *va.* Notch; scratch; scrape against.

**kerubi** (·—ˈ) Cherub. **~yun,** Cherubim.

**kervan** Caravan. **~a katılmak,** to join the procession, to go with the rest. **~başı** (·ˈ··), **–nı,** *or* **~cı,** leader of a caravan. **~kıran,** the planet Venus *when a morning star.* **~saray** (···ˈ), caravanserai; inn with a large courtyard.

**kerye** Iron hoop *tightened by bolts*; connecting part of a handcuff.

**kes** Person, individual.

**kesad** A not being saleable; dullness (of a market). **~lık,** condition of dullness in a market; time of scarcity *or* unemployment.

**kesafet** (·—ˈ), **–ti** Density; thickness, opacity; coarseness.

**kesb** Acquisition; gain; earning. **~etmek** (ˈ··), **–eder,** to earn; gain; acquire.

**kese** Purse; small bag; case; cyst; hair-glove *for rubbing the body*; *formerly* a sum of 500 piastres; the power of the purse, wealth. **~den eklemek,** to be out of pocket: **~sine* güvenmek,** to be able to afford: **~ sürmek,** to rub the body with a hair-glove (*as in a Turkish bath*).

**kese·dar,** Keeper of the Purse; treasurer. **~kâğıdı, –nı,** paper bag. **~lemek,** to rub the body with a hair-glove. **~li,** having a bag *etc.*: **~ kurd,** tapeworm cyst.

**kesek** Clod; a turf; a turf of peat *for fuel.*

**kesel** Slackness; indolence.

**kesenkes** (·ˈ·) Decisive; categorical; definite.

**keser** Adze.

**kesici** That *or* who cuts; slaughterman. **~ diş,** incisor tooth: **yol ~,** highwayman.

**kesif** Dense; thick; opaque; thickly populated.

**kesik** Cut; broken; spoilt; castrated; curdled; weary. Skim-milk cheese. **~ kalmak,** to be interrupted *or* abruptly ended: **~ sulama usulü,** system of supplying irrigation for certain periods only. **~lik,** state of being cut *or* broken; lassitude.

**kesilmek** *vn. pass of* **kesmek.** Be cut *etc.*; be cut off; cease; be exhausted; be turned into, become; pretend to be; be curdled. **···den ~,** to cease from, to be unable to do any longer: **ayaklarım buz kesildi,** my feet are like ice: **dizlerim kesiliyor,** my knees are giving way: **göz ~,** to be 'all eyes': **kuvvetten ~,** to be exhausted: **nefesi ~,** to be out of breath: **taş ~,** to become as hard as stone; to be petrified: **ümid kesilecek bir hal,** a hopeless state: **yağmur kesildi,** the rain has stopped: **yemekten içmekten ~,** to lose one's appetite.

**kesim** Act of cutting; slaughter (of an animal); cut, shape, form; make; fashion; abstention; vacation; agreed price *or* rent. **~ vakti,** agreed time of payment: **~e vermek,** to put up a farm for rent: **su ~i,** waterline *of a ship.* **~ci,** contractor who undertakes to farm a branch of the revenue.

**kesin** Definite; certain. **~ olarak,** for certain, certainly.

**kesinti** Clipping; cutting; chip; deduction *from a sum.* **~ye almak** (*sl.*), secretly to make fun of *s.o.*

**kesir, –sri** Fraction.

**kesişmek** *va.* Conclude *an agreement*; settle *an account*; draw *a game*; fix *a price.* **söz ~,** to come to an agreement.

**kesitaşı** (·ˈ··), **–nı** A flat stone *at a river's edge upon which washerwomen beat out their washing.*

**kesken** Small rodent; cockchafer grub.

**keski** Bill-hook; coulter. **~ kalemi,** cold chisel; parting-tool (lathe).

**keskin** Sharp; keen; pungent; severe; decided; peremptory. Edge *of a cutting instrument.* **~lik,** sharpness; pungency; incisiveness; shrewdness; cutting edge. **keskinliğine koymak,** to set edgewise.

**kesme** Cut; that can be cut; decided, definite. Shears; Turkish delight; a kind of flat macaroni; soft rock; apostrophe; a kind of openwork embroidery. **~ şeker,** lump sugar. **~ce,** with the right to cut for examination (*in buying a water-melon etc.*); *bought* in a lump lot; for a lump sum.

**kesmek** *va.* Cut; cut off; interrupt; intercept; cut down, diminish; determine, decide, agree upon; cut the throat of, kill; castrate; issue (a warrant). *vn.* Cut well, be sharp; cost; (*sl.*) talk boringly; exaggerate. **kesip atmak,** to destroy root and branch; settle offhand; settle once and for all, to be dogmatic: **aklı* ~,** to understand; realize; think possible: **ağrıyı ~,** to stop pain: ⌜**başını kes!**⌝, 'duck your head!': **···den elini ~,** to cease from doing *stg.*: **gözü* ~,** to deem possible; **gözüm kesmiyor,** I dare not: **hesabı ~,** to settle an account; to cut off relations: **kısa ~,** to be brief: **···in önünü ~,** to bar the way of; to prevent: **paha ~,** to fix a price: **para (akçe) ~,** to mint money: ⌜**sesini kes!**⌝ 'shut up!': **söz ~,** to arrange a marriage: **sütten ~,** to wean: **ümidini* ~,** to despair: **yol ~,** to hold up, to rob; to bar the way: **yolunu ~,** to stop *s.o.*; (of a ship) to diminish speed.

**kesmelik** Quarry.

**kesmez** Blunt.

**kesmik** Chaff mixed with broken straw.

**kesret, -ti** Multitude; great quantity.

**kesretmek** (ʹ··), **-eder** *va.* Break; subdue, defeat; abate.

**kestane** (·–ʹ) Chestnut. **~ dorusu,** chestnut-bay colour: **~ kargası,** jay: ⌜**~ kabuğundan çıkmış da kabuğunu beğenmemiş**⌝, 'the chestnut emerged from its shell and did not like the look of it', *said of one who is ashamed of his origin*: **deniz ~si,** sea-urchin. **~lik,** chestnut grove.

**kestirme** *verb. n. of* **kestirmek.** Snooze. Definite, decisive; approximate. **~den gitmek,** not to beat about the bush: **~ yol,** short cut.

**kestirmek** *va. caus. of* **kesmek.** Cause to cut *etc.*; shorten; cause to cease; appreciate; estimate; cause to curdle, turn *milk etc.* sour; decide; perceive; clearly understand; take a bearing of (*mil.*). *vn.* Have a snooze. **kestirip atmak,** to destroy utterly: **gözüne* ~,** *v.* **göz**: **oğlunu*~,** to have one's son circumcised.

**keş[1]** Skim-milk cheese; cheese made from yoghourt. ⌜**ağzının tadını bilen ~ yer**⌝, ''tis caviare to the general', people like you (him) can hardly be expected to appreciate so good a thing.

**keş[2]** Gullible; foolish.

**keşf·etmek** (ʹ··), **-eder** *va.* Uncover; discover; reconnoitre; examine carefully and estimate cost *or* value; guess. **~i,** *v.* **keşif. ~iyat** (··ʹ), **-tı,** discoveries.

**keşide** (·–ʹ) A drawing in a lottery. **~ etm.,** to draw (a lot in a lottery).

**keşif, -şfi** Discovery; scrutiny; investigation, valuation; estimate of cost; reconnaissance; forecasting, divination. **~ kolu,** reconnoitring patrol. **~name** (··–ʹ), written estimate of costs.

**keşiş** Christian priest *or* monk. **~hane** (··–ʹ), monastery; convent.

**Keşişdağı** (·ʹ··), **-nı** *The old name of* Mount Olympus (near Bursa). **keşişleme,** South-east wind (*as blowing from the Keşişdağı to Istanbul*); sirocco (*in the Mediterranean*).

**keşke, keşki** Would that ...! **~ bilsem,** would that I knew!; **~ bilseydim,** would that I had known!

**keşkek** Dish of wheat boiled with minced meat.

**keşkül** Beggar's bowl; *abb. for* **~fukara,** a sweet made of milk and pistachio nuts.

**keşmekeş** Great confusion.

**ket[1], -ti** Obstacle. **···e ~ vurmak,** to stand in the way of ... .

**ket[2], -ti** Starch. **~al,** starched and glazed cotton *or* linen stuff.

**kete** Cake made of rice-flour.

**ketebe[1]** *pl. of* **kâtib.** Writers; clerks.

**ketebe[2]** Diploma for excellence in calligraphy; inscription at the end of a book (*taking the place of the modern title-page*).

**keten** Flax; linen. **~ helvası,** sweetmeat made of honey and oil: **~ kuşu,** linnet: **~ tohumu,** linseed. **~cik,** grass-wrack (seaweed).

**ketim, -tmi** A concealing *or* keeping secret. **ketmetmek** (ʹ··), **-eder,** to conceal; not to divulge; dissimulate.

**ketum** (·ʹ) Discreet; reticent; keeping a secret. **~iyet, -ti** *or* **~luk,** discretion; reticence; keeping one's mouth shut.

**kevgir** Perforated skimmer.

**kevser** A river in Paradise; nectar.

**keyfemayeşa** (ʹ·–·–) As it pleases your fancy; arbitrarily; just anyhow.

**keyf·etmek** (ʹ··), **-eder** *vn.* Amuse oneself; enjoy oneself. **~i,** *v.* **keyif. ~î** (·ʹ), arbitrary; capricious.

**keyfiyet, -ti** Condition; circumstance; affair. ⌜~ böyle böyle!⌝, 'well, that's how the matter stands!'

**keyif, -yfi** Health; bodily and mental condition; merriment, fun, good spirits; pleasure, amusement; inclination, whim, fancy; slight intoxication. Hilarious; tipsy. **keyfince,** as he pleases, arbitrarily: ⌜**keyfin bilir**⌝, 'well, as you please!': ~ **çatmak,** to make merry: ~ **etm.,** to amuse *or* enjoy oneself: **keyfi* gelmek.,** to be delighted, to feel in a good humour: ~ **halinde,** intoxicated: **keyfim iyi değil,** I am not feeling very well: **keyfi kaçtı,** he is out of spirits: **keyfiniz nasıl?,** how are you?: ~ **olmak,** to be tipsy: ~**i sıra,** arbitrarily: ~ **sormak,** to inquire after *s.o.'s* health, to say 'how are you?': ~ **vermek,** to intoxicate: **keyfim yerinde,** I am well: ~ **yetiştirmek,** to make merry with drink: **ehli ~,** sensualist: **gel keyfim gel,** *expression of comfort and contentment.*

**keyif·lenmek,** to enjoy oneself; be tipsy. ~**li,** merry; happy; tipsy. ~**siz,** indisposed; gloomy, not cheerful. ~**sizlik,** indisposition; ailment; depression.

**kez** Time. **bu ~,** this time: **her ~,** always.

**keza** Thus; in like manner; too. ~ **ve ~,** and so forth.

**kıbal, -li** Shape; manner; style.

**kıble** *Direction to which a Moslem turns when praying, i.e. towards Mecca*; south; south wind; place *or* person towards which *or* whom everyone turns. ~**nüma,** compass which indicates the direction of the **kıble**; any compass.

**Kıbrıs** Cyprus. ~**lı,** Cypriot. ···**taşı, -nı,** crystal cut like a diamond; paste.

**kıcı** Any bitter pungent herb.

**kıç, -çı** Hinder part; behind; buttocks, rump; stern *of a ship.* ~ **atmak,** (of a horse) to lash out with both feet: ···**e ~ attırmak** (*sl.*), to get the better of *s.o.*: ~**tan kara etm.,** to moor by the stern: ~ **üstü oturmak,** to remain helpless: **baş ~ yok,** there is neither leader nor led. ~**ın ~ın,** backwards; astern. ~**lı,** *ship* down by the stern.

**kıdem** Antiquity; priority; precedence; seniority. ~**en** (·˙·), by right of priority *or* seniority. ~**li,** senior; earliest. ~**siz,** without seniority; junior. ~**sizlik,** juniority.

**kıkır** *In ~ ~,* giggling. ~**tı,** giggling.

**kıkır·dak** Cartilage; gristle; crackling. ~**damak,** to make a crackling noise; (*sl.*) to die.

**kıl** Hair; bristle; hair's-breadth. Made of hair. ~ **çadır,** hair tent: ~**ına bile dokunulmaz,** a hair of whose head must not be touched, sacrosanct: ~ **kadar,** the smallest degree *or* quantity: ···**e ~ kaldı,** he (it) was within a hair's-breadth of ... : ~ **kalem,** camelhair brush: ~**ını kıpırdatmadan,** without turning a hair: ~**ı kırk yarmak,** to split hairs; ~**ı kırk yaran,** too meticulous, hair-splitting: ~ **şaşmadan,** with scrupulous care.

**kılağı** Wire-edge. ~**sını almak,** *or* ~**lamak,** to put a fine edge *on a tool.*

**kılaptan** Gilt-copper wire; gold wire wound on silk; trimming of false gold thread. ~ **işleme,** worked with imitation gold thread.

**kılavuz** Guide; pilot; go-between *in arranging a marriage*; leader of a file of animals (*usually a donkey*); gimlet point of an auger; screw-tap; **adit** (*mining*); corncrake. ⌜~**u karga olanın (burnu boktan ayrılmaz)**⌝, 'who takes a crow for his guide will never have his nose far from dung'; if you take bad advice you will regret it. ~**luk,** profession of guide *or* pilot.

**kılbarak** A kind of Shetland pony *bred in parts of Turkey.*

**kılburun** Narrow promontory.

**kılçık** Fish-bone; awn; string *of a bean-pod.*

**kılefte** (*sl.*) Theft.

**kılıbık** Man ruled by his wife; henpecked man.

**kılıc** Sword. Curved. ~**ına,** on edge: ~**dan geçirmek,** to put to the sword: ~**ının hakkı olarak,** by right of conquest: ~ **oynatmak,** to be the ruler, to dominate. ~**bacak,** bandylegged. ~**balığı, -nı,** swordfish. ~**lama,** edgewise; set on edge; slung from the shoulder; crosswise.

**kılıf** Case; sheath. ⌜**minareyi çalan ~ını hazırlar**⌝, 'he who steals a minaret will prepare a case for it'; he who undertakes a risky business will take care to provide for the consequences; the thief will take care to cover up his tracks.

**kılık** Shape; appearance; cut; costume; aspect. ~ **kıyafet,** one's dress: ⌜~ **kıyafet köpeklere ziyafet**⌝, very dirty and untidy. ~**lı,** having *such* *and such* an appearance *or* dress; well-shaped; well-dressed. **çoban ~,** dressed like a shepherd. **~sız,** deformed; ugly; badly dressed.

**kılınmak** *vn. pass. of* **kılmak.** Be done *or* performed (*used as an auxiliary verb in place of* **edilmek** *in formal or official language*).

**kılkıran** Alopecia; baldness.

**kılkuyruk** Pintail duck; tatterdemalion; shifty-looking fellow.

**kıllanmak** *vn.* Become hairy; (of a youth) to begin to show a beard.

**kıllı** Hairy.

**kılmak** Do; perform. **namaz ~,** to perform the service of worship of Islam.

**kılsız** Hairless; beardless.

**kılükal** (−·⌐), **–li** Tittle-tattle; gossip.

**kılyakı** (⌐··) Seton.

**kımıl** An insect pest of cereals (*Aelia*).

**kımılda·mak, ~nmak** *vn.* Move. **~tmak** *va.* Move; shake.

**kımız** Koumiss (fermented mare's milk).

**kın** Sheath (of a sword *etc.*).

**kına** Henna. **~ yakmak,** to apply henna *to a part to be dyed*; to be overjoyed. **~cık,** rust *of plants.* **~çiçeği** (·⌐···), **–ni,** a species of balsam. **~gecesi, –ni,** the night two days before a wedding (*observed as a night of entertainment and on which, formerly, the bride had her fingers and toes freshly dyed with henna*). **~lı,** dyed with henna: **~** *or* **~ keklik,** the Greek partridge *or* chukor.

**kınak** Claw; finger-joint.

**kınakına** (··⌐·) Cinchona bark.

**kınamak** *va.* Reproach; make sarcastic remarks to; taunt, mock.

**kınap, kınnap** Yarn; twine.

**kıp kıp, kıpır kıpır** *Expresses constant rapid movement.*

**kıpık** Half-closed eye; winking.

**kıpır** *v.* **kıp. ~damak, ~danmak,** to move slightly; start; quiver; vibrate. **~datmak,** to cause to start *or* move quickly; agitate: ⌜**işleri biraz kıpırdattık**⌝, 'we've got a move on'. **~dı, ~tı,** slight quick movement; start; quiver.

**kıpıştırmak** *va. Only in* **göz ~,** to wink.

**kıp·kırmızı, ~kızıl** Bright red; very red.

**kıpma** Wink; twinkling of an eye.

**kıpmak** *va.* Wink *or* blink *the eye. v. also* **kırpmak.**

**Kıpt, Kıptı** Copt; Coptic; gipsy. **~ça,** the Coptic language.

**kır**[1] Country (*as opp. to town*); uncultivated land; wilderness. **~ koşusu,** cross-country race.

**kır**[2] Grey. Greyness; grey horse.

**kıraat, –ti** Reading; reading-lesson. **~ kitabı,** reading-book (for teaching). **~hane** (·—·—⌐), public reading-room; coffee-house where newspapers are kept.

**kırac** Uncultivated land. Parched; sterile.

**kıracak** Nut-crackers.

**kırağı** Hoar-frost.

**kıral** (·⌐), **kral** King. **~î** (·—⌐), royal. **~içe,** queen. **~iyet** (·—·⌐), **–ti, ~lık,** kingdom; kingship.

**kıran**[1] Edge; shore; horizon.

**kıran**[2] Breaking; destructive. Epidemic, murrain. *As suffix,* breaking-, destroying-, *e.g.* **dalgakıran,** breakwater. **~ girmek,** for an epidemic to break out; for some kind of plague *or* disease to cause damage: ... **kıtlığına ~ girmedi,** there is no shortage of ...

**kırân** Conjunction of planets. **sahib ~,** 'lord of a fortunate conjunction', *title given to a victorious monarch.*

**kıranta** Man whose hair is beginning to turn grey. Grizzled.

**kırasıya** Bone-breaking; violent.

**kırat, –tı** Carat; quality; value.

**kırba** Water-skin; leather bottle; wine-bibber.

**kırbaç** Whip; scourge. **~lamak,** to whip.

**kırçıl** Sprinkled with grey.

**kırçıla** Marline.

**kırdonlu** Having a grey coat (horse).

**kırgın**[1] Disappointed, hurt.

**kırgın**[2], **kırçın** Murrain.

**kırık** Broken; cracked; milder (weather). Break, fracture; fragment, splinter. **~ dökük,** metal scrap *etc.*; broken-down; odds and ends of; broken (French *etc.*): **vücudüm ~,** I am not feeling up to much: **~ numara almak,** to get bad marks: **~tahtası,** splint. **~çı,** bone-setter. **~lık,** state of being broken; physical weariness *or* weakness: **hayal kırıklığı,** disappointment.

**kırılmak** *vn. pass. of* **kırmak.** Break, be broken; be ruined *or* killed; suffer heavy casualties; be hurt *or* offended; die of laughter; be mitigated, become milder. **kırılıp dökülmek,** to be constantly broken; (of a woman) to be coquettish: **beli*** ~, to be exhausted; to be discouraged: **burnunun* direği** ~, to be nearly knocked down by a stench: **bir işe eli** ~, to be an adept at a thing: **gülmeden** ~, to have one's sides ache with laughter: **sular kırıldı,** the weather is warmer.

**Kırım**[1] The Crimea.

**kırım**[2] Wholesale slaughter; crease *in clothes.*

**kırıntı** Fragment; crumb; crushed stone; fissure.

**kırışık** Wrinkle; crack, small fissure.

**kırış·mak** *vn.* Become wrinkled; kill one another; mutually break. **~tırmak** (*sl.*) (of a woman) to 'carry on' with a man.

**kırıt·ış** Coquetry. **~kan,** coquettish; flighty. **~mak,** to behave in a coquettish manner.

**kırk, -kı** Forty; *used especially to denote a large indefinite number.* ~ **anahtarlı,** very wealthy; a man of property: **~ı çıkmak,** (of a woman) to have completed forty days after childbirth: ~ **defa,** countless times: ~ **ikindi,** the rainy season (*in parts of Anatolia*): ⌜**~ından sonra saz çalmak**⌝, to start doing *stg.* later in life than usual: ~ **yılda bir,** 'once in a blue moon', on very rare occasions: ⌜**biz ~ kişiyiz, birbirimizi biliriz**⌝, we know too much about him (you) to be taken in.

**kırk·ambar** general store; general dealer; person of encyclopedic knowledge; ship carrying a large and varied cargo. **~ar,** forty each; forty at a time. **~ayak,** centipede; crab-louse. **~bayır,** third stomach of ruminants. **~geçid,** very winding river.

**kırkı** Shears *for shearing sheep.* **~cı,** sheep-shearer. **~m,** the shearing season; the clip *of wool.*

**kırk·ıncı** Fortieth. **~lamak,** to make forty; to do *stg.* forty times; complete forty days *after an event*: **hastayı** ~, to cure a sick person by spells. **~lar,** the Forty Saints of Islam: **~a karışmak,** to disappear. **~lı,** having forty parts; being born within forty days of *s.o.* else: **onunla ~dır,** he was born soon after him. **~lık,** forty years old.

**kırkma** Shearing; hair cut so as to cover the forehead.

**kırkmak** *va.* Shear; clip.

**kırlağan** The bubo of plague; plague.

**kırlangıc** Swallow; martin; gurnard; (*formerly*) a light swift galley. ~ **dönümü,** the time of the swallows' southward migration, early October: ~ **fırtınası,** storms occurring about the end of March: **dağ ~ı,** sand-martin: **kılıc ~ı,** the swift: **pencere ~ı,** house martin. **~otu, -nu,** the greater celandine; turmeric.

**kırma** Pleat; fold; a variety of handwriting; crushed barley; half-breed. Broken; folding (gun *etc.*). **~cı,** seller of small grains; corn-merchant; folder (bookbinding).

**kırmak** *va.* Break; split; kibble (corn); kill, destroy; offend; fold, crease; discount (bill); change (money); lower (price); mitigate, abate (cold *etc.*). **kırıp dökmek** to destroy; to keep on breaking: **kırıp geçirmek,** to destroy; to tyrannize: **dümeni sola (sağa)** ~, to steer to port (starboard); **dümeni** ~ (*sl.*), to 'make off': **koz (pot)** ~, to make a faux pas, to 'drop a brick': **maaş** ~, to borrow money on salary due but not paid: **mezadda bir şeyi** ~, to make a successful bid for a thing at an auction: **numarasını** ~, to cut off *a schoolchild's* marks, to give bad marks to him: **para** ~, to coin money (grow rich).

**kırmalı** Pleated; creased.

**kırmız** Cochineal. ~ **madeni,** kermes mineral (trisulphide of antimony).

**kırmızı** Red. **~msı, ~mtrak,** reddish. **~lık,** redness.

**kırp·ık** Clipped. **~ıntı,** clippings.

**kırpıştırmak** *va.* Blink *the eyes.*

**kırpmak** *va.* Clip; trim; shear; cut down (expenses); wink. **gözünü kırpmadan,** without batting an eyelid, without turning a hair: ⌜**ben senin karanlıkta göz kırptığını ne bileyim?**⌝, how was I to know?

**kırtas·î** (·–́) Pertaining to stationery. ~ **işler,** office work. **~iye** (·–·́), office expenses; stationery; 'red tape'. **~iyeci,** stationer; bureaucrat; one given to 'red tape'. **~iyecilik,** trade of a stationer; bureaucracy, 'red tape'.

**kırtipil** Wearing old clothes; common; insignificant.

**kıs kıs** *Imitates the sound of suppressed laughter.* ~ ~ **gülmek,** to laugh under one's breath; to snigger.

**kısa** Short. ~ **geçmek,** to refer

briefly to a subject: ⌜~ günün kârı⌝, (*iron.*) it's better than nothing; what more did you expect?: **aklı ~**, of limited intelligence: **sözün ~sı**, in short. **~ca** (·ˈ·), shortly, briefly. **~cık**, very short. **~lık**, shortness.

**kısac** Pincers; pliers; claw *of a crab.*

**kısal·mak** *vn.* Become short; shrink. **~tma**, abbreviation.

**kısas** (·ˈ) Retaliation; lex talionis. **~a ~**, an eye for an eye: **~ etm.**, to put to death for manslaughter.

**kısık** Pinched; squeezed up; screwed up (eyes); hoarse *or* choked (voice). Groin. **~lı**, flowing in a mere trickle.

**kısılmak** *vn. pass. of* **kısmak.** Be pinched; be squeezed; be in difficulties; (of voice) to become hoarse.

**kısım**[1], **–smı** Part; portion; piece; kind, sort. **Arab kısmı**, the Arabs: **kadın kısmı**, the female sex, womenfolk.

**kısım**[2] Handful.

**kısır** Barren, sterile. **~lık**, sterility.

**kısırğanmak** *va.* **birini bir şeyden ~**, to grudge someone something.

**kıskac** Pair of folding steps; vice; pincers.

**kıskanc** Jealous; envious. **~lık**, jealousy; envy.

**kıskanmak** *va.* Envy; be jealous of. **gözünden ~**, to be very jealous of, to regard as the apple of one's eye.

**kıskıvrak** (ˈ··) Tightly bound *or* squeezed; tightly coiled up; neat and tidy. **~ yakalamak**, to catch so that escape is impossible.

**kısma** *verb. n. of* **kısmak**, **istim ~ cihazı**, throttle.

**kısmak** *va.* Squeeze; tighten; pinch; cut down, diminish; be stingy with; (of a horse *or* dog) to put the tail between the legs. **dilini* kısıp oturmak**, to remain silent: **kulak ~**, (of an animal) to fold back the ears.

**kısmen** (ˈ·) Partly, partially.

**kısmet, –ti** Destiny; lot; fate; luck. **~ !**, perhaps!, it may be!: **~i açık**, fortunate; (of a girl) sought after: **~i* açılmak**, to be successful, to be in luck; **~i* ···den açılmak**, to have success in ...; **~i* ···den açılmamak**, to have the bad luck to ...: **~ ise**, if fate so decrees: **~ olmamak**, not to be possible; not to succeed: **kızın ~i çıktı**, the girl's luck has come, *i.e. s.o.* has asked for her hand. **~li**, fortunate, lucky.

**kısm·i** *v.* kısım. **~î**, partial.

**kısrak** Mare. **kız~**, womenfolk.

**kıssa** Story, tale, fable. **~dan hisse**, the moral of a tale. **~han**, storyteller.

**kıstas** Large pair of scales; criterion.

**kıstelyevm** Deduction from wages *for absence or lateness.*

**kıstırmak** *va. caus. of* **kısmak.** Cause to be pinched *etc.*; crush (a finger); catch by driving into a narrow place; corner.

**kış**[1] Winter; winter cold. **~ı etm.**, to reach the winter; to stay *somewhere* till winter comes: **~ta kıyamette**, in the depth of winter: **kara ~**, the depth of winter: ⌜**Allah dağına göre verir ~ı**⌝, 'God tempers the wind to the shorn lamb'.

**kış**[2] *Noise made to scare away birds etc.* **~alamak**, to shoo away *birds etc.*

**kışın** (ˈ·) In the winter.

**kışkırt·mak** *va.* Incite; excite. **~ı**, incitement. **~ıcı**, provocative; inciting; inciter.

**kışla** (ˈ·) Barracks.

**kış·lak** Winter quarters for animals, nomads *or* an army. **~lamak**, to become wintry *or* cold; to pass the winter. **~lık**, suitable for the winter: winter residence.

**kışt** *v.* **kış**[2].

**kıt, –tı** Little; few; scarce; deficient; rarely. **~ı ~ına hesablamak**, to cut it fine: **~ı ~ına idare etm.**, just to be able to make both ends meet: **~ kanaat**, *v.* **kıtkanaat**: **~ı ~ına** *or* **~a ~ yetişmek**, to be barely sufficient.

**kıt'a** Portion; piece; continent; district; detachment *of troops*; segment; size, dimension. **üç ~ gemi**, three ships. **~at, –tı**, *pl. of* **kıt'a.**

**kıtal**[1] (·ˈ), **–li** A killing; battle; massacre.

**kıtal**[2] A kind of landing-net for catching fish.

**kıtık** Refuse of flax; tow; stuffing *of a mattress etc.*

**kıtıpiyos** Common; poor, trifling, insignificant.

**kıtır** Maize grains cracked over a fire; a lie. **~ ~**, *imitates a crackling sound, the crunching of stg. by the teeth; also indicates ferocity as in gnashing the teeth*: **~ atmak**, to lie, to make an impudent exaggeration: **~ ~ kesmek**, to kill in cold blood. **~cı**, liar. **~ damak**, to make a crunching sound.

**kıtkanaat** (ˈ···) Having to be satisfied

with little; in scarcity. ~ **geçinmek,** barely to be able to live: ~ **yetişmek,** barely to suffice.

**kıtlama** Drinking tea while holding a piece of sugar in the mouth.

**kıtlaşmak** *vn.* Become scarce.

**kıtlık** Scarcity; dearth; famine. **insan kıtlığında,** *in such phrases as*: ⌜**insan kıtlığında buraya beni tayin ettiler**⌝, (*apologetically*) 'as they couldn't get anyone else they appointed me here'.

**kıtmır** *The name of the dog of the Seven Sleepers*; dog.

**kıvam** (·ʼ) Proper degree of consistency *or* maturity; the right moment to do anything. **~ına gelmek** *or* **~ını bulmak,** to come to the right consistency *or* degree; to be at the best possible moment *for some action*: **~ında olm.,** to be in its prime.

**kıvanc** Legitimate pride.

**kıvanca** Transhipment; change of watch *on a ship*.

**kıvanmak** Be legitimately proud.

**kıvılcım** Spark.

**kıvır, ~ ~** In curls, wriggling, writhing. **~cık,** curly; crisp. A kind of small-tailed, curly-haired sheep; the mutton therefrom: ~ **lahana,** curly kale: ~ **salata,** cabbage lettuce: ~ **zıvır,** trifling, insignificant.

**kıvırmak** *va.* Curl; twist; coil; turn in an edge of cloth and sew it round; invent (a lie); put in order; execute with success; (*sl.*) eat greedily; earn. *vn.* Succeed; turn; dance gracefully. **burun ~,** to turn up one's nose.

**kıvrak** Supple; brisk; dexterous; agile; coquettish.

**kıvranmak** *vn.* Writhe; wriggle; be agitated.

**kıvr·ık** Curled, twisted; curly; hemmed; folded. Fold. **pantolonun kıvrığı,** the turn-up of trousers. **~ılmak,** *pass. of* **kıvırmak,** to be twisted *etc.*; to be squeezed into a tight place. **~ım,** twist; curl; fold: ~ ~ **olm.** (**kıvrılmak**), to be doubled up with pain. **~ıntı,** coil; a winding; turn; twist.

**kıyafet** (·–ʼ), **-ti** The general appearance and dress *of a person*; aspect; physiognomy. ~ **düşkünü,** wretchedly clothed: **tebdili ~ etm.,** to disguise oneself. **~li,** in *such-and-such* a shape *or* dress. **~siz,** ill-looking; untidy.

**kıyak** (*sl.*) Pretty; elegant; smart.

**kıyamamak** *va. impotential of* **kıymak.** Spare the life of; have pity on; grudge (expense); be unable to bring oneself *to do stg.*; not to have the heart to.

**kıyamet** (·–ʼ), **-ti** The Resurrection of the Dead; the end of the world; great disaster; tumult. ~ **alâmeti!,** how dreadful!: ~ **gibi (kadar), heaps** of: ~ **koparmak,** to create an uproar, to 'raise hell': ~ **koptu, a great** disaster has occurred; hell has broken loose; ~ **mi kopar?,** 'what the devil does it matter!': **kış ~,** intense cold: **kızılca ~,** the 'hell of a row'.

**kıyas** (·ʼ) Comparison, analogy; reasoning, syllogism; rule; opinion. ~ **etm.,** to compare: ~ **ile, by analogy**: ~ **kabul etmez,** incomparable: **bu ~ üzere,** by analogy with this; at this rate: '**var ~ et!**', 'draw your own conclusions!' **~en** (·ʼ·), by comparison; by analogy; by rule. **~î** (·–ʼ), in accordance with rule; analogous; regular.

**kıyasıya** Murderous; merciless.

**kıyı** Edge; shore; bank; corner; extremity. **~dan ~dan,** '(creeping) along the shore', very cautiously: **~da bucakta (köşede),** in holes and corners, in out-of-the-way places: **~ya çekilmek,** to withdraw, to get out of the way: **~ya inmek (çıkarmak),** to land *from a ship*: ~ **sıra,** *v.* **kıyısıra.**

**kıyıcı**[1] Frequenter of the shore; shore fisherman.

**kıyıcı**[2] Who cuts up; cruel, pitiless. Tobacco-cutter.

**kıyık** Minced; chopped up.

**kıyılmak** *vn. pass. of* **kıymak.** Be minced *etc.*; have a feeling of debility, ache. **bakmağa kıyılmaz bir manzara,** a view one is never tired of looking at, a marvellous view.

**kıyım** Act *or* manner of mincing *or* chopping; a single quantity of mincing. **iri ~,** heavily-built. **~lı,** minced; chopped up: **ince ~ tütün,** finely cut tobacco.

**kıyıntı** Anything chopped up; a griping of the stomach; aching of the limbs; languor.

**kıyısıra** (·ʼ··) Along the shore; coastwise.

**kıyma** Minced meat. Minced, finely chopped. ~ **tahtası,** chopping-board. **~lı,** with minced meat.

**kıymak** *va.* (*with acc.*) Mince, chop up fine; slaughter, massacre; decide on; (*with dat.*) bring oneself to do an injury; not to spare; sacrifice; *v. also* **kıyamamak. canına*** ~, to have no pity on and kill: **nikâh** ~, to perform the marriage ceremony.

**kıymet, -ti** Value; price; esteem. ~ **bilmek,** to appreciate the value of; to appreciate (show gratitude for). ~**lendirmek,** to bring into profitable use, utilize. ~**li,** valuable, precious. ~**siz,** worthless.

**kıymık** Splinter.

**kız** Girl; daughter; virgin; queen (cards). ~ **alıp vermek,** to intermarry: ⌜~**ını dövmiyen dizini döver**⌝, keep your daughter in order or you will regret it later: ~ **gibi,** new, untouched; beautiful: ~ **ismi,** maiden name: ~ **oğlan** ~, virginal; young and fresh: ~ **tarafı,** the bride's relatives: ⌜**beğenmiyen** ~**ını vermesin!**⌝, 'anyone who does not like me needn't give me his daughter', I don't care what others think; who does not like it may lump it!: **hanım** ~, a shy little girl.

**kızak** Sledge; slide; slipway; ways *for launching a ship.* **kızağa çekilmek,** (of a ship) to be drawn up on a slipway; (of a man) to be put on the shelf: ~**tan indirmek,** to launch *a ship*: ~ **kaymak,** to slide on ice, to sledge: ~ **yapmak,** to slide.

**kızaklık** Joist.

**kızalak** Wild poppy.

**kızamık** Measles. ~ **çıkarmak,** to have measles. ~**çık,** German measles.

**kızan** Youth, lad; sturdy country lad. **karı** ~, wives and children; the whole population.

**kızar·mak** *vn.* Turn red; blush; ripen; be roasted *or* toasted. **kızarıp bozarmak,** to grow red and pale by turns. ~**tma,** roasted; roast meat *etc.* ~**tmak,** to make red; cause to blush; roast, grill, fry: **yüz (yüzünü*)** ~, to overcome one's natural reluctance *in asking a favour etc.*

**kızdırmak** *va.* Heat; anger, annoy.

**kızgın** Hot; red-hot; angry; excited; feverish; on heat (sexually). ~**lık,** great heat; fury; feverish activity; sexual excitement.

**kızıl** Red; red-hot; golden. Scarlet fever. ~ **buğday,** spelt: ~**deli,** a raving madman: ~ **şap,** light purple.

**kızıl·ağac,** alder; *also (comm.)* Brazil wood. **K**~**ay,** Red Crescent (Turkish Red Cross). ~**baş,** *a Shiite sect; a military class in the army of Shah Ismail; used to describe* a person of easy morals. ~**ca,** reddish: ~**kıyamet,** a fearful uproar. ~**cık,** cornelian cherry; cornel wood: ~ **sopası,** 'a rod in pickle'. ~**elma,** *v.* **elma.** ~**kanad,** rudd, roach. ~**lık,** redness; red colour; rouge; scarlet fever. ~**ötesi, -ni,** infra-red. ~**tı,** redness; red spot.

**kızışmak** *vn.* Get angry *or* excited; become heated; increase in fury *or* violence.

**kız·kardeş** Sister. **K**~**kulesi, -ni,** the Maiden's Tower at the entrance to the Bosphorus. ~**kuşu, -nu,** peewit. ~**larağası, -nı** *formerly* the chief black eunuch *in the Imperial household.* ~**lık,** maidenhood; virginity. ~**memesi, -ni,** a kind of small bitter orange.

**kızmak** *vn.* Get hot; glow; get red-hot; get angry; get excited; (of an animal) to be on heat. **gözü*** ~, to be beside oneself *with anger or passion.*

**ki** That; as; in order that; *sometimes,* seeing that, since, *e.g.* **işitmedi** ~ **cevab versin,** how can he answer since he didn't hear?: *expresses surprise, e.g.* ⌜**bilmiyor mu** ~**?**⌝, 'doesn't he know then?'; ⌜**geldim** ~ **kimseler yok**⌝, 'I came and found there was no one there!': *it often expresses stg. unsaid, e.g.* **öyle yağmur yağdı** ~, it rained so that ...; **öyle şaştım** ~, I was so surprised that ...: *used with* **bilmem** *in such cases as*: **bilmem** ~ **ne yapmalı,** I don't know what one should do!; **bilmem** ~ **kime sorsam!,** I don't know who to ask!: **ola** ~, it may be that ...: **ta** ~ until.

**-ki** *suffix. When placed after a noun in the locative or genitive it forms a pronoun or adjective.*

**kibar** (·⌄) Noble; rich; belonging to the upper class; distinguished. **orman** ~**ı,** bear; coarse, uncouth man. ~**lık,** greatness; nobility; gentle birth: ~ **taslamak,** to pretend to be a gentleman.

**kibir, -bri** Pride, haughtiness; contempt. **-li,** proud, haughty; contemptuous.

**kibrit, -ti** Sulphur; match. ~ **suyu,** dilute sulphuric acid: **köküne** ~ **suyu dökmek (ekmek),** to exterminate. ~**-çi,** matchseller.

**kifayet** (·—⌄), **-ti** Sufficiency; ability, capacity. ~ **etm.,** to suffice: **···le** ~ **etm.,** to be contented with ... . ~**li,**

having sufficient capacity *or* ability; adequate. **~siz,** inadequate.

**kik, -ki** Ship's gig; skiff.

**kikirik** *In* **Con ~,** John Bull.

**kil** Fuller's earth; clay.

**kile** *Measure of capacity just over a bushel.* **dipsiz ~,** spendthrift who is always borrowing: ⌜**dipsiz ~ boş ambar**⌝, 'it's a sheer waste of time'; 'there is no end to it'.

**kiler** Store-room; pantry; larder. **~ci,** a kind of butler *or* housekeeper.

**kilermeni** Kind of red clay, *formerly used in medicine*; bole armeniac.

**kilid** Lock. **~ altında,** under lock and key: **···e ~ vurmak,** to lock: **asma ~,** padlock. **~lemek,** to lock. **~lenmek,** to be locked; (of teeth *etc.*) to be clenched. **~li,** furnished with a lock; locked.

**kilim** Woven matting; carpet without pile. **~i kebeyi sermek,** to settle in a place.

**kilis** *v.* **kils.**

**kilise** (·ʾ·) Church.

**kiliz** Reed. **~ balığı,** tench.

**killi** Clayey.

**kilo** Kilogram. **~ almak,** to put on weight.

**kils** Limestone. **~î,** calcareous.

**kilükal** (—·ʾ) Gossip, tittle-tattle.

**kim** Who?; whoever. **~iniz,** some of you; **~isi,** some people: ⌜**~ ~e**⌝, 'nobody will notice it'; 'nobody knows anything about it': **~i ..., ~i ...,** some ..., others ...: **~e ne!,** what does it matter to anyone!: ⌜**~ ~inle, o da benimle**⌝, everyone has his *bête noire* and I am his: **her ~,** whoever: **sen ~ oluyorsun?,** 'who are you to *say etc.* ?'.

**kimse** Someone; anyone; (*with neg.*) nobody, no one. **~den korkmam,** I am afraid of no one: **~si yok,** he has no one belonging to him, no friends: **hiç bir ~,** not a soul, no one. **~cikler,** *only in* **~ yok,** there's not a soul there. **~siz,** without relations or friends; destitute. **~sizlik,** a being without anyone to look after one; destitution.

**kimya** (·ʾ) Chemistry; rare and precious thing. **bil ~,** chemically. **~ğer,** chemist. **~ğerlik,** science *or* profession of chemistry.

**kimyevî** (··ʾ) Chemical.

**kimyon** Cummin. **~i,** sage-green coloured.

**kin** Malice; grudge; hatred. **~ tutmak (bağlamak),** to cherish a grudge, to nurse a hatred. **~ci, ~dar, ~li,** vindictive; nourishing a grudge.

**kinaye** (·–ʾ) Allusion; hint; innuendo. **~li,** allusive; sarcastic.

**kinin** Quinine.

**kir** Dirt; uncleanliness. **~ götürmek,** (of a cloth *etc.*) not to show the dirt: **~ tutmak,** to show the dirt.

**kira** (·ʾ) A hiring; hire; rent. **~ ile tutmak,** to rent *or* hire: **~ya vermek,** to let out on hire, to let: ⌜**ağzını ~ya mı verdın?**⌝, *said to one who is silent on a matter which concerns him*: **ayak ~sı,** payment to *s.o.* for going somewhere: ⌜**ayağına ~ mı istiyorsun?**⌝, 'why can't you take the trouble to come (go)?': **diş ~sı,** *v.* **dişkirası**: ⌜**her gördüğünden göz ~sı istemek**⌝, to wish to imitate everything one sees. **kira·cı,** who rents *or* hires; tenant. **~lamak,** to rent, hire; let; let out on hire. **~li,** rented; let. **~lık,** for hire, to let.

**kiram** (·ʾ) Noble, honourable (*in such phrases as* **ulemayı ~, azayı ~**). **~en,** *in* **~ kâtibin,** the Recording Angels.

**kiraz, kirez** Cherry. **~ ayı,** May. **~ elması,** a very small kind of apple.

**kirde** A kind of maize bread.

**kirec** Lime. **~ gibi,** very white: **sönmemiş ~,** quicklime: **~ ocağı,** lime-kiln: **~ sütü,** whitewash: **~ taşı,** limestone. **~kaymağı, -nı,** slaked lime. **~li,** containing lime; whitewashed.

**kiremit** Tile.

**kirez** Cherry.

**kiriş** Catgut; bowstring; violin string; rafter. **~i kırmak** (*sl.*), to run away: **~li köprü,** girder bridge: **kulakları ~te,** all ears. **~leme,** board set on edge; set on edge. **~lemek,** to string (a bow).

**kirizma** Trenching of land. **~etm.,** to double-trench.

**kir·lemek, ~letmek** *va.* Dirty, soil; slander, violate. **~lenmek,** to become dirty; become morally soiled; (of a woman) to have her periods. **~li,** dirty; soiled: dirty linen, the washing: **~ye atmak,** to put *clothes etc.* aside for washing: ⌜**~ çamaşırlarını ortaya çıkarmak**⌝, to show up *s.o.*'s misdeeds '(*not* 'to wash one's dirty linen in public'). **~lilik,** dirtiness; canonical uncleanness.

**kirpi** Hedgehog.

**kirpik**[1] Eyelash. **kirpiğimi kırpmadım (kavuşturmadım),** I didn't sleep a wink.
**kirpik**[2], **kerpik** Carbuncle (stone).
**kirş** Rumen; paunch.
**kisbî** Acquired (*as opp. to natural*); who earns.
**kispet, -ti** Costume; wrestler's shorts.
**Kisra** (·-́) Chosroes; one of the Sassanian dynasty.
**kisve** Garment; costume; costume of a special class; wrestler's shorts.
**kiş** Check! (at chess).
**kişi** Person; human being; *indef. pron.*, one, you. ⌜**~ ettiğini bulur**⌝, as you sow, so will you reap. **~lik,** special to *so many* persons; **yüz ~,** (*stg.*) for a hundred people. **~oğlu, ~zade** (··-̀), of gentle birth. **~zadelik,** being of a good family.
**kişlemek** *va.* Shoo away (birds).
**kişmiş** A small kind of raisin.
**kişnemek** *vn.* Neigh.
**kişniş** Coriander.
**kitab** (·̀) Book. **~a el basmak,** to take an oath on a sacred book; to be quite certain about *stg.*: **~a uydurmak,** to do *stg.* dishonest in an apparently honest way; to get round a law, an agreement *etc.*
**kitab·cı,** bookseller. **~e** (·-̀), inscription *on a monument etc.* **~eli** (·-·̀), having an inscription; having an ornamental pattern. **~et** (·-̀), **-ti,** art of writing; writing lesson; essay-writing; literary style; office and duties of a clerk *or* secretary. **~evi, -ni,** bookshop. **~hane** (··-̀), library; bookshop. **~î** (·--́), pertaining to books; bookish: librarian; believer in a sacred book. **~iyat** (·-·̀), **-tı,** bibliography. **~iye** (·-·̀), bast. **~sız,** not possessing a book; not believing in a sacred book, pagan; **hesabsız ~,** without due consideration; without counting the cost.
**kitara** (·̀·) Guitar.
**kitle** *v.* **kütle.**
**kit·lemek** *va.* Lock. **~li,** furnished with a lock; locked.
**kitre** Gum tragacanth.
**kiyanus** Cyanogen.
**kiyaset** (·-̀), **-ti** Shrewdness; sagacity.
**kiyotin** Guillotine.
**kizib, -zbi** Lie.
**klâkson** Motor horn.
**klâsör** (*Fr. classeur*) File.
**klâviye** (*Fr. clavier*) Keyboard.
**klefte** (̀··) Theft. **~ci,** thief.
**kliket** (*Fr. cliquetis*) Pinking *of an engine.* **~leşmek,** to pink, knock.
**klor** Chlorine.
**klüb** Club.
**kobay** (*Fr. cobaye*) Guinea-pig, cavy.
**koca** Old, ancient; large, great; famous. Husband; old man; elder. **~ baş,** cattle; *v. also* **kocabaş**: **~ya varmak,** (of a woman) to marry: **karı ~,** a married couple; **karı ~ kavgası,** domestic squabble.
**koca·baş,** hawfinch (?); large beet. **~başı,** village headman. **~karı,** old woman: **~ soğuğu** (= **berdelacuz**), cold spell at the end of March. **~lamak, ~laşmak, ~lmak, ~mak,** to grow old. **~lı,** having a husband. **~lık,** old age; state of being a husband. **~man,** huge, enormous. **~sız,** unmarried (woman); widow. **~tmak, ~ltmak,** to cause to grow old. **~yemişi** (·̀···), **-ni,** arbutus.
**kocunmak** *vn.* Take offence; sulk; be scared. ⌜**(al kaşağıyı, gir ahıra) yarası olan kocunsun**⌝, 'go into the stable with a curry-comb, the horse with a sore will be scared' ('let the galled jade wince'); a guilty conscience betrays itself; *hence* ⌜**yaran yok ne kocunuyorsun?**⌝, 'you've done nothing wrong, why are you scared?'
**koç, -çu** Ram; sturdy and plucky young man. **~ burunlu,** Roman-nosed: **yabani ~,** wild sheep. **~başı** (̀··), **-nı,** battering-ram. **~boynuzu, -nu,** cleat. **~katımı, -nı,** ramming season. **~kar,** ram used for fighting. **~lanmak,** to become a ram; to act violently *or* bravely.
**koçak** Brave; generous.
**koçan** Corncob; stump; heart *of a vegetable*; stump of counterfoils.
**koçu** Granary; cattle-shed; bullock-cart.
**kod** Code.
**kodaman** Large; clumsy; slow of movement; notable, influential. Magnate, 'bigwig'.
**kodes** (*sl.*) Prison, 'clink'.
**kodeş, kodoş** Pimp.
**kof** Hollow; rotten; weak; stupid; ignorant. **~luk,** hollowness; ignorance.
**kofana** Large **lüfer** fish.
**kofti** (*sl.*) Lie; trick.
**koğlamak, koğmak** *etc. v.* **kovlamak, kovmak** *etc.*
**koğuş, kovuş** Large room; dormi-

tory; ward. **~ağacı, -nı,** baulk of timber; beam, joist.

**kok, -ku** Coke. **~laştırmak,** to carbonize.

**kokar** That smells; fetid, stinking. **akarı ~ı yok,** in perfect order. **~ca,** polecat.

**kokla·mak** *va.* Smell; nuzzle; get the wind of *some coming event.* **koklayanın burnu düşmek,** to be unpleasant. **~şmak,** to smell one another; caress and kiss one another. **~tmak,** *caus. of* **koklamak** *q.v.*; to give just a whiff of, *i.e.* give in very minute quantity.

**kok·mak** *vn.* Smell; stink; go bad; be at hand, give signs of being about to happen. ⌜**ne kokar ne bulaşır**⌝, harmless but useless. **~muş,** putrid; rotten (egg *etc.*); very lazy; dirty (man).

**kokona, kokana** (·ˈ·) Elderly Greek woman.

**kokoreç** A dish of sheep's lungs.

**kokoroz** Ear of maize; the maize plant; any pointed ill-shaped thing.

**kokoz** Very poor; (*sl.*) hard up. **~luk,** destitution.

**kokteyl** Cocktail.

**koku** Smell; scent; indication *or* inkling of *stg.* as yet unseen. **···in ~sunu almak,** to perceive the smell of; to get an inkling *or* have a presentiment of *stg.*: **~su çıkmak,** (of a secret) to be divulged; (of anything) to give signs of its approach: **ben herkesin ağzının ~sunu çekemem,** I can't put up with other people's whims *or* moods.

**koku·lu** Having *a certain* smell; perfumed. **~tmak,** to give out a smell; smell (badly): make *a place* smell; cause *s.o.* to smell; sicken, disgust.

**kol** Arm; foreleg; neck; team; troupe; patrol; wing *of an army*; column *of troops*; branch, subdivision; handle; bar; strand *of rope.* **~ ~a,** arm-in-arm: **~ atmak,** to send forth branches, ramify, extend, develop: **~ gezmek,** to go the rounds: **···in ~una girmek,** to take *s.o.* by the arm: **~ vurmak,** to patrol: **~ yürütmek,** to splice (rope).

**kola** Starch; starch paste. **~lamak,** to starch. **~lı,** starched.

**kolaçan** A walking about (*esp. with an eye to pilfering*). **~a çıkmak,** to wander about seeing what can be picked up: **~ etm.,** to rummage about, poke about.

**kolağası** (ˈ···), **-nı** (*Formerly*) adjutant-major.

**kolan**[1] Broad band *or* belt; girth; binding round the bottom of a tent; rope of a swing. **~ vurmak,** to girth a horse; to swing while standing up. **~yeri, -ni,** the part of a horse where the girth goes.

**kolan**[2] Young foal; wild ass.

**kolay** Easy. Easy way to do *stg.*; means. **~ına bakmak,** to look about for the easiest way of doing *stg.*: **~ını bulmak,** to find an easy way: ⌜**~ gele!**⌝, 'may it be easy!' (*said to s.o. at work*): **dile ~,** easy to say *but not so easy to do.*

**kolay·ca,** pretty easy; quite easy. **~lamak,** to facilitate; to have nearly finished (a job, money, food *etc.*); to 'break the back' of *a task.* **~laşmak, ~lanmak,** to become easy; to be nearly finished. **~lık,** easiness; facility in working; means; easy circumstances, comfort: **~ göstermek,** to give facilities, to make things easy.

**kol·cu** Watchman; custom-house guard; agent for servants. **~çak,** gauntlet; mitten; cuff-protector; armlet. **~demiri** (ˈ···), **-ni,** iron bar for barring a door. **~düğmesi, -ni,** cuff-link.

**kolej** College.

**kolera** (·ˈ·) Cholera.

**kolkola** (ˈ··) Arm-in-arm.

**kollamak** *va.* Search; keep under observation; look after, protect. **sıra (fırsat) ~,** to watch for an opportunity.

**kolleksiyon** Collection.

**kollu** Having arms *or* sleeves; having *so many* contingents (military force); having *so many* strands (rope). **dört ~ya binmek** (*sl.*), *lit.* to mount the four-handled (bier), *i.e.* to go to the grave.

**kolluk** Cuff.

**kolon** Colonist; column; colon.

**kolona** (·ˈ·) Mullion.

**kolonya** (·ˈ·) Eau-de-Cologne.

**kolordu** (ˈ··) Army Corps.

**kolsaati** (ˈ···), **-nı** Wrist-watch.

**koltuk** Armpit; arm-chair; out-of-the-way spot; small wine-shop; hawker; dealer in old clothes; flattery; *the ceremony during a wedding when the bridegroom gives his arm to the bride, hence* **koltuğa girmek,** to marry; **koltuğuna* girmek (tak-**

mak), to give one's arm to *s.o.*, to put one's arm through his: ~ **değneği,** crutch; ~ **değneğiyle,** with the help of others: ~ **halatı,** mooring-rope: **koltuğu* kabarmak,** to swell with pride: ~**ta olm.**, to be another's guest, to go to an entertainment at another's expense: ···**in koltuğuna sığınmak,** to be under the wing of ...: ~ **vermek,** to flatter.

**koltuk·altı** (·ˋ··), **-nı,** armpit; space under the arm. ~**çu,** old-clothes man; keeper of a small out-of-the-way tavern; flatterer, hypocrite. ~**lamak,** to support by the arm; to take *stg.* under the arm. ~**lu,** having arms (chair); arm-chair.

**kolye** (*Fr. collier*) Necklace.

**kolyoz** Spanish mackerel.

**koma¹** (ˋ·) Apostrophe.

**koma²** Fellow wife; second wife of two.

**koma³** Act of putting. Put; set down; (matter) brought forward. *Adv. expressing haste*: ~ **koş!,** run hard!

**komak** *v.* **koymak.**

**komalika** (··ˋ·) Shellac.

**kombina** (·ˋ·) Combine.

**komiser** Superintendent of police.

**komisyoncu** Commission-agent.

**komita** (·ˋ·) Revolutionary committee; secret society; member of such a society. ~**cı,** member of a secret revolutionary society; rebel (*esp. applied to Bulgarian, Greek, and Serbian revolutionary plotters at the end of the 19th and beginning of the 20th centuries*).

**kompas** *v.* **kumpas.**

**komplo** (*Fr. complot*) Plot.

**komposto** (·ˋ·) Stewed fruit.

**komşu** Neighbour. ⌜~**nun tavuğu** ~**ya kaz görünür**⌝, one always envies another's possessions: ⌜**ev alma** ~ **al**⌝, one's neighbours matter more than one's house: ⌜**gülme** ~**na gelir başına**⌝, don't laugh at another's misfortune, it may happen to you one day. ~**luk,** a being a neighbour; neighbourly deed.

**komuta** Command; order.

**komutan** Commander; commandant.

**komün·ism** Communism. ~**ist,** communist.

**konak¹** Scurf. ~**lı¹**, scurfy.

**konak²** Halting-place; stage; day's journey; mansion; government house. ~ **etm.**, to make a stop on a journey. ~**lamak,** to stay for the night *when on a journey*; (*mil.*) to be billeted. ~**lı²**, who lives in a big house; gentleman.

**konc** Leg of a boot *or* stocking.

**konca** Bud.

**koncolos** Vampire; werewolf.

**kondurmak** *va. caus. of* **konmak** *q.v.*; find lodgings for *s.o.*; quarter (troops); retort. ···**e** ~, to put *or* place on ...; to attribute to, to charge *s.o.* with *stg.*: **hastalığı kendine kondurmuyor,** he will never admit his illness.

**konferans** Conference; lecture.

**konfor** Comfort.

**koni** Cone. ~**k,** conic.

**konişmento** (··ˋ·) Bill of lading.

**konmak** *vn.* Alight; settle; perch; camp; make a night's halt *during a journey*; have a piece of good luck. **konup göçmek,** to lead a nomadic life: **mirasa** ~, to come into an inheritance.

**konser** Concert.

**konsey** (*Fr. conseil*) Council.

**konsol** (*Fr. console*) Bracket; corbel.

**konsolos** Consul. ~**hane** (···–ˋ), consulate.

**konsolto** (·ˋ·), **konsültasyon** Medical consultation.

**kont, -tu** Count.

**kontak** Contact (*elect.*); short circuit. **kafadan** ~ (*sl.*), a bit touched in the head.

**kontenjan** (*Fr. contingent*) Quota.

**kontes** Countess.

**kontluk** County.

**kontra** (ˋ·) Against. ~ **flok,** flying-jib: ~**lar sancaktan (iskeleden) seyretmek,** to be on the starboard (port) tack.

**kontrat** Contract.

**kontrol** Control.

**konu** Subject (matter).

**konuk** Guest. ~ **lamak,** to entertain; put up (a guest); give a feast.

**konukomşu** The neighbours; all the neighbourhood.

**konulmak** *vn. impers. of* **konmak.** Be put, placed, set.

**konuşkan** Loquacious.

**konuş·mak** *vn.* Converse, talk. *va.* Talk about. **birisile konuşmamak,** to have no more to do with a person. ~**turmak** *caus. of* **konuşmak;** to introduce *s.o. to another.*

**konyak** Brandy.

**koparmak** *va.* Pluck; break off; take by force; get *stg.* out of *s.o.*; set up (an outcry). ⌜**domuzdan kıl ~**⌝, to succeed in getting *stg.* difficult (*e.g. money from a miser*): **izin ~**, to manage to get leave: **kıyamet ~**, to raise hell, to make a great fuss *or* row: **ödünü ~**, to terrify *s.o.*: **toz ~**, to raise a great dust: **tuttuğunu koparır**, he sees a thing through; he is resolute, dogged: **zincirini ~**, to become furious; to go raving mad.

**kopasıca** (··ꞌ·) (*Lit. may it break off*) **kafası ~!**, damn him!

**kopay** *v.* **kopoy.**

**kopça** Hook-and-eye. **dişi ~**, eye; **erkek ~**, hook. **~lamak**, to fasten with hooks-and-eyes.

**kopil** Small Greek boy; rascal.

**kopmak** *vn.* Break in two; snap; set out *or* start off on an action; break out, begin (of any violent commotion or natural disturbance); ache violently; (*dial.*) run. **bora koptu**, the storm burst: **gönülden ~**, (of an act) to proceed from kindness of heart; **gönülden ne koparsa**, whatever one feels inclined to give: **patırdı koptu**, a great noise arose.

**kopoy** Sporting dog; hound.

**kopuk** Broken off, torn; penniless; vagabond.

**kopuz** Kind of guitar.

**kopya, kopye** (ꞌ·) Copy. **~cı**, who copies; who cribs *at examinations.*

**kor** Red-hot cinder.

**koramiral** (ꞌ···), **-li** Vice-admiral.

**kordele, kordela** (·ꞌ·) Ribbon.

**kordon** Cord; watch-chain; cordon; strand of rope *when separated.*

**korgeneral** (ꞌ···), **-lı** Corps Commander, Lieut.-General.

**korınga** Sainfoin.

**korindon** Corundum.

**korkak** Timid; cowardly. Coward. **~lık**, cowardice; timidity.

**korkmak** *vn.* Be afraid. **···den ~**, to be afraid of, to fear: **···den gözü* ~**, to dread, be terrified of.

**korku** Fear; alarm; danger. ⌜**~ dağları bekler**⌝, of course he is (you *etc.* are) afraid: **can (baş) ~su**, fear for one's life. **~lmak**, *impers. of* **korkmak**: **korkulur bir şey**, a thing to be afraid of. **~lu**, frightening, dangerous: **~ ruya**, nightmare; ⌜**~ ruya görmektense uyanık durmak hayırlıdır**⌝, it is better to remain awake than to have a nightmare, better to be safe than sorry. **~luk**, scarecrow; banister; parapet; guard of a sword-hilt; mere figurehead. **~nc**, terrible, fearful. **~suz**, fearless, intrepid; safe. **~suzluk**, fearlessness; safety. **~tmak**, to frighten, threaten.

**korluk** Fire of red-hot embers; brazier.

**korna, korne** (ꞌ·) Motor horn.

**korno** (ꞌ· Horn; powder-horn; oil-can.

**koro** (ꞌ·) Chorus.

**korsan** Pirate; corsair.

**korse** Corset.

**koru** Small wood, copse. **~cu**, rural guard; forest watchman.

**koruk** Unripe grape. **~ lüferi**, medium-sized **lüfer**, caught in August.

**koru·mak** *va.* Defend; watch over; cover (expenses). **~nma**, defence: **millî ~ mahkemesi**, special tribunal set up for emergencies: **pasif ~**, air-raid precautions. **~nmak**, to defend oneself; to take shelter: **···den ~**, to avoid. **~yucu**, protective: defender: **~ tababet**, preventive medicine.

**korunga** *v.* **korınga.**

**koskoca** (ꞌ··) Enormous; very eminent.

**koşma** A kind of popular ballad.

**koş·mak** *vn.* Run. *va.* Harness; give as escort *or* companion; put to work; attribute; lay down (conditions). **baş ~**, to be obstinate, to insist: **işe ~**, *va.* to put to work: *vn.* to rush to work. **~turmak**, *va. caus. of* **koşmak**, cause to run *etc.*; *also* to run about and tire oneself in *doing stg.*; to dispatch.

**koşu** Race. **~ atı** racehorse: **~ yolu**, racecourse, track: **bir ~** (*pronounced* **bikoşu**), quickly, with a dash. **~cu**, runner. **~lu**, harnessed; (*mil.*) horse-drawn.

**koşuk** Ballad; folk-song.

**koşulmak** *vn. pass. and impers. of* **koşmak. böyle koşulur**, one runs like that.

**koşum** Act of harnessing; harness. **~ hayvanı**, carriage-horse, draught-horse: **~ kayışı**, trace.

**koşuşmak** *vn.* Run together; crowd in; make a concerted rush.

**kota** Quota.

**kotarmak** *va.* Dish up *food*; serve out. **pişirip ~**, to cook and serve up *food*; to settle *a question*; to finish off *a job.*

**kotra** (ꞌ·) Cutter (boat). Pen for small animals.

**kova** Bucket.
**kovalamaca** The game of 'touch-last' or 'catch'.
**kovalamak** *va.* Pursue; endeavour to obtain; wait for. **arkasından atla kovalar gibi,** 'as though pursued by horsemen', in great haste.
**kovan** Hive; cartridge-case; shell-case. **torpito ~ı,** torpedo-tube: **~lı anahtar,** box spanner.
**kovcu** Informer.
**kovlamak** *va.* Denounce.
**kovmak, koğmak** *va.* Drive away; turn back; repel; persecute; denounce, slander.
**kovucu** Who chases; *v. also* **kovcu.**
**kovuk** Hollow. Cavity. **dişinin kovuğuna bile gitmedi,** 'not enough to fill the hollow of a tooth', *i.e.* a very exiguous portion of food.
**kovulamak** *va.* Denounce.
**kovuş** *v.* **koğuş.**
**kovuşturma** Legal proceedings, prosecution.
**koy** Small bay *or* inlet; nook.
**koyak** Valley.
**koyar** Confluence of two streams.
**koymak, komak** *va.* Put; let go; leave; permit; suppose. **⌜koydunsa bul!⌝,** 'I can't find it anywhere': **koyup gitmek,** to leave a thing and go away: **araya ~,** to use *s.o.* as intermediary: **bahis ~,** to lay a wager: **(bir yola) baş ~,** to be ready to sacrifice oneself for *stg.*: **el ~,** to requisition: **elden koymamak,** not to grudge; not to neglect: **ortaya (meydana) ~,** to produce; to bring forward *an argument etc.*; to prove: **üstüne ~,** to add: **yanına koymamak,** to leave unpunished: **yola ~,** to send *s.o.* off on a journey; **yoluna ~,** to put right; to set *a business* going.
**koynu** v. **koyun[1].**
**koyu** Thick; dense (liquid, darkness); deep, dark (colour); true, genuine; fervent, extreme. **~lanmak, ~laşmak,** to become dense; become dark (colour). **~lmak[1],** to become dense *etc.* **~ltmak,** to render dense; thicken (soup *etc.*); darken (colour). **~luk,** density; deepness *of colour.*
**koyulmak[2]** *vn. pass. of* **koymak.** Be put, *etc.*; be poured: **···e ~,** be busied with; set to *work etc.*, begin; fall on; attack.
**koyun[1], -ynu** Bosom; breast pocket. **~ koyna,** in each other's arms: **koynuna* almak,** to take to bed with one: **birinin koynuna girmek,** to go to bed with *s.o.*: **elleri koynunda,** helpless, not knowing what to do: **yüzü ~,** face downwards.
**koyun[2]** Sheep; mild, spiritless person; simpleton. **dağ ~u,** mouflon. **~-gözü** (·ˋ··), **-nu,** feverfew; camomile. **~yılı, -nı,** 'the year of the sheep' (*the eighth year in the old Turkish cycle of years*).
**koyuvermek, koyvermek** *va.* Let go; just to put down; allow. **altına ~,** to soil one's clothes involuntarily: **kahkahayı ~,** to burst into laughter: **kendini kapıp ~,** to cease to take an interest in oneself, in one's ordinary life, in one's business *etc.*; to let oneself drift; to feel quite free: **sakal ~,** to let the beard grow.
**koz** Walnut; trump *at cards.* **~ helvası,** nougat: **~ kaybetmek,** to lose one's case *at law or in an argument*: **~ kırmak,** to play a trump; to commit an indiscretion, make a *faux pas.* **~unu pay etm.,** to reach a settlement: **~ paylaşmak,** to go shares, to come to an agreement; **···le ~unu paylaşmak,** to settle accounts with ...; **'sizinle paylaşacak ~um var',** 'I have a bone to pick with you'; **⌜biz ~umuzu kendimiz paylaşırız⌝,** we'll settle our differences without your help.
**koza** Silk cocoon; a kind of bean; any small round thing.
**kozak** Cone of a coniferous tree; round thing.
**kozalak** Cypress cone; small stunted thing. **~lı,** coniferous.
**köçek** Dancing-boy; camel foal.
**köfte** Meat rissole.
**köftehor** Cunning rogue (*half affectionately*); **~!** lucky dog!
**köhne** Old; worn; antiquated; second-hand.
**kök[1], -kü** Tuning-key of a stringed instrument. **sazı ~ etm.,** to tune a stringed instrument.
**kök[2], -kü** Root; base; fang *of a tooth*; origin; lode. **~ünden koparmak,** to eradicate: **~ünü kurutmak,** to exterminate: **~ salmak,** to be deeply rooted: **işi ~ünden kesip atmak,** to reject *or* settle a matter once and for all: **⌜hepsinin ~üne kibrit suyu!⌝,** 'to hell with the lot of them!'
**kök·boyası** (ˋ···), **-nı,** madder. **~çü,** herbalist. **~lemek,** to uproot; to clear roots from the ground. **~lenmek, ~leşmek,** to take root;

become firmly established. **~lü,** having roots; rooted.

**köken** Branch of a melon *or* marrow plant.

**köknar** Fir.

**köle** Slave. **~niz,** your very humble servant: **~ doyuran,** a filling *food.*

**kölemen** Mameluke; *formerly* a corps of military slaves.

**kömür** Charcoal; *also used for* coal (*properly* **maden** *or* **taş ~ü**). Coal-black. **~ ocağı,** coal-mine. **~cü,** charcoal burner; coal-dealer; stoker. **~lük,** coal-hole; coal-cellar; bunker (*naut.*).

**köpek** Dog; vile man. **~ balığı,** shark. **~dişi** (·ˈ··), **-ni,** canine tooth. **~lemek, ~leşmek,** to cringe like a beaten dog. **~lik,** low-down action; baseness. **~memesi, -ni,** tumour which comes under the armpit.

**köpoğlu, -nu** Scoundrel. **~ köpek!,** (dog son of a dog!) *term of violent abuse.* **~luk,** a dirty trick.

**köprü** Bridge; hasp (of a lock). **~nün gözleri,** the arches of a bridge.

**köprücük** Collar-bone.

**köpük** Froth; foam; scum; lather.

**köpürmek** *vn.* Froth; foam; foam at the mouth.

**kör** Blind; without foresight; careless; blunt (knife); small-meshed (net). **~ ~üne,** blindly; carelessly: **~ boğaz,** appetite (*contemptuously*): **~ düğüm,** knot that can't be undone, tangle; deadlock: ⌜**~ kadıya ~sün demek**⌝, to 'call a spade a spade', *hence* **~ kadı,** outspoken, downright: **~ kandıl,** blind drunk: **~ kaya,** submerged rock: **~ünü kırmak,** to humble the pride of *s.o.*: **~ kütük,** dead drunk: ⌜**~ ~ parmağım gözüne**⌝, as plain as a pikestaff: **~ olası herif,** the cursed fellow: ⌜**~ ölür badem gözlü olur (kel ölür sırma saçlı olur)**⌝, 'when the blind man dies, they say he had almond eyes (when the bald one dies, they say he had golden hair)', *exaggerated praise of the dead or the past*: **~ talih,** bad luck, evil destiny: **~ tane,** 'bunted' grain; smut-ball: **elinin ~ü!,** you can go to hell!: ⌜**kavga elinin ~ünden çıkar**⌝, a mere trifle may cause a quarrel: **gözün ~ olsun (olası)!,** curse you! **kör·barsak,** caecum, appendix. **~boğaz,** gluttonous. **~döğüşü, -nü,** confusion, muddle. **~ebe** (ˈ··), blind man's buff; the blindfolded player.

**körfez** Gulf.

**kör·körüne** (ˈ···) Blindly; at random; carelessly. **~lemeden, ~lemesine,** blindly; at random. **~lemek,** *v.* körletmek. **~lenmek,** to become blind; become blunt; to get 'rusty' or stale (*fig.*). **~letmek, ~leştirmek,** to blind; to blunt; damp, discourage; bring to nought: **nefsini ~,** to 'take the edge off' one's appetite *or* desire; to have a snack to satisfy one's hunger for the moment. **~lük,** blindness; bluntness; lack of foresight; blundering.

**köroğlu** (ˈ··), **-nu** *The hero of a popular legend*; wife, 'the missus'. **~nun ayvazı,** an inseparable companion: **bir ~ bir ayvaz,** just a wife and husband without children.

**körpe** Fresh; tender; very young and fresh. **~lik,** freshness; tenderness; youth.

**körük** Bellows; folding hood (of a car). **yangına ~le gitmek,** to add fuel to the flames. **~çü,** bellows-maker; one who fans the flame, instigator, agitator. **~lemek,** to fan a flame with bellows; encourage, incite. **~lü,** having bellows: **~ bavul,** expanding suit-case.

**kör·yılan** (ˈ··) Blind-worm. **~yol** (ˈ·), railway branch line coming to a dead end.

**kös**[1] Big drum. ⌜**~ dinlemiş (davulun sesi vız gelir)**⌝, 'who has heard the big drum takes no notice of the sound of the kettle-drum', too sophisticated to be impressed; callous, insensitive.

**kös**[2] *In* **~ ~ yürümek,** to walk in a pensive and dejected manner.

**köse** With little or no beard; sparsely timbered. ⌜**her şey olur biter, ~nin sakalı bitmez**⌝, (*a pun*) 'everything comes to an end (**biter**), but a beardless man's beard does not grow (**bitmez**)', all things come to an end: ⌜**vay benim ~ sakalım!**⌝, I am nonplussed. **~lik,** scantiness of beard.

**köseğen** Sensitive plant (a kind of mimosa).

**köseği** Poker; piece of wood burnt at the end.

**kösele** (·ˈ·) Stout leather *used for soles.* **~ suratlı,** shameless. **~taşı, -nı,** sandstone *used for polishing marble.*

**kösem, kösemen** Ram *or* goat that leads the flock; bell wether; ram trained to fight; daredevil.

**köskötürüm** (ˈ...) Completely paralysed.
**köstebek** Mole. ~ **illeti**, a kind of scrofula *which a mole was supposed to cure*.
**köstek** Watch-chain; fetter, hobble; brake. **köstеği kırmak**, to break one's fetters, run away.
**kösümek, kösünmek** *vn.* Be on heat.
**köşe** Corner; angle; nook; retreat. ~ ~ *or* ~ **bucak**, every hole and corner: ~ **başı**, street-corner: ~ **kadısı**, stay-at-home: ~**ye oturmak**, (of a girl) to get married: ~ **sarrafı**, street-corner money-changer: **baş (üst)** ~**ye çıkmak**, to place oneself at the top of the table, *i.e.* in the post of honour: **bir** ~**ye çekilmek**, to go into retirement, to withdraw from public life.
**köşebent** Angle-tie.
**köşek** Camel foal.
**köşe·kapmaca** Puss-in-the-corner. ~**leme**, angular; having angles *or* corners. ~**li**, having corners *or* angles: **üç** ~, three-cornered, triangular. ~**taşı, -nı**, corner-stone.
**köşk, -kü** Pavilion; summer-house; villa; after deck-cabin. ~**lü**, *a man formerly employed to give warning of a fire*.
**kötek**[1] A beating. ~ **atmak (çekmek)**, to give a beating: ~ **yemek**, to get a beating.
**kötek**[2] The Umbra (fish) (*Umbrina cirrhosa*).
**kötü** Bad. ~ **kadın**, prostitute: ~ **kişi olm.**, to become a bad man in the eyes of *s.o.* ~**lemek** *va.*, to speak ill of, slander: *vn.* to become a wreck *from illness*. ~**leşmek**, to become bad; deteriorate. ~**lük**, badness; bad action; harm.
**kötüm·semek** *va.* Think ill of. ~**ser**, pessimistic; derogatory.
**kötürüm** Paralysed; crippled. ~**lük**, paralysis.
**köy** Village; country (*as opp. to town*). ~**lü**, belonging to a village; peasant; fellow villager; rough, bucolic. ~**lülük**, a being born in *or* belonging to a vıllage.
**köz** Embers.
**kral** *v.* **kıral**.
**krank** *In* ~ **mili**, crankshaft.
**kredi** Credit.
**kremayer** (*Fr. cremaillère*) Rack (*mech.*).
**krep** Crêpe; pancake. ~**döşin**, crêpe-de-Chine.
**kriko** (ˈ.) Jack (for lifting).
**kriz** (*Fr. crise*) Crisis.
**kroki** (*Fr. croquis*) Sketch.
**krom** Chromium.
**kron** Crown (coin).
**kropi** *In* ~ **bağı**, figure-of-eight knot.
**kruvazör** Cruiser.
**kubbe** Dome; cupola; vault *of heaven*. **habbeyi** ~ **yapmak**, to make a mountain out of a molehill: **bir yalanın** ~**sini yapmak**, to make a lie seem true by telling another.
**kubur** Holster; quiver; horse pistol; hole in old-fashioned latrine. ~**luk**, quiver.
**kucak** Breast; embrace; armful; lap. ~ ~, by armfuls: ~ **kucağa**, in one another's arms: **kucağına almak**, to embrace; to take on one's lap: ~ **(ta) çocuk**, a child in arms. ~**lamak**, to embrace; surround; include. ~**laşmak**, to embrace one another.
**kuçu kuçu** Bow-wow; *call to a dog*.
**kudema** (..ˈ) *pl. of* **kadim**. The ancients; eminent people; elders.
**kudret, -ti** Power; strength; capacity; the omnipotence of God; wealth; nature. ~**ten**, natural *phenomena etc.*: ~**im yetişmez**, I am not strong enough; I can't afford it: ~**helvası, -nı**, manna. ~**li**, powerful; capable. ~**siz**, powerless; feeble incapable.
**kud·sal** *v.* **kutsal**. ~**sî** (.ˈ), sacred; divine. ~**siyat, -tı**, sacred things. ~**siyet, -tı**, sanctity.
**kudum** (.ˈ) Arrival.
**kudur·mak** *vn.* Be attacked by rabies; go mad. ~**muş**, mad (dog): ⌜**alışmış** ~**tan beterdir**⌝, a habit is a curse. ~**tmak**, to infuriate.
**kuduz** Hydrophobia, rabies. Suffering from hydrophobia; furious. ~**böceği, -ni**, cantharides.
**Kudüs** Jerusalem.
**kufa** Round wickerwork coracle *used on the Tigris*.
**kûfî** (–ˈ) Cufic characters for Arabic writing.
**kuğu** Swan.
**kûhi** Tumble-down.
**kuka** Ball (of wool *etc.*); coconut-wood.
**kukla** (ˈ.) Doll; puppet; very small man.
**kuku** Cuckoo.

**kukulete** (··̀·) Hood; cowl.

**kukulya** (·̀·) Silkworm cocoon. **~cı,** gipsy fortune-teller: **~ fırtınası,** storm occurring in mid-April.

**kukumav** The Little Owl. **~ gibi,** sitting apart by himself.

**kul** Slave; creature; man (*in relation to God*); (*formerly*) Janissary. **~unuz,** your servant, I: **~ hakkı,** one's duty to one's neighbour (*as opp. to one's duty to God*): **···e ~ kurban olm.,** to be devoted to ...: **~ olm.,** to be the slave of *s.o.*, to do everything asked of one: **~ yapısı,** man-made, perishable: **Allahın· ~u,** human being: ⌜**Allahın bildiğini ~dan ne saklıyalım**⌝, I may as well say it: ⌜**Allah ne verir de ~ kaldırmaz (götürmez)**⌝, human beings can bear whatever burdens Fate decrees.

**kula** (̀··) Russet; dun (horse).

**kulaç** Fathom. **~ ~,** in full measure, freely: **~ atmak,** to take soundings; to swim overarm: **çift ~ yüzmek,** to swim double overarm. **~lamak,** to fathom; to measure with the extended arms; to walk swiftly.

**kulağakaçan** Earwig.

**kulak** Ear; attention; ear-shaped projection *or* handle; flap; peg *of a violin*; mouldboard (plough); guard *round a key-hole etc.*; slip of paper *attached to a letter*; branch pipe; small ball of forcemeat *put in soup*. **~tan kulağa,** (news *etc.*) secretly passed on: **kulağı* ağır,** hard of hearing: **~ asmak,** to lend an ear, to pay attention; '**~ asma!**', 'pay no heed!': **kulağını* bükmek,** to warn *s.o.* secretly about *stg.*: **kulağı delik,** alert, intelligent: **~ demiri,** mouldboard *of a plough*: **kulağını* doldurmak,** to persuade *or* prime *s.o.*: **~ dolgunluğu,** hearsay; knowledge acquired by listening to others: **~tan dolma,** hearsay: **kulağına girmedi,** he paid no attention: **~ kesilmek,** to be all ears, to listen attentively: **~ kiri,** ear-wax: **kulağına* koymak,** to prime *s.o.*, to drop him a hint: **~ memesi,** lobe of the ear: **~ misafiri,** one who overhears: **bir kulağı sağır olm.,** to turn a deaf ear to *stg.*, to wink at it: **kulağına* sokmak,** to force *stg.* on *s.o.'s* attention: **~ tozu,** the sensitive spot behind the ear; *also sometimes used for* ear-drum: **~ zarı,** ear-drum: **eli kulağında,** *stg.* that may happen any moment, imminent: **can kulağı ile dinlemek,** to listen with rapt attention: **göz ~ olm.,** to be all eyes and ears.

**kulak·çın** Ear-flap. **~lık,** ear-flap; headphone.

**kule** Tower; turret.

**kulis** Coulisse.

**kullanış** Method of using. **~lı,** serviceable; handy. **~sız,** unhandy; not practical.

**kullanmak** *va.* Use; employ; treat; deal tactfully with, humour: direct; drive (a car *etc.*); take habitually (a food, drink, tobacco *etc.*); appoint (to an office *etc.*).

**kulluk** Slavery, servitude; worship.

**kullukçu** (*Formerly*) Janissary stationed at a guard-house; subaltern in the Janissaries.

**kuloğlu** (̀···), **-nu** (*Formerly*) member of a military force consisting of the sons of slaves.

**kulp, -pu** Handle (of a jug *etc.*); pretext. **bir ~una getirmek,** to seize an opportunity *to say stg.*: **~takmak,** to invent a pretext; to find an excuse *for blame or ridicule*; **(ucunu) ~unu bulmak,** to find a way of settling a matter: **(ucunu) ~unu kaybetmek,** to be at a loss to know what to do: **yumurtaya ~ takmak,** to give the most absurd pretexts.

**kulplu** Having a handle.

**kuluçka** Broody hen. **~devri,** incubation period (*also of a disease*): **~ makinesi,** incubator: **~ya oturmak,** (of a bird) to sit: **~ya yatmak,** to incubate.

**kulumbur** Swivel-gun; carronade.

**kulunc** Colic; cramp; lumbago. **~ kırmak,** to cure lumbago by massage.

**kulüb, klüb** Club.

**kulübe** Hut; shed; sentry-box.

**kum** Sand; gravel; gravel (disease). **~ balığı,** sand-eel: **~ gibi kaynamak,** to swarm in countless numbers: **~a oturmak,** (of a ship) to run on to a sandbank: **~ saati,** hour-glass: ⌜**denizde ~ onda para**⌝, he is immensely wealthy.

**kuma** The second wife (of two).

**kumanda** Military command (order); command (authority). **~ etm.,** to command; to give a command. **~n,** military commander; major (in some armies).

**kumandarya** (··̀·) The Comanderia wine of Cyprus.

**kumanya** (·̀·) Ship's provisions;

portable rations *of a soldier*; small stern locker of a boat.

**kumar** Gambling. **~baz, ~cı,** gambler. **~bazlık,** gambling. **~hane** (··–ˈ), gambling-den.

**kumaş** Tissue; fabric, stuff; cloth; texture; quality.

**kumbara** (ˈ··) Bomb; money-box. **~ çivisi,** hobnail: **el ~sı,** hand-grenade. **~cı,** bombardier.

**kumkuma** (·ˈ·) Narrow-necked vase *or* bottle; ink-bottle; *used in such expressions as*: **esrar ~sı,** one full of secrets; **malûmat ~sı,** a mine of information.

**kum·lu** Sandy; gravelly; gritty; speckled with small spots (cloth *etc.*). **~luk,** sandy place; sands; sandy.

**kumpanya** (·ˈ·) Company.

**kumpas** Callipers; composing-stick; consideration and calculation; trick, plot. **~ kurmak,** to calculate, plot, take counsel: **~ı iyi kurdu,** he laid his plans well. **~lı,** arranged; concerted; plotted; secretly *or* treacherously planned.

**kumral** Reddish-yellow; light-brown (hair); light chestnut (horse); darkish (complexion).

**kumru** Turtle-dove.

**kumsal** Sandy. Sandy place; sand beach. **~lık,** sand-pit; gravel-pit.

**kundak** Bundle of rags; swaddling clothes; bun *of hair*; bundle of oily rags for incendiary purposes; stock *of a gun*; gun-carriage. **~tan beri,** from the cradle: **~taki çocuk,** a baby in arms: **···e ~ sokmak,** to set fire to; sabotage. **~çı,** gun-stock maker; incendiary; one who wrecks *a project etc.* **~çılık,** arson; wrecking (*fig.*). **~lamak,** to swaddle; set fire to; wreck (*fig.*), sabotage; to do the hair up in a bun. **~lı,** swaddled; filled with combustibles.

**kundura** (ˈ··) Shoe. **~cı,** shoemaker.

**kunduz** Beaver; (*err.*) otter. **~ böceği,** cantharides.

**kunt** Strong; thick; solid.

**kuntrat, kunturat, -tı** Contract. **-lı,** let *or* sold by contract.

**kupa** (ˈ·) Cup; wine-glass; cupful; hearts (cards); coupé.

**kupkuru** (ˈ··) Bone-dry.

**kupon** Coupon; sufficient cloth to make a suit.

**kur**[1] (*Fr. cours*) Course *of studies etc.*; rate of exchange.

**kur**[2] (*Fr. cour*) Courtship.

**kura kura** *v.* **kurmak.**

**kur'a** The drawing of lots; military conscription; year (class) of conscripts. **~ çekmek,** to draw lots: **~ isabet etm.,** for the lot to fall to one; to be recruited by drawing lots: **~ya girmek,** to reach military age: **~ neferi,** conscript: ⌜**kaçın ~sıyız**⌝, 'I'm too old a bird to be caught by chaff'. **~cı,** officer *or* committee charged with the drawing of lots for military service.

**kurabiye** (·–·ˈ) Cake made with almonds *or* nuts.

**kurada** Shrivelled; decrepit; worn out.

**kurak** Dry, arid. **~lık,** drought.

**kuran** (*Fr. courant*) Current *of air etc.* **~der** (*Fr. courant d'air*), draught.

**Kur'an** (·ˈ) The Koran.

**kurbağa** Frog; the Star-gazer (*Uranoscopus scaber*). **kara ~,** toad. ⌜**yaptığı hayır ürkütüğü ~ya değmez**⌝, 'the good done will not make up for the frog which was frightened', *said about doing stg. which entails more fuss than it is worth.* **~cık,** little frog; tumour on the tongue; handle of a window-frame; wire-cutters. **~lama,** frogwise: the breast stroke in swimming.

**kurban** (·ˈ) Sacrifice; victim. **~ bayramı,** the Moslem Festival of Sacrifices: **~ kesmek,** to kill as a sacrifice: **~ olm.,** to sacrifice oneself, to be a victim: ⌜**~ olayım!**⌝, 'I beseech you!': **~ payı,** part of the sacrificed sheep given to the poor: **can ~,** (i) a thing one would give one's life for; (ii) 'I'd be only too glad!' **~lık,** animal destined for sacrifice: **~ koyun,** sheep for sacrifice; mild uncomplaining man.

**kurca** Irritation; itching. **~ çıbanı,** an irritable ulcer.

**kurcalamak** *va.* Scratch; rub; irritate; meddle with; fiddle about with, tamper with. **zihnini* ~,** to cause one to 'scratch one's head', to worry.

**kurcata** Crosstrees (*naut.*).

**kurd** *v.* **kurt. ~ağzı** (ˈ··), **-nı,** dovetail; (*naut.*) fairlead. **~ayağı, -nı,** club-moss.

**kurdela** (·ˈ·) Ribbon.

**kurdeşen** Rash (measles, nettle-rash, harvest-bug bites).

**kurena** (··ˈ–) *pl. of* **karîn.** Associates, companions; *as sing.* a chamberlain of the Sultan.

**Kureyş** The Koreish (the Prophet's tribe).

**kurgan** Castle; fortress.

**kuriye** Courier.

**kurlağan** Whitlow.

**kurma** An erecting *etc.* Portable; *toy etc.* that winds up.

**kurmak** *va.* Set up; establish; organize; plan, meditate; set (trap); cock (gun); pitch (tent); wind (clock *etc.*); prime (a person); make (a pickle). *vn.* Brood over *stg.* **kura kura,** by brooding over *stg.*: **bağdaş ~,** to sit cross-legged in oriental fashion: **sofra ~,** to lay a table.

**kurmay** Staff (*mil.*). **Genel ~ Başkanı,** Chief of the General Staff.

**kurna** Basin of a bath *or* fountain; sink.

**kurnaz** Cunning; shrewd. **~lık,** cunning, shrewdness.

**kuron** (*Fr. couronne*) Crown *of a tooth*; crown, kroner (coin).

**kurs**[1] (*Fr. cours*) Course *of lessons etc.*

**kurs**[2] Disk; lozenge; pastille of incense.

**kursak** Crop *of a bird*; stomach; dried bladder *or* its membrane. **~lı,** greedy; 'full of guts'; goitrous person.

**kurşun** Lead; bullet; lead seal. **~ atmak,** to fire a rifle *etc.*; **~a dizmek,** to execute by shooting: **~ dökmek,** *to perform the superstitious custom of melting lead and pouring it into cold water over the head of a sick person*; **~cu kadın,** the woman who performs this ceremony: **~ kâğıdı,** tinfoil: **~ sirkesi,** solution of subacetate of lead, Goulard water: **~ tavası,** ladle for melting lead: **~ tuzu,** subacetate of lead: ⌜**şeytan kulağına ~!**⌝, (may the Devil's ears be plugged with lead!), 'touch wood!'

**kurşun·î** (··–), lead-coloured. **~kalem,** lead pencil. **~lamak,** to cover *or* seal with lead.

**kurt, -du** Wolf; worm, maggot. **~ dökmek,** to pass a worm; **~larını* dökmek,** (i) to sow one's wild oats; (ii) to achieve a long desired object: **~kapanı,** pit for trapping wolves; a wrestling trick: **~ masalı,** *v.* **kurtmasalı**: ···**in kurdu olm.,** to be an old hand at ...; to be a hard-bitten ...: **eski ~,** old hand, old stager: ⌜**hangi dağda ~ öldü?**⌝, *said when stg. pleasant happens unexpectedly*: **içine* ~ düşmek,** to have a misgiving; **içini* ~ gibi yemek,** to be consumed by anxiety, to be very worried: **kafasına* ~ sokmak,** to put an idea into *s.o.'s* head.

**kurt·ağzı** (–··), **-nı** Dovetail; (*naut.*) fair-lead. **~bağrı, -nı,** privet. **~boğan,** aconite. **~çuk,** grub. **~lanmak,** to become maggoty *or* worm-eaten; become agitated *or* impatient; fidget. **~lu,** maggoty, wormy; uneasy, suspicious; fidgety: **~ kaşar (peynir),** a fidgety child. **~ mantarı, -nı,** puff-ball. **~masalı, -nı,** a story told to explain away *stg.*; 'the same old story': **~ okumak,** to invent all sorts of pretexts in order to get out of doing *stg.* **~pençesi, -ni, ~tırnağı, -nı,** bistort. **~yeniği, -ni,** worm-hole in wood; *also used for* **bityeniği,** *stg.* 'fishy'.

**kurtarmak** *va.* Save, rescue; redeem *stg. pawned*; recover *one's losses at a game*; *also used err. for* **kotarmak** *q.v.* ⌜**daha (bundan) aşağısı kurtarmaz**⌝, (i) I can't sell it for less; (ii) nothing less will do for him, he must always have the best (*iron.*).

**kurtul·mak** *vn.* Escape; be saved; slip out; (of a pregnant woman) to be delivered; ···**den ~,** to be rid of, be free from; get out of; lose one's grip of. **~uş,** liberation; escape; way of escape.

**kuru** Dry; dried; withered; emaciated; bare; mere. Dry land; dry part *of anything.* **~ ~ya,** uselessly, in vain; without good reason; mere: **~ ekmek,** dry bread (bread and nothing else): **~ gürültü,** mere clamour; just rumour: **~ iftira,** sheer calumny: **~ kafa,** skull; stupid; **bir ~ kafa kalmak,** (of a widow *etc.*) to be left all alone: **~ kahve,** roasted *or* ground coffee-beans: **~ kalabalık,** an aimless crowd; **~ kalabalık etm.,** to hang around and do nothing: **~ oda,** unfurnished room: **~ sandalye,** non-upholstered arm-chair: **~ sıkı,** blank shot; empty threat: **~ tahtada kalmak,** to lose one's furniture; to be destitute: ⌜**~nun yanında yaş ta yanar**⌝, 'the green burns along with the dry', *i.e.* the innocent suffer with the guilty: **~ yerde (toprakta),** on the bare earth: **dut ~su,** dried mulberries: **kara ~,** dark and skinny: **piç ~su,** a tiresome naughty child.

**kurucu** Founder.

**kurulamak** *va.* Wipe dry; dry.

**kurulmak** *vn. pass. of* **kurmak.** Be founded *etc.*; pose; swagger; settle oneself comfortably. **dünya kurulalıdan beri,** since the beginning of the world.

**kurultay** Assembly; congress.

**kurulu** Established; set up; strung (bow); ready to fire (gun).

**kuruluk** Dryness.

**kuruluş** Foundation; structure; (*mil.*) distribution of forces.

**kurum**[1] Soot. ~ **tutmak,** to be full of soot.

**kurum**[2] Pose; conceit. ~**undan geçilmiyor,** his conceit is intolerable: ~ ~ **kurulmak,** to be exceedingly puffed-up: ~ **satmak,** to give oneself airs.

**kurum**[3] Association; society.

**kurumak** *vn.* Dry; wither up; become thin; become paralysed. ⌜**dilin kurusun!**⌝, 'may your tongue be withered!' (*a curse*): **kanı*** ~, to be worried out of one's life.

**kurum·lanmak** *vn.* Be puffed-up, give oneself airs. ~**lu,** conceited, puffed-up. ~**suz,** without conceit; modest.

**kurun** (·⁎) *pl. of* karin. Ages. ~**u ûlâ,** ancient times: ~**u vusta,** the Middle Ages: ~**u vustaî,** medieval.

**kurunmak** *vn.* Dry oneself.

**kuruntu** Strange fancy; unfounded suspicion; illusion; melancholy. ~**lu,** afflicted with unfounded fears *or* suspicions.

**kuruş** Piastre. ~**luk,** piastre piece; piastre's worth.

**kurut** Dried milk product.

**kurutma** Action of drying. ~ **kağıdı,** blotting-paper.

**kurutmak** *va.* Dry; cause to shrivel. **kanını*** ~, to persecute, vex, exasperate: **kökünü*** ~, to eradicate, utterly destroy.

**kuruyası** *Optative form of* **kurumak.** ⌜**ağzı** ~⌝, 'may his tongue be withered!' (*a curse*).

**kuskun** Crupper; stern-cable. ~**u düşük,** broken-down (horse); down and out, too wretched to bother about his dress. ~**suz,** without a crupper; free, unbridled; neglected, broken-down.

**kuskus** Dough in small pellets *used for pilaf*; semolina.

**kusmak** *va. & vn.* Vomit; (of cloth after being dyed *or* cleaned) to show up an old stain. **kusacağım geliyor,** I feel like being sick; I am utterly disgusted: **kan** ~, to be in great pain, to suffer greatly.

**kus·muk, –uk, –untu** Vomit. ~ **turucu,** emetic.

**kusur** (·⁎) Failure to do one's duty; defect, fault; remainder *of a sum of money.* ~**a bakmamak,** to overlook an offence, to forgive: ~ **etmemek,** to spare no effort: **analar** ~**u,** a poor sort of mother, a parody of a mother. ~**lu,** faulty; incomplete, defective. ~**suz,** without defect; complete; innocent.

**kuş** Bird. ~**a benzetmek,** to spoil *stg.* by trying to improve it: ~ **beyinli,** of limited intelligence: ~**a dönmek,** to look 'something like' (*iron.*): ~ **gibi,** very light *or* agile: ⌜~ **uçmaz, kervan geçmez**⌝, a desolate, deserted spot: ~ **uçurmaz,** very alert and capable: ~ **uykusu,** a sleep from which one awakes at the slightest sound: ⌜**ağzıyle*** ~ **tutsa* faydası yok**⌝, even if he were to perform a miracle it would be no use now: ⌜**her** ~**un eti yenmez**⌝, everyone is not at your service: **ona** ~**um kondu,** I took to him.

**kuşak** Sash; girdle; cummerbund; supporting beam; generation. **ipsiz** ~**sız,** vagabond. ~**lama,** diagonally. ~**lamak,** to brace *or* tie (a wall *etc.*).

**kuşam** *In* **giyim** ~, dress: **giyimli** ~**lı,** smartly dressed.

**kuşane** *v.* **kuşhane.**

**kuşan·mak** *vn. & va.* Gird oneself; put on a sash; gird on *a sword*; dress. **giyinmiş kuşanmış,** all dressed up: ⌜**iş becerenin, kılc kuşananın**⌝, success comes to those who know their job. ~**tı,** *in* **giyinti** ~, clothes.

**kuşatmak** *va.* Wind round the waist; gird on; surround, envelop; besiege.

**kuş·bakışı** (⁎···), **–nı** Bird's-eye view. ~**başı,** in small pieces; (of snow) in big flakes: ~ **et,** pieces of meat the size of a walnut. ~**baz,** bird-fancier; bird-catcher. ~**burnu, –nu,** beak; hip *of the dog-rose.* ~**çu,** falconer; bird-fancier. ~**dili, –ni,** thieves' slang; childish language *with an f or other letter added to each syllable.* ~**gömü,** *in* ~ **et,** the fillet of meat on each side of the back-bone. ~**hane** (·–⁎), place where hawks used to be kept; small saucepan. ~**konmaz,** asparagus.

**kuşe** (*Fr. couché*) *In* ~ **kâğıd,** art paper.
**kuşku** Suspicion. ~**lanmak,** to feel nervous *or* suspicious.
**kuş·lokumu** (ˈ···), **-nu** Kind of sweet cake *sold in the street to children.* ~**luk,** aviary; forenoon; lunch. ~**-palazı, -nı,** diphtheria. ~**sütü,** ~**südü, -nü,** (*lit.* bird's milk); a non-existent *or* unobtainable thing: ˹~**-nden gayri her şey vardı**˺, there was every conceivable thing to eat: ˹~ **ile beslemek**˺, to look after *s.o.* with every possible care. ~**tüyü, -nü,** feather: ~ **yatak,** feather-bed, ~**üzümü, -nü,** dried currants. ~**yemi, -ni,** canary seed; bird-seed.
**kut, -tu** Luck; prosperity.
**kut·lamak** *va.* Celebrate; congratulate. ~**lu,** lucky; auspicious; happy. ~**lulamak,** to offer congratulations to *s.o. on a feast-day etc.*
**kutnu** Kind of silk and cotton cloth. **Mısır** ~**su,** fustian.
**kut·sal** *Invented word to replace* **kutsî** (*properly* **kudsî** *q.v.*); sacred.
**kutsuz** Unlucky. ~**luk,** bad luck.
**kutub, -tbu** Pole (of the Earth); pole (*elect.*); most eminent person; the axis around which a business revolves. **kutbu şimalî (cenubî),** North (South) Pole. ~**eyn,** the two poles. ~**yıldızı, -nı,** the Pole Star.
**kutur, -tru** Region; diameter.
**kuva** *pl. of* **kuvvet.** Powers; forces. ~**yı külliye,** (*mil.*) main force.
**kuvafür** (*Fr. coiffeur*) Hairdresser.
**kuvars** Quartz.
**kuvve** *As* **kuvvet** *q.v.*; *also* faculty; quality; potency; possibility. ~**den fiile getirmek (çıkarmak),** to put a project into execution.
**kuvvet, -ti** Strength; force; power; vigour. ~**le,** strongly: ~**ten düşmek,** to weaken, to lose strength: **paraya** ~ **muvaffak oldu,** he succeeded thanks to money: **var** ~**i ile,** with all his might. ~**lenmek,** to become strong; to be strengthened. ~**li,** strong, powerful. ~**siz,** weak, without strength. ~**sizlik,** weakness.
**kuyd** *Only used to emphasize* **kayıd.**
**kuyruk** Tail; appendix; follower; queue; train (of a dress *or* a great personage); corner *or* tail *of the eye*; breech *of a gun.* **kuyruğuna baka baka,** very dejectedly: **kuyruğuna basmak,** to provoke *s.o.*: **kuyruğu* (kapana) kısılmak** *or* **ele vermek,** to be caught by the tail, to be in great straits: **kuyruğunu* kısmak,** to put the tail between the legs (*also fig.*): ~ **sallamak,** to wag the tail; to fawn and flatter: **kuyruğunu* tava sapına çevirmek,** to thrash *s.o.*: **kuyruğuna* teneke bağlamak,** to make a laughing-stock of *s.o.*: **kuyruğu titretmek** (*sl.*), to die: **çekiver kuyruğunu!,** forget about him! he's not worth worrying about.
**kuyruk·acısı, -nı,** rancour; desire for vengeance for some wrong suffered. ~**lu,** having a tail: ~ **piyano,** grand piano: ~ **saat,** grandfather clock: ~ **sürme,** eyelid stain (antimony) slightly overdone: ~ **yalan,** a 'whopping' lie: ~ **yıldız,** comet. ~**sallıyan,** wagtail. ~ **sokumu, -nu,** coccyx. ~**yağı, -nı,** fat melted down from the tail of the fat-tailed sheep.
**kuytu** Sheltered from the wind; snug; dark; hidden. Sheltered nook; remote spot.
**kuyu** Well; pit; borehole; mine-shaft. ~ **fındığı,** a kind of hazelnut *which is buried to give it a special flavour*: ···**in** ~**sunu kazmak,** to lay a trap for *s.o.*: **kar** ~**su,** snow-pit *for keeping snow for summer use.* ~**cu,** well-sinker.
**kuyud** (·ˈ) *pl. of* **kayıd.** Bonds *etc.* ~**at, -tı,** registrations.
**kuyumcu** Jeweller; goldsmith.
**kuzahiye** Iris *of the eye.*
**kuzey** North.
**kuzgun** Raven. ˹~**a yavrusu şahin görünür**˺, 'all his geese are swans': ˹**ya devlet başa ya** ~ **leşe**˺, either good fortune or a miserable death; there is no saying what may happen to one; ˹neck or nothing˺. ~**cuk,** grille in a prison door. **K**~ **denizi, -ni,** the Caspian Sea. ~**î** (··ˈ), black as a raven.
**kuzu** Lamb; mild man. ~**m,** my dear chap!: ~ **sarması,** lamb chitterlings, *v.* **canciğer**: **anasının körpe** ~**su,** mother's little darling. ~**cuk,** little *or* pet lamb. ~**çıbanı, -nı,** small boil. ~**dişi, -ni,** milk-tooth. ~**-kestanesi, -ni,** a small variety of chestnut *eaten raw.* ~**kulağı, -nı,** sheep's-sorrel: ~ **tozu,** potassium oxalate. ~**lamak,** to lamb. ~**laşmak,** to become as mild as a lamb. ~**lu,** big with young *or* with a lamb at side. ~**mantarı, -nı,** morel.
**kübik** Cubic.

**küçücük** (ʾ··) Tiny; darling.

**küçük** Small; young; insignificant. Child; young animal. **~ten beri,** from childhood: **~ düşmek,** to look small, to feel ashamed: **~ düşürmek,** to make *s.o.* feel small: **kendimizi ~ düşürmiyelim,** do not let us demean ourselves: **~ zabıt,** non-commissioned officer.

**küçük·dil,** uvula; *v.* **dil**[1]. **~lü,** intermixed with small: **~ büyüklü,** some large, some small; young and old. **~lük,** smallness; childhood; pettiness; indignity. **~semek,** to despise; belittle.

**küçül·mek** *vn.* Become small; be reduced; wane; feel insignificant. **~tmek,** to make small; diminish; reduce; belittle. **~tücü,** humiliating.

**küçümsemek** *va.* Belittle.

**küçürek** Rather small.

**küf** Mould, mouldiness. **~ bağlamak (tutmak),** to become mouldy: **~ tadı,** mouldy taste.

**küfe** Large deep basket *usually carried on the back.* **~ ile getirilmek,** to be so drunk that one has to be carried home in a basket: ⌜**arkasında yumurta ~si yok ya, dönüverir**⌝, there's nothing to stop him changing his mind, his plans *etc.* **~ci,** basket-maker; porter *who carries goods in a large basket on his back.* **~li,** who carries goods in a **küfe.** **~lik,** basketful; a man so drunk that he has to be carried home in a basket.

**küfeki** *v.* **kefeki.**

**küffar** *pl. of* **kâfir.** Unbelievers.

**küf·lenmek** *vn.* Turn mouldy; suffer from neglect. **~lü,** mouldy; perished from neglect; out-of-date: **~ para,** hoarded money.

**küfran** (·ʾ) Ingratitude (*also* **~ı nimet**).

**küfür, -frü** Unbelief; blasphemy; cursing, swearing. **~ etm.,** to curse and swear. **~baz,** swearing; foul-mouthed.

**küfür küfür** *Imitates the rustling of the wind.*

**küfüv, -fvü** An equal in rank *or* social status (*mainly in connexion with marriage*).

**küheylân** Pure-bred Arab horse.

**kükremek** *vn.* Become infuriated with rage, (lion) roar.

**kükürt** Sulphur. **~lü,** sulphurous.

**kül**[1] Ashes. Ash-coloured; ruined. **~ etm., to ruin: ~ kedisi,** one who feels the cold, who likes the fire: **~ kesilmek,** to turn pale: **~ (kömür) olm.,** to be utterly ruined: **···in ~ünü savurmak,** to ruin *s.o.*: **~ yutmak,** to be duped.

**kül**[2], **-llü** The whole; all. **~ halinde,** as a whole.

**külâh** Conical hat; anything of that shape; trick, deceit. **~ giydirmek,** to play a trick on *s.o.*: **~ kapmak,** to secure some advantage for oneself by cunning: **~ sallamak,** to flatter, toady: ⌜**Alinin ~ını Veliye, Velinin ~ını Aliye giydirmek**⌝, to make one's way *or* earn one's living by little tricks: ⌜**bunu (gecelik) ~ıma anlat!**⌝, 'tell me another!': ⌜**sonra ~ları değişiriz**⌝, *do this or* we shall fall out: ⌜**şeytana ~ını ters giydirmek**⌝, to be very cunning.

**külâhçı** Trickster.

**külbastı** Grilled cutlet.

**külbütör** (*Fr. culbuteur*) Rocker-arm.

**külçe** Metal ingot; heap; pile; bunch *of keys.* **~ gibi oturmak,** to collapse from fatigue.

**küldür** *v.* **paldır.**

**külek** Tub *with handles.*

**külfet, -ti** Trouble; inconvenience; great expense; ceremonious behaviour. **~ etm.,** to put oneself to inconvenience. **~li,** troublesome; laborious; expensive; ceremonious; forced, unnatural. **~siz,** easy; without inconvenience; unceremonious; natural; spontaneous; informal; not involving great expense.

**külhan** Stoke-hole of a bath. **~beyi** (·ʾ··), **-ni,** a rough, a rowdy; a young blood of the lower classes. **~i** (·–ʾ), urchin; young scamp; merry fellow.

**külkedisi** *v.* **kül.**

**külleme** Mildew of vines.

**kül·lemek** *va.* Cover with ashes. **~lenmek,** to be turned to ashes; smoulder; cool down, die down.

**küll·î** (·ʾ) Total; universal; abundant. **~iyat, -tı,** complete works of an author. **~iyen** (·ʾ·), totally; entirely; (*with neg.*) not at all, absolutely not. **~iyet, -ti,** abundance: **~le,** in great quantity: **~iyetli,** abundant.

**küllü**[1] Containing *or* mixed with ashes. **~ su,** lye.

**küllü**[2] *v.* **küllî.**

**külot, -tu** (*Fr. culotte*) Knickerbockers; riding-breeches.

**külrengi** (ʾ··) Ash-coloured.

**kültür** Culture.
**külünk** Pick; mace; crow-bar.
**külüstür** Shabby; out-of-date; poor in quality.
**kümbet, -ti** Vault; dome; projection; (*sl.*) the behind. **~li,** projecting.
**küme** Heap; mass; mound; hill; straw *or* reed hut; hide *for shooting.*
**kümeç** *v.* **gümeç.**
**kümes** Poultry-house; coop; hut. **~ hayvanları,** poultry.
**künbed** *v.* **kümbet.**
**künde** Fetter, hobble; trap; ambush. **~ye almak (düşürmek),** to throw by a trick *in wrestling*: **~den atmak,** to trip *s.o.* up (*fig.*).
**künh** Essence; reality. **~üne varmak,** to get to the bottom of a matter, to learn thoroughly.
**künk** Earthenware water-pipe. **~ döşemek,** to lay down water *or* drainage pipes.
**künye** Patronymic. **~ (defteri)** register of names, *esp.* Army list: **~si bozuk,** who has a bad record.
**küp, -pü** Large earthenware jar. **~lere binmek,** to get into a rage: **~ gibi,** enormously stout: **altın ~ü,** a Croesus: **~ünü doldurmak,** to grow rich; to feather one's nest.
**küpe** Ear-ring; dewlap. ⌜**kulağında* ~ olsun**⌝, take that piece of advice, let that be a warning to you; **bu benim kulağıma ~ oldu,** that was a lesson to me; I never forgot it. **~çiçeği** (·ˈ···), **-ni,** fuchsia. **~li,** wearing ear-rings; having a dewlap.
**küpeşte** (·ˈ·) Gunwale (boat); bulwarks (ship); rail *of banisters.*
**kür** Health cure.
**Kürd** Kurd; Kurdish.
**kürdan** (*Fr. cure-dents*) Tooth-pick.
**küre** Globe; sphere. **~iarz,** terrestrial sphere; the Earth.
**kürek** Shovel; oar; hard labour, penal servitude. **~çekmek,** to row: **akıntıya ~ çekmek,** to row against the current, to struggle in vain: ⌜**tek ~le mehtaba çıkmak**⌝, (to go for a moonlight row with only one oar), (i) to undertake *stg.* with insufficient means; (ii) to make a feeble attempt at mocking *s.o.* **~çi,** oarsman, rower. **~kemiği, -ni,** shoulder-blade. **~li,** having *so many* oars; **iki çifte ~ sandal,** a two pair-oar boat.
**kürelmek, küremek** *va.* Shovel up; clear away.
**kürevî** (··ˈ) Spherical. **~yat** (···ˈ), **-tı,** spherical trigonometry.
**kürk, -kü** Fur; fur-coat. ⌜**ye ~üm, ye!**⌝ *or* ⌜**buyurun ~üm!**⌝, *used when s.o. is judged by his outward appearance only.* **~çü,** furrier. **~lü,** of fur; adorned with fur; fur-bearing (animal); wearing a fur-coat.
**kürsü** Throne; sofa; chair; footstool; pulpit; dais; tribune; the upper heaven supporting the throne of God. **~ taşı,** pedestal stone.
**kürtaj** Curetting.
**kürtün** Large and clumsy pack-saddle.
**küs** Easily offended; sulky.
**küskü** Crow-bar; fire-dog; iron wedge; half-burnt piece of wood.
**küskün** Disgruntled; offended. **talih ~ü,** unlucky. **~lük,** vexation; a being in the sulks.
**küskütük** Helplessly drunk.
**küsmek** *vn.* Be offended; sulk. ⌜**bahtına (talihine) küssün**⌝, it's just a bit of bad luck: **talihe (tecelliye) ~,** to be tired of life; to curse one's fate: ⌜**tavşan dağa küsmüş, dağın haberi olmamış**⌝, the hare was offended with the mountain, but the mountain never noticed it.
**küspe** Residue of crushed seeds; oil-cake.
**küstah** Insolent. **~lık,** insolence, effrontery.
**küstere** Jack-plane; grindstone. **~ye tutmak,** to sharpen on a grindstone.
**küstümotu** (·ˈ··), **-nu** Mimosa.
**küsuf** Solar eclipse.
**küsur** *pl. of* **kesir,** Fractions. **iki bin ~,** two thousand odd: **bir ~ yıldır,** it's a year and a bit since .... **~at** (··ˈ), **-tı,** *pl. of pl. of* **kesir:** groups of fractions; fractions (*arith.*).
**küşad** Opening; inauguration; opening *at chess or backgammon.* **~ etm.,** to open (exhibition, hospital *etc.*): **resmi ~,** official opening *or* inauguration. **~e** (·—ˈ), open; cheerful.
**küşayiş** (·—ˈ) Cheerfulness.
**küt**[1] Blunt; not pointed; paralysed.
**küt**[2] *Imitates the noise of knocking on a door or of the heart beating etc.* **~ diye vurdu,** he gave it a sharp blow: **para ~ cebe,** he popped the money into his pocket.
**kütah** (·ˈ) *Vulg. form of* **kûtah** *only used in* **ömrünü (gencliğini) ~ etm.,** not to have enjoyed one's life (youth).
**kütle** Heap; block; mass; great

quantity; aggregate. ~**vî** (··ˈ), massive; in the bulk or mass.

**kütlemek** *vn.* Give out a thudding noise.

**kütlü** *Cotton* with the seed in it.

**küttedek** (ˈ··) With a bang.

**kütüb** *pl. of* **kitab.** Books. ~**hane,** ~**ane** (··—ˈ), library; book-shop.

**kütük** Tree-stump; baulk; log; stock *of a vine*; ledger, register; cartridge pouch. ~ **gibi,** dead drunk: **cehennem kütüğü,** a hardened sinner: **eski** ~, a seasoned log; an experienced old man. ~**lük,** belt with cartridge pouches attached.

**kütür** Crisp, fresh (fruit). ~ ~, the noise made when eating such; a crunching sound. ~**demek,** to give out a crashing *or* crunching sound. ~**dü,** the sound of *stg.* cracking *or* of an apple *or* a cucumber being eaten.

**küvet, -ti** (*Fr. cuvette*) Wash-hand basin; developing dish.

# L

**lâakal** (ˈ··) At least; not less than.

**lâalettayin** (ˈ··—·) At random; whosoever; whatsoever.

**lâbada** Dock (plant).

**lâbirent, -ti** Labyrinth.

**lâciverd** (—·ˈ) Lapis-lazuli; dark-blue colour. ~**î** (—··ˈ), dark-blue.

**lâçka** (ˈ·) Let go!; slacken off (rope)! Play *or* slack *in machinery.* ~ **etm.,** to slacken *or* cast off *a rope*; to get slack: **demiri** ~ **etm.,** to let go the anchor.

**lâden** (—ˈ) Laudanum; resin of Cistus plants.

**lâdes** (—ˈ) A game *or* bet with the wishbone of a fowl. ~ **tutuşmak,** to pull a wishbone with one another: ⌜**bile bile** ~!⌝, 'I know I shall be worsted', *said when one is obviously going to be cheated.*

**lâdin** Spruce.

**lâdinga** (·ˈ·) Cartridge-belt.

**lâdinî** (——ˈ) Not connected with religion; lay.

**lâf** Word; talk; empty words; boasting; [*in many phrases* **lâf** *is interchangeable with* **söz** *q.v.*]. ~ ~**fı açar,** one topic leads to another: ~ **altında kalmamak,** to be quick to retort, to give as good as one gets: ⌜~ **anlayan beri gelsin**⌝, no one seems to see the point *of what I am saying*: ~ **atmak,** to chatter; to make insolent remarks to a woman in the streets: ~**ını bilmek,** to weigh one's words: ~ **değil,** it's no trifle, it's important: ~ **ebeşi,** a great talker; quick at repartee: ···**i** ~ **etm.,** to gossip about (unfavourably); ···**in** ~**ını etm.,** to talk about *s.o. or stg.*: ~**ü güzaf,** empty words, brag: ~ **işitmek,** to be rebuked: ⌜~**kıtlığında asmalar budayım**⌝ *or* ⌜~ **söyledi balkabağı**⌝, 'what's that got to do with it?; 'don't talk rot!': ⌜~ **ola beri gele**⌝, that's nothing to do with the question; that is beside the mark: ~ **olsun** (*or* **ola**) **diye,** just for *stg.* to say: '**Ahmedin yanında onun** ~**ı olur mu?**', he cannot be spoken of in the same breath with Ahmed: '**böyle yapmıyalım,** ~ **olur**', 'don't let's do this or people will talk': ~**a tutmak,** to engage *s.o.* in conversation *thereby preventing him from working etc.*: **uzun** ~**ın kısası,** in short.

**lâfazan** Braggart; windbag. ~**lık,** chatter; boasting.

**lâfetmek** (ˈ··), **-eder** *vn.* Speak, talk; gossip; grumble.

**lâfız, -fzı** Word. **lâfzı murad,** (i) said but not meant; (ii) person *or* thing of no account.

**lâf·zan** (ˈ·) Literally; *v. also* **lâfazan.** ~**î** (·ˈ), literal.

**lâğar** (—ˈ) Thin and weak; weedy (*of animals only*).

**lâğım, -ğmı** Underground tunnel; sewer; explosive mine; adit. ~ **açmak,** to dig a drain; to tunnel for a mine: ~ **atmak,** to fire a mine. ~**cı,** sewerman; sapper.

**lâğıv, -ğvı** Cancellation, annulment; suppression. **lağvetmek** (ˈ··), **-eder,** to abrogate, cancel; abolish.

**lahana** (·ˈ·) Cabbage. ~ **turşusu,** pickled cabbage: ⌜**bu ne perhiz, bu ne** ~ **turşusu?**⌝, *said of two opposite extremes*: **frenk** ~**sı,** Brussels sprouts.

**lâhavle** (—ˈ·) *The use of the expression* '**lâ havle ve lâ kuvvete illâ billâh**' (there is no power nor strength but in God) *used to express anger or impatience*: ~ **çekmek,** to use this expression.

**Lahey** The Hague.

**lâhid, -hdi** Tomb.

**lâhik** (—ˈ), **-ki** Joined on to; added; succeeding to *or* newly appointed to a post; present holder of an office. ~a (–·ˈ), appendix; suffix; codicil; additional note.

**lâhin, -hni** Note; tone; melody.

**lâhur·aki** (··ˈ·) A fine merino *imitating Lahore shawls*. ~**î** (··ˈ), the stuff of which Lahore shawls are made.

**lâhza, lâhze** The twinkling of an eye; instant.

**lâik** (*Fr. laïque*) Lay, secular.

**lâin** (–ˈ) Who curses. ~ (·ˈ), accursed; godforsaken; execrable.

**lâk, lâka** Lacquer.

**lâkab** Family name; cognomen; nickname.

**lâkayd** (–ˈ) Indifferent; nonchalant. ~**i** (–·ˈ), nonchalance; indifference.

**lâke** Lacquered.

**lâkerda** (·ˈ·) Salted tunny.

**lâkırdı** Word; talk; promise. ~ **ağzından* dökülmek,** to talk unwillingly: ~ **altında kalmamak,** to give as good as one gets, to be quick to retort: ~**ya boğmak,** purposely to obscure *or* divert the conversation by a lot of irrelevant talk: ~ **etm.,** to talk; to gossip: ···**in** ~**sını etm.,** to mention: ~ **karıştırmak,** to draw a red herring across the trail: ~ **taşımak,** to repeat to *s.o.* other people's gossip about him: **bu çocuk** ~ **anlamıyor,** this child is incorrigible. ~**cı,** loquacious; chatterbox.

**lâkin** (–ˈ) But; nevertheless.

**lâklâk,** ~**a** The clacking noise made by storks; senseless chatter. ⌜**leyleğin ömrü** ~ **ile geçer**⌝, *said of people who just talk and do nothing.* ~**ıyat, -tı,** twaddle.

**lâl** (–), **-lı** Ruby; garnet; ruby lips; red ink.

**lâla** (ˈ·) Servant placed in charge of a boy; tutor; pedagogue. ⌜~ **paşa eğlencesi değilim**⌝, 'it's not my job to keep people amused *or* flattered'.

**lâlanga** (·ˈ·) Kind of pancake.

**lâle**[1] (–ˈ) Iron ring *formerly put round the neck of convicts and lunatics*; forked stick *for picking figs*.

**lâle**[2] (–ˈ) Tulip. ~ **devri,** the early eighteenth century (*when tulips were greatly in vogue*). ~**gün,** red.

**lâm**[1] The Arabic letter L. ~ **elif çevirmek,** to take a stroll: ~ **cim istemez,** *or* ~**ı cimi yok,** it must be done; there's no question about it!; *also used of a very critical situation*, 'it's all up!'

**lâm**[2] (*Fr. lame*) Thin plate; microscope slide.

**lâma** Sheet of metal. ~ **demiri,** sheet-iron.

**lâmba**[1] (ˈ·) Cornice; mortise; rebate.

**lâmba**[2] (ˈ·) Lamp; radio valve. ~ **gömleği,** incandescent mantle.

**lâmel** (*Fr. lamelle*) Cover-glass (microscope).

**lâmı** (–ˈ) Shining.

**lândo, lândon** Landau.

**lânet** (ˈ·), **-ti** Curse, imprecation. Damnable; peevish, cross-grained. ~ **etm.,** to curse: ~ **olsun!,** a curse upon him (it *etc.*). ~**leme,** act of cursing *or* of pronouncing an anathema against *s.o.*: accursed, anathematized. ~**lemek,** to curse; to pronounce a formal anathema against *s.o.*

**lângır lûngur** Who speaks in a loud voice and with a vulgar accent; random and tactless (talk); lumbering along.

**lâp lâp** Flop! flap! ~ ~ **yemek,** to eat greedily, smacking the lips.

**lâpa** Rice pudding; any moist dish; poultice. Soft, flabby. ~ ~ **kar yâğmak,** to snow in large flakes. ~**cı,** fond of sloppy dishes; languid, flabby (person); milksop. ~**msı,** flabby.

**lâpçin** Sort of indoor boot *laced at the side* ~ **ağızlı,** windbag.

**lâpina** (·ˈ·) Wrasse.

**Lâpon** Laplander. ~**ya** (·ˈ·), Lapland.

**lâppadak, lârpadak** (ˈ··) Suddenly; with a flop.

**lârmo** (ˈ·) *In* ~ **yakası,** luff *of a sail.*

**lâsta** (ˈ·) Maximum load a ship can carry.

**lâstik** Rubber; galoshes; tyre. ~**li,** made of rubber; elastic.

**lâşe** Corpse, carcass; putrid thing.

**lâşka** (ˈ·) *v.* **lâçka.**

**lâta** (ˈ·) Lath; gown *formerly worn by ulema*; frock *of a priest.*

**lâtarna, lâterna** (·ˈ·) Barrel-organ.

**lâteşbih** (ˈ··) Anything but ...; far from being ... .

**lâtif** Fine; slender; pleasant; elegant; light; subtle; witty. ~**e** (·–ˈ), Joke, witticism; anecdote.

**lâtilokum** Turkish delight.

**lâtin**[1] (ˈ·) Eastern Catholic; Latin. ~ **çiçeği,** nasturtium. ~**ce,** Latin language.

**lâtin**[2] (ˈ·) Lateen sail.

**lâübali** (–·–ˈ) Free-and-easy; too familiar *or* intimate; careless; offhand. **~leşmek,** to be too free-and-easy; to take liberties. **~lik,** too free-and-easy behaviour; an offhand manner.

**lâv** Lava.

**lâva** (ˈ··) Pull!; hoist away! **~etm.,** to pull a boat *etc.*

**lâvanta** (·ˈ·) Lavender water; perfume. **~ çiçeği,** lavender (flower).

**lâvha** *v.* **levha.**

**lâvta**[1] (ˈ·) Obstetric forceps; doctor *or* midwife.

**lâvta**[2] (ˈ·) Lute.

**lâyık** (–ˈ), **-kı, -ğı** Suitable; worthy. That which one deserves. **~ile,** as it should be; in a worthy manner: **~ını bulmak,** to get one's deserts: **~ görmek,** to deem worthy *or* suitable: **···e ~ olm.,** to be worthy of, to merit: **~ı vechile,** properly, adequately: ⌜**ağzına ~**⌝, (worthy of your palate); that was a really good meal; I wish you could have been there.

**lâyiha** (–·ˈ) Explanatory document; project; bill (proposed law).

**Lâz** The Laz people *of the SE. coast of the Black Sea.* **~ca** (ˈ·), in the Laz way; the Laz language.

**lâza** Tray *or* small trough.

**lâzım** (–ˈ) Necessary; requisite; neuter (verb). **~ gelmek,** to be necessary; to be a necessary consequence; **~ gelenlere,** to whom it concerns: **~ olm.,** to be necessary; to be needed: **neme (nesine, nenize) ~?,** what's that to me (him, you)?; *also sometimes*: but still; all the same: **senin ne üstüne ~?,** what's that got to do with you? **~lı,** necessary; unavoidable. **~lık,** chamber-pot.

**lâzime** (–·ˈ) A necessary thing; natural consequence; corollary; obligation; requisite *for a journey etc.*; ship's stores. **~ci,** ship-chandler.

**leb** Lip; edge. **~i derya,** sea-shore. **~aleb** (ˈ··), brimful.

**leblebi** Roasted chick-peas; bullet. **demir ~,** a very difficult task; a hard nut to crack; *stg.* hard to stomach: ⌜**leb demeden ~yi anlamak**⌝, to understand instantly. **~ci,** seller of roasted chick-peas.

**lef, -ffi** An enclosing. **~fen** (ˈ·), enclosed (in a letter *or* parcel). **~fetmek** (ˈ··), **-eder,** to enclose *in a letter etc.*

**leğen, liğen** Bowl, basin. **~ ibrik,** bowl and ewer (*formerly handed round after a meal for washing the hands*): ⌜**her şey bitti, işimiz bir ~ örtüsüne kaldı**⌝, 'all is complete save for a covering for the basin', *said ironically when an unnecessary thing is provided while essentials are missing.*

**leh** For him *or* it; in favour of him. **~imde,** in my favour: **~ ve aleyh,** for and against.

**Leh** Pole; Polish. **~çe**[1] (ˈ·), Polish (language). **~istan,** Poland. **~li,** Polish, Pole.

**lehçe**[2] (·ˈ) Dialect.

**lehim** Solder. **~lemek,** to solder. **~li,** soldered.

**leh·tar** Supporter; in favour of. **~te,** in his (its) favour: **~ ve aleyhte,** pro and con.

**lehülhamd** (ˈ··) Thank God!

**leke** Spot of dirt; stain; mark. **~ çıkarmak,** to remove a stain: **···e ~ getirmek,** to dishonour, to stain the character of: **~ olm.,** to become stained *or* spotted: **~ sürmek,** to besmirch *s.o.'s name.* **~ci,** a cleaner of clothes: **~ toprağı,** fuller's earth. **~lemek,** to spot, stain; cast aspersions upon. **~li,** spotted, stained; dishonoured: **~ humma,** typhus fever; spotted fever. **~siz,** spotless; immaculate.

**leken** A kind of snowshoe.

**lenduha** (·–ˈ) Enormous; clumsy.

**lenger** Anchor; large deep copper dish.

**lento** Lintel.

**lep** *v.* **leb.**

**lepiska** (·ˈ·) Flaxen (hair).

**leş** Carcass. **~ bağı,** running bowline: **gemi ~i,** wreck.

**letafet** (·–ˈ), **-ti** Charm; grace; elegance; amiability.

**levanten** Levantine.

**levazım** (·–ˈ), **levazımat, -tı** *pl. of* **lâzime.** Necessities; materials; munitions; supplies, provisions; commissariat department, Quartermaster-General's department.

**leve** (*Fr. levée*) Trick (at cards).

**levend** *Formerly* an irregular military force; a gay young spark; a fine well-set-up lad.

**levha** Signboard; inscribed card; framed inscription; metal plate; slab; picture.

**leviye** (*Fr. levier*) Gear lever.

**levrek** Bass (fish). **tatlı su levreği,** perch.

**leylâk** Lilac. **~î** (··ˊ), lilac-coloured.

**leylek** Stork.

**leylî** (·ˊ) Boarder *at a school.*
**leziz** Tasty; delicious; delightful.
**lezzet, -ti** Taste; flavour; pleasure, enjoyment. ~ **duymak (almak),** to enjoy the taste of a thing; to find pleasure in *stg.* ~**li,** pleasant to the taste; delightful. ~**siz,** tasteless; insipid.
**lık·ırdamak** *vn.* Gurgle. ~**lık,** gurgling.
**libade** (·–ˋ) Short quilted coat.
**libas** (·ˋ) Garment.
**libre** (ˋ·) Pound (weight).
**lider** (ˊ·) Leader. ~**lik,** leadership.
**lif** Fibre; loofah; bunch of palm fibres *used for scrubbing oneself in a bath.* ~ **gibi,** rough, scratchy. ~**î** (–ˊ), fibrous.
**lifti** Mortise chisel.
**liftinuskuru** (ˋ····) (Lifting screw); turn-buckle.
**lig** League; union.
**liğen** *v.* **leğen.**
**liken** Lichen.
**likorinos** (··ˋ·) Smoked mullet.
**liman** Harbour. ~ **odası,** harbour-master's office: ~ **reisi,** harbour-master. ~**lamak,** to come into harbour; (of the wind *or* sea) to die down. ~**lık,** place serving as a harbour; calm sea; calm (sea); suitable for a harbour: **süt** ~, dead calm; absolute quiet.
**limba** A kind of barge.
**limbo**[1] (ˋ·) Mortise.
**limbo**[2] (ˋ·) Salvage.
**lime** (–ˋ) Strip. ~ ~, in strips; in tatters.
**limon** Lemon. Lemon-coloured; made of lemon-tree wood. ~ **gibi sararmak,** to turn pale: ~ **küfü,** bluish-green. ~**ata** (··ˋ·), fresh lemonade. ~**atacı,** lemonade seller. ~**î** (··ˊ), pale yellow; capricious, touchy. ~**lu,** flavoured with lemon juice; sour. ~**luk,** conservatory. ~**tuzu** (·ˋ··), **-nu,** citric acid.
**linç** Lynching. ~ **usulü,** lynch law. ~**etmek** (ˋ··), **-eder,** to lynch.
**linyit, -ti** Lignite.
**lira** (ˋ·) Lira; pound; Turkish lira of 100 piastres. ~**lık,** of the value of a lira.
**liret, -ti** Italian lira.
**lisa** (ˋ·) Sheet of a sail. ~ **etm.,** to hoist sail.
**lisan** (·ˋ) Language; talk. ~**aşina** (·–·ˊ), linguist. ~**ıhal** (·–ˋ), the conveyance of meaning without words; a giving to understand *stg.* without saying it. ~**î** (·–ˊ), lingual; linguistic. ~**iyat** (·–·ˋ), **-tı,** linguistics.
**lisans** (*Fr. licence*) Diploma *or* examination *for passing out of High School or University*; licence.
**lise** (ˋ·) (*Fr. lycée*) Grammar School; High School. ~**li,** student at a Grammar or High School.
**liste** (ˋ·) List.
**litre** (ˋ·) Litre (= $1\frac{3}{4}$ pints). ~**lik,** holding *or* amounting to one litre.
**liva** (·ˊ) Brigade (*mil.*); *formerly an administrative district governed by a Mutasarrif.*
**livar** Enclosure connected with the sea *for keeping fishes alive.*
**livre** Livery.
**liyakat** (·–ˋ), **-tı** Merit; suitability.
**liyme** *v.* **lime.**
**liynet, -ti** Looseness of the bowels; mild diarrhoea.
**lobut, -tu** Cudgel; Indian clubs.
**loca** (ˋ·) (*Fr. loge*) Box *at the theatre*; Masonic lodge; small room, cell.
**loça** (ˋ·) Hawse-pipe.
**loda** (ˋ·) Heap of straw *etc.* covered with earth.
**lodos** South-west wind; south-west; south-westerly gale. ~ **poyraz,** blowing hot and cold: ~**a tutulmak,** to sway about like a drunken man. ~**lamak,** (of the wind) to blow from the south-west; (of the weather) to become mild. ~**luk,** exposed to the south-west.
**lofça** Very large nail *used in building construction.*
**loğ** Stone roller.
**loğusa, lohusa** Woman after childbirth. ~ **otu,** aristolochia: ~ **şekeri,** a red-coloured sugar *used in* ~ **şerbeti,** a drink offered to visitors to a newly confined woman.
**loka** (ˋ·) Untidy.
**lokanta** (·ˋ·) Restaurant. ~**cı,** restaurant keeper.
**lokavut, lokavt** Lock-out.
**lokma** Mouthful; morsel; a kind of sweet fritter *distributed to the poor on the death of a relation*; rounded head of a bone, condyle; screw-die. ~ ~, piece by piece: ~ **dökmek,** to distribute lokma: ~ **göz,** pop-eyed: ⌜**ağzına vur** ~**sını al!**⌝, 'you can take the very bread out of his mouth', *said of a very mild person*: **ağzından** ~**sını almak,** to take from *s.o.* what right-

fully belongs to him: ⌈büyük ~ yut, büyük söz söyleme!⌉, 'don't boast about the future!': gözleri ~ gibi fırlamış, his eyes started out of his head.

**lokmacı**, a maker of **lokma**; sponger.

**Lokman** *Name of two legendary sages, one regarded as the father of medicine, the other as a famous storyteller.* ⌈~ hekimin ye dediği⌉, 'advised by the doctor as good to eat', *said of an attractive woman or delicious food.* **~ruhu** (·ˋ··), **-nu**, sulphuric ether.

**lokum** Turkish Delight; diamonds (cards).

**lololo** Nonsense; empty words *used to deceive or put off s.o.* ⌈**bize de mi ~?**⌉, 'do you really think I can be taken in by that?'

**lombar** Port *in a ship's side.*

**lomboz** Port-hole; dead-light; scuttle for air.

**lonca** Tradesmen's guild *or* corporation; meeting-place of such.

**lop** Round and soft (that can be swallowed at a mouthful). **~ ~**, *describes the falling of a soft round thing or the swallowing of such*: **~ et**, boneless meat: **~ ~ yutmak** *or* **~latmak**, to bolt (food).

**loppadak** (ˋ··) *v.* **lop lop.**

**lopur lopur** *In* **~ ~ yutmak (yemek)**, to swallow greedily in large mouthfuls.

**lor** Curd of goat's milk. **~peyniri** (ˋ··), **-ni**, cheese of goat's milk.

**lord, lort** Lord. **~lar Kamarası**, the House of Lords.

**lorta** (ˋ·) Shoemaker's last.

**lostarya** (·ˋ·) Small tavern.

**lostra** (ˋ·) Shoe polish. **~cı**, bootblack.

**lostromo** (·ˋ·) Boatswain.

**losyon** (*Fr. lotion*) Eau de Cologne; scent.

**loş** Dark, gloomy; slack, weak. **~luk**, darkness; slackness.

**lotarya** (·ˋ·) Lottery.

**löğusa** *v.* **loğusa.**

**lök** Awkward; clumsy; sluggish. Male camel. **~ gibi oturmak**, to sit in an awkward and lazy manner.

**lûbiyat** (··ˋ), **-tı** Games; amusements.

**lûgat, -tı** Word; dictionary. **~ paralamak**, to use learned *or* pedantic language. **-çe**, vocabulary; glossary.

**lululu** *v.* **lololo.**

**lûmbar, lûmboz** *v.* **lombar, lomboz.**

**lûtf·en** (ˋ·) As a favour; please! **~etmek** (ˋ··), **-eder**, to do the favour; to have the kindness; to kindly send.

**lûtuf, -tfu** Kindness, goodness; favour. **···mek lûtfunda bulunmak**, to be so kind as to .... **~dide** (··–ˋ), who has received a favour. **~kâr** (··ˋ), kind; gracious. **~name** (··–ˋ), *polite form for* your letter.

**lüb, -bbü** Kernel; marrow; essence; heart.

**Lübnan** (ˋ·) Lebanon.

**lüfer** *Name of a delicious fish caught in the Bosphorus,* blue-fish (*Temnodon saltator*).

**lüks** (*Fr. luxe*) Luxury. **~ lâmbası**, vaporized oil lamp.

**lüle** Pipe; bowl of a tobacco pipe; spout; paper cone; curl; fold; a kind of water measure. **~ci**, maker of pipe-bowls: **~ çamuru**, red clay from which pipe-bowls are made. **~taşı** (·ˋ··), **-nı**, meerschaum.

**lüp**[1] *v.* **lüb.**

**lüp**[2], **-pü** *Stg.* got without cost trouble. **~ diye yutmak**, to gulp down: **~e konmak**, to get *stg.* gratis *or* without effort. **~çü**, one who lives by his wits; parasite.

**lüzucet** (·–ˋ), **-ti** Viscosity. **-li**, viscous.

**lüzum** Necessity; need. **~lu**, necessary; needed; useful: **~ lüzumsuz yere**, even when not needed. **~suz**, unnecessary; useless.

# M

**maada** (–·ˋ *or* –ˋ) Besides; in addition to; except. Rest; remainder. **bundan ~**, furthermore, besides this: **~sı**, the rest of it.

**maahaza** (ˋ··–) In spite of this; nevertheless.

**maaile** (ˋ···) Together with the family.

**maalesef** (ˋ···) Unfortunately; with regret.

**maal·iftihar** (ˋ····) With pride; with pleasure. **~memnuniye** (ˋ··–··), with pleasure.

**maamafih** (ˋ···) Nevertheless.

**maarif** (·–ˋ) *pl. of* **marifet.** Branches of science; education. **~ Vekâleti**, Ministry of Education. **~çi**, educationalist.

**maaş** (·ˋ) Salary; allowance *to widows etc.* **açık ~**, half-pay, payment while

unemployed. ~lı, receiving a salary *etc.*
**maattessüf** (ˋ....) With regret, unfortunately.
**maazallah** (·ˊ··) God preserve us!; Heaven forfend!
**mabaid** (–·ˋ), **–badi** Sequence, continuation; remainder. **mabadi var,** to be continued.
**mâbed** (–ˋ) Place of worship, temple.
**mabeyin, –yni** Interval; room between the women's quarters and the men's quarters *in a large house*; the private apartments of the Palace, *where the Sultan usually received*; relations between two people. **mabeynde,** between them; **mabeynimizde,** between us: **mabeynleri bozuk,** they are on bad terms. **~ci,** Court chamberlain.
**mablak** Spatula; putty knife.
**mabud** (–ˋ) Worshipped. God; idol. ⌜**vermedi ~, ne yapsın Mahmud?**⌝, 'man proposes, God disposes'; it can't be helped. **~e** (––ˋ), Goddess.
**macar** Hungarian; (*sl.*) louse. **~ca,** Hungarian language. **M~istan,** Hungary. **~lı,** Hungarian.
**macera** (–·ˊ) Adventure. **~lı,** adventurous; hazardous. **~perest,** adventurous.
**macun** (–ˋ) Putty; paste; cement; electuary; fruit paste. **diş ~u,** toothpaste. **~lamak,** to stop up with putty *or* cement.
**maç, –çı** Match (football *etc.*).
**maça** (ˋ·) Spade *at cards*; core *of a moulding*. **~ bey,** the knave of spades; ostentatious and conceited person.
**Maçin** (–ˋ) Southern China.
**maçuna** (·ˋ·) Crane (machine).
**madalya** (·ˋ·) Medal.
**madam** Madame; Mrs. (*only of non-Turkish women*).
**madampol, –lü** Madapollam; calico.
**madd·e** Matter; substance; material; subject; article *or* paragraph *of a regulation or law*. **~eci,** materialist. **~ecilik,** materialism. **~eten** (ˋ··), materially. **~î** (·ˊ), *a.* material. **~iyat, –tı,** material things; *as sing.* materialism.
**madem** (–ˋ), **mademki** While; since; as.
**maden** (–ˋ) Mine; mineral; metal; a mine *of learning etc.* **~direği,** pit-prop: **~ kömürü,** coal: **~ ocağı,** mine (coal *etc.*): **~ suyu,** mineral water. **~ci,** miner; mining expert; metallurgist; mine owner. **~î** (–·ˊ), *a.* metal, mineral.
**madik** Trick, ruse. **~ atmak,** to cheat.
**madrabaz** Middleman; cheat, impostor.
**madud** (–ˋ) Counted; limited. **···den ~,** comprised in ...; considered as ... .
**madum** (–ˋ) Non-existent. **~iyet, –ti,** non-existence; absence.
**madun** (–ˊ) Inferior; subordinate.
**mafa** (ˋ·) Bolt *or* screw with a ring at the end.
**mafevk** (–ˋ), **–kı** A superior. **~ında,** above him *or* it.
**mafiş** (–ˋ) Finished!; nothing left! A kind of very light fritter.
**mafsal** Joint; articulation.
**mağara** (·ˋ·) Cave; pit.
**mağaza** Large store, shop; storehouse.
**mağdur** (·ˋ) Unjustly treated; wronged. Sufferer, victim. **~iyet** (·–·ˋ), **–ti,** a being unjustly treated; oppression; a suffering loss.
**mağf·iret, –ti** Remission of sins, forgiveness; grace. **~ur,** whose sins are forgiven; deceased.
**mağlûb** (·ˋ) Defeated, overcome. **hırsına ~ olarak,** losing his temper. **~iyet,** (·–·ˋ), **–ti,** defeat.
**mağmum** (·ˋ) Sad; anxious; gloomy; overclouded. **~iyet** (·–··), **–ti, ~luk,** sadness, gloominess.
**magnez·ya** (·ˋ·) Magnesia. **~yum** (·ˋ·), magnesium.
**mağrib** West; sunset. **Mağrib,** Morocco. **~î** (··ˊ), Moroccan; Moor: ⌜**mal bulmuş ~ye dönmek**⌝, to be overjoyed.
**mağrur** (·ˋ) Self-confident; proud, conceited. **~en** (·ˊ), proudly; confidently; **···e ~,** trusting in ... . **~luk,** conceit, over-confidence.
**mağşuş** (·ˋ) Alloyed; adulterated; base (coin).
**mah** Month; moon. **~ gelmemek,** to be insufficient, to have no effect.
**mahafil** (·–ˋ) *pl. of* **mahfil.** Resorts *etc.*
**mahal, –lli** Place; post; occasion. **~linde,** at his post; opportune: **···e ~ bırakmamak** *or* **vermemek,** not to give occasion for ...: **~line masruf (emek, para),** (efforts, money) well spent: **bu sözlere ~ yoktur,** there is no occasion for using such words.
**mahalâkallah** (ˊ....) 'What God

has created', (all creation); very crowded; a great crowd.
**mahalle** Quarter *of a town*; ward. ~ **çocuğu,** street urchin, guttersnipe: ~ **karıları,** common women: ~ **tavrı,** vulgar manners. ~**eli,** person belonging to a quarter *or* ward; neighbour. ~**î,** local.
**mahallebi** Sweet dish made with rice and milk. ~**ci,** maker and seller of milk dishes: ~ **çocuğu,** mother's darling; milksop.
**mahalsiz** Inopportune; out-of-place.
**maharet** (·—ʹ), **-ti** Skill, proficiency. ~**li,** skilful, proficient. ~**siz,** unskilful; clumsy.
**mahaşerallah** (ʹ····) (Whom God has assembled together), very crowded.
**mahbes** Prison.
**mahbub** (·ʹ) Beloved.
**mahbus** (·ʹ) Imprisoned; prisoner.
**mahcub** (·ʹ) Ashamed; bashful. ~**iyet** (·—·ʹ), **-ti,** bashfulness; shame; modesty.
**mahcur** (·ʹ) Under interdiction; not allowed to dispose of his property.
**mahcuz** (·ʹ) Sequestrated.
**mahdud** (·ʹ) Limited; definite; bounded.
**mahdum** (·ʹ) Son.
**mahfaza** Case; casket; sheath. ~**lı,** having a case; kept in a case.
**mahfe** A frame across a camel's back with seats on either side.
**mahfi** (·ʹ) Hidden; secret; clandestine. ~**ce** (·ʹ·), secretly.
**mahfil** Place of resort; circle; club; private pew in a mosque; masonic lodge.
**mahfuz** (·ʹ) Protected; looked after; committed to memory; reserved. ~**en** (·ʹ·), under guard *or* protection; in custody. ~**iyet** (·—·ʹ), **-ti,** protection; conservation; safeguard; reservation.
**mahıv, -hvı** Destruction; annihilation; abolition.
**mahir** (—ʹ) Skilful. ~**lik,** skill.
**mahiyet** (—·ʹ), **-ti** Reality; the true nature of a thing; character. ... **bir** ~ **almak,** to assume *such and such* a character *or* form. ~**li,** having the character of: **ultimatom** ~ **bir nota,** a note bearing the character of an ultimatum.
**mahkeme** Court of Justice. **bidayet** ~**si** *or* ~**i asliye,** Court of First Instance: **ceza** ~**si,** Criminal Court: **hukuk** ~**si,** Civil Court: **sulh** ~**si,** court of summary jurisdiction, Police Court. ~**lik,** a matter for the courts: ~ **olm.,** to have a dispute which can only be settled in a court of law.
**mahkûk** (·ʹ)- **-kü** Engraved; scratched; erased. Erasure. ~**ât** (·—ʹ), **-tı,** engravings; inscriptions.
**mahkûm** (·ʹ) Sentenced; condemned; judged; subject to. The condemned. ~**iyet** (·—·ʹ), **-ti,** condemnation; sentence.
**mahlâs** Surname; pseudonym.
**mahleb** Mahaleb (cherry).
**mahlû** (·ʹ) Dethroned, deposed.
**mahlûk** (·ʹ), **-ku** Created. Creature. ~**at, -tı,** created things; creatures.
**mahlûl** (·ʹ), **-lü** Dissolved; vacant; escheated (property). Solution, lotion. ~**at** (·—ʹ), **-tı,** property lapsing to the State on the owner's death because there are no heirs.
**mahlût** (·ʹ), **-tu** Mixed; adulterated. Mixture.
**mahmudiye** (·—·ʹ) Gold coin of 25 piastres *coined in the reign of Mahmud II.*
**mahmur** (·ʹ) Heavy after a drunken sleep; pleasantly torpid (*as a baby after a good sleep*); sleepy, languid (eye); lackadaisical. ~ **bakış,** a soft, tender look. ~**luk,** heaviness after a drunken sleep, 'hang-over'; *also* a pleasant sleepy feeling; dreaminess *or* tenderness of look: ~ **bozmak,** to take a pick-me-up to cure a 'hang-over' (a hair of the dog that bit one).
**mahmuz** (·ʹ) Spur; ram *of a ship*. **çavdar** ~**u,** ergot: **direk** ~**ları,** climbing irons. ~**lamak,** to spur.
**mahrama** *v.* **makrama.**
**mahrec** Outlet; origin, source; vocal organs; pronunciation of a letter *or* sound; denominator (*arith.*); specialized school for a profession *or* trade.
**mahrek, -ki** Orbit *of a planet etc.*; trajectory.
**mahrem** Confidential; secret; intimate. ~**ane** (··—ʹ), confidential (speech *etc.*); confidentially, as a secret. ~**iyet, -ti,** the condition of being **mahrem**; a being a confidant.
**mahruk** (·ʹ) Burnt. Fuel; combustible. ~**at** (·—ʹ), **-tı** Combustibles; fuel.
**mahrum** (·ʹ) Deprived; disappointed. ~ **kalmak,** to be disappointed; to remain deprived (of = ···**den**). ~**iyet** (·—·ʹ), **-ti,** deprivation;

destitution: ~e katlanmak, to suffer privation; to do without.

**mahrut** (·ꜜ), **-tu** Cone. **kara ~u subay**, a naval officer on shore duty *or* an air-force officer on ground duty. **~î** (·–ꜝ), conical: **~ çadır**, bell tent.

**mahsub** (·ꜜ) Counted; calculated. **~ etm.**, to count, to reckon in an account. **~en** (·ꜝ·), on account; to the account of (= *dat.*).

**mahsul** (·ꜜ), **-lü** Product; produce; crop; result. **~at** (·–ꜜ), **-tı**, *pl. of* mahsul. **~dar** (··ꜝ), **~lü**, productive, fertile.

**mahsur** (·ꜜ) Confined; limited; besieged; cut off *by floods etc.*

**mahsus** (·ꜜ) Special; peculiar to; proper; particular; private; reserved; not seriously meant. (ꜝ·), specially, expressly, on purpose.

**mahşer** The Last Judgement; great crowd; great confusion.

**mahud** (·ꜜ) Well-known, notorious; (*contemptuously*) your.

**mahun** Mahogany.

**mahunya** (·ꜝ·) Barberry (*Berberis*).

**mâhur** (–ꜝ) A mode of Oriental music.

**mahv·etmek** (ꜝ··), **-eder** *va.* Destroy; abolish. **~olmak** (ꜝ··), to be destroyed, ruined, abolished.

**mahviyet, -ti** Modesty; unobtrusiveness.

**mahya** Texts *or* figures made by lamps suspended between minarets during Ramazan; festoon.

**mahz** Pure; unmixed; mere. **~ı riya**, sheer hypocrisy. **~a** (ꜝ–), merely; only; entirely.

**mahzen** Underground store-house; granary; cellar.

**mahzun** (·ꜜ) Sad, grieved. **~iyet** (·–·ꜝ), **-ti**, grief, sadness.

**mahzur** (·ꜜ) Something to be guarded against; objection; inconvenience; danger. **~u şer'î**, a religious objection.

**mahzuz** (·ꜜ) Pleased, happy. **~iyet** (·–·ꜝ), **-ti**, joy; pleasure.

**mai** (–ꜝ) Light blue.

**mail** (–ꜝ) Leaning; inclined; oblique; inclined to; tending towards. **~i inhidam**, likely to collapse, in a dangerous condition (of a building). **~e**, slope.

**maişet** (·–ꜝ), **-ti** Means of subsistence; livelihood.

**maiyet, -ti** Suite; following. **~inde**, in his suite; accompanying him: **~ vapuru**, 'stationnaire' of an ambassador.

**makabil** (–·ꜝ), **-bli** That which goes before. **makabline şamil olm.**, (of a law *etc.*) to be retrospective.

**makad** Covering (of a sofa *etc.*); cushion; the behind.

**makale** (·–ꜝ) Article *in a newspaper*.

**makam** (·ꜜ) Place; abode; post; rank; office; executive; tomb of a saint; tune. **~ında**, in token of, after the manner of, *e.g.* **tezyif ~ında**, by way of derision, in contempt: **~ tutturmak**, to strike up a tune; to annoy by constant repetition: **alâkalı ~**, the competent authority: **her ~dan söylemek**, to talk on all kinds of subjects. **~at, -tı**, high offices; authorities. **~lı**, harmonious. **~sız**, inharmonious, discordant.

**makar, -rrı** Seat; centre. **~rı saltanat**, the Capital.

**makara** (·ꜝ·) Pulley; reel; spool; drum. **~ gibi söylemek**, to chatter incessantly: **~ları salıvermek** *or* **koyuvermek**, to burst into roars of laughter: **üç dilli ~**, a three-sheaved block.

**makarna** (·ꜝ·) Macaroni. **~cı**, a macaroni maker *or* seller; an Italian.

**makas** Scissors; shears; claw *of a lobster etc.*; switch *or* points of a railway; steering-rods; anything in the form of scissors; a cross-leg throw in wrestling. **~ ateşi**, cross-fire: **~ gülle**, chain-shot: **~ hakkı**, cuttings left over after cutting out a suit. **~çı**, a railway pointsman. **~lamak**, to cut with scissors; to tweak with the first and second fingers; to rob; plagiarize. **~lı**, having the form of scissors: **~ dürbün**, stereo-telescope. **~tar**, tailor's cutter.

**makber, makbere** Tomb; cemetery.

**makbul** (·ꜜ) Accepted; acceptable; liked. **~e geçmek**, to be received with pleasure: **~ümdür**, I accept gladly.

**makbuz** (·ꜜ) Received. Receipt *for payment*. **yüz lira ~um olmuştur**, I acknowledge receipt of 100 lira. **~at** (·–ꜜ), **-tı**, receipts.

**maket, -ti** (*Fr. maquette*) Sketch; outline; model.

**makferlân** (··ꜝ·) (*Macfarlane*) Inverness cape (woman's cloak with cape).

**makhur** (·ꜜ) Overwhelmed; defeated.

**maki** Lemur.

**Makidonya** (··ꜝ·) Macedonia.

**makina, makine** (ˋ··) Machine; engine. **~ye verilerken,** 'Stop Press' (in a newspaper): **~ zabıtı,** engineer officer (*naut.*): **yazı ~si,** typewriter. **~ci,** mechanic; engine-driver. **~li** (·ˋ··), driven by *or* fitted with a machine: **~ tüfek,** machine-gun.
**makinist, -ti** Engine-driver; engineer; mechanic.
**makiyaj** (*Fr. maquillage*) Making-up, painting the face.
**makkab** *For* **matkab.** Drill.
**makrama** Napkin; kerchief; handkerchief; bedspread; face-towel.
**maksad** Aim, purpose, intention. **~ile,** with a view to, with the intention of: ... **demekten ~,** ... means ... .
**maksur** (·⁎) Shortened; limited; reduced (military service). **~e** (·–ˋ), private pew in a mosque; private grounds. **hizmeti ~,** reduced military service.
**makta** (·–́), **-aı** Place of cutting; section; pause; cutting (in a wood).
**maktel** Place of execution *or* murder.
**maktu** (·⁎), **-uu** Cut off; interrupted; fixed (price). **~a,** cutting *of a newspaper etc.* **~an** (·–́·), at a fixed price; in the lump.
**maktul** (·⁎) Killed.
**mâkul** (–⁎) Reasonable; wise; prudent. **~ görmek,** to deem reasonable, to approve. **~ât** (––⁎), **-tı,** conceivable and comprehensible things; a thing based on reason. **~iyet** (·–·ˋ), **-ti,** reasonableness.
**makule** (·–ˋ) Kind, sort; category. **bu ~ adamlar,** this sort of men.
**makûs** (–⁎) Inverted; inverse; reversed; reflected; opposed; perverse (fate *etc.*); unlucky. **~en** (––́·), inversely.
**mal** Property; possession; wealth; goods; an animal owned (sheep, cows *etc.*); scamp, scoundrel; loose woman. **~ canlısı,** covetous, avaricious: **~ edinmek,** to become rich; to appropriate; *v. also* **maletmek**: **~ edinmemek** *or* **üzerine ~ etmemek,** to take no account of, not to worry about: **~ etm.,** *v.* **maletmek**: **~ın gözü,** rascal, ne'er-do-weel: **~ meydanda,** the proof is here, *i.e.* it's obviously bad: **~ mülk** *or* **~ menal,** property; goods: **~ olm.,** *v.* **malolmak**: **baba ~ı,** patrimony: **hepsi bir ~,** one's as bad as the other: **ne ~ olduğu anlaşıldı,** it is clear now what a scoundrel he is.
**mala** (ˋ·) Bricklayer's trowel. **~lamak,** to work *or* smooth with a trowel.
**malafa** Arbor; mandrel.
**malakof** Crinoline.
**malak** Buffalo calf.
**malâmal** Brimful.
**malârya** (·ˋ·) Malaria.
**malâyani** (–––́) Meaningless; useless. Nonsense; futile thing.
**mal·dar** Wealthy. **~edilmek** (ˋ···), *pass. of* **maletmek,** to be appropriated *etc.*; to belong. **~en** (–́·), in goods; financially.
**maletmek** (ˋ··), **-eder** *va.* Take possession of, appropriate; attribute to, ascribe to. **bir şeyi üstüne** *or* **kendine ~,** to appropriate *stg.* that does not belong to one: **üstüne ~,** to enter as a debit against: **···i ucuza ~,** to get *stg.* cheap.
**malgama** Amalgam.
**malıtaşı, -ni** Large stone used as an anchor.
**malî** (––́) Pertaining to property; financial.
**malihulya** (···–́) Melancholy; whim, fancy.
**malik** (–ˋ), **-ki** Owning, possessing; (*with dat.*) owner of. **···e ~ olm.,** to possess: **kendine ~ olmamak,** to lose one's self-control; to be unconscious. **~âne** (–·–ˋ), State lands held in fief by a private owner; large estate. **~iyet, -ti,** ownership; rights of ownership.
**maliye** (–·ˋ) *Fem. of* **malî.** Financial. Finance; Ministry of Finance. **~ci,** financier.
**maliyet** (–·ˋ), **-ti** Cost; **~ fiati,** cost price.
**malmüdürü** (ˋ···), **-nü** Financial officer of a district.
**malolmak** (ˋ··) *vn.* (*with dat.*). Cost. **kaça maloldu?,** how much did it cost?
**mal·perest** Who worships money. **~sahibi** (ˋ···), **-ni,** proprietor. **~sandığı** (ˋ···), **-nı,** government financial department.
**Malta** (ˋ·) Malta. **~ eriği (muşmulası),** loquat. **~ palamudu,** the Pilot Fish (*Naucrates Ductor*). **~taşı** (ˋ···), **-nı,** a soft building stone.
**maltız** Maltese; a kind of brazier; a kind of goat.
**malûl** (–⁎), **-lü** Ill; invalid; defective. **~ gazi,** disabled soldier: **···le ~,** tainted with. **~in** (––́), disabled soldiers; war victims. **~iyet** (–́··), **-ti,** infirmity; defect.

**malûm** (–⸌) Known; active (verb). Yes; true! **~u ilâm,** a telling *stg.* that is known to all: **~unuzdur ki,** you know that.
**malûmat** (––⸌), **-tı** Information; knowledge. **~ına müracaat etm.,** to ask for information about: **~ sahibi,** a man of learning: **ondan ~ım yok,** I have no knowledge of, *or* no information about, that. **~furuş,** who poses as learned; pedant. **~lı,** learned; educated; well-informed. **~sız,** ignorant; uneducated. **~tar,** informed.
**malzeme** Necessaries; materials.
**mama** (*inf.*) Food. **~ bezi,** bib.
**mamelek** (–·⸌) All that one possesses.
**mamul** (–⸌) Made; manufactured. **···den ~,** made of ... . **~ât** (––⸌), **-tı,** manufactures; manufactured goods.
**mamur** (–⸌) Prosperous, flourishing. **dört başı ~,** flourishing in every way; first-class. **~e** (––⸌), a prosperous and cultivated place. **~iyet** (––·⸌), **-ti, ~luk** (––⸌), a flourishing condition; prosperity.
**mamut, -tu** Mammoth.
**mâna** (–⸌) Meaning; sense; motive; essence; dream; reason. **···den ~ çıkarmak,** to put a false interpretation on ...; to read into *a remark* an insinuation which was not there: **~ vermek,** to interpret, to translate (orally); **···e ~ vermemek,** to be unable to explain *stg.*; to be rather suspicious about it: **âlemi ~da görmek,** to see in a dream: **tam ~sile,** in the fullest sense of the word. **mâna·lı** (––⸌), significant; having the meaning of; suggestive; allusive. **~sız** (––⸌), senseless; without significance.
**manastır** Monastery.
**manav** Fruiterer.
**manca** (⸌·) (*vulg.*) Food.
**mancana** (·⸌·) Large water-cask *on board ship.*
**mancınık** Catapult; ballista.
**manda**[1] Water buffalo; very fat person.
**manda**[2] (*Fr. mandat*) Mandate.
**mandal** Latch; catch; tumbler, pawl; cleat; clothes-peg; tuning-peg *of violin etc.* **dış kapının ~ı,** a distant relative: **kırk kapının ~ı,** one who goes everywhere and pokes his nose into everything. **~lamak,** to shut with a latch; to hang up washing with clothes-pegs.
**mandapost, -tu** Postal money-order.
**mandar** Small pulley-block.
**mandater** (*Fr. mandataire*) Mandatory.
**mandepsi** (*sl.*) *In* **~ya basmak (gelmek),** to be taken in, to be cheated.
**mandıra** (⸌··) Small cow-shed; sheep-pen; cheese dairy. **~ köpeği,** cattle dog; a brutal man *in the service of others.*
**mandoz** Block; pulley.
**mandren** (Drill) chuck.
**mânen** (⸌·) Morally (*as opp. to* materially).
**manend** (–⸌), **menend** Resembling; similar.
**maneska** (·⸌·) A pulley with two double blocks.
**manevî** (–·⸌) Moral (*as opp. to* material); spiritual. **~ evlâd,** another's child who is, to all intents and purposes, a son: **şahsı ~,** juridical person. **~yat, -tı,** spiritual and moral matters; morale.
**manevra** (·⸌·) Manœuvres; a manœuvre, trick. **~ çevirmek,** to manœuvre: **~ fişeği,** blank cartridge: **~ yapmak,** to manœuvre; to shunt *a train.* **~lı,** (*sl.*) foul, diabolic: **~ iş,** stratagem, ruse.
**manga** (⸌·) Squad (*mil.*); mess (*mil.*).
**mangal** Brazier.
**mangır** *An obsolete* copper coin; money; small disk of charcoal dust *placed on the bowl of a narghile to keep it alight.*
**mangiz** (*sl.*) Money.
**mani**[1] (–⸌) Song; ballad.
**mani**[2] (–⸌) (*Fr. manie*) Mania.
**mâni** (–⸌), **-ii** Preventing; hindering. Obstacle; impediment. **~ olm.,** to prevent, hinder.
**mania** (–·⸌) Obstacle; difficulty. **~lı,** presenting obstacles *or* difficulties: **~ yarış,** obstacle race.
**manidar** (––⸌) Significant; expressive.
**manifatura** (··⸌··) Textiles. **~cı,** draper.
**manika** (·⸌·) Windsail *for ventilating a ship.*
**manipülatör** Signalling-key.
**manita** Swindle.
**manivela** (··⸌·) Lever; crank.
**mankafa** (⸌··) Stupid; dazed; awkward; big; suffering from chronic glanders (horse). **~lık,** stupidity, thick-headedness; glanders.
**manke** (*Fr. manqué*) Miss.

**manken** Mannequin; scarecrow; ninny.
**manolya** (·ˋ·) Magnolia.
**mansab** River-mouth.
**mansıb** High office.
**Manş** (*Fr. Manche*) The English Channel.
**manşet, -ti** (*Fr. manchette*) Cuff; headline.
**mantar** Mushroom; cork; inner sole of a shoe: lie, invention. ~ **atmak,** to tell lies: ~**a basmak,** to be duped: ~ **meşesi,** cork-oak: ~ **pabuc,** shoe with cork soles *or* high cork heels. ~**cı,** mushroom-seller; liar.
**mantı**[1] Large pulley-block (on a ship).
**mantı**[2] A kind of meat pasty.
**mantık** Logic. ~**î** (··–́), logical.
**mantin** A thick silk tissue.
**mantinato** (··ˋ·) A kept woman; mistress.
**manto** (*Fr. manteau*) Cloak; mantle.
**mantol** Menthol.
**manya** (ˋ·) Mania.
**manyat** A kind of fishing-net.
**manyet·ize** Magnetized; mesmerized. ~**izma** (··ˋ·), magnetism; mesmerism. ~**o** (·ˋ·), magneto.
**manyezi** Magnesia.
**manzar** Aspect; appearance. ~**a,** view; spectacle; panorama; perspective. ~**alı,** having a fine view.
**manzum** (·ˋ) Written in rhyme and metre. ~**e** (·–ˋ), row; system; series; poem.
**manzur** (·ˋ) Seen; considered. ... ~**unuz oldu mu?,** have you seen ... ?
**mapa** Eye-bolt.
**maraba** Shepherd's assistant.
**marabet, -ti** Marabout.
**maral** Doe.
**marangoz** Joiner; cabinet-maker. ~ **balığı,** saw-fish. ~**luk,** joinery; cabinet-making.
**mâraz** Place of an occurrence; occasion.
**maraz** Disease, illness; worry; pain; evil. ⌜**merhametten** ~ **çıkar**⌝, misplaced pity may be the cause of evil. ~**î** (··–́), pertaining to disease; pathological; morbid.
**maraza** *Vulg. for* **muaraza.** Controversy; tumult.
**mareşal** Marshal (*mil.*).
**marifet** (–·ˋ), **-ti** Knowledge; skill; talent; craft, skilled trade; clever thing; contrivance; curiosity; means; intervention. **onun** ~**iyle oldu,** it happened by means of him, *i.e.* he did it *or* it occurred with his knowledge: **ehli** ~, talented, able: '**gördün mü yaptığın** ~**i?**', 'see what a mess you've made of it!' ~**li,** skilful; talented; cleverly made.
**mariz**[1] (·–́) Sick, ailing; depressed.
**mariz**[2] A beating.
**marka** (ˋ·) Mark; trademark; initials on clothing. ~**lı** (ˋ··), marked; bearing the mark of.
**markacı** (*sl.*) Swindler.
**marki** Marquess. ~**z,** marchioness.
**Marko** *Prop. n.* ~ **Paşa,** *Abdul Hamid's chief physician, renowned for his patience in listening to his patients, hence*: ⌜**derdini** ~ **Paşaya anlat!**⌝, 'nobody wants to hear about your troubles!'
**maroken** Morocco leather.
**marpuc** Tube of a narghile.
**mars** Grand slam; (at backgammon) a game lost without taking a piece. ~ **olm.,** to be badly beaten; to be dumbfounded.
**marsık** Imperfectly burnt charcoal *giving off poisonous fumes.* ~ **gibi,** black and ugly (woman).
**marsıvan (eşeği)** Ass.
**marş** March (tune); treadle *of a loom etc.* March! ~ **motörü,** (car) starter.
**marşandiz** Goods train.
**mart, -tı** March (month). ~ **havası,** changeable weather.
**martaval** Nonsense; lies. ~ **okumak,** to talk nonsense; to tell lies.
**martı** Gull.
**martika** (·ˋ·) A kind of two-masted sailing-vessel used along the Black Sea coast.
**martin** Martini rifle. **at** ~**i!,** *v.* **Debreli.**
**martoloz** Sailor from the Danube in the pay of the Turks.
**mâruf** (–ˋ) Known; well known; proper, usual. ... **demekle** ~, *or* ... **namile** ~, known as ... .
**marul** Cos lettuce. ~**cuk,** Christmas rose.
**mâruz** (–ˋ) Exposed to (*with dat.*); presented; submitted. ~**at** (––ˋ), **-tı,** representations; matters submitted *by an inferior to a superior.*
**marya** (ˋ··) Ewe; female animal; young fish, fry.
**masa** (ˋ·) Table; office desk; department in a government office; bankrupt's effects.
**masad** Steel *for sharpening knives.*

**masal** Story, tale; myth; silly tale. ~ **kabilinden,** fabulous: ~ **okumak,** to read *or* tell a tale; to romance: ~ **söylemek,** to tell idle tales. ~**cı,** story-teller.

**mâsara** Oil *or* wine press.

**masarif** (·–ˋ) *pl. of* **masraf.** Expenses; disbursements.

**masdar** Infinitive.

**mâsebak** (–·ˋ), **–kı** That which has gone before; the past; precedent.

**mâsiva** (–·ˊ) The vanities of this world. ~**dan geçmek,** to renounce all but God.

**masiyet** (–·ˋ), **–ti** Disobedience; sin.

**maskara** Buffoon; laughing-stock; droll child. Ridiculous; dishonoured. ~ **etm.** *or* ~**ya almak,** to make a laughing-stock of. ~**lanmak,** to make oneself a laughing-stock; to play the buffoon. ~**lık,** buffoonery; making oneself ridiculous; shame; dishonour: ~ **etm.,** to play the fool.

**maske** (ˋ·) Mask. ~**yi indirmek** *or* **atmak,** to throw off the mask: ~**sini* indirmek,** to unmask *s.o.* ~**lemek,** to mask; to camouflage. ~**li,** masked.

**maslahat, –tı** Business; affair; the proper course. **idarei** ~ **etm.,** to manage with what one has. ~**güzar,** Chargé d'Affaires.

**maslak** Stone trough *for watering animals*; running tap; water tower.

**maslub** (·ˋ) Hanged. ~**en** (·ˊ·), by hanging.

**masmavı** (ˋ··) Very blue.

**masnu** (·ˊ) Manufactured; artificial, false. ~**at** (·–ˋ), **–tı,** things made by skill; manufactures; created things; false news.

**masraf** Expense; outlay. ~**ını çıkarmak,** to pay fori tself: ~ **kapısı,** the expenditure side of an account; ~ **kapısını açmak,** to cause expense: ···**i** ~**a sokmak,** to put *s.o.* to expense. ~**lı,** expensive.

**masruf** (·ˊ) Spent; expended. Expense.

**mass·edici** Absorbent. ~**etmek** (ˋ··), **–eder,** to suck; to absorb.

**mastar** *v.* **masdar.**

**mastara** (·ˋ·) Index (of a sextant *etc.*); cursor.

**mastelâ** Small tub.

**mastı** Small short-legged dog. ~**bacak,** *v.* **bastıbacak.**

**mastika** (·ˊ·) Resin; raki with resin in it.

**mastur** (*sl.*) Drunk; drowsy.

**masturi** The broadest part of a ship.

**mâsum**( –ˋ) Innocent. Child. ~**iyet** (––·ˋ), **–ti,** innocence; infancy.

**masun** (·ˋ) Preserved, guarded; safe; inviolable. ~**iyet** (·–·ˋ), **–ti,** security; inviolability; immunity.

**masura** (·ˋ·) Small reed; weaver's shuttle; bobbin; spout; measure of water (= ¼ of a **lûle**).

**maş** Indian pulse.

**maşa** Tongs; pincers. **emniyet** ~**sı,** split-pin. ~**lı,** having pincers *or* tongs: **eli** ~, truculent, malevolent: ~ **gözlük,** pince-nez. ~**lık,** a being the tool of *s.o.*: ···**e** ~ **etm.,** to be the catspaw of ... .

**maşallah** (ˊ·–) (*Lit.* what (wonders) God hath willed!); wonderful!; *used to express admiration or wonder; to admire a child etc. without saying it would incur the risk of the evil eye.* Charm *worn by children to avert the evil eye.* ~**ı var!** *has much the same meaning as* **maşallah**: **kırk bir kere** ~, to say ~ over and over again in admiration.

**maşatlık** Non-Moslem (*esp.* Jewish) cemetery.

**maşlah** A loose open-fronted cloak without sleeves.

**maşraba** Metal pot *or* mug.

**maşrık, –kı** The East. ~**ı âzam**, Grand Lodge (Freemasonry).

**mâşuk** (–ˋ), **–ku** (*fem.* ~**a**) Beloved.

**mat[1], –tı** Checkmate (chess).

**mat[2]** Matt.

**matafora** (··ˋ·) Davit.

**matafyon** Eyelet.

**matah** *A form of* **meta,** goods, *only used contemptuously.* '**bu** ~ **değil ya!**', it's not as precious as all that!

**matara** (·ˋ·) Leather *or* tin water-bottle. ~**cı,** water-carrier attached to caravans.

**matbaa** Printing-press. ~**cı,** printer.

**matbah, mutfak** Kitchen.

**matbu** (·ˊ) Printed. ~**a** (·–ˋ), printed matter. ~**at** (·–ˋ), **–tı,** the papers; the Press.

**matem** (–ˋ) Mourning. ~ **etm.,** *or* ~ **tutmak,** to mourn; to go into mourning: ~ **havası** *or* **marşı,** funeral march. ~**li,** *a.* mourning; in mourning.

**matetmek** (ˋ··), **–eder** Checkmate; defeat.

**matiz[1]** Drunk.

**matiz[2]** Long splice.

**matkab** Drill; auger.
**matlab** Demand; wish; question.
**matlub** (·-̍) Desired; demanded. Debt due; desideratum. **~a kaydetmek,** to pass to *s.o.'s* credit: **~a muvafık çıkmadı,** it was not up to standard, not up to expectation: **~ vechile,** as desired; as it should be. **~at, -tı,** demands; debts due.
**matmazel** Mademoiselle.
**matolmak** (̍··) *vn.* Be checkmated *or* defeated.
**matrak, -kı** Mace; cudgel; (*sl.*) joking.
**matrud** (·-̍) Driven away; expelled.
**matruş** (·-̍) Shaven.
**mâtuf** (–̍) Directed; imputed; aiming at (= *dat.*).
**maun** Mahogany.
**maval** Lie; story.
**mavera** (–·-̍) That which is beyond. **~î** (–·–-̍), which lies beyond; transcendental. **~ünnehir,** Transoxiana.
**mavi** (–̍) Blue. **~lik,** blue colour. **~msi, ~mtrak,** bluish. **~ş,** blue-eyed; blond.
**mavna** (̍·), **mavuna** (·̍·) Barge; lighter. **~cı,** lighterman; bargee.
**mavzer** (̍·) Mauser.
**maya** Ferment; yeast; essence, essential; origin; stock; talent; brood mare *or other female animal.* **bu ~ daki** ..., this sort of ...; **~sı bozuk,** a 'bad lot'. **~lanmak,** to ferment; increase; be accumulated. **~lı,** fermented, leavened. **~lık,** anything serving as a ferment.
**mayasıl** Chilblain; (*pop.*) piles. **~-otu** (··̍··), **-nu,** scrofularia (?); *name given to various medicinal plants.*
**mayasız** Unfermented; unleavened; worthless (person).
**maydonoz** Parsley. ⌜**neler neler de ~lu köfteler**⌝, 'ah, I could tell you a lot of stories' *or* 'wouldn't you like to know!'
**mayhoş** Slightly acid; tart; bitter-sweet. **~luk,** a slightly sour taste.
**mayın** Floating mine.
**mayıs**[1] May (month). **~böceği, -ni,** cockchafer.
**mayıs**[2] Fresh cow *or* sheep dung; dried cow-dung *used as a fuel.*
**mayi** (–̍) Liquid; fluid. **~ mahruk,** liquid fuel. **~at** (–·̍), **-tı,** liquids. **~iyet, -ti,** liquidity.
**mayistra** (·̍·) Mainsail. **~ sereni,** mainyard.
**maymun** Monkey. **~ iştahlı,** inconstant, capricious: **~ suratlı,** hideous, repulsive. **~cuk,** little monkey; picklock. **~luk,** drollery.
**mayn** Floating mine.
**mayna** (̍·) Down sails! **~ etm.,** to down sails; to cease work: **~ olm.,** to come to a stop; to calm down.
**mayo** (*Fr. maillot*) Bathing costume. **~lu,** wearing a bathing costume.
**maytab** (*corr. of* **mehtab** *q.v.*) Small firework; Bengal fire. **~a atmak,** *or* **~ etm.,** to make fun of: **çanak ~ı,** Bengal fire.
**mazarrat, -tı** Injury, harm; detriment. **~ı dokunmak,** to harm, to be detrimental.
**mazbata** (·̍·) Official report; protocol; minutes *of a meeting.*
**mazbut** (·̍) Recorded; well protected *against rain etc.*; solid, well-built (house); fixed *in the mind*; decided; neat, compact; correct (style *etc.*); level-headed; decent.
**mazeret** (–·̍), **-ti** Excuse; apology.
**mazğal** Embrasure; loop-hole; slit in a machine-gun shield; *err. for* **mıskal** *q.v.*
**mazhar** Who acquires; honoured; distinguished; the object of (favours, honour *etc.*). **~iyet, -ti,** attainment; distinction; acquisition; success.
**mazi** Gall-nut; gall-bearing oak; arbor vitae; thuja.
**mazî** (–-̍) Past; bygone. The past; the past tense. **~ye karışmak,** to belong to bygone days, to be a thing of the past.
**mazlum** (·̍) Victim of cruelty; oppressed; mild, inoffensive. **~iyet** (·–·̍), **-ti, ~luk,** mildness; tractability.
**maznun** (·̍) Suspected; accused (of = **ile** *or* ···**den**).
**mazot, mazut, -tu** Crude oil.
**mazul** (–̍) Dismissed *from a post*; out of office. **~en** (–-̍·), *adv.* in retirement; discharged. **~in** (––-̍), dismissed officials. **~iyet** (––·̍), **-ti,** condition of being dismissed: **~ maaşı,** salary paid to an official dismissed from his post.
**mazur** (–̍) Excused; excusable. ···**in bir şeyini ~ görmek,** to pardon *s.o.* for *stg.*: **~ tutmak,** to hold *s.o.* excused.
**mazut** *v.* **mazot.**
**meal** (·-̍), **-li** Meaning, purport. **şu ~de bir şayia,** a rumour to this

effect. ~**en** (·⸌·), having regard to the meaning.
**mebde, -ei** Beginning; origin; first principle; starting-point.
**meblâğ** Sum of money; amount.
**mebni** (·⸌) Built, erected. ···**e** ~, based on.
**mebus** (·⸌) Deputy; member of Parliament. ~**luk,** quality of a deputy.
**mebzul** (·⸌) Abundant, lavish; cheap. ~**iyet** (·—·⸌), **-ti,** abundance; lavishness.
**mecal** (·⸌), **-li** Power; ability; possibility. ~**î* kalmamak,** to have no power left, to be no longer able. ~**siz,** powerless, exhausted.
**mecaz** (·⸌) Metaphor; figurative expression. ~**en** (·⸌·), figuratively. ~**î** (·—⸌), figurative; metaphorical.
**mecbur** (·⸌) Compelled. ~ **etm.,** to compel. ~**en** (·⸌·), by force, compulsorily. ~**î** (·—⸌), obligatory; forced. ~**iyet** (·—·⸌), **-ti,** compulsion; obligation; necessity.
**meccan·en** (·⸌·) Gratis; gratuitously. ~**î** (·—⸌), gratuitous.
**mecelle** The old Turkish Civil Code.
**mechul** (·⸌) Unknown; passive (verb).
**mecidiye** Silver coin of 20 piastres.
**meclis** Sitting; assembly; council; scene of a play; social gathering. **Millet** ~**i** *or* ~**i millî,** the National Assembly, Chamber of Deputies. ~ **kurmak,** to sit in council: **sözüm** ~**ten dışarı,** excuse the term!
**mecmu** (·⸌), **-uu** Collection; heap; total. ···**in** ~**u,** all of .... ~**a,** review, periodical, magazine. ~**an** (·⸌·), in all; wholly, totally.
**mecnun** (·⸌) Mad; madly in love. **M**~, *name of the hero of Eastern romance, the lover of Leyla.* ~**iyet** (·—·⸌), **-ti,** ~**luk,** madness; passion.
**mecra** (·⸌) Watercourse, conduit, canal; course.
**mecruh** (·⸌) Wounded; confuted, untenable.
**Mecuc** (—⸌) Magog; fabulous race of dwarfs. **Yecuc ve** ~, Gog and Magog.
**mecus, mecusî** (·—⸌) Magians; fire-worshippers. Pertaining to fire-worshippers; pagan; *also loosely used for* Hindu. ~**ilik** (·——⸌), ~**iyet, -ti,** fire-worship; Zoroastrianism; paganism.
**meczub** (·⸌) Attracted; crazy. One possessed by devils; ecstatic dervish; crazy fellow.

**meç, -çi** Rapier; foil.
**med, -ddi** Prolongation, extension; flow of the tide; high tide.
**medar** (·⸌) Centre of movement; orbit; tropic; point on which a question turns; means; help. ~**ı maişet,** means of subsistence: ···**e** ~ **olm.,** to help.
**meddah** Eulogist; public story-teller *or* mimic. ~**lık,** quality and occupation of such a story-teller; sycophancy, toadying.
**meded** Help, aid. ~!, help!: ~ **Allah,** 'only God can help!' (*used about a desperate situation or terrible event*).
**medenî** (··⸌) Civilized; civil; civic. ~ **cesaret,** moral courage. ~**ye,** *fem. of* **medenî.** ~**yet, -ti,** civilization.
**medfun** (·⸌) Buried.
**medhal, -li** Entrance; beginning; introductory principles; connexion, influence. ~**i olmamak,** to have no connexion with, to have nothing to do with. ~**dar,** participating; involved in.
**medh·etmek** (⸌··), **-eder** *va.* Praise. ~**i,** *v.* **medih.** ~**iye,** eulogy. ~**üsena** (···⸌), ~ **etm.,** to praise and extol.
**medih, -dhi** Praise.
**medine** (·⸌·) Town, city. **Medine,** Medina: ~ **fıkarası,** in rags and tatters.
**medlûl** (·⸌), **-lü** Sense, meaning.
**medrese** (·⸌·) *Formerly* Moslem theological school. ~**li,** educated at a Moslem school
**medyun** (·⸌) Indebted; in debt. Debtor. (*with acc.*) Owing ... .
**mefahir** (·—⸌) *pl. of* **mefharet.**
**mefhar** Glory. ~**et, -ti,** cause *or* object of glory; pride.
**mefhum** (·⸌) Understood. Sense; significance; concept, idea.
**mefkûre** (·—⸌) Ideal. ~**ci,** idealist. ~**vî** (·—·⸌), ideal.
**meflûc** (·⸌) Paralysed.
**mefruk** (·⸌) Separated; disjoined.
**mefruş** (·⸌) Spread (carpet *etc.*); furnished. ~**at** (·—⸌), **-tı,** carpets, mats *etc.*; furniture.
**mefruz** (·⸌) Hypothetical.
**mefsuh** (·⸌) Annulled; abrogated.
**meftun** (·⸌) (*with dat.*) Madly in love with; admiring. ~**iyet, -ti,** a being madly in love; intense admiration.
**meftûr** (·⸌) Languid; lukewarm.
**mef'ul, meful** (·⸌) Made; done; passive. **ismi** ~, passive participle.

**meğer** (ʹ·), **meğerse** (·ʹ·) But; however; only; it seems that. **~ki** (ʹ··), provided that.
**meh** *v.* mah.
**mehabet** (·–ʹ), **–ti** Awe; majesty. **~li,** majestic; awe-inspiring.
**meharet** (·–ʹ), **–ti** Skill; proficiency.
**mehaz** (–ʹ) Source from which something is taken; authorities *used in writing a book.*
**Mehdi** The Moslem Messiah, *who will appear in due time to deliver the faithful*; Mahdi.
**mehel** *For* **mahal** *in* **~dir,** it serves him right. **sana ~dir,** it serves you right.
**mehengir** Carpenter's gauge; surface gauge. **mafsallı ~,** scribing-block.
**mehenk** *v.* **mihenk.**
**mehmaemken** (·ʹ··) As far as possible.
**Mehmedcik** The Turkish 'Tommy Atkins'.
**mehtab** Moonlight; *v. also* **maytab.**
**mehter** Band which played to a great man. **~hane** (··–ʹ), a military band in the suite of a Vizier *and* the place where that band lived.
**mekân** (·ʹ) Place; site; abode. **~ tutmak,** to establish oneself.
**mekanik** Mechanics.
**mekik** Weaver's shuttle. **~ dokumak,** to be moved about from pillar to post.
**mekkâre** (·–ʹ) Pack-horses of an army. **~ci,** soldier in charge of military pack-animals.
**Mekke** (ʹ·) Mecca.
**mekruh** (·ʹ) Abominable, disgusting; not prohibited but frowned upon by religious law. **~at** (·–ʹ), **–tı,** disgusting things.
**mekşuf** (·ʹ) Uncovered; discovered; manifest.
**mekteb** School. **~i asmak,** to play truant: **~ görmüş,** who has been educated at a school: **ana ~i,** kindergarten. **~li,** school-child; who has a school diploma; officer who has been at a military school.
**mektub** (·ʹ) Written. Letter. **~cu,** Chief Secretary of a Ministry. **~laşmak,** to correspond by letter.
**melâike** (·–·ʹ) *pl. of* **melek.** Angels; *as sing.* angel.
**melâl** (·ʹ), **–li** Melancholy; depression.
**melâmet** (·–ʹ), **–ti** Blame, censure.
**melâmî** (·–ʹ) *Member of a sect of dervishes who disregard the outward rites of religion.*
**mel'anet, –ti** An execrable act.
**melâs** Molasses.
**melce, –ei** Refuge, asylum.
**melek** Angel. **~ otu,** angelica. **~üssiyane** (·····–ʹ), guardian angel.
**melek·ât** (··ʹ), **–tı** *pl. of* **meleke.** Innate faculties; *as sing.* natural capacity. **~e,** proficiency, skill; natural faculty.
**melekût, –tu** God's spiritual dominion; creation.
**melemek** *vn.* Bleat.
**melengeç** *v.* merlengeç.
**melez** Cross-bred; half-bred. Mulatto; a cross (in breeding). **~ ağacı,** larch. **~lemek,** to cross *in breeding*: **ilk melezleme,** first cross.
**melfuf** (·ʹ) Wrapped up; enclosed *in a letter.* Enclosure. **~at** (·–ʹ), **–tı,** enclosures. **~en** (·ʹ·), as an enclosure.
**melhem** Ointment; salve. **⌈kelin ~i olsa kendi başına sürer⌉,** 'don't expect help from one who needs help himself'.
**melhuz** (·ʹ) Anticipated; probable. **~at** (·–ʹ), **–tı,** things occurring to the mind; anticipated *or* probable events.
**melik, –ki** King. **~e** (·–ʹ), queen.
**melisa** Lemon verbena.
**melon** Bowler-hat.
**meltem** Breeze that blows off the shore every day in summer.
**melûl** (·ʹ) Low-spirited; vexed. **~ ~ bakmak,** to wear a piteous expression.
**mel'un** (·ʹ) Accursed. Accursed man.
**memat** (·ʹ), **–tı** Death. **hayat ~ meselesi,** a matter of life or death.
**memba** (·ʹ), **–aı** Spring; source; origin.
**meme** Teat; nipple; lobe *of the ear*; tumour; burner *of a lamp.* **~den kesmek,** to wean. **~li,** having teats; mammiferous: **~ hayvan,** mammal. **~lik,** cover to protect a sore teat *or* to prevent sucking.
**memişhane** (··–ʹ) Privy.
**memleha** Salt-pit; saltworks.
**memleket, –ti** Dominion; country; town; a man's home district. **~li,** inhabitant; fellow countryman.
**memlûk** (·ʹ), **–kü** Possessed. Slave; Mameluke. **~iyet** (·–·ʹ), **–ti,** slavery.

**memnu** (·⊥) Forbidden. **~at, -tı,** forbidden things. **~iyet, -ti,** prohibition.
**memnun** (·ˋ) Pleased; glad; happy; grateful; under an obligation. **~ etm.,** to please; to make happy. **~en** (·⊥·), gladly, with pleasure. **~iyet** (·–·ˋ), **-ti,** pleasure; gratitude.
**memşa** (·⊥), **memşane** Privy.
**memul** (–ˋ), **-lü** Hoped; expected; desired. Thing hoped for; a hope. **~ etm.,** to hope, expect.
**memun** (–ˋ) Secure; trusted.
**memur** (–ˋ) (*with dat.*) Charged with; ordered to. Official; agent; employee. **~ olm.,** to be charged with *some duty*: **devlet ~u,** a government official: **sıhhiye ~u,** health officer. **~en** (–⊥·), officially. **~iyet** (––·ˋ), **-ti,** official duty; appointment; office; charge; post. **~luk** (––ˋ), quality and duties of an official; official post.
**men, -n'i** A preventing *or* prohibiting; prohibition.
**menafi** (·–⊥), **-ii** *pl. of* **menfaat.** Benefits *etc.*
**menakib** (·–ˋ) *pl. of* **menkibe.** Eulogies; legends; epic deeds. **~name,** life of a saint.
**menba** *v.* **memba.**
**mendebur** Idle; good-for-nothing; disagreeable, disgusting.
**mendil** Handkerchief.
**mendirek** Breakwater; artificial harbour.
**menekşe** (·ˋ·) Violet.
**menend** Resembling, like. **misli ~i yok,** he has no peer.
**menengüş** *v.* **merlengec.**
**menetmek** (ˋ··), **-eder** *va.* Forbid; prevent.
**meneviş** Fruit of the terebinth; wavy appearance of shot silk; blueing of steel. **~li,** wavy; watered (silk); blued (steel).
**menfa** (·⊥) Place of exile.
**menfaat, -ti** Use; advantage; profit. **~ görmek,** to experience a benefit. **~li,** useful; advantageous; beneficial. **~perest,** self-seeking; always looking for gain.
**menfez** Hole; air-hole; vent.
**menfi** (·⊥) Exiled, banished; negative; contrary, perverse, antagonistic; adverse. **~lik,** negation; denying; contrariness. **~yen** (·⊥·), in exile; negatively; perversely.
**menfur** (·ˋ) Loathed, abhorred. herkesin **~u,** detested by everyone.
**mengec** Large shuttle for wool.
**mengel** Bangle.
**mengene** (ˋ··) Press; vice; clamp; mangle.
**menhus** (·ˋ) Ill-omened, inauspicious; cursed.
**menî** Semen.
**men'i** (ˋ·) *v.* **men.**
**menkibe, menkabe** Epic; panegyric; recital of a great man's deeds *or* virtues; legend. **~vî** (···⊥), legendary.
**menkûb** (·ˋ) Disgraced. **~iyet** (·–·ˋ), **-ti,** a being disgraced.
**menkul** (·ˋ) Transported, conveyed; traditional. **~dur ki,** it is traditionally related that: **kendinden ~,** that which, according to himself, he possesses: **gayri ~,** real (estate). **~ât, -tı,** movables; traditions.
**menolunmak** (ˋ···) *vn.* Be prevented *or* forbidden.
**mensub** (·ˋ) Related to, connected with. **~at** (·–ˋ), **-tı,** things belonging *or* attributed; one's relatives. **~iyet** (·–·ˋ), **-ti,** relationship; connexion; membership (of a society *etc.*).
**mensuc** (·ˋ) Woven. **~at** (·–ˋ), **-tı,** textiles.
**mensur** (·ˋ) In prose. **~e** (·–ˋ), prose.
**menşe** Place of origin. **~li,** originating from; exported from.
**menşur** Prism; royal patent. ... **~undan geçirmek,** to see *stg.* in the light of ... .
**menteşe** (ˋ··) Hinge.
**menus** (–ˋ) To which one is accustomed; familiar; in common use, current.
**menzil** Halting-place; stage, day's journey; house; inn; range (of a gun); transport branch of an army; lines of communication. **~e,** degree; rank; social status. **~li,** having a range of ... : **uzun ~,** long-range (gun *etc.*).
**menzul** (·ˋ), **-lü** Stricken with apoplexy; paralysed.
**mer'a** Pasture.
**merak, -kı, -ğı** Curiosity; whim; passion *for stg.*, great interest; anxiety; depression, melancholy. **birini ~ta bırakmak,** to cause anxiety to *s.o.*: **birinin ~ına dokunmak,** (of a matter) to make *s.o.* uneasy: **···e ~ etm.,** to have a passion for, *or* to be

very interested in *stg.*; **bir şeyi ~ etm.**, to be curious about a thing: **birini ~ etm.**, to be anxious about a person: **~ halini almak,** (of *stg.*) to become a passion: **bir şeye ~ı* olm.**, to have a passion for, to make a hobby of *stg.* **~aver** (··–ˈ), causing anxiety. **~lanmak,** to be anxious: ···**e ~,** to be curious about *or* have an interest in *stg.* **~lı,** curious; interested in, fond of; anxious: **futbol ~sı,** a football devotee. **~sız,** free from anxiety; uninterested, indifferent.

**meral, -li** Roe; doe.

**mer'alık** (·–ˈ) Well-pastured. Pasturage.

**meram** Desire; intention; aim. **~ etm.**, to wish, intend, strive: **~ anlamaz,** unreasonable; who can't be made to understand: **~ını* anlatmak,** to be able to express what one wants: ⌜**~ anlıyan beri gelsin**⌝, what's the good of talking; nobody takes any notice!: ⌜**~ın elinden bir şey kurtulmaz**⌝, 'where there's a will there's a way': ⌜**~ etmiye görsün**⌝, if he wants a thing he gets it: **~ına* nail olm.**, to attain one's object.

**meraret** (·–ˈ), **-ti** Bitterness.

**merasim** (·–ˈ) Ceremonies; established usages; *as sing.* ceremony; commemoration. **~ geçişi,** ceremonial march-past: **~le karşılamak,** to give a ceremonious welcome to ... .

**meratib** (·–ˈ) *pl. of* **mertebe.** Ranks; degrees.

**merbut** (·ˈ) Attached; appended; dependent; captive (balloon), **~en** (·ˈ·), as an attachment *or* appendage. **~iyet** (·–·ˈ), **-ti,** dependence; devotion.

**mercan** Coral. **~ balığı,** red sea-bream: **~ kayaları,** coral reef, atoll: **~ terlik,** red leather slippers. **~köşk,** marjoram.

**merci** (·ˈ), **-ii** Place to which recourse is had; source to which a thing is referred; recourse; reference; competent authority.

**mercimek** Lentil. ···**le mercimeği fırına vermek,** to flirt with; to come to terms with: **su mercimeği,** duckweed.

**mercu** (·ˈ) Requested.

**merd** Man; brave, manly man. Manly; brave; fine *in character.* **~ane**[1] (·–ˈ), manly, virile: in a manly way. **~ce** (ˈ·), bravely; as becomes a man. **~lik,** manliness, courage.

**merdane**[2] (·–ˈ) Inking cylinder; rolling-pin; roller.

**merdiven** Ladder; steps; stairs. **kırkına ~ dayamak,** to be nearing forty *years of age.*

**merdüm** Man; human being. **~ek,** little man; manikin; pupil of the eye. **~giriz,** misanthropic; unsociable.

**meret, -ti** The damned fellow!; the cursed thing!

**merhaba** (ˈ·· *or* ˈ·–) Good-day!; how are you? **~laşmak,** to greet one another.

**merhale** A day's journey; stage. **~ ~,** by stages.

**merhamet, -ti** Mercy, pity; **~li,** merciful; tender-hearted. **~siz,** merciless; cruel.

**merhem** Ointment; salve.

**merhum** (·ˈ) Deceased; 'the late ...' (*correctly only used of Moslems*). **~ olm.**, to die.

**merhun** (·ˈ) Pledged; pawned; contingent.

**mer'i** (·ˈ) Observed; in force (of a law *etc.*).

**Meriç** (ˈ·) The river Maritza.

**Merih** Mars.

**merinos** Merino.

**mer'iyet, -ti** A being valid *or* in force; validity. **~e geçmek,** to come into force.

**merkad** Resting-place; bed; grave.

**merkeb** Mount (horse *etc.*); donkey; ship. **~ci,** donkeyman.

**merkez** Centre; administrative centre; central office; condition, manner. **polis ~i,** police-station. **siklet ~i,** centre of gravity: **'düşünceniz ne ~dedir?'**, 'what do you think about it?' **~î** (··ˈ), central. **~iyet, -ti,** centralization; a being central. **~lenmek, -ileşmek,** to be centralized.

**merkum** (·ˈ) The said (person), the above-mentioned (*contemptuously*).

**merlanos** Whiting.

**merlengec** Mastic-tree (Lentiscus).

**mermer** Marble. **~ kaymağı** *or* **su ~i,** alabaster: **~ kireci,** lime from burnt marble. **~lik,** marble paving. **~şahi** (··–ˈ), book muslin.

**mermi** Projectile; shell.

**mersin** Myrtle. **~ balığı,** sturgeon: **~ morinası,** the largest of the sturgeons (*Acipenser huso*): **çiga (çuka) ~,** sterlet (*Acipenser ruthenus*).

**mersiye** Elegy.

**mersörize** Mercerized.

**mert** *v.* **merd.**
**mertebe** Degree; rank; grade. **mümkün ~,** as far as possible.
**mertek** Squared baulk of timber; beam. **elifi ~ zannetmek,** to be very ignorant.
**Meryem** Miriam; Mary. **~ana** (·˙··), the Virgin Mary.
**merzengûş** Marjoram.
**mes** *v.* **mest**[2].
**mesabe** (·–˙) Degree; quality; nature. ... **~sinde,** of the nature of ..., like ... .
**mesafe** (·–˙) Distance; space. **kat'ı ~ etm.,** to traverse a distance.
**mesağ** (·⊥) Sanction; lawfulness.
**mesaha** (·–˙) The measurement of land; the measure *of a field etc.* **~ şeridi,** tape-measure: **~ zinciri,** measuring-chain.
**mesam·e** (·–˙) (*pl.* **~at**) Pore.
**mesane** (·–˙) Bladder.
**mesavi** (·–⊥) Evil conditions *or* acts; misdeeds, vices.
**mesbuk** (·˙) Preceded; surpassed; having a precedent. **~ hizmetlerine,** for services rendered: **gayri ~,** unprecedented.
**mescid** Mosque (*esp.* a small mosque).
**mesel** Proverb; parable; instance. **~â** (˙·–), for instance, for example.
**mesele** Question; problem; thesis; a matter of concern. **hiçten bir ~ çıkarmak,** to make a fuss about nothing: **bir şeyi ~ yapmak,** to make a to-do about *stg.*
**meserret, -ti** Joy; rejoicing.
**meshetmek** (˙··), **-eder** *va.* Stroke; rub lightly; wipe the shoes with the palm of the hand to replace ritual ablution.
**mesih**[1], **-shi** A touching lightly with the hand.
**Mesih**[2] (·⊥), **~a** The Messiah.
**mesina** (·˙·) Silkworm gut.
**mesire** (·–˙) Promenade; excursion spot.
**mesken** Dwelling.
**meskenet, -ti** Lack of spirit; sluggishness.
**meskûkât** (·–˙), **-tı** Coins.
**meskûn** (·˙) Inhabited.
**meskût** (·˙) Silenced; passed over in silence.
**meslek, -ki, -ği** Career, profession; mode of acting *or* thinking; moral character; principle. **~ten,** by profession: **~ sahibi,** who has a profession; a man of sound principles: **~ten yetişme,** professional, 'de carrière'. **~daş,** one of the same profession; colleague. **~î** (··⊥), professional. **~siz,** without a career; unprincipled.
**mesmu** (·⊥) Heard, audible; valid. **~at** (·–˙), **-tı,** rumours; hearsay.
**mesned** Place of support; fulcrum; post, office of dignity.
**mesnevi** (··⊥) A poem in rhymed couplets.
**mesrur** (·˙) Glad; contented. **~iyet** (·–·˙), **-ti,** happiness.
**mest**[1], **-ti** Light soleless boot, *worn in the house or with overshoes.*
**mest**[2] Drunk. **~ etm.,** to intoxicate; to enchant: **~ olm.,** to be intoxicated; to be enraptured. **~ane** (·–˙), in a drunken manner. **~edici** (˙···), intoxicating.
**mestur** (·˙) Covered; veiled; secret. **~ endaht,** indirect fire (*mil.*). **~e** (·–˙), *for* **tahsisatı ~,** secret funds.
**mes'ud** (·˙) Happy; fortunate. **~iyet** (·–·˙), **-ti,** happiness.
**mesul** (·˙) Responsible, answerable (for = ···**den**). **~iyet** (·–·˙), **-ti,** responsibility. **~iyetli,** involving responsibility.
**meşahir** (·–˙) *pl. of* **meşhur.** Famous men.
**meşak** (·⊥), **-kkı** *pl. of* **~kat, -tı,** hardship; trouble. **~katlı,** troublesome; exhausting, wearisome.
**meşale** Torch. **~ci,** torch-bearer.
**meşbu** (·⊥) Satiated; saturated.
**meşe** Oak. Oaken. **~ odunu,** blockhead. **~cik,** germander. **~lik,** wood of oak trees.
**meşgale** Business; occupation; preoccupation; pastime.
**meşgul** (·˙) Occupied; busy; preoccupied. **~iyet** (·–·˙), **-ti,** preoccupation; occupation; a being busy.
**meşhed** Place of martyrdom; battlefield; tomb of a **şehid,** *q.v.* **~i,** inhabitant of Meshed: **~ mubalâğası,** a gross exaggeration.
**meşhud** (·˙) Seen, witnessed.
**meşhur** (·˙) Famous; well-known.
**meşihat** (·–˙), **-tı** Office of the Sheikh-ul-Islam.
**meşin** Leather. **~ suratlı,** thick-skinned.
**meşk, -ki** Model for writing; copybook; musical exercise. **~ olm.,** to serve as a model. **~etmek** (˙··),

**–eder,** *va.* to take as a model; *vn.* to practise writing; to practise music.
**meşkûk** (·⁎) Doubted; doubtful.
**meşkûr** (·⁎) Acknowledged with thanks; deserving thanks.
**meşreb** Natural disposition; character. (*As a compound*) having *such and such* a character, *e.g.* **hafif∼,** flighty, frivolous.
**meşru** (·–) Legal; legitimate. **∼iyet** (·—·), **–ti,** **∼luk,** legitimacy; legality.
**meşrub** (·⁎) Drinkable. Beverage. **∼at** (·–⁎), **–tı,** drinks.
**meşruh** (·⁎) Commented on; explained. **∼at** (·–⁎), **–tı,** marginal notes; written comments: **∼ta bulunmak**, to add comments *or* explanations.
**meşrut** (·⁎) Stipulated; bound by conditions. **∼a** (·–), property left by will with conditions attached. **∼i** (·–), constitutional. **∼iyet** (·—·), **–ti,** constitutional government.
**meşum** (·⁎) Inauspicious, ill-omened; sinister.
**meşveret, –ti** Consultation; council. **ehli ∼,** member of a council; competent adviser.
**meta** (·–), **–ai** Merchandise; goods. **ne ∼ olduğunu öğrendik,** we know now what sort of fellow he is (*i.e.* a scoundrel).
**metafora** (··–·) Davit. **borda ∼sı,** boat-boom.
**metalib** (·–) *pl. of* **matlab.** Requests; things demanded.
**metanet** (·–), **–ti** Firmness; solidity; tenacity; toughness.
**metazori** (··–·) (*sl.*) By force and threats.
**metbu** (·–), **–uu** Obeyed. Sovereign. **devlet ∼ası,** the state of which he is a subject. **∼iyet** (·—·), **–ti,** sovereignty.
**metelik** Obsolete coin of 10 paras. **∼ etmez,** not worth **a** 'sou': **meteliğe kurşun atmak,** to be penniless: **···e ∼ vermemek,** not to care a damn for .... **∼siz,** without a 'sou', penniless.
**meteris** Trench (*mil.*).
**methal** *etc. v.* **medhal** *etc.*
**metin, –tni** Text.
**metîn** (·–) Solid; firm; tough; trustworthy; strong (in character).
**metot** Method.
**metre** (–·) Metre. **∼ kare,** square metre: **∼ küp,** cubic metre.
**metres** (*Fr. maîtresse*) Mistress; kept woman.
**metris** Trench (*mil.*).
**metropolit** Greek bishop; Metropolitan. **∼lik,** title *or* office of a Metropolitan.
**metrûk** (·⁎) Left; abandoned; deserted; neglected; obsolete. **∼ât** (·–⁎), **–tı,** effects left by a deceased person. **∼e** (·–) *fem. of* **metruk;** divorced woman.
**mevad** (·–), **–ddı** *pl. of* **madde.** Matters; materials; objects; articles, paragraphs.
**mevali** (·—) *pl. of* **mevlâ.** Masters.
**mevc** Wave. **∼e,** a single wave *or* ripple.
**mevcud** (·⁎) Existing; present. The number present (at a meeting *etc.*); stock; available force (*mil.*). **∼at** (·–⁎), **–tı,** all existing things; creation. **∼iyet** (·—·), **–ti,** existence; presence: **bir ∼ göstermek,** to make oneself noticed.
**mevdu** (·–) Entrusted. **∼at** (·–⁎), **–tı,** things entrusted; deposits (with a bank *etc.*).
**mevhum** (·⁎) Imaginary, fancied; fictitious. **∼at** (·–⁎), **–tı,** imaginations; imaginary fears.
**mevki, –ii** Place; position; situation; post; seat *in a theatre etc.*; class *on a train etc.* **∼ine göre,** according to circumstances. **∼î** (··–), local. **∼li,** having a first-class compartment, seat *etc.*
**mevkib** Procession.
**mevkuf** (·⁎) Stopped; arrested; detained; dependent. Arrested person, prisoner. **∼en** (·–·), under arrest. **∼iyet** (·–·), **–ti,** arrest; detention; a being held in mortmain.
**mevkut** (·⁎) Fixed for a certain period; periodical.
**Mevlâ** (·–) The Lord God. **mevlâ,** master. **∼sını* bulmak,** to get what one deserves: ⌜**arayan ∼sını da bulur belâsını da**⌝, he got what he was asking for; it serves him right. **∼na** (·––), 'our lord', *title formerly given to great religious personages*; *dervishes' mode of addressing a man.*
**mevlevî** (··–) *A member of the order of 'Whirling Dervishes'.*
**mevlid** Time *or* place of birth; birthday.
**mevlûd** Born. Child; birthday of Mohammed; poems in honour of that birthday; memorial service held forty days after a death. **∼ kandili,** evening of the Feast of the Prophet's birthday.

**mevrud** (·⁎) Arrived; touched upon. ~**at** (·–⁎), **-tı,** things arrived (letters *etc.*).
**mevrus** (·⁎) Inherited; hereditary.
**mevsim** Season; proper time *for anything*; the between seasons, spring and autumn. ~**lik,** suitable for spring *or* autumn; seasonal. ~**siz,** unseasonable; premature; out-of-place.
**mevsuf** (·⁎) Endowed; qualified by *such and such an adjective.* Substantive.
**mevsuk** (·⁎) Trusted; reliable; authentic; documented. ~**an** (·⁎·), authentically, reliably. ~**iyet** (·–·⁎), **-ti,** authenticity; reliability.
**mevt, -ti** Death. ~**a** (·⁎), the dead.
**mev'ud** (·⁎) Promised; appointed; predestined. **eceli ~,** the appointed time of death.
**mevzi** (·⁎), **-ii** Place; position. ~**î** (··⁎), local.
**mevzu** (·⁎), **-uu** Placed; laid; situated; instituted; subject (to); conventional; customary. Subject; proposition. **bahis ~u** *or* ~**ubahis, -hsi,** in question; subject under discussion: **para ~ değildir,** it is not a question of money. ~**a** (·–⁎), *fem. of* **mevzu**: postulate; convention. ~**at** (·–⁎), **-tı,** subjects; rules, dispositions, regulations; conventions; legislation.
**mevzun** (·⁎) Weighted; balanced; symmetrical.
**mey** Wine.
**meyal** *In* **hayal ~,** *v.* **hayalmeyal.**
**meyan**[1] *n.* Middle; midst; interval. ~**larında,** in the midst of them; between them: **bu ~da,** amongst these; including ...; in the meantime.
**meyan**[2] Liquorice. ~**balı** (·⁎··), **-nı,** liquorice (sweet, medicine). ~**kökü, -nü,** liquorice root.
**meyane** (·–⁎) Middle; interval; correct degree of cooking (jam *etc.*). Middling, moderate. ~**de,** between us (them *etc.*): ~**sini bulmak,** to find the means; to reconcile; to reach just the right moment *for stg. to be done*: ~**si gelmek,** to reach the right moment; (of a dish) to reach the right consistency: ~**ye girmek,** to get between, as an obstacle *or* as mediator. ~**ci,** middle-man.
**meydan** Open space; public square; arena, ring, ground; the open; opportunity. ~**da,** in the open; houseless; exposed; manifest, obvious; 'evidently': **~ aramak,** to seek space; to seek an opportunity: ~**a atmak,** to put forward, to suggest: **~ bulmak,** to find an opportunity: ~**a çıkmak,** to come forth, to show oneself; (of a child) to grow up: ~**a çıkarmak,** to expose to view, publish, discover; to bring up *a child* to maturity: ~**a gelmek,** to come into the open; to become celebrated; to reach maturity: ~**a getirmek,** to bring into view; to form *or* create: **harb ~ı,** battlefield: **~ muharebesi,** pitched battle: **~ okumak,** to challenge: **~ vermek,** to give an opportunity; to give encouragement: **At ~ı,** the Hippodrome in Istanbul: '**vaziyet ~da!**', 'you know very well what the situation is!'
**meydan·cı,** man employed to clean public places; caretaker. ~**lık,** open flat space.
**meyhane** (·–⁎) Wine-shop, tavern. ~**ci,** tavern-keeper, publican.
**meyil, -yli** Inclination; slope; propensity; affection, liking. **···e ~ göstermek,** to have an inclination for ... . ~**li,** inclined.
**meyl·etmek** (⁎··), **-eder** *vn.* Be inclined. **···e ~,** to have a propensity *or* liking for. ~**i,** *v.* **meyil.**
**meymenet, -ti** A being lucky *or* auspicious. ~**li,** auspicious. ~**siz,** inauspicious; unsympathetic, disagreeable (person).
**meyus** (–⁎) Hopeless, despairing. ~**iyet** (––·⁎), **-ti,** hopelessness, despair.
**meyva** Fruit. **~ yaprağı,** fruit-bud. ~**cı,** fruiterer. ~**lı,** made of fruit. ~**lık,** fruit-garden; receptacle for fruit.
**meyval** (*with dat.*) Very inclined towards; very fond of.
**meyzin** *v.* **müezzin.**
**mezad** Auction; auction-place. ~**a koymak (vermek, çıkarmak),** to sell by auction: **~ malı,** goods bought at an auction; cheap trifles; bargain; cheap, tawdry: **~ olm.,** to be sold by auction. ~**cı,** auctioneer.
**mezalim** (·–⁎) Cruelties.
**mezamir** (·–⁎) Psalms.
**mezar** Grave, tomb. **~ kaçkını,** a person with one foot in the grave. ~**cı,** grave-digger. ~**lık,** cemetery. ~**taşı** (·⁎··), **-nı,** tombstone.
**mezbaha** Slaughterhouse.
**mezbele** Refuse-heap.

**mezbuhane** (·——ˈ) Desperate; suicidal.
**mezbur** (·ˈ) Aforesaid.
**mezc** A mixing *or* blending. **~etmek** (ˈ··), **-eder,** to mix, combine, blend.
**meze** Appetizer; snack; hors-d'œuvre. **~ci,** seller of snacks. **~lik,** anything used as an appetizer *or* to accompany a drink.
**mezeborda** (·ˈ··) Broadside.
**mezellet, -ti** Abjectness; baseness.
**mezestre** Half-mast.
**mezevolta** Half-hitch.
**mezğerdek, mezğeldek** The Little Bustard.
**mezheb** Religion, creed; doctrine; sect. **~i geniş,** too tolerant in matters of morals.
**mezin** *v.* **müezzin.**
**mezit, -ti, mezitbalığı, -nı** Whiting.
**meziyet, -ti** Virtue; talent; value.
**mezkûr** (·ˈ) Aforementioned; the said ... .
**mezmum** (·ˈ) Blamed; ill-spoken of.
**mezru** (·ˈ) Sown. **~at** (·—ˈ), **-tı,** sowings; seeded fields; crops.
**mezun** (—ˈ) Authorized; on leave; having a school diploma; graduate; excused *from performing some duty.* **~en** (—ˈ·), *adv.* on leave. **~iyet** (——·ˈ), **-ti,** leave, furlough; permission, authorization.
**mıcır** Grit used in road surfacing; coal-dust; small ashes. **~ık,** crushed, squashed.
**mıh** Nail. **~ gibi kapalı,** tightly shut. **~î** (·ˈ), nail-shaped: **hattı ~** *or* **~ yazı,** cuneiform writing. **~lamak,** to nail. **~lanmak,** to be nailed; to be nailed to the spot. **~lı,** nailed.
**mıhladız** *v.* **mıknatıs.**
**mıknatıs** Magnet. **~î** (·——ˈ), magnetic. **~iyet, -ti,** magnetism.
**mıncık** *v.* **cıcık.**
**mıncıklamak** *va.* Claw about (as a cat); tease apart; handle too much; over-caress.
**mıntaka** Zone; district. **~vî** (···ˈ), regional.
**mırdar** *v.* **murdar.**
**mırıl·damak, ~danmak** *vn.* Mutter to oneself; grumble. **~tı,** muttering; grumbling.
**mırın** *Only in* **~ kırın etm.,** to show vague disapproval; to appear unwilling; to boggle.
**mırmır** *Imitates the sound of muttering.*
**mırnav** Miaow.
**Mısır**[1] (ˈ·) Egypt. **mısırlı,** Egyptian.
**mısır**[2] Maize; parched maize.
**mıskal** Burnisher. **~a vurmak,** to burnish.
**mısra** (·ˈ), **-aı** Hemistich; line of poetry.
**mışıl mışıl** *Imitates the sound of heavy breathing.* **~ ~ uyumak,** to sleep soundly.
**mıymıntı** Weak; slack; useless (man).
**mızık** Shilly-shallying, indecision; a not obeying the rules of a game. **~çı,** an untrustworthy, querulous, dithering sort of fellow; one who does not obey the rules of a game. **~lanmak,** to spoil a game by not obeying the rules.
**mızıka** (·ˈ·) Military band. **ağız ~sı,** mouth-organ.
**mızmız** Hesitant, unable to make up his mind; lazy, slow; querulous.
**mızrab** Instrument for striking; plectrum.
**mızrak, -kı** Lance. ⌜**~ çuvala sığmaz**⌝, 'the spear will not go into the sack', *used to deride an obvious falsehood*: ⌜**tek (yek) at tek mızrakı**⌝, all alone; without family. **~lı,** armed with a lance; lancer: **~ ilmihal,** (*formerly*) a standard elementary school-book.
**mi, mı, mu, mü** 1. *Interrogative particle.* 2. *Adds emphasis.* **cahil mi cahil,** ignorant beyond words: **yapar mı yapar !,** he's quite capable of doing that; of course he will do that.
**miad** (—ˈ) Fixed place *or* time; fixed period *for the renewal of clothes etc. issued to soldiers or pupils, or for the renewal of tyres on a car etc.*
**mibzer** Drill for sowing.
**micmer** Censer.
**miço** (ˈ·) Cabin-boy; boy waiter.
**mide** (—ˈ) Stomach; good taste. **~si* bulanmak,** to be nauseated; to feel suspicious: **~si kaldırmak (kabul etm.,)** to feel like eating *stg.*; to swallow an insult: **~ye oturmak,** to lie heavy on the stomach, be indigestible. **~ci,** who thinks only of his belly; self-seeker. **~siz,** having bad taste; eating anything. **~vî** (—·ˈ), good for the stomach.
**Midilli** (·ˈ·) Mitylene. **midilli,** small shaggy pony: **mahşer ~si,** short person; mischief-maker.
**midye** (ˈ·) Mussel.
**miftah** Key.

**miğfer** Helmet.
**miğra, miğri** (ˈ·) Conger eel.
**miğren** Migraine, headache.
**mihanik** Mechanics. **~î,** mechanical.
**mihber** Test-tube.
**mihenk, mihek** Touchstone; test; standard. **mihenge vurmak,** to test. **~çi,** silversmith who tests metals.
**mihir, -hri** Sun; affection, love.
**mihman** Guest. **~dar,** host; official charged with offering hospitality to distinguished visitors. **~nüvaz,** hospitable.
**mihnet, -ti** Trouble; affliction. **~keş, ~zede,** afflicted; disconsolate.
**mihrab** Niche in a mosque indicating the position of Mecca (and corresponding to the altar in a church). ⌜**cami yıkılmış ama ~ yerinde**⌝, in spite of damage the essential part is unharmed; a woman still beautiful though no longer young.
**Mihrace** (·–ˈ) Maharajah.
**mihver** Pivot; axle; axis.
**mika** (ˈ·) Mica.
**mikdar** Quantity; amount; value. **bir ~,** a small amount.
**mikrob** Microbe; evil person.
**mikyas** (·ˈ) Measuring instrument; proportion; scale; standard. **altın ~ından ayrılmak,** to come off the Gold Standard.
**mil**[1] Style; probe; obelisk; pivot. **ana ~i,** lead-screw of a lathe: **menteşe ~i,** gudgeon: **piston ~i,** gudgeon-pin: **göze ~ çekmek,** to blind *with an instrument.*
**mil**[2] Silt.
**mil**[3] Mile.
**milâd** (–ˈ) Birthday; birth of Christ. **~î** (––ˈ), Anno Domini, A.D.
**milel** *pl. of* **millet.** Peoples; nations.
**milim** Thousandth; 'millième'.
**milis** Militia.
**millet, -ti** Nation; people; people united by a common faith; a class of people; crowd. **Millet Meclisi,** the National Assembly. **~vekili** (·ˈ···), **-ni,** deputy; member of the Turkish Parliament.
**millî** (·ˈ) National. **~leştirmek,** to nationalize. **~yet, -ti,** nationality; religious community. **~yetçi, ~yetperver,** nationalist.
**milyar** Thousand million.
**milyon** Million.
**mim** The Arabic letter **mim** (m); tick, mark. **~ koymak,** to tick off, to make a mark against *stg.*
**mimar** (–ˈ) Architect. **~i** (––ˈ), **~lık,** architecture. **~î** (––ˈ), architectural.
**mimber** Pulpit in a mosque.
**mimik** Mimic.
**mim·lemek** *va.* To mark down (as suspect *etc.*). **~li,** marked off; politically suspect; a marked *man.*
**mina** (ˈ–) Enamel; glass; sky. Blue.
**minakop** The Umbra (fish) (*Umbrina cirrhosa*).
**minare** (·–ˈ) Minaret. **~ boyu,** 30 or 40 feet high: **şeytan ~si,** whelk.
**minder** Cushion; mattress; wrestling ring. **~ çürütmek,** (of a visitor *etc.*) to show no signs of going away.
**mine** (ˈ·) Enamel; dial (of a clock). **~çiçeği** (ˈ····), **-ni,** verbena. **~lemek,** to enamel. **~li** (ˈ··), enamelled.
**minelgaraib** *In* **ve ~,** how odd!
**mingayrihaddin** (·ˈ···) Although it is not for me to say; without wishing to lay down the law.
**minha** (ˈ·) *In* **anha ~,** *in such expressions as* **anha ~ razı oldu,** after much humming and hawing he agreed.
**mini·cik** Tiny. **~k,** small and sweet. **~mini** (·ˈ··), tiny.
**minimom** Minimum.
**minkale** Protractor.
**minnacık** (ˈ··) Tiny.
**minnet, -ti** Obligation *for a favour received*; taunt *about a former kindness.* **~ etm.,** to put oneself under an obligation *to s.o.*; to ask a favour; to bow: **canına* ~ (bilmek),** *expresses great pleasure, especially a pleasure mingled with relief*; **canıma ~!,** so much the better! *or* I accep twith pleasure (and relief): ⌜**ne sakala ~ ne bıyığa**⌝, avoid being under an obligation to anyone.
**minnettar,** grateful; indebted. **~lık,** gratitude.
**mintan** Sort of waistcoat with sleeves; shirt.
**mintarafillah** (ˈ···) By divine dispensation; fortunately; thank Heaven!
**minval, -li** Method; manner. **bu ~ üzere,** in this manner.
**mir** Chief; commander.
**mira** (ˈ·) Surveyor's rod.
**mirac** (–ˈ) Mohammed's ascent to heaven.
**mirahor** (*Formerly*) Master of the Horse.

**miralay** (*Formerly*) Colonel; Captain (Navy).

**miras** Inheritance. **~a konmak (yemek)**, to come into an inheritance: ⌜**ölüm hak, ~ helâl**⌝, there is nothing wrong in coming into an inheritance. **~çı,** inheritor. **~yedi,** spendthrift. **~yedilik,** extravagance, squandering.

**mirî** (–⁻́) Belonging to the State. The State treasury. **~ ambarı,** a government storehouse: **~ için,** on government account: **~ malı,** public money *or* property.

**mirikelâm** (–··⁻́) Eloquent man; good orator.

**mirim** (–̀) My dear!

**mirliva** (··⁻́) (*Formerly*) Brigadier-general.

**mis** Musk. **~ gibi,** sweetly scented; delicious; in a perfect manner: **kavga ~ gibi kokar,** a quarrel seems certain.

**misafir** (·–̀) Guest; visitor; a speck in the eye. **~ konağı,** guest-house for travellers in a village: **~ odası,** guest-chamber, reception room: **~ tohumu,** natural child. **~eten** (·–̀··), as a guest. **~hane** (·–·–̀), public guest-house for travellers. **~lik, ~et, –ti,** a being a guest; visit: **misafirliğe gitmek,** to pay a visit. **~perver,** hospitable. **~perverlik,** hospitality.

**misak** (–̀), **–kı** Solemn promise; pact.

**misal** (·̀), **–li** Model; precedent; like, match. **~ getirmek,** to give an example.

**misbah** (·̀) Fin; bladder; float.

**misil, –sli** A similar; an equal amount; as much again. **iki misli,** the double: **misli yok,** matchless. **~li, ~lü,** like, similar: **bu ~ adamlar,** these sort of people: **o ~,** like that, similar.

**misina** (̀··) Gut (trace).

**misk, –ki** *v.* mis.

**miskab** Drill; augur.

**miskal**[1] (·̀), **–li** Weight of 1½ drams (for precious stones). **~ ile,** a tiny quantity: ⌜**alışveriş ~le**⌝, 'business is business'.

**miskal**[2] (·̀), **–li** Pan-pipe.

**misket**[1], **–ti** Scented fruit; muscatel grape; Lady apple; grape-shot.

**misket**[2], **–ti** Musket.

**miskin** Poor; wretched; lazy; abject; poor-spirited; leprous. **~ illeti,** leprosy. **~hane** (··–̀), leper hospital. **~lenmek,** to become poor *or* wretched; become indolent. **~lik,** poverty; abjectness; incompetence; leprosy.

**misli** *v.* misil.

**misvak, –kı** Piece of wood beaten into fibres at one end for use as a toothbrush. **başı ~lı,** fanatic.

**mit, –ti** Myth.

**mitil** A kind of light quilt.

**miting** Meeting.

**miyan** Middle; midst; interval; loins. **~larında,** between them. **~cı,** go-between; intercessor.

**miyane** *v.* meyane.

**miyar** (·⁻́) Standard *of weight or measure*; chemical reagent.

**miyavlamak** *vn.* Miaow.

**miyop** Short-sighted.

**mizac, –ci** Temperament; disposition; state of health; constitution; mood, whim. **~gir,** obsequious; sycophantic. **~li,** having *such and such* a temperament *or* constitution: **tez ~,** hasty, passionate: **zayıf ~,** of weak constitution. **~siz,** unwell. **~sizlik,** indisposition.

**mizah** (·̀) Jest, joke. **~î** (·–⁻́), humorous.

**mizan** (–̀) Balance; pair of scales.

**mizana** (·̀·) Mizzen.

**mizansen** (*Fr. mise en scène*) Staging of a play (*also fig.*).

**mizitra** (·̀·) Fresh cheese from goat's milk.

**mizmar** (·̀) Pipe, flute; windpipe.

**mobilya** (·̀·) Furniture.

**moda** (̀·) Fashion.

**model** Pattern; model. **~ci,** pattern-maker.

**Moğol** Mongolian. **~istan,** Mongolia.

**mola** (̀·) Rest; pause; act of letting go *or* slacking off.

**molla** Theological student; chief judge; doctor of Moslem law.

**moloz** Rough stone; rubble. Useless. **~ duvarı,** wall of rough stones.

**monden** (*Fr. mondain*) Mundane; worldly.

**montaj** (*Fr. montage*) Mounting; fitting.

**mor** Purple, violet. ⌜**alı alına ~u ~una**⌝, flushed and out-of-breath: **~ salkım,** wistaria: **~ tavuk,** Purple Gallinule. **~armak,** to become purple; become bruised; be red with weeping. **~luk,** a being purple *or* violet. **~umtrak, ~umsu,** purplish.

**morg** (*Fr. morgue*) Mortuary.

**morina** (·⁻́·) Cod. **mersin ~sı,** the largest sturgeon *or* beluga.

**morto** (*sl.*) Dead. ~**yu çekmek,** to die.
**moruk** (*sl.*) Dotard; old fogey; 'the old man', 'the governor'.
**moskof** Russian; ruthless. ~**toprağı,** holystone.
**mosmor** (ˋ··) Bright purple.
**mostra** (ˋ·) Pattern; sample. ~ **olm.** (*sl.*), to make an exhibition of oneself. ~**lık,** a thing that is a sample only and not for sale, *hence* a person who puts in an appearance but does nothing.
**motör** Motor; (*generally*) motorboat. ~**lü,** having a motor; motorized.
**motris** (*Fr. motrice*) Electric locomotive.
**mozalak** Stunted.
**muaddel** Corrected; modified.
**muad·ele** (·—·ˋ) Equation. ~**elet, -ti,** a being equivalent. ~**il** (·—ˋ), equivalent: an equivalent, a similar one.
**muaf** Pardoned; excused; exempt; immune (to = **den**). ···**den** ~ **tutulmak,** to be exempted from ... . ~**iyet** (·—·ˋ), **-ti,** a being excused; exemption; immunity.
**muahede** (·—·ˋ) Pact; treaty. ~**name** (·—··—ˋ), document containing an agreement *or* treaty.
**muaheze** (·—·ˋ) Censure; criticism. ~ **etm.,** to blame *or* criticize.
**muahhar** Posterior; deferred; subsequent. ~**en** (·ˋ··), subsequently.
**muahid** (·—ˋ) Contracting. Signatory *of a treaty etc.*
**muakkib** Follower; pursuer.
**muallâ** (··ˊ) Exalted; sublime.
**muallâk** Suspended; in suspense. ~**ta kalan meseleler,** outstanding questions, matters in suspense.
**muallim** Teacher. ~**ime,** female teacher.
**muamele** (·—·ˋ) (*pl.* **muamelât**) A dealing with another; treatment; conduct; transaction; procedure; interest on money; sexual relations. ···**e fena** (**iyi**) ~ **etm.,** to treat *s.o.* badly (well): ~ **görmüş,** approved and marked 'for action' (of an application *or* decision): ···**i** ~**ye koymak,** to take the necessary official steps to carry out *an approved application etc.*: ~ **vergisi,** tax on business transactions: ···**le** ~ **yapmak,** to have dealings with ... . **muamele·ci** (·—··ˋ), broker. ~**li,** which is under consideration: (of an application *etc.*) approved and marked 'for action'.
**muamma** (··ˊ) Enigma.
**muammer** Long-lived.
**muannid** Obstinate, unyielding.
**muaraza** (·—·ˋ) Controversy.
**muarefe** (·—·ˋ) A being acquainted.
**muariz** (·—ˋ) Opposing; hostile; objecting. Opponent, antagonist.
**muasir** (·—ˋ) Contemporary.
**muaşaka** (·—·ˋ) A loving one another; love-making.
**muaşeret** (·—·ˋ), **-ti** Social intercourse. ~ **etm.,** to live together: **adabı** ~, the rules of social behaviour, code of manners.
**muateb** (·—ˋ) Reproved.
**muattal** Vacant; abandoned; disused; (factory) idle.
**muattar** Perfumed.
**muavaza** (·—·ˋ) Exchange; compensation.
**muavenet** (·—·ˋ), **-ti** Help, assistance. ···**e** ~ **etm.,** to help.
**muavin** (·—ˋ) Assistant, *esp.* an assistant official, assistant headmaster *etc.*; half-back (football). ~**lik,** post of an assistant official.
**muayede** (·—·ˋ) Visit of congratulation on a feast-day; wishing one another the compliments of the day.
**muayene** (·—·ˋ) Inspection; examination; scrutiny. ~ **etm.,** to inspect, examine, scrutinize. ~**ci,** customhouse inspector. ~**hane** (·—··—ˋ), doctor's consulting-room.
**muayyen** Definite; determined; known.
**muazzam** Great; huge; important. **düveli** ~**a,** the Great Powers.
**muazzeb** Tormented, pained.
**muazzez** Cherished; esteemed; honoured. ~**en** (·ˋ··), in an honoured way; with great honours.
**mubah** Neither commanded nor forbidden by religious law; lawful, permissible.
**mubassır** Superintendent; usher at a school.
**mubayaa** (·—·ˋ) (*pl.* ~**t**) Purchase; commercial transaction; wholesale buying.
**mucib** (—ˋ) Rendering necessary, causing. Cause; motive; requirement; necessary consequence. **bir** ~ **çekmek,** to pass a petition *etc.* as 'approved' (*v.* **mucibince**): ~**den**

çıkmak, (of a decree *etc.*) to be approved: ~i ibret, a means of warning to others, an example: ~ olm., to cause: esbabı ~e, determining cause, preamble to a Bill. ~ince (··̀·), according to requirements; as necessary; (*formula used by a superior official to denote that stg. is approved and passed for action*).
**mucid** (–̀) Inventor. Inventing.
**mucir** (–̀) Who lets *or* hires out.
**muciz** (–̀) Laconic, concise.
**mûciz** (–̀) Overpowering; perplexing.
**mucize** (–·̀) Miracle; wonder. ~ **kabîlinden,** by a miracle. ~**vî** (–··́), miraculous.
**mucur** Scoriae; dross; slag.
**muço** (̀·) Cabin-boy; boy waiter.
**mudil** (–̀) Difficult; arduous. ~**e,** difficult, complicated matter.
**mufarakat** (·–·̀), **-tı** Separation. ~ **etm.,** to part; to say goodbye to one another.
**mufarik** (·–̀) Separate; separable. **lâzımı gayri** ~, indispensable.
**mufassal** Detailed; lengthy. ~**an** (·̀··), at length and in detail.
**muğ** Magian; fire-worshipper; tavern-keeper.
**mugaddi** (··́) Nutritious.
**mugalâta** (·–·̀), **maglata** Misleading argument; sophistry; fallacy.
**muganni** (··́) Singer. ~**ye,** female singer.
**mugayeret** (·–·̀), **-ti** A not conforming; difference; opposition.
**mugayir** (·–̀) Opposed; contrary; adverse. ~**i edeb,** contrary to good manners; uncivil.
**muğber** Hurt, offended.
**muğfil** Deceitful; deceptive.
**muğlak** Abstruse; obscure; complicated.
**muhabbet, -ti** Love; affection; friendship; friendly chat. ~ **çiçeği,** mignonette: ~ **etm.,** to have a friendly chat. ~**li,** friendly; affectionate. ~**name,** friendly letter.
**muhab·ere** (·–·̀) Correspondence by letter *etc.*; signals service (*mil.*). ~**erat** (·–·̀), **-tı,** correspondence; communication. ~**ir,** correspondent.
**muhaceret** (·–·̀), **-ti** Emigration.
**muhacim** (·–̀) Assailant; raider; forward (football).
**muhacir** (·–̀) Emigrant; refugee (*esp.* Moslem).
**muhaddeb** Convex.
**muhaddis** One who studies and hands on the traditions of Islam.
**muhafaza** (·–·̀) Protection; conservation; preservation. ~ **etm.,** to protect, take care of, keep. ~**kâr** (·–··́), conservative.
**muhaff·ef** Lightened; alleviated. ~**if,** lightening, alleviating: **esbabı** ~**e,** extenuating circumstances.
**muhafız** (·–̀) Guarding. Guard; defender; commander of a fort. ~ **kıtası,** bodyguard.
**muhakeme** (·–·̀) Hearing of a case in court; trial; judgement, discernment. ~ **etm.,** to judge, decide. ~**li,** of sound judgement.
**muhakk·ak** Certain; well-known. Without doubt, certainly. ~**ik, -kı,** who verifies *or* scrutinizes.
**muhal** (·́) Impossible; inconceivable. **farzı** ~, if the impossible were to happen; in the unlikely event of.
**muhalefet** (·–·̀), **-ti** Opposition; contrariness. ···**e** ~ **etm.,** to oppose; to disagree with: ···**e** ~**ten dolayı,** for contravention of ...: **havanın** ~**i yüzünden,** on account of unfavourable weather.
**muhalif** (·–̀) Opposing; contrary; contradictory.
**muhalled** Eternal. ~**at, -tı,** immortal works; classics.
**Muhammed** Mohammed. ~**î** (···́), pertaining to Mohammed; Moslem.
**muhammen** Estimated. ~ **fiat,** estimated price.
**muhammes** Pentagon. Pentagonal· having five (verses *etc.*).
**muhammin** Who estimates. Valuer.
**muhar·ebe** (·–·̀) Battle; war. ~**ib** (·–̀), belligerent; warrior; combatant.
**muharref** Altered; falsified.
**muharrem** The month Muharrem, the first month of the Arabic lunar year.
**muharrer** Written. ~**at, -tı,** writings; correspondence.
**muharrik, -ki** Causing to move; motive; stirring up. Motor; instigator, agitator.
**muharrir** Writer; editor; author.
**muharriş** Irritating; itching.
**muhasama** (·–·̀) Hostility; contention; dispute. ~**at, -tı,** hostilities; war.
**muhasara** (·–·̀) Siege. ~ **etm.,** to besiege.
**muhaseb·at** (·–·́), **-tı** Accounts.

**~e,** accountancy, book-keeping; the accounts office of a business: **~ görmek,** to audit the accounts. **~eci,** accountant; Chief Accountant; auditor.

**muhasım** (·–·) Hostile; opponent, adversary.

**muhasır** (·–·) Besieging, investing.

**muhasib** (·–·) Who reckons. Accountant. **~lik,** book-keeping; accountant's office.

**muhassala** Resultant.

**muhassenat, -tı** Good *or* beautiful things; virtues; advantages.

**muhat** (·–), **-tı** Surrounded; contained. **~ı ilim,** within the bounds of knowledge, known.

**muhatab** (·–·) One addressed in speech; second person (*gram.*).

**muhatara** (·–·) Danger. **~lı,** dangerous.

**muhavere** (·–·) Conversation; dialogue. **~ etm.,** to converse together.

**muhavv·el** Changed; transformed; turned over to. **~ile,** transformer.

**muhayyel** Imagined; imaginary. **~ât, -tı,** imaginations; fancies.

**muhayyer** Who has a choice *or* option; optional; on approval.

**muhayyir** Bewildering.

**muhbir** Who gives information; newspaper correspondent, reporter.

**muhdes** Invented; created; newly created (post *etc.*); added (parts).

**muhib, -bbi** Who loves; friend.

**muhik, -kki** True, right; justifiable.

**muhit** (·–), **-ti** Surrounding; comprehending. Circumference; surroundings; the circle in which one moves, milieu.

**muhkem** Firm; sound; strong; tight.

**muhlis** Sincere. Sincere friend. **halis ~,** unadulterated; pure-blooded.

**muhrib** Destructive. Destroyer (*naut.*).

**muhtac** (·–) In want; indigent; in need. **···e ~ olm.,** to be in need of. **~lık,** a being in need; neediness, poverty.

**muhtar** Chosen, elected; free to choose, independent, autonomous. Headman *of a village or quarter*. **faili~,** a free agent. **~iyet** (·–·), **-ti,** freedom of action, autonomy. **~lık,** the post and duties of a headman.

**muhtas, -ssı** Special; exclusively used for (= *dat.*).

**muhtasar** Shortened, abridged; frugal, unpretentious. **~ ve müfid,** brief but to the point; pithy. **~an** (·–··), in brief; concisely.

**muhtekir** Profiteer.

**muhtel, -lli** Spoilt; injured; disturbed. **şuuru ~,** mentally afflicted.

**muhtelif** Diverse, various.

**muhtelis** Embezzler.

**muhtelit** Mixed; composite. **~ tedrisat,** co-education.

**muhtemel** Possible; probable.

**muhterem** Respected; honoured.

**muhteri** (··–), **-ii** Who invents. Inventor.

**muhteris** Covetous; desirous.

**muhteriz** Cautious; reserved; timid, hesitating.

**muhteşem** Magnificent; majestic.

**muhtev·a** (··–) Contents. **~i** (··–), containing. **~iyat** (···–), **-tı,** contents.

**muhtıra** Note; memorandum. **~ defteri,** note-book; diary.

**muhzir** Process-server; bailiff *of a court*.

**muin** Who helps; assisting. **Allah ~in olsun,** God help you!

**muit** (*Formerly*) Usher *or* supervisor in a school

**mukaar** Concave.

**mukabele** (·–·) A confronting; a reciprocating; reward; retaliation; retort, reply; a collating *or* comparing; recitation of the Koran. **~ etm.,** to confront; collate; retaliate; retort; reciprocate; resist: **···le ~ görmek,** to be received with (applause *etc.*): **~ okumak,** to recite the Koran by heart. **~bilmisil, -sli,** reprisal, tit-for-tat. **~ci,** an official who collates documents; reciter of the Koran. **~ten** (·–···), in return; reciprocally.

**mukabil** (·–·) Facing, opposite; corresponding; equivalent. The opposite; equivalent; thing given in return; in return; in compensation. In return for (= *dat.*). **~inde,** opposite; in return: **~ hücum,** counterattack: **buna ~,** on the other hand.

**mukaddem** Previous; preferable; first. Before, ago. **~a** (·–·–), previously, in the past. **~e** (*pl.* **~at, -tı**), preface; preliminary; forerunner of an event; premiss.

**mukadder** Decreed by Providence; predestined, fated; inevitable. Fate. **~at, -tı,** preordained things; destiny fate.

**mukaddes** Sacred, holy. **~at, -tı,** sacred things.

**mukaffa** (··⌄) Rhymed.
**mukallid** Mimicking. Mimic.
**mukannen** Fixed; regular.
**mukarenet** (·—·⌄), **–ti** A drawing nearer; *rapprochement*; association; conjunction *of stars*.
**mukarin** (·—⌄) Joined; associated. **~i hakikat,** close to the truth, true.
**mukarrer** Established, fixed; decided; certain. **~at, –tı,** decisions.
**mukaseme** (·—·⌄) A sharing out; apportionment.
**mukassi** Oppressive, stuffy.
**mukataa** A farming out of public revenue; sale of a business for a lump sum; rent paid to the **evkaf** for cultivated land turned into building land *or* gardens. **~lı,** subject to the **mukataa** rent.
**mukavele** (·—·⌄) Agreement; contract. **~ etm..** to agree mutually to do *stg*. **~li,** that has been settled by an agreement. **~name** (·—··—⌄), written agreement; pact.
**mukav·emet** (·—·⌄), **–ti** Resistance; endurance. **~koşusu,** long-distance race: **yarı ~ koşusu,** medium distance race. **~im** (·—⌄), resisting; resistant; strong; enduring.
**mukavva** (·⌄·) Cardboard.
**mukavves** Curved; arched.
**mukavvi** (··⌄) Strengthening; tonic.
**mukayese** (·—·⌄) Comparison. **~ etm.,** to compare.
**mukayyed** Bound; restricted; registered; diligent, attentive. **~ olm.,** to be registered; to attend diligently (to *stg*.).
**mukayyid** Registrar.
**mukim** Who dwells *or* stays; stationary.
**mukni** (·⌄) Satisfying; convincing. **delâili ~a.** convincing proofs.
**muktaza** (··⌄) Required; necessary. Need; exigency; requirement. **~sınca,** as required by, according to the requirements of ... .
**muktazi** (··⌄) Necessary. **~yat, –tı,** requirements; requisites.
**muktebes** Acquired from another; quoted.
**muktedir** Capable; powerful. **···e ~ olm.,** to be able to, be capable of.
**muktesid** Economical; careful in spending.
**mum** Wax; candle. **~la aramak,** to search for *stg*. very difficult to find; to crave for; to miss bitterly: **~a çevirmek,** to render disciplined and obedient: **⌜~ dibine ışık vermez⌝**, 'a candle does not light its own bottom', helping others without helping oneself: **~ gibi,** (i) very upright; stiff; like new; (ii) docile: **~ hala gibi,** (i) awake; (ii) restless, always on the move: **~ olm.** *or* **~a dönmek,** to become disciplined *or* compliant: **⌜~ söndü⌝**, 'the candle was extinguished', *alluding to a Bektashi ceremony, in which the lights were put out and which was popularly supposed to be immoral, hence* immoral, improper.
**mumaileyh** (⌄—··) Afore-mentioned (*fem.* **~a,** *pl.* **~im**).
**mum·cu** Tallow-chandler. **~hala** *v.* **mum. ~lamak,** to wax; to attach a seal to. **~luk,** candle-power. **~yağı** (⌄··), **–nı,** tallow.
**mumya** (⌄·) Mummy; a shrivelled sallow man.
**munakkid** *v.* **münekkid.**
**mundar** Dirty; *v.* **murdar.**
**munhasır** Restricted; limited. **···e ~,** limited to, exclusively for the use of. **~an** (·⌄··), exclusively.
**munis** (—⌄) Companionable, sociable; tame; friendly; familiar.
**munkatı** Cut off; interrupted; come to an end; separated.
**munsab** Poured; flowing. **···e ~ olm.,** to flow into. (*err. for* **mansab**) Estuary.
**munsif** Just, equitable, fair. **~ane** (··—⌄), just, reasonable (act); in an equitable manner, fairly.
**muntazam** In a line; regular; well-arranged, tidy, orderly; regular (army). **~an** (·⌄··), in an orderly manner; regularly.
**muntaz·ar** Awaited, expected. **~ır,** who waits expectantly: **~ olm.,** to be ready and waiting; to await (*with dat.*).
**munzam** Added, extra, additional; appended.
**murabaha** (·—·⌄) Usury. **~cı,** usurer.
**murabba**[1] (··⌄), **–aı** Square; squared.
**murabba**[2] (··⌄) Marmalade.
**murabit** (·—⌄), **–ti** Moslem hermit, marabout.
**murad** Wish; intention, aim. **~ da o!,** that's exactly what's wanted: **~ edinmek,** to hope, desire: **~ına ermek,** to attain one's desire: **~ etm.,** to desire, to propose to oneself: **bunu demekten ~,** by that is meant.
**murafaa** Recourse to a court; a

pleading before a court; a summoning *s.o.* to appear in court.
**murahhas** Delegated. Delegate; plenipotentiary. **heyeti ~a,** delegation with full powers.
**murak·abe** (·–·ʾ) Vigilance; control; supervision; religious meditation. ···i **~ altında bulundurmak,** to put under control. **~ıb,** controller; auditor.
**murassa** (··–ʾ), **–aı** Bejewelled.
**murdar, mırdar** Dirty; unclean. **~ ilik,** spinal marrow. **~lık,** dirt, filth.
**Musa** Moses.
**musab** (·ʾ) Stricken by illness *or* calamity. Victim.
**musadd·ak** Confirmed; certified. **~ik,** who confirms *or* certifies.
**musademe** (·–·ʾ) Collision; encounter; skirmish; percussion (*mil.*). **~ iğnesi,** firing pin: **~ kıtası,** detachment of shock troops: **~ tapası,** percussion fuse; **tavikli ~ tapası,** delayed percussion fuse: ···**le ~ etm.,** to collide with, to encounter.
**musadere** (·–·ʾ) Confiscation. **~ etm.,** to confiscate.
**musaf** *v.* Mushaf.
**musafaha** (·–·ʾ) A shaking hands.
**musahabe, musahabet** (·–·ʾ), **–ti** Company; conversation. **~ etm.,** to associate *or* to have conversation with another.
**musahh·ah** Corrected. **~ih,** corrector; proof-reader.
**musahib** (·–ʾ) Companion; gentleman-in-waiting.
**musalâha** (·–·ʾ) A making peace with another; reconciliation.
**musalla** (··–ʾ) Public place for prayer. **~ taşı,** stone on which the coffin is placed during the funeral service.
**musallat, –tı** Worrying; attacking. ···**e ~ olm.,** to worry, pester, fall upon; (of robbers *etc.*) to haunt a place.
**musammem** Decided upon; intended.
**musandıra** (·ʾ··) Large wardrobe for storing mattresses *etc.*; fixed slab at the end of a sofa frame; sideboard.
**musanna** (··–ʾ), **–aı** Skilfully *or* artistically made.
**musann·efat** (···–ʾ), **–tı** Literary works. **~if,** compiler of a book; classifier.
**musarrah** Openly and clearly set forth. **~an** (·ʾ··), clearly and explicitly.
**musavv·er** Imagined; depicted, illustrated. **~ir,** who depicts; artist, designer.
**Musevi** (–·–ʾ) Jew.
**Mushaf** The Koran.
**musır, –rrı** Persevering; persistent. **~rane** (··–ʾ), pertinacious, insistent (acts); in a persistent manner.
**musib** (·–ʾ) That hits the mark; right, appropriate.
**musibet** (·–ʾ), **–ti** Calamity, evil; tiresome person. Ill-omened, foul. **~!**, you pest!: ⌜**bir ~ bin nasihatten yeğdir**⌝, experience is the best teacher. **~li,** disastrous; ill-omened.
**musik·ar** (–·ʾ) Pandean pipe **~i** (–·–ʾ), music (*in general*); (**müzik** *is European music as opp. to* **saz,** *Oriental music, while* **mızıka** *is a military band*).
**muska** Amulet; charm. **şirinlik ~sı,** a charm worn by women to gain the affections of a man.
**muslih** Who puts right; conciliatory. **~ane** (··–ʾ), improving, setting right, peaceable (*not used of persons*); in a conciliatory manner.
**muslin** Muslin.
**musluk** Tap; spigot. **~taşı** (·ʾ··), **–nı,** stone basin under a tap.
**muson** (*Fr. mousson*) Monsoon.
**mustalah** Too technical and pedantic.
**mustatil** (··–ʾ) Rectangle.
**muş** (*Fr. mouche*) Steam launch.
**muşamba** (·ʾ·) Oiled silk; tarpaulin; linoleum; mackintosh, waterproof. **~ gibi,** very dirty (clothes *etc.*).
**muşer** Curved saw.
**muşmula** (ʾ··) Medlar; any shrivelled thing; 'old fogey'.
**muşta** Fist; blow with the fist; iron ball *used by shoemakers for pounding seams.* **~lamak,** to thump, pound.
**mut, –tu** Luck; fortune; happiness.
**muta** Robin.
**mûta** (––ʾ) Given. Datum.
**mutaassıb** Fanatical; bigoted. Fanatic.
**mutaazzım** Proud, arrogant.
**mutab·akat** (·–·ʾ), **–tı** Conformity; agreement. ···**le ~ etm.,** to agree with. **~ık** (·–ʾ), conforming; agreeing.
**mûtad** (––ʾ) Customary; habitual. Custom, habit. **~ım değil,** it is not my habit: **~ hilafına,** contrary to custom.
**mutalebat** (·–·–ʾ), **–tı** Things demanded; demands.

**mutallâka** Divorced woman.
**mutantan** Accompanied by noise and pomp; ostentatious; gorgeous.
**mutarıza** (–··ˋ) Bracket; parenthesis.
**mutasallif** Boastful; presumptuous.
**mutasarrıf** Owning; having the disposal of *a thing*; (*formerly*) governor of a sanjak (province). ···e ~ **olm.**, to be absolute master *or* owner of ... . **~lık**, post and jurisdiction of a mutasarrıf.
**mutasavver** Imagined; contemplated; projected.
**mutasavvıf** A Sufi; who studies mysticism.
**mutatabbib** Quack (doctor).
**mutavaat** (·–·ˋ), **–tı** Submission; obedience; compliance; reflexive form of a verb.
**mutazallim** Who complains of injustice.
**mutazammin** Comprising, containing. ···**i** ~ **olm.**, to comprise, contain.
**mutazarrır** Injured; suffering loss.
**mutbah** Kitchen.
**muteber** (–·ˋ) Esteemed; of good repute; solvent; enjoying credit; valid. Notable (*pl.* **~an** (–··ˊ)).
**mutedil** (–·ˋ) Temperate; moderate; mild.
**mutekid** (–·ˋ) Who firmly believes; religious.
**mutekif** (–·ˋ) Who retires for fasting and prayer.
**mutemed** (–·ˋ) Relied on; reliable. Fiduciary; man entrusted with the finances of a department.
**mutena** (–·ˊ) Important; carefully attended to; select, refined, 'distingué'.
**muterif** (–·ˋ) Confessing, acknowledging.
**muteriz** (–·ˋ) Opposing; objecting. ~ **olm.**, to make objections.
**mutfak** Kitchen; cuisine.
**muti** (·ˊ) Obedient.
**mutlak, –kı** Absolute; autocratic; unconditional. Absolutely; certainly. ~ **gelecek,** he is sure to come: ~ **gelmeli,** he *must* come: **vekili** ~, a plenipotentiary. **~a** (ˋ·–), absolutely; without fail; certainly. **~iyet, –ti,** absolutism; autocracy.
**mutlu** Lucky, fortunate. **ne ~!**, what luck! how fortunate!: **ne ~ ona!**, how lucky for him! **~luk,** luck, good fortune.
**mutmain** Tranquil; satisfied, assured.
**muttaki** Pious; devout.
**muttali** (··ˊ) Informed; aware.
**muttarid** In succession; regular; uniform. **~en** (·ˋ··), regularly; in regular succession.
**muttasıf** Endowed (with = **ile**); having the quality of ... .
**muttasıl** (··ˋ) Joined to another; continuous. (ˋ··), continuously; all the time.
**muvacehe** (·–·ˋ) A being face to face; confrontation. ···**in ~sinde,** in the presence of, vis-à-vis ...; with regard to ...; ~ **etm.,** to confront.
**muvafakat** (·–·ˋ), **–tı** Agreement, consent. ~ **etm.,** to agree, consent.
**muvaffak** Successful. ~ **olm.,** to succeed: **bunu yapmağa ~ oldu,** he succeeded in doing this. **~iyet, –ti,** success; victory (*pl.* **~iyat**). **~iyetli,** successful. **~iyetsizlik,** lack of success, failure.
**muvafık** (·–ˋ) Agreeable; suitable; conformable; favourable.
**muvakkat** Temporary; provisory. **~en** (·ˋ··), temporarily.
**muvakkit, –ti** Time-keeper at a mosque; chronometer. **~ane** (···–ˋ), clock-room of the **muvakkit.**
**muvasala** (·–·ˋ) Communication. ~ **hatları,** lines of communication. **~t, –tı,** arrival.
**muvazaa** (·–·ˋ) Collusion; a pretending to agree with one another for some ulterior motive; dissimulation. **~tan** (·–ˋ··), in pretence, in collusion.
**muvazat** (·–ˊ), **–tı** A being parallel; a drawing a parallel.
**muvazene, muvazenet** (·–·ˋ), **–ti** Equilibrium, balance. **umumî** ~, general budget (of a government). **~li,** in equilibrium; balanced; levelheaded; well-judged. **~siz,** unbalanced.
**muvazi** (·–ˊ) Parallel.
**muvazin** (·–ˋ) Equal in weight; wel balanced.
**muvazzaf** Having a duty; charged with; salaried; regular (army); on the active list (officer).
**muylu** Trunnion *of a cannon*; hub *of a car*.
**muz** Banana.
**muzaffer** Victorious. **~en** (·ˋ··), victoriously. **~iyet, –ti,** victory, triumph.
**muzari** (·–ˋ), **–ii** Aorist tense.
**muzır** Harmful, detrimental.

**muzib** (–·) Plaguing, tormenting, teasing; mischievous. **~lik,** teasing; mischievous behaviour; practical joke.
**muzlim** Dark, gloomy; sinister.
**muzmahil** Dispersed; annihilated. ~ **olm.,** to come to nought; to be annihilated.
**muzmer** Secret (thought, intention); implied but not expressed. **~at, –tı,** hidden thoughts *or* desires.
**muztar** Forced, compelled.
**muztarib** Agitated, worried, disturbed; suffering (from = ···**dan**).
**mübad·ele** (·–·) Exchange; barter. **~il** (·–·), exchanging; subject to exchange (of a population *etc.*); Turkish immigrant from Greece after the 1921–22 war, settled on land of an exchanged Greek.
**mübahase** (·–·) Discussion. ~ **etm.,** to discuss.
**mübalâğa** (·–·) Exaggeration. ~ **etm.,** to exaggerate: **ismi** ~, the superlative adjective. **~cı,** given to exaggeration. **~lı,** exaggerated.
**mübarek** (·–·) Blessed; sacred; bountiful; auspicious. ~!, bless it!: ~ **herif,** the blessed fellow (*iron.*): ~ **ağzını* açmak,** to start speaking evil.
**mübar·eze** (·–·) Combat between champions; contest; duel. **~iz,** champion; duellist; wrestler.
**mübaşir** (·–·) Process-server; official who conveys the order of a department; agent.
**mübay·enet** (·–·), **–ti** Divergence; conflict (of statements *etc.*). **~in** (·–·), opposed; different.
**mübdi** (·–), **–ii** Innovator; creator (aesthetically).
**mübeccel** Honoured; reverenced.
**müberra** (··–) Absolved; free.
**müberrid** Cooling. Refrigerator.
**mübeşş·er** Who receives good news. **~ir,** announcing good news.
**mübeyyin** Declaring, stating, certifying; making clear.
**mübhem** Vague, indefinite. **~iyet, –ti,** vagueness; lack of precision.
**mübin** Evident, manifest. **dini** ~, Islam.
**mübrem** Inexorable (decree *etc.*); inevitable; urgent.
**mübtelâ** (··–) (*with dat.*) Suffering from; addicted to; having a passion for; in love with. **kumara** ~ *or* **kumar ~sı,** addicted to gambling.
**mübtezel** Abundant; common; of no account; vulgar.
**mücadele** (·–·) Dispute; struggle.
**mücah·ede** (·–·) Endeavour; a fighting for Islam. **~id** (·–·), champion (of Islam *or* of some ideal).
**mücaseret** (·–·), **–ti** A daring. ···**meğe** ~ **etm.,** to dare to ... .
**mücazat** (·––), **–tı** Punishment; retribution. ···**e** ~ **etm.,** to punish.
**mücbir** Compelling; coercive.
**mücedd·ed** Renewed; new. **~en** (·'··), newly, recently; afresh, anew. **~id,** who renews, renovates *or* innovates.
**mücehh·ez** Equipped; furnished. **~iz,** who equips; shipowner.
**mücellâ** (··–) Polished; shining.
**mücell·ed** Bound (book). Volume. **~id,** bookbinder.
**mücerreb** Proved; tested. **~at, –tı,** things proved by experience.
**mücerr·ed** Isolated; simple, mere; pure; abstract; incorporeal; unmarried. Merely, simply. **~id,** isolating; insulating; insulator.
**mücessem** Having a body; corporeal; solid; personified. **kurei ~e,** globe: **namusu** ~, honour personified.
**mücevher** Bejewelled. Jewel. **~at, –tı,** jewellery.
**mücmel** Succinct, concise. Summary. **~en** ('··), concisely, summarily.
**mücrim** Culpable, guilty. Culprit; criminal; (*pl.* **mücrimin; iadei** ~, extradition). **~iyet, –ti,** guilt, culpability.
**müctehid** Expounder of Islamic laws.
**mücver** Croquette.
**müdaf·aa** (·–·) Defence; resistance. **~i** (·–·), **–ii,** defender; back (football).
**müdah·ele** (·–·) Interference; intervention. ~ **etm.,** to meddle, interfere (in *or* with = *dat.*). **~il** (·–·), intervening; intervener.
**müdah·ene** (·–·) Flattery; sycophancy. **~eneci, ~in** (·–·), flatterer; sycophant.
**müdara** (·––) Dissimulation; feigned friendship.
**müdavat** (·––), **–tı** Medical treatment.
**müdavele** (·–·) A causing to circulate. **~i efkâr,** exchange of views.
**müdavemet** (·–·), **–ti** Assiduity; unremitting attention (to work); a

frequenting (school *etc.*). ···e ~ **etm.,** to be assiduous in, to frequent.
**müdavi** (·–˗̀) Who treats *or* cures.
**müdavim** (·–˗̀) Who frequents; assiduous, persevering. Habitué, regular visitor, customer *etc.*
**müddea** (··˗̀) Claim; accusation; thesis; subject of a claim before a court. **~aleyh,** defendant.
**müddei** (··˗̀) Who asserts *or* claims; claimant; plaintiff; prosecutor. **~ umumi** (··–·–˗̀), Public Prosecutor.
**müddet, -ti** Space of time; period; interval. **~i ömür,** lifetime: **~ tayin etm.,** to fix a time *or* period of delay: **~siz olarak,** indefinitely.
**müdebbir** Prudent; far-sighted. Efficient manager.
**müdekkik, -ki** Who investigates minutely; research student. **~ane** (···–˗̀), minute (researches *etc.*): minutely, meticulously.
**müderris** Professor; *a grade in the hierarchy of the* **ulema.**
**müdevver** Round, circular, spherical; transferred (to a new balance sheet).
**müdhiş** Terrible, fearful; enormous; extraordinary; excessive.
**müdir** *v.* müdür. **~an,** *pl. of* **müdir.**
**müdrik** Perceiving, comprehending.
**müdrir** Diuretic.
**müdür** Who manages *or* superintends. Director; administrator; official governing a **nahiye** (sub-district). **~lük, ~iyet, -ti,** office and functions of a müdür; directorate; head office.
**müebbed** Eternal; perpetual. **~en** (·˗̀··), eternally; in perpetuity.
**müekkil** Client *of a lawyer.*
**müell·ef** Composed, compiled. **~at, -tı,** written works; literary compositions. **~if,** author: **~ hakkı,** author's rights.
**müellim** Painful; grievous.
**müemmen** Assured, safeguarded.
**müennes** Female; feminine.
**müesses** Founded; established. **~e,** foundation; establishment; institution (*pl.* **~at, -tı**; **~ı hayriye,** pious foundations, benevolent institutions).
**müessif** Sad; regrettable.
**müessir** Effective; influential; touching. **~ fiili,** assault and battery: **ibreti ~e,** an effective warning.
**müessis** Who establishes; founder (*pl.* **~in, ~an**).
**müeyy·ed** Strengthened; corroborated; aided. **~id,** strengthening; corroborating. **~ide,** corroborative *or* confirming statement *or* action; sanction (*jur.*).
**müezzin, meyzin** Muezzin; he who calls Moslems to prayer.
**müfessir** Who expounds. Commentator of the Koran.
**müfettiş** Who examines *or* investigates. Inspector. **~lik,** inspectorship; inspectorate.
**müfid** Useful; advantageous.
**müflis** Bankrupt; penniless.
**müfred** Separate; single; isolated. Singular (*gram.*). **~at** (··˗̀), **-tı,** details; particulars; enumeration; detailed inventory: **~ programı,** programme giving items (of a school course *etc.*): **~ı tıb,** materia medica.
**müfrez** Separated; detached. **~e,** detachment of troops.
**müfrit** Excessive; beyond bounds; exaggerated. Extremist.
**müfsid** Disturber of the peace. Seditious; mischief-making; subversive.
**müftehir** (*with* **ile**) Who glories in; who is proud of.
**müfteri** (··˗̀) Calumniator, slanderer. **~yat, -tı,** calumnies.
**müfteris** Predatory; rapacious; that tears its prey.
**müfti, müftü** Moslem jurist; mufti, senior Moslem priest. **~lik,** office and rank of a mufti.
**mühendis** Engineer. **~hane** (···–˗̀), school of engineering; (*formerly*) school of gunnery. **~lik,** profession of engineering.
**müheykel** Giant; huge; clumsily built.
**müheyya** (··˗̀) Ready, prepared.
**müheyyic** Stirring, exciting, rousing.
**mühim** Important; urgent. **~mat** (··˗̀), **-tı,** munitions of war; ammunition; important matters. **~me,** important *or* urgent affair.
**mühlet, -ti** Respite; delay; grace. **~ vermek,** to grant a delay.
**mühlik** Dangerous; destructive; deadly.
**mühm·el** Neglected; abandoned; meaningless. **~at, -tı,** meaningless words. **~il,** negligent.
**mühre** Burnisher; glass ball *or* cowry shell *or* agate *used for polishing paper or parchment*; a vertebra; bead; artificial bird, decoy (shooting). **~lemek,** to polish with a **mühre. ~li,** polished (paper).

**mühtedi** (··˗́) Converted to Islam.
**mühür, -hrü** Seal; signet-ring; impression of a seal; the whorl of hair at the top of the head. **~ünu* basmak,** to put one's seal to *stg.*; to guarantee the truth of *stg.*: **~ünü* yalamak,** to go back on one's word. **~dar,** seal-keeper; private secretary to a Minister. **~lemek,** to stamp with a seal. **~lü,** sealed.
**müjde** Good news; present given to a bearer of good news. **~ etm.,** *or* **~lemek,** to bring good news. **~ci,** who brings *or* announces good news. **~lik,** gift to the bringer of good news.
**mükâfat** (·–̀), **-tı** Reward; prize. **~en** (·˗́–·), as a reward.
**mükâleme** (·–·̀) Conversation; dialogue; (pl. **~at,** *diplomatic* negotiations).
**mükedder** Sad, grieved; turbid.
**mükellef** Bound; obliged; liable; taxed; conscripted; highly adorned, sumptuous. **~iyet, -ti,** obligation (to pay taxes, perform services *etc.*), liability.
**mükemmel** Complete; perfect; excellent. **~!,** splendid! **~en** (·̀··), perfectly.
**mükerrer** Repeated. **~en** (·̀··), repeatedly.
**mükeyyifat** (···̀), **-tı** Intoxicants; stimulants; narcotics.
**mükrim** Who treats with honour; hospitable; kind.
**müktes·eb** Acquired; earned. **~ hak,** vested interest. **~at, -tı,** acquisitions; attainments. **~ib,** who acquires *or* earns.
**mülâhaza** (·–·̀) Consideration; observation; reflection. ... **~sile,** in consideration of ...: ... **~sındayım,** I am of the opinion that ...: **~ etm.,** to observe; to consider: (*pl.* **~at, -tı, ~ hanesi,** column for remarks). **~sızlık,** thoughtlessness; rashness.
**mülâk·at** (·–̀), **-tı** Meeting; interview; audience. **~ etm.,** to have an audience *or* interview. **~î** (·–˗́), who meets *or* interviews: **...e ~ olm.,** to meet with ..., to have an interview with ... .
**mülây·emet** (·–·̀), **-ti** A suiting *or* fitting; mildness, gentleness; softness; freedom of the bowels. **~im,** suitable; mild, gentle; soft, pliant; free in the bowels: **~ gelmek,** to seem reasonable, to appeal *to one*.
**mülâz·emet** (·–·̀), **-ti** A holding tenaciously to *stg.*; persistence, assiduity; a serving as an unpaid beginner in an official post; novitiate. **~ etm.,** to serve as a novice *or* candidate for a post. **~im,** novice; assistant functionary; (*formerly*) lieutenant.
**mülemma** (··˗́) Soiled, smeared; variegated. Poem written in two languages.
**mülevves** Soiled, dirty.
**müleyyin** Softening; laxative.
**mülga** (·˗́) Suppressed; abolished.
**mülhak** Added, appended; annexed; attached; dependent (on = *dat.*). **~at, -tı,** added things; places subordinate *to a centre of government*.
**mülhem** Inspired; suggested. **...den ~ olm.,** to be inspired by ... .
**mülhid** Atheist. Irreligious; heretic. **~lik,** unbelief; irreligion.
**mülk, -kü** Possession; property; dominion. **~ almak,** to acquire landed property: **~ sahibi,** landed proprietor. **~i** (·˗́), belonging to the state; civil; civilian. **~iye,** civil service; school for civil servants. **~iyet, -ti,** the quality of being freehold.
**mültefit** Attentive, courteous, kind.
**mülteka** (··˗́) Junctions (of rivers *or* roads); place of meeting.
**mültez·em** Favoured; considered necessary. **~im,** one who farms a branch of the revenue: regarding as necessary; favouring.
**mümanaat** (·–·̀), **-tı** Opposition; prevention. **~ etm.,** to oppose, prevent.
**mümarese** (·–·̀) Skill acquired by practice; dexterity; practice, training.
**mümas** (·˗́), **-ssı** Touching. Tangent.
**mümaşat** (·–˗́), **-tı** A complying; feigned approval; flattery: **...e ~ etm.,** to make concession to ...; to approve in order to flatter.
**mümbit** Fertile; productive.
**mümessil** Representative; editor; actor.
**mümeyyiz** (*Legally*) capable of distinguishing between right and wrong; distinguishing; distinctive. Chief clerk; examining official *at a school*.
**mümin** (–̀) Believer in Islam (*pl.* **~in**).
**mümkün** Possible. **~ olduğu kadar** *or* **~ mertebe,** as far as possible.

**mümtaz** (·˙) Distinguished; privileged; autonomous. **~iyet** (·—·˙), **-ti, ~lık,** distinction; autonomy.
**münacat** (·—˙), **-tı** Prayer to God; inward prayer; prayer in poetry.
**münadi** (·—˙) Herald; public crier.
**münaferet** (·—·˙), **-ti** Mutual aversion.
**münafık** (·—˙), **-kkı** Hypocrite; mischief-maker; double-dealing.
**münafi** (·—˙) (*with dat.*) Opposed *or* contrary to.
**münakale** (·—·˙) Transport; transfer; (*pl.* münakalat, transport; communications).
**münakasa** (·—·˙) A giving of a contract to the lowest bidder; a putting up to tender; adjudication. **~ya konulmak,** to be put up to tender.
**münakaşa** (·—·˙) Dispute. **~ etm.,** to dispute: **~ götürmez,** indisputable.
**mün'akid** Concluded; ratified; assembled, convoked.
**mün'akis** Reflected; inverted; reflex.
**münakkah** Polished (style); carefully revised.
**münakkaş** Ornamented, decorated.
**münakkid** Critic.
**münasebet** (·—·˙), **-ti** Fitness; proportion; reason; relation, connexion; pretext, motive; opportunity; (*pl.* münasebat, relations *between nations or people*). ... **~iyle,** in connexion with ... : **~ almaz,** it is not seemly: **~ ~ aramak,** to seek an opportunity: **bir ~ düşürmek,** to find a suitable occasion: **bir ~ ile,** on a fitting occasion; under a suitable pretext: **bu ~le,** in this connexion: '**ne ~!**', 'by no means!'; 'not a bit of it!'; 'not likely!'; 'what's that got to do with it?'
**münasebet·li,** reasonable; suitable; opportune. **~siz,** inopportune; unsuitable; unseemly; unreasonable. **~sizlik,** unseemly action; silly act; *stg.* done at the wrong moment. **~tar,** connected with.
**münasib** (·—˙) Suitable; proper; opportune. **~ görmek,** to think proper *or* opportune.
**münavebe** (·—·˙) Alteration; a taking turns; rotation. **~ten** (·—˙··), in turns, alternately.
**münazaa** (·—·˙) Dispute; quarrel. **~lı,** about which there is a dispute; controversial.
**münazara** (·—·˙) Discussion; debate.
**münbit** Fertile.
**müncer** Drawn; attracted; ···**e ~,** resulting in, leading to. **~ olm.,** to be drawn *or* attracted; to result in; to have as consequence.
**münci** (·˙) Saving. Rescuer, deliverer.
**mündefi** (··˙) Repulsed, driven away; removed.
**mündemic** Entering in; contained in.
**münderic** Inserted; written; published. **~at, -tı,** contents (of a letter *or* book).
**münebbih** Rousing; awakening. Stimulant; excitant (coffee *etc.*).
**müneccim** Astronomer; astrologer.
**münekkid** Criticizing. Critic.
**münevver** Enlightened; educated. Educated person, intellectual.
**münezzeh** Kept free (from = **den**); exempt; pure.
**münfail** Annoyed; offended.
**münferid** Separated; isolated; insulated. **~en** (·˙··), separately; singly; alone.
**münfesih** Abrogated, annulled; abolished.
**münhal** Loosened; solved; dissolved; vacant.
**münhani** Bent, curved. Curve; arc.
**münhat** Fallen down; depressed, low-lying (place); low; degraded.
**münhemik** (*with dat.*) Given up **to;** absorbed in; indulging in.
**münhezim** Routed, defeated. **~en** (·˙··), in rout, in disorderly flight.
**münkad** (·˙) Docile, submissive, obedient.
**münkalib** Transformed.
**münkariz** Extinct (of a dynasty *etc.*); exterminated; perished.
**münkasim** Divided into parts.
**münker** Denied; unacceptable; not allowed by religious law. *Name of one of the two angels who question the dead* (*the other is* **Nekir**).
**münkesir** Broken; annoyed; brokenhearted.
**münkir** Who denies *or* disbelieves.
**münşerih** Cheerful, in good humour.
**münşi** (·˙) Secretary. **~yane** (·——˙), written in a bombastic style.
**müntah·ab** Selected; elected; privileged. **~abat, -tı,** selected passages (of an author); anthology. **~ib,** who selects; elector.
**müntahil** Plagiarist.
**müntakil** Migrating; transferred; passed on, inherited.

**müntakim** Avenging.
**münteha** (··–̀) Limit; extreme. Final.
**müntehab** *etc. v.* müntahab *etc.*
**müntehi** (··–̀) Final; last. ~ **olm.**, to end (in = *dat.*).
**müntehir** Suicide (person).
**müntesib** (*with dat.*) Connected with; having relations with; belonging to.
**münteşir** Disseminated; diffused; published.
**münzevi** (··–̀) Retiring to a solitary place. Recluse, hermit. ~ **yaşamak,** to live a secluded life.
**müphem** *v.* mübhem.
**müpteda, müptelâ** *etc. v.* mübteda *etc.*
**mür, -rrü** Myrrh.
**müracaat** (·–·̀), **-tı,** (*pl.* ·–·–̀) Recourse; application; reference. ···**e** ~ **etm.**, to have recourse to, refer to, consult. ~**gâh** (·–··–̀), place *or* authority to which recourse must be had.
**mürai** (·–̀) Hypocritical. Hypocrite. ~**lik,** hypocrisy.
**mürdesenk** Litharge, peroxide of lead.
**mürdüm,** ~**eriği** Damson.
**mürebbi** (··–̀) Who brings up *or* educates. Tutor; trainer. ~**ye,** governess.
**mürеccah** Preferable; preferred.
**müreffeh** Prosperous; well-to-do. ~**en** (·̀··), comfortably; in easy circumstances.
**mürekkeb** Compound; composed (of = **den**). Ink. **cehli** ~, the ignorance of one who is sure he knows: **oldukça** ~ **yalamış,** somewhat educated. ~**balığı, -nı,** squid, cuttlefish. ~**lemek,** to ink in; to cover with ink. ~**li,** inky; filled with ink: ~ **kalem,** fountain-pen.
**müressem** Designed; decorated with designs *or* illustrations.
**müretteb** Set in order, arranged; prepared; invented, concocted; incumbent (duty); allotted; destined for; 'ear-marked' for. ~**at, -tı,** *pl. used as sing.*, appropriation; crew (of a ship); troops allotted to *or* destined for some place *or* duty.
**mürettib** Compositor. ~ **hatası,** misprint.
**mürevvic** Giving currency to; propagating (ideas *etc.*); who pushes *or* facilitates *stg.*
**mürid** Novice in an order of dervishes; disciple. ⌜**şeyh uçmamış** ~**i uçurmuş**⌝, the sheikh could not fly but his disciple said he did; the prodigies of a great man are often only the invention of his admirers.
**mürnel** Three-stranded twine.
**mürşid** Who shows the right way; guide; spiritual teacher.
**mürteci, -ii** Reactionary.
**mürted, -ddi** Apostate, renegade.
**mürtefi** (··–̀) Elevated; raised; removed.
**mürtekib** Corrupt; taking bribes; who commits a sin.
**mürtesem** Delineated. Projection.
**mürteşi** (··–̀) Who accepts bribes, corrupt.
**mürur** (·–̀) Passage (of a person); lapse of time. ~ **tezkeresi,** permit to pass, pass. ~**iye,** toll; permit to pass. ~**uzaman,** prescription, limitation (time after which an action cannot be taken): ~**la sakıt olm.,** (of an action at law) to fail under the Statute of Limitations.
**mürüvvet, -ti** Generosity; munificence; blessing; family feast *for a birth, marriage etc.* **evlâdının** ~**ini görmek,** to see one's child grow up and get married. ~**li,** generous; considerate. ~**siz,** ungenerous; inconsiderate.
**mürver** Elder (tree).
**müsaade** (·–·̀) Permission; favour. ···**e** ~ **etm.**, to permit, consent to. ~**kâr,** tolerant.
**müsab·aka** (·–·̀) Competition; race. ~**ya girmek,** to compete: ~ **imtihanı,** competitive examination. ~**ık** (·–̀), competitor.
**müsademe, müsadere,** *see* musademe *etc.*
**müsaid** (·–̀) Permitting; favourable; convenient. ~ **davranmak,** to show oneself favourably disposed.
**müsakkafat** (··–̀), **-tı** Roofed buildings; house property; Vakif income derived from house property.
**müsalâha** (·–·̀) A making peace, reconciliation.
**müsalemet** (·–·̀), **-ti** Tranquillity; a living in peace, harmony.
**müsamaha** (·–·̀) Indulgence; tolerance; forbearance; negligence. ~ **etm.**, to shut one's eyes *to an impropriety.* ~**kâr,** ~**cı,** indulgent; non-censorious; tolerant.
**müsamere** (·–·̀) Evening entertainment; soirée; concert.
**müsaraat** ·–·̀), **-tı** Haste.

**müsav·at** (·–⌐), **–tı** Equality. **~i** (·–⌐), equal: equivalent; the sign =.
**müsb·et** Proved; established; positive; positive (*elect.*). The positive sign (*elect.*). **~ cevab,** an answer in the affirmative: **~ kafalı,** realistic, hard-headed. **~it,** proving: **evrakı ~e,** documents in proof of *stg.*
**müsebbib** Cause, motive; author, instigator.
**müseccel** Officially registered; matriculated; notorious (thief *etc.*).
**müseddes** Having six parts; hexagon; stanza of six lines.
**müsekkin** Quieting, calming. Anodyne, sedative.
**müsellâh** Armed. **gayri ~,** unfit to bear arms, fit only for non-combatant duties (*mil.*): **kafadan gayri ~,** off his head, having 'a screw loose'. **~an** (·⌐··), with arms in their hands.
**müsellem** Admitted by all, incontestable. **~at, –tı,** admitted truths, truisms.
**müselles** Triple; triangular. Triangle.
**müselsel** Connected; consecutive; linked.
**müsemma** (··⌐) Named, called.
**müsevvid** Secretary who drafts letters *etc.*
**müshil** Purgative.
**müskir** Intoxicating. **~at, –tı,** intoxicants.
**müskit** Who silences; who reduces to silence by his arguments, persuasive.
**müslim** Moslem.
**müslüman** Moslem; pious; honest. **~ca** (··⌐·), in a Moslem way; honestly; honourably.
**müsmir** Fruitful; productive; successful.
**müspet** *v.* **müsbet.**
**müsrif** Spendthrift; extravagant; prodigal.
**müstacel** (·–⌐) Urgent. **~en** (·⌐··), urgently, in haste. **~iyet, –ti,** urgency.
**müstafi** (·–⌐) Who resigns.
**müstağni** (··⌐) Having no need (of = **den**); independent; satisfied; disdainful. **bir şeyden ~ olmamak,** to be unable to do without a thing: **izahtan ~dir,** it has no need of explanation.
**müstağrak** Submerged; overwhelmed. **···e ~,** plunged into; inundated by.
**müstahak, –kkı** Deserts; due reward *or* punishment. **~kını bulmak,** to get one's deserts: **···e ~ olm.,** to deserve, to be entitled to: **'Allah ~kını versin!',** 'may Heaven punish him!'
**müstahdem** Employed. Employee.
**müstahkar** Despised; contemptible.
**müstahkem** Fortified.
**müstahsil** Productive. Producer. **kuvvei ~e,** productive capacity.
**müstahzar** Made ready, prepared. Ready-made drug. **~at, –tı,** prepared drugs, medicinal preparations.
**müstaid** Clever, capable. **···e ~** clever at; capable of; inclined, disposed to (an illness).
**müstakar, –rrı** Settled place *or* time; *used for* **müstakır** *q.v.*
**müstakbel** Future. The future; the future tense.
**müstakır** Settled; stationary; stable.
**müstakil** Independent; apart. Absolutely; solely; expressly. **~len** (··⌐·), independently; absolutely; solely; expressly.
**müstakim** Straight; upright, honest.
**müstamel** (·–⌐) Used; employed; not new.
**müstantık** Examining magistrate.
**müstatil** Oblong.
**müstear** (··⌐) Temporarily borrowed; temporary. **namı ~,** pseudonym.
**müsteb'ad** Remote; improbable; far-fetched.
**müstebid** Despotic, tyrannical. Despot, tyrant. **~ane** (···–⌐), despotic (action *etc.*); despotically, tyrannically.
**müstecir** (·–⌐) Who rents; tenant. **~en** (·⌐··) on lease; as a tenant.
**müstefid** (··⌐) Who profits. **···den ~ olm.,** to profit by; to learn from; to enjoy (rights *etc.*).
**müstefreşe** Concubine.
**müstehab** Laudable; recommended but not enjoined by religious law.
**müstehase** (··–⌐) Fossil (*pl.* **müstehasat**).
**müstehcen** Loathsome; obscene.
**müstehlik, –ki** Consuming. Consumer.
**müstehzi** (··⌐) Jeering, mocking; sarcastic, ironical.
**müstekim** Straight; upright, honest.
**müstekreh** Loathed; disgusting.
**müstelzim** Necessitating; involving; implying.
**müstemleke** Colony (*pl.* **müstemlekât**).

**müstenid** (*with dat.*) Relying *or* based on. **~en** (·⁀··) based on, relying on.
**müstenkif** Who abstains *or* refuses.
**müstensih** Who makes a copy. Copying press; hectograph.
**müsterham** Implored, begged.
**müsterih** (··⁀) At rest; at ease. **~ etm.**, to set at ease.
**müstesna** (··⁀) Excluded, exceptional; extraordinary. **~lık,** a being exceptional.
**müsteşar** Councillor; Secretary of State. **~lık,** under-secretaryship.
**müsteşrik, -ki** Orientalist.
**müstetir** Veiled; hidden. **tahtında ~,** implied but not expressed, 'sous-entendu'.
**müstevli** (··⁀) Invading; predominant; prevalent; epidemic. **···e ~ olm.**, to invade.
**müstezad** (··⁀), **-dı** Increased; supplemented. Poem having a rhymed supplement to each hemistitch.
**müsvedde** Draft, rough copy. **~lik,** paper for rough copies.
**müşaare** (·—·⁀) Poetic contest.
**müşab·ehet** (·—·⁀), **-ti** Resemblance. **~ih** (·—⁀) resembling: **~ olm.**, to resemble.
**müşahede** (·—·⁀) A seeing *or* witnessing; observation (*med.*). **~ altına alınmak,** to be placed under observation: **~ etm.**, to see, witness.
**müşahhas** Personified; identified; concrete (*as opp. to* **mücerred,** abstract).
**müşahid** (·—⁀) Observer.
**müşareket** (·—·⁀), **-ti** Participation; partnership; association; reciprocal form of the verb.
**müşarileyh** (·—·⁀), **müşarünileyh** Afore-mentioned; the said; (*fem.* **~a,** *pl.* **~im**).
**müşateme** (·—·⁀) Mutual abuse.
**müşav·ere** (·—·⁀) Consultation; deliberation. **~ir** (·—⁀) who consults *or* is consulted; counsellor; consultant (*med.*): **hukuk ~i,** legal adviser.
**müşekkel** Of imposing form; huge.
**müşerref** Honoured. **~ olm.**, to be honoured, to feel it an honour: **~ oldum,** I am glad to meet you.
**müşevveş** Confused; dubious.
**müşevvik, -ki** Inciting; encouraging. Instigator.
**müşfik** Tender; compassionate.
**müşir** *v.* **müşür.**
**müş'ir** Marking; pointing out; informing. Index, pointer.
**müşiriyet, -ti** Rank of a field-marshal.
**müşkül** Difficult. Difficulty; doubt; (*pl.* **müşkülât, -tı**: **~ çıkarmak,** to raise difficulties: **~ çekmek** *or* **düçarı ~ olm.**, to meet with difficulties). **~pesend,** hard to please; fastidious; exacting.
**müşrik, -ki** Polytheist; pagan; (*pl.* **~in,** people who do not believe in the unity of God).
**müştak, -kkı** Derived. Derivative; (*pl.* **~kat, -tı**).
**müştâk** (·⁀) Filled with desire (of, to = *dat.*).
**müştehi** (··⁀) Appetizing.
**müştehir** Well-known; renowned.
**müşteki** (··⁀) Who complains. Complainant.
**müştemilat** (··⁀), **-tı** Contents; annexes; out-houses.
**müşterek** Common; shared; collective; joint. **faslı ~,** ratio, intersection (*math.*). **~en** (·⁀··), in common; jointly.
**müşteri**[1] Customer; purchaser; client.
**müşteri**[2] (··⁀) The planet Jupiter.
**müşür** Field-marshal. **~lük,** rank and position of a field-marshal.
**mütaa-** *v.* **mütea-.**
**mütabaat** (·—·⁀), **-tı** A conforming *or* following; conformity; obedience. **~ etm.**, to follow, conform.
**mütalâa** (·—·⁀) A studying; observation, remark; opinion. **... ~sında bulunmak,** to be of the opinion that ... : **~ etm.**, to read, study.
**mütareke** (·—·⁀) Armistice.
**müteaccib** Astonished, amazed.
**müteaddi** (···⁀) Aggressive; transitive (verb).
**müteaddid** Numerous; several.
**müteaffin** Putrid; stinking.
**müteahhid** Contractor; purveyor.
**müteakıb** (··—⁀) Successive; subsequent. **birbirini ~,** one after the other. **~en** (··⁀··) subsequently; successively.
**müteallik** (*with dat.*) Dependent on; relative to; concerning; connected with.
**müteammid** Acting deliberately *or* with premeditation. **~en** (··⁀··), with premeditation; deliberately.
**müteammim** General; in common use.
**mütearife** (···—·⁀) Axiom.

**mütearrız** Attacking; aggressive. Aggressor.
**müteazzım** Proud, arrogant.
**mütebadil** (··–˙) Taking each other's place; interchangeable; alternate.
**mütebahhir** Very erudite. Learned scholar.
**mütebaki** (··–˙) Remaining; outstanding. Remainder; balance; surplus.
**mütebariz** (··–˙) Prominent; outstanding.
**mütebasbıs** Fawning; cringing.
**mütebayin** (··–˙) Distinct; contrasting; incommensurable (*arith.*).
**mütebellir** Crystallized; as clear as crystal; very evident.
**mütebessim** Smiling.
**mütebeyyin** Manifest; proved.
**mütecanis** (··–˙) Homogeneous; of the same kind.
**mütecasir** (··–˙) Audacious; presumptuous. **~ olm.**, to dare (impudently).
**mütecaviz** (··–˙) Exceeding; transgressing; exorbitant; presumptuous. Aggressor. **bini** *or* **binden ~**, exceeding a thousand.
**müteceddid** Innovator; modernist; following the latest fashion.
**mütecelli** (···˙) Becoming manifest. **~ olm.**, to become manifest; to show oneself.
**mütecellid** Daring, courageous; challenging. **~ane bir hareket**, a brave defiant action.
**mütecessim** Appearing solid; corporeal; personified.
**mütecessis** Inquisitive, curious.
**mütedahil** (··–˙) Entering each other; intermixed; overlapping; commensurable (*arith.*); in arrears (of payments *etc.*).
**mütedair** (··–˙) Concerning, relative.
**mütedavil** (··–˙) Current; in common use. **~ sermaye**, working capital.
**mütedeyyin** Religious.
**müteehhil** Married.
**müteellim** Suffering; grieved.
**müteemmil** Reflecting; meditative.
**müteenni** Slow and cautious; circumspect.
**müteessif** Grieved, sorry, regretful.
**müteessir** Hurt; grieved; touched; affected. **~ olm.**, to be grieved, touched, affected, impressed. **~en** (··˙··), **··den ~**, grieving that ..., in sorrow for ... .
**mütefavit** Dissimilar; various.
**mütefekkir** Thoughtful, pensive. Thinker. **kuvvei ~e**, the thinking faculty.
**mütefennin** Versed in science *or* art. Scientist.
**müteferri** (···˙), **-ii** Having ramifications; derived; accessory; subordinate. **~yat, -tı**, offshoots; ramifications; derivatives.
**müteferrid** Isolated; sole, unique.
**müteferrik** Separated; dispersed; various, miscellaneous. **~a**, money for miscellaneous expenses; petty cash; sundries; the department of a police station dealing with petty offences, licences *etc.*
**mütefessih** Putrefied; degenerate.
**mütefevvik** Superior.
**mütegallib** Usurping; tyrannical. Tyrant. **~ane** (····–˙), tyrannically; in a cruel and violent manner. **~e**, oppressors; usurpers.
**mütegayyir** Changed; spoilt.
**mütehakkim** Despotic; domineering.
**mütehalif** (··–˙) Diverse; mutually opposed.
**mütehalik** (··–˙) Precipitate; enthusiastic.
**mütehallik** (*with* **ile**) Endowed with; possessing.
**mütehammil** (*with dat.*) Supporting; enduring; capable of.
**müteharrik** Moving; movable; portable.
**mütehassıl** Produced; resulting.
**mütehassıs** Specialist.
**mütehassir** Who regrets an absent person *or* thing desired; longing; disappointed.
**mütehassis** Moved (by emotion). **··le ~**, animated by (a desire *etc.*).
**mütehavvil** Changing; variable.
**mütehayyir** Amazed; bewildered.
**mütehevvir** Impetuous; rash; furious.
**müteheyyic** Excited.
**mütekabil** (··–˙) Opposite each other, reciprocal; mutual. **~en** (··˙··), mutually; reciprocally. **~iyet, -ti**, reciprocity.
**mütekaddim** Preceding; former; ancient; (*pl.* **~in**, the ancients; men of old).
**mütekaid** (··–˙) Retired on a pension. Pensioner (*pl.* **~in**).
**mütekallis** Shrunken; contracted.
**mütekâmil** (··–˙) Arrived at perfection, perfected; developed by evolution.

**mütekâsif** (··–˙) Thick; condensed; concentrated.
**mütekebbir** Proud, haughty.
**mütekeffil** (*with acc.*) Standing surety for; responsible for; guaranteeing.
**mütekellim** Speaking. Speaker; first person (*gram.*).
**mütelevvin** Of various colours; variegated; fickle.
**mütelezziz** Enjoying the taste; relishing. ···**den** ~ **olm.**, to relish, to enjoy.
**mütemadi** (··–˙) Continuing; continuous. ~**iyen** (··–˙) continuously; continually.
**mütemarız** (··–˙) Feigning illness, malingering.
**mütemayil** (··–˙) Inclined; leaning.
**mütemayiz** (··–˙) Distinguished (for = **ile**).
**mütemeddin** Civilized.
**mütemekkin** Settled (in a place); established.
**mütemenna** Desired; asked for. Wish; request.
**mütemmim** Completing, perfecting; supplementary. Supplement. **cüz'i** ~, integral part.
**mütenakız** (··–˙) Contradictory.
**mütenasib** (··–˙) Proportional; symmetrical; well-proportioned; well-built. **mebsuten (makûsen)** ~, directly (indirectly) proportional.
**mütenazır** (··–˙) Facing one another; corresponding; symmetrical.
**mütenebbih** Warned; vigilant; on his guard *as the result of unpleasant experience*.
**müteneffir** Feeling aversion (*for* = **den**).
**müteneffiz** Influential; (*pl.* ~**an**, influential people).
**mütenekkir** Disguised; unrecognizable; incognito. ~**en** (··˙··), incognito (*adv.*).
**mütenevvi** (···˙) Of various kinds; diverse.
**müteradif** (··–˙) Synonymous.
**müterakim** (··–˙) Accumulated.
**müterakki** (···˙) Progressive.
**müterassıd** Observer; lying in wait.
**müterc·em** Translated. ~**im,** translating; translator.
**mütereddi** (···˙) Depraved; degenerate.
**mütereddid** Hesitating; undecided.
**müterekkib** Consisting of parts; composed (of = **den**).
**müterennim** Singing; trilling.
**müterettib** Arranged in order. ···**den** ~, resulting from: **uhdemize** ~ **vazife,** the duty which it falls on us to perform.
**mütesanid** (··–˙) Mutually supporting; solidarity; joint (responsibility).
**müteselli** (···˙) Who consoles himself; comforted.
**mütesellim** Who takes delivery of *stg.* from another; (*formerly*) a deputy lieutenant-governor and collector of taxes.
**müteselsil** Forming a chain; in continuous succession; uninterrupted (sequence). ~ **mesuliyet,** joint liability. ~**en** (··˙··), in continuous succession; one after the other.
**müteşabih** (··–˙) Resembling one another; similar.
**müteşar** (··–˙) Who professes to be a poet; poetaster.
**müteşebbis** Who has initiative; enterprising; who starts an enterprise. ···**e** ~ **olm.**, to set to work on.
**müteşekkil** Formed. ···**den** ~, formed *or* composed of.
**müteşekkir** Thankful; grateful.
**mütetabbib** A tyro in medicine; quack.
**mütetebbi** (···˙) Who investigates *or* researches.
**mütevahhiş** Frightened; scared.
**mütevakkıf** At a stop, standing still; dependent (on = *dat.*).
**mütevali** (··–˙) Consecutive; successive; continuous. ~**yen** (··˙··), successively, consecutively, continuously.
**mütevassıt** Medium, intermediary; mean.
**mütevazı** (··–˙) Humble; modest.
**mütevazi** (··–˙) Parallel. ~**yüladla** (··–···˙), parallelogram.
**mütevazin** (··–˙) Equal in weight; balancing one another.
**müteveccih** (*with dat.*) Turned towards; facing; aimed at; favourably disposed to. ~**en** (··˙··), in the direction of.
**müteveffa** (···˙) Deceased; the late.
**mütevehhim** Who has imaginary fears *or* suspicions.
**mütevekkil** Who puts his trust in God; resigned.
**mütevelli** (···˙) Administrator, *esp.* the trustee of a pious foundation.
**mütevellid** Born; caused; resulting.
**müteverrim** Consumptive.
**müteyakkız** Wide awake, vigilant.

**mütezellil** Humiliated; debased; mean, contemptible, servile.
**müthiş** *v.* **müdhiş.**
**müttakî** (··⊥) God-fearing; devout.
**müttefik, -ki** Agreeing; unanimous; allied. Ally. **~an** (·ˋ··), unanimously; in agreement.
**müttehem** Accused *or* suspected (of = **ile**).
**müttehid** United; unanimous. **~en** (·ˋ··), unanimously; unitedly. **~ül-meal,** having the same purport.
**müttehim** Accusing; *used also err. for* **müttehem.**
**müttekâ** (··⊥) Anything leaned upon; bolster, cushion.
**müvacehe** (·—·ˋ) A being face to face. **~sinde,** in the presence of.
**müverrih** Historian; chronicler; (*pl.* **~in**).
**müvesvis** Apprehensive; suspicious; troubled by scruples.
**müvezzi** (··⊥), **-ii** Distributing. Newsvendor; paper boy; postman.
**müyesser** Facilitated; practicable; helped by God. **~ olm.,** to be granted by God.
**müzah·eret** (·—·ˋ), **-ti** Help, support. **~ir** (·—ˋ), who supports *or* assists.
**müzahrefat** (···ˋ), **-tı** Excrement; filth.
**müzakere** (·—·ˋ) Discussion; conference; negotiation; the rehearsing of their lesson by schoolboys amongst themselves; (*pl.* **müzakerat,** negotiations): **~ etm.,** to talk over, discuss *stg.* **~ci,** master who hears boys their lessons; tutor.
**müzayaka** (·—·ˋ) Hardship; straits. **~ ile geçinmek,** to subsist with difficulty.
**müzayede** (·—·ˋ) Auction.
**müze** (ˋ·) Museum.
**müzehh·eb** Gilded, gilt. **~ib,** gilder.
**müzehher** In flower.
**müzekker** Male; masculine.
**müzekkere** Memorandum; note; warrant. **tevkif ~si,** warrant of arrest.
**müzevvir** Who falsifies; trickster; sneak; mischief-maker. **~lik,** tale-telling; knavery.
**müzeyyel** Having an appendix *or* addendum *or* postscript; (document) having an answer appended below. **~en** (·ˋ··), as an appendix *or* addendum.
**müzeyyen** Embellished; decorated.
**müz'ic** Annoying; vexatious.
**müzik** Music (*European as opp. to* **saz,** *oriental music*).
**müzmin** Chronic. **~leşmek,** to become chronic.

# N

**na** There!; there it is!; take it! **~ kafa!,** what a fool I (he) was!; what am I thinking about!: **~ sana!,** so much for you!; there, take that!
**na-, nâ-** (—) *Pers. negative prefix.*
**nabemahal** (—··ˋ) Out-of-place; untimely.
**nabemevsim** (—··ˋ) Premature; out-of-season.
**nabız, -bzı** Pulse. **nabzına bakmak** *or* **nabzını yoklamak,** to feel *s.o.'s* pulse: ⌜**nabza göre şerbet vermek**⌝, to use tact with a person. **~gir,** tactful, diplomatic.
**nabzı** *v.* **nabız.**
**nacak** Large axe with a hammer at the back. ⌜**oldu olacak kırıldı ~**⌝, there's nothing more to be done about it; the die is cast.
**nacar** Carpenter.
**naçar** (—ˋ) Who has no remedy; forced by necessity; helpless, in distress. Reluctantly; of necessity.
**naçiz** (—ˋ) Of no account, insignificant; modest, humble. **~ane** (——ˋ), humble (petition *etc.*) belonging to one's humble self; humbly.
**nadan** (—ˋ) Tactless; unmannerly; uneducated.
**nadas** The preliminary ploughing of land for cleaning before preparing the seed-bed; fallowing: **yeşil ~,** green manuring of land.
**nadide** (——ˋ) Never seen before; curious; rare.
**nadim** (—ˋ) Regretful; contrite. **···e ~ olm.,** to regret, to be sorry for having done.
**nadir** (—ˋ) Rare; unusual. **~at, -tı,** rarities; rarity. **~e** (—·ˋ), rarity; amusing anecdote. **~en** (⊥··), rarely.
**naehil** (—·ˋ) Unworthy; incapable; inexpert, unqualified.
**nafaka** Means of subsistence, livelihood; maintenance allowance; alimony. **···e ~ bağlamak,** to assign a subsistence allowance to: **~sını temin etm.,** to earn one's living.
**nafi** (—ˋ) Useful; profitable; beneficial. **~a** (—·ˋ), Public Works; Ministry of Public Works.

**nafile** (–·˙) Useless; in vain. A supererogatory act (of prayer *etc.*). ~!, it's no use!; don't persist!: ~ **yere**, uselessly, in vain.
**nafiz** (–˙) Penetrating; influential.
**nagant** An old-fashioned kind of pistol.
**nağme** Tune; song. **ara ~si**, intermezzo, interlude.
**nah** *v.* **na.**
**nahak** (–˙) Unjust; iniquitous. ~ **yere**, unjustly, unfairly.
**nahif** Thin, emaciated; weak, fragile.
**nahiv, -hvi** Syntax.
**nahiye** (–·˙) Region; sub-district *of one or more villages, forming an administrative unit under a* **müdür.** ~**vî** (–··˙), regional; local (*med.*).
**nahoş** (–˙) Disagreeable; unpleasant; unwell. ~**luk**, unpleasantness; indisposition.
**nahvet, -ti** Pride; conceit.
**nahvî** (·˙) Connected with syntax.
**naib** (–˙) One who acts for others; substitute; judge; regent; vice-regent.
**nail** (–˙) Who obtains. ···**e** ~ **olm.**, to obtain, attain. ~**iyet, -ti**, acquisition; attainment (of an object *etc.*).
**naim** (·˙) *Only in* **nazü** ~ **içinde**, in comfort and luxury.
**nakarat, -tı** Refrain; tiresome repetition; harping.
**nakavt** (˙·) Knock-out.
**nakd·en** (˙·) In cash; for ready money. ~**î** (·˙), cash, in ready money.
**nakıl** *corr. of* **nahil.** Palm branch. ~ **çiçeği**, phlox.
**nâkıl** (–˙) *v.* **nâkil.**
**nakır, -krı** Sculpture; carving.
**nâkıs** (–˙) Deficient; defective; minus; below zero.
**nakış, -kşı** Design; drawing; picture; embroidery; decoration. ~ **işlemek**, to embroider.
**nakız, -kzı** Annulment; violation (of a treaty *etc.*).
**nakibuleşraf** (*Formerly*) The representative at Istanbul of the Sherif of Mecca.
**nakid, -kdi** Cash; money; ready money.
**nâkil** (–˙) Transporting; transferring; narrating. Narrator; conductor (*elect.*). ~**iyet**, ~**ti**, a being a transporter; conductivity.
**nakil, -kli** Transport; removal; transfer; narration; translation. ~ **vasıtaları**, means of transport.
**nakisa** (·–˙) Defect; shame. ···**e** ~ **getirmek**, to bring dishonour upon.
**nakkaş** Artist (*esp.* an illuminator of manuscripts); decorator.
**naklen** (˙·) By tradition; by transfer.
**nakletmek** (˙··), **-eder** *va.* Transport; transfer; relate, narrate; translate. *vn.* Move, change one's abode.
**nakl·i** *v.* **nakil.** ~**î** (·˙), pertaining to transport; traditional. ~**iyat, -tı**, things handed down by tradition; means of transport; transport (*mil.*). ~**iye**, transport expenses; means of transport: ~ **gemisi (tayyaresi)**, transport ship, (plane).
**nakris** Gout.
**nakş** *v.* **nakış.** ~**berâb**, writing on water, building on sand. ~**etmek** (˙··), **-eder**, to decorate; to design.
**nakten, naktî** *v.* **nakden, nakdî.**
**nakz·en** (˙·) By annulment; in violation. ~**etmek** (˙··), **-eder**, to annul, quash; violate; contradict. ~**ı**, *v.* **nakız.**
**nal** Horseshoe. ~**lari dikmek**, (of a horse and contemptuously of a man) to die: ~ **döken**, stony road: **dört ~a gitmek**, to gallop at full speed: ⌜**hem ~ına hem mıhına**⌝, to act impartially, to try to please both parties; to hit out right and left regardless of persons: ⌜**iş üç ~la bir ata kaldı**⌝, 'all that is wanted now is three shoes and a horse', *said ironically of stg. that has only just begun.*
**nalân** (––˙) Moaning, lamenting. ~ **olm.**, to moan *or* lament.
**nalband** Shoeing-smith, farrier; horse-doctor. ~**lık**, horse-shoeing, farriery.
**nalbur** Man who makes *or* sells horseshoes *or* small articles of hardware, ironmonger.
**nalça** Iron tip *or* heel on a boot.
**nâle, nâliş** (–˙) Moan, groan.
**nâlet** (*corr. of* **lânet**) Cross-grained, peevish. ~ **olsun!**, damn the fellow!
**nalın** Pair of pattens *or* clogs. ~**cı**, clog-maker: ~ **keseri**, clog-maker's adze; egoist; ⌜~ **keseri gibi kendine yontmak**⌝, to think only of one's own advantage.
**nallamak** *va.* Shoe (a horse).
**nam** (–) Name; renown; reputation; quality. Named. ~**ına**, in the name of; by way of *in such phrases as*: **akraba ~ kimsesi yok**, he has nothing in the way of relations; **para ~ bir şey yok**, there isn't a farthing, I

haven't a 'sou': **~ında,** of the name of: **~ ile,** under the name of: **~ kazanmak,** to make a name for oneself: **~ ve hesabına,** for, on behalf of: **~ ve nişanı kalmadı,** it has left no trace, it has perished utterly: **dünyaya ~ vermek (salmak),** to acquire a world-wide reputation.

**namaglûb** (–·ꞌ) Unconquered; invincible.

**namahrem** (–·ꞌ) Not related *or* intimate; not having access to the harem.

**namaz** (·ꞌ) Ritual worship; prayer. **~ kılmak,** to perform the ritual prayers of Islam; **~ı kılındı,** his burial service has been read; (*sl.*) he's as good as dead. **~gâh,** open space devoted to prayer.

**namdar** (––ꞌ) Famous, celebrated.

**name** (–ꞌ) Letter; love-letter; document; *used in compounds for any written agreement or document, e.g.* **kanunname,** code of laws; **nizamname,** regulations; **sulhname,** treaty of peace.

**namerd** (–ꞌ) Unmanly; cruel; cowardly; vile. Vile, despicable person. **~e muhtac olm.,** to be obliged to ask help from one whom one despises: **~e muhtac olmamak,** to depend on no one for one's living, to be under obligation to no one. **~ce,** cowardly; unmanly; contemptible. **~lik,** cruelty; cowardice; vileness.

**namlı** Renowned, famous.

**namlu** Barrel (of a gun); blade (of a sword). **~ matkabı,** D-bit.

**namus** (–ꞌ) Honour; good name; rectitude, honesty. **~una dokunmak,** to affect one's honour; to hurt one's pride. **~kâr, ~lu,** honourable; honest, upright. **~suz,** without honour; shameless; dishonest.

**namütenahi** (–··–ꞌ) Unending; infinite.

**namzed** Nominated, designated; betrothed. Candidate; betrothed person; military cadet. **~lik,** candidature.

**nan** Bread; livelihood. **~ü nimet,** benevolence, generosity.

**nane** (–ꞌ) Mint; peppermint. **~ yemek,** to commit a blunder; to say *stg.* silly: ⌜**bak yediği ~!**⌝, what a silly thing to say (do)!

**nanemolla** (–ꞌ··) A timid and useless fop.

**nanik** Long nose, snook. **~ yapmak,** to cock a snook.

**nankör** Ungrateful. **~lük,** ingratitude.

**nanpare** (·–ꞌ) A piece of bread; livelihood. **~ye muhtac olm.,** to be destitute.

**nar** Pomegranate.

**nâr** (–) Fire; hell-fire; pain, injury. **~a yakmak,** to injure: **birinin ~ına yanmak,** to suffer for another's misdeeds: **başı ~a yandı,** he has burnt his fingers.

**nara** (–ꞌ) Cry, shout. **~ atmak,** to yell at the top of the voice.

**narcıl** Coconut.

**nardenk** Syrup made of pomegranate *or* damson juice.

**nargile** Water-pipe, hookah, narghile.

**narh** Price officially fixed.

**narin** (–ꞌ) Slim, slender; tender, delicate.

**nark, -kı** *v.* **narh.**

**nas, -ssı** Dogma.

**nasb** Nomination; appointment. **~ etmek** (ꞌ··), **-eder,** to nominate, appoint.

**nâsıh** (–ꞌ) Adviser, counsellor.

**nasıl** (ꞌ·) How?; what sort?; whatever sort. **~sınız?,** how are you?: **~sa,** in any case; somehow or other: **her ~sa,** in whatever way; somehow or other: **~ ki,** just as; as a matter of fact: **hem de ~!,** *emphasizes what has gone before, e.g.* **çocuk isterdim ... hem de ~ !,** I longed for a child—oh! how I longed!; **şimdi nezle oldum ... hem de ~!,** now I've got a cold, and what a cold!

**nasır** Wart; corn; callosity. **~lanmak,** to get warts *or* corns; to become calloused *or* callous. **~lı,** warty; with corns; calloused; callous.

**nasib** (·ꞌ) Lot, share, portion; one's lot in life. **~ almak,** to be initiated *into a dervish order*: **...den ~ almak,** to enjoy: **... ~im olmadı, ...** did not fall to my lot; it was not vouchsafed to me to ...: **~ olursa,** if destiny should will it: **insanlıktan ~i yok,** there is nothing human about him.

**nasihat** (·–ꞌ), **-ti** Advice; admonition. **~ etm.,** to advise: **birinin ~ini tutmak,** to follow *s.o.'s* advice.

**nasir** (–ꞌ) Prose-writer.

**nassı** *v.* **nas.**

**nasuh** *In* **~ nusuh tövbesi,** never again!

**naş** (–) Bier *or* coffin with a corpse; corpse.
**nâşir** (–ꜝ) Divulging. Publisher.
**natamam** (–·ꜝ) Incomplete.
**natık** (–ꜝ) Speaking; expressing *or* setting forth. **~a,** the faculty of speech; eloquence. **~alı,** eloquent.
**natır** Servant in a women's bath; watchman *of a garden or vineyard.*
**natuk** (·–ꜝ) Eloquent.
**natura** (·ꜝ·) Nature; constitution.
**navlun** Sum paid for chartering a ship; freight. **~ mukavelesi,** charter-party.
**nay, nayzen** *v.* **ney, neyzen.**
**naz** Mincing air; coquetry; coyness; whims; disdain; smirking; endearments. **~ etm., ~yapmak** *or* **~a çekmek,** to show coyness *about doing stg.*; to pretend not to be keen *about stg. when one really is*: **~ını* çekmek,** to tolerate *s.o.*, to put up with his peculiarities: **~ı* geçmek,** for one's whims to be tolerated; to be a 'persona grata': **~u nimet içinde büyümek,** to grow up amid fondlings and favours; to be spoilt: ⌜**bu ne ~?**⌝, why this coyness?; why so reluctant?: ⌜**çok ~ âşık usandırır**⌝, excessive coyness annoys the lover; 'oh well! if you don't want to, don't!'
**nazar** Look; regard; consideration; the evil eye. **~ımda,** in my view, as for me: **~ında,** according to him, in his view: ... **~iyle bakmak,** to regard as ...: **~ boncuğu,** bead worn to avert the evil eye: **birisine ~ı değmek,** to overlook *s.o.*, to cause illness *etc.* by the evil eye: **çocuğa ~ değdi** *or* **çocuk ~a geldi,** the evil eye has been cast upon the child, he has been overlooked: **···in ~ı dikkatini celbetmek,** to attract the attention of: **~ı itibara almak,** to take into consideration: **sarfı ~ etm.,** to give up, to renounce.
**nazaran** (ꜝ··) (*with dat.*) According to; with regard to; in proportion to; seeing that.
**nazare** *In* **~ye almak,** to make fun of.
**nazari** (··–ꜝ) Theoretical; visual. **~yat** (···ꜝ), **–tı,** *pl. of* **nazariye. ~yatçı,** theorist. **~ye** (···ꜝ), theory.
**nazarlık** Charm against the evil eye.
**nazenin** (–·ꜝ), **nazende** Graceful; delicate; amiable, nice; petted, spoilt; (*sarcastically*) whippersnapper.
**nazım** (·ꜝ), **–zmı** Versification, verse.
**nâzım** (–ꜝ) Regulator (*mech.*); who arranges; versifier.
**nazır** (–ꜝ) Who watches; overlooking, facing. Minister; spectator. **denize ~,** overlooking the sea: **Hariciye ~ı,** Minister for Foreign Affairs: **hazır ve ~,** all-present and all-seeing (*an attribute of the Almighty*). **~lık,** Ministry.
**nazik** (–ꜝ) Delicate; courteous. **~lik,** delicacy; courtesy.
**nazil** (–ꜝ) Descending; alighting. **~ olm.,** to descend, alight.
**nazir** (·–ꜝ) Match, like. **~e** (·–ꜝ), a similar thing; a poem written to resemble another poem in form and subject.
**nazlanmak** *vn.* Behave coquettishly; be coy; feign reluctance; behave in an affected manner.
**nazlı** Coquettish; coy; wayward; spoilt; reluctant; ticklish (job *etc.*).
**nazm·en** (ꜝ·) In verse. **~ı,** *v.* **nazım.**
**ne**[1] **~ ... ~ ...,** neither ... nor ... .
**ne**[2] What?; what; whatever; how. **~den,** for what reason?, why?: **~ye,** what for?, why?: **~yin nesi?,** *lit.* what relation to whom?, *i.e.* who on earth is that person?: **o adam sizin ~niz?,** what relation to you is that man?: **~yleyim** *etc. v.* **neyleyim**: **~ de olsa,** still; all the same; after all: **~ demek?,** what does it mean?; *v.* **demek**: **~ler gördüm!,** what didn't I see!; I saw all sorts of things: **~ güzel!,** how nice!: **~ ise!,** fortunately; well never mind; anyway: **~ ... ise,** whatever, *e.g.* **~ işitse inanır,** he believes whatever he hears: **~me lâzım,** *v.* **lâzım**: **~ olur ~ olmaz,** just in case: **~den sonra,** shortly afterwards: **~ vakit (zaman)?,** when?: **~ var ~ yok?,** what's the news?, how are things getting on?: **bu ~ hal!,** what's all this?: **hususî tren ~sine*?,** who is *he* to have a special train?
**nebat** (·ꜝ), **–tı** Vegetation; plant (*pl.* **~at**; **ilmi ~,** botany; **~ bahçesi,** botanical garden). **~î** (·–ꜝ), vegetable; botanical.
**nebevî** (··–ꜝ) Pertaining to a prophet.
**nebi** (·–ꜝ) Prophet.
**nebze** Particle; bit. **bir ~,** a little bit.
**necabet** (·–ꜝ), **–ti** Nobility. **~li,** noble.
**necaset** (·–ꜝ), **–ti** Impurity; excrement.
**Necaşi** (·––ꜝ) Negus; Emperor of Abyssinia.

**necat** (·ˋ), **-tı** Salvation; safety. ~ **bulmak,** to escape.
**nece** (ˋ·) In what language?
**Necef** Nejef. ~ **taşı,** rock-crystal.
**neci** Of what trade? **~dir?,** what is his trade?, what does he do?
**necib** Noble, of high lineage.
**Necid, ~cdi** Nejd.
**nedamet** (·—ˋ), **-ti** Regret; remorse. ~ **etm., (getirmek),** to regret.
**nedense** (·ˋ·) ~ *or* **her ~,** somehow or other; for some reason or other.
**nedim** (·ˋ) Boon companion; courtier; court buffoon. **~e** (·—ˋ), Lady of the Court.
**nedret, -ti** Rarity.
**nefais** (·—ˋ) Rare and exquisite things.
**nefaset** (·—ˋ), **-ti** A being exquisite, beautiful *or* rare.
**nefer** Individual; person; private soldier.
**nefes** Breath; moment; spell. ~ **aldırmamak,** to give no rest, not to give any respite: ~ **almak,** to breathe; to breathe freely again: ~ **çekmek,** (i) to take a whiff (of tobacco *etc.*); (ii) to smoke hashish: ~ **darlığı,** asthma: ~ **etm.,** to cure by breathing on *s.o.* and casting a spell: ~ **~e olm.,** to be out of breath: **~i* tutulmak,** to be unable to breathe; to have an attack of asthma: ~ **tüketmek,** to talk oneself hoarse: ~ **vermek,** to breathe out: **geniş ~ almak,** to breathe freely again.
**nefes·darlığı, -nı,** asthma. **~lenmek,** to breathe; take a short rest; breathe again (with relief). **~lik,** ventilator; vent-hole; the time passed in taking a breath: **bir ~ canı kalmış,** he's quite worn out; he looks wretched.
**nefi, -f'i** Advantage; profit. **nef'ine,** to one's own advantage: ~ **hazine,** the maxim that the first consideration in official matters must be the Public Purse.
**nefis, -fsi** Soul; life; self; essence; concupiscence. **nefsine,** in himself (itself): **nefsini* beğenmek,** to think a lot of oneself: **nefsine düşkün,** self-indulgent: **nefsine* mağlûb olm.,** to be overcome by one's desires: **nefsine* yedirememek,** to be unable to bring oneself *to do stg.*: **nefsini* yenmek,** to master oneself.
**nefîs** (·ˋ) Excellent; exquisite; rare. **sanayii ~e,** the fine arts. **~e** (·—ˋ), an exquisite *or* beautiful object.
**nefiy, -fyi** Banishment; exile; negation. ~ **edatı,** the negative particle.
**nefret, -ti** Disgust; loathing. **···den ~ etm.,** to detest, to feel an aversion for.
**nefrit, -ti** Nephritis.
**nefsan·î** (·—ˋ) Sensual, carnal; rancorous, malignant. **~iyet, -ti,** sensuality; spite. **~iyetçi,** one who bears malice, a spiteful man. **~iyetli,** spiteful.
**nefs·î** (·ˋ) Pertaining to the soul *or* to self; sensual. **~i,** *v.* **nefis.**
**neft, -ti** Naphtha. **~î,** of a dark brownish-green colour. **~yağı** (ˋ··), **-nı,** naphtha oil.
**nefy·etmek** (ˋ···), **-eder** *va.* Banish, exile; deny. **~i,** *v.* **nefiy.**
**nehale** (·—ˋ) Mat *to put under plates.*
**nehari** (·—ˋ) Pertaining to the day, diurnal; who is a day pupil, home boarder.
**nehir, -hri** River.
**nehy·etmek** (ˋ··), **-eder** *va.* Prohibit. **~i,** *from* **nehiy,** prohibition.
**nekadar** (ˋ··) How much?; what a lot; however much. ~ **olsa,** after all.
**nekahet** (·—ˋ), **-ti** Convalescence.
**nekes** Mean, stingy. **~lik,** stinginess.
**Nekir** *One of the angels who question men in their graves (the other is* **Münker**).
**nekre** Who makes odd and witty remarks.
**nelik** (ˋ·) The character *or* nature *of a thing.*
**nem** Moisture; damp. ~ **kapmak,** to absorb damp: **buluttan ~ kapmak,** to be unduly touchy *or* suspicious.
**nema** (·ˋ) Growth, increase; interest on money; profit. **~landırmak,** to make profitable.
**Nemçe** (ˋ·) Austria; Austrian.
**nemelâzımcılık** *The attitude of one who says* **neme lâzım,** *'what's that to do with me?'*; indifference, unconcern.
**nem·lenmek** *vn.* Become damp. **~li, ~nâk,** damp, humid.
**nemrud** Nimrod (*an impious Chaldean king, who cast Abraham into the flames*); cruel; very obstinate; very contrary; unmanageable. **~luk,** cruelty, tyranny; obstinacy: **~luğu* tutmak,** to have a fit of obstinacy.
**Nemse** (ˋ·) *Old word for* Austria. ~ **arpası,** pearl barley: ~ **böreği,** a kind of meat patty.

**nerde, nerden** *v.* nerede, nereden *under* nere.

**nere** (\`·) What place?; what part?; whatsoever place. **~m?**, what part of me?: **~n?**, what part of you?: **burası ~si?**, what place is this? **~de** (\`··), where?; wherever: **~ ise**, before long: **~ kaldı ki ...**, how much less ...; let alone that ...: **bu ~, o ~!**, there is no comparison between the two. **~den** (\`··), from where?; whence?; wheresoever: **~ geldi?**, where did he come from?; *also* **~ geldi!**, I wish to goodness he hadn't come!: **~ söyledim!**, why on earth did I say that!: **~ nereye**, for some reason or other; I don't quite know why. **~deyse** (\`····), before long. **~li** (\`··), coming from what place?: **~siniz?**, where do you hail from? **~ye** (\`··), to what place?, whither?; to whatever place: **~ giderse gitsin**, let him go where he will; wherever he goes.

**nergis** Narcissus; the eye of a beauty.

**nesc·etmek** (\`··), **-eder** *va.* Weave. **~i**, *v.* **nesic.**

**neseb** Family; genealogy. **~i sahih olmıyan çocuk**, illegitimate child. **~en** (\`··), by descent; by family. **~î** (··\_\`), relating to one's family; genealogical.

**nesic, -sci** Weaving; web; tissue.

**nesih, -shi** Abolition; abrogation; effacement; a kind of Arabic script.

**nesil, -sli** Descendants; generation; family. **~i münkarız oldu**, his race is extinct.

**nesir, -sri** Prose. **~ ci**, prose-writer.

**nesne** Thing; anything.

**nesr·en** (\`·) In prose. **~i**, *v.* **nesir.**

**nesturi** (·–_\`) Nestorian.

**neşat, -tı** Cheerfulness, gaiety; alacrity. **~lı**, in good spirits.

**neş'e** Slight intoxication; gaiety, merriment; joy. **~si yerinde**, he is in good humour. **~lendirmek**, to render merry; put in a good humour; intoxicate slightly. **~lenmek**, grow merry; become slightly drunk. **~li**, merry; in good humour. **~siz**, in bad humour; sad.

**neş'et, -ti** A coming into existence; origin; adolescence. **~ etm.**, to originate; to grow up; to pass out of a school.

**neşide** A popularly recited poem.

**neşir, -şri** A spreading broadcast; a publishing; publication; promulgation.

**neşr'etmek** (\`··), **-eder** *va.* Spread abroad; publish; diffuse. **~iyat, -tı**, publications.

**neşter** Lancet.

**neşvünema** (···_\`) Growth. **~ bulmak**, to grow and flourish.

**neta** (*naut.*) Clear; properly stowed; shipshape.

**netameli** Ill-omened; sinister; *thing or person* best avoided.

**netekim** *v.* nitekim.

**netice** (·–\`) Consequence; effect; result; conclusion. ... **~si**, as the result of ...: **~i kelâm**, in short; in conclusion.

**neuzübillah** (·_\`···) *Lit.* 'we take refuge with God'; God help us!

**neva** (·_\`) Tune, melody. **soğuk ~**, antipathetic, dour.

**nevale** (·–\`) Portion; food; meal. **~ düzmek**, to provide food.

**nev'ama** (··_\`) In a certain manner; to a certain extent; a kind of; so to speak.

**nevazil** (·–_\`) Cold in the head. **~ olm.**, to have a cold.

**nevcivan** Youth; young man.

**nevha** Lament; wail for the dead.

**nevheves** Frivolous; capricious.

**nevi, -v'i** Species; sort, variety, kind. **~ ~**, of various kinds: **~ beşer**, the human race: **nev'i şahsına münhasır**, of its own kind; 'sui generis': **cins ve ~**, genus and species.

**nev'ima** *v.* **nevama.**

**nevmid** Without hope; in despair. **~ane** (··–\`), desperate. **~i** (·–\_\`), despair, hopelessness.

**nevr** *In* **~i* dönmek**, for one's mood to change; to become moody *or* angry.

**nevralji** Neuralgia.

**nevruz** The Persian New Year's Day (22 March). **~iye** (·–·\`), *a kind of sweetmeat offered as a present on the Persian New Year's Day.* **~otu, -nu**, toad-flax.

**nevzad** Newly-born (child).

**ney** Reed; flute. **~zen**, flute player.

**neyleyim** (\`··) *etc. for* **ne eyleyim** *etc.* What can I do? **neylersiniz?**, what is one to do about it?

**nezafet** (·–\`), **-ti** Cleanliness. **~i fenniye**, the Hygiene Department.

**nezahet** (·–\`), **-ti** Purity; cleanness; decency, decorum.

**nezaket** (·–\`), **-ti** Delicacy; refinement; good breeding; a matter requiring delicacy in its treatment. **~li**, refined, delicate.

**nezamandır** (ˈ···) For a long time past.
**nezaret** (·–ˈ), **–ti** Prospect, view; inspection; supervision, superintendence; administration; direction; Ministry. ···**e** ~ **etm.**, to superintend, direct, inspect: **Hariciye N~i**, Ministry of Foreign Affairs.
**nezd** Vicinity of a person. **~inizde**, near you; in your opinion: **hükûmet ~inde**, in the view of the Government; with the Government; 'auprès du gouvernement'.
**nezetmek** (ˈ··), **–eder** *va.* Tear away; remove.
**nezf·î** (·ˈ) Haemorrhagic. **~i**, *v.* **nezif.**
**nezif, –zfi** Haemorrhage. **nezfi dimağî**, cerebral haemorrhage.
**nezih** (·ˈ) Pure in life and character; quiet, pleasant (place).
**nezir, –zri** A vowing *or* devoting; vow; thing vowed.
**nezle** Cold in the head. ~ **olm.**, to get a cold in the head.
**nezolunmak** (ˈ···) *vn.* Be removed.
**nezretmek** (ˈ··), **–eder** *va.* Vow; promise to give (as a vow).
**nıkrıs** Gout.
**nısab** (·ˈ) *The minimum income above which the Moslem tax of* **zekât** *becomes payable*; the number necessary for a quorum; the proper condition of a thing. **~ını bulmak**, to acquire the proper degree *or* condition.
**nısfıye** Small flute.
**nısıf, –sfı** A half.
**nışadır** Salammoniac; ammonia.
**nışasta** (·ˈ·) Starch.
**nice** (ˈ·) How many!; many a ...; how?; however many; howsoever. ~ **adamlar**, how many men ...!; many a man: ~ **olur?**, how will it be?; what will happen?: ⌜~ **senelere!**⌝, *a formal greeting on feast-days like* 'many happy returns of the day!'
**nicelik** (ˈ··) State; quantity.
**niçin** (ˈ·) Why?
**nida** (·ˈ) Cry; shout. **harfi ~**, interjection (*gram.*).
**nifak** (·ˈ), **-kı** Discord, enmity, strife.
**nihaî** (·–ˈ) Final.
**nihale** *v.* **nehale.**
**nihayet** (·–ˈ), **–ti** End; extremity; extreme. (ˈ–·) At last; at most. ~ **bulmak**, to come to an end: ~ **derecede**, extremely: ···**e** ~ **vermek**, to bring to an end, to put an end to. **~siz**, endless; infinite; countless.
**nikâb** (·ˈ) Veil with two holes for the eyes; mask.
**nikâh** Betrothal; marriage. ~ **düşmek**, for a marriage to be possible, *i.e. for the parties not to be within the prohibited degrees of relationship*: ~ **etm.**, to betroth, to marry: ~ **kıymak**, to perform the ceremony of marriage: ⌜**anasının ~ını istemek**⌝, to ask an absurd price for a thing.
**nikâh·lanmak**, to become betrothed *or* married. **~lı**, betrothed; married (*v.* **nikâh**). **~sız**, unmarried. ~ **yaşamak**, to live together without being married.
**nikbet, –ti** Misfortune; disgrace.
**nikbin** (·ˈ) Optimistic. Optimist. **~lik**, optimism.
**nikel** Nickel.
**nilüfer** Water-lily.
**nimet** (–ˈ), **–ti** Blessing; good fortune; benefaction; favour; food (*esp.* bread).
**nimresmî** (ˈ·–) Semi-official.
**nine** (ˈ·) Granny; mummy.
**ninni** Lullaby.
**nirengi** (··ˈ) Triangulation. ~ **noktası**, trigonometrical point; landmark, reference mark, guide mark.
**nisa** (·ˈ) Women. **~iye** (·–·ˈ), gynaecology.
**nisan** (–ˈ) April.
**nisb·et, –ti** Relation; proportion, ratio; comparison; relationship; spite. **~ine**, out of spite; in defiance: ~ **etm.**, to attribute; to compare; to act spitefully: ~ **tutmak**, to compare: ~ **vermek**, to say *stg.* out of spite: **imkân ~inde**, within the limits of possibility. **~etçi**, spiteful; defiant. **~eten** (ˈ··), relatively; in comparison; in proportion; in order to spite, spitefully. **~î** (·ˈ), proportionate; proportional. **~iyet, –ti**, relativity.
**nişadır** *v.* **nışadır.**
**nişan** (·ˈ) Sign; mark; indication; scar; target; order, decoration; engagement, betrothal; token given on betrothal. ~ **almak**, to take aim *at stg.*: **~a atmak**, to shoot at a target: **~dan dönmek**, to break off an engagement (of marriage): ~ **koymak**, to make a mark: ···**den** ~ **vermek**, to bear a resemblance to: ~ **yapmak**, to arrange an engagement: ~ **yüzüğü**, engagement ring: **namü ~ı kalmadı**, no trace of him remains.
**nişan·cı**, marksman. **~e** (·–ˈ), sign, mark. ~ **gâh** (··ˈ), butt, target;

backsight (of a gun): ~ **dürbünü,** telescopic sight. ~**lamak,** to sign; to aim at; to become engaged to, betroth. ~**lı,** engaged to be married.

**nişasta** (·ˈ·) Starch.

**nitekim** (ˈ··) Even as; just as; thus; e.g.; as a matter of fact.

**nitelik** Quality.

**niyabet** (·–ˈ), **-ti** An acting as substitute; office and functions of a **naib** *q.v.* ~**i saltanat,** regency.

**niyaz** (·ˈ) Entreaty, supplication. ~ **etm.,** to ask as a favour, to entreat for: **naz ~ile yapmak,** to do *stg.* only after much entreaty. ~**kâr,** ~**mend,** supplicant.

**niye** Why?

**niyet, -ti** Resolve; intention; formal resolve to perform some religious act. ... ~ **ile,** with the intention of ...: ~**i bozmak,** to change one's mind: ~ **çekmek,** to have one's fortune told *by drawing slips of paper bearing various possible (or impossible) prophecies*: ~ **etm.,** to resolve: ~ **kuyusu,** well at which vows are made.

**niyetli,** who has an intention; who has resolved to fast: **bu gün ~yim,** I am fasting today.

**niza** (·ˈ), **-aı** Quarrel; dispute. ~ **etm.,** to contend, dispute. ~**cı,** contentious; quarrelsome. ~**lı,** about which there is a dispute.

**nizam** (·ˈ) (*pl.* ~**at**) Order; regularity; law; regulation; system. ~**a koymak (getirmek),** to put in order. ~**en** (·ˈ·), according to law; legally. ~**î** (·–ˈ), legal; regularized. ~**iye** (·–·ˈ), the Regular Army: ~ **kapısı,** the main entrance to a barracks. ~**lı,** in order; regular; legal. ~**name** (··–ˈ), regulation. ~**sız,** in disorder; irregular; illegal. ~**sızlık,** disorder; irregularity; illegality.

**nobran** Arrogant; discourteous, churlish, ill-bred.

**noelbaba** (·ˈ··) Father Christmas.

**nohut** Chick-pea ⌜~ **oda bakla sofa**⌝, *said of a small-roomed house.*

**noksan** Deficient; defective; missing. Deficiency, defect, shortcoming. ~ **görmek,** to suffer a falling-off. ~**lık,** deficiency, defect.

**nokta** Point, dot; spot; speck; full stop; centre punch; dot *in Morse* (*dash* = **çizgi**); isolated sentry; military *or* police post. ~**i nazar,** point of view: ~**sı** ~**sına,** exactly, in every way. ~**lamak,** to dot, punctuate, mark with a centre punch. ~**lı,** dotted; punctuated.

**Norveç** (ˈ·) Norway.

**not, -tu** Note; memorandum; mark (in school). ~**etm.,** to make a note: ~ **tutmak,** to take notes: ···**e** ~ **vermek,** to give marks to; to pass judgement on.

**nota** (ˈ·) Note (musical, diplomatic); bill; memorandum; music (book *or* score).

**noter** Notary.

**nöbet, -ti** Turn (of duty *etc.*); watch (of a sentry *etc.*); access (of fever); set performance of a military band. ~**le,** in turn, by turns. ~ **beklemek,** to mount guard; to await one's turn: ~ **çalmak,** (of a band) to play before a sovereign *or* governor: ~**e çıkmak (girmek),** to mount guard; to go on sentry duty: ~ **şekeri,** sugar candy: **bir isteri** ~**i,** an attack of hysterics.

**nöbet·çi,** on guard; on duty; sentry; watchman: ~ **eczane,** pharmacy whose turn it is to be open at night: ~ **zabiti,** officer of the watch; orderly officer. ~**leşme,** a taking turns; rotation of crops. ~**leşmek,** to take turns; to take turn and turn about.

**nufus** *v.* **nüfus.**

**Nuh** Noah. ~**'un gemisi,** Noah's Ark: ~ **nebiden kalma,** antediluvian: ⌜~ **der peygamber demez**⌝, he is very obstinate: ~ **teknesi,** merrythought.

**nuhuset** (·–ˈ), **-ti** A being unlucky; evil omen.

**nukre** Ingot of silver.

**numara** (ˈ··) Number; marks; trick; performance; item, event (in an entertainment). ~ **on,** full marks: ~ **yapmak,** to play a part, to act: **iyi (fena)** ~ **vermek,** to give good (bad) marks *to s.o., i.e.* to think well (ill) of him: **tam** ~ **almak,** to get full marks: 100 **(yüz)** ~, W.C.: ~**cı,** tall-talker; charlatan. ~**lı,** numbered.

**numune** *v.* **nümune.**

**nur** (–) Light; brilliance; halo, the spiritual light of saintliness; glory. ⌜~ **içinde yatsın!**⌝, *said when mentioning a beloved dead one*: ~ **yüzlü,** benevolent looking (old man): **göz** ~**u dökmek,** to try one's eyes with work: **gözümün** ~**u,** my darling: **mezarına** ~ **inmek (yağmak),** for a light to descend upon his tomb, *i.e.* to be very holy.

**nuranî** (––ˈ), **nurlu** Luminous, shining; majestic; of blessed aspect.

~**yet, -ti,** a having a saintly aspect.

**nusayri** (··⅃) Of the Nusayriyye sect (inhabiting North Syria).

**nusha, nuska** (*corr. of* **muska**). Amulet, charm.

**nusret, -ti, nusrat** Help, *esp.* Divine help; victory.

**nutuk, -tku** The faculty of speech; speech; discourse. **nutka gelmek,** to begin to speak: **nutku* tutulmak,** to be tongue-tied, to be confused and silent.

**nüfus** (·ɂ) *pl. of* **nefis.** People; souls; inhabitants; *as sing.* person; inhabitant; (*for* ~ **tezkeresi**) identity papers. ~ **cüzdanı (kâgıdı),** identity book: ~ **kütüğü,** state register of persons: ~ **memurluğu,** Registry of Births *etc.*: ~ **tahriri (sayımı),** census: **beher** ~**a,** for each inhabitant, per head.

**nüfus·ca** (·⅃·), as regards persons: ~ **zayiat yokmuş,** there was no loss of life. ~**lu,** having ... inhabitants: **otuz milyon** ~ **bir devlet,** a country of 30 million inhabitants.

**nüfuz** (·ɂ) Penetration; permeation; insight; influence. ~ **etm.,** to penetrate; to go into; to influence. ~ **sahibi,** an influential person: ~**u nazar sahibi,** a person of insight. ~**lu,** influential. ~**suz,** without influence.

**nüks·etmek** (ɂ··), **-eder** *vn.* (Of a disease) to return and cause a relapse. ~**ü,** *v.* **nüküs.**

**nükte** Subtle point; nicety *of language*; witty remark; epigram. ~ **saçmak,** to make witty remarks. ~**ci,** witty; witty person. ~**dan,** witty. ~**li,** witty; subtle (of speech).

**nüküs, -ksü** Relapse (in illness).

**nümayiş** (·—ɂ) Show; pomp; simulation; demonstration (political *etc.*). ~**çi,** demonstrator. ~**kâr,** who demonstrates *or* makes a show; who simulates.

**nümune** (—·ɂ) Sample; pattern; instance, example; model. ~ **çiftliği,** model farm. ~**lik,** pattern, sample; (*sl.*) ridiculous, absurd.

**nüvaziş** (·—ɂ) A caressing *or* petting; kind treatment. ~ **etm.,** to caress, to show kindness. ~**kâr** (·—·⅃), caressing; kindly, attentive.

**nüve** Focus; centre; nucleus.

**nüzul** (·ɂ), **-lü** A descending *or* alighting; apoplexy. ~**etm.,** to descend, alight: **ona** ~ **isabet etti,** he had an apoplectic stroke.

# O

**o**[1] That; those. ~ **adam,** that man: ~ **çocuklar,** those children: ~ **bir,** that other one, the other: ~ **bir gün,** the other day, several days ago: **yarın değil** ~ **bir gün,** the day after tomorrow: ~ **bu,** whether this or that: ~ **gün bu gün,** from that day onwards.

**o**[2], **-nu** (**onun, ona, ondan,** *pl.* **onlar**). He; she; it.

**oba** Large nomad tent *in several compartments*; nomad family.

**obur** Gluttonous, greedy. ~**luk,** gluttony.

**obüs** Shell; howitzer.

**ocak**[1] January.

**ocak**[2] Furnace, kiln, hearth, fireplace, oven, range; quarry, mine: bed (garden); fraternity, guild, club; family, dynasty; home. ~ **çekirgesi,** cricket: ···**in ocağına düşmek,** to seek the protection of; to implore; to be at the mercy of: ~ **kaşı,** stone stand for saucepans *etc.* in front of a fireplace: **ocağı söndü,** his line has died out: ···**in ocağını söndürmek,** to destroy the family of, to ruin: **ev** ~ **kurmak,** to set up house, to start a family: **kömür ocağı,** coal-mine: **yeniçeri ocağı** *or* ~ **halkı,** the Corps of Janissaries: **Türk Ocağı,** the Turkish Nationalist Club.

**ocak·çı,** chimney-sweep; stoker. ~**çılık,** profession of a chimney-sweep. ~**lı,** having a fire-place; belonging to the **Türk Ocağı.** ~**lık,** fire-place; hearth-stone; chimney; (*formerly*) a family estate given by the sovereign; baulk of timber *serving as base for a superstructure*: ~ **demiri,** sheet-anchor.

**od** Fire. ⌜~ **yok ocak yok**⌝, no fire, no hearth', *used to imply great poverty*: **yüreğine*** ~ **düşmek,** to be deeply grieved.

**oda** Room; office; (*formerly*) Janissary barracks. **ticaret** ~**sı,** Chamber of Commerce: **yatak** ~**sı,** bedroom; **yemek** ~**sı,** dining-room. ~**başı** (·ɂ··), **-nı,** man in charge of the rooms of an inn *or* caravanserai. ~**cı,** man employed to clean and watch the rooms of a public building *or* an office; servant at an inn.

**odalık** Concubine; odalisque.

**odun** Firewood; log; cudgel; stupid coarse fellow. **~cu,** wood-cutter; seller of firewood. **~luk,** wood-shed; forest *or* tree suitable for cutting for firewood.

**of!** *Interjection expressing disgust, grief, or annoyance.* **~lamak.** to ejaculate 'ugh!': **oflayıp poflamak,** to say 'ugh!' from heat and weariness.

**ofsayd** Offside (football *etc.*).

**oğalamak** *va.* Rub and press with the hand; crumble.

**oğlak** Kid.

**oğlan** Boy; servant; catamite; tongue (carpentry). **kız ~ kız,** young virgin. **~cık,** little boy.

**oğlu** *v.* **oğul.** *Also in such expressions as*: **yok ~ yok,** it simply wasn't there: **varoğlu var,** everything was there.

**oğmaç, oğmak**[1] Freshly made **tarhana** (curd soup).

**oğmak**[2] *va.* Rub and press with the hand; massage; polish.

**oğul, -ğlu** Son; swarm of bees. **~dan oğla,** from father to son: **~ arısı,** young bee: **er oğlu er,** a fine man. **~balı** (·ˋ··), **-nı,** honey from a fresh swarm. **~duruk,** womb. **~luk,** sonship; duty of a son; adopted son. **~otu, -nu,** Lemon verbena.

**oğulmak** *vn.* Be rubbed, massaged, polished.

**oğuşturmak** *va.* Rub against each other.

**oğuz** *Name of a legendary Turkish king; name given to that part of the Turkish race inhabiting S. W. Asia.*

**oh!** *Interjection expressing satisfaction etc.* **~ olsun!,** serve you right!; I'm so glad! (*malignantly*): **~ çekmek,** to gloat over another's misfortunes: **~ démek,** (i) to breathe a sigh of satisfaction; (ii) to rest.

**ok, -ku** Arrow; pole *of a carriage*; *any long straight piece of wood at right angles to another part.* ⌜**~ yaydan fırlamış**⌝, 'the arrow has sped from the bow', *i.e.* the deed is done and can't be undone: **tatar ~u,** crossbow. **~çu,** bowman, archer; bow-maker.

**okadar** *v.* **kadar.**

**okka** Oke (*a measure of weight = 400 dirhems = 2·8 lb.*). **~ altına gitmek,** to bear the brunt, to be the chief victim: **~ çekmek,** to weigh heavy: **~ tutmak,** to be heavy: **kara ~,** the old **okka** *as opposed to the modern use = kilogram*: ⌜**nereye gitsen ~ dörtyüz dirhem**⌝, men are the same everywhere: ⌜**tam ~ dörtyüz dirhem bir adam**⌝, a first-class thoroughly reliable man: **yeni ~,** 1 kilogram. **okka·lı,** weighing *so many* okes; heavy, weighty, important: **~ kahve,** large cup of coffee. **~lık,** an oke of *stg.*

**oklava** Rolling-pin.

**okluk** Quiver.

**oksijen** Oxygen.

**okşamak** *va.* Caress, fondle; flatter; faintly resemble, remind one of. **zevkini* ~,** to be to one's taste.

**okşayış** Caress; petting.

**oktruva** Octroi.

**okul** School. **ilk ~,** primary school. **~lu,** who has been to school.

**okumak** *va.* Read; learn; study; sing; recite; say a prayer; invite; call; exorcize; (*sl.*) curse. **kendi bildiğini ~,** to go one's own way: **···in canına ~,** to harass, ruin: **düğüne ~,** to invite to a wedding-feast: **···e lânet ~,** to curse *s.o.*: **masal ~,** to romance: **meydan ~,** to defy, to challenge: **pek ~,** to read aloud; to sing loudly: **···e rahmet ~,** to pray for the soul of, to bless.

**okumamışlık** Illiteracy.

**okumuş** Educated; learned.

**okun·aklı** Legible. **~mak,** *pass. of* **okumak** *q.v.* Be prayed over *or* exorcized. **okunmuş su,** holy water.

**okut·mak** *va. caus. of* **okumak.** Cause to read *or* learn; instruct; educate; sell. **başını* ~,** to get oneself exorcized; ⌜**kendini okut!**⌝, go and get yourself exorcized!, *i.e.* you're crazy!: ⌜**gelen gidene rahmet okuttu**⌝, the new-comer caused his predecessor to be blessed, *i.e.* regretted. **~turmak,** *caus. of* **okutmak**: cause to be taught; have educated; cause to be sold.

**okuyucu** Reader; singer; exorcist; one who recites incantations.

**okyanus** Ocean. **~ya** (··ˋ·), Oceania.

**okyılanı** (ˋ···), **-nı** A small poisonous snake.

**ola** *3rd pers. sing. opt. of* **olmak. ~ ki,** it may be *or* happen that: **geçmiş ~,** (i) *said to one who has just been ill or met with an accident,* 'I hope you are all right now' *or* 'I'm glad you're better!'; (ii) 'it's too late, you've missed your chance!'

**olabilir** Possible; it may be.

**olacak** *3rd pers. sing. fut. and fut. part. of* **olmak.** Which may *or* will happen; ~!, so Fate has willed!: ~ **olur,** what is fated will happen: ~ **iş değil!,** that's impossible!; it's absurd!: **olacağı nedir?,** what's its lowest price?: **iş olacağına varır,** things must take their course: **o kâtib** ~ **adam,** that confounded clerk.

**olağan** That commonly happens; frequent; possible; probable; everyday. ~**üstü** (··⸌··), extraordinary.

**olagelmek** (·⸌··) *vn.* Happen now and again; happen frequently.

**olamak** *va.* Remove the suckers of *a vine.*

**olan** *pres. part. of* **olmak.** Being; becoming; that which is *or* happens. ~ **biten,** event: ~ **oldu,** what's done is done; it's too late now!: **insan** ~, a decent fellow, a real man: ⌈**sana** ~ **olmuş**⌉, 'what's come over you?'; you must be mad *to do or speak thus*!: ⌈**vay bize** ~**lar!**⌉, how disappointing!; what a nuisance!

**olanca** (·⸌·) Utmost; all possible. ~ **kuvvetiyle,** with all his strength.

**olarak** (·⸌·) *gerund. of* **olmak.** Being; *often used to make an adverb from an adjective, e.g.* **mevziî** ~, locally; **kat'î** ~ **yasak,** definitely forbidden; *other uses can best be shown by examples*: **ilk defa** ~, for the first time; **bunun neticesi** ~, as the result of this; **sarhoş** ~, in a drunken condition.

**olası** *optative form of* **olmak.** May it be!; which may be, possible, probable. **gözü kör** ~ **(olasıca),** 'may his eye be blind!', 'curse the fellow!': **oldum** ~**ya,** ever since I can remember.

**olay** Event; fact.

**olaydım, olaydı** (·⸌·) *etc. In such phrases as*: **söylemez olaydım!,** would to God I hadn't said it!; **gelmez olaydı!,** I wish to Heaven he hadn't come!

**oldu** *3rd pers. sing. perf. of* **olmak. olan** ~, what's done is done; it's too late now!: ~ **bitti,** 'fait accompli': ⌈~ **da bitti maşallah!**⌉, 'well, that's done the trick!': ⌈**ne** ~ **ne olmaz**⌉, just in case: ~ **olacak,** the inevitable has happened, *or* (*since one has got so far or things have taken this turn*) one may as well ..., *or* it's no use worrying any more about it.

**oldukça** (·⸌·) Rather. ~ **zengin,** pretty rich.

**oldurmak** *va. caus. of* **olmak.** Cause to be *or* become; cause to ripen; bring to perfection; raise (plants).

**olduysa** *3rd pers. sing. past cond. of* **olmak.** *In such phrases as*: **bu işte ne** ~ **bana oldu,** it is I who have had to bear the brunt; I am the one who suffered most.

**olgun** Ripe; mature. ~**luk,** ripeness; maturity: ~ **imtihanı,** examination *corresponding to the Higher School Certificate.*

**olmadık** (⸌··) *neg. past part. of* **olmak.** That has not happened; without precedent; inacceptable. ~ **bahaneler ileri sürürek,** putting forward all sorts of pretexts: ~ **bir şey değil,** it may well happen; it is not out of the common: ~ **olmaz,** anything may happen; nothing is impossible.

**olmak, olur** [*This verb is so important and its many parts so frequently used in idiomatic ways that these have been largely put under special headings, v.* **ola, olacak** *etc., etc.*] *vn.* Be; become; happen; be suitable; fit; ripen, mature; be cooked *or* prepared; (*sl.*) get drunk. ···**den** ~, to lose *stg.*; **işinden oldu,** it cost him his job: ~ **bitmek,** (*both verbs declined*) to happen, *e.g.* **oldu bitti,** it happened; **olup bitenler,** events; **ne olup bitiyor?,** what is happening?: **bana her şeyi olduğu gibi söyle,** tell me everything as it is, *i.e.* the whole truth: **çocuğu olmuyor,** she can't have a child.

**olmamış** (⸌··) *neg. past part. of* **olmak.** That did not happen *etc.* Not ripe, immature.

**olmasına** *In such phrases as*: '**genc** ~ **genc, amma tembel**', 'yes, he's young, it's true, but he's lazy' (*in reply to a question*).

**olmaz** *3rd pers. sing. aor. neg. of* **olmak.** That does not happen; impossible; wrong; unsuitable. ~!, it's impossible; you mustn't do it!: ~ ~, there is nothing that does not happen, *i.e.* anything may happen: **hiç olmazsa,** at least, at any rate: **ne oldu (olur) ne** ~, just in case.

**olmıya** (⸌··) *In* ~ **ki** ..., beware lest ... .

**olmıyacak** *neg. fut. part. of* **olmak.** Unlikely; unseemly; unsuitable.

**olmıyarak** *neg. gerund of* **olmak. kendisinde** ~, beside himself (with anger, joy, *etc.*).

**olmuş** *past part. of* **olmak.** That has

happened; completed; ripe, mature. ~ **bitmiş iş,** 'fait accompli'.

**olsa** *3rd pers. sing. cond. of* **olmak.** ~ ~, at the very most; in the last resort; ~ ~ **olabilir,** the worst that can happen is ...: ⌜~ **da olur, olmasa da**⌝, it's all the same whatever happens.

**olsun** *3rd pers. sing. imperative of* **olmak.** So be it!; even; if only. ~!, all right!, so be it!, I don't mind: ~ ~, at the very most: **bir kuruş** ~, even a piastre; if only one piastre: **iyi saatte** ~**lar,** spirits, the unseen: **iş diye** ~, *v.* **iş**: **oh** ~!, *v.* **oh.**

**olta**[1] Fishing-line. ~ **iğnesi,** fish-hook: ~ **yemi,** bait.

**olta**[2] *v.* **volta.**

**oluk** Gutter pipe; groove. ~ ~, in streams. ~**lu,** grooved: ~ **kalem,** gouge: ~ **sac,** corrugated iron.

**olunmak** *vn. As an auxiliary verb with Arabic verbal nouns it forms the passive of verbs, the active forms of which are formed with* **etmek,** *e.g.* **zikrolunmak,** *pass. of* **zikretmek**; **tatbik** ~, *pass. of* **tatbik etmek.**

**olupbitti** (·ˋ··) 'Fait accompli'.

**olur** *3rd pers. sing. aor. of* **olmak.** It is; it becomes *etc.* Possible; it may be; permissible. ~!, all right!; you may do it!: ~ **mu** ~, of course it's possible!; it may happen so; one never knows: ~**una bağlamak,** to make the best of a matter: ~**u ile iktifa etm.,** to make the best of things: ~ **olmaz,** any, whatever; anybody; whoever: ~ **şey değil!,** it's incredible!: **bir gün** ~ **ki,** the day will come that ...: **hiç** ~ **mu?,** can such a thing be!; can such a thing be done!: **işi** ~**una bırakmak,** to leave things to take their own course: **ne** ~ **ne olmaz,** just in case.

**oluş** *verbal noun of* **olmak.** State *or* manner of being *or* becoming. Nature; condition; event. ~**unda,** in itself, in reality.

**omaca** Stump of a tree.

**ombra** (ˋ·) Umber.

**omlet, -ti** Omelette.

**omurga** (·ˋ·) Backbone; keel. **iç** ~, keelson. ~**lı,** vertebrate.

**omuz** Shoulder. ~**una almak,** to shoulder; to undertake: ···**e** ~ **çevirmek,** to give the cold shoulder to: ~ **ile öpüşmek,** to be almost equal to; to be neck and neck with: ~ **silkmek (kaldırmak),** to shrug the shoulders: ~**da taşımak,** to honour, to hold in high esteem: ~ **vermek,** to push *or* hold with the shoulder: ···**e** ~ **vermek,** to give *s.o.* a shoulder up: **bir lâfa** ~ **vermemek,** to ignore what *s.o.* says: ~**a vurmak,** to shoulder: ~ ~**a yürümek,** to walk shoulder to shoulder.

**omuz·başı, -nı,** point of the shoulder. ~**daş,** companion, pal; fellow-**tulumbacı** (*q.v.*): ~ **pantolonu,** bell-bottomed trousers. ~**lamak,** to shoulder, give a shoulder to, assist. ~**lu,** having *such-and-such* shoulders. ~**luk,** epaulet; quarter (*naut.*).

**on** Ten. ~ **bir,** eleven; ~ **iki,** twelve, *etc. etc.*: ~ **numara,** full marks.

**ona** *dat. of* **o.** To him, her, it.

**onar** Ten each; ten at a time.

**onarmak** *va.* Repair; put in order.

**onbaşı** (ˋ··), **-nı** Corporal.

**onbeş** (ˋ·) Fifteen. ~ **gün,** fortnight.

**onca** According to him (her); in his (her) opinion; as far as he (she) is concerned.

**oncuk** *Only in* ~ **boncuk,** trashy jewellery.

**ondabir** A tenth.

**ondalık** A tenth; tithe; ten per cent. ~**çı,** one who works on a ten per cent. commission.

**ondan** *abl. of* **o.** From him, her, it.

**ondurmak** *va. caus. of* **onmak.** Cure, heal; improve; better the condition of. ⌜**ne öldürür ne ondurur**⌝, 'he neither kills nor heals', *said of a rather half-hearted man, who is unwilling to do what is expected of him.*

**ongun** Flourishing, prosperous.

**oniki** Twelve. **O**~ **ada,** the Dodecanese. ~**parmak,** duodenum. ~**telli,** a 12-stringed guitar.

**onlar** *pl. of* **o.** They.

**onlu** Having ten parts. The ten *of a suit of cards.* ~**k,** of ten parts; worth ten (piastres *etc.*); ten piastre *or* ten para piece.

**on·madık** (ˋ··) Not cured; not healed; unfortunate. ~**maz,** that will not heal. ~**mak** *vn.,* heal up; mend, improve.

**ons** Ounce.

**onsuz** Without him, her, it.

**onul·mak** *vn.* Be healed; be cured. ~**maz,** incurable.

**onun** *gen. of* **o.** Of him, her, it; his, hers, its. ~**la beraber,** together with him, her, it; for all that; at the same time.

**onuncu** Tenth.

**onur** Dignity; honour; self-respect. **~una* dokunmak,** to hurt *a person's* pride. **~lamak,** to do honour to.

**operatör** Surgeon.

**or** Dyke; earthwork protected by a ditch.

**ora** (ˈ·) That place (*only used with suffixes*). **~sı,** that place; that affair: **~sı öyle,** that is so: **~ları,** those places; those circumstances: ⌜**~sı sizin, burası benim dolaşmak**⌝, to saunter about. **~da** (ˈ··), there. **~dan** (ˈ··), from there, thence. **~ya** (ˈ··), to that place, thither, there.

**orak** Sickle, reaping-hook; harvest. **~böceği, -ni,** grasshopper. **~çı,** reaper. **~lamak,** to reap.

**oralı** Of that place; native of that place. **~ olmamak,** to pay no attention; to feign indifference.

**oramiral** Vice-admiral.

**oran** Measure; scale; proportion; symmetry; estimate.

**orantı** Ratio.

**orası** *v.* **ora.**

**oraya** (ˈ··) To that place, thither, there.

**ordinaryüs** (··ˈ·) Senior professor holding a 'chair' in a university.

**ordino** (ˈ··) Certificate of ownership; delivery order; licence; order.

**ordu** Army; camp. **~ kurmak,** to encamp. **~bozan,** spoil-sport. **~gâh,** military camp: **açık ~,** bivouack.

**org** Organ.

**orgeneral** (ˈ···) Full general; Army Commander.

**orkinos** (ˈ··) Large tunny-fish.

**orman** Forest; wood. **~ horozu,** blackcock. **~cı,** forester; forest-guard. **~cık,** little wood, copse. **~lık,** thickly wooded; woodland.

**orospu** Prostitute, whore.

**orostopoğlu** (··ˈ··), **-nu** Son of a bitch; scoundrel.

**orsa** (ˈ·) Direction of the wind as regards a ship's sailing. **~ alabanda!,** down with the helm!; **~ alabanda yatmak (eğlendirmek),** to be hove to: **~ boca etm.,** to tack and veer, to cruise about; to struggle along: **~ etm.,** to luff: **~sına seyretmek (gitmek),** to be close-hauled, to hug the wind. **~lamak,** to hug the wind.

**orta** Middle; centre; mean; the space around one; the high sea; a regiment of the Janissaries. Middle; central; medium; middling; public. **~da,** in the middle; in sight; **~da bir şey yok,** there is nothing to be seen; **~da bir çok sözler geçti,** many words passed between those present: **~ya almak,** to put in the middle, to set in the midst: **~da bırakmak,** to leave in the lurch: **~boylu,** of medium height: **~sını bulmak,** to find a middle course, to come to a compromise: **~ya çıkmak,** to arise, to come into being: **~ya dökülmek,** (of a secret) to be revealed: **~dan gidermek (kaldırmak),** to remove, to do away with: **~ hizmetçisi,** housemaid: **~ işi,** housework: **~dan kalkmak,** to be removed, to disappear, to be destroyed: **~dan kaybolmak,** to be lost to view, to disappear: **~ya koymak,** to produce; to put forward; **canı* ~ya koymak,** to stake one's life: **~ malı,** common to all; common possession; prostitute: **ikisi ~sı,** the middle *or* mean between the two; **ikisi ~sı yok,** there is no middle course: **varını* yoğunu* ~ya atmak,** to be ready to sacrifice one's all.

**orta·çağ,** the Middle Ages. **~elçi,** Minister Plenipotentiary. **~halli,** of moderate means; moderate.

**ortak** Partner; associate; accomplice; fellow wife in a polygamous household. **~ olm.,** to be a partner. **~laşa,** as a partner; in common. **~laşmak,** to enter into partnership with one another. **~lık,** partnership; being a wife in a polygamous family: ⌜**öküz öldü ~ ayrıldı**⌝, 'the ox has died, the partnership is dissolved', *i.e.* the reason which united the two parties no longer exists.

**ortalama** Medium; average. **~ bir hesabla,** on an average. **~k,** to divide in the middle; split the difference; put in the midst; reach the middle.

**ortalık** One's immediate surroundings; the world around; the face of nature; people, the public. **~ ağarmak,** for the dawn to break: **ortalığı birbirine katmak,** to make a mess, to turn the place upside down; to cause alarm and confusion: **~ hizmetçisi,** housemaid: **~ kararmak,** for night to fall: **~ karışmak,** for rebellion *or* disturbances to break out: **~ta kimse yok,** there is no one about: **'~ pahalılık!',** everything is dear!: **~ süpürmek,** to sweep up; **gayet iyi ~ süpürür,** she is a good housemaid:

**ortalığı toplamak,** to tidy up, to put a room straight.
**orta·malı** *v.* **orta malı.** **~mekteb,** secondary school. **~oyunu, -nu,** the old Turkish theatrical show. **~zaman** (·⸌··), the Middle Ages.
**ortanca**[1] Middle; middling; middle son *of three.*
**ortanca**[2] (·⸌·) Hydrangea.
**ortıkolana** Water-rail.
**ortodoks** Greek Orthodox (Christian).
**oruc** Fasting; fast. **~ açmak (bozmak),** to break *or* violate the fast: **~ tutmak,** to fast: **~ yemek,** purposely not to observe the fast: ⌜**gâvura (papaza) kızıp ~ bozmak**⌝, 'to quarrel with a Christian and break the fast (to spite him)', *i.e.* ⌜to cut off one's nose to spite one's face⌝. **~lu,** fasting. **~suz,** not fasting.
**orun** Place; post, employment.
**osmanlı** Ottoman. **~ca** (··⸌·), the Ottoman Turkish language, *i.e.* Turkish before the Revolution.
**osur·mak** *vn.* Break wind, fart. **~uk,** fart.
**oş, oşt** *Cry used to drive away a dog.*
**ot**[1] *v.* **od.**
**ot**[2], **-tu** Grass; any small plant *or* herb; weed; fodder. **~ tutunmak,** to remove hair by means of a depilatory.
**otak, otağ** Large nomad tent. **otağı hümayun,** the Imperial tent.
**otarmak** *va.* Pasture.
**otel** Hotel. **~ci,** hotel-keeper.
**otlak** Grassland, pasture. **~ lüferi,** medium-sized **lüfer,** caught in August. **~çı,** sponger. **~iye,** pasture tax; rent on pasture.
**otla·mak** *vn.* Be out to pasture; graze; lead a bovine existence, vegetate; (*sl.*) sponge. **hala koyduğum yerde otluyor,** he is just where he was; he's made no progress. **~nmak,** be at pasture; be fed off by beasts; be overgrown with grass.
**otlubağa** (·⸌··) Toad.
**otluk** Pasture; hayrick; hay-barn; priming powder.
**oto** Motor-car. **~büs,** motor-bus. **~kar,** large motor-car. **~mobil,** motor-car: **~ idare etm.,** *or* **kullanmak,** to drive a car.
**oto-** Auto-.
**oturak** Seat; thwart; residence; halting-place; place on which a thing stands (foot, stand, bottom); the posterior; chamber-pot. Seated; resting; halting; sedentary (*as opp. to nomadic*). **~ âlemi,** a drinking-party with dancing women: **yarım ~,** stretcher (in a rowing-boat). **~lı,** well settled; solidly based; sound solid (man); well-chosen, dignified (language).
**oturmak** *vn.* Sit; sit and enjoy oneself; rest; fit well; (of a ship) to be stranded, to run aground; (of a building) to settle *or* sink; (of a liquid) to settle. ⌜**otur oturduğun yerde!**⌝ *or* ⌜**otur oturmana bak!**⌝, 'don't meddle!'; 'mind your own business!': **tahta ~,** to ascend the throne: **···in üstüne ~,** to appropriate what does not belong to one; to embezzle: **···in üzerine ~,** not to restore *a borrowed thing,* to stick to *it*: **yerine ~,** to sit down; to fit into its place.
**oturtma** Set, mounted (gem). A dish of minced meat with vegetables.
**oturtmak** *va. caus. of* **oturmak.** Cause *or* allow to sit *etc.*; seat; place; run *a ship* aground; set, mount (jewel).
**oturulmak** *impers. of* **oturmak. bu evde oturulmaz,** one can't live in this house.
**oturum** Sitting (of a court *etc.*).
**oturuşmak** *vn.* Become calm *or* pacified.
**otuz** Thirty. **~ar,** thirty each; thirty at a time. **~luk,** of thirty ...; worth thirty *piastres etc.*; thirty years old. **~uncu,** thirtieth.
**ova** Grassy plain; meadow. **~lık,** level country, plain.
**ovalamak, ovmak** *etc. v.* **oğalamak** *etc.*
**oy** Opinion; vote. **beyaz ~,** vote in favour; **kırmızı ~,** vote against; **yeşil ~,** neutral vote.
**oya** Pinking; embroidery on the edges of a garment. **~lamak,** to pink *or* embroider, *v.* **oya**; put *s.o.* off to gain time; distract *s.o.'s* attention; make *one* waste one's time; keep *a child* quiet by amusing it. **~landırmak,** to put *s.o.* off by false promises. **~lanmak,** to loiter; be put off by frivolous pretexts; distract oneself. **~lı,** pinked; embroidered on the edge.
**oydu** (⸌··) It was he.
**oygalamak** *v.* **oyulgalamak.**
**oyluk** Thigh.
**oylum** Excavation; pit; pock-mark. Hollowed out; holed; carved. **~ ~,** (of smoke *etc.*) curling.
**oyma** Decoration by hollowing out; sculpture, carving, engraving *etc.*

~**cı,** sculptor, engraver. ~**lı,** carved, sculptured *etc.*

**oymak¹** *va.* Excavate; scoop out; engrave; carve; cut out *paper etc.* in decorative designs.

**oymak²** Subdivision of a people; tribe; troop of boy-scouts.

**oynak** Playful; frisky; mobile; unstable; flirtatious; loose, having much play, shifting (*mech.*); handy (airplane). ~**lık,** playfulness, friskiness; lightness of character, frivolity; looseness, play (*mech.*). ~**yeri, -ni,** articulation.

**oynamak** *vn.* & *va.* Play; move; dance; skip; jump about; be loose, have too much play (*mech.*); palpitate; dawdle; slip (of the ground in a landslide). **oynama!,** don't joke!, be serious! *or* don't dawdle!: **oynıya oynıya,** joyfully: **aklı*** ~, to go off one's head: **içe** ~, to keep losing *in a game*: **yerinden bile oynamıyor,** he does not even budge from his position: **yüreği*** ~, to be startled: **zilsiz** ~, to be overflowing with joy: **yer yerinden oynasa,** if the heavens fall, come what may.

**oynanılmak** *impers. of* **oynamak. burada oynanılmaz,** no playing here.

**oynaş** Playfellow; sweetheart, lover. ~**mak,** to play with one another; joke together.

**oynatmak** *va. caus. of* **oynamak.** Cause to play, dance *etc.*; move; perform (a play); dupe; trifle with. **aklını*** ~, to go off one's head: **at** ~, to be of some account, to have a say in matters: **yüzünü gözünü** ~, to make a wry face.

**oysaki** Yet; however.

**oyuk** Hollowed out; gouged out. Hollow part of a thing; grotto, cave; run *of a mole etc.*

**oyulga** Tacking; sewing loosely together. ~**lamak,** to tack *cloth etc.* together; baste. ~**nmak,** to be tacked.

**oyum¹** A scooping *or* digging out; pit; hollow; tap-root. ~**lamak,** to take root.

**oyum²** (ˋ·) I am he.

**oyun** Game; play; spectacle; jest; dance. ~ **almak,** to win a game: ~ **etm. (yapmak),** to play a trick, deceive: ~**a gelmek,** to be deceived: ~ **oyuncak,** an easy matter, a mere trifle: ~ **vermek,** to lose a game: **can** ~**u,** a venture where life is at stake.

**oyun·baz,** playful; deceitful; swindler. ~**bozan,** spoil-sport, kill-joy; quarrelsome; who keeps making fresh difficulties; who goes back on his word at the last minute. ~**cak,** toy, plaything; trifle, easy job; laughing-stock. ~**cu,** player; actor; gambler; dancer; comedian; trickster.

**ozan** Wandering minstrel.

**öbek** Heap; mound; group. ~ ~, scattered groups.

**öbür** The other. ~**ü,** the other one; that one.

**öc** Revenge; bet. ···**den** ~ **almak,** to take revenge on; ~**ünü almak,** to avenge: ~ **tutmak,** to lay a wager.

**öd** Gall; bile; courage, 'guts'. ~**ü* kopmak (patlamak),** to be frightened to death.

**ödağacı** (ˋ···), **-nı** Aloe wood.

**ödek** Indemnity; compensation.

**öde·mek** *va.* Pay; indemnify. ~**şmek,** to pay one another; settle accounts.

**ödev** Duty.

**ödlek** Cowardly.

**ödül** Reward; prize.

**ödünc** Loan. ~ **almak,** to borrow: ~ **vermek,** to lend: **deniz** ~**ü,** bottomry.

**öf!** *Expresses disgust*; 'ugh!'

**öfke** Anger, rage. ~ **topuklarına çıkmak,** to fly into a rage. ~**ci,** irascible. ~**lenmek,** to grow angry, get into a rage. ~**li,** choleric; hot-headed; impetuous.

**öğle, öğlen** Noon. ~**den sonra,** afternoon: ~ **yemeği,** the midday meal. ~**yin** (·ˋ·), at noon, about midday.

**öğmek, övmek** *va.* Praise; commend.

**öğren·ci** Pupil, student. ~**im,** education.

**öğrenmek** *va.* Learn; become accustomed to *or* familiar with.

**öğret·ici** Instructive, didactic. ~**im,** instruction; lessons.

**öğretmek** *va.* Teach (a thing); suggest; admonish, punish. **söz** ~, to put words into the mouth, to prompt.

**öğretmen** Teacher; instructor.

**öğücü, övücü** Who praises.

**öğün, övün** Share; portion; meal. ~**lük,** enough for *so many* meals,

**öğüngen, övüngen** Boastful, vainglorious.
**öğünmek, övünmek** *vn.* Praise oneself; boast. ~ **olmasın,** without wishing to boast.
**öğür·demek** *vn.* Retch; sob. **~mek,** to low, bellow; retch; belch. **~tmek,** to cause to bellow *etc.*; break in (an animal). **~tü,** a lowing *or* bellowing; a retching; belch.
**öğüş, övüş** Praise, commendation.
**öğüt** Advice. **~lemek,** to advise.
**öğüt·mek, övütmek** *va.* Grind; eat heartily. **~ücü,** that grinds; molar (tooth).
**öhö** *Expresses contempt or derision.*
**ökçe** Heel *of a boot.* **~li,** having heels; **yüksek ~,** high-heeled. **~siz,** without heels (shoe).
**ökse** Birdlime. **~ye basmak,** to fall into a trap. **~otu** (·ˈ··), **-nu,** mistletoe.
**öksemek, öksümek** *va.* Long for; miss.
**öksü** Half-burnt piece of wood.
**öksür·mek** *vn.* Cough; be at the last gasp. **~ük,** cough. ~ **otu,** coltsfoot.
**öksüz** Orphan (without mother); without relations *or* friends; the Piper fish (*Trigla lyra*). ~ **babası,** charitable man: ~ **sevindiren,** a common tawdry thing.
**öküz** Ox; heavy, stupid person. ⌜**~ün altında buzağı aramak**⌝, 'to search for a calf under an ox', *i.e.* to hunt for *stg.* in the most unlikely place: ⌜**~ü bacaya çıkarmak**⌝, to undertake an impossibility: ⌜**~ü bıçağın yanına götürmek**⌝, to make things unduly difficult: ~ **damı,** ox-stall, cow-shed. **öküz·gözü, -nü,** leopard's bane, arnica. **~lük,** a being bovine; dullness, stupidity.
**ölçek** *A measure of capacity for grain* $=\frac{1}{4}$ of a **kile**; measure; scale.
**ölçer** Poker; fire-rake.
**ölç·mek** *va.* Measure. **ölçüp biçmek (tartmak),** to decide after full consideration; to plan carefully. **~ü,** a measure; dimensions; quantity: ~ **vermek,** to be measured *for a suit of clothes*: **ağzının* ~sünü vermek,** to put *s.o.* in his place with a snub *or* reproof: **boyunun* ~sünü almak,** to learn one's lesson: **göz ~sü,** an estimate made by eye. **~ülü,** temperate, moderate, well-balanced. **~üm,** the measure of a thing; appraisal, estimate; air, manner: **~ünü bozmak,** to alter one's behaviour for the worse: ~ **etm.,** to give oneself airs. **~üsüz,** unmeasured; immeasurable; immoderate, excessive. **~üşmek,** *only in* ···**le boy ~,** to measure oneself against *s.o.*
**öldüresiye** With the intention of killing; ruthlessly; to death.
**öldür·mek** *va. caus. of* **ölmek.** Kill; kill time; render soft *or* tender. **körünü (nefsini) öldürmez,** he will not give way. **~ücü,** mortal, fatal, deadly.
**ölgün** Faded; withered; enervated; calm (sea).
**ölker** Nap, pile; down *on fruit.* **~siz şeftalı,** nectarine.
**öllük** Child's diaper.
**ölmek, -ür** *vn.* Die; fade, wither, lose freshness; suffer great grief *or* anxiety. ⌜**ölür müsün öldürür müsün**⌝, *said when very angry with s.o., but unable to do anything about it*: ⌜**~ var dönmek yoktur**⌝, he has (we *etc.* have) burnt his (our *etc.*) boats.
**ölmez** He *etc.* does not die. Undying, immortal; hard-wearing, resistant. **~leştirmek,** to immortalize. **~oğlu,** hard-wearing, resistant.
**ölmüş** Dead.
**ölü, ölük** Dead; feeble, lifeless; faded, withered. Corpse. ~ **deniz,** swell: **yarın gelmezse ~mü öp!,** I'll eat my hat if he doesn't come tomorrow!, *but* **oraya gidersen ~mü öp!,** I adjure you not to go there.
**ölücü** Mortal.
**ölüm** Death. ~ **dirim,** life-and-death: ~ **döşeği,** death-bed: ⌜**~ var dirim var**⌝, one must be prepared for all eventualities: **bu ~lü dünya,** this transitory world. **~lük,** *in such phrases as*: **~, dirimlik, 50 liram var,** all I have in the world is 50 liras.
**ömür, -mrü** Life; existence; enjoyment of life, happiness; age. Odd, amusing. **ömrü* oldukça,** for the rest of one's life, as long as one lives: ⌜**ömrünüz çok olsun!**⌝, 'may your life be long', *i.e.* may you be rewarded!: ~ **sürmek,** to live happily: **ömrü yokmuş,** said of one who died when very young: ⌜**Allah ~ler versin!**⌝, 'may God prolong your life' (*an expression of gratitude*): **çok ~ adamdır,** he's an odd fellow (but one can't help liking him): **ne ~ şey!,** how wonderful!; how beautiful!: **sizlere ~!,** may you live!, *euphemism for* 'he is dead'.

**ömür·lü,** having a ... life: **kısa ~,** short-lived. **~süz,** short-lived.
**ön** Front; space in front; breast, chest; the future. Front; foremost; *as prefix,* forefront. **~ümde, ~ümüzde, ~lerinde** *etc.*, in front of me, us, them *etc.*: **~ümüzdeki hafta,** the coming week: **···in ~ ünü almak,** to prevent, to avoid: **~ü ardı,** the beginning and the end; **···ünü ardını bilmek,** to be tactful and considerate: **~ünü ardını düşünmek,** to think well before acting, to act circumspectly: **~üne arkasına bakmadan,** thoughtlessly, without regard for the consequences: **~üne bak!,** look out!, mind!, *or* mind your own business!: **~üne bakmak,** to hang one's head: **···in ~üne düşmek,** to place oneself at the head *or* in front of ...; **düş ~üme!,** come along with me!: **···in ~üne geçmek,** to avoid, to take measures against, to combat: **···in ~üne gelmek,** to come before ..., to present oneself to ...; **~üne gelenle kavga eder,** he quarrels with everyone he meets: **···in ~ ünü kesmek,** to bar the way of, to waylay: **~ünde sonunda,** sooner or later: **başı ~ünde,** quiet, unassuming: **göz ~ünde tutmak,** not to lose sight of, to take into consideration.
**önayak** Pioneer; the first to do *stg.*; promoter; ringleader.
**önce** A little in front; first: **···den ~,** before. **~den,** beforehand; first of all: **~leri,** at the beginning; at first; formerly: **ilk ~,** first of all; in the very first place.
**öncü** Vanguard.
**önder** Leader. **~lik,** leadership.
**önem** Importance. **~li,** important; notable.
**önerge** Proposal; motion.
**önle·mek** *va.* Resist; face; stop, prevent. **~yici,** preventive.
**önlü** Pertaining to the front. **~ ve arkalı bir sahife,** a page written both on the back and the front.
**önlük** Apron.
**önsöz** (ˈ·) Preface.
**önüsıra** (·ˈ··) Before him.
**öpmek** *va.* Kiss. ⌜**öp babanın elini!**⌝, 'well, this is a mess!'; 'what's to be done now?': ⌜**öp de başına koy!**⌝, 'be thankful for small mercies!': ⌜**annemin ölüsünü öpeyim ki**⌝, (*an oath*) 'may my mother die if ...': ⌜**ağzını öpeyim!**⌝, *said to one who has given good news*: **el etek ~,** humbly to beg a favour: **gözlerinizden öperim,** (*in a letter*) I send you my love: '**sen bu evi elini öpene beş bin liraya satarsın**', 'you could sell this house for £5,000 and the buyer would be grateful to you'; 'many would be only too glad to give you £5,000 for it'.
**öp·ücük** Kiss. **~üş,** act *or* manner of kissing; kiss. **~üşmek,** to kiss one another; be reconciled.
**ör** Fence; artificial barrier.
**örçin** Rope ladder.
**ördek** Duck; urinal *for use in bed.* **~ balığı,** striped wrasse. **~başı,** greenish-blue.
**örek** Network of a tissue; build of a wall.
**öreke** Distaff; midwife's stool.
**örf** Common usage; extra-judicial civil usage. **~en** (ˈ·), according to common usage. **~î,** conventional, customary: **~idare,** martial law: **idarei ~ye,** state of siege.
**örgü** Plaited *or* knitted thing; plait; tress of hair; rush mat; tissue.
**örme** Plaiting; knitting; darning. Plaited; knitted; interwoven; darned; built with mortar.
**örmek** *va.* Plait; knit; interlace; darn; build; lay *bricks or stones in building.*
**örnek** Specimen, sample; model; pattern. **tıpkı örneğini yapmış,** he has copied the model exactly. **~lik,** model; sample: serving as a model *or* sample.
**örs** Anvil.
**örselemek** *va.* Handle roughly, misuse; wear out; spoil; rumple; exhaust, weaken.
**örtbas** A hushing up *or* concealment. **~ etm.,** to hush up, to endeavour to conceal *some unpleasant fact or defect.*
**ört·mek** *va.* Cover; wrap; veil; conceal; shut. **~türmek,** *caus. of* **örtmek.**
**örtü** Cover; wrap; roof; blanket. **masa ~sü,** tablecloth: **baş ~sü,** scarf worn over the head. **~cü,** covering, concealing. **~lü,** covered; wrapped up; concealed; shut; roofed; obscure (of speech): **üstü ~ söz,** equivocal, dubious speech; words implying more than they say. **~nmek** *vn.*, to cover oneself; to veil oneself: *va.* to cover oneself with. **~süz,** uncovered; open; bare.

**örü** Texture; web; darn; wall, building; barrier, division; enclosed space. **~cü,** mender, darner. **~lü,** plaited, knitted; enclosed *by a wall or hurdles.*

**örümcek** Spider; cobweb. **~ almak,** to sweep away cobwebs: **~ kafalı,** old-fashioned, incapable of accepting new ideas: **~ kuşu,** shrike: **~ tutmak,** to be covered with cobwebs: **şeytan örümceği,** gossamer. **~ağı, -nı,** spider's web; cobweb. **~lenmek,** to become covered with cobwebs.

**öşür** *v.* **üşür.**

**öte** The farther side; what is on the farther side. Other; farther; being farther away *or* on the other side. **~de, ~ye,** over there, farther on: **~si,** the rest; what follows; **~si berisi,** this side and that; here and there: **~sine varmak,** to surpass, to exceed what is usual: **daha ~,** farther still: **denizin ~si,** overseas: **bir kaç adım ~si,** a few paces away: **bundan ~sine karışmam,** further than that I am not concerned.

**öte·beri** (*both* **öte** *and* **beri** *are declined, v.* **öte**), this and that; various things; this side and that side; here and there. **~deberide,** here and there: **~denberi** (··ˈ··), from of old; heretofore. **~ki,** the other; the farther; farther away: **~ beriki,** this one and that, the one and the other. **~yanda,** on the other side.

**ötleğen** Warbler.

**ötlek** Cowardly; timid.

**ötmek** *vn.* Sing (of birds); resound, echo; crow (of a cock); talk foolishly, chatter. **borusu* ~,** to have a say in a matter; for one's words to carry weight.

**öttürmek** *va. caus. of* **ötmek.** Cause to sing *etc.*; talk airily; swank. **öttürme!,** don't talk through your hat!: **onun yapacağı yok, sade öttürüyor,** he can't do anything, he's only swanking.

**ötürü** (*with abl.*) By reason of, on account of.

**ötüş** Singing (of birds); resonance. **~mek,** (of birds) to sing together; sound together.

**övmek, övücü, övünmek** *etc. v.* **öğmek** *etc.*

**övünc** Boasting.

**övüş, övütmek** *etc. v.* **öğüş** *etc.*

**öyle** So; in that manner. Such; like that. **~mi ?,** is that so?: **~ ya!.** isn't it so! **~ ise, öyleyse,** if so, in that case: **~ bir adam,** such a man: **⌜~ de battık, böyle de⌝,** we're done for anyhow: **bana ~ geliyor ki,** it seems to me that.

**öyle·ce,** in that manner; somewhat so. **~lik,** that manner; such a way: **~le,** in such a manner. **~si,** such. **~yse,** if so, in that case.

**öyük**[1] Scarecrow.

**öyük**[2], **höyük** Artificial hill; mound.

**öyün** *v.* **öğün.**

**öz** Own; real; genuine; essential. Marrow; kernel; essence; pith; cream; self. **~ü sözü bir,** genuine, sincere: **~ türkçe,** pure Turkish: **az konuştu fakat ~ konuştu,** he spoke little but to the point: **⌜açık ~ tok söz⌝,** sincerely and frankly: **sözü ~üne uymak,** to be sincere; to be consistent: **⌜az olsun ~ olsun⌝,** it's not the quantity but the quality that matters.

**Özbek** The Uzbek people *of Central Asia.*

**özbeöz** (ˈ··) Essentially. Real, true.

**özdek** Stem, trunk; matter.

**özel** Personal; private; special. **~ ad,** proper name. **~lik,** peculiarity.

**özen** Pains; a taking pains. **~ bezen,** trinkets, ornaments; ceremony; ruse: **~e bezene,** painstaking, with particular care.

**özengi** *v.* **üzengi.**

**özen·mek** *vn.* Take pains, be painstaking, try hard; (*with dat.*) take pains about *stg.*; desire ardently; try to imitate *others*; have a passing fancy *to do stg.*: **özenip bezenmek,** to go to great trouble, take great pains. **~siz,** careless; not elaborate; superficial. **~ti,** aping; pseudo-, counterfeit; mock; swanking: **~ şair,** a pseudo poet.

**özle·mek** *va.* Wish for; long for. **~nmek**[1], *pass. of* **özlemek.**

**özlenmek**[2], **özleşmek** *vn.* Become of a pasty consistency; acquire pith, kernel *etc.*

**özleştirmek** *va.* Purify (a language); render authentic and unadulterated.

**özleyiş** Longing; home-sickness.

**özlü** Having a kernel, pith *etc. v.* **öz**; sappy, pulpy; of a sticky *or* pasty consistency; pithy terse; substantial; fertile. **~ toprak,** potter's clay.

**özlük** Selfishness; essence; pith; substance of a thing; stickiness.

**özür, -zrü** Defect; impediment; apology, pardon. **~ dilemek,** to ask pardon: **özrü meşru** a legitimate excuse: ⌜**özrü kabahatinden büyük**⌝, his excuse is worse than his fault. **~lü,** defective; having an excuse. **~süz pürüzsüz,** free from any defect.

# P

**pa** (–) Foot; leg.
**pabuc, papuc** Shoe; slipper; tag (of an electric wire). ⌜**pabucumu alırsın!**⌝, 'nothing doing!': **~dan aşağı,** contemptible: **…in ayağının papucu olamamak,** not to be fit to hold a candle to …: ⌜**…e ~ bırakmamak**⌝, not to be intimidated by …: **~larını çevirmek,** to give *s.o.* a hint that it is time he went: ⌜**pabucu dama atılmak**⌝, to fall into discredit: **pabucumda değil,** it doesn't concern me, it doesn't matter to me: **…in ~larını eline vermek,** to send *s.o.* away; to give the sack *to s.o.*: **~ eskitmek (paralamak),** to display great energy in following up a matter: **~suz kaçmak,** to flee in haste; to be very frightened: ⌜**~ kadar dili var**⌝, 'his tongue is as long as a shoe's', *said of one who answers back rudely*: ⌜**~ pahalı**⌝, (*on recognizing that s.o. is too much for one*) 'time to quit', 'time to throw up the sponge!': ⌜**…e pabucunu ters giydirmek**⌝, (i) to teach *s.o.* a lesson; (ii) to play a trick on *s.o.*; (iii) to cause *s.o.* to escape hurriedly: ⌜**iki ayağı bir ~da**⌝, hustled, perplexed: ⌜**…in iki ayağını* bir pabuca sokmak**⌝, to hustle *or* confuse *s.o.*: ⌜**sen kendini pabucu büyüğe okut!**⌝, 'get yourself exorcized by one with big shoes (a hodja)!', *i.e.* you must be crazy *to do such a thing*!
**pabuççu** Shoemaker; attendant who looks after people's shoes in a public building. **~ kölesi,** a shoemaker's iron last.
**paça** Sheep's *or* pig's trotters; dish made from trotters; lower part of the leg; lower part of the trouser leg. **~sını çekecek hali olmamak,** to be hopelessly clumsy and incapable: **~sı dürük,** slovenly, untidy: **~ günü,** the day after a wedding *when a dish of trotters is eaten*: **~sını kurtarmak,** to escape: **~ları sıvamak,** to tuck up one's trousers; to 'roll up one's sleeves', to get down to a job; ⌜**çayı görmeden ~ları sıvamak**⌝, to count one's chickens before they are hatched: **baş ~sı,** jelly made of sheep's head: **yaka ~ götürmek,** to collar *s.o.* and take him away.
**paçal** The proportion of various grains legally permitted in bread.
**paçalı** Having trouser legs; made of trotter jelly; (fowl *etc.*) having feathered legs. **dar ~,** having narrow trousers.
**paçarız** Crosswise; intricate. **~ düşmek,** to fall foul of a thing and get entangled in it.
**paçavra** (·ˈ·) Rag; miserable rag (of a newspaper). **~ etm.** *or* **~ya çevirmek,** to botch, to make a mess of *stg.*: **~ hastalığı,** influenza. **~cı,** ragman.
**paçuz, paçoz** Small grey mullet; (*sl.*) prostitute.
**padavra** (·ˈ·) Shingle; thin board *used under tiles*. **~ gibi** *or* **~sı çıkmış,** so thin that his ribs show.
**padişah** Ruler, sovereign, *esp.* the Sultan of Turkey. **~lık,** Sultanate.
**pafta** (ˈ·) Metal decoration on horses' harness; screw-plate; each section of a large map; large coloured spot. **~ ~,** decorated with metal plates; covered with spots: **~ kolu,** die stock: **~ lokması,** screw-die.
**pağurya** Crab.
**pah** Bevel. **~lamak,** to bevel.
**paha, baha** Price. **~ biçilmez,** priceless: **~ya çıkarmak,** to raise in price: **~ya çıkmak,** to rise in price: **her ne ~sına olursa olsun,** at whatever cost: **yok ~sına,** for a mere nothing. **~cı,** who sells at a high price. **~lanmak,** to become dear. **~lı,** dear, high-priced: **~ya oturmak,** to be very expensive.
**pahıl** Uncharitable; dog-in-the-manger.
**pak** (–) Clean, pure, untarnished; holy. **pirü ~,** spotlessly clean.
**paket, -ti** Packet; parcel. **~lemek,** to make into a parcel; pack up.
**pak·lamak** *va.* Clean; clear *of an accusation*, acquit; take away; use up; kill. **~lık,** purity, cleanliness.
**pakt, -tı** Pact.
**pala** Scimitar; blade of an oar *or* similarly shaped thing, paddle; thin wide plank set on edge. **~ çalmak,** to brandish a scimitar; to swagger about; to strive. **~bıyık,** large curved moustache.
**palaçka** Loot.

**palalık** Edge of a rafter. **palalığına koymak,** to set up on edge.
**palamar** Hawser. **~ı çözmek (koparmak),** to slip the cable; to make off. **~gözü, -nü,** hawse-hole.
**palamud**[1] The Short-finned Tunny *or* Pelamid (*Pelamys Sarda*).
**palamud**[2] Valonia (dried acorns of *Quercus aegilops*).
**palan** Broad soft saddle without frame.
**palandız** Stone in which the tap of a public fountain is fixed.
**palanga, palanko** (·ˋ·) Tackle, pulley-block.
**palanka** (·ˋ·) Redoubt of a fortress.
**palas** Coarse textile; rag.
**palas pandıras** Hastily; abruptly; brusquely.
**palaska** (·ˋ·) Cartridge-belt; bandolier.
**palaspare** (··–ˋ) Rags.
**palavra** (·ˋ·) Main-deck; idle talk, boast. **~cı,** braggart, boasting.
**palaz** Duckling, gosling *etc.* Fat. **~lamak,** to grow plump; (of a child) to grow up; to become noticeably better off.
**paldım** Crupper.
**paldırküldür** (·ˋ··) With great noise.
**palet, -ti** (*Fr. palette*) Mat; palette; articulated track of a caterpillar vehicle. **~li,** tracked (vehicle).
**palikarya** (··ˋ·) Greek youth; Greek rowdy.
**palto** (ˋ·) Overcoat.
**palûze** (––ˋ) *A kind of blancmange or jelly made of starch and sugar.*
**palyaço** (·ˋ·) Harlequin; buffoon.
**pamuk** Cotton. **~ atmak,** to card cotton: **~ gibi atmak,** to throw into utter confusion: **~ balı,** white honey: **~ balığı,** Blue shark: **~ barutu,** gun-cotton: **~ ipliği,** cotton thread; **~ ipliğile bağlamak,** to make a temporary *or* unsatisfactory arrangement: **kulaktan pamuğunu çıkarmak,** to become attentive; to listen to advice. **pamuk·akı** (··ˋ·), *a kind of* cotton thread *used for embroidery.* **~çuk,** aphtha, thrush (*med.*). **~lu,** of cotton; wadded.
**pan** (*Fr. panne*) Breakdown (of a car).
**panayır** Fair.
**pancar** Beet. **~ kesilmek,** to turn red in the face.
**pancur** Outside shutter, *esp.* slatted shutter.
**pandantif** Pendant (necklace).
**pandispanya** (··ˋ·) *A kind of cake resembling Bath Bun.*
**pandomima** (··ˋ·) Pantomime.
**pansiyon** Pension, boarding-house.
**pantolon** Trousers.
**pantufla** (·ˋ·) Felt slipper.
**panya** (ˋ·) Stern painter of a boat.
**panzehir** Antidote.
**Papa** (ˋ·) Pope. **~lık,** papacy.
**papağan** Parrot.
**papak** High Persian lambskin cap.
**papara** (·ˋ·) Cheese soup; any insipid thing. **~ yemek,** to get a severe scolding.
**papatya** (·ˋ·) Camomile; daisy.
**papaz** Priest; monk; king (cards). **~a dönmüş,** in need of a hair-cut: **~ kaçtı (uçtu),** *a kind of card game*: ⌜**~a kızıp perhiz bozmak**⌝, 'to get angry with the priest and break one's fast', *i.e.* to ⌜cut off one's nose to spite one's face⌝: **~ uçurmak,** to have a drinking party: **~ yahnisi,** fish *or* meat stewed with vegetables: ⌜**her gün ~ pilav yemez**⌝, you can't expect a good thing to go on for ever; one mustn't take things for granted. **~lık,** office and duties of a priest, priesthood.
**papazi** (·–ˋ) A kind of gauze.
**papel** (*sl.*) A Turkish pound note. **~ci,** cardsharper.
**papiyon** (*Fr. papillon*) Bow-tie. **~ somunu,** winged nut: **~ vidası,** thumb screw.
**papuc** *v.* **pabuc.**
**papura** (·ˋ·) Heavy plough *drawn by two yoke of oxen.*
**papye·büvar** (*Fr. papier buvard*) Blotting-paper. **~kuşe** (*Fr. papier couché*), surface-coated paper, art paper.
**para** Money; a para (the fortieth part of a piastre). **~ bozmak,** to change money: **~ çekmek,** (i) to withdraw money *from a bank etc.*; (ii) to squeeze money *out of s.o.*: **~ ~yı çeker,** money breeds money: **~dan çıkmak,** to be obliged to spend money: **~ ile değil,** very cheap: **~ etm.,** to be worth *stg.*: **~ etmemek,** to be worthless; to be in vain, to have no effect: **para farkı,** rate of exchange: ⌜**~na geçer hükmün**⌝, 'your jurisdiction goes with your money', *said to one who has been paid for stg. but still wishes to concern himself with it*: **~yı sokağa atmak,** to throw money away: **~ tutmak,** to

have money; to save money: ~ **yemek,** to spend money on pleasure: **dünyanın ~sı,** a 'mint of money': **kaç para?,** what's it worth?; what's the use!, it's in vain, it's too late.

**paraçol** A one-horse light carriage without springs; knee *of timber*; bracket.

**paradi** Gallery (theatre).

**paraf** Flourish *at the end of a signature.* **~e etmek,** to initial (agreement *etc.*).

**parafin** Paraffin wax.

**parafudr** (*Fr. parafoudre*) Spark-gap arrester.

**paragöz(lü)** Money-grubber.

**parakete** (··ˈ·), **parageta, paragat** Ship's log (instrument); fishing-line with several hooks. ~ **hesabı,** dead reckoning.

**para·lamak** *va.* Tear *or* cut to pieces. **lûgat ~,** to use learned words *in order to show off.* **~lanmak,** to be torn *or* broken to pieces; become rich; strain every nerve; to do everything possible to help *s.o.* **~latmak,** *caus. of* **paralamak.**

**paralel** Parallel. ~ **bağlanmış,** shunt (motor *etc.*), connected in parallel.

**paralı** Having money; rich; requiring payment.

**paralık** Of *so many* paras. **bir ~ etm.,** to vilify, to ruin *s.o.'s* reputation.

**paramparça** (·ˈ··) All in bits. ~ **etm.,** to tear to pieces, to break to bits.

**parantez** Parenthesis; inverted commas.

**parasız** Without money; penniless; gratis. ~ **tellal,** one who blazes abroad news that is no concern of his.

**parasvana** *v.* **prazvana.**

**paraşol** *v.* **paraçol.**

**paraşüt** Parachute. **~çü,** parachutist.

**paratoner** (*Fr. paratonnerre*) Lightning-conductor.

**paravana** (··ˈ·) Folding screen.

**parazit** Parasite; atmospherics (W/T.).

**parça** Piece; bit; segment; length *of cloth.* ... **~sı,** *used contemptuously of a person of a certain category, e.g.* **kâtib ~sı,** a very ordinary clerk, some sort of clerk: ~ ~, in bits, in pieces; by instalments; ~ ~ **etm.,** to break *or* tear to pieces; ~ ~ **olm.,** to be broken to bits, be torn to pieces: ~ **başına,** per piece; piecework: ~ **tesiri bomba,** fragmentation bomb: **bir ~,** a bit, a little, a moment: **elmas ~sı,** enchanting (child *etc.*). **parça·cı,** seller of piece goods. **~k purçak,** in rags and tatters. **~lamak,** to break *or* cut into pieces. **~lanmak,** *as* **paralanmak.**

**pare** (–ˈ) Piece. ~ ~, all in pieces: 101 ~, salute of 101 guns.

**parele** (·ˈ·) Mortise-joint.

**parıl** ~ ~, very brightly, brilliantly, flashing. **~damak,** to gleam, glitter, twinkle. **~tı,** glitter, gleam, flash.

**parke** Parquet; small paving stones.

**parlak** Bright; brilliant; shining; successful, influential. **~lık,** brilliance; beauty; influence; ability.

**parla·mak** *vn.* Shine; burn and flare up; flare up in anger; acquire influence, become distinguished. *va.* Fly *a falcon* at its quarry. **~tmak,** *caus. of* **parlamak,** to cause to shine *etc.*; (*sl.*) to drink. **~yış,** brilliance; a shining.

**parmak** Finger; toe; spoke *of a wheel*; bar, rod, single piece of a railing *etc.* (*v.* **parmaklık**); peak *of a hill*; *measure = about* $1\frac{1}{4}$ *in.*; a touch *or* taste *of stg.* **parmağı* ağzında kalmak,** to be lost in admiration; to be astounded: **parmağımı basıyorum,** I'll take my oath on it: **~la gösterilmek,** to be a person of distinction: ~ **hesabı,** counting on the fingers; syllabic metre *in Turkish poetry*: ~ **izi,** finger-print: ~ **kadar çocuk,** a tiny child: **···e parmağını* koymak,** to take a hand in *a matter*: **···i parmağında* oynatmak,** to twist *s.o.* round one's little finger: **···i parmağına sarmak (dolamak),** to have 'a bee in one's bonnet' about *stg.*: **~la sayılmak,** to be counted on one's fingers, to be rare: **···e parmağını* sokmak,** to meddle with *stg.*: ~ **üzümü,** a long kind of grape: **~larını yemek,** to find a dish delicious: **baş ~,** thumb, big toe: ⌜**beş ~ bir değil**⌝, everyone is different: ⌜**on parmağında on kara**⌝, *said of one who calumniates everyone*: ⌜**on parmağında on hüner (marifet)**⌝, he is very skilful *or* versatile.

**parmak·çı,** maker of banisters, spokes *etc.* **~lamak,** to finger; meddle with; stir up, incite. **~lık,** railing; balustrade; banisters; grating; grill.

**parmıcan** Parmesan cheese.

**parola** (·ˈ·) Password.

**parpa** (ˈ··) Young turbot.
**parpar** Brightly; gleamingly.
**pars** Leopard; cheetah.
**parsa** (ˈ·) Collection of money. **~ toplamak,** to make a collection, to hand the hat round: **~yı başkası topladı,** *s.o.* else got the benefit: **···den ~ toplamak,** to profit by ..., to exploit.
**parsal, partal** Old, worn clothes. In tatters.
**parsel** (*Fr. parcelle*) Plot *of land.*
**parşömen** Parchment.
**parti** Political party; party (social); game, match; consignment *of goods.* **~yi vurmak,** to do a good stroke of business.
**parya** (ˈ·) Pariah; outcast.
**pas¹** Rust; tarnish, dirt. **~ açmak,** to clean off rust; (*sl.*), to have a drink: **~ tutmak,** to rust: **buğday ~ı,** rust of wheat: **dil ~ı,** fur on the tongue: **kulakların ~ını gidermek,** to hear good music *after having been deprived of it, v.* **paslanmak.**
**pas²** Stake (at cards); pass! (at bridge); pass (football).
**pasak** Dirty and untidy clothes. **~lı,** dirty *or* untidy in dress, slovenly.
**pasaparola** (···ˈ·) Verbal command passed along the ranks (*mil.*).
**pasaport, -tu** Passport.
**paskal** Comic. Buffoon.
**paskalya** (·ˈ·) Easter. **~ yumurtası gibi,** very much made-up woman.
**pas·lanmak** *vn.* Become rusty; (of the tongue) to be furred. **kulakları* ~,** not to have heard good music for some time: **~lanmaz,** stainless (steel). **~lı,** rusty; dirty; pale, faded.
**paso** (ˈ·) Pass (on a railway *etc.*). **benden ~!,** 'I've had enough of it!'; 'don't count on me!'
**paspas** Doormat.
**pasta** (ˈ·) Sweet cake; pastry; tart; fold, pleat. **~hane** (··–ˈ), pastry-shop.
**pastal, pastav** Bundle of tobacco leaves.
**pastırma** Preserve of dried salt meat. **~sını* çıkarmak,** to give *s.o.* a good thrashing: **balık ~sı,** dried *or* salted fish: **domuz ~sı,** ham *or* bacon: **~ yazı,** a spell of warm sunny weather early in December.
**paşa** Pasha (*formerly the highest title of Turkish civil and military officials*). **Paşa eli,** (*formerly*) European Turkey: **~ kapısı,** government office *in the provinces.* **~lık,** title and rank of pasha. **~zade** (··–ˈ), son of a pasha.
**pat¹, -tı** Aster; diamond star.
**pat²** Suddenly. *Onom. expressing the noise of a blow with a flat thing.*
**pat³** Flat; snub (nose).
**pata¹** A drawn game; stalemate; deadlock. 'All square'.
**pata²** Cringle (*naut.*).
**patadak, patadan** (ˈ··) All of a sudden.
**patak** A whacking. **~lamak,** to give a whacking to *s.o.*; beat *a carpet or washing.*
**patalya** (·ˈ·) Small ship's-boat.
**patates** (·ˈ·) Potato.
**patavatsız** Who talks at random without reflection; tactless.
**paten** Skate.
**patenta** (·ˈ·) Patent; licence; bill of health; letters of naturalization. **~lı,** possessing papers of naturalization; having a patent *or* licence.
**patır** A tapping sound. **~ ~,** *imitates the noise of footsteps*: **~ kütür,** with the noise of footsteps; noisily. **~damak,** to make a knocking noise; make the noise of footsteps. **~dı,** noise; row: **~ çıkarmak,** to make a row, to cause a commotion: **ayak ~sı,** the noise of footsteps; a mere threat; a false alarm. **~dılı,** noisy; rowdy.
**patik** Child's shoe.
**patika** (·ˈ·) Foot-path; track.
**patinaj ~ yapmak,** to slip, to skid.
**patiska** (·ˈ·) Cambric.
**patküt** (ˈ·) Noise of repeated blows.
**patlak** Explosion; bursting. Burst; torn open. **~ gözlü,** goggle-eyed: **~ vermek,** to burst; to burst out (rebellion *etc.*); to be discovered *or* divulged.
**patla·mak** *vn.* Burst; explode; burst open; (of winter *etc.*) to arrive suddenly. **patlama!,** keep calm!; wait a moment!: **beş yüz liraya patladı,** it cost 500 lira: **tabanları* ~,** for the feet to become blistered, to be very tired from walking. **~ngıc,** pop-gun; cracker; paper bag blown up and burst. **~tmak,** to cause to burst *or* explode: make *s.o.* furious: **kafa ~,** to work very hard (mentally); to rack one's brains to try to understand *stg.*; to give s.o. a headache by noise *or* worrying.
**patlıcan** Aubergine, egg-plant. ⌜**acı ~ı kırağı çalmaz**⌝, the bad one escapes; *or* you can't spoil what is

already spoilt: ⌜**seninki (onunki) can da herkesinki ~ mı?**⌝, 'why should you (he) be the only one?', 'why should you be treated differently from others?' **~giller,** solanaceae.

**patra** *v.* **çetra.**

**patrik** Patriarch. **~hane** (··–⌄), Patriarchate (house). **~lik,** patriarchate, the office of a patriarch.

**patron** Head *of a firm or business*; model, pattern.

**pattadak** *v.* **patadak.**

**pavurya** (·⌄·) Large edible crab. **ayı ~sı,** spider crab.

**pay¹** Share; lot, portion; reproach, blame; margin, tolerance (*mech.*). **···den ~ biçmek,** to take as an example, to judge *or* deduce from; ⌜**kendinden ~ biç!**⌝, what would *you* do in the circumstances?; put yourself in his place!: **~ etm.,** to share, to go shares: **ağzının* ~ını almak,** to get snubbed in return for an insolent remark: **ağzının* ~ını vermek,** to snub *s.o.* severely, to give *s.o.* 'what for' for insolence: **ihtiyat ~ı bırakmak,** to leave a margin for safety: **kardeş ~ı,** equally, half and half: **kedi ~ı,** cat's-meat.

**pay²** Foot. **~a ~,** completely.

**pâyan** (–⌄) End, extremity. **neş'esine ~ yoktu,** his joy was unbounded.

**payanda** (·⌄·) Prop, support. **···e ~ vermek,** to prop up.

**paydos** Break, rest. **~!,** enough!, pax!: **~etm.,** to cease work, to knock off.

**pâye** (–⌄) Grade, rank; degree; dignity. **···e ~ vermek,** to show deference to, to esteem unduly.

**payi·dar** (··⌄) Firm, stable; enduring. **~mal,** trodden under foot, oppressed, despised: **~ etm.,** to trample on; destroy; despise. **~taht, -tı,** residence of the sovereign; capital.

**pay·lamak** *va.* Assign a share *to*; give one his due, scold. **~laşmak,** to share, divide up; go shares. **~latmak,** to scold, reprove.

**paytak** Knock-kneed; bandy-legged. Pawn (chess).

**payvant** Fetter; hobble.

**pazar¹** Sunday.

**pazar²** Bazaar; market; marketplace; bargaining. **~ etm.,** to bargain: **~ kesmek,** to conclude a bargain; to settle a price: **can ~ı,** situation where life is at stake: ⌜**evdeki ~ çarsıya uymaz**⌝, things don't turn out as one expects; 'counting one's chickens before they are hatched'.

**pazar·cı,** dealer in a market. **~laşmak,** to bargain, chaffer; settle a price. **~lık,** bargaining; deal; agreement as to price: **pazarlığını bozmak,** to break one's bargain *or* agreement: **~etm.,** to bargain, to do a deal: **~ uymamak,** for no agreement to be reached in bargaining: **içten ~,** mental reservation. **~lıklı,** clever at bargaining: **içten (içinden) ~,** cunning, insincere, hypocritical.

**pazartesi** (·⌄··) Monday.

**pazen** *v.* **bazen.**

**pazı¹** Mountain spinach.

**pazı², bazu** Arm; muscle; strength. **~bend,** armlet; amulet worn round the arm.

**pazvant** Watchman.

**pec** *v.* **bec.**

**peçe** Black veil *worn by Moslem women.* **~li,** veiled.

**peçeta** (·⌄·) Peseta.

**peçete** (·⌄·) Napkin.

**peçiç** *A game in which sea-shells are used in place of dice.*

**peçuta, peçota** (·⌄·) Largest sized bonito (palamut).

**pedavra** *v.* **padavra.**

**peder** Father. **~ane** (··–⌄), fatherly, paternal. **~şahî** (··–⌄), patriarchal.

**pehlivan** Wrestler; hero. **yalancı ~,** not quite such a fine fellow as he makes out. **~lık,** wrestling; bravery.

**pehpeh** Bravo! **~lemek,** to applaud; flatter.

**pehriz, perhiz** Abstinence; continence; diet (*med.*). **~ tutmak,** to fast; to observe a diet: *v.* **lahana.** **~li, ~kâr,** fasting; on a diet.

**pejmürde** Withered, faded; decayed; shabby.

**pek¹** Hard; firm; unyielding; violent; tight. **~ gözlü** *or* **gözü ~,** bold, courageous: **~ yürekli,** hard-hearted: **~ yüzlü** *or* **yüzü ~,** brazen-faced; thick-skinned; callous; blunt; with no respect for the feelings of others: **ağzı ~,** who can keep a secret, discreet: **eli ~,** stingy: ⌜**karnı tok, sırtı ~**⌝, well-fed, well set-up, one who leads a happy life.

**pek²** Very; very much; very often; violently; loudly. **~ ~,** at most: **~ iyi,** very good; very well; all right! **~âlâ** (⌄––), very good; all right! **~i,** *v.* **pek iyi.**

**pekçe** *dim. of* **pek**[1]. Rather hard *etc.*; somewhat loudly *or* violently.
**pekgözlü** *v.* **pek**[1].
**pek·işmek** *vn.* Become hard *or* firm *or* tight. **~iştirmek,** *caus. of* **pekişmek, yüzü ~,** to become **pek yüzlü,** *v.* **pek**[1]. **~itmek,** to make hard *or* firm; strengthen. **~leştirmek,** *v.* **pekiştirmek. ~lik,** hardness; firmness; stinginess; constipation.
**pekmez** Boiled grape-juice. **şeker ~i,** molasses.
**peksimet** Hard biscuit; ship's biscuit.
**pek·yürekli, ~yüzlü** *v.* **pek**[1].
**pelesenk**[1] Balsam; balm.
**pelesenk**[2] Common bittern; *v. also* **persenk.**
**pelin** Wormwood; absinthe.
**pelit** Acorn; valonia.
**pelte** Jelly. Jelly-like, flabby. **~lenmek, ~leşmek,** to become jellified *or* flabby.
**peltek** Lisp. **~lemek,** to lisp. **~lik,** lisping.
**pelüş** Plush.
**pembe** Rose colour. Rosy, pink. **toz ~si,** light pink. **~lik,** rose colour, rosiness. **~msi,** rather rosy, pinkish.
**penc** *In* **~ tavuğu,** guinea-fowl.
**pencere** (˙··) Window.
**pencik** Title-deed to a slave. **~li,** (slave) sold with a title-deed.
**pençe** The whole hand; paw; talon; strength; violence; ancient official signature of the Sultans; large stain; sole *for shoe repair*; tuber, crown (of asparagus *etc.*). **···e ~ atmak,** to lay hands on, seize: **~ vurmak,** to re-sole (a shoe): **~ ~ yanaklar,** fresh pink cheeks: **el ~ durmak,** to stand in an attitude of respect *with the right hand grasping the left.*
**pençe·lemek,** to grasp, seize; claw. **~leşmek,** to lock fingers with another and have a test of strength; be at grips with (= **ile**); engage in contest. **~li,** having claws; formidable; repaired, re-soled (boots).
**peni** Penny.
**pens** (*Fr. pince*) Pincers; pleat.
**pepe, pepeme** Stammer. Stammerer. **~lemek,** to stammer, stutter. **~lik, ~melik,** a stammering.
**perakende** (·—˙·) Dispersed, scattered; disjoined; confused; retail. **~ci,** retailer.
**perçem** Tuft of hair; long lock on the top of the head; forelock (of a horse). ⌜**tut kelin ~inden**⌝, 'it's like trying to hold a bald man by his hair', *said of one from whom due reparation for a wrong cannot be expected or who cannot be held responsible.*
**perçin** Clenching of a nail; riveting a bolt; putting a nut on the protruding end of a bolt; rivet. **~ etm.** *or* **~lemek,** to rivet, clench. **~li,** riveted.
**perdah** Polish; gloss; finishing shave. **~lamak,** to polish, burnish; shave a second time. **~lı,** polished; shining. **~sız,** unpolished; dull; matt.
**perde** Curtain; screen; partition shutting off the women's quarters; veil; membrane; cataract (eye); act (of a play); fret of a stringed instrument: musical note, pitch of the voice; modesty, chastity. **···e ~ çekmek,** to veil: **~si* yırtılmak,** to be shameless: **göz ~si,** cataract: **tiz ~den,** on a high note; violently: **üst (yüksek) ~den,** on a high note, in a high pitch: **üst ~den başlamak,** to start threatening *or* cursing: **üst ~den atıp tutmak,** to talk big, to give oneself airs.
**perde·ci** Doorkeeper. **~dar,** doorkeeper. **~lenmek,** to be veiled *or* curtailed; (eye) to have cataract. **~li,** veiled, curtained; webbed; membraneous; having cataract. **~siz,** without veil; unscreened; shameless, immodest. **~sizlik,** immodesty, unchastity.
**pereme** Greek two-oared boat.
**perende** (·˙·) Somersault. **~ atmak,** to turn a somersault: **···in önünde ~ atmak,** to dupe, to play tricks on.
**perese** Mason's plumb-line; level; direction, bearing; state, condition. **···i ~ye almak** to weigh (a matter *etc.*): **bir ~ye geldi ki,** it came to such a point that: **···i ~sine getirmek,** to choose the right moment for *stg.*
**-perest** *Pers. suffix.* Worshipping ..., *e.g.* **zevk~,** a worshipper of pleasure; **put~,** an idolator. **~iş,** a worshipping, adoration: **···e ~ etm.,** to worship. **~kâr,** adoring, worshipping; adorer.
**pergel** Pair of compasses. **iğneli ~,** dividers. **~lemek,** to measure *or* scribe with compasses; think out.
**perhiz** *v.* **pehriz.**
**peri** Fairy; good genius; beautiful person. **Ahmed'in perisi Mehmed'den hoşlanmamış,** Ahmed dislikes Mehmed: **~ illeti,** epilepsy; hysteria: **~si pis,** he likes dirt, he is never clean.

~**cik,** little fairy; bolt of a door; epilepsy. ~**li,** haunted; sinister; possessed.

**perişan** Scattered; disordered, in confusion; routed; perplexed; wretched, ruined. ~ **etm.,** to scatter, to ruin, to rout. ~ **olm.,** to be scattered, routed; to be in a state of misery *or* ruin. ~**lık,** a state of disorder, ruin *or* wretchedness.

**perki** A voracious freshwater fish.

**perkitmek, perkişmek** *etc. v.* **pekitmek** *etc.*

**permeçe** (·꞉·) Small hawser; tow-rope.

**persenk** Refrain; word continually repeated *like 'you know', 'you see'.*

**perşembe** Thursday. ⌜~**nin gelişi çarşambadan belli olur**⌝, 'it's as clear as that night follows day' *or* 'one can feel it coming'.

**pertav** Jump; run taken before a jump; bowshot. ~ **almak,** to take a run before a jump: ~ **etm.,** to shoot (an arrow *etc.*).

**pertavsız** Magnifying-glass; burning-glass.

**peru** A single pear-shaped gem.

**peruka** (·꞉·) Wig. **perukâr,** hair-dresser.

**perva** (·∸) Heed, attention; fear. ~**sız,** fearless; caring for no one. ~**sızlık,** fearlessness; a not caring for anyone.

**pervane** (−−꞉) Moth; fly-wheel; propeller; paddle-wheel; sails of a windmill; sun-fish. **üstüne** ~ **olm.,** to look after and protect.

**pervaz** Ornamental border; cornice; moulding; fringe. ~**lık,** anything suitable *or* sufficient for a border.

**-perver** *Pers. suffix.* Nourishing ...; caring for ..., *e.g.* **misafir**~, hospitable; **hayâl**~, nourishing vain hopes.

**pes**[1] Low, soft (voice).

**pes**[2] *In such phrases as:* ~ **dedir(t)mek,** to make *s.o.* cry small, give in: ~ **demek,** to submit, to give in: ~ **etm.,** to cry small: **yanında** ~ **demek,** to recognize the superiority of, to give in to: **buna** ~**!,** 'it's beyond me!' *i.e.* it passes my comprehension; *or* 'it's the limit!' (of *s.o.'s* behaviour *or* speech).

**pesek** Tartar *of the teeth.*

**pespaye** (·−꞉) Common, vulgar.

**pest** *v.* **pes.**

**pestenkerani** (···−꞉) Idiotic; nonsensical.

**pestil** Fruit pulp pressed into thin layers and dried. ~**ini* çıkarmak,** to beat *s.o.* to a jelly: ~**i* çıkmak,** to be beaten *or* crushed: ~ **gibi yatmak,** to lie exhausted.

**peş** The space behind; edging *of a garment.* ~**imde,** ~**inde** *etc.* after *or* behind me, him *etc.*: ~**inden,** following ..., in company with ...: ~**ini bırakmak,** to cease following: ~ **inde dolaşmak (gezmek),** to go around with *s.o.*; to pursue *a matter*: ~**ine düşmek,** to follow: ~**inden koşmak,** to run after: ~**i sıra,** behind him, following him, with him; afterwards: **···i** ~**ine* takmak,** to bring along with one: **···in** ~**ine takılmak,** to tack oneself on to *s.o.*

**peşiman** (·−꞉) *v.* **pişman.**

**peşin** Former; first; paid in advance, ready (money). First, in the first place; in advance. ~ **almak,** to buy for cash: ~ **söylemek,** to tell in advance, prognosticate: **berveçhi** ~, in anticipation; ready money being paid down. ~**ci,** one who deals for cash.

**peşisıra** *v.* **peş.**

**peşkeş** Gift. ~ **çekmek,** to make a present of a thing that does not belong to one.

**peşkir** Napkin.

**peşrev** Overture, prelude. (*mus.*); grape-shot; preliminary movements of wrestlers. ⌜**zurnada** ~ **olmaz (ne çıkarsa bahtına)**⌝, you mustn't expect too much: *v. also* **pişrev.**

**peştahta** Small desk; counter; money-changer's board.

**peştamal** Large bath-towel. ~ **kuşanmak,** to finish one's apprenticeship and become a master workman. ~**cı,** dealer in bath-towels. ~**iye,** (*formerly*) a fee paid by an apprentice to his master on leaving his apprenticeship to work on his own. ~**lık,** money paid for the goodwill of a business: ~ **vermek,** to buy the goodwill.

**petalides** (··꞉·) Limpet.

**petek** Honeycomb; any circular disk; jar made of clay. ~ **gözü,** cell in honeycomb. ~**göz,** compound *or* mosaic eye *of insects.*

**petkir** Hair sieve.

**petrol** Petroleum (*not* petrol, *which is* **benzin**).

**pey** Earnest-money; money on account; deposit; bid *at an auction.* ~ **sürmek,** to run up the bidding: ~

tutulmaz, that cannot be taken as serious, not to be relied upon: ~ **vermek,** to pay a deposit: ~ **vurmak,** to make a bid.

**peyda** (·⌐) Existent; produced; manifest; born. ~ **etm.,** to procure, beget, create; acquire (a habit *etc.*): ~ **olm.,** to come into being, be manifested, appear.

**peydahlamak** *va. see* **peyda etm., çocuk** ~, (of an unmarried woman) to be with child.

**peyderpey** (ˋ··) One after the other; in succession; step by step.

**peygamber** Prophet *esp.* the prophet Mohammed. ~ **ağacı,** guaiacum, lignum vitae: ~ **çiçeği,** cornflower: ~ **devesi,** the praying mantis: ~ **kuşu,** wagtail.

**peyk, -ki** Lackey; messenger; follower; satellite.

**peyke** (ˋ·) Wooden bench.

**pey·lemek** *va.* Pay a deposit on *stg.* (to seal a bargain); make sure of getting *stg.*; book, engage. ~**leşmek,** to conclude a bargain by paying a deposit.

**peynir** Cheese. ~ **dişi,** the last remaining tooth of an old man; ~ **dişli,** toothless: ~ **ekmekle yemek,** to do *stg.* easily, as a matter of course; ⌜**aklını** ~ **ekmekle mi yedin?**⌝, 'have you gone out of your senses?': **kirli hanım** ~**i,** a soft rich white cheese: ⌜**lafla** ~ **gemisi yürümez**⌝, 'fine words butter no parsnips'. ~**ci,** cheesemonger. ~**lenmek,** to become cheesy; (of milk) to coagulate; to be flavoured *or* sprinkled with cheese.

**peyrev** Who follows in another's footsteps; subordinate; follower, imitator.

**pezevenk** Pimp; *a term of general abuse.*

**pıhtı** Coagulated liquid. Coagulated, clotted. **kan** ~**sı,** clot of blood. ~**lanmak,** ~**laşmak,** to become coagulated, to clot. ~**latmak,** to cause to coagulate.

**pılıpırtı** (*pl.* ~**lar,** *but acc.* **pılıyı pırtıyı**). Old rubbish; belongings. **pılıyı pırtıyı toplamak,** to pack up one's belongings.

**pınar** Spring, source **göz** ~**ı,** inner corner of the eye.

**pır** A whizzing *or* whirring. ~ ~ **etm.,** to whiz, to whirr: **lâmba** ~ ~ **edip söndü,** the lamp spluttered and went out.

**pırasa, pırazvana** *etc. v.* **prasa** *etc.*

**pırıl** *etc. v.* **parıl** *etc.*

**pırlak** Lure, decoy.

**pırlamak** *vn.* Flutter. *v. also* **parlamak. pırlayıp gitmek,** to fly away.

**pırlangıc** Musical top.

**pırlanmak** (Of a young bird) to flutter and try to fly.

**pırlanta** (·ˋ·) A brilliant. Set with brilliants.

**pırnal, pırnar** Ilex, holm-oak.

**pırpı** *A kind of stone used as an antidote to snake-bites*; monkshood.

**pırpırı** Shabbily dressed. A dissolute rake; (*formerly*) a cloak of red felt *worn by constables.*

**pırpıt** Coarse home-made cloth. Old, worn-out, shabby.

**pırtı** *v.* **pılıpırtı.**

**pırtık** Torn; ragged; in rags.

**pırt·lak** Bulging. ~**lamak,** to bulge out.

**pısırık** Shy, diffident; weak, incapable.

**pıt** *In* ~ **yok,** there's not a sound.

**pıtır** ~ ~, *imitates the sound of rapid footsteps.* ~**damak,** to make a tapping sound, to crackle. ~**dı,** light tapping *or* crackling sound.

**pıtrak** Plant bearing innumerable burrs (such as burdock). ~ **gibi,** covered with fruit.

**piç**[1] Bastard; offshoot, sucker; cuticle of the nail; the small, incomplete *or* deficient replica of anything. ~ **etm.,** to spoil (sleep *etc.*) by frequent interruption: ~ **kurusu,** tiresome, naughty child.

**piç**[2] Pitch *of a propeller.*

**piçota** Largest sized palamut (bonito).

**pide** *Kind of* bread *baked in thin flat strips.*

**pikap, -pı** Pick-up (gramophone).

**pike**[1] *The cotton material* piqué, quilting.

**pike**[2] (*Fr. piqué*) ~ **uçağı,** dive-bomber: ~ **etm.,** (of an aircraft) to dive.

**pil** (*Fr. pile*) Electric battery.

**pilâki** Stew *of fish or beans with oil and onion, eaten cold.*

**pilâv** Pilaf. ⌜~**dan dönenin kaşığı kırılsın**⌝, 'one does not refuse a good offer', *also* 'I won't fail', 'I'll abide by my decision': **iç** ~, pilaf with meat and raisins.

**piliç** Chick. ~ **çıkarmak,** to hatch eggs.

**pineklemek** *vn.* Slumber; doze.

**pines** A kind of sea mollusc, pinna.
**pinpon** Dotard.
**pinti** Miserly, stingy; shabby. **~lemek,** to become dirty and shabby from miserliness. **~lik,** stinginess, meanness; shabbiness.
**pipo** Tobacco pipe.
**pir** Old man; founder of an order of dervishes; patron saint. **~ aşkına,** just 'for love', without expecting anything in return: **~i fani,** a decrepit old man: ⌜**~ ol!**⌝, 'bravo!': ⌜**bir oldu amma ~ oldu**⌝, it was a great success: **bir kızarsa ~ kızar,** he doesn't often lose his temper, but when he *does* ...!
**pire** Flea; aphis. ⌜**~yi deve yapmak**⌝, to exaggerate grossly: **~ gibi,** very agile, lively: ⌜**~ için yorgan yakmak**⌝, 'to burn a blanket to get rid of a flea', *i.e.* to cause great damage to oneself in order to avenge a trifle: ⌜**~yi nallamak**⌝, 'to shoe a flea', to attempt the impossible, *but it sometimes means* 'to be very cunning'.
**pire·lenmek,** to become infested with fleas; to hunt for fleas on oneself; to be worried by suspicion; to feel uneasy; to be in a bad temper. **~li,** full of fleas; suspicious; uneasy. **~otu, -nu,** pyrethrum; insect powder.
**pirina** (·˙·), **pirin** Cake of crushed olives from which the oil has been extracted.
**pirinc**[1] Rice. **~ örgüsü,** moss-stitch (knitting): **~ su kaldırmamak,** (of rice) to be unable to absorb much water; **~i su kaldırmaz,** he is very touchy.
**pirinc**[2] Brass.
**pirohu, piruhi** *A dish of* stewed dough with cheese.
**pirüpak** (–·˙–) Spotlessly clean.
**piryol** Cask wide at the bottom and narrow at the top; cylindrical iron vessel; 'turnip' watch.
**pirzola** (·˙· *or* ˙··) Cutlet, chop.
**pis** Dirty; foul; obscene. **~i ~ine,** uselessly, in vain: **~ ~ düşünmek,** to brood, to appear distraught and worried: **~ ~ gülmek,** to laugh in a scoffing manner. **~boğaz,** greedy; so greedy that he eats anything at any time.
**pisi** (*Childish language*) Cat; pussy-cat. **~ ~!,** puss! puss!: **~ balığı,** plaice; **dere ~ balığı,** flounder.
**pisi pisine** *v.* **pis.**
**pisin** (*Fr. piscine*) Fish-pond; bathing-pool.
**piskopos** Bishop. **~luk,** bishopric; rank of bishop; bishop's house.
**pis·lemek** *va.* Dirty; soil; spoil; relieve oneself; make a mess of; (of an animal *or* baby) to make a mess on *stg.* **~lenmek, ~leşmek,** to become dirty; to be soiled. **~letmek,** to make dirty, to soil. **~lik,** dirtiness; dirt; mess; obscenity.
**pist**[1] *Noise made to drive away cats.*
**pist**[2] (*Fr. piste*) Running track; runway.
**piston** Piston; (*sl.*) backing, influence.
**pişdar** Vanguard.
**pişik** Inflamed sore *in the groin or armpit resulting from heat or sweat.*
**pişim** Act of being cooked; amount cooked at one time.
**pişirim** Act of cooking; amount that one will cook at one time.
**pişirmek** *va.* Cook; bake; ripen; plan (a course of action); hatch (a plot); (of sweat) to cause a sore, *v.* **pişik;** to learn *stg.* extremely well. **işi ~,** to concoct a plan; to have an amorous adventure.
**pişkin** Well-cooked; well-baked; ripe, mature; 'hard-baked', hardened, experienced. **~lik,** a being well cooked *or* ripened; maturity; experience, knowledge of the world.
**pişman** Sorry (for = *dat.*); regretful; penitent. ⌜**geldiğine geleceğine ~ etm.**⌝, to give *s.o.* a hot reception. **~lık,** regret; penitence.
**piş·mek** *vn.* Be cooked *or* baked; ripen, mature; be perfected; become experienced; be overcome by the heat; (of the skin) to be chafed *or* inflamed by heat *or* sweat. **~memiş,** uncooked; immature; inexperienced.
**pişrev** Scout; forerunner; prelude; preamble. **el ~i,** feints made by wrestlers before coming to grips. *v. also* **peşrev.**
**pişti** *A kind of* ball game.
**piştov** Pistol.
**pitalinez** *corr. of* **petalides.** Limpet.
**piyade** (·–˙) Foot-soldier; infantry; pedestrian; pawn (chess); a single-pair boat; a man of small capacity *or* knowledge. **deniz ~leri,** marines.
**piyale** (·–˙) Cup; cup of wine.
**piyan** Seizing (of a rope).
**piyango** (·˙·) Lottery; raffle. **~su çıkmak,** to win a lottery.

**piyano** (·ˋ·) Piano.
**piyasa** (·ˋ·) Place of passage; public place; open space; rate of exchange; current price. **~ya çıkmak,** to go out for a walk: **~ya düşmek,** (of a woman) to go on the streets: **~ etm.,** to walk about.
**piyata** (·ˋ·) Oval dish.
**piyaz** Stew with onions. **fasulye ~ı,** salad of beans cooked with onions, oil, vinegar, eggs *etc.*: **~ı vermek** *or* **~ları basmak,** to applaud, to flatter. **~cı,** flatterer. **~lamak,** to marinade; to sing the praises of, to flatter.
**piyes** (*Fr. pièce*) Piece, theatrical play.
**plâçka** (ˋ·) Spoil, booty. **~cı,** raider, freebooter.
**plâk** (*Fr. plaque*) Plate; disk; gramophone record; clay pigeon. **···i ~a almak,** to make a record of (song *etc.*).
**plâka** Plate; name-plate.
**plân** Plan. **arka ~,** background; **ön ~,** foreground.
**plânçete** (·ˋ·) Plane-table.
**plânör** (*Fr. planeur*) Glider.
**plânya** (ˋ·) Carpenter's plane.
**plâsman** (*Fr. placement*) Investment *of money*.
**plâtin** Platinum.
**podösüet** (*Fr. peau de suède*) Suede leather.
**pof** *Imitates a popping noise.*
**poğaça** (·ˋ·) Flaky pastry.
**poğama** (·ˋ·) Crane; derrick.
**pohpoh** Bravo! Applause; flattery. **~lamak,** to applaud; to flatter.
**poliçe** (·ˋ·) Bill of exchange; insurance policy.
**poligon** (*Fr. polygon*) Artillery range.
**polis** Police.
**politika** (··ˋ·) Politics; policy, cunning *or* flattery in conversation. **~cı,** politician; one who knows when to use flattery *or* cunning in his talk.
**Polon·ya** (·ˋ·) Poland. **~yalı, ~ez,** Polish.
**Pomak** Bulgarian Moslem.
**pomata** (·ˋ·) Pomade.
**pompa** (ˋ·) Pump.
**ponza** (ˋ·) Pounce, pumice; rottenstone.
**porselen** Porcelain; porcelain insulator.
**porsuk**[1] Badger. **~ ağacı,** yew.
**porsuk**[2], **porsumak** *v.* **pörsük, pörsümek.**
**porsun** Boatswain.
**portakal** Orange.
**Portekiz** (ˋ··) Portugal. **~li,** Portuguese.
**portmanto** (·ˋ·) Coat-stand; coat-hanger; (*not* portmanteau *which is* **bavul**).
**portolon** Chart; plan.
**posa** (ˋ·) Sediment; dregs. **~lanmak,** to deposit a sediment; to settle as dregs.
**posbıyık** (ˋ··) Large bushy moustache; having such a moustache.
**post, -tu** Skin; hide; tanned skin with the fur on, *esp. when used as a rug*; rind, shell; office, post. **~ elden gitmek,** to lose one's skin (life); to lose one's post: **~ kalpak,** sheepskin cap: **~ kapmak,** to get an office (*with a suggestion of fraud or cunning*): **~ kavgası,** a quarrel about getting a post: **~u kurtarmak,** to save one's skin (life): **~una oturmak,** to take possession of one's official post; to assume airs: **~u sermek,** to settle oneself down in a place *or* a post *with the intention of staying*: **⌜(inanma dostuna) saman doldurur postuna⌝,** *said when a friend has betrayed one.*
**posta** (ˋ·) The post (letters); postal service; mail steamer; mail train; military post; shift, gang, relay. **~ etm.,** (of a policeman) to take *s.o.* up, to take to the police-station: **~yı kesmek,** to cease frequenting a place *or* doing stg. **~ polisi,** policeman on a beat.
**posta·cı,** postman. **~hane** (···—ˋ), post-office. **~l**[1], postal. **~ne** (·—ˋ), post-office. **~pulu,** postage stamp.
**postal**[2] Heavy army shoe; loose woman.
**postnişin** Established in an office of dignity, *esp.* as head of a religious order.
**poşu** *v.* **puşu.**
**pot, -tu** Crease, fold, pleat; punt, raft; wooden floor of a cattle-shed. Puckered; too full (dress). **~ gelmek,** to go wrong, to turn out badly: **~ kırmak,** to make a 'faux pas', to 'drop a brick': **~ yeri,** defect.
**pota** (ˋ·) Crucible.
**potin** Boot.
**pot·lanmak** *vn.* Be creased *or* puckered. **~lu,** puckered; in folds. **~uk,** puckered; full (dress); in pleats: young camel.
**potur** Pleat; fold; corrugation; full pleated knee-breeches. Puckered;

pleated. **~lu,** wearing the breeches known as **potur**: peasant.
**poyra** (\`·) Hub; axle-end (of a car). **~luk,** log of hard wood out of which hubs are made.
**poyraz** North-East wind; NE. point of the compass. **~ kuşu,** oyster-catcher: **ağzını* ~a açmak,** to be disappointed; to get nothing.
**poz** (*Fr. pose*) Pose; exposure (photography).
**pörçük pörçük** In bits and pieces.
**pörs·ük** Shrivelled up; withered. **~-ümek,** to shrivel up, become withered *or* wrinkled.
**pösteki** Sheepskin; sheepskin rug used to sit on. **~ olm.,** to become limp: **~ saymak,** (to count the hairs of a sheepskin), to be engaged on a useless and tedious task: **~sini* sermek,** to flay *s.o.*, to give *s.o.* a severe thrashing: **deliye ~ saydırmak,** to give *s.o.* a tiresome but useless job.
**prafa** (\`·) **~ oyunu,** a card game for three people.
**pranga** (\`·) Fetters attached to the legs of criminals; penal servitude.
**prasa** (\`·) Leek. **~ bıyıklı,** with a very long moustache: ⌜**~ olsa yemem**⌝, 'I couldn't eat anything'.
**pratik** Practical. Practice.
**pratika** (·\`·) Pratique, clean bill of health.
**prazvana** (·\`·) Shank *of a blade*; metal socket; ferrule.
**prens** Prince. **~lik,** principality. **~es,** princess.
**prese** (\`·), **preze** *v.* **perese.**
**prevantoryum** Sanatorium for the treatment of tuberculosis in its early stage.
**prim** Premium; prize.
**priz** (*Fr. prise*) Electric connexion, plug; drive (*mech.*).
**profilli** Sectional (iron).
**prostela** ·\`·) Apron.
**protesto** (·\`·) Protest. **~ etm.,** to protest: **~ çekmek,** (of a government) to make a formal protest.
**prova**[1] (\`·) Trial; printer's proof. **~ etm.,** to try on (a dress).
**prova**[2] (\`·) Prow, bow *of a ship*.
**provadifortuna** (\`···\`·) Protest of average (*comm.*).
**Prusya** Prussia. **~lı,** Prussian.
**pruva** *v.* **prova**[2].
**puc** Cleft between the buttocks.
**puding** Pudding; conglomerate (*geol.*).
**puduk** Newly-born camel foal.
**puf** Puff. **~böreği, -ni** (\`···), pastry puff stuffed with meat and cheese. **~la** (\`·), puffed out; soft; down (eiderdown). **~lamak,** to blow, to puff; to become puffed out *or* soft: **oflayıp ~,** to puff and blow, pant.
**puhu** Eagle owl.
**pul** Thin round disk; scale (*as* fish scale); spangle; washer; stamp (postage *etc.*); piece (in draughts and backgammon); nut (of a bolt); small coin, mite. **bir ~ etmemek,** to be worthless: **para ~,** money; **parası ~u yok,** he has no money.
**pulad** Steel.
**pul·lamak** *va.* Stamp. **allayıp ~,** to ornament with spangles; to overdecorate. **~lu,** scaly; spangled; bearing stamps; spotted.
**pulluk** Heavy plough.
**pund** Position of a ship. **~unu bulmak** *or* **~una getirmek,** to find a suitable opportunity *to do stg.*: **~ tâyini,** finding a ship's position by astronomical calculations.
**punta** (\`·) (Lathe) centre. **fener ~,** live centre: **gezer ~,** back centre; **gezer ~ gövdesi,** tailstock, poppet.
**puntal** Stanchion.
**punto** (\`·) Size of type.
**pupa** (\`·) Stern *of a ship*; following wind. **~ gitmek,** to sail before the wind: **···e ~ yelken açmak,** to take advantage of ..., to profit by ... .
**puro** Cigar.
**pus**[1] Mist, haze; condensation (on a cold glass); bloom *on fruit*; blight, mildew; moss *on trees and plants*; web *made by insects on leaves*; crust *formed on the nipples of ewes*.
**pus**[2] (*Fr. pouce*) Inch.
**pusad, -dı** Arms; equipment; instruments of war. **~cı,** clown in the **ortaoyunu** having a slapstick *or* wooden sword. **~lanmak,** to be armed; to put on armour. **~lı,** armed; clad in armour.
**pusarık** Haze; hazy weather; mirage.
**puse** (–\`) Kiss.
**pusla**[1] (\`·), **pusula** Note; memorandum; list. **~ etm.,** to make a note of, to make a list.
**pusla**[2] (\`·), **pusula** *v.* **pusula**[2].
**pus·lanmak** *vn.* (Of a cold glass) to be misty with condensation; (of fruit) to have the bloom on it. **~lu,** hazy, misty; having the bloom on it.
**puslu** *In* **uslu ~,** quiet and shy.
**pusmak** *vn.* Crouch down; lie in

ambush; descend; become misty; be grieved *or* offended.

**pusu** Ambush. ~ **kurmak,** to lay an ambush: ~**ya yatmak,** to lie in wait.

**pusula**[1] *v.* **pusla**[1].

**pusula**[2] (ˋ··) Compass. ~**yı şaşırmak,** to lose one's bearings; to be bewildered.

**puşide** (——ˋ) Covered; hidden. Covering; quilt; a kind of scarf worn round the head.

**puşt, -tu** Catamite.

**puşu** Light turban *formerly worn by soldiers*; kerchief worn round the head.

**put, -tu** Idol; beautiful boy *or* girl. ~ **gibi durmak,** to stand like a graven image. ~**kıran,** iconoclast, ~**laştırmak,** to idolize. ~**perest,** idolater. ~**perestlik,** idolatry.

**putlamak** *vn.* (Of a camel) to foal.

**putrel** (*Fr. poutrelle*) Iron beam *or* post *used in building.*

**puvan** Point; score.

**püf** *Imitates a puff of wind or breath.* ~ **noktası,** a weak spot. ~**etmek** (ˋ··), **-eder,** ~**lemek,** ~**kürmek,** to blow out, blow on; puff.

**püfür püfür** *Imitates the noise of a gentle breeze.*

**pünez** (*Fr. punaise*) Drawing-pin.

**pür-** *Pers. prefix.* Full of ..., *e.g.* **pürhiddet,** full of fury; **pürümid,** full of hope.

**pürçek** Curl.

**pürçük** *In* **bölük** ~, in fragments; incomplete.

**pürmelâl** *In* **hali** ~, a hopeless state.

**pürnakıl** Full of ... , covered with ... .

**pürtük** Uneven, rough, knobbly. Knob, small protuberance.

**pürüz** Shagginess; roughness; unevenness, irregularity; roughness on the surface of castings; fluff; hitch, difficulty. ~ **ayıklamak** *or* **temizlemek,** to settle a matter; to clear away difficulties. ~**lenmek,** to become rough *or* shaggy; to be beset with difficulties. ~**lü,** rough, shaggy; covered with ink stains *or* splashes of paint; uneven, irregular; beset with difficulties. ~**süz,** even, smooth; without defects.

**püskü** *v.* **eskipüskü.**

**püskül** Tuft; tassel; difficulties. ~**lendirmek,** to tease out into a tassel; to attach a tassel; to cause to become difficult *or* complicated. ~**lü,** tasselled; difficult, complicated: ~ **belâ,** a serious calamity; a damnable nuisance.

**püskür·geç** Atomizer. ~**me,** scattered about (of water *etc.*): anything splashed about: ~ **ben,** beauty spots scattered about over the face: ~ **memesi,** injection nozzle: ~ **pompa,** fuel-pump (Diesel). ~**mek,** to blow out water from the mouth; to spray (liquid); to foam at the mouth; splutter; scatter, drive away: **ateş** ~, to be furiously angry. ~**tmek,** *caus. of* **püskürmek**; to scatter (an enemy).

**püsür** *Used in conjunction with words expressing filth*; unpleasant additions and accessories. **bu işin bir çok boku ~ü var,** there are a lot of complications in this business.

# R

**Rab, -bbi** The Lord God. **Rabbim!,** my God!: **Rabbena,** our Lord, God: **ya Rabbi!,** oh! my God! ~**bani** (·—ˊ), divine: **ilhami** ~, divine inspiration.

**rabıt**[1], **-btı** A binding; bond; connexion; grammatical construction.

**rabıt**[2] (—ˋ) Binding, connecting. ~**a** (—·ˋ), tie, bond; connexion; orderly arrangement; copula; logical *or* grammatical conformity. ~**alı,** in good order; regular; decorous, well-conducted, decent. ~**asız,** disordered; irregular; incoherent; rambling; disorderly. ~**asızlık,** disorder; irregularity; bad behaviour.

**rabt·etmek** *va.* Bind; fasten; connect. **bir karara** ~, to decide *or* settle *a question etc.* ~**iye,** tether, tie, bond; fastener.

**raca** (ˋ·) Rajah.

**raci** (—ˋ) Returning; concerning, relating to. ···**e** ~ **olm.,** to concern, to fall on.

**racon** (*sl.*) Custom, rule.

**radansa** (·ˋ·) Thimble *or* cringle *of a rope.*

**radde** Degree. ~**sinde,** approximately: **saat on** ~**lerinde,** about ten o'clock: **bayılma** ~ **lerine gelmiştim,** I came near to fainting.

**radike** Dandelion.

**radyo** (ˋ·) Radio, wireless.

**raf** Shelf. ~**a koymak,** to shelve *a matter.*

**rafadan** Very lightly boiled egg.

**rafızî** (–·⊥) Heretic; Shiite.

**rağbet, –ti** Desire, inclination. ···**e** ~ wish for *stg.*; demand for *goods.* ~**ten düşmek,** to be no longer in demand, to be out of favour: ···**e** ~ **etm.** to wish for, to esteem: ~ **görmek,** to be in demand, to be esteemed: ···**e** ~**i var,** he has an inclination for: ···**e** ~**i yok,** he has no inclination for. **rağbet·li,** desirous; having an inclination *for stg.*; in demand, sought after, liked: ~**si çok,** there is a great demand *for some kind of goods*: ~**si yok,** there is no demand for it. ~**siz,** who feels no inclination *for stg.*; unesteemed, not sought after. ~**sizlik,** a being without inclination; lack of esteem, a not being in demand.

**rağm** Spite. ~**ına,** out of spite: ···**in** ~**ına,** to spite .... ~**en** (ˈ·), in spite of (*dat.*).

**rahat, –tı** Rest; ease; comfort; quiet. At ease; tranquil; comfortable; easy. ~**ına bakmak,** to look after one's own comfort; ~**ınıza bakın!,** (i) please don't worry; it really doesn't matter; (ii) don't bring trouble upon yourself by unnecessary interference: ~ **mı battı?,** *lit.* 'did comfort disturb you?', *i.e.* 'why did you give up your comfortable job?': ~ **etm.,** to be at ease, to make oneself comfortable, to rest: ~ **durmak,** to keep quiet, not to fidget; ~ **dur!,** don't fidget!: **bir gün** ~ **yüzü görmedim,** I have not had a day's peace. **rahat·lanmak,** to rest; take one's ease; calm oneself. ~**lık,** ease; comfort; quiet: **Allah** ~ **versin,** sleep well! ~**sız,** unquiet; uneasy; indisposed; uncomfortable: ~**etm.,** to disturb: ~ **olmayınız!,** don't be uneasy, don't worry! ~**sızlık,** uneasiness; discomfort; indisposition.

**rahib** (–ˈ) Monk; ~**e,** nun.

**rahim, –hmi** Womb.

**rahle** Low reading-desk.

**rahmet, –ti** Mercy; God's compassion; rain. ···**e** ~ **okumak,** to pray for the soul of ...; to regret *stg. lost*: ···**e** ~ **okutmak,** to cause one to regret ...; to make one long for ... (in place of *stg.* else): ···**e Allah** ~ **eylesin!,** God's blessing on ...! ~**li,** ~**lik,** the deceased, the late.

**rahvan** Amble; ambling horse. Ambling.

**raiyye** (–·ˈ) Flock at pasture; subject community, people under a ruler.

**rakam** Figure; arithmetic.

**rakı** Raki, arrack. **sakız** ~**sı,** raki with mastic in it. ~**cı,** maker *or* seller of raki; raki addict.

**rakîb** (·⊥) Rival. ~**siz,** unrivalled.

**rakkas** Dancer; pendulum. ~ **çarkı,** balance-wheel of a watch.

**raks** Dance, dancing. ~**an** *a.*, dancing. ~**etmek** (ˈ··), **–eder,** to dance; oscillate.

**râm** Tame, gentle, submissive. ~ **etm.,** to subjugate, to cause to yield.

**ramak, –kı** Smallest possible quantity of anything. **bu kazanın hayatına mal olmasına** ~ **kaldı,** this accident all but cost him his life.

**ramazan** Ramazan, *the ninth month of the Moslem year, during which Moslems fast between dawn and sunset.* ~ **tiryakisi,** a quick-tempered man (*orig. from the bad temper caused by fasting*). ~**lık,** suitable for Ramazan.

**rampa** (ˈ·) Boarding (*naut.*); a ship's going alongside another *or* a wharf; incline; loading-platform. ~ **etm.,** to board *an enemy's ship*; (*sl.*) to accost (a woman); to join *a party* uninvited.

**randevu** (ˈ··) Rendezvous.

**randıman** (*Fr. rendement*) Yield; profit; output. ~**lı,** profitable.

**ranza** (ˈ·) Bunk (*naut.*).

**rap** *Imitates the sound of marching.*

**rapor** Report; medical certificate.

**rapt-** *v.* **rabt-.**

**ras** *v.* **rast.**

**rasad** A watching; astronomical *or* meteorological observation. ~**hane** (··–ˈ), observatory; meteorological station.

**ras·gele, rastgele** Chance, met by hazard. By chance; at random. **rasgele!,** good luck! ~**gelmek** (ˈ··) (*with dat.*), to meet by chance, come across; hit *the mark*; turn out right, succeed. ~**getirmek** (ˈ···), to succeed in meeting; to cause to succeed; cause to hit the mark: **Allah işinizi rasgetirsin!,** may God grant you success! ~**gitmek,** *in* **iş** ~, (of an affair) to succeed.

**râsıd** (–ˈ) Observer.

**raslamak, rastlamak** (*with dat.*) Meet by chance; coincide with.

**raspa** (ˈ·) Scraper; grater. ~ **etm.,** to scrape: ~ **taşı,** pumice-stone, holystone.

**rast, –tı, ras** Straight; right; proper;

straightforward; in order; successful. ~ **gitmek,** to turn out well, prosper: ~ **makamı,** *a mode in Oriental music.*

**rastık** Cosmetic used for blackening the eyebrows. ~ **çekmek,** to blacken the eyebrows: ~ **mürekkeb,** black indelible ink: ~ **taşı,** antimony.

**raşit·ik** Rachitic, rickety. ~**izm,** rickets.

**râtıb** (–') Damp.

**ravak** Run honey.

**râvend** Rhubarb.

**ravnt** Round (boxing *etc.*).

**ray** Rail (railway *etc.*).

**râya** (–'), **reaya** *pl. of* **raiyye.** Subjects, *esp.* the non-Moslem subjects of the Ottoman Empire.

**rayic** (–') Current; in demand; saleable; in common use. Market price; current value. ~ **usulü,** ad valorem.

**rayiha** (–·') Smell, odour. ~**lı,** scented.

**razakı** *A variety of* white grape, *from which the best raisins are made.*

**razı** (–') Satisfied, contented; willing; pleased. ~ **etm.,** to satisfy (a person), to obtain the acquiescence of: ~ **olm.,** to be willing, to consent; to approve; to be pleased. ⌜**alan** ~, **satan** ~⌝, the buyer satisfied, the seller satisfied; since those concerned are content, no one else should interfere: **Allah** ~ **olsun,** may God reward you (him *etc.*).

**reaya** (·–') *v.* **raya.**

**rebab** A sort of guitar. ~**î,** lyrical.

**rebi** (·'), **–ii** Spring. ~**î** (·–'), pertaining to spring, vernal. ~ **yülâhır,** *fourth month of the Moslem year.* ~**yülevvel,** *third month of the Moslem year.*

**receb** *Seventh month of the Moslem year.*

**recmetmek** ('··), **–eder,** to pelt; stone to death.

**reçel** Fruit preserve; jam.

**reçete** (·'·) Prescription; recipe.

**reçina, reçine** (·'·) Resin.

**red, –ddi** A repelling *or* rejecting; refutation; repudiation. ~**dile cevab vermek,** to answer with a refusal: ~**di hâkim talebi,** a demand for other judges. ~**detmek** ('··), **–eder,** to reject; repel; return (an answer); repudiate; refute.

**redif** (·') Reserve. Reservist; word repeated at the end of every line of a poem.

**refah** (·') Easy circumstances; comfort; luxury.

**refakat** (·–'), **–ti** Accompaniment; companionship. ···**in** ~**inde,** accompanied by, in the company of...: ···**e** ~ **etm.,** to accompany *s.o.*

**ref'etmek** ('··), **–eder** *va.* Raise, heighten; promote; remove; annul.

**refik** (·'), **–ki** Companion, associate. ~**a** (·–'), wife.

**regaib** (·–') **leylei** ~, the 12th of Rejeb, *anniversary of the conception of Mohammed.*

**rehavet** (·–'), **–ti** Softness; limpness; slackness, lethargy.

**rehber** Guide; guide-book. ~**lik,** a guiding: ~ **etm.,** to guide.

**rehin, –hni** Pawn, pledge, security; hostage. **rehne koymak** *or* **bırakmak,** to pawn *or* pledge, to give as security. ~**e** (·–'), object given as pledge; hostage.

**reis** Head; chief; president; captain (of a merchant ship). *a.* Principal, chief. ~ **vekili,** Vice-president. ~**icumhur,** President of the Republic. ~**ülküttab,** (*formerly*) Minister of Foreign Affairs.

**reji** ('·) Regie (former administration of the Tobacco Monopoly).

**rejisör** (*Fr. regisseur*) Stage-manager.

**rekabet** (·–'), **–ti** Rivalry; competition.

**rekâket** (·–'), **–ti** Defect in speech; stammer; incoherence; defect in style.

**rekât, –tı** Complete act of worship with the prescribed postures.

**reklâm** (*Fr. réclame*) Advertisement.

**rekolta, rekolte** (·'·) Harvest; crop.

**rekor** Record. ~**cu,** ~**tmen,** record-breaker.

**rekz** A setting up *or* planting; erection. ~**etmek** ('··), **–eder,** to set up, erect; plant.

**remil, –mli** Sand; geomancy. ~ **atmak,** to tell the future by geomancy.

**remiz, –mzi** Sign; nod; wink; allusion; symbol. ~**lendirmek,** to symbolize.

**remz·etmek** ('··), **–eder** *vn. & va.* Make a sign; allude, hint; symbolize, typify. ~**î** (·'), allusive; symbolical.

**Ren**[1] The Rhine.

**ren**[2] (*Fr. renne*) Reindeer.

**rencber** Labourer; workman; farm-hand; day-labourer. ~**başı, –nı,** labourers' foreman. ~**lik,** occupation of labourer *or* farm-hand.

**rencide** (·–') Pained, hurt. ~ **etm.,** to hurt.

**rende** Carpenter's plane; grater. **~lemek,** to plane; shave; grate.
**rengârenk** Multi-coloured; variegated.
**renk** Colour; colour *or* complexion of a thing. **~ atmak,** to lose colour, fade: **~ten renge girmek,** to change colour; to be inconstant; to blush with shame: **rengi* kaçmak,** to turn pale: **~ ~ olm.,** to change colour *from emotion*: **~ vermemek,** to appear unmoved, to conceal one's feelings: **rengi yok,** colourless; unreliable: **bin renge girmek,** to keep changing colour; to be inconstant; to use every kind of subterfuge.
**renk·lemek,** to colour. **~li,** coloured; of *such and such* a colour. **~siz,** colourless; pale; without personality.
**resçekmek** *v.* **rest.**
**re'sen** (ˋ·) On one's own account *or* initiative; directly.
**resif** (*Fr. recif*) Reef.
**re'sikâr** Supreme direction *of a business etc.*; highest post.
**resim, -smi** Design; drawing; picture; ceremony; due, tax, toll. **~ çekmek,** to take a photograph: **resmini* çıkarmak,** to have one's photo taken: **resmi geçid,** military review: **resmi kabul,** official reception: **resmi küşad,** inauguration: **mahvolduğumuzun resmidir,** our ruin is certain.
**resim·ci,** draughtsman; artist; photographer; picture-dealer. **~li,** illustrated.
**resm·en** (ˋ·) Officially; ceremoniously; as a mere matter of form. **~etmek** (ˋ··), **-eder,** to draw; describe (a circle *etc.*). **~i,** *v.* **resim. ~î** (·ˊ), official; ceremonious; formal; done as a matter of form. **~iyet, -ti,** official character *of stg.*; formality: **işi ~e dökmek,** to become formal, to adopt an official manner.
**ressam** Designer; artist. **~lık,** the art *or* profession of painting.
**rest** (*Fr. reste*) *In* **~ çekmek,** to stake all; to act boldly, to dare to do *stg.*: **bir şeye ~ çekmek,** to stake a thing: **şerefine ~ çekti,** he staked his honour.
**resto** (ˋ·) *Cry used by waiters to cancel an order.*
**resul, -lü** Envoy; apostle; prophet.
**reşid** Who has attained his majority (18 years old).
**reşme** Chain *or* leather noseband *for a horse.*
**ret** *v.* **red.**
**retuş** (*Fr. retouche*) Retouch.
**reva** (·ˊ) Proper, suitable. **~ görmek,** to deem proper.
**revac** A being in demand; a being current. **~landırmak,** to cause *stg.* to be in demand. **~lı,** in demand; current. **~sız,** not in demand; not current.
**revak, -kı** Porch; pavilion; arbour. **~î** (·—ˊ), stoic.
**revan** Going; flowing. **~ olm.,** to go, to flow.
**revani** (·—ˋ) *A kind of sweet made with semolina.*
**revanş** (*Fr. revanche*) **~ maçı,** return match.
**revir** Small military hospital; infirmary *at a factory or school*; sick-bay. **~ âmiri,** district inspector.
**reviş** A going; gait; trend; conduct.
**revnak, -kı** Brightness, splendour; brilliance; sparkle; beauty. **~lı,** brilliant; splendid; beautiful; in full prime. **~sız,** lustreless, dull.
**rey** Opinion; judgement; vote. **~ beyan etm.,** to set forth one's opinion: **~e koymak,** to put to the vote: **~ vermek,** to vote.
**rezakı** *v.* **razakı.**
**rezalet** (·—ˋ), **-ti** Vileness, baseness; scandal; scandalous behaviour. **~ çekmek,** to suffer disgrace: **~ çıkarmak,** to cause a scandal.
**reze** Hinge.
**rezene** Fennel.
**rezil** Vile, base; disreputable; disgraced. Scoundrel. **~ etm.,** to hold up to scorn, to disgrace: **~ olm.,** to be disgraced.
**rıfk, -kı** Gentleness; suavity.
**rıh** Sand *used to dry writing.* **~dan,** sand-box.
**rıhtım** Quay, wharf.
**rıza** (·ˊ) Consent, acquiescence; resignation. **~sı oluyor,** he gives his consent, he approves: **Allah ~sı için,** for the love of God!: **kazaya ~!,** it can't be helped!: **kazaya ~ göstermek,** to resign oneself to one's fate.
**rızk, -kı** One's daily food; sustenance; the necessaries of life.
**riayet** (·—ˋ), **-ti** A respecting; observance; respect, esteem; respectful treatment; consideration, regard, kind attention. **···e ~ etm.,** to treat with respect, to pay attention to. **~en** (·ˊ··), (*with dat.*) out of respect for ...; in consideration of ... . **~kâr,** respectful; considerate. **~siz,** dis-

respectful; irreverent. **~sizlik,** disrespect; irreverence.
**rica** (·–ˈ) Request, prayer. **~ etm.,** to request, to desire: **size bir ~m var** *or* **sizden bir ~da bulunacağım,** I have a request to make of you: **~ minnet,** after much beseeching. **~cı,** who makes a request: intercessor.
**rical, -li** Men; men of importance, high officials.
**ricat, -ti** Return; retreat. **~ etm.,** to retreat.
**rik'a** Cursive Arabic script *formerly used chiefly by Turks.*
**rikkat, -ti** Slenderness; delicacy; tenderness, compassion. **~amiz** (··–ˈ), **~engiz,** piteous, deplorable. **~li,** compassionate, pitiful.
**rimel** A cosmetic for the eyelashes.
**rind** A jolly, unconventional, humorous man. **~ane** (·–ˈ), in the manner of a **rind. ~lik,** the quality of a **rind. ~meşreb,** having the temperament of a **rind.**
**ringa** (ˈ·) Herring.
**risale** (·–ˈ) Treatise; pamphlet.
**risalet** (·–ˈ), **-ti** Mission of a prophet; apostleship.
**rivayet** (·–ˈ), **-ti** Tale; rumour. **~ etm.,** to tell, relate: **~ olunmak,** to be narrated, to be rumoured.
**riya** (·–ˈ) Hypocrisy. **~kâr** (·–ˈ), hypocritical. **~kârlık,** hypocrisy. **~kârane** (·–––ˈ), hypocritical (conduct *etc.*): in a hypocritical manner.
**riyal** Silver dollar.
**riyaset** (·–ˈ), **-ti** Presidency; chairmanship. **~icumhur,** Presidency of the Republic.
**riyasız** (·–ˈ) Without hypocrisy, genuine, sincere.
**riyazet** (·–ˈ), **-ti** Asceticism; mortification of the flesh.
**riyazî** (·–ˈ) Mathematical. **~yat, -tı,** mathematics.
**rizofora** Mangrove.
**robdöşambr** (*Fr. robe de chambre*) Dressing-gown.
**roda** (ˈ·) Coil of rope.
**rodaj** (*Fr. rodage*) **~ yapmak,** to grind in (a valve).
**roka** (ˈ·) Rocket (plant used for salad).
**roket, -ti** Rocket.
**rol, -lü** Role, part.
**rom** Rum.
**Romalı** (ˈ··) Roman.
**roman** Novel. **~cı,** novelist.
**Romanyalı** (·ˈ··) Roumanian.
**romatizma** (··ˈ·) Rheumatism. **~lı,** rheumatic.
**rondelâ** Washer.
**rosto** (ˈ·) Roasted. Roast meat.
**rota** (ˈ·) Ship's course.
**roza** (ˈ·) Rose diamond.
**rozet, -ti** Rosette.
**römork, -ku** (*Fr. remorque*) Trailer (vehicle). **~ör** (*remorqueur*), tug.
**Rönesans** Renaissance.
**rönons** (*Fr. renonce*) Revoke (cards).
**rötar** (*Fr. retard*) Delay.
**ruam** Glanders.
**ruba** (·–ˈ) Clothing, clothes. **~lık** (ˈ··), suitable for making clothes.
**rubab** *v.* **rebab.**
**ruberu** (ˈ·–) Face to face; opposite.
**rubu** (·–ˈ), **-b'u** Quarter.
**rubya** (ˈ·) Rupee.
**rugan** Varnish; grease; patent leather. **~lamak,** to varnish; polish. **~lı,** varnished; polished; patent leather.
**ruh**[1] (ˈ) Soul; spirit; essence; energy, activity. **~ haleti,** psychological condition; morale; mood: **~u bile duymadı,** he didn't even notice.
**ruh**[2] Rook *or* castle (chess).
**ruhan·î** (–ˈ) Spiritual; clerical; immaterial. Angel; ghost. **~iyet, -ti,** spirituality; saintliness.
**ruhban** *pl. of* **rahib.** Clergy. **~iyet** (·–·ˈ), **-ti,** monastic life.
**ruh·î** (–ˈ) Psychic; psychological. **haleti ~ye,** the psychological condition. **~iyat** (–·ˈ), **-tı,** psychology.
**ruh·lanmak** *vn.* Become animated; revive. **~lu,** animated; having *such and such* a spirit; lively; vivid, full of feeling. Spirit level.
**ruhsad, -dı** Permission; permit; leave; dismissal. **~ vermek,** to give leave; to dismiss. **~iye,** licence. **~lı,** authorized; on leave. **~name** (··–ˈ), permit.
**ruhsuz** Inanimate, lifeless; spiritless. **~luk,** lifelessness; lack of spirit.
**Ruhulkudüs** (*To Moslems*) the Angel Gabriel; (*to Christians*) the Holy Ghost.
**rulo** (*Fr. rouleau*) Roller; roll.
**Rum** Byzantine Greek; Greek of Turkish nationality. (*Formerly*) Turkey. **~ca** (ˈ·), modern Greek language. **~î** (–ˈ), belonging to the ancient Romans *or* to the Byzantine Greeks; Gregorian (calendar).
**rumuz** *pl. of* **remiz.** Signs, nods *etc.*; *as sing.* symbol; abbreviation; initial. **~at** (··–ˈ), **-tı,** formulae.

**rupye** Rupee.
**Rus** Russian. **~ça,** the Russian language. **~laşmak,** to become Russian.
**Rusya** (ʹ·) Russia. **~lı** (ʹ··) Russian.
**rutubet** (·–ʹ), **-ti** Dampness, humidity. **~li,** damp.
**ruy** Face; surface.
**ruya** (–ʹ) Dream. ~ **görmek,** to have a dream: ~ **tabiri,** the interpretation of dreams.
**ruz** Day. **~u ceza** *or* **~u hesab,** the Day of Judgement.
**ruzname** (·–ʹ) Diary; agenda.
**rüchan** Preponderance; preference; advantage. ~ **hakkı,** priority. **~ıyet** (·–·ʹ), **-tı,** a being preferable; a having priority.
**rücu** (·ʹ), **-uu** Return; a going back on one's word. ~ **etm.,** to return; to go back on *one's word etc.*
**rüesa** (··ʹ) *pl. of* **reis.** Heads; chiefs.
**rüfai** (·–ʹ) *A sect of dervishes, renowned for their spectacular performances.* ⌜**bu işe ~ler karışır**⌝, what happens next is beyond our ken.
**rüfeka** (··ʹ) *pl. of* **refik.** Companions; partners.
**rükû, -ûu** A bowing down in prayer.
**rükün, -knü** Pillar, column; prop, support; fundamental principle.
**rüküş (hanım)** A comically dressed, helpless little woman.
**rüsum** (·ʹ) *pl. of* **resim.** Usages; rites; dues, taxes. **belediye ~u,** municipal taxes: **80** ~ **tonluk motör,** a motorboat of 80 tons registered tonnage. **~at** (·–ʹ), **-tı,** dues, taxes; *as sing.* Customs Administration.
**rüsvay** Publicly disgraced; object of scorn.
**rüşd** Majority, coming of age. **~ünü* isbat etm.,** to come of age: **sinni ~,** the age of reason, majority. **~iye,** high school, secondary school.
**rüşeym** Embryo, germ.
**rüşvet, -ti** Bribe; ship's spare rigging stores; place where these are kept. ~**i kelâm,** complimentary words before criticizing: ~ **yemek,** to accept a bribe: ~ **yedirmek (vermek),** to give a bribe. **~çi,** taker of bribes.
**rütbe** Degree; grade; rank. ~ **almak,** to rise in rank.
**rüya** *v.* **ruya.**
**rüyet, -ti** A seeing; visibility. **dâvayı** ~ **etm.,** to hear a case *at law.*
**rüzgâr** Wind. ~ **ile gitmek,** to sail with the wind; to trim one's sails to suit the occasion: ~ **gülü,** compass rose: ~ **payi,** wind allowance in shooting: ~ **üstü,** the windward side. **~altı,** the lee: **~ya düşme,** leeway: ~ **yakası,** the leech of a sail.

# S

**-sa** *v.* **ise.**
**saadet** (·–ʹ), **-ti** Happiness; prosperity. **~le,** good-bye!. **~li, ~lû,** happy; fortunate; *official title formerly given to generals etc.*
**saat, -ti** Hour; time; time of day; one hour's journey on foot; watch, clock. ~ **altıda,** at six o'clock: ~ **besaat,** from hour to hour: ~ **kaç?,** what's the time?: **~ler olsun!,** *popular form of* **sıhhatler olsun!,** *q.v.*: ~ **tutmak,** to time (a race *etc.*): ~ **vurmak,** to strike the hour: **bir** ~ **evvel,** as soon as possible: **iyi ~te olsunlar,** djinns, evil spirits: **o ~,** at that moment; at once. **saat·başı, -nı,** the end of an hour; a general pause in the conversation. **~çi,** watchmaker. **~çilik,** the trade of a maker, seller *or* repairer of watches. **~lik,** lasting *so many* hours.
**saba**[1] (·ʹ) Gentle wind from the east.
**Saba**[2] *v.* **Seba.**
**sabah** Morning. In the morning; tomorrow morning: **~ları,** in the morning; every morning: **~tan,** from early morning; from tomorrow: **~ı bulmak (etmek),** to stay awake all night, to work *etc.* through the night: **~a çıkmak,** (of a sick person) to live till the morning: **~ınız hayır olsun!,** good morning!: ~ **yıldızı,** the planet Venus *as a morning star.*
**sabah·çı** Early riser; working *or* remaining till daybreak. **~ki,** this morning's .... **~lamak,** to sit up all night; to become morning. **~leyin** (·ʹ··), in the morning, early. **~lık,** woman's morning dress.
**saban** Plough. ~ **bıçağı (keskisi, kılıcı),** coulter: ~ **kulağı,** mouldboard: ~ **oku,** pole of a plough: ~ **ökçesi,** heel of a plough: ~ **uc demiri,** ploughshare. **~kıran,** rest-harrow.
**sabık** (–ʹ) Former, previous, preceding; foregoing. **~a** (–·ʹ), former misdeed; previous conviction. **on ~sı bulunan bir yankesici,** a pickpocket with ten previous convictions. ~-

**alı**, previously convicted; recidivist. **~an** (–̀··), formerly; previously.
**sabır, -brı** Patience; aloes; *v. also* **sabur. ~lı**, patient. **~sız**, impatient. **~sızlanmak**, to grow impatient.
**sabi** (·–̀) Male child, boy.
**sâbih** Swimming; floating. **~ havuz**, floating dock.
**sabit** (–̀) Fixed; stationary; firm; proved. **~ kadem**, steadfast: **~ mürekkeb**, indelible ink: **~ olm.**, to be fixed *or* firm; to be proved *or* confirmed: **~ seviye kabı**, float-chamber (carburettor).
**sabretmek** (̀··), **-eder** *vn.* Be patient; (*with dat.*) endure *stg.*
**sabuk** Astray; senseless. **abuk ~ söylemek**, to talk nonsense.
**sabun** Soap. **⌈suya ~a dokunmamak⌉**, to avoid anything likely to cause trouble. **~cu**, soap-maker, soap-seller. **~hane** (··–̀) Soap factory. **~lamak**, to soap. **~lu**, soapy.
**sabur** Patient. **ya ~ çekmek**, to call on God (the Patient One) *to give one patience*, to grow angry and impatient.
**sabura** (·–̀·) Ballast.
**sac**[1] Sheet-iron; iron plate; thin iron plate for cooking, girdle. Made of sheet-iron.
**sac**[2] Teak.
**sacayağı** (̀···), **-nı** Trivet; trio. **~ yürümek**, to march with one company in front and two in the rear separated by an interval.
**saç**[1], **-çı** Hair. **~ ~a baş başa gelmek**, to come to blows: **⌈~ı bitmedik yetim⌉**, an orphan while still a baby: **⌈~ına sakalına bakmadan⌉**, forgetting his age: **⌈~ sakal biribirine karışmış⌉**, dishevelled and upset.
**saç**[2] *v.* **sac**[1].
**saçak** Eaves of a house; fringe. **⌈alev saçağı sardı⌉**, 'the flames have reached the eaves', the matter has got out of control: **salkım ~**, hanging about in disorder *or* in rags. **~lı**, having eaves; fringed; dishevelled, untidily dressed.
**saçbağı** (̀··), **-nı** Hair ribbon.
**saçı** (*Formerly*) wedding present given to a bride; coins, sweets, rice *etc.* thrown over a bride.
**saçık** Disordered; scattered. **açık ~**, untidily dressed; immodest, indecent.
**saçılmak** *vn. pass. of* **saçmak.** Be scattered; be sprinkled. **açılıp ~**, to be scattered; to dissipate one's forces; to be immodestly dressed (*also* **dökülüp ~**).
**saçıntı** Things thrown and scattered about.
**saçıştırmak** *va.* Sprinkle; dredge; sow.
**saçkıran** Baldness; alopecia.
**saçlı** Having hair; hairy. **~ sakallı**, of mature age.
**saçma** Act of scattering; anything scattered *or* sprinkled; small shot; cast-net; nonsense. Nonsensical.
**saçmak** *va.* Scatter; sprinkle, dredge; sow broadcast. **saçıp savurmak**, to spend money prodigally: **ateş ~**, to fire in all directions; to be in a furious rage: **yaraya tuz ~**, (to sprinkle salt on the wound), to add fuel to the flames: **yaş ~**, to weep copiously.
**saçma·lamak** *vn.* Talk nonsense; say incongruous things. **~sapan**, incongruous nonsense.
**saçula** (·̀·) Wooden mould *for casting metals.*
**sada, seda** (·–̀) Sound; voice.
**sadaka** Alms; charity. **~ya muhtac olm.**, to be reduced to penury: **baş* göz* ~sı**, a sacrifice *made to avert some mishap*: **⌈verilmiş ~dan varmış⌉**, you've had a lucky escape!
**sadakat** (·–̀), **-ti** Faithful friendship; fidelity; devotion. **~li**, faithful; devoted.
**sadakor** Raw silk.
**sadalı** (·–̀) Sounding; with a voice; talking (film). **~ harf**, vowel.
**sadaret** (·–̀), **-ti** Grand Vizierate.
**sadasız** (·–̀) Without sound; silent. **~ harf**, consonant.
**sade** (–̀) Mere; simple; unmixed; pure; single (flower); plain, unadorned; unsweetened (coffee); unstuffed (pastry). Simply, merely, just. **~ güzel**, beautiful in itself without ornament: **~ suya çorba**, soup made without fat, *hence* **~ suya**, mere, plain; unimportant. **~ce** (–̀··), simply, merely.
**saded** Point *or* object in view; intention; scope. **~e gelelim**, let us come to the point under discussion: **~den haric**, extraneous to the question, off the point.
**sadedil** (–·̀) Ingenuous; guileless; naïve. **~lik**, ingenuousness, simpleness of heart.
**sadef** *v.* **sedef.**
**sade·leştirmek**, to simplify. **~lik**, simplicity, plainness. **~yağ**, butter.
**sadık** (–̀) True; sincere; faithful; honest; devoted.

**sadır, -drı** Breast, chest; heart; front; prominence; post of honour; most important person. **sadra geçmek,** to take the chief seat in an assembly; to become Grand Vizier: **sadra şifa verecek,** satisfactory.

**sâdır** (–ˈ) Emanating. ~ **olm.,** to emanate (of a decree *etc.*); to happen.

**sadme** Collision; sudden blow.

**sadra** *v.* **sadır.**

**sadrazam** (·–ˈ) Grand Vizier.

**saf, -ffı** Row; line; rank. ~ ~, in rows *or* ranks: ~ **düzmek,** to draw up a line of battle.

**sâf** (–) Pure; unadulterated; limpid; sincere, unfeigned; ingenuous, simple, naïve.

**safa, sefa** (·ˈ–) Freedom from anxiety; peace, ease; enjoyment, pleasure. ~ **geldiniz!,** welcome! (*the reply to which is* ~ **bulduk**): ~**yi hatır,** peace and quiet, ease and comfort: ~ **sürmek,** to live a quiet and happy life; to enjoy oneself.

**safahat** (··ˈ), **-tı** *pl. of* **safha.** Phases.

**saf·derun** Simple; sincere; silly. ~**derunluk,** simplicity; credulity. ~**dil,** simple-hearted, naïve, credulous.

**safer** *Second month of the Moslem year.*

**saffet, -ti** Purity, sincerity; ingenuousness.

**saffıharb** Line of battle. ~ **gemisi,** capital ship.

**safha** Surface, face; plate; leaf, page; phase.

**sâfi** (ˈ–·) Clear; pure; sincere; net. ~ **su,** mere water and nothing else: **gayri** ~, gross (profits *etc.*).

**safir** Sapphire.

**saflık** Simplicity; ingenuousness.

**safra[1]** Ballast. ~ **atmak,** to get rid of useless persons *or* things.

**safra[2]** Bile; gall. ~ **bastırmak,** to have a snack: ~**sı bulanmak,** to be nauseated: ~**m kabarıyor,** I feel like being sick. ~**lı,** bilious; feeling sick *or* giddy. ~**vî** (·–ˈ–), biliary; bilious.

**safran** Saffron.

**safsata** False reasoning; sophistry; quibbling. ~**ya düşmek,** to use silly arguments. ~**cı,** sophist.

**safşiken, safzen** Who breaks the enemy's ranks; valiant.

**sağ[1]** Alive; safe; sound in body; trustworthy; strong. ~ **akçe (para),** good money, genuine coin: ~ **kalanlar,** the survivors (after a disaster *or* battle): ~ **kurtulmak,** to escape with one's life; to come out safe and sound: ~ **ol! (olsun!),** *a phrase for thanking*: ~ **salim** *or* **selâmet,** safe and sound: ⌜**başınız** ~ **olsun!**⌝, *formal phrase for offering condolences on the loss of a relative*; **başın** ~ **olsuna gitmek,** to pay a visit of condolence: ⌜**sen** ~ **ben selâmet**⌝, 'well, there's none left' (*regretfully and perhaps reproachfully*); *also in such phrases as*: '**bu işi bir ayda bitirdik mi, sen** ~ **ben selâmet!**', 'if we get this job done in a month, we shall have seen the last of it, thank goodness!': **siz** ~ **olun!,** it doesn't matter, don't worry!

**sağ[2]** Right, right-hand. The right-hand side. ~**a bak!,** eyes right!: ~**ına soluna bakmak,** to look about one: ~**ına soluna bakmamak,** to act without consideration: ~**ı solu yok,** tactless; eccentric; devil-may-care; unpredictable: ~**ını solunu şaşırmak,** to be bewildered, not to know what to do.

**sağaçık** Outside right (football).

**sağalmak** *vn.* Be cured, become well.

**sağanak, sağnak** Heavy rainstorm, downpour; squall; sudden loss *or* damage.

**sağcı** Right-wing sympathizer.

**sağdıç** Intimate friend of the bride *or* bridegroom; intimate friend *in general*; godfather. ~ **emeği,** useless efforts.

**sağı** Bird excrement; fowl dung.

**sağılmak** *vn. pass. of* **sağmak.** Be milked *etc.*; be frayed; (of a snake) to uncoil itself.

**sağım** Single act of milking; quantity milked; quantity of honey taken. ~**lı,** kept for milking; in milk. ~**lık,** beast kept for milking.

**sağır** Deaf; giving out a dull sound (*as a full cask*); opaque (glass); closed-up, sham (door *etc.*); having a low conductivity of heat (kettle *etc.*); sound-proof. ⌜~ **sultanın bile duyup bildiği**⌝, a thing that is known and obvious to all: **bir kulağı** ~ **olm.,** to condone, to take no notice of *a fault*: **hem** ~ **hem sığır,** stupid and obstinate. ~**lık,** deafness.

**sağış** *n.* Milking.

**sağiç** (ˈ·) Inside right (football).

**sağlam** Sound; whole; healthy; trustworthy; wholesome. ~**a** *or* ~ **kazığa bağlamak,** to make safe *or* sure: ~ **rüzgâr,** a steady wind. ~-

**lamak, sağlamak,** to make safe, secure *or* certain; put in working order; secure, ensure. **~laşmak,** to become sound *or* firm; be put right (broken machine *etc.*); prove to be true; become safe, escape from danger. **~lık,** soundness, wholeness; safety; health; trustworthiness; wholesomeness; truthfulness (being true).

**sağlı** Connected with the right hand; having a right-hand side.

**sağlıcak** *Used in expressions like* **~la gidiniz!,** go in good health!, good journey!

**sağlık** A being alive; life; good health; health in general (*instead of the proper Turkish* sıhhat). **~!,** all is well!: **sağlığında,** in his lifetime, while he was alive: **~ almak,** to ask the way; to have *stg.* recommended: **~ olsun!,** never mind!: **~ selâmetle,** in health and safety: **~ vermek,** to recommend (a shop, a doctor *etc.*): **can sağlığı,** health: ⌜**her şeyin başı ~ (can sağlığı)**⌝, health above all!: **bundan iyisi ~ (can sağlığı),** the best one can have: **eline ~!,** well done!; thank you!: ⌜**üstüme iyilik ~!**⌝, (i) that's the limit!; (ii) I've never heard of such a thing!; (iii) God forbid!

**sağmak** *va.* Milk; fleece (*fig.*); take honey; (of a cloud) to pour out rain; unwind.

**sağmal** (Cow *etc.*) kept for milking.

**sağnak** *v.* **sağanak.**

**sağrı** Rump; leather made from the rump of a horse; mountain ridge.

**sağyağ** (ˈ·) Cooking butter.

**sah, -hhı** Official flourish on a document *to show that it has been examined or registered.*

**saha** (–ˈ) Open space; courtyard; field (*also fig.*); area; ground (football *etc.*). **hayat ~sı,** 'Lebensraum'.

**sahabe** (·–ˈ) The disciples of Mohammed.

**sahabet** (·–ˈ), **-ti** Support, protection; patronage. **···e ~ etm.,** to support and protect.

**sahaf** Seller of secondhand books.

**sahan** Large copper food dish; a dish of food. **~lık,** marble-topped stand *to put dishes on*; landing *on a staircase*; platform *of a tram etc.*; amount of food in a sahan.

**sahhaf** *v.* **sahaf.**

**sahi** (–ˈ) *v.* **sahih.**

**sahib** (–ˈ) Possessing; endowed with. Owner, possessor; master. **···e ~ çıkmak,** to claim ownership of *stg.*; to stand as protector *or* patron of *s.o.*: **namus ~i,** a man of honour: **mal ~i,** owner of a property, rich man: **tabiat ~i,** a man of good taste: **ev ~i,** the master *or* the owner of a house.

**sahib·e,** female owner *etc.* **~lik,** ownership; protection: **···e ~ etm.,** to protect, to champion. **~siz,** unowned, ownerless; without a protector; abandoned.

**sahi·ci** (ˈ··) Real, genuine, true. **~den** (–·ˈ), really, truly.

**sahife** (··ˈ) Page; leaf.

**sahih** (·ˈ) Sound; true, correct. **~ mi?,** is that really so?

**sahil** (–ˈ) Shore, coast; bank. **~i selâmet,** safety from danger. **~hane** (–·–ˈ), house on the sea-shore *or* the bank of a river. **~topu** (–ˈ··), **-nu,** coastal gun.

**sahileş·mek** *vn.* Turn out to be true; be confirmed. **~tirmek,** to verify, show to be true; fulfil.

**sahne** Scene (theatre); stage.

**sahra** (·ˈ) Open plain; open country; wilderness, desert: field (*mil.*). **~yı kebir,** the Sahara: **~ topu,** field-gun.

**sahte** False; spurious; counterfeit, sham. **~kâr,** who counterfeits *or* forges. **~kârlık,** forgery; counterfeiting, false coining. **~lik,** falsity, spuriousness. **~vakar,** who assumes an air of dignity; who pretends to be imposing.

**sahtiyan** Morocco leather.

**sahur** Meal before dawn during the Ramazan fast.

**saik** (–ˈ), **-ki** Driving, impelling. Factor. **···e ~ olm.,** to cause, to induce.

**sair** (–ˈ) That remains; the rest of; other; that walks *or* goes; well-known, current. **ve ~e** (*abb. to* **v.s.**) etcetera. **~filmenam,** somnambulist.

**saka** Water-carrier. **~kuşu, -nu,** goldfinch.

**sakağı** Glanders; farcy.

**sakak** Double chin; dewlap.

**sakal** Beard; whiskers. **~ bırakmak** *or* **salıvermek,** to let the beard grow: **~ı ele vermek,** to allow oneself to be led by the nose; to allow all one's secrets to be guessed: **···in ~ına gülmek,** to make a fool of ...; to ridicule, to deceive: **~ımı uzatsam değecek,** very close: **ak ~ kara ~,**

high and low, rich and poor: **ak ~dan yok ~a kadar,** old and young alike. **sakal·lanmak** *vn.*, to grow a beard. **~lı,** bearded: ⌜**her ~yı baban mı sandın?**⌝, people are not always what you think they are.

**sakamet** (·–ˌ), **–ti** Defect; fault; harm.

**sakamonya** (··ˌ·) Scammony (*a purgative*).

**sakangur** Skink; coarse book-muslin.

**sakar** White blaze on a horse's forehead. Ill-omened, unlucky; sinister; (servant *etc.*) who always breaks things. **~ meki,** coot. **~ca,** white-fronted goose. **~lı,** having a white blaze (horse); unlucky, ill-omened. **~lık,** clumsiness.

**sakat, –tı** Unsound, defective; disabled, invalid; broken, damaged. Offals. **~at, –tı,** offals. **~çı,** seller of offals. **~lamak,** to injure, damage, mutilate. **~lık,** infirmity; defect; mistake.

**sakıf, –kfı** Roof.

**sakın** Beware!, take care!, mind!; don't! **~gan,** timid; cautious; retiring. **~ıcı,** cautious man who avoids risks. **~ma,** a looking after oneself; an avoidance of risks: **~sı olmamak,** not to care a straw for anyone.

**sakınmak** *vn.* Take care of oneself; be cautious. *va.* Protect *oneself or one's property.* **···den ~,** to guard oneself from ..., be on one's guard against ...: **gözünü budaktan sakınmaz,** dare-devil: **sözünü sakınmaz,** he doesn't mince words: **yankesicilerden sakınınız!,** beware of pickpockets!

**sakır ~ ~,** shivering *or* trembling. **~damak,** to shiver with cold *or* fear. **~dı,** a shivering with cold *or* fear.

**sakırga** Tick.

**sakıt** (–ˌ) Falling; dethroned. **~ olm.,** to fall, to lapse, to become of no effect.

**sakız** Mastic; chewing-gum. **~** *or* **~ gibi,** very white and clean: **âlemin ağzında ~ olm.,** to be a matter of common talk: **birinin ağzının ~ı olm.,** to be an object of *s.o.'s* ill-natured gossip; *stg.* that *s.o.* never ceases talking about.

**sâki** (–ˍ) Cupbearer; distributor of water.

**sakil** Heavy; ugly.

**sakim** Faulty, defective; harmful; wrong.

**sakin** (–ˌ) Quiet; motionless; stationary; calm; allayed; quiescent (letter); who inhabits *or* dwells.

**sakiname** (–––ˌ) Lyric addressed to a cupbearer.

**sâkit** (–ˌ) Silent; taciturn.

**sakiye** Female cupbearer.

**saklamak** *va.* Hide; keep secret; keep, store, save for future use; preserve from danger. **Allah saklasın!,** God preserve us!: **halktan ~,** to keep a secret from the people; to preserve from the rabble.

**saklambac** Hide-and-seek.

**sakla·nmak** *vn.* Hide oneself; be concealed; be kept *or* stored; be preserved from harm. **~yıcı,** protecting, preserving; concealing, secretive.

**saklı** Hidden; secret; put aside; preserved.

**saklıya** *v.* **saklamak.**

**saksağan** Magpie. ⌜**dam üstünde ~vur beline kazmayı**⌝ (*a nonsensical rhyme*), 'what's that got to do with it?'; 'what rot you are talking!'

**saksı** Flower-pot; vase. **~lık,** shelf for flower-pots; place for keeping flowers in pots during the winter.

**Saksunya** (·ˌ·) Saxony. **s~,** Dresden china.

**sakulta** (·ˌ·) Case-shot.

**sal**[1] (*Fr. salle*) Hall.

**sal**[2], **–lı** Raft; stretcher; bier.

**salâ** Call to prayer *or* to a funeral.

**salâbet** (·–ˌ), **–ti** Firmness; solidity. **~li,** firm; strong *in faith or character.*

**salacak** Slab on which corpses are placed for washing.

**salâh** Goodness, righteousness; improvement. **~a doğru,** improving, getting better. **~ kesbetmek,** (of conditions) to improve: **sulhu ~,** peace and amity.

**salâhiyet** (·–·ˌ), **–ti** Authority *or* right to do *stg.*; competence. **~li,** authorized; competent; (the department *or* authority) concerned. **~name,** credentials; written authority.

**salak** Silly, doltish.

**salam** *A kind of* sausage.

**salamandra** (··ˌ·) Salamander.

**Salamon** Solomon *when used to mean Jews in general* (*cf. our 'Moses'*); (*King Solomon is* **Süleyman peygamber**).

**salamura** (··ˌ·) Brine for pickling; anything pickled in brine.

**salapurya** (··ˌ·) Small lighter.

**salaş** Booth; market stall; temporary shed.
**salaşpur** A loosely woven cotton fabric *used for linings*.
**salât** (·–ˆ), **–tı** Moslem ritual prayer. **~ü selâm getirmek,** to pronounce the formula calling God's benediction on the Prophet (*during prayer or in times of peril*).
**salata** (·ˈ·) Salad; lettuce. **frenk ~sı,** endive. **~lık** (·ˈ··), cucumber.
**salâtüselâm** (·–··–ˆ) Prayer in moments of danger, *v.* **salât.**
**salâvat** (··–ˆ), **–tı** *pl. of* **salât.** Prayers. **yanına ~la varılır,** approached only with fear and trembling (*said of a proud inaccessible man or a cruel or angry man*).
**salb, –bi** Execution by hanging. **~etmek** (ˈ··), **–eder,** to hang. **~en** (ˈ·), by hanging.
**salça** (ˈ·) Sauce; tomato sauce.
**saldır·gan** Aggressor. Aggressive. **~ım,** aggression, attack. **~ma,** act of attacking; large knife.
**saldır·mak** *va. caus. of* **salmak,** *q.v.*; rush *s.o.* to a place. *vn.* Hurl oneself; make an attack (on = *dat.*) **~mazlık,** non-aggression.
**salep** (–ˈ) Salep (*root of Orchis mascula*); the hot drink made of this root.
**salgın** Aggressive; savage (animal); contagious. Epidemic (*also fig.*); contagion; general tax levied on a community; annual tribute.
**salhane** (·–ˈ) Slaughter-house.
**salı** Tuesday.
**salık** Information; indication as to one's way. **~ vermek,** to give directions as to route.
**salıncak** Swing; hammock. **~ sandalye,** rocking-chair: **kayık ~,** swing-boat.
**salınmak**[1] *vn. pass. of* **salmak.** Be thrown *etc.*; (of water, electricity) to be turned on. '**onun suyuna pirinc salınmaz**', he is not to be relied upon.
**salın·mak**[2] *vn.* Sway; oscillate; loiter along, swaying from side to side. **~tı,** swell (at sea); a swaying about. **~tılı,** running with a swell (sea); swaying; tottering.
**salıvermek** (·ˈ··) *va.* Let go; set free; release; allow to grow (beard *etc.*). **kahkaha ~,** to burst out laughing.
**salib** Cross. **~iahmer** (·–··ˈ), the Red Cross.
**salih** (–ˈ) Good; serviceable; suitable, proper; pious.
**sâlik** (–ˈ), **–ki** Who follows *a profession*. Class of dervishes above the novitiate; devotee.
**salim** (–ˈ) Safe, sound; free from (defects *etc.*). **~en** (–ˆ··), safely and soundly.
**salis** (–ˈ) Third. **~e,** a third; one-sixtieth of a second. **~en** (–ˆ··), thirdly.
**salkım** Hanging bunch *of grapes or flowers*; any tree bearing hanging flowers (acacia, wistaria *etc.*); cluster. Hanging, pendent. **~ ateş,** firework ending in a shower of stars: **~ saçak,** hanging about untidily *or* in rags: **~ söğüt,** weeping-willow. ⌜**halka verir talkını, kendi yutar ~ı**⌝, he preaches one thing and practises another.
**sallabaş** Afflicted with an involuntary shaking of the head.
**sallamak** *va.* Swing; rock; shake; wag; put off; leave in suspense. **baş ~,** to nod the head; to listen without paying attention; **her şeye baş ~,** to agree to everything, to raise no objection: **el ~,** to wave the hand; ⌜**elini sallasan ellisi, kolunu sallasan çifte tellisi**⌝, you have only to wave your hand and people will come in crowds; no dearth of candidates: **hiç sallamamak** (*sl.*), to pay no attention; not to care a brass farthing: **··e kuyruk ~,** to flatter.
**sallan·dirmak** *va.* Swing; shake; hang (execute); put off, delay; put off by false promises. **~mak** *vn.*, swing about; rock; oscillate; totter; be about to fall; loiter, lounge about, waste time.
**sallapatı** Without reflection; suddenly. Careless; tactless.
**sallasırt etm.** *va.* Hoist on the shoulders, shoulder.
**salma** *verb. n. of* **salmak.** Untethered, let out to pasture; running continuously (water). Local rate levied on villages; (formerly) a kind of police force; long hanging sleeve; a kind of stew containing rice. **güvercin ~sı,** pigeon-loft.
**salmak** *va.* Throw; let go; send in haste; insert; spread out; let hang down; send forth (shoots, smoke); postpone; impose (tax); give out (shade); cast (shadow); lay (foundation); throw into a saucepan. *va.* Be aggressive; (*with dat.*) hurl one-

self at, attack. **atını ~**, to let one's horse go at full speed: **boy ~**, (of young trees *etc.*) to grow: **su ~**, to add water to *a thing when cooking*: **suya ~**, to waste.
**salmastra** (·ˈ·) Cord wound round anything to protect it against chafing; gasket; packing. **~ kutusu**, stuffing-box.
**salname** (——ˈ) Almanac.
**salon** Guest-room; dining-room; hall.
**salpa** Loose; slack; untidy; slovenly dressed; up and down (anchor).
**salt** Mere, simple. Merely, solely.
**salta**[1] *A kind of* short jacket.
**salta**[2] (ˈ·) **etm.** *vn.* Stand on the hind legs (dog *etc.*). *va.* Slacken off (a tight rope).
**saltanat, -tı** Sovereignty, dominion; authority, rule; pomp, magnificence. **~ sürmek**, to rule as Sultan; to live in great splendour. **~lı**, regal, pompous, magnificent.
**salya** (ˈ·) Saliva.
**salyangoz** Snail. ⌜**Müslüman mahallesinde ~ satmak**⌝, 'to sell snails in the Moslem quarter', to do *stg.* that 'isn't done'.
**Sam**[1] Shem.
**sam**[2], **-mmi** Poisonous wind, simoom; blight. **~ vurmak (çalmak)**, for the wind to scorch *or* the blight to injure.
**sâman** (—ˈ) Wealth; well-being.
**saman** Straw. ⌜**~ altından su yürütmek**⌝, to do *stg.* in an underhand way, to intrigue covertly: **~ gibi**, insipid. **~ tozu (çöpü)**, chaff: ⌜**sakla ~ı, gelir zamanı!**⌝, don't waste anything, it may come in useful some day. **saman·i** (·—ˈ), straw-coloured. **~kâğıdı, -ni**, tracing-paper. **~kapan**, amber. **~lık**, straw-rick; place for storing straw: ⌜**iki gönül bir olunca ~ seyran olur**⌝, where two hearts beat as one the surroundings make no difference. **~uğrusu, -nu, ~yolu, -nu**, the Milky Way.
**Samî** (—ˈ) Semitic.
**sami** (—ˈ), **-ii** Who hears. Listener. **~in** (—·ˈ), listeners, the audience.
**samim** (·ˈ) Inmost *or* essential part of a thing; essence. **~î** (·——ˈ) sincere; cordial. **~iyet** (·—·ˈ), **-ti**, sincerity.
**samsa** *A kind of* pastry *sweetened with syrup.*
**samsun** Mastiff. **~cu**, soldier in charge of mastiffs, *formerly used against the enemy.*
**samur** Sable. **~ kürk**, sable-skin coat; **~ kürkü sırtına almak**, to take the blame, to bear the responsibility: **~ kürkü ···in sırtına giydirmek**, to lay the blame on ...: **~ kaş**, thick black eyebrows: **sarı ~**, pine-marten: **su ~u**, otter.
**samyeli** (ˈ··), **-ni** Hot poisonous wind, simoom.
**san**[1] Reputation, esteem; surname. **adı ~ı yok**, he is of no account, of no repute: **adımız ~ımız**, our good name.
**san**[2] A disease of corn crops, *causing yellowness.*
**sana** *dat. of* **sen.** To you. **o ~ ne?**, what's that to you?
**san'at, -ti** *or* **-tı** Trade, calling, craft; industry; art; skill, ability. **~le**, artistically. **~çı, ~kâr**, artisan; artist; actor. **~kârlık**, profession of an artisan; profession of art; skill, artistic ability.
**sanayi** (·—ˈ), **-ii** *pl. of* **san'at.** Industries *etc.* **~ci**, industrialist. **~leştirmek**, to industrialize.
**sancak** Flag, standard; starboard side; (*formerly*) a subdivision of a vilayet. **sancağı şerif**, the flag of the Prophet, *only unfurled for a Holy War.* **~tar**, standard-bearer.
**sancı** Stomach-ache, colic, gripes, stitch. **~lanmak**, to have a stomach-ache *or* similar internal pain. **~lı**, having a stomach-ache. **~mak**, to ache, to be griped.
**sancılmak** *vn. pass. of* **sançmak.**
**sandal**[1] Rowing-boat. **~cı**, boatman.
**sandal**[2] Sandalwood.
**sandal**[3] Sandal (shoe).
**sandal**[4] Brocade. **Sandal Bedesteni**, the Municipal Auction-rooms at Istanbul.
**sandalye** (·ˈ·) Chair; office, post. **~ kavgası**, struggle for a post: **~siz nazır**, Minister without portfolio: **koltuklu ~**, arm-chair.
**sandık** Chest, coffer, box; coffer-dam; cash-box; cash department *of a government or business*; old-fashioned fire-pump; box for measuring sand *etc.* **~ emini**, cashier: **~ eşyası**, clothes *etc.* forming part of a bride's dowry: **~ sepet**, bag and baggage: **tasarruf sandığı**, Savings Bank. **~lı**, furnished with a box: thin board *used for veneer*: **~ saat**, grandfather clock.
**sanduka** (·ˈ·) Sarcophagus.
**sandviç** Sandwich.

**sanı** Idea; imagination; surmise. **~cı,** who thinks *or* meditates; who imagines *or* surmises.

**sanık** Suspected; accused.

**sani** (ˊ–) *a.* Second.

**sanih** (–ˋ) Occurring to the mind. **~ olm.,** to occur to one; for an inspiration to come to one. **~a** (–·ˋ), sudden thought; inspiration.

**saniye** (–·ˋ) Second, moment; seconds-hand. **~si ~sine,** dead on time, to the very second: **tabiati ~,** second nature. **~li,** (watch) with seconds-hand: **~ tapa,** time-fuse. **~n** (ˊ··), secondly.

**sank, sankı** Bird excrement.

**sankı** (ˋ·) As if, as though; supposing that.

**sanlı** Well-known; esteemed. **adlı ~,** well-known and talked about.

**sanmak, -ır** *va.* & *vn.* Think, suppose; deem.

**sansar** Pine-marten; polecat. **~ gibi,** stealthy, sly.

**sansür** (*Fr. censure*) Censorship.

**santimetre** Centimetre.

**santral, -li** (*Fr. centrale*) Telephone exchange; power-house.

**santur** Dulcimer.

**sanzatu** (*Fr. sans atout*) No trumps.

**sap** Thread; stem; handle; stalk. ⌜**~ derken saman demek**⌝, to talk twaddle: ⌜**~ demeden samanı anlamak**⌝, to be very 'quick in the uptake': **~ına kadar,** to the core, utterly.

**sapa** Off the road; out-of-the-way, secluded. **~ düşmek,** to be off the main road; to be remote *or* inaccessible: **~ yol,** by-road, side-street.

**sapan** Sling; catapult; sling *for hoisting heavy articles.* **direk ~ı,** strop (*naut.*): ⌜**peşinden ~ taşı yetişmez**⌝, 'a stone from a sling couldn't catch him up', *said of one running away very hard.* **~balığı, -nı,** thresher shark.

**saparna** (·ˋ·) Sarsaparilla.

**saparta** (·ˋ·) Broadside; severe scolding.

**sapasağlam** (ˋ···) *v.* **sapsağlam.**

**sapık** Gone astray; eccentric; crazy.

**sapılmak** *vn. impers. of* **sapmak, buradan sapılır,** here one turns off *in another direction.*

**sapır sapır** *Imitates the noise of continuously falling things.* **~ ~ dökülmek,** to fall continuously (*e.g.* grass being mown).

**sapıtmak** *va.* Cause to go astray. *vn.* Go off one's head; talk nonsense. **sözü ~,** to lose the thread of an argument: **yolu ~,** to take the wrong turning.

**sapkın**[1] Astray; off the right road.

**sapkin**[2] Harpoon; fish-spear.

**saplama** Stud (*mech.*).

**sapla·mak** *va.* Thrust into; pierce; skewer. **~nmak,** (*with dat.*) to sink into, penetrate; get an idea fixed into the mind; get a handle *or* stem.

**saplı** Having a handle *or* stem; sticking into a thing. Bowl *or* pot with a handle; scoop, ladle. **eski politikaya ~ kaldılar,** they stuck to their old policy.

**sapmak** *vn.* Swerve; deviate; diverge; turn off into a different direction; go astray; fall into error. **çıkmaza ~,** to get into a blind alley: **yalana ~,** to have recourse to lying.

**sapsarı** (ˋ··) Bright yellow; very pale.

**sapsız** Without a handle *or* stem. **~ balta,** one without backing *or* influence: **ipsiz ~,** without a tie, vagabond; disconnected, incoherent.

**saptırmak** *va. caus. of* **sapmak,** *q.v.*

**sar'a** Epileptic fit. **~sı var (tutuyor),** he is in a fit.

**sarac** Saddler; leather-worker. **~lık,** the trade of a saddler, saddlery. **~hane, ~ane** (··–ˋ), saddlery market; saddlery workshop.

**sarahat** (·–ˋ), **-ti** Clearness; explicitness. **~en** (·ˊ··), clearly, explicitly.

**saraka** Pillory; ridicule. **~ya almak** (**sarmak**), to deride, ridicule.

**saralı** Subject to epileptic fits.

**sararmak** *vn.* Turn yellow *or* pale.

**sarasker** *v.* **serasker.**

**saray** Palace; mansion; government house *or* office. **~ Burnu,** Seraglio Point: **~ lokması,** a sweetmeat *made of dough, eggs and sugar.* **~lı,** attached to a palace; brought up in a palace *or* mansion: palace servant *or* slave. **~patı,** China Aster.

**sardalya** (·ˋ·) Sardine, pilchard; packed like sardines.

**sardırmak** *va. caus. of* **sarmak,** *q.v.* **~** *or* **merak ~,** to start to have a passion for … .

**Sardunya** (·ˋ·) Sardinia; geranium (pelargonium).

**sarf** Expenditure; use; grammar. **~etmek** (ˋ··), **-eder,** to spend, expend. **~iyat** (··ˊ), **-tı,** expenses.

**sarfınazar** Putting aside. **her şeyden**

~, apart from everything else: ~ **etm.**, to disregard; to relinquish.
**sargı** Bandage.
**sarhoş** Drunk. **içmeden ~**, 'drunk without drinking', *said of one who behaves oddly.* **~luk**, drunkenness.
**sarı** Yellow; fair-haired; pale; haggard. Yellow colour; yolk. **~ altın**, pure gold; a Turkish lira piece: **~çam**, Scotch pine: ⌜**(orduda) ~çizmeli (Mehmed Ağa)**⌝, an unknown person; just anybody; a nobody; *said also of an incomplete address.* **sarı·ağız**, the Meagre (*Sciaena aquila*). **~asma**, golden oriole. **~ca**, yellowish: **~ arı**, wasp. **~çalı**, barberry. **~göz**, a kind of sea-bream.
**sarığıburma** A kind of sweet pastry.
**sarık** Turban. **~ sarmak**, to put on *or* wear a turban. **~lı**, wearing a turban.
**sarı·kanad** Medium-sized **Lüfer** (fish). **~lı**[1], coloured yellow. **~lık**, yellowness; jaundice; blight.
**sarılı**[2] Wound; fastened; surrounded.
**sarılmak** *vn. pass. of* **sarmak.** Be surrounded; be enveloped; be entangled; be bandaged; (*with dat.*) be wound *or* wrapped round; be wound *or* wrapped up in; clasp; embrace; throw oneself upon; be absorbed in; give oneself up to *work etc.*; (of a plant) to entwine itself round, to climb. **ayağa ~**, to clasp the legs in entreaty: **boynuna* ~**, to throw one's arms round the neck of *s.o.*: ⌜**denize düşen yılana sarılır**⌝, a desperate man will face any risk: **···e dört elle ~**, to give one's utmost to ..., to show the greatest zeal in ...: **silâha ~**, to take up arms; to seize a weapon *with the intention of using it.*
**sarım**[1] Sharp; trenchant; decided.
**sarım**[2] Bandage; turn of winding (of elect. coil *etc.*).
**sarımsak** *v.* **sarmısak.**
**sarımsı, sarımtrak** Yellowish.
**sarınmak** *va. & vn.* Wrap oneself (in = *dat.*); gird oneself.
**sarıot** (·˙·) Small pine plank.
**sarısabır** (·˙··), **-brı** The finest quality of aloes.
**sarısalkım** (·˙··) Laburnum.
**sarışın** Fair-haired, blond.
**sarih** Clear; explicit.
**sark·ık** Pendulous; hanging loosely; flabby. **~ılmak**, to hang down; be suspended. **~ınmak**, to hang down; lean over; (*with dat.*) molest, worry.
**sarkıntı** Robbery, spoliation; molestation. **~lık**, act of robbery *or* molestation; importunate *or* insulting behaviour to a woman.
**sarkıtmak** *va.* Let hang down; suspend; hang. **dudak ~**, to pout, sulk.
**sarkmak** *va.* Hang down; lean out (of a window *etc.*); (*with dat.*) come down on, attack suddenly. **bir yere kadar ~**, to go as far as a place.
**sarma** Act of winding *or* enveloping; embrace; a thing wrapped up in *stg.* else; meat and rice wrapped up in vine leaves (**yaprakdolması**); trip (with the leg). **~ya almak**, to get a hold round one's adversary's leg in wrestling: **kuzu ~sı**, lamb chitterlings; ⌜**can ciğer kuzu ~sı**⌝ a very intimate friendship: **hedefi ~ kabiliyeti**, the pattern made by a shot-gun.
**sarmak** *va.* Wind *or* wrap round; bandage; embrace; cling to; surround; wind (wool *etc.*); comprehend, take in; captivate, interest. *vn.* (Of a vine *etc.*) to climb; busy oneself about *stg.* ⌜**ateş saçağı** (*or* **bacayı**) **sardı**⌝, things are pretty desperate: **başına* ~**, to wind *stg.* round the head; (of wine *etc.*) to go to the head; **bu işi (adamı) nereden başıma sardın?**, why did you saddle me with this matter (man)?: **işe ~**, to go to work assiduously on a matter: **bu adam beni hiç sarmadı**, I didn't take to that man at all.
**sarmalamak** *va. Only in* **sarıp ~**, to pack tightly.
**sarmaşdolaş** (·˙··) Close embrace. In a close embrace; inextricably intertwined. **~ olm.**, to embrace one another; to be very close friends.
**sarmaşık** Intertwined. Ivy. **çit sarmaşığı**, the Greater Bindweed.
**sarmaşmak** *vn.* Embrace one another; be intertwined.
**sarmısak** Garlic. **bir diş ~**, a clove of garlic.
**sarnıc** Cistern; tank; **~lı vapur** *or* **~ gemisi**, tanker.
**sarp, -pı** Steep; hard, difficult; inaccessible; intractable. **iş ~a sarıyor**, the matter becomes very complicated *or* serious. **~laşmak**, to become steep; become difficult *or* impracticable.
**sarpa** (˙·) Sea-bream.
**sarpon** Pit for storing grain; silo; baker's dough-tub.
**sarraf** Money-changer; banker. **insan ~ı**, a good judge of men: **koltuk**

(köşe) ~ı, street money-changer. ~**lık,** profession of money-changer; money-changer's fee.
**sarsak** Palsied; shaking from feebleness; quivering; idiot. ~**lık,** palsy; clumsiness.
**sarsık** Walking with a quivering gait; shaky.
**sars·ılış** A being shaken; shock; jolt. ~**ıntı,** a being shaken; shock; concussion; earthquake; disaster. ~**ma,** a shaking (*active*); shake; joggle.
**sarsmak** *va.* Shake; agitate; joggle; give a shock to; upset.
**sart, -tı** Rope made of reeds.
**sası** Mouldy smell. Smelling mouldy.
**sataşmak** *vn.* Become aggressive; seek a quarrel; (*with dat.*) annoy; interfere with; tease.
**saten** Satin.
**sath·an** (ˋ·) Superficially. ~**ı,** *v.* **satıh.** ~**î** (·ˊ), superficial. ~**ice** (·ˊ·), superficially.
**satı** Sale. ~**cı,** salesman, seller; hawker. **ayak** ~**sı,** street hawker.
**satıh, -thı** Upper surface of a thing; face; superficies; plane. **sathı mail,** inclined plane.
**satılık** On sale; for sale.
**satım** Sale. ~**lık,** commission on sale.
**satın** Sale. ~ **almak,** to buy. ~**alıcı,** purchaser.
**satır**[1] Large knife *for cutting meat*; tobacco cutter; executioner's sword. ~ **atmak,** to exterminate.
**satır**[2] Line *of writing.* ~ **başı,** paragraph. ~**lık,** containing *so many* lines.
**satış** Manner of selling; sale.
**satlıcan** Pleurisy.
**satmak** *va.* Sell; make a false show of; pretend to be; (*sl.*) get rid of *s.o.* **ağız** ~, to boast; to blow one's own trumpet: **kendini** ~, to give oneself airs; **kendini satmasını bilmek,** to know how to make the best of oneself, to be able to display one's abilities.
**satranc** Chess; a check pattern; chequered. ~**lı,** with a check pattern.
**satvet, -ti** Spring, rush; attack; force, vigour.
**savab** Right action; correct judgement; *also used for* **sevab** *q.v.* **hata** ~ **cedveli,** list of corrections, errata.
**savak** Cistern from which water is distributed; hatch to a mill-pond; sluice.
**savaş** Struggle, fight; battle; war. ~**çı,** combatant. ~**mak,** to struggle, fight; dispute: ···meğe ~, to struggle to do ..., to work hard to ....
**savat**[1], **-tı** Engraving in black on silver; niello, Tula work. ~**lamak,** to engrave thus. ~**lı,** engraved with black.
**savat**[2], **-tı** Watering-place for cattle; place for fattening cattle. ~**lamak,** ~**mak,** to fatten cattle at pasture.
**savla** (ˋ·) Signal-halyards.
**savma** *verb. noun of* **savmak.** Sent away; got rid of. **baştan** ~, superficial, careless, perfunctory: **baştan** ~ **cevab,** evasive answer.
**savmak** *va.* Drive away; dismiss; get rid of; avoid, escape from; get over (an illness). *vn.* Pass away, come to an end. **başından*** ~, to get rid of *s.o.*: **baştan** ~, to do *stg.* superficially *or* carelessly: **satıp** ~, to sell all one has; **sata sava geçinmek,** to be reduced to such straits that one has to sell one's belongings to live: **sırasını*** (**nöbetini**)~, to have done one's turn; to have 'done one's bit'.
**savruk** Awkward, clumsy; too hasty.
**savrulmak** *vn. pass. of* **savurmak.**
**savsak** Negligent, dilatory. ~ ~, prowling about. ~**lamak,** to put *s.o.* off with excuses *or* pretexts; to put off doing *stg.*
**savulmak** *vn.* Stand aside; get out of the way. **savul!,** stand clear!, get out of the way!
**savunma** Defence. **Millî** ~ **Bakanlığı,** Ministry of National Defence. ~**k,** to defend oneself.
**savurmak** *va.* Toss about; throw into the air; blow about; winnow; brandish (a sword). *vn.* Blow violently; bluster, brag. **atıp** ~, to bluster: **küfür** ~, to let out an oath: ···**in külünü** ~, to vow vengeance on ...: **sirkeyi** ~, to brew vinegar.
**savuş·mak** *vn.* Pass; cease; slip away. ~**maz,** that will not go away; incurable. ~**turmak,** to cause to go away *or* cease; ward off; escape *or* avoid *some disagreeable thing or person.*
**saydırmak** *va. caus. of* **saymak.** ···**i yerinde** ~, to make *s.o.* mark time, to prevent *s.o.* making any progress.
**saye** (–ˋ) Shadow. ... ~**sinde,** thanks to ...: **bu** ~**de,** in this manner. ~**ban** (–·ˋ), canopy, tent: **sayenizde** ~ **oluyorum,** thanks to you I am all right (*slightly ironical*).

**sayfiye** Country house: **~ye gitmek,** to go into the country.
**sayfa** *v.* **sahife.**
**saygı** Respect, esteem; thoughtfulness, consideration. **~larımı sunarım,** I present my respects. **~lı,** respectful; considerate; well-mannered. **~sız,** inconsiderate; disrespectful. **~sızlık,** disrespect; lack of consideration.
**sayı** Number; reckoning. ⌜**~m suyum yok**⌝, (*orig. from a children's game*) 'pax!'; I'm not in this; I'm taking no part! **~cı,** official teller of sheep for taxation.
**sayıklama** Delirium; talking in one's sleep; dreaming of *stg.* longed for.
**sayıklamak** *vn.* Talk in one's sleep *or* in delirium; rave; dream of *stg.* longed for.
**sayılı** Counted; limited in number; numbered; marked, special. **~ gün,** a red-letter day.
**sayılmak** *vn. pass. of* **saymak.** Be counted *or* numbered; be esteemed; be taken into account. **hatırı sayılır,** who counts, respected; considerable.
**sayım** A counting; census. **~ vergisi,** tax on the number of animals.
**sayın** Esteemed; excellent. **~ Bay ...,** Dear Mr. ... (*in a letter*).
**sayısız** Innumerable.
**sayışmak** *vn.* Settle accounts with one another. **sövüşüp ~,** to swear at one another.
**Sayıştay** The Exchequer and Audit Department.
**saymak** *va.* Count; number, enumerate; regard, count as; esteem, respect; deem, suppose. **sayıp dökmek,** to recount at length: '**bunu saymayız, sizi yine bekleriz**', 'I don't call this a visit, you must come again': **···in hatırını ~,** to have consideration for, to respect: **hiçe ~,** to account as nothing, to hold of no account: **···e sövüp ~,** to swear at: **yerinde ~,** to mark time; to make no progress.
**saymamazlık** Disrespect; irreverence.
**sayvan** Flounce, fringe; umbel; the external ear; tent.
**saz**[1] Rush; reed. **~ benizli,** pale.
**saz**[2] Musical instrument; band; Oriental music. **~ söz,** music and conversation; party: **~a söze düşkün,** fond of parties: **~ şairi,** minstrel; one who improvises songs and music: **ince ~,** Turkish orchestra of stringed instruments: **meydan ~ı,** a kind of large guitar. **~cı,** player of a musical instrument; musician.
**sazan** Carp. **yeşil ~,** tench.
**sazende** (–·⌄) Player of a musical instrument; lute player.
**sazlık** Reed-bed; place covered with rushes.
**se** Three (at dice).
**-se** *v.* **ise.**
**Seba** Saba (ancient town in Arabia). **~ kraliçesi,** the Queen of Sheba.
**sebat** (·⌄), **-tı** Stability; firmness; perseverance; constancy. **~kâr,** enduring; persistent. **~lı,** enduring; stable; persevering. **~sız,** unstable, fickle; lacking perseverance.
**sebeb** Cause, reason; source; means; occasion. **... ~ile,** on account of ..., owing to ...: **~ aramak,** to search for a pretext: **e ~ olm.,** to cause, to occasion: **bu ~le,** for this reason, therefore: **her hangi bir ~le,** for some reason or other.
**sebeb·iyet, -ti,** a being the cause *or* motive: **···e ~ vermek,** to cause. **~lenmek,** to earn one's living; to get a small profit *out of stg.*; to get *stg.* out of *a thing.* **~li,** having a reason *or* excuse: **~ sebebsiz,** without any reason. **~siz,** without a cause *or* reason: ⌜**sebeb olanlar ~ kalsın**⌝, (*a curse*) 'may those responsible suffer for it!'
**sebil** (·⌄) Road; public fountain; free distribution of water. **~ etm.,** to spend lavishly, squander. **~ci,** man who distributes water gratis. **~hane** (··–⌄), public fountain: **~ bardağı gibi,** all in a row (*slightly derogatory*).
**sebk, -kı, sebkat, -tı** A going before; precedence. **~etmek** (⌄··), **-eder** *vn.*, to precede, happen before.
**sebze** Green plant; vegetable. **kuru ~,** pulse. **~vat, -tı,** *pl. of* **sebze.**
**seccade** (·–⌄) Prayer rug. **~ci,** maker *or* seller of prayer rugs; servant whose duty it is to look after the prayer rug of a great man.
**secde** Act of prostrating oneself in worship. **~ etm.,** *or* **~ye varmak,** to perform the ritual prostrations in prayer: **~ye yatmak,** to prostrate oneself, to fall down and worship (*dat.*).
**seci, -c'i** Rhyme in prose.
**seciye** Moral quality; character; natural disposition. **~li,** of high

moral character, good. **~siz,** untrustworthy; vicious.
**seçici** Who chooses *or* distinguishes.
**seçik** Clear; distinct.
**seç·ilmek** *vn. pass. of* **seçmek.** **~ilmiş,** picked, choice; what is left after the best has been taken. **~im,** choice; election. **~kin,** choice; outstanding. **~me,** a choosing; choice; election; perception: selected, choice. **~mece,** allowing the purchaser to pick and choose (of a commercial transaction).
**seç·mek** *va.* Choose, select; elect; perceive; distinguish, make out: **yemek ~,** to be particular about one's food. **~men,** elector.
**sed, -ddi** Barrier; obstacle; dam; bank; fence; rampart; obstruction *to an action.* **Çin Seddi,** the Great Wall of China. **~detmek** (ˈ··), **-eder,** to bar, obstruct, barricade.
**seda** (·ˈ) *v.* **sada.**
**sedef** Mother-of-pearl; shell producing mother-of-pearl. Made of mother-of-pearl. **~ hastalığı,** psoriasis. **~li,** made of *or* decorated with mother-of-pearl. **~otu, -nu,** rue.
**sedir**[1] Platform at the top of a room with a divan on it; divan, sofa. **baş ~e geçmek,** to take the top seat; to take the first place, to be very important: **erkân ~i,** a kind of high sofa.
**sedir**[2] Cedar.
**sedye** (ˈ·) Sedan-chair; stretcher. **~ci,** stretcher-bearer. **~lik,** a stretcher case.
**sefa** (·ˈ) *v.* **safa.**
**sefahet** (·—ˈ), **-ti** Foolish squandering; dissipation.
**sefalet** (·—ˈ), **-ti** Poverty; misery. **~ çekmek,** to suffer privation: **~e düşmek,** to be reduced to poverty.
**sefaret** (·—ˈ), **-ti** Ambassadorship; embassy, legation. **~hane** (·—·—ˈ), embassy (building), legation.
**sefer** Journey; voyage; campaign; time. **~ açmak,** to start hostilities: **~ etm.,** to go on a journey: **~e gitmek,** to go to war, to start a campaign: **bu ~,** this time.
**sefer·ber,** mobilized for war. **~berlik,** mobilization. **~î,** pertaining to travel *or* to a campaign: **heyeti ~ye,** expeditionary corps: **ordunun ~ mevcudu,** the war strength of an army. **~tası, -nı,** travelling food box *with several metal dishes fastened together.*
**sefih** (·ˈ) Spendthrift, prodigal; dissolute.
**sefil** (·ˈ) Poor; miserable; destitute.
**sefine** (·—ˈ) Ship.
**sefir** (·ˈ) Ambassador.
**segâh** *Name of a musical cadence.*
**seğirdim** Recoil *of a gun*; distance run in a race. **~ yapmak,** to recoil.
**seğirme** Vibration; tremor; nervous twitch. **~k,** to tremble; to twitch nervously: **gözü ~,** for the eye to twitch (*supposed to be a good omen*).
**seğirtmek** *vn.* Run; hasten.
**segman** (*Fr. segment*) Piston-ring.
**seğmen** *Formerly a division of soldiers incorporated with the Janissaries*; servant in charge of dogs; young men, armed and in national costume, *who take part in a procession.*
**seher** Time just before dawn; early morning.
**sehim, -hmi** Share, portion; Treasury bond.
**sehiv, -hvi** Mistake; inadvertence.
**sehpa** (·ˈ) Tripod; easel; gallows. **atlama ~sı,** vaulting-horse.
**sehv·en** (ˈ·) Inadvertently. **~etmek** (ˈ··), **-eder,** to make a slip.
**sek** (*Fr. sec.*). Dry (wine); champagne.
**sekalet** (·—ˈ), **-ti** Ugliness; eyesore.
**sekban** *v.* **seğmen.**
**seki** Pedestal; stone seat; white sock on a horse.
**sekiz** Eight. ⌜**Allah ~de verdiğini dokuzda almaz**⌝, what is fated will happen. **~er,** eight each; eight at a time. **~inci,** eighth. **~li,** having eight; the eight *of a suit of cards.* **~lik,** worth *or* containing eight ... .
**sek·me** Hop; ricochet. **~mek,** to hop; to run in a series of jumps; to ricochet; to miss: **bir gün bile sekmez (sektirmez),** he never misses a single day. **~sek,** hop-scotch.
**seksen** Eighty. **~er,** eighty each; eighty at a time. **~inci,** eightieth. **~lik,** worth *or* containing eighty ...; octogenarian.
**sekte** Pause; interruption; apoplexy. **~ vermek,** to relax, to give a respite: **kalb ~si,** heart failure. **~lenmek,** to be hindered *or* interrupted.
**sektir·me** A causing to rebound *or* ricochet. **~mek,** to cause to hop, rebound, ricochet; *v.* **sekmek.**
**sel** Torrent; inundation; flood.
**selâ** Saddle *of a bicycle.*
**selâm** Salutation; greeting; salute.

~ **almak,** to acknowledge a salute *or* another's bow: ~ (*or* ~**a**) **durmak,** to rise respectfully to receive the salute of a superior; to present arms: ~ **dur!,** (*formerly*) present arms!: ~ **etm.,** to send one's compliments: ···**le** ~**ı sabahı kesmek,** to cease relations with ...: ~ **vermek,** to greet, to salute: ⌜**ne** ~ **ne sabah**⌝, without as much as saying good-morning; ignoring everyone: **resmi** ~, military salute: ···**in size** ~**ı var,** ... sends you his greetings: **yerden** ~, a very deferential bow.

**selâmet** (·–ˋ), **-ti** Safety; security; freedom from danger *or* illness; soundness; liberation; successful result; (sentence) free from defect *or* error. ~**le!,** good-bye and good luck!: ~ **bulmak** *or* ~**e çıkmak,** to gain safety; to turn out well: **Allah** ~ **versin!** *said when mentioning an absent friend* (*often when criticizing him*): **sahili** ~, the shore of safety, *i.e.* escape from danger: ⌜(**şuradan şuraya**) ~**le çıkmayayım**⌝, (*an oath*) 'if it be not true may I die!'

**selâmetlemek,** to see *s.o.* off; to wish *s.o.* God-speed.

**selâm·lamak** *va.* Salute; greet. ~**lık,** the part of a large Moslem house reserved for males; (*formerly*) the public procession of the Sultan to a mosque at noon on Fridays. ~**ünaleyküm,** 'peace be on you' (*the formal greeting of Moslems to each other, the reply being* **aleyküm selâm**): ~ **demeden,** without so much as by your leave; brusquely and tactlessly. ~**ünkavlen,** *euphemism for* paralysis, stroke.

**Selânik** (·ˋ·) Salonica. **yürek** ~, a coward. ~**li,** Jewish convert to Islam.

**selâtin** *pl. of* **sultan.** Sovereigns. Imperial; grand.

**selb** A depriving; negation. ~**etmek** (ˋ··), **-eder,** to deprive forcibly.

**Selçuk** Seljuk. ~**î** (··ˋ), ~**lu,** Seljukian.

**sele** Flattish wicker basket.

**selef** Predecessor; ancestor; man of old.

**selis** Fluent; easy-flowing (style *etc.*).

**sellemehüsselâm** Without ceremony; without being announced; unexpectedly; rudely.

**selliseyf etm.** *va.* Draw the sword.

**sellüloit** Celluloid.

**selpik** Languid, lazy.

**selvi** Cypress. ~**lik,** place abounding in cypresses.

**sema**[1] (·ˋ) Sky; heaven.

**sema**[2], **-ai** Dervishes' dance. ~**hane** (·––ˋ), dervish conventicle for religious music and dancing.

**semahat** (·–ˋ), **-ti** Generosity, munificence.

**semaî** (·–ˋ) A solemn dance tune.

**sembol, -lü** Symbol.

**semender** Salamander.

**semer** Pack-saddle; pad *used by porters for carrying heavy weights.* ~**ci,** maker of pack-saddles.

**semere** Fruit; profit; result.

**semih** Liberal; generous.

**semir·gin** Fat because he is lazy and lazy because he is fat. ~**mek,** to grow fat. ~**tmek,** to fatten; to manure (the ground).

**semiz** Fat; fleshy. ~**ce,** rather fat. ~**lik,** fatness. ~**otu** (·ˋ··), **-nu** purslane.

**sempati** Sympathy.

**semt, -ti** Direction; region, neighbourhood; quarter in which one lives. ~ ~, in certain places; in every quarter.

**semum** (·ˋ) Simoom.

**sen** Thou, you. ~ ~ **ol!,** (*a warning*) *in such phrases as*: ~ ~ **ol, bir daha bunu yapma!,** now don't forget, you mustn't do this again!: '**biz bu işi gizli tutmağa çalışırken** ~ **git her kese söyle!**', 'just when we were trying to keep the matter quiet, what must you (he) do but go and tell everyone!'

**sena** (·ˋ) Praise; eulogy.

**sendelemek** *vn.* Totter; stagger.

**sendere** Thin board; shingle.

**sendika** Trade union.

**sene** Year. ~**lik,** lasting *so many* years; of *so many* years.

**sened** Written proof of a transaction *etc.*; document; title-deed; voucher; receipt. ~**siz sepetsiz,** without giving *or* demanding any written proof: **nakliye** ~**i,** bill-of-lading, way-bill: **tasarruf** ~**i,** title-deed.

**senevî** (··ˋ) Annual.

**senfoni** Symphony.

**seni** *acc. of* **sen.** Thee; you; *sometimes used as a term of abuse usually in conjunction with* **gidi** *q.v.*

**senin** *gen. of* **sen.** Thine, your, of you. *In such phrases as*: **o sokak** ~ **bu sokak benim gezip duruyor,** he spends all his time wandering about this

street and that (the streets generally). **~ki,** yours.
**senli** *In* ~ **benli,** hail-fellow-well-met; unpretentious, free-and-easy. ~ **benli konuşmak,** to have a confidential talk; to be intimate.
**sentetik** Synthetic.
**sepek** Pivot of a millstone.
**sepet, -ti** Basket; basketful. Wickerwork. ~ **havası çalmak,** to show a person that his room is preferable to his company, *v.* **sepetlemek**: **sandığı ~i kaldırmak,** to remove all one's belongings. **~kulpu, -nu,** broad arch. **~leme,** woven like a basket. **~lemek,** to get rid of a tiresome person; to sack, dismiss. **~li,** having a basket; ~ **motosiklet,** motor-cycle having a side-car.
**sepi** Dressing for hides; tanning; dyeing of furs. **~ci,** tanner. **~lemek,** to tan; to prepare furs. **~li,** tanned (hide); dyed and prepared (fur).
**sepken** Anything sprinkled; shower of rain; slight fall of snow.
**sepmek** *va.* Sip noisily; gobble; sprinkle. *vn.* Drizzle.
**septik** Sceptical.
**sepya** Sepia.
**ser**[1] Head; chief; top; end. ⌜**~den mi geçmeli yardan mı?**⌝, 'must one give up one's life or one's love?', *i.e.* which of two difficult alternatives?: **~de genclik var!,** what else do you expect of a young man?: **~de müdürlük var!,** well, after all he is the manager: ⌜**~ vermek sır vermemek**⌝, better die than give away a secret.
**ser**[2] (*Fr. serre*) Greenhouse; hothouse.
**serab** (·⏗) Mirage.
**serapa** (·—⏗) From head to foot; utterly.
**serasker** (*Formerly*) Commander-in-chief; Minister of War. ~ **kapısı,** offices of the Minister of War.
**serazad** (·—⏗) Free; independent.
**serbest, serbes** Free; independent; unreserved, frank. **~çe** (·⏗·), freely. **~î,** liberty. **~lik,** freedom; frankness.
**serçe** Sparrow. **~giller,** passerines. **~parmak,** little finger *or* toe.
**serdar** (·⏗) Military chief; general.
**serdengeçti** Troops selected for a desperate enterprise, forlorn hope, suicide squad. **~lik,** mad enterprise; foolhardiness.
**serdetmek** (⏗··), **-eder** *va.* Set forth; expound.
**serdümen** Quartermaster (*naut.*).
**sere** Span between the thumb and first finger.
**seren** Yard (*naut.*); boom, spar.
**Serendib** Ceylon.
**sereserpe** (·⏗··) Free and unrestrained.
**sergerde** Chief; leader of a band *of bandits or irregulars.*
**sergi** Anything spread; mat *or* carpet *on which goods are exposed for sale*; temporary stall; shop-front; exhibition.
**sergüzeşt, -ti** Adventure.
**serhad, -ddi** Frontier. ~ **narası,** war-cry.
**serhademe** Head servant *in an office etc.*
**seri**[1] Series. ~ **halinde imal,** mass production.
**seri**[2] (·⏗) Quick, swift. **~an** (·⏗·), quickly.
**seril·i** Stretched out *or* spread on the ground. **~mek** *vn. pass. of* **sermek**; to lie at full length on the ground; fall ill; drop in a faint: **yatağa ~,** to be seriously ill.
**serin** Cool. Cool weather *or* air. **~-lemek, ~lenmek, ~leşmek,** to become cool; to cool oneself. **~lik,** coolness.
**serkeş** Unruly, rebellious. **~lik,** disobedience, rebelliousness.
**serlâvha, serlevha** Title, heading.
**sermaye** (·—⏗) Capital; stock; first cost; acquired knowledge; (*sl.*) prostitute. ~ **komak,** to invest capital. **~dar,** capitalist. **~li,** having a capital of .... **~siz,** without capital; without attainments.
**sermek** *va.* Spread out on the ground; beat down to the ground; spread over; neglect *one's job.* **ipe un ~,** to plead vain excuses; to offer empty assurances: **sere serpe,** *v.* **sereserpe.**
**sermest, -ti** Drunk; intoxicated *with joy etc.* **~i** (··⏗), intoxication.
**ser·muharrir** Editor-in-chief. **~-mürettib,** chief compositor.
**serpantin** (*Fr. serpentin*) Paper streamer.
**serp·ilmek** *vn. pass. of* **serpmek.** Fall as if sprinkled; stretch oneself out to rest; (of a child) to grow apace. **yüreğine* su ~,** to be relieved *or* comforted. **~inti,** drizzle; spray from a falling liquid; traces left behind of a thing; repercussion: **~-sine uğramak,** to feel the effects of

some remote cause. **~iştirmek,** to sprinkle in small quantities; distribute *or* scatter small amounts of money; (of rain) just to begin to drop. **~me,** sprinkled about; sprinkled with; a sprinkling; cast-net.

**serpmek** *va.* Sprinkle slightly; scatter with the hand. *vn.* Fall in a sprinkle. **saçlarına* kır ~,** for one's hair to have a sprinkling of grey.

**serpuş** Headgear.

**sersem** Stunned; bewildered; scatter-brained; foolish. **~ sepet,** stupid; stupidly. **~lemek,** to be stunned *or* stupefied; to lose one's head; become silly *or* absent-minded *or* forgetful. **~lik,** stupefaction; confusion; stupidity; wool-gathering.

**serseri** (··-́) Vagabond; tramp. Loose (mine *etc.*). **~ce** (··-́·), in the manner of a vagabond. **~lik,** vagabondage, vagrancy.

**sert** Hard; harsh, severe; violent; potent. **~lenmek, ~leşmek,** to become hard, severe, violent. **~lik,** hardness; harshness; violence; potency.

**serteser** (\`··) Utterly, entirely.

**serüven** *v.* **sergüzeşt.**

**servet, -ti** Riches, wealth. **~li,** wealthy.

**servi** Cypress. **~ boylu,** of a slender and graceful build. **Lübnan ~si,** cedar.

**servis** Service.

**serzeniş** Reproach; reprimand. **~kâr,** reproachful.

**ses** Sound; noise; voice; cry. **~ çıkarmak,** to speak; to blab: **~ çıkarmamak,** to say nothing; to condone (*with dat.*): **~ çıkmak,** to be heard; to become known; to be rumoured: **~i çıkmaz,** taciturn: **~ini* kesmek** *or* **~i* kesilmek,** to cease speaking; to be reduced to silence; **~ini kes!,** shut up!: **~ seda yok,** not a sound to be heard: **~ vermek,** to give out a sound; to say *stg*: **~ vermemek,** not to answer *when called.*

**ses·lemek,** to hearken, to give ear. **~lenmek,** to call out to *s.o.* (*dat.*); to reply to one calling. **~li,** voiced, having *such* a voice; noisy; talking (film): **kaba ~,** gruff-voiced. **~siz,** voiceless; quiet; silent; meek; having a poor tone (mus. instrument). **~sizlik,** quietness, silence; meekness.

**set, -ti** *coll. form of* **sed** *q.v.*

**seten** Satin.

**setir, -tri** A covering, veiling *or* hiding.

**setre** Old-fashioned form of frock-coat.

**setretmek** (\`··), **-eder** *va.* Cover; hide.

**sevab** (·\`) God's reward for a pious act *or* good conduct on earth; good deed, meritorious action. **~a girmek (nail olm.),** to acquire merit in God's sight: **~ etm.,** to do a good deed; to live virtuously. **~lı,** meritorious.

**sevda** (·-́) Melancholy; spleen; passion, love; intense longing; scheme, project. **bu ~dan vazgeç!,** give up this idea!: **kara ~,** hypochondria. **~lı,** madly in love; enamoured. **~vî** (·—-́), atrabilious; melancholic; amorous. **~zede,** stricken with passion; smitten with desire.

**sevgi** Love; affection; compassion. **~li,** lovable; beloved; darling.

**sevil·ir** Lovable; amiable; desirable. **~mek,** *pass. of* **sevmek** *q.v.*; be lovable *or* amiable.

**sevim** Love; affection; affability. **~li,** lovable; genial. **~siz,** unattractive; unsympathetic (not likeable).

**sevin·c** Joy; delight. **~cli,** joyful. **~mek,** to be pleased *or* happy.

**sevişmek** *vn.* Love *or* caress one another.

**seviye** Equality; level; rank; degree. **yaşayış ~si,** the standard of living.

**sevk, -kı** A driving, urging, inciting; dispatch (of troops *etc.*). **~etmek** (\`··), **-eder,** to drive, impel; urge, incite; send: **tekaüde ~,** to pension off. **~itabiî,** instinct. **~ıyat, -tı,** dispatch of troops; consignments *of goods.* **~ulceyş,** strategy. **~ulceyşî,** strategic.

**sevmek** *va.* Love; like; fondle. **yerini ~,** (of plants) to prosper, to grow well.

**seyahat** (·—\`), **-ti** Journey; travelling; expedition (polar, scientific). **~name** (·—·—\`), book of travels.

**seyelân** A flowing; flood. **~ etm.,** to stream, pour.

**seyid** Master, lord, chief; Seyyid (descendant of the Prophet).

**seyif, -yfi** Sword.

**seyir, -yri** Movement; progress; travel; excursion; voyage; a looking on at a thing; spectacle. **~e çıkmak,** to go for a walk *or* ride, to make an excursion: ~ **jurnalı**, log-book: ~ **kılavuz kitabi**, sailing-directions: ~ **tecrübesi**, trial trip *of a ship*: **piston seyri**, piston stroke. ~**ci**, spectator; one who merely looks on.

**seyirmek** *etc. v.* **seğirmek** *etc.*

**seyis** Groom.

**Seylân** Ceylon. ~ **taşı**, garnet.

**seylâb** Flood; torrent.

**seyran** Outing; pleasure trip; excursion. ~ **etm.** *or* ~**a çıkmak**, to go for a trip, to make an excursion.

**seyrek** Wide apart; few and far between; at infrequent intervals *of space or time*; rare; rarely; loosely woven; sparse. ~**çe**, somewhat infrequent. ~**leşmek**, to become infrequent; to be at wide intervals. ~**lik**, distance of intervals; rarity of occurrence, infrequency; looseness of texture.

**seyrelmek** *v.* **seyrekleşmek.**

**seyretmek** (ˋ··), **–eder** *vn.* Move; go along; behave; look on. *va.* Look at; look on at (without interfering). **seyret!**, now you'll see (what's going to happen).

**seyr·gâh** *v.* **seyrangâh.** ~**üsefer**, traffic (*movement of people and vehicles*): ~ **memuru**, traffic policeman.

**seyyah** Traveller, tourist.

**seyyal, –li** *a.* Fluid; liquid.

**seyyar** Mobile; portable. ~ **satıcı**, street hawker. ~**e** (·–·), planet.

**seyyiat, –tı** *pl. of* **seyyie.**

**seyyie** Evil thing; evil deed; vice; evil consequence. ~**sini çekmek**, to suffer the consequences of an evil act.

**seza** (·ˊ) Meet, fit, suitable; worthy (of = *dat.*). Merited punishment *or* reward.

**sezaryen** Caesarian (operation).

**sez·i** Feeling; intuition. ~**inlemek**, ~**inmek**, ~**insemek**, to be aware of, be conscious of; have an inkling of. ~**inti**, perception; inkling.

**sezmek** *va.* Perceive; feel; discern, make out. **evvelden** ~, to have a presentiment of.

**sıcacık** Warm; pleasantly hot.

**sıcak** Hot. Heat; hot place; hot bath. **sıcağı sıcağına**, 'while the iron is hot', at once: ~ **kanlı**, warm-hearted; amiable (*not* hot-blooded): ⌜**elini** ~ **sudan soğuk suya sokmamak**⌝ (of a woman) to be unwilling to do housework. ~**lık**, heat.

**sıçan** Rat; mouse. ~**dişi** (·ˋ··), **–ni**, a kind of fine edging to linen. ~**kırı**, **–nı**, mouse-coloured horse. ~**kulağı**, **–nı**, chickweed. ~**kuyruğu**, **–nu**, rat-tailed file. ~**otu**, **–nu**, arsenic. ~**yolu**, **–nu**, underground passage, *esp.* a gallery for a landmine.

**sıç·ılmak** *vn.* Be fouled with one's own excrement; be filthy; (of a thing) to be damaged. ~**mak**, to open the bowels; to go to the rear; befoul *or* spoil *stg.*

**sıçra·ma** Act of jumping. ~ **tahtası**, spring-board. ~**mak**, to jump, spring; start; spurt out. ~**yış**, a jumping *or* springing.

**sıdık, –dkı** Truth; sincerity. ~**ile çalışmak**, to put one's heart into one's work.

**sıfat, –tı** Attribute; adjective; mien. ... ~**ile**, in the capacity of ...: ~**ından anladım**, I could see from his expression.

**sıfır** Zero; nought; cipher. ⌜~**a** ~ **elde var bir**⌝, nought and carry one (*arith.*), *i.e.* the barest minimum: ~ **numara makine**, No. O hairdresser's clipper, *to cut as closely as possible*: ~**ı tüketmek**, to exhaust one's means, to be reduced to the last extremity: **solda** ~, 'a nought on the left-hand side', a mere cypher.

**sığ** Shallow. Shoal; sandbank. ~**a oturmak**, (of a ship) to go aground.

**sığamak** *va.* Tuck *or* roll up (skirts, shirt-sleeves *etc.*); rub with the hand; smooth; massage; *v. also* **sıvamak.**

**sığanmak** *see* **sıvanmak.**

**sığın** Moose, elk.

**sığınak** Shelter (alpine, air-raid *etc.*).

**sığın·mak** *vn.* Squeeze into *a narrow place*; take shelter *or* refuge. **Allaha sığındık**, I trust in God. ~**tı**, one who takes refuge *or* to whom shelter has been given (*derogatory*); parasite.

**sığır** Ox; bull; cow; buffalo. ~ **eti**, beef: **kara** ~, buffalo: **su** ~**ı**, water buffalo: ⌜**ben** ~ **yüreği yutmadım**⌝, I'm sick of telling you! ~**cık**, starling. ~**dili**, **–ni**, ox-tongue; bugloss. ~**gözü**, **–nü**, corn marigold. ~**kuyruğu**, **–nu**, mullein. ~**lık**, bovine stupidity; boorishness. ~**tmaç**, herdsman, drover.

**sığışmak** *vn.* Go *or* fit into a confined space with difficulty.

**sığla** Liquidambar tree (*L. orientalis*).

**sığlık** Shallow; sandbank.

**sığmak** *vn.* (*with dat.*) Go into; be contained by. **akla sığar,** plausible; admissible: **ele avuca sığmaz bir çocuk,** an unruly child: **içi içine sığmamak,** to be unable to contain oneself *from joy etc.*: **kabına sığmamak,** to be uncontrollably impatient.

**sıhhat, -ti** Health; truth. **~te bulunmak,** to be in good health: **~ler olsun!,** 'good health to you!', *said to one having had a bath or a shave.*

**sıhhî** (·–́) Pertaining to health; hygienic. **~ye,** Health Department.

**sık** Close together; dense; closely woven; frequent; tight. Frequently. **~boğaz,** urgently: **~ etm.,** to take by the throat; to force *s.o.*

**sıkı**[1] *a.* Tight; firmly driven in; strict, severe; heavy (gale); tight-fisted. **~ bas!,** hold tight!; stand firm!: **~ durmak,** to 'sit tight', to stick fast: **~ esmek,** to blow a gale: **~ yürümek,** to walk briskly: **ağzı ~,** secretive: **ayağına ~,** a good walker.

**sıkı**[2] *n.* Pressing necessity; trouble, straits; severe menace *or* reprimand; wad *for a firearm.* **~ya dayanmak,** to stand hard work, to brave trouble: **~ya gelmek,** to meet with great difficulty, to be hard put to it: **~yı görünce,** when pressed, compelled *or* threatened: **~ya koymak,** to press *s.o.* hard, to try to force *s.o.* to do *stg.*: **~yı yemek,** to receive a severe threat *or* reprimand.

**sıkı·fıkı,** close together; intimate: **~ konuşmak,** to have a confidential chat. **~lamak,** to tighten; to press with questions.

**sıkıcı** Tiresome; boring.

**sıkılgan** Easily embarrassed; awkward; shy. **~ olmıyan,** unconstrained.

**sıkılma** *verb. noun of* **sıkılmak.** A being bored *etc.*; a being ashamed. **bu adamda hiç ~ yok,** this man has no sense of shame.

**sıkılmak** *vn. pass. of* **sıkmak.** Be pressed *or* squeezed; be in difficulties; be bored, annoyed, uneasy, ashamed. **başı* ~,** to be in straits: **canı* sıkılmak,** to be bored.

**sıkımlık** *In* ⌜**bir ~ canı var**⌝, you could knock him down with a feather (of a very weak person).

**sıkıntı** Annoyance; boredom; embarrassment; discomfort; distress; weariness; worry; financial straits. **~ çekmek,** to suffer annoyance *or* inconvenience: **~da olm.,** to be in straits: **~ vermek,** to cause vexation *or* inconvenience.

**sıkış·ık** Closely pressed together; crowded; congested. **~mak,** to be closely pressed together; be crowded together; be in straits; become urgent; be 'taken short'. **~tırmak,** to press, squeeze; tighten; force; oppress; question closely; press *money etc.* into another's hand.

**sıkkın** Annoyed, disgusted; in difficulty, in need.

**sıklamak** *vn. In* **ağlamak ~,** to weep and lament.

**sık·lanmak, ~laşmak** *vn.* Be frequent (in time *or* space); be closely woven; be close together. **~laştırmak, ~latmak,** to bring close together; render frequent: **adımlarını* ~,** to hasten one's pace.

**sıklet, -ti** Heaviness, weight; uneasiness, languor. **~ çekmek,** to be bored: **~ vermek,** to annoy, bore.

**sıklık** Density of texture; frequency. Densely populated.

**sıkma** *verb. noun of* **sıkmak.** Act of squeezing *etc.*; a kind of tightly fitting trousers.

**sıkmak** *va.* Press; squeeze; tighten; put pressure on; dun *for money*; cause annoyance, embarrassment *or* discomfort; discharge (firearm). **···in canını* ~,** to annoy *or* bore: **dış ~,** to set the teeth *for perseverance or endurance.*

**sıksık** *v.* **sık.**

**sıla** Reunion with friends *or* family; visit to one's native country; relative pronoun. **~ya gitmek,** to visit one's native country, to go home: **~ hastalığı,** home-sickness. **~cı,** who sets off to visit his home; (soldier) on leave.

**sımsıkı** Very tight; squeezed; narrow.

**sınai** (·–́) Industrial.

**sınamak** *va.* Try, test.

**sınav** Examination.

**sındı** Large cutting-out scissors.

**sınd·ık, ~ırgı** Scene of a defeat. **~ırmak,** to rout.

**sınıf** Class; sort; category. **~ı(nı) geçmek,** to be promoted to another class: **~ta kalmak,** to fail to get promotion (in a school).

**sınık** *v.* **sıngın.**

**sınır** Frontier. **~ dışından,** from abroad. **~daş,** having a common frontier; bordering. **~lamak,** to limit, determine.

**sınmak** *vn.* Break; be routed; be scattered; be broke (bankrupt).

**sıpa** Year-old donkey foal; year-old fawn.

**sır**[1] Glaze (of pottery); silvering (mirror). **~ vermek,** to give *pottery* a glaze, *v. also* sır[2].

**sır**[2], **-rrı** Secret; mystery. **~ açmak,** to reveal *or* confide a secret: **~ra kadem basmak** *or* **~ olup gitmek,** to disappear: **~ saklamak,** to keep secret: **~ tutmak,** to keep secret; to keep a secret: **~ vermek,** to betray a secret, *v. also* sır[1]: **akıl ~ erecek gibi değil,** unintelligible, inexplicable, mysterious.

**sıra** Row; file; rank; order, sequence; series; regularity; turn; opportune moment; bench; desk; line *of writing.* In a row *or* line *or* layer. Along; by. **~ ~,** in rows, courses, *or* layers: **~da,** in a row; ... **~da,** just at the moment that ..., as ...: **~sında,** in his (its) turn; when necessary: **~ ile,** in rows; in turn; in order: **~sı ile,** respectively: **~dan bir adam,** any ordinary man (*i.e.* not specially selected): **~ya bakmak,** to pay attention to time *or* turn: **~ beklemek,** to await one's turn, to queue: **~sı düştü,** the right moment for it has come: **~ düşürmek,** to find a favourable opportunity: **~ evler,** houses in a row: **~sı gelmişken,** by the way, apropos: **~sına getirmek,** to await a favourable opportunity: **~sına göre,** according to circumstances: **~ gözetmek,** to wait for a suitable moment; to pay regard to *s.o.'s* turn *or* seniority: **~sına koymak,** to put into its proper place; to set to rights: **~ malı,** ordinary goods (not specially made): **aklı ~,** according to him (*implying disbelief or contempt*): **arada ~da,** here and there: **kıyı ~,** along the shore: **sözün ~sı,** the context of a word; the course of a discourse: **söz ~sı bende,** now it's my turn to speak: **yanı ~,** by his side; together with him.

**-sıra** *Suffix forming an adverb of place or time.* **ardısıra,** after him; one after the other: **önüsıra,** in front of him.

**sıraca** Scrofula. **~ otu,** mullein; (?) scrophularia. **~lı,** scrofulous.

**sıra·dağ** Mountain chain. **~lamak,** to arrange in a row; set up in order; enumerate a series (of complaints *etc.*); (of a child) to begin to walk holding on to one thing after another. **~lanmak,** to queue up. **~lı,** in a row; in due order; at the right moment: **~ sırasız,** in and out of season, at all sorts of times. **~sız,** out of order; ill-timed; improper.

**sırat** (·ˈ) *In* **~köprüsü, -nü,** a bridge on the road to heaven *very narrow and difficult to pass*; a steep and dangerous road.

**sıravari** (··–ˈ) In a line *or* row.

**Sırbistan, Sırbiya** Serbia.

**sırça** Glass; paste (false diamond); spun glass; glass bead.

**sırdaş** Confidant, intimate. **~lık,** intimate friendship.

**sırf** Pure; mere; sheer.

**sırık** Pole; stick *for climbing plants.* **~la atlama,** pole-jump: **~ gibi boy büyütmek,** to grow in size but not in sense: **~ gibi durmak,** to stand aside and do nothing.

**sırıklamak** *va.* Carry off, steal.

**sırım** Leathern thong; strap. **~ gibi,** wiry (person).

**sırıt·kan** Given to grinning. **~mak,** to grin; (of a defect) to show up; be a fiasco. **sırıta kalmak,** to remain grinning like a dead person.

**sır·lamak** *va.* Glaze (pottery); silver (mirror). **~lı,** glazed; silvered (mirror).

**sırma** Lace *or* embroidery of silver *or* silver-gilt thread. Golden (hair *etc.*). **sarı ~,** gold-thread; gold-lace. **~lı,** embroidered with gold *or* silver thread. **~keş,** maker of gold *or* silver thread; embroiderer in same.

**sırnaş·ık** Tiresome; pertinacious, importunate. **~mak,** to worry, annoy.

**Sırp, ~lı** Serb, Serbian. **~ça,** Serbian (language).

**sırr·a, ~ı** *v.* sır. **~en** (ˈ·), secretly. **~olmak,** to disappear.

**sırsıklam** (ˈ··) Very wet; wet to the skin.

**sırt, -tı** Back; ridge. **~a almak,** to shoulder; to undertake: **~ından* atmak,** to get rid of, free oneself from: **~ından* çıkarmak,** to get *stg.* at another's expense: **···in ~ından geçinmek,** to live at *s.o.'s* expense: **~ı* kaşınıyor,** he's itching for a beating: **···in ~ını yere getirmek,** to overcome, get the better of: **bıçak ~ı kadar,** very little difference: ⌜**karnı tok ~ı pek**⌝, well fed and well clad, in easy circumstances.

**sırtar** A kind of lizard *with thick skin on the back*. **~balığı, -nı,** freshwater bream.
**sırtarmak** *vn.* Grin.
**sırtlamak** *va.* Take on one's back; back, support.
**sırtlan** Hyæna.
**sıska** Dropsical; rickety; thin and weak, puny. **~lık,** dropsy; rickets.
**sıtkı** *v.* sıdık.
**sıtma** Fever; malaria. **~ görmemiş ses,** a rich deep voice: **gizli ~,** masked *or* intermittent fever; underhand intriguer. **~lı,** malarial.
**sıva** Plaster (*building*). **~ harcı,** stucco. **~cı,** plasterer: **~ kuşu,** nuthatch. **~lı¹,** plastered, stuccoed. **~lı²,** with sleeves rolled up. **~ma,** laid on *like plaster*; covered with; washed over with: **~ kel,** bald all over. **~mak¹,** to plaster; daub; cover over with; soil. **~mak²,** *v.* sığamak. **~nmak¹,** *pass. of* sıvamak¹: to be plastered with. **~nmak²,** to roll up *one's trousers or sleeves*; to get to work.
**sıvaşmak** *vn.* Become sticky; adhere, stick to; be dirtied with some sticky thing.
**sıvazlamak** *va.* Stroke; caress.
**sıvı** Liquid.
**sıvık, sıvış** Semi-fluid; sticky; bedaubed.
**sıvışık, sıvışkan** Sticky; importunate, boring (person). **~lık,** stickiness; importunity.
**sıvışmak** *v.* sıvaşmak *or* sivişmek.
**sıyanet** (·–¹), **-ti** Preservation.
**sıyga** Mood, tense. **~ya çekmek,** to cross-examine *s.o.*
**sıyır·mak** *va.* Tear *or* peel off; strip off; skim off; graze; draw *a sword etc.*; polish off (finish up). **···den yakasını* ~,** to get out of (a difficulty *etc.*); to escape from … .
**sıyrık** Peeled; skinned; abraded; brazen-faced. Abrasion.
**sıyr·ılmak** *vn. pass. of* sıyırmak. Be skinned *or* peeled; be scraped *or* rubbed off; slip off, sneak away; get out of a difficulty. **···den ~,** to be stripped of; to get rid of …: **ondan sıdkım sıyrıldı,** I've lost all confidence in him. **~ıntı,** scrapings (from a kitchen utensil *etc.*); peelings; scratch.
**sızdır·mak** *va. caus. of* sızmak. Cause to ooze out; squeeze (money) out of; cause to drop into a drunken sleep. **~maz,** (water-)tight.

**sızı** Ache, pain; grief.
**sızıltı** Lamentation; complaint; discontent. **~ çıkarmak,** to utter murmurings of discontent; to give rise to murmurings of discontent: **~ya meydan vermemek,** not to give anybody cause to complain.
**sızıntı** Oozings, tricklings; an oozing out *of secrets, information etc.*
**sızırmak** *v.* sızdırmak.
**sızla·mak** *vn.* Suffer sharp pain. **burnunun* direği ~,** to suffer acute pain (*also fig.*). **~nmak,** to moan with pain; lament; complain. **~tmak,** to pain; cause to groan *or* lament.
**sızmak** *vn.* Ooze; trickle; leak; (of a secret) to leak out; infiltrate (*mil.*); drop into a drunken slumber. **aralarından su sızmamak,** to be closely tied *by mutual interest etc.*; to be very close friends.
**sicil, -lli** Register; judicial record. **~li ahval,** register of service *etc.* of an employee: **~ etm.** *or* **~le kaydetmek,** to enter into the register of a court of record. **~li,** registered; previously convicted.
**sicim** String; cord. **~ gibi yağmur,** pelting rain.
**sidik** Urine. **~ damlaması,** incontinence of urine: **~ tutulması,** retention of urine: **~ yarışı,** futile rivalry; dispute about trifles, *but* ⌜**sen onunla ~ yarışına çıkamazsın**⌝, it's hopeless for you to think of competing with him: **~ zoru,** difficulty in passing water.
**sidik·kavuğu, -nu,** bladder. **~li,** soiled with urine; suffering from incontinence of urine. **~şekeri, -ni,** diabetes. **~yolu, -nu,** urethra.
**sidre** Lotus tree.
**sif** C.i.f. (*comm.*).
**sifon** Siphon.
**siftah** First stroke of business; first sale of a new commodity. For the first time. **~ etm.,** to do the first stroke of business of the day; to begin. **~lamak,** to make the first sale of the day; to begin.
**siftinmek** *vn.* Wriggle about and scratch oneself; approach a person in a cringing, fawning manner; guzzle.
**siga** *v.* sıyga.
**sigar** Cigar. **~a** (·¹·), cigarette: **~ iskemlesi,** small three-legged table *or* stand. **~alık** (·¹··), cigarette case.
**siğil** Wart; green scum on water.

**sigorta** (·ˋ·) Insurance; insurance company. **~ya yatırmak,** to cover by insurance. **~cı,** insurance agent. **~lı,** insured.

**sihir, -hri** Magic; sorcery, witchcraft; charm; fascination. **~baz,** who practises magic; magician, sorcerer. **~bazlık,** magic, sorcery. **~lemek,** to bewitch, enchant. **~li,** bewitched.

**sihrî** (·ˋ) Magical.

**sikke**[1] Coin.

**sikke**[2] Dervish's cap.

**siklon** Cyclone.

**silâh** Weapon, arm. **~ başı,** call to arms, alarm: **~ başına!,** to arms!: **~a davranmak,** to take up and prepare to use a weapon. **~çı,** armourer. **~endaz,** fusilier; marine. **~hane** (··–ˋ), armoury; arsenal. **~lı,** armed. **~lık,** belt for carrying weapons. **~sız,** unarmed: **~a ayırmak,** to allot to non-combatant duties. **~sızlanma,** disarmament. **~şor,** musketeer; armed guards of a palace. **~şorluk,** knighthood, skill in the use of arms.

**silecek** Large bath-towel.

**silgi** Duster for cleaning a blackboard; sponge for wiping a slate; eraser.

**silici** Professional cleaner *or* polisher; one who planes boards for building.

**silik** Rubbed out; worn; indistinct; insignificant; second-rate.

**silindir** Cylinder; roller. **~ şapka,** top-hat: **el ~i,** garden roller.

**silinmek** *vn. pass. of* **silmek.** Be scraped *or* rubbed down *for polishing.*

**silinti** Wipings; anything wiped off.

**silk·elemek** *va.* Shake off (dust *etc.*). **~inmek,** to shake oneself; shake off the effects of *stg.*; shake oneself free. **~inti,** shake; a shaking *or* trembling; anything shaken off.

**silkme** *Dish of finely chopped aubergines or gourds and meat.*

**silkmek** *va.* Shake; shake off; pounce (make a copy of a drawing *etc.* by perforating holes in it and shaking through a coloured powder). **toz ~,** to shake off dust, to beat: **yaka ~,** to shake the collar *as a sign of disgust.*

**sille** Box on the ear; slap.

**silme** *verb. noun of* **silmek.** Wiped; scrubbed; planed; shaven; levelled to the brim (of a measure of corn). **~ tahtası,** board for levelling off a measure of grain.

**silmek** *va.* Wipe; scrub; plane; rub down; polish; erase; remove the excess of anything (skim foam off beer, level off a heap of grain *and so forth*). **silip süpürmek,** to make a clean sweep of: **burnunu* ~,** to blow the nose.

**silsile** Chain; line, series; pedigree; dynasty; a chain of promotions through seniority. **~i meratib,** a hierarchy. **~name** (···–ˋ), pedigree, genealogical tree.

**silyon** *In* **~ feneri,** navigation light on mast.

**sim** Silver. **~ şerid,** imitation silver braid.

**sima** (–ˋ) Face, features; figure, personage.

**simit** Roll of bread in the shape of a ring; lifebuoy.

**simsar** (·ˋ) Broker; middleman; commission agent. **~lık,** profession of broker.

**simsiyah** (ˋ··) Jet black.

**simya** Alchemy.

**sinameki** (·–·ˋ) Senna.

**sinare** (·ˋ·) Large fish-hook.

**sinarit, -ti, sinagrid** A kind of sea-perch (*Dentex vulgaris*).

**sincab** (·ˋ) Grey squirrel; fur of grey squirrel. **~i** (·–ˋ), of grey squirrel fur; dark grey.

**sinderus** Gum sandarac.

**sindirmek** *va. caus. of* **sinmek.** Digest; swallow; assimilate; terrify, cow. **sindire sindire,** permeating; very thoroughly: **içine* sindirmemek,** to spoil one's pleasure, *v.* **sinmek.**

**sine**[1] (–ˋ) Bosom, breast; projection. **~ye çekmek,** to put up with, to resign oneself to.

**sine**[2], **sinece** *gerund of* **sinmek. ~ ~,** so as to sink into; **~ ~ yağmur yağmak,** to rain in a soaking manner.

**sinek** Fly. **~ avlamak,** to potter about, idle: **~ kaydı tıraş,** a very smooth shave. **~kâğıdı** (·ˋ···), **-nı,** fly-paper. **~lik,** fly-whisk.

**sinema** Cinema. **~sı oynanmak,** (of a book *or* play) to be filmed.

**sini** Round metal tray, *used as a small table.*

**sinir** Sinew; nerve; fibre; rib (*bot.*). **~leri* ayakta olmak,** for one's nerves to be on edge. **~ hastalığı,** neuralgia: **~ tutulması,** cramp. **~lemek,** to free from sinews, to hamstring. **~lenmek,** to become irritated; to have one's nerves set on edge; to be hamstrung. **~li,** on edge, irritable;

sinewy, wiry. **~lilik,** a state of nerves, irritability; wiriness.

**sinmek** *vn.* Be absorbed, swallowed, digested; sink into the ground; penetrate; be hidden; crouch down *to hide oneself*; 'sing small', be humiliated; be cowed. **içime* sinmiyor,** my pleasure is spoilt.

**sinmez** Indigestible.

**sinsi** Stealthy; slinking; sneaking; insidious. **~lik,** stealthiness, subtlety; underhand dealing.

**sintine** (·꜠·) Bilge *of a ship.*

**sinyal** Signal.

**sipahi** (·—꜠) (*Formerly*) cavalry soldier.

**sipariş** (·—꜠) Order (*comm.*); commission; allotment of pay *made by a soldier etc. to relatives.*

**siper** Shield; shelter; trench; rampart; peak of a cap; anything acting as a protection *or* guard; top-slide (lathe). ~ **almak,** to parry a blow; to take shelter behind *stg.*; **~e almak,** to take under one's protection: ~ **olm.,** to shield with one's own body: **elini** ~ **etm.,** to shade the eyes with one's hand.

**sipsi** Boatswain's pipe; reed *of a clarinet etc.*

**sipsivri** (꜠··) *In* ~ **kalmak,** to be suddenly deserted by everyone; to be destitute: ~ **çıkagelmek,** to appear unexpectedly (when not wanted).

**sirayet** (·—꜠), **-ti** A spreading *or* propagating itself; contagion, infection. ~ **etm.,** (of a disease) to spread.

**siret** (—꜠), **-ti** Moral quality; conduct; character. **~i suretine uymaz,** whose appearance belies his character. *v.* **siyer.**

**sirk, -ki** Circus.

**sirke**[1] Vinegar. ⌜**bedava** ~ **baldan tatlıdır**⌝, a gift *or* an undeserved honour will often work wonders: ⌜**keskin** ~ **küpüne zarar verir**⌝, sour vinegar harms its jar, *said of one whose bad temper does him harm*: **talaş ~si,** wood-vinegar (acetic acid).

**sirke**[2] Nit.

**sirkengebin, sirkencebin** (···꜡) Drink made of vinegar and honey; oxymel.

**sirküler** Circular.

**siroko** (·꜠·) Sirocco.

**sirto** (꜠·) A kind of dance.

**sis** Fog, mist. **~lenmek,** to become damp *or* foggy; (of a glass *etc.*) to be covered with dew. **~li,** foggy, misty, hazy.

**Sisam** Samos.

**sistire** (·꜠·) Instrument for scraping the dough off a kneading-trough; carpenter's scraper.

**sitayiş** (·—꜠) Eulogy. ~ **etm.,** to eulogize.

**sitem** Reproach. **~dide** (··—꜠), unjustly treated; ill-treated. **~kâr,** reproachful.

**sitteisevir** 'The six of the Ox'; (*six days in April when the sun is in Taurus, reputed to be a time of bad weather*).

**sittin** Sixty. **~sene** (·꜠··), for a very long time.

**Sivas** Sivas. ⌜**göründü ~ın bağları**⌝, 'the vineyards of Sivas appeared', an unexpected and unpleasant situation.

**sivil** Civilian; in mufti; plain-clothes policeman; (*sl.*) naked.

**sivilce** Pimple.

**sivişik** *v.* **sıvışık.**

**sivişmek** *vn.* Disappear, decamp.

**sivri** Sharp-pointed; tapering; tall and slim. ~ **akıllı,** eccentric, odd and self-opinionated: ~ **kafalı,** obstinate. **~burun,** with a pointed nose. **~lmek,** to become pointed *or* prominent; to make rapid progress in one's career. **~ltmek ~tmek,** to make pointed *or* sharp at the end. **~sinek,** mosquito: ⌜**anlayana** ~ **saz, anlamayana davul zurna az**⌝, a whisper is enough for a wise man, a shout won't make a fool understand.

**siya** (꜠·) Reversing oars and rowing backwards. **~!,** back oars!

**siyah** Black. **~i** (·—꜡), negro. **~lık,** blackness; a figure in the dark.

**siyaset** (·—꜠), **-ti** Politics; policy; diplomacy. **~en** (·꜡··), politically; diplomatically.

**siyasi** (·—꜡) Political; diplomatic. **~yat** (·—·꜠), **-tı,** diplomacy; politics.

**siymek** *vn.* (Of a dog) to urinate. *v.* **cami.**

**siz** *pron. 2nd pers. pl.* You (*also used politely for sing.*).

**-siz, -sız, -suz, -süz** *suffix.* Without ...; ...less. **sensiz,** without you; **parasız,** without money: **tüysüz,** hairless: **susuz,** without water.

**-sizin, -sızın** *suffix used with the infinitive.* Without ...; before ... . **görmeksizin,** without seeing: **yazmaksızın,** without *or* before writing.

**sizinki** Yours.

**-sizlik** *Suffix forming a noun of the adjective formed with* **-siz** (*with due alteration of vowels for euphony*): **parasızlık,** lack of money: **susuzluk,** thirst, aridity.
**smokin** Dinner-jacket.
**soba** (ˈ·) Stove; hothouse. **~cı,** maker *or* installer of stoves.
**sodyum** (ˈ·) Sodium.
**sof** Wool; cloth made from the hair of goats, camels *etc.*; camlet; mohair; alpaca.
**sofa** (ˈ·) Hall; ante-room; stone bench; sofa.
**sofî** Sufi, mystic; devotee.
**sofra** Dining-table; meal. **~ başına geçmek,** to sit down to a meal: **~ bezi,** tablecloth: **~yı kaldırmak,** to clear away (after a meal): **~ kurmak,** to lay the table: **~ takımı,** a table service: **yer ~sı,** tray laid on a low stool to serve as table. **~cı,** butler; parlourman.
**softa** Moslem theological student; bigot, fanatic. Behind the times, old-fashioned.
**sofu** Religious, devout; fanatic. **kaba** (**ham**) **~,** an intolerant bigot. **~luk,** religious devotion.
**soğan** Onion; bulb. **kaba ~,** a 'rotter', a coward: **soyup ~a çevirmek,** to pillage, to strip. **~cık,** pickling-onion; shallot. **~lı,** prepared with onions. **~lık,** onion bed *or* garden; place for storing onions. **~zarı** (·ˈ··), **-nı,** onion-skin.
**soğuk** Cold; frigid, unfriendly; out-of-place, in bad taste. Cold weather. **~lar,** the cold weather, winter: **~ algınlığı,** a cold, chill: **~ almak,** to catch cold: **~ damga,** embossed stamp: **~ davranmak,** to behave coldly: **~ durmak,** to look on coldly: **~ kaçmak** (**düşmek**), (of a deed *or* word) to be out-of-place, in bad taste: **~ söz,** an unfriendly word. **soğuk·bez,** cotton cloth, jaconet. **~ça,** somewhat cold. **~kanlı** (·ˈ··), calm, cool-headed; cold, unsympathetic; (*not* cold-blooded =**merhametsiz** *or* **hissiz**). **~lamak,** to catch a chill. **~luk,** cold, coldness; chilliness of manner; cooling room in a hammam; cold sweet: **~ etm.,** to act stupidly *or* in bad taste.
**soğumak** *vn.* Become cold; catch cold; be chilly in one's relations; cease to care (for = **den**). **araları ~,** for the relations between two people to be strained: **arası ~,** for an interval to pass and interest in *stg.* to be lost, *or* it forgotten: **···den buz gibi ~,** to take a great dislike to ... .
**soğutmak** *va. caus. of* **soğumak.** Render cold; cool; alienate. **bir işin arasını ~,** to neglect an affair: **iki kişinin arasını ~,** to sow discord between two people: **ziyaretlerin arasını ~,** to curtail their visits to one another.
**sohbet, -ti** Friendly intercourse; chat, conversation. **~ etm.,** to have a chat: **helva ~i,** an evening party where the guests make and eat **helva** together.
**sok** (*Fr. soc*) Ploughshare. **üç ~lu pulluk,** three-furrow plough.
**sokak** Road, street; *often means* 'not in the house, outside', *e.g.* **sokağa çıktı,** he went out; he is not at home: **yemeği ~ta yemek,** to eat out (at a restaurant *etc.*): **~ süpürgesi,** a loose woman: ⌜**alt yanı** (**tarafı**) **çıkmaz ~**⌝, 'this leads nowhere' (of a discussion *etc.*).
**sokma** *verb. noun of* **sokmak.** Introduced from outside; imported.
**sokmak** *va.* Thrust into, insert; introduce; involve, entail; drive into; let in; (of an insect *or* snake) to sting *or* bite; injure; calumniate. **başını* sokacak bir yer,** a dwelling however modest: **···e burnunu* ~,** to poke one's nose into ...: **canına*** (*or* **canının* içine**) **sokacağı* gelmek,** to be extremely fond of, to adore: **birinin gözüne ~,** to show *a thing* in a rude *or* reproachful manner: **birini işe ~,** to put *s.o.* into a job: **bir işe parmak ~,** to meddle with a matter.
**sokra** Butt end *of a plank joining it to another.*
**sokulgan** Sociable, quick to make friends.
**sokulmak** *vn. pass. of* **sokmak. ···e ~,** to insinuate oneself into; push into; cultivate friendly relations with.
**sokum** Act of inserting; place where a thing is inserted; a kind of cheese sandwich made with **yufka** *q.v.* **kuyruk ~u,** where the tail of an animal joins the body; coccyx.
**sokur** Mole. Blind; one-eyed.
**sokuş·mak** *vn.* Push oneself gently *into a place or amongst others.* **~turmak,** to push gently *or* secretly *into stg.* **···in eline ~,** to slip (a tip *etc.*) into *s.o.'s* hand: **araya ~,** to try to

find time for *some job* amid other occupations.
**sol** Left. Left-hand side. **~dan geri dönmek,** to turn left about, to retire: **~ tarafından kalmak,** to get out of bed the wrong side. **~ak,** left-handed. **~cu,** left-winger (politician). **~iç** (ˈ·), inside left (football *etc.*).
**solgun** Pale; faded; withered.
**solmak** *vn.* Fade; wither; become pale.
**solucan** Worm; ascaris, round worm. **~ gibi,** pale and thin; unpleasant (person).
**solugan** Short of breath; asthmatic. Shortness of breath; asthma.
**soluk**[1] Faded; withered; pale.
**soluk**[2] Breath; breathing; panting. **~ soluğa,** panting, out-of-breath: **~ almak,** to take breath, to recover oneself; **derhal soluğu telefonda almıştı,** he at once hurried off to the telephone; **soluğu Bagdadda almak,** 'to hasten to Baghdad', to fly the country, to escape: **soluğu dar almak,** hardly to be able to breathe; to escape narrowly: **~ aldırmak,** to give one time to take breath, to give a respite: **bir ~ta,** in a flash. **soluk·lanmak,** to take a long and easy breath. **~suz,** without breath; without respite.
**solumak** *vn.* Breathe heavily; pant. **burundan ~,** to snort with anger; to pant heavily.
**som**[1] Solid, not hollow; massive.
**som**[2] Ivory from a fish's tooth, rhinoceros horn *etc.*
**som**[3] **~ balığı,** salmon.
**somak** Sumach.
**somaki** Porphyry.
**somun** Loaf; nut (female screw).
**somur·mak, ~tmak** *vn.* Pout; frown; sulk. **~tkan,** sulky.
**somye** (ˈ·) (*Fr. sommier*) Spring mattress.
**son** End; result; afterbirth. Last; latter; final. **~unda,** in the end, finally: **~dan bir evvelki (ikinci),** penultimate, last but one. **~ defa,** the last time: **~ derece,** (to) the last degree; the uttermost: **~unu düşünmek,** to think of the result (of an action): **~unu getirememek,** to fail to achieve *stg.*: ⌜**~a kalan dona kalır**⌝, 'the Devil take the hindmost': **eni ~u,** the ultimate end, the long and the short of it: **bunun ~u yoktur,** no good will come of this.
**sonbahar** (ˈ··) Autumn.
**sonda** (ˈ·) Probe; bore; catheter.
**sondaj** (*Fr. sondage*) Sounding.
**sondurmak** *va.* Stretch.
**sonek** Suffix.
**sonra** In future; afterwards; by and by. The future; a later time; consequence. **···dan ~,** after ...; **bundan ~,** after this; **ondan ~,** after him (it), after that; then: **~dan, ~ları,** later; recently: **~sı,** its result, its sequel: **~ya bırakmak,** to put off till another time: **~dan görme,** a parvenu, an upstart: **~dan görmemiş,** who is not a parvenu *or* upstart: **~dan olma,** comparatively recent: **en ~,** at the very end, last of all: **aklı* ~dan gelmek,** to think of *stg.* too late. **sonra·cık,** *only in* **ondan sonracığıma,** after that. **~ki** (ˈ··), who comes later; that happens later: **~ler,** successors; posterity; those who come later. **~sız,** eternal.
**sonsuz** Endless; eternal; infinite; useless, without results.
**sontraş** Farrier's clippers for paring hoofs.
**sonuç** End; result.
**sonuncu** Last; final.
**sop, -pu** *Only in* **soy ~,** family and relations, *v.* **soy.**
**sopa** (ˈ·) Thick stick, cudgel; beating; stripe *in cloth.* **···e ~ atmak,** to give a beating to: **~ yemek,** to get a beating. **~lı,** armed with a stick; having broad stripes (cloth *etc.*): **eli ~,** a man of violence, bully.
**sorgu** Question; interrogation. **~ hâkimi,** interrogating magistrate ('juge d'instruction'): **···i ~ya çekmek,** to cross-examine.
**sorguç** Plume; crest; aigrette.
**sorgun** Ben-tree; moringa.
**sormak**[1] *va.* Ask; inquire; inquire about. **sora sora,** by dint of repeated questions: **···i** *or* **···in hatırını ~,** to inquire about a person's health, to ask news of *s.o.*: ⌜**ne sen sor ne de ben söyleyim**⌝, 'I think we had better not go into that'; ⌜ask no questions and you'll be told no lies⌝.
**sormak**[2] *va.* Suck.
**sorti** (*Fr. sortie*) Point (*elec.*).
**soru** Question; interrogation. **~ günü,** the Day of Judgement.
**sorulmak** *vn. pass. of* **sormak. soruluyor,** it is asked: **benden sorulmaz,** I am not responsible.
**sorumak** *va.* Suck noisily.

**sorumlu** Responsible.

**soruş·mak** *vn.* Question one another. **~turmak,** to make investigations: **birine bir şeyi ~,** to ask *s.o.* about *stg.*

**sorut·kan** Sulky. **~mak,** to be cross, sulky.

**soy** Family; race; lineage; ancestors; descendants; kind, sort. Pure-blooded; noble. **~a çekmek,** to take after one's family: ⌜**~dur** (*or* **~ ~a**) **çeker**⌝, heredity is strong: **~ sop** (*acc.* **soyu sopu**), one's family and relations; **~u sopu belli,** he comes of a good family. **~adı** (ˈ··), **-nı,** family name; surname: **~ kanunu,** the law requiring every Turk to have a surname. **~ca** (ˈ·) as a family; as regards family. **~lu,** pure-bred; of good family.

**soygun** Pillage, spoliation. Undressed; stripped; robbed. **~ vermek,** to be plundered. **~cu,** plunderer, pillager. **~culuk,** plundering; fleecing.

**soymak** *va.* Strip; undress; peel; flay; rob; sack.

**soysop** *v.* **soy.**

**soysuz** Of bad race; degenerate; good-for-nothing. **~luk,** degeneracy; worthlessness. **~laşmak,** to degenerate.

**soytarı** Clown, buffoon. **~lık,** buffoonery.

**soyucu** Footpad; brigand.

**soyun·mak** *vn.* Undress oneself; change one's clothes; take off clothes (in order to work). **soyunup dökünmek,** to change into comfortable clothes. **~tu,** peel; bark; anything stripped off.

**söbü** Oval; conical; cylindrical.

**söğmek** *v.* **sövmek.**

**söğüş** Boiled meat; cold meat.

**söğüt** Willow. **~ yaprağı,** willow leaf; a very thin kind of dagger.

**sökmek** *va.* Pull up; tear down; rip open; dismantle; undo; break through (an obstacle), surmount (a difficulty); break up (land); decipher. *vn.* Succeed; (of a purge) to take effect; appear, come out; (of mucus) to flow; (of dawn) to break. **bu böyle sökmez,** you won't get any further like this; this won't do: **burada zorbalık sökmez,** if you think you can succeed by force, you're wrong.

**sökük** Unstitched. Dropped stitch *in knitting*; rent. **~ örmek,** to repair a rent.

**sökülmek** *vn. pass. of* **sökmek.** (*sl.*) be forced to give *or* pay. **sökülüp atılmak,** to be utterly eradicated.

**sökün etm.** *vn.* Appear suddenly; crop up; come one after the other.

**söküntü** Rent in a seam; place where knitting has unravelled; sudden rush of a crowd.

**söküotu** (·ˈ··), **-nu** Bird's-foot trefoil.

**sölpü·k** Lax; flabby. **~mek,** to hang flabbily; to be flabby *or* sluggish.

**sömürge** Colony.

**sömürmek** *va.* Gobble down; devour; nose about and eat (as a cow).

**söndürmek** *va.* Extinguish; slake; deflate.

**sönmek** *vn.* Be extinguished, go out (of a fire); (of a sail) to become slack, to flap; be deflated.

**sönük** Extinguished; dim; tarnished; slack (sail); washed out; deflated; obscure, undistinguished.

**sör** (*Fr. sœur*) Nun, sister *in a hospital.*

**söve** Door *or* window frame; posts on a wagon for holding up the load.

**söv·mek** *vn.* Curse and swear (at = *dat.*). **···e sövüp saymak,** to swear at *s.o.* **~üş,** cursing, vituperation. **~üşmek,** to swear at one another.

**söylemek** *va. & vn.* Speak; say; tell; explain. **arkasından ~ gibi olmasın amma ...,** I don't like to say it behind his back, but ...: **benden söylemesi,** well, I've warned you!, *or* I felt I ought to say *stg.* (but I needn't go any further): **büyük ~,** to boast: **şarkı ~,** to sing a song: **yalan ~,** to lie.

**söyle·nilmek,** to be said; to be pronounced. **~nmek,** to be spoken or said; to speak to oneself; mutter; grumble: **söylendiğine göre,** according to what is said. **~nti,** rumour. **~şmek,** to talk over, discuss *stg.*; to converse; to consult with one another. **~tmek,** *va. caus. of* **söylemek**; ⌜**beş para ver söylet on para ver sustur**⌝, 'if you pay him five paras to talk you will give him ten to shut up' (*said of a bore*). **~yiş,** manner of speaking: **~ine göre,** according to what he said; from his manner of speaking.

**söz** Word; speech; talk; rumour; gossip; promise; agreement. **~de,** in word only; so-called; as though, as if; supposing that: ~ **açmak,** to start a conversation about *stg.*; ~ ~**ü açar,** one topic leads to another: ~ **altında kalmamak,** not to remain silent when attacked *or* insulted: ~ **anlamak,** *v.* **anlamak**: ~ **anlatmak,** to persuade: ~ **aramızda,** between ourselves: ~ **arasında,** in the course of conversation; by the way: ~**ünü bilmek,** to be tactful and considerate in one's talk: ⌜~ **bir Allah bir**⌝, I am a man of my word: ···**in** ~**ünden çıkmak,** to disregard the advice of ... : ~**ünü*** **değiştirmek,** to change one's tone: ~**ünden*** **dönmek,** to go back on one's word: ~**ünde*** **durmak,** to keep one's word: ~ **ebesi,** quick at repartee; a good talker: ~**ünün eri,** a man of his word: ~ **etm.,** to talk, gossip *unfavourably about s.o.*: ~**ü geçen,** aforesaid; ~**ü geçer,** what he says 'goes', influential;···**in** ~**ü geçti,** mention was made of ... : ~ **geçirmek,** to make *s.o.* listen to one, to make one's influence felt: ~ **gelişi,** for example: ~**ün gelişi,** in the course of conversation: ···**e** ~ **gelmek,** to be gossiped about: ~ **götürmez,** it is beyond question: ~ **işitmek,** to 'be told off': ~ **kaldırmak,** not to take offence at a joke *or* a contradiction; ~ **kaldırmaz,** who cannot stand being contradicted; who cannot take a joke against himself: ~ **kesmek,** to decide *or* agree; to conclude a marriage agreement: ~**ünü*** **kesmek,** to cut *s.o.* short, to interrupt him: ~ **kesimi,** agreement to marry, engagement: ~ **olsun diye,** just for *stg.* to say; without meaning it: ~ **olur,** people will talk about it: ~**üm ona,** so-called, alleged: ~ **onun,** what he says is right; it is his turn to speak: ~ **sahibi,** a master of words, eloquent; who has a say *in a matter*: ~ **ünü*** **tutmak,** to keep one's word; ···**in** ~**ünü tutmak,** to take the advice of, to obey: ~ **vermek,** to give one's word, to promise: ~ **yok,** there's nothing to say to that; quite true!: ~**üm yabana,** *as* ~**üm ona**; *also* 'pardon the expression!'; *v. also* **yaban.**

**söz·başı, -nı,** heading *of a chapter etc.* ~**birliği, -ni,** unanimity, agreement. ~**cü,** spokesman. ~**leşmek,** to agree together. ~**lü,** agreed together; having promised; engaged to be married; in words, verbal, oral. ~**lük,** dictionary.

**spiker** Announcer (radio).

**spor** Sports; games; (*not quite the English* 'sport'). ~**cu,** ~**tmen,** athlete, games player, sportsman.

**staj** (*Fr. stage*) Apprenticeship; course of instruction; probation. ~**yer** (*Fr. stagiaire*), apprentice; probationer.

**stepne** (Stepney) Spare wheel *for car.*

**stok, -ku** Stock.

**stor** (*Fr. store*) Blind.

**su, -yu** Water; fluid; stream; sap; broth; temper *of steel*; running pattern *or* decoration. ~**dan,** worthless; insignificant: ~ **almak,** (of a ship) to make water, leak: ~ **başı,** source; spring; fountain; waterside; *v. also* **subaşı**; ~**yun başı,** source; centre; most important part *of a business, etc.*; ~**yu başından kesmek,** to cut off at its source; to nip in the bud: ~ **bendi,** reservoir: ~ **böreği,** a kind of pastry with flaked crust: ~**yunu çekmek,** to be exhausted, to be used up; **para** ~**yunu çekti,** the money is exhausted: **bu şehrin** ~**yu mu çıktı?,** what's wrong with this town, why don't you like it?: ~ ~ **dökmek,** to make water, to pump ship: ~**ya düşmek,** to fail, to come to nought: ~ **etm.,** (**yapmak**), (of a ship) to make water, to leak: ~**geçmez** (**geçirmez**), waterproof: ⌜~ **gibi aziz ol!**⌝, *a form of thanks to one who offers water*: ~ **gibi bilmek,** to know perfectly; ~ **gibi okumak,** to read fluently: ···**in** ~**yuna** (~**yunca**) **gitmek,** not to go counter to *s.o.*, to treat with tact, to flatter when necessary: ~**ya göstermek,** to wash *stg.* lightly: ~ **götürür,** that will bear further inspection: ~ **götürmez,** indisputable: ~ **içinde,** easily; certainly: ~ **kaldırmak,** to absorb water: ~**lar kararmak,** for darkness to fall: ~ **katılmamış,** unadulterated: ~**yu** (**tokmakla**) **kesiyor,** (of a knife) it is very blunt: ~**yu sert,** hard-tempered (steel); harsh (man): ~ **vermek,** to temper *steel*: **bir geminin çektiği** ~, the water a ship draws: **saat iki** ~**larında,** about two o'clock: ~**yunun suyu,** a distant relationship; only a remote connexion: (**ağaclara**) ~ **yürümek,** for the sap of trees to begin to rise, for them to begin to shoot.

**sual, -li** Question; inquiry; request. **kabir** (**ahret**) ~**i,** endless interrogation. ~**li,** containing a question: ~

cevablı, in the form of question and answer.
**sualtı** (ˋ··) Underwater.
**subaşı** (ˋ··), **–nı** (*Formerly*) police superintendent; *v. also* **su başı.**
**subay** Officer.
**subye** (ˋ··) (*Fr. sous-pieds*) Trouser-straps.
**sucu** Water-seller.
**sucuk** Sausage; sweetmeat made of grape-juice and nuts. **sucuğunu çıkarmak,** to give a good thrashing to; to tire out *or* exhaust: ~ **gibi ıslanmak,** *or* ~ **olm.** *or* ~ **kesilmek,** to be wet through.
**suç, –çu** Fault; offence; crime; sin. ···**e** ~**atmak,** to attribute an offence to ...: ~ **etm.,** to commit an offence, to sin: **suçundan geçmek,** to overlook *s.o.'s* offence: ···**e** ~ **yükletmek,** to lay the blame on ... .
**suçiçeği** (ˋ···), **–ni** Chicken-pox.
**suç·lamak** *va.* Accuse. ~**landırmak,** to find guilty. ~**lu,** guilty: ···**den** ~ charged with ...: ~ **çıkarmak,** to find guilty: ~ **durmak,** to stand with the air of a guilty person. ~**suz,** not guilty, innocent. ~**üstü** (ˋ··), red-handed, 'in flagrante delicto'.
**sudak** Pike-perch (*Lucioperca*).
**sudan**[1] *abl. of* **su.** Insignificant; worthless.
**Sudan**[2] (ˊ·) Sudan. ~**lı,** Sudanese.
**sudolabı** (ˋ···), **–nı** Wheel for raising water.
**sudur** (·ˊ) Emanation; issuing (of a decree *etc.*). ~ **etm.,** to be issued; to take place.
**sufî** (–ˊ) Moslem mystic.
**suflör** (*Fr. souffler*) Prompter (theatre).
**sugeçirmez** Waterproof.
**sui–** (–·) *prefix.* Evil-. ~**istimal** (–··–ˋ), abuse; misuse: ~ **etm.,** to misuse, abuse: **emniyeti** ~, breach of confidence. ~**kasd,** criminal attempt; malice aforethought. ~**kasdcı,** one who makes an attempt upon life; conspirator. ~**niyet, –ti,** evil intention. ~**tefehhüm,** misunderstanding. ~**tefsir,** misinterpretation. ~**zan, –nnı,** suspicion, distrust.
**su·kabağı,** (ˋ···) **–nı** Gourd *used for holding water.* ~**kamışı, –nı,** bulrush.
**sukbe** Small hole; puncture.
**su·kerevizi** (ˋ····), **–ni** Water-hemlock. ~**kesimi, –ni,** waterline *of a ship.* ~**kuşu, –nu,** moorhen.
**sukut** (·ˋ) **–tu** Fall; lapse; abortion. **hayal** ~**u,** disappointment.
**sulak** Watery; marshy. Water-trough; water-bowl.
**sula·mak** *va.* Water; dilute with water; irrigate; pay cash, pay in advance. ~**ndırmak,** *caus. of* **sulanmak; ağzını*** ~, to make one's mouth water. ~**nmak,** to become wet *or* watery; be provided with water; be irrigated; flirt; become silly *or* too familiar: ···**e** ~, to manifest interest in *stg.* one hopes to get; to envy: **ağzım sulandı,** my mouth watered: **beyni*** ~, to have a softening of the brain, to become senile: **kanı*** ~, to be anaemic.
**sulb**[1], **–bü** The loins; vertebral column; offspring of one's loins, descendants.
**sulb**[2] Hard; solid.
**suleha** (··ˊ) *pl. of* **salih.** Pious people.
**sulh** Peace; reconciliation; accord. ~ **hakimi,** justice of the peace, police-court magistrate: ~ **mahkemesi,** minor court for petty offences (= *magistrate's police-court*): ~ **olm.,** to come to an amicable agreement: ~**a sübhana yatmak,** to become docile *or* amenable.
**sulh·cu,** peace-loving; pacifist. ~**en** (ˋ·), peaceably.
**sulta** Sovereignty; power; authority.
**sultan** Sultan; Sultan's daughter, Sultana. ~**î** (·–ˊ), pertaining to a sultan; imperial; (*formerly*) secondary school; a kind of pipless grape. ~**lık,** sultanate: great happiness.
**sulu** Watery; moist; juicy; silly; importunate; too familiar. ~ **boya,** water-colour (paint): ~ **sepken,** sleet. ~**ca,** somewhat watery; rather too familiar. ~**luk,** a being importunate, silly, too familiar.
**suluk** Skin disease affecting a baby's head; aigrette; water-bowl *in a bird-cage.* ~ **zinciri,** curb-chain.
**sumak** Sumach.
**sumen** (*Fr. sous-main*) Writing-pad; blotting-pad.
**sumercimeği** (ˋ····), **–ni** Duckweed.
**sundurma** Open shed; lean-to roof.
**sungur** Falcon.
**sun'î** (·ˊ) Artificial; false; affected.
**sunmak** *va.* Put forward; offer; present.
**sunturlu** Severe (scolding); resounding (oath).
**sunu** (·ˊ), **–n'u** Act; creation. **bunda**

ne ~m var?, what part have I played in this?, how am I to blame?
**supap, -pı** (*Fr. soupape*) Valve.
**sûr** City wall, rampart. ~ **dahilinde,** within the city's walls.
**sur** (–) Trumpet, *esp.* the trumpet of the Day of Judgement.
**sura** A kind of soft silk fabric.
**surat, -tı** Face; mien; sour face. ~ **asmak,** to frown, to make a sour face: ⌈**~a bak süngüye davran**⌉, 'look at his face and get your bayonet ready', *said when speaking of a very ugly or disagreeable looking man*: ⌈**~ından düşen bin parça olur**⌉, very bad-tempered: ~ **düşkünü,** very ugly: ~ **etm.,** to look sulky: **yüz ~ davul derisi (hak getire, mahkeme duvarı),** brazen-faced, shameless.
**surat·lı** Sulky. **~sız,** sulky; ugly.
**sure** (–·) Chapter of the Koran.
**suret** (–·), **-ti** Form, shape; aspect; manner; picture; copy; case; supposition. ~ **çıkarmak,** to take a likeness; to make a copy: **~ine girmek,** to assume the form of: **~i haktan görünmek,** to appear sincere: **bir ~le,** in some way or other; to such a degree that: **bu ~le,** in this way: **hüsnü ~le,** in a proper manner.
**sureta** (–··–) Outwardly, in appearance; simulated; as a matter of form.
**surî** (––) Apparent; feigned.
**Suriye** (–··) Syria. **~li,** Syrian.
**susam** Sesame; (*for* **süsen**) iris.
**susa·mak** *vn.* Be thirsty. **···e ~,** to thirst for, long for: **canına (ölümüne) ~,** to be extremely foolhardy; to court death: **···in canına ~,** to act without pity towards: **dayağa ~,** to 'ask for' a beating. **~mış,** thirsty; longing for (*dat.*).
**susamuru** (·...), **-nu** Otter.
**susa·nmak** *impers. of* **susamak. ~tmak** *caus. of* **susamak,** to make thirsty, to cause to long for.
**susığırı** (·...), **-nı** Water-buffalo.
**susmak** *vn.* Be silent; cease speaking.
**suspus** Reduced to silence; silent and cowering.
**susta**[1] (··) Safety-catch. **~lı çakı,** clasp-knife.
**susta**[2] (··) *In* ~ **durmak,** (of a dog *etc.*) to stand on its hind legs; to be very obsequious.
**susuz** Waterless; arid; thirsty. ⌈**birini suya götürüp ~ getirmek**⌉, to out-do *s.o.*, to make short work of *s.o.*
**sutaş** (*Fr. soutache*) Braid. **~lı,** braided.
**suteresi** (·...), **-ni** Watercress.
**sutyen** (*Fr. soustien*) Brassière.
**suüstü** (·..) Above-water. **bir denizaltının ~ sürati,** a submarine's surface speed.
**suvare** (*Fr. soirée*) Evening party.
**suvar·ım** Manner of watering; amount of water given at one irrigation. **~mak,** to water (an animal).
**suyol·cu** (·..) Man responsible for upkeep and repair of water conduits. **~u, -nu,** water conduit; aqueduct; urinary passage; watermark *in paper*.
**suyunca** *v.* **su.**
**suz·inâk** (–·–), **~nâk** *Name of a sad mode of music.* **~iş** (–·), a burning; suffering; great sorrow.
**sübek** Urinal attached to a baby's cradle.
**sübhan** (·–) *In* ⌈**sulha ~a yatmaz**⌉, 'he is adamant, uncompromising'. **~allah,** *lit.* 'glory be to God'; oh, my God!
**sübut** (·–), **-tu** A being proved. ~ **bulmak,** to be proved; ~ **delili,** certain proof.
**sübye**[1] (··) Emulsion; sweet drink *made with pounded almonds and melon seeds.* **~leştirmek,** to emulsify.
**sübye**[2] (··) Cuttle-fish.
**sübye**[3] *In* ~ **armalı,** fore-and-aft rigged *ship*.
**süfera** (··–) *pl. of* **sefir.** Ambassadors.
**süflî** Low; common, low-down; shabby. **~yet, -ti, ~lik,** lowness; shabbiness.
**sühulet** (·–·), **-ti** Facility, ease.
**sühunet** (·–·), **-ti** Temperature.
**süklüm püklüm** In a crestfallen manner; hanging the head; sheepishly.
**sükûn** (·–) Calm, quiet; repose. **~et** (·–·), **-ti,** quiet, calm; rest.
**sükût** (·–), **-tu** Silence. ~ **etm.,** to be silent. ~ **hakkı,** hush-money. **~î** (·––), taciturn.
**sülale** (·–·) Family; line; descendants.
**Süleyman** Suleyman; Solomon. ~ **Peygamber,** King Solomon: ⌈**mühür kimdeyse ~ odur**⌉, 'who holds the seal is Solomon', *i.e.* he who can prove his authority is the rightful person.
**süline** (·..) Razor-fish, solen.
**sülûk** (·–), **-kü** A following a career; a belonging to a religious order;

contemplative life. ···e ~ **etm.**, to follow the career of … .
**sülüğen** Red lead.
**sülük** Leech; tendril *of a vine etc.*
**sülümen** Corrosive sublimate. **tatlı** ~, calomel.
**sülün** Pheasant. ~ **gibi,** tall and graceful: ~ **Bey,** a brisk, lively little man, Mr. Mouse.
**sülüs, -lsü** A third; a style of Arabic script with large letters. ~**an** (··ʾ) two-thirds.
**sümbük, -kü** Hoof.
**sümbül** Hyacinth. ~**i** (··–́), cloudy, overcast.
**sümkürmek** *vn.* Expel mucus from the nose.
**sümmettedarik** (ʾ··—·) *Said of stg. entered on without preparation*; on the spur of the moment.
**sümsük** Imbecile; uncouth.
**sümük** Mucus, *esp. that of the nose.* ~**lü,** covered with mucus; slimy; snivelling. ~**lüböcek,** slug.
**sümürmek** *v.* **sömürmek.**
**sündüs** Silk brocade.
**sünepe** Slovenly; sluggish.
**sünger** Sponge. ~**le silmek,** to wipe off the slate: **ev** ~**i,** dry rot: ···**in üzerinden** ~ **geçirmek,** to pass the sponge over, to cancel. ~**kâğıdı, -nı,** blotting-paper. ~**taşı, -nı,** pumice-stone.
**süngü** Bayonet; spine. ~**sü ağır,** slow-moving: ~**sü düşük,** subdued, depressed, crestfallen: **ocak** ~**sü,** poker. ~**lemek,** to bayonet.
**sünnet, -ti** Moslem practices and rules *not laid down in the Koran but due to the Prophet's own habits and words*; ritual circumcision. ~ **düğünü,** circumcision feast: ~ **etm.,** to circumcise; to amend: ~ **olm.,** to be circumcised: **ehli** ~, orthodox, Sunni Moslems. ~**ci,** circumciser. ~**leme,** *like Mohammed's beard, i.e.* round and short (beard). ~**lemek** (*sl.*), to eat up entirely, to finish off. ~**li,** circumcised. ~**siz,** uncircumcised.
**sünnî** (·–́) Orthodox; Sunnite.
**sünuh** (·ʾ) A *matter's* occurring to the mind. ~**at** (·—ʾ), **-tı,** thoughts that occur to the mind; inspiration.
**süphan** *etc. v.* **sübhan.**
**süprü·lmek** *vn. pass. of* **süpürmek.** Be swept. ~**ntü,** sweepings; rubbish. ~**ntücü,** dustman. ~**ntülük,** dust-heap, rubbish-heap.
**süpürge** Broom; brush. **saçını** ~ **etm.**, (of a woman) to work hard in the house: ~ **darısı,** sorghum: ~**otu,** heather: ~ **sapı yemek,** to 'get the stick'. ~**ci,** maker *or* seller of brooms; street-sweeper.
**süpür·mek** *va.* Sweep; brush; sweep away; clear the dish. ~**ücü,** sweeper.
**süpya** (ʾ·) Squid.
**sürahi** (·—ʾ) Decanter, waterbottle.
**sürat, sür'at, -ti** Speed, velocity; haste. ~**le,** quickly: ~**i intikal,** quick-wittedness, perspicacity. ~**li,** quick; hurried.
**sürc** Stumble; slip; mistake. ~**mek,** to stumble; to make a mistake. ~**ülisan,** slip of the tongue, 'lapsus linguae'.
**süre** Period: extension.
**sürek** A drove *of cattle*; duration; fast driver. ~ **avı,** drive (shooting). ~**li,** lasting, prolonged. ~**siz,** transitory.
**sürfe** Caterpillar; maggot; teredo.
**sürgü** Harrow; roller; bolt *of a door*; plasterer's trowel; cursor; sliding bar; till. ~**lemek,** to harrow (a field); bolt (door); roll (a road); smooth (plaster); ~**lü,** bolted; sliding: ~ **pergel,** beam-compasses.
**sürgün** Banishment; place of exile; an exile; shoot, sucker; diarrhoea. Exiled. ~ **avı,** drive, battue: ~ **etm.,** to banish. ~**lük,** exile; place of exile; diarrhoea; purgative.
**sürme** Anything drawn; bolt; drawer; aperient; kohl; move (chess *etc.*); smut *of wheat etc.* Sliding, drawing in and out. ~ **pencere,** sash window: ⌜**adamın gözünden** ~**yi çalar**⌝, 'he would steal the salve off a man's eye' (*of a very crafty person*). ~**dan,** pot for eye-salve.
**sürmek** *va.* Drive in front; drive away; drive (a vehicle); banish; push along; push forth (buds *etc.*); rub on, smear; plough; sell; spend (time, life). *vn.* Push on, go on; continue, extend; (of time) to pass; germinate. **boy** ~, to shoot up, grow tall: **boya** ~, to lay on paint: **çift** ~, to plough: **çok sürmedi,** it did not last long; it was not long before … : ···**e el** ~, to touch, to meddle with; **işe el sürmemek,** not to lift a finger *to do stg.*: **hüküm** ~, to prevail: **içi** ~, to have diarrhoea: **ileri** ~, to put forward (suggestion, proposal): **öne** ~, to bring up *a matter etc.* (*with an innuendo of cunning*); to put *s.o.* forward

to meet a difficult situation: **saltanat ~,** to reign: **safa ~,** to lead a pleasant life: **yağ ~,** to spread butter, to rub on oil: **yüz ~,** to prostrate oneself humbly.

**sürme·lemek** *va.* Bolt (a door); put on eye-salve. **~li,** having a bolt; bolted; sliding; tinged with eye-salve: **~ kumpas,** slide gauge, sliding calipers. **~taşı** (.ˈ..), **-nı,** antimony (*used for blackening the eye-brows*).

**sürre** Purse; gifts *formerly sent to Mecca annually by the Sultan.* **~ alayı,** the procession which accompanied the **sürre**: **~ devesi,** an oddly-dressed person; **~ devesi gibi dolaşmak,** to loaf about with an air of being busy: **~ emini,** the official entrusted with the delivery of the **sürre.**

**sürtmek** *va.* Rub *one thing* against another; rub with the hand; wear down by friction. *vn.* Wander about aimlessly. ⌜**sürt Allah kerim**⌝ (*said of vagabonds*): **sokak sokak ~,** to lounge about through the streets: **taban ~,** to walk incessantly.

**sürtük** (Woman) always walking the streets.

**sürt·ülmek** *vn. pass. of* **sürtmek.** Be rubbed *etc.* **~ünmek,** to rub oneself against *stg.*; drag oneself along; creep; toady; seek a quarrel, behave in a provocative manner.

**sürur** (.ˈ) Joy; pleasure.

**sürü** Herd; flock; crowd, gang; a lot of. **~ ~,** in droves, in flocks: **~den ayrılmamak** *or* **~ye katılmak,** to join the throng, to do as others do; (of a child) to cease to be a baby and go in company with others: **~süne bereket** (*sl.*), heaps of: **~ sepet,** all together; all the lot; in great numbers. **~cü,** drover; driver *of a vehicle*; man in charge of post-horses, *esp.* those carrying the mail.

**sürük·lemek** *va.* Drag; drag along the ground; drag *s.o.* somewhere against his will; carry *one's audience, one's readers* with one; involve, lead to, entail. **atalete ~,** to bring to a standstill, to render useless. **~lenmek,** to drag oneself; to be dragged; to drag on, be protracted. **~leyici,** fascinating, attractive.

**sürülmek** *vn. pass. of* **sürmek.** Be rubbed *etc.*

**sürüm** Act of driving *etc.*, *v.* **sürmek**; rapid sale, great demand *for some article.* **~ ~,** *used to strengthen the meaning of* **sürünmek,** *q.v.* **~lü,** finding a ready sale; in great demand. **~süz,** hard to sell; not in demand.

**sürümek** *va.* Drag along the ground; procrastinate. **ayak ~,** to shuffle along; to be very reluctant about *stg.* one has agreed to do: **ipini ~,** to lead a life of crime and deserve the gallows.

**sürünceme** Delay; a matter's dragging on. **~de kalmak,** to drag on, to be long drawn out. **~li,** dilatory.

**süründürmek** *va.* Bring into utter misery.

**sürüngen** Reptile.

**sürünmek** *va.* Rub in *or* on.. *vn.* Drag oneself along the ground; grovel; live in misery.

**sürür** (.ˈ) Sulphide of mercury; vermilion varnish.

**sürüş·mek** *vn.* Rub together. **~türmek** *va.,* to rub in slowly and gently; massage.

**sürütme** Drag-net, trawl; a kind of fish-hook.

**sürütmek** *va. caus. of* **sürmek.** Cause to drag *etc.* ⌜**iti öldürene sürütürler**⌝, one must clear up the mess one has made.

**Süryani** (.—ˈ) Syrian Christian (of Northern Irak); Syriac language. **~ce,** Syriac.

**süs** Ornament, decoration; elegance of dress; toilet; luxury. **~ saltanat,** great luxury: **kendisine ... ~ünü vermek,** to play the part of ...: **kaza ~ü vermek,** to pass off as an accident. **süs·lemek, ~lendirmek** *va.* Adorn; embellish. **~lenmek,** to adorn oneself, deck oneself out. **~lü,** ornamented; decorated; carefully dressed; luxurious.

**süsmek** *va.* Butt, toss; gore.

**süt, -tü, -dü** Milk; milk-like juice. **~ ağzı,** beestings (colostrum): **~ beyaz,** milk-white: **~ü bozuk,** base; without character; unprincipled: ···**in ~üne havale etm.,** to leave *stg. to another's* sense of honour: **~ kardeş,** foster-brother *or* sister: **~ kesildi,** the milk has turned sour: **~ kesimi,** weaning; **~ten kesmek,** to wean: **~ kırı,** milk-white horse: **~ kuzusu,** sucking lamb; baby: ⌜**ağzı* ~ kokuyor**⌝, he behaves like a child: **aslan ~ü,** raki. **kuş ~ü,** 'bird's milk',

*stg.* impossible to find; **kuş ~ünden gayrı her şey var,** nothing is lacking.
**süt·ana, ~anne** (ˈ··), wet-nurse; foster-mother. **~baba,** foster-father. **~başı, -nı,** cream. **~çü,** milkman, dairyman. **~çülük,** dairying, trade of a milkman. **~dişi, -ni,** milk-tooth. **~hane** (·–ˈ), dairy. **~laç,** rice-pudding. **~leğen,** spurge, euphorbia. **~liman** (ˈ··), dead calm: **ortalık ~,** everything is perfectly quiet; 'the coast is clear!' **~lü,** milky; in milk. **~lüce,** petty spurge. **~lük,** dairy. **~nine, ~ne,** wet-nurse, foster-mother. **~oğul, -ğlu,** foster-child. **~otu, -nu,** milkwort, polygala. **~süz,** without milk; base, ignoble. **~vurgunu, -nu,** (child) suffering from bad milk; rickety.
**sütun** (·ˈ) Column; pillar; column (in a newspaper); beam *of light.*
**süvari** (·–ˈ) Cavalryman; captain *in the navy*; mounted police; cavalry. **~ askeri,** cavalryman. **~lik,** career and duties of a cavalryman *or* a captain.
**süve** *v.* **söve.**
**süveter** Sweater.
**Süveyş** Suez.
**süz·geç** Filter; strainer; rose *of a watering-can.* **~gü,** fine filter. **~gün,** languid; half-closed (eye); grown thin. **~me,** filtered; strained; run (honey).
**süzmek** *va.* Strain, filter; examine closely; half-close the eyes, look attentively at *stg.* through half-closed eyes. **birine göz ~,** to cast amorous glances at *s.o.*
**süzük** Drawn, strained (face *etc.*).
**süzülmek** *vn. pass. of* **süzmek.** Be strained *or* filtered; become thin; become languorous; be very closely examined; slip *or* creep away; glide along swiftly and silently. **gözleri* ~,** for the eyes to be half-closed; to be sleepy.

# Ş

**şa** (*abbrev. for* **yaşa**) Long live ...!; hurrah!
**şaban** (–ˈ) *The eighth month of the Moslem year.*
**şablon** Pattern (*mech.*).
**şabrak** Horse-cloth; saddle-cover.
**şad, şadan, şaduman** Joyful, happy. **~i** (–ˈ), **~umani,** joy, happiness.
**şadırvan** Tank of water *with a jet in the middle*; tank *attached to mosques for ablutions.*
**şaduman** (–·ˈ) *v.* **şad.**
**şafak, -kı** (*Originally*) twilight, dusk; (*now*) dawn. **~ atmak,** for dawn to break; **birisinde ~ atmak,** for *s.o.* to turn pale with fright: **~la beraber,** at dawn: **~ sökmek,** for dawn to break.
**şâfiî** (–·ˈ) The Shafi sect of Islam. **~ köpeği,** a dirty-faced fellow.
**şaft, -tı** Shaft.
**şah**[1] (*) Shah. ⌜**ben ~ımı bukadar severim**⌝, I can't risk any more; don't count on me any further.
**şah**[2] Fork; horn. **~a kalkmak,** (of a horse) to rear.
**şahadet** (·–ˈ) *v.* **şehadet.**
**şahane** (––ˈ) Royal, imperial; regal, magnificent.
**şahbaz** Royal falcon; champion, hero. Fine. ⌜**şahtı (şah idi) ~ oldu**⌝, 'finer and finer!' (*only used ironically*).
**şahdamar** (ˈ··), **şahdamarı, -nı** Aorta.
**şaheser** (–·ˈ) Masterpiece.
**şahıs**[1], **-hsı** Person; individual; personal features. **~a mahsus,** personal. **~landırmak,** to personify.
**şahıs**[2] (–ˈ), **-hsı** Surveyor's rod.
**şahid** (–ˈ) Witness; example (*gram.*); control (in an experiment); example *from a well-known work.* **~ tutmak,** to call on one to witness, to accept as witness: **kırk ~ lâzım,** one can hardly believe it: **···e ~ olm.,** to witness. **~lik,** a giving evidence, testimony; a being witness *to stg.*
**şâhika** (–·ˈ) Summit.
**şahin** (–ˈ) Peregrine falcon. **~ bakışlı,** with fierce and piercing eyes. **~ci,** falconer in charge of the peregrines.
**şahlanmak** *vn.* (Of a horse) to rear; become angry and threatening; get out of hand.
**şahmerdan** Battering-ram; pile-driver; beetle (heavy hammer).
**şahnişin** Bay-window on an enclosed balcony.
**şahrem** *Only in* **~ ~ çatlamak (yırtılmak),** (of the skin) to be covered with cracks; (of cloth) to be much torn.
**şahs·an** (ˈ·) Personally; in person. **~ı,** *v.* **sahıs.** **~î** (·ˈ), personal,

private. **~iyat** (..⁄), **–tı**, personalities; personal matters: **~a dökülmek,** (of a discussion) to descend to personalities. **~iyet, –ti,** personality; important person.

**şaibe** (–.⁄) Stain; defect.

**şair** (–⁄) Poet; minstrel. **~ane** (–.–⁄), poetical; in a poetical manner. **~lik,** the quality of a poet *or* minstrel.

**şak, –kı** A clacking noise (*as of wood against wood, the crack of a whip, a box on the ear*).

**şaka** Fun; joke; jest. **~ya boğmak,** to try to pass off *a matter* with a joke: ⌜**~ derken kaka olur**⌝, 'a joke may easily become a serious matter': **~ya gelmez,** (a matter) not to be joked about; (a man) who can't take a joke; who is not to be joked with, severe: **~ götürmez,** serious, not a joking matter: **~ kaldırmak,** to be able to stand a joke: **~ya vurmak,** to pretend to take *stg.* as a joke: **~sı yok,** in earnest; not to be trifled with: **el ~sı,** practical joke: **hoyrat ~sı,** horseplay.

**şakacıktan** (i) As a joke; (ii) (to do *stg.* serious) under the pretence of a joke.

**şakak** Temple (of the head).

**şakalaşmak** *vn.* Joke with one another.

**şakavet** *v.* **şekavet.**

**şakayık, –kı, –ğı** Peony.

**şakımak** *vn.* (Of a nightingale *or* canary) to sing loudly.

**şakır** *In ~ ~, imitating the noise of rain, of splashing, of a continuous rattle or jingle.* **~ şakur** (**şukur**), *imitates a hollow rattling and banging noise.* **~damak,** to make the noises expressed by **şakır**; rattle, jingle; sing vociferously (nightingale). **~tı,** continuous clatter *or* rattle, *v.* **şakır.**

**şakıt, –tı** Lamprey.

**şakî** (.⁄) Brigand; rebel; outlaw.

**şakird** (–⁄) Pupil; apprentice. **~lik,** quality of ; pupil *or* apprentice; apprenticeship; wages paid to an apprentice.

**şakketmek** (⁄. ), **–eder** *va.* Cleave; split.

**şaklaban** Mimic; jester, buffoon; amusing fellow; charlatan. **~lık,** mimicry; buffoonery.

**şak·lamak** *vn.* Make a loud cracking noise, *v.* **şak.** **~latmak,** to crack (whip); cause a loud cracking noise to be made.

**şakrak** Noisy; mirthful; vivacious; chatty. **~ kuşu,** bullfinch.

**şakşak** Slap-stick; large castanet; applause; toadying. **~çı,** toady; 'yes-man'. **~çılık,** base adulation.

**şakul** (–⁄) Plumb-line; plummet. **~î** (––⁄ *or* –.⁄), perpendicular. **~lamak,** to set up with a plumb-line; plan; measure.

**şal** Shawl, *esp.* a Cashmere shawl; pall.

**şalgam** Turnip.

**şali** (–⁄) Alpaca; camlet; bunting.

**şalter** Switch (*elec.*).

**şalupa** (.⁄.) Sloop.

**şalvar** Baggy trousers.

**Şam¹** Damascus. **~ fıstığı,** pistachio nut.

**şam²** Evening.

**şama** Wax taper. **~lı kibrit,** vesta match.

**şamandıra** (.⁄..) Buoy; float *for a wick*; burner *of a paraffin lamp*; ball *of a ball-cock.*

**şamar** Slap; box on the ear.

**şamata** Great noise, uproar, hubbub. **~cı,** a noisy, uproarious person. **~lı,** noisy.

**şambabası** (⁄...), **–nı** A kind of sweet pastry.

**şambrnuvar** (*Fr. chambre noire*) Dark-room.

**şamdan** Candlestick.

**şamfıstığı** (⁄...), **–nı** Pistachio nut.

**şamil** (–⁄) Comprising; including; comprehensive.

**şampanya** (.⁄.) Champagne.

**şampanze** Chimpanzee.

**şampiyon** Champion.

**şampuvan** Shampoo.

**şan** Fame, renown; glory; reputation; state, quality, aspect; display; importance. **~ına* düşmek** (**yakışmak**) *or* **~ından* olm.,** to befit one's station *or* dignity; **~ından* olm.,** *also* to be peculiar to, to be characteristic of.

**şangır** *etc. v.* **şıngır** *etc.*

**şanlı** Glorious; famous; fine-looking. **~ şöhretli,** fine and imposing.

**şano** (⁄.) Stage (theatre).

**şans** Chance; luck. **~lı,** lucky. **~sız,** unlucky.

**şansız** Without renown, unknown; undistinguished. **~ şöhretsiz,** insignificant in appearance.

**şantaj** (*Fr. chantage*) Blackmail.

**şantöz** (*Fr. chanteuse*) Female singer.

**şap¹, –pı** Alum; coral reef. **~ gibi**

dönmek *or* ~ kesilmek, to become bitter: ~ illeti, foot-and-mouth disease: ~a oturmak, to be in a hopeless dilemma; to be greatly disconcerted: kızıl ~, light purple colour: ⌜ne ~ ne şeker⌝, neither one thing nor the other.

**şap² şap** *Imitates the sound of kissing.*

**Şapdenizi** (ˈ···), **-ni** The Red Sea.

**şapır** *In* ~ ~ *or* ~ **şupur,** *imitates the smacking of lips.* **~damak,** to make a smacking noise (in kissing *or* eating). **~tı,** smacking noise of the lips.

**şapka** (ˈ·) Hat; truck *of a mast*; cowl *of a chimney.* **supap ~sı,** valve head. **~cı,** hatter. **~lı,** wearing a hat; having a circumflex.

**şaplak** Smack on the face.

**şap·lamak** *vn.* Make a smacking noise *with the lips or hand.* **~latmak,** to cause to make a smacking noise: **bir tokat ~,** to give a resounding slap.

**şaprak** Saddle covering.

**şapşal** Untidy; slovenly.

**şar şar** *Imitates the sound of slashing.*

**şarab** Wine. **~cı,** wine-merchant. **~hane** (··–ˈ), wine factory; storehouse for wine.

**şarampol** Palisade; stockade.

**şarapnel** Shrapnel.

**şarbon** (*Fr. charbon*) Anthrax; smut *of wheat etc.*

**şarıl** Sound of running water. **~ ~ akmak, ~damak,** to flow with a splashing noise. **~tı,** a gurgling, splashing noise.

**şarih** (–ˈ) Commentator; annotator.

**şark, -kı** The East.

**şarkadak** (ˈ··) *Imitates the noise of a thing falling.* **~ bayılmak,** to fall down in a faint.

**şark·an** (ˈ·) In *or* from an easterly direction. **~î** (·ˈ–), eastern. **~ıyat, -tı,** the study of Oriental languages and literature.

**şarkı** Song. **~ söylemek,** to sing. **~cı,** song-writer; singer.

**şarlamak** *v.* şarıldamak.

**şarpa** (ˈ·) Scarf.

**şart, -tı** Condition, stipulation; article of an agreement; conditional clause; oath *with the words* '~ olsun!' (*may I be divorced from my wife if this be not true*). **~ koşmak,** to lay down a condition. **~lamak,** to wash clothes *or* vessels in accordance with the requirements of canon law. **~laşmak,** mutually to agree to conditions. **~lı,** having a condition attached; who takes the oath called şart *q.v.* **~name** (·–ˈ), list of conditions; specification; contract.

**şasi** Chassis.

**şâşaa** (–·ˈ) Glitter, sparkle; splendour. **~lanmak,** to sparkle; to be magnificent *or* pompous. **~lı,** sparkling; resplendent; gorgeous.

**şaşalamak** *vn.* Be bewildered *or* confused.

**şaşı** Squinting, squint-eyed.

**şaşılacak** Surprising; wonderful.

**şaşı·lamak** *vn.* Squint. **~lık,** squint, squinting.

**şaşır·mak** *va.* Be confused about *stg.*; lose *the way etc. vn.* Become bewildered *or* embarrassed; lose one's head. **~tma,** a misleading *or* confusing; zigzag; tongue-twister; transplanting. **~tmaca,** tongue-twister; puzzle. **~tmak,** to confuse, bewilder; mislead; transplant (seedlings *etc.*).

**şaşkaloz** Cross-eyed.

**şaşkın** Bewildered, confused; stupid. **~lık,** bewilderment; stupidity.

**şaşmak** *vn.* Miss one's way, go astray; deviate; be surprised *or* bewildered. **şaşa kalmak,** to be bewildered: **şaşacak şey,** a surprising thing: **şaşmamak,** not to go astray; not to deviate; not to go wrong; to be punctual, to keep exact time (of a clock).

**şat¹** Flat-bottomed boat; lighter.

**şat², -ttı** Large river, *esp.* the united Tigris and Euphrates, *also called* **Şattülarab.**

**şatafat, -tı** Luxury; ostentatious living. **~lı,** pretentious, showy.

**şathiyat** (··ˈ), **-tı** Flippant and satirical writings.

**şatır** (–ˈ) Gay, vivacious; agile.

**şatranc** Chess.

**şavul** *v.* şakul.

**şayak** Serge.

**şayan** (–ˈ) Fitting; suitable; worthy; deserving. **~ı dikkat,** worth attention, notable: **~ı takdir,** praiseworthy.

**şayed** (ˈ–·) If perchance; lest; perhaps.

**şayi** (–ˈ), **-ii** Divulged; commonly known; shared in common. **~a** (–·ˈ), news spread about, rumour. **~at, -tı,** *pl. of* şayia.

**şayka** (ˈ·) *A kind of* boat *used in the Black Sea.*

**şaz** Exception.

**şeamet** (·–˼), **–ti** A being inauspicious; evil omen. **~li,** inauspicious, ill-omened.
**şebboy** Wallflower; stock. **sarı ~,** wallflower; **kırmızı ~,** stock.
**şebek** Baboon.
**şebek·e** Net; lattice-work; grating; network *of railways etc.*; band *of robbers etc.*; ring *of dealers etc.*
**şecaat** (·–˼), **–ti** Bravery, courage. ⌈**~ arz ederken**⌉, when boasting of one's qualities to reveal one's defects.
**şecer** Tree. **~e,** genealogical tree.
**şeci** (·–˼) Brave, bold.
**şedid** (·˼) Hard; strong; violent; severe.
**şef** Chief, leader. **~garson,** head waiter. **~lik,** duties of a chief; chief office: **istasyon şefliği,** stationmaster's office.
**şefaat** (·–˼), **–ti** Intercession. **~ci,** intercessor.
**şeffaf** Transparent; diaphanous.
**şefkat, –ti** Compassion; affection. **~li,** compassionate; affectionate.
**şeftali** (·–˼) Peach; kiss.
**şeftren** (*Fr. chef de train*) Guard *of a train.*
**şehadet** (·–˼), **–ti** A witnessing; testimony; evidence; a thing witnessed; a testifying to Islam; death in battle (*of a Moslem only*); martyrdom. **~ etm.,** to bear witness (to = *dat.*): **~ getirmek,** to pronounce the formula 'there is no God but God, Mohammed is the apostle of God': **~e nail olm.,** to gain martyrdom, to die on the field of battle: **~ parmağı,** the index finger. **~name** (·–·–˼), testimonial; diploma; certificate.
**şehamet** (·–˼), **–ti** Boldness coupled with efficiency.
**şehbender** Turkish *or* Persian consul. **~hane** (···–˼), consulate. **~lik,** rank and duties of a consul.
**şehid** Martyr; (*of Moslems only*) one who dies in battle *or* in the service of his country. **~lik,** martyrdom, death in battle *etc.*, *v.* **şehid**: cemetery *or* monument for those who died in battle.
**şehir**[1], **–hri** Month.
**şehir**[2], **–hri** Large town, city. **~dışı, –nı,** suburb. **~li,** townsman, citizen.
**şehîr** (·–˼) Celebrated.
**şehislâm** *v.* **şeyhülislâm.**
**şehlâ** Having a slight cast in the eye.
**şehrayin** (·–˼) Illumination of a town *for a festival etc.*
**şehr·emaneti** (˼·–·˼), **–ni** Prefecture of a large town. **~emini, –ni,** prefect of a large town.
**şehriye** Vermicelli. **arpa ~si,** pearl-barley.
**şehvanî** (·–˼) Lustful; sensual.
**şehvet, –ti** Lust; sensuality. **~li,** sensual; voluptuous. **~perest,** sensual; enslaved by lust.
**şehzade** (·–˼) Prince, *esp.* a Sultan's son.
**şek, –kki** Doubt; uncertainty. **~ ve şübhe,** doubt and misgiving.
**şekavet** (·–˼), **–ti** Brigandage.
**şeker** Sugar; a sweet; darling. Sweet. **~im,** my darling: **~ bayramı,** *name of the feast during the first three days after the Ramazan fast*: **~ illeti,** diabetes: **~ kamışı,** sugar-cane.
**şeker·ci,** sweet-seller; sugar-merchant; confectioner: **~ boyası,** poke-weed (*Phytolacca*). **~leme,** candied fruit; sugar-plum; doze, nap. **~lemek,** to sugar, to preserve in sugar. **~li,** sugared; sweetened with sugar. **~renk,** whitish brown; uncordial, cool (relations): **araları ~,** their relations are somewhat cool, they are not on very good terms.
**şekil, –kli** Form; shape; figure; plan; diagram; kind; manner; features. **bir şekle girmek,** to take a form *or* shape: **bir şekle koymak,** to manage somehow or other: **şekle riayeten,** for form's sake.
**şekil·lendirmek** to give a form *or* shape to; to form. **~siz,** shapeless; uncouth; without diagrams.
**şekl·en** (˼–) In form; in appearance. **~i,** *v.* **şekil.**
**şekva** (·–˼) Complaint.
**şelâle** Waterfall.
**şema** (˼·) Outline; sketch; plan, diagram.
**şemail** (·–˼) Features.
**şemm·e** A single sniff; a whiff; a very small quantity; hint (of reproach *etc.*). **~î** (·–˼), pertaining to smell; olfactory.
**şems** Sun. **~e,** ornamental figure of the sun. **~î** (·–˼), solar. **~iye,** parasol, umbrella.
**şen** Joyous, cheerful.
**şenaat** (·–˼), **–ti** Foulness; wickedness.
**şenelmek** *v.* **şenlenmek.**
**şeni** (·–˼) Disgraceful, vile, immoral.

**~a** (·–˘), vile *or* immoral act.
**şe'nî** (·–́) Real. **~yet, -ti,** reality.
**şen·lenmek** *vn.* Become cheerful, gay, joyful; become inhabited and prosperous. **~lik,** gaiety, cheerfulness; public rejoicings, illuminations *etc.*; prosperity, increase in amenities.
**şer, -rri** Evil; wickedness; harm; quarrel. Bad, wicked. şerrine* **lânet,** a curse upon his wickedness! may God curse him!
**-şer, -şar** *Suffix to numerals making them distributive.* **ikişer,** two each, two at a time.
**şerait** (·–˘), **-ti** *pl. of* **şart.** Conditions.
**şer'an** (˘·) In accordance with canon law.
**şerbet, -ti** Sweet drink; sherbet; infusion; medicinal *or* aperient draught. **~** *or* **gübre ~i,** liquid manure: **~ almak,** to take an aperient. **~ci,** seller of sherbet. **~çiotu, -nu,** hop (plant). **~lenmek,** to be rendered immune *to a disease.* **~li,** with sherbet; infused; immune, bewitched *against snake-bites etc.*; notorious, hardened. **~lik,** anything used for making sherbet.
**şeref** Honour; glory; excellence; legitimate pride, exaltation; superiority, distinction. **~ bulmak,** to be honoured; to increase in value. **~bahş,** that honours, who honours (by his presence *etc.*). **~e,** gallery of a minaret *whence the call to prayer is made.* **~iye,** tax on the increase in land value due to building. **~lenmek,** to acquire honour, be honoured; increase in value. **~li,** honoured, esteemed; favoured *or* distinguished (district). **~yab** (··–́), honoured.
**şerh** Explanation; commentary. **~etmek** (˘··), **-eder,** to explain, comment on.
**şerha** Cut. **~ ~ doğramak,** to cut to pieces.
**şer·i, -r'i** The Moslem religious law, sheri. **~'î** (·–́), pertaining to the religious law: **helei ~ye,** a legal way of getting round the law. **~at** (·–˘), **-ti,** canonical obligations; the religious law: ⌜**~in kestiği parmak acımaz**⌝, 'just punishment is not resented'. **~ci,** an upholder of the Religious Law.
**şerid** Ribbon; tape; band, belt; film (cinema); tapeworm. **~ arşın,** tape-measure. **~lemek,** to bind *or* ornament with ribbon *or* tape. **~li,** beribboned.
**şerif** Noble; descended from Mohammed; *formerly a title of the Governor of Mecca.*
**şerik, -ki** Partner; shareholder; companion.
**şerir** Bad, wicked; rebellious. Scoundrel.
**şeş** *Pers. num.* Six, *only used in backgammon and in*: **~i beş görmek,** to squint; to be thoroughly confused: **sana ~i beş gösteririm!,** I'll knock stars out of you!
**şetaret** (·–˘), **-ti** Merriment, gaiety.
**şev** Slope; glacis; bevel. Sloping. **~ ~ine,** sloping; not at right angles.
**şevk**[1], **-ki** Desire; ardent yearning; eagerness; mirth. **~e gelmek,** to become eager; to grow merry. **~î**[1] (·–́), **~li,** eager, desirous; gay. **~siz,** without eagerness; cold; dull.
**şevk**[2], **-ki** Thorn. **~î**[2] (·–́), thorny; spinal.
**şevket, -ti** Majesty, pomp. **~li, ~lû, ~meab,** majestic (*title of a sovereign*).
**şev·lenmek** *vn.* Slope; incline. **~li,** sloping; bevelled.
**şevval, -li** *The tenth month of the Moslem lunar year.*
**şey** Thing; *often used when one cannot find the right word or name,* what d'you call it, what's his name. **bir ~, a** thing, something: **bir şeyler,** something or other: **bir ~ değil,** it is nothing; it doesn't matter; not at all! (*in reply to thanks etc.*): ⌜**bir ~dir oldu**⌝, 'I'm sorry it happened, but it can't be helped now': **çok ~ !,** really!; how strange!: **hiç bir ~,** nothing.
**şeyda** (·–́) Madly in love.
**şeyh** Old man; elder; sheikh; head of a family *or* tribe; head of a religious order. **~ülharem,** the Governor of the town and province of Medina. **~ülislâm,** Sheikhulislam.
**şeytan** Satan; devil; crafty man; sharp little devil (child). **~ arabası,** floating thistledown *or* dandelion seeds; hand-driven trolley *used by railwaymen*; bicycle: **~ın ard ayağı** *or* **kıç bacağı,** a limb of Satan, a crafty fellow: ⌜**~ azabda gerek**⌝, it serves you (him *etc.*) right!: **~ın bacağını kırmak,** to overcome temptation: **~ boku (tersi),** asafoetida: ⌜**~a külâhını ters giydirir**⌝, 'he could cheat the Devil himself, he is very cunning': **~ diyor ki,** I am very

tempted *to do stg. foolish*: ⌜~ın işi yok⌝, 'by pure bad luck': **~ları tepesine çıktı,** he lost his temper: **~ tırnağı,** a hangnail: **~ tüyü,** a talisman *supposed to give personal attraction*: **~a uymak,** to let oneself be led astray; to yield to temptation: **aksi ~,** as bad luck would have it; 'how provoking!'.

**şeytan·bezi, -ni,** velveteen. **~et, -ti,** act of devilry; malice; craftiness. **~î** (·–⌐), diabolical. **~lık,** devilry; slyness; malicious cunning.

**şık[1], -kkı** One of two alternatives.

**şık[2]** (*Fr. chic*) Chic; smart. **~lık,** smartness.

**şıkır** *In ~ ~, imitates a jingling or clinking noise; also used of dazzling light.* **~damak,** to rattle, jingle: **cebleri* ~,** to have plenty of money. **~tı,** a jingling noise.

**şıldır** *In* **~ ~ bakmak,** (of a baby) to stare with wide-open eyes.

**şıllık** Gaudily dressed (woman). Loose woman.

**şımar·ık** Spoilt (child); saucy, impertinent; stuck-up. **~mak,** to be spoilt by indulgence; to get above oneself; to lose one's self-control. **~tmak,** to spoil *a child*.

**şınanay** Refrain, fol-de-rol, tra-la-la.

**şıngıl** *v.* **çıngıl.**

**şıngır** *In ~ ~, imitates the noise of breaking glass.* **~damak,** to crash, to make the noise of breaking glass. **~tı,** the noise of breaking glass.

**şıp, -pı** Noise of a drop falling; a sudden slight noise. Easily, quickly, at once. **~ diye geldi,** all of a sudden he came: **~ın işi,** very quickly and easily.

**şıpıdık** (⌐··) Low-heeled shoe *or* slipper.

**şıpıldamak** *vn.* (Of water) to make a lapping noise, to lap.

**şıpır** *In ~ ~,* falling in drops. **~tılı,** splashing: **~ hava,** rainy weather.

**şıpka** (⌐·) Rope *or* wire net *used on a ship*; torpedo-net.

**şıppadak** (⌐··), **şıpsak** Quickly, at once.

**şıpsevdi** (⌐··) (*sl.*) Quick to fall in love; susceptible.

**şıpşıp** Slipper without any back.

**şıra** (⌐·) Must, unfermented grape-juice. **~ lı,** juicy.

**şırak** *Imitates a sudden sharp noise.*

**şırfıntı** *A term of contempt for women*; bitch (*of a woman*).

**şırıl** *In ~ ~, imitates the noise of gently running water.* **~damak,** to make such a noise. **~tı,** noise of running water, splashing, gurgling.

**şırınga** (·⌐·) Syringe; enema; hypodermic syringe; injection.

**şırlağan** Oil of sesame.

**şırlamak** *v.* **şırıldamak.**

**şırlop** Eggs fried in yoghourt.

**şırp, şırpadak** *v.* **şıp, şıpadak.**

**şıvgar** *v.* **şivgar.**

**şiar** Characteristic, trait.

**şiddet, -ti** Violence; severity. **~li,** violent; vehement; severe.

**şifa** (·⌐) Restoration to health; healing. **~ bulmak,** to recover health: **~yı bulmak (kapmak),** (*iron.*) to fall ill; to turn out badly: **~ olsun!,** 'may it bring you health!', *said to one having a drink or taking medicine*: **~ vermek,** to restore health. **~lı,** healing; wholesome.

**şifah·en** (·⌐·) Orally, verbally. **~î** (·–⌐), oral, verbal.

**şifre** (⌐·) Cipher; code. **~ açmak (çözmek),** to decode, decipher. **~li,** in cipher; with initials.

**Siî** (–⌐) Shiite. **~lik,** a belonging to the Shiite sect, Shiism.

**şiir** Poetry; poem.

**şikâr** (·⌐) The chase; hunting; game *killed in the chase*; prey, victim; anything rare and much sought after; booty. **~ bir şey mi?,** is it such a rarity?.

**şikâyet** (·–⌐), **-ti** Complaint. **~ etm.,** to complain: **···i ~ etm.,** to complain about. **~çi,** complainant. **~name** (·–·–⌐), written complaint.

**şikem** Belly; womb. **~perver,** glutton; gluttonous.

**şikest** Break, fracture; destruction. **~e,** broken; the Persian style of Arabic writing.

**şile** Wild marjoram.

**şilep** Tramp steamer; cargo boat.

**Şili** Chili.

**şilin** Shilling.

**şilte** Thin mattress *or* quilt.

**şimal, -li** The North. **~ yıldızı,** the Pole Star. **~en** (·⌐·), to *or* from the north. **~î** (·–⌐), northern, north.

**şimden** *v.* **şimdiden.**

**şimdi** (⌐·) The present. Now. **~den** henceforth; already: **~den tezi yok,** the sooner the better; with all speed: **~ye kadar (değin),** up till now. **~cik** (⌐··), this very moment, now at once. **~ki,** the present; the actual. **~lik,** for the present.

**şimşek** Lightning. **~li fener,** flashing light (lighthouse).
**şimşir** Box (tree).
**şin** Disgrace.
**şinik** *Measure for cereals equalling a quarter bushel (10 litres).*
**şinşile** Chinchilla.
**şip** A kind of coarse gauze.
**şiraze** (−−ˋ) Head-band of a bound volume; a thing that holds other things together; order, regularity. **~den çıkmak,** to lose one's mental balance: **~sinden çıkmak,** *or* **~si bozulmak,** (of a matter) to deteriorate beyond recovery.
**şirden** The second stomach of ruminants.
**şirin** Sweet; affable; charming. **~lik,** sweetness; amiability.
**şirk, -ki** Polytheism. **Tanrıya ~ koşmak,** to attribute a partner to God, *i.e.* to be a polytheist.
**şirket, -ti** Partnership; joint ownership; joint-stock company. **~ vapuru,** (*formerly*) Bosphorus ferry steamer.
**şirlan** *v.* **şırlağan.**
**şirpençe** Carbuncle; anthrax.
**şirret, -ti** Evilly disposed, malicious. Malicious person; tartar, shrew, virago. **~lik,** malice; evil disposition.
**şist, -ti** Schist.
**şiş**[1] Spit; skewer; rapier; knitting-needle; axle. **~e geçirmek,** to skewer: ⌜**ne ~ yansın ne kebab**⌝, so that neither party suffers damage.
**şiş**[2] Swelling; tumour. Swollen.
**şişane** (·−ˋ) Rifling *of a gun*; rifled gun. ⌜**altı kaval üstü ~**⌝, 'smoothbore below and rifled above', a paradoxical absurdity; wearing clothes that do not go with one another.
**şişe** Bottle; lamp-glass; cupping-glass; moulded *or* planed lath.
**şişek** Year-old lamb; teg.
**şişirmek** *va.* Cause to swell; inflate; exaggerate; (*sl.*) do *stg.* hastily and carelessly; 'cram' for an exam.: stab.
**şişkin** Swollen; puffed up. **~lik,** distension, puffiness, swelling.
**şişko** (ˋ·) Very fat (man).
**şişlemek** *va.* Spit, skewer; stab.
**şişman** Fat. **~lık,** fatness, obesity.
**şişmek** *vn.* Swell; become inflated; grow fat; become swollen; be distended; (of a runner) to be unable to continue for want of breath. **kafası ~,** to be distraught *or* dazed *by noise etc.*
**şiv** *v.* **şev.**
**şive** (−ˋ) Accent, pronunciation; idiom; gracefulness; style.
**şivgar** Whipple-tree of a gun-carriage.
**şoför** Chauffeur; driver of a motor-car.
**şom** Inauspicious; sinister, gloomy. Evil omen. **~ağızlı** (ˋ···), who always predicts misfortune.
**şont** Shunt (dynamo *etc.*).
**şorolop** *Imitates the sound of a thing gulped down.*
**şose** (ˋ·) (*Fr. chaussée*) Macadamized road.
**şoven** (*Fr. chauvin*) Chauvinistic.
**şöhret, -ti** Fame; reputation; name by which a man is known, pseudonym. **~ hastası,** one who seeks notoriety. **~li,** famous; notorious.
**şölen** Feast.
**şömine** (*Fr. cheminée*) Fireplace.
**şövaliye** (*Fr. chevalier*) Knight. **~ yüzük,** ring with a crest on it. **~lik,** chivalry.
**şöyle** In that manner; so; just. Of that sort, such. **~ bir baktı,** he just glanced at, *or* he looked with contempt: **~ böyle,** so so; not too well; roughly speaking: **~ dursun,** ... let alone ...: **~ ki,** in such a manner that; as follows; that is to say. **~ce** (ˋ··), in this manner; in such wise.
**şu, -nu** (*pl.* **şunlar**) This; that (**şu** *is between* **o** = that there *and* **bu** = this here); this person; this thing. **~nu bunu bilmem,** I won't hear of any excuses *or* objections: ⌜**~nun bunun ~su busu ile alâkadar olmıyan**⌝, not interested in other people's private affairs: **~na buna,** to this person and that, to all sorts of people: **şundan bundan (şuradan buradan) konuşmak,** to talk about this and that: **~ kadar,** *v.* **şukadar.**
**şubat, -tı** February.
**şube** (−ˋ) Branch; section; branch office. **~lenmek,** to branch out, ramify.
**şuh** Lively; full of fun; coquettish; pert. **~luk,** liveliness, playfulness; coquettishness.
**şukadar** (ˋ··) So much; so many; so; this much. **~ (var) ki,** moreover, it remains to be said: **bundan ~ sene evvel,** many years ago.
**şule** (−ˋ) Flame.
**şum** *v.* **şom.**

**şuncağız** *dim. of* **şu.** This little one.
**şundan, şunu, şunun** *v.* **şu.**
**şûra** (–⌄) Council. **Şûrayı Devlet,** the Council of State.
**şura** (⌄·) This place; that place. **~m,** this part of me: **~da,** here; **~da burada,** here and there: **~ya,** hither: **~dan,** hence: **~ları,** these places: **~larda,** in these parts: **~sı muhakkaktır,** this much is certain. **~cık** (⌄··), just here; close by. **~lı** (⌄··), belonging to this place; inhabitant of this place. **~sı** (⌄··), **-nı,** this place; this fact: **~nı unutmamalı,** one must not forget this point.
**şurdan** *for* **şuradan,** *v.* **şura.**
**şurub** Syrup; sweet medicine.
**şuur** (·⌄) Comprehension; intelligence; conscience; mind. **~u bozuk,** out of his senses. **~altı,** subconscious. **~lu,** intelligent; conscious (will, *etc.*); sensible, judicious. **~suz,** unconscious (deed *etc.*); callous, heedless; unreasonable.
**şübhe** Doubt; suspicion; uncertainty. **···den ~ etm.,** to suspect: **~ye düşmek,** to begin to suspect, to have a suspicion: **ne ~ ?,** what doubt can there be?, most certainly!
**şübhe·lenmek,** to have a suspicion *or* doubt (about = **den**). **~li,** doubtful; causing suspicion.
**şükr·an** Thankfulness; gratitude. **~etmek** (⌄··), **-eder,** to feel grateful; be thankful (for = **dat.**): ⌜**bir yiyip bin ~**⌝, to be very thankful for one's circumstances. **~ü,** *v.* **şükür.**
**şükür, -krü** Thanks; gratitude. **~ ki** or **Allaha ~,** thanks to God :**···in şükrünü bilmek,** to be grateful for ... .
**şümul** (·⌄), **-lü** An including *or* comprehending; comprehensiveness. **~ü olm.,** to include, cover, embrace. **~lendirmek,** to amplify, extend, generalize. **~lü,** comprehensive.
**şüphe** *v.* **şubhe.**
**şürekâ** (··⌄) *pl. of* **şerik.** Partners *etc.*
**şüru** (·⌄), **-uu** Commencement. **...e ~ etm.,** to commence ... .
**şüt, -tü** Shot (football). **~ çekmek,** to shoot (football).
**şüyu**[1] (·⌄), **-uu** Publicity; divulgation. **~ bulmak,** to be noised abroad, to become common gossip: ⌜**bir şeyin ~u vukuundan beterdir**⌝, the news of an event is often more harmful than its happening.
**şüyu**[2] (·⌄) Undivided shares *in a property.* **izalei ~,** the dividing up of a property amongst various owners.

# T

**ta**[1] And, *v.* **da**[1].
**ta**[2] (–) Even; even as far as; even until. **~ ki,** so that, in order that: **~ kendisi,** his very self: **hakikatin ta kendisidir,** it is the very truth.
**taaccüb** Wonder, astonishment. **···e ~ etm.,** to marvel, be astonished at.
**taaddüd** A multiplying; plurality. **~ etm.,** to multiply, to become frequent. **~üzevecat, -tı,** polygamy.
**taaffün** Putrefaction; stink.
**taahhüd** Undertaking *or* engagement *to do stg.*; registration *of letters etc.* **~ etm.,** to undertake, engage to do. **~at,** (··⌄), **-tı,** *pl. of* **taahhüd: âza ~ı,** membership fees. **~lü,** registered (letter). **~name** (···–⌄), written undertaking; contract.
**taalâllah** (·⌄··) *In* ⌜**tevekkel ~**⌝, *pop. pronunciation of* **tevekkeltu alâllah,** I resign myself to God, *i.e.* I resign myself to my fate; I need not do anything about it.
**taallûk, -ku** Connexion; relation; attachment. **···e ~ etm.,** to have connexion with, to concern. **~at** (···⌄), **-tı** Relations; family connexions.
**taammüd** An acting intentionally; premeditation. **~en** (·⌄··), intentionally; with premeditation.
**taammüm** A becoming general. **~ etm.,** to become general, to spread.
**taannüd** Obstinacy; persistence.
**taarruz** Attack; aggression; assault; molestation. **···e ~ etm.,** to attack, assault; violate (a woman); **···le ~ etm.,** to be in conflict with. **~î** (···⌄), aggressive; offensive (attacking).
**taassub** Bigotry; fanaticism; zeal, earnestness. **~kâr,** fanatical.
**taayyün** A being manifest. **~ etm.,** to become clear *or* manifest; to be defined *or* determined.
**taazzum** A pretending to be great; arrogance.
**tab** (–) Strength. **~ütüvanı kalmadı,** he has no strength left.
**tababet** (·–⌄), **-ti** The science of medicine.
**tabak**[1] Plate; dish. **~ gibi,** quite flat.

**tabak²** Tanner. ~**hane** (··–⌄), tannery. ~**lık,** trade of a tanner.

**tabaka¹** Layer; stratum; sheet (of paper); fold; class.

**tabaka²** (·⌄·) Tobacco *or* cigarette box.

**taban** Sole (of a foot *or* a shoe); heel; firmness, pluck; girder, wall-plate; floor; base; plateau; fine steel; agricultural roller; bed of a river. ~ **inciri,** small sweet figs ripening last: ~**ı*** **kaldırmak,** to take to one's heels: ~ **kılıc,** sword of Damascus steel: ~**a kuvvet,** by dint of hard walking: ~**ı*** **sızlamak,** for one's soles to ache from fatigue: ~ **tepmek,** to walk a long way; to tire oneself by walking: ~**ları*** **yağlamak,** to prepare to go to a distant place on foot: ~ ~**a zıd,** diametrically opposed.

**tabanca** (·⌄·) Pistol.

**taban·lı** Soled; brave, firm. ~**sız,** soleless; cowardly, weak. ~**vayla,** on foot (*comic variation of* **tramvay**).

**tabasbus** A cringing *or* fawning. ···**e** ~ **etm.,** to fawn and flatter.

**tabelâ** (·⌄·) Table (list); soldier's ration-book; list of food *in schools, hospitals etc.*; card of treatment *hung on a patient's bed in hospitals*; sign *of a shop or firm*. ~**cı,** sign-painter.

**tab'etmek** (⌄··), **–eder** *va.* Print.

**tabı, –b'ı** Natural quality, disposition: a printing *or* stamping; an edition.

**tabıl** Drum; tympanum.

**tâbi** (–⌄), **–ii** Following; dependent; imitating; subject; conforming; submissive. Subject *of a state*; tributary *of a river*. ···**e** ~ **olm.,** to follow; to be dependent on; to be subject to; to imitate.

**tabi** (–⌄), **–ii** Printer; editor; publisher.

**tabiat** (·–⌄), **–ti** Nature; character, natural quality, disposition; habit; regularity *of the bowels*; taste, refinement. ~**ile,** naturally, of itself: ···**i** ~ **etm.,** to make a habit of ...: ~ **sahibi,** a man of taste. ~**li,** having *such and such* a nature; possessing good taste. ~**siz,** devoid of good taste. ~**üstü,** supernatural.

**tabib** (·⌄) Doctor, physician.

**tabiî** (·–⌄) Natural; normal. Naturally, of course. **def'i** ~, a going to stool. ~**yeci,** teacher of *or* specialist in natural history. ~**yun** (·–·⌄), naturalists; natural philosophers.

**tâbiiyet, –ti, tâbilik** Nationality; a conforming; dependence.

**tâbir** (–⌄) Word; phrase; expression; explanation; interpretation of dreams. ~ **olunmak,** to be called, to have the name of. ~**at, –tı,** *pl. of* tâbir. ~**ci,** interpreter of dreams. ~**name,** book explaining the interpretation of dreams.

**tâbiye** (–·⌄) Tactics. ~**vî** (–··⌄), tactical.

**tabla** (⌄·) Circular tray; ash-tray; scale of a balance; flat surface; disk.

**tablo** (*Fr. tableau*) Picture.

**tabu** Taboo.

**tabur** Battalion. ~**cu,** discharged from hospital, *esp.* soldier passed fit for service after an illness: ~ **edilmek,** to be sent back to his battalion.

**tabure** (*Fr. tabouret*) Footstool.

**tabut, –tu** Coffin; bier; large egg-box. ~**luk,** place in a mosque where coffins are laid.

**tabya** (⌄·) Bastion; redoubt; fort.

**tac** Crown; diadem; crest *of a bird*. ~ **giymek,** ~**lanmak,** to be crowned: **baş** ~**ı,** loved and honoured above all (*mainly of a person*). ~**dar** (–⌄), crowned head, sovereign.

**Tacik** Persian inhabitant of Turkestan.

**tâcil** (–⌄) **etm.** *va.* Hasten; accelerate.

**tacir** (–⌄) Merchant.

**tâciz** (–⌄) A bothering *or* worrying. ~ **etm.,** to annoy, worry, disturb.

**tad·almak** *and* ~**ı** *v.* tat. ~**ıcı,** taster. ~**ım,** the faculty of taste. ~**ımlık,** just a taste.

**tadil** (–⌄) Adjustment; rectification; modification. ~ **etm.,** to adjust, modify, moderate. ~**ât** (––⌄), **–tı,** modifications. ~**en** (–⌄·), by way of modification *or* rectification.

**taflan** Cherry laurel. Dark green (eyes). **frenk** ~**ı,** Portugal laurel.

**tafra** Conceit; pride. ~ **satmak,** to give oneself airs. ~**cı,** conceited. ~-**furuş,** conceited; boastful.

**tafsil** (·⌄) Detailed explanation; detail. ~ **etm.,** to explain in detail. ~**ât, –tı,** details, particulars. ~**en** (·⌄·), in detail.

**tafta** (⌄·) Taffeta.

**tagallüb** Usurpation; domination; tyranny; mastery. ~ **etm.,** to usurp power.

**taganni** (··⌄) A singing.

**tagayyüb** Disappearance.

**tagayyür** Change, variation; deterioration.
**tağdiye** A feeding; feed, input.
**tağşiş** (·ʹ) Adulteration.
**tağyir** (·ʹ) Change; deterioration.
**tahaddüs** Occurrence.
**tahaffuz** A guarding oneself.
**tahakkuk, -ku** A proving to be true; a being realized; verification. ~ **etm.**, to prove true; to be realized; to come into existence: ~ **ettirmek,** to certify, verify; realize (a desire *etc.*): ~ **memuru,** official charged with the final assessment of a tax: ~ **tarihi,** date when a tax becomes due.
**tahakküm** Arbitrary power; dictation; oppression. ~ **etm.**, to dominate, tyrannize, oppress.
**tahallûs** Adoption of a pseudonym.
**tahammül** Endurance, patience. ~ **etm.**, to endure, support, put up with: ~ **olunmıyacak derecede,** to an intolerable degree.
**tahammür** Fermentation.
**tahan** *v.* **tahin.**
**taharet** (·—ʹ), **-ti** Cleanliness; canonical purification.
**taharri** (··∸) Search; research; investigation. ~ **(memuru),** plainclothes police, detective. ~**yat, -tı,** *pl. of* taharri.
**tahassür** Longing.
**tahassüs** Feeling; sensation. ~**at, -tı,** impressions, feeling (being touched).
**tahaşşüd** Concentration.
**tahattur** An occuring to the mind.
**tahayyül** Imagination; fancy. ~ **etm.**, to imagine, fancy. ~**ât, -tı,** ideas; fancies.
**tahayyür** Amazement; bewilderment.
**tahdid** (·ʹ) Limitation. ~ **etm.**, to limit. ~**at** (·—ʹ), **-tı,** limitations.
**tahfif** (·ʹ) Alleviation; assuagement. ~ **etm.**, to lighten, relieve, mitigate.
**tahın, -hnı** A grinding.
**tahin** (—ʹ) Flour, *esp.* sesame flour; sesame oil. ~ **helvası,** sweetmeat *made of sesame seed with honey or sugar.* ~**î** (—·ʹ), of a yellowish-grey colour.
**tahkik** (·ʹ), **-ki** Verification; investigation. ~**etm.**, to verify, ascertain, investigate. ~**at** (·—ʹ), **-tı,** investigations; research; inquiry.
**tahkim** (·ʹ) Act of strengthening *or* fortifying. ~ **etm.**, to strengthen; to fortify. ~**at, -tı,** fortifications.
**tahkir** (·ʹ) Insult.
**tahkiye** Narration, story-telling.
**tahlif** (·ʹ), **etm.**, *va.* Administer an oath to; adjure.
**tahlil** (·ʹ) Analysis. ~ **etm.**, to analyse. ~**î** (·—∸), analytical.
**tahlis** (·ʹ) A liberating *or* rescuing. ~ **etm.**, to free, to rescue. ~**iye** (·—·ʹ), lifeboat service: ~ **sandalı,** lifeboat: ~ **simidi,** lifebuoy.
**tahliye** An emptying; discharge (of cargo); evacuation. ~ **etm.**, to discharge (cargo); empty; set free. ~**ci,** stevedore.
**tahmil** (·ʹ) A loading *or* imposing. ~**etm.**, to load (ship); to impose (a burden); to impute.
**tahmin** (·ʹ) Estimate; conjecture, guess. ~ **etm.**, to estimate, conjecture, calculate: **göz** ~**ile,** by eye (*in judging distance, weight etc.*). ~**en** (·∸·), approximately. ~**î** (·—∸), approximate; conjectural.
**tahmis** (·ʹ) A parching *or* roasting; coffee roasting and grinding establishment. ~**çi,** a roaster and grinder of coffee.
**tahnit** (·ʹ), **-ti** Embalming *of the dead*; stuffing *of animals or birds.*
**tahra** Pruning hook.
**tahrib** (·ʹ) Destruction. ~ **etm.**, to destroy. ~ **maddesi,** explosive: ~ **tanesi (mermisi),** high-explosive shell: ~**at** (·—ʹ), **-tı,** destructions. ~**kâr,** destructive, deadly.
**tahrif** (·ʹ) Distortion; falsification *of a document by erasure or addition*; fraudulent alteration. ~ **etm.**, to falsify, distort, misrepresent, alter fraudulently.
**tahrik** (·ʹ), **-ki** Incitement; instigation; provocation. ~**amiz** (·——∸), subversive, provocative. ~**kât** (·—∸), **-tı,** movements, instigations; subversive acts.
**tahrilli** *In* ~ **göz,** an eye that naturally looks as if the eyelids had been artificially darkened.
**tahrim** (·ʹ) Prohibition. ~ **etm.**, to prohibit, to declare unlawful.
**tahrir** (·ʹ) A writing; a setting forth in words; essay; a registering. ~**at** (·—ʹ), **-tı,** documents; dispatches; *as sing.* official letter; circular note: ~ **kalemi,** secretariat: ~ **müdürü,** secretary-general. ~**en** (·∸·), in writing. ~**î** (·—∸), written.
**tahrirli** *v.* **tahrilli.**
**tahriş** (·ʹ) Irritation.

**tahsil** (·ˋ) Production; acquisition; collection of taxes; study, education. ~ **etm.**, to produce; acquire; collect (dues *etc.*); study. ~**i emval kanunu,** law concerning the collection of taxes and legal penalties: ~ **görmek,** to study: **hasılı** ~ **etm.**, to produce what already exists, to 'carry coals to Newcastle': **ilk** ~, elementary (primary) education. ~**at** (·–ˋ), **–tı,** moneys collected; dues, taxes. ~**dar,** tax-collector; agent.

**tahsin** (·ˋ) Admiration; certificate *formerly given as a prize in schools.*

**tahsis** (·ˋ) Assignment; appropriation. ~ **etm.**, to assign *to a special purpose or person.* ~**at** (·–ˋ), **–tı,** allowance; grant.

**tahşid** (·ˋ) Concentration. ~ **etm.**, to assemble, to concentrate (*esp.* troops). ~**at** (·–ˋ), **–tı,** concentrations.

**tahşiye** Annotation; marginal note.

**taht**[1], **–tı** Throne. ~**a geçmek,** to succeed to the throne: ~**tan indirmek,** to dethrone: ~**a oturtmak,** to enthrone.

**taht**[2], **–tı** Under surface; space under; *as prefix,* under-, sub-.

**tahta** Board, plank; sheet (of metal); bed *in a garden*; wood. Wooden. ~**ya çamaşıra gitmek,** to go as charwoman to a house: ~**ya kaldırmak,** to make a pupil go to the blackboard: ʹ**bir** ~**sı eksik**ʹ, having 'a screw loose'; half-witted: **bostan** ~**sı**, garden-bed: **iman** ~**sı**, the breast: **yaş (çürük)** ~**ya basmak,** to fall into a trap: **yazar bozar** ~**sı**, child's slate: **yek** ~**da,** at one go.

**tahta·biti, –ni,** bed-bug. ~**boş,** raised platform on a roof *with posts for clothes-lines.* ~**kurusu, –nu,** bed-bug. ~**lı,** planked, boarded; wood-pigeon: ~ **köy** (*sl.*), cemetery: ~ **köye gitmek,** to die. ~ **pabuc,** slippers. ~**perde,** fence *or* partition of boards. ~**revalli** (···ˋ·), see-saw.

**taht·elârz** (*accent on this and following words is on the first syllable*) Subterranean. ~**elbahir,** submarine. ~**elcild,** subcutaneous. ~**elhıfz,** under escort. ~**essıfır,** below zero. ~**eşşuur,** subconscious.

**tahtıravan** Litter; palanquin.

**tahvil** (·ˋ) A transforming *or* converting; conversion *of debt*; draft, security. ~ **etm.**, to convert, transmute. ~**ât, –tı,** securities; debentures.

**taife** (–·ˋ) Body of men, crew, gang; class; tribe.

**tak**[1] *Imitates a knocking sound.* **canına*** ~ **demek (etmek),** to become intolerable.

**tak**[2] (–), **–kı** Arch; vault. ~**ı zafer,** triumphal arch.

**tak**[3] *imp. of* **takmak. süngü** ~ **vaziyetinde,** with bayonets fixed.

**taka** (ˋ·) Small sailing-boat *used in the Black Sea.*

**takaddüm** Precedence; priority. ···**e** ~ **etm.**, to precede, to anticipate.

**takallûs** Contraction *of muscles etc.*

**takarrüb** Approach; proximity.

**takarrür** A being established; a being decided. ~ **etm.**, to be decided *or* confirmed.

**takas** A setting off claims against each other; clearing (*fin.*); exchange of goods; compensation. ~ **etm. (olm.)**, to balance off (mutual claims, debts *etc.*): ~ **odası,** clearing-house. ~**lamak,** to compensate. ~ **tukas olm.**, to be all square (of claims, accounts *etc.*).

**takat** (–ˋ), **–ti** Strength; power. **buna** ~ **getirmez,** this is beyond his strength.

**takatuka** (··ˋ·) Noise; tumult; a kind of roller *used in printing.*

**takav** Horseshoe. ~**cı,** farrier.

**takayyüd** Attention; care. ···**e** ~ **etm.**, to pay attention to, to take care of. ~**at, –tı,** precautions, precautionary measures.

**takaza** (·–ˋ *but coll.* ··ˋ) Taunt.

**takbih** (·ˋ) Disapproval; blame.

**takbil** (·ˋ) Kissing.

**takdim** (·ˋ) A giving precedence; presentation; offer. ~ **etm.**, to give precedence *or* preference; to present, offer; introduce *one person to another*: ~ **ve tehir,** transposition of terms: **ehemmi mühimme** ~ **etm.**, to do the most important thing first. ~**e** (·–ˋ), offering to a superior, presentation.

**takdir** (·ˋ) Predestination, fate; appreciation; supposition; case. ~ **etm.**, to foreordain, prearrange; know the value of, appreciate; understand; suppose; take for granted; allot (a task): ~ **böyle imiş,** it was so fated: **o** ~**de,** in that case.

**takdirkâr** Appreciative; admirer.

**takdis** (·ˋ) Sanctification; consecration; veneration.

**takılış** Banter, raillery.

**takılmak** *vn. pass. of* **takmak.** Be

affixed; be stuck; attach oneself *to a person*. *va.* (*with dat.*) Deride; banter; ridicule; worry. ···e gözü*~, for one's eye to be caught by *stg.*: ···e zihni*~, to apply one's mind to *a problem etc.*

**takım** A set, lot, *or* number *of things*; tea (dinner *etc.*) service; suit *of clothes*; suit *of cards*; squad of men, crew, gang, team; class *of people*; cigarette-holder. ~ ~, in sets, in lots; in classes: **av ~ı**, shooting *or* fishing tackle: **ayak ~ı**, the common people: **bir ~**, a lot of: **çalgı ~ı**, orchestra: **yatak ~ı**, bed-clothes.

**takımadalar** (·ˈ···) Archipelago.

**takın·mak** *va.* Attach to oneself; put on, wear; assume (attitude), put on, affect. ⌜**tel(ler) takınsın**⌝, 'well, let him rejoice (but *I* certainly shall not)!': ⌜**terbiyeni takın!**⌝, behave yourself. **~tı**, small debt; temporary failure to pass an exam.; connexion *with a person*; affair *with a woman*.

**takır** ~ ~, *imitates the noise of horses' hooves etc.* ~ **tukur**, *imitates an alternate tapping and knocking noise.* **~damak**, to make a tapping *or* knocking noise. **~tı**, a tapping *or* knocking noise.

**takıştırmak** *va.* Fasten on neatly; wear as an ornament; dress up. **takıp ~**, to ornament oneself.

**takızafer** (—··ˈ) Triumphal arch.

**takib** (—ˈ) Pursuit; persecution. ~ **etm.**, to follow, pursue, follow up. **~at** (——ˈ), **-tı**, (**kanunî**) ~, legal proceedings, prosecution.

**takke** Skull-cap; night-cap. **~sini havaya (göğe) atmak**, to throw one's hat into the air for joy: ⌜**al ~ ver külâh**⌝, familiar joking and talk; a being on very intimate terms.

**takla, taklak** Somersault. ~ **atmak**, to turn a somersault: ~ **attırmak**, to twist *s.o.* round one's little finger: ⌜**kırk paraya dokuz ~ atar**⌝, he'd do anything for money. **~cı**, tumbler, acrobat.

**taklavat** *In* **takım** ~, bag and baggage; the whole family; the whole lot of them.

**taklib** (·ˈ) Inversion; reversal.

**taklid** (·ˈ) Imitation, counterfeiting. Counterfeit, sham. ~ **etm.**, to follow blindly; imitate; feign, sham: ···**in ~ini yapmak**, to mimic *s.o.*

**takma** *verb. noun of* **takmak.** Act of attaching *etc.* Stuck on; attached; false (beard *etc.*); that can be fitted together; pre-fabricated (house). ~ **aletleri**, make-up appliances.

**takmak** *va.* Affix, attach; put on; give (a name *etc.*); give as a present *to a bride*; (*sl.*) incur a debt; (*sl.*) surpass; (*sl.*) *abbrev. for* **kancayı ~**, *v.* **kanca**: **adam ~**, to appoint a man to accompany another: ···**e çelme ~**, to trip up, to play a dirty trick on.

**takmamak** *neg. of* **takmak.** (*sl.*) Take no notice of.

**takoz** Wooden wedge; prop *used to shore up a ship on the ways.*

**takrib** (·ˈ) Approximation. **~en** (·ˈ·), approximately, about. **~î** (·—ˈ), approximate.

**takrir** (·ˈ) A confirming; deposition, statement, memorandum; report; official note; official notification of transference of real property; delivery (manner of speaking of lecturer *etc.*); motion (in an assembly). ~ **etm.**, to establish, confirm; depose, state, set forth: ~ **sahibi**, the mover of a motion: **ders ~ etm.**, to give a lesson: **sual ~i**, interpellation.

**takriz** Eulogy of a book; appreciatory preface by an important literary man to another's book.

**taksa** (ˈ·) Postage due. **~pulu, -nu**, postage-due stamp.

**taksi** (ˈ·) Taxicab.

**taksim** (·ˈ) Division; partition; distribution; instrumental solo. ~ **etm.**, to divide into parts; share out; distribute; divide (*arith.*): **~at** (·—ˈ), **-tı**, compartments of a building: **idarî (mülkî) ~**, administrative divisions (vilayets *etc.*).

**taksir** (·ˈ) Fault. **aç ~**, fasting, hungry. **~at** (·—ˈ), **-tı** *pl. of* **taksir** *also used as sing.*; fault; sin: ⌜**Allah ~ını affetsin!**⌝, *said when criticizing a dead person.*

**taksit** (·ˈ), **-ti** Instalment.

**taktak** Wooden instrument for beating the washing.

**taktuk** *Imitates the sound of knocking.*

**takunya** (·ˈ·) Clog; sabot.

**takuş** *In* ~ **tukuş**, with a clatter.

**takva** (·ˈ) Fear of God; piety.

**takvim** (·ˈ) Almanac, calendar.

**takviye** Reinforcement.

**takyid** (·ˈ) Restriction. **~at** (·ˈ), **-tı**, restrictions (*also used as sing.*).

**talâk, -kı** Divorce. **~ı selâse**, final and irrevocable divorce.

**talâkat** (·—ˈ), **-ti** Eloquence.

**talan** Pillage. **alan ~,** in utter confusion.
**talaş** Wood shavings; sawdust; filings. **~ kebabı,** a kind of meat patty.
**talaz** Wave, billow; (of silk *etc.*) a being ruffled up.
**taleb** Request; demand. **~name** (··–̀), written request.
**talebe** Student; pupil.
**tali**[1] (–̀), **–i'i** Pollen.
**tali**[2] (–̀) *v.* **talih.**
**tâli** (–́) Secondary (education); subordinate.
**talib** (–̀) Applicant; suitor; customer. **···e ~ olm.,** to seek after, aspire to, strive for.
**talih** (–̀) Good fortune; luck. **birinin ~ine bakmak,** to tell *s.o.'s* fortune. **~li,** lucky. **~siz,** unlucky.
**talik** (–*), **–kı** Suspension; Persian style of Arabic writing. **~ etm.,** to suspend; put off, defer.
**talika** (·̀·) Four-wheel cart suspended on straps.
**talim** (–*) Instruction; practice; drill. **~ ve terbiye,** instruction, training: **~ fişeği,** blank cartridge. **~at** (––*), **–tı,** instructions. **~atname** (––·–̀), book of instructions. **~hane** (–·–̀), parade ground; drill-hall. **~li** (–·̀), instructed, practised, drilled. **~name** (–·–̀), drill-book.
**talimar** Cutwater *of a ship.*
**talisiz** Unfortunate.
**talk, –kı** Talc.
**talkın** *Coll. form of* **telkin,** *v.* **salkım.**
**tallahi** (̀–·) *Usually in conjunction with* **vallahi,** by God!
**taltif** (·*) Favour; recompense. **···le ~ etm.,** to confer (a rank *etc.*).
**tam** Complete; perfect. Completely; exactly. **~ yol,** full speed.
**tamah, tama** Greed, avarice. **···e ~ etm.,** to covet, desire. **~kâr,** greedy; avaricious; stingy.
**tamam** (·*) Whole. Complete; finished; ready; 'pat', just right; true, correct. **~!,** that's right!, that's it!, *or used to express unpleasant surprise,* 'there you are!': **~ile,** wholly, in its entirety: **···e ~ gelmek,** to just suit.
**tamam·en** (·́·), completely, entirely. **~lamak,** to complete, finish; make good *a defect.*
**tambur** A kind of guitar. **~a** (·̀·), a small **tambur: ağız ~sı,** Jews'-harp. **~acı,** player of the **tambura. ~î** (·–́), player of the **tambur.**
**tamim** (–*) Generalization; circular. **~ olunmak,** to be circulated. **~en** (–́·), by circular.
**tamir** (–*) Repair; restoration. **~at** (––̀), **–tı,** repairs.
**tampon** Buffer; wad, plug (*med.*).
**tamtakır** (̀··) Absolutely empty.
**tamters** (̀·) The exact opposite.
**tan** Dawn.
**tanassur** Conversion to Christianity.
**tandır** Oven *made in a hole in the earth.* **~ ekmeği,** bread baked in an earth oven. **~name** (··–̀), an old wives' tale.
**tane** (–̀) Grain; seed; pip; piece; bullet, cannon-ball; a single thing. **~ ~,** in separate grains; one by one; **~ ~ söylemek,** to speak each word distinctly: **~ye gelmek,** to form seed *or* berries: **kaç ~ ?,** how many?: **üç ~ elma,** three apples: **ver bir ~,** give me one.
**tane·cik** (–·̀), unique, only: granule. **~lemek,** to separate into grains; granulate. **~lenmek,** to produce grains *or* berries; to be separated into grains. **~li,** having grains *or* berries; in separate grains.
**tanen** Tannin.
**tangır tungur** Loud clanging noise.
**tango** (̀·) Tango; loudly-dressed woman.
**tanık** Witness. **~lık,** evidence.
**tanı·mak** *va.* Know, be acquainted with; recognize; acknowledge; listen to. **tanımamak,** to pay no attention to.
**tanı·mak, ~lmak,** *vn. pass. of* **tanımak. ~şıklık,** mutual acquaintance. **~şmak,** to make acquaintance with one another; be acquaintances. **~tmak,** *caus. of* **tanımak,** to introduce *one person to another.*
**tan·lamak** *vn.* Dawn. **~layın,** at dawn.
**tannan** Resounding; ringing.
**Tanrı** God. **~ dağı,** the Tien Shan range: **~nın günü,** every blessed day.
**tansiyon** (*Fr. tension*) Blood-pressure.
**tantana** Pomp, display; magnificence.
**tantuna** *In* **~ gitmek** (*sl.*), to be lost, ruined, dismissed.
**tanyeri** (̀··), **–ni** East; dawn. **~ ağarıyor,** dawn is breaking.
**tanzifat** (·–*), **–tı** Town scavenging service.
**tanzim** (·*) A putting in order; an

organizing. **~at** (·–⁎), **–tı,** reforms; reorganization; the political reforms of Abdulmejid in 1839. **~atçı,** reformer.

**tanzir** (·⁎) An imitating a poem. ~ **etm.,** to imitate; to produce a similar.

**tapa** (⁎·) Fuse; *v.* **tıpa.**

**tapan** Harrow.

**tapın·ak** Temple. **~mak,** to worship, adore (*with dat.*).

**tapkur** *In* ~ **kolanı,** girth, surcingle.

**tapmak** *va.* (*with dat.*) Worship.

**tapon** Discarded; common, second-rate.

**taptaze** (⁎··) Absolutely fresh.

**tapu** (⁎·) Title-deed.

**taraça** (·⁎·) Terrace.

**taraf** Side; direction; district; part; end; party *to a cause or dispute*; protector. ~ ~, on this side and that: ... **~ına,** towards ... : ... **~ından,** on the part of ..., by ..., from the direction of ... : **o ~a,** in that direction: **bu ~lar,** these parts, this neighbourhood: **...in ~ını tutmak** (*or* **···den ~a çıkmak**), to take the part of, to side with: **ona ~ çıktı,** he became his protector.

**taraf·dar,** partisan; adherent: **···e ~ olm.,** to be in favour of. **~darlık,** partiality, partisanship. **~lı,** supporter. **~sız,** neutral, impartial. **~sızlık,** neutrality, impartiality.

**tarak** Comb; rake; harrow; weaver's reed; crest (of a bird); drag; gills *of a fish*: instep; scallop; serrated pattern *on cloth etc.* ~ **dubası,** dredger: **ayak tarağı,** tarsus: **kar ~ makinesi,** snow-plough: ⌜**kırk (bin) ~ta bezi var**⌝, 'he has cloth on a thousand looms', he looks after a host of things, he has many irons in the fire: **vida tarağı,** screw chaser.

**tarak·çı,** maker *or* seller of combs. **~lamak,** to comb; rake; harrow; dredge; paint with zigzag lines. **~lı,** crested (bird); broad-footed. **~otu** (·⁎··), **–nu,** teazle.

**tarama** Act of combing *or* dredging; search; soft roe; red caviare. ~ **resim,** shaded drawing, hachure: **arama ~,** search *by police etc.*

**tara·mak** *va.* Comb; rake; harrow; dredge; search minutely. **arayıp ~,** to search carefully for *stg.*; to make minute inquiries about *s.o.'s* whereabouts: **demir ~,** to drag anchor. **~nmak,** to comb oneself; be combed, raked *etc.*

**tarassud** Observation. ~ **altında,** under surveillance.

**tarator** Sauce made with vinegar and walnuts.

**taravet** (·–⁎), **–ti** Freshness; juiciness.

**tarayıcı mayn ~ gemi,** minesweeper.

**taraz** Combings; fibres combed out. ~ ~ **(turaz),** dishevelled. **~lanmak,** to become rough by combing *or* friction; be frayed; be dishevelled.

**tarçın** Cinnamon.

**tard** Expulsion, repulsion. ~ **etm.,** to drive away; expel; degrade (an officer).

**tarh**[1] Flower-bed; garden border.

**tarh**[2] Subtraction; imposition *of taxes.*

**tarhana** Preparation of dried curds and flour; soup made of this.

**tarhun** Tarragon.

**tarım** Agriculture.

**tarınmak** *In* **arınıp ~,** to be thoroughly cleaned.

**tarif** (–⁎) Description; definition; recipe.

**tarife** (–·⁎) Tariff; time-table.

**tarih** (–⁎) History; date; epoch; chronogram. ~ **atmak,** to put the date: **~a karışmak,** to be a thing of the past.

**tarih·çe,** short history. **~çi,** historian. **~en** (⁎··), historically. **~î** (–·⁎), historical. **~li,** dated ... . **~siz,** undated.

**tarik** (·⁎) Way, road; method. ~ **bedeli,** road-tax. **~at** (·–⁎), **–tı,** way, path; religious order. **~atçı,** member of a religious order, dervish.

**târiz** (–⁎) Allusion; hint; innuendo.

**tarla** (⁎·) Field. **~kuşu, –nu,** lark.

**tartaklamak** *va.* Harass; manhandle.

**tartı** Act of weighing; weight; scale. **~ya gelmez,** imponderable. **~lı,** weighed; balanced; well-pondered. **~lmak,** *vn. pass. of* **tartmak**; be weighed. **~şmak,** to argue, dispute.

**tartmak** *va.* Weigh; ponder well; estimate; weigh up (a person).

**tartura** Turner's wheel; wheel of a spinning-wheel.

**tarumar** (–·⁎) Scattered; in disorder. ~ **etm.,** to rout.

**tarz** Form, shape, appearance; manner; method; sort; demeanour. **~ı hâl,** method of solution (of a difficulty *etc.*): **~ı hareket,** behaviour.

**tarziye** Apology. ~ **vermek,** to make an apology.

**tas** Bowl *or* cup *with a rounded bottom.* **~ı tarağı toplamak,** to pack up and go: başı ~, bald-headed: ⌜**eski hamam eski** ~⌝, the same old story: hamam ~ı, metal bowl *used for throwing water over oneself in a bath.*

**tasa** Worry. ~**sını çekmek,** to suffer grief *or* regret for *s.o. or stg.*; ⌜~**mın on beşi**⌝, 'I don't care a hang!': ⌜~**sı sana mı düştü?**⌝, 'why worry, it doesn't concern you!': ~ **vermek,** to cause anxiety. ⌜**o** ~ **bu** ~⌝, we've got enough to worry about as it is.

**tasallût, -tu** Aggression; outrage.

**tasallüb etm.** *vn.* Become hard.

**tasallüf** A being vain and presumptuous.

**tasannu** (··⊥), **-uu** Artifice.

**tasar** Project; draft. ~**ı,** bill, draft law. ~**lamak,** to plan, project.

**tasarruf** Possession; economy, saving. ~**umda,** at my disposal: ···**e** ~ **etm.** to have the use and disposal of; to possess: ···**in** ~**unda olm.,** to be in the possession and at the disposal of: ~ **sandığı,** savings-bank.

**tasarsız** Extempore; improvised.

**tasasız** Free from care; light-hearted.

**tasavvuf** Mysticism; sufism. ~**î** (···⊥), mystical; sufic.

**tasavvur** Imagination; idea; conception. ~**unda olm.,** to propose to.

**tasdi** (·⊥) ~ **etm.,** to pay a visit to *s.o.*

**tasdik** (·⩘), **-ki** Confirmation; ratification. ~**li,** certified. ~**name** (··–⩘), certificate.

**tasfiye** Liquidation; clearance; elimination. **hesabları** ~ **etm.,** to wind up accounts. ~**hane** (···–⩘), refinery.

**tashih** (·⩘) Correction; reading of proofs.

**tasımlamak** *va.* Plan, project; imagine.

**taskebabı** (⩘···), **-nı** Meat cut in small pieces and roasted.

**taslak** Anything in the rough; draft; sketch; model. **adam taslağı,** a boor, clown: **hekim taslağı,** an incompetent, bungling doctor. ~**çı,** pattern-maker.

**taslamak** *va.* Pretend to *stg.* one does not possess; make a show of.

**tasma** Collar (of a dog *etc.*); strap *of clogs.*

**tasni** (·⊥), **-ii** Fabrication; invention. ~**at, -tı,** inventions, falsehoods.

**tasnif** (·⩘) Classification.

**tasrif** (·⩘) Declension; conjugation.

**tasrih** (·⩘) ~ **etm.,** to make clear, specify. ~**at** (·–⩘), **-tı,** explanations. ~**en** (·⊥·), by way of clarification.

**tastamam** (⩘··) Quite complete; perfect.

**tasvib** (·⩘) Approval. ~**en** (·⊥·), with approval.

**tasvir** (·⩘) Picture; description. ~ **etm.,** to depict, describe.

**taş** Stone; rock; precious stone; piece *or* man (chess, draughts *etc.*); allusion, innuendo. Stone; hard as stone. ···**e** ~ **atmak,** to direct an unpleasant allusion at *s.o.*: ~ **çatlasa,** whatever happens, under any circumstances: ···**e** ~ **çıkar(t)mak,** to give points to, to be greatly superior to *s.o.*: ~ **düşürmek,** to pass a gall- (bladder-) stone: ~**ı gediğine (yerine) koymak,** to give as good as one gets; to make a clever retort: ~ **kesilmek,** to be petrified: ⌜~**ı sıksa suyunu çıkarır**⌝, he is incredibly strong; he succeeds in whatever he undertakes: ~**a tutmak,** to pelt with stones: ⌜~ **yerinde ağırdır**⌝, the value of anything depends on its proper use: **bağıra** ~ **basmak,** to resign oneself, to suffer without complaint: **başını** ~**tan** ~**a vurmak,** to repent bitterly: **dağ** ~, a whole heap of: ⌜**hangi** ~**ı kaldırsan altından çıkar**⌝, he always turns up everywhere; he has a finger in every pie (*in a good sense*).

**taşak** Testicle.

**taş·bademi** (⩘···), **-ni** Wild almond. ~**balığı, -nı,** non-migratory fish. ~**basması, -nı,** lithograph. ~**çı,** stonemason; quarryman: ~ **kalemi,** stonemason's chisel.

**taşım** A coming to the boil.

**taşıma** *verb. n. of* taşımak. Carried. ⌜~ **su ile değirmen dönmez**⌝, an enterprise cannot succeed with inadequate means.

**taşımak** *va.* Carry; transport; bear; pass on *another's words*; spread *gossip.*

**taşınmak** *vn. pass. of* **taşımak.** Be carried *etc.*; move one's belongings to another place; change one's abode; go too often *to a place.* **düşünüp** ~, to ponder deeply.

**taşırmak** *va. caus. of* **taşmak.** Cause to overflow. *vn.* Go beyond the bounds of decency, go too far. ⌜**bardağı taşıran damla**⌝, the 'last straw'.

**taşıt** Transport; vehicle.

**taşıtmak** *va. caus. of* taşımak.

**taşkın** Overflowing; overlapping; excessive; exuberant; insolent. ~**lık,** excess; impetuosity; insolence.

**taş·lama** Stoning; grinding (*mech.*). **~lamak,** to stone; to pave with stones; to grind. **~lı,** stony. **~lık,** stony, rocky place; paved courtyard; stone threshold; gizzard.
**taşmak** *vn.* Overflow; boil over; go too far.
**taşocağı** (ˋ···), **-nı** Stone quarry.
**taşra** (ˋ·) The provinces. **~lı,** provincial. **~lık,** conditions of provincial life; the provinces.
**taş·tahta** (ˋ··) Slate *for writing.* **~-yürekli,** stony-hearted, cruel.
**tat** Taste; flavour; relish; charm. **tadını almak,** to acquire a taste for; to enjoy: **tadında bırakmak,** not to overdo *stg.*: **tadını çıkarmak,** to get the utmost enjoyment out of *stg.*: **tadı damağında* kalmış,** the flavour of it still lingers on his palate: **tadını (tuzunu) kaçırmak (bozmak),** to spoil the enjoyment (of a party *etc.*); **artık tadını kaçırıyorsun!,** you are going a bit too far!: **tadı (tuzu) kalmadı,** it no longer gives any pleasure: **tadına varmak,** to get the full flavour of, to enjoy: **tadından yenmez,** delicious; (*iron.*) insipid: **ağız tadile,** with zest: **ağzının tadını bilmek,** to be a gourmet: ⌜**ağza ~ boğaza feryad**⌝, 'the mouth enjoys the taste, the belly cries for more', *said of a very small amount of delicious food*: ⌜**ağzımın tadını bozma!**⌝, don't spoil my enjoyment!: **evin tadı,** domestic peace.
**tatar** Tartar; courier (*also* ~ **ağası**). **~böreği, -ni,** a kind of dough pasty *made with minced meat and yaourt.*
**tatarcık** Sandfly; midge.
**tatar·ı** Undercooked (food). **~sı,** rather like a Tartar.
**tatbik** (·ˋ) Adaptation; application; comparison. ~ **edilmek,** to be brought into application, to come into force (of a law *etc.*): ~ **sahasına koymak (çıkarmak),** to put into practice. **~an** (·ˋ·), conformably (to); according to. **~at** (·—ˋ), **-tı,** applications (of a law *etc.*); putting into practice; manuœvres (*mil.*): **~ta,** in practice. **~î** (·—ˋ), comparative; practical.
**tatık, tatım** A taste; small bit eaten to try the taste.
**tatil** (—ˋ) Holiday; rest. Closed (of an office *etc.*) for a holiday. **~etm.,** to suspend, to cause to cease: ~ **olm.,** to be closed (for a holiday): ~ **yapmak,** to take a holiday.
**tatlamak** *va.* Sweeten; flavour.
**tatlı** Sweet; agreeable. Sweetmeat; sweet. ~ **belâ,** 'a sweet curse', *term of endearment for a child*: ~ **dil,** soft words: ···**i ~ya bağlamak,** to settle *a matter* amicably: **canı ~,** who avoids any personal discomfort *or* labour. **~ca,** sweetish; rather agreeable. **~cı,** maker *or* seller of sweetmeats; sweet-toothed. **~lık,** sweetness; kindness. **~msı,** sweetish. **~su** (·ˋ·), drinking water; fresh water (*as opp. to* salt water): ~ **Frengi,** Levantine.
**tatmak, -dar** *va.* Taste; try; experience.
**tatmin** (·ˋ) A tranquillizing reassurance. ~ **etm.,** to satisfy (a curiosity, a desire); to calm, reassure. **~kâr,** satisfactory.
**tatsız** Tasteless; disagreeable; insipid. ~ **tuzsuz,** insipid; stupid and dull. **~lık,** insipidity, dullness; disagreeable behaviour. **~lanmak,** to become insipid; to behave in a disagreeable manner.
**taun** (—ˋ) Pest; plague; epidemic.
**tav** Proper heat *for tempering or hammering a metal*; opportune moment; well-nourished condition, fatness; water sprinkled *on paper, tobacco, etc., before pressing*; a doubling of the stakes *at gambling.* **~ını bulmak,** to acquire the right condition *for working*: ~ **fırını,** tempering furnace: **~a getirmek** *or* **~ını vermek,** to bring to the correct heat: **~ına getirmek,** to bring to the right condition (*fig.*): **~ı kaçırmak,** to miss the right moment; to let slip a suitable opportunity: ~ **sürmek,** to double the stakes: ~ **vermek,** to bring to the requisite degree of dampness.
**tava** Frying-pan; fried food; trough *for slaking lime.*
**tavaf** The ceremony of going round the Kaaba at the pilgrimage to Mecca. ···**i** ~ **etm.,** to keep going round *a place.*
**tavan** Ceiling; highest point reached *by an aircraft or a projectile.* ~ **arası,** garret.
**tavassut, -tu** Intervention; mediation.
**tavattun etm.** *vn.* Settle *in a place.*
**tavazzuh etm.** *vn.* Become clear.
**tavcı** Accomplice in a swindle.
**tavır, -vrı** Mode, manner; kind attitude; arrogant manner. ~ **sat-**

**mak,** to give oneself airs. **~lı,** having *such and such* a manner; arrogant.

**taviz** (–ʹ) Substitution; compensation; concession; compromise. **~at** (– –ʹ), **–tı,** compensations *etc.*: **~ta bulunmak,** to make concessions.

**tavla**[1] (ʹ·) Backgammon. **~ pulu,** a piece at backgammon. **~cı,** backgammon player.

**tavla**[2] (ʹ·) Stable; place where horses are tethered. **~ halatı,** tether-rope: **~ uşağı, ~ci,** stable-boy.

**tav·lamak** *va.* Bring *a thing* to its best condition, *v.* **tav**; sprinkle water on *paper, tobacco etc. before pressing*; deceive, swindle, *v.* **tavcı.** **~lanmak,** *pass. of* **tavlamak**; *also* (of an animal) to get fat. **~lı,** at its best condition, *v.* **tav**; damped (paper, tobacco); red-hot (iron); in prime condition (animal).

**tavrı** *v.* **tavır.**

**tav·samak** *vn.* Lose its **tav** *q.v.*; fall away from its prime; cool down. **pazar ~,** for the market to become dull.

**tavsif** (·ʹ) Description.

**tavsit** (·ʹ) **etm.** *va.* Cause to mediate; use as an intermediary.

**tavsiye** Recommendation.

**tavşan** Hare. **ada –ı,** rabbit: ⌜**~ boku gibi (ne kokar ne bulaşır)**⌝, 'like a hare's droppings (neither smelly nor messy)', harmless but useless: **~ dudağı,** hare-lip: ⌜**~a kaç tazıya tut (demek)**⌝, ⌜to run with the hare and hunt with the hounds⌝: **~ kanı,** bright carmine colour: ⌜**~ın suyunun suyu**⌝, a very distant connexion: **~ uykusu,** sleep with the eyes half-closed; inattention; pretended sleep: **~ yürekli,** timid.

**tavşan·cıl,** eagle, vulture. **~kulağı, –nı,** cyclamen. **~lamak,** to become as thin as a hare. **~yemez,** *nickname given to the Shiite sect called* **Kızılbaş.**

**tavuk** Hen. ⌜**benim başına gelen pişmis tavuğun başina gelmedi**⌝, you can't think what I've been through!: **~ balığı,** *a local name for* the whiting.

**tavuk·göğsü** (·ʹ··), **–nü,** sweet dish *made with milk and the pounded breast of a fowl.* **~götü, –nü,** wart. **~karası, –nı,** night-blindness, nyctalopia. **~kanadı, –nı,** fan *for fanning a fire.*

**tavus** Peacock. **~ kuyruğu,** peacock's tail; (*sl.*) violent vomiting.

**tavzif** (·ʹ) **etm.** *va.* Entrust *s.o.* with a duty.

**tavzih** (·ʹ) Clear explanation; explanatory correction. **~en** (·ʹ·), by way of explanation.

**tay** Foal.

**taya** Child's nurse.

**tayfa** Band; troop; crew; gang; sailor.

**tayfun** Typhoon.

**tayın** Ration, *esp.* small loaf issued to soldiers (*also* **~ ekmeği**).

**tâyin** (–ʹ) Appointment; designation. **~ etm.,** to appoint; decide, fix, settle.

**Taymis** (ʹ·) Thames; *The Times* newspaper.

**tayyar** Flying; volatile. **~e** (·–ʹ), aeroplane: **~ böceği,** dragon-fly. **~eci,** airman, aviator. **~ecilik,** aviation, flying.

**tayyör** (*Fr. tailleur*) Tailor-made costume.

**tazallüm** A complaining of wrong.

**tazammun** A comprising; implication.

**tazarru** (··ʹ), **–uu** Supplication.

**taze** (–ʹ) Fresh; new, recent; young; tender. Young girl. **~ yaprak,** fresh vine-leaves. **~lemek,** to freshen up; renew. **~lik,** freshness; tenderness; youth.

**tazı** Greyhound; sleuth. **~ gibi,** thin, peaked, pinched, **~lamak,** to become thin.

**tâzim** (–ʹ) An honouring; respect. **~at** (– –ʹ), **–tı,** honours; homage. **~en** (–ʹ·), out of respect; as an honour.

**taziye** (–·ʹ) Condolence. **~ etm.,** to offer condolence.

**tazmin** (·ʹ) Indemnification; indemnity. **~at** (·–ʹ), **–tı,** indemnities; reparations.

**tazyik** (·ʹ), **–ki** Pressure.

**te** And; also; *v.* **da**[1].

**teadül** (·–ʹ) A being equal.

**teahhur** Delay.

**teakub** (·–ʹ) A following one after the other.

**teali** (·–ʹ) Elevation; loftiness.

**teamül** (·–ʹ) Custom, practice.

**teati** (·–ʹ) Exchange.

**teavün** (·–ʹ) Mutual assistance.

**tebaa** Subject *of a state.*

**tebaan** (ʹ··) In conformity with (*dat.*).

**tebaiyet, –ti** Allegiance; submission.

**tebarüz** (·–ʹ) **etm.** *vn.* Become manifest *or* prominent. ~ **ettirmek,** to show clearly, to point out.

**tebayün** (·—ˡ) A being mutually different; inconsistency.

**tebcil** (·ˡ) Veneration; honouring.

**tebdil** (·ˡ) Change, alteration; conversion; exchange. In disguise, incognito. **~i hava,** change of air.

**tebeddül** A being changed; alteration. **~ât** (···ˡ), **-tı,** changes.

**tebelleş olm.** Pester, worry.

**tebelluğ etm.** *vn.* Receive an official communication; be informed.

**tebellür** Crystallization. ~ **etm.,** to crystallize; to become crystal clear.

**teberru** (··ˡ), **-uu** Charitable gift; donation. **~at, -tı,** donations.

**teberrüken** (·ˡ··) Counting *stg.* as a blessing *or* as a good omen; as a compliment.

**tebessüm** Smile. ~ **etm.,** to smile.

**tebeşir** Chalk.

**tebeyyün etm.** *vn.* Become clear; be evident; be proved.

**tebhir** (·ˡ) Fumigation; disinfection. **~hane** (··—ˡ), fumigating station.

**teb'id** (·ˡ) Banishment.

**tebliğ** (·ˡ) Transmission; communication. **resmî** ~, official communiqué. **~at** (·—ˡ), **-tı,** communications; reports.

**tebrik** (·ˡ), **-ki** Congratulation. **~ât** (·—ˡ), **-tı,** congratulations.

**tebriye** Acquittal; exoneration. **~i zimmet etm.,** to prove one's innocence.

**tebşir** (·ˡ) ~ **etm.,** to bring good news, to cheer with glad tidings.

**tebyiz** (·ˡ) ~ **etm.,** to make a fair copy.

**tecahül** (·—ˡ) ~ **etm.,** to pretend ignorance (of = **den**): **~ü arifane,** making use of an assumed ignorance for the purpose of satire *or* irony.

**tecanüs** (·—ˡ) Homogeneity.

**tecasür** (·—ˡ) ~ **etm.,** to dare, presume.

**tecavüz** (·—ˡ) Excess; aggression; attack.

**Teccal** Antichrist; imposter; fearing neither God nor man.

**tecdid** (·ˡ) Renewal. ~ **etm.,** to renew.

**teceddüd** Renovation; reform.

**tecelli** (··ˡ) Manifestation; destiny; luck.

**tecemmülât, -tı,** odds and ends; a lot of junk.

**tecennün** A becoming insane.

**tecerrüd** Isolation.

**tecessüm** ~ **etm.,** to assume, *or* appear like, a solid body; to be personified.

**tecessüs** Inquisitiveness; curiosity.

**techil** (·ˡ) **etm.** *va.* Show up *s.o.'s* ignorance.

**techiz** (·ˡ) **etm.,** to equip, fit out; to lay out *a corpse.* **~at** (·—ˡ), **-tı,** *pl. of* **techiz**; *as sing.,* equipment, *esp.* the rigging out of a ship.

**tecil** (—ˡ) ~ **etm.,** to defer, postpone (*esp.* military service *or* punishment).

**tecrid** (·ˡ) ~ **etm.,** to strip, to isolate.

**tecrübe** Trial; test; experiment; experience. **~sini etm.,** to try *stg.* out, to experiment on: ~ **tahtası,** *stg.* that can be experimented on with impunity; 'guinea-pig': **kalem ~si,** essay.

**tecrübe·li,** experienced; proved by trial. **~siz,** inexperienced; not yet tested. **~î** (··ˡ), experimental.

**tecvid** (·ˡ) Art of reading *or* reciting the Koran with proper rhythm.

**tecviz** (·ˡ) ~ **etm.,** to declare lawful, permit.

**tecziye** Punishment. ~ **etm.,** to punish.

**tedabir** (·—ˡ) *pl. of* **tedbir.** Plans; dispositions; measures.

**tedafuî** (·—·ˡ) Defensive.

**tedahül** (·—ˡ) Arrears; interaction. **~de,** in arrears.

**tedai** (·—ˡ) Association of ideas.

**tedarik** (·—ˡ) ~ **etm.,** to procure, obtain, prepare, provide. **~ât, -tı,** preparations. **~li,** prepared; fitted out: ~ **bulunmak,** to be prepared with all requirements.

**tedavi** (·—ˡ) Medical treatment; cure.

**tedavül** (·—ˡ) Circulation; currency.

**tedbir** (·ˡ) Plan; measure; course of action; foresight. **~li,** provident, thoughtful; cautious. **~siz,** improvident; thoughtless.

**tedenni** (··ˡ) Retrogression; decline.

**tedfin** (·ˡ) ~ **etm.,** to bury.

**tedhiş** (·ˡ) A terrifying. ~ **etm.,** to terrify.

**tedib** (—ˡ) ~ **etm.,** to correct, punish.

**tedirgin** Restless, discontented. ~ **etm.,** to disturb, upset.

**tediye** (—·ˡ) Payment. ~ **etm.,** to pay. **~li,** cash on delivery (C.O.D.).

**tedkik** (·ˡ), **-ki** Close examination; scrutiny. ~ **etm.,** to investigate carefully, to examine closely, to go into a matter. ~**at** (·—⁎), **–tı,** investigations; researches.

**tedric** (·⁎) An advancing by degrees. ~**en** (·—̀·), by degrees, gradually. ~**î** (·——̀), gradual.

**tedris** (·⁎) Instruction, teaching. ~**at** (·—⁎), **–tı,** instruction; course of lessons.

**tedvin** (·⁎) **etm.** *va.* Codify; register.

**tedvir** (·⁎) ~ **etm.,** to cause to revolve; to administer *a business.*

**teeddüb** A refraining (out of good manners) from some action. ~**en** (·ˋ··), out of politeness.

**teehhül** A marrying. ~ **etm.,** to marry.

**teehhür** Postponement; delay.

**teemmül** ~ **etm.,** to reflect, deliberate.

**teenni** (··—̀) Circumspection; caution.

**teessüf** Regret; a being sorry.

**teessür** A being affected; emotion; grief.

**teessüs** A being founded *or* established.

**teeyyüd** ~ **etm.,** to be strengthened, confirmed.

**tef** Tambourine. ···**i** ~**e koymak** (~**e koyup çalmak**), publicly to speak ill of, to hold up to public ridicule.

**tefahür** (·—ˋ) Boasting; arrogance. ~ **etm.,** to boast (about = **ile**).

**tefe** Machine for winding silk; hank of spun silk; flap of a saddle.

**tefeci** Usurer. ~**lik,** usury.

**tefecik** *In* **ufacık** ~, tiny; dainty.

**tefehhüm** **etm.** *va.* Understand gradually; perceive.

**tefek** *Only in* **ufak** ~, small; insignificant: **ufak** ~ **şeyler,** various trifles.

**tefekkür** Reflection.

**tefeli** Of close texture (cloth *etc.*).

**teferruat, –tı** Details; accessories.

**teferrüd** **etm.** *vn.* Distinguish oneself.

**tefessuh** Decomposition; putrefaction.

**tefevvuk, –ku** Superiority.

**tefeyyüz** ~ **etm.,** to profit (morally).

**tefne** (ˋ·) Bay-tree.

**tefrik** (·⁎), **–kı** Separation; distinction. ~ **etm.,** to separate, to distinguish; to discriminate.

**tefrika** Discord; serial *in a newspaper.* ···**i** ~**ya düşürmek,** to sow discord between.

**tefriş** (·⁎) ~ **etm.,** to spread (carpet *etc.*); to furnish; to cover (a floor). ~**at** (·—⁎), **–tı,** furnishings.

**tefrit** (·⁎), **–ti** A doing less than one's duty; remissness; deficiency. **birincisi ifrattı, ikincisi** ~ **oldu,** the first was excessive, the second not enough: **ifrattan** ~**e,** from one extreme to the other.

**tefsir** (·⁎) Commentary; interpretation.

**teftiş** (·⁎) Investigation; inspection.

**tegafül** (·—ˋ) Feigned ignorance *or* inattention.

**teğelti** Saddle-pad; numdah.

**teğmen** Lieutenant.

**tehacüm** (·—ˋ) A rushing *or* crowding together; concerted rush *or* attack.

**tehalüf** (·—ˋ) Difference, dissimilarity.

**tehalük** (·—ˋ), **–kü** Keenness, zeal.

**tehassür** Longing; regret.

**tehcir** (·⁎) **etm.** *va.* Deport.

**tehdid** (·⁎) Threat. ~ **etm.,** to threaten.

**tehevvür** Sudden outburst of anger, fury.

**teheyyüc** Excitement; emotion.

**tehir** (—⁎) Delay; postponement.

**tehlike** Danger. ~**ye atılmak,** to court danger: ~**ye atmak,** to risk: ~**ye sokmak,** to endanger. ~**li,** dangerous. ~**siz,** without danger; inoffensive.

**tehlil** (·⁎) The recitation of the Moslem formula 'there is no God but God'.

**tehyic** (·⁎) **etm.** *va.* Excite.

**tehzil** (·⁎) **etm.** *va.* Ridicule; mock.

**tek, –ki** A single thing; odd number. Single; unique; alone, solitary; odd (not even). Only, merely; once; as long as, provided that. ~ ~, odd ones, not a pair: ~ **başına,** apart; on one's own: ~ **durmak** (**oturmak**), to sit by oneself; to be quiet: ~ **elden,** under one management *or* command; from one centre: ~**e** ~ **kavga,** single combat: ~ **tük,** here and there, now and then: **bir** ~ **atmak** (*sl.*), to have a drink: **çift mi** ~ **mi?,** odd or even?

**tekabül** (·—ˋ) ···**e** ~ **etm.,** to correspond to, to be proportional to.

**tekâmül** (·—ˋ) Evolution. ~ **etm.,** to mature, develop.

**tekâpu** (·——̀) Sycophancy; base flattery.

**tekâsüf** Condensation; density.

**tekaüd** (·—ˋ) Retirement; pension. Retired; pensioned. ~ **maaşı,** half-pay; pension: ~**e sevketmek,** *or* ~ **etm.,** to pension off. ~**iye,** pension; deduction from salary for pension.

~**lük**, a being pensioned; retirement on a pension.
**tekbaşına** (ꜝ···) All by himself.
**tekbir** (·ꜝ) A proclaiming the greatness of God in the formula '**Allahu ekber**' (God is most great). ~ **almak** (**getirmek**), to pronounce that formula.
**tekdir** (·ꜝ) Scolding; reprimand.
**teke** He-goat. ~**den süt çıkarmak**, to be very skilful in getting what one wants: **deniz** ~**si**, prawn.
**tekebbür** Haughtiness.
**tekeffül** A becoming bail *or* surety. ···**i** *or* ···**e** ~ **etm.**, to stand surety for, to guarantee.
**tekel** Monopoly.
**tekellüf** Ceremoniousness, formality. ~**lü**, elaborate; bombastic. ~**süz**, plain; not overdone.
**tekemmül** ~ **etm.**, to be perfected.
**teker**[1] One at a time. ~ ~, one by one.
**teker**[2], **tekerlek** Wheel. Circular, round. ~ **arası**, width of track; ~ **tabanı**, tyre of a cart-wheel.
**tekerleme** Rigmarole; stereotyped formal way of speaking; the use of similarly sounding *or* rhyming words to produce a meretricious effect in speech *or* writing; roller (wave).
**teker·lemek** *va.* Roll; let slip out inadvertently; blurt out. ~**lenmek**, to roll round (*vn.*); turn head over heels; fall over; (*sl.*) die; (*sl.*) to be sacked. ~**meker** (·ꜝ··), head over heels.
**tekerrür** Repetition.
**teketek** (ꜝ··) *In* ~ **harb** (**kavga**), single combat.
**tekevvün** A coming into existence.
**tekfin** (·ꜝ) **etm.** *va.* Wrap in a winding-sheet.
**tekfir** (·ꜝ) *va.* Accuse *a Moslem* of heresy *or* blasphemy.
**tekfur** Prince of the Byzantine Empire.
**tekgözlük** (ꜝ··) Monocle.
**tekid** (–ꜝ) Confirmation; corroboration; repetition (of a message *or* order). ~**en** (–ꜝ·), in confirmation; as a repeat.
**tekin** ~ **değil**, inauspicious, haunted, ill-omened; dangerous; (man) with whom it is best to have nothing to do. ~**siz**, taboo.
**tekir** Tabby (cat). ~ **balığı**, striped red mullet.
**tekke** Dervish Convent. ⌜~**yi bekliyen çorbayi içer**⌝, he who serves (a party *etc.*) patiently will be rewarded eventually.
**teklif** (·ꜝ) Proposal; offer; motion *before an assembly*; etiquette; ceremony; custom; tax, obligation. ~ **etm.**, to propose; offer formally; submit; move (a motion); bid, tender: ~ **sahibi**, mover *of a motion or bill*; bidder; one who submits a tender: ~ **ve tekellüf**, the rules of etiquette and decorum: ~ **yok**, there is no need for ceremony.
**teklif·at** (·–ꜝ), **–tı**, proposals; formalities. ~**li**, with whom one must stand on ceremony; formal. ~**siz**, without ceremony; free-and-easy. ~**sizlik**, unceremoniousness.
**tekme** Kick. ~ **atmak**, to give a kick: ~ **yemek**, to get a kick, receive a blow; to fall into disgrace. ~**lemek**, to kick (*va.*).
**tekmil** (·ꜝ) All; the whole of. ~ **etm.**, to complete, finish: ~ **olm.**, to be completed.
**tekne** Trough; hull; craft. ~ **kazıntısı**, 'the last scraping of dough from the trough', *used about the youngest child of a numerous family*: **kuru** ~, the bare hull.
**tekn·ik** Technique. ~ **isyen**, technician.
**tekrar** Repetition; recurrence. Again. ~ ~, over and over again: ~ **etm.**, to repeat: ~ **olm.**, to happen again, to recur. ~**lamak**, to repeat.
**tekrir** (·ꜝ) Repetition.
**teksif** (·ꜝ) ~ **etm.**, to condense, compress; concentrate.
**tektük** (ꜝ·) Here and there; now and again. One or two; occasional.
**tekvin** (·ꜝ) Creation; production.
**tekzib** (·ꜝ) A declaring to be a lie.
**tel** Wire; fibre; a single thread *or* hair; string *of a musical instrument*; silver *or* gold thread used to decorate a bride's hair; telegram. Made of wire. ~ **çekmek**, to draw wire; to enclose with wire; to send a wire (*also* ~ **vurmak**): ~ **kafes**, iron cage: **her** ~**den çalmak**, to know a bit of everything, to be a jack-of-all-trades: **birinin hassas** ~**ine dokunmak**, to touch *s.o.* on his tender spot.
**telâ** (ꜝ·) Horsehair stiffening *of coat collars etc.*
**telâffuz** Pronunciation.
**telâfi** (·–ꜝ) ~ **etm.**, to make up for, compensate: ~**si imkânsız**, irreparable.

**telâki** (·–⁄) A meeting one another.
**telâkki** (··⁄) Reception; mode of receiving *or* regarding; interpretation, view.
**telâş** Flurry, confusion; alarm; hurry. **~a düşmek,** to be flurried, alarmed. **~lı,** flurried; agitated. **~sız,** calm, composed.
**telâtin** Russia leather.
**teldolab** (⁄··) Meat-safe.
**telebbüs** A putting on of clothes. **~ etm.,** to put on clothes, to dress.
**telef ~ etm.,** to destroy, ruin, kill: **~ olm.,** to be destroyed *or* ruined, to be killed. **~at, -tı,** casualties (*mil.*); losses of life *in an accident etc.*: **~ vermek,** to suffer losses *in battle*: **~ verdirmek,** to inflict losses.
**telefon** Telephone.
**teleme** *In* **~ peyniri,** fresh unsalted cheese. **~ peyniri gibi,** soft, flabby.
**teles** Threadbare. **~imek,** to become threadbare.
**teleskop, -pu** Astronomical telescope.
**televvün** A changing colour; fickleness.
**telezzüz ~ etm.,** to enjoy the taste of, to take pleasure in.
**telgraf** Telegraph. **~hane** (··–⁄), telegraph office. **~lamak,** to telegraph. **~name** (··–⁄), telegram.
**telhis** (·⁄) **~ etm.,** to summarize, make an abstract of.
**telif** (–⁄) A reconciling; composition; compromise. **~ etm.,** to reconcile, to square (conflicting facts); to write *or* compile: **~ hakkı,** copyright: **···le kabili ~,** compatible with ... .
**tel'in** (·⁄) A cursing. **~ etm.,** to curse.
**telkadayıfı** (⁄···) Sweet dish *of thin shreds of batter baked with butter and syrup.*
**telkâri, -ni** Woven of gold *or* silver thread. Stuff thus woven; filigree.
**telkih** (·⁄) Grafting; inoculation.
**telkin** (·⁄) Suggestion; inspiration; inculcation; final rites at a funeral. **~at** (·–⁄), **-tı,** harmful suggestions: **~a kapılmak,** to be influenced by suggestions. **~li,** *or* **~le tedavi,** faith-healing.
**tellâk, -kı** Bath attendant; shampooer.
**tellâl** Town-crier; broker; middleman. **parasız ~,** (unpaid crier), one who spreads news about matters that are no concern of his. **~iye** fee paid to a **tellâl. ~lık,** profession of a **tellâl.**
**tellemek** *va.* Adorn with gold wire *or* thread; deck out; embellish (a story); praise extravagantly; wire (telegraph). **telleyip pullamak,** to deck out with gold thread *etc.*, to cover with decorations; embroider *a narrative with exaggerations.*
**tellendirmek** *vn.* (*sl.*) Enjoy a smoke.
**telli** Wired; decorated with gold *or* silver wire *or* thread. **~ bebek,** extravagantly dressed; dandy: **~ pullu,** decked out: **~ ~ turna,** the Crowned Crane.
**telmih** (·⁄) Allusion; hint.
**telörgü** (⁄··) Barbed wire fence *or* entanglement; wire-netting.
**telsiz** Wireless. **~lemek,** to radio.
**teltik** Defect; small balance of an account. **teltiği temizlemek,** to pay off a small balance of an account. **~li,** *account* with a small balance owing; *sum of money* with a fractional amount, not a round sum. **~siz,** round *sum*; without fractions; complete, whole; fully paid.
**telüre, telöre** Casement.
**telve** Coffee-grounds. **~ falı,** fortune-telling by the appearance of coffee-grounds.
**telvin** (·⁄) A colouring.
**telvis** (·⁄) A soiling. **~ etm.,** to defile, soil.
**temadi** (·–⁄) A continuing uninterruptedly.
**temaruz** (·–⁄) A feigning sickness.
**temas** Contact. **···e ~ etm.,** to touch; to touch on (a subject *etc.*): **···le ~ etm.,** to make contact with, to get in touch with.
**temaşa** (·–⁄) Spectacle, show, scene; the theatre. **~ etm.,** to look on at, to enjoy *the scene.*
**temayül** (·–⁄) Inclination; bias; tendency.
**temayüz** (·–⁄) **etm.** *vn.* Be distinguished *or* privileged.
**tembel** Lazy. Lazy man. ⌈**~e iş buyur sana akıl öğretsin**⌉, if you ask a lazy man to do something, he will give you better advice, *i.e.* he will try to get out of it. **~hane** (··–⁄), leper-house; office where work is neglected. **~lik,** laziness.
**temcid** A glorifying God; canticle intoned from minarets before dawn. **~ pilâvı,** a thing that grows wearisome from repetition.
**temdid** (·⁄) **~ etm.,** to prolong, extend, stretch.

**temdin** (·*) **etm.** *va.* Civilize.
**temeddün** A becoming civilized.
**temel** Foundation; base. **~inden,** fundamentally; at bottom: **~ atmak,** to lay a foundation: **~ taşı,** foundation-stone: **~ çivisi çakmak (kakmak),** to have every intention of staying in a place *or* in the world: **~ tutmak,** to become firm in its place; to settle down permanently.
**temel·lenmek,** to be firmly settled *or* based. **~leşmek,** to become firmly established. **~li,** well-founded; permanent; fundamental: **~ gitti,** he went for good. **~siz,** without foundation; baseless.
**temenna** (··*), **temennah** Oriental salute *bringing the fingers of the right hand to the lips and then to the forehead.* **~ etm. (çakmak),** to salute *as above.*
**temenni** (··*) Desire, wish.
**temerküz** Concentration.
**temerrüd** Obstinacy; perverseness.
**temessük, -kü** A taking firm hold; bill acknowledging a debt *or* claim; title-deed.
**temessül** A being assimilated.
**temettü, -üü** Profit, gain; dividend.
**temevvüc** Fluctuation, undulation.
**temeyyüz** **~ etm.,** to become distinguished.
**temhir** (·*) A sealing.
**temin** (—*) **~ etm.,** to assure, ensure. **~at** (——*), **-tı,** security; guarantee; assurance.
**temiz** Clean; pure; honourable. **~e çekmek,** to make a fair copy *of a writing*: **kendini ~e çıkarmak,** to clear oneself (of a charge *etc.*): **~e çıkmak,** to be cleared, to prove innocent: **~ bir dayak,** a sound thrashing: **~ giyinmek,** to dress respectably: **~e havale etm.,** to clean up; to kill: **~ konuşmak,** to talk in a correct and polished manner: **~ para,** net sum of money (after deductions): **~ yemek,** to eat good food in good places.
**temiz·ce,** fairly clean: cleanly, nicely. **~leme,** act of cleaning. **~lemek,** to clean; clean up; clear away; despoil, rob; (*sl.*) to kill. **~leyici,** cleansing: cleaner. **~lik,** cleanliness; purity; honesty; act of cleaning; purge: **~ işleri,** scavenging service.
**temkin** (·*) Self-possession; dignity; composure. **~li,** grave, dignified; self-possessed.
**temlik** (·*) A putting in possession.
**temmuz** July.
**tempo** Time (*mus.*).
**temren** Head of an arrow *or* spear.
**temrin** Exercise *given to a pupil*; practice.
**temriye** A skin disease; lichen.
**temsil** (·*) Representation (agency); performance *of a play*; assimilation; parable; symbol; saying, maxim: **~ bürosu,** Press Bureau of a Ministry: **~ etm.,** to represent; present (play); assimilate; compare: **söz ~i,** for instance.
**temsil·en** (·*·), representing. **~î** (·—*), imaginary; representative.
**temyiz** (·*) A separating *or* distinguishing; discernment; soundness of judgement; appeal (law). **~ etm.,** to distinguish; to appeal (*jur.*): **~ mahkemesi,** Supreme Court of Appeal. **~en** (·*·), on appeal (legal).
**ten** The body; flesh. **~ fanilâsı,** vest: **~ rengi,** flesh-colour.
**tenakus** (·—*) Decrease, diminution. **~ etm.,** to decrease, diminish.
**tenakuz** (·—*) Contradiction.
**tenasüb** (·—*) Proportion; symmetry. **~î** (·—·*), proportional.
**tenasüh** (·—*) Metempsychosis.
**tenasül** (·—*) Reproduction, generation. **~ aleti,** the organ of generation. **~î** (·—·*), genital; sexual.
**tenazur** (·—*) A being symmetrical.
**tenbih** (·*) Warning; order, injunction; stimulation. **~ etm.,** to warn; enjoin; excite (nerves). **~at** (·—*), **-tı,** warnings; orders.
**tencere** (*··) Saucepan. ⌜**~ ~ye dibin kara demiş**⌝, 'the pot called the kettle black': ⌜**~de pişirip kapağında yemek**⌝, to live very economically: ⌜**~ tava hepsi bir hava**⌝, everyone goes their own way; nobody cares a straw: ⌜**~ yuvarlandı kapağını buldu**⌝, ⌜birds of a feather flock together⌝.
**tenef** Guy rope of a tent.
**teneffür** Aversion; disgust.
**teneffüs** Respiration. **~hane** (···—*), recreation room; place for resting.
**teneke** Tin; tinplate; a tin (*esp.* a paraffin tin); amount held by a paraffin tin. **(arkasından) ~ çalmak,** to boo *s.o.* publicly: **~sini eline vermek,** to give *s.o.* the sack: **~ mahallesi,** shanty-town: **(yüzü) ~ kaplı,** brazen-faced: **sarı ~,** brass.

~**ci,** tinsmith. ~**li,** tinned: **geçmişi** ~, cursèd.

**teneşir** The bench on which the corpse is washed (*also* ~ **tahtası**). ~**e gelesi!,** (*a curse*) 'may he die!': ~**e sürmek,** (of a bad habit *etc.*) to last till death: ⌜**onu ancak** ~ **paklar**⌝, 'only death can cleanse him', *said of a great scoundrel.* ~**lik,** place for washing corpses *in the courtyard of a mosque.*

**tenevvü, -üü** Variation. ~ **etm.,** to vary.

**tenevvür** Enlightenment.

**tenezzüh** Pleasure walk, excursion, jaunt.

**tenezzül** A coming down; decline; condescension. ···**e** ~ **etm.,** to deign to ...; to deign to accept ... . ~**en** (·̀··), condescendingly; graciously.

**tenha** Deserted (place). ~**laşmak,** to become deserted *or* empty (of places). ~**lık,** solitude; deserted *or* lonely place.

**tenkid** (·̀) Criticism. ~**ci,** critic.

**tenkil** (·̀) ~ **etm.,** to repress *a rebellion.*

**tenkis** (·̀) A diminishing; diminution. ~ **etm.,** to curtail, diminish.

**tenkiye** Clyster.

**tennure** (·–̀) Dervish's skirt.

**tenperver** Who takes great care of himself; fond of comfort; soft.

**tensib** (·̀) Approval. **bunu sizin** ~**inize bırakıyorum,** I leave this to your discretion: ~ **etm.,** to approve.

**tensik** (·̀), **-kı** ~ **etm.,** to reorganize, reform. ~**at** (·–̀), **-tı,** reforms; a combing out of inefficient officials.

**tente** (̀·) Awning. ~**li,** with an awning.

**tentene** (̀··) Lace. ~**li,** ornamented with lace.

**tentürdiyod** Tincture of iodine.

**tenvir** (·̀) Illumination. ~ **etm.,** to illuminate; to make clear: ~ **fişeği,** flare (*mil.*): ~ **mermisi,** star shell: ~ **tabancası,** Very pistol. ~**at** (·–̀), **-tı,** lighting *of a street etc.*

**tenzih** (·̀) A considering free from defect.

**tenzil** (·̀) A lowering *or* diminishing; reduction (of prices). ~**ât** (·–̀), **-tı,** reductions (of prices *etc.*). ~**âtlı,** reduced in price: ~ **satış,** bargain sale.

**tepe**[1] *v.* **tepmek.**

**tepe**[2] Hill; summit; crown of the head; crest *of a bird.* ~**den,** in a superior manner, condescendingly: ~ **aşağı gitmek,** to fall headlong; (of a business) to go downhill: ~**si* atmak,** to be infuriated: ···**e** ~**den bakmak,** to look down on, to despise: ~ **camı,** skylight: ~**sine* çıkmak,** to presume on *s.o.'s kindness etc.*: ~**sine* dikilmek,** to worry, to insist: ~**den inme,** sudden, unexpected; from above, from a higher authority: ~**den tırnağa,** from head to foot: ~ **taklak,** on one's head, upside-down, head foremost: ~ **üstü,** upside-down, head first, headlong: **acısı** ~**mden çıktı,** I felt very sore about it: **aklı** ~**sinden yukarı,** thoughtlessly, absent-mindedly.

**tepe·cik,** little hill. ~**göz,** with an upward squint; whose forehead is so narrow that his eyes seem near his hair; star-gazer (fish, *Uranoscopus*); a legendary monster in Turkish epics (*cf.* Cyclops). ~**leme,** brimful; mound, heap; a killing; sound thrashing. ~**lemek,** to knock on the head; kill; thrash unmercifully. ~**li,** crested (bird).

**tepengi** Thick pad of a pack-saddle; broad girth.

**tepinmek** *vn.* Kick and stamp; dance *with joy or in anger.*

**tepir** Hair sieve. ~**lemek,** to pass through a fine sieve.

**tepişmek** *vn.* Kick one another; quarrel violently. ⌜**atlar tepişir eşekler ezilir**⌝, 'the weak go to the wall'.

**tepme** Kick; relapse (in illness). ~ **atmak,** to kick: **geri** ~, recoil.

**tepmek** *vn.* Kick; recoil; (of an illness) to recur. *va.* Kick; spurn; underestimate. **tepe tepe kullanmak,** to wear *a garment* continuously and give it rough usage: **at** ~, to spur on a horse: **bir fırsatı** ~, to spurn an opportunity: **nimeti** ~, to spurn a piece of luck.

**tepremek, teprenmek** *vn.* Struggle; bestir oneself.

**tepreşmek** *vn. As* **tepremek**; (of an illness) to recur.

**tepsi** Small tray.

**ter** Sweat, perspiration. ~**e batmak,** to sweat heavily: ~ **dökmek (basmak),** to sweat; to labour hard; to sweat with terror *or* anxiety: **soğuk** ~**ler döktürmek,** to cause to break out in a cold sweat *from terror*: **kan** ~, profuse sweat; **kan** ~ **içinde kalmak,** to be in a profuse sweat; to undergo great exertion.

**terakki** (··⊥) Advance, progress; increase. ~ etm., to make progress, to advance; increase. **~perver,** progressive.

**teraküm** (·—⊥) Accumulation.

**terane** (·—⊥) Tune; refrain; yarn, concocted story; subterfuge.

**teravi** (·—⊥), **teravih** Prayer special to the nights of Ramazan.

**terazi** (·—⊥) Balance; pair of scales.

**terbiye** A bringing up; education; training; good manners; a teaching manners; correction, punishment; sauce; rein. **~sini vermek,** to teach *s.o.* his manners; to reprimand: **beden ~si,** physical training, gymnastics. **terbiye·ci,** one who educates; pedagogue; trainer. **~li,** well brought up; good-mannered; educated; flavoured *with a sauce etc.* **~siz,** badly brought up; ill-mannered; uneducated; without a sauce *or* seasoning. **~sizlik,** bad manners; rudeness; lack of education. **~vî** (···⊥), educational.

**tercih** (·⊥) Preference; priority. ~ etm., to prefer. **~an** (·⊥·), preferably, in preference.

**terciibend** (·—·⊥) Poem in which each stanza ends with the same couplet.

**tercüm·an** Interpreter; translator; dragoman. **~anlık,** office *or* profession of an interpreter: ~ etm., to act as interpreter. **~e,** translation: ···**e** ~ etm., to translate into ... . **~eihal, -li,** biography; memoirs.

**terdöşeği** (⊥···), **-ni** Childbed.

**tere** Cress. **~ci,** seller of cress: ⌜**~ye tere satmak**⌝, 'to teach your grandmother to suck eggs': **~ye tere satma!,** don't try to take me in!

**tereddi** (··⊥) Degeneration; deterioration.

**tereddüd** Hesitation; indecision.

**tereffü, -üü** Elevation.

**tereke** Estate of a deceased person; heritage; sale of a dead man's effects.

**terekküb** ~ etm., to be composed *or* compounded.

**terelelli** (··⊥·) Feather-brained; frivolous.

**terementi** (··⊥·) Terebinth; turpentine.

**terennüm** ~ etm., to sing, hum, warble.

**tereotu** (·⊥··), **-nu** Dill.

**teres** Cuckold; pimp; scoundrel.

**teressüb** Sediment, precipitation.

**teressüm etm.** *vn.* Be pictured; become evident.

**terettüb etm.** *vn.* Be incumbent (upon = *dat.*).

**tereyağ** (·⊥·), **tereyağı, -nı** Fresh butter. ⌜**~ından kıl çeker gibi**⌝, skilfully and easily.

**terfi** (·⊥), **-ii** Promotion; advancement. ~ etm., to be promoted. **~an** (·⊥·), by way of promotion; on promotion.

**terfih** (·⊥) A causing to live in prosperity. ~ etm., to bring prosperity to; to better the condition of.

**terfik** (·⊥) **etm.** *va.* Send as escort *or* companion. **~an** (·⊥·), as companion; by way of escort.

**terhin** (·⊥) **etm.** *va.* Pawn; pledge.

**terhis** (·⊥) Discharge *of a soldier after serving his time.* ~ etm., to authorize; to discharge (soldier); to demobilize: ~ **tezkeresi,** discharge papers *of a soldier.*

**terim** Term; technical term.

**terk, -ki** Abandonment; renouncement; omission. **~etmek** (⊥··), **-eder,** to abandon; renounce; relinquish; neglect; leave (an inheritance, a post *etc.*).

**terkeş** Quiver.

**terki** Anything strapped to the back of a saddle. ~ **bağı,** strap for fastening things to the back of a saddle: **~ye almak** *or* **~mek,** to take as pillion rider.

**terkib** (·⊥) A joining *or* compounding; composition; compound; structure; compound word; phrase; synthesis. ~ etm., to compose, compound; constitute. **~î** (·—⊥), composite; synthetic. **~ibend,** a poem, the stanzas of which are connected by a refrain.

**terkos** (⊥·) Mains water-supply; water laid on (in Istanbul).

**ter·lemek** *vn.* Sweat, perspire; be covered with dew (of a glass); start growing (of a moustache); be very tired; be embarrassed. **~letici,** sudorific; hard, fatiguing (work). **~letmek,** to cause to sweat; greatly fatigue. **~li,** sweating, perspiring: ~ **su içmek,** to drink water when perspiring.

**terlik** Slipper.

**termometre** Thermometer.

**termos** Thermos flask.

**ternöv** (*Fr. Terre-neuve*) Newfoundland dog.

**ters** Back *or* reverse *of a thing*; wrong *or* reverse direction; excrement.

Reverse; wrong; opposite; inside out; peevish, contrary; unfortunate, ill-timed. ~ ~ **bakmak,** to look sourly: ~**i* dönmek,** to lose one's bearings: ~ **gelmek,** to be the wrong way about; to be in the opposite direction: ~ **gitmek,** to go wrong, to turn out badly: ~**ine* olm.,** to happen contrary to one's wishes: ~**inden okumak,** (i) to misunderstand; (ii) to be very quick-witted: ~ **pers,** very inverted; all wrong; disconcerted: ~**ine yazmak,** to write the reverse way: ~ **yüz,** empty-handed, disappointed; ~ **yüz etm.,** to turn (a suit of clothes): **elin ~i,** the back of the hand.

**tersane** (·–ˈ) Dockyard; maritime arsenal (*esp. that at Istanbul*). ~**li,** attached to the maritime arsenal; (*formerly*) naval officer *or* rating.

**tersim** (·ˈ) **etm.** *va.* Picture; design, draw.

**ters·lemek** *vn. va.* Scold; answer harshly; snub; dung; befoul with dung. ~**lenmek,** to be in a bad temper; to behave in a peevish contrary way: to be snubbed. ~**lik,** a turning out in the reverse of what was hoped; a being reversed *or* wrong; contrariness, vexatiousness. ~**pers,** ~**yüz,** *v.* **ters.**

**tertemiz** (ˈ··) Absolutely clean.

**tertib** (·ˈ) Arrangement; order; disposition; plan, project; recipe; medical prescription; composition; setting-up of type; format *of a book.* **hafif ~,** slightly, just a little: **tam ~,** fully, thoroughly.

**tertib·at** (·–ˈ), **–tı,** arrangements; dispositions; apparatus; installations. ~**ci,** good at organizing; who is always planning. ~**lemek,** to organize, arrange. ~**lenme,** arrangement; disposition (*mil.*). ~**li,** well-organized; tidy. ~**siz,** without system; ill-prepared; badly planned. ~**sizlik,** bad organization.

**terütaze** Quite fresh.

**tervic** (·ˈ) ~ **etm.,** to make current; to advocate, encourage.

**terzi** Tailor. ~ **sabunu,** French chalk (soapstone). ~**hane** (··–ˈ), tailor's shop; clothing factory. ~**lik,** tailoring.

**terzil** (·ˈ) **etm.** *va.* Treat with ignominy; humiliate before others; ill-treat; insult.

**tesadüf** (·–ˈ) A meeting by chance; chance event; coincidence. ···**e** ~ **etm.,** to come across; to coincide with; to happen by chance: ~**e bak ki!,** what a strange coincidence! ~**en** (·ˈ··), by chance; by coincidence. ~**î** (·–·ˈ), chance (event *etc.*), fortuitous.

**tesahüb** (·–ˈ) ···**e** ~ **etm.,** to claim to be the owner *or* the author of.

**tesalüb** (·–ˈ) Cross-breeding.

**tesamüh** (·–ˈ) Condonation.

**tesanüd** (·–ˈ) Solidarity.

**tesavi** (·–ˈ) Parity.

**tesbih** (·ˈ) Rosary. ~ **ağacı,** Indian lilac, bead-tree (*Melia Azedarach*): ~ **çekmek,** to tell one's beads. ~**böceği, –ni,** woodlouse.

**tesbit** (·ˈ), **–ti** ~ **etm.,** to establish, stabilize; confirm; (*mil.*) to tie down *an enemy*: **hesabı** ~ **etm.,** to make up an account.

**tescil** (·ˈ) ~ **etm.,** to register, record.

**teselli** (··ˈ) Consolation. ~ **bulmak,** to console oneself: ~ **vermek,** to console.

**tesellüm** ~ **etm.,** to take delivery: ~ **tecrübesi,** full-power trials *of a ship before delivery.*

**teselsül** Continuous succession; train of ideas; sequence of events. ~ **etm.,** to follow in an uninterrupted series.

**tesemmüm** A being poisoned.

**tesettür** A being veiled.

**teseyyüb** Negligence, slackness.

**teshil** (·ˈ) ~ **etm.,** to facilitate. ~**ât** (·–ˈ), **–tı,** facilities.

**teshin** (·ˈ) A heating.

**teshir**[1] (·ˈ) Fascination; enchantment.

**teshir**[2] (·ˈ) Conquest; subjugation.

**tes'id** (·ˈ) Celebration, festival.

**tesir** (–ˈ) Effect; impression; influence. ~**at** (––ˈ) **–tı,** impressions; effects; influences. ~**li,** touching, moving; impressive; efficacious. ~**siz,** ineffective; without influence; free, without being influenced. ~**sizlik,** inefficacy.

**tesis** (–ˈ) A laying a foundation, a basing *a matter on stg.*; foundation. ~ **etm.,** to found, establish, institute, base. ~**at** (––ˈ), **–tı,** institutions; establishments; plants (industrial).

**teskere** (ˈ··) Litter; stretcher; bier; hand barrow. ~**ci,** stretcher-bearer.

**teskin** (·ˈ) **etm.** *va.* Pacify, calm; assuage.

**teslih** (·ˈ) An arming. ~ **etm.,** to arm. ~**at** (·–ˈ), **–tı,** armament(s).

**teslim** (·⁄) A handing over; delivery. ~ **almak**, to take delivery of: ~ **etm.**, to hand over, deliver; to give up, surrender; to pay over *money*; to admit *an argument*: ~ **olm.**, to surrender, give oneself up: **~i ruh etm.**, to give up the ghost: ~ **ve tesellüm**, the handing over of an office to a successor. **~at** (·–⁄), **–tı**, instalments (of money *etc.*). **~iyet** (·–·⁄), **–ti**, submission; resignation: **arzı** ~ **etm.**, to surrender.
**teslis** (·⁄) The Trinity.
**tesliye** Consolation.
**tesmim** (·⁄) A poisoning.
**tesmiye** A naming.
**tesniye** Dual (*gram.*).
**tespih, tespit** *v.* **tesbih, tesbit.**
**tesri** (·⁄), **–ii** A hastening; acceleration.
**testere** (⁄··) Saw. **kayış ~si**, bandsaw: **kıl** ~, fretsaw: **kollu** ~, hacksaw. **~balığı, –nı**, sawfish. **~burun**, goosander.
**testi** Pitcher; jug. ⌈**~yi kıran da bir, suyu getiren de**⌉, *said reproachfully when a deserving person is no better treated than an undeserving one*: ⌈**su ~si su yolunda kırılır**⌉, one must take the risks of one's occupation.
**tesvid** (·⁄) Rough draft.
**tesviye** Arrangement; payment, settlement (of an account); adjustment; smoothing; fitting; free pass (railway *etc.*) *given to travelling soldiers*. ~ **atölyesi**, fitting-shop; ~ **hududu**, contour line: ~ **ruhu**, spirit-level: **toprak ~si**, levelling of the ground: **kum ~si**, ballast *of a railway or road*. **~ci**, fitter; one who levels the ground. **~cilik**, fitting (*mech.*).
**teşahhus** Personification.
**teşaur** (·–⁄) A pretending to be a poet.
**teşbih** (·⁄) Comparison; simile. ⌈**~te hata olmaz**⌉, only for the sake of comparison; let it not be misunderstood.
**teşci, –ii** Encouragement.
**teşdid** (·⁄) Intensification; doubling a consonant.
**teşebbüs** Enterprise; effort; initiative. ···**e** ~ **etm.**, to set to work at ... , to undertake, start *an enterprise etc.* **~at, –tı**, enterprises; efforts: **~ta bulunmak**, to take steps towards *doing stg.*
**teşekkül** A being formed; formation; organization; association. ~ **etm.**, to be formed *or* constituted. **~ât, –tı**, formations, associations.
**teşekkür** A giving thanks. ~ **etm.**, to thank: ~ **ederim**, thank you!
**teşerrüf** A being honoured. ~ **ettim**, I am honoured (to meet you).
**teşeüm etm.** *vn.* Draw an evil omen. ···**den** ~ **etm.**, to consider ... as inauspicious.
**teşevvüş** Confusion.
**teşhir** (·⁄) A making public *or* notorious; an exposing to the public view; exhibition.
**teşhis** (·⁄) Recognition; identification; diagnosis. ~ **etm.**, to identify, diagnose.
**teşkil** (·⁄) Formation; organization. **~ât** (·–⁄), **–tı**, organization: **~ı esasiye kanunu**, the constitution. **~âtçı**, organizer. **~âtlandırmak**, to organize. **~âtlı**, organized.
**teşmil** (·⁄) Extension; generalization. ~ **etm.**, to extend to, to include.
**teşne** Thirsty; longing (for=*dat.*).
**teşri** (·⁄), **–ii** Legislation. **~î** (·–⁄), legislative: ~ **kuvvet**, the legislative power: ~ **masuniyet**, the immunity of legislators.
**teşrif** (·⁄) ~ **etm.**, to honour *by one's presence*: ···**e** ~ **etm.**, to come *or* go to: ~ **mi?**, must you be going so soon?
**teşrif·at** (·–⁄), **–tı** Ceremonial; protocol. **~atçı**, Master of the Ceremonies.
**teşrih** (·⁄) Dissection; anatomy. ~ **etm.**, to dissect; to examine minutely. **~hane** (··–⁄), dissecting-room.
**teşrik** (·⁄), **–ki** A making *s.o.* partner. **~i mesai**, joint effort, co-operation.
**teşrini·evvel** (·–··⁄) October (*now* **Ekim**). **~sani** (·–·–⁄), November (*now* **Kasım**).
**teşvik** (·⁄), **–ki** Encouragement; incitement. **~kâr**, encouraging.
**teşviş** (·⁄) ~ **etm.**, to confuse, disorder.
**teşyi** (·⁄), **–ii** A seeing *s.o.* off. ~ **etm.**, to see *s.o.* off; to follow a funeral.
**tetabuk** (·–⁄), **–ku** ···**e** ~ **etm.**, to correspond to, to conform with.
**tetanos** Tetanus.
**tetebbü, –üü** Study; research. **~at, –tı**, studies, researches.
**tetevvüc** Coronation. ~ **etm.**, to be crowned.
**tetik** Trigger. Vigilant; agile, quick. **tetiğini bozmamak**, to keep a cool

head: ~ **bulunmak** *or* ~ **üzerinde olm.**, to be vigilant, to be on the 'qui-vive': **adımı ~ almak,** to proceed with caution: **alt ~te,** at half-cock: **üst ~te,** at full cock. **~lik,** a being very alert, promptness, agility.
**tetimmat** (··ꜜ), **–tı** Accessories.
**tetkik** *v.* tedkik.
**tetre** Sumach.
**tetvic** ~ **etm.,** to crown.
**tevabi** (·—ꜜ), **–ii** Followers; dependants; dependencies.
**tevafuk** (·—ꜜ), **–ku** Agreement; compatibility; conformity.
**tevakki** (··ꜜ) ···**den ~ etm.,** to beware of.
**tevakkuf** A stopping. ~ **etm.,** to stop, to stay: ···**e ~ etm.,** to depend on, to require (thought, time *etc.*).
**tevali** (·—ꜜ) Uninterrupted succession.
**tevarüd** (·—ꜜ) Coincidence; unintentional composition of the same verse by different poets.
**tevarüs** (·—ꜜ) An inheriting.
**tevatür** (·—ꜜ) Hearsay; generally current report. **~en** (·ꜜ··), by common report: ~ **sabıt,** known to all.
**tevazu** (·—ꜜ), **–uu** Modesty.
**tevazün** (·—ꜜ) Equilibrium.
**tevbih** (—ꜜ) Rebuke, reprimand.
**tevcih** (·ꜜ) ~ **etm.,** to turn towards (*va.*); to direct *one's words or looks* towards; to confer (an office *or* rank); to nominate: ~ **olunmak,** to be awarded (a decoration *or* post of honour).
**tevdi** (·ꜜ), **–ii** ~ **etm.,** to entrust (an affair *or* secret); to deposit (money); to commit *to the charge of* another; to present, tender. **~at** (·—ꜜ), **–tı,** deposits (in a bank *etc.*).
**teveccüh** A turning towards; favour, kindness, goodwill. **~ünüz efendim!,** it's kind of you to say so (*in reply to a compliment*): ···**e ~ etm.,** to turn towards (*vn.*); to turn one's attention to; to face; (of a duty) to fall to one's lot.
**tevehhüm** ~ **etm.,** to imagine, to fancy.
**tevekkeli** (*Followed by a neg.*), it was not without reason that ...; it was not for nothing that ... .
**tevekkül** A putting one's trust in God; resignation.
**tevellüd** Birth. **~lü,** born *in such and such a year.*
**teverrüm** ~ **etm.,** to become consumptive.
**tevessü, –üü** Extension, expansion.
**tevessül etm.** *vn.* Have recourse to.
**tevettür** Tension.
**tevezzü, –üü** Distribution.
**tevfik** (·ꜜ), **–ki** Guidance (*esp.* divine); adaptation; success. **~an** (·ꜜ·), in accordance *or* conformity (with = *dat.*).
**tevhid** (·ꜜ) Unification; consolidation; monotheism. ~ **etm.,** to unite.
**tevil** (·ꜜ) An explaining away; forced interpretation.
**teviye** *In* **bir ~,** continuously.
**tevkif** (·ꜜ) Detention; arrest. ~ **etm.,** to detain, arrest, stop; to deduct (a sum of money). **~at** (·—ꜜ), **–tı,** deductions (from wages *etc.*); stoppages of pay; arrests. **~hane** (··—ꜜ), place of custody of arrested persons.
**tevkil** (·ꜜ) **etm.,** *va.* Appoint as representative *or* deputy.
**tevlid** (·ꜜ) ~ **etm.,** to create; cause.
**tevliyet, –ti** Appointment of a **mütevelli** *q.v.*; office of **mütevelli.**
**Tevrat, –tı** The Pentateuch. **~î** (··ꜜ), biblical.
**tevriye** Use of an ambiguous word.
**tevsi** (·ꜜ), **–ii** Enlargement.
**tevsik** (·ꜜ), **–ki** ~ **etm.,** to confirm, prove; prove by documentary evidence.
**tevsim** (·ꜜ) **etm.** *va.* Name.
**tevzi** (·ꜜ), **–ii** Distribution; delivery (of letters *etc.*). **~at, –tı,** distributions; postal deliveries.
**tevzin** (·ꜜ) **etm.** *va.* Balance.
**teyakkuz** Vigilance.
**teyel** Coarse sewing, tacking. **~lemek,** to sew coarsely, tack. **~li,** tacked.
**teyelti** *v.* **teğelti.**
**teyemmüm** Ritual ablution with sand *or* earth in default of water; a looking longingly at *stg.* one cannot have.
**teyemmün** A regarding as lucky. **~en** (·ꜜ··), as a token of good luck; as an act of piety.
**teyid** (—ꜜ) Confirmation.
**teyze** (ꜜ·) Maternal aunt. **hanim ~,** *polite term of address for any elderly lady.* **~zade** (··—ꜜ), cousin (child of a maternal aunt).
**tez**[1] Quick. Quickly, promptly. ~ **elden,** without delay, in haste: **bugünden ~i yok,** this very day, immediately: **yarından ~i yok, işe başlamalı,** tomorrow at the latest

the work must be begun: **canı (içi) ~,** hustler, energetic; impatient.

**tez**[2] (*Fr. thèse*) Thesis; question.

**tezad, -ddı** Contrast; contradiction; incompatibility.

**tezahür** (·–¹) Manifestation. **~ etm.,** to become manifest, to appear. **~at, -tı,** public demonstration; ovation.

**tezayüd** (·–¹) An increasing; growth.

**tezcanlı** (¹··) Hustling; energetic; impatient of delay.

**tezebzüb** Confusion; disorder.

**tezek** Dried dung *used as fuel.* **yer tezeği,** peat.

**tezekkür** **~ etm.,** to discuss, consider.

**tezelden** (¹··) Without delay; in haste.

**tezellül** Abasement.

**tezelzül** **~ etm.,** to be shaken.

**tezene** Plectrum.

**tezevvüc** A taking a wife; matrimony.

**tezeyyün** An adorning oneself.

**tezgâh** Loom; work-bench; counter; shipbuilding yard; workshop; machine-tool. **~ başı yapmak** (*sl.*), to have a drink at a bar: **dikiş ~ı,** bookbinder's stitching-frame. **~dar,** one who serves at a counter; shop-assistant.

**tezhib** (·¹) A gilding *or* inlaying with gold.

**tezkâr** Remembrance; mention.

**tezkere** (¹··) Note; memorandum; official certificate *or* receipt; soldier's discharge papers; biographical memoir. **~sini eline vermek,** to give *s.o.* 'the sack': **av ~si,** shooting licence: **nüfus ~si,** identity papers: **unvan ~si,** trade-licence: **yol ~si,** permit to travel. **~ci** (¹···), discharged soldier; reservist. **~lik,** paper used for official notes; soldier due for discharge.

**tezkiye** Purification; praise. **~sini* düzeltmek,** to reform oneself; **~si bozuk,** who has a bad reputation: **~ etm.** to clear *s.o.'s* character.

**tez·lenmek** *vn.* Make haste; be impatient. **~lik,** speed; haste; impatience.

**tezlil** (·¹) A humiliating.

**tezvic** (·¹) **etm.** *va.* Unite in matrimony.

**tezvir** (·¹) Wilful misrepresentation; falsehood, deceit; malicious instigation. **~atta bulunmak,** to engage in malicious misrepresentations.

**tezyid** (·¹) Augmentation, increase.

**tezyif** (·¹) Derision; mockery.

**tezyil** (·¹) **etm.** *va.* Add as a supplement.

**tezyin** (·¹) **~ etm.,** to adorn, embellish. **~at** (·–¹), **-tı,** adornments. **~î** (··¹), decorative.

**tıb, -bbı** The science of medicine; therapeutics. **~ben** (¹·), medically. **~bî** (·¹), medical. **~biye,** medical faculty *or* school. **~biyeli,** medical student.

**tığ** Crochet-needle; bodkin; awl; knitting-needle; plane-iron. **~ gibi,** slender but strong and active, wiry: **~ örgüsü,** crochet.

**tığala** (·¹·) Gum euphorbium.

**tığlamak** *va.* Lance; pierce with a needle; (*sl.*) slaughter (an animal). *vn.* (Of a wound) to give a piercing pain.

**tıhal** Spleen.

**tık tık** *Imitates a ticking noise.*

**tıka** *In* **~ basa,** crammed full. **~ç,** plug; stopper; gag. **~lı,** stopped up; plugged.

**tıkamak** *va.* Stop up; plug; gag. **burnunu* ~,** to hold the nose: **iştahı ~,** to spoil the appetite: **sözü ağza ~,** to interrupt *s.o.* speaking, to shut *s.o.* up.

**tıkan·ık** Stopped up; choked. **~ıklık,** choking; suffocation; a being stopped up; interruption of communications (*mil.*); lack of appetite. **~mak,** *v.n. pass. of* **tıkamak,** to be stopped up; to choke; be suffocated; to lose one's appetite.

**tıkınmak** *vn.* Stuff oneself; eat in haste; gulp down one's food.

**tıkır** *Imitates a clinking or rattling or tapping noise*; (*sl.*) money, 'chink'. **~ ~,** with a clinking *or* rattling noise: **~ ~ işlemek,** (of a clock *etc.*) to go perfectly; to go 'like clockwork': **işler ~ında gidiyor,** business is going well: **keyfim ~ında,** I am in the best of spirits.

**tıkır·damak** *vn.* Make the sounds indicated by **tıkır** *q.v.* **~dı, ~tı,** a rattling *or* clinking sound: **~lı telgraf âleti,** sounder (telegraphy).

**tıkış·mak** *vn.* Be crammed *or* squeezed together. **~tırmak,** to cram into a small space; to bolt (food).

**tıkız** Fleshy; hard.

**tıklım tıklım** Brimful.

**tıkmak** *va.* Thrust, squeeze *or* cram into. **deliğe (hapse) ~,** to clap into jail: **lâkırdıyı ağza ~,** to cram down *s.o.'s* throat an assertion he has made.

**tıknaz** Plumpish; stout.
**tıknefes** (ˈ··) Short of breath; asthmatic.
**tıksır·ık** A suppressed sneeze. **~mak,** to sneeze with the mouth shut.
**tılsım** Talisman; charm; spell. **~ bozmak,** to break a spell: **~ı bozuldu,** the spell is broken; he (it) no longer has any influence. **~lı,** having a charm *or* spell; spell-binding.
**tım** *In* ⌜**kim kime ~ ~a** (*or* **dum duma**)⌝, nobody will take any notice; nobody knows (knew) anything about it!
**tımar** Grooming a *horse*; *formerly* a kind of fief granted by the Sultan to soldiers. **~ etm.,** to groom. **~cı,** holder of a fief.
**tımarhane** (··–ˈ) Lunatic asylum. **~ kaçkını,** escaped lunatic: ⌜**~cinin gözü kör olsun!**⌝, 'damn the asylum warder (for letting you escape)!, you ought to be shut up!
**tın tın** *Imitates metallic sounds.*
**tınaz** Stack of hay *or* corn. **~ gibi,** a whole heap of ... .
**tıngır** *or* **~sız** (*sl.*) Stony-broke. **~ ~** *or* **~ mıngır,** *imitates the sound of metallic things knocking together*; cash down. **~damak,** to tinkle, clink, clang. **~tı,** clinking, clanking noise.
**tınlamak** *vn.* Tinkle, ring (of metal *etc.*).
**tınmak** *vn.* Make a sound; *usually only in a negative form.* **tınmamak,** not to utter a sound; to take no notice; to pretend not to see *or* hear: **tınmayıvermek,** simply to take no notice: **tınmaz,** who pays no attention, who takes no notice; who says nothing; **tınmaz melâike** (*iron.*), quiet, aloof; non-committal.
**tıp** *v.* **tıb.**
**tıpa, tapa** Stop; plug; cork; mop *for oiling guns*; fuse.
**tıpatıp** (ˈ··) Exactly; absolutely.
**tıpır·damak** *vn.* Make a noise as of drops falling; walk with little noise; (of the heart) to go pit-a-pat. **~dı,** sound of drops falling *or* of light footsteps.
**tıpış tıpış** *Imitates the noise of the small steps of a child.* **~ ~ gitmek,** to walk with small steps; to go willy-nilly.
**tıpkı** (ˈ·) Exactly like; in just the same way. **~sı,** exactly like it; the very image of ...: **~ basım,** facsimile.
**tırabzan** Hand-rail; banister. **~ babası,** newel, post with knob at the end of a banister; a father who has no influence over his children.
**tıraş** Shaving; the hair which grows between shaves; boring talk; bragging. **~ etm.,** to shave; to cut; to tell lies to, to take *s.o.* in; to bore: **~ı gelmiş** (**uzamış**), needing a shave: **~ olm.,** to shave oneself, to get a shave: **~a tutmak,** to detain *s.o.* by idle talk; to buttonhole *s.o.*
**tıraş·çı,** boring talker; braggart; swindler. **~lamak,** to prune; to thin out trees. **~lı,** needing a shave; shaved, clean-shaven, unbearded (*the context will usually show which of these contradictory meanings is right*): **~ geldi,** he came unshaved.
**tırfıl** Trefoil; clover.
**tırhalli** *v.* **turhalli.**
**tırık** *Imitates the noise of two hard things striking against each other; usually used with* **tırak.**
**tırıklamak** *va.* (*sl.*) Steal.
**tırıl** Naked; thinly clad; 'stony-broke'. **~lamak,** to shiver with cold; to be 'broke'. **~lık,** a being 'stony-broke'.
**tırıs** Trot. **~ gitmek,** to trot.
**tırkaz** Bar behind a door to keep it shut.
**tırma·lamak** *va.* Scratch; worry, annoy; offend (the ears, the taste). **~nmak,** to cling with the claws *or* the finger-tips: **···e ~,** to climb (a tree, a mountain *etc.*).
**tırmık** Scratch; rake; harrow; drag-hook. **~lamak,** to scratch, rake, harrow.
**tırnak** Finger-nail; toe-nail; claw; hoof; ejector *of a gun*; fluke *of an anchor*; catch (*mech.*). **~ çekici,** claw-hammer: **~ işareti,** inverted commas: ⌜**tırnağının kiri bile olamaz**⌝, he's not fit to lick his boots: **tırnağını* sökmek,** to torture: **birine ~ takmak,** to have one's knife into a person: **dişinden tırnağından artırmak,** to pinch and scrape: **dişi tırnağı döküldü,** he worked himself to death: **tepeden tırnağa,** from head to foot.
**tırnak·çı,** pickpocket. **~lamak,** to scratch with the nails. **~lı,** having nails *or* claws; spiked (wheel).
**tırpan** Scythe; trepan. **···e ~ atmak,** to exterminate.
**tırpana** (·ˈ·) Skate (fish).
**tırtık** Unevenness. **~ ~,** uneven, jagged. **~çı** (*sl.*), pickpocket; rogue.

~**lamak,** to pull to pieces; pluck; rob. ~**lı,** rough; jagged.
**tırtıl** Caterpillar; milling *of a coin*; knurl; perforation *of a stamp*. ~**lı,** having a milled edge: ~ **tekerlek,** caterpillar wheels.
**tıs** Goose's hiss. ~!, hush!: ~ **dememek,** not to make the slightest noise; not to raise the slightest objection. ~**lamak,** to hiss like a goose; to spit like a cat.
**tıynet, -ti** Natural character. ~**siz,** of low character.
**ti** *In* ~ **işareti,** a bugle call.
**ticaret** (·–ˈ), **-ti** Trade, commerce; profit. ~ **etm.,** to engage in commerce; to earn, make a profit: **T~ Odası,** Chamber of Commerce. ~**hane** (·–·–ˈ), business house; firm; place where a man has his business. ~**li,** profitable.
**ticarî** (·–ˈ) Commercial.
**tifo** (ˈ·) Typhoid fever.
**tiftik** Mohair; fine soft wool *clipped from sheep in spring*. ~ ~ **etm.,** to unravel, to pull into threads: ~ **gibi,** very soft: ~ **keçisi,** Angora goat: **keten tiftiği,** lint. ~**lenmek,** to become unravelled *or* frayed.
**tifüs** (ˈ·) Typhus.
**tik**[1], **-ki** Teak.
**tik**[2], **-ki** (*Fr. tic*) Twitching.
**tike** (ˈ·) Piece, patch. ~ ~, patched.
**tiksin·mek** *vn.* (*with abl.*) Be disgusted with; loathe. ~**ti,** disgust; loathing.
**tilâvet** (·–ˈ), **-ti** Religious reading *or* chanting.
**tilki** Fox; cunning fellow. ⌜~**nin dönüp dolaşıp geleceği yer kürkçü dükkânıdır**⌝, (i) you (he) will end up there anyhow; (ii) you (he) can't help coming back here (*or* to the same job) however hard you try to avoid it: ~ **tırnağı,** the Spotted Orchis *from which the drink* **salep** *is made*. ~**leşmek,** to become crafty. ~**lik,** craftiness.
**tilmiz** Disciple; pupil.
**tim** Team.
**timsah** (·ˈ) Crocodile.
**timsal** (·ˈ), **-li** Symbol; model, example.
**tin** *In* ~ ~ **yürümek,** (of a baby) to toddle; (of a very old man) to toddle along briskly; to move with unexpected agility.
**tip, -pi** (*Fr. type*) Type; queer specimen.
**tipi** Blizzard, snow-storm. ~**lemek,** to blow a blizzard, snow heavily.
**tipik** Typical.
**tir** *As* **tiril** *q.v.*
**tiraj** (*Fr. tirage*) Circulation *of a newspaper*.
**tiramola** (··ˈ·) *Naut. interj.* Haul (on a rope)! A kind of capstan. ~ **etm.,** to tack ship; to go about.
**tirandaz** Popular form of **tirendaz,** *but meaning* trim, well-dressed; dexterous, skilful.
**tirbuşon** (*Fr. tire-bouchon*) Corkscrew.
**tire**[1] Sewing cotton. Cotton (*a*).
**tire**[2] (*Fr. tiret*) Dash, hyphen.
**tirendaz** Archer; *v. also* **tirandaz.**
**tirenti** (·ˈ·) Fall (of a pulley); boat-falls.
**tirfillenmek** *vn.* Become threadbare.
**tirid** Bread soaked in gravy; feeble old man. **suyuna** ~, without substance; perfunctory: **suyuna ~ geçinmek,** to live on next to nothing. ~**leşmek,** to become old and feeble.
**tiril** *In* ~ ~ **titremek,** to tremble like an aspen leaf. ~**demek,** to shiver.
**tirit** *v.* **tirid.**
**tiriz** Lath, batten; moulding (*architecture*); piping (*dressmaking*).
**tirkeş** Quiver.
**tirlin** (*Fr. tire-ligne*) Drawing-pen.
**tirpidin, tirpit.** Small mattock.
**tirsi balığı, -nı** Shad.
**tirşe** Vellum. Pale green.
**tiryak** Theriac; antidote to poison. ~**i** (·–ˈ), addicted to alcohol, tobacco, opium *etc.*; tiresome, 'difficile'. ~**ilik,** a being addicted to *stg.*; the thing to which one is addicted; obsession; smoking.
**titiz** Peevish, captious, hard to please, 'difficile'; fastidious, sensitive; meticulous, particular, extremely careful. ~**lenmek,** to be tiresome and hard to please; to become annoyed.
**titre·k** Trembling. ~ **kavak,** aspen. ~**mek,** to shiver; to tremble: ··· **in üzerine** ~, to love *s.o.* so tenderly that one is always on tenterhooks about him.
**tiyatro** (·ˈ·) Theatre. ~**cu,** theatre owner; actor.
**tiz** High-pitched. **yarım ton** ~, sharp (*mus.*).
**tohum** Seed; grain; semen; eggs (of insects). ~**a kaçmak** *or* **bağlamak,** to go to seed; ~**a kaçmış,** *also* aged,

past its prime: **fesad ~u saçmak,** to sow sedition *or* discontent. **~luk,** suitable for seed; kept for breeding; garden bed: **~ buğday,** seed wheat.

**tok** Satiated; full; deep (voice); closely-woven, thick (cloth). ⌜**~ evin aç kedisi**⌝, well off but still not content: **~ gözlü,** contented; not covetous: **~ sözlü,** who does not mince his words, outspoken: **~ tutmak,** (of a food) to be filling: **gözü ~,** contented, free from greed: **karnım ~,** I am not hungry: I can't be taken in *by that sort of talk.*

**toka**[1] Buckle. **~lı,** having a buckle; buckled.

**toka**[2] **~ etm.,** to shake hands; to clink glasses; (*sl.*) to give, pay. **~laşmak,** to shake hands.

**tokaç** Mallet; bat *for beating out washing.*

**tokat** Cuff, box on the ears. **···e ~ aşketmek (atmak)** *or* **···i ~lamak,** to give *s.o.* a box on the ears.

**toklu** Having pendent glands (goat); yearling lamb.

**tokluk** Satiety; thickness *or* density *of cloth.* **boğaz tokluğuna çalışmak,** to work in return for one's board.

**tokmak** Mallet; beetle (*implement*); door-knocker; clapper *of a bell*; wooden pestle: **~ gibi,** chubby (baby).

**tokurcun** Stook *of wheat etc.*; *a game played with pebbles or marbles.*

**tokur·datmak** *va.* Make *a hookah* bubble. **~tu,** the bubbling noise of a hookah.

**tokuş** *v.* **dokuş.**

**tokuş·mak** *vn.* Butt one another; collide. **~turmak,** to cause to collide; to clink *glasses*; to cannon (billiards).

**tokuz** Thick, closely-woven (cloth).

**tolga** Helmet.

**toloz** Arched vault.

**tomar** Roll *or* scroll (of paper *etc.*). Cylindrical. **top ~ı,** rammer *or* swab *for a gun.*

**tombak** Gold-plated copper; copper zinc alloy.

**tombalak** Round as a ball; plump.

**tombaz** Barge; punt; pontoon. **~lardan yapılmış sal,** ferry (*mil.*); **zincirli ~,** chain-ferry.

**tombola** (ˈ··) Tombola; lotto.

**tombul** Plump.

**tomruk** Bud; heavy log. **~lanmak,** to put forth buds.

**tomurcuk** Bud.

**ton**[1] (*Fr. thon*) Preserved tunny.

**ton**[2] Ton. **~aj,** tonnage.

**tondura** Tundra.

**tonel** *Vulg. for* **tünel,** *only in* **~ geçmek** (*sl.*), to be absent-minded, to be wool-gathering.

**tonga** Trap, trick. **~ya basmak (düşmek),** to fall into a trap, to be deceived.

**tonilâto** (··ˈ·) Tonnage; ton. **~luk,** having a tonnage of ...; a ton's weight of ... .

**tonluk** Of *so many* tons.

**tonoz**[1] Vault.

**tonoz**[2] **~ etm.,** to warp *a ship*: **~ balığı,** tunny: **~ demiri,** kedge-anchor.

**tonton** Darling.

**top**[1], **-pu** Round; collected together. Ball; any round thing; a whole; roll (of cloth *or* paper). **~ ~,** in groups, in lumps; **~u ~u,** in all: **~tan,** wholesale; as a whole: **~unuz,** all of you: **~unu birden,** one and all: **~ gibi,** willy-nilly, without question.

**top**[2], **-pu** Gun, cannon. **~ atmak,** to fire a gun: **~** *or* **~u atmak,** to 'go bust'; to be 'ploughed' *in an exam.*: **~un ağzında,** the one most in danger: **~a tutmak,** to bombard *a place.*

**topaç** Top (plaything), teetotum; the thick rounded part of a Turkish oar. **~ gibi,** sturdy (child).

**topak** Roundish lump. Short and fat.

**topal** Lame. Cripple. ⌜**~ eşekle kervana karışmak**⌝, to undertake *stg.* with inadequate means. **~lamak,** to limp. **~lık,** lameness.

**toparlak** Round. Limber (*mil.*).

**topar·lamak** *va.* Collect together; pack up; roll up (*mil.*). **kendini ~,** to pull oneself together. **~lanmak,** *vn. pass. of* **toparlamak,** to be collected together *etc.*; to pull oneself together.

**topatan** An early oblong kind of melon.

**topçeker** Large gunboat; tractor for pulling guns.

**topçu** Artilleryman; gunner; the artillery; (*sl.*) one who looks like being 'ploughed' in an exam. **~luk,** gunnery; duties and profession of a gunner.

**topla** Three-pronged winnowing fork.

**toplamak** *va.* Collect together; convene; gather; sum up; fold up (clothes); tidy up; clear away; put on weight. **ağzını topla!**, shut up! *said to a person who is insulting one*: **aklını başına ~**, to collect one's wits, to pull oneself together: **kendini ~**, to recover *from an illness*.

**toplan·ılmak** *vn. impers. form of pass. of* **toplamak**; *as* **toplanmak**. **~mak** *vn.*, to collect, assemble, come together; regain one's health; put on flesh; (of a boil) to come to a head. **~tı**, assembly, gathering, meeting.

**toplaşmak** *vn.* Gather together.

**toplu** Having a knob *or* rounded head; compact; collected together, in a mass; well-arranged, tidy; plump; collective; (*for* **~ iğne**) pin. **~ ateş**, concentrated fire (*mil.*): **~ iğne**, pin. **~luk**, compactness; community; gathering.

**toprak** Earth; soil; land. Earthen. **~ altı**, the subsoil: ⌜**toprağı bol olsun!**⌝, 'may he rest in peace! (*referring to s.o. dead*): **~ boya**, paint in powder form. **~ sokak**, unpaved street: **eski ~**, old but well-preserved (person): **kara ~**, the grave. **toprak·bastı**, tax on goods *or* beings entering a town. **~lamak**, to cover with earth.

**toptan** Wholesale; in the mass. **~cı**, wholesaler. **~cılık**, wholesale trading.

**topu topu** In all; all told.

**topuk** Heel; ankle; fetlock; bar *of a river*; heel *of a shoe*; heel *of a mast*.

**topuz** Mace; knob (on a stick *etc.*); knot of hair. Short, thick (man).

**topyekûn** (ˈ··) Total.

**tor** Net; tissue. **~ağ**, fine-meshed fishing-net.

**torak** Charcoal pit; kiln; dried skim milk.

**toraman** Young, wild and untamed. Robust young man; pet animal.

**torba** Bag; scrotum; cyst. **~ yoğurdu**, yaourt strained in a bag: ⌜**ağzında ~ mı var?**⌝, have you lost your tongue?: **yem(lik) ~sı**, nose-bag.

**torik** Large palamut.

**torina** (·ˈ·) Grampus.

**torlak** Unbroken colt; wild youth.

**torluk** Mud hut; charcoal kiln.

**torna** (ˈ·) Lathe. **~ etm.**, to turn *on a lathe*: **~ aynası**, potter's wheel; lathe chuck. **~cı**, turner.

**tornavida** (··ˈ·) Screwdriver.

**tornistan**[1] (ˈ··) **etm.** *vn.* Go astern; give up a project *etc.*

**tornistan**[2] (ˈ··) Turned (suit of clothes).

**Toros** Taurus.

**torpil** Mine (explosive); torpedo; (*sl.*) a 'friend at court'. **~lemek**, to torpedo.

**torpito, torpido** Torpedo; torpedo-boat. **~ kovanı**, torpedo-tube.

**tortop** (ˈ·) Quite round.

**tortu** Deposit; dregs; sediment. **~lu**, having sediment; turbid.

**torun** Grandchild; two-year-old camel.

**tos** A blow with the head. **~ vurmak**, to butt. **~lamak**, to butt; (of a ship) to have a slight collision *or* (*sl.*) to pitch; (*sl.*) to pay.

**tostoparlak** (ˈ···) Quite round.

**tosun** Young bull; fine robust young man.

**toy**[1] Raw, inexperienced, 'green'.

**toy**[2] Banquet.

**toygar** A kind of lark.

**toyluk** Inexperience; rawness.

**toz** Dust; powder. Like dust; in powder form. **~ almak**, to dust: **~u dumana katmak**, to raise clouds of dust; to make a great ado; to create confusion: **~ etm.**, to raise the dust: **~ koparmak**, to kick up a dust: **~ silkmek**, to beat out the dust, to dust; **~unu silkmek**, to give *s.o.* 'a dusting': **ayağının ~u ile**, at the moment of arrival, without delay: **ortalığı ~ pembe görmek**, to see the world through rose-coloured spectacles: **···in üzerine ~ kondurmamak**, not to allow anything to be said against ... , not to hear any criticism of ... . **toz·amak**, to raise the dust: **~armak**, to become dust; to go to powder. **~koparan**, place exposed to strong winds. **~lanmak**, to become dusty. **~lu**, dusty. **~luk**, gaiter, dusty place.

**tozmak** *In* **gezip ~**, to saunter about and enjoy oneself.

**toz·pembe** (ˈ··) *v.* **toz.** **~untu**, any fine thing like dust. **~utmak**, to raise a dust; to go too far, become unreasonable; to go mad.

**töhmet, -ti** Imputation; guilt. **~li**, under suspicion; guilty. **~siz**, not suspected; innocent.

**tökesimek, tökezlemek** *vn.* Stumble.

**tömbeki** The Persian tobacco *smoked in hookahs.*
**töre** Custom; rule; law.
**tören** Ceremony; celebration.
**törpü** Rasp; file. **ömür ~sü,** an exhausting task; a trying person. **~lemek,** to rasp, file.
**törü** *v.* **töre.**
**tövbe** Repentance. **~!,** pax!; enough!: **···e ~ etm.,** to repent having done *stg.* and vow not to do it again: **~ istiğfar,** to repent and ask God's pardon: **~ler olsun!,** I'll never do it again!: **~ler ~si,** *as* **~ler olsun: yedi ceddine ~ etm.,** to swear one will give up *stg.* for good and all.
**tövbe·kâr,** penitent: **~ (kadın),** reformed prostitute. **~li,** penitent; under a vow not to sin again *or* not to do *stg.* again.
**Trablus** (ˈ·) Tripoli. **~ugarb,** Tripoli in Africa: **~ uşŞam,** Tripoli in Syria.
**Trabzon** (ˈ·) Trebizond. **~ hurması,** date-plum: *v. also* **tırabzan.**
**trahom** Trachoma.
**trahoma** (·ˈ·) Dowry (of a non-Moslem).
**trak** *v.* **trank.**
**Trakya** (ˈ·) Thrace.
**trampa** (ˈ·) Barter; exchange.
**trampete** (·ˈ·) Side drum.
**tramplen** (*Fr. tremplin*) Spring-board.
**tramvay** Tram. **~cı,** tram-driver.
**trank** *Imitates the clanking of metal (usually in connexion with the payment of cash).*
**transit, -ti** Passage of goods without paying custom dues; through (traffic).
**traş** *v.* **tıraş.**
**travers** (*Fr. traverse*) Railway sleeper. **~e çıkmak,** to sail to windward by tacking.
**tren, tiren** Train.
**trinketa** (·ˈ·) Foresail.
**trişin** Trichina.
**trup, -pu** Troupe; theatrical company.
**tu** *Interj. expressing disgust.*
**tufan** (–ˈ) Violent rainstorm; flood; the Flood.
**tufeyli** (··ˈ) Parasite; sponger; toady.
**tuğ** Horse-tail *attached to a helmet or flagstaff as a sign of rank.*
**tuğ·amiral** Rear-Admiral. **~ay,** brigade. **~bay,** brigadier. **~general,** brigadier-general.
**tuğla** (ˈ·) Brick. **~ harmanı,** brick-yard. **~cı,** brickmaker.
**tuğlu** Wearing a crest of horsehair *v.* **tuğ.**
**tuğra** *v.* **tura.** The Sultan's monogram.
**tuğyan ~ etm.,** to overflow; to rebel.
**tuhaf** Uncommon; curious, odd; comic, amusing. **~!** *or* **~ şey!,** that's odd!, how curious!: **~ıma gitti,** it seemed odd to me: **işin ~ı,** the odd thing about it is .... **~çı** *v.* **tuhafiyeci.** **~iye** (·–·ˈ), millinery, drapery. **~iyeci,** milliner, draper, fancy-dealer. **~lık,** a being odd *or* funny: **~ yapmak,** to make funny remarks.
**tul** (–), **-lü** Length; longitude. **~ dairesi,** meridian.
**tulga** Helmet.
**tulû** (·ˈ), **-ûu** Rising *of the sun or a star*; birth *of an idea.* **~at, -tı,** sudden ideas; improvisations; *as sing.* popular theatre *where the actors improvise.* **~atçı,** actor who improvises.
**tulum, tuluk** Skin made into a bag *to hold water etc., or used as a float for a raft* (**kelek**); overalls: **~ gibi** *or* **yağ ~u,** as fat as a pig: **~ peyniri,** a kind of cheese made in a skin.
**tulumba** (·ˈ·) Pump; fire-engine; waterspout. **~ tatlısı,** a sweetmeat *made of dough soaked in syrup.* **~cı,** a member of the old independent fire brigades; a rough, rowdy; an unmannerly youth: **~ koğuşu,** a dormitory for firemen; an assembly of roughs.
**tuman** Long wide drawers *or* trousers.
**tumba** (ˈ·) Tilting-truck; a turning upside down; (of children) a tumbling into bed.
**tumturak** Bombast; pompous speech. **~lı,** bombastic; high-flown.
**Tuna** The Danube.
**tunc** Bronze.
**Tunus** (ˈ·) Tunis. **~ gediği,** a thing got without trouble; a rich woman married for her money. **~lu,** Tunisian.
**tur**[1] (*Fr. tour*) Tour; promenade.
**tur**[2] *In* **~ yağı,** a vegetable oil.
**tura** (ˈ·), **tuğra** The Sultan's monogram; the imperial cipher; skein (of silk *etc.*); a game *played with a knotted handkerchief.* **yazı mı ~ mı?** heads or tails?; **yazı ~ atalım,** let's toss for it.

**turac** Francolin.
**Turanî** (– – ⌐) Turanian.
**turb(a)** (*Fr. tourbe*) Peat (for fuel).
**turfa** Unclean; not fresh (food). **~ olm.,** to fall into disesteem, to be despised. **~lamak,** to despise, treat with contempt.
**turfanda** Early fruit *or* vegetables; (*jokingly*) novice, new. **son ~,** the last fruit *etc.* of the season. **~lık,** garden for growing early fruit *etc.*
**turhalli** (⌐··) *Only in* ⌜**her halli ~**⌝ *or* ⌜**~ bir halli**⌝, everyone is in as bad a state as his neighbour; all in confusion.
**Turisina** Mount Sinai.
**turna** Crane (bird). **~ balığı,** pike; garfish: ⌜**~yı gözünden vurmak**⌝, to hit the mark (*fig.*); to do a good stroke of business: **~ katarı,** a flock of cranes; a procession of people: **~ kırı,** ashen grey colour.
**turne** (*Fr. tournée*) Tour.
**turnike** (*Fr. tourniquet*) Turnstile.
**turnsol** (*Fr. tournesol*), **turnusol, -lü** Dyer's croton; turnsole (purple dye); litmus paper.
**turnuva** (⌐··) (*Fr. tournoi*) Tourney; tournament.
**turp, -pu** Radish. **~ gibi,** robust, 'as sound as a bell': ⌜**aklına ~ sıkayım**⌝, (*sl.*), 'what rot!'
**Tursina** Mount Sinai.
**turşu** Pickle. **~ gibi,** very weary: **~ kesilmek (olm.),** to turn sour; (of food) to go bad; to become slack, to have no energy: **~ kurmak,** to pickle; ⌜**~sunu mu kuracaksın?**⌝, 'do you want to pickle it?', *or* ⌜**~sunu kur!**⌝, 'pickle it!', *angry retorts to one who has refused to give stg.*; 'all right, keep it then!': **~ suratlı (yüzlü),** sour-faced. **turşu·cu,** maker and seller of pickles. **~luk,** material suitable for pickling.
**turunc** Seville orange. **~u,** orange colour.
**tuş** (*Fr. touche*) Key (of piano, typewriter *etc.*).
**tutam** Small handful.
**tutamak** Handle; habitual custom; proof, evidence; means of livelihood. **~ bulmak,** to find a pretext: **~ noktası,** *stg.* to catch hold of: **~ vermek,** to afford a pretext. **~lı,** provided with a handle *or* a grip; high-principled, to be depended on. **~sız,** without a grip *or* handle *or* support; without principles, not to be depended on.
**tutar** Epilepsy, seizure, fit; total, sum. Holding, seizing, *v.* **tutmak.**
**tutarak** Seizure, fit; kindling wood, tinder. **tutarağı tuttu,** he had a seizure; he has a fit of obstinacy.
**tutkal** Glue; size. **~ gibi,** *used of a person one can't get rid of*: **balık ~ı,** isinglass. **~lamak,** to glue.
**tutkun** (*with dat.*) Affected by; given to; in love with.
**tutma** Act of holding; grip, hold.
**tutmak** 1. *va.* Hold; hold on to; keep, retain; preserve; take, catch, seize; stop, detain; esteem, account, reckon, suppose; contain, hold; agree with, tally with; take the part of, favour; listen to *advice etc.*; engage (servant); hire (house *etc.*); spread to, reach (a district); (of a ship) to touch at *a port*; (*sl.*) have as wife. 2. *vn.* Take root; (of seeds, grafts, vaccination *etc.*) to take; succeed; (of a curse) to take effect; hold on, endure, last; (of a sum of money, an account *etc.*) to amount to, to reach; stick, adhere; come into one's head, occur to one; (of rain *etc.*) to begin; (of an illness *or* pain) to come on, to attack one; (of wine *etc.*) to go to the head. **tutalım ki,** let us suppose that: **tutuyor musun?** (*sl.*), 'have you any cash?': **tuttu, İstanbula gitti,** it occurred to him to go to Istanbul: **atıp ~,** to blame, to criticize; to rant; to talk boastfully *or* airily: **ayağını çabuk tut!,** 'get a move on!': **başı* ~,** for one's head to ache *from noise or worry*: **birbirini tutmaz şeyler,** incompatible things; things that don't make sense: **bu listeler birbirini tutmuyor,** these lists do not tally: **deniz (tren** *etc.*) **~,** for the sea (a train *etc.*) to cause one to feel sick: **···den tutun da ... kadar,** starting from ... to ...: **dişim tuttu,** my tooth has begun to ache again: **dümen** (*etc.*) **tutmuyor,** the rudder (*etc.*) is not working, is having no effect: **elinden ~,** to give *s.o.* a helping hand: **elini çabuk tut!,** get on with the job!: **ev ~,** to rent a house: **iş ~,** to work, to have a job: **kar tutmadı,** the snow did not lie: **kendimi tutmadım,** I could not restrain myself, I could not help ...: **···i lâfa ~,** to detain *s.o.* by talking to him: **bir işi sağlam ~,** to

start *stg.* on a sound basis: **sözünü*** ~, to keep one's word: **···in sözünü** ~, to listen to ..., to take *s.o.'s* advice: **şahid** ~, to bring forward as a witness: **hiç bir yerim tutmuyor,** I feel rotten all over: **yolu** ~ (*sl.*), to make off, to clear out: **··· e yüzü* tutmamak,** not to have the face to ..., not to dare to ...: **o akşam sinemaya gideceğimiz tuttu,** we took it into our heads to go to the cinema that evening: **bunu görenin güleceği tutar,** anyone seeing this would feel like laughing: **herifin edebsizliği tutunca da tutar ha!,** when the fellow is rude, he *is* rude.

**tutsak** Prisoner, captive. **~lık,** captivity.

**tutturmak** *va. caus. of* **tutmak.** Cause to hold *etc.*; begin; cause to catch on *or* succeed. *vn.* Begin and continue; hit the mark; keep bothering *s.o.* to do *stg.* (like a spoilt child). **tuttura bildiği fiatı istemek** *or* **tuttura bildiğine satmak,** (of a shopkeeper *etc.*) to ask a price *or* to sell at a price without regard to real value.

**tutuk** Paralysed; having had a stroke; stopped up; impeded; embarrassed; tongue-tied; stuttering; slow, hesitant; (*sl.*) 'gone on', in love with. **~luk,** *any of the states described by* **tutuk**; breakdown, stoppage: **göğüs tutukluğu,** shortness of breath.

**tutulma** *verb. n. of* **tutulmak.** Act of being held *etc.*; a falling in love. **güneş ~sı,** an eclipse of the sun: **dil ~sı,** an impediment of speech; a being tongue-tied.

**tutulmak** *vn. pass. of* **tutmak.** Be held *etc.*; be struck with, fall in love with; be angered; succeed, catch on; be rented *or* hired; be seized with illness; stutter, be tongue-tied. **···e** ~, to fall in love with, to be mad about ...; to be furious with ...: **ay tutuldu,** the moon is in eclipse: **boynu*** ~, to have a stiff neck: **dili*** ~, to be tongue-tied; to be struck dumb *by fear etc.*: ⌜**hay dilin tutulasıca!**⌝, 'curse your tongue!', *said to one who has said what he ought not to*: **elle tutulur,** tangible: **kendi ağzile tutuldu,** is given away by his own words *or* he contradicts himself: **sesi** ~, to become hoarse: **tifoya** ~, to catch typhoid.

**tutum** Manner, conduct, procedure; economy, thrift. **~lu,** thrifty.

**tutun** *v.* **tutmak** *and* **tutunmak.**

**tutunmak** *va.* Apply to oneself; wear. *vn.* Hold on, cling, take a hold. **bu moda çabuk tutundu,** this fashion caught on quickly: **örtü** ~, to put on a wrap: **sıkı tutun!,** hold tight!: **sülük** ~, to apply leeches to oneself.

**tutuş·mak** *vn.* Catch hold of one another; quarrel; catch fire, be on fire; flare up *in anger.* **bahse** ~, to bet: **etekleri*** *or* **eli* ayağı** ~, to be in a great fright; to be in desperation: **kavgaya** ~, to quarrel; to come to blows. **~turmak,** *va. caus. of* **tutuşmak,** to set on fire; set *persons* by the ears, cause to quarrel; press into a person's hand (*esp.* a bribe): **··· için yanıp** ~, to be violently in love with ....

**tuvalet** (*Fr. toilette*), **-ti** Toilet; articles of toilet; toilet table; lavatory.

**tuz** Salt. **···e** ~ **ekmek,** to add salt to ...: ~ **biber ekmek,** to make things worse; to be the last straw; **bu o işin ~u biberidir,** this is a necessary addition; **~u biberi yerinde,** properly seasoned; nothing lacking, all right: **~la buz olm.,** to be smashed to bits; to be utterly routed: ~ **ekmek hakkı,** gratitude towards a benefactor; ~ **ekmek haini,** ungrateful to a benefactor: **~u kuru,** well-off, in easy circumstances; without worries: **ingiliz ~u,** Epsom salts: **tadı ~u yok,** tasteless, insipid.

**tuzak** Trap. **···e** ~ **kurmak,** to lay a trap for *s.o.*

**tuz·la** (ˈ·), **~lak** Saltpan; salt-mine. **~lama,** act of salting; salted; pickled in brine. **~lamak,** to salt, pickle in brine: ⌜**tuzlayım da kokma**⌝, *stg. like* 'utter nonsense!'; 'you must be off your head!' **~lu,** salt, salted; pickled; expensive: ~ **oturmak** *or* **~ya mal olm.,** to cost dear. **~luca,** oversalted; rather expensive. **~luk,** salt-cellar. **~luluk,** saltiness, salinity. **~ruhu** (ˈ··), **-nu,** hydrochloric acid; spirit of salt. **~suz,** unsalted; insipid.

**tüb** Tube.

**tüccar** *pl. of* **tacir** *used as sing.* Merchant.

**tüfek** Gun, rifle. ~ **çatmak,** to pile arms. **dolu** ~, 'a loaded gun', *said of a choleric person.* **~çi,** gun-maker; armourer; (*formerly*) guard at the

Imperial Palace. ~**hane** (··–!), armoury.

**tüh** *Interj. expressing regret or annoyance.* ~ **sana!**, shame on you!

**tük·enmek** *vn.* Be exhausted; come to an end; give out. ~**enmez,** inexhaustible; a kind of syrup *made of fruit juice, to which water is constantly added.* ~**etmek,** to exhaust; use up; spend: **nefes** ~, to wear oneself out in trying to explain *stg.*: **sıfırı** ~, to have come to the end of one's resources.

**tükür·mek** *va. & vn.* Spit; spit out. **tükürdüğünü yalamak,** to eat one's words: ⌜**aşağı tükürsem sakalım, yukarı tükürsem bıyığım**⌝, I'm between the devil and the deep blue sea: ⌜**birbirinin ağzına tükürmüş gibi**⌝, 'as though they had spit into each other's mouths', *said of two people who give exactly the same version of an event.*

**tükürük, tükrük,** spittle, saliva: ~ **bezleri,** salivary glands: ~ **otu,** Star of Bethlehem, ornithogalum: **ağzında tükürüğü kurudu,** he talked himself hoarse.

**tül** Tulle.

**tülbend** Muslin; gauze. ~ **kuruyuncaya kadar,** in a very short time.

**tüm·amiral** Vice-Admiral. ~**general,** Major-General.

**tümen** Great number; great heap; ten thousand; a Persian gold coin; division (*mil.*). ~ ~, in vast numbers.

**tümör** Tumour.

**tümsek** Small mound; protuberance.

**tümselmek** *vn.* Rise *out of the ground*; become round; be protuberant.

**tün** Night. ~**aydın!** good evening!

**tünek** Perch *in a hen-house etc.* ~**lemek, tünemek,** to perch (of birds).

**tünel** Tunnel. ~ **geçmek,** *v.* **tonel.**

**tüp** Tube.

**türâbâ** (·–!) *Only in* **haraben** ~, tumbledown, utterly ruined.

**türbe** Tomb, grave, mausoleum. ~**dar,** keeper of a tomb.

**türe·di** Upstart; parvenu. ~**mek,** to spring up suddenly.

**Türk, -kü** Turk. Turkish. ~**çe,** the Turkish language; in Turkish: ~**si,** in plain Turkish, *i.e.* the long and the short of it (*cf. 'in plain English'*). ~**çü,** admirer and supporter of Turkish language and culture. ~**iyat, -tı,** the study of things Turkish. ~**iye** (!··), Turkey. ~**iyeli,** belonging to *or* originating from Turkey. ~**leşmek,** to adopt Turkish habits; become like a Turk. ~**lük,** the quality of being a Turk. ~**men,** Turcoman.

**türkü** Folk-song. ~ **söylemek,** to sing: ⌜**kimin arabasına binerse onun** ~**sünü çağırır**⌝, he changes the tune to suit the occasion.

**türlü** Sort, kind, variety; a dish of meat with various kinds of vegetables. ~ ~, of all sorts, various: ~**sünü görmek,** to have varied experiences of *stg.*: **bir** ~, in some way, somehow; (*with neg.*) in no way whatever, not at all; **her** ~, every sort of.

**tüs** *Only with* **tüy.** Down. **tüyü** ~**ü yok,** the hair has not sprouted on his cheeks.

**tütmek** *vn.* (Of a chimney *etc.*) to smoke. **gözünde*** (**burnunda***) ~, to be longed for.

**tütsü** Fumigation; fumigant; incense; the use of incense in sorcery. ~**lemek,** to fumigate; to cure *meat etc.* by smoking: **kafayı** ~, to become tipsy. ~**lü,** fumigated; smoked: **kafası** ~, befuddled, tipsy.

**tütün** Smoke; tobacco. ~ **balığı,** smoked fish: ~ **içmek,** to smoke (tobacco). ~**cü,** tobacconist; grower of tobacco. ~**lemek,** to cure by smoking.

**tüy** Feather; down; hair. ~ **gibi,** as light as a feather, agile: ~**lerim ürperdi,** my hair stood on end: (**dilde**) ~ **bitmek,** to be weary of repeating *stg.*: **bu her şeye** ~ **dikti,** this was the last straw. ~**kalem,** quill-pen. ~**lenmek,** to grow feathers; to start to grow a beard; to be well-feathered, to be rich. ~**lü,** feathered; well-to-do. ~**süz,** unfeathered; unfledged; young. ~**tüs,** *v.* **tüs.**

**tüymek** *vn.* (*sl.*) Slip away.

**tüzük** Rules and regulations *of a society etc.*

# U

**ubudiyet** (·–·!), **-ti** Loyal devotion to God; slavery, servitude.

**ubur** (·!) A passing; passage; crossing. ~ **etm.**, to pass, cross: **müruru ~**, street traffic. ;

**uc** Tip, point; extremity; end, frontier; top; direction, course; cause, motive; steel pen; ploughshare. **~ ~a,** end to end; point to point: **~dan ~a,** from one extreme to the other: **~ bölüğü,** leading company (*mil.*): **~u dokunmak,** to involve *or* affect *one*; to entail: **~ ~a gelmek,** to be just enough: **~u ~una getirmek** *or* **~ ~a karşılamak,** to make both ends meet: **···in ~unu göstermek,** to drop a hint of *stg. advantageous as an inducement*: **~unu kaçırmak,** to lose the thread of a matter: **~unu ortasını bulmak,** to get to the bottom of a matter: **bu işin ~u ortası belli değil,** one doesn't know how to tackle this matter: **işin ~unda para var,** there is money to be made out of it: **···in ~unda bir şey olm.,** for there to be some secret purpose behind *stg.*: **~ vermek,** to appear; to sprout, grow; (of a boil) to come to a head: **ayakların ~una basarak,** on tiptoe: **başı ~unda,** by his side: **burnunun ~unu görmemek,** to be blind with pride: **dilinin* ~una gelmek,** for *stg.* to be on the tip of one's tongue (and which one refrains with difficulty from saying): **dilinin* ~unda olm.,** to be on the tip of one's tongue (of *stg.* one is trying to recollect): **dünyanın bir ~unda,** at the other end of the world: **iki ~unu bir araya getirmek,** to make both ends meet; to make a success of a business *etc.*

**uclu** Pointed; having a nib in it (penholder).

**ucra** (·⊥) Remote; out-of-the-way; solitary.

**ucsuz** Without a point; with no nib (penholder). **~ bucaksız,** endless, vast.

**ucube** (––⊥) Strange thing; a wonder; curiosity; abortion; monstrosity.

**ucun ucun** Partially; bit by bit.

**ucuz** Cheap. **~ kurtulmak,** to get off lightly: **~ satmak,** to sell cheaply: **sudan ~,** dirt cheap. **~cu,** who sells cheaply; bargain hunter. **~lamak,** to become cheap. **~luk,** cheapness; place where living is cheap.

**uçak** Climbing plant; aeroplane. **~savar** (·⊥··), anti-aircraft.

**uçar** Flying; volatile. **~ı,** who flies in the face of all decency; dissolute: **~ çapkın,** debauchee, rake: **~ takımından,** an arrant scoundrel.

**uçkun** Spark. Flying.

**uçkur** Band for holding up trousers. **harama ~ çözmek,** to have illegitimate sexual relations. **~luk,** seam for passing band through top of trousers.

**uçmak** *vn.* Fly; evaporate; fall *from a great height*; fade away, disappear; act outrageously, go beyond all bounds; be wild *with joy etc.* **uçan kuşa borclu,** in debt to everybody: **benzi uçtu,** he turned pale.

**uçucu** Flying; volatile.

**uçuk**[1] *In* **rengi (benzi) ~,** whose colour has fled, pale.

**uçuk**[2] Blain on the lips; vesicle; herpes. **~lamak,** to have vesicles break out on the lips.

**uçur·mak** *va. caus. of* **uçmak.** Cause to fly *etc.*; fly (an aeroplane); exaggerate; boast about; praise excessively. **kellesini ~,** to cut off a person's head: **kuş uçurmamak,** to not let a living thing pass, to be very vigilant. **~tma,** kite (paper). **~tmak,** *va. caus. of* **uçurmak,** *but used as caus. of* **uçmak,** to cause *or* allow to fly *etc.* **~um,** precipice; abyss.

**uçuş** Flight *of a plane.*

**uçuşmak** *vn.* Fly together, fly in a flock; fly about; flap the wings and fly noisily.

**ud** Lute. **~î** (–⊥), lute player.

**uf** *Interj. expressing boredom, annoyance, fatigue.* **~ puf demek,** to sigh, to express annoyance.

**ufacık** (⊥··) Very small, tiny.

**ufak** Small. **~ tefek,** small; of no account; small and short (man); trifles, small things: **ekmek ufağı,** crumbs: **un ~,** as fine as flour. **~lı,** *in* **irili ~,** large and small together. **~lık, ~para,** small change; lice.

**ufal·amak** *va.* Crumble; break up. **~mak,** to become smaller. **~tmak, ufatmak,** to reduce in size. **~anmak, ufanmak,** to be broken up small; crumble.

**ufarak** Smaller; very small.

**ufk·an** (⊥·) Horizontally. **~î** (·⊥), horizontal. **~u,** *v.* **ufuk.**

**uflamak** *vn.* Say 'oof' *expressing boredom or annoyance.*

**ufuk, -fku** Horizon.

**ufunet** (·–⊥), **-ti** Inflammation. **~lenmek,** to putrefy; become inflamed and putrid. **~li,** putrefying; putrid.

**uğalamak, uğmak** *etc., v.* **oğalamak** *etc.*

**uğrak** Place through which one passes. **yol uğrağı,** place lying on one's road.

**uğramak** *va.* (*with dat.*) Stop *or* touch at *a place*; meet with (an accident); suffer (an illness); look in on *s.o.*; undergo (a change *etc.*). *vn.* Be possessed by an evil spirit: **ahali sokağa uğramış,** everybody has come out into the street: **gözleri dışarıya uğramış,** his eyes started out of his head: **···in semtine uğramamak,** not to go near … .

**uğraş** Struggle, fight. **~mak,** to struggle; to strive hard. **···le ~,** to be busy with *stg.*; to have a down on *s.o.* **~tırmak,** *caus. of* **uğraşmak**; *also,* to cause annoyance to, to disturb.

**uğratmak** *va. caus. of* **uğramak.** Cause to stop at, *etc.*; cause to encounter: expose *s.o. to a danger etc.*; send away.

**uğul·damak** *vn.* Hum; buzz; (of the wind) to howl. **~tu,** humming *or* buzzing noise; a singing in the ears.

**uğur, -ğru** Good luck; good omen. **~lar olsun!,** good luck!; good journey! (*said to one departing*): **···in uğruna (uğrunda),** for the sake of …; on account of … . **~lamak,** to wish *s.o.* good luck; see *s.o.* off *on a journey.* **~lu,** lucky; auspicious: **~ kademli,** bringing good luck, a happy omen: **ayağı ~ gelmek,** to bring luck. **~suz,** bringing bad luck; ill-omened: rascal: **ayağı ~ gelmek,** to bring bad luck. **~suzluk,** ill omen; a being unlucky.

**uğuşturmak, uğuz** *v.* **oğuşturmak, oğuz.**

**uhde** Obligation; responsibility. **···in ~sinde olm.,** to be entrusted to … , to be in the charge of …: **···in ~sinden gelmek,** to carry out the task of … , to discharge the duty of …: **~sine geçirmek,** to charge *s.o.* with the duty *etc.* of … .

**uhrevî** (··–̲) Pertaining to the next world.

**uhuvvet, -ti** Brotherhood.

**ukalâ** (··–̲) *pl. of* **âkil.** Sages, wise people; *as sing.* wiseacre, know-all, prig. **~dümbeleği, -ni,** pretentious quack; wiseacre. **~lık,** quality of a wiseacre; pretence of being clever; conceitedness; wisecrack.

**ukba** (·–̲) The next world.

**ukde, ukte** Knot; ganglion; *stg.* that sticks in the throat. **adamın içine ~ olur,** it's a thing one can't get over; it rankles.

**ukubet** (·–̲), **-ti** Retribution; punishment (*esp.* in the next world); torture.

**Ulah** Wallachian.

**ulak** Courier. **el ulağı,** messenger-boy, page: **bana el ulağı oluyor,** he helps me in small jobs.

**ulama** Consecutive, uninterrupted. Appendix. **~k,** to join *one thing to another*; to bring into contact.

**ulan** My good fellow!; man alive!

**ulaş·ık** Reaching. **~mak,** (*with dat.*) reach, arrive at; meet. **~tırma,** a causing to reach; communication: **~ Bakanı,** Minister of Communications (Transport).

**ulema** (··–̲) *pl. of* **âlim.** Learned men; ulema, doctors of Moslem theology.

**ulu** Great. Great man. **~larımız,** our great men. **~geyik,** red deer. **~lamak,** to extol, honour. **~lanmak,** to be honoured; to be puffed up. **~luk,** greatness.

**ulûm** *pl. of* **ilim.** Sciences.

**uluma** The howling *of dogs.* **~k,** to howl.

**uluorta** (·–̲··) Openly; without reserve; recklessly. Unfounded, gratuitous (assertion).

**ulus** Nation.

**uluşmak** *vn.* Howl together in packs (wolves *etc.*).

**ulvî** (·–̲) High; sublime.

**umacı** Ogre *or* bogy man (to frighten children).

**umde** Principle.

**umma** Hope; expectation. **···i ~ya uğratmak,** to disappoint *s.o.* **~dık** (–̲··), unexpected.

**ummak** *va.* Hope; expect.

**umman** (·–̲) Ocean; the Indian Ocean.

**umran** (·–̲) Prosperity.

**umud** Hope.

**umulmak** *vn. pass. of* **ummak.** Be hoped *or* expected.

**umum** (·–̲) General; universal; all. The public; people in general. **umum·en** (·–̲·) generally; universally. **~hane** (··–̲) brothel. **~î** (·––̲) general; universal; public: **efkâri ~ye,** public opinion. **~iyet**

(·—·ˈ), **-ti,** generality; universality: **~le,** in general: **~ itibarile,** on the whole; generally speaking.
**umunmak** *va.* Set one's hopes on.
**umur** (·ˈ) *pl. of* **emir.** Affairs; matters; *as sing.* a matter of importance; concern. **~ etmemek,** not to trouble about *stg.*: **~unda* bile olmamak,** not to care; to be indifferent: **ne ~un?,** what's that to you?: **ne ~unda!,** what did he care! **~samak,** to be concerned about: to consider important.
**un** Flour; meal. **has ~,** fine white flour: ⌜**~umu eledim, eleğimi duvara astım**⌝, I've finished with this sort of thing; I'm too old; I've done it all before: **~ ufak,** as fine as flour. **~cu,** flour merchant.
**un·lamak** *va.* Sprinkle with flour. **~lu,** prepared with flour; farinaceous.
**un·madık** (ˈ··) Incurable; that will not heal. An unlucky person. **~mak,** to heal; to get well. **~maz,** that will not heal; incorrigible; unlucky.
**unsur** Element; component part.
**unulmaz** *v.* unmaz.
**unut·kan** Forgetful. **~kanlık,** forgetfulness. **~mabeni** (·ˈ···) forget-me-not.
**unut·mak** *va.* Forget. ⌜**(havada bulut) sen onu unut!**⌝, you'll never see it again. **~ucu,** forgetful.
**unvan** (·ˈ) Title; superscription. **~lı,** bearing the title of ...; entitled; who gives himself airs.
**upuzun** (ˈ··) Extremely long.
**ur** Wen; tumour; goitre; excrescence.
**urba** *Pop. form of* **ruba,** dress, robe.
**urban** Bedouin Arabs.
**urğan** Rope.
**uruc** (·ˈ) **~ etm.,** to ascend.
**uruk, -ku** *pl. of* ırk Races.
**urup** *An old measure,* ⅛ of an **arşın,** *i.e. approx.* 3 in.
**usan** Boredom; disgust. **~c,** boring, tedious; boredom, tedium: ···**den ~ gelmek,** to be bored by ... . **~dırıcı,** boring, tedious. **~dırmak,** to bore, sicken, disgust. **~mak,** (*with abl.*) to become sick of, bored *or* disgusted with.
**uskumru** (·ˈ·) Mackerel.
**uskunca** (·ˈ·) Sponge *for a gun.*
**uskur** (*Eng. screw*) Screw, propeller; screw-driven ship.
**usla·ndırmak** *va.* Bring *s.o.* to his senses; make *s.o.* behave. **~nmak,** to become sensible; come to one's senses.
**uslu** Well-behaved; good (child); quiet (horse); sensible. **~ oturmak,** to sit still, keep quiet. **~luk,** a being well-behaved *or* sensible.
**usta** Master *of a trade or craft*; master workman; craftsman; foreman; overseer; *formerly* a woman superintendent of servants and slaves. Skilled; clever; experienced. **~ olm. (çikmak),** to finish one's apprenticeship and become a master workman. **~lık,** mastery *of a trade or craft*; proficiency; master-stroke. **~lıklı,** masterly; cleverly made; cunningly devised.
**ustunc** Portable case of instruments.
**ustura** (ˈ·· *or* ·ˈ·) Razor. **~ tutmak,** to use a razor, to shave.
**usturlab** Astrolabe.
**usturmaça** (··ˈ·) Fender, collision-mat.
**usturpa** (·ˈ·) Mop of rope ends; scourge.
**usturuplu** (*sl.*) Striking, impressive; right, decent, 'comme il faut'.
**usul** (·ˈ), **-lü** Method; system; manner; procedure; time (*mus.*). Gently, carefully, quietly. **~la,** carefully, gently: **~ü dairesinde,** in the recognized way: **~ü muhakeme,** judicial procedure: **~ vurmak,** to beat time (in Turkish music).
**usul·lacık** (·ˈ··), very slowly, gently *or* quietly. **~süz,** unmethodical; irregular; contrary to rules.
**uşak** Male servant; boy, youth; shop assistant. **~kapan,** lammergeier. **~lık,** childhood; profession of a man-servant.
**uşkun** A kind of wild rhubarb.
**uşşak** (·ˈ), **-kı** *pl. of* **âşık** Lovers; dervishes; *name of a mode in Turkish music.*
**ut** *v.* ud.
**utan·acak** Shameful. **~c,** shame; modesty; bashfulness. **~dırmak,** to make ashamed; put to shame; cause to blush; cause to look foolish. **~ğac, ~ğan, ~ık,** bashful; shy; shame-faced. **~ış,** a being ashamed. **~ma,** act of being ashamed.
**utan·mak** *vn.* Be ashamed; be shy *or* bashful; blush with shame *or* embarrassment. **~maz,** shameless; impudent. **~mazlık,** impudence, shamelessness.

**Utarid** The planet Mercury.
**uvalamak, uvmak, uvuşturmak** *v.* oğalamak *etc.*
**uyan·dırıcı** Awakening; arousing. **~dırmak,** to awaken; arouse; revive (a fire *etc.*); stir *a country into activity and prosperity.* **~ık,** awake; vigilant, wide awake, smart. **~ıklık,** a being awake; vigilance, smartness. **uyanmak** *vn.* Awake; wake up; (of a fire) to burn up: (of plants) to start growth; revive, come to life.
**uyar** Like. **~ı yok,** incomparable.
**uyarmak** *v.* uyandırmak.
**uydurma** Made-up (story, excuse, *etc.*).
**uydurmak** *va.* Cause to conform *or* agree; make to fit; adapt; invent, make up. **anahtar ~,** to use a false key: **···e ayak ~,** to keep in step with ...: **işi kitaba ~,** to manage an affair cleverly: **pazarlığı ~,** to make a good bargain: '**yine işini uydurdu!**', he's brought it off again! (*of a person who is always successful*).
**uydurmasyon** (*sl. and jokingly*). *An invented word for* invention, fable; made-up.
**uygun** Conformable; in accord; fitting; cheap, reasonable (price); just right. **işimiz ~,** our affairs are all right. **~luk,** a being appropriate *or* fitting. **~suz,** not conforming; unsuitable; unseemly: **~ bir kadın,** an immoral woman. **~suzluk,** impropriety: bad behaviour.
**uyku** Sleep. **~m açıldı (dağıldı),** my sleepiness has passed off: **~sunu* açmak,** to shake off one's drowsiness: **~sunu* almak,** to have a good night's rest: **~ basmak,** for sleep to overcome one: **~ çekmek,** to sleep: **~ya dalmak,** to fall asleep: **~m geldi (var),** I am sleepy: **bütün gece gözüme ~ girmedi,** I haven't slept a wink all night: **~ ilâcı,** sleeping-draught: **~m kaçtı,** I can't get to sleep: **~ kestirmek,** to have a nap: **~ sersemliği,** drowsiness: **~ya varmak,** to go to sleep: **~ya yatmak,** to lie down to sleep. **uyku·cu,** fond of sleep. **~suz,** sleepless. **~suzluk,** sleeplessness, insomnia.
**uyluk** Thigh. **~ kemiği,** femur.
**uymak** *vn.* Conform; agree; fit; suit; answer; harmonize; adapt oneself; follow; listen *to*; be fitting *or* seemly; be arranged *or* settled. **aklına* ~,** to yield to some temptation: ⌈**dakikası dakikasına uymaz**⌉, he changes his mood every five minutes; hard to get on with: **o kimseye uymaz,** he goes his own way, listening to no one: **iyi ki size uymadım,** it's a good thing I didn't listen to you: **şeytana ~,** to be tempted and led astray.
**uysal** Conciliatory; easy-going.
**uyuklamak** *vn.* Dose.
**uyumak** *vn.* Sleep; go to sleep; be negligent *or* slothful; come to a standstill, make no progress; clot, coagulate.
**uyun·mak** *impers. of* **uyumak. burada uyunur mu?,** can one get any sleep here? **~tu,** half-asleep, lazy.
**uyur** Sleepy; lethargic; dormant; still.
**uyuşmak**[1] *reciprocal of* **uymak** Come to a mutual understanding.
**uyuş·mak**[2] *reciprocal of* **uyumak.** Become numb *or* insensible; (of pain *etc.*) to relax, slacken. **~turmak,** to benumb; assuage; deaden (pain *etc.*). **~turucu,** that benumbs *or* deadens: **~ ilâc,** anodyne, narcotic. **~uk,** numbed; insensible; indolent. **~ukluk,** numbness; laziness.
**uyut·mak** *va. caus. of* **uyumak.** Send to sleep; bore; keep quiet *or* put off *s.o. by vague promises etc.*; let *a matter* become dormant, put off indefinitely. **~ucu,** soporific.
**uyuz** Itch; mange; scab. Having the itch; mangy; scabby. **~ otu,** scabious.
**uz**[1] *In* **az gittik ~ gittik,** we went on and on and on.
**uz**[2] Good; seemly. **az olsun ~ olsun,** let it be little but let it be good.
**uzak** Distant, remote, far off. Distant place; the distance. **~tan,** from far off: **~tan akraba,** a distant relative: **bir şeye ~tan bakmak,** to look on at *stg.* from a distance, *i.e.* to take no part in it: **uzağa düşmek,** to be far *from one another*: **~tan merhaba,** a distant greeting: **~tan tanımak,** to know *s.o.* by sight: **gözden ~ tutmamak,** to keep a close eye upon. **uzak·ça,** rather far off. **~laşmak,** to retire to a distance; to be far away. **~lık,** distance; remoteness.
**uza·mak** *vn.* Stretch; grow longer; extend; be prolonged. **dili* ~,** to criticize *or* attack presumptuously. **~nmak,** to be prolonged *or* extended;

to stretch oneself out; expand: ···e ~, to extend to, to go as far as. ~**tmak,** to extend, stretch out; postpone; prolong; make tedious *or* tiresome; allow to grow long (hair, beard): to be prolix; be importunate: **uzatmıyalım,** 'to come to the point!', in short: **dil** ~, to be forward; to criticize presumptuously: **el** ~, to seize, to lay hands on; to meddle.

**uzlaşma** Agreement, understanding. ~**k,** to come to an agreement *or* understanding.

**uzman** Expert.

**uzun** Long; long-winded; in detail. ~ **boylu,** tall, long, lengthy: ~ **etm.,** to hold forth at great length; ~ **etme!,** that's enough!, don't keep on about it!; *or* oh! come on!, don't maintain that attitude!: ~ **oturmak,** to sit with outstretched legs: ···**i** ~ **tutmak,** to drag out *a job etc.*: ~ **uzadıya,** at great length: **eli** ~, pilferer.

**uzun·ca,** somewhat long *or* tall. ~**kulaklı,** long-eared. ~**luk,** length, lengthiness, height.

**uzuv, -zvu** Organ; limb.

**uzv·î** (·⁻) Organic. ~**iyet, -ti,** organism.

# Ü

**ücra** (·⁻) Remote, out-of-the-way.

**ücret, -ti** Pay; wage; fee; cost (of postage, telegram *etc.*); price (of a railway ticket). ~**li,** receiving pay. ~**siz,** unpaid; without payment; gratis.

**üç** Three. ~ **aşağı beş yukarı,** after some haggling: ~ **aşağı beş yukarı dolaşmak,** to walk up and down aimlessly *or* anxiously: ~**te bir,** a third: ~ **buçuk atmak** (*sl.*), to be very frightened: ~ **buçuk cahil** (**serseri** *etc.*), a handful of ignoramuses (vagabonds *etc.*).

**üç·ambarlı** (⁻···), three-decker (man-of-war). ~**aylar,** the three sacred months of Islam, *viz.* Rejeb, Shaban, and Ramazan. ~**er,** three each; three at a time. ~**gen,** triangle. ~**köşeli,** triangular, three-cornered. ~**leme,** triple; three-stranded *rope etc.* ~**lemek,** to make three: **evlerini üçledi,** he has bought a third house. ~**leşmek,** to amount to three. ~**lü,** consisting of three; marked with the number three. ~**lük,** of the value of three *piastres etc.*; for three persons (coffee *etc.*). Trinity. ~**üncü,** third. ~**üncülük,** a being third; third place. ~**üz,** triplets; one of triplets. ~**üzlü,** triple: ~ **bir taret,** turret with three guns.

**üdeba** (··⁻) *pl. of* **edib.** Literary men.

**üflemek** *va.* Blow out *a candle etc.*; blow upon; blow up (toy balloon *etc.*); blow *a musical instrument. vn.* Blow, puff, pant. ⌜**sütten (çorbadan) ağzı yanan ayranı üfliyerek içer**⌝, ⌜a scalded cat fears cold water⌝.

**üful** (·⁻), **-lü** Extinction; death.

**üfür·mek** *vn.* Blow, puff. *va.* Blow upon; blow up *with the breath*; cast a spell upon *or* cure by breathing on. **canına** ~ (*sl.*), to ruin. ~**ük,** a breathing on a sick person to cure him. ~**ükçü,** sorcerer who claims to cure by such breathing. ~**üm,** puff; blast.

**üğendire,** ~**ğendirek** Ox-goad.

**ülema** *v.* **ulema.**

**üleş** Portion, share. ~**mek,** to go shares. ~**tirmek,** to share out.

**ülfet, -ti** Familiarity; friendship.

**ülke** Country; province.

**Ülker** The Pleiades.

**ülûhiyet** (·–·⁻), **-ti** Divinity.

**ümera** (··⁻) *pl. of* **emir.** Chiefs; commanders.

**ümid** (·⁻) Hope; expectation. ⌜~ **dünyası bu**⌝, hope never dies: ~ **etm.,** to hope; to expect: ~ **etmediği felâket başına geldi,** an unexpected catastrophe befell him: ···**den** ~**ini kesmek,** to give up hope of ...: ~**im var,** I hope: **kat'ı** ~ **etm.,** to abandon hope.

**ümid·bahş,** hopeful; that gives hope. ~**gâh,** person *or* thing on which one's hopes are fixed. ~**lendirmek,** to make hopeful, to fill with hope. ~**lenmek,** to be hopeful; to conceive hopes. ~**siz,** without hope; hopeless; desperate: **âtiden** ~, despairing of the future. ~**sizlik,** hopelessness, desperation. ~**var,** hopeful.

**ümmet, -ti** Community *of the same religion*; people, nation. ~**i Muhammed,** Moslems; *often used to mean* people, folk *in general.*

**ümmî** (·⁻) Illiterate.

**ümniye** Hope; desire; purpose.

**ün** Voice, cry; fame, reputation. ~ **salmak,** to become famous: **kuru** ~, a mere name, an empty sound. ~**lemek,** to cry out; sing. ~**lü,** famous; honoured.

**üniforma** (··˙·) Uniform.

**ünsiyet, -ti** Familiarity; a being on friendly terms with *s.o.* ···**le** ~ **etm.** (**peyda etm.**), to be intimately acquainted with.

**ürcufe** (·—˙) False rumour.

**Ürdün** Transjordan.

**üre** Urea.

**üre·ğen** Productive; prolific. ~**m,** increase. ~**me,** reproduction, procreation. ~**mek,** to multiply, increase (*vn.*). ~**tmek,** to cause to multiply; breed.

**ürgün** Pond *or* backwater *caused by the overflow of a river or the sea.*

**ürkek** Timid, fearful. ~**lik,** timidity.

**ürk·mek** *vn.* Start with fear; be frightened; (of a horse) to shy. ~**üntü,** sudden fright; panic. ~**ütmek,** to startle, scare.

**ürper·mek** *vn.* (Of the hair) to stand on end; shiver. ~**tmek,** to make the hair stand on end: **tüyleri ürperten,** hair-raising.

**ürümek** Howl (of dogs *etc.*).

**ürün** Product.

**üryan** Naked; bare. ~**i** (·—˙), a kind of thin-skinned plum; plum *or* prune skinned and dried.

**üs, -ssü** Base; basis; **deniz** ~**sü,** naval base: **hareket** ~**sü,** base of operations (*mil.*): ~**sü mizanı doldurmak,** to reach the required standard (in an exam.).

**üsera** (··˙) *pl. of* **esir.** Prisoners-of-war. ~ **karargâhı (kampı),** prisoners-of-war camp.

**Üsküdar** Scutari. ⌜**atı alan** ~**ı geçti**⌝, too late, nothing to be done!

**üslûb** (·˙) Style.

**üssü** *v.* üs.

**üst, -tü** *v. also* **üzeri.** Upper *or* outside surface; the top *of a thing*; the space over *a thing*; clothing; superior (*n.*); remainder; change; address *of a letter*. Upper, uppermost. (*with poss. suffix*) On; over; on the top of. ~ ~**e,** *v.* **üstüste:** ~**te,** above; on top: ~**ten,** superficially: ~**ü açık,** obscene, 'smutty': ~**üne* almak,** to take upon oneself; to put on *clothes etc.*; to take (a remark *etc.*) as being directed against oneself; to row ahead: ~**ten almak,** to talk *or* behave in a superior manner: ~**ünden* atmak,** to try to avoid *or* get rid of: ~**üne basmak,** to hit the nail on the head: ~**ü başı,** *v.* **üstbaş:** ~**üne bırakmak,** to quit, to give up: ···**in** ~**üne çevirmek (geçirmek),** to turn over to ... , to transfer to ...: ~ **çıkmak,** to win: ~**e çıkmak,** to pretend to be innocent: ~**üne çıkmak,** to come to the top; to get the better of: ~ **fırçası,** clothes-brush: ~**ünden geçmek,** to violate (a woman): **bunun** ~**ünden beş sene geçti,** five years elapsed after this: ~ **gelmek,** to surpass, to prevail: ~**üne gelmek,** to turn up, *to appear when one is doing stg.*: ~**ü temiz,** he is cleanly dressed: ~**üne* varmak,** to keep on at *s.o.*; to put on an extra price; to bid higher *at an auction*: ~**üme varma!,** don't keep on at me!; don't insist!: ~**e vermek,** to give in addition: to suffer loss in a business transaction: ~ **yan,** next door (in a street); a little farther on: ···**in** ~**üne yapmak,** to make over *stg.* to *s.o.*: **adı** ~**ünde,** as the name implies; as befits the name: **baş** ~**ünde tutmak,** to honour, revere: **kış** ~**ü,** at the coming of winter: **senin ne** ~**üne lâzım?,** what business is that of yours?

**üstad** (·˙) Master; teacher; expert. ~**ane** (·——˙), masterly.

**üstbaş** (*both words declined*) Dress, attire. **üstüne başına etm.,** (i) (of a child) to foul its clothing; (ii) to abuse *s.o.* violently: ~ **kalmamak,** to have no clothes left.

**üstderi** (˙··) Epidermis.

**üste** Further; in addition. **işin** ~**sinden gelmek,** to succeed, to cope with a matter. ~**leme,** relapse. ~**lemek,** (of an illness) to recrudesce; to dwell on *stg. with regret or desire*; to persist: **gelmek istemiyor, fazla üsteleme!,** he doesn't want to come, don't press the matter! ~**lik,** furthermore; in addition.

**üsteğmen** (˙··) First lieutenant.

**üstinsan** (˙··) Superman.

**üstübec** White lead.

**üstün** Superior. ~ **gelmek,** to come out superior; to be victorious: ~ **tutmak,** to consider superior, to prefer: ⌜**el elden** ~**dür**⌝, there is always one superior. ~**körü** (·˙··), superficial. ~**lük,** superiority.

**üstüne** *v.* üst *and also* üzerine.
**üstüpü** Oakum, tow; mop *for cleaning guns.*
**üstüste** (ʹ··) One on top of the other. iki sene ~, two years in succession.
**üstüvan·e** (··–ʹ) Cylinder. **~î** (··–ʹ), cylindrical.
**üşen·gec, üşengen** Lazy, slothful. **~iklik,** laziness, sloth.
**üşenmek** *vn.* (*with dat.*) Be too lazy to do *stg.*; do with reluctance.
**üşmek** *vn.* Flock to a place.
**üşne** A kind of moss *used as a perfume.*
**üşniye** Algae.
**üşümek** *vn.* Feel cold; catch cold.
**üşüntü** A flocking together; crowd, mob.
**üşür, -şrü** Tenth; tithe.
**üşürmek** *va. caus. of* **üşmek.** Collect together; cause to make a concerted attack.
**üşüşmek** *vn.* Crowd together; make a concerted attack.
**üşütmek** *va. caus. of* **üşümek.** Cause to feel *or* catch cold.
**ütmek** *vn. dial. for* **yutmak.** Win in a game (children).
**ütü** Flat-iron; crease *made by ironing.* **~cü,** ironer; laundress. **~lemek,** to iron; to singe the hair off *sheep's trotters etc.*: **kafa ~** (*sl.*), to bore *s.o.* stiff. **~lü,** ironed, singed.
**ütüme** Roasted fresh wheat.
**üvendire** Ox-goad.
**üvey-.** Step-. **~ana,** stepmother: **~baba,** stepfather, *etc.*
**üvey·k** Turtle dove.
**üvez** Rowan tree and berry.
**üye** Member (of a council *etc.*).
**üzek** Pole of an ox-cart.
**üzengi** Stirrup. **çanak ~,** cylindrical stirrup. **~lemek,** to spur *a horse* with the stirrup (*oriental stirrups having pointed corners for this purpose*).
**üzenmek** *v.* **özenmek.**
**üzere, üzre** On, upon; according to; about; on the subject of; on condition of; for the purpose of; at the point of, just about to.
**üzeri** Upper *or* outer surface *of a thing*; space above *a thing*; remainder, change (money). On; over; about. **üzeri** *is practically interchangeable with* **üst** *and many phrases under* **üst** *can equally well be used with* **üzeri. ~ne* almak,** to take it upon oneself *to do stg.*: **~nde para yok,** I have no money on me: **~nde* kalmak,** to remain on one as a debt *or* a charge: **~ temiz,** he is cleanly dressed: **kabahati kendi ~nden atmak,** to exculpate oneself: **kabahati birisinin ~ne atmak,** to throw the blame on *s.o.*
**üzerlik** Rue seeds *used as a fumigant.*
**üzgeç** Rope ladder.
**üzgü** Oppression; cruelty.
**üzgün** Weak, invalid, ill; anxious, worried. **~ balığı,** Dragonet.
**üzlet, -ti** A retiring into seclusion; solitude; isolation.
**üzmek** *va.* Treat harshly; cause to break down *from grief or anxiety.*
**üzre** *v.* **üzere.**
**üzülmek** *vn. pass. of* **üzmek.** Be worn out; be sorry *or* worried; regret having been unable to do *stg.*
**üzüm** Grape. ⌜**~ ~e baka baka kararır**⌝, 'evil communications corrupt good manners': ⌜**~ünü ye de bağını sorma**⌝, if you get a pleasure *or* a benefit, there is no need to worry about its source. **~suyu, -nu,** grape-juice; wine.
**üzüntü** Anxiety; dejection; fatigue. **~lü,** tedious.

# V

**vabeste** (–·ʹ) Dependent; depending (on = *dat.*).
**vacib** (–ʹ) Incumbent; necessary.
**vade** (–ʹ) Fixed term *or* date; maturity *of a bill etc.* **~si gelmek,** to fall due: **~si geçmiş,** of which the date of payment has expired. **~li,** maturing at a certain date.
**vadetmek** (ʹ··), **-eder** *va.* Promise.
**vadi** (–ʹ) Valley; sense, tenor. **bu ~de,** in this sphere, in this line: **her ~den,** on every subject.
**va'di** (–ʹ) *v.* **vaid.**
**vafi** (–ʹ) Abundant. ⌜**kâfi ve ~**⌝, enough and more than enough.
**vaftiz** Baptism. **~ anası,** godmother: **~ babası,** godfather: **~ etm.,** to baptize.
**vagon** Railway wagon. **yataklı ~,** sleeping-car. **~li** (*Fr.* *wagon-lit*), sleeping-car. **~lu,** having so many wagons *or* trucks. **~luk,** truck load.
**vah** *Interj. expressing pity or regret.*
**vaha** (–ʹ) Oasis.

**vahamet** (·—ꜝ), **-ti** Gravity, seriousness (of a situation). **~li**, grave.
**vahdaniyet** (·—·ꜝ), **-ti** The Unity of God.
**vahdet, -ti** Unity. **~i vücud**, pantheism.
**vahi** (—ꜝ) Futile, silly.
**vâhid** (—ꜝ) One. **~ikıyasi** (—···—ꜝ), unit of measurement.
**vahiy, -hyi** Divine inspiration; God's revelation *to a prophet*.
**vahlanmak** *vn.* Say 'alas!', 'what a pity!'
**vahş·et, -ti** Wildness, savageness. **~i** (·ꜝ), wild; savage; shy, afraid of man. **~ice** (·ꜝ·), **~iyane** (·——ꜝ), wild; brutal; savage; in a wild, brutal *or* savage way. **-ilik** (·—ꜝ), wildness; savageness; bestiality, brutality.
**vahy·etmek** (ꜝ··), **-eder** *va.* Inspire; (of God) to reveal. **~i**, *v.* **vahiy.**
**vaız, -a'zı** Admonition; sermon.
**vaid, -a'di** Promise. **va'dinde durmak,** to abide by one's promise. **~leşmek,** to make promises to one another.
**vaiz** (—ꜝ) One who admonishes; preacher.
**vak** The Lesser Bittern.
**vak'a** Event, occurrence.
**vakaâ** *v.* **vakıâ.**
**vakar** *v.* **vekar.**
**vakayi** (·—ꜝ), **-ii** *pl. of* **vakıa.** Events; calamities; battles. **~name** (··——ꜝ), chronicle.
**vaketa** (·ꜝ·) Calf-skin leather.
**vakfe** Stop; pause; interval.
**vakf·etmek** (ꜝ··), **-eder** *va.* Devote *property* to a pious foundation; devote *or* dedicate *oneself* (*or time*) *to some purpose*. **~iye,** deed of trust of a pious foundation. **~ı,** *v.* **vakıf.**
**vakfon** Nickle-plated copper *or* iron.
**vakıa** (—·ꜝ) Fact; dream, vision. **bu bir ~dır, inkâra mahal yok,** this is a fact which cannot be denied.
**vakıâ** (ꜝ·—) In fact, actually; it is true that.
**vakıf, -kfı** Pious foundation; Wakf. **~name** (··—ꜝ), deed of trust of a pious foundation.
**vâkıf** (—ꜝ) Aware, cognizant; who devotes property to a pious foundation. **···e ~ olm.,** to be aware of ..., to be cognizant of ... .
**vaki** (—ꜝ), **-ii** Happening, taking place; actual. **~ olm.,** to happen, befall.
**vakit, -kti** Time; (*with a participle*) when. **~ile,** in times past: **~inde,** at the right time: **~ ~,** at times: **~ler hayır olsun!,** good day!: **bir ~,** once, once upon a time: **her ~,** always: **hiçbir ~,** never: **ne ~?,** when?: **o ~,** then, in that case: **o ~ bu ~,** from then till now: **hali vakti yerinde,** well-off.
**vakit·li,** done at the right time; in due season. **~ vakitsiz,** at all sorts of times. **~siz,** untimely.
**vakom** *v.* **vakum.**
**vakt·aki** (·—ꜝ) When, at the time that ... . **~i,** *v.* **vakit.** **~ihal,** financial circumstances: **~ yerinde,** in easy circumstances. **~ile** (·ꜝ·), in the past; at one time: **~ görmüş geçirmiş,** who has seen better days.
**vakum** (ꜝ·) Vacuum; (·ꜝ), Vacuum oil.
**vakur** Grave, dignified. **~ane** (·——ꜝ), grave *or* dignified (behaviour *etc.*).
**valâ** (—ꜝ) High; eminent. **alayi ~ ile,** with great pomp and ceremony (*iron.*).
**valf** Valve.
**vali** (—ꜝ) Vali; Governor of a province.
**valide** (—·ꜝ) Mother.
**vallah** (ꜝ·), **vallahi** (·ꜝ·) By God!; I swear it is so!
**vanilya** (·ꜝ·) Vanilla.
**vantilâtör** Ventilator; fan.
**vantuz** (*Fr. ventouse*) Cupping-glass; sucker *of an octopus etc.*
**vapur** Steamer. **kara ~u** (*vulg.*), train.
**var**[1] Existent; present; at hand; available. Belongings, possessions; wealth. There is, there are. **~ etm.,** to cause to be present, to create: **yoktan ~ etm.,** to make *stg.* out of nothing: **~ kuvvetile,** with all his strength: **~ mısın?,** I dare you to do it; I bet you can't do it: **~ olm.,** *v.* **varolmak: ~ı yoğu,** all that he has: **~sa ... yoksa ...,** he thinks of nothing else than ... , *e.g.* **~sa oğlu yoksa oğlu,** he just lives for his son: **altmışında ~ yok,** he must be about sixty: **bir ~ bir yok,** transitory, uncertain: **ne ~ ?,** what's the matter?: **ne ~ ne yok?,** what's the news?: **ne ~ ne yok,** whatever there is; whatever one possesses.
**var**[2] *v.* **varmak.**
**varak, -kı** Leaf; petal; sheet of paper. **altın ~,** gold-leaf: **bakır ~,** copper-foil.

**varak·a,** a single leaf; note, letter, document. **–çı,** a worker in gold-leaf, gilder. **~lamak,** to ornament with gold-leaf. **~lı,** gilded. **~pare** (··–ˈ), scrap of paper; worthless document; humble request; 'rag' (newspaper).
**varan** *In such phrases as*: ~ **dört!,** that makes four!, that's number four!
**varda** (ˈ·) Look out!; keep clear!; make way!
**vardakosta** (··ˈ·) Fat but imposing person.
**vardatopu** (·ˈ··), **–nu** Signal gun.
**vardavelâ** (··ˈ·) Rail of a ship. **kıç ~sı,** taffrail.
**vardırmak** *va. caus. of* **varmak.** Allow to reach; cause to arrive.
**vardiya** (ˈ··) Watch (on board ship).
**vardula** (·ˈ·), **vardolos** Welt of a shoe.
**varek** Seaweed, kelp.
**vâreste** (–·ˈ) Free; exempt.
**varetmek** (ˈ··) *v.* **var.**
**vargel** ~ **tezgâhı,** shaper.
**varılmak** *impers. of* **varmak. oraya iki saatte varılır,** one reaches there in two hours.
**varış** *verb. n. of* **varmak.** Arrival *etc.*; *also* quickness of perception.
**–vari** *Pers. suffix.* Similar to ..., like ... .
**varid** (–ˈ) That which arrives *or* happens; probable; admissible. **bu ~ değildir,** it is unlikely to happen. **~at** (–·ˈ), **–tı,** revenues, income.
**varil** Small cask.
**varis** (·ˈ) Varicose veins.
**vâris** (–ˈ) Inheriting. Heir.
**variyet** (–·ˈ), **–ti** Wealth, riches; income. **~li,** well-to-do, wealthy.
**varlık** Existence; presence; personality; possessions, wealth. **~ göstermek,** to make one's presence felt; to achieve *stg.*: **böyle işler ~la olur,** one must be well-off to do such things. **~lı,** well-to-do.
**varmak, –ır** *vn.* (*with dat.*) Go towards; arrive; reach, attain; approach; result, end in; (of a woman) to marry. **varsın,** *v.* **varsın**: **var istediğini yap!,** do whatever you like! (a challenge): **varıncaya kadar,** up to, to: **dili*** ~, to dare to say; **dilim varmıyor,** I can hardly bring myself to say it: **eli*** ~, to dare to do: **···in farkına** ~, to perceive, to notice: **fenaya** ~, to turn out badly: **kocaya** ~, to marry a husband: **bu iş neye varacak?,** what will be the end of this?.
**varmış** *past of* **var** *and of* **varmak.** ⌜**bir ~ bir yokmuş**⌝, 'once upon a time'.
**varolmak** (ˈ··) *vn.* Exist. **varol!,** 'may you live long!' well done!, bravo!: **varolsun!,** 'long may he live!'
**varoş** Suburb.
**varsağı** A special form of song.
**varsam** Lesser Weaver *or* Sting-fish.
**varsın** *In such phrases as*: ~ **gelsin,** let him come if he likes: ~ **gelmesin,** it doesn't matter whether he comes or not: ~ **o da keman öğrensin,** well, if he wants to learn the violin, let him (it's not a bad idea).
**varta** Abyss; great peril. **~yı atlatmak,** to escape great danger.
**varyemez** Miserly.
**varyos, balyos** Sledge-hammer. **~çu,** a smith's striker *or* hammerman.
**vasat, –tı** Middle. Mediocre. **~î** (··ˊ), central; mean. Mean, average. **···in ~sini almak,** to take the mean *or* average of ... .
**vasf·etmek** (ˈ··), **–eder** *va.* Describe.
**vasıf, –sfı** Quality; description; epithet. **~landırmak,** to qualify; describe.
**vâsıl** (–ˈ) Arriving, joining. ~ **olm.,** to arrive.
**vasıta** Means; intermediary; vehicle. **···in ~sile,** by means of; **sizin ~nızla,** by means of you, through you. **~sız,** direct.
**vasi** (–ˈ) Executor; trustee; guardian.
**vâsi** (–ˊ) Extensive, wide; abundant.
**vasiyet, –ti** Will, testament; last request of a dying person. **~name** (····–ˈ), written will.
**vasletmek** (ˈ··), **–eder** *va.* Unite; join.
**vaşak** Lynx; fur of the lynx.
**vat, –tı** Watt.
**vatan** One's native country, motherland. **~daş,** compatriot; fellow countryman. **~daşlık,** a being a compatriot. **~î** (··ˊ), pertaining to one's native land. **~perver, ~sever,** patriotic; patriot. **~perverlik,** patriotism, love of one's country.
**vatoz** Thornback Ray. **iğneli ~,** Sting-ray.
**vaveylâ** (–·ˊ) Alas! Cry of horror *or* lament. **~yı koparmak,** to raise a cry of horror *or* lament.

**vay** *Interj. expressing surprise or regret.* ~ **sen misin!**, hullo, is that you!: ~ **başım**, oh, my poor head!: ~ **başıma!**, woe is me!: ~ **canına!**, how amazing!

**va'zetmek, vâzetmek** (–́··), **–eder** *va. & vn.* Admonish; preach.

**vaz'etmek** (ˋ··), **–eder** *va.* Put; place; lay (foundation); impose (tax).

**vazgeçirmek** (ˋ···) *va. caus. of* **vazgeçmek.**

**vazgeçmek** (ˋ··) *va.* (*with abl.*) Give up; cease from; abandon (a project).

**va'zı** (–ˋ) *v.* **vaız.**

**vazı, –z'ı** Act of putting down *or* laying; an imposing (a tax). ~**yed,** a laying hands on; seizure: ···**e** ~ **etm.**, to seize, confiscate; to take up *a matter.*

**vâzı** (–ˋ), **–ıı** Who lays down *or* institutes. ~**ıkanun,** legislator.

**vazıh** (–ˋ) Clear, manifest. ~**an** (–́··), clearly.

**vazife** (·–ˋ) Duty; obligation. ~ **etm.**, *or* ~**sinden olm.**, to care, to mind, to be interested: **ne** ~**n** *or* **ne üstune** ~**?**, what's that to you? **vazife·dar,** charged with an official duty. ~**siz,** without an official duty; who neglects his work, careless, slack. ~**şinas,** dutiful; conscientious.

**vaziyet, –ti** Position; situation; attitude. ~ **almak,** to stand to attention.

**vazo** (ˋ·) Vase.

**vazolunmak** (ˋ···) *vn. pass. of* **va'zetmek.**

**ve** And. ~ **saire,** etcetera (*abbrev.* **v.s.**).

**veba** (·–́) Plague, pestilence.

**vebal, –li** Sin *that will be punished in the next world*; evil consequence. ~**i boynuna,** on his (your) head be it!

**vebalı** Stricken with the plague.

**veballi** That will be punished in the next world.

**veca** (·–́), **–aı** Pain; colic. ~**lı,** painful; in pain.

**vecaib** (·–ˋ) *pl. of* **vecibe.** Obligations.

**vecd, vecid, –cdi** Ecstasy; rapture.

**vech·e** Direction; side. ···**e** ~ **vermek,** to direct *s.o.* ~**i,** *v.* **vecih.**

**vecih, –chi** Manner; means. **hiçbir vechile,** in no wise, by no means: **lâyıkı vechile,** as it should be done:

**veciz** Laconic, terse. ~**e** (·–ˋ), terse saying; aphorism.

**veda** (·–́), **–aı** Farewell. ···**e** ~ **etm.**, to bid farewell to ... . ~**laşmak,** to bid each other farewell. ~**name** (·––ˋ), letter of farewell.

**vedia** (·–ˋ) A thing deposited *or* given into safe-keeping.

**vefa** (·–́) Fidelity; loyalty; faithfulness. ···**e ömrü** ~ **etmedi,** he did not live long enough to ... . ~**kâr,** ~**lı,** faithful, loyal, constant. ~**sız,** faithless; disloyal. ~**sızlık,** faithlessness; disloyalty.

**vefat** (·ˋ), **–tı** Death; decease.

**vefiyat** (··ˋ), **–tı** *pl. of* **vefat.** Deaths; mortality.

**Vehabî** (·––́) Wahhabi; Wahhabite.

**vehham** Given to forebodings; apprehensive; suspicious.

**vehim, –hmi** Foreboding; groundless fear; surmise; illusion; delusion.

**vehle** Moment, instant. ~**ten** (ˋ··), for the first moment, just at first.

**vehm·etmek** (ˋ··), **–eder** *va.* Forebode; fear; surmise; have the illusion that ... . ~**i,** *v.* **vehim.**

**vekâlet** (·–ˋ), **–ti** A being agent *or* representative of another; attorneyship; Ministry. ···**e** ~ **etm.**, to represent *s.o.*; act as agent *or* attorney for *s.o.* ~**en** (·–́··), as representative *or* deputy of another; by proxy. ~**name** (·–·–ˋ), power of attorney.

**vekar** Staidness, gravity; dignity; dignified calmness. ~**lı,** grave, dignified; calm. ~**sız,** lacking in dignity *or* seriousness. ~**sızlık,** lack of dignity.

**vekil** (·ˋ) Agent; representative; deputy; attorney; proxy; Minister of State. ~ **etm.**, to appoint as one's representative: ···**e** ~ **olm.**, to represent *or* act as deputy for *s.o.* ~**harc,** steward *or* majordomo *in a great house.* ~**lik,** quality *or* duties of a **vekil**; agency; attorneyship.

**velâdet, –ti** Birth.

**velâkin** (·–́·) But still.

**velâyet** (·–ˋ), **–ti** Saintship.

**veled** Child; rascal.

**velena** (·ˋ·) Staysail.

**velense** (·ˋ·) Kind of thick blanket; horse-rug.

**velev, velevki** (·ˋ·) Even if; even though.

**velfecri** *In* ⌜**gözleri** ~ **okumak**⌝, to give the impression of being very astute and wide awake (*slightly derogatory*).

**velhasıl** (ˈ–·) In fine, in short.
**veli** (·ˈ–) Guardian *of a child etc.*; friend of God, saint. ~**ahd,** heir to the throne. ~**lik,** qualities and duties of a guardian; quality of a saint. ~**nimet, -ti,** benefactor, patron.
**velûd** (·ˈ–) Prolific; productive. ~**iyet, -ti,** prolificacy; productivity.
**velvele** Outcry; clamour, hubbub. ~**li,** tumultuous, noisy.
**velyetmek** (ˈ··), **-eder** *vn.* Follow in succession.
**Venedik** (·ˈ·) Venice. ~**li,** Venetian.
**ventil** Valve; cock.
**vento** (ˈ·) Topping-lift; guy.
**veraset** (·–ˈ), **-ti** Inheritance; heritage.
**vere** (ˈ·) Capitulation, surrender.
**verecek** Debt.
**verem** Tuberculosis. Tuberculous, consumptive. ~**li,** tuberculous.
**verese** *pl. of* **varis.** Heirs.
**veresi, veresiye** On credit.
**veretmek** (ˈ··), **-eder** (*sl.*) *va.* (*with dat.*) Keep on hitting; give several blows to.
**verev** Oblique; diagonal; slanting.
**vergi** Gift; tax. **Allah** ~**si,** gift of God, talent: **Allaha** ~, only God can do that: ···**e**~ **olm.,** to be special to, to be the speciality of … .
**verici** Who *or* which gives.
**verile** *Lit.* 'let it be given'. ~ **emri,** order for payment of government money.
**verim** Produce; return; profit; output. ~**li,** profitable; productive.
**veriş** Mode *or* act of giving. **alış** ~, trade, commerce. ~**mek,** to give to one another, exchange.
**veriştirmek** *vn.* Utter abuse; swear.
**vermek, -ir** *va.* Give; deliver; pay; offer; sell; attribute; undergo (losses); teach. *As aux. verb implies easiness and quickness, e.g.* **anlayıverdi,** he readily understood; *it sometimes implies a courteous request, e.g.* **kapıyı açıver !,** would you mind opening the door? **ver elini Boğaziçi (dedik gittik),** we decided to go up the Bosphorus and off we went: **açık** ~, to have a deficit: **alıp verememek,** to disagree: **ara** ~, to suspend, to interrupt: **başbaşa verdiler,** they laid their heads together: ···**i çocukluğuna*** ~, to overlook *stg.* as due only to youthfulness: **ele** ~, to hand over, to betray: ···**i talihsizliğe** ~, to put *stg.* down to bad luck: **ver yansın,** *v.* **veryansın.**
**vernik** Varnish. ~**lemek,** to varnish.
**veryansın** (ˈ··) **etm.** *vn.* Squander; get excited *or* enthusiastic; let oneself go.
**vesaik** (·–ˈ), **-kı** *pl. of* **vesika.** Documents.
**vesait** (·–ˈ), **-ti** *pl. of* **vasıta.** Means. ~**i nakliye,** means of transport.
**vesatet** (·–ˈ), **-ti** Mediation.
**vesayet** (·–ˈ), **-ti** Trusteeship.
**vesika** (·–ˈ) Title-deed; document; ration card. ~ **ile,** by coupon (rationed). ~**lı,** licensed (prostitute).
**vesile** (·–ˈ) Means; pretext; opportunity. ~ **buldukça,** on any pretext: **bir** ~ **ile,** under some pretext: **bu** ~ **ile,** by this means, taking this opportunity.
**vesselâm** (ˈ··) So that's that!; so that's an end of the matter! **anlaşılmaz insansın** ~**!,** you're an odd creature and that's all about it!
**vestiyer** (*Fr. vestiaire*) Cloak-room; coat-peg.
**vesvese** Anxiety; secret fear; scruple. ~ **etm.,** to be inwardly unhappy *or* anxious, to have misgivings.
**veya** (·ˈ–), **veyahud** Or.
**veyselkarani** *In* ⌜**yemen ellerinde** ~⌝, completely lost; utterly at a loss.
**vezaret** (·–ˈ), **-ti** Vizierate.
**vezin, -zni** Metre (poetry).
**vezir** (·ˈ) Vizier, Minister; queen (chess). ~**iazam** (·–·–ˈ), the Grand Vizier. ~**lik,** office and duties of a Vizier.
**vezne** Pay-office. ~**dar,** treasurer; cashier.
**vıcık** Half liquid half solid; sticky; dirty. **ortalık** ~ ~, the whole place was a quagmire.
**vık** *In* **içi*** ~ ~ **etm.,** to be very impatient.
**vınlamak** *vn.* Buzz; hum.
**vır** *In* ~ ~, *imitating continuous and exasperating talk.* ~ ~ **başının* etini yemek,** to keep on nagging at *s.o.* ~**ıldamak,** to talk incessantly; to keep complaining querulously. ~**ıltı,** tiresome talk; nagging; querulousness. ~**lamak,** to nag.
**vız** Buzz. ~ **gelmek,** to be a matter of indifference: ⌜~ **gelir tırıs gider**⌝, I don't care two hoots. ~**ıldamak,** to buzz, hum; keep on complaining.

**~ıltı,** a buzzing *or* whirring noise; querulous complaining. **~ır ~ır,** *imitates the whirring of a machine and used to describe stg. that can easily and quickly be done.* **~lamak,** *v.* **vızıldamak.**
**vicah** (·ˇ) Personal presence. **~en** (·ˊ·), in the presence of. **~î** (·ˊ–), done in the presence of *s.o.*
**vicdan** (·ˇ) Conscience. **~azabı,** the pangs of conscience. **~en** (·ˊ·), in accordance with one's conscience. **~î** (·–ˊ), pertaining to conscience. **~lı,** conscientious; honest. **~sız,** without a conscience; unscrupulous. **~sızlık,** lack of a conscience; unscrupulousness.
**vida** (ˊ·) Screw. **~lamak,** to screw; to screw down. **~lı** (ˊ··), screwed: **~ cıvata,** screw-bolt.
**vidala, videle** (·ˊ·) Box-calf; calf-skin.
**viğle** (ˊ·) Observation post at a **dalyan,** *q.v.*
**vikaye** (·–ˊ) Protection; prophylaxis.
**vilâyet** (·–ˊ), **-ti** Province governed by a Vali. **~li,** of *such and such a* vilayet; fellow countryman.
**vinç** Crane; winch.
**vira** (ˊ·) *Interj. giving the* order to set a crane going. (ˊ·) Continuously.
**viraj** (*Fr. virage*) Curve of a road. **bir ~ı dönmek,** to go round a bend.
**viran** (–ˇ) Devastated, ruined. **~e** (––ˊ), ruin. **~elik** (––·ˊ), a place of ruins. **~lık,** a being a ruin; ruin.
**vird** Constantly repeated saying. **~-etmek** (ˊ··), **-eder,** to repeat constantly. **~izeban,** *stg.* repeated constantly.
**vire** *v.* **vira.**
**virgül** (*Fr. virgule*) Comma. **noktalı ~,** semicolon.
**visal** (·ˇ), **-li** Meeting; lover's union.
**viski** (ˊ·) Whisky.
**vişne** (ˊ·) Morello cherry. **~çürüğü, -nü,** purplish-brown colour.
**vites** (*Fr. vitesse*) Gear; gears. **~ kolu,** gear-lever: **~ kutusu,** gear-box.
**vitrin** (*Fr. vitrine*) Shop window.
**viya** (ˊ·) *Order to steer straight after altering course.*
**viyak** *In* **~ ~,** squawking.
**viza, vize** (ˊ·) Visa. **~lı** (ˊ··), with a visa.
**vizita** (·ˊ·) Medical visit; doctor's fee.
**vizon** (*Fr. vison*) Mink.
**volan** (*Fr. volant*) Fly-wheel.
**voli** (ˊ·) Space covered by a cast of a circular fishing-net; (*sl.*) successful stroke of business. **~ çevirmek,** to cast a net.
**volkan** Volcano.
**volta** (ˊ·) A round turn (knot); tack; a turn *in walking up and down.* **~ vurmak (etmek),** to go to windward by tacking; to cruise about; to walk up and down; to beat about the bush: **meze ~,** half-hitch.
**v.s.** *Abbrev. for* **ve saire.** Etcetera.
**vuku** (·ˊ), **-uu** Occurrence, event. **~ bulmak,** to happen, to take place. **~at** (·–ˇ), **-tı,** events, incidents.
**vukuf** (·ˇ) Knowledge; information. **ehli ~,** expert. **~lu,** well-informed. **~suz,** ignorant; badly informed. **~suzluk,** lack of information, ignorance.
**vurdumduymaz** Insensitive; thick-skinned.
**vurğu** Accent, stress.
**vurğun** Struck; (*with dat.*) in love with; 'gone on'. Good stroke of business; profiteering. **~ vurmak,** to make a successful speculation. **~cu,** speculator; profiteer.
**vurmak, -ur** *va.* (*with acc.*) Hit *stg.* (*acc.*) against *stg.* (*dat.*); put *one thing* (*acc.*) on *another* (*dat.*); hit and kill, shoot dead; apply (paint *etc.*); chafe, gall, blister; (of pickpocket) to steal; swindle. *va.* (*with dat.*) Strike, hit, knock; take (a road *or* direction); (of the sun) to beat down on; (of wind, light *etc.*) to penetrate into; to pretend to be. ⌜**vur deyince öldürmek**⌝, to exceed one's orders *or* advice; **vur dedikse öldür demedik ya!,** I told you to hit him, not to kill him! (*used fig.*): ⌜**vur patlasın çal oynasın**⌝, squandering money on pleasure; going 'on the bust': **···i açığa ~,** to divulge (secret *etc.*): **adam ~,** to kill a man; **adamı ~,** to kill the man; **adama ~,** to strike a man: **baş ~,** (of a ship) to pitch: **···e baş ~,** to have recourse to ...: **başına* ~,** (of wine) to go to the head; (of fumes *etc.*) to give one a headache: **boynunu* ~,** to decapitate: **deliliğe ~,** to feign madness: **dışarı (içeri) ~,** (of a disease) to affect one externally (internally): **harbde haylı vurmuştu,** he is said to have made a pile of money during the war

(*with a hint of profiteering*): **ormana ~**, to take the forest road: **para ~**, (i) to steal by pickpocketing; (ii) to make money by dubious means, to profiteer: **yağmur içeriye vuruyor,** the rain is coming in (to a room *etc.*): **yakı ~**, to apply a blister: **kendini yerden yere ~**, to roll on the ground in agony: **zavallı adam kim vurduya gitti,** the poor fellow was killed, but nobody knows by whom.

**vurulmak** *vn. pass. of* **vurmak.** Be struck *etc.* **…e ~**, to be in love with *s.o.*: **beyninden vurulmuşa döndü,** he was thunderstruck.

**vuruş** Blow. **~mak,** to fight.

**vuslat, -tı** Union *with one's beloved.*

**vusta** (·–) **kuruna ~**, the Middle Ages.

**vusul** (·–), **-lü** Arrival. **~ünde,** on his arrival.

**vuzuh** (·–) Clearness; clarity *of speech etc.* **~suzluk,** lack of clearness, obscurity.

**vücud, -dü** Existence, being; the human body. **~ bulmak** *or* **~e gelmek,** to come into existence, to arise: **~e getirmek,** to bring into being, to produce. **~ca** (·–·), bodily; physically. **~lü,** large in body, heavily built. **~süz,** bodiless; non-existent; small, weak.

**vükelâ** (··–) *pl. of* **vekil.** Ministers. **~ heyeti,** Council of Ministers, Cabinet. **~lık,** post of Minister.

**vürud** (·–), **-dü** Arrival.

**vüsat, vüs'at, -ti** Spaciousness.

**vüsuk** (·–), **-ku** Authenticity.

**vüzera** (··–) *pl. of* **vezir,** Ministers.

# Y

**ya**[1] Ah indeed!; oh!; then!; so!; especially; don't forget! …; after all; yes, of course! **~ ben ne yapayım?,** well then, what shall I do?: **~ duyarsa?,** yes, but what if he hears?: **~ öyle mi?,** ah! is that so?: **gördün mü ~!,** there! you see what happens!: **istediğin oldu ~!,** after all you got what you wanted!: **'Niçin söylemedi?' 'Söyledi ~!',** 'Why didn't he say?' 'He *did* say!': **öyle ~!,** yes, indeed!

**ya**[2] O…; oh!; hi!; **~ rabbi,** oh!, my God!

**ya**[3] Or. ~ … ~ … , either … or … .

**yaba** Wooden fork *with three to five prongs for winnowing, carrying hay, etc.* **~lamak,** to winnow *or* carry hay with a **yaba.**

**yaban** Desert; wilderness. *In compounds (the other word having the 3rd pers. possessive suffix)* wild-, *e.g.* **~keisi** , wild goat. **~ın köpeği,** an outcast: **sözüm ~a,** excuse the expression!; *also* the 'so-called …'.

**yaban·cı,** stranger; foreigner; foreign: **…in ~sı,** a stranger to … . **~cılık,** a being a stranger *or* foreigner. **~i** (·–·), belonging to the desert *or* wilds; untamed, wild; boorish, unmannerly. **~ilik,** wildness; boorishness. **~lık,** visiting *clothes*; one's best clothes; wild place; wildness.

**yad**[1] Strange; alien. Stranger. **~ elde,** in a foreign land; away from home.

**yad**[2] (–) Remembrance; mention. **…i ~a getirmek,** to call to mind. **~etmek** (–··). **-eder,** to recollect; mention. **~igâr** (–·–), keepsake; souvenir; scoundrel; notorious.

**yadırgamak** *va.* Regard as a stranger; find *stg.* strange *or* odd; (of a child) to cry at a stranger.

**Yafa** (–·) Jaffa.

**yafta** (–·) Label; placard.

**yağ** Oil; fat; grease; ointment. **~ bağlamak,** to put on fat: **~ bal olsun!,** I hope you'll enjoy it (food): **araları ~ bal,** all is well between them; ⌜**~dan kıl çeker gibi**⌝, with the greatest of ease: **ekmeğine ~ sürüldü,** that was an unexpected bit of luck; *v. also* **ekmek: içi* ~ bağlamak,** to feel pleased and relieved: ⌜**kendi ~ile kavrulmak**⌝, to manage by oneself; to live very modestly: **yüreğim ~ bağladı,** I was overjoyed: **yüreğimin ~ı eridi,** I was prostrated with grief.

**yağ·cı,** maker *or* seller of butter *or* oil; an unctuous person. **~cılık,** trade of a dealer in fats *or* oils. **~dan, ~danlık,** grease-pot; oil-can. **~hane** (·–·), oil-mill; butter factory; creamery. **~ımsı,** oily; greasy.

**yağdırmak** *va. caus. of* **yağmak.** Rain down *bombs etc.*; pour out *money etc.*

**yağı** Enemy.

**yağır** Withers *of a horse*; saddle-gall.

**yağış** Manner of raining; rain. **~lı,** rainy.

**yağız** Black (horse); very dark (man).
**yağ·lamak** *va.* Grease; oil; butter; flatter; grease the palm of; cause to make money. **yağlayıp ballamak,** to paint in glowing colours. **~lanmak,** *vn. pass. of* **yağlamak,** to be greased *etc.*; to become dirty with grease; to be benefited, to gain a profit. **~lı,** fat; greasy; oily; dirty with grease; rich, free with money; profitable: **~ ballı olm.,** to be on the sweetest of terms: **~ kapı,** a rich employer: **~ kuyruk,** the fat tail of a sheep; a 'milch-cow'; a profitable business: **~ lokma,** a rich windfall: **~ müşteri,** a profitable customer; a 'milch-cow'; (*iron.*) a hard bargainer. **~lık,** napkin, handkerchief. **~lıkçı,** (i) dealer in handkerchiefs *etc.*; (ii) man who hires out wedding garments.
**yağma** Booty; loot. ⌜**~ Hasan'ın böreği**⌝, *phrase used to describe an irresponsible waste of other people's* (*or public*) *money*: **~ etm.,** to plunder: ⌜**~yok!**⌝, 'nothing doing!', 'you can't get away with that!' **~cı,** plunderer, pillager. **~cılık,** pillage.
**yağmak** *vn.* Rain; be poured out in abundance. **dolu ~,** to hail: **kar ~,** to snow: **yağmur ~,** to rain.
**yağmur** Rain. ⌜**~dan kaçarken doluya tutulmak**⌝, 'out of the frying-pan into the fire'; ⌜**~ olsa kimsenin tarlasına yağmaz**⌝, 'if he were rain itself he would not rain on anyone's field', *used to describe an uncharitable and disobliging person.*
**yağmur·ca,** fallow deer. **~cın, ~-kuşu,** plover. **~lamak,** to become rainy. **~lu,** rainy. **~luk,** raincoat; transom *of a door, etc.*
**yağsız** Without fat *or* oil; skim milk *or* cheese.
**yahey** (–\`) *Interj. expressing a feeling of comfort.*
**yahni** A meat stew with onions. ⌜**ucuz etin ~si tatsız olur**⌝, you get what you pay for; if a thing is cheap you can't expect much of it.
**yahşi** Pretty; agreeable; good. ⌜**anan ~ baban ~**⌝, using the most honeyed words.
**yahu** (\`·) *Interj. calling s.o. or emphasizing a remark* (*often rather reproachfully*). **bu da geçer, ~!,** don't worry, this will pass!: **~ dün niçin bize gelmedin?,** why didn't you come to us yesterday? (*reproachfully*).
**yahud** (\`·) Or.
**yahudi** (·–\`) Jew. Jewish. **~ baklası,** lupin: **~ pazarlığı,** hard bargaining. **yahudi·ce** (·–\`·), Hebrew language. **~lik,** quality of a Jew; Jewish method of business; stinginess.
**yaka** Collar; bank, shore, edge; edge *or* corner of a sail. **~sı açılmadık küfür,** an unusual and obscene oath, *hence* **~sı açılmadık,** unusual, unheard-of: **~dan atmak,** to get rid of: **...in ~sını bırakmamak** not to let *s.o.* go; **~mı bırak!,** leave me in peace!: **~yı ele vermek,** to be caught *or* arrested: **~yı kurtarmak (sıyırmak),** to escape: **~sından** (*or* **-sını**) **tutmak,** to hold responsible; to keep hold of, not to let escape: **~ silkmek,** to be disgusted, 'fed up': **...in ~sına yapışmak,** to hold responsible; to force *s.o.* to do *stg.*: ⌜**iki ~sı bir araya gelmemek**⌝, to fail to get on; to make a mess of things (financially): ⌜**iki elim ~nda!**⌝, I won't let you get out of this!: **öte ~,** the other side *or* shore: **ruzgâri ~ya almak,** to luff (a boat).
**yaka·lamak,** to collar, seize; find; hold responsible. **~lık,** stuff suitable for collars; collar of a shirt.
**yakamoz** Phosphorescence in the sea.
**yakarış** Entreaty, prayer.
**yakarmak** *Only in conjunction with* **yalvarmak,** to implore.
**yakaza** Wakefulness.
**yakı** Cautery; blister; plaster. **~ açmak,** to apply a cautery: **pehlivan ~sı,** actual cautery (burning) *or* cautery by applying a caustic substance.
**yakıcı** Burning, smarting; biting *to the taste.*
**yakılmak** *vn. pass. of* **yakmak.** Be burnt *etc.*; pour out one's woes.
**yakın** Near (in place *or* time). Nearby place, neighbourhood; recent time; near future. From near at hand; closely, thoroughly. **~ımızda,** close to us: **~da, ~larda,** near by; in the near future; recently: **~ akraba,** near relation: **~dan,** closely: **~ amir,** immediate superior (*mil.*): **~a getirmek,** to bring near; (of a telescope) to magnify: **~ muharebe,** close action (*mil.*): **~ zamanda,** not long ago: ⌜**akıl var ~ var**⌝, it's only common sense.

**yakın·laşmak,** to draw near, approach. **~lık,** nearness, proximity.
**yakıotu** (.ˋ..), **-nu** Willow-herb.
**yakışık** Suitability; most suitable way; apparent truth, plausibility; beauty. **~ almak,** to be suitable: **~ almaz,** it is 'not done'. **~lı,** suitable, becoming; comely, handsome. **~sız,** unsuitable, unbecoming; ugly.
**yakış·mak** *vn.* Be suitable *or* becoming; be proper *or* fit: look well, be pretty *or* handsome (of a thing or dress). **size yakışmaz,** it does not become you; it is not worthy of you. **~tırmak,** *va. caus. of* **yakışmak,** to think *stg.* becoming *to a person*; to expect *stg. of a person*: **bunu sana yakıştıramadım,** I would not have expected that of you: **yakıştırıp takıştırmak,** to embellish, decorate: **yakıştırıp uydurmak,** to invent *stg.* suitable to the occasion: **giydiğini yakıştırır,** he dresses well.
**yakıt, -tı** Fuel.
**yakin, yakîn** (.ˊ) Certainty. **~en** (.ˊ.), for certain, positively.
**yaklaşılmak** *vn.* Be approached; be approachable.
**yaklaşmak** *vn.(with dat.)* Draw near; approach; approximate; resemble.
**yakmak** *va.* Burn; set on fire; scorch; light; inflame with love; apply (poultice, henna); dupe. **···e abayı ~,** to fall in love with ...: **···in başını ateşe ~,** to bring misfortune on *s.o.*: **can ~,** to oppress; to hurt: ⌜**dışarısı seni yakar içerisi beni yakar**⌝, he (she, it) looks attractive to those who don't know about him (*etc.*): **elektriği ~,** to turn on the electric light: **türku ~,** to compose a song: **'yaklaşma, yakarım!',** 'don't come near or I'll fire!': ⌜**yere bakar yürek yakar**⌝, he's not as innocent as he looks.
**Yakubi** (−−ˋ) Jacobite (Christian sect in Irak).
**yakut** (−ˋ), **-tu** Ruby. **gök ~,** sapphire: **lâl ~,** garnet: **mor ~,** amethyst: **sarı ~,** topaz.
**yalab·ık** Shining; gorgeous. Glitter, sparkle. **~ımak,** to shine, sparkle.
**yalak** Trough; drinking-basin *at a fountain.*
**yalama** Licking; sore; erosion, abrasion; wear, play (*mech.*). **~ resim,** wash-drawing.
**yalamak** *va.* Lick; graze; (of artillery fire *etc.*) to sweep over *a place.* ⌜**bal tutan parmağını yalar**⌝, one connected with a profitable business is bound to get pickings: **çanak ~,** to toady: **imzayı ~,** to dishonour one's signature: **mürekkeb yalamış,** having some education: **tükürdüğünü ~,** to swallow one's words.
**yalan** Lie, falsehood. False. **~dan,** not seriously; **~dan yıkamak,** to wash superficially: **birinin ~ını çıkarmak,** to show up *s.o.'s* lies; to give the lie to *s.o.*: **~ çıkmak,** to turn out untrue: **birinin ~ını tutmak,** to catch *s.o.* out in his lying: **~ yanlış,** carelessly, superficially: **~ yere yemin etm.,** to perjure oneself.
**yalan·cı,** liar; false; deceitful; imitation: **~ çıkarmak,** *v.* **çıkarmak**: **~ dolma,** vine-leaves *or* vegetable stuffed with rice: ⌜**~nın mumu yatsıya kadar yanar**⌝, a lie has only a short life: **başkasının ~sıyım,** '(... so I was told), if it be not correct it is not I that lie'. **~cıktan,** superficially; not meaning it; in pretence. **~cılık,** lying, mendacity. **~lamak,** to deny, contradict.
**yalanmak** *vn.* Lick oneself; get a little profit *out of stg.*
**yalap yalap** In a sparkling manner.
**yalatmak** *va. caus. of* **yalamak.** Graze; let *s.o.* get a small profit.
**yalayıcı** Licking. **~ ateş** (*mil.*), grazing fire, fire with a flat trajectory.
**yalaz** Flame. **~lanmak,** to flame up, blaze up.
**yalçın** Rugged; steep; bare; slippery.
**yaldırak** Shining, brilliant.
**yaldız** Gilding; false decoration. **~dan ibaret,** superficial. **~cı,** gilder. **~lamak,** to gild; to put a false finish to. **~lı,** gilt; lacquered; falsely adorned.
**yalelli** (.ˋ.) Arab song; orgy; amusement. **~ gibi,** unending, monotonous.
**yalı** Shore; beach; waterside residence. **~ çapkını,** kingfisher: **~ kazığı,** tall thin person: **~ mevsimi,** the summer season: **~ uşağı,** one born and bred by the seaside.
**yalım**[1] Blade *of a sword etc.*; stock; kind; nature. **dağ ~ı,** steep slope, cliff, headland.
**yalın**[1], **yalım**[2] Flame; glitter; lightning. **~lamak,** to blaze; glitter.
**yalın**[2] Single; bare, stripped; naked (sword). **~ayak** (.ˋ..) barefooted:

⌜~ başı kabak⌝, bareheaded and barefooted, in rags. **~kat** (·ˋ·), singlefold; superficial, shallow. **~kılıc** (·ˋ··), naked sword, drawn sword.

**yallah** (ˋ·) *interj.* (*abbrev. of* **ya Allah!**) Come!, go! **~ ~**, at most.

**yalman** The pointed cutting part of a weapon.

**yalnız** (ˋ·) Alone. Only. **~ başına***, by oneself, single-handed. **~ca** (·ˋ·), alone; by oneself. **~lık**, solitude; loneliness.

**yalpa** The rolling of a ship. **~ vurmak,** (of a ship) to roll; (of a drunken man) to sway about, to lurch. **~lık,** gimbals: **~ omurga,** bilge-keel.

**yalpak** Friendly.

**yaltak** Fawning; sycophantic; cringing. **~lanmak,** to fawn, flatter obsequiously.

**yalvar·ıcı** Entreating, imploring. **~mak,** to entreat, beg, implore.

**yama** Patch. **~ gibi durmak,** to look out of place: ⌜**~ küçük delik büyük**⌝, the means are insufficient for the end: **~ vurmak,** to put on a patch. **~cı,** patcher; repairer *of clothes, boots etc.*; skinflint.

**yamac** Slope of a hill; side.

**yamak** Assistant, mate. **aşcı yamağı,** under-cook; kitchen-maid.

**yamalak** *Only in* **yarım ~,** defective, half *done etc.* **yarım ~ bir türkçe ile,** in broken Turkish.

**yama·lamak** *va.* Patch. **~lı,** patched; with scars on the face. **~mak,** to patch; stick on; impose *s.o. or stg.* on to *s.o.*

**yaman** Strong, violent, cruel; capable, smart, efficient.

**yamanmak** *vn. pass. of* **yamamak.** Be patched on; instal oneself, get a footing.

**yamçı** Thick rough cape; felt saddlecover.

**yamrı** *Only in* **~ yumru,** uneven and lumpy; gnarled.

**yamyam** Cannibal. **~lık,** cannibalism.

**yamyassı** (ˋ··) Quite flat.

**yan** Side; flank; vicinity *of a thing*; presence *of a person*; direction, bearing. **~a,** *v.* **yana: ~dan,** from the side; sideways; in profile: **~ıma,** to my side, towards me; **~ımda,** at my side; by me; **~ımdan,** from my side; from me: **~ında,** at his (your) side; by him (you); in his (your) opinion: **~ ~,** sideways, sidelong: **~ ~a,** side by side: **···in ~ında,** in comparison with ...: **bir ~dan bir ~a,** from one side to the other: **~ına* almak,** to take into one's service: **~ ateşi** flanking fire, enfilade: **~ bakmak,** to look askance; ⌜**varmı bana ~ bakan?**⌝, 'is there anyone who dares to quarrel with me?': ⌜**~ baksam kabahat**⌝, every little error is brought up against me: **~ basmak,** (*sl.*), to have one's hopes dashed; to be deceived: **~ı başında,** at his side: **~ına* bırakmamak (koymamak),** not to leave unpunished: **~ çizmek,** to sneak off; pay no heed to: **~ gelmek,** to make oneself comfortable: **~ gözle bakmak,** to look at *s.o.* out of the corner of the eye; to look at *s.o. or stg.* with evil intentions: **~ına kalmak,** to remain unpunished: **bunu ~ına koymam!,** I shan't forget that, you'll pay for it!: **~ında oturmak,** to sit down by him; to lodge with him: **···in alt ~ını sormak,** to inquire further into *a matter*: **dört ~a bakmak,** to look in every direction: **ellerim ~ıma gelecek,** I shall die one day (so I must tell the truth); **hilâfım varsa iki elim ~ıma gelsin!,** may I die if I am telling a lie!: **o zamandan bu ~a,** since that time.

**yana**[1] *dat. of* **yan,** *also used as a prep.* (*with dat.*) Towards; (*with abl.*) as regards, concerning; in favour of, on the side of. **···den ~ olm.,** to be on the side of ..., to take the part of ....

**yana**[2] *gerund of* **yanmak. ~ ~** *or* **~ yakıla,** in a moving way; pouring out one's sorrows.

**yanak** Cheek. **yanağından kan damlıyor,** his cheeks are ruddy with health.

**yanar** *v.* **yanmak.** Inflammable. **~dağ** (·ˋ·), volcano. **~döner,** shot (silk *etc.*).

**yanaş·ık** Adjacent. **~ nizam** (*mil.*), close order. **~ılmak,** to be approached; be approachable: **yanaşılmaz,** unapproachable; inaccessible. **~ma,** act of approaching; casual labourer, hireling.

**yanaşmak** *vn.* (*with dat.*) Draw near, approach (of a ship) to come alongside: accede *to a request*; incline, seem willing.

**yanayakıla** *v.* **yana**[2].

**yançizmek** *v.* **yan.**

**yandık** Camel-thorn.

**yangelmek** (ˈ··), **–ir** *vn.* Take one's ease; enjoy oneself lazily.

**yangın** Burnt; burning; suffering. Conflagration, fire; victim *of a fire etc.* (*also fig.*). ~ **bombası,** incendiary bomb: ~**dan çıkmış gibi,** destitute: ~**a gitmek,** to go in great haste: ~ **var!,** fire! fire!

**yanıbaşı** (·ˈ··), **–nı** Place by the side of a person. ~**nda*,** by the side of ..., close by ... .

**yanık** Burn, scald; blight; stinking smut; bunt. Burnt, scorched; tanned; blighted; lighted, turned on (electric light *etc.*); doleful, piteous, touching, pathetic. ~ ~, in a moving way: ~ **kokmak,** to smell of burning; to have a smoky flavour. ~**kara,** bubo of plague. ~**yağı** (·ˈ··), **–nı,** ointment for burns.

**yanılmak**[1] *vn.* Complain *or* fret *at a lost opportunity.*

**yanıl·mak**[2] *vn.* Make a mistake; go wrong. ~**maz,** unfailing; faultless. ~**tmac,** tongue-twister. ~**tmak,** to cause to make a mistake; lead into error.

**yanısıra** (·ˈ··) By the side of.

**yani** (ˈ·) That is to say; i.e.; namely.

**Yani** *Greek prop. name.* ⌜**kırk yıllık ~ olur mu Kâni?**⌝, ⌜can the leopard change his spots?⌝

**yankesici** (ˈ···) Pickpocket.

**yankı** Echo; reaction. ~**lamak,** to echo.

**yanlış** Error, blunder. Wrong, incorrect; formidable (man). ~ **herif,** a man not to be trifled with: ~ **kapıyı çalmak,** to be out in one's reckoning; to make a bad shot: ~ **yere,** wrongly, falsely. ~**lık,** mistake, blunder.

**yanma** Burning, combustion.

**yanmak** *vn.* Burn; be alight; catch fire; be burnt *or* over-roasted; (of a plant) to be blighted *by heat or cold*; be painful, hurt; be very thirsty; be ruined, be 'done for'; feel grieved *or* sorry; recount one's woes; become invalid *or* forfeited; lose one's turn *in a game.* **yana yana,** complaining bitterly: **yanıp yakılmak,** to pour out one's woes piteously: ···**e** ~, to be consumed with passion for ...: ···**den ağzı** ~, to 'burn one's fingers' with *stg.*: **başkasının derdine** ~, to suffer on account of, *or* for the sake of, another: **canım yandı,** I have suffered; I was hurt: **derd** ~, to pour out one's sorrows: ···**in elinden** ~ to suffer wrong at the hands of ...: **haline** ~, to complain of one's circumstances: **haydi yandı!,** time's up! (a turn on a roundabout *etc.*): **içi** ~, to suffer: **yüreği yanmış,** distressed, afflicted.

**yanpırı, yanpiri** Leaning to one side; distorted; awkward; crabwise.

**yansı** Lop-sided.

**yanyana** (ˈ··) Side by side; contiguous.

**yap yap** Gently; slowly; by degrees.

**yapak, yapağı** Wool, *esp.* the wool of a sheep shorn in the spring (*as opp. to* **yün,** *the shorter wool of an autumn-shorn sheep*). ~**çı,** wool-merchant. ~**lı,** well-fleeced (sheep); woollen.

**yapayalnız** (ˈ···) Absolutely alone.

**yapı** Building, edifice; build *of the body.* ~**cı,** maker; builder; creative; constructive.

**yapı·lı** Made; of *such and such a* construction *or* build: **genc** ~, of youthful build; **hafif** ~, lightly constructed. ~**lış,** method of construction; process of building. ~**lmak** *pass. of* **yapmak,** to be made *etc.*; to become rich; to be drunk.

**yapıncak**[1] Rug *to protect a horse from rain*; a kind of shaggy coat.

**yapıncak**[2] A variety of white grape from the village of that name.

**yapınmak** *va.* Have made for oneself (clothes *etc.*).

**yapış** *In* ~ ~, sticky. ~**ık,** stuck on, attached: ~ **kulaklı,** with closely adhering ears. ~**kan,** sticky; adhesive; pertinacious, importunate. ~**kanlık,** stickiness; pertinacity. ~**kanotu, –nu,** pellitory.

**yapış·mak** *vn.* Stick, adhere; stick to one (as a bore); set about *stg.* (*dat.*). ···**in eteğine** ~, to hang on to *s.o.'s* coat *in entreaty.* ~**tırma,** act of sticking on; a thing stuck on. ~**tırmak,** to stick on, attach; to say *stg.* in quick reply: **tokat** ~, to give a box on the ear.

**yapıtaşı, –nı** Building stone.

**yapma** Act of doing *or* making. False, imitation, sham. **yerden** ~, very short, squat. ~**cık,** artificial; feigned; false: affectation.

**yapmak** *va.* Do; make; construct; arrange; set to rights; make ready. **yapma!,** incredible!: **yapacağı ben**

**bilirim,** you just see what I'll do! (a threat): **yapacağını yap!,** do your worst!: **bana yapacağını yaptı,** he did me all the harm he could: ⌜**amma yaptın** (ha)!⌝, is it possible!; you don't say so!: **geçen kış çok soğuk yaptı,** last winter was very cold: **gülmeden yapamadım,** I couldn't help laughing: **gönül** (**hatır**) ~, to satisfy, to content: **ne yapıp yapıp,** in some way or other; by every possible means: ⌜**sanki iş yaptın değil mi?**⌝, a fine lot of use you are!

**yaprak** Leaf; vine-leaf; sheet of paper; flake; layer. ~ **aşısı,** budding: ~ **biti,** aphis: ~ **dolması,** stuffed vine-leaves: ~ **kurmak,** to pickle vine-leaves: ~ **sıgara,** cigar: ~ **tütün,** tobacco in the leaf.
**yaprak·dökümü, -nü,** the fall of the leaf; autumn. ~**lanmak,** to come into leaf; become flaky; (of a sail *or* flag) to flap in the wind. ~**lı,** leafed; leafy; flaky; having *so many* leaves. ~**sız,** leafless; bare (tree).

**yapyalnız** (\`..) Absolutely alone.

**yar**[1] Precipice; abyss. ~**dan uçmak,** to fall down a precipice: ⌜**deveyi ~dan uçuran bir tutam ottur**⌝, a small incident may cause a great disaster.

**yar**[2], **yâr, -ri** Friend; lover. ~**ü ağyar,** friend and foe, all the world: ~**i gar,** intimate friend: ~ **olm.,** to be a helping friend, to assist.

**yara** Wound; sore, cut; boil. ~ **açmak,** to wound: ~**yı deşmek,** to open a wound; to touch a sore spot: ···**in** ~**sına dokunmak,** to touch *s.o.* on his tender spot: ~ **toplamak,** for a boil to come to a head (*also fig.*): ~**ya tuz ekmek,** to put salt on a wound, to increase another's pain by one's words: **dil** ~**sı,** a wounding of the feelings by a bitter word.

**Yarabbi** (\`..) My God!

**yaradan** Creating. The Creator. ~**a kurban olayım!,** *expression of admiration on seeing a beautiful child*: ~**a sığınırım!,** God help me!: ~**a sığınıp vurmak,** to strike a violent blow.

**yaradılış** Creation; nature, temperament; constitution.

**yara·lamak** *va.* Wound; hit *a ship with a shell.* ~**lı,** wounded: **kalbden** ~, afflicted, grieved.

**yara·mak** *vn.* Be serviceable *or* useful; be of use, be suitable. **işe yarar,** serviceable, useful for the purpose in hand: **işe yaramaz,** useless: **sana iyilik yaramaz ki,** it's no good treating you kindly. ~**maz,** good-for-nothing; naughty. ~**mazlık,** naughtiness; bad behaviour.

**yâran** *pl. of* **yar.** Friends; lovers.

**yaranmak** *vn.* (*with dat.*) Make oneself serviceable; curry favour. **ne yaptımsa kendisine yaranamadım,** I did everything I could to please, but all in vain.

**yarar** Serviceable, useful; capable. ~**lık,** courage.

**yarasa** Bat.

**yaraşık** Pleasing appearance; suitability.

**yaraş·mak** *vn.* Be suitable; be pleasing in appearance; harmonize, go well with (*dat.*). **güzele ne yaraşmaz?,** a beauty looks well in any thing. ~**tırmak,** to make suit; to deem suitable *or* becoming; invent (a lie *etc.*).

**yaratan** *v.* **yaradan.**

**yaratıcı** Creative; creating.

**yaratılış** *v.* **yaradılış.**

**yaratmak** *va.* Create. ⌜**Allah yaratmış dememek**⌝, to be merciless: ⌜**küçük dağları ben yarattım diyor**⌝, 'he says "I created little hills"', he is incredibly conceited.

**yarbay** Lieutenant-colonel.

**yarda** (\`.) Yard. ~**lık,** *so many* yards long.

**yardak** Mate; accomplice. ~**çı,** helper; accomplice; lickspittle. ~**çılık,** aid; complicity.

**yardım** Help, assistance. ···**e** ~ **etm.,** to help, succour: ···**ın** ~**ına yetişmek,** to come to the aid of ... . ~**cı,** helper, assistant; auxiliary: ~ **fiil,** auxiliary verb. ~**cılık,** quality of a helper; assistance. ~**cısız,** without a helper; without assistance. ~**laşmak,** to help one another.

**yaren** Friend. ~**lik,** friendly conversation *or* joking.

**yargı** Judgement in a court of law. ~**c,** judge. ~**lamak,** to hear a case; try; judge.

**yargıtay,** Supreme Court of Appeal.

**yarı** Half. ~ ~**ya,** on a fifty-fifty basis, taking equal shares: ~ **çekili bayrak,** half-masted flag: ~**da kalmak,** to be left half-finished, to be broken off in the middle: ~ **yolda,** half-way; ~ **yolda bırakmak,** to leave half-finished, to give up before completion.

**yarıbuçuk** Small; insufficient; 'only half a ...': ~ **askerle,** with only a handful of soldiers.

**yarıcı**[1] Who splits; wood-chopper; who breaks through the enemy's line.

**yarıcı**[2] One who works another's land for half the profit. ~**lık,** the working of land on that basis.

**yarıgece** (·ˈ··) Midnight. At midnight.

**yarık** Split, cleft, cracked. Crack, fissure.

**yarılamak** *va.* Be half-way to *a place*; be half-way through *a job or a period.*

**yarılmak** *vn. pass. of* **yarmak.** Be split *etc.*; crack. ⌜**yer yarılıp içine girmiş**⌝, 'the earth must have cracked and it dropped into the crack', *said of stg. that can't be found.*

**yarım** Half. ~ **ağızla,** not seriously meant (*as of an invitation etc.*): ⌜~ **elma gönül alma**⌝, a very small kindness may win a heart: ~ **kan,** half-bred: ~ **saat,** half an hour. **yarım·ada** (·ˈ··), peninsula. ~**ca** (·ˈ·), megrims, severe headache. ~**lamak,** to be half-way through *stg.*; to half finish. ~**pabuc** (·ˈ··), pauper; vagabond. ~**sağ,** half right (*mil.*). ~**sol,** half left (*mil.*). ~**yamalak,** perfunctory; only half *done, learnt etc.*: incompletely.

**yarın** (ˈ·) Tomorrow. ~ **değil öbür gün,** the day after tomorrow. ~**ki** (ˈ··), of *or* belonging to tomorrow.

**yarış** Manner of splitting; race; competition. **çene** ~**ı,** empty chatter.

**yarış·mak** *vn.* Race; compete. ~**tırmak,** cause to race *etc.*: **çene** ~, to wag the jaw incessantly, to chatter nonsense.

**yarıyarıya, yarıyolda** *v.* **yarı.**

**yarlığ** Command, edict.

**yarlıgamak** *va.* (Of God) to pardon sins.

**yarlık**[1] A place of precipices.

**yarlık**[2] *v.* **yarlığ.**

**yarlık**[3] Friendship; kindness.

**yarma** Act of splitting; cleft, fissure; break-through (*mil.*); railway cutting; large coarse man. Cleft, split; coarsely ground (wheat *etc.*). ~ **muharebesi,** battle aiming at a break-through: ~ **şeftali,** freestone peach.

**yarmak** *va.* Split; cleave; cut through; break through (*mil.*). **başını gözünü** ~, to smash *s.o.'s* face; to make a mess of, to mangle (a lesson, a language *etc.*): **kafa göz** ~, to be very tactless: **kılı kırk** ~, to split hairs.

**yas** Mourning. ~ **tutmak,** to be in mourning.

**yasa** Law; code of laws.

**yasak** Prohibition; interdict. Forbidden, prohibited. ~ **etm.,** to forbid: ~ **savmak,** to serve in case of need, to 'do' when nothing better is available; to do *stg.* merely to comply with a rule: ~ **savar,** (*stg.*) that will do for the time being.

**yasakçı** (*Formerly*) a man who went in front to clear the way for a great person; guard for an ambassador *or* consul.

**yasemin** (–·ˈ) Jasmine.

**yasla·mak** *va.* Support; bolster up. ~**nmak,** lean against *stg.*; support oneself.

**yaslı** In mourning.

**yassı**[1] *v.* **yatsı.**

**yassı**[2] Flat and wide. ~**kadayıf** (·ˈ···), small cakes of batter soaked in syrup. ~**lanmak,** ~**laşmak,** ~**lmak,** to become flat and wide.

**yastağac** Bream.

**yastığac** Pastry-board.

**yastık** Bolster, pillow; cushion; pad; nursery-bed (garden). ~ **yüzü,** pillow-case: **bir yastığa baş koymak,** to get married: **bir** ~**ta kocamak,** to have a long married life: **başı*** ~ **görmemek,** not to sleep a wink: **yüz yastığı,** pillow.

**yaş**[1] Wet; damp; fresh. Wetness, moisture; tears. ~**a bastırmak,** to deceive, cheat: ~ **dökmek,** to shed tears: ~**ını içine akıtmak,** to hide one's grief: ~ **odun,** green wood: ~ **tahtaya basmak,** to be cheated, to be taken in: ⌜**ağac** ~**ken eğilir**⌝, 'as the twig is bent the tree is inclined'.

**yaş**[2] Age. ~**ını başını almış,** of mature years; ~**ına başına bakmadan,** regardless of his (your) age: ~**ında değil** *or* **daha** ~**ı yok,** he is not yet one year old: ~**ı ne başı ne!,** he's too young! ⌜**akıl** ~**ta değil baştadır**⌝, age is no guarantee of wisdom: **ben** ~**ta,** of my age: ⌜**bir** ~**ıma daha girdim**⌝, I am astonished: **kaç** ~**ındasın?,** how old are you?: **on** ~**ına girdi,** he is in his tenth year.

**yaşa** Long live ...!
**yaşama** Living; the art of living.
**yaşamak** *vn.* Live; know how to live. **yaşadınız!**, you are in luck!
**yaşanmak** *impers. of* **yaşamak. burada iyi yaşanıyor,** one lives well here.
**yaşanmış** True to life.
**yaşarmak** *vn.* Become wet; become fresh.
**yaşasın** Long may he live!; long live ...!
**yaşatmak** *va. caus. of* **yaşamak.** Cause *or* allow to live; keep alive; represent as alive (on the stage).
**yaşayış** Method of living; life; livelihood.
**yaşdaş, yaşıt** Of the same age.
**yaşlanmak**[1] *vn.* Become wet.
**yaşlanmak**[2] *vn.* Grow old.
**yaşlı**[1] Wet; suffused with tears.
**yaşlı**[2] Aged. **~başlı,** of mature years. **~ca,** getting on in years. **~lık,** old age; advanced years.
**yaşlık** Wetness; damp weather; juiciness.
**yaşmak** Veil *worn by oriental women.*
**yat, -tı** Yacht.
**yatağan** Heavy curved knife; yataghan.
**yatak** Bed, couch; lair; anchorage, berth; bearing *of a shaft*; chamber *of a gun*; receiver of stolen goods; screen *for illicit enterprises*; place of congregation, mart; river-bed; ore-bed. **yatağa düşmek,** to take to one's bed: **av yatağı,** place frequented by game: **bilye yatağı,** ball-bearing: **zengin yatağı,** the residential quarter of a town.
**yatak·hane** (··—ᐟ), dormitory. **~lı,** furnished with a bed; having *so many* beds; deep-channelled (river); 'wagon-lit', sleeping-car: **~lık,** bedstead; place for storing beds; a being a receiver of stolen goods; for *so many* beds: **~ hasta,** ill enough to have to go to bed. **~odası, -nı,** bedroom.
**yatalak** Bedridden.
**yatı ~ mektebi,** boarding-school: **gece ~sı,** staying the night as a guest. **~lı,** where one sleeps: **~ mekteb,** boarding-school.
**yatık** Leaning to one side; gently rising (ground). **~ yollu top,** gun with a flat trajectory: **30 derece ~, (ship) with a list** of 30 degrees: *v. also* **yatkın.**
**yatır** Place where a saint is buried. **~ çeşme,** saint's tomb with a fountain.
**yatırım** Deposit; investment.
**yatırmak** *va. caus. of* **yatmak.** Lay down; cause to lie down; make lean *or* slope; throw to the ground; overthrow; deposit *money in a bank etc.*
**yatısız** Home-boarder (at a school).
**yatış·mak** *vn.* Calm down; become quiet. **~tırmak,** to calm, tranquillize.
**yatkı** Crease; fold.
**yatkın** Leaning to one side; inclined. **eli ~** (*with dat.*), accustomed to, fairly skilled at *some job.*
**yatmak** *vn.* Lie down; go to bed; be in bed; pass the night; be bedridden; become flat; (of a ship) to lie at anchor; be broken in, be used to *work etc.*; yield, consent; stay in prison. **···e ~,** to lie on; lean towards; agree to: **eli* yatmak,** to be used (to work *etc.*), to be fairly skilful: **bir yerde yatıp kalkmak,** to live *or* lodge at a place: **sulha (subhana) ~** *or* **yola ~,** to be (with some difficulty) brought to reason *or* to agreement: **bir şeyin üstüne ~** (*sl.*), to 'hang on to' *stg. that does not belong to one*: **yan ~,** to lean over to one side.
**yatsı** Time of going to bed; **~ (namazı)** prayer said by Moslems two hours after sunset. ⌜**~dan sonra ezan okumak**⌝, to do *stg.* too late *or* at an unsuitable moment.
**yavan** Plain, dry *food*; without oil *or* fat; tasteless.
**yavaş** Slow; gentle, mild; soft (sound); docile. **~ ~!,** gently!; steady!; don't be in a hurry!
**yavaş·ça, ~çacık** Gently; slowly. **~lamak,** to become slow *or* mild; to slow down; (of rain) to slacken.
**yave** (—ᐟ) Foolish talk. Commonplace.
**yaver** (—ᐟ) Helping. Aide-de-camp.
**yavru** The young *of an animal or bird*; cub; chick; *affectionate term for a child.* **~ çıkarmak,** (of a bird) to hatch out chicks: **konak ~su,** a little mansion. **~cuk,** *dim. of* **yavru. ~lamak,** (of an animal) to bring forth young.
**yavşak** Young louse, nit.
**yavşan** Thorny, spiny. **~ otu,** *Artemisia fragrans* (?).
**yavuklu** Betrothed, engaged.
**yavuz** Resolute; efficient; good

excellent. ⌜~ **hırsız ev sahibini bastırır**⌝, *said of one who is in the wrong but who carries the day by bluff*: ~ Selim, Selim the First.

**yay** Bow (of an archer); bow (violin). ~ **burcu,** the constellation of Sagittarius: **araba ~ları,** the springs of a carriage: ⌜**ok ~dan fırladı**⌝, the thing has happened, there can be no recall.

**yaya, yayan** On foot; pedestrian; of little account; without skill. ~ **kalmak,** to be compelled to go on foot *for want of a horse or vehicle*; to be in a difficult situation, to get oneself into a pretty pass: ⌜~ **kaldın tatar ağası**⌝, now you're stranded!; you're on the wrong track!

**yayakaldırımı, -nı** Foot pavement.

**yayanlık** A going on foot.

**yaygara** Shout; outcry, clamour. **~yı basmak (koparmak),** to raise an outcry; to make a great to-do about nothing. **~cı,** noisy; brawling.

**yaygı** Something spread out as a covering.

**yayık**[1] Spread out; broad, wide. ~ ~, in a dawdling manner; drawling one's words.

**yayık**[2] Churn.

**yayılma** Act of being spread out *etc.*; deployment (*mil.*). **~k,** *vn. pass. of* **yaymak,** to be spread out; spread, be disseminated; be stretched out on the ground *by a blow or fainting.*

**yayın**[1] Publication.

**yayın**[2], **~balığı** Sheat-fish, silurus.

**yayla** High plateau; summer camping-ground. **~k,** summer pasture on high ground. **~kiye,** rent paid for a **yaylak.**

**yaylı** Armed with a bow; having springs. Carriage with springs *used for long journeys.* ~ **gözlük,** pince-nez.

**yaylım** *In* ~ **ateş,** volley; drum-fire.

**yayma** Act of spreading *etc.*; small dealer's stall. **~cı,** small dealer *whose goods are spread out on a stall.*

**yaymak** *va.* Spread; scatter; disseminate.

**yayvan** Broad; spreading out; slack. ~ ~ **gülmek,** to laugh uproariously: ~ ~ **konuşmak,** to drawl: **ağzı ~,** garrulous.

**yaz** Summer.

**yazboz tahtası** School slate.

**yazdırmak** *va. caus. of* **yazmak.** Cause to write *or* be written; have inscribed *or* registered.

**yazı**[1] Plain; flat place.

**yazı**[2] Writing; calligraphy; manuscript; inscription; destiny; a written article. ~ **makinesi,** typewriter: ~ **tahtası,** blackboard: ~ **taşı,** slate. **yazı·cı,** writer; clerk; secretary. **~hane** (··–ˈ), writing-table, desk; office.

**yazık** A pity; a shame; deplorable; what a pity!; what a shame! **~lar olsun!,** shame!: **adama ~ oldu,** it was a pity that this happened to the man.

**yazı·lı** Written; inscribed, registered; decreed by Fate, destined. **~lış,** method of writing; spelling. **~lmak,** *vn. pass. of* **yazmak,** to be written *etc.*; be registered *or* enrolled; be entered *at a school or university, etc.*

**yazın** (ˈ·) In summer.

**yazış** Manner of writing. **~mak,** to write to one another, correspond.

**yaz·lamak** *vn.* Pass the summer *in a place.* **~lı,** *only in* ~ **kışlı,** summer and winter alike. **~lık,** suitable for the summer; summer clothing; rent for the summer.

**yazma** Act of writing; hand-painted *or* hand-printed kerchief *or* bed-spread. Written, manuscript; hand-painted. **~cı,** one who paints *or* prints fine stuffs (muslin *etc.*).

**yazmak** *va.* Write; inscribe; register; enrol; (*auxiliary verb, used in the past only*) to have been on the point of ...; **düşeyazdı,** he all but fell. **yazıp bozmak,** to give an order and then a counter-order; to be capricious: ~ **çizmek,** to compose in writing: **yüz ~,** to decorate the face of a bride.

**ye** *v.* **ya**[3].

**Yecuc** (–ˈ) *In* ~ **Mecuc,** Gog and Magog.

**yedek** Halter; tow-rope; led animal; reserve horse; reserve; spare part. Spare; in reserve. **yedeğe almak,** ~ (**~te** *or* **yedeğe**) **çekmek,** to take in tow, to tow: ~ **parçaları,** spare parts: ~ **subay,** supplementary officer, reserve conscript officer: **yedeğe vermek,** to have *a horse* led *or a boat* towed: **gıda ~leri,** reserves of food.

**yedek·çi,** man who leads a spare horse; man who tows a boat: ~ **yolu,** towpath. **~li,** in lead; in tow; having a spare horse; provided with a spare part. **~lik,** serving as a reserve *or* spare part.

**yedi** Seven. ~ **adalar,** the Ionian Islands: ~ **başlı** (**yılan**), a seven-headed hydra; venomous woman; dangerous man: ~ **canlı,** having seven lives, invincible.
**yedirme** *verb. noun of* **yedirmek.** A preparation of hemp, lime, and oil *used to make a watertight joint in pipes.*
**yedirmek** *va. caus. of* **yemek.** Cause to eat *or* be eaten; feed; cause to swallow; let *oil etc.* be absorbed; expend, lose *money*: **nefsine** ~, to swallow *an insult etc.*: ···**e para** (**rüşvet**) ~, to bribe *s.o.*
**yedişer** Seven each; seven at a time.
**yediveren** Any prolific plant; plant producing several crops a year.
**yedmek** *va.* Lead *or* tow with a rope.
**yegân** One by one. ~**e** (·—ˈ), sole, unique.
**yeğ** *v.* **yey.**
**yeğen** Nephew; niece.
**yek**[1] *v.* **yey.**
**yek**[2] One. ~**ten,** all at once; without any reason: ~ **kalem,** straightway. ~**dığer,** one another; each other.
**yeke** Tiller.
**yekinmek** *vn.* Make a great effort.
**yek·nasak** Uniform; monotonous. ~**nasaklık,** uniformity; monotony. ~**pare** (·—ˈ), in a single piece.
**yek·renk** All of one colour. ~**san,** one with; level; **yerle** ~, levelled to the ground. ~**ta** (·ˈ), unique. ~**ten,** all at once; without any reason.
**yekûn** (·ˈ) Total; sum.
**yel** Wind; flatulence; rheumatism. ~ **gibi gelmek,** to slip in unobserved: ⌜~ **üfürdü su götürdü**⌝, easily and without effort: ⌜**yerinde** ~**ler esiyor**⌝, it no longer exists.
**yeldeğirmeni, -ni** Windmill.
**yeldirme** *A kind of* light cloak *formerly worn by women.*
**yele** Mane. ~**li,** maned: **uzun** ~, having a long mane.
**yelek** Waistcoat, vest; wing-feather, pinion; feather of an arrow.
**yelim** Mucilage; gum.
**yelken** Sail. ~ **açmak,** to hoist sails: ~ **gemisi,** sailing-ship: ~**leri indirmek** *or* **mayna etm.,** to lower sails: ~**i suya indirmek,** to humble oneself, to knuckle under: ⌜(**yel yeperek**) ~ **kürek**⌝, in a great hurry.
**yelken·bezi, -ni,** sailcloth, canvas. ~**ci,** sailor *on board a sailing vessel*; sail-maker. ~**li,** fitted with sails: sailing-boat.
**yelkovan** Minute-hand of a watch; weather-cock; eaves-board; smoke-cowl; the Bosphorus shearwater (*Puffinus yelkovanus*).
**yel·lemek** *va.* Blow upon; fan. ~**lenmek,** break wind, fart. ~**li,** windy; flatulent; rheumatic.
**yellim** *In* ~ **yeperek** (**yepelek**), running swiftly; in great haste.
**yelloz** Whore.
**yelpaze** (·—ˈ) Fan. ~**lemek,** to fan. ~**lenmek,** to fan oneself.
**yeltenmek** *va.* (*with dat.*) Strive *or* dare to do *stg.* beyond one's powers *or* rights.
**yelve** Woodcock; (*in some districts*) snipe.
**yelyeperek** *v.* **yel.**
**yem** Food; a ration of food; fodder; bait; priming of a muzzle-loading gun. ~ **borusu,** bugle-call for horse fodder; ~ **borusunu çalmak,** to put *s.o.* off by empty promises: ~ **kestirmek,** to stop and feed the horses.
**yemek**[1] (**yer, yiyor, yiyecek, yiyen**) *va. & vn.* Eat; feed; consume; spend; dissipate; suffer (an infliction, a beating *etc.*); take *a bribe.* ⌜**yeme de yanında yat!**⌝, *said of a delicious food almost too good to touch*: **yiyip bitirmek,** to consume utterly, to exhaust: ···**in başını** ~, to ruin, to kill; **başının* etini** ~, to worry the life out of ... , to nag at: **ceza** ~, to receive punishment: ···**in hakkını** (**parasını**) ~, to do *s.o.* out of his rights (money): **içi* içini*** ~, to be consumed with impatience *or* anxiety: **kendi kendini** ~, to worry oneself to death: **kumarda** ~, to gamble away: **miras** ~, to come into an inheritance: **yağmur** ~, to be wetted through by rain.
**yemek**[2] *n.* Food; a meal; a dish *or* course *of food.* ~ **borusu,** bugle-call for food. ~**hane** (··—ˈ), dining-hall. ~**li,** with food; with a meal; of *so many* courses *or* dishes. ~**lik,** serving as food; edible: a thing destined as food; money for food. ~**siz,** without food; without a meal.
**Yemen** (ˈ·) The Yemen. ~ **zafranı,** dyer's rocket, yellow-seed. ~**li,** native *or* inhabitant of the Yemen.
**yemeni** A kind of light shoe *worn by peasants*; coloured cotton handkerchief. ~**li,** wearing a **yemeni.**

**yemin** The right hand; the right side *or* direction. ~ **bozmak,** to violate an oath: ~ **etm.,** to swear, to take an oath: ~ **ettirmek,** to administer an oath: ⌜~ **etsem başım ağrımaz**⌝, I can say with a clear conscience: **bir şeyin üzerine** ~ **etm.,** to swear by a thing. ~**li,** bound by oath.

**yemiş** Fruit. ~ **vermek,** to bear fruit. ~**çi,** fruiterer. ~**lik,** fruit-garden; fruit-dish.

**yemleme** Bait; alluring words. ~**k,** to bait *a hook or trap*; prime *a gun*; entice.

**yemlik** Suitable for food *for animals*. Trough, manger; nose-bag; bribe; an easy prey (in gambling).

**yemyeşil** (ˈ··) Very green.

**yen** Sleeve; cuff.

**yenge** (ˈ·) A woman's sister-in-law *or* aunt-in-law; elderly woman who helps and attends a bride.

**yengec** Crab.

**yeni** New; recent; raw, inexperienced. Recently. ~**den,** afresh, anew; ~**den** ~**ye,** always over again, ever anew: ~ **çıkma,** newly brought out; new-fangled.

**yeni·bahar,** pimento. ~**baştan** (·ˈ··), anew, afresh. **Y~cami,** *name of a famous mosque in Istanbul*: ~ **traşı,** a rough-and-ready haircut. ~**ce,** fairly new *or* recent.

**yeniçeri** (·ˈ··) Janissary; the Corps of Janissaries. ~**ağası, -nı,** the Commander-in-chief of the Janissaries.

**yenidünya** (·ˈ··) The New World, America; the Japanese medlar, loquat; glass ball *used as an ornament in a room*.

**yenik** Place nibbled *or* gnawed by insects *etc.*; moth-eaten place.

**yeni·lemek** *va.* Renew; renovate. ~**lik,** newness; novelty; rawness, inexperience.

**yenilmek¹** *vn. pass. of* **yemek.** Be eaten; be edible. **bu yenilir yutulur şey değil!,** this is intolerable!

**yenilmek²** *vn. pass. of* **yenmek².** Be overcome; lose *at a game*.

**yenir** Edible.

**yenirce** Canker of trees.

**yenişmek** *vn.* Try to beat one another; wrestle; fight. **yenişememek,** to tie, to dead-heat.

**yenli** Having sleeves.

**yenmek¹, -ir** *vn. pass. of* **yemek.** Be eaten *etc.*; be edible; be worn *or* frayed.

**yenmek², -er** *va. vn.* Overcome, conquer; be victorious; win *at a game*.

**yepyeni** (ˈ··) Brand new.

**yer¹** *v.* **yemek.**

**yer²** The earth; ground; place; space; room; landed property; situation, employment; mark *left behind by a thing*. ~**de,** on the ground; on the Earth: ~**inde,** in its place; suitable; to the point, well put; correct, right: ~**e bakmak,** to cast one's eyes to the ground *from modesty or shame*; ⌜~**e bakar yürek yakar**⌝, not so innocent as he looks; artful dodger: ~**e batmak,** to perish: ~ **bulmak,** to find a place, to get a situation: ~**ini bulmak,** (of an order *or* a request) to be carried out; (of a man) to find his right job: ~ **etm.,** to leave a mark; to impress, make an impression: ~**e geçmek,** to feel ready to sink into the earth for shame; ~**e geçsin!,** may he perish!: ~**ine*** **geçmek,** to replace *s.o.*: ~**ine gelmek,** to come into place, to come all right: ~**ine getirmek,** to carry out (an order *etc.*): ~**de kalmak,** not to be appreciated: ~**ine koymak,** to replace *stg.*: ⌜~**in kulağı var**⌝, 'walls have ears': ~**e vurmak,** to defeat; to discredit: ~**den** ~**e vurmak,** to throw violently to the ground: ~**i yok,** it is uncalled for, it is out of place: **başı*** ~**ine gelmek,** to recover oneself: **baş üzerinde** ~**i var,** he is esteemed and respected: **bunu yapsa** ~**idir,** if he does this he will be quite right.

**yer·altı** (ˈ··), **-nı,** ground below the surface; tunnel; underground chamber: ~ **kablosu,** subterranean cable. ~**çamı** (ˈ··), **-nı,** a kind of heath. ~**elması** (ˈ···), **-nı,** Jerusalem artichoke. ~**eşeği** (ˈ···), **-ni,** woodlouse. ~**fesleğeni** (ˈ····), **-ni,** dog's mercury. ~**fıstığı** (ˈ···), **-nı,** peanut. ~**göçkeni** (ˈ···), **-nı,** mole.

**yerinmek** *va.* (*with dat.*) Feel regret for; be sorry about.

**yer·kabuğu** (ˈ···), **-nu** The Earth's crust. ~**katı** (ˈ··), **-nı,** ground-floor.

**yerleş·ik** Settled; established. ~**mek,** to settle down; become established; get into an employment *or* office. ~**tirmek,** *caus. of* **yerleşmek,** to put into place; arrange in proper order; put into an employment; settle *s.o. into a place of residence.*

**yerli** Local; indigenous; native. ~ **~si**, a local inhabitant: ~ **dolab**, a fixed cupboard: ~ **mal** *or* **malı**, local produce: ~ **yerinde**, in its (his) proper place: ~ **yerine**, each to his post.

**yermek** *va.* Loathe; blame; criticize. **aş ~**, to turn from one's proper food and long for *stg.* else (*used esp. of a pregnant woman*).

**yer·siz** Without a home; out of place. **~sıçanı** (ˈ···), **-nı**, mole. **~yüzü** (ˈ··), **-nü**, the face of the earth; the world.

**yesar** The left; left hand.

**ye'si** *v.* **yeis.**

**yeşermek** *vn.* Become green; bloom.

**yeşil** Green, verdant; fresh. **Y~ay**, the Green Crescent (Turkish Temperance Society). **~bağa**, tree-frog. **~baş**, the mallard drake. **~imtrak**, greenish. **~lenmek**, to become green; be freshened: **···e ~**, to be amorously excited by ... . **~li**, mixed with green. **~lik**, greenness, verdure; meadow; salad; green vegetables, greens.

**yeşim** Jade.

**yeter** Sufficient. Enough! **~lik**, competence. **~sizlik**, insufficiency, inadequacy.

**yetim** (·ˈ) Orphan; fatherless child. **~hane** (··–ˈ), orphanage. **~lik**, quality of an orphan; orphanage.

**yetinmek** *vn.* Be contented (with = **ile**).

**yetiş·kin** *v.* **yetişmis.** **~me**, *verb. noun of* **yetişmek**: arrived at full growth; ripe; perfected. **alaydan ~**, risen from the ranks; not professional.

**yetişmek** *vn. & va.* (*with dat.*) Reach; attain; suffice; attain maturity, grow up; be brought up; be ready *or* on hand in time; catch (a train *etc.*); arrive to the help of ; cope with; live to see *a certain event etc.*; have lived long enough to have seen *a person or event*; (of plants) to grow. **yetişin!**, help!: **yetişme (yetişmiyesi)**, 'may you (he) never grow up!' (*a curse*): **ben size yetişirim**, I'll catch up with you.

**yetiş·miş**, arrived; reached maturity; grown up. **~tirici**, breeder *of animals.* **~me**, **···in ~si**, brought up by ... . **~tirmek**, *caus. of* **yetişmek**, cause to reach *etc.*; bring up, educate; breed (animals); convey (news), send (information); be unduly precipitate in passing on unnecessary information.

**yetki** Competence; qualification. **~li**, competent; qualified.

**yetmek**[1] *v.* **yedmek.**

**yetmek**[2] *vn.* Suffice; reach, attain. **canına* ~**, to become intolerable; **artık canıma yetti**, I've had enough of it: **···e gücü ~**, to be capable of, to have the strength to ... .

**yetmiş** Seventy. **~er**, seventy each. **~inci**, seventieth. **~li**, containing seventy. **~lik**, of the value of seventy *piastres etc.*; weighing seventy *kilos etc.*; seventy years old; person seventy years old.

**yevm** Day. **~î** (·ˈ–), daily. **~iye**, daily pay; day-book.

**yey, yeğ** Better; preferable.

**yezid, yezit** Impious; cruel; vile fellow.

**yığılı** Heaped, piled up.

**yığılmak** *vn. pass. of* **yığmak.** Be heaped up; crowd together; fall in a faint; collapse.

**yığın** Heap; pile; crowd. ~ **~**, in heaps: **~la** ... . heaps of ... . **~ak** (*mil.*), concentration. **~mak**, to be concentrated. **~lık**, crowd; mass. **~tı**, accumulation; heap; crowd.

**yığışmak** *vn.* Crowd together.

**yığmak** *va.* Collect in a heap; pile up; accumulate; mass (troops).

**yıkama** Act of washing.

**yıka·mak** *va.* Wash. **~nmak**, to wash oneself; have a bath; be washed.

**yık·ıcı** Destructive. Breaker-up (of old ships *or* buildings); junk dealer. **~ık**, demolished; fallen down; ruined. **~ılmak**, *pass. of* **yıkmak**, be demolished *etc.*; fall down; become decrepit; take oneself off, clear out: **yıkıl git!**, clear out! **~ım**, ruin; crash (bankruptcy). **~ıntı**, heap of ruins; debris.

**yıkmak** *va.* Pull down; demolish; ruin, overthrow; unload (an animal). **ev ~**, to bring ruin to a home; to sow domestic discord: **yakıp ~**, to destroy utterly.

**yıl** Year. ~ **başı**, New Year's Day: ⌜~ **uğursuzundur**⌝, 'the times belong to the rascals', said of undeserving people who get money *or* position: ⌜**ar ~ı değil kâr ~ı**⌝, one can't be squeamish about money matters these days: **ayda ~da bir,** very rarely: **kırk ~da bir,** very seldom; just for once.

**yılan** Snake. **~ gibi,** treacherous; repulsive: **~ gömleği,** snake-skin: **~ hikâyesi,** a long-winded story; *stg.* that never ends: **~ın kuyruğuna basmak,** to arouse the spite of a venomous and powerful person: ⌜**uyuyan ~ın kuyruğuna basma!**⌝, 'let sleeping dogs lie!' **yılan·balığı, –nı,** eel. **~cı,** snake-charmer. **~cık,** erysipelas. **~cıl,** the Sacred Ibis. **~kavı** (··–ˈ), spiral, winding. **~yastığı, –nı,** arum.

**yılbaşı** (ˈ··), **–nı** The New Year; New Year's Day.

**yıldırak** Bright, shining.

**yıldırıcı** Terrifying; one who causes fear and anxiety.

**yıldırım** Lightning, thunderbolt. **~ cezası,** fine levied on the spot, automatic fine: **~ telgrafı,** urgent telegram: **~la vurulmuşa döndü,** he was thunderstruck *with terror.*

**yıldırmak** *va.* Frighten, daunt, cow.

**yıldız** Star; Pole star; north; destiny. **~ı ···ile barışmak,** to be on good terms with ..., to get on with ...: **~ı düşkün,** ill-fated: **~ı parlak,** whose star is in the ascendant, lucky. **~böceği, –ni,** firefly; glow-worm, **~çiçeği, –ni,** dahlia.

**yıldönümü** (ˈ···), **–nü** Anniversary.

**yılgın** Cowed, daunted. **~lık,** a being cowed: **göz yılgınlığı,** terror.

**yılışık** Sticky; importunate; grinning unpleasantly.

**yılışmak** *vn.* Grin unpleasantly; smile impudently.

**yıl·lamak** *vn.* Be a year *in one place*; become a year old. **~lanmak,** to become a year *or* several years old; remain several years; grow old; (of a matter) to drag on for a long time.

**yıllık** One year old; *so many* years old. One year's rent; a year's salary; year-book, annual. **bunca ~ dost,** a very old friend. **~çı, ~lı,** who receives a yearly salary; who pays a yearly rent; servant paid by the year.

**yıl·mak** *vn.* Be afraid. **···den ~,** to dread: **···den gözü ~,** to be terrified of ... . **~maz,** undaunted.

**yıpranma** Wear and tear.

**yıpranmak** *vn.* Wear out; be worn *by friction or use*; grow old prematurely.

**yıprat·ıcı** That wears out; exhausting; toilsome. **~ma,** act of wearing out: **~ harbi,** war of attrition. **~mak,** to wear out (*va.*); wear *by friction etc.*

**yırt·ıcı** Tearing, rending. **~ hayvan (kuş),** beast (bird) of prey. **~ık,** torn, rent; ragged; brazen-faced: **~ pırtık,** in rags.

**yırtılmak** *vn. pass. of* **yırtmak.** Be torn *or* rent; (**yüzünün perdesi**) **~,** to become insolent *or* shameless.

**yırtınmak** *vn.* Shriek in desperation; be fearfully perturbed; struggle hopelessly.

**yırtmac** Slit in a sleeve *etc.*

**yırtmak** *va.* Tear, rend; burst; tear to pieces; break in *a horse.* **···in yüzünü ~,** to render insolent; to encourage *s.o.* to be too forward.

**yısa** *v.* **yisa.**

**yiğit, –ti** Young man; fine manly youngster; hero. Brave, stout-hearted. ⌜**her ~in kârı değil**⌝, it's not a thing anybody can do; it's not as easy as all that. **~lik,** courage, pluck, heroism: **yiğitliğe leke sürmemek,** to save one's face.

**yine** (ˈ·) *v.* **gene.**

**yirmi** Twenty. **~ yaş dişi,** wisdom-tooth. **~lik,** of the value, weight, length of twenty ...: twenty para *or* twenty piastre piece: **~ bir genc,** a twenty-year-old youth. **~nci,** twentieth. **~şer,** twenty each; twenty at a time.

**yisa** (ˈ·) Hoist away!; pull!

**yitik** Lost.

**yit·irmek** *va.* Lose; cause to go astray, ruin. **~mek,** to be lost; go to ruin.

**yiv** Groove; rifling *of a gun*; chamfer; hem; stripe. **~li,** grooved; chamfered; rifled.

**yiy·ecek** Edible; to be eaten. Food. *v.* **yemek.** **~en,** *v.* **yemek.** **~ici,** who eats; glutton; greedy; corroding (sore). **~inti,** anything edible; edibles; food-stuff. **~ip,** *v.* **yemek.** **~iş,** manner of eating: ⌜**her yiğitin bir yoğurt ~i var**⌝, everyone has his own way of doing things.

**yobaz** Religious fanatic.

**yoğrulmak** *vn. pass. of* **yoğurmak.**

**yoğun** Thick; stout; coarse; big; gross, unmannerly.

**yoğuna** *v.* **yok.**
**yoğurmak** *va.* Knead.
**yoğurt** Yaourt (a kind of sour milk). ~ **çalmak,** to make yaourt: ⌜**hiç kimse yoğurdum kara demez**⌝, nobody runs down his own handiwork, 'nobody cries "stinking fish!" ': ⌜**sütten ağzı yanan yoğurdu üfler de yer**⌝, 'a scalded cat fears cold water'. **~çu,** maker *or* seller of yaourt.
**yok** Non-existent; absent. Non-existence; nothing. No. There is not; **yok** *is used with possessive suffixes to make the negative of the verb 'to have', e.g.* **evi ~,** he has no house. **yoğuna,** vainly; for nothing: ~ **değildir,** there are (it is) not wanting, there are (is): ~ **olm.,** to be annihilated, to cease to exist: ~ **pahasına,** for a mere song: ~ **yere,** without reason; uselessly: **Londrada ~ ~,** there is nothing that is not to be found in London: **hiç ~tansa ona da razı olduk,** we agreed to that as better than nothing: **ne var ne ~ ?,** what's the news?: **var ~,** there is very little if any: **varı yoğu,** all he possesses, his all.
**yoketmek** (ʹ··), **-eder** *va.* Render non-existent; annihilate.
**yoklama** Examination; inspection; roll-call; call-up *of recruits.* **~cı,** military inspector.
**yoklamak** *va.* Feel with the fingers *or* hand; examine, inspect; search; try, test; visit *a sick person.* **yoklaya yoklaya gitmek,** to feel one's way along: **···in ağzını ~,** to sound *a person,* to try to learn his opinion.
**yokluk** Absence; non-existence; lack; poverty. ⌜**adam yokluğunda**⌝, 'for want of a man'.
**yoksa** (ʹ·) If not; otherwise; or; if there be not; but not; I wonder if. ~ **gelmiyecek mi ?,** perhaps he isn't coming after all.
**yoksul, yoksuz** Possessing nothing; destitute; in need of; lacking. **dayak ~u,** who deserves a beating: **fırsat ~u,** one who always looks out for an opportunity *to do stg. wrong*; **namus ~u,** devoid of honour. **~luk,** destitution.
**yokuş** Rise; ascent; slope. Rising (ground). **~lu,** rising (ground); sloping upwards.
**yokyere** (ʹ··) In vain, uselessly.

**yol** Road; way; street; channel, canal; means, medium; manner, method; behaviour; rule, law; journey; career; rate of speed; stripe. ~ ~, striped; in lines: **~una,** for, for the sake of: **~unda,** for the sake of; in order, going as it should, *e.g.* **bütün işler ~undadır,** all goes well: **~unda, ~undaki,** in the sense that, to the effect that: **~iyle (~u ile),** via; by way of; properly, duly: ~ **almak,** to get up speed; to advance: **···in ~una bakmak,** to await *s.o.'s* arrival: **···in ~unu beklemek** to lie in wait for ...: ~ **bilmek,** to know the way; to know how to do *stg.*; to know how to behave: **~unu bulmak,** to find a way of doing *stg.*: **~undan çıkarmak,** to mislead, pervert: **~dan çıkmak,** to go off the rails; to go to the bad: **···e ~u* düşmek,** for one's road to lead one to ... , *e.g.* **oraya ~unuz düşerse,** if you are going that way: **~a düşmek,** to set out, to set forth on a journey: **~lara düşmek,** to set out in search of *s.o.* (*implies an emergency*): ~ **erkân,** social conventions: **~a gelmek,** to come round, to think better of *stg.*; to come to reason: **~a getirmek,** to bring *s.o.* round to the right view *or* course, to bring to reason, to persuade: **~una girmek,** (of a matter) to come right: ~ **iz bilmek,** to know the rules of social behaviour: **~dan kalmak,** to be kept back, to be detained: **~larda kalmak,** to be delayed on the road: ~ **kesmek,** to waylay a road, to stage a hold-up; **~unu kesmek,** to stop *s.o.*: **~una koymak,** to set right; **~undan koymak,** to prevent, detain, delay; **~-üstü,** on the road: ~ **vermek,** to give passage, to make way; to give *a horse* his head; to let go his own way; to discharge dismiss from service: **···e ~ vermek,** to open the way to, to cause: ~ **vurmak,** to waylay a road, to stage a hold-up: **~a vurmak,** to see *s.o.* off *on a journey*: **bir ~** (*provincial*), once: ⌜**bize ~ göründü**⌝, we must be going: **bu vapurun ~u yok,** this steamer has no speed: ⌜**(sıçanın geçtiğini aramam ama) ~ olur**⌝, I don't mind it once, but it might become a precedent.
**yol·bilir,** well-mannered; adroit. **~-bilmez,** unmannerly, awkward. **~-cu,** traveller; passenger; child about

to be born; one at the point of death; woman of the streets: ~ **geçirmek,** to accompany *s.o.* on the start of a journey: ⌜**~ yolunda gerek**⌝, a traveller's place is on the road; one should stick to what one is doing: **yine ~ olduk,** I must start on my travels again. **~culuk,** travelling, travel: **~ ne zaman?,** when do you set out? **~daş,** fellow traveller; comrade: **can -ı,** a very intimate friend, a faithful friend; *s.o.* to keep one company: **kapı ~ı,** fellow servant; in the same employment. **~daşlık,** a being a fellow traveller; comradeship, fellowship.

**yoldurmak** *va. caus. of* **yolmak.** Cause to pluck *etc.* **···e ot ~,** to make *s.o.* work hard.

**yolgeçen** *In* **~ hanı,** *used of a place which is much frequented.*

**yol·harcı** (˙..), **-nı** Travelling expenses; journey-money. **~kesici,** highwayman, brigand.

**yollama** Act of sending; Goods Department of a railway; military transport service; large beams of building timber. **~ müdürü,** railway official in charge of the Goods Department.

**yol·lamak** *va.* Send; dispatch. **~lanmak,** be sent off; set off *on a journey etc.*; advance.

**yollu** Having *such and such* roads; striped; having *such and such* a way *or* manner; proper, correct, regular; fast (ship *etc.*). Of the nature of, by way of. **~ yolsuz işler,** irregular actions: **nasihat ~,** of the nature of advice.

**yolluk** Provisions for a journey; journey-money.

**yolmak** *va.* Pluck; tear out; strip; despoil. **saçını başını ~,** to tear one's hair; to tear out *s.o.'s* hair.

**yolsuz** Roadless; irregular; without speed (ship, motor-car *etc.*). **~luk,** irregularity; impropriety.

**yoluk** Plucked; hairless.

**yolunda** *v.* **yol.**

**yolunmak** *vn. pass. of* **yolmak.** Be plucked; be robbed; lose one's money *at gambling*; tear one's hair *with grief.*

**yolüstü** (˙..) (Place) lying on one's road; (window *etc.*) looking on the road.

**yoma** Cable-laid (rope). **~ bağı,** carrick bend.

**yonca** Clover; trefoil; lucerne.

**yonga** Chip, chipping; kindling. ⌜**mal canın ~sıdır**⌝, it is hard to part with anything one owns.

**yonma, yonmak** *v.* **yontma, yontmak.**

**yontma** Act of chipping. Chipped, cut. **~ taş,** dressed stone.

**yont·mak** *va.* Cut, chip into shape; dress (stone); pare (nails); sharpen (pencil). **kendinden yana ~,** to turn to one's own advantage. **~ulmamış** (.˙..), not chipped *or* hewn; rough, uneducated.

**yordam** Agility; dexterity. **el ~ı,** feeling with the hand: **el ~ile,** by touch: **yol ~ bilmek,** to know the proper way to do *stg.*; **yolunu ~ını şaşırmak,** to lose one's bearings; not to know where one is. **~lı,** *in* **yollu ~,** in the proper way, 'comme il faut'.

**yorga** Jog-trot. Going at a jog-trot.

**yorgan** Quilt. ⌜**~ gitti kavga bitti**⌝, 'the dispute is ended', *said ironically when the subject of a dispute no longer exists*: ⌜**ayağını ~a göre uzatmak**⌝, 'to cut one's coat according to one's cloth'. **~cı,** quilt-maker; upholsterer. **~iğnesi, -ni,** quilting-needle. **~lık,** suitable for making quilts. **~yüzü, -nü,** outer covering of a quilt.

**yorgun** Tired, weary. **~ argın,** dead tired. **~luk,** weariness, fatigue: **~ almak,** to rest from one's fatigue: **~ kahvesi,** coffee to revive one when tired.

**yorma**[1] Act of tiring.

**yorma**[2] Interpretation of a dream *or* an omen.

**yormak**[1] *va.* Tire, fatigue. **ağız ~,** to talk *or* plead in vain: **çene ~,** to chatter, talk nonsense.

**yormak**[2] *va.* Interpret a dream *or* omen; presage; attribute. **hayra ~,** to interpret favourably, to regard as auspicious: **üstüne ~,** to take *a remark etc.* as directed against oneself.

**yortu** Christian feast.

**yorucu** Fatiguing, wearisome.

**yorulmak**[1] *vn. pass. of* **yormak**[1]**.** Be tired; tire oneself out in vain.

**yorulmak**[2] *vn. pass. of* **yormak**[2]**.** Be interpreted *etc.*

**yorum** Interpretation of a dream *or* omen; a counting as auspicious. **~cu,**

commentator. **~lamak,** to comment on, explain.
**yosma** Pretty; graceful; attractive. Pretty and attractive person; coquette. **~m,** my pet! **~lık,** gracefulness; charm.
**yosun** Moss. ~ **bağlamak,** to become covered with moss: **deniz ~u,** seaweed: **taş ~u,** lichen. **~lu,** mossy, covered with moss.
**yoz** Untrained, wild. **~laşmak,** to become wild; to lose qualities that have been acquired.
**yön** Direction; quarter; regard, relation. **gıda ~ünden,** as regards food: **ne ~den?,** in what respect? **~elme,** direction, course. **~eltmek,** to direct; turn towards. **~etim,** direction, administration, management: **sıkı ~,** martial law. **~lü,** directed *or* turned towards.
**yöre** Side; neighbourhood; suburb.
**yörük** *v.* **yürük.**
**yörünge** Course taken *by a thing*; orbit; trajectory.
**yudum** Mouthful; draught *of water etc.*
**yuf** *Interj. expressing disgust.* ~ **borusu çalmak,** to boo: ~ **sana!,** shame on you!
**yufka** Thin; weak; poor. Thin layer of dough; wafer; unleavened bread in thin sheets. ⌜**arkası ~**⌝, (*jokingly*) there's nothing more (to eat)!: **cebi ~,** penniless: **sırtı ~,** scantily clad. **~lık,** flour suitable for flake pastry; thinness; poverty; soft-heartedness.
**yuğurmak** *v.* **yoğurmak.**
**yuha** (ˈ.) *A shout of contempt or derision.* **~ya tutmak** *or* **···e ~ çekmek,** to hold up to derision; to hoot.
**yukar·da** On high; above; overhead. **~dan,** from above: ~ **almak,** to behave in a condescending manner.
**yukarı** High; upper; top. Above; upwards; on high; up. Upper *or* topmost part. **~ya yığmak,** to pile up high; to exaggerate.
**yulaf** Oats.
**yular** Halter. ⌜**başında ~ı eksik**⌝, he's an ass.
**yumak**[1] *va.* Wash.
**yumak**[2] Ball (of wool, string *etc.*). **~lamak,** to wind into a ball.
**yummak** *va.* Shut; close (eye, fist). ⌜**ağzını açıp gözünü ~**⌝, to lose one's temper and say bitter things without reflection: **···e göz ~,** to shut the eyes to, to wink at ...; **göz yummamak,** not to wink at *or* condone; not to sleep a wink.
**yumru** Round. Round thing; boil; tubercle (*bot.*). ~ **burun,** bottlenose: **boyun ~su,** goitre.
**yumruk** Fist; blow with the fist. **~larını sıkmak,** to clench the fists; to threaten with the fists. **~lamak,** to hit with the fist; pound with the fist, knead. **~luk,** *in* ⌜**bir ~ canı var**⌝, one blow and he's a dead man.
**yumru·lmak, ~lanmak** *vn.* Become swollen; get a boil. **~luk,** roundness; swelling (of a boil *etc.*).
**yumuk** Closed by swelling (eye); half-shut (eye); plump; soft.
**yumulmak** *vn.* (Of the eye) to become closed *by swelling.*
**yumurcak** Bubo; plague; pestilential child.
**yumurta** Egg; darning mushroom. ~ **akı,** the white of an egg: ⌜**~ kabuğunu beğenmemiş**⌝, *said of one who runs down those who brought him up or educated him*: ⌜**~ kapıya gelince**⌝, at the very last minute; *or* when the situation has become serious: ~ **ökçeli ayakkabı,** shoes with oval high heels, *worn by town roughs*: ~ **sarısı,** the yolk of an egg.
**yumurta·lık,** ovary; egg-cup. **~lamak,** to lay eggs; spawn; invent *a story*; blurt out, say *stg.* indiscreet.
**yumuşacık** Rather soft *or* mild.
**yumuşak** Soft; mild; yielding. Soft part *of anything.* ~ **başlı,** docile: **yüzü ~,** too kind to refuse. **~lık,** softness; mildness.
**yumuşamak** *vn.* Become soft; become pliant *or* yielding; calm down.
**yunak, yunaklık** Wash-house; place for washing clothes on a river bank.
**Yunan, ~istan** Greece. **~ca** (.ˈ.), ancient Greek *or* modern Greek *as spoken in Greece* (*modern Greek as spoken in Turkey is* rumca). **~lı,** Greek.
**yunmak** *vn.* Wash oneself.
**Yunus** Jonah. **~balığı, -nı,** porpoise; dolphin.
**yurak** Spoke *of a wheel.*
**yurd** Native country; home; habitation; estate. **yeri ~u yok (bellisiz),** homeless, vagabond. **~daş,** fellow countryman, compatriot. **~lanmak,** to settle *in a place.* **~luk,** estate, domain. **~sever,** patriotic.

**yusufçuk** A small kind of turtle-dove; dragonfly.
**yusyumru** (ˈ··) Quite round; very swollen.
**yusyuvarlak** (ˈ···) Quite round.
**yutkunmak** *vn.* Swallow one's spittle; gulp in suppressing one's emotions. **bir şey karşısında ~,** to resign oneself to doing without *stg.*
**yutmak** *va.* Swallow; gulp down; swallow *an insult*; endure *an injury* in silence; believe *a lie*, swallow *a tall story*; win at cards; appropriate wrongfully; fail to see *a joke*. **kazık (baston) yutmuş gibi,** as stiff as a poker: **riyaziyeyi yutmuş,** he knows mathematics from A to Z.
**yut·turmak** *va. caus. of* **yutmak.** Cause to swallow *etc.*, *esp.* cause to swallow *a lie*, swindle. **~turulmak,** *pass. of* **yutturmak,** to be taken in by *a lie etc.* **~ucu,** who swallows *or* devours; who wins *at cards*. **~ulmak,** *pass. of* **yutmak,** be swallowed *etc.*; lose *at cards*: ⸢**ekmek bile çiğnenmeden yutulmaz**⸣, no reward without effort. **~um,** *v.* **yudum.**
**yuva** Nest; home; socket; seating *of a valve*. **~ kurmak,** to build a nest; to set up a home: **···in ~sını* yapmak,** to give *s.o.* 'what for', to teach him a lesson: ⸢**garib kuşun ~sını Allah yapar**⸣, God cares for all. **~lamak,** to nest.
**yuvarlak** Round, spherical. Ball; marble; limber of a gun; cylinder of a printing-press. **~lık,** roundness.
**yuvar·lamak** *va.* Rotate; roll; roll up; swallow greedily; utter *a lie*; toss off (a glass of wine). **~lanmak** *vn.*, to revolve, turn round; roll; topple over; be tormented *by grief or worry*; lose one's job, be sacked; (*sl.*) die suddenly: **yuvarlanıp gitmek,** to worry along somehow. **~latmak,** to cause to rotate *etc.*; make round.
**yüce, yücel** High; exalted *person or position*. High place. **~lenmek, ~lmek,** to become high; rise. **~lik,** height; loftiness; exalted rank.
**yük, -kü** Load; burden; heavy task *or* responsibility; cargo; large cupboard for bedding. **~ünü almak,** to take all it can hold: **~ün altından kalkamamak,** to find one's duties too much for one: **~ arabası,** wagon: ⸢**~te hafif pahada ağır**⸣, small in bulk but valuable: **~ hayvanı,** pack-animal: **~ünü tutmak,** to become rich: **~ü üzerinden atmak,** to decline *or* shift a responsibility: **···e ~ vurmak,** to load *an animal*.
**yük·lemek,** to load; place a load on; throw the blame on; impute; attribute. **~lenmek** *pass. of* **yüklemek,** to be loaded *etc.*; take a load upon oneself: **···e ~,** to throw oneself upon, attack; keep on at *s.o.* **~letmek,** to place a load on; load (a ship *etc.*); impose *a duty, an expense etc.*; impute, attribute. **~lü,** loaded (with = *nom.*); pregnant; drunk; overburdened with work; in debt; rich: **kömür ~,** loaded with coal. **~lük,** large cupboard *or* closet for bedding.
**yüksek** High; loud (voice). High altitude. **~ten atmak,** to boast: **··· e ~ten bakmak,** to look down upon: **~lerde dolaşmak,** to aim high, to be ambitious: **~ten kopmak,** to set out with great pretensions: **~ sesle okumak,** to read aloud: **~ten uçmak,** to be ambitious *or* presumptuous. **~lik,** height, elevation.
**yüksel·mek** *vn.* Mount; rise; be promoted; (of a ship) to gain the open sea. **~tmek,** to raise; praise excessively.
**yüksük** Thimble. **~le ölçmek,** to dole out stingily: **~ otu,** foxglove.
**yüksünmek** *va. & vn.* Regard as burdensome; grudge giving *or* fulfilling a promise.
**yün** Wool. Woollen. **~lenmek,** to become woolly; to be frayed. **~lü,** woollen.
**yürek** Heart; stomach; centre of the affections; courage, boldness. [*In many idioms* **yürek** *is used instead of* **iç** *or* **gönül**; *if the phrase cannot be found under* **yürek** *consult* **iç** *and* **gönül**]. **~ler acısı,** a heartbreaking event *or* condition: **yüreği* atmak,** for one's heart to palpitate from emotion: **yüreği* (içi) bayılmak,** to be very hungry: **~ çarpıntısı,** palpitation of the heart; misgivings: **yüreği dar,** impatient: **~ dayanmaz,** unbearable; heartbreaking: **yüreğine* inmek,** to be struck with great fear, agitation *or* shame: **yüreği* kabarmak,** to be nauseated; to suffer trouble *or* pain: **yüreğini* kaldırmak,** thoroughly to upset *or* excite *s.o.*: **yüreği* kalkmak,** to be alarmed; to be very upset **yüreği katı,** obstinate, obdurate

**yüreği* katılmak,** to suffer greatly: **yüreği* oynamak,** to have a fluttering of the heart; to have misgivings: **yüreği pek,** stout-hearted: **yüreği* tükenmek** *or* **~ tüketmek,** to wear oneself out *in trying to explain stg.*: **~ vermek,** to hearten, to give courage: **yüreği* yanmak,** to be grieved; to feel pity; to meet with disaster: ... **···se yüreğim yanmaz,** it wouldn't have been so bad if ...: **yüreği yufka,** easily moved; compassionate.

**yürek·lenmek,** to take heart, be emboldened. **~li,** stout-hearted, plucky; -hearted.

**yürük** Fast, fleet. Nomad.

**yürüme** *verb. n. of* **yürümek. kan ~,** congestion of blood.

**yürümek** *vn.* Walk; advance, make progress; hurry along; (*sl.*) die; come into force, have effect. ⌜**Allah yürü ya kulum demiş**⌝, 'Allah said "advance, my slave!"', *said, rather enviously, of one who has made rapid progress*: ... **modası aldı yürüdü,** the fashion of ... has grown: **···in üstüne ~,** to 'go for' *s.o.* (physically): **üzerine ~,** to march against, to attack.

**yürünmek** *vn. impers. of* **yürümek.** Be walked. **on saat yüründü,** a march of ten hours was made.

**yürürlük** A being in force; validity.

**yürütmek** *va. caus. of* **yürümek.** Cause to walk *etc.*; put forward (an idea, a proposal *etc.*); (*sl.*) 'walk off with', pilfer; (*sl.*) 'sack', dismiss; put into force. **faiz ~,** to reckon up and pay interest.

**yürüyüş** Gait; march; assault. **~ kapısı,** wicket-gate: **mevzun ~,** marching in step.

**yüz[1]** Face; surface; the face *or* right side (of cloth *etc.*); the outer covering of a thing; motive, cause; boldness; effrontery. **~ünde,** on, on the surface of: **~ünden,** for the sake of, on account of: **~ akı,** *v.* **yüzakı**: **~ ~e bakamam,** I can't look him in the face (*because of some past injury I have done him*): **~üne bakılır** *or* **~üne bakılacak gibi,** good-looking: **~üne bakmağa kıyılmaz,** one can't help looking at her (him), very beautiful: **~üne bakılmaz,** ugly, horrible: **~ bulmak,** to be emboldened; to become presumptuous; to be spoilt *by kind treatment*: **~ünden* çekmek,** to suffer at the hands of ...: **···den ~ çevirmek,** to turn away from: **~e çıkmak,** to come to the surface; to be insolent: **~ etm.,** to hand *stg.* over; to face up *two boards that are to be joined together*: **~e gelen,** select, superior: **~ ~e gelmek,** to come face to face with, to meet; ⌜**bir gün olur ~ ~e gelirsin**⌝, don't make an enemy of anyone, you may meet him again some day: **~ geri,** *v.* **yüzgeri**: **~ göre,** just to please *or* flatter: **~ü görmek,** *in such phrases as*: **bir az rahat ~ü görmek,** to find a little peace; **para ~ü görmek,** to get a bit of money: **~ göz olm.,** to be too intimate *or* familiar: **~ünü gözünü açmak,** to talk to a child about unseemly things, *esp.* sex: **~ü* gülmek,** to be happy *or* delighted; **~ü güler,** with smiling face, happy: **~üne* gülmek,** to smile at, to be friendly towards (*with a hint of insincerity*): **her taraf insanın ~üne gülüyor,** everything is tidy and clean: **~üne* kan gelmek,** to recover one's health: **~ü kara,** one who in the past has done *stg.* to be ashamed of: **~ karası,** dishonour, disgrace: **~ kızartıcı,** shameful, disgusting: **~ünden okumak,** to read (**okumak** *alone often means 'to recite'*): **~ü* olmamak,** not to dare, not to have the face to; to be unable to refuse; to be unable to face: **~ü pek,** brazen-faced: **~ü soğuk,** dour: **~ suyu,** honour, self-respect; **babasının ~ü suyu hürmetine,** out of respect to his father; thanks to his father: **~ tutmak,** to begin; to take a turn towards: **ona bunu söylemeğe ~üm tutmaz,** I can't bring myself to tell him this: ⌜**~ ~den utanır**⌝, it is difficult to say such things to a person's face; it is hard to refuse personally: **~ üstü,** *v.* **yüzüstü**: **~ vermek,** to be indulgent *to,* to spoil; to give encouragement: **···e ~ vurmak,** to have recourse to: **~üne* vurmak,** to cast *stg.* in a person's teeth; to reproach *s.o.* with *stg.*: **~ yazısı,** decorations stuck on to the face of a village bride: **~ü* yok,** he has not the face, he dare not; one can't hold out against; **sıcağa ~üm yok,** I can't face the heat; **kumara ~ü yok,** he can't resist a gamble: **ne ~le geldin?,** how have you the face to come?: **ne ~le söyleceğim?,** how shall I bring myself to tell him?.

**yüz²** Hundred; one hundred. **~lerce**, by hundreds; hundreds of ...: **~de beş**, five per cent.

**yüzakı** (ˈ..), **-nı** Personal honour; *both words declined in such phrases as*: **işten yüzümün akile çıktım**, I came out of the affair with unblemished honour; **yüzünün akile buradan çekil git!**, clear out of here before I show you up!

**yüzbaşı** Captain (army). **~lık**, rank of captain.

**yüzde** Per cent. Percentage; rate per cent. **~lik**, percentage: **tesbit edilen kâr ~leri**, fixed percentages of profit.

**yüzdürme** *verb. n. of* **yüzdürmek. ~ kuvveti** (**kabiliyeti**), buoyancy. **~k**, *caus. of* **yüzmek**, to float (a sunken ship *etc.*).

**yüzer** A hundred each; a hundred at a time.

**yüzgec** Swimming; floating; knowing how to swim. Fin; float *of a seaplane*.

**yüzgeri** *In* **~ dönmek**, to face about, to turn around; to retreat.

**yüzgörümü** (ˈ...), **-nü** Present given by the bridegroom *on first seeing the face of the bride*. **yüzgörümlüğü**, the thing given as such a present.

**yüzlemek** *va.* Accuse *or* reproach *a person* to his face, *producing some positive proof in support of the accusation.*

**yüzlerce** (.ˈ.) In hundreds; in great numbers.

**yüzleş·mek** *vn.* Meet face to face; be confronted with one another. **~tirmek**, to bring face to face; confront.

**yüzlü¹** With *such and such* a face *or* surface. **güler ~**, smiling: **pek ~**, brazen-faced: **iki ~**, double-faced.

**yüzlü²** Having a hundred ... .

**yüzlük¹** Cover *or* protection for the face.

**yüzlük²** Costing, worth *or* weighing one hundred ... . A hundred para *or* piastre piece.

**yüzlülük** State of having *such and such* a face. **iki ~**, a being double-faced, duplicity.

**yüzmek¹** *vn.* Swim; float.

**yüzmek²** *va.* Flay; skin; despoil. ⌜**yüzdük yüzdük kuyruğuna geldik**⌝, the main job is done, the lesser one will easily be done; *or* 'a little patience, we're nearly there'.

**yüzsüz** Brazen-faced, shameless. **~lük**, effrontery.

**yüzük** Ring. **yüzüğü geri çevirmek**, to break off an engagement.

**yüzükoyun** (.ˈ..) Face downwards, lying on one's face, upside-down.

**yüzüncü** Hundredth.

**yüzüstü** (ˈ..) Face downwards; as things are. **~ bırakmak**, to leave things as they are *or* incomplete: **beni buralarda ~ bırakma!**, don't leave me to my fate in these regions!: **~ kapanmak**, to be prostrate with one's face on the ground.

**yüzyıl** Century.

**yüzyüze** (ˈ..) Face to face; *v.* **yüz.**

# Z

**zabıt, -btı** A holding firmly; a taking possession; conquest; forcible seizure; comprehension; a taking down in writing, a recording; minutes; legal proceedings. **~ kâtibi**, clerk of the court; clerk charged with the duty of recording the proceedings of any assembly: **~ tutmak**, to take legal proceedings.

**zabıta** (–.ˈ) Police. **~lık**, a matter for the police; a person whom only the police can deal with.

**zabıtname** (..–ˈ) Minutes *of a meeting etc.*; proceedings; protocol.

**zabit** (––ˈ), **-ti** Officer. Capable of command, who can keep discipline. **~ namzedi**, gentleman cadet. **~an**, officers. **~lik**, rank, status and duties of an officer; commission.

**zabt·etmek** (ˈ..), **-eder** *va.* Hold firmly; seize; take possession of; restrain, master; take down in writing. **~iye**, gendarmerie, gendarme. **~ı** *v.* **zabıt.**

**zac** Vitriol; sulphate of iron. **ak ~**, sulphate of zinc. **~yağı** (ˈ..), **-nı**, sulphuric acid.

**zade** (–ˈ) Born; noble. Son; (*in compounds*) son of ... . **~gân** (–.ˈ), nobles: noble: **~ sınıfı**, the nobility.

**zâf** *v.* **zaıf.**

**zafer** Success; victory.

**za'fı** *v.* **zaıf.**

**zafiyet** (–.ˈ), **-ti** Weakness; thinness.

**zağ** Keenness of edge *of a sword etc.*

**zağanos** Large owl *trained to hunt*; *v. also* çağanoz.
**zağar** Hound.
**zağlamak** *va.* Give a keen edge to; polish (sword, knife).
**zâhib** (–ꜝ) ···**e** ~ **olm.**, to incline *to an idea*; to surmise.
**zâhir** (–ꜝ) Apparent. Outside. Clearly, evidently. ~ **olm.**, to be *or* become evident. ~**en** (ꜝ–··), outwardly; to outward appearance. ~**î** (–·ꜝ), external, outward.
**zahire** (·–ꜝ) Store of grain *or* provisions; provisions. ~ **ambarı**, granary.
**zahmet, -ti** Trouble; difficulty; distress; fatigue. ~ **buyurdunuz**, you have put yourself to great trouble: ~ **çekmek**, to suffer trouble *or* fatigue: ~**ine değdi**, it was worth the trouble: ~ **etm.**, to give oneself trouble, to put oneself to inconvenience; ~ **etmeyiniz**, don't trouble yourself!: ~**e sokmak** *or* ~ **vermek**, to cause pain *or* trouble.
**zahmet·li**, troublesome; difficult. ~**siz**, free from trouble; easy. ~**sizce** (··ꜝ·), easily, without trouble.
**zaıf, -a'fı** Weakness.
**zaid** (–ꜝ) Redundant, superfluous.
**zakkum** Oleander. Very bitter.
**zalim** (–ꜝ) Unjust; tyrannical; cruel. Tyrant. ~**lik**, cruelty, oppression, tyranny.
**Zaloğlu** *In* ~ **Rüstem,** *a hero in Persian legend*; *type of a very strong man.*
**zam, -mmı** Addition.
**zaman** (·ꜝ) Time; period. (*with participle*) When. ~**ında**, in his day; in its season; at the right time: ~**la**, in the course of time: ~ **adamı**, opportunist: ~**a uymak**, to conform to the requirements of the time: **aman** ~ **yok**, without respite: **bir** ~ *or* **bir** ~**lar**, at one time; formerly, once: **gel** ~ **git** ~, after a certain time; in due course: **hiçbir** ~, never: **ne** ~**dır**, how long ago it is that ...: **o** ~, then.
**zamane** (·–ꜝ), the age; the present time; fortune. ~ **adamı**, opportunist: ~ **çocukları**, the children of today.
**zambak** Lily.
**zamin** (–ꜝ) Who stands surety. ~ **olm.**, to be guarantee, to make oneself responsible for.
**zamir** (·ꜝ) Personal pronoun.
**zamk, -mkı** Gum. ~**lamak**, to gum. ~**lı**, gummed.
**zamm·etmek** (ꜝ··), **-eder** *va.* Add; increase. ~**ı**, *v.* **zam.**
**zampara** Womanizer, rake. ~**lık etm.**, to run after women.
**zan, -nnı** Opinion; surmise; suspicion. ~**nımca** *or* ~**nıma göre,** in my opinion: ... ~**nındayım**, I am of the opinion that ... .
**zanaat, -tı** Craft; handicraft.
**zangır** *In* ~ ~, trembling; with the teeth chattering; making a clanking *or* rattling noise. ~**damak**, to have the teeth chatter through fear; tremble; clank; rattle. ~**tı**, a clanking *or* rattling noise.
**zangoc** Verger of a church.
**zanka** (ꜝ·) Horse-drawn sleigh.
**zann·etmek** (ꜝ··), **-eder** *va.* Think; suppose. ~**ı**, *v.* **zan.**
**zaparta** (·ꜝ·) Broadside; severe scolding.
**zaptetmek** *etc. v.* **zabtetmek** *etc.*
**zar**[1] Membrane; film; thin skin *of an onion etc.* **kulak** ~**ı**, ear-drum: **beyin** ~**ı veremi**, cerebral meningitis.
**zar**[2] Dice. ~ **atmak**, to throw dice: ~ **tutmak**, to cheat in throwing dice.
**zâr** (–) Weeping bitterly; miserable; thin. **ahu** ~, bitter lamentation.
**zarafet** (·–ꜝ), **-ti** Elegance; grace; delicacy.
**zarar** Damage, injury; loss; harm ~ **etm.**, *or* **görmek**, to suffer harm ~**ına satmak**, to sell at a loss: ~ **vermek**, to cause harm *or* loss: ~**ı yok**, it doesn't matter!; never mind!: **ne** ~, what does it matter!
**zarar·dide** (··–ꜝ), who has suffered injury *or* loss. ~**lı**, harmful; who suffers harm: ~ **çıkmak**, to come out the loser. ~**sız**, harmless; innocent; safe, unhurt; not so bad; pretty good.
**zarf** Receptacle; envelope; cover; case; cupholder *in which a cup of hot coffee is placed*; adverb. **bir ay** ~**ında**, during a month; for a month; within a month.
**zarfçı** Crook; confidence trickster.
**zarı** *In* ~ ~ **ağlamak**, to weep bitterly.
**zarif** Elegant, graceful; delicate; witty. ~**lik**, elegance; delicacy.
**zarta** (ꜝ·) Fart.
**zartzurt** *Imitates loud and domineering words.* ~ **etm.**, to give orders *or*

to talk in a loud and blustering manner.

**zaruret, -ti** Need; want; necessity; poverty. ~ **halinde,** in case of necessity.

**zarurî** (·–˗) Necessary; unavoidable.

**zarzor** (˗·) Willy-nilly; barely.

**zat** (*), **-tı** Individual. ~**ı aliniz (alileri)** your exalted person, *polite for* you: ~ **işleri,** personnel department: ~**ı mesele,** the essence of the matter, the real point: ~**a mahsus,** personal, not transferable: **haddi ~ında,** in itself, in its essence.

**zaten** (˗·) In any case; as a matter of fact.

**zati** (–˗) *Vulg. for* **zaten.**

**zatî** (–˗) Essential; original; personal.

**zatürrie** Pneumonia.

**zavahir** (·–˗) *pl. of* **zâhir.** Outside; visible parts; appearance. ~**e kapılmak,** to be taken in by the outward appearance of a thing *or* man: ~**i kurtarmak,** to save appearances.

**zavallı** (˗··) Unlucky; miserable. **ah ~!,** poor chap!

**zaviye** (–·˗) Corner; cell *of a recluse*; angle.

**zayetmek** (˗··), **-eder** *va.* Lose.

**zayıf** Weak; thin; weakly; of little weight *or* authority. ~**lamak,** to become enfeebled; become thin. ~**lık,** weakness, debility; emaciation.

**zayi** (–˗), **-ii** Lost; destroyed. ~ **etm.,** to lose: ~ **olm.,** to be lost, to perish: ~**inden vermek,** to issue a duplicate of a lost document (*identity papers etc.*). ~**at** (–·˗), losses: ~ **vermek,** to suffer casualties: ~ **verdirmek,** to inflict losses.

**zayolmak** (˗··) *vn.* Be lost; perish.

**zeamet** (·–˗), **-ti** A large fief.

**zebani** (·–˗) Demon of hell; cruel monster.

**zebellâ** Huge; thick-set. Huge man.

**zeberced** Chrysolite; beryl; topaz. Pale bluish-green.

**zebil** *v.* **sebil.**

**zebun** (·˗) Weak, helpless; exhausted. **ihtiraslara ~ olanlar,** those who are powerless against their desires: **yanlış bir düşüncenin ~u,** the dupe of an erroneous idea. **zebun·küş,** who oppresses the weak *or* defenceless; cruel and cowardly.

**zebur** The Psalms of David.

**zecir, -cri** Restraint; violence; a compelling to labour unwillingly.

**zecr·en** (˗·) With violence; forcibly. ~**î** (·˗), violent; compulsory; forcible.

**-zede** *Pers. suffix.* Stricken by ... , *e.g.* **kazazede,** overtaken by disaster, shipwrecked.

**zedelemek** *va.* Damage by striking; maltreat; bruise.

**zefir**[1] Deep sigh; expiration.

**zefir**[2] Zephir (tissue and garment).

**zehab** (·˗) Belief; imagination.

**zehir, -hri** Poison; anything very bitter. **beyaz ~,** cocaine, heroin *etc.*: ⌜**içime ~ (zemberek) oldu**⌝, it spoilt my enjoyment. ~**lemek,** to poison. ~**li,** poisonous, venomous; poisoned (food).

**zehri** *v.* **zehir.**

**zekâ** (·˗) Intelligence.

**zekât** (·˗), **-tı** Alms *prescribed by Islam* (*one fortieth of income*); tax for the relief of the poor. ⌜**malın kadar ~ın olsun!**⌝, *said as a reproach to s.o. who gives very little.*

**Zekeriya** (··˗·) Zachariah. ~ **sofrası kurmak,** to prepare a dinner of various sorts of dried fruit *to which guests are invited and at which the host makes some solemn wish.*

**zeki** Intelligent.

**zelil** Low, base, contemptible.

**zelzele** Earthquake.

**zem, -mmi** Blame; censure.

**zeman** *v.* **zaman.**

**zemberek** Spring (of a watch *etc.*); any very bitter thing. **zehir ~,** bitter as poison; a very cantankerous person. ~**li,** fitted with a spring.

**zembil** Basket *woven of rushes or palm-leaves.* ⌜**gökten ~le inmemiş ya!**⌝, he's nothing out of the way; he's just the same as anybody else.

**zemherir, zemheri** Extreme cold; the depth of winter. ~ **zürafası,** one who wears very inadequate clothes in winter.

**zemin** (·˗) The earth, the world; ground; ground *of a design*; subject-matter *of a discourse*; meaning, sense. ~ **katı,** ground-floor: ~**ü zaman,** conditions of time and space.

**zemmetmek** (˗··), **-eder** *va.* Censure; denigrate, slander.

**zemzem** Name of the sacred well at Mecca. ⌜~ **kuyusuna işemek**⌝, to do stg. monstrous merely to acquire notoriety: **çoklarına nisbetle ~le yıkanmış gibidir,** he is a perfect paragon compared with most of them.

**zenaat** *v.* zanaat.
**zencefil** Ginger.
**zenci** (·–́) Negro. ~ **ticareti,** the slave-trade.
**zencifre** Vermilion; cinnabar.
**zencir** *v.* zincir.
**zendost** Fond of women; rake.
**Zengibar** Zanzibar.
**zengin** Rich. Wealthy man. ~**lik,** riches, wealth.
**zer** *v.* zeri.
**zerdali** (·–́) Wild apricot.
**zerdava** Beech marten.
**zerde** Dish of sweetened rice coloured with saffron *served at weddings.*
**zerdeçal, zerdeçav** Turmeric.
**Zerdüşt, -tü** Zoroaster. ~**i,** Zoroastrian; fire-worshipper.
**zer·'etmek** (́··), **-eder** *va.* Sow. ~**i** (·–́), **-r'i,** a sowing; sown seed; seed for sowing. ~**iyat** (··–́), **-tı,** sowings; crops; cultivation.
**zerk, -ki** Injection. ~**etmek** (́··), **-eder,** to inject, give an injection of.
**zerre** Atom; mote. ~ **kadar,** in the slightest degree: ~**nin** ~**si,** absolutely nil.
**zerzevat, -tı** Vegetables. ~**çı,** greengrocer.
**zevahir** *v.* zavahir.
**zeval** (·́), **-li** Noon; decline; decadence; adversity. ~ **bulmak,** (of a nation *etc.*) to decline: ⌜**elçiye** ~ **olmaz**⌝, an ambassador cannot be blamed for his mission.
**zeval·î** (·–́), reckoned from noon: ~ **saat ikide,** two p.m.; two by European time. ~**siz,** unfading; everlasting, permanent.
**zevat** (·́), **-tı** *pl. of* zat. Persons.
**zevc** Husband. ~**e,** wife.
**zevk, -kı, -ki** The sense of taste; taste, flavour; appreciation *of a thing;* good taste; enjoyment, pleasure. ···**i** ~**e almak,** to make fun of ...: ~**ine bak!,** enjoy yourself!: ~**ini* okşamak,** to please: ~**inde* olm.,** to be enjoying *or* amusing oneself: ~ **ve safa,** amusement, pleasures: ~**ı selim,** good taste; ~**ı selim sahibi,** a man of taste: ~**ine varmak,** to appreciate *stg.* ~**lenmek,** to amuse oneself: ···**le** ~, to mock at, to make fun of ... . ~**li,** pleasant; amusing. ~**siz,** tasteless; ugly; in bad taste. ~**sizlik,** bad taste.
**zevzek** Silly; giddy; talkative. ~**lenmek,** to behave in a silly manner; to say stupid things. ~**lik,** silly flighty behaviour; senseless chatter.
**zeybek** *A class of the population in the Izmir and Aydin districts; formerly* a kind of light infantryman. ~ **oyunu,** the Zeybek folk-dance.
**zeyç, -ci** Astronomical tables.
**zeyl, zeyil, -yli** Appendix; addendum, postscript.
**zeytin** Olive. ~**ci,** dealer in olives. ~**lik,** olive grove. ~**yağı, -nı,** olive-oil: ~ **gibi üste çıkmak,** to come off best; to get the better of an argument.
**zeytuni** (·–́) Olive-green.
**zıbarmak** *vn.* Become torpid from drink; (*contemptuously*) to go to bed, to sleep; to die.
**zıbın** Wadded jacket for a baby.
**zıcret, -ti** Distress, oppression (*med.*).
**zıd, -ddı** The contrary; the opposite; opposition; detestation. ···**in** ~**dına* basmak,** to do *stg.* to spite *s.o.*: ···**in** ~**ına* gitmek,** to act contrary to the wishes of ...; to oppose. ~**diyet, -ti,** opposition; repugnance, antipathy; contrast.
**zıh** Edging; border; fillet; moulding. ~**lamak,** to put a border *or* edging to.
**zıkkım** The food of the damned; any bitter *or* unpleasant food. ~**lanmak,** to stuff oneself with food (*only used in anger or contempt*).
**zılgıt** (*sl.*) Threat; scolding. ~ **vermek,** to scold: ~**ı yemek,** to get a scolding.
**zımba** Drill; punch. ~**lamak,** to drill, punch; (*sl.*) stab. ~**lı,** perforated; with a hole punched in it.
**zımbır·datmak** *va.* Twang; strum *on a stringed instrument.* ~**tı,** a twanging *or* strumming noise.
**zımn·en** (́·) By implication; tacitly. ~ **anlatmak,** to imply. ~**î** (·–́), implied, indirectly *or* tacitly understood. ~**ında,** with a view to; for the purpose of.
**zımpara** Emery. ~ **kâğıdı,** emery paper; sandpaper.
**zındık** Misbeliever; atheist.
**zıngır** *In* ~ ~, *imitates the noise of violent trembling.* ~**damak,** to tremble violently; rattle. ~**tı** A rattling *or* trembling noise.
**zınk** *Imitates the noise of a moving thing brought to an abrupt standstill.* **otomobil** ~ **diye durdu,** the motor-car came to an abrupt stop.

**zıp** Suddenly; pop! ~ **diye çıkmak,** to pop up all of a sudden: ~ ~ **sıçramak,** to jump about wildly. ~-**çıktı,** a foppish bounder; upstart.
**zıpır** Hare-brained; madcap.
**zıpka** (ˈ·) A kind of tight-fitting breeches.
**zıpkın** Fish-spear; harpoon.
**zıplamak** *vn.* Jump, skip *or* bounce about.
**zıppadak** (ˈ··) Suddenly, unexpectedly; with one bound.
**zıpzıp, -pı** Marble (plaything).
**zır zır** *Imitates a continuous and tiresome noise.*
**zırdeli** (ˈ··) Raving mad.
**zırh** Armour; crease; braid down the side of trousers. ~**lı,** armoured; armour-plated; braided (trousers): battleship. ~**sız,** unarmoured; without braid (trousers).
**zırıl** *In* ~ ~, in streams. ~**damak,** to keep up an incessant chatter *or* clatter. ~**tı,** continuous chatter *or* clatter; squabble; (*sl.*) dirty, silly, useless: **kaynana** ~**sı,** rattle: **kocakarı** ~**sı,** the cackle of old women.
**zırlak** Senselessly yelling; bawling.
**zırlamak** *vn.* Keep up a continuous noise; (*contemptuously*) weep.
**zırnık** Orpiment; yellow arsenic. ~ **bile alamazsın,** you won't get a farthing *out of him*: **adama** ~ **vermez,** not a red cent will he give.
**zırt zırt** Frequently; at unexpected *or* unsuitable times.
**zırtapoz** Crazy.
**zırva** Silly chatter. ⌜~ **tevil götürmez**⌝, it's no use trying to make sense out of foolish talk. ~**lamak,** to talk twaddle.
**zırzop, -pu** Silly ass.
**zıt** *v.* zıd.
**zıvana** (·ˈ·) Inner tube; tenon (mortise); mouthpiece of a cigarette. ~**dan çıkmak,** to be befuddled; to be in a rage: ~ **testeresi,** tenon-saw. ~**lı,** having a tube at the end (cigarette). ~**sız,** crazy.
**zıya** (·ˈ), **-aı** Loss.
**zibidi** Oddly dressed; eccentric, crazy; upstart.
**zifaf** (·ˈ) Entry of the bridegroom to the nuptial chamber. ~ **gecesi,** the first night of wedded life.
**zifir** Deposit in a pipe stem. Bitter; dark. ~**i,** pitch-black.
**zifos** Splash of mud. Useless, in vain. ~ **yemek,** to be bespattered with mud.
**zift, -ti** Pitch. Pitch-black; very bitter. ~ **yesin** *or* ~**in pekini yesin,** he may starve for all I care. ~**lemek,** to daub with pitch. ~**lenmek,** to be daubed with pitch; overeat; *used contemptuously for* **yemek,** eat, consume, squander; to put to one's own use (not too honestly). ~**li,** daubed with pitch.
**zihin, -hni** Mind; intelligence; memory. **zihnini* bozmak (bulandırmak),** to make one suspicious: **zihnim durdu,** my mind ceased to work, I couldn't take anything in: **zihni* karışmak,** to be confused: ···**in zihnini* karıştırmak,** to confuse *s.o.*: ···**e zihni* takılmak,** for one's attention to be caught *by some striking fact or knotty point*: ~**de tutmak,** to bear in mind: ~ **(zihnini*) yormak,** to think hard, to rack one's brains: **sarfı** ~ **etm.,** to apply the mind to *stg.*
**zihn·en** (ˈ·) Mentally; in one's mind. ~ **hesab etm.,** to reckon in one's head. ~**î** (·ˈ), mental; intellectual. ~**iyet, -ti,** mentality.
**zikir, -kri** Remembrance, recollection; mention; recitation of the attributes of God; dervish religious service.
**zikr·etmek** (ˈ··), **-eder** *va.* Mention; intone religious formulae *or* prayers. ~**i,** *v.* **zikir.** ~**icemil,** praise: '~**iniz geçti**', (*politely*) you were mentioned. ~**olunmak,** to be mentioned.
**zikzak** Zigzag. ~**vari** (··–ˈ), zigzagging.
**zil** Cymbals; bells on a tambourine; gong; bell. ~ **takınmak,** to make merry: **etekleri** ~ **çalıyor,** he is in transports: **karnım** ~ **çalıyor,** I feel peckish.
**zil·hicce** *The twelfth month of the Moslem year.* ~**kade** (·–ˈ), *the eleventh month of the Moslem year.*
**zillet, -ti** Abasement, degradation.
**zilli** With cymbals *or* bells; having a bell; badly behaved. ~ **bebek,** doll which, when squeezed, strikes cymbals: ~ **maşa,** jingle, jingling Johnny.
**zilzurna** (ˈ··) Blind drunk.
**zimmet, -ti** Duty; obligation; charge; debt; debit side *of an account*. ~**ine iki bin lira geçirmiş,** he has misappropriated (embezzled) 2,000 lira: ~**te kalmak,** to be owing.

**zina**[1] Bumble-bee.

**zina**[2] (·⊥) Adultery; fornication. **~kâr,** adulterous.

**zincifre** Vermilion; cinnabar.

**zincir, zencir** Chain; fetters; succession, series. **~ini koparmak,** (of a madman) to run amok: ···i **~e vurmak,** to put *s.o.* in chains: **gemi ~i,** anchor chain.

**zincir·leme,** in a continuous series; act of chaining; a proceeding continuously: **temsil ~ usulüne göre devam ediyor,** the show goes on continuously (as in a cinema): **perakendeciler birbirlerine ~ usulile mal satıyorlar,** the retailers sell goods to one another in succession, *each time taking the legally allowed profit and selling to the public at a high price, thus evading the anti-profiteering laws.* **~lemek,** to chain; to connect in a series. **~li,** provided with a chain; chained; in a continuous manner: **~ han,** old-fashioned inn. **~lik,** chain locker (*naut.*).

**zindan** (·ˈ) Dungeon; dark place. Very dark. ⌈**dünya başına* ~ olm.**⌉, for the world to become a place of darkness, *i.e.* suddenly to feel very depressed. **~cı,** warden of a dungeon; jailer. **~delen,** medium-sized bonito (**palamut**).

**zinde** Alive; active, energetic. **~gi** (··⊥), **~lik,** life; animation, activity.

**zinhar** Beware!; take care!; by no means.

**zir** (–) Lowest *or* under part; top string of a lute; (*with* **izafat**) under. ···**in ~i idaresinde,** under the direction of … .

**zira** (⊥·) Because.

**zirâ** (·⊥), **-âı** Cubit.

**zira·at** (·–ˈ), **-ti** Agriculture. **~î** (·–⊥), agricultural.

**zirde** Underneath, below.

**ziruh** (–ˈ) Alive. Animate object.

**zirüzeber** (⊥···) Upside-down.

**zirve** Summit; peak.

**zivoma** (·ˈ·) Carpenter's square.

**ziya** (·⊥) Light. **~dar,** luminous; well lighted (room *etc.*).

**ziyade** (·–ˈ) Increase; more; surplus; excess. More; much; too much; excessive, superfluous. Too; very. ···**dan ~,** more than …: **~sile,** to a great degree; largely: ⌈**Allah ~ etsin!**⌉, *a form of thanks for a meal*: **hadden ~,** beyond measure: **pek ~,** extremely, excessively. **~ce,** rather more, somewhat. **~leşmek,** to increase (*vn.*).

**ziyafet** (·–ˈ), **-ti** Feast, banquet; dinner-party. **~ çekmek,** to give a feast.

**ziyan** (·ˈ) Loss; damage. **~ına,** at a loss: **~ çekmek (görmek),** to suffer loss *or* damage, to suffer prejudice: **~ sebil olm.,** to be wasted: **~ı yok!,** no matter!

**ziyan·cı, ~kâr,** prejudicial, injurious. **~lı,** injured; who suffers loss: **bu işten ben ~ çıktım,** it is I who came out the loser in this business. **~sız,** harmless; pretty good.

**ziyaret** (·–ˈ), **-ti** Visit; pilgrimage. **~ etm.,** to pay a visit; to perform a pilgrimage: ⌈**hem ~ hem ticaret**⌉, to combine a visit with business: **yarın ~inize geleceğim,** I will pay you a visit tomorrow. **~çi,** visitor; pilgrim. **~gâh,** a much visited place; place to which a pilgrimage is made.

**ziynet, -ti** Ornament; decoration. **~lemek,** to adorn, embellish. **~li,** ornamented; embellished.

**zoka** (ˈ·) Artificial bait; spinner. **~yı yutmak,** to be duped.

**zom** (*sl.*) Drunk.

**zonklamak** *vn.* Throb with pain.

**zor** Strength; violence; difficulty; compulsion. Difficult, hard; fatiguing; forced (marriage). With difficulty; only just. **~la,** by force; with difficulty: **~ belâ,** by great efforts, after great trouble: **~a gelmek,** *or* **~u altında kalmak,** to be forced *or* constrained: ⌈**~la güzellik olmaz**⌉, no good can be achieved by force: **~u* ne?** *or* **ne ~u*?,** what's the matter with him; what does he want?; why should he do it (he is not obliged to)?: **~un ne?,** what's the trouble?, what's the matter with you?; **~ları ne imiş?,** what did they want?, *etc.*: … **~unda olm. (kalmak),** to be obliged to …: **aklından ~u var,** he's off his head.

**zoraki** Forced; involuntary; under compulsion; by force. **~lik,** a forced thing.

**zorba** Who uses force; rebel; bully. Violent; brutal. **~lık,** the use of force; violence; bullying.

**zor·lamak** *va.* Force; exert one's strength; handle roughly; misuse; urge strongly. **kendini ~,** to force oneself; to exert oneself. ~**lan,** *v.* **zorla** (*under* **zor**). ~**lanmak,** to force oneself; be forced; be roughly handled. ~**laşmak,** to grow difficult, become harder. ~**lu,** strong; violent; powerful, influential. ~**luk,** difficulty; arduousness. ~**zoruna** (ˋ···), with great difficulty.

**zuafa** (··ˊ) *pl. of* **zayıf.** Weak ones.

**zuhur** (·ˋ) ~ **etm.,** to appear; come to pass; come into existence: ~**a gelmek,** to happen: ~**a getirmek,** to cause to happen, to bring to pass: **sahib** ~, a man who rises from obscurity to power.

**zuhur·at** (·–ˋ), **-tı,** sudden occurrences; chance events; unexpected events; the turn of events; unexpected expenses: ~**a tâbi olm.,** to depend on events. ~**i** (·–ˋ), a clown *in the old Turkish theatre*: ~**ye çıkar gibi,** dressed in a ludicrous manner: ~ **kolu,** a band of clowns.

**zulmet, -ti** Darkness.

**zulmetmek** (ˋ··), **-eder** *va.* (*with dat.*) Do a wrong to; treat unjustly *or* cruelly.

**zulumba** (·ˋ·), **zulumpad** Zedoary (*aromatic root*).

**zulüm, -lmü** Wrong; oppression; cruelty. ~**kâr,** oppressive; tyrannical; cruel.

**zurna** A kind of shrill pipe *usually accompanied by a drum*; the Saury Pike (*Scombresox saurus*). **çatlak** ~, a garrulous man: **davul** ~ **ile,** publicly; ostentatiously. ~**cı,** ~**zen,** player of the **zurna.**

**zücaciye** (·–·ˋ) Glassware.

**züğürd** Destitute; bankrupt; 'stony-broke'. ~ **tesellisi,** cold comfort: **akıl züğürdü,** of poor intelligence: **çehre züğürdü,** lacking in good looks: **terbiye züğürdü,** uneducated. ~**lük,** indigence; bankruptcy; a being 'stony-broke'. ~**lemek,** to become destitute; go bankrupt.

**Zühal, -li** Saturn.

**zühd** Pious asceticism.

**zühre** The planet Venus. ~**vi** (··ˊ), venereal.

**zühul** (·ˋ), **-lü** Error; omission; forgetfulness.

**zül, -llü** Degradation; humiliation.

**zülfikar** The two-bladed *or* cleft sword of Ali.

**zülfüyar** (The beloved one's curl), *only in* ~**a dokunmak,** to touch a tender spot; to 'put one's foot in it'.

**zülûbiye, zülbiye** A cake made with honey and almonds.

**zülüf, -lfü** Side-lock of hair; love-lock; tassel. ~**lü,** having love-locks; wearing a tasselled cap: ~ **baltacı,** *formerly a class of Palace guards.*

**zümre** Party; body; set of people; group; class. ~ **dersi,** a group of studies embracing different subjects: ~**i düveliye,** a group of states *or* Powers: **halk** ~**si,** the common people. ~**vi** (··ˊ), belonging to *such and such* a class *or* group.

**zümrüd** Emerald; emerald green. ~**i** (··ˊ), ~**in,** emerald green. ~**üanka,** a fabulous bird *said to inhabit the Caucasus*; phoenix; a will-o'-the-wisp: ⌜~ **gibi ismi var cismi yok**⌝, 'like the phoenix it has a name but no body', *said of stg. that does not really exist.*

**zünnar** Rope girdle *formerly worn by Christians in Turkey*; (*med.*) shingles.

**zünüb, -nbü** Sins.

**züppe** Fop; coxcomb; affected person. Affected; snobbish. ~ **münevver,** one who affectedly pretends to be an intellectual. ~**lik,** foppishness; affectation; snobbery.

**zürafa** (·–ˊ) Giraffe.

**zürefa** (··ˊ) *pl. of* **zarif.** Witty people.

**zürra** (·ˊ), **-aı** Cultivators, farmers.

**zürriyet, -ti** Issue, progeny; descendants. ~**i kesildi,** his family has died out for want of heirs.

**züvvar** *pl. of* **zair.** Visitors; pilgrims.

**züyuf** (·ˊ) Base money; spurious coins.

# ENGLISH–TURKISH

# ENGLISH-TURKISH DICTIONARY



**a** [êy], A harfi. **not to know A from B,** kara cahil olmak.
**a** [ə ; êy]. *Sesli harfle başlıyan bir kelimeden evvel* an *olur. Gayrı muayyen harfi tarif.* Bir. **~ man and ~ woman,** bir erkekle bir kadın. (*Umumileştirici mânada*): **~ woman takes life too seriously,** kadınlar hayatı fazla ciddiye alırlar. (*Tevziî mânada*): **apples at fivepence a pound,** libresi beş peni'ye elma: **I said *a* [êy] potato,** bir tane patates dedim (yani iki, üç değil).
**a–** [ə–]. *Bu harfile başlıyan bazı kelimeler için, onların köklerine bakınız.*
**A.A** (*kıs.*) **anti-aircraft,** Uçaksavar.
**aback** [əˈbak]. **to be taken ~,** şaşalamak.
**abaft** [əˈbâft]. (*den.*) Geminin kıç tarafında.
**abandon** [ˈəbandən] *n.* Serbestlik, kendini bırakma. *v.* Terketmek, bırakmak, vazgeçmek. **to ~ oneself to ...,** ···e kapılmak. **~ment,** Terk, bırak(ıl)ma.
**abase** [əˈbêys]. Alçaltmak, tezlil etm. **~ment,** alçaltma, tezlil.
**abash** [əˈbaş]. Utandırmak.
**abate** [əˈbêyt]. İn(dir)mek, azal(t)-mak. **~ment,** azalma; tenzilât.
**abb·ess** [ˈabes]. Baş rahibe. **~ey** [ˈabi]. Manastır. **~ot** [ˈabot]. Başkeşiş.
**abbreviat·e** [əˈbrîviêyt]. İhtisar etm.; kısaltmak. **~ion,** [–ˈêyşn], ihtisar; kısaltma.
**abc** [ˌêyˌbîˈsî] Alfabe; harf sırasile tertib edilmiş rehber vs.; evveliyat.
**abdicat·e** [ˈabdikêyt]. Saltanatı, tahtı terketmek; (hakkından) vazgeçmek; çekilmek. **~ion,** [–ˈkêyşn], tahttan çekilme; hakkından vazgeçme.
**abdomen** [abˈdoumen]. Karın, batın.
**abduct** [abˈdʌkt]. (Birisini) kaçırmak. **~ion,** kaçırma. **~or,** kaçıran.
**abeam** [əˈbîm]. Bordada.
**abed** [əˈbed]. Yatakta.
**aberration** [abəˈrêyşn]. İnhiraf; anormallik. **in a fit of ~,** dalgınlıkla: **mental ~,** delilik.
**abet** [əˈbet]. **to aid and ~,** suç ortağı olmak. **~tor,** suç ortaği.
**abeyance** [əˈbêyəns]. **in ~,** cari değil, muteber olmıyan.
**abhor** [əbˈhô(r)]. Nefret etmek. **~-rence** [–ˈhorəns], nefret, tiksinme. **~rent,** nefret verici, tiksindirici.
**abide** [əˈbâyd] (**abode** [əˈboud]). Kalmak. **to ~ by one's word,** sözünde durmak: **I can't ~ him,** ona tahammül edemem.
**abili·ty** [əˈbiləti]. İktidar, ehliyet; kanunî salâhiyet: *pl.* **~ties,** meziyetler. **to the best of one's ~ty,** yapabildiği kadar.
**abject** [ˈabcekt]. Sefil, düşkün.
**abjure** [abˈcuə(r)]. Bir kanaat, yemin ederek vazgeçmek; bir yere bir daha dönmiyeceğine yemin etmek.
**ablative** [ˈablətiv]. Mefulünanh.
**ablaze** [əˈblêyz]. Tutuşmuş, alev almış.
**able** [ˈêybl]. Muktedir, yapabilen, meharetli, elverişli. **to be ~ to,** *fiillerde iktidarî mâna ifade eder, mes.* **to be ~ to do,** yapabilmek: **an ~ piece of work,** usta işi, mükemmel eser. **~-bodied,** sağlam, dinc: **~ seaman,** bahriye onbaşışı.
**ablution** [əˈbluşn]. Abdest. **~s,** alelâde yıkanma.
**abnorm·al** [abˈnôml]. Anormal; istisnaî; tabiate aykırı; fevkelâde. **~-(al)ity,** [–ˈm(al)əti], anormallik, tabiate aykırılık.
**aboard** [əˈbôd]. Gemide; gemiye. **all ~ !,** herkes (gemiye vs.) girsin!.
**abode**[1] [əˈboud] *n.* Oturulan yer, ikametgâh. **~**[2] *p.p. v.* **abide.**
**aboli·sh** [əˈboliş]. İlga etm.; lâğvetmek; kaldırmak. **~tion** [–ˈlişn]. İlga, lâğv; kaldırılma. **~tionist,** bir şeyin ilgası tarafdarı.
**abominable** [əˈbominəbl]. İğrenc, nefret verici; berbad.
**abominat·e** [əˈbominêyt]. Nefret etm., iğrenmek. **~ion** [–ˈneyşn], nefret, iğrenme: **that ~ of a ...,** o mel'un....
**aborigin·al** [ˌabəˈricinl]. Yerli, aslî. **~es** [–îz], asıl yerliler.
**abort** [əˈbôt]. Çocuk düşürmek. **~ion** [əˈbôşn], çocuk düşürme; pek çirkin, biçimsiz şey. **~ive,** vakitsiz ve ölü doğmuş; akım kalmış.

**abound** [əˈbaund]. Bol olm.; mebzulen bulunmak. **the forest ~s in tigers,** ormanda kaplan boldur.
**about** [əˈbaut]. Etrafında, civarda; aşağı yukarı, hemen hemen; hakkında, dair; üzere. *Bir fiille kullanıldığı zaman* sağa sola *ve* etrafa *mânalarını ifade eder.* **he is ~ again,** kalktı, iyileşti: **there is a lot of influenza ~,** ortada pek çok grip var: **~ to do stg.,** bir şeyi yapmak üzere: **to go ~,** dolaşmak; geriye dönmek: **he doesn't go ~ much,** pek ortaya çıkmıyor: **~ here,** bu taraflarda: **how [what] ~ going to the cinema?,** sinemaya gidelim mi? (ne derseniz?): **to know what one is ~,** işini bilmek; ne yaptığını bilmek: **I haven't a penny ~ me,** üstümde on para yok: **to move ~,** dolaşıp durmak: **to order s.o. ~,** birisine emir vermek: **you must do something ~ it!,** bunun bir çaresini bulmalısınız!: **there is something ~ him I don't like,** her nedense bu adamdan hoşlanamıyorum: **there's something wrong ~ it,** bunun bozuk bir tarafı var: **it's ~ time to go,** artık gitmeliyiz: **it's ~ time the post came,** (i) posta neredeyse gelecek; (ii) posta nerede kaldı?: **~ turn!,** geriye dön!: **to walk ~ the streets,** sokaklarda gezmek [dolaşmak]: **what are you ~?,** neler yapıyorsunuz (bakalım)? **~** *v.* **to ~ ship,** geriye dönmek.
**above** [əˈbʌv]. Yukarı, üst, üstün; fazla, fevkinde; yukarıda, üstünde; gökte, göklerde. **~ all,** her şeyden evvel, bilhassa: **to be ~ (all) suspicion,** (her türlü) şübheden azade olm.: **~ oneself,** haddini bilmez; coşkun: **I am ~ doing that,** bunu yapmağa tenezzül etmem: **over and ~ that,** fazla olarak. **~-board,** açıkça, dürüst. **~-mentioned,** yukarıda zikredilen; ismi geçen.
**abras·ion** [əˈbrêyjn]. Aşındırma; sıyrık. **~ive,** [-siv] aşındırıcı(madde).
**abreast** [əˈbrest]. Yanyana; geride kalmıyan. **to be ~ of the times,** zamana uygun olm.; zamanın adamı olm.: **to march two ~,** ikişer ikişer yürümek.
**abridge** [əˈbric]. Telhis etm., kısaltmak; tahdid etm. **~ment,** icmal, hulâsa, kısaltma.
**abroad** [əˈbrôd]. Yabancı memlekette; haricde; her tarafa. **there is a rumour ~ that,** etrafta dolaşan şayialara nazaran.
**abrogate** [ˈabrogêyt]. Feshetmek, ilga etm.
**abrupt** [əˈbrʌpt]. Anî; sert ve kısa; kesik; dik.
**abscess** [ˈabses]. Apse, çıban.
**abscond** [əbˈskond]. Sıvışmak; kaçmak.
**absen·ce** [ˈabsəns]. Yokluk; noksan. **~ of mind,** dalgınlık: **to be conspicuous by one's ~,** bulunmayışile göze çarpmak: **leave of ~,** izin: **sentenced in his ~,** gıyaben mahkûm. **~t**[1] [ˈabsnt] *a.* Yok; hazır bulunmıyan. **~t-minded,** dalgın. **~t**[2] [əbˈsent] *v.* **to~ oneself,** gelmemek. **~tee** [ˌabsnˈtî] (Yoklamada vs.) bulunmıyan kimse. **~tee landlord,** malikânesinde yaşamıyan arazi sahibi. **~teeism,** arazi sahibinin haricde yaşaması; devamsızlık.
**absolute** [ˈabsəlyût]. Mutlak, kat'î; tam, mükemmel; otoriter. **~ly,** mutlaka, kat'î olarak, tamamen.
**absol·ution** [absoˈlyûşn]. Günahların kilise tarafından affı. **~ve** [əbˈsolv]. Günahlarını affetmek; beraet ettirmek; serbest bırakmak.
**absor·b** [əbˈsôb]. Massetmek, emmek, içine çekmek; hızını kesmek; tamamen meşgul etmek. **to become ~ed in stg.,** bir şeye dalmak. **~bent,** emici, idrofil. **~bing,** kavrayıcı, meraklı. **~ption** [əbˈsôpşn]. Emilme, massedilme; zihnî meşguliyet, dalma.
**abst·ain** [əbˈstêyn]. Çekinmek, geri durmak. **~ainer,** içki içmiyen; **total ~,** içkiye tövbeli. **~emious** [əbˈstîmyəs]. (Bilhassa yemek ve içmekte) azla kanaat eden; kanaatkâr; mütevazı (sofra). **~ention** [əbˈstenşn]. Çekinme, geri durma; istinkâf. **~inence** [ˈabstinəns]. Zevkten geri durma; perhiz; imsâk. **~inent,** yemede içmede mümsik, perhizkâr.
**abstract**[1] [ˈabstrakt] *n.* Hülâsa; icmal. **~**[2] *a.* Mücerred; nazarî; muğlak. **~**[3] [əbˈstrakt] *vb.* Çıkarmak; ayırmak; tecrid etm.; hulâsa etm.; aşırmak. **~ed** [əbˈstraktid]. Dalgın. **~ion** [əbˈstrakşn]. Mücerred fikir: hayâl; dalgınlık; ayırma; tecrid; aşırma.
**abstruse** [əbˈstrûs]. Muğlak; derin.
**absurd** [əbˈsö(r)d] Gülünc; mânasız, saçma. **~ity,** gülünçlük, mânasızlık; saçmalık.
**abund·ance** [əˈbʌndəns]. Bolluk, bereket, zenginlik. **to live in ~,**

refah içinde yaşamak. **~ant,** bol, mebzul, bereketli; çok.
**abus·e** [ə'byûs] *n.* Suiistimal, kötüye kullanma; fena kullanma; küfür. *v.* [ə'byûz]. Suiistimal etm., kötüye kullanmak; küfretmek. **~ive** [əb'yûsiv]. Ağzı bozuk; yolsuz.
**abut** [ə'bʌt]. **~ on [against],** ···e bitişik olm.; dayanmak. **~ment,** mesned, dayanak; köprü ayağı.
**abys·mal** [ə'bizməl]. Dipsiz; sonsuz. **~s** [ə'bis]. Yerin dibi; uçurum.
**academic** [akə'demik]. Eflâtun felsefesine mensub; üniversiteye aid; âlimane; nazarî. **~al,** bir üniversite veya koleje aid. **~als,** üniversite elbisesi. **~ian** [ə'kade'mişn]. Akademi âzası.
**academy** [ə'kademi]. Muayyen bir fen veya san'atin tahsil edildiği yer; (İskoç) orta mekteb. **Royal Academy,** İngiliz güzel san'atler akademisi.
**accede** [ak'sîd]. Razı olm.; tahta çıkmak.
**accelerat·e** [ək'selərêyt]. Hızlan(dır)mak. **~or,** akseleratör; hızlandırıcı.
**accent** ['aksnt] *n.* Vurgu, aksan; şive, ağız. *v.* [ək'sent]. Vurgulamak, aksan koymak. **~uate** [ək'sentyuêyt]. Vurgulamak.
**accept** [ak'sept] Kabul etm., razı olmak. **~able,** kabul edilebilir; münasib. **~ance,** kabul; iyi karşılanma; (poliçe) bittavassut kabul. **~or,** bir poliçeyi kabul eden kimse.
**access** ['akses]. Giriş, girme; nöbet. **To have ~ to,** bir yere [birisinin yanına] girebilmek: **within easy ~ of,** ···e kolayca gidilebilir. **~ibility,** yaklaşılabilme, girilebilme, erişilebilme. **~ible,** erişilebilir; tesir edilebilir.
**accession** [ak'seşn]. Tahta çıkış; zam, ilâve; varış.
**accessory** [ak'sesəri]. Fer'î; teferrüat; yardımcı; (suçta) ortak.
**accidence** ['aksidəns]. Tasrifler bahsi; morfoloji.
**accident** ['aksidənt]. Kaza; tesadüf; ârıza. **~s will happen,** kazanın önüne geçilmez. **~al** ['dentl], tesadüfî, ârızî; kazaen.
**accla·im** [ə'klêym]. Alkışlamak; beğenmek. **~mation** [akla'mêyşn]. Alkış(larla kabul etme).
**acclimatize** [ə'klâymətâyz]. Yabancı iklime, muhite alıştırmak. **to get [become] ~d,** yeni iklime, muhite, alışmak.
**accommod·ate** [ə'komədêyt]. Uydurmak, telif etm.; yerleştirmek. **to ~ s.o. with ...,** birine ... vermek, tedarik etmek. **~ating,** müşkülât çıkarmıyan; uygun. **~ation,** yatacak veya kalacak yer; ödünç parası; uydurma; tatbik.
**accompan·iment** [ə'kʌmpənimnt]. Refakat; tetimmat; (*mus.*) sesle beraber çalınan parça. **~ist,** (*mus.*) refakat eden. **~y** [ə'kʌmpəni]. Arkadaşlık etm., refakat etm., beraber bulunmak.
**accomplice** [ə'komplis]. Suç ortağı.
**accomplish** [ə'kompliş]. Yapıp bitirmek; meydana getirmek; başarmak. **~ed,** başarılmış; hünerli; incelmiş. **~ment,** yapıp bitirme, meydana getirme; başarılmış eser; başarı, muvaffakiyet; marifet.
**accord** [ə'kôd] *n.* Muvafakat, razı olma; anlaşma; uygunluk. *v.* Uymak; (lûtfen) vermek. **of one's own ~,** kendiliğinden. **in ~ance with,** göre, mucibince. **~ing** [ə'kôdin(g)], **~ to,** ···e göre, ... nazaran. **~ingly,** bundan dolayı, gereğince.
**accost** [ə'kost]. Yanaşmak; yaklaşıp söz açmak; sarkıntılık etmek.
**account**[1] [ə'kaunt] *n.* Hesab; ehemmiyet, itibar; rapor, hikâye, anlatış. **by all ~s,** herkesin söylediğine göre: **to call s.o. to ~,** birisinden hesab sormak: **to go to one's (last) ~,** ölmek: **to keep the ~s,** hesab tutmak: **of no ~,** ehemmiyetsiz: **on ~ of,** ···den dolayı: **on every ~,** her bakımdan: **on his own ~,** kendiliğinden, kendi başına: **on no ~ [not on any ~],** hiç bir suretle, kat'iyen: **to pay on ~,** mahsuben ödemek: **to take into ~,** göz önünde tutmak: **to turn [put] to ~,** bir şeyden istifade temin etmek. **~**[2] *v.* Addetmek, tutmak. **to be ~ed of,** sayılmak: **to ~ for,** hesab vermek; mesul tutulmak; izah etm.: **there is no ~ing for tastes,** herkesin zevkine karışılmaz; bu zevk meselesidir. **~able,** mes'ul; izah edilebilir. **~ancy** [ə'kauntənsi]. Defter tutma, muhasebecilik. **~ant,** muhasib, muhasebeci.
**accoutrement** [ə'kûtrimnt], Askerî eşya.
**accredit** [a'kredit]. İtimadnameli elçi göndermek. **~ed,** resmen tanınmış (şahıs).
**accretion** [a'krîşn]. İlâve; gelişerek birleşme.
**accrue** [ə'krû]. Hasıl olm.; eklenmek.

**accumul·ate** [əˈkyûmyulêyt]. Birik(tir)mek, topla(n)mak, art(tır)mak; teraküm etmek. **~ation,** toplama, biriktirme; yığın. **~ative,** biriktirici; artan, biriken. **~ator** [–tə(r)]. Akümülatör.

**accura·cy** [ˈakyurəsi]. Doğruluk, kat'ilik, sıhhat. **~te** [ˈakyurit]. Doğru, sahih, sadık; dikkatli.

**accus·ation** [akyuˈzêyşn]. İttiham; suçlandirma. **~ative** [əˈkyûzətiv]. Mef'ülünbih. **~e** [əˈkyûz]. Suçlandırmak, ittiham etmek.

**accustom** [əˈkʌstəm]. Alıştırmak. **~ed,** *a.* alışık, alışkan; mutad; her zamanki.

**ace** [êys]. (İskambil) birli; (zak) yek; çok düşman uçağı düşüren havacı. **within an ~ of,** kıl kaldı.

**acerbity** [əˈsö(r)biti]. Burukluk; (söz) acılık.

**ache** [êyk]. Ağrı(mak), acı(mak). **all ~s and pains,** inliye sıklıya: **it makes my heart ~,** içim parçalanıyor.

**achieve** [əˈçîv]. Yapıp bitirmek, meydana çıkarmak; başarmak. **~ment,** muvaffakiyet, başarı.

**Achilles** [aˈkilîz]. Aşil. **The heel of ~,** insanın en zayif noktası.

**acid** [ˈasid]. Ekşi; hamızî. **~ity** [əˈsiditi], ekşilik, asidlik.

**acknowledge** [akˈnolic]. İtiraf etm., kabul etm.; tasdik etm.; tanımak; mukabele etmek. **to ~ receipt of,** alındığını bildirmek: **to ~ books consulted,** müracaat edilen kitabları zikretmek. **~ment,** kabul, itiraf; tasdik; zikretme: **by way of ~,** karşılık olarak; teşekkür için.

**acme** [ˈakmi]. En yüksek yer, zirve, evc.

**acolyte** [ˈakolâyt]. Kilisede küçük memur.

**acorn** [ˈêykôn]. Meşe palamudu.

**acoustic** [əˈkûstik]. Sese aid. **~s,** akustik; ses tertibatı.

**acquaint** [əˈkwêynt]. Tanıtmak, bildirmek, haber vermek, malûmat vermek. **to become [make oneself] ~ed with . . .,** ···le tanışmak, ahbab olm.; öğrenmek, istinas peyda etmek. **~ance,** tanıma, malûmat; tanıdık, bildik: **he improves upon ~,** tanıdıkça daha iyidir: **to have a wide ~ with French,** fransızcadaki bilgisi derin olmak. **~anceship,** tanışıklık.

**acquiesce** [ˌakwiˈes]. Muvafakat etm.; kabul etm. **~nce,** muvafakat, kabul; teslimiyet.

**acqui·re** [akˈwâyr]. Elde etm.; edinmek; sahib olmak. **~rement,** edinme; elde edilmiş (malûmat vs.), bilgi, hüner; müktesebat. **~sition** [ˌakwiˈzişn]. Elde etme; iktisab; elde edilen şey; kazanc. **~sitive** [əkˈwizitiv], haris, dünya malına düşkün.

**acquit** [əˈkwit]. Beraet ettirmek; suçsuz çıkarmak; (borcunu) ödemek. **to ~ oneself,** davranmak, hareket etmek. **~tal,** beraet; ödeme.

**acre** [ˈêykə(r)]. İngiliz dönümü (0·4 hektar). **God's ~,** mezarlık. **~age** [ˈêykric], dönüm mikdarı; saha.

**acrid** [ˈakrid]. Kekre; (uslûb) acı, zehirli.

**acrimon·ious** [akriˈmọunyəs]. Haşin, hırçın, sert. **~y** [ˈakrimoni], hırçınlık, sertlik.

**acrobat** [ˈakrobat]. Cambaz. **~ic** [–ˈbatik], cambaza aid, cambazca. **~ics,** cambazlık.

**across** [əˈkros]. Ortasından, bir yandan bir yana, karşıdan karşıya; çapraz; genişliğine; öbür tarafa. **We shall soon be ~,** bir az sonra karşıda olacağız: **to come ~ s.o.,** birisile karşılaşmak: **to go ~ a bridge,** bir köprüden geçmek: **the river is a mile ~,** nehrin genişliği bir mildir.

**act**[1] [akt] *n.* Yapılan şey, iş, fiil; hareket; kanun; vesika; (piyes) perde. **~ of God,** tabiî sebeb: **to catch s.o. in the (very) ~,** birisini suçüstü yakalamak. **~**[2] *v.* Hareket etm., davranmak; vazife görmek; işlemek, tesir etm.; rol oynamak, temsil etmek. **to ~ for s.o.,** vekili olm.: **he is only ~ing,** (i) numara yapıyor; (ii) sade vekillik yapıyor: **to ~ up to one's principles,** prensiplerine uygun olarak yaşamak: **to ~ upon instructions,** talimata göre hareket etmek. **~ing** [ˈaktin(g)] *a.* Vekil. **~ for . . .,** ···in vazifesini gören. *n.* (Tiyatro) temsil, oyun. **~ion** [ˈakşn]. Fiil, hareket, iş; tesir; muharebe. **~ at law,** dava: **to bring an ~ against s.o.,** birisi aleyhine dava açmak: **to go into ~,** muharebeye girişmek: **killed in ~,** muharebede ölmüş: **a man of ~,** faal adam: **out of ~,** saf harici; (makine) işlemiyen: **to put out of ~,** işlemiyecek hale koymak: **to take ~,** harekete geçmek. **~ionable** [ˈakşnəbl]. Davaya sebeb olabilir. **~ivate** [ˈaktivêyt]. Faal bir hale getirmek. **~ive** [ˈaktiv]. Faal, hareketli, canlı; çevik; faaliyette, vazifede; tesirli. **Verb ~,** müteaddi fiil: ~ **service,** cebhe hizmeti. ~**ivity** [–ˈtiviti], faaliyet, hareketlilik, canlılık, çeviklik. ~**or** [ˈaktə(r)]. Aktör; rol oynayan. ~**ress** [ˈaktris]. Aktris. ~**ual** [ˈakçuəl]. Gerçek, hakikî, fiilî; elle tutulabilir; asıl. ~**ually,** gerçek, hakikaten; hattâ (çok garib, ama). ~**uality** [–ˈaliti], gerçek, hakikat, hakikî vaziyet. ~**uary** [ˈakçuəri]. İstatistikçi. ~**uate** [ˈakçuêyt]. İşletmek, tahrik etmek.

**acumen** [əˈkyûmən]. Zekâ keskinliği.

**acute** [əˈkyût]. Keskin, sivri; keskin zekâlı; şiddetli; dar (zaviye).

**A.D.** [ˈêyˈdî]. **Anno Domini,** milâddan sonra.

**adage** [ˈadic]. Mesel, atasözü.

**adamant** [ˈadəmənt] *n.* Son derece sert bir şey. *a.* Eğilmez, müsamahasız. ~**ine** [–ˈmantâyn] elmas gibi sert; çok inadcı.

**adapt** [əˈdapt]. İntibak ettirmek, uygun bir hale getirmek; tadil etm.; iktibas etmek. ~**ability,** intibak kabiliyeti. ~**able,** uyabilir, intibak edilebilir. ~**ation,** uyma, intibak; iktibas.

**A.D.C.** [êyˈdîˈsî]. (*abb.*) **Aide-de-camp,** yaver.

**add** [ad]. Katmak, eklemek, ilâve etm., zammetmek, üstüne koymak. **to ~ to my work,** işim kâfi değilmiş gibi, üstelik bir de . . .: **to ~ up,** toplamak; baliğ olm., yekûn tutmak: **it all ~s up to this,** bunun neticesi . . . dir; hulâsa.

**addict**[1] [ˈadikt] *n.* (Afyon, vs.ye) düşkün, tiryaki, mübtelâ. ~[2] [əˈdikt] *v.* **to ~ oneself to** [**to be ~ed to**], ···e kendini vermek, düşkün olm. ~**ion,** düşkünlük, tiryakilik, ibtilâ.

**addition** [əˈdişn]. İlâve, zam; toplama. **in ~,** bundan başka, ilâve olarak. ~**al,** eklenen, ilâve, fazla.

**addled** [ˈadld]. Kokmuş (yumurta); sersem olmuş (baş). ~**-headed,** ~**-brain,** beyinsiz, kuş beyinli.

**address** [əˈdres] *n.* Adres; hitabe, nutuk; hal ve tavır; meharet. *v.* Hitab etm.; adres yazmak; nutuk söylemek. **form of ~,** hitab şekli, ünvan: **to ~ oneself to s.o.,** birisine hitab etm.: **to ~ oneself to a task,** bir işe girişmek, koyulmak. ~**ee** [–ˈsîy], kendisine mektub yazılan.

**adduce** [əˈdyûs]. Delil olarak ileri sürmek.

**adept** [ˈadept]. Üstad; usta, mahir.

**adequa·cy** [ˈadikwisi]. Yeterlik, kifayet; münasib olma. ~**te,** kâfî, elverişli; uygun.

**adhe·re** [ədˈhiə(r)]. Yapışmak; iltihak etm., bağlı kalmak; israr etmek. ~**rence,** yapışma, iltihak. ~**rent,** *a.* yapışık; *n.* tarafdar. ~**sion** [ədˈhîjn]. Yapışma, iltihak etme. ~**sive** [–siv], yapışkan.

**ad hoc** [ˈadˈhok]. (*Lât.*) Bilhassa bunun için.

**ad infinitum** [ˈad infiˈnâytum]. (*Lât.*) Ebediyen, sonsuz olarak.

**ad interim** [ˈad ˈintərim]. (*Lât.*) Muvakkaten.

**adjacent** [əˈcêysnt]. Bitişik, komşu, yakın.

**adjectiv·e** [ˈaciktiv]. Sıfat. ~**al** [–ˈtâyvl], sıfata aid.

**adjoin** [əˈcôyn]. Bitişik olmak.

**adjourn** [əˈcö(r)n]. Tehir etm.; başka yere gitmek. ~**ment,** tehir, tâlik.

**adjudge** [əˈcʌc]. Hüküm vermek, karar vermek; (mükâfat vs.) vermek.

**adjudicate** [əˈcûdikêyt]. Karar vermek, hüküm vermek; ihale etmek.

**adjure** [aˈcuə(r)]. İnkisarla tehdid ederek taleb etmek.

**adjust** [əˈcʌst]. Düzeltmek, tadil etm.; halletmek; ayar etm.; tanzim etmek. **to ~ oneself to,** ···e intibak etmek: ~**able,** tanzim edilebilir; ayarlanabilir. ~**ment,** tatbik etme; düzeltme; ayarlama; (bir ihtilâfı) bertaraf etme.

**adjutant** [ˈacutənt]. Emir subayı. ~**-general (A.D.G.),** ordunun bütün zat işlerine bakan yüksek rütbede subayı.

**ad lib.** [adˈlib]. (*Lât.*) Arzu olunduğu kadar.

**administ·er** [ədˈministə(r)]. İdare etm.; (ilac vs.) vermek; (yemin) verdirmek. ~**ration** [ədˌminisˈtrêyşn]. İdare; hükûmet; (ilac) verme; (yemin) verdirme. ~**rative** [ədˈministrətiv], idarî. ~**rator** [ədˈministrêytə(r)], müdür, idareci.

**admir·able** [ˈadmrəbl]. Takdire lâyık; hayran olmağa değer. ~**ation** [ˌadməˈrêyşn]. Hayranlık, takdir. **to be the ~ of everyone,** herkesin hayranlığını çekmek: **to be struck with ~,** hayran kalmak. ~**e** [ədˈmâyə(r)]. Hayran olm.; çok beğenmek. ~**er,** hayran olan kimse, meraklı; âşık.

**admiral** [ˈadmərəl]. Amiral. ~**ty,** Amirallik Dairesi (İngiliz Bahriye Nezareti): **First Lord of the ~,**

Bahriye Nazırı: **Court of ~,** deniz mahkemesi.

**admiss·ible** [əd'misəbl]. Kabul olunabilir. **~ion** [əd'mişn]. İtiraf; ikrar; kabul; giriş, dühûl; duhuliye.

**admit** [əd'mit]. İtiraf etm., ikrar etm.; kabul etm.; içeri almak; imkân vermek. **let it be ~ted that,** itiraf edelim ki. **~tedly,** herkesin itiraf edeceği gibi, muhakkak, vakıa. **~tance,** (bir yere) kabul; giriş; duhuliye: **no ~,** girilmez.

**admonish** [əd'moniş]. İhtar etm., tenbih etm., azarlamak.

**ad nauseam** ['ad 'nôsiəm]. (*Lât.*) Bıktırıncaya kadar.

**ado** [ə'dû]. Telâş, gürültü, patırdı. **much ~ about nothing,** küçük bir şeyi mesele yapmak.

**adolescen·t** [ado'lesənt]. 15 ile 20 yaş arasında olan; çocukça. **~ce,** genclik.

**adopt** [ə'dopt]. Benimsemek; kabul etm.; evlâd edinmek; seçmek. **~ed, ~ive child,** manevî evlâd. **~ion,** evlâd edinme; benimseme, kabul etme; seçme.

**ador·e** [ə'dôə(r)] Tapmak. **~able,** tapılacak kadar güzel, iyi. **~ation** [adə'rêyşn], tapma.

**adorn** [ə'dôn]. Süslemek, tezyin etmek. **~ment,** süs, ziynet.

**Adrianople** [ˌêydryə'noupl]. Edirne.

**adrift** [ə'drift]. Rüzgâra ve akıntıya tâbi. **to cut (a boat) ~,** palamarı çözmek: **to cut oneself ~ from,** ···le münasebeti kesmek: **to turn s.o. ~,** birisini ortada bırakmak.

**adroit** [ə'drôyt]. Meharetli, usta, becerikli.

**adult** ['adʌlt]. Büyük (çocuğun aksi olarak), büluğa varmış, kâhil.

**adulter·ate** [ə'dʌltərêyt]. İçine fena şeyler karıştırmak; kalitesini düşürmek, bozmak. **~ation,** tağşiş. **~er** [ə'dʌltərə(r)]. Zani. **~ess,** zaniye. **~ous,** zina yapan. **~y,** zina.

**ad valorem** ['ad va'lôrem]. (*Lât.*) Değere göre. **~ duty,** kıymet üzerinden resim.

**advance**[1] [əd'vâns] *n.* İlerleme: gelişme; terakki. **~ guard,** öncü, pişdar: **in ~,** önce(den); peşin: **to book in ~,** önceden yer tutmak: **to make an ~,** (i) avans vermek; (ii) ilerlemek: **to make ~s,** ilk adımı atmak; işvebazlıkla cezbetmek: **~**[2] *v.* İlerle(t)mek; terakki etm.; yürütmek; ileri sürmek; art(ır)mak; avans vermek. **~d,** ileri, yüksek. **~ment,** ilerleme, terakki; terfi.

**advantage** [əd'vântic] *n.* Menfaat, fayda, kâr, kazanç; üstünlük. *v.* Faydalı olm., bir fayda, üstünlük, kazanc temin etmek. **to show to ~,** en müsaid, en iyi, şekilde göstermek veya görünmek: **to turn out to s.o.'s ~,** birisinin işine yaramak. **~ous** [ˌadvən'têycəs], faydalı, kârlı; üstünlük temin eden.

**advent** ['advənt]. Gelme, baş gösterme; İsa'nın zuhuru. **the second ~,** hıristiyan inanışına göre İsa'nın dünyaya ikinci defa gelişi.

**adventur·e** [əd'vençə(r)] *n.* Macera, sergüzest; tehlikeli iş. *v.* Tehlikeye koymak, riske etmek. **~er,** sergüzestçi, macera meraklısı; dolandırıcı. **~ous,** macera arıyan; atılgan; tehlikeli.

**adverb** ['advö(r)b]. (*gram.*) Zarf. **~ial** [ad'vö(r)byəl], zarfa mensub: **~ phrase,** zarf gibi kullanılan tabir.

**advers·ary** ['advö(r)səri]. Muhalif, düşman. **~e** ['advö(r)s]. Zıd; karşı; müsaid olmıyan. **~ity** [əd'vö(r)siti]. Felâket, talihin ters gitmesi; düşkünlük; sıkıntı.

**advertise** ['advətâyz]. Ilân etm., reklâmını yapmak. **~ment** [əd'vö(r)tizmənt], ilân, reklâm.

**advi·ce** [əd'vâys]. Öğüt, nasihat, tavsiye. **~s,** haber, malûmat. **to take s.o.'s ~,** birisinin tavsiyesine göre hareket etm.: **to take medical ~,** doktora sormak: **~ note, letter of ~,** ihbarname. **~sable** [əd'vâyzebl]. Tavsiye edilir; uygun, makul, münasib. **~se** [əd'vâyz]. Tavsiye etm.; fikir vermek. **to ~ s.o. of stg.,** birisini bir şey hakkında ikaz etm., haber vermek: **you would be well ~d to obey him,** ona itaat etseniz daha iyi olur. **~ser,** öğüt veren, tavsiye eden; müşavir. **~sory,** istişari.

**advoca·cy** ['advəkasi]. Tarafını tutma, müdafaa. **~te** ['advokət] *n.* Bir şeye taraflı, müdafaasını yapan; avukat. **Lord ~,** (İskoçya'da) baş savcı. *v.* ['advokêyt] Tarafını tutmak, tavsiye etmek.

**advt.** (*abb.*) **advertisement.**

**adze** [adz]. Keser.

**Aegean** [î'ciən]. Eğe, Adalar Denizi.

**aegis** ['îcis]. Zeus'in kalkanı; himaye.

**aeon** ['îən]. Kâinatın yaşı; ebediyet.

**aerate** ['eərêyt]. Havalandırmak.

**aerial** [ˈe̯əriəl] *a.* Havaî. *n.* Telsiz anteni.
**aero-** [ˈe̯əro-]. Hava-. **~batics** [–ˈbatiks], (uçakla) havada cambazlık. **~drome** [–ˈdro̯um], hava meydanı. **~nautic(al),** uçuculuğa aid. **~nautics,** havacılık. **~plane** [ˈe̯əroplêyn]. Uçak, tayyare.
**aesthet·e** [ˈîsθît]. Estetikçi. **~ic** [îsˈθetik], iyi zevke uygun. **~ics,** estetik, güzellik ilmi.
**afar** [əˈfâ(r)]. Uzak. **~ off,** uzakta.
**affable** [ˈafəbl]. Nazik, lûtufkâr.
**affair** [əˈfe̯ə(r)]. İş, mesele. **~s,** iş. **~ of honour,** şeref meselesi: **love ~,** aşk macerası: **that is my ~,** bu mesele bana aid.
**affect**[1] [aˈfekt] *v.* Tesir etm.; işlemek; müteessir etmek. **~ed with a disease,** bir hastalığa mübtelâ. **~ing,** dokunaklı, tesirli, içli. **~**[2], Yalancıktan yapmak; gibi görünmek; hoşlanmak. **to ~ ignorance,** bilmemezlikten gelmek: **~ation** [–ˈtêyşn], yapmacık; sun'ilik; naz. **~ed,** yapmacıklı, sun'i.
**affection** [əˈfekşn]. Şefkat, muhabbet, sevgı; ârıza, hastalık. **~ate,** şefkatli, kalbden bağlı; muhabbetli: **yours ~ly,** (mektub sonunda) sevgilerle.
**affidavit** [ˈafiˈdêyvit]. Yeminli ve yazılı ifade.
**affiliat·e** [əˈfilyêyt]. **be ~ed to,** (bir cemiyet hakkında) başka bir cemiyete bağlanmak: **~ed firms,** birbirine bağlı firmalar. **~ion,** (i) başka bir cemiyete bağlanması; (ii) bir çocuğun nesebini isbat: **~ order,** bir çocuğun nesebini isbat eden mahkeme kararı.
**affinity** [aˈfiniti]. Yakınlık; benzeşme; cisimlerin birleşme temayülü; sıhriyet.
**affirm** [aˈfö(r)m]. İddia etm., tasdik etm. **~ation,** iddia, tasdik, teyid. **~ative,** müsbet; tasdik eden.
**affix** [aˈfiks]. Bağlamak; takmak; (mühür) basmak; (pul) yapıştırmak.
**afflict** [aˈflikt]. Istırab vermek; acı vermek. **to be ~ed by,** ···e duçar olm., mübtelâ olmak. **~ion,** ıstırab, derd, keder.
**affluen·t** [ˈaflu̯ənt] *a.* Zengin; refah içinde; bol. **~ce,** zenginlik, bolluk.
**afford** [aˈfôd]. Vermek, vesile olm.; (paraca vs.) gücü yetmek, muktedir olm., hali ve vaziyeti müsaid olmak. **to be unable to ~,** harcı olmamak: **can you ~ the time?,** vaktiniz var mı, müsaid mi?
**afforest** [aˈforest]. Ağaçlandırmak, ormanlaştırmak.
**affray** [əˈfrêy]. Arbede, kavga.
**affront** [əˈfrʌnt] *n.* (Alenî) hakaret, tahkir. *v.* **~,** hakaret etm.: **to suffer an ~,** hakarete uğramak.
**afield** [əˈfîld]. **to be ~,** iş başında olm.: **to go far [farther] ~,** çok (daha) uzağa gitmek.
**aflame** [əˈflêym]. Alev içinde, tutuşmuş.
**afloat** [əˈflo̯ut]. Su üzerinde; denizde. **to keep ~,** su üzerinde durmak: **to set a ship ~,** bir gemiyi yüzdürmek.
**afoot** [əˈfut]. Yaya. **there is something ~,** bir şeyler dönüyor.
**afore-** [əˈfô(r)]. Önceden; yukarıda. **~thought** [əˈfôˌθôt]. **with (of) malice ~,** taammüden.
**afraid** [əˈfrêyd]. Korkmuş. **to be ~ of,** ···den korkmak: **I am ~ so, that,** maalesef, yazık ki.
**afresh** [əˈfreş]. Yeniden, yeni baştan, tekrar.
**African** [ˈafrikən]. Afrika'ya aid; Afrikalı.
**aft** [âft]. Geminin kıç tarafında.
**after** [ˈâftə(r)]. Sonra, bundan sonra; bunun üzerine; sonradan; ertesi; arkasından, peşi sıra, peşinde; dununda; göre, nazaran; tarzında, üslûbunda. **~ all (is said and done),** sonunda, netice itibarile; ne de olsa. **~ you,** (önce) siz buyurunuz: **in ~ days,** ileride, gelecekte: **in ~ life,** yaşlandıkça; sonradan: **the day ~ tomorrow,** öbürgün: **it is ~ one o'clock,** saat biri geçiyor: **on and ~,** ···den itibaren: **time ~ time,** kırk defa: **for years ~,** bundan sonra senelerce: **what are you ~ ?,** ne arıyorsunuz?; maksadınız nedir? **~-care** (hastalık vs.den) sonraki bakım. **~-deck,** arka güverte. **~-effects,** neticeleri, tesirleri; serpentileri. **~-life, (i) the ~,** ahret; (ii) ömrün sonraki vakitleri. **~-taste,** ağızda bir şeyin asıl tadı gittikten sonra hissedilen tad. **~crop** [ˈâftəˈkrop]. İkinci mahsûl. **~glow** [ˈâftəglo̯u]. Batan güneşin son ışıkları. **~math** [ˈâftəˌmâθ]. Netice, âkıbet. **~noon** [ˌâftəˈnûn]. Öğleden sonra, ikindi üstü. **this ~,** bugün öğleden sonra. **~thought** [ˈâftəθôt]. Sonradan gelen fikir. **~wards** [ˈâftəwədz]. Sonra, sonradan.
**again** [əˈgen, əˈgêyn]. Yine, tekrar, yeniden, daha; diğer taraftan. **as much ~,** iki misli: **half as much ~,**

bir buçuk misli: **(every) now and ~,** ara sıra: **time and ~,** tekrar tekrar, kaç defa: **what's his name ~?,** ismi ne idi bakayım?

**against** [ə'genst, ə'gêynst]. Karşı, mukabil; aleyhte, muhalif; karşısında; rağmen. **to buy provisions ~ the winter,** kış için (kışlık) erzak almak: **to run up ~ s.o.,** birisile tesadüfen karşılaşmak.

**age** [êyc] *n.* Yaş; çağ; devir; uzun zaman. *v.* Yaşlanmak, ihtiyarlamak. **at his ~,** o yaşta: **to come of ~,** reşid olm.: **it will last for ~s,** çok zaman sürer [dayanır]: **I haven't seen him for ~s,** onu hanidir görmedim: **the Middle Ages,** ortaçağ: **of an ~,** aynı yaşta: **of an ~ to marry,** evlenecek yaşta: **over ~,** yaşını geçirmiş: **under ~,** şakird, küçük. **age-long,** asırlık. **~d** (i) ['êycid]. Yaşlı, yaşlanmış; **the ~,** yaşlılar, ihtiyarlar; (ii) [êycd]. Yaşında. **~ing,** ihtiyarlıyan. **~less,** ihtiyarlamaz; herdemtaze.

**agen·cy** ['êycənsi]. Vasıta, delâlet; vekillik, acentelik; daire. **news ~,** ajans. **~t** ['êycənt]. Vasıta, âmil; vekil; acente, ajan. **to be a free ~,** hareket serbestisine malik olmak.

**agenda** [a'cendə]. Ruzname, gündem.

**aggravat·e** ['agrəvêyt]. Daha fenalaştırmak; vahim bir hale getirmek; sabrını tüketmek, kızdırmak. **~ing circumstance,** şiddetlendirici sebeb [hal]. **~ion,** şiddetlendirme; kızdırma; hiddet; şiddetlendirici sebeb.

**aggregate** ['agrigit] *n.* Bütün; küme; toplanmış; kütle; (betonda) çakıllı kum, taşkırığı vs. **in the ~,** bir bütün olarak. *v.* Birleştirmek, toplamak; yekûn tutmak.

**aggress·ion** [ə'greşn]. Saldırma, tecavüz. **~ive,** tecavüzkâr, kavgacı. **~or,** saldıran, mütecaviz.

**aggrieve** [ə'grîv]. Keder vermek, mağdur etmek. **~d,** mağdur, mazlum.

**aghast** [ə'gâst]. Dehşet içinde.

**agil·e** ['acâyl]. Çevik, atik, faal. **~ity** [ə'ciliti], çeviklik, atiklik, faaliyet.

**agitat·e** ['acətêyt]. Sallamak; rahatsız etm., heyecan vermek; tahrik etm.; karıştırmak. **to ~ for stg.,** bir şeyi elde etmek için israrla uğraşmak. **~ion,** sallama; heyecan; tahrik; kaynaşma; israrla taleb etme. **~or,** ['acitêytə(r)], heyecana getiren; tahrikçi, körükçü.

**aglow** [ə'glou]. Parlıyan; ateş içinde.

**agnostic** [ag'nostik]. Allahın ve hakikatın bilinemiyeceğine inanan.

**ago** [ə'gou]. Önce, evvel. **a little while ~,** bir az evvel: **long ~,** çok eskiden: **so (as) long ~ as 1880,** daha 1880 de: **no longer ~ than last week,** daha geçen hafta.

**agog** [ə'gog]. **to be ~ for,** şiddetle ummak: **to be (all) ~,** telaş içinde olmak.

**agonize** ['agonâyz]. İşkence etm.; işkence görmek.

**agony** ['agəni]. Istırabdan kıvranma; şiddetli acı, ağrı. **~ column,** şahsî ilânlar sütunu.

**agrarian** [ə'greəriən]. Ziraate dair.

**agree** [ə'grî]. Uyuşmak; hak vermek; razı olm.; **~ to,** kabul etm.; ···e razı olm., tasdik etm.; **~ with,** ···e uygun olm., ···le bir fikirde olm. **it does not ~ with him,** ona dokunuyor. **~able** [a'griəbl]. Hoş, nazik; uygun, münasib; razı. **~ment** [ə'grîmnt]. Uyuşma, anlaşma; mukavele; mutabakat: **to be in ~,** ayni fikirde olm.; mutabık olm.; razı olm.: **to enter into an ~ with s.o.,** birisile bir mukavele yapmak.

**agricultur·e** ['agrikʌlçə(r)]. Ziraat. **~al** [ˌagri'kʌlçərəl]. ziraate aid, ziraatçi (millet). **~ist,** ziraatçi.

**aground** [ə'graund]. Karaya oturmuş. **to run ~,** karaya otur(t)mak.

**aha** [a'ha, a'hâ]. Ha!; tamam!

**ahead** [ə'hed]. İleride; önde. **to draw ~,** daha öne gitmek: **to go ~,** önden gitmek; ilerlemek; yol almak: **go ~!,** devam ediniz; siz buyurunuz!: **~ of one's time,** zamanına göre ileri.

**ahoy** [ə'hôy]. (*den.*) Hey!, ey!. **boat ~!,** hey gemidekiler!

**aid** [êyd] *n.* Yardım; imdad. *v.* Yardım etm., imdad etm. **in ~ of,** menfaatine.

**aide(s)-de-camp,** ['êydəkâ(n)]. Yaver(ler).

**ail** [êyl]. Rahatsız etm.; keyifsiz olmak. **what ~s you?,** neniz var? **~ment** ['êylmnt]. Hastalık, illet.

**aileron** ['êylərə(n)]. Uçak kanadının hareket eden kısmı.

**aim** [êym] *n.* Nişan alma; hedef; maksad, gaye. *v.* (Bir silahı) nişan vaziyetine getirmek; nişan almak. **~ at,** kasdetmek; gayret etm.; istihdaf etmek. **~less,** maksadsız, gayesiz; neticesiz.

**air**[1] [eə(r)] *v.* Havalandırmak;

(çamaşır vs.) ısıtmak. **to ~ one's grievances,** derdini ortaya dökmek: **to ~ one's knowledge,** malûmatını satmak. ~² *n.* Hava; tavır, hal. **to beat the ~,** akıntıya kürek çekmek: **there is an ~ of comfort in this house,** insan bu evde kendini rahat hissediyor: **to give oneself [to put on] ~s,** bir takım haller, tavırlar takınmak, numara yapmak: **to have an [the] ~ of,** gibi görünmek: **he has an ~ about him,** onun tesirli bir hali var: **it is all in the ~ as yet,** daha ortada bir şey yok: **there are lots of rumours in the ~,** etrafta bir çok rivayetler dolaşıyor: **our left flank was in the ~,** (*mil.*) sol kanadımız açıkta kaldı: **there's something in the ~,** ortada bir şeyler oluyor: **to be on the ~,** radyoda konuşmak: **to put on the ~,** radyo ile neşretmek: **to melt (vanish) into thin ~,** yok olmak: **to walk on ~,** etekleri zil çalmak. **~-base,** hava üssü. **~-cooled,** hava ile soğutulan. **~-line,** hava hattı [yolu]. **~-liner,** muayyen bir hatta tahsis edilmiş yolcu uçağı. **~-lock,** bir su borusunda vs. hava boşluğu. **~-mail,** hava postası. **~-minded,** havacılığın ehemmiyetine inanan. **~-pocket,** hava boşluğu. **~-port,** hava limanı. **~-tight,** hava geçmiyen. **~craft** [ˈeəkrâft]. Uçaklar, tayyareler. **~man** [ˈeəmən]. Uçakçı, tayyareci. **~plane** [ˈeəplêyn]. Uçak. **~ship** [ˈeəşip]. Sevki kabil balon, zeplin. **~worthy** [ˈeəwö(r)ði]. (Uçak hakkında) havada durabilen, uçmıya müsaid.

**air·y** [ˈeəri]. Havalı, havadar; hafif; havaî. **~ily, to talk ~,** dem vurmak.

**aisle** [âyl]. Bir kilisede sütunlarla ayrılmış koridor.

**aitch** [êyç]. H harfinin ingilizce adı. **to drop one's ~es,** h harfini telaffuz etmemek (ki İngiltere'de iyi tahsil ve terbiye görmemiş tabakadan mânasına gelir).

**ajar** [əˈcâ(r)]. Yarı açık, aralık (kapı).

**akin** [əˈkin]. **~ to,** akraba; yakın, benzeyen.

**alabaster** [ˈaləbâstə(r)]. Ak mermer.

**alack** [əˈlak]. Yazık, eyvah, heyhat.

**alacrity** [əˈlakriti]. Çeviklik, canlılık.

**alarm** [əˈlâm] *n.* Silah başı işareti; tehlike işareti; endişe, telaş. *v.* Telaşa düşürmek. **to be ~ed at stg.,** ···den telaşa düşmek: **to raise [give] the ~,** tehlikeyi haber vermek: **to take ~,** telaşa düşmek. **~ing,** endişe veren. **~ist,** ortalığı her zaman telaşa düşüren; çabuk telaşlanan. **~-clock,** uyandıran saat.

**alas** [aˈlâs]. Vay!, yazık!; maalesef.

**Albania** [alˈbêynyə]. Arnavutluk. **~n,** arnavut.

**alb·ino** [alˈbînou]. Derisi, saçları ve kaşları beyaz insan veya hayvan. **~umen** [ˈalbyumən]. Albümin; yumurta akı.

**alchem·ist** [ˈalkəmist]. Simyager. **~y,** eski kimya, simya.

**alcohol** [ˈalkəhol]. Alkol; alkollu içki. **~ic,** [alkəˈholik] *a.* alkollu; *n.* ayyaş. **~ism** [ˈalkəholizm], ayyaşlık.

**alcove** [ˈalkouv]. Bir odanın içerlek kısmı; cumba; bir odada duvar içindeki yuva gibi girinti.

**alderman,** *pl.* **-men** [ˈôldəmən, -men]. Belediye meclisinin kıdemli âzası.

**ale** [êyl]. **~-house,** birahane.

**Aleck** [ˈalek]. **smart ~,** kurnaz; bilmediği yok.

**alert** [əˈlö(r)t]. Uyanık, dikkatli: **to be on the ~,** tetikte olmak

**alfresco** [alˈfresko]. Açık havada.

**algebra** [ˈalcebra]. Cebir. **~ic(al)** [alcəbrêəkl], cebre aid.

**Algeria** [alˈcîryə]. Cezayir. **~n,** Cezayire aid; Cezayirli. **Algiers** [alˈcîəz]. Cezayir şehri.

**alias** [ˈêylyas]. Takma ad; öteki adı, yahud.

**alibi** [ˈaləbây]. Suç işlendiği zaman başka yerde bulunulduğu iddiası.

**alien** [ˈêyliən]. Ecnebi, yabancı; uymıyan. **~ate** [ˈêyliənêyt]. Uzaklaştırmak, soğutmak; (bir malı) ferağ ve temlik etm., devretmek. **~ation,** uzaklaşma, soğuma; ferağ ve temlik: **mental ~,** cinnet. **~ist** [ˈêyliənist]. Akliyeci.

**alight¹** [əˈlâyt] *v.* İnmek; konmak. ~² *a.* Ateş içinde. **to catch ~,** tutuşmak: **to set ~,** tutuşturmak.

**align** [əˈlâyn]. Sıraya koymak, dizmek. **~ment,** dizilme, sıraya girme, hiza.

**alike** [əˈlâyk]. Benzer; aynı, müsavi. **summer and winter ~,** hem yaz hem kış.

**aliment** [ˈalimənt]. Yiyecek, gıda. **~ary,** yiyeceğe dair; besleyici. **~ation,** gıda verme, besleme.

**alimony** [ˈaliməni]. Nafaka.

**alive** [əˈlâyv]. Canlı, diri, hayatta; kaynaşan. **to be ~ to,** farkında olm.:

to keep stg. ~, yaşatmak, muhafaza etm.: **look ~!**, çabuk ol: **man ~!**, nasıl olur!: **the kindest man ~**, dünyada en iyi insan.

**alkal·i** [ˈalkəlây]. Kalevi, kali. **~ine,** alkalinli. **~oid,** şibih kalevi.

**all** [ôl]. Bütün, hep, hepsi, her. *Sıfatların başında gelirse tekid ifade eder, mes.* **all-powerful,** tam kuvvetli. **~ of us,** hepimiz: **~ alone,** yapyalnız: **~ at once,** birdenbire: **do you see him at ~?**, onu hiç görüyor musunuz? **if you go there at ~**, eğer oraya gidecek olursanız: **he will come tomorrow if at ~**, şayed gelirse yarın gelecek: **if it is at ~ cold,** bir parça soğuk olsa: **~ the better,** çok daha iyi; iyi ya!, isabet!: **he ~ but died,** az kaldı ölüyordu: **it's ~ but done,** yapıldı [bitti] sayılır: **~ by oneself [myself, etc.]**, kendi başına [başıma vs.]; yapyalnız: **to be ~ for ...**, ···in tarafdarı olm.: **for ~ he may say,** ne söylerse söylesin: **for ~ his talent,** (bütün) kabiliyetine rağmen: **his son was ~ in ~ to him,** oğlu onun için hayatta her şey idi: **taking it ~ in ~**, heyeti umumiye itibarile: **he lost his ~**, varını yoğunu kaybetti: **my ~**, varım yoğum: **not at ~**, hiç; hiç değil; bir şey değil: **~ of a thousand pounds,** en aşağı bin lira: **~ over,** her tarafta; bitmiş: **~ right,** pek iyi; hay hay: **he's not so stupid as ~ that,** artık bu kadar da abtal değildir: **he's ~ there,** sen onu hiç merak etme, o açıkgözdür: **he's not ~ there,** pek aklı başında değil, terelelli: **~ the way,** yolun sonuna kadar. **~ Fools' Day,** bir nisan. **~-in,** yüzde yüz her şey dahil: **~ wrestling,** serbest güreş. **~-purpose,** her şeye yarıyan. **~-round,** (atlet vs.) her sahada mükemmel; her bakımdan. **~ Saints' Day,** azizler günü (1 kasım). **~-weather,** her havaya elverişli.

**allay** [əˈlêy]. Yatıştırmak, teskin etmek.

**allegation** [aliˈgêyşn]. İleri sürme, iddia.

**allege** [əˈlec]. İleri sürmek, iddia etm.

**allegiance** [əˈlîcəns]. Biat, sadakat; tabiiyet.

**allegor·y** [ˈaligəri]. Remzî hikâye. **~ical** [–ˈgorikl], remzî, kinaye yolile.

**alleviate** [əˈlîviêyt]. Azaltmak, hafifletmek, yatıştırmak.

**alley** [ˈali]. (Şehirde) dar yol, aralık; (parkta) ağaçlıklı yol. **blind ~**, çıkmaz sokak.

**alli·ance** [əˈlâyəns]. İttifak, birlik. **~ed** [aˈlâyd] *a.* Müttefik; yakın; münasebeti olan.

**alligator** [ˈaligêytə(r)]. Amerika timsahı.

**alliteration** [əˌlitəˈrêyşn]. Bir kelime grupunda baş harflerin kafiyeli olması.

**allocat·e** [ˈaləkêyt]. Tahsis etm. **~ion,** tahsis etme; tahsis edilen mikdar; hisse.

**allot** [əˈlot]. Tahsis etm.; hisselere ayırmak. **~ment,** hisselere ayırma; hisse, pay; kiraya verilen küçük bahçe.

**allow** [əˈlau]. Müsaade etm.; bırakmak; kabul etm.; hoş görmek; vermek; tahsis etmek. **to ~ for,** hesabetmek: **~ing for the circumstances,** şartları hesaba katarak: **the matter ~s of no delay,** işin beklemeğe tahammülü yok. **~able,** kabul edilebilir; caiz, meşru, mubah. **~ance,** tahsisat; istihkak; tenzilât; ihtiyat payı: **to make ~(s) for,** hesaba katmak: **to put s.o. on short ~**, istihkakını kısmak: **travelling ~**, harcırah. **~edly,** herkesin kabul [itiraf] ettiği üzere.

**alloy** [ˈalôy] *n.* Halita. [əˈlôy] *v.* Halita yapmak; değerini bozmak.

**allu·de** [əˈlûd]. **~ to,** İma etm.; taş atmak; kısaca zikretmek. **~sion** [əˈlûjn]. İma; tas; zikretme.

**allur·e** [əˈlyuə(r)]. Çekmek, cezbetmek; kandırmak. **~ing,** çekici, cazib.

**alluv·ial** [əˈlûviəl]. Taşan suların bıraktığı toprağa aid. **~ium,** taşan suların biraktığı toprak.

**ally** [ˈalây] *a.* Müttefik; dost; arkadaş. [əˈlây] *v.* Birleştirmek, ittifak et(tir)mek.

**alma mater** [ˈalma ˈmâtə(r)]. (*Lât.*) Bir kimsenin tahsilini görüp yetiştiği üniversite.

**almanac** [ˈolmənak]. Takvim; almanak.

**almighty** [ôlˈmâyti]. Her şeye kaadir; Kaadiri Mutlak; (*coll.*) tam; büyük.

**almond** [ˈâmənd]. Badem. **sugar ~**, badem şekeri.

**almost** [ˈôlmoust]. Hemen hemen, aşağı yukarı; az kaldı.

**alms** [âmz] *n. pl.* Sadaka. **~house,** fakirlere mahsus imaret.

**aloft** [əˈloft]. Yukarıda; gemi direğinde.

**alone** [əˈloun]. Yalnız; tek başına;

sade. **let ~,** şöyle dursun; **he can't talk his own language, let ~ English,** İngilizçe şöyle dursun kendi lisanını bile konuşamaz: **to let ~,** kendi haline bırakmak; rahat bırakmak: **let well ~,** iyidir, fazla kurcalama: **with that courtesy which is his ~,** kendine mahsus nezaketile.

**along** [ə'lon(g)]. Boyunca; uzunluğuna; ileriye. **all ~,** başından itibaren: **come ~!,** haydi gel!: **~ with,** beraber: **to get ~ with,** ···le geçinmek, uyuşmak: **get ~ with you!,** (*coll.*) haydi canım!. **~shore,** sahil boyunca, kıyı sıra. **~side,** yanyana; borda bordaya; bordada; **~ the quay,** rıhtım yanında.

**aloof** [ə'lûf]. Uzakta; sokulmaz. **to hold [keep, stand] ~,** uzak durmak, sokulmamak. **~ness,** çekingenlik, sokulmayış.

**aloud** [ə'laud]. Yüksek sesle.

**alpha** ['alfa]. Yunan alfabesinin birinci harfi. **~ and Omega,** ilk ve son; başı ve sonu.

**alphabet** ['alfəbet]. Alfabe. **~ical** [–'betikl], alfabe sırasile.

**alpin·e** ['alpâyn]. Alp dağlarına aid ve dair. **~ist** ['alpinist], dağcı, alpinist.

**already** [ôl'redi]. Daha; daha şimdiden; şimdiye kadar; kadar. **I've been here an hour ~,** buraya geleli tam bir saat oldu: **two o'clock ~!,** saat ne çabuk iki olmuş!

**also** ['ôlsou]. De; dahi; aynı zamanda; bir de.

**altar** ['oltə(r)]. Mabedin en mukaddes yeri; mihrab; mabedde kurban kesilen yer. **to lead s.o. to the ~,** birile evlenmek.

**alter** ['oltə(r)]. Değiş(tir)mek, tadil etm., başka şekle sokmak [girmek]. **that ~s matters (the case),** o başka, o zaman iş değişir: **to ~ for the better [worse],** daha iyi [fena] olmak. **~ation** [–'rêyşn], değiştirme, değişiklik; tadilât.

**altercation** ['oltə'kêyşn]. Münakaşa.

**alternat·e** ['oltənêyt] *v.* Nöbetleşe değiş(tir)mek; birbirini takib etmek. **~ing,** mütenavib. **~e** [ôl'tö(r)nit] *a.* Nöbetleşe değişen, mutenavib. **on ~ days,** gün aşırı. **~ive** [ol'tö(r)netiv]. İki hal veya şıktan birisi (olan). **there is no ~,** tek çare [şekil] budur: **the ~ plan would be,** yegâne mümkün olan diğer şekil budur.

**although** [ôl'ðou]. Her ne kadar.

**alti·meter** ['altimîtə(r)]. İrtifa ölçen alet. **~tude** ['altityûd]. Deniz sathından irtifa.

**altogether** [ôltə'geðə(r)]. Hep beraber; tamamen, büsbütün; hepsi; umumiyetle.

**altru·ism** ['altruizm]. Başkalarını düşünme. **~ist,** diğergâm. **~istic** [–'istik], başkalarını düşünen.

**aluminium** [alyu'miniəm]. Alüminyom.

**alumnus** [ə'lʌmnʌs]. Bir kolej mezunu.

**always** ['ôlwêyz, –wiz]. Her zaman, daima. **there is ~ your car,** olmazsa [sıkışınca] otomobiliniz var.

**am** [am]. *1st. p. sing. pres. of* **be. I am here,** buradayım.

**a.m.** ['êy 'em]. Öğleden evvel.

**amalgam** [ə'malgəm]. Cıva ile bir madenin halitası. **~ate,** [ə'malgəmêyt], mezcetmek, terkib etm.; birleşmek.

**amass** [ə'mas]. Yığmak, toplamak.

**amateur** ['amətö(r) (–tyuə(r))]. Meraklı; amatör, heveskâr. **~ish** [–tyuəriş], (fena mânada) amatör işi.

**amatory** ['amatori]. Aşıkane.

**amaze** [ə'mêyz]. Hayrette bırakmak. **I was ~d,** ağzım açık kaldı. **~ment,** şaşkınlık.

**Amazon** ['amazən]. Eski Yunan mitolojisinde muharib kadın; erkek yapılı kadın.

**ambassad·or** [am'basədə(r)]. Büyük elçi, sefir. **~ress,** büyük elçinin eşi; sefire.

**amber** [ambə(r)]. Kehriba (rengi). **~gris** [–'grîs], amber.

**ambidextrous** ['ambi'dekstrəs]. İki elini de kullanabilen.

**ambigu·ity** [ambi'gyuiti]. Başka mânaya da gelebilme; iltibas. **~ous** [am'bigyuəs], başka mânaya da gelebilen; iltibaslı, mübhem; şübheli.

**ambiti·on** [am'bişn]. Yükselme ihtirası; şiddetle arzu edilen şey. **~ous,** gözü ileride, büyük emeller peşinde; ihtiras sahibi; ikbal perest.

**amble** ['ambl] *n.* Rahvan *v.* Rahvan gitmek; rahat ve sakin yürümek.

**ambulance** ['ambyuləns]. Seyyar hastahane; hasta arabası.

**ambush** ['ambuş]. Pusu. Pusuya düsürmek [yatırmak]. **to lay an ~,** pusu kurmak: **to lie in ~,** pusuya yatmak.

**ameliorate** [ə'mîliərêyt]. İyileş(tir)mek.

**amen** [ˈâˈmen, ˈêyˈmen]. Amin.
**amenable** [əˈmînibl]. Tabi; makul, uysal. ~ **to reason,** sulha subhana yatar.
**amend** [əˈmend]. Düzeltmek; tadil etm., islah etmek. **to ~ one's ways,** hal ve hareketini düzeltmek. **~s,** tazminat; **to make ~s,** tazmin etm. **~ment,** düzeltme; tadil.
**amenity** [əˈmîniti, –ˈmen–]. Hoşluk, lâtiflik. **the amenities of life,** hayatın zevki, ğüzel tarafı.
**amethyst** [ˈameθist]. Mor yakut.
**amiable** [ˈêymiəbl]. Kendini sevdirir, hoş.
**ami·cable** [ˈamikəbl]. Dostça. **~ty** [ˈamiti]. Dostluk.
**amid(st)** [əˈmid(st)]. Ortasında.
**amidships** [əˈmidşips]. Geminin ortasında.
**amiss** [əˈmis]. Yanlış; bozuk. **to take stg. ~,** bir şeyi fenaya almak: **don't take it ~!,** hatırınız kalmasın!: **a glass of beer wouldn't be [come] ~,** şimdi bir bardak bira olsa fena olmaz.
**ammeter** [ˈamitə(r)]. Ampermetre
**ammonia** [əˈmounyə]. Nişadır. **~c,** amonyaka aid. **~ted** [–ˈyêyted], amonyaklı.
**ammunition** [ˌamyuˈnişn]. Cebhane; mühimmat.
**amnesia** [amˈnîziə]. Hafızayı kaybetme hastalığı.
**amnesty** [ˈamnesti]. Umumî af (ilân etm.).
**among(st)** [əˈmʌn(g)(st)]. Arasında; içinde. ~ **you,** aranızda.
**amoral** [aˈmorəl]. Ahlakla ilişiği olmıyan.
**amorous** [ˈamərəs]. Âşık; aşka aid.
**amorphous** [aˈmôfʌs]. Şekilsiz.
**amount** [əˈmaunt] *n.* Mikdar; meblağ; yekûn. *v.* (para vs.) baliğ olm., varmak, tutmak. **it ~s to the same thing,** aynı hesaba gelir, aynı şeydir: **he will never ~ to much,** ondan mühim bir şey beklenmez.
**amp** [amp]. Amper.
**amphib·ian** [amˈfibyən]. Hem karada hem suda yaşıyan (hayvan) veya giden (uçak vs.). **~ious,** hem karada hem de suda yaşıyabilen.
**ampl·e** [ˈampl]. Geniş, bol; kâfi. **~ify** [ˈamplifây]. Büyütmek, genişletmek; kuvvetini artırmak; tafsil veya ilâve etmek. **~ifier,** amplifikatör. **~itude** [ˈamplityûd]. Genişlik; bolluk.
**amputat·e** [ˈampyutêyt]. (Bir uzvu) kesmek. **~ion,** ampütasyon.

**amuck** [əˈmʌk]. **to run ~,** Kudurmuş gibi etrafa saldırmak.
**amulet** [ˈamyulet]. Muska, tılısım.
**amus·e** [əˈmyûz]. Eğlendirmek; oyalamak; güldürmek. **to ~ oneself (by, with),** eğlenmek. **~ement,** eğlence. **~ing,** eğlenceli, güldürücü: **the ~ing thing about it,** işin tuhafı.
**an** [an] *v.* **a.**
**anachronism** [aˈnakronizm]. Bir şahıs, hadise veya şeyi aid olmadığı zamana koyma.
**anaem·ia** [aˈnîmyə]. Kansızlık. **~ic,** kansız; zayıf.
**anaesth·esia** [anisˈθîziə]. Hissi uyuşturma. **~etic** [–ˈθetik], hissi uyuşturan madde. **~etist** [aˈnîsθətist], hissi ibtal eden, anastezi veren kimse. **~etize** [aˈnîsθətâyz], hissi uyuşturmak.
**anagram** [ˈanagram]. Başka bir kelime veya cümlenin harflerile meydana getirilen kelime veya cümle.
**analog·ous** [aˈnaləgəs], benzer, kıyas edilebilir. **~y** [aˈnaləci], kıyas; münasebet.
**analy·se** [ˈanəlâyz]. Tahlil etmek. **~sis** [əˈnaləsis], tahlil, analiz. **~st** [–list], tahlilci. **~tic(al)** [–ˈlitik(l)], tahlilî.
**anarch·ic(al)** [aˈnâkik(l)]. (Vaziyet vs.) karmakarışık, anarşik. **~ist** [ˈanakist], anarşist. **~y** [ˈanəki], anarşi, karmakarışık vaziyet.
**anathema** [əˈnaθəma]. Lânet(leme). **to be ~ to s.o.,** nefretini mucib olmak. **~tize** [anaˈθîmətâyz], lânetlemek.
**anatom·ical** [anaˈtomikl]. Anatomiye aid, teşrihî. **~ist** [aˈnatəmist], teşrihci. **~y,** teşrih, anatomi.
**ancest·or** [ˈansestə(r)]. Ata, ced. **~ral** [anˈsestrəl], atadan kalma, atalara aid, irsî. **~ress,** nine, kadın ced. **~ry,** soy.
**anchor** [ˈan(g)kə(r)] *n.* Gemi demiri, çapa. *v.* Demirlemek, demir atmak; iyice tesbit etmek. **to come to ~,** demirlemek: **to drag ~,** demiri sürüklemek: **to let go the ~,** demiri funda etm.: **to lie (ride) at ~,** demirli olm.: **to weigh ~,** demiri vira etmek. **~age,** demirleme yeri, ücreti.
**anchorite** [ˈan(g)kərâyt]. Münzevi, tariki dünya.
**anchovy** [anˈçouvi]. **(fresh)** hamsi; **(preserved)** ançüviz.
**ancient** [ˈêynşənt]. Eski, kadim; ihtiyar. **the ~s,** eski Yunan gibi kadim milletler.

**ancillary** [anˈsiləri]. Tabi; fer'î; bağlı.
**and** [and]. Ve, ile, bir de, de, daha. **better ~ better,** gittikçe daha iyi: **to walk two ~ two,** ikişer ikişer yürümek.
**anecdot·e** [ˈanikdout]. Fıkra, lâtife, hikâye.
**anemometer** [ˌanəˈmomətə(r)]. Rüzgâr ölçen alet.
**anemone** [əˈnemoni]. Mânise lâlesi. **sea ~,** deniz inciri.
**aneroid** [ˈanərôyd]. **~ (barometer),** mayisiz işleyen barometre.
**anew** [əˈnyû]. Yeniden, tekrar.
**angel** [ˈêyncəl]. Melek. **be an ~ and . . .,** ne olursun . . .: **ʳfools rush in where ~s fear to treadˈ,** akıllının ayağını atmıyacağı yere deliler koşar. **~ic** [anˈcelik], melek gibi; meleklere mahsus. **~us** [ˈancəlʌs]. (Katoliklerce) sabah, öğle ve akşam okunan bir dua.
**ang·er** [ˈan(g)gə(r)]. Hiddet, öfke. Kızdırmak, öfkelendirmek. **easily ~-ed,** pek çabuk kızar. **~ry** [ˈan(g)gri]. Hiddetli, öfkeli; iltihablı. **to be (get) ~,** darılmak, gücenmek: **to be ~ about stg.,** bir şeye kızmak: **to be ~ with s.o.,** birisine kızmak: **to get ~,** kızmak: **to make ~,** kızdırmak.
**angle**[1] [ˈan(g)gl]. Zaviye, köşe; açı. **~**[2]**.** Olta. Olta ile balık tutmak. **to ~ for,** tutmağa çalışmak. **~r,** balık avcısı.
**anglicism** [ˈan(g)glisizm]. İngilizceye has tabirler kullanma.
**anglo-** [ˈan(g)glou]. **~-Indian,** İngilizle Hindli melezi; Hindistanda doğmuş İngiliz. **~phil,** İngiliz muhibbi.
**anguish** [ˈan(g)gwiş]. Müdhiş acı veya keder.
**animadver·sion** [ˌanimadˈvö(r)şn]. Tenkid; çekiştirme. **~t (on),** tenkid etm.
**animal** [ˈaniml]. Hayvan. Hayvana aid.
**animat·e** [ˈanimət] *a.* Canlı. [ˈanimêyt] *v.* Canlandırmak; harekete getirmek. **~ion** [–ˈmêyşn], canlılık, hareket, heyecan.
**animosity** [ˌaniˈmositi]. Düşmanlık, nefret.
**ankle** [ˈan(g)kl]. Topuğun yan kemiği.
**annal·ist** [ˈanalist]. Vak'anüvis, tarihçi. **~s,** vakayiname, tarih.
**Anne** [ˈan]. **Queen ~'s dead!,** bu (ta) ne zamanın hikâyesi: bu artık bayatladı: **Queen ~'s Bounty,** fakir rahiblerin için kurulmuş vakf.
**annex** [əˈneks] *v.* İlhak etm., eklemek, katmak. **~e** [ˈaneks] *n.* İlâve kısım.
**annihilat·e** [əˈnâyhilêyt]. Yok etm., imha etmek. **~ion** [–ˈlêyşn], mahvetme.
**anniversary** [ˌaniˈvö(r)səri]. Yıldönümü.
**annotate** [ˈanotêyt]. (Kitabın kenarına) haşiye yazmak; (metni) şerh ve izah etmek.
**announce** [əˈnauns]. Haber vermek, bildirmek. **~ment,** haber verme; ilân. **~r,** haber veren; (radyoda) haberleri okuyan.
**annoy** [əˈnôy]. Taciz etm., rahatsız etm., kızdırmak. **~ance,** kızıp canı sıkılma, üzülme; taciz eden kimse, şey.
**annu·al** [ˈanyuəl]. Yıllık, senelik; bir sene süren veya her sene vukubulan. **~ity** [əˈnyuiti], senelik tahsisat, ücret.
**annul** [əˈnʌl]. Feshetmek, ibtal etmek.
**annunciation** [əˈnʌnsiˈêyşn]. Haber.
**anode** [ˈanoud]. (*elect.*) Müsbet kutub.
**anoint** [əˈnôynt]. Yağlamak; (bir kralı vs. merasim icabi) yağlıyarak takdis etmek.
**anomal·ous** [əˈnoməl əs]. Kaideye uymıyan; müstesna. **~y,** usulsüzlük; istisna.
**anon.** *v.* **anonymous.**
**anonym·ity** [ənoˈnimiti]. Anonimlik. **~ous** [əˈnoniməs], isimsiz, imzasız, anonim.
**another** [əˈnʌðə(r)]. Başka; bir başka; (bir) daha. **one ~,** birbirini: **in ~ ten years,** bundan on sene sonra: **that's (quite) ~ matter,** o zaman iş değişir: **science is one thing, art is ~,** ilim başka, san'at başka: **taking one thing with ~,** sonunda; vasatî olarak.
**answer** [ˈânsə(r)] *n.* Cevab; karşılık. *v.* Cevab vermek; karşılamak. **in ~,** cevab olarak: **to ~ to a description,** bir tasvir veya tarife uymak: **to ~ the door,** gelip kapıyı açmak: **to ~ for s.o. (stg.),** birinin namına söz söylemek; tekeffül etm.; mesul olm.: **to ~ the helm** (gemi) dümeni dinlemek: **he has a lot to ~ for,** yaptığı bir çok şeylerden mesul tutulacak: **to ~ the purpose,** maksada kâfi gelmek: **his scheme didn't ~,** projesi muvaffak olmadı. **~able** [ˈânsərəbl]. Mesul, sorumlu; cevab verilebilir.

**ant** [ant]. Karınca.
**antagon·ism** [an'tagənizm]. Düşmanlık; zıddiyet; rekabet. **~ist,** muhalif, düşman. **~istic** [-'istik], zıd, muhalif. **~ize** [an'tagənâyz], düşman etmek.
**antarctic** [an'tâktik]. Cenub kutbuna aid; cenub kutbu bölgesi.
**ante-** ['anti] *pref.* ···den evvel. **~cedent** [ˌanti'sîdnt]. Önceki, önce olan (şey); mukaddem. **the ~s of s.o.,** birisinin geçmişi, cemaziyelevveli. **~date** [ˌanti'dêyt]. Önceki tarihi koymak; daha evvel gelmek. **~ diluvian** [ˌantidə'lûviən]. Tufandan önceye aid; çok eski veya yaşlı. **~-natal** [anti'nêytl]. Doğumdan evvel. **~rior** [an'tîriə(r)]. Ön, önceki. **~-room** ['antirum]. İçinden başka odaya geçilen oda, bekleme odasi.
**antelope** ['antiloup]. Ceylan.
**anthem** ['anθəm]. Dinî şarkı. **National ~,** millî marş.
**anthology** [an'θoloci]. Seçme yazılar kitabı, antoloji.
**anthropo·id** ['anθropôyd]. İnsana benzer. **~logy** [anθro'polici]. İnsanın menşei ve gelişmesi ile ırkları, örf ve âdetler ve itikadları tedkik eden ilim.
**anti-** ['anti] *pref.* ···e karşı muhalif. **~-aircraft** ['anti 'eəkrâft]. **~ gun,** uçaksavar top.
**antics** ['antiks]. Maskaralık, soytarılık.
**anticipat·e** [ən'tisipêyt]. Merakla beklemek; başkalarından evvel davranmak; önceden yapmak; önlemek. **~ion** [-'pêyşn], umma, merakla bekleme; tahmin: **in ~,** peşin olarak: **in ~ of stg.,** ilerdeki bir şeyi düşünerek. **~ory** [-pətəri], ilerisini düsünerek.
**anticlerical** [ˌanti'klerikl]. Rahib sınıfı aleyhinde.
**anticlimax** [ˌanti'klâymaks]. Yüksek, güzel, bir düşünce, söz vs.den, birdenbire gülünç değersize düşme; fena tezad.
**antidote** ['antidout]. Panzehir çare.
**Antioch** ['antiok]. Antakya.
**antipath·y** [an'tipəθi]. Hoşlanmama. **~etic(al)** [-'θetik] sevimsiz.
**antipodes** [an'tipədîz]. Yer yüzünde her hangi bir noktanın mukabili olan nokta; bir şeyin taban tabana zıddı. **The ~,** Avusturalya ve Yeni Zelanda.
**antiqu·arian** [anti'kweriən]. Antikaya aid; antika meraklısı; antikacı. **~ated** ['antikwêytid], Eskimiş, modası geçmiş. **~e** [an'tîk]. Çok eski; **~ dealer,** antikacı. **~ity** [an'tikwiti]. Eskilik; eski çağ; eski zamana aid şey.
**anti-semit·e** [anti'sîmâyt]. Yahudi düşmanı. **~ism** [-'semitizm]. Yahudi düsmanlığı.
**antiseptic** [anti'septik]. Taafünü önliyen.
**antisocial** [anti'souşəl]. Cemiyete karşı (hareket vs.).
**antithesis,** *pl.* **-es** [an'tiθəsis, -sîz]. Tezad.
**antler** ['antlə(r)]. Geyik boynuzu.
**antonym** ['antonim]. Zıd.
**anvil** ['anvil]. Örs.
**anxi·ety** [an(g)'zâyəti]. Endişe; şiddetli arzu. **~ous** ['an(g)kşəs], endişeli; üzüntülü; istekli: **to be ~ about,** ···i merak etm.: **to be ~ to,** arzu etm. ehemmiyet vermek.
**any** ['eni]. Bir; her hangi; her hangi bir; rastgele; her; bir az, bir mikdar, bir dereceye kadar. **not ~,** hiç: [*Menfi ve sualli cümlelerde* **some (one)** *yerine kullanılır*]. **~ longer [~ further],** artık, daha fazla: **~ more,** daha fazla, başka: **have you ~?,** (ondan) sizde var mı?: **I haven't ~,** bende hiç yok: **I'm not having ~,** (*coll.*) yağma yok: **there is little if ~,** olsa bile [varsa da] pek az: **he knows English if ~ man does,** İngilizceyi bilse bilse o bilir: **he may come ~ minute,** her an gelebilir: **not ~ too well,** pek o kadar iyi değil: **is the patient ~ better?,** hasta bir az daha iyi mi?: **come ~ time,** ne zaman istersen gel. [*Bazı hallerde* **any**'*nin mânası cümle içinde vurgulu olup olmadığına göre değişir, mes.*: **a good dentist is to be found in any *town* of any *size*,** her hangi büyük bir şehirde iyi bir dişçi bulunabilir; *fakat*: **a dentist of some sort is to be found in *any* town of *any* size,** her hangi büyüklükte rastgele bir şehirde şöyle böyle bir dişçi bulunabilir.]
**anybody, anyone** ['enibodi 'eniwʌn]. Biri, bir kimse, kim; herkes; kim olsa; rastgele. (*Menfi ve sualli cümlelerde*) hiç kimse. **~ but he,** ondan başka kim olsa: **is he ~?,** hatırı sayılır [mühim bir adam] mı?: **he isn't just ~,** o rastgele bir adam değildir: **he will never be ~,** onun mühim bir adam olmasına imkân

yok: **everybody who is ~ was there,** bellibaşlı herkes orada idi: **to look at him ~ would think that he was an old man,** onu gören ihtiyar zanneder: **he is a poet if ~ is,** şair diye ona denir.

**anyhow** [ˈenihau̯]. Nasıl olursa, nasıl olsa, nasılsa; hoş . . . ya!; olsun, her halde, ne ise, yine. **things are (going) all ~,** işler karmakarışık: **you can't do this ~,** (i) bu işi siz nasıl olsa yapamazsınız; (ii) bu işi rastgele yapamazsınız.

**anyone** *v.* **anybody.**

**anything** [ˈeniθin(g)]. Bir şey, her hangi bir şey, her şey; ne olsa. (*Menfi ve sualli cümlelerde*) hiçbir şey. **~ but that,** (tek) bu olmasın de ne olursa olsun: **like ~,** şiddetle: **do you ever see ~ of him?,** onu gördüğünüz var mı?: **if he is ~ of a gentleman he will apologize,** efendi adamsa özür diler: **are you ~ of a musician?,** musikiden anlar mısınız.: **he's ~ but a fool,** hiç te abdal değildir: **it's as easy as ~,** bundan kolay bir şey yok.

**anyway** [ˈeniwêy] *v.* **anyhow.**

**anywhere** [ˈeniwe̯ə(r)]. Her yerde, nerede olursa olsun. (*Menfi ve sualli cümlelerde*) hiç bir yerde. **~ but there,** oradan başka her yerde: **this won't get you ~,** bu işin sonu yok (size bir faydası yok).

**anywise** [ˈeniwâyz]. Her hangi bir şekilde; (*menfi cümlelerde*) hiç bir suretle.

**apart** [əˈpât]. Ayrı ayrı; ayrı, bir tarafta; ayrılmış. **~ from,** ···den başka, ... bir tarafa: **joking ~,** şaka bertaraf: **it is difficult to tell them ~,** onları birbirinden ayırmak [ayırd etm.] güçtür: **to move ~,** ayrılmak: **to take a machine ~,** bir makineyi sökmek.

**apartment** [əˈpâtmnt]. Oda; salon. **~s,** daire.

**apath·etic** [ˌapaˈθetik]. Hissiz; alâkasız. **~y** [ˈapəθi] hissizlik; alâkasızlık.

**ap·e** [êyp] *n.* (Kuyruksuz iri) maymun. *v.* Taklid etm., taklidini yapmak. **~ish** [ˈêypiş]. Maymun gibi, maymunca.

**aperture** [ˈapətyu̯ə(r)]. Aralık, delik, açık.

**apex,** *pl.* **-es, apices** [ˈêypiks, -iz, ˈêypisîz]. Zirve, doruk, tepe.

**aphorism** [ˈafərizm]. Vecize; hikmet.

**aphrodisiac** [afroˈdiziak]. Cinsî iştahı artıran mukavvi.

**apiece** [əˈpîs]. Her biri; adam başına.

**apocryphal** [əˈpokrifəl]. Hakikati şübheli; uydurma.

**apolog·etic** [əˌpoləˈcetik] *a.* Özür dileyen: müdafaa eden (söz veya yazı ile). **~ist,** [əˈpolocist] *n.* bir fikir veya davayı müdafaa eden. **~gize,** [–ˈcâyz] *v.* özür dilemek, itizar etm., af dilemek. **~y** [əˈpoloci] *n.* özür dileme, itizar, af dileme, tarziye; müdafaa: **to make an ~,** özür dilemek: **to demand an ~,** tarziye istemek: **this ~ for a letter,** bu mektub kusuru; bu mektub bozması.

**apople·ctic** [ˌapoˈplektik]. İnmeli; felce aid. **~xy** [ˈapopleksi]. İnme, felc.

**apost·asy, -cy** [aˈpostəsi] Dininden dönme, irtidad. **~ate,** dininden dönmüş, mürted.

**apost·le** [əˈposl]. Havari; din lideri veya misyoner; bir hareketin lideri. **~olic** [–ˈtolik], havarilere aid; papaya aid.

**apostrophe** [əˈpostrəfi]. Virgül, apostrof.

**appal** [əˈpôl]. Dehşet içinde bırakmak. **~ling,** müdhiş.

**apparatus,** *pl.* **-us, -uses** [apəˈrêytəs, -əsiz]. Alet, cihaz, vasıta.

**apparel** [əˈparəl]. Elbise, üst baş. Giydirip kuşatmak; süslemek.

**apparent** [əˈparənt, əˈpe̯ərənt]. Besbelli; ortada; aşikâr; görünüşteki. **heir ~,** veliahd. **~ly,** görünüşe göre; her halde (olmalı).

**apparition** [apəˈrişn]. Hayalet; görünme.

**appeal** [əˈpîl]. Başvurma; yalvarma; daha yüksek mahkeme (makama) müracaat; çağırma; cazibe. Başvurmak, yalvarmak; daha yüksek makhemeye müracaat etm., istinaf etm.; cezbetmek, hoşa gitmek. **this song does not ~ to me,** bu şarkı beni sarmıyor: **Court of ~,** istinaf mahkemesi. **~ing,** (bakış vs.) yalvaran; dokunaklı; cazib.

**appear** [əˈpi̯ə(r)]. Görünmek, çıkmak, meydana çıkmak; gibi görünmek. **it ~s,** öyle görünüyor, anlaşıldığına göre; (*bazen Türkçedeki naklî mazi,* -miş, *yerine kullanılır*).

**appearance** [əˈpi̯ərens]. Görünme, zuhur; (sahneye vs.) çıkma; görünüş, zevahir; manzara; (kitab vs.) intişar. **to [by] all ~s,** görünüşe nazaran: **for the sake of ~s [to save ~s],** zevahiri kurtarmak için: **his ~ is against**

him, zevahir onun lehine değil: **to put in an ~,** isbatı vücud etm.
**appease** [əˈpîz]. Yatıştırmak, teskin etm.
**appella·nt** [aˈpelənt]. Başvuran; istinaf eden. **~tion** [ˌapəˈlêyşn]. İsim (verme); ünvan.
**append** [əˈpend]. Eklemek, ilâve etm.; (imza) basmak. **~aĝe,** bir şeyin tali kısmı.
**appendi·citis** [əˌpendiˈsâytis]. Apandisit. **~x** [əˈpendiks]. İlâve, zeyl; körbağırsak.
**appertain** [ˌapö(r)ˈtêyn]. Aid olmak.
**appeti·te** [ˈapitâyt]. İştah; istek. **~zer** [ˌapitâyzə(r)]. İştah açan şey; meze; aperitif. **~ziĝ,** iştah açan; nefis.
**applau·d** [əˈplôd]. Alkışlamak. **~se,** alkış.
**apple** [ˈapl]. Elma. **~ of the eye,** gözbebeği: **~ of discord,** hakkında münazaa edilen şey: **to upset the ~-cart,** bir çuval inciri berbad etmek. **~-pie,** üstü hamurlu elma tortası: **~ bed,** muziplik için karmakarışık edilen yatak: **in ~ order,** gayet muntazam.
**appl·iance** [əˈplâyəns]. Alet; vasıta; tertibat. **~icable** [ˈaplikəbl]. Tatbik edilebilir; uygun. **~icant** [ˈaplikənt], müracaat sahibi; namzed; istida veren. **~ication** [–ˈkêyşn], tatbik etme; başvurma; müracaat; istida, taleb; çok dikkat ve gayret; (ilac vs.) sürme, koyma: **on ~,** taleb vukuunda. **~y** [əˈplây]. Üstüne koymak; tatbik etm., kullanmak; tahsis etm.; başvurmak, taleb etm.; uygun düşmek. **to ~ oneself to,** kendini bir işe vermek; çok dikkat ve gayret sarfetmek: **to ~ to s.o.,** ···e müracaat etm.: **to ~ for s.o.,** birisini taleb etm.: **to ~ for a post,** bir iş için müracaat etm.
**appoint** [əˈpôynt]. Tayin etm.; kararlaştırmak. **~ed,** tayin edilen: **well ~,** iyi döşenmiş (ev vs.). **~ment,** tayin; memuriyet; randevu: **~s,** döşeme techizat.
**apportion** [əˈpôşn]. Taksim etm.; paylaşmak.
**apposite** [ˈapozit]. Uygun, münasib.
**apprais·e** [əˈprêyz]. Kıymet takdir etmek. **~al** [**~ement**], kıymet takdiri.
**appreci·able** [əˈprîşəbl]. Takdir ve tahmin edilebilir; farkedilir (derecede). **~ate** [–iêyt], kıymet biçmek; takdir etm.; takdir ve teşekkür etm.; beğenmek; farketmek; kıymeti artmak. **~ation** [–siˈêyşn], kıymet biçme; takdir; kıymetin yükselmesi: **to write an ~ of ...** hakkında bir tenkid yazmak. **~ative** [**~atory**], takdir eden, kadirşinas; anlıyarak beğenen; **to be ~ative of** ...den anlamak (zevk duymak).
**apprehen·d** [ˌapriˈhend]. Yakalamak, tevkif etm.; kavramak; korkmak. **~sion** [ˌapriˈhenşn]. Yakalama, tevkif; kavrama; vehim. **~sive,** endişe eden, vehimli.
**apprentice** [əˈprentis]. Çırak; acemi talebe.
**apprise** [əˈprâyz]. Haber [malûmat] vermek.
**approach** [əˈprouç]. Yaklaşma, yanaşma; bir yere getiren yol. Yaklaşmak, yanaşmak; müracaat etmek. **to make ~es to s.o.,** birine avans yapmak: **I'll ~ him on the matter,** bu meseleyi kendisine açacağım: **to ~ a question,** bir meseleyi ele almak. **~able,** yaklaşılabilir, yanaşılabilir.
**approbation** [ˌaproˈbêyşn]. Kabul, tasvib, beğenme. **on ~,** muhayyer (mal); tecrübe edilen (hizmetçi vs.).
**appropriat·e¹** [əˈproupri·it] *a.* Uygun, münasib. **~²** [əˈproupriêyt] *v.* Kendine mal etm.; tasarruf etm.; tahsis etmek. **~ion** [–ˈêyşn], kendine mal etme; tasarruf; tahsis(at).
**approv·al** [əˈprûvl]. Muvafık bulma, tasvib; kabul; tasdik. **goods on ~,** muhayyer mal. **~e** [əˈprûv]. Muvafık bulmak, tasvib etm.; kabul etm.; tasdik etmek.
**approximat·e¹** [əˈproksimit] *a.* Takribî, yaklaşan; tahminî. **~ely,** takriben, tahminen. **~e²** [əˈproksimêyt] *v.* Çok yaklaşmak; yaklaştırmak. **~ion,** yaklaşma; takrib; hakikate yakın tahmin.
**appurtenance** [əˈpö(r)tənəns]. İlâve edilen tali şey. **~s,** teferrüat.
**apricot** [ˈêyprikot]. Kayısı.
**April** [ˈêypril]. Nisan. **~-fool's-day,** nisanın birinci günü: **to make an ~-fool of s.o.,** nisanın birinci günü birisine muziblik yapmak.
**apron** [ˈêyprən]. Önluk. **to be tied to one's mother's ~-strings,** ağzı süt kokmak.
**apt** [apt]. Uygun, elverişli; yerinde; meyyal; muhtemel; zeki, istidadlı. **glass is ~ to break,** cam kırılır: **this train is ~ to be late,** bu tren gecikirse şaşmam [geciktiğini çok gördüm].

**~itude** [ˈaptityûd], istidad, kabiliyet. **~ness,** uygunluk; istidad.
**aqu·arium** [əˈkweəriəm]. Suda yaşıyan hayvanlar müzesi, cam dolabı. **~atic** [əˈkwatik]. Suda yaşıyan; suda yapılan. **~educt** [aˈkwidʌkt]. Su kemeri.
**aquiline** [ˈakwilâyn]. Kartal (gagası) gibi.
**Arab** [ˈarəb]. Arab. **~ia** [aˈrêybiə], Arabistan. **~ian,** Arablara, Arabistan'a aid: **~ nights,** bin bir gece masalları. **~ic,** Arablara, Arabistan'a aid; arabca.
**arable** [ˈarəbl]. Sürülebilir; ziraate elverişli.
**arbit·er** [ˈâbitə(r)]. Hakem. **~ral,** hakeme aid. **~rament** [âˈbitrimnt], hakem kararı. **~rary,** keyfî, indî. **~rate** [ˈâbitrêyt]. Hakem olarak karar vermek; hakeme müracaat etmek. **~ration** [–ˈtrêyşn], hakem usulü; hakem kararı ile hal.
**arbor·eal** [âˈboriəl]. Ağaclara aid. **~iculture** [ˈâboriˌkʌltyuə(r)], ağaççılık.
**arbour** [ˈâbə(r)]. Çardak.
**arc** [âk]. Kavis. **electric ~,** elektrik yayı. **~-lamp,** elektrik yayı lâmbası.
**arcade** [âˈkêyd]. Bir sıra kemer; kemeraltı.
**arch**[1] [âç]. Kemer; tak. Kemer yapmak; kıvırmak. **to ~ the back,** (kedi) sırtını kamburlaştırmak: **triumphal ~,** takızafer. **~way** [ˈâçwêy]. Kemer; kemerli geçid. **~**[2] (*Kadınlar ve çocuklar*) açıkgöz, şeytan. **~**[3] *pref.* Baş, en baş; . . . şahı. **~angel** [ˈâkˌêyncəl]. En büyük melek. **~-bishop** [ˈâçˌbişop]. Baş piskopos.
**archa·eology** [ˌâkiˈoloci]. Arkeoloji. **~ic** [âˈkêy·ik]. Çok eski; modası geçmiş; kullanılmaz olmuş. **~ism,** çok eskilik; çok eski ve unutulmuş tabir vs.
**archer** [ˈâçə(r)]. Okçu. **~y,** okçuluk.
**archipelago** [âkiˈpelago]. Takım adalar.
**architect** [ˈâkitekt]. Mimar. **Naval ~,** deniz inşaat mühendisi. **~ure** [ˈâkitektyuə(r)], mimarlık.
**archiv·es** [ˈâkâyvz]. Hazinei evrak, arşiv; vesikalar, siciller. **~ist** [ˈâkivist], evrak veya sicil memuru.
**arctic** [ˈâktik]. Şimal kutbuna aid; şimal kutbu; kutub gibi.
**ard·ent** [ˈâdənt]. Ateşli; coşkun; çok gayretli. **to wish ~ly,** şiddetle arzu etmek. **~our** ˈâdə(r)]. Ateşlilik, gayret; heyecan; şiddetli sıcak.
**arduous** [ˈâdyuəs]. Zahmetli, güç; sarp, dik (yol).
**are** [â(r)] *v.* **be. we ~ here,** biz buradayız.
**area** [ˈeəriə]. Mesaha; yüz ölçüsü; saha; bölge; avlu.
**arena** [əˈrînə]. Oyun sahası, spor meydanı; faaliyet sahası.
**Argentin·a** [ˈâcəntînə]. Arjantin. **~e** [ˈâcənˌtâyn), Arjantinli: **the ~ (Republic)** Arjantin (Cumhuriyeti).
**argosy** [ˈâgosi]. (*Şair*) büyük tüccar gemisi.
**argu·e** [ˈâgyû]. Münakaşa etm.; delil (sebeb) göstermek, bir fikir ileri sürmek. **to ~ down,** ilzam etm., (delil üstünlüğü ile) susturmak: **to ~ s.o. into doing stg.,** birisini bir şeyi yapmağa ikna etmek. **~fy** [–fây], ukalâlık etmek. **~ment** [–mnt], münakaşa; muhakeme; delil, fikir, tez; hülâsa. **~mentative** [–ˈmentətiv], münakaşa meraklısı; (eser) tenkid ve muhakemeli.
**arid** [ˈarid]. Kurak; kavruk; kuru.
**arise** [əˈrâyz]. Kalkmak; çıkmak; zuhur etm., hasıl olm., doğmak. **to ~ from the dead,** (ölüler) dirilmek: **should the occasion ~,** fırsat zuhurunda; icab ederse.
**aristocra·cy** [arisˈtokrəsi]. Asiller sınıfı. **~t** [ˈaristokrat], asilzade.
**arithmetic** [əˈriθmətik]. Hesab. **~al** [–ˈmetikl], hesaba aid.
**ark** [âk]. Tahta sandık. **Noah's ~,** Nuh'un gemisi.
**arm**[1] [âm] *n.* Kol. **to carry in one's ~s,** kucakta taşımak: **to carry stg. at ~'s length,** bir şeyi kolunu uzatarak taşımak: **to give one's ~ to s.o.** [**to take s.o.'s ~**], birinin koluna girmek: **to keep s.o. at ~'s length,** birini yanına yaklaştırmamak. **~-badge,** pazubend. **~-chair,** koltuk. **~-in-~,** kolkola. **~-rest,** kol dayanacak yer. **-armed,** kollu, *mes*: **one-armed,** tek kollu. **~**[2]**.** Silâh. Silâhlandırmak. **to ~s!,** silâh başı!: **the fourth ~,** hava kuvvetleri: **to lay down one's ~s,** teslim olm.: **a nation in** [**under**] **~s,** silâhlanmış millet: **to take up** [**to rise up in**] **~s,** silâha sarılmak: **to be up in ~s,** ayaklanmak. **~ed,** silâhlı. **~ful** [ˈâmful]. Kucak (dolusu); **in ~s, by the ~,** kucak kucak. **~let** [ˈâmlit]. Kol halkası; pazubend. **~pit** [ˈâmpit]. Koltuk altı.
**armada** [âˈmâdə]. Büyük donanma.

**armament** [ˈâməmənt]. Silâhlar; techizat; teslihât.
**armature** [ˈâmətyu̯ə(r)]. Zırh; bir hayvan, nebatı koruyan kabuk; armatür.
**Armenia** [âˈmîni̯ə]. Ermenistan. **~n,** ermeni, ermenice.
**armistice** [ˈâmistis]. Mütareke. **~ day,** 1918 mütarekesinin yıldönümü (11 kasım).
**armorial** [âˈmori̯əl]. Armaya aid. **~ bearings,** arma.
**armour** [ˈâmə(r)] Zırh. Zırhla kaplamak. **in full ~,** tepeden tırnağa zırhlı (silâhlı). **~er,** zırhcı; silâhcı; tüfekçi. **~y,** silâhhane.
**army** [ˈâmi]. Ordu; kalabalık. **to be in the ~,** askerde olm.; (meslekçe) asker olm.: **to join the ~,** askere gitmek, asker olmak. **~-corps,** kolordu. **~-list,** ordu subay listesi.
**aroma** [əˈrou̯mə]. Güzel koku; baharlı koku. **~tic** [–ˈmatik], güzel kokulu.
**around** [əˈrau̯nd]. Çevresinde, etrafında; civarında; sularında. **all ~,** çepeçevre, dört taraftan.
**arouse** [əˈrau̯z]. Uyandırmak; canlandırmak, teşvik etm.; sebeb olm.
**arraign** [əˈrêyn]. İttiham etm., mahkemeye vermek.
**arrange** [əˈrêync]. Sıraya koymak; düzeltmek; tertib etm., hazırlık yapmak; tanzim etm.; yoluna koymak; kararlaştırmak; anlaşarak halletmek; (*mus.*) tatbik etm. **one cannot ~ for everything,** insan her şeyi (önceden) tertib edemez. **~ment,** sıraya koyma, tanzim; düzeltme; hal; tertib; sıra; hazırlık; anlaşma: **~s,** tertibat.
**arrant** [ˈarənt]. Tamamen, son derece, katıksız. **~ nonsense,** deli saçması.
**array** [əˈrêy]. Sıra, saf; teşhir; gösterişli manzara; süslü elbise. Sıraya koymak, tanzim etm.; giydirmek, süslemek. **in battle ~,** muharebe nizamında.
**arrear(s)** [əˈri̯ə(z)]. Geri kalan (iş vs.); ödenmemiş (borc vs.); bekaya. **to get [fall] into ~,** (tediye vs.) gecikmek, geri kalmak.
**arrest** [əˈrest]. Yakalama, tevkif; durdurma. Yakalamak, tevkif etm.; tutmak, alıkomak; durdurmak.
**arriv·al** [əˈrâyvl]. Varış, gelme, vasıl olma. **a new ~,** yeni gelen şahis. **~e** [əˈrâyv]. Vasıl olm., varmak; muvaffak olmak. **to ~ upon the scene,** çıkagelmek.
**arrogan·ce** [ˈarəgəns]. Küstahca gurur; nahvet. **~t,** Küstahca mağrur; mağrurane.
**arrogate** [ˈarogêyt]. **to ~ to oneself,** (haksız yere) benimsemek, kendine maletmek.
**arrow** [ˈarou̯]. Ok. **~-head,** ok başı, temren.
**arsenal** [ˈâsənəl]. Tersane; silâhhane; cebhanelik; silâh fabrikası.
**arson** [ˈâsən]. Kundakçılık.
**art**[1] [ât] *v.* be. **~**[2]**.** San'at; hüner, maharet; ictimaî ilimler; resim ve heykeltraşlık. **~ school,** ressamlık ve heykeltraşlık mektebi: **the black ~,** büyü: **the fine ~s,** güzel san'atler: **the noble ~ (of self-defence),** boks: **I had no ~ or part in it,** ben işin içinde değildim: **he is studying ~,** resim tahsil ediyor. **~ful** [ˈâtfəl]. Kurnaz; hilekâr; usta. **~fulness,** kurnazlık. **~isan** [âtiˈzan]. Zanaatkâr, el işçisi. **~ist** [ˈâtist]. San'atkâr; ressam; artist. **~istic** [–ˈtistik], san'ate aid, san'atkârca, san'atkâr ruhlu. **~less** [ˈâtlis]. Sade, tabiî, saf; oyunsuz; işlenmemiş.
**arter·ial** [âˈti̯əri̯əl]. Kan damarına aid; damara benziyen. **~ road,** ana yol. **~y** [ˈâtəri]. Kırmızı kan damarı; ana cadde, mühim yol.
**arthritis** [âˈθrâytis]. Mafsal iltihabı.
**artichoke** [ˈâtiçou̯k]. Enginar. **(Jerusalem) ~,** yer elması.
**article** [ˈâtikl] *n.* Makale; madde; eşya, parça, nesne; harfitarif. *v.* Birisini (talebe olarak) bir mimar vs.nin yanına vermek. **~d clerk,** staj gören kâtib: **~s of apprenticeship,** usta ile çırak arasında mukavele: **~ of faith,** akide, iman edilen şey: **~s of war,** askerî ceza kanunu ve nizamnamesi.
**articulate** [âˈtikyulit] *a.* Mafsallı; seçkin, tane tane söylenen; lâkırdısı anlaşılır. *v.* [–lêyt] Mafsal ile birleştirmek; mafsal teşkil etm.; tane tane telâffuz etm., seçkin konuşmak.
**artifice** [ˈâtifis]. Hile, oyun, kurnazlık, desise; ustalık, marifet. **~r** [âˈtifisə(r)]. Usta işçi; ordu veya bahriyede bazı zanaatlerde mütehassıs olan er.
**artificial** [âtiˈfişəl]. Yapma, sun'î. **~ity** [–ˈaliti], sun'ilik.
**artillery** [âˈtiləri]. Toplar; topçuluk; topçu (sınıfı). **~man,** topçu eri.
**Aryan** [ˈe̯əri̯ən]. Hind-Avrupa grupuna mensub. Arî.

**as** [az]. Gibi; kadar; olarak; sıfatla; için; ···diği için; ···dikçe; ···ince; ···diğinden: ~ **if** [~ **though**], sanki. ~ **I do not know,** bilmediğim için: ~ **a nation the English,** İngiliz milleti umumiyetle: **A is to B** ~ **C is to D,** A'nın B'ye nisbeti ne ise C'nin D'ye nisbeti de odur: ~ **big** ~ **...,** ··· kadar büyük; **twice** ~ **big** ~ **this,** bunun iki misli büyüklükte: **is it** ~ **difficult** ~ **that?,** bu kadar güç mü?: ~ **far** ~ **I can,** elimden geldiği kadar: **he is** ~ **industrious** ~ **he is intelligent,** zeki olduğu kadar çalışkandır: **by day** ~ **well** ~ **by night,** hem gece hem gündüz: ~ **you like,** nasıl isterseniz: ~ **for that** [~ **regards that,** ~ **to that**], buna gelince: ~ **to** [**for**] **you,** size gelince: ~ **a child,** çocukken: ~ **often happens,** ekseriya olduğu gibi: **he came** ~ **I left,** ben giderken o geldi: ~ **he spoke,** bunu söyler söylemez [söylediği anda]: ~ **the season advances so the days get longer,** mevsim ilerledikçe günler uzuyor: ~ **you treat others so will they treat you,** başkalarına nasıl muamele edersen onlar da sana öyle muamele ederler: **leave it** ~ **it is!,** olduğu gibi bırakınız!: **I should like to come but** ~ **it is I cannot,** gelmek isterdim fakat bu vaziyette mümkün değil: **we should like more pupils but,** ~ **it is, the school is full up,** daha fazla talebe isterdik fakat mekteb daha şimdiden doldu: **cold** ~ **it is I'll have a swim,** soğuğa rağmen yüzmeğe gideceğim: **if I had seen it,** ~ **I did not, I would have told the police,** eğer ben görseydim—ki görmedim—polise haber verirdim: **I had him** ~ **a partner,** o benim ortağımdı: **I remember him** ~ **having been a good tennis player,** ben onun iyi bir tenisçi olduğunu hatırlıyorum: **will you be so kind** ~ **to tell me?,** lûtfen bana söyler misiniz?: **this child is lazy** ~ **lazy,** bu çocuk tembel mi tembel: ~ **you were!,** *evvelce verilen bir emri hükümsüz bırakamak için verilen emir*: **the train is at 6.30;** ~ **you were! 7.30,** tren altı buçukta, pardon!, yedi buçukta.

**asbestos** [as<sup>ı</sup>bestos]. Yanmaz kâğıd, asbest.

**ascend** [ə<sup>ı</sup>send]. Çıkmak, yükselmek; tırmanmak.

**ascendan·cy** [ə<sup>ı</sup>sendnsi]. Üstünlük; hakimiyet; nüfuz. **~t,** yükselen, hakim olan, nüfuz kazanan: **to be in the** ~, yıldızı parlamak; talih ve itibarı artmak.

**ascen·sion** [ə<sup>ı</sup>senşən]. Yükselme. **~t** [ə<sup>ı</sup>sent], (dağa vs.) çıkış, tırmanma; yokuş; yükselme.

**ascertain** [<sub>ı</sub>asö(r)<sup>ı</sup>têyn]. Doğrusunu öğrenmek; anlamak; tahkik ve tayin etmek.

**ascetic** [ə<sup>ı</sup>setik]. Zevklerden el çekmiş, zahid. **~ism** [-sizm], zahidlik, riyazet.

**ascribe** [as<sup>ı</sup>krâyb]. Atfetmek; irca etm., isnad etm.

**asep·sis** [a<sup>ı</sup>sepsis]. Mikrobsuzluk, asepsi. **~tic,** ilac kullanmadan mikrobların önüne geçen, aseptik.

**ash**[1] [aş]. Dişbudak ağacı. **mountain** ~, üvez ağacı. **~**[2]. Kül. **a person's ~es,** yakılan cesedin külleri: **~-bin,** çöp tenekesi. **~-tray,** sigara tablası. **~en** [<sup>ı</sup>aşən]. Kül gibi, kül renginde olan.

**ashamed** [ə<sup>ı</sup>şêymd]. Utanmış, mahcub. **to be** ~, utanmak.

**ashore** [ə<sup>ı</sup>şôə(r)]. Sahilde, karada. **to go** ~, karaya çıkmak: **to run** ~, karaya otur(t)mak.

**Asia** [<sup>ı</sup>êyşə]. Asya. **~tic** [<sub>ı</sub>êyşi<sup>ı</sup>atik], Asya'ya aid, Asyalı.

**aside** [ə<sup>ı</sup>sâyd]. Bir tarafa; köşeye; kendi kendine; alçak sesle söylenen söz.

**ask** [âsk]. Sormak; davet etmek. ~ **for,** istemek; dilemek: **I ~ed the director,** müdüre sordum: **I ~ed for the director,** müdürü görmek istedim: **I ~ed about the director,** müdürü sordum: **to** ~ **after s.o.** [**s.o.'s health**], hatırını sormak: **to** ~ **s.o. in,** birini eve [içeriye] davet etm.: **to** ~ **s.o. out,** birini bir lokanta vs. gibi dışarıda bir yere davet etm.: **to** ~ **s.o. up,** birini apartıman vs. gibi yukarıda bir yere davet etm.: **to** ~ **back,** geri istemek: **to** ~ **s.o. back,** mukabele olarak davet etm.: **he ~ed for it (he brought it on himself),** çanak tuttu: **he's ~ing for a beating,** (dayak yemek için) kaşınıyor. **~ing** [<sup>ı</sup>âskin(g)] *n.* **you may have it** [**it's yours**] **for the** ~, dilediğin zaman senindir.

**askance** [ə<sup>ı</sup>skans]. **to look** ~ **on,** ···e şübhe ve itimadsızlıkla bakmak.

**aslant** [ə<sup>ı</sup>slânt]. Bir yana doğru, iğri.

**asleep** [ə<sup>ı</sup>slîp]. Uykuda, uyumuş. **to be** ~, uyumak, uyukuda olm.: **my foot is** ~, ayağım uyuşmuş.

**asparagus** [əs<sup>ı</sup>parəgʌs]. Kuşkonmaz.

**aspect** [ˈaspekt]. Görünüş; manzara; yüz; cebhe, taraf. **a south ~,** cenuba bakış.
**aspen** [ˈaspən]. Titrek kavak. **to tremble like an ~ leaf,** tir tir titremek.
**asperity** [əsˈperiti]. Sertlik, haşinlik.
**aspers·e** [əsˈpö(r)s]. İftira etm. **~ion,** iftira: **to cast ~s upon s.o.,** birisinin üzerine iftira atmak.
**asphyxiat·e** [asˈfiksiêyt]. (Havagazı vs.) boğmak. **~ion** [–ˈêyşn], havagazı vs. ile boğulma.
**aspir·ant** [ˈaspərant]. Tâlib; şiddetle heves ve arzu eden. **~e** [əsˈpâyə(r)]. **~ to,** emel beslemek; talib olm., peşinde olmak. **~ation,** emel, iştiyak.
**aspirat·e** [ˈaspirêyt] *v.* (Bir harfi) solukla telâffuz etm.; (havayı, suyu) içine çekmek, emmek. *a. & n.* [–rit], solukla telâffuz edilen (harf); h harfi. **to drop his ~s,** h harfini telâffuz etmemek (ki halk tabakasına mensubiyeti gösterir). **~ion,** (hava vs.yi) içine çekme, emme; (bir harfi) solukla telâffuz etme.
**ass** [as]. Eşek; abdal, enayi. **to make an ~ of oneself,** abdallık etm., enayilik etm., gülünc olmak. [*İngilizcede bu kelimenin mecazî mânası türkçedeki kadar ağır değildir ve şaka olarak kullanılır.*]
**assail** [əˈsêyl]. Saldırmak, hücum etmek. **~ ant,** saldıran kimse.
**assassin** [əˈsasin]. Kaatil. **~ate,** katletmek, öldürmek. **~ation,** taammüden katil.
**assault** [əˈsôlt]. Hücum; (birdenbire) taarruz; tecavüz. Hücum etm., tecavüz etmek.
**assay** [aˈsêy]. Bir maden halitasını, bir külçe ayarını, tayin etme(k); deneme(k), tecrübe etme(k).
**assembl·age** [əˈsemblic]. Toplantı; toplama; birleştirme. **~e** [əˈsembl]. Topla(n)mak, birleş(tir)mek; (makine vs.) kurmak. **~y,** Meclis; ictima, toplantı; (makine) kur(ul)ma.
**assent** [əˈsent]. Rıza, tasvib. **to ~ (to),** Razı olm., tasvib etm.; kabul etm.: **to receive the Royal Assent,** (kanun hakkında) kıral tarafından tasdik edilmek.
**assert** [əˈsö(r)t]. İleri sürmek; ısrar etmek. **to ~ oneself,** otoritesini göstermek. **~ion,** [əˈsö(r)şn] ileri sürme, iddia; ısrar; hakkını kullanma. **~ive,** fazla iddialı; kendine fazla güvenir.
**assess** [əˈses]. (Vergi vs. hakkında) mikdarını tayin etm.; kıymet biçmek; **~able,** zarar ve ziyan keşif ve takdir olunabilir; vergi tarh olunabilir; kıymeti takdir olunabilir. **~ment,** keşif ve takdir etme; takdir edilen mikdar (meblağ). **~or,** muhammin; tahakkuk memuru.
**asset** [ˈaset]. Fayda temin eden şey; kâr; aktif. **~s,** mallar; borcları ödeyecek karşılık.
**assidu·ity** [asiˈdyûiti]. Dikkat ve devamla çalışma. **~ous** [əˈsidyuəs], dikkatli ve devamlı (çalışan, çalışma).
**assign** [əˈsâyn]. Ayırmak, tahsis etm.; (bir malı) ferağ etm.; atfetmek. **~ment,** tayin, tahsis; taksim; ferağ. **~ation** [ˌasigˈnêyşn]. Buluşma için zaman ve yer tayini; (*jur.*) nakil ve ferağ.
**assimilate** [əˈsimilêyt]. Hazmetmek; massetmek; benzetmek.
**assist** [əˈsist]. Yardım etm.; iştirak etmek. **to ~ at,** ···de hazır bulunmak. **~ance,** yardım, muavenet. **~ant,** yardımcı, muavin; (bir mağaza vs.de) satıcı.
**assizes** [əˈsâyzəs]. İngiltere'deki kontluklarda zaman zaman kurulan ve seyyar hâkimler tarafından teşkil edilen muvakkat mahkemeler.
**associat·e** [əˈsouşiêyt] *v.* Birleştirmek; ortak olm.; düşüp kalkmak; iştirak etm.; akla getirmek. *n.* [–şi·it], Ortak, şerik; yarı âza, muhabir âza; arkadaş. **~ion** [–ˈêyşn], birleşme, ortaklık, iştirak, münasebet; hatıra; birlik, cemiyet: **~ football,** yalnız ayakla oynanan futbol.
**assort** [əˈsôt]. Cinslere veya çeşitlere ayırmak; uymak. **~ed,** çeşitli. **~ment,** cinslere, çeşitlere, ayırma; muhtelif çeşitlerden mürekkeb takım.
**assuage** [əˈswêyc]. Yumuşatmak; gidermek.
**assum·e** [əˈsyûm]. Üstüne almak; takınmak; halini almak; (iktidarı vs.) eline almak; sahib olm.; farzetmek, sanmak. **~ption** [əˈsʌmşn]. Üstüne alma; takınma; farzetme, sanma.
**assur·ance** [əˈşuərəns]. Temin; kendine güvenme; teminat; pişkinlik; kendini beğenmişlik; sigorta.
**assure** [əˈşuə(r)]. Temin etm.; teminatta bulunmak; sigorta etmek. **~d,** müemmen. **~dly,** muhakkak, şübhesiz.

**asterisk** [ˈastərisk]. Yıldız işareti (*).
**astern** [əˈstö(r)n]. Geminin kıç tarafında; arkada. **to go ~,** geri gitmek.
**asthma** [ˈasθmə]. Nefes darlığı. **~tic** [ˌasθˈmatik], nefes darlığına aid; nefes darlığı çeken.
**astir** [əˈstö(r)]. Harekette; heyecanlı. **to set ~,** harekete getirmek, kımıldatmak.
**astonish** [əˈstoniş]. Hayrette bırakmak, şaşırtmak. **~ed,** şaşkın, şaşmış, hayran. **~ment,** hayret, şaşkınlık, şaşırma.
**astound** [əˈstaund]. Hayretten dondurmak, şaşkınlıktan serseme çevirmek. **~ing,** sersemletici; müdhiş.
**astray** [əˈstrêy]. Yolunu şaşırmış, doğru yoldan çıkmış. **to go ~,** yolunu şaşırmak; baştan çıkmak: **to lead ~,** baştan çıkarmak.
**astride** [əˈstrâyd]. Bacakları ayrılmış, ata biner gibi.
**astringent** [əˈstrincənt]. Kanı durduran (ilac); kaabız; ağız buruşturucu.
**astrolo·ger** [ˌasˈtroləcə(r)]. Müneccim. **~gy,** yıldızlardan talihi okuma, ilmi nücum, müneccimlik.
**astrono·mer** [əˈstronəmə(r)]. Hey'et âlimi, felekiyatçı. **~mic(al)** [–ˈnomikl], hey'et ilmine aid; **prices are ~,** fiatler akla durgunluk verir. **~my,** hey'et ilmi, felekiyat.
**astute** [əˈstyût]. Keskin zekâli; kurnaz.
**asunder** [əˈsʌndə(r)]. (Birbirinden) ayrı, ayrılmış; parçalara ayrılmış. **to tear ~,** ikiye ayırmak. **to put ~,** ayırmak.
**asylum** [əˈsâyləm]. Sığınak, melce. **(lunatic) ~,** tımarhane.
**asymmetr·ic** [ˌasiˈmetrik]. Tenazursuz, nisbetsiz. **~y** [əˈsimətri], tenazursuzluk.
**at** [at]. …de, …ye. **to be ~ s.o.,** başının etini yemek: **to be (always) ~ stg.,** bir şey ile (daima) meşgul olm.: **he is ~ it again,** (işte) yeni başladı: **while we are ~ it,** hazır bu iş üzerinde iken: **~ night,** geceleyin, gece: **~ your request,** talebiniz üzerine: **~ them!,** ileri!, hücum!
**atavis·m** [ˈatavizm]. Atalara çekiş. **~tic** [–ˈvistik], atalara çeken.
**ate** [et, êyt] *v.* eat.
**atheis·m** [ˈêyθi·izm]. Allahı inkâr etme, dinsizlik. **~t,** Allahı inkâr eden; dinsiz.
**Athen·ian** [əˈθîniən]. Atina'ya aid, Atinalı. **~s** [ˈaθənz], Atina.
**athirst** [əˈθö(r)st]. **~ for,** …e susamış, teşne.
**athlet·e** [ˈaθlît]. Atlet, sporcu. **~ic** [əˈθletik], gürbüz ve çevik; spora aid. **~ics,** atletik sporlar.
**at-home** [ətˈhoum]. Misafir kabulü. **~ day,** kabul günü.
**Atlantic** [atˈlantik]. **the ~ (Ocean),** Atlas okyanusu.
**atlas** [ˈatlas]. Atlas; harita.
**atmospher·e** [ˈatməsfiə(r)]. Atmosfer; muhit, hava. **~ic(al)** [–ˈferik(l)], atmosfere aid, havaya aid.
**atoll** [ˈatol]. Ortasında bir göl teşkil eden halka biçiminde mercan adası.
**atom** [ˈatəm]. Atom; zerre. **~ic(al)** [əˈtomik(l)], atoma aid. **~ize** [ˈatəmâyz], mayii toz haline getirmek, püskürmek.
**atone** [əˈtoun]. **~ for,** kefaret vermek, **~ment,** kefaret, telâfi.
**atroci·ous** [əˈtrouşəs]. Vahşice; gaddar; çok çirkin; berbad. **~ty** [əˈtrositi], vahşet; zulüm, gaddarlık.
**atrophy** [ˈatrəfi]. Dumur; zafiyet. Dumura uğratmak; zayıf düş(ür)mek.
**attach** [əˈtaç]. Bağlamak, birleştirmek; (*jur.*) haczetmek; bağlı olm., merbut olmak. **to ~ oneself to,** …e iltihak etm., takılmak: **to be ~ed to,** …e bağlı olm., merbut olm.; sevmek. **~ment,** bağlılık; dostluk; rabıta; bir şeyin takılabilir parçası; (*jur.*) tevkif, haciz.
**attaché** [əˈtaşêy]. Ataşe. **~-case,** evrak çantası.
**attack** [əˈtak] *n.* Saldırış, hücum, taarruz; nöbet. *v.* Saldırmak, hücum etm.; (bir işe) girişmek. **to return to the ~,** tekrar hücuma geçmek.
**attain** [əˈtêyn]. Varmak, ermek, erişmek, kazanmak. **~ment,** varma, erişme, elde etme; hüner, malûmat: **~s,** müktesebat marifetler.
**attar** [ˈatə(r)]. **~ of roses,** Gülyağı.
**attempt** [əˈtempt]. Teşebbüs (etm.), çalışma(k); tecrübe (etmek). **to make an ~ on s.o.'s life,** birinin hayatına kasdetmek: **~ed murder,** cinayet teşebbüsü: **I'll do it or perish in the ~,** ne olursa olsun onu yapacağım.
**atten·d** [əˈtend]. Gidip hazır bulunmak; beraber olm.; tedavi etm.; dikkat etm. **to ~ to,** dinlemek; bakmak; …e hizmet etm.; …le meşgul olm.: **to ~ on s.o.,** refakatinde bulunmak. **~dance** [əˈtendəns]: Hazır bulunma; devam; hizmet; refaket; tedavi. **there was a small ~,** gelenler

azdı. ~**dant,** *a.* beraber olan, refaket eden; hazır ve mevcud olan; devam eden; *n.* Hizmetçi; (mağaza, müze, tiyatro vs. de) memur: **medical ~,** bir şahısın hususî doktoru. ~**tion** [əˈtenşn]. Dikkat; ihtimam; itina; meşgul olma; alâka gösterme; nezaket. ~**!,** hazırol!: **to pay ~,** dikkat etm.: **to pay one's ~s to,** ···e kur yapmak: **to stand at ~,** hazırol vaziyetinde durmak. ~**tive** [eˈtentiv]. Dikkatli; itinalı.

**attenuate** [əˈtenyuêyt]. İnceltmek.

**attest** [əˈtest]. Tasdik etmek.

**attic** [ˈatik]. Tavan arası, çatı odası.

**attire** [əˈtâyə(r)]. Elbise. Giydirmek.

**attitude** [ˈatityûd]. Davranış, tavır, duruş. **to strike an ~,** poz almak.

**attorney** [əˈtö(r)ni]. Vekil, mümessil; avukat. **power of ~,** vekâletname. ~**-general,** baş müddeiumumî.

**attract** [əˈtrakt]. Çekmek, cezbetmek. ~**ion,** çekme, cazibe; cezbeden şey; alım. ~**ive,** çeken, cazib; alımlı, göz alıcı.

**attribut·e** [ˈatribyût] *n.* Vasıf, sıfat, hassa. *v.* [əˈtribyût]. Atfetmek; yormak; maletmek. ~**ion** [–ˈbyûşn], atfetme, hamletme; yorma; hassa. ~**ive** [əˈtribyutiv], vasıflandıran; sıfat.

**attrition** [əˈtrişn]. Aşınma; yenme. **war of ~,** aşındırma harbi.

**auburn** [ˈôbən]. Kestane rengi.

**auction** [ˈôkşn] *n.* Mezad, açık arttırma ile satış. *vb.* Mezada çıkarmak. ~**eer** [–ˈniə(r)], mezad tellalı.

**audac·ious** [ôˈdêyşəs]. Cür'etli; atılgan. ~**ity** [ôˈdasiti], cür'et, cesaret.

**audible** [ˈôdəbl]. İşitilebilir, işitilir.

**audience** [ˈôdyəns]. Dinleyiciler; seyirciler; dinleme; huzura kabul.

**audit** [ˈôdit]. Hesabları teftiş; teftiş etme(k); murakabe.

**audit·ion** [ôˈdişn]. Dinleme; bir şarkıcı vs.nin sesini tecrübe için dinleme. ~**or,** dinleyici; murakib. ~**orium** [–ˈtôriəm], dinleyicilere mahsus yer.

**augment** [ôgˈment]. Arttırmak, çoğaltmak.

**augur** [ˈôgə(r)]. Kâhin, falcı. (Fala bakarak vs.) haber vermek. **it ~s no good,** hayra alâmet değil: **it ~s well,** hayra alâmettir. ~**y** [ˈôgyuri], kehanet, fal; alâmet; falcılık: **to take the ~ies,** fala bakmak.

**august** [ôˈgʌst]. Mübeccel; muhteşem.

**August** [ˈôgəst]. Ağustos.

**aunt** [ânt]. Teyze; hala; yenge (dayı veya amca karısı). ~ **Sally,** herkesin takıldığı kimse.

**aural** [ˈôrəl]. Kulağa aid.

**Aurora** [ôˈrôrə]. Şafak tanrısı; fecir. ~ **borealis,** şimal kutbu fecri.

**auspic·e** [ˈôspis]. Fal. **under the ~s of,** ···in himayesi altında; sayesinde. ~**ious** [ôsˈpişəs]. Müsaid, uygun; uğurlu.

**auster·e** [ôsˈtiə(r)]. Sert; haşin; müsamahasız; süssüz. ~**ity** [ôsˈteriti], süssüzlük; imsâk.

**Australia** [ôsˈtrêylyə]. Avustralya. ~**n,** Avustralyalı.

**Austria** [ˈôstriə]. Avusturya. ~**n,** Avusturyalı.

**authentic** [ôˈθentik]. Mevsuk; hakikî. ~**ate** [–kêyt], tevsik etm.; hakikî olduğunu göstermek; müellifini tesbit etmek. ~**ity** [–ˈtisiti], mevsukiyet; hakikî olma.

**author** [ˈôθə(r)]. Müellif; fail; sebeb olan.

**author·itarian** [ôˌθoriˈteriən]. Otoriter. ~**itative** [ôˈθoritətiv], resmî; âmirane; salâhiyet ve ihtisas sahibi; mevsuk. ~**ity** [ôˈθoriti]. Salâhiyet; nüfuz; otorite; salâhiyet ve hüküm sahibi şahıs; müsaade. **the authorities,** makamlar, idare: **to act on s.o.'s ~,** birinin müsaadesile hareket etm.: **to be an ~ on stg.,** bir meselede ihtisas sahibi olm.: **on good ~,** mevsuk kaynaktan: **under s.o.'s ~,** birinin emrinde. ~**ize** [ˈôθərâyz]. Salâhiyet vermek, mezuniyet vermek; müsaade etmek. ~**ized,** salâhiyetli, resmî: **the ~ Version (of the Bible),** İncilin 1611 de yapılan İngilizce resmî tercümesi.

**autobiography** [ˌôtəbâyˈogrəfi]. Hatırat, müellif tarafından yazılan tercümeihal.

**autocra·cy** [ôˈtokrəsi]. İstibdad. ~**t(ic)** [ˌôtəˈkrat(ik)], müstebid.

**autograph** [ˈôtəgrâf]. Birinin kendi elyazısı, imzası(nı atmak).

**automatic** [ˌôtəˈmatik]. Otomatik. ~ **machine,** para atınca bilet vs. veren makine.

**autono·mous** [ôˈtonəməs]. Kendi kendini idare eden, muhtar. ~**my,** kendi kendini idare salâhiyeti, muhtariyet.

**autopsy** [ôˈtopsi]. Fethimeyyit, otopsi.

**autumn** [ˈôtəm]. Sonbahar. ~**al** [–ˈʌmnel], sonbahara aid.

**auxiliary** [ôg'zilyəri]. Yardımcı: tâli.
**avail** [ə'vêyl] *n.* **of no ~,** beyhude. *v.* Faydası olm. **to ~ oneself of,** ···den istifade etm. **~able,** mevcud, elde edilebilir; (bilet vs.) muteber; mer'î.
**avalanche** ['avəlânç]. Çığ.
**avaric·e** ['avəris]. Tamah; para hırsı. **~ious** [ˌavə'rişəs], tamahkâr, paraya haris.
**avenge** [ə'venc]. Öc almak, intikam alma.
**avenue** ['avənyu]. Ağaçlıklı cadde; geniş cadde; (*fig.*) bir yere eriştiren yol.
**aver** [ə'və(r)]. Doğru olduğunu ileri sürmek.
**average** ['avəric]. Orta, vasat. Vasatî(sini almak). **general ~,** avarya: **on an ~,** ortalama: **to strike an ~,** vasatîsini çıkarmak.
**avers·e** [ə'vö(r)s]. Muhalif, zıd; hazzetmez. **~ion** [ə'vö(r)şn]. Nefret; hoşlanmayış; zıddiyet; nefret edilen şey. **pet ~,** en çok nefret edilen kimse veya şey.
**avert** [ə'vö(r)t]. Önüne geçmek; bertaraf etmek.
**aviary** ['êyviəri]. Kuşhane.
**aviat·ion** [ˌêyvi'êyşn]. Uçakçılık, havacılık. **~or** ['êyviêytə(r)], uçakçı, havacı.
**avid** ['avid]. Haris, açgözlü; doymaz.
**avoid** [ə'vôyd]. İctinab etm.; ···den sakınmak; kurtulmak, önlemek. **~able,** kaçınabilir, önlenebilir. **~ance,** kaçınma; önleme.
**avoirdupois** [ˌavədə'pôyz, avwadyu'pwa]. İngiliz ağırlık ölçüsü sistemi.
**avow** [ə'vau]. İtiraf etm. **~al,** itiraf. **~edly,** açıkça.
**awake** (*p.* **awoke**) [ə'wêyk, ə'wouk]. Uyan(dır)mak. Uyanmış; uyanık. **to be ~ to a danger,** bir tehlikenin farkında olm.: **wide ~,** tamamen uyanmış; gözü açık. **~n,** Uyandırmak; birinin gözünü açmak. **~ning,** uyanma; kendine gelme; gözü açılma: **a rude ~,** acı bir sukutu hayal.
**award** [ə'wôd]. Hüküm; karar; tazminat, mükâfat. (Mükâfat tazminat olarak) vermek; ihale etm.; karar vermek.
**aware** [ə'weə(r)]. Haberdar, farkında, bilir, agâh. **to become ~ of,** öğrenmek: **not that I am ~ of,** benim bildiğime göre (böyle) değil; benim haberim yok.
**awash** [ə'woş]. Su seviyesinde.
**away** [ə'wêy]. Uzağa, uzakta; ötede, öteye; bir tarafa; bir düziye. [*Bir fiil sonuna gelince şu mânaları ifade eder*:—(i) uzaklaşma *veya* değiştirme, *mes.* **to walk ~,** yürüyüp uzaklaşmak; (ii) devam, *mes.* **to write ~,** yazmağa devam etm.; (iii) tüketmek, *mes.* **the sea is eating ~ the rocks,** deniz kayaları aşındırıyor; **to waste ~,** eriyip bitmek.] **~ back in 1900,** daha [tâ] 1900 de: **~ with you!,** defol! : **~ with it!,** kaldır!, götür!: **he is ~,** seyahattedir: **far ~,** çok uzakta: **he is far and ~ the cleverest,** fersah fersah en zekisidir: **I must ~,** gitmeliyim: **right ~,** hemen: **to stay ~,** (i) orada bulunmamak; (ii) başka yerde kalmak.
**awe** [ô]. Hürmet, hayranlıkla karışık korku, huşu (telkin etm.). **to stand in ~ of,** huşula telâkki etm., **to strike with ~,** korku vermek. **~some** ['ôsəm], dehşet veren. **~-inspiring,** huşu telkin eden.
**awful** ['ôfl]. Korkunc, müdhiş; heybetli; (*coll.*) berbad. **thanks ~ly,** sonsuz teşekkürler: **he's ~ly kind,** son derece mültefittir.
**awhile** [ə'wâyl]. Bir müddet, bir az.
**awkward** ['ôkwəd]. Beceriksiz; hantal; aksi; sıkıntılı. **the ~ age,** çocukluktan çıkma çağı: **an ~ customer,** tehlikeli adam.
**awl** [ôl]. Biz.
**awning** ['ônin(g)]. Tente.
**awoke** [ə'wouk] *v.* **awake.**
**awry** [a'rây]. Çarpık. **to go ~,** ters gitmek.
**axe** [aks]. Balta. **to have an ~ to grind,** bir işte çıkarı olmak.
**axi·al** ['aksiəl]. Mihvere aid. **~s** ['aksis]. Mihver.
**axiom** ['aksiəm]. Mütearife. **~atic,** besbelli.
**axle** ['aksəl]. Dingil; mil.
**ay(e)**[1] [ây]. Evet; hayhay; kabul reyi. **the ~s have it,** kabul edilmiştir. **~**[2]. Daima.
**azure** ['ajuə(r)]. Havaî mavi; mavi gök.

# B

**B** [bî]. B harfi; (*mus.*) si.
**B.A.** ['bî'êy]. (*abb.*) **Bachelor of Arts.**
**baa** [bâ]. Melemek.
**babble** ['babl] (Çocuk gibi) manasız sesler **çıkarma(k)**; saçma sapan

konuşma(k); şırıldamak; şırıltı. **to ~ out stg.**, boşboğazlık etmek. **~er,** boşboğaz, geveze.

**babe** [bêyb]. Küçük çocuk, bebek.

**Babel** [ˈbêybl]. Babil. **a ~ of talk,** gürültülü konuşma.

**baboon** [baˈbûn]. Kısa kuyruklu iri bir cins maymun.

**baby** [ˈbêybi]. Bebek, küçük çocuk; küçük. **~ grand,** kısa kuyruklu piyano: **he left me to carry the ~,** her şeyi benim üstüme bıraktı (o işin içinden sıyrıldı): **to have a ~,** doğurmak: **she is going to have a ~,** hamiledir. **~hood,** bebeklik, çocukluk. **~ish,** çocukça, bebekçe. **~-farm,** çocuk bakımevi (ücretli). **~-linen,** çocuk takımı.

**bacchan·alia** [bakəˈnêyliə], Baküs festivali; içki âlemi. **~alian,** işrete aid.

**baccy** [ˈbaki]. (*abb.*) **tobacco,** tütün.

**bachelor** [ˈbaçələ(r)]. Bekâr. **~ of Arts,** Edebiyat Fakültesi mezunu: **~ of Science,** Fen Fakültesi mezunu. **~hood,** bekârlık.

**back¹** [bak]. Arka; sırt; ters; (futbol) bek, müdafi. Arkada bulunan; karşı. Geri, geride, geriye; mukabeleten; evvel. **~current,** ters akıntı; **~ to front,** ters: **to answer ~,** karşılık vermek: **when will he be ~?,** ne zaman dönecek?: **to be on one's ~,** (i) arka üstü yatmak; (ii) hasta yatmak: **to break one's ~,** belini kırmak: **to break the ~ of the work,** bir işin çoğunu bitirmek: **(ship) to break her ~,** (gemi) omurgasını kırmak: **it all comes ~ to me now,** şimdi her şeyi tekrar hatırlıyorum: **to get at the ~ of stg.,** işin içyüzünü anlamak: **the idea at the ~ of one's mind,** asıl maksad: **to get (a bit of) one's own ~,** acısını çıkarmak: **to put one's ~ into stg.,** var kuvvetile çalışmak: **to put [get] s.o.'s ~ up,** birini kızdırmak: **to sit with one's ~ to,** arkası ···e dönük oturmak: **a house standing ~ from the road,** içerlek ev: **there's something at the ~ of it,** bunun içinde bir iş var: **to turn one's ~ on s.o.,** birine arkasını çevirmek; birinden yüz çevirmek: **with one's ~ to the wall,** (çarpışma vs.) ricat hattı kesilmiş olarak; mezbuhane. **~-breaking,** insanın belini kıran, çok yorucu. **~-chat,** karşılık verme. **~-fire,** (*otom.*) alevin geri tepmesi. **~-hand,** (tenisde) soldan vurma. **~ handed,** elinin tersiyle; **a ~-handed compliment,** hoşa gitmiyen kompliman. **~-number,** günü geçmiş; eski; modası geçmiş. **~-pay,** tediyesi gecikmiş maaş. **~-scratching,** piyaz, karşılıklı pohpoh. **~-sight,** gez. **~-tooth,** azı dişi. **~-yard,** bir evin arka avlusu. **~²** *v.* Arka olm.; muzaheret etm., lehinde söylemek; himaye etm.; geri yürütmek; geri geri gitmek; arkadan desteklemek. **(wind) to ~,** (rüzgâr) sağdan sola değişmek: **to ~ a bill,** aval vermek: **to ~ a horse,** (yarışta) bir ata para yatırmak. **~ down,** iddiasından vazgeçmek. **~ out,** geri geri gitmek; **to ~ out of a promise,** vadinden dönmek. **~ up,** lehinde söylemek. **~bencher** [ˌbakˈbençə(r)]. Parlamentonun kabinede yahud ön safta bulunmıyan âzası. **~bite** [ˈbakbâyt]. Birini giyabında zemmetmek; arkasından çekiştirmek. **~bone** [ˈbakbo͡un]. Belkemiği; (*fig.*) karakter. **English to the ~,** sapına kadar İngiliz. **~er** [ˈbakə(r)]. Bir yarışta para koyan; tarafdar. **~ground** [ˈbakgra͡und]. Arka plân; geri; fon. **~gammon** [bakˈgamən]. Tavla. **~side** [bakˈsâyd]. Kıç. **~slide** [bakˈslâyd]. Fena yola sapmak; tekrar hataya düşmek. **~stair(s)** [bakˈste͡ər(s)]. Arka merdiven. **~ gossip,** hizmetçi dedikodusu. **~ward** [ˈbakwəd]. Geri; gecikmiş. **~wards,** geriye doğru; geri; tersine. **~wardness,** gerilik. **~wash** [ˈbakwoş]. Geminin akıntısı. **~water** [bakwôtə(r)]. Faaliyetsiz bir yer. **~woodsman,** [bakˈwudzmən]. Balta görmemiş ormanda yerleşmiş kimse.

**bacon** [ˈbêykən]. Tuzlanmış, tütsülenmiş domuz eti. **to save one's ~,** yakayı kurtarmak.

**bad** [bad]. *a.* Kötü, fena; münasebetsiz, uygunsuz; bozuk, kokmuş; kusurlu; kalp; değersiz; berbad; (delil vs.) kifayetsiz. *n.* Fenalık; kötülük. **to be ~ at stg.,** bir şeyde iyi olmamak, becerememek: **~ debt,** tahsili kabil olmıyan alacak: **to go ~,** bozulmak, kokmak, çürümek: **to go to the ~,** fena yola sapmak: **to go from ~ to worse,** gittikçe fenalaşmak: **to call s.o. ~ names,** birine karşı fena sözler sarfetmek: **my ~ leg,** sakat bacağım: **it is ~ly needed,** buna çok büyük ihtiyac var: **it is very ~ of you to . . .,** ···diğiniz için çok kabahat-

lisiniz: **to be taken ~,** birdenbire hasta olmak: **I am £100 to the ~,** yüz lira kaybım var: **to want stg. ~ly,** bir şeyi şiddetle arzu etmek. **~-looking,** çirkin. **~-tempered,** huysuz, aksi. **~ness,** fenalık, kötülük.
**bade** [bad, bêyd]. *v.* **bid.**
**badge** [bac]. Alâmet, işaret; plâka; rozet.
**badger**[1] [ˈbacə(r)] *n.* Porsuk. ~[2] *v.* Başının etini yemek.
**baffle** [ˈbafl]. Ses vs.nin hareketini kontrol eden bir levha. Şaşırtmak. **to be ~d,** apışıp kalmak: **to ~ description,** tarifi imkânsız olmak.
**bag** [bag] *n.* Torba; çuval; kese kâğıdı; çanta; kese. *v.* Torbaya koymak; (elbise) şişmek; (*coll.*) aşırmak. **~ and baggage,** her şeyi ile.
**bagatelle** [ˌbagəˈtel]. Ehemmiyetsiz şey.
**baggage** [ˈbagic]. Bagaj; yolcu eşyası; ordu ağırlığı; yüzsüz, şımarık kadın.
**baggy** [ˈbagi]. Üstten sarkan (elbise).
**bagpipe** [ˈbagpâyp]. Gayda. **~s,** İskoçların tulum çalgısı.
**bail**[1] [ˈbêyl]. Kefalet (senedi); kefil. **to ~ out,** kefalet vererek tahliye ettirmek. **to go ~ for s.o.,** birine kefil olm. ~[2] *v.* Sandalın suyunu boşaltmak. **to ~ out,** paraşütle atlamak. **~ey** [ˈbêyli]. **The Old ~,** Londra cinayet mahkemesi. **~iff** [ˈbêylif]. Çiftlik kâhyası; icra memuru.
**bairn** [beərn]. (İskoç lehçesinde) çocuk.
**bait** [bêyt]. (Olta vs. için) yem; çekici ve aldatıcı şey. Eziyet etm.
**baize** [bêyz]. Kabaca dokunmuş yün kumaş (bilârdo ve oyun masaları için kullanılır).
**bake** [bêyk]. (Fırında) pişirmek. **~r,** ekmekçi: **~'s dozen,** on üç. **~house,** fırın, ekmekçi dükkânı. **~ry,** ekmekçilik; ekmekçi dükkanı.
**baking-powder.** Sun'î maya.
**baksheesh** [bakˈşîş]. Bahşiş.
**balance** [ˈbaləns]. Terazi; rakkas; muvazene; bakiye; hesab farkı. Denketmek; muvazene kurmak (bulmak); karşılaştırmak. **to hang in the ~,** muallakta, nazik bir vaziyette bulunmak: **to strike a ~,** bilanço çıkarmak: **to throw s.o. off his ~,** birinin muvazenesini kaybettirmek, şaşırmak: **to turn the ~,** vaziyeti değiştirmek. **~d,** muvazeneli. **~-sheet,** bilanço: **to get out a ~,** bir bilanço tanzim etmek.
**bald** [bôld]. Saçları dökülmüş, kel. **~ness,** saçsızlık. **~-headed,** kel: **to go at it ~,** düşüncesizce atılmak.
**balderdash** [ˈbôldədaş]. Saçma sapan söz.
**bale** [bêyl]. Balya (yapmak), denk (bağlamak).
**baleful** [ˈbêylfl]. Zararlı; uğursuz.
**balk** [bôlk]. Kiriş; engel; (iki tarla arasında) sürülmemiş kısım. Mani olm.: **to ~ at stg.,** bir şey karşısında durmak; tereddüd etmek.
**ball**[1] [bôl]. Top; yuvarlak; küre; yumak; mermi. **to ~ up,** top haline gelmek: **to have the ~ at one's feet,** eline fırsat geçmek: **to keep the ~ rolling,** işi devam ettirmek: **to start the ~ rolling,** işi açmak. **~-bearing,** bilye; bilyeli. **~-cartridge,** kurşunlu fişek. **~-cock,** yüzen top ile işliyen kapama valfı. ~[2]. Balo. **~room** [ˌbôlrum]. Dans salonu.
**ballad** [ˈbaləd]. Basit şarkı; tahkiyevî şiir.
**ballast** [ˈbaləst]. Safra (koymak); çakıl (döşemek); muvazenesini temin etmek. **to have ~,** temkinli olmak.
**balle·rina** [ˌbaləˈrînə]. Bale dansözü. **~t** [ˈbalêy]. Bale.
**ballistics** [baˈlistiks]. Atış ilmi.
**balloon** [bəˈlûn]. Balon. Balon gibi şişmek; balonla çıkmak.
**ballot** [ˈbalət] *n.* Kur'a; rey atma; reylerin yekûnu. *v.* Kur'a çekmek. **to ~ for,** rey ile seçmek: **to take a ~,** kur'a çekmek. **~-box,** rey kutusu. **~-paper,** rey kâğıdı.
**balm** [bâm]. Oğul otu; teskin eden şey; merhem. **~y,** güzel ve ağır kokulu; sıcak ve lâtif; (*coll.*) kaçık.
**balustrade** [ˌbaləˈstrêyd]. Tırabzan parmaklığı.
**bamboo** [bamˈbû]. Bambu (ağacı).
**bamboozle** [bamˈbûzl]. Aldatmak; kafese koymak.
**ban** [ban]. Aforoz; umumî efkâr tarafından mahkûmiyet; yasak. Yasak etm.; aforoz etmek. **to put a ~ on,** yasak etm.
**banal** [bəˈnâl]. Adî, mübtezel; basmakalıb. **~ity** [ˈnaliti] basmakalıb söz vs.
**banana** [bəˈnânə]. Muz.
**band**[1] [band]. Bağ; şerid; kuşak, kayış. Bağlamak. ~[2]. Takım; güruh. **to ~ together,** birleşmek; bir araya gelmek. ~[3]. Bando; çalgıcılar heyeti. **~master** [ˈbandˈmâstə(r)]. Bando şefi. **~sman,** *pl.*

**-men** [ˈbansmən]. Mızıkacı, bando çalgıcısı.
**bandage** [ˈbandəc]. Sargı, bağ. Sarmak, sargıya koymak.
**bandit** [ˈbandit]. Haydud, şaki.
**bandolier** [bandoˈliə(r)]. Omuz kayışı; fişeklik.
**bandy**[1] [ˈbandi]. **to ~ words,** ağız kavgası etmek. **~**[2]. **~(-legged),** İğri bacaklı.
**bane** [bêyn]. Afet, felâket; zehir. **~ful,** mahvedici, öldürücü; zehirli.
**bang** [ban(g)] *n.* (*ech.*). Bum, çat; vuruş; gürültü; patlama. *v.* Gürültü ile vur(ul)mak, kapa(n)mak; çarpmak, patlamak. **~ in the face,** tam yüzüne: **~ on time,** dakikası dakikasına: **to go ~,** patlamak.
**bangle** [ˈban(g)l]. Bilezik; halka.
**banish** [ˈbaniş]. Sürmek, koğmak. **~ment,** sürgün.
**banister** [ˈbanistə(r)]. Tırabzan.
**banjo** [banˈcou]. Kitaraya benzer bir çalgı.
**bank**[1] [ban(g)k] *n.* Sed; kıyı, kenar; sığlık; *v.* Sed çekmek; (uçak) bir tarafa yatmak. **to ~ up,** yığ(ıl)mak. **~**[2]. Kürekçi yeri, kürek sırası; org klâvyesi. **~**[3]. Banka; (oyunda) banko. Bankaya koymak. **to ~ on stg.,** bir şeye ümid bağlamak. **~ holiday,** resmî tatil günü. **~er,** bankacı. **~-note,** kâğıd para.
**bankrupt** [ˈban(g)krʌpt]. Müflis; meteliksiz. İflâs ettirmek: **to go ~,** iflâs etmek. **~cy,** iflâs; mahvolma.
**banner** [ˈbanə(r)]. Bayrak, sancak.
**banns** [bans]. Evlenme ilânı. **to forbid the ~,** evlenmelerine itiraz etm.: **to put up the ~,** evlenme ilânını resmî teşhir etmek.
**banquet** [ˈban(g)kwit]. Ziyafet.
**bantam** [ˈbantəm]. Küçük cins tavuk, ispenc. **~ weight,** horoz sıklet.
**banter** [ˈbantə(r)]. Şaka, lâtife, alay (etmek).
**bapti·sm** [ˈbaptizm]. Vaftiz. **~smal** [–ˈtizməl], vaftize aid. **~ze,** [–ˈtâyz], vaftiz etm., isim vermek.
**bar**[1] [bâ(r)] *n.* Çizgi; çubuk; (tahta veya madenden) kol; engel; çıta; kalıb (sabun); (mahkemede) suçlu yeri; baro; mahkeme; içki dağıtılan veya satılan tezgâh veya oda; nehir ağzında kum seddi. *v.* Kapamak; demirlemek; engel olm.; yasak etmek. **to be called to the ~,** baroya yazılmak: **colour ~,** beyaz ırkla renkli irklar arasında fark gözetme: **horizontal ~,** barfiks. **~maid** [ˈbâˌmêyd]. İçki tezgâhında çalışan kiz **~man** [ˈbâmən], içki tezgâhında garson. **~**[2], **~ring** [bâ(r), ˈbârin(g)]. Maada, müstesna. **~ none,** istisnasız.
**barb** [bâb]. Ok ucu; olta kancası, diken. **~ed** [bâbd]. Dikenli, kancalı: **~ wire,** dikenli tel: **~-wire entanglement,** tel örgü.
**barbar·ian** [bâˈberiən]. Barbar, vahşi; yabancı. **~ic** [–ˈbarik], barbarca, vahşice. **~ism** [ˈbâbərizm], barbarlık; alışılmamış kelime veya tabir. **~ity** [–ˈbariti], vahşet. **~ous,** barbarca, vahşice; zalim.
**barbecue** [ˈbâbəkyu]. Bütün olarak kızartılmış hayvan çevirmesi; bunun yendiği toplantı.
**barber** [ˈbâbə(r)]. Berber.
**bard** [bâd]. Saz şairi; şair.
**bare** [beə(r)]. Çıplak; açık; çorak; boş; sade; süssüz; ancak kâfi. Soymak; açmak. **~ly,** hemen hemen, ancak: **a ~ majority,** zayıf bir ekseriyet: **to earn a ~ living,** ancak hayatını kazanabilmek: **to lay ~,** açmak, açığa vurmak; yakıp yıkmak. **~** *v.* **bear.** **~back** [ˈbeəbak]. **to ride ~,** ata eğersiz binmek. **~faced** [ˈbeəfêysd]. Yüzsüz, hayasız. **~foot(ed)** [ˈbeəfut, –ˈfutid]. Yalınayak.
**bargain** [ˈbâgin]. Ticarî anlaşma, iş; pazarlık; kelepir. Ticarî bir işe girişmek; pazarlık etm. **It's a ~,** (i) uyuştuk!; (ii) kelepirdir: **into the ~,** üstelik, caba: **I didn't ~ for that,** bunu hiç beklemedim: **he got more than he ~ed for,** başına belâyı satın aldı: **to make a good ~,** kârli bir iş yapmak: **to strike [drive] a ~,** birile pazarlığı uydurmak, anlaşmak.
**barge**[1] [bâc] *n.* Mavna; yük dubası. **admiral's ~,** amirale mahsus sandal: **state ~,** saltanat kayığı. **~-pole,** uzun avara gönderi: **I wouldn't touch it with a ~,** kırk yıl görmesem aramam. **~e** [bâˈcî]. Mavnacı, dubacı. **~**[2] *v.* **~ into [against],** ···e çarpmak; şiddetle tos vurmak: **to ~ in,** yersiz medahale etmek.
**bark**[1] [bâk]. Ağac kabuğu. **to ~ one's shins,** incik kemiği sıyrılmak. **~**[2]. Havlama(k); haşin bir sesle söylemek. ⌜**his ~ is worse than his bite**⌝, sen onun bağırıp çağırmasına bakma: **to ~ up the wrong tree,** yanlış yerde aramak. **~**[3]. (*Şair.*) Sandal, gemi.
**barley** [ˈbâli]. Arpa. **~corn,** arpa tanesi.

**barm** [bâm]. Bira mayası. **~y,** (*sl.*) zıpır, kaçık.
**barn** [bân]. Zahire ambarı; ahır. **~yard,** çiftlik avlusu; kümes.
**barnacle** [ˈbânikl]. Kayalara, gemi diplerine yapışan bir nevi midye.
**barque** [bâk]. Üç direkli yelken gemisi.
**barrack(s)**[1] [ˈbarək(s)] *n.* Kışla; kışla gibi bina. ~[2] *v.* Sarakaya almak; yuhaya tutmak.
**barrage** [ˈbarâj]. Nehir barajı; bend; baraj atışı.
**barrel** [ˈbarəl]. Fıçı, varil; namlu. Fıçıya koymak. **~-organ,** laterna.
**barren** [ˈbarən]. Kısır; çorak.
**barricade** [ˌbarəˈkêyd]. Mania; siper. Sokak siperleri kurmak.
**barrier** [ˈbariə(r)]. Mania; engel; çit.
**barrister** [ˈbaristə(r)]. (Yüksek mahkemelere çıkabilen) avukat.
**barrow**[1] [ˈbarou]. Elarabası. ~[2]. Höyük; mezar tümseği.
**barter** [ˈbâtə(r)]. Mübadele, trampa. Trampa etmek. **to ~ away,** şerefini vs. satmak.
**basal** [ˈbêysl]. Kaideye veya esasa aid.
**base**[1] [bêys] *n.* Esas; temel, kaide; dib; üs; kök. *v.* Kurmak; esasını koymak; temelini atmak; istinad etmek. **~ment** [bêysmnt]. Bodrum katı. ~[2] *a.* Aşağılık, alçak; mağşuş, hileli. **~less,** esassız; temelsiz: **~ness,** adilik, alçaklık: **~-born,** soysuz; gayrimeşru.
**bash** [baş]. Yumruk (vurmak); çökertme(k).
**bashful** [ˈbaşfl]. Utangaç, mahcub; kızarıp bozaran.
**basic** [ˈbêysik] *a.* Esas, temel.
**basilica** [baˈzilikə]. Bir kilise.
**basin** [ˈbêysn]. Çanak, tas, leğen; aptesane çukuru; havuz; havza.
**basis,** *pl.* **-es** [ˈbêysis, -îz]. Esas, temel; prensip; menşe.
**bask** [bâsk]. Ateşe, güneşe, karşı oturmak (yatmak).
**basket** [ˈbâskit]. Sepet, küfe; sepet gibi örülü şey.
**bas-relief** [ˈbasrəlîf]. Kabartma.
**bass** [bêys]. Baso.
**bassoon** [baˈsûn]. Çifte kamışlı bir çalgı.
**bastard** [ˈbastəd]. Piç; sahte. **~y,** piçlik.
**baste**[1] [bêyst]. Teyellemek, iliştirmek. ~[2]. Et kızartırken üzerine erimiş yağ dökmek; (*coll.*) ıslatmak, dayak atmak.
**bastinado** [ˌbastəˈnâdou]. Falaka (çekmek).
**bastion** [ˈbastiən]. Kale burcu; müdafaa.
**bat**[1] [bat]. Yarasa. ~[2]. Kriket vs. sopası; raket. Kriket oyununda vurmak. **to carry one's ~** krikette (*and fig.*) yenilmemek; **to do stg. off one's own ~,** bir işi kendiliğinden veya yalnız başına yapmak: **without ~ting an eyelid,** göz kırpmadan. **~sman** [ˈbatsmən]. Kriket oyununda sopa ile topa vuran. ~[3]. (*coll.*) Sür'at. **at a rare ~,** rüzgâr gibi.
**batch** [baç]. Bir ağız (fırın) ekmek; takım alay, yığın.
**bate** [bêyt] = **abate. With ~d breath,** nefesi kesilerek.
**bath** [bâθ, *pl.* bâðz]. Banyo, hamam. Banyo yapmak.
**bath·e** [bêyð]. (Nehir veya denizde) banyo (etm., yapmak); yüzme(k); ıslatmak, sulamak. **~ing,** deniz banyosu: **~ costume,** mayo.
**bathos** [ˈbêyθos]. Üslûbda yüksekten gülünce düşme.
**batman** [ˈbatmən]. Emir eri.
**baton** [ˈbatən]. Asâ; değnek.
**battalion** [bəˈtalyən]. Tabur.
**batten**[1] [ˈbatən]. Tiriz (çekmek). **to ~ down,** (*naut.*) lombar kapaklarını sımsıkı kapatmak. ~[2]. Semirmek, tüylenmek, gelişmek.
**batter**[1] [batə(r)] *n.* Sulu hamur. ~[2] *v.* Durmadan vurmak, dövmek; hırpalamak. **~y** [ˈbatəri]. Batarya; pil, akümülatör. (*jur.*) Döğme, darb.
**battle** [ˈbatl]. Muharebe; mücadele. Döğüşmek, mücadele etm. **to fight s.o.'s ~,** birinin tarafını tutmak: **that's half the ~,** mücadele yarı kazanıldı sayılır. **~ments** [ˈbatəlmnts]. Kale burcunda mazgallı siperler; (*şair.*) kale. **~ship** [ˈbatlˌşip]. Zırhlı; muharebe gemisi.
**bauble** [ˈbôbl]. Gösterişli fakat değersiz şey.
**bawdy** [ˈbôdi]. Açıksaçık, müstehcen. **~-house,** umumhane.
**bawl** [bôl]. (*ech.*) Haykırma(k), barbar bağırma(k); feryad (etm.).
**bay**[1] [bêy]. Defne. **~-rum,** bir saç suyu. ~[2]. Koy, körfez. ~[3]. Cumba; duvar bölmesi. **~-window,** cumba (penceresi). ~[4]. (Köpek) havlama(k). **to be [stand] at ~,** mezbuhane mücadeleye girişmek. ~[5]. Doru (at).
**bayonet** [ˈbeənit]. Süngü(lemek).
**bazaar** [bəˈzâ(r)]. (Şark memleket-

lerinde) çarşı; iane toplamak için muhtelif eşya satılan yer.

**B.C.** [ˈbîˈsî] **before Christ,** Milâddan evvel.

**be** [bî] (*pres. ind.* **am, art, is,** *pl.* **are;** *past ind.* **was, wast, was,** *pl.* **were;** *pres. sub.* **be;** *past sub.* **were, wert, were;** *pres. part.* **being,** *p.p.* **been;** *imp.* **be.** *Bu kelimelere bakınız*). Olmak; imek; var olmak, mevcud olmak; bulunmak; hazır olmak; kâin olmak. *Mechul fiil teşkiline yarıyan yardımcı fiil, mes.*: **to love,** sevmek; **to be loved,** sevilmek. **so ~ it,** pek iyi; öyle olsun.

**be-** [bi-]. *Bu harflerile başlıyan bazı kelimeler için onların köklerine bakınız.*

**beach** [bîç]. Kumsal, sahil; plaj. Karaya oturtmak; karaya çekmek.

**beacon** [ˈbîkən]. Yüksek bir yerde yakılan işaret ateşi, kulesi.

**bead** [bîd]. Tesbih veya gerdanlık tanesi; boncuk; tane; arpacık. **to draw a ~ on,** ···e nişan almak.

**beadle** [ˈbîdl]. Bir kilise memuru; mübaşir.

**beagle** [ˈbîgl]. Küçük av köpeği.

**beak** (bîk]. Gaga; (*sl.*) sulh hâkimi; (*mekteb argosu*) muallim.

**beaker** [ˈbîkə(r)]. Büyük bardak.

**beam**[1] [bîm]. Şua, ışık. Işık saçmak; (neşeden vs.) gözleri parlamak. **a ~ of delight,** geniş tebessüm. **~**[2]. Kiriş, hatıl, putrel, kemere; terazi kolu; geminin eni. **a ~ wind,** yandan gelen rüzgâr: **on the starboard [port] ~,** sancak [iskele] tarafında: **broad in the ~,** geniş kalçalı (kimse): **to be on one's ~ ends,** (kimse) sıfırı tüketmek.

**bean** [bîn]. Bakla ve fasulya gibi nebatların umumî adı. **broad ~,** bakla: **French ~s,** taze fasulya: **haricot ~s,** taze fasulya; kuru fasulya: **runner ~s,** çalı fasulyası: **he hasn't a ~,** meteliği yok: **to be full of ~s,** kanlı canlı olm.: **to give s.o. ~s,** (*coll.*) dünyanın kaç bucak olduğunu göstermek: **to spill the ~s,** baklayı ağzından çıkarmak.

**bear**[1] [beə(r)] *n.* Ayı; borsada fiatları düşürerek hava oyunu oynayan kimse (aksi = **bull**). **polar ~,** beyaz ayı. **~-fight,** itişip kakışma, kargaşalık. **~-garden,** pis, intizamsız, bir yer. **~ish** [ˈbeəriş]. Hamhalat; kaba: huysuz. **~skin** [ˈbeəskin]. Ayı postu. **~**[2] **(bore (bare), borne)** [bô(r) (beə(r)), bôn] *v.* Taşımak; kaldırmak; tahammül etm.; dayanmak; doğurmak; (meyva) vermek; malik olm.; teveccüh etm., cihetine dönmek: **to ~ oneself well,** (i) vücudunu dik tutmak; (ii) hal ve hareketi iyi olm.: **she has borne him many children,** ondan bir çok çocuğu oldu: **it was gradually borne in upon him that...,** yavaş yavaş şuna kani oldu ki ...: **to ~ with,** karşı sabırlı olm.: **to bring to ~,** kullanmak: **to bring all one's guns to ~ on...,** bütün toplarını ···in üzerine çevirmek: **the road ~s north,** yol şimale teveccüh eder: **the lighthouse ~s due west,** fenerin kerterizi tam garbdedir: **how does the land ~?,** kara hangi cihette?: **here we ~ to the right,** burada bir az sağa meyledeceğiz. **~ away (off),** götürmek: **to ~ away the prize,** mükâfatı kazanmak. **~ down,** bastırmak, ezmek: **to ~ down upon s.o.,** (gemi vs.) hücum etm. için yaklaşmak; heybetle üzerine gelmek. **~ on,** üstüne basmak: **to bring a gun, etc., to ~ on stg.,** top vs.yi bir şeyin üzerine meylettirmek: **to bring all one's strength to ~ on stg.,** bir şeyin üzerine bütün kuvvetiyle basmak. **~ out,** dışarıya götürmek: **to ~ s.o. out,** birinin sözünü tasdik etm.: **this ~s out what I said,** bu benim söylediğimi teyid ediyor. **~ up,** sabır ve tahammül etm.: **~ up!,** cesaret! **~able** [ˈbeərəbl]. Taşınabilir; dayanılabilir. **~er** [ˈbeərə(r)]. Taşıyan; hâmil; getiren. **~ing** [ˈbeərin(g)] *n.* Kerteriz; münasebet; mâna; şümul; doğurma; tavır; (*mech.*) yatak. **beyond all ~,** dayanılmaz: **to be in full ~,** (meyva ağacı) tam kıvamında olm.: **to lose one's ~s,** nerede bulunduğunu bilmemek: **to take one's ~s,** cihet tayin etmek.

**beard** [biəd]. Sakal; başak dikeni. Sakalından yakalamak; meydan okumak. **to ~ the lion in his den,** bir şey taleb etm. için korkulan birinin yanına gitmek.

**beast** [bîst]. (Dört ayaklı) hayvan; canavar; sığır; kaba ve iğrenc adam; hınzır. **~ly,** hayvanca; iğrenc; berbad; hınzırca (hareket vs.).

**beat**[1] [bît] *n.* Vurma, çarpma; (davul vs.) çalma; tempo; (polis vs.) devir. **~**[2] *v.* Döğmek; vurmak; (kalb) çarpmak; (kanad) çırpmak; (davul vs.) çalmak; yenmek, mağlub etm.; (yumurta vs.) çalkamak; hayrette bırakmak. that ~s me!, aklım ermez!: that ~s everything!, bu tüy dikti! now then, ~ it!, haydı, çık! çek arabanı!: strawberries ~ cherries (any day), çilek kirazdan kat kat üstündür: to ~ to arms, silâh başı çalmak: to ~ the record, rekor kırmak: to ~ a retreat, ric'at etm.: to ~ the retreat, (trampetle) ric'at emri vermek: to ~ time, tempo tutmak: to ~ a wood, (av için) ormanı taramak: to ~ to windward, (*naut.*) rüzgâr karşı yol almak. ~ **back,** püskürtmek, geri çevirmek. ~ **down,** indirmek, pazarlıkla indirmek; ezmek: the sun ~ down upon our heads, güneş başımızda kaynıyordu. ~ **in,** to ~ in a door, kapıyı kırıp girmek. ~ **off,** püskürtmek. ~ **out,** vurup çıkarmak: to ~ out s.o.'s brains, birinin beynini patlatmak: to ~ up game, ormanda avlanırken kuşları havalandırmak. ~**en** [ˈbîtn] *p.p. v.* beat. *a.* the ~ track, çiğnenmiş yol; (*fig.*) herkesin gittiği yol. ~**er** [ˈbîtə(r)]. Döğen (kimse); sürek avında kuşları vs. yerinden çıkaran adam; tokmak. ~**ing** [ˈbîtin(g)]. Dövme, dayak. to give a ~ to, ···e dayak atmak.

**beatif·ic** [ˌbiəˈtifik]. Pürneşe; mes'ud eden. ~**y,** [bîatifây] (papa) birini azizler sırasına idhal etmek.

**beatitude** [biˈatityûd]. Ahiret saadeti.

**beau,** *pl.* **-s, -x,** [bou, -z]. Şık, züppe; âşık.

**beaut·iful** [ˈbyûtəfl]. Çok güzel; lâtif. ~**ify** [-ˈfây], güzelleştirmek. ~**y** [ˈbyûti]. Güzellik; güzel kimse; (*coll.*) mükemmel, nefis şey. the ~ of it is . . ., işin en güzel tarafı . . . ~**y-spot,** (yüzdeki) ben; güzel manzaralı yer.

**beaver** [ˈbîvə(r)]. Kunduz, kastor.

**becalm** [biˈkâm]. to be ~ed, (yelkenli) rüzgârsızlıktan kımıldanamamak.

**became** [biˈkêym] *v.* become.

**because** [biˈkoəz]. Çünkü, zira; ···den dolayı; için, sebebile.

**beck** [bek]. to be at s.o.'s ~ and call, birinin eli altında (emrine hazır) bulunmak.

**beckon** [ˈbekən]. (Birine) gelmesi için işaret etmek.

**becom·e** [became] [biˈkʌm, biˈkêym]. Olmak; gittikçe . . . olmak; yakışmak; *sıfatlardan fiil teşkil etmeğe yarar, mes.*: to ~ old, ihtiyarlamak: to ~ ill, hastalanmak. what has ~ of him?, o ne oldu?: that hat does not ~ you, o şapka size yakışmaz. ~**ing** [biˈkʌmin(g)] *a.* Uygun, yakışık.

**bed**[1] [bed] *n.* Yatak, yatacak yer; bahçe tarhı; nehir yatağı; tabaka, kat; temel. ~ and board, yatmak ve yiyip içmek: ~ of roses, rahat vaziyet: to be brought to ~ of a child, bir çocuk dünyaya getirmek: to die in one's ~, eceliyle ölmek: to get out of ~ on the wrong side, (o gün) huysuz olm.: to go to ~, (uyumak üzere) yatmak: to keep to one's ~, yatakta hasta olm.: to make a ~, yatağı düzeltmek: ⌜as you make your ~ so you must lie on it⌝, kendi yaptığını çekmeli: spare ~, misafir yatağı: spring ~, somye: to take to one's ~, yatağa düşmek. ~**-bug,** tahtakurusu. ~**chamber** [ˈbedçêymbə(r)]. Yatak odası: Gentleman of the ~, kıralın şahsî hizmetinde bulunan asilzade. ~**clothes** [ˈbedklouðz]. Yatak takımı. ~**ding** [ˈbedin(g)]. Yatak takımı. ~**fellow** [ˈbedfelou]. Yatak arkadaşı. ~**-linen,** yatak çarşafı ve yastık kılıfı. ~**post** [ˈbedpoust]. Karyola direği: between you and me and the ~, söz aramızda. ~**ridden** [ˈbedridn]. Yatalak. ~**-rock price,** (fiat) olacağı: to get down to ~, bir işin esasına gitmek. ~**room** [ˈbedrum]. Yatak odası: spare ~, misafir yatak odası. ~**side** [ˈbedsâyd]. Yatak yanı, başucu: ~ manner, (doktorlar hakkında) hastalara muamele şekli. ~**sore** [ˈbedsô(r)]. Çok yatmaktan vücudun soyulması. ~**-spread** [ˈbedspred]. Yatak örtüsü. ~**stead** [ˈbedsted]. Karyola (kereveti). ~[2] *v.* Bir tarh içine dikmek. to ~ down a horse, at için ahıra samandan yatak yapmak: to ~ down a machine, makineyi sağlam bir kaide üzerine yerleştirmek.

**Bedlam** [ˈbedləm]. Tımarhane; gürültü; 'Toptaşı'. ~**ite,** kaçık.

**Bedouin** [ˈbeduin]. Bedevî.

**bee** [bî]. Arı. to have a ~ in one's bonnet, bir şeyle bozmak. ~**hive** [ˈbîhâyv]. Arıkovanı. ~**-keeper,** arıcı. ~**-line,** en kısa yol. ~**swax** [ˈbîzwaks]. Balmumu.

**beech** [bîç]. Kayın ağacı. ~**-mast,** ~**-nut,** kayın palamudu.

**beef** [bîf]. Sığır eti; (*coll.*) kuvvet, iri yarılık. corned ~, konserve sığır eti. ~**eater** [ˈbîfîte(r)]. Londra kalesi

bekçisi. **~steak** [ˌbîfˈstêyk]. Biftek. **~-tea,** sığır eti suyu. **~y** [ˈbîfi]. (*coll.*) Etli butlu, iri yarı.
**Beelzebub** [biˈelzəbʌb]. Şeytan, iblis.
**been** [bîn] *p.p. of* **be.** Olmuş, imiş; gitmiş. **where have you ~?,** nerede idiniz?, nereye gittiniz? : **I have ~ ill,** (şimdiye kadar) hasta idim : **he has ~ punished,** cezalandırıldı : **I have ~ to London,** Londraya gittim; Londrada bulundum.
**beer** [ˈbiə(r)]. Bira. **small ~,** ehemmiyetsiz şeyler. **to think no small ~ of oneself,** küçük dağları ben yarattım demek. **~y,** bira kokan; çakırkeyf.
**beet** [bît]. Pancar. **~root** [ˈbîtrut]. (Yenilen) pancar.
**beetle** [ˈbîtl]. Bokböceği; böcek.
**befall** [biˈfôl]. Zuhur etm., vuku bulmak; başa gelmek.
**befit** [biˈfit]. Uyumak, münasib olm., yakışmak.
**before** [biˈfô(r)]. Önce, evvel; önde, önünde; bir önceki. **the day ~,** bir gün evvel : **~ Christ,** Milâddan evvel : **it ought to have been done ~ now,** şimdiye kadar yapılmış olmalıydı : **it was long ~ he came,** (i) gelmesi uzun sürdü, uzadı; (ii) o buraya gelmeden çok evveldi : **I will die ~ I give in,** teslim etmekten ölmeyi tercih ederim. **~hand** [biˈfô·hand]. Önceden, daha evvel.
**befriend** [biˈfrend]. ···e dostça hareket etm.; yardım etmek.
**beg** [beg]. Dilenmek; istemek; dilemek; yalvarmak; (köpek) salta durmak. **I ~ to . . .,** hürmetle . . . : **I ~ of you,** rica ederim : **I ~ to differ,** müsaadenizle ben bu fikirde değilim : **to ~ the question,** dava delil almak : **to ~ s.o. off,** birini affettirmek : **these jobs go (a-) ~ging,** bu işlere pek talib yok. **~gar** [ˈbegə(r)]. Dilenci; çok fakir kimse. Dilenciye çevirmek; iflâs ettirmek. **lucky ~!,** köftehor! : **poor ~!,** zavallı adamcağız! : ⌜**~s cannot be choosers**⌝, (i) 'dilenciye hıyar vermişler, eğridir diye beğenmemiş'; (ii) oluru ile iktifa etmeli *kabilinden* : **the beauty of the scene ~s description,** manzaranın güzelliği tarife sığmaz. **~garly,** dilenciye verir gibi; gülünc (mikdarda az). **~gary,** dilencilik.
**began** *v.* **begin.**
**begin** (**began, begun**) [biˈgin, -gan, -ʌn]. Başlamak. **to ~ with,** evvelâ, ilk önce. **~ning,** başlangıc; baş; esas. **~ner,** başlayıcı, mübtedi; acemi.
**begone** [biˈgon]. Defol!, yıkıl!, çekil!
**begrudge** [biˈgrʌc]. Vermek istememek.
**beguile** [biˈgâyl]. Aldatmak; baştan cıkarmak; eğlendirmek. **to ~ the time,** vakit geçirmek.
**begun** *v.* **begin.**
**behalf** [biˈhâf]. **on ~ of s.o.** [**on s.o.'s ~**], biri namına; tarafından; biri lehinde.
**behav·e** [biˈhêyv]. Davranmak; hareket etmek. **~ yourself!,** uslu otur!; terbiyeni takın! : **well ~d,** uslu, terbiyeli. **~iour** [biˈhêyvyə(r)]. Tavır, hareket; muaşeret; işleyiş.
**behead** [biˈhed]. Başını kesmek.
**beheld** *v.* **behold.**
**behest** [biˈhest]. Emir, irade.
**behind** [biˈhâynd]. Arkada, arkasında; arkadan, geride. Kıç; arka. **to be ~ the times,** eski kafalı olm. **~hand,** gecikmiş; geride.
**behold** (**beheld**) [biˈhould, -held]. Bakmak, görmek. **~er** [biˈhouldə(r)]. Seyirci; şahid.
**being** [ˈbî·in(g)] *pres. part. of* **be.** *n.* Mevcudiyet; varlık; mahluk. **human ~,** insan : **to come into ~,** meydana çıkmak, vücud bulmak. **for the time ~,** şimdilik : **you are ~ obstinate,** (şu anda) inad ediyorsunuz.
**belch** [belç]. Geğirme(k); püskürtme(k).
**beleaguer** [biˈlîgə(r)]. Kuşatmak, muhasara etmek.
**belfry** [ˈbelfri]. Çan kulesi (sahanlığı). **to have bats in the ~,** bir tahtası eksik olmak.
**Belgi·um** [ˈbelcəm]. Belçika. **~an** [-cən], Belçika'ya aid; Belçikalı.
**belie·f** [biˈlîf]. İnanma; iman; kanaat. **to the best of my ~,** benim bildiğime göre. **~ve** [biˈlîv]. İnanmak; iman etm., zannetmek. **to ~ in . . .,** ···e itimad etm. : **to make ~ to do stg.,** bir şeyi yapıyor gibi görünmek. **~ver,** inanan; mümin.
**belittle** [biˈlitl]. Küçültmek, alçaltmak.
**bell** [bel]. Çan; kampana; çıngırak; zil. (Etek vs.) şişmek, havalanmak. **to ~ the cat,** kimsenin yanaşamadığı tehlikeli bir işi üzerine almak : ⌜**~, book, and candle**⌝, resmî lânetleme : **there's (a ring at) the ~,** zil çalıyor. **~-pull,** çekerek çalınan zil kordonu.

~-push, elektrikli zil dügmesi. ~-wether, kösemen.
**belle** [bel]. Güzel kadın, dilber.
**bellicos·e** [ˈbelikoṷz]. Harbci, harbi seven. **~ity** [–ˈkositi], harbcilik; kavgacılık.
**belligeren·t** [biˈlicərənt]. Muharib; muhasım. **~cy,** muharib hali.
**bellow** [ˈbeloṷ]. Böğürme(k); kükreme(k).
**bellows** [ˈbeloṷz]. Körük.
**belly** [ˈbeli]. Karın. Şiş(ir)mek. **~ful,** karın dolusu: **to have a ~,** tıkabasa, bıkmak.
**belong** [biˈlon(g)]. Aid olm., ···nin olm., mensub olm.; sakinlerinden olmak. **I ~ here,** buralıyım. **~ings,** aid olan şeyler.
**beloved** [bıˈlʌvd] *p.p.* Sevilen. *a. and n.* [biˈlʌvid], sevgili, aziz; canım.
**below** [biˈloṷ]. Aşağı; aşağıda; aşağısında; altında.
**belt** [belt]. Kemer; kuşak; bel kayışı; bölge. Kayışla döğmek. **to hit below the ~,** (boksta) kemerden aşağı (usulsüz) vurmak; (*fig.*) alçakça (kahbece) hareket etmek. **~ed,** kemerli.
**bemoan** [biˈmoṷn]. (Bir şeyden) inliyerek şikâyet etm.; ah ve figan etmek.
**Ben¹** [ben]. İskoçya'da dağ tepesi. **~².** *abb.* **Benjamin. Big ~,** Londra'da Parlamento Sarayının saat kulesi çanı.
**bench** [benç]. Sıra; tezgâh; hâkim kürsüsü; mahkeme. **Front ~,** Avam Kamarasında nazırlara ve eski nazırlara tahsis edilen ön sıra; **back ~es,** diğer mebusların oturduğu sıralar: **Queen's ~,** en yüksek İngiliz mahkemesi. **~er,** kıdemli avukat.
**bend¹** [bend] *n.* İğilme, inhina; dirsek; köşe; dönemeç; düğüm; (arma üzerindeki) şerid, çizgi. **~ sinister,** (armada) gayrımeşruluk alâmeti olan muvazi çizgiler: **to take a ~,** virajdan dönmek. **~²** (**bent**) *v.* İğ(il)mek; bük(ül)mek; kavislen(dir)mek; kıvırmak; kıvrılmak; germek; bir tarafa çevirmek (çevrilmek); tevcih etm.; (*naut.*) bağlamak.
**beneath** [biˈniθ]. Altında, altta; dununda. **it is ~ him to . . .,** ···e tenezzül etmez.
**benediction** [beniˈdikşn]. Hayır dua; takdis.
**benefact·ion** [beniˈfakşn]. İyilik, ihsan, hayır. **~or** [ˈbeniˌfaktə(r)], iyilik eden; velinimet; hayır sahibi.
**benefic·e** [ˈbenifis]. Aidatlı papazlık mesnedi. **~ence** [biˈnefisəns]. İyilik, lûtuf. **~ent,** iyi, lûtufkâr; mubarek. **~ial** [beniˈfişl]. Faydalı; yarar. **~iary,** aidat alan; faydalanan.
**benefit** [ˈbenifit]. Fayda; istifade, kâr, menfaat; tazminat. Yaramak, faydalı olm., fayda etmek. **the ~ of the doubt,** (*jur.*) şübhe halinde maznun lehine karar: **~ society,** karşılıklı yardım cemiyeti.
**benighted** [biˈnâytəd]. (Mecburen) geceye kalmış; karanlıkta kalmış; cahil.
**benign** [biˈnâyn]. Yumuşak huylu; (*med.*) tehlikesiz. **~ant** [biˈnignənt], iyi kalbli; yumuşak huylu; müsaid. **~ity** (biˈnigniti], yumuşaklık; (*med.*) zararsızlık.
**bent** [bent] *p.p.* **bend.** *a.* İğilmiş, bükülmüş, aklına koymuş. *n.* Meyil; temayül. **to be ~ on doing stg.,** bir seyi yapmağa azmetmek: **to have a ~ for,** ···e istidadı olm.: **to be homeward ~,** eve doğru yolunda olmak.
**benzine** [benˈzîn]. Yağ lekesini çıkarmak kullanılan bir nevi benzin.
**beque·ath** [biˈkwîð, –îθ]. Vasiyet etmek. **~st** [biˈkwest]. Vasiyetle bırakılan şey.
**bereave** (**bereaved** *veya* **bereft**) [biˈrîv, –rîvd, –reft]. (Ölüm hakkında) birisini bir sevdiğinden mahrum etm. **~ment,** büyük kayıb (ölüm), matem.
**berry** [ˈberi]. Çekirdeksiz sulu küçük meyva (çilek, frenküzümü, böğürtlen vs.); tane.
**berth** [bö(r)θ]. (Vapur, tren veya uçakta) yatak, ranza; (geminin) demir yeri, rıhtımdaki yeri; yer, memuriyet. Demir yeri vermek; yanaştırmak; rıhtıma yanaşmak. **to give a wide ~ to,** ···den uzak durmak, çekinmek; alarga durmak.
**beseech** (**-ed, besought**) [biˈsîç, –sôt]. Yalvarmak, istirham etmek.
**beset** (*past.* **beset**) [biˈset]. Kuşatmak; hücum etmek. **~ting sin,** insanın daima düştüğü hata.
**beside** [biˈsâyd]. Yanında, yanına; dışında, haricinde. **to be ~ oneself,** (hiddetten vs.) kendini kaybetmek.
**besides** [biˈsâydz]. Bundan başka; bundan maada; zaten; bir de.
**besiege** [biˈsîc]. Muhasara etmek.
**besom** [ˈbîzm]. Çalı süpürgesi.
**besought** *v.* **beseech.**
**bespeak** (**-spoke, -spoken**) [biˈspîk,

–spǫuk, –spǫukən]. Ismarlamak; önceden almak, tutmak.

**best** [best]. En iyi. **~ man,** sağdıc: **at (the) ~ he is not generous,** en hafif tabirle cömert değildir: **to get [have] the ~ of it [to come off ~],** üstün olm., galib gelmek: **to do one's ~,** elinden gelen her şeyi yapmak: **in one's ~,** en iyi elbisesile: **she is not looking her ~ today,** bugün her zamanki gibi güzel değil: **he looks his ~ in uniform,** ona en çok yakışan üniformadır: **to make the ~ of it,** oluru ile iktifa etm.: **to the ~ of my knowledge,** benim bildiğime göre: **the ~ part of the year,** (i) senenin mühim bir kısmını; (ii) en güzel mevsimi: **he can lie with the ~,** yalancılıkta eşsizdir.

**bestial** [ˈbestyəl]. Hayvanî, hayvanca. **~ity** [ˌbestiˈaliti], hayvanlık.

**bestow** [biˈstǫu]. Vermek, ihsan etmek.

**bet** [bet]. Bahis, iddia. Bahse girmek, bahis tutuşmak. **to make [lay] a ~,** bahse girmek: **to take (up) a ~,** bahsi kabul etm.: **you ~ your life!,** elbette, ne zannettiniz! **~ting,** bahis, bahse girme.

**bête-noire** [ˈbêytˈnwâ(r)]. Nefret edilen adam veya şey.

**betimes** [biˈtâymz]. Erken, erkenden.

**betray** [biˈtrêy]. Hiyanet etm., aldatmak, ihanet etm.; yanlış yola sevketmek; göstermek, ifşa etmek. **~al,** ihanet, düşmana teslim; ifşa; açığa vurma.

**betroth** [biˈtrǫuð]. Nişanlamak. **~al,** nişanlama, nişan. **~ed,** nişanlı.

**better**[1] [ˈbetə(r)] *a. adv. & n.* Daha iyi; üstün; üst; mafevk. **~ and ~,** gittikçe daha iyi: **~ still . . .,** daha iyisi . . .: **that's ~!,** hah şöyle; işte şimdi oldu; bu çok daha iyi: **a change for the ~,** iyileşme, düzelme: **he has seen ~ days,** şimdiki hali fena fakat ne günler görmüştür: **to do stg. for ~ or worse,** bir şeyi (neticesi) ne olursa olsun yapmak: **to take s.o. for ~ or worse,** birini olduğu gibi (iyi ve fena taraflarile) kabûl etm.: **to get the ~ of s.o.,** birini mağlub etm.: **to go one ~,** arttırmak; (birini bir şeyde) bastırmak: **you had ~ tell him,** ona söyleseniz daha iyi olur: **so much the ~!,** daha iyi ya!; olsun!: **for the ~ part of the year,** senenin yarısından fazlası, mühim bir kısmında: **you are ~ off than I am,** sizin vaziyetiniz benimkinden daha müsaiddir: **to think ~ of it,** fikrini değiştirmek, vazgeçmek. **~**[2] *v* Daha iyi yapmak (olmak), düzeltmek; iyileş(tir)mek. **~**[3], **bettor** *n.* Bahse giren; bahis tutan.

**between** [biˈtwîn]. Ara, arasında, arada. **you must choose ~ them,** ikisinden birini seçmeniz lâzım

**betwixt** [biˈtwikst] = **between. ~ and between,** ikisinin ortası.

**bevel** [ˈbevl]. Pah(lamak), şev (vermek).

**beverage** [ˈbevəric]. İçilen şey; meşrubat.

**bevy** [ˈbevi]. Küme, grup, takım, sürü.

**beware** [biˈwẹə(r)]. **~ (of),** (···den) sakınmak, korunmak.

**bewilder** [biˈwildə(r)]. Şaşırtmak.

**bewitch** [biˈwiç]. Büyülemek, teshir etmek.

**beyond** [biˈyond]. İleri; daha uzak; öte, öteye, ötede, ötesinde; sonra; dışında; üstünde; aşarak. **the ~,** öte, mavera: **~ belief,** inanılmıyacak: **~ doubt,** şübhe götürmez: **~ words,** tarif edilmez: **at the back of ~,** dünyanın öteki ucunda: **it's ~ me,** buna aklım ermez: **that's (going) ~ a joke,** iş şaka olmaktan çıkıyor.

**bias** [ˈbâyəs]. Meyil, temayül; peşin hüküm; bir tarafı tercih; çapraz, çapraz kesme. Bir tarafa tesir etm., tarafsızlığını bozmak. **to be ~ed against s.o.,** birinin aleyhinde olmağa meyletmek.

**bib** [bib]. Çocuk göğüslüğü; önlüğün göğse gelen kısmı. **to put on one's best ~ and tucker,** iki dirhem bir çekirdek olmak.

**Bibl·e** [ˈbâybl]. İncil. **~ class,** din dersi. **~ical** [ˈbiblikl]. İncile aid.

**biblio·graphy** [ˌbibliˈogrəfi]. Bibliyografya. **~maniac** [–ˈmêyniak], kitab meraklısı; kitab delisi. **~phile** [ˈbibliofil], kitab seven.

**bicentenary** [ˌbâysenˈtînəri]. İki yüzüncü yıldönümü.

**biceps** [ˈbâyseps]. Pazı.

**bicker** [ˈbikə(r)]. Atışmak, çekişmek. **~ing,** ağız dalaşı.

**bicycl·e** [ˈbâysikl]. Bisiklet. Bisikletle gitmek. **~ist,** bisiklete binen.

**bid**[1] [bid]. Fiat teklif (etm.), pey sürme(k). **to make a ~ for power,** iktidarı ele geçirmeğe teşebbüs etmek. **~ding,** mezatta artırma. **~der,** pey süren, arttıran. **~**[2] *v.* Emretmek, kumanda etm.; davet etm.; temenni etm. **he ~s fair to be a great doctor,** büyük bir doktor olacağa benziyor. **~ding,** emir; davet: **to be at s.o.'s ~,** birinin emrinde olmak.

**bide** [bâyd] = **abide. To ~ one's time,** zamanını [fırsatını] beklemek.

**biennial** [bâyˈlenyəl]. İki senelik; iki senede bir olan.

**bier** [ˈbi̯ə(r)]. Cenaze teskeresi (*bazan* tekerlekli).

**big** [big]. İri, büyük, büyümüş; mühim. **~ with child,** hamile: **~ with consequences,** ağır neticeler doğurabilir: **~ end,** biyel başı: **to talk ~,** yüksekten atmak. **~ger,** daha büyük.

**bigam·ist** [ˈbigəmist]. Evli iken üstüne evlenen kimse. **~ous,** iki karılılığa veya iki kocalılığa aid. **~y,** evli iken üstüne evlenme.

**bigot** [ˈbigət] *n.* **~ed** *a.* Müteassıb, darkafalı(kimse). **~ry,** taasub, darkafalılık.

**bigwig** [ˈbigwig]. Kodaman.

**bike** [bâyk]. (*coll.*) **bicycle.**

**bile** [bâyl]. Safra; huysuzluk, öfke. **to stir s.o.'s ~,** birinin damarına basmak. **bilestone,** safra kesesinde bulunan taş.

**bilge** [bilc]. (*naut.*) Sintine; fıçı karnı; (*coll.*) herze. **~-keel,** yalpalık omurga.

**bilingual** [ˈbâyˈlin(g)gwəl]. İki dilli; iki lisan konuşan.

**bilious** [ˈbilyəs]. Safralı; huysuz.

**bilk** [bilk]. Para vermeden sıyrılmak.

**bill**[1] [bil]. Gaga; ağız; (*esk.*) teber. **to ~ and coo,** sevişip koklaşmak. **~**[2]. Hesab, fatura; sened; poliçe; afiş, ilân; kanun lâyihası; kâğıd para. Faturasını yapmak; ilân yapıştırmak. **~ of exchange,** poliçe: **~ of fare,** yemek listesi: **~ of lading,** konişmento, yük senedi. **~-poster,** ilân yapıştırıcı.

**billet** [ˈbilit]. (Seferber askerin) konak tezkeresi; konak yeri; iş, vazife. Konak etm.; yerleştirmek, yerleşmek. **every bullet has its ~,** kaderin önüne geçilmez.

**billiards** [ˈbilyədz]. Bilârdo.

**Billingsgate** [ˈbilin(g)sgêyt]. Londra balık pazarı; ağızbozukluğu (Kasımpaşa ağzı).

**billion** [ˈbilyən]. Trilyon; (*Amer.*) milyar.

**billow** [ˈbilo̯u]. Büyük dalga. Dalgalanmak.

**bi-monthly** [ˈbâyˈmʌnθli]. İki ayda bir.

**bin** [bin]. Kab; kutu; sandık; ambar.

**bind** (**bound**) [bâynd, ba̯und]. Bağlamak, rabtetmek; sarmak; mukayyed etm.; mecbur etm.; kabız vermek; cildlemek; (çimento vs.) tutmak, donmak; (makine vs.) sıkışmak. **to be bound to do stg.,** bir şeyi yapmağa mecbur olmak. **~ down,** mecbur etm., mukayyed etmek. **~ over,** (*jur.*) birinin cezasını tecil etmek. **~er** [ˈbâyndə(r)]. Bağlayıcı; mücellid; (defter vs.) kab; biçer bağlar makine. **~ing** [ˈbâyndin(g)] *a.* Bağlayıcı; yapıştırıcı; tutucu: muteber, cari; vâcib; kabız verici. *n.* Cildleme; cild.

**binge** [binc]. (*sl.*) Âlem, cümbüş.

**binocular** [biˈnokyulə(r)]. İki gözle kullanılan. **~s,** dürbün.

**biochemistry** [ˈbâyoˈkemistri]. Biyoşimi.

**biograph·er** [bâyˈogrəfə(r)]. Birinin tercümeihalini yazan. **~ical** [–ˈgrafikl], tercümeihale aid. **~y** [–ˈogrəfi], tercümeihal; hayat.

**biolog·y** [bâyˈoloci]. Biyoloji. **~ical** [–ˈlocikl], biyolojiye aid.

**biped** [ˈbâyped]. İki ayaklı hayvan.

**biplane** [ˈbâyplêyn]. Çift satıhlı uçak.

**birch** [bö(r)ç]. Huş ağacı; huş dallarından kamçı. Huş dalı ile döğmek.

**bird** [bö(r)d]. Kuş; (*sl.*) herif. **hen ~,** dişi kuş: **song ~,** ötücü kuş: **~ of passage,** muhacir kuş; bir yerde muvakkaten kalan kimse. **~-fancier,** kuşçu. **~'s-eye,** kuş bakışı; umumî. **~'s-nest,** kuş yuvası (nı aramak).

**birth** [bö(r)θ]. Doğum, doğma. **to give ~ to,** doğurmak; meydana çıkarmak. **~-control,** doğumun tahdidi. **~-mark,** doğuşta mevcud yüz lekesi. **~-rate,** doğum nisbeti. **~day,** doğum günü; doğum yıldönümü. **in one's ~ suit,** çırçıplak. **~right,** kıdem hakkı; doğum dolayısile hak; doğuşta kazanılan hak.

**Biscay** [ˈbiskêy]. **the Bay of ~,** Gaskonya körfezi.

**biscuit** [ˈbiskit]. Bisküi. **that takes the ~,** (*sl.*) artık bu kadarı da fazla.

**bisect** [bâyˈsekt]. İkiye biçmek; iki müsavi kısma bölmek.

**bishop** [ˈbişəp]. Piskopos; satranc oyununda fil. **~ric,** piskoposluk.

**bit**[1] [bit]. Gem. **to champ the ~,** gemini ısırmak; öfkeden veya sabırsızlıktan kudurmak: **to take the ~ between its [one's] teeth,** gemi azıya almak. **~²**. Parça; lokma, kırıntı; matkab. **a ~,** bir parça, bir az: **a good ~,** oldukça: **not a ~,** hiç değil; estağfurullah!: **not a ~ of it!** ne gezer: **a ~ of luck,** talih; devlet kuşu. **~³** *v.* **bite.**

**bitch** [biç]. Dişi köpek; kahpe.

**bit·e (bit, bitten)** [bâyt, bit, bitən]. Isırmak, dişlemek; (balık) oltaya vurmak; (biber, soğuk) yakmak; (rüzgâr) kesmek. Isırış; ısırma; dişleme; (balık) oltaya vurma; lokma. **to ~ the dust,** harbde veya bir mücadelede ölmek: ⌜**once bitten twice shy**⌝, ⌜çorbadan ağzı yanan ayranı üfler de içer⌝: **to be bitten with a desire to do stg.,** bir şey yapmak arzusile yanmak [kıvranmak]: **to ~ off,** ısırıp koparmak: **to ~ s.o.'s head off,** birine ters ve şiddetle cevab vermek. **~er** [ˈbâytə(r)]. Isırıcı. **the ~ bit,** men dakka dukka. **~ing** [ˈbâytin(g)] *a.* Acı, keskin, zehirli.

**bitter** [ˈbitə(r)]. Acı, keskin, sert, şiddetli, meraretli, amansız. **~ enemies,** can düşmanları: **to be ~ [to feel ~ly] about stg.,** bir şey için kin beslemek; kendini mağdur hissetmek: **to the ~ end,** en sonuna kadar. **~ness** [ˈbitənis]. Acılık, sertlik; kin.

**bitum·en** [ˈbityumən]. Zift. **~inous** [-ˈtyûminəs], ziftli.

**bivouac** [ˈbivuak] *n. & v.* Açıkta (çadırsız) ordugâh (kurmak); açıkta yatmak.

**bizarre** [biˈzâ(r)]. Garib, acayib; biçimsiz.

**blab, ~ber** [blab, blabə(r)]. Geveze(lik etm.); boşboğaz(lık etm.). **to ~ out,** ağzından kaçırmak.

**black** [blak]. Kara; siyah; zenci. Karartmak; karalamak; (ayakkabı) boyamak. **to be ~ and blue (all over),** vücudu mosmor olm.: **to beat s.o. ~ and blue,** birinin pestilini çıkarmak: **the Black Country,** İngiltere'nin Staffordshire ve Warwickshire kontluklarının sanayî bölgesi: **the Black Death,** büyük veba: **to have a ~ eye,** gözü şişmek, morarmak: **to be ~ in the face,** (hiddet vs. den) morarmak: **~ ingratitude,** tuz ekmek hainliği, nankörlük: **to look ~,** surat asmak: **to look as ~ as thunder,** yüzü karmakarışık (tehdidkâr) olm.: **~market,** karaborsa: **to set stg. down in ~ and white,** yazıya geçirmek: **to work in ~ and white,** çini mürekkebi yahud karakalemle resim yapmak. **~-coated workers,** kâtib vs. gibi elişi yapmıyan memur sınıfı. **~-letter,** gotik harf. **-out,** karartma. **~amoor** [ˈblakəmô(r)]. Zenci. **~ball** [ˈblakbôl]. (Bir klübde âza seçilirken) aleyhte rey vermek. **~beetle** [blakˈbîtl]. Karafatma; hamam böceği. **~bird** [ˈblakbö(r)d]. Karatavuk. **~-board** [ˈblakbôd]. Yazı tahtası; kara tahta. **~en** [ˈblakən]. Karartmak, karalamak; lekelemek, iftira etmek. **~guard** [ˈblagâd]. Edebsiz, rezil. **~ing** [ˈblakin(g)]. Ayakkabı boyası; kurşun tozu. **~lead** [blakˈled]. Grafit; kurşun tozu. **~leg** [ˈblakleg]. Amele grev halinde iken çalışan işçi. **~list** [ˈblaklist] *v.* Kara listeye koymak. **~mail** [blakˈmêyl]. Şantaj; para koparmak için ıskandalla tehdid etme. Şantaj yapmak. **~shirt** [ˈblakşö(r)t]. Karagömlekli; Faşist. **~smith** [ˈblaksmiθ]. Demirci; nalbant.

**bladder** [ˈbladə(r)]. Mesane; kabarcık; futbol topunun lâstik kısmı.

**blade** [blêyd]. Bıçak vs. ağzı; kılıç; kürek palası; pervane kanadı; arpa vs. nin ince yaprağı; bir şeyin yassı ve geniş tarafı. **razor ~,** jilet bıçağı.

**blame** [blêym]. Kabahat; mesuliyet; ···e kabahat bulma(k), ayıblama(k); mesul tutmak. **to bear the ~,** kabahati üzerine almak: **to lay [put] the ~ on s.o.,** kabahati birinin üzerine atmak: **they ~ each other,** kabahati birbirinin üzerine atıyorlar: **to ~ stg. for an accident,** kazayı bir şeye atfetmek: **you have only yourself to ~,** kabahati başkasında arama. **~less,** lekesiz; tertemiz; kusursuz, masum. **~worthy,** tekdire lâyik, ayıblanmağa müstahak.

**blanch** [blânç]. Ağar(t)mak; beyazlatmak, beyazlanmak; sararmak.

**blancmange** [bləˈmônj]. Sütlü pelte.

**bland** [bland]. Yumuşak, tatlı; mülâyim, nazik; (*ekseriya bir az sun'î ve müstehzi mânasına gelir*).

**blank** [blank]. Boş; yazısız; manasız; şaşkın. Boşluk. **~ cartridge [shot],** kuru sıkı. **~ cheque,** açık bono: **~ verse,** kafiyesiz şiir, serbest nazım: **to draw a ~,** (piyangoda) boş çekmek: **to look ~,** şaşkın şaşkın bakmak. **~ly,** şaşkın bir vaziyette: **he ~ denied it,** tamamen ve katiyetle inkâr etti.

**blanket** [ˈblankit]. Yün yorgan, battaniye. (*Amer.*) umumî. Yorgan gibi örtmek.
**blare** [blêyr]. Boru sesi; şiddetli ve sert ses. (Boru) ötmek.
**blarney** [ˈblâni]. Dil dökme(k).
**blasphem·e** [blasˈfîm]. (Mukaddes şeylere) hürmetsizlikte bulunmak; küfretmek. **~ous** [ˈblasfəməs], mukaddes şeylere hürmetsiz; dinsiz, imansız. **~y,** [ˈblasfəmi], mukaddes şeylere hürmetsizlik; küfür.
**blast** [blâst]. Şiddetli ve anî hava cereyanı; boru, düdük sesi; patlama, infilak. (Dinamitle) atmak; patlatmak; kavurmak; kırıp geçirmek; mahvetmek; (*coll.*) ···e lânet etmek. **~ you!,** Allah belânı versin!: **~ it!,** lânet olsun!: **to be in ~,** (yüksek fırın) yanmak, faaliyette olmak. **~ing,** *n.* patlama; berhava etme. **~ing-powder,** lağım barutu. **~-furnace,** yüksek fırın.
**blatan·cy** [ˈblêytənsi]. Göze çarpma. **~t,** göze çarpan; şamatalı: **~ injustice,** göz göre göre büyük haksızlık.
**blaze**[1] [blêyz]. Alev (parıltısı); ateş; parlama; ışık bolluğu; alevlenme. Alevlenmek, tutuşmak; parlamak, parıldamak. **in a ~,** alevler içinde; tutuşmuş: **go to ~s!,** cehennem ol!, defol!: **what the ~s!,** ne haltetmeğe . . .: **like ~s,** çılgınca. **~ away,** sürekli bir ateş atmak. **~ up,** birdenbire parlamak. **~**[2]**.** Bir ağacı kabuğunu keserek işaretlemek. **to ~ a trail,** yol çizmek, yol açmak; çığır açmak. **~**[3]**.** İlân etm., yaymak. **~r** [ˈblêyzə(r)]. Spor caketi.
**blazon** [ˈblêyzən]. Arma. Arma çizmek; işaret koymak. **to ~ forth [out],** davul zurna ile ilân etmek.
**bleach** [blîç]. Ağartan, beyazlatan şey. Ağartmak, beyazlatmak. **~ing-powder,** kumaşı ağartan klorlu toz.
**bleak** [blîk] *a.* Çıplak, rüzgâra maruz; soğuk; (*fig.*) ümidsiz.
**bleary** [bliə̯ri]. Sulanmış ve kızarmış (göz).
**bleat** [blît]. Meleme(k).
**bleed (bled)** [blîd, bled]. Kan almak; kanamak; kanını dökmek. **to ~ s.o.,** (*coll.*) birinin parasını sızdırmak: **to ~ s.o. white,** birinin varını yoğunu elinden almak: **my heart ~s,** içim paralanıyor.
**blemish** [ˈblemiş]. Kusur; leke. Hafifçe bozmak; dokunmak.
**blend** [blend]. (Çay vs.) harman; halita. Karış(tır)mak; harman etm. (olm.); (renkler) uy(dur)mak.
**bless (blessed, blest)** [bles, blest]. Takdis etm., hayır dua etm.; (Allaha) hamdetmek. **God ~ you!,** (veda ederken) Allaha emanet ol!: **(God) ~ me [you]!, I'm blest, ~ my soul; ~ the boy!,** *hayret veya hiddet ifade eden tabirler*: **to be ~ed with stg.,** nasib olm.: **I ~ my stars that . . .,** çok şükür olsun ki . . .: **I'm blest if I know!,** hiç bilmiyorum; nereden bileyim. **~ed** [blesid] *a.* Mubarek; mes'ud; Allahlık. **the Blessed Virgin,** Meryem ana: **every ~ day,** Allahın günü. **~ing** [blesin(g)] *n.* Hayır dua. **the ~s of civilization,** medeniyetin nimetleri. **that's a ~!,** çok şükür; hamdolsun!, isabet!
**blether** [bleðə(r)]. Manasız, boş sözler (söylemek).
**blew** *v.* **blow.**
**blight** [blâyt]. Nebatlara ârız olan hastalık; her hangi bir kötü tesir. (Güneş, rüzgâr vs.) yakmak, kavurmak; bozmak, mahvetmek. **to ~ s.o.'s hopes,** birinin ümidlerini boşa çıkarmak. **~er** [blâytə(r)]. (*sl.*) Herif; Allahın belâsı. **poor ~!,** zavallı; **you lucky ~!,** seni köftehor! **~y** [blâyti]. (*Asker argosu; aslı* 'bilâdi') İngiltere; yurd.
**blimp** [blimp]. Keşif balonu.
**blind**[1] (blâynd) *a.* Kör; iyi görünmez; deliksiz (duvar); çıkmaz (yol). *v.* Kör etm., körleştirmek; gözünü kamaştırmak. **the ~,** körler: ⌜**can the ~ lead the ~?**⌝ ⌜kelden köseye yardım olur mu?⌝: **to go at a thing ~(ly),** bir işe körükörüne girişmek: **to turn a ~ eye to stg.,** göz yummak. **~fold,** gözleri bağlı; körükörüne; gözlerini bağlamak. **~-man's-buff,** körebe. **~**[2] *n.* Abajur; stor. **roller ~,** yaylı perde.
**blink** [blin(g)k]. Gözlerini kırpıştırma(k); kesik kesik parıldamak. **to ~ the facts,** hakikate gözlerini yummak: **to ~ the question,** meseleye yanaşmamak. **~ers** [ˈblin(g)kö(r)s]. Atın göz siperi.
**bliss** [blis]. Saadet, bahtiyarlık. **~ful,** mes'ud, bahtiyar.
**blister** [blistə(r)]. Kabarcık; su kapma; yakı. Kabartmak, su kapmasına sebeb olm.; yakı koymak. **~ing (language,** *etc.*), zehirli.
**blithe(some)** [blâyð(-səm)]. Neş'eli.

**blithering** [ˈbliðərin(g)]. ~ **idiot,** hebenneka.
**blizzard** [blizəd]. Tipi.
**bloat** [bloͧut]. Şişirmek, kabartmak. **~ed** *a.* göbeği yağ bağlamış: **~er,** tuzlanmış ve tütsülenmiş ringa balığı.
**blob** [blob]. Benek, leke, su damlası.
**block** [blok] *n.* Kütük; kaya vs. parçası; sokak ortasında kalan ada; (*Amer.*) sokak; dairelerden mürekkeb bina; arsa parçası; bir bütün teşkil eden şeyler; klişe, kalıb; makara; tıkanma. *v.* Tıkamak; kapamak; mani olm.; kalıblamak. ~ **capitals,** büyük harfler: **to go to [perish on] the ~,** (*tarih*) başı kesilerek idam edilmek: **traffic ~,** yolun tıkanması. ~ **out,** kaba taslak çizmek; (sansör vs.) karalamak, çıkarmak. ~ **up,** bir kapı veya pencereyi örmek; (delik vs.) tıkamak. **~head** [ˈblokhed]. Mankafa, ahmak. **~-house** [ˈblokhaͧus]. Gözetleme kulesi.
**blockade** [bloˈkêyd]. Abluka (etm.); kuşatma(k). **to run the ~,** ablukayı yarmak. **~-runner,** ablukayı yaran gemi.
**bloke** [bloͧuk]. (*coll.*) Herif, adam.
**blond** [blond]. Sarışın.
**blood** [blʌd]. Kan; soy, asalet. Av köpeğini kana alıştırmak. **there is bad ~ between them,** aralarında husumet var: **to cause bad ~,** aralarını bozmak: **blue ~,** hanedandan, aristokrat: **it made my ~ boil,** tepem attı; **in cold ~,** önceden hesablı: **his ~ ran cold,** tüyleri diken diken oldu: **to draw ~,** kanatmak: **fresh [new] ~,** (bir cemiyet veya işe alınan) yeni unsurlar: **he is out for ~,** kana susamış; (*fig.*.) yanına yanaşılmaz: **it runs in the ~,** soyunda vardır: ⌜**to get ~ out of a stone**⌝, merhametsizden merhamet beklemek: ~ **will tell,** asalet bellidir: ⌜**~ is thicker than water**⌝, kan rabıtası her şeyden kuvvetlidir: **his ~ is up,** kızıştı: **young ~,** (i) genc unsur; (ii) genc ve kibar delikanlı; köyün kabadayısı. **~-curdling,** tüyler ürpertici. **~-pressure,** tansiyon. **~-vessel,** ev'iye; kan damarı. **~hound** [ˈblʌdhaͧund]. Kaçan bir kaatıl vs.yi yakalamakta kullanılan bir cins zağar; (*fig.*) detektif. **~shed** [ˈblʌdşed]. Kan dökme. **~shot** [ˈblʌdşot]. ~ **eye,** kanlanmış göz. **~sucker** [ˈblʌdsʌkə(r)]. Sülük; (*fig.*) insanın kanını emen tefeci vs. **~y** [ˈblʌdi]. Kanlı, kanlanmış; (*sl.*) mel'un, Allahın belâsı.
**bloom** [blûm]. Çiçek; tazelik, genclik; meyva dumanı. Çiçek açmak, çiçekte olm.; gelişmek. **to burst into ~,** çiçek açmak: **in full ~,** tamamen çiçek açmış; (*fig.*) tam gelişme halinde. **~ing,** çiçek açmış; bereketli, mamur; taze; (*sl.*) (*şaka*) Allahın belâsı.
**bloomer** [ˈblûmə(r)]. (*coll.*) Gaf.
**blossom** [ˈblosəm]. Ağac çiçeği. (Ağac) çiçek açmak. **to ~ out,** açılmak, gelişip güzelleşmek.
**blot** [blot]. Mürekkeb lekesi; leke. Lekelemek; kirletmek; mürekkebini kurutmak. **to ~ one's copybook,** küçük bir hata ile şöhretine halel getirmek. ~ **out,** silmek; örtmek.
**blotch** [bloç]. Mürekkeb, boya vs. lekesi; deri üzerinde kırmızı leke. Büyük lekelerle kaplamak. **~y,** lekeli.
**blotting paper** [ˈblotin(g)]. Kurutma kâğıdı.
**blouse** [blaͧuz]. Bluz; işçi gömleği.
**blow**[1] [bloͧu] *n.* Darbe; vuruş. **to come to ~s,** yumruk yumruğa (kavgaya) tutuşmak: **to strike a ~ for liberty,** hürriyet için bir hamle yapmak. **~**[2] (**blew, blown**) [bloͧu, blû, bloͧun] *v.* Esmek; solumak; üflemek; hohlamak; (ampul, sigorta) yanmak; fırlatmak; (boru vs.) çalmak; çalınmak; körüklemek; (sinek) yumurtlamak; (balina) su fışkırtmak; (sır vs.yi) ifşa etm.; (çiçek) açmak. **to be ~n,** soluğu kesilmek: **you be ~ed!,** (*sl.*) sen vız gelirsin!; artık senden bıktım!: **I'll be ~ed if I will [do]!,** ···sam bana da adam demesinler!: **well I'm ~ed!,** deme, Allah aşkına!: **to ~ a boiler,** bir kazandan istim boşaltmak: **to ~ the tanks of a submarine,** bir denizaltının su haznelerini boşaltmak (çıkış yapmak): ~ **the expense [the expense be ~ed!],** (*coll.*) masrafa aldırma!: **it is ~ing a gale,** fırtına var: **to let the horses ~,** atlara soluk aldırmak: **to ~ hot and cold,** (bir şey hakkında) bir dediği bir dediğine uymamak: **to ~ a kiss,** işaretle buse göndermek: **to ~ one's nose,** burnunu silmek: **the door blew open,** kapı rüzgârla açıldı. ~ **about,** oraya buraya uç(ur)mak. ~ **down,** devirmek, yere yatırmak. ~ **in,** (*coll.*) çıkagelmek; uğramak. ~ **off,** rüzgârdan uçmak: ~ **off steam,** istim

salıvermek. ~ **out,** üfleyip söndürmek; üfleyip çıkarmak; (rüzgârdan) sönmek: **to ~ out one's cheeks,** avurtlarını şişirmek. ~ **over,** (rüzgâr) devirmek: **the storm has ~n over,** fırtına geçti: **the scandal soon blew over,** rezalet çabucak unutuldu. ~ **up,** berhava etm.; atmak; şişirmek; berhava olm., patlamak; (*coll.*) haşlamak: **it's ~ing up for rain,** rüzgâr yağmur getirecek: **to be ~n up with conceit,** kibirden kabarmak. **~-fly,** et sineği. **~-lamp,** kaynaklama lâmbası, prümüs lâmbası. **~-pipe,** kuyumcu lâmbası; üfliyerek zehirli ok atmak için boru. ~[3] *n.* (Rüzgâr) üfleme; sümkürme; (boru vs.) çalma. **to go for a ~,** hava almak için bir gezinti yapmak. **~er** [ˈblǫuə(r)]. Üfleyici; esici; hava üfleyen makine. **~y** [ˈblǫui]. Rüzgârlı, fırtınalı.

**blowzy** [ˈblaụzi]. Kırmızı yüzlü; saçları karışık (kadın).

**blubber**[1] [ˈblʌbə(r)] *n.* Balina yağı. ~[2] *v.* Gürültü ile ağlamak, zırlamak.

**bludgeon** [ˈblʌcən]. Kalın ve kısa sopa, matrak. Matrak ile vurmak.

**blue** [blû]. Mavi; mavi renk; çivit; Mavileştirmek; çivitlemek. ~ **water,** deniz: **the ~s,** cansıkıntısı: **Cambridge ~,** açık mavi: **to get one's ~,** Oxford veya Cambridge üniversitesini sporda temsil etm.: **to go ~,** morarmak: **to look ~,** neş'esiz görünmek; bozulmak: **things are looking ~,** vaziyet fena görünüyor: **to ~ one's money,** parasını çarçur etm.: **out of the ~,** damdan düşer gibi; **Oxford ~,** koyu mavi: **true ~,** sadık: **you may talk till you are ~ in the face,** konuşabildiğin kadar konuş! (faydası yok). **~-black,** mavi yazıp kuruyunca siyah olan (mürekkeb). **~-pencil,** mavi kalemle işaret etm. veya karalamak. **~-print,** mavi zemin üzerine beyazla çizilmiş makine vs. projesi, mavi kopya. **~beard** [ˌblûˈbiạd]. Karılarını öldüren adam; mavi sakal. **~bell** [ˈblubel]. Yabani sümbül. **~bottle** [ˈblubotl]. Büyük mavi sinek. **~jacket** [ˈblucakit]. Bahriye neferi. **~stocking** [ˈblustokin(g)]. Okumuş kadın.

**bluff**[1] [blʌf]. Dik, sarp; toksözlü, açık. Sarp kayalık. ~[2]. Blöf, kuru sıkı tehdid. Blöf yapmak. **to call s.o.'s ~,** blöfe aldırmamak.

**blunder** [ˈblʌndə(r)]. Büyük (ahmakça) bir hata (yapmak). **to ~ against [into] s.o.,** birine çarpmak: **to ~ through,** iyi kötü işi başarmak: **to ~ upon the truth,** hakikati tesadüfen keşfetmek: **to ~ one's way along,** çarpa çarpa ilerlemek.

**blunt** (blʌnt). Kör (bıçak vs.); lâfını sakınmaz (kimse); açık, pervasız (söz). Körleştirmek; hassasiyetini gidermek.

**blur** [blö(r)]. Hayal meyal görülen şey; leke, bulaşık. (Yazı vs.) bulaştırmak. ~ **out,** bulandırıp gizlemek.

**blurt** [blö(r)t]. ~ **out,** birdenbire söylemek; düşünmeden söylemek.

**blush** [blʌş]. Kızarma(k); utanma(k). **at the first ~,** ilk bakışta: **in the first ~ of youth,** gencliğin ilk çağlarında: **to ~ for s.o.,** biri namına utanmak.

**bluster** [ˈblʌstə(r)]. Kabadayılık. Sert esmek; bağırıp çağırmak; yüksekten atmak. **to ~ out threats,** tehdid savurmak. **~er,** yüksekten atan kabadayı.

**boar** [bô(r)]. Erkek domuz. **wild ~,** yaban domuzu.

**board** [bôd]. Tahta; levha; karton; sofra, yemek; pansiyon; idare meclisi; daire. Tahta ile kaplamak; döşemek; birine yemeğini vermek; (gemiye vs.) binmek; (gemiye) hücüm edip girmek; pansiyon olmak. **the ~s,** tiyatro sahnesi: **on ~,** gemide, trende vs.de: ~ **of directors,** idare meclisi: **bed and ~ [~ and lodging],** tam pansiyon: **Board of Trade,** Ticaret Nezareti: ~ **wages,** yemek parası da dahil olarak verilen ücret: **above ~,** açıkça, dürüst: **to go by the ~,** (*naut.*) direk vs. güverteden denize düşmek; (*fig.*) büsbütün elden çıkmak: **to sweep the ~,** kumarda masadaki bütün parayı kazanmak; bir müsabaka vs.de bütün mükâfatları kazanmak. ~ **out,** pansiyona yerleştirmek. ~ **up,** tahta ile kapatmak. **~er** [ˈbôdə(r)]. Pansiyon kiracısı; yatılı (talebe). **~ing** [ˈbôdin(g)]. Tahta perde; tahta kaplama. **~ing-house,** pansiyon. **~ing-school,** yatılı mekteb.

**boast** [bǫust]. Öğünme(k), iftihar (etm.). **to ~ of [about],** ···le böbürlenmek: **without wishing to ~,** öğünmek gibi olmasın ama. . . . **~ful,** öğüngen palavracı.

**boat** [bǫut]. Sandal, kayık, gemi.

Sandalla gezmek. **all in the same ~,** aynı halde: **to burn one's ~s,** gemilerini yakmak; ⌜ölmek var dönmek yok⌝ *kabilinden.* **~-deck,** filika güvertesi. **~-hook,** kayık kancası. **~-house, ~shed,** kayıkhane. **~er,** (sert) hasır şapka, kanotye. **~man,** kayıkçı. **~swain** [ˈbo̯usən]. Porsun, lostromo. **bosun's mate,** lostromo muavini: **~'s chair [cradle],** gemiyi boyayan işçinin oturduğu asılı iskemle.

**bob**[1] [bob]. Şakul; saç lülesi; kesik kuyruk. **to ~ one's hair,** (kadın) saçlarını kısa kestirmek. **~**[2]**.** Baş hareketi. Oynama(k), kımıldama(k); hafifçe iğilme(k). **~ down,** sakınmak için birdenbire iğilmek. **~ up,** birdenbire meydana çıkmak. **~**[3]**.** (*sl.*) Şilin.

**bobbin** [ˈbobin]. Makara; pamuk iği.

**bobby** [ˈbobi] **Robert**'in kısaltmış şekli; polis.

**bode** [bo̯ud]. **it ~s no good,** hayra alâmet değil.

**bodice** [ˈbodis]. Korsaj.

**bod·ied** [ˈbodid]. Vücudlü. **able ~,** sağlam. **~ily** [ˈbodili]. Vücude aid; cismanî; tamamile, kâmilen. **~y** [ˈbodi]. Vücud; gövde; beden; cesed; cisim; şahıs; grup, heyet, cemiyet; takım, mecmu; esas. **~ and soul,** bütün mevcudiyetile: **to keep ~ and soul together,** kıtakıt yaşamak: **heavenly ~,** semavî cisim. **~y-colour,** kesif sulu boya. **~yguard,** hassa askeri. **~y-snatcher,** cesed hırsızı. **~y(work),** (otomobil vs.) karoseri.

**bodkin** [ˈbodkin]. Şerid geçirmek için şiş.

**bog** [bog]. Batak(lık). Batağa saplamak.

**bogey** [ˈbo̯ugi]. Umacı.

**boggle** [ˈbogl]. **to ~ at doing stg.,** bir işe yanaşmamak.

**bogie** [ˈbo̯ugi]. Vagon veya lokomotifin ön kısmının dayandığı çifte dingilli kademe.

**bogus** [ˈbo̯ugəs]. Sahte, taklid.

**Bohemian** [boˈhîmi̯ən]. Bohemyalı; derbeder, kalender; çingene.

**boil**[1] [bôyl] *n.* Çıban. **~**[2] *n.* Kaynama. *v.* Kayna(t)mak; haşlamak. **~ed egg,** rafada yumurta: **hard-~ed egg,** hazırlop yumurta: **to be on the ~,** kaynamak: **to come to the ~,** kaynamağa başlamak: **to go off the ~,** kaynaması durmak: **to keep the pot ~ing,** aileyi geçindirmek. **~ away,** kaynayıp buhar olmak. **~ down,** kaynayıp suyunu çekmek; hülâsa etm.: **it all ~s down to this,** hülâsası budur. **~ over,** kaynayıp taşmak: **to ~ over with rage,** hiddetten köpürmek. **~ up,** (süt vs.) kaynayıp kabarmak. **~er** [ˈbôylə]. Kazan. **~ing-point** [ˈboilin(g)]. Kaynama derecesi.

**boisterous** [ˈbôystərəs]. Şiddetli; gürültülü; taşkın.

**bold** [bo̯uld]. Cesur; kendinden emin; küstah. **as ~ as brass,** çok yüzsüz: **to make ~,** cüret etm.: **to put a ~ face on the matter,** haklı imiş gibi cesaret takınmak. **~-faced,** yüzsüz, küstah. **~ type,** kalın harf.

**bole** [bo̯ul]. Ağac gövdesi.

**boll** [bol]. Pamuk ve keten tohumlarını örten mahfaza. **~-weevil,** pamuk kurdu.

**bollard** [ˈbolö(r)d]. (*naut.*) Baba; bite.

**bolster** [ˈbo̯ulstə(r)]. Uzun yastık, alt yastık. **~ up,** yastık koymak; desteklemek.

**bolt**[1] [bo̯ult]. Sürme(lemek); cıvata (ile bağlamak); tüfek mekanizması; yıldırım. **~ upright,** dimdik: **a ~ from the blue,** bulutsuz havada şimşek kadar umulmadık ve anî vak'a: **he has shot his last ~,** son kurşununu attı. **~**[2]**.** Kaçma(k), firar; atılma. Birdenbire fırlayıp gitmek; çiğnemeden acele yemek. **to make a ~ for it,** tabanları kaldırmak: **to make a ~ for stg.,** bir şeye doğru atılmak (koşmak).

**bomb** [bom]. Bomba. (Bilhassa uçaktan) bombalamak. **~er** [ˈbomə(r)]. Bombacı; bomba uşağı.

**bombard** [bomˈbâd]. Bombardıman etmek. **~ment,** bombardıman.

**bombastic** [ˈbombastik]. Tumturaklı.

**bona fide** [ˈbo̯unəˈfâydi]. Hakikî. **~s,** hüsnüniyet.

**bonanza** [bo̯uˈnanzə]. Zengin bir maden damarı; büyük bir kazanc membaı.

**bond** [bond] *n.* Bağ; kayıd; sened; tahvilat; kefalet; münasebet. **to be in ~,** (eşya) gümrükte olm.: **to take goods out of ~,** eşyayı gümrükten çıkarmak: **his word is as good as his ~,** sözünün eridir. **~holder** [ˈbondho̯uldə(r)]. Tahvil hamili.

**bond·age** [ˈbondəc]. Kölelik, esirlik, serflik. **~sman** [ˈbonsmən]. Serf, köle; kefil.

**bone** [boun]. Kemik; kılçık. Kemiklerini (kılçıklarını) ayıklamak; (*sl.*) aşırmak. **to the ~,** iliklerine kadar: **a bag of ~s [skin and ~],** bir deri bir kemik: **to feel stg. in one's ~s,** (niçin olduğunu bilmeden) emin olm.: ⌜**hard words break no ~s**⌝, sert sözler insanın bir yerini kırmaz: **I have a ~ to pick with you,** seninle paylaşılacak kozum var: **to make no ~s about . . .,** ···de hiç tereddüd etmemek. **~setter,** çıkıkçı. **~-dry,** kupkuru. **~ idle,** çok tembel. **~-shaker,** köhne otomobil veya bisiklet.

**bonfire** [ˈbonfâyə(r)]. Şenlik ateşi; bahçe süprüntüsünü ortadan kaldırmak için açıkta yakılan ateş.

**bonnet** [ˈbonit]. Çocuk ve kadınların giydiği başlık; iskoç beresi; motor kapağı.

**bonny** [ˈboni]. (*İskoç*) güzel, hoş.

**bonus** [ˈbounəs]. İkramiye. **~share,** temettü hissesi: **cost-of-living ~,** hayat pahalılığı zammı.

**bony** [ˈbouni]. Kemikleri görünen; kılçıklı.

**boo** [bû]. Yuha. **~ (at),** ···e yuha diye bağırmak; ıslık çalmak. ⌜**he wouldn't say ~ to a goose**⌝, ⌜ağzına vur lokmasını al⌝ *kabilinden*.

**booby** [ˈbûbi]. Ahmak; şaşkın. **~ prize,** sonuncu gelene verilen mükâfat. **~-trap,** kapıdan girenin başına düşecek şekilde (şaka için) asılan şey.

**boohoo** [buˈhu]. Şımarıkça ağlama(k).

**book**[1] [buk] *n.* Kitab; defter; bab. **the Book,** incil: **to be in s.o.'s good [bad] ~s,** birinin gözünde ol(ma)mak: **to bring s.o. to ~,** birini hesab vermeğe mecbur tutmak: **to keep the ~s of a firm,** bir firmanın defterini tutmak: **to make a ~,** at yarışında bahse girmek: **on the ~s,** bir klüb vs. âzası olarak ismi defterde yazılı olm.: **to speak like a ~,** kitabî konuşmak: **it won't suit my ~,** bu işime gelmez. **~-end,** kitab dayayacağı. **~-keeper,** defter tutan, muhasib. **~-keeping,** muhasebe. **~-learning,** kitabî malûmat. **~-maker,** at yarışlarında bahis defteri tutan adam. **~**[2] *v.* Hesaba yazmak; (yer vs.) tutmak; bilet almak. **to be ~ed up,** (doktor, iş adamı vs.) vakti dolu olm.; (otel vs.) bütün yerler tutulmuş olmak. **~bind·er** [ˈbukbâyndə(r)]. Mücellid. **~case** [ˈbukkêys]. Kitab dolabı, kütübhane. **~ing-clerk.** Gişedeki biletçi. **~ing-office,** gişe. **~ish** [ˈbukiş]. Okumağa meraklı. **~let** [ˈbuklit]. Küçük kitab; risale. **~-mark(er)** [ˈbukmâkə(r)]. Okunan sahifeyi göstermek üzere konulan şey. **~seller** [ˈbukselə(r)]. Kitabcı. **~shop** [ˈbukşop]. Kitabhane. **~stall** [ˈbukstôl]. İstasyonda kitabcı dükkânı.

**boom**[1] [bûm]. Bumba; seren; liman ağzındaki mania. **~**[2]. (*ech.*) (Top vs.) gürleme; (rüzgâr) uğultu. Gürlemek; uğuldamak. **~**[3]. Fiatlarda anî yükselme; büyük rağbet. Reklam yapmak; yükselmek.

**boomerang** [ˈbûməran(g)]. Avustralya yerlilerinin bükülmüş sert tahtadan bir silâhı ki fırlatıldığı yere avdet eder; yapanın üstüne dönen kötü hareket.

**boon**[1] [bûn] *n.* Nimet; lûtuf, iyilik. **~**[2] *a.* (*şair*) Şen, neş'eli. **~ companion,** içki, eğlence, arkadaşı; çok yakın arkadaş.

**boor** [ˈbuə(r)]. Kaba, yontulmamış adam. **~ish,** kaba, hoyrat.

**boost** [bûst]. Yardım için itme(k); reklamını yapmak; (*elect.*) cereyan kuvvetini artırmak.

**boot**[1] [bût]. Potin, ayakkabı; çizme; otomobil arkasındakı eşya yeri. Tepmek. **the ~ is on the other leg,** vaziyet tam aksi: **to die in one's ~s,** eceli kaza ile ölmek: **to get the ~,** pabucu eline verilmek, sepetlenmek: **to give s.o. the (order of the) ~,** birinin pabucunu eline vermek: **to have one's heart in one's ~s,** ödü kopmak. **~-black,** ayakkabı boyacısı. **~ed** [ˈbûtid] **and spurred,** harekete hazır. **~lace** [ˈbûtlêys]. Ayakkabı bağı. **~leg** [ˈbûtleg]. (*Amer.*) içki kaçakçılığı etmek. Kaçak (içki). **~less** [ˈbûtlis], Ayakkabısız; faydasız, lüzumsüz. **~-polish,** ayakkabı boyası. **~s** [bûts]. Otel hademesi; ayakkabı temizleyici. **~**[2]. **to ~,** üstelik, fazla olarak. **~**[3]. **what ~s it to ...?,** neye yarar?

**booth** [bûð]. Baraka; külübe.

**booty** [ˈbûti]. Ganimet; çapul; kazanc.

**booze** [ˈbûz]. İçki; çakıntı. Kafayı tütsülemek. **to be on the ~,** kafayı çekmek.

**boracic** [boˈrasik]. Borakslı, borik.

**border** [ˈbôdə(r)]. Kenar; pervaz; tiriz; hudud. Etrafını çevirmek; kenar yapmak. **to ~ on,** bitişik olm.;

(*fig.*) çok yaklaşmak: **the Border,** İskoçya hududu: **colour ~ing on red,** kırmızıya çalan renk. **~er,** hududda oturan. **~land,** sınırdaş memleket; iki şey arası.

**bore**[1] [bô(r)]. Çap; boru kutru; sondaj deliği. Delmek; sondaj yapmak. **~hole** [ˈbôhoul]. Sondaj deliği; kuyu. **~r** [ˈbôrə(r)]. Delici; burgu, matkab. **~**[2]. Can sıkıcı kimse; usandırıcı şey. Can sıkmak, usandırmak. **to be ~d stiff,** can sıkıntısından patlamak. **~dom** [ˈbôdəm]. Can sıkıntısı. **~**[3]. Bazı nehirlerin ağızlarının birdenbire daralmasından husule gelen med dalgası. **~**[4] *v.* **bear**[2].

**boring** [ˈbôrin(g)]. Can sıkıcı. Delici. Delik, sondaj.

**born** [bôn]. Doğmus. **to be ~,** doğmak: **a ~ poet,** doğuşta şair: **in all my ~ days,** bütün ömrümde: **her latest ~,** son doğan çocuğu: **London ~,** Londra'da doğmuş: **a Londoner ~ and bred,** doğma büyüme Londralı: **~ tired,** daima yorgun; fitraten tembel ve uyuşuk.

**borne** *v.* **bear**[2].

**borough** [ˈbʌrə]. Kasaba; belediyesi olan şehir.

**borrow** [ˈborou]. Ödünc almak; borc almak; iğreti almak.

**Borstal** [ˈbôstəl]. İngiltere'de suçlu çocuklara mahsus islahhane.

**bosh** [boş]. Boş şey, saçma.

**bosom** [ˈbuzəm]. Göğüs; kucak. **~ friend,** candan dost.

**Bosphorus** [ˈbosforəs]. Boğaziçi.

**boss**[1] [bos]. Yumru; çıkıntı; meme. **~**[2] (*coll.*) Patron; sözü geçen. (Bir daireyi vs.) idare etm.; birisine hâkim olmak. **she's the ~,** evde karısının sözü geçer. **~y** [ˈbosi]. (*coll.*) Zorba; mütehakkim.

**bosun** *v.* **boatswain.**

**botan·ic(al)** [boˈtanik(l)]. Nebatat ilmine aid. **~ist,** [ˈbotənist], nebatatçı. **~ize** [ˈbotənâyz], nebatat nümuneleri toplamak. **~y** [ˈbotəni], nebatat ilmi.

**botch** [boç]. Biçimsiz yama; kaba iş. Baştan savma yapmak; kabaca yamamak.

**both** [bouθ]. İkisi de; her ikisi de; iki, her iki. **~ ... and ...,** hem ... hem ...: **~ of you,** her ikiniz: **you can't have it ~ ways,** ikisinin ortası olmaz, ya öyle ya böyle.

**bother** [ˈboðə(r)]. Canını sıkma(k), taciz etme(k); üzüntülü iş. Üzmek; tasdi etmek. **~ it!,** Allah müstahakkını versin!: **I can't be ~ed!,** bana ne?: **he couldn't ~ [be ~ed] to do it,** yapmağa üşendi: **don't ~ me!,** beni rahat bırak! **don't ~ about me,** beni düşünme: **he doesn't ~ about anything,** hiç bir şeye aldırmıyor. **~ation, ~!,** Allah müstahakkını versin! **~some,** can sıkıcı, müz'ic, üzüntülü.

**bottle** [ˈbotl]. Şişe; sürahi. Şişeye, kavanoza, koymak. **to ~ up one's feelings, anger,** *etc.*, hislerini, hiddetini vs. tutmak: **to ~ up a fleet,** filoyu çıkma yolunu tutarak sıkıştırmak. **~-green,** çok koyu yeşil. **~-neck,** şişe boğazı; nakil vasıtalarının izdihamdan saplanıp kalması; bir işi çıkmaza sokan [duraklatan] şey. **~-nosed,** patlıcan burunlu. **~-washer,** her işe bakan adam.

**bottom** [ˈbotəm] *n.* Dib; aşağı taraf; alt; kıç; asıl, esas; (*esk.*) gemi. *a.* Alt; asgarî. *v.* Dibe dokunmak. **at ~,** hakikatte, esas itibarile: **he is at the ~ of all this,** bütün bunların arkasında o var: **I bet my ~ dollar,** nesine istersen bahse girerim: **the ~ has fallen out of the market,** piyasa çöktü: **to get to the ~ of a matter,** bir meselenin içyüzünü öğrenmek: **from the ~ of one's heart,** çok samimî: **to knock the ~ out of an argument,** bir muhakemeyi cerhetmek, yıkmak: **to touch ~,** (i) (gemi) dibi karaya dokunmak; (ii) en aşağı dereceye varmak. **flat ~ed** [ˈbotomd]. Dibi düz. **~less** [ˈbotəmlis]. Dibsiz; çok derin; sonsuz. **the ~ pit,** gayya. **~most** [ˈbotəmmoust]. En aşağı.

**boudoir** [ˈbudwa]. Kadın odası, buduvar.

**bough** [bau]. Dal.

**bought** *v.* **buy.**

**boulder** [ˈbouldə(r)]. Münferid kaya parçası.

**boulevard** [ˈbûlvâd]. Bulvar.

**bounc·e** [bauns]. Zıplama, sıçrama; (*coll.*) yüksekten atma, ögünme. Zıpla(t)mak, sıçra(t)mak: **~ in [out],** birdenbire girmek [çıkmak]. **~ing** [baunsin(g)] *a.* Gürbüz, canlı ve neş'eli (çocuk).

**bound**[1] [baund]. Had, hudud, sınır. Sınır teşkil etm., hudud çizmek. **to break ~s,** (bir asker, talebe) yasak edilen yere girmek: **the town is out of ~s,** şehre gitmek yasaktır: **to keep within ~s,** hududu aşmamak: **to set**

**~s,** tahdid etmek. **~ary** [ˈbaundəri]. Hudud, sınır. **~er** [ˈbaundə(r)]. Âdî, adam; centilmenin aksi. **~less** [ˈbaundlis]. Hududsuz; sonsuz. **~²**. Atlama(k), sıçrama(k). **to advance by leaps and ~s,** şaşırtıcı bir süratle ilerlemek: **to ~ away,** zıplayarak gitmek. **~³**. **ship ~ for ...,** ···e gitmek üzere olan gemi. **whither ~ ?,** nereye? **~⁴** *v.* **bind.** *a.* Mecbur; kayıdlı; cildli.

**bount·iful** [ˈbauntəful]. Cömert, alicenab; bol. **~y** [ˈbaunti]. Cömertlik; ihsan; ikramiye, prim.

**bouquet** [ˈbûkêy]. Çicek demeti; buket; (şarab) koku.

**bout** [baut]. Girişme; nöbet, sıra.

**bovine** [ˈbovâyn]. Öküze aid; ahmak; ağır.

**bow¹** [bou]. Yay; ilmek; fiyonga; kavis. **to have two strings to one's ~,** ikinci bir imkâna malik olm. **~-legged** [bouˈlegd]. İğrı bacaklı, paytak. **~line** [ˈboulin]. İzbarça bağı: **running ~,** müteharrik izbarça bağı: **~man,** *pl.* **-men** [ˈbouman, -men]. Okçu, tirendaz. **~shot** [ˈbouşot]. Ok atımı. **~sprit** [ˈbousprit]. Civadra. **~string** [ˈboustrin(g)]. Ok kirişi; kemend. *v.* (**bowstrung**), Kemend ile boğmak. **~-window** [ˈbouˈwindou]. Kavisli pençere; cumba. **~²** [bau]. İğilme, baş eğme. İğilmek, başını eğmek. **to ~ and scrape to s.o.,** birine kandilli temenna etm.: **to ~ to the inevitable,** kaderi olduğu gibi kabul etm. **~³** [bau] *n.* Geminin başı, prova; provaya en yakın kürekçisi. **to cross the ~s of a ship,** bir geminin önüne geçmek.

**bowdlerize** [ˈboudlərâyz]. Bir eserin müstehcen görülen yerlerini çıkarmak.

**bowel** [ˈbauəl]. Barsak. **~s,** barsaklar; iç. **to empty the ~s,** abdeste çıkmak.

**bower** [ˈbauə(r)]. Çardak.

**bowl¹** [boul] *n.* Kâse, tas, çanak; pipo ağzı. **~²**. Tahta top, yuvarlak. **~s,** tahta toplarla oynanan bir nevi oyun. Topu yerde yuvarlamak; (krikette) topu atmak. **~ along,** (araba vs.) hızla gitmek. **~ out,** (krikette) birini oyun harici yapmak; (*fig.*) temizlemek. **~ over,** devirmek, düşürmek; şaşırtmak. **~er,** (krikette) topu atan oyuncu.

**bowler (hat)** [ˈboulə(r) (hat)]. Melon şapka.

**box¹** [boks] *n.* Şimşir ağacı. **~wood,** şimşir tahtası. **~²** *n.* Kutu, sandık; loca; ahırın bölmesi; arabacı yeri; mahkemede şahid veya maznun yeri. *v.* Kutuya koymak. **to ~ the compass,** pusula kertelerini saymak: **to be in the same ~,** aynı vaziyette bulunmak. **~-office,** tiyatro vs. gişesi: **~ receipts,** tiyatro geliri. **~³**. Boks yapma(k). **to ~ s.o.'s ears,** birine tokat atmak. **~er,** boksör. **~ing,** boks. **~ing-day,** Noelin ertesi günü (26 aralık; İngiltere'de resmî tatil günü ki postacı, çöpçü vs.nin **Christmas-box** denilen hediye veya bahşişleri o gün verilir).

**boy** [bôy]. (On sekiz yaşına kadar erkek) çocuk; oğul; çocuk garson; talebe; şarkta yerli uşak. **~s will be ~s,** çocuktur yapacak: **old ~ !,** ahbab!: **an old ~,** bir mektebin eski mezunu; ihtiyarın biri: **one of the ~s** eğlenceye düşkün. **from a ~,** çocukluğundan beri. **~hood** [ˈbôyhud]. Çocukluk çağı. **~ish** [ˈbôy·iş]. Erkek çocuk gibi.

**boycott** [ˈbôykot]. Boykot (yapmak).

**Bp.** (*abb.*). **bishop.**

**brace** [brêys]. (Mimarî) bağ; destek; rabıt işareti ({); makkab sapı. Destek vurmak; kuvvet vermek. **~s,** pantalon askısı. **~ and bit,** el makkabı: **a ~ of birds,** iki kuş.

**bracelet** [ˈbrêyslit]. Bilezik.

**bracing** [ˈbrêysin(g)]. Canlandırıcı.

**bracken** [ˈbrakn]. Büyük serhas.

**bracket** [ˈbraket]. Kol, destek; parantez. Parantez içine almak; bir rabıt işareti ile birleştirmek; bir tutmak; (topçulukta) hedefi makas içine almak. **square ~s,** köşeli parantez.

**brackish** [ˈbrakiş]. Tuzlumsu, acı (su).

**bradawl** [ˈbradôl]. Biz.

**Bradshaw** [ˈbradşô]. (İngiltere'de) tren tarifesi.

**brag** [brag]. Yüksekten atma(k); öğünme(k). **~ of [about],** ···le böbürlenmek. **~gart,** kendini öven kimse.

**braid** [brêyd]. Örgülü şerid, vs. (Saç vs.yi) örmek, kurdele ile bağlamak. **gold ~,** sırma kordon.

**braille** [brêyl]. Körlere mahsus kabartma yazı.

**brain** [brêyn]. Beyin, dimağ; kafa. Beynini patlatmak. **he has ~s,** çok kafalıdır: **to get [have] stg. on the ~,** bir şeyi aklından çıkaramamak. **~y,**

(*coll.*) kafalı. **~-fag,** beyin yorgunluğu. **~ pan,** kafatası. **to have a ~-storm,** bir şey beynine vurmak. **~-wave,** birdenbire gelen parlak fikir.

**brake** [brêyk]. Fren; büyük yolcu arabası. **~** veya **put on** [**apply**] **the ~,** fren yapmak. **~ van,** şefdötren furgonu: **150 ~ horsepower,** 150 bremze beygir kuvveti. **~sman,** frenci. **~-drum,** fren makarası.

**bramble** [ˈbrambl]. Böğürtlen çalısı.

**bran** [bran]. Kepek. **~-mash,** su ile karıştırılmış kepek.

**branch** [brânç]. Dal; kol; şube. Dallanmak. **~ out,** dalbudak salmak; şubelere ayrılmak; kol teşkil etmek. **~ off,** çatallaşıp ayrılmak.

**brand** [brand] *n.* Yanık odun parçası; meş'ale; kızgın demir; damga; marka, cins. *v.* Dağlamak; damgalamak. ⌜**a ~ from the burning**⌝, cehennemde yanmaktan kurtulmuş kimse. **~-new,** yepyeni.

**brandish** [ˈbrandiş]. Sallamak, savurmak.

**brandy** [ˈbrandi]. Konyak.

**brass** [brâs]. Pirinç; (*sl.*) para. **the ~,** pirinçten yapılmış musiki aletleri. **~y,** pirinç gibi.

**brassière** [brasiˈe̯ə(r)]. Sutiyen.

**brat** [brat]. Yumurcak, veled.

**bravado** [brəˈvâdo̯u]. Kurusıkı kabadayılık; palavra.

**brave** [brêyv]. Cesur, yiğit. Muharib. Karşı gelmek; meydan okumak. **~ry,** Cesaret, kahramanlık.

**bravo** [brâˈvo̯u]. Aferin!

**brawl** [brôl]. Gürültülü kavga (etmek).

**brawn** [brôn]. Kaba et; adale, kuvvet; paça gibi dondurulmuş domuz eti. **~y,** kuvvetli, adaleli.

**bray** [brêy]. Anırma(k).

**brazen** [ˈbrêyzən]. Pirinçten, tunçtan; yüzsüz, utanmaz. **to ~ it out,** yüzsüzlükle bastırmak, pişkinlik etmek.

**brazier** [ˈbrêyzi̯ə(r)]. Mangal.

**Brazil** [brəˈzil]. Brezilya. **~ian,** brezilyalı.

**breach** [brîç]. Yarık; kırık; aralık; riayetsizlik; kırma(k), bozma(k). **~ of faith,** vadini tutmama: **~ of promise,** evlenme vadini bozma: **~ of the peace,** sukûnu ihlâl.

**bread** [bred]. Ekmek. **~ and butter,** tereyağlı ekmek; rızık, geçim yolu: **he knows on which side his ~ is buttered,** menfaatinin hangi tarafta olduğunu biliyor. **~-winner,** aile geçindiren.

**breadth** [bredθ]. Genişlik, en, vüs'at.

**break** (**broke, broken**) [brêyk, bro̯uk, bro̯ukən]. Kırık, kırıklık; kırılma; açıklık; ara, fasıla; teneffüs; sesi değişme. Kırmak; bozmak; tutmamak; kesmek; alıştırmak; mahvetmek, işini bitirmek; kırılmak; topun seyri değişmek. **the ~ of day,** şafak: **a ~ in the weather,** havanın değişmesi: **to ~ a blow, a fall,** *etc.*, bir darbe vs.nin şiddetini azaltmak: **to ~ ground,** toprak sürmek; ilk adımı atmak: **to ~ fresh ground,** çığır açmak: **to ~ s.o. of a habit,** birini bir âdetten vazgeçirmek: **to ~ s.o.'s heart,** birine keder ve ıstırab vermek: **to ~ one's journey,** uzun bir seyahatte bir yerde kalmak: **to ~ the news (gently),** münasib bir lisanla haber vermek: **to ~ an officer,** bir zabıtın rütbesini indirmek: **to ~ a set** (gümüş vs.) takımını bozmak. **~ away,** koparmak; kaçıp kurtulmak; ayrılmak. **~ down,** yıkmak, inhilal ettirmek; sakatlanmak; büyük bir teessürden dolayı ağlamak. **~ in,** zorla girmek; alıştırmak; söze karışmak: **to ~ in (up)on a company,** bir grupa birdenbire karışmak. **~ into,** zorla girmek; birdenbire bir şeye başlamak: **to ~ into one's reserves,** ihtiyattan sarfetmek. **~ off,** koparmak, kesmek; ayrılmak: **to ~ off an engagement,** nişanı bozmak. **~ out,** patlamak; kaçmak: **(of the face) to ~ out into pimples,** yüzü sivilce ile kaplanmak: **to ~ out into a sweat,** ter basmak. **~ through,** yarmak. **~ up,** dağıtmak; toprağı sürmek; parçala(n)mak; (mekteb) devre sonunda tatil olm.; (hava) bozulmak. **to ~ with s.o.,** birisile bozuşmak, alâkasını kesmek. **~able** [ˈbrêykəbl]. Kırılacak (şey). **~age** [ˈbrêykəc]. Kırılma. **~down** [ˈbrêykda̯un]. Bozulma; (sıhhat) çökme. **~er** [ˈbrêykə(r)]. Sahile çarpan dalga. **~fast** [ˈbrekfəst]. Sabah kahvaltısı. Kahvaltı etmek. **~neck** [ˈbrêyknek]. Çok tehlikeli. **~water** [ˈbrêykwôtə(r)]. Dalga kıran.

**breast** [brest]. Göğüs; meme; (*fig.*) iç, kalb. Göğüslemek; (tepeye) tırmanmak. **to make a clean ~ of it,** her şeyi itiraf etmek. **~work,** düzlükte kurulan siper.

**breath** [breθ]. Soluk, nefes alma; hafifçe esme; buğu; püfürtü. **all in the same ~,** aynı zamanda: **to have a bad ~,** ağzı kokmak: **to draw ~,** nefes almak: **it is the very ~ of life to me,** bu benim için canım kadar azizdir: **to lose one's ~,** nefesi kesilmek: **out of ~,** nefesi kesilmiş bir halde: **to save one's ~,** nefes tüketmemek: **to speak below [under] one's ~,** alçak sesle konuşmak.

**breath·e** [brîð]. Nefes almak; teneffüs etm.; mırıldanmak; şoylemek. **to ~ courage into s.o.,** birine cesaret vermek: **to ~ freely again,** rahat bir nefes almak; ferahlanmak: **to ~ heavily,** zahmetle nefes almak: **to ~ one's last,** son nefesini vermek: **to ~ forth [out] threats,** tehdid savurmak. **~ing-space,** nefes alacak zaman veya yer; istirahat fırsatı.

**bred** *v.* **breed. well [ill] ~,** terbiyeli [terbiyesiz]: **well-~ horse,** cins at.

**breech** [brîç]. Kıç; kaynak yeri; top kuyruğu. **~es,** dizin altından bağlanan kısa pantalon; (*coll.*) pantalon: **to wear the ~,** kocasına hâkim olmak. **~-block,** kama. **~-loading,** kuyruktan dolma.

**breed** [brîd]. Soy, cins. (**bred** [bred]). Doğurmak; yetiştirmek, beslemek; çoğalmak. **~ing,** soy; yetiştirme, besleme; terbiye.

**breez·e** [brîz]. Hafif rüzgâr, (*coll.*) atışma. **~y,** hafif rüzgârlı; canlı.

**brevity** [ˈbreviti]. Kısalık.

**brew** [brû]. Kaynatıp tahammur ettirerek bira vs. yapmak; (çay) demlemek. Hazırlanmış meşrub. **there is something ~ing,** bir şeyler dönüyor. **~ery,** bira fabrikası.

**briar** [ˈbrâyə(r)]. Yabani gül.

**bribe** [brâyb]. Rüşvet (vermek). **~ry,** rüşvet verme, alma.

**bric-a-brac** [ˈbrikəbrak]. Biblolar.

**brick** [brik]. Tuğla; kalıp. Tuğla döşemek. **to ~ up,** tuğla ile örmek: **to drop a ~,** (*coll.*) çam devirmek: **he's a ~,** (*coll.*) dört yüz dirhem adamdır. **~layer,** duvarcı.

**bridal** [ˈbrâydl]. Geline aid; evlenmeğe aid.

**bride** [brâyd]. Gelin. **~groom** [ˈbrâydgrûm], güvey. **~smaid,** düğünde gelinin yanında bulunan genc kız.

**bridge** [bric]. Köprü; briç (kâğıd oynu). Üzerine köprü kurmak. **~-head,** köprübaşı.

**bridle** [ˈbrâydl]. At başlığı, yular. Gem vurmak, dizgin takmak. **~-path,** atlılara mahsus dar yol.

**brief**[1] [brîf] *a.* Kısa, mücmel. **in ~,** hülâsa olarak: **to be ~,** sözün kısası. **~**[2] *n.* Dava dosyası. *v.* **to ~ a barrister,** bir davayı bir avukata vermek: **to ~ a case,** bir davanın dosyasını tanzim etm.: **to hold a ~ for s.o.,** birini mahkemede müdafaa etm.: **I don't hold any ~ for him, but ...,** onu müdafaa etmek benim vazifem değil, fakat .... **~-case,** evrak çantası.

**brig.** (*abb.*) **brigadier.**

**brigad·e** [briˈgêyd]. Liva; tugay; müfreze. **fire-~,** itfaiye. **~ier** [brigəˈdiə(r)], mirliva; tuğbay, tuğgeneral.

**brigand** [ˈbrigənd]. Haydud, eşkıya. **~age,** haydudluk, eşkıyalık.

**bright** [brâyt]. Parlak; aydınlık; (hava) açık; uyanık, zeki; neş'eli. **~en** [ˈbrâytən]. Parla(t)mak; neş'elen(dir)mek.

**brillian·ce** [ˈbrilyəns]. Fevkalâde parlaklık, şa'saa. **~t,** fevkalâde parlak; pırlanta.

**brim** [brim]. Bardak vs. ağzı, kenar. **to ~ over,** taşmak. **~-full,** ağız ağza dolu.

**brimstone** [ˈbrimstoun]. Kükürt.

**brin·e** [brâyn]. Tuzlu su, salamura. **~y,** tuzlu: **the ~,** (*coll.*) deniz.

**bring** (**brought**) [brin(g), brôt]. Getirmek; irca etm.; bir hale getirmek. **to ~ into action [play],** ortaya koymak; harekete geçirmek: **to ~ guns to bear on stg.,** topları bir şey üzerine çevirmek: **he brought it on himself,** bunu başına kendi getirdi: **I could not ~ myself to tell him,** ona söylemeğe dilim varmadı: **he couldn't ~ himself to,** ...mağa gönlü razı olmadı. **~ about,** vukua getirmek; sebeb olm. **~ along,** yanında getirmek. **~ down,** düşürmek, indirmek: **to ~ down the house,** alkış tufanı koparmak. **~ forth,** doğurmak; meydana çıkarmak; mahsul vermek. **~ forward,** ileri getirmek; hesab yekûnunu nakletmek. **~ in,** içeri getirmek; idhal etmek. **to ~ off a success,** başarmak; **to ~ it off,** (*coll.*) muvaffak olmak. **~ on,** sahneye koymak; sebeb olm., hasıl etm.: **this warm weather will ~ on the strawberries,** bu sıcak havada çilekler çabuk olacak. **~ out,** meydana koymak; belli etm.; göstermek; neşretmek. ~ **round,** ayıltmak; kandırmak; yola getirmek. **to ~ a patient through,** bir hastayı kurtarmak. ~ **to,** ayıltmak. ~ **under,** rametmek. ~ **up,** yaklaştırmak; yanaşmak; büyütmek, terbiye etm.; ileri sürmek: **to ~ up one's food,** istifra etm.: **to ~ stg. up against s.o.,** birinin aleyhine bir şeyi ileri sürmek: **to be brought up short by stg.,** bir şeye çarpıp birdenbire durmak.

**brink** [brin(g)k]. Kenar (nehir, uçurum). **on the ~ of,** hemen hemen, üzere.

**brisk** [brisk]. Faal; canlı; uyanık; işlek; canlandırıcı; sert (rüzgâr).

**bristl·e** [ˈbrisl]. Domuz kılı; sert kıl veya tüy. Tüylerini kabartmak; (*fig.*) kabarıp kavgaya hazır olmak. **the whole question ~s with difficulties,** mesele baştan başa güçlüklerle kaplıdır. **~y,** kıllı, diken gibi.

**Brit·ain** [ˈbritən]. Britanya: **Great ~,** Büyük Britanya. **~annia,** Büyük Britanya ve İngiliz İmperatorluğunun sembolü olan kadın; İngiltere. **~annic,** Britanya'ya aid. **~ish,** Britanyalı, İngiliz: **the ~,** İngiliz, İskoç, Galyalı ve Şimalî İrlandalıların umumî adı. **~on,** Britanyalı; İngiliz.

**Brittany** [ˈbritəni]. Fransadaki Bretanya.

**brittle** [ˈbritl]. Kolay kırılır; gevrek.

**broach** [brouç]. Boşaltma tığı. Delik açmak; söze girişmek.

**broad** [brôd]. Geniş; vüs'atli; müsamahalı; serbest, liberal. ~ **accent,** kaba telâffuz: **~ly speaking,** umumiyetle: **the (Norfolk) ~s,** Norfolk kontluğunun göller ve bataklıklar bölgesi. **~-minded,** geniş fikirli, müsamahalı. **~en** [ˈbrôdən]. Genişletmek. **~sheet** [ˈbrôdşît]. Yalnız bir tarafı basılmış büyük kâğıd. **~side** [ˈbrôdsâyd]. Borda; borda topları; borda ateşi; şiddetli hücum. ~ **on,** yanlamasına.

**broadcast** [ˈbrôdkâst]. Yayılmış, dağıtılmış. Radyo yayını. Etrafa serpmek; radyo ile neşretmek. **to sow ~,** saçarak ekmek.

**brocade** [broˈkêyd]. İşlemeli kumaş; kılaptanlı çatma. Kılaptanla işlemek.

**brochure** [ˈbrouşuə(r)]. Risale.

**brogue** [broug]. İrlandalı şivesi.

**broil**[1] [brôyl] *n.* Kavga; gürültü. ~[2]. Iskara (et). Iskarada kızartmak. **~ing hot,** şiddetli sıcak.

**broke** [brouk] *v.* **break**; (*coll.*) Cebi delik, meteliksiz.

**broken** [ˈbroukən] *v.* **break**; kırılmış; kolu kanadı kırılmış. ~ **down,** çökük; bozuk; bitkin; kurada.

**broker** [ˈbroukə(r)]. Komisyoncu; acenta. **~age,** komisyon parası.

**bronchia** [ˈbronkia]. Ciğer kasabaları.

**bronchitis** [bronˈkâytis]. Bronşit.

**bronco** [ˈbronko]. Yabani, yarı ehlî at.

**bronze** [bronz]. Tunç. Tunçlaş(tır)mak; (güneş vs.) deriyi yakmak.

**brooch** [brouç]. Ziynet iğnesi, broş.

**brood** [brûd]. Bir defada yumurtadan çıkan civcivler veya kuş yavruları. Kuluçkaya yatmak; kara kara düşünmek. **to ~ over [on] stg.,** bir şeyi kurmak: **~ mare,** damızlık kısrak. **~y,** kuluçka tavuk; dalgın (kimse).

**brook**[1] [bruk] *n.* Dere, çay. ~[2] *v.* (*Menfi cümlelerde*) tahammül etm., dayanmak. **the matter ~s no delay,** meselenin beklemeğe tahammülü yoktur.

**broom** [brûm]. Süpürge; ʳ**a new ~ sweeps clean**ˀ, yeni memur vs. iyi iş görür.

**broth** [broθ]. Etsuyu.

**brothel** [ˈbroðəl]. Umumhane.

**brother** [ˈbrʌðə(r)]. Erkek kardeş; bir cemiyet veya teşkilat azâsı (*bu mânada cem'i* **brethren** [ˈbreðren]). **Older ~,** ağabey. **~hood,** kardeşlik; arkadaşlık; (ahilik gibi) cemiyet, teşkilat. **~-in-arms,** silâh arkadaşı. **~-in-law,** enişte, bacanak.

**brought** *v.* **bring.**

**brow** [brau]. Alın; kaş; bayırın sırtı. **~beat** [ˈbraubît]. Sert bakarak korkutmak; kuru sıkı tehdid etm.

**brown** [braun]. Esmer, kahve rengi, kestane rengi. Esmerletmek; (güneş) yakmak. **to be ~ed off,** (*sl.*) usanmak, bıkmak.

**brownie** [ˈbrauni]. Hizmet veya iyilik eden peri; (8–11 yaşlarında) küçük izci kız.

**browse** [brauz]. Otlamak. ~ **on,** (yaprak vs.) yemek: **to ~ among books,** kitab karıştırmak.

**bruise** [brûz]. Bere, çürük. Berelemek, çürütmek; (*fig.*) yaralamak. **~r,** zorba.

**brunette** [bruˈnet]. Esmer; esmer kadın.

**brunt** [brʌnt]. En şiddetli darbe veya kısım. **to bear the ~ of,** acısına katlanmak.

**brush** [brʌş]. Fırça; süpürge; tüylü kuyruk; hafif çarpışma. Fırçalamak; süpürmek, hafifçe dokunup geçmek. **sweeping ~,** süpürge. **~ aside [away],** bertaraf etm., itibara almamak. **~ down,** üstünü fırçalamak, (atı) tımar etmek. **~ out,** fırça ile temizlemek. **~ up,** fırçalamak: **to ~ up one's French,** fransızcasını tazelemek. **~wood** [ˈbrʌşwud]. Çalılık.

**brusque** [brʌsk, brûsk]. Haşin; sert.

**brut·al** [ˈbrûtl]. Hayvanca; zalim; canavarca. **~ality** [–ˈtaliti], vahşilik. **~alize,** hayvanlaştırmak, kabalaştırmak. **~e** [brût]. Hayvan; canavar; zalim. **by ~ force,** sırf kuvvete dayanarak; zorbalıkla. **~ish,** hayvan gibi, pek kaba.

**B.Sc.** (*abb.*) **Bachelor of Science,** Fen fakültesi mezunu.

**bubble** [ˈbʌbl]. Hava kabarcığı; hayal, olmıyacak şey. Köpürmek. **to ~ over,** taşmak; coşmak: **to prick the ~,** birinin kurduğu hayali yıkmak.

**buccaneer** [bʌkəˈniə(r)]. Korsan(lık etmek).

**buck** [bʌk]. Erkek karaca, geyik, keçi veya tavşan; şık adam, züppe; (*Amer. sl.*) dolar. (At) sıçrayıp kıç atmak. **to pass the ~,** (*Amer. sl.*) bir mesuliyeti vs. üzerinden atmak. **~ up,** (*coll.*) birine kuvvet ve cesaret vermek; cesaret bulmak; harekete gelmek.

**bucket** [ˈbʌkit]. Kova.

**buckle** [ˈbʌkl]. Toka(lamak), kopça(lamak). **~ to** girişmek, çok çalışmak. **~ up,** (maden) sıcak vs. den bükülmek.

**bucolic** [byuˈkolik]. Kır hayatına dair.

**bud** [bʌd]. Tomurcuk(lanmak), konca (vermek); (bir ağacı) aşılamak. **~ding,** yaprak aşısı: **a ~ poet,** yetişmekte olan şair.

**budge** [bʌc]. Kımılda(t)mak; fikrini değiştir(t)mek.

**budget** [bʌcit]. Bütçe; mecmua, paket. **to ~ for,** bütçeye koymak.

**buff** [bʌf]. Meşin; deri asker caketi. Deve tüyü rengi. Deri ile parlatmak. **in the ~,** çırılçıplak. **blindman's ~,** körebe oyunu.

**buffalo** [ˈbʌfəlou]. Manda.

**buffer** [ˈbʌfə(r)]. Müsademe yayı; tampon; eski kafalı adam; beceriksiz adam.

**buffet**[1] [ˈbʌfit]. Yumruk(la vurmak); vurmak, çarpmak. **~**[2] [ˈbufêy]. Tabak dolabı; büfe, yemek tezgâhı.

**buffoon** [bʌˈfûn]. Soytarı, maskara. **~ery,** soytarılık, maskaralık.

**bug** [bʌg]. Tahtakurusu; böcek.

**bugbear** [ˈbʌgbeə(r)]. Korkulan şey; nefret edilen adam veya şey.

**buggy** [ˈbʌgi]. Bir kişilik hafif araba.

**bugle** [ˈbyûgl]. Boru (çalmak). **~r,** borazan.

**build** (built) [bild, bilt]. Yapı yapmak, inşa etm.; kurmak; yapmak. Yapı, biçim. **I am built that way,** ben böyleyim: **I am not built that way,** bu bana gelmez. **~ing** *n.* yapı; bina, ev. **~ up,** takviye etm.; kurmak: **built-up area,** meskûn bölge: **well-built man,** biçimli adam.

**bulb** [bʌlb]. (Lâle vs. hakkında) soğan; elektrik ampulu. **~ous,** soğanlı.

**bulg·e** [bʌlc]. Bel verme(k); şiş (yapmak). **~ing,** pırtlak; fırlayan.

**bulk** [bʌlk]. Hacim, cüsse; büyük kısım. **to ~ large,** çok yer tutmak. **in ~,** toptan. **~y,** hacimli, kocaman.

**bulkhead** [ˈbʌlkhed]. Gemi bölmesi.

**bull**[1] [bul]. Boğa; (fil vs.) iri hayvanların erkeği; boğa gibi, çok kuvvetli. **a ~ in a china shop,** patavatsız adam: **to take the ~ by the horns,** bir tehlikeyi önlemek için birden ve cesaretle atılmak: **John Bull,** İngiltere ve İngilizlere verilen isim. **~dog,** buldok köpeği; cesur ve inadcı adam. **~-dozer** [–douzə(r)], engel, enkaz vs.yi temizlemek için kullanılan büyük makine. **~'s-eye** [ˈbulzây]. Hedefin ortası; tam vuruş; bir nevi nane şekeri. **~**[2]**.** Fiatların yükseleceğini hesab ederek tahvilat alan kimse, hosye. **to ~ the market,** borsada fiatları yükselterek hava oyunu yapmak. **~**[3]**.** Papanın emirnamesi veya piskopos tayin beratı.

**bullet** [ˈbulit]. Kurşun, mermi.

**bulletin** [ˈbulətin]. Kısa haber; tebliğ; mecmua.

**bullion** [ˈbʌlyən]. Altın, gümüş külçesi.

**bully** [ˈbuli]. Zorba, kabadayı. Korkutmak, tehdidle zorlamak; küçük veya zayıf a fena muamele etmek.

**bully-beef** [ˈbulibîf]. Konserve sığır eti.

**bulrush** [ˈbulrʌş]. Saz, hasır sazı.

**bulwark** [ˈbulwək]. Siper, istihkâm; mendirek; küpeşte.

**bum** [bʌm]. (*pop.*) Kıç; (*Amer. coll.*) tembel, işe yaramaz, sarhoş.
**bumble-bee** [ˈbʌmblˈbî]. Hezen arısı, zina.
**bump** [bʌmp]. Çarpma, vuruş, sarsıntı; yumru. Çarpmak; vurmak, çarpışmak, sars(ıl)mak. **~er,** otomobilin önündeki müsademe tamponu; çok büyük. **~y** [ˈbʌmpi]. Yamrı yumru; çıkıntılı.
**bumptious** [ˈbʌmşəs]. Kendini beğenmiş; küstahça ukalâ.
**bun** [bʌn]. Çörek; saç topuzu. **that takes the ~ !**, artık bu kadar olur!
**bunch** [bʌnç]. Salkım; deste; demet; sürü; takım. Bir araya toplamak. **he's the best of the ~,** içlerinde en iyisi odur.
**bundle** [ˈbʌndl]. Bohça; demet, deste. Demet vs. yapmak; çıkın yapmak. **to ~ s.o. off,** birini savmak.
**bung** [bʌn(g)]. Fıçı tapası, tıkaç. Tapasını kapamak, tıkamak. **to be ~ed up,** tıkanmak.
**bungalow** [ˈbʌngalo̯u]. Tek katlı kır evi.
**bungle** [ˈbʌn(g)gl]. Bozma(k), berbad etme(k). **to make a ~ of stg.,** bir şeyi yüzüne gözüne bulaştırmak.
**bunion** [ˈbʌnyən]. Ayak parmağının üzerinde hasıl olan şiş.
**bunk**[1] [bʌn(g)k]. Kabine yatağı; ranza. **~**[2] (*sl.*) Saçma, palavra. **~**[3] (*sl.*) **to do a ~,** sıvışmak.
**bunker** [ˈbʌn(g)kə(r)]. Vapur kömürlüğü; golf sahasında mania.
**bunkum** [ˈbʌn(g)km]. Saçma, palavra.
**bunny** [ˈbʌni]. Tavşan.
**bunting** [ˈbʌntin(g)]. Bayrak kumaşı, şali; bayraklar.
**buoy** [bôy]. Şamandıra (koymak); cankurtaran. **to ~ up,** su üzerinde tutmak; (*fig.*) ümid, cesaret vermek. **~ancy** [ˈbôyənsi]. Yüzme kabiliyeti; izafî ağırlık; (fiat vs.) yükselme kabiliyeti; neş'e. **~ant,** yüzebilir, hafif; neş'eli, canlı.
**burden** [ˈbö(r)dən]. Yük, ağırlık; elem; geminin yük kabiliyeti; ana fikir. Yüklemek; ağırlaştırmak; sıkmak. **to be a ~ to s.o.,** birine eziyet vermek: **to make s.o.'s life a ~,** birini doğduğuna pişman ettirmek: **the ~ of proof,** beyyine külfeti.
**bureau** [byuro̯u]. Yazı masası; yazıhane. **~cracy** [byu̯əˈrokrəsi]. Hükûmet dairelerile idare; merkeziyetçilik; kırtasiyecilik. **~crat** [ˈbyu̯əro-krat], merkeziyet tarafdarı; memur; kırtasiyeci.
**burgh** [ˈbʌrə]. Kasaba.
**burgl·ar** [ˈbö(r)glə(r)]. (Eve giren) hırsız. **~ary,** hırsızlık. **~e** [ˈbö(r)gl]. Çalmak.
**burial** [ˈberi̯əl]. Gömme, defnetme. **~-ground,** mezarlık.
**burlesque** [bö(r)ˈlesk]. Komik şekilde taklid.
**burly** [ˈbö(r)li]. İri yarı, kuvvetli.
**Burm·a** [ˈbö(r)mə]. Birmanya. **~an, ~ese,** Birmanyalı.
**burn** (**burnt**) [bö(r)n, bö(r)nt]. Yakmak, yanmak. Yanık. **to be ~t to death,** diri diri yanmak: **I hope his ears are ~ing,** (birinden bahsederken) kulakları çınlasın!: **to ~ one's fingers,** başını belâya sokmak: **money ~s a hole in his pocket,** cebinde para durmaz: **he has money to ~,** denizde kum onda para. **~ down,** (bir şehri vs.) yakmak. **to be ~t out,** evi tamamen yanmak. **~-up,** tamamen yakmak; (ateş) canlanmak. **~ing** [ˈbö(r)nin(g)] *a.* Yanan. **a ~ question,** hayatî mesele; pek mühim mesele. **~-ing-glass,** pertavsız.
**burnish** [ˈbö(r)niş]. Mıskal ile parlatmak.
**burrow** [ˈbʌro̯u]. Tavşan vs. yuvası. İn açmak, delik kazmak.
**bursar** [ˈbö(r)sə(r)]. (İngiliz üniversite ve mekteblerinde) muhasebeci; (İskoçya'da) burslu talebe. **~y,** kolej muhasebesi; burs.
**burst** (**burst**) [bö(r)st]. Patla(t)ma(k); çatla(t)ma(k); birdenbire çıkma(k). **to ~ asunder,** kırmak: **to ~ forth,** birdenbire çıkmak, söylemek: **I was ~ing with impatience,** sabırsızlıktan içim içimi yiyordu: **to be ~ing with laughter,** gülmekten katılmak: **to ~ a door open,** bir kapıyı kırıp açmak: **to ~ out,** birdenbire bağırmak: **to ~ out laughing,** kahkaha koparmak: **to ~ into tears,** gözünden yaşlar boşanmak: **the truth ~ (in) upon me,** birdenbire hakikati anladım.
**bury** [ˈberi]. Gömmek; saklamak; unutmak.
**bus** [bʌs]. Otobüs. **to miss the ~,** (*fig.*) fırsatı kaçırmak. **~man** [ˈbʌsmən]. Otobüs şoförü veya biletçisi: **to take a ~'s holiday,** tatil veya istirahat zamanında da mesleğine aid iş yapmak.
**bush** [buş]. Çalı, çalılık; fidan. **to beat about the ~,** sözü döndürüp

dolaştırmak; ⌜**good wine needs no ~**⌝, iyi mal için reklama hacet yok. **~y** [ˈbuşi]. Çalılık; sık, fırça gibi.

**bushel** [ˈbuşəl]. İngiliz kilesi (takriben 36 litre).

**Bushman** [ˈbuşman]. Cenubî Afrika'nin bir yerli kabilesine mensub kimse.

**business** [ˈbiznis]. İş, san'at, meşguliyet, vazife; ticaret; firma; husus, mesele. **~ is ~**, alışveriş miskalle; iş başkadır: **the ~ end (of a tool,** *etc.*), bir alet vs.nin sivri, keskin tarafı: **he is in ~ for himself,** kendi hesabına çalışıyor: **he means ~**, ciddidir: **to have no ~ to ...**, hakkı olmamak: **it's none of your ~**, size aid bir şey değil: **to be out of ~**, işten çekilmek: **good ~!**, hele şükür!: **to send s.o. about his ~**, birini defetmek. **~like,** işe elverişli; pratik.

**bust**[1] [bʌst]. Yarım heykel; göğüs. **~**[2]. (*sl.*) **to go ~**, İflâs etmek.

**bustle** [ˈbʌsl]. Telâşlı faaliyet. Telâş ve acele etm.; telâşa vermek.

**busy** [ˈbizi]. Meşgul, işi olan; faal, işlek, hareketli; işgüzar. Meşgul etmek. **the ~ hours,** en çok faaliyet olan saatler: **to get ~**, işe girişmek. **~body,** işgüzar, her işe burnunu sokan.

**but** [bʌt]. Fakat, ama, lâkin, ancak; başka; yalnız. **~ for,** olmasaydı: **~ for that,** bu olmasa: **~ yet,** böyle olmakla beraber: **anyone ~ me,** benden başka herkes: **anything ~ that,** bu olmasın da ne olursa olsun: **he is anything ~ a hero,** o kahramandan başka her şeydir: **I cannot ~ believe,** inanmamak mümkün değildir ki: **had I ~ known,** eğer bilseydim: **he knows ~ little,** pek az bilir: **never a year passes ~ he comes to visit us,** bir sene yoktur ki o bizim ziyaretimize gelmesin.

**butcher** [ˈbuçə(r)]. Kasab; kaatil. (Kasablık) hayvan kesmek; doğramak. **~y,** kasablık; kesip doğrama; katliâm.

**butler** [ˈbʌtlə(r)]. Sofra ve kiler işlerine bakan erkek kâhya, sofracı.

**butt**[1] [bʌt]. Büyük fıçı, varil. **~**[2]. Hedef. **the ~s,** atış sahası. **~**[3]. Tos. Tos vur(dur)mak; sümmek. **to ~ in,** sellemehüsselâm girişmek.

**butter** [ˈbʌtə(r)]. Tereyağı. Üzerine tereyağı sürmek. **to ~ up,** müdahene etm.; çok medhetmek: **he looks as though ~ wouldn't melt in his mouth,** *zahiren pek uslu ve masum görünen bir kimse hakkında kullanılır.* **~cup,** Düğün çiçeği. **~fly,** kelebek. **~milk,** yayık ayranı. **~y,** kiler. **~-fingers,** daima elinden bir şey düşüren kimse (*esp.* atılan topu tutamıyan).

**buttock** [ˈbʌtək]. But, kaba et. **~s,** kıç.

**button** [ˈbʌtən]. Düğme. Düğmele(n)mek; ilikle(n)mek. (**boy in**) **~s,** otel veya klüpte çocuk garson. **~-hole,** *n.* ilik; yakaya takılan çiçek; *v.* yakalayıp zorla dinletmek.

**buttress** [ˈbʌtris]. Destek(lemek), payanda (vurmak).

**buxom** [ˈbʌksəm]. Toplu, sıhhatli ve neş'eli (kadın).

**buy** (**bought**) [bây, bôt]. Satın almak, almak. **a good** [**bad**] **~**, kârlı [zararlı] alışveriş. **~ in,** (mezadda) satıcı namına satın almak; istok etmek. **~ off,** ···e para vererek kurtulmak. **~ out,** hissesini vs. satın almak. **~ over,** rüşvetle birini satın almak. **~ up,** piyasa mevcudunun hepsini satın almak. **~er** [ˈbâyə(r)]. Satın alıcı.

**buzz** [bʌz]. (*ech.*) Vızıltı; gürültü. Vızıldamak; (kulak) çınlamak. **to ~ about,** öteye beriye telâşla gidip gelmek: **~ off!,** (*sl.*) çek arabayı! **~er** [ˈbʌzə(r)]. Buhar düdüğü. Telgraf işareti.

**by**[1] [bây]. Vasıtasile; ···ile; ···den, tarafından; ···de, yanında, yakınında, nezdinde; ···e kadar; önünden, yakınından; geçmiş; bir tarafa; göre, nazaran; suretile. **~ and ~**, bir azdan: **~ day,** gündüz: **~ doing that,** bunu yapmak suretile: **~ error,** yanlışlıkla: **~ far,** büyük bir farkla: **~ God,** Allah hakkı için: **~ now** [**~ this time**], şimdiye kadar: **~ oneself,** kendi kendine: **~ rights,** usülen, usüle göre: **~ sea,** deniz yolile: **~ the sea,** deniz kenarında: ⌜**do as you would be done ~**⌝, sana nasıl muamele etmelerini istiyorsan başkalarına öyle muamele et!: **to do one's duty ~ s.o.**, birine karşı olan vazifesini yapmak: **one ~ one,** birer birer: **three feet ~ two,** boyu üç eni iki kadem: **~ 1500,** 1500 tarihlerinde: **the meeting will be over ~ 5 o'clock,** toplantı her halde saat beşte biter. **~**[2], [**bye**] *pref.* Talî; ikinci derecede. *n.* Spor musabakalarında oyuncuların tasnifinde tek kalan oyuncu; istitrad. **~ the ~**, istitraden; hatırıma gelmişken. **~-**

election, ara seçimi. ~(e)-law, mahalli idarelerce konan nizam. ~-pass, *n.* talî yol. *v.* uğramamak, yanından geçmek; (manilerden) ictinab etm.: ~ road, bir ana yol üzerinde kalabalık bölgelerden kaçınmak için yapılan talî yol. ~-product, talî mahsul. ~-word, darbımesel olmuş.
**bye-bye** [bâybây]. Allaha ısmarladık!, güle güle!
**bygone** [ˈbâygon]. Geçmiş, eski. let ~s be ~s, geçmişi unutalım.
**bystander** [ˈbâystandə(r)]. Seyrici.
**Byzantium** [bâyˈzantyəm]. Bizans.

# C

**C** [sî]. C harfı; (*mus.*) do; büyük C Roma rakamlarında 100 sayısını gösterir.
**cab** [kab]. Kira arabası; taksi; lokomotifin makinist yeri; kamyonda şoför yeri. ~by [ˈkabi]. (*coll.*) Arabacı. ~man [ˈkabmən]. Kira arabacısı.
**cabal** [kəˈbal]. Gizli komite. ~istic, [kabəˈlistik] gizli.
**cabaret** [ˈkabərêy]. Dansedilen çalgılı ve içkili lokanta.
**cabbage** [ˈkabic]. Lahana.
**cabin** [ˈkabin]. Kamara; kulübe. ~-boy, muço.
**cabinet** [ˈkabinət]. (Camlı veya çekmeceli) dolab; kabine; küçük hususî oda. ~-maker, ince iş yapan marangoz. ~-work, ince marangozluk.
**cable** [ˈkêybl]. Kalın halat; kablo; telgraf. Kablo ile telgraf çekmek. ~'s length, 185·2 metrelik mesafe ölçüsü.
**caboodle** [kəˈbûdl]. (*coll.*) the whole ~, sürü sepet; cümbür cemaat.
**ca' canny** [ˈkâˈkani]. Yavaş yavaş. to ~ [to go ~], kendini pek sıkmadan çalışmak.
**cache** [kaş]. Erzak konan gizli yer.
**cachet** [ˈkaşêy]. Mühür, damga; kapsül; hususiyet, alâmet.
**cackle** [ˈkakl]. Gıdaklama(k); gürültü ile gülme(k), konuşma(k).
**cacophony** [kaˈkofoni]. Tenafür, ahenksizlik.
**cactus** [ˈkaktʌs]. Frenk inciri.
**cad** [kad]. Aşağılık adam; âdî mahluk. ~dish [ˈkadiş]. Aşağılık, âdî.
**cadaverous** [kaˈdavərəs]. Cesed gibi; zayıf ve sapsarı.
**cadence** [ˈkêydəns]. Ahenk; ses perdesi.
**cadet** [kəˈdet]. Küçük kardeş veya oğul; askerî (*esp.* deniz) mekteb talebesi. ~ Corps, mekteb taburu.
**cadge** [kac]. Dilenmek; dilencilikle, madrabazlıkla elde etm.; otlamak. ~r, dilenci; otlakçı.
**café** [ˈkafêy]. Kahvehane; lokanta. ~teria [–ˈtiəriə], müşterilerin kendi hizmetlerini kendileri gördükleri lokanta.
**cage** [kêyc]. Kafes; asansör odası. Kafese koymak.
**Cairo** [ˈkâyrou]. Kahire.
**caisson** [ˈkêysən]. Cebhane sandığı veya vagonu; su altı temel işlerinde kullanılan sandık; batan gemileri yüzdürmek için kullanılan duba.
**caitiff** [ˈkêytif]. Korkak, alçak, aşağılık.
**cajole** [kəˈcoul]. Kandırmak; güzel sözlerle aldatmak. to ~ s.o. out of stg., birini kandırıp bir şey koparmak. ~ry, kandırma.
**cake** [kêyk]. Pasta, kek; kalıb, parça. Katılaşmak, kabuk bağlamak. ~s and ale, hayatın zevkleri: ⌜you can't eat your ~ and have it too⌝, bir şeyi hem sarfedip hem de ona malik olmak imkânsızdır: they are selling like hot ~s, kapışılıyor.
**calamit·ous** [kaˈlamitəs]. Felâketli; âfet gibi. ~y, âfet, felâket, beliyye.
**calc·ify** [ˈkalsifây]. Kireçlendirmek. ~ium, kils.
**calcula·te** [ˈkalkyulêyt]. Hesablamak; tahmin etm.; hesab etm. to ~ on s.o., birine bel bağlamak. ~tion, hesab(lama).
**calculus** [ˈkalkyuləs]. Hesabı asgarı namütenahi.
**Caledonia** [ˌkaliˈdounyə]. Kaledonya; İskoçya.
**calendar** [ˈkalendə(r)]. Takvim; liste; kayıd defteri.
**calends** [ˈkalends]. Eski Roma takviminde ayın ilk günü. on the Greek ~, balık kavağa çıkınca.
**calf**[1] *pl.* **calves** [kâf, kâvz]. Buzağı; dana; (fil vs.) yavrusu; dana derisi; beceriksiz. to kill the fatted ~, aile efradından birinin dönüşünü kutlamak. ~-love, ilk aşk.
**calf**[2], *pl.* **calves.** Bacağın dizden aşağı arka kısmı; baldır.
**calib·rate** [ˈkalibrêyt]. Çapını tayin etm.; derecesini ayarlamak. ~re [ˈ–bə(r)], çap; değer; ehliyet.

**calico** [ˈkaləkou]. Pamuk bezi.
**caliph** [ˈkêylif]. Halife.
**call**[1] [kôl] *n.* Çağırma; bağırma; davet; ziyaret; telefon etme; yoklama; sebeb. **at ~,** emre muhavvel: **to have a close ~,** dar kurtulmak: **to give a ~,** seslenmek; telefon etm.: **there's no ~ to blush,** utanacak bir sebeb yok: **there's no ~ for rejoicing,** ortada sevinecek bir şey görmüyorum: **to pay a ~ on s.o.,** birini ziyaret etm.: **port of ~,** geminin uğradığı liman: **to put a ~ through,** uzak mesafeye telefon etm.: **within ~,** çağırılınca işidebilecek bir mesafede. **~**[2] *v.* Çağırmak; bağırmak; isim vermek; davet etm.; uyandırmak; ziyaret etm.; telefon etm. **London ~ing!** (radyo) burası Londra: **he is ~ed Tom,** ismi Tom'dur: **he ~ed him a liar,** yalancı olduğunu (yüzüne karşı) söyledi: **to ~ s.o.** *or* **stg. after s.o.,** bir kimseye veya bir şeye birinin adını vermek: **to ~ back,** cevabını vermek; tekrar uğramak: **to ~ s.o. back,** birisini geri çağırmak. **~ for,** birini çağırtmak; icabetmek; taleb etm.; uğrayıp almak: '**to be (left till) ~ed for**', gelinip alınacak. **~ forth,** sebeb olm.; ortaya çıkarmak. **~ in,** (içeri) çağırmak; (kâğıd para vs.) yeni para çıkarmak için geri çekmek: **to ~ off a strike,** bir greve nihayet verilmesini emretmek: **to ~ off a deal,** bir anlaşmayı ibtal etm.: **to ~ on s.o.,** birini ziyaret etm. **~ out,** bağırmak; duelloya davet etmek. **~ over,** yoklama yapmak. **~ together,** toplamak; ictimaa davet etmek. **~ up,** yukarıya çağırmak; askere çağırmak; hatıra vs. uyandırmak: **to ~ upon God,** *etc.*, Allahtan veya azizlerden istemek: **to ~ upon s.o. to do stg.,** birinden bir şeyi yapmasını taleb etm.: **to ~ upon s.o. for help,** yardım için birine başvurmak: '**I now ~ upon Mr. B.**', şimdi sözü Mr. B.'ye veriyorum. **~er** [ˈkôlə(r)]. Ziyaretçi; telefon eden kimse. **~ing** [ˈkôlin(g)] *n.* Meslek; iş.
**calligraphy** [kaˈligrəfi]. Hattatlık.
**callipers** [ˈkalipö(r)z]. Çap pergeli; kumpas.
**callous** [kaˈləs]. Nasırlı; sertleşmiş; katı; hissiz.
**calm** [kâm]. Sakinlik; soğukkanlılık. Sakin; durgun; heyecansız; soğukkanlı. Teskin etm., yatıştırmak. **to ~ down,** yatış(tır)mak.
**calumn·iate** [kəˈlʌmniêyt]. İftira etmek. **~y** [ˈkaləmni], iftira.
**Calvary** [ˈkalvəri]. İsa'nın çarmıha gerildiği yer; büyük ıstırab.
**camarilla** [ˌkaməˈrilə]. Gizli komite.
**camber** [ˈkambə(r)]. Muhaddeb olma(k).
**came** *v.* **come.**
**camel** [ˈkaməl]. Deve.
**cameo** [ˈkamiou]. İşlemeli akik.
**camera** [ˈkamərə]. Fotograf makinesi. **in ~,** hususî; gizli olarak. **~man,** fotografçı.
**camouflage** [ˈkamuflâj]. Kamuflaj (yapmak); örtmek, gizlemek.
**camp** [kamp]. Karargâh; kamp. Kamp kurmak.
**campaign** [kamˈpêyn]. (Askerî) sefer; seferberlik. Mücadele etmek.
**camphor** [ˈkamfə(r)]. Kâfur.
**can**[1] [kan]. Madenî kab; maşrapa; teneke kutu. Konserve yapmak; kutuya koymak. **~**[2], **(could).** *v. Yardımcı fiil; (masdar halinde kullanılmaz; bunun yerine* **to be able** *kullanılır).* Muktedir olm., ···bilmek; *fiilerin başına gelerek iktidarî sıygasını teşkil eder*: **I can do,** yapabilirim: **I could not do,** yapamazdım: **it cannot be done,** bu mümkün değildir: **what ~ it be?,** ne olabilir?, nedir acaba?: **~ you swim?,** yüzmek bilir misiniz?: **I will do what I ~,** elimden geleni yaparım.
**canal** [kəˈnal]. Kanal; geçid. **~ize** [ˈkanəlâyz], kanal açmak.
**canary** [kəˈneəri]. Kanarya.
**cancel** [ˈkansl]. Silmek, feshetmek; kaldırmak; hükümsüz bırakmak; vazgeçmek; ibtal etmek. **~lation,** silme; feshetme.
**cancer** [ˈkansə(r)]. Kanser. **~ous,** kanserli.
**candelabra** [ˌkandəˈlabra]. Kollu şamdan.
**candid** [ˈkandid]. Samimî, açık; tarafsız.
**candida·te** [ˈkandidit]. Namzed. **~ture,** namzedlık.
**candle** [ˈkandl]. Mum. **to burn the ~ at both ends,** başka başka bir çok işler yaparak kuvvetini tüketmek: **the game is not worth the ~,** astarı yüzünden pahalı: **he cannot hold a ~ to you,** o senin eline su dökemez; senin ayağının pabucu olamaz. **~stick,** şamdan.
**candour** [ˈkandə(r)]. Açık kalblilik; tarafsızlık.

**candy** [ˈkandi]. Şekerleme (yapmak); şeker gibi olmak.
**cane** [kêyn]. Baston, değnek; kamış. Dayak atmak. **to get the ~,** dayak yemek.
**canine** [ˈkanâyn]. Köpeğe aid, gibi.
**canister** [ˈkanistə(r)]. Teneke kutu.
**canker** [ˈkankə(r)]. Ağız yarası, karha; ağaç kanseri; (*fig.*) çürütücü tesir. Kemirmek.
**cannibal** [ˈkanibl]. Yamyam. **~ism,** yamyamlık.
**cannon** [ˈkanən]. Top; (bilardoda) karambol (yapmak); çarpmak. **~-ball,** top güllesi. **~-fodder,** harbde malzeme gibi harcanan adam. **~ade,** top ateşi.
**cannot** = **can not.**
**canny** [ˈkani]. Açıkgöz, cinfikirli, hesabî.
**canoe** [kəˈnû]. Hafif sandal, kano.
**canon** [ˈkanən]. Kilise kanunu, dinî nizam, dinî liste; esas; bir rahib rütbesi. **~ical** [kəˈnonikl], kilise kanununa göre; incile göre; meşru, kabul edilmiş; rahiblere mahsus (elbise vs.). **~ize** [ˈkanənâyz]. (Bir kimseyi) azizler sırasına koymak.
**canopy** [ˈkanəpi]. Gölgelik; sayeban; kubbe; saçak. Gölgelik veya saçakla örtmek.
**can't** [kânt] = **cannot.**
**cant**[1] [kant]. Müraice söz. **a ~ phrase,** tekerleme. **~**[2]**.** Meyil. Meyil vermek; iğilmek; yan yat(ır)mak.
**cantankerous** [kənˈtankərəs]. Ters, aksi, dirliksiz, huysuz.
**cantata** [kanˈtâta]. Bir koro tarafından okunan bestelenmiş manzume.
**canteen** [kanˈtîn]. Kantin; portatif sofra takımı.
**canter** [ˈkantə(r)]. Eşkin (gitmek). **to win in a ~,** zahmetsiz kazanmak.
**Canterbury** [ˈkantəbəri]. **Archbishop of ~,** İngilitere'nin başpiskoposu.
**cantilever** [ˈkantiˌlîvə(r)]. Dirsek, destek.
**cantonment** [kanˈtûnmənt]. Askerî konak yeri.
**canvas** [ˈkanvəs]. Yelken [çadır] bezi; kanaviçe; yelken; yağlı boya resim.
**canvass** [ˈkanvəs]. Köy köy vs. dolaşarak rey toplama(k) veya sipariş kaydetme(k); ahalinin fikirlerini sorma(k).
**canyon** [ˈkanyən]. İki tarafı uçurum dere.
**cap**[1] [kap] *n.* Kasket; kepi; başlık; kapak. **~ and bells,** çıngıraklı soytarı külâhi: **~ and gown,** üniversite hoca ve talebesine mahsus şapka ve cüppe: **~ in hand,** tazimkâr bir tavırla: ˹**if the ~ fits wear it!**˺, yarası olan kocunsun!: **black ~,** İngiltere'de idam kararı verirken hâkimin giydiği şapka: **to get one's ~,** (sporda) birinci takım için seçilmek: **to set one's ~ at a man,** (kadın) bir erkeği avlamak. **~**[2] *v.* **that ~s all!,** bu hepsine tüy dikti: **to ~ a story,** bir hikâye vs.yi bastırmak (daha iyisini söylemek).
**capab·le** [ˈkêypəbl]. Muktedir, kabiliyetli, ehliyetli. **~ility** [–ˈbiliti], kabiliyet, iktidar.
**capaci·ous** [kəˈpêyşəs]. İstiablı; geniş, vasi, büyük. **~ty** [–ˈpasiti], istiab, alış kabiliyeti; verim; iktidar, kabiliyet; sıfat: **full to ~,** tamamen dolu.
**caparison** [keˈparisən]. Haşe (örtmek).
**cape**[1] [kêyp]. Pelerin. **~**[2]**.** Burun. **the C~ (of Good Hope),** Ümid Burnu.
**caper**[1] [ˈkêypə(r)] *n.* Gebre. **~**[2]**.** Sıçrama(k), zıplama(k).
**capital** [ˈkapitl]. Baş, büyük; son derece mühim. Büyük harf; devlet merkezi; sermaye; sütun başlığı. **~!,** fevkalâde! **~ punishment,** ölüm cezası: **to make ~ out of stg.,** bir şeyden istifade etm. **~ism** [ˈkapitəlizm, kəˈpitəlizm]. Hususî mülkiyete dayanan iktisadî sistem. **~ist,** sermayesini işleten kimse; zengin kimse. **~ize,** sermayeye çevirmek; sermaye olarak kullanmak; (bir şeyi) kendi istifadesi için kullanmak; bir gelirin sermayesini hesab etmek.
**capitulate** [kəˈpityulêyt]. (Şartla) teslim olmak.
**capric·e** [kəˈprîs]. Anî heves; sebatsızlık. **~ious** [–ˈprişəs], maymun iştahlı; sebatsız.
**capricorn** [ˈkaprikôn]. Keçi burcu. **Tropic of C~,** medarı cedî.
**capstan** [ˈkapstən]. Bocurgat.
**capsule** [ˈkapsyul]. Şişe kapağı; ilâc kapsülü.
**captain** [ˈkaptin]. Yüzbaşı; baş; reis; (deniz) albay; süvari. Bir seferi idare etm. **Group ~,** (hava) albay. **~cy,** yüzbaşılık; (deniz) albaylık; reislik.
**caption** [ˈkapşn]. (Makale vs.) başlık; (resim, filim vs.) yazı, izahat.
**captious** [ˈkapşəs]. Kusur bulan; titiz.
**captivate** [ˈkaptivêyt]. Teshir etm., gönlünü kapmak; bendetmek.

**captiv·e** [ˈkaptiv]. Esir. ~ **balloon,** karaya bağlı balon. ~**ity** [–ˈtiviti], esirlik.
**captor** [ˈkaptə(r)]. Esir eden. ~**ure** [ˈkaptyuə(r)]. Esir alma(k); yakalama(k); zabtetme(k); zorla, hile, ile alma(k); esir; ganimet.
**car** [kâ(r)]. İki tekerlekli araba; otomobil; vagon; (balon vs.de) oda.
**carafe** [kəˈrâf]. Sürahi.
**carat** [ˈkarət]. Kırat.
**caravan** [karəˈvan]. Kervan; araba şeklinde küçük ev; çingene arabası. Böyle bir arabada seyahat etmek.
**carbolic** [kâˈbolik]. ~(**acid**), fenol.
**carbon** [ˈkâbon]. Karbon; kopya kâğıdı. ~**aceous,** karbonlu, karbon gibi. ~**ize** [ˈkâbonâyz], karbonlaştırmak.
**carbuncle** [ˈkâbʌnkl]. Şirpençe çıbanı.
**carcass, -case** [ˈkâkəs]. Leş; cesed; (istihfaf) vücud; enkaz.
**card**[1] [kâd] *n.* Kart; karton; kâğıd; iskambil; mukavva; bilet; fiş; kartvizit. ~ **index,** fiş usulü dosya, liste: **he's a knowing** ~, o ne tilkidir: **it is (quite) on the** ~**s that,** olabilir; haritada var: **to play one's** ~**s well,** elindeki kozu iyi oynamak: **he's a queer** ~, antikanın biridir: **he has a** ~ **up his sleeve,** daha son kozunu oynamadı: **to lay one's** ~**s on the table,** gizlisi kapaklısı olmamak. ~**board** [ˈkâdbôd]. Mukavva, karton. ~**-table,** oyun masası. ~[2] *v.* Yün vs.yi tel tarakla taramak. ~**er,** tarak makinesi.
**cardiac** [ˈkâdiak]. Kalbe aid; kalbe kuvvet verici (ilâc), kordiyal.
**cardigan** [ˈkâdəgan]. Yün örgüsü ceket.
**cardinal**[1] [ˈkâdənal] *n.* Kardinal. ~ **red,** al (renk). ~[2] *a.* Baş, esaslı, en mühim. ~ **numbers,** adî sayılar: ~ **points,** dört ana cihet.
**care**[1] [keə(r)] *n.* Dikkat; endişe, keder; kaygı; himaye, muhafaza. ~ **of (c/o),** vasıtasile, elile: ⌈~ **killed the cat**⌉, kendini fazla üzme!: ~**s of State,** devletin mesuliyetleri: **to take** ~, dikkat etm., ihtimam ve itina etm.: **that matter will take** ~ **of itself,** o iş kendi kendine düzelir: **want of** ~, bakımsızlık. ~**free** [ˈkeəfrî]. Kaygısız; kayıdsız. ~**ful** [ˈkeəful]. Dikkatli, ihtimamlı, itinalı, ihtiyatlı, ö çülü. **be** ~ **!,** dikkat et!, sakın! ~**less** [ˈkeəlis]. Dikkatsiz, düşüncesiz; ihmalkâr; kayıdsız; mübalâtsız. ~**taker** [ˈkeətêykə(r)]. Kapıcı, bekçi. ~**worn** [ˈkeəwôn]. Kaygı ve kederden bitkin; muztarib. ~[2] *v.* Endişe etm., düşünmek; keder etm.; umursamak. ~ **for,** beğenmek; bakmak, ihtimam etm.: **I don't** ~**!,** (i) bence aynı şey; (ii) bana ne?: **I don't** ~ **for him,** o hoşuma gitmiyor: **who is caring for him?,** ona kim bakıyor?: **I don't** ~ **to be seen in his company,** onun yanında görülmek istemem: **what do I** ~**?,** bana ne?: **I don't** ~ **what he says,** ne söylerse söylesin (aldırmam): **I didn't** ~ **for that novel,** o roman beni sarmadı: **I don't** ~ **two hoots [a red cent, a brass farthing],** bana vız gelir: **to** ~ **for nothing,** alâkadar olmamak; pervasız olm.: **for all I** ~, bana kalırsa: **not that I** ~, bana vız gelir: **that's all he** ~**s about,** bütün düşündüğü [ehemmiyet verdiği] bu: **if you** ~ **to ...,** ... arzu ederseniz: **would you** ~ **to read this paper?,** gazeteyi okumak ister misiniz?
**careen** [kəˈrîn]. (Gemiyi) tamir için yan yatırmak.
**career**[1] [kəˈriə(r)] *n.* Meslek; meslek hayatı; hayat. **to take up a** ~, bir mesleğe girmek. ~[2]. **to** ~ **about,** delice ve başıboş koşma(k): **in full** ~, tam hızla.
**caress** [kəˈres]. Okşama(k), öpme(k).
**caret** [ˈkarət]. Yazıda çıkıntı işareti (‸).
**cargo** [ˈkâgou]. Yük, hamule.
**caricature** [ˈkarikəˈtyuə(r)]. Karikatür(ünü yapmak). Tehzil etmek.
**carillon** [kaˈrilyən]. Muayyen havalar çalmak üzere tertib edilmiş çanlar.
**carmine** [ˈkâmâyn]. Lâl (rengi), açık ve parlak kırmızı.
**carnage** [ˈkânic]. Kan dökme, kıtal.
**carnal** [ˈkânəl]. Cinsî; şehvanî; dünyevî.
**carnation** [kâˈnêyşn]. Karanfil; açık pembe.
**carnival** [ˈkânəval]. Cümbüş; âlem.
**carnivor·a** [kəˈnivəra]. Et yiyen hayvanlar. ~**e,** [ˈkânivô(r)], et yiyen hayvan veya nebat. ~**ous,** et yiyen.
**carol** [ˈkarəl]. Neş'eli şarkı; dinî şarkı. Şarkı söylemek.
**carous·al** [kəˈrauzl]. İçki âlemi. ~**e,** âlem yapmak, işret etmek.
**carp** [kâp] *v.* Kusur bulmak; şikâyet etmek. **to** ~ **at,** tutturmak, çekiştirmek.

**carpent·er** [ˈkâpəntə(r)]. Doğramacı(lık yapmak). **~ry,** doğramacılık.
**carpet** [ˈkâpit]. Halı; örtü. Halı döşemek; kaplamak. **to be on the ~,** azarlanmak. **~-bag,** halı torba. **~-bagger,** maceraperest ve prensipi olmıyan politikacı. **~-knight,** salon zabiti.
**carriage** [ˈkaric]. Araba; vagon; top arabası; taşıma, nakil; taşıma ücreti; tavır.
**carrier** [ˈkariə(r)]. Taşıyan kimse, şey; nâkil; hastalık taşıyan. **aircraft ~,** uçak gemisi: **~ pigeon,** posta güvercini.
**carrion** [ˈkariən]. Leş, laşe; pis, kokmuş.
**carrot** [ˈkarət]. Havuc. **~-headed,** kızıl saçlı.
**carry**[1] [ˈkari] *n.* Menzil; taşıma. **~**[2] *v.* Taşımak; götürmek; nakletmek; geçirmek; tutmak; (kale vs.) zabtetmek; (faiz) getirmek; (top) menzili ... olmak; (ses) ···den işidilmek; ... kadar gitmek. **~ two** (cemederken) elde var iki: **to ~ all before one,** bütün rakibleri yenmek: **to ~ the day,** muvaffak olm.: **to ~ one's liquor well,** içkiye dayanmak: **the motion was carried,** teklif kabul edildi: **to ~ oneself well,** vücudunu dik tutmak: **to ~ one's point,** kendi fikrini kabul ettirmek: (**shop**) **to ~ a certain article,** (dukkân) bir eşya üzerinden muamele yapmak: **his word carries no authority,** sözünün değeri yoktur. **~ away,** alıp götürmek; coşturmak. **~ forward,** bir yekûnu vs. nakletmek. **~ off,** alıp götürmek; kazanmak; (hastalık) öldürmek: **to ~ it off well,** becermek. **~ on,** yapmak, devam ettirmek: devam etm.; vazifesinden ayrılan birinin işine bakmak: **~ on!,** devam ediniz!; siz işinize bakınız!: **I don't like the way she carries on,** gidişini pek beğenmiyorum: **she carried on dreadfully,** (*sl.*) kıyamet kopardı: **to ~ on with s.o.,** (*sl.*) birisile mercimeği fırına vermek: **to ~ on a correspondence with s.o.,** birisile mektublaşmak. **~ out,** yerine getirmek; icra etm. **~ through,** muvaffakiyetle yerine getirmek; başarmak. **~ing** [ˈkari·in(g)] *n.* Taşıma; nakletme. **~ capacity,** istiab: **I don't like your ~s on,** (*sl.*) ben senin gidişatını beğenmiyorum.
**cart** [kât]. İki tekerlekli araba; elle itilen araba. Araba ile taşımak. **to ~ about,** taşıyıp durmak: **to be in the ~,** hapı yutmak: **to put the ~ before the horse,** bir işi tersinden yapmak. **~age,** araba ile taşıma; nakliye parası. **~-wheel,** araba tekerliği: **to turn ~s,** elleri üzerinde havada dönerek takla atmak. **~er** [ˈkâtə(r)]. Yük arabacısı.
**cartel** [kâˈtel]. Fabrikalar arasında anlaşma.
**cartilage** [ˈkâtilic]. Kıkırdak.
**cartograph·er** [kâˈtogrəfə(r)]. Haritacı. **~y,** haritacılık.
**carton** [ˈkâtən]. Mukavva (kutu).
**cartoon** [kâˈtûn]. Karikatür(ünü yapmak). **~ist,** karikatürcü.
**cartridge** [ˈkâtric]. Hartuc; fişek.
**carve** [kâv]. (Sofrada) eti kesip dağıtmak; oymak. **~r,** eti kesen; büyük et bıçağı; oymacı.
**cascade** [kasˈkêyd]. Çağlayan, şelâle.
**case**[1] [kêys]. Hal, vaziyet, durum, keyfiyet; mesele, vak'a, hadise; dava; tasrif hali; hastalık vak'ası, hasta. **in any ~,** her halde: **that alters the ~,** o zaman vaziyet değişir: **as in the ~ of ...,** ... için olduğu gibi: **in ~ it rains,** belki yağmur yağar: **in ~ he does not come,** gelmemesi ihtimaline karşı: **in ~ he comes give him this letter,** gelecek olursa bu mektubu ona ver: **in no ~,** hiç bir suretle: **just in ~,** ne olur ne olmaz: **you have no ~,** davaya hakkınız yok: **there is no ~ against you,** bu meselede aleyhinize dava açılamaz: **to make out a ~,** davayı isbat etm.: **that is not the ~,** vaziyet böyle değildir: **should the ~ occur,** icab ettiği takdirde: **the ~ in point,** bahis mevzuu olan mesele. **~**[2]. Kılıf; mahfaza, kutu, çanta; kasa. Kaplamak, örtmek; kutuya koymak. **display ~,** eşya teşhir edilen camekân. **~-hardened,** dışı sertleştirilmiş; (*fig.*) nasırlanmış; sıkıntıya alışmış.
**casemate** [ˈkêysmêyt]. Mazgallı siper.
**casement** [ˈkêysmənt]. Menteşeli pencere.
**cash** [kaş]. Para; nakid; peşin para. Paraya çevirmek; (çeki) bozmak. **~ account,** kasa hesabı: **~ down,** peşin para; derhal ödenen para: **~ price,** peşin fiat: **to ~ in,** paraya çevirmek: **to ~ in on stg.,** ···den kâr temin etm., istifade etm.: **to be out of ~,** (yanında) parası olmamak: **terms ~,** peşin ödenir. **~-book,** kasa defteri. **~ box,** para kutusu. **~-register,**

dukkân vs.de alınan parayı kaydeden makine. **~ier¹** [ka'şiə(r)] *n.* Kasadar, veznedar. **~ier²** *v.* Hizmetten çıkarmak, tardetmek.

**cashmere** ['kaşmîr]. Keşmir şalı.

**cask** [kâsk]. Fıçı, varil.

**casket** ['kâskit]. Mücevher çekmecesi.

**Caspian** ['kaspiən]. **the ~ (Sea),** Hazer denizi.

**casserole** ['kasəroul]. Güvec; saplı tencere.

**cassock** ['kasək]. Cüppe.

**cast¹** [kâst] *v.* Atmak; saçmak; (demir) dökmek; salmak; rol tevzi etm.; **to ~ (up) figures,** rakamları toplamak: **to ~ the lead,** iskandil etmek. **~-net,** serpme ağ. **~-off,** çıkarılıp atılmış; eski (elbise vs.). **~ about,** etrafa atmak: **to ~ about for stg.,** aramak; çare aramak. **~ away,** atmak: **to be ~ away,** (gemi) karaya düşmek; kazaya uğramak. **~away** ['kâstəwêy]. Deniz kazasına uğrayıp hücra bir ada vs.de kalmış. **~ down,** aşağı atmak; indirmek: **to be ~ down,** neş'esiz, keyifsiz, kederli olmak. **~ off,** çıkarıp atmak; reddetmek; (örme işinde) ilmik geçirmek: **~ off!,** (*naut.*) alarga! **~²** *n.* Atma; atış; dökme; voli; oltanın kancalı ucu; bir piyesin şahısları; rol tevzii; şehlâ. **a man of his ~,** bu karakterde bir adam. **~³** *a.* Atılmış; dökme. **~ing** ['kâstin(g)] *n.* Dökme; döküm. **~-iron,** dökme demirden yapılmış: **he has a ~ constitution,** bünyesi demir gibi sağlam.

**caste** [kâst]. Hind cemiyetinde sınıf; sınıf farkı. **to lose ~,** cemiyetteki yerini ve itibarını kaybetmek.

**castellated** ['kastelêytid]. Mazgallı ve kuleli.

**castigate** ['kastigêyt]. Cezalandırmak; şiddetle tenkid ve tekdir etmek.

**castle** ['kâsl]. Kale, hisar; şato; (satranc) ruh. (Satranc) şah ile ruhu aşırmak. **to build ~s in the air,** hayal kurmak.

**castor** ['kâstə(r)]. Koltuk vs. tekerleği. **~oil,** hintyağı.

**castrate** [kas'trêyt]. Hadım etmek.

**casual** ['kajyuəl]. Tesadüfî, rastgele; plansız, dikkatsiz; lâübali; muvakkat. Muvakkat işçi. **~ty** ['kajyulti]. Kaza; (*mil.*) zayiat, kayıb; ölü, yaralı. **fatal ~,** ölüm.

**casuist** ['kazuist]. Ahlak meselelerinde doğru ile yanlışı inceden inceye araştıran adam; safsatacı. **~ry,** safsata.

**cat¹** [kat] *n.* Kedi. **to let the ~ out of the bag,** bir sırrı ağzından kaçırmak: ⌜**a ~ may look at a king**⌝, hepimiz insanız *kabilinden bir tabir*: **to be like a ~ on hot bricks,** diken üstünde oturmak: **to see which way the ~ jumps,** rüzgârın nereden eseceğini beklemek: **there is not room to swing a ~,** kımıldanacak yer yok: **enough to make a ~ laugh,** insan gülmekten bayılır. **~-burglar,** kapı vs.yi kırmadan tırmanarak eve girip soyan hırsız. **~call** ['katkôl]. (Tiyatroda vs.) ıslık (çalmak); yuha(lamak). **~-o'-nine-tails,** dokuz kamçılı kırbac. **~'s-cradle,** çocukların iki el parmaklarına ip geçirerek oynadıkları bir oyun. **~'s-eye,** bir nevi değerli taş. **~'s-paw,** başkası tarafından alet olarak kullanılan kimse. **~ty** ['kati]. Sinsi ve müstehzi. **~²** *v.* **to ~ the anchor,** demiri grivaya vurmak.

**catacomb** ['katakûm]. Yer altında dehliz şeklinde mezarlık.

**catafalque** ['katafalk]. Cenaze merasiminde üzerinde tabutun teşhir ediliği kaide.

**catalepsy** ['katalepsi]. Nöbet sırasında vücudun kaskatı kesildiği bir hastalık.

**catalogue** ['katalog]. Katalog (yapmak).

**catapult** ['katapʌlt]. Mancınık (ile atmak); lâstikli sapan.

**cataract** ['katərakt]. Dik şelâle; sel; (gözde) perde.

**catarrh** [kə'tâ(r)]. Muhatî gışanın iltihabı; nezle, akıntı.

**catastroph·e** [kə'tastrəfi]. Felâket; musibet; feci netice. **~ic** [katə'strofik], felâketli; müdhiş; facialı.

**catch** [kaç] *n.* Tutuş, kapış; sürgü, kol vs. gibi tutan şey; tutulan şey; hile. **a good ~,** (*coll.*) (bir kız bakımından) kelepir koca: **he [it] is no great ~,** bulunmaz hind kumaşı [matah] değil: **there's a ~ in it,** 'bir püf noktası vardır'; 'ucuzdur illeti var' *kabilinden*: **a ~ question,** öyle bir sual ki hatıra gelebilecek ilk cevabı yanlıştır. **~²** (**caught**) [kôt] *v.* Yakalamak, tutmak, almak, ele geçirmek; kapmak; yetişmek; çarpmak; (ateş) tutuşmak. **to ~ at,** tutmağa çalışmak: **~ me (doing such a thing)!,** bunu yapmak mı? ne münasebet!; bunu yaparsam arab olayım!: **you don't ~ me!**, ben faka basmam; yağma yok!: **to ~ s.o. a blow,** birine bir yumruk eklemek: **a sound caught my ear,** kulağıma bir ses çarptı: **a picture caught my eye,** bir resim gözüme ilişti: **you'll ~ it!**, paparayı yiyeceksin!: **I didn't quite ~ your name,** isminizi iyi işidemedim. **~-as-~-can,** serbest güreş. **~ on,** alıp yürümek, rağbeti olmak. **~ out,** gafil avlamak; (yaparken) yakalamak. **~ up,** yetişmek: **I'll ~ up with you,** size yetişirim. **~ing** [ˈkaçin(g)] *a.* Bulaşık, sari. **~ment** [ˈkaçmənt]. **~ basin** [**~ area**], bütün suları bir nehre akan bölge. **~word** [ˈkaçwö(r)d]. Bir siyasî parti vs.nin daima tekrar edilen vecizesi; şiar. **~y** [ˈkaçi]. Kolay hatırda kalan; cazib.

**catechi·sm** [ˈkatəkizm]. Sual ve cevab ile öğretme; bu suretle yazılmış din kitabı. **~ze** [–kâyz], sual ve cevab usulü ile öğretmek.

**categor·ical** [ˌkatəˈgorikl]. Kat'î; şartsız; kesenkes. **~y** [ˈkatigəri]. Sınıf, sıra, tabak.

**cater** [ˈkêytə(r)]. **to ~ for,** ... için yemek tedarik etm.; zevk, eğlence vs. temin etmek. **~er,** yiyecek tedarik eden, vekilharc. **~ing,** yemek tedariki.

**caterpillar** [ˈkatəpilə(r)]. Tırtıl.

**caterwaul** [ˈkatəwôl]. Miyavlama(k).

**catgut** [ˈkatgʌt]. Kiriş.

**cathedral** [kəˈθîdrəl]. İçinde piskopos kürsüsü bulunan büyük kilise.

**catholic** [ˈkaθəlik]. Âlemşümul, her şeyi ihtiva eden, her kese yarayan veya herkesle alâkadar olan; geniş fikirli; bütün Hıristiyanları içine alan. **~ism** [kaˈθolisizm], katoliklik. **~ity** [–ˈlisiti], âlemşümullük; geniş fikirlilik; liberallik.

**cattle** [ˈkatl]. *collective n.* Sığır. **~-show,** ziraat sergisi. **~-truck,** öküz nakline mahsus vagon.

**Caucas·ian** [kôˈkêyjyən]. Kafkasya'ya aid, Kafkasyalı. **~us** [ˈkôkəsəs], Kafkaslar, Kafkas dağları.

**caucus** [ˈkôkʌs]. Siyasî parti liderlerinin toplantısı.

**caught** *v.* **catch.**

**cauldron** [ˈkôldrən]. Kazan.

**cauliflower** [ˈkoliflauə(r)]. Karnibahar.

**caus·al** [ˈkôzl]. Sebebe dair; sebeb ifade eden. **~ality** [–ˈzaliti], sebeb ve netice münasebeti. **~ative** [–ˈzêytiv], sebeb olucu; (*gram.*) müteaddi. **~e** [kôz]. Sebeb, illet; münasebet; dava; büyük mesele; tarafdarlık. ···e sebeb olm., mucib olm.; doğurmak; **to ~ s.o. to do stg.,** birine bir şey yaptırtmak. **to make common ~ with s.o.,** bir gaye uğrunda birisile birleşmek.

**causeway** [ˈkôzwêy]. Islak bir sahada yapılan yüksekçe yol.

**caustic** [ˈkôstik]. (Eti) yakıcı; acı ve dokunaklı, yaralayıcı (söz). Yakıcı ilâc.

**cauter·ize** [ˈkôtərâyz]. Dağlamak, yakmak. **~y,** dağlama, yakma (aleti).

**caut·ion** [ˈkôşn]. İhtiyat, basiret; ihtar. İhtar etm.; tehdid etmek. **~!**, dikkat!: **a ~,** (*sl.*) antika, numara. **~ionary** [–əri], ihtar aid, ihtiyatî. **~ious,** ihtiyatlı; ölçülü; bir az korkak; çekingen.

**caval·cade** [kavlˈkêyd]. Süvari alayı; atlılar ve arabaların geçişi. **~ier** [kavəˈliə(r)]. Atlı, süvari; kadınlara karşı çok nazik adam; bir hanıma refaket eden erkek. Serbest, lâübali. **~ry** [ˈkavlri]. Süvari sınıfı; süvari.

**cave** [kêyv]. Mağara, in. **to ~ in,** Çökmek, yıkılmak; (*sl.*) teslim olm., razı olmak.

**caviar** [ˈkaviâ(r)]. Havyar.

**cavil** [ˈkavil]. Daima kusur bulmak; manasız şekilde itiraz etmek.

**cavity** [ˈkaviti]. Çukur, oyuk, boşluk. **~ wall,** arayeri boş bırakılan duvar.

**caw** [kô]. (Karga) bağırma(k).

**C.B.** (*abb.*), **Companion of the Bath,** *v.* **order.**[2]

**C.B.E.** (*abb.*) **Commander of the British Empire,** *v.* **order**[2].

**cease** [sîs]. Durmak, bitmek, kesilmek, dinmek; vazgeçmek. **~ fire!** ateş kes!: **he has ~d to write,** artık yazmıyor: **without ~,** durmadan. **~less,** durmadan; sürekli.

**cedar** [ˈsîdə(r)]. Sedir, Lübnan servisi.

**cede** [sîd]. Teslim etm., devretmek; feragat etmek.

**ceiling** [ˈsîlin(g)]. Tavan; uçağın âzami irtifaı; fiatların âzamı haddi.

**celebr·ate** [ˈselibrêyt]. Kutlamak; ruhanî icra etm.; bayram yapmak; şöhretini yaymak. **~ated,** meşhur. **~ation** [–ˈbrêyşn], kutlama; tes'id; ruhanî âyin icrası. **~ity** [siˈlebriti], meşhur şahıs; şöhret.

**celerity** [siˈleriti]. Hız, sür'at.

**celery** [ˈseləri]. Kereviz.
**celestial** [siˈlestyəl]. Göğe aid, semavî; ilâhi.
**celiba·cy** [ˈselibəsi]. Bekârlık. **~te,** bekâr.
**cell** [sel]. Hücre; küçük oda; manastır veya zindan odası; petek gözü; elektrik pili.
**cellar** [ˈselə(r)]. Mahzen, yeraltı kiler; şarab mahzeni.
**'cell·ist** [ˈçelist]. Viyolonselci. **~o** [ˈçelou], viyolonsel.
**cellul·ar** [ˈselyulə(r)]. Hücreye aid; hücrelerden mürekkeb; göz göz. **~e,** hüceyre.
**Celtic** [ˈkeltik, ˈseltik]. Keltlere aid; keltçe.
**cement** [siˈment]. Çimento. Tutkal, birleştiren şey. Çimentolamak; iyice birleştirmek, yapıştırmak.
**cemetery** [ˈsemetəri]. Mezarlık.
**cenotaph** [ˈsenoutâf]. Başka yerde gömülü biri için dikilen abide.
**censor** [ˈsensə(r)]. Sansör; kontrol memuru; tenkidci. Sansür etm.; yasak etmek. **~ial** [–ˈsôriəl], sansöre aid. **~ious** [–sôriəs], tenkidci; daima kusur bulan. **~ship,** sansür, sansür idaresi.
**censure** [ˈsenşuə(r)]. Tenkid etme(k); kabahat bulma(k), tevbih etme(k).
**census** [ˈsensʌs]. Nüfus sayımı. **to take a ~,** nüfus sayımı yapmak.
**cent** [sent]. Doların yüzde biri, sent. (*coll.*) metelik, zerre. **per ~,** yüzde.
**centen·arian** [ˌsentəˈneriən]. Yüz yaşında veya daha fazla olan. **~ary** [senˈtînəri]. Yüzüncü yıldönümü. **~nial,** yüz yıla veya yüzüncü yıldönümüne aid.
**centimeter** [ˈsentimîtə(r)]. Santimetre.
**centipede** [ˈsentipîd]. Kırkayak.
**central** [ˈsentrəl]. Orta; merkez, merkezî. Santral. **~ize** [–âyz], merkezileştirmek. **~ization** [–âyˈzêyşn], merkezileşme.
**centre** [ˈsentə(r)]. Orta, merkez. Ortaya koymak; bir noktaya toplamak; temerküz ettirmek. **~-board,** kontra omurga(lı tekne).
**centri·c** [ˈsentrik]. Ortaya aid, merkezî. **~fugal** [–ˈtrifyugl], merkezden uzaklaşan. **~petal** [–ˈtripətl], merkeze yaklaşan.
**century** [ˈsençəri]. Asır, yüzyıl.
**ceramics** [siˈramiks]. Çinicilik.
**cereal** [ˈsiəriəl]. Hububat nevinden; zahire, hububat.
**cerebral** [ˈserebrəl]. Beyine aid.
**ceremon·ial** [seriˈmounyəl]. Resmî; merasimli; âyine aid; merasim, âyin. **~ious,** merasim ve teşrifatlı; tekellüflü. **~y** [ˈserəməni]. Merasim, teşrifat: **no need to stand upon ~ here,** burada teklif tekellüfe lüzum yok.
**cert** [sö(r)t]. (*coll.*) = **certainty. It's a (dead) ~,** hiç şübhe yok, elde bir.
**certain** [ˈsö(r)tn, –tin]. Kat'î, muhakkak; emin; bazı, bir, bir az. **a~,** ismi zikredilmek istenmiyen bir (kimse vs.), adı lâzım değil: **a lady of a ~ age,** yaşlıça bir hanım: **a ~ Mr. Brown,** Mr. Brown isminde bir adam: **he will come for ~ [he is ~ to come],** muhakkak gelir: **to make ~ of stg.,** tahkik etm., temin etm., hakkında emin olmak. **~ly,** elbette, şübhesiz; hay hay. **~ty,** katiyet, kat'î şey: **for a [of a] ~,** muhakkak.
**certi·fiable** [ˈsö(r)tifâyəbl]. Tasdik edilebilir; doktor raporile tımarhaneye gönderilecek kadar deli. **~ficate** [sö(r)ˈtifikit]. Tasdikname, vesika; şehadetname, ehliyetname. Bir vesika veya şehadetname vermek. **~fy** [ˈsö(r)tifây]. Tasdik etm.; resmiyet vermek; şehadetname, vesika vermek. **~tude** [ˈsö(r)tityûd]. Kat'iyet; şübhesiz olma.
**cessation** [seˈsêyşn]. Durma, inkıta.
**cession** [ˈseşn]. Terk, devir, ferağ.
**cess·pit, ~pool** [ˈsespit, –pûl]. Lağım çukuru.
**Ceylon** [siˈlon]. Seylan, Serendib.
**cf.** [ˈsîˈef] (*abb. Lât.* confer) Karşılaştırınız.
**chafe** [çêyf]. Sürt(ün)mek; sürterek berelemek; ısıtmak için oğmak; sinirlen(dir)mek.
**chaff**[1] [çâf] *n.* Saman tozu; hububat kabuğu ve saman kırıntısı. **~**[2]**.** Şaka (etm.), takılma(k).
**chaffinch** [ˈçafinç]. İspinoz.
**chagrin** [ˈşagrin, şaˈgrîn]. Üzüntü, infial. Ümidini kırmak; canını sıkmak.
**chain** [çêyn]. Zincir; silsile: 66 kademlik ölçü, ölçme aleti. Zincirlemek. **~ up,** zincire bağlamak: **to put in ~s,** zincire vurmak. **~-armour [~mail],** örme zırh elbise. **~-gang,** zincire vurulmuş olarak çalışan mahkûmlar. **~-store,** büyük bir magazanın şubesi.
**chair** [çeə(r)]. Sandalye, iskemle;

kürsü; sedye; makam; reislik makamı. Omuzda taşımak. **to take a ~,** oturmak: **to take the ~,** bir toplantıya başkanlık etm.: **folding ~,** açılır kapanır iskemle. **~man,** reis, başkan; sedyeci.
**Chaldean** [kal'dîən]. Kildanî.
**chalet** ['şalêy]. Dağ kulübesi; köşk.
**chalice** ['çalis]. Âyinde kullanılan kadeh.
**chalk** [çôk]. Tebeşir. **to ~ (up),** tebeşirle yazmak: **to ~ out,** tebeşirle tasarlamak: **French ~,** talk: **as different as ~ from cheese,** arasında dağlar kadar fark var: **A. is better than B. by a long ~,** A. B.'den fersah fersah daha iyidir: **'will Ahmed win the race?' 'Not by a long ~',** 'Ahmed yarışı kazanacak mı?' 'Ne münasebet (tam aksi)'. **~y,** tebeşirli, tebeşir gibi.
**challenge** ['çalinc]. Meydan okuma(k); düelloya davet etme(k); 'alnını karışlamak'; itiraz etm.; parola sormak.
**chamber** [çêymbə(r)]. Oda; yatak odası; toplantı vs. salonu; meclis; kısım, bölme; depo. **~(-pot),** lâzımlık: **~s,** avukat yazıhanesi: **~ of Deputies,** millet meclisi. **~lain** [–lin], mabeynci. **~maid** [–mêyd], oda hizmetçisi.
**chameleon** [kə'mîlyən]. Bukalemun.
**champ** [çamp]. (At gemini) ısırmak.
**champagne** [şam'pêyn]. Şampanya.
**champion** ['çampyən]. Şampiyon; müdafaacı, kahraman; mübariz. Müdafaa etm., tarafını tutmak.
**chanc·e**[1] [çâns] *n.* Kısmet, talih, kader; fırsat; imkân; ihtimal. *a.* Tesadüfî. **by ~,** tesadüfen: **will you be there by any ~?,** orada olmak (oraya gitmek) ihtimaliniz var mı?: **the ~s are that,** çok muhtemeldir ki: **the ~s are against his coming,** gelmesi ihtimali zayıftır: **the ~s are against me,** vaziyet aleyhimedir: **to have an eye to the main ~,** daima kendi çıkarına bakmak: **to give s.o. a ~,** (i) birine kendini göstermek için imkân vermek; (ii) imkân vermek, müsaade etm.; insaflı olm.: **there is an off ~ that ...,** ... muhtemel değil fakat olabilir: **he stands a ~ of winning,** kazanmak ihtimali yok değildir: **to take a ~,** bir kere denemek; **to take one's ~,** talihini denemek: **to take a long ~,** muvafakkiyet ihtimali pek zayıf olan bir işe girişmek. **~y,** talihe bağlı. **~**[2] *v.* Bir kere denemek; tesadüfe bırakmak. **to ~ to do stg.,** bir şeyi tesadüfen yapmak: **if you ~ to see him,** onu görecek olursanız: **to ~ upon s.o.,** birine rastlamak.
**chancel** ['çânsəl]. Kilisede mihrabın etrafında rahiblere ve koroya mahsus yer.
**chance·llery** ['çânsələri]. Elçilik kançılaryası. **~llor** ['çânsələ(r)]. Büyük rütbeli devlet memuru; bir üniversitenin fahrî rektorü. **Vice-~,** üniversite rektorü: **~ of the Exchequer,** (İngiltere'de) Maliye Nazırı: **Lord ~,** Lordlar Kamarası reisi ve Adliye Nazırı. **~ry** ['çânsəri]. **Court of ~,** Lordlar Kamarasından sonra İngiltere'de en yüksek mahkeme: **to put one's head in ~** kapana kısılmak.
**chandelier** [şande'liə(r)]. Avize.
**change** [çêync]. Değişme, değişiklik; tahavvül; tebdil; bozukluk (para), paranın üstü; değişik elbise; aktarma. Degiş(tir)mek; tahvil etm., tahavvül etm.; başkalaşmak; boz(ul)mak; aktarma yapmak. **to ~ into,** ···e tahvil etm., tahavvül etm.; çevirmek: **for a ~,** değişiklik olsun diye: **to ~ from bad to worse,** gittikçe daha beter olm.: **to ~ for the better [worse],** iyiliğe [kötülüğe] yüz tutmak: **a ~ of front,** (*mil.*) cebhe değiştirme; (*fig.*) yüzgeri etme: **the ~ of life,** (kadın) adetten kesilme: **you won't get much ~ out of me!,** (*coll.*) benden yana ümidini kes!: **'no ~ given',** para bozulmaz: **to ~ hands,** (bir şeyin) sahibi değişmek: **to ring the ~s,** mevcud imkânları denemek: **I couldn't wish it ~d,** bu hale (vaziyete) razıyım. **~ about,** değişip durmak. **~ down,** vitesi küçültmek. **~ over,** bir usulden bir usule geçmek; tarz değiştirmek. **~ up,** vitesi büyültmek. **~able** ['çêyncəbl]. Kararsız; ittiradsız; değişebilir. **~ling** ['çêynclin(g)]. Gizlice değiştirilen çocuk. **changing-room** ['çêyncin(g) 'rûm]. Soyunma odası.
**channel** ['çanl]. Nehir yatağı; boğaz, kanal; yol, geçid; mecra, mahrec. Kanal açmak; oymak. **the (English) ~,** Manş denizi: **the ~ Islands,** Jersey ve Guernsey adaları.
**chant** [çânt]. Şarkı, dinî şarkı; gülbank. Makam ile okumak; gülbank çekmek. **to ~ the praises of, ...** göklere çıkarmak.

**chanty** [ˈşanti]. Heyamola şarkısı, gemici şarkısı.

**chao·s** [ˈkêyos]. Büyük, karışıklık. **~tic** [keˈotik], karmakarışık; intizamsız.

**chap**[1] [çap]. Deride çatlak. Deri çatlamak. **~**[2]. (*coll.*) Arkadaş, adam; çocuk. **old ~**, azizim; (bizim) ahbab.

**chapel** [ˈçapl]. Küçük kilise; mekteb vs.nin kilisesi. **Nonconformist** mezhebinin kilisesi.

**chaperon** [ˈşaperoun]. Bir genc kıza refakat eden yaşlı kadın. Bir genc kıza hâmilik etmek.

**chaplain** [ˈçaplin]. Bir aile, alay vs.de dinî âyin icra etmekle vazifeli rahib.

**chapter** [ˈçaptə(r)]. Bab, fasıl; bahis; kısım; şube; bir katedrale bağlı rahibler. **to give ~ and verse,** kaynaklarını zikretmek: **~ house,** bir katedrale bağlı rahiblerin binası.

**char**[1] [çâ(r)] *v.* **to go out ~ring,** gündelikle ev hizmeti yapmak. *n.* = **charwoman. ~**[2] *v.* Yakarak kömür haline getirmek.

**char-a-banc** [ˈşarəban(g)]. Büyük gezinti veya seyahat araba veya otobüsü.

**character** [ˈkaraktə(r)]. Sıfat, alâmet, mahiyet, cins, vasıf; ahlakî vasıf, seciye; huy, ahlak; şöhret, hususiyet; harf; piyes ve romanda şahıs; aktörün oynadığı rol. **to be in ~ [out of] with,** uy[ma]mak: **to clear s.o.'s ~,** temize çıkarmak; **to give s.o. a good ~,** birine iyi numara vermek; bonservis vermek: **he's a ~,** o bir âlemdir, kimseye benzemez: **in his ~ of,** ... sıfatile: **a public ~,** tanınmış bir şahsiyet. **~istic** [ˌkariktəˈristik]. Hususiyet; başkalarından ayıran şey. Hususiyetine uygun; ayırd edici; kendisine mahsus. **~ize** [ˈ–râyz], tavsif etm.; temayüz etmek.

**charcoal** [ˈçâkoul]. Odun, mangal, kömürü; karakalem. **~-burner,** kömürcü.

**charge**[1] [çâc] *n.* Süvari veya süngü hücumu; (futbol) şarj. *v.* Şiddetli ve anî bir şekilde hücum etm.; şarj etmek. **to ~ down upon s.o.,** birinin üzerine doğru kat'î ve tehdidkâr bir şekilde ilerlemek: **to ~ into stg.,** ···e çarpmak: **to return to the ~,** tekrar hücuma geçmek; tekrar başlamak. **~r** [ˈçâcə(r)]. Muharebe atı. **~**[2] *n.* Bir defada doldurulan mikdar; barut hakkı; dolu; hamule; (*elect.*) şarj; harc, masraf; vazife, memuriyet; mesuliyet; emanet; nasihat; ittiham. **at a ~ of ...,** ... ücretle, masrafla: **to be a ~ on s.o.,** birine yük olm.: **to bring [lay] a ~ against s.o.,** birini ittiham etm.: **free of ~,** bedava: **to give s.o. ~ of [over], to put s.o. in ~ of,** birine bir şeyi tevdi etm.: **to be in ~ of,** bakmak, vazifeli olm., mesul olm.: **to give s.o. in ~,** polise teslim etm.: **to take s.o. in ~,** tevkif etm.: **to take ~ of,** birini veya bir şeyi üstüne almak: **list of ~s,** tarife: **to make a ~ for stg.,** bir şey için para almak. **~**[3] *v.* Yükletmek; doldurmak; para istemek; masraf yazmak; ittiham etm.; iş, vazife vermek; tenbih etm., emretmek. **~able** [ˈçâcəbl]. İtham olunabilir; hesabına yazılabilir; yükletilebilir.

**chariot** [ˈçariət]. Eski harb veya yarış arabası; saltanat arabası. **~eer** [–ˈtiə(r)], zafer veya yarış arabası sürücüsü.

**charit·able** [ˈçaritəbl]. Hayırsever, merhametli, cömert. **~ institution,** hayır müessesesi veya cemiyeti. **~y** [ˈçariti]. Merhamet; iyilik, sadaka, fukaraya yardım; muhabbet: **~ ball,** bir hayır cemiyeti menfaatine balo: ˹**~ begins at home**˺, insan herkesten evvel kendi ailesi halkına yardım etmelidir: **~ school,** yetimler yurdu: **~-boy, -girl,** yetimler yurdunda yetiştirilen çocuk.

**charm** [çâm]. Sihir; cazibe; tılısım. Teshir etm.; cezbetmek; son derece hoşa gitmek. **~ing,** çok hoş, pek cazib.

**chart** [çât]. Deniz haritası; istatistik vs. grafiği; şema. Haritaya almak; grafiğini çıkarmak; haritada göstermek.

**charter** [ˈçâtə(r)]. Berat, imtiyaz, patenta (vermek); kiralamak. **~ed accountant,** mütehassıs muhasib. **~-party,** navlun mukavelesi.

**charwoman** [ˈçâwumən], *pl.* **-women** [–wimen], tahtaya çamaşıra gelen gündelikçi kadın.

**chary** [ˈçeəri]. İhtiyatlı. **to be ~ of doing stg.,** bir şeyi yapmağa çekinmek.

**chase** [çêys]. Kovalama(k); takib etme(k); avlama(k); koşma(k). **the ~,** av: **to ~ away,** koğmak: **to ~ (off) after stg.,** bir şeyin peşinden gitmek: **to give ~ to s.o.,** birini kovalamak: **wild goose ~,** olmıyacak bir şeyin peşinden gitme. **~r,** avcı.

**chasm** [ˈkazm]. Büyük yarık; uçurum.
**chassis** [şasî]. Otomobil iskelet kısmı.
**chast·e** [çêyst]. İffetli. **~en** [ˈçêysn]. Gururunu kırmak. **~ise** [çasˈtâyz]. Ceza vermek, dögmek. **~ity** [ˈçastiti]. İffet.
**chat** [çat]. Sohbet, hoşbeş. Sohbet etm.
**chattel** [ˈçatl]. Menkul eşya. **goods and ~s,** her türlü menkul eşya.
**chatt·er** [ˈçatə(r)]. Gevezelik (etm.); çene çalma(k). **~box,** geveze, boşboğaz. **~y,** sohbet meraklısı.
**cheap** [çîp]. Ucuz; âdi, pespaye. **dirt ~,** sudan ucuz: **on the ~,** ucuza, ucuz olarak: **to feel ~,** keyifsiz olm.: (*bazan*) mahcub olm.: **to hold stg. ~,** bir şeye ehemmiyet vermemek: **to make oneself ~,** kendini küçük düşürmek. **~-jack,** seyyar satıcı. **~en,** ucuzla(t)mak; değerini düşürmek.
**cheat** [çît]. Dolandırıcı; hilekâr. Dolandırmak; aldatmak.
**check**[1] [çek] *n.* (Satranc) keş; durdurma; engel; muvaffakiyetsizlik; kontrol; (vestiyer vs.de) fiş. *v.* (Satranc) keş etm.; durdurmak; menetmek; önlemek; bakmak, yoklamak; duraklamak. **to hold [keep] in ~,** durdurmak: **to keep a ~ on stg.,** kontrol etmek. **~**[2] *a.* Satranclı (kumaş). **~**[3]. (*Amer.*) = **cheque. ~ers** [ˈçekö(r)z]. (*Amer.*) Dama oyunu. **~mate** [çekˈmêyt]. Satrancda mat etme(k); yenme(k); işini bozmak.
**cheek** [çîk]. Yanak; (*coll.*) yüzsüzlük, küstahlık. ···e küstahlık etm. **~ by jowl,** haşır neşir: **to have the ~ to,** ... küstahlığında bulunmak. **~y,** yüzsüz.
**cheep** [çîp]. Cıvıltı. Cıvıldamak.
**cheer** [çi̯ə(r)]. Neş'e; teşvik; alkış; bravo; yaşa! sesi. Alkışlamak. **~ (up),** neş'elen(dir)mek; içi(ni) açmak: **~ on,** teşvik etm.; **~ up!,** üzülme!: **what ~?,** ne var, ne yok? **~ful,** neş'eli; güler yüzlü; ferah. **~io!,** (*coll.*) Güle güle!; eyvallah!, sıhhatinize!. **~less,** kasvetli; hüzün verici. **~y,** neş'eli.
**cheese** [çîz]. Peynir. **~monger,** peynirci. **~-paring,** hesabî; hesabilik; hasislik. **cheesy,** peynir gibi.
**chef** [şef]. Aşçı başı.
**chemical** [ˈkemikl]. Kimyevî, şimik. **~s,** ecza: kimyevî maddeler.
**chemist** [ˈkemist]. Eczacı; kimyager. **~ry,** kimya.
**cheque** [çek]. Çek. **crossed ~,** çizgili çek. **~-book,** çek defteri.
**chequer** [çekə(r)]. Damalı yapmak; ittiradı bozmak. **a ~ed career,** dalgalı hayat.
**cherish** [ˈçeriş]. Aziz tutmak; beslemek.
**cherry** [ˈçeri]. Kiraz. ˹**to make two bites at a ~**˺, bir işi lüzumsuz yere ağır almak.
**cherub** [ˈçerʌb]. Melek; nur topu gibi çocuk; masum yüzlü kimse. **~ic** [–ˈrûbik], melek gibi; masum yüzlü.
**chess** [çes]. Satranc oyunu. **~men,** satranc taşları.
**chest** [çest]. Sandık, kutu; göğüs. **~ of drawers,** çekmeceli dolab: **to get it off one's ~,** (*coll.*) içini dökmek.
**chestnut** [ˈçesnʌt]. Kestane; (*coll.*) bayat şaka, hikâye.
**chew** [çû]. Çiğnemek. **~ing-gum,** sakız.
**chicanery** [şiˈkêynəri]. Hile.
**chick** [çik]. Civciv. **~-pea,** nohut.
**chicken** [ˈçikin]. Piliç, civciv; tavuk eti. ˹**don't count your ~s before they are hatched**˺, çayı görmeden paçaları sıvama: **she is no ~,** pek körpe denmez. **~-hearted,** korkak. **~-pox,** su çiceği hastalığı.
**chide** [çâyd]. Azarlamak.
**chief** [çîf]. Baş; reis; büyük. **commander-in-~,** başkumandan. **~tain** [ˈçîftən]. Kabile reisi.
**chilblain** [ˈçilblêyn]. Mayasıl.
**child** [çâyld]. Çocuk. **be a good ~!,** uslu dur!: **from a ~,** küçükten beri: **to be with ~,** hamile olmak. **~birth,** çocuk doğurma. **~hood,** çocukluk: **second ~,** bunaklık. **~ish,** çocukça: **to grow ~,** bunamak. **~like,** çocuk gibi. **~'s-play,** kolay iş: **it's mere ~,** işten bile değil.
**chill** [çil]. Soğuk algınlığı; soğukluk. Soğutmak; üşütmek; (madeni) suya batırmak; şevkini [ümidini] kırmak. **~ed meat,** dondurulmuş et: **to cast a ~ over the company,** meclise soğukluk getirmek: **a cold ~ came over him,** tüyleri ürperdi: **to take the ~ off stg.,** hafifçe ısıtmak. **~y,** (nahoş şekilde) serin; soğuk; hep üşüyen.
**chime** [çâym]. Muhtelif havalar çalabilir çanlar; ahenkli çan sesi. (Çanlar) çal(ın)mak. **to ~ the hour,** (saat) saat vurmak: **to ~ in,** söze karışmak.
**chimer·a** [kiˈmi̯ərə]. Ejderha; kor kunç hayal; imkânsız fikir. **~ical** [–ˈmerikl], hayalî; imkânsız.

**chimney** [ˈçimni]. Baca; lâmba şişesi. **~-corner,** ocak başı. **~-piece,** ocak rafı. **~-pot,** baca külâhı. **~-sweep,** baca temizleyici.

**chin** [çin]. Çene. **to wag one's ~,** çene çalmak. **~-strap,** çene kayışı.

**China** [ˈçâynə]. Çin; **china,** çini, porselen. **~man** (*pl.* **-men**), Çinli.

**Chinese** [çâyˈnîz]. Çinli, Çine aid; çince.

**chink**[1] [çink]. Yarık; çatlak. **~**[2]. Çinlama(k). **~**[3]. (*sl.*) Çinli.

**chip** [çip]. Yonga; çentik; küçük parça; kırıntı; kızarmış patates; (kâğıd vs. oyunlarında saymak için) fiş. Yontmak; çentmek. **to ~ in,** söze karışmak: **a ~ of the old block,** babasına benziyen oğul.

**chiropody** [kâyˈropodi]. Nasırcılık.

**chirp** [çö(r)p]. Cıvıltı. Cıvıldamak. **~y,** (*sl.*) neş'eli, şen.

**chisel** [ˈçizl]. Çelik kalem; marangoz, taşçı, kalemi. Kalemle oymak, yontmak; (*sl.*) dolandırmak. **cold ~,** demirci kalemi.

**chival·rous** [ˈşivəlrəs]. Şövalye gibi; alicenab. **~ry,** şövalyelik (ki kahramanlık, şeref, kadına hürmet, zayıfı himaye, cömertlik vasıflarını ihtiva eder).

**chivy** [ˈçivi]. Koğalamak. **to ~ s.o. about,** birine nefes aldırmamak.

**chlorine** [ˈklorîn]. Klor.

**chlorophyll** [ˈklorofil]. Nebatlara yeşillik veren madde.

**chock** [çok]. Takoz (koymak). **~-a-block,** hıncahınc; tıkabasa.

**chocolate** [ˈçoklit]. Çikolata.

**choice** [çôys] *n.* Seçme, intihab, tercih; tercih hakkı; çeşit; seçilen şey. *a.* Seçkin. **for ~,** tercihen.

**choir** [kwâyə(r)]. Koro; kilisede koro mahalli.

**choke** [çouk]. Tıkanma; nefes alamama; boğulma; (*otom.*) jikle. Tıka(n)mak; nefesini tıkamak; boğ(ul)mak; nefes alamamak. **~ back,** (gözyaşı vs.) tutmak. **~ off,** (*sl.*) vazgeçirmek.

**cholera** [ˈkolera]. Kolera.

**choose** (**chose, chosen**) [çûz, çouz, çouzn]. Seçmek, intihab etm.; ihtiyar etm.; tercih etm.; karar vermek. **he cannot ~ but accept,** kabul etmekten başka bir şey yapamaz: **there is nothing to ~ between them,** aralarında hiç fark yoktur: **I do not ~ to do so,** bunu yapacak değilim: **when I ~,** istediğim zaman.

**chop**[1] [çop]. Balta veya satır darbesi; kotlet, pirzolalık; deniz şıpırtısı. (Balta veya satırla) kesmek; doğramak, kıymak; (dalgalar) çırpınmak. **~ off,** kesip koparmak: **~ up,** doğramak, kıymak. **~-house,** kebabcı dükkânı. **~per** [ˈçopə(r)]. Satır. **~py** [ˈçopi]. Hafifçe dalgalı; (rüzgâr) değişen, mütehavvil. **~**[2]. **to lick one's ~s,** yalanmak. **~**[3]. Anî değişiklik; rüzgârın değişmesi. **to ~ and change,** bir saati bir saatine uymamak: **~s and changes,** mütemadî değişmeler.

**choral** [ˈkôrəl]. Koroya aid.

**chord** [ˈkôd]. Kiriş, tel; (*geom.*) veter; ahenk, akort. **to touch the right ~,** can alacak noktaya dokunmak.

**chore** [ço(r)]. Zor ve zevksiz iş.

**choreography** [ˌkôriˈogrəfi]. Bale danslarını tanzim etme.

**chorister** [ˈkoristə(r)]. Koro şarkıcısı.

**chortle** [ˈçôtl]. Hafif sesle gülmek.

**chorus** [ˈkôrəs]. Koro; nakarat. **in ~,** hep beraber ve aynı zamanda. **~-girl,** müzikhol dansözü.

**chose, chosen** *v.* **choose.**

**Christ** [krâyst]. İsa. **~en** [ˈkrisn], vaftiz etm., isim vermek. **~ening,** vaftiz. **~endom** [ˈkrisəndəm], Hıristiyanlık âlemi. **~ian** [ˈkristyən], Hıristiyan: **the ~ era,** Milâdî sene. **~ianity** [kristiˈaniti], Hıristiyanlık.

**Christmas** [ˈkrisməs], Noel. **a merry ~!,** Noeliniz kutlu olsun!. **~-box** *v.* **boxing-day.** **~-card,** Noel tebriki. **~-day,** Noel günü (25 aralık). **~-eve,** Noel arifesi. **~-tide,** Noel zamanı.

**chronic** [ˈkronik]. Müzmin, devamlı.

**chronicle** [ˈkronikl]. Vakayiname, tarih. Vakayii, tarihini, yazmak. **~r,** vak'anüvis, tarihçi.

**chronolog·ical** [ˌkronoˈlocikl]. Tarih sırasına göre. **~y** [kroˈnoloci]. Zamanın devirlere ayrılması; hadiseleri tarih sırasına göre veren cedvel.

**chrysalis** [ˈkrisəlis]. Koza.

**chubby** [ˈçʌbi]. Tombul.

**chuck**[1] [çʌk]. Atma(k), fırlatma(k); kaldırıp atmak. **to ~ about,** saçmak, savurmak; **to ~ one's weight about,** azamet satmak: **to ~ out,** kapı dışarı etm.: **to ~ s.o. under the chin,** çenesini okşamak: **to get the ~,** (*sl.*) işinden çıkarılmak: **to give s.o. the ~,** (*sl.*) birine yol vermek: **to ~ up,** (işini) bırakmak; vazgeçmek. **~**[2]

*n.* Torna aynası. **drill ~,** matkab mandreni.
**chuckle** [ˈçʌkl]. Hafif sesle kendi kendine gülme(k).
**chum** [çʌm]. Ahbab. **to ~ up with s.o.,** birile ahbab olmak. **~my,** ahbabca.
**chump** [çʌmp]. Kütük; (*sl.*) kafa; ahmak.
**chunk** [çʌnk]. İri parça (odun vs.).
**church** [çö(r)ç]. Kilise. **the ~ of England,** Anglikan kilisesi: **to go into the ~,** rahib olm.: **High ~,** protestan kilisenin âyinlere çok ehemmiyet veren bir kolu (*bunun aksine* **Low ~** *denir*). **~warden,** kilise mütevellisi; uzun saplı toprak pipo. **~yard,** kilise avlusu, mezarlığı. **~-worker,** kilisenin hayır işlerine yardım eden kimse.
**churl** [çö(r)l]. Kaba ve terbiyesiz adam. **~ish,** kaba, terbiyesiz.
**churn** [çö(r)n]. Yayık. Sütü yayıkta çalkamak; yağ çıkarmak; suyu vs. köpürtmek.
**chute** [şût]. Oluk; kızak; çağlıyan.
**chutney** [ˈçʌtni]. Baharlı bir hind salçası.
**cicada** [siˈkâda]. Ağustos böceği.
**cicatrice** [ˌsikətris]. Yara izi, kabuğu.
**cicerone** [ˈsisərǫun]. Seyyah rehberi.
**C.I.D.** *v.* **criminal.**
**cider** [ˈsâydə(r)]. Elma şırası.
**cigar** [siˈgâ(r)]. Yaprak sigara, puro. **~ette** [ˈsigəˈret], sigara: **~-holder,** ağızlık.
**cinch** [sinç]. (*sl.*) Elde bir.
**cinder** [ˈsində(r)]. Kor; kül.
**Cinderella** [ˌsindəˈrela]. Ihmal edilmiş, takdir edilmiyen kimse.
**cinema** [ˈsinima]. Sinema. **~tograph** [–ˈmatəgrâf], sinema makinesi.
**cipher** [ˈsâyfə(r)]. Sıfır; solda sıfır (kimse); şifre; gizli yazı; şifre anahtarı; marka. Şifre ile yazmak; hesab yapmak.
**circle** [ˈsö(r)kl]. Daire; halka; (tiyatroda) balkon; grup; meclis; mahfil. Etrafını dönmek, dolaşmak. **to square the ~,** imkânsız bir şeye girişmek.
**circuit** [ˈsö(r)kit]. Etrafını dönme, dolaşma; devri; dolaşılan saha; muhit; (*elect.*) devre; İngiltere'de seyyar mahkeme. **to go on ~,** (hâkim, avukat) seyyar mahkeme ile dolaşmak.
**circulat·e** [ˈsö(r)kyulêyt]. Dolaş(tır)mak; tedavül et(tir)mek, yay(ıl)mak; neşretmek, neşredilmek. **~ion** [–ˈlêyşn], dolaşma; cereyan; deveran; tevdavül; (gazete vs.) satış mikdarı.
**circumcis·e** [ˈsö(r)kəmsâyz]. Sünnet etmek. **~ion** [–ˈsijn], sünnet.
**circumference** [səˈkʌmfərəns]. Muhit, çevre.
**circumflex** [ˈsö(r)kʌmfleks]. (ˆ) işaretli vurgu.
**circumspect** [ˈsö(r)kəmspekt]. İhtiyatlı, basiretli; düşünceli; müteenni.
**circumstance** [ˈsö(r)kəmstans]. Hal; şart, vaziyet; hadise; keyfiyet; vak'a; tafsilat; merasim. **~s,** malî vaziyet. **in [under] the ~s,** bu vaziyette; bu ahvalde: **in no ~s,** hiç bir şekilde: **my worldly ~s,** malî vaziyetim: **with pomp and ~,** büyük merasimle. **~d, well ~d,** hali vakti yerinde: **as I was ~d,** içinde bulunduğum vaziyette.
**circumstantial** [ˌsö(r)kəmˈstanşəl]. Mufassal; arizî, talî. **~ evidence,** emare.
**circus** [ˈsö(r)kəs]. Sirk, cambazhane; bir kaç caddenin birleştiği meydan.
**cistern** [ˈsistö(r)n]. Sarnıc.
**citadel** [ˈsitədel]. Kale; hisar.
**cite** [sâyt]. Celbetmek, mahkemeye davet etm.; zikretmek; şahid göstermek.
**citizen** [ˈsitizn]. Hemşeri; vatandaş; sivil şahıs; şehirli. **~ship,** vatandaşlık, vatanperverlik.
**city** [ˈsiti]. Büyük şehir. **the City,** Londra'nın iş merkezi: **he is in the ~,** Londra'da iş adamıdır.
**civic** [ˈsivik]. Şehre aid; medenî; vatandaşlığa aid. **the ~ authorities,** belediye makamları. **~s,** yurtbilgisi; vatandaşlık bilgisi.
**civil** [ˈsivil]. Vatandaşlara aid; devlete, millete, aid; medenî; nazik, terbiyeli. **~ engineering,** köprü, yol, liman vs. mühendisliği: **~ law,** medenî kanun: **~ servant,** mülkiye memuru: **~ service,** mülkiye hizmeti, devlet memurluğu: **~ list,** Kıralın ve sarayın tahsisatı: **~ war,** dahilî harb.
**civili·an** [siˈvilyən]. Sivil; başıbozuk. **~ty** [siˈvılıti], nezaket, terbiye. **~zation** [ˌsivilâyˈzêyşn], medeniyet. **~ze** [ˈsivilâyz], medenileştirmek: **to become ~d,** medenileşmek.
**clack** [klak]. (*ech.*) Tıkırtı. Tıkırdamak.
**clad** *v.* **clothe.**
**claim** [klêym]. İddia etm.; istemek; taleb etmek. İddia; taleb; taleb edilen şey. **to stake out a ~,** bir arsanın hududunu çizmek: **to jump a**

~, sahibinin kullanmadığı bir arsayı işgal etmek. **~ant,** davacı; taleb sahibi; hak iddia eden.

**clairvoyan·ce** [klęəˈvoyəns]. İstikbali görmek kabiliyeti. **~t,** gaibi gören.

**clamber** [ˈklambə(r)]. Güçlükle tırmanma(k).

**clammy** [ˈklami]. Soğuk ve ıslak; yapışkan.

**clamo·rous** [ˈklamərəs]. Gürültülü. **~ur,** gürültü; yaygara. **to ~ for,** yaygara ile istemek.

**clamp** [klamp]. Mengene; kenet; köşebent; yığın. Mengeneye kıstırmak; kenetlemek: yığmak.

**clan** [klan]. Kabile. **~nish** [ˈklaniş]. Kabileye aid; biribirine bağlı ve yabancıları sevmiyen. **~sman,** *pl.* **-men,** bir kabileye veya klana mensub adam.

**clandestine** [klanˈdestin]. Gizli, el altından.

**clang** [klan(g)]. (*ech.*) Tınlama(k), çınlama(k). **~our,** bir sürü madenî sesler.

**clank** [klank]. (*ech.*) Şıkırtı, şıkırdatmak.

**clap** [klap]. El çırpma; (tek) gök gürleme. El çırpmak, alkışlamak; vurmak; ansızın ve şiddetle koymak. **to ~ s.o. on the back,** tebrik için vs. birinin sırtına vurmak: **to ~ eyes on s.o.,** birini birdenbire görmek; **I've never ~ped eyes on him,** onu hayatımda hiç görmedim: **to ~ on one's hat,** şapkasını başına geçirmek: **to ~ a pistol to s.o.'s head,** birinin başına tabanca dayamak: **to ~ s.o. in prison,** birini hapse tıkmak. **~per** [ˈklapə(r)]. Çan tokmağı. **~trap** [ˈklaptrap]. Gösteriş için boş söz veya hareket; safsata, palavra.

**clarify** [ˈklarifây]. Tasfiye etm.; berrak bir hale getirmek; aydınlatmak, izah etmek.

**clash** [klaş]. (*ech.*) Çarpma veya çarpışma sesi; çarpışma(k), çatışma(k); şiddetli ihtilaf (halinde olm.); uyuşmamak.

**clasp** [klâsp]. Toka; bağlamağa yarıyan şey; el sıkma. Sıkmak; kucaklamak; bağlamak. **to ~ one's hands,** el kavuşturmak. **~-knife,** sustalı çakı.

**class** [klâs]. Sınıf; tabaka; nevi, cins. Tasnif etmek. **no ~** (*coll.*) aşağı sınıf veya derecede. **~y** [ˈklâsi]. (*sl.*) Mükemmel, ekâbir.

**classic** [klasik]. Mükemmel; klâsik; klâsik muharrir. **the ~ s,** eski Yunan ve Lâtin edebiyatı. **~al,** klâsik: **~ scholar,** Yunan ve Lâtin edebiyatları âlimi.

**classif·ication** [ˈklasifiˈkêyşn]. Sınıflara ayırma, tasnif. **~y,** sınıflara ayırmak, tasnif etmek.

**clatter** [ˈklatə]. (*ech.*) Takırtı; takırdamak. **to ~ downstairs,** merdivenden gürültü ile inmek: **to come ~ing down,** paldır küldür düşmek.

**clause** [klôz]. Madde; şart; hüküm; cümlenin bir kısmı.

**claustrophobia** [ˌklôstroˈfǫubyə]. Kapalı yerlerden korkma.

**claw** [klô]. Hayvan pençesi(ndeki iğri tırnak); istakoz kıskacı; çekiçin çivi söken tarafı. Tırmalamak, pençelemek.

**clay** [klêy]. Balçık, kil; insan vücudu. **~ey,** balçıklı.

**clean** [klîn] *a.* Temiz, saf; biçimli. *adv.* Tamamen, iyice. *v.* Temizlemek. **as ~ as a new pin,** tertemiz: **to give stg. a ~,** bir şeyi temizlemek: **to have ~ hands,** rüşvete el sürmemiş olmak. **~ out,** boşaltıp temizlemek; boşaltmak; **to be ~ed out,** meteliksiz kalmak. **~ up,** süprüntü vs.yi toplıyarak temizlemek; düzeltmek; (*coll.*) bitirmek. **~ly** *a.* [ˈklenli]. (Tabiat itibarile) temiz. *adv.* [ˈklînli], temiz bir surette. **~se** [klenz]. Temizlemek.

**clear¹** [ˈklįə(r)] *a.* Açık, aydınlık; berrak, şeffaf, temiz; saf; net; tam; aşikâr; engelsiz, serbest. **to be [get] ~ of . . .,** ···den kurtulmak: **a ~ majority,** tam ekseriyet: **a ~ profit,** safi kâr: **a ~ thinker,** vazıh fikirli: **'all ~!',** yol açık!: **to sound the 'all-~',** bir hava hücumunun sonunda 'tehlike geçti' işareti vermek: **the coast is ~,** 'ortalık sütliman': **to stand [keep] ~,** açık durmak, çekilmek: **the moment the train was ~ of the station,** tren istasyondan ayrılır ayrılmaz: **my conscience is ~,** vicdanen müsterihim: **to make oneself ~,** maksadını açıkça izah etm.: **are you quite ~ about that?,** (i) bunu iyice anladınız mı?: (ii) bu hususta tamamen emin misiniz?: **to send a message in ~,** açık (şifresiz) bir haber göndermek. **~-cut,** düzgün, biçimli; kat'î. **~-headed,** anlayışlı, açık kafalı. **~²** *v.* Açık hale getirmek; berrak hale getirmek; temizlemek;

ayıklamak; kurtarmak; engelleri kaldırmak; boşaltmak; tebriye etm.; ···den ayrılmak; gümrükten çıkarmak; (hava) açılmak: (gemi) hareket etmek. **to ~ the air,** (i) havayı temizlemek; (ii) vaziyeti tavzih etm.: **to ~ one's conscience,** vicdanını müsterih kılmak: **to ~ the decks for action,** (i) muharebe için güverteyi neta etm.; (ii) bir iş için hazırlık yapmak: **to ~ the ground (for),** (···e) yol açmak, zemin hazırlamak: **to ~ £100 [10 per cent.],** yüz lira [yüzde on] safi kâr temin etm.: **(jumping) to ~ an obstacle,** atlarken dokunmadan bir maniadan aşmak: **to ~ a ship,** (i) gemiyi tahliye etm.; (ii) bir geminin bütün masraflarını vererek hareket müsaadesini almak: **to ~ the table,** sofrayı kaldırmak. **~ away,** kaldırmak; temizlemek, derleyip toplamak. **~ off,** (borc vs.) ödemek; temizlemek; kaçmak, sıvışmak. **~ out,** boşaltmak; çekilip gitmek: **~ out!,** çek arabanı! **~ up,** temizlemek; aydınlatmak; halletmek. **~ing** [ˈkliərin(g)] *n.* Ormanda açık saha; bankalar arasında çek ve sened mübadelesi suretile hesablama; kliring. **~ing-house,** takas odası.

**cleavage** [ˈklîvic]. Yarılma; ayrılma, ayrılık.

**cleave¹ (-d, cleft)** [klîv, -d, kleft]. Yarmak, yarılmak; çatlamak. **a cleft stick,** çatallı değnek: **to be in a cleft stick,** çıkmaza girmek. **~². to ~ to . . .,** ···e bağlı olm., yapışmak.

**clef** [klef]. (*mus.*) Nota anahtarı.

**cleft** *v.* **cleave.**

**clemen·cy** [ˈklemensi]. Merhamet: mülayimlik. **~t,** mülayim.

**clench** [klenç]. (Dişlerini, yumruklarını) sıkmak; sımsıkı yakalamak.

**cler'gy** [ˈklö(r)ci]. Rahibler, rahib sınıfı. **~man,** *pl.* **-men,** rahib, papaz.

**cleric** [ˈklerik]. Rahib. **~al** [ˈklerikl]. Kâtiblere, rahiblere, aid. **~alism,** siyasette rahiblerin nüfuzu veya buna tarafdarlık.

**clerk** [klâk]. Kâtib, yazıcı; evrak memuru; (*jur.*) rahib. **Town Clerk,** İngiliz belediyelerinde tahrirat kâtibi ile evrak müdürü vazifelerine yakın iş gören memur: **~ of the weather,** hava işlerini idare ettiği farzolunan şahıs: **~ of works,** işbaşı. **~ly,** kâtibce.

**clever** [klevə(r)]. Zeki; maharetli; becerikli. **he was too ~ for us,** bizden daha kurnaz çıktı.

**cliché** [ˈklışêy]. Basmakalıb söz, klişe.

**click** [klik]. (*ech.*) Sert ve kesik ses; şıkırtı. Şıkırdamak; (topuklarını) çarpmak; (*sl.*) uyuşmak.

**client** [ˈklâyənt]. Müşteri; müekkil. **~ele** [klîonˈtel], müşteriler.

**cliff** [klif]. Yar, uçurum, kayalık.

**climat·e** [ˈklâymit]. İklim. **~ic** [–ˈatik], iklime aid.

**climax** [ˈklâymaks]. En buhranlı nokta; en yüksek derece.

**climb** [klâym]. Tırmanma(k), çıkma(k); yükselme(k); yokuş. **to ~ down,** tırmanarak inmek; yelkenleri suya indirmek. **~er,** tırmanan; dağcı; sarmaşık nebat; her ne pahasına olursa olsun sosyetede muvaffak olmak isteyen adam.

**clinch** [klinç]. Perçin(lemek); (boks) girift olma(k). **to ~ an argument,** bir münakaşada karşısındakini (kuvvetli bir cevabla) susturmak: **to ~ a bargain,** pazarlığı uydurmak.

**cling (clung)** [klin(g), klʌn(g)]. Sımsıkı sarılmak; vazgeçmemek; bağlanmak. **to ~ together,** birleşik olm.; 'anca beraber kanca beraber' olmak.

**clinic** [ˈklinik]. Klinik. **~al,** kliniğe aid: **~ thermometer,** doktor termometrosu, derece.

**clink¹** [klin(g)k]. (*ech.*) Şıkırtı. Şıkırda(t)mak. **to ~ glasses,** kadeh tokuşturmak. **~².** (*sl.*) Kodes.

**clip** [klip]. Kırpma; kırkım. Kırpmak; kırkmak; kesmek; zimbalamak. **to ~ s.o.'s claws,** (*fig.*) birinin tırnaklarını sökmek: **to ~ s.o.'s wings,** birinin hareketini, faaliyetini, tahdid etm.: **to ~ one's words,** kelimelerin sonunu yutmak. **~pers** [ˈklipö(r)z]. Kırpma aleti; saç traş makinesi.

**cloak** [klouk]. Harmaniye, pelerin; (*fig.*) behane. Örtmek, gizlemek. **~-room,** emanet, gardırob: **ladies' ~,** 'kadınlara'.

**clock** [klok]. Büyük saat (duvar saati vs.). **to ~ in [out],** (işe gelen [işini bitiren] amele vs.) saati çevirmek: **grandfather ~,** dolablı saat: **what o'~ is it?,** saat kaç?: **it is three o'~,** saat üç: **like one o'~,** (*sl.*) mükemmel: **twenty minutes by the ~,** tam yirmi dakika: **to sleep the ~ round,** on iki saat uyumak. **~wise** saat akrebinin döndüğü istikamette **~work,** saat makinesiyle işliyer

makine: **like ~**, saat gibi, muntazam.
**clod** [klod]. Toprak parçası; ahmak. **~-hopper,** hödük.
**clog** [klog]. Takunya. Tıka(n)mak.
**cloister** [ˈklôystə(r)]. Manastır; avlu etrafındaki kemerli yolu. Manastıra kapamak; tecrid etmek.
**close**[1] [klous] *a.* Yakın, bitişik; dikkatli; sık; mahdud; kapalı, havasız; hasis; fazla ketum; samimî (dost). **~ by** [**~ at hand**], yakında; civarda: **a ~ win,** pek küçük farkla kazanma: **to lie ~**, bir kenara saklamak: **a ~ prisoner,** sıkı nezaret altında olan mahbus: **at ~ quarters,** çok yakından; (*naut.*) borda bordaya: **a ~ thing,** (*coll.*) uc uca, müşkülatla: **~ time** [**season**], muayyen hayvanların avlanmasının yasak olduğu müddet [mevsim]: **a ~ translation,** aslına çok yakın tercüme. **~-cropped** [**~-cut**], kısa kesilmiş. **~-fisted,** cimri. **~-fitting,** iyi oturan (elbise), dar. **~-set,** (**eyes**) birbirine yakın (gözler). **~-up,** çok yakından çekilen resim. **~**[2] [klouz] *n.* Nihayet; kapanma. *v.* Kapatmak; kapanmak; bit(ir)mek; sona er(dir)mek; (gemi) yaklaşmak. **to ~ the books,** hesabı kapatmak: **to ~ the ranks,** safları şıkıştırmak; (*fig.*) tehlike karşısında birleşmek: **to ~ with s.o.,** birile göğüs göğüse gelmek; birile anlaşmaya varmak: **to ~ with a bargain,** pazarlığı uydurmak. **~ about** [**~ round**], kuşatmak, etrafını çevirmek. **~ down,** büsbütün kapatmak, kapanmak; (radyo) neşriyatını bitirmek. **~ in, night is closing in,** karanlık basıyor: **the days are closing in,** günler kısalıyor: **to ~ in on s.o.,** etrafını çevirerek yaklaşmak. **~ up,** kapatmak; kapanmak; örtülmek; yaklaşmak, sıkışmak. **~d** [klouzd]. Kapanmış. **~ shop,** yalnız bir sendika azâsını kullanan fabrika vs. **~ness** [ˈklousnis]. Yakınlık; sıkılık; havasizlık; hasislik; ketumiyet.
**closet** [ˈklozit]. Küçük yahud hususî oda; dolab; helâ. Birile bir yerde kapanmak.
**clos·ing** [ˈklouzin(g)] *n.* Kapatma; kapanma; tatil. *a.* Son, sonuncu. **~ure** [ˈkloujuə(r)]. Kapatma. **to move the ~**, müzakerenin kifayetine karar vermek.
**clot** [klot]. Pıhtı(laşmak); (süt) kesilmek.
**cloth** [kloθ]. Kumaş. **the ~**, rahiblik. **American ~** [**oil-~**], muşamba: **to lay the ~**, sofrayı kurmak: '**the respect due to his ~**', mensub olduğu mesleğe gereken hürmet. **~e** (**-d** veya **clad**) [klouð, -d, klad]. Giydirmek; örtmek, kaplamak. **~es** [klouðz]. Elbise, esvab; çamaşır. **suit of ~**, kostüm, takım elbise: **~-line,** çamaşır ipi: **~-peg,** mandal. **~ier** [klouðiə(r)]. Kumaşçı; elbiseci. **~ing** [ˈklouðin(g)]. Giyim, giyecek.
**cloud** [klaud]. Bulut. Bulutlamak; bulandırmak; (**brow**) kaşı çatılmak. **in the ~s,** dalgın: **to be under a ~**, şübhe altında olm.; gözden düşmüş olm.: ⌈**every ~ has a silver lining**⌉, (ne kadar fena olursa olsun) her işte bir hayır vardır. **~ed,** bulutlu; bulanık. **~y,** bulutlu; bulanık; (fikir) mübhem.
**clout** [klaut]. (*sl.*) Darbe. Vurmak.
**cloven** [ˈklouvən] *v.* **cleave. ~-footed** [**~-hoofed**], çatal ayaklı tırnaklı (*Şeytan umumiyetle çatal ayaklı olarak gösterilir*).
**clover** [ˈklouvə(r)]. Yonca, tırfıl. **to be in ~**, çok talihli olmak.
**clown** [klaun]. Soytarı, palyaço; hödük. Soytarılık etmek.
**club**[1] [klʌb]. Çomak; ıspatı. Sopa ile vurmak. **~-foot,** yumru ayak. **~**[2]. Klüb, cemiyet. **to ~ together,** muayyen bir maksadla bir araya gelmek, paralarını birleştirmek. **~-house,** spor klübü binası.
**cluck** [klʌk]. (Tavuk) guluklama(k).
**clue** [klû]. İpucu; miftah.
**clump**[1] [klʌmp]. Ağac veya çiçek kümesi. **to ~ together,** yığmak. **~**[2]. (*ech.*) **to ~ about,** ağır basarak yürümek.
**clumsy** [ˈklʌmzi]. Beceriksiz; hantal; acemi; biçimsiz.
**clung** *v.* **cling.**
**cluster** [ˈklʌstə(r)]. Demet; salkım. **to ~ together,** toplanmak.
**clutch** [klʌç]. Kavrama; ambreyaj. **~es,** pençe. Kavramak, yakalamak, tutmak. **to fall into s.o.'s ~es,** birinin pençesine düşmek: **to make a ~ at stg.,** bir şeyi yakalamak için anî bir hareket yapmak: **to let in the ~**, ambreyaj yapmak: **to throw out the ~**, debreyaj yapmak.
**clutter** [ˈklʌtə(r)]. Darmadağınlık. **to ~ up,** darmadağın şeylerle doldurmak.
**C.M.G.** [ˈsîˈemˈcî]. (*abb.*) **Companion of St. Michael and St. George,** *v.* **order**[2].

**co-** [kou] *pref.* *Bazı kelimelerin başına gelerek* iştirak *ifade eder.*
**Co.** [*abb.*) **Company,** şirket.
**C.O.** [ˈsîˈou]. (*abb.*) **Commanding Officer,** kumandan.
**c/o** (*abb.*) **Care of ..., ...** vasıtasi ile; ... eliyle.
**coach**[1] [kouç]. Büyük yolcu arabası; vagon; otobüs. ~ **and six,** altı atlı araba. ~**man,** *pl.* **-men,** arabacı. ~**work** [ˈkouçwö(r)k]. Karoseri. ~[2]. İmtihana hazırlayan hoca; spor antrenörü. Hususî ders vererek imtihana hazırlamak; (sporcuya) antrenman yaptırmak.
**coagul·ant** [kouˈagyûlənt]. Pıhtılaştırıcı (madde). ~**ate,** koyulaş(tır)mak; pıhtılaş(tır)mak.
**coal** [koul]. Maden kömürü. Kömür almak; kömür vermek. ⌜**to carry** ~**s to Newcastle**⌝, denize su taşımak: **to haul s.o. over the** ~**s,** birini haşlamak: **to heap** ~**s of fire on s.o.'s head,** kötülüğe iyilikle mukabele ederek mahcub etm.: **live** ~**s,** kor. ~**-bearing,** kömür ihtiva eden. ~**-black,** simsiyah, kapkara. ~**-cellar,** kömürlük. ~**-field,** maden kömürü havzası. ~**-gas,** hava gazı. ~**-pit,** kömür ocağı. ~**-tar,** katran.
**coalesce** [kouəˈles]. Birleşmek, kaynaşmak. ~**nce,** birlesme, kaynaşma.
**coalition** [kouəˈlişn]. Birleşme; siyasî partilerin vs. muvakkat anlaşması.
**coarse** [kôs]. Kaba; âdî, aşağılık; sert. ~**n,** kabalaş(tır)mak.
**coast** [koust]. Denizkenarı, sahil. Sahil boyunca gitmek. ~**er,** sahil gemisi. ~**line,** sahil boyu. ~**wise,** sahil boyunca.
**coat**[1] [kout] *n.* Ceket; manto; at vs. tüyü; yağlı boya tabakası. ~ **of arms,** arma: ~ **of mail,** zırh elbise: ~ **and skirt,** kostüm, tayör: **dress-**~, frak: **morning-**~, jaketatay: (**over**)~ [(**top**)~], pardesü, palto: ⌜**to cut one's** ~ **according to one's cloth**⌝, ayağını yorganına göre uzatmak: ~**-hanger,** elbise askısı. ~[2] *v.* (Bir şeyin üstünü) boya vs. tabakasile kaplamak; kaplamak. ~**ed tongue,** paslı dil. ~**ing,** boya vs. tabakası.
**coax** [kouks]. Dil dökerek ikna etm.
**cobble**[1] [ˈkobl]. ~(**stone**), arnavut kaldırım taşı. ~[2] *v.* Yamamak. ~**r,** ayakkabı tamircisi.
**cobra** [ˈkoubrə]. Çok zehirli bir yılan.
**cobweb** [ˈkobweb]. Örümcek ağı.

**blow away the** ~**s,** başını dinlendirmek.
**cock**[1] [kok] *n.* Horoz; musluk; ventil; tetik; (*sl.*) erkeğin tenasül aleti. ~**bird,** erkek kuş: **at full** ~, tam kurulu (silâh): **at half** ~, yarı kurulu: **old** ~!, (*sl.*) azizim: ~ **of the walk,** bir yerde borusu öten kimse. ~**-a-doodle-doo!,** kokoriko! ~**-a-hoop,** çok sevinen ve övünen. ~**-and-bull story,** inanılmaz hikâye. ~**-crow,** sabah karanlığında. ~**-eyed,** şaşı; eğri. ~**-fight,** horoz dövüşü. ~**-sure,** (*cont.*) kendinden fazla emin. ~**pit** [ˈkokpit]. Horoz dövüşü meydanı; dövüş yeri; uçakta pilot yeri. ~**roach** [ˈkokrouç]. Hamamböceği. ~**scomb** ˈ[kokskoum]. Horozibiği. ~**tail** [ˈboktêyl]. Kokteyl. ~**y** [ˈkoki]. Kendini beğenmiş, kurumlu; ispenç horozu gibi. ~[2] *v.* Dikmek; horozu tetiğe almak. **to** ~ **the eye,** göz ucu ile bakmak: **to** ~ **the ears,** kulaklarını dikmek: **to** ~ **one's hat,** şapkasını yan giymek.
**cocked** [kokt]. ~ **hat,** kenarları kalkık resmî şapka. **to knock s.o. into a** ~ **hat,** birinin pastırmasını çıkarmak; birini mahvetmek.
**cockney** [ˈkokni]. Londra'nın fakir halkından; bu halkın konuştuğu lehçe.
**coco·a** [ˈkoukou]. Kakao. ~**nut,** büyük Hindistan cevizi; (*sl.*) kafa.
**cocoon** [koˈkûn]. Koza.
**cod** (*sl.*) Aldatmak.
**coddle** [ˈkodl]. Nazla büyütmek. **to** ~ **oneself,** sıhhatine fazla itina etmek.
**code** [koud]. Kanun mecmuası; düstur; şifre. Şifre ile yazmak. **morse** ~, mors işareti.
**codger** [ˈkocə(r)]. Tuhaf ve garib adam.
**codicil** [ˈkoudəsil]. Vasiyetnameye ek.
**codify** [ˈkoudifây]. Bir sisteme göre tanzim etmek.
**co-education** [ˈkouedyuˈkêyşn]. Muhtelit tedrisat.
**coefficient** [ˈkoueˈfişnt]. (*mat.*) Emsal rakamı.
**coequal** [kouˈîkwəl]. Eşit; müsavi.
**coerc·e** [kouˈö(r)s]. Zorlamak; itaate mecbur etmek. ~**ion** [-ˈö(r)şn], zorlama, tazyik. ~**ive,** zorlayan, mecburî.
**coeval** [kouˈîvl]. Yaşıt, muasır.
**coffee** [ˈkofi]. Kahve. **black** ~, sütsüz kahve: **white** ~, sütlü kahve: ~ **beans,** çekirdek kahve: ~ **coloured,**

sütlü kahve renginde: ~ **grounds,** kahve telvesi: ~ **room,** (otelde) salon.
**coffer** [ˈkofə(r)]. Sandık, kasa. **~s,** hazine. **~dam,** su geçmez sandık.
**coffin** [ˈkofin]. Tabut. Tabuta koymak.
**coğ** [kog]. Çark dişi. **~-wheel,** dişli (çark).
**coğent** [ˈkoucənt]. İkna ve ilzam edici.
**coğitate** [ˈkocitêyt]. Düşünüp taşınmak.
**coğnizan·ce** [ˈkognizəns]. Malûmat; (*jur.*) salâhiyet. **to take ~ of stg.,** (*jur.*) göz önüne almak. **to be ~t of a fact,** bir şey hakkında malûmat sahibi olmak.
**cohabit** [kouˈhabit]. Karı koca gibi yaşamak. **~ation** [–ˈtêyşn], cinsî münasebet.
**cohe·re** [kouˈhiə(r)]. Yapışmak, tutmak; (mana) birbirini tutmak; insicamlı olmak. **~rent,** yapışık; birbirini tutan, insicamlı. **~ sion** [kouˈhîjn]. Yapışma, birleşme. **~sive** [–hîsiv], yapışıcı, yapışık.
**coil** [kôyl]. Kangal; roda; saç buklesi; (*elect.*) bobin. Kangal etm.: ~ **up,** kıvrılıp yatmak.
**coin** [kôyn]. Maden para, sikke. Para basmak, (yeni bir kelime vs.) uydurmak. **to ~ money,** para kırmak: **to pay s.o. back in his own ~,** birine aynı şekilde mukabele etmek. **~age,** [ˈkôynəc], para basma; bir memleketin para sistemi. **~er,** para basan; kalpazan.
**coincide** [ˌkouinˈsâyd]. Tesadüf etm., mutabık olm.; uymak. **~nce** [–ˈinsidəns], tesadüf; uygunluk. **~nt,** tesadüf; tesadüfî.
**coke** [kouk]. Kok (yapmak).
**col.** (*abb.*) = **colonel,** albay.
**cold** [kould] *n.* Soğuk; nezle. *a.* Soğuk; donuk (renk); serinlik veren (renk). **to be [feel] ~,** üşümek; (hava) soğuk olm.: **a ~ in the head,** nezle: **to catch ~,** soğuk almak: **to catch a ~,** nezle olmak, birinden nezle almak: **you will catch your death of ~,** fena halde soğuk alacaksın: **to get [grow] ~,** soğumak: **to leave s.o. out in the ~,** birini açıkta bırakmak (*fig.*): **that leaves me ~,** bu beni hiç alâkadar etmez: **to give s.o. the ~ shoulder [to cold-shoulder s.o.],** birine omuz çevirmek: **~ store [storage],** soğutma tertibatlı depo. **~-blooded,** kanı soğuk olan (hayvan); soğuk, hissiz; merhametsiz (insan); taamüden yapılmış, merhametsizce (fiil, iş). **~-hearted,** merhametsiz, duygusuz.
**collàborat·e** [koˈlabərêyt]. İş birliği yapmak. **~ion** [–ˈrêyşn], işbirliği.
**collaps·e** [kəˈlaps]. Çökme(k), göçme(k), yıkılma(k); suya düşmek; birdenbire düşme(k). **~ible** [kəˈlapsəbl]. Katlanır (iskemle).
**collar** [ˈkolə(r)]. Yaka(lık); tasma; halka. Yakalamak; yaka takmak. **~-bone,** köprücük kemiği.
**collat·e** [kəˈlêyt]. (İki metin vs.yi) dikkatle mukayese etm. **~ion** [kəˈlêyşn]. (Metinleri) karşılaştırma; hafif yemek.
**collateral** [kəˈlatərəl]. Yanyana; muvazi. **~ security,** munzam teminat.
**colleağue** [ˈkolîg]. İş arkadaşı; meslekdaş.
**collect** [kəˈlekt] *v.* Toplamak, bir araya getirmek; biriktirmek; koleksiyon yapmak; (uğrayıp) almak. **~ion** [kəˈlekşn], toplama, bir araya getirme; koleksiyon; para toplama; cibayet; yığın; toplanan para; posta kutularından mektubların toplanması. **~ive,** müşterek, hep bir olarak; çokluk ifade eden (kelime). **~or** [kəˈlektə(r)], toplayan, toplayıcı; koleksiyoncu. **tax-~,** tahsildar: **ticket ~,** biletçi.
**colleğe** [ˈkolic]. Bir üniversiteye bağlı yatılı yurd; yüksek mekteb, kolej.
**colli·de** [kəˈlâyd]. Şiddetle çarpmak, çarpışmak. **~sion** [kəˈlijn]. Çarpışma, çarpma. **to come into ~ with,** …e çarpışmak.
**collier** [ˈkoliə(r)]. Kömür gemisi; maden kömürü işçisi. **~y,** maden kömürü ocağı.
**colloqu·ial** [kəˈloukwiəl]. Konuşma diline aid. **~ialism,** konuşma dilinde kullanılan. **~y** [ˈkolokwi], konuşma, sohbet.
**collusiˈon** [kəˈlûjn]. Muvazaa, gizli anlaşma. **~ve** [–lûsiv], muvazaa nevinden.
**colon** [ˈkoulən]. İki nokta (:).
**colonel** [ˈkö(r)nl]. Albay, miralay.
**colon·ial** [kəˈlouniəl]. Müstemlekeye aid; müstemleke halkından. **~ist** [ˈkolonist], bir müstemlekede yerleşen; bir müstemlekeyi ilk kuranlardan. **~ize** [ˈkolonâyz], bir yeri müstemleke haline getirmek. **~y** [ˈkoləni]. Müstemleke.

**coloration** [ˌkoloˈreyşn]. Renklerin vaziyeti, renk.

**coloss·al** [kəˈlosl]. Çok büyük, muazzam. **~us,** dev gibi şahıs, şey.

**colour** [ˈkʌlə]. Renk; boya; canlılık. Renk vermek, boyamak; başka bir şekil vermek; kızarmak. **~s,** bayrak, bandıra; askerlik hizmeti; yarış atı sahibinin işareti olan renkler. **~ bar [line],** beyazlarla diğer ırklara mensub insanlar arasındaki ictimaî, siyasî vs. fark: **to change ~,** rengi uçmak: **to put a false ~ on things,** hadiseleri yanlış bir şekilde göstermek: **to sail under false ~s,** sahte bandıra ile çıkmak; (*fig.*) sahte hüviyet takınmak: **with ~s flying,** bayraklar dalgalanarak: **with flying ~s,** büyük muvaffakiyetle: **to get one's ~s,** bir kolej vs.nin birinci takım oyuncusu olm.: **to give [lend] ~ to a rumour,** bir rivayeti takviye etm.: **to join the ~s,** askere gönüllü yazılmak: **local ~,** mahallî renk: **to lose ~,** rengi atmak: **to lower one's ~s,** teslim bayrağı çekmek: **I should like to see the ~ of his money before ..,** ···den evvel parasının yüzünü görmek isterim: **to nail one's ~s to the mast,** ölünceye kadar çarpışmak: **to be off ~,** keyifsiz olm.: **oil ~(s),** yağlı boya: **the ~ problem,** zenci (veya sarı ırka mensub milletler) meselesi: **to stick to one's ~s,** kanaatlerine bağlı kalmak: **to show oneself in one's true ~s,** yüzünden maskeyi indirmek: **under ~ of,** ... behanesile: **water ~,** sulu boya. **~-bearer,** bayraktar. **~-blind,** bazı renkleri ayıramıyan. **~ed** [ˈkʌlö(r)d]. Renkli; zenci; tesir altında kalmış. **~ful** [ˈkʌləful]. Renkli, canlı. **~less,** renksiz; soluk.

**colt** [koult]. Tay.

**column** [ˈkoləm]. Sütun, direk; kol. **spinal ~,** belkemiği.

**coma** [ˈkoumə] (*med.*) Derin baygınlık. **~tose** [ˈkomətouz], koma halinde; pek uyuşuk.

**comb** [koum]. Tarak; tarama; horoz ibiği. Taramak. **~ out,** (karışık saçı) taramak; temizlemek.

**combat** [ˈkombat, ˈkʌmbət]. Çarpışma; mücadele. ···le çarpışmak, mücadele etm. **~ant,** muharib. **~ive,** kavgacı.

**combin·ation** [ˌkombiˈnêyşn]. Birleş(tir)me; mezcetme; birlik; tertib; terkib. *pl.*, don ve gömleği birleşik iç çaşmaırı. **~ lock,** şifreli kilid. **~e**[1] [ˈkombâyn] *n.* Ticaret ve sanayide birleşme, kartel; biçer döver makinesi. **~e**[2] [kəmˈbâyn] *v.* Birleş(tir)mek; mezcetmek; birlik olm.; tertib etm.

**combust·ible** [kəmˈbʌstibl] *a.* Yanabilir, (*fig.*) tutuşup parlamağa hazır. *n.* Yakıt. **~ion** [kəmˈbʌsçən], yanma, ihtirak, tutuşma.

**come** (**came**) [kʌm, kêym]. Gelmek, varmak; olmak; netice itibarile olmak. **let them all ~,** varsın hepsi gelsin: **~, ~!** [**~ now!**], haydi canım!, haydi bakalım!: **~ what may,** ne olursa olsun: **a week ~ Sunday,** pazar günü haftası olacak: **he will be three ~ April,** nisanda üç yaşında olacak: **what will ~ of it ?,** neticesi ne olacak ?: **don't try to ~ it over me!,** bana hükmetmeğe kalkma!: **in the time to ~,** istikbalde: **for six weeks to ~,** gelecek altı hafta içinde: **how did you ~ to do that ?,** nasıl oldu da bunu yaptınız?: **now that I ~ to think of it,** şimdi (bu meseleyi) tekrar düşününce: **it (all) ~s to this that ...,** neticesi [hülâsası] şudur ...: **'Tom is very lazy.' 'Well, if it ~s to that, so are you',** 'Tom pek tembel!'. 'Ona bakarsan sen de tembelsin!': **I have ~ to believe that,** şu kanaate vardım ki: **I came to like [hate] him,** sonunda ondan hoşlandım [nefret ettim]: **what does the total ~ to?,** yekûn ne tutuyor?: **what are things [we] coming to ?,** nereye gidiyoruz ?; bunun sonu ne olacak ?. **~ about,** olmak, vukubulmak: (gemi) volta etmek. **~ across,** geçmek; rast gelmek; tesadüfen bulmak. **~ against,** karşı gelmek; çarpmak. **~ along,** ilerlemek: **~ along!,** cabuk ol!: **he's coming along nicely with his Turkish,** (*coll.*) türkçesi epeyce ilerliyor. **~ away,** ayrılıp gelmek, bırakıp gelmek; (bir şey) yerinden çıkmak, sökülmek. **~ between,** aralarına girmek. **~ by,** önünden geçmek; elde etmek; eline geçirmek: **all his money was honestly ~ by,** bütün parasını namusile kazanmıştır. **~ down,** inmek; düşmek: **to ~ down from the University,** üniversiteyi bitirmek: **to ~ down in the world,** (maddî vaziyet bakımından) düşmek: **to ~ down handsomely,** cömert davranmak: **to ~ down (up)on s.o.,** şiddetle azarlamak. **~ in,** girmek; (moda, meyva vs.) çıkmak: **the tide is coming in,**

med yükseliyor: **and where do I ~ in?,** ya, ben ne olacağım? (*infial ifade eder*). ~ **into,** girmek; (bir şeye) varis olmak. ~ **off,** çıkmak; ayrılmak; düşmek; vaki olm.; muvaffak olm.: **to ~ off badly,** altta kalmak. ~ **on,** ilerlemek; terakki etm.: ~ **on!,** haydi bakalım!: **night is coming on,** karanlık basıyor: **if it ~s on to rain we shall get wet,** yağmur yağacak olursa ıslanırız: **your case ~s on tomorrow,** yarın sizin davanızın sırası gelecek. ~ **out,** çıkmak; (çiçek) açılmak; belirmek: **to ~ out (on strike),** grev yapmak: **(of a girl) to ~ out,** (genc kız) ilk defa toplantılara gitmek: **to ~ out in a rash, spots,** *etc.*, kızıl lekeler vs. dökmek: **to ~ out with a remark,** birdenbire söze karışarak bir şey söylemek. ~ **round,** dolaşıp gelmek; (yakın bir yerden) gelmek; etrafına toplanmak; kendine gelmek; kanmak: **the time has ~ round to ...,** ···zamanı yine geldi: **you will soon ~ round to my way of thinking,** yakında benim dediğime gelirsin. ~ **through,** geç(ir)mek; kurtulmak. ~ **to,** ayılmak. ~ **up,** yukarı gelmek, çıkmak; yetişmek: **to ~ up to the University,** üniversiteye başlamak. **to ~ up to s.o.,** birine yanaşmak: **to ~ up with s.o.,** birine yetişmek: **the play did not ~ up to my expectations,** piyes umduğum gibi çıkmadı: **to ~ up against,** ···le karşılaşmak. ~ **upon,** rasgelmek; üzerine gelmek. **~ly** [ˈkʌmli]. Yakışıklı, güzel. **~r** [ˈkʌmə(r)]. Gelen. **all ~s,** kim gelirse, her gelen.

**comed·ian** [kəˈmîdiən]. Komedi aktörü; komedi muharriri; komik kimse. **~y** [ˈkomədi]. Komedi.

**comet** [ˈkomit]. Kuyruklu yıldız.

**comfort** [ˈkʌmfət]. Teselli; ferahlık, rahat. Teselli etm. **be of good ~!,** metin olunuz!: **cold ~,** züğürt tesellisi. **~able** [ˈkʌmfətəbl], rahat; sıkıntısız; keyifli; kâfi: **to be ~ably off,** hali vakti yerinde olmak. **~er,** teselli eden kimse; emzik.

**comic** [ˈkomik]. Tuhaf. **~al,** tuhaf, güldürücü.

**coming** [ˈkʌmin(g)] *a.* Gelecek. *n.* Gelme; varış. **a ~ man,** istikbali açık adam.

**comma** [ˈkomə]. Virgül. **inverted ~s,** tırnak işareti.

**command** [kəˈmând]. Emir; idare; kumanda altında bulunan ordu vs.; hakimiyet. Emretmek; kumanda etm.; kumandanlık etm.; hâkim olmak. **to have ~ of several languages,** bir kaç lisana vakıf olm.: **to ~ respect,** hürmet telkin etm.: **second in ~,** kumandan muavini: **by royal ~,** Kıraliçenin emrile (*baz.* davetile): **word of ~,** emir, kumanda. **~ant** [ˌkomanˈdant]. Kumandan, âmir. **~eer** [ˌkomanˈdiə(r)]. El komak; zorla almak. **~er** [kəˈmândə(r)]. Kumandan, âmir; (deniz) yarbay. **~er-in-chief,** başkumandan. **~ment** [kəˈmândmənt]. On emirden biri; Allahın emri.

**commemorat·e** [kəˈmemərêyt]. Tes'id etm.; hatırasını kutlulamak. **~ion,** tes'id.

**commence** [kəˈmens]. Başlamak. **~ment,** başlangıc.

**commend** [kəˈmend]. Medhetmek; emanet etm., tevdi etmek. **this did not ~ itself to me,** ben bunu uygun bulmadım: **highly ~ed,** bir müsabaka vs.de kazanandan sonra gelene verilen sıfat. **~ation** [–ˈdêyşn], takdir etme; öğme.

**commensur·able** [kəˈmensyurəbl]. Ölçülebilir; mütenasib. **~ate,** müsavi.

**comment** [ˈkomənt]. Mülahaza; mütalaa; izahat; tenkid. Mütalaa serdetmek. ~ **on,** şerhetmek, izah etm., tenkid etmek. **~ary,** şerh, izahat: **running ~,** bir hadise devam ederken başka biri tarafından verilen izahat, *ve bundan,* bir maç vs. devam ederken radyo ile verilen izahat. **~ator** [–ˌtêytə(r)], tefsirci; radyo muhabiri.

**commerc·e** [ˈkomö(r)s]. Ticaret, alış veriş; münasebet. **~ial** [–mö(r)şl], ticarî.

**commissariat** [komiˈseəriət]. Askerî levâzim dairesi.

**commission** [kəˈmişn] *n.* Vazife, memuriyet; vazife verme; heyet; sipariş; yüzdelik; yapma, irtikab. *v.* Vazife veya memuriyet vermek; salâhiyet vermek; hizmete koymak. ~ **agent,** komisyoncu: **~ed officer,** subay, zabit: **to get one's ~,** subay tayin olunmak: **Royal Commission,** Parlamento kararile kurulan tahkikat vs. heyeti. **~aire** [kəˌmişəˈneə(r)]. Otel, sinema kapısında bekliyen uniformalı memur. **~er** [kəˈmişənə(r)]. Komisyon âzası; murahhas; müdür; komiser.

**commit** [kəˈmit]. Teslim etm., tevdi etm.; işlemek; irtikab etmek. **to ~ to**

memory, ezberlemek: **to ~ oneself,** kendini geri çekilemiyecek bir vaziyete sokmak: **without ~ting myself,** ihtiyat kaydı ile: **to ~ to paper [writing],** yazmak: **to ~ to prison,** habse mahkûm etmek. **~ment** [kə'mitmənt]. Taahhüd; vaid; bağlantı. **~tal** [kə'mitl]. Tevdi etme; habsetme; irtikab; taahhüd.

**committee** [kə'miti]. Heyet, meclis, komisyon, encümen.

**commodious** [kə'moudyəs]. Ferah, geniş.

**commodity** [kə'moditi]. Mal; ticaret eşyası; faydalı şey.

**commodore** ['komodô(r)]. Komodor; yat klübü fahrî reisi. **air ~,** Hava tuğgeneral.

**common**[1] ['komən] *n.* Umumî arazi; çayır. **~**[2] *a.* Müşterek; umumî; hep bilinen; hergünkü; alelâde; adî, kaba saba. **~ courtesy [honesty],** en ibtidaî nezaket [dürüstlük] kaidesi: **~ knowledge,** malûmu ilâm: **nothing out of the ~,** fevkelâde bir şey değil: **to be ~ talk,** herkesin ağzında olmak: **to have stg. in ~ with s.o.,** birile müşterek bir şeyi olmak; birile benzeyen bir tarafı olmak. **~-room,** muallimlere mahsus umumî oda. **~er** ['komənə(r)]. Asalet unvanı taşımıyan kimse; Avam kamarası âzası. **~place** ['komənplêys] *a.* Umumî, alelâde; kaba saba. *n.* Basmakalıb şey. **~s** ['komənz]. Halk, asîl olmıyanlar. **the (House of) Commons,** Avam Kamarası: **to be on short ~,** yiyeceği kıt olmak. **~weal** ['komənwiəl]. **the ~,** umumî menfaat. **~wealth** ['komənwelθ]. Devlet, cumhuriyet; umumî menfaat. **the Commonwealth,** İngiltere'de Cromwell zamanında (1649–60) cumhuriyet devri.

**commotion** [kə'mouşn]. Karışıklık, heyecan; hayuhuy.

**commun·al** ['komunəl]. Müşterek, bir cemaate aid. **~e**[1] ['komûn] *v.* Birile konuşmak. **to ~ with oneself,** istiğraka dalmak. **~e**[2] *n.* (Fransa'da) nahiye. **~ion** [kə'myûnyən]. Münasebet; fikir ve his iştiraki; komünyon âyini.

**communica·ble** [kə'myunikəbl]. Nakledilebilir, tebliğ edilbilir; bulaşık (hastalık). **~te** [kə'myunikêyt]. Nakletmek; tebliğ etm.; haber vermek; irtibatı olm. **~tion** [–'kêyşn], nakil; tebliğ; muvasala; temas: **~s,** münakalât. **~tive** [–'myunikətiv], havadis vermeğe hevesli; konuşkan.

**communis·m** ['komyunizm]. Komünizm. **~t,** komünist.

**community** [kə'myûniti]. Halk, cemiyet, cemaat; iştirak. **the ~,** devlet, halk.

**commute** [kə'myût]. Hafifletmek; (cezayı) tahvil etm.; (*Amer.*) mevsim abonman karnesi ile gidip gelmek.

**compact**[1] ['kompakt] *n.* Anlaşma. **~**[2] [kəm'pakt] *a.* Sıkı, kesif; veciz. **~**[3] ['kompakt]. Pudralık.

**compan·ion** [kəm'paniən]. Arkadaş; eş; refik; refakat eden kimse; (*naut.*) kaporta. **~ionable,** arkadaşlığı hoş. **~y** ['kʌmpəni]. Refakat; arkadaşlık; cemiyet, şirket; (*mil.*) bölük; trup; arkadaşlar; misafirler; tayfa; ortaklar. **he is good ~,** arkadaşlığı iyidir: **to keep s.o. ~,** birine arkadaşlık etm.: **to keep bad ~,** fena insanlarla düşüp kalkmak: **to part ~ (with s.o.),** (birinden) ayrılmak.

**compar·able** ['kompərəbl]. Mukayese edilebilir. **~ative** [kəm'parətiv], nisbî, kıyasî. **~e** [kəm'peə(r)]. Mukayese (etm.), kıyas etm.; karşılaştırmak. **beyond ~,** emsalsiz: **this ~s favourably with that,** onunla mukayesesi, bunun lehine netice verir: **John can't ~ with him,** John onunla mukayese edilemez: **nobody can ~ with him in French,** fransızcada hiç kimse ona çıkışamaz. **~ison** [kəm'parisn]. Mukayese, kıyas; karşılaştırma. **there is no ~ between them,** onlar birbirile mukayese edilemez.

**compartment** [kəm'pâtmənt]. Bölme; daire; kompartiman; göz.

**compass**[1] ['kʌmpəs] *n.* Muhit; hacim; ihata; pusula; pergel. **that's beyond my ~,** bu benim iktidarımın haricindedir: **in small ~,** küçük hacimde. **~-card,** pusula kartı. **~**[2] *v.* Bir şeyin etrafını dolaşmak; ihata etm.; kavramak.

**compassion** [kəm'paşn]. Merhamet, acıma. **~ate,** merhametli, şefkatli.

**compatible** [kəm'patəbl]. Beraber olabilir; uygun; telifi kabil.

**compatriot** [kəm'patriət]. Vatandaş, hemşeri.

**compel** [kəm'pel]. Zorlamak, mecbur etm.

**compensat·e** ['kompənsêyt]. Bedelini ödemek; tazmin etm.; telâfi etmek. **~ion** ['sêyşn], Tazmin, telâfi; karşılık; bedel.

**compete** [kəmˈpît]. Rekabet etm., müsabakaya girmek.
**competen·ce** [ˈkompitəns]. Salâhiyet; maharet; kifâyet; yeterlik. **~cy,** geçinecek kadar gelir. **~t,** salâhiyetli; muktedir; kâfi.
**competit·ion** [ˌkompəˈtişn]. Rekabet; müsabaka. **~ive** [kəmˈpetitiv], rekabete aid: **~ examination,** müsabaka imtihanı: **~ price,** rekabet edebilecek fiat. **~or** [kəmˈpetitə(r)], Müsabakaya giren kimse; rakib.
**compil·ation** [ˌkompiˈlêyşn]. Seçip toplama. **~e** [kəmˈpâyl]. Seçip toplamak; muhtelif eserlerden derlemek.
**complacen·cy** [kəmˈplêysənsi]. Kayıdsızlık; kendini beğenmişlik. **~t,** kayıdsız. **(self-) ~,** kendini beğenmiş.
**complain** [kəmˈplêyn]. Şikâyet etm.; ah ve vah etmek. **~ant,** (*jur.*) şikâyetçi; davacı. **~t,** şikâyet: **to lodge a ~ against s.o.,** biri hakkında şikâyette bulunmak.
**complement** [ˈkomplimənt] *n.* Tamamlayıcı şey; tam takım; mürettebat. *v.* [–ˈment], Tamamlamak. **~ary** [–ˈmentəri], tamamlayıcı.
**complet·e** [kəmˈplît] *a.* Tam, tamam; mükemmel. *v.* Tamamlamak, doldurmak. **~ion** [–ˈplîşn], tamamlama, yerine getirme.
**complex** [ˈkompleks] *a.* Mürekkeb; karışık. *n.* Ruhî anormallik. **inferiority ~,** aşağılık hissi. **~ity** [kəmˈpleksiti]. Karışıklık.
**complexion** [kəmˈplekşn]. Ten, cild, yüzün rengi; mahiyet. **that puts another ~ on the matter,** o zaman vaziyet değişir.
**complian·ce** [kəmˈplâyəns]. Razı olma; uysallık. **in ~ with,** ···e uygun olarak, göre: **to refuse ~ with an order,** bir emre itaat etmemek. **~t,** uysal; evetefendimci.
**complicat·e** [ˈkomplikêyt]. Karıştırmak; bir işi karmakarışık etm. **~ion** [–ˈkêyşn], karışıklık; müşkülât; (*med.*) ihtilât.
**complicity** [kəmˈplisiti]. Suç ortaklığı.
**compliment** [ˈkomplimənt]. Öğme, cemile. Tebrik etm.; medhetmek. **~s,** hürmet; tebrik. **~ary** [–ˈmentəri], cemilekâr, taltifkâr; medihli.
**comply** [kəmˈplây]. (**~ with,** ···e) razı olm., muvafakat etm.; imtisal etm.
**component** [kəmˈpounənt] *n.* Esaslı parça. *a.* Terkib eden.
**comport** [kəmˈpôt]. Uymak. **to ~ oneself,** hareket etm. **~ment,** davranış, hareket tarzı.
**compos·e** [kəmˈpouz]. Tertib etm., tanzim etm.; düzeltmek; yazmak; bestelemek. **~r,** bestekâr. **~ite** [ˈkompəzit,-zâyt]. Mürekkeb. **~ition** [ˈkompəˈzişn]. Terkib (etme); tertib (etme); tahrir; beste. **~itor** [kəmˈpositə(r)]. Mürettib.
**compound**[1] [ˈkompaund] *n.* Mahlut. *a.* Mürekkeb; mahlut. **~**[2] [kəmˈpaund] *v.* Terkib etm., taksitler toptan tediye etmek. **to ~ debts,** bir borc üzerinde alacaklı ile anlaşmak.
**comprehen·d** [ˌkompriˈhend]. Anlamak, kavramak; ihtiva etmek. **~sible** [ˌkompriˈhensibl]. Anlaşılabilir. **~sion** [–ˈhenşn], anlama; kavrayış; şümul. **~ive,** şümullü; geniş, idrake aid.
**compress**[1] [ˈkompres] *n.* (*med.*) Islak bez. **~**[2] [kəmˈpres] *v.* Sıkmak, tazyik etm.; hülâsa etmek. **~ion** [–ˈpreşn], sıkıştırma, tazyik; teksif.
**comprise** [kəmˈprâyz]. İhtiva etm.
**compromise** [ˈkomprəmâyz]. Uzlaşma; ikisi ortası. (Şerefini vs.) tehlikeye koymak; bir uzlaşmaya varmak.
**compuls·ion** [kəmˈpʌlşn]. Cebir, zorlama. **under ~,** zorla, mecbur kalarak. **~ory** [–ˈpʌlsəri], mecburî.
**compunction** [kəmˈpʌnkşn]. Vicdan azabı.
**comput·e** [kəmˈpyût]. Hesab etm.; tahmin etmek. **~ation** [–têyşn], hesab; tahmin.
**comrade** [ˈkomrid]. Arkadaş; yoldaş.
**con** [kon]. (*yal.*) **the pros and ~s of a question,** bir meselenin lehinde ve aleyhinde olan noktalar.
**concav·e** [ˈkonkêyv]. Mukaar, çukur. **~ity** [–ˈkaviti], mukaarlık, çukur(luk).
**conceal** [kənˈsîl]. Gizlemek, saklamak, örtmek. **~ment,** gizle(n)me, sakla(n)ma.
**concede** [kənˈsîd]. Teslim etm.; itiraf etm.
**conceit** [kənˈsît]. Kibir, kendini beğenme. **~ed,** kibirli, kendini beğenmiş.
**conceiv·able** [kənˈsîvəbl]. Düşünülebilir, akıl alabilir, makul; akla gelecek (her şey). **~e** [kənˈsîv]. Düşünmek; aklı almak; tasavvur etm.; gebe kalmak. **to ~ a dislike for s.o.,** birinden birdenbire nefret etm.

**concentrat·e** [ˈkonsəntrêyt]. Bir yere topla(n)mak; temerküz et(tir)mek; bir hedefe çevirmek. **~ion** [–ˈtrêyşn], bir yere toplama; temerküz; bir hedefe çevirme. **~ed,** koyu kesif; kuvvetli.
**concentric** [kənˈsentrik]. Müşterek merkezli.
**concept** [ˈkonsept]. Fikir, mefhum. **~ion** [–ˈsepşn], idrak etme; fikir; gebe kalma.
**concern** [kənˈsö(r)n] *n.* Alâka, münasebet; mesele; iş; endişe; şirket. *v.* Aid olm., dair olm., münasebeti (alâkası) olm.; raci olm.; **to be ~ed [~ oneself] with,** …e karışmak. **as ~s,** …e gelince, . . . itibarile: **as far as I am ~ed,** bana gelince; bence: **of ~,** mühim: **it's no ~ of mine,** bu beni alâkadar etmez: **to be ~ed about stg.,** bir şeyden endişe etm.: **his honour is ~ed,** şerefi mevzuubahistir: **I was very ~ed to hear,** …duyunca çok müteessir oldum. **~ing,** dair, hakkinda.
**concert**[1] [ˈkonsö(r)t] *n.* Ahenk; konser. **to act in ~ with s.o.,** birile elbirliğiyle hareket etm.: **~ pitch,** yüksek seviyede tutmak. **~**[2] [kən ˈsö(r)t] *v.* Müşavere ve müzakere edip tanzim etm.
**concession** [kənˈseşn]. Teslim; imtiyaz.
**conciliat·e** [kənˈsiliêyt]. Gönlünü almak; yatıştırmak; uzlaştırmak. **~ion** [–ˈêyşn], uzlaş(tır)ma; barış(tır)ma. **~ory** [–ˈsiliətəri], gönül alıcı.
**concise** [kənˈsâys]. Muhtasar, mücmel; veciz.
**conclave** [ˈkonklêyv]. Hususî toplantı.
**conclu·de** [kənˈklûd]. Bitirmek; akdetmek; bir neticeye varmak. **to be ~d (in our next),** sonu gelecek sayıda. **~sion** [kənˈklûjn]. Netice; akdetme; nihayet verme; netice çıkarma. **to try ~s with s.o.,** birile boy ölçmek. **~sive** [–ˈklûsiv], ikna edici; kat'î.
**concoct** [kənˈkokt]. Uydurmak.
**concord** [ˈkonkôd]. Uygunluk; ahenk. **~ance** [–ˈkôdəns], uygunluk; indeks.
**concourse** [ˈkonkôs]. Toplantı; birleşme.
**concrete**[1] [ˈkonkrît] *a.* Hakikî; müspet; müşahhas; elle tutulur; muayyen. **~**[2] *n.* Beton. *v.* Beton kaplamak, betonlamak. *a.* Betondan yapılmış.
**concubine** [ˈkonkyubâyn]. Odalık.
**concur** [kənˈkö(r)]. Uymak; razı olm.: uyuşmak; kabul etmek. **~rence** [–ˈkʌrəns], muvafakat; tesadüf. **~rent** [–ˈkʌrənt], aynı zamanda vakı olan; uygun.
**concussion** [kənˈkʌşn]. Sadme; sarsıntı; beyin sarsılması.
**condemn** [kənˈdem]. Mahkûm etm.; ayıplamak; kullanmağa uygun bulmamak. **~ed cell,** idam mahkûmu hücresi. **~ation** [–ˈnêyşn], mahkûm etme.
**condens·e** [kənˈdens]. Teksif etm., koyulaştırmak; (yazıyı) hülâsa etmek. **~ation** [–ˈsêyşn], teksif, kesif mayi. **~er,** mükessif; imbik; kondansatör.
**condescen·d** [ˌkondiˈsend]. Tenezzülde bulunmak. **~ding,** himayekâr. **~sion,** tenezzül; lûtufkârlık.
**condition** [kənˈdişn]. Hal; şart; vaziyet; medenî hal. Şarta bağlamak; iyi bir hale getirmek. **on ~ that** …, … şartı ile: **to keep (oneself) in ~,** idmanlı olm.: **to be out of ~,** hamlamak: **to be ~ed by stg.,** bir şeye bağlı olmak. **~al,** şarta bağlı; (*gram.*) şart sıygası.
**condol·e** [kənˈdoul]. **~ with,** …e taziyede bulunmak. **~ence** [–ˈdouləns], taziye.
**condone** [kənˈdoun]. Affetmek; göz yummak.
**conduc·e** [kənˈdyûs]. Mucib olm.; yardım etm. **~ive,** mucib olan.
**conduct** [ˈkondʌkt] *n.* Davranış; hareket; idare. *v.* [–ˈdʌkt] Sevketmek; idare etm.; rehberlik etm.; nakletmek; orkestrayı idare etmek. **certificate of good ~,** hüsnühal kâğıdı: **to ~ oneself,** (iyi, fena) hareket etm.: **~ed tours,** bir rehberin idaresinde gezinti. **~ing** (*elect.*), nâkil. **~ivity** [–ˈtiviti], (sıcak vs.yi) nakil kabiliyeti. **~or** [kənˈdʌktə(r)], sevk eden, idare eden; orkestra şefi; biletçi; nakleden, nâkil.
**con·e** [koun]. Mahrut; kozalak; koni. **~ical,** mahrutî.
**confection** [kənˈfekşn]. Tertib etme; tertib edilmiş şey; şekerleme. **~er,** Şekerlemeci, pastacı. **~ery,** pasta, şekerleme vs.
**confedera·cy** [kənˈfedərəsi]. Devletler birliği; ittifak. **~te**[1] [–ˈfedərit] *a.* & *n.* birleşmiş, müttefik; suç ortağı. **~te**[2] [–ˈfedərêyt] *v.* (devletler) birleş(tir)mek; birlik olmak. **~tion** [–ˈrêyşn], devletler birliği; ittifak.

**confer** [kənˈfö(r)]. Müzakere etm., görüşmek; (unvan vs.) tevcih etm. **~ence** [ˈkonfərəns], kongre; toplantı; müzakere. **~ment** [–ˈfö(r)mənt], tevcih, verme.
**confess** [kənˈfes]. İtiraf etm.; günah çıkar(t)mak. **~ion** [–ˈfeşn], itiraf; ikrar; günah çıkarma: **~ of faith,** imanını ikrar. **~or,** günah çıkaran papaz; itiraf eden kimse.
**confid·ant** [ˈkonfidant]. Sırdaş; yakın dost. **~e** [kənˈfâyd]. (Sır vs.) tevdi etmek. **~ in,** ···e emniyet etm., itimad etmek. **~ence** [ˈkonfidəns]. İtimad, emniyet; mahremlik. **to make a ~ to s.o.,** birine bir sır söylemek: **~ man,** dolandırıcı, tavcı: **~-trick,** kandırarak dolandırma, tavlama. **~ent** [ˈkonfidənt]. Emin; kendine pek güvenir. **~ential** [–denşiəl], mahrem; gizli; itimad edilebilir.
**confine** [kənˈfâyn] *v.* Tahdid etm.; kuşatmak, habsetmek. **to be ~d,** (i) tahdid edilmek; habsedilmek; (ii) çocuk doğurmak: **to ~ oneself to,** ···le iktifa etm. **~ment,** habsedilme, mahbusluk; loğusalık. **~s** [ˈkonfâyns]. Hudud.
**confirm** [kənˈfö(r)m]. Teyid etm., tasdik etm.; takviye etm. **~ation,** teyid, tasdik; hıristiyan çocuğunun büluğ zamanı kilise camiasına kabulü âyini. **~ative,** tasdikî. **~ed** *a.* kök salmış; ıslah olmaz.
**confiscat·e** [ˈkonfiskêyt]. Musadere etmek. **~ion** [–ˈkəyşn], musadere.
**conflict** [ˈkonflikt] *n.* İhtilâf, mücadele. *v.* [kənˈflikt], Birbirini tutmamak. **~ with,** ···e zıd olm., muhalif olmak. **~ing,** zıd; birbirini tutmaz.
**confluence** [ˈkonfluəns]. Beraber akma.
**conform** [kənˈfôm]. Tatbik etm.; intibak etm. **~able** [–ˈfôməbl], uygun; benzer. **~ation** [–ˈmêyşn], şekil, bünye. **~ity,** uygunluk: **in ~ with,** mucibince.
**confound** [kənˈfaund]. Bozmak; şaşırtmak; karıştırmek. **~ it!,** lânet olsun!. **~ed,** lânetleme: **it was ~ly cold,** Allahın belâsı bir soğuk vardı.
**confront** [kənˈfrʌnt]. Karşılaş(tır)mak, yüzleştirmek. **~ation,** muvacehe, yüzleştirme.
**confus·e** [kənˈfyûz]. Karıştırmak, şaşırtmak. **to get ~d,** şaşırmak. **~d,** *a.* Mahcub; şaşırmış; karışık. **~ion** [kənˈfyûjn]. Karışıklık, intizamsızlık; karıştırma; şaşkınlık, utanma. **in ~,** karmakarışık: **to be covered with ~,** fena halde mahcub olm.: **to put s.o. to ~,** birini mahcub etm.: **to fall into ~,** karmakarışık olm.: **~ worse confounded,** karışıklığın danıskası.
**congeal** [kənˈcîl]. Don(dur)mak; katılaş(tır)mak; pıhtılaş(tır)mak.
**congenial** [kənˈcîniəl]. Cana yakın; uygun; hoş.
**congenital** [kənˈcenitl]. Doğuşta olan.
**congest** [kəncest]. Topla(n)mak; kan toplamasına sebeb olmak. **~ed area,** fazla nüfuzlu [pek kalabalık] bölge. **~ion** [–ˈcesçən], izdiham, kalabalık; kan birikmesi, ihtikan.
**congratulat·e** [kənˈgratyulêyt]. Tebrik etmek. **~ion** [–ˈlêyşn], tebrik.
**congregat·e** [ˈkongrigêyt]. Toplanmak, birleşmek. **~ion** [–gêyşn], cemaat; toplanma.
**congress** [ˈkongres]. Birleşme; kongre; (Amerika Birleşik Devletlerinde) Senato ile Temsilciler meclisinden mürekkeb Milli Meclis. **~ional** [–ˈgreşənəl], Kongre'ye aid. **~man,** (Amerika'da) Millet Meclisi âzası.
**congruen·ce** [ˈkongruəns]. Uygunluk. **~t** [**congruous**], uygun; münasib.
**conifer** [ˈkounifö(r)]. Kozalaklı ağac.
**conjecture** [kənˈcektyuə(r)]. Zan, tahmin (etmek).
**conjugal** [ˈkoncugl]. Evliliğe aid.
**conjugat·e** [ˈkoncugêyt] (*gram.*) Tasrif etmek. **~ion,** tasrif.
**conjunc·tion** [kənˈcʌn(g)kşn]. Birleşme; atıf edatı; iktiran. **in ~ with,** ···le beraber. **~ture** [kəncʌn(g)kçə(r)]. Durum. **at this ~,** ahval böyle iken.
**conjur·e** [ˈkʌncə(r)]. Hokkabazlık yapmak. **~ up,** sihirbazlıkla davet etm.; hatıra getirmek: **a name to ~ with,** sihirli isim. **~er** [**~or**], hokkabaz.
**conk** [konk]. **~ out,** (*sl.*) makine birdenbire durmak.
**connect** [kəˈnekt]. Bağlamak; birleştirmek; münasebet tesis etm.; (tren vs.) aktarması olmak. **to be well ~ed,** iyi aileye mensub olmak. **~ing** [**-tive**] [kəˈnektin(g), -tiv]. Bağlıyan, rabtedici. **~ion** [**-nexion**] [kəˈnekşn]. Bağ; irtibat; münasebet; alâka; yakınlık; sıhriyet, sıhrî ak-

raba; aktarma. **in ~ with,** münasebetile: **in this ~,** bu hususta.
**conning-tower** [ˈkonin(g)ˈtaṷə(r)]. Kumanda kulesi.
**conniv·e** [kəˈnâyv]. Göz yummak. **~ance,** göz yumma, müsamaha.
**connoisseur** [ˌkoniˈsö(r)]. Ehil, mütehassıs.
**conque·r** [ˈkon(g)kə(r)]. Fethetmek, zabtetmek. **~ror,** fetheden, fatih. **the Conqueror,** 1066 'da İngiltere'ye çıkan Norman kıralı William. **~st** [ˈkonkwest]. Fethetme.
**conscien·ce** [ˈkonşəns]. Vicdan. **in all ~,** doğrusu: **with a clear ~,** vicdanı müsterih olarak: **~ money,** vicdan azabı yüzünden yerine iade eden para: **~ stricken,** vicdan azabına kapılmış: **I would not have the ~ to do it,** bunu yapmağa vicdanım razı olmaz. **~tious** [ˌkonşiˈenşəs]. Vicdanlı, dürüst. **~ objector,** vicdanına karşı olduğunu söyliyerek askerlik yapmak istemiyen: **~ scruple,** vicdan endişesi.
**conscious** [ˈkonşəs]. Kendine malik; ayılmış; şuurlu; kasdî. **to be ~ of stg.,** bir şeyin farkında olm.: **to become ~ of stg.,** bir şeyin farkına varmak. **~ness,** kendine malik olma; şuur; his: **to lose ~,** kendini kaybetmek: **to regain ~,** kendine gelmek.
**conscript** [ˈkonskript] *n.* Askere alınmış adam, kur'a askeri. *v.* [kənˈskript], Askere almak. **~ion** [–ˈskripşn], mecburî askerlik; askere alma.
**consecrate** [ˈkonsikrêyt]. Takdis etm.
**consecutive** [kənˈsekyutiv]. Müteakib; üstüste olan.
**consensus** [kənˈsensəs]. Umumî muvafakat.
**consent** [kənˈsent]. Razı olma(k); muvafakat etme(k); müsaade. **by common ~,** herkesin muvafakatile.
**consequence** [ˈkonsikwəns]. Netice; akibet; ehemmiyet. **it is of no ~,** ehemmiyeti yok: **in ~,** bu sebeble: **in ~ of,** …in neticesinde: **do this or take the ~s!,** ya bu işi yap yahud akibetine [mesuliyetine] katlan.
**consequent** [ˈkonsikwənt]. Tabi, bağlı; neticesi olan. **~ly,** neticesi olarak. **~ial** [–ˈkwenşiəl], netice olarak husule gelen; azametli.
**conserv·ancy** [kənˈsö(r)vənsi]. Muhafaza, koruma. **~ation** [ˌkonsö(r)ˈvêyşn]. Muhafaza, himaye. **~ative** [kənˈsö(r)vətiv]. Muhafazakâr; mutedil: **on ~ lines,** eski usulde. **~ator** [kənˈsö(r)vətə(r)]. Muhafız. **~atory** [kənˈsö(r)vətri]. Limonluk. **~e** [kənˈsö(r)v] *v.* Muhafaza etm., korumak, himaye etm.; (meyva vs.) konserve yapmak. *n.* Konserve; reçel.
**consider** [kənˈsidə(r)]. Düşünüp taşınmak; nazarı itibara almak; düşünmek; riayet etm.; addetmek, saymak. **all things ~ed,** her şeyi göz önünde tutarak. **~able** [kənˈsidərəbl]. Mühim, büyük. **~ate** [kənˈsidərit]. Başkalarına karşı saygılı. **~ation** [kənˈsidəˈrêyşn]. Mütalaa; saygı; düşünce; itibar; dikkat; ehemmiyet; göz önünde tutulacak şey; bedel, karşılık. **for a ~,** ivaz mukabilinde: **in ~ of,** düşünerek, göz önünde tutarak; dolayısile; karşılık olarak: **money is no ~,** para mevzuu bahis değil; **on no ~,** hiç bir sebeble: **the question under ~,** tedkikedilmekte olan mesele: **to take into ~,** gözönünde tutmak. **~ing** [kənˈsidərin(g)]. Göre, dolayı, bakınca; göz önünde tutulursa.
**consign** [kənˈsâyn]. Göndermek; sevketmek; teslim etm. **~ee,** [–ˈnî], (kendisine) gönderilen. **~ment** [–ˈsâynmənt], gönderme, sevk, teslim; gönderilen bir mikdar eşya: **for ~ abroad,** harice sevkedilecek: **on ~,** satıldığı zaman tediye edilmek üzere teslim. **~or** [–ˈnô(r)], gönderen; teslim eden.
**consist** [kənˈsist]. Mürekkeb olm., ibaret olmak. **~ence [-ency]** [kənˈsistəns, -si]. Koyuluk; birbirini tutma. **~ent** [kənˈsistənt]. Aynı prensiplere, hareket tarzına vs. uyan (şey, kimse), uygun; birbirini tutan.
**consol·e** [kənˈsoṷl]. Teselli etmek. **~ation** [–ˈlêyşn], teselli.
**consolidate** [kənˈsolidêyt]. Takviye etm., pekiştirmek; birleştirmek.
**consonant** [ˈkonsənənt]. *n.* Sessiz harf. *a.* Uyar, uygun; ahenkli.
**consort**[1] [ˈkonsôt]. Zevc(e), eş. **to act in ~ with,** …le birlik (olarak) hareket etm.: **Prince Consort,** bir kadın hükümdarının kocası. **~**[2] [kənˈsôt] *v.* **~ with,** …le düşüp kalkmak.
**conspicuous** [kənˈspikyuəs]. Göze çarpan.
**conspir·acy** [kənˈspirəsi]. Fena maksadla gizli ittifak; suikasd. **~ator,**

suikasdcı. ~e [kənˈspâyə(r)]. (Biri aleyhine) fesad kurmak; suikasd hazırlamak; (hadiseler) yardım etmek.
**constab·le** [ˈkʌnstəbl]. Polis memuru. **Chief ~**, polis komiseri: **special ~**, yardımcı polis. **~ulary** [kənˈstabyuləri], zabıta polis.
**constan·cy** [ˈkonstənsi]. Sebat, devamlılık; sadakat. **~t**, değişmez, daimî; devamlı; sadık, vefalı.
**constellation** [ˌkonstəˈlêyşn]. Takımyıldız.
**consternation** [ˌkonstəˈnêyşn]. Donup kalma; haşyet.
**constipat·e** [ˈkonstipêyt]. Kabız yapmak, peklik vermek. **~ion** [–ˈpêyşn], kabız.
**constitu·ency** [kənˈstituənsi]. İntihab dairesi. **~ent** [kənˈstituənt] *a.* Teşkil eden. *n.* Bir intihab dairesindeki müntehiblerden biri. **~te** [ˈkonstityût]. Teşkil etm., terkib etm.; tayin etmek. **~tion** [–ˈtyûşn], terkib, teşkil; tesis etme; bünye; teşekkül; meşrutiyet; kanunu esasî, ana yasa: **he has an iron ~**, bünyesi demir gibidir. **~tional**, bünyevî; meşrutî.
**constrain** [kənˈstrêyn]. Zorlamak, icbar etm., tazyik etm.; zorla tutmak, alıkoymak. **~ed** *a.* sıkıntılı. **~t**, zorlama, icbar, cebir, tazyik; sıkıntı; kendini tutma: **to put s.o. under ~**, birini nezaret altında bulundurmak: **to speak without ~**, serbestçe konuşmak.
**constrict** [kənˈstrikt]. Sık(ıştır)mak; daraltmak. **~ion**, sık(ıştır)ma, daraltma; tazyik; darlık. **~or**, sıkıştırıcı; çekici.
**construct** [kənˈstrʌkt]. İnşa etm.; yapmak, kurmak. **~ion**, inşa etme; yapı, inşaat; cümle yapısı; mana (verme): **to put a wrong ~ on stg.**, bir şeye yanlış mana vermek. **~ional**, inşaata aid. **~ive**, yapıcı; yaratıcı; faydalı. **~or**, inşaatçı.
**construe** [ˈkonstrû]. İzah etm.; mana vermek; tercüme etmek.
**consul** [ˈkonsl]. Konsolos; (*tarih*) konsül. **~ar** [–syulə(r)], konsolosluğa aid. **~ate** [–syulit], konsoloshane; konsüllük.
**consult** [kənˈsʌlt]. Danışmak: sormak; bakmak. **to ~ one's own interest**, kendi menfaatini düşünmek: **to ~ s.o.'s feelings**, birinin hislerine hürmet etm.: **to ~ together**, birbirine danışmak; (bir meseleyi) beraber konuşmak. **~ant**, müşavir doktor; mütehassıs. **~ation** [–ˈtêyşn], danışma; sorma; (lûgata) bakma; konsültasyon. **~ative** [–ˈsʌltətiv], istişarî. **~ing**; **~ hours**, muayene saatleri: **~ room**, muayenehane.
**consum·e** [kənˈsyûm]. Yakıp kül etm.; yiyip bitirmek, sarfetmek, istihlâk etm. **to be ~ed with boredom**, can sıkıntısından patlamak: **to be ~ed with jealousy**, kıskançlıktan içi içini yemek: **to be ~ed with thirst**, susuzluktan yanmak. **~er**, müstehlik; sarfeden; (havagazi vs.) kullanan: **~ goods**, istihlâk eşyası. **~ption** [kənˈsʌmpşn]. İstihlâk, sarfetme; verem. **~ptive**, veremli.
**consummate**[1] [kənˈsʌmit] *a.* Tam, mükemmel; usta. **~**[2] [ˈkonsəmêyt] *v.* Tamamlamak. **to ~ the marriage**, karı kocalık işini tamamlamak.
**contact** [ˈkontakt] *n.* Temas; münasebet; (*elect.*) kontak. *v.* [kənˈtakt] Temas etm. **~-breaker**, otomatik şalter.
**contagi·on** [kənˈtêycən]. Sirayet, bulaşma; bulaşık hastalık. **~ous** sarî, başkalarına geçer.
**contain** [kənˈtêyn]. İhtiva etm.; tutmak. **to ~ oneself**, kendini tutmak: **to be unable to ~ oneself**, içi içine sığmamak. **~er**, (kutu, şişe vs.) kab.
**contaminat·e** [kənˈtaminêyt]. (Temasla) bulaştırmak, kirletmek, ifsad etmek. **~ion** [–ˈnêyşn], bulaştırma, telvis, ifsad.
**contemplat·e** [ˈkontəmplêyt]. Seyretmek; uzun uzadıya düşünmek; mülahaza etm.; ümid etm.; niyet etm. **~ion** [–ˈplêyşn], seyretme; uzun uzadıya düşünme; mülahaza; istiğrak; niyet. **~ive** [ˈkontəmplêytiv], mülahazaya aid.
**contemporary** [kənˈtempəri]. Muasır, cağdaş.
**contempt** [kənˈtempt]. İstihfaf; küçük görme. **~ of court**, adliye nizamlarına riayetsizlik: **beneath ~**, son derece aşağılık: **to bring into ~**, küçük düşürmek: **to hold in ~**, hakir görmek. **~ible**, aşağılık, zelil, rezil. **~uous** [–tyuəs], istihfafkâr.
**conten·d** [kənˈtend]. Çarpışmak; müsabakaya girmek; yarışmak; çekişmek; iddia etm. **~tion** [kənˈtenşn]. Münakaşa; iddia. **to be a bone of ~**, münazaa mevzuu olm. **~tious**, kavgacı, münakaşalı.

**content¹** [ˈkontent] *n.* Muhteva, içindeki. **~s,** muhteva, içindekiler. **~²** [kənˈtent] *n.* Hoşnudluk; tatmin edilme. *a.* Hoşnud; memnun. *v.* Hoşnud etm., tatmin etm., memnun etmek. **to be ~ with,** ···le iktifa etm.: **to your heart's ~,** canınızın istediği kadar. **~ed** [kənˈtentid]. Halinden memnun. **~ment** [kənˈtentmənt]. Halinden memnun olma.

**contest** [ˈkontest] *n.* Müsabaka; çarpışma; mücadele. *v.* [kənˈtest] Muhalefet etm.; (bir şey için) mücadele etm.; müsabakaya girmek. **to ~ a seat,** bir intihab dairesi için namzedliğini koymak. **~ant** [–ˈtestənt], Mübariz; müsabakaya giren.

**context** [ˈkontekst]. Siyak; siyaku sibak.

**contiguous** [konˈtigyu̯əs]. Bitişik, yakın.

**continen·ce** [ˈkontinəns]. İmsâk. **~t¹,** imsâkli; perhizkâr. **~t²** [ˈkontinənt], *n.* kıt'a. **the C~,** Avrupa kıt'ası (İngiltere haric). **~tal,** [–ˈnentl], kıt'aya aid; Avrupa kıt'asına aid.

**contingen·cy** [kənˈtincənsi]. İhtimal, tesadüf; beklenmedik hal. **to provide for ~cies,** her ihtimali düşünmek. **~t** [kənˈtincənt]. *a.* Arizî, tesadüfî; şarta bağlı. *n.* Bir mikdar asker vs.; **to be ~ upon stg.,** bir hadiseye tâbi olmak.

**continu·al** [kənˈtinyu̯əl]. Devamlı; sık sık olan. **~ance,** devam (müddeti). **~ation** [–ˈêyşn], devam (etme): **~ course,** mektebi bitirenler için kur. **~e** [kənˈtinyû], devam et(tir)mek; imtidad etm.: (gazete) '**to be ~d**', 'devamı var'. **~ity** [–ˈnyûiti], devamlılık, süreklilik. **~ous** [–ˈtinyu̯əs], devamlı, fasılasız.

**contort** [kənˈtôt]. Burmak, kıvırmak. **~ion** [–ˈtôşn] burma, kıvrık. **~ionist** [–ˈtôşənist], vücudünü türlü şekillere sokan cambaz.

**contour** [ˈkontu̯ə(r)]. Dış hatları. **~ (-line),** (harita) tesviye hududu. **~-map,** tesviye haritası.

**contra** [ˈkontra, –ə] *pref.* Karşı, aksi, zıd *manalarına gelen ön ek.*

**contraband** [ˈkontrəband]. Kaçak (çılık).

**contract¹** [ˈkontrakt] *n.* Mukavele; taahhüd. **the ~ for the bridge was placed with ...,** köprünün inşası filan şirkete ihale edildi: **to put work out to ~,** bir işi müteahhide vermek. **~²** [kənˈtrakt] *v.* Büzmek; daraltmak, kısaltmak; (hastalık vs.ye) tutulmak; taahhüde girmek. **~ing parties,** âkid taraflar. **~ible,** çekilebilir. **~ion** [–ˈtrakşn], çekilme; takallüs; kısalma. **~or,** müteahhid. **~ual** [–ˈtraktyu̯əl], mukaveleye müteallik.

**contradict** [ˌkontrəˈdikt]. Aksini söylemek; yalanlamak. **~ion** [–ˈdikşn], tekzib: **~ in terms,** mütenakız tabir: **in ~ to,** aksine olarak. **~ory** [–ˈdiktəri], birbirini nakzeden, mütenakız.

**contraption** [kənˈtrapşn]. (*coll.*) Garib bir alet; tertibat.

**contrar·iness** [kənˈtre̯ərinis]. Aksilik, inadcılık. **~iwise** [ˈkontrəriwâyz, kənˈtre̯əriwâyz], aksine olarak; diğer taraftan; inad için. **~y** [ˈkontrəri]. Muhalif; aksi; uygun olmıyan; [kənˈtre̯əri], aksi (şahıs). **on the ~,** bilâkis, tersine: **I have nothing to say to the ~,** buna karşı hiç bir diyeceğim yoktur.

**contrast** [ˈkontrâst] *n.* Tezad. *v.* [kənˈtrâst], Tezad teşkil etm.; karşılaştırmak.

**contraven·e** [ˌkontrəˈvîn]. Karşı gelmek; muhalefet etm.; (kanun vs.ye) tecavuz etm. **~tion** [–venşn], hilâfına hareket.

**contribut·e** [kənˈtribyût]. Yardım etm., iane vermek; (işe vs.) iştirak etm.; (gazete vs.ye) yazı vermek. **~ion** [–ˈbyûşn], yardım, iane, iştirak; (gazete vs.ye) verilen yazı; vergi. **~or,** yardım eden, iştirak eden, yazı veren. **~ory,** yardım edici: **~ negligence,** kazaya uğrıyanın kısmî mesuliyeti.

**contriv·ance** [kənˈtrâyvəns]. Hususî tertibat. **~e** [kənˈtrâyv]. İcad etm.; bir usul kurmak; bir çaresini bulmak.

**control** [kənˈtro̯ul]. Mürakebe; iktidar, nüfuz, idare. İdare etm.; tanzim etm.; hâkim olm; zabtetmek; bastırmak. **circumstances beyond our ~,** elimizde olmıyan sebebler: **~ yourself!,** kendini kaybetme!: **to lose ~ of oneself,** kendini kaybetmek: **to lose ~ (of a business,** *etc.*), ipin ucunu kaçırmak: **to be out of ~,** idare edilmez bir hale gelmek. **~lable** [kənˈtro̯uləbl]. İdare edilebilir. **~ler,** mürakib; idare eden.

**controver·sial** [ˌkontroˈvö(r)şi̯əl]. İhtilaflı; münakaşacı. **~sialist,** münakaşacı. **~sy** [ˈkontrəvö(r)si], ihtilaf, münakaşa: **to carry on a ~ with i.o.,** birile münakaşaya girişmek.

beyond ~, su götürmez. ~t [ˈkontrovö(r)t]. Tekzib etm.
**convalesce** [ˌkonvəˈles]. Nekahatte olmak. **~nce,** nekahat. **~nt,** nekahatte olan.
**convene** [kənˈvîn]. Toplantıya davet etmek.
**convenien·ce** [kənˈvînyəns]. Müsaidlik; rahatlık; kolaylık. **at your ~,** müsaid zamanınızda: **at your earliest ~,** mümkünse bir an evvel: **public ~,** helâ: **all modern ~s,** bütün modern konfor. **~t** [kənˈvînyənt]. Elverişli; münasib. **to make it ~ to do stg.,** bir şeyi yapmağı kolaylaştırmak.
**convent** [ˈkonvənt]. (Rahibelerin) manastır.
**convention** [kənˈvenşn]. Toplantı; muahede; âdet, mevzua. **social ~s,** ictimaî mevzuat. **~al** [–ˈvenşənl], âdete uygun; basmakalıb.
**converge** [kənˈvö(r)c]. Aynı noktaya yaklaşmak.
**convers·ation** [ˈkonvö(r)ˈsêyşn]. Konuşma, sohbet. **to hold a ~ with s.o.,** birisile konuşmak. **~ational,** konuşmaya aid. **~ationalist** [–ˈsêyşnəlist], hoşsohbet kimse. **~e**[1] [kənˈvö(r)s] *v.* Konuşmak. **~**[2] [ˈkonvö(r)s] *n.* Zıd, aksi. **~ely,** diğer taraftan, buna karşılık olarak.
**conver·sion** [kənˈvö(r)şn]. Tahvil etme, değiştirme; dinini vs. değiştirme. **~t**[1] [ˈkonvö(r)t] *n.* Dinini vs. değiştirmiş, dönme. **~t**[2] [kənˈvö(r)t]. *v.* Tahvil etm.; dinini vs. değiştirmek. **to ~ funds to one's own use,** para ihtilas etmek. **~ter,** muhavvile, transformatör. **~tible,** tahvil edilebilir.
**convex** [ˈkonveks]. Muhaddeb; kabarık.
**convey** [kənˈvêy]. Taşımak, nakletmek; ifade etm.; ferağ etmek. **~ance** [–ˈvêyəns], nakil, sevk; taşıma (vasıtası); ifade; ferağ (senedi). **~or,** taşıyan.
**convict**[1] [ˈkonvikt] *n.* Kürek mahkûmu. **~**[2] [kənˈvikt] *v.* Suçunu isbat etm., suçlandırmak.
**conviction** [kənˈvikşn]. Mahkûm etme [olma]; kanaat. **to carry ~,** ikna etmek.
**convince** [kənˈvins]. İkna etm.
**convivial** [kənˈviviəl]. Cümbüşlere düşkün; şen.
**convo·cation** [ˌkonvəˈkêyşn]. İctima, meclis. **~ke** [kənˈvouk]. (Bir meclisi) toplamak.
**convoy** [ˈkonvôy] *n.* Muhafaza altında giden kafile. *v.* [kənˈvôy], (Korumak için) refakat etmek.
**convuls·e** [kənˈvʌls]. Şiddetle sarsmak. **~ion**·[–ˈvʌlşn], ihtilac; ihtilâl. **~ive,** ihtilaclı.
**coo** [kû]. Güvercin ve kumru ötüşü. **billing and ~ing,** öpüşüp koklaşma.
**cook** [kuk]. Aşçı. Piş(ir)mek. **to ~ accounts,** hesabı tahrif etm.: **to ~ s.o.'s goose,** birinin yuvasını yapmak: **he is ~ed,** (koşucu) kesildi. **~er,** yemek pişirilen ocak, kab. **~ery,** aşçılık: **~-book,** yemek kitabı.
**cool** [kûl]. Serin(lik), soğuk; soğukkanlı. Soğu(t)mak, serinle(t)mek. **to ~ down,** serinlemek, dinlenmek; yatış(tır)mak: **to ~ off,** (heyecan vs.) sönmek.
**coolie** [ˈkûli]. (Hindistanda). hamal.
**coop** [kûp]. **to ~ up,** dar bir yere kapamak.
**cooper** [ˈkûpə(r)]. Fıçıcı.
**co-operat·e** [kouˈopərêyt]. İşbirliği yapmak; birlikte çalışmak. **~ion** [–ˈrêyşn], işbirliği. **~ive** [–ˈopərətiv], kooperatif; yardım etmeğe razı.
**co-ordinate** [kouˈôdənêyt] *v.* Tanzim etm., tertib etmek. *n.* [–nit], (*gram.*) Aynı mertebede olan (cümle vs.); (*math.*) vaz'îkemiyet. *a.* İnsicamlı, muntazam.
**cop** [kop]. (*sl.*) Aynasız (polis). **if you are late you'll ~ it,** geç kalırsan görürsün.
**copartner** [kouˈpâtnə(r)]. Kâra ortak.
**cope** [koup]. **~ with,** başa çıkmak.
**copious** [ˈkoupyəs]. Bol, mebzul.
**copper** [ˈkopə(r)]. Bakır; kazan; bakır para, peni; (*sl.*) polis. **~plate,** bakır levha: **~ writing,** basma gibi muntazam el yazısı. **~smith,** bakırcı.
**copse** [kops]. Küçük koru.
**copy** [kopi]. Kopya, nüsha; taklid; suret; örnek. Kopya etm.; taklid etmek. **fair ~,** temiz(e çekilmiş): **rough ~,** müsvedde. **~-book,** meşk defteri. **~right** [ˈkopirâyt]. Telif hakkı(nı almak).
**coquet** [koˈket]. Naz etm. **~ry** [ˈkokətri], cilve. **~te,** nazlı. **~tish,** nazlı işveli.
**coral** [ˈkorəl]. Mercan.
**cord** [kôd]. İp, bağ, şerid.
**cordial** [ˈkôdiəl]. Samimî; kalbe kuvvet veren. **~ity** [–ˈaliti], samimiyet.

**corduroy** [ˈkôdyurôy]. Pamuklu kadife. **~s**, pamuklu kadife pantalon.
**core** [kô(r)]. İç, göbek, esas; (dökme) maça. **to ~**, göbeği çıkarmak.
**co-respondent** [kou·resˈpondənt]. (Bir boşanma davasında) zinada ortak olan.
**cork** [kôk]. Mantar (tıpa). Mantarla tıkamak. **~screw**, mantar burgusu, tirbuşon.
**corn**[1] [kôn]. Ekin, zahire; tane; (İngiltere'de) buğday; (Amerika'da) mısır. **~-cob**, mısır koçanı. **~ crops**, hububat: **Indian ~**, mısır. **~ed beef**, konserve sığır eti. **~**[2] *n.* Nasır. **to tread on s.o.'s ~s**, birinin bamteline basmak.
**corner**[1] [ˈkônə(r)] *n.*. Köşe (başı), dönemec. **to drive s.o. into a ~**, birini sıkıştırmak: **to put a child in the ~**, bir çocuğu cezaya dikmek: **round the ~**, köşeyi dönünce: **to rub the ~s off s.o.**, birini bir az yontmak: **to turn the ~**, tehlike vs.yi geçiştirmek: **to take a ~**, viraj yapmak: **to make a ~ in wheat**, buğday ihtikârı yapmak. **~wise**, köşeyi başa getirerek. **~-stone**, köşe [temel] taşı. **~**[2] *v.* Sıkıştırmak; çevirmek; viraj yapmak; ... ihtikârı yapmak.
**corollary** [kəˈroləri]. Netice.
**coronation** [ˌkorəˈnêyşn]. Tac giyme.
**coroner** [ˈkorənə(r)]. (İngiliz hukuku) şüpheli ölüm vak'alarını tahkik eden mahallî idare memuru.
**corporal**[1] [ˈkôpərəl] *a.* Vücude aid, bedenî. **~**[2] *n.* Onbaşı.
**corporate** [ˈkôpərêyt]. Birleşmiş. **body ~ [~ body]**, hükmî şahıs; cemiyet.
**corporation** [ˈkôpəˈrêyşn]. Teşekkül, cemiyet; lonca; belediye meclisi; (*sl.*) göbek. **to develop a ~**, göbek bağlamak.
**corps** [kô(r)]. Hey'et; kolordu.
**corpse** [kôps]. Cesed.
**corpulent** [ˈkôpyulənt]. Şişman.
**correct** [kəˈrekt] *v.* Düzeltmek, tashih etm. *a.* Doğru, dürüst, münasib. **to stand ~ed**, hatasını kabul etm.: **it's the ~ thing**, usul budur. **~ion** [–ˈrekşn], düzeltme, tashih; cezalandırma: **I speak under [subject to] ~**, belki yanılıyorum: **house of ~**, ıslahhane.
**correlate** [ˈkorəlêyt]. İki şey arasında münasebet tesis etm.; birbirine taallûku olmak.
**correspond** [ˌkorisˈpond]. Tekabül etm.; muhabere etmek. **~ence**, mektublaşma; tekabül: **~ school**, muhabere usulü ile tedrisat yapan mekteb. **~ent**, muhabir. **~ing**, uygun, karşılık olan.
**corridor** [ˈkoridô(r)]. Koridor; geçid.
**corroborat·e** [kəˈrobərêyt]. Teyid etm., tasdik etmek. **~ion** [–ˈrêyşn], teyid, tasdik.
**corrode** [kəˈroud]. Aşındırmak; (pas) yemek.
**corros·ion** [kəˈroujn]. Aşın(dır)ma, paslan(dır)ma. **~ive**, [–ˈrousiv], aşındırıcı, paslandırıcı: **non- ~**, paslanmaz.
**corrugat·e** [ˈkorugêyt]. Dalgalı bir hale getirmek. **~ed iron**, oluklu sac.
**corrupt** [kəˈrʌpt] *a.* Bozulmuş, çürümüş. *v.* Bozmak, çürütmek; ayartmak; (birini) rüşvetle elde etmek. **~ion** [–ˈrʌpşn], bozulma; tefessüh; irtikab.
**corsair** [ˈkôseə(r)]. Korsan (gemisi).
**cosmetic** [ˈkozmetik]. Makyaj levazımı.
**cosmo·politan** [ˌkozmoˈpolitən]. Bütün dünyaya şamil. **~s** [ˈkozmos]. Kâinat.
**Cossack** [ˈkosak]. Kazak.
**cost** [kost]. Fiat; ücret; masraf; zarar. Mal olm., fiatı ... olm.; fiat biçmek. **~ of living**, hayat pahalılığı: **at all ~s [at any ~]**, her ne pahasına olursa olsun: **at little [great] ~**, az [çok] masrafla: **at the ~ of one's life**, hayatı pahasına: **at s.o.'s ~**, birinin zararına: **~ price**, maliyet fiatı: **with ~s**, mahkeme masrafları suçluya aid olarak. **~ly** [ˈkostli]. Çok değerli, pahalı.
**coster(monger)** [ˌkostəˌmʌn(g)gə(r)]. Seyyar satıcı.
**costume** [ˈkostyûm]. Kıyafet; elbise; tayör; tarihî elbise. **bathing ~**, mayo.
**cosy** [kouzi] *a.* Sıcak rahat ve kuytu; keyifli.
**cot** [kot]. Çocuk yatağı.
**cottage** [ˈkotic]. Küçük kır evi, kulübe. **~ hospital**, küçük şehir hastahanesi.
**cotton** [ˈkotn]. Pamuk (bezi, ipliği). **absorbent ~**, eczalı pamuk: **printed ~**, pamuklu basma: **sewing ~**, tire. **to ~ on to s.o.**, (*sl.*) birinden hoşlanmak. **~-wool**, kaba pamuk: **to bring up a child in ~**, bir çocuğu pek nazlı büyütmek.

**couch** [kauç]. Yatak; kanape.
**cough** [kof]. Öksürük. Öksürmek. **to give a ~**, manalı manalı öksürmek: **to ~ out stg.**, öksürerek ağzından çıkarmak.
**could** *v.* **can.**
**council** [ˈkaunsl]. Meclis, divan, şura. **County ~**, kontluk (vilâyet) meclisi. **~lor**, meclis âzası.
**counsel** [ˈkaunsl] *n.* Danışma; istişare, müşavere; fikir; nasihat; dava vekili, avukat. *v.* Tavsiye etm., öğüt vermek. **~ in chambers**, Müşavir avukat: **~ of perfection**, erişilmesi güç ideal: **to keep one's own ~**, kendi düşüncelerini kendine saklamak: **Queen's Counsel**, İngiltere'de en yüksek avukat derecesi: **to take ~ with s.o.**, birine danışmak. **~lor**, müşavir; elçilik müsteşarı.
**count**[1] [kaunt] *n.* Kont. **~ess**, kontes. **~**[2] *n.* Sayma, sayı. *v.* Saymak; sayılmak; (reyleri) tasnif etmek. **to keep ~ of**, sayısını hatırlamak: **to lose ~ of**, sayısını hatırlıyamamak: **to ~ on doing stg.**, bir şey yapmağı hesablamak: **to ~ (up)on s.o.**, birine güvenmek. **~ in**, hesaba katmak. **~ out**, birer birer saymak: **to be ~ed out**, nakavt olm: **you can ~ me out of this**, bu işte beni saymayın. **~ up**, yekûnunu hesab etmek. **~er**[1] [ˈkauntə(r)] *n.* Sayıcı âlet; (oyunda) sayı fişi; tezgâh. **under the ~**, el altından. **~er**[2]. *a.* Karşı, mukabil, aksi. *v.* Karşılamak; önlemek.
**countenance** [ˈkauntənəns]. Yüz, çehre; tasvib. Tasvib etmek. **to give [lend] ~ to**, teşvik etm.: **to keep one's ~**, temkinini bozmamak: **to stare s.o. out of ~**, dik dik bakarak birisini bozmak.
**counter-** *pref.* *Mürekkeb kelimelerde hemen daima* karşı, mukabil *ile tercüme edilir.* **~-clockwise**, saat akrebinin dönüş cihetine aykırı olarak. **~act** [kauntərakt]. Mukabele etm., önlemek. **~action** [–ˈrakşn], mukabil hareket. **~balance** [ˈkauntəˈbaləns]. Muvazene karşılık. Tevzin etm. **~feit** [ˈkauntəfît] *a.* Sahte, taklid. *v.* Taklid etm. **~feiter**, kalpazan; sahtekâr. **~foil** [ˈkauntəfôyl]. Defter koçanı. **~mand** [ˈkauntəˈmând]. Bir emri geri almak ibtal etmek. **~part** [ˈkauntəpât]. Suret; karşılık. **~sign** [ˈkauntəsâyn]. Birinin imzaladığı bir şeyi tasdik için imza etmek.
**country** [ˈkʌntri]. Memleket, yurd; kır; sayfiye; taşra; köy. **~ cousin**, dışarlıklı: **to go up ~**, memleketin içerilerine doğru gitmek: **open ~**, kırlar; ormansız ova: **to strike [go] across ~**, bir yere tarlalar arasından gitmek. **~-house**, sayfiye. **~man** [ˈkʌntrimən]. Vatandaş; kırda yaşıyan adam. **~side** [ˌkʌntriˈsâyd]. Kır. **~woman**, *pl.* **-men** [ˈkʌntriwumən, -wimin]. Kadın vatandaş; köylü kadın.
**county** [ˈkaunti]. İngiltere'de vilâyet, kontluk. **~ town**, kontluk merkezi.
**coupl·e** [ˈkʌpl]. Çift; iki eş. Birleştirmek, eş yapmak. **to go [hunt, run] in ~s**, bir şeyi daima iki kişi beraber yapmak. **~ing** [ˈkʌplin(g)]. Birleştirme cihazı.
**coupon** [ˈkûpon]. Sened vs. koçanı.
**courage** [ˈkʌric]. Cesaret. **to take [pluck up, muster up] ~**, cesaretini toplamak. **~ous** [kʌˈrêycəs], cesaretli.
**course**[1] [kôs] *n.* Akış; istikamet; rota; yol; hareket; devam müddeti; dersler; öğün; pist. **of ~**, elbette; **in the ~ of**, esnasında: **in due ~**, sırası gelince: **evil ~s**, fena hareketler: **to hold (on) one's ~**, tuttuğu rotadan [yoldan] ayrılmamak: **as a matter of ~**, tabiî olarak: **that is a matter of ~**, bu mevzuu bahis değildir: **in the ordinary ~ of things**, normal olarak: **there was no ~ open to me but to run away**, benim için kaçmaktan başka yapacak bir şey yoktu: **to set the ~**, rotayı tayin etm.: **to take one's own ~**, kendi bildiği gibi hareket etm.: **in the ~ of time**, zamanla: **~ of treatment**, (*med.*) tedavi yolu. **~**[2]. Akmak.
**court**[1] [kôt] *n.* Avlu; oyun sahası; mahkeme (âzaları); saray (halkı). **~ dress**, saray elbisesi: **~ room**, mahkeme salonu: **Ambassador to the ~ of St. James**, İngiliz kraliçenin nezdinde elçi: **the Law ~s**, Adliye sarayı: **to pay ~ to s.o.**, birine kur yapmak: **in open ~**, alenî muhakemede. **~-house**, mahkeme binası. **~ martial**, harb divanı(na vermek). **~**[2] *v.* Hulûs göstermek; kur yapmak; aramak. **~eous** [ˈkö(r)tyəs]. Nazik; çelebi. **~esan** [ˈkôtizan]. Fahişe. **~esy** [ˈkö(r)təsi]. Nezaket; düşünceli hareket. **by ~ [as a matter of ~]**, nezaketen: **~ title**, nezaketen verilen ünvan. **~ier**

[ˈkôtiə(r)]. Saray mensubu; nedim. **~ly** [ˈkôtli]. Kibar, nazik. **~ship** [ˈkôtşip]. Kur yapma, muaşaka. **~yard** [ˈkôtˈyâd]. Avlu.
**cousin** [ˈkʌzn]. Amca [dayı, hala, teyze] çocuğu; uzak akraba.
**cove**[1] [kouv]. Küçük körfez. ~[2] (*sl.*) Herif.
**covenant** [ˈkʌvənənt]. Mukavele, ahid(name); anlaşma. Mukavele yapmak.
**Coventry** [ˈkovəntri]. **to send to ~,** ···le alâkayı kesmek; boykot etmek.
**cover**[1] [ˈkʌvə(r)] *n.* Örtü; zarf; kab; kapak; sığınak; himaye; behane; (*comm.*) karşılık: **to give s.o. ~,** birini barındırmak: **outer ~,** dış lastiği: **to read a book from ~ to ~,** bir kitabı baştan başa okumak: **to take ~,** sığınmak: **under separate ~,** ayrıca. ~[2] *v.* Örtmek; kaplamak; sarmak; gizlemek; kapamak; ihtiva etm.; (aygır) çiftleşmek; (masrafı) karşılamak. **to ~ a deficit,** bir para açığını kapatmak: **to ~ a distance,** bir mesafe almak: **to ~ s.o. with ridicule,** birini gülünc bir hale düşürmek: **to ~ s.o. with a weapon,** silâhı birine dikmek: **to ~ up,** örtmek. **~ing** [ˈkʌvərin(g)] *n.* Örtü. *a.* **~ letter,** gönderilen bir şeyi izah eden mektub. **~t** [ˈkʌvö(r)t] *a.* Gizli.
**covet** [ˈkʌvit]. Hased etm.; şiddetle arzu etm. **~ous,** hased eden, haris.
**cow**[1] [kau] *n.* İnek. **~-** ..., dişi ···: **wait till the ~s come home,** çıkmaz ayın son çarşambasına kadar bekle. **~herd** [ˈkauˈhö(r)d]. Sığır çobanı. **~hide** [ˈkauhâyd]. Sığır derisi. ~[2] *v.* Korkutmak. **~ed look,** dayak yemiş gibi bir hal.
**coward** [ˈkauəd]. Korkak, alçak. **~ice,** korkaklık. **~ly,** korkak.
**cower** [ˈkauə(r)]. Korkudan sinmek.
**cowl** [kaul]. (Keşişlere mahsus) başlıklı cüppe; baca başlığı.
**cox** [koks] *v.* Dümen kullanmak. *n.* **~(swain)** [ˈkoks(n)]. Dümenci.
**coxcomb** [ˈkokskoum]. Horoz ibiği; züppe.
**coy** [kôy]. Nazlı; çekingen.
**crab** [krab] *n.* Yengec; Seretan burcu; kusur bulmak. **to catch a ~,** (kürek çekerken) yanlış bir hareketle küreği kımıldamaz bir hale getirmek. **~wise,** yan yan yürüyen. **~bed** [krabd]. Huysuz; karışık.
**crack**[1] [krak] *n.* Çatırtı; (kamçı) şaklama; çatlak, yarık; aralık; şiddetle vurma; sohbet. *a.* (*coll.*) Yaman, mükemmel. **to have a ~ at,** bir kere denemek. **~-brained,** çatlak kafalı. ~[2] *v.* Çatla(t)mak; çatırda(t)mak; şakla(t)mak; yar(ıl)mak; kır(ıl)mak; (ses) çatallaşmak. **to ~ a bottle,** bir şişe içmek: **to ~ a crib,** eve girip hırsızlık etmek: **to ~ a joke,** nükte yapmak. **~ up,** parçalanmak; (*coll.*) öğmek, medhetmek. **~ed,** *a.* çatlak, yarık; (*sl.*) kaçık. **~er** [ˈkrakə(r)]. Kıracak âlet; patlangıc; ince gevrek bisküvit. **~le** [ˈkrakl]. Çıtırdı. Çıtırdamak. **~sman** [ˈkraksmən]. Ev soyan hırsız.
**cradle** [ˈkrêydl]. Beşik. Beşiğe yatırmak. **~ song,** ninni.
**craft**[1] [krâft]. Hüner, marifet; san'at; hile; kurnazlık. **~sman,** san'at erbabı. **~smanship,** hüner. ~[2]. Gemi; uçak. **~y** [ˈkrâfti]. Kurnaz. **~iness,** kurnazlık.
**cram** [kram] *n.* Ziyade kalabalık. *v.* Doldurmak; tık(ın)mak; alelâcele bir imtihana hazırla(n)mak.
**cramp** [kramp] *n.* Sinir büzülmesi. **to feel ~ed for room,** yeri dar olm., sıkışık olm.: **to ~ one's style,** birinin maharet vs. sini bozmak: **~ed handwriting,** sıkışık yazı: **~ed style,** sıkıntılı üslûb.
**crane** [krêyn]. Turna kuşu; maçuna. **to ~ one's neck,** boynunu uzatmak: **to ~ forward,** başını ileri uzatmak.
**crank**[1] [kran(g)k] *n.* Manivelâ. *v.* Manivelâ ile işletmek. ~[2] *n.* Meraklı eksantrik kimse. **~y** [kranki]. Huysuz; garib.
**crash** [kraş]. (*ech.*) *n.* Tarraka; gürültü; şangır şungur; gürültü ile düşme, çarpma; (otomobil, uçak vs.) kazası; iflas; çözülme. *v.* Çatırdamak; gürültü ile çarp(tır)mak, düş(ür)mek, kır(ıl)mak. **~-dive,** (bir denizaltı) hızla dibe dalma.
**crass** [kras]. **~ ignorance,** kara cehalet.
**crate** [krêyt]. Büyük ambalaj sandığı. (Eşyayı) sandıklamak.
**crater** [ˈkrêytə(r)]. Yanardağ ağzı.
**crav·e** [krêyv]. Şiddetle arzu etm.; yalvararak istemek. **~ing** [ˈkrêyvin(g)]. Şiddetli arzu, doymak bilmez iştah.
**crawl** [krôl]. Yerde sürünme(k); krol yüzme(k); yavaş yürümek.
**crayon** [ˈkrêyon]. Renkli kalem.
**craz·e** [krêyz]. Şiddetli merak. Çıldırtmak. **~iness** [ˈkrêyzənis]. Çıl-

gınlık, delilik. **~y** [ˈkrêyzi] *n.* Deli, çılgın. *a.* **to be ~ about [over], ...** için deli olm., çıldırmak: **to drive [send] s.o. ~,** birini çıldırtmak.
**creak** [krîk]. (*ech.*) Gıcırtı. Gıcırdamak.
**cream** [krîm]. Kaymak; bir şeyin en iyi kısmı. **~y,** kaymaklı.
**crease** [krîs]. Kırma; kat; pli. Buruşturmak. **well-~d trousers,** ütülü pantalon.
**creat·e** [krîˈêyt]. Yaratmak; bir memuriyete tayin etm.; vücude getirmek: icad etm. **~ion** [–ˈêyşn], yaradılış; hilkat; icad; kâinat; san'at eseri. **~ive,** yaratıcı. **~or,** yaratıcı. **~ure** [ˈkrîçə(r)]. Mahluk; kul; başkasının aleti. **dumb ~s,** hayvanlar.
**cred·ence** [ˈkrîdens]. İnanma, itimad. **to give ~ to,** inanmak. **~entials** [kriˈdenşlz]. İtimadname, hüviyet vesikası. **~ible**[ˈkredibl]. İnanılabilir. **~it** [ˈkredit] *n.* İtibar; kredi; itimad; şeref; şöhret. *v.* İnanmak; itibar etm.; itimad etm.; bir krediyi hesabına geçirmek. **on ~,** veresiye: **I ~ed him with [I gave him ~ for] more intelligence,** ben onu daha zeki zannediyordum: **it does him ~,** bu ona şeref verir; **to gain ~,** gittikçe inanılmak; şeref kazanmak: **he gained ~s in French and Latin,** (imtihanda) fransızca ve lâtinceden iyi derece ile muvaffak oldu: **to give ~,** kredi açmak; inanmak. **to lend ~ to,** takviye etm. **~itable** [ˈkreditəbl]. Şerefli. **~itor** [ˈkreditə(r)]. Alacaklı. **~ side** (muhasebe defterinin) alacak kısmı. **~ulity** [kreˈdyûliti]. Saf(dil)lik. **~ulous** [ˈkredyuləs], saf(dil).
**creed** [krîd]. İman; itikad.
**creek** [krîk]. Nehir ağzi; vadi.
**creep** (**crept**) [krîp, krept]. Sürünmek, ağır ağır ilerlemek, sokulmak. Sürünme. **to give s.o. the ~s,** birini ürpertmek: **to ~ along,** gizlice ilerlemek: **to ~ away,** sessizce sıvışmak: **old age is ~ing on,** ihtiyarlık çöküyor. **~er,** yerde sürünerek yayılan nebat. **~y, to feel ~,** ürpermek: **~ story,** tüyler ürpertici bir hikâye: **~ crawly feeling,** tüyler ürperme.
**cremat·e** [kriˈmêyt]. (Ölüyü) yakmak. **~ion** [–ˈmêyşn], ölüyü yakma. **~orium** [–məˈtôriəm], ölülerin yakıldığı müessese.
**crept** *v.* **creep.**
**crescendo** [kreˈşendou]. (*mus.*) Gittikçe artarak.
**crescent** [ˈkresənt]. Hilâl, yeni ay; yarım daire şeklinde sokak. **the ~,** İslamın remzi. **Red ~,** Kızılay.
**crest** [krest]. Tepelik; sorguc; yele; (dağ, dalga vs.) tepesi; arma başlığı. **~fallen,** süngüsü düşük.
**Cret·an** [ˈkrîtən]. Giridli. **~e,** Girid.
**cretin** [ˈkretin]. İri kafalı ve doğuştan ebleh.
**crevasse** [kriˈvas]. Uçurum.
**crevice** [ˈkrevis]. Yarık, çatlak.
**crew** [krû]. Tayfa; mürettebat.
**crew** *v.* **crow**[2].
**crib** *n.* İntihal (yapmak); mektebde başkasından kopya etmek.
**crick** [krik]. Hafif burkulma.
**cricket**[1] [ˈkriket]. Cırcır böceği. **~**[2]. Kriket oyunu. **that's not ~,** bu doğru değil, bu yapılmaz. **~er,** kriketçi.
**cried** *v.* **cry,**
**crier** [ˈkrâyə(r)]. Tellâl, ilân eden.
**crime** [krâym]. Cinayet, cürüm, suç.
**Crimea** [krâyˈmiə]. Kırım. **~n,** Kırıma aid.
**criminal** [ˈkriminl]. Cürme aid; mücrim, cani. **the ~ Investigation Department (C.I.D.),** Emniyet Cinayet Dairesi: **habitual ~,** cürmü itiyad haline getiren kimse.
**crimson** [ˈkrimzn]. Fesrengi.
**cringe** [krinc]. Köpekleşme(k).
**crinkl·e** [ˈkrin(g)kl]. Kıvrım, kat. Kıvırmak, buruşturmak. **~d paper,** krep.
**cripple** [ˈkripl]. Sakat; malûl. Sakat etm.; bozmak. Zarar vermek.
**crisis** [ˈkrâysis]. Buhran, kriz.
**crisp** [krisp]. Gevrek; serin (hava); keskin.
**criss-cross** [ˈkrisˈkros]. Çapraz çizgili.
**critic** [ˈkritik]. Münekkid, tenkidci. **arm-chair ~,** oturduğu yerden tenkid eden. **~al,** tenkide aid; tenkidci; tehlikeli; can alıcı. **~ism** [ˈkritisizm], tenkid. **~ize** [–ˈsâyz], tenkid etm.; kusur bulmak.
**croak** [krouk]. (*ech.*) (Kurbağa) vakvak etm.; (karga) gaklama(k); çirkin bir sesle bağırmak. (*sl.*) ölmek. **~er,** şikâyetçi.
**crock. to ~ up,** çökmek: **to be ~ed,** sakatlanmak.
**crockery** [ˈkrokəri]. Tabak takımı.
**crocodile** [ˈkrokədâyl]. Timsah; ikişer kişilik yürüyüş sırası. **~ tears,** sahte gözyaşları.
**crocus** [ˈkroukəs]. Çiğdem.

**croft** [kroft]. Küçük çiftlik.
**crony** [ˈkrọuni]. Ahbab, kafadar.
**crook** [kruk] *n.* Kanca; çoban değneği; dolandırıcı, hilekâr. *v.* Kıvırmak. **~ed** [–id], iğri; dalavereci.
**croon** [krûn]. Mırıldıyarak şarkı söyleme(k).
**crop**[1] [krop] *n.* Mahsul; ekin; kısa kesilmiş saç. *v.* Kırpmak; kısaltmak; (hayvan) yemek; mahsul vermek. **to give s.o. a close ~**, birinin saçını dibinden kesmek: **Eton ~**, alâgarson kesilmiş. **~ up**, birdenbire zuhur etm., ara sıra meydana çıkmak. **~**[2]. Kamçı sapı; kısa kamçı.
**cropper** [ˈkropə(r)]. **to come a ~**, (*sl.*) fena halde düşmek; fiyasko yapmak.
**cross**[1] [kros] *n.* Haç; keder; melez. *a.* Çapraz; ters; aksi; dargin. **to be ~**, darılmak: **fiery ~**, halkı isyana teşvik eden işaret: **to be at ~ purposes**, birbirlerinin maksadını yanlış anlayıp muhalefet etmek: **to make the sign of the ~**, haç çıkarmak: **to sign with a ~**, (ümmi) imza yerine haç çizmek. **~**[2]. *v.* Karşıdan karşıya geçmek; (hayvanları) karıştırmak, karşılaşmak; çapraz koymak. **to ~ one's legs**, bacak bacak üstüne atmak: **to ~ oneself**, haç çıkarmak: **to ~ s.o.('s plans)**, birinin işini bozmak. **~ out**, çizmek. **~ over**, karşıya geçmek. **~**[3] *pref.* Bir yandan bir yana *veya* çaprazlama *manalarını ifade eder.* **~-bar**, kol demiri. **~-bow**, Tatar [mancınık] oku. **~-bred**, melez. **~-breed**, melez (yetiştirmek). **~-country race**, kırkoşusu. **~-examination**, sorgu. **~-examine**, istintak etm. **~-eyed**, şaşı. **~-fire**, çatal ateşi. **~-grained**, damarı ters (ağac); huysuz (kimse). **~-legged**, bacak bacak üzerine. **~-patch**, huysuz densiz çocuk. **~-question**, istintak etmek. **~-roads**, dört yol ağzı: **we are at the ~**, kat'î karar zamanı geldi. **~-section**, makta. **~-talk**, karşılıklı münakaşa. **~-wind**, muhalif rüzgâr. **~ing** [ˈkrosin(g)] *n.* Geçid; geçiş; tesalub. **~wise** [ˈkroswâyz]. Çapraz, ters. **~word** [ˈkroswö(r)d]. **~ (puzzle)**, bulmaca.
**crouch** [krạuç]. Çömelme(k); iğilme(k).
**crow**[1] [krọu] *n.* Karga. **as the ~ flies**, dümdüz, dosdoğru. **~'s-feet**, göz kenarı daki kırışık. **~'s-nest**, (*naut.*) çanaklık. **~**[2] (**crew**, veya **crowed**) [krọu, krû, krọud] *v.* (Horoz) ötmek; sevincinden bağırmak. *n.* Horoz ötüşü. **to ~ over s.o.**, (mağlub edilen birine karşı) fazla sevinc göstermek. **~bar** [ˈkrọubâ(r)]. Demir manivelâ kolu.
**crowd** [krạud]. Kalabalık; halk, kütle. Bir araya toplamak, tıka basa doldurmak; sıkışmak. **to ~ in**, kalabalık halinde girmek: **to ~ out**, ···e yer bırakmamak: **to ~ on sail**, bütün yelkenlerini açmak. **~ed**, kalabalık; dolu.
**crown**[1] [krạun] *n.* Tac; hükümdarlık, krallık; hükümdar, kıral vs.; çelenk; şeref; şapka tepesi; tepe, baş; **~ lands**, kırala aid arazi (mirî): **~ lawyer**, hükümetin müdafaa vekili: **~ prince**, veliahd: **half a ~**, iki buçuk şilin. **~**[2] *v.* Tac giydirmek; şereflendirmek; mükâfatlandırmak; dişe kuron takmak. **to ~ all**, üstelik en fenası [iyisi]. **~ing**, en son, en yüksek.
**crucial** [ˈkrûşịəl]. Kat'î; çok mühim.
**cruci·fix** [ˈkrûsifiks]. Çarmıh. **~fixion** [–ˈfikşn], çarmıha germe. **~fy** [ˈkrûsifây]. Çarmıha germek; işkence etmek.
**crude** [krûd]. Ham, çiğ; kaba.
**cruel** [krụəl]. Zalim, insafsız. **~ty**, zulüm; işkence.
**cruise** [krûz]. Deniz gezintisi (yapmak). **~r**, [ˈkrûzə(r)], kruvazör: **~ weight**, yarı-ağır sıklet.
**crumb** [krʌm]. Ekmek kırıntısı; küçük parça. **a ~ of comfort**, küçücük bir teselli. **~le** [ˈkrʌmbl]. (Ekmek vs.) ufal(t)mak; parçala(n)mak; yıkılmak. **~ly**, kırıntılı.
**crumple** [ˈkrʌmpl]. Buruşturmak. **to ~ up**, çökmek.
**crunch** [krʌnç]. (*ech.*) Kıtır kıtır yemek; (kar vs.) gıcırdamak.
**crusade** [krûˈsêyd]. Haçlı seferi. **~r**, Haçlı.
**crush** [krʌş] *v.* Ezmek; buruşturmak; sıkmak; tazyik etm. *n.* Kalabalık. **please ~ up a little**, lûtfen bir az sıkışınız.
**crust** [krʌst]. Ekmek kabuğu; sert kabuk. **~y** [ˈkrʌsti]. Kabuklu (ekmek); haşin.
**crutch** [krʌç]. Koltuk değneği; destek.
**crux** [krʌks]. Esas, en mühim, nokta.
**cry** (**cried**) [krây, –d] *v.* Bağırmak; ağlamak. *n.* Nara, feryad; bağırma;

ağlama. to ~ one's eyes [heart] out, hüngür hüngür ağlamak; it's a far ~ to, çok uzaktır: the pack is in full ~, (avda) şikârın kokusunu alan köpekler bağrışıyorlar: the crowd was in full ~ after the thief, kalabalık hırsızın arkasından bağrışarak koşuyordu: to have a good ~, doya doya ağlamak: within ~, çağırınca duyulabilecek mesafede. ~ing, bağrıyan; ağlıyan; pek göze çarpan; rezalet teşkil eden. ~ off, sözünü geri almak. ~ out, bağırarak söylemek.
**crypt** [kript]. Kilise bodrumu. ~ic [ˈkriptik]. Esrarlı.
**crystal** [ˈkristəl]. Billur. ~ clear, billur gibi duru: ~ gazing, billûra bakarak falcılık. ~line [–lâyn], billûr gibi. ~lize [–lâyz], tebellûr etm. ~lization [–zêyşn], tebellür.
**cub** [kʌb]. Hayvan yavrusu. (unlicked) ~, yontulmamış delikanlı.
**cubby-hole** [ˈkʌbihoul]. Küçük göz.
**cub·e** [kyûb]. Mikâb, küb. Mikâbını bulmak. ~ic(al), mikâb şeklinde; kübik. ~icle [kyûbikl]. Küçük yatak odası; hücre.
**cuckoo** [ˈkukû]. Guguk kuşu. ~-clock, guguklu saat.
**cucumber** [ˈkyûkʌmbə(r)]. Hıyar, salatalık. cool as a ~, fevkalâde soğukkanlı.
**cud** [kʌd]. to chew the ~, geviş getirmek.
**cuddle** [kʌdl]. Kucaklama(k), okşama(k); kucaklaşmak.
**cudgel** [ˈkʌcl]. Sopa, kalın değnek. Dayak atmak, döğmek. to take up the ~s for s.o., birini şiddetle müdafaa etm.: to ~ one's brains, zihnini yormak.
**cue**[1] [kyû]. İşaret. to give s.o. the ~, birine (bir şey hakkında) işaret vermek: to take the ~ from ..., birinden işaret almak. ~[2]. Bilârdo istekası.
**cuff** [kʌf]. Kolluk. Hafif tokat (atmak).
**cul-de-sac** [ˈkuldəˈsak]. Çıkmaz.
**culminat·e** [ˈkʌlminêyt]. En son noktaya varmak; neticelenmek. ~ion [–ˈnêyşn], son nokta; en yüksek derece.
**culp·ability** [ˈkʌlpəˈbiliti]. Suçluluk. ~able [ˈkʌlpəbl], suçlu. ~rit [ˈkʌlprit]. Suçlu.
**cult** [kʌlt]. Mezheb; ibadet.
**cultivat·e** [ˈkʌltivêyt]. Çift sürmek; toprağı işlemek; yetiştirmek; geliştirmek, terbiye etmek. ~ed, tahsil görmüş; işlenmiş. ~ion [–ˈvêyşn], toprağı işleme; cift sürme; terbiye. ~or, ciftçi.
**cultur·al** [ˈkʌltyurəl]. Kültüre aid. ~e, yetiştirme; kültür. ~ed, kültürlü.
**cumber** [ˈkʌmbə(r)]. Yük olm. ~some, havaleli.
**cumulative** [ˈkyûmyulətiv]. Müterakim.
**cunning** [ˈkʌnin(g)] *n.* Kurnazlık, hile: maharet. *a.* Kurnaz; meharetli.
**cup** [kʌp] *n.* Fincan; kupa. in one's ~s, sarhoş iken. ~-bearer, saki. ~-final, futbol şampiyonluğu finali. ~ful [ˈkʌpful]. Fincan dolusu. ~-tie, futbol şampiyonluk eleme maçı.
**cupboard** [ˈkʌbəd]. Dolab. ~ love, yiyecek bir şey verileceği için gösterilen muhabbet.
**cupidity** [kyûˈpiditi]. Hırs, açgözlülük.
**cur** [kö(r)]. Âdi köpek; terbiyesiz.
**cur·able** [ˈkyûrəbl]. Tedavi edilebilir. ~ative [ˈkyûrətiv]. İyileştirici, şifa verici. ~ator [kyûˈrêytə(r)]. Müze müdürü. ~e [kyuə(r)]. Tedavi etmek; tütsülemek. Tedavi; şifa, çare. ~e-all, her derde deva.
**curb** [kö(r)b]. Gem zinciri. Ata gem vurmak; hiddetini tutmak. to put a ~ on one's passions, ihtiraslarına gem vurmak.
**curd** [kö(r)d]. Kesilmiş süt. ~le [ˈkö(r)dl]. (Süt) kesilmek; (kan) pıhtılaşmak. enough to ~ one's blood, tüylerini ürpertecek derecede.
**curfew** [ˈkö(r)fyû]. Fevkalâde hallerde halkın evinden dışarı çıkması yasak olduğu zaman.
**curio** [ˈkyûriou]. Nadir şey. ~sity [ˌkyuəriˈosəti], merak; antika: old ~ shop, antikacı mağazası. ~us [ˈkyuəriəs], meraklı, hevesli; garib: ~ly enough, garibi şu ki.
**curl** [kö(r)l]. Kıvrım, büklüm. Kıvırmak, kıvrılmak; bukle yapmak. to ~ oneself up, dertop olm. ~ing irons, saç maşası. ~y [ˈkö(r)li]. Kıvırcık; dalgalı.
**currant** [ˈkʌrənt]. Kuşüzümü.
**currency** [ˈkʌrənsi]. Revac; rayic..
**current** [ˈkʌrənt] *a.* Cari; hali hazıra aid, bugünkü. *n.* Cereyan, akıntı. ~ number, bir mecmua son çıkan nüshası: it is ~ly reported that, umumiyetle söylenildiğine göre: in ~ use, umumiyetle kullanılan.

**curriculum** [kʌˈrikyuləm]. Müfredat programı.
**curry** [ˈkari]. Atı kaşağılamak. **to ~ favour with s.o.**, müdahene ile birinin gözüne girmeğe çalışmak.
**curse** [kö(r)s]. Lânet; belâ; inkisar. Lânet etm. **to ~ one's fate**, bahtına küsmek: **what ~d weather!**, hava da Allahın belâsı!
**cursory** [ˈkö(r)səri]. Acele; sathî.
**curt** [kö(r)t]. Kısa, kuru; nezaketsizce kısa.
**curtail** [kö(r)ˈtêyl]. Kısaltmak, kısmak. **~ment**, kısaltma.
**curtain** [ˈkö(r)tən]. Perde. **~ lecture**, zevcenin kocasını yatakta azarlaması.
**curts(e)y** [ˈkö(r)tsi]. Diz kırarak reverans (yapmak).
**curv·ature** [ˈkö(r)vətyuə(r)]. Kavislenme. **~e** [kö(r)v]. Kavis (çizmek).
**cush·ion** [ˈkuşən]. Yastık (koymak); sademeyi hafifletmek. **~y** [ˈkuşi]. (*sl.*) Kolay ve rahat (iş), otlak.
**cuss** [kʌs]. (*sl.*) Küfür; herif. **~edness**, inadcılık, aksilik; nisbet.
**custard** [ˈkʌstəd]. Yumurtalı, sütlü krema.
**custod·ian** [kʌsˈtoudiən]. Muhafız. **~y** [ˈkʌstədi], muhafaza; nezaret; tevkif: **to take into ~**, tevkif etmek.
**custom** [ˈkʌstəm]. İtiyad, âdet, örf. **~s**, gümrük. **~ary**, âdet olan. **~er**, müşteri: **a queer ~**, garib bir adam.
**cut**[1] [kʌt] *n.* Kesme; kesik, yara; darbe; yarma; kesip çıkarma; kesik parça; biçki, biçim; (fiat vs.) indirme; kader darbesi; tanımamazlıktan gelme. **to be a ~ above ...**, ···e tenezzül etmemek: **to make a clean ~ with**, ···le alâkayı tamamen kesmek: **a prime ~**, kasablık etin en seçme parçası: **short ~**, kestirme yol: **an unkind ~**, dokunaklı ve kırıcı söz, hareket: **the unkindest ~ of all**, en fecii. **~**[2] *a.* Kesik, kesilmiş. **~ and dried**, hazır (fikir vs.); kat'î şekilde tesbit edilmiş (plân vs.): **~ glass**, billur: **low-~ dress**, dekolte elbise: **~ price**, tenzilatlı fiat: **~ and thrust**, göğüs göğüse kavga. **~**[3] (cut) *v.* Kesmek, biçmek, yontmak, yarmak; (fiat) indirmek. **to ~ s.o. (dead)**, birini görmemezlikten gelmek: **that ~s both ways**, bu iki yüzlü bir kılıcdır: **to ~ a corner**, köşeyi dönmeyip kestirmeden gitmek; (*otom.*) köşeye sürünerek viraj yapmak: **to ~ across country**, kırdan kestirme gitmek: **to ~ a lecture, etc.**, (*coll.*) bir ders vs.yi asmak: **to ~ and run**, (*fig.*) sür'atle sıvışmak: **to ~ the whole concern**, bir işle alâkasını kesmek. **~ away**, kesip çıkarmak. **~ back**, yontmak; kısaltmak; (*coll.*) sür'atle dönüp geri gitmek. **~ down**, kesip devirmek; kısmak; biçmek. **~ in**, söze karışmak; (yarışta) rakibinin yolunu kesmek. **~ into**, yarmak; bir parça kesmek; söze karışmak. **~ off**, kesip koparmak, ayırmak: **to be ~ off**, ölmek. **~ out**, kesip çıkarmak; biçmek; oymak: **to ~ s.o. out**, birinin bir işte yerini almak: **he is ~ out for this job**, bu iş onun için biçilmiş kaftandır. **~ up**, doğramak; bozmak: **to be ~ up**, kendini üzmek: **to ~ up nasty [ugly]**, (*sl.*) hiddete kapılmak. **~lery** [ˈkʌtləri]. Çatal bıçak takımı. **~ter** [ˈkʌtə(r)]. Kesici, biçici; kotra. **~-throat** [ˈkʌtθrout]. Katil; insafsız. **~ting** [ˈkʌtin(g)] *a.* Keskin; tesirli. *n.* Yarma; gazete maktuası.
**cute** [kyût]. (*coll.*) Açıkgöz; zeki; (*Amer.*) zarif, hoş.
**C.V.O.** (*abb.*) **Companion of the Royal Victorian Order**, *v.* **order.**[2]
**cycl·e** [ˈsâykl]. Devir, devre; bisiklet. Bisikletle gitmek. **~ic(al)**, devrî. **~ist**, bisikletçi.
**cyclone** [ˈsâykloun]. Kasırga, siklon.
**Cyclopean** [sâyˈkloupiən]. Dev gibi.
**cylind·er** [ˈsilində(r)]. Üstüvane, silindir. **~rical** [–ˈlindrikl], üstüvanî.
**cynic** [ˈsinik]. Her şeyi kötü gözle gören. **~al**, kelbî, müstehzi. **~ism** [ˈsinisizm], Kelbiyun felsefesi.
**cypher** [ˈsâyfə(r)]. Şifre.
**cypress** [ˈsâypris]. Selvi.
**Cypr·ian** [ˈsipriən]. Kıbrıslı. **~iot**, Kıbrıslı. **~us** [ˈsâyprəs], Kıbrıs.
**cwt.** (*abb.*) **hundredweight.**
**Czar** [zâ]. Çar. **~evitch** [ˈzârəviç] eski Rus veliahdı. **~ina** [zâˈrînə] Çariçe.
**Czech** [çek]. Çek. **~oslovakia** Çekoslovakya.

# D

**d** [dî]. D harfı; (*mus.*) re; peni'nin kısaltması; Roma sayılarında 500.
**dab**[1] [dab]. Hafifçe vurma(k); yumuşak ve ıslak bir şey(le bastırmak)

~². (*coll.*) **to be a ~ (hand) at stg.**, bir şeyi yaman bilmek.
**dabble** [ˈdabl]. **to ~ in stg.**, bir az meşgul olm.
**dad, daddy** [dad, ˈdadi]. (*coll.*) Baba.
**daffodil** [ˈdafədil]. Yabani nergis.
**daft** [dâft]. Kaçık, sapık.
**dagger** [ˈdagə(r)]. Kama, hançer; (†) işareti. **to be at ~s drawn**, birbirinin kanına susamak: **to look ~s at s.o.**, bir kaşık suda boğacakmış gibi bakmak.
**daily** [ˈdêyli]. Günlük, gündelik. Her gün; gün geçtikçe. Gündelik gazete.
**dainty** [ˈdêynti]. Zarif ince ve nazik; nazlı. Nefis yiyecek.
**dairy** [ˈdeəri]. Süthane; sütçü dükkânı. **~ farm**, süt istihsal edilen çiftlik.
**daisy** [ˈdêyzi]. Papatya.
**dally** [ˈdali]. Haylazlık etm., vakit geçirmek. **to ~ with**, bir şeyle oynamak: **to ~ with s.o.**, birini oynatmak.
**dam** [dam]. Bend (suyu). Bendle kapamak; zabtetmek.
**damage** [ˈdamic] *n.* Zarar, ziyan; (*sl.*) fiat, masraf. **~s**, tazminat. *v.* Zarar vermek.
**Damas·cus** [dəˈmâskəs]. Şam. **~k** [ˈdaməsk, dəˈmâsk]. Şam kumaşı; Şam işi.
**dame** [dêym]. Hanım; yaşlı kadın.
**damn** [dam]. Küfür; Allah belâsını versin! Lânetlemek; mahkûm etmek. **~ it!**, hay Allah müstehakkını versin! **well I'm ~ed!**, artık çok oluyor: **I'll see him ~ed first**, dünyada olmaz; **do your ~edest!**, elinden geleni arkana koyma: **it's not worth a (tuppenny) ~**, on para etmez. **~able**, lânet ve nefrete lâyık. **~ation** [–ˈnêyşn], lânet, tel'in; Allahın belâsı. **~ed**, lânetleme.
**Damocles** [ˈdamoklîz]. **the sword of ~**, İnsanın başında daimî tehlike.
**damp** [damp] *n.* Rutubet; buğu. *a.* Nemli, ıslak. *v.* ~ veya **~en**, hafifçe ıslatmak; (ateş, ses) bastırmak; (heyecan vs.) soğumak, sönmek. **to ~ s.o.'s ardour**, birinin hevesini kırmak. **~er**, (sobada) ateş tanzim kapağı; sesi kısma cihazı: **to put a ~ on the company**, toplantıya soğuk bir hava getirmek.
**damson** [ˈdamzn]. Mürdüm eriği.
**dance** [dâns]. Dans; balo. Dansetmek. **to ~ attendance on s.o.**, birinin etrafında dört dönmek: **to ~ for joy**, evincinden takla atmak: **to ~ with rage**, hiddetten tepinmek: **to lead s.o. a ~**, birinin başına iş açmak: **I'll make him ~ to a different tune**, ben ona gösteririm. **~r**, danseden kimse; çengi, rakkase.
**dandelion** [ˈdandilâyən]. Karahindiba.
**dandruff** [ˈdandrəf]. (Saçta) kepek.
**dandy** [ˈdandi]. Fazla şık; iki dirhem bir çekirdek; (*Amer.*) yaman, mükemmel.
**Dan·e** [dêyn]. Danimarkalı. **~ish** [ˈdêyniş]. Danimarkaya aid; danimarkaca.
**danger** [ˈdêyncə(r)]. Tehlike. **~ous** [ˈdêyncərəs], tehlikeli.
**Darby** [ˈdâbi]. **~ and Joan**, Arzu ile Kanber.
**dar·e** [ˈdeə(r)]. Kalkışmak, cesaret etm. **how ~ you!**, bu ne cesaret, küstahlık!: **I ~ say**, olabilir: **to ~ s.o. to do stg.**, birine bir şeyi 'yapamazsın' diye meydan okumak: **don't you ~ touch him!**, ona dokunayım deme! **~e-devil**, gözünü çopten sakınmaz. **~ing** [ˈdeərin(g)] *n.* Cesaret, yiğitlik. *a.* Cesur, atılgan.
**dark** [dâk]. Karanlık. Koyu; esmer; gizli. **the Dark Ages**, Ortaçağın ilk yarısı: **to be in the ~**, haberi olmamak: **after ~**, ortalık karardıktan sonra: **it is getting ~**, ortalık kararıyor: **a ~ horse**, hakkında bir şey bilinmiyen yarış atı [rakib]: **to keep stg. ~**, bir şeyi gizli tutmak. **~-eyed**, kara gözlü. **~en**, karar(t)mak, koyulaş(tır)mak. **~ness**, karanlık, (renk) koyuluk. **~y**, (*coll.*) zenci.
**darling** [ˈdâlin(g)]. Sevgili.
**darn** [dân]. Örerek tamir etme(k).
**dart** [dât]. Ok, hafif mızrak, cirid: birdenbire atılma. At(ıl)mak. **~s**, küçük okları içiçe daire şeklinde bir hedefe atmaktan ibaret bir oyun.
**dash¹** [daş] *n.* Seğirtme; anî ve hızlı koşuş; atılma; hamle; atılganlık; damla; çizgi (—). **to cut a ~**, gösteriş yapmak: **to make a ~ at**, ···e saldırmak: **to make a ~ for [to]**, ···e doğru atılmak. **~²** *v.* Şiddetle atmak, fırlatmak, çarptırmak; seğirtmek. **to ~ at s.o.**, birinin üzerine atılmak: **to ~ s.o.'s hopes [spirits]**, ···in ümidlerini [cesaretini] kırmak: **all my hopes were ~ed to the ground**, bütün ümidlerim suya düştü: **to ~ to pieces**, fırlatarak parça parça etmek. **~ along**, hızla gitmek. ~

**away,** hızla ayrılmak. ~ **in,** paldır küldür girmek. ~ **off,** (i) sür'atle uzaklaşmak; (ii) sür'atle karalamak. ~ **out,** dışarı fırlamak: **to ~ out s.o.'s brains,** birinin beynini patlatmak. **~board** [ˈdaşbôd]. Âlet tablosu. **~ing** [ˈdaşin(g)] *a.* Atılgan, yaman; göze çarpan.

**data** [ˈdêytə]. Mûtalar.

**date**[1] [dêyt]. Hurma. **~-palm,** hurma ağacı. **~**[2]. Tarih. Tarihini atmak; tarihini [eskiliğini, yaşını] belli etmek. ~ **of a bill,** bir senedin vadesi: **six months after ~ [at six months' ~],** altı ay sonunda: **interest to ~,** bugüne kadar olan faiz: **to have a ~ with s.o.,** birisile bir sözü (randevusu) olm.: **out of ~,** modası geçmiş: **under the ~ of May 9th,** 9 mayıs tarihinde: **to be up to ~,** zamana uygun olm.; modern [yeni fikirli] olm.; işini günü gününe yetiştirmek.

**dative** [ˈdêytiv]. Mef'ulünbih.

**daughter** [ˈdôtə(r)]. Kız (evlâd). **~-in-law,** gelin.

**daunt** [dônt]. Korkutmak. **~less,** yılmaz.

**dawdle** [ˈdôdl]. Ağır davranmak.

**dawn** [dôn]. Şafak; başlangıc. Gün ağarmak, şafak sökmek. **at length it ~ed on me that ...,** nihayet anladım ki....

**day** [dêy]. Gün, gündüz; zaman; günlük. ~ **after ~,** arka arkaya her gün: ~ **by ~,** günden güne: **all ~ long,** bütün gün akşama kadar: **the ~ before yesterday,** evvelki gün: **before ~,** güneş doğmadan evvel: **break of ~,** şafak: **by ~,** gündüz: **it was broad ~,** güneş doğalı çok olmuştu: **to carry the ~,** kazanmak: **the ~ is ours,** kazandık: **the ~ was going badly for the English,** muharebe İngilizlerin aleyhine gidiyordu: **from that ~ to this,** o gün bugündür: **the good old ~s,** hey gidi günler: **in the good old ~s [in the ~s of old],** eski zamanda: **he has had his ~, his ~is done [over],** onun zamanı geçti: **in my ~,** benim zamanımda: **it's many a long ~ since ...,** ne zamandan beri ...: **to ask a girl to name the ~,** (*coll.*) bir kıza evlenme teklifi yapmak: **one of these (fine) ~s,** (ikaz veya tehdid makamında) günün birinde: **the other ~,** geçen gün: **he has seen better ~s,** kibar düşkünüdür; o ne günler görmüştür: **some ~,** bir gün: **this ~ week,** gelecek hafta bu gün: **it is three years ago to a~,** günü gününe üç sene evvel: **to this very ~,** bu gün bile, hâlâ: ⌈**it's all in the ~'s work**⌉, bu işe giren buna katlanır (beklenmedik bir şey değil). **~-boy,** neharî. **~-dream,** hülya. **~-labourer,** gündelikçi. **~light** [ˈdêylâyt]. Gündüz; açıklık: **in broad ~,** güpe gündüz: **by ~,** gündüz(ün): **to begin to see ~,** bir işin içyüzünü anlamağa başlamak; üzüntülü bir işin sonuna yaklaştığını sezmek.

**daze** [dêyz]. Sersemlik. Sersemletmek.

**dazzle** [ˈdazl]. Gözlerini kamaştırma(k).

**D.C.L.** (*abb.*) **Doctor of Civil Law,** Hukuk Doktoru.

**D.D.** (*abb.*) **Doctor of Divinity,** İlahiyat Doktoru.

**de-** [dî-]. *pref. Şu manaları taşır:*—(i) *Bir şeyin tam aksini yapmak, mes.* **mobilize,** seferber etm.; **demobilize,** terhıs etm.: (ii) aşağıya, *mes.* **ascend,** çıkmak; **descend,** inmek: (iii) uzak, ayrı, *mes.* **rail,** ray; **derail,** raydan çıkarmak: (iv) tam, *mes.* **despoil,** tamamen soymak.

**dead** [ded]. Ölü, ölmüş; ölü gibi; kat'î, tam(amen). **I am ~ against it,** ben bunun tamamen aleyhindeyim: ~ **beat,** ibresi sallanmıyan (alet): **a ~ cert(ainty),** elde bir: ~ **and done for,** onun işi bitti: **in ~ earnest,** son derece ciddî: **to go ~,** (bir uzuv) uyuşmak: ~ **heat,** yarışta başbaşa varış: **the ~ hours,** gece yarısı: ~ **letter,** (i) sahibine teslim edilmiyen mektub; (ii) mer'i olmıyan kanun: **a ~ loss,** tam kayb: ~ **march,** cenaze marşı: ⌈~ **men tell no tales**⌉, ölüler konuşmaz (*bir sırrı ifşa etmemesi için öldürülen kimse hakkında kullanılır*): **at ~ of night,** gece yarısı: ~ **on time,** tam vaktinde: ~ **reckoning,** gemi mevkiinin parakete ve pusula vasıtasile, rasadsız tayini: **the Dead Sea,** Lut Denizi: ~ **secret,** son derece gizli: **a ~ shot,** keskin nişancı: **to come to a ~ stop,** anî olarak ve tam durmak: ~ **to ...,** ...e karşı hissiz: ~ **white,** mat beyaz boya: **in the ~ of winter,** karakışta: ~ **wire,** elektrik cereyanı geçmiyen tel: ~ **to the world,** son derece bitkin, sarhoş. **~-alive,** ölü gibi. **~-beat,** bitkin bir halde. **~-centre,** ölü nokta. **~-end,** çıkmaz. **~-weight,** (i) kesilmiş hayvanın ağırlığı; (ii) borc vs. hakkında:— ağır yük. **~en** [ˈdedn]. Hafifletmek; ses geçmez hale getirmek. **~ly** [ˈdedli]. Öldürücü; tehlikeli; ölü gibi; müdhiş. **in ~ earnest,** şakası yok: **~ sin,** kebair. **~ness** [ˈdednis]. Uyuşukluk; durgunluk.

**deaf** [def]. Sağır. **~ as a post,** duvar gibi sağır: **to turn a ~ ear to,** reddetmek: ⌜**none so ~ as those who won't hear**⌝, işitmek istemiyen kadar sağır olmaz. **~-mute,** sağır ve dilsiz. **~en,** sağırlaştırmak. **~ness,** sağırlık.

**deal**[1] [dîl] *n.* (Çok) mikdar; çok. **a good ~,** çok: **a great ~,** pek çok. **~**[2] *n.* Ticarî muamele; pazarlık; oyun kâğıdını dağıtma, el. *v.* Muamele etm.; pazarlık etm.; iş yapmak; dağıtmak; (darbe) indirmek; oyun kâğıdını dağıtmak. **well, that's a ~,** pek iyi uyuştuk: **to ~ in ...,** ... ticareti yapmak: **to ~ out,** tevzi etm.: **to ~ with (a matter),** (bir mesele) ile meşgul olm.: **to ~ with s.o.,** birisile ticaret yapmak; birisile meşgul olm.: **I'll ~ with him!,** onu bana bırak! **~**[3] *n.* Çam tahtası. **~er** [ˈdîlə(r)]. Satıcı, tüccar: (oyunda) kâğıd dağıtan. **double-~er,** iki yüzlü.

**dean** [dîn]. Dekan; İngiliz katedral başrahibi.

**dear** [diə(r)]. Aziz; sevgili; pahalı. **~ ~!** [**~ me!**], aman yarabbi!: **oh ~!,** yazık!: **my ~ fellow,** azizim: **to get ~** [**~er**], pahalılaşmak: **you shall pay ~(ly) for this!,** bu size pahalıya mal olacak: **to run for ~ life,** var kuvvetile koşmak.

**death** [deθ]. Ölüm, vefat. **you'll be the ~ of me!,** (i) benim ölümüme sebeb olacaksın; (ii) beni gülmekten öldüreceksin: **to be in at the ~,** *v.* **kill**: **the Black Death,** ortaçağdaki veba: **to do to ~,** zulm ederek öldürmek: **meat done to ~,** fazla pişirmiş et: **(fashion, story) done to ~,** (moda, hikâye vs.) insanı bıktıracak derecede yayılmış, tekrarlanmış: **to drink oneself to ~,** kendini işretle öldürmek: **to put to ~,** idam etm.: **sick to ~ of,** ···den son derece bıkmış: **war to the ~,** ölesiye harb. **~-agony,** can çekişme. **~-blow,** öldürücü darbe. **~-duties,** veraset ve intikal vergisi. **~-less** [ˈdeθlis].Ölmez. **~ly,** ölü gibi. **~-mask,** bir ölünün yüzünün kalıbı. **~-rate,** ölüm nisbeti. **~-trap,** ölüm tehlikesi olan yer.

**debar** [diˈbâ(r)]. **~ s.o. from stg.,** birini bir şeyden mahrum etm.; **to ~ s.o. from doing stg.,** birini bir şey yapmaktan menetmek.

**debase** [diˈbêys]. Alçaltmak; âdileştirmek; ayarını bozmak. **~ment,** alçalma.

**debat·able** [diˈbêytəbl]. Münakaşası kabil; kat'î olmıyan. **~e,** (bir meseleyi) müzakere (etm.).

**debauch** [diˈbôç]. İşret. Ahlâkını bozmak. **~ee** [–ˈçî], sefih. **~ery** [–ˈbôçəri], sefahat.

**debit** [ˈdebit]. Zimmet (hanesi). Zimmet kaydetmek: **~ balance,** (bütçede) açık.

**debonair** [deboˈneə(r)]. Şen, hoş ve nazik.

**debris** [ˈdêyˈbrî]. Enkaz; kırıntı.

**debt** [det]. Borc. **I shall always be in your ~,** size karşı daima borclu olacağım: **to be head over ears [up to the eyes] in ~,** uçan kuşa borclu olmak. **~or,** borclu.

**debunk** [dîˈbʌnk]. (*coll.*) 'Putları kırmak'.

**debut** [ˈdêybyû]. (Aktör) sahneye ilk çıkış; (genc kız) sosyeteye ilk giriş.

**decade** [diˈkêyd]. On yıl. **~nce** [ˈdekədens]. Tereddi, gerileme. **~nt,** inhitat eden.

**decamp** [diˈkamp]. Gizlice sıvışmak, savuşmak.

**decapitate** [diˈkapitêyt]. Başını kesmek.

**decay** [diˈkêy]. Çürüme(k); bozulma(k).

**decease** [diˈsîs]. Ölüm. Vefat etm. **~d,** merhum.

**decei·t** [diˈsît]. Aldatma, hile, yalan. **~tful,** aldatıcı, hilekâr. **~ve** [diˈsîv]. Aldatmak, hile yapmak.

**December** [diˈsembə(r)]. Aralık ayı.

**decen·cy** [ˈdîsnsi]. Edeb; nezahet. **the (common) decencies,** edeb. **~t** [ˈdîsnt]. Edebli; münasib; kâfi; (*coll.*) rabıtalı.

**decentralize** [dîˈsentrəlâyz]. Ademi merkezileştirmek.

**decept·ion** [diˈsepşn]. Aldatma, hile. **~ive** [–ˈtiv], aldatıcı.

**decide** [diˈsâyd]. Karar vermek; (hakkında) hüküm vermek; (birine bir şey hakkında) karar verdirmek. **~d,** kararlaştırılmış; kat'î; kat'î fikirli. **~dly,** muhakkak.

**deciduous** [diˈsiduəs]. Dökülür.

**decimal** [ˈdesiml]. Ondalı, âşarî.

**decimate** [ˈdesimêyt]. Büyük bir kısmını öldürmek.

**decipher** [dîˈsâyfə(r)]. Şifreyi okumak.

**decis·ion** [di'sijn]. Karar; kat'î fikirlilik. **~ive** [–'sâysiv], kat'î.
**deck**[1] [dek] *n.* Güverte. *v.* **~ over,** güverte koymak. **the lower ~,** alt güverte; bahriye erleri. **double-~er bus,** iki katlı otobüs. **~-chair,** açılıp kapanır sandalye. **~-hand,** âdi gemici. **~-house,** güverte kamarası. **~**[2] *v.* Süslemek, donatmak.
**declar·ation** [ˌdeklə'rêyşn]. Beyan, ifade. **~e** [di'kleə(r)]. Açıkça söylemek; beyan etm., ifade etm. **to ~ for [against] stg.,** bir şeyin lehinde [aleyhinde] olduğunu söylemek: **have you anything to ~?,** (gümrükte) gümrüğe tabi bir şeyiniz var mı? **~ed,** alenî.
**decline**[1] [di'klâyn] *n.* İnme; inhitat; zeval. **to be on the ~,** azalmağa yüz tutmak; rağbetten düşmek: **to go into a ~,** vereme tutulmak. **~**[2] *v.* Nazikâne reddetmek; kabul etmemek; meyletmek; zayıflamak; azalmak; tasrif etmek. **in one's declining years,** hayatın sonuna doğru.
**decode** [dî'koud]. Şifreyi çözmek.
**decompos·e** [ˌdîkəm'pouz]. Tefessüh etm., çürü(t)mek; (unsurlara) ayr(ıl)mak. **~ition** [pe'zişn], (unsurlara) ayrılma; çürüme.
**decorat·e** ['dekərêyt]. Süslemek, donatmak; nişan vermek. **~ion** [–'rêyşn], süsleme, donatma; nişan (verme). **~ive,** süsleyici. **~or,** mefruşatçı.
**decorous** ['dekərəs]. Terbiyeye uygun.
**decoy** [di'kôy]. Tuzak; yem; çağırtkan (kuş). Tuzağa düşürmek; hile ile cezbetmek.
**decrease** *n.* ['dîkrîs] azalma. *v.* [di'krîs] azal(t)mak. **to be on the ~,** gittikçe azalmak.
**decree** [di'krî]. İrade, hüküm. İrade etm.
**decrepit** [di'krepit]. İhtiyar ve dermansız; bitkin; köhne.
**dedicat·e** ['dedikêyt]. Vakfetmek; takdis etm.; ithaf etmek. **~ion** [–'kêyşn], vakıf, takdis; ithaf.
**deduc·e** [di'dyûs]. Netice çıkarmak, istidlâl etmek. **~t** [di'dʌkt]. Hesabdan tenzil etm.; çıkarmak. **~tion,** tenzil edilen mikdar; istidlâl. **~tive,** istidlâlî.
**deed** [dîd]. Fiil; hareket; sened, mukavelename, hüccet. **in ~,** hakikatte: ⌜**~s not words**⌝, ⌜ayinesi iştir kişinin lâfa bakılmaz⌝. **~-box,** evrak kutusu.
**deep** [dîp] *a.* Derin; gür (ses); koyu (renk). *n.* Derin yer. **the ~,** deniz: **to go off the ~ end,** (*coll.*) hiddetlenmek: **~ into the night,** gecenin ilerlemiş saatlerinde: **in the ~ of winter,** karakışta: **to commit a body to the ~,** bir ölüyü denize gömmek: **two [four] ~,** iki [dört] sıra: ⌜**still waters run ~**⌝, derin düşünen insanlar çok konuşmaz. **~-chested,** geniş göğüslü. **~-laid plan,** gizlice ve meharetle hazırlanmiş plân. **~-seated,** köklü. **~-set eyes,** çukur göz. **~en** ['dîpn]. Derinleş(tir)mek; koyulaş(tır)mak.
**deer** [diə(r)]. Geyik. **~skin,** ceylan derisi.
**deface** [di'fêys]. Görünüşünü bozmak.
**defam·ation** [ˌdîfa'mêyşn]. İftira. **~atory** [–'famətəri], iftiralı. **~e** [di'fêym], iftira etm.
**default** [di'fôlt] *n.* (Bir şeyi yapmakta vs.) kusur; gıyab; noksan. *v.* (Bir şeyi yapmakta) kusur etm.; mahkemede hazır bulunmamak; borclarını ödeyememek. **judgement by ~,** gıyabî hüküm: **in ~ of,** hazır bulunmadığı için. **~er,** borclarını ödemiyen; (*jur.*) gaib; (*mil.*) suçlu.
**defeat** [di'fît]. Mağlubiyet, bozgun. Mağlub etm., yenmek. **~ism,** bozgunculuk.
**defect** [di'fekt]. Noksan, kusur; sakatlık. **~ion** [di'fekşn]. Mensub olduğu parti, ordudan çekilme. **~ive** [di'fektiv]. Kusurlu, noksan, sakat; (*gram.*) eksik sıygalı.
**defen·ce** [di'fens]. Müdafaa, himaye, koruma; müdafaaname. **~ces,** müdafaa siperleri. **counsel for the ~,** müdafaa vekili. **~d** [di'fend]. Müdafaa etm., himaye etm., korumak; tarafını tutmak. **~dant,** dâva edilen. **~der,** müdafi. **~sible** [di'fensibl]. Müdafaa edilebilir. **to be [stand] on the ~sive,** müdafaada kalmak.
**defer**[1] [di'fö(r)]. Tehir etm. **~red payment,** tehir edilen ödeme; taksitle tediye. **~ment,** tehir. **~**[2]. **to ~to,** ···e hürmet etmek. **~ence** ['defərəns], riayet: **in [out of] ~,** hürmeten: **with all due ~ to you,** hatırınız kalmasın! **~ential** [ˌdefə'renşiəl]. hürmetkâr.
**defian·ce** [di'fâyəns]. Meydan okuma. **to bid ~ to s.o. [to set s.o. at ~],** birisine meydan okumak. **in ~ of the law,** kanunu hiçe sayarak. **~t,** meydan okuyan.

**defici·ency** [diˈfişnsi]. Noksan; açık. **~ent,** kusurlu, noksan. **~t** [ˈdefisit]. (Bütçede vs.) açık; noksan.
**defile**[1] [ˈdîfâyl] *n.* Pek dar geçid. **~**[2] [diˈfâyl] *v.* Pisletmek.
**defin·able** [diˈfâynəbl]. Tarif edilebilir. **~e** [diˈfâyn]. Tarif etm.; tesbit ve tayin etmek. **~ite** [ˈdefinit]. Kat'î, muayyen: **~ly,** muhakkak. **~ition** [–ˈnişn], tarif; izah, tayin. **~itive** [–ˈfinitiv], nihaî; kat'î.
**deflat·e** [dîˈflêyt]. Havasını boşaltarak indirmek; azaltmak. **~ion** [–ˈflêyşn], (lâstik vs.) havasını boşaltarak indirme; fiatları indirmek için piyasadaki para mikdarını azaltma.
**deflect** [diˈflekt]. İnhiraf et(tir)mek.
**deforest** [dîˈforest]. Bir bölgenin ağaclarını kesmek.
**deform** [diˈfôm]. Şeklini bozmak. **~ed,** biçimsiz, çirkin. **~ation,** şeklini bozma. **~ity** [–ˈfômiti], biçimsizlik; sakatlık.
**defraud** [diˈfrôd]. Dolandırmak. **to ~ s.o. of stg.,** birini hile ile bir şeyden mahrum etm.
**defray** [diˈfrêy]. (Masraf vs.yi) ödemek.
**deft** [deft]. Mahir, usta.
**defy** [diˈfây]. Meydan okumak. **I ~ you to do so!,** yap da göreyim!: **to ~ description,** tasviri imkânsiz olmak.
**degenerate** [dîˈcenərit]. *a.* Soysuzlaşmış; *v.* [–rêyt], tereddi etm., soysuzlaşmak. **~tion** [–ˈrêyşn], tereddi.
**degrad·ation** [ˌdegrəˈdêyşn]. Rütbe indirme; zillet, şerefsizlik. **~e** [dîˈgrêyd], rütbesini indirmek; tezlil etm., haysiyetini kırmak.
**degree** [diˈgrî]. Derece; rütbe; ünvan; paye. **by ~s,** yavaş yavaş: **to some ~,** bir dereceye kadar: **to take one's ~,** bir üniversiteden mezun olm.
**deify** [ˈdî·ifây]. Tanrılaştırmak.
**deign** [dêyn]. Tenezzül etmek.
**deity** [ˈdî·iti]. İlâh, tanrı.
**deject·ed** [diˈcektid]. Kederli, keyfi kaçmış; süngüsü düşük. **~ion** [–şn], meyusluk, melâl, keyifsizlik.
**delay** [diˈlêy]. Gecik(tir)me(k) tehir (etm.). **~ed-action,** tavikli.
**delegat·e** [ˈdeligit] *n.* Mürahhas. *v.* [–gêyt], Mürahhas olarak göndermek. **~ion** [–şn], mürahhas heyeti.
**delet·e** [diˈlît]. Silip çıkarmak. **~ion** [–şn], silip çıkarma.
**deliberat·e** [diˈlibərit] *a.* Kasdî; mahsus; düşünceli; temkinli. *v.* [–rêyt] Uzun uzadıya düşünmek, müzakere etmek. **~ion** [–rêyşn], teemmül; müzakere; dikkat ve itina. **~ive** [–tiv], müzakereye aid.
**delica·cy** [ˈdelikəsi]. Zarafet, incelik, nezaket; hassaslık; nahiflik. **table delicacies,** nefis yiyecek. **~te** [ˈdelikit], ince, zarif, nazik, hassas; nahif: **to tread on ~ ground,** nazik bir meseleye dokunmak.
**delicious** [diˈlişəs]. Nefis, hoş, tatlı, leziz.
**delight** [diˈlâyt]. Zevk, safa; neş'e. Çok zevk vermek; sevin(dir)mek; zevk almak, bayılmak. **~ful,** pek hoş.
**delimit** [dîˈlimit]. Hududunu çizmek.
**delinquen·cy** [diˈlinkwənsi]. Suçluluk. **~t,** suçlu.
**deliri·ous** [diˈliriəs]. Sayıklıyan; çılgın. **~um,** sayıklama; **~ tremens,** hezeyanı mürteiş.
**deliver** [diˈlivə(r)]. Kurtarmak; teslim etm.; (mektub vs.) tevzi etmek. **to ~ a message,** başkasına aid bir haberi vermek: **to ~ s.o., stg. (up, over) to s.o.,** birine bir şeyi teslim etm.: **to ~ a woman (of a child),** bir kadını doğurtmak. **~ance** [diˈlivrəns]. Kurtarış, ifade etme. **~y,** teslim; tevzi; konuşma verme tarzı; doğurma: **to accept ~ of,** teslim almak: **to take ~ of,** tesellüm etm.: **on ~,** tesliminde.
**delu·de** [diˈlyûd]. Aldatmak; kandırmak. **~sion** [diˈlûjn]. Aldatma; aldanma; hayal. **under the ~,** vehminde. **~sive** [–siv], aldatıcı.
**deluge** [ˈdelyuc]. Tufan; şiddetli yağmur. Sel basmak, tufana boğmak; çok ıslatmak.
**demand** [diˈmând]. Taleb; isteme; ihtiyac. Taleb etm., istemek; icabetmek. **to be in great ~,** çok rağbette olm.: **I have many ~s upon my time,** vaktim doludur: **payable on ~,** ibrazında tediye olunacak.
**demean** [diˈmîn]. **~ oneself,** kendini alçaltmak. **~our,** hal, tavır.
**demented** [diˈmentid]. Deli, çılgın.
**demi-** [ˈdemi]. *pref.* Yarı ....
**demilitarize** [dîˈmilitərâyz]. Gayrıaskerî hale getirmek.
**demobilize** dîˈmoubilâyz]. Terhis etmek.
**democra·cy** [diˈmokrəsi]. Halk hükûmeti. **~t** [ˈdeməkrat], halkçı. **~tic** [–ˈkratik], halk hükûmetine aid.

**demol·ish** [di'moliş]. Yıkmak, tahrib etmek. **~ition** [demə'lişn], yıkma, tahrib.
**demon** ['dîmən]. Şeytan, iblis. **~iac** [dî'mouniak], şeytanî, cılgın.
**demonstra·ble** [di'monstrəbl]. İsbat veya izah edilebilir. **~te** ['demənstrêyt]. Tecrübe, tatbikat ile isbat etm.; izah etm., iyice göstermek; nümayişte bulunmak. **~tion** [–'strêyşn], tatbikat ile isbat, izah; nümayiş. **to make a ~**, tezahuratta bulunmak, nümayiş yapmak. **~tive** [–'monstrətiv], coşkun; hislerini saklıyamayıp açıkça gösteren: **~ adjective,** işaret sıfatı. **~tor** ['demənstrêytə(r)], nümayişçi; profesörün asistanı.
**demoraliz·e** [di'morəlâyz]. Ahlâkını bozmak; yeis vermek. **~ation** [–'zêyşn], ahlâkını bozma, cesaretini kırma.
**demure** [di'myuə(r)]. Uslu, çekingen.
**den** [den]. İn; sığınak; ufak oda.
**denationalize** [dî'naşənəlâyz]. Devlet inhisarından çıkarmak.
**denia·ble** [di'nâyəbl]. İnkâr edilebilir. **~l** [di'nâyəl]. İnkâr; red; feragat. **a ~ of justice,** ihkakı haktan imtina: **I will take no ~,** muhakkak ···melisiniz.
**Denmark** ['denmâk]. Danimarka.
**denominat·e** [di'nominêyt]. **to ~ s.o. (stg.) as ...,** ... adını vermek, tavsif etmek. **~ion** [–'nêyşn], isim; ad verme; zümre, nevi, cins. **~or** [–'nominêytə(r)] (*math.*) mahrec.
**denote** [di'nout]. Göstermek.
**denounce** [di'nauns]. Alenen itham etm.; şiddetle aleyhinde bulunmak; muahede vs.nin bittiğini haber vermek.
**dens·e** [dens]. Sık, kesif, koyu; abdal. **~ity,** sıklık, kesafet, koyuluk.
**dent** [dent]. Çentik. Çentmek.
**dent·al** ['dentl]. Dişe aid. **~ surgeon,** diş doktoru. **~ifrice** [–ifris], diş macunu, tozu. **~ist** [–ist], dişçi. **~istry,** dişçilik. **~ure** ['dençə(r)], takma diş; dişler.
**denud·ation** [dînyu'dêyşn]. Çıplak bırakma. **~e** [dî'nyûd], çıplak bırakmak; mahrum etme.
**denunciat·ion** [di'nʌnsiêyşn]. Alenen itham etme; şiddetle aleyhinde bulunma; (bir muahede vs.nin) yenilenmiyeceğini haber verme.
**deny** [di'nây]. İnkâr etm.; tanımamak. **to ~ stg. to s.o.,** birine bir şeyi vermemek: **to ~ oneself stg.,** kendini bir şeyden mahrum etm.: **there's no ~ing that,** ···dığı inkâr edilemez: **he is not to be denied,** ona red cevabı verilemez.
**depart** [di'pât]. Gitmek, ayrılmak. **~ from,** terketmek. **~ed,** gitmiş, ayrılmış: **the ~,** merhum. **~ure** [di'pâçə(r)]. Gitme, ayrılma, mufarakat, azimet, kalkış. **a new ~,** yeni bir temayül, âdet.
**department** [di'pâtmnt]. Daire, şube, kalem; kısım. **~ store,** büyük mağaza. **~al** [–'mentl], şube, daireye aid.
**depend** [di'pend]. Asılı olm., asılmak. **~ on,** ···e bağlı olm., tâbi olm.; güvenmek: **that ~s [it all ~s],** belli olmaz: **that ~s on you,** bu size bağlıdır: **to ~ on s.o.,** geçimi birine bağlı olm.: **to ~ (up)on s.o.,** birine güvenmek; birinden emin olmak. **~able,** güvenilir, emin. **~ant,** başkasının himaye, yardımına muhtac olan kimse. **~ence** [di'pendens]. Bağlılık; bel bağlama; güvenme. **to place ~ on s.o.,** birine güvenmek. **~ency,** müstemleke, tâbi yer; bağlı olma. **~–encies,** müştemilat. **~ent,** birine bağlı, tâbi; asılı.
**depict** [di'pikt]. Tasvir etm., göstermek.
**deplet·e** [di'plît]. Tüketmek. **~ion** [–plîşn], tüketme.
**deplor·e** [di'plô(r)]. Acımak; ···e müteessir olm.; fena bulmak. **~able,** acınacak, merhamete değer; berbad.
**depopulate** [dî'popyulêyt]. Nüfusunu boşaltmak, azaltmak.
**deport[1]** [di'pôt]. Memleketten dışarı tardetmek, sürmek. **~ation** [dîpô'têyşn], sürgün. **~[2]. to ~ oneself,** davranmak. **~ment,** tavır.
**depos·e** [di'pouz]. Azletmek, hal'etmek. **to ~ to a fact,** şehadet etmek. **~ition** [dîpou'zişn]. Hal'etme; (*jur.*) yazılı ifadesi.
**deposit** [di'pozit] *n.* Tortu; tabiî yığıntı; emanet, mevduat; pey. *v.* Koymak; tevdi etm.; yatırmak; pey vermek. **on ~,** emanette; faize yatırılan para. **~ary,** emanetçi. **~or,** emanet eden; para yatıran. **~ory,** ambar.
**depot** ['depou]. Ambar; ardiye.
**deprav·e** [di'prêyv]. İfsad etm. **~ed,** ahlâkı bozuk. **~ity** [–'praviti], ahlâk bozukluğu.
**depreciat·e** [di'prîşiêyt]. Kıymetini

düşürmek; kıymetten düşmek. **~ion,** kıymetini düşürme; (*comm.*) aşınma.

**depress** [di'pres]. Bastırmak, indirmek; neş'esini kırmak. **~ed,** kederli, süngüsü düşük. **~ing,** kasvetli. **~ion** [–'preşn], kasvet; (*comm.*) durgunluk; çukur.

**depriv·e** [di'prâyv]. Mahrum etm., zorla elinden almak. **~al,** mahrumiyet. **~ation** [ˌdepri'vêyşn], mahrum etme (olma).

**depth** [depθ]. Derinlik; boy; derin yer; tam ortası. **to get out of one's ~,** (suda) ayağı yerden kesilmek; (*fig.*) salâhiyeti haricine çıkmak: **in the ~s of despair,** tam bir ümıdsizlik içinde: **in the ~ of winter,** karakışta. **~-charge,** su bombası.

**deput·e** [di'pyût]. Vekil tayin etm. **~ation** [ˌdepyu'têyşn], murahhas heyeti; vekil tayin etme. **~ize** ['depyutâyz], birine vekâlet etmek. **~y** ['depyuti]. Vekil; meb'us; murahhas: **~-chairman,** reis vekili. **~-governor,** vali muavini.

**derail** ['dî'rêyl]. (Treni) yoldan çıkarmak.

**derange** [di'rêync]. (Sırasını vs.) bozmak, aklına dokunmak.

**Derby** ['dâbi]. **the ~,** 1780'denberi Epsom'da yapılan meşhur at yarışı.

**derelict** ['derəlikt]. Terkedilmiş; metrûk (gemi). **~ion** [–'likşn], terketme: **~ of duty,** vazifenin ihmali.

**deri·de** [di'râyd]. Alay etm. **~sion** [di'rijn]. İstihza, alay. **to hold s.o. in ~,** birile alay etmek.

**deriv·e** [di'râyv]. Çık(ar)mak; müştak olm. **to ~ pleasure,** zevk bulmak. **~ation** [ˌderi'vêyşn], iştikak; menşe. **~ative** [–'rivətiv], müştak.

**derogat·e** ['derogêyt]. **~ from,** azaltmak. **~ion** [–'gêyşn], ihlâl, dokunma. **~ory** [–'rogətəri], itibar kırıcı.

**descen·d** [di'send]. İnmek; alçalmak; (bir aileden) çıkmak; (babadan oğula) geçmek. **to ~ on s.o.,** birinin üzerine çullanmak: **well ~ded,** iyi aileye mensub. **~dant,** bir aileden gelen kimse, hafid. **~t** [di'sent]. İnme; nesil, şecere.

**descri·be** [di'skrâyb]. Tarif etm., tasvir etm.; anlatmak. **~ption** [dis'kripşn]. Tarif; anlatma; tasvir; çeşit. **to answer to s.o.'s ~,** birinin tasvirine uymak. **~ptive,** tasvirî; **~ of,** …i tasvir eden.

**desecrate** ['desikrêyt]. (Mukaddes bir şeye karşı) hürmetsizlik etm.

**desert**[1] [di'zö(r)t] *n.* Lâyik olan şey. **to get one's ~s,** lâyığını bulmak. **~**[2] ['dezət] *n.* Çöl; çorak. **~**[3] [di'zö(r)t] *v.* Bırakıp kaçmak; askerden kaçmak. **~ed,** terk edilmiş; hali, tenha. **~er,** asker kaçağı. **~ion** [–'zö(r)şn], bırakıp kaçma; firar.

**deserv·e** [di'zö(r)v]. Hak etm., lâyik olmak. **~edly,** haklı olarak. **~ing,** lâyik: **this is a ~ case,** bu adam yardıma lâyıktır.

**design** [di'zâyn] *n.* Plan; taslak, resim; model; maksad. *v.* Çizmek, planını yapmak; tertib etm.; niyet etmek. **by ~,** kasden: **with this ~,** bu maksadla. **~ing,** kurnaz. **~edly** [di'zâynidli]. Kasden. **~er** [di'zâynə(r)]. (Kumaş desenleri veya elbise modelleri çizen) ressam.

**designat·e** ['dezignêyt] *v.* Tayin etm., tahsis etm. *a.* Tayin edilmiş. **~ion,** tayin, tahsis; ünvan, sıfat.

**desir·e** [di'zâyə(r)]. Arzu, istek. Arzu etm., istemek. **it leaves much to be ~ed,** mükemmel olmaktan uzaktır. **~ability** [–ə'biliti], hoşa gitme: **the ~ of stg.,** bir şeyin faydalı olup olmadığı. **~able** [–ebl], makbul, hoş. **~ous** [–rəs], arzu eden, talib.

**desk** [desk]. Yazı masası; kasa; mekteb sırası.

**desolat·e** ['desəlit] *a.* Issız, tenha; viran; perişan; kimsesiz. *v.* ['desolêyt], Harab etm., perişan etm. **~ion,** [desə'lêyşn], harablık; ıssızlık.

**despair** [dis'peə(r)]. Yeis, çaresizlik. Ümidini kesmek.

**despatch** *v.* **dispatch.**

**desperat·e** ['despərit]. Ümidsiz; çok tehlikeli; şiddetli. **~ion** [–'rêyşn], ümidsizlik, çaresizlik.

**despicable** [dis'pikəbl]. Alçak.

**despise** [dis'pâyz]. Hakir görmek.

**despite** [dis'pâyt]. **~ [in ~ of],** …e rağmen.

**despoil** [dis'pôyl]. Soymak.

**despond** [dis'pond]. Yeis. Ümidsizliğe düşmek. **~ency,** ümidsizlik; bedbinlik. **~ent,** ümidsiz.

**despot** ['despot]. Müstebid. **~ic** [–'potik], müstebid. **~ism** despətizm], istibdad.

**dessert** [di'zö(r)t]. (Yemeğin sonunda) meyva.

**destin·ation** [ˌdesti'nêyşn]. Gönderilen veya gidilecek yer. **~e** ['destin]. Tahsis etm.; nasib etmek. **to be ~ed,** tahsis edilmek: **I was ~d to see all this,** kaderimde bütün

bunları görmek de varmış. **~y** [ˈdestini]. Kader; talih.
**destitut·e** [ˈdestityût]. Yoksul; parasız. **~ion** [–ˈtyûşn], yoksulluk; mahrumiyet.
**destroy** [disˈtrôy]. Yıkmak, tahrib etm. **~er,** muhrib; tahrib eden.
**destruct·ible** [disˈtrʌktəbl]. Tahribi mümkün. **~ion** [disˈtrʌkşn], tahrib; mahvolma; harabe. **~ive,** tahrib edici. **~or,** tahrib edici.
**detach** [diˈtaç]. Ayırmak, çözmek: (*mil.*) hususî bir vazife ile göndermek. **~ed,** ayrı, müstakil; tarafsız. **~ment,** ayırma; müfreze; tarafsızlık.
**detail** [ˈdîtêyl]. Tafsilat; ayrı parça; hususî bir vazife için seçilen grup. Tafsil etm. **in ~,** mufassalan: **in every ~,** her noktada.
**detain** [diˈtêyn]. Alıkoymak; geciktirmek; hapsetmek.
**detect** [diˈtekt]. Meydana çıkarmak, keşfetmek, farketmek. **~ion** [–tekşn], meydana çıkarma, keşif: **to escape ~,** gözden kaçmak. **~ive,** polis hafiyesi: **~ story,** cinaî roman. **~or,** meydana çıkaran.
**detention** [diˈtenşn]. Alıkoyma; tevkif.
**deter** [diˈtö(r)]. Vazgeçirmek. **nothing will ~ him,** hiç bir şey onu durduramaz. **~rent** [diˈterənt]. Vazgeçiren, önleyici.
**deteriorat·e** [diˈti̯əri̯ərêyt]. Fenalaş(tır)mak. **~ion** [–ˈrêyşn], fenalaşma, kıymetten düşme.
**determin·e** [diˈtö(r)min]. Kat'î karar vermek; tesbit etm. **~able,** tayini mümkün. **~ate,** mahdud. **~ation** [–ˈnêyşn], azim, karar, tesbit: **an air of ~,** azimli tavır.
**detest** [diˈtest]. Nefret etm. **~able,** iğrenc; berbad. **~ation** [dîtesˈtêyşn], nefret: **to hold stg. in ~,** bir şeyden nefret etmek.
**dethrone** [diˈθro̯un]. Tahttan indirmek.
**detonat·e** [ˈdetənêyt]. Patla(t)mak.
**detour** [ˈdêytu̯ə(r)]. Sapa yol. **to make a ~,** başka yoldan dolaşmak.
**detract** [diˈtrakt]. **~ from,** azaltmak. **~or,** başkalarını çekiştiren.
**detriment** [ˈdetrimənt]. Zarar.
**deuce**[1] [dyûs]. (Oyun kâğıdı) ikili; (tenis) 40 sayı ile beraber vaziyet. **~**[2]. (*coll.*) Şeytan. **go to the ~!,** cehennem ol!: **we are in the ~ of a mess,** ayıkla pirincin taşını!: **he is the ~ of a liar,** sunturlu yalancıdır: **to play the ~ with stg.,** bir şeyi berbad etmek. **~d** [–sid], berbad; bir çok.
**devaluation** [dîvalyûˈêyşn]. Kıymetini düşürme.
**devastat·e** [ˈdevəstêyt]. Tahrib etm., mahvetmek. **~ion** [–ˈstêyşn], tahrib.
**develop** [diˈveləp]. İnkişaf et(tir)mek; genişle(t)mek, tedricen meydana çık(ar)mak; (itiyad) peyda etm.; (*fot.*) banyo ile izhar etmek. **let us see how things ~,** hadiselerin inkişafını bekliyelim. **~ment,** inkişaf; ilerleme; netice; hadise.
**deviat·e** [ˈdîviêyt]. Sapmak. **~ion** [–ˈêyşn], sapma, inhiraf.
**device** [diˈvâys]. Cihaz, alet; tedbir; hüner; hile; arma üzerindeki cümle. **to leave s.o. to his own ~s,** işine karışmamak.
**devil** [ˈdevl] *n.* Şeytan, iblis; habis ruh; Allahın belâsı; zalim; hain. **~ a one** [**bit**], hiç mi hiç: ⌜**between the ~ and the deep blue sea**⌝, ⌜aşağı tükürsem sakalım yukarı tükürsem bıyığım⌝: **the blue ~s,** iç sıkıntısı: **to give the ~ his due,** kötü adamın bile hakkını vermek: **go to the ~!,** cehennem ol!: **he has gone to the ~,** mahvoldu: **how the ~ do you know that?,** bunu da nereden biliyorsun?: **(to do stg.) like the ~,** alabildiğine: **there'll be the ~ to pay,** bunun acısı sonra çıkar: **to play the ~ with,** berbad etm.: **the poor ~,** zavallı: **to raise the ~,** kıyamet koparmak: ⌜**talk of the ~ (and he's sure to appear)**⌝, *kendisinden bahsedilirken çıkagelen biri hakkında kullanılır*: **what the ~ are you doing?,** ne halt ediyorsun?: **The ~ he is!,** 'Yok canım! deme!' **~-may-care,** hiç kimseye aldırmaz. **~ish** [ˈdeviliş]. Habis; şeytanî; berbad. **~ment** [**~ry**], habaset; şeytanlık; çılgınlık.
**devious** [ˈdîvi̯əs]. Sapa; iğri.
**devise** [diˈvâyz]. İcadetmek; kurmak.
**devoid** [diˈvôyd]. **~ of,** ···den arî, mahrum.
**devol·ution** [ˌdîvoˈlyûşn]. Miras yolu ile geçme; vazifenin başkasına devri. **~ve** [diˈvolv]. **to ~ on,** ···e devretmek; geçmek.
**devot·e** [diˈvo̯ut]. Vakfetmek; tahsis etm. **~ed,** candan bağlı. **to be ~ed to sport,** kendini spora vermek. **~ee** [devo̯uˈtî], düşkün; hayran; sofu

~ion [di'vouşn], sadakat; fedakârlık; dindarlık. ~ional, dua, ibadete aid. ~ions, dua, ibadet.
**devour** [di'vauə(r)]. Hayvan gibi yemek.
**devout** [di'vaut]. Dindar; çok bağlı, candan.
**dew** [dyû]. Çiğ. ~y, çiğ ile kaplı. ~-drop, çiğ damlası.
**dexter·ity** [deks'teriti]. Hüner, ustalık. ~ous ['dekstrəs], becerikli.
**D.F.C.** ['dî'ef'sî]. Distinguished Flying Cross. *Havacılara verilen bir nişan.*
**diabetes** [,dâyə'bîtîz]. Şeker hastalığı.
**diabolical** [,dâye'bolikl]. Mel'unca.
**diadem** ['dâyədem]. Tac.
**diagnos·e** ['dâyəgnouz]. (Hastalığı) teşhis etmek. ~is [–'nousis], teşhis.
**diagonal** [dây'agənl]. Çapraz (hat).
**diagram** ['dâyəgram]. Şekil, şema. ~matic [–'matik], şema halinde.
**dial** ['dâyəl] *n.* Saat minesi; taksimatlı daire; güneş saati; rakamları ihtiva eden daire. *v.* Otomatik telefonun numaralarını çevirmek.
**dialect** ['dâyəlekt]. Lehçe.
**dialogue** ['dâyəlog]. Muhavere.
**diamet·er** [dây'amətə(r)]. Kutur. ~rical [–'metrikl], kutrî: ~ly opposed, taban tabana zıd.
**diamond** ['dâyəmənd]. Elmas; baklava şekli; (iskambil) karo: ⌜~ cut ~⌝, ⌜dinsizin hakkından imansız gelir⌝: a rough ~, kaba fakat iyi kalbli: ~ wedding, bir düğünün altmışıncı yıldönümü.
**diaper** ['dâyəpə(r)]. Kundak bezi.
**diary** ['dâyəri]. Ruzname; muhtıra defteri.
**dice** [dâys]. Oyun zarları. Zar oynamak.
**dickens** ['dikənz] (*Nezaketen* devil *yerine kullanılır*) şeytan. what the ~ are you doing here?, burada ne halt ediyorsun?
**dicky** ['diki]. (*sl.*) Kırık, çürük; sarsak.
**dictat·e** [dik'têyt]. Söyleyip yazdırmak; zorla kabul ettirmek. I won't be ~ed to, ben emre gelemem. ~ion, imlâ; emretme. ~or, emreden kimse. ~orship, diktatörlük.
**diction** ['dikşn]. Konuşma şekli. ~ary ['dikşnri]. Lûgat kitabı; sözlük.
**did** *v.* do.
**diddle** ['didl]. (*coll.*) Aldatmak; yutturmak.
**die**[1] [dây] *n.* Oyun zarı; kalıb; ıstampa. the ~ is cast, ok yaydan çıktı. ~[2] (died, dying) [dây, dâyd, dây·in(g)] *v.* Ölmek, vefat etmek. to be dying to do stg., bir şeyi yapmağı şiddetle arzu etm.: 'never say ~!', cesaretini kaybetme!: to ~ away [down], gittikçe hafifleyip kaybolmak: to ~ off, birer birer ölmek: to ~ out, yavaş yavaş ortadan kalkmak.
**diet** ['dâyət] *n.* Yiyecek, gıda; perhiz yemeği. *v.* Perhize koymak. ~ary, gıda rejimi.
**differ** ['difə(r)]. Farketmek, benzememek, farklı olm. I beg to ~, müsaadenizle ben bu fikirde değilim. ~ence ['difərəns]. Fark; ihtilaf. ~s arose, münakaşa çıktı: that made all the ~, bu her şeyi değiştirdi: it makes no ~, hepsi bir: settle your ~s, anlaşınız: to split the ~, farkı paylaşmak. ~ent, farklı; muhtelif; başka: I feel a ~ man, kendimi bambaşka hissediyorum: that's a ~ matter, o başka mesele. ~entiate [difə'renşiêyt], ayırd etm., fark etm.
**difficult** ['difiklt]. Güç, zor, müşkül. ~ (to get on with), titiz, huysuz. ~y, zorluk, müşkûlat; sıkıntı.
**diffiden·ce** ['difidəns]. Çekingenlik. ~t, çekingen.
**diffus·e**[1] [di'fyûs] *a.* Pek tafsilatlı; yayılmış. ~[2] [di'fyûz] *v.* Yaymak, dağıtmak. ~ion [–fyûjn], yayılma.
**dig**[1] [dig] *n.* Kazma; dürtme. to give s.o. a ~ in the ribs, birini dürtmek: to have a ~ at s.o., (*fig.*) birini dürtmek. ~[2] (dug) [dig, dʌg] *v.* Kazmak; çukur açmak; dürtmek. to ~ away at stg., (*coll.*) çok çalışmak: to ~ in, siper kazmak: to ~ into [through], (kazıp) delmek: to ~ one's toes in, direnmek: to ~ out, kazıp çıkarmak: to ~ up, meydana çıkarmak. ~ging, *n.* kazma. ~gings (*abb.* digs), pansiyon.
**digest**[1] ['dâycest] *n.* İcmal, hulâsa. ~[2] [di'cest, dây'cest] *v.* Hazmetmek; sindirmek. ~ible, hazmı kolay. ~ion [–'cesçən], hazım. ~ive, midevî.
**digit** ['dicit]. Parmak; rakam.
**digni·fied** ['dignifâyd]. Vakur, ağır başlı. ~fy ['dignifây]. Yükseltmek, şeref vermek. ~tary ['dignitəri]. Yüksek rütbeli olan. ~ty ['digniti]. Vakar, haysiyet; yüksek makam, rütbe. to be [stand] on one's ~, yukarıdan almak: it is beneath your ~ to accept it, bunu kabul etmeğe tenezzül edemezsiniz.
**digress** [dây'gres]. Mevzudan ayrıl-

mak. ~**ion** [–ˈgreşn], sadedden ayrılma.

**dilapidat·e** [diˈlapidêyt]. Harab etm.; kırıp dökmek. ~**ed,** harab, köhne. ~**ion** [–ˈdêyşn], harabolma.

**dilat·e** [dâyˈlêyt]. İnbisat ettirmek; (gözler) büyümek. **to ~ upon stg.**, bir mevzuu uzun uzadıya anlatmak. ~**ion** [–ˈlêyşn], inbisat.

**dilatory** [ˈdilətri]. Bati; sürüncemeli.

**dilemma** [dâyˈlemə]. İki şıklı [müşkül] vaziyet; çıkmaz.

**diligen·ce** [ˈdilicens]. Gayretli çalışma. ~**t,** gayretli, çalışkan.

**dilly-dally** [ˈdiliˈdali]. Boş vakit geçirmek.

**dilute** [dâyˈlyût] *a.* Sulandırılmış. *v.* Sulandırmak, hafifletmek.

**dim** [dim] *a.* Donuk, loş, hafif. *v.* Karartmak, donuklaştırmak. ~**mer** [ˈdimə(r)]. Işığı azaltan tertibat.

**dimension** [dâyˈmenşn]. Eb'ad, ölçü. ~**al,** buudlu.

**dimin·ish** [diˈminiş]. Azaltmak, indirmek. ~**ution** [ˌdimiˈnyûşn]. Azal(t)ma, in(dir)me. ~**utive** [diˈminyutiv], ufak; tasgir.

**dimple** [ˈdimpl]. Yanak [çene] çukuru.

**din** [din] *n.* Gürültü, patırdı. *v.* **to ~ stg. into s.o. [s.o.'s ears]**, mütemadiyen söyliyerek bir şeyi birinin kafasına sokmak.

**din·e** [dâyn]. Akşam yemeğini yemek. **to ~ out,** dışarıda yemek. ~**ing-car,** yemek vagonu. ~**ing-room,** yemek odası. ~**ner** [ˈdinə(r)]. Akşam yemegi; ziyafet. ~**ner-jacket,** smokin.

**ding-dong** [ˈdin(g)ˈdon(g)]. Çan sesleri. **a ~ struggle,** kâh bir tarafın, kâh öbür tarafın lehine inkişaf eden mücadele.

**dinghy** [ˈdin(g)i]. Pek küçük sandal.

**dingy** [ˈdinci]. Rengi solmuş, kirli.

**dint** [dint]. **by ~ of,...** kuvvetile, vasıtasile: **by ~ of working,** çalışa çalışa.

**diocese** [ˈdâyəsis]. Piskoposluk sahası.

**dip** [dip] *n.* Dal(dır)ma; içine bir şey daldıran madde; çukur. *v.* Dal(dır)mak, batırıp çıkarmak; (yol) iniş olm. **to ~ a flag,** bayrağı arya etm.: **to ~ into a book,** bir kitabı gözden geçirmek: **to ~ into one's purse,** çok masrafa girmek: **to ~ sheep,** koyunları ilâclı suya daldırmak.

**diphthong** [ˈdifθon(g)]. Bir hece teşkıl eden iki sesli harf.

**diploma** [diˈplǫumə]. Şehadetname.

**diploma·cy** [diˈplǫuməsi]. Diplomatik; diplomasi; maharet. ~**tic,** [ˌdipləˈmatik], maharetli, usûl bilen: ~ **service,** hariciye hizmeti. ~**t(ist)** [-ˈmat, –ˈplǫumətist], diplomat.

**dipsomania** [ˌdipsǫuˈmêyniə]. İçki ibtilası. ~**c,** ayyaş.

**dire** [dâyə(r)]. Dehşetli.

**direct**[1] [dâyˈrekt, di-] *v.* Sağlık vermek; idare etm.; tevcih etm.; emir vermek. ~[2] *a.* Dosdoğru; tam; tok sözlü. ~ **object,** (*gram.*) sarih mef'ul. ~**ly,** doğrudan doğruya; hemen: **I will come ~ I've finished,** bitirir bitirmez gelirim. ~**ion** [diˈrekşn, dây-]. İdare; talimat; emir; tarif; adres; cihet. **to lose one's sense of ~,** nerede olduğunu bilememek. ~**ional,** istikamete aid. ~**or** [diˈrektə(r), dây-]. Müdür; şirketin idare âzası; cihet verici alet. ~**orate** [diˈrektərit], müdürlük, idare meclisi. ~**ory** [diˈrektəri]. Rehber.

**dirge** [dö(r)c]. Cenaze şarkısı.

**dirt** [dö(r)t]. Kir, pislik; çamur. **to eat ~,** (*coll.*) tarziye vermeğe mecbur olm.: **to throw ~ at s.o.,** birini çamura bulamak: **to treat s.o. like ~,** birine köpek muamelesi etmek. ~**-cheap,** sudan ucuz. ~**-track,** kül dökülmüş yarış yolu. ~**y** [ˈdö(r)ti] *a.* Pis, kirli; berbad; aşağılık, alçak. *v.* Kirletmek; kirlenmek. **to have a ~ mind,** aklı daima müstehcen şeylerde olm.: **to play a ~ trick on s.o.,** birine âdi bir oyun oynamak: **do your own ~ work!,** beni bu şübheli işe sokma!

**dis-** [dis] *pref. Şu manalara gelir:*—(i) aksi: **contented,** memnun; **discontented,** gayrimemnun; (ii) yapılan bir şeyi bozma: **hearten,** cesaret vermek; **dishearten,** cesaretini kırmak; (iii) uzaklaştırma: **to disperse,** dağıtmak.

**disab·ility** [ˌdisəˈbiliti]. Sakatlık; kabiliyetsizlik. ~**le** [disˈêybl]. Sakat etm.; hasara uğratmak. ~**led,** malûl.

**disabuse** [ˌdisəˈbyûz]. Gözünü açmak.

**disadvantage** [ˌdisədˈvântic]. İnsanın aleyhine olan vaziyet vs. **to be at a ~,** (başkalarına nisbetle) daha zayıf bir vaziyette olmak. ~**ous** [–ədvanˈtêycəs], aleyhine olan, gayri müsaid.

**disaffected** [ˌdisəˈfektid]. Hükûmete karşı gayrimemnun; asi.

**disagree** [ˌdisəˈgrî]. İhtilâf etm.; farklı olm. ~ **with,...** ile uyuşmamak,

fikri başka olm.; (sıhhat vs.) dokunmak. I ~, ben bu fikirde değilim. **~able** [–ˈgriəbl], nahoş; huysuz. **~ment,** ihtilâf, kavga.
**disappear** [disəˈpiə(r)]. Gözden (ortalıktan) kaybolmak. **~ance** [ˈpiərəns], gözden kaybolma.
**disappoint** [disəˈpôynt]. Ümidini boşa çıkarmak; vadini tutmamak. **I am ~ed in him,** o beklediğim gibi çıkmadı: **how ~ing!,** ne aksilik! **~ment,** ümidi boşa çıkma.
**disapprov·e** [ˌdisəˈprûv]. Tasvib etmemek. **~al,** takbih.
**disarm** [disˈâm]. Silâhını almak; şübhe hislerini gidermek. **~ament** [–məmənt], silâhsızlama.
**disarrange** [ˌdisəˈrêync]. Tertibini bozmak.
**disast·er** [diˈzâstə(r)]. Felâket. **~rous,** feci.
**disband** [disˈband]. Terhis etm. dağılmak.
**disbelie·f** [ˈdisbəˈlîf]. İmansızlık. **~ve,** inanmamak. **~ver,** iman etmiyen.
**disc** [disk]. Daire. **identity ~,** (*mil.*) künye.
**discard** [disˈkâd]. Bertaraf etm.
**discern** [diˈsö(r)n]. Farketmek. **~ing,** zeki. **~ible,** farkedilebilir. **~ment,** anlayış.
**discharge**[1] [disˈçâc] *n.* Boşaltma; boşalan şey; cerahat; (silâh) atış; işten çıkarılma; tahliye; ödeme; ifa. **in the ~ of his duties,** vazifesinin ifası sırasında. ~[2] *v.* Boşaltmak, serbest bırakmak; ateş etm.; ödemek; ifa etm.; işten çıkarmak; tahliye etm.; cerahat akmak.
**disciple** [diˈsâypl]. Şakird.
**disciplin·e** [ˈdisiplin]. İnzibat. Terbiye etmek. **~arian** [–ˈneəriən], sert amir. **~ary** [–əri], disipline aid.
**disclaim** [disˈklêym]. Feragat etm.; inkâr etm.
**disclose** [disˈklouz]. İfşa etmek.
**discolour** [disˈkʌlə(r)]. Rengini bozmak.
**discomfort** [disˈkʌmfət]. Rahatsızlık. Rahatsız etm.
**discompose** [ˌdiskəmˈpouz]. Bozmak.
**disconcert** [ˌdiskənˈsö(r)t]. Şaşırtmak. **~ing,** şaşırtıcı; daima beklenmedik şekilde hareket eden.
**disconnect** [ˌdiskəˈnekt]. Birbirinden ayırmak. **~ed,** rabıtasız.
**disconsolate** [disˈkonsəlit]. Kederli.
**discontent** [ˌdiskənˈtent]. Hoşnudsuzluk. **~ed,** hoşnudsuz.
**discontinu·e** [ˌdiskənˈtinyû]. Vazgeçmek, kesmek. **~ance,** kesilme. **~ity** [–kontıˈnyûiti], devamsızlık.
**discord** [ˈdiskôd]. İhtilaf; ahenksizlik. **to sow ~,** aralarını bozmak. **~ant** [–ˈkôdənt], ahenksiz.
**discount** [ˈdiskaunt] *n.* İskonto. *v.* [–ˈkaunt], İskonto etm.; (sened) kır(dır)mak. **at a ~,** iskonto ile.
**discourage** [disˈkʌric]. Cesaretini kırmak.
**discourse** [ˈdiskôs] *n.* Hitabe. *v.* [–ˈkôs], Konuşmak.
**discover** [disˈkʌvə(r)]. Keşfetmek; anlamak; meydana çıkarmak. **~y,** keşif.
**discredit** [disˈkredit]. İtibarsızlık; itimadsızlık. *v.* İtibardan düşürmek; inanmamak. **it is to his ~ that, ...** onun aleyhine kaydedilecek bir şeydir. **~able,** haysiyet kırıcı.
**discre·et** [disˈkrît]. Ketum. **~tion** [disˈkreşn]. Ketumluk, teenni. **to reach years of ~,** (*jur.*) mümeyyiz olm.: **to use ~,** teenni ile hareket etmek. **~tionary,** ihtiyarî: **~ power,** takdir salâhiyeti.
**discrepan·cy** [disˈkrepənsi]. Uymama; fark. **~t,** uymıyan, farklı.
**discriminat·e** [disˈkriminêyt]. Ayırd etm.; tefrik etmek. **to ~ between people,** farklı muamele etm. **~ing,** ehil; titiz. **~ion** [–ˈnêyşn], ayırd etme; muhakeme; **no ~!,** ayrı seçi yok!
**discuss** [disˈkʌs]. ...i görüşmek; müzakere etm. **~ion** [–ˈkʌşn], müzakere; bahis: **the subject under ~,** bahis mevzuu olan mesele.
**disdain** [disˈdêyn]. İstihfaf (etm.). **~ful,** istihfafkâr.
**disease** [diˈzîz]. İllet. **~d,** illetli.
**disembark** [ˌdisemˈbâk]. Karaya çık(ar)mak. **~ation** [–ˈkêyşn], karaya çık(ar)ma.
**disengage** [ˌdisenˈgêyc]. Rabıta ve alâkasını kesmek. **~d,** meşgul olmıyan.
**disfavour** [disˈfêyvə(r)]. **to fall into ~,** gözden düşmek.
**disfigure** [disˈfigə(r)]. Biçimini bozmak.
**disgorge** [disˈgôc]. Kusmak; zorla geri vermek.
**disgrace** [disˈgrêys]. Gözden düşme; rezalet; ayıb. Gözden düşürmek; rezil etmek. **to be in ~,** gözden düşmüş olm. **~ful,** rezil, çok ayıb.
**disguise** [disˈgâyz]. Kıyafet tebdili. Kıyafetini değiştirmek; gizlemek.

**disgust** [dis'gʌst]. Tiksinme; nefret. Çok canını sıkmak. **~ing,** iğrenc.
**dish** [diş]. Büyük yemek tabağı; yemek. (*sl.*) haklamak, işini bozmak: **~ oneself,** kendi kendini mahvetmek: **to ~ up,** kotarmak. **~-cloth,** bulaşık bezi. **~-water,** bulaşık suyu; pek sulu ve tadsız çorba.
**dishearten** [dis'hâtn]. Cesaretini kırmak.
**dishon·est** [dis'onist]. Namussuz. **~esty,** namussuzluk. **~our** [dis'onə(r)]. Şerefsizlik; leke. Şeref ve haysiyetini kırmak. **to ~ a bill,** bir poliçeyi kabul etmemek: **to ~ one's word,** sözünde durmamak. **~ourable** [–'onrəbl], namussuz; rezil.
**disillusion** [ˌdisi'lûjn]. Gözünü açma(k).
**disinclination** [ˌdisinkli'nêyşn]. İsteksizlik.
**disinfect** [ˌdisin'fekt]. Dezenfekte etmek. **~ant,** antiseptik (madde).
**disinherit** [ˌdisin'herit]. Mirastan mahrum etmek.
**disintegrate** [dis'intigrêyt]. Küçük parçalara ayır(ıl)mak.
**disinterested** [dis'intərestid]. Menfaat düşünmiyen.
**dislike** [dis'lâyk]. Beğenmeme(k). **to take a ~ to s.o.,** birinden soğumak.
**dislocate** ['disləkêyt]. Yerinden çıkarmak; altüst etmek. **~d,** çıkık; altüst.
**dislodge** ['disloc]. Yerinden oynatmak.
**disloyal** [dis'lôyəl]. Sadakatsiz. **~ty,** vefasızlık.
**dismal** ['dizml]. Kederli; sönük.
**dismantle** [dis'mantl]. Sökmek.
**dismay** [dis'mêy]. Korku, hayretten donup kalma. Cesaretini kırmak.
**dismiss** [dis'mis]. Savmak; gitmeğe izin vermek; kovmak; (davayı) reddetmek. **~al,** işten çıkarma; reddetme.
**dismount** [dis'maunt]. (Attan) in(dir)mek; (makineyi) sökmek.
**disobliging** [ˌdisə'blâycin(g)]. Aksi.
**disorder** [dis'ôdə(r)]. İntizamsızlık. Karıştırmak. **~ed,** bozuk, intizamsız. **~ly,** karmakarışık; itaatsiz.
**disown** [dis'oun]. Tanımamak.
**dispatch** [dis'paç]. Gönderme; rapor; acele; öldürme. Göndermek; tamamlamak; öldürmek. **with all possible ~,** mümkün olan süratle. **~-box,** evrak çantası. **~-rider,** (*mil.*) haberci.
**dispel** [dis'pel]. Gidermek.
**dispens·e** [dis'pens]. Tevzi etm.; ilâc yapıp vermek. **to ~ with,** ···siz yapabilmek. **~ary,** dispanser. **~ation** [–'sêyşn], tevzi etme; muafiyetname. **~er,** dispanser eczacısı.
**dispers·e** [dis'pö(r)s]. Dağıtmak; dağılmak. **~al, ~ion** [–sl, –şn], dağıtma.
**displace** [dis'plêys]. Yerinden çıkarmak; yerini almak. **~ment,** yerinden çıkarılma; mai mahrec.
**display** [dis'plêy]. Gösteriş; nümayiş; şatafat. Teşhir etm.
**displeas·e** [dis'plîz]. Hoşuna gitmemek; gücendirmek. **to be ~ed with** [**at**], ···den memnun olmamak. **~ing,** nahoş. **~ure** [–'plejə(r)], gücenme; iğbirar.
**disposal** [dis'pouzl]. Tertib; kullanış; bertaraf etme; tasarruf. **at the ~ of,** ···in tasarrufunda: **at your ~,** emrinize hazır: **for ~,** satılık.
**dispose** [dis'pouz]. Tanzim etm., tertib etm. **to ~ of,** bertaraf etm.; halletmek; satmak; **to be ~d of,** bertaraf edilmek: **to be ~d to,** mütemayil olm.: **'give what you feel ~d to!',** gönlünden ne koparsa ver!: ⌜**man proposes, God ~s**⌝, ⌜takdir tedbiri bozar⌝. **~d,** hazır: **well ~,** hayırhah: **ill ~,** bedhah.
**disposition** [ˌdispə'zişn]. Tertib; tasarruf; tabiat; meyil, istek; istidad.
**dispossess** [ˌdispou'zes]. Mahrum etmek.
**disproportion** [ˌdisprə'pôşn]. Nisbetsizlik. **~ate,** nisbetsiz.
**disprove** [dis'prûv]. Cerhetmek; aksini isbat etm.
**disput·e** [dis'pyût]. Münakaşa; niza. Münakaşa etm., itiraz etm. **~ation** [–pyu'têyşn], münazara.
**disquiet** [dis'kwâyət] *n.* Rahatsızlık; endişe. *v.* **~** [**~en**], huzurunu kaçırmak. **~ing,** endişe verici.
**disregard** [ˌdisri'gâd]. Hürmetsizlik. Aldırmamak; itibar etmemek.
**disreput·e** ['disri'pyût]. Fena şöhret. **to bring into ~,** itibardan düşürmek: **to fall into ~,** itibardan düşmek. **~able** [–'repyutəbl], sui şöhret sahibi; rezil; (*coll.*) kılıksız, külhan ey kılıklı.
**disrespect** [ˌdisri'spekt]. Hürmetsizlik.
**disrupt** [dis'rʌpt]. Kesmek. **~ive,** zorla ayırıcı.
**dissatis·fy** ['dis'satisfây]. Memnun edememek. **~faction** [–'fakşn], hoşnudsuzluk. **~fied,** gayrı memnun.
**dissect** [di'sekt]. Teşrih etm.

**dissemble** [di'sembl]. Hakikî hislerini gizlemek.
**dissen·sion** [di'senşn]. Niza; tefrika. **~t** [di'sent]. Aynı fikirde olmama(k); Anglikan kilisesinden ayrılma(k). **~ter,** muhalif.
**dissident** ['disident]. Ekseriyetten ayrılan.
**dissimulate** [di'simyulêyt]. Hislerini gizlemek.
**dissipat·e** ['disipêyt]. İsraf etm.; sefahat etmek. **~ed,** sefih. **~ion** [–'pêyşn], sefahat.
**dissociate** [di'soușiêyt]. **to ~ oneself from,** alâkasını kesmek.
**dissol·ute** ['disəlyût]. Sefih, ahlaksız. **~ution** [ˌdisə'lûşn]. İnhilâl; erime; sona erme; ölüm. **~ve** [di'zolv]. Eri(t)mek; inhilâl et(tir)mek; sona ermek.
**dissua·de** [di'swêyd]. Vazgeçirmek. **~sion** [–swêyjn], vazgeçirme.
**distan·ce** ['distəns]. Mesafe; uzaklık; fasıla. **~ lends enchantment to the view,** uzaktan davulun sesi hoş gelir: **to keep s.o. at a ~,** birine soğuk davranmak: **to keep one's ~,** haddini bilmek. **~t** ['distənt]. Uzak mesafede; soğuk.
**distaste** [dis'têyst]. Hoşlanmayış. **~ful,** hoş olmıyan.
**distemper** [dis'tempə(r)]. Tutkallı boya (ile boyamak).
**distend** [dis'tend]. Ger(il)mek.
**distil** [dis'til.] İmbikten çekmek, taktir etm.; damıtmak. **~lation** [–'lêyşn], taktir (edilmiş). **~lery,** taktirhane.
**distinct** [dis'tin(g)kt]. Ayrı; vazıh, kat'î. **~ly,** vazıh olarak. **~ion** [dis'tin(g)kşn]. Ayırma, fark; şöhret; nişan. **to gain ~,** temayüz etm.: **a man of ~,** seçkin adam: **without ~,** fark gözetmeden. **~ive,** hususî.
**distinguish** [dis'tin(g)gwiş]. Ayırmak; tefrik etm.; meşhur yapmak. **to ~ oneself by,** …le temayüz etmek. **~ed,** seçkin; meşhur. **~able,** ayırd edilebilir.
**distort** [dis'tôt]. Bükmek, bozmak. **~ion** [–'tôşn], bükülme; tahrif.
**distract** [dis'trakt]. (Dikkati) başka tarafa çekmek; işgal etm.; çıldırtmak. **~ed,** çılgın. **~ion** [–şn], işgal etme; eğlence; çılgınlık: **to drive s.o. to ~,** birini çıldırtmak.
**distress** [dis'tres]. Istırab (vermek); sıkıntıya sokmak. **~ed,** mustarib; sefalet içinde. **~ful,** elemli.
**distribut·e** [dis'tribyût]. Dağıtmak, tevzi etm.; taksim etmek. **~ion** [–'byûşn], dağıtma, tevzi; taksim; hisse; dağılma, tevezzü. **~or,** distribütör.
**district** ['distrikt]. Mıntaka, bölge; mahalle.
**distrust** [dis'trʌst] *n.* İtimadsızlık. *v.* …e inanmamak. **~ful,** itimadsız.
**disturb** [dis'tö(r)b]. Rahatsız etm.; bozmak; endişe vermek. **~ance** [–əns], rahatsız etme (olma), sıkıntı.
**disunite** [ˌdisyu'nâyt]. Aralarını açmak.
**disuse** [ˌdis'yûs]. Kullanılmayış.
**ditch** [diç]. Hendek. Hendeğe yuvarlamak. **to die in the last ~,** sonuna kadar dayanmak. **as dull as ~water,** son derece ruhsuz ve sıkıcı.
**dither** ['diðə(r)]. Şaşırıp duralamak.
**ditto** ['ditou]. Aynı şey. **to say ~,** 'evet efendim' demek.
**ditty** ['diti]. Küçük şarkı.
**divan** ['dâyvan, di'van]. Minder; divan.
**div·e** [dâyv]. Dalma(k); pike uçuşu (yapmak); (*Amer.*) meyhane. **gambling ~,** kumarhane. **~er,** dalgıç.
**diverg·e** [dây'vö(r)c]. Ayrılmak; sapmak. **~ent,** farklı.
**diver·se** [dây'vö(r)s]. Muhtelif, farklı. **~sion** [dây'vö(r)şn, di-]. Başka tarafa çevir(il)me; eğlence. **~sity** [dây'vö(r)siti]. Başkalık, tenevvü. **~t** [dây'vö(r)t]. Başka tarafa çevirmek, yolunu değiştirtmek; eğlendirmek.
**divide** [di'vâyd]. Bölmek, taksim etm., paylaşmak; ayırmak; ayrılmak. **~ up,** parçalamak. **to ~ the House,** (İngiliz Parlamentosunda) rey verdirmek. **~nd** ['dividend]. Temettü hissesi.
**divin·e**[1] [di'vâyn]. *v.* Keşfetmek; kehanette bulunmak. **~**[2] *a.* İlâhi; mükemmel. **~ity** [–'viniti], Allahlık; ilâhiyat.
**divis·ible** [di'vizibl]. Taksimi kabil. **~ion** [–'vijn], taksim; kısım; parça; tümen; tefrika, (İngiliz Parlamentosunda) meb'usların rey vermek için ayrılmaları.
**divorce** [di'vôs]. Boşanma. Boşatmak; boşa(n)mak.
**dizz·y** ['dizi]. Başı dönen. **to feel ~,** başı dönmek. **~iness,** başdönmesi.
**do**[1] **(did, done)** [dû, did, dʌn]. (*Yardımcı fiil için bk.* **do**[2]). Yapmak,

etmek; bitirmek; başarmak; tanzim etm.; (*sl.*) aldatmak; uygun gelmek. **to be done,** yapılmak; tamamlanmak; (et) kâfi pişirilmek; bitkin bir hale gelmek; (*sl.*) aldanmak: **I am done,** mahvoldum: **I've been done,** aldatıldım: **have done!,** yetişir!, sus!: ⌜**~ as you would be done by**⌝, sana karşı nasıl hareket edilmesini istiyorsan sen de başkalarına karşı öyle hareket et: **I can't ~ on £500 a year,** senede beş yüz lira ile idare edemem: **how ~ you ~?,** nasılsınız? *yalnız birbirine takdim olunan kimseler tarafından kullanılır ve cevab olarak aynı tabir tekrar edilir*: **what can I ~ for you?,** ne emiriniz var?: **he is ~ing law [medicine],** hukuk [tıb] tahsil ediyor: **this isn't very suitable but I will make it ~ [make ~ with it],** bu pek elverişli değil fakat idare edeceğim: **no you don't!,** öyle yağma yok!: **this sort of thing isn't done,** böyle şey yapılmaz: **this meat is not done,** bu et iyi pişmemiş: **what's ~ing here?,** burada neler oluyor?: **there's nothing ~ing here,** burada hiç bir şey olmuyor: **to ~ a town,** (seyyah) bir şehri gezmek: **what are you ~ing?,** ne yapıyorsunuz?: **what ~ you ~?,** (i) ne yaparsınız?; (ii) işiniz [mesleğiniz] nedir?: **well done!,** aferin!: **he did well [badly] in his examination,** imtihanda muvaffak oldu [olamadı]: **he does himself very well,** rahatına iyi bakar: **they ~ you very well at this restaurant,** bu lokantanın yemekleri çok iyidir: **potatoes ~ very well in this district,** bu mıntakada patates iyi yetişir: **the patient is ~ing well,** hasta iyileşiyor: **that will ~,** (i) bu olur; (ii) artık yeter!: **that won't ~,** bu olmaz: **to ~ away with s.o.,** birini öldürmek: **to ~ away with stg.,** bir şeyi lâğvetmek. **to ~ s.o. in,** birini öldürmek. **~ up,** tamir etm., süslemek. **~ with, he had a lot to ~ with the success of the scheme,** plânın muvaffak olmasında onun büyük hissesi var: **I've had a lot to ~ with horses,** atla çok meşgul oldum: **I could ~ with a bit more help,** bana bir az daha yardım eden olsa fena olmaz: **I could ~ with another £100 a year,** senede yüz lira daha alsam fena olmaz: **what ~ you ~ with yourself all day long?,** bütün gün vaktinizi nasıl geçiriyorsunuz?. **~².** *aux. v. Yardımcı fiil olarak* **do**: (1) *Sual teşkiline yarar.* **~ you know?,** biliyor musunuz?: **does he speak English?,** İngilizce bilir mi?: **did you see him?,** onu gördünüz mü? (2) *Menfi fiil teşkiline yarar.* **I ~ not [don't] know,** bilmiyorum: **he does not come,** gelmez: **we did not hear,** işitmedik. (3) *Tekid için kullanılır.* **I <u>do</u> know him,** onu vallahi tanıyorum: **he <u>did</u> say so,** o vallahi böyle dedi: **<u>do</u> come tomorrow,** ne olur yarın gel. (4) *Bir fiili tekrar etmemek için onun yerine kullanılır.* **'Who knows this?' 'I ~',** bunu kim biliyor? —Ben (biliyorum): **'He went to Paris'. 'Did he?',** Parise gitti.—Ya, öyle mi? (5) *'Değil mi' yerine, tasrif edilerek kullanılır.* **You see him every day, don't you?,** siz onu her gün görürsünüz, değil mi?: **he speaks English, doesn't he?,** o İngilizce bilir, değil mi?: **they came last week, didn't they?,** geçen hafta geldiler, değil mi? **~³.** (*abb.*) = **ditto** Aynı şey.

**docil·e** [ˈdousâyl]. Uslu. **~ity** [–ˈsiliti], uysallık.

**dock¹** *n.* Havuz; rıhtım. *v.* Bir gemiyi havuza sokmak. **dry ~,** sabit havuz: **naval ~s,** tersane. **~er,** liman amelesi. **~².** (Mahkeme) maznun yeri. **~yard** [ˈdokyâd]. Havuz fabrikaları, tersane.

**doctor** [ˈdoktə(r)]. Doktor; âlim. Tedavi etm.; tahrif etmek. **~ate,** doktora.

**doctrine** [ˈdoktrin]. Mezheb, nazariye.

**document** [ˈdokyumnt]. Vesika; evrak. Tevsik etmek. **~s pertaining to the case,** dava dosyası. **~ary** [–ˈmentəri], vesikaya dayanan: **~evidence,** yazılı delil.

**dodder** [ˈdodə(r)] (İhtiyarlıktan) titremek. **~er,** sarsak ihtiyar.

**Dodecanese** [douˈdekənîz]. Oniki Ada.

**dodge** [doc]. Oyun, kurnazlık. Yana kaçınmak. **~r,** hilekâr.

**dodo** [ˈdoudou]. ⌜**as dead as the ~**⌝, ortadan kalkmış.

**doe** [dou]. Dişi geyik, tavşan.

**doer** [ˈdûə(r)]. Yapan, eden, fail.

**does, doest** *v.* **do.**

**dog** *n.* Köpek. *v.* (Birinin) peşinden gitmek. **dirty ~,** alçak herif: ⌜**every ~ has his day**⌝, herkesin sırası gelir: **gay ~,** çapkın: ⌜**give a ~ a bad name and hang him**⌝, bir adamın 'adı çıkacağına canı çıksın' *kabilinden*: **to**

go to the ~s, sefalete düşmek: to take ⌜a hair of the dog that bit you⌝, bir içki âleminin ertesi günü mahmurluğunu gidermek için bir bardak daha içmek: to ⌜help a lame ~ over a stile⌝, çaresiz kalmış birini müşkülattan kurtarmak: to lead a ~'s life, başı derdden kurtulmamak: you lucky ~!, seni gidi köftehor seni!: ⌜you can't teach an old ~ tricks⌝, yeni bir şeyi oğrenemem: to throw to the ~s, israf etm.: to be top ~, üstün gelmek. ~-collar, (*sl.*) rahiblere mahsus yuvarlak yakalık. ~-ear, kitabın sahifesini kıvırmak. ~-fox, erkek tilki.

**dogged** [ˈdogid]. Azimli. ⌜it's ~ as does it⌝, sebat etmeli. **~ness**, sebat.

**doggo** [ˈdogou]. to lie ~, ölmüş gibi yapmak.

**dogma** [ˈdogmə]. Nas, akide. **~tic** [–ˈmatik], kat'î. **~tize** [–tâyz], kat'î söylemek.

**doings** [ˈdûin(g)z]. Faaliyet; yapılan şeyler.

**dole** [doul]. Sadaka; işsizlere hükûmetçe verilen haftalık. to ~ out, (tamahkârca) azar azar tevzi etmek. to go on the ~, haftalık yardımını almak.

**doleful** [ˈdoulfəl]. Mahzun.

**doll** [dol]. Bebek (kukla).

**dolt** [doult]. Alık. **~ish**, ahmak.

**domain** [dəˈmêyn]. Malikâne; mülk; saha. it does not come within my ~, bu benim saham haricindedir.

**dome** [doum]. Kubbe. **~d**, kubbeli.

**domestic** [dəˈmestik] *a.* Eve aid, aile hayatını seven; ehlî; memlekete aid; yerli. *n.* Hizmetçi. **~ate**, ehlileştirmek.

**domicile** [ˈdomisâyl]. Mesken, ikametgâh.

**domin·ance** [ˈdominəns]. Hakimiyet, nüfuz. **~ant**, hâkim, nüfuzlu. **~ate** [ˈdominêyt]. Hâkim olm. **~ation** [–ˈnêyşn], hakimiyet, tahakküm. **~eer** [ˌdomiˈniə(r)]. Tahakküm etmek. **~eering**, mütehakkim.

**dominion** [dəˈminyən]. Hakimiyet; memleket; dominyon.

**don**[1] [don]. Oxford veya Cambridge üniversitelerinde hoca. **~**[2] *v.* Giydirmek.

**donate** [douˈnêyt]. Teberru etm.; vermek.

**done** *v.* do.

**donkey** [ˈdonki]. Eşek. she would talk the hind leg off a ~, beş para ver söylet on para ver sustur.

**donor** [ˈdounô(r)]. Hibe eden, veren.

**don't** [dount]. = do not.

**doodle** [ˈdûdl]. Dalgın iken kâğıd üzerine şekiller çizmek.

**doom** [dûm]. Feci akıbet; kader; hüküm. Mahkûm etmek. the day of ~, kıyamet günü: his ~ is sealed, o mahvolmuş demektir. **~sday** [ˈdûmsdêy]. Kıyamet günü: ~ Book, İngiltere'de Norman istilâsindan sonra kıralın emrile tertib olunan emlâk defteri.

**door** [dôə(r)]. Kapı. the fault lies at my ~, kabahat benimdir: to keep within ~s, sokağa çıkmamak: out of ~s, açık havada: to show s.o. the ~, birini kapı dışarı etm.; to show s.o. to the ~, birini kapıya kadar teşyi etm.: to turn s.o. out of ~s, birini dışarı çıkarmak: next ~, yandaki kapı. **~keeper**, kapıcı. **~mat**, paspas. **~way**, kapı yeri. dead as a ~-nail, ölmüş gitmiş.

**dope** [doup] *n.* Çiriş; (*sl.*) afyon gibi sersemletici ilâc; haber. *v.* İlâcla sersemletmek. ~ habit, morfin gibi ilâc ibtilâsı.

**dormant** [ˈdômənt]. Uyuyan; haretsiz.

**dormitory** [ˈdômitəri]. Yatakhane.

**dose** [dous]. Doz (vermek), ilâc vermek.

**doss** [dos]. (*sl.*) to ~ (down), yatmak.

**dot** [dot]. Nokta; benek. Noktasını koymak; he arrived on the ~, elifi elifine geldi.

**dot·e** [dout]. Bunamak. to ~ upon, ···e mübtelâ olmak. **~age** [ˈdoutic], bunama. **~ing**, bunak; bir şeye gülünc bir ibtilâsı olan.

**dotty** [ˈdoti]. (*sl.*) Sapık.

**double**[1] [ˈdʌbl] *n.* Çift; iki misli, iki kat; iki porsiyon; benzer. at the ~, koşar adım: to lead a ~ life, biri herkese malûm biri hiç kimsenin bilmediği iki türlü hayat sürmek: to play a ~ game, iki tarafı da idare etm. **~-barrelled**, çift namlulu. **~-breasted**, kruaze (ceket). **~-cross**, iki tarafla da anlaşmış görünüp her ikisini de aldatma(k). **~-dealing**, iki yüzlülük. **~-decker**, iki ambarlı gemi; iki katlı otobüs. **~-dutch**, anlaşılmaz lâkırdı [lisan]. **~-quick**, çok çabuk, koşarak. **~**[2] *v.* İki misli yapmak, olm.; aynı olın.; katlamak; ikiye bükmek. ~ **back,** katlamak; sür'atle geri dönmek. ~ **over,** ikiye bükmek. ~ **up,** ikiye bük(ül)mek.

**doubt** [dąut]. Şübhe. …den şübhe etm. **no** ~, şübhesiz, muhakkak: ~**ing Thomas,** her şeyden şübhe eden: **I ~ whether he will come,** geleceğini pek zannetmem. ~**ful,** şübheli. ~**less,** şübhesiz.

**douche** [dûş]. Duş (yapmak).

**dough** [dou̧]. Hamur; (*sl.*) para.

**doughty** [ˈdąuti]. Cesur.

**dour** [du̧ə(r)]. Soğuk, sert; inadcı.

**dove** [dʌv]. Kumru. ~**cot,** güvercinlik. ~**tail** [ˈdʌvtêyl]. Kurtağzı ile eklemek; (*fig.*) iki şeyi telif etm.

**dowager** [ˈdąuəcə(r)]. Ünvan sahibinin dul kalan karısı. **Queen** ~, valide kraliçe.

**dowdy** [ˈdąudi]. Kılıksız, fena giyinmiş.

**down**[1] [dąun] *n.* Alçak tebeşirli tepe. ~[2] *n.* Kuşun ana tüyü. ~**y,** tüylü. ~[3] *v.* İndirmek, yenmek. **to ~ tools,** grev yapmak. ~[4]. *adv. prep.* Aşağı, aşağıya. **to be [feel]** ~, keyfi yerinde olmamak: **to be ~ with influenza,** gripten yatmak: **to be ~ for stg.,** bir listeye ismini yazdırmış olmak: **just let me get that** ~, dur, yazayım: **to have** ~, yıktırmak, kestirmek; **to have a ~ [be ~] on s.o.,** birine kancayı takmak: **to hit a man when he is** ~, zebunküşlük etm.: **~ on one's luck,** talihsiz: **I am ten pounds ~ on this,** bu işte on lira açığım var: **~ and out,** sefalet içinde: **I should like that ~ on paper,** bunu yazı ile tesbit edelim: **~ to the beginning of the 19th century,** 19uncu asrın başına kadar: **this tyre is** ~, bu lastik sönmüş: ~ **under,** Avustralya ve Yeni Zelanda: **up and** ~, bir aşağı bir yukarı; bazan iyi bazan fena: **~ with …!,** … kahrolsun! ~**-at-heel,** topuğu çiğnemiş (ayakkabı); düşkün, sefalet içinde (adam). ~**-stream,** bir nehirde akıntı istikametinde olan. ~**cast** [ˈ–kâst]. Süngüsü düşük, keyifsiz; (gözler) inik. ~**fall** [ˈ–fôl]. Düşme; yağma. ~**hearted** [ˈ–ˈhâtid]. Meyus. ~**hill** [ˈ–ˈhil]. İnişli; yokuş aşağı. **to go** ~, (*fig.*) gittikçe fenalaşmak. ~**most** [ˈ–mou̧st]. En aşağı. ~**pour** [ˈ–po̧ə(r)]. Şiddetli yağmur. ~**right** [ˈ–râyt]. Dobra dobra söyliyen; kat'î; tamamen. ~**stairs** [ˈ–ˈstȩəz]. Merdivenin alt başında; aşağı, aşağıda. ~**trodden** [ˈ–trodən]. Ayaklar altında çiğenen; zulüm gören. ~**ward** [ˈ–wəd]. Aşağı doğru olan. ~**wards** [–wədz], aşağı doğru: **from** … ~, …den itibaren: **face** ~, yüzükoyun.

**dowry** [ˈdąuri]. Drahoma.

**doze** [dou̧z]. Uyuklama(k), kestirme(k).

**dozen** [ˈdʌzn]. Düzine. **to sell by the** ~, düzine ile satmak: **to talk nineteen to the** ~, çene çalmak.

**Dr.** = **doctor.**

**drab** [drab] *n.* Pasaklı kadın; sürtük. *a.* Renksiz; zevksiz.

**Draconic** [draˈkonik]. Merhametsiz.

**draft** [drâft]. Taslak, müsvedde; müfreze, kıt'a; tediye emri. Taslağını yapmak; kıt'ayı ayırmak, göndermek.

**drag**[1] [drag] *n.* Çekmeğe yarıyan şey; tarama ağı; çengel; engel, mâni. ~[2] *v.* Sürü(kle)mek; yere sürünmek. ~ **about [along],** sürüklemek. ~ **away,** zorla alıp götürmek. ~ **on,** sürüklenmek. ~ **out,** sürükleyip çıkarmak; bir işi uzatmak: **to ~ out a wretched existence,** sürünerek yaşamak. ~ **up,** (çocuğu) gelişi güzel terbiye etm.

**draggle** [ˈdragl]. Çamurda sürüyerek kirletmek.

**dragoman** [ˈdragəman]. Şarkta tercüman.

**dragon** [ˈdragən]. Ejderha; çok sert ve müsamahasız kimse.

**dragoon** [drəˈgûn]. Ağır süvari. (Halka) zulmetmek.

**drain** [drêyn]. Lağım, su yolu; devamlı masraf, yük. Suyunu boşaltmak; süzmek; son damlasına kadar içmek; (arazi vs.) kurutmak; (para) tüketmek. **to throw money down the** ~, parayı sokağa atmak: **to ~ s.o. dry,** birinin parasını son santimine kadar almak. ~**age** [ˈdrêynic]. Suların akması; bir bataklığın kurutulması; lağım tesisatı. ~**er** [ˈdrêynə(r)]. Süzgeç.

**drama** [ˈdrâmə]. Dram, piyes; tiyatro san'ati. ~**tic** [drəˈmatik], dram gibi; tiyatroya aid; heyecanlı. ~**tist** [ˈdramətist], tiyatro müellifi. ~**tize** [ˈdramətâyz], dram haline getirmek; heyecanlı bir hale sokmak; bir romanı piyese çevirmek.

**drank** *v.* **drink.**

**draper** [ˈdrêypə(r)]. Manifaturacı, tuhafiyeci.

**drastic** [ˈdrastik]. Kat'î, şiddetli.

**drat** [drat]. (*coll.*) ~ **the child!**, aman buyumurcak! ~**ted,** hınzır.
**draught** [drâft]. Çek(il)me; içme, yudum; geminin çektiği su; hava cereyanı; ilâc. **at a** ~, bir yudumda. ~**s,** dama oyunu. ~**y** [ˈdrâfti]. Cereyanlı. **it is** ~, cereyan yapıyor.
**draughtsman** [ˈdrâftsmən]. Teknik ressam, desinatör. ~**ship,** ressamlik.
**draw**[1] [drô] *n.* Kur'a vs. çekme; çok rağbet gören şey; berabere kalmış oyun. ~[2] (**drew, drawn**) [drô, drû, drôn] *v.* Çek(il)mek; keşide etm.; celbetmek; resim yapmak; çizmek. **the battle was** ~**n,** muharebe neticesiz kaldı: **to** ~ **a game,** bir oyunda berabere kalmak: **to be hanged, drawn, and quartered,** (eskiden) asılıp, barsakları çıkartılıp, parçalanmak: **to** ~ **near,** yaklaşmak: **the train drew into the station,** tren istasyona girdi: **to** ~ **round the table,** masanın etrafına toplanmak: **to try to** ~ **s.o.,** söyletmek: **I won't be** ~**n,** ağzımdan lâf alamazsın: **I had to** ~ **upon my savings,** tasarruf ettiğim paradan alıp sarfetmeğe mecbur oldum: **to** ~ **upon one's memory,** hafızasını yoklamak. ~ **apart,** ayrılmak. ~ **aside,** bir tarafa çekmek, bir kenara çekilmek. ~ **away,** uzaklaşmak. ~ **back,** geri çekmek; (perde) açmak. ~ **in,** içine çekmek: **the day is** ~**ing in,** akşam oluyor: **the days are** ~**ing in,** günler kısalıyor. ~ **on,** giymek; geçirmek: **the evening was** ~**ing on,** akşam yaklaşıyordu. ~ **out, the days are** ~**ing out,** günler uzuyor: **to** ~ **s.o. out,** birini konuşturmak. ~ **to,** çekip kapatmak. ~ **up,** (plân vs.yi) tertib etm.; (otom. vs.) durmak: **to** ~ **oneself up,** dik durmak: **to** ~ **up to the table,** masaya yaklaşmak: **to** ~ **up with s.o.,** birine yetişmek. ~**back** [ˈdrôbak]. Mahzur; ihracat primi. ~**bridge** [ˈdrôbric]. İndirilip kaldırılabilen veya bir kenara çekilebilen köprü. ~**er** [ˈdrô(r)]. Çekmece, göz; çeken, çekici. **chest of** ~**s,** çekmeceli dolab. ~**ers** [ˈdrôz]. Don. ~**ing** [ˈdrôin(g)] *n.* Yağlı ve sulu boyadan gayrı resim san'atı. ~**ing-room,** salon, misafir odası. ~**ing-pin,** raptiye.
**drawl** [drôl]. Kelimeleri uzatarak konuşma(k).
**drawn** [drôn] *v.* **draw.** *a.* (Çehre) süzük; (muharebe) neticesiz; (oyun) berabere; (kılıc) çekilmiş.
**dray** [drêy]. Ağır yük arabası.
**dread** [dred] *n.* Korku, dehşet; haşyet. *a.* Korkunc. *v.* ···den yılmak. ~**ful,** müdhiş, korkunc: ~**ly,** müdhiş, son derece.
**dream** [drîm]. Ruya, hûlya. Ruya görmek; hûlyaya dalmak; ruyada görmek. **day** ~, hûlya: **little did I** ~, hayalimde bile yoktu: **I should not** ~ **of doing that,** bunu kat'iyen yapmam. ~**er,** ruya gören; hayalperest. ~**y,** hulyalı; dalgın.
**dreary** [ˈdriəri]. Kasvetli; ıssız.
**dredge** [drec]. Tarak makinesi (ile taramak).
**dregs** [dregs]. Tortu, telve.
**drench** [drenç]. İyice ıslatmak. ~**ed to the skin,** iliklerine kadar ıslanmış.
**dress**[1] [dres] *n.* Elbise; üstbaş; kıyafet; tuvalet; kostüm. **full** ~, merasim elbisesi. ~**-circle,** (tiyatro) balkon. ~[2] *v.* Giydirmek; giyinmek; süslemek; pansıman yapmak; sarmak; yontmak; düzeltmek; askeri hizaya getirmek. **to** ~ **for dinner,** akşam tuvaleti, frak vs. giymek: **do we** ~**?**, smokin vs. mecburî mi?: **to** ~ **ship,** gemiyi bayraklarla donatmak: **right** ~!, (*mil.*) sağdan hizaya gel! ~ **down,** şiddetle azarlamak; dayak atmak. ~ **up,** (çocuklar) büyük adam kıyafetine girmek; (büyükler) giyinip kuşanmak. ~**-maker** [ˈdresmêykə(r)]. Kadın terzisi. ~**y** [ˈdresi]. Gösterişli giyinen.
**drew** *v.* **draw.**
**dribble** [ˈdribl]. Damlama(k); salya (sı akmak); (futbol) topu sürmek.
**dried** [drâyd] *a.* Kuru; kurumuş. ~**-up,** kavrulmuş.
**drier, driest** *v.* **dry**[1].
**drift** [drift]. Temayül; kar yığıntısı; su, havada cereyanla sürüklenme(k); rastgele sürüklenme(k). **to let oneself** ~, kapıp koyuvermek: **to let things** ~, işleri kendi haline bırakmak. ~**-wood,** suda yüzen tahta parçası. ~**er** [ˈdriftə(r)]. Ağ çeken bir nevi balıkçı gemisi.
**drill**[1] [dril] *n.* Matkab. *v.* Delmek, (kuyu) açmak. ~[2] *n.* (*mil.*) Talim. *v.* Talim et(tir)mek.
**drink** (**drank, drunk**) [drin(g)k, dran(g)k, drʌn(g)k]. İçmek. İçecek şey; içki. **to have a** ~, bir şeyi içmek: **give me a** ~ **of water,** bana bir az su ver: **to be the worse for** ~, sarhoş olm.: **to** ~ **oneself to death,** işretten kendini öldürmek: **the** ~ **question,**

alkolizm meselesi: **soft ~**, alkolsüz içki: **strong ~**, alkollü içki: **to take to ~**, kendini içkiye vermek: **to ~ the waters**, içmelere gitmek. **~ away, to ~ away one's fortune**, varını yoğunu içkide bitirmek: **to ~ away one's sorrows**, kederini içki ile dağıtmak. **~ in**, emmek; **he drank it all in**, hepsini yuttu. **~ up**, içip bitirmek. **~able** [ˈdrin(g)kəbl]. İçilir.

**drip** [drip]. Damlama(k), su sızma(k). **to be ~ping wet**, sırsıklam olmak.

**drive**[1] [drâyv] *n.* (Araba ile) gezinti; (bahçede) araba yolu; enerji ve sürükleme kabiliyeti; sürek avı; makineyi işleten vasıta. **belt ~**, (makine) kayışla işleme. **~**[2] **(drove, driven)** [drâyv, drouv, drivn] *v.* Önüne katıp sürmek; sevketmek; sıkıştırmak; (araba vs.) kullanmak: **to ~ a nail**, bir çiviyi çakmak: **the rain was driving in our faces**, yağmur yüzümüze çarpıyordu: **to ~ s.o. to the station**, birini (araba vs. ile) istasyona götürmek: **to ~ a tunnel**, tünel açmak: **what are you driving at?**, maksadınız nedir?, ne demek istiyorsunuz? **~ away**, koğmak; uzaklaş(tır)mak. **~ in**, içeri sürmek; sokmak: **to ~ in [home] a point**, bir noktayı hiç tereddüdde mahal kalmıyacak şekilde anlatmak. **~ off**, koğmak; (araba ile) ayrılmak. **~ on**, ileri sürmek; (araba ile) durmadan ilerlemek. **~ out**, koğmak. **~n** [ˈdrivn] *a.* **~ snow**, yığılmış kar. **~r** [ˈdrâyvə(r)]. Sürücü; arabacı, şoför.

**drivel** [ˈdrivl]. Salya(sı akmak); saçma(lamak).

**driving** [ˈdrâyvin(g)] *n.* (Araba vs.yi) kullanma. *a.* Yürütücü; sevkedici. **~ licence**, şoförlük ruhsatiyesi: **~ rain**, rüzgârlı yağmur: **~ test**, şoförlük imtihanı. **~-belt**, çark kayışı.

**drizzle** [ˈdrizl]. Çiseleme(k).

**droll** [droul]. Tuhaf, komik; garib.

**dromedary** [ˈdromədəri]. Hecin devesi.

**drone** [droun]. Erkek arı; tembel adam. Vızıldamak.

**droop** [drûp]. Sarkma(k); (göz) inik olmak.

**drop**[1] [drop] *n.* Damla; damla şeklinde bir şey; düşme, inme; yukarıdan aşağıya mesafe. **~-forge**, kalıp ile dövmek. **~**[2] *v.* [*Bu fiil umumiyetle tesadüfî ve gelişi güzel bir hareketi ifade eder.*] Düşürmek; damla(t)mak; bırakmak; indirmek; inmek; yere düşmek. **to ~ behind**, tedricen geri(de) kalmak: **to ~ into the habit of**, …i adet edinmek: **to ~ a habit**, bir adetten vazgeçmek: **I've ~ped £100 over this**, bu işte yüz lira kaybettim: **to ~ a letter [word]**, (yazıda) bir harfi [kelimeyi] atlamak; (konuşurken) bir harfi telaffüz etmemek: **to ~ s.o. a line**, birine iki satır bir şey yazmak: **there the matter ~ped**, mesele öylece kaldı: **I am ready to ~**, (yorgunluktan) ayakta duramıyorum: **to ~ a remark**, bir şeyi söyleyivermek: **where shall I ~ you?**, sizi nerede bırakayım? **~ in**, rasgele uğramak; çökmek. **~per** [ˈdropə(r)]. Damlalık.

**drought** [draut]. Kuraklık.

**drove** [drouv]. (Yürüyüş halinde) sürü. **~r**, sürücü. *v.* **drive.**

**drown** [draun]. Suda boğ(ul)mak; su basmak. **to ~ oneself**, suda intihar etm.: **to ~ one's sorrows in drink**, kederini içki ile avutmak.

**drows·e** [drauz]. Uyuklamak. **~y**, uykusu basmış.

**drudge** [drʌc]. Ağır ve zahmetli işler görmek. **~ry**, ağır, zahmetli, zevksiz iş.

**drug** [drʌg]. İlâc. Uyuşturucu ilâc (vermek). **~-addict [~-fiend]**, kokain gibi zehirli ilâclara mübtelâ. **~-store** (*Amer.*), eczane. **~gist** [ˈdrʌgist]. Eczacı.

**Druid** [ˈdrûid]. Eski Kelt rahibi.

**drum** [drʌm] *n.* Davul; kulak zarı; davul şeklinde kutu; kayış çemberi. *v.* Davul çalmak. **to ~ stg. into s.o.**, bir şeyi birine tekrar ede ede öğretmek: **to bang the big ~**, reklam yapmak: **~ major**, askerî bando şefi: **to ~ out**, (*mil.*) ordudan tardetmek. **~-head court martial**, harb zamanında vaka mahallinde fevkalâdeden olarak kurulan askerî mahkeme. **~stick** [ˈdrʌmstik]. Davul tokmağı.

**drunk** [drʌnk] *v.* **drink**; *a.* sarhoş. **~ard**, *n.* ayyaş, sarhoş. **~en**, *a.* sarhoş.

**Druse** [drûs]. Dürzü.

**dry**[1] [drây] *a.* Kuru; kurak; susuz; suyu çekilmiş; (şarab) tatlı olmıyan; içki yasak olan (yer). **to feel ~**, susamak: **to go ~**, kurumak; (bir memleket) içkiyi yasak etm.: **~ humour**, gülmeden ve tabiî olarak yapılan nüktecilik: **to run ~**, kurumak: **a ~ smile**, zoraki gülümseme: **~ work**, insanı susatan iş. **~-**

**as-dust,** kupkuru (muharrir vs.). **~-clean,** yıkamadan temizlemek. **~-dock,** havuz(lamak). **~-eyed,** ağlamıyan. **~-shod,** ayaklarını ıslatmadan. ~[2] *v.* Kurutmak; kuru(la)mak, yaşını silmek. **to ~ up,** suyu çekilmek; silmek; **oh, ~ up!,** aman, kes sesini! **~er** [ˈdrâyə(r)]. Kurutucu şey. **~ing-cupboard,** nemli çamaşırların kurutulduğu dolab.

**D.S.O.** [ˈdî ˈes ˈou] = **Distinguished Service Order,** bir askerî nişan.

**dual** [ˈdyuəl] *a.* Çift, ikili.

**dubious** [ˈdyûbyəs]. Şübheli; kat'î olmıyan.

**duc·al** [ˈdyûkl]. Dukaya aid. **~hess** [ˈdʌçis]. Düşes. **~hy** [ˈdʌçi]. Dukalık.

**duck**[1] [dʌk] *n.* Ördek; *muhabbet ifade eden tabir.* **~'s egg,** sıfır: **like water off a ~'s back,** hiç tesirsiz: **~s and drakes,** suda taş sektirme oyunu: **to take to French like a ~ to water,** fransızcadan hoşlanmak ve onu kolay bulmak. ~[2] *v.* Dal(dır)mak; birdenbire başını eğmek. **to get a ~ing,** suya düşüp ıslanmak. **~ling** [ˈdʌklin(g)]. Ördek yavrusu.

**duct** [dʌkt]. Boru; su yolu.

**ductile** [ˈdʌktâyl]. Tel haline konabilir.

**dud** [dʌd]. İşe yaramaz (şey); patlamamış mermi. **~ cheque,** karşılıksız çek.

**due** [dyû] *a.* Uygun, münasib; icab eden; gelmesi vs. beklenen; borcu ödeme vakti gelmiş. *adv.* Tam. *n.* Hak, istihkak. **~s,** resim, vergi, borc. **~ to ...,** ...den dolayı, ...in sayesinde: **to be ~ to,** sebebi ... olmak; ...in hakkı olmak: **it is ~ to him,** onun sayesinde: **it is due to him,** onun hakkıdır: **after ~ consideration,** iyice düşünüp taşındıktan sonra: **to fall ~,** (bir borcun vs.) vadesi gelmek: **~ south,** tam cenuba doğru.

**duffer** [ˈdʌfə(r)]. Kalınkafalı kimse; beceriksiz adam.

**dug** [dʌg]. *v.* **dig. ~-out,** bir ağac kütüğü oyularak yapılmış kayık; bir tepe vs. içine oyularak yapılmış sığınak.

**duke** [dyûk]. Duka. **~dom,** dukalık.

**dull** [dʌl] *a.* Sıkıntılı; ağır; gabi; donuk; kesad; ruhsuz; kör, kesmez. *v.* Sersemletmek; (ağrı) hafifleş(tir)mek; körleş(tir)mek. **I feel ~,** içim sıkılıyor: **to be ~ of hearing,** ağır işitmek.

**duly** [ˈdyûli]. Hakkile; uygun bir şekilde; beklendiği gibi.

**dumb** [dʌm]. Dilsiz, sessiz. **~ show** pandomima. **~found** [dʌmˈfaund]. Şaşkına çevirmek.

**dummy** [ˈdʌmi]. Taklid şey.

**dump** [dʌmp]. (*ech.*) *v.* Atmak; (bir yükü) boşaltmak; rekabet için piyasaya çok ucuz fiatla mal çıkarmak. Süprüntü yığını; cebhane deposu.

**dumps** [dʌmps]. Hüzün, keyifsizlik. **to be down in the ~,** kederli olm.

**dumpy** [ˈdʌmpi]. Bodur.

**dun** [dʌn]. Israrla alacağını istemek.

**dunce** [dʌns]. Kalınkafalı.

**dune** [dyûn]. Kum tepeceği.

**dung** [dʌn(g)]. Gübre.

**dungarees** [dʌn(g)əˈrîz]. Mavi işçi pantalonu.

**dungeon** [ˈdʌncən]. Zindan.

**dupe** [dyûp]. Aldatılmış kimse. Aldatmak.

**duplicat·e** [ˈdyûplikit] *n.* İki nüshadan biri, bir şeyin aynı. *a.* Çift. *v.* [-kêyt]. Suretini çıkarmak, iki misli yapmak; kopya etmek. **in ~,** iki nüsha olarak: **~ing machine,** teksir makinesi.

**duplicity** [dyûˈplisiti]. İki yüzlülük.

**durable** [ˈdyûrəbl]. Devamlı, sürekli, dayanıklı.

**duration** [dyûˈrêyşn]. Devam, müddet.

**during** [ˈdyûrin(g)]. Esnasında, zarfında.

**durst** [dö(r)st] = **dared.**

**dusk** [dʌsk]. Akşam karanlığı. **it is growing ~,** hava kararıyor: **at ~,** akşam üstü. **~y,** karanlık; esmer.

**dust** [dʌst]. Toz (almak). **to bite the ~,** bozguna uğramak: **to ~ s.o.'s jacket,** birine dayak atmak, 'tozunu almak'; **to kick up a ~,** toz etmek; mesele yapmak: **to shake the ~ of ... off one's feet,** lânet olsun diye alâkasını kesmek: **to throw ~ in s.o.'s eyes,** aldatmak. **~bin** [ˈdʌstbin]. Çöp tenekesi. **~-cart,** çöp arabası. **~er** [ˈdʌstə(r)]. Toz bezi. **~man** [ˈdʌstmən]. Çöpçü. **~-pan,** faraş. **~y** [ˈdʌsti]. Tozlu.

**Dutch** [dʌç]. Felemenkli; felemenkçe; Holanda'ya aid. **~ courage,** içkiden gelen cesaret: **double ~,** anlaşılmaz lisan. **~man,** Felemenkli; (*Amer.*) Alman.

**duti·able** [ˈdyûtiəbl]. Gümrük resmine tabi. **~ful** [ˈdyûtifl]. Hürmetkâr ve itaatli.

**duty** [ˈdyûti]. Vazife; hürmet; vergi,

resim. ~ **call,** nezaket ziyareti: **to be on ~,** vazifede olm., nöbetçi olm.: **to do ~ for,** …in yerini tutmak: **liable to ~,** gümrük resmine tabi: **from a sense of ~** [as in ~ bound], (manevî) vazife icabı. **~-free,** gümrük resminden muaf.

**dwarf** [dwôf] *n.* & *a.* Cüce, bodur. *vb.* Gelişmesine mani olm.; cüce göstermek. **~ish,** cüce gibi.

**dwell** [dwel]. İkamet etm.; kalmak; ısrarla durmak. **to ~ on stg.,** bir şey üzerinde durmak. **~ing,** mesken, ikametgâh.

**dwindle** [ˈdwindl]. Tedricen azalmak.

**dye** [dây]. Boya. Boya(n)mak. **~d in the wool,** (kumaş) yün halinde iken boyanmış; esaslı; koyu: **a liar of the deepest ~,** sunturlu bir yalancı: **this material ~s well,** bu kumaş boyaya gelir. **~r,** boyacı. **~-works,** boyahane.

**dying** *v.* die.

**dynamic** [dâyˈnamik]. Son derece cevval ve enerjik.

**dynasty** [ˈdinəsti]. Hanedan, sülâle.

# E

**e** [î]. E harfi: (*mus.*) mi.

**each** [îç]. Herbir; her; başına, beheri. **~ one of us,** herbirimiz; **~ one of them,** herbiri: **~ other,** birbiri: **we see ~ other every day,** birbirimizi her gün görüyoruz: **we ~ have our own room,** herbirimizim bir odası var.

**eager** [ˈîgə(r)]. Hevesli; istekli; haris; gayretli. **to be ~ for,** çok istemek. **~ness,** şiddetli arzu; gayret.

**eagle** [ˈîgl]. Kartal; karakuş. **~-eyed,** keskin gözlü.

**ear** [i̯ə(r)]. Kulak; başak. **to be all ~s,** kulak kesilmek: **last night your ~s must have burnt,** dün gece her halde kulaklarınız çınlamıştır: **to have a good [no] ear,** musikide kulağı hassas ol[ma]mak: **to play by ~,** (musiki) ezberden çalmak: **to bring a storm about one's ~s,** başına belâ açmak: ⌜**walls have ~s**⌝, yerin kulağı var: **to be up to the ~s in work,** işi başından aşmak: **a word in your ~!,** kulağınıza bir şey söyliyeceğim. **~-drum,** kulak zarı. **~-mark,** inek vs.nin kulağına yapılan marka; muayyen bir maksad için tahsis etmek. **~phone,** kulaklık. **~-ring,** küpe. **~-shot, within ~,** işitilecek mesafede. **~-splitting,** kulakları patlatan. **~-trumpet,** sağır borusu. **-~ed** [i̯əd] … kulaklı.

**earl** [ö(r)l]. (İngiltere'de) kont.

**early** [ˈö(r)li]. Erken; önce; eski; ilk; başlangıcında; turfanda. **as ~ as 1700,** daha 1700 senesinde: **~ closing day,** dükkânların öğleden sonra kapalı olduğu gün: **at an ~ date,** yakında: **in ~ days,** eskiden: **~ enough,** zamanında: **to keep ~ hours,** erken yatıp erken kalkmak: **~ in the list,** listenin baş tarafında.

**earn** [ö(r)n]. Kazanmak; kesbetmek. **~ings** [ˈö(r)nin(g)s]. Kazanc, kâr; gelir.

**earnest** [ˈö(r)nest] *a.* Ciddî; hakikî. *n.* Pey akçesi. **an ~ of one's goodwill,** hüsnüniyetinin delili olarak: **in (deadly) ~,** ciddî olarak: **he is very much in ~,** işi çok ciddiye alıyor: **I ~ly hope,** kuvvetle ümid ederim.

**earth** [ö(r)θ] *n.* Toprak; arz, dünya; yeryüzü; (*elect.*) toprak. *v.* (*elect.*) Toprağa bağlamak [iletmek]. **~ over,** toprak ile örtmek: **to come back to ~,** hülyayı bırakmak: **to go to ~,** (tilki) inine girmek; (*fig.*) saklanmak: **to go to the ends of the ~,** dünyanın öteki ucuna gitmek: **where on ~ have you been?,** neredeydin yahu?: **why on ~ …?,** ne halt etmeğe …?: **to run to ~,** (tilkiyi) inine kaçırmak; (hırsız vs.yi) takib ederek bulmak. **~en** [ˈö(r)θən]. Topraktan yapılmış: **~enware,** çanak çömlek: **glazed ~,** kaba çini. **~ly** [ˈö(r)θli]. Dünyevi. **there is no ~ reason for …,** … için hiç sebeb yoktur. **~quake** [ˈö(r)θkwêyk]. Zelzele. **~worm** [ˈö(r)θwö(r)m]. Solucan. **~y** [ˈö(r)θi]. Topraklı; toprak gibi; maddî.

**ease**[1] [îz] *n.* Rahat; huzur; refah; kolaylık. **to be at one's ~,** rahat olm.: **to be [feel] at ~,** içi rahat olm.: **ill at ~,** huzursuz, endişeli: **to live a life of ~,** işsiz ve rahat yaşamak: **to set s.o. [s.o.'s mind] at ~,** birinin içini rahat ettirmek: **stand at ~!,** (*mil.*) rahat!: **to take one's ~,** yangelmek. **~**[2] *v.* Hafifletmek; yatıştırmak; gevşetmek: **~ up [off],** yavaşla(t)mak; gevşe(t)mek.

**easi·er** [ˈîzi̯ə(r)]. Daha kolay. **~ly** [ˈîzili] kolayca; yavaş yavaş: **he is ~ the best player,** o fersah fersah en iyi oyuncudur. **~ness** [ˈîzinis] kolaylık; selaset.

**east** [îst]. Şark, doğu, gündoğusu. **~-end,** Londra'nın fakir ve kalabalık kısmı. **~erly** [ˈîstəli], **~ wind,** doğudan esen rüzgâr: **~ course,** doğuya doğru rota. **~ern** [–tən], Şarka aid. **~ward** [–wəd], doğuya doğru giden: **~s,** doğuya doğru.
**Easter** [ˈîstə(r)]. Paskalya.
**easy** [ˈîzi]. Kolay; rahat; müreffeh. **~ (ahead)!,** yavaş ileri!: ⌜**~ come ~ go**⌝, ⌜haydan gelen huya gider⌝: **~ to get on with,** kolay geçinilir: ⌜**easier said than done**⌝, dile kolay: **to go ~ with stg.,** bir şeyi idare ile kullanmak: **honours ~,** berabere kalma: **the market was ~,** piyasa durgundu: **by ~ payments,** küçük taksitlerle: **within ~ reach of, ...** kolaylıkla erişilebilir: **stand ~!,** (*mil.*) yerinde rahat!: **to take it ~,** yangelmek; mola vermek; kendini fazla yormamak; çubuğunu tellendirmek: **to take life ~,** hayatta bir şeye aldırmayıp keyfine bakarak yaşamak. **~-chair,** koltuk. **~-going,** babacan.
**eat** (**ate, eaten**) [ît, et, ˈîtn]. Yemek. **he ~s out of my hand,** (i) (hayvan) elimden yem yiyiyor; (ii) (insan) bir dediğimi iki etmez: **to ~ its head off,** (at) iş görmiyerek semirmek: **to ~ one's heart out,** içi içini yemek: **he is ~ing me out of house and home,** onun boğazına para yetiştiremiyorum: **to ~ one's words,** tükürdüğünü yalamak. **~ away,** aşındırmak. **~ up,** yiyip bitirmek. **to ~ up the miles,** (otomobil) çok hızlı gitmek. **~able** [ˈîtəbl]. Yenir: **~s,** gıda, yiyecek. **~er** [ˈîtə(r)]. **small [great] ~,** az [çok] yiyen kimse. **~ing** [ˈîtin(g)]. **~-house,** ahçı dükkânı.
**eaves** [îvz]. Dam saçağı. **~drop** [ˈîvzdrop]. Gizlice dinlemek.
**ebb** [eb]. (Deniz) cezir. Cezir olm.; azalmak. **to ~ and flow,** (i) med ve cezir olm.; (ii) azalıp çoğalmak; (iii) (muharebe vs.) kâh bir tarafın kâh diğer tarafın lehine gelişmek: **to ~ away,** tedricen tükenmek: **the patient is at a low ~,** hastanın vaziyeti çok fenadır.
**ebony** [ˈebəni]. Abanoz.
**eccentric** [ekˈsentrik]. Dişmerkezli; acayib. **~ity** [–ˈtrisiti], garabet.
**ecclesiastic** [iˈklîziˈastik]. Rahib; ulema. **~al,** kiliseye aid.
**echelon** [ˈeşelon]. Kademe nizamı.
**echo** [ˈekou]. Aksisada, yankı. Ses aksettirmek; ses geri gelmek.
**éclat** [êyˈklâ]. Parlak muvaffakiyet.
**eclipse** [iˈklips]. Ay veya güneş tutulması. Husuf, küsufa uğratmak; gölgede bırakmak.
**econom·ic** [ˌîkəˈnomik]. İktisadî. **~ical,** idareli. **~ics,** iktisad ilmi. **~ist,** [–ˈkonəmist], iktisadcı. **~ize,** [–ˈkonəmâyz], idareli kullanmak, tasarruf etmek. **~y,** [îˈkonəmi], iktisad; tasarruf, tutum, idare: **political ~,** iktisad ilmi.
**ecstasy** [ˈekstəsi]. Vecd; coşkunluk.
**eddy** [ˈedi]. Anafor (yapmak).
**Eden** [ˈîdn]. **the Garden of ~,** Aden, İrembağı.
**edge**[1] [ec] *n.* Bıçak vs. ağzı; kenar; sırt. **on ~,** (i) kenar üstünde; (ii) sinirli: **to put an ~ on to (knife,** *etc.*), kılağını vermek: **not to put too fine an ~ upon it,** kılı kırk yarmadan: **to set the teeth on ~,** diş kamaştırmak: **straight ~,** cedvel: **to take the ~ off,** körletmek. **~**[2] *v.* Kenar çekmek; kenarında bulunmak. **~ away,** yavaş yavaş uzaklaşmak: **~ one's way in,** yavaş yavaş sokulmak. **~d** [ˈecd], keskin; ağızlı. **~ways, ~wise** [ˈedcwêyz, –wâyz], palalığına; yandan; yan yan. **I couldn't get a word in ~,** ağzımı açıp bir söz söyliyemedim.
**edible** [ˈedibl]. Yenir.
**edict** [ˈîdikt]. İrade; ferman.
**edify** [ˈedifây]. Mânen yükseltmek; öğretip tenvir etm.
**edit** [ˈedit]. (Bir yazıyı) neşre hazırlamak; (bir gazeteyi) idare etmek. **~ion** [iˈdişn], tabı, basım. **~or** [ˈeditə(r)], gazete müdürü; neşreden kimse. **~orial** [–ˈtôriəl], başmakale.
**educat·e** [ˈedyukêyt]. Talim ve terbiye etm.; tahsil ettirmek; alıştırmak. **he was ~ed in England,** İngiltere'de tahsil etti. **~ed,** okumuş, münevver. **~ion** [–ˈkêyşn], talim ve terbiye; tahsil; maarif. **Ministry of ~,** Maarif Nezareti. **~ional,** talim ve terbiye aid. **~ive** [ˈedyukêytiv], terbiye edici.
**eel** [îl]. Yılan balığı.
**e'er** [eə(r)]. (*şair.*) = **ever.**
**eerie** [ˈiəri]. Tüyler ürpetici.
**efface** [iˈfêys]. Silmek. **to ~ oneself,** bir tarafa çekilmek.
**effect**[1] [iˈfekt] *n.* Tesir; nüfuz; netice; meal. **~s,** mal, eşya. **to carry into ~,** tatbik etm.: **for ~,** tesir yapmak için: **in ~,** filhakika; doğrusu: **to give ~ to,** yerine getirmek: **of no ~,**

tesirsiz: **to no ~**, boşuboşuna: **it had no ~**, tesir etmedi; **it had no ~ whatever,** bana mısın demedi: **personal ~s,** şahsî eşya: **to take ~**, tesir yapmak; mer'î olm.; (aşı vs.) tutmak: **words to that ~**, o mealdaki sözler. ~² *v.* Tesir etmek; netice vermek; yerine getirmek; istihsal etmek. **to ~ an entrance,** bir yere zorlıyarak girmek: **to ~ a policy of insurance,** sigorta mukavelesi akdetmek. **~ive** [iˈfektiv]. Tesirli, işe yarar; hakikî; elverişli; mevcud. **~iveness,** tesirlilik, fayda. **~ual** [iˈfektyu̯əl], tesirli; istenilen neticeyi hasıl eden.

**effeminate** [iˈfeminit]. Kadın gibi.

**effervesce** [ˌefəˈves]. Kaynayıp köpürmek.

**efficac·ious** [ˌefiˈkêyşəs]. Tesirli, faydalı. **~y** [ˈefikəsi], fayda, tesir.

**efficien·cy** [iˈfişnsi]. Ehliyet; iktidar; verim. **~t,** ehliyetli; muktedir; elverişli.

**effigy** [ˈefici]. İnsan modeli. **to burn [hang] in ~**, bir adamı tahkir için resmini yakmak [asmak].

**effort** [ˈefət]. Çabalama, gayret; (*coll.*) eser. **to exert every ~**, her gayreti sarfetmek: **to make an ~**, bir cehd sarfetmek. **~less,** cehidsiz ve kolayca.

**effrontery** [iˈfrʌntəri]. Yüzsüzlük, küstahlık.

**effus·ion** [iˈfyûjn]. Dök(ül)me; coşkunluk; (*cont.*) değersiz eser. **~ive** [–siv], coşkun; bol.

**e.g.** [ˈî ˈcî]. = **exempli gratia,** meselâ.

**egg¹** [eg]. Yumurta; (balık) tohum. **boiled ~**, rafadan yumurta: **fried ~**, sahanda yumurta: **poached ~**, suda pişmiş kabuksuz yumurta: **hard-boiled ~**, hazırlop: **scrambled ~s,** tereyağile çalkalanıp pişirilen yumurta: **bad ~**, kokmuş yumurta; değersiz adam. **~-cup,** rafadan yumurta kabı. **~-plant,** patlıcan. **~-shaped,** yumurta biçimde. ~² *v.* **to ~ s.o. on,** (*sl.*) tahrik etm.

**egois·m** [ˈego̯uizm]. Hodbinlik; kendini beğenme. **~t,** hodbin; kendini beğenen.

**egotis·m** [ˈegətizm]. Hep kendini düşünme benlik davası. **~t,** yalnız kendini düşünen kimse.

**Egypt** [ˈîcipt]. Mısır. **~ian** [iˈcipşn], mısırlı. **~ology** [–ˈtoləci], mısriyyat.

**eider** [ˈâydə(r)]. **~down,** kuştüyü yorgan.

**eight** [êyt]. Sekiz. **figure of ~ knot,** kropi bağı. **~h** [êytθ], sekizinci; sekizde bir. **~een** [êyˈtîn], on sekiz. **~y** [ˈêyti], seksen. **~ieth** [ˈêyti̯əθ], sekseninci.

**either** [ˈâyðə(r)]. İkisinden biri. **~ I or you,** ya ben ya sen: **~ this or that,** ya bu ya şu: **nor that ~**, ne de bu: **I won't go ~**, (sen gitmezsen) ben de gitmem.

**ejaculate** [iˈcakyulêyt]. Birdenbire söylemek; fışkırtmak.

**eject** [iˈcekt]. Dışarı atmak; koğmak; fışkırtmak. **~ion** [–kşn], fışkırtma.

**eke** [îk]. **~ out,** idare ile kullanarak yetiştirmek. **to ~ out a living,** kıt kanaat geçinmek.

**elaborate** [iˈlabərit] *a.* Dikkatle hazırlanmış; inceden inceye. *v.* [–rêyt], Özenerek tertib etm. **to ~ upon,** mufassalan izah etmek.

**elapse** [iˈlaps]. (Vakit) geçmek.

**elastic** [iˈlastik] *a.* Elâstiki. *n.* Lâstikli şerid, ip. **~ity** [–ˈtisiti], elâstikiyet.

**elat·e** [iˈlêyt]. Sevindirmek. **to be ~d,** haz ve gurur duymak. **~ion** [–ˈlêyşn], büyük sevinc; gurur.

**elbow** [ˈelbo̯u]. Dirsek. Dirsekle dürtmek. **at one's ~**, yanında: **to be out at ~s,** (ceket) dirsekleri delinmek; (insan) düşkün ve çapaçul olm.: **to ~ one's way through**; itip kakarak yol açmak. **~-grease,** el emeği. **~-room,** kollarını kımıldatacak yer.

**elder** [ˈeldə(r)]. İki kişinin en yaşlısı; yaşlı ve mühim adam; ced. **~ brother,** ağabey: **obey your ~s,** büyüklerinize itaat ediniz. **~ly,** yaşlı.

**eldest** [ˈeldist]. En yaşlı.

**elect** [iˈlekt] *v.* Seçmek, intihab etm.; karar vermek. *a.* Seçkin. **the ~**, cennete gidecek olanlar. **~ion** [–kşn], intihab, seçim. **~or,** müntehib, seçmen. **~oral** [–tərəl], seçime aid. **~orate,** müntehibler.

**electric** [iˈlektrik]. Elektrikî. **~al,** elektrikî, elektriğe aid. **~ally,** elektrik vasıtasile. **~ian** [–ˈtrişn], elektrikçi. **~ity** [ˈtrisiti], elektrik.

**electri·fy** [iˈlektrifây]. Elektriklendirmek; heyecanlandırmak. **~fication** [–fiˈkêyşn], elektriklendirme. **~fying,** heyecandırıcı.

**electro-** [iˈlektro̯u]. Elektrikli veya elektrik vasıtasile yapılmış. **~-magnet,** elektrik mıknatısı. **~-plate,** elektrik vasıtasile kaplamak. **~cute**

[iˈlektrəkyût]. Elektrikle öldürmek, idam etmek. ~**cution** [–ˈkyûşn], elektrikle ölüm.
**elegan·ce** [ˈelegəns]. Zarafet, letafet. ~**t,** zarif, nazik.
**elegy** [ˈeleci]. Mersiye.
**element** [ˈelimənt]. Eleman; unsur, esas, cevher; âmil. ~**s,** mebadi, başlangıc. **the** ~**s,** tabiatin kudretleri: **to brave the** ~**s,** tabiat kudretlerine meydan okumak: **to be in one's** ~, kendi muhitinde olm.: **to be out of one's** ~, kendi muhitinde olmamak. ~**al** [–ˈmentl], unsura aid; esaslı; tabiate aid. ~**ary** [–ˈmentəri], ibtidaî; basit; ilk: ~ **school,** ilkmekteb.
**elephant** [ˈelifənt]. Fil. **a white** ~, lüzümsüz ve masraflı mülk. ~**ine** [–ˈfantâyn], fil gibi; dev gibi; ağır ve hantal.
**elevat·e** [ˈelivêyt]. Yükseltmek. ~**ed,** yüksek; (*coll.*) çakırkeyif: ~ **railway,** bir şehir içinde sütunlar üzerinde yapılmış demiryolu. ~**ing,** yükseltici. ~**ion** [–ˈvêyşn], yükseltme; yükseklik; topun menzil zaviyesi; yapı maktaı. ~**or** [–tə(r)], asansör; hububat ambarı.
**eleven** [iˈlevn]. Onbir. ~**th,** onbirinci: **at the** ~ **hour,** son dakikada.
**elf** [elf]. Cüce ve muzib peri.
**elicit** [iˈlisit]. (Hakikatı) meydana çıkarmak.
**eligible** [ˈelicibl]. İntihab edilebilir; elverişli, münasib.
**eliminat·e** [iˈliminêyt]. Kaldırmak, bertaraf etm. ~**ion** [–ˈnêyşn], bertaraf etme, ifna.
**Elizabethan** [iˌlizəˈbîθən]. Birinci kıraliçe Elizabeth devrine aid.
**ell** [el]. ˹**if you give him an inch he will take an** ~˺, ˹yüz verirsen astarını da ister˺.
**ellip·se** [iˈlips]. Kat'ı nakıs. ~**tical** [–ˈliptikl], kat'ı nakısa aid.
**elm** [elm]. Karaağac.
**elocution** [ˌeləˈkyûşn]. Konuşma sanati.
**elongat·e** [ˌîlon(g)ˈgêyt]. Gerip uzatmak. ~**ed,** ince ve uzun.
**elope** [iˈlo̤up]. Aşıkı ile gizlice kaçmak.
**eloquen·ce** [ˈeləkwəns]. Belâğat. ~**t,** beliğ.
**else** [els]. Yoksa; başka. **anyone** ~, başkası, başka biri: **either this or** ~ **that,** ya bu ya şu: **come in or** ~ **go out,** ya içeri gir ya dışarı çık: **can I see somebody** ~**?,** başka birini görebilir miyim?: **he eats little** ~ **than bread,** ekmekten başka pek bir şey yemez: **he thinks of little** ~ **but money,** paradan başka pek bir şey düşünmez: **no one** [**nobody**] ~, başka hiç kimse: **nothing** ~, **thank you,** başka bir şey istemem. ~**where** [elsˈwe̤ə(r)]. Başka yer(ler)de.
**elucidate** [iˈlyûsidêyt]. Aydınlatmak.
**elude** [iˈlyûd]. Ustalıkla başından savmak; ···den kaçamak yapmak. **to** ~ **a blow,** bir darbeden kaçınmak.
**elusive** [iˈlyûsiv]. Tutulmaz, bulunmaz. **he is a most** ~ **person,** bu adamı ele geçirmek çok güç: **an** ~ **reply,** kaçamaklı cevab.
**Elysian** [iˈlisi̤ən]. **the** ~ **fields,** cennet bahçeleri.
**'em** [əm] = **them.**
**emaciate** [iˈmêysiêyt]. Bir deri bir kemik yapmak.
**emanat·e** [ˈîmənêyt]. Çıkmak; sadır olm. ~**ion** [–ˈnêyşn], sudur; çıkan gaz vs.
**emancipat·e** [iˈmansipêyt]. Azad etm., esaretten kurtarmak. ~**ion** [–ˈpêyşn], esaretten kurtulma, kurtarma; azadlık; hürriyet.
**embalm** [emˈbâm]. Tahnit etmek.
**embank** [emˈban(g)k]. Topraktan sed çekmek. ~**ment,** toprak sed; rıhtım.
**embargo** [emˈbâgo̤u]. Menetme. **to lay an** ~ **upon,** ···e ambargo koymak: **to put an** ~ **on public meetings,** umumî toplantıları yasak etmek.
**embark** [emˈbâk]. Gemiye bin(dir)mek: **to** ~ **upon,** ···e girişmek ~**ation** [–ˈkêyşn], gemiye bin(dir)me
**embarrass** [imˈbaras]. Sıkıntıya sokmak; rahatsız etm. ~**ed,** sıkılgan; paraca sıkıntıda. ~**ment,** sıkıntı; engel.
**embassy** [ˈembasi]. Sefaret; sefarethane.
**embellish** [emˈbeliş]. Süslemek.
**ember** [ˈembə(r)]. Kor.
**embezzle** [emˈbezl]. İhtilâs etm.; zimmetine para geçirmek
**embitter** [emˈbitə(r)]. Ekşitmek; dünyadan nefret ettirmek; ters ve huysuz yapmak. **to** ~ **a quarrel,** bir kavgayı körüklemek. ~**ed,** dünyadan nefret etmiş.
**emblem** [ˈembləm]. Remiz; timsal; işaret.
**embod·iment** [imˈbodimnt]. Tecessüm. ~**y** [imˈbodi]. Tecessüm ettirmek: bir bütün halinde toplamak.

**embrace** [im'brêys] *v.* Kucaklamak, benimsemek; memnuniyetle kabul etm.; şamil olm. *n.* Kucaklaşma. **to ~ a career,** bir mesleğe intisab etmek.
**embroider** [im'brôydə(r)]. İğne ile nakış işlemek; işkembeden atmak. **~y** [-dəri], nakış.
**embryo** ['embriou]. Rüşeym; tohum. **in ~,** rüşeym halinde; gelişmemiş: **a doctor in ~,** istikbalin doktoru.
**emend** [i'mend]. Tashih etm.; hatasını düzeltmek.
**emerald** [emərəld]. Zümrüd; yemyeşil. **the Emerald Isle,** İrlanda.
**emerge** [i'mö(r)c]. Suyun yüzüne çıkmak; ortaya çıkmak; birdenbire zuhur etm.; netice olarak anlaşılmak. **~nce** [-əns], ortaya çıkma.
**emergency** [i'mö(r)cənsi]. Derhal harekete geçme [bir çrae bulmayı] icab ettiren hadise; fevkalâde hal, anî tehlike; buhran. **in case of ~,** zaruret halinde: **to provide for emergencies,** beklenmedik vaziyete karşı hazırlıklı bulunmak: **a state of ~,** harb hali için icabeden hazırlıkları yapma emri: **~ repairs,** mübrem tamirat. **~-brake** [**~-exit,** *etc.*] ihtiyac zamanında kullanılan fren, çıkış vs.
**emery** ['eməri]. Zımpara. **~ paper,** zımpara kâğıdı.
**emetic** [i'metik]. Kusturucu (ilâc).
**emigr·ant** ['emigrənt]. (Kendi memleketinden) hicret eden. **~ate,** hicret etmek. **~ation** [-'grêyşn], muhaceret; göçme.
**eminen·ce** ['eminəns]. Yükseklik; yüksek yer. **~t,** yüksek; mümtaz; meşhur. **~tly,** fevkalâde.
**emiss·ary** ['emisəri]. Gönderilen kimse; casus. **~ion** [i'mişn]. Salıverme; neşredilen şey; banknotun tedavüle ihracı.
**emit** [i'mit]. Dışarıya yaymak; salıvermek; neşretmek.
**emolument** [i'molyumənt]. Maaş.
**emotion** [i'mouşn]. Heyecan; his. **~al,** içli; heyecanlı; müteessir edici.
**emperor** ['empərə(r)]. İmparator.
**emphas·is** ['emfəsis]. Tekid; vurgu. **~ize** [-sâyz], tebarüz ettirmek; ···e ehemmiyet vermek; vurgulamak.
**emphatic** [im'fatik]. Kat'î; ehemmiyetle söylenen. **~ally,** kat'î olarak.
**empire** ['empâyə(r)]. İmparatorluk; saltanat; hakimiyet. **~ day,** Britanya İmparatorluğun millî bayramı (24 mayıs).
**employ** [im'plôy]. Kullanmak; istihdam etmek. Hizmet. **to ~ oneself with** [**in**], ···le meşgul olm.: **to keep ~ed,** meşgul etmek. **~ee** [-î], müstahdem; memur; amele. **~er,** patron, iş veren kimse. **~ment,** iş, hizmet; meşguliyet; kullanma; kullanış.
**empress** ['empres]. İmparatoriçe.
**emptiness** ['emtinis]. Boşluk.
**empty** ['empti] *a.* Boş; aç; nafile; kuru (tehdid vs.). *n.* İçi boş şey. *v.* Boşaltmak; tahliye etmek. **empties,** boş kutular, şişeler vs.: **'to be taken on an ~ stomach',** açkarnına alınacak: **to go away ~,** eli boş gitmek. **~-handed,** eli boş. **~-headed,** akılsız.
**emulat·e** ['emyulêyt]. Rekabet etm.; gıbta etm.; taklid etmek. **~ion** [-'lêyşn], rekabet, gıbta.
**enable** [i'nêybl]. İktidar vermek; muktedir kılmak; imkân vermek.
**enact** [i'nakt]. (Kanun) vazetmek; irade etm.; icra etm.; (bir rolu) oynamak. **~ment,** kararname; irade; vaz'etme.
**enamel** [i'naml]. Mine, mineli iş. **~ paint,** vernikli boya. **-ware,** emaye.
**enamour** [i'namə(r)]. **to be ~ed of,** ···e âşık olm.
**encamp** [in'kamp]. Ordugâh kurmak. **~ment,** ordugâh; kamp.
**enchant** [in'çânt]. Teshir etm., sihirlemek. **~ing,** sihirli, cazibeli. **~ment,** sihir; cazibe. **~ress** [-tris], sihirli kadın, dilber.
**encircle** [in'sö(r)kl]. Kuşatmak.
**enclos·e** [in'klouz]. Kuşatmak; leffetmek; kapatmak. **~ed,** kuşatılmış; kapanmış; leffen gönderilen. **~ure** [in'kloujuə(r)]. Kapatma; leffen gönderilen şey; duvarla çevrilmiş arsa; kuşatan duvar. **the Royal ~,** Kıraliçeye mahsus mahal.
**encompass** [in'kʌmpəs]. Tamamen etrafını çevirmek. **to ~ s.o.'s death,** kumpas kurarak birinin ölümüne sebeb olmak.
**encounter** [in'kauntə(r)]. ···le karşılaşmak; ···e tesadüf etm.; uğramak. Çarpışma; karşılaşma; mücadele.
**encourag·e** [in'kʌric]. Cesaret vermek; teşvik etm. **~ement,** teşvik; yüz verme. **~ing,** cesaret [ümid] verici.
**encroach** [in'krouç]. **to ~ upon,** ···e tecavüz etmek. **to ~ on s.o.'s time,** birinin vaktini almak.

**encumb·er** [inˈkʌmbə(r)]. ···e yük olm.; mani olm.; (yol) tıkamak. (property) to be ~ed, (mülk) ipotekli olmak. ~**rance** [–brəns], yük, engel.
**encyclopaedia** [inˌsâykləˈpîdyə]. Ansiklopedi.
**end**[1] [end] *n.* Son, nihayet; uc; akibet; bakiye; gaye. **at an ~**, bitmiş, tükenmiş: **to come to a bad ~**, sonu [akibeti] fena olm.: **to make both ~s meet**, iki ucunu bir araya getirmek: **to bring to an ~**, nihayet v rmek: **by the ~ of the day**, uzun bir günun sonunda: **to change ~s**, (futbolda) haftaymda sahada yer değiştirmek: **the ~s of the earth**, dünyanın bir ucu: **from ~ to ~**, baştan başa: **in the ~**, sonunda: ⌜**the ~ justifies the means**⌝, gaye vasıtayı mubah kılar: **to keep one's ~ up**, dayanmak: **to make an ~ of**, bitirmek; mahvetmek: **to meet one's ~**, eceli gelmek: **no ~ of**, sonsuz: **to no ~**, boşuna, nafile: **to think no ~ of**, çok sevmek: **to think no ~ of oneself**, kendini çok beğenmek: **he's no ~ of a fellow**, yaman bir adamdır: **to put an ~ to**, ···e nihayet vermek: **~ on**, kirişleme: **to meet ~ on**, burun buruna çarpışmak: **to stand** [**set**] **on ~**, kirişlemesine koymak: **five hours on ~**, beş saat mütemadiyen: **three days straight** [**right**] **on ~**, üstüste üç gün: **the ~ of time**, kıyamet günü: **to the ~ that** ..., ... maksadıyle: **and that's an ~ of it!**, işte bu kadar!: **to begin at the wrong ~**, tersinden başlamak. ~[2] *v.* Bitirmek, nihayet vermek; bitmek, sona ermek. **to ~ in a point**, sivri bir ucla nihayet bulmak: **to ~ in smoke**, suya düşmek: **he ~ed by saying** ..., sonunda ... dedi. ~**ed** [ˈendid]. ... uclu. ~**ing** [ˈendin(g)]. Son; son ek. ~**less** [ˈendlis]. Sonsuz; bitmez tükenmez.
**endear** [inˈdiə(r)]. Sevdirmek. ~**ments**, muhabbetli sözler.
**endeavour** [inˈdevə(r)]. Çalışma(k), gayret (etm.).
**endorse** [inˈdôs]. Tasdik etm.; (çeki muhatab) imzalamak. ~**ment**, ciro; tasdik etme.
**endow** [inˈdau]. Bir hayır müessesesi için irad temin etmek. ~**ed with**, malik, haiz. ~**ment**, irad temin etme.
**endur·e** [inˈdyuə(r)]. Tahammül etm.; dayanmak; daimî olmak. ~**able**, çekilir, tahammül olunur. ~**ance**, tahammül; dayanıklılık. ~**ing**, dayanıklı; devamlı.
**enemy** [ˈenəmi]. Düşman. **how goes the ~?**, saat kaç?
**energ·etic** [ˌenəˈcetik]. Faal; kuvvetli. ~**ize** [ˈenö(r)câyz], faaliyet vermek. ~**y** [ˈenəci], faaliyet; çalışkanlık; kuvvet.
**enfold** [inˈfould]. Sarmak; kucaklamak.
**enforce** [inˈfôs]. İnfaz etm.; teyid etm. **to ~ obedience**, itaat ett rmek: **to ~ one's will upon s.o.**, arzusunu birine zorla kabul ettirmek. ~**able**, infazı kabil. ~**d**, mecburî. ~**ment**, infaz etme, icra.
**enfranchise** [inˈfrançâyz]. Seçim hakkı vermek.
**engag·e** [inˈgêyc]. Vaadetmek; taahhüd etm.; üzerine almak; hizmetine almak; cezbetmek; (dikkatini) çekmek; ···le mucadeleye girişmek. **to be engaged**, meşgul olm.; nişanlı olm.; tutulmuş olmak: **to ~ s.o. in conversation** [**to ~ in conversation with s.o.**], birisile mükâlemeye girişmek: **to ~ in battle**, muharebeye girmek: **to ~ in politics**, siyasete girişmek. ~**ed** [inˈgêycd]. Nişanlı; meşgul. ~**ement** [inˈgêycment]. Nişanlanma; taahhüd; çarpışma. ~**-ring**, nişan yüzüğü: **to meet one's ~s**, taahüddünü tutmak, borclarını ödemek: **owing to a previous ~**, daha evvel başka yere söz vermiş olduğum için: **social ~s**, davet gibi meşguliyetler. ~**ing** [inˈgêycin(g)]. Hoş, alımlı.
**engine** [ˈencin]. Makine; lokomotif. ~**-driver**, makinist. ~**-room**, makine dairesi. ~**-shed**, lokomotif garajı. ~**er** [ˌenciˈniə(r)]. Mühendis; istihkâm subayı. Mühendislik yapmak; kurnazca, ustaca, tertib etm. **the Engineers**, istihkâm sınıfı: **civil ~**, yol, su vs. mühendisi. ~**ering**, mühendislik.
**Eng·land** [ˈin(g)glənd]. İngiltere.
**English** [ˈin(g)gliş]. İngiliz; ingilizce. **in plain ~**, açıkçası. ~**man**, İngiliz. ~**woman**, İngiliz kadını.
**engrav·e** [inˈgrêyv]. Hâkketmek. ~**er**, hâkkâk. ~**ing**, mahkûk resim.
**engross** [inˈgrous]. Tamamen zabt ve işgal etm. **to be ~ed in**, ···e kapanmak.
**enhance** [inˈhâns]. Artırmak.
**enigma** [inˈigmə]. Muamma. ~**tic(al)** [–ˈmatik(l)], muammalı, esrarengiz.

**enjoy** [in'côy]. ···de zevk almak; ···in tadını almak. **to ~ oneself,** eğlenmek: **to ~ the confidence of s.o.,** birinin itimadını kazanmış olm.: **to ~ doing stg.,** bir şeyi yapmaktan zevk almak: **to ~ good health,** sıhhati iyi olmak. **~able,** hoş; zevk verici; eğlenceli. **~ment,** zevk; eğlence; tad.

**enlarge** [in'lâc]. Büyü(t)mek; genişlemek. **to ~ upon,** hakkında sözü uzatmak. **~ment,** büyütme; dahame; agrandisman.

**enlighten** [in'lâytn]. Aydınlatmak, tenvir etm. **~ed,** münevver. **~ment,** tavzih; tenevvür.

**enlist** [in'list]. (Gönüllü) asker olm.; asker kaydetmek. **to ~ the services of,** ···in yardımını temin etmek.

**enliven** [in'lâyvn]. Canlandırmak.

**enmity** ['enmiti]. Husumet, adavet. **at ~ with,** ···le arası açık.

**enormous** [i'nôməs]. Muazzam, iri; pek büyük.

**enough** [i'nʌf]. Kâfi; yetişir; kâfi derecede; oldukça. **~ and more than ~,** kâfi ve vafi. **to be ~,** kâfi gelmek, yetmek: **~ said,** fazla söze ne hacet?: **curiously ~,** işin tuhafı: **I've had ~ of you,** senden illâllah!: **more than ~,** lüzumundan fazla: **he writes well ~, but ...,** yazısı fena değil, amma ....

**enquire, -y** *v.* **inquire, -y.**

**enrol** [in'ro͜ul]. Askere yazmak; asker olmak; deftere kaydetmek; aza kaydetmek, olmak.

**en route** ['ân'rût]. Yolda.

**ensign** ['ensâyn]. Sancak, bayrak; alâmet; bayrakdar. **the White Ensign,** İngiliz bahriyesinin bayrağı: **the Red Ensign,** İngiliz ticaret filosunun bayrağı.

**enslave** [in'slêyv]. Köle yapmak.

**ensu·e** [in'syû]. Sonradan gelmek. **~ing,** sonradan gelen; gelecek.

**ensure** [in'şô(r), –şu͜ə(r)]. Temin etmek.

**entail** [in'têyl]. Sebeb olm.; icabetmek.

**entangle** [in'tan(g)gl]. Dolaştırmak; karmakarışık etm. **~ment,** mânia: **barbed wire ~,** dikenli tel mânia.

**entente** [ân'tânt]. Anlaşma; itilâf. **~ cordiale,** İngiltere ile Fransa arasındaki dostluk: **the Triple ~,** İtilâfi müselles.

**enter** ['entə(r)]. Girmek; ···e dahil olm.; girişmek; binmek; içeriye girmek; kaydetmek. **~ that to me!,** bunu hesabıma yazınız: **to ~ into an agreement,** bir mukavele akdetmek: **to ~ into the spirit of the game [thing],** bir oyunun [şeyin] ruhuna nüfuz etm.: **to ~ for a race,** bir yarışa yazılmak: **to ~ a horse for a race,** bir atı yarışa kaydettirmek. **~ on, upon,** başlamak: **to ~ upon one's twentieth year,** yirmi yaşına basmak.

**enterpris·e** ['entə(r)prâyz]. Teşebbüs; iş; acarlık. **~ing,** müteşebbis, acar, atılgan.

**entertain** [ˌentə'têyn]. Misafirliğe kabul etm.; eğlendirmek; kabul etm. **they ~ a lot,** misafirleri eksik olmaz. **~ing,** eğlenceli; hoşsohbet. **~ment,** eğlenti; eğlence: **~ allowance,** ziyafet tahsisatı.

**enthrone** [in'θro͜un]. Tahta oturtmak.

**enthuse** [in'θyûz]. (*coll.*) Coşmak. **to ~ about,** bir şeyi göklere çıkarmak.

**enthusias·m**[ in'θyûziazm]. Coşkunluk; büyük heyecan; şevk ve hayret. **~t,** bir şeyin hayranı. **~tic** [–'astik], coşkun; hayran.

**entic·e** [in'tâys]. Tatlılıkla cezbetmek; ayartmak. **~ing,** cazibeli; ayartıcı. **~ement,** ayartma; cezbetmek için kullanılan tatlı söz.

**entire** [in'tâyə(r)]. Tam, bütün; tamam. **~ly,** büsbütün, tamamen. **~ty,** bütünlük: **in its ~,** tamamı ile, bütünü ile.

**entitle** [in'tâytl]. İsim vermek; hak ve salâhiyet vermek. **to be ~d to do stg.,** bir şeyi yapmağa salâhiyeti olm.: **you are not ~d to say such a thing,** böyle bir şeyi söylemeğe hakkınız yok.

**entity** ['entiti]. Varlık; mevcudiyet.

**entrails** ['entrêylz]. Barsaklar.

**entrance**[1] ['entrəns]. Giriş, girme, duhul; medhal. **~-fee [-money],** duhuliye. **~**[2] [in'trâns]. Vecde getirmek; hayran etmek.

**entreat** [in'trît]. Yalvarmak. **~y,** yalvarma; ısrarla rica etme.

**entrench** [in'trenç]. Metris yapmak; siper kazmak. **to ~ oneself,** yerleştirmek: **to ~ upon,** ···e tecavüz etmek. **~ment,** metris, istihkâm, siper.

**entre nous** ['ântr 'nû]. (*Fr.*) Söz aramızda.

**entrepôt** ['ântrəpo͜u]. Ardiye; ticaret merkezi.

**entrust** [in'trʌst]. Tevdi etmek.

**entry** ['entri]. Medhal; girme, giriş; kayıd; kaydedilen şey, kimse. **~-form,** kayıd varakası.

**enumerate** [iˈnyûmərêyt]. Birer birer saymak.
**enunciat·e** [iˈnʌnsiêyt]. Vuzuhla beyan etm.; ileri sürmek; telâffuz etmek. **~ion** [–ˈêyşn], konuşma tarzı; ileri sürme.
**envelop** [inˈveləp]. Sarmak; kaplamak. **~e** [ˈenvəloup, on-]. Zarf.
**envenom** [inˈvenəm]. Zehir katmak; kızıştırmak.
**envi·able** [ˈenviəbl]. Gıbta edilecek. **~ous** [ˈenviəs]. Hasud; kıskanc; gıbta eden. **to be ~ of,** imrenmek, gıbta etmek.
**environ** [inˈvâyrən]. Kuşatmak. **~ment,** muhit; kuşatma; etraf. **~s** [inˈvâyrənz], civar, etraf.
**envisage** [inˈvizic]. Zihninde canlandırmak; tasavvur etm.
**envoy** [ˈenvôy]. Elçi, murahhas.
**envy** [ˈenvi]. Gıbta; hased, kıskanc. Gıbta etm., kıskanmak. **to be the ~ of all,** herkesin gıbta ettiği bir şey olm.: **to be green with ~,** çok kıskanmak.
**ephemeral** [iˈfemərəl]. Bir günde yaşıyan; fanî; gelip geçici.
**epic** [ˈepik]. Destan tarzında.
**epicentre** [ˈepisentə(r)]. Zelzele merkezi.
**epicure** [ˈepikyuə(r)]. Yemek ve içki meraklısı.
**epidemic** [epiˈdemik]. Salgın.
**epigram** [ˈepigram]. Hicivli şiir; iğneli söz; vecize. **~matic** [–ˈmatik], vecize tarzında.
**epilep·sy** [ˈepilepsi]. Sar'a illeti. **~tic** [–ˈleptik], sar'alı.
**epilogue** [ˈepilog]. Hatime.
**Epiphany** [eˈpifəni]. Üç şarklı âlimin İsa'yı ziyaretleri yıldönümü ki Hıristıyanlar tarafından 6 ocakta tes'id olunur (Ortodokslarca haçı suya atma).
**episcopal** [eˈpiskəpl]. Piskoposa aid.
**episode** [ˈepisoud]. Vak'a, hadise; fıkra.
**epistle** [iˈpisl]. Mektub; name.
**epitaph** [ˈepitâf]. Mezar kitabesi.
**epithet** [ˈepiθet]. Vasıf; sıfat; lâkab.
**epitom·e** [eˈpitoumi]. Hulâsa; zübde. **~ize** [iˈpitəmâyz], icmal etmek.
**epoch** [ˈîpok]. Devir.
**eponymous** [iˈponiməs]. Adını veren.
**equab·le** [ˈekwəbəl]. Yeknesak, az değişen; mutedil (iklim); ölçülü. **~ility** [–ˈbiliti], itidal, yeknesaklık, değişmezlik.
**equal** [ˈîkwəl] *a.* Müsavi; denk; aynı seviyede. *n.* Küfüv; eş. *v.* Müsavi olm., eş olm.; yetişmek. **~s,** akran; emsal; aynı rütbede olanlar. **to be ~ to the occasion [to a task],** bir işin uhdesinden gelmek: **to get ~ with s.o.,** birisinden acısını çıkarmak: **I don't feel ~ to it,** bunu yapacak halim yok: **the ~ sign,** müsavi işareti (=). **~ly,** aynı derecede; müsavi olarak. **~itarian** [ˌîkwoliˈteəriən]. Umumî müsavat tarafdarı. **~ity** [îˈkwoliti]. Müsavat. **on an ~ with,** ···le müsavi olarak. **~ize** [ˈîkwəlâyz]. Müsavi etm.; (maçta) beraberlik temin etmek.
**equanimity** [ˌîkwəˈnimiti]. Temkin. **to recover one's ~,** kendini toplamak.
**equat·e** [iˈkwêyt]. Müsavi yapmak; tadil etmek. **~ion** [iˈkwêyşn]. Muadele; tadil etme. **simple ~,** basit muadele: **quadratic ~,** ikinci dereceden muadele. **~or** [iˈkwêytə(r)]. Hattıistiva. **~ial** [ˌekwaˈtôriəl], hattıistivaya aid.
**equi-** [îkwi] *pref.* Müsavi mânasını ifade eder. **~distant** [–ˈdistənt]. Aynı uzaklıkta olan. **~lateral** [–ˈlatərəl]. Dılıları müsavi. **~librium** [–ˈlibriəm]. Muvazene. **~nox** [ˈîkwənoks]. Gündönümü.
**equip**. [iˈkwip]. Donatmak; techiz etmek. **~ment,** techiz etme; techizat: takım; levazım.
**equitable** [ˈekwitəbl]. İnsaflı, âdil.
**equity** [ˈekwiti]. Adalet, insaf.
**equivalen·ce** [iˈkwivələns]. Teadül. **~t,** muadil; müsavi; bedel; karşılık.
**equivoca·l** [iˈkwivəkl]. İki mânalı; mübhem; iltibaslı; şübheli. **~te** [iˈkwivəkêyt]. Kandırmak için iki mânalı sözler kullanmak.
**era** [ˈiərə]. Çağ; tarihin devrelerinden biri; hicrî, milâdî vs. tarih başlangıcı.
**eradicate** [iˈradikêyt]. Kökünden sökmek.
**eras·e** [iˈrêyz]. Silmek, çizmek. **~er,** lâstik (silgi). **~ure** [iˈrêyjyə(r)], silme; silinti.
**ere** [eə(r)]. Evvel, ···den önce. **~ now,** bundan evvel: **~ long,** neredeyse.
**erect** [iˈrekt] *v.* Dikmek, kurmak, inşa etm. *a.* Dimdik; kaim. **~ion** [iˈrekşn], bina; dikme, inşa etme; kurma.
**ergo** [ˈö(r)gou]. Bu sebebden.
**Erin** [ˈerin]. İrlanda.
**ermine** [ˈö(r)min]. Kakım (kürkü).

**erode** [iˈrǫud]. Kemirmek, aşındırmak.
**Eros** [ˈeros]. Aşk ilâhı.
**eros·ion** [iˈrǫujn]. Kemirme; aşın(dır)ma. **~ive** [–siv], kemirici, aşındırıcı.
**erotic** [eˈrotik]. Aşka aid; şehvanî.
**err** [ö(r)]. Yanılmak; dalâlete düşmek.
**errand** [ˈerənd]. Bir iş için gönderme. **a fool's ~**, neticesiz olacağı önceden bilinen iş vs. **~-boy**, bakkal çırağı.
**erratic** [iˈratik]. Hareketi intizamsız; devamlı olmıyan.
**erra·tum**, *pl.* **-ta** [ˈerêytəm, -ta]. Yanlışlık. **~ta**, düzeltme cedveli.
**erroneous** [iˈrǫunyəs]. Hatalı, yanlış.
**error** [ˈerə(r)]. Hata, yanlışlık; kabahat; dalâlet. **clerical ~**, istinsah hatası: **printer's ~**, tertib hatası: **to be in ~**, yanılmak: '**~s and omissions excepted**' (**e. & o.e.**), (bir hesabda) 'muhtemel yanlış ve noksanlar müstesna'.
**erstwhile** [ˈö(r)stwâyl]. Vaktiyle; sabık.
**erudit·e** [ˈerudâyt]. Âlim. **~ion** [–ˈdişn], âlimlik; büyük vukuf.
**erupt** [iˈrʌpt]. Fışkırmak; indifa etm. **~ion** [–ˈrʌpşn], indifa; fışkırma; (kızamık vs.) dökme.
**escalator** [ˈeskəlêytə(r)]. Müteharrik merdiven.
**escap·ade** [ˌeskəˈpêyd]. Genclik çılgınlığı. **~e** [isˈkêyp]. Kaçmak, kurtulmak; (gaz) sızmak. Kaçış, firar; kurtulma; sızma. **he ~d with a fright**, korkmaktan başka bir zarar görmedi: **to ~ notice**, gözden kaçmak: **to have a narrow ~**, dar kurtulmak: **to make one's ~**, kaçıp kurtulmak: **not a word ~d him**, (i) kelime kaçırmadı; (ii) ağzından bir söz çıkmadı.
**escort** [ˈeskôt] *n.* Muhafız; maiyet alayı; kavaliye. *v.* [isˈkôt], Maiyet, muhafız, sıfatile refaket etmek. **under ~**, muhafaza altında.
**esoteric** [esoˈterik]. Batınî (felsefe); gizli.
**especial** [isˈpeşl]. Mahsus; hususî; *v. keza* **special**: **~ly**, bilhassa: **my ~ friend**, en iyi arkadaşım.
**espionage** [ˈespiənâj]. Casusluk.
**espous·al** [isˈpauzl]. Nikâh; kabul. **~e**, ···le evlenmek; tarafdarı olm.
**esquire** [isˈkwâyə(r)]. (Eskiden) genc asilzade; (şimdi) bir 'gentleman'in fahrî ünvanı ki isminden sonra yazılır, *mes.* **P. Jones, Esq.**
**essay** [ˈesêy] *n.* Deneme; kalem tecrübesi. *v.* [eˈsêy], Tecrübe etm.; denemek; çabalamak. **~ist** [ˈesêyist], denemeler yazan muharrir.
**essen·ce** [ˈesəns]. Öz, asıl; künh. **the ~ of the matter**, işin esası: **meat ~**, et hulâsası. **~tial** [iˈsenşl] *a.* Aslî, esaslı; elzem. *n.* Elzem şey; en mühim nokta.
**establish** [isˈtabliş]. Tesis etm., kurmak; yerleştirmek; tahakkuk ettirmek; tasdik etmek. **to ~ oneself in business**, ticaret hayatına girmek: **to ~ oneself in a place**, bir yerde yerleşmek. **~ed**, yerleşmiş; sabit; sağlam: **an ~ fact**, tesbit edilmiş bir vakıa. **~ment**, tesis etme, kurma; müessese; teşkilat; **to keep up a big ~**, büyük bir evi ve bir çok hizmetçisi olmak.
**estate** [isˈtêyt]. Malikâne; emlâk; bir adamın menkul ve gayrimenkul emlâki; miras; hal, vaziyet. **~ duty**, intikal vergisi: **the ~s of the realm**, İngiltere'de üç siyasî sınıf (asilzadeler, ruhban sınıfı ve avam): **the fourth ~**, matbuat: **personal ~**, menkul mallar: **real ~**, gayrimenkul mallar: **of high** [**low**] **~**, ictimaî mevkii yüksek [aşağı]. **~-agency**, emlâk acentası (daire). **~-agent**, (i) emlâk acentası (kimse); (ii) büyük emlâk idare eden memur.
**esteem** [isˈtîm]. İtibar (etm.), hürmet (etm.); takdir etm.; saymak. **self ~**, (i) izzetinefis; (ii) kendini beğenme.
**estima·ble** [ˈestiməbl]. Değerli. **~te¹** [ˈestimit] *n.* Tahmin; kıymet takdiri. **the ~s**, bütçe. **~te²** [ˈestimêyt] *v.* Tahmin etm., kararlamak: **~d value**, muhammen kıymet. **~tion** [–ˈmêyşn], takdir, tahmin; rey; itibar: **in my ~**, bence.
**estrange** [isˈtrêync]. Soğutmak. **to become ~d from s.o.**, birisinden soğumak.
**estuary** [ˈestyuəri]. Nehir mansabı; halic.
**etern·al** [iˈtö(r)nl]. Ezelî ve ebedî; sonsuz. **~ity** [–niti], ebediyet: **the eternities**, ebedî hakikatler.
**ether** [ˈîθiə(r)]. Eter. **the ~**, esîr; sema. **~eal** [iˈθiəriəl], esirî; gayet hafif ve nazik; semavî; ruhanî.
**ethic(al)** [ˈeθik(l)]. Ahlâk ilmine aid; ahlâkî. **~s** [ˈeθiks], ahlâk ilmi.
**Ethiopia** [îθiˈǫupyə]. Habeşistan. **~n**, habeş.
**etiquette** [ˌetıˈket, ˈetiket]. Âdabı muaşeret. **not to stand upon ~**, teklifsiz olm.

**etymology** [ˌetiˈmoləci]. İştikak ilmi.
**eucharist** [ˈyûkərist]. Hıristiyanlarca İsa'nın etini ve kanını temsil eden ekmekle şarabın yenmesi âyini.
**eugenic** [yûˈcenik]. İnsan ırkını ıslâh ilmine aid.
**eulog·ist** [ˈyûləcist]. Medhiyeci. **~istic** [–ˈcistik], medhedici. **~ize** [ˈyûləcâyz], medhüsena etmek. **~y** [ˈyûləci], medih; kaside.
**eunuch** [ˈyûnək]. Haremağası; hadım.
**euphony** [ˈyûfəni]. Ses ahengi.
**Eurasian** [yuəˈrêyziən]. Avrupalı ile Asyalı melezi.
**Europe** [ˈyuərəp]. Avrupa. **~an** [–ˈpiən], Avrupalı.
**evacuat·e** [iˈvakyuêyt]. Tahliye etm., boşaltmak; ifraz etmek. **~ion** [ˈêyşn], tahliye, boşaltma, ifraz.
**evade** [iˈvêyd]. İctinab etm., savmak; …den kurtulmak.
**evaluate** [iˈvalyuêyt]. Takdir etmek.
**evanescent** [ˌîvaˈnesənt]. Çabuk kaybolur; süreksiz, fanî.
**evangel·ic(al)** [ˌîvanˈcelik(l)]. İncile aid; protestanlığa aid. **~ism** [iˈvancəlizm], incili neşretme. **~ist,** incil muharriri; incili neşreden kimse. **~ize,** hıristiyan etmek.
**evaporat·e** [iˈvaperêyt]. Buharlaş(tır)mak; buhar gibi uçmak. **~ion** [–ˈrêyşn], tebahhur; buhar olma; buğu.
**evas·ion** [iˈvêyjn]. Kaçınma; baştan savma. **~ive** [–siv], kaçamaklı: **to take ~ action,** (uçak) zikzak yaparak ateşten kaçınmak.
**Eve**[1] [îv]. Havva. **~**[2]**.** Arife; bir gün önce. **on the ~ of,** arifesinde.
**even**[1] [îvən]. *a.* Düz, ârızasız; bir hizada; muntazam; müsavi; çift (tek değil). **to be ~,** (oyunda) berabere kalmak: **~ bet,** müşterek bahiste müsavi risk: **to get ~ with s.o.,** birisinden acısını çıkarmak: **to lay ~ odds,** müsavi şartlarla bahse girmek: **an ~ sum,** yuvarlak hesab. **~**[2] *adv.* Bile; hattâ. **~ if** [**~ though**], …se bile: **~ now it is not too late,** hattâ şimdi bile geç sayılmaz: **~ so,** hattâ, böyle olsa bile: **~ then,** (i) o zaman bile; (ii) buna rağmen: **if ~ I could see him,** bari onu görebilsem. **~**[3] *v.* Tesviye etm., düz etmek. **to ~ out,** müsavileştirmek. **~ly** [ˈîvənli]. Muntazaman; müsavi olarak. **~ matched,** uygun ve müsavi. **~ness** [ˈîvənnis] Düzlük; müsavilik; intizam; ittirad.
**evening** [ˈîvnin(g)]. Akşam(lık). **in the ~,** akşamlayın: **~ party,** suare. **~-dress** [**-wear**], tuvalet; smokin veya frak.
**event** [iˈvent]. Vak'a, hadise; hal; numara. **at all ~s,** her halde: **in the ~ of,** … takdirde: **in the course of ~s,** neticede; zamanla: **in either ~,** her iki halde de: **to be wise after the ~,** iş işten geçtikten sonra akıl öğretmek. **~ful,** vak'alarla dolu; maceralı. **~ual** [iˈventyuəl]. Son olarak; netice olarak; nihaî. **~ually,** en sonunda, neticede; akibet. **~uality** [–ˈaliti], ihtimal; takdir. **~uate** [iˈventyuêyt]. Vukubulmak; çıkmak.
**ever** [ˈevə(r)]. Daima; bir vakitte; her hangi bir vakitte. **~ after,** ondan sonra hep: **~ and anon** [**~ and again**], arasıra: **as cold a winter as ~ you saw,** hiç görülmemiş derecede soğuk bir kış: **as quick as ~ you can,** nekadar çabuk olmak mümkünse: **he is as idle as ~,** eskisi gibi hep tembeldir: **for ~,** ebediyen; daima: **for ~ and ~,** ebediyen: **he went for ~,** bütün bütün gitti: **England for ~!,** yaşasın İngiltere!; **not ~,** hiç bir zaman: **if ~ you see him,** onu görecek olursanız: **now, if ~, is the time,** bu işin bir zamanı varsa işte şimdidir: **I seldom, if ~, go there,** oraya gitsem bile pek seyrek giderim: **he is a poet if ~ there was one,** ben şair diye buna derim: **~ since,** işte o zamandan itibaren: **~ so easy,** o kadar kolay ki: **I waited ~ so long,** o kadar bekledim ki: **thank you ~ so much,** pek çok teşekkür ederim: **what ~ is the matter?,** Allah Allah ne oldu?: **who ~ heard of such a thing?,** bu hiç işidilmiş şey midir?: **we are the best friends ~,** biz fevkalâde iyi dostuz. **~green** [ˈevəgrîn]. Yaz kış yeşil olan. **~-lasting** [ˈevəˈlâstin(g)]. Daimî; pek dayanıklı; bitmez. **from ~,** ezelden beri. **~more** [ˈevəˈmô(r)]. Ebediyen; daima.
**every** [ˈevri]. Her, herbir. **~ bit as good as,** tıpkı … kadar iyi: **~ few minutes,** her bir kaç dakika: **I expect him ~ minute,** onu bekliyorum, neredeyse gelir: ⌜**~ man for himself!**⌝, herkes başının çaresine baksın: **~ now and again,** arasıra: **~ one,** …den her biri (**~one**=herkes): **~ other one,** iki kişide bir: **~ other day,** gün aşırı: **~ third man,** üç kişide bir. **~body** [–bədi, –bodi], herkes. **~day,** *a.* hergünkü; günlük. **~one** [ˈevriwʌn],

herkes. ~**thing,** herşey. ~**where** [ˈevriweə(r)], her yerde: ~ **you go,** her gittiğiniz yerde.
**evict** [iˈvikt]. Hükmen tahliye ettirmek; bir yerden çıkarmak. ~**ion,** çıkartma, çıkarılma; tahliye ettirme.
**eviden·ce** [ˈevidəns]. Şahadet; delil; delâlet; beyyine. ···e delil olmak. **to be in** ~, göze çarpmak: **to bear [give]** ~ **of,** göstermek, delâlet etm.: **to give** ~, şahadet etm.: **to turn Queen's** ~, suç ortakları aleyhine şahadet etm. ~**t** [ˈevident]. Aşikâr; vazıh. ~**tly,** aşikâr olarak; her halde.
**evil** [ˈîvl] *a.* Fena, kötü. *n.* Kötülük, fenalık; zarar; belâ; derd. **the** ~ **eye,** nazar: **the Evil One,** şeytan: **of** ~ **omen,** uğursuz: ~ **spirit,** habis ruh: **to speak** ~ **of,** ···e iftira etmek. ~-**doer,** günahkâr. ~-**minded,** kötü niyetli.
**evince** [iˈvins]. Göstermek.
**evoca·tion** [ˌîvoˈkêyşn]. Davet, çağırma; hatırlatma. ~**tive,** hatırlatan, andıran.
**evoke** [iˈvouk]. Davet etm., çağırmak; andırmak, hatırlatmak.
**evolution** [ˌivəˈlyûşn]. Tekâmül, gelişme; manevra. ~**ary,** tekâmüle aid. ~**ist,** tekâmülcü.
**evolve** [iˈvolv]. Tekâmül et(tir)mek.
**ex**[1] [eks]. ···den. ~ **ship [works],** gemi [fabrika]dan teslim. ~-[2] *pref.* Sabık ....
**exact**[1] [igˈzakt] *a.* Kat'i, tam; doğru; aynen. **or, to be more** ~, yahud, daha doğrusu. ~**ly,** kat'i olarak, aynen; ~!, çok doğru, hakkınız var. ~[2] *v.* Cebren para almak; icabettirmek. ~**ing,** müşkülpesend; çok şey istiyen; zahmetli. ~**ion** [igˈzakşn], müfrit taleb; keyfî ve ölçüsüz vergi. ~**itude** [igˈzaktityûd]. Kat'iyet; doğruluk.
**exaggerat·e** [igˈzacərêyt]. Mubalağa etm., ifrat etm. ~**ed,** müfrit, mubalağali. ~**ion** [–ˈrêyşn], mubalağa, ifrat.
**exalt** [igˈzôlt]. Yükseltmek; göklere çıkarmak. ~**ation,** coşkunluk, büyük heyecan. ~**ed,** yüksek, âli; coşkun.
**exam** [igˈzam]. (*abb.*) **examination.** ~**ination** [igˈzamiˈnêyşn]. İmtihan; muayene; teftiş; istintak. **to sit for an** ~, bir imtihana girmek: **under** ~, tedkik edilmekte; muayene neticesinde. ~**ine** [igˈzamin]. Tedkik etm., muayene etm.; istintak etm.; imtihan etmek. ~**iner,** mümeyyiz. ~**inee** [–î], imtihan edilen kimse.
**example** [igˈzâmpl]. Misal; örnek; nümune; nüsha; ibret. **for** ~, meselâ: **to make an** ~ **of s.o.,** başkalarına ibret olsun diye birini cezalandırmak: **a practical** ~, müşahhas bir misal: **to set an** ~, örnek olm.: **without** ~, emsalı görmemiş.
**exasperat·e** [igˈzâspərêyt]. Sabrını tüketmek. ~**ing,** insan (öfkeden) çıldırtan. ~**ion,** şiddetli öfke.
**excavat·e** [ˈekskavêyt]. Kazmak, hafriyat yapmak. ~**ion,** hafriyat, kazı. ~**or,** hafriyatcı; kazma makinesi.
**exceed** [ikˈsîd]. Aşmak. ~**ingly,** ifrat derecede.
**excel** [ikˈsel]. Üstün olm.
**excellen·ce** [ˈeksəlens]. Mükemmellik; mümtazlık. ~**cy, His [Your]** ~, sefir, nazır vs.ye verilen ünvan. ~**t,** mükemmel; nefis
**except**[1] [ikˈsept] *v.* İstisna etm., haric tutmak. **present company** ~**ed,** sizden iyi olmasın. ~[2] *prep.* ···den başka, müstesna, haric. ~**ion** [ikˈsepşn]. İstisna. **to take** ~ **to,** kabul etmemek: **with the** ~ **of,** ... müstesna olarak: **without** ~, istisnasız. ~**al,** istisnaî; fevkalâde; nadir.
**excerpt** [ˈeksö(r)pt]. Bir kitab vs.den alınmış parça.
**excess** [ikˈses]. İfrat; fazla; aşırı hareket. ~**es,** taşkınlıklar, mezalim; zevk ve eğlencede ifrat. ~ **fare,** (bilet) mevki farkı vs. için zam: ~ **luggage,** nizamî ağırlığı aşan eşya: ~ **profits tax,** fazla kazanc vergisi. ~**ive,** müfrit; ölçüden aşırı: ~**ly,** ifrat derecede.
**exchange**[1] [iksˈçêync] *v.* Mübadele etm., trampa etm. **to** ~ **stg. for stg.,** bir şeyi bir şeyle değiştirmek: **to** ~ **greetings,** selâmlaşmak: **to** ~ **hats,** birbirinin şapkasını almak: **to** ~ **posts,** becayiş etmek. ~[2] *n.* Mübadele, trampa, teati, becayiş; borsa; kambiyo. **bill of** ~, poliçe: ⌜~ **is no robbery**⌝, mübadele meşrudur: **foreign** ~, döviz: **in** ~ **for,** ···e bedel: **to give in part** ~, bir şey satın alırken ücretin bir kısmı yerine bir eşya vermek: (**rate of**) ~, kambiyo rayici: **telephone** ~, telefon santralı. ~**able,** mübadele edilebilir.
**exchequer** [iksˈçekə(r)]. Devlet hazinesi. **the Exchequer,** Maliye Nezareti: **the Chancellor of the** ~, Maliye Nazırı: ~ **bill,** hazine bonosu: **my** ~ **is empty,** kesem boş.

**excis·e**[1] [ikˈsâyz] *n.* İspirto, tütün vs. üzerine konan vergi (istihlâk vergisi); bu vergileri toplıyan daire. ~[2] *v.* Kesip çıkarmak. **~ion** [–sijn], kesip çıkarma.

**excit·able** [ikˈsâytəbl]. Çabuk heyecanlanan; muvazenesiz. **~ability,** çabuk heyecanlanma. **~e** [ikˈsâyt]. Heyecanlandırmak; tahrik etm.; sebeb olmak. **~ed,** heyecanlanmış. **~ing,** heyecanlı. **~ement,** heyecan; coşkunluk.

**exclaim** [iksˈklêym]. Nida etm.; birdenbire demek. **to ~ at [against],** ···e karşı protesto etm.

**exclamat·ion** [ˌeksklәˈmêyşn]. Sevinc, hayret nidası; nida; âni sôz. **~ mark,** nida işareti (!). **~ory** [-ˈklamәtәri], sevinc, hayret ifade eden.

**exclud·e** [iksˈklûd]. İçeri almamak; haric tutmak; kabul etmemek. **this ~s all possibility of doubt,** bu hiç bir şübhe bırakmıyor. **~ing,** ... haric.

**exclusˈion** [iksˈklûjn]. İçeri almama; kabul etmeme. **to the ~ of,** ···i haric tutarak. **~ive** [–ˈklûsiv], başkalarını dahil etmemek üzere; pek hususî; hesaba dahil olmıyan. **~ of ,** ... haric: **the Joneses are very ~,** Jones'lar pek kibar geçinirler: **this is a very ~ club,** burası pek seçkin bir klübdür: **'plant' and 'animal' are ~ terms,** nebat ve hayvan tabirlerinin telifi kabil değildir.

**excommunicate** [ˌekskәˈmyûnikêyt]. Aforoz etmek.

**excruciating** [iksˈkrûşiêytin(g)]. Dayanılmaz derecede eziyet edici.

**excursion** [iksˈkö(r)şn]. Gezinti; istitrad. **~ists,** bir gün için gezintiye gidenler.

**excus·able** [iksˈkyûzәbl]. Mazur görülebilir. **~e**[1] [iksˈkyûz] *v.* Mazur görmek; affetmek; muaf tutmak. **~ me,** affedersiniz: **if you will ~ the expression,** haşa huzurdan: **to ~ s.o. from doing stg.,** birini bir şeyden muaf tutmak. **~e**[2] [iksˈkyûs] *n.* Mazeret; vesile, bahane. **⌜ignorance of the law is no ~⌝,** kanunu bilmemek mazeret değildir: **to make ~s,** mazeret göstermek; özür dilemek.

**execr·able** [ˈeksikrәbl]. Menfur, pek çirkin. **~ate** [ˈeksikrêyt], nefret etm. **~ation** [–ˈkrêyşn], lânet; nefret; küfür.

**execut·e** [ˈeksikyût]. İfa etm., icra etm.; yerine getirmek; infaz etm.; idam etmek. **to ~ a deed,** bir senedi imza vs. ile tamamlamak. **~ant** [igˈzekyutәnt], icra edici. **~ion** [ˌeksiˈkyûşn], ifa, icra, ikmal; idam: **to do great ~,** cok zarar vermek. **~ioner,** cellâd. **~ive** [igˈzekyutiv] *a.* İcra ve tenfizle mükellef; icrai idare eden. *n.* İcra eden idare, faal idare. **~ duties,** (*mil.* vs.) idare vazifeleri: **~ power,** icra kuvveti. **~or** [igˈzekyutә(r)]. Vasiyeti tenfiz memuru. **~rix,** vasiyeti tenfize memur kadın.

**exempl·ary** [igˈzemplәri]. Örnek olarak. **~ify** [igˈzemplifây], temsil etm.; ... örneği olmak.

**exempt** [igˈzempt]. Muaf tutmak. Muaf; ârî. **~ion** [–şn], muafiyet.

**exercise**[1] [ˈeksәsâyz] *n.* Kullanma; icra; talim, alıştırma; mekteb vazifesi; beden terbiyesi. **in the ~ of one's duties,** vazife esnasında. **mental ~,** zihni işletme: **to take ~,** vücudü işletmek. **~-book,** mekteb defteri. ~[2] *v.* Kullanmak; sarfetmek; yapmak; talim etm., edilmek; düşündürmek. **to ~ oneself,** vücudünü işletmek: **to ~ a horse,** bir atı gezdirmek: **to ~ an influence upon,** ···e tesir etm.: **to ~ one's mind,** zihnini işgal etm.

**exert** [igˈzö(r)t]. Kullanmak, sarfetmek; göstermek. **to ~ oneself,** gayret sarfetmek, uğraşmak. **~ion** [igˈzö(r)şn], cehid, uğraşma; meşakkat.

**exeunt** [ˈeksiʌnt]. (Tiyatro) sahneden çıkarlar.

**exhal·e** [eksˈhêyl]. Nefesi dışarı vermek. **~ation** [–ˈlêyşn], tebahhur.

**exhaust**[1] [igˈzôst] *v.* Tüketmek; sarfedip bitirmek; bitkin bir hale getirmek. **~ing,** pek yorucu. **~ion,** bitkinlik. **~ive,** pek tafsilatlı, etraflı. ~[2] *n.* Egsoz; boşaltılmış gazler. **~-pipe,** egsoz borusu.

**exhibit** [igˈzibit]. Teşhir etm., göstermek; izhar etmek. Teşhir edilen şey; mahkemede delil olarak ibraz edilen şey. **~ion** [ˌeksiˈbişn], sergi; teşhir etme, gösterme; üniversitede küçük burs: **to make an ~ of oneself,** kendini rezil etm.: **an ~ of temper,** birdenbire öfkelenme gösterme. **~or** [igˈzibitә(r)], bir sergide teşhir eden kimse.

**exhilarat·e** [igˈzilәrêyt]. Keyif ve neş'e vermek. **~ing,** canlandırıcı. **~ion** [–ˈrêyşn], canlanma.

**exhort** [igˈzôt]. Tenbih etmek.

**exhume** [eks'hyûm]. Mezardan çıkarmak.
**exigen·cy** [ek'sicənsi]. Mübremlik; zaruret. **~t,** mübrem.
**exile** ['eksâyl]. Nefyetmek; sürgüne göndermek. Sürgün; mülteci.
**exist** [ig'zist]. Mevcud olm., varolmak, yaşamak. **~ence,** mevcudiyet; hayat; varlık: **to be in ~,** mevcud olm.: **to come into ~,** doğmak. **~ent** [**~ing**], mevcud; var.
**exit** ['eksit]. Çıkış, çıkma. **to make one's ~,** sahneden çıkmak.
**ex(-)officio** ['eks o'fişio]. Memuriyeti dolayısile.
**exonerate** [ig'zonərêyt]. Suçsuz çıkarmak.
**exorbitant** [ig'zôbitənt]. Aşırı, fahiş (fiat).
**exotic** [ek'sotik]. Başka iklime aid; ecnebi memleketten gelen.
**expand** [iks'pand]. Genişle(t)mek; imbisat et(tir)mek; (kanadları) açmak. **~ing,** genişliyen, gelişen.
**expans·e** [iks'pans]. Büyük saha. **~ible,** genişliyebilir. **~ion** [iks'panşn], imbisat; genişleme; çoğalma. **~ionist policy,** genişleme siyaseti. **~ive** [iks'pansiv], imbisat edici; geniş; yayvan; çok konuşur.
**expatriate** [eks'patriêyt]. Sürgün etmek.
**expect** [iks'pekt]. Muhtemel kılmak; ümid etm., beklemek. **as one might ~,** pek tabiî olarak: **I ~ so,** her halde: **don't ~ me till you see me,** beni bekleme, gelirsem gelirim: **to ~ s.o. to do stg.,** birinin bir şeyi yapmasını beklemek. **~ant mother,** gebe kadın. **~ation** [ˌekspek'têyşn]. Bekleme, ümid etme. **the ~ of life,** yaşanılacağı ümid edilen müddet: **to come up to [fall short of] ~s,** beklendiği gibi çık[ma]mak: **contrary to all ~s,** bütün beklenilenlerin aksine olarak.
**expedien·cy, -ce** [iks'pîdyənsi, -əns]. Münasebet, uygunluk; şahsî menfaat. **~t,** *a.* münasib, muvafık, uygun; *n.* çare, tedbir.
**expedit·e** ['ekspidâyt]. Tacil etm., kolaylaştırmak. **~ion**[1] [ˌekspi'dişn]. Sefer heyeti. **~ionary force,** kuvvei seferiye. **~ion**[2]. Acele. **~ious** [-işəs], aceleci, çabuk.
**expel** [iks'pel]. Tardetmek, kovmak.
**expend** [iks'pend]. Sarfetmek, harcetmek. **~iture** [-ityuə(r)], masraf; sarfiyat.
**expens·e** [iks'pens]. Masraf. **~es,** masraflar, sarfiyat. **at the ~ of,** pahasına: **we had a laugh at his ~,** hepimiz onun bu haline güldük: **to go to great ~,** çok masraf etmek: **to put s.o. to ~,** birini masrafa sokmak: **travelling ~s,** harcırah. **~ive** [iks'pensiv], pahalı: **to live ~ly,** lüks yaşamak.
**experi·ence** [iks'piəriəns]. Tecrübe; görgü. Tecrübe etm., görmek. **~enced,** tecrübeli; görmüş geçirmiş; malûmatlı. **~ment** [iks'perimənt]. Tecrübe (etm.), denme(k). **~mental** [-mentl], tecrübî.
**expert** ['ekspö(r)t]. Mütehassıs, ehil, usta.
**expiat·e** ['ekspiêyt]. Kefaret vermek; cezasını çekmek. **~ion** [-'êyşn], kefaret. **~ory** ['ekspiətəri], kefaret olarak.
**expir·e** [iks'pâyə(r)]. Ölmek; sona ermek; nefes vermek. **~ation** [ˌekspâyə'rêyşn], sona erme; müddetin hitamı; nefes verme. **~y** [iks'pâyəri], müddetin hitamı; sona erme; ölme.
**explain** [iks'plêyn]. Tavzih etm.; anlatmak; izah(at) vermek. **to ~ oneself,** meramını anlatmak: **to ~ away,** tevil etmek.
**explana·tion** [ˌeksplə'nêyşn]. İzah; anlatma; tevil. **to give an ~ of one s conduct,** hareketini izah etm. **~tory** [iks'planətəri], izahat verici.
**explicable** [iks'plikəbl]. İzah edilebilir.
**explicit** [iks'plisit]. Vazıh; kat'î; aşikâr. **~ly,** kat'î surette.
**explode** [iks'ploud]. Patla(t)mak; infilak et(tir)mek; tekzib etmek.
**exploit**[1] ['eksplôyt] *n.* Kahrahmanlık. **~**[2] [iks'plôyt] *v.* İşletmek; …den istifade etmek. **~ation** [-'têyşn], işletme; istismar etme.
**explorat·ion** [ˌeksplô'rêyşn]. Keşif ve araştırma; tedkik. **~ory** [-'plorətəri], istikşafa aid.
**explore** [iks'plô(r)]. Hakkında tedkik yapmak, araştırmak; …de tedkik için seyahate çıkmak. **to ~ for,** araştırmak. **~r,** kâşif.
**explos·ion** [iks'ploujn]. İnfilak, patlama. **~ive** [-'plousiv], patlayıcı; infilakî madde.
**export** ['ekspôt] *n.* İhrac etme; ihrac edilen mal. *v.* [eks'pôt]. İhrac etm., ihracat yapmak. **~ation** [-'têyşn], ihrac. **~er,** ihracatçı.
**expos·e** [iks'pouz]. Açıkta, maruz

bırakmak; teşhir etm. **to ~ oneself to danger,** kendini tehlikeye maruz bırakmak. **~ed,** açıkta kalmış; keşfedilmiş; rüzgâra maruz. **~ure** [iksˈpoʒə(r)]. Ortaya koyma; teşhir; rezalet; maruz etme; açık havada soğuğa maruz kalma; (*fot.*) poz.

**expound** [iksˈpaund]. Tefsir etm.

**express**[1] [iksˈpres] *a.* Mahsus; kat'î; sür'atli. *n.* (**-train**), ekpres. **~ly,** bilhassa: **~ for this purpose,** bilhassa bu maksad için. **~**[2] *v.* Sıkıp (suyunu) çıkarmak; ifade etm.; göstermek. **~ion** [iksˈpreşn]. Tabir; izhar; mana; yüz ifadesi; sıkıp çıkarma. **you could tell by the ~ of his voice,** sesinin tonundan belli idi: **he wore a very serious ~,** yüzünün ifadesi çok ciddî idi: **try to read with more ~,** daha manalı okumağa çalış. **~ionless,** hiç bir şey ifade etmiyen. **~ive,** manalı.

**expropriate** [iksˈprouprieyt]. İstimlâk etm.

**expuls·ion** [iksˈpʌlşn]. Tardetme; kovma; çıkarma. **~ive** [–siv], defedici.

**expurgate** [ˈekspö(r)gêyt]. Tenkih etm.; bir kitabdan ahlâk vs.ye aykırı kısımları tayyetmek.

**exquisite** [ˈekskwizit]. Enfes, latif; gayet ince ve nazik; keskin.

**ex-service-man** [eksˈsö(r)vismən], *pl.* **-men.** Terhis edilmiş asker.

**extant** [ˈekstant]. Hâla baki ve mevcud.

**extempor·aneous** [ˌekstempəˈrêynyəs]. İrticalî. **~e** [eksˈtempəri], irticalen; hazırlıksız. **~ize,** irticalen söylemek; hazırlık yapmaksızın söylemek, yazmak, çalmak.

**extend** [iksˈtend]. Uzatmak; uzanmak; temdid etm.; takdim etm. **to ~ a welcome to,** ···e 'hoş geldiniz!' demek.

**extens·ible** [iksˈtensibl]. Uzanabilir. **~ion** [iksˈtenşn] uzanma, temdid; büyütme; zam, ilâve: **university ~,** üniversite derslerinin harice teşmili usulü. **~ive,** vâsi, geniş. **~ively,** büyük mikdarda.

**extent** [iksˈtent]. Mertebe, derece; mesaha; mikdar. **to a certain [to some] ~,** bir dereceye kadar: **to such an ~ that,** o derecede ki.

**extenuat·e** [iksˈtenyuêyt]. Hafifletmek. **~ing circumstances,** cezayı hafifletici sebebler.

**exterior** [eksˈtiəriə(r)]. Dış; dış taraf; haricî; görünüş.

**exterminate** [eksˈtö(r)minêyt]. İmha etm.

**external** [eksˈtö(r)nl]. Haricî; dış; zahirî. **to judge by ~s,** zavahire göre hükmetmek: **Minister for External Affairs,** Hariciye Nazırı.

**extinct** [iksˈtin(g)kt]. Sönmüş; münkariz; hâlen mevcud olmıyan. **~ion** [–kşn], inkıraz; söndürme; itfa.

**extinguish** [iksˈtin(g)gwiş]. Söndürmek; itfa etm.; ilga etmek. **~er,** yangın söndürme aleti.

**extort** [iksˈtôt]. Başkasından zor, tehdid ile almak. **to ~ a promise from s.o.,** birinden bir vaid koparmak. **~ion** [–ˈtôşn], zor, tehdid ile alma; gasbetme. **~ionate,** gaddar; fahiş. **~ioner,** zorla alan adam; fahiş fiat istiyen kimse.

**extra** [ˈekstrə] *a.* Fazla, ziyade; esas masraftan haric; ilâve olarak, munzam. *n.* İlâve, zam; gazetenin fevkalâde tab'ı. **~s,** ilâve masraf: **little ~s,** ufak tefek ilâve masraf.

**extra-** *pref.* Ziyade ...; ···den haric; ···den dışı olan.

**extract** [ˈekstrakt] *n.* Hulâsa; esans. *v.* [iksˈtrakt]. Çıkarmak; sökmek. **~ion** [–ˈtrakşn], çıkarma, sökme; soy. **~or,** söküçü âlet.

**extradit·e** [ˈekstrədâyt]. Bir mücrimi kendi memleketine iade etmek. **~able** [–ˈdâytəbl], mücrimin iadesini icabettiren (cürüm). **~ion** [–ˈdişn], mücrimi iade etme.

**extrajudicial** [ˈekstrəˈcûdişəl]. Mahkemeden haric olan.

**extramural** [ˈekstrəˈmyûrəl]. Üniversiteye mensub olmıyan (ders, muallim vs.).

**extraneous** [eksˈtrêyniəs]. Dışarıdan gelen.

**extraordinary** [iksˈtrôdnəri]. Fevkalâde; garib.

**extra-special** [ˈekstrəˈspeşl]. Fevkalâde; çok hususî.

**extravagan·ce** [iksˈtravəgəns]. İsraf, müsriflik; itidalsizlik. **~t,** müsrif; çok masraflı; müfrit, itidalsiz. **~za,** fantezi.

**extrem·e** [iksˈtrîm] *a.* Son derece; çok şiddetli; en uzak; müfrit; itidalsiz. *n.* En uzak nokta, had. **~ely,** son derece; pek çok: **an ~ case,** müfrit, müstesna hal: **to drive s.o. to ~s,** birini ifrata sevketmek: **to go to ~s,** ifrata varmak: **to go from one ~ to the other,** ifrattan tefrite geçmek: **in the ~,** son derecede: **the ~**

penalty, ölüm cezası. ~**ist** [iks'trîmist]. İfratçı. ~**ity** [iks'tremiti]. Uc; nihayet; son derece; son çare; büyük tehlike. **the extremities (body)**, eller ve ayaklar: **to be in great ~**, son derece sefalet içinde olmak.

**extricate** ['ekstrikêyt]. Kurtarmak. **to ~ oneself**, işin içinden çıkmak.

**exuberan·ce** [ig'zyûbərəns]. Coşkunluk; bolluk. ~**t**, coşkun; bol.

**exult** [ig'zʌlt]. **to ~ at [in]**, ···den çok sevinmek; iftihar etm.: **to ~ over s.o.**, mağlub edilen rakib karşısında 'oh olsun' diye sevinmek. ~**ant**, mesrur; çok sevinen. ~**ation** [–têyşn], büyük sevinc; iftihar.

**eye**[1] [ây] *n.* Göz; iğne deliği; budak; ~**s front!**, ileri bak!: ~**s right!**, sağa bak! **an ~ for an ~**, kısas usulu: **'that's all my ~ (and Betty Martin)!'**, bütün bunlar kuru lâf: **to be all ~s**, dikkat kesilmek: **he has ~s at the back of his head**, onun görmediği yoktur: **to give s.o. a black ~**, gözünü morartmak: **to cast an ~ over**, ···e göz gezdirmek: **to cast down one's ~s**, yere bakmak: **to catch s.o.'s ~**, dikkatini çekmek: **there was not a dry ~ in the room**, odada ağlamıyan yoktu: **to give an ~ to stg.**, bir şeye bakmak: **to have an ~ for stg.**, bir şeyin iyisini seçebilmek hassası: **just keep an ~ on this child**, bu çocuğa göz kulak oluver: **to keep one's ~s skinned**, gözünü dört açmak: **to make ~s at**, ···e göz etm.: **mind your ~!**, dikkat et!: **in the mind's ~**, hayalinde: **to do stg. with one's ~s open**, bir şeyi göz göre göre yapmak: **to be in the public ~**, halkın gözünde olm. **to see ~ to ~ with s.o.**, birisile aynı fikirde olm.: **you can see that with half an ~**, bu aşikârdır: **I set ~s on England for the first time**, İngiltere'yi ilk defa gördüm: **I am up to the ~s in work**, işten başımı kaşıyacak vaktim yok: **with an ~ to**, ... maksad ile. ~**-opener**, dersi ibret olan şey. ~**-strain**, göz yorgunluğu. ~**-wash**, göz ilâcı; göz boyası. ~[2] *v.* ···e göz atmak; bakmak; süzmek. ~**ball** ['âybôl]. Göz küresi. ~**brow** ['âybrau]. Kaş: **to knit the ~s**, kaşları çatmak. ~**glass** ['âyglâs]. Tek gözlük, monokl; ~**glasses**, kelebek gözlük. ~**lash** ['âylaş]. Kirpik. ~**less** ['âylis]. Gözsüz; kör. ~**lid** ['âylid]. Gözkapağı. **to hang on by the ~s**, pamuk ipliğine bağlı olmak. ~**shot** ['âyşot]. **within [out of] ~**, gözle görülebilecek [görülemiyecek] mesafede. ~**sight** ['âysâyt]. Görme. **to have good ~**, gözleri iyi görmek. ~**sore** ['âysô(r)]. Göze batan şey. ~**witness** ['âywitnis]. Gözü ile gören; şahid.

**eyrie** [eə'ri]. Kartalın yuvası.

# F

**f** [ef]. F harfi. (*abb.*) **Fellow.**

**fable** ['fêybl]. Masal; efsane.

**fabric** ['fabrik]. Bina, yapı; kumaş. ~**ate** ['fabrikêyt]. Yapmak, imal etm.; uydurmak. ~**ation** [–'kêyşn], uydurma; yapma; icad.

**fabulous** ['fabyuləs]. Efsanevî; şayanı hayret. **at a ~ price**, ateş pahasına: ~**ly rich**, Karun kadar zengin.

**façade** [fə'sâd]. Bina cebhesi; zavahir.

**face**[1] [fêys] *n.* Yüz, çehre; saat minesi. **in the ~ of all men**, âleme karşı: **in the ~ of danger**, tehlike karşısında: **to fly in the ~ of facts**, hakikate aldırmamak: **to fly in the ~ of Providence**, kadere karşı mücadele etm.: **to keep a straight ~**, gülmemek: **to make a ~**, yüzünü gözünü oynatmak: **on the ~ of it**, görünüşte: **to pull a long ~**, suratını bir karış asmak: **to put a bold ~ on it**, korktuğu halde cesur görünmek: **to save one's ~**, zavahiri kurtarmak için: **to set one's ~ against**, ···e karşı cebhe almak. ~**-value**, zahirî kıymet. ~[2] *v.* Karşılamak; yüzüne bakmak; karşı olm.; dayanmak; katlanmak; kaplamak; nâzır olmak. **to be ~d with**, ···le kaplanmak; ... karşı karşıya bulunmak: **~ this way!**, bu tarafa dönünüz!: **the house ~s south**, ev cenuba bakar: **I can't ~ another winter here**, bir kış daha burada kalmayı göze alamam: **the difficulties that ~ us**, karşımızdaki güçlükler. **~ about**, yüzgeri etmek. ~**d** [fêysd]. Kaplanmış; yüzlü.

**facetious** [fe'sîşəs]. Olur olmaz her şey hakkında nükte yapan.

**facil·e** ['fasâyl]. Kolay; uysal. ~**itate** [fe'silitêyt]. Kolaylaştırmak. ~**ity**, kolaylık; istidad: ~**ities**, imkân ve vasıtalar.

**facing** ['fêysin(g)] *a.* Nâzır, karşı olan; müteveccih. *n.* Kaplama; dış astarı.

**facsimile** [fak<sup>l</sup>simili]. Tıpkısı, tam kopya.
**fact** [fakt]. Fiil; vak'a; hakikat; keyfiyet. ~ **and fiction,** hakikat ve hayal: **the ~ is that,** hakikat şudur ki: **an accomplished ~,** emrivaki: **apart from the ~ that ...,** ···den başka: **in ~,** hakikaten hattâ: **in point of ~,** aslını ararsan: **to look ~s in the face,** hadiseleri olduğu gibi görmek: **to stick to ~s,** vak'aları göz önünde tutmak; **owing to the ~ that ...,** ···den dolayı.
**factio·n** [ˈfakşn]. Nifak; parti içinde ayrılık. **~us** [ˈfakşəs]. Mücadeleci; fitneci; geçimsiz.
**factitious** [fakˈtişəs]. Sun'î.
**factor** [ˈfaktə(r)]. Âmil; komisyoncu; çiftlik kâhyası; kasım. **greatest common ~,** kasımı müştereki âzam.
**factory** [ˈfaktəri]. İmalâthane, fabrika. **~-hand,** fabrika amelesi.
**factotum** [fakˈtoutəm]. Her işi gören adam.
**factual** [ˈfaktyuəl]. Vak'aya aid; hakikî.
**faculty** [ˈfaklti]. İstidad, kabiliyet; fakülte.
**fad** [fad]. Şahsî bir âdet; gelip geçici bir moda. **full of ~s,** bir takım garib adetleri olan.
**fade** [fêyd]. Sol(dur)mak, rengi atmak; zail olm., yavaş yavaş gözden kaybolmak. **to ~ away,** gözden kaybolmak: **to ~ one scene into another,** (filim) sahneyi tedricen değiştirmek. **~-out,** (radyo) sesin kaybolması.
**fag** [fag]. Yor(ul)mak. Yorucu iş; (*sl.*) sigara. **it's too much ~!,** (*sl.*) zahmete değmez: **what a ~!,** bu da bir angarya! **~-end,** izmarit; son.
**faggot** [ˈfagət] *n.* İnce odun demeti.
**fail** [fêyl]. Becerememek; muvaffak olmamak; olmamak; ihmal etm.; tükenmek; iflâs etm.; yardım etmemek; vücuddan düşmek. **to ~ to do stg.,** bir şeyi yapmamak: **to ~ in one's duty,** vazifesinden kusur etm.: **whatever you do, don't ~ me!,** ne olursa olsun aman beni atlatma!: **his heart ~ed him,** cesaret edemedi: **I ~ to see why, ...** sebebini anlamıyorum: **~ing payment,** tediye edilmediği halde: **without ~,** mutlaka, muhakkak olarak: **words ~ me,** kelime bulamıyorum: **this will do ~ing all else,** başka hiç bir şey bulunmazsa bu olur. **~ing** [ˈfêylin(g)] *n.* Eksiklik. *a.* Zayıflıyan; eriyip giden. **~ this,** bu bulunmadığı halde. **~ure** [ˈfêylyə(r)]. Muvaffakiyetsizlik; iflâs; noksan.
**faint**[1] [fêynt] *v.* Bayılmak. *n.* Baygınlık. *a.* Bayılacak. **a dead ~,** ölü gibi baygın olma: **a ~ing fit,** baygınlık. **~**[2] *a.* Hafif, zayıf; soluk; hayal meyal; mübhem. ⌜**~heart never won fair lady**⌝, yüreksiz adam aşkta muvaffak olmaz.
**fair**[1] [feə(r)] *n.* Panayır; sergi. **~**[2] *a.* İnsaflı, adalete uygun; hilesiz; güzel; sarışın; (hava) iyi; fena değil, şöyle böyle. ⌜**all's ~ in love and war**⌝, aşkta ve harbde her şey caizdir: **given a ~ chance,** (adilane) imkân verildiği halde: **there is a ~ chance that we shall win,** kazanmamız oldukça muhtemeldir: **~ copy,** temize çekilmis nüsha: **~ game,** (*fig.*) meşru hedef: **'by ~ means or foul',** hangi vasıta ile olursa olsun; **~ to middling,** şöyle böyle: **one's ~ name,** lekesiz nam: **~ play,** dürüst hareket: **to put s.o. off with ~ promises,** birini güzel vaidlerle oyalamak: **the ~ sex,** cinsi latif: **~ and square,** dürüst, insaflı: **to hit stg. ~ and square,** bir şeyin tam ortasına vurmak: **he is in a ~ way to lose his job,** işinden olması kuvvetle muhtemeldir: **a ~ wind,** müsaid rüzgâr: **~ words,** güzel sözler: **he ~ly beamed with delight,** sevincinden adeta ağzı kulaklarına vardı. **~-haired,** sarışın. **~-minded,** munsif. **~-sized,** büyükçe. **~-spoken,** nezaketli. **~-weather sailor,** yalnız iyi havada denize çıkan gemici: **~ friend,** iyi gün dostu. **~ness** [ˈfeənis]. İnsaflılık, adalet. **in all ~,** munsif olmak için. **~way** [ˈfeəwêy]. Nehir vs.de gemilerin geçmesine ayrılan yer.
**fairy** [ˈfeəri]. Peri. **~-cycle,** oyuncak bisiklet. **~-like,** peri gibi. **~-tale,** peri masalı; masal. **~land** [ˈfeəriland]. Periler diyarı.
**fait accompli** [fêyt akomplî]. Emrivaki.
**faith** [fêyθ]. İman; inan; itimad; vefa. **bad ~,** suiniyet; ihanet: **to die in the ~,** imanlı olarak ölmek: **in good ~,** hüsnüniyetle: **to have [put] ~ in,** inanmak, itimad etm.: **to keep ~,** vadinde durmak: **to lose ~ in s.o.,** birisinden sıdkı sıyrılmak. **~- healing,** telkinle tedavi. **~ful** [ˈfêyθfəl]. Vefalı. **the ~,** müminler: **to promise ~ly,** katiyetle vadetmek: **yours ~ly,** (iş mektublarının sonun-

da) hürmetlerimi sunarım. **~less** [ˈfêyθlis]. Vefasız; hain; imansız.
**fake** [fêyk]. Yapma şey; sahte şey; taklid. Taklid etmek. **to ~ up,** uydurmak.
**falcon** [ˈfolkn]. Doğan, şahin. **~er,** şahinci. **~ry,** av kuşları terbiyesi.
**fall**[1] [fôl] *n.* Sukut, düşme; inkıraz; çökme; şelâle; yaprak dökümü, sonbahar; yağış: **the Fall,** ilk insanın (Adem'in) günahı: **to have a ~,** düşmek: **to ride for a ~,** (*fig.*) başının belâsını aramak. **~**[2] *v.* **(fell, fallen)** [fel, ˈfôlən]. Düşmek, sukut etm.; dökülmek; çökmek; azalmak; tesadüf etm.; olmak. **to ~ for,** ···e abayı yakmak; ···den aldanmak: **his eye fell upon me,** gözü bana ilişti: **his face fell,** suratı asıldı: **to ~ into a habit,** bir şeyi adet edinmek: **night is ~ing,** hava kararıyor: **to ~ into temptation,** iğvaya kapılmak. **~ away,** terketmek; dininden dönmek: **the profits fell away to nothing,** kâr gitgide sıfıra düştü. **~ back,** geri çekilmek: **to ~ back upon,** ···e başvurmak. **~ behind,** geride kalmak. **~ down,** yere düşmek; çökmek. **~ in,** çökmek; içeri düşmek; (kira mukavelesi vs.) müddeti bitmek; (*mil.*) sıraya girmek: **to ~ in with,** tesadüf etm.; kabul etmek. **~ off,** ···den düşmek; azalmak; evvelki gibi olmamak. **~ out,** dışarıya düşmek; dökülmek; vukua gelmek; külâhları değiştirmek; (*mil.*) sıradan ayrılmak: **to ~ out with s.o.,** birisile bozuşmak. **~ over,** devrilmek: **to ~ over an obstacle,** bir maniaya çarpıp düşmek: **people were ~ing over one another to buy the book,** halk bu kitabı kapışıyordu. **~ through,** geçip düşmek; suya düşmek; vazgeçilmek. **~ to,** başlamak; yemeğe saldırmak: **now then, ~ to!,** haydi! işinize! **~en** [ˈfôlən], düşmüş; dökülmüş; düşük; günahkâr. **the ~,** muharebede ölenler, şehidler.
**fallac·ious** [fəˈlêyşəs]. Aldatıcı; safsatalı. **~y** [ˈfaləsi], mugalata, safsata; yanlışlık.
**fallible** [ˈfalibl]. Yanılabilir.
**fallow** [ˈfalou]. Nadas. Nadasa bırakmak; dinlendirmek. **to lie ~,** (toprak) nadas halinde kalmak; (*fig.*) işlenmemiş olmak.
**fals·e** [fols, fôls]. Sahte, taklid, yapma(cık); asılsız, yanlış; hileli, hain. **~ alarm,** yersiz telaş; **~ imprisonment,** haksız yere hapis: **to raise ~ hopes,** beyhude ümid uyandırmak. **~ehood** [–hud], yalan. **~ify** [ˈfolsifây], tahrif etm.; yanlış çıkarmak. **~ity,** yanlışlık; yalancılık; sahtelik.
**falter** [ˈfôltə(r)]. Tereddüd göstermek; bocalamak; kekelemek.
**fame** [fêym]. Şöhret, nam. **house of ill ~,** umumhane. **~d,** şöhretli.
**familiar** [fəˈmilyə(r)] *a.* Mutad, alışmış; senli benli, teklifsiz. *n.* **~ (spirit),** bir büyücü emrindeki cin; hiç ayrılmaz arkadaş. **I am not ~ with Turkish,** Türkçe bilmem: **to be too ~,** lâübali olmak. **~ity** [–ˈariti], senli benli olma; teklifsizlik; alışıklık: ⌜**~ breeds contempt**⌝, alışkanlık her şeyin ehemmiyetini düşürür. **~ize** [–ˈmilyərâyz], alıştırmak.
**family** [ˈfamili]. Aile; soy; cins. **~ likeness,** bir ailede birbirine benzeme: **~ man,** ev bark sahibi: **it runs in the ~,** bütün aile halkı böyledir: **in the ~ way,** gebe.
**fami·ne** [ˈfamin]. Kıtlık, umumî açlık. **~sh** [ˈfamiş]. Aç kalmak; aç bırakmak. **~shed** [**~shing**], çok aç; açlıktan ölen.
**famous** [ˈfêyməs]. Meşhur. **~ly,** (*coll.*) pek iyi, mükemmel.
**fan** [fan]. Yelpaze; vantilatör; (*sl.*) meraklı.
**fanatic** [fəˈnatik]. Kaba sofu. **~al,** mutaassıb, koyu. **~ism** [–sizm], taassub.
**fanc·ier** [ˈfansiə(r)]. ... meraklısı. **bird ~,** kuşbaz. **~iful** [ˈfansiful]. Hayal mahsulü, tuhaf; hava ve hevesine tâbi.
**fancy**[1] [ˈfansi] *n.* Hayal; heves. *a.* Süslü, tuhaf. **I have a ~ that,** bana öyle geliyor ki: **to take a ~ to,** ···den (nedense) hoşlanmak. **~-dress ball,** kıyafet balosu. **~-free,** kimseye gönül vermemiş. **~-goods,** fantezi eşya. **~-work,** ince el işi. **~**[2]. *v.* Tasavvur etm.; zannetmek; gözü tutmak, beğenmek. **~ now!, ~that!,** çok şey!, acayib!: **to ~ oneself,** kendini beğenmek: **he rather fancies his French,** fransızcasını bir şey zannediyor.
**fanfare** [ˈfanfeə(r)]. Merasim borusu.
**fang** [fan(g)]. Sivri uzun diş.
**fanlight** [ˈfanlâyt]. Kapı üstü penceresi.
**fantas·tic** [fanˈtastik]. Hayalî; akla hayret veren; gülünc. **~y** [ˈfantəsi]. Hayal, hayalî resim; acayib fikir.

**far** [fâ(r)]. Uzak; ötedeki; bir hayli. ~ **away** [**off**], uzak, uzakta: ~ **and away the best** [**the cheapest,** *etc.*], fersah fersah daha iyi [ucuz vs.]: ~ **and wide,** yurdun [dünyanın] dört köşesinde: **as** ~ **as,** ···e kadar: ~ **better** [**worse**], çok daha iyi [fena]: ~ **from it,** bilâkis: **he is** ~ **from well,** hiç iyi değildir: **at the** ~ **end of the street,** caddenin öbür ucunda: **to go too** ~, fazla ileri gitmek. ~ **into the night,** gece geç vakte kadar: **to make one's money go** ~, parasını yetiştirmek: **that is going too** ~, bu kadarı da fazla: **so** ~, şimdiye kadar: **so** ~ **so good,** şimdiye kadar hoş: **so** ~ **as I know,** benim bildiğim kadar: **so** ~ **from ...,** ... şöyle dursun: **the night was** ~ **spent,** gece ilerlemişti. ~**-away look,** uzaklara dalmış bakış. **few and** ~**-between,** pek nadir. ~**-famed,** dünyaca tanınmış. ~**-fetched,** zoraki. ~**-flung,** çok uzaklara yayılmış. ~**-off,** çok uzak. ~**-reaching,** uzaklara erişen. ~**-seeing** [**-sighted**], durendiş.

**farc·e** [fâs]. Âmiyane komedi. ~**ical** [–ikl], gülünc.

**fare**[1] [fe͜ə(r)] *n.* Nakliye ücreti; kira arabası müşterisi. **single** [**return**] ~, yalnız gitme [gidip gelme] ücreti: ~**s, please!,** biletler, beyler! ~[2] *n.* Yiyecek ve içecek. ~[3] *v.* Seyahate [sefere] çıkmak; (iş, hal) iyi [fena] olmak. **to** ~ **forth,** yola çıkmak: **how did you** ~**?,** nasıl oldu?: **how** ~**s it?,** işler nasıl gidiyor?: ~ **thee well!,** elveda!: **it** ~**d ill with him,** muvaffak olmadı: **if you do that it will** ~ **ill with you,** bunu yaparsan, vay haline! ~**well** [ˈfe͜əˈwel]. Allahaısmarladık! **to bid** ~ **to** [**to take** ~ **of**], ···e veda etmek.

**farm** [fâm]. Çiftlik, Çiftçilik etmek. **to** ~ **out,** iltizama vermek. ~**er,** çiftçi. ~**ing,** çiftçilik: **stock** ~, hayvan yetiştirme. ~**-hand,** çiftlik amelesi. ~**-yard,** çiftlik avlusu.

**farrier** [ˈfari͜ə(r)]. Nalband.

**farthe·r** [ˈfarðə(r)]. **far**'*dan ismi tafdil.* Daha uzak; *v.* **further.** ~ **back,** daha geride: ~ **off,** ondan uzak: ~ **on,** daha ileride. ~**most,** en uzak. ~**st** [ˈfâðist]. En uzak.

**farthing** [ˈfâðin(g)]. En küçük ingiliz parası; bir peninin dörtte biri; metelik. **not worth a brass** ~, on para etmez.

**fascinat·e** [ˈfasinêyt]. Teshir etm., büyülemek. ~**ing,** teshir edici. ~**ion** [–ˈnêyşn], cazibe, sihir.

**fashion** [ˈfaşn]. Moda; kılık; tarz; âdet; görenek. Şekil vermek. **after a** ~, şöyle böyle: **in the** ~, modada: **out of** ~, modası geçmiş: **to lead the** ~, modaya örnek olm.; **to set the** ~, moda çıkarmak: **a man of** ~, son moda giyinen adam. ~**-plate,** moda resimleri. ~**able** [ˈfaşənəbl], modaya, âdete, uygun; şık.

**fast**[1] [fâst] *n.* Oruc; perhiz. *v.* Oruc tutmak. **to break one's** ~, orucunu bozmak: **to be taken** ~**ing,** açkarnına alınacak. ~[2] *a.* Sıkı; ayrılmaz; sabit; rengi uçmaz. ~ **asleep,** derin uykuda: ~ **by,** yakında: **they are** ~ **friends,** sıkıfıkı dostturlar: **to make** ~, sıkı bağlamak: **to play** ~ **and loose,** iki yüzlülük yapmak. ~[3] *a.* Süratlı; hafifmeşreb; (saat) ileri. **the** ~ **set,** sefihler: **as** ~ **as I do this,** ben bunu eder etmez. ~**ness** [ˈfâstnis]. Sürat; yanaşılmaz yer; kale.

**fasten** [ˈfâsn]. Bağlamak, iliştirmek; sıkıca kapatmak. **to** ~ **a crime on s.o.,** bir cürmü birine yükletmek: **to** ~ **down,** yapıştırmak: **to** ~ **on to,** ···e takılmak. ~**er** [~**ing**], bağ, kilid; rabtiye.

**fastidious** [fasˈtidi͜əs]. Titiz; müşkülpesend.

**fat** [fat] *a.* Şişman; yağlı. *n.* Yağ. *v.* Semirtmek. ⌜**the** ~ **is in the fire**⌝, işte şimdi kıyamet kopacak: ~ **land,** bereketli toprak: ⌜**to live on the** ~ **of the land**⌝, tam bir refah içinde yaşamak: **a** ~ **salary,** (*coll.*) dolgun maaş: **'that's a** ~ **lot of use!',** (*coll.*) (*cont.*) Maşallah! ne kadar faydalı şey! ~**ten** [ˈfatn]. Şışmanla(t)mak. ~**ty** [ˈfati]. Yağlı.

**fatal** [ˈfêytl]. Öldürücü; meş'um; mühlik; mukadder. **a** ~ **mistake,** vahim bir hata. ~**ism** [ˈfêytəlizm]. Kaderiye. ~**ist,** kaderci. ~**istic,** kadere inanan. ~**ity** [fəˈtaliti]. Ölümü mucib olan kaza; felâket.

**fate** [fêyt]. Kader, kısmet, alınyazısı; akibet; ecel. **the Fates,** ecel perileri. ~**d,** mukadder. ~**ful,** kadere bağlı; mühim.

**father** [ˈfâðə(r)]. Baba; ced; katolik papaz. Evladlığa kabul etmek. **the Holy Father,** Papa: **to** ~ **a child upon s.o.,** babası budur diye isnad etm.: **to** ~ **stg. on s.o.,** bir şeyi birine atfetmek: **to talk to s.o. like a** ~, birini azarlamak: **to play the heavy** ~, çok ciddî nasihat vermek. **~hood,** babalık. **~land,** anavatan. **~less,** yetim. **~ly,** baba gibi. **~-in-law,** kayınpeder.

**fathom** [ˈfaðəm]. Kulac (= 1·829 m.) İskandil etm.; içyüzünü anlamak. **~less,** dibsiz.

**fatigue** [fəˈtîg]. Yorgunluk; (*mil.*) angarya. Yormak. **to be on ~,** angaryada çalışmak: **~ party,** angaryacılar.

**fatuous** [ˈfatyu̯əs]. Akılsız ve beyhude.

**fault** [folt]. Kusur, kabahat; eksiklik; hata; (geol.) fay. **to be at ~,** kabahatli olm.: **to find ~ with,** tenkid etm., ···de kusur bulmak: **generous to a ~,** ifrat derecede cömert: **through no ~ of his,** kendi taksiri olmadan. **~iness,** eksiklik. **~less,** mükemmel. **~y,** kusurlu; eksik; kabahatli; sakim. **~-finder,** daima kusur bulan.

**fauna** [ˈfôna]. Bir mıntakanın hayvanları.

**faux pas** [ˈfo̯uˈpâ]. Pot; hata. **to make a ~,** pot kırmak.

**favour¹** [ˈfêyvə(r)] *n.* Lûtuf, inayet; iltimas, himaye; tarafdarlık; kurdele. **as a ~,** bir lûtuf olarak: **to ask a ~ of s.o.,** birinden bir ricada bulunmak: **to be in ~ of doing stg.,** bir şeyi yapmağa tarafdar olm.: **to be in ~ with s.o.,** birinin gözünde olm.: **to be out of ~,** gözden düşmek: **by ~ of,** ···in delâletile: **to decide in ~ of,** lehine karar vermek: **to find ~ with s.o.** [**to gain s.o.'s ~**], birinin gözüne girmek: **without fear or ~,** kimseden korkmadan ve kimseye minnet etmeden. **~²** *v.* Tarafdarı olm.; tercih etm.; musaid olm.; kolaylaştırmak; iltimas etm. **he ~s his father,** babasına benziyor. **~able** [ˈfêyvərəbl]. Müsaid; uygun, muvafık. **to look upon stg. with a ~ eye** [**to regard stg. ~ably**], bir şeyi tasvible karşılamak. **~ed** [ˈfêyvə(r)d]. Tercih edilen. **ill ~,** çirkin: **well ~,** yakışıklı: **the ~ few,** talihli bir avuc adam. **~ite** [ˈfêyvrit] *a.* En çok beğenilen; müreccah, makbul. *n.* Sevgili; gözde; ikbal. **he is a general ~,** herkes onu sever. **~itism** [ˈfêyvəritizm]. İltimasçılık.

**fawn¹** [fôn] *n.* Geyik yavrusu. *a.* Açık kahverengi. **~²** *v.* **~ upon,** müdahene etmek. **~ing,** müdaheneci.

**fealty** [ˈfiə̯lti]. Biat; sadakat.

**fear** [fiə̯(r)]. Korku, endişe. Korkmak; endişe etm.; yılmak. **to ~ for s.o.,** birisi için endişe etm.: **to go** [**be, stand**] **in ~ of,** ···den korkmak: **for ~ that,** ···den korkarak: **in mortal ~** (**of one's life**), ölüm tehlikesile: **no ~!,** ne münasebet!: **to put the ~ of God into s.o.,** birine haddini bildirmek. **~ful** [ˈfiə̯ful]. Müdhiş, korkunc; korkak, endişeli. **~less,** pervasız; yılmaz. **~some** [ˈfiə̯səm], korkunc.

**feasib·le** [ˈfîzibl]. Yapabilir, mümkün. **~ility,** imkân.

**feast** [fîst]. Ziyafet; bayram. Ziyafette yiyip içmek. **to ~ on stg.,** bir şeyi büyük zevk ile yemek: ⌜**enough is as good as a ~**⌝, her şeyin fazlası fazla.

**feat** [fît]. Hayret verici iş; büyük maharet icab ettiren şey; kahramanlık.

**feather¹** [ˈfeðə(r)] *n.* Kuş tüyü; ok yeleği. ⌜**birds of a ~ flock together**⌝, ⌜tencere yuvarlandı kapağını buldu⌝: **to make the ~s fly,** kıyamet kopmasına sebeb olm.: **that's a ~ in his cap,** bu onun için övünülecek bir şeydir: **in full ~,** tam tüylü; keyfi yerinde: ⌜**you could have knocked me down with a ~**⌝, hayretten küçük dilimi yuttum: **to show the white ~,** korkaklık etmek. **~y,** kuş tüyü gibi. **~²** *v.* (Kuş) tüylenmek; (kürekçi) pala çevirmek. **to ~ one's nest,** küpünü doldurmak. **~ed,** kuş tüylü.

**feature** [ˈfîçə(r)] *n.* Yüz uzuvlarından biri; hususiyet. *v.* Tavsif etm., göstermek. **~s,** yüzün hatları.

**February** [ˈfebru̯əri]. Şubat.

**fecund** [ˈfîkʌnd]. Velûd. **~ity** [—ˈkʌnditi], velûdiyet; mahsullü olma.

**fed** *v.* **feed.**

**federat·e** [ˈfedərêyt]. Birleştirmek. **~tion** [—ˈrêyşn], birlik; dahilî istiklallerini muhafaza eden devlet vs.den mürekkeb birlik.

**fee** [fî]. Ücret; hak; bahşiş; (**doctor's**) vizita. Ücret vermek. **entrance ~,** duhuliye.

**feeble** [ˈfîbl]. Kuvvetsiz, zayıf; yavan. **~-minded,** ebleh.

**feed** (**fed, fed**) [fîd, fed] *v.* Yemek yemek; otlamak; yem vermek; beslemek; (lâzım olan maddeyi) temin etm.; tağdiye etmek. *n.* Yem; gıda; (makinede) tağdiye: **forced ~,** tazyik ile tağdiye: **to be off one's ~,** iştahsız olmak: **to ~ on ...,** ···le beslemek: **to ~ s.o. on,** birini ···le beslemek: **to ~ out of one's hand,** (hayvan) yemini avucdan almak; (*fig.*) birinin avucunun içinde olm.: **to ~ up,** bol gıda ile kuvvetlendirmek. semirtmek: **to**

be fed up, bıkmak: I am fed up with you, senden illallah! ~er [ˈfîdə(r)]. Besleyici; yiyen; çocuk önlüğü; emzik. ~ing [ˈfîdin(g)]. Yem(ek) verme, besleme; tağdiye: forcible ~, zorla yedirme: ~-bottle, emzik.

**feel**[1] [fîl] *n.* Dokunma hissi; el yordamı. rough to the ~, teması kaba. ~[2] (felt) [felt] *v.* Hissetmek; duymak; el ile dokunmak; el yordamiyle bulmak. to ~ about for [after] stg., bir şeyi el yordamiyle aramak: to ~ cold, üşümek: to ~ for s.o., birine müteessir olm.: to ~ hot, harareti olm.: to ~ like doing stg., canı istemek: to ~ in one's pockets for stg., ceblerini yoklamak: I don't ~ quite myself, kendimi o kadar iyi hissetmiyorum: I don't ~ up to it, bunu yapacak halim yok: to ~ one's way, yolunu el yordamiyle bulmak; (*fig.*) yavaş yavaş ve ihtiyatla ilerlemek: to ~ well, keyfi yerinde olm.: to ~ unwell, keyifsiz olmak. ~er [ˈfîlə(r)]. Böceğin lems âleti. to throw [put] out a ~, iskandil etm. ~ing [ˈfîlin(g)]. His; hassasiyet. I have a ~ that, bana öyle geliyor ki: the general ~ is that, umumiyetle zannediliyor ki: to have no ~s, hissiz olm.: I speak with ~, samimi olarak söylüyorum.

**feet** *v.* foot.

**feign** [fêyn]. Yalandan yapmak; uydurmak. to ~ sick [illness], temarüz etm.: to ~ ignorance, bilmemezlikten gelmek. ~ed, sahte.

**feint** [fêynt]. (*mil.*) Hile(li hareket yapmak).

**felicit·ate** [fiˈlisitêyt]. Tebrik etmek. ~ous, pek yerinde olan; mes'ud. ~y, saadet; uygunluk.

**feline** [ˈfîlâyn]. Kedi cinsinden.

**fell**[1] [fel] *v.* fall. ~[2] *v.* Yere (kesip) indirmek. ~ *a.* Meş'um; merhametsiz. ~ *n.* Kayalık tepe.

**fellow** [ˈfelou]. Herif; adam; eş; aynı dereceden kimse; üniversite hocası; bir ilim cemiyetinin âzası. a good ~, iyi adam: poor ~, zavallı: you might let a ~ speak, bırak da anlatayım. ~-being [-creature], hemcins. ~-countryman, vatandaş. ~-feeling, birinin halinden anlama. ~-servant, kapı yoldaşı. ~ship [ˈfelouşip]. Arkadaşlık; üniversitede hocalık.

**felon** [ˈfelən]. Mücrim, cani; habis. ~ious [ˈlounyəs], caniyane. with ~ intent, cürüm işlemek maksadile. ~y, cürüm, cinayet

**felt**[1] [felt] *v.* feel. ~[2] *n.* Keçe; kebe.

**female** [ˈfîmêyl]. Dişi.

**feminine** [ˈfeminin]. Kadına aid; kadın gibi; müennes.

**fen** [fen]. Bataklık mıntaka.

**fenc·e**[1] [fens] *n.* Tarla veya bahçe etrafındaki tahta perde; parmaklık; hırsız yatağı. *v.* Etrafını parmaklık ile çevirmek. to ~ off a field, bir tarlayı tel vs. ile ayırmak; to sit on the ~, suya sabuna dokunmadan tarafsız kalmak. ~ing[1], çit, tahtaperde. ~[2] *v.* Eskrim yapmak. ~ing[2], eskrim.

**fend** [fend]. to ~ for oneself, kendi yağı ile kavrulmak: to ~ for s.o., birinin ihtiyaclarına bakmak: to ~ off, defetmek, savmak. ~er [ˈfendə(r)]. Usturmaca; ocak siperi; (*otom.*) siper.

**ferment** [ˈfö(r)ment] *n.* Maya; kaynama; büyük heyecan. *v.* [fö(r)ˈment]. Mayala(n)mak; heyecanlan(dır)mak. ~ation [–êyşn], mayalanma.

**fern** [fö(r)n]. Eğreltiotu.

**feroci·ous** [fəˈrouşəs]. Vahşî, canavarca. ~ty [–ˈrositi], canavarlık, vahşilik.

**-ferous** [–fərəs] *suff.* ... hamil, husule getiren, ihtiva eden; *mes.* auriferous, altın ihtiva eden.

**ferret** [ˈferit]. Kır sansarı (ile avlamak). to ~ about, araştırmak: to ~ out, arkasını bırakmıyarak bulup çıkarmak.

**ferro-concrete** [ˈferouˈkonkrît]. Betonarme.

**ferry** [ˈferi]. Nehir vs.nin kayık vs. ile geçilen yeri; sahiller arasında işliyen kayık vs. to ~ across, nehrin bir sahilinden öbür sahiline geçirmek: aerial ~, nehrin üzerinde sahilden sahile geçen asma vagon.

**fertil·e** [ˈfö(r)tâyl]. Mümbit; bereketli; ilkah edilmiş. ~ity [–ˈtiliti], mümbitlik; kuvvei inbatiye. ~ize [–lâyz], ilkah etm.; mümbitleştirmek; gübrelemek. ~izer, gübre *esp.* sun'î gübre.

**ferv·ency** [ˈfö(r)vensi]. Hararetlilik; coşkunluk. ~ent, hararetli, coşkun. ~id [ˈfö(r)vid]. Hararetli, ateşli, hiddetli. ~our [ˈfö(r)və(r)]. Hararet, büyük gayret; coşkunluk.

**fester** [ˈfestə(r)]. İrinlenmek. ~ing, cerahatlenen.

**festiv·al** [ˈfestivl]. Bayram; yortu; eğlenceli toplantı. ~e, neşeli, bay-

rama aid. **~ity** [–ˈtiviti], bayram (eğlentileri); şenlik, cümbüş.
**festoon** [fesˈtûn]. Çiçek, bayrak vs.yi 'mahya' gibi asarak süslemek.
**fetch** [feç]. Gidip getirmek; (filan fiatla) satılmak; (*sl.*) cezbetmek. **to ~ and carry,** süflî işler yapmak: **to ~ s.o. a blow,** birine bir tokat aşketmek. **~ up,** alıp yukarıya getirmek; kusturmak: **he'll ~ up in prison,** hapsi boylıyacak.
**fête** [fêyt]. Bayram. Ziyafet ile ağırlamak.
**fetid** [ˈfîtid]. Pis kokulu; ufunetli.
**fetish** [ˈfîtiş]. İbtidai kavimlerin taptıkları şey; put. **to make a ~ of stg.,** bir şeye yersiz olarak pek fazla değer vermek.
**fetter** [ˈfetə(r)]. Köstek, bukağı. Bukağılamak; zincire vurmak. **to burst [throw off] one's ~s,** köstegi kırmak.
**fettle** [ˈfetl]. Hal. **in fine ~,** keyfi yerinde.
**feud** [fyûd]. Aile veya ferdler arasında düşmanlık. **blood ~,** kan gütme. **~al** [ˈfyûdl]. Derebeyliğe aid. **~alism,** derebeylik.
**fever** [ˈfîvə(r)]. Humma; sıtma; hararet. **~ish,** hummalı; ateşi olan.
**few** [fyû]. Az. **a ~,** bir kaç: **a good ~, quite a ~,** bir çok: **some ~,** bazıları: **every ~ days,** birkaç günde bir. **~er,** daha az. **~est,** en az. **~ness,** azlık.
**fiancé**(**e**) [fiˈonse]. Nişanlı.
**fib** [fib]. Ehemmiyetsiz bir yalan (söylemek).
**fibr·e** [ˈfâybə(r)] Lif; seciye. **~ous,** lifi, lifli.
**fickle** [ˈfikl]. Bir dalda durmaz, gelgeç.
**fict·ion** [ˈfikşn]. Hayal; hikâye; masal; roman nevi. **legal ~,** hukukî mevhume. **~itious** [fikˈtişəs], hayalî; uydurma; aslı olmıyan.
**fiddle** [ˈfidl]. Keman (çalmak). **as fit as a ~,** turp gibi (sıhhatte): **to play second ~,** ikinci derecede bir rol oynamak: **to ~ (about) with stg.,** kurcalamak: **to ~ away one's time,** oyalanarak vaktini israf etm.: **to ~ over a job,** bir iş üzerinde oynamak. **~dedee** [–diˈdî], saçma. **~r,** kemancı (*köt.*); köy kemancısı; bir iş üzerinde oyalanan kimse. **~sticks!,** saçma! **~-stick,** keman yayı. **~-string,** keman kirişi.
**fiddling** [ˈfidlin(g)]. Ufak tefek ehemmiyetsiz (iş vs.).

**fidelity** [fiˈdeliti, fây-]. Vefa; doğruluk.
**fidget** [ˈficit]. Rahat oturmamak; bir türlü rahat oturmıyan kimse. **to have the ~s,** çocuk gibi bir türlü rahat durmamak. **~y,** yerinde durmıyan.
**fiduciary** [fâyˈdyûsəri]. İtimada bağlı. **~ money,** kâğıd para.
**fie** [fây]. Ayıb!, utanmaz mısın!
**fief** [fîf]. Tımar, zeamet.
**field** [fîld] *n.* Tarla; saha; harb meydanı; muharebe; (dürbün vs.) rüyet sahası; bir yarışa iştirak edenler. *v.* (Kriket) vurulan topu kapmak. **beasts of the ~,** tabiat halinde yaşıyan hayvanlar: ⌜**a fair ~ and no favour**⌝, müsavi şartlar altında: **to hold the ~,** üstünlüğü muhafaza etm.: **to leave s.o. in possession of the ~,** meydanı birine bırakmak: **to take the ~,** harbe girmek. **~-artillery,** sahra topları. **~-day,** (*mil.*) manevra günü; muvaffakiyetli gün. **~-glass(es),** çifte dürbün. **~-hospital,** (*mil.*) seyyar hastahane. **~-marshal,** mareşal, müşür. **~-sports,** kır eğlenceleri (avcılık vs.).
**fiend** [fînd]. Zebani; habis ruh; gaddar ve zalim adam; bir şeyin müfrit derecede tiryakisi. **the Fiend,** İblis. **~ish,** şeytanî.
**fierce** [fiəs]. Vahşi, azgın, hiddetli, şiddetli.
**fiery** [ˈfâyəri]. Ateşli, alevli; atılgan.
**fift·een** [fifˈtîn]. On beş. **~eenth,** on beşinci. **~h** [ˈfifθ]. Beşinci; beşte bir. **~ieth** [fiftiəθ]. Ellinci; ellide bir. **~y,** elli: **~-~,** yarı yarıya; **he is in the fifties,** ellisini geçti.
**fig** [fig]. İncir (ağacı). **green ~s,** taze incir: **not to care a ~, for, ...** vızgelmek: **in full ~,** giyinmiş kuşanmış.
**fight** (**fought**) [fâyt, fôt]. Döğüşmek; harbetmek; kavga etm.; savaşmak; ···le uğraşmak; aleyhine dava açmak. Döğüş, kavga; muharebe; mücadele. **to ~ down,** mücadele ede ede mağlub etm.: **free ~,** kalabalık arasında çıkan kavga: **to ~ off,** büyük bir gayretle defetmek: **to ~ it out,** sonuna kadar mücadele etm.: **to ~ one's way out,** bir kalabalığın içinden dövüşe dövüşe kurtulmak: **to show ~,** kavga edecek olmak: **to ~ shy of,** ictinab etm.; sakınmak: **stand-up ~,** usulü dairesinde kavga. **~er** [ˈfâytə(r)]. Muharib; mücadeleci; avcı uçağı. **~ing** [ˈfâytin(g)] *a.* Muharib. *n.* Harb;

döğüş. the ~ line, muharebe hattı: he has a ~ chance of recovery, (hastalıkla) mücadele edebilirse iyileşir. ~ing-cock, döğüş horozu.
**figment** [ˈfigmənt]. **a ~ of the imagination,** hayal mahsulü.
**figurative** [ˈfigyurətiv]. Mecazî.
**figure**¹ [ˈfigə(r)] *n.* Şekil, biçim; vücud; rakam; fiat; mikdar; şahsiyet. **~s,** hesab; **I am no good at ~s,** hesabım çok fenadır: **~ of eight,** 8 şeklinde: **a ~ of speech,** mecaz: **to cut [make] a fine [poor] ~,** parlak [zavallı] bir tesir bırakmak: **what a ~ of fun!,** ne gülünc manzara!: **to go into ~s,** (hesab işine) rakamlara gelmek: **to keep one's ~,** vücudunun biçimini muhafaza etm.: **in round ~s,** yuvarlak hesab: **to work out the ~(s),** hesablamak. **~²** *v.* Hesab etm.; tasavvur etmek. **to ~ as,** kendine ... süsünü vermek: **to ~ stg. to oneself,** tasavvur etm.: **to ~ out the expense,** masrafını hesabetmek. **~head** [ˈfigəhed]. Gemi arslanı. **he's only a ~,** o orada mostralıktır.
**filament** [ˈfiləmnt]. İnce tel.
**filch** [filç]. Aşırmak.
**file**¹ [fâyl]. Eğe. Eğelemek. **~ down,** eğeleyip düzeltmek: **~ off [away],** eğeleyip gidermek. **~².** Sıra; dosya; fihrist; sıra ile dizilmiş eski gazeteler. Dosyaya koymak; tasnif etmek. **in ~,** çift sıra: **in single [Indian] ~,** tek sıra: **rank and ~,** subaylar ve erler: **to ~ a petition,** mahkemeye istida vermek: **to ~ one's petition (in bankruptcy),** mahkemeye muracaat ederek iflâsını bildirmek.
**filial** [ˈfilyəl]. Evlâd vazifesine aid.
**fill** [fil]. Dolmak; doldurmak; doyurmak. **to ~ the bill,** lazımgeleni yapmak: **to eat one's ~,** tıkabasa doymak: **to have one's ~ of stg.,** ···e doymak: **to ~ the part,** rolunu yapmak: **to ~ requirements,** ihtiyacları karşılamak. **~ in,** (çek, liste, çukur vs.yi) doldurmak; tamamlamak. **~ out,** şiş(ir)mek. **~ up,** doldurmak; tamamlamak; (kapı) örmek. **~ing** [ˈfilin(g)] *n.* Doldurma; dolma. **~-station,** benzin istasyonu.
**fillet** [ˈfilit]. Fileto; dilim. **to ~ fish,** balığın kılçığını çıkarıp ikiye bölmek.
**filly** [ˈfili]. Dişi tay.
**film** [film]. Zar; pelikül; filim. Zarla kaplamak; filime çekmek. **the ~s,** sinema: **(eye) to ~ over,** gözü bulanmak. **~y,** bulanık; pek hafif; şeffaf.
**filter** [ˈfiltö(r)]. Süzgeç. Süzgeçten geçirmek; süzülmek.
**filth** [filθ]. Pislik. **~y,** pis; müstehcen.
**fin** [fin]. Balık kanadı.
**final** [ˈfâynl]. Son, sonuncu, nihaî; kat'î. Son imtihan; (musabaka) son yarış; son maç. **~e** [fiˈnâli], final: **the grand ~,** tantanalı bitiş. **~ist,** (musabaka) sonuna kadar kalan rakiblerin biri. **~ity** [fâyˈnaliti], kat'ilik; son olma; gaiyet. **~ly** [ˈfâynəli], nihayet.
**financ·e** [fiˈnans, ˈfâynans]. Maliye; maliyecilik. Parasını temin etm., sermaye tedarik etmek. **~ial** [–ˈnanşl], malî, paraya aid. **~ier** [–ˈnansiə(r)], maliyesi; sermayedar.
**find** [fâynd] *n.* Buluş; bulunmuş şey; keşif. **~** *v.* **(found)** [fâynd, faund]. Bulmak; keşfetmek; rastgelmek; farketmek; addetmek; öğrenmek. **to ~ oneself,** kendi kabiliyetini keşfetmek: **to ~ s.o. in clothes, food,** *etc.*, birine elbise, gıda vs. tedarik etm.: **I couldn't ~ it in my heart to ...,** içim götürmedi: **it has been found that,** tesbit edilmiştir ki: **he is not to be found,** (aradık) bulmak mümkün değil: **wages £5 all found,** iaşe ve ibate ile beş lira haftalık: **the court found the prisoner guilty,** mahkeme cürmü sabit gördü: **the judge found for the plaintiff,** hakim davacı lehine karar verdi. **~ out,** keşfetmek: **to ~ s.o. out,** birinin ne mal olduğunu anlamak. **~ing** [ˈfâyndin(g)] *n.* Bulma; varılan netice.
**fine**¹ [fâyn]. Para cezası(na çarpmak). **~². in ~,** hulâsa. **~³.** İnce; nazik; hâlis; güzel, mükemmel. **to cut [run] it ~,** zaman [para]yı kıtakıt hesablamak: **prices are cut very ~,** fiatlar asgari hadde indirilmiş: **one ~ day,** günün birinde: **one of these ~ days,** günün birinde: ⌜**~ feathers make ~ birds**⌝, ⌜süslü elbiseler insanı kibar gösterir⌝: **a ~ looking man,** kelle kulak yerinde: **the ~r points of stg.,** ···in incelikleri: **not to put too ~ a point on it,** ince eleyip sık dokumadan. **~ness** [ˈfâynnis]. İncelik; zarafet; güzellik. **~ry** [ˈfâynəri]. Güzel elbise. **~sse** [fiˈnes]. Maharet; kurnazlık.
**finger** [ˈfin(g)gə(r)]. Parmak (ile dokunmak). **first ~,** şehadet parmağı: **second [middle] ~,** ortaparmak:

third ~, adsız [yüzük] parmağı: **little** ~, serçe parmak: **to burn one's ~s over stg.**, bir şeyden ağzı yanmak: **to have green ~s**, bahçede her şeyi kolayca yetiştirme mahareti olm.: **to have stg. at one's ~ ends**, bir işin girdisini çıktısını bilmek: **to lay one's ~ on the cause**, meselenin esasına parmağını basmak: **I won't let anyone lay a ~ on him**, onun kılına dokundurtmam. **~-bowl**, sofrada el yıkmağa mahsus tas. **~-print**, parmak izi.

**finicky** [ˈfiniki]. Ehemmiyetsiz şeyler üzerinde gayet titiz olan.

**finish** [ˈfiniş]. Son, nihayet; bitirme; varış. Bit(ir)mek; sona er(dir)mek; hitam bulmak. **he's ~ed**, işi bitti: **to ~ third**, (yarışta) üçüncü gelmek: **we ~ed up all square**, berabere kaldık: **to ~ off**, tamamen bitirmek: **I've ~ed with him!**, onunla alâkam kalmadı: **wait till I've ~ed with him!**, ben ona dünyanın kaç bucak olduğunu gösteririm. **~ed** [ˈfinişd] *a.* tamamlanmış; hazır; mükemmel. **~ing** *a.* bitirici; tamamlayıcı. **~ line**, varış hattı: **~ touch**, tamamlayıcı ameliye.

**finite** [ˈfâynâyt]. Sonu var; mahdud.

**Fin·land** [ˈfinlənd]. Finlandiya. **~lander** [**~n**, **~nish**], Finlandiyalı.

**fiord** [ˈfiˈôd]. Dar ve derin körfez.

**fir** [fö(r)]. Köknar.

**fire**[1] [ˈfâyə(r)] *n.* Ateş; yangın. **to be between two ~s**, her iki taraftan hücuma uğramak: ⌜**a burnt child fears the ~**⌝, ⌜sütten ağzı yanan ayranı üfler de içer⌝: **on ~**, tutuşmuş: **to be on ~**, yanmak: **to catch ~**, ateş almak: **to get on like a house on ~**, (i) sür'atle terakki etm.; (ii) çok iyi anlaşmak: **to set ~ to stg.** [**to set stg. on ~**], tutuşturmak: **to be under ~**, düşman ateşine uğramak: **to go through ~ and water for ...**, ... için her şeyi göze almak. **~-alarm**, yangın işareti. **~-brigade**, itfaiye. **~-damp**, grizu. **~-eater**, yiğit. **~-engine**, yangın tulumbası. **~-escape**, yangın merdiveni. **~-extinguisher**, yangın söndürme aleti. **~-fighting**, yangın söndürme (teşkilatı vs.). **~-policy**, yangın sigortası. **~-proof**, ateşe dayanır, yanmaz **~-raising**, kundakçılık. **~-ship**, kundakçı gemi. **~-station**, itfaiye merkezi. **~-worship**, ateşe tapma. **~**[2] *v.* Tutuşturmak; ateş etm.: (*sl.*) koğmak. **to ~ bricks**, tuğla fırınlamak: **to ~ an engine, boiler**, kazanı yakmak: **to ~ a question at s.o.**, birine birdenbire bir sual sormak. **~arm** [ˈfâyrâm]. Ateşli silâh. **~brand** [ˈfâyəbrand]. Fesadcı; kundakçı. **~man** [ˈfâyəmən]. İtfaiyeci; gemi ateşçisi. **~side** [ˈfâyəsâyd]. Ocak başı. **~wood** [ˈfâyəwud]. Ateşlik odun; çıra. **~works** [ˈfâyəwö(r)ks]. Hava fişekleri; (*fig.*) heyecanlı nutuklar vs.

**firm**[1] [fö(r)m] *n.* Firma; ticarethane. **~**[2] *a.* Muhkem; sabit; sağlam; katı; kat'î. **~ friends**, sıkı dostlar: **to be ~ about stg.**, bir şey üzerinde ısrar etm.: **to hold stg. ~ly**, bir şeyi sımsıkı tutmak: **to stand ~**, dayanmak. **~ness** [ˈfö(r)mnis]. Metanet; istikrar.

**firmament** [ˈfö(r)məmənt]. Sema, feza, gök.

**first** [fö(r)st]. Birinci; ilk; evvel; başta; en önce; evvelâ; ilk defa olarak. **~ of all**, en evvel: **at ~**, ilkönce: ⌜**~ come ~ served**⌝, ilk gelen sıraya girer: **~ and foremost**, her şeyden evvel: **from the ~**, başlangıcdan beri: **at ~ hand**, doğrudan doğruya: **~ and last**, bir kere...: **~ or last**, er geç: **from ~ to last**, başlangıcdan sonuna kadar: **in the ~ place**, evvelâ: **I will go there ~ thing tomorrow**, yarın ilk iş olarak oraya gideceğim: **~ things ~**, ehemmi mühimme takdim. **~-aid**, ilk sıhhî imdad. **~-born**, ilk, en büyük çocuk. **~-class**, mükemmel; birinci sınıf, mevki. **~-fruits**, ilk mahsul; (*fig.*) ilk semeresi. **~-rate**, mükemmel.

**fiscal** [ˈfiskl]. Devletin hazinesine aid.

**fish** [fiş]. Balık (avlamak). ⌜**all is ~ that comes to his net**⌝, bir çıkarı olan her şey onun makbûlüdür: **to ~ for compliments**, kendini medhettirmek için bahane aramak: **to drink like a ~**, içkide yüzmek: **to feed the ~es**, boğulmak. ⌜**there's as good ~ in the sea as ever came out of it**⌝, ⌜Amasya'nın bardağı biri olmazsa biri daha⌝: ⌜**neither ~, flesh nor good red herring**⌝, hiç bir şeye benzemiyor: **I've other ~ to fry**, daha mühim işlerim var: **a ~ out of water**, karaya vurmuş balık gibi: **he's a queer ~**, o bir âlem: **to ~ in troubled waters**, bulanık suda balık avlamak: **to ~ up stg.**, bir şeyi suyun dibinden çıkarmak. **~-cake,** balık köftesi. **~-bone,** kılçık. **~-hook,** olta iğnesi. **~erman** [ˈfişəmən]. Balıkçı. **~ery** [ˈfişəri]. Balıkçılık; balık tutulan yer. **~ing** [ˈfişin(g)] *n.* Balıkçılık, balık avı. **~ing-line,** olta ipliği. **~ing-rod,** olta sırığı. **~ing-smack,** küçük balıkçı gemi. **~-monger** [ˈfişmʌn(g)gə(r)]. Balık satıcısı. **~wife** [ˈfişwâyf]. Kadın balık satıcısı. **~y** [ˈfişi]. Balık gibi; balık kokulu; şübheli.

**fissure** [ˈfişuə(r)] *n.* Yarık; *v.* Yar(ıl)mak.

**fist** [fist]. Yumruk. **~icuffs** [ˈfistikʌfs], yumruk kavgası.

**fit**[1] [fit] *n.* Sara; hastalık vs. nöbeti. **~ of anger,** hiddet galeyanı: **fainting ~,** baygınlık: **to fall into a ~,** sarası tutmak: **to be in ~s of laughter,** gülmekten katılmak: **he will have a ~ when he knows,** bunu duyarsa adama inme iner: **to work by ~s and starts,** rastgele çalışmak: **he had a ~ of idleness,** tembellik damarı tuttu: **to frighten s.o. into ~s,** birinin ödünü koparmak. **~**[2] *a.* Münasib, uygun, lâyık; muktedir; sıhhati iyi; idmanlı. *n.* Elbise vs.nin uyması. **this coat is a good [bad] ~,** bu ceket uyuyor [uymuyor]: **~ to drink [eat],** içilir [yenir]: **I feel ~ to drop,** ayakta duracak halim yok: **he is not ~ for the post,** bu yerin ehli değildir: **he is ~ for nothing,** bir işe yaramaz: **he is not ~ to be seen,** âlem içine çıkmaz: **to think [see] ~ to do stg.,** bir şeyi uygun bulmak: **a tight ~,** pek dar, sıkışık. **~**[3] *v.* Uy(dur)mak; takmak; donatmak; (elbise) prova etm. **to ~ s.o. for a career,** birini bir mesleğe hazırlamak: **to ~ together,** birbirine geçmek. **~ in,** birbirine geçirmek; uymak: **I'll ~ it in somehow (meeting, engagement,** *etc.*), her halde sıkıştırmağa çalışırım. **~ on,** (elbise) prova etm.; (lâstik) takmak. **~ out,** donatmak. **~ up,** kurmak. **~ful** [ˈfitfəl]. İntizamsiz; devamlı olmıyan; kaprisli. **~ness** [ˈfitnis]. Uygunluk; kabiliyet; sağlık. **~ter** [ˈfitə(r)] *n.* Tesviyeci. **~ting**[1] [ˈfitin(g)] *a.* Uygun, münasib, lâyık, yakışır. **~ting**[2] *n.* Tesviyecilik; elbisenin provası. **~s,** mobilya; tertibat.

**five** [fâyv]. Beş. **~r** (*coll.*) beş ingiliz liralık banknot. **~fold,** beş misli.

**fix**[1] [fiks] *n.* Güç bir vaziyet. **~**[2] *v.* Tesbit etm.; sokmak; bağlamak; kurmak; yerleştirmek; kararlaştırmak. **I'll ~ him!,** (i) onun icabına bakarım; (ii) onunla anlaşırım: **to ~ the blame on s.o.,** kabahatın birinde olduğunu isbat etm.: **to ~ s.o. with one's eye,** birine dik dik bakmak. **~ up(on),** seçmek; kararlaştırmak. **~ up,** kurmak; tanzim etm.: **I'll ~ you up,** sizin için icabeden hazırlık vs.yi yaparım. **~ative** [ˈfiksətiv]. Tesbit edici (madde). **~ed** [fikst]. Sabit; bağlı; muayyen; maktu. **~ity** [ˈfiksiti]. Sabitlik; değişmezlik. **~ture** [ˈfiksçə(r)]. Sabit demirbaş; bir yere bağlı olan kimse, şey. **I seem to be a ~ here,** buraya bağlandım kaldım.

**fizz** [fiz]. (*ech.*) Gazoz gibi fışıldamak; (*coll.*) şampanya. **~le** [fiˈzl], **to ~ out,** boşa çıkmak, suya düşmek. **~y** [ˈfizi]. Gazoz gibi.

**flabbergast** [ˈflabəgâst]. Şaşırtmak; hayrette bırakmak.

**flabby** [ˈflabi]. Lâpamsı; gevşek; yumuşak.

**flag**[1] [flag]. Erinmek; dermansız kalmak. **~**[2] *n.* Bayrak, sancak. **to keep the ~ flying,** milletinin şerefini muhafaza etm.: **to lower (strike) one's ~,** teslim için bandıra indirmek: **black ~,** korsan bandırası: **yellow ~,** karantina işareti olan sarı bayrak. **~-captain,** Amiral gemisinin kumandanı. **~-day,** rozet dağıtılan gün. **~-officer,** Amiral rütbesinde olan bahriye zabiti. **~ship** [ˈflagşip]. Amiral gemisi. **~staff** [ˈflagstâf]. Bayrak direği.

**flagellate** [ˈflacəlêyt]. Kamçılamak.

**flagon** [ˈflagən]. Büyük şişe.

**flagran·t** [ˈflêygrənt]. Göze batan (fenalık); rezalet nevinden; aşikâr (günah). **~cy,** kabahatın aşıkârlığı. **~te delicto** [flêyˈgranti diˈliktou]. Cürmümeşhud; suçüstü.

**flag(stone)** [ˈflagstoun]. Yassı kaldırım taşı.

**flail** [flêyl]. Harman döveni.

**flair** [fleə(r)]. **to have a ~ for stg.,** bir şeye hususî bir istidadı olm.

**flak·e** [flêyk]. İnce tabaka; kuşbaşı kar. Tabaka tabaka ayrılmak. **~y** [ˈflêyki]. İnce tabakadan mürekkeb. **~ pastry,** yufkalı hamur işi.

**flamboyant** [flamˈbôyənt]. Parlak renkli.

**flame** [flêym]. Alev; aşk ateşi. Alevlenmek. **to ~ up,** birdenbire alevlenmek; öfkelenmek: **an old ~ of mine,** benim eski bir göz ağrısı.

**flange** [flanc]. Kalkık kenar.

**flank** [flan(g)k]. Böğür; yan; ordu cenahı. Yanında olm.; yandan tehdid etmek. **to take the enemy in the ~,** düşmanın cenahına hücum etm.: **to turn s.o.'s ~,** yandan hücum etm.

**flannel** [ˈflanl]. Fanilâ. **~ette,** fanilâ taklidi kumaş.

**flap** [flap]. (*ech.*) Kuş kanadının vuruşu; sarkık parça; menteşeli kenar; kapak. Kanadlarını çırpmak; salla(n)mak. (*sl.*) meraklanmak. **~per** [ˈflapə(r)]. Kaplumbağa vs.nin kolu; (*sl.*) saçını daha yapmıyan genc kız.

**flare** [fleə(r)]. Alev aydınlığı; işaret fişeği. Alev gibi parlamak. **to ~ up,** birdenbire alevlenmek. **~-up,** ansızın alevlenme; birdenbire hiddetlenme.

**flash**[1] [flaş] *n.* Anî ışık; şimşek; lem'a. *v.* Şimşek çakmak; ışıldamak. **a ~ in the pan,** (*fig.*) saman alevi: **to ~ past,** şimşek gibi geçmek: **the truth ~ed upon me,** kafamda bir şimşek çaktı. **~-lamp,** ceb feneri. **~-light,** işaret feneri. **~-point,** iştial noktası. **~²,** **-y** *a.* Sahte gösterişli.

**flask** [flâsk]. Yassı ceb şişesi.

**flat**[1] [flat] *n.* Apartıman (dairesi). **~²** *a.* Düz(lük), müstevi, yassı; yavan; (*mus.*) bemol; (renk) mat; (bira) köpüğü dağılmış; *n.* Düz satıh; el ayası; münhat ova. **he ~ly insulted me,** bana bayağı hakaret etti: **to fall ~ on one's face,** pat diye yüzükoyun düşmek: **(joke) to fall ~,** (nükte) muvaffak olmamak: **to lie down ~ on the ground,** boylu boyuna yere yatmak: **to go ~ out,** alabildiğine koşmak: **~ race,** maniasız yarış: **a ~ rate of pay,** muayyen bir ücret ödeme: **a ~ refusal,** tam bir red: **that's ~!,** işte o kadar!: **~ tyre,** sönük lâstik. **~-fish,** yassı balık. **~-footed,** (*med.*) düztaban. **~ness** [ˈflatnis]. Düzlük; yassılık. **~ten** [ˈflatn]. **to ~ oneself against a wall,** kendini duvara doğru yapıştırmak: **to ~ down [out],** yassılatmak; ezmek: **to ~ out,** (uçak) pikeden sonra doğrulmak.

**flatter** [ˈflatə(r)]. Fazla övmek; aslından daha güzel göstermek. **to ~ oneself,** övünmek; boş ümide düşmek. **~y,** yaltaklanma.

**flatulen·ce, -cy** [ˈflatyuləns]. Midede gaz toplanması. **~t,** midede gaz hasıl edici; sözü boş ve tumturaklı.

**flaunt** [flônt]. Gösteriş yapmak; azametle teşhir etm.

**flavour** [ˈflêyvə(r)]. Çeşni, lezzet. Çeşni vermek. **to ~ of stg.,** ···e çalmak. **~ing,** çeşni verici şey.

**flaw** [flô]. Kusur, eksiklik; çatlak. **~less,** kusursuz.

**flax** [flaks]. Keten. **~en,** lepiska.

**flay** [flêy]. Derisini yüzmek; merhametsizce tenkid etmek.

**flea** [flî]. Pire. **to send s.o. away with a ~ in his ear,** birini ters bir cevabla koğmak. **~-bite,** pire yeniği: **a mere ~,** devede kulak. **~-bitten,** pire yeniklerile dolu.

**fleck** [flek]. Benek(lemek).

**fled** *v.* **flee.**

**fledge·d** [flecd]. Tüylenmiş. **~ling,** yavru kuş; toy kimse.

**flee** **(fled)** [flî, fled]. Kaç(ın)mak; (zaman) uçup gitmek.

**fleec·e** [flîs]. Yünlü post; koyundan kırpılan yün mikdarı. Kırkmak; soymak. **~y,** yünlü; **~ clouds,** tırtık tırtık bulutlar.

**fleet**[1] [flît] *n.* Donanma, filo. **~²** *a.* Sür'atli, hızlı giden. **~ing,** süreksiz, fani.

**Flemish** [ˈflemiş]. Flaman.

**flesh** [fleş]. Et; vücud. **one's own ~ and blood,** aynı etten ve kandan olanlar: **it's more than ~ and blood can stand,** buna can dayanmaz: **to make s.o.'s ~ creep,** birinin tüylerini ürpertmek: **the lusts of the ~,** şehvanî arzular: **to put on ~,** şişmanlamak: **to lose ~,** zayıflamak: **to go the way of all ~,** ölmek: **~ wound,** hafif bir yara. **~-colour,** ten rengi: **~-eating,** et yiyici. **~-pots,** yiyecek bolluğu; lüks hayat. **~y** [ˈfleşi]. Etli canlı, şişmanca.

**flew** *v.* **fly.**

**flex** [fleks]. Bükmek, eğmek. Kordon teli. **~ible** [ˈfleksibl], bükülebilir; uysal. **~ibility,** eğilebilme; uysallık.

**flick** [flik]. (*ech.*) Fiske (vurmak). **a ~ of the wrist,** sür'atli bir bilek hareketi: **the ~s,** (*sl.*) sinema. **~er** ˈflikə(r)]. (Alev, ışık) oynamak.

**flier** [ˈflâyə(r)]. Uçan kimse.

**flight**[1] [flâyt]. Uçma; mahrek; mesafe; kuş sürüsü. **a ~ of fancy,** hayal oyunu: **in the first ~,** ön safta: **a ~ of planes,** uçakların grup halinde uçuşu: **a ~ of stairs,** merdiven. **~-lieutenant,** hava yüzbaşısı. **~²,**

Kaçış; firar. **to take ~**, uçmak: **to take to ~**, kaçmak: **to put to ~**, hezimete uğratmak. **~y** [ˈflâyti]. Hafif mizaclı.
**flimsy** [ˈflimzi]. İnce; sağlam olmıyan. (*coll.*) ince daktilo kâğıdı.
**flinch** [flinç]. Korkup sakınmak. **without ~ing**, hiç sakınmadan.
**fling** (**flung**) [flin(g), flun(g)]. Hızlı atmak; atılmak, seğirtmek. Fırlatma; bir iskoç dansı. **to have one's ~**, genclik çılgınlıkları yapmak: **to have a ~ at stg.**, (*coll.*) bir şeyi şöyle bir denemek: **to ~ one's arms about**, kollarını savurmak: **to ~ out at s.o.**, birine birdenbire küfür etm.: **to ~ open**, şiddetle açmak: **to ~ up a job**, (*coll.*) bir işten birdenbire çıkmak.
**flint** [flint]. Çakmaktaşı. **with a heart of ~**, katı yürekli: **to skin a ~**, sineğin yağını hesab etmek. **~y**, taş gibi katı.
**flip** [flip]. Fiske (vurmak).
**flippan·t** [ˈflipənt]. Hiç bir şeyi ciddiye almıyan. **~cy**, hürmetsizlik.
**flipper** [ˈflipə(r)]. Kaplumbağa vs.nin kolu.
**flirt** [flö(r)t]. İşvebaz. Flört etmek. **~ation** [–ˈtêyşn], flört etme.
**flit** [flit]. Gölge gibi geçmek; şuraya buraya uçmak; göçmek. **~ter** [ˈflitə(r)]. Hafifçe ve çabuk şuraya buraya uçmak. —
**float**[1] [flout] *n*. Sal; şamandıra; ağ mantarı; olta mantarı. **~**[2] *v*. Yüz-(dür)mek; batmamak; havada durmak. **to ~ a loan**, bir istikraz çıkarmak: **to ~ a company**, bir şirket kurmak. **~ing** [ˈfloutin(g)] *a*. Yüzen. **~ capital**, mütedavil sermaye: **~ debt**, mütehavvil borc: **~ population**, sabit olmıyan nüfus.
**flock** [flok]. Hayvan sürüsü. **to ~ (together)**, koyun sürüsü gibi toplanmak: **~s and herds**, koyun ve sığır: **a pastor and his ~**, rahib ile cemaatı.
**floe** [flou]. Yüzen buz parçası.
**flog** [flog]. Kamçılamak. **to ~ a dead horse**, tarihe karışmış bir şeyi canlandırmağa çalışmak. **~ging**, dayak.
**flood** [flʌd]. Tufan; su taşması, sel; med. Sel basmak; taşmak. **to be ~ed with letters**, mektub yağmuruna tutulmak. **~-gate**, bend kapağı. **~-light**, projektörle aydınlatma(k). **~-lit**, projektörlerle aydınlatılmış. **~-tide**, med. **~ing** *n*. sel basma.
**floor** [flô(r)]. Oda zemini; döşeme; kat. Ev zeminini tahta döşemek. **to ~ s.o.**, birine cevab verilemiyecek sual sormak. **to take the ~**, söylemek için ayağa kalkmak: **house with two ~s**, iki katlı ev: **to wipe the ~ with s.o.**, (*coll.*) birini tam manası ile mağlûb etmek. **~-cloth**, tahta bezi. **~-polish**, mobilya cilâsı. **~-space**, döşeme sahası.
**flop** [flop] (*ech.*) Yere düşen ağırca bir şeyin sesi. Cop diye yere düşmek. **to go ~**, ansızın düşmek; (*fig.*) suya düşmek. **~py**, (şapka) sarkık.
**flor·a** [flôra]. Bir mıntakada yetişen bütün nebatat. **~al** [ˈflôrəl]. Çiçeklere aid. **~escence** [flôˈresens]. Çiçek açması. **~ist** [ˈflorist]. Çiçekçi.
**florid** [ˈflorid]. Fazla süslü; kırmızı yüzlü.
**florin** [ˈflorin]. İki şilin kıymetinde gümüş para.
**flotation** [flouˈtêyşn]. Yüzdürme: istikrazı çıkarma; hisse senedlerini satarak bir şirket kurma.
**flotilla** [flouˈtilə]. Küçük filo.
**flotsam** [ˈflotsəm]. Denizde sahibsiz olarak yüzen eşya. **~ and jetsam**, denizde yüzen ve karaya vuran enkaz.
**flounce** [flauns]. Kadın elbiselerinde volan. **to ~ about**, öfkeli hareket etmek.
**flounder** [ˈflaundə(r)] *v*. Çamura veya suya bata çıka yürümek.
**flour** [ˈflauə(r)]. Un. **~y**, unlu. **~-mill**, değirmen.
**flourish**[1] [ˈflʌriş]. (Kılıc) öteye beriye savurmak. Savurma; gösterişli hareket. **a ~ of trumpets**, merasim borusu. **~**[2]. Mamur olm.; gelişmek; tıkırında gitmek. **~ing** *a*. mamur; iyi.
**flout** [flaut]. İstihfaf ile aldırmamak.
**flow** [flou]. Akmak; cereyan etm. Akış; akıntı; med. **a ready ~ of language**, çok selis konuşma. **~ing** [ˈflouin(g)]. Akan; selis; seyyal; (elbise) gevşek ve sarkık.
**flower** [ˈflau(r)]. Çiçek. Çiçeklenmek. **to burst into ~**, birdenbire çiçeklenmek: **the ~ of the army**, ordunun en seçkin kısmı: **in the ~ of one's youth**, gencliğinin en parlak çağında. **~y**, çiçekli; fazla süslü. **~-girl**, çiçekçi kadın. **~-pot**, saksı.
**flown** *v*. **fly**.
**flu** [flû]. (*abb.*) **influenza.** Grip.
**fluctuat·e** [ˈflʌktyuêyt]. Dalga gibi inip kalkmak; kararsız olm.; değişmek. **the temperature ~s from day**

**to day,** hararet her gün değişir. **~ion** [–ˈêyşn], temevvüc; tereddüd; değişme.

**fluen·cy** [ˈflûənsi]. Selâset. **~t,** selis: **to speak a language ~ly,** bir lisanı kolay konuşmak.

**fluff** [flʌf]. Tüy ve hav döküntüsü. **to ~ up its feathers,** (kuş) tüyleri kabartmak. **~y,** yumuşak ve kaba tüylü.

**fluid** [ˈflûid]. Mayi; su. Seyyal; akar; şekli kolay değişen. **~ity** [–ˈiditi], mayilik.

**fluk·e** [flûk]. Baht işi. Baht işi becermek. **~y,** baht işi elde edilmiş; talihe bağlı.

**flummox** [ˈflʌməks]. Şaşırtmak.

**flung** *v.* **fling.**

**flunkey** [ˈflʌn(g)ki]. (*cont.*) Uşak, peyk.

**flurry** [ˈflʌri]. Anî bir bora, kar; telâş, heyecan. Telâşa düşürmek. **to get flurried,** telâşa düşmek.

**flush**¹ [flʌş]. Hızlı akıtmak; (yüz) kızar(t)mak. Sür'atlı su akıntısı; galeyan. **to ~ out a drain,** gerizi bol su ile temizlemek: **to be in the full ~ of health,** yanağından kan damlamak: **in the first ~ of victory,** zafer sarhoşluğu ile. **~**² *a.* Taşarcasına dolu; ayni seviyede; bir hizada. **~ with ...,** ···in yüzünden dışarı taşmıyan.

**fluster** [ˈflʌstə(r)]. Telâş, heyecan. Telâşa düşürmek. **to be ~ed** [**to be all in a ~**], çırpınmak, telâş etm.

**flut·e**¹ [flût]. Flâvta. **~**². Yiv açmak. **~ed,** yivli. **~ing,** yiv şeklinde süs.

**flutter** [ˈflʌtə(r)]. Çırpınma(k); telâşa düşme(k); şuraya buraya uçma(k), yürek çarpıntısı. **to have a little ~,** az para koyarak kumar, bahise, girmek.

**fluvial** [ˈflûviəl]. Nehre aid.

**flux** [flʌks]. Akma; vücuddan gayritabiî sızıntı; değişiklik; lehim suyu. **to be in a state of ~,** sık sık değişmek.

**fly**¹ [flây] *n.* Uçuş; çadır perdesi; pantalonun ön yırtmacı. **the flies,** tiyatro sahnesinin üstü. **~-button,** pantalonun ön düğmesi. **~-leaf,** bir kitabın başında ve sonundaki boş yaprak. **~ wheel,** volan. **~-by-night,** gece hayatı mübtelâsı; borcunu vermeden gece sıvışan. **~**² *n.* Sinek; balık tutmak için sun'î sinek. **they died like flies,** yığın yığın öldüler: **there are no flies on him,** çok açıkgözdür: **to rise to the ~,** (balık) sun'î sineğe doğru sıçramak; (insan) kendisini tahrik etmek için mahsus söylenen söze kanarak kızmak. **~-blown,** sinek tersile lekelenmiş. **~**³ (**flew, flown**) [flây, flû, flǫun] *v.* Uçmak; kaçmak; sür'atle koşmak; uçak ile seyahat etm.; uçarak geçmek; uçurmak. **to ~ to arms,** (bir millet) silâha sarılmak: **to ~ at,** ···e saldırmak: **the bird has flown,** aranılan kimse kayıblara karıştı: **to ~ high,** yüksekte uçmak; gözü yükseklerde olm.: **to let ~ at s.o.,** birine ağzına geleni söylemek; birine bir tokat aşketmek: **to make the money ~,** har vurup harman savurmak: **the door flew open,** kapı şırak diye açıldı: **to ~ to pieces,** parça parça olm.: **to send ~ing,** kaçırtmak: **to send things ~ing,** ortalığı darmadağın etmek. **~ back,** geri uçmak; mümkün mertebe çabuk avdet etm. **~ off,** uçup ayrılmak; (düğme) kopmak. **~ing** [ˈflây·in(g)]. Uçan; uçma; tayyarecilik. **~ column,** seyyar kıta: **the ~ of a flag,** bayrak asma: **to pay a ~ visit to London,** Londra'ya şöyle bir uğramak: **~ start,** başlama noktasını son hızla geçme: **~ squad,** (polis) yıldırım kıt'ası. **~ing-boat,** tekneli deniz uçağı. **~ing-buttress,** kemerli payanda.

**foal** [fǫul]. Tay; sıpa. (Kısrak, eşek) doğurmak.

**foam** [fǫum]. Köpük. Köpürmek.

**fob**¹ [fob] *n.* Saat cebi. **~**² *v.* **~ off,** aldatmak: **to ~ stg. off on s.o.,** birine hile ile değersiz bir şey satmak.

**foc·al** [ˈfǫukl]. Mihraka aid. **~ length,** mihrak mesafesi. **~us** [ˈfǫukəs]. Mihrak. Mihraka getirmek; ayar etm.; temerküz etmek. **in ~,** ayarlı; vazıh: **out of ~,** ayarsız; vuzuhsuz: **all eyes were ~ed on ...,** bütün gözler ···e dikilmişti. **~using,** bir noktaya toplanma; temerküz.

**foc's'le** [ˈfǫuksl]. *abb.* **forecastle.**

**fodder** [ˈfodə(r)]. (Kuru ot) yem (vermek).

**foe, ~man** [fǫu, ˈfǫumən]. Düşman.

**fog** [fog]. Sis. Sis gibi kuşatmak, karartmak; şaşırtmak. **I am in a complete ~,** nerede bulunduğumu bilmiyorum; vaziyeti hiç anlamıyorum. **~-bound,** sis yüzünden hareket etmiyen; sisle kaplı. **~-horn,** sis düdüğü. **~-signal,** sis işareti. **~gy** [ˈfogi]. Sisli. **I haven't the foggiest idea,** hiç haberim yok.

**fog(e)y** [ˈfọugi]. **old ~**, eski kafalı adam.
**foible** [ˈfôybl]. Zayıf taraf.
**foil**[1] [fôyl] *n.* Pek ince safiha; bir şeyi iyice tebarüz ettirmek için kullanılan şey. **~**[2] *n.* Talim meçi. **~** *v.* İşini bozmak, önüne geçmek.
**foist** [fôyst]. Kandırıp yutturmak, hile ile sokuşturmak. **to ~ oneself on s.o.**, birine yamanmak.
**fold**[1] [fọuld]. Koyun ağılı. Ağıla kapatmak. **~**[2]. Kıvrım, kat; çukur. Katlamak; kuşatmak. **to ~ the arms**, kollarını kavuşturmak. **~er** [ˈfọuldə(r)]. Dosya zarfı. **~ing** [ˈfọuldin(g)] *a.* Katlanabilir, katlanır. **~-ing-chair (-ladder)**, açılıp kapanır iskemle (merdiven).
**foliage** [ˈfọulyəc]. Ağac yaprakları.
**folio** [ˈfọuliọu]. Büyük kıtada kitab; bir kere katlanmış kâğıd tabakası.
**folk** [fọulk]. Halk; millet. **country ~**, kırda yaşıyan kimseler. **~-dance**, halk oyunu. **~-song**, halk türküsü. **~lore** [ˈfọuklô(r)]. Halk bilgisi.
**follow** [ˈfolọu]. Takib etm., peşinden gitmek veya gelmek; ···den sonra gelmek; halef olm.; neticesi olm., ···den çıkmak; gözden kaybetmemek; tâbi olm., tarafdarı olm.; (söylenilen bir şeyi) anlamak. **as ~s**, aşağıdaki gibi: **to ~ s.o. about**, birinin peşine takılmak: **it ~s that ...**, bundan şu netice çıkar ki ...: **it does not ~ that ...**, bundan ... neticesi çıkarılmaz: **to ~ the plough**, çiftçi olm.: **to ~ a profession**, bir mesleğe mensub olm.: **to ~ the sea**, gemici olmak. **~ on**, ara vermeden devam etm.; sonra gelmek. **~ out**, nihayete kadar takib etm.; yerine getirmek. **~ up**, peşini bırakmamak; devam etm.; **to ~ up a clue**, bir ipucunu takib etm.: **to ~ up a victory**, bir zaferi sonuna kadar getirmek. **~er** [ˈfolọuə(r)]. Maiyet erkânından biri; peyk; tarafdar; halef. **~ing** [ˈfolọuin(g)] *a.* Müteakıb; aşağıdaki. *n.* Maiyet; tarafdarların mecmuu. **a ~ sea**, kıça doğru gelen dalgalar.
**folly** [ˈfoli]. Akılsızlık, ahmaklık.
**foment** [fọuˈment]. (Şişi) sıcak su ile ısıtmak; kışkırtmak. **~ation** [–ˈtêyşn], sıcak su tatbiki.
**fond** [fond]. Çok seven; şefkatli; gayrımakul derecede seven; saf. **~ of**, ... düşkünü, çok seven. **~ly**, safca. **~le** [ˈfondl]. Okşamak.
**font** [font]. Vaftiz kurnası.
**food** [fûd]. Yiyecek şey, yemek, gıda; yem. **to give ~ for thought**, düşündürmek. **~stuffs**, gıda maddeleri.
**fool** [fûl]. Ahmak, budala; soytarı. Maskaralık etm.; aldatmak. **to ~ about [around]**, avare avare dolaşmak: **to ~ away the time**, vaktini boş geçirmek: **to make a ~ of oneself**, kendini gülünç etmek: **to make a ~ of s.o.**, birini maskaraya çevirmek: **to play the ~**, maskaralık etm. **~'s cap**, deli külâhı. **~-proof**, pek kolay, pek emin. **~ery** [ˈfûləri]. Ahmaklık; maskaralık. **~hardy** [ˈfûlhâdi]. Delice cesur. **~ish** [ˈfûliş]. Akılsız, ahmak. **to feel ~**, kendini gülünç (olmuş) hissetmek: **to look ~**, gülünç görünmek.
**foolscap** [ˈfûlskap]. Esericedid kâğıdı.
**foot**[1], *pl.* **feet** [fut, fît]. Ayak; kaide; dib; kadem, 12 pusluk (= 30·479 sant.); şiirde vezin; piyade askeri. **to be carried off one's feet**, dalga ile sürüklenmek; heyecana kapılmak: **to be carried out feet foremost**, cenazesi çıkmak: **to find one's feet**, muhite alışmak: **to have [get] cold feet**, (*fig.*) korkmak: **~ and horse**, piyade ve süvari askerleri; **20,000 ~**, 20,000 piyade askeri: **to keep one's feet**, düşmemek: **to knock s.o. off his feet**, birini yere sürmek: **to light on one's feet**, ayakları üstüne düşmek: **on ~**, yaya: **to set on ~**, kurmak: **to go at a ~'s pace**, yaya sür'atile gitmek: **to put one's best ~ foremost**, acele etm.; gayret göstermek: **to put one's ~ down**, ayak diremek: **to put one's ~ in it**, pot kırmak: **to set s.o. on his feet**, bir adama muayyen bir para vererek müstakil bir iş kurmasını temin etm.: **it will need a lot of money to set this business on its feet again**, bu işi tekrar yoluna koymak için çok para lâzım: **to sit at s.o.'s feet**, birinin talebesi olmak. **~-and-mouth disease**, şap hastalığı. **~-bridge**, yayalara mahsus dar köprü. **~-hills**, dağ eteği. **~-passenger**, yaya. **~-plate**, lokomotifin sahanlığı. **~-rule**, 12 pusluk ölçü; kadem. **~-slogger**, (*sl.*) piyade neferi. **~**[2] *v.* **to ~ it**, yaya gitmek: **to ~ the bill**, hesabı ödemek. **~ball** [ˈfutbôl]. Futbol. **~baller**, futbolcu. **-~ed** [ˈfutid]. ... ayaklı. **~fall** [ˈfutfôl]. Ayak sesi. **~gear** [ˈfutgiə(r)]. Ayakkabı. **~hold** [ˈfut-

hǫuld]. Ayak basacak yer. **to get a ~,** ayağile tutunmak: **to lose one's ~,** tırmanırken ayağı kurtulmak. **~ing** [ˈfutin(g)]. *v.* **foothold**; seviye, vaziyet. **on a war ~,** seferî vaziyette. **~lights** [ˈfutlâyts]. Tiyatro sahnelerinin önlerindeki bir sıra ışık. **~man** [ˈfutmən]. Perdeci; uşak. **~mark** [ˈfutmâk]. Ayak izi. **~note** [ˈfutnǫut]. Sahife dibindeki not. **~pad** [ˈfutpad]. Yol kesen haydud. **~path** [ˈfutpâθ]. Keçi yolu; patika. **~print** [ˈfutprint]. Ayak izi. **~sore** [ˈfutsô(r)]. Ayakları pişmiş. **~step** [ˈfutstep]. Ayak sesi; adım. **to follow [walk, tread] in the ~s of,** …in izinden gitmek. **~stool** [ˈfutstûl]. Tabure. **~way** [ˈfutwêy]. Kaldırım; yayalara mahsus geçid. **~wear** [ˈfutweə(r)]. Ayakkabı.

**footl·e** [ˈfûtl]. **to ~ about,** (*sl.*) boş gezmek. **~ing,** ehemmiyetsiz; manasız.

**fop** [fop]. Kendini beğenmiş; züppe.

**for** [fô(r); fə(r)]. İçin; zarfında; müddetle; mesafe dahilinde; olarak; …den; …den dolayı; …e; çünkü. **~ all [aught] I care,** bana vız gelir: **~ all that,** söylenen (yapılan) her şeye rağmen: **~ all his wealth he is unhappy,** bütün servetine rağmen mes'ud değildir: **~ all I know,** benim bildiğime göre: **to be all ~,** …in lehinde olm.: **~ ever,** ebediyen: **~ the first time,** ilk defa olarak: **in ~,** *v.* **in**: **he's ~ it,** göreceği var!: **to weep [dance] ~ joy,** sevincden ağlamak [oynamak]: **to leave England ~ France,** İngiltere'den Fransa'ya hareket etm.: **~ life,** ölünceye kadar; kaydı hayat ile: **~ miles and miles,** kilometrolarca: **~ myself, I would rather ...,** bana gelince ... tercih ederim: **I am doing this ~ myself,** bunu kendim için yapıyorum: **I can do it ~ myself, thank you,** teşekkür ederim, ben kendim yapayım: **oh ~ peace!,** ah! bir sulh olsa!: **~ years and years,** senelerce.

**forage** [ˈforic]. Hayvanlara ve bilhassa atlara verilen ot, saman vs. Yem aramak; araştırarak elde etmek. **~-cap,** yumuşak asker kasketi.

**forasmuch** [ˈfôrasˈmʌç]. **~ as,** madem ki.

**foray** [ˈforêy]. Akın (etmek).

**forbade** *v.* **forbid.**

**forbear**[1] [ˈfôbeə(r)] *n.* Ced, dede, ata. **~**[2] **(forbore, forborne)** [fəˈbeə(r), –ˈbô(r), –ˈbôn]. Çekinmek, sakınmak; sabırlı olmak. **to ~ to say [to ~ from saying],** söylemekten çekinmek. **~ance,** sabır; çekinme.

**forbid (forbade, forbidden)** [fəˈbid, –ˈbad, –ˈbidn]. Yasak etm.; menetmek. **God ~!,** Allah göstermesin!; **to ~ s.o. to do stg.,** bir şeyi yapmağı birine yasak etm.: **to ~ s.o. the house,** birini evine sokmamak. **~ding,** korkunc; ekşi yüzlü.

**forbore, forborne** *v.* **forbear.**

**force**[1] [fôs] *n.* Kuvvet; zor; asker veya polis kıtası. **this law is still in ~,** bu kanun hâlâ mer'idir: **this law will come into ~ tomorrow,** bu kanun yarın meriyete giriyor: **the police arrived in ~,** büyük mikdarda polis kuvveti geldi: **to put in ~,** infaz etm.: **there is ~ in what he says,** söylediği boş değil: **~ majeure,** esbabı mücbire. **~**[2] *v.* Zorlamak; mecbur etm.; sun'î hararetle yetiştirmek. **my hand was ~d,** zora getirdiler: **to ~ the pace,** bir şeyi hızlandırmak; **I tried to ~ a smile,** gülmeğe çalıştım.. **~ back,** defetmek; geri itmek. **~d** [fôsd] *a.* Zoraki; mecburî. **~ful** [ˈfôsfəl]. Girgin; tesirli.

**forceps** [ˈfôseps]. Kerpeten; pens.

**ford** [fôd]. Nehir geçidi. Bir nehirden yaya geçmek. **~able,** yürüyerek geçilebilir.

**fore** [fô(r)]. Ön (taraf). **at the ~,** pruva direğinde: **to the ~,** ileride mevcud ve hizmete hazır: **to come to the ~,** başa gelmek. **~-** *pref.* Önceden; ön, önde. **~-and-aft,** baştan kıça kadar. **~arm**[1] [ˈfôrâm] *n.* Kolun ön kısmı. **~arm**[2] [fôrˈâm] *v.* Önceden silâhlan(dır)mak. *v.* **forewarned. ~boding** [fôˈbǫudin(g)]. Bir felâketin geleceğini önceden hissetme. **~cast** [ˈfôkâst] *n.* Olacak bir şeyin keşfi; tahmin. *v.* [–ˈkâst]; İstikbalde vukubulacak şey (bilhassa hava) hakkında haber vermek. **weather ~,** hava raporu. **~castle** [ˈfoksl] (*um.* **foc's'le** *yazılır*). Tayfa kamarası. **~close** [fôˈklǫuz]. İpotekli bir malı haczetmek. **~father** [ˈfôfâðə(r)]. Ced, ata. **~finger** [ˈfôfin(g)gə(r)]. Şehadet parmağı. **~gather** [fôˈgaðə(r)]. Toplanmak. **~go**[1] *v.* **forgo. ~**[2] [fôˈgǫu]. Takaddüm etmek. **~going,** yukarıdaki. **~gone** [fôˈgon]. **a ~ conclusion,** önceden belli olan netice. **~ground**

[ˈfôgraund]. Ön plân. ~**head** [ˈfored]. Alın. ~**leg** [ˈfôleg]. Hayvanın ön ayağı. ~**lock** [ˈfôlok]. Perçem: '**to take time by the** ~', fırsat kaçırmamak. ~**man,** [ˈfômən]. Kalfa; jüri reisi. ~**most** [ˈfômoust]. En önde olan. **first and** ~, ilkönce. ~**runner** [ˈfôrʌnə(r)]. Haber verici; müjdeci; alâmet, delil. ~**see (foresaw, foreseen)** [fôˈsî, –sô, –sîn]. Önceden sezmek; geleceğini anlamak. ~**shadow** [fôˈşadou], İstikbale aid bir şeye delâlet etm. ~**shore** [ˈfôşô(r)]. Sahilin med ve cezir işaretleri arasındaki kısmı. ~**shorten** [fôˈşôtn]. Perspektif kaidelerine nazaran kısaltmak. ~**sight** [ˈfôsâyt]. Basiret. ~**stall** [fôˈstôl]. Bir işte (başka birisinden) evvel davranmak; önune geçmek. ~**taste** [ˈfôtêyst]. Sonra gelecek bir saadet [felâketin] nümunesi. ~**tell (foretold)** [fôˈtel, –tould]. Önceden haber vermek. ~**thought** [ˈfôθôt]. Dûrendişlik. ~**warn** [fôˈwôn]. Önceden haber verip ikaz etmek. ⌜~**ed is forearmed**⌝, önceden haberi olan hazır olur. ~**woman** [ˈfôwumən]. Kadın kalfa. ~**word** [ˈfôwö(r)d]. Önsöz.

**foreign** [ˈforin]. Ecnebi; yabancı; haricî. ~ **trade,** dış ticaret: **Foreign Office,** İngiliz Hariciye Nezareti: **Foreign Secretary,** İngiliz Hariciye Nazırı: **Minister of Foreign Affairs,** Hariciye Vekili. ~**er,** ecnebi; yabancı adam.

**forensic** [fəˈrensik]. Mahkemeye aid. ~ **medicine,** adlî tıb: ~ **skill,** avukatlık mahareti.

**forest** [ˈforist]. Orman. ~**er,** orman memuru. ~**-guard,** orman korucusu. ~**ry** [ˈforestri]. Ormancılık.

**forever** [fəˈrevə(r)]. Ebedî olarak; daima. ~**more,** ebedî olarak.

**forfeit** [ˈfôfit]. Hata veya ihmaldan dolayı kaybedilen (şey); hakkın sukutu. Hata veya ihmaldan dolayı bir şeyi [hakkı] kaybetmek. ~ **money,** cayma tazminatı. ~**ure,** hakkın sukutu.

**forgave** *v.* **forgive.**

**forge**[1] [fôc]. Demirci ocağı; nalband dükkânı. Demir dövmek; uydurmak, taklidini yapmak; kalpazanlık etmek. ~**d,** sahte, kalp. ~**r,** kalpazan. ~**ry,** kalpazanlık; tahrif. ~[2] *v.* **to** ~ **ahead,** (gemi) ağır ağır ileri gitmek; (yarışta) yavaş yavaş önüne geçmek; (işte) gittikçe terakki etmek.

**forget (forgot, forgotten)** [fôˈget, –ˈgot, ~ˈgotn]. Unutmak; ihmal etmek. ~ **(about) it!,** onu artık düşünme!: '**and don't you** ~ **it!**', bunu unutayım deme!: **never to be forgotten,** unutulmaz. ~**ful** [fôˈgetfəl]. Unutkan; ihmalci.

**forgiv·able** [ˈfôˈgivəbl]. Affedilebilir. ~**e (forgave, forgiven)** [fôˈgiv, –ˈgêyv, –ˈgivn]. Affetmek; bağışlamak. ~**ing,** *a.* müsamahalı. ~**eness,** af, affetme.

**forgo (forwent, forgone)** [fôˈgou, –ˈwent, –ˈgon]. Vazgeçmek; ···den kendini mahrum etmek.

**forgot, forgotten** *v.* **forget.**

**fork** [fôk]. Çatal; çatallı bel; yaba; apış. Çatallaşmak; bel ile eşelemek. **to** ~ **out [up],** (*sl.*) uçlanmak. ~**ed,** çatallı.

**forlorn** [fôˈlôn]. Kimsesiz; metrûk; ümidsiz. ~ **hope,** fedailer; ümidsiz teşebbüs.

**form**[1] [fôm] *n.* Şekil, biçim; endam; tarz; suret; kalıb; usul, âdet; âdabımuaşeret; formüler; forma; sınıf; peyke; (atlet) form. **as a matter of** ~ [**for** ~**'s sake**], âdet yerini bulsun diye: **it is bad** ~, **it is not good** ~, yapılmaz: **to fill up a** ~, bir formüleri doldurmak: **to be in good** ~, tam kıvamında olm.: **in** ~ [**out of** ~], (i) formunda [formunda değil]; (ii) derste [ders haricinde]. ~**-room,** sınıf, dershane. ~[2] *v.* Şekil vermek; teşkil etm.; terkib etm.; kurmak; hâsıl olm.; şekil almak. **to** ~ **fours,** (*mil.*) dörder olm.: **to** ~ **a habit,** âdet edinmek. ~**al** [ˈfôməl]. Resmî; teklifli; soğuk tavırlı. ~**ality** [––ˈmaliti], merasim, tekellüf, usul: **as a mere** ~, âdet yerini bulsun diye. ~**at** [ˈfôma]. Kitab forması. ~**ation** [fôˈmêyşn]. Teşkil, teşekkül; tertib, nizam; kurma, ihdas etme. ~**less** [ˈfômlis]. Biçimsiz, şekilsiz.

**former** [ˈfômə(r)]. Evvelki, sabık, eski. **the** ~ ..., **the latter** ..., evvelki ..., sonuncu .... ~**ly,** eskiden, vaktiyle, eski zamanlarda.

**formidable** [ˈfômidəbl]. Yenmesi güç; korkulacak; heybetli.

**formula** [ˈfômyulə]. Formül; reçete. ~**te** [ˈfômyulêyt]. Kat'î ve vazıh bir tarzda ifade etm.; formül haline koymak.

**forsake (forsook, forsaken)** [fôˈsêyk, –ˈsuk, –ˈsêykn]. Terketmek: vazgeçmek; yüzüstü bırakmak.

**forsw·ear** (forswore, forsworn) [fôˈswęə(r), –ˈswô(r), –ˈswôn]. Tövbe etmek. **to ~ oneself,** yalan yere yemin etmek. **~orn** [fôˈswôn] *a.* Yalan yere yeminli
**fort** [fôt]. Kale; tabya.
**forte** [ˈfôte]. Bir kimsenin kuvvetli tarafı.
**forth** [fôθ]. İleri; dışarı; açığa; sonra. **to go ~,** çıkmak: **and so ~,** vesaire: **back and ~,** (bir) ileri (bir) geri: **from this time ~,** şimdiden sonra. **~-coming** [fôθˈkʌmin(g)]. Gelecek, önümüzdeki; hazır. **no help was ~,** yardımdan eser yoktu: **he is not very ~,** o pek kapalı. **~right** [fôθˈrâyt]. Açıkça. **~with** [fôθˈwiθ]. Derhal, hemen.
**fortieth** [ˈfôtiəθ]. Kırkıncı; kırkta bir.
**forti·fiable** [ˈfôtifâyəbl]. Tahkim edilebilir. **~fication** [ˈfôtifiˈkêyşn]. İstihkâm; tahkim etme. **~fications,** tahkimat. **~fy** [ˈfôtifây]. Tahkim etm.; takviye etm., kuvvetlendirmek. **~tude** [ˈfôtityûd]. Şecaat; cesaret; metanet.
**fortnight** [ˈfôtnâyt]. On beş günlük müddet; iki hafta. **this day ~,** iki hafta sonra bu gün. **~ly,** on beş günde bir olan.
**fortress** [ˈfôtris]. Müstahkem yer.
**fortuitous** [fôˈtyûitəs]. Tesadüfî, ârızî.
**fortun·ate** [ˈfôtyunêyt]. Bahtiyar, talihli. **~ately,** bereket versin. **~e** [ˈfôtyən]. Baht, talih, kışmet; servet. **it has cost me a ~,** bu bana dünya kadar paraya maloldu: **to come into a ~,** büyük bir servete varis olm.: **by good ~,** bereket versin: **to make a ~,** büyük bir servet toplamak: **a man of ~,** pek zengin adam: **to marry a ~,** zengin bir kadınla evlenmek: **to tell ~s,** fala bakmak. **~e-hunter,** evlenmek için zengin arıyan kimse. **~e-teller,** falcı.
**forty** [ˈfôti]. Kırk.
**forward** [ˈfôwəd]. İleri, ileriye doğru; ilerdeki; ileriye giden, müterakki; şımarık; haddinden bir az fazla serbest; (futbol) muhacim. **from that day ~,** o günden itibaren: **to look ~ to stg.,** bir şeye önceden sevinmek: **'to be ~ed'** ['please ~!'], (mektub zarfında) lûtfen yeni adresine gönderiniz! **~s,** ileriye doğru.
**fossil** [ˈfosl]. Müstehase (halinde). **an old ~,** eski kafalı ihtiyar. **~ize,** müstehase haline koymak.
**foster**[1] [ˈfostə(r)]. Teşvik etm.; beslemek. **~-**[2]. Süt···. **~-mother,** sütnine.
**fought** *v.* **fight.**
**foul** [fau̯l]. Pis; iğrenc; ufunetli; tıkanmış; dolaşık; bozuk (hava). Favul. Kirletmek, kirlenmek; dolaştırmak. **to fall ~ of s.o.,** birisile çatışmak: **to fall ~ of the law,** kanunun pençesine düşmek: **~ play is suspected,** bir suikasddan şübhe ediliyor: **to run ~ of another ship,** başka gemiye çarpmak. **~ness,** pislik; habaset; bozukluk.
**found**[1] [fau̯nd]. Kurmak, tesis etmek. **~**[2]. (Demir vs.) dökmek. **~**[3] *v.* **find.** **~ation** [fau̯nˈdêyşn]. Kurma, tesis; müessese; temel; vakıf. **~ation-stone** temel taşı. **~er**[1] [ˈfau̯ndə(r)] *n.* Kurucu, müessis. **~er**[2] *n.* Dökmeci. **~er**[3] *v.* (Gemi) batmak; (at) yıkılmak. **~ling** [ˈfau̯ndlin(g)]. Sokakta terkedilen kimsesiz çocuk. **~ry** [ˈfau̯ndri]. Dökümhane.
**fount**[1] [fau̯nt]. Pınar; memba, menşe. **~**[2]. Aynı puntoda hurufat takımı. **~ain** [ˈfau̯ntin]. Fıskıye; pınar, memba: **~-head,** asıl menşe: **~-pen,** dolma kalem.
**four** [fô(r)]. Dört. **to be on all ~s with ...,** ···ile müsavi olm.: **to go [run] on all ~s,** ellerile dizlerinin üstünde yürümek: **to form ~s,** (*mil.*) dörder olm.: **open to the ~ winds,** her yana açık. **~-handed,** dört elli; dört kol (iskambil). **~-in-hand,** dört atlı araba. **~-master,** dört direkli gemi. **~-square,** muhkem. **~fold** [ˈfôfou̯ld]. Dört misli. **~score** [ˈfôskô(r)]. Seksen. **~teen** [fôˈtîn]. On dört. **~teenth,** on dördüncü. **~th** [fôθ]. Dördüncü; dörtte bir.
**fowl** [fau̯l]. Tavuk; kuş. **to keep ~s,** kümes hayvanları beslemek: **wild ~,** avlanacak su kuşları. **~-house,** tavuk kulübesi. **~er,** kuş avcısı. **~ing,** kuş avcılığı.
**fox** [foks]. Tilki; kurnaz adam. Kurnazlık etm., aldatmak. **~glove,** yüksükotu. **~y,** tilki gibi; kurnaz; (saç) kızıl. **~-hound,** tilki avına mahsus bir nevi zağar. **~-hunting,** tilki avı. **~-terrier,** bir cins küçük köpek.
**foyer** [ˈfwaye]. (Tiyatroda) medhal.
**fraction** [ˈfrakşn]. Küçük parça; kır(ıl)ma; kesir. **in the ~ of a second,** bir anda. **~al,** kesirlere aid.
**fractious** [ˈfrakşəs]. Ters, huysuz; (at) harın; dikkafalı.

**fracture** [ˈfraktyə(r)]. Kırma; kırık. Kır(ıl)mak; çatla(t)mak. **to set a ~,** kırığı (yerine oturtup) sarmak.
**fragil·e** [ˈfracâyl]. Kolay kırılır; nazik, cılız; gevrek. **~ity** [–ˈciliti], kolay kırılabilme, gevreklik.
**fragment** [ˈfragmənt]. Küçük parça, kıta. **~ary** [–ˈmentəri], parça halinde.
**fragran·ce** [ˈfrêygrəns]. Güzel koku. **~t,** güzel kokulu.
**frail** [frêyl]. Çelimsiz; kolay kırılır; günaha kolay girer. **~ty,** çelimsizlik; manevî zaaf.
**frame**[1] [frêym] *n.* Çerceve; çatı; yapı; iskelet; vücud; süve; usul; nebatat camekânı. **~ of mind,** ruhî halet. **~**[2] *v.* Çerçevelemek; tertib etm., kurmak. **how is the new apprentice framing?,** yeni çırak işe alışıyor mu? **~-up,** danışıklı dövüş; birini suçlu göstermek için kurulan kumpas. **~work** [ˈfrêymwö(r)k]. Çatı; bina kafesi; iskelet.
**France** [frâns]. Fransa.
**franchise** [ˈfrançâyz]. İntihabatta seçim hakkı.
**francophile** [ˈfrankǫufil]. Fransız tarafdarı.
**frank** [frank] *a.* Açık sözlü, samimî. **~ly,** açıkça: **~ (speaking),** açıkçası. **~ness,** açık sözlülük, samimiyet.
**frantic** [ˈfrantik]. Çılgınca heyecanlanmış.
**fratern·al** [frəˈtö(r)nl]. Kardeşçe; kardeşler arasındaki. **~ity,** kardeşlik; uhuvvet; dostların cemiyeti. **~ize** [ˈfratənâyz], kardeş gibi görüşmek.
**fratricide** [ˈfratrisâyd]. Kardeşini öldürme; kardeş kaatili.
**fraud** [frôd]. Hile(-kârlık); sahtekâr; müzevir; (*coll.*) katakulli; umulduğu gibi çıkmıyan. **a pious ~,** sahte dindar. **~ulent** [ˈfrôdyulənt]. Hileli, düzenbaz: **~ conversion,** ihtilâs. **~ulence,** hilekârlık.
**fraught** [frôt]. **~ with,** ···le yüklü. **~ with danger,** pek tehlikeli.
**fray**[1] [frêy] *n.* Arbede; muharebe. **eager for the ~,** kavgaya hazır. **~**[2] *v.* Tarazlan(dır)mak; aşınmak. **~ed nerves,** yıpranmış sinirler.
**frazzle** [ˈfrazl]. (*coll.*) **to beat to a ~,** birini (oyunda) adamakıllı yenmek.
**freak** [frîk]. Acibei hilkat; garibe; çılgınlık. **~ish,** acayib, garibe nevinden.
**freckle** [ˈfrekl]. Çil(lendirmek). **~d,** çilli.
**free**[1] [frî] *v.* Serbest bırakmak; azad etm.; kurtarmak; tahliye etm. **~**[2] *a.* Serbest; hür; bağlı olmıyan; azade; ücretsiz; ihtiyarî: meşgul olmıyan; serbest ve kolayca hareket eden. **~ly,** ihtiyarî olarak; bolca; açıkça, serbestçe: **~ from [of],** ···den muaf, ···siz: **~ on board (f.o.b.),** vapurda teslim: **to break ~ from,** kendini ···den kurtarmak: **~ and easy,** teklifsiz: **as a ~ gift,** hediye olarak: **to have a ~ hand,** istediği gibi harekette serbest olm.: **to give s.o. a ~ hand,** tam serbestlik vermek: **to be ~ of s.o.'s house,** birinin evine serbestçe girip çıkmak: **~ imports,** gümrük resmine tâbi olmıyan ithalât: **~ love,** nikâhsız yaşama: **to make ~ with s.o.,** birisile laübali olm.: **to be ~ with one's money,** eli açık olm.: **~ on rail (f.o.r.),** trende teslim: **to set ~,** serbest bırakmak; azadetmek; kurtarmak: **~ trade,** serbest mübadele: **~ will,** iradei cüz'iye: **to do stg. of one's own ~ will,** bir şeyi kendi iradesile yapmak. **~-born,** hür doğmuş. **~-hand,** serbest el ile çizilen (kroki). **~-handed,** eli açık. **~-lance,** müstakil gazeteci. **~-thinker,** serbest fikirli. **~-thought,** serbest fikir. **~-wheel,** serbest tekerlek(li); bisiklet üzerinde pedal çevirmeden gitmek. **~booter** [ˈfrîbûtə(r)]. Korsan. **~dom** [ˈfrîdəm]. Hürriyet; serbestlik, azadelik; istiklâl; muafiyet; açıklık. **~ of speech,** söz hürriyeti: **~ of a city,** bir şehrin fahrî hemşeriliği. **~hold** [ˈfrîhǫuld], mülk (olan). **~holder,** mülk sahibi. **~man** [ˈfrîmən], hür adam (köle olmıyan); hemşeri.
**freez·e** (**froze, frozen**) [frîz, frǫuz, frǫuzn]. Don(dur)mak. **It is ~ing,** hava pek soğuk; donuyor: **to ~ the blood (in one's veins),** tüylerini ürpertmek: **to ~ on to stg.,** (*sl.*) bir şeyi kendine mal etm.: **to ~ out,** (*sl.*) istiskal ederek savmak. **~ing,** çok soğuk; donma; don. **~ing-point,** donma noktası.
**freight** [frêyt]. Navlun; yük. (Gemiyi) yükletmek; yük için kiralamak. **~age,** navlun (ücreti). **~er,** şilep. **~-car,** (*Amer.*) yük vagonu.
**French** [frenç]. Fransız(ca). **~ chalk,** terzi sabunu: **to take ~ leave,** izinsiz gitmek: **~ window,** pencereli kapı. **~-polish,** alkol ile cilâ. **~man** [frençmən]. Fransız.

**frenz·ied** [ˈfrenzid]. Çılgın; pek öfkeli. **~y**, çılgınlık.
**frequen·cy** [ˈfrîkwənsi]. Sık sık olma. **~t**[1] [ˈfrîkwənt] *a.* Mükerrer; sık sık olan. **~tly**, çok defa. **~t**[2] [friˈkwent] *v.* (Bir yere) sık sık gitmek. **~ted**, ayaküstü.
**fresco** [ˈfreskou]. Renkli duvar resmi.
**fresh** [freş]. Taze; taravetli; yaş; başka; (at) azgın; (su) tatlı; acemi; yeni gelen; biraz küstah. **to let some ~ air into a room**, odayı havalandırmak: **to get some ~ air**, bir az hava almak: **a ~ breeze**, serin rüzgâr: **as ~ as a daisy**, terütaze: **a youngster (just) ~ from school**, çiçeği burnunda bir lise mezunu: **~ butter**, tuzsuz tereyağı: **in the ~ of the morning**, sabahın serinliğinde. **~-coloured**, renkli, sıhhatli (yüz). **~en** [ˈfreşn]. Tazeleştirmek; taravet vermek; (rüzgâr) sertleşmek. **~man** [ˈfreşmən]. Üniversitede:- birinci sınıf talebesi. **~ness** [ˈfreşnis]. Tazelik; taravet; acemilik. **~water** [freşˈwôtə(r)]. Tatlı suya aid.
**fret**[1] [fret]. Köşeli yunan nakşı (ile süslemek). **~-saw**, oyma testeresi. **~wòrk** [ˈfretwö(r)k]. Kafes gibi oyma. **~**[2]. Kemirmek; kendini yemek. Endişe, merak. **to ~ and fume**, sabırsızlanıp öfkelenmek. **~-ful**, şikâyetçi; huysuz (çocuk).
**friable** [ˈfrâyəbl]. Gevrek.
**friar** [ˈfrâyə(r)]. Katolik keşişi. **~y**, manastır.
**friction** [ˈfrikşn]. Sürtünme; oğma; uyuşmamazlık.
**Friday** [ˈfrâydi]. Cuma. **Good Friday**, paskalyadan evvelki cuma.
**fried** *v.* **fry. ~ eggs**, sahanda yumurta.
**friend** [frend]. Dost, ahbab. **~less** [ˈfrendlis]. Ahbabsız; kimsesiz. **~ly** [ˈfrendli]. Kanısıcak; dostça; dosta yakışır. **~ society**, yardımlaşma cemiyeti: **~ wind**, müsaid rüzgâr. **~ship** [ˈfrendşip]. Dostluk, ahbablık.
**frigate** [ˈfrigit]. Firkateyn.
**fright** [frâyt]. Âni korku; (*sl.*) gülünç veya çirkin kimse; korkuluk. **to be in a ~**, korku içinde olm.: **to give s.o. a ~**, korkutmak: **to take ~**, ürkmek: **what a ~ she looks in that hat!**, o şapka ile ne gülünç görünüyor! **~en** [ˈfrâytn]. Korkutmak; ürkütmek. **to ~ away [off]**, korkutup kaçırmak: **to ~ s.o. into doing stg.**, birini korkutup bir şey yaptırmak: **to ~ s.o. out of his wits**, birinin ödünü koparmak. **~ened**, korkmuş: **to be ~ of [at]**, ···den korkmak: **to be ~ to death**, korkudan ödü patlamak. **~ful** [ˈfrâytfəl]. Korkunc; müdhiş, dehşetli: **I'm ~ly sorry**, (*coll.*) aman affedersiniz. **~fulness**, tedhiş, dehşet.
**frigid** [ˈfricid]. Soğuk. **~ity**, soğukluk.
**frill** [fril]. Kırmalı yakalık; (*coll.*) faydasız süs.
**fringe** [frinc]. Saçak (takmak).
**frippery** [ˈfripəri]. Değersiz süs.
**frisk** [frisk]. **~ (about)**, sıçrayarak oynamak. **~y**, oynak. **~iness**, şetaret.
**fritter**[1] [ˈfritə(r)] *n.* Dilim dilim kızartılmış elma vs. **~**[2] *v.* **~ away**, parça parça doğramak; azar azar israf etmek.
**frivol** [ˈfrivl]. Değersiz şeylerle vakit geçirmek. **~ous**, hafifmeşreb; havai. **~ity** [–ˈvoliti], havailik.
**frizz·le**[1] [ˈfrizl]. Saç kıvırmak. **~**[2]. Cızırdatarak kızartmak. **~y** [ˈfrizi]. Kıvırcık.
**fro** [frou]. **to and ~**, şuraya buraya.
**frock** [frok]. Kadın veya çocuk elbisesi. **~-coat**, redingot.
**frog**[1] [frog]. Kurbağa. **~-march**, dört kişi bir adamı yüzükoyun taşımak. **~-spawn**, kurbağa yumurtası. **~**[2]. Toka; çapraz.
**frolic** [ˈfrolik]. Gülüp oynama(k). **~some**, oynak.
**from** [from]. ···den. **~ my father**, babamdan: **~ ignorance**, cehaletten: **~ what I heard**, işittiğime göre: **~ childhood**, çocukluktan beri.
**front**[1] [frʌnt] *v.* Karşısında durmak; teveccüh etm. **~**[2] *n.* Ön; cebhe; saha; yüz. *a.* Öndeki; cebheye aid. **~ to ~**, yüz yüze: **in ~**, önde: **in ~ of**, önünde; karşısında: **to put a bold ~ on it**, cesur görünmek: **to come to the ~**, ön plâna gelmek: **a house on the ~**, denize karşı ev: **~ room**, sokak üstündeki oda: **the sea ~**, deniz kenarı. **~-door**, sokak kapısı; ön kapı. **~-view**, önden görünüş. **~-line soldiers [trenches]**, cebhe hattı askeri [siperleri]. **~age** [ˈfrʌntic]. (Bina vs.) cebhe, yüz. **~al** [ˈfrʌntl]. Alna aid. **~ attack**, cebheden taarruz. **~ispiece** [ˈfrʌntispîs]. Kitabın ilk sahifesine konan resim.
**frontier** [ˈfrʌntiə(r)]. Hudud; sınır. **~sman**, hudud üzerinde yaşıyan kimse.

**frost** [frost]. Ayaz, kırağı, don; (*sl.*) muvaffakiyetsizlik. Ayazlatmak; şekerle kaplamak; (cam) buzlu yapmak. **white ~**, kırağı: **black ~**, kırağısız şiddetli soğuk: **Jack Frost**, şahıslaştırılan kırağı. **~bite**, bir uzvun donması. **~bitten**, (uzuv) donmuş. **~y**, ayazlı: **a ~ reception**, soğuk bir kabul.

**froth** [froθ]. Köpük; manasız sözler. Köpürmek. **to ~ over**, köpürerek taşmak. **~y**, köpüklü.

**frown** [fraun]. Kaş çatma(k). **to ~ upon stg.**, hoş görmemek. **~ing**, asık suratlı.

**frowsty** [frausti]. Sıcak, havasız ve pis kokulu.

**frowzy** [frauzi]. Pasaklı; küf kokulu.

**frozen** [ˈfrouzn]. *v.* **freeze.** *a.* Donmuş. **~ assets**, nakte kolay çevrilemiyen mal.

**fructify** [ˈfrʌktifây]. Meyva vermek.

**frugal** [ˈfrûgl]. Tutumla, idareli; sade; bol olmıyan. **~ity** [–ˈgaliti], tutum; (yemek) sade ve az olma.

**fruit** [frût]. Meyva, yemiş; netice; kâr. **the ~s of the earth**, toprağın mahsulleri: **stone ~**, çekirdekli meyvalar. **~erer**, manav, yemişçi. **~ful** [ˈfrûtfəl]. Meyvalı; müsmir; verimli. **~ion** [fruˈişn]. Nail olma; mazhar olma. **to come to ~**, semere vermek. **~less** [ˈfrûtlis]. Neticesiz; beyhude. **~y** [ˈfrûti]. Meyva lezzetli.

**frump** [frʌmp]. Kılıksız eski zaman kadını. **~ish**, eski moda giyinmiş.

**frustrat·e** [frʌsˈtrêyt]. İşini bozmak; engel olm. **~ion**, menedilme; hüsran.

**fry**[1] [frây]. Tavada kızartma(k). **~**[2]. Yavru balık. **small ~**, ehemmiyetsiz kimseler. **~ing-pan** [ˈfrây·in(g)ˈpan]. Tava. ⌜**out of the ~ into the fire**⌝, ⌜yağmurdan kaçarken doluya tutulmak⌝.

**fuddle** [ˈfʌdl]. Zihnini bulandırmak. **~d**, çakırkeyf.

**fuel** [fyuəl]. Yakılacak şey; mahrukat; yakıt; benzin; tahrik edecek şey. Mahrukat almak, tedarik etmek. **to add ~ to the flames**, yangına körükle gitmek: **liquid ~**, akar yakıt. **~-pump**, benzin pompası.

**fug** [fʌg]. Sıcaktan ve havasızlıktan hâsıl olan kasvetli hava. **~gy** [ˈfʌgi]. Kasvetli olan.

**fugitive** [ˈfyûcitiv]. Kaçak; fanî; mülteci; kaçıcı.

**fulcrum** [ˈfʌlkrəm]. Manivelâ mesnedi.

**fulfil** [fulˈfil]. Yerine getirmek, icra etm.; tamamlamak; gidermek. **~ment**, yerine getirme, icra.

**full** [ful] *a.* Dolu, dolgun; tam; bütün; tok; bol. **of ~ age**, yirmi bir yaşına girmiş: **~ brother [sister]**, öz kardeş: **to have one's hands ~**, başında çok iş olm.: **he is very ~ of himself**, hep kendinden bahseder: **I waited two ~ hours**, tam iki saat bekledim: **in ~**, tamamen: **name in ~**, (ismin ilk harfleri değil) tam isim: **~ lips**, dolgun dudaklar: **it's a ~ three miles**, en aşağı üç mil uzaktır: **the ~ moon**, dolun ay: **everyone is ~ of the news**, bu havadisten başka bir şeyden bahsetmiyorlar: **~ session [meeting]**, umumî ictima: **~ stop**, nokta: **to come to a ~ stop**, tamamen durmak: **to the ~**; tamamile: **~ up**, dop dolu: **~ well**, epeyi. **~-back**, (futbol) müdafi. **~-blooded**, (i) öz; (ii) dinc. **~-blown**, tamamen açılmış (çiçek): **a ~ doctor**, tahsilini bitirmiş doktor. **~-dress**, büyük üniforma. **~-length**, tam boy. **~-page**, bütün sahifeyi kaplıyan. **~-sized**, tabiî büyüklükte. **~-time job**, (insanın) bütün vaktini alan iş. **~er** [ˈfulə(r)]. Çırpıcı: **~'s earth**, kil. **~ness** [ˈfulnis]. Doluluk; bolluk. **in the ~ of time**, zamanı gelince. **~y** [ˈfuli]. Tamamen; büsbütün. **~y-fashioned**, ayağa uyan bir şekilde örülmüş (çorab).

**fulminate** [ˈfʌlminêyt]. Şimşek gibi çakmak; (*fig.*) ateş püskürmek. Patlayıcı madde.

**fulsome** [ˈfulsəm]. Mide bulandıracak kadar.

**fumble** [ˈfʌmbl]. Beceriksizce yapmak. **to ~ for stg.**, el yordamıyla aramak: **to ~ for words**, kekelemek.

**fume** [fyûm]. Pis kokulu ve muzir duman; öfke. Tütsülemek; fena gaz çıkarmak; hiddetlenmek.

**fumigate** [ˈfyûmigêyt]. Tütsülemek; gaz ile dezenfekte etmek.

**fun** [fʌn]. Eğlence, zevk; neş'e; şaka. **in ~ [for ~]**, şaka olarak: **~ fair**, bayram yeri: **he is full of ~**, çok neş'eli ve tuhafdır: **it was great ~**, çok eğlenceli oldu: **like ~** (*sl.*) delicesine: **to make ~ of [to poke ~ at]**, ···ile alay etm.: **it's poor ~ to ...**, eğlenceli bir şey değil: **to do stg. for the ~ of the thing**, bir şeyi zevk için yapmak.

**function** [ˈfʌn(g)şn]. Asıl iş, vazife; maksad; resmî merasim; toplantı.

İşlemek; iş görmek. **~al,** iş görür; amelî: **~ disease,** vücud azasından birinin intizamsızlığı. **~ary** [ˈfʌn(g)şənəri]. Memur.

**fund** [fʌnd] *n.* Bir şeye tahsis edilen meblağ; sermaye; stok. **~s,** para, sermaye; sandık: **the Funds,** devlet eshamı: **to be in ~s,** hazır parası olm.: **he has a ~ of knowledge,** derin malûmatı var: **public ~s,** devlet parası: **to start a ~,** iane açmak: **sinking ~,** amortisman.

**fundament** [ˈfʌndəmənt]. Esas; temel. **~al** [–ˈmentl], esasî; başlıca. **~als,** esaslar.

**funer·al** [fyûnərəl]. Cenaze alayı. **at a ~ pace,** ağır ağır yürüyerek: **that's your ~!,** (*coll.*) sen bilirsin, keyfine!. **~eal** [fyuˈnîriəl]. Mateme aid; kasvetli.

**fung·us,** *pl.* **-uses, -i** [ˈfʌn(g)gəs, -gəsəs, -gây]. Mantar. **~ous,** mantara aid.

**funicular** [fyûˈnikyulə(r)]. Kablo ile işliyen (demiryolu).

**funk** [fʌnk]. (*coll.*) Korku; korkak. ···den korkmak. **to be in a (blue) ~,** çok korkmak.

**funnel** [ˈfʌnl]. Huni; vapur bacası.

**funny** [ˈfʌni]. Tuhaf; eğlenceli; güldürücü; gülünç; mizahî; garib. **he's a ~ fellow,** tuhaf [acayib] bir adamdır: **none of your ~ tricks!,** dalavere istemem!: **there is something ~ about this,** bu iş bir az tuhaf. **I feel rather ~,** (kendimde) bir tuhaflık hissediyorum: **funnily enough,** işin tuhafı. **~-bone,** dirsek kemiğinin hassas noktası.

**fur** [fö(r)]. Kürk; hayvan postu; kedinin tüyü; dil pası. (Dil) paslanmak. **~ and feather,** tavşanlar ve av kuşları: **to make the ~ fly,** şiddetli bir kavgaya sebeb olm. **~-lined,** içi kürkülü. **~-trade,** kürkçülük. **~red** [fö(r)d]. Kürklü; (dil) paslı; (kazan) kilsî tabaka ile kaplı. **~rier** [ˈfʌriə(r)]. Kürkçü. **~ry** [ˈfʌri]. Kürk kaplı (hayvan); kürk gibi.

**furbish** [ˈfö(r)biş]. **~ (up),** Silmek; parlatmak; tazelemek.

**furious** [ˈfyuəriəs]. Kızgın, azgın, kudurmuş. **to be ~ with s.o.,** birine fena halde kızmak.

**furl** [fö(r)l]. (Yelkeni) sarmak.

**furlong** [ˈfö(r)lon(g)]. Bir milin sekizde bir (220 yarda = 201 m.).

**furlough** [ˈfö(r)lou]. Sıla, izin. **to go on ~,** sılaya gitmek.

**furnace** [ˈfö(r)nis]. Ocak, fırın; cehennem gibi yer. **blast ~,** yüksek fırın.

**furnish** [ˈfö(r)niş]. Tedarik etm.; döşemek; donatmak; vermek. **~ed house,** döşeli ev: **to be let ~ed,** mobilyası ile kiralık. **~ings,** döşeme, mefruşat.

**furniture** [ˈfö(r)nityə(r)]. Mobilya. **~-remover,** ev nakleden müteahhid. **~-van,** döşeme nakliye kamyonu.

**furrow** [ˈfʌrou]. Sapan izi; tekerlek izi; gemi izi; yüz kırışığı. İz açmak; kırıştırmak.

**further** [ˈfö(r)ðə(r)]. *v.* **farther;** daha; fazla; ötedeki; daha çok; yeni; bundan başka. İlerletmek; kolaylaştırmak. **I did not pursue the matter ~,** bunun üzerinde fazla durmadım: **'awaiting your ~ orders',** (*comm.*) yeni siparişlerinizi bekliyerek: **I'll see you ~ first!,** cehennem ol!: **until ~ notice,** iş'arı ahire kadar: **without ~ ado,** hemen. **~ance** [ˈfö(r)ðərəns]. Muavenet, ilerletme. **for the ~ of [in ~ of],** ... kolaylaştırmak için.

**furthest** [ˈfö(r)ðist]. *v.* **farthest.**

**furtive** [ˈfö(r)tiv]. Sinsi; hırsızlama; el altından. **to cast ~ glances at,** göz ucuyla bakmak.

**fury** [ˈfyûri]. Kızgınlık; hiddet; kudurma. **in a ~,** gayet öfkeli: **to work like ~,** domuzuna çalışmak.

**fuse**[1] [fyûz] *n.* (Mermi) tapa; (*elect.*) sigorta; fitil. **~-box,** sigorta kovanı. **~-wire,** sigorta teli. **~**[2] *v.* Eri(t)mek; birleştirmek. **the light has ~d,** elektriğin sigortası yandı.

**fuselage** [ˈfyûzəlâj]. Uçak çatısı.

**fusilier** [fyûzəˈliə(r)]. (*esk.*) Tüfekli piyade.

**fusillade** [fyuzəˈlêyd]. Devamlı silâh ateşi. **a ~ of questions,** sual yağmuru.

**fusion** [ˈfyûjn]. Eri(t)me; birleşme.

**fuss** [fʌs]. Lüzumsuz gürültü, faaliyet; sebebsiz telâş. Lüzumsuz yere telaşlanmak. **to make [kick up] a ~,** mesele çıkarmak: **to make a ~ about stg.,** bir şeyi mesele yapmak: **to make a ~ of s.o.,** birinin üzerine çok düşmek. **to ~ about [around],** sağa sola titizlenmek: **to ~ over s.o.,** birinin üzerine titremek. **~y,** ince eleyip sık dokunan; fazla meraklı.

**fusty** [ˈfʌsti]. Küf kokulu; köhne.

**futil·e** [ˈfyûtâyl]. Beyhude; vâhi. **~ity** [–ˈtiliti], beyhudelik.

**futur·e** [ˈfyûtyə(r)]. İstikbal, gelecek zaman. Müstakbel, gelecek. **~ism** [ˈfyûtyərizm]. San'at vs.dea nanevî usullere aykırı bir yol tutan bir hareket. ~**ity** [fyuˈtyûriti]. İstikbal.

**fuzz** [fʌz]. Hafif kıl. ~**y**, kıvırcık; bulanık; hayal meyal.

# G

**G** [cî]. G harfi; (*mus.*) sol notası.

**gab** [gab]. Palavra; ağız. **to have the gift of the** ~, çenebaz olmak.

**gabble** [ˈgabl]. (*ech.*) Kaz gibi ses çıkarma(k); çabuk ve anlaşılmıyacak şekilde konuşma(k).

**gable** [ˈgêybl]. ~ (**-end**), damın müselles şeklinde olan yanı.

**gad** [gad]. **to** ~ **about**, eğlence peşinde gezmek. ~**about** [ˈgadəbaut]. Avare; serseri.

**Gad. By** ~! Allah hakkı için!

**gadget** [ˈgacit]. Hünerli küçük bir alet.

**Gael** [gêyl]. İskoçyalı Kelt. ~**ic**, iskoç dili.

**gaff** [gaf]. Balıkçı kancası (ile tutmak). **to blow the** ~, (*sl.*) sırrı meydana çıkarmak.

**gaffer** [ˈgafə(r)]. İhtiyar köylü; amele başı.

**gag** [gag]. Bağırmaktan men için ağıza tıkılan şey; (parlamentoda) müzakereye nihayet verme; tuluat kabilinden söz, hareket. Susturmak.

**gage** [gêyc]. Rehin, teminat. Rehin olarak vermek. **to throw down the** ~ **to s.o.**, birine meydan okumak.

**gaiety** [ˈgêyəti]. Şenlik; neş'e; eğlenti.

**gaily** *v.* **gay.**

**gain** [gêyn]. Kazanc, kâr; artırma. Kazanmak; kâr etm.; ···e varmak. **to** ~ **ground**, ilerlemek: **to** ~ **on a competitor**, rakibine yaklaşmak: **a bad habit** ~**s on one**, fena bir itiyad gittikçe kökleşir: **to** ~ **on one's pursuers**, kendisini takib edenlerden uzaklaşmak: **to** ~ **s.o. over**, birini ikna ederek kazanmak: **I have** ~**ed five pounds this month**, (i) bu ay beş lira kazandım; (ii) bu ay 2 kilo aldım: **to** ~ **strength**, kuvvetlenmek: **to** ~ **the upper hand**, üstün gelmek: **this watch** ~**s ten minutes a day**, bu saat her gün on dakika ileri gider: **to** ~ **weight**, kilo almak; ehemmiyet kazanmak. ~**ful** [ˈgêynfl]. Kârlı.

**gainsay** (**gainsaid**) [gêynˈsêy, –ˈsed]. İnkâr etmek.

**gait** [gêyt]. Yürüyüş.

**gaiter** [ˈgêyte(r)]. Getr; tozluk.

**gala** [ˈgâlə]. Büyük şenlik.

**galaxy** [ˈgalaksi]. Kehkeşan; seçkin zevattan mürekkeb toplantı.

**gale** [gêyl]. Kuvvetli rüzgâr.

**gall**[1] [gôl]. Safra; çok acı şey; kin. **pen dipped in** ~, zehirli kalem: **it was** ~ **and wormwood to him**, bu ona çok acı geldi. ~**-bladder**, safra kesesi. ~**-stone**, safra kesesinde taş. ~[2]. Ağaclarda böceklerden hasıl olan şiş. ~[3]. Yağır (yapmak); sürünerek yara etm.; incitmek. ~**ing** [ˈgôlin(g)]. İncitici.

**gallant** [galənt]. Yiğit; muhteşem. [gəˈlant] Kadınlara karşı fazla nazik. ~**ry** [ˈgaləntri], yiğitlik.

**gallery** [ˈgaləri]. Dehliz; üstü kapalı balkon; (tiyatro) en üst ve ucuz mevki; yeraltı yolu. **to play to the** ~, şöhret kazanmak için avama hoş görünmek.

**galley** [ˈgali]. Kadırga; büyük sandal; gemi mutfağı. ~**-proof**, tashih provası. ~**-slave**, (*esk.*) kürek mahkûmu.

**Gallic** [ˈgalik]. Eski Galya'ya aid; fransız. ~**ism** [ˈgalisizm], fransızcaya mahsus tabir.

**gallivant** [galiˈvant]. Eğlence peşinde koşmak.

**gallon** [ˈgalən], Galon = 4·54 litre.

**gallop** [ˈgaləp]. Dörtnala koşma(k); pek acele yapmak. ~**ing consumption**, çabuk ilerliyen verem.

**gallows** [ˈgalouz]. Darağacı; idam sehpası. ~**-bird**, ipten kazıktan kurtulmuş.

**galore** [gəˈlô(r)]. Bol; çok.

**galvanize** [ˈgalvənâyz]. Galvanize etm.; canlandırmak.

**gambit** [ˈgambit]. (Satranc ve *fig.*) daha iyi bir vaziyet kazanmak için kaybedilen el.

**gambl·e** [ˈgambl]. Kumar; baht işi. Kumar oynamak. **to** ~ **away stg.**, bir şeyi kumarda kaybetmek. ~**er**, kumarbaz: ~ **on the Stock Exchange**, acyocu. ~**ing**, kumar: **-den**, kumarhane.

**gambol** [ˈgambəl]. Zıplama(k).

**game**[1] [gêym]. Oyun; eğlence; hile; alay; (*yln. müfred*) av hayvanı; av eti. Kumar oynamak. **do you like** ~**s?**, oyunları sever misiniz?: **it's all in the** ~, bir işin hem iyi hem fena tarafına razı olmalı: **big** ~, arslan gibi büyük av: **fair** ~, avlanması caiz olan av: **a politician is fair** ~

for everyone, politikacılar herkes için meşru bir hedeftir: **to fly at higher ~**, gözü daha yüksekte olmak: **to have the ~ in one's hands**, vaziyete hâkim olacağından emin olmak: **to have a ~ with s.o.**, birile (birisine) oyun oynamak: **what's his little ~?**, ne dolab çeviriyor acaba?: **now, none of your little ~s!**, bana oyun oynıyamazsın!: **to beat s.o. at his own ~**, düşmanı kendi oyunile yenmek: **to make ~ of**, alaya almak: **to play the ~**, dürüst hareket etm.: **to play s.o.'s ~**, birinin ekmeğine yağ sürmek: **a ~s shop**, spor levazımatı mağazası: **to spoil s.o.'s ~**, birinin plânını bozmak: **two can play at that ~!**, bu oyunu başkaları da bilir!: **the ~ is up!**, hapı yuttuk. **~-bag**, av çantası. **~-cock**, dövüştüren horoz. **~keeper** [ˈgêymkîpə(r)]. Av bekçisi. **~-licence**, av ruhsatiyesi. **~-preserve**, hususî av korusu. **~²** *a.* Cesaretli. **he is ~ for anything**, hiç bir şeyden yüksünmez: **to die ~**, sonuna kadar sebat göstermak. **~³** *a.* Sakat. **he has a ~ leg**, bacaklarının biri sakat.

**gamp** [gamp]. İhtiyar hastabakıcı; battal şemsiye.

**gamut** [ˈgamət]. En pesten en tize kadar bütün ses perdeleri; tam vüs'at.

**gander** [ˈgandə(r)]. Erkek kaz. ⌜**what's sauce for the goose is sauce for the ~**⌝, ⌜seninki can da benimki can değil mi?⌝.

**gang¹** [gan(g)]. Avene; çete; ekip; sürü. **the whole ~**, bütün güruh. **~²** *v.* (İskoç). Gitmek. **~-plank** (vapurdan yolcu çıkarmak için) iskele. **~er** [ˈgan(g)gə(r)]. Amele ekipinin başı **~ster** [ˈgan(g)stə(r)]. Haydud; Alikıran baş kesen. **~way** [ˈgan(g)wêy]. Geçid; sıralar arasındaki aralık; gemiden rıhtıma geçen köprü. **~!**, destur!

**gangrene** [ˈgan(g)grîn]. Yaranın çürümesi.

**gaol** [cêyl]. Hapishane. Hapsetmek. **to break ~**, hapishaneden kaçmak. **~-bird**, hapishane kaçkını; sabıkalı. **~er**, zindancı.

**gap** [gap]. Aralık; gedik; açık yer; eksiklik. **to fill [stop] a ~**, gedik kapamak; bir eksiği tamamlamak.

**gap·e** [gêyp]. Hayretten ağzı açık kalmak; esneme(k); ağız gibi açmak. **the ~s**, mütemadi esneme. **~ing** [ˈgêypin(g)] *a.* Ağzı açık; dev ağzı gibi açık (uçurum); geniş ve açık (yara).

**garage** [ˈgarâj]. Garaj. Garaja koymak.

**garb** [gâb]. Kılık, kıyafet. Giydirmek.

**garbage** [ˈgâbic]. Süprüntü, pislik.

**garble** [ˈgâbl]. Tahrif etmek.

**garden** [ˈgâdn]. Bahçe (-ye mahsus). Bahçivanlık etmek. **common or ~**, alelâde: **kitchen ~**, sebze bahçesi: **winter ~**, kış bahçesi; büyük limonluk. **~er** [ˈgâdnə(r)], bahçivan.

**gargantuan** [gâˈgantyuən]. Dev gibi.

**gargle** [ˈgâgl]. Gargara (etmek).

**gargoyle** [ˈgâgôyl]. İnsan, hayvan, başına benziyen oluk ağzı.

**garish** [ˈgeəriş]. Çiğ parlak.

**garland** [ˈgâlənd]. Çiçekten çelenk (ile süslemek).

**garlic** [ˈgâlik]. Sarmısak.

**garment** [ˈgâmənt]. Elbise parçası.

**garner** [ˈgânə(r)]. Zahire ambarı. Biriktirmek.

**garnet** [ˈgânit]. Lâltaşı.

**garni·sh** [ˈgâniş]. (Yemek) süslemek. **~ture** [ˈgânitye(r)]. Süs, ziynet.

**garret** [ˈgarit]. Tavanarası.

**garrison** [ˈgarisn]. Garnizon. Bir kaleye asker yerleştirmek. **~ town**, askerî birliklerin daimî olarak bulunduğu şehir.

**garrotte** [gaˈrot]. Boğarak öldürmek.

**garrul·ity** [gaˈrûliti]. Çenebazlık. **~ous** [ˈgaryuləs], geveze.

**garter** [ˈgâtə(r)]. Çorab bağı, dizbağı. **the Garter**, Dizbağı nişanı.

**gas** [gas]. Gaz; havagazi; (*sl.*) boş lâkırdi; (*Amer. coll.*) otomobilin benzini. Gazlemek. **coal ~**, havagazi: **laughing ~**, nitrojenli oksid gazi: **to have ~**, bayıltılmak: **to step on the ~**, (*Amer.*) otomobili hızlandırmak. **~-bag**, gaz zarfı; geveze. **~-burner**, gaz ibiği. **~-fire**, gaz ocağı. **~-fittings**, havagazi aletleri. **~-main**, ana gaz borusu. **~-man**, havagazi memuru. **~-proof**, gazden masun. **~-ring**, tek ateşli havagazi ocağı. **~-works**, gazhane. **~eous** [ˈgêysiəs]. Gaz halinde. **~ometer** [gaˈsomətə(r)]. Havagazi deposu. **~sy** [ˈgasi]. Gazli; gaz gibi.

**gash** [gaş]. Uzunca ve derince yara yapmak. Bıçak vs. yarası.

**gasket** [ˈgaskit]. Conta.

**gasp** [gâsp]. Hayret, acıdan nefesini tutma(k); zorlukla solumak. **to ~**

for breath, nefesi kesilmek: **to give a ~**, korku, hayretten nefesi kesilmek: **to be at one's last ~**, ölüm halinde olm.: **to fight to the last ~**, son nefesine kadar dövüşmek.

**gastr·ic** [ˈgastrik]. Mideye aid. **~itis** [–ˈtrâytis], mide iltihabı.

**gastrono·me** [ˈgastrənom]. Yemek meraklısı. **~my**, şikemperverlik; yemek ihtisası.

**gate** [gêyt]. Kapalı olmıyan yerler arasında kapı; mânia; giriş; bir maça para ile giren seyirciler; bunlardan alınan para mikdarı. Bir talebeyi izinsiz bırakmak. **~-crasher**, (*sl.*) biletsiz davetsiz giren kimse. **~-house**, kapı bekçisinin evi. **~-keeper**, dış kapı bekçisi. **~-legged** (table), kanadlı (masa). **~-money**, maç, duhuliye parasının mecmuu. **~way** [ˈgêytwêy]. Giriş yeri; kapı.

**gather** [ˈgaðə(r)]. Toplamak, biriktirmek; birleşmek; artırmak; anlamak; hükmetmek; cerahat bağlamak; (alın) buruşturmak; (elbise) pli yapmak. **to ~ oneself together (for a spring)**, (sıçramak için) gerilmek: **to ~ round**, etrafına toplanmak: **to ~ speed**, gittikçe hızlanmak: **to ~ strength**, (hasta) kuvvetlenmek: **as will be ~ed from the enclosed letter**, ilişik mektubdan anlaşılacağı üzere: **a storm is ~ing**, bulutlar toplanıyor: **in the ~ing darkness**, gittikçe basan karanlıkta: **to be ~ed to one's fathers**, ölmek. **~ed** [ˈgaðəd] *a.* Buruşuk, çatık. **to have a ~ finger**, parmağı iltihablanmak. **~ing** [ˈgaðərin(g)]. Toplantı; ictima; iltihab.

**gauche** [gouş]. Savruk; patavatsız.

**gaud·y** [ˈgôdi]. Çiğ renkli; zevksizce süslenmiş.

**gauge** [gêyc]. Ölçü; mikyas; çap; demiryol rayları arasındaki açıklık; mehengir. Ölçmek; tahmin etmek.

**Gaul** [gôl]. Eski Galya(lı).

**gaunt** [gônt]. Zayıf ve çökük yanaklı; (dağ) yalçın, korku veren.

**gauntlet** [ˈgôntlit]. Zırhlı eldiven; kolçak. **to throw down the ~**, meydan okumak: **to take up the ~**, meydan okuyanın davetini kabul etm.: **to run the ~**, her taraftan gelen hücumlara maruz olmak.

**gauz·e** [gôz]. Bürümcük, gaz. **wire ~**, eleklik tel örgüsü kumaş. **~y**, bürümcük gibi.

**gave** *v.* give.

**gawk** [gôk]. Uzun boylu hantal kimse.

**gay** [gêy]. Şen; zevk düşkünü; parlak renkli. **a ~ dog**, çapkın adam: **to lead a ~ life**, zevk ve eğlence içinde yaşamak: **to talk gaily about stg.**, bir şeyden dem vurmak: **a ~ woman**, hafifmeşreb kadın.

**gaze** [gêyz]. Gözünü dikerek bakmak. Devamlı bakış. **to ~ at [on, upon]**, ···e dikkatle uzun uzun bakmak: **a dreadful sight met his ~**, korkunc bir manzara gözüne ilişti.

**gazette** [gəˈzet]. Resmî gazete (ile ilân etmek). **~er** [gazəˈtiə(r)], coğrafya lûgatı.

**gear** [giə(r)]. Şahsî eşya; takım; levazımat; cihaz; çark tertibatı; (*otom.*) vites. Çark dişleri birbirine geçmek. **to ~ up**, vites artırmak: **to ~ down**, vites azaltmak: **in ~**, dişler birbirine geçmiş: **out of ~**, dişler çıkmış: **to throw out of ~**, çark dişlerini birbirinden çıkarmak; bozmak. **~-box**, vites kutusu. **~-lever**, vites kolu. **~-wheel**, dişli çark. **~ing** [ˈgiərin(g)]. Dişli çark tertibatı.

**gee** [cî]. **~ up!**, deh! **~-~**, (çocuk lisanında) at.

**geese.** goose'*un cemi.*

**geld** [geld]. İğdiş etmek.

**gem** [cem]. Kıymetli taş; mücevher; seçme ve kıymetli şey.

**gendarme** [jonˈdâm]. Jandarma.

**gender** [ˈcendə(r)]. Cins; ismin cinsi.

**genealog·y** [ˌcîniˈaloci]. Silsile; soy; şecere. **~ist**, şecereci. **~ical** [–ˈlocikl], şecereye aid: **~ tree**, şecere.

**general** [ˈcenrəl]. Umumî; ekseriyetle olan. General. **in ~**, umumiyetle: **as a ~ thing**, umumî olarak: **~ post**, memuriyetler arasında umumî değişiklik: **the ~ reader**, okuyucu kütlesi. **~ly**, ekseriya. **~issimo** [ˌcenərəˈlisimou]. Baş kumandan. **~ity** [ˌcenəˈraliti]. Umumilik; ekseriyet. **~ize** [ˈcenrəlâyz], tamim etmek. **~ization** [–ˈzêyşn], tamim. **~ship** [ˈcenrlşip]. Generallık; kumanda kabiliyeti.

**generat·e** [ˈcenərêyt]. Tevlid **etm.**; husule getirmek; hasıl etmek. **~ion** [–ˈrêyşn], tenasül; hasıl etme [edilme]; nesil: **the rising ~**, yeni nesil. **~ive** [ˈcenərətiv], tenasülî; hasıl edici. **~or** [ˈcenərêytə(r)], müvellid; hasıl edici cihaz.

**generic** [cəˈnerik]. Nev'e aid; **umumî.**

**gener·osity** [cenəˈrositi]. Cömerdlik; bolluk. **~ous** [ˈcenrəs], cömerd; bol; alicenab.

**genesis** [ˈcenisis]. Tevellüd; tekvin; başlangıc.

**genial** [ˈcînyəl]. Hoş; güler yüzlü; mülayim; müsaid (iklim). **~ity** [–ˈaliti], tatlılık.

**genie** [ˈcîni]. Cin.

**genital** [ˈcenitl]. Tenasüle aid. **the ~s** [**the ~ organs**], tenasül âleti.

**genitive** [ˈcenitiv]. Muzafünileyh.

**genius**¹, *pl.* **-es** [ˈcînyəs, -iz]. Deha; hususiyet. **to have a ~ for doing stg.**, bir şeyde hususî bir kabiliyet göstermek. **~²**, *pl.* **genii** [ˈcînyəs, –yây]. Cin, ruh. **s.o.'s evil ~**, birinin habis ruhu.

**genocide** [ˈcenosâyd]. Bir milleti öldürme.

**gent** [cent]. (*abb.*) = **gentleman.** (*Aşağı tabaka*) Kişi, şahıs; (*ticarette*) erkek: **gents' underclothing**, erkek çamaşırı. **~eel** [cenˈtîl]. Kibar; sahte kibar. **~ile** [ˈcentâyl]. Yahudi olmıyan. **~ility** [cenˈtiliti]. Kibarlık. **shabby ~**, düşkün kibarın hali. **~le** [ˈcentl]. Yavaş; nazik; hafif. **the ~ art**, olta ile balık avı: **of ~ birth**, kibar: **~ exercise**, hafif idman: **~ reader!**, aziz okuyucu!: **the ~(r) sex**, cinsi lâtif: **gently does it!**, yavaş! **~lefolk(s)** [ˈcentlfouk(s)]. Kibar terbiyeli kimseler. **~leman** [ˈcentlmən]. Kibar sınıfından kimse; centilmen; efendi. **~ in waiting**, kıralın hususî hizmetinde bulunan asilzade: **he is not a ~**, o kibar sınıfından değildir: **he's no ~**, o adam değildir: **~'s agreement**, kontratsız anlaşma. **~lemanly** [ˈcentlmənli]. Bir centilmene yakışır surette. **~leness** [ˈcentlnis]. Yumuşaklık, naziklik. **~lewoman** [ˈcentlwumən]. Kibar sınıfından kadın. **~ly** [ˈcentli] *v.* **gentle.** **~ry** [ˈcentri]. Asillerden sonra gelen tabaka.

**genuine** [ˈcenyuin]. Hakikî; sahih; samimî; su katılmadık.

**genus**, *pl.* **genera** [ˈcînəs, cenəra]. Cins.

**geograph·y** [ciˈogrəfi]. Coğrafya. **~er**, coğrafyacı.

**geolog·y** [ciˈoləci]. Jeoloji. **~ist** [–ˈoləcist], jeoloji mütehassısı.

**geometr·y** [ciˈometri]. Hendese. **~ician** [–ˈtrişn], hendeseci.

**Georgia** [ˈcôcyə]. Gürcistan; Birleşik Amerika'nın Georgia eyaleti. **~n**, Gürcü; gürcüce; I–IV üncü George devrine aid.

**germ** [cö(r)m]. Tohum; mikrob; esas. **~-carrier**, mikrob taşıyan. **~-killer**, mikrob öldürücü ilâc. **~icide** [ˈcö(r)misâyd]. Mikrob öldürücü şey. **~inate** [ˈcö(r)minêyt]. Filizlenmek; çimlen(dir)mek; neşvünema bulmak.

**german**¹ [ˈcö(r)mən]. Ebeveyni aynı olan. **cousin ~**, amca vs. çocukları. **~²**. Alman; almanca. **~y**, Almanya.

**germane** [cö(r)mêyn]. Aid, müteallık.

**gerrymander** [ˈcerimandə(r)]. Seçimlerde hile yapmak.

**gerund** [cerənd]. İngilizcede, '-ing' ile nihayetlenen, fiilden yapılma isim.

**gestation** [cesˈtêyşn]. Gebelik müddeti.

**ges·ticulate** [cesˈtikyulêyt]. Söz söylerken çok el hareketleri yapmak. **~ure** [ˈcestyə(r)]. Eli şareti; hareket. Maksadını âza hareketlerile ifade etmek

**get** (**got, got(ten)**) [get, got(n)]. Almak, elde etm.; olmak; bulmak; tutmak, yakalamak; vurmak; alıp getirmek; kazanmak; varmak; malik olm.; (**get**'*ten sonra bir ismi meful gelirse umumiyetle müteaddi ifade eder*: **to ~ a house built**, bir ev yaptırmak: **to ~ one's hair cut**, saçını kestirmek). **we are not ~ting anywhere** [**we are ~ting nowhere**], bundan bir netice çıkmaz: **to ~ breakfast**, kahvaltı etmek: **to ~ the breakfast**, kahvaltıyı hazırlamak: **to ~ one's arm broken**, kolu kırılmak: **what's that got to do with it?**, bunun onunla ne münasebeti var?: **~ going!**, haydi!: **to ~ a tree to grow**, bir ağacı yetiştirmeğe muvaffak olm.: **I have got to go to London**, Londra'ya gitmeliyim: **to become a diplomat you have got to learn French**, diplomat olabilmek için fransızca öğrenmeğe mecbursunuz: **to ~ s.o. home**, birini evine götürmek: **the play didn't really ~ me**, (*coll.*) piyes beni sarmadı: **I don't ~ you** [**your meaning**], anlamıyorum: **to ~ s.o. into a place**, birini bir yere kayırmak: **where has that book got to?**, o kitab nereye gitti?: **where did you ~ to know that?**, nasıl oldu da bunu öğrendiniz?: **I got to know him during the war**, kendisini tesadüfen harb esnasında tanıdım: **later I got to know him better**, sonraları onu iyice [daha iyi] tanıdım:

**what's got you?**, (*coll.*) sana ne oldu?. ~ **about**, dolaşmak; yayılmak: (**invalid**) **to be able to ~ about again**, (hasta) yataktan kalkıp dolaşabilmek. ~ **across**, bir taraftan öbür tarafa geç(ir)mek; asmak; (*coll.*) (piyes) muvaffak olmak. ~ **along**, ilerlemek; geçinmek: **to ~ along with s.o.**, birisile geçinmek: **to ~ along without stg.**, bir şeysiz de olabilmek: ~ **along with you!**, (i) haydi git!; (ii) amma yaptın ha!. ~ **at**, ermek; kavramak; yetişmek: **difficult to ~ at**, gitmesi güç: **what are you ~ting at?**, maksadınız nedir?: **if I can ~ at him he'll be sorry**, bir elime geçerse hali yamandır: **to ~ at a witness**, bir şahidi ayartmak: **he's been ~ting at you**, size dil dökmüş. ~ **away**, kaçıp kurtulmak; ayrılmak; koparmak; kapıp götürmek: ~ **away with you!**, haydi canım!: **there's no ~ting away from it**, bunu kabul etmeliyiz: **he'll never ~ away with that**, kimse yutturamaz. ~ **back**, evine dönmek; geri almak; telâfi etm.: **to ~ one's own back**, acısını çıkarmak: **to ~ stg. back into its box**, bir şeyi tekrar kutusuna koymak. ~ **by**, geçmek. ~ **down**, yere inmek; indirmek; yazmak; yutmak: (**to a dog**) ~ **down!**, in aşağı!: **to ~ down to one's work** [**to ~ down to it**], işe iyice girişmek: **to ~ down to facts**, vakıalara gelmek. ~ **in**, girmek; vasıl olm.; içeri almak; toplamak; sokmak; ekmek: **to ~ in with s.o.**, birinin gözüne girmek: **to ~ in a supply of coal**, *etc.*, kömür vs. alıp depo etm.: **to ~ in the harvest**, mahsulü toplamak: **to ~ in s.o. to see to the gas**, *etc.*, birini çağırıp havagazini vs. göstermek: **to ~ in for a constituency**, mebus seçilmek: **to ~ a blow in**, bir darbe indirmek: **I couldn't ~ a word in**, ağzımı açıp bir kelime söyliyemedim: **to ~ one's hand in**, elini alıştırmak. ~ **into**, girmek; sokmak; giyinmek: **to ~ into a club**, bir klübe girebilmek: **to ~ into bad habits**, fena itiyadlara alışmak: **to ~ into the way of doing stg.**, bir şeyi âdet edinmek: **to ~ s.o. into the way of doing stg.**, birini bir şeye alıştırmak: **to ~ into a temper**, hiddetlenmek: **to ~ stg. into one's head**, bir fikir edinmek; hatırlamak. ~ **off**, bir şeyden inmek, ayrılmak; yola çıkmak; kurtulmak; çıkarmak, soyunmak; kurtarmak, beraet ettirmek: **to ~ off a duty**, bir işten muaf olm.: **to ~ off a stranded ship**, karaya oturmuş bir gemi (i) -den çıkmak; (ii) -yi yüzdürmek: (**girl**) **to ~ off with a man**, (kız) birisile evlenmeğe muvaffak olm.: **to ~ stg. off one's hands**, bir şeyi başından atmak; bir şeyden kurtulmak: **to ~ one's daughter off one's hands**, kızını evlendirmek. ~ **on**, binmek; giyinmek; terakki etm., muvaffak olm.; yaklaşmak; birbirile geçinmek: **how are you ~ting on?**, nasılsınız?; işleriniz [sihhatiniz] nasıl?: **to ~ on in life**, muvaffak olm.: **to be ~ting on for fifty**, ellisine merdiven dayamak: **it is ~ting on for ten**, saat ona yaklaşıyor: **I can't ~ these shoes on**, bu ayakkabları giyemiyorum (dar geliyor): ~ **on with you!**, haydi canım!: **to ~ on with s.o.**, birisile geçinmek: **to ~ on without s.o.** [**stg.**], bir kimsesiz [şeysiz] yapabilmek: **to ~ on with the job** [**to ~ on with it**], bir işe devam etm.: **how did you ~ on with your exam.?**, imtihanınız nasıl geçti? ~ **out**, kurtulmak; çık(ar)mak; çözmek, halletmek; kazanmak; sız(dır)mak; tertib etm.: ~ **out!**, defol!: ~ **out with you!**, haydi canım!: **to ~ out without loss**, zarar etmeden bir işin içinden çıkmak: **to ~ out of doing stg.**, bir işten kurtulmak: **to ~ out of the habit of doing stg.**, bir itiyaddan kurtulmak: **to ~ out of the way of doing stg.**, (iyi) bir alışkanlığı kaybetmek: ~ **out of my** [**the**] **way!**, yolumdan (önümden) çekil: **I shall ~ nothing out of it**, bundan benim elime bir şey geçmiyecek: **to ~ out a scheme**, bir plân hazırlamak. ~ **over**, aşmak, üzerinden geç(ir)mek; atlamak: **I can't ~ over it**, (i) geçemem; (ii) hazmedemiyorum; (iii) hâlâ şaşıyorum: **he can't ~ over his loss**, kayıbını unutamıyor: **I shall be glad when I ~ it over**, bu işi bitirsem de kurtulsam: **to ~ over one's shyness**, çekingenlikten kurtulmak. ~ **round**, gidivermek; dolaşıp geçmek; yayılmak: **as you ~ round the corner**, köşeyi dönünce: **to ~ round s.o.**, birini dil dökerek kandırmak: **to ~ round a difficulty**, bir müşkülü yolunu bulup halletmek: **to ~ round the law**, hilei şer'iyesini bulmak: **I'll ~ round this evening if I can**, imkân olursa bu akşam giderim. ~ **through**, geçmek; bitirmek; içinden geçirmek;

yetişmek: to ~ **through to s.o.,** (telefon) birisile irtibat temin etm.: to ~ **a bill through Parliament,** bir kanunu meclisten geçirmek. ~ **together,** biriktirmek; topla(n)mak. ~ **under,** altına girmek; altından geçmek; hakkından gelmek: to ~ **a fire under,** yangın söndürmek. ~ **up,** (ayağa) kalkmak; (rüzgâr) artmak; yükselmek; yukarısına çıkmak; ayağa kaldırmak; bindirmek: **to** ~ **oneself up,** süslenmek: **to** ~ **oneself up as ...,** kendine ... süsü vermek: **to** ~ **up a hill,** bir tepeye çıkmak: **to** ~ **up to mischief,** yaramazlık yapmak: **to** ~ **up a play,** bir piyes tertib etm.: **to** ~ **up to s.o.,** birine yetişmek: **got up (woman),** fazla makiyajlı (kadın).

**gewgaw** [ˈgyûgô]. Cicili bicili şey.

**geyser** [ˈgêyzə(r), gîzə(r)]. Sıcak su fışkırtan pınar; banyo için havagazile işliyen ısıtma cihazı.

**ghastly** [ˈgâstli]. Korkunc, müdhiş: **a** ~ **light,** meş'um aydınlık: **a** ~ **smile,** zoraki sırıtma.

**gherkin** [ˈgö(r)kin]. Turşuluk hıyar.

**ghost** [go̤ust]. Hayalet, hortlak. **the Holy Ghost,** Ruhülkudüs: **to raise a** ~, ruh çağırmak: **to lay a** ~, bir cin, ruh, koğmak: **to be the mere** ~ **of one's former self,** iğne ipliğe dönmek: **not to have the** ~ **of a chance,** en küçük bir ümidi olmamak: **to give up the** ~, ruhunu teslim etm.: **I haven't the** ~ **of an idea,** zerre kadar haberim yok. ~**like,** hayalet gibi. ~**ly,** hayalet gibi; ruhanî.

**ghoul** [ga̤ul]. Leş yediği sanılan gulyabani; zebani; hortlak; iğrenc şeyleri seven kimse. ~**ish,** hortlak gibi; iğrenc.

**giant** [ˈcâyənt]. Dev; iriyarı. ~**ess,** dişi dev.

**gibber** [ˈcibə(r)]. Maymun gibi sesler çıkarma(k); anlaşılmaz tarzda konuşmak. **a** ~**ing idiot,** ebleh. ~**ish,** abuksabuk sözler.

**gibbet** [ˈcibit]. Darağacı.

**gibe** [câyb]. İstihzalı alay, sözle yaralamak.

**gidd·y** [ˈgidi]. Başı dönmüş; başdöndürücü; hoppa. **to play the** ~ **goat,** maskaralık etmek. ~**iness,** başdönmesi; hoppalık.

**gift** [gift]. Hediye, armağan; hüner. **to have a** ~ **for languages,** lisana istidadı olm.: **I would not have it at a** ~, bedava verseler almam: ⌜**one should not look a** ~ **horse in the mouth**⌝, hediye atın dişine bakılmaz. ~**ed,** hünerli, istidadlı.

**gigantic** [câyˈgantik]. Devasa, kocaman.

**giggle** [ˈgigl]. (*ech.*) Kıkır kıkır (şımarık) gülme(k).

**gild** [gild]. Yaldızlamak. **to** ⌜~ **the lily**⌝, mükemmel bir şeyi lüzumsuz yere süslemek.

**gills** [gilz]. Galsame; mantarın altındaki safihalar. **to look rosy** [**green**] **about the** ~, sıhhatli [keyifsiz] görünmek.

**gilt** [gilt]. Yaldız(lı). ~**-edged,** kenarı yaldızlı (kitab): ~ **securities,** itimada şayan tahvilat.

**gimcrack** [ˈcimkrak]. Mezad malı; derme çatma.

**gimlet** [ˈgimlit]. Burgu. ~**-eyed,** keskin gözlü.

**gin** [cin]. Ardıc suyu, cin. ~ ². Tuzak; çırçır. Pamuğu çırçır ile tohumdan ayırmak.

**ginger** [cincə(r)]. Zencefil; gayret. Kızıl (saç). **to** ~ **up,** gayret vermek, canlandırmak. ~**-beer,** zencefil şurubu. ~**bread** [ˈcincəbred]. Zencefilli kurabiye: ⌜**to take the gilt off the** ~⌝, bir şeyin en cazibeli tarafını çıkarmak. ~**ly** [ˈcincəlı]. İhtiyatla; çekinerek. **to tread** [**walk**] ~, pek dikkatli yürümek.

**gipsy** [ˈcipsi]. Kıptı, çingene.

**giraffe** [ˈciˈrâf]. Zürafe.

**gird** [gö(r)d]. Sarmak, kuşatmak. **to** ~ **oneself for the fray,** mücadeleye hazırlanmak.

**girder** [ˈgö(r)də(r)]. Taban; kiriş.

**girdle** [ˈgö(r)dl]. Kuşak (bağlamak); kuşatmak.

**girl** [gö(r)l]. Kız. **one's best** ~ [~ **friend**], sevgili. ~**hood,** kızlık çağı. ~**ish,** genc kız gibi.

**girth** [gö(r)θ]. Kolan; (ağac) muhit ölçüsü.

**gist** [cist]. Meal; hulâsa.

**give**¹ [giv] *n.* Esneklik. ~² **(gave, given)** [giv, gêyv, givn] *v.* Vermek; bağışlamak; nasib etm.; esnemek; çözülmek; çökmek; eğilmek; **(a laugh, shout,** *etc.*) gülmek, bağırmak vs. **to** ~ **stg. to s.o.** veya **to** ~ **s.o. stg.,** birine bir şey vermek: **to** ~ **it (to) s.o.,** (*sl.*) birini haşlamak: **to** ~ **as good as one gets,** taşı gediğine koymak: **the frost is giving,** don çözülüyor: **I** ~ **you our host,** (kadeh kaldırırken) ev sahibinin şerefine!: **we must** ~ **ourselves an hour to get**

there, oraya kadar yolu bir saat hesab etmeliyiz: **to ~ one to think,** düşündürmek: **the window ~s upon the road,** pencere sokağa bakıyor: **to ~ way,** kopmak; teslim olm.: **to ~ way to,** ···e teslim olm.; ···in fikrini kabul etm.: ···e yol vermek: **to ~ s.o. what for,** (*sl.*) birine dünyanın kaç bucak olduğunu göstermek. **~-and-take,** karşılıklı fedakârlık. **~ away,** bağışlamak; elinden çıkarmak; ifşa etm.: **to ~ away the bride,** nikâh merasiminde kızı resmen kocasına vermek: **to ~ s.o. away,** birini ele vermek: **to ~ oneself away,** foya vermek: **to ~ the show away,** bir sırrı ifşa etmek. **~ back,** geri vermek; iade etmek. **~ forth,** çıkarmak; yaymak. **~ in,** teslim olmak. **~ off,** çıkarmak, yaymak. **~ out,** işaa etm.; neşretmek; dağıtmak; tükenmek. **~ over,** teslim etm.; vazgeçmek. **~ up,** terketmek; teslim etm.; vazgeçmek; bırakmak: **I ~ it up!,** benden pes: **I had given you up,** geleceğinizden umidi kesmiştim: **to ~ oneself up,** teslim olm.; kendini polise teslim etm.: **to ~ oneself up to sport,** kendini spora vermek: **to ~ up a game,** bir oyunu bırakmak: **to ~ up the game [the struggle],** mücadeleden vazgeçmek: **to ~ up a patient,** bir hastadan ümid kesmek. **~n** [ˈgivn]. *v.* **give.** *a.* Muayyen; malûm; mübtelâ. **at a ~ time,** muayyen bir zamanda: **in a ~ time,** muayyen bir zaman zarfında: **~ to drink,** içkiye mübtelâ: **I am not ~ that way,** ben böyle bir adam değilim.

**glaci·al** [ˈglêysiəl]. Glâsyeye aid; buz gibi. **~er** [ˈglasyə(r)]. Cümudiye; glâsye.

**glad** [glad]. Memnun. **I am very ~ of it,** ondan pek memnunum: **I shall be only too ~ to help you,** size memnuniyetle yardım ederim: **I should be ~ of some help,** bir az yardım eden olursa memnun olurum. **~den** [ˈgladn]. Sevindirmek.

**glade** [glêyd]. Orman açıklığı.

**gladiator** [ˈgladiêytə(r)]. Eski Romada:- insan, vahşi hayvanlarla, dövüşen pehlivan.

**glamo·ur** [ˈglamə(r)]. Sihir, cazibe; parlaklık. **~rous,** cazibeli.

**glance** [glâns]. Kısa bakış. Bakıvermek; hafifçe vurup sekmek. **to ~ at,** ···e bir bakmak: **to ~ aside [off],** (kurşun) sekmek; (kılıc) sıyırmak: **to ~ through [over] a document,** bir yazıya şöyle bir göz gezdirmek.

**gland** [gland]. Gudde. **~ular** [ˈglandyulə(r)]. Guddevî.

**glar·e** [gleə(r)]. Kamaştırıcı ışık; dargın bakış. Ters ters bakmak; parıldamak. **in the full ~ of the sun,** güneşin alnında: **in the ~ of publicity,** âlemin gözü önünde. **~ing,** göz kamaştırıcı; çiğ renkli; inkâr edilemez.

**glass** [glâs]. Cam, sırça; bardak; dürbün; barometre. **~es,** gözlük; çifte dürbün. **~ eye,** sun'î göz: **cut ~,** billûr: **frosted [ground] ~,** buzlu cam: **plate ~,** ayna camı: **the ~ is falling,** barometre düşüyor: **to have a ~ too much,** çakırkeyif olm.: **grown under ~,** limonlukta yetiştirilmiş: ⌜**those who live in ~ houses should not throw stones**⌝, sırça evde oturan komşusuna taş atmaz: **to wear ~es,** gözlük kullanmak. **~-blower,** şişeci. **~-case,** camekân. **~-cutter,** camcı kalemi. **~-house,** limonluk. **~-works,** cam fabrikası. **~ful** [ˈglâsful]. Bardak dolusu. **~y** [ˈglâsi]. Cam gibi; ayna gibi.

**glaz·e** [glêyz]. Sır. Cam geçirmek; sırlamak; (göz) bulanmak. **~ed,** camlı; sırlı; bulanık. **~ier** [ˈglêyziə(r)]. Camcı: **~'s diamond,** elmastıraş.

**gleam** [glîm]. Muvakkat ve hafif parıltı; şua. Parıldamak. **a ~ of hope,** bir ümid lem'ası.

**glean** [glîn]. Hasaddan sonra yerde kalan başakları toplamak.

**glee** [glî]. Sevinc. **~ful,** sevincli.

**glen** [glen]. Küçük vâdi.

**glib** [glib]. (*köt.*) Cerbezeli; dil döken.

**glide** [glâyd]. Kaymak; (kuş) kanadlarını kımıldatmadan uçmak; (uçak) mötörü işletmeden inme(k). **~r,** planör.

**glimmer** [ˈglimə(r)]. Donuk ışık (yaymak); görünür görünmez aydınlık. **a ~ of hope,** ümid lem'ası.

**glimpse** [glimps]. Bir an için görmek; süreksiz bakış. **to catch a ~ of,** bir an için görmek.

**glint** [glint]. Parlaklık; parıltı. Parıldamak.

**glisten** [ˈglisn]. Parlamak.

**glitter** [ˈglitə(r)]. Parıltı. Parıldamak. ⌜**all is not gold that ~s**⌝, her parlıyan altın değildir.

**gloaming** [ˈglo͡umin(g)]. Akşamın alaca karanlığı.

**gloat** [glo͡ut]. Şeytanî bir haz göstermek. **to ~ on [over],** ···i şeytanî bir hazla seyretmek.

**glob·e** [gloub]. Küre; dünya; lâmba karpuzu. **~-trotter,** devriâlem seyyahı. **~ular** [ˈglobyulə(r)]. Küre şeklinde.

**gloom** [glûm]. Karanlık; hüzün. **to cast a ~ over the company,** toplantıya kasvet vermek. **~iness,** karanlık; mahzunluk. **~y,** loş; kapanık; kederli; kasvetli; hüzün verici: **to see the ~ side of things,** her şeyi fena tarafından görmek.

**glor·ify** [ˈglôrifây]. Tebcil etm.; müfrit derecede medhetmek. **~ification** [–fiˈkêyşn], tebcil; mübalâğalı medhetme. **~ified,** tebcil edilmiş; (*coll.*) gözde büyütülmüş. **~ious** [ˈglôriəs]. Şanlı; parlak. **a ~ day,** günlük güneşlik bir gün: **to have a ~ time,** fevkalâde eğlenmek. **~y** [ˈglôri]. Şan; şeref; debdebe; parlaklık. **to ~ in,** ···le iftihar etmek: **to cover oneself with ~,** şan kazanmak: **~ be to God!,** hamdolsun!: **~ be!,** maşallah!: **to go to ~,** (*coll.*) mahvolmak: **Old Glory,** Birleşik Amerika bayrağı: **~-hole,** (*sl.*) karmakarışık oda veya dolab vs.

**gloss** [glos]. Şerh; cilâ. Cilâlamak. **to put a ~ on the truth,** hakikatı örtmek: **to ~ over a fault,** bir kusuru örtmek. **~y** [ˈglosi]. Parlak, perdahlı.

**glossary** [ˈglosəri]. Eski, nadir, sözleri tefsir eden küçük sözlük.

**glove** [glʌv]. Eldiven (giydirmek). **to fit like a ~,** tıpatıp uymak: **to put on the ~s,** boks etm.: **to throw down the ~,** meydan okumak: **to take up the ~,** mücadeleyi kabul etm.: **to handle s.o. with the ~s off,** merhametsizce davranmak.

**glow** [glou]. Kızıl parıltı; hararet; yüzü yanma. Parıldamak; yüzü yanmak; içine ateş basmak. **in a ~,** vücudü [yüzü] hararetlenmiş: **he ~ed with pleasure,** sevincden gözleri parladı: **in the first ~ of enthusiasm,** ilk heyecanın verdiği ateşle: **~ing with health,** yanaklarından kan damlıyarak: **to speak in ~ing terms of ...,** birini göklere çıkarmak.

**glower** [ˈglauə(r)]. Yiyecekmiş gibi bakmak.

**glue** [glû]. Tutkal(lamak). **~y,** tutkallı.

**glum** [glʌm]. Somurtkan; süngüsü düşük.

**glut** [glʌt]. Fazla bolluk. Ziyadesile doyurmak. **~ton** [ˈglʌtn]. Obur. **he is a ~ for work,** inek gibi çalışıyor. **~tonous,** obur gibi. **~tony,** oburluk.

**G.M.T.** [ˈcî ˈem ˈtî]. **Greenwich Mean Time,** Greenwich saati.

**gnarled** [nâld]. Boğumlu; çarpık çurpuk.

**gnash** [naş]. **to ~ the teeth,** dişlerini gıcırdatmak.

**gnat** [nat]. Sivrisinek. ⌜**to strain at a ~ and swallow a camel**⌝, küçük bir kabahatı mesele yaptığı halde büyük bir kusura göz yummak.

**gnaw** [nô]. Kemirmek. **the ~ings of hunger,** açlıktan kıvranma.

**gnome** [noum]. Bir nevi cüce cin.

**go** **(went, gone)** [gou, went, gon]. Gitmek; yürümek; olmak; geçmek. [...**ing** *ile biten fiil şekillerinin başında* 'gidip yapmak' *manasına gelir. mes.*: **to ~ shopping,** alış verişe gitmek: **to ~ hunting,** avlanmağa gitmek.] **who ~es there?,** kimdir o?: **it ~es without saying that ...,** ... bedihidir: **to make things ~,** işleri yürütmek: **now don't ~ thinking that I am your enemy,** benim sana düşman olduğum fikrini aklından çıkar: **I can't make it any better, so let it ~,** bundan iyisini yapamam, olduğu gibi kalsın: **six months gone with child,** altı aylık hamile. **~-ahead,** müteşebbis. **~-as-you-please,** serbest; rastgele. **~-between,** arabulucu. **~-by, to give s.o. the ~,** birini atlatmak. **~ about,** dolaşmak; dönmek: **to ~ about one's work,** işine gücüne devam etm.: **I'll show you how to ~ about the job,** bu işin nasıl yapılacağını size gösteririm. **~ against,** karşı gitmek: **his appearance ~es against him,** zevahiri onun lehinde değildir. **~ back,** geri gitmek; dönmek: **his family ~es back to the Conquest,** ailesi fethe (Norman istilâsına) kadar çıkar: **to ~ back on a promise [on one's word],** sözünden dönmek: **to ~ back on a friend,** bir arkadaşına ihanet etmek. **~ before,** önünde gitmek; takaddüm etmek. **~ behind,** arkada gitmek; içyüzünü aramak. **~ by,** geçmek: **don't let this chance ~ by!,** bu fırsatı kaçırma!: **to ~ by the directions,** talimata göre hareket etm.: **to ~ by appearances;** zevahire göre hükmetmek: **that's nothing to ~ by,** buna istinaden bir şey yapılamaz. **~ down,** inmek; batmak;

zeval bulmak: **the tyre has gone down,** lâstik söndü: **the swelling has gone down,** şiş azaldı: **to ~ down from the University,** üniversiteyi bitirmek: **that won't ~ down with me,** ben bunu yutmam: **the speech went down well,** nutuk iyi tesir etti: **he has gone down in the world,** vaktile ne günler görmüştür: **to ~ down to posterity,** ebediyete intikal etmek. **~ for,** gidip aramak; hücum etm.: **someone ought to ~ for the doctor,** birisi doktor çağırsın. **~ forward,** ileri gitmek: geminin ön kısmına gitmek: **what is ~ing forward?,** ne oluyor? **~ in,** (eve) girmek; (güneş) örtülmek: **to ~ in for stg.,** bir şeye meraklı olm.; bir mesleğe girmek: **to ~ in for a car,** bir otomobil alıp kullanmak: **to ~ in for an exam.,** bir imtihana girmek: **to ~ in with s.o. for stg.,** birisile beraber bir işe girişmek: **~ in and win!,** haydi bakalım! **~ into,** girmek; girişmek; tedkik etm.: **to ~ into second gear,** ikinci vitese girmek. **~ off,** ayrılmak; fenalaşmak: **everything went off well,** her şey iyi geçti: (horse) **to ~ off its feed,** (at) iştahı kapanmak. **~ on,** devam etm.; ileri gitmek; vukubulmak: **~ on!,** devam et; haydi canım!; ileri git!: **to ~ on with stg.,** bir şeye devam etm.: **to ~ on at s.o.,** birinin başının etini yemek: **what's ~ing on here?,** burada ne oluyor?: **I have enough to ~ on with,** şimdilik yanımdaki kâfidir: **he is ~ing on for fifty,** ellisine yaklaşıyor: **I don't like the way he is ~ing on,** gidişini beğenmiyorum: **this has been ~ing on for years,** bu senelerce devam edegelmiştir. **~ out,** dışarıya gitmek; (sokağa) çıkmak; sönmek: **he ~es out teaching,** evlerde husuŝi ders veriyor. **~ over,** üzerinden geçmek; aşmak: **to ~ over an account,** hesabın üzerinden geçmek: **to ~ over a house,** bir evi gezmek: **to ~ over the ground,** araziyi keşfetmek; bir sahada çalışmak: **to ~ over to the enemy,** düşman tarafına geçmek: **to ~ over stg. in one's mind,** zihninden geçirmek. **~ round,** dolaşmak; dönmek; deveran etm.: **there is not enough to ~ round,** bu herkese yetişmez. **~ through,** içinden geçip ötesine çıkmak; okuyup teftiş etm.: **the bill has gone through,** kanun layihası kabul edildi: **the deal did not ~ through,** pazarlık uymadı: **the book has gone through three editions,** kitab üç defa basıldı: **to ~ through a fortune,** bir serveti yiyip bitirmek: **you don't know what I've gone through!,** başıma geleni sorma!: **to ~ through s.o.'s pockets,** birinin ceblerini aramak: **to ~ through with stg.,** bir şeyi bitirinceye kadar ayrılmamak: **we've got to ~ through with it,** sonuna kadar dayanmalıyız. **~ under,** altından geçmek; batmak; yenilmek; mahvolmak. **~ up,** çıkmak; (fiatı) yükselmek; patlamak: **to ~ up a form,** sınıfını geçmek: **to ~ up to the University,** üniversiteye gitmek: **to ~ up in flames,** tutuşup mahvolmak: **to ~ up to a person,** birine yaklaşmak.

**goad** [go͡ud]. Üvendire (ile yürütmek). **to ~ s.o. on,** dürtmek, teşvik etmek.

**goal** [go͡ul]. Hedef; kale. **~-keeper,** kaleci.

**goat** [go͡ut]. Keçi; ahmak. **billy ~,** teke: **nanny ~,** dişi keçi: **to get s.o.'s ~,** (*sl.*) birini sinirlendirmek: **to play the ~,** budalaca davranmak. **~ee,** çene ucundaki küçük sakal.

**gobble**[1] [ˈgobl]. Çabuk yemek. **~**[2]. (*ech.*) Hindi gibi ses çıkarmak. **~r,** baba hindi.

**goblet** [ˈgoblit]. Kadeh.

**goblin** [ˈgoblin]. Çirkin ve cüce cin.

**G.O.C.** [ˈcîˈo͡uˈsî]. **General Officer Commanding,** başkumandan.

**God** [god]. Allah; Tanrı; **god,** ilâh. **a feast fit for the ~s,** pek mükellef bir ziyafet: **to make a little (tin) ~ of s.o.,** birine, nerede ise, tapmak: **~ forbid!,** haşa!: **~ willing,** inşallah: **thank ~!,** çok şükür: **would to ~,** Allah vere de: **~'s acre,** mezarlık. **~-fearing,** dindar. **~-forsaken,** Allahın belâsı: **a ~ spot,** cehennemin dibi. **to wish s.o. ~-speed,** birini uğurlamak. **~child.** Vaftiz çocuğu. **~dess,** ilâhe. **~father,** vaftiz babası. **~less,** imansız. **~like,** ilâhî. **~liness,** dindarlık. **~ly,** dindar. **~mother,** vaftiz annesi. **~send,** beklenmiyen nimet. **~son,** vaftiz oğlu.

**goer** [ˈgo͡uə(r)]. **good [bad] ~,** iyi [fena] giden (at).

**goggle** [ˈgogl]. (Gözler) fırlamak. **~s,** (*sl.*) gözlük. **~-eyed,** pırtlak gözlü.

**going** [ˈgo͡uin(g)]. Gidiş; gitme. Giden, işliyen; faaliyette olan. **that's good ~,** iyi sayılır: **it is rough ~ on that road,** bu yol pek sarsar: **while**

the ~ is good, vaziyet müsaid iken. **goings-on**, gidişat.
**gold** [gould]. Altın (rengi.). **old ~**, donuk altın rengi: **to sell s.o. a ~ brick**, birini dolandırmak. **~-bearing**, altın madeni ihtiva eden (toprak). **~-digger**, altın arayıcı; (*Amer. coll.*) erkeklerden para sızdıran kadın. **~-diggings**, sathî altın madeni. **~-field**, altın bulunan mıntaka. **~-leaf**, altın varak. **~ mine**, altın madeni; pek kârlı iş. **~-plated**, altın kaplamalı. **~-rimmed**, altın çerçeveli (gözlük). **~-rush**, altın bulunan mıntakaya üşüşme. **~en** [ˈgouldn]. Altından mamul; altın renkli. **the ~ rule**, en iyi kaide: **the ~ age**, insanların saadet ve sulh içinde yaşadıkları hayalî bir devir; bir memleketin en parlak devri: **a ~ opportunity**, bulunmaz fırsat: **the ~ Horn**, Haliç: **~ wedding**, bir izdivacın ellinci yıldönümü. **~fish** [ˈgouldfiş]. Kırmızı balık. **~smith**[ ˈgouldsmiθ]. Kuyumcu.
**golly** [ˈgoli]. (*sl.*) Allah! Allah!
**gone** *v.* go.
**gong** [gon(g)]. Haber çanı.
**good** [gud]. İyi, güzel; faydalı; uslu; müstakim; çok, büyük. İyilik, hayır; faide, menfaat. **~s**, *v.* **goods. a ~ deal** [**a ~ many**], bir çok: **~ day, ~ morning, ~ afternoon, ~ evening, ~ night**, *günün muhtelif zamanlarında kullanılan selâm şekli*: **he's as ~ as dead**, namazı kılındı: **he didn't say it but he as ~ as said it**, demedi amma dedi sayılır: **to be ~ enough to ..., ...** lûtfunda bulunmak: **that's not ~ enough**, bu olmaz, bu uygun değil: **to go for ~**, temelli olarak gitmek: **for ~ and all**, temelli olarak: **~ for you!**, aferin!: **fruit is ~ for one**, meyva faydalıdır: **~ for nothing**, hiçe yaramaz: **the ticket is ~ for two months**, bu bilet iki ay için muteberdir: **this horse is ~ for another five years**, bu at beş sene daha dayanır: **to give as ~ as one gets**, birinin ağzının payını vermek: **to hold ~**, muteber olm.: **your ~ lady**, refikanız: **it's no ~, don't persist!**, nafile!, ısrar etme!: **that's very ~ of you**, çok lûtufkârsınız: **~ old James!**, yaşa J.!: **your ~ selves**, (ticarî muhaberelerde) siz: **it's too ~ to be true**, inanılmıyacak kadar iyi. **~-fellowship**, iyi arkadaşlık. **~-for-nothing**, yaramaz; serseri. **~-humoured**, güler yüzlü. **~-looking**, güzel yüzlü, yakışıklı. **~-bye** [ˈgudˈbây]. Allaha ısmarladık. [gudˈbây], Veda. **~ for the present**, şimdilik Allaha ısmarladık: **~ to all that!**, artık bütün bunlara elveda. **~ish** [ˈgudiş]. İyice. **it's a ~ step from here**, buradan epeyce uzaktır. **~ly** [ˈgudli]. Güzel. **a ~ inheritance**, dolgun bir miras. **~ness** [ˈgudnis]. İyilik; cevher. **have the ~ to ...**, lûtfen ...: **~ gracious!** [**my ~!**], aman Yarabbi!: **thank ~**, hamdolsun: **~ knows**, Allah bilir. **~s** [gudz]. Eşya, emtia. **~ train**, marşandiz: **to deliver the ~**, (i) eşyayı teslim etm.; (ii) vadini tutmak: **that's the ~!**, maşallah! **~will** [gudˈwil]. İyi niyet; peştemallık. **to retain s.o.'s ~**, birinin teveccühünü muhafaza etmek. **~y-~y** [ˈgudi], fazla iyi, (*istihza*) melek.
**goose**, *pl.* **geese** [gûs, gîs]. Kaz; budala. ⌜**all his geese are swans**⌝, 'kargaya yavrusu şahin görünür'. **~-flesh**, tüyleri ürpermiş; soğuktan titriyen insanın derisi. **~-step**; (*mil.*) kaz adımı. **~berry** [ˈguzbri]. Bektaşı üzümü. **to play ~**, iki sevgiliye refakat etmek.
**Gordian** [gôdiən]. **to cut the ~ knot**, müşkül bir vaziyetten anî ve kat'î bir çare ile kurtulmak.
**gor·e**[1] [gô(r)]. Pıhtılaşmış kan. **~y** [ˈgôrı]. Kanlı. **~**[2]. Birini boynuzlarıyle yaralamak.
**gorge** [gôc]. Boğaz, gırtlak; dar geçid. Tıkabasa doyurmak. **to ~ oneself**, doyuncaya kadar yemek: **my ~ rose**, midem bulandı.
**gorgeous** [ˈgôcəs]. Pek parlak, debdebeli. **we had a ~ time**, (*coll.*) fevkalâde vakit geçirdik.
**gorse** [gôs]. Karaçalı.
**gosh** [goş]. *Hayret nidası.*
**gosling** [ˈgozlin(g)]. Kaz palazı.
**gospel** [ˈgospl]. İncil. **to take stg. for ~**, mutlaka doğru olarak kabul etmek.
**gossamer** [ˈgosəmə(r)]. İnce ve hafif bürümcük; pek hafif ve ince şey.
**gossip** [ˈgosip]. Dedikodu(cu). Dedikodu yapmak. **to have a ~**, yarenlik etmek.
**got, gotten** *v.* **get.**
**gouge** [gauc]. Oluklu marangoz kalemi (-yle oymak). **to ~ out s.o.'s eye**, birinin gözünü parmakla çıkarmak.
**gourd** [gûəd]. Sukabağı; ondan yapılmış su kabı.

**gour·mand** [ˈguəmând]. Obur kimse. **~met** [ˈguəmêy]. Yemeğine titiz.
**gout** [gaut]. Damla illeti. **~y,** damlalı.
**govern** [gʌvən]. Hâkim olm.; idare etm., hükûmet sürmek; zabtemek; tanzim etmek. **this verb ~s the dative,** bu fiil mef'ulünileyh alır. **~ess** [ˈgʌvənis]. Mürebbiye. **~ment** [ˈgʌvənmənt]. Hükûmet; idare (etme); rejim. **to form a ~,** hükûmet teşkil etm.: **the ~ party,** iktidar partisi. **~mental** [–ˈmentl], hükûmete aid. **~or** [ˈgûvənə(r)]. Vali; idare heyeti âzası; regülatör. **the ~,** (*coll.*) babam; iş sahibi.; beyim, paşam. **~or-general,** umumî vali.
**gown** [gaun]. Kadın robu; profesörlerinin kisvesi. Bu kisveyi giydirmek. **town and ~,** bir üniversite şehirindeki halk ile üniversite talebeleri.
**G.P.** [ˈcîˈpî]. **General Practitioner,** ihtisas yapmamış doktor.
**grab** [grab]. Çabuk bir hareketle kapma(k); ele geçirmek; gasbetme(k). Kazma makinesinin kıskaçlı kovası. **policy of ~,** gasıb siyaseti. **~ber,** gasbedici.
**grace**[1] [grêys] *n.* Zarafet, cazibe; inayet: lûtuf; rahmet; mühlet; yemekten önce ve sonra şükür duası. **the Graces,** (*mit.*) insanlara ve tabiate güzellik ve letafet veren üç ilâhe: **His Grace [Your ~],** başpiskopos ve düklere verilen lâkab: **Act of Grace,** umumî af kanunu: **as an act of ~,** bir lûtuf olarak: **to be in the good [bad] ~s of,** ···in gözünde olmak [olmamak]: **to do stg. with a good ~,** hoşlanmadığı bir şeyi memnuniyetle yapmak: **to do stg. with a bad ~,** bir şeyi söylene söylene yapmak: **days of ~,** tediyesi için verilen üç gün mühlet: **by the ~ of God,** Allahın inayetile: **he had the ~ to be ashamed,** hiç olmazsa utandı: **it has the saving ~ that ...,** kendisini affetiren tarafı ... dir: **to say ~,** yemekten önce ve sonra dua etm.: **in this year of ~,** bu sene. **~**[2] *v.* Tesrif etm.; tezyin etmek. **~ful** [ˈgrêysfəl]. Zarif, lâtif, nazik; endamlı. **~fulness,** zarafet, letafet. **~less** [ˈgrêyslis]. Hayırsız; haylaz.
**gracious** [ˈgrêyşəs]. Nazik, mültefit, lûtufkâr; (Allah) merhametli, inayetkâr; tenezzül edici. **~ (me)! [goodness ~!],** Aman yarabbi!
**grad·ation** [grəˈdêyşn]. Tedricî değişme; derece. **~e** [grêyd]. Rütbe; derece, sınıf; meyil derecesi. Tasnif etm.; derecelere ayırmak; bir yolun meylini tanzim etmek. **to be on the down [up] ~,** [kötüleşmek] iyileşmek. **~ient** [ˈgrêydiənt]. Yolun meyili. **upward ~,** yokuş: **downward ~,** iniş. **~ual** [ˈgradyuəl]. Tedricî.
**graduate** [ˈgradyuit] *n.* Üniversite mezunu. *v.* [–uêyt] üniversiteden mezun olm.; derece derece taksim etm.
**graft**[1] [grâft]. Ağac aşısı; aşı kalemi. Aşılamak. **skin ~ing,** deri yamama. **~**[2]**.** Resmî işlerinde nufuzunu kullanarak suiistimal; rüşvet alıp verme. **~er,** rüşvet veren.
**grain**[1] [grêyn]. Arpa vs. tanesi; hububat; zerre. **~**[2]**.** Ağac damarı (taklidle boyamak). **it goes against the ~ for me to do it,** bunu istemiyerek yapıyorum. **~ed,** damar damar.
**gramma·r** [ˈgramə(r)]. Gramer (kitabı): **~-school,** hususî lise. **~tical** [grəˈmatikl]. Gramer kaidesine göre.
**gramophone** [ˈgraməfoun]. Gramafon.
**grampus** [ˈgrampəs]. **puff like a ~,** manda gibi solumak.
**granary** [ˈgranəri]. Zahire ambarı; zahiresi bol olan ve ihracat yapan memleket.
**grand** [grand]. Büyük ve muhteşem; kibar; en mühim, baş; (*coll.*) âlâ, mükemmel. **they are rather ~ people,** pek tantanalı ve azametli kimsedirler: **he's a ~ fellow,** bulunmaz adamdır: **I am not feeling very ~ today,** bugün bir parça keyifsizim: **~stand,** (spor sahalarında) tribün: **~ total,** umumî yekûn. **~child,** *pl.* **-ren** [ˈgrançâyld, -çildrən]. Torun. **~-dad,** büyük baba. **~-daughter,** kız torun. **~eur** [ˈgrandyə(r)]. Azamet; heybet; debdebe. **~father** [ˈgranfâðə(r)]. Büyük baba. **~iloquent** [granˈdiləkwənt]. Tumturaklı. **~iose** [ˈgrandiouz]. Muazzam; pek gösterişli. **~(mam)ma** [gran(mə)mâ] Büyük anne, nine. **~mother** [ˈgranmʌðə(r)]. Büyük valide. **'teach your ~ to suck eggs!',** 'sen giderken ben geliyordum'; babana akıl öğret! **~ly,** fazla ihtimamlı. **~pa(pa)** [ˈgran(pə)pa]. Büyük baba; dede. **~parent** [ˈgranpeərənt]. Dede veya nine. **~son** [ˈgran(d)sʌn.] Erkek torun.

**grange** [grêync]. Köşk ile çiftlik.
**grannie, granny** [ˈgrani]. Nine.
**grant** [grânt]. İhsan etm.; imtiyaz vermek; bağışlamak; kabul etm. İmtiyaz; tahsisat; ihsan. ~**ed that...**, kabul edelim ki ...: **to receive a state ~**, devletten tahsisat almak: **to take stg. for ~ed**, bir şeyi hakikî gibi kabul etm.: **you take too much for ~ed**, her şeyi olmuş bitmiş gibi farzediyorsun.
**granula·r** [ˈgranyulə(r)]. Tanecikli. ~**te** [ˈgranyulêyt]. Tanelenmek, taneletmek: ~**ted sugar**, toz şeker. ~**tion** [–ˈlêyşn], tanelenme.
**grape** [grêyp]. Üzüm. **'sour ~s!'**, ⌜kedi uzanamadığı ciğere pis der⌝. ~**-fruit**, grepfrut.
**graph** [graf]. Grafik. **-~**, yazılmış, yazan, yazıcı. ~**ic** [ˈgrafik]. Yazı, sanatlerine aid; grafik ile gösterilen; çizilen; canlı.
**grapp·le** [ˈgrapl]. Tutmak, yakalamak. **to ~ with s.o.**, birisile göğüs göğüse dövüşmek: **to ~ with a difficulty**, bir güçlükle pençeleşmek. ~**ling-iron**, bir şeyi tutmak için çengelli bir demir.
**grasp** [grâsp]. Kavrama(k); sımsıkı tutma(k); tahakküm, pençe. **to ~ at stg.**, bir şeyi kapmağa çalışmak: **to have stg. within one's ~**, bir şey elinin altında olm.: **to escape from s.o.'s ~**, elinden kurtulmak: **beyond one's ~**, erişilemez; kavranılmaz: **to have a good ~ of a subject**, bir mevzua iyice vâkıf olm. ~**ing**, açgözlü.
**grass** [grâs]. Çayır otu, çimen. Çimen ile kaplamak. **not to let the ~ grow under one's feet**, vakit kaybetmemek: **'keep off the ~!'**, çimene basmayınız!: **to put a horse out to ~**, atı çayıra çıkarmak. ~**-snake**, âdi zararsız yılan. ~**-widow**, kocası bir yere gittiği için yalnız yaşıyan kadın. ~**hopper** [ˈgrâshopə(r)]. Âdi ufak çayır çekirgesi. ~**land** [ˈgrâsland]. Otluk yer; çayır. ~**y** [ˈgrâsi]. Çimenli.
**grate**[1] [grêyt] *n.* Önü açık ingiliz ocağı. ~[2] *v.* Rendelemek; (diş) gıcırda(t)mak. **to ~ on the ear**, kulakları tırmalamak. ~**r** [ˈgrêytə(r)]. Rende.
**grateful** [ˈgrêytfəl]. Minnettar.
**gratif·ication** [ˌgratifiˈkêyşn]. Memnuniyet; bahşiş. ~**y** [ˈgratifây], memnun etm.; gidermek.
**grating**[1] [ˈgrêytin(g)]. Parmaklık; boru süzgeçi; gezinti ıskarası. ~[2]. Gıcırtı. Kulakları tırmalayıcı.
**gratis** [ˈgrêytis]. Bedava; caba.
**gratitude** [ˈgratityûd]. Minnet(tarlık); şükran.
**gratuit·ous** [grəˈtyûitəs]. Bedava; sebebsiz; uluorta. ~**y** [grəˈtyûiti]. Bahşiş; vazifesini ikmal eden asker vs.ye verilen para mükâfatı.
**grave**[1] [grêyv] *a.* Mühim; ciddî; ağır başlı; vahim. ~[2] *n.* Mezar, kabir. **to have one foot in the ~**, bir ayağı çukurda olm.: **he must have turned in his ~**, (bunu duysa vs.) mezarında rahatsız olur: **beyond the ~**, kabrin ötesi. ~**-clothes**, kefen. ~**-digger**, mezarcı. ~**stone** [ˈgrêyvstoun]. Mezar taşı. ~**yard** [ˈgrêyvyâd]. Mezarlık. ~[3] *v.* Hakketmek. ~**n image**, put. ~**r** [ˈgrêyvə(r)]. Hakkâk (kalemi). ~[4] *v.* Geminin dibini temizlemek.
**gravel** [ˈgravl]. Çakıllı kum; kum hastalığı. **to be ~led**, (*sl.*) apışıp kalmak. ~**ly**, çakıllı. ~**-path**, çakıllı kum döşeli yol. ~**-pit**, çakıl ocağı.
**gravit·ate** [ˈgravitêyt]. Cazibe kuvvetile düşmek; cezbolunmak. **to ~ towards**, (bir noktaya) meyletmek. ~**ation** [–ˈtêyşn], cazibe kuvveti. ~**y** [ˈgraviti]. Ağırlık; sıklet; cazibe kuvveti; ciddilik; ağırbaşlılık; ehemmiyet: **centre of ~**, sıklet merkezi: **specific ~**, izafî sıklet.
**gravy** [ˈgrêyvi]. Pişirilirken etten akan yağ ve su; bununla yapılan salça.
**gray** *v.* **grey.**
**graze**[1] [grêyz]. Otlamak. ~[2]. Sıyırmak. Sıyrık.
**greas·e** [grîs]. Yenmiyecek her türlü yağ. Yağlamak. **to ~ s.o.'s palm**, birine rüşvet vermek. ~**er**, yağlayıcı. ~**y**, yağlı; kayıcı. ~**e-cup**, gres kutusu. ~**e-gun**, gres pompası. ~**e-paint**, makiyaj yağı.
**great** [grêyt]. Büyük; muazzam; iri; şöhretli; mühim. ~**s**, Oxford'da edebiyat mezunu olmak için son imtihan. **a ~ many**, pek çok: **there are not a ~ many**, çok yok: **a ~ deal**, çok: **he is ~ on dogs**, (*coll.*) köpek meraklısıdır. ~**ness**, büyüklük; çokluk; şöhret. ~**-coat**, palto. ~**-grandchild**, torunun çocuğu. ~**-grandfather**, dedenin babası. ~**-~-grandfather**, dedenin dedesi. ~**-hearted**, âlicenab.

**Gre·cian** [ˈgrîşyən]. Eski Yunanistana aid. **~ece** [grîs]. Yunanistan. **~ek** [grîk]. Yunanlı; rum(ca); eski yunanca. **the ~ church,** Ortodoks kilisesi: **it's all ~ to me,** buna hiç aklım ermez.

**greed** [grîd]. Oburluk; hırs; açgözlülük; iştah. **~y,** obur; açgözlü, haris. **~iness,** oburluk; hırs.

**green** [grîn]. Yeşil (renk); taze; çiğ; acemi, toy. Çimenlik. **he is not as ~ as he looks,** göründüğü kadar toy değil: **to go** [**turn**] **~,** sararmak: **to keep s.o.'s memory ~,** hatırasını canlı tutmak: **a ~ old age,** dinc ihtiyarlık. **~ness,** yeşillik; toyluk. **~ery** [ˈgrînəri]. Yeşillik. **~fly** [ˈgrînflây]. Yaprak biti. **~gage** [grînˈgêyc]. Ufak yeşil. **~grocer** [ˈgrîngroușə(r)]. Manav. **~horn** [ˈgrînhôn]. Tecrübesiz genc, toy. **~house** [ˈgrînhaus]. Limonluk; ser. **~stuff** [ˈgrînstʌf]. Yenir yeşillik.

**Greenwich** [ˈgriniç]. **~ Mean Time** (**G.M.T.**), Greenwich (vasatî) saati.

**greet** [grît]. Selâmlamak; istikbal etmek. **to ~ the eye** [**ear**], göze [kulağa] çarpmak: **to ~ a speech with cheers,** bir nutku alkişlarla karşılamak. **~ing,** selâm: **New-Year ~s,** yılbaşı tebrikleri.

**gregarious** [griˈgeəriəs]. Sürü halinde yaşıyan.

**grenad·e** [griˈnêyd]. El bombası. **~ier** [ˌgrenəˈdiə(r)]. (*esk.*) Humbaracı asker.

**grew** *v.* **grow.**

**grey** [grêy]. Kurşunî, gri; kır (saçlı). **to turn** [**go**] **~,** kırlaşmak. **~ matter,** sincabî madde: **grown ~ in the service,** saçları hizmette ağartmış. **~beard** [ˈgrêybiəd]. Aksakallı. **~hound** [ˈgrêyhaund]. Tazı. **ocean ~,** süratli yolcu vapuru.

**grid** [grid]. Izgara; tel kalbur; şebeke. **the ~,** millî elektrik şebekesi. **~-map,** kareli harita.

**grie·f** [grîf]. Keder, hüzün; esef. **to come to ~,** suya düşmek; belâsını bulmak; kazaya uğramak: **to my great ~ I had no son,** heyhat ki oğlum olmadı. **~vance** [ˈgrîvəns]. Derd, şikâyeti mucib hal. **to air one's ~s,** derdleşmek: **pet ~,** derdi günü. **~ve** [grîv]. Kederlenmek; esef etm.; keder vermek. **~ved,** kederli: **I am ~ to hear that ...,** ... işitmekle çok müteessirim. **~vous** [ˈgrîvəs]. Keder verici, acı, feci; ağır.

**grill** [gril]. Izgara(da pişmiş et). Izgarada pişirmek.

**grille** [gril]. Kapı, pencere parmaklığı.

**grim** [grim]. Haşin; ekşi yüzlü; meş'um; tehdidkâr; merhametsiz. **~ humour,** acı nükte: **to hang on like ~ death,** (bir şeye) mezbuhane sarılmak. **~ness** [ˈgrimnis]. Sertlik. **~ace** [griˈmêys]. İşmizaz (göstermek); yüzünü ekşitme(k).

**grim·e** [grâym]. Deriye işliyen kir, toz vs. **~iness** [ˈgrâyminis]. Kirlilik. **~y** [ˈgrâymi]. Kirli.

**grin** [grin]. Sırıtma(k). **to ~ and bear it,** güler yüzle tahammül etmek.

**grind** (**ground**) [grâynd, graund]. Öğütmek; toz haline getirmek; ezmek; (dişlerini) gıcırda(t)mak; sürterek parıldamak; bilemek. Gıcırtı; meşakkatli iş. **~er,** öğütme makinesi; bileyici. **~ing,** gıcırtı; öğütme; ezme: **~ poverty,** ezici sefalet. **~stone** [ˈgrâynstoun], değirmen taşı; bileği taşı: **to keep one's nose to the ~,** durmadan çalışmak.

**grip** [grip]. Sıkı tutmak; ···e pençe atmak; kavramak. Sıkı tutma; pençe; seyahat çantası. **to be at ~s** [**come to ~s**] **with the enemy,** düşmanla kapışmak; **to get a good ~ on** [**of**] **stg.,** bir şeye iyice tutunmak: **to have** [**get**] **a good ~ of the situation** [**of a subject**], vaziyeti [bir mevzuu] iyice kavramak: **to lose one's ~ on affairs,** işlerin yakasını bırakmak.

**gripe** [grâyp]. Kulunc tutmak; şiddetli karınağrısı vermek. **the ~s,** kulunc.

**grisly** [ˈgrizli]. Korkunc, ürpertici.

**grist** [grist]. Öğütülecek zahire. **that brings ~ to the mill,** kâr kârdır: ⌜**all is ~ that comes to his mill**⌝, her şeyden kâr çıkarır.

**grit** [grit]. İri parçalı kum; sebat, cesaret. Gıcırdatmak. **~stone,** kumtaşı. **~ty,** kumlu.

**grizzl·ed** [ˈgrizld]. Kır düşmüş. **~y** [ˈgrizli]. Kır: **~-bear,** Şimalî Amerikada büyük bir ayı.

**groan** [groun]. İnlemek; inilti; sızlanma(k).

**grocer** [ˈgrousə(r)]. Bakkal. **~y,** bakkal dükkânı. **~ies,** bakkaliye.

**groggy** [ˈgrogi]. Sendeliyen. **to feel ~,** halsizlik hissetmek.

**groin** [grôyn]. Kasık.

**groom** [grûm]. Seyis; = **bridegroom,** güvey. (Atı) timar etmek. **well ~ed,** (at) iyi tımar edilmiş; (insan) çeki

düzen verilmiş. **~sman** [ˈgrûmsmən]. Güveyin sağdıcı.
**groove** [grûv]. Yiv. Yiv açmak. **to get into a ~,** eski âdetlerine bağlı olm.
**grope** [grou̯p]. Elleri ile yoklamak, yolunu bulmak. **to ~ for [after] stg.,** bir şeyi el yordamı ile araştırmak.
**gross**[1] [grou̯s]. On iki düzine; grosa. **~**[2]. Kaba; şişko; hantal; kalın; gayri safi; toptan. **~ ignorance,** kara cahillik: **a ~ mistake,** pek büyük hata: **~ tonnage,** gayri safi tonilâto.
**grotesque** [grou̯ˈtesk]. Güldürecek biçimde; garib.
**grotto** [ˈgrotou̯]. Ufak, güzel mağara.
**ground**[1] [grau̯nd] *v.* **grind.** Öğütülmüş. **~ glass,** buzlu cam. **~**[2] *n.* Yer, zemin; toprak; arsa; saha; sebeb; plân. **~s,** bir evin hususî arazisi; esbabı mucibe; telve. **above ~,** yer yüzünde. **to break ~,** toprağı kazmak: **to break new [fresh] ~,** çığır açmak: **this report covers a great deal of new ~,** bu rapor bir çok yeni noktalara temas ediyor: **to cut the ~ from under s.o.'s feet,** birinin dayandığı noktayı çürütmek: **down to the ~,** tamamen: **to fall to the ~,** yere düşmek; (*fig.*) suya düşmek: **to gain ~,** ilerlemek; ehemmiyeti artmak: **to give [lose] ~,** gerilemek; tedricen ehemmiyetini kaybetmek: **to give ~ for ...,** ···e mahal vermek: **to go to ~,** inine girmek; gizlenip gözden kaybolmak: **on the ~ that, ...** ileri sürerek: **firm ~,** sağlam temele dayanarak: **to stand one's ~,** mevkiini muhafaza etmek. **~-plan,** temel plânı. **~-sheet,** yere serilen su geçmez yaygı. **~**[3] *v.* Karaya otur(t)mak; tesis etmek. **to ~ one's arguments on,** delillerini ···e istinad ettirmek. **~ed** [ˈgrau̯ndid]. (Uçak) meydânda kalmağa mecbur olan: **well ~ in Latin,** lâtincenin esaslarını kavramış. **~ing** [ˈgrau̯ndin(g)]. **to have a good ~ in stg.,** bir şeyin esaslarını iyice bilmek. **~less** [ˈgrau̯ndlis]. Asılsız; sebebsiz. **~sman** [ˈgrau̯nzmən]. Oyun sahalarına bakan adam. **~work** [ˈgrau̯ndwö(r)k]. Zemin; temel, esas.
**group** [grûp]. Küme; öbek; manzume. Grup halinde topla(n)mak. **in ~s,** öbek öbek.
**grouse**[1] [graus]. (*coll.*) Homurdanma(k). **~**[2]. Bir nevi keklik.
**grove** [grou̯v]. Koru, ormancık.
**grovel** [ˈgrovl]. Yerde sürünmek; alçakcasına yalvarmak. **~ling,** alçak dalkavuk nevinden.
**grow** (**grew, grown**) [grou̯, grû, grou̯n]. Büyümek, boy atmak; çoğalmak; olmak, ···lemek, ···lenmek, ···leşmek; gittikçe ... olm.; yetiş(tir)mek; (sakal) salıvermek. **to ~ into a woman,** kadın olmak: **this picture ~s (up)on one,** bu resim insanı gittikçe sarıyor: **he will ~ out of it,** büyüdükçe ondan vazgeçer: **to ~ out of one's clothes,** çocuk büyüdükçe elbiseleri dar gelmek: **one ~s to like it,** insan gittikçe ondan hoşlanıyor: **to ~ up,** büyümek. **~ing** [ˈgrou̯in(g)] *a.* Büyüyen, yetişen; artan. *n.* Yetiş(tir)me, artma. **~n** [grou̯n] *v.* **grow.** *a.* Büyümüş; yetiştirilmiş. **tower ~ over with ivy,** sarmaşık kaplı kule: **well ~,** boyu bosu yerinde. **~-up,** büyük, yetişmiş: **the ~s,** (çocukların aksi olarak) büyükler. **~th** [grou̯θ]. Büyüme, inkişaf; artma; şiş, ur. **a week's ~ of beard,** bir haftalık tıraş.
**growl** [grau̯l]. Hırlamak; homurdanmak. Hırıltı.
**groyne** [grôyn]. Deniz kenarında kazıklar vs. ile yapılan sed.
**grub**[1] [grʌb]. Sürfe, kurd; (*sl.*) manca. **~**[2] *v.* Eşelemek; (*sl.*) çok çalışmak. **to ~ up,** (toprağı) hafifçe kazmak; (kökler) sökmek. **~by** [ˈgrʌbi]. Kirli, pis.
**grudg·e**[1] [grʌc] *n.* Kin, hınc. **to bear s.o. a ~,** birine kin beslemek. **~**[2] *v.* Esirgemek; çok görmek; hased etm. **~ing,** gönülsüz; esirgeyici: **he is ~ in his praise,** medhi cömerdce değildir: **~ly,** istemiyerek.
**gruel** [ˈgruəl]. Sulu yulaf lâpası.
**gruelling** [ˈgruəlin(g)]. Meşakkatli, çok yorucu. Meşakkatli ve takat kesen iş, maç vs.
**gruesome** [ˈgrûsəm]. Ürkütücü, ürpertici.
**gruff** [grʌf]. Hırçın; boğuk sesli.
**grumbl·e** [ˈgrʌmbl]. Mırıldamak; şikâyet etm. Şikâyet. **to ~ at,** ···den şikâyet etmek. **~ing** [ˈgrʌmblin(g)]. Mırıltı, şikâyet. Homurdanarak.
**grumpy** [ˈgrʌmpi]. Somurtkan.
**Grundy** [ˈgrʌndi]. **Mrs. ~,** ahlâk hususunda pek titiz ve müteassıb kimse.
**grunt** [grʌnt]. Domuz gibi hırıldamak. Hırıltı.
**guano** [ˈgwânou̯]. Kurutulmuş kuş gübresi.
**guarant·ee** [garanˈtî]. Kefil, teminat.

Kefalet etm.; temin etm.: **to stand ~ for s.o.,** birine kefil olmak. **~or** [garanˈtô(r)]. Kefil; zâmin.
**guard** [gâd]. Muhafız; nöbet(çi); bekçi: muhafaza; dikkat, teyakkuz. Korumak, muhafaza etmek. **the Guards,** muhafız alayları: **to ~ against,** önlemek: **to be on ~,** nöbet beklemek: **to be on one's ~,** tetik durmak: **to be caught off one's ~,** gafil avlanmak: **to come off ~,** nöbeti bitmek: **to go on [mount] ~,** nöbete çıkmak: **~ of honour,** ihtiram kıt'ası: **to keep ~,** nöbet beklemek: **'one of the old ~',** eskilerden (asker, politikacı): **to put s.o. on his ~,** birini ikaz ederek ihtiyatlı olmasını söylemek: **to throw s.o. off his ~,** birini gaflete sevketmek. **~-house [-room],** askeri karakol; nöbetçi odası. **~-rail,** korkuluk. **~ed** [ˈgâdid] *a.* İhtiyatlı, muhteriz. **~ian** [ˈgâdyən]. Vasi; bekçi; muhafız. **~ angel,** koruyan melek: **Board of Guardians,** belediye iane heyeti. **~ianship,** vesayet. **~ship** [ˈgâdşip]. Muhafız gemi. **~sman** [ˈgâdsmən]. İngiliz Kıralının hassa alaylarından birine mensub asker.
**guer(r)illa** [gəˈrilə]. Çete (harbi).
**guess** [ges]. Tahmin (etm.); bilme(k); keşfetmek; zannetmek. **I give you three ~es,** üç defada bilirsen ne iyi: **by ~ (-work),** rastgele, tahminî olarak: **to keep s.o. ~ing,** birini şaşırtarak aldatmak: **it's pure ~-work,** tahminden ibaret: **~ whom I met!,** kime rastgelsem beğenirsiniz!
**guest** [gest]. Misafir; davetli; pansiyoner. **~-house,** misafirhane; pansiyon.
**guffaw** [gʌˈfô]. Kaba, gürültülü, kahkaha (salıvermek).
**guid·e** [gâyd]. Rehber, kılavuz; delâlet; talimat; örnek. Yol göstermek, delâlet etm.; irşad etm.; idare etmek. **~ance,** rehberlik; yol gösterme; delâlet; nasihat. **~ing,** yol gösterici.
**guild** [gild]. Esnaf loncası. **~hall,** lonca salonu: **the Guildhall,** Londra belediye dairesi.
**guile** [gâyl]. Kurnazlık; hilekârlık. **~ful,** hilekâr. **~less,** dürüst; saf.
**guilt** [gilt]. Suçluluk; mücrimlik. **~y,** suçlu; mücrim: **to find s.o. ~,** birini suçlu olduğunu tesbit etm.: **to plead ~,** suçunu (mahkemede) itiraf etm.: **~ conscience,** suçlu olduğunu bilme: **~ look,** suçlu bakış.
**guinea** [ˈgini]. Eski İngiliz lirası; (*şim.*) bir lira bir şilin.
**guise** [gâyz]. Kıyafet; şekil.
**guitar** [giˈtâ(r)]. Kitara.
**gulf** [gʌlf]. Körfez; büyük aralık.
**gull** [gʌl]. Martı; safderun. Aldatmak. **~ible** [ˈgʌləbl]. Kolayca aldanan.
**gullet** [ˈgʌlit]. Boğaz.
**gully** [ˈgʌli]. Sel çukuru; dere.
**gulp** [gʌlp]. Yudum; yutma sesi (çıkarmak). **to ~ down,** yutmak.
**gum**[1] [gʌm]. Diş eti. **~boil** [ˈgʌmbôyl]. Diş etinde çıban. **~**[2]. Zamk; göz çapağı. Zamklamak: **to ~ down,** zamk ile yapıştırmak: **to ~ up,** yapışıp kımıldamamak. **~-boots,** lâstik çizme. **~-tree,** zamk ağacı: **up a ~,** müşkül bir vaziyette. **~my** [ˈgʌmi]. Zamklı, yapışkan.
**gumption** [ˈgumşn]. Akılselim.
**gun** [gʌn]. Top; tüfek; (*Amer.*) tabanca. **to blow great ~s,** fırtına kopmak: **to stick to one's ~s,** teslim olmamak; iddiasından vazgeçmemek: **spray ~,** püsküren alet. **~-carriage,** top kundağı. **~-running,** silâh kaçakçılığı. **~-shy,** (köpek) tüfek sesinden korkar. **~-cotton** [ˈgʌnkotn]. Pamuk barutu. **~fire** [ˈgʌnfây(r)]. Top ateşi. **~man** [ˈgʌnmən]. Silâhlı cani. **~metal** [ˈgʌnmetl]. Top bronzu. **~ner** [ˈgʌnə(r)]. Topçu. **~nery,** topçuluk fenni. **~powder** [ˈgʌnpaudə(r)]. Barut. **~shot** [ˈgʌnşot]. Top, tüfek atışı: **within [out of] ~,** tüfek menzili içinde [haricinde]. **~smith** [ˈgʌnsmiθ]. Silâhcı; tüfekçi.
**gurgle** [ˈgö(r)gl]. (*ech.*) Lıklık. Lıkırdamak.
**gush** [gʌş]. Fışkırmak; fazla hassasiyetle konuşma(k). **~ing,** fışkıran: **~ woman,** coşkun, taşkın kadın. **~er,** petrol fışkırtan kuyu.
**gusset** [ˈgʌsit]. Genişletmek için ilâve olunan kısım.
**gust** [gʌst]. Rüzgâr, yağmurun anî bir savruntusu. **a ~ of anger,** hiddet dalgası. **~y** [ˈgʌsti]. Arasıra şiddetle esen.
**gusto** [ˈgʌstou]. Zevk; ağıztadı.
**gut** [gʌt]. Barsak; çalgı kirişi. İçini dışına çıkarmak. **~s,** barsaklar; (*coll.*) cesaret. **the fire ~ted the house,** yangın evin içini tamamen tahrib etti.
**gutter** [ˈgʌtə(r)]. Oluk; su yolu. (Mum) akıp gitmek. **born in the ~,**

en aşağı halk tabakasından. **~-press,** âdi gazeteler. **~-snipe,** afacan.
**Guy**[1] [gây]. Bir erkek ismi. **~ Fawkes,** 5 Kasım 1605'da İngiliz Parlamento binasını barut ile berhava etmeğe teşebbüs eden adam: **~ Fawkes' Day,** 5 Kasım. **~**[2]. 5 kasımda dolaştırılan ve Guy Fawkes'u temsil eden manken; korkuluk gibi. (*Amer. coll.*) adam, herif. Takılmak. **~**[3]. **~(-rope),** direk, çadır vs. yerini tutmak için halat. Bir iple tesbit etmek.
**guzzle** [ˈgʌzl]. Tıkınmak. **~r,** obur.
**gymnas·ium** [cimˈnêyziəm]. Jimnastikhane. **~tics,** jımnastik, idman.
**gyrat·e** [câyˈrêyt]. Deveran etmek. **~ion** [–ˈrêyşn], deveran, devir. **~ory,** dönücü; deveran eden.

# H

**H** [êyç]. H harfi. **to drop one's h's,** h harfini telaffuz etmemek.
**ha** [hâ]. *Hayret, sevinc, şübhe veya muvafakkiyet ifade eder.* **~, ~,** *gülme sesini taklid eder.*
**Habeas Corpus** [ˈhabias kôpʌs]. **~ ~ Act,** haksız tevkifi meneden meşhur ingiliz kanunu.
**haberdasher** [ˈhabəˌdaşə(r)]. İğne, kurdelâ gibi ufak tefek satan dükkâncı; tuhafiyeci.
**habit** [ˈhabit]. İtiyad, âdet; huy; mutad. **contrary to ~,** mutad hilafına; **to get [grow] into the ~ of,** …e alışmak: **to make a ~ of,** âdet edinmek: **out of [from force of] ~,** alışkanlıkla. **~ual** [həˈbityuəl]. Mutad; itiyada bağlı; daimî; alışmış. **~ drunkard,** ayyaş. **~uate** [həˈbityuêyt]. Alıştırmak. **~ué** [həˈbityuêy]. Gedikli (müşteri vs.).
**hâbit·able** [ˈhabitəbl]. Oturulabilir. **~at** [ˈhabitat]. Bir hayvan yetiştiği yer. **~ation** [habiˈtêyşn]. İkametgâh. **fit for ~,** ikamete elverişli.
**hachure** [ˈhaşuə(r)]. Tarama (hatları ile göstermek).
**hack**[1] [hak]. Kabaca, intizamsızca kesmek; incik kemiğine tekme atmak. Kaba yara. **to ~ one's way through,** yolunu yarıp açmak. **~-saw,** kollu testere. **~**[2]. Kira beygiri; âdi işler gören adam. **literary ~,** gündelik muharrir. **~-work,** bir gazetedeki gündelik alelâde yazı işleri.
**hackle** [ˈhakl]. Horozun boynundaki uzun tüyler. **with his ~s up,** dövüşmeğe hazır.
**hackney** [ˈhakni]. Kira beygiri veya arabası. **~ed,** çok kullanılıp bayağılaşmış.
**had** *v.* **have.**
**Hades** [ˈhêydîz]. Cehennem.
**haema-, -mo** [ˈhîmə, –mou]. Kana aid. **~rrhage** [ˈheməric]. Kan kaybetme. **~rrhoids** [ˈhemərôydz]. Basur.
**haft** [hâft]. Sap (takmak).
**hag** [hag]. Acuze, cadaloz.
**haggard** [ˈhagəd]. Istırab, açlık veya acıdan dolayı benzi sararmış.
**haggle** [ˈhagl]. Çekişe çekişe pazarlık etmek.
**Hague** [hêyg]. **The ~,** Lahey.
**hail**[1] [hêyl]. Dolu (yağmak). **~-stone,** dolu tanesi. **~**[2]. Çağırmak; seslemek; (geçen bir gemiye) işaret vermek. Çağırma, seslenme. **where does this ship ~ from?,** bu gemi hangi limandan geliyor?: **where do you ~ from?,** siz ne taraftansınız?: **within ~,** seslenebilecek bir mesafede.
**hair** [heə(r)]. Saç; kıl; tüy. **against the ~,** tüyün tersine: **to do one's ~,** saç tuvaletini yapmak: **a fine head of ~,** gür ve güzel saç: **keep your ~ on!,** (*sl.*) öfkelenme!: **to lose one's ~,** saçı dökülmek; (*sl.*) öfkelenmek: **to put one's ~ up,** (kız) saçlarını topuz yapmak: **to split ~s,** kılı kırk yarmak: **for one's ~ to stand on end,** tüyleri ürpermek: **not to turn a ~,** istifini bozmamak. **~-cut,** saç tıraşı. **~-raising,** tüyler ürpertici. **~-restorer,** saç ilâcı. **~'s-breadth, he escaped drowning by a ~,** boğulmasına kıl kaldı: **to be within a ~ of death,** ölmesine kıl kalmak. **~-shirt,** (riyazet) kıl gömlek. **~-splitting,** kılı kırk yarma. **~-spring,** ince helezoni yay. **~dresser** [ˈheədresə(r)]. Berber, kuaför. **~dressing,** berberlik. **-~ed** [heəd] *suff.* …saçlı; …tüylü. **~less** [ˈheəlis]. Saçsız; tüysüz; dızlak. **~pin** [ˈheəpin]. Firkete. **a ~ bend,** birdenbire tam aksi istikamette kıvrılan yolun dönemeci. **~y** [ˈheəri]. Kıllı; tüylü.
**halcyon** [ˈhalsiən]. **~ days,** sakin ve mes'ud günler.
**hale** [hêyl] *a.* Gürbüz ve dinc. **to be ~ and hearty,** dinc ve canlı olmak.

**half**, *pl.* **halves** [hâf, hâvz]. Yarım, nısıf; yarı; buçuk; mektebin yarım senelik devre müddeti; (futbol) devre. **three and a ~**, üç buçuk: **~ as big again**, bir buçuk misli: **he only got ~ as many marks as I did**, o benim aldığım numaraların yarısını aldı: **I was ~ afraid that ...**, ... diye bir az korktum: **to make a good start is ~ the battle**, iyi başlanan iş yarı bitmiş demektir: **a man's better ~**, bir adamın karısı: **to do stg. by halves**, bir işi yarım yamalak yapmak: **too clever by ~**, lüzumundan fazla akıllı: **~ a crown**, iki buçuk şilin: **to cut in ~**, ikiye bölmek: **to go halves**, müsavi olarak paylaşmak: **not ~ bad**, hiç te fena değil: **not ~!**, (*coll.*) pek çok: **not ~ enough**, yarısı bile değil: **he didn't ~ swear**, (*coll.*) öyle bir küfür etti ki: **to lean ~ out of the window**, pencereden yarı beline kadar sarkmak: **return ~**, dönüş bileti. **~-and-~**, (kahve) yarısı süt yarısı kahve; (viski) yarısı viski yarısı soda. **~-back**, muavin. **~-baked**, yarı pişmiş; abtal. **~-breed** *n.* melez. **~-brother**, üvey kardeş. **~-caste**, melez adam. **~-fare**, yarım bilet. **~-hearted**, isteksiz, gevşek. **~-holiday**, yarım azad. **~-hour**, yarım saat: **~ly**, yarım saatte bir. **~-mast**, yarıya indirilmiş (bayrak). **~-measure**, yarım yamalak tedbir. **~-price**, yarı fiat; yarım duhuliye. **~-seas-over**, yarı sarhoş, çakırkeyif. **~-term**, mekteb devresinin ortasındaki tatil. **~-time, to work ~**, yarım gün çalışmak. **~-tone**, açıkla koyu arası. **~-turn**, yarım devir. **~-way**, yarı yolda: **to meet s.o. ~**, (*fig.*) karşısındaki ile uyuşmak: **to get ~ through a book**, bir kitabı yarılamak: **there's no ~ house**, ikisi ortası olmaz. **~-witted**, ebleh, safdil. **~-year**, altı aylık müddet: **~ly**, altı ayda bir. **~penny** [ˈhêypeni]. (½*d.*); yarım peni: **~-worth**, yarım penilik.

**halitosis** [haliˈtǫusis]. (*med.*) Fena kokulu nefes.

**hall** [hôl]. Büyük salon; büyük sayfiye; üniversitelerde küçük kolej. **servant's ~**, hizmetçilere mahsus yemekhane: **~ porter**, (otel) kapıcı.

**hallmark** [ˈhôlmâk]. Altın ile gümüşten yapılan şeylere basılan resmî ayar damgası. **the ~ of genius**, dehanın alâmeti farikası.

**hallo** [həˈlǫu]. Allo!; yahu!

**halloo** [həˈlû]. Yüksek sesle birini çağırma(k).

**hallow** [ˈhalǫu]. Takdis etmek.

**hallucination** [ˌhalûsiˈnêyşn]. Kuruntu, vehim, birsam.

**halo** [ˈhêylǫu]. Hâle (ile kuşatmak).

**halt**[1] [hôlt]. Duruş; durak. Duraklamak; durdurmak. **~!**, dur!: **to come to a ~**, birdenbire durmak: **to call a ~**, nihayet vermek. **~**[2]. Topal; aksak. Aksaklık. Topallamak; bocalamak.

**halter** [ˈhôltə(r)]. Yular; idam ipi. Yular takmak.

**halve** [hâv]. İki müsavi kısma bölmek; yarıya indirmek. **~s** *v.* **half.**

**ham** [ham]. Domuzun tuzlanmış ve tütsülenmiş budu; jambon.

**hamlet** [ˈhamlit]. Küçük köy.

**hammer** [ˈhamə(r)]. Çekic; tüfek horozu; mezadcı tokmağı. Çekicle vurmak. **to ~ away at stg.**, bir şeye ısrarla çalışmak: **to come under the ~**, mezadda satılmak: **to give s.o. a good ~ing**, birini iyice dövmek; birini bir oyunda kolayca yenmek: **~ and tongs**, var kuvvetle. **~ out**, çekic ile yassılaştırmak; (*coll.*) uydurmak; çok zahmetle (eser) yazmak.

**hamper**[1] [ˈhampə(r)] *n.* Büyük kapaklı sepet. **~**[2] *v.* Serbest hareketine mâni olm.

**hamstring** [ˈhamstrin(g)]. Hayvanın ard ayak veteri (kesmek); (*fig.*) baltalamak; mecalsız bırakmak.

**hand**[1] [hand] *n.* El; saat akrebi; işçi, tayfa; el yazısı; imza. **on all ~s**, her taraftan [–ta]: **all ~s on deck!**, (*naut.*) herkes güverteye!: **the ship was lost with all ~s**, gemi bütün tayfası ve yolcularile beraber kayboldu: **at ~**, yanında, hazırda: **to ask for the ~ of ...**, (bir kıza) talib olmak: **to bear a ~**, yardım etm.: **to bring up an animal by ~**, bir hayvanı kendi elile yetiştirmek: **at first ~**, doğrudan doğruya: **at second ~**, ikinci elden: **to get stg. off one's ~s**, (i) bir şeyi başından atmak; (ii) bir şeyden kurtulmak: **to give a ~**, yardım etm.: **to go ~ in ~ with**, başa baş gitmek: **to be a good ~ at doing stg.**, bir şeye eli yatmak: **to be in good ~s**, iyi insanların elinde olm.: **the work is now in ~**, iş derdesttir: **the situation is now in ~**, hükûmet vaziyete hâkim: **these children need taking in ~**, bu çocukları yola getirmeli: **to**

lay ~s on ..., (bir şeyi) bulmak; ···i yakalamak: **living from ~ to mouth,** günü gününe yaşama: **~s off!** dokunma!: **on ~,** mevcudda: **I have a lot of work on ~,** üzerimde bir çok iş var: **on the one ~ ..., and on the other ~ ...,** bir taraftan ..., diğer taraftan ...: **out of ~** *adv.* derhal; lemeden; *a.* haşarı, çığırından çıkmış: **to let stg. get out of ~,** ipin ucunu kaçırmak: **my ~ is out,** itiyadı kaybetmişim: **~ over ~ [fist],** ipe tırmanır gibi el hareketi ile; (*fig.*) muntazam ve sür'atle ilerliyerek: **for one's own ~,** kendi çıkarına: **to put one's ~ to stg.,** bir şeyi ele almak: **to throw in one's ~,** vazgeçmek: **to have time on one's ~s,** bol vakti olm.: **to ~,** el altında; **yours to ~,** (ticarette) mektubunuz alınmıştır. **~-made,** el işi. **~-pick,** elle ayırmak. **~-to-~ fight,** göğüs göğüse muharebe. **~²** *v.* El ile vermek, uzatmak. **~ down,** el ile tutup indirmek; babadan oğula miras bırakmak; nesilden nesle nakletmek. **~ in,** vermek; teslim et.: **to ~ in one's resignation,** istifa etmek. **~ on,** elden ele geçirmek; nesilden nesle nakletmek. **~ out,** el ile dışarıya vermek; dağıtmak. **~ over,** devrü teslim etm.; havale etm.; ele vermek. **~ round,** elden ele dolaştırmak. **~bill** [ˈhandbil]. El ilânı. **~book** [ˈhandbuk]. Elkitabı; rehber. **~cuff** [ˈhandkʌf]. Kelepçe (-ye vurmak). **~ful** [ˈhandful]. Avuc dolusu; bir avuc; ele avuca sığmaz bir çocuk. **~hold** [ˈhandhould]. Tutunacak yer. **~icap** [ˈhandikap]. Handikap; (*fig.*) engel, yük; müvaffakiyeti veya ilerlemeği güçleştiren her şey. Handikap vermek; engel olmak: **he is ~ped by his poverty,** fakirliği onun elini kolunu bağlıyor. **~icraft** [ˈhandikrâft]. El hüneri, sanati: **~sman,** el işçisi. **~iness** [ˈhandinis]. Kullanışlılık; marifet; el işleri yapabilmek kabiliyeti. **~iwork** [ˈhandiwö(r)k]. El işi; eser; marifet. **~kerchief** [ˈhan(g)kəçîf]. Mendil. **~le** [ˈhandl]. Sap; kulp; vasıta. Ellemek, dokunmak, kullanmak; idare etmek. **to have a ~ to one's name,** asalet unvanına malik olm.: **he is hard to ~,** onu idare etmek güçtür: **to ~ a lot of money,** elinden çok para geçmek: **to ~ a situation,** bir vaziyeti idare etmek: **~-bar,** gidon. **~ling** [ˈhandlin(g)]. Kullanma, idare etme; dokunma: **rough ~,** fena muamele. **~shake** [ˈhandşêyk]. El sıkma. **~some** [ˈhansəm]. Güzel; cömerd; (*coll.*) büyük: ⌜**~ is as ~ does**⌝, 'dışarısı seni yakar içerisi beni yakar'. **~-work** [ˈhandwö(r)k]. El işi. **~-writing** [ˈhandrâytin(g)]. El yazısı. **~y** [ˈhandi]. Eli hünerli; becerikli; kullanışlı; faydalı; el altında. **it will come in ~ some day,** bir gün işe yarar: **I always keep some ~,** daima bir az bulundururum. **~yman,** el ulağı.

**hang¹** [han(g)] *n.* Giyilen şeyin duruşu. **I don't care a ~,** bana vızgelir: **to get the ~ of stg.,** yolunu, yordamını bulmak. **~²** *v.* **(hung, hung** *veya* **hanged)** [han(g); hʌn(g), han(g)d]. As(ıl)mak; sarkmak. **~ it (all)!,** Allah belâsını versin!: **to ~ a door,** kapı yerine takmak: **~ the expense!,** masraf ne olursa olsun: **to ~ fire,** gecikmek, sürüklenmek: **the streets were hung with flags,** sokaklara bayraklar asılmıştır: **I'll kill him if I ~ for it,** beni asacaklarını bilsem onu gine öldürürüm: **to let things go ~,** işleri ihmal etm.: **to ~ one's head,** başını önüne eğmek: **~ed if I know!,** ben ne bileyim?: **to ~ meat,** eti yumuşatmak için bir müddet asmak: **to ~ by a thread [hair],** kıl kalmak; çok tehlikede olmak. **to have a ~-dog look,** süngüsü düşük bir hali olmak. **~-over,** (hastalık) bakiyesi; (sarhoşluk) mahmurluk. **hang about,** avare dolaşmak; bir şeyin etrafında dönüp dolaşmak. **~ back,** geri durmak; tereddüd göstermek. **~ down,** sarkmak. **~ on,** dayanmak; ···e bağlı olm.: **to ~ on to stg.,** bir şeye asılmak; sıkı tutmak: **to ~ on s.o.'s words,** birini ağzı açık dinlemek. **~ out,** dışarıya asmak; sarkmak; (*coll.*) ikamet etmek. **~ over,** eğilmek, sarkmak; üzerine asılı durmak. **~ together,** birbirine uymak; anca beraber kanca beraber olmak. **~ up,** asmak; geciktirmek. **~ar** [ˈhan(g)ə(r)]. Baraka. **~er** [ˈhan(g)ə(r)]. Çengel, askı: **~er-on,** çanak yalayıcı; peyk. **~ing** [ˈhan(g)in(g)]. Asılı; asma. **this is a ~ matter,** bu insanı darağacına götürür. **~ings,** sarkık kumaşlar. **~man** [ˈhan(g)man]. Cellâd.

**hank** [han(g)k]. Çile; kangal.

**hanker** [ˈhan(g)kə(r)]. **to ~ after, to**

have a ~ing for ..., hasretmek; (bir şey) gözünde tütmek.
**hanky-panky** [ˈhan(g)kiˈpan(g)ki]. Dalavere, el altından iş; göz boyası.
**ha'·penny,** *pl.* **ha'pence** [ˈhêypeni, –pens]. Yarım peni. **three ~,** bir buçuk peni. **~p'orth** [ˈhêypəθ]. **halfpenny-worth,** yarım penilik. **he hasn't a ~ of sense,** on paralık aklı yoktur.
**hap·hazard** [hapˈhazəd]. Rastgele; gelişi güzel. **~less** [ˈhaplis], Talihsiz, bedbaht. **~ly** [ˈhapli]. Tesadüfen. **~pen** [ˈhapən]. Olmak, vukuagelmek; olup bitmek; rastgelmek; tesadüfen vukubulmak. **don't let it ~ again!,** bir daha yapayım deme!: **it so ~ed that ...,** tesadüfen ...: **do you ~ to know?,** acaba biliyor musunuz?: **to ~ upon stg.,** rastgele bulmak: **if anything should ~ to me,** şayed bana bir hal olursa. **~pening,** vak'a: **~s,** olup bitenler.
**happ·iness** [ˈhapinis]. Saadet; memnuniyet; bahtiyarlık. **~y** [ˈhapi]. Mesud; memnun; bahtiyar; isabetli. **as ~ as the day is long [as a king, as a sandboy],** son derece mes'ud: **many ~ returns of the day!,** nice seneler!: **of ~ memory,** cennetmekân: **~ thought!,** ne güzel buluş. **~y-go-lucky,** düşüncesiz; gelişi güzel.
**harangue** [həˈran(g)]. Palavralı, gürültülü bir hitabe(de bulunmak).
**harass** [ˈharəs]. Taciz etm.; mükerrer hücumlarla yormak.
**harbinger** [ˈhâbincə(r)]. Müjdeci; haberci.
**harbour** [ˈhâbə(r)]. Liman; melce. Limanda demirlemek; beslemek. **~-dues,** liman rüsumu. **~-master,** liman reisi.
**hard** [hâd]. Katı; sert; pek; çetin; zor, müşkül; ağır; şefkatsız; cimri. Gayretle, şiddetle; güçlükle. **he is always ~ at it,** mütemadiyen çalışıp çabalıyor: **to be ~ at work,** harıl harıl çalışmak: **~ by,** yakında: **~ cash,** peşin para: **to be ~ on one's clothes,** elbisesini çabuk eskitmek: **to die ~,** kolay kolay ölmemek: **~ drinker,** çok içen kimse: **~ drinks,** alkollü içkiler: **~ facts,** inkâredilemez hakikatler: **~ and fast,** çok kat'î: **~ of hearing,** ağır işitir: **he got two years ~ labour,** iki sene küreğe mahkûm oldu: **~ lines [luck],** aksi talih: **to be ~ on s.o.,** birine karşı insafsızlık etm.: **it is ~ on him,** bu onun talihsizliğidir: **~ to please,** müşkülpesend: **to be ~ put to it,** akla karayı seçmek; hal gelmek: **~ swearing,** okkalı küfürler: **to have a ~ time of it,** çok sıkıntı çekmek: **to try one's ~est,** elinden geleni yapmak: **~ up,** parasız: **we are ~ up for sugar,** şekerimiz kıt: **~ upon (his heels),** tam peşinden: **~ water,** kireçli su: **it will go ~ with us if ...,** ···se halimiz yamandır: **~ worker,** çok çalışan kimse. **~-baked,** sertleşinceye kadar pişirilmiş; pişkin. **~-bitten,** pişkin. **~-boiled,** pişkin: **~ egg,** hazırlop yumurta. **~-earned [-won],** çok çalışarak kazanılmış. **~-fisted,** cimri. **~-hearted,** merhametsiz. **~-mouthed,** ağzı sert. **~-set,** donmuş; katılaşmış. **~-wearing,** dayanıklı (kumaş). **~-working,** çok çalışkan. **~en** [ˈhâdn]. Pekleştirmek; katılaş(tır)mak; sertleş(tir)mek; meşakkate alıştırmak. **prices are ~ing,** fiatlar yükseliyor. **~ened,** katılaşmış; meşakkate alışkın: **~ criminal,** sabıkalı. **~ihood [~iness]** [ˈhâdihud, –nis]. Cesaret; yüz; meşakkate dayanma. **~ness** [ˈhâdnis]. Katılık, peklik, sertlik; güçlük: **~ of hearing,** ağır işitme: **~ of heart,** taşyüreklilik. **~ship** [ˈhâdşip]. Meşakkat. **~ware** [ˈhâdweə(r)]. Madenî eşya. **~wood** [ˈhâdwud]. Kozalaklı ağaclardan gayrı bütün diğer ağacların odunu. **~y** [ˈhâdi]. Meşakkate dayanır; cesur; (nebat) soğuğa dayanır: **~ annual,** (*fig.*) her sene ortaya çıkan mevzu.
**hare** [heə(r)]. Tavşan. (*coll.*) Tavşan gibi koşmak. **run like a ~!,** var kuvvetile koş!: **~ and hounds,** *v.* paperchase: ⌜**first catch your ~ (and then cook it)**⌝, ⌜ayıyı vurmadan postunu satma!⌝: ⌜**to run with the ~ and hunt with the hounds**⌝, ⌜tavşana kaç tazıya tut⌝. **~-brained,** deli, sersem, çılgın. **~-lip,** tavşandudağı.
**haricot** [ˈharikou]. Kuru fasulye.
**hark** [hâk]. Dinlemek, kulak vermek. **to ~ back to stg.,** aynı mevzua dönmek.
**harlequin** [ˈhâləkwin]. Palyaço.
**harlot** [ˈhâlət]. Orospu.
**harm** [hâm]. Zarar (vermek); dokunma(k); fenalık (yapmak). **you will come to ~,** size zarar gelir: **out of ~'s way,** emin bir yerde. **~ful,** zararlı, dokunur. **~less,** zararsız, dokunmaz.
**harmon·y** [ˈhâməni]. Ahenk (ilmi);

tesanüd; mutabakat. **~ious** [–ˈmǫunyəs], ahenkli. **~ium** [-ˈmǫunyəm], küçük seyyar org. **~ize** [ˈhâmənâyz], telif etm., (ahenge) uydurmak; birbirine uymak.

**harness** [ˈhânis]. Koşum takımı. Koşmak; bir nehir elektrik istihsali için kullanmak. **to die in ~**, ölünceye kadar mesleğinde çalışmak: **to get back into ~**, iş başına avdet etmek. **~-maker**, sarac.

**harp** [hâp]. **to ~ on**, (bir mevzua) mütemadiyen dönmek.

**harpoon** [hâˈpûn]. Zıpkın (-lamak).

**harpy** [ˈhâpi]. Tamahkâr ve hasis kimse.

**harrow** [ˈharǫu]. Sürgü (geçirmek). **to ~ s.o.'s feelings**, birinin yüreğini parçalamak. **~ing**, yürek parçalayıc.

**harry** [ˈhari]. Yıkıp yakmak; eziyet vermek.

**Harry.** (*abb.*) **Henry. Old ~**, Şeytan: **to play Old ~ with ...**, berbad etmek.

**harsh** [hâş]. Haşin; merhametsiz; kekre; kulakları tırmalıyan (ses).

**harum-scarum** [ˈheərəmˈskeərəm]. Delişmen.

**harvest** [ˈhâvist]. Hasad (mevsimi); ekin biçme(k); mahsul (toplamak). **~er**, orakçı (makinesi).

**has** *v.* **have.**

**has-been** [ˈhazbîn]. Zamanında değerli olan fakat artık hükmü kalmamış bulunan kimse veya şey.

**hash** [haş]. Doğramak, kıymak; ikinci defa pişirip yahni yapmak. Kıymalı yemek; karışık şey; temcid pilavı. **to make a ~ of stg.**, yüzünü gözünü bulaştırmak: **to settle s.o.'s ~**, icabına bakmak.

**hast** [hast]. (*2nd pers. sing. pres.*) **have.**

**hast·e** [hêyst]. Acele (etmek). **to do stg. in ~**, bir şeyi acele yapmak; üstünkörü yapmak: **to make ~**, acele etm.: ⌜**more ~ less speed**⌝, ⌜tiz reftar olanın payine dâmen dolaşır⌝. **~en** [ˈhêysn]. Acele et(tir)mek; hız vermek. **to ~ to a place**, bir yerde soluğu almak. **~y** [ˈhêysti]. Aceleci; üstünkörü; atılgan. **~iness**, aceleci olma; tezcanlılık.

**hat** [hat]. Şapka. **top ~ [silk ~]**, silindir şapka. **I'll eat my ~ if ..., ...** arab olayım!: **keep it under your ~!**, kimseye söyleme!: **to raise one's ~ to s.o.**, birine şapka çıkarmak: **to take off one's ~ to**, (*coll.*) birinin üstünlüğünü itiraf etmek. **~-peg**, şapka asacak kanca. **~-stand**, şapkalık. **~band** [ˈhatband]. Şapka kurdelâsı. **~less**, şapkasız. **~ter**, şapkacı.

**hatch**[1] [haç]. (*naut.*) Kaporta, ambar ağzı; dam geçidi. **service ~**, servis penceresi: **under ~es**, güverte altında. **~**[2]. Yumurtadan civciv çık(ar)mak. **to ~ a plot**, kumpas kurmak. **~way** [ˈhaçwêy]. Kaporta ağzı.

**hatchet** [ˈhaçit]. Küçük balta. **to bury the ~**, barışmak, sulh yapmak. **~-faced**, dar ve sivri yüzlü.

**hate** [hêyt]. Nefret (etm.), kin (beslemek). **to ~ s.o. like poison**, birini bir kaşık suda boğacak kadar nefret etm.: **I should ~ to be late**, katiyen geç kalmak istemem: **I ~ your going away**, sizin gitmenize çok müteessirim. **~ful**, nefret edilen.

**hath** [haθ]. (*esk.*) (*3rd pers. sing. pres.*) **have.**

**hatred** [ˈhêytrid]. Nefret; kin.

**haught·y** [ˈhôti]. Mağrur; azamet satan. **~iness**, gurur; kurum.

**haul** [hôl]. Kuvvetle çekmek; sürüklemek; yedek çekmek. Çekme; balık ağını çekme (mikdarı); (*fig.*) elde edilen şeylerin mikdarı. **to ~ s.o. over the coals**, birini tekdir etm. **~ down**, indirmek. **~ up**, yukarıya çekmek; (kayık) karaya çekmek. **~age** [ˈhôlic]. Nakletme; kamyonla eşya nakli: **~ contractor**, kamyonla nakleden müteahhid. **~ier** [ˈhôliə(r)]. Kamyoncu; (madenlerde) maden vagonlarını götüren adam.

**haunch** [hônç]. Kalça; sağrı.

**haunt** [hônt]. Birinin sık sık gittiği yer. Sık sık uğramak; tayf halinde sık görünmek; (fikir) musallat olmak. **~ed**, perili. **~ing**, insanın aklından çıkmıyan.

**have**[1] **(had)** [hav, had]. Malik olm., sahib olm.; almak; tutmak. *Yardımcı fiil, (gramere bk.); müteaddi teşkiline yarar, mes.* **to ~ one's boots repaired**, ayakkablarını tamir ettirmek: **to ~ a house built**, bir ev yaptırmak: **I ~ a house**, evim var: **he has a horse**, atı var: **to ~ to do stg.**, bir şeyi yapmağa mecbur olm.: **to be had**, (*sl.*) aldatılmak: **I had him there**, (*coll.*) onu burada kıstırdım: **let him ~ it!**, (i) ona ver!; (ii) vur!; (iii) ağzının payını ver!: **he will ~ it that ...**, iddia ediyor ki: **as Plato has it**, Eflâtunun dediği gibi: **rumour has it that**, rivayete göre: **I won't ~ such behaviour**, böyle harekete müsaade edemem: **I won't ~ anything said against him,** onun aleyhinde söz söyletmem: **we had a lot of rain last week,** geçen hafta çok yağmur yağdı: **to ~ dinner,** *etc.* akşam yemeği yemek vs.: **he had his leg broken,** ayağı kırıldı. **~ at,** ···e hücum etmek. **~ in,** to **~ s.o. in,** birini eve davet etm.: **to ~ the doctor in,** doktor çağırmak. **~ on, to ~ stg. on,** (i) bir şey giymek; (ii) bir işi olmak: **to ~ nothing on,** (i) çıplak olm.; (ii) azade olm.: **to ~** s.o. **on,** (*sl.*) birini aldatmak. **~ out, to ~ a tooth out,** dişini çıkartmak: **to ~ a matter out,** bir meseleyi münakaşa ve halletmek. **~ up,** (*coll.*) mahkemeye celbetmek. **~**[2] *n.* (*coll.*) Hile, dolab. **the ~s and the have-nots,** zenginler ve fakirler.

**haven** [ˈhêyvn]. Liman; melce, sığınak.

**haversack** [ˈhavəsak]. Arka çantası.

**havoc** [ˈhavək]. Büyük zarar; tahribat. **to make ~ of** [**to work ~, to play ~ with**], berbad etmek.

**hawk**[1] [hôk] *n.* Atmaca gibi bir çok yırtıcı kuşlara verilen ad. *v.* Doğan ile avlamak. **~**[2]. Ayak satıcılığı etmek. **~er,** ayak satıcısı.

**hawser** [ˈhôzə(r)]. Palamar.

**hay** [hêy]. Kuru ot. **to make ~,** biçilmiş otu güneşe yayıp kurutmak: **to make ~ of stg.,** karmakarışık etm.: ⌜**to make ~ while the sun shines**⌝, bir işi fırsat varken yapmak. **~making,** ot biçme ve kurutma (mevsimi). **~stack,** kuru ot yığını.

**hazard** [ˈhazəd]. Baht; kaza; tehlike. Talihe bırakmak; tehlikeye koymak. **at all ~s,** her şeyi göze alarak. **~ous,** tehlikeli; baht işi.

**haz·e** [hêyz]. Hafif sis. **~y** [ˈhêyzi]. Puslu; mübhem; bulanık. **I am a bit ~ about the date,** tarihini pek hatırlamıyorum. **~iness,** sis; mübhemlik.

**hazel** [ˈhêyzl]. Fındık ağacı. **~ eyes,** elâ gözler. **~-nut,** fındık.

**H.E.** [ˈêyçˈî]. (*abb.*) (i) **His Excellency,** Hazretleri; (ii) **high explosive,** yüksek infilaklı (madde).

**he** [hî]. (İsmi geçen erkek adam) o. **~-** (*pref.*) Erkek ...; **~-goat,** teke. **~-man,** erkek adam.

**head**[1] [hed] *n.* Baş, kafa; tepe; reis; zekâ; tane; bab, madde; (futbol) kafa vuruşu. **sixpence a ~** [**per ~**], adam başına altı peni: **taller by a ~,** bir baş boyu daha uzun: **to win by a ~,** bir baş boyu farkla kazanmak: **to win by a short ~,** pek küçük farkla kazanmak: **~ first,** baş aşağı: **to give a horse his ~,** atın başını serbest bırakmak: **to go to one's ~,** (içki vs.) başına vurmak: **to keep one's ~,** kendini kaybetmemek: **to keep one's ~ above water,** su üzerinde durmak; (*fig.*) borca girmeden geçinmek: **to lay ~s together,** baş başa vermek: **to lose one's ~,** (i) pusulayı şaşırmak; (ii) idam edilmek: **to make ~,** ilerlemek: **to make ~ against stg.,** mukavemetini kırmak: **he is off his ~,** aklından zoru var: **to stand on one's ~,** baş aşağı durmak: **on your own ~ be it!,** günahı boynuna!: **~ on to the wind,** rüzgâra karşı: **he gives orders over my ~,** bana sormadan emirler veriyor: **to talk over s.o.'s ~,** birine anlayamıyacağı şeylerden bahsetmek; **under separate ~s,** ayrı fasıllarda: **~s or tails?,** yazı mı tuğra mı?: ⌜**~s I win, tails you lose**⌝, ne olursa olsun ben kazanırım: **I cannot make ~ or tail of this,** bundan hiç bir mana çıkaramıyorum: **to take it into one's ~ to ..., ...** aklına esmek: **now matters are coming to a ~,** işte şimdi dananın kuyruğu kopacak: **to bring matters to a ~,** bir işi kat'î bir neticeye bağlamak: ⌜**two ~s are better than one**⌝, ⌜akıl akıldan üstündür⌝. **~-dress** [**~-gear**], başlık. **~-hunter,** insan avcısı. **~-light,** projektör. **~-on,** baş başa (çarpışma vs.). **~-phone,** kulaklık. **~-splitting,** başını catlatan (ağrı). **~-wind,** pruva rüzgârı. **~-work,** zihin işi. **~**[2] *v.* Başa geçmek; başta olmak; başa koymak. **to ~ the ball,** (futbol) kafa vurmak: **the country is ~ing for disaster,** memleket felâkete doğru sürükleniyor: **to ~ the list,** listenin başında gelmek. **~ off,** yolunu kesmek. **~ache** [ˈhedêyk]. Başağrısı. **-~ed** [ˈhedid] *suff.* ···başlı; ···saçlı. **~er** [ˈhedə(r)]. Başaşağı dalış. **~ing** [ˈhedin(g)]. Serlevha; başlık. **to come under the ~ of ..., ...** faslına girmek. **~land** [ˈhedlənd]. Burun. **~line** [ˈhedlâyn]. Başlık. **~long** [ˈhedlon(g)]. Başaşağı; paldır küldür; düşüncesiz. **~man** [ˈhedmən]. Kabile reisi; köy muhtarı. **~quarters** [hedˈkwôtəz]. (*abb.* **H.Q.**) Karagâh; merkez. **General ~ (G.H.Q.),** umumî karargâh. **~sman** [ˈhedzmən]. Cellâd. **~stone** [ˈhedstoun]. Mezar taşı. ~**strong** [ˈhedstron(g)]. Dikkafalı; inadcı. ~**way** [ˈhedwêy]. İlerleme; hız. **to gather ~**, hız almak; **to make ~**, ilerlemek. ~**y** [ˈhedi]. Delişmen.

**heal** [hîl]. Şifa vermek; iyileş(tir)mek: ~ **up**, (yara) kapanmak: **to ~ the breach**, barıştırmak. ~**th** [helθ]. Sıhhat, sağlık. **to drink s.o.'s ~**, birinin sıhhatine içmek: **Bill of ~**, pratika. ~**thful**, sıhhî; şifa veren. ~**thy**, sıhhati yerinde; sağlam; sıhhate yarar: ~ **appetite**, iyi iştah: ~ **criticism**, iyi niyetle, faydalı tenkid. ~**thiness**, sıhhatlı olma.

**heap** [hîp]. Yığın; küme(lemek). **a ~ of** [~**s of**], bir çok: **to fall in a ~**, düşüp yığılmak: **to be struck all of a ~**, (*sl.*) hayretten küçük dilini yutmak: **to ~ praises on s.o.**, birini sitayişlere boğmak. ~**ed**, *a.* ağız ağza dolu.

**hear (heard)** [hiə(r), hö(r)d]. İşitmek; duymak; dinlemek; haber almak. ~**!** ~**!**, bravo!: **he likes to ~ himself talk**, çok konuşmaktan hoşlanıyor: **I have ~d it said, ...** söylendiğini işittim: **to ~ from s.o.**, birinden mektub almak: **I won't ~ of it!**, dünyada bunu kabul edemem. ~**er** [ˈhiərə(r)]. Dinleyici. ~**ing** [ˈhiərin(g)] *n.* Dinleme; işitme duygusu. **within ~ (out of ~)**, işitil(miy)ecek mesafede: **to condemn s.o. without a ~**, birini dinlemeden aleyhinde hüküm vermek: **to be quick of ~**, kulağı keskin olmak: **it was said in my ~**, kulaklarımla duydum. ~**say** [ˈhiəsêy]. Şayia; kulaktan dolma.

**heart** [hât]. Kalb; gönül; can; yürek; merkez; merhamet; cesaret; gayret; (iskambil) kupa. **after one's own ~**, tam istediği gibi: **at ~**, içinden: **to have s.o.'s welfare at ~**, birinin saadetile candan alâkadar olmak: ~ **attack**, kalb krizi: **to have one's ~ in one's boots**, meyüs olmak: **from the bottom of my ~**, en candan: **to break one's ~ over stg.**, bir şeyden dolayı içi içini yemek: **to break one's ~**, mahvetmek: **by ~**, ezberden: **to learn by ~**, ezberlemek: **a change of ~**, hislerin değişmesi: **to one's ~'s content**, canının istediği kadar: ~ **failure**, kalb sektesi: **not to find it in one's ~ [not to have the ~ to]**, kıyamamak: **in good ~**, keyfi yerinde: **to have one's ~ in one's work**, işini sevmek: **in my ~ of ~s**, kalbimin derinliklerinde: **to lose ~**, ye'se düşmek: **to lose one's ~ to**, ···e kalbini kaptırmak: **the ~ of the matter**, meselenin esası: **to have one's ~ in one's mouth**, canı ağzına gelmek: **to put new ~ into s.o.**, birine yeniden cesaret vermek: **his ~ is in the right place**, iyi niyetlidir: **to set one's ~ on**, ···i aklına koymak: ~ **and soul**, can ve gönülden: **to take stg. to ~**, bir şeyi kendine derd etm.: **a sight that goes to the ~**, yürek parçalıyan bir manzara. ~**-ache**, keder. ~**-breaking**, son derece keder verici. ~**-broken**, kederden kolu kanadı kırılmış. ~**-felt**, samimî; içten. ~**-searching**, vicdanı araştıran. **to feel a tug at the ~-strings**, kalbinin en hassas teline dokunulmak. ~**-to-~**, samimî. ~**-whole**, kimseye gönül vermemiş. ~**burn** [ˈhâtbö(r)n]. Mide ekşimesi. ~**ed** [ˈhâtid]. ···kalbli, ···yürekli, (*fig.*) **stony-~**, taşyürekli. ~**en** [ˈhâtn]. Cesaret vermek. ~**less** [ˈhâtlis]. Merhametsiz. ~**rending** [ˈhâtrendin(g)]. Yürek parçalayıcı. ~**y** [ˈhâti]. Candan; dinc; fazla ne'şeli ve gürültülü: ~ **appetite**, iyi iştah: ~ **meal**, bol bir yemek.

**hearth** [hâθ]. Ocak; ocağın önü; ocağın başı. **without ~ or home**, od yok ocak yok. ~**stone**, ocaktaşı; kil. ~**-rug**, ocağın önüne serilen kilim.

**heat** [hît]. Hararet, sıcaklık; kızgınlık. ~ **(up)**, Isıtmak; ısınmak. **on ~**, (dişi hayvan) kızgın: **to get ~ed**, hiddetlenmek. ~**ed**, ısınmış; hararetli. ~**er**, ısıtıcı şey; **electric ~**, elektrik sobası. ~**-stroke**, güneş çarpması. ~**-wave**, sıcak dalgası. ~**ing** [ˈhîtin(g)]. Isıtıcı; ısıtma cihazı. **central ~**, kalorifer.

**heath** [hîθ]. Fundalık; çorak arazi. ~**er** [ˈheðə(r)]. Süpürge otu; funda.

**heathen** [ˈhîðn]. Putperest; kâfir. ~**ish**, putperestliğe aid, kâfir.

**heave (heaved veya hove)** [hîv, -d, houv]. (Ağır bir şeyi) atmak, kaldırmak; deniz gibi kabarıp inmek. Şiddetle atma, itme, çekme. **to ~ on a rope**, palamar zorla çekmek: **to ~ the lead**, iskandil atmak: **to ~ a sigh**, derin derin iç çekmek: **to ~ in sight**, birdenbire görünüvermek: **to ~ to** (gemi) dur(dur)mak: ~ **ho!**, yisa!.

**heaven** [ˈhevn]. Gök, sema; cennet Allah. **the ~s**, sema; kubbe: **in ~** cennette; öbür dünyada: **to go to ~** cennete gitmek: **good ~s!**, Allah

Allah!: **thank ~!**, bereket versin; çok şükür: **for ~'s sake,** Allah aşkına: **would to ~**, ah keşki: **to move ~ and earth to get stg.**, bir şeyi elde etmek için yapmadık bir şey bırakmamak. **~-sent**, Allah tarafından gönderilen. **~ly** [ˈhevənli]. Semavî; ilâhî; cennete aid; Allahtan; (*coll.*) pek nefis: **Our ~ Father**, Allah.

**heav·iness** [ˈhevinis]. Ağırlık, sıklet; kasvet; uyuşukluk. **~ of heart,** keder. **~y** [ˈhevi]. Ağır; sakil; iri yapılı; kalın; kasvetli; şiddetli; cansız. **~ crop,** zengin mahsul: **a ~ day,** (i) kasvetli bir gün; (ii) çok çalışılan gün: **~ fire,** şiddetli top ateşi: **to have a ~ heart,** kederli olmak: **~ meal,** bol yemek: **to be a ~ sleeper,** uykusu ağır olmak: **a ~ step,** tok ayak sesi: **time hangs ~ on his hands,** yapacak bir şey olmadığı için cansıkıntısı içindedir. **~-eyed,** gözleri (uykusuzluktan) çakmak. **~-handed,** beceriksiz; zalim. **~-hearted,** kederli.

**Hebrew** [ˈhîbru]. Yahudi; ibranice.

**hecatomb** [ˈhekətûm]. Katliâm.

**heckle** [ˈhekl]. Birini nutuk söylerken müşkül suallerle sıkıştırmak.

**hectic** [ˈhektik]. Pek heyecanlı; telâşlı.

**hecto-** [ˈhektou̯]. Yüz ....

**hector** [ˈhektə(r)]. Kabadayı. Yüksekten atmak.

**he'd** [hîd] = **he had; he would.**

**hedge** [hec]. Çit; mania. Çit çevirmek; kaçamaklı davranmak. **~d about with difficulties,** müşkülâtla çevrilmiş. **~hog** [ˈhechog]. Kirpi.

**hedonism** [ˈhîdənizm]. Hayatın maksadını zevk telâkki eden felsefe.

**heed** [hîd]. Aldırmak; kulak vermek; dikkat etmek. Aldırış; ehemmiyet verme; dikkat. **to pay [take] no ~,** aldırmamak; dikkat etmemek. **~ful,** dikkatli; ihtiyatlı. **~less,** dikkatsiz; aldırış etmiyen.

**hee-haw** [ˈhîˈhô]. Anırma(k).

**heel**[1] [hîl]. Topuk; ökçe; bir şeyin ara kısmı. Ökçe vurmak. **to be at [on] one's ~s,** tam peşinde olm.: **to bring s.o. to ~,** birini yola getirmek: **to come to ~,** (köpek) çağırılınca sahibinin peşinden gelmek; (*fig.*) itaat etmek: **to be down [out] at ~s,** ökçeleri ezilmiş olm.; sefalette olm.: **(horse) to fling out its ~s,** (at) çifte vurmak: **head over ~s,** tepe taklak: **to kick [cool] one's ~s,** işsiz, sabırsızlanarak beklemek: **to show a clean pair of ~s,** kaçıp gözden kaybolmak: **to take to one's ~s,** tabanları yağlamak: **to tread upon s.o.'s ~s,** birini yakından takib etmek: **to turn on one's ~,** birden bire dönmek: **to be under the ~ of the invader,** müstevlinin çizmesi altında olmak. **~-tap,** kadehin dibindeki son yudum. **~**[2] *v.* **~ (over),** (gemi) bir yana yat(tır)mak.

**hefty** [ˈhefti]. İri yarı.

**hegemony** [hîˈceməni]. Tahakküm.

**Hegira** [ˈhecra, həˈci̯əra]. Hicret.

**heifer** [ˈhefə(r)]. Doğurmamış genc inek.

**heigh** [hêy]. *Dikkat celbi nidası.*

**height** [hâyt]. İrtifa, yükseklik; yüksek yer; zirve; boy; son derece. **in the ~ of the battle,** muharebenin en civcivli zamanında: **the ~ of folly,** deliliğin son mertebesi. **~en,** yükseltmek; artırmak.

**heinous** [ˈhêynəs]. İğrenc; affolunmaz.

**heir** [e̯ə(r)]. Vâris, mirasçı. **~ to the throne,** veliahd. **~-apparent,** mirastan iskat edilemiyen meşru mirasçı. **~ess** [ˈe̯əris]. Kadın vâris. **~loom** [ˈe̯əlûm]. Babadan kalma kıymetli bir şey.

**held** *v.* **hold.**

**heliotrope** [ˈhelyotro̯up]. Açık mor.

**hell** [hel]. Cehennem. **a ~ upon earth,** Allahın belâsı bir yer: **~ let loose,** cehennemden nümune: **to make the ~ of a noise,** çok gürültü yapmak: **to raise ~,** kıyamet koparmak: **to ride ~ for leather,** (at) dörtnala koşmak: **to work like ~,** domuzuna çalışmak: **what the ~ do you want?,** ne istiyorsun, be adam? **~-fire,** cehennem azabı. **~ish** [ˈheliş]. Cehenneme aid; müdhiş.

**he'll** [hîl] = **he will.**

**Hellen·e** [ˈhelîn]. Helen. **~ic** [–ˈlînik], Yunanlılara aid.

**hello** *v.* **hallo.**

**helm** [helm]. Dümen (yekesi). **at the ~,** dümende; idare eden: **to answer the ~,** dümeni dinlemek. **~sman** [ˈhelmsmən]. Dümenci.

**helmet** [ˈhelmit]. Miğfer; tulga. **sun ~,** kolonyal şapka.

**helot** [ˈhelət]. Esir, köle.

**help** [help]. Yardımcı, muavin. Yardım (etm.), muavenet (etm.), imdad (etm.); kolaylaştırmak; sofrada yemek dağıtmak. **~!,** imdad!: **I can't ~ it,** elimde değil: **I can't ~**

thinking, bence muhakkak: **it can't be ~ed [there's no ~ for it]**, çare yok: **don't be longer than you can't ~**, mümkünse fazla gecikme: **I couldn't ~ laughing**, gülmekten kendimi alamadım: **to cry for ~**, 'imdad' diye bağırmak: **to ~ s.o. down [in, out, up]**, birine inerken, vs. yardım etm.: **so ~ me God!**, Allah şahid olsun!: **mother's ~**, çocuğa bakmağa gelen hizmetçi: **past ~**, ümid yok: **to ~ s.o. on [off] with a coat**, *etc.*, palto vs. giyerken (çıkarırken) birine yardım etm.: **~ yourself!**, (yemek) siz kendiniz buyurun! **~er**, yardımcı, muavin. **~ful**, kolaylaştırıcı; yardım eden. **~ing**, *n.* bir tabak yemek. **~less**, âciz; kimsesiz; çaresiz.

**helter-skelter** [ˈheltə(r)ˈskeltə(r)]. Kaçan kaçana.

**hem¹** [hem]. Dikilmiş kenar; etek. Kenarını kıvırıp dikmek. **~ in**, kuşatmak. **~².** Manalı manalı öksürmek.

**hemi-** [ˈhemi] *pref.* Yarım ...; *mes.* **~sphere** [ˈhemisfiə(r)]. Yarımküre.

**hemp** [hemp]. Kenevir; benk. **~en**, kenevirden yapılan.

**hen** [hen]. Tavuk; dişi kuş. **to set a ~**, bir tavuğu kuluçkaya yatırmak. **~-coop**, tavuk kafesi. **~-house**, tavuk kulübesi. **~-party**, (*coll.*) yalnız kadınlardan mürekkeb toplantı. **~-pecked**, kılıbık.

**hence** [hens]. Buradan; bu sebebden; binaenaleyh. **ten years ~**, bundan on sene sonra. **~forth** [ˈhensˈfôθ], [**~forward** [–ˈfôwəd]], bundan sonra.

**henchman** [ˈhençmən]. Sadık yardımcı; uşak.

**henna** [ˈhennə]. Kına.

**her** [hö(r)]. (Dişi hakkında) onu; onun;. ona.

**herald** [ˈherəld]. Haberci, müjdeci. İlân etm.; geleceğini haber vermek. **~ic**, armacılığa aid. **~ry**, armacılık.

**herb** [hö(r)b]. Ot, nebat, *esp.* baharat makamında kullanılan otlar. **~alist** [ˈhö(r)bəlist]. İlâclık satan kimse.

**herculean** [hö(r)kyuˈliən]. Herkül'e aid; beşerî kuvvetin fevkinde olan.

**herd** [hö(r)d]. Hayvan sürüsü. Sürüye katmak; hayvan sürüsüne bakmak. **to ~ together**, sürü halinde toplanmak: **the common ~**, ayaktakımı: **the ~ instinct**, sürü hissi. **~sman**, sığırtmaç.

**here** [hiə(r)]. Burada; buraya. **about ~**, bu civarda: **~ and there**, şurada burada; arasıra: **~ and now**, derhal: **~ there and everywhere**, her tarafta: **that's neither ~ nor there**, bunun mesele ile bir alâkası yok: **~'s your book**, işte kitabınız!: **~!**, gel!: **~ goes!**, haydi bakalım!. **~about(s)** [ˈhiərabaut(s)]. Bu civarda. **~after** [hiərˈâftə(r)]. İstikbalde; bundan sonra; aşağıda: **the ~**, ahret. **~by** [ˈhiəˈbây]. *Bir mukavele başında:* **I ~** ..., ben bu vesika ile .... **~in** [ˈhiərin]. Bunun içinde. **~tofore** [ˈhiətoˈfô(r)]. Şimdiye kadar; evvelce. **~upon**, bunun üzerine. **~with**, bununla; ilişik olarak.

**heredit·y** [həˈrediti]. İrsiyet; irs. **~ary**, miras ile intikal eden; mevrus.

**here·sy** [ˈherəsi]. Bir akideye muhalif olan mezheb; itizal; yanlış fikir. **~tic** [ˈherətik]. İtizalci. **~tical** [–ˈretikl], itizale aid.

**herita·ble** [ˈheritəbl]. Miras ile intikal edebilen. **~ge** [ˈheritêyc]. Miras; tereke.

**hermaphrodite** [hö(r)ˈmafroudâyt]. Hünsa.

**hermetic** [hö(r)ˈmetik]. Hava geçmez şekilde kapalı.

**hermit** [ˈhö(r)mit]. Tariki dünya. **~age**, zaviye.

**hernia** [ˈhö(r)nyə]. Fıtık.

**hero** [ˈhiərou]. Kahraman. **~ic** [hiəˈrouik]. Kahramanca. **~ics**, mübalâğalı ve tumturaklı sözler. **~ine** [ˈherouin]. Kadın kahraman. **~ism** [ˈherouizm]. Kahramanlık.

**Herod** [ˈherod]. **to out-~ ~**, zulümde Firavuna taş çıkartmak.

**heroin** [ˈherouin]. Morfin hulâsası, eroin.

**heron** [ˈherən]. Balıkçıl.

**herring** [ˈherin(g)]. Ringa balığı. **red ~**, tütsülenmiş ringa: (*fig.*) **to draw a red ~ across the trail**, bir söz söyliyerek muhavereyi kasden sadedden çevirmek. **~-bone**, ringa kılçığı gibi. **~-pond**, (*şak.*) Atlantiğin şimal kısmı.

**hers** [hö(r)z]. (Kadın) onun, onunki. **a friend of ~**, dostlarından biri. **~elf** [hö(r)ˈself]. (Kadın) kendisi.

**he's** [hîz] = **he is.**

**hesita·ncy** [ˈhezitənsi]. Tereddüd. **~nt**, mütereddid. **~te** [ˈhezitêyt]. Tereddüd etmek. **~tion** [–ˈtêyşn], tereddüd: **without the slightest ~**, hiç tereddüd etmiyerek.

**heterodox** [ˈhetərodoks]. Müesses dinlere, fikirlere, aykırı olan. **~y,** böyle bir aykırılık.
**heterogeneous** [ˌhetəroˈcenəs]. Birbirine uymıyan.
**hew** (**hewed, hewn**) [hyû, hyûd, hyûn]. Balta ile vurarak kesmek, yontmak. **to ~ coal,** kömür kazmak: **to ~ out a career for oneself,** çalışıp çabalıyarak meslek hayatını yapmak: **to ~ one's way through,** vurarak kendine yol açmak.
**hexagon** [ˈheksəgon]. Altı köşeli şekil.
**heyday** [ˈhêyˈdêy]. ···in en parlak devri.
**H.H.** =His Highness. *v.* highness.
**hi** [hây]. Yahu!; bana bak!
**hibernate** [ˈhâybənêyt]. Kış mevsimini uykuda geçirmek.
**hiccough, hiccup** [ˈhikʌp]. Hıçkırık (tutmak).
**hid(den)** [ˈhidn] *v.* **hide**; *a.* Saklı, gizli.
**hide**[1] (**hid, hidden**) [hâyd, hid, hidn]. Sakla(n)mak; gizle(n)mek. **to ~ one's head,** utancdan sinmek. **~-and-seek,** saklambac. **~**[2]**.** Deri; post. **~-bound,** dar ve değişmez fikirli.
**hideous** [ˈhidyəs]. Son derece çirkin, iğrenc.
**hiding** [ˈhâydin(g)] *n.* Sakla(n)ma; (*coll.*) dayak.
**hierarchy** [ˈhâyərâki]. Memurların derece ve rütbe silsilesi.
**hieroglyph** [ˈhâyərouglif]. Eski Mısırlıların yazısı; anlaşılmaz yazı.
**higgledy-piggledy** [ˈhigəldiˈpigəldi]. Karmakarışık.
**high** [hây]. Yüksek; yüksekteki; ulu; asil; (et) hafifçe bozulmuş; (*mus.*) tiz; (deniz, rüzgâr) sert. **to go as ~ as £100,** yüz liraya kadar vermek: **~ day,** bayram: **~ and dry,** (gemi) tamamen karada: **to leave s.o. ~ and dry,** birini yüzüstü bırakmak: **from on ~,** yukarıdan; Allahtan: **with a ~ hand,** keyfî; karakuşî: **to ride the ~ horse,** yukarıdan almak: **~ latitudes,** kutba yakın olan mıntakalar: **~ lights,** bir resim ışıklı tarafları; en göze çarpan şeyler: **~ life,** yüksek sosyete hayatı: **to hunt ~ and low for stg.,** fellek fellek aramak: **~ and mighty,** (*fig.*) mağrur: **~ noon,** tam öğle vakti: **on ~,** gökte: **feelings ran ~,** pek hiddetli münakaşalar oldu: **the ~ seas,** açık denizler: **in ~ spirits,** neş'eli: **the ~ table,** kolejde profesörlere mahsus sofra: **to have a ~ time,** çok eğlenmek: **it's ~ time we went,** artık gitmek zamanı geldi: **~ water,** med: **~ words,** hiddetli sözler, kavga. **~-born,** asil. **~-class,** iyi cinsten; yüksek sınıftan. **~-flown,** tumturaklı. **~-flyer,** gözü yüksekte olan. **~-grade,** iyi cinsten. **~-handed,** mütehakkim. **~-minded,** âlicenab. **~-sounding,** tantanalı. **~-spirited,** cesur; atılgan; ateşli. **~brow** [ˈhâybrau]. Fikir ve sanatte ince zevk sahibi. **~land** [hâyland]. Dağlık araziye aid. **~lands,** dağlık; Iskoçya'nın şimali. **~lander,** dağlı. **~ly** [ˈhâyli]. Son derece; pek çok. **~ness** [ˈhâynis]. Yükseklik; irtifa. **His Highness** [**Your ~**], prenslere verilen unvan. **~way** [ˈhâywêy]. Ana yol. **the King's ~,** devlet yolu. **~wayman,** yolkesici.
**hike** [ˈhâyk]. (*coll.*) Yaya olarak seyahat etmek.
**hilari·ous** [hiˈleəriəs]. Neş'eli ve gürültücü. **~ty** [-ˈlariti], gürültülü neş'e.
**hill** [hil]. Tepe; yokuş; küme. **up ~ and down dale** [**over ~ and dale**], dere tepe. **~ock,** tepecik. **~side,** tepe yamacı. **~top,** tepe üstü. **~y,** inişli yokuşlu.
**hilt** [hilt]. Kabza. **up to the ~,** tamamen.
**him** [him]. (Erkek) onu, ona, ondan, onunla: **it's ~,** odur: **that's ~,** işte odur. **~self,** kendisi.
**hind** [hâynd]. Arka. **to get on one's ~ legs,** (*şak.*) ayağa kalkmak. **~most** [ˈhâyndmoust]. En arkadaki. ⌜**the devil take the ~**⌝, ⌜sona kalan donakalır⌝.
**hinder** [ˈhində(r)]. Engel olm.; mâni olmak.
**Hind·i** [ˈhindi]. Hindce. **~u** [hinˈdû]. Hindu, Hindli; hindce. **~ustan,** Hindistan. **~ustani,** hindce.
**hindrance** [ˈhindrəns]. Engel, mânia. **without let or ~,** mâni olmadan.
**hinge** [hinc]. Menteşe; esas noktası. Menteşe üzerinde dönmek. **to ~ on ...,** (*fig.*) ···e bağlı olmak.
**hint** [hint]. İma; üstü kapalı anlatma; fikir verme. İma etm.; çıtlatmak. **a broad ~,** açık bir ima: **to give** [**drop**] **a ~,** birine çıtlatıvermek: **to give ~s about stg.,** bir şey hakkında birine bir fikir vermek: **not the slightest ~ of ...,** ···den en küçük bir iz bile yok.
**hinterland** [ˈhintəland]. Sahilin gerisindeki arazi.

**hip**[1] [hip]. Kalça. **to smite ~ and thigh,** bozguna uğratmak. **~-bath,** badya şeklinde banyo. **~-pocket,** pantalonun arka cebi. **~**[2]. **~! ~! hurrah!,** şa! şa! şa!

**hippo-** [ˈhipou] *pref.* Ata aid. **~-drome** [ˈhipoudroum]. At meydanı; sirk. **~potamus** [ˌhipouˈpotəməs]. Su aygırı; (*abb.* **hippo.**).

**hire** [hâyə(r)]. Kira (ücreti). Kiralamak. **on ~;** kiralık: **to ~ out** [**to let out on ~**], kiraya vermek. **~-purchase,** taksitle satın alma.

**his** [hiz]. (Erkek) onun; onunki.

**hiss** [his]. S harfi gibi ses; tıslama. Tıslamak; ıslıkla tezyif etmek.

**histor·ian** [hisˈtôriən]. Müverrih; tarihçi. **~ic** [hisˈtorik]. Tarihî. **~ical,** tarihe aid, tarihî. **~y** [ˈhistəri]. Tarih (kitabı). **that's ancient ~,** (i) o tarihe karıştı; (ii) onu herkes bilir: **the inner ~ of stg.,** bir meselenin içyüzü: **s.o.'s past ~,** birinin geçmişi: **a patient's ~,** bir hastanın sıhhî tercümei hali.

**histrionic** [histriˈonik]. Aktörlüğe aid.

**hit** [hit]. Vurmak, dövmek, çarpmak; isabet etm.; ararken tesadüfen bulmak. Darbe; vurma; isabet; muvafakkiyet. **to have a sly ~ at s.o.,** birine taş atmak: **that was a ~ at you,** bu taş sana: **to be hard ~,** büyük bir sarsıntıya uğramak: **to make a ~ (of stg.),** muvaffak olm.: **~ or miss,** rastgele. **~ back,** karşılığını vermek. **~ off, that ~s him off exactly [to a T],** tam o!, tıpkı o!: **we don't ~ it off,** geçinemiyoruz. **~ on,** tesadüfen bulmak: **I've ~ upon a good plan,** iyi bir plân aklıma geldi. **~ out,** darbeler savurmak.

**hitch** [hiç]. Anî çekiş, engel. İliştirmek, takmak. **there is a ~ somewhere,** bir yerde bir ârıza var: **without a ~,** pürüzsüz. **~-hike** [ˈhiçˈhâyk]. Kısmen yaya ve kısmen de yoldan geçen otomobillere parasız binerek seyahat etmek.

**hither** [ˈhiðə(r)]. Buraya, bu tarafa. **~ and thither,** şuraya buraya. **~to,** şimdiye kadar.

**hive** [hâyv]. Arıkovanı; bir kovanın arılarının mecmuu.

**H.M.S.** [êyçˈemˈes] = **Her Majesty's Ship,** İngiliz donanmasına mensub gemi.

**ho** [hou]. *Hayret, zafer, istihza celbetme nidası.* ···e doğru. **westward ho!,** batıya doğru!

**hoar** [hô(r)]. Ağarmış, kır. **~-frost,** kırağı. **~y** [ˈhôri]. Ağarmış; ak saçlı; asırdide. **~iness,** ak saçlılık; pek eski olma.

**hoard** [ˈhôd]. **~ (up),** biriktirmek; istifçilik etmek. Define; istif edilen şey. **~ing** [ˈhôdin(g)], tahta perde: **advertisement ~,** afiş tahtası.

**hoarse** [hôs]. Boğuk, kısık (sesli). **to shout oneself ~,** sesi kısılıncaya kadar bağırmak.

**hoax** [houks]. Muziblik; şaka olarak aldatma(k); kafese koymak.

**hobble** [ˈhobl]. Köstek(lemek); topallamak. **to ~ along,** topallıyarak yürümek.

**hobbledehoy** [ˌhobldiˈhôy]. Gelişme çağında biçimsiz hareketli ve utangac çocuk.

**hobby**[1] [ˈhobi]. İş haricinde zevk için meşguliyet; merak. **~-horse,** değnekten at: **to ride one's ~,** çok sevdiği bir mevzua dönmek.

**hobnail** [ˈhobnêyl]. Kabara.

**hob-nob** [ˈhobˈnob]. **to ~ with s.o.,** birisile senli benli olmak.

**hobo** [ˈhoubou]. (*Amer.*) Serseri, aylak.

**Hobson** [ˈhobsən]. **~'a choice,** tercih imkânı yok.

**hockey** [ˈhoki]. Çomakla oynanan bir top oyunu.

**hocus** [ˈhoukəs]. Aldatmak. İçeceğe ilâc katarak uyutmak. **~-pocus,** göz boyası.

**hod** [hod]. Duvarcı arkalığı.

**Hodge** [hoc]. İngiliz köylüsünün temsili.

**hoe** [hou]. Çapa(lamak). **a hard row to ~,** uzun ve meşakkatlı iş.

**hog** [hog]. Domuz; obur kimse. **to go the whole ~,** bir işte sonuna kadar gitmek. **~'s-back,** balık sırtı tepe. **~shead** [ˈhogshed]. Büyük fıçı.

**hoist** [hôyst]. Hisa etm.; yükseltmek; (bayrak) çekmek. Yükseltme cihazı. **to give s.o. a ~ up,** kalkarken veya ata binerken birine yardım etmek.

**hold**[1] [hould]. Gemi ambarı. **~**[2] *n.* Tutunacak yer; tutma. **to get ~ of stg.,** bir şeyi ele geçirmek: **I can't get a ~,** tutunamıyorum: **to have a ~ over s.o.,** birine hükmü geçmek: **to keep ~ of stg.,** bir şeyi salıvermemek: **to lose one's ~,** (tutunduğu yerden) eli kurtulmak; birinin üzerindeki nüfuzunu kaybetmek. **~**[3] **(held)** [hould, held] *v.* Tutmak; işgal etm.; saymak; içine almak, ihtiva etm.; sabit durmak. **to ~ s.o. to his promise,** birine vadini tutturmak. **~-all,** bir nevi hurc; öteberi doldurmak için bir nevi bavul. **~ back,** geri tutmak; saklamak; alıkoymak; çekinmek. **~ down,** yerinde tutmak; yerde tutmak; (başını) eğmek; inkiyad altında tutmak. **~ forth,** bir mevzuda yüksekten atmak; kandırmak için teklif etmek. **~ in,** zabtetmek. **~ off,** yaklaştırmamak: **if the rain ~s off,** yağmur yağmıyacak olursa. **~ on,** bırakmamak; devam etm.: **~ on a bit!,** yavaş!, biraz dur! **out,** uzatmak; kandırmak için teklif ileri sürmek; teslim olmamak; dayanmak. **~ over,** geri bırakmak; tehir etmek. **~ together,** bir arada tutmak; çözülmemek; (anca beraber, kanca beraber) ayrılmamak. **~ up,** yukarı tutmak; kaldırmak; durdurmak; yolunu kesip soymak; düşmemek: **to ~ s.o. up to ridicule,** sözlerile birini gülünç etm.: **to ~ s.o. up as an example,** birini örnek göstermek: **to ~ up one's head again,** başını bir daha doğrultmak: **~ up!,** aman düşme! **~ with, I don't ~ with these new ideas,** bu yeni fikirlerle başım hoş değil. **~er** [ˈhǫuldə(r)]. Tutan adam; tutacak şey; ... sahibi. **~fast** [ˈhǫuldfâst]. Tutucu alet; kenet. **~ing** [ˈhǫuldin(g)] *n.* Tutma; arazi parçası: **I have a small ~ in X. Company,** X. şirketinde bir kaç hissem var.

**hole** [hǫul]. Delik; çukur; in; fena vaziyet; nahoş bir yer. Delik açmak. **to be in a ~,** fena vaziyette bulunmak: **to get s.o. out of a ~,** birini fena bir vaziyetten kurtarmak: **in ~s** [**full of ~s**], delik deşik: **to make a ~ in ...,** ···de bir delik açmak; ···de büyük bir gedik açmak: **to pick ~s,** kusur bulmak: **to search every ~ and corner,** bütün kıyı bucağı araştırmak.

**holiday** [ˈholidêy]. Tatil günü, müddeti; azad; bayram. **the ~s,** mekteb tatili: **to be on ~** [**on one's ~s**], izinli olm.; seyahatte olm.: **half ~,** yarım azad. **~maker,** tenezzüh yerlerine iznini geçirmek için gelen kimse.

**holiness** [ˈhǫulinis]. Kudsiyet; mubareklik; dindarlık. **His H~.** Papa.

**Holland** [ˈholənd]. Holanda, Felemenk.

**hollo(a)** [ˈholǫu]. Bağırmak. Bağırış.

**hollow** [ˈholǫu]. Kof, içi boş; çukur (yer); gayrı samimî. **~ out,** içini oymak: **the ~ of the hand,** avuc içi: **~ excuse,** boş mazeret: **~ sound,** içi boş bir şeyden gelen ses: **to beat s.o. ~,** (*coll.*) birini kolayca yenmek.

**holocaust** [ˈholokôst]. Büyük insan telefatına sebeb olan yangın; katliâm.

**holster** [ˈholstə(r)]. Tabanca kılıfı.

**holy** [ˈhǫuli]. Mukaddes, kudsî, mubarek; şerif. **the H~ Ghost,** ruhülkudüs: **the H~ Land,** Filistin: **H~ of Holies,** musevi mabedinin en iç kısmı; harim: **~ orders,** papazlık: **to take ~ orders,** papaz olm.: **that child is a ~ terror,** (*coll.*) bu çocuk Allahın belâsıdır.

**home** [hǫum]. Aile ocağı; yuva; ev; yurd, vatan; melce. **to be at ~,** evde bulunmak; misafirleri kabul etm.: **to be at ~ with a subject,** bir mevzuu iyice bilmek: **to feel at ~ with s.o.,** birisile hiç yabancılık hissetmemek: **to make oneself at ~,** misafir gibi durmamak; muhitine alışmak: **to bring stg. ~ to s.o.,** (i) birinin suçlu olduğunu meydana çıkarmak; (ii) bir şeyi birine iyice anlatmak: **starvation has now been brought ~ to us,** açlığın ne olduğunu yakından görüp anladık: **the H~ Counties,** Londra'ya bitişik kontluklar: **to drive a point ~,** bir şeyi birinin zihnine yerleştirmek: **England is the ~ of freedom,** İngiltere hurriyetin yurdudur: **to feel quite at ~,** kendi muhitinde gibi hissetmek: **it's a ~ from ~,** burası insanın kendi evi sayılır: **to go ~,** (i) eve gitmek; sılaya gitmek; (ii) isabet etm.: **to go to one's last ~,** ölmek: **it is a ~ match today,** bugünkü maç bizim sahamızda oynanıyor: **to take an example nearer ~,** daha yakın bir misal getirmek: **the H~ Office,** Dahiliye Nezareti: **the H~ Secretary,** Dahiliye Nazırı: **the ~ side,** sahası üzerinde maç yapılan takım: **~ trade,** memleketin iç ticareti: **a ~ truth,** nazik noktaya dokunan hakikat. **~-bound,** vatana giden. **~-coming,** sıla; memleketine gelme. **~-grown,** kendi bahçesinde yetiştirilen. **H~-Guard,** yurd müdafaası için siviller arasında yapılan askerî teşkilat. **~-made,** evde yapılmış. **~-work,** (mekteb) ev vazifesi. **~less** [ˈhǫumlis]. Evsiz; kimsesiz. **~ly** [ˈhǫumli]. Basit; gösterişsiz; (*Amer.*) güzel olmıyan (yüz). **~sick** [ˈhǫumsik]. Vatan hasreti çeken. **~sickness,** dâüssıla. **~spun** [ˈhǫum-

spʌn]. Ev dokuması; basit. **~stead** [ˈhoumsted]. Çiftlik ve müştemilâtı. **~ward** [ˈhoumwəd]. Vatana, evine doğru giden. **~wards,** evine doğru.
**homicid·e** [ˈhomisâyd]. Adam öldürme; kaatil. **~al,** adam öldürmeğe mütemayil.
**homily** [ˈhomili]. Vaız. **to read s.o. a ~,** uzunuzadıya nasihat vermek.
**homing** [ˈhoumin(g)]. **~ pigeon,** posta güvercini.
**homo** [ˈhoumou]. (*Lât.*) İnsan.
**homo-** *pref.* Aynı; müşterek. **~-geneous** [houmouˈcînyəs]. Mütecanis.
**homonym** [ˈhomounim]. İmlâları bir olduğu halde muhtelif manalar ifade eden kelimeler; *mes.* yüz (100), yüz (çehre).
**honest** [ˈonist]. Dürüst; itimada şayan; doğru; samimî. **to make an ~ woman of s.o.,** baştan çıkardığı kadınla evlenmek. **~y** [ˈonisti]. Dürüstlük, doğruluk, samimilik.
**honey** [hʌni]. Bal: tatlılık. **my ~,** sevgilim. **~comb,** bal peteği; delik deşik etm.: **the army was ~ed with discontent,** ordu hoşnudsuzluk yüzünden için için çürüyordu. **~ed,** ballı; tatlı. **~moon,** balayı.
**honk** [hon(g)k]. (*otom.*) Korna sesi (çıkarmak).
**honor·arium** [onəˈreəriəm]. Meslek adamına hizmeti mukabilinde verilen ücret. **~ary** [ˈonərəri]. Fahrî. **~ific** [onəˈrifik]. Hürmet ifade eden (tabir).
**honour** [ˈonə(r)]. Şeref; tâzim; namus (kârlık); fazilet; rütbe; (iskambil) en yüksek beş koz. Hürmet göstermek, şeref vermek; tebcil etmek. **~s (degree),** ihtisas imtihanı: **the Honours List,** Kıraliçenin doğum yıldönümünde ve yılbaşında Kıraliçe tarafından tevcih edilen rütbe vs. listesi. **I am ~ed,** müşerrefim: **I am in ~ bound to do this,** bu işi yapmak benim namus borcumdur: **to ~ a bill,** bir poliçeyi tediye etm.: **to do the ~s (of the house),** ev sahibi vazifesini görmek: **to receive with full ~s,** büyük merasimle kabul etm.: **His H~ [Your H~],** hâkim vs.lere verilen unvan: **to leave stg. to s.o.'s sense of ~,** bir şeyi birinin sütüne havale etm.: **to put s.o. on his ~,** bir şeyi birinin namusuna havale etm.: **to ~ one's signature,** taahhüdünü tutmak: **to take ~s,** ihtisas imtihanını vermek. **~able** [ˈonərəbl]. Namuslu; müstakim; muhterem. **the H~,** *bir Lord'un çocuklarına verilen unvan*: **the Right H~,** *İngiliz kıralının hususî meclis azalarına verilen unvan.* **~ed** [ˈonəd] *a.* Şerefli; müşerref.
**hood** [hud]. Kukulete; omuzluklu başlık; (otom. vs.) körük; atmaca başlığı. **~ed,** kukuleteli.
**hoodlum** [ˈhûdləm]. (*Amer.*) Külhanbeyi.
**hoodwink** [ˈhudwin(g)k]. Aldatmak; göz boyamak.
**hoof** (*pl.* **hooves**) [hûf, hûvz]. At vs.nin tırnağı; (*sl.*) tepme. **to ~ it,** (*sl.*) sıvışmak: **to ~ s.o. out** (*sl.*) birini tepip dışarıya atmak.
**hook** [huk]. Çengel; kanca; olta iğnesi; kıvrıntı. Kanca ile tutmak; çengele asmak; (balığı) oltaya takmak; (*sl.*) (kız) koca avlamak. **'by ~ or by crook',** her hangi bir şekilde: **to ~ it,** (*sl.*) sıvışmak. **~ed** [hukd] *a.* Kanca gibi; kancalı. **~ nose,** gaga burun.
**hooligan** [ˈhûligən]. Azgın külhanbeyi. **~ism,** azgınlık.
**hoop** [hûp]. Kasnak; fıçı, oyuncak çemberi; halka. Çemberlemek.
**hoot** [hût]. (*ech.*) Baykuş gibi ötmek; (*otom.*) boru çalmak. Baykuş sesi; boru veya düdük çalması. **to ~ s.o. down,** birini yuhalarla susturmak.
**hooves** *v.* **hoof.**
**hop**[1] [hop]. Bir ayak üstünde sıçramak; seke seke yürümek. Sıçrama; (*coll.*) pek kısa bir mesafe; (*coll.*) raks. **~, skip and jump,** üç adım atlama: **~ it!,** (*sl.*) çek arabanı!: **to catch s.o. on the ~,** birini zayıf bir vaziyette yakalamak: **a flight in three ~s,** üç menzilde yapılan tayyare seferi. **~**[2]. Şerbetçiotu.
**hope** [houp]. Ümid; umma; emel. Ümid etm.; emel beslemek; itimad etmek. **to ~ for,** vukubulmasını arzu etm.: **to ~ against ~,** olmıyacak bir şeyi ümid etm.: **in the ~ of ...,** ümidile: **past (all) ~,** ümidsiz. **~d-for,** umulan ve istenilen. **~ful** [ˈhoupfəl], Ümidli; ümid verici. **~less** [ˈhouplis]. Ümidsiz; meyus. **that boy is ~,** o çocuk ıslah kabul etmez.
**hopper** [ˈhopə(r)]. Sıçrıyan (böcek); (değirmen) oluğu.
**hopscotch** [ˈhopskoç]. Seksek oyunu.
**horde** [hôd]. Nizamsız kalabalık; istilâ eden sürü.
**horizon** [həˈrâyzn]. Ufuk. **~tal** [horiˈzontl], ufkî; satıh.

**horn** [hôn]. Boynuz; boru; klakson. **to draw in one's ~s,** iddialarını kısmak: **the ~s of the moon,** hilâlin ucları: **the Golden H~,** Haliç. **~ed** [hônd]. Boynuzlu. **~y** [ˈhôni]. Boynuz gibi katı; nasırlanmış.

**hornet** [ˈhônit]. Büyük sarı arı. **to bring a ~'s nest about one's ears** [**to stir up a ~'s nest**], başına belâ açmak.

**horo·logy** [hoˈrolədji]. Saatçilik; zaman ölçme sanatı. **~scope** [ˈhorəskoup]. Zayiçe. **to cast one's ~,** zayiçesine bakmak.

**horr·ible** [ˈhorəbl]. Müdhiş; iğrenc; (*coll.*) berbad. **~id** [ˈhorid]. (*esk.*) Müdhiş; iğrenc; (*şim.*) çirkin; pis; nahoş. **don't be so ~ to each other!,** birbirinizin gözünü oymayın!: **you ~ thing!,** seni utanmaz seni! **~ific** [hoˈrifik]. Dehşetli, korkunc. **~ify** [ˈhorifây]. Dehşet vermek; ürpertmek. **~or** [ˈhorə(r)]. Dehşet; nefret; iğrenme; ürperme; iğrenc, korkunc bir şey. **to have a ~ of,** ···den iğrenmek, şiddetle nefret etmek. **~or-stricken,** [**~struck**], dehşete kapılmış.

**hors-d'œuvre** [ôˈdö(r)vr]. Çerez.

**horse** [hôs]. At, beygir; süvari; aygır; atlama sehpası. Ata bindirmek. **~ and foot,** süvari ve piyade: ⌜**don't look a gift ~ in the mouth**⌝, hediyede kusur aranmaz: **to mount** [**ride**] **the high ~,** yüksekten atmak: ⌜**wild ~s wouldn't drag it from me!**⌝, öldürseler söylemem: ⌜**you can take a ~ to the water but you can't make him drink**⌝, ⌜Nuh der peygamber demez⌝: **white ~s,** köpüklü dalgalar. **~-box,** at nakline mahsus vagon. **~-coper** [**-dealer**], at cambazı. **~-drawn,** atla çekilen (araba vs.). **H~ Guards,** hassa süvari alayı. **~-laugh**(**ter**), kaba kahkaha. **~-play,** eşek şakası. **~-power,** beygir kuvveti. **~-show,** at sergisi. **~back** [ˈhôsbak], **on ~,** ata binmiş. **~hair,** at kılı(ndan yapılmış). **~man,** atlı, binici: **~ship,** binicilik. **~shoe,** at nalı (şeklinde). **~woman,** kadın binici.

**horsy** [ˈhôsi]. Seyisler gibi giyinen ve konuşan.

**horticultur·e** [ˈhôtikʌltyə(r)]. Bahçıvanlık. **al** [-ˈkʌltyərəl], bahçıvanlığa aid.

**hose** [houz]. Hortum; (*comm.*) çorab. **half-~,** kısa çorab. **~-pipe,** hortum.

**hosier** [ˈhouziə(r)]. Çorabcı. **~y,** çorab ve iç çamaşırı gibi eşya.

**hospi·ce** [ˈhospis]. İmarethane. **~table** [hosˈpitəbl]. Misafirperver; mükrim. **~tal** [ˈhospitl]. Hastahane. **to walk the ~s,** (tıb talebesi) hastahanelerde çalışmak. **~tality** [ˌhospiˈtaliti]. Misafirperverlik. **to show s.o. ~,** birini evine kabul etm.

**host¹** [houst]. Misafir kabul eden ev sahibı; ziyafet veren kimse; otelci. '**to reckon without one's ~**', ⌜evdeki hesab çarşıya uymaz⌝. **~ess** [ˈhoustis]. Misafir kabul eden ev sahibesi. **~².** Kalabalık; çokluk; ordu. **the Heavenly ~s,** melekler; yıldızlar: **he is a ~ in himself,** bir çok adama bedeldir: **a (whole) ~ of servants,** bir sürü hizmetçi: **Lord God of H~s,** ordulara zafer veren Allah.

**hostage** [ˈhostic]. Rehîne.

**hostel, -ry** [ˈhostl(ri)]. (*esk.*) Han; (*şim.*) talebelerin ikametine mahsus ev. **youth ~s,** bazı memleketlerde yayan seyahat eden gencler için tesis edilen yurdlar.

**hostil·e** [ˈhostâyl]. Düşmanca; düşmana aid; hasmane; muhalif; aleyhdar. **~ity** [–ˈtiliti], düşmanlık, husumet, muhalefet. **~ities,** harb hali.

**hot** [hot]. Sıcak, hararetli: kızgın; baharlı. **to ~ up,** ısıtmak: **~ air,** (*fig.*) boş lâkırdı: **to get all ~ and bothered,** fazla telaşa düşmek: **burning ~,** yakacak kadar sıcak: **to go ~ and cold all over,** ürpermek: **~ dog,** (*Amer.*) sıcak sucuklu sanduviç: **a ~ favourite,** halkın çok tuttuğu (at vs.): **to give it s.o. ~,** ağzının payını vermek: **news ~ from the press,** gazetelerden dumanı üstünde bir havadis: **to make the place too ~ for s.o.,** birini bulunduğu yerden kaçırmak: **~ and strong,** pek şiddetli bir tarzda: **to be ~ stuff at stg.,** (*coll.*) ···de yaman olm.: **to be ~ on s.o.'s tracks,** takib edilen kimseye çok yaklaşmak: **to be in ~ water,** gözden düşmüş olm.: **to get into ~ water,** başına iş açmak. **~-blooded,** sıcak kanlı; hiddetli; atılgan. **~-foot,** çok acele. **~-tempered,** çabuk öfkelenen. **~-water-bottle,** soğuk havada kullanılan lâstikten sıcak su şişesi. **~bed** [ˈhotbed]. **a ~ of sedition,** isyan yatağı. **~head** [ˈhothed]. Mütehevvir kimse; ateşli genc. **~headed,** mütehevvir. **~house** [ˈhothaus]. Limonluk; ser.

**hotchpotch** [ˈhoçpoç]. Karmakarışıklık.

**hotel** [hou̯ˈtel]. Otel. **~-keeper,** otelci.
**hound** [hau̯nd]. Köpek *esp.* av köpeği; alçak herif. **to ~ s.o. down,** bilâfasıla birini takib ederek kovmak: **to ~ s.o. on,** birini tahrik etm.: **to ride to ~s,** ata binip köpek sürüsile tilki avına iştirak etmek.
**hour** [ˈau̯ə(r)]. Saat (60 dakika). **what ~ is it?,** saat kaç?: **~ by ~,** saatten saate, her saat: **after ~s,** iş saatlerinden sonra: **to keep late ~s,** geç yatmak: **office ~s,** çalışma saatleri: **questions of the ~,** zamanın meseleleri: **in the small ~s (of the morning),** sabaha karşı gece yarısından sonra: **the ~ has struck (to do stg.),** zamanı geldi: **to take ~s over stg.,** bir iş üzerinde saatlerce durmak. **~ly** [ˈau̯əli]. Saatte bir olan: **we expect him ~,** neredeyse gelir, onu bir iki saate kadar bekliyoruz: **he lives in ~ fear of death,** her an ölüm korkusu içinde yaşıyor.
**house** [hau̯s]. Ev, hane, mesken; hanedan; ticarî müessese; tiyatro seyircileri. *v.* [hau̯z]. Bir eve koymak; iskân etm.; muhafaza içine koymak. **H~ of Commons,** Avam kamarası: **country ~,** sayfiye: **~ full!,** (tiyatro) yer yok!: **~ of God,** kilise: **~ and home,** ev bark: **~ of ill fame,** umumhane: **to keep ~,** ev idare etm.: **to keep to the ~,** dışarı çıkmamak: **to keep open ~,** evini misafirlere açık tutmak: **H~ of Lords,** Lordlar kamarası: **to get on like a ~ on fire,** mükemmelen ilerlemek; (birisile) çok iyi geçinmek: **to set up ~,** yuva kurmak: **to set one's ~ in order,** işlerini tanzim etm.: **town ~,** konak. **~-agent,** ev simsarı. **~-party,** bir kaç gün için evde toplanan misafirler. **~ physician,** bir hastahanenin daimî doktoru. **~-property,** akarat. **~-surgeon,** bir hastahanenin daimî operatörü. **~-top,** evin dam ve bacaları: **to proclaim from the ~s,** âleme tellâl etmek. **~-warming,** yeni bir eve yerleşmek münasebetiyle verilen ziyafet. **~-boat** [ˈhau̯sbou̯t]. Üzerinde ev kurulan duba. **~breaker** [ˈhau̯sbrêykə(r)]. Ev hırsızı. **~ful** [ˈhau̯sfəl]. Ev dolusu. **~hold** [ˈhau̯shou̯ld]. Ev halkı. Eve aid; ev idaresine aid. **~ gods,** evin en sevilen eşyası: **~ troops,** hassa alayı: **~ word,** harcıâlem kelime. **~holder,** ev sahibi. **~-keep·er** [ˈhau̯skîpə(r)], Evi idare eden kadın. **my wife is a good ~,** refikam evi iyi idare eder. **~keeping,** ev idaresi. **~maid** [ˈhau̯smêyd]. Ortalık hizmetçisi: **~'s knee,** dizkapağı iltihabı. **~master** [ˈhau̯smâstə(r)]. Büyük İngiliz yatı mekteblerini teşkil eden evlerden birini idare eden muallim. **~wife,** *pl.* **-wives** [ˈhau̯swâyf, –wâyvz]. Ev kadını. **~wifery** [–ˈwiferi], ev idaresi.
**housing** [ˈhau̯zin(g)]. İskân etme; mahfaza.
**hove** *v.* **heave.**
**hover** [ˈhovə(r)]. (Kuş) az hareket ederek bir yerin üstünde uçmak. **to ~ about a place,** çok ayrılmıyarak bir yerin etrafında dolaşmak.
**how** [hau̯]. Nasıl; ne (kadar.) **~ is it (that) ···?,** nasıl oluyor da ···?: **do you know ~ to do it?,** nasıl yapılacağını biliyor musunuz?: **~ do you do?,** (ilk tanışıldığı zaman) memnun oldum: **~ long?,** uzunluğu ne kadar?; ne kadar (müddet)?: **~ nice!,** ne güzel: **~ old is he?,** o kaç yaşında? **~beit** [hau̯ˈbî·it]. Ne ise. **~ever** [hau̯ˈevə(r)]. Maamafih; her nekadar; ... ise de. **~ much,** her nekadar.
**howl** [hau̯l]. (*ech.*) Uluma(k); bağırma(k); (bebek) bağırarak ağlamak. **to ~ with laughter,** yüksek sesle kahkaha atmak: **to ~ a speaker down,** bir hatibi yuhalarla susturmak. **~er** [ˈhau̯lə(r)]. Güldürecek hata; büyük gaf.
**hoyden** [ˈhôydn]. Gürültücü erkeksi kız.
**H.P.** = **horsepower,** beygir kuvveti.
**H.Q.** = **headquarters,** karagâh.
**H.R.H.** = **His [Her] Royal Highness,** *pren[ses]lere verilen unvan*; fehametlu.
**hub** [hʌb]. Tekerlek göbeği. **the ~ of the universe,** dünyanın en mühim yeri.
**hubbub** [ˈhʌbʌb]. (*ech.*) Velvele: gürültü.
**hubby** [ˈhʌbi]. (*coll.*) Kocacağım.
**huddle** [ˈhʌdl]. Karışık ve sık bir sürü. Sıkı bir halde toplanmak.
**hue**[1] [hyû]. Renk. **~**[2]. **(to raise) a ~ and cry,** hırsız vs.yi tutmak için telaşla çağrışma(k).
**huff** [hʌf]. Dargınlık. Küstürmek. **to be in a ~,** küsmek. **~y,** dargın.
**hug** [hʌg]. Sıkıca kucaklama(k); sarılmak. **to ~ the shore,** (gemi) sahile sokulmak.

**huge** [hyûc]. Kocaman, lenduha.
**hugger-mugger** [ˈhʌgəmʌgə(r)]. İntizamsızlık.
**hulk** [hʌlk]. Gemi iskeleti. **~ing,** büyük ve hantal.
**hull** [hʌl]. Gemi teknesi; (fındık) dış kabuğu. Dış kabuğunu soymak. **~ down,** ufukta kaybolan tekne.
**hullabaloo** [ˌhʌləbəˈlû]. Velvele.
**hullo** [hʌlou]. Yahu!; allo!; vay!; merhaba!
**hum** [hʌm]. (*ech.*) Vızıldamak; mırıldamak; arı gibi çalışmak. Vızıltı, mırıltı. **to ~ and haw,** kemküm etm.: **to make things ~,** harıl harıl çalıştırmak.
**human** [ˈhyûmən]. İnsanî, beşerî. İnsan, beşer. **~ beings,** insanlar; âdem oğlu: **if it is ~ly possible,** beşerî imkân dahilinde ise: **~ly speaking,** beşerî bakımdan. **~e** [hyuˈmêyn]. İnsaniyetli, merhametli. **~eness,** merhamet. **~itarian** [hyumaniˈteəriən]. İnsaniyetperver; merhametli. **~ity** [hyuˈmaniti]. Beşeriyet, insaniyet; merhamet. **the humanities,** edebî ilimler ve *esp.* Lâtin ve Yunan klâsikleri. **~ize** [ˈhyûmənâyz]. İnsanlaştırmak. **~kind** [ˈhyûmənˈkâynd]. İnsanlık, beni âdem.
**humble** [ˈhʌmbl]. Alçakgönüllü. Kibrini kırmak. **of ~ birth,** mütevazı bir aileye mensub: **in my ~ opinion,** fikri acizaneme göre: **to eat ~ pie,** yanıldığını itiraf etm.: **your ~ servant,** 'aciz bendeleri' *resmî mektubların sonunda*: **to ~ oneself,** baş eğmek; küçülmek.
**humbug** [ˈhʌmbʌg]. Şarlatan(lık); riyakârlık. **that's all ~!,** bu hep palavradır: **there's no ~ about him,** içi dışı birdir.
**humdrum** [ˈhʌmdrʌm]. Cansıkıcı. **a ~ existence,** yeknesak bir hayat.
**humid** [ˈhyûmid]. Rutubetli, nemli. **~ity** [–ˈmiditi], rutubet.
**humili·ate** [hyuˈmiliêyt]. Terzil etm.; küçük düşürmek; kibrini kırmak. **~ation** [–ˈêyşn], tezlil, zillet, küçük düşme. **~ty** [hyuˈmiliti]. Tevazu, alçakgönüllülük.
**hummock** [ˈhʌmək]. Tümsek, tepecik.
**humo·rist** [ˈhyûmərist]. Mizah muharriri; nükteci adam. **~rous** [ˈhyûmərəs]. Nükteli; mizahî; güldürücü. **~ur** [ˈhyûmə(r)]. Mizah; huy; tabiat. Keyfine hizmet etm.; gönlünü almak. **in a good ~,** neş'esi yerinde: **out of ~,** ters, suratlı; keyfi kaçmış: **lacking in ~,** nükteden anlamaz: **a sense of ~,** gülecek tarafı görme kabiliyeti.
**hump** [hʌmp]. Hörgüc; kambur; tümsek. Kambur etmek. **to ~ up one's shoulders,** omuzunu kamburlaştırmak: **to have the ~,** (*sl.*) canı sıkılmak: **that gives me the ~,** (*sl.*) bu canımı sıkıyor. **~back** [**~-ed**], kambur.
**hunch** [hʌnç]. Kambur(laştırmak). **to be ~ed up,** dertop büzülmek: **to have a ~,** içine doğmak. **~back,** kambur.
**hundred** [ˈhʌndrəd]. Yüz. **in ~s,** yüzlerce: **a ~ per cent.,** yüzde yüz: **to have five ~ a year,** senede beş yüz lira geliri olmak. **~fold,** yüz misli. **~th,** yüzüncü; yüzde bir. **~weight,** (*Eng.*) 112 libre; (*Amer.*) 100 libre.
**hung** *v.* hang.
**Hungar·y** [ˈhʌngəri]. Macaristan. **~ian** [–ˈgeəriən], macar(ca).
**hung·er** [ˈhʌngə(r)]. Açlık; kıtlık; şiddetli istek. **to ~ for,** çok arzu etm. **~ry** [ˈhʌngri]. Aç, acıkmıs; müştak; çok gübre istiyen (toprak). **to go ~,** açlık çekmek.
**hunk** [hʌnk]. İri parça.
**hunt** [hʌnt]. Avlamak. Av; tilki avı cemiyeti. **to ~ (about) for stg.,** bir şeyi araştırmak. **~ down,** yakalayıncaya kadar peşini bırakmamak. **~ out [up],** arayıp meydana çıkarmak. **~er** [ˈhʌntə(r)]. Avcı; arayıcı; av atı. **~ing** [ˈhʌntin(g)]. Avcılık. **to go hunting,** tilki avına gitmek: **to go house ~,** ev aramak. **~ing-ground,** av sahası: **a happy ~ for collectors,** antika meraklıları için müsaid yer. **~ress** [ˈhʌntris]. Avcı kadın. **~sman** [ˈhʌntsmən]. Avcı.
**hurdle** [ˈhö(r)dl]. Mani; mania; yarış maniası. Manialarla kuşatmak; manialı koşu yapmak. **~r,** manialı koşuya giren.
**hurl** [hö(r)l]. Fırlatmak.
**hurly-burly** [hö(r)liˈbö(r)li]. Karışıklık.
**hurrah, -ray** [huˈrâ, -rêy]. Yaşa!, Hura!
**hurricane** [ˈhʌrikən]. Kasırga. **~-lamp,** rüzgârda sönmeyen fener.
**hurr·ied** [ˈhʌrid] *a.* Telaşla yapılmış; aceleci. **~y** [ˈhʌri]. Acele, istical. Acele et(tir)mek; hareket ettirmek. **~ up!,** çabuk ol!: **to be in a ~,** acelesi olmak.

**hurt** [ˈhö(r)t]. İncitmek; canını yakmak; acıtmak; rencide etm.; zarar vermek. Yara; zarar. **to do s.o. a ~,** birine zarar vermek; birine haksızlıkta bulunmak: **to get ~,** yaralanmak, incinmek: **to ~ oneself,** bir yerini acıtmak. **~ful,** zararlı, muzir.
**hurtle** [ˈhö(r)tl]. Fırlamak; şiddet ve gürültü ile hareket etmek.
**husband**[1] [ˈhʌsbənd] *n.* Koca, zevc. **~**[2] *v.* Tasarrufla idare etm.; iktisad etmek. **~ry,** ciftçilik.
**hush** [hʌş]. Sus!; sükûn. Sakin olmak; sus(tur)mak; teskin etmek. **to ~ up,** örtbas etm.: ⌈**the ~ before the storm**⌉, fırtınayı haber veren durgunluk. **~-~,** çok gizli. **~-money,** birinin ağzını kapatmak için verilen rüşvet.
**husk** [hʌsk]. Kabuk; kılıf.
**husky**[1] [ˈhʌski]. (Ses) kısık. **~**[2]**.** Dinc. Eskimo kızak köpeği.
**hussy** [ˈhʌzi]. Edebsiz kız.
**hustle** [ˈhʌsl]. Acele; itip kakma(k); sıkışmak; acele et(tir)mek. **I won't be ~d,** dara gelemem. **~r,** işini yürüten.
**hut** [hʌt]. Kulübe; baraka. **~ments** [ˈhʌtmənts]. Asker barakaları.
**hutch** [hʌç]. (Tavşan) kafesli sandık.
**hyacinth** [ˈhâyəsinθ]. Sümbül.
**hybrid** [ˈhâybrid]. Melez. **~ize,** melez olarak yetiştirmek.
**hydrant** [ˈhâydrənt]. Hortum takılan su borusu.
**hydraulic** [hâyˈdrôlik]. Su kuvveti ile işliyen.
**hydro-** [ˈhâydrou] *pref.* Suya aid. **~,** kaplıca. **~-electric** [ˈhâydroueˈlektrik]. Su kuvveti ile istihsal edilen elektrik .... **~phobia** [hâydrouˈfoubyə]. Kuduz; sudan korkma illeti.
**hygien·e** [hâyˈcîn]. Hıfzısıhha, sıhhat bilgisi. **~ic,** sıhhî.
**hymn** [him]. İlâhi; millî marş. Temcid etmek.
**hypercritical** [ˈhâypəˈkritikl]. Müfrit tenkidci.
**hyphen** [ˈhâyfn]. (İki kelime arasındaki) çizgi. **~ated,** çizgi ile ayrılarak yazılan (kelime).
**hypno·sis** [hipˈnousis]. Sunî uyutma. **~tic** [–ˈnotik], (sunî olarak) uyutucu. **~tism** [ˈhipnəˈtizm], sunî uyutma usulü. **~tist,** ipnotizma mütehassısı. **~tize,** ipnotizma etmek.
**hypocri·sy** [hiˈpokrisi]. Riyakârlık. **~te** [ˈhipəkrit], riyakâr, ikiyüzlü adam. **~tical** [–ˈkritikl], ikiyüzlü.
**hypothe·sis** [hâyˈpoθesis]. Faraziye. **~tical** [–ˈθetikl], farazî.
**hyster·ia** [hisˈtiəriə]. Sinir bozukluğu. **~ical** [–ˈterikl], isterik. **~ics** [–ˈteriks], sinir buhranı: **to go into ~,** sinir buhranına kapılmak.

# I

**I** [ây]. I harfi. **~,** ben.
**ib., ibid.** (*abb.*) (*Lât.*) **ibidem.** Ayni yerde.
**ice** [âys]. Buz; dondurma. Buz ile kaplamak; soğutmak; keyki erimiş şeker ile kaplamak. **to break the ~,** resmiyeti kaldırmak: **to cut no ~,** hiç tesir etmemek. **~-bound,** buz ile kuşatılmış (gemi). **~-cream,** dondurma. **~berg** [ˈâysbö(r)g]. Buz adası. **~d** [âysd]. Buz ile soğutulmuş; buzlu; şeker ile kaplanmış (keyk vs.). **Iceland** [ˈâysland]. İzlanda: **~er, ~ic,** İzlandalı.
**ic·icle** [ˈâysikl]. Buz parçası; sarkık uzun buz. **~iness** [ˈâysinis]. Buz gibilik. **~ing** [ˈâysin(g)]. Bir keykin şeker kaplaması; uçağın üzerini kaplıyan buz. **~y** [ˈâysi]. Buzlu; pek soğuk.
**iconoclast** [âyˈkonoklast]. Putkıran.
**idea** [âyˈdiə]. Fikir; tasavvur; niyet, maksad. **the ~!,** amma yaptın ha!: **to get ~s into one's head,** olmıyacak şeyler beklemek: **he has some ~ of how to row,** bir parçacık kürek çekmesini bilir: **I have an ~ that ...,** bana öyle geliyor ki: **I have no ~,** hiç bilmiyorum: **I had no ~ that ...,** hiç haberim yoktu: **a man of ~s,** buluş sahibi: **what an ~!,** hiç olur mu?
**ideal** [âyˈdiəl]. Mefkûre; ülkü. **~ism,** mefkûrecilik. **~ist,** mefkureci. **~ize,** idealleştirmek.
**identical** [âyˈdentikl]. Tamamile aynı; farksız; mutabık.
**identi·fy** [âyˈdentifây]. Hüviyetini tesbit etm.; teşhis etmek. **to ~ oneself with ...,** ···e iştirak etm. **~fication** [-fiˈkêyşn], hüviyet tesbiti: **~ papers,** hüviyet varakası. **~ty** [âyˈdentiti]. Hüviyet; ayniyet: **~ card,** hüviyet varakası: **mistaken ~,** yanlış hüviyet tesbiti.
**ideology** [âydiˈoləci]. Siyasî, iktisadî bir nazariye etrafındaki fikirler sistemi.
**idiocy** [ˈidiəsi]. Belâhet; ahmaklık.

**idiom** [ˈidyəm]. Şive; dil; tabir. **~atic** [–ˈmatik], hususî tabirlere aid.
**idiot** [ˈidyət]. Anadan doğma ebleh; bön safdil adam. **~ic** [idiˈotik], ahmakça, budalaca.
**idle** [ˈâydl]. Tembel; işsiz; âtıl; beyhude. Vaktini beyhude geçirmek; iş görmemek. **~ capital,** âtıl sermaye: **to run ~,** (makine) boşuna işlemek: **the ~ rich,** işsiz zenginler: **out of ~ curiosity,** sırf tecessüs sevkile. **~ness** [ˈâydlnis]. Tembellik; işsizlik; avarelik; beyhudelik. **~r** [ˈâydlə(r)]. Haylaz.
**idol** [ˈaydl]. Put; tapılan kimse. **~-ater** [âyˈdolətə(r)], putperest. **~atrous,** puta *tapan. **~atry** [âyˈdolətri], putperestlik. **~ize** [ˈâydəlâyz], putlaştırmak; perestiş etm.
**idyll** [ˈidil]. Kır hayatına aid ve ekseriyetle âşıkane küçük manzume. **~ic,** bir idilin mevzuu olmağa lâyik olan; sâf, zarif (aşk, manzara).
**i.e.** [ˈâyˈî]. (*abb.*) (*Lât.*) **id est,** Yani, demek ki.
**if** [if]. Eğer; şayed; ···sa, ···se. **ask ~ he is at home,** sor bakalım evde mi: **do you know ~ he is at home?,** onun evde olup olmadığını biliyor musunuz?: **~ I were you,** sizin yerinizde olsam: ⌜**~ifs and ans were pots and pans**⌝, ⌜olsa ile bulsa ile iş olmaz⌝: **oh ~ he could only come!,** ah!, bir gelebilse!: **~ only to please me,** benim hatırım için bile olsa: **see ~ you can open this,** şunu açabilir misiniz acaba: **it is only worth £50, ~ that,** ancak elli lira eder, o da şübheli: **I wonder ~ ...,** acaba ....
**ignit·e** [igˈnâyt]. Tutuş(tur)mak; iştial et(tir)mek. **~ion** [igˈnişn]. Tutuşturma; işal; kontak.
**ignoble** [igˈnoubl]. Alçak; rezil.
**ignomin·ious** [ˌignoˈminiəs]. Rezil; yüz kızartıcı. **~y** [ˈignəmini]. Rezalet; hacalet.
**ignor·amus** [ˌignəˈrêyməs]. Kara cahil kimse. **~ance** [ˈignərəns]. Cahillik; haberi olmamazlık. **~ant,** cahil; tahsilsiz; bihaber: **to be ~ of stg.,** bir şeyi bilmemek; bir şeyden haberi olmamak. **~e** [igˈnô(r)]. Aldırmamak; kulak asmamak.
**il-** =**in-** *pref. menfi mana ifade eder.*
**ill** [il]. Fenalık; kötülük; zarar; belâ. Hasta; fena, kötü; muzır. **to be ~ with (measles),** (kızamık) -tan yatmak: **to fall [be taken] ~,** hastalanmak: **although I can ~ afford it,** benim pek harcım değil amma: **I can ~ afford to offend that man,** o adamı darıltmak pek işime gelmez: **it ~ becomes you,** sana yakışmaz: **~ at ease,** huzursuz, meraklı: **~ fitted,** uymaz: **~ luck,** talihsizlik: **~ humour,** huysuzluk. **~ness** [ˈilnis]. Hastalık. **~ provided,** techizat noksan: **to take stg. ~ [in ~ part],** bir şeyi fena karşılamak. **~-** *pref.* Fena ...; *menfi mana ifade eder, mes.*: **~-deserved,** mustahak olmıyan; **~-pleased,** memnun olmıyan. **~-advised,** ihtiyatsız: **you would be ~ to ...,** ... yapmakla ihtiyatsızlık etmiş olursunuz. **~-assorted,** birbirine uymaz. **~-bred,** terbiyesiz. **~-considered,** düşüncesiz. **~-deserved,** mustahak olmıyan. **~-fame,** fena şöhret: **house of ~,** umumhane. **~-fated,** bedbaht; uğursuz. **~-favoured,** çirkin. **~-feeling,** kin: **no ~!,** kimsenin hatırı kalmasın! **~-founded,** asılsız. **~-gotten,** gayri meşru surette kazanılmış: ⌜**~ gains never prosper**⌝, haram mal sahibine hayretmez. **~-natured,** huysuz, aksi. **~-omened,** meşum. **~-luck,** aksilik; talihsizlik: **as ~ would have it,** aksi gibi. **~-pleased,** gayrimemnun. **~-starred,** bahtı kara. **~-timed,** vakitsiz; münasebetsiz. **~-treat [~-use],** fena muamele etm.; hırpalamak. **~treatment. ~-usage,** fena muamele. **~-will,** adavet; kindarlık: **to bear s.o. ~,** birine garaz beslemek.
**illegitim·acy** [ˌiləˈcitəməsi]. Gayrimeşru olma; piçlik. **~ate,** gayrimeşru; piç.
**illicit** [iˈlisit]. Kanuna aykırı.
**illiter·acy** [iˈlitərəsi]. Okumamışlık. **~ate,** okuma yazma bilmiyen.
**illumin·ate** [iˈlyûminêyt]. Aydınlatmak; üzerine ışık saçmak; renkli resimler ve harflerle süslemek. **~-ation** [iˈlyûmiˈnêyşn]. Tenvir; aydınlatma. **~ations,** donanma.
**illusion** [iˈlûjn]. Aldatıcı görüş; hayal. **optical ~,** görme hatası. **~ist,** hokkabaz.
**illustrat·e** [ˈiləstrêyt]. Resim vs. ile süslemek; resim ve misallerle izah etm.; misal getirerek anlatmak. **~ed,** resimli. **~ion** [–ˈstrêyşn], resim; izah: **by way of ~,** (izah için) misal alarak. **~ive** [ˈiləstrêytiv], izah verici; misal getirici.

**illustrious** [iˈlʌstriəs]. Şöhretli.
**im-, in-, ir-** *pref.* (i) *Menfilik ifade eder, mes.*: **possible,** mümkün; **impossible,** imkânsız; (ii) *idhal ifade eder, mes.*: **immerse,** daldırmak; **ingress,** giriş. *Sadece menfilik ifade eden bu gibi kelimelerinin çoğu lûgate alınmamıştır. Bunların manalarını müsbet şekillerine bakarak çıkarmak kolaydır.*
**I'm** = **I am.**
**image** [ˈimic]. Resim; şekil; aynı; put. **~ry** [ˈimicəri], tahayyülât; teşbih ve tasvir.
**imagin·e** [iˈmacin]. Tasavvur etm.; tahayyül etm.; sanmak. **~able,** tasavvur edilebilir. **~ary,** hayalî; muhayyel. **~ation** [–ˈnêyşn], tasavvur; muhayyele; icad kudreti.
**imbecile** [ˈimbisîl]. Ahmak, budala, ebleh.
**imbibe** [imˈbâyb]. Massetmek; içmek.
**imbue** [imˈbyû]. İlham etm.; işba etmek. **~d with superstitions,** hurafelerle meşbu.
**imitat·e** [ˈimitêyt]. Taklid etm.; birini örnek tutmak; eserine uymak. **~ion** [–ˈtêyşn], taklid; sahte eser: **in ~ of ...,** ···in eserine uyarak. **~ive,** taklidî.
**immaculate** [iˈmakyulit]. Lekesiz.
**immaterial** [ˌiməˈtiəriəl]. Ehemmiyetsiz.
**immature** [ˌiməˈtyuə(r)]. Olmamış, kemale ermemiş.
**immeasurable** [iˈmejərəbl]. Ölçülmez; hadsiz.
**immediate** [iˈmîdyət]. Derhal olan; müstacel, mübrem; akabinde vukubulan; doğrudan doğruya. **in the ~ future,** yakın istikbalde: **my ~ neighbour,** bitişik komşum. **~ly,** derhal: **~ you hear me shout, come to me,** bağırdığımı işitir işitmez bana gel.
**immemorial** [ˌiməˈmôriəl]. Zamanı bilinmiyecek kadar eski. **from time ~,** ezeldenberi.
**immens·e** [iˈmens]. Ucu bucağı olmıyan; hadsiz; kocaman. **~ely,** (*coll.*) son derece, pek çok. **~ity,** sonsuz büyüklük; hadsizlik.
**immerse** [iˈmö(r)s]. Daldırmak. **to be ~d in one's work,** işine dalmak.
**immigr·ant** [ˈimigrənt]. (Gelen) muhacir. **~ate,** bir memlekete göçmek. **~ation** [–ˈgrêyşn], bir memlekete muhaceret.
**imminen·ce** [ˈiminəns]. Vukuu yakın olma. **~t,** vukuu yakın ve muhakkak.
**immobil·e** [iˈmoubâyl]. Hareketsiz; kımıldanmaz; sabit. **~ity** [–ˈbiliti], hareketsizlik. **~ize** [iˈmoubilâyz], kımıldanamaz hale getirmek; durdurmak.
**immoderate** [iˈmodərit]. İtidâlsiz; ifrat derecede.
**immodest** [iˈmodist]. İffetsiz; açık saçık. **~y,** iffetsizlik; açıksaçıklık.
**immoral** [iˈmorəl]. Ahlâkı bozuk.
**immortal** [iˈmôtəl]. Ölmez. **the ~s,** mitolojik ilâhlar; pek meşhur şairer; ölmezler. **~ity** [–ˈtaliti], ölmezlik. **~ize** [–ˈmôtəlâyz], ölmezleştirmek.
**immovable** [iˈmûvəbl]. Kımıldanamaz; değiştirilemez; gayri menkul.
**immun·e** [iˈmyûn]. Muaf; masun. **~ity,** muafiyet; masuniyet. **~ize** [ˈimyunâyz], muaf kılmak; muafiyet vermek.
**imp** [imp]. Küçük şeytan; afacan.
**impact** [ˈimpakt]. Çarpma; musademe.
**impair** [imˈpeə(r)]. Bozmak.
**impart** [imˈpât]. Vermek; bahsetmek.
**impartial** [imˈpâşl]. Tarafsız; munsıf. **~ity** [–şiˈaliti], tarafsızlık.
**impass·able** [imˈpâsəbl]. Geçilmez. **~e** [ˈimpâs]. Çıkmaz; içinden çıkılmaz vaziyet.
**impassioned** [imˈpaşənd]. Müteheyyic.
**impassive** [imˈpasiv]. Teessürsüz.
**impatien·ce** [imˈpêyşns]. Sabırsızlık. **~t,** sabırsız; tezcanlı: **~ of control,** hüküm altına girmez.
**impeach** [imˈpîç]. Hakkında suizan beslemek; şübhelenmek.
**impeccable** [imˈpekəbl]. Kusursuz.
**impecunious** [impəˈkyûniəs]. Parasız.
**imped·e** [imˈpîd]. Mâni olm., engel çıkarmak. **~iment** [–ˈpedimənt], mania, engel: **~ of speech,** pelteklik.
**impel** [imˈpel]. Sevketmek; zorlamak.
**impend** [imˈpend]. Vukubulmak üzere olm. **~ing,** yakında vukubulacak.
**impenetrable** [imˈpenitrəbl]. Girilemez; sırrına erişilemez; kapalı.
**impenitent** [imˈpenitent]. Tövbe etmez.

**imperative** [imˈperətiv]. Emir sıygası. Zarurî; mübrem.
**imperceptible** [ˌimpəˈseptəbl]. Sezilemez.
**imperfect** [imˈpö(r)fikt]. Hikâyei hal. Tamamlanmamış; eksik; kusurlu. **~ion** [–ˈfekşn], eksiklik; kusur.
**imperi·al** [imˈpiəriəl]. İmparatora aid; şahane. **~ous** [imˈpiəriəs]. Mütehakkim; zarurî.
**imperil** [imˈperil]. Tehlikeye koymak.
**imperishable** [imˈperişəbl]. Zevalsiz; ebedî; bozulmaz.
**impermeable** [imˈpö(r)miəbl]. Su geçmez.
**impersonal** [imˈpö(r)sənl]. Muayyen bir şahsa aid olmıyan.. Gayrişahsî (fiil).
**impersonate** [imˈpö(r)sənêyt]. Bir şahıs rolünü yapmak; temsil etm.
**impertinen·ce** [imˈpö(r)tinəns]. Haddini bilmezlik; küstahlık. **~t,** haddini bilmez; küstah.
**impervious** [imˈpö(r)viəs]. Su vs. geçirmez; nüfuz ettirmez.
**impetu·ous** [imˈpetyuəs]. Şiddetli; coşkun. **~osity** [–ˈositi], coşkunluk.
**impetus** [ˈimpitəs]. Hız; şiddet.
**impi·ety** [imˈpâyəti]. Dine karşı hürmetsizlik. **~ous** [ˈimpiəs]. Allahtan korkmaz; dine karşı hürmetsiz.
**implacable** [imˈplakəbl]. Amansız, yavuz.
**implement**[1] [ˈimpləmənt] *n.* Alet; vasıta. **~s,** takım. **~**[2] [ˈimpliment] *v.* Tamamlamak; yerine getirmek.
**implicat·e** [ˈimplikêyt]. Sokmak; medhaldar etm.: tazammun etmek. **to be ~ed,** medhaldar olm.: **without ~ing anyone,** kimseyi karıştırmadan. **~ion** [–ˈkêyşn], medhaldar olma; tazammun: **by ~,** zımnen.
**implicit** [imˈplisit]. Kat'î; zımnî. **~ obedience,** itirazsız itaat.
**implied** [imˈplâyd]. Zımnî.
**implore** [imˈplô(r)]. Yalvarmak.
**imply** [imˈplây]. Delâlet etm.; tazammun etm.; kasdetmek; ima etmek.
**impolite** [impəˈlâyt]. Nezaketsiz.
**import**[1] [ˈimpôt]. İdhal. **~s,** idhalât. *v.* [imˈpôt], İdhal etmek. **~ation** [impôˈtêyşn]. İdhal etme; idhal edilen şey. **~er** [imˈpôtə(r)]. İdhalatçı. **~**[2] [ˈimpôt]. Mâna; ehemmiyet. **~ance** [imˈpôtəns]. Ehemmiyet; nüfuz. **of ~,** mühim. **~ant** [imˈpôtənt]. Ehemmiyetli; nüfuzlu. **to look ~,** mühim bir adam tavrı takınmak.
**importun·ate** [imˈpôtyunit]. İsrarla taleb eden.
**impos·e** [imˈpouz]. Üzerine koymak, yüklemek; zorla yaptırmak. **to ~ on,** aldatmak: **to ~ oneself on,** takılmak. **~ing** [imˈpouzin(g)]. Heybetli; kellifelli. **~ition** [ˌimpəˈzişn]. Yükletme; insafsız yük, vergi vs.; hile; yazı cezası.
**impossib·ility** [imˌposəˈbiliti]. İmkânsızlık; imkânsız şey. **a physical ~,** maddeten imkânsız. **~le** [imˈposibl]. İmkânsız; mümkün değil; muhal. **an ~ person,** tahammül edilmez kimse.
**impostor** [imˈpostə(r)]. Sahtekâr; düzme.
**impoten·ce** [ˈimpətəns]. Kudretsizlik; âcizlik; (cinsî) iktidarsızlık. **~t,** kudretsiz; âciz; (cinsî) iktidarsız.
**impound** [imˈpaund]. Müsadere etm.
**impoverish** [imˈpovəriş]. Fakirleştirmek.
**impracticable** [imˈpraktikəbl]. Yapılamaz; icra edilemez.
**impress**[1] *n.* [ˈimpres]. Damga; nişane. *v.* [imˈpres]. Basmak; üzerine iz bırakmak; derin tesir bırakmak; dikkatini celbetmek. **I was favourably ~ed by the youth,** genç üzerimde iyi bir tesir bıraktı. **~**[2] [imˈpres]. Zorla askere almak; bir maksad için kullanmak. **~ion** [imˈpreşn]. İntiba; tabı; tesir; (damga) nişanı. **to make a good [bad] ~,** iyi [fena] tesir bırakmak: **I am under the ~ that ...,** bana öyle geliyor ki. **~ionable,** kolay müteessir olan. **~ive** [imˈpresiv]. Tesir edici; unutulmaz.
**imprint** *n.* [ˈimprint]. Marka. *v.* [imˈprint]. Marka basmak; zihnine sokmak.
**imprison** [imˈprizn]. Hapsetmek. **~ment,** hapsetme; hapsedilme: **a month's ~,** bir ay hapis.
**improba·ble** [imˈprobəbl]. Muhtemel olmıyan. **~bility** [–ˈbiliti], muhtemel olmayış; inanılmazlık.
**impromptu** [imˈpromtyû]. Hazırlıksız; irticalen.
**improper** [imˈpropə(r)]. Yersiz; açık saçık.
**impropriety** [ˌimprouˈprâyəti]. Yakışıksızlık; uygunsuzluk.
**improve** [imˈprûv]. İyileş(tir)mek; ıslah etm. **to ~ the occasion [the shining hour],** ele geçirdiği fırsattan azami istifade etm.: **to ~ (up)on stg.,** ilâve, değiştirme ile bir şeyi

daha iyi hale koymak: **he ~s on acquaintance,** tanıdıkça insan onu o kadar fena bulmuyor: **to ~ on s.o.'s offer,** birinin teklif ettiğinden fazlasını vermek. **~ment** [im'prûvmənt]. İyileş(tir)me; ıslah; terakki; tekâmül; faydalı ilâve. **to be an ~ on stg.,** bir şeyden daha iyi olmak: **open to ~,** ıslaha muhtac.
**improviden·ce** [im'providəns]. Basiretsizlik; israf. **~t,** müsrif; basiretsiz.
**improvis·e** ['imprəvâyz]. İrticalen söylemek; hazırlık yapmadan muvakkaten tedarik etm. **~ation** [–'zêyşn], irtical; muvakkat tedbir.
**impruden·ce** [im'prûdəns]. Düsüncesizlik; tedbirsizlik. **~t,** düşüncesiz.
**impuden·ce** ['impyudəns]. Yüzsüzlük. **~t,** yüzsüz; saygısız.
**impuls·e** ['impʌls]. Sevk; itme; düşünmeden yapılan anî hareket. **~ive** [im'pʌlsiv]. Atılgan; düşünmeden derhal harekete geçen: **~ force,** itici kuvvet.
**impunity** [im'pyûniti]. **to do stg. with ~,** cezasını çekmeden yapmak.
**impur·e** [im'pyuə(r)]. Pis; iffetsiz; mağşuş. **~ity** [–'pyûriti], pislik; iffetsizlik; mahlutluk.
**impute** [im'pyût]. İsnad etm.
**in** [in]. ···de; içinde; ···e; içine. **we are ~ for a storm,** muhakkak fırtına olacak: **we are ~ for trouble,** başımıza iş çıkacak: **now we're ~ for it!,** şimdi hapı yuttuk!: **to write ~ ink,** mürekkeble yazmak: **is the fire still ~?,** ateş hâlâ yanıyor mu?: **is the train ~ yet?,** tren geldi mi?: **apples are now ~,** elma çıktı.
**in-** *v.* im-.
**inability** ['inəbiliti]. Kabiliyetsizlik.
**naccessible** [ˌinak'sesəbl]. Varılamaz; yanına girilemez.
**inaccura·cy** [in'akyurəsi]. Yanlışlık. **~te,** sahih olmıyan.
**inaction** [in'akşn]. Hareketsizlik; atalet.
**inactiv·e** [in'aktiv]. Faaliyet göstermiyen; âtıl. **~ity** [–'tiviti], faaliyetsizlik; atalet: **masterly ~,** basiretli hareketsizlik.
**inadequate** [in'adikwət]. Kâfi olmıyan.
**inane** [in'êyn]. Akılsız; mânasız.
**inanimate** [in'animit]. Cansız.
**inapt** [in'apt]. Meharetsiz; yersiz.
**inarticulate** [ˌinâ'tikyulêyt]. Sözsüz; gayrı natık; mafsalsız.
**inasmuch** [inaz'mʌç]. **~ as,** mademki; ... binaen; ···e göre.
**inaudible** [in'ôdibl]. İşidilemez.
**inaugura·l** [in'ôgyurəl]. Açma merasime aid. **~te,** açmak; başlamak. **~tion** [–'rêyşn], açma merasimi.
**inborn** ['inbôn]. Fıtrî; doğuştan.
**incalculable** [in'kalkyuləbl]. Hesab edilemez. **his temper is ~,** huyu hiç belli olmaz.
**incandescent** ['inkan'desənt]. Beyaz hararette olan.
**incapable** [in'kêypəbl]. İktidarsız. **~ of proof,** isbat edilemez: **~ of appreciating,** takdirden âciz.
**incapacit·ate** [inkə'pasitêyt]. İktidardan mahrum etm.; âciz bırakmak. **~y** [inkə'pasiti]. İktidarsızlık.
**incarcerate** [in'kâsərêyt]. Hapsetmek.
**incarnat·e** [in'kânit]. Beşer şeklinde olan. Tecessüm ettirmek. **a devil ~,** şeytanın ta kendisi. **~ion,** canlı timsal.
**incendiar·ism** [in'sendyərizm]. Kundakçılık. **~y,** kundakçı, yangın çıkarıcı.
**incense**[1] ['insens] *n.* Tütsü; buhur. **~**[2] [in'sens] *v.* Öfkelendirmek.
**incentive** [in'sentiv]. Teşvik edici şey.
**inception** [ın'sepşn]. Başlangıc.
**incertitude** [in'sö(r)tityûd]. Kararsızlık.
**incessant** [in'sesənt]. Fasılasız.
**incest** ['insest]. Yakın akraba arasında cinsî münasebet. **~uous** ['sestyuəs], buna aid.
**inch** [inç]. Pus = 25·4 mm. **to die by ~es,** yavaş yavaş ölmek: **to escape by ~es,** ···e kıl kalmak: **he knows every ~ of the place,** buraları avucunun içi gibi bilir: **to flog s.o. within an ~ of his life,** birinin dayaktan canını çıkarmak: **every ~ a soldier,** iliklerine kadar asker: **not to yield an ~,** bir karış gerilememek.
**incident** ['insidənt]. Hadise. Aid. **~al** [–'dentl], ârızî; tesadüfî; ufak tefek (masraf). **~ally,** tesadüfen.
**incis·e** [in'sâyz]. Deşmek; oymak. **~ion** [in'sijn], yarma; bıçak ile açılmış yer. **~ive** [–'sâysiv], keskin.
**inclin·ation** [ˌinkli'nêyşn]. Meyil; istek. **~e** *n.* ['inklâyn]. Meyil; yokuş. *v.* [in'klâyn]. Meylet(tir)mek; eğ(il)mek; müsaid olm.; çalmak. **~ed,** mütemayil; müsaid: **to be [to feel] ~ (to do stg.),** canı istemek:

not to feel ~, canı istememek: **he is ~ that way,** onun huyu böyledir.
**inclu·de** [inˈklûd]. Şamil olm., dahil etm.; ihtiva etm. **up to and ~ding Dec. 31st,** 31 aralığa kadar (31 aralık dahil). **~sion** [inˈklûjn]. Dahil etme; dahil bulunma. **~sive** [–siv], dahil: **~ terms,** her şey dahil olarak ücret.
**incognito** [inˈkognitou]. (*abb.* **incog.**) Mütenekkiren; takma adla. Takma ad.
**income** [ˈinkʌm]. Gelir; kazanc. **private ~,** şahsî gelir.
**incom·er** [ˈinkʌmə(r)]. Giren. **~ing,** giren: **the ~ tenant,** yeni kiracı.
**incomparable** [inˈkompərəbl]. Eşsiz.
**incompatible** [ˌinkəmˈpatəbl]. Biribirine uymaz; uyuşamaz.
**incompeten·t** [inˈkompətənt]. Beceriksiz; salâhiyetsiz. **~ce,** beceriksizlik.
**inconceivable** [ˌinkonˈsîvəbl]. Tasavvur olunamaz; hayret verici.
**inconclusive** [ˌinkonˈklûsiv]. Neticesiz.
**incongruous** [inˈkon(g)gruəs]. Birbirine uymaz.
**inconsisten·t** [ˌinkənˈsistənt]. Telif edilemez; kararsız. **~cy,** telif edilemeyiş.
**inconsolable** [ˌinkonˈsouləbl]. Teselli kabul etmez.
**inconspicuous** [ˌinkənˈspikyuəs]. Göze çarpmaz; ehemmiyetsiz.
**inconvenien·t** [ˈinkənˈvîniənt]. Zahmet verici; vakitsiz; münasib olmıyan; rahatsız. **~ce,** zahmet; rahatsızlık. Zahmet, güçlük vermek.
**incorporat·e** [inˈkôpərêyt]. Birleş(tir)mek; tevhid etm.; şirket teşkil etmek. **~ed,** *a.* birleşik: **~ company,** anonim şirket. **~ion** [–ˈrêyşn], birleş(tir)me; dahil edilme.
**incorrigible** [inˈkoricəbl]. Islah kabul etmez.
**incorruptible** [ˌinkoˈrʌptəbl]. Rüşvet kabul etmez; dürüst; çürümez.
**increase** *n.* [ˈinkrîs]. Art(tır)ma; çoğalma. *v.* [inˈkrîs]. Art(tır)mak; çoğal(t)mak; büyü(t)mek. **to be on the ~,** artmakta olmak.
**incred·ible** [inˈkredibl]. İnanılmaz. **~ulity** [ˌinkrəˈdyûliti]. İnanmazlık. **~ulous** [–ˈkredyuləs], güç inanır.
**increment** [ˈinkrimənt]. Artma; zam. **unearned ~,** şerefiye.
**incriminate** [inˈkriminêyt]. Kabahatli saydırmak.
**incubat·e** [ˈinkyubêyt]. (Yumurta) kuluçkaya yatırmak. **~ion** [–ˈbêyşn], kuluçka devresi. **~or,** kuluçka makinesi.
**incur** [inˈkö(r)]. Uğramak; başına getirmek. **to ~ expense,** masrafa girmek.
**incurable** [inˈkyûrəbl]. Tedavi edilemez; ıslah kabul etmez.
**indebted** [inˈdetid]. Borclu; minnettar.
**indecent** [inˈdîsnt]. Açık saçık.
**indecis·ion** [ˌindiˈsijn]. Kararsızlık, tereddüd. **~ive** [–ˈsâysiv], kat'î olmıyan; kararsız.
**indeclinable** [ˌindiˈklâynəbl]. (*gram.*) Tasrif edilmiyen.
**indecor·ous** [inˈdekərəs]. Muaşeret âdabına aykırı
**indeed** [inˈdîd]. Hakikaten; çok. **~!,** ya!
**indefatigable** [ˌindiˈfatigəbl]. Yorulmak bilmez.
**indefensible** [ˌindiˈfensəbl]. Müdafaa edilemez; mazur görülemez.
**indefin·able** [ˌindiˈfâynəbl]. Mübhem. **~ite** [inˈdefinit]. Muayyen olmıyan; mübhem; (*gram.*) **past ~,** gayrimuayyen mazi: **~ article,** gayrimuayyen harfitarif: **~ pronoun,** mübhem zamir. **to postpone ~ly,** müddetsiz olarak tehir etmek.
**indelible** [inˈdelibl]. Silinmez. **~ pencil,** kopya kalemi.
**indelicate** [inˈdelikət]. Nezaketsiz.
**indemni·fy** [inˈdemnifây]. Tazmin etmek. **~ty** [inˈdemniti]. Tazminat.
**indent**[1] [inˈdent]. Diş diş kesmek. Çukur; satır başı boşluğu. **~ation,** çukur. **~**[2]. Asker vs.ye lâzım olan bir şeyi resmen istemek. **~ure** [inˈdentyə(r)]. Resmî sened. **~ures,** hizmet mukavelesi.
**independen·ce** [ˌindiˈpendəns]. İstiklâl; kimseye muhtac olmamak için kâfi gelir. **~t,** müstakil; kendi başına olan.
**indescribable** [ˌindiˈskrâybəbl]. Tarifi imkânsız.
**indestructible** [ˌindiˈstrʌktəbl]. Yıkılmaz.
**index,** *pl.* **indexes** [ˈindeks, -iz]. Fihrist, cedvel; şahadet parmağı. *pl.* **indices** [ˈindisîz], rakam, üs. İndeks tertib etmek.
**India** [ˈindyə]. Hindistan. **~man,** İngiltere ile Hindistan arasında sefer yapan gemi. **~n,** Hindli: **Red ~,** Şimalî Amerika yerlisi, kırmızı derili:

~-ink, çini mürekkebi. ~**rubber** [ˌindyəˈrʌbə(r)]. Lâstik.
**indicat·e** [ˈindikêyt]. Göstermek; delâlet etmek. ~**ion** [–ˈkêyşn], alâmet; emare; delâlet. ~**ive** [inˈdikətiv], (*gram.*) ihbar tasrifi: ~ **of,** …e delâlet eden. ~**or,** müşir, ibre; sürat vs.yi gösteren cihaz.
**indict** [inˈdâyt]. İttiham etm.; mahkemeye vermek. ~**able,** ~ **offence,** cürüm, suç. ~**ment,** ittihamname.
**Indies** [ˈindîz]. **East** ~, Şarkî Hind: **West** ~, Antil denizindeki adalar.
**indifferen·ce** [inˈdifərəns]. Kayıdsızlık, fütursuzluk; bigânelik. ~**t,** kayıdsız, fütursüz; bigâne; iyi olmıyan.
**indigenous** [inˈdicənəs]. Yerli.
**indign·ant** [inˈdignənt]. Haklı olarak infial duyan. ~**ation** [–ˈnêyşn], haksız muamele hasıl ettiği his; infial. ~**ity** [inˈdigniti]. Hakaret, rezalet.
**indigo** [ˈindigou]. Çivit.
**indirect** [ˌindiˈrekt, –dâyˈrekt]. Vasıtalı; doğrudan doğruya olmıyan; dolaşık. ~ **speech,** (*gram.*) nakledilen söz ('bilmem' dedi, yerine 'bilmediğini söyledi' gibi).
**indiscipline** [inˈdisiplin]. İtaatsizlik.
**indiscre·et** [indisˈkrît]. Boşboğaz; düşüncesiz. ~**tion** [ˌindisˈkreşn]. Boşboğazlık; düşüncesiz hareket; pot kırma.
**indiscriminate** [ˌindisˈkriminit]. Fark gözetmeden yapılmış; körükörüne.
**indispensable** [ˌindisˈpensəbl]. Elzem.
**indispos·e** [indisˈpouz]. Soğutmak; arzusunu kırmak. ~**ed,** keyifsiz; isteksiz. ~**ition** [ˌindispəˈzişn], keyifsizlik.
**indisputable** [indisˈpyûtəbl]. Muhakkak.
**indissoluble** [ˌindiˈsolyubl]. Erimez.
**indistinct** [ˌindisˈtin(g)kt]. Kolayca seçilemez; belli belirsiz.
**individual** [ˌindiˈvidyuəl]. Ferdî; hususî; şahsî; tek. Ferd; adam. ~**ist,** ferdiyetçi. ~**ity** [–ˈaliti], şahsiyet; hususiyet; ferdiyet.
**Indo-China** [ˈindouˈçâynə]. Hindiçini.
**indolen·ce** [ˈindələns]. Tembellik. ~**t,** tembel, gevşek.
**indomitable** [inˈdomitəbl]. Yılmaz.
**indoor** [ˈindô(r)]. Ev içinde olan, yapılan. ~**s** [inˈdôz], ev içinde: ~**s and out,** ev içinde ve dışarıda: **to go** ~**s,** eve girmek.
**indubitable** [inˈdyûbitəbl]. Muhakkak.
**induce** [inˈdyûs]. Kandırıp bir şeyi yaptırmak; imale etm.; teşvik etm.; (*elect.*) endüksiyon cereyanı hasıl etmek. **nothing will** ~ **me to do it,** dünyada onu yapmam. ~**ment,** teşvik edici (verilen) şey: **to hold out** ~**s to s.o. to do stg.,** birine bir şeyi yaptırmak için cazib vaidlerde bulunmak.
**induct** [inˈdʌkt]. Birini memuriyetine resmen oturtmak. ~**ion** [–ˈdʌkşn], resmen oturtma; (*elect.*) endüksiyon; (mantık) kıyas. ~**ive,** istikraî.
**indulge** [inˈdʌlc]. (Birinin) isteklerine razı olm.; şımartmak; (ümide) kapılmak. **to** ~ **(oneself) in,** …e mübtelâ olm.: **to** ~ **in a cigar,** masrafa bakmayıp bir puro içmek: **to** ~ **freely in drink,** içkiye fazla düşkün olmak. ~**nce** [inˈdʌlcəns]. Müsamaha; ibtilâ. ~**nt,** müsamahakâr; göz yuman.
**industr·ial** [inˈdʌstriəl]. Sınaî. **the** ~ **revolution,** 18 ve 19 uncu asırlarda sanayideki bir çok icadlar ve sür'atli inkişaf. ~**ialism,** sanayicilik. ~**ialist,** sanayici. ~**ialize,** sınaîleştirmek. ~**ious** [inˈdʌstriəs]. Çalışkan. ~**y** [ˈindəstri]. Çalışkanlık; sanayi.
**inebriate** [inˈîbriêyt]. Sarhoş (etm.).
**ineffect·ive, -ual** [ˌiniˈfektiv, -tyuəl]. Tesirsiz; nafile; beceriksiz.
**inefficien·t** [ˌiniˈfişənt]. (İnsan) ehliyetsiz; (makine, tedbir) iyi işlemiyen; tesirsiz. ~**cy,** ehliyetsizlik; iyi işlemeyiş.
**inept** [inˈept]. Yersiz; ahmakça.
**inequality** [ˌinîˈkwoliti]. Müsavatsızlık; pürüzlülük; değişiklik.
**inert** [inˈö(r)t]. Cansız, hareketsiz; âtıl; kimyevî tesiri haiz olmıyan. ~**ia** [inˈö(r)şyə], atalet.
**inestimable** [inˈestiməbl]. Paha biçilmez; pek kıymetli.
**inevitable** [inˈevitəbl]. Kaçınılamaz; ictinab edilemez.
**inexact** [inigˈzakt]. Yanlış.
**inexhaustible** [ˌinigˈzôstəbl]. Bitmez.
**inexpedient** [ˌineksˈpîdiənt]. Münasib olmıyan.
**inexperience** [ˌiniksˈpiəriəns]. Tecrübesizlik. ~**d,** tecrübesiz, acemi.
**inexplicable** [inˈeksplikəbl]. İzah edilemez.
**inexpressible** [ˌiniksˈpresəbl]. Tarif edilemez; sözle söylenmez.
**infallib·le** [inˈfaləbl]. Hiç yanılmaz;

muhakkak: **an ~ remedy,** birebir ilâc. **~ility** [–ˈbiliti], yanılamazlık.

**infam·ous** [ˈinfəməs]. Kötülüğü meşhur. **~y** [ˈinfəmi]. Fena şöhret; rezalet.

**infan·cy** [ˈinfənsi]. (5 yaşına kadar) çocukluk; küçük olma hali; (bir teşebbüsün) ilk devresi. **~t** [ˈinfənt]. Pek küçük çocuk; bebek; küçük. **~ticide** [–ˈfantisâyd], küçük çocuk katilliği. **~tile** [ˈinfəntâyl], küçük çocuğa aid; çocukça.

**infantry** [ˈinfəntri]. Piyade askeri.

**infatuat·e** [inˈfatyuêyt]. Aşktan çılgın bir hale getirmek. **~ion** [–ˈêyşn], çılgınca âşık olma.

**infect** [inˈfekt]. Sirayet ettirmek, bulaştırmak; aşılamak. **~ion** [–ˈfekşn], sirayet, bulaşma. **~ious,** sari, bulaşık.

**infer** [inˈfö(r)]. İstidlâl etm.; zımnen delâlet etm. **~ence** [ˈinfərəns], istidlâl: **to draw an ~,** istidlâl etmek. **~ential** [–ˈrenşl], istidlâlî.

**inferior** [inˈfiərio(r)]. Madun; aşağıda bulunan; alt; âdi. **to be in no way ~ to s.o.,** her cihetten biri kadar iyi olmak. **~ity** [–ˈoriti], madunluk; aşağılık, iyi olmama: **~ complex,** aşağılık duygusu.

**infern·al** [inˈfö(r)nl]. Cehenneme aid; (*coll.*) Allahın belâsı. **~ noise,** müdhiş gürültü. **~o** [inˈfö(r)nou]. Cehennem.

**infertil·e** [inˈfö(r)tâyl]. Mahsulsuz; kısır.

**infest** [inˈfest]. (Muzır bir şey) etrafı sarmak. **this house is ~ed with rats,** bu evi fareler istilâ etmiş.

**infidel** [ˈinfidl]. Kâfir; imansız. **~ity** [ˌinfiˈdeliti]. Sadakatsizlik; imansızlık.

**infiltrate** [ˈinfiltrêyt]. Sızıp girmek; gizlice ve tedricen sokulmak.

**infinit·e** [ˈinfinit]. Sonsuz; hududsuz. **the ~,** feza: **the I~,** Allah: **to take ~ pains,** son derece itina etmek. **~ely,** son derecede. **~esimal** [ˌinfiniˈtesiml]. Son derecede küçük. **~ive** [inˈfinitiv]. Mastar. **~y** [inˈfiniti]. Sonsuzluk.

**infirm** [inˈfö(r)m]. Malûl; hastalıklı; kararsız. **~ary,** [–əri], Darülâceze; hastane. **~ity,** malûllük, dermansızlık; kararsızlık.

**inflam·e** [inˈflêym]. Alevlendirmek; iltihablandırmak. **~ed,** iltihablı; alevlenmiş. **~mable** [–ˈflaməbl], çabuk ateş alır. **~mation** [–ˈmêyşn], iltihab.

**inflat·e** [inˈflêyt]. Şişirmek; artırmak. **to ~ the currency,** enflâsyon yapmak. **~ed,** şiş(iril)miş; fahiş (fiat): **~ with pride,** kibirden kabarmış. **~ion** [–ˈflêyşn], şişirme; enflâsyon.

**inflec·t** [inˈflekt]. (Sesi) tadil etm.; (*gram.*) tasrif etmek. **~tion** [**~xion**] [inˈflekşn]. Tasrif.

**inflexib·le** [inˈfleksəbl]. Eğilmez; kararından dönmez. **~ility** [–ˈbiliti], eğilmezlik; sertlik.

**inflict** [inˈflikt]. Birinin başına nahoş bir şey getirmek. **to ~ a punishment on s.o.,** birini cezaya çarptırmak: **to ~ pain,** canını acıtmak; ıstırab vermek: **to ~ a wound on s.o.,** birini yaralamak: **to ~ one's company on s.o.,** (istenmediği halde) birini ziyaret etmek. **~ion** [inˈflikşn]. (Ceza vs.) verme; ···e duçar etme; eza.

**inflow** [ˈinflou]. İçeriye doğru akış.

**influen·ce** [ˈinfluəns]. Tesir (yapmak); nüfuz(u altında tutmak). **to have ~,** (i) nüfuzlu olm.; (ii) arkası olm.: **under the ~ of drink,** sarhoşluk esnasında. **~tial** [influˈenşl]. Nüfuzlu; tesirli.

**influenza** [ˌinfluˈenzə]. Grip.

**inform** [inˈfôm]. Bildirmek; ···e haber vermek. **to ~ s.o. about stg.,** birine bir şey hakkında malûmat vermek: **to ~ against s.o.,** birini jurnal etmek. **~ant** [inˈfômənt]. Haber veren kimse. **~ation** [ˌinfəˈmêyşn]. Malûmat, haber, istihbarat: **for your ~,** bilgi edinmeniz için. **~ative** [–ˈfômətiv], malûmat verici. **~ed** [inˈfômd] *a.* Haberdar. **well ~,** malûmatlı. **~er** [inˈfômə(r)]. Birini ihbar, şikâyet, eden, jurnalci; gammaz, münafık.

**informal** [inˈfôml]. Gayri resmî. **~ity** [–ˈmaliti], teklifsizlik.

**infra dig.** [ˈinfrəˈdig]. Vekara uymaz.

**infra-red** [ˈinfraˈred]. Kızılötesi.

**infrequent** [inˈfrîkwənt]. Nadir.

**infringe** [inˈfrinc]. İhlâl etm.; bozmak. **to ~ a patent,** ihtira beratının hakkına tecavüz etm.: **to ~ upon s.o.'s rights,** birinin haklarına tecavüz etmek. **~ment,** ihlâl: **~ of copyright,** telif hakkına tecavüz.

**infuriate** [inˈfyuəriêyt]. Kudurtmak. **~d,** kudurmuş, fena hiddetlenmiş.

**ingen·ious** [inˈcîniəs]. Hünerli; mucidin hünerini gösteren. **~uity** [ˌinciˈnyuiti]. Hüner, marifet.

**ingenuous** [inˈcenyu̯əs]. Sadedil, masum; samimî.
**ingot** [ˈingət]. Külçe.
**ingrained** [inˈgrêynd]. Kökleşmiş; çıkarılmaz.
**ingratiat·e** [inˈgrêyşiêyt]. **to ~ oneself with s.o.**, gözüne girmeğe çalışmak. **~ing**, sokulgan.
**ingratitude** [inˈgratityûd]. Nankörlük.
**ingredient** [inˈgrîdi̯ənt]. Bir şeyin terkibine giren madde.
**inhabit** [inˈhabit]. İçinde oturmak; ···de ikamet etmek. **~able**, oturulabilir. **~ant**, oturan kimse: **the ~s of the village**, köy ahalisi. **~ed**, meskûn.
**inhale** [inˈhêyl]. Nefes çekmek.
**inherent** [inˈhi̯ərənt]. Cibillî, tabiî; aslî: **~ defect**, esasta olan kusur.
**inherit** [inˈherit]. Miras olarak almak. **~ance**, veraset, miras. **~or**, vâris.
**inhibit** [inˈhibit]. Yasak etm.; (hisler) tutmak. **~ion** [–ˈbişn], yasak etme; nehiy.
**inhospitable** [ˌinhosˈpitəbl, –ˈhos–]. Misafir sevmez; dağ başı gibi.
**inhuman** [inˈhyûmən]. Gayri insanî; zalim. **~e** [inhyuˈmêyn], zalim. **~ity** [–ˈmaniti], insaniyetsizlık; zalimlik.
**inimical** [inˈimikl]. Düşman; gayrı müsaid.
**inimitable** [inˈimitəbl]. Taklid edilemez; eşsiz.
**iniquit·ous** [inˈikwitəs]. İnsafsız. **~y**, günah.
**initial** [iˈnişl]. İlk; başlangıcta bulunan. Bir kelimenin (şahıs adının) ilk harfi. Parafe etmek.
**initiat·e** [iˈnişiêyt]. Başlamak; sırlarını oğreterek bir cemiyete kabul etm.; ilimde ilk adımını attırmak; (birine) bir şeyin esaslarını öğretmek. **~ion** [iˈnişiêyşn]. Başlama, ilk adımını atma; girme. **~ive** [iˈnişyətiv], şahsî teşebbüs: **he has no ~**, müteşebbis değildir: **to take the ~**, bir iş için ilk adımı atmak: **on one's own ~**, kendi teşebbüsile. **~or**, önayak olan kimse.
**inject** [inˈcekt]. Zerketmek, enjeksiyon yapmak; içeri sokmak. **~ion** [–ˈcekşn], enjeksiyon: **to give an ~**, şırınga yapmak.
**injudicious** [ˌincyuˈdişəs]. Tedbirsiz.
**injunction** [inˈcʌnkşn]. Kat'î emir; mahkeme tarafından verilen ihtar. **to give s.o. strict ~s**, birine kat'i surette ihtar etmek.
**injur·e** [ˈincə(r)]. Zarar vermek; dokunmak; bozmak; sakat etmek. **~ed**, yaralanmış; zarar görmüş: **in an ~ tone of voice**, yaralı bir sesle: **the ~ party**, (*jur.*) zarara uğrıyan kimse. **~ious** [–ˈcu̯əri̯əs], zararlı. **~y** [ˈincəri], zarar; haksızlık; yara: **to do s.o. an ~**, birine haksızlık etm.
**injustice** [inˈcʌstis]. İnsafsızlık. **to do s.o. an ~**, (i) birine karşı insafsızlık etm.; (ii) birinin günahına girmek.
**ink** [in(g)k]. Mürekkeb (sürmek). **~ in [over]**, kurşun kalemile yazılmış bir şey üzerinden mürekkeb ile geçmek: **printers' ~**, matbaa mürekkebi: **indian ~**, çini mürekkebi. **~y**, mürekkebli; kapkara. **~ pot** [ˈin(g)k]pot]. Mürekkeb hokkası. **~stand**, yazı takımı.
**inkling** [ˈin(g)klin(g)]. İma; hafif şübhe. **to get [have] an ~ of stg.**, bir şeyin kokusunu almak: **I hadn't an ~ of what was to happen**, ne olacağından zerre kadar haberim yoktu.
**inlaid** [inˈlêyd]. Üzerine altın vs. kakarak nakışlar yapılmış.
**inland** [ˈinlənd]. Bir memleketin denizden uzak iç kısmı. Dahilî. **~ trade**, memleketin iç ticareti. **I~ Revenue.** Tahsilat İdaresi.
**inlay** (**inlaid**) [inˈlêy, -lêyd]. Bir maddenin üzerinde açılan yuvalara başka bir madde kakıp oturtmak.
**inlet** [ˈinlet]. Medhal; giriş yolu; körfezcik.
**inmate** [ˈinmêyt]. Bir ev, odada oturan kimse.
**inmost** [ˈinmo̯ust]. En içerideki; derunî.
**inn** [in]. Han; küçük otel; meyhane. **I~s of Court**, Londra'da avukatlık stajını yapmak hakkını veren cemiyetler ve onlara aid binalar. **~-keeper** [ˈinkîpə(r)]. Hancı, meyhaneci.
**innate** [iˈnêyt]. Fıtrî; doğuştan.
**inner** [ˈinə(r)]. Dahilî; iç; derunî. **to look after the ~ man**, karnını doyurmak. **~most**, en içerindeki.
**innings** [ˈinin(g)z]. **he has had a long ~**, (*fig.*) bir makam vs.de çok uzun müddet kaldı; çok yaşadı: **well, he has had a good ~**, (ölen birisi hakkında) maşallah çok yaşadı: **my ~ now!**, şimdi sıra bende.
**innocen·ce** [ˈinəsəns]. Masumiyet;

saflık. **~t,** masum; hilesiz. Masum çocuk. **~ of clothes,** cıplak.
**innocuous** [iˈnokyuəs]. Zararsız.
**innovat·ion** [inəˈvêyşn]. Bid'at; yenilik. **~or** [ˈinəvêytə(r)], yenilik tarafdarı; bid'at ehli.
**innuendo** [ˌinyuˈendou]. İma. **to make ~s against s.o.,** birine taş atmak.
**innumerable** [iˈnyûmərəbl]. Sayısız; pek çok.
**inoculat·e** [iˈnokyulêyt]. Aşılamak; telkih etmek. **~ion** [–ˈlêyşn], aşıla(n)-ma.
**inoffensive** [ˌinəˈfensiv]. Zararsız.
**inopportune** [inˈopətyûn]. Münasib olmıyan.
**inordinate** [inˈôdinit]. Hadden fazla.
**inorganic** [ˌinôˈganik]. Gayriuzvî; camid.
**in-patient** [ˈinˈpêyşənt]. Hastahanede kalan hasta.
**input** [ˈinput]. Tağdiye.
**inquest** [ˈinkwest]. Tahkik; bir ölümün sebebini araştıran adlî tahkikat.
**inquir·e** [inˈkwâyə(r)]. Sual sormak. **to ~ about s.o. [stg.],** birisi [bir şey] hakkında malûmat edinmek: **to ~ after s.o.,** birinin hatırını sormak: **to ~ into stg.,** bir şeyi tahkik etmek. **~ing,** araştırıcı. **~y** [inˈkwâyəri]. Sual; sorgu; tahkikat; tedkik; istifsar. **to make inquiries about s.o.,** birinin hakkında tahkikat yapmak: **to make inquiries after s.o.,** birinin hatırını sormak: **Court of ~,** tahkikat heyeti.
**inquisit·ion** [ˌinkwiˈzişn]. Resmî tahkik ve tedkik. **the I~,** enkizisyon mahkemesi. **~ive** [inˈkwizitiv]. (Yersiz olarak) mütecessis.
**inroad** [ˈinroud]. Akın; tecavüz. **to make ~s upon one's capital,** sermayesinde rahneler açmak.
**inrush** [ˈinrʌş]. İçeriye doğru şiddetli akın.
**insan·e** [inˈsêyn]. Mecnun; deli. **~itary** [inˈsanitəri]. Sıhhate muzır. **~ity** [inˈsaniti]. Akıl hastalığı; delilik.
**insatiable** [inˈsêyşəbl]. Doymak bilmez; açgözlü.
**inscri·be** [inˈskrâyb]. Yazmak; kaydetmek; hâkketmek. **~ption** [inˈskripşn]. Kitabe; yazı; kaydetme; tescil etme.
**inscrutable** [inˈskrûtəbl]. Sırrına erişilemez.
**insect** [ˈinsekt]. Böcek, haşere. **~-icide** [inˈsektisâyd], böcek öldürücü (ilâc).
**insecur·e** [ˌinsiˈkyuə(r)]. Emin olmıyan; tehlikeye maruz. **~ity** [–ˈkyûriti], emniyetsizlik.
**insens·ate** [inˈsensêyt]. Hissiz; çılgınca. **~ible** [inˈsensibl]. Hissiz; baygın; hissolunmaz; duymaz. **to be knocked ~,** bir darbe ile kendinden geçmek. **~ibility** [–ˈbiliti], hissizlik; bayılma. **~itive** [inˈsensitiv]. Hassas olmıyan.
**inseparable** [inˈsepərəbl]. Ayrılmaz.
**insert** [inˈsö(r)t]. Dercetmek; sokmak; ilâve etmek. **~ion** [–ˈsö(r)şn], derc(etme); sokma; ilâve (etme).
**inset** [ˈinset]. Dercedilen şey; büyük resmin kenarındaki küçük resim.
**inshore** [ˈinşô(r)]. Sahilde; sahile yakın.
**inside** [ˈinˈsâyd]. İç; orta yer; karın; iç tarafındaki; dahilî; ev içindeki; içeride, içeriye; ···in içinde, ···in içine. **~s,** karın ve barsaklar. **the ~ of an affair,** işin içyüzü: **to have ~ information,** bir şeyi yerinden, kaynağından öğrenmek: **~ right,** (futbol) sağiç: **~ left,** soliç: **~ out,** tersyüz: **to turn everything ~ out,** ortalığı altüst etm.: **to know stg. ~ out,** bir şeyin içini dışını bilmek: **~ of a week,** bir haftadan az.
**insidious** [inˈsidyəs]. Gizli sokulur.
**insight** [ˈinsâyt]. Feraset; içyüzü, huyu çabuk kavramak kabiliyeti.
**insignia** [inˈsigniyə]. Bir makamın veya bir rütbenin resmî alâmetleri.
**insignifican·ce** [ˌinsigˈnifikəns]. Ehemmiyetsizlik. **~t,** cüzi; dikkate değmez.
**insincer·e** [ˌinsinˈsiə(r)]. Gayrı samimî; ikiyüzlü; sahte. **~ity** [–ˈseriti], samimiyetsizlik; sahtelik.
**insinuat·e** [inˈsinyuêyt]. İma etm.; yavaşça ve kurnazca sokmak. **to ~ oneself,** sokulmak. **~ion** [–ˈêyşn], ima; üstü kapalı itham.
**insipid** [inˈsipid]. Yavan; lezzetsiz.
**insist** [inˈsist]. **~ (on)** Israr etm.; ayak diremek; ısrarla tasdik etm. **I ~ on obedience,** muhakkak itaat isterim. **~ence,** ısrar (etme); ayak direme. **~ent,** musir; muannid; müz'ic.
**insobriety** [ˌinsoˈbrâyəti]. Ayyaşlık; itidalsizlik.
**insolen·ce** [ˈinsələns]. Küstahlık. **~t,** küstah.
**insoluble** [inˈsolyubl]. Erimez; halledilemez.

**insolven·cy** [inˈsolvənsi]. İflâs. **~t,** müflis.
**insomnia** [inˈsomnyə]. Uykusuzluk.
**insomuch** [ˈinsouˈmʌç]. **~ as [that],** hattâ; o kadar ki.
**inspect** [inˈspekt]. Teftiş etm.; muayene etmek. **~ion** [–ˈspekşn], teftiş, muayene. **~or,** müfettiş; muayene memuru. **~orate** [–ərit], müfettişlik; teftiş heyeti.
**inspir·ation** [ˈinspiˈrêyşn]. Telkin; ilham; nefes alma. **~e** [inˈspâyə(r)]. İlham etm.; telkin etm.; teşvik etm.; nefesi içeri çekmek. **~ed,** mülhem; ilhamlı. **an ~ article in a paper,** nüfuzlu bir kimse tarafından gizlice telkin edilen makale. **~ing,** heyecanlandırıcı; teşvik edici.
**instability** [ˌinstəˈbiliti]. Sebatsızlık.
**instal·l** [inˈstôl]. Yerleştirmek; ik'adetmek. **to ~ central heating,** kalorifer tesisatı yapmak. **~lation,** yerleştirme; ik'ad; tesisat. **~ment** [inˈstôlmənt]. Taksit; kurma. **the ~ system,** taksitle satın alma.
**instance** [ˈinstəns]. Misal; defa; ısrar. Misal olarak iradetmek. **for ~,** meselâ: **in the first ~,** ilkönce: **at the ~ of,** ···in ısrarile: **an isolated ~,** tek bir misal: **in many ~s,** çok kere: **in the present ~,** bu defa: **Court of First I~,** bidayet mahkemesi.
**instant** [ˈinstənt]. An, lâhza. Mübrem; anî. **come this ~!,** derhal gel!: **I expect ~ obedience,** derhal itaat isterim: **on the ~,** derhal: **the ~ I hear from him,** ondan haber alır almaz: **on the 4th ~ (inst.),** bu ayın dördünde. **~aneous** [ˌinstənˈtêynyəs]. Anî; enstantane.
**instead** [inˈsted]. Yerde; **~ of ...,** ···in yerine; bedel olarak. **I told him to come; ~ he ran away,** gelmesini söyledim, halbuki o kaçıp gitti.
**instep** [ˈinstep]. Ayağın üst kısmı.
**instigat·e** [ˈinstigêyt]. Tahrik etm.; kışkırtmak. **~ion** [–ˈgêyşn], tahrik; kışkırtma. **~or,** kışkırtan adam.
**instil** [inˈstil]. Zihnine yerleştirmek.
**instinct** [ˈinstin(g)kt]. Sevkitabiî; insiyak. **~ive** [inˈstin(g)ktiv], insiyakî.
**institut·e** [ˈinstityût]. Tesis etm.; başlamak. Müessese; enstitü. **~ion** [ˌinstiˈtyûşn]. Tesis (etme); müessese; cemiyet; müesses âdet.
**instruct** [inˈstrʌkt]. Öğretmek; talimat vermek; emretmek. **~ion** [–kşn], Öğretme; emir: **~s,** talimat; malûmat; emirler: **~ book,** rehber: **~ for use,** (ilâc vs.) kullanma şekli. **~ive,** ders verici; ibret verici. **~or** muallim; usta.
**instrument** [ˈinstrumənt]. Alet; çalgı; (*jur.*) sened. **~al** [–ˈmentl], çalgı ile çalınan: **~ in (doing stg.),** yardım eden. **~alist,** çalgıcı. **~ality** [–ˈtaliti], **through the ~ of,** ···in vasıtasile, yardımıyla.
**insubordinate** [ˌinsəˈbôdinit]. İtaatsiz.
**insufferable** [inˈsʌfrəbl]. Tahammülfersa.
**insufficient** [ˌinsəˈfişənt]. Kâfi olmıyan; eksik.
**insular** [ˈinsyulə(r)]. Adaya aid; fazla darfikirli. **~ity** [–ˈlariti] fazla darfikirlilik.
**insulat·e** [ˈinsyulêyt]. (*elect.*) Tecrid etmek. **~ing,** tecrid edici. **~ion** [–ˈlêyşn], tecrid (etme): **~or,** mücerrid, izolatör.
**insult** *v.* [inˈsʌlt]. Tahkir etm.; şerefine dokunmak. *n.* [ˈinsʌlt]. Hakaret. **to add ~ to injury,** bir fenalığa başka bir fenalığı katmak. **~ing,** tahkir edici.
**insuperable** [inˈsyûprəbl]. Aşılmaz.
**insur·e** [inˈşô(r)]. Sigorta etm.; *v.* ensure. **~ance** [–ˈşôrəns], sigorta.
**insurrection** [ˌinsəˈrekşn]. İsyan.
**intact** [inˈtakt]. Tamam.
**intake** [ˈintêyk]. İçeriye alınmış şeyin mikdarı; (su vs.nin) içeriye girdiği yer.
**intangible** [inˈtancəbl]. Cisimsiz; maddî olmıyan.
**integra·l** [ˈintegrəl]. Tam; bütün; tamamî. **to be an ~ part of stg.,** bir şeyin tamamlayıcı cüz'ü olmak. **~te** [ˈintigrêyt]. Tamamlamak.
**integrity** [inˈtegriti]. Doğruluk; tamamlık.
**intellect** [ˈintilekt]. Akıl; idrak kabiliyeti. **~ual** [–ˈlektyuəl], akla, zekâya mensub; yüksek zekâ sahibi; münevver.
**intelligen·ce** [inˈtelicəns]. Zekâ, akıl; istihbarat; haber: **~ officer,** istihbarat subayı. **~t,** zeki. **~tsia** [–ˈcentsyə], münevverler.
**intelligible** [inˈtelicəbl]. Kolay anlaşılır.
**intempera·nce** [inˈtempərəns]. İtidalsizlik; bekrilik; sertlik. **~te,** itidalsiz; müfrit; çok içer; sert.
**intend** [inˈtend]. Niyet etm., maksadı olm.; kasdetmek. **I ~ed no**

harm, hiç bir fenalık kasdetmedim: **was his remark ~ed ?**, o sözü kasden mi söyledi?: **that remark was ~ed for you,** o sözü sizi kasdederek söyledi: **this portrait is ~ed for me,** bu, gûya benim resmim: **this watch was ~ed for you,** bu saat sizin içindi. **~ed** [inˈtendid] *a.* Kasden yapılmış veya söylenmiş. **my words had the ~ effect,** sözlerim istediğim tesiri yaptı. *n.* (*coll.*) Yavuklu.

**intens·e** [inˈtens]. Keskin; son derecede olan; derin. **~ely,** gayet; büyük alâka ile. **~ify** [inˈtensifây]. Şiddetini arttırmak; kuvvetlendirmek. **~ity,** şiddet; kuvvet. **~ive,** derin; gayretli, kesif.

**intent** [inˈtent]. Kasıd, maksad; meal. Dikkatli. **to all ~s and purposes,** esas itibarile: **to be ~ on stg.,** zihni bir şeyle meşgul olm.: **to be ~ on doing stg.,** bir şeyi yapmağa kasdetmek. **~ness,** fevkalâde dikkat. **~ion** [inˈtenşn]. Niyet, kasıd. **with the ~ of ···ing, ...** maksadiyle: **I have not the slightest ~ of ···ing [to ...],** ···e hiç niyetim yok: **to do stg. with the best ~s,** bir şeyi hiç bir fena niyetle yapmamak. **~ional,** kasdî. **~ionally,** kasden. **~ed, ...** niyetli: **well-~,** iyi niyetli.

**inter** [inˈtö(r)]. Defnetmek, gömmek.

**inter-** [ˈintə(r)] *pref.* Beyn···; ···arası(nda); mütekabilen; *mes.* **~connected,** biribirine bağlı.

**interact** [ˌintərˈakt]. Birbirine müessir olmak.

**interallied** [ˌintərˈalâyd]. Müttefikler arasındaki.

**intercede** [ˌintəˈsîd]. **to ~ (with s.o.) for s.o.,** biri için şefaat etm.

**intercept** [ˌintəˈsept]. Yolunu kesmek; tutmak. **~ion** [–ˈsepşn], kesme; tutulma.

**intercession** [ˌintəˈseşn]. Şefaat.

**interchange** [ˌintəˈçêync]. Mübadele (etm.); becayiş (etm.). **~able,** birbirinin yerine konulabilir.

**intercommunication** [ˌintəkomyuniˈkêyşn]. Biribirile haberleşme.

**intercourse** [ˈintəkôs]. Muaşeret; (cinsî) münasebet.

**interdependent** [ˈintədiˈpendənt]. Birbirine bağlı.

**interdict** *n.* [ˈintədikt]. Yasak emri; *v.* [–ˈdikt]. Yasak etm.

**interest**[1] [ˈintrist, ˈintərest] *n.* Alâka; ilgi; merak; faiz; menfaat. **it is in your ~,** sizin menfaatinizedir: **to take an ~ in ...,** ···e alâka göstermek: **to have an ~ in a business,** bir işe para yatırmış olm.: **to bear ~ at 5%,** 5% faiz getirmek: **questions of public ~,** herkesi alâkadar eden meseleler: **this is not in the public ~,** bu halkın menfaatine değildir: **the shipping ~,** deniz ticareti ile alâkalı olanlar. **~**[2] *v.* Alâkasını uyandırmak; alâkadar etm.; sarmak; ilgilendirmek. **to ~ oneself [be ~ed] in stg.,** bir şeye meraklı olm. **~ed** *a.* alâkadar, ilgili: **~ in stg.,** bir şeyin meraklısı: **~ motives,** menfaatperestlik: **he is an ~ party,** bu işte şahsî menfaatı var. **~ing** [ˈintərestin(g)]. Alâka uyandırıcı; meraklı; mühim.

**interfere** [ˌintəˈfiə(r)]. Müdahale etm., karışmak. **to ~ with,** burnunu sokmak; dokunmak; engel olm.: **this tree ~s with the view,** bu ağac manzaraya mâni oluyor. **~nce** [–əns], müdahale; karışma; engel; (radyo) parazit.

**interim** [ˈintərim]. Muvakkat. Aralık vakti. **ad ~,** muvakkaten: **~ dividend,** ara temettü: **in the ~,** bu aralıkta.

**interior** [inˈtiəriə(r)]. İç; dahilî. Iç; memleketin iç tarafı.

**interjection** [ˌintəˈcekşn]. (*gram.*) Nida.

**interlock** [ˌintəˈlok]. Birbirine bağla(n)mak.

**interloper** [ˈintəloupə(r)]. Hakkı olmadığı halde bir yere, işe giren.

**interlude** [ˈintəlyûd]. Fasıla; (tiyatro) aralık.

**intermarry** [ˌintəˈmari]. Muhtelif aileler vs. arasında evlenmek.

**intermediate** [ˌintəˈmîdyət]. Mütevassıt; iki şey arasında bulunan; orta.

**interminable** [inˈtö(r)minəbl]. Cansıkacak kadar uzun.

**intermittent** [ˌintəˈmitənt]. Durup yine işliyen; fasılalı.

**intern** [intö(r)n]. Kalebent etm.; enterne etmek. **~ee** [ˌintö(r)ˈnî]. Hükümet tarafından bir yere kapatılan kimse. **~ment** [inˈtö(r)nment]. Çıkması mümkün olmıyan bir yere kapatma; enterne etme.

**internal** [inˈtö(r)nl]. Dahilî; içe aid. **~-combustion engine,** dahilî ihtiraklı makine.

**international** [ˌintəˈnaşnl]. Beynelmilel; milletlerarası (oyuncu).

**interplay** [ˈintəplêy]. Karşılıklı tesir.

**interpose** [ˌintəˈpo̱uz]. Arasına koymak; müdahale etmek.
**interpret** [inˈtö(r)prit]. Mânasını izah etm.; tercümanlık etmek. **~ation** [–ˈtêyşn], izah; tercüme. **~er** [–ˈtö(r)prita(r)], tercüman.
**interrogat·e** [inˈterəgêyt]. Sorguya çekmek. **~ion** [–ˈgêyşn], istifham; sorgu: **~ mark,** istifham işareti. **~ive** [–ˈrogətiv], sorgu ifade eden; istifhamlı.
**interrupt** [ˌintəˈrʌpt]. Fasılaya uğratmak; (sözünü) kesmek. **~ion** [–ˈrʌpşn], kesme; sözün kesilmesi: **without ~,** durmadan.
**intersect** [ˌintəˈsekt]. Tekatu etm. **~ion** [–ˈsekşn], tekatu (yeri); faslı müşterek.
**interval** [ˈintəvl]. Ara, fasıla. **at ~s,** arasıra.
**interven·e** [ˌintəˈvîn]. Araya girmek; müdahele etmek. **~tion** [–ˈvenşn], araya girme; müdahele; tavassut.
**interview** [ˈintəvyû]. Görüşme(k); (···le) mülâkat.
**intestate** [inˈtestət]. Vasiyetsiz ölmüş.
**intestine** [inˈtestin]. Barsak.
**intima·cy** [ˈintiməsi]. Sıkı dostluk; mahremlik; cinsî münasebet. **~te**[1] [ˈintimit]. İçli dışlı; kafadar; mahrem. **~te**[2] [ˈintimêyt] *v.* Üstü kapalı anlatmak; bildirmek. **~tion** [––ˈmêyşn], haber.
**intimidat·e** [inˈtimidêyt]. Korkutmak. **~ion** [–ˈdêyşn], korkutma.
**into** [ˈintu]. İçine doğru; ···e, ···in içeriye; [*bir fiil ile birlikte olduğu zaman o fiile bak*]. **to change stg. ~ stg.,** bir şeyi bir şeye tahvil etm.: **to grow ~ a man,** büyüyüp adam olm.: **to work far ~ the night,** gece yarılarına kadar çalışmak: **5 ~ 12 goes 2 and 2 over,** 12 de 5 iki defa var, 2 artar.
**intolera·ble** [inˈtolərəbl]. Tahammül edilmez. **~nce** [inˈtolərəns]. Müsamahasızlık; taassub. **~nt,** müsamahasız; mütaassıb: **he is ~ of this drug,** bu ilâc ona dokunur.
**intonation** [ˌintoṷˈnêyşn]. Sesin ahengi.
**intoxic·ant** [inˈtoksikənt]. Sarhoş edici. **~ate,** sarhoş etm.; (sevinç) çılgın bir hale getirmek. **~ated,** sarhoş. **~ation** [–ˈkêyşn], sarhoşluk.
**intransitive** [inˈtrânsitiv]. (*gram.*) Lâzım (fiil).
**intrepid** [inˈtrepid]. Cesur; pervasız. **~ity** [–ˈpiditi], cesaret.
**intrica·cy** [ˈintrikəsi]. Karışıklık; giriftlik. **~te** [ˈintrikit], girift; muğlak, karışık.
**intrigue** [inˈtrîg]. Entrika (yapmak); desise. Tecessüsünü tahrik etmek.
**intrinsic** [inˈtrinsik]. Zatî. **~ally,** aslen.
**introduc·e** [ˌintrəˈdyûs]. Takdim etm.; sokmak; idhal etm.; ileri sürmek. **~tion** [ˌintrəˈdʌkşn]. Mukaddeme; başlangıc; içeri sokma; takdim etme. **letter of ~,** tavsiye mektubu: **to give s.o. an ~ to s.o.,** birisi için birine tavsiye (mektubu) vermek. **~tory** [ˌintrəˈdʌktəri]. Mukaddeme kabilinden; takdim edici.
**introspec·tion** [ˌintroṷˈspekşn]. Kendi ruhunu tedkik etme. **~tive,** kendi içini tedkik eden.
**introver·sion** [ˌintroṷˈvö(r)şn]. Kendi içine dönme. **~t** [ˈintroṷvö(r)t], içe dönük.
**intru·de** [inˈtrûd]. Zorla içeri sokmak; karışmak; münasebetsizce sokulmak. **to ~ on,** ···e tecavüz etm.: **I hope I'm not ~ing,** inşallah münasebetsiz zamanda gelmedim. **~der,** hakkı olmadığı bir yere giren kimse. **~sion** [inˈtrûjn]. İçeri sok(ul)ma; zorla dühul; tecavüz; münasebetsizce girme. **~sive** [–siv], sokulgan (*köt.*); mütecaviz.
**intuit·ion** [ˌintyuˈişn]. Seziş; hads. **~ive** [–ˈtyûitiv], seziş ile olan.
**inundat·e** [ˈinʌndêyt]. Suya basmak. **~ion** [–ˈdêyşn], su basması, feyezan.
**inure** [iˈnyuə(r)]. (Meşakkat) alıştırmak.
**invade** [inˈvêyd]. İstilâ etm. **~r,** müstevli.
**invalid**[1] [inˈvalid]. Hükümsüz; batıl. **~ity** [–ˈliditi], hükümsüzlük, butlan. **~ate** [inˈvalidêyt]. İbtal etm. **~**[2] [ˈinvəlîd]. Hasta; illetli. **to be ~ed out of the army,** çürüğe çıkarılmak.
**invaluable** [inˈvalyuəbl]. Son derece kıymetli; pek faydalı.
**invariable** [inˈveəriəbl]. Değişmez; sabit.
**invasion** [inˈvêyjn]. İstilâ; tecavüz.
**invective** [inˈvektiv]. Acı söz; sövme.
**inveigh** [inˈvêy]. **to ~ against ...,** aleyhinde şiddetli söz söylemek; tenkid etmek.
**inveigle** [inˈvêygl]. Hile ile kandırmak.
**invent** [inˈvent]. İcadetmek; ihtira

etm.; uydurmak. ~**ion** [–ˈvenşn], icad; ihtira; uydurma (havadis). ~**ive**, icad kabiliyeti olan. ~**or**, mucid, muhteri.
**inventory** [ˈinvəntri]. Müfredat defteri. **to make an** ~, ···in müfredatını tanzim etmek.
**inver·se** [inˈvö(r)s]. Makûs; ters; zıd. ~ **ratio**, makûs nisbet. ~**ion** [–ˈvö(r)şn], makûs yapma, olma. ~**t** [inˈvö(r)t]. Makûs yapmak; tersine çevirmek. ~**ted commas**, tırnak işareti.
**invertebrate** [inˈvö(r)tibrit]. Fıkrasız.
**invest** [inˈvest]. Faize para yatırmak; muhasara etm.; birini makamına oturtmak; birine salâhiyet vermek. ~**iture** [inˈvestityuə(r)]. Memuriyetin resmen tevcihi; Kıral tarafından nişan ve unvanların resmen tevcihi merasimi. ~**ment** [inˈvestmənt]. Para yatırma; yatırılan para; muhasara etme. **steel shares are a good** ~, çelik hisse senedleri kârlıdır. ~**or** [inˈvestə(r)]. Sermayesini yatıran.
**investigat·e** [inˈvestigêyt]. Tedkik etm., tahkik etmek. ~**ion**, tedkik; tahkikat; teftiş.
**inveterate** [inˈvetərit]. Zamanla kökleşmiş.
**invidious** [inˈvidyəs]. Hatır kırıcı; kıskandırıcı.
**invigilate** [inˈvicilêyt]. Yazılı imtihanda nezaret etmek.
**invigorate** [inˈvigərêyt]. Dincleştirmek.
**invincible** [inˈvinsibl]. Yenilemez.
**inviolable** [inˈvayələbl]. Nakzedilemez.
**invisi·ble** [inˈvizibl]. Gözle görülemez. ~ **ink**, gizli mürekkeb.
**invit·ation** [inviˈtêyşn]. Davet. ~**e** [inˈvâyt]. Davet etm., çağırmak. **to** ~ **trouble**, belâyı satın almak. ~**ing** [inˈvâytin(g)]. *a.* Cazib; lezzetli; hoş.
**invocation** [ˌinvoˈkêyşn]. İstimdad; dua.
**invoice** [ˈinvôys]. Fatura (çıkarmak). **as per** ~, fatura mucibince.
**invoke** [inˈvouk]. İstimdad etm.; yalvarmak.
**involuntar·y** [inˈvoləntəri]. İhtiyarî olmıyan. ~**ily**, istemiyerek.
**involve** [inˈvolv]. Sarmak; istilzam etm.; karıştırmak; mucib olm.; sokmak. **to be** ~**d**, medhaldar olm.; alâkalı olmak: **his honour is** ~**d**, şerefi mevzuu bahistir. ~**d** *a.* girift, muğlak; karışık; borclu.

**invulnerable** [inˈvʌlnrəbl]. Yaralanmaz.
**inward** [ˈinwəd]. Dahilî; içe doğru. ~**ly**, içte. ~**ness**, esas; içyüz. ~**s**, içeriye doğru.
**iodine** [ˈayədîn]. Tentürdiyod.
**iota** [ayˈoutə]. Yota. **not one** ~, zerre kadar.
**I.O.U.** [ay ouˈyû]. (*abb.*) **I owe you** (size borcum var). Borc senedi.
**ir-** *v.* **im-**.
**irascible** [iˈrasibl]. Çabuk öfkelenir.
**ir·ate** [ayˈrêyt]. Hiddetlenmiş; öfkeli. ~**e** [ayə(r)]. Öfke, hiddet.
**Ireland** [ˈayələnd]. İrlanda.
**iris** [ˈayəris]. Kuzahiye.
**Irish** [ˈayəriş]. İrlandalı; irlandaca. ~**man**, İrlandalı.
**irk** [ö(r)k]. Üzmek; usandırmak. ~**some** [ˈö(r)ksəm], usandırıcı.
**iron** [ˈây(r)ən]. Demir(den yapılmış). Ütü(lemek). **old** ~, hurda demir: **to be in** ~**s**, zincirli olm.: ⌜**to have many** ~**s in the fire**⌝, ⌜kırk tarakta bezi olmak⌝: ⌜**strike while the** ~ **is hot!**⌝, demiri tavında iken dövmeli: **to put s.o. in** ~**s**, birini prangaya vurmak. ~**clad**, zırhlı. ~**-foundry**, dökümhane. ~**-shod**, demir uclu. ~**ing** [ˈayənin(g)]. Ütüleme. ~**master** [ˈayənmâstə(r)]. Demir imalatçısı. ~**monger**, hırdavatçı. ~**mongery**, hırdavat. ~**worker** [ˈayənwö(r)kə(r)]. Demirci.
**iron·ic(al)** [ayˈronik(l)]. İstihzalı; kaderin bir cilvesi gibi. ~**y** [ˈayrəni]. İstihza. ~ **of fate**, kaderin cilvesi.
**irradiat·e** [iˈrêydiêyt]. Şualar neşretmek; parlatmak.
**irrational** [iˈraşənl]. Gayrı makul; akla uymaz.
**irreconcilable** [ˌirekənˈsâyləbl]. Telif edilemez.
**irrecoverable** [ˌiriˈkʌvərəbl]. Geri alınamaz; istirdadı imkânsız.
**irrefutable** [iˈrefyutəbl]. Reddedilemez.
**irregular** [iˈregyulə(r)]. Gayrı muntazam; usûl ve nizama aykırı; arızalı; (*gram.*) gayrı kıyasî. ~ **life**, sefih hayat: ~ **troops**, başıbozuk asker. ~**ity** [–ˈlariti], intizamsızlık; usule aykırılık; gayrıkıyasîlik.
**irrelevan·t** [iˈreləvənt]. Yersiz; münasebeti olmıyan. ~**ce** [~**cy**], yersiz olma; münasebetsizlik.
**irreligious** [ˌiriˈlicəs]. Dindar olmıyan; dine mugayir.

**irremediable** [ˌiriˈmîdyəbl]. Şifa bulmaz; çaresiz.
**irreparable** [iˈrepərəbl]. Telâfisi mümkün olmıyan.
**irrepressible** [ˌiriˈpresəbl]. Zabtolunmaz.
**irreproachable** [ˈiriˈproụçəbl]. Kusursuz.
**irresistible** [ˌiriˈzistibl]. Önüne durulmaz; dayanılmaz.
**irresolute** [iˈrezolyut]. Kararsız.
**irrespective** [ˌiriˈspektiv]. ~ **of ...,** ...e bakmaksızın.
**irresponsible** [ˌiriˈsponsibl]. Gayrı mes'ul; sersem.
**irresponsive** [ˌiriˈsponsiv]. Mukabele etmez.
**irretrievable** [ˌiriˈtrîvəbl]. Telâfi edilemez; bir daha ele geçmez.
**irreveren·t** [iˈrevərənt]. Hürmetsiz. **~ce**, hürmetsizlik.
**irreversible** [ˌiriˈvö(r)səbl]. Tersine çevirilemez.
**irrevocable** [ˌiriˈvokəbl]. Değiştirilemez.
**irrigat·e** [ˈirigêyt]. Sulamak; iska etmek. **~ion** [–ˈgêyşn], sulama, iska.
**irrita·ble** [ˈiritəbl]. Çabuk kızar; çabuk iltihablanır. **~bility** [–ˈbiliti], öfkelilik; tahriş edilebilme. **~te** [ˈiritêyt]. Gücendirmek; sinirlendirmek; tahriş etm. **~tion,** dargınlık; taharrüş.
**irruption** [iˈrʌpşn]. İstilâ; içeriye üşüşme.
**is** [iz]. *3rd pers. sing. pres. ind. of* **be.** Dır, dir vs.
**-ish** [iş]. *Şu manalari ifade eden son ek*: (i) Oldukça; **greenish,** yeşilimsi; **oldish,** yaşlıca. (ii) gibi, benzer; **foolish,** deli gibi. (iii) Mütemayil; **bookish,** kitabî.
**Islam·ic** [isˈlamik]. İslâmiyete aid.
**island** [ˈâylənd]. Ada; (kalabalık caddelerde) yayalara mahsus adacık. **~er,** adalı.
**isle** [âyl]. Ada. **the British Isles,** Britanya ve İrlanda.
**isolat·e** [ˈâysəlêyt]. Tecrid etm.; ayırmak. **~ed,** ücra; münferid; yalnız. **~ion** [ˌâysoˈlêyşn]. Tecrid; infirad; ücralık; yalnızlık. **~ hospital,** tecrid hastahanesi. **~ionism,** infiradcılık. **~ionist,** infiradcı.
**Israel** [ˈizrêyəl]. Beni İsrail. İsrail devleti. **~ite** [–lâyt], yahudi.
**issue**[1] [ˈisyû] *v.* Çıkmak, neş'et etm.; çıkarmak; neşretmek; tedavüle çıkarmak; tevzi etm. **to ~ passports,** pasaport vermek: **to ~ a warrant of arrest,** bir tevkif müzekkeresi çıkarmak. **~**[2] *n.* Çıkış; çıkarma; mahrec; netice; evlâd; neşretme; nüsha; dağıtma; bir defada tedavüle çıkarılan para. **to bring a matter to an ~,** meseleyi bir neticeye bağlamak: **to die without ~,** bilâ veled vefat etm.: **to evade the ~,** asıl mevzudan kaçmak: **in the ~,** neticede: **to join ~ with s.o.,** münakaşa etm.: **to obscure the ~,** asıl mevzuu kaybettirmek: **the point at ~,** münakaşa edilen nokta.
**isthmus** [ˈismʌs]. Berzah.
**it** [it]. *Cansızlar hakkında kullanılan şahıs zamiri.* O; onu; ona. **~ is raining,** yağmur yağıyor: **~ is getting late,** geç oluyor: **~ seems to me,** bana öyle geliyor ki: **he thinks he's *it*,** küçük dağları ben yarattım diyor: **this book is absolutely *it*,** bu kitab yamandır: **he hasn't got ~ in him to do that,** o bu işin adamı değildir.
**ital.** (*abb.*) **italics** [iˈtaliks]. İtalik harfler.
**Italy** [ˈitəli]. İtalya.
**itch** [iç]. Kaşınmak. Kaşıntı. **the ~,** uyuz illeti: **to ~ to do stg.,** bir şey yapmak için içi gitmek: **to have an ~ for writing,** yazmağı şiddetle arzu etm.: **he's ~ing for trouble,** başının belâsını arıyor. **~y,** kaşıntılı, uyuz.
**item** [ˈâytem]. Kalem; madde; dahi, keza. **~s,** müfredat. **to give the ~s,** madde madde saymak: **~s on the agenda,** ruznamedeki meseleler: **news ~s,** muhtelif haberler: **the last ~ on the programme,** programdaki son numara.
**itiner·ant** [âyˈtinərənt]. Seyyah; seyyar. **~ary** [–rəri], (seyahat) program.
**its** [its]. *Gen. of* **it.** Onun.
**it's = it is; it has.**
**itself** [itˈself]. Kendi. **by ~,** kendi kendine; **in ~,** haddizatında.
**I've = I have.**
**ivory** [ˈâyvəri]. Fildişi (rengi); fildişinden yapılmış. **black ~,** (*esk.*) zenci köleler.
**ivy** [ˈâyvi]. Sarmaşık.

# J

**J** [cêy]. J harfi.
**jab** [cab]. Sivri bir şeyile dürtme(k).
**jabber** [ˈcabə(r)]. (*ech.*) Maymun gibi konuşma(k).
**Jack**[1] [cak]. (*abb.*) John; (iskambil)

bacak. ~ Tar, bahriyeli nefer (*cf.* Mehmedcik): ⌜**before you could say ~ Robinson**⌝, kaşla göz arasında: **every man ~,** bilâistisna herkes. **~-in-the-box,** açılınca içinden bir yaya bağlı bebek fırlayan kutu. **~-knife,** büyük sustalı çakı. **~².** Kriko. **~ up,** kriko ile kaldırmak. **~³.** Cıvadra sancağı. **Union J~,** İngiliz bayrağı.

**jackal** [ˈcakôl]. Çakal.

**jackanapes** [ˈcakənêyps]. Küçük çapkın.

**jackass** [ˈcakas]. Eşek; ahmak.

**jackdaw** [ˈcakdô]. Küçük karga.

**jacket** [ˈcakit]. Caket; dış örtü; kitab zarfı; kabuk.

**Jacobean** [ˌcakoˈbîən]. İngiliz kıralı birinci James'in devrine aid.

**jade¹** [cêyd]. Yeşimtaşı. **~².** Lağar beygir; oynak kız. **~d** [ˈcêydid]. Bitkin, mecalsız.

**jag** [cag]. Sivri ve pürüzlü çıkıntı. Pürüzlü bir tarzda kesmek. **~ged,** sivri ve pürüzlü.

**jaguar** [ˈcagyua(r)]. Cenubî Amerika kaplanı.

**jail** [cêyl]. Hapishane. Hapsetmek.

**jam¹** [cam]. Reçel. **~².** Sıkışıklık; sıkışık kalabalık. Sıkıstırıp basmak; yuvarlanmasına, yürümesine mani olm.; (radyo) neşriyatı bozmak; (makine) sıkışmak.

**Jamaica** [cəˈmêykə]. Camayika.

**jamboree** [camboˈrî]. Eğlenceli toplantı, cümbüş (*gen.* izcilerin büyük toplantısı).

**jangle** [ˈcan(g)gl]. (*ech.*) Ahenksiz ses (çıkar(t)mak).

**janissary** [ˈcanisəri]. Yeniçeri.

**janitor** [ˈcanitə(r)]. Kapıcı.

**January** [ˈcanyuəri]. Ocak ayı.

**Japan¹** [cəˈpan], Japonya. **~anese** [–əˈnîz], japonyalı; Japonca. **~²** [cəˈpan]. Pek katı ve parlak bir nevi vernik(le kaplamak).

**jar¹** [câ(r)]. Kavanoz; küp. **~².** Gıcırtı; sarsıntı; nifak. **~ (on),** gıcırdamak; sarsmak; dokunmak; uymamak. **that's a bit of a ~!,** (*coll.*) bu pek tepeden inme oldu! **~ring** [cârin(g)]. Ahenksiz; uygunsuz; sarsan.

**jargon** [ˈcâgn]. Bozuk şive; meslekî argo.

**jaundice** [ˈcôndis]. Sarılık hastalığı. **~d,** safravî: **to take a ~ view of the world,** dünyayı karanlık görmek.

**jaunt** [cônt]. Kısa gezinti. **~y** [ˈcônti]. Şen; havaî.

**javelin** [ˈcavəlin]. Kısa mızrak.

**jaw** [cô]. Çene; ağız; (*coll.*) cansıkıcı nasihat. (*coll.*) Çene çalmak; cansıkıcı nasihat vermek. **hold your ~!,** çeneni tut!: **the ~s of death,** ölümün pençesi.

**jay** [cêy]. Alakarga. **~-walker** [ˈcêywôkə(r)]. Sokakta dalgın ve ihtiyatsızca yürüyen kimse.

**jazz** [caz]. Caz (müziği ile dansetmek).

**jealous** [ˈceləs]. Kıskanc. **to be ~ of [for],** ···den kıskanmak. **~y,** kıskanclık.

**jeans** [cînz]. İş için mavi pantalon.

**jeer** [ciə(r)]. İstihza, alay. Yuha çekmek. **to ~ at,** alaya almak.

**jejune** [cəˈcûn]. Kıt; yavan.

**jelly** [ˈceli]. Pelte; donmuş etsuyu. Pelteleş(tir)mek. **to pound s.o. to a ~,** pestilini çıkarmak. **~-fish,** denizanası.

**jemmy** [ˈcemi]. Kısa bir manivelâ.

**jeopard·ize** [ˈcepədâyz]. Tehlikeye koymak. **~y,** tehlike.

**Jericho** [ˈcerikou]. **go to ~!,** cehenneme git!

**jerk** [cö(r)k]. Anî sarsıntı. Birdenbire sarsarak hareket etmek. **~y** [ˈcö(r)ki]. İntizamsız ve anî hareketlerle yapılan.

**jerry** [ˈceri]. (*sl.*) Alman askeri; (*sl.*) lâzımlık. **~-builder,** ucuz, âdi ev yapıcısı. **~-built,** derme çatma.

**jersey** [ˈcö(r)zi]. Kazak; jerse.

**jest** [cest]. Alay (etm.), şaka (etmek). **to say stg. in ~,** bir şeyi şaka olarak söylemek. **~er,** soytarı.

**Jesuit** [ˈcezyuit]. Cizvit. **~ical** [–ˈitikl], hilekâr.

**Jesus** [ˈcîzəs]. İsa peygamber.

**jet¹** [cet]. **~-black** simsiyah. **~².** Fıskıye; (*otom.*) ciglör; (havagazi) alevi. Fışkırmak.

**jet·sam** [ˈcetsəm]. Deniz enkazı. **~-tison** [ˈcetisən]. Gemiyi kurtarmak için eşyayı denize atma(k).

**jetty** [ˈceti]. İskele; dalgakıran.

**Jew** [cû]. Yahudi. **~ess** [ˈcues]. Yahudi kadın. **~ish** [–iş], yahudi, musevi. **~ry,** yahudiler.

**jewel** [cuəl]. Kıymetli taş; mücevher. **a ~ of a servant,** bulunmaz bir hizmetçi. **~led,** mücevherle süslü. **~ler,** kuyumcu. **~ry** [**~lery**], mücevherat.

**jib¹** [cib]. **to ~ at doing stg.,** bir şeyi yapmaktan imtina etm. **~²** (*naut.*) Flok: **~-boom,** baston.

**jibe** *v.* **gibe.**

**jiffy** [ˈcifi]. (*coll.*) An, lâhza.
**jig** [cig]. Pek canlı bir raks (yapmak). **to ~ up and down,** dans eder gibi sıçramak.
**jiggered** [ˈcigəd]. (*sl.*) **well I'm ~!,** olur şey değil!: **I'm ~ if I'll do it!,** yaparsam Arab olayım!
**jigsaw** [ˈcigsô]. Makineli oyma testeresı. **~ puzzle,** muhtelif parçalar halinde kesilen bir resmi tekrar birleştirmekten ibaret bir oyun.
**jilt** [cilt]. Fındıkçı kız. Bir adama evlenme vadedip sözünde durmamak.
**jingle** [ˈcin(g)gl]. (*ech.*) Çıngırtı. Çıngırdamak.
**jingo** [ˈcin(g)go̱u]. Şoven. **by ~,** vallahi! **~ism,** şovenlik.
**jinks** [cin(g)ks]. **high ~,** cümbüş.
**job**[1] [cob] *n.* İş, vazife; götürü iş; hizmet; meşguliyet; zor bir iş; iltimaslı muamele. **that's a bad ~!,** aksilik!: **to make the best of a bad ~,** fena şartlardan azamî istifade temin etm.: **to give s.o. up as a bad ~,** bundan hayır gelmez diye bırakmak: **to do work by the ~,** götürü iş yapmak: **to be paid by the ~,** götürü ücret almak: **this is not everybody's ~,** bu iş her babayiğitin kârı değil: **it's a good ~ that ...,** bereket versin ki: **to make a good ~ of stg.,** bir işi iyi yapmak: **we had a ~ to get there,** oraya gidinceye kadar hal olduk: **he knows his ~,** işini biliyor: **a ~ lot,** ucuz alınan müteneyvi eşya: **to be out of a ~,** işsiz olmak. **~**[2] *v.* İş görmek; komisyonculuk etmek. **~ber,** borsada bir nevi simsar; resmî işlerde suiistimal yapan kimse. **~bery,** resmî işlerde suiistimal. **~bing,** komisyonculuk: **a ~ workman,** götürü iş yapan işçi.
**Job** [co̱ub]. Eyub; sabrın timsali. **~'s comforter,** insanın yarasına tuz biber eken.
**jockey** [ˈcoki]. Cokey. Manevra yapmak. **to ~ s.o. into doing stg.,** kurnazlıkla kandırıp birine bir işi yaptırmak.
**jocular** [ˈcokyulə(r)]. Şakacı.
**jog** [cog]. Hafifce dürtme(k). **to ~ along,** iyi kötü yuvarlanıp gitmek: **to ~ s.o.'s memory,** birinin bir şeyi hatırlamasına yardım etmek. **~-trot,** ağır ve rahat lenk gidiş.
**johnny** [ˈconi]. (*sl.*) Adam, herif.
**join** [côyn]. Birleş(tir)mek; kat(ıl)mak; bitiş(tir)mek; ...e iltihak etm.; bir araya gelmek. İki şeyin birleştiği yer. **to ~ battle,** muharebeye başlamak: **to ~ forces [hands] with s.o.,** birisile işbirliği yapmak: **to ~ a club,** bir klübe âza yazılmak: **to ~ one's ship,** (denizci) gemiye dönmek: **to ~ up,** *va.* birleştirmek; *vn.* askere yazılmak. **~er** [ˈcôynə(r)]. Doğramacı: **~ery,** doğrama işi. **~t**[1] [côynt] *n.* İki şeyin birleştiği yer; ek yeri; mafsal; büyük parça et. **to find a ~ in s.o.'s armour,** birinin zayıf damarını bulmak: **the time is out of ~,** ortalık altüst oldu: **his nose was put out of ~,** burnu kırıldı. **~t**[2] *a.* Birleşik; müşterek: **~ly liable,** müştereken mesul: **~ly and severally liable,** müteselsilen ve münferiden mesul. **~ted** [ˈcôyntid]. Mafsallı.
**joke** [co̱uk]. Şaka; lâtife. Şaka yapmak; şakadan söylemek. **joking apart,** şaka bertaraf: **the ~ of it is that ...,** tuhafı şu ki: **practical ~,** el şakası. **~r,** şakacı; (iskambil) bazı oyunlarda en kıymetli sayılan kâğıd.
**joll·ity** [ˈcoliti]. Neş'elilik, cümbüş. **~y** [coli]. Şen; güler yüzlü, güzel; (*coll.*) çok. **I'll take ~ good care not to go there again,** bir daha oraya gidersem bana da adam demesinler: **'I won't do it!' 'You ~ well will!',** 'Ben bunu yapmam.' 'Top gibi yaparsın!'
**jolt** [co̱ult]. Sarsıntı. Sarsmak.
**Jonah** [co̱una]. Yunus peygamber; uğursuz adam.
**Jordan** [ˈcôdən]. Erdün (nehri). **this side of ~,** bu dünyada.
**jostle** [ˈcosl]. İtip kakmak; birbirine sürtünmek.
**jot** [cot]. Zerre. **to ~ stg. down,** bir şeyi yazıvermek: **I don't care a ~,** bana vız gelir.
**journal** [ˈcö(r)nl]. Gazete; yevmiye defteri; ruzname. **~ese** [–ˈlîz], (*sl.*) fena gazeteci üslûbü. **~ism** [–lizm], gazetecilik. **~ist,** gazeteci.
**journey** [ˈcö(r)ni]. Seyahat (etm.). **~man** [ˈcö(r)nimən]. (*esk.*) Kalfa; (*şim.*) para ile tutulan adam.
**Jove** [co̱uv]. Jüpiter. **by ~!,** Vallahi.
**jovial** [ˈco̱uvyəl]. Şen; keyifli; güler yüzlü. **~ity** [–ˈaliti], guleryüzlülük, şenlik.
**jowl** [ca̱ul]. Çene. **cheek by ~,** haşır neşir.
**joy** [côy]. Sevinc; neş'e; zevk. **oh, ~!,** aman, ne güzel!: **I wish you ~ (of it),** (i) güle güle!; (ii) (*istihzalı olarak*) Allah versin! **~ful** [ˈcôyfl], sevindirici; memnun edici; sevincli; neş'-

eli. ~**less** [–lis], kederli. ~**ous** *v.* **joyful.** ~-**bells,** düğün vs. için çalınan çanlar. ~-**ride,** sahibinin müsaadesini almıyarak otomobilinde yapılan gezinti.

**J.P.** [ˈcêyˈpî]. (*abb.*) **Justice of the Peace,** sulh hâkimi.

**jubil·ant** [ˈcûbilənt]. Büyük neş'e içinde; pek memnun. ~**ation** [--ˈlêyşn], pek çok sevinme; bayram etme. ~**ee** [ˈcûbilî]. Mühim bir vakanın ellinci yıldönümü. **silver** ~, 25 inci yıldönümü: **diamond** ~, 60 ıncı yıldönümü.

**Judas** [ˈcûdas]. Yehuda; hain. ~**kiss,** pek haince bir hareket.

**judge** [cʌc]. Hâkim; hakem; bir şeyden iyi anlıyan. Mahkemede karar vermek; hüküm vermek; tahmin etmek. **judging by** ..., ...e bakılırsa; ...e nazaran: ~ **of my surprise!,** hayretimi düşün!: **a good** ~ **of men,** insan sarrafı. ~**ment** [ˈcʌcmənt]. Muhakeme kararı; hüküm; fikir. **the day of** ~ [**the Last** ~], kıyamet günü: **it's a** ~ **on you,** bu sana Allahın cezasıdır.

**judic·ature** [ˈcûdikətyuə(r)]. Hâkimlik. **the** ~, hâkimler. ~**ial** [cûˈdişyəl]. Mahkemeye aid; hâkimliğe aid; hükmî; adlî; bitaraf. ~ **murder,** mahkeme kararile fakat haksız olan idam. ~**iary** [cûˈdişyəri]. Adliye idaresi; hâkimler. ~**ious** [cûˈdişəs]. Müdebbir; makul.

**jug** [cʌg]. Küçük testi; kulplu su kabı; (*sl.*) kodes(e tıkmak).

**juggle** [ˈcʌgl]. Hokkabazlık yapmak. **to** ~ **with figures,** *etc.*, rakamlar vs. ile oynamak. ~**r,** hokkabaz.

**juic·e** [cûs]. Meyvanın suyu; (*sl.*) benzin; elektrik cereyanı. ~**y,** sulu.

**July** [cuˈlây]. Temmuz.

**jumble** [cʌmbl]. Karmakarışık yığın. ~-**sale,** bir hayır cemiyeti menfaatine ufak tefek eşya satışı.

**jump** [cʌmp]. Atlama; sıçrayış; irkilme; mania; (fiat) ansızın yükseliş. (Üzerine) atlamak; sıçramak; zıplamak; irkilmek. **to** ~ **to a conclusion,** acele hüküm vermek: **to** ~ **down s.o.'s throat,** birini şiddetle terslemek: **to** ~ **up** [**to** ~ **to one's feet**], sıçrayıp ayağa kalkmak: **he would** ~ **at it,** dünden hazır. ~**er**[1] [ˈcʌmpə(r)]. Atlıyan kimse, at. ~**er**[2]. Örgülü triko. ~**iness** [ˈcʌmpinis]. Sinirlilik. ~**y** [ˈcʌmpi]. Sinirli; her sesten korkan.

**junct·ion** [ˈcʌn(g)kşn]. Birleşme (yeri); ittisal; kavşak. ~**ure** [ˈcʌn(g)ktyə(r)]. Birleşme yeri; vaziyet: **at this** ~, bu (buhranlı) anda.

**June** [cûn]. Haziran.

**jungle** [ˈcʌn(g)gl]. Cengel.

**junior** [ˈcûnyə(r)]. Daha genc; en genc; kıdem itibarile madun. **the** ~**s,** gencler: ~ **counsel,** avukat yardımcısı: **the** ~ **school,** mektebin ilk kısmı: **he is ten years my** ~, o benden on sene küçüktür: **Thomson** ~, Thomson'un oğlu.

**junk.** Eski halat parçaları.

**junket** [ˈcʌn(g)kit]. Bir nevi yoğurt; âlem (yapmak).

**juris·diction** [ˌcuərisˈdikşn]. Kaza hakkı (cari olan daire). ~**prudence** [-ˈprûdəns], hukuk ilmi. ~**t** [ˈcûrist]. Hukukçu.

**jur·or** [ˈcuərə(r)]. Jüri âzası. ~**y** [ˈcuəri]. Jüri heyeti. ~**man,** jüri âzası. ~-**box,** mahkemede jüri mevkii.

**just**[1] [cʌst] *a.* Âdil; müstahak; adalete uygun; haklı. **the sleep of the** ~, deliksiz bir uyku. ~**ness** [ˈcʌstnis]. Haklı olma; doğruluk. ~[2] *adv.* Hemen; henüz şimdi; tam, tamamen; şöyle bir; daradar. ~ **as you say,** tıpkı dediğiniz gibi: ~ **as he spoke,** konuşur konuşmaz: **I am** ~ **coming,** hemen şimdi geliyorum: ~ **for a joke,** sadece şaka olsun diye: **I was** ~ **going out, when** ..., tam sokağa çıkacağım sırada ...: **that's** ~ **it!** (i) işte mesele burada; (ii) işte tam bu!: ~ **listen!,** bir az dinle!: ~ **listen to him!,** şuna bak, nasıl saçmalıyor!: ~ **a moment!,** bir dakika!; ~ **a moment!,** dur bakalım!: **I saw him** ~ **now,** onu şimdi gördüm: **business is bad** ~ **now,** şu sırada işler kötü: **it is** ~ **four o'clock,** saat tam dört: ~ **once,** yalnız bir defa: **this book is** ~ **out,** bu kitab yeni çıktı: ~ **so,** (i) doğru; (ii) tam öyle: ~ **then,** tam o anda: ~ **there,** tam orada: ~ **think of it!,** hayret!: ~ **take a seat, will you!,** şöyle bir az oturur musunuz: **I can't go** ~ **yet,** şimdilik henüz gidemem: **'Is it raining?' 'Just'.** 'Yağmur yağıyor mu?' 'Serpiştiriyor': **'Was he angry?' 'Wasn't he** ~**!'**, 'Kızdı mı?' 'Hem de nasıl!': **'Well, I'll do as you say.'** 'I should ~ **think you will!'**, 'Pek iyi, sizin dediğiniz gibi yaparım.' 'Elbette öyle yapacaksın!': **if I get this job**

won't I ~ **work**!, bu iş olursa öyle bir çalışacağım ki: '**Do you like** ...?' '**Don't I** ~!' '... sever misiniz?' 'Hem de nasıl!'

**justice** [ˈcʌstis]. Adalet, hak; hâkim. ~ **of the Peace**, sulh hâkimi: **the Lord Chief J**~, İngiltere'de yüksek mahkemenin (Queen's Bench) dairesi reisi: **to do** ~ **to stg.**, bir şeyin hakkını vermek: **in** ~ **to him**, onun hakkını vermiş olmak için: **to bring to** ~, mahkemeye vermek: **he did not do himself** ~, kendini gösteremedi.

**justif·y** [ˈcʌstifây]. Haklı çıkarmak; mazur göstermek. ~**iable** [–ˈfâyəbl] hak verilebilir; mazur görülebilir. ~**ication** [ˌcʌstifiˈkêyşn], haklı sebeb; mazur gösterme.

**jut** [cʌt]. ~ **out**, çıkıntı halinde bulunmak.

**juvenile** [ˈcûvənâyl]. Gencliğe mahsus.

# K

**K** [kêy]. K harfi.

**Kalends** [ˈkalends]. ⌜**on the Greek** ~⌝, ⌜balık kavağa çıkınca⌝.

**K.B.E.** [ˈkêy ˈbî ˈî]. **Knight Commander of the British Empire**; *v.* **order²**.

**K.C.** [ˈkêyˈsî]. **King's Counsel**; *v.* **king.**

**K.C.B., K.C.I.E.**, vs. *v.* **order²**.

**keel** [kîl]. Geminin omurgası. Karina etmek. **on an even** ~, ufkî: **to** ~ **over**, albura olmak.

**keen** [kîn]. Keskin; sert; hassas; gayretli. **to be** ~ **on stg.**, bir şeye hevesli olm.: **I am not very** ~ **on it**, bundan pek hoşlanmam: **as** ~ **as mustard**, pek gayretli.

**keep¹** [kîp] *n.* Bir hisarın iç kalesi. ~² *n.* Yiyecek; gıda. **to work for one's** ~, boğaz tokluğuna çalışmak: **he isn't worth his** ~, yediği ekmeği hak etmiyor. ~³ *v.* (**kept**) [kept]. Tutmak; alıkoymak; bırakmamak; saklamak; beslemek; idare etm.; riayet etmek. Kendini tutmak; devam etm.: ~ (**good**) bozulmamak. **you may** ~ **this**, bu sizde kalsın: (in **a shop**) **do you** ~ **soap**?, (dükkânda) sizde sabun bulunur mu?: **to** ~ **at it**, çalışmak: **to** ~ **s.o. at it**, bir işi yapması için birinin üstüne düşmek: **to** ~ **at work**, işe devam etm.: **to** ~ **one's bed**, (hastalıktan) yataktan çıkmamak: **to** ~ **s.o. in clothes**, birinin giyimini temin etm.: **don't let me** ~ **you!**, sizi alıkoymayayım: **to** ~ **stg. from s.o.**, birisinden bir haberi gizlemek: **to** ~ **s.o. from doing stg.**, birini bir şey yapmaktan alıkoymak: **how are you** ~**ing?**, nasılsınız?: **to** ~ **to the left**, soldan gitmek: **fish won't** ~ **in summer**, balık yazın çabuk bozulur: **this meat will be all the better for** ~**ing**, bu et bir kaç gün saklanırsa daha iyi olur: **to** ~ **stg. to oneself**, gizleyip kimseye söylememek: **they** ~ **to themselves**, başkalarına sokulmuyorlar: **to** ~ **s.o. to his promise**, birine vadini tutturmak: **to** ~ **one's seat**, yerinden kalkmamak; at üzerinden düşmemek: **to** ~ **s.o. waiting**, birini bekletmek. ~ **away**, yaklaş(tır)mamak. ~ **back**, alıkoymak; durdurmak; ihtiyat olarak tutmak; saklamak, ketmetmek; geri kalmak; **you are** ~**ing stg. back!**, dilinin altında bir şey var. ~ **down**, bastırmak; aşağıda tutmak; inkıyad ettirmek; yükselmesine mani olm.; aşağıda kalmak, büzülüp saklanmak: **to** ~ **expenses down**, fazla masrafı önlemek. ~ **in**, içeride tutmak; salıvermemek; evden çıkmasını menetmek; izinsiz bırakmak; evde kalmak: **to be kept in**, izinsiz kalmak: **to** ~ **the fire in**, ocağı söndürmemek: **to** ~ **one's hand in**, alışkanlığını kaybetmemek: **to** ~ **in with s.o.**, birisile iyi münasebetlerini muhafaza etmek. ~ **off**, defetmek; uzak kalmak: **if the rain** ~**s off**, yağmur yağmazsa; ~ **your hands off!**, dokunma! ~ **on**, çıkarmamak; devam etm.: **to** ~ **on at s.o.**, birinin başının etini yemek: **don't** ~ **on about it!**, fazla ısrar etme!: **to** ~ **on doing stg.**, bir şeyi yapıp durmak. ~ **out**, içeri bırakmamak; dışında kalmak: **to** ~ **s.o. out of his rights**, birini hakkından mahrum etm.: **to** ~ **out of a quarrel**, bir kavgaya karışmamak. ~ **together**, bir arada tutmak, kalmak; dağılmamak. ~ **under**, zabtetmek; inkıyad altına almak; bastırmak: **to** ~ **a fire under**, bir yangının büyümesini menetmek. ~ **up**, düşürmemek; devam etm.; idame etm.; muhafaza etm.; geri kalmamak: **he couldn't** ~ **up**, daima geri kalıyordu: **to** ~ **up one's courage**, cesaretini kaybetmemek: ~ **it up!**, dayan!: **to** ~ **s.o**

up at night, birinin yatmasına mani olm.: **to ~ up with the times,** zamana uymak. **~er** [ˈkîpə(r)]. Bekçi; muhafız; av bekçisi; sahib. **~ing** [ˈkîpin(g)]. **to be in s.o.'s ~,** birinin muhafazası altında olm.: **in ~ with** ..., ...e uygun olarak: **out of ~ with** ..., ...e uymaz. **~sake** [ˈkîpsêyk]. Hatıra.

**ken** [ken]. Tanımak. Birinin bildiği veya gördüğü saha.

**kennel** [ˈkenl]. Köpek kulübesi.

**kept** [kept] *v.* **keep; ~ woman,** kapatma.

**kerb** [kö(r)b]. Kaldırımın kenar taşı.

**kernel** [ˈkö(r)nel]. Çekirdek içi; esas.

**kerosene** [ˌkeroˈsîn]. Petrol, gaz.

**ketchup** [ˈkeçəp]. Domates salçası.

**kettle** [ˈketl]. (Su) ibrik. **to put the ~ on,** su kaynatmak: ⌜**here's a pretty ~ of fish!**⌝, ⌜ayıkla şimdi pirincin taşını⌝.

**key**[1] [kî] *n.* Anahtar; tuş; kurma sapı; ses perdesi. **in the ~ of C,** do perdesi: **~ map,** ana harita: **~ industry,** ana sanayi: **~ point,** mühim nokta: **to touch the right ~,** tam yerinde söylemek. **~board,** klâviye. **~ hole,** anahtar deliği. **~stone,** kilid taşı; (*fig.*) esas. **~-note,** baş perde; esas. **~-ring,** anahtar halkası. **~-word,** bir şifrenin anahtarı. **~**[2] *v.* **to be ~ed up,** meraklı heyecanda olmak.

**K.G.** [ˈkêy ˈcî]. **Knight of the Garter**; *v.* **order**[2].

**khaki** [ˈkâki]. Hâki renk(li). **to get into ~,** asker olmak.

**kibosh** [ˈkâyboş]. Saçma. **to put the ~ on,** matetmek.

**kick** [kik]. Tepmek; çifte atmak; seğirdim yapmak. Tepme; çifte; seğirdim; **to ~ the bucket,** (*sl.*) nalları dikmek: **this drink has a ~ in it,** bu içki oldukça kuvvetli: **to get a ~ out of stg.,** bir şeyin zevkini çıkarmak: ⌜**to get more ~s than ha'pence**⌝, takdirden çok tenkide uğramak: **to ~ up a fuss,** [**rumpus**], mesele çıkarmak: **to ~ one's heels,** sabırsızlanarak beklemek: **I felt like ~ing myself,** yaptığıma çok pişman oldum: **he has no ~ left in him,** mecalsizdir: **to ~ off,** bir futbol maçına başlamak: **don't leave your things ~ing about,** eşyanı ötede beride bırakma!: **to be ~ed out,** kovulmak. **~-back,** seğirdim.

**kid**[1] [kid]. *n.* Oğlak (derisi); (*coll.*) çocuk. **to handle s.o. with ~ gloves,** birini nezaketle idare etm.: **not a job for ~ gloves,** (i) kirli bir iş; (ii) bu meselede merhametsizce davranmalı. **~**[2] *v.* (*coll.*). Aldatmak; takınmak. **you can't ~ me!,** öyle yağma yok!: **I was only ~ding,** şaka söyledim.

**kidnap** [ˈkidnap]. Cebren kaçırmak; çocuk çalmak.

**kidney** [ˈkidni]. Böbrek; huy.

**kill** [kil]. Öldürme(k), katletmek. Öldürülmüş hayvan. **to be in at the ~,** bir teşebbüsde muvaffakiyet anında hazır bulunmak: **~ or cure remedy,** şiddetli ve tehlikeli ilâc, tedbir: **to ~ with kindness,** lüzumundan fazla ihtimamla vermek. **~-joy,** neş'e bozan. **~ off,** imha etm.; kökünü kazımak. **~er** [ˈkilə(r)]. Kaatil; öldürücü. **humane ~,** hayvanları eziyetsizce öldüren tabanca. **~ing** [ˈkilin(g)] *a.* Öldürücü. **too ~ for words,** (*coll.*) dayanılmaz derecede tuhaf.

**kiln** [kiln]. Kireç yakmak için ocak.

**kilt** [kilt]. İskoçyalılar giydikleri kısa eteklik.

**kin** [kin]. Akraba. **next of ~,** en yakın akraba: **to inform the next of ~,** ailesine haber vermek. **~dred** [ˈkindrid]. Akrabalar; akrabalık. **a ~ spirit,** kafaca yakın bir adam. **~sfolk** [ˈkinzfouk]. Akrabalar. **~-ship,** hısımlık. **~sman,** akraba.

**kind**[1] [kâynd] *n.* Cins, nevi, çeşid; makule; keyfiyet. **payment in ~,** aynen verilen ücret: **to repay s.o. in ~,** aynı ile mukabele etm.: **these ~ of men,** bu gibiler: **he is the ~ of man who always succeeds,** bu her zaman muvaffak olan adamlardandır: **in a ~ of way,** şöyle böyle: **nothing of the ~,** hiç değil!: **something of the ~,** öyle bir şey: **I ~ of expected it,** bunu âdeta bekliyordum. **~**[2] *a.* Müşfik; hayırhah; sevimli; dostane; nazik. **be so ~ as to ...,** lûtfen ...: **to be ~ to s.o.,** birine iyi muamele etm.: **it is very ~ of you,** çok naziksiniz: **give him my ~ regards,** hürmetlerimi söyle!: **~ly ...** [**will you ~ly ...**], lûtfen ...: **he didn't take it very ~ly,** pek hoşuna gitmedi: **to take ~ly to ...,** ...e ısınmak. **~-hearted,** şefkatli. **~ly** [ˈkâyndli] *adv.* *v.* **kind**[1]. *a.* İyi kalbli. **~ness** [ˈkâyndnis]. İyilik, iyi muamele; iltifat.

**kindergarten** [ˈkindəgâtn]. Ana mektebi.

**kindle**[1] [ˈkindl]. Tutuş(tur)mak.

**kindle²**. **to be in ~,** (tavşan vs.) gebe olmak.
**king** [kin(g)]. Kıral; (dama) dama olan taş; (satranc) şah. **the ~ of beasts,** aslan: **the ~ of birds,** kartal: **the ~ of ~s,** Allah. **~dom** [ˈkin(g)dəm]. Kıraliyet; devlet. **the United K~,** Büyük Britanya. **~ly** [ˈkin(g)li]. Kıral gibi; şahane. **~ship** [ˈkin(g)şip]. Kırallık.
**kink** [kin(g)k]. Kıvrım; (düşüncede) gariblik.
**kiosk** [ˈkiosk]. Gazete satılan kulübe.
**kipper** [ˈkipə(r)]. Tütsülenmiş ringa balığı.
**kirk** [kö(r)k]. (İskoçya'da) kilise.
**kiss** [kis]. Buse; öpücük. Öpmek; öpüşmek. **to ~ the book,** yemin ederken İncili öpmek: **to ~ the dust,** mağlûb olmak: **to ~ hands,** (büyük bir memuriyete tayin olunca) kıralın elini öpmek.
**kit** [kit]. Asker eşyası; pılıpırtı; avadanlık. **to pack up one's ~,** pılıpırtıyı toplamak. **~-bag,** asker hurcu.
**kitchen** [ˈkiçin]. Mutfak. **~maid,** mutfak hizmetçisi. **~-dresser,** mutfak tabaklığı. **~-garden,** sebze bahçesi. **~-range,** mutfak sobası.
**kite** [kâyt]. Çaylak; uçurtma.
**kitten** [ˈkitn]. Kedi yavrusu.
**knack** [nak]. Hüner; işin sırrı. **to have the ~ of doing stg.,** işin sırrını bilmek.
**knav·e** [nêyv]. Alçak herif, dolandırıcı; (iskambil) bacak. **~ery,** alçaklık. **~ish,** hilekâr.
**knead** [nîd]. Yuğurmak; oğmak.
**knee** [nî]. Diz. **to bend the ~ to,** ...e boyun eğmek: **to bring s.o. to his ~s,** (*fig.*) diz çöktürmek: **to go down on one's ~s to s.o.,** birinin ayaklarına kapanmak. **~-cap,** diz kapağı. **-~d,** ... dizli. **~-deep,** diz boyu derinliğinde olan. **~-hole,** yazıhanede diz boşluğu. **~-joint,** diz mafsalı. **~l (knelt)** [nîl, nelt]. Diz çökmek; dizüstü oturmak.
**knell** [nel]. Matem çanı çalınma.
**knelt** *v.* **kneel.**
**knew** *v.* **know.**
**knickers** [ˈnikəz]. Kısa pantalon; kadın donu.
**knick-knack** [ˈniknak]. Küçük süslü şey.
**knife,** *pl.* **knives** [nâyf, nâyvz]. Bıçak; çakı. Bıçaklamak. **the ~,** neşter, ameliyat: **to have one's ~ into s.o.,** birine kancayı takmak: **war to the ~,** kıyasıya kavga. **~-edge,** bıçak ağzı; pek dar dağ sırtı; terazi kolunun asılı durduğu ince çelik parçası. **~-grinder,** bıçak bileyici.
**knight** [nâyt]. (*esk.*) Şövalye; silâhşor; (*şim.*) şövalye rütbesini ve isminin önüne 'Sir' lâkabı koyma hakkını haiz olan kimse; (satranc) fers. Şövalye rütbesi vermek. **~ errant,** orta çağda diyar diyar dolaşan şövalye. **~hood,** şövalyelik. **~ly,** şövalyeye yakışır.
**knit** [nit]. Örmek; triko yapmak; birbirine birleştirmek. **to ~ the eyebrows,** kaşlarını çatmak. Örülmüş. **loosely ~ frame,** gevşek yapılı (kimse). **~ted,** örülmüş. **~ eyebrows,** çatık kaşlar. **~ting,** örme işi: **~-needle,** örgü şişi.
**knob** [nob]. Topuz; (*sl.*) baş, kelle. **~b(l)y,** topuzlu.
**knock** [nok]. Vurma, darbe; kapı çalınması; (makine) kliket. Vurmak; çarpmak; kapıyı çalmak; (makine) kliketleşmek. **to ~ s.o. on the head,** öldürmek: **to ~ one's head against stg.,** başını bir şeye çarpmak. **~-about,** gürültülü (soytarı); kaba işe elverişli (elbise). **~-down, a ~ blow,** sersemletici vuruş: **~ price,** en aşağı fiat. **~-kneed,** dizleri bitişik ve baldırları ayrık. **~-out,** nakavt. **~ about,** hırpalamak: **~ about (the world),** feleğin çemberinden geçmek. **~ down,** yere vurmak; yıkmak: **to be ~ed down to s.o.,** (mezadda) birinin üzerinde kalmak. **~ in,** vurup kakmak; kırmak. **~ off,** yerinden fırlatmak; paydos etm.: **to ~ something off the price,** fiatını kırmak. **~ out,** vurup çıkartmak; bir yumrukla sersemletmek. **~ over,** devirmek. **~ up,** vurup yukarıya fırlatmak; (*coll.*) derme çatma yapmak; bitkin bir hale koymak: **he is quite ~ed up,** bitkin bir haldedir: **to ~ up against s.o.,** birine tesadüf etm.: **to ~ up against stg.,** bir şeye çarpmak. **~er** [nokə(r)]. Kapı tokmağı.
**knoll** [noul]. Yuvarlak tepe.
**knot** [not]. Düğüm; bağ; budak; gemi sür'at ölçüsü (1 **knot** = bir saatte bir deniz mili); (ahali) ufak küme. Düğüm haline bağlamak. **~ty,** düğüm düğüm; budaklı: **a ~ point,** çatallı bir mesele.
**know¹** [nou] *n.* **to be in the ~,** işin içyüzünü bilmek. **~-how,** bir şeyi yapma sırrı. **~²** **(knew, known**

[nou, nyû, noun] *v.* Bilmek; tanımak; haberi olm.; ayırd etm. **I don't ~ about that,** Vallahi, orasını bilemem: **he worked all he knew,** alabildiğine çalıştı: **I ~ better than that,** (i) bu kadarcık şeyi bilirim; (ii) ben bundan iyisini bilirim: **he ~s better than to do that,** artık bu kadarını da bilir!: **you ought to ~ better!,** bu kadarcık şeyi bilmeliydiniz!: **a man is ~n by his friends,** insan ahbabından bellidir: **don't I ~ it!,** bilmez miyim?: **to get to ~ s.o.,** birini zamanla tanımak; tesadüfen tanımak: **I would have you ~ that ...,** şunu bilmiş ol ki ...: **to ~ how,** yolunu bilmek: **'wouldn't you just like to ~!',** 'neler neler de maydonozlu köfteler!': **to make oneself ~n to s.o.,** kendini birisine tanıtmak: **it has never been ~n to snow here,** buralara kar yağdığı hiç görülmemiştir: **not if I ~ it!,** dünyada yap(tır)mam!: **to ~ of s.o.,** birini bilmek (fakat şahsan tanımamak): **not that I ~ of,** benim bildiğime göre değil: **to ~ what one is talking about,** bahsettiği şeyi iyi bilmek: **I ~ not what,** bilmem ne: **to ~ what's what,** bir işten anlamak. **~able** [ˈnouəbl]. Tefrik edilebilir. **~ing** [ˈnouin(g)] *a.* Açıkgöz; cin fikirli. **a ~ smile,** çok anlayışlı bir tebessüm. **~ingly,** kasden. **~ledge** [ˈnolic]. Bilgi; ilim; malûmat; haber. **common ~,** herkesce malûm şey: **to (the best of) my ~,** benim bildiğime göre; **to my certain ~,** iyice biliyorum ki: **without my ~,** benim haberim olmadan: **you have grown out of all ~,** tanınmaz şekilde büyümüşsün. **~ledgeable,** malümâtlı. **~n** [noun] *v.* **know²**. *a.* Belli; tanınmış. **~ as ...,** ... olarak tanınmış: **this is what is ~ as ...,** buna ... denilir: **a ~ thief,** herkesce malûm bir hırsız: **well ~,** meşhur; herkesce malûm.

**knuckle** [ˈnʌkl]. Parmak mafsalı. **to ~ under,** teslim olm.: **to rap s.o. over the ~s,** parmaklarının üzerine vurmak; (*fig.*) birini hafifçe haşlamak. **~-bone,** aşık kemiği: **~s,** aşık oyunu. **~-duster,** demir muşta.

**kosher** [ˈkoşə(r)]. Kaşar; turfa olmıyan.

**kowtow** [ˈkauˈtau]. **to ~ to s.o.,** (*fig.*) birinin karşısında elpençe divan durmak.

**Kt.** = **Knight.**

**kudos** [ˈkyûdos]. İtibar, şeref.

# L

**L** [el]. L harfi.

**label** [ˈlêybl]. Yafta (yapıştırmak); marka.

**laboratory** [ləˈborətri]. Laboratuvar.

**laborious** [ləˈbôriəs]. Yorucu; zahmetli; çalışkan. **~ly,** zahmetle.

**labour¹** [ˈlêybə(r)] *n.* Çalışma; zahmet; iş(çilik); İngiliz İşçi Partisi; el emeği; doğurma (sancısı). **a ~ of love,** merak saikasile yapılan iş: **forced ~,** angarya: **hard ~,** ağır hapis cezası: **skilled ~,** usta işi. **~-saving,** işten tasarruf (yolu). **~²** *v.* Çalışmak; çabalamak; uğraşmak. **to ~ a point,** bir meselenin üzerinde lüzumundan fazla durmak: **to ~ along,** güçlükle ilerlemek: **to ~ under a delusion,** bir hayale kapılmak. **~ed,** (üslûb) ağır; (nefes) zahmetle alma. **~er** [ˈlêybrə(r)], rençper, işçi. **the ~ing class,** işçi sınıfı.

**labyrinth** [ˈlabərinθ]. Dolambaçlı yer.

**lace** [lêys]. Dantelâ; (ayakkabının) bağı. Bağlarını bağlamak; dövmek. **milk ~d with rum,** romlu süt.

**lacerate** [ˈlasərêyt]. Yırtmak.

**lack** [lak]. Eksiklik; ihtiyac. Eksik olm.; mahrum olmak. **for ~ of ...,** ...sizlik yüzünden. **~-lustre,** fersiz. **~ing** [ˈlakin(g)]. Eksik; muhtac. **he is ~ in courage,** kâfı derecede cesur değil.

**lackey** [ˈlaki]. Uşak; peyk.

**laconic** [ləˈkonik]. İcazlı; kısa ve kestirme.

**lacquer** [ˈlakə(r)]. Lâke (ile kaplamak).

**lad** [lad]. Genc, delikanlı, erkek çocuk. **a regular ~,** çapkın.

**ladder** [ˈladə(r)]. El merdiveni. **I've a ~ in my stocking,** çorabim kaçtı: **to be at the top of the ~,** yüksek mevkide bulunmak.

**lad·e** [lêyd]. Yükletmek **~en,** yüklü. **~ing** [ˈlêydin(g)]. Bir geminin yükü. **bill of ~,** konişmento.

**ladified** [ˈlêydifâyd]. Hanımefendilik taslıyan.

**ladle** [ˈlêydl]. Kepçe. **~ (out),** kepçe ile dağıtmak.

**lady** [ˈlêydi]. Hanım (efendi); kibar hanım; **lord** ve **knight**'*ların zevcelerine verilen unvan.* **'Ladies',** hanımlara (mahsus helâ): **~ doctor,** kadın doktor: **she looks a ~,** hanımefendiye benziyor: **a ladies' man,** kadınlardan hoşlanan ve onların hoşlandığı erkek: **my ~,** 'Lady' unvanı taşıyana hitab tarzı: **Our L~,** Meryemana. **~like,** bir hanımefendiye yakışır. **L~-chapel,** kilisede Meryemanaya tahsis edilen kısım. **~-killer,** kadınlar nezdinde muvaffakiyet iddiasında olan.

**lag**[1] [lag]. Gecikme; sebeble netice arasındaki zaman farkı. **to ~ behind,** geri kalmak. **~**[2]. Gayrinâkil madde ile sarmak. **~**[3]. **an old ~,** (*sl.*) sabıkalı. **~gard** [ˈlagö(r)d]. Tembel; bati.

**lagoon** [ləˈgûn]. Denize bağlı olan sığ göl.

**laicize** [ˈlêy·isâyz]. Lâikleştirmek.

**laid** *v.* **lay.**

**lain** *v.* **lie.**

**lair** [ˈleə(r)]. Vahşi hayvan ini; haydud yatağı.

**laird** [leəd]. İskoçya'da emlâk sahibi.

**laity** [ˈlêy·iti]. Ruhanî sınıfa mensub olmıyanlar.

**lake** [lêyk]. Göl. **the L~s,** şimalî İngiltere'de göller bölgesi.

**lam** [lam]. (*sl.*) **to ~ into s.o.,** dayak atmak.

**lamb** [lam]. Kuzu(lamak). **~s' tails,** fındık çiçekleri.

**lame** [lêym]. Topal; sakat. Topallatmak. **a ~ excuse,** zayıf bir mazeret.

**lament** [ləˈment]. İnilti; şikâyet; ölüye ağlama(k). İnlemek; şikâyet etmek. **the late ~ed,** merhum. **~able** [ˈlaməntəbl], ağlanacak, acınacak.

**lamp** [lamp]. Lâmba, fener. **safety ~,** madenci feneri: **standard ~,** ayaklı lâmba. **~light,** lâmba ışığı. **~lighter,** sokak lâmbalarını yakan adam. **~-black,** lâmba isi. **~-glass,** lâmba şişesi. **~-post,** fener direği. **~-shade,** abajur.

**lampoon** [lamˈpûn]. Hiciv. Hicvederek tahkir etmek.

**lance** [lâns]. Mızrak. Deşmek. **~r** [ˈlânsə(r)]. Mızraklı süvari. **~t** [ˈlânsit]. Neşter.

**land**[1] [land] *n.* Kara; toprak; memleket; emlâk; arsa; arazi. **L~'s End,** cenubu garbî İngiltere'nin Atlantiğe uzanan burnu: **from L~'s End to John o' Groats,** İngiltere'nin bir başından bir başına: **back to the ~,** toprağa dönüş: **by ~,** karadan: **dry ~,** kara (yani deniz değil): **the Holy L~,** Arzı Mukaddes, Filistin: **the ~ of the living,** bu dünya: **to make [sight] ~,** karayı görmek: **native ~,** anavatan: **to see how the ~ lies,** vaziyeti anlamak. **~-agent,** çiftlik kâhyası; emlâk tellalı. **~-girl,** ziraatte çalışan kadınlar teşkilatına mensub kadın. **~-mine,** kara maynı. **~**[2] *v.* Karaya çıkmak [inmek]; karaya çıkarmak [indirmek]; götürmek; sokmak. **he always ~s on his feet,** daima dört ayak üstüne düşer: **I was ~ed with this big house,** bu berhane başıma kaldı: **to ~ s.o. one,** birine tokat aşketmek: **that will ~ you in prison,** bu yüzden hapse girersen: **to ~ on the sea,** (uçak) denize inmek: **this will ~ us in trouble,** bu bizim başımıza iş açacak. **~ed** [ˈlandid] *a.* **~ property,** arazi, mülk: **~ proprietor,** arazi sahibi. **~fall** [ˈlandfôl]. Karanın görünmesi. **~ing** [ˈlandin(g)]. *n.* Karaya çıkma, inme; sahanlık: **forced ~,** mecburî iniş. **~ing-ground,** iniş sahası. **~ing-stage,** iskele. **~lady** [ˈlandlêydi]. Emlâk, ev sahibesi. **~locked** [ˈlandˌlokt]. Hemen her taraftan kara ile kuşatılmış (liman). **~lord** [ˈlandlôd]. Emlâk, ev sahibi; hancı. **~lubber** [ˈlandlʌbə(r)]. Denizciliğe alışık olmıyan adam. **~mark** [ˈlandmâk]. Hudud işareti; alâmet; mühim hadise. **~owner** [ˈlandounə(r)]. Arazi sahibi. **~scape** [ˈlanskêyp]. Manzara (resmi). **~slide** [ˈlanslâyd]. Heyelân; siyasî hezimet. **~sman** [ˈlansmən]. Denize alışık olmıyan adam. **~ward** [ˈlandwəd]. Kara tarafına bakan: **~wards,** karaya doğru.

**lane** [lêyn]. Dar yol; okyanus gemilerinin seyrettikleri muayyen yol; iki sıra halk arasındaki geçid.

**language** [ˈlan(g)gwic]. Lisan, dil; konuşma tarzı. **bad ~,** küfür.

**langu·id** [ˈlan(g)gwid]. Gevşek; cansız. **~ish** [ˈlan(g)gwiş]. Gevşemek; zayıf düşmek; mecalsiz kalmak. **to ~ after [for] stg.,** bir şeyin arzusile erimek. **~or** [ˈlan(g)gwə(r)]. Gevşeklik; baygınlık; fütur; cansızlık; mahmurluk. **~orous,** gevşek, baygın, cansız.

**lank** [lan(g)k]. Uzun ve zayıf; (saç) uzun, ince ve düz. **~y,** uzun bacaklı ve zayıf.

**lantern** [ˈlantö(r)n]. Fener; projektör. **Chinese ~,** kâğıd feneri: **magic ~,** hayalcı feneri: **~ lecture,** projeksiyonlu konferans: **~ slide,** projeksiyon camı.

**lap**[1] [lap] *n.* Diz üstü; bir yarışta bir devir. **in the ~ of the gods,** Allahın elinde: **the last ~,** bir yarışın son devri; (*fig.*) çoğu gitti azı kaldı. **~ful** [ˈlapful]. Kucak dolusu. **~**[2] *v.* Yalıyarak içmek; (deniz) şıpıldamak. **to ~ the course,** pisti bir defa dönmek: **~ped in luxury,** lükse gark olmuş: **to ~ over,** üst üste bindirmek: **to ~ up,** şapırtı ile içmek; (*fig.*) inanmak.
**lapel** [ˈlapl, ləˈpel]. Yaka devrimi.
**lapse**[1] [laps] *n.* Sehiv; mürurruzaman. **~ of duty,** vazifede kusur: **~ of memory,** birdenbire unutma: **with the ~ of time,** zamanla. **~**[2] *v.* (Zaman) geçmek; düşmek; hata yapmak; battal olm.; (hak) başkasına intikal etmek. **to ~ into silence,** birdenbire susmak. **~d,** battal.
**larceny** [ˈlâsəni]. **petty ~,** küçük hırsızlık.
**larch** [lâç]. Melez çamı.
**lard** [lâd]. Domuz yağı.
**larder** [ˈlâdə(r)]. Kiler.
**large** [lâc] Büyük, iri. **as ~ as life,** tabiî büyüklükte; sapa sağlam: **at ~,** serbest: **taking it by and ~,** heyeti mecmuası itibariyle. **~ly,** ekseriyetle. **~ness,** büyüklük. **~-sized,** büyük boy (şey).
**lark**[1] [lâk]. Tarla kuşu. **to get up with the ~,** şafakla beraber kalkmak. **~**[2]. Şaka. **to ~ about,** çocukça eğlenmek: **for a ~,** sırf şaka için.
**larva** [ˈlâvə]. Sürfe; tırtıl.
**lascar** [ˈlaska(r)]. Avrupa gemilerinde çalışan Hindli gemici.
**lascivious** [ləˈsivyəs]. Şehvete düşkün.
**lash** [laş]. Kamçı ucu, darbesi; (*fig.*) acı hiciv. Kamçılamak; hicivli sözlerle tezyif etmek. **to ~ stg. to ...,** bir şeyi ···e iple bağlamak: **to ~ oneself into a fury,** hiddetten kudurdukça kudurmak: **to ~ its tail,** kuyruğunu hiddetle oynatmak: **to ~ out,** çifte atmak. **~ing,** *n.* kırbaçlama; bağlayan ip: **~s of (beer),** (*sl.*) bolluk.
**lass** [las]. Kız. **~ie,** kızcağız.
**lassitude** [ˈlasityûd]. Yorgunluk, kesiklik.
**last**[1] [lâst] *n.* Kundura kalıbı. **~**[2] *a.* Son(uncu); geçen. **at ~,** nihayet: **at long ~,** en sonunda: **the ~ but one,** sondan ikinci: **the L~ Day,** kıyamet günü: **we shall never hear the ~ of it,** bundan kurtuluş yok: **we haven't heard the ~ of it,** daha başımıza neler gelecek: **in my ~,** son mektubumda: '**~ but not least**', son fakat mühim: **to look one's ~ on,** son defa görmek: **that was the ~ we saw of him,** gidiş o gidiş: **the ~ thing in hats,** en son moda şapka: **that is the ~ thing to frighten me,** hele bundan hiç korkmam: **this day ~ week,** geçen hafta bugün: **to have the ~ word,** (i) münakaşada altta kalmamak; (ii) son söz kendisinde olm.: **the ~ word has been said on that,** ictihad kapısı kapandı. **~**[3] *v.* Dayanmak; kalmak; sürmek; devam etmek. **it's too good to ~,** (hava) böyle güzel devam etmez; (talih) bu böyle sürüp gitmez: **the journey ~ed two months,** seyahat iki ay sürdü: **how long does your leave ~?,** izniniz ne kadardır?: **our food will only ~ a month,** yiyeceğimiz bir ay dayanır: **this overcoat will ~ me the winter,** bu palto beni yaza çıkarır. **~ing** [ˈlâstin(g)] *a.* Dayanıklı; devamlı; sürekli. **~ly** [ˈlâstli]. Sonunda; nihayet; hulâsa.
**latch** [laç]. Mandal(lamak). **~-key,** sokak kapısı için ceb anahtarı.
**late** [lêyt]. Geç; gecikmiş; eski; merhum: **~r,** daha geç, daha sonra: **~st,** en geç, en son. **at the ~st,** en geç: **to be ~ (for stg.),** (bir şey için) gecikmek: **I was ~ in going to bed,** geç yattım: **it's a bit ~ in the day to ...,** ... için çok geç kaldınız: **early and ~,** bütün gün: **it is getting ~,** geç oluyor: **~ in life,** ilerlemiş bir yaşta: **~ at night,** gece geç vakit; **~ into the night,** gece geç vakitlere kadar: **in the ~ nineties,** bin dokuz yüz senesine doğru: **Mr. J. ~ of Bristol,** bundan evvel Bristol'da ikamet eden Mr. J.: **the ~ Mr. J.,** merhum Mr. J.: **in ~ summer,** yazın sonuna doğru: **to arrive too ~,** yetişememek: **before it is too ~,** iş işten geçmeden: **to stay up ~,** geç vakte kadar yatmamak: **of ~ years,** son seneler zarfında. **~ly** [ˈlêytli]. Geçenlerde. **as ~ as yesterday,** daha dün. **~ness** [ˈlêytnis]. Geçlik.
**latent** [ˈlêytnt]. Gizli olarak mevcud.
**lateral** [ˈlatərəl]. Yandaki.
**lathe** [lêyð]. Torna tezgâhı.
**lather** [ˈlaðə(r)]. Sabun köpüğü. Sabunlamak; (*sl.*) dayak atmak.
**Latin** [ˈlatin]. Lâtince. **~ characters,** Lâtin harfleri.
**latitud·e** [ˈlatityûd]. Genişlik; ser-

bestlik. Arz derecesi. **in these ~s,** bu iklimlerde.
**latrine** [lə'trîn]. Abteshane.
**latter** ['latə(r)]. Sonraki; son zikredilen. **~ end,** ölüm. **~ly,** geçenlerde.
**lattice** ['latis]. Kafes (şeklinde).
**Latvia** ['latvya]. Letonya. **~n,** Letonyalı; let dili.
**laud** [lôd]. Övmek. **~able,** medhe lâyik.
**laugh**[1] [lâf] *n.* Gülme; gülüş; alay. **to have the ~ of s.o.,** birini bozmak: **with a ~,** gülerek. **~**[2] *v.* Gülmek. **to ~ at s.o.,** birisi ile alay etm.: **to ~ at stg.,** bir şeye gülmek: **there's nothing to ~ at,** gülecek bir şey yok: **to get (oneself) ~ed at,** kendisine güldürmek: **to ~ stg. off,** şakaya vurmak: **to ~ s.o. out of a thing,** alay ede ede vazgeçirtmek: **to ~ over stg.,** bir şeye gülmek: **to ~ s.o. to scorn,** birisini alay ederek küçük düşürmek: **I'll make him ~ on the other side of his face,** ben ona gülmeyi gösteririm. **~able** ['lâfəbl]. Gülünç; gülecek. **~ing** ['lâfin(g)]. Gülen: **~ly,** gülerek. **to burst out ~,** kahkaha atmak: **it's no ~ matter,** mesele ciddidir. **to make a ~-stock of oneself,** âleme maskara olmak **~ter** ['lâftə(r)]. Gülme, kahkaha. **to split with ~,** katılırcasına gülmek: **to roar with ~,** kahkaha koparmak.
**launch**[1] [lônç] *n.* **motor ~,** motör. **~**[2] *v.* Suya indirmek; atmak. **to ~ an offensive,** taarruza başlamak: **to ~ forth on an enterprise,** bir teşebbüse girişmek: **once he is ~ed on this subject ...,** bir kere bu mevzua girişti mi ....
**laund·er** ['lôndə(r)]. Çamaşır yıkamak. **~ress** [-dris], çamaşırcı kadın. **~ry,** çamaşır(hane).
**laureate** ['lôryêyt]. Başı defneli. **the Poet ~,** İngiltere'nin resmî baş şairi.
**laurel** ['lorəl]. Taflan, defne ağacı. **to look to one's ~s,** şöhretini başkasına kaptırmamak: **to rest on one's ~s,** bir muvaffakiyetle iktifa etmek.
**lava** ['lâva]. Lav.
**lavatory** ['lavətri]. Yıkanma yeri; tuvalet.
**lavender** ['lavində(r)]. Lâvanta otu.
**lavish** ['laviş]. Bol bol sarf; müsrif. Sahavetle sarfetmek. **~ness,** sahavet; bolluk.
**law** [lô]. Kanun; hukuk; adalet; kaide. **to be a ~ unto oneself,** bildiğini okumak: **to be at ~,** dâvalı olm.: **court of ~,** mahkeme: **to go to ~ with s.o.,** birisinin aleyhine dâva açmak: **to keep the ~,** kanuna riayet etm.: **to lay down the ~,** ahkâm kurmak: **without wishing to lay down the ~,** haddim olmıyarak: **to practise ~,** hukukçuluk etm.: **to take the ~ into one's own hands,** bizzat ihkakı hak etmek. **~-abiding,** kanuna riayetkâr. **~-breaker,** kanun tanımıyan. **~-lord,** Lordlar kamarasının hukukçu azası. **~-maker,** kanun vâzıı. **~ful** ['lôfl]. Meşru; kanuna uygun; mubah. **~giver** ['lôgivə(r)]. Kanun vâzıı. **~less** ['lôlis]. Kanun tanımaz: **~lessness,** nizamsızlık. **~suit** ['lôsyût]. Dâva. **~yer** ['lôyə(r)]. Dâva vekili; avukat.
**lawn** [lôn]. Çimenlik. **~-mower,** çimen kırpma makinesi. **~-tennis,** tenis oyunu.
**lax** [laks]. Gevşek. **~ in morals,** hafifmeşreb: **~ attendance,** devamsızlık.
**laxative** ['laksətiv]. Liynet verici (ilâc).
**lay**[1] [lêy] *n.* Şarkı. **~**[2] *a.* Lâik; ruhanî sınıftan olmıyan. **~**[3] *v.* **lie.** **~**[4] **(laid)** [lêy, lêyd] *v.* Koymak; bırakmak; yatırmak; sermek; kurmak. **to ~ about one,** sağa sola vurmak: **to ~ a bet,** bahse girmek: **to ~ a complaint,** şikayette bulunmak: **to ~ (dinner) for three,** üç kişilik sofra kurmak: **to ~ eggs,** yumurtlamak: **to ~ the fire,** ocağı hazırlamak: **to ~ s.o. to rest,** birini gömmek: **to ~ the table,** sofrayı kurmak: **to ~ that ...,** bahse girmek. **~ aside,** bir tarafa koymak. **~ by,** (para) bir kenara koymak. **~ down,** bir yere bırakmak; (halı) döşemek: **to ~ down one's arms,** teslim olm.: **to ~ down conditions,** şartları tayin etm.: **to ~ down one's hand,** elini göstermek: **to ~ down one's life,** hayatını feda etm.: **to ~ down one's office,** memuriyetini terketmek: **to ~ oneself down,** yatmak: **to ~ down that ...,** ... şart koşmak. **~ in,** tedarik etm.; dayak atmak. **~ off, ~ off workmen,** işçiler muvakkaten yol vermek. **~ on,** (boya) sürmek; (vergi) kesmek: **to ~ on water,** su tesisatını yapmak: **to ~ it on thick,** ballandırdıkça ballandırmak: **to ~ on with a will,** kamçı vs. yapıştırmak. **~ out,** dizmek ve yaymak: **to ~ s.o. out,** birisini yere

sermek: **to ~ out a camp,** ordugâh plânını çizmek: **to ~ out a corpse,** ölüyü techiz ve tekfin etm.: **to ~ out money,** para harcamak: **to ~ oneself out to please,** göze girmeğe çalışmak. **~ up,** toplamak: **to ~ up a ship,** *etc.*, gemi vs. işletmemek: **to be laid up (by illness),** yatak hastası olmak. **~man** [ˈlêymən]. Ruhanî sınıftan olmıyan kimse; bir meslek, ilmin yabancısı. **~-out** [lêyˈaut]. Plân; düzen; tertib.

**layer**[1] [ˈlêyə(r)]. Kat, tabaka. Kat kat koymak. **~**[2]. **a good [bad] ~,** çok [az] yumurtlayan.

**laz·e** [lêyz]. Tembelleşmek. **to ~ about,** tembelce vakit geçirmek. **~iness** [ˈlêyzinis]. Tembellik, haylazlık. **~y** [ˈlêyzi]. Tembel, haylaz. **~y-bones,** tembel çocuk.

**lead**[1] [led]. Kurşun; iskandil kurşunu. Kurşun ile kaplamak. **to heave the ~,** iskandil atmak: **to swing the ~,** temaruz etm. **~**[2] [ˈlîd] *n.* Önde bulunma; rehberlik; delâlet; köpek kayışı; (*elect.*) tel. **to follow s.o.'s ~,** birini takib etm.: **to give the ~,** örnek olm.: **to give s.o. a ~,** yol göstermek: **to take the ~,** önayak olm. **~**[3] **(led)** [lîd, led]. *v.* (Elinden tutup) alıp getirmek; yedmek; önlerine geçmek; müncer olm.; idare etm.; emrinde olm.; götürmek, çıkmak; çıkarmak. **to ~ (at cards),** açmak: **to ~ a happy life,** bahtiyar bir hayat sürmek: **to ~ s.o. a miserable life,** birine sefil bir hayat yaşatmak: **to ~ a movement,** bir harekete önayak olm.: **to ~ to nothing,** hiç bir neticeye çıkmamak: **to ~ the way,** yol göstermek, öne geçmek: **easily led,** kolayca başkasına tabi olan: **I am led to the conclusion that ...,** şu neticeye vardım ki .... **~ away,** elinden tutup alıp götürmek; ayartmak: **to be led away,** başkasının tesirine kapılmak. **~ off,** elinden tutup alıp götürmek; başlamak. **~ on,** önüne geçmek; yol göstermek; teşvik etm.: **to ~ s.o. on to say stg.,** birini bir şeyi söylemeğe sevketmek: **that ~s on to what I was going to say,** bu asıl söyliyeceğim şeye götürür. **~ up,** (elinden tutup) yukarı çıkarmak: **to ~ up to a subject,** muhavereyi bir bahse götürmek. **~en** [ˈledn]. Kurşundan yapılmış: **~-footed,** ağır yürüyen. **~er** [ˈlîdə(r)]. Reis; kumandan; önayak; baş makale: **~ership,** sevkü idare kudreti; reislik: **under the ~ of,** ···in idaresi altında. **~ing** [ˈlîdin(g)] *a.* En mühim, başta gelen: **~ article,** baş makale: **~ cases,** emsal teşkil eden dâvalar: **~ part,** baş rol: **~ question,** istenilen cevaba götüren sual: **the ~ question of the day,** günün en mühim meselesi.

**leaf,** *pl.* **leaves** [lîf, lîvz]. Yaprak; varak; lâma. **~ of a table,** masanın uzatma eki: **in ~,** yeşermiş: **the fall of the ~,** yaprakdökümü: **to take a ~ out of s.o.'s book,** örnek almak: **to turn a ~ down,** kitabın sahifesini kıvırmak: **to turn over a new ~,** yaşayışını ıslah etmek. **~let,** risale. **~y,** yapraklı. **~-mould,** yaprak gübresi.

**league**[1] [lîg]. Fersah. **~**[2]. Cemiyet; birlik. **to ~ together,** birleşmek: **L~ of Nations,** Milletler Cemiyeti: **to be in ~ with,** ile birlik etm.: **to form a ~ against s.o.,** birinin aleyhine birleşmek.

**leak** [lîk]. Sızıntı; delik. Sızmak, akmak. **to ~ out,** dışarı sızmak: **to spring a ~,** (gemi) su edecek kadar yarılmak: **to stop a ~,** deliği tıkamak. **~age** [–ic], sızıntı; ifşa. **~y,** sızar; delik.

**lean**[1] [lîn] *a.* & *n.* Zayıf; lağar; kıt. Yağsız et. **a ~ diet,** perhiz, bol olmıyan yemek: **~ years,** kıtlık yılları. **~**[2] *v.* **(leant)** [lîn, lent]. Meyletmek; temayül etm.; eğilmek. **~ against,** ···e dayanmak: **~ back,** arkasına dayanmak: **~ forward,** ileriye eğilmek: **~ out,** sarkmak: **~ over,** abanmak: **~ towards,** temayül etm., meyletmek: **~ upon,** ···e dayanmak. **~-to shed,** sundurma.

**leap** (-ed *veya* leapt) [lîp, -t, lept]. Atlamak; sıçramak. **to ~ at an offer,** bir teklifi tehalükle kabul etm.: **to ~ for joy,** sevincinden sıçramak. **~-frog,** birdirbir. **~-year,** kebise sene.

**learn** (-ed *veya* learnt) [lö(r)n, –d, –t]. Öğrenmek; haber almak. **to ~ from s.o.,** (i) birisinden haber almak; (ii) birisine imtisal etm.: **to ~ a lesson,** iyi bir ders almak: ⌜**it's never too late to ~**⌝, *veya* ⌜**live and ~**⌝, insan her yaşta öğrenebilir. **~ed** [ˈlö(r)nid] *a.* âlim. **~er,** talebe, acemi: **a quick ~,** çabuk öğrenir. **~ing** *n.* öğrenme; bilgi.

**lease** [lîs]. Uzun müddetle kiralama(k). **to take on a new ~ of life,** yeniden hayata doğmak. **~hold,** kiralanmış.

**leash** [lîş]. Köpeklere takılan sırım.

**least** [lîst]. En az; en küçük; asgarî. **at ~,** hiç olmazsa: **ten days at the ~,** en aşağı on gün: **he deserves it ~ of all,** o buna herkesten daha az layıktır, müstahaktır: **not in the ~,** estağfurullah!: **not the ~ bit,** hiç bir şekilde: **it doesn't matter in the ~,** hiç ehemmiyeti yok: **to say the ~ of it,** en hafif tabirile. **~ways,** (*pop.*) hiç olmazsa.

**leather** [ˈleðə(r)]. Kösele; meşin; deri. Kösele ile **kaplamak**; (*sl.*) dayak atmak. **morocco ~,** sahtiyan: **patent ~,** rugan. **~y** [ˈleðəri]. Kösele gibi; (et) sahtiyan gibi.

**leave**[1] [lîv] *n.* Müsaade; izin; mezuniyet. **by your ~!,** müsaadenizle!: **on ~,** izinli: **to take one's ~,** veda etm.: **to take French ~,** sıvışmak; izinsiz gitmek. **~-taking,** veda. **~**[2] *v.* (left) [lîv, left]. Bırakmak, terketmek; bir yerden çıkmak; tevdi etm.; hareket etmek. **let us ~ it at that,** (bu bahsi) burada keselim: **he is still too ill to ~ his bed,** henüz yataktan çıkamıyacak derecede hastadır: **~ him to himself!,** onu kendi haline bırak!: **to ~ hold of stg.,** salıvermek: **four from six ~s two,** altıdan dört çıktı iki kaldı: **he ~s at 5 o'clock,** saat beşte çıkar: **he left school at eighteen,** mektebi on sekiz yaşında bitirdi: **to ~ the table,** sofradan kalkmak: ⌜**take it or ~ it!**⌝, ⌜ister beğen, ister beğenme!⌝: **left to oneself,** yalnız kalınca: **he was left £1,000,** kendisine bin lira miras kaldı: **when the father died they were left very badly off,** babaları ölünce darda kaldılar: **there is only one bottle left,** yalnız bir şişe kaldı: **there was nothing left but to ~ the country,** memleketi terketmekten başka çare yoktu: **there was nothing left to me but to sign it,** bana yalnız imza etmek kaldı. **~ about,** ortada bırakmak. **~ alone,** kendi haline bırakmak; rahat bırakmak. **~ behind,** unutmak; geride bırakmak. **~ off,** vazgeçmek; bırakmak. **~ out,** dışarıda bırakmak; etmemek; atlamak. **~ over,** tehir etm.: **to be left over,** kalmak.

**leaven** [ˈlevn], Maya(landırmak); tesir etm.

**leaves** [lîvz], *pl. of* **leaf.**

**leaving** [ˈlîvin(g)] *n.* Bırakma; ayrılma; hareket etme. **~ certificate,** orta tahsil diploması. **~s,** artık.

**Lebanon** [ˈlebənən]. Lübnan.

**lecherous** [ˈleçərəs]. Şehvete düşkün.

**lecture** [ˈlekçə(r)]. Umumi ders, konferans, (vermek). **to give s.o. a ~,** birine va'zetmek; tekdir etmek. **~r,** konferans veren adam; doçent: **~-ship,** doçentlik.

**led** [led] *v.* **lead.** *a.* **a ~ horse,** yedek at.

**ledge** [lec]. Düz çıkıntı; kaya tabakası.

**ledger** [ˈlecə(r)]. Ana defter.

**lee** [lî]. Rüzgâraltı. **under the ~ of ...,** ... muhafazalı tarafından. **~-ward** [ˈlîwəd, ˈluəd]. Rüzgâraltı tarafına aid. **~way** [ˈlîwêy]. **to make up ~,** kaybedilen vakti telâfi etmek.

**leech** [lîç]. Sülük; (*esk.*) hekim.

**leek** [lîk]. Pırasa.

**leer** [liə(r)]. Şehvetle yan bakmak.

**lees** [lîz]. Tortu.

**left**[1] *v.* **leave. ~**[2]. Sol; soldan; sola doğru. Sol taraf. **the ~,** sol parti. **~-hand,** sol tarafdaki. **~-handed,** solak: **a ~ compliment,** kompliman yaparken çam devirme: **~ marriage,** küfüv olmıyan izdivac.

**leg** [leg]. Bacak; ayak; but; konc. **I am on my ~s all day,** bütün gün ayaktayım: **to get on one's ~s again,** iyileşmek: **to give s.o. a ~ up,** birisine yardım etm.: **to be on his [its] last ~s,** ölmek [sönmek] üzere olm.: **not to leave s.o. a ~ to stand on,** iddialarını birer birer çürüterek diyecek bir şey bırakmamak: **to pull s.o.'s ~,** birile alay etm.: **to run as fast as one's ~s will carry one,** var kuvvetile koşmak: **to get one's sea ~s,** geminin hareketine alışıp ayakta durabilmek: **to set s.o. [stg.] on his [its] ~s again,** tutup kaldırmak; diriltmek: **to show a ~,** yataktan kalkmak: **to stand on one's own ~s,** kendi yağile kavrulmak. **~gings** [ˈlegin(g)gz]. Tozluk, getr. **~gy** [ˈlegi]. Uzun bacaklı.

**legacy** [ˈlegəsi]. Vasiyet ile birine bırakılan şey. **a ~ of the past,** mazinin mirası.

**legal** [ˈlîgl]. Kanunî, meşru, hukukî. **to take ~ advice,** adlî istişarede bulunmak. **~ity** [-ˈgaliti], kanuniyet. **~ize** [ˈlîgəlâyz], tecviz etm.

**legat·e** [ˈleget]. Elçi; Papa elçisi. **~ion** [liˈgêyşn]. Orta elçilik; sefarethane.

**legend** [ˈlecənd]. Menkibe; hikâye; kitabe. **~ary** [-əri], efsanevî.

**legible** [ˈlecibl]. Okunaklı.
**legion** [ˈlîcən]. Alay; bir çok. **their name is ~,** sayısızdırlar. **~ary,** lejiyoner.
**legislat·e** [ˈlecislêyt]. Kanun yapmak. **~ion** [–ˈlêyşn], kanun yapma; mevzu kanunlar. **~ive** [–iv], kanun yapan, kanun vazıına aid. **~or** [–ə(r)], kanun vazıi. **~ure** [–yə(r)], kanun yapan heyetlerin mecmuu.
**legitima·cy** [liˈcitiməsi]. Meşru olma. **~te** *a.* [liˈcitimət]. Meşru, helâl. **~ pride,** haklı gurur: **~ child,** meşru çocuk.
**leisure** [ˈlejə(r)]. Boş vakit; rahat. **to be at ~,** serbest olm.: **to do stg. at one's ~,** bir şeyi acelesiz ve müsaid bir zamanda yapmak: **people of ~,** meşguliyetsiz kimseler. **~d** [ˈlejəd], vakti bol. **~ly,** acelesiz: **a ~ journey,** geze geze seyahat.
**lemon** [ˈlemən]. Limon(lu). **~ade** [ˌleməˈnêyd], limonata; gazoz.
**lend** [lend]. İare etm.; borc vermek. **to ~ an ear,** kulak asmak: **to ~ itself to ~,** ···e müsaid olm.: **to ~ help [a hand] to ...,** yardım etm.: **to ~ out books,** kira ile kitab vermek: **~ing library,** kira ile kitab veren kütübhane.
**length** [len(g)kθ]. Uzunluk: **~ of time,** müddet: **~ of service,** kıdem. **at ~,** (i) nihayet; (ii) uzun uzadıya; (iii) baştan sona kadar: **a dress ~,** bir elbiselik kumaş: **to fall all one's ~ [full ~] on the ground,** yere boylu boyuna düşmek: **to go to the ~ of, ...** dereceye kadar varmak: **to go to any ~(s),** her çareye başvurmak: **overall ~,** tam uzunluk: **over the ~ and breadth of the country,** memleketin dört yanında: **of some ~,** oldukça uzun: **to turn in its own ~,** olduğu yerde dönmek: **to win by a ~,** (yarışı) bir kayık, at boyu farkla kazanmak. **~en** [ˈlen(g)kθn]. Uza(t)mak; temdid etm. **his face ~ed,** surat astı. **~ways, -wise** [ˈlen(g)kθwêyz, -wâyz]. Uzunluğuna. **~y** [ˈlen(g)kθi]. Uzun uzadıya.
**lenien·ce, -cy** [ˈlîniəns(i)]. Müsamaha. **~t,** müsamahakâr.
**lens** [lenz]. Adese; objektif; pertavsız.
**lent**[1] [lent] *v.* **lend.** **~**[2]. Büyük perhiz. **~en,** ona aid.
**leonine** [ˈlîyənâyn]. Aslan gibi.
**leopard** [ˈlepəd]. Pars. ⌜**can the ~ change his spots?**⌝, ⌜kırk yıllık Yani olur mu Kâni?⌝
**lep·er** [ˈlepə(r)]. Miskin adam; cüzamlı adam. **~rosy** [ˈleprəsi]. Cüzam; miskinlik.
**less** [les]. Daha az; daha küçük; eksik. **to grow ~,** azalmak: **~ and ~,** gittikçe azalıyor: **in ~ than no time,** bir anda: **London is no ~ expensive than Paris,** Londra pahalılıkta Paris'ten aşağı değildir: **he writes with no ~ knowledge than clarity,** malûmatlı olduğu kadar vuzuh ile de yazıyor: **he has no ~ than a thousand a year,** senelik geliri tam bin liradir: **he has not ~ than a thousand a year,** senede en aşağı bin lira geliri var: **no ~ a person than the King himself,** bizzat kıral: **none the ~,** buna rağmen: **he can't speak his own language, still ~ Turkish,** türkçe şöyle dursun kendi lisanını konuşamaz: ⌜**the ~ said the better**⌝, nekadar az söylenirse o kadar iyidir. **-~** *suff.* ···siz, *mes.* **harm~,** zararsız. **~en** [ˈlesn]. Küçül(t)mek; azal(t)mak. **~er** [ˈlesə(r)]. Daha küçük; daha az. **the ~ of two evils,** ehveni şer.
**lesson** [ˈlesn]. Ders; ibret. **to draw a ~ from stg.,** bir şeyden ibret almak: **let that be a ~ to you,** bundan ibret al: **to learn one's ~ (by bitter experience),** boyunun ölçüsünü almak.
**lest** [lest]. Olmasın diye; belki. **~ we forget,** unutmıyalım diye.
**let**[1] [let]. Kiraya vermek. Kiralama. **house to ~,** kiralık ev. **~**[2]. Bırakmak; müsaade vermek. let *fiili başka bir fiilin başına geldiği zaman onu müteaddi yapar, mes.* **to *let* fall,** duşürmek: **to *let* know,** bil*dir*mek. **to ~ alone, ~ be,** kendi haline bırakmak: **he can't even walk, ~ alone run,** koşmak şöyle dursun, yürüyemez bile: **~ the cost be ever so high,** ne kadar pahalı olursa olsun: **to ~ go,** salıvermek: **to ~ things go,** işin peşini bırakmak: **he ~ himself go,** veryansın etti: **~ me hear what happened,** olanı biteni anlat. **~ down,** indirmek; düşürmek; yarı yolda bırakmak: **he ~ me down badly,** beni çok fena sukutu hayale uğrattı: **to ~ s.o. down gently,** birini hafifçe cezalandırmak: **to ~ down a coat,** *etc.*, bir elbiseyi uzatmak: **to ~ the fire down,** ateşi kendi kendine sönmeğe bırakmak: **to ~ one's hair down,** saçını çözmek: **to ~ a tyre down,** lâstiği söndürmek: **the chair ~ him down,** iskemle çökerek onu

düşürdü. ~-**down,** sukutu hayal. ~ **in,** içeri almak; eklemek: **this will ~ me in for a lot of work,** bu bir sürü iş demektir: **I've been ~ in for a thousand pounds,** bu bana bin liraya maloldu: **I didn't know what I was ~ting myself in for,** başıma gelecekleri bilmedim: **to ~ s.o. in on a secret,** bir sırrı birine açmak. ~ **off,** (silâh) atmak; salıvermek; affetmek: **to ~ s.o. off lightly,** hafif bir ceza ile salıvermek: **to be ~ off with a fine,** bir para cezasıyle yakayı kurtarmak: **to ~ off steam,** içini boşaltmak. ~-**off,** ucuz kurtulma. ~ **on, to ~ on about stg.,** bir şeyi ifşa etmek. ~ **out,** çıkarmak; salıvermek; boşaltmak; uzatmak: **to ~ out at s.o.,** çifte atmak: **to ~ boats out (on hire),** kayık kiraya vermek: **to ~ out (a secret),** (bir sırrı) ifşa etm.: **to ~ a strap out one hole,** kayışı bir delik açmak. -~ *suff. Küçültme eki. mes.* **stream~,** küçük çay.

**lethal** [ˈlîθl]. Öldürücü.

**letharg·ic** [liˈθâcik]. Rehavetli. ~**y** [ˈleθəci], uyuşukluk.

**letter, lettuce** See p. 807.

**Levant** [liˈvant]. **the ~,** Akdenizin şark sahilindeki memleketler. ~**ine** [levnˈtîn, ˈlevntâyn], tatlısu frengi.

**level**[1] [ˈlevl] *n.* Seviye; hiza; tesviye aleti. **to find one's ~,** kendi muhit ve seviyesini bulmak: **on a ~ with ...,** ile bir hizada: **on the ~,** düzlükte; dürüst. ~[2] *a.* Düz; ufkî. ~ **with,** bir hizada: ~ **crossing,** demiryolu geçid: **to do one's ~ best,** elinden geleni yapmak: **to keep a ~ head,** soğukkanlılığını muhafaza etm. ~-**headed,** muvazeneli; soğukkanlı. ~[3] *v.* Tesviye etmek. **to ~ a blow at s.o.,** birisine bir darbe indirmek: **to ~ a gun at s.o.,** tüfeği bir kimseye çevirmek: **to ~ up,** bir hizaya getirmek.

**lever** [ˈlîvə(r)]. Manivelâ; (*fig.*) vasıta. Manivelâ ile kaldırmak. ~**age,** manivelâ kudreti.

**leviathan** [leˈvâyəθən]. Ejderha.

**levity** [ˈleviti]. Hiffet.

**levy** [ˈlevi]. Toplama; vergi; asker toplama. Para, asker toplamak. **to ~ blackmail,** birine şantaj yapmak: **to ~ a fine on s.o.,** birisinden para cezası almak: **to ~ a tax,** vergi tarhetmek: **to ~ a tribute on ...,** ...den haraç almak: **to ~ war on s.o.,** birisile harb etmek.

**lewd** [lyûd]. İffetsiz; fuhşa aid.

**lexicon** [ˈleksikən]. Lûgat.

**liab·ility** [ˌlâyəˈbiliti]. Zimmet; mesuliyet; temayül; (hastalığa) istidad. **assets and liabilities,** aktif ve pasif. ~**le** [ˈlayəbl]. Mesul; tâbi; maruz; muhtemel.

**liaison** [liˈêyzon]. İrtibat, münasebet.

**liar** [ˈlâyə(r)]. Yalancı.

**libel** [ˈlâybl]. İftira. Aleyhine iftirada bulunmak. **to bring an action for ~ against s.o.,** birisi aleyhine zem ve kadih dâvasi açmak. ~**lous,** iftiralı.

**liberal** [ˈlibərəl]. Cömerd; bol; hürriyet tarafdarı, serbest fikirli: ~ **education,** ictimaî ilimler tahsili. ~**ism,** ahrarlık. ~**ity** [ˌlibəˈraliti], cömerdlik; fikir genişliği.

**liberat·e** [ˈlibərêyt]. Serbest bırakmak; kurtarmak. ~**or,** halâskâr, kurtarıcı.

**libertin·e** [ˈlibətîn]. Çapkın; sefih.

**liberty** [ˈlibəti]. Hürriyet; serbestlik. **to be at ~ to do stg.,** bir şeyi yapmakta serbest olm.: ~ **man,** izinli bahriyeli: **to set at ~,** serbest bırakmak: **to take the ~ of (doing stg.),** ictisar etm.: **to take liberties with a person,** başına çıkmak.

**librar·y** [ˈlâybrəri]. Kütübhane. ~-**ian** [lâyˈbreəriən], kütübhaneci.

**lice,** *pl. of* louse.

**licen·ce** [ˈlâysns]. Ruhsat; izin tezkeresi; ehliyetname; hadden aşırı serbestlik. ~**se** [ˈlâysns]. *v.* Ruhsat, izin tezkeresi, salâhiyet vermek. ~**see** [lâysənˈsî]. Ruhsatlı kimse. ~**tious** [lâyˈsenşəs]. Çapkın.

**licit** [ˈlisit]. Meşru.

**lick**[1] [lik] *n.* Yalama. **at full ~,** (*coll.*) alabildiğine bir süratle. ~[2] *v.* Yalamak; (*sl.*) üstün gelmek. **to ~ s.o.'s boots,** yaltaklanmak: **he is not fit to ~ that man's boots,** o adamın kestiği tırnak olamaz: **to ~ the dust,** kahrolunmak: **to ~ one's lips,** yalanmak: **this ~s me,** buna aklım ermez: **to ~ into shape,** adam etmek. ~**ing,** yalama; (*sl.*) yenilme.

**lid** [lid]. Kapak; gözkapağı. **that puts the ~ on it !,** (*sl.*) bir bu eksikti.

**lie**[1] **(lied, lying)** [lây, lâyd, lây·in(g)]. Yalan söylemek. Yalan. **to give s.o. the ~ (direct),** tekzib etmek: **a pack of ~s,** yalan dolan: **a white ~,** iş bitiren yalan. ~[2] *n.* Durum. **the ~ of the land,** arazi vaziyeti; (*fig.*) vaziyet hal, şerait. ~[3] *v.* **(lay, lain)** [lây, lêy, lêyn]. Yatmak; uzanmak; durmak; bulunmak; vâki olm.; düşmek. **no action would ~ in this case,**

bu halde dâva mesmu olamaz: **to ~ at the point of death,** ölmek üzere olm.: **as far as in me ~,** elimden geldiği kadar: **here ~s ..., ...** burada medfundur: **the snow never ~s there,** orasını kar tutmaz: **to ~ under suspicion,** şübhe altında kalmak: **time ~s heavy on my hands,** işsizlikten sıkılıyorum. **~ about,** meydanda kalmak. **~ down,** yatmak: **to take stg. lying down,** ses çıkarmadan kabul etm.: **he won't take it lying down,** kolay kolay kabul etmiyecek. **~ in,** loğusa olmak. **~ off,** (gemi) açıkta yatmak. **~ over,** tehir edilmek. **~ to,** geminin başını rüzgâra çevirip durmak. **~ up,** yatakta kalmak.

**liege** [lîc]. Eski zamanda bir derebeyine tâbi kimse.

**lien** [ˈliən]. İhtiyatî haciz.

**lieu** [lyû]. **in ~ of,** yerine, bedel olarak.

**lieutenant** [lefˈtenənt]. Vekil; (deniz) yüzbaşı; (kara) teğmen: **~-colonel,** (asker) yarbay: **~-commander,** (deniz) kıdemli yüzbaşı: **~-general,** korgeneral: **~-governor,** bir umumî valiye tâbi vali.

**life,** *pl.* **lives** [lâyf, lâyvz]. Hayat; ömür. Hayat boyunca. **to bring to ~,** diriltmek: **to carry one's ~ in one's hands,** kelleyi koltuğa almak: **the ~ to come,** ahret: **to come to ~,** canlanmak: **to do stg. for dear ~,** bir şeyi var kuvvetile yapmak: **to draw from ~,** bir modele bakarak resim yapmak: **to escape with one's ~,** postu kurtarmak: **to fly for one's ~,** can korkusile kaçmak: **for ~,** kaydi hayat şartile: **I can't for the ~ of me understand,** buna hiç aklım ermiyor: **to be a good ~,** sigortaya pek elverişli olm.: **not on your ~!,** aslâ!: **run for your lives!,** kaçabilen kaçsın!: **he has seen ~,** görmüş geçirmiş: **a ~ sentence,** müebbed hapis: **one's position in ~,** insanın cemiyetteki mevkii: **still ~,** natür mort: **to take s.o.'s ~,** birini öldürmek: **to take one's own ~,** kendini öldürmek: **at my time of ~,** benim yaşımda: **to the ~,** tıpkı: **true to ~,** yaşanmış: **without accident to ~ or limb,** kılına halel gelmeden: **upon my ~,** başım hakkı için. **~-blood,** kan, can. **L~-guards,** İngiliz kıraliçenin süvari muhafızları. **~-insurance,** hayat sigortası. **~-interest,** kaydi hayat şartile intifa hakkı. **~-line,** cankurtaran halatı. **~-saving,** can kurtaran. **~belt** [ˈlâyfbelt]. Cankurtaran kemeri. **~boat** [ˈlâyfbout]. Cankurtaran sandalı. **~buoy** [ˈlâyfbôy]. Cankurtaran simidi. **~less** [ˈlâyflis]. Cansız; ruhsuz. **~like** [ˈlâyflâyk]. Canlı mahluk gibi. **~long** [ˈlâyflon(g)]. Hayat müddetince. **a ~ friend,** çok eski dost. **~size** [ˈlâyfsâyz]. Tabiî büyüklükte. **~time** [ˈlâyftâym]. Hayat müddeti. **in his ~,** hayatta iken: **we shan't see it in our ~,** bizim ömrümüzde görmiyeceğiz.

**lift¹** [lift] *n.* Asansör; yükseltme. **to give s.o. a ~,** birini arabasına almak: **to give s.o. a ~ up,** birine yardım etm.: **to get a ~ up in the world,** mevkiini yükseltmek. **~²** *v.* Kaldırmak; yükseltmek; (*sl.*) çalmak. Kalkmak. **the clouds are ~ing** bulutlar yükseliyorlar: **to ~ something down,** indirmek: **to ~ up one's head,** başını kaldırmak; (*fig.*) herkesin yüzüne bakabilmek: **to ~ potatoes,** patates toplamak: **to ~ s.o. up,** birini tutup kaldırmak: **~ing power,** kaldırma kudreti.

**light¹** [lâyt]. *n.* Işık, aydınlık, nur; gündüz; far, lâmba; bakım. **to act according to one's ~s,** kendi ölçülerine göre hareket etm.: **to bring to ~,** meydana çıkarmak: **to come to ~,** meydana çıkmak: **the ~ of day,** gündüz;: **to get [stand] in s.o.'s ~,** birisine karanlık etm.; mâni olm.: **to give s.o. a ~,** birine ateş (kibrit) vermek: **to place stg. in a good ~,** bir şeyi müsaid bir zaviyeden göstermek: **a leading ~,** mümtaz ve mutena kimse: **to be in one's own ~,** kendi kendisine karanlık etm.: **~s out,** ışıkları söndürme: **to begin to see ~,** anlamak: **to set ~ to stg.,** tutuşturmak: **~s and shades of expression,** ifade incelikleri. **~²** *v.* (**-ed** veya **lit**) [ˈlâytid, lit]. Yakmak; tutuşturmak; aydınlatmak; ışık vermek. Tutuşmak; parıldamak. **to ~ a fire,** ateş yakmak: **the match will not ~,** kibrit yanmıyor: **to ~ up,** ışık vermek; aydın olm.: **lit up,** aydınlanmış; (*sl.*) çakır keyf: **his face lit up,** yüzü güldü: **to ~ the way for s.o.,** birine ışık tutmak. **~³** (**-ed** veya **lit**). Konmak. **to ~ on one's feet,** ayakları üstüne düşmek: **to ~ upon,** raslamak. **~⁴** *a.* Hafif; açık (renk); **~ hair,** sarı saç. ⌜**~ come, ~ go**⌝, ⌜haydan gelen huya gider⌉: **with a ~ heart,** neş'e ile: **to make ~ of,** yabana atmak: **to be a ~ sleeper,** uykusu hafif olm.: **to travel ~,** az eşya ile seyahat etmek. **~-fingered,** yankesici gibi. **~-handed,** eli hafif; becerikli. **~-headed,** sayıklıyan; hoppa. **~-hearted,** gamsız, şen; **in a ~ manner [~ly],** düşüncesiz. **~-minded,** hoppa. **~-weight,** (boks) hafifsıklet; (kumaş) hafif. **~en**[1] [ˈlâytn]. Hafifletmek. **~en**[2]. Aydınlatmak; (renk) daha açık yapmak. Şimşek çakmak; aydınlanmak, açılmak. **~er**[1] [ˈlâytə(r)] *a.* Daha hafif, açık. **~er**[2] *n.* Mavna. **~er**[3] *n.* Yakıcı alet: **petrol ~,** benzinli çakmak. **~house** [ˈlâythaus]. Fener kulesi: **~-keeper,** fenerci. **~ing** [ˈlâytin(g)]. Tenvirat; aydınlatma: **~-up time,** (oto) ışıkları yakma zamanı. **~ly** [ˈlâytli]. Hafifçe. **to get off ~,** ucuz kurtulmak: **to speak ~ of s.o.,** birisini istihfaf etm.: **to take stg. ~,** hafiften almak. **~ness** [ˈlâytnis]. Hafiflik, hiffet. **~ning** [ˈlâytnin(g)]. Şimşek, yıldırım; şimşek gibi; **as quick as ~ [with ~ speed, like greased ~],** şaşırtıcı bir hızla: **~-conductor,** siperisaika. **~s** [lâyts]. Akciğer. **~ship** [ˈlâytşip]. Fener gemisi.

**like**[1] [lâyk] *a. & prep.* Benzer; aynı; gibi. **to be ~,** benzemek: **as ~ as not [~ enough],** belki de: **fellows ~ you,** senin gibiler: **he ran ~ anything,** alabildiğine koştu: **women are ~ that,** kadınlar böyledir: **that's just ~ a woman,** bir kadından bundan başka ne beklersin?: ⌈**~ master, ~ man**⌉, efendi nasılsa uşak öyledir: **today is nothing ~ so hot as yesterday,** bugünkü sıcak dünkü sıcağın yanında hiç bir şey değil: **that's something ~!,** ha şöyle!: **that's something ~ a ship!,** işte gemi diye buna derler: **it will cost something ~ ten pounds,** aşağı yukarı on liraya malolacak: **what's he ~?,** nasıl bir adam? **~**[2] *n.* Benzeri: **to do the ~,** aynını yapmak: **I never saw the ~ of it,** böylesini hiç görmedim: **we shall never see his ~ again,** onun gibisini bir daha göremeyiz: **the ~s of you and me,** sizin ve benim gibiler. **~**[3] *n.* Çok tercih edilen şey. **~s and dislikes,** hoşlandığı ve hoşlanmadığı şeyler **~**[4] *v.* Hoşlanmak; beğenmek; istemek. **I ~ him,** o hoşuma gidiyor: **I ~ that!,** maşallah!: **I should ~ to have seen him,** keşki onu görmüş olsaydım: **I can do as I ~ with him,** o avucumun içindedir: **how do you ~ him?,** onu nasıl buluyorsunuz?: **how do you ~ your tea?,** çayınız nasıl olsun?: **as much as (ever) you ~,** ne kadar istersiniz: **just as you ~!** siz bilirsiniz!: **I should ~ time to ...,** ... için zamana ihtiyacım var: **you may say what you ~,** siz ne derseniz deyiniz: **whether you ~ it or not,** isteseniz de istemeseniz de. **-~** *suff.* -varı; benzer; -ca, -ce. **~able** [ˈlâykəbl]. Sevimli. **~lihood** [ˈlâyklihud]. İhtimal. **in all ~,** pek muhtemeldir ki. **~ly** [ˈlâykli]. Muhtemel; ···cak gibi; münasib; yakışıklı. **as ~ as not,** muhtemeldir ki: **not ~!,** ne münasebet!: **that's a ~ story!,** olacak şey değil! **~-minded.** Hemfikir. **~n** [ˈlâykn]. Benzetmek. **~ness** [ˈlâyknis]. Benzerlik; insan resmi. **it is a good ~,** (resmi) çok benziyor. **~wise** [ˈlâykwâyz]. Aynı vechile; hem.

**liking** [ˈlâykin(g)]. Beğenme; hoşlanma; zevk; meyl. **to have a ~ for,** ···den hoşlanmak: **I have taken a ~ to him,** ona ısındım: **is this to your ~?,** bu zevkinize göre midir?

**lilac** [ˈlâylək]. Leylâk. Açık mor.

**Lilliputian** [liliˈpyûşn]. Ufacık; cüce.

**lily** [ˈlili]. Zambak.

**limb** [lim]. Uzuv; dal. **~ed,** ... uzuvlu.

**limbo** [ˈlimbou]. Cehennem dibi.

**lime**[1] [lâym]. Ihlamur ağacı. **~**[2]. Misket limonu. **~**[3]. Kirec. (Toprağa) kirec serpmek. **slaked ~,** sönmüş kirec. **~-kiln,** kirec ocağı. **~light** [ˈlâymlâyt]. **to be in the ~,** göz önünde olm., halkın dilinde dolaşmak.

**limerick** [ˈlimərik]. Beş mısralık mizahî manzume.

**limit**[1] [ˈlimit] *n.* Had, sinir. **he's the ~,** o artık fazla oluyor: **that's the ~!,** bu kadar olur!: **within a five-mile ~,** beş mil dahilinde: **it is true within ~s,** muayyen bir hadde kadar doğrudur: **without ~,** hadsiz. **~less,** hududsuz, sonsuz. **~**[2] *v.* Tahdid etm.; kısmak. **to ~ oneself to ...,** ···le iktifa etmek. **~ation** [–ˈtêyşn], tahdidat; mühlet: **he has his ~s,** zayıf noktaları var: **Statute of ~s,** müruruzaman kanunu.

**limp**[1] [limp]. Topallama(k); aksaklık. **~**[2] *a.* Gevşek; yumuşak. **to feel as ~ as a wet rag,** suyu çıkmış limon gibi bitkin olmak.

**limpet** [ˈlimpit]. **to stick like a ~,** sülük gibi yapışmak.
**limpid** [ˈlimpid]. Berrak.
**line**[1] [lâyn] *n.* Çizgi; hat; yol; sera; çubuk; satır; saf; sülâle; olta. **what is his ~ of business?,** işi nedir?: **to cross the L~,** istiva hattından geçmek: **in direct ~,** babadan oğula: **one must draw the ~ somewhere,** her şeyin bir haddi var: **I draw the ~ at ...,** ...ya kadar gitmem: **to drop s.o. a ~,** birine bir iki satır göndermek: **to fall into ~ with ...,** ...e uymak: **to fall out of ~,** ayırmak: **to give s.o. a ~ on stg.,** ipucu vermek: **~ of goods,** mal çeşidi: **it's hard ~s on him,** ona çok yazık: **in a ~,** bir sıra halinde: **the ~s of a ship,** geminin şekli: **shipping ~,** deniz nakliye kumpanyası: **something in that ~,** aşağı yukarı bu neviden: **that is not in my ~,** bu benim işim değil: **railway ~,** demiryolu hattı: **~ of thought,** fikir silsilesi: **the ~ to be taken,** tutulacak yol: **to read between the ~s,** üstü kapalısını kavramak: **on the right ~s,** doğru yol üzerinde. **~**[2] *v.* Astar koymak; kaplamak. **~age** [ˈlini·ic]. Soy; sülâle; nesil. **~al** [ˈliniəl]. Hattî. **~ descendant,** doğrudan doğruya torun. **~ament** [ˈliniəmənt]. Yüz hattı. **~s,** çehre. **~ar** [ˈliniə(r)]. Hattî. **~ measurement,** uzunluk ölçüsü. **~d** [lâynd]. Çizgili; buruşuk; astarlı. **a well-~ purse,** dolgun kese. **~r** [ˈlâynə(r)]. Büyük yolcu vapuru. **~sman** [ˈlâynsmən]. Telgraf, hattı amelesi; (futbol) hat hâkemi.
**linen** [ˈlinən]. Keten bezi(nden yapılmış eşya); çamaşır. **⌜to wash one's dirty ~ in public⌝,** (aile) kendi içyüzünü ortaya dökmek.
**linger** [ˈlin(g)gə(r)]. Bir yerden ayrılmak istemiyerek kalmak; hastalığı uzayıp ölmemek. **to ~ on a subject,** bir mevzuu uzatmak. **~ing, a ~ death,** uzun bir can çekişme: **a ~ doubt,** kurtulunmaz bir şübhe: **a ~ look,** gözünü ayırmadan bir bakış.
**lingo** [ˈlin(g)gou]. Yabancı dil.
**lingu·a franca** [ˈlin(g)gwaˈfranka]. Müşterek anlaşma vasıtası olan bir lisan. **~al** [ˈlin(g)gwəl]. Dile aid. **~ist,** lisan mütehassısı; çok yabancı dil bilen. **~istic** [–ˈistik], dil bilgisine aid.
**lining** [ˈlâynin(g)]. Astar; iç kaplaması. **⌜there's a silver ~ to every cloud⌝,** her şeyde bir hayır vardır.
**link** [lin(g)k]. Zincir halkası; rabıta. Bağlamak. **to ~ arms,** kol kola girmek: **missing ~,** noksan halka; insanla maymun arasındakı mahluk.
**lino** [ˈlâynou] *abb. for* **~leum** [liˈnoulyəm]. Muşamba.
**linseed** [ˈlinsîd]. Keten tohumu.
**lint** [lint]. Keten tiftiği.
**lion** [ˈlâyən]. Aslan; herkesin merakını uyandıran kimse. **to make a ~ of [~ize] s.o.,** muvafakkiyet göstermiş bir kimseyi toplantılarda nazarı dikkati ona celbetmek: **the ~'s mouth,** çok tehlikeli bir yer: **the ~'s share,** aslan payı: **to twist the ~'s tail,** kasden İngilizleri kızdıracak neşriyatta bulunmak. **~ess,** dişi aslan.
**lip** [lip]. Dudak, ağız; (*sl.*) yüzsüzlük. **to hang on s.o.'s ~s,** birisinin ağzına bakmak: **to keep a stiff upper ~,** korku, keder göstermemek: **none of your ~!,** yüzsüzlüğün lüzumu yok! **~ped,** dudaklı. **~stick,** dudak boyası. **~-read,** (sağırlar) sözü dudak hareketlerinden anlamak. **~-service,** samimî olmıyan cemile.
**liqu·efy** [ˈlikwifây]. Su haline getirmek, gelmek. **~id** [ˈlikwid]. Mayi; seyyal; berrak; nakte kolay çevrilir; elde bulunan (para). Mayi, sulu olan cisim. **~idate** [ˈlikwidêyt]. Tasfiye etmek. **~idation** [–ˈdêyşn], tasfiye. **~or** [ˈlikə(r)]. İçki; mahlul. **to be the worse for ~,** çakırkeyf olmak.
**liquorice** [ˈlikəris]. Meyankökü.
**lisp** [lisp]. **S** ve **Z** harflerini th veya **dh** gibi telâffuz (etm.); peltek konuşma(k).
**list**[1] [list]. Fihrist; cedvel; liste; kadro. Bir deftere yazmak. **on the active ~,** faal hizmette: **alphabetical ~,** alfabe sırasile cedvel: **on the danger ~,** (hasta) ölüm tehlikesinde. **~**[2]. (Gemi) yan yatma(k). **~less** [ˈlistlis]. Kayıdsız; gevşek; melûl.
**listen** [ˈlisn]. Kulak vermek. **~ to ...,** ...i dinlemek; ...e dikkat etm.: **to ~ in,** kulak misafiri olm.; radyoyu dinlemek. **~er,** dinleyici: **to be a good ~,** başkasının sözlerini sabırla dinlemek: **⌜~s never hear good of themselves⌝,** kulak misafiri kendisi hakkında iyi bir şey işitmez.
**lit** *v.* **light.**
**literacy** [ˈlitərəsi]. Okur yazarlık.
**literal** [ˈlitrəl]. Harfî; lafzî; harfi harfine; mecazî değil; hakikî. **he ~ly**

has to beg his food, geçinmek icin tam manasile dileniyor.

**litera·ry** [ˈlitərəri]. Edebiyata aid. ~ **man**, edib. ~**te** [–rit], okur yazar. ~**ture** [ˈlitrətyə(r)]. Edebiyat (mesleği, eserleri); kitabiyat; rehber. **I wish I had some ~**, keşki yanımda okunacak bir şey bulunsaydı.

**lithe** [lâyð]. Eğilir bükülür; çevik.

**litho·graph** [ˈliθograf]. Taşbasması resim (yapmak). ~**graphy** [–ˈlogrəfi], taşbasması.

**Lithuania** [liθyûˈêynyə]. Litvanya. ~**n**, litvanyalı, litvanyaca.

**litig·ant** [ˈlitigənt]. Dâvacı. ~**ate**, mahkemeye müracaat etmek. ~**ation** [–ˈgêyşn], dâva etme.

**litter**[1] [ˈlitə(r)]. Sedye; teskere. ~[2]. Çörçöp; karmakarışıklık. Dağıtmak. ~[3]. Domuz vs.nin bir batında doğurduğu yavrular.

**little** [ˈlitl]. Küçük, ufak; az. **a ~**, biraz: ~ **by ~**, tedricen: **every ~ helps**, ne kadar az olursa olsun işe yarar: **for a ~ (while)**, bir müddet: **so that's your ~ game!**, demek kurduğun kumpas buydu!: **he ~ knows what fate awaits him**, başına gelecekten haberi yok: ~ **or nothing**, hiç denilecek kadar: **the ~ ones**, yavrular: **the ~ people**, periler: **to think ~ of stg.**, bir şeyi değersiz tutmak: **to think ~ of others**, (i) herkesi küçük görmek; (ii) başkalarını düşünmemek: **he thinks ~ of walking 20 miles**, onun için 20 mil işten bile değil: **I know his ~ ways**, ben onun acayibliklerini bilirim: **he did what ~ he could**, elinden gelen azıcık yardımı esirgemedi. ~**-Englander**, imparatorluk siyaseti aleyhdarı İngiliz.

**liturgy** [ˈlitö(r)ci]. Cemaatle ibadete mahsus dualar.

**live**[1] [lâyv] *a.* Diri, canlı, hayatta. ~ **cartridge**, hakikî mermi: ~ **coals**, kor halinde kömür: ~ **wire**, elektrikli tel; pek faal adam. ~**-bait**, canlı balık yemi. ~[2] [liv] *v.* Yaşamak; geçinmek; ikamet etmek. **to ~ down a scandal**, bir rezaleti unutturmak: **one has got to ~**, geçim dünyası bu!: ⌜**~ and learn!**⌝, 'bir yaşıma daha girdim': ⌜**~ and let ~**⌝, herkesin yaşamağa hakkı var: **to ~ in**, (hizmetçi) çalıştığı yerde yatmak ve yiyip içmek: **long ~!**, yaşasın!: **to ~ on**, yaşamağa devam etm.: **to ~ on very little**, az para ile geçinmek: **he has very little to ~ on**, geçinecek parası çok az: **to ~ on s.o.**, birisinin parasile geçinmek: **to ~ on vegetables**, sebze yiyerek yaşamak: **I can't ~ up to my wife**, karımın gidişine uyamıyorum. ~**lihood** [ˈlâyvlihud]. Maişet; geçinme. ~**liness** [ˈlâyvlinis]. Şetaret; canlılık. ~**long** [ˈlivlon(g), ˈlâyv–]. **the ~ day**, bitmek tükenmek bilmez gün. ~**ly** [ˈlâyvli]. Canlı; neş'eli; civelek. ~**n** [ˈlâyvn]. ~ **up**, canlan(dır)mak. ~**r** [livə(r)]. (Filan tarzda) yaşıyan. **evil [loose] ~**, sefih, ahlaksız.

**liver** [ˈlivə(r)]. Karaciğer. **to have a ~**, karaciğeri bozuk olm. ~**ish** [ˈlivəriş]. Karaciğeri bozuk; safralı.

**liver·ied** [ˈlivərid]. Resmî elbiseli (uşak). ~**y** [ˈlivəri]. Büyük konaklarda uşakların giydikleri hususî elbise. ~**-company**, Londra'da büyük esnaf cemiyeti.

**lives** *n. pl. of* **life**.

**livestock** [ˈlâyvstok]. Bir çiftlikteki canlı hayvanlar.

**livid** [ˈlivid]. Bere renginde olan. ~ **with anger**, hiddetten mosmor.

**living**[1] [ˈlivin(g)] *a.* Diri, hayatta. **a ~ death**, ölümden beter bir hayat: **in the land of the ~**, yaşıyanlar arasında: **no ~ man could do better**, hayatta bulunanlar arasında hiç kimse daha iyi yapamaz. ~[2] *n.* Hayat tarzı; geçinme; yaşama; bir yere papaz tayin etme. **to earn one's own ~**, eli ekmek tutmak: **to make a ~**, hayatını kazanmak: **a ~ wage**, asgarî geçinme ücreti. ~**-room**, oturma odası.

**lizard** [ˈlizəd]. Kertenkele.

**load**[1] [loud] *n.* Yük; hamule. **I have a ~ on my mind**, zihnimi meşgul eden bir şey var: **to take a ~ off one's mind**, rahat nefes aldırmak: ~**s of**, pek çok. ~[2] *v.* Yükletmek; (silâh) doldurmak. **to ~ favours on s.o.** [**to ~ s.o. with favours**], birine lûtuf yağdırmak. ~**ed**, yüklü; dolu: ~ **dice**, cıvalı zar.

**loaf**[1], *pl.* **loaves** [louf, louvz]. Bir ekmek, somun. **a ~ of sugar**, şeker kellesi: ⌜**half a ~ is better than no bread**⌝, bu hiç yoktan iyidir. ~[2] *v.* **to ~ about**, haylazca vakit geçirmek. ~**er**, haylaz herif.

**loan** [loun]. İstikraz; ödünc verme (alma); ariyet. Ariyet vermek; ikraz etmek. **to have the ~ of**, ariyeten almak: **to raise a ~**, istikrazda bulunmak.

**loath** [louθ]. İsteksiz. **to be ~ to do stg.**, bir şeyi yapmağa gönlü olmamak: **he did it nothing ~**, memnuniyetle yaptı.
**loath·e** [louð]. Nefret etm. **~ing** [louðin(g)], nefret; istikrah. **~some** [ˈlouθsəm], iğrenc.
**loaves.** *pl. of* **loaf.**
**lob** [lob]. Havaya atmak.
**lobby** [ˈlobi]. Koridor. Meclis koridorlarında mebusların reylerini taleb etm.
**lobster** [ˈlobstə(r)]. İstakoz. **~-pot**, istakoz sepeti.
**local** [ˈloukl]. Mahallî; oralı, buralı. **the ~**, (*sl.*) mahalledeki birahane: **~ government**, mahallî idare. **~ity** [louˈkaliti], yer; civar; semt: **to have a bump of ~**, kolaylıkla cihet tayin edebilmek. **~ize** [ˈloukəlâyz], mevziileştirmek; tahdid etm.; tecrid etmek. **~ly**, kendi mahallinde; civarında.
**locat·e** [louˈkêyt]. Yerini tayin, keşfetmek. **to be ~ed in a place**, bir yerde yerleştirilmek. **~ion** [louˈkêyşn], yerini tayin etme, keşfetme; yerleştirme; mevki.
**locative** [ˈlokətiv]. Mefulünfih.
**lock**[1] [lok] *n*. Lüle; perçem; kâkül. **~**[2] *n*. Kilid; tüfek çakmağı; mütehharrik kapılı kanal seddi. **~-gate**, bu seddin kapısı: **~, stock and barrel**, ne var ne yok, hepsi: **under ~ and key**, kilid altında. **~**[3] *v*. Kilidle(n)mek; kenetlenmek; (makine parçaları) iç içe geç(ir)mek. **to ~ s.o. in**, birinin üzerinden kapıyı kilidlemek: **to ~ out**, kapıyı kilidleyip birini dışarıda bırakmak; fabrikayı kilidleyip işçileri dışarıda bırakmak: **to ~ up**, kilid altında bulundurmak; hapsetmek. **~-out**, lokavt. **~-up**, mahpuslar nezarethanesi. **~jaw** [ˈlokcô]. Tetanos, küzaz. **~smith** [ˈloksmiθ]. Çilingir.
**locket** [ˈlokit]. Madalyon.
**locomot·ion** [ˌloukəˈmouşn]. Bir yerden kalkıp başka bir yere gitme. **~ive** [ˈloukəmoutiv], hareket ettirici; lokomotif.
**locum-tenens** [ˈloukəmˈtînenz]. Muvakkaten başkasının vazifesini üzerine alan kimse; vekil.
**locust** [ˈloukʌst]. Şarka mahsus büyük ve tahribkâr çekirge.
**locution** [loˈkyûşn]. Tabir; ifade tarzı.
**lodge**[1] [loc] *n*. Kapıcı bahçevan evi; farmason locası. **shooting ~**, av köşkü. **~-keeper**, bahçe kapıcısı. **~**[2] *v*. Muvakkaten bir evde otur(t)mak. **to ~ a complaint**, bir şikâyette bulunmak: **to ~ with s.o.**, birinin evinde oturmak. **~r**, başkasının evinde bir iki oda kiralıyan kimse: **to take in ~s**, kendi evinde bir kaç oda kiraya vermek.
**lodging** [ˈlocin(g)]. Barınacak yer. **~s**, pansiyon. **~-house**, pansiyon: **common ~**, (düşkünler için) misafirhane.
**loft** [loft]. Tavanarası. **~y** [ˈlofti]. Yüksek; âlî.
**log**[1] [log]. Kütük; (*naut.*) parakete. **to heave the ~**, parakete atmak. **~-book**, gemi jurnalı. **~-cabin**, kütüklerden yapılmış kulübe. **~**[2] *v*. Gemi jurnalına kaydetmek.
**loggerheads** [ˈlogəhedz]. **at ~**, araları açık.
**logic** [ˈlocik]. Mantık ilmi. **~al** [–kl], mantıkî.
**loin** [lôyn]. Fileto. **~s**, bel altı: **to gird up one's ~**, etekleri sıvamak. **~-chop**, pirzola. **~-cloth**, peştamal.
**loiter** [ˈlôytə(r)]. Boş gezmek, dolaşmak.
**loll** [lol]. **to ~ about**, tembelce oturmak.
**lollipop** [ˈlolipop]. Şekerleme.
**lone** [loun]. Tenha; kimsesiz. **to play a ~ hand**, bir işte yalnız başına kalmak. **~ly** [ˈlounli], **~some**, yalnız; tenha.
**long**[1] [lon(g)] *v*. **to ~ to do stg.**, bir şeyi yapmağı çok arzu etmek: **to ~ for**, ···in hasretini çekmek: **he ~s for Istanbul**, İstanbul burnundan tütüyor. **~**[2]. (*abb.*) **longitude. ~**[3] *a.* & *n.* Uzun (müddet). **to be ~ in the arm**, kolları uzun olm.: **as ~ as I live**, ömrüm oldukça: **it will take as ~ as three years**, üç sene kadar sürer: **you can play there as ~ as you don't go near the river**, nehrin yanına gitmemek şartile orada oynayabilirsiniz: **a week at the ~est**, en fazla bir hafta: **I had only ~ enough to eat a sandwich**, yalnız bir sanduvic yiyecek vakit buldum: **how ~ will it take?**, ne kadar sürecek?: **how ~ have you been here?**, buraya geleli ne kadar oldu?: **he hasn't ~ to live**, fazla yaşayamaz: **a ~ purse**, dolu kese: **the ~ and the short of it**, işin hulâsası: **the best by a ~ way**, çok büyük bir farkla en iyisi: **to go a ~ way**, uzak gitmek; büyük bir tesir yapmak: **he**

will go a ~ **way,** bu adam çok ilerler: **to take the ~est way round,** en uzak yoldan gelmek: **he is a ~ time in coming,** geç kaldı. **~-ago,** eski zaman(a aid). **~-bow,** eski İngiliz ve çok kuvvetli bir yay. **~-drawn-out,** uzun uzadıya. **~-lost friend,** çoktandır görülmiyen bir ahbab. **~-sighted,** presbit; müdebbir; uzağı gören. **~-standing,** müzmin; eski. **~-suffering,** sabir(lı); müsamaha(kâr). **~-winded,** uzun uzadıya konuşan. **~evity** [lonˈceviti], uzun ömürlülük. **~hand** [ˈlon(g)hand], âdi yazı (*yani istenografya değil*). **~itude** [ˈloncityûd], tûl. **~shoreman** [lon(g)ˈşômən], gemi boşaltıcı. **~ways, -wise** [ˈlon(g)wêyz, –wâyz]. Uzunluğuna.

**look**[1] [luk] *n.* Bakış; görünüş; manzara. **to have a ~ at,** gözden geçirmek: **to take a good ~ at,** iyice bakmak: **to have [take] a ~ round (the town),** (şehri) dolaşmak: **one could see by his ~ that he was angry,** kızdığı yüzünden belli idi: **I don't like his ~s,** bu adamın yüzünü hiç beğenmiyorum: **I don't like the ~ of the thing,** bana bu iş şübheli görünüyor: **from the ~ of him,** görünüşüne nazaran: **he has the ~ of his father,** o babasını andırıyor: **good ~s,** güzellik. **~-in,** teklifsiz bir ziyaret: **not to have a ~,** kazanması hiç muhtemel olmamak. **~-out,** gözetleme (yeri); gözcü; manzara: **to keep a ~,** dikkat etmek: **to be on the ~ for,** kollamak: **it is a poor ~ for him,** istikbali karanlık görünüyor: **that's his ~,** bu onun bileceği şey. ~[2] *v.* Bakmak; görünmek. **to ~ like,** benzemek: **it ~s like raining,** yağmur yağacağa benziyor: **she ~s her age,** yaşını gösteriyor: **you ~ well,** iyi [sıhhatli] görünüyorsun: **you ~ well in that hat,** o şapka size yakışıyor. **~ about,** **to ~ about one,** etrafına bakmak: **to ~ about for s.o.,** gözlerile birisini araştırmak: **to ~ about for a job,** bir iş aramak. **~ after,** bakmak; çekip çevirmek; idare etm.: **he is old enough to ~ after himself,** artık kendini idare edecek yaştadır. **~ at,** bakmak: **to ~ at him one would think he was starved,** yüzüne bakan onu açlıktan ölüyor zanneder: **what sort of man is he to ~ at?,** (şeklen) nasıl bir adam?: **he's not much to ~ at, but ...,** görünüşte bir şeye benzemiyor, amma ...: **the way of ~ing at things,** görüş tarzı. **~ back,** arkaya [geriye] bakmak: **to ~ back upon the past,** maziye dönüp bakmak: **what a day to ~ back to!,** bu günü daima zevkle hatırlayacağız! **~ down,** aşağıya bakmak: **to ~ down a list,** bir listeyi gözden geçirmek: **to ~ down upon,** hor görmek. **~ for,** aramak; beklemek: **he is ~ing for trouble,** belâsını arıyor. **~ forward,** ileri bakmak: **I am ~ing forward to seeing him,** onu göreceğim zamanı zevkle bekliyorum. **~ in, to ~ in at the window,** pencereden içeriye bakmak: **to ~ in upon s.o.,** geçerken birisine uğramak. **~ into,** tedkik etm.; göz önünde tutmak. **~ on,** bakmak; boş durup seyirci olm.; *v.* **~ upon.** **~ out,** dışarı bakmak; nazır olm.; dikkat etm.; aramak: **~ out (for yourself)!,** dikkat et!: **everyone must ~ out for himself,** herkes başının çaresine bakmalı: **to ~ out a train in the time-table,** tarifede trene bakmak. **~ over,** göz gezdirmek; nazır olm.: **to ~ over a house,** bir evi gezmek: **to ~ s.o. all over,** birisini baştan aşağı süzmek. **~ round,** etrafına bakmak; dönüp bakmak, gezmek. **~ through,** gözden geçirmek: **to ~ s.o. through and through,** birine içini okur gibi bakmak. **~ to, to ~ to stg.,** bir şeye bakmak: **to ~ to the future,** istikbali düşünmek: **to ~ to s.o. to do stg.,** bir iş için birisine güvenmek: **I ~ to going to Scotland this autumn,** bu sonbahar İskoçya'ya gideceğimi umuyorum. **~ up,** yukarıya bakmak; başını kaldırmak: **to ~ up to s.o.,** birine itibar etm.; **business is ~ing up,** işler canlanıyor: **to ~ up a word in the dictionary,** bir kelime için lûğate bakmak: **to ~ s.o. up,** birisini gidip görmek. **~ upon,** bakmak, saymak: **nice to ~ upon,** güzel. **~er-on** [ˈlukərˌon], seyirci. **-~ing** [ˈlukin(g)] *suff.* ... görünüşlü; yüzlü. **~ing-glass** [ˈlukin(g)ˌglâs], ayna.

**loom**[1] [lûm] *n.* Dokuma tezgâhı. ~[2] *v.* Karaltı gibi gözükmek. **to ~ large,** vukuu pek yakın görünmek: **dangers ~ing ahead,** tehdid eden tehlikeler.

**loony** [ˈlûni]. Meczub, kaçık.

**loop** [lûp] *n.* İlmik; (yol)un dirseği. *v.* İlmiklemek. **to ~ the ~,** (uçak) takla atmak. **~-hole,** mazgal; kaçamak. ~-**line,** ana hattan ayrılıp sonra tekrar kavuşan demiryolu.

**loose**[1] [lûs] *a.* Gevşek; ayrı; sallanan, başıboş; sarkık. **to become** [**get**] ~, gevşemek; ayrılmak; yerinden çıkmak; çözülmek: **to break** ~, boşamak; kurtulmak: **to cast** ~, geminin halatlarını salıvermek: ~**ly clad,** bol elbiseli: **to be at a** ~ **end,** işsiz olm.: **to go on the** ~, hovardalık etm.: **to let** [**set**] ~, serbest bırakmak: ~ **living,** sefahat: **to carry money** ~ **in one's pocket,** parayı cebinde taşımak: **a** ~ **translation,** serbest tercüme: **word** ~**ly employed,** tam yerinde kullanılmıyan kelime: ~**ly woven,** seyrek dokunmuş. ~-**fitting,** gevşek; (elbise) bol. ~-**leaf,** yaprakları ayrı ayrı olan (defter). ~[2] *v.* Salıvermek; serbest bırakmak; çözmek. **to** ~ **one's hold,** sıkı tutmamak: **to** ~ **hold of stg.,** salıvermek. ~**n** [ˈlûsn], gevşe(t)mek; çözmek. **to** ~ **a cough,** öksürüğü söktürmek: **to** ~ **s.o.'s tongue,** dilini çözmek.

**loot** [lût]. Çapul; ganimet, yağma malı. Yağma etm., çapullamak.

**lop**[1] (**lopped**) [lop, lopt]. **to** ~ (**off**), budamak; ucunu kesmek. ~[2]. **to** ~ **over,** sarkmak. ~-**sided,** bir tarafa yatkın.

**lope** [loup]. Uzun adımlarla yürümek. Bu yürüyüş.

**loquacious** [loˈkwêyşəs]. Geveze.

**lord** [lôd]. Efendi; sahib. **to** ~ **it over,** tahakküm etmek istemek: **the L**~, Cenabı Hak: **the L**~**'s Day,** pazar günü: **good L**~ !, Allah! Allah!: **in the year of our L**~, milâdın ... senesinde. ~**ly,** muhteşem; kibirli. ~**ship,** lordluk: **your** ~, zati asilâneleri.

**lore** [loə(r)]. Bilgi, ilim.

**lorry** [ˈlori]. Kamyon.

**los·e** (**lost**) [lûz, lost]. Kaybetmek. **to be lost** [~ **oneself**] **in** ..., ···e dalmak: **the joke was lost on him,** nükteyi anlamadı: **the affair lost nothing in the telling,** bu mesele anlatılırken dallandı budaklandı: **this doctor has lost several patients,** (i) hastalarından bir çoğu öldu; (ii) müşterilerinden bir çoğunu kaybetti: **to** ~ **one's reason,** aklını kaçırmak: **I lost most of what he said,** söylediğinin çoğunu işidemedim: **I have lost sight of him,** (i) onu gözden kaybettim; (ii) onu gördüğüm yok: **to** ~ **one's strength,** kuvvetten düşmek: **my watch** ~**s ten minutes a day,** saatim günde on dakika geri kalıyor. ~**er** [ˈlûzə(r)]. Mağlub; kaybeden kimse. **I am the** ~ **by it,** bu işte kaybeden benim: **to be a bad** ~, oyunda kaybedince kendine hâkim olmamak. ~**ing** [ˈlûzin(g)] *a.* **a** ~ **game,** kaybedileceği muhakkak olan oyun: **a** ~ **concern,** kârlı olmıyan bir iş.

**loss** [los]. Kayıb; zayi olma; zarar. **to be at a** ~, şaşırmak: **to cut one's** ~**es,** zarardan kâr etm.: **he** [**it**] **is no** ~, bu kayıb sayılmaz.

**lost** [lost] *v.* **lose.** *a.* Kaybolmuş. ~ **property office,** kayıb eşya için müracaat yeri: **to look** [**seem**] ~, yadırgıyor gibi görünmek.

**lot** [lot]. Kur'a; piyango; kısmet; takım, mikdar; çok; kısım; müzayedeye çıkarılan malların beheri; arsa. **a** ~ **of** [~**s of**], bir sürü, çok: **all the** ~, sürü sepet: **a bad** ~, sağlam ayakkabı değil: **to cast** [**draw**] ~**s,** kur'a çekmek: **it did not fall to my** ~, bana nasib olmadı: **in** ~**s,** takım halinde: **to make a** ~ **of s.o.,** birini başına çıkartmak: **to buy in one** ~, hepsini toptan almak: **that's the** ~, hepsi bu kadar: **to think a** ~ **of oneself,** kendini bir şey zannetmek: **to throw in one's** ~ **with** ..., ···le mukadderatını birleştirmek: **what a** ~ **of people!,** ne kadar kalabalık!: **the whole** ~ [**all the** ~] **of you,** hepiniz.

**lottery** [ˈlotəri]. Piyango; kısmet meselesi.

**loud** [laud]. Gürültülü; (ses) yüksek; (renk) çiğ. **to be** ~ **in one's praises of,** fazla medhetmek: **out** ~, cehren, yüksek sesle. ~**ly,** yüksek sesle: ~ **dressed,** gösterişli giyinmiş. ~**speaker** [ˈlaudˈspîkə(r)], hoparlör.

**lounge** [launc]. Hol; teneffüs salonu. **to** ~ (**about**), hiç bir şey yapmıyarak tembelce oturmak, gezmek. ~-**chair,** şezlong. ~-**suit,** günlük elbise.

**lous·e,** *pl.* **lice** [laus, lâys]. Bit; kehle. ~**y** [ˈlauzi], bitli; alçak.

**lout** [laut]. Hantal, ayı gibi. ~**ish,** mankafa; hantal.

**love**[1] [lʌv] *v.* Sevmek; âşık olm.; ···den hoşlanmak. **'Will you come?' 'I should** ~ **to',** 'Gelir misiniz?' 'Memnuniyetle'. ~[2] *n.* Aşk; sevgi, muhabbet; sevgili; aşk mabudu; (oyunda) pata. **to be in** ~, âşık olm.: **to fall in** ~ **with s.o.,** birine gönlünü kaptırmak: **first** ~, ilk gözağrısı: **to do stg.**

for the ~ of it, bir şeyi merak saikasile veya zevk için yapmak: **it cannot be had for ~ or money,** bu ne para ile ne de hatır için bulunur: **to work** (just) **for ~,** fisebilillah [pir aşkına] çalışmak: **give him my ~!,** gözlerinden öperim!: **there is no ~ lost between them,** biribirlerinden hoşlanmazlar: **an old ~ of mine,** eski sevgililerimden biri: **he sends you his ~,** size selâm söyledi. **~-child,** gayrimeşru çocuk. **~-match,** aşk izdivacı. **~ly** [ˈlʌvli], gayet güzel, lâtif; hoş. **~r** [ˈlʌvə(r)], âşık; sevgili. **a ~ of stg.,** mübtelâ, düşkün. **…-lover,** … seven, … meraklısı. **~sick** [ˈlʌvsik], sevdazade; mecnun.

**loving** [ˈlʌvin(g)] *a.* Muhabbetli; seven. **~-kindness,** şefkat, hayırhahlık.

**low**[1] [lou] *v.* Böğürmek. **~**[2] *a.* Alçak, yüksek olmıyan; düşük; münhat; kaba; (ses) pes; (fiat) ucuz. **of ~ birth** [**~-born**], aşağı tabakadan: **a ~ bow,** derin reverans: **to bring** [**lay**] **s.o. ~,** yere sermek: **to lie ~,** bir köşede saklanmak: **in ~ spirits** [**~-spirited**], süngüsü düşük: **so ~ had he sunk,** bu dereceye sukut etmiş. **~-bred,** soysuz, kaba. **~-brow,** fikir meselelerine alâkasız. **~-class,** âdi; aşağı tabakadan. **~-down,** rezil, alçak: **to give s.o. the ~,** (*sl.*) birine bir mesele hakkında fikir, malûmat vermek. **~-level,** aşağı hizada olan. **~-necked,** dekolte (elbise). **~-pitched,** (dam) alçak; (ses) pes perdeli; (tavan) basık. **~-water mark,** (deniz) cezrin en aşağı seviyesi. **~er** [ˈlouə(r)] *a. comp. of* low, daha aşağı vs.; madun. **the ~ classes,** aşağı tabaka. *v.* İndirmek; azaltmak; (ses) yavaşlatmak. **~ermost,** en aşağı. **~land** [ˈloulənd], münhat (arazi): **the Lowlands,** cenub İskoçya: **~lander,** cenub İskoçyalı. **~ly** [ˈlouli], mütevazı.

**loyal** [ˈlôyəl]. Samimî; (saltanata) sadık. **to drink the ~ toast,** kıraliçenin sıhhatine kadeh kaldırmak. **~ist,** hanedana sadık kimse. **~ty,** samimiyet; (saltanata) sadakat.

**lozenge** [ˈlozinc]. Main; baklava şekli; pastil.

**lubber** [ˈlʌbə(r)]. Hantal, beceriksiz adam.

**lubric·ant** [ˈlyûbrikənt]. Yağlıyan. Yağ. **~ate** [–kêyt], yağlamak: well **~d,** (*sl.*) kafayı çekmiş.

**lucid** [ˈlûsid]. Vazıh, iyi anlaşılır; berrak; parlak. **to have ~ intervals,** (deli, sayıklıyan hasta) arasıra kendisine gelmek. **~ity** [–ˈsiditi], vuzuh; berraklık.

**lucifer** [ˈlûsifə(r)]. Sabah yıldızı; şeytan. **as proud as ~,** gayet kibirli.

**luck** [lʌk]. Baht, talih, şans. **as ~ would have it,** tesadüfen: **bad ~,** aksilik; talihsizlik: **better ~ next time!,** inşallah gelecek defa daha iyi olur: **a bit** [**piece, stroke**] **of ~,** düşeş: **to be down on one's ~,** talihi ters gitmek: **good ~,** talih: **hard ~!,** yazık!: **yes, worse ~!,** maalesef!. **~ily,** bereket versin ki. **~y,** talihli; bahtiyar; uğurlu: **~ dog!,** köftehor!: **~ hit** [**shot**], tesadüfen hedefi isabet ettirme: **how ~!,** ne âlâ!: **to make a ~ shot,** (*fig.*) boş atıp dolu tutmak: **thank your ~ stars!,** bir yiyip bin şükret!: **~-dip,** piyango torbası.

**lucr·ative** [ˈlyûkrətiv]. Kazanclı, kârlı. **~e** [ˈlûkə(r)], para: **for filthy ~,** sırf para için.

**ludicrous** [ˈlyûdikrəs]. Gülünç.

**luff** [lʌf]. Yelkenin rüzgâr yakası. Orsa etm.

**lug** [lʌg]. Şiddetli çekiş. Sürüklemek.

**luggage** [ˈlʌgic]. Yol eşyası, bagaj. **~-van,** bagaj vagonu.

**lukewarm** [ˈlûkwôm]. Ilık; gevşek, gayretsiz. **to be rather ~ about stg.,** bir şeye karşı alâkasız, meraksız olmak.

**lull** [lʌl]. Muvakkat sükûnet. Uyuşturmak, teskin etmek. **~aby** [ˈlʌləbây]. Ninni.

**lumb·ago** [lʌmˈbêygou]. Belağrısı. **~ar** [ˈlʌmbar], arka alt tarafına aid.

**lumber** [ˈlʌmbə(r)]. Lüzumsuz eşya; kesilmiş kereste; kabuklu kereste. **to ~ up,** lüzumsuz eşya ile doldurmak: **to ~ along,** hantal hantal yürümek. **~-yard,** kereste deposu.

**lumin·ary** [ˈlûminəri]. Işık veren bir cisim; büyük âlim. **~ous** [–nəs], ışık saçıcı, aydınlatıcı.

**lump**[1] [lʌmp] *n.* Büyük parça; topak, şiş; ahmak. **a big ~ of a boy,** iri yarı bir çocuk: **to have a ~ in the throat,** boğazı düğümlenmek: **in the ~,** toptan: **~ sugar,** kesme şeker: **a ~ of sugar,** bir şeker tanesi: **~ sum,** toptan. **~ish** [ˈlʌmpiş], hantal; ahmak. **~y,** pıhtılı; topaklı. **~**[2] *v.* Yığmak; **to ~ along,** hantal hantal yürümek. ⌜**if you don't like it you may ~ it!**⌝, beğenmezsen beğenme.

**lunacy** [ˈlûnəsi]. Cinnet, delilik.

**lunar** [ˈlûnə(r)]. Aya aid, kamerî.
**lunatic** [ˈlûnətik]. Deli, mecnun. ~ **asylum**, tımarhane.
**lunch** [lʌnç]. Öğle yemeği(ni yemek).
**lung** [lʌn(g)]. Akciğer.
**lunge** [lʌnc]. Hamle, saldırış; meç, kılıc, ile hamle. **to ~ out at s.o.**, birine yumrukla vurmağa çalışmak: **to ~ forward**, birdenbire kendisini ileri atmak.
**lurch** [lö(r)ç]. (Gemi) ânî yalpa; (otomobil) ansızın sıçrama; (sarhoş) sendelme. Böyle ânî bir hareket yapmak. **to leave s.o. in the ~**, birini yüzüstü bırakmak.
**lure** [lyuə(r)]. Cezbetmek; ayartmak; vaid, yalan, ile cezbetmek. Cazibe; hile; tuzak. **the ~ of the deep**, denizin cazibesi.
**lurid** [ˈlyuərid]. Korkunc bir kızıllıkta; korkunc bir şekilde tasvir eden (üslûb).
**lurk** [lö(r)k]. Kötü niyetle gizlenmek. **to ~ about [be on the ~]**, gizli gizli dolaşmak. **~ing**, gizlenmiş: **a ~ suspicion**, mübhem bir şübhe.
**luscious** [ˈlʌşəs]. Pek tatlı, usareli (meyva); çok süslü (üslûb); ağzı sulandıran (tasvir).
**lush** [lʌş]. Mebzul; usareli.
**lust** [lʌst]. Şehvet; hırs. **~ after**, şiddetle istemek; hırs beslemek. **~ful**, şehvanî.
**lustr·e** [ˈlʌstə(r)]. Parlaklık; cilâ, perdah; avize. **to shed ~ on**, ···e şöhret vermek. **~eless**, donuk. **~ous** [–strəs], parlak; cilâlı.
**lusty** [ˈlʌsti]. Gürbüz, dinc, kuvvetli.
**lute** [lyût]. Lâvta; ud.
**luxe** [lüks] (*yal.*) **de ~**, lüks.
**luxuri·ance** [lʌkˈsyuəriəns]. Mebzuliyet, bolluk. **~ant**, mebzul, bol; bereketli. **~ate**, (nebat) mebzul olm.; (insan) bolluk ve zevk içinde yaşamak. **~ous**, tantanalı; pek süslü; zevk ve sefaya dalmış: **to live a ~ life**, lüks yaşamak.
**luxury** [ˈlʌkşəri]. Lüks; her zaman tadılmıyan zevk. **I gave myself the ~ of a cigar**, fevkalâdeden olarak bir puro alıp içtim.
**lying**[1] [ˈlây·in(g)]. *v.* **lie**[1]; Yalan söyleme, yalancılık. Yalancı. **~**[2] *v.* **lie**[3]. Yatan; uzanmış; bulunan; vâki. **~-in**, loğusalık: **~ hospital**, doğumevi.
**lynch** [linç]. Linçetmek.
**lynx** [links]. Vaşak. **~-eyed**, keskin nazarlı.
**lyric** [ˈlirik]. Lirik. **~al**, liriğe aid; heyecanlı.

# M

**M** [em]. M harfi.
**M.A.** [ˈemˈêy]. (*abb.*) Master of Arts, üniversite mezuniyet diploması ile Doktora arasında bir derece.
**Ma** [mâ]. (*abb.*) Mamma.
**Ma'am** [mam]. (*abb.*) Madam. **School-~**, (*Amer.*) muallime.
**macabre** [maˈkâbr]. Ürpertici, meş'um.
**macadam** [maˈkadəm]. Şose. **~ize**, kırılmış taşları ile şose yapmak.
**macaroni** [ˌmakəˈrouni]. Makarna.
**macaroon** [ˌmakəˈrûn]. Bademli kurabiye.
**mace** [mêys]. Gürz; topuz (şeklinde merasim asâsı).
**macerate** [ˈmasərêyt]. Suda ıslatıp yumuşatmak.
**machiavellian** [ˌmakiəˈveliən]. Gayet sinsi ve hilekâr.
**machination** [makiˈnêyşn]. Entrika; kumpas kurma.
**machin·e** [məˈşîn]. Makine; bisiklet. Bir makine ile şekil vermek, tekâmül ettirmek; makine ile dikmek. **~-gun**, makinalı tüfek. **~-made**, fabrika işi. **~-shop**, atölye. **~-tool**, makineli alet. **~ery**, makineler, mekanizma; vasıtalar: **the ~ of government**, idare makinesi. **~ist**, makinist.
**mackerel** [makrəl]. Uskumru. **~-sky**, kapalı havada görünen top top bulutlar.
**mackintosh** [ˈmakintoş]. Yağmurluk.
**macro-** [ˈmakrou] *pref.* Uzun ..., büyük .... **~cosm** [–kozm], kâinat.
**mad** [mad]. Deli, mecnun; kuduz; öfkeli. **as ~ as a hatter [as a March hare]**, zırdeli: **~ about [on] football, *etc.***, futbol vs. delisi: **~ for revenge**, intikama susamış: **to be ~ with s.o.**, birine hiddetinden deli olm.: **to drive s.o. ~**, çıldırtmak: **to go ~**, aklını bozmak, çıldırmak: **socialism gone ~**, müfrit [delice] sosyalizm. **~cap** [ˈmadkap], delişmen, zıpır. **~den** [ˈmadn], çıldırtmak. **~ding** [ˈmadin(g)], **the ~ crowd**, büyük şehrin velvelesi. **~house** [ˈmadhaus], tımarhane. **~man**, deli. **~ness**, delilik.
**made** *v.* **make. made-up**, uydurma; makiyajlı.
**Madeira** [məˈdiərə]. Mader **adası**; mader şarabı. **~ cake**, pandispanya.

**Madonna** [maˈdonə]. Meryemana; Meryemana tasviri. ~ **lily,** beyaz zambak.

**madrigal** [ˈmadrigl]. Aşka dair kısa manzume; çalgısız bir şarkı.

**maelstrom** [ˈmêylstrom]. Büyük girdab.

**magazine** [ˌmagəˈzîn]. Cebhanelik; fişek hazinesi; mecmua. **powder ~,** barut deposu: ~ **rifle,** mükerrer ateşli tüfek.

**magdalen** [ˈmagdəlin]. Tövbekâr fahişe.

**magenta** [məˈcentə]. Kırmızı ile eflâtun arasında bir renk.

**maggot** [ˈmagət]. Kurd. **~y,** kurdlu.

**Magi** [ˈmêycây]. İsa yeni doğduğu zaman hediye getiren üç şarklı âlim.

**magic** [ˈmacik]. Sihir(li), büyü(lü), sihirbazlık. **as if by ~,** mucize kabilinden: **black ~,** büyü: ~ **lantern,** hayal feneri: **to work like ~,** mucize gibi tesir etmek. **~al,** sihirli, sehhar: **to have a ~ effect,** bir büyü tesiri yapmak. **~ian** [məˈcişn], sihirbaz, büyücü.

**magisterial** [macisˈtiəriəl]. Mütehakkim, hâkimane; sulh hâkimine aid.

**magistra·te** [ˈmacistrit, –rêyt]. Sulh hâkimi. **~cy** [**~ture**], sulh hâkimliği.

**Magna Charta** [ˈmagnaˈçâta]. İngiltere'de 1215'de şahsî ve siyasî hurriyeti temin eden kanun.

**magnanim·ity** [ˌmagnəˈnimiti]. Âlicenablık. **~ous** [magˈnaniməs], âlicenab.

**magnate** [ˈmagnêyt]. Eşraf. **industrial ~,** sanayi kodamanlarından.

**magnet** [ˈmagnit]. Mıknatis; cazibeli kimse, şey. **bar ~,** çubuk mıknatis: **horseshoe ~,** at nalı mıknatis. **~ic** [–ˈnetik], mıknatisli; cazibeli. **~ism** [–tizm], mıknatisiyet; manyatizma. **~ize,** mıknatislemek; manyatizma yapmak. **~o** [–ˈnîtou], manyeto.

**magnification** [ˌmagnifiˈkêyşn]. Büyütme.

**magnificen·ce** [magˈnifisəns]. İhtişam; azamet. **~t,** muhteşem; mükemmel.

**magnif·ier** [ˈmagnifâyə(r)]. Pertavsız; mübalağacı. **~y,** büyütmek; hakkında mübalağa etm.: **~ing glass,** pertavsız: **~ing power,** büyütme kuvveti.

**magnitude** [ˈmagnityûd]. Büyüklük, azamet; ehemmiyet.

**magnolia** [magˈnoulyə]. Manolya.

**magnum** [ˈmagnəm]. ~ **opus,** şaheser.

**magpie** [ˈmagpây]. Saksağan.

**Magyar** [ˈmagyâ(r)]. Macar.

**maharajah** [ˌmahaˈrâca]. Mihrace.

**Mahdi** [mâdi]. Mehdi.

**mahogany** [məˈhogəni]. Maun.

**Mahomet** [maˈhomit]. Muhammed. **~an,** Müslüman.

**maid** [mêyd]. Kız; kadın hizmetçi. ~ **of honour,** damdonör: **old ~,** evlenmemiş yaşlı kız. **~-of-all-work,** her işe bakan hizmetçi. **~servant** [ˈ––sö(r)vənt], kız hizmetçi.

**maiden** [ˈmêydn]. Kız; bakire. Yeni, kullanılmamış. ~ **name,** bir kızın evlenmeden evvelki soyadı: ~ **speech,** bir mebusun ilk nutku: ~ **voyage,** bir geminin ilk seferi. **~hood,** kızlık, bekâret. **~like** [**~ly**], kız gibi, kıza yakışır; afif.

**mail**[1] [mêyl]. Zırh elbise. **~ed,** zırhlı: **the ~ fist,** kuvvet ile tehdid. **~**[2]. Posta (ile göndermek). **~-boat,** posta vapuru. **~-cart,** posta arabası; çocuk arabası. **~-order,** posta ile gönderilen sipariş. **~-packet,** posta vapuru. **~-van,** posta furgonu.

**maim** [mêym]. Sakatlamak. **~ed,** sakat.

**main**[1] [mêyn] *n.* Kuvvet; (*esk.*) okyanus; su, havagazi, elektrik ana boru, kablosu. **with might and ~,** var kuvvetile: **in the ~,** alelekser: **to take one's power from the ~s,** elektriği ana hattan almak. **~**[2] *a.* Ana; baş; başlıca. **all ~ services,** ana hizmetler (su, havagazi, elektrik): **the ~ force (army,** *etc.*), ana kuvveti: **by ~ force,** cebren. **~ly,** başlıca. **~land,** ada olmıyan kara. **~mast,** ana direk. **~sail** [ˈmêynsl], mayistra yelkeni. **~spring,** ana yay; başlıca âmil. **~stay** [–stêy], ana istralya; istinad noktası. **he is the ~ of the business,** o işin temelidir.

**maintain** [mêynˈtêyn]. Tutmak; muhafaza etm., bakmak; müdafaa etm.; beslemek; masrafını görmek; iddia etmek. **~able,** tutulabilir; müdafaası kabil.

**maintenance** [ˈmêyntənəns]. Bakım; nafaka; iaşe; idame. **in ~ of this contention,** bu iddianın isbatı için: ~ **order,** nafaka kararı.

**maize** [mêyz]. Mısır (buğdayı).

**majestic** [məˈcestik]. Muhteşem.

**majesty** [ˈmacisti]. Haşmet; şevket.

His M~, zati şahane; haşmetlû; Kıral Hazretleri: **Your ~,** Haşmetmeab.

**major**[1] [ˈmêycə(r)] *n.* Binbaşı. **~-general,** tümgeneral. **~**[2] *a.* Daha büyük; pek büyük; reşid. **~ity** [məˈcoriti], ekseriyet; reşid olma; binbaşılık.

**make**[1] [mêyk] *n.* Cins, çeşid; marka; imal; yapı; fıtrat. **to be on the ~,** ne yapıp yapıp zengin, muvaffak, olmak için çalışmak. **~**[2] *v.* (**made** [mêyd]). Yapmak, etmek, kılmak; yaratmak, imal etm.; husule getirmek; teşkil etm.; kazanmak. **make** *fiili ingilizcede müteaddi teşkilinde de kullanılır.* **~ for** [**towards**], ···e doğru gitmek, kapağı atmak: **he is not so stupid as you ~ him,** zannettiğiniz kadar abdal değildir: **he is as dishonest** [**honest**] **as they ~ them,** son derece namussuz [namuslu] dur: **he made as if** [**though**] **to get up,** kalkacak gibi oldu: **this book made him,** onu adam eden bu kitabdır: **what made you do that?,** bunu ne diye yaptın?: **we must ~ do with it,** bununla idare etmeliyiz: **don't ~ a fool of yourself!,** kendini gülünç etme!: **he was made for this job,** tam bu işin adamı: **idleness does not ~ for wealth,** zenginliğin yolu tembellik değildir: **to ~ good,** muvaffak olm.: **to ~ stg. good,** bir şeyi telâfi etm.: **he will ~ a good doctor** [**soldier**], iyi bir dotor [asker] olur: **we shan't ~ it,** (tren vs.ye) yetişemiyeceğiz: **we'll ~ the village today,** köyü bügün çıkarırız: **show what you are made of!,** kendini göster!: **I don't know what to ~ of it** [**I can ~ nothing of it**], hiç bir şey anlamıyorum: **will you ~ one of the party?,** siz de bizimle beraber gelir misiniz?: **this book ~s pleasant reading,** bu kitab zevkle okunuyor: **these stones ~ hard walking,** bu taşlar üzerinde zahmetle yürünüyor: **that ~s ten,** bununla on oldu: **what do you ~ the time?,** sizin saatinize göre saat kaç? **~ away, to ~ away with stg.,** kaldırmak; mahvetmek; aşırmak: **to ~ away with s.o.,** birini öldürmek, yok etm.: **to ~ away with oneself,** intihar etmek. **~ off,** kaçmak, sıvışmak: **to ~ off with stg.,** alıp götürmek; bir şeyi yürütmek. **~ out,** anlamak, çözmek; sökmek: **to ~ out an account** [**list**], fatura, hesabı [liste] yapmak: **I can just ~ out stg. in the distance,** uzakta hayal meval bir şey seçiyorum: **how do you ~ that out?,** bunu nereden çıkarttınız?, buna nasıl hükmediyorsunuz?: **he made himself out to be a rich man,** kendisinin zengin olduğunu söyledi: **he is not such a villain as people ~ out,** herkesin söylediği kadar fena bir insan değildir. **~ over,** havale etm.: **he made over his farm to his son,** çiftliğini oğlunun üstüne yaptı, *fakat*: **he made over £500 a year,** senede beş yüz liradan fazla kazandı. **~ up,** uydurmak; tamamlamak; makiyaj yapmak; telâfi etm.: **to ~ it up,** barışmak: **to ~ up to s.o.,** gönlünü almak; ···e yaranmak; yüzüne gülmek: **we must ~ it up to him,** ona bunu ödemeliyiz, telâfi etmeliyiz: **to ~ up an account** [**a list**], hesabı [listeyi] yapmak, tamamlamak: **to ~ up the books,** hesabı kapatmak: **to ~ up the fire** [**stove**], ateşi [sobayı] canlandırmak: **to ~ up material into a dress,** kumaştan elbise yapmak: **to ~ up a lie** [**story**], yalan [hikâye] uydurmak: **to ~ up lost ground** [**time**], geri kalan işi (kaybedilen vakti) telâfi etm.: **to ~ up one's mind,** karar vermek: **to ~ up a prescription,** bir reçete yapmak. **~-believe,** yalancıktan. **~-do** [**-shift**], yasak savan; iğreti. **~-up,** düzgün, makiyaj; yaradılış. **~-weight** [ˈ–wêyt], vezni tamamlamak için teraziye konan şey. **~r** [ˈmêykə(r)], yapıcı; fabrikatör; Halik. **to go to one's ~,** Allahına kavuşmak.

**making** [ˈmêykin(g)] *n.* İmal, yapma. **~s,** küçük kazanclar. **this event was the ~ of him,** onu adam eden bu vakadır: **he has the ~s of a poet,** onda şairlik hamuru var: **this quarrel was none of my ~,** bu kavgayı ben çıkarmadım.

**mal-** *pref.* Fena .... **~adjustment** [ˌmaləˈcʌstmənt], intibaksızlık. **~administration** [ˌmalədminisˈtrêyşn], fena idare, idaresizlik; vazifeyi suiistimal. **~nutrition** [ˌmalnyuˈtrişn]. Gıdasızlık; fena tagaddi. **~odorous** [malˈoudərəs]. Fena kokulu. **~practice** [malˈpraktis]. Kanun veya ahlâka aykırı hareket; irtikâb. **~treat** [malˈtrît], hırpalamak; fena muamele etm.; hor kullanmak.

**maladroit** [ˈmalədrôyt]. Beceriksiz; münasebetsiz.

**malady** [ˈmalədi]. Hastalık, illet.

**malaise** [maˈlêyz]. Keyifsizlik; sıkıntı.
**malapropism** [ˈmaləpropizm]. Bir kelime veya tabirin yanlış yerde ve şekilde kullanılması.
**malaria** [məˈleərię]. Sıtma. **~l,** sıtmalı.
**Malay** [məˈlêy]. Malezyalı.
**malcontent** [ˈmalkəntent]. (Hükümetten) gayrimemnun.
**male** [mêyl]. Erkek.
**malediction** [ˌmaliˈdikşn]. Lânet.
**malefactor** [ˈmalifaktə(r)]. Cani.
**maleficent** [maˈlefisənt]. Muzır, zararlı.
**malevolen·t** [məˈlevələnt]. Kötü niyetli, kindar. **~ce,** kindarlık.
**malic·e** [ˈmalis]. Kin; kötü niyet; hiyanet; habaset. **with ~ aforethought,** kasden. **~ious** [məˈlişəs], şirret; habis; kindar.
**malign**[1] [məˈlâyn] *v.* İftira etm.; günahına girmek. **~er,** iftiracı. **~**[2] *a.* Menhus; muzır. **~ant** [məˈlignənt], muzip; şerir; habis; vahim (hastalık vs.).
**malinger** [məˈlin(g)gə(r)]. Temaruz etm., yalancıktan hastalanmak.
**mallard** [ˈmaləd]. Yaban ördeği.
**malleable** [ˈmaliębl]. Çekiçle dövülerek kırılmadan biçime konabilir; uysal.
**mallet** [ˈmalit]. Tokmak, tokaç.
**malt** [molt]. Malt. **~-house,** malt fabrikası.
**Maltese** [molˈtîz]. Maltız; maltız dili.
**Mameluke** [ˈmaməlûk]. Memlûk.
**mammal** [ˈmaml]. Memeli hayvan. **~ia** [məˈmêylyə], memeli hayvanlar sınıfı.
**mammoth** [ˈmaməθ]. Mamut. Dev gibi, kocaman.
**mammy** [ˈmami]. Anneciğim; ihtiyar zenci kadın.
**man**[1] [man] *v.* **to ~ a fort,** kaleye kuvvet koymak: **to ~ a ship,** gemiye tayfa koymak: **fully ~ned,** tam kadrolu. **~**[2], *n. pl.* **men** [man, men]. Adam; insan; erkek; er; kimse; amele, işçi; insanoğlu, beşer; (dama vs.) taş. **be a ~!,** cesur ol!: **(as) ~ and boy,** çocukluktan beri: **a ~'s ~,** erkek adam: **officers and men,** subaylar ve erler: **the ~ in the street,** alelâde kimse, orta adam: **her young ~,** (kız hakkında) erkek arkadaş (yavuklu). **~-at-arms,** (ortaçağda) asker, *esp.* zırhlı süvari askeri. **~-eater,** adam yiyen kaplan vs. **~-of-war,** harb gemisi. **~ful** [ˈmanful], merd, cesur. **~handle** [ˈmanhandl], elle (insan kuvvetile) hareket ettirmek; hırpalamak. **~hole** [ˈmanhǫul], (büyük kazan) tamircinin gireceği delik; yollarda gaz vs. tamircisinin çalıştığı çukur. **~hood** [ˈmanhud], beşeriyet; erkeklik; büluğ; yiğitlik; bir memleketin erkekleri. **~kind** [manˈkâynd]. Ademoğlu, insanlık. **~ly** [ˈmanli]. Merd, merdane, yiğitçe. **~liness,** merdlik, yigitlik, erkeklik. **~nish** [ˈmaniş], erkek gibi; erkeksi. **~power** [ˈmanpaųə(r)], el emeği; işçiler; bir memleket askerinin adedi. **~servant** [ˈmansö(r)vənt], *pl.* **menservants** [ˈmensö(r)vənts], uşak. **~slaughter** [ˈmanslôtə(r)]. Kasden olmıyarak adam öldürme.
**manacle** [ˈmanəkl]. Kelepçe (takmak).
**manag·e** [ˈmanic]. İdare etm.; kullanmak; çekip çevirmek; becermek; muvaffak olm.; geçinmek. **we'll ~ it somehow,** elbette bir yolunu buluruz: **how on earth did you ~ to break that vase?,** nasıl yaptın da o vazoyu kırdın?: **he tried to mount his horse but couldn't ~ it,** ata binmeğe çalıştı, beceremedi: **he can't ~ this horse at all,** bu atı hiç zaptedemez: **I ~d to escape,** bir kolayını bulup kaçtım: **'what day shall I come?' 'Well, can you ~ Saturday?',** 'ne gün geleyim?' 'Cumartesi nasıl? [cumartesi gelebilir misiniz]?' **I can't ~ more than £100,** yüz liradan fazla sarfedemem: **I can't ~ all that meat,** bu etin hepsini yiyemem: **if we go away, how will we ~ about the dog?,** gidersek köpeği ne yaparız? **~eable** [ˈmanicibl], idare edilebilir; kullanışlı. **~ement** [ˈmanicmənt], idare; müdürlük. **~er** [ˈmanicə(r)], müdür; idareci. **~eress,** müdire. **~erial** [ˌmaniˈciəriəl], idareye, müdüre aid. **~ing** [ˈmanicin(g)], idareci; becerikli; işgüzar ve mütehakkim.
**mandarin**[1] [ˈmandərin]. Mandaren. **~**[2]**(e),** mandalina.
**mandat·e** *n.* [ˈmandêyt]. Emir; vekillik; manda; vesayet; Papa iradesi; iktidardaki partiyi seçen müntehiblerin verdiği talimat. *v.* [manˈdêyt], manda altına koymak. **~ory** [ˈmandətəri], mandaya aid; vekil. **~ state,** manda sahibi devlet.

**mandible** [ˈmandibl]. Alt çene; kuş gagalarının parçalarından her biri; böcek ağzının çıkıntılı kısmı.
**mane** [mêyn]. Yele.
**mange** [mêync]. Uyuz.
**manger** [ˈmêyncə(r)]. Yemlik. ⌜**a dog in the ~**⌝, kendi kullanmadığı bir şeyden başkasının istifadesini istemiyen kimse.
**mangle**[1] [ˈman(g)gl]. Çamaşır mengenesi(nden geçirmek). **~**[2]. Parçalamak, yırtmak, delik deşik etmek. **to ~ a language**, bir lisanı ezmek: **to ~ a quotation**, yarım yamalak iktibas etmek.
**mango** [ˈman(g)gou]. Hind kirazı.
**mangy** [ˈmêynci]. Uyuz; (*sl.*) pintice.
**mania** [ˈmêynyə]. Cinnet; mani. **~c** [ˈmêynyak], tehlikeli deli; manyak.
**manicure** [ˈmanikyuə(r)]. Manikür (yapmak).
**manifest**[1] [ˈmanifest]. Zahir, belli, aşikâr. Açıkça göstermek; izhar etmek. **to ~ itself**, tecelli etm.; belli olmak. **~ation** [–ˈtêyşn], tecelli; izhar; tezahürat. **~**[2]. Manifesto. **~o** [maniˈfestou], beyanname.
**manifold** [ˈmanifould]. Türlü türlü, katmerli; çok. Müstensihle yapılan yazı (bir çok suretini çıkarmak).
**manikin** [ˈmanikin]. Ufacık adam; kukla.
**manil(l)a** [məˈnilə]. Manila kendirinden yapılan ip.
**manipulate** [məˈnipyulêyt]. El ile işlemek; idare etm.; suiistimal etmek.
**manna** [ˈmana]. Kudret helvası; balsıra.
**mannequin** [ˈmanikin]. Manken.
**manner** [ˈmanə(r)]. Tarz, tavır, usul, yol; âdet. **~s**, terbiye; muaşeret; âdet: **good ~s**, muaşeret adabı: **bad ~s**, görgüsüzlük, terbiyesizlik. **all ~ of people [things]**, her türlü, her cins, halk [eşya]: **as (if) to the ~ born**, sanki böyle [bu iş için] doğmuş: **to forget one's ~s**, terbiyesini bozmak; kendini unutmak: **in a ~ of speaking**, tabir caizse, söz gelişi: **in like ~**, aynı tarzda: **no ~ of doubt**, hiç şübhe yok: **to teach s.o. ~s**, birine terbiye dersi vermek: **what ~ of man is he?**, nasıl bir adam?. **~ed** [ˈmanö(r)d], sahte, müfrit: **-~**, (filan tarzda) hareket eden: **bad-~**, terbiyesiz: **well-~**, terbiyeli: **coarse ~**, kaba. **~ism** [ˈmanərizm], tasannu; (bir muharrire aid) hususiyet. **~less**, görgüsüz, terbiyesiz. **~ly**, terbiyeli.
**manœuvre** [məˈnûvə(r)]. Manevra; hile; tertibat. Manevra yapmak; (gemi) kullanmak. **~s**, manevra, askerî tatbikat.
**manor** [ˈmanə(r)]. Malikâne; tımar. **lord of the ~**, malikane sahibi. **~-house**, malikâne sahibinin köşkü. **~ial** [məˈnôriəl], 'manor'a aid.
**mansion** [ˈmanşən]. Kâşane, büyük konak. **The M~ House**, Londra belediye reisinin dairesi.
**mantelpiece, -shelf** [ˈmantlpîs, –şelf]. Ocak rafı.
**mantilla** [manˈtila]. İspanyol kadınlarının kullandığı başörtüsü.
**mantle** [ˈmantəl]. Harmani; manto; lâmba gömleği. Harmani ile örtmek; yayılıp renk vermek.
**manual**[1] [ˈmanyuəl]. El ile yapılan; elişi. **~ labour**, el emeği, el işi. **~**[2]. Risale; dua kitabı; orgun el ile çalınan tuşları.
**manufact·ory** [ˌmanyuˈfaktəri]. Fabrika. **~ure** [–ˈfaktyə(r)], imal; mamul şey; yapmak, imal etm.; uydurmak: **~s**, (sinaî) mamulât. **~urer**, fabrikacı, fabrikatör, imalâtçı.
**manur·e** [məˈnyuə(r)]. Gübre(lemek).
**manuscript** [ˈmanyuskript]. Yazma; el yazması; müsvedde.
**Manx** [manks]. **Man** adasına aid. **~ cat**, kuyruksuz kedi.
**many** [ˈmeni]. Çok, bir çok; o kadar; müteaddid; türlü, muhtelif. **as ~ again** [**twice as ~**], bir bu kadar daha: **there were as ~ as a hundred people there**, orada yüz kişi kadar vardı: **as ~ as you like**, ne kadar isterseniz: **a good ~**, oldukça, bir çok: **a great ~**, pek çok, bir hayli: **how ~?**, kaç tane?: **~ of us**, çoğumuz, içimizden çoğu: **so ~**, o kadar çok; şu kadar: **I told him in so ~ words**, (açıkça söylemeden) münasib şekilde anlattım: **three too ~**, üç tane fazla. **~-sided**, çok taraflı.
**map** [map]. Harita(sını yapmak). **to draw a ~**, harita yapmak.
**maple** [ˈmêypl]. (*Amer.*) Akça ağac; isfendan çınarı.
**mar** [mâ(r)]. Bozmak; ihlal etmek. **to make or ~**, ya (iyi bir şey) yapmak ya bozmak.
**maraud** [məˈrôd]. Plâçkaya çıkmak; yağma etmek. **~er**, plâçkacı.
**marbl·e** [ˈmâbl]. Mermer; bilye. Mermerden yapılmış. **~ed**, mermer döşeli; ebru.

**March**[1] [mâç]. Mart. **~**[2]. Hudud, serhad. **to ~ with,** ile hemhudud olmak. **~**[3]. (Asker hakkında) yürümek. Askerî yürüyüş; marş; terakki, ilerleme. **~ past,** geçid resmi: **quick ~!,** ileri arş!: **to give s.o. ~ing orders,** birine yol vermek.

**marchioness** [ˈmâşənes]. Markiz.

**mare** [meə(r)]. Kısrak. ⌜**the grey ~ is the better horse**⌝, karısı kendisine üstündür. **~'s-nest,** asılsız bir haber, hulya.

**margarine** [mâcəˈrîn, ˈmâgərîn]. Margarin.

**margin** [ˈmâcin]. Kenar; zırh; pay; mesafe; ara; tolerans. **to allow s.o. some ~,** bir dereceye kadar hareket serbestisi vermek: **to allow a ~ for mistakes,** hatayı hesaba katmak: **to allow a ~ for safety,** ihtiyat payı bırakmak: **by a narrow ~,** daradar, az bir farkla. **~al,** sahifenin kenarında bulunan; deniz kıyılarında bulunan. **~ note,** haşiye.

**marguerite** [mâgəˈrît]. Papatya.

**Maria** [məˈrâya]. **Black ~,** hapishane arabası.

**marine** [məˈrîn]. Denize aid, bahrî. Deniz silahendazı. **merchant ~,** ticaret filosu: ⌜**tell it to the ~s!**⌝, külahıma dinlet! **~r** [ˈmarinə(r)], gemici: **master ~,** ticaret gemisinde ehliyetnameli kaptan.

**marionette** [ˌmariəˈnet]. Kukla.

**marital** [ˈmaritl]. Kocalığa veya evlilik hayatına aid.

**maritime** [ˈmaritâym]. Denizciliğe aid; denize aid, denize yakın.

**mark**[1] [mâk] *n.* Alâmet, işaret; çizgi, çetele; marka, damga; nişan, hedef; numara; mark. **as a ~ of (my) esteem,** takdir nişanesi olarak: **below the ~ [not up to the ~],** (i) keyifsiz; (ii) tam ehil değil; (iii) istenilen kalitede değil, kifayetsiz: **a man of ~,** ehemmiyetli bir adam: **to make one's ~,** (i) temayüz etm.; (ii) (yazma bilmiyenler hakkında) imza yerine işaret koymak: **to be (a bit) off the ~,** tahminde biraz yanılmak; hedefi tutmamak: **to get off the ~ quickly,** (bir yarışta) derhal hareket etm.; (bir işe) derhal girişmek: **'save the ~!',** tövbeler olsun!; sözüm ona: **to be wide of the ~,** hedefi tutmamak; yanlış tahmin etmek: **to toe the ~,** hizaya girmek; herkese uymak; usule riayet etm.; yola gelmek. **~**[2] *v.* Çizmek; işaret koymak; nişan yapmak; berelemek; marka yapmak; dikkat etm.; göstermek; numara vermek. **~ my words!,** sözüme mim koy!; duvara yazıyorum: **to ~ time,** yerinde saymak. **~ed** [mâkd] *a.*, damgalı; işaretli; mimli; göze çarpan: **~ card,** işaretli kâğıd: **a ~ man,** mimli bir adam. **~er** [ˈmâkə(r)], işaret eden kimse. **~ing** [ˈmâkin(g)] *n. gen.* **~ings,** benekler, işaretler. **~ing-ink,** (çamasır için) sabit mürekkeb. **~sman,** nişancı; atıcı.

**market** [ˈmâkit]. Çarsı; pazar; hal. Pazarda satmak. **to be in the ~ for stg.,** bir şeyi satın almak istemek: **to be on [come into] the ~,** satışa çıkmak: **to go ~ing,** pazara gitmek: **to find a ready ~,** revac görmek: **the ~ has risen,** fiatlar yükseldi. **~able,** satılabilir. **~-garden,** bostan. **~-house,** hal. **~-price,** satış fiatı, pazar fiatı.

**marline** [ˈmâlin]. İki kollu ince halat.

**marmalade** [ˈmâməlêyd]. Portakal reçeli.

**maroon**[1] [məˈrûn]. Vişne çürüğü rengi. **~**[2]. Karaya çıkarıp ıssız bir adada bırakmak. **to be ~ed,** haricle münasebeti kesilmek.

**marplot** [ˈmâplot]. Bir tedbir fuzulî bir müdahale ile bozan kimse.

**marque** [mâk]. **letters of ~,** (vaktile) verilen korsanlık fermanı.

**marquee** [mâˈkî]. Büyük çadır.

**marquis, -quess** [ˈmâkwis]. Marki.

**marriage** [ˈmaric]. Evlenme, izdivac. **to give s.o. in ~,** kocaya vermek: **to seek s.o. [s.o.'s hand] in ~,** bir kıza talib olm.: **~ lines,** evlenme kâğıdı: **relative by ~,** sıhrî: **~ settlement,** bir kız evlenirken babası tarafından yatırılan meblağ. **~able,** evlenecek çağda, gelinlik.

**married** [ˈmarid]. Evlenmiş: **~ man,** ev bark sahibi: **a ~ couple,** karıkoca: **to get ~,** evlenmek.

**marrow** [ˈmarou]. İlik; öz; sakız kabağı. **~bone,** ilikli kemik.

**marry** [ˈmari]. Evlen(dir)mek; nikâhla vermek; kocaya vermek. **to ~ beneath one,** küfüv olmıyanla evlenmek: **to ~ into a family,** evlenme yolu ile bir aileye girmek: **to ~ money,** zengin bir kimse ile evlenmek.

**Mars** [mâz]. Merih yıldızı; harb ilâhı.

**marsh** [mâş]. Bataklık. **~ gas,** durgun sudan intişar eden karbonlu hidrojen. **~land,** bataklık yer. **~y,** bataklık, sulak.

**marshal**[1] [ˈmâşəl] *n.* Mareşal, müşir; teşrifat memuru. ~[2] *v.* Dizmek, sıralamak. **to ~ facts,** vakaları toplayıp mantıkî bir tertibe koymak. **~ling, ~ yard,** trenlerin tasnif ve tertib edildiği istasyon.
**marsupial** [mâˈsyûpiəl]. Keseli (hayvan).
**mart** [mât]. Çarşı; ticaret merkezi.
**martial** [ˈmâşl]. Harbe aid, harbî, cengâver; harbe elverişli. **court ~,** divanı harb, askerî mahkeme: **~ law,** örfî idare.
**Martian** [ˈmâşən]. Merih yıldızına aid.
**martin** [ˈmâtin]. **house ~,** pencere kırlangıcı.
**martinet** [mâtiˈnet]. Müfrit disiplinci.
**martyr** [ˈmâtə(r)]. Şehid; din uğrunda ölen adam; fikir kurbanı; mağdur. Şehid etmek. **a ~ to rheumatism,** romatizma kurbanı: **to make a ~ of oneself,** şöhret kazanmak için fedakârlık eder görünmek. **~dom,** şehidlik; büyük ıstırab.
**marvel** [ˈmâvəl]. Mucize; acibe. Şaşmak, taaccüb etmek. **it's a ~ to me that** ..., ... beni hayrette bırakıyor: **to work ~s,** mucize gibi tesir etmek. **~lous,** acayib, fevkalâde; şaşılacak.
**marzipan** [ˌmâziˈpan]. Badem ezmesi.
**mascot** [ˈmaskot]. Uğur getiren adam, şey; tılsım.
**masculine** [ˈmaskyulin]. Erkeğe aid, erkeğe benziyen; müzekker.
**mash** [maş]. Ezilmiş ve sulu bir madde; lâpa; ezme. Lâpa haline koymak; ezmek.
**masher** [ˈmaşə(r.]. Kadın avcılığı taslıyan züppe.
**mask** [mâsk]. Maske; nikab; alçıdan yüz kalıbı. Maskelemek; örtmek; örtbas etm., gizlemek. **~ed ball,** maskeli balo: **to drop [throw off] the ~,** maskeyi yüzünden indirmek: **under the ~ of,** ··· perdesi altında.
**mason** [ˈmêysn]. Taşçı; duvarcı; farmason. **~ic** [maˈsonik], masonluğa aid. **~ry** [ˈmêysnri], duvarcılık; duvarcı, taşçı, işi; masonluk.
**masque** [mask]. Amatör bir temsil.
**masquerade** [ˌmaskəˈrêyd]. Maskeli balo; kıyafet tebdili; taslama, gibi görünme. **to ~ as** ..., taslamak, ...gibi görünmek, geçinmek.
**mass**[1] [mas]. Katolik kilise âyini; bunun için yazılan musiki. ~[2] *n.* Kütle; hacım; yığın, küme; mecmu: **the ~es,** avam takımı: **a ~ of people,** büyük kalabalık: **people in the ~,** umumiyetle halk: **the great ~ of the people,** halkın ekseriyeti: **~ meeting,** kalabalık miting: **~ production,** seri halinde imal: **~ executions,** toptan idamlar: **~ rising,** bütün memleketin ayaklanması. ~[3] *v.* Yığmak; toplamak; cemetmek; bir araya getirmek. Kütle halinde toplanmak. **~ive** [ˈmasiv], ağır ve kalın; kocaman; som; lenduha; kütlevî.
**massacre** [ˈmasəkə(r)]. Katliam (etmek). Kılıcdan geçirmek.
**mass·age** [maˈsâj]. Uğma, uğuşturma(k), masaj (yapmak). **~eur** [maˈsö(r)], tellak; masajcı. **~euse,** kadın masajcı.
**mast**[1] [mâst]. Gemi direği. **to sail before the ~,** tayfa olarak hizmet etmek. **~ed,** direkli. **~**[2]. Meşe ve kayın ağaclarının palamudu.
**master**[1] [ˈmâstə(r)] *n.* Âmir; sahib; reis; efendi; usta, üstad; muallim, hoca; tüccar gemisi kaptanı; üst gelen. **Master of Arts** (**M.A.**), içtimaî ilimlerden mezun (İngiliz üniversitelerinde mezuniyet diploması ile Doktora arasında bir derece): **~ of Ceremonies,** teşrifat memuru: **~ of the Horse,** Mirahor: **~ of Hounds,** bir tilki (geyik) avını idare eden adam: **to be one's own ~,** müstakil olm.: **to be ~ in one's own house,** kendi evinin efendisi olm.: **you are not ~ of yourself,** iradeniz elinizde değil: **the young ~,** küçük bey. **~-at-arms,** harb gemisinde inzibat cavuşu. **~-key,** ana anahtar. **~-stroke,** üstadca bir tedbir. ~[2] *v.* Zabtetmek; âmirce idare etm.; itaat ettirmek; hâkim olmak. **to ~ a difficulty,** bir güçlüğü yenmek: **to ~ a subject,** bir mevzua hâkim olmak. **~ful** [ˈmâstəful], mütehakkim; inadcı; iradesi kuvvetli. **~ly,** üstadca. **~piece,** şaheser. **~y** hâkimiyet; galebe; üstünlük; üstadlık.
**mastic** [ˈmastik]. Sakız. **~ate** [ˈmastikêyt], çiğnemek.
**mastiff** [ˈmâstif]. Samsun.
**mastoid** [ˈmastôyd]. Halemî.
**mat**[1] [mat]. Hasır; paspas; keçe; nihale; karmakarışık yığın. Hasır döşemek, örmek; saç ve emsalini birbirine yapıştırıp top etmek. **col-**

lision ~, usturmaca. ~². Donuk; mat. Donuk hale koymak.

**match¹** [maç]. Kibrit. **to strike a ~**, kibrit çakmak. ~². Maç; oyun. ~³. Misal, misil, eş. **to be a ~ for**, ···e denk olm., eş olm.: **to make a good ~**, fevkalâde bir eş bulmak: **a good ~ of colours**, birbirini tutan renkler: **to meet one's ~**, dengine raslamak. ~⁴ *v*. Birbirine uymak; mütenasib olmak. Birbirine uydurmak; eşini bulmak. **to ~ s.o. against another**, boy ölçüştürmek: **to be well ~ed**, uyuşmak; birbirinin dengi olmak. **~board(ing)** [ˈmaçbôd(ing)], birbirine geçme tahta. **~less** [ˈmaçlis], misli yok, emsalsiz. **~lock** [ˈmaçlok], çakmaklı tüfek. **~maker** [ˈmaçmêykə(r)], çöpçatan. **~wood** [ˈmaçwud], **made of ~**, çörden çöpten: **to burn like ~**, çıra gibi yanmak: **smashed to ~**, parça parça edilmiş.

**mate¹** [mêyt] *n*. Eş; arkadaş; iş ortaği; yamak; tüccar gemilerinde ikinci kaptan. ~² *v*. Çiftleş(tir)mek; evlen(dir)mek; eş olm.; eşini bulmak. ~³. Satrancda mat (etmek).

**mater** [ˈmêytə(r)]. Anne. **Alma Mater** [ˈmâtə(r)], bir kimsenin okuduğu mekteb veya üniversite.

**material** [məˈti̯əri̯əl]. Madde; kumaş; bir kitab için lâzım gelen malzeme. Maddî; elzem. **~s**, levâzım, malzeme: **raw ~**, ham madde. **~ism**, maddicilik. **~ist**, maddici. **~istic** [–ˈlistik], maddî, maddicilik mesleğine aid. **~ize**, maddileştirmek; gerçekleşmek; tahakkuk etm.

**matern·al** [məˈtö(r)nl]. Anaya ve analığa aid. **~ uncle**, dayı. **~ity**, analık: **~ home**, doğumevi.

**mathematic·s** [ˌmaθəˈmatiks]. Riyaziye. **applied ~**, tatbikî riyaziye: **pure ~**, nazarî riyaziye. **~al**, riyazî. **~ian** [–ˈtişn], riyaziyeci.

**matins** [ˈmatinz]. Sabah ibadeti.

**matriarch** [ˈmêytriâk]. İbtidai bir kabilede hem ana hem hâkime sayılan kimse. **~al** [–ˈâkl], ana hâkimiyetine aid.

**matric** [məˈtrik]. (*abb.*) **matriculation.**

**matricide** [ˈmatrisâyd]. Ana kaatil(liğ)i.

**matriculat·e** [məˈtrikyulêyt]. Üniversiteye girmek için lâzım gelen imtihanı vermek. **~ion** [–ˈlêyşn], üniversiteye kaydolunma ve bunun için lazım gelen imtihan.

**matrimon·y** [ˈmatriməni]. Evlilik, evlenme. **~ial** [–ˈmo̯unyəl], evliliğe aid.

**matrix** [ˈmêytriks]. Dölyatağı; harf kalıbı; kalıb dişi; kıymetli taş parçasını ihtiva eden kaya.

**matron** [ˈmêytrən]. Yaşlı ve muhterem evli kadın; ana kadın; hatun; bir hastahane vs.nin âmiri olan kadın; bir mektebde çocukların sıhhatine bakan kadın. **~ly**, ana kadına yakışan.

**matter¹** [ˈmatə(r)] *n*. Madde; cevher; mesele, dâva, iş; ehemmiyet; eser mevzuu; cerahat. **that's (quite) another ~**, o başka bir mesele [bahis]: **as a ~ of course**, tabiî olarak; hiç düşünmeden: **[a] ~ of fact**, vakıa, hakikat: **as a ~ of fact**, zaten; doğrusu: **for that ~**, ona gelince: **it is a ~ for rejoicing that ...**, sevinmeye değer bir meseledir ki: **in the ~ of ...**, ···in hususunda: **it makes no ~**, zarar yok; fark yok; ehemmiyeti yok: **no great ~**, bir şey değil: **it will be a ~ of two months**, bu iki aylık bir meseledir: **to settle ~s**, meseleyi halletmek, kapatmak: **it's a ~ of taste**, bu bir zevk meselesidir: **what's the ~?**, ne var?, ne oldu?: **what's the ~ with you?**, neniz var?, size ne oldu?: **well, what ~!**, ne çıkar?: **is there anything the ~ with you?**, size bir şey mi oldu?. **~-of-fact**, maddî, hissiz, kuru, pratik. ~² *v*. Ehemmiyeti olmak. **it does not ~**, ehemmiyeti yok: **it ~s a good deal to me**, benim için ehemmiyeti var: **it doesn't ~ to me whether he comes or not**, gelse de gelmese de bence müsavi.

**matting** [ˈmatin(g)]. Hasır örgüsü; keçe. **coco-nut ~**, koko yol keçesi.

**mattress** [ˈmatris]. Minder.

**matur·e** [məˈtyûə(r)]. Olgun; kemale ermiş; yaşını başını almış; geçkin; vadesi gelmiş (bono). Olgunlaştırmak; pişirmek; kemale ermek; ödenme zamanı gelmek. **after ~ consideration**, düşünüp taşındıktan sonra. **~ity** [məˈtyûriti], olgunluk, kemal; tediye vadesi.

**maudlin** [ˈmôdlin]. Sarhoşluktan cıvık ve ağlamalı bir halde olan.

**maul** [môl]. Dövüp berelemek; hırpalamak.

**maunder** [ˈmôndə(r)]. İnsicamsız konuşmak; dalgın dalgın gezinmek.

**Maundy** [ˈmôndi]. **~ Thursday**, paskalyadan önceki perşembe günü.

**mausoleum** [môsou'liəm]. Türbe.
**mauve** [mouv]. Leylâk rengi; leylâkî.
**maw** [mô]. Karın; hayvan ağzı.
**mawkish** ['môkiş]. Tiksindirici; yavan.
**maxim** ['maksim]. Vecize; darbımesel; düstur; şiar.
**maximum** ['maksiməm]. En yüksek derece, azamî had; gaye; narh. Azamî, en büyük.
**May** [mêy] *n.* Mayıs; akdiken. ~ **Day**, mayısın birinci günü. ~**pole** ['mêypoul], mayısın birinde bayram yapmak için dikilen direk.
**may** (**might**) [mêy, mâyt] *v.* (*Yardımcı fiil. bk. dahi* **might**). *Bu fiil birleştiği fiile ihtimal veya müsaade manalarını ilâve eder.* ~ **I come in?**, girebilir miyim?: ~ **they be happy!**, mes'ud olsunlar: **be that as it** ~, her ne olursa olsun: **it** ~ **be that ...**, olabilir ki: **he** ~ **come tonight**, (i) bu akşam belki gelir; (ii) bu akşam gelebilir; bu akşam gelmesine müsaade var: **it** ~ **rain**, yağmur yağabilir; yağmur yağması muhtemeldir: **we** ~ **as well stay where we are**, bulunduğumuz yerde kalsak daha iyi: **work as he** ~, nekadar çalışırsa çalışsın: **if I** ~ **say so**, kusura bakmayın amma ...: **I hope we** ~ **meet again**, ümid ederim ki yine görüşürüz: **who** ~ **you be?**, siz kimsiniz?, siz kim oluyorsunuz? ~**be** ['mêybî], belki; ihtimal ki.
**mayor** [meə(r)]. Belediye reisi. ~**al**, belediye reisine aid. ~**alty**, belediye reisliği. ~**ess**, belediye reisinin karısı; kadın belediye reisi.
**maze** [mêyz]. Lâbirent. Sersemletmek. **to be in a** ~, ne yapacağını bilmemek.
**M.D.** ['em'dî]. **Doctor of Medicine**, doktor, hekim.
**Mdlle.** (*abb.*) **Mademoiselle.**
**me** [mî]. Beni; (*coll.*) ben. **to** ~, bana: **from** ~, benden: **ah** ~!, eyvah!: **dear** ~!, yok canım!; ne yazık!
**meadow** ['medou]. Çayır, otlak.
**meagre** ['mîgə(r)]. Zayıf; kıt; yavan.
**meal**[1] [mîl]. Yulaf arpa veya mısır unu. ~**y** ['mîli], unlu, un gibi. ~**y-mouthed**, yapmacıktan tatlı dilli; yaltak. ~[2]. Yemek. **at** ~**s**, yemeklerde: **to make a** ~ **of**, yiyip bitirmek; yemek yerine yemek.
**mean**[1] [mîn] *n. & a.* Orta; vasat(î); *pl.* ~**s**, *v.* **means**. **the golden [happy]** ~, ne ifrat ne tefrit; ikisi ortası. ~[2] *a.* Cimri; alçak, aşağı. **to take a** ~ **advantage of s.o.**, bir fırsatı âdice kullanarak birisine üstün gelmek: **the** ~**est Frenchman expects good cooking**, en aşağı bir Fransız bile yemeğin iyi pişirilmesini ister: **he has no** ~ **opinion of himself**, kendini epeyi beğenmiştir: **he is no** ~ **scholar**, o mühim bir âlimdir: **to think** ~**ly of s.o.**, (i) birini pek gözü tutmamak; (ii) küçümsemek. ~**ness** ['mînnis], cimrilik; aşağılık, çingenelik. ~[3] (**meant**) [ment] *v.* Demek istemek, kasdetmek; manası olm.; muradetmek; kararlaştırmak; ifade etmek. **I didn't** ~ **to be rude**, bu nezaketsizliği kasden yapmadım: **he** ~**s well**, (···e rağmen) hüsnüniyet sahibidir: **he** ~**s no harm**, fenalık kasdetmiyor: **I** ~ **to write a book**, bir kitab yazmak niyetindeyim: **I** ~ **what I say**, bu hususta ciddiyim: **I** ~ **to be obeyed**, bana itaat edilmesini isterim yoksa ...: **that remark was** ~**t for you**, bu sözü sizi kasdederek söyledim: **I** ~**t this necklace for you**, bu gerdanlığı size vermeği düşünüyordum: **the name** ~**s nothing to me**, bu ismi hiç hatırlamıyorum: **this portrait is** ~**t to be me**, bu gûya [sözde] benim resmim: **what do you** ~ **by behaving like that?**, bu hareketinizle ne demek istiyorsunuz?; ne cesaretle böyle hareket ediyorsunuz?
**meander** [mi'andə(r)]. Yılankavı olm., sağda solda dolaşmak.
**meaning** ['mînin(g)] *n.* Mana; meal; kasıd. Manalı. **what's the** ~ **of this?**, (i) bunun manası nedir? (ii) bu ne demek?, bu ne!, bu nasıl şey!: **well-**~, iyi kalbli; (aslında) iyi niyetli. ~**less**, manasız, abes.
**means** [mînz]. (*um. müfred olarak kullanılır*); Vasıta; vesile; yol, suret; imkânlar; servet, para. **a** ~ **to an end**, gaye için vasıta: **by all** ~**s**, (i) hayhay, elbette; (ii) ne yapıp yapıp: **by all** ~**s let him learn Turkish**, varsın türkçe öğrensin: **by any** ~ **you can**, her ne suretle olursa olsun: **he is not by any** ~ **a rich man**, hiç te zengin bir adam değildir: **it is beyond my** ~, benim harcım değil; bu benim için imkânsız: **to live beyond one's** ~, gelirinden fazla sarfetmek: **a man of** ~, varlıklı bir adam: **by no (manner of)** ~, hiç bir suretle; katiyen: **there is no** ~ **of doing it**, bunu yapmağa imkân yok: **private** ~, bir şahsın

kendi geliri (bir vazifeden aldığı ücret haricinde): **by some ~ or other,** her hangi bir şekilde: **by ~ of ...,** ... vasıtasile: **he has been the ~ of ...,** onun vasıtasile ...: **without ~,** geliri olmıyan; fakir: **ways and ~,** türlü türlü vasıta *esp.* malî vasıta: **Committee of Ways and M~,** bütçe encümeni.

**meant** *v.* **mean**³.

**meantime** [mîn'tâym], **~while** [–'wâyl]. Bu arada. **in the ~,** bu müddet zarfında; bununla beraber.

**measles** ['mîzlz]. Kızamık. **German ~,** kızamıkçık.

**measly** ['mîzli]. (*sl.*) Değersiz, sefil.

**measur·e** ['mejə(r)]. Ölçü; mikyas; ölçme; had; vezin; nizam; tedbir. Ölçmek; mesaha etm.; tartmak; ölçüsü ... kadar olmak. **beyond [out of all] ~,** çok fazla, sonsuz: **to ~ one's length (upon the ground),** boylu boyuna yere serilmek: **made to ~,** ısmarlama: **to ~ out,** ölçerek tayin etm., dağıtmak: **in some ~,** bir dereceye kadar: **to take ~s,** tedbir almak. **~able,** ölçülebilir; yakın. **~ed,** ölçülü. **~ing, ~ chain,** mesaha zinciri: **~-glass,** dereceli bardak. **~ment,** ölçme, ölçü.

**meat** [mît]. Et; lüb; esas. **⌜as full of ~ as an egg⌝,** özlü, esaslı: **music is ~ and drink to him,** musiki onun için gıda gibidir: **that book is rather strong ~ for the young,** bu kitab çocuklar için çok ağırdır. **~y,** etli; özlü. **~-safe,** tel dolab.

**mechan·ic** [me'kanik], **–ician** [–kə'nişn]. Makineci; makine işçisi; **~s,** mihanik. **~ical,** mihanikî: makineye aid, makine ile işlenen. **~ism** ['mekənizm], makine tertibatı, mekanizma. **~ize** ['mekənâyz], makineleştirmek.

**medal** ['medəl]. Madalya. **~lion** [mə'dalyən], madalyon. **~list,** madalya kazanan.

**meddle** ['medl]. Karışmak; lüzumsuz yere müdahale etmek. **to ~ in,** ···e burnunu sokmak: **to ~ with,** ···e karışmak, dokunmak, kâhyalık etmek. **~r,** müdahaleci. **~some** ['medlsəm], her şeye burnunu sokan.

**Medes** [mîdz]. Medler. **⌜a law of the ~ and Persians⌝,** asla değişmez âdet, anane.

**medi(a)eval** [medi'îvəl]. Orta çağlara aid.

**mediat·e** ['mîdiêyt]. Tavassut etm., araya girmek. **~ion** [–'êyşn], tavassut: **~or,** mütevassıt; arabulan.

**medical** ['medikəl]. Tıbbî; hekimliğe aid. **~ board,** heyeti sıhhiye: **~ man,** doktor: **~ Officer of Health,** sıhhat memuru.

**medicament** ['medikəment]. İlâc.

**medicin·e** ['medisn, 'medsin]. İlâc; hekimlik, tıb. **to give s.o. a dose of his own ~** birine mukabelebilmisil yapmak **~al** [–'disinəl], ilâc gibi kullanılabilir; şifalı.

**mediocr·e** ['mîdi'oukə(r)]. Orta, vasat; aşağı; pek parlak değil. **~ity** [–'okriti], orta, vasat olma; alelâde insan.

**meditat·e** ['meditêyt]. Düşünceye dalmak, düşünmek; zihninde bir şey tertib etm., niyet etm. **~ion** [–'têyşn], düşünme, düşünüp taşınma; dalgınlık; murakebe. **~ive** ['meditətiv], dalgın; düşünceli.

**Mediterranean** [ˌmeditə'rêynyən]. Akdeniz(e aid).

**medium** ['mîdyəm]. Orta. Orta. vasat; vasıta, çare, yol; delâlet; medyom. **happy ~,** tam karar, ne ifrat ne tefrit.

**meed** [mîd]. Mükâfat. **one's ~ of praise,** müstahak olduğu medih.

**meek** [mîk]. Alçak gönüllü; halim; uysal. **~ness,** alçak gönüllülük.

**meerschaum** ['miəşəm]. Lületaşı.

**meet**¹ [mît] *a.* Münasib; lâyik; yakışır. **~**² (**met** [met]) *v.* Rasgelmek; yüzyüze gelmek; bir araya gelmek; buluşmak; görüşmek; karşılamak. **to ~ s.o.,** (i) birine rasgelmek; (ii) uyuşmağa hazır olm.: **to ~ the case,** vaziyete, hal ve şartlara uymak: **to ~ one's death,** bir kaza ile ölmek: **to make both ends ~,** geçinebilmek: **to ~ one's eyes,** göze ilişmek: **there is more in it than ~s the eye,** pek göründüğü gibi değil: **to ~ with an accident,** *etc.*, kaza vs.ye uğramak: **pleased to ~ you!,** (*Amer.*) müşerref oldum: **to ~ a train,** bir treni karşılamak. **~ing** ['mîtin(g)] *n.*, ictima, toplantı; miting; (nehirler) kavuşma.

**mega-** *pref.* Büyük ..., ... büyüten. **~lomania** [ˌmegəlou'mêynyə], azamet hastalığı. **~phone** ['megəfoun], megafon.

**melanchol·y** ['melən(g)kəli]. Malihulya; karasevda; melâl Gamlı, meraklı; hüzün verici. **~ia** [–'koulyə], malihulya. **~ic** [–'kolik], karasevdalı; mahzun.

**mêlée** [ˈmelêy]. Kördövüşü, arbede.
**mellifluous** [məˈliflu̯əs]. Bal gibi, tatlı.
**mellow** [ˈmelo̯u]. Olgun; (ses, renk) tatlı; (şarab) yumuşak; (*coll.*) çakırkeyif. **to** ~ *veya* **to grow** ~, olgunlaşmak; yaş ve tecrübe ile müsamahakâr ve temkinli olmak.
**melod·y** [ˈmelədi]. Ezgi, nağme; melodi. **~ious** [məˈlo̯udyəs], ahenkli.
**melon** [ˈmelən]. Kavun.
**melt** [melt]. Eri(t)mek; yumuşamak. ~ **away,** eriyip kaybolmak: ~ **down,** eritip birleştirmek: **to** ~ **into tears,** gözlerinden yaş boşanmak. **~ing-pot,** pota: **to be in the** ~, inhilâl etm.; baştan başa değiştirilmek.
**member** [ˈmembə(r)]. Uzuv; âza. **M~ of Parliament** (**M.P.**), meb'us, milletvekili. **~ship,** âzalık; âzaların adedi.
**membrane** [ˈmembrêyn]. Zar; gışa.
**memento** [məˈmento̯u]. Yadigâr; hatıra.
**memo** [ˈmîmo̯u] *v.* **memorandum.**
**memoir** [ˈmemwâ(r)]. İlmî muhtıra; hatıra. **~s,** hatırat.
**memor·able** [ˈmemərəbl]. Hatırlamağa değer; unutulmaz; mühim. **~andum,** *pl.* **–da** [ˌmeməˈrandəm, –da], muhtıra, nota: **to make a** ~ **of stg.,** not almak: **~-book,** muhtıra defteri, karne. **~ial** [məˈmôri̯əl], abide; yadigâr; muhtıra, arzuhal. Hatırlatıcı (abide vs.). **~ize** [ˈmemərâyz], ezberlemek. **~y** [ˈeməri], hafıza; hatıra. **to the best of my** ~, hatırladığıma göre: **of blessed** ~, rahmetli: **in** ~ **of** ..., ···in hatırasına: **within living** ~, hatırlıyanlar vardır.
**mend** [mend]. Tamir etm., iyi hale koymak; ıslah etm.; yamamak. İyileşmek. **to** ~ **one's ways,** halini ıslah etmek: ⌜**least said, soonest** ~**ed**⌝, fazla kurcalama!
**mendaci·ous** [menˈdêyşəs]. Yalancı.
**mendic·ant** [ˈmendəkənt]. Dilenci.
**menfolk** [ˈmenfo̯uk]. Ailenin erkekleri.
**menial** [ˈmînyəl]. Süflî; aşağı; (*cont.*) hizmetçi.
**menstruation** [ˌmenstruˈêyşn]. Hayız, (kadınların) aybaşı.
**mental** [ˈmentəl]. Akla aid; zihnî. **~ly afflicted,** şuuru muhtel: ~ **hospital** [**home**], tımarhane: ~ **specialist,** akliyeci: **to make a** ~ **reservation,** içinden pazarlık etmek. **~ity** [mənˈtaliti], zihniyet.
**mention** [ˈmenşn]. Anma; zikir. Anmak; zikretmek; adını anmak; bahsetmek. **don't** ~ **it!,** estağfurullah!: **we need hardly** ~ **that** ..., zikre lüzum yoktur ki ...: **~ed in dispatches,** harbde kumandan raporunda adı zikredilmiş olan: **to receive an honourable** ~, bir müsabaka vs.de derece almayıp sadece zikre lâyık görülmek: **to make** ~ **of,** zikretmek: **not to** ~, bundan başka: **nothing worth** ~**ing,** zikretmeğe değmez bir şey: **to** ~ **s.o. in one's will,** vasiyetnamede varisler arasında zikretmek.
**mentor** [ˈmentô(r)]. Müşavir; akıl hocası.
**menu** [ˈmenyû]. Yemek listesi.
**mercantile** [ˈmö(r)kəntâyl]. Ticarete aid; esnafça. ~ **marine,** deniz ticaret filosu.
**mercenary** [ˈmö(r)sənəri]. Ücretli (asker), para hırsı ile yapılmış; menfaatperest.
**merchandize** [ˈmö(r)çəndâyz]. Ticarî eşya; mal.
**merchant** [ˈmö(r)çənt]. Tüccar. ~ **prince,** pek mühim ve zengin tüccar: ~ **service,** deniz ticareti mesleği: ~ **ship,** yük gemisi. **~man,** yük gemisi.
**merci·ful** [ˈmö(r)sifəl]. Merhametli; **~ly,** merhametli; çok şükür. **~less,** merhametsiz; amansız.
**Mercur·y**[1] [ˈmö(r)kyuri]. Ticaret tanrısı; Utarid seyyaresi. ~[2]. Cıva. **~ial** [–ˈkyûri̯əl], cıvalı, cıva gibi; sebatsız.
**mercy** [mö(r)si]. Merhamet; aman; rahmet. ~ **(on us)!,** aman!, ya Rabbi!: **to be at the** ~ **of** ..., mukadderatı ···in elinde olm.: **at the** ~ **of the waves,** dalgaların keyfine bağlı: **I am at your** ~, boynum kıldan ince: **for** ~**'s sake!,** Allah aşkına: **left to the tender mercies of** ..., ···in eline düşmüş: ⌜**be thankful for small mercies!**⌝, ⌜öp de başına koy!⌝: **to throw oneself on s.o.'s** ~, birinin ocağına düşmek: **sister of** ~, rahibe: **what a** ~**!,** ne âlâ!
**mere**[1] [mi̯ə(r)] *n.* Küçük ve sığ göl. ~[2] *a.* Saf, sade; sadece, sırf; ···dan başka bir şey değil: **~ly,** mahza, yalnız; âdeta: **he is a** ~ **boy,** daha çocuktur: **a** ~ **nobody** [**nothing**], pek ehemmiyetsiz bir adam [şey]: **the** ~ **sight of him,** onu görmek bile.
**merge** [mö(r)c]. Birleş(tir)mek; dahil olm.; içinde kaybolmak; (renk) çalmak. **~r,** birleşme.
**meridian** [məˈridyən]. Nısfınnehar

dairesi; tul hattı. ~ **altitude,** gayeti irtifa.

**meringue** [məˈran(g)]. Yumurta akından yapılan bir nevi tatlı.

**merino** [məˈrînou]. Merinos koyunu, yünü.

**merit** [ˈmerit]. Değer, liyakat; meziyet. Müstahak olm., değmek. **to judge stg. on its ~s,** değerine göre hüküm vermek. **~orious** [–ˈtôriəs], değerli; medhe lâyık.

**merlin** [ˈmö(r)lin]. Çakır doğan.

**mermaid** [ˈmö(r)mêyd]. Deniz kızı.

**merriment** [ˈmerimənt]. Neş'e; keyif; cümbüş.

**merry** [ˈmeri]. Şetaretli; şen, neş'eli; güleryüzlü; çakırkeyif. **to make ~,** cümbüş etm.: **to make ~ over stg. [s.o.],** bir şeyi [kimseyi] alaya almak. **~-go-round,** atlıkarınca. **~maker,** cümbüş eden.

**meseems** [miˈsîmz]. (*esk.*) Bana öyle geliyor ki.

**mesh** [meş]. Ağ gözü. (Çark dişleri) birbirine geçmek.

**mesmer·ism** [ˈmezmərizm]. Manyatizma. **~ize** [–âyz] …e manyatizma yapmak.

**Mesopotamia** [ˌmesopoˈtêymyə]. Irak.

**mess**[1] [mes] *n.* Karışıklık; pislik; askerî sofra(da oturanlar). **to get into a ~,** üstünü başını kirletmek; başını belâya sokmak; karmakarışık olm.: **to make a ~ of things,** berbad etm.: **to make a ~ of stg.,** bir şeyi kirletmek: **here's a pretty ~!,** ayıkla pirincin taşını! **~**[2] *v.* Kirletmek; askerî sofrada yemek yemek. **to ~ about [around],** sinek avlamak: **to ~ stg. [s.o.] about,** karıştırmak; dokunmak: **to ~ things up,** işleri berbad etm. **~mate,** sofra arkadaşı. **~-tin,** aş kabı. **~-up,** karışıklık. **~y** [ˈmesi], kirli; karmakarışık.

**message** [ˈmesic]. Hususî haber. **to leave a ~ for s.o.,** birisi için hususî bir haber bırakmak: **to run ~s,** ufak tefek işlere koşmak.

**messenger** [ˈmesəncə(r)]. Haberci. **~ boy,** hususî haber vs. götüren çocuk: **Queen's ~,** kuriye.

**Messiah** [meˈsâyə]. Mesih.

**Messrs.** [ˈmesö(r)z]. Mr.*'in cemi; um. firmalar hakkında kullanılır.*

**met** *v.* **meet.**

**metal** [ˈmetəl]. Maden(î). Yolu kırık taşlarla tesviye etmek. **to leave the ~s,** (lokomotif vs.) raydan çıkmak. **~lic,** [məˈtalik], madenî, madenden yapılmış. **~lurgy** [meˈtalö(r)ci], maden ilmi.

**metamorphosis** [ˌmetəmôˈfousis]. İstihale.

**metaphor** [ˈmetəfə(r)]. Mecaz; istiare. **mixed ~,** birbirini tutmıyan mecazlar. **~ical** [–ˈforikl], mecazî: **~ly,** istiare suretile.

**meteor** [ˈmîtyô(r)]. Şahab. **~ic** [mîtiˈorik], şimşek gibi parlayıp geçen. **~ite** [ˈmîtiərâyt], haceri semavî. **~ology** [ˌmîtyəˈroləci], hava bilgisi.

**meter** [ˈmîtə(r)]. Zaman, sürat vs. ölçmeğe mahsus alet; saat, kontör.

**methinks** [miˈθinks]. (*esk.*) Bana öyle geliyor ki.

**method** [ˈmeθəd]. Usul; kaide; tarz; metod; nizam. **⌜there's ~ in his madness⌝,** göründüğü kadar deli değil. **~ical** [məˈθodikl], muntazam, tertibli; usulü dairesinde çalışan.

**methylated** [ˈmeθilêytid]. **~ spirit,** odun alkolü ile karıştırılmış ispirto.

**meticulous** [məˈtikyuləs]. Titiz; çok dikkatli; dakik.

**metr·e** [ˈmîtə(r)]. Metre; şiirde vezin. **~ic** [ˈmetrik], **~ system,** aşarî usul; sistem metrik.

**metropol·is** [məˈtropəlis]. Devlet merkezi; büyük şehir. **~itan** [––ˈpolitn], devlet merkezine aid; başpiskopos.

**mettle** [ˈmetl]. Ataklık, cesaret. **to put s.o. on his ~,** göreyim seni diye teşvik etm.: **to show one's ~,** kendini göstermek. **~some,** atılgan, atak; sert başlı.

**mew** [myû]. Miyavlamak.

**mews** [myûz]. Bir şehirde sıra ahırlar. **The Royal ~,** Kıraliçe ahırları.

**miaow** [mîˈyau] (*ech.*) Miyavlamak.

**mickle** [ˈmikl]. **⌜many a ~ makes a muckle⌝, ⌜**damlıya damlıya göl olur**⌝.**

**micro-** [ˈmâykrou] *pref.* Küçüklük *ifade eder.*

**microbe** [ˈmâykroub]. Mikrob.

**microscop·e** [ˈmâykroskoup]. Mikroskop. **~ic** [–ˈskopik], yalnız mikroskop ile görünür; çok küçük.

**mid** [mid]. Orta: arasında. **in ~ air,** havada. **~day** [ˈmidˈdêy], öğle üzeri. **~land** [ˈmidlənd], bir memleketin iç kısmı. **the M~s,** Orta İngiltere. **~most** [ˈmidmoust], en ortadaki. **~night** [ˈmidnâyt], gece yarısı: **to burn the ~ oil,** geç vakitlere kadar çalışmak. **~shipman** [ˈmidşipmən], deniz asteğmeni. ~**ships** [ˈmidşips], geminin ortasında. ~**st** [midst], orta arada. **in the ~ of all this,** tam bu arada: **in our ~,** içimizde: **in the ~ of them,** ortalarında. ~**stream** [midˈstrîm], nehrin ortası. ~**-summer** [ˈmidsʌmə(r)], yaz ortası: **M~ Day,** 21 Haziran: **~ madness,** zırdelilik. ~**way** [ˈmidwêy], yarı yolda. ~**wife** [ˈmidwâyf], ebe: ~**ry** [–ˈwifəri], ebelik. ~**winter** [midˈwintə(r)], kış ortası; karakış.

**middle** [ˈmidl]. Orta; bel. Ortadaki. **in the ~ of it all,** tam ortasında: **to be in the ~ of doing stg.,** bir şeyle meşgul olm.: **there is no ~ way,** ikisinin ortası yoktur. ~**man,** kabzımal; komisyoncu. ~**most,** en ortadaki. ~**-aged,** orta yaşlı. ~**-class,** orta sınıf(a mensub).

**middling** [ˈmidlin(g)]. İyice; ne iyi ne kötü.

**midge** [mic]. Tatarcık.

**midget** [ˈmicit]. Cüce; minimini.

**mien** [mîn]. Eda, surat.

**might**[1] [mâyt] *v. past of* **may.** **~ I see him?,** acaba kendisini görebilir miyim?: **we ~ as well go a little further,** (oldu olacak) biraz daha uzağa gidebiliriz: **one ~ as well throw money away as give it to him,** buna para vermek parayı sokağa atmak demektir: **it ~ be better to tell him,** kendisine söylemek belki de daha iyidir: **he ~ have been thirty,** otuz yaşında ya var ya yoktu: **he ~ have been a bit more generous,** biraz daha cömerd olamaz mıydı: **you ~ shut the door when you come in,** kapıyı kapayamaz mıydın?: **you ~ shut the door, will you?,** kapıyı kapar mısınız: **strive as they ~ they could not move it,** ne kadar çabaladılarsa da onu kımıldatamadılar. ~[2] *n.* Kudret; kuvvet. **with all one's ~** [**with ~ and main**], var kuvvetile. ~**y,** kudretli; muazzam; (*sl.*) son derece.

**migra·nt** [ˈmâygrənt]. Göçücü; göçebe. ~**te** [mâyˈgrêyt], göçmek. ~**tory** [ˈmâygrətəri], göçücü, göçebe.

**Mike** [mâyk]. (*abb.*) **Michael. for the love of ~,** Allah aşkına.

**mild** [mâyld]. Mülâyim; halim; mutedil; uslu; hafif. **draw it ~!,** mübalağa etme!, atma!: **to put it ~ly,** en hafif tabirle.

**mildew** [ˈmildyû]. Küf(lenmek).

**mile** [mâyl]. Mil. **to be ~s ahead of s.o.,** birini fersah fersah geçmek. ~**age** [ˈmâylic], mil hesabile mesafe: **car with low ~,** az kullanılmış otomobil. ~**stone,** mil gösteren taş: **~s in one's life,** bir insanın hayatındaki mühim vakalar.

**milit·ant** [ˈmilitənt]. Uğraşan, savaşan, muharib. ~**arism** [ˈmilitərizm], harbculuk. ~**arist,** harbcu. ~**ary** [ˈmilitəri], askerî: **the ~,** asker sınıfı; ordu. ~**ate** [ˈmilitêyt], **to ~ against stg.,** bir şeye müsaid olmamak, engel olmak. ~**ia** [məˈlişə], milis; redif askeri.

**milk** [milk]. Süt. Sağmak; süt vermek. **new ~,** henüz sağılmış süt: **to come home with the ~,** sabaha karşı eve dönmek: ⌜**a land of ~ and honey**⌝, pek bolluk ve mamur memleket: ⌜**it's no use crying over spilt ~**⌝, ⌜oldu olacak kırıldı nacak⌝. ~**er,** sağıcı. ~**maid,** sağıcı kız. ~**man,** sütçü; inekçi. ~**pail,** süt gerdeli. ~**sop,** lâpacı. ~**y,** sütlü; süt gibi: **the M~ Way,** kehkeşan. ~**-and-water,** yavan, gayretsiz. ~**-can,** süt kabı; güğüm.

**mill** [mil]. Değirmen; kumaş ve iplik fabrikası. Öğütmek; frezelemek; tırtıllamak; dövüşmek: **to ~ around,** (koyun vs.) kaynaşmak: **to go** [**pass**] **through the ~,** (*fig.*) büyük meşakkatler çekip tecrübelerle hayatta pişmek. ~**-hand,** kumaş veya iplik fabrikası amelesi. ~**-pond,** değirmen havuzu. ~**-race,** değirmen arkı. ~**ed** [mild], tırtıllı. ~**er** [ˈmilə(r)], değirmenci. ~**stone** [ˈmilstǫun], değirmentaşı. ⌜**to be between the upper and the nether ~**⌝, örs ile çekic arasında kalmak: ⌜**a ~ round one's neck**⌝, insanın hayatta muvaffak olmasına engel olan şey.

**millen·ary** [miˈlenəri]. Bininci yıldönümü. ~**nium** [–ˈlenyəm], bin yıllik müddet: **the ~,** tam bir sulh ve saadet devri.

**millet** [ˈmilit]. Darı.

**milliner** [ˈmilinə(r)]. Kadın şapkacısı; tuhafiyeci. ~**y,** tuhafiyeci eşyası.

**million** [ˈmilyən]. Milyon. **he is one in a ~,** eşi yoktur. ~**aire** [–ˈe̱ə(r)], milyoner. ~**th,** milyonuncu.

**mimic** [ˈmimik]. Taklid (etmek). ~**ry,** mukallidlik.

**minaret** [ˈminəret]. Minare.

**minatory** [ˈminətəri]. Tehdidkâr.

**minc·e** [mins]. Kıyma(k); doğramak. **he does not ~ his words,** sözünü esirgemez. ~**emeat,** kıyma; içyağı kuru üzüm şeker vs. ile yapılan bir tatlı: **to make ~ of s.o.,** parça parça etm. ~**e-pie,** ~**emeat** ile doldurulmuş börek. ~**ing** [ˈminsin(g)], nazlı, kırıtkan.

**mind**[1] [mâynd] *n.* Akıl; beyin; hatıra; fikir; niyet. **to bear in ~,** hatırda tutmak: **to call to ~,** hatırlamak: **to give s.o. a piece of one's ~** [**to tell s.o. one's ~**], ağzına geleni söylemek: **I have a good [half a] ~ to,** şeytan diyor: **he knows his own ~,** o ne yapacağını bilir: **to make up one's ~,** karar vermek: **we must make up our ~ to the fact that ..., ...** fikre kendimizi alıştırmalıyız: **to my ~,** bana göre: **that's a weight off my ~,** yüreğime su serpildi: **to take s.o.'s ~ off his troubles,** derdlerini unutturmak: **to have stg. on one's ~,** zihnini işgal eden, insana derd olan, bir şeyi olmak: **to be out of one's ~,** aklını kaçırmak: **he puts me in ~ of his father,** babasını andırıyor: **put me in ~ of it tomorrow!,** yarın bana hatırlat!: **to be in one's right ~,** aklı başında olm.: **to set one's ~ on stg.,** bir şeyi aklına koymak: **to take it into one's ~ (to do stg.),** (bir şeyi yapmak) aklına esmek: **time out of ~,** oldum olasıya; ezelden beri: **to be in two ~s about stg.,** karar verememek, bocalamak. ~[2] *v.* Dikkat etm.; bakmak; aldırış etmek. **~ (yourself)!,** dikkat et!: **~ you don't break it!,** sakın kırma!, aman, kırarsın!: **~ what you are about!,** ne yaptığınıza dikkat edin: **I don't ~ him,** (i) hiç fena adam değildir; (ii) ona hiç aldırmam!: **I don't ~ your going there sometimes,** arasıra oraya gitmenize bir diyeceğim yok: **if you don't ~ my saying so,** sözüme gücenmezseniz: **never ~!,** zarar yok!; **never ~ the expense!,** masrafın ehemmiyeti yok: **never ~ what he says,** söylediğine bakma!: **I shouldn't ~ a drink,** susadım, bir şey içsek fena olmaz: **would you ~ shutting the door!,** lûtfen kapıyı kapar mısınız. ~**ed** [ˈmâyndid], niyetli, istekli. -~, ... düşünceli, ... fikirli. **if you are so ~,** eğer böyle istiyorsanız: **mechanically ~,** makineden anlar. ~**ful** [ˈmâyndful], düşünerek, hatırlayarak. **~ of one's health,** sıhhatine dikkat eden.

**mine**[1] [mâyn]. Benim; benimki. ~[2]. Maden ocağı (kazmak); lağım (açmak); mayn (koymak). **to ~ for gold,** altın araştırmak. ~**r,** madenci. ~**ral** [ˈminərəl], maden. Madenî. ~**ralogy** [ˈˈraləci], madenler bilgisi. ~**sweeper** [ˈmâynswîpə(r)], arama tarama gemisi.

**mingle** [ˈmin(g)gl]. Karış(tır)mak; katmak.

**miniature** [ˈminiatyə(r)]. Minyatür. Küçücük.

**minim·ize** [ˈminimâyz]. Küçümsemek; azaltmak. ~**um** [–əm], asgarî; en az, en küçük. En küçük mikdar.

**mining** [ˈmâynin(g)]. Madencilik; madenciliğe aid.

**minion** [ˈminyən]. (*cont.*) Gözde; köle, peyk. **the ~s of the law,** polis, vs.

**minister**[1] [ˈministə(r)] *n.* Vekil, nazır, bakan; orta elçi; papaz. ~[2] *v.* **to ~ to s.o. [s.o.'s needs],** birinin ihtiyacını temin etmek. ~**ial** [ˌminisˈtiəriəl], vekil veya nazıra aid; hükümet idaresine aid; iktidarda bulunan partiye aid.

**ministr·ation** [ˌminisˈtrêyşn]. Hizmet; ihtimam. ~**y** [ˈministri], vekillik, nezaret, bakanlık; papazlık. Kabine.

**mink** [min(g)k]. Vizon(un kıymetli kürkü).

**minnow** [ˈminou]. Pek küçük bir tatlısu balığı.

**minor** [ˈmâynə(r)]. Daha küçük; küçükçe; ehemmiyetsiz; tâli. Reşit olmamış kimse. **Asia M~,** Anadolu: **Smith ~,** iki Smith kardeşlerin küçüğü. ~**ity** [mâyˈnoriti], akalliyet; azlık; küçüklük: **to be in a ~ of one,** fikrinde yalnız kalmak.

**minster** [ˈminstə(r)]. Büyük kilise.

**minstrel** [ˈminstrəl]. Ortaçağda saz şairi.

**mint**[1] [mint]. Darbhane. Para basmak. **in ~ condition,** yepyeni: **to have a ~ of money,** para kesmek. ~[2]. Nane.

**minus** [ˈmâynəs]. Eksik; ...siz: çıkartma işareti. **a ~ quantity,** sıfırdan aşağı mikdar: **ten ~ two equals eight,** on nakıs iki müsavi sekiz.

**minute**[1] [ˈminit]. Dakika; lâhza; muhtıra. ~**s (of a meeting),** mazbata. **I'll come in a ~,** şimdi gelirim: **he may come any ~ (now),** şimdi neredeyse gelir: **to make [write] a ~,** derkenar yazmak: **to make a ~**

of stg., not tutmak: on [to] the ~, dakikası dakikasına: wait a ~!, biraz bekle! ~² [mây'nyût]. Ufacık, minimini. ~ly, inceden inceye.

**minx** [min(g)ks]. Haspa; kurnaz kız.

**mirac·le** ['mirəkl]. Keramet; mucize; harika. **by a ~**, mucize kabilinden: **~ play**, dinî piyes: **to work ~s**, mucize yapmak, keramet göstermek; mucize gibi tesir etmek. **~ulous** [mi'rakyuləs], harikulâde.

**mirage** [mi'râj]. Serab.

**mire** [mây(r)]. Çamur, bataklık; pislik.

**mirror** ['mirə(r)]. Ayna. Aksettirmek.

**mirth** [mö(r)θ]. Neş'e; gülme. **~ful**, şen, eğlenceli, gülüşen.

**mis-** *pref.* *Bir kelimeye* fena, zıd, eksik *vs. gibi menfi manalar veren bir ek; mes.* **misadvise**, fena nasihat vermek: **misjudge**, yanlış hüküm vermek.

**misadventure** [ˌmisəd'ventyə(r)]. Kaza; aksi tesadüf.

**misalliance** [ˌmisa'lâyəns]. Küfüv olmıyan kimse ile evlenme; münasebetsiz birleşme.

**misanthrop·e, -ist** ['misanθroup, mi'sanθrəpist]. Merdümgiriz, adamcıl.

**misapplication** [ˌmisapli'kêyşn]. Yanlış tatbik; suiistimal.

**misapply** [ˌmisə'plây]. Yerinde kullanmamak; beyhude yere sarfetmek.

**misapprehend** [ˌmisapri'hend]. Yanlış anlamak; yanılmak.

**misappropriation** [ˌmisə'proupriêyşn]. Zimmete geçirme.

**misbehav·e** [ˌmisbi'hêyv]. Edebsizlik etm., yaramazlık etmek. **~iour**, yaramazlık.

**misbegotten** [misbi'gotn]. Piç; alçak.

**misbelie·f** [ˌmisbi'lîf]. Yanlış itikad; imansızlık. **~ver**, imansız; kâfir.

**miscall** [mis'kôl]. Yanlış isim vermek.

**miscarr·iage** [mis'karic]. Çocuk düşürme; (proje) suya düşme. **~ of justice**, adlî hata. **~y** [mis'kari], çocuk düşürmek; boşa çıkmak; suya düşmek; (mektub vs.) yerine varmayıp kaybolmak.

**miscellan·eous** [ˌmisə'lêynyəs]. Türlü türlü; çesidli, muhtelif. **~y** [mi'seleni], muhtelif mevzulara aid eserlerden mürekkeb mecmua; türlü türlü eşyanın toplanması.

**mischance** [mis'çâns]. Kaza; aksilik; talihsizlik.

**mischie·f** ['misçif]. Yaramazlık; afacanlık; zarar; fesad; muziblik; hainlik. **to be up to some ~**, yaramazlık yapmak; kumpas kurmak: **to keep s.o. out of ~**, yaramazlıktan alıkoymak için bir çocuğa iş vermek: **to make ~**, fesad karıştırmak: **to make ~ between people**, aralarını bozmak: **out of pure ~**, sırf şeytanlıktan. **~f-maker**, fesadcı; ara bozucu. **~vous** ['misçivəs]. Muzib; yaramaz; zararlı; hain; ara bozucu.

**misconception** [ˌmiskən'sepşn]. Yanlış anlama, hata.

**misconduct** *n.* [mis'kondʌkt]. Fena hareket, idare. *v.* [miskən'dʌkt], fena idare etm.: **to ~ oneself**, fena harekette bulunmak.

**misconstru·ction** [ˌmiskən'strʌkşn]. Yanlış anlama. **~e** [–'strû], yanlış anlamak; ters mana vermek.

**miscount** [mis'kaunt]. Yanlış sayılma. Yanlış hesab etm., yanlış saymak.

**miscreant** ['miskriənt]. Suçlu; habis.

**misdeal** [mis'dîl]. Kâğıd oyunlarında kâğıdı yanlış dağıtma(k).

**misdeed** [mis'dîd]. Kabahat; fenalık.

**misdemean** [misdə'mîn]. **to ~ oneself**, kötü harekette bulunmak. **~our** ['–mînə(r)], kabahat, kusur.

**misdirect** [ˌmisdây'rekt]. Yanlış nasihat vermek; adresini yanlış yazmak; (işi) fena idare etmek.

**miser(ly)** ['mâyzə(r)(li)]. Hasis, cimri.

**miser·able** ['mizrəbl]. Sefil; bedbaht; miskin; berbad; feci. **I am feeling pretty ~**, pek fenayım: **a ~ salary**, pek cüzi maaş: **~ weather**, berbad hava. **~y** ['mizəri], sefalet, perişanlık; ıstırab; acı. **to put an animal out of its ~**, bir hayvanı öldürüp eziyetten kurtarmak: **you little ~!**, (*coll.*) seni hınzır yumurcak!

**misfire** [mis'fâyə(r)]. Ateş almama(k); (nükte) anlaşılmamak; yerinde olmamak.

**misfit** [mis'fit]. Uymıyan elbise; yerinin adamı olmıyan kimse; cemiyete intibak etmiyen kimse.

**misfortune** [mis'fôtyûn]. Bedbahtlık, talihsizlik; kaza.

**misgiv·e** (-**gave**) [mis'giv, -gêyv]. Şübheye düşürmek. **~ing**, şübhe, endişe, korku. **with some ~ [not without ~s]**, biraz korkarak.

**misguided** [misˈgâydid]. Dalâlete düşmüş; yanlış yola sapmış.
**mishandle** [misˈhandl]. Hor kullanmak; fena muamele, idare etmek.
**mishap** [misˈhap]. Kaza; aksilik.
**misinformed** [misinˈfômd]. Yanlış haber almış.
**misjudge** [misˈcʌc]. Yanlış hüküm vermek; hatalı fikir edinmek.
**mislay** (**mislaid**) [misˈlêy, -lêyd]. Yanlış veya hatırlanamıyacak bir yere koymak.
**mislead** (**misled**) [misˈlîd, -led]. Baştan çıkarmak; aldatmak.
**mismanage** [misˈmanic]. Fena idare etmek.
**misogynist** [miˈsocinist]. Kadın düşmanı.
**misplace** [misˈplêys]. Yanlış yere koymak. **~d**, yerinde değil; yersiz.
**misprint** [misˈprint]. Tabı hatası.
**mispronounce** [ˌmisproˈnauns]. Yanlış telâffuz etmek.
**misrepresent** [ˌmisrepriˈzent]. Yanlış tarif etmek.
**misrule** [misˈrûl]. Fena idare (etmek).
**miss**[1] [mis]. *Evli olmıyan kadın unvanı*; kız; hanım kız. **~**[2]. İsabet etmeme; manke; (atışta) boşa gitme. İsabet etmemek; kaçırmak; aramak; göreceği gelmek; eksik olmak. **to give stg. a ~**, atlamak, vazgeçmek: **we shall all ~ him**, onu çok arıyacağız: **he's no great ~**, yokluğunu pek hissetmiyoruz: **it will never be ~ed**, eksikliğini kimse farketmez: **you haven't ~ed much**, mühim bir şey kaçırmış olmadınız: **it's hit or ~**, rasgele; ne olursa olsun diye: **to ~ the market**, piyasa fırsatı kaçırmak: **to ~ the point**, bir şeyin esasını anlamamak: **to ~ one's way**, yolu şaşırmak: **who is ~ing?**, kim eksik?: ⌜**a ~ is as good as a mile**⌝, muvaffakiyetsizliğin küçüğü de büyüğü de farksızdır (*mes. bir treni bir veya on dakika farkla kaçırmak ayni şeydir*). **~ing** *a.*, eksik, noksan; namevcud; kaybolmuş; (harbde) kayıb.
**misshapen** [misˈşêypn]. Çelimsiz, biçimsiz.
**missile** [ˈmisâyl]. Atılan şey; mermi.
**mission** [ˈmişn]. Hususî memuriyet, vazife; elçilik; heyeti mahsusa. **foreign ~s**, dışarıya gönderilen misyoner heyetleri: **his ~ in life**, kendisi için tayin ettiği hayat vazifesi. **~ary** [ˈmişənəri], misyoner, dai.
**missis** [ˈmisiz]. *Evli kadının unvanı* (*daima* Mrs. *yazılır*).
**missive** [ˈmisiv]. Mektub.
**misspent** [misˈspent]. Yanlış yere sarfedilmiş. **a ~ youth**, tembelce veya çapkınca geçirilen genclik.
**missus** [ˈmisəs]. *Hizmetçi tarafından ev sahibine verilen unvan*; hanımefendi. **the ~** [**my ~**], köroğlu; bizimki.
**mist** [mist]. Sis. **to ~ over**, buğula(n)mak. **~y** [ˈmisti], sisli; hayal meyal.
**mistake**[1] [misˈtêyk]. Hata; yanlışlık. **by ~**, yanlışlıkla: **make no ~ about it!**, anlamadım deme!: **she's a pretty woman and no ~**, o kadın güzel mi güzel. **~**[2] (**mistook, mistaken**) [misˈtuk, –ˈtêykn]. Yanlış anlamak. **to be ~n**, yanılmak: **to ~ s.o. for s.o. else**, birini başkasına benzetmek. **~n** *a.* Yanlış, hatalı.
**mister** [ˈmistə(r)]. (*Daima* Mr. *yazılır*). *Erkeğin ismi önünde kullanılan unvan.* Bey, Bey efendi; Bay. *Bazı memurlara verilen unvan. mes.* **Mr. President**, Reis bey.
**mistime** [misˈtâym]. Vaktini yanlış hesab etm.; bir şeyi vakitsiz yapmak.
**mistletoe** [ˈmisltou]. Burc, ökseotu.
**mistress** [ˈmistris]. *Evli kadınlara hitab unvanı* (Mrs.); muallime; sahibe; metres.
**mistrust** [misˈtrʌst]. İtimadsızlık. İtimad etmemek. **~ful**, itimad etmiyen.
**misunder·stand** (**-stood**) [misʌndəˈstand, -stud]. Yanlış anlamak. **~standing**, anlaşamamazlık; geçimsizlik. **~stood** *a.*, yanlış anlaşılmış; kıymeti takdir edilmemiş.
**misuse** *n.* [misˈyûs]. Hor, yanlış kullanma; suiistimal; kullanma. *v.* [–ˈyûz], hor kullanmak; yanlış maksada hasretmek; suiistimal etmek.
**mite** [mâyt]. Ufacık şey *esp.* çocuk; peynir kurdu; pek ufak bir sikke. **not a ~ left**, zerresi kalmadı: ⌜**the widow's ~**⌝, ⌜çok veren maldan, az veren candan⌝.
**mitigate** [ˈmitigêyt]. Yumuşatmak; tadil etm.
**mitre** [ˈmâytə(r)]. Piskopos tacı; şev gönye.
**mitten** [ˈmitn]. Kolçak. Parmaksız eldiven.
**mix** [miks]. Karıştırmak; tahlit etm.; karmak; birleşmek. **~up**, birbirinden ayırt edememek, karıştırmak: **to get**

~ed up, zihni karışmak. ~ed *a.*, karışık; mahlut; muhtelit: ~ bathing, kadın ve erkek bir arada yüzmek: ~ marriage, muhtelif ırklardan kimselerin evlenmesi: to get ~, karışmak. ~er, karıştırıcı alet; a good ~, iyi geçinen kimse. ~-up, kargaşalık; karışıklık. ~ture [ˈmikstyə(r)]. Karıştırma; karışık şey; halita; terkib.
**Mme.** (*abb.*) Madame.
**mnemonic** [nîˈmonik]. Hatırlayıcı.
**moan** [mǫun]. İnilti; figan. İnlemek.
**moat** [mǫut]. Kale hendeği ki ekseriya su ile doludur.
**mob** [mob]. Kalabalık; güruh. Saldırmak; birine hücum etmek. ~ law, linç kanunu.
**mobil·e** [ˈmǫubâyl]. Müteharrik; oynak; değişken; seyyar. ~ity [–ˈbiliti], müteharriklik; oynaklık: the ~ of an army, ordunun hareket kabiliyeti. ~ize [ˈmǫubilâyz], seferber etmek (edilmek). ~ization [–ˈzêyşn], seferberlik.
**mocassin** [ˈmokəsin]. Geyik derisinden yapılan çarık.
**mock** [mok]. ~ (at), alay etm.; ehemmiyet vermemek; alay için taklid etmek. Eğlenmek; şaka etmek. Yapma, sahte, taklid. to make a ~ of s.o., birini maskara etmek. ~ery [ˈmokəri], alay; gülünc bir şey; oyun: the trial was a mere ~, muhakeme komediden ibaretti.
**mod·e** [mǫud]. Tavır, tarz; moda; usul; minval; (*mus.*) makam. ~ish *a.*, son moda.
**model** [ˈmodl]. Örnek, nümune; tip. Örnek olan, mükemmel; model olan. Modelini yapmak; modellik etmek.
**moderat·e** *a.* [ˈmodərit]. Mutedil, orta, alelâde; aşırı derece veya mikdarda olmıyan. *v.* [–rêyt], tadil etm.; azal(t)mak. ~ion, itidal.
**modern** [ˈmodö(r)n]. Asrî; yeni; şimdiki zamana aid. ~ languages, bugün konuşulan lisanlar. ~ism, yenilik tarafdarlığı; yeni düşünce, kelime. ~ize [–nâyz], asrileştirmek; yenileştirmek.
**modest** [ˈmodist]. Mütevazı, alçak gönüllü; afif. ~y, tevazu, alçak gönüllülük; sadelik; iffet, utangaclık. with all due ~, övünmek gibi olmasın amma.
**modicum** [ˈmodikəm]. Cüzi, bir az.
**modif·y** [ˈmǫdifây]. Tadil etm.; azaltmak; şeklini değiştirmek. ~i cation [–fiˈkêyşn], tadil; değiştirme.
**modulate** [ˈmodyulêyt]. Sesini tadil etm.; gamını değiştirmek.
**mohair** [ˈmǫuhęə(r)]. Tiftik (kumaş).
**Mohammedan** [mǫuˈhamidn]. Müslüman.
**moist** [môyst]. Yaş, nemli. ~en [ˈmôysn], hafifçe ıslatmak. ~ure [ˈmoistyə(r)], rutubet, nem.
**mole**[1] [mǫul]. Ben. ~[2]. Köstebek. ~ hill, köstebek yuvası: ⌜to make a mountain out of a ~⌝, ⌜habbeyi kubbe yapmak⌝. ~skin, köstebek kürkü. ~[3]. Mendirek, dalgakıran.
**molecule** [ˈmǫulikyûl]. Zerre; molekül.
**molest** [mǫuˈlest]. Taciz etm.; dokunmak.
**mollify** [ˈmolifây]. Yumuşatmak; gönül almak.
**mollycoddle** [ˈmolikodl]. Hanım evladı; lâpacı. Nazlı büyütmek.
**molten** [ˈmǫultən]. Erimiş.
**moment** [ˈmǫumənt]. An, lâhza; ehemmiyet; vezniyet. to be of ~, mühim olm.: not for a ~!, asla!: one ~ [half a ~]!, bir dakika!, biraz dur!: the ~ I saw him, onu gördüğüm anda: this ~, derhal. ~ary, âni. ~ous [mǫuˈmentəs], çok ehemmiyetli. ~um, vezniyet; hız.
**monarch** [ˈmonək]. Hükümdar, kıral. ~ist, kırallık tarafdarı. ~y, saltanat, kırallık.
**monast·ery** [ˈmonəstri]. Manastır. ~ic [moˈnastik], keşişliğe ve manastır hayatına aid.
**Monday** [ˈmʌndi]. Pazartesi.
**monetary** [ˈmʌnətəri]. Paraya aid, nakdî.
**money** [ˈmʌni]. Para; nakid; servet. it's a bargain for the ~, o fiata kelepirdir: it will bring in big ~, bu işte çok para var: to come into ~, paraya konmak: to throw good ~ after bad, zararlı bir işe devamda inad etm.: you've had your ~'s worth, masrafını bol bol çıkarttın. ~ed, zengin, paralı. ~lender, tefeci. ~-box, kumbara. ~-changer, sarraf. ~-grubber, para düşkünü.
**-monger** [–mʌn(g)gə(r)]. *suff.* ···ci, ... satıcı, ... çıkaran *manasında bir son ek, mes.* cheese~, peynirci; war~, harbi körükliyen.
**mongrel** [ˈmʌn(g)grəl]. Melez; soyu karışık her şey *esp.* köpek.
**monitor** [ˈmonitə(r)]. Vâiz; mektebde küçük sınıflara nezaret eden yüksek sınıf talebesi. Yabancı telsiz neşriyatını kontrol için takib etmek.

**monk** [mʌn(g)k]. Keşiş; rahib. **~ish,** keşiş gibi.
**monkey** [ˈmʌn(g)ki]. Maymun; kazık kakma makinesi. **to ~ about with stg.,** kurcalamak, haltetmek. **~ tricks,** açıkgözlük: **to put one's ~ up,** öfkelendirmek: **you young ~!,** seni çapkın seni!
**mono-** [ˈmonou̯]. *pref.* Tek-. **~cle** [ˈmonokl], tekgözlük. **~gamy** [moˈnogəmi], tek karılılık. **~gram** [ˈmonogram], bir şahıs isimlerinin ilk harflerinden yapılan tek şekil halinde marka. **~graph** [ˈmonou̯graf], yalnız bir şey yahud şahış hakkında bir eser. **~pol·y** [məˈnopəli], inhisar; tekel: **~ist,** inhisarcı: **~ize,** inhisar altına almak. **~syllable** [ˌmonəˈsiləbl], tek heceli kelime. **~ton·e** [ˈmonətou̯n], yeknesak ahenk: **~ous** [məˈnotənəs], yeknesak; cansıkıcı: **~y** [məˈnotəni], yeknesaklık; can sıkıntısı.
**monsoon** [monˈsûn]. Hind okyanusunda mevsim rüzgârı; muson.
**monst·er** [ˈmonstə(r)]. Hilkat garibesi; canavar. Dev gibi, azman. **~rous** [ˈmonstrəs], azman; inanılmaz derecede; müdhiş. **it's perfectly ~,** olur rezalet değil. **~rosity** [–ˈstrositi], hilkat garibesi, azman, galatitabiat.
**month** [mʌnθ]. Ay. **a ~ of Sundays,** kırk yıl; çok uzun müddet: **this day ~,** gelecek ay bugün. **~ly,** ayda bir olan; aylık: **monthlies,** aylık mecmualar.
**monument** [ˈmonyumənt]. Abide; anıt. **~al** [–ˈmentl], muazzam, heybetli.
**moo** [mû]. Böğürme(k).
**mood**[1] [mûd]. Fiil sıygası. ~[2]. Ruh haleti. **he is in one of his bad ~s,** yine aksiliği üzerinde: **in a good ~,** keyfi yerinde: **to be in the ~ to …,** …i canı istemek: **not to be in the ~ to,** içinden gelmemek: **he is in a generous ~ this morning,** bu sabah cömerdliği tuttu: **he is in no laughing ~,** yüzü hiç gülmüyor: **a man of ~s,** günü gününe uymaz. **~y,** huysuz, dargın; günü gününe uymaz.
**moon** [mûn]. Ay, kamer. **to ~ about,** dalgın dalgın etrafta gezinmek: **full ~,** dolunay: **new ~,** hilâl: **the changes [phases] of the ~,** ayın safhaları: **to cry for the ~,** olmıyacak şey istemek: **the man in the ~,** aydede: **I know no more about it than the man in the ~,** ben nereden bileyim? **~beam,** ayın şuaı. **~light,** mehtab, ay ışığı. **~shine,** saçma sapan söz.
**moor**[1], **~land** [mu̯ə(r), ˈmu̯ələnd]. Yüksek, ağacsız ve fundalıklı boş arazi. ~[2] *v.* Gemiyi karaya bağlamak; demirlemek. **~ing,** demir yeri: **ship at her ~s,** (gemi) dubaya bağlanmış; **to cast off [cut] one's ~s,** acele uzaklaşmak için palamarı kesmek. **M~**[3] *n.*, **~ish** [ˈmu̯əriş] *a.* Mağribî.
**moose** [mûs]. Kanada geyiği.
**moot** [mût]. Münakaşalı. İleri sürmek. **a ~ point,** su götürür bir mesele.
**mop** [mop]. Saplı tahta bezi; bulaşık için sicim fırça. Islatarak silmek. **to ~ up,** temizlemek; yok etmek: **to ~ one's brow,** mendille alnını silmek: **to ~ the floor with,** (*sl.*) haklamak.
**mope** [mou̯p]. Süngüsü düşük olmak.
**moraine** [məˈrêyn]. Glasiyelerin eteklerinde biriken kaya parçaları; buzultaş.
**moral** [ˈmorəl]. Manevî; ahlâkî; iyi ahlâklı, dürüst. Kıssadan hisse. **~s,** ahlâk. **~ courage,** medeni cesaret. **~e** [moˈrâl], maneviyat. **~ist,** ahlâkçı. **~ity** [məˈraliti], ahlâk(îlik); fazilet. **~ize** [ˈmorəlâyz], ahlâktan dem vurmak.
**morass** [məˈras]. Bataklık.
**moratorium** [ˌmorəˈtôri̯əm]. Borcların tecili.
**morbid** [ˈmôbid]. Marazî; iğrenc; nahoş şeylere marazî bir alâka gösteren.
**mordant** [ˈmôdənt]. Isırıcı; iğneli.
**more** [mô(r)]. Daha; daha çok; daha ziyade. **~ and ~,** gittikçe: **all the ~,** haydi haydi: **as many ~ (again),** bu kadar daha: **I'll have as many ~ as you can spare,** bundan fazla ne kadar verebilirseniz alırım: **the ~ the better,** ne kadar çok olursa o kadar iyi: **~ than enough,** yeter de artar: **~ or less,** aşağı yukarı: **the ~ you talk the less you think,** ne kadar çok konuşursan o kadar az düşünürsün: **a little ~ and I should have killed him,** az kaldı onu öldürüyordum: **never (no) ~!,** Allah göstermesin!: **it is no ~,** artık ortada yok: **he is no ~ Russian than I am,** kim demiş onu rus diye?: **'I can't understand it.' 'No ~ can I.'** 'Ben bunu anlamıyorum.'—'Benden de al, o kadar.': **'He said you couldn't speak Spanish.' 'No ~ I can.'** 'Sizin ispanyolca bilmediğinizi söyledi.'—'Bilmem, ya!'

one or ~, bir veya bir kaç: **the ~'s the pity!**, çok daha yazık: **the ~ so as ...**, bilhassa şunun için ki ...: **what ~ could you want!**, bundan iyisi can sağlığı: **what is ~**, dahası var. **~over** [mô'rouvə(r)], bundan başka; şu da var ki.
**morganatic** [môgə'natik]. Bir prensin asîl olmıyan bir kadınla evlenmesine aid.
**moribund** ['moribʌnd]. Ölüm halinde.
**morning** ['mônin(g)]. Sabah; öğleden evvel. Sabahlık. **the first thing in the ~**, sabahleyin erkenden: **Good ~!**, sabahlarınız hayır olsun; günaydın!: **it is my ~ off**, bugün öğleden evvel izinliyim. **~coat**, jaketatay.
**Morocc·o** [mə'rokou]. Fas. **~an** [-kən], faslı: **~ leather**, maroken, sahtiyan.
**moron** ['môron]. Ebleh, ahmak.
**morose** [mə'rous]. Abus, huysuz, gülmez.
**morphia** ['môfiə]. Morfin.
**morrow** ['morou]. Yarın.
**morsel** ['môsl]. Lokma; parça.
**mortal** ['môtl]. İnsan, beşer. Öldürücü; fani. **~ combat**, ölünceye kadar mücadele: **a ~ enemy**, can düşmanı: **~ fear**, can acısı: **~ sins**, büyük günahlar. **~ity** [–'taliti], vefiyat; ölüm nisbeti.
**mortar**[1] ['môtə(r)]. Harc. **~-board**, harc teknesi. **~**[2]. Havan; dibek; havantopu.
**mortgage** ['môgic]. İpotek; rehin. İpotek etmek.
**morti·fy** ['môtifây]. Nefse eza etm.; riyazet yapmak; tezlil etmek. Gangren olm. **~fication** [–fi'kêyşn], riyazet; zillet; kendisini zelil hissetme; (*med.*) nesiclerin gangren hali.
**mortuary** ['môtyuəri]. Morg.
**Mosaic** [mou'zêyik]. Musa peygambere aid; **mosaic**, mozaik.
**Moslem** ['mozlem]. Müslüman.
**mosque** [mosk]. Cami, mescid.
**mosquito** [mos'kîtou]. Sivrisinek. **~-net**, cibinlik.
**moss** [mos]. Yosun. **~y**, yosunlu.
**most** [moust]. **much** *veya* **many**' *nin tafdili*. En, en ziyade; pek çok; son derecede; ziyadesile; en büyük mikdar vs. **at ~**, en fazla; nihayet: **at the very ~ £100**, topu topu yüz lira: **to make the ~ of oneself**, kendini göstermek: **to make the ~ of one's hair**, saç tuvaletini kendisine en yakışan şekilde yapmak: **to make the ~ of a story**, bir hikâyeyi ballandıra ballandıra anlatmak: **it is a lovely day; let's make the ~ of it**, bugün hava çok güzel, aman ziyan etmiyelim: **we haven't much petrol; we must make the ~ of it**, cok benzinimiz yok fakat olanile idarei maslahat etmemiz lâzım: **he is more enterprising than ~**, pek çok kimselerden daha müteşebbisdir. **-most**. *suff. Bazı kelimelerin sonuna gelerek en çok, en ziyade manasını ifade eder, mes.*, **inner ~**, en iç.
**moth** [moθ]. Pervane; güve. **~-eaten**, güve yemiş; köhne.
**mother** [mʌðə(r)]. Anne, ana, valide. Analık etm.; evlâd gibi beslemek. **~ country** [**~ land**], anavatan: **~'s darling**, hanım evlâdı: **~ earth**, tabiat, toprak: **every ~'s son**, istisnasız her ferd: **~ wit**, feraset. **~hood**, analık. **~ly**, ana(lığ)a aid; ana gibi. **~-of-pearl**, sedef.
**moti·on** ['mouşn]. Hareket; kımıldanma; mekanizma; def'i tabiî; teklif; işaret. **to ~ s.o. to do stg.**, bir işaretle birini bir şeyi yapmağa davet etmek. **to put forward a ~**, bir teklif vermek: **the ~ was carried**, teklif kabul edildi: **to set in ~**, hareket ettirmek, işletmek. **~ve** ['moutiv], hareket ettirici. Sebeb, âmil, saik.
**motley** ['motli]. Renk renk; çeşid çeşid. **a ~ crowd**, her çeşid halktan kalabalık.
**motor** ['moutə(r)]. Motör; otomobil. Hareket ettirici; âmil. Otomobil ile gitmek. **~ist**, otomobil kullanan kimse. **~ize**, motör ile techiz etm. **~-boat**, motör (bot). **~-bus**, otobüs. **~-car**, otomobil. **~-cycle**, motosiklet.
**motto** ['motou]. Şiar; arma rümuzu.
**mould**[1] [mould]. Küf. **~y**, küflü; bayat; (*sl.*) yavan, sıkıntılı. **~**[2]. Kalıb; dökme kalıb; şekil; yaradılış. Kalıba dökmek; biçim vermek. **~er**[1], dökmeci. **~er**[2], cürüyüp toz haline gelmek. **~ing**, silme; pervaz.
**moult** [moult]. Tüy, saç vs.sini dökme(k).
**mound** [maund]. Küme; tepecik; höyük.
**mount**[1] [maunt] *v.* Bin(dir)mek; üzerine çıkmak; mukavva yahud beze yapıştırmak; yukarı çıkmak; yükselmek. **to ~ guard**, nöbetçi olm.: **to ~ the throne**, cülus etm.: **this ship ~s ten guns**, bu geminin on tane topu var: **to ~ up**, artmak;

çok olmak. **~ed** *a.*, atlı; mukavvaya yapışmiş (foto vs.): **silver-~**, gümüş geçirilmiş. **~²** *n.* Dağ, tepe; binek.
**mountain** [ˈmauntin]. Dağ. **~eer** [–ˈniə(r)], dağlı; dağcı; dağcılık yapmak. **~ous,** dağlık.
**mountebank** [ˈmauntiban(g)k]. Sokak cambazı; sahte doktor.
**mourn** [môn]. Matemini tutmak; ölümüne ağlamak; hasret çekmek. **~er,** matemli; matemci. **~ful** [ˈmônfəl], hüzünlü; ağlamış; yanık. **~ing,** matem(li).
**mouse** [maus]. Fındık faresi; küçük fare. Sıçan avlamak. **~trap,** fare kapanı.
**moustache** [musˈtâş]. Bıyık.
**mouth** *n.* [mauθ]. Ağız. *v.* [mauð], **to ~ one's words,** kelimeleri resmî bir eda ile ve tane tane telâffuz etmek. **useless ~,** fuzulî şahıs: **by word of ~,** şifahen. **~ful,** ağız dolusu. **~piece,** ağızlık: **to be the ~ of s.o.,** başkasının namına konuşmak. **~-organ,** ağız mızıkası.
**movable** [ˈmûvəbl]. Müteharrik; zamanı değişen. **~s,** mobilya; menkul eşya.
**move¹** [mûv] *n.* Kımıldanma; hareket (ettirme); taşınma; göç; tedbir; (satranc vs.) sürme. **to be always on the ~,** bir türlü yerleşememek; daima hareket halinde olmak: **I'm always on the ~,** dur yok otur yok: **to get a ~ on,** (*coll.*) çabuk olm.: **we must make a ~,** artık biz kaçalım; bir şey yapmalıyız: **he is up to every ~ (in the game),** o ne kurttur. **~²** *v.* Tahrik etm., işletmek; kımıldatmak; yürütmek; yer değiştirmek; nakletmek; kaldırmak; tesir etm., müteessir etm.; teklif etmek. Kımıldamak, yer değişmek; hareket etm.; yürümek; taşınmak, göçmek. **to ~ that ...,** ... teklif etm.: **to be ~d (by emotion),** mütehassis olm.: **keep moving!,** durmayınız!: **when the spirit ~s me,** canım istediği zaman: **it is time we were moving,** artık biz gidelim: **to ~ in high society,** yüksek sosyeteye devam etmek. **~ about,** dolaşmak; mütemadiyen taşınmak; bir yerden başka yere kaldırmak. **~ in,** taşınılan eve girmek. **~ off,** hareket etm., gitmek. **~ on,** başka bir yere göçmek; (polis) bir kalabalık dağıtmak. **~d** *a.*, müteessir. **~ment** [ˈmûvmənt], hareket, kımıldanış; saatin mekanizması.
**mov·ie** [ˈmûvi]. (*sl.*) Sinema filimi. **the ~s,** sinema. **~ing** *a.*, müteharrik, hareket halinde; müessir, acıklı.
**mow** [mou]. Ot biçmek; çimen kırpmak. **to ~ down the enemy,** düşmanı biçmek. **~er,** tırpancı; kırpma makinesi.
**M.P.** [ˈemˈpî]. **Member of Parliament,** milletvekili.
**Mr.** (*abb.*) Mister. (*Unvan*) ... efendi.
**Mrs.** (*abb.*) Mistress. (*Unvan*) ... hanımefendi.
**MS.** *pl.* **MSS.** (*abb.*) Manuscript, elyazması.
**much** [mʌç]. Çok; kadar. **~ of an age [size],** aşağı yukarı aynı yaşta [boyda]: **~ as I should like to ...,** çok isterdim ama ...: **~ as I like him ...,** kendisini çok severim, ama ...: **~ to my astonishment he did not come,** ne dersiniz gelmedi; **as ~ again,** bu kadar daha, bir misli daha: **I thought as ~,** bunu bekliyordum: **it's as ~ as saying he is a liar,** bu ona yalancı demeğe gelir: **it is as ~ as we can do to feed ourselves,** halbuki kendimizi besliyebilirsek ne mutlu!: **he looked at me as ~ as to say ...,** söylemek ister gibi baktı: **ever so ~ richer,** çok daha fazla zengin: **'What will it cost?' 'Ever so ~',** 'Kaça gelecek?' 'Pek pahalıya': **how ~?,** ne kadar?: **~ he knows about it!,** (*cont.*) tamam, şimdi bildi!: **to make ~ of s.o.,** (i) birisi için bayram yapmak; (ii) başının üstünde gezdirmek: **to make ~ of stg.,** izam etm., büyütmek; mübalağa etm.: **I can't make ~ of it,** ondan pek anlamam: **(pretty) ~ the same,** hemen hemen aynı: **not ~!,** (*sl.*) ne münasebet!: **not ~ of a doctor,** adamakıllı bir doktor değil: **so ~ the better!,** daha iyi ya!: **so ~ so that ...,** o derecede ki ...: **so ~ for your promise!,** nerede kaldı senin vaidin?: **he would not so ~ as answer,** cevab bile vermedi: **so ~ for that question, now for the next,** işte bu mesele böyle, şimdi ötekine geçelim: **this ~ is certain,** şurası muhakkaktır ki: **one can have too ~ of a good thing,** ⌜her şeyin fazlası fazla⌝: **you can't have too ~ of a good thing,** ⌜fazla mal göz çıkarmaz⌝: **that's (a bit) too ~ of a good thing,** bu kadarı da bir az fazla. **~ness** [ˈmʌçnis], **it's much of a ~,** ha öyle olmuş ha böyle (ikisi de bir).

**muck** [mʌk]. Gübre; pislik; çamur. **to ~ about,** (*coll.*) sürtmek: **to ~ up a job,** (*coll.*) bir işi berbat etmek. **~y,** pis; çamurlu.

**mud** [mʌd]. Çamur. **~dy,** çamurlu. **~guard,** tekerlek çamurluğu. **~-lark,** afacan. **~-slinging,** biribirine çamur atma (*fig.*).

**muddle** [ˈmʌdl]. Karışıklık; arabsaçı. Şaşırtmak; karıştırmak. **to be in a ~,** zihni karışmak: **to get in a ~,** işleri karışmak; belâya çatmak: **to ~ things up,** karıştırmak: **to ~ through,** yapılan hatalara rağmen işin içinden muvaffakiyetle çıkmak. **~-headed,** zihni karışık; sersem; kalın kafalı.

**muff** [mʌf]. Manşon.

**muffin** [ˈmʌfin]. Yassı pide.

**muffle** [ˈmʌfl]. Büründürmek; sarıp sarmalamak; (çan vs. sesini boğmak için) sarmak. **to ~ oneself up,** kendini sarıp sarmalamak. **~r,** boyun atkısı.

**mufti** [ˈmʌfti]. Sivil elbise.

**mug**[1] [mʌg]. Maşraba; bardak; (*sl.*) çehre. **~**[2], (*coll.*) Safdil, bön. **~**[3], (*sl.*) Çok çalışmak. **to ~ up a subject,** bir imtihan için bir mevzua çok çalışmak. **~gy** [ˈmʌgi], rutubetli ve ağır (hava).

**mulberry** [ˈmʌlberi]. Dut (ağacı).

**mul·e** [myûl]. Katır; inadcı adam; masura makinesi. **~eteer,** katırcı. **~ish,** katır gibi.

**mullet** [ˈmʌlit]. **red ~,** tekir balığı; barbunya: **grey ~,** kefal balığı.

**multi-** [ˈmʌlti] *pref.* Çok ..., *mes.* **~form,** çok şekilli; **~-engined,** çok makineli. **~farious** [ˌmʌltiˈfe̤əri̤əs]. Çeşid çeşid; muhtelif.

**multipl·e** [ˈmʌltipl]. Muhtelif, müteaddit, katmerli. (Riyaziye) misil; mazrub. **lowest common ~,** misli müşteriki asgar: **~ store,** bir çok şehirde şubhesi olan büyük mağaza. **~ication** [ˌmʌltipliˈkêyşn], zarb; çoğalma; çarpma: **~ table,** kerrat cedveli. **~icity** [–ˈplisiti], çokluk, kesret. **~y,** [ˈmʌltiplây], zarbetmek; çoğaltmak; çarpmak.

**multitude** [ˈmʌltityûd]. Kalabalık.

**mum** [mʌm]. Sessiz. **to keep ~,** ses çıkarmamak: **~'s the word !,** kimse duymasın !

**mumble** [ˈmʌmbl]. Mırıldanmak; anlaşılmaz tarzda konuşmak.

**mummy**[1] [ˈmʌmi]. Mumya. **~**[2]. Anneciğim.

**mumps** [mʌmps]. Kabakulak.

**munch** [mʌnç]. (*ech.*) Çiğnemek.

**mundane** [ˈmʌndêyn]. Dünyevi.

**municipal** [myuˈnisipl]. Belediyeye aid. **~ity** [–ˈpaliti], belediye idaresi.

**munificent** [myuˈnifisənt]. Cömerd.

**munitions** [myuˈnişənz]. Mühimmat; askerî levazım.

**mural** [ˈmyûrəl]. Duvara aid; duvara asılı.

**murder** [ˈmö(r)də(r)]. Taammüden katil. Katletmek. ⌜**~ will out**⌝, hakikat (kabahat) sonunda meydana çıkar. **~er,** kaatil. **~ous,** katletmeğe niyetli.

**murky** [ˈmö(r)ki]. Kararmış; isli. **a ~ past,** şübheli bir mazi.

**murmur** [ˈmö(r)mə(r)]. Mırıltı; ses; homurtu. Mırıldanmak; homurdanmak. **to ~ at** [**against**] **stg.,** bir şeye karşı homurdanarak söylenmek.

**musc·le** [ˈmʌsl]. Adale; kuvvet. **not to move a ~,** kılını kıpırdatmamak. **~ular** [ˈmʌskyulə(r)], adalî, adaleli; kuvvetli.

**muse**[1] [myûz] *n.* Esatirde dokuz güzel sanat ilâhesinden her biri; şiir perisi: **to invoke the ~,** ilham davet etmek. **~**[2] *v.* Düşünceye dalmak.

**museum** [myûˈzi̤əm]. Müze. **a ~ piece,** bir müzede teşhir edilmeğe lâyık; (*cont.*) müzelik.

**mush** [mʌş]. Pelte; lâpa.

**mushroom** [ˈmʌşrum]. Mantar; türedi; birdenbire zuhur eden.

**music** [ˈmyûzik]. Musiki; müzik. **to face the ~,** tenkidcileri cesaretle karşılamak: **to set to ~,** bestelemek. **~al** [–kl], müziğe aid, müziği sever; çalgılı; ahenkli. **~ian** [myuˈzişn], musikişinas; çalgıcı.

**musk** [mʌsk]. Mis; misk otu.

**musket** [ˈmʌskit]. Misket tüfeği. **~eer** [–ˈti̤ə(r)], tüfekçi; silahşor. **~ry,** küçük silâhlarla ateş.

**muslin** [ˈmʌzlin]. Müslin(den yapılmış).

**mussel** [ˈmʌsl]. Midye.

**must**[1] [mʌst] *n.* Şira. **~**[2] *v.* [*Tasrifte hiç değişmiyen bir yardımcı fiil; umumiyetle vücubî sıygası ifade eder.*] Mecbur olm., lâzım olm.; icabetmek, her halde ... olmak. **I ~ go,** gitmeliyim, gitmem lâzım: **I ~ not** [**mustn't**] **go,** gitmemem lâzım, gitmem yasaktır; [gitmem lâzım değil, **I need not go**]: **you ~ learn Turkish,** (i) türkçe öğrenmeniz lâzım; (ii) türkçe öğrenseniz iyi olur: **you <u>must</u> learn T.,**

muhakkak türkçe öğrenmelisiniz: **you ~ have forgotten me,** beni unutmuşsunuzdur: **you must know him,** onu tanımamanıza imkân yok: **I ~ be going,** artık gitmeliyim: **I am going because I must,** mecbur olduğum için gidiyorum (yoksa kalırdım): **do so if you ~,** icabediyorsa [zarurî ise] yapınız: **England, you ~ know, is not all factories,** şurasını söyleyim ki İngiltere fabrikadan ibaret değil: **it ~ be ten o'clock,** saat, her halde, on vardır: **I ~ have made a mistake,** her halde bir hata yaptım: **if you go that way you must meet him,** oradan giderseniz muhakkak ona rastlarsınız: **just as we were starting on our journey he ~ break his leg,** tam seyahate çıkacağımız sırada aksi gibi bacağı kırıldı.

**mustard** [ˈmʌstəd]. Hardal.

**muster** [ˈmʌstə(r)]. Asker toplama; toplantı. Topla(n)mak. **to pass ~,** teftişten geçirilmek; kabul edilebilmek.

**musty** [ˈmʌsti]. Küf kokulu; küflü; köhne.

**mute** [myût]. Sessiz; dilsiz. Dilsiz insan; okunmıyan harf; ücretli matemli. Bir çalgı vs.nin sesini kısmak. **deaf ~,** sağır ve dilsiz adam.

**mutilate** [ˈmyûtilêyt]. Bir uzvunu kesmek; sakat etm.; bozmak, kırmak.

**mutin·eer** [ˌmyûtiˈniə(r)]. Âsi (asker). **~ous** [ˈmyûtinəs], isyan halinde. **~y** [ˈmyûtini], isyan (askerî). İsyan etm.; ayaklanmak.

**mutter** [ˈmʌtə(r)]. Mırıldamak; homurdanmak. Mırıltı.

**mutton** [ˈmʌtn]. Koyun eti. **~-chop,** koyun pirzolası.

**mutual** [ˈmyûtyuəl]. Karşılıklı; mütekabil. **a ~ friend,** müşterek dost.

**muzzle** [ˈmʌzl]. Hayvan burnu; top ağzı; burunsalık. Burunsalık takmak; (*fig.*) ağzına gem vurmak. **~-loader,** ağızdan dolma top.

**my** [mây]. Benim. **~!,** aman! **~self** [mâyˈself], ben kendim.

**myriad** [ˈmiriəd]. Çok büyük aded. **~s of,** binlerce.

**myster·ious** [misˈtiəriəs]. Esrarengiz; gizli kapaklı. **~y** [ˈmistəri], esrarlı, anlaşılmaz şey; hikmet; muamma.

**mystic** [ˈmistik]. Tasavvufî; sufî. **~ism** [ˈmistisizm], tasavvuf, mistisizm.

**mystify** [ˈmistifây]. Esrarlı bir oyunla aldatmak; esrarengiz görünmek, göstermek.

**myth** [miθ]. Esatir hikâyesi; masal; hurafe; ismi var cısmı yok. **~ical,** esatiri. **~ology** [mâyˈθoləci], esatir; mitoloji.

# N

**N** [en]. N harfı.

**nab** [nab] (*sl.*) Kapmak; aşırmak; yakalamak.

**nabob** [ˈnêybob]. Nevvab; çok zengin adam.

**nadir** [ˈnêydə(r)]. Semtikadem; en aşağı nokta veya safha.

**nag¹** [nag] *n.* Ufak beygir. **~²** *v.* Dırlamak. **to ~ at s.o.,** birinin başının etini yemek.

**nail** [nêyl]. Çivi; tırnak; pençe. Çivilemek. **~ down,** çivi ile kapatmak: **~ up,** çiviliyerek kapatmak (asmak). **as hard as ~s,** çok sıhhatli ve dayanıklı: **as right as ~s,** dosdoğru: **to hit the ~ on the head,** tam üzerine basmak: **to ~ a lie to the mast,** yalanını meydana koymak: **that's another ~ in his coffin,** bu onun sonunu (ölümünü) biraz daha yaklaştırır: **on the ~,** (*sl.*) derhal, peşin para ile: **he stood ~ed to the spot,** donakaldı: **to ~ s.o. (down) to his promise,** birine vaidini tutturmak.

**naïve** [naˈîv, nêyv]. Saf; sadedil; bön. **~ty** [–iti], sadedillik.

**naked** [ˈnêyked]. Çıplak. **~ sword,** yalın kılıc: **visible to the ~ eye,** gözle görülebilir (dürbünsüz).

**namby-pamby** [ˈnambiˈpambi]. Mahallebici hanım evlâdı ve yapmacıklı (genc).

**name¹** [nêym] *n.* İsim, ad, nam; şöhret. **to get a bad ~,** adı çıkmak: **by ~,** isminde: **to know by ~,** gıyaben tanımak: **to call (each other) ~s,** birbirine fena sözler söylemek: **'to mention [~] no ~s',** isim tasrih etmek istemiyorum: **I'll do so or my ~ is not ...,** bunu yapmazsam bana da adam demesinler: **to send in one's ~,** ismini içeri haber vermek; bir müsabakaya ismini yazdırmak: **to put one's ~ down,** ismini yazdırmak: **a king in ~ only,** yalnız adı kıral: **what in the ~ of goodness are you doing?,** ne yapıyorsun Allah aşkına? **~²** *v.*

İsim koymak; ad takmak; zikretmek; tayin etmek. **to ~ s.o. after** [*Amer.* for] **s.o.**, birine birinin ismini vermek. **~less** [ˈnêymlis], isimsiz; mechul; anlatılamaz. **a person who shall be ~**, ismini söylemiyeceğim bir zat. **~ly**, yani; şöyle ki. **~sake** [ˈnêymsêyk], adaş.

**nanny** [ˈnani]. Dadı. **~-goat**, dişi keçi.

**nap**[1] [nap]. Hafif uyku. **to take** [**have**] **a ~**, hafif uykuya dalmak; kestirmek. **~ping, to catch s.o. ~**, birini gafil avlamak. **~**[2], hav.

**nape** [nêyp]. **the ~ of the neck**, ense.

**naphtha** [ˈnafθə]. Neft.

**napkin** [ˈnapkin]. Peşkir; peçete.

**narcissus** [nâˈsisəs]. Nerkis.

**narcotic** [nâˈkotik]. Uyutucu; uyuşturucu.

**narrat·e** [nəˈrêyt]. Nakletmek; hikâye anlatmak. **~ive** [ˈnarətiv], nakil; hikâye; ifade; fıkra. Nakil ve rivayete aid. **~or** [–ˈrêytə(r)], nakleden; hikâyeci.

**narrow** [ˈnarǫu]. Dar; mahdud. Daraltmak; darlaşmak. **the ~s**, boğaz, dar liman ağzı. **~-gauge railway**, dar hatlı demiryolu. **~-minded**, darkafalı; muteassıb.

**nasal** [ˈnêyzl]. Buruna aid; enfî; gunneli. Genizden okunan harf.

**nascent** [ˈnasənt]. Doğan; vücude gelen.

**nasty** [ˈnâsti]. Hoşa gitmiyen; nahoş; pis; iğrenc. **a ~ sea**, dalgalı deniz: **to turn ~**, hiddete kapılmak; tehdidkâr olm.: **that's a ~ one !**, (*coll.*) *lâfı ağza kapayan bir cevab hakkında kullanılır; nahoş bir haber alındığı zaman söylenir.*

**nation** [ˈnêyşn]. Millet; devlet; memleket. **~al** [ˈnaşnl]. millî; tebaa. **~alist**, milliyetçi; milliyetperver. **~ality** [ˌnaşəˈnaliti], milliyet; tabiiyet. **~alize** [ˈnaşnəlâyz], millileştirmek; devletleştirmek.

**nativ·e** [ˈnêytiv]. Yerli; doğma; memlekete aid; memlekette yetişen. **~ language**, ana dil: **~ land**, anavatan. **~ity** [nəˈtiviti], İsa'nın doğumu.

**natty** [ˈnati]. Zarif, süslü.

**natural** [ˈnatyurəl]. Tabiî; tabiate aid; anadan doğma; cibillî; sun'î değil; halis; normal. Anadan doğma budala. **~ child**, piç: **~ father**, piçin babası: **~ history**, tabiiye: **to die a ~ death**, eceliyle ölmek: **it comes ~ to him**, ona çok kolay gelir. **~ist** [ˈnatyurəlist], hayvanları ve nebatları tedkik eden meraklı. **~ize** [ˈnatyurəlâyz], tâbiiyete kabul etm.; hayvan vs. yeni iklime alıştırmak: yerleştirmek. **~ly** [ˈnatyurəli], tabiî surette, kolayca: **~!**, tabiî !

**nature** [ˈnêytyə(r)]. Tabiat; hilkat; yaradılış; mahiyet. **by ~**, fıtraten: **from ~**, canlı bir modelden, tabiî manzaradan: **something in the ~ of** ..., ...in kabilinden. **-~d**, ...huylu, ... tabiatli.

**naught** [nôt]. Hiç, sıfır. **to bring to ~**, akamete uğratmak: **to come to ~**, suya düşmek: **to set at ~**, hiçe saymak.

**naught·y** [ˈnôti]. Yaramaz, huysuz. **~iness**, yaramazlık.

**nause·a** [ˈnôsįə]. Bulantı; iğrenme. **~ate**, mide bulandırmak; iğrendirmek; bıktırmak: **~ous**, iğrenc, bıktırıcı.

**nautical** [ˈnôtikl]. Gemiciliğe aid. **~ mile**, deniz mili.

**naval** [ˈnêyvl]. Harb gemilerine ve bahriyeye aid. **~ architecture**, gemi mühendisliği: **~ battle**, deniz muharebesi.

**nave** [nêyv]. Kiliselerin ortasında bulunan geniş yer.

**navel** [ˈnêyvl]. Göbek.

**navigable** [ˈnavigəbl]. Seyrüsefere müsaid (sular); denize dayanabilir (gemi).

**nàvigat·e** [ˈnavigêyt]. Gemi idare etm., kullanmak. **~ion** [–ˈgêyşn], deniz yolculuğu; denizcilik ilmi: **~ lights**, bir gemide bulunması icabeden ışıklar. **~or**, gemici; gemi veya uçağın rotasından mesul olan zabit.

**navvy** [ˈnavi]. Toprak tesviyesi vs. ağır işlerde çalışan amele. **steam ~**, kazma makinesi.

**navy** [ˈnêyvi]. Bahriye. **merchant ~**, ticaret filosu: **~ blue**, lâciverd.

**nay** [nêy]. Hayır. **I cannot say him ~**, ona hayır diyemem: **he will not take ~**, menfi cevab kabul etmez.

**N.B.** [ˈenˈbî]. (*Lât.*) **nota bene**, ihtar.

**near** [nįə(r)]. Yakın; karib; takribî; civarında; (*coll.*) cimri. Yaklaşmak. **~by**, yanında: **as ~ as I can remember**, hatırımda kaldığına göre: **I am nowhere ~ as rich as you are**, ben sizin kadar zengin olmaktan uzağım: **to come** [**draw**] **~**, yaklaşmak: **I came ~ to being drowned**, az kaldı boğuluyordum: **those who are ~ and**

dear to us, yakınlarımız: ~ at hand, yanında: ~ing forty, kırkına merdiven dayamış: to go by the ~est road, kestirme yoldan gitmek: the ~ side, sol taraf: it was a ~ thing, dar kurtuldum: ~ upon a hundred, yüz ya var ya yok. ~ly [ˈniəli], hemen hemen; takriben; yakın. I ~ fainted, az kaldı bayılıyordum: he is not ~ as rich as you, o sizin kadar zengin olmaktan uzaktır. ~-sighted [ˌniəˈsâytid], miyop.

**neat** [nît]. Muntazam; zarif; temiz; basit ve iyi; tertibli; (rakı vs.) susuz.

**nebul·a**, *pl.* **-ae** [ˈnebyulə, –î]. Küçük yıldızlar kümesi. **~ous**, sisli; mübhem.

**necess·ary** [ˈnesəseri]. Lâzım; gereken. Lâzım olan şey, eşya. **the ~**, bir şey için lâzım olan para: **the necessaries of life**, zarurî ihtiyaclar: **to be ~**, lâzım olm., icabetmek: **to do the ~**, icabeden şeyleri yapmak: **if ~**, icab ederse. **~itate** [niˈsesitêyt], icabetmek. **~itous**, muhtac, fakir. **~ity** [niˈsesəti], zaruret; lüzum; mecburiyet: **of ~**, zarurî olarak: ⌜**~ knows no law**⌝, muztar kalınca her şey yapılır.

**neck** [nek]. Boyun; (şişe) boğaz. **a ~ of land**, küçük berzah: **~ and crop**, tamamen: **to fall on s.o.'s ~**, birinin boynuna sarılmak: **to get it in the ~**, (*sl.*) şiddetli bir darbeye uğramak: **~ and ~**, başabaş: ⌜**~ or nothing**⌝, ⌜ya devlet başa ya kuzgun leşe⌝: **to save one's ~**, postu kurtarmak: **to be up to one's ~ in debt**, uçan kuşa borclu olm.; **to be up to one's ~ in work**, işi başından aşmak. **~cloth**, boyunbağı. **~lace** [ˈneklis] gerdanlık. **~-tie**, boyunbağı.

**nectar** [ˈnekta(r)]. Kevser.

**need**¹ [nîd] *n.* İhtiyac; lüzum; zaruret. **in ~**, muhtac; fakir: **to have [be in, stand in] ~ of ...**, ...e muhtac olm.: **in case of ~**, icabında: **my ~s are few**, ihtiyaclarım mahduddur: **in times of ~**, müşkül zamanda. **~**² *v.* Muhtac olm.; icabetmek; istemek. **he ~s a lot of asking**, yalvartmadan bir şey yapmaz: **why ~ he have come tonight (of all nights)?**, ne diye tutup da bu akşam geldi?: **you only ~ed to ask**, sormanız kâfi idi: **he ~ not go, ~ he?**, onun gitmesi lâzım değil, değil mi?: **you ~ not have done it**, yapmıyabilirdiniz: **you ~n't have been so rude**, (yaptığınız) bu nezaketsizliğe hiç lüzum yoktu. **~ful** [ˈnîdfəl] *v.*, **necessary**. **~less** [ˈnîdlis], lüzumsuz; beyhude. **~s** [nîdz] *adv.* *Yal.* **must** *ile kullanılır*; **if ~ must**, mutlak lâzımsa; zarurî ise: ⌜**~ must when the devil drives**⌝, mutlak yapmalıyım.

**needle** [ˈnîdl]. İğne; tığ; örgü şişi; ibre; çuvaldız. ⌜**it's like looking for a ~ in a haystack**⌝, saman yığınında iğne aramak gibi (*yani* bulmak hemen hemen imkânsız): **as sharp as a ~**, şeytan gibi zeki. **~woman**, dikişçi kadın. **~work**, dikişçilik; işleme.

**ne'er** [neə(r)]. =**never**. **~-do-well [-weel]**, adam olmaz; serseri.

**nefarious** [niˈfeəriəs]. Çirkin; şeni.

**negat·ion** [niˈgêyşn]. İnkâr; nefi. **the ~ of ...**, ...in zıddı. **~ive** [ˈnegətiv], nefi (edatı); negatif; menfi. Reddetmek.

**neglect** [niˈglekt]. İhmal etm.; iyi bakmamak. Bakımsızlık; ihmal. **~ed**, bakımsız; metrûk. **~ful**, ihmalci; dikkatsiz.

**neglig·ence** [ˈneglicəns]. İhmal; gaflet; itinasızlık. **~ent**, ihmalci, kayıdsız, dikkatsiz. **~ible** [–əbl], ehemmiyetsiz; sayılmaz.

**negoti·ate** [niˈgouşiêyt]. Müzakereye girişmek; havale ve ciro etmek. **~able**, havale ve ciro edilebilir; cirolu. **~ation** [niˌgouşiˈêyşn], müzakere; akdetme. **~ator** [niˈgouşiêytə(r)], müzakereye memur kimse; murahhas.

**negr·o** [ˈnîgrou]. Zenci. **~ess**, zenci kadın. **~oid**, zenciye benzer.

**neigh** [nêy]. Kişneme(k).

**neighbour** [ˈnêybə(r)]. Komşu. **one's duty towards one's ~**, insanlara karşı vazifelerimiz. **~hood** [–hud], civar; cihet. **~ing**, komşu, bitişik. **~ly**, iyi komşu gibi.

**neither** [ˈnâyðə(r)]. Hiç birisi. **~ ... nor ...**, ne ... ne .... **if you don't go ~ shall I**, siz gitmezseniz ben de gitmem.

**nem. con.** [ˈnemˈkon]. (*Lât.*) = **nemine contradicente.** Kimse muhalefet etmiyerek.

**neo-** [ˈnîou] *pref.* Yeni .... **~logism**, yeni kelime; eski kelimelerin yeni manası.

**nephew** [ˈnefyu]. Yeğen (erkek).

**nepotism** [ˈnepətizm]. Akraba kayırma.

**Neptune** [ˈneptyûn]. Deniz ilâhı; Neptün seyyaresi.

**nerv·e** [nö(r)v]. Sinir, asab; cesaret. Cesaret vermek. **to get on s.o.'s ~s,** birinin sinirine dokunmak: **to have the ~ to,** cüret etm.: **to lose one's ~,** cesaretini kaybetmek: **a man of ~,** pek soğukkanlı: **you have got a ~!,** ne cesaret!: **to be in a state of ~s,** sinirli olm.: **to strain every ~,** alabildiğine çabalamak. **~e-racking,** sinirleri bozan. **~ous** [ˈnö(r)vəs], ürkek; sıkılgan; evhamlı; sinirli; asabî; sinirlere aid. **to be ~ about doing stg.,** bir şeyi yapmaktan çekinmek: **to be ~ about s.o.,** birini merak etm. **~-ousness,** ürkeklik, sıkılganlık, çekingenlik. **~y,** sinirli; ürkek.

**nest** [nest]. Yuva; (haydud) yatağı; küçük içiçe kutular vs. Yuva yapmak. **~ling,** kuş yavrusu. **~-egg,** fol; ihtiyaten bir tarafa konulan küçük sermaye. **~le** [ˈnesl], kendi yuvasını yapmak; sığınmak; saklanmak.

**net**[1] [net]. Ağ şebeke. Ağ ile tutmak, örtmek. **~**[2]. Sâfi; darası alınmış. (Bir iş) sâfi kâr temin etmek. **he ~ted a nice little sum,** bu işte epeyi para vurdu. **~ted** [ˈnetid], ağ halinde; ağ gibi. **~ting,** ağ örgüsü. **~work** [ˈnetwö(r)k], şebeke.

**nether** [ˈneðə(r)]. Alttaki. **The N~-lands** [ˈneðələndz], Holanda, Felemenk.

**nettle** [ˈnetl]. Isırgan. (*fig.*), kızdırmak.

**neur·algia** [nyu̯əˈralcə]. Nevralji. **~itis** [–ˈrâytis], sinir iltihabı. **~-ologist** [–ˈroləcist], asabiyeci. **~osis** [–ˈrou̯sis], sinir hastalığı. **~otic** [–ˈrotik], asabî.

**neuter** [ˈnyûtə(r)]. Cinsiyetsiz; (*gram.*) müzekker olmıyan, bitaraf.

**neutral** [ˈnyûtrəl]. Tarafsız: (renk) kurşunî: (otom.) ölü nokta. **~ity** [–ˈtraliti], tarafsızlık. **~ize** [ˈnyutrəlâyz], tesirsiz bırakmak; akamete uğratmak.

**never** [ˈnevə(r)]. Hiçbir zaman; aslâ; kat'iyyen. **~ a one,** hiç biri bile değil: **~ again will I go there,** bir daha oraya gitmem: **be he ~ so angry,** ne kadar kızarsa kızsın: **you surely ~ said that!,** nasıl oldu da bunu söyledin?: **I have ~ yet seen it,** onu daha hiç görmedim: **well I ~!,** Allah! Allah! **~-ending,** bitmez; fasılasız. **~-failing,** tükenmez. **~-more** [ˈnevəmô(r)], bir daha hiç. **~theless** [ˌnevəðəˈles], bununla beraber; maamafih.

**new** [nyû]. Yeni; taze; acemi. **N~ World,** Amerika: **N~ Year,** yeni sene, sene başı: **to see the N~ Year in,** yıl başını kutlamak: **to wish s.o. a Happy N~ Year,** birinin yeni yılını tebrik etmek. **~-comer,** yeni gelen. **~-fangled,** yeni çıkma. **~-fashioned,** yeni moda. **~-laid, ~ egg,** günlük yumurta.

**Newcastle** [ˈnyûkâsl]. ⌜**to carry coals to ~**⌝, dereye su taşımak.

**Newfoundland** [ˌnyûfəndˈland]. Ternöv; [nyûˈfau̯ndlənd], ternöv köpeği.

**news** [nyûz]. Havadis; haber. **to be in the ~,** herkesin ağzında olm.: **to break the ~ gently,** alıştıra alıştıra söylemek: **a good [bad] piece of ~,** iyi [fena] haber: **this is ~ to me,** bunu işitmemiştim: **what's the ~?,** ne var ne yok? **~agent,** gazeteci (dükkânı). **~monger,** havadis kumkuması. **~paper,** gazete. **~-agency,** haber alma acentası. **~-print,** gazete kâğıdı.

**next** [nekst]. Gelecek; en yakın; ertesi; sonraki; öteki; önümüzdeki; bitişik; yanında; bundan sonra. **to be continued in our ~,** devamı gelecek sayıda: **he lives ~ door to us,** bitişiğimizde oturur: **the thing ~ my heart,** en çok arzu ettiğim şey: **the ~ largest,** ondan sonra en büyüğü: **~ to nothing,** hemen hemen hiç: **to wear flannel ~ the skin,** fanilayı tenine giymek: **the ~ time I see him,** bir daha onu gördüğüm zaman: **what ~!,** olur şey değil!: **what ~, please?,** (dükkânda) başka ne istiyorsunuz?: **who comes ~?,** sıra kimde?: **the year after ~,** öbür sene. **~-door neighbour,** bitişik komşu.

**nib** [nib]. Kalemucu.

**nibble** [ˈnibl]. Kemirmek; (koyun) otlatmak. Ufacık lokma.

**nice** [nâys]. Hoş, sevimli; tatlı; cazib; nefis; ince; titiz. **a ~ distinction,** ince bir fark: **to be ~ to s.o.,** birine iyi muamele etm.: **it is ~ of you to ...,** ···mekle nezaket gösterdiniz: **it is ~ and cool,** hava çok tatlı ve serin: **this is a ~ mess,** işler arabsaçına döndü. **~ty** [ˈnâysiti], ince nokta, incelik. **to a ~,** tam karar; tamamile: **the niceties of a language,** bir lisanın gavamızı.

**niche** [niç]. Duvarda hücre.

**nick** [nik]. Çentmek. Çentik. **in the**

~ **of time,** tam zamanında. **Old N~,** Şeytan. **~name** [ˈniknêym], lâkab (takmak).

**nickle** [ˈnikl]. Nikel; beş sentlik Amerikan parası. **~-plated,** nikel kaplama.

**niece** [nîs]. Yeğen (kız).

**niggard** [ˈnigö(r)d]. Cimri adam. **~ly,** cimri, pinti.

**nigger** [ˈnigə(r)]. Zenci. **to work like a ~,** domuzuna çalışmak: ⌜**a ~ in the woodpile**⌝, çapanoğlu.

**niggling** [ˈniglin(g)]. Lüzumsuz teferruatlı.

**nigh** [nây]. Yakın; aşağı yukarı.

**night** [nâyt]. Gece; karanlık. **at** [**by**] **~,** geceleyin: **good ~!,** geceniz hayırlı olsun!: **to have a good** [**bad**] **~** (**'s rest**), iyi [fena] uyumak: **in the ~,** geceleyin: **to make a ~ of it,** geceyi zevk ile geçirmek; sabahlamak: **a ~ out,** zevk ile geçirilen gece; bir hizmetçinin izinli olduğu gece: **first ~,** piyesin ilk temsili. **~dress** [**~gown**], gecelik. **~fall,** akşam üzeri. **~ly,** her gece yapılan. **~mare,** kâbus. **~shirt,** gecelik entari. **~cap,** gecelik takke; yatarken içilen (alkollü) içki. **~-light,** idare lâmbası. **~-shift,** gece nöbeti; gece çalışan ekip. **~ingale** [ˈnâytingêyl], bülbül.

**nil** [nil]. Hiç; sıfır.

**nimble** [ˈnimbl]. Çevik; tetik; tez.

**nimbus** [ˈnimbəs]. Yağmur bulutu; hale.

**nincompoop** [ˈnin(g)kəmpûp]. Alık, avanak.

**nine** [nâyn]. Dokuz. **to have ~ lives (like a cat),** yedi canlı olm.: **dressed up to the ~s,** iki dirhem bir çekirdek: **a ~ days' wonder,** birdenbire meşhur olup kısa zamanda unutulan şey: **~ times out of ten,** hemen her defa. **~pins,** şişe şekilli dikili dokuz tahtayı uzaktan bir tahta topla devirmekten ibaret olan oyun: **to go down like ~,** iskambil gibi devrilmek. **~teen,** on dokuz: **to talk ~ to the dozen,** makine gibi konuşmak. **~ty,** doksan: **the ~ties,** 1890 ile 1900 seneleri arasında.

**ninny** [ˈnini]. Alık.

**ninth** [nâynθ]. Dokuzuncu.

**nip**[1] [nip]. Çimdik; hafif ısırma. Çimdiklemek; hafifçe ısırmak; (kırağı) haşlamak. (*sl.*) Çabuk gitmek. **there is a ~ in the air,** hava sert: **to ~ in the bud,** bir kötülüğün vs. daha başlangıcda önüne geçmek. **to ~ off,** ucunu dişleyip koparmak.

**~**[2]. Bir yudum içki.

**nipper** [ˈnipə(r)]. (*sl.*) Çocuk.

**nippers** [ˈnipö(r)z]. Ufak kerpeten.

**nipple** [ˈnipl]. Memebaşı.

**nippy** [ˈnipi]. (*coll.*) Çevki, atik; (hava) keskin. **look ~ about it!,** haydi çabuk ol!

**nitr·ate** [ˈnâytrêyt]. Azotiyet. **~ic acid,** asid nitrik. **~ogen** [ˈnâytrocən], azot, nitrojen.

**nitwit** [ˈnitwit]. Budala.

**no** [nou̯]. Hayır; yok; öyle değil; hiç; menfi cevab. **~es,** bir teklif aleyhine rey verenler. **it is ~ distance,** uzak değil: **friend or ~ (friend) he can't behave like that,** dost olsun böyle hareket edemez: **he is ~ genius,** elbette dâhi değil: **in less than ~ time,** pek az sonra: **~ man** [**one**], hiç bir kimse: **make ~ mistake!,** duydum duymadım deme!: **I made ~ reply,** cevab vermedim: **there is ~ saying what he will do next,** bundan sonra ne yapacağı bilinmez: **~ smoking!,** tütün içilmez: **at ~ time,** hiç bir vakit: **whether he comes or ~,** gelse de gelmese de: **tell me whether you are coming or ~,** gelip gelmiyeceğinizi haber veriniz.

**Noah** [ˈnou̯ə]. Nuh peygamber. **~'s ark,** Nuh gemisi.

**nob** [nob]. (*sl.*) Baş, kelle; kibar.

**nob·ility** [nou̯ˈbiliti]. Asalet. **the ~,** asılzadeler. **~le** [ˈnou̯bl], asîl; ulvi. Asılzade. **~man,** asılzade.

**nobody** [ˈnou̯bodi]. Hiç kimse. Değersiz adam. **he's a mere ~,** o solda sıfırdır.

**nocturnal** [nokˈtö(r)nl]. Geceleyin olan.

**nod** [nod]. Baş sallama; başla işaret. Kabul, selâm vs. ifade etmek için başını sallamak; pineklemek. **to have a ~ding acquaintance with s.o.,** birile pek az tanışmak.

**nohow** [ˈnou̯hau̯]. Hiç bir suretle.

**nois·e** [nôyz]. Gürültü; velvele; ses. **to ~ abroad,** (bir haberi) yaymak: **to be a big ~,** borusu ötmek: **to make a ~ in the world,** meşhur olmak. **~y** [ˈnôyzi], gürültülü, patırdılı.

**noisome** [ˈnôysəm]. Muzır; iğrenc.

**nolens volens** [ˈnou̯lenzˈvou̯lenz]. (*Lât*). İster istemez.

**nomad** [ˈnou̯mad]. Göçebe; bedevi. **~ic** [noˈmadik], göçebe.

**nomin·al** [ˈnominl]. İsmi olup cismi olmıyan; itibarî; isim veya isimlere

aid. ~ **value**, itibar: **he is the ~ head,** adı reis. **~ate,** tayin etm.; bir memuriyet için teklif etm.; namzedliğe seçmek. **~ative,** mücerred hali. **~ee** [nomiˈnî], namzed; mansub.
**non-** [non] *suff. Nefi edatı*; gayri-; ademi-.
**non-aggression** [nonəˈgreşn]. Ademi tecavüz; saldırmazlık.
**nonce** [nons]. **for the ~,** bu kere.
**nonchalan·t** [ˈnonşalənt]. Kayıdsız; kaygısız. **~ce,** kayıdsızlık.
**non-combatant** [nonˈkombətənt]. Gayri muharib.
**non-commissioned** [nonkəˈmişnd]. **~ officer (N.C.O.),** küçük zabit.
**non-committal** [nonkəˈmitl]. Suya sabuna dokunmaz; mübhem.
**nonconformist** [nonkənˈfômist]. Anglikan kilisesinden itizal eden protestan.
**nondescript** [ˈnondiskript]. Tasnif ve tarif edilemez; tuhaf.
**none** [nʌn]. Hiç; hiç kimse; hiç bir. ⌜**~ so blind as those who won't see**⌝, en fena kör görmek istemiyendir: **~ but he knows the secret,** bu sırrı ondan başka kimse bilmiyor: **you know, ~ better,** siz herkesten iyi bilirsiniz: **any food is better than ~ (at all),** her hangi bir yemek hiç yoktan iyidir: **~ of your cheek!,** yüzsüzlüğün lüzumu yok!: **~ of them are coming,** hiç biri gelmiyecek: **~ of this is suitable,** bunun hiç bir kısmı uygun değil: **I was ~ too soon,** tam zamanında yetiştim: **he is ~ too well off,** (malî) vaziyeti pek o kadar iyi değil: **I like him ~ the worse for that,** bundan dolayı onu daha az seviyor değilim: **he is ~ the worse for his illness,** hastalık geçirdiği halde sıhhatine hiç tesir etmedi.
**nonentity** [nǫuˈnentiti]. Ehemmiyetsiz, solda sıfır bir adam.
**non-essential** [noneˈsenşl]. Feri; tâli.
**non-existent** [onegˈzistənt]. Gayri mevcud.
**nonplus** [ˈnonplʌs]. Şaşırtmak. **to be ~sed,** apışık kalmak.
**non-resident** [nonˈrezidənt]. İkamet etmiyen. **~ landowner,** kendi emlâkinde oturmıyan mülk sahibi: **~ student,** nehari talebe.
**nonsens·e** [ˈnonsəns]. Saçma; boş sözler; ahmaklık. **~ical** [–ˈsensikl], saçma sapan.
**non-stop** [ˈnonˈstop]. **~ train,** hiç bir yerde durmıyan (doğru giden) tren: **~ flight,** doğrudan doğruya uçuş.
**noodles** [ˈnûdlz]. Erişte.
**nook** [nuk]. Bucak, köşe.
**noon** [nûn]. Öğle vakti. **~day** [**~tide**], öğle.
**noose** [nûs]. İlmik, kemend. İlmikle tutmak.
**nor** [nô(r)]. Ve ne de. **neither ... ~ ...,** ne ... ve ne de ....
**nor'.** (*abb.*) **north.**
**norm** [nôm]. Nümune; alelâde tip. **~al** [ˈnôml], alelâde; tabiî. **~ality** [–ˈmaliti], alelâdelik.
**Norseman** [ˈnôsmən]. Eski Norveçli.
**north** [nôθ]. Şimal; (*naut.*) yıldız. **the ~ Country,** İngiltere'nin şimali: **the ~ star,** kutub yıldızı. **~erly** [ˈnôð-ö(r)li], **a ~ wind,** şimalden gelen rüzgâr: **a ~ course,** şimale doğru rota. **~ern** [ˈnôðən], şimale aid: **~ lights,** şimal fecri. **~ward(s),** şimale dogru. **~-east (NE.),** şimali şarkı; poyraz: **~-west (NW.),** şimaligarbi; karayel.
**Nor·way** [ˈnôwêy]. Norveç. **~wegian** [nôˈwîcịən], norveçli.
**nor'wester** [nôˈwestə(r)]. Şiddetli karayel rüzgârı; muşamba şapka.
**nose** [nǫuz]. Burun; uc; koku hassası. **~ about,** kolaçan etm.; **to ~ her way through fog,** *etc.*, (gemi) sis vs.de yolunu arayıp bulmak: **to ~ a thing out,** ısrarla arayarak meydana çıkarmak: **to blow one's ~,** burnunu silmek: ⌜**to cut off one's ~ to spite one's face**⌝, ⌜gâvura kızıp oruc bozmak⌝: **to follow one's ~,** dosdoğru gitmek; aklı selimini kullanmak: **to have a good ~,** (köpek) iyi koku almak: **to lead s.o. by the ~,** birini parmağında çevirmek: **to pay through the ~,** ateş pahasına almak: **to put s.o.'s ~ out of joint,** birinin pabucunu dama atmak: **to speak through the ~,** genizden konuşmak: **to turn up the ~,** burun kıvırmak. **~bag,** yem torbası. **~gay,** çiçek demeti. **~-dive,** pike. **~-ring,** burun halkası.
**-nosed** [nǫuzd] *suff.* ... burunlu.
**nostalg·ia** [nosˈtalcịə]. Dâüssıla; vatan hasreti. **~ic,** sıla hastası.
**nostril** [ˈnostril]. Burun deliği.
**nostrum** [ˈnostrəm]. Şarlatan ilâcı.
**nosy** [ˈnǫuzi]. Her şeye burnunu sokan: **~ Parker,** böyle bir adam.
**not** [not]. *Nefi edatı*. Değil. **I do ~ know [I don't know, I know ~],** bilmiyorum: **~ at all,** hiç, asla; bir şey değil: **~ everybody can do this,**

değme adam bunu yapamaz: ~ **a few,** az değil: '**Are you going?**' '~ **I**', 'Gidecek misin?' 'Ben mi?, ne münasebet!': ~ **that,** maamafih: ~ **that I know of,** benim bildiğime göre hayır: **I think** ~, zannetmem.
**notab·le** [ˈnǫutəbl]. Tanınmış; şayanı dikkat; muteber; zikre değer. Eşraftan biri. **~ility** [–ˈbiliti], meşhur kimse; ilerigelen. **~ly,** bahusus; epeyce.
**notary** [ˈnǫutəri]. Noter.
**notch** [noç]. Çentik. Çentmek.
**note**[1] [nǫut] *n.* Not; muhtıra; tezkere; nota; banknot; ehemmiyet. **credit** ~, kredi senedi: **to compare ~s,** karşılıklı fikir ve intibalarını söylemek: ~ **of hand,** borc senedi: **a man of** ~, mühim kimse: **to take** ~, dikkat etm.: **to take ~s,** not etm.: **there was a** ~ **of anger in what he said,** sözlerinde hiddet kokusu vardı: **he struck just the right** ~, sözleri çok uygun düştü: **it is worthy of** ~ **that** ..., dikkate değer ki. ~[2] *v.* Dikkat etm.; not etm. **I'll just** ~ **that down,** bunu kaydedivereyim. **~book** [ˈnǫutbuk], muhtıra, not defteri. **~d** [ˈnǫutid], meşhur; maruf; görülmüş. **~paper** [ˈnǫutpêypə(r)], mektub kâğıdı. **~-worthy** [ˈnǫutwö(r)ði], dikkate değer; mühim.
**nothing** [ˈnʌθin(g)]. Hiç; hiç bir şey; sıfır. ~ **but the best,** en iyisinden aşağı olmaz: **to come to** ~, suya düşmek: ~ **doing!,** yağma yok!: **there's** ~ **doing there,** orada iş yok: **that is** ~ **to do with me,** o bana aid değil: **to have** ~ **to do with** ..., ... ile hiç münasebeti olmamak: **I will have** ~ **to do with him,** onun yüzünü bile görmek istemem: **it is** ~ **less [else] than cheating,** dolandırıcılıktan başka bir şey değil: **all that goes for** ~, bütün bunlar hiçe sayılıyor: **there is** ~ **for it but to** ..., ...mekten başka yapacak bir şey yok: **he is** ~ **if not generous,** o da cömerd değilse kim cömerddir?: ~ **like so much,** hiç te o kadar değil: **only a hundred pounds, a mere** ~, ata deveye değil ya yüz liranın içinde: ~ **near so pretty as her sister,** hiç te kardeşi kadar güzel değil: **it is not for** ~ **that** ..., tevekkeli (değil): **he is** ~ **of a scholar,** hiç âlim değil: **to say** ~ **of** ..., ... üstelik: **it's** ~ **to me whether you do it or not,** ister yap ister yapma bana göre hava hoş.
**notice** [ˈnǫutis]. İhbar, ilân; dikkat. Farkında olm.; aldırış etmek. **to avoid** ~, göze çarpmamak için: **it has come to my** ~ **that,** öğrendiğime göre: **until further** ~, işari ahire kadar: **to give s.o.** ~, önceden haber vermek; (hizmetçiyi) savmak; (hizmetçi, kiracı) çıkacağını haber vermek: **I gave him a week's** ~, (savmadan) bir hafta önce haber verdim: **I must have** ~, önceden haberim olmalı: **at a moment's** ~, (önce haber vermeden) birdenbire: **at short** ~, kısa mühletle: **to take** ~ **of,** nazarı itibara almak: **to take no** ~, aldırış etmemek. **~able** [ˈnǫutisəbl], dikkate değer; göze çarpan.
**notif·y** [ˈnǫutifây]. Haberdar etm.; tebliğ etm.; bildirmek. **~iable** [–ˈfâyəbl], haber verilmesi elzem. **~ication** [–fiˈkêyşn], ihbar; tebliğ.
**notion** [ˈnǫuşn]. Fikir; mefhum; zan. **that's a good** ~!, çok iyi bir fikir!: **you have no** ~ **how dull it was,** ne kadar sıkıcı olduğunu tasavvur edemezsiniz: **I have a** ~ **he is going to resign,** istifa edeceğini hissediyorum: '**I haven't a** ~!', 'Hiç bir fikrim yok.'
**notori·ety** [ˌnǫutəˈrâyəti]. Şöhret (*gen.* fena); dile düşmüşlük. **to seek** ~, göze çarpmak istemek. **~ous** [nǫuˈtôri̯əs], mahud; adı çıkmış: **~ly,** cümleye malum olduğu üzere.
**notwithstanding** [ˌnotwiθˈstandin(g)]. (Buna) rağmen; her nekadar.
**nought** [nôt]. Sıfır.
**noun** [na̯un]. (*gram.*) İsim.
**nourish** [ˈnʌriş]. Beslemek. **~ing,** besleyici. **~ment,** gıda, yemek.
**nous** [na̯us]. Akıl, zekâ.
**Nova Scotia** [ˈnǫuvəˈskǫuşə]. Yeni İskoçya.
**novel**[1] [ˈnovl] *n.* Roman. **~ist,** romancı. ~[2] *a.* Yeni; yeni usul, çıkma; taptaze. **~ty,** yenilik; yeni moda.
**November** [nǫuˈvembə(r)]. İkinciteşrin; kasım.
**novice** [ˈnovis]. Acemi; papaz çömezi.
**now** [na̯u]. Şimdi; bu anda; işte; bu halde; henüz; artık. Şimdiki zaman. ~ ... ~ ..., gâh ... gâh ...: ~, **what's the trouble?,** ne var bakalım?: ~ **then!,** sakın ha!: **the train ought to be here by** ~, tren şimdiye kadar gelmiş olmalıydı: **between** ~ **and then,** o zamana kadar: **oh come** ~!, haydi canım!: **(every)** ~ **and then [again],** arasıra: ~ **for it!,** haydi bakalım!: **from** ~ **on,** şimdiden sonra:

just ~, hemen şimdi: **they won't be long ~,** nerede ise gelirler (fazla gecikmezler): **~ or never [~ if ever],** ya şimdi ya hiç!: **until ~ [up to ~],** şimdiye kadar: **well ~!,** Allah! Allah! **~adays** [ˈnauədêyz], bu günlerde; şimdiki zamanlarda.

**nowhere** [ˈnouweə(r)]. Hiç bir yerde.

**nowise** [ˈnouwâyz]. Hiç bir surette.

**noxious** [ˈnokşəs]. Muzır, zararlı; mühlik.

**nozzle** [ˈnozl]. Ağızlık; hortumbaşı.

**nucle·ar** [ˈnyûkliə(r)]. Nüveye aid. **~us,** *pl.* **~i** [ˈnyûkliəs, -kliây], nüve; çekirdek.

**nud·e** [nyûd]. Çıplak (insan resmi, heykeli). **~ism,** çıplak gezenlerin mesleği. **~ity,** çıplaklık.

**nudge** [nʌc]. El ile dürtme(k).

**nugget** [ˈnʌgit]. İşlenmemiş küçük külçe.

**nuisance** [ˈnyûsəns]. Sıkıntı veren veya taciz eden şey, hareket; derd; başbelâsı. **a little ~,** (çocuk) başağrısı: **the man's a ~,** bu adam da başbelâsı: **what a ~!,** yazık!.

**null** [nʌl]. Hükümsüz, battal. **~ and void,** keenlemyekün. **~ify,** ibtal etm. **~ity,** hükümsüzlük; hiçlik; nüfuz olmıyan adam.

**numb** [nʌm]. Uyuşuk. Uyuşturmak.

**number**[1] [ˈnʌmbə(r)] *n.* Aded; sayı; rakam; numara; takım; nüsha. **~s,** çok. **a ~ of,** bir kaç: **great ~s of,** bir çok: **a small ~ of,** bir kaç, bir az: **even ~,** cift aded: **odd ~,** tek aded: **whole ~,** tam aded: **among the ~,** aralarında: **any ~ of,** çok mikdarda: **they were ten in ~,** sayıları on kadardı: **to look after ~ one,** kendi çıkarına bakmak: **to be overcome by ~s,** sayı üstünlüğüne yenilmek: **one of their ~,** onlardan biri: **his ~ is up,** (*sl.*) yandı, mahvoldu. **~**[2] *v.* Saymak; hesab etm.; numara koymak; numaralamak. **their army ~s ten thousand,** ordularının yekûnu on bindir: **to ~ off,** (*mil.*) numara saymak: **~ off!,** (emir) sağdan say! **~less** [ˈnʌmbəlis], sayısız; hesabsız.

**numer·al** [ˈnyûmərəl]. Rakam, aded. Adedî. **cardinal ~s,** adî adedler: **ordinal ~s,** sıra adedleri. **~ical** [–ˈmerikl], sayı ve rakama aid: **~ superiority,** sayıca üstünlük. **~ous** [ˈnyûmərəs], çok; kalabalık.

**numskull** [ˈnʌmskʌl]. Mankafa.

**nun** [nʌn]. Rahibe, sör. **~nery,** rahibe manastırı.

**nuncio** [ˈnʌnsyou]. **Papal ~,** Papa elçisi.

**nuptial** [ˈnʌpşl]. Düğüne aid; zifafe aid. **~s,** düğün. **~ chamber,** zifaf odası.

**nurse** [ˈnö(r)s]. Dadı; hastabakıcı. Emzirmek; hastaya bakmak; kucağında tutmak; idare ile kullanmak. **wet ~,** sütnine: **to ~ a grudge,** *etc.*, kin vs. beslemek. **~maid,** dadı kız. **~ry** [ˈnö(r)səri], çocuk odası: **~ garden,** fidanlık. **~ryman,** fidanlık bahçevanı.

**nurture** [ˈnö(r)tyə(r)]. Beslemek; büyütmek.

**nut** [nʌt]. Fındık; cıvata somunu; (*sl.*) kafa. Fındık toplamak. **hard [tough] ~ to crack,** demir leblebi: **to be dead ~s on stg.,** (*sl.*) bir şeye mübtelâ olm.: **he can't run [write,** *etc.*] **for ~s,** (*sl.*) ne yapsan koşamaz [yazamaz vs.]: **off his ~,** (*sl.*) bir tahtası eksik. **~brown,** fındık renginde. **~crackers,** fındık kıracağı. **~meg,** küçük hindistancevizi. **~shell** [ˈnʌtşel], fındık kabuğu. **to put the matter in a ~,** kısaca anlatmak.

**nutri·ment** [ˈnyûtrimənt]. Gıda. **~tion** [–ˈtrişn], besleyiş. **~tious** [–ˈtrişəs], besleyici.

**nuzzle** [ˈnʌzl]. Burunla eşelemek.

**nymph** [nimf]. Su, orman perisi.

# O

**O** [ou]. O harfi. *v.* **oh**; *hitab nidası* (*pek az kullanılır*): **O God!,** ya Allah!

**oaf** [ouf]. Budala, beceriksiz çocuk.

**oak** [ouk]. Meşe. **heart of ~,** aslan yüreklilik: **hearts of ~,** meşe tahtasından yapılmış eski İngiliz harb gemileri. **~en,** meşe odunundan yapılmış.

**oar** [ô(r)]. Kürek. **to pull a good ~,** iyi kürekçi olm.: **to put in one's ~,** münasebetsizce müdahale etm.: **to rest on one's ~s,** dinlenmek. **-~ed,** ... kürekli. **~sman,** kürekçi.

**oasis** [ouˈêysis]. Vaha.

**oat** [out]. Yulaf tanesi. **~s,** yulaf. **to sow one's wild ~s,** genclik çılgınlıkları yapmak. **~meal,** yulaf unu.

**oath** [ouθ]. Yemin; küfür. **to let out an ~,** küfür savurmak: **to put s.o. on his ~,** birine yemin ettirmek: **to take the ~,** yemin etmek.

**obdura·cy** [ˈobdyurəsi]. İnadcılık. **~te,** inadcı; tövbe etmez.

**obedien·ce** [oˈbîdyəns]. İtaat. **~t,** itaatli; söz anlar: **your ~ servant** [**yours ~ly**], *resmî veya ticarî yazılan mektubun sonunda kullanılır.*

**obeisance** [oˈbêysəns]. Baş eğme. **to make ~,** baş eğmek; biat etmek.

**obelisk** [ˈobəlisk]. Sütun; dikili taş.

**obey** [ouˈbêy, əˈbêy]. İtaat etm.; söz dinlemek; imtisal etmek.

**obituary** [oˈbityuəri]. **~ notice,** (gazetede) ölüm ilânı; ölünün tercümeihali: **~ column,** gazetede ölüm ilânlarına mahsus sütun.

**object**[1] [ˈobcikt] *n.* Şey, nesne; hedef, maksad; (*gram.*) mefulünbih. **an ~ of pity** [**ridicule**], acınacak [gülünecek] şey: **money no ~,** paranın ehemmiyeti yok: **an ~ for study,** tedkik mevzuu: **what is the ~ of all this?,** bütün bundan maksad ne?: **with the ~ of ..., ...** maksadile. **~ive** [–ˈcektiv], afakî; mefule aid; hedef, maksad; dürbünün büyük adesesi; (*fot.*) adese. **~ivity** [–ˈtiviti], afakilik. **~**[2] [obˈcekt] *v.* İtiraz etm., razı olmamak. **to ~ to,** münasib görmemek, beğenmemek; reddetmek: **to ~ that ..., ...** diye itiraz etm., protesto etmek. **~ion** [obˈcekşn], itiraz, protesto; mahzur, mâni: **if you have no ~,** mahzur görmezseniz: **to make no ~ to,** ···e razı olm.: **to raise an ~,** bir mahzur ileri sürmek: **to take ~ to,** ···e itiraz etmek. **~ionable,** mahzurlu; nahoş; tahammül edilmez. **~or,** itirazcı, protesto eden.

**oblig·ation** [obliˈgêyşn]. Mecburiyet; mükellefiyet; taahhüd; minnet. **to meet one's ~s,** borclarını ödemek: **to put oneself under an ~ to s.o.,** birine karşı minnet altında kalmak: **I am under no ~ to ..., ...** mecburiyetinde değilim. **~atory** [oˈbligətəri], mecburî; zarurî. **~e** [oˈblâyc], mecbur etm.; mükellef etm.; minnet altında bırakmak. **to be ~d,** mecbur olm.; mükellef olm.; minnettar olmak: **to ~ a friend,** hatır için yardım etm.: **can you ~ me with a light?,** ateşinizi lûtfedermisiniz?: **~ me by shutting the door,** kapıyı kapatmak lûtfunda bulunur musunuz?: **I should be much ~d if you would write to him,** ona lûtfen yazarsanız çok minnetdar olurum: **you will ~ me by not doing this again,** bunu bir daha yapmazsanız çok memnun olurum. **~ing** [oˈblâycin(g)], lûtufkâr; yardım etmeğe hazır.

**oblique** [oˈblîk]. Eğri; dolambaçlı: **~ case,** (*gram.*) isimlerin tasrif hali: **~ oration** [**narrative**], naklî ifade.

**obliterate** [oˈblitərêyt]. Silmek; yok etmek.

**obliv·ion** [oˈblivyən]. Unut(ul)ma. **to pass into ~,** unutulup gitmek. **~ious,** unutkan; haberdar olmıyarak: **~ of the fact that ..., ...** kâmilen unutarak.

**oblong** [ˈoblon(g)]. Boyu eninden fazla.

**obnoxious** [obˈnokşəs]. Menfur; nahoş.

**obscen·e** [obˈsîn]. Açık saçık; müstehcen. **~ity,** müstehcenlik.

**obscur·e** [obˈskyuə(r)]. Karartmak; gizlemek; örtmek. Karanlık; vazıh değil; mechul. **~ity,** karanlık; mechulluk.

**obsequious** [obˈsîkwiəs]. Alçak derecede mütevazı.

**observ·ance** [obˈzö(r)vəns]. Dine yahud kanuna riayet. **religious ~s,** dinî âyinler. **~ant,** dikkatli; her şeyi düşünen; dine ve kanuna riayetkâr. **~ation** [ˌobzö(r)ˈvêyşn], gözetleme; rasad; tedkik; müşahede; ihtar; söz. **to escape ~,** görülmemek (için): **to keep under ~,** göz hapsine almak; tarassud altında bulundurmak. **~atory** [–ˈzö(r)vətəri], rasadhane. **~e** [obˈzö(r)v], riayet göstermek; gözetlemek, tarassud etm.; dikkatle bakmak; söylemek; ihtarda bulunmak; farkında olmak. **to ~ silence,** ağzını açmamak: **he never ~s anything,** hiç bir şeyin farkına varmaz.

**obsess** [obˈses]. Zihnine musallat olm. **to be ~ed by an idea,** bir fikir aklından çıkmamak. **~ion** [–ˈseşn], musallat olan fikir; sabit fikir; daimî endişe.

**obsolete** [ˈobsəlît]. Kullanılmaz olmuş, modası geçmiş.

**obstacle** [ˈobstəkl]. Engel; mâni; mania.

**obstetric** [obˈstetrik]. Ebeliğe aid.

**obstina·cy** [ˈobstinəsi]. İnadcılık; dikkafalılık. **~te** [–nit], inadcı; dikkafalı; (hastalık) tedavisi zor.

**obstreperous** [obˈstrepərəs]. Haşarı.

**obstruct** [obˈstrʌkt]. Tıkamak; engel olm.; menetmeğe çalışmak. **to ~ the traffic,** yolu tıkamak: **to ~ the view,** manzarayı kapamak. **~ion** [–kşn], engel, mâni; tıkama. **~ive,** hail.

**obtain** [obˈtêyn]. Elde etmek, ele

geçirmek; istihsal etm. Âdet olm.; hüküm sürmek. **his ability ~ed him a good post,** kabiliyeti ona iyi bir mevki kazandırdı. **~able,** elde edilebilir; bulunur.

**obtru·de** [obˈtrûd]. İleri sokmak; sokulmak. **to ~ oneself upon s.o.,** birine münasebetsizce sokulmak. **~sive** [–ˈtrûsiv], göze batar.

**obtuse** [obˈtyûs]. Sivri olmıyan; kalınkafalı. **~ angle,** münferic zaviye.

**obverse** [obˈvö(r)s]. (Para vs.) yüz tarafı.

**obviate** [ˈobviêyt]. Önüne geçmek; çaresini bulmak.

**obvious** [ˈobviəs]. Aşikâr; meydanda.

**occasion**[1] [oˈkêyjn]. *n.* Fırsat; vesile; vaziyet; vak'a; lüzum. **as ~ requires,** vaziyete göre; icabında: **should the ~ arise,** icabında: **to celebrate the ~,** hadiseyi tesid etmek için: **on ~,** bazan; **on one ~,** bir defa: **on several ~s,** bir çok defa: **on such an ~,** böyle bir halde [vaziyette]: **to rise to the ~,** lâyıkı ile başarmak: **on the ~ of his marriage,** düğünü münasebetiyle. **~**[2] *v.* Sebeb olm., mucib olmak. **~al** [oˈkêyjənl], ara sıra olan.

**occident** [ˈoksidənt]. Garb. **~al** [–ˈdentl], garbî.

**occult** [oˈkʌlt] *a.* Gizli; gaibe aid. **the ~ [the ~ sciences],** sihirbazlık.

**occup·ant** [ˈokyupənt]. İşgal eden; bir evin sahib veya kiracısı. **the ~s of the car,** otomobildeki kimseler. **~ation** [okyuˈpêyşn]. İşgüç, meşgale; işgal. **to be in ~ of a house,** bir evde oturmak. **~ier** [ˈokyupâyə(r)] *v.* **occupant.** **~y** [oˈkyupây], işgal etm., ···de oturmak; (bir şehri) zabtetmek; iş vermek. **to be occupied in ..., ...** ile meşgul olm.: **to ~ one's time in doing stg.,** bir şeyi yapmakla vakit geçirmek.

**occur** [oˈkö(r)]. Vukubulmak; ara sıra meydana çıkmak; bulunmak; hatırına gelmek. **this must not ~ again,** bu bir daha tekerrür etmemeli. **~rence** [oˈkʌrəns], vak'a; hadise; vukubulma. **everyday ~,** günlük hadise: **to be of frequent ~,** sık sık vuku bulmak.

**ocean** [ˈouşn]. Okyanus, deniz. **-~ going** (ship), okyanusta sefer eden (gemi).

**o'clock** [əˈklok]. **what ~ is it?** saat kaç?: **two ~,** saat iki.

**octagon** [ˈoktagən]. Sekizdılılı şekil.

**octo-** [ˈoktou] *pref.* Sekiz ....

**October** [okˈtoubə(r)]. Birinciteşrin, ekim.

**octogenarian** [ˌoktoucəˈneəriən]. Seksen yaşında.

**octopus** [ˈoktəpʌs]. Ahtapot.

**ocul·ar** [ˈokyulə(r)]. Göze aid; gözle görülen. **~ist,** göz hekimi, gözcü.

**odd** [od]. Tek (çift değil); eşsız; seyrek; tuhaf. **in ~ corners,** kıyıda bucakta: **employed on ~ jobs,** öteberi işlerde çalışan: **forty ~,** kırk küsür: **to make up the ~ amount [money],** bir meblağın üstünü tamamlamak: **at ~ moments [times],** boş vakitlerde: **to strike one as ~,** garibine gitmek: **well that's ~!,** tuhaf şey! **~ity** [ˈoditi], acayib (adam, şey). **~ments** [ˈodmənts], ufak tefek şeyler. **~s** [odz], **the ~ are against him,** ihtimaller aleyhinedir: **to fight against great ~,** büyük üstünlüğe karşı çarpışmak: **to be at ~ with s.o.,** araları açık olm.: **the ~ are that,** muhtemeldir ki: **~ and ends,** ufak tefek şeyler: **it makes no ~,** zarar yok; hepsi bir: **~ on [against] a horse,** (at yarışında) bir atın lehine [aleyhine] olan ihtimaller: **what's the ~!,** ne zarar var? ne çıkar?

**ode** [oud]. Bir nevi lirik şiir.

**odi·ous** [ˈoudyəs]. Nefret verici, menfur. **~um** [ˈoudyəm], nefret; muhitçe sevilmeme. **to incur ~,** herkesin nefretine uğramak.

**odour** [ˈoudə(r)]. Koku. **to be in good ~,** gözde olmak: **to be in bad ~,** gözden düşmek.

**Odyssey** [ˈodisi]. Odise; heyecanlı sergüzeşt; destan.

**o'er** [ô(r)]. *v.* **over.**

**of** [ov, əv]. ···in; ···den: **~ this,** bunun, bundan. **~ itself,** kendi kendine: **a child ~ five,** beş yaşında bir çocuk: **the city ~ Paris,** Paris şehri: **a fool ~ a man,** abdalın biri: **the love ~ God,** Allahın kullarına olan sevgisi; kulun Allah sevgisi.

**off** [of]. Uzakta; uzağa; dışarıya; ···den; ···den uzak; sol taraf; sol tarafdaki. [*Bir fiilin yanında olduğu zaman manası o fiil ile verilmiştir; umumiyetle bir yerden bir yere hareket veya bir hareketin durması manasını ifade eder.*] **be ~!,** çek arabanı!: **to be badly [well] ~,** hali vakti yerinde olmamak [olmak]: **to be badly ~ for sugar [coffee],** şekeri [kahvesi] az kalmak: **how are we ~ for coal?,** kömürümüz ne kadar kaldı?: **to break-**

fast ~ **bread and cheese,** kahvaltıyı peynir ekmekle yapmak: **you are better ~ where you are,** şimdiki vaziyetiniz daha iyi: **come ~ the grass!,** çimenden çık [çekil]: **the concert is ~,** konser verilmiyecek: **the deal is ~,** pazarlık bozuldu: ~ **day,** izinli gün; çalışılmıyan gün; insanın her zamanki gibi muvafakkiyetli olmadığı gün: ~ **duty,** vazifesi bitmiş; serbest: **to be ~ one's food,** canı yemek istememek: **hats ~!,** şapkaları çıkarın!: **I'm ~,** ben gidiyorum: **the meat is a bit ~,** et biraz ağırlaşmış: ~ **and on,** ara sıra: **to allow five per cent. ~ for ready money,** peşin para için yüzde beş tenzilât yapmak: **the house is ~ the main road,** ev caddeden sapa düşer: **in the ~ season,** mevsimi olmadığı zaman: **I have very little ~ time,** çok az serbest vaktim var.

**offal** [ˈofl]. Kasablık hayvanların baş, işkembe, ciğer vs.gibi kısımları.

**offen·ce** [oˈfens]. Cünha, kabahat; ihlâl; alınma; hücum, taarruz. **to give ~,** gücendirmek: **no ~ meant,** kimsenin hatırı kalmasın: **to take ~,** alınmak, hatırı kalmak. **~d** [oˈfend], gücendirmek, hatırını kırmak; kabahat işlemek, kusur etmek. **to be ~ed,** alınmak; hatırı kalmak: **to ~ against** (**the law,** *etc.*), ihlâl etm. **~der,** kabahat işliyen: **first ~,** ilk defa olarak suçlu: **old** [**hardened**] **~,** sabıkalı suçlu. **~sive** [oˈfensiv], taarruza aid; hatır kırıcı; tiksindirici. Taarruz. **to take the ~,** taarruza geçmek.

**offer** [ˈofə(r)]. Takdim (etm.), teklif (etm.); sunma(k); arz(etmek); göstermek; vermek; ileri sürmek. Zuhur etmek. **to ~ battle,** muharebeye davet etm.: **he ~ed to strike me,** baną vuracak gibi oldu. **~ing,** feda edilen şey; kurban. **~tory** [ˈofətəri], kilisede zekât toplama ve toplanan para.

**off-hand** [ˈofhand]. Hazırlıksız; irticalen; ha deyince; ~ *veya* **~ed** [–ˈhandid], teklifsiz; soğuk tavırla.

**offic·e** [ˈofis]. Yazıhane; idarehane; bakanlık; memuriyet, vazife; âyin. **~s,** ticarî daire; delâlet. **to be in ~,** (parti) iktidarda bulunmak: **Foreign** [**Home, War**] **~,** Hariciye [Dahiliye, Harbiye] bakanlığı: **to take ~,** (parti) iktidara geçmek; (bakan) makama geçmek: **through the good ~s of,** …in delâletile. **~-boy,** odacı (oğlan). **~er** [ˈofisə(r)], zabit, subay; memur. **~ial** [oˈfişəl], memur. Resmî. **~ialdom,** bürokrasi. kırtasiyecilik. **~iate** [oˈfişiêyt], resmî vazife ifa etm.; dini âyin icra etmek. **to ~ as host,** ev sahibi vazifesini görmek. **~ious** [oˈfişəs], işgüzar; çalmadan oynar.

**offing** [ˈofin(g)]. **the ~,** karadan uzak fakat görülebilen açık deniz: **in the ~,** açıkta: **a job in the ~,** muhtemel olan iş.

**offset** [ˈofset]. (*Bazan* **outset** *yerinde kullanılır.*) Bedel, karşılık. **to serve as an ~ to stg.,** bir şeyin güzelliğini belirtmek.

**offshoot** [ˈofşût]. Filiz; şube; torun.

**offshore** [ˈofşô(r)]. Karadan gelen; karadan biraz uzakta. ~ **wind,** meltem.

**offside** [ofˈsâyd]. Ofsayd.

**offspring** [ˈofsprin(g)]. Zürriyet; çoluk çocuk.

**oft, often** [oft, ofn]. Çok defa; sık sık; ekseriya. **as ~ as,** her vakit ki: **as ~ as not** [**more ~ than not**], ekseriya, çok defa: **how ~,** kaç defa: **it cannot be too ~ repeated,** ne kadar tekrar edilse yeridir.

**ogle** [ˈougl]. Cilveli bakışlarla süzmek.

**ogre** [ˈougə(r)]. İnsan eti yiyen dev; canavar. **~ss** [ˈougris]. ogre'nin dişi.

**O.H.M.S.** = **On Her Majesty's Service.** Devlet hizmetinde (*resmî evrak üzerine yazılır*).

**oil** [ôyl]. Yağ; petrol; gaz. Yağlamak. **to burn the midnight ~,** göz nuru dökmek: **to pour ~ on the flames,** körüklemek: **to pour ~ on troubled waters,** fırtınayı yatıştırmak: **to paint in ~s,** yağlı boya ile resim yapmak: **to strike ~,** petrol keşfetmek; vurgun vurmak: **to ~ s.o.'s palm,** rüşvet yedirmek: **to ~ the wheels,** işi kolaylaştırmak. **~cloth,** Amerikan bezi; muşamba. **~skin,** gamsele. **~s,** gamseleden ceket ve pantalon. **~stone,** bileği taşı. **~y,** yağlı; yağlanmış; mütebasbıs. **~-bearing,** yağ veren (nebat); petrollu. **~-cake,** küspe. **~-can,** yağdanlık. **~-colour,** yağlı boya resim.

**ointment** [ˈôyntmənt]. Melhem. ⌜**a fly in the ~**⌝, ⌜sinek küçüktür ama mide bulandırır⌝.

**old** [ould]. İhtiyar, yaşlı; eski, kadim; külüstür; modası geçmiş. ~ **age,** ihtiyarlık: **any ~ thing,** (*coll.*) ne olursa olsun: **an ~ friend,** ihtiyar bir dost; eski dost: **to grow ~,** yaşlanmak: **an ~ hand,** tecrübeli kimse: **as ~ as the hills,** çok eski: **how ~ are you?,** kaç yaşındasın?: **five years ~,** beş yaşında: **~ maid,** ihtiyar kız: **~ man [chap]!,** ahbab! **the ~ man,** babam; patron; (gemide) kaptan: **the same ~ thing [story],** ⌜eski hamam eski tas⌝: **that's an ~ trick,** bu hileyi herkes bilir: **~ woman,** koca karı; (erkek hakkında) lâpacı; titiz: **my ~ woman,** bizimki (karım): **the ~ year,** hemen bitmiş, bitmekte olan yıl. **~-clothes-man,** eskici. **~-fashioned,** modası geçmiş; eski kafalı. **~-world,** eski zamana aid: **the ~,** Şarkî Yarımküre. **~en** [ˈouldn], **in ~ times,** eski zamanlarda.

**olfactory** [olˈfaktəri]. Koku hissine aid.

**oligarchy** [ˈoligâki]. Küçük ve zengin bir zümre hâkim olduğu idare, hükümet.

**olive** [ˈoliv]. Zeytin (ağacı); zeytunî. **~branch,** barış alâmeti olan zeytin dalı: **to hold out the ~,** barış için ilk teşebbüsü yapmak. **~-green,** zeytunî renk.

**Olympi·an** [ouˈlimpyən]. İlâhların mekânı olan Olimpos dağına aid; lâhuti; pek muhteşem ve azametli. **~c, the ~ games,** Olimpiyadlar.

**om·en** [ˈoumən]. Fal. **to ~ well [badly],** istikbal için iyi [fena] alâmet olm.: **of good [bad] ~,** uğurlu [uğursuz]: **to regard as a good [bad] ~,** …i iyiye [fenaya] yormak. **~inous** [ˈominəs], meşum, uğursuz; tehdidkâr.

**omi·ssion** [oˈmişn]. Zühul, ihmal; unutulmuş şey; atlanmış kelime; kusur. **~t** [ouˈmit, oˈmit], yanlışlıkla unutmak; ihmal etm.; atlamak.

**omni-** *pref. Bir kelımeye* her kes, hepsi, bütün (için) *manalar veren bir ek.* **~bus** [ˈomnibəs], omnibüs. **~potent** [omˈnipətənt], mutlak bir kudret sahibi. **The ~,** Allah, Kadiri mutlak. **~potence,** tam ve mutlak kudret. **~scient** [omˈnişənt], âlimi kül; her şeyi bilir. **~vorous** [omˈnivərəs], her şeyi yiyen. **an ~ reader,** ne bulursa okuyan.

**on** [on]. Üzerinde; üzerine; üstünde; …de. Temas halinde olarak; civarında; kuşatarak; cihetinde; ilerisinde. [*Fiil ile olduğu zaman o fiile bak; o zaman* ilerleme, devam *veya* bağlantı *mânalarını ifade eder.*] **~ my entering the room,** ben odaya girince: **~ hearing this,** bunu işitince: **~ Tuesday,** salı günü: **~ June 10th,** on Haziranda: **~ with your coat!,** caketini giy!: **~ with the work!,** işe devam et!: **the brakes are ~,** frenlidir: **from that day ~,** o günden beri: **the examination is now ~,** imtihan başladı veya devam ediyor: **just ~ a year ago,** takriben bir sene evvel: **later ~,** daha sonra: **the police are ~ to him,** zabıta onun peşindedir: **what's ~ at the cinema?,** sinemada ne oynuyor?: **what's ~ today?,** bu gün ne var?: **without anything ~,** çırçıplak: **well ~ in years,** yaşı ilerlemiş.

**once** [wʌns]. (Yalnız) bir defa; eskiden: **at ~,** derhal; aynı zamanda: **to do too much at ~,** (i) bir çok şeyi birden yapmak; (ii) bir şeyle fasılasız meşgul olm.: **all at ~,** birdenbire; hepsi birlikte: **~ (and) for all,** ilk ve son defa; (tehdidle) son defa olarak: **for ~,** bir defaya mahsus olarak: **for this ~,** bir defalık: **~ more [again],** bir kere daha, tekrar: **~ upon a time,** bir varmış bir yokmuş: **~ a week,** haftada bir defa: **~ in a while [way],** nadiren.

**one** [wʌn]. Bir; tek; [*Fr. 'on', almanca 'man' gibi fiillerin meçhul halini teşkil etmeğe yarar*; insan; *mes.* **when ~ thinks,** düşünüldüğü zaman, insan düşündüğü zaman]: **~ and all,** istisnasız hepsi: **~ after the other,** arka arkaya: **any ~ of you,** içinizden her hangi biri: **the next but ~,** daha sonraki: **~ by ~,** birer birer: **it makes ~ angry,** bu insanı kızdırır: **~ John Smith,** J. S. isminde biri: **I like good plays, but loathe bad ~s,** iyi piyesleri severim, fenalardan nefret ederim: **a duck and her young ~s,** ördek ve yavruları: **our dear ~s,** sevdiklerimiz: **he's a knowing ~,** çok bilmişin biridir: **I am not much of a ~ for football,** futbol bana gelmez: **I am not the ~ to …,** (onu) yapacak adam değilim: **you can have ~ or the other, but not both,** ya birini ya ötekini alabilirsiniz, fakat ikisi birden olmaz: **~'s own house,** insanın kendi evi: **the old ~s,** ihtiyarlar: **~ and sixpence,** bir buçuk şilin: **~ so wise,** böyle akıllı bir adam: **that's ~ (up) to us,** (*sl.*) bununla biz bir sayı kazandık: **that's ~ way of doing it,** bu böyle de yapılabilir: **to be at ~ with s.o.,** birisile hemfikir olmak. **~-armed,** tek kollu, çolak. **~-eyed,**

tek gözlü. **~-sided,** tek taraflı; insafsız; müsavi değil; bir cihetli. **~-way street,** nakil vasıtalarının yalnız bir istikametten gittiği sokak. **~self** [wʌnˈself], kendisi.

**onerous** [ˈonərəs]. Ağır, külfetli.

**onion** [ˈʌnyən]. Soğan.

**onlooker** [ˈonˌlukə(r)]. Seyirci.

**only** [ˈo̤unli] *a.* Tek; yegâne. *adv.* Yalnız, sade. *conj.* Fakat. **~ you can do it,** sizden başka bunu kimse yapamaz: **I ~ came here today,** buraya daha bugün geldim: **he is ~ rich because he is dishonest,** ancak dürüst olmamak suretile zengin oldu: **if ~ I could see him!,** ah onu bir görebilsem!: **if ~ I could see him, I could persuade him,** onu bir görebilsem ikna ederdim: **my one and ~ hope,** yegâne ümidim: **it is ~ too true,** maalesef hakikat budur: **he'd be ~ too glad [pleased],** o dünden hazır, canına minnet.

**onomatopoeic** [ˌonoˌmato̤uˈpî-ik]. Taklidî ahenkle yapılmış (kelime), *mes.* **bow-wow** (köpek); **bang!** (top sesi).

**on·rush** [ˈonrʌş]. Hücum; saldırış. **~set** [ˈonset], hücum, hamle: **from the ~,** başlangıcından. **~slaught** [ˈonslôt]. Hücum, saldırış.

**onus** [ˈo̤unəs]. Yük; mesuliyet. **the ~ of proof lies with the plaintiff,** isbat dâvacıya düşer.

**onward** [ˈonwəd]. İlerliyen. **~s,** ileri. **from now ~,** bundan böyle [sonra]: **from tomorrow ~,** yarından itibaren.

**ooze** [ûz]. Balçık; sızıntı. Sız(dır)mak. **my courage is oozing,** cesaretim kesiliyor: **he ~s conceit,** baştan başa kibir.

**opaque** [o̤uˈpêyk]. Kesif; şeffaf olmıyan.

**open**[1] [ˈo̤upn] *a.* Açık; kilidlenmemiş; açılmış; meydanda; duçar. *n.* Açık (ev haricinde); açık deniz; saha, meydanlık. **~ boat,** güvertesiz gemi. **break ~,** kırıp açmak: **to cut ~,** kesip açmak: **~ to doubt,** su götürür: **half ~,** aralık: **to keep ~ house,** evinin kapısını açık tutmak (*fig.*): **to keep an ~ mind,** (fikren) tarafsız kalmak: **~ race, *etc.*,** umuma açık yarış vs.: **the ~ sea,** engin, açık deniz: **an ~ secret,** herkese malûm bir sır: **to be ~ to advice,** fikir, tavsiye vs.yi kabule hazır olm.: **it is ~ to you to object,** itiraz etmekte serbestsiniz. **~**[2] *v.* Açmak; başlamak. **~ out,** yaymak; açılmak: **to ~ out a hole,** bir deliği genişletmek. **~ up,** açmak; başlamak; açıp genişletmek. **~ing** [ˈo̤upənin(g)], aç(ıl)ma; başlangıc; ilk hareket; delik, ağız; açıklık. **a good ~ for a young man,** bir delikanlı için iyi bir imkân (iş vs.): **an ~ for trade,** ticaret için mahrec, pazar. **~ly** [ˈo̤upənli], açıkça; açıktan açığa; el âleme karşı.

**opera** [ˈopərə]. Opera. **~-glass,** opera dürbünü. **~tic** [–ˈratik], operaya aid.

**opera·ble** [ˈopərəbl]. Ameliyat edilebilir. **~te** [ˈopərêyt], ameliyat yapmak; işlemek; tesir etm.; işletmek. **to ~ on s.o. for appendicitis,** birine apandisit ameliyatı yapmak. **~ting-table,** ameliyat masası. **~ting-theatre,** ameliyat odası. **~tion** [opəˈrêyşn], ameliyat; işle(t)me; tesir; faaliyet; meriyet; (*mil.*) tatbikat. **to come into ~,** meriyete girmek. **~tive** [ˈopərətiv], ameliyata aid; âmil; amele. **~tor,** makine işleten adam; (borsada) acyocu: **wireless ~,** telsizci.

**opiate** [ˈo̤upiət]. Uyutucu, narkotik.

**opinion** [oˈpinyən]. Kanaat; fikir; zan; mutalaa. **in my ~,** bence, kanaatimce: **to have [hold] a high ~ of,** takdir etm.: **to have no [a poor] ~ of,** ···e fazla kıymet vermemek: **to be of the ~ that,** ... kanaatinde bulunmak: **I am entirely of your ~,** fikrinize tamamen iştirak ediyorum: **public ~,** efkâri umumiye: **to take another ~,** (*med.*) bir başka doktora da sormak. **~ated** [oˈpinyənêytid], fikrinden dönmez; inadcı.

**opium** [ˈo̤upyəm]. Afyon. **~ den,** afyonkeşler kahvesi. **~-eater [~-fiend],** afyon tiryakisi.

**opponent** [oˈpo̤unənt]. Muhalif; rakib.

**opportun·e** [ˈopôtyûn]. Müsaid zamanda olan; tam vaktinde gelen; muvafık, uygun. **~ist** [–ˈtyûnist], zamane adamı; fırsat düşkünü. **~ity** [ˌopəˈtyûniti], fırsat; vesile. **to give an ~,** meydan vermek: **to seek an ~,** vesile aramak: **a golden ~,** ele geçmez fırsat, kelepir.

**oppos·e** [oˈpo̤uz]. Karşı koymak; muhalefet etm.; önüne geçmek; aleyhinde olmak. **~ed,** karşısında; aksi: **country as ~ to town,** kır, şehrin aksine olarak .... **~ite** [ˈopəzit], mukabil; karşıkarşıya; karşısında; aksi; karşı. **the exact ~,** taban tabana zıd: **one's ~ number,** karşı tarafta aynı rütbe veya vazifede olan kimse: **the ~ sex,** öteki cins. **~ition** [ˌopəˈzişn], muhalefet; mukavemet; itiraz; muhalif parti; zıddiyet; rekabet.

**oppress** [oˈpres]. Zulmetmek, ezmek, tazyik etm. **~ion** [oˈpreşn], zulüm, tazyik; sıkıntı, kasvet. **~ive,** zalim; ezici; ağır; can sıkıcı, kasvetli. **~or,** zalim, gaddar.

**opprobri·ous** [oˈprǫubriəs]. Hakaret edici.

**opt** [opt]. Seçmek. **to ~ for,** tercih etmek. **~ative** [opˈtêytiv], temenni sıygası. **~ion** [ˈopşn], hakkı hıyar; intihab; seçme. **to have an ~ on stg.,** bir şey üzerinde tercih hakkı olm. **~ional,** ihtiyarî.

**optic** [ˈoptik]. Görmeğe aid, basarî. **~ nerve,** göz siniri. **~al,** görmeğe aid: **~ illusion,** görme hatası: **~ instruments,** dürbün, mikroskop gibi aletler. **~ian** [–ˈtişn], gözlükçü. **~s,** basariyat ilmi, optik.

**optim·ism** [ˈoptimizm]. Nikbinlik. **~ist,** nikbin. **~istic** [–ˈmistik], nikbin(ce).

**opulen·t** [ˈopyulənt]. Zengin; bol. **~ce,** bolluk: **to live in ~,** refah içinde yaşamak.

**or** [ö(r)]. Yahud; veya; yoksa. **either ... ~ ...,** ya ... ya ...: **not either ... ~ ...,** ne ... ve ne ...: **will he go ~ not?,** gidecek mi, gitmiyecek mi?

**orac·le** [ˈorəkl]. Gaibden haber; kehanet. **to consult an ~,** fala bakmak: **to work the ~,** piston işletmek; iltimas temin etmek. **~ular** [oˈrakyulə(r)], iki manalı; kehanet kabilinden.

**oral** [ˈôrəl]. Şifahî; sözlü; ağza aid.

**orange** [ˈorənc]. Portakal (rengi). **Seville ~,** turunc.

**orat·ion** [oˈrêyşn]. Nutuk, hitabe; cansıkıcı nutuk. **~or** [ˈorətə(r)], hatib. **~ory** [ˈorətri], hatiblik; belâgat.

**orb** [ôb]. Küre; göz küresi. **~it** [ˈôbit], mahrek; göz çukuru.

**orchard** [ˈôçəd]. Yemiş bahçesi.

**orchestra** [ˈôkestrə]. Orkestra; tiyatroda orkestra yeri. **~tion** [–ˈtrêyşn], notaların orkestra aletlerine göre tertibi.

**orchid** [ˈôkid]. Orkide.

**ordain** [ôˈdêyn]. İrade etm.; (Allah) takdir etm., mukadder kılmak; ruhanî rütbe tevcih etmek. **to be ~ed,** papazlığa tayin olunmak.

**ordeal** [ôˈdîl]. Ateşten gömlek; çetin bir tecrübe; mihnet.

**order**[1] [ˈôdə(r)] *n.* İntizam, nizam; tertib; usul, yol; usul ve âdab; sıra, saf; mertebe derece; tabaka, sınıf; tarikat; emir, ihtar; sipariş. **~! ~!,** [**to call s.o. to ~**], *meclis vs.de bir azayı müzakere usulüne davet için nida*: **in ~ of age** [**seniority,** *etc.*]. yaş [kıdem vs.] sırasile: **~ on a bank,** banka havalesi: **~ of battle,** muharebe nizamı: **to call for ~s,** sipariş almak için uğramak: **close ~,** (*mil.*) yanaşık nizam: **~ of the day,** günlük emir: **Holy Orders,** papazlık; **to take Holy ~s,** papaz olm.: **all in ~,** her şey yerli yerinde; nizama uygun: **in good ~,** düzgün; işliyen; iyi bir halde: **to put in ~,** düzeltmek; sıraya koymak: **in ~ to do stg.,** bir şey yapmak için: **in ~ that stg. may be done,** bir şey yapabilmesi için: **to keep ~,** intizamı temin etm.; **to keep children in ~,** çocukları uslu tutmak: **~ of knighthood,** şövalyelik rütbesi, *v.* **order**[2]: **made to ~,** ısmarlama: **in marching ~,** (*mil.*) yürüyüş techizatile: **the old ~ of things,** eski devir, eski nizam ve usul: **out of ~,** bozuk, işlemez; usulsüz: **out of its ~,** sıradan çıkmış. **~**[2] *n.* Nişan. **~ of knighthood,** şövalyelik rütbesi. *Başlıca İngiliz nişanları şunlardır*:—(i) **Order of the Garter (G.)**; (ii) **~ of the Bath (B.)**; (iii) **~ of St. Michael and St. George (M.G.)**; (iv) **~ of the British Empire (B.E.)**; (v) **Royal Victorian Order (V.O.).** *Garter Nişanının yalnız bir rütbesi vardır*: **Knight of the Garter (K.G.)**: *ötekilerin dört veya beş rütbesi vardir*:—1. **Grand Commander,** *mes.* **~ of the Bath (G.C.B.)**; 2. **Knight Commander,** *mes.* **~ of the Bath (K.C.B.)**; 3. **Commander,** *mes.* **~ of St. Michael and St. George (C.M.G.)**; 4. **Officer,** *mes.* **~ of the British Empire (O.B.E.)**; 5. **Member,** *mes.* **~ of the Victorian Order (M.V.O.).** **~**[3] *v.* Emretmek, emir vermek; tanzim etm., idare etm.; tayin etm.; sipariş etm., ısmarlamak. **~ arms!,** hazırol!. **~ about,** (sağa sola) emretmek. **~ off,** ayrılmasını emretmek: **to ~ a player off the field,** (hakem) oyuncuyu sahadan çıkarmak. **~ed** [ˈôdö(r)d] *a.,* muntazam. **well ~,** nizamlı, iyi idare edilen. **~ly** [ˈôdəli], tertibli; toplu;

usullü; uslu; muntazam. Emir eri. ~ **officer,** nöbetçi subayı. ~ **room,** kışlalarda kalem odası.
**ordinal** [ˈôdinəl]. ~ **number,** sıra sayısı.
**ordinance** [ˈôdənəns]. Talimatname; ihtar; ferman; âyin.
**ordinar·y** [ˈôdinəri]. Alelâde, mutad, her zamanki; tipik; âdi. **above the** ~, alelâdenin üstünde: **a very** ~ **kind of man,** kendi halinde bir adam: **out of the** ~, müstesna, fevkalâde: ~ **seaman,** (bahriyede) üçüncü sınıf gemici: ~ **share,** âdi hisse: **physician-in-**~ **to the Queen,** Kıraliçenin hususî doktoru. ~**ily,** bermutad, alelâde, umumiyetle.
**ordination** [ˌôdiˈnêyşn]. Papazlığa kabul edilme.
**ordnance** [ˈôdnəns]. Top; askerî levazım ve techizat dairesi. ~ **Survey,** İngiltere'nin harita dairesi.
**ordure** [ˈôdyuə(r)]. Pislik; müzahrefat.
**ore** [ô(r)]. Maden filizi.
**organ** [ˈôgən]. Uzuv; vasıta olan şey, kimse; cihaz; organ; gazete; org. **barrel** ~, lâterna: **mouth** ~, ağız mızıkası: **the vocal** ~**s,** ses cihazı. ~**-grinder,** lâternacı. ~**ic** [ôˈganik], uzvî; esasî. ~**ist** [ˈôgənist], orgcu.
**organiz·e** [ˈôgənâyz]. Teşkil et.; tanzim ve tertib etmek. ~**ation** [–ˈzêyşn], teşkil; teşekkül; teşkilat; bünye. ~**er,** tanzim ve tertib eden; teşkilatçı.
**orgy** [ˈôci] *pl.* **orgies.** Sefahat âlemi; cümbüş.
**orient** [ˈôriənt]. Şark; doğu. ~ **pearl,** en iyi cinsden inci. ~**al** [–ˈentl], şarkî, şarka mahsus. ~**alist** [–ˈentəlist], müsteşrik. ~**ate** [ˈôriəntêyt], bir şeyin mevkiini tayin etmek. **to** ~ **oneself,** kendi vaziyetini tayin etmek, takdir etmek. ~**ation** [–ˈtêyşn], cihet tayini.
**orifice** [ˈorifis]. Delik; ağız.
**origin** [ˈoricin]. Mebde, menşe, menba; asıl; nesil. ~**al** [oˈricinl], esasî, asıl; ilk; kopya olmıyan; ilk defa meydana konmuş; yepyeni; yeni fikirler meydana getirmek iktidarı olan. Bir yazının aslı; aslî nüsha; metin: ~ **sin,** bütün insanların doğuşta mevcud olan (?) günahı. ~**ality** [–ˈnaliti], yepyenilik; bambaşka bir tarzda olma; kimseye benzememezlik. ~**ate** [oˈricinêyt], icad etm., ihtira etm.; meydana gelmek.
**Orion** [oˈrâyən]. Cebbar burcu.
**ornament** [ˈônəmənt]. Süs, ziynet. Süslemek, tezyin etmek. ~**al** [—ˈmentl], süslü; güzel; gösterişli.
**ornate** [ôˈnêyt]. Fazla süslenmiş.
**ornithology** [ˌôniˈθoləci]. Kuşlar bahsi.
**orphan** [ˈôfən]. Öksüz, yetim. ~**age** [–ic], yetimler yurdu; öksüzlük. ~**ed,** yetim kalmış.
**ortho-** [ˈôθou]. *pref.* Doğru ....
**orthodox** [ˈôθoudoks]. Akidesi sahih; umumiyetle kabul edilen bir hakikate veya bir fikre uygun; usul ve erkâna uygun, ortodoks; Ortodoks kilisesine mensub. ~**y,** akidenin sıhhati; ortodoksluk; bir insanın fikirlerinin örfe uygunluğu.
**oscillat·e** [ˈosilêyt]. Saat rakkası gibi hareket etm., sallanmak; sars(ıl)mak. ~**ion** [–ˈlêyşn], salınma.
**osprey** [ˈosprêy]. Balık kartalı (?); sorguc.
**ossify** [ˈosifây]. Kemikleş(tir)mek.
**ostensible** [osˈtensibl]. Surî, zahirî.
**ostentat·ion** [ˌostənˈtêyşn]. Gösteriş, çalım. ~**ious,** gösterişçi, cakalı.
**osteopath** [ˈostiopaθ]. Kırıkçı.
**ostler** [ˈoslə(r)]. Han seyisi.
**ostracize** [ˈostrəsâyz]. Cemiyet harici ilân etm.; birile her türlü münasebeti kesmek.
**ostrich** [ˈostric]. Devekuşu.
**other** [ˈʌðə(r)]. Başka, diğer; sair; öbür; o bir, öteki. **the** ~**s,** ötekiler: **every** ~, her ikinci: **the** ~ **day,** geçenlerde: **some** ~ **day,** başka bir gün: **some ...,** ~**s ...,** bazısı ..., bazısı ...: **one or** ~ **of us,** aramızdan biri: **the** ~ **world,** öbür dünya: **I could not do** ~ **than ... [I could do no** ~ **than...],** benim için ···den başka yapacak bir şey yoktu: **fancy coming this day of all** ~**s!,** başka günleri bırakıp sen tut da bu gün gel! ~**wise** [ˈʌðəwâyz], başka türlü; yoksa; aksi takdirde. **he could not do** ~ **than ...,** ···den başka bir şey yapamadı: **he is rather mean,** ~ **he is pleasant,** bir az hasistir, yoksa hoş adamdır. ~**worldly** [ˈʌðəwö(r)ldli], dünyevî olmıyan; ahret adamı; bu dünyadan değil.
**otter** [ˈotə(r)]. Su samuru, lutr.
**Ottoman**[1] [ˈotəmən]. Osmanlı. ~[2]. Sedir.
**ought**[1] [ôt]. *Tasrif edilmez yardımcı fiil; umumiyetle vücubî fiil ile tercüme edilir.* Lâzım, elverişli veya uygun olmak. **I** ~ **to go,** gitmem

lâzım; gitmeliyim: I ~ to have gone, gitmeliydim: **I ~ to know, but I don't,** bilmem lâzım amma bilmiyorum: **to behave as one ~,** icab ettiği gibi hareket etm.: **you ~ to read this book,** bu kitabı her halde okuyunuz: **she ~ to be married soon,** (i) (yaşı geçiyor) bir an evvel evlenmesi lâzım; (ii) (cazibeli olduğu icin) bu kızın her halde çabuk kısmeti çıkar. **~².** Hiç; sıfır.

**ounce¹** [auns]. İngiliz tartı ölçüsü. **avoirdupois ~,** librenin $\frac{1}{16}$ sı = 28·35 gram: **troy ~,** librenin $\frac{1}{12}$ sı = 31·1 gram. **he hasn't an ~ of sense,** on paralık aklı yok. **~²,** kar parsı.

**our** [auə(r)]. Bizim, bize aid. **~s,** bizimki. **this house is ~,** bu ev bizimdir: **a friend of ~,** ahbablarımızdan biri. **~selves** [-ˈselvz], **we**'*yi tekid için kullanılır*: **we ~ prefer to live in London,** biz kendimiz Londra'da oturmağı tercih ediyoruz.

**oust** [aust]. Yerinden çıkarmak; birini yerinden edip kendisi o yeri almak.

**out** [aut]. Dışarı; haricde; ···den; ···den dışarı; evde değil. [*Fiille birlikte* dışarı *veya* tamamlama *mânalarını ifade eder. mes.*:– **to run ~,** (i) dışarıya koşmak; (ii) tükenmek: **to think stg. ~,** bir şeyi düşünüp taşınmak.] **~ and ~,** tamamen, son derece: **all ~,** alabildiğine (koşmak, çalışmak vs.): **my wife is ~,** refikam evde değil: **he is ~ and about again,** (bir hasta hakkında) artık kalktı: **the secret is ~,** sır ifşa edildi: **the miners are ~ again,** madenciler yine grev yapıyorlar: **to be ~ in one's calculations,** hesablarında yanılmak: **he is ten pounds ~ in his accounts,** hesabında on liralık hata var: **day ~,** (hizmetçi) izinli gün: **the fire is ~,** ateş sönmüş: **to know the ins and ~s of stg.,** bir şeyin içini dışını bilmek: **to be ~ of it,** (i) bir muhiti yadırgamak; (ii) bir iş vs. ile alâkasını kesmek; (iii) (yarış vs.de) kaybedeceği muhakkak olm.: **to be ~ of sugar,** şekeri kalmamak: **to feel ~ of it,** bir muhitte kendini yabancı hissetmek: **~ of politeness,** nezaketen: **you will be well ~ of the whole business,** bu işten yakayı sıyırsanız sizin için çok iyi olur: **my patience is ~,** sabrım tükendi: **put him ~! [~ with him!],** onu dışarı at!: **there is no other way ~,** başka çıkar yol yok: **~ with it,** (i) ver bakalım!; (ii) ağzından baklayı çıkar! **~-of-date,** modası geçmiş. **~-of-place,** yersiz, mevsimsiz. **~-of-the-way,** hücra, sapa; garib. **~-of-pocket,** cebden. **~-patient,** hastahanede yatmadan tedavi edilen hasta: **~s' department,** hastahanenin dispanseri. **~-relief,** fakir ailelere yapılan para yardımı.

**out-** *pref. Hemen hemen her fiilin ve bir çok isimlerin başına getirilebilir ve* üstünlük *veya* fazlalık *ifade eder, mes.*:– **to ~-row s.o.,** başkasından daha iyi kürek çekmek: **an ~size in shoes,** en büyük boyda ayakkabı.

**outbid** [autˈbid]. **to ~ s.o. at an auction,** müzayedede birisinden fazla pey sürmek.

**outboard** [ˈautbôd]. Dıştan takma (deniz motörü).

**outbreak** [ˈautbrêyk]. Zuhur; çıkış; vuku; kıyam. **~ of temper,** hiddet galeyanı.

**outburst** [ˈautbö(r)st]. Patlama, infilak; fışkırma; kopma; taşkınlık.

**outcast** [ˈautkâst]. Kimsesiz; serseri.

**outclass** [autˈklâs]. (Başkalarına) pek üstün olmak.

**outcome** [ˈautkəm]. Netice; son; akibet.

**outcrop** [ˈautkrop]. Arz tabakasının yer yüzüne çıkması.

**outcry** [ˈautkrây]. Feryad; haykırma; şikâyet sesi.

**outdistance** [autˈdistəns]. Birini geçmek.

**outdo (-did, -done)** [autˈdu, -did, -dʌn]. Fevkinde olm., geçmek; ···den galebe çalmak. **not to be outdone,** altta kalmamak.

**outdoor** [autˈdô(r)]. Ev dışında, açık havada yaşıyan, bulunan, vukubulan vs. **~s,** ev dışında, açık havada.

**outer** [ˈautə(r)]. Dış tarafta bulunan; haricî; en uzak. **~most,** en dışta, en uzakta.

**outfit** [ˈautfit]. Techizat; takım; levazimat; sefer levazimatı; elbise. **first-aid ~,** ilk yardım kutusu: **the whole ~,** takım taklavat. **~ter,** hazırcı; erkek elbisesi satıcısı.

**outflank** [autˈflan(g)k]. (*mil.*) Cenah taşmak.

**outflow** [ˈautflou]. Dışarıya akan (su, havagaz vs.) nin mikdarı; mahrec.

**outgeneral** [autˈcenrəl]. Düşmandan daha iyi manevra yapmak.

**outgoing** [ˈautgouin(g)]. Çıkan, kalkan. **~s,** masraf, sarfiyat.

**outgrow** [au̯tˈgrou̯]. **to ~ s.o.**, birisinden daha çabuk büyümek: **to ~ clothes**, (büyüdükçe) elbisesi dar gelmek.
**outhouse** [ˈau̯thau̯s]. Mülhak bina.
**outing** [ˈau̯tin(g)]. Gezinti.
**outlandish** [au̯tˈlandiş]. Pek garib; vahşi.
**outlast** [au̯tˈlâst]. Daha çok dayanmak.
**outlaw** [ˈau̯tlô]. Kanun harici (kimse). Kanunî haklardan mahrum etm.; yasak etm.
**outlay** [ˈau̯tlêy]. Sarfiyat; bir teşebbüse başlamak için lâzımgelen masraf.
**outlet** [ˈau̯tlet]. Mahrec; menfez; (*comm.*) pazar.
**outline** [ˈau̯tlâyn]. ~ *veya* **~s**, dış hatları; şekil; taslak. Taslağını çizmek. **main [general, broad] ~s of a plan**, bir plan vs.nin ana hatları: **in ~**, kabataslak.
**outlive** [au̯tˈliv]. (Başkaları öldüğü halde) artakalmak.
**outlook** [ˈau̯tluk]. Manzara; görünüş; ihtimal. **the ~ is not cheerful**, vaziyet pek parlak değil.
**outlying** [ˈau̯tlây·in(g)]. Etrafta olan; uzakça.
**outmost** [ˈau̯tmou̯st]. En dıştaki.
**outnumber** [au̯tˈnʌmbə(r)]. Sayıca üstün olmak.
**outpost** [ˈau̯tpou̯st]. İleri karakol.
**output** [ˈau̯tput]. Verim; istihsal.
**outrage** [ˈau̯trêyc]. Tecavüz; zorbalık; suikasd; rezalet. Zorlamak; tecavüz etm.; din, iffet, kanun vs.ye karşı hareket etmek. **~ous** [–ˈrêycəs], son derece mütecaviz; müdhiş; rezilane.
**outright** [au̯tˈrâyt] *adv.* Tamamile, büsbütün; açıkça; kat'i bir surette; derhal. *a.* [ˈau̯trâyt], Açık; müsbet; dobra dobra; kat'i. **to buy stg. ~**, derhal ve peşin para ile almak: **to kill ~**, hemen öldürmek.
**outrun** [au̯tˈrʌn]. ···den daha çabuk koşmak. **his ambition ~s his ability**, ihtirası kabiliyetinden üstün.
**outset** [ˈau̯tset]. Başlangıc, iptida. **at the ~**, ilkönce: **from the ~**, başlanğıcdanberi.
**outshine** [au̯tˈşâyn]. ···den daha parlak olm., gölgede bırakmak.
**outside** [au̯tˈsâyd]. Dış, haricî; dış taraf, haric; dışarı, haricde, dışında. **at the (very) ~**, olsa olsa: **an ~ chance**, küçük bir ihtimal: **to get an ~ opinion**, dışarıdan birinin fikrini almak: **~ porter**, serbest hamal: **~ price**, en yüksek fiat: **~ work**, (i) işçinin kendi evinde yaptığı iş; (ii) evin dış kısmında tamirat ve tezyinat; (iii) vazife veya iş saatleri haricinde yapılan iş. **~r** [au̯tˈsâydə(r)], bir meslek, parti, mahfil vs.ye mensub olmıyan kimse; kibar olmıyan kimse; (at yarışlarında) muteber olmıyan bir at.
**outskirts** [ˈau̯tskö(r)ts]. Civar, kenar.
**outspoken** [au̯tˈspou̯kn]. Açık, tok sözlü; kör kadı.
**outstanding** [au̯tˈstandin(g)]. Çıkıntılı; göze çarpan; mümtaz; kalbur üstü; muallakta kalan; tedahülde kalan.
**outstay** [au̯tˈstêy]. ···den fazla kalmak. **to ~ one's leave**, iznini geçirmek: **to ~ one's welcome**, bir evde çok kalarak kendisini istiskal ettirmek.
**outstrip** [au̯tˈstrip]. Geçmek.
**outward** [ˈau̯twəd]. Dış, zahirî. **the ~ journey**, dışarıya sefer: **~bound ship**, limandan çıkan gemi. **~s**, dışarıya doğru.
**outweigh** [au̯tˈwêy]. Daha mühim olmak.
**outwit** [au̯tˈwit]. ···den daha kurnazca davranmak.
**ova** [ˈou̯va]. *pl. of* **ovum**.
**oval** [ˈou̯vl]. Beyzî.
**ovary** [ˈou̯vəri]. Mebiz; tohumluk.
**ovation** [ou̯ˈvêyşn]. Halkın bir kimseyı çılgınca alkışlaması.
**oven** [ʌvn]. Fırın.
**over** [ˈou̯və(r)]. Üstünde, üzerinde; yukarısında; fevkinde; ···den fazla; öbür tarafına (-da); karşı yakasına (-da); bütün sathına; baştan başa; hakkında; nazaran; bitmiş. **~ again**, bir daha: **all ~ again**, yeni baştan: **~ against**, karşısında: **all ~**, her tarafında: **it's all ~**, bitti: **to be wet all ~**, tepeden tırnağa ıslanmak: **to be all ~ dust**, elbise toz içinde kalmak: **all ~ the place**, her tarafa (dağılmış vs.): **~ and ~**, tekrar tekrar; yuvarlanarak: **men of twenty and ~**, yirmi yaşında ve yirmiden yukarı olanlar: **he is ~ eighty**, sekseni geçkindir: **~ and above this**, bundan başka: **that is ~ and done with**, oldu bitti: **5 into 13 goes twice and 3 ~**, 13te 5 iki defa var, 3 kalır: **~ the border**, hududun ötesine: **what's come ~ you?**, sana ne oluyor?: ~

there [yonder], karşıda, orada: **several times ~**, üstüste bir kaç defa: **~ the last five years**, son beş sene zarfında: **~ the way**, karşı tarafta.
**over-** *pref. Bir fiile ve bazan bir sıfat veya isme eklendiği zaman şu mânaları ihtiva eder:*— (i) lüzumundan fazla; icab ettiğinden fazla; *mes.* **to overeat** = fazla yemek; **to overload** = fazla yükletmek: (ii) en üst; üstünlük; örtme, *mes.* **overcoat** = palto; **overlord** = metbu; **overcome** = hakkından gelmek. **Over** *edatı hemen her kelimenin başına konarak yenı bir kelime icad olunabilir.* **Over** *ile başlayıp lûgatte bulunmıyan kelimeleri asıl kelimede arayıp* **over** *ile aldıkları yeni mânayı yukardaki izahattan çıkarmak mümkündür.*
**overall** [ˈouvərôl]. Arkadan ilikli göğüslük. **~s,** çekme, tulum. **overall length,** tam boy.
**overawe** [ouvərˈô]. Korkutup hareket ve muhalefetten menetmek.
**overbalance** [ouvəˈbaləns]. Müvazenesini kaybetmek, devrilmek.
**overbearing** [ouvəˈbeərin(g)]. Mütehakkim.
**overboard** [ˈouvəbôd]. Gemiden denize düşmüş. **man ~!**, denize adam düşmüş!
**overcast** [ˈouvəkâst]. Bulutlu; endişeli. **face ~ with fear,** korkulu kaplı yüz.
**overcharge** [ouvəˈçâc] *v.* Fahiş fiat istemek. *n.* [ˈouvə–]. Fahiş fiat; zam.
**overcoat** [ˈouvəkout]. Palto.
**overcome** (**-came**) [ouvəˈkʌm, –kêym]. Hakkından gelmek; zabtetmek. **to be ~ by grief,** *etc.*, keder vs.ye kapılmak.
**overdo** [ouvəˈdû]. İfrat etm.; çok pişirmek. **to ~ oneself** [**it**], kendini çok yormak.
**overdraft** [ˈouvədrâft]. Açık itibar.
**overdraw** (**-drew, -drawn**) [ouvəˈdrô, –drû, –drôn]. Bankadaki mevduatından fazla para çekmek.
**overdue** [ouvəˈdyû]. Vadesi geçmiş; gecikmiş.
**overflow** *v.* [ouvəˈflou]. Taşmak. *n.* [ˈouvə-], havuz vs.de fazla suların dökülmesine mahsus oluk. **~ meeting,** toplantı yerine sığmıyan halka mahsus bir toplantı. **~ing,** taşma; bol.
**overgrow** [ouvəˈgrou]. (Nebat) bir yer vs.yi kaplamak. **~n,** otlarla kaplanmış: **~ child,** yaşından fazla büyümüş çocuk.
**overhaul** *v.* [ouvəˈhôl]. Muayene, tamir etm.; arkadan yetişmek. *n.* [ˈouvəhôl]. Dikkatli muayene; tamir.
**overhead** *adv.* [ouvəˈhed]. Başın üstte; havada. *a.* [ˈouvəhed]. Yukarıda olan. **~ cable,** havaî kablo. **~s,** umumi masraflar.
**overhear** [ouvəˈhiə(r)]. (Bir şeye) kulak misafiri olm.; tesadüfen işitmek.
**overjoyed** [ouvəˈcôyd]. Ziyadesile memnun; etekleri zil çalıyor.
**overland** [ˈouvəland]. Karadan.
**overlap** *v.* [ouvəˈlap]. Üst üste katlanmak. *n.* [ˈouvə–]. Sarkan kısım.
**overleaf** [ouvəˈlîf]. Bir kâğıdın arkasında.
**overlook** [ouvəˈluk]. Yüksekten bakmak; ···e nazır olm.; ···e göz yummak; unutmak; nezaret etm. **~ it this time!,** bu sefer bağışla, hoşgör!: **~ing the sea,** denize nazır.
**overlord** [ˈouvəlôd]. Metbu; âmir.
**overpower** [ouvəˈpauə(r)]. Hakkından gelmek; mağlub etmek. **~ing,** tahammül edilmez.
**overrate** [ouvəˈrêyt]. Fazla kıymet vermek.
**overreach** [ouvəˈrîç]. Hile ile yenmek. **to ~ oneself,** haddini aşırmak.
**override** (**-rode, -ridden**) [ouvəˈrâyd, –roud, –ridn]. Çok binerek yormak; üstün gelmek. **to ~ one's authority,** salâhiyetini aşmak: **this decision ~s all others,** bu karar ötekileri ibtal eder.
**overrule** [ouvəˈrûl]. Cerhetmek; hükümsüz bırakmak.
**overrun** (**-ran**) [ouvəˈrʌn, –ran]. Haddini tecavüz etmek; her tarafına yayılmak: (bir makineyi) çok işletmek.
**oversea** [ˈouvəsî]. Denizaşırı. **~s** [ouvəˈsîz], denizaşırı memleketlerde -e, -den.
**oversee** [ouvəˈsî]. Gözetmek, nezaret etmek. **~r** [ˈouvəsiə(r)], müfettiş; nezaretçi.
**overshoes** [ˈouvəşûz]. Galoş.
**oversight** [ˈouvəsâyt]. Dikkatsizlik; sehiv; nezaret. **by** [**through**] **an ~,** dikkatsizlikle.
**oversleep** (**-slept**) [ouvəˈslîp, –slept]. **to ~ oneself,** uyuya kalıp gecikmek.
**overstate** [ouvəˈstêyt]. Mübalağa etmek.

**overstep** [quvəˈstep]. Haddini aşmak.
**overt** [ˈquvö(r)t]. Meydanda; açık.
**overtake** (-took, -taken) [quvəˈtêyk, -tuk, -têykn]. Arkadan yetişmek; yetişip geçmek; başına gelmek. **to ~ arrears of work,** geri kalan işi yetiştirmek.
**overtax** [quvəˈtaks]. Ağır vergilerle ezmek. **to ~ s.o.'s patience,** birinin sabrını taşırmak: **to ~ one's strength,** kendini yıpratmak.
**overthrow** *v.* (-threw, -thrown) [quvəˈθrqu, -θrû, -θrqun]. Devirmek; yıkmak; yenmek. *n.* [ˈquvəθrqu]. İnkiraz; devrilme.
**overtime** [ˈquvətâym]. Vazife harici çalışılan zaman.
**overture** [ˈquvətyuə(r)]. Müzakere teklifi; uvertür. **to make ~s to s.o.,** bir iş hakkında ilk teklifi yapmak.
**overturn** [quvəˈtö(r)n]. Devirmek. Devrilmek.
**overwhelm** [quvəˈwelm]. Kahretmek; ezmek. **to be ~ed with joy,** sevincinden kendini kaybetmek: **I am ~ed by your generosity,** cömerdlikle beni son derece mahcub ediyorsunuz. **~ing,** kahir; ezici.
**overwork** [quvəˈwö(r)k]. Fazla çalıştırmak. **~ (oneself),** sıhhatini bozacak derecede çalışmak. Fazla çalışma.
**overwrought** [quvəˈrôt]. Fazla heyecan veya çalışmaktan bitkin.
**ovum,** *pl.* **ova** [ˈquvəm, ˈquva]. Yumurta; tohum.
**ow·e** [qu]. Borcu olm., borclu olmak. **~ing,** borc olarak; bakiye; tediyesi lâzım: **~ to ...,** ... yüzünden, ...den dolayı.
**owl** [aul]. Baykuş. **~ish,** baykuş gibi.
**own**[1] [qun] *v.* Tasarruf etm., malik olm., sahib olm.; tanımak, itiraf etmek. **he ~s three houses,** üç evi var: **to ~ up (to),** itiraf etmek. **~**[2] *a.* & *n.* Kendi, kendinin, kendisinki. **my ~ house,** kendi evim: **the house is my ~,** ev kendime aid: **~ brother,** öz kardeş: **to come into one's ~,** kendi malını elde etm.; lâyik olduğu mevkii almak: **I do my ~ cooking,** yemeğimi kendim pişiriyorum: **to get one's ~ back,** kuyruk acısını çıkarmak: **to hold one's ~,** mevkiini tutmak: **he has money of his ~,** kendi parası var: **on one's ~,** kendi başına: **I am all on my ~ today,** bugün kendi kendimeyim: **to love truth for its ~ sake,** hakikati hakikat için sevmek: **my time is my ~,** vaktimi istediğim gibi kullanabilirim. **~er** [ˈqunə(r)], sahib; mal sahibi; tasarruf eden. **~ership,** sahiblik: **'under new ~',** sahibi değişmiştir.
**ox,** *pl.* **oxen,** [oks, ˈoksn]. Öküz.
**oxid·e** [ˈoksâyd]. Humuz, oksid. **~ize** [ˈoksidâyz], tahammuz et(tir)mek; paslanmak.
**Oxonian** [okˈsqunıən]. Oxford üniversitesine mensub.
**oyster** [ˈôystə(r)]. İstridye. ⌜**as close as an ~**⌝, çenesini bıçak açmıyor.
**oz.** (*abb.*) **ounce(s)**[1].

# P

**P** [pî]. P harfi.
**pa** [pâ]. (*coll.*) Baba.
**pace**[1] [pêys] *n.* Adım; yürüyüş; sürat. **to go the ~,** çabuk koşmak; sefahat içinde yaşamak: **the ~ is too hot for me,** ben onlarla yarış edemem: **to keep ~ with ...,** ...e adım uydurmak: **to put s.o. through his ~s,** birini yoklamak: **to quicken one's ~,** adımlarını açmak. **~**[2] *v.* Adımlamak, arşınlamak; (bir koşucu) idman ettirmek. **to ~ up and down,** bir aşağı bir yukarı gezmek.
**pacif·ic** [pəˈsifik]. Sulhperver, barışsever; yumuşak başlı. **the P~,** Büyük Okyanus. **~ist** [ˈpasifist], sulh tarafdarı. **~y** [ˈpasifây], teskin etm., yatıştırmak.
**pack**[1] [pak] *v.* (Eşyayı) bavula koymak; denk etm.; paket yapmak; sıkı tıkmak; kıtık ile tıkamak. Sürü halinde toplanmak: **~ (up),** eşyasını toplayıp sandıklarına koymak; gitmeğe hazırlanmak. **~ up,** (*sl.*) vazgeçmek; kaçmak: **to ~ a meeting [a jury,** *etc.*], bir meclis, bir jüri vs.de kendi tarafdarlarının ekseriyetini temin etm.: **the room is ~ed,** oda hıncahınc: **~ed like sardines,** balık istifi: **to send s.o. ~ing,** (*coll.*) birini kovmak: **to ~ together,** sıkı sıkı toplanmak. **~**[2] *n.* Takım; sürü; bohça, arkaçantası; (*med.*) sargı. **a ~ of cards,** iskambil destesi: **~ of hounds,** av köpeği sürüsü: **(ice-)~,** buz birikintisi: **a ~ of lies,** yalan dolan. **~-animal,** yük hayvanı **~-saddle,** semer. **~age** [ˈpakic]

bohça; paket; küçük deste. ~et [ˈpakit], paket; deste. ~et-boat, yolcu ve posta gemisi. ~ing [ˈpakin(g)], ambalaj; denk bağlama; tıkaç. ~ing-needle, çuvaldız.

**pact** [pakt]. Misak; mukavele.

**pad**[1] [pad] *n.* Küçük yastık; (yara vs.için) pamuk yastık; sumen; zımbalı not defteri; istampa; palan, belleme; bazı hayvanların yumuşak tabanı; tilki ve tavşan pençesi. ~[2] *v.* İçini yün vs. ile doldurmak; fodra etm.; bir yazı haşviyat katmak. **to ~ along,** kurt gibi sessizce koşmak. **~ding,** fodra; haşviyat.

**paddle** [ˈpadl]. Kısa kürek; vapurun yan çark tahtası; su değirmeninin kanadı; su kaplumbağasının ayağı. Kanoyu kısa kürek ile yürütmek; çıplak ayaklarını suya sokup oynamak. **~-boat,** yandan çarklı gemi. **~-wheel,** yan çarkı.

**paddock** [ˈpadək]. Ahır civarında küçük çayır.

**paddy** [ˈpadi]. Çeltik.

**padlock** [ˈpadlok]. Asma kilid (ile kilidlemek).

**padre** [ˈpâdrêy]. Orduya mensub papaz.

**paean** [ˈpîən]. Zafer türküsü.

**pagan** [ˈpêygən]. Putperest; ehli kitab olmıyan; kâfir. **~ism,** putperestlik.

**page**[1] [pêyc]. Sahife. **~**[2], iç oğlanı; peyk.

**pageant** [ˈpacənt]. Debdebeli alay.

**pagoda** [pəˈgo̯udə]. Çinde mabed kulesi.

**paid** [pêyd] *v.* **pay** *a.* Ödenmiş; tediye edilmiş. **to put ~ to ...,** temizlemek.

**pail** [pêyl]. Gerdel; helke.

**pain** [pêyn]. Acı, ağrı; elem. **~s,** zahmet, meşakkat. *v.* **to ~ [give ~ to],** acıtmak, ağrıtmak, canını yakmak; kederlendirmek. **to be in ~,** acı duymak: **on ~ of death,** ölüm cezasile: **it ~s me to say this,** bunu söylemek bana elem veriyor: **~s and penalties,** kanunî cezalar: **to put a wounded animal out of its ~,** yaralı bir hayvanı öldürüp eziyetten kurtarmak: **to take ~s [be at great ~s] to do stg.,** bir şeyi yapmak için çok uğraşmak: **to take ~s over stg.,** bir şeye son derece itina etmek. **~ed,** canı sıkılmış, kederli. **~ful,** acı veren; ıstırab veren; cansıkıcı, müteessir edici. **~staking** [ˈpêynztêykin(g)], itinalı.

**paint** [pêynt]. Boya. Boyamak; boya ile resmetmek; tarif ve tasvir etm.; düzgün vs. sürmek. **to ~ the face,** makiyaj yapmak: **he ~s,** ressamdır: **to ~ the town red,** sokaklarda cümbüş yaparak ortalığı altüst etmek. **~er,** boyacı; ressam. **~ing,** boyalı resim; ressamlık.

**pair**[1] [pe̯ə(r)] *n.* Çift; iki aded; karı koca; eş. **carriage and ~,** iki atlı araba: **a ~ of scissors,** makas: **a ~ of steps,** üç ayaklı merdiven: **a ~ of trousers,** pantalon. **~**[2] *v.* Çift çift tertib etm.; eşini bulmak; eş olm.: **to ~ off,** ikişer ikişer ayırmak.

**pal** [pal]. (*coll.*) Arkadaş; kafadar. **to ~ up with,** ···le kafadar olmak.

**palace** [ˈpaləs]. Saray.

**palat·e** [ˈpalət]. Damak. **to have a fine ~,** ağzının tadını bilmek: **to have no ~ for ...,** ···e karşı iştahı olmamak. **~able,** lezzetli.

**palatial** [pəˈlêyşl]. Saray gibi, muhteşem.

**palaver** [paˈlâvə(r)]. Vahşi kabilelerde müzakere; palavra; boş lakırdı.

**pale**[1] [pêyl] *n.* Kazık; (*esk.*) hudud. **beyond the ~,** (parya gibi) cemiyete kabul edilemez. **~**[2] *a.* Solgun; sönük; açık (renk). *v.* **turn ~,** sapsarı kesilmek, rengi uçmak.

**pali·ng** [ˈpêylin(g)]. Parmaklık. **~sade** [paliˈsêyd], ağac veya demir kazıklarla yapılmış çit.

**pall**[1] [pôl] *n.* Tabut örtüsü; (*fig.*) örtü. **~**[2] *v.* Yavanlaşmak; tadını kaybetmek. **it never ~s on one,** insan ona hiç doyamaz.

**palliat·e** [ˈpaliêyt]. Hafifletmek; muvakkaten dindirmek.

**pall·id** [ˈpalid]. Solgun. **~or** [ˈpalə(r)], solgunluk.

**palm**[1] [pâm]. Hurma ağacı. **to carry off the ~,** galebe çalmak; mükâfat kazanmak. **P~ Sunday,** Paskalyadan önceki pazar günü. **~**[2]. Avuc, aya. **to ~ stg. off on s.o.,** birine (değersiz) bir şeyi yutturmak: **to ~ a card,** kâğıd oyununda elçabukluğu ile bir kâğıd elde etmek: **to grease the ~,** rüşvet vermek. **~ist,** el falcısı. **~istry,** el falı. **~-oil,** hurma yağı; rüşvet. **~y** [ˈpâmi], **~ days,** (geçmişteki) mesud günler.

**palpable** [ˈpalpəbl]. Belli.

**palpitat·e** [ˈpalpitêyt]. Titremek; (yürek) oynamak. **~ion** [–ˈtêyşn], yürek oynaması.

**paltry** [ˈpôltri]. Miskin, değersiz.

**pamper** [ˈpampə(r)]. Şımartmak.
**pamphlet** [ˈpamflit]. Risale. **~eer** [–ˈtiə(r)], risale muharriri; hicviyeci.
**pan**[1] [pan] *n*. Yassı kab; çanak; leğen; tava; terazi gözü. ⌜**a flash in the ~**⌝, neticesiz bir hamle. **~**[2] *v*. Altınlı toprak demir tavada yıkayıp altını ayırmak. **to ~ out,** yıkayıştan bir mikdar altın çıkmak; neticelenmek: **it didn't ~ out as we expected,** umduğumuz gibi çıkmadı.
**Pan.** Eski Yunanlıların kır ilâhi; tabiat.
**pan-** *pref*. Tam ..., hep ....
**panacea** [panaˈsîə]. Her derde deva.
**pancake** [ˈpankêyk]. Gözleme. (Uçak) ufkî vaziyette yere şiddetle çarpmak: **as flat as a ~,** yamyassı.
**pandemonium** [ˌpandiˈmouniəm]. Bütün şeytanların toplandığı yer; velvele. **~ broke out,** kıyamet koptu.
**pander** [ˈpandə(r)]. Pezevenk(lik etmek). **to ~ to s.o.,** birisine yüz vermek: **to ~ to some vice,** bir kötülüğü teşvik etmek.
**pane** [pêyn]. Pencere camının bir tek parçası.
**panegyric** [paniˈcirik]. Medhiye.
**panel** [ˈpanəl]. (Kapı) ayna; kaplama tahtası; levha; memur vs. listesi. Tahta kaplama ile kaplamak. **~ling,** tahta kaplama.
**pang** [pan(g)]. Âni ve şiddetli sancı. **the ~s of death,** ölüm ihtilâcları: **the ~s of remorse,** vicdan azabı.
**panic** [ˈpanik]. Ansızın ve yersiz telaş ve korku. Ansızın ve çok defa esassız bir telaşa düşmek. **~ky,** kolayca telaşa düşen.
**panjandrum** [panˈcandrʌm]. *Şatafatlı bir rütbe ve memuriyet ifade eden uydurma bir unvan, bilh.* **the Great ~.**
**pannier** [ˈpaniə(r)]. Küfe.
**panoply** [ˈpanopli]. Tam takım zırh ve silâh. **in full ~,** tamamen mücehhez.
**panorama** [panoˈrâmə]. Geniş manzara.
**pant** [pant]. Sık sık nefes almak. **to ~ for breath,** nefes nefese olm.: **to ~ for [after] stg.,** şiddetle arzu etmek.
**pantechnicon** [panˈteknikən]. Ev eşyası taşımağa mahsus büyük araba.
**pantheism** [ˈpanθi·izm]. Vahdeti vücud.
**panther** [ˈpanθə(r)]. Pars.
**pantomime** [ˈpantoumâym]. (*esk.*) Sessiz tiyatro; (*şim.*) bir peri masalına dayanan piyes.
**pantry** [ˈpantri]. Sofra takımının muhafaza edildiği ve yıkandığı oda.
**pants** [pants]. Don; pantalon.
**pap** [pap]. Meme; lâpa.
**papa** [pəˈpâ]. Baba.
**papa·cy** [ˈpêypəsi]. Papalık (idaresi ve hükümeti). **~l** [ˈpêypl], Papaya aid.
**paper** [ˈpêypə(r)]. Kâğıd; gazete; imtihan sualleri; (bir kongre vs. için yazılan) tedkik. Kâğıddan yapılmış. Kâğıd ile kaplamak. **~s,** evrak; hüviyet cüzdanı; gemi evrakı. **brown ~,** paket kâğıdı: **to put to [down on] ~,** yazmak: **~ profits,** kâğıd üzerine (nazarî) kâr: **to read a ~,** gazete okumak; konferans vermek: **to send in one's ~s,** istifa etm.: **to set a ~,** imtihan sualleri hazırlamak. **~-hanger,** duvar kâğıdcısı. **~-knife,** kâğıd bıçağı. **~-mill,** kâğıd fabrikası. **~-weight,** prespapye.
**papier mâché** [ˌpapyeˈmaşe]. Kartonpat.
**papist** [ˈpêypist]. Katolik.
**par** [pa(r)]. Müsavilik; başabaş olma. **to be on a ~ with,** ···e müsavi olm.; ···le aynı seviyede olm.: **at ~,** başabaş: **above ~,** başabaştan yukarı: **below ~,** başabaştan aşağı: **to feel below ~,** bir az keyifsiz olmak.
**par.** (*abb.*) **paragraph.**
**parable** [ˈparəbl]. Manevî bir hakikati göstermek için anlatılan hikâye. **to speak in ~s,** kinayeli konuşmak.
**parabola** [pəˈrabola]. Kat'ı mükâfi.
**parachute** [ˈparəşût]. Paraşüt.
**parade** [pəˈrêyd]. Nümayiş; merasim; teftiş nizamı; piyasa yeri. Gösteriş yapmak; teşhir etm.; (askeri) ictimaa çağırmak; (asker) ictimaa çıkmak.
**paradise** [ˈparədâys]. Cennet: **bird of ~,** cennet kuşu: **an earthly ~,** yeryüzü cenneti: **to live in a fool's ~,** hakikatten gafil olarak yalancı bir saadet içinde yaşamak.
**paraffin** [ˈparəfîn]. Gaz, petrol. **liquid ~,** mayi halinde vazelin: **~ wax,** parafin.
**paragon** [ˈparəgən]. Fazilet örneği.
**paragraph** [ˈparəgrâf]. Satırbaşı; küçük fıkra.
**parallel** [ˈparəlel]. Muvazi; mütevazi; musalih. Muvazi hat; haritadaki arz dairelerinden biri; mukayese. Benzemek; mukayese etmek. **without ~,** emsalsiz; hiç görülmemiş.
**paraly·se** [ˈparəlâyz]. Felce uğrat-

mak. **to be ~d with fear,** korkudan donakalmak. **~sis** [pəˈralisis], inme; felc. **~tic** [–ˈlitik], mefluc, kötürüm: **to have a ~ stroke,** felc gelmek.

**paramount** [ˈparəmaunt]. Üstün; faik.

**paramour** [ˈparəmuə(r)]. Gayri meşru sevgili.

**parapet** [ˈparəpit]. Korkuluk.

**paraphernalia** [ˌparəfəˈnêylyə]. Her hangi techizat takım; cihaz.

**paraphrase** [ˈparəfrêyz]. Bir yazının mânasını başka kelimelerle ifade etme(k).

**parasit·e** [ˈparəsâyt]. Başkasının sırtından geçinen. **~ic** [–ˈsitik], tufeyli halinde olan.

**parasol** [ˈparəsol]. Şemsiye (güneşe karşı).

**parboil** [ˈpâbôyl]. Yarı kaynatmak.

**parcel** [ˈpâsl]. Paket; takım, yığın. **~ out,** hisse hisse ayırmak; ifraz etmek. **part and ~ of,** ···le hallihamur olan.

**parch** [pâç]. Kavurup kurutmak. **to be ~ed with thirst,** susuzluktan yanmak.

**parchment** [ˈpâçmənt]. Tirşe, parşömen.

**pardon** [ˈpâdn]. Af. Affetmek. **~!** [**I beg your ~!**], affedersiniz!; efendim? **~ me!,** affedersiniz! **~able,** affedilebilir; mazur görülebilir.

**pare** [peə(r)]. Yontmak; kabuğunu soymak.

**parent** [ˈpeərənt]. Baba veya anne. Ana; esaslı. **~s,** ebeveyn, ana baba. **~age** [–tic], nesil, soy. **~al** [pəˈrentl], ebeveyne aid. **~hood,** babalık *veya* analık.

**parenthe·sis,** *pl.* **-ses** [pəˈrenθəsis, –sîz]. Parantez. **~tic** [–ˈθetik], istitrad kabilinden.

**pariah** [ˈpâriə]. Parya. **~-dog,** sokak köpeği.

**pari passu** [ˈpeərâyˈpasyû]. (*Lât.*) Müsavi adımla.

**parish** [ˈpariş]. Bir papazın dinî bölgesi; mahalle. **to go on the ~,** mahalle tarafından beslenmek: **the whole ~,** bütün mahalle. **~ioner** [pəˈrişənə(r)], bir papazın dinî bölgesinde oturan kimse.

**parity** [ˈpariti]. Başa baş olma.

**park** [pâk]. Etrafı çevreli ağaclık geniş yer, koru; arabaları vs.nin toplu bulunduğu yer. **car ~,** otomobillerin muvakkaten bırakıldığı yer: **public ~,** park: **to ~ a car in a street,** otomobili muvakkaten bir sokakta bırakmak: **'no ~ing here',** buraya otomobil bırakılmaz.

**parlance** [ˈpâləns]. Konuşma tarzı. **in common ~,** konuşma dilinde: **in legal ~,** hukuk lisanında.

**parley** [ˈpâli]. Müzakere, (*esp.* mütareke hakkında). Müzakereye girmek.

**parliament** [ˈpâləmənt]. Parlamento; millet meclisi. **the Houses of ~,** Londrada parlamento sarayı: **both houses of ~,** ingiliz parlamentosunun Avam ve Lordlar kamaraları. **~ary,** parlamentoya aid.

**parlour** [ˈpâlə(r)]. Küçük salon, oturma odası. **bar ~,** bir birahanenin arka odası: **~ games,** toplantılara mahsus eğlenceli oyunlar: **~ tricks,** bir toplantıya gelenleri eğlendirici marifetler. **~maid,** sofracı.

**parlous** [ˈpâləs]. Tehlikeli; telâş verici.

**parochial** [pəˈroukyəl]. **Parish** veya mahalleye aid; mahallî; darkafalı.

**parody** [ˈparədi]. Ciddi bir eserin gülünç bir şekilde taklidi. Bu tarzda taklid etm.

**parole** [pəˈroul]. Bir mahpus veya esirin kaçmıyacağı hakkında verdiği söz. **to be on ~,** böyle namus üzerine verilen sözle serbest bırakılmak.

**paroxysm** [ˈparoksizm]. Şiddetli nöbet.

**parquet** [ˈpâkêy]. Parke.

**parricide** [ˈparisâyd]. Anasını, babasını veya yakın akrabasını öldürme.

**parrot** [ˈparət]. Papağan.

**parry** [ˈpari]. Darbeyi çelme(k). **to ~ a question,** sıkıcı bir suali baştan savma bir cevab ile atlatmak.

**parse** [pâz]. Bir kelime veya cümleyi gramer kaidesine göre tahlil etmek.

**Parsee** [pâˈsî]. Parsî; zerdüştî.

**parsimon·y** [ˈpâsimoni]. Cimrilik. **~ious** [–ˈmounyəs], ifrat derecede cimri.

**parsley** [ˈpâsli]. Maydonoz.

**parsnip** [ˈpâsnip]. Yabani havuc. ⌜**fine words butter no ~s**⌝, ⌜lâfla peynir gemisi yürümez⌝.

**parson** [ˈpâsn]. İngiliz papazı. **~'s nose,** pişmiş tavuğun kıçı. **~age,** papaz evi.

**part**[1] [pât] *n.* Kısım; parça; cüzü; fasıl; hisse; rol; taraf, cihet. *adv.* Kısmen. **for my ~,** bence, bana kalırsa: **in ~,** kısmen: **in ~s,** bazı yerlerde: **in these ~s**, bu taraflarda: **he looks the ~**, tam işinin adamı görünüyor: **for the most ~**, ekseriya: **I had no ~ in it**, ben dahil değildim: **on the one ~ ... and on the other ~**, bir taraftan ..., öbür taraftan da ...: **on the ~ of s.o.**, birisinin tarafından: **a man of ~s**, değerli bir adam: **orchestral ~s**, muhtelif aletleri çalanlara düşen kısımlar: **to play a ~**, rol oynamak: **to play the ~ of, ...** süsü vermek: **~s of speech**, kelime aksamı: **to take ~ in**, ···de iştirak etm.: **to take the ~ of**, (i) ···süsü vermek; (ii) tarafdarı olm.: **to take stg. in good [bad] ~**, bir şeyi iyi [fena] karşılamak. **~-owner**, tasarruf ortağı; mal ortağı. **~-time job**, günün bütün iş saatlerini doldurmıyan vazife. ~² *v.* Ayırmak; tefrik etm.; kırmak. Ayrılmak; kırılmak. **to ~ with stg.**, vermek, terketmek: **to ~ company with s.o.**, birisinden ayrılmak: ⌈**the best of friends must ~**⌉, hiç bir şey ebedî değildir.

**partake** (**-took, -taken**) [pâˈtêyk, -tuk, -têykn]. **~ of [in]**, iştirak etm., dahil olmak. **to ~ of a meal**, yemek yemek.

**Parthian** [ˈpâθiən]. **a ~ shot**, eski Partlar gibi ayrılırken söylenen dokunaklı söz.

**partial** [ˈpâşl]. Kısmî; umumî olmıyan; insafsızca tarafgir. **to be ~ to stg.**, bir şeye düşkün olm.; hoşlanmak. **~ity** [–ˈaliti], tarafgirlik; beğenme.

**participa·nt** [pâˈtisipənt]. İştirak eden. **~te** [–pêyt], **to ~ in stg.**, iştirak etm.

**participle** [ˈpâtisipl]. *Fiilin tasrifiile yapılan sıfat.*

**particular**¹ [pəˈtikyulə(r)] *a.* Mahsus; has; hususî, şahsî; tek; pek dikkatli; titiz. **in ~**, bilhassa: **to be ~ about one's food**, yemek seçmek: **to be ~ about one's dress**, giyim hususunda titiz olm.: **I don't like this ~ one**, hoşuma gitmiyen yalnız bu: **I am ~ly fond of this one**, bilhassa bundan hoşlanıyorum. ~² *n.* Nokta; husus. **~s**, tafsilat. **alike in every ~**, her hususta aynı: **in this ~**, bu hususta. **~ity** [–ˈlariti], hususiyet. **~ize** [pâˈtikyulərâyz], tayin ve tahsis etm.; birer birer zikretmek.

**parting** [ˈpâtin(g)] *n.* Ayrılma; veda; saç elifi. *a.* Taksim edici; ayrılırken yapılan. **a ~ kiss**, ayrılık busesi: **a ~ shot**, ayrılırken söylenen dokunaklı söz: **to be at the ~ of the ways**, dört yol ağzında olmak.

**partisan** [pâtiˈzan]. Tarafgir; bitaraf olmıyan. **~ship**, tarafgirlik.

**partition** [pâˈtişn]. Bölme (duvar); taksim etme. Bölmek; taksim etm.; paylaşmak. **to ~ off**, bölmelere ayırmak.

**partly** [ˈpâtli]. Kısmen.

**partner** [ˈpâtnə(r)]. Ortak, şerik; dans arkadaşı. Ortak olm. **sleeping ~**, komanditer. **~ship**, ortaklık: **to go [enter] into ~ with s.o.**, birine ortak olm.: **to take s.o. into ~**, birini ortaklığa almak.

**partridge** [ˈpâtric]. Çil.

**party** [ˈpâti]. Fırka; cemiyet; grup; takım, ekip; (bir kontrat vs.de) taraf; davet; şahıs. **~ dress**, bir davete gitmek için elbise: **dinner ~**, ziyafet: **evening ~**, suvare: **firing ~**, kurşuna dizen müfreze: **rescue ~**, kurtarma ekip: **third ~**, üçüncü şahıs: **to be a ~ to a crime**, bir cinayete iştirak etm.: **I will be no ~ to such a step**, ben böyle bir tesebbüşe iştirak edemem: **to give a ~**, dans, ziyafet vs. için davet etm.: **will you join our ~?**, bizimle beraber gelir misiniz?: **he was one of the ~**, o da grupa dahildi: **a ~ of the name of Smith**, S. adında birisi: **a funny old ~**, antika bir ihtiyar: **~ spirit**, fırkacılık zihniyeti. **~-wall**, ara duvarı.

**parvenue** [ˈpâvənyu]. Sonradan görme; zıpçıktı.

**pasha** [ˈpâşa]. Paşa.

**pass**¹ [pâs] *n.* Geçid; iki dağ arası. **to hold the ~**, en mühim mevkii müdafaa etm.: **to sell the ~**, ihanet etmek. ~² *n.* Paso; yol tezkeresi; hal; imtihanda geçme, geçecek derece; (futbol) pas. **to come to ~**, vukubulmak: **things have come to a pretty ~**, işler şimdi tam karıştı: **things have come to such a ~ that ...**, işler öyle bir vaziyete girdi ki: **conjuror's ~**, hokkabazın elçabukluğu: **free ~**, (tren vs. için) paso. **~-book**, (banka) hesab cüzdanı. **~port** [ˈpâspôt]. Pasaport. **~word** [ˈpâswö(r)d]. Parola. ~³ *v.* Geçmek, aşmak; tecavüz etm.; tasdik etm., kabul etm.; geçirmek; sayılmak; rayic olm.; pas vermek. **~ friend (all's well)**, (parola soran nöbetçinin cevabı) geç!: **he ~es for a great writer**, büyük bir muharrir sayılır: to ~ a law, bir kanunu kabul etm.: to let s.o. ~, birini geçirmek: let me say in ~ing, istitraden söyleyim ki: to ~ a motion, (i) bir teklifi kabul etm.; (ii) defi hacet etm.: that won't ~!, bu olmaz! ~ away, ölmek. ~ by, önünden, yanından geçmek. ~ off, the pain has ~ed off, ağrı geçti: everything ~ed off without a hitch, her şey ârızasız geçti: to ~ oneself off for ..., kendine ... süsü vermek: to ~ off a false coin on s.o., birine sahte para sürmek: to ~ stg. off as a joke, bir şeyi şakaya vurmak. ~ on, geçip devam etm.: read this and ~ it on!, bunu okuduktan sonra dolaştırınız. ~ out, dışarıya çıkmak; bayılmak; dışarıya geçirmek. ~ over, öbür tarafa geçmek; aşmak; geçirmek; göz yummak: to ~ over to the enemy, düşmana iltihak etmek. ~ round, etrafını dolaş(tır)mak. ~able [ˈpâsəbl], şöyle böyle; geçilir.

**passage** [ˈpasic]. Geçme; geçid; koridor; (bir kitabda) parça. a ~ of arms, sert sözler teatisi: to have a good [bad] ~, deniz yolculuğu iyi geçmek [geçmemek]: bird of ~, göçücü kuş: to work one's ~, gemide çalışarak yol ücretini ödemek.

**passenger** [ˈpasincə(r)]. Yolcu.

**passer-by** [ˈpâsəˈbây]. Tesadüfen geçen kimse. the ~s-by, gelip geçenler.

**passim** [ˈpasim]. Her yerinde.

**passing** [ˈpâsin(g)]. Geçen; geçici; fâni. Pek çok, son derece. ~ events, günün meselesi: he made a ~ remark, tesadüfen bir mütalâa ileri sürdü. ~-bell, matem çanı.

**passion** [ˈpaşn]. İhtiras; aşk; şiddetli istek; hiddet. to be in a ~, şiddetle öfkelenmek: a fit of ~, hiddet galeyanı: to have a ~ for s.o., birine vurgun olm.: to have a ~ for stg., bir şeye son derece düşkün olm.: ruling ~, bir kimsenin en büyük merakı: the P~, İsa'nin ıstırabı. ~ate, heyecanlı, ihtiraslı; çabuk öfkelenen.

**passiv·e** [ˈpasiv]. Muti; itiraz etmiyen; pasif; (*gram.*) meful; mechul fiil. to take up a ~ attitude, pasif davranmak. ~ity [–ˈsiviti], hareketsizlik.

**passover** [ˈpâsovə(r)]. Yahudilerin hamursuz bayramı.

**past** [pâst]. Geçmiş, sabık; geçerek; sonra; öbür tarafına. the ~, geçmiş, mazi; mazi sıygası: in the ~, eskiden: for some time ~, bir müddettenberi: he walked ~ the house, evinin önünden yürüyerek geçti: ten minutes ~ two, ikiyi on geçe: he is ~ seventy, yetmişini geçmiştir: ~ all understanding, insanın aklı almıyan: ~ endurance, tahammül edilmez: to be ~ caring for stg. or s.o., aldırmamak: a thing of the ~, tarihe karışmış: a town with a ~, tarihî bir şehir: a woman with a ~, mazide maceraları olan kadın. ~-master, maharet sahibi.

**paste** [pêyst]. Hamur; macun. ~-board [–bôd]. Mukavva.

**pasteurize** [ˈpastərâyz]. Pastörize etm.

**pastime** [ˈpastâym]. Eğlence, oyun.

**pastor** [ˈpâstə(r)]. Papaz; mürşid. ~al [ˈpâstərəl]. Çobanlara aid; kırlara ve köylere aid; papazlara *esp.* piskoposlara aid; otlamağa mahsus. Çoban şiiri. ~ people, hayvancılıkla yaşıyan halk.

**pastry** [ˈpêystri]. Hamur işi; pasta. ~-cook, pastacı.

**pastur·age** [ˈpâstyuric]. Otlak, mera. ~e [–tyə(r)], otlak, çayır. Otla(t)mak.

**pasty** [ˈpêysti] *a.* Hamur gibi; yapışkan. ~-faced, uçuk benizli.

**pat**[1] [pat]. (*ech.*) El ile çabuk ve hafif vuruş. Okşamak. a ~ on the back, sırtını okşama: to ~ s.o. on the back, teşvik, tebrik etmek, cesaret vermek için birinin sırtını okşamak: to ~ oneself on the back, kendi yaptığı bir şeyi beğenmek. ~[2]. Tamamile uygun; tam zamanında.

**patch** [paç]. Yama; küçük arazi parçası; yara üzerine yapıştırılan bez parçası. Yamamak. a ~ of blue sky, bulutlar arasında bir parça mavi gök: cabbage ~, lahana tarhı: not to be a ~ on s.o., birinin eline su dökememek: to strike a bad ~, talihi muvakkaten ters gitmek. ~ up, kabaca yamamak: to ~ things up [to ~ up a quarrel], barışmak: a ~ed-up peace, derme çatma bir sulh. ~work [ˈpaçwö(r)k], muhtelif renk ve büyüklükte parçalardan dikilmiş şey. ~y [ˈpaçi], muntazam olmıyan. his work is ~, yaptığı iş kısmen iyi kısmen fenadır.

**pate** [pêyt]. Kelle, kafa.

**patent**[1] [ˈpêytənt]. İhtira beratı;

icad. Beratlı; hünerli. Patenta almak. ~ **leather,** rugan. **~ee** [–ˈtî], berat sahibi. ~² *a.* Aşikâr, besbelli.

**pater** [ˈpêytə(r)]. **the** ~, babam. **~familias** [–fəˈmilias], ev bark sahibi. **~nal** [pəˈtö(r)nl], babaya aid. **~nity,** babalık: **of doubtful** ~, babası şübheli. **~noster** [ˈpatənostə(r)], hazreti İsa'ya aid lâtince meşhur bir dua.

**path** [pâθ]. Patika, keçi yolu: (yıldız) mahrek; meslek. **to leave the beaten** ~, herkesin takib ettiği yoldan ayrılmak: **to cross s.o.'s** ~, birinin önüne çıkmak; birinin arzusuna karşı gelmek. **~finder,** çığır açan kimse. **~less,** yolsuz.

**pathetic** [pəˈθetik]. Acıklı, dokunaklı.

**patholog·ical** [ˌpaθəˈlocikl]. Hastalığa aid, marazî. **~y,** hastalıklar ilmi.

**pathos** [ˈpêyθos]. Dokunaklı ve keder verici hassa.

**pathway** [ˈpâθwêy]. İnce yol, patika.

**patien·ce** [ˈpêyşəns]. Sabır; tahammül; yalnız oynanan iskambil oyunu. **to have** ~ **with s.o.,** birine karşı sabırlı davranmak: **to try s.o.'s** ~, birinin sabrını tüketmek. **~t** [ˈpêyşnt], sabırlı; mütehammil; hasta: **a doctor's ~s,** bir doktorun müşterileri.

**patriarch** [ˈpêytriâk]. İbrahim gibi ilk resüllerin biri; Ortodoks kilisesinin piskoposu, patrik; aile, kabile reisi; muhterem ihtiyar. **~al** [–ˈâkl], patriklere aid; pederşahi; yaşlı ve muhterem. **~ate,** patriklik.

**patrician** [pəˈtrişn]. Asılzade; aristokrat.

**patrimony** [ˈpatriməni]. Ecdaddan kalma miras; kilise vakfı.

**patriot** [ˈpatriət]. Vatanperver. **~ic** [–ˈotik], vatanperver(ane). **~ism** [ˈpatriətizm], vatanperverlik.

**patrol** [pəˈtroul]. Devriye; kol. Devriye gezmek; etrafını dolaşmak.

**patron** [ˈpêytrən]. Velinimet; hâmi; müşteri. ~ **saint,** koruyucu aziz. **~age** [ˈpatrənic], himaye; müşterilik; hâmi sıfatı takınma. **~ize** [ˈpatrənâyz], himaye etm.; bir dükkân vs.nin müşteri olm.; hâmi sıfatı takınmak: **a ~ing air,** yukarıdan alma.

**pattern** [ˈpatö(r)n]. Nümune; örnek; mostra; döküm kalıbı; resim.

**paucity** [ˈpôsiti]. Azlık, kıtlık.

**paunch** [pônç]. Karın; iri göbek.

**pauper** [ˈpôpə(r)]. Fakir, yoksul adam. **~ism,** yoksulluk. **~ize,** yoksul düşürmek.

**pause** [pôz]. Müvakkaten durma; duraklama; durak; fasıla. Duraklamak; tereddüd etmek. **to make s.o.** ~, birini düşündürmek.

**pav·e** [pêyv]. Kaldırım döşemek. **to** ~ **the way for s.o.,** birinin işini kolaylaştırmak. **~ement,** kaldırım. **~ing,** taş döşeme: **~-stone,** kaldırım taşı.

**pavilion** [pəˈvilyən]. Büyük sivri çadır.

**paw** [pô]. Pençe; (*sl.*) el. (At) yeri deşmek; (*sl.*) ellemek.

**pawn**¹ [pôn]. (Satrancda) paytak. **to be s.o.'s** ~, birinin aleti olm.: **to be a mere** ~ **in the game,** bir işte ehemmiyetsiz bir alet olmak. ~². Rehin. Rehne koymak. **in** ~, rehin olarak verilmiş: **to take out of** ~, rehinden çıkarmak. **~broker,** rehinci. **~-ticket,** rehin makbuzu.

**pax** [paks]. ~!, tövbe!

**pay**¹ [pêy] *n.* Maaş; ücret; aylık. **to be in s.o.'s** ~, birinin hizmetinde olmak. ~² *v.* (**paid**) [pêyd]. Ödemek; borcunu vermek; kârlı olm.; arzetmek, göstermek; etmek. **to** ~ **attention,** dikkat etm.: **to** ~ **no attention,** boş vermek: **it will** ~ **for itself,** masrafını çıkarır: **the business does not** ~, bu iş kâr getirmez: **to** ~ **s.o. to do stg.,** birine para ile bir şeyi yaptırmak: **it will** ~ **you to do this,** bunu yapmakta faydanız var: **to** ~ **money into a bank,** bankaya para yatırmak: **to** ~ **respect to s.o.,** birine hürmet göstermek: **to** ~ **one's respects,** saygılarını sunmak: **to** ~ **one's way,** normal bir hayat yaşıyacak kadar kazanmak. ~ **away,** sarfetmek. ~ **back,** geri vermek; karşılığını vermek: **to** ~ **s.o. back** (i) birinden alınan parayı iade etmek, (ii) birinden acısını çıkarmak. ~ **down,** peşin vermek. ~ **for,** .. için para vermek: **to** ~ **for a mistake** *etc.*, bir hata vs.nin cezasını çekmek: **I'll make you** ~ **for this,** ben bunun acısını senden çıkarırım. ~ **in,** tediye etm.: **to** ~ **in a cheque,** bankaya çek yatırmak. ~ **off,** (borc, hesab) temizlemek: **to** ~ **off a servant,** hizmetçiye ücretini verip yol vermek. ~ **out,** harcetmek; (**rope**), halat kaloma etm.: **I'll** ~ **you out for that** ben bunun acısını senden çıkarırım

**I'll ~ him out for that,** onun da benden alacağı olsun! **~ up,** borcunu ödemek: **~ up!,** parayı ver bakalım! **~able** [ˈpêyəbl], ödenmesi lâzım olan. **accounts ~,** pasif borclar: **~ at sight,** görüldüğünde [**on demand,** ibrazında; **to bearer,** hâmiline; **to order,** emre] ödenecek. **~ee** [pêyˈî], alıcı. **~ing** [ˈpêy·in(g)], ödeme; tediye. **a ~ concern,** kârlı bir iş: **~ guest,** hususî bir evde pansiyoner. **~master** [ˈpêymâstə(r)], (ordu) mutemed; (bahriye) levazım memuru: **~ General,** İngilterede muhasebe işlerine bakan devlet dairesinin başı. **~ment** [ˈpêymənt], tediye, ödeme; ücret; hizmet mukabili. **~ in full,** tasfiye.

**p.c.** (*abb.*) **per cent.,** yüzde.

**pea** [pî]. Bezelye. **sweet ~,** ıtırşahi: **green ~s,** taze bezelye: **as like as two ~s,** bir elmanın bir yarısı biri bir yarısı biri: **as easy as shelling ~s,** çok kolay. **~-soup,** kuru bezelye çorbası.

**peace** [pîs]. Sulh; barış; rahat. **to break** [**disturb**] **the ~,** asayişi bozmak: **~ establishment of the army,** ordunun hazarî kadrosu: **to hold one's ~,** susmak: **justice of the ~,** sulh hakimi: **to keep the ~,** asayişi muhafaza etm.: **to leave s.o. in ~,** birini rahat bırakmak: **to live in ~,** kavgasız yaşamak: **to make one's ~ with s.o.,** birisile barışmak: **to have ~ of mind,** başı dinc olmak. **~able,** asayişli; sakin. **~ful,** rahat, sakin; sulhperver. **~time,** sulh zamanındaki.

**peach** [pîç]. Şeftali. **what a ~ of a child!,** maşallah altıntopu gibi çocuk!

**pea·cock** [ˈpîkok]. Tavus. **~hen,** dişi tavus.

**peak** [pîk]. Zirve, şahika; bir şeyin en yüksek noktası. **~ hours,** yol, dükkân vs.nin en işlek saatleri: **~ load,** (bir dinamonun) azamî yükü.

**peaky** [ˈpîki]. Zayıf ve solgun (çocuk).

**peal** [pîl]. Kilise kulesinin çan tertibatı; birçok çanların sesi. Çanları çalmak; (gök) gürlemek. **~ of laughter,** kahkaha.

**peanut** [ˈpînʌt]. Yer fıstığı.

**pear** [pe̯ə(r)], Armud.

**pearl** [pö(r)l]. İnci (avlamak); (su, ter) inci gibi taneler hasıl etm. **~ button,** sedef düğme. **~y,** inci gibi. **~-barley,** frenk arpası. **~-diver,** inci avcısı. **~-grey,** açık külrengi.

**peasant** [ˈpezənt]. Köylü. **~ry,** köylüler.

**peat** [pît]. Yer tezeği, turb. **~-bog,** yer tezeğinin teşekkül ettiği bataklık.

**pebble** [ˈpebl]. Çakıltaşı. ⌜**not the only ~ on the beach**⌝, ⌜gökten zembille inmemiş ya!⌝.

**peccadillo** [pekəˈdilo̯u]. Küçük suç.

**peck¹** [pek]. Gagalama(k); keskin bir şeyle yarmak. **to ~ at,** gaga ile vurmak, ucuyla çiğnemek. **~er** [ˈpekə(r)], (*sl.*) **to keep up one's ~,** yılmamak, ümidsizliğe kapılmamak. **~ish** [ˈpekiş], (*sl.*) **to feel ~,** karnı zil çalmak. **~².** Kuru şeyler ölçüşü; kilenin dörtte biri. **a ~ of troubles,** bir yığın derd.

**pectoral** [ˈpektərəl]. Göğüse aid, sadrî.

**peculate** [ˈpekyulêyt]. İhtilâs etm.

**peculiar** [piˈkûlyə(r)]. Has, mahsus; ferdî, zatî; acayib, tuhaf, garib. **~ity** [–liˈariti], hususiyet; gariblik, tuhaflık.

**pecuniary** [piˈkyûnie̯ri]. Paraya aid. **~ embarrassment,** para sıkıntısı.

**pedal** [ˈpedəl]. (Piyano, bisiklet vs.) ayak basamağı. Pedal ile hareket ettirmek.

**pedant** [ˈpedənt]. Ukalâ; malûmatfüruş. **~ic** [–ˈdantik], ukalâca. **~ry** [ˈpedəntri], ukalâlık; ilim sahasında titizlik.

**peddl·e** [ˈpedl]. Seyyar sokak satıcılığı yapmak.

**pedestal** [ˈpedistl]. Kaide; heykel ayaklığı. **to put s.o. on a ~,** birini mükemmel addetmek.

**pedestrian** [pəˈdestri̯ən]. Yaya yürüyen, piyade; (üslûb) yavan, cansız.

**pedigree** [ˈpedigrî]. Şecere; nesil. Safkan, cins.

**pedlar** [ˈpedlə(r)]. Ayak esnafı, seyyar satıcı.

**peel** [pîl]. Yemişin kabuğu(nu soymak). **~ (off),** soyulmak; (deri) pul pul dökülmek. **~ing,** soyulma: **~s,** soyuntu.

**peep¹** [pîp]. (*ech.*). Civciv gibi ötme(k). **~².** Gizlice bakıverme; azıcık bir bakış. **to ~ at** [**to take a ~ at**], hırsızlama bakıvermek; bir aralıktan bakmak. **at ~ of day,** şafakta. **~-hole,** gözetleme deliği.

**peer¹** [pi̯ə(r)]. Eş; akrandan biri; İngiltere'de asalet rütbesini haiz olan kimse, lord. **~age,** lordlar sınıfı; lordluk: asılzadelerin salnamesi. **~less,** eşsiz. **~²** *v.* Karanlıkta veya hayalmeyal olan bir şeye dikkatle bakmak.

**peev·ed** [pîvd]. Küskün. **~ish,** titiz, hırçın.

**peg**[1] [peg] *n.* Ağac çivi; küçük kazık; mandal; akort vidası; bir yudum (viski): **a ~ to hang a grievance on,** şikâyet vesilesi: ⌜**a square ~ in a round hole**⌝, yerinin adamı değil: **to take s.o. down a ~ or two,** birinin kibrini kırmak. **~**[2] *v.* Ağac çivi ile mıhlamak. **to ~ away at stg.,** bir işte azim ve sebatla çalışmak: **to ~ clothes on the line,** çamasırı ipe mandallamak: **to ~ down,** kazığa bağlamak: **to ~ the exchange,** kambiyoyu tesbit etm.: **to ~ out,** (*sl.*) kuyruğu titretmek, ölmek: **to ~ out a claim,** yeni keşfedilen altın vs.yi ihtiva eden bir arazide muayyen bir parçayı kazıklarla sınırlandırarak üzerinde hak iddia etm.

**Pegasus** [ˈpegəsəs]. (*mit.*) Kanadlı at; (burc) Feresi âzam; şairin ilhamı.

**pejorative** [piˈcoritiv]. Bir kelimeye fena bir mâna verme.

**pelf** [pelf]. İrtikâb ile alınan para; vurgun.

**pelican** [ˈpelikən]. Kaşıkçı kuşu.

**pellet** [ˈpelit]. Ufak top; tane; hap.

**pel-mell** [ˈpelˈmel]. Karmakarışık bir halde.

**Peloponnesus** [ˈpelopoˈnîsəs]. Mora.

**pelt**[1] [pelt] *n.* (Kürklü) deri; pösteki. **~**[2] *v.* **to ~ s.o. with stones,** *etc.*, birine taş vs. yağdırmak. (Yağmur) bardaktan boşanmak: **~ing rain,** sicim gibi yağmur: **at full ~,** alabildiğine koşarak.

**pen**[1] [pen]. Ağıl; kümes. Ağıla koymak. **to ~ up,** kapatmak. **~**[2]. Kalem. Yazmak. **~-name,** (muharrir) takma ad. **~holder** [ˈpenhouldə(r)], demir kalem sapı. **~knife** [ˈpennâyf], çakı. **~manship** [ˈpenmənşip], hattatlık.

**penal** [ˈpînl]. Cezaya aid; ceza icab ettiren. **~ code,** ceza kanunu: **~ servitude,** kürek cezası. **~ize,** cezalandırmak; eziyet etmek. **~ty** [ˈpenəlti], ceza; para cezası; kefaret: (sporda) penaltı: **~ clause,** (bir mukavelede) zamanında teslim edilmiyen sipariş hakkında tazminat hükmü: **on ~ of death,** ölüm cezası ile: **to pay the ~ of a mistake,** *etc.*, bir hata vs.nin cezasını çekmek.

**penance** [ˈpenəns]. Kefaret.

**pence** [pens] *v.* **penny.**

**penchant** [ˈpo(n)ˈşo(n), ˈpençənt]. İstidad, temayül.

**pencil** [ˈpensl]. Kurşun kalem (ile yazmak). **~ of light,** şua.

**pend·ant** [ˈpendənt] *n.* Askı; pandantif. **~ent** [ˈpendənt] *a.*, sarkık; muallak. **~ing** [ˈpendin(g)], muallak. **~ his arrival,** gelinceye kadar. **~ulous** [ˈpendyuləs], sarkık; sallanan. **~ulum** [ˈpendyuləm], rakkas.

**penetrat·e** [ˈpenitrêyt]. Delip girmek, nüfuz etm.; sinmek; işlemek; hulûl etmek. **~ion** [–ˈtrêyşn], hulûl; nüfuz etme; sokuluş.

**penguin** [ˈpengwin]. Penguen.

**peninsula** [pəˈninsyulə]. Yarımada. **~r,** yarımadaya aid: **the ~ War** 1808–14 İspanya harbi.

**peniten·ce** [ˈpenitəns]. Pişmanlık, tövbe. **~t,** pişman, tövbekâr, nadim. **~tiary** [–ˈtenşəri], ıslahhane; (*Amer.*) hapishane.

**penn·ant, ~on** [ˈpenənt]. Flâma, flândra.

**penniless** [ˈpenilis]. Meteliksiz; yoksul.

**penny** [ˈpeni]. (*pl.* **pennies** *olursa sikke sayısını ve* **pence** *olursa para mikdarını gösterir*). Peni; bir şilinin $\frac{1}{12}$ kısmı. **he hasn't a ~ (to bless himself with),** meteliksizdir: **~ dreadful,** ucuz ve değersiz cinaî roman: **to earn [turn] an honest ~,** namusile ufak tefek para kazanmak: ⌜**in for a ~ in for a pound**⌝, ⌜(nasıl olsa) öyle de battık böyle de⌝: **to look twice at every ~,** pek tutumlu olm.: **it will cost a pretty ~,** epeyiceye malolacak: ⌜**a ~ for your thoughts**⌝, ⌜binin yarısı beş yüz (o da sende yok)⌝: ⌜**take care of the pence and the pounds will look after themselves**⌝, küçük masraflara dikkat edersen, büyükleri kendiliğinden gözetilmiş olur: ⌜**~ wise pound foolish**⌝, küçük işlerde hasis büyük işlerde müsrif: **not a ~ the worse,** hiç bir zarar görmeden. **~weight,** bir ölçü (*takriben* $1\frac{1}{2}$ *gram*). **~worth,** penilik. **~-in-the-slot machine,** para atınca bilet vs. veren makine. **~-piece,** bir penilik para: **I haven't a ~,** meteliğim yok. **~-whistle,** teneke flüt.

**pension**[1] [ˈpo(n)syo(n)]. Pansiyon. **~**[2] [ˈpenşn]. Tekaüd maaşı. **to ~ (off),** emekliye ayırmak: **to retire on a ~,** tekaüd edilmek. **~er,** mütekaid, emekli.

**pensive** [ˈpensiv]. Dalgın, düşünceli.

**pent** [pent]. ~ **up** [in], kapanmış; zabtedilmiş fakat taşmak üzere olan (his, hiddet vs.). ~**house** [ˈpenthaus], sundurma.
**penta-** [ˈpenta]. *pref.* Beş-. ~**gon** ˈpentəgon], beş dılılı. ~**meter** [penˈtamətə(r)], beş heceli mısra. ~**teuch** [ˈpentatyûk], tevratın ilk beş kitabı.
**Pentecost** [ˈpentəkost]. Yahudilerde:–gülbayramı: Hıristiyanlarda:–hamsin yortusu.
**penultimate** [peˈnʌltimit]. Sondan önceki.
**penumbra** [peˈnʌmbra]. Esas gölge etrafında hasıl olan hafif gölge.
**penur·y** [ˈpenyuri]. Yoksulluk; kıtlık. ~**ious** [–ˈnyûriəs], kıt; yoksul; cimri.
**people** [ˈpîpəl], (*pl.*). Halk; insan; aile ve akraba. (*sing with pl.* ~**s**). Millet. İskân etmek. ~ **at large,** umumiyetle herkes: ~ **say,** diyorlar: **the common** ~, halk tabakası: **how are your** ~**?**, sizinkiler ne halde?: **my** ~ **came from Ireland,** biz aslen İrlandalıyız: **the King and his** ~, Kıral ve tebaası: **old** [**young**] ~, ihtiyarlar [gencler]: **what are you** ~ **going to do?**, sizler ne yapacaksınız?
**pep** [pep]. (*sl.*) Hararet, gayret. **full of** ~, girişken, gayretli.
**pepper** [ˈpepə(r)]. Biber. Biber ekmek. ~**corn,** biber tanesi. ~**mint,** nane; nane ruhu; nane şekeri. ~**y,** biberli; tez mizaclı, hemen parlar (adam). ~**-pot,** biberlik.
**per** [pö(r)]. Vasıtasile, delâletile. ~ **cent.,** yüzde: ~ **week,** haftada: ~ **head,** adam başına: **as** ~ **invoice** [**sample**], fatura [nümune] ye uygun olarak.
**peradventure** [pöradˈventyə(r)]. Belki, şayed; kazara.
**perambulat·e** [pəˈrambyulêyt]. Etrafını dolaşmak, gezmek. ~**or,** çocuk arabası (*abb.* **pram**).
**perceive** [pəˈsîv]. Farketmek; kavramak.
**percentage** [pəˈsentic]. Yüzdelik. **only a small** ~ **of plague victims recovered,** vebalıların pek azı kurtuldu.
**percept·ible** [pəˈseptibl]. Duyulur, hissedilir; farkına varılır. **barely** [**hardly**] ~, belli belirsiz. ~**ion,** his, duyma; idrak; (verginin) tahsili.
**perch** [pö(r)ç]. Tünek. Tüneklemek. **to knock s.o. off his** ~, tünekten indirmek.
**perchance** [pəˈçâns]. Şayed; belki.
**percolate** [ˈpəkəlêyt]. Süz(ül)mek; filtreden geçirmek.
**percussion** [pəˈkʌşn]. Şiddetle vuruş; çarpma; müsademe. ~ **cap,** kapsol: ~ **instruments,** vurularak çalınan çalgılar (davul vs.).
**perdition** [pəˈdişn]. Cehennem azabına uğrama; mahvolma. **to consign to** ~, lânet etmek.
**peregrination** [ˌperigriˈnêyşn]. Dolaşma.
**peremptory** [pəˈremptəri]. Kat'i, müsbet, kestirme; mütehakkimane.
**perennial** [pəˈrenyəl]. Daimî, sürekli, mütemadi. Bir kaç yıl yaşıyan nebat.
**perfect** *a.* [ˈpö(r)fikt]. Mükemmel, kusursuz; tam, bütün; (*gram.*) mazi. *v.* [–ˈfekt] İkmal etm., mükemmelleştirmek; tamamlamak. ~**ion** [pəˈfekşn], mükemmellik, ikmal: **to** ~, mükemmelen; tamamen.
**perfid·ious** [pəˈfidyəs]. Hain, haince; vefasız. ~**y** [ˈpö(r)fidi] hainlik; vefasızlık.
**perforat·e** [ˈpö(r)ferêyt]. Delmek; içinden geçmek. ~**ion** [–ˈrêyşn] delme; ufak delik: ~ **of a stamp,** pulun tırtılı.
**perforce** [pəˈfôs]. İster istemez; zorla.
**perform** [pəˈfôm]. İcra etm., ifa etm., yerine getirmek. ~**ing dogs,** numara yapan köpekler: **to** ~ **on a musical instrument,** bir çalgı çalmak: **to** ~ **in a play,** bir piyeste bir rol oynamak. ~**ance,** icra, ifa, yerine getirme; işleme; eser; temsil, numara: **to put up a good** ~, başarmak: **a sorry** ~, muvaffakiyetsiz bir iş. ~**er,** aktör; icra edici: **a good** [**poor**] ~, işini iyi [fena] yapan kimse.
**perfume** *n.* [ˈpö(r)fyum]. Güzel koku; parföm. *v.* [pəˈfyûm]. Güzel koku yaymak. ~**ry** [–ˈfyûməri], ıtriyat.
**perfunctory** [pəˈfʌnktəri]. Mühmel; yarım yamalak; iş olsun diye.
**pergola** [ˈpö(r)golə]. Çardak.
**perhaps** [pə(r)ˈhaps]. Belki; olabilir ki; şayed.
**peril** [ˈperil]. Tehlike, muhatara. **touch him at your** ~**!**, ona dokunursan vay sana!: **you do this at your** ~, bunu yaparsan günahı boynuna!
~**ous,** tehlikeli.
**perimeter** [pəˈrimitə(r)]. Çevre, muhit.
**period** [ˈpiəriəd]. Devir; müddet;

devre; nöbet; cümle. Muayyen bir devre aid. **the ~,** şimdiki zaman: **monthly ~s,** (kadının) aybaşı. **~ic(al),** muntazam devrelerde yapılan; (*yal.* **~ical**) mecmua.

**periphery** [pə'rifəri]. Dairenin muhiti.

**perish** ['periş]. Telef olm.; can vermek; harab olm.; çürümek, bozulmak. **I'll do it or ~ in the attempt,** ben bunu ölürüm de yine yaparım: **~ the thought!,** Allah göstermesin! **~able,** fâni; bozulabilen; çabucak çürüyen: **~s,** çabuk cürüyen [bozulan] mallar (et, yemiş gibi). **~ing cold,** öldürücü soğuk. **~ed,** çürümüş, bozulmuş: **to be ~ with cold,** soğuktan donmak.

**perjur·e** ['pö(r)cə(r)]. **to ~ oneself,** yalan yere yemin etm.; yemininden dönmek. **~y,** yalan yere yemin; yemininden dönme; yalancı şahidlik.

**perk** [pö(r)k]. **to ~ up,** (kuş gibi) çevikçe başını kaldırmak; (kulaklarını) dikmek; canlanmak: **to ~ s.o. up,** birini canlandırmak. **~y,** canlı açıkgöz ve biraz yüzsüz.

**permanen·ce** ['pö(r)mənens]. Daimilik, devam. **~t,** daimî; devamlı; değişmez. **~ way,** demiryolu döşeli yol.

**permea·bility** [ˌpö(r)miə'biliti]. Su ve gaz geçebilme. **~te** ['pö(r)miêyt]. Süzmek.

**permiss·ible** [pə'misibl]. Caiz; helâl. **~ion** [–'mişn], izin, ruhsat, müsaade. **~ive,** izin verici, müsaade eden.

**permit** *n.* ['pö(r)mit]. Müsaade, ruhsat (tezkeresi). *v.* [pə'mit]. Müsaade etm., izin ve ruhsat vermek; yol vermek; bırakmak; kabul etmek.

**permutation** [pö(r)mu'têyşn]. Tebadül.

**pernicious** [pə'nişəs]. Muzır; menhus; zararlı.

**pernickety** [pə'nikəti]. Titiz; (iş) nazik.

**peroration** [pero'rêyşn]. Nutkun hatimesi.

**perpendicular** [ˌpö(r)pen'dikyulə(r)]. Amudî, şakulî; kaim; dik. Kaim hat, amud. **out of the ~,** şakulî olmıyan.

**perpetrat·e** ['pö(r)pitrêyt]. İrtikab etm., yapmak. **~or,** mürtekib.

**perpetu·al** [pə'petyuəl]. Daimî, fasılasız; mütemadi. **~ate** [–tyuêyt], devam ettirmek; ebedileştirmek. **~ity** [–'tyûiti], daimilik: **in [for, to] ~,** ebediyen.

**perplex** [pə'pleks]. Şaşırtmak; zihnini karıştırmak. **~ity,** şaşkınlık; tereddüd, teşevvüş.

**perquisite** ['pö(r)kwizit]. Aylığa ilâveten alınan para. **to get ~s,** çimlenmek.

**persecut·e** ['pö(r)sikyût]. Zulmetmek, eziyet etmek. **~ion** [–'kyûşn], zulüm, eziyet: **~ mania,** herkesin kendi aleyhinde bulunduğu vehminden ibaret bir hastalık.

**persever·e** [pö(r)si'viə(r)]. Sebat etm.; devam etmek. **~ance,** sebat, ısrar.

**Persia** ['pö(r)şə]. İran. **~n** [–şn], acem, iranlı; acemce, farisî: **~ cat,** Van kedisi: **~ Gulf,** Basra körfezi.

**persist** [pə'sist]. Israr etm., sebat etm.; devam etm.; üstelemek. **~ence,** ısrar; sebat. **~ent,** musir; inadcı; devamlı.

**person** ['pö(r)sn]. Kişi, kimse; şahıs, ferd, zat; vücud; (*gram.*) şahıs. **in ~,** bizzat, şahsen: **to have a commanding ~,** vücudu heybetli olm.: **to be no respecter of ~s,** hem nalına hem mıhına vurmak. **~able,** yakışıklı. **~age,** mühim adam; şahıs. **~al,** şahsî; hususî; ferdî; kendine aid; menkul: **~ column,** (bir gazetede) kücük şahsî ilânlar: **don't let us be ~,** şahsiyata girmiyelim. **~ality,** şahsiyet; varlık; hal, vaziyet; mühim zat; *pl.* **~ities,** şahsiyat. **~ally,** şahsen; bizzat; fikrimce: **~ I don't like beer,** ben kendim bira sevmem. **~alty,** şahsî servet. **~ate,** bir şahıs rolünü yapmak. **~ification** [pəˌsonifi'kêyşn], teşahhus, tecessüm. **~ify** [–'sonifây], tecessüm etm., teşahhus ettirmek: **he is virtue ~ified,** mücessem fazilettir. **~nel** [–'nel], bir iş veya bir müesseseye mensub memurlar, müstahdemler; geminin tayfası.

**perspective** [pə'spektiv]. Tenazur; perspektiv; manzara. **in ~,** tenazura göre: **to see stg. in its true ~,** bir şeyi hakiki çehresile görmek.

**perspicac·ious** [pö(r)spi'kêyşəs]. Anlayışlı; sürati intikal sahibi. **~ity** [–'kasiti], anlayış, feraset, sürati intikal.

**perspir·ation** [pö(r)spə'rêyşn]. Terleme, ter. **to break into ~,** terlemeğe başlamak. **~e** [pə'spâyə(r)], terlemek.

**persua·de** [pə'swêyd]. İkna etm.; razı etmek. **to be ~d that ...,** **···e** kail olmak. **~sion** [–'swêyjn], ikna;

kandırma; inandırma; kanaat; itikad. ~**sive** [–ˈswêysiv], ikna edici; inandırıcı; saik; teşvik eden (hareket vs.).
**pert** [pö(r)t]. Şımarık; arsız.
**pertain** [pəˈtêyn]. Aid olm.; raci olmak.
**pertinac·ious** [pö(r)tiˈnêyşəs]. Musir; inadcı. ~**ity** [–ˈnasiti], sebat; ısrar; inad.
**pertinent** [ˈpö(r)tinənt]. Münasib; uygun.
**perturb** [pəˈtö(r)b]. Rahatsız etm.; endişeye düşürmek. ~**ation** [–ˈbêyşn], heyecan; endişe, merak.
**perus·al** [peˈrûzl]. Mütalaa. ~**e,** mütalâa etm.
**perva·de** [pö(r)ˈvêyd]. Nüfuz ve istilâ etm.; her tarafa yayılmak. ~**ding** [~**sive**], her tarafa yayılan.
**perver·se** [pəˈvö(r)s]. Ters, aksi; huysuz, titiz; bozuk. ~**sion** [–ˈvö(r)şn], tahrif; bozma; bozukluk. ~**ity,** terslik, aksilik; huysuzluk. ~**t** *v.* [pəˈvö(r)t], tahrif etm.; bozmak; baştan çıkarmak; dininden döndürmek. *n.* [ˈpö(r)vö(r)t], Mürted; cinsî dalâlete düşmüş.
**pervious** [ˈpö(r)viəs]. Geçilebilir.
**pessimis·m** [ˈpesimizm]. Bedbinlik. ~**t,** bedbin. ~**tic** [–ˈmistik], bedbin(ce).
**pest** [pest]. Veba, taûn; başbelâsı. ~**iferous** [pesˈtifərəs], veba neşreden; iğrenc. ~**ilence** [ˈpestiləns], sarî ve öldürücü hastalık; veba, taûn. ~**ilent,** öldürücü, mühlik; menfur. ~**ilential** [–ˈlenşl], bulaşıcı, sarî; melun ve menfur.
**pester** [ˈpestə(r)]. İzac etm.; çullanmak; başının etini yemek; tebelleş olmak.
**pestle** [ˈpesl]. Havaneli.
**pet**[1] [pet]. (Kedi, köpek gibi) evde zevk için beslenen hayvan; cici; herkese tercih edilen. Nazlı büyütmek; okşamak. **my** ~!, cicim!, canım!: ~ **aversion,** en çok nefret edilen adam, şey: ~ **grievance,** derdi günü: ~ **name,** sevilen bir kimseye takılan ad. ~[2]. Küskünlük, gücenme. **to be in a** ~ küsmek.
**petal** [ˈpetl]. Tüveyc yaprağı.
**petard** [peˈtâd]. ⌜**to be hoist with one's own** ~⌝, kendi kazdığı kuyuya düşmek.
**Peter** [ˈpîtə(r)]. Bir erkek ismi. **Blue** ~, hareket flâması: ⌜**to rob** ~ **to pay Paul**⌝, başkasına vermek için birinden gasbetmek; eski borcunu ödemek için yeniden borc almak. ~ **out** [ˈpîtəˈraut]. Tükenmek; yavaşça yok olm.; suya düşmek.
**petition** [pəˈtişn]. İstida, dilekçe (vermek).
**petrel** [ˈpetrel]. **storm(y)** ~, fırtına kuşu; (*fig.*) gelince ortalığı birbirine karıştıran kimse.
**petrif·y** [ˈpetrifây]. Taş haline koymak veya gelmek. **to be** ~**ied,** taş kesilmek.
**petrol** [ˈpetrol]. Benzin. ~**eum** [pəˈtroulyəm], gaz, petrol. ~**iferous** [–ˈlifərəs], petrollü (toprak).
**petticoat** [ˈpetikout]. İç eteklik. ~ **government,** evde kadının hakimiyeti.
**pettifogg·er** [ˈpetifogə(r)]. Aşağılık avukat; safsatacı. ~**ing,** kılı kırk yaran.
**pett·iness** [ˈpetinis]. Ehemmiyetsizlik; dar düşüncelilik; miskinlik. ~**y** [ˈpeti], ehemmiyetsiz; küçük; darkafalı; miskin; adî: ~ **cash,** müteferrika: ~ **larceny,** ufak hırsızlık: ~ **officer,** bahriyede küçük zabit.
**pettish** [ˈpetiş]. Hırçın; çabuk küser.
**petulant** [ˈpetyulənt]. Tez mizaclı; küseğen.
**pew** [pyû]. Kilisede ibadet edenlere mahsus arkalıklı peyke; mahfil.
**pewter** [ˈpyûtə(r)]. Kalay ile başka bir maden halitası. ~ **ware,** bu halitadan yapılan kablar.
**phalanx** [ˈfalanks]. Mızraklı alay, falanj.
**phant·asm** [ˈfantazm]. Hayalet; tayf; vehim. ~**om** [ˈfantəm], tayf, hayal(et); heyulâ; ismi var cismi yok.
**Pharaoh** [ˈfeərou]. Firavun.
**pharis·ee** [ˈfarisî]. Fariz; riyakâr. ~**aical** [fariˈsêy·ikl], riyakâr.
**pharma·ceutical** [ˌfâməˈkyûtikl]. Eczacılığa aid. ~**copoeia** [–koˈpîyə], ilâclar kitabı. ~**cy** [ˈfâməsi], eczacılık; eczahane.
**phase** [fêyz]. Safha.
**pheasant** [ˈfezənt]. Sülün.
**phenomen·al** [fiˈnominl]. Garib, fevkalâde; hayret verici. ~**on,** *pl.* ~**a,** şuura akseden vaka; tabiî hadise; cilve; fevkalâde hadise, şahıs vs.
**phew** [fyû]. *Yorgunluk, hayret vs. nidası.*
**phial** [ˈfâyəl]. Küçük şişe.
**philander** [fiˈlandə(r)]. Flört yapmak.

**philanthrop·ic** [filanˈθropik]. Hayırsever; hamiyetli; şefkatlı. **~ist** [fiˈlanθrəpist], insansever; hayır sahibi. **~y** [fiˈlanθrəpi], hayırseverlik.

**philately** [fiˈlatəli]. Posta pulculuk.

**philharmonic** [filhâˈmonik]. **~ Society,** musiki sevenler cemiyeti.

**philippic** [fiˈlipik]. Sert ve acı nutuk.

**Philistine** [ˈfilistâyn]. Cahil, zevksiz adam.

**philology** [fiˈloləci]. Filoloji; dil bilgisi.

**philosoph·er** [fiˈlosəfə(r)]. Feylosof; kalender. **the ~'s stone,** haceri felsefî. **~ical** [–ˈsofikl], felsefî; kalenderane: **to take things ~ly,** başına geleni sabırla ve kalenderane kabul etmek. **~y** [–ˈlosəfi], felsefe.

**philtre** [ˈfiltə(r)]. Aşk iksiri.

**phlegm** [flem]. Balgam; soğukluk. **~atic** [flegˈmatic], heyecanlanmaz; lenfavî.

**phobia** [ˈfoubyə]. Sebebsiz korku; fobi. **-~,** *suff.* ···den nefret etme.

**phoenix** [ˈfîniks]. Yandıktan sonra kendi külünden tekrar vücud bulan efsanevi bir kuş.

**phone** [foun]. (*abb.*) **telephone. ~tic** [fouˈnetik]. Sese aid; fonetik: **~s,** fonetik ilmi.

**phosphor·us** [ˈfosfərəs]. Fosfor. **~-escence** [-ˈresəns], yakamoz. **~ic acid,** hamızı fosfor. **~ous,** fosforlu.

**photo** [ˈfoutou]. (*abb.*) **photograph. ~-** *pref.* Ziyaya aid; foto-. **~graph** [ˈfoutougrâf], fotograf; resim çekmek. **~er** [–ˈtogrəfə(r)], fotografcı. **~y** [–ˈtogrəfi], fotografya. **~gravure** [ˌfotograˈvyuə(r)], fotografya hakkâklığı.

**phrase** [frêyz]. İbare; ifade tarzı; cümle; tabir. Bir fikri ifade etmek için kelime ve cümleler seçmek. **as the ~ goes,** meşhur tabir ile.

**phut** [fʌt]. **to go ~,** (*sl.*) suya düşmek.

**physic** [ˈfisik]. İlâc *esp.* müshil. İlâc, müshil vermek. **~s,** fizik ilmi. **~al** [ˈfizikl], fizikî; maddî; bedenî: **~ training,** beden terbiyesi. **~ian** [fiˈzişn], hekim, doktor. **~ist** [ˈfizisist], fizik mütehassısı.

**physio·gnomy** [fiziˈonəmi]. Yüz, çehre. **~logy** [fiziˈoləci], fiziyoloji.

**physique** [fiˈzîk]. Vücudun bünye ve kuvveti. **of poor ~,** cılız, kuvvetsiz.

**pian·o** [piˈyanou]. Piyano. **grand ~,** kuyruklu piyano: **upright ~,** dik piyano. **~ist** [ˈpîyənist], piyanist.

**piastre** [piˈâstə(r)]. Kuruş.

**piazza** [piˈatsa] (İtalyada) meydan; (Amerikada) ev etrafındaki kapalı taraça.

**piccaninny** [pikəˈnini]. Zenci çocuk; yavru.

**pick**[1] [pik]. Sivri kazma. **~**[2] *n.* Seçme; güzide şey. **the ~ of the bunch,** en iyisi. **~**[3] *v.* Sivri bir aletle delmek, kazmak; ayıklamak; toplamak; seçmek. **to ~ acquaintance with s.o.,** birisile dostluk tesisine vesile aramak: **to ~ a bone,** kemiğin etini ayıklamak: **to have a bone to ~ with s.o.,** birisile paylaşılacak kozu olm.: **to ~ s.o.'s brains,** birisinden malûmat koparmak: **to ~ and choose,** titizce seçmek: **to ~ holes in stg.,** bir şeyin tenkid edilecek taraflarını bulmak: **to ~ a lock,** kilidi maymuncukla açmak: **to ~ to pieces,** didiklemek: **to ~ pockets,** yankesicilik yapmak: **to ~ a quarrel,** kavga etmek için vesile aramak: **to ~ sides,** bir oyunda (iki kaptan) oyuncularını seçmek: **to ~ one's steps [way],** dikkatle yürümek: **to ~ one's teeth,** dişlerini karıştırmak. **~ at one's food,** iştahsız yemek. **~ off,** el ile tutup kaldırmak; dikkatle nişan alıp vurmak. **~ on [upon], I don't know why he ~ed on me,** bilmem neden buldu buldu da beni buldu. **~ out,** el ile tutup çıkarmak; seçmek: **to ~ out s.o. in a crowd,** kalabalıkta birini görmek. **~ over,** (meyva vs.yi) birer birer muayene edip iyilerini ayırmak. **~ up,** eğilip bir şeyi yerden almak; rasgele bulmak; yolda durup birini otomobil vs.ye almak: (hasta) iyileşmek: **to ~ up stg. cheap,** kelepir olarak bulmak: **to ~ up courage,** cesaretini toplamak: (**searchlight**) **to ~ up a plane,** (bir ışıldak) uçağı yakalamak: **to ~ up a language,** *etc.*, bir lisan vs.yi çabuk öğrenmek: **to ~ up a livelihood,** hayatını oradan buradan kazanmak: **to ~ up speed,** hızını arttırmak: **to ~ up one's strength,** (hasta) kendini toplamak. **~-up,** pikap. **~-a-back** [ˈpikəbak], **to carry s.o. ~,** birini sırtta taşımak. **~axe** [ˈpikaks], kazma. **~ing** [ˈpikin(g)], toplama, devşirme. **~ and stealing,** aşırma: **~s,** ufak tefek kâr: **to get ~s,** çöplenmek. **~pocket** [ˈpikpokit], yankesici.

**picket** [ˈpikət]. Kazık; ileri karakol;

(grev esnasında) çalışmak isteyen işçilere mâni olmak için nöbet bekliyen amele. (Atı) kazığa bağlamak. **to ~ a factory,** (grev esnasında) bir fabrikaya 'picket' dikmek.

**pickle** [ˈpikl]. Turşu; salamura; sıkıntılı vaziyet; yaramaz çocuk. Salamuraya yatırmak; turşu kurmak. **to be in a nice [sorry] ~,** müşkül bir vaziyette olm.: **you little ~!,** (çocuğa) seni gidi seni!

**picnic** [ˈpiknik]. Piknik.

**pictorial** [pikˈtôriəl]. Resimli.

**picture** [ˈpiktyə(r)]. Resim; tasvir; levha; timsal; pek güzel şey; filim. Tasvir etmek. **the ~s,** sinema. **~ to yourself ...!,** ... tasavvur et!: **he is the ~ of his father,** tıpkı babası: **he is the ~ of health,** aslan gibi sıhhatli: **that child is a perfect ~,** o çocuk bebek gibi güzel: **to come into the ~,** sahaya girmek, mevzuubahis olm.: **to be very much in the ~,** bir meselede çok mühim rolü olm.: **to be out of the ~,** sayılmamak. **~-gallery,** resim müzesi. **~-house,** sinema. **~-rail,** duvara tablo asmak için çıta. **~sque** [ˌpiktyəˈresk], resmi yapılacak kadar güzel.

**piddle** [ˈpidl]. Çişini etmek.

**pidgin** [ˈpicin]. **~ English,** yarı bozuk ingilizce yarı çince bir lisan, ki, tüccarlar ile hizmetçiler tarafından kullanılır.

**pie** [pây]. Etli veya meyvalı börek. **fruit ~,** torta: **to have a finger in the ~,** bir işe karışmak.

**pie·bald, ~d** [ˈpâybôld]. Alacalı.

**piece** [pîs]. Parça; kısım; tane; sikke; yama; (satranc, dama) taş; piyes. **to ~ together,** birleştirmek, yan yana koymak: **a ~ of advice,** bir tavsiye: **all of one ~,** yekpare: **all to [in] ~s,** paramparça: **a ~ of artillery,** bir top: **by the ~,** tane ile; parça başına: **to fall [go] to ~s,** parça parça olm.; harab olm.: **a ~ of (good) luck,** talih eseri: **to take to ~s,** sökmek: **a ~ of water,** bir küçük göl vs.: **this is a ~ of my work,** bu benim işimden bir örnektir. **~-goods,** mensucat. **~-work,** işe göre ücret; parça başına ücret. **~meal** [ˈpîsmîl]. Bölük pürçük, parça parça.

**pied-à-terre** [piêydâˈteə(r)]. Baş sokacak yer.

**pier** [piə(r)]. Denize uzanmış iskele; rıhtım; payanda, destek.

**pierc·e** [piəs]. Delmek; delip geçmek, açmak. **~ing,** acı, keskin; iliğe işliyen (soğuk).

**pierrot** [ˈpiərou]. Palyaço.

**piet·ism** [ˈpâyətizm]. Müfrit dindarlık. **~y** [ˈpâyəti], dindarlık: **filial ~,** ana babaya karşı muhabbet ve hürmet.

**piffle** [ˈpifl]. (*sl.*) Saçma(lamak).

**pig** [pig]. Domuz; pis herif; obur; maden külçesi. **to ~ it,** ahırda gibi yaşamak: **don't be a ~!,** (i) oburluk etme!; (ii) insaf et! **~gery** [ˈpigəri], domuz ahırı; çok pis yer. **~gish,** pis; pisboğaz. **~-headed,** dikkafalı, inadcı. **~-iron,** pik demir. **~let,** domuz yavrusu. **~man,** domuz çobanı. **~skin** [ˈpigskin], domuz derisi. **~sticking,** mızrak ile yaban domuzu avcılığı. **~sty,** domuz ahırı; pek pis ev, oda. **~tail,** uzun ve arkaya sarkık saç örgüsü. **~wash,** domuz yemi; mutfak süprüntüsü.

**pigeon** [ˈpicən]. Güvercin; safdil, aval. **clay ~** (spor) sunî güvercin. **~-chested,** çıkık göğüslü. **~-hole,** yazıhane gözü: (evrakı vs.) gözlere koymak; hasır altı etmek. **~-toed,** ayaklarını içeri basarak yürüyen.

**pigment** [ˈpigmənt]. Hayvan [nebat] nesiclerine renk veren madde; boya maddesi.

**pigmy** [ˈpigmi]. Cüce.

**pike** [pâyk]. Kargı; zirve. **~staff,** kargı sapı. ⌜**as plain as a ~**⌝, ⌜kör kör parmağım gözüne⌝.

**pile**[1] [pâyl]. Yığın, küme; büyük ve muhteşem bina. **~ (up),** yığmak; biriktirmek. Yığılmak, birikmek. **to ~ arms,** tüfek çatmak: **to make one's ~,** küpünü doldurmak: **to ~ it on,** (*coll.*) mübalağa etm.: **to ~ on the agony,** acı veya korkunc tafsilat vermek. **~**[2]. Büyük kazık. **~-driver,** şahmerdan. **~**[3]. Hav. **~**[4]. **~s,** basur.

**pilfer** [ˈpilfə(r)]. Aşırmak.

**pilgrim** [ˈpilgrim]. Hacı; seyyah. **The P~ Fathers,** 'Mayflower' gemisinde Amerikaya muhaceret eden İngilizler. **~age,** hacca gitme; uzun seyahat.

**pill** [pil]. Hap. **~-box,** hap kutusu; (*mil.*) beton kule şeklinde makineli tüfek yuvası.

**pillage** [ˈpilic]. Yağma, çapul. Yağma etm., soymak.

**pillar** [ˈpilə(r)]. Direk; sütun. **to be driven from ~ to post,** mekik dokumak. **~-box,** (üstüvane şeklinde) posta kutusu.

**pillion** [ˈpilyən]. Terki. **to ride ~,** terkiye binmek.
**pillory** [ˈpiləri]. Teşhir direği; teşhir cezası. (Bir suçluyu ceza olarak) teşhir etmek.
**pillow** [ˈpilou]. Baş yastığı. **~-case,** yastık yüzü.
**pilot** [ˈpâylət]. Kılavuz; rehber; pilot. Gemiye kılavuzluk etm.; uçak kullanmak; yol göstermek. **~age,** kılavuzluk (ücreti).
**pimp** [pimp]. Pezevenk(lik etmek).
**pimpl·e** [ˈpimpl]. Ergenlik; sivilce. **~y,** ergenlikli, sivilceli.
**pin**[1] [pin] *n.* Toplu iğne; mil. **~s,** (*sl.*) bacaklar. **I don't care a ~!,** bana vız gelir: **you could have heard a ~ drop,** sinek uçsa işitilirdi: **~s and needles,** karıncalanma: **to be on ~s and needles,** çok meraklanmak: **for two ~s I'd box your ears,** benden tokat yemediğine şükret!: **safety ~,** çengelli iğne: **split ~,** gupilya. **~**[2] *v.* İğnelemek; iğne ile bağlamak. **to ~ s.o.'s arms to his side,** birinin kollarını arkasından kıskıvrak yakalamak: **to be ~ned against a wall,** duvara sıkıştırılmak: **to be ~ned under a fallen beam,** düşen bir kalasın altında sıkışmak: **to ~ s.o. down to facts,** birini sırf vakiaları söylemeğe mecbur etm.: **to ~ one's hopes on ...,** ümidini ···e bağlamak.
**pinafore** [ˈpinəfô(r)]. Çocuk önlüğü.
**pince-nez** [ˈpansnêy]. Kelebek gözlüğü.
**pincers** [ˈpinsö(r)z]. **(pair of) ~,** kerpeten.
**pinch**[1] [pinç] *n.* Çimdik; tutam; sıkıntı, darlık: **it will do at a ~,** zaruret halinde yasak savar: **it was a close ~,** bıçak sırtı kadar bir şey kaldı: **to feel the ~,** zaruret içinde kalmak: **to give s.o. a ~,** birini çimdiklemek: **the ~ of hunger [poverty,** *etc.*], açlık, fakirlik vs.nin ıstırabı. **~**[2] *v.* Çimdiklemek; sıkıştırıp sıkıntı vermek; (ayakkabı) sıkmak; fazla tutumlu olm.; (*sl.*) aşırmak, yakalamak. **to be ~ed for money,** para sıkıntısı çekmek: **to ~ and scrape,** dişinden tırnağından artırmak: **that's where the shoe ~es,** işte derd burada: ⌈**everyone knows best where his own shoe ~es**⌉, herkes kendi derdini herkesten iyi bilir.
**pinchbeck** [ˈpinçbek]. Altın taklidi; sahte.
**pine**[1] [pâyn] *n.* Çam. **Austrian ~,** kara çam: **Scotch ~,** sarı çam: **stone ~,** fıstık çamı. **~apple** [ˈpâynapl], ananas. **~-cone,** çam kozalağı. **~**[2] *v.* **to ~ for,** ···in hasretini çekmek: **he ~s for London,** Londra gözünde tütüyor: **to ~ away,** yavaş yavaş kuvvetten düşmek.
**ping** [pin(g)]. (*ech.*) Kurşun vızıltısı.
**pinion** [ˈpinyən]. Kanad (tüyü); çark feneri. Kuş kanadının ucunu kesmek; kollarını bağlamak.
**pink**[1] [pin(g)k]. Küçük karanfil. Pembe. **in the ~ of condition,** mükemmel idmanlı; (meyva vs.) tam olgun: **the ~ of perfection,** mükemmelliğin en yüksek derecesi. **~**[2], (*ech.*) (Otomobil) kliketleşmek.
**pinnace** [ˈpinəs]. Kürekli filika.
**pinnacle** [ˈpinəkl]. Zirve.
**pint** [pâynt]. Bir galon'un sekizde biri (= *0·586 litre*). **to have a ~,** bir bardak bira çekmek.
**pioneer** [pâyoˈniə(r)]. Yol açmak için önden giden kimse; önayak olan; (*esk.*) baltacı neferi. **~ work,** bakir, el değmemiş bir mevzu veya iş.
**pious** [ˈpâyəs]. Dindar, sofu. **~ fraud,** riyakâr.
**pip**[1] [pip]. Ufak çekirdek; (iskambil vs. üzerindeki) sayı, benek. **~**[2], **to have [get] the ~,** (*coll.*) sıkılmak, üzülmek: **to give s.o. the ~,** birinin keyfini bozmak. **~**[3]. (Askerlikte ve telfonda) P harfi: **~ emma = p.m.,** öğleden sonra.
**pipe** [pâyp]. Pipo; çubuk; künk; kaval. Çocuk gibi ince bir sesle söylemek: (*naut.*) düdük ile kumanda vermek: **to ~ on board,** gemiye çıkan bir kimseyi porsun düdüğü ile selâmlamak. **the ~s,** gayda: **to smoke the ~ of peace,** barışmak: ⌈**put that in your ~ and smoke it!**⌉, kulağında küpe olsun!: **to ~ up,** taganniye başlamak. **~-line,** petrol vs.yi nakletmek için yere döşenen boru. **~r** [ˈpâypə(r)], gaydacı; kavalcı. **to pay the ~,** icabeden masrafı ödemek: ⌈**who pays the ~ calls the tune**⌉, ⌈parayı veren düdüğü çalar⌉.
**piquan·t** [ˈpîkənt]. Acı, yakıcı, keskin. **~cy,** (tad hakkında) dokunaklılık, keskinlik; bir şeyin meraklı tarafı.
**pique** [pîk]. Güceniklik, küskünlük. Küstürmek; izzeti nefsine dokunmak. **to ~ oneself on stg.,** bir şeyle öğünmek.
**pira·cy** [ˈpâyrəsi]. Korsanlık; telif

hakkı gasıblığı. ~**te** [ˈpâyərət], korsan; telif hakkına tecavüz edip bir kitabı neşretmek. ~**tical** [–ˈratikl], korsancasına.

**pirouette** [piruˈet]. Bir ayak üzerinde tam çark etme(k).

**piscatorial** [ˌpiskəˈtôriəl]. Balıkçılığa aid.

**pish** [piş]. (*ech.*) *Hoşnudsuzluk nidası.*

**piss** [pis]. Sidik. İşemek.

**pistachio** [pisˈtâşyou]. Şam fıstığı.

**pistol** [ˈpistl]. Tabanca.

**piston** [ˈpistən]. Piston. ~**-ring,** segman.

**pit**[1] [pit] *n.* Çukur; yerde kazılan tuzak; çopur; tiyatroda parter; maden ocağı. **the bottomless ~,** gayya, cehennem: **the ~ of the stomach,** göğüs çukuru. ~**-head,** maden ocağının ağzı. ~**-prop,** (maden ocağında) direk, destek. ~**man,** madenci. ~[2] *v.* (Pas) madeni karıncalandırmak; (çiçek illeti) yüzde çopur bırakmak. **to ~ oneself against s.o.,** boy ölçüşmek.

**pit-a-pat** [ˈpitəˈpat]. (*ech.*) Hafif hafif çarpma. (**heart**) **to go ~,** yürek tıp tıp atmak.

**pitch**[1] [piç]. Zift. Zift ile kaplamak. ~**-black,** simsiyah. ~**-dark,** zifiri karanlık. ~[2] *n.* Mertebe, derece; yükseklik; meyil; ses perdesi; yalpa; sokak [pazar] satıcısının muayyen yeri; (oyun) meydan; pervane piçi; (vida) adım. **full ~,** (top vs.) atılan bir şeyin yere çarpmadan başka bir şeye vurması: **to the highest ~,** son dereceye kadar: **to such a ~ that ...,** öyle bir mertebede ki.... ~[3] *v.* Atmak; yere dikmek; (çadır) kurmak; bir şeyi atıp muayyen bir yere düşürtmek; sesin perdesini ayarlamak. Konmak; yere inmek; (gemi) baş vurmak. **to ~ one's hopes very high,** gözü yükseklerde olmak. **~ in(to),** yumruk [dil] ile tecavüzde bulunmak; tehalükle girişmek; bir yere başaşağı düşmek. **~ on** [**upon**], ···e konmak; seçmek, karar vermek: **to ~ on one's head,** tepe üstü düşmek. ~**-and-toss** [ˈpiçəndˈtos], yazı mı tura mı oyunu. ~**ed battle,** meydan muharebesi. ~**fork** [ˈpiçfôk], diğeren (ile savurmak); fırlatmak.

**pitcher** [ˈpiçə(r)]. Testi. ʳ**little ~s have long ears**ˀ, ʳçocuktan al haberiˀ.

**piteous** [ˈpitiəs]. Acınacak halde.

**pitfall** [ˈpitfôl]. Tuzak gizli tehlike. **there are a lot of ~s in this business,** bu işte ayağını denk almalı.

**pith** [piθ]. (Nebatta) öz; cevher; kudret; esaslı kısım. ~**iness,** vecizlik. ~**y,** özlü; veciz; kısa ve keskin (üslûb).

**piti·able** [ˈpitiəbl]. Acınacak. ~**ful,** merhametli; acınacak; miskin. ~**-less,** amansız, merhametsiz.

**pittance** [ˈpitəns]. Pek cüzî ücret. **a mere ~,** ölmiyecek kadar kazanc.

**pitted** [ˈpitəd] *a.* Çiçek bozuklu; (maden asit veya pastan) karıncalanmış.

**pity** [ˈpiti]. Merhamet; acıma; şefkat, rikkat. ~ [**take ~ on**], acımak; merhamet etmek. **for ~'s sake!,** Allah aşkına!: **what a ~!,** yazık!

**pivot** [ˈpivət]. Mil; mihver. Bir mihver etrafında dönmek. ~**al,** merkezî; en mühim.

**pixy, -ie** [ˈpiksi]. Peri, cin.

**placard** [ˈplakâd]. Yafta. Üzerine yafta yapıştırmak.

**placate** [pləˈkêyt]. Teskin etm., yatıştırmak; hatırını yapmak.

**place**[1] [plêys] *n.* Yer; mevki; mahal; meydan; memuriyet; konak, ev. **in ~ of,** yerine, bedel olarak: **in the first ~,** ilkönce; evvela; **it is not my ~ to do it,** benim vazifem değil: **to four ~s of decimals,** dördüncü kesre kadar: **to know one's ~,** haddini bilmek: **to lay a ~** (**at table**), sofrada bir kişilik yer kurmak: **in the next ~,** sonra; bundan başka: **out of ~,** kendi yerinde olmıyan; münasebetsiz: **to feel out of ~,** yadırgamak: **to look out of ~,** yama gibi durmak: **in its proper ~,** yerli yerinde: **to put s.o. in his ~,** birine haddini bildirmek: **to take ~,** vukubulmak: **to take a ~,** bir işe girmek; bir ev kiralamak; bir yerde oturmak; bir mevkii zabtetmek. ~[2] *v.* Koymak; yerleştirmek; vazifeye yerleştirmek (para) yatırmak. **to be awkwardly ~d,** müşkül bir vaziyette olm.: **to ~ a book with a publisher,** bir kitabı naşire kabul ettirmek: **I can't ~ him,** kim olduğunu tamamen tayin edemiyorum: **to ~ confidence in s.o.,** birine itimad etm.: **to ~ an order,** bir sipariş vermek: **to ~ a matter in s.o.'s hands,** bir işi birinin eline vermek: **to be well ~d in a class,** sınıfta derecesi iyi olmak.

**placid** [ˈplasid]. Sakin; durgun; halim.

**plagiarism** [ˈplêyciərizm]. İntihal.
**plagu·e** [plêyg]. Veba; taun; belâ. Taciz etm., üzmek; kasıp kavurmak. **a ~ on him!**, kör olasıca!: **to ~ s.o.'s life out,** birinin başının etini yemek. **~y,** baş ağrıtıcı.
**plain¹** [plêyn] *n.* Ova; sahra. **~²** *a.* Vazih, aşikâr; sade; güzel değil. **(person) to be ~,** (insan) güzel olmamak: **to be ~ with s.o.,** birisine dobra dobra söylemek: **~ dealing,** dürüst hareket: **in ~ clothes,** sivil elbiseli: **in quiet ~ clothes,** sade giyinmiş: **in ~ English,** açıkçası; ingilizcesi: **in ~ words,** açıkçası. **~-spoken,** tok sözlü; dobra dobra söyliyen. **~ness** [ˈplêynnis], sadelik; vuzuh; toksözlülük; çirkinlik.
**plaint** [plêynt]. Şikâyet. **~iff** [ˈplêyntif], davacı, müddei. **~ive** [ˈ–iv], iniltili, ağlamış, sızlamalı.
**plait** [plat]. Örgü, saç örgüsü. Saç, hasır örmek.
**plan** [plan]. Tedbir; tasavvur; niyet; fikir; plân, resim, harita. Tertib etm.; plânını çizmek; tasavvur etm.; niyet etm. **everything went according to ~,** her şey plâna uygun olarak cereyan etti: **the best ~ would be to ...,** yapılacak en iyi şey ... dir: **to ~ to do stg.,** bir şeyi yapmağa niyet etmek.
**plane¹** [plêyn]. Çınar. **~².** Rende(lemek). **~³.** Düz satıh; seviye; uçağın kanadı; (*abb.*) **aeroplane,** uçak. **~-table,** plânçete.
**planet** [ˈplanit]. Seyyare. **~ary system,** seyyareler manzumesi.
**plank** [plan(g)k]. Uzun tahta, kalas. Tahta döşemek. **to ~ down money,** parayı nakden ödemek: **to ~ oneself down,** pat diye bir yere oturmak: **to walk the ~,** gemiden denize doğru uzatılmış bir kalas üzerinde gözleri bağlı olarak yürümek ki korsan gemilerinde idam şekli idi.
**plant¹** [plânt] *n.* Demirbaş eşya; sanayide kullanılan her türlü aletler vs. **~²** *n.* Nebat; ot; fide: (*sl.*) hile. **~³** *v.* (Fidan vs.) dikmek. **to ~ a blow,** bir darbe indirmek: **to ~ a field with barley,** bir tarlaya arpa ekmek: **to ~ an idea in s.o.'s mind,** birinin aklına bir fikir koymak: **to ~ oneself in front of s.o.,** birinin karşısında dikilmek: **to ~ out,** (fide) saksıdan çıkarıp dikmek. **~ation** [plânˈtêyşn], fidanlık; şekerkamışı, pamuk, çay vs. edilen tarla. **~er** [ˈplântə(r)], bu tarlalarını idare eden kimse.
**plaque** [plak]. Maden safhası; tabelâ.
**plash** [plaş]. (*ech.*) Çağıltı; suya çarpma sesi. Çağıldamak; suda çırpınmak.
**plasm(a)** [ˈplazm(a)]. Kan ve lenfa suyu.
**plaster** [ˈplâstə(r)]. Sıva; alçı; yakı. Sıvamak; yakı yapıştırmak. **adhesive ~,** yakı bezi: **court ~,** ingiliz yakısı: **~ of Paris,** alçı: **a ~ saint,** sahte veli. **~er,** sıvacı.
**plastic** [ˈplastik]. Yuğurulabilir; istenilen şekle sokulabilir. Heykeltraşlık vs. **the ~ arts,** heykeltraşlık, çömlekçiliği gibi sanaatler.
**plate** [plêyt]. Tabak; madenî levha; plâk; fotograf camı; gümüş (kaplamalı) sofra takımı; takma diş dizisi; basma resim. Madenî levha ile kaplamak. **hot-~,** soba vs.nin yemek ısıtmağa mahsus yeri: **number ~,** numara plâkası. **~ful,** tabak dolusu. **~-armour,** gemi zırhı. **~-glass,** ayna camı. **~-rack,** tabak rafı. **~layer** [ˈ–leə(r)], kay tamircisi.
**platen** [ˈplatən]. Makine tahtası; yazı makinesinde kâğıd silindiri.
**platform** [ˈplatfôm]. Düz çatı; sahanlık; peron; top temeli; platform; siyasî partinin programı. **a good ~ speaker,** iyi bir siyasî hatib.
**plating** [ˈplêytin(g)]. Kaplama(cılık).
**platinum** [ˈplatinəm]. Platin.
**platitud·e** [ˈplatityûd]. Bayağılık; basmakalıb, beylik. **~inize** [–ˈtyûdinâyz], tatsız tuzsuz konuşmak. **~inous** [–ˈtyûdinəs], basmakalıb ve yavan (söz veya kimse).
**Plato** [ˈplêytou]. Eflâtun. **~nic** [pləˈtonik], Eflâtun felsefesine aid; nazariyeden ibaret; zararsız: **~ love,** ideal, manevî aşk.
**platoon** [pləˈtûn]. Askerî müfreze, takım.
**platter** [ˈplatə(r)]. Ağacdan yapılmış tabak.
**plaudits** [ˈplôdits]. Alkış.
**plausible** [ˈplôzibl]. Akla yakın, makul; zahiren makul hakikatte değil.
**play¹** [plêy] *n.* Oyun; oynama; eğlence; şaka; kumar; hareket (serbestisi); faaliyet; piyes; lâçka. **to call into ~,** meydana çıkarmak; davet etm.: **to come into ~,** ortaya çıkmak, rol oynamak: **in ~,** alay için: **in full ~,** tam faaliyette: **to give full ~ to**

one's abilities, *etc.*, birinin istidadının vs. gelişmesine tam imkân vermek: **to make much ~ of stg.**, bir hadise vs.yi mütemadiyen büyüterek bir maksad için kullanmak: **the ~ runs high**, büyük çapta kumar oynanıyor: **~ on words**, cinas. **~²** *v.* Oynamak; eğlenmek; kumar oynamak; (*mus.*) çal(ın)mak. **to ~ the fine lady**, kibar hanım rolü oynamak: **to ~ the man**, erkekçe hareket etm.: **I'll ~ you for drinks**, sizinle içkisine oynarım (kaybeden içkileri ısmarlıyacak): **I'll ~ you for five shillings**, sizinle beş şilinine oynarım: **to ~ for one's own hand**, kendine yontmak: **to ~ into the hands of s.o.** [**to ~ s.o.'s game**], birinin ekmeğine yağ sürmek: **to ~ the game**, namuslu davranmak: **to ~ a part**, rol oynamak: **to ~ for time**, vakit kazanmak için oyalamak. **~ at**, (filan oyunu) oynamak: **(children) to ~ at being soldiers**, *etc.*, çocuklar askerlik vs. oynamak: **what are you ~ing at?**, ne kumpas kuruyorsun?: **to ~ s.o. at chess**, birisile satranc oynamak. **~ away**, kumarda (servetini vs.) kaybetmek: **we are ~ing away tomorrow**, yarın karşı tarafın sahasında oynuyoruz. **~ off, to ~ off s.o. against s.o. else**, kendi menfaati için birini başkasına karşı kullanmak. **~ on**, oynamağa devam etm.: **to ~ on** [**upon**] **s.o.'s feelings**, birinin merhamet vs. hissinden istifade etm.: **the fire-engine ~ed on the house**, itfaiye hortumları eve tevcih etti. **~ out**, oyunu sonuna kadar oynamak: **the organ ~ed the people out**, halk kiliseden çıkıncaya kadar org çaldı: **to be ~ed out**, takati kalmamak; modası geçmiş olm. **~ up**, gayretle oynamak: **to ~ up to s.o.**, birine yaranmak. **~er** [ˈplêyə(r)]. Oyuncu; aktör. **~fellow**, oyun arkadaşı. **~ful**, şakacı, neşeli. **~goer**, tiyatro meraklısı. **~ground**, oyun sahası; eğlence yeri. **~house**, tiyatro. **~mate**, oyun arkadaşı. **~thing**, oyuncak. **~time**, oyun zamanı; teneffüs vakti. **~wright**, tiyatro müellifi.

**plea** [plî]. Müdafaaname; bahane; vesile. **~ for mercy**, aman talebi: **on the ~ of ..., ...** bahanesile. **~d**, dava etm.; bir davayı ileri sürmek yahud bir davaya karşı müdafaada bulunmak; bir mahzuru ileri sürmek; öne sürmek; yalvarmak. **to ~ guilty or not guilty**, suç [mesuliyeti] kabul veya redd etm.: **to ~ illness**, *etc.*, hastalığını vs. ileri sürmek (bahane etm.): **to ~ with s.o.**, birine yalvarmak: **to ~ s.o.'s cause with s.o.**, birisi için başka birine şefaat etmek. **~ding**, yalvarıcı; müdafaa sanati; ithamname ve müdafaaname; yalvarma: **special ~**, mugalata. **~der**, avukat.

**pleas·ant** [ˈplezənt]. Hoş; şirin, canayakın. **to make oneself ~ to s.o.**, birinin yüzüne gülmek; birinin gözüne girmeğe çalışmak; iltifat etmek. **~antry**, latife; şaka. **~e** [plîz]. Hoşuna gitmek; memnun etm.; göze girmek. **~!** [**if you ~**], lûtfen, rica ederim: **do as you ~!**, istediğinizi yapınız; siz bilirsiniz!: **to do as one ~s**, istediğini yapmak: **hard to ~**, müşkülpesend: **~ God!**, inşallah: **there is no ~ing him**, onu memnun etmek mümkün değil: **~ yourself!**, siz bilirsiniz, nasıl isterseniz. **~ed**, memnun, razı: **to be ~ to do stg.**, bir şeyi memnuniyetle yapmak; (resmî) buyurmak: **he is very well ~ with himself**, kendini beğenmiş; yaptığından memnun. **~ing** [ˈplîzin(g)]. Hoş; canayakın; sempatik. **~urable** [plejərəbl]. Memnun edici; hoş. **~ure** [ˈplcjə(r)]. Haz, zevk, memnuniyet; eğlence; keyf; tenezzüh. **at ~**, istenildiği zaman: **at the ~ of ...**, ...in keyfine göre: **without consulting my ~**, bana danışmadan: **office held during ~**, birinin arzusuna bağlı vazife: **I have much ~ in informing you that ...**, size bildirmekle memnunum: **~ resort**, eğlence şehri veya yeri: **~ ground**, eğlence meydanı: **~ trip**, tenezzüh seyahati: **what is your ~?**, emriniz nedir?

**pleat** [plît]. Kırma, plise (yapmak).

**pleb·eian** [pliˈbîən]. Ayak takımına aid; halk tabakasından. **~s** [plebz], **the ~**, ayak takımı; avam.

**plebiscite** [ˈplebisâyt]. Plebisit.

**pledge** [plec]. Rehin; vaid, söz; teminat; tövbe; sıhhatine içme. Rehin [teminat] olarak vermek; sıhhatine içmek. **to ~ oneself to ...**, vadetmek: **~ of good faith**, hüsnüniyet teminatı: **to put stg. in ~**, bir şeyi rehine koymak: **to take** [**sign**] **the ~**, içki içmeğe tövbeli olm.: **I am under ~ of secrecy**, bu sırrı söylememeğe söz verdim.

**Pleiades** [ˈplâyədîz]. **the ~**, Süreyya.

**plen·ary** [ˈplînəri]. Tam; umumî;

mutlak. ~ **assembly,** umumî heyet ictimaı: ~ **powers,** tam salâhiyet. **~ipotentiary** [ˈplînipəˈtenşəri]. Tam salâhiyetli (elçi vs.). **~itude** [ˈplînityûd]. Tamamlık; bolluk.

**plent·y** [ˈplenti]. Bolluk; çokluk. ~ **of,** çok. **~eous** [**~iful**], bol, mebzul; bereketli.

**plethora** [ˈpleθorə]. Çokluk; fazlalık.

**plia·ble** [ˈplâyəbl]. Eğilebilir, bükülür; yumuşak, uysal. **~nt,** eğilip bükülür.

**pliers** [ˈplâyö(r)z]. Kıskaç; kargaburun.

**plight**[1] [plâyt] *n.* Hal, vaziyet. **a sorry** ~, müşkül veya acıklı bir vaziyet. ~[2] *v.* **to** ~ **one's word** [**troth**], söz vermek, *esp.* evlenmeğe söz vermek.

**Plimsoll** [ˈplimsol]. ~ **line,** gemilerin müsaade olunan yükü aldığı zaman suya batacağı kısmı gösteren hat. **~s,** üstü bez tabanı lâstik ayakkabı.

**plinth** [plinθ]. Sütun, heykel ayaklığı.

**plod** [plod]. Ağır ağır veya yorgun gibi yürümek. **to** ~ **along,** ağır ağır fakat sebatla yürümek, çalışmak. **~der,** sürekli gayretle çalışan kimse.

**plop** [plop]. (*ech.*) *Ağır bir şeyin suya düşme sesi*; cumburlop. Lop diye suya düşmek.

**plot**[1] [plot]. Arsa; tarh; küçük arazi parçası. ~[2]. Entrika; gizli tertib; suikasd; bir piyes veya hikâyenin plânı; grafik. Suikasd tertib etm.; kumpas kurmak; haritasını veya grafiğini çizmek.

**plough** [plau̯]. Saban, pulluk. Çift sürmek. **the** ~, Büyük Ayı: **to** ~ **s.o. in an exam.,** (*coll.*) birini imtihanda çaktırmak: **to follow the** ~, çiftçilik yapmak: **to put one's hand to the** ~, bir işe gayretle girişmek. ~ **in,** saban ile gömmek. ~ **through, to** ~ **one's way through the mud,** çamurda güçlükle ilerlemek: (**ship**) **to** ~ **through the waves,** (gemi) dalgaları yarmak. ~ **up,** bir çayırı sabanla sürmek. **~-boy,** çiftçi yamağı. ~ **land** [ˈplau̯land], sabanla sürülmüş arazi. **~share,** sabanın uc demiri.

**plover** [ˈplʌvə(r)]. Kızkuşu.

**plow.** (*Amer.*) *v.* **plough.**

**pluck**[1] [plʌk] *v.* Yolmak; soymak; koparmak; (*sl.*) imtihandan döndürmek. **to** ~ **out** [**off**], çekip koparmak: **to** ~ **s.o. by the sleeve,** birini yeninden çimdikler gibi çekmek: **to** ~ **up courage,** cesaretini toplamak. ~[2] *n.* Cesaret; kabadayılık. **~y,** yiğit, gözü pek.

**plug** [plʌg]. Tapa; tıkaç; (*elect.*) fiş; (helâda) su haznesi kolu: (*sl.*) yumruk darbesi. Tıkamak; (*sl.*) yumruklamak. **sparking** ~, buji: **wall** ~, priz, dişi fiş: ~ **tobacco,** ağız tütünü. ~ **away,** sebatla çalışmak. ~ **in,** (*elect.*) fişi prize sokmak.

**plum** [plʌm]. Erik; (*sl.*) en iyisi: **the** **~s,** en iyi memuriyetler vs. **~-cake,** kuru üzümlü kek. **~-pudding,** *esp.* Noel'de yenilen kuru üzümlü meşhur ingiliz puding'i.

**plum·age** [ˈplûmic]. Kuşun tüyleri. **~e** [plûm], büyük ve gösterişli tüy; sorguç. **to** ~ **itself,** (kuş) tüylerini düzeltmek: **to** ~ **oneself on stg.,** bir şeyle övünmek: **borrowed** **~s,** karganın tavus tüyleri giymesi gibi sahte unvan vs.

**plumb** [plʌm]. Şakul; iskandil kurşunu; şakulî vaziyet. Şakulî; (*sl.*) tastamam. İskandil etm.; inceden inceye tedkik etm.; evin su tertibatını kurmak. **~er,** kurşuncu. **~ing,** evin boru tertibatı.

**plump**[1] [plʌmp]. Semiz; tombul. ~[2], (*ech.*) *Ağır düşme sesi*; cumburlop; ansızın. **to** ~ **for,** ···e rey vermek, tercih etmek.

**plunder** [ˈplʌndə(r)]. Yağma (etm.); soymak.

**plunge** [plʌnc]. Daldırmak, batırmak; sokmak. Atılmak, dalmak; (gemi) baş vurmak. Suya dalma; saldırış. **to take the** ~, geri dönülmesi imkânsız bir işe girişmek ('ok yaydan çıktı' *kabilinden*). **~r** [ˈplʌncə(r)]. Azgın kumarbaz vs.; tulumba pistonu.

**pluperfect** [plûˈpö(r)fikt]. Hikâyei mazi sıygası, *mes.* **he had seen,** görmüştü.

**plural** [ˈplu̯ərəl]. Cemi. ~ **vote,** birden fazla rey kullanma hakkı. **~ity** [–ˈraliti], çokluk; bir adamda birkaç vazifenin birleşmesi.

**plus** [plʌs]. Zaid işareti (+); ilâvesile; müsbet. ~ **side of an account,** hesabın alacak hanesi. **~-fours,** golf pantalonu.

**plush** [plʌş]. Pelüş.

**Pluto** [ˈplûtou̯]. Mitolojide cehennem hükümdarı; *dahi* zenginliğin ilâhı.

**plutocra·cy** [pluˈtokrəsi]. Zenginler hakimiyeti. **~t** [ˈplûtəkrat], nüfuzlu zengin; pek zengin adam.

**ply**[1] [plây] *n.* Katmer, kat; plise; ip kolu. **five-~ wood,** beş katlı kontrplak; **three-~ rope,** üç kollu halat. **~wood,** kontrplak. ~[2] *v.* Kuvvetle işletmek, kullanmak; sıkıştırmak. Muayyen bir şekilde sefer yapmak. **to ~ the oars,** çala kürek kürek çekmek: **to ~ a trade,** bir sanat icra etm.: **to ~ s.o. with drink,** birine mütemadiyen içki vermek: **to ~ s.o. with questions,** birini suallerle sıkıştırmak: **car ~ing for hire,** kira otomobili.

**p.m.** [ˈpîˈem]. (*abb. Lât.*) **post meridiem,** öğleden sonra.

**pneumatic** [nyuˈmatik]. Hava ile işliyen: **~ tyre,** lâstik.

**pneumonia** [nyûˈmounyə]. Pnömoni.

**P.O.** (*abb.*) **Post Office,** Postahane.

**poach**[1] [pouç]. Yumurtayı kabuksuz olarak suda pişirmek. ~[2]. Ruhsatsiz avlamak; başkasının malını haksız olarak almak. **to ~ on s.o.'s preserves,** başkasının sahasına tecavüz etmek. **~er,** ruhsatsız avlanan avcı.

**pocket** [ˈpokit]. Ceb; kese; torba; (havacılık) hava boşluğu. Derceb etm.; cebe koymak; (hislerini) zabtetmek; (tahkir vs.yi) hazmetmek. **to be in ~,** (bir işten) kâr etm., kârlı çıkmak: **to be out of ~,** cebden [keseden] eklemek; zararlı çıkmak: **to have s.o. in one's ~,** birini avucunda tutmak: **always to have one's hand in one's ~,** mütemadiyen para vermeğe mecbur olm.: **to line one's ~s,** kesesini doldurmak: **~s under the eyes,** gözlerin altındaki sarkık etler. **~-book,** muhtıra defteri. **~-handkerchief,** mendil. **~-knife,** çakı. **~-money,** ceb harclığı.

**pock-mark** [ˈpokmâk]. Çiçekbozuğu.

**pod** [pod]. (Bakla vs.) kabuk. Kabuğunu soymak.

**podgy** [ˈpoci]. Şişko; bodur.

**poem** [ˈpouim]. Şiir, manzume.

**poet** [ˈpouit]. Şair. **~ess,** kadın şair. **~ic(al)** [–ˈetikl], şairane; şiire aid. **~ry** [ˈpouitri], şiir sanaati, nazım.

**poignant** [ˈpôynənt]. Keskin; tesirli; ıstırab verici.

**point**[1] [pôynt] *n.* Nokta; derece; kerte; uc; burun; cihet; mesele; sadet; maksad; hususiyet; (*elect.*) sorti; (oyun) puvan. **at all ~s,** her cihetle, her bakımdan: **to be on the ~ of doing stg.,** bir şeyi yapmak üzere olm.: **to be to the ~,** (söz) isabetli olm., yerinde olm.: **beside the ~,** sadedden haric; yersiz: **to come to the ~,** sadede gelmek: **at the ~ of death,** ölmek üzere iken: **to give ~s to ...,** ···e taş çıkarmak: **figures that give ~ to his argument,** iddiasını takviye eden rakamlar: **the ~s of a horse,** *etc.*, at vs.nin bedenî vasıfları: **the case in ~,** bahis mevzuu olan mesele: **in ~ of fact,** hakikatte: **in ~ of numbers,** sayıca: **the ~ of a joke,** bir nüktenin inceliği: **to make a ~,** bir noktayı isbat etm.: **to make a ~ of ...,** ···e bilhassa dikkati çekmek: **I make a ~ of being in bed by eleven,** saat on birde muhakkak yatarım (buna ehemmiyet veririm): **off the ~,** sadedden haric: **rash to the ~ of madness,** çılgınlık derecesinde atılgan: **railway ~s,** demiryolunu makası: **what's the ~ of doing this?,** bunu yapmakta ne mana var?: **two ~ five,** iki virgül beş. **~-blank,** ufkî ateş edilmiş: **to fire at s.o. ~,** birine çok yakından ateş etm.: **to ask s.o. ~,** ağzında gevelemeden birdenbire sormak: **to refuse ~,** kat'i olarak reddetmek. **~-duty, policeman on ~,** muayyen bir yerde vazife gören polis. **~-to-~ (race)** kırda yapılan manialı at koşusu. ~[2] *v.* Sivriltmek; bir noktaya çevirmek, tevcih etm.; göstermek; delâlet etm.; (av köpeği) ferma etmek. **to ~ a moral,** (kıssadan) hisse çıkarmak. **to ~ a wall,** duvar derzetmek. **~ at,** parmak ile göstermek: **to ~ one's stick at stg.,** bir şeyi değnekle işaret etmek. **~ out,** dikkati çekmek; ihtar etm.; belirtmek; göstermek: **may I ~ out that ...?,** şu noktayı hatırlatabilir miyim ki? **~ed,** sivri uclu; dokunaklı, iğneli. **~er** [ˈpôyntə(r)], fermacı av köpeği; iğne, dilcik. **~less** [ˈpôyntlis], manasız; faydasız; beyhude. **~sman,** makasçı.

**poise** [pôyz]. Müvazene, temkin; duruş. Müvazenede tutmak. **to be ~d,** asılmak.

**poison** [ˈpôyzən]. Zehir(lemek). **to take ~,** kendini zehirlemek: **he hates me like ~,** elinden gelse beni bir kaşık suda boğar: ⌜**one man's meat is another man's ~**⌝, birisi için zehir olan şey başkası için iksir olabilir. **~ous,** zehirli; öldürücü; menhus; muzır.

**poke**[1] [pouk]. ⌜**to buy a pig in a ~**⌝, bir şeyi görmeden satın almak. ~[2]. Dürtüş; dirsek vurma. Parmak vs.

ile dürtmek; dürterek sokmak. **to ~ fun at s.o.,** birisile alay etm.: **to give s.o. a ~ in the ribs,** şaka için birinin kaburgalarını parmakla dürtmek: **to ~ a hole in stg.,** dürterek delik açmak: **to ~ one's nose into ...,** ...e burnunu sokmak. **~ about,** kurcalamak, karıştırmak. **~ out, to ~ s.o.'s eye out,** dürterek birinin gözünü çıkarmak: **to ~ one's head out of the window,** başını pencereden uzatmak.

**poker**[1] [ˈpǫukə(r)]. Küskü. ⌜**as stiff as a ~**⌝, baston yutmuş gibi. **~**[2]. Poker. **~-face,** (poker oyuncusu gibi) hislerini hiç belli etmiyen yüz.

**poky** [ˈpǫuki]. (Oda) dar ve âdi; (ev) nohut oda bakla sofa: (iş) ehemmiyetsiz, hakîr.

**Poland** [ˈpǫulənd]. Polonya.

**polar** [ˈpǫulə(r)]. Kutbî. **~ bear,** beyaz ayı: **~ circle,** kutub medarı.

**pole**[1] [pǫul]. Kutub. **~s apart,** aralarında dağ var. **~**[2]. Sırık; direk. **under bare ~s,** (gemi) yelkenler inik olarak. **~-jump,** sırıkla atlama. **P~**, Polonyalı. **~-axe** [ˈpǫulaks], mezbahalarda hayvanları öldürmeğe mahsus topuz. Bu aletle öldürmek.

**polemic** [poˈlemik]. Kalem münakaşası.

**police** [pəˈlîs]. Zabıta, polis. İnzibat altına almak; asayaşi temin etmek. **~man** [–mən], polis, zabıta memuru. **~woman,** kadın polis. **~-court,** sulh mahkemesi. **~-station,** karakol. **~-van,** hapishane arabası.

**policy**[1] [ˈpolisi]. Siyaset; hareket tarzı; tedbir. **a matter of public ~,** umumun [halkın] menfaatini alâkadar eden şey. **~**[2]. Sigorta mukavelesi; poliçe. **to take out a ~,** (bir şeyi, bir yere) sigorta ettirmek.

**polish** [ˈpoliş]. Cilâ, perdah; kundura boyası; parlaklık; nezaket. Parlatmak; cilâlamak, perdahlamak; kundura boyamak; kabalığını gidermek. **~ed,** cilâlı; parlak: **~ manners,** terbiyeli, nazik hal ve tavır: **~ style,** zarif üslûb. **~ off,** silip süpürmek; çabuk bitirmek. **~ up,** parlatmak: **to ~ up one's English,** ingilizcesinin pasını silmek.

**Polish** [ˈpǫuliş]. Polonyalı; leh; lehçe.

**polite** [pəˈlâyt]. Nazik, kibar, terbiyeli. **to do the ~,** vazife vs. icabi nezaket göstermek: **~ society,** terbiyeli muhit. **~ness,** nezaket, terbiyelilik; iltifat.

**politic** [ˈpolitik]. İhtiyatlı, akıllı; kurnaz. **the body ~,** devlet, siyasî cemiyet. **~al** [pəˈlitikl], siyasî; politikaya aid. **~s,** politika, siyasiyat: **to go into ~,** siyasî hayata atılmak: **what are your ~?,** siyasî kanaatleriniz nedir?

**poll** [pǫul]. Baş; rey verme; verilen reylerin sayısı. Seçimde rey vermek. Rey almak; öküz vs.nin boynuzlarını kesmek; ağacın tepesini kesmek. **to declare the ~,** seçimlerin neticesini ilân etm.: **to go to the ~,** seçimde rey vermek: **to head the ~,** seçimde kazanmak. **~ed** [pǫuld], boynuzları kesilmiş (hayvan); tepesi budanmış (ağac).

**pollard** [ˈpoləd]. Tepesi budanmış ağac; boynuzları kesilmiş hayvan.

**poll·en** [ˈpolən]. Tali, gubar. **~inate** [ˈpolinêyt], (çiçek) tenasül tozu yaymak.

**pollute** [poˈlyût]. Kirletmek, telvis etm.

**polo** [ˈpǫulǫu]. Çevgene benziyen bir oyun.

**poltroon** [polˈtrûn]. Korkak adam. **~ery,** korkaklık.

**poly-** [ˈpoli-]. *pref.* Çok .... **~andry** [ˈpoliˌandri], çok kocalılık. **~gamy** [poˈligəmi], çok karılılık. **~glot** [ˈpoliglot], çok dil bilen. **~syllabic** [ˌpolisiˈlabik], çok heceli. **~technic** [poliˈteknik], muhtelif ilimlere şamil; sanat mektebi; mühendis mektebi. **~theism** [poliˈθî·izm], müşriklik.

**pomade** [poˈmâd]. Merhem.

**pomegranate** [pomˈgranit]. Nar.

**pommel** [ˈpʌml]. Yumruklamak.

**pomp** [pomp]. Debdebe, alayiş. **with ~ and circumstance,** büyük merasimle. **~ous** [ˈpompəs], mutantan, debdebeli; azametli; sahte vakar; şatafatlı. **~osity** [–ˈpositi], kendini beğenmişlik, sahte vakarlık; tumturaklı saçma.

**pom-pom** [ˈpompom]. Ufak seri ateşli top.

**pond** [pond]. Havuz; gölcük. **~ life,** durgun su içinde yaşıyan mahluklar.

**ponder** [ˈpondə(r)]. Düşünceye dalmak; düşünüp taşınmak. **~able** [ˈpondərəbl], tartılır. **~ables,** maddî şeyler. **~ous,** hantal; havaleli; can-sıkıcı.

**poniard** [ˈponyəd]. Hançer(lemek).

**pontif·f** [ˈpontif]. Ruhanî reis. **the sovereign ~,** Papa. **~ical** [–ˈtifikl], Papaya mensub; kurumlu. **~icate,**

papalık; ruhanî reislik etm.; tumturaklı cafcaflı sözler söylemek.
**pontoon** [pon'tûn]. Duba; köprü dubası.
**pony** ['pouni]. Küçük at, midilli.
**poodle** ['pûdl]. Kıvırcık tüylü köpek.
**pooh** [pû]. *İstihfaf edatı.* **~-~**, istihfaf etm.; istihfaf ile reddetmek.
**pool**[1] [pûl]. Gölcük; bir nehrin derin ve sakin kısmı; bahçe havuzu. **~**[2]. Kumar masalarında pay sandığı. **~**[3]. Bir merkezde toplamak, birleştirmek.
**poop** [pûp]. Pupa. **to be ~ed,** geminin kıçı büyük bir dalga altında kalmak.
**poor** [puə(r), pô(r)]. Fakir, yoksul; zavallı; değersiz; noksan; kısır; silik; zayıf; bakımsız. **the ~,** fakirler: **the ~ chap,** zavallı; adamcağız: **in my ~ opinion,** benim aciz kanaatime göre: **to have a ~ opinion of s.o.,** birisine pek kıymet vermemek: **a ~ sort of mother,** analar kusuru: **~ you!,** vah zavallı! **~-law,** fakirlere mahallî idarelerce yardım hakkında kanun. **~-relief,** fakirlere yardım. **~house** ['pôhaus], darülaceze. **~ly** ['puəli], çok iyi olmıyarak, hasta. **~ness** ['puənis, pônis], fakirlik; mahsûlsüzlük; eksiklik, değersizlik.
**pop** [pop]. (*ech.*) Pat!, çat!, güm! Patlama sesi. Hafif sesle patlamak: (*sl.*) rehine vermek. **to be in ~,** rehinde olm.: **to go ~,** pat diye patlamak: **to ~ the question,** (*sl.*) evlenme teklifi yapmak. **~-eyed,** pırtlak gözlü. **~-gun,** oyuncak tüfek. **~ in,** girivermek. **~ off,** (*sl.*) nalları dikmek. **~ over [round],** gidivermek. **~ up,** sipsivri çıkmak.
**pop·e** [poup]. Papa; ortodoks papazı. **~ish** ['poupiş], (*köt.*) Papalığa aid; katolik dinine aid. **~ery,** (*köt.*) Papa tarafdarlığı, katoliklik.
**popinjay** ['popincêy]. Züppe.
**poplar** ['poplə(r)]. Kavak.
**poppy** ['popi]. Gelincik; haşhaş.
**popul·ace** ['popyuləs]. Ayak takımı, avam. **~ar** ['popyulə(r)], halka aid; halkın hoşuna gider; herkesin anlıyabileceği tarzda; makbul. **~ error [belief],** umumî hata [kanaat]: **a ~ boy,** herkes tarafından sevilen çocuk: **~ government,** halk hükûmeti. **~arity** [–'lariti], umumî muhabbet; rağbet. **~arize** [–ərâyz], halka sevdirmek; halkın seviyesine indirmek. **~ate** ['popyûlêyt], iskân ettirmek; mamur etmek: **thickly ~ated,** nüfusu kesif. **~ation** [–'lêyşn], nüfus; ahali. **~ous** ['popyuləs], nüfusu çok; kalabalık.
**porcelain** ['pôslin]. Porselen.
**porch** [pôç]. Kapı saçaklığı.
**porcupine** ['pôkyupâyn]. Oklu kirpi.
**pore**[1] [pô(r)]. Mesame. **~**[2], **to ~ over a book,** bir kitaba dalmak: **to ~ over a subject,** bir mevzu üzerinde uzun uzadıya düşünmek.
**pork** [pôk]. Domuz eti. **~er,** genc besili domuz. **~y,** semiz. **~-pie,** domuz etile yapılmış kıymalı börek: **~ hat,** yuvarlak yassı şapka.
**pornography** [pô'nogrəfi]. Açık saçık yazılar.
**porous** ['pôrəs], mesameli, süzgeç gibi.
**porridge** ['poric]. Yulaf unundan yapılmış lâpa.
**port**[1] [pôt]. Liman. **home ~,** menşe limanı: **~ of registry,** geminin kayıdlı olduğu liman: **to put into ~,** limana girmek: ⌜**any ~ in a storm**⌝, başı sıkışan adam ince eleyip sık dokumaz. **~**[2]. Geminin iskele tarafı(na aid). **to go to ~,** iskeleye doğru gitmek. **~**[3]. Lûmbar. **~-hole,** lûmbuz. **~**[4]. Tavır, hal. **~**[5]. Porto şarabı.
**port·able** ['pôtəbl]. Taşınabilir, portatif. **~age** ['pôtic], taşıma, taşınma; nakliye ücreti. **~er** ['pôtə(r)], kapıcı; hamal: **~erage,** hamaliye.
**portal** ['pôtl]. Cümle kapısı.
**portcullis** [pôtkʌlis]. Yukarıdan aşağı kapanan tarak şeklinde kale kapısı.
**Porte** [pôt]. **the Sublime ~,** Babıâli.
**porten·d** [pô'tend]. Delâlet etmek. **~t** ['pôtənt]. İstikbalı gösteren alâmet; şer alâmeti; manalı alâmet. **~tous** [–'tentəs], uğursuz; mucize kabilinden.
**portfolio** [pôt'fouliou]. Evrak çantası; portföy; vekillik. **minister without ~,** devlet bakanı.
**portion** ['pôşn]. Hisse, pay; nasib; mikdar, parça; bir tabak yemek. **to ~ out,** paylaştırmak, taksim etmek.
**portly** ['pôtli]. Şişman; heybetli.
**portmanteau** [pôt'mantou]. Bavul.
**portra·it** ['pôtrêyt]. İnsan resmi, tasvir. **to have one's ~ taken,** fotograf çektirmek: **to sit for one's ~,** resmini yaptırmak. **~y** [pô'trêy], birinin resmini yapmak; tasvir etm. **~yal,** tasvir etme.
**Portug·al** ['pôtyugəl]. Portekiz. **~uese** [·-'gîz], Portekizli, portekizce.

**pose** [pǫuz]. Tavır, hal; sahte tavır. Muayyen bir vaziyet aldırmak; ileri sürmek, sormak. Sahte bir tavır takınmak. **to ~ as a doctor,** doktorluk taslamak: **without ~,** samimî. **~r,** müşkül bir sual; poz alan: **to give [set] s.o. a ~,** birine çetin bir sual sormak. **~ur** [pǫuˈzö(r)], sahte tavırlı.

**posh** [poş]. (*sl*). Pek şık, gösterişli. **to ~ oneself up,** giyinip kuşanmak.

**position** [poˈzişn]. Vaziyet, hal; mevzi; yer; durum; ictimai seviye; mevki; vazife. Yerini tayin etmek.

**positive** [ˈpozitiv]. Müsbet; kat'i; aşikâr; muhakkak; emin; fazla emin. Pozitif. **~ sign,** zaid, toplama işareti (+): **a ~ miracle,** tam bir mucize.

**posse** [ˈposi]. Müfreze, takım.

**possess** [pǫuˈzes]. ···e malik olm., ···in sahibi olm., tasarruf etmek. **all I ~,** varım yoğum: **to be ~ed by fear,** korkuya kapılmak: **to be ~ed with an idea,** (yanlış) bir fikre kapılmak: **to ~ oneself,** kendini tutmak: **to ~ oneself of,** zabtetmek, ele geçirmek: **to ~ one's soul in peace,** başını dinlemek. **~ed,** *a.*, perili; mecnun. **~ion** [pǫuˈzeşn], tasarruf; mülk, mal; müstemleke. **~s,** mal, servet; zatî eşya. **to be in ~ of,**··· e malik olm.: **to be in the ~ of s.o.,** (bir şey) birinin elinde olm.: **in full ~ of his faculties,** aklî melekelerine tamamen hâkim: **to get [take] ~ of stg.,** bir şeyi elde etm., zabtetmek: **to remain in ~ of the field,** muharebe meydanına hâkim olm.: **house to be sold with vacant ~,** derhal tahliye edilecek satılık ev. **~ive** [pǫuˈzesiv], tesahübkâr. **~ case,** muzafıileyh: **~ pronoun,** mülki zamir. **~or** [pǫuˈzesə(r)], mal sahibi; tasarruf eden.

**possib·ility** [ˌposiˈbiliti]. İmkân; ihtimal; kabil olma; mümkün şey. **to allow for all ~ilities,** her ihtimalı düşünmek: **if by any ~ I do not come,** eğer her hangi bir sebeble gelmiyecek olursam: **I cannot by any ~ be there in time,** vaktinde orada olmama imkân yoktur: **the proposal has ~ies,** teklifin muvaffakiyeti ihtimalı yok değildir. **~le** [ˈposibl], mümkün; muhtemel; olur, olabilir; belki. **as far as ~,** mümkün mertebe: **it is just ~ I may not come,** gelmemem de imkânsız değildir: **it is just ~ to live on £200 a year,** senede 200 lira ile kıtakıt yaşamak mümkündür: **what ~ reason have you to refuse this post?,** ne diye bu vazifeyi reddettin, Allah aşkına?

**possum** [ˈposəm]. **to play ~,** ölmüş gibi yapmak.

**post**[1] [pǫust] *n.* Kısa direk; kazık; sütun. **to go to the ~,** yarışa girmek: **to be left at the ~,** (yarışın başında) geride kalmak: **to win on the ~,** yarışın son dakikada [at başı farkla] kazanmak: **winning ~,** (bir yarışta) bitiriş direği. **~**[2] *v.* **~ (up),** direk üzerine veya umumî bir yere dikmek; yapıştırmak: '**~ no bills!**', buraya ilân yapıştırılmaz! **~**[3]. Posta; postahane. Posta ile göndermek; posta kutusuna atmak; posta arabası ile [menzil ile] seyahat etmek. **there has been a general ~ among the staff,** memurlar arasında esaslı bir değişiklik oldu: **to open one's ~,** mektublarını okumak: **by return of ~,** gelecek posta ile. **~man,** postacı. **~mark,** posta damgası. **~master** posta müdürü: **~ General,** İngiltere Posta ve Telgraf Bakanı. **~mistress,** kadın posta müdürü. **~-boy,** tatar, postiyon. **~-chaise,** tatar arabası. **~-free,** posta ücreti olmıyarak. **~-haste,** alelacele, müstacelen. **~**[4]. **~ (up),** yevmiye defterindeki hesabları ana deftere geçirmek; tam malûmat vermek. **to ~ oneself up in a matter,** bir mesele hakkında malûmat edinmek: **to keep s.o. ~ed up,** birini vaziyet vs.den daima haberdar etmek. **~**[5]. Memuriyet, vazife; (*mil.*) nokta, karakol. Bir yere koymak, yerleştirmek; tayin etmek. **to die at one's ~,** vazife başında ölmek: **to take up one's ~,** vazifeye başlamak. **~**[6]. **Last ~,** yat borusu. **to sound the last ~ (over the grave),** bir askerin cenazesinde mezar başında merasim icabı 'yat borusu' çalmak.

**post-** *pref.* Sonraki; ···den sonra gelen.

**post·age** [ˈpǫustic]. Posta ücreti. **~-age-stamp,** posta pulu. **~al,** postaya aid. **~card,** kartpostal.

**post-date** [ˈpǫustˈdêyt]. (Çek vs.ye) muahhar tarih atmak.

**poster** [ˈpǫustə(r)]. Duvar ilâni; yafta.

**posteri·or** [posˈtiəriə(r)]. Sonraki; muahhar; arkadaki. Kıç; arka. **~ty** [posˈteriti], nesil; gelecek nesiller.

**postern** [ˈpostö(r)n]. Arka veya yan kapı.

**posthumous** [ˈpostyuməs]. Ölümünden sonra; babasının ölümünden sonra doğmuş; müellifin ölümünden sonra neşredilmiş.
**post-mortem** [ˌpoustˈmôtəm]. Fethimeyyit.
**postpone** [pousˈpoun]. Tehir etm., tecil etm.; başka zamana bırakmak. **~ment,** tehir, tecil.
**postscript** [ˈpouskript]. (*abb.* **P.S.**), Mektub haşiyesi; derkenar.
**postulate** [ˈpostyulêyt]. Mevzu, lâzım olan şart. Şart koymak.
**posture** [ˈpostyuə(r)]. Tavır. Vücude vaziyet vermek. **to ~ as ...,** ... takınmak, taslamak.
**post-war** [ˈpoustˈwô(r)] *a.* Harb sonrası.
**posy** [ˈpouzi]. Çiçek demeti.
**pot** [pot]. Çömlek; kavanoz; kab; saksı; testi; lâzımlık; kanyot; (*sl.*) mükâfat kupası. Kavanoza koymak; (*sl.*) tüfek ile vurmak. **a big ~,** (*sl.*) kodaman: **to go to ~,** suya düşmek, iflâs etm.: **to have a ~ [~s] of money,** altın babası olm.: **to keep the ~ boiling,** (i) geçimini kazanmak; (ii) sohbetin soğuyup tavsamasını önlemek: ⌜**the ~ called the kettle black**⌝, ⌜tencere tencereye dibin kara demiş⌝: **~s and pans,** kab kacak. **~-bellied,** karnı şişkin. **~-boiler,** sırf para kazanmak için yazılan yazı. **~-hole,** yol çukuru. **~-house,** meyhane. **~-hunter,** sırf mükâfat kazanmak için müsabakalara giren oyuncu. **~-luck, come and take ~ with us,** bize yemeğe buyurun, ne çıkarsa bahtınıza. **~-shot, to take a ~ at stg.,** bir av vs.ye rasgele ateş etm.; bir kere talihini denemek.
**potato,** *pl.* **-oes** [poˈtêytou]. Patates.
**poten·cy** [ˈpoutənsi]. Kuvvet; tesir. **~t,** kuvvetli, dokunaklı. **~tate,** hükümdar.
**potential** [pəˈtenşl]. Bilkuvve mevcud; muhtemel; (*gram.*) iktidarî. **~it·y** [–ˈaliti], bilkuvve mevcud kuvvet: **situation full of ~ies,** her türlü imkânlara müsaid vaziyet.
**pother** [ˈpoθə(r)]. Karışıklık.
**potion** [ˈpouşn]. İçilecek ilâc; şerbet.
**potsherd** [ˈpotşö(r)d]. Kırık çömlek parçası.
**pottage** [ˈpotic]. Koyu çorba.
**potter**[1] [ˈpotə(r)]. Çömlekçi. **~y,** çömlekçilik; çömlek fabrikası. **the Potteries,** Staffordshire kontluğunun başlıca çömlekçilik ile meşgul mıntakası. **~**[2] *v.* **~ (about),** ufak tefek şeylerle uğraşmak; sinek avlamak. **to ~ along,** acele etmeden yürümek.
**potty** [ˈpoti]. (*sl.*) Değersiz; kolay; kaçık.
**pouch** [pauç]. Torba, kese; (gözlerin altında) sarkık et. Derceb etm.
**poult·erer** [ˈpoultərə(r)]. Tavuk vs. satıcısı. **~ry** [ˈpoultri], kümes hayvanları: **~-yard,** kümes.
**poultice** [ˈpoultis]. Lâpa (koymak).
**pounce** [pauns]. **~ on [upon],** üstüne atılmak, saldırmak; pençelemek.
**pound**[1] [paund]. Lira; ingiliz lirası; libre, funt [= 0·454 gram.]. **~**[2] *v.* Havanda dövmek; ufalamak; mütemadiyen vurmak; çarpmak. **the boat was ~ing on the rocks,** gemi kayalara çarpıyordu: **to ~ along,** (insan) güm güm basarak yürümek. **~age** [ˈpaundic], lira başına komisyon, gümrük resmi vs. **-~er** [paundə(r)], ... librelik.
**pour** [pô(r)]. Dökmek; yağdırmak; boşaltmak. Şiddetle yağmur yağmak; sel gibi akmak. **~ off,** bir mayii bir kabdan başka bir kaba dökmek. **~ out,** dökmek; boşaltmak: **to ~ out one's heart,** kalbini açmak. **~ing** [ˈpôrin(g)] *a.*, **~ rain,** sel gibi yağmur: **a ~ wet day,** çok yağmurlu gün.
**pout** [paut]. Dudak bükme(k).
**poverty** [ˈpovəti]. Fakirlik; yoksulluk; kıtlık. **~-stricken,** sefalet içinde.
**powder** [ˈpaudə(r)]. Toz; pudra; barut. Toz haline getirmek; pudra sürmek. **to keep one's ~ dry,** her ihtimale karşı hazır bulunmak: **to smell ~ for the first time,** ilk defa muharebeye girmek. **~y,** toz gibi.
**power** [ˈpauə(r)]. Kudret; kuvvet; iktidar; devlet; salâhiyet; kabiliyet. **a ~ of,** (*sl.*) çok: **the ~s that be,** âmir vaziyetinde bulunanlar: **it is beyond my ~,** elimde değil; buna muktedir değilim: **his ~s are failing,** (ihtiyarlıktan vs.) melekeleri zayıflıyor: **to exceed one's ~s,** salâhiyetini aşmak: **to come into ~,** (bir parti) iktidara geçmek: **to fall into s.o.'s ~,** birinin eline düşmek: **to give s.o. full ~s,** birine tam salâhiyet vermek: **the Great P~s,** Büyük Devletler: **to have s.o. in one's ~,** birini avucunun içinde tutmak: **more ~ to him [to his elbow]!,** Allah gücünü arttırsın! **-~ed,** ... kudretli; ... takatli. **~less,**

kuvvetsiz, âciz: **they are ~ in the matter,** bu hususta bir şey yapamazlar. **~-house,** kuvvet merkezi; elektrik santralı.

**pow-wow** [ˈpauwau]. Müzakere (etmek), toplantı (yapmak).

**pp.** (*abb.*) **pages,** sahifeler.

**p.p.** (*abb. Lât.*) **per pro(curationem),** vekâleten.

**practica·ble** [ˈpraktikəbl]. Tatbiki, icrası mümkün; geçilir; makul. **~l** [ˈpraktikl], amelî; becerikli; hesabını kitabını bilir; kullanışlı; elverişli; tatbikî. **~ joke,** el şakası: **~ example,** müşahhas misal. **~lly,** tatbikat itibarıyle; hemen hemen: **~ none,** hemen hiç.

**practice** [ˈpraktis] *n.* Nazariye karşılığı; amel, tatbikat; ameliyat; âdet, usul; tecrübe, idman, alışma; hareket tarzı; (hekim, avukatın) müşterilerinin mecmuu ve çalıştığı mıntaka. **in ~,** bilfiil, hakikatte, tatbikatta: **to do stg. for ~,** bir işi alışmak için yapmak: **to make a ~ of doing stg.,** bir şeyi âdet edinmek: **to talk a language well needs a lot of ~,** bir lisanı iyi konuşmak çok pratiğe bağlıdır: **out of ~,** idmansız; pratiğini kaybetmiş: **⌜~ makes perfect⌝,** yapa yapa (boza) öğrenilir: **this doctor [lawyer] has a large ~,** bu hekimin [avukatın] çok müşterisi var.

**practise** *v.* Yapmak, icra etm., tatbik etm.; meşketmek. İdman etmek; doktorluk, avukatlık etmek. **to ~ a deceit,** hile yapmak: **⌜~ what you preach!⌝,** amelin kavline uysun! **~d** *a.* tecrübeli; mahir.

**practitioner** [prakˈtişənə(r)]. **medical ~,** doktor: **general ~,** ihtisası olmıyan doktor.

**pragmat·ic(al)** [pragˈmatik(l)]. Hodbin; mütehakkim. **~ism** [ˈpragmətizm], maddilik; ukalâlık.

**prairie** [ˈpreəri]. Vâsi düz ve ağacsız çayırlık.

**praise** [prêyz]. Öğme; medih. Medhetmek; tazim etmek. **beyond all ~,** ne kadar medhetsem azdır: **to sing the ~s of,** göklere çıkarmak: **to sing [sound] one's own ~s,** kendini öğmek: **to speak in ~ of s.o.,** birinden sitayişle bahsetmek, öğmek. **~worthy** [–wö(r)ði], medhe lâyik.

**pram** [pram]. (*abb.*) **perambulator** *q.v.*: küçük bot.

**prance** [prâns]. Hoplamak. **to ~ about,** şuraya buraya sıçramak.

**prank** [pran(g)k] *n.* Yaramazlık, şeytanlık. **to play ~s on s.o.,** birine azizlik yapmak.

**prate** [prêyt]. Dem vurmak.

**prattle** [ˈpratl]. Çoçuk gibi konuşma(k).

**prawn** [prôn]. Büyük karides.

**pray** [prêy]. Dua etm.; çok rica etm., yalvarmak. **I ~ he may soon return,** inşallah yakında yine döner: **~ be seated!,** buyurunuz oturunuz!: **he's past ~ing for,** (i) artık ümid yok; (ii) ıslâh kabul etmez: **and what do you want, ~?,** siz ne istiyorsunuz, Allah aşkına? **~er** [ˈpreə(r)], dua; yalvarma; istida: **to say one's ~s,** duasını etmek.

**pre-** [prî] *pref.* Önceden.

**preach** [prîç]. Va'zetmek. **to ~ to s.o.,** birine uzun uzadıya nasihat vermek: **to ~ the Gospel,** İncili [hıristyanlığı] neşretmek. **~er,** vaiz. **~ify** [–ifây], uzun uzadıya nasihat vermek.

**preamble** [ˈprîambl]. Mukaddeme, önsöz.

**precarious** [priˈkeəriəs]. Güvenilemez; tehlikeli. **a ~ living,** kifayetsiz kazanc.

**precaution** [priˈkôşn]. İhtiyat(lı tedbir). **as a ~,** ne olur ne olmaz. **~ary,** ihtiyatî.

**precede** [priˈsîd]. Önünden gitmek, öne geçirmek; takaddüm etm.; takaddüm hakkı olmak. **~nce** [ˈpresidəns], takaddüm (hakkı): **to have [take] ~ of s.o.,** birinin önüne geçme [takaddüm] hakkına malik olm.: **this matter takes ~ of all others,** bu mesele hepsinden daha mühimdir. **~nt** [ˈpresidənt] *n.* misal, emsal; teamül: **according to ~,** emsali gibi, usule göre: [priˈsîdənt] *a.* evvelki, mukaddem.

**precept** [ˈprîsept]. Kaide, usul, düstür.

**precinct** [ˈprîsin(g)kt]. Mukaddes bir yerin çevresi. **~s,** etraf, daire.

**precious** [ˈpreşəs]. Kıymetli; nadide; yapmacıklı: (*sl.*) çok; canım. **he took ~ good care not to go there again,** bir daha oraya gitmemeğe son derece dikkat etti.

**precipi·ce** [ˈpresipis]. Uçurum, yar, **~tous** [priˈsipitəs]. Uçurum gibi; sarp, dik.

**precipit·ance, -cy** [priˈsipitəns, -si]. Acelecilik; atılma; düşünmeden davranma. **~ate** *a.* [–tit], acul; atılgan;

düşünmeden yap(ıl)an; a ~ **retreat,** paldır küldür ricat. *n.* Tortu; çöküntü. *v.* [–têyt], Tacil etm., pek çabuk neticeye vardırmak; çöktürmek. ~**ation** [–ˈtêyşn], acele, atılma; tortulanma; yağış.

**precis** [ˈprêysî]. Hulâsa; fezleke.

**precis·e** [priˈsâys]. Kat'i; tam; muayyen; titiz. **at two o'clock ~ly,** tam (elifi elifine) saat ikide: ~**ly so!,** tamam! ~**ion** [priˈsijn], kat'iyet; vuzuh, açıklık. ~ **instruments,** hassas aletler.

**preclude** [priˈklûd]. Menetmek; meydan vermemek.

**precoc·ious** [priˈkoușəs]. Vaktinden evvel yetişmiş; çabuk inkişaf etmiş. ~**ity** [–ˈkositi], vakitsiz olgunluk.

**preconce·ive** [ˌprîkənˈsîv]. Önceden ve tedkik etmeden bir fikir edinmek. ~**d idea,** peşin hüküm. ~**ption** [ˌprîkənˈsepşn], peşin hüküm.

**preconcert** [ˌprîkənˈsö(r)t]. Önceden müşavere edip karar vermek.

**precursor** [priˈkö(r)sə(r)]. Öncü; haberci, işaret.

**predatory** [priˈdêytəri]. Yırtıcı; yağmacı; tamahkâr.

**predecease** [ˌprîdiˈsîs]. ···den evvel ölme(k).

**predecessor** [ˈprîdisesə(r)]. Selef.

**predestin·e** [priˈdestin]. Takdir etmek. ~**ation** [–ˈnêyşn], kaza ve kader; takdir.

**predicament** [priˈdikəmənt]. Fena hal; müşkül vaziyet. **to be in an awkward ~,** 'aşağı tükürsem sakalım, yukarı tükürsem bıyığım' vaziyetinde olmak.

**predicate** [ˈpredikət] *n.* Müsned; (*gram.*) haber. *v.* [–ˈkêyt]. Kaziyede hüküm ve isnad etmek.

**predict** [priˈdikt]. Önceden haber vermek; kehanette bulunmak. ~**ion** [–ˈdikşn], kehanet, önceden haber verme: **your ~ came true,** dediğiniz çıktı. ~**or** [–tə(r)], uçaksavar toplarının ateş edecekleri noktayı tayin eden alet.

**predilection** [ˈprîdiˈlekşn]. Tercih; meyil. **to have a ~ for stg.,** bir şeyi tercih etm.

**predispos·e** [ˌprîdisˈpouz]. Müsaid bir hale getirmek. **to be ~ed to do stg.,** önceki bir vaziyeti göze alarak bir şeyi yapmağa mütemayil olmak. ~**ition** [–ˈzişn], istidad, kabiliyet; (bir hastalığa) müsaid olma.

**predomin·ant** [prîˈdominənt]. Üstün, galib; ekseriyeti olan. ~**ate,** üstün olm.; ekseriyet teşkil etmek.

**pre-eminent** [prîˈeminənt]. Üstün; mümtaz.

**preen** [prîn]. Kuş gibi gagasıyla tüylerini taramak. **to ~ oneself,** kendine çeki düzen vermek.

**preface** [ˈprefis]. Mukaddeme (yapmak), önsöz (yapmak).

**prefect** [ˈprîfekt]. Büyük ingiliz mekteblerinde bazı imtiyaz ve salâhiyetler verilen kıdemli çocuk; (Fransa vs.de) vali.

**prefer** [priˈfö(r)]. Tercih etm., hoşlanmak; tayin etm., terfi ettirmek; ileri sürmek. ~**able** [ˈprefrəbl], tercih edilir. ~**ence** [ˈprefrəns], tercih: **to have a ~ for stg.,** bir şeyi tercih etm.: **I have no ~,** bence hepsi bir: **in ~,** tercihen: ~ **share,** imtiyazlı hisse. ~**ential** [prefəˈrenşl], tercih hakkı olan, tercih edilen; imtiyazlı. ~**ment** [priˈfö(r)mənt], terfi; bir memuriyete tayin edilme.

**prefix** [ˈprîfiks] *n.* Ön ek. *v.* [priˈfiks]. Önüne koymak.

**pregnan·t** [ˈpregnənt]. Gebe; yüklü; manidar. ~ **with consequences,** neticeler doğuracak olan. ~**cy,** gebelik.

**prehistoric** [ˌprîhisˈtorik]. Tarihten önceki.

**prejud·ge** [prîˈcʌc]. Tedkik etmeden hüküm vermek. ~**ice** [ˈprecudis], peşin hüküm; sebebsiz beğen(me)me; tarafgirlik; zarar. Zarar vermek. **to have a ~ against,** ···e karşı peşin hükmü olm.: **without ~,** bütün hakları mahfuz kalarak: **without ~ to anyone,** kimseye zarar vermeden. ~**iced,** peşin hüküm besliyen; zarar görmüş. ~**icial** [–ˈdişl], zararlı, muzır.

**prelate** [ˈprelət]. Büyük rütbeli rahib.

**preliminar·y** [priˈliminəri]. Mukaddeme olarak; başlangıc kabilinden. Başlangıcda yapılan şey. ~**ies,** başlangıc.

**prelude** [ˈprelyud]. Asıl hadiseye başlangıc mahiyetinde olan şey; mukaddeme tarzında olan vaka. Mukaddeme gibi bir şey söylemek, yapmak.

**premature** [ˈprematyuə(r), ˌprîməˈtyuə(r)]. Mevsimsiz; vaktinden evvel; (**child**) vaktinden evvel doğmuş. **to be ~,** vaktini beklemeden yapmak; acele etmek.

**premeditate** [prîˈmeditêyt]. Önceden

kararlaştırmak; önceden kasdetmek, taammüden yapmak. **~d,** kasdî.

**premier** [ˈpremyə(r)]. Kıdemli; baştaki. Başvekil.

**premise** [ˈpremis] *n.* Kaziye.

**premises** [ˈpremisiz]. Bir evin yahud dükkânın odaları ve arazisi. **no one allowed on the ~,** buraya girilmez: **to see s.o. off the ~,** birini kapı dışarı etmek.

**premium** [ˈprîmyəm]. Mükâfat; ikramiye; sigorta ücreti. **to be at a ~,** çok aranılmak: **to put a ~ on (idleness,** *etc.***),** (tembelliği vs.) teşvik etmek.

**premonit·ion** [ˌprîmoˈnişn]. Önceden hissetme. **~ory** [-ˈmonitəri], ihtar eden: **~ symptoms,** hastalığı haber veren âraz.

**prenatal** [prîˈnêytl]. Doğumdan evvel.

**preoccup·y** [prîˈokyupây]. Zihnini işgal etm. **~ation** [-ˈpêyşn], zihin meşguliyeti: **one's chief ~,** en büyük endişe. **~ied,** fikri dağınık.

**prepaid** [ˈprîpêyd]. Peşin olarak ödenmiş.

**prepar·ation** [ˌprepəˈrêyşn]. Hazırlama; tertib; (mektebde) hususî olarak çalışma. **to make (one's) ~s for stg.,** bir şey için tertibat almak. **~tory** [priˈparətəri], hazırlayıcı, ihzarî: **~ to doing stg.,** başlamadan evvel: **~ school,** 'public school' lere hazırlayan ilk mekteb. **~e** [priˈpeə(r)], hazırlamak, tertib etm. Hazırlanmak. **to ~ s.o. for a bad piece of news,** birini fena bir habere alıştırmak. **~ed,** hazır: **to be ~ to ...,** ...göze almak.

**preponder·ant** [priˈpondərənt]. Üstün. **~ate,** üstün gelmek, daha nüfuzlu olmak.

**preposition** [prepəˈzişn]. Atıf veya cer edatı.

**prepossess** [ˌprîpəˈzes]. Celbetmek, gönlünü çekmek. **~ing,** cazibeli.

**preposterous** [priˈpostərəs]. Akıl almaz.

**prerequisite** [prîˈrekwizit]. Önceden lâzımgelen (şey).

**prerogative** [priˈrogətiv]. İmtiyaz; hususî hak.

**presage** [ˈpresic] *n.* İstikbale aid alâmet; fal. *v.* [priˈsêyc]. Alâmet olm.; istikbalden haber vermek.

**presbyterian** [ˌprezbiˈtiəriən]. İskoçya protestan kilisesine aid.

**prescri·be** [prisˈkrâyb]. Emretmek; tenbih etm.; (*med.*) tavsiye etm., reçete vermek. **~bed, ~ task,** verilen muayyen vazife: **within the ~ time,** tayin edilen müddet zarfında. **~ption** [priˈskripşn], reçete. **~ptive,** örf ve adetle yerleşmiş.

**presence** [ˈprezəns]. Hazır bulunma, huzur; vekar. **in the ~ of,** huzurunda, hazır bulunduğu halde: **to have a good ~,** vakur ve heybetli olm.: **he makes his ~ felt,** varlığını etrafına hissettiriyor: **~ of mind,** soğukkanlılık: **saving your ~,** haşa huzurdan: **your ~ is requested,** hazır bulunmanız rica olunur.

**present**[1] [ˈprezənt] *n.* & *a.* Hazır; mevcud; şimdiki. Şimdiki zaman; halihazır. **at ~,** simdiki halde; bu anda: **at the ~ time,** bu zamanda: **to be ~,** hazır bulunmak: **for the ~,** şimdilik: **up to the ~,** şimidiye kadar: **the ~ writer,** bu satırların muharriri. **~**[2] *n.* Hediye, armağan. **~**[3] [priˈzent]. Takdim etm., sunmak; hediye etm.; prezante etm.; arzetmek; ibraz etmek. **to ~ arms,** (*mil.*) selâm vaziyeti almak: **to ~ oneself,** isbatı vücud etm.: **to ~ a pistol at s.o.'s head,** birinin başına tabanca dayamak: **to ~ a play,** bir piyes temsil etm.: **to ~ stg. to s.o.** [**~ s.o. with stg.**], birine bir şeyi hediye etm. **~able** [priˈzentəbl], takdim edilebilir; yakışıklı. **do make yourself ~,** kendine çeki düzen ver. **~ation** [ˌprezenˈtêyşn], arzetme; sunma; hediye; ibraz; huzura çıkma; (bir dava vs.yi) anlatış. **~ copy,** hediyelik nüsha.

**presentiment** [priˈzentimənt]. Önceden hissetme.

**preserv·e** [priˈzö(r)v]. Muhafaza etm., korumak; kurtarmak; idame etm.; konservesini yapmak. Konserve, reçel. **game ~,** av hayvanlarını korumak için ayrılan koru vs. **~ation** [ˈprezəˈvêyşn], muhafaza, koruma; **in a good state of ~,** iyi muhafaza edilmiş. **~ed,** muhafaza edilmiş; konserve halinde: **well** [**badly**] **~ building,** iyi bakıl[ma]mış bina: **well ~ (man),** yaşına göre genc.

**preside** [priˈzâyd]. Riyaset etm., başkanlık yapmak. **to ~ at** [**over**] **a meeting,** bir cemiyete riyaset etmek. **~ncy** [ˈprezidənsi], riyaset, başkanlık. **~nt** [ˈprezidənt], reis; başkan. **~ of the Board of Trade,** İngiliz Ticaret Bakanı. **~ntial** [preziˈdenşl], riyasete aid.

**press**[1] [pres] *n.* Baskı; cendere; mengene; matbaa, basımevi; matbuat, basın; izdiham. **~ of business,** işin çokluğu: **to have a good ~,** gazetelerde iyi geçmek: **in the ~,** tabedilmekte olan. **~**[2] *v.* Bas(tır)mak; sık(ıştır)mak; tazyik etm.; ısrar etm.; (elbise) ütülemek; zorla askerliğe almak. **to ~ forward** [**on**], tacil etm.; acele etm.: **time ~es,** vakit dardır: **to be ~ed for time,** vakti dar olm.: **to ~ into service,** askerliğe almak; zorla kullanmak. **~-box,** gazeteciler locası. **~-cutting,** gazete maktuası. **~-gallery,** Parlamentoda matbuat locası. **~-gang,** eskiden donanma için cebren adam toplamağa memur olan kol. **~-stud,** çıtçıt. **~ed** [prest], sıkışmış; ütülenmiş; daralmış. **to be hard ~,** sıkışık vaziyette olmak. **~ing** [ˈpresin(g)], mübrem; müstacel; mühim; mecburî; sıkıcı. **~man** [ˈpresmən], gazeteci. **~ure** [ˈpreşə(r)], basma; sıkma; baskı; tazyik; cebir. **to bring ~ to bear on s.o.,** birinin üzerinde tazyik icra etm.: **~ of business,** işlerin çokluğu dolayısı ile: **under ~ of necessity,** zaruret sevki ile: **to work at high ~,** hummalı bir şekilde çalışmak. **~-cooker,** buhar çıkarmıyan sıkı kapaklı tencere, düdüklü tencere.

**prestige** [presˈtîj]. Nüfuz, şöhret.

**presto** [ˈprestou]. **hey ~!,** (hokkabazın nidası) 'oldu da bitti maşallah!'; haydi bakalım.

**presum·ably** [priˈzyûməbli]. İhtimal ki; galiba; her halde. **~e,** tahmin etm., ihtimal vermek; addetmek, saymak; haddini tecavüz etm. **I hope I am not ~ing on your kindness,** umarım ki iyiliğinizi suiistimal etmiyorum: **Mr. Smith, I ~!,** yanılmıyorsam siz Mr. S. siniz. **~ption** [priˈzʌmşn], tahmine dayanan hüküm; ihtimal; haddini bilmeyiş. **~ptive,** veraset itibariyle tayin edilmiş: **heir ~,** muhtemel varis: **~ evidence,** ahval ve şeraitten çıkarılan delil. **~ptuous** [–ˈtyuəs], haddini bilmiyen.

**preten·ce** [priˈtens]. Yalandan yapma; bahane; iddia. **false ~s,** sahtekârlık: **he makes no ~ to ...,** hiç bir ... iddiası yoktur: **under the ~ of,** bahanesile. **~d** [priˈtend], yalandan yapmak; ... gibi yapmak; uydurmak; iddiada bulunmak. **to ~ illness** [**~ to be ill**], yalancıktan hastalanmak: **(children) let's ~ to be soldiers!,** askerlik oynıyalım: **he does not ~ to be clever,** zekilik iddiasında değildir: **he's only ~ing,** aslı yok. **~er,** bir tahta hak iddia eden. **~sion** [priˈtenşn], iddia; hak davası; ehliyet iddiası. **a man of no ~s,** iddiasız bir adam. **~tious** [–şəs], iddialı, gösterişli.

**preterite** [ˈpretərit]. Mazi sıygası.

**preternatural** [ˌprîtəˈnatyurəl]. Gayri tabiî.

**pretext** [ˈprîtekst] *n.* Bahane, vesile. **under** [**on**] **the ~ of,** bahanesile. *v.* [–ˈtekst]. Vesile yapmak.

**prettiness** [ˈpritinis]. Güzellik.

**pretty** [ˈpriti]. Güzel (*kadın veya çocuk; erkek ise istihfaf*) hoş, sevimli; oldukça. **here's a ~ mess** [**pass, state of affairs**]**!,** ayıkla pirincin taşını!: **~ much the same,** hemen hemen eskisi gibi. **~-~,** cicili bicili.

**preva·il** [priˈvêyl]. Hüküm sürmek; adet olm.; yenmek. **to ~ upon s.o. to do stg.,** birini ikna etm. **~iling,** hüküm süren; galib; cari. **~lent** [ˈprevələnt]. Hüküm süren, cari.

**prevaricate** [priˈvarikêyt]. Kaçamaklı söz söylemek; yalanla kandırmak.

**prevent** [priˈvent]. Önlemek, menetmek, mâni olm.; meydan vermemek; bırakmamak. **~able,** men'i mümkün. **~ion** [–ˈvenşn], önleme, menetme, mâni olma. **~ive,** önleyici, menedici; mania; hastalığı meneden ilâc veya tedbir.

**previous** [ˈprîvyəs]. Önceki, evvelki, sabık. **~ to ...,** ···den evvel: **~ly,** evvelce, şimdiye kadar.

**prey** [prêy]. Av, şikâr. **to ~ upon,** avlayıp yemek; soymak. **bird of ~,** yırtıcı kuş: **an easy ~,** dişe gelir: **to be a ~ to ...,** ···e duçar olm.: **to ~ upon one's mind,** zihnini kurcalamak.

**price** [prâys]. Fiat, paha. Fiat koymak; kıymetini tahmin etmek. **at any ~,** ne pahasına olursa olsun: **not at any ~,** dünyada, kati'yen hayır: **you can buy it at a ~,** parayı gözden çıkarırsanız alabilirsiniz: **beyond ~,** paha biçilmez: **cash ~,** peşin fiat: **cost ~,** maliyet fiatı: **to set a ~ on s.o.'s head,** birinin başı için para koymak: (*sl.*) **what ~ my new car?,** yeni otomobilime ne dersin bakalım! **~d,** fiat konulmuş: **high-** [**low-**] **~,** yüksek [alçak] fiatlı. **~less,**

paha biçilmez; bulunmaz; (*sl.*) **he's a ~ fellow!**, ömürdür.

**prick** [prik]. (İğne) batmak, sokmak; iğnelemek. İğne, diken batması; hafif yara. **to ~ off names on a list,** bir listedeki isimlere işaret etm.: **to ~ up the ears,** (at) kulaklarını dikmek; (insan) kulak kabartmak: **to kick against the ~s,** beyhude yere kafa tutarak kendine zarar getirmek. **~ing, the ~s of conscience,** vicdan azabı.

**prick·le** [ˈprikl]. Diken, iğnelemek. **~ly,** dikenli; kirpi gibi; (mesele) çapraşık; (insan) titiz, huysuz: **~ heat,** sıcak memleketlerde terden hasıl olan ısılık: **~ pear,** frenk inciri.

**pride** [prâyd]. Kibir, gurur, azamet; iftihar. **to ~ oneself upon stg.,** bir şeyle öğünmek: ⌜**~ comes before a fall**⌝, kibirin sonu fenadır: **false ~,** yersiz izzetinefis: **he is the ~ of his family,** ailesi onunla iftihar ediyor: **proper ~,** izzetinefis: **to put one's ~ in one's pocket [to pocket one's ~, to swallow one's ~],** gururunu yenmek: **to take an empty ~ in doing stg.,** bir şeyden yok yere gurur duymak: **to take ~ in doing stg.,** bir şeyi iyi yapmakla gurur duymak.

**priest** [prîst]. Rahib, papaz. **~hood,** rahiblik, papazlık: **the ~,** papazlar: **to enter the ~,** papaz olmak. **~ly,** papazlığa aid; rahib gibi. **~-ridden,** rahib sınıfının nüfuzunun altında olan (memleket).

**prig** [prig]. Ukalâ, kendini beğenmiş. **~gish,** fazilet taslayan.

**prim** [prim]. Pek soğuk resmî ve tertibli; sunî şekilde pek muntazam (bahçe). **~ and proper,** söz ve kıyafet hususunda pek fazla titiz.

**prima·cy** [ˈprâyməsi]. Başpiskoposluk; üstünlük. **~te** [ˈprâymêyt], başpiskopos.

**prima facie** [ˈprâyma ˈfêyşi·î]. **a ~ case,** ilk delillere göre haklı görünen dava.

**primary** [ˈprâyməri]. Asıl, ana, esaslı, başlıca, en mühim. **~ colours,** ana renkler: **~ education,** ilk öğretim: **~school,** ilkmekteb.

**prime**[1] [prâym]. Birinci; baş(lıca); esaslı; en iyi, seçme. En iyisi; kıvam, tav, olgunluk çağı. **in ~ condition,** mükemmel bir halde: **of ~ importance,** en ehemmiyetli: **in the ~ of life [in one's ~],** hayatın en olgun devresinde: **~ necessity,** en zarurî ihtiyac: **~ minister,** başbakan: **~ number,** aslî sayı: **to be past one's ~,** zamanı geçmiş olmak. **~**[2] *v.* (Tüfek) ağız otunu koymak; boyanın ilk astarını sürmek: (insanın) kulağını doldurmak. **~r** [ˈprâymə(r)], ilk okuma kitabı.

**primeval** [prâyˈmîvl]. En eski, ilk; ibtidaî; dünyanın en eski devirlerine aid.

**primitive** [ˈprimitiv]. İbtidaî; en eski; pek basit, kaba; eski usul.

**primogeniture** [ˌprâymoˈcenityuə(r)]. Ekberiyet. Aynı ana babadan doğan çocuklar arasında en büyüklük.

**primrose** [ˈprimrouz]. Yaban çuhaçiçeği.

**prince** [prins]. Prens; hükümdar; şehzade. **~ of the Church,** kardinal: **~ consort,** kadın hükümdarın kocası olan prens: **~ of Wales,** ingiliz veliahdi: **the ~ of darkness,** İblis. **~ly,** prense aid, lâyık: **a ~ gift,** şahane bir hediye. **~ss,** prenses.

**principal** [ˈprinsipl]. Bellibaşlı; başlıca; en mühim. Baş; müdür; reis; müekkil; sermaye. **~ity** [ˌprinsiˈpaliti], prenslik.

**principle** [ˈprinsipl]. Prensip, kaide; umde. **~s,** erkân; ahlâk kaideleri. **to act up to one's ~s,** prensiplerine sadık kalmak: **first ~,** mebde: **on ~,** prensip itibarile. **~d, high-~,** yüksek prensip sahibi: **low-~,** aşağı prensipli.

**print** [print]. Damga; matbu şey; basma kumaş; emprime; (*fot.*) klişeden basılmış resim; nüsha. Tabetmek, basmak: (*fot.*) klişeden kâğıda geçirmek. **blue-~,** mavi teknik resim kopyası; proje: **in ~,** matbu, basılmış: **he likes to see himself in ~,** yazısını neşredilmiş görmekten hoşlanıyor: **large [small] ~,** büyük [küçük] harfler: **out of ~,** mevcudu kalmamış (kitab): **to rush into ~,** olur olmaz bir şeyi gazetede neşretmek. **~er,** matbaacı; matbaa işçisi: **~'s ink,** matbaa mürekkebi: **~'s reader,** musahhih. **~ing,** matbaacılık; tabetme; basma. **~ing-press,** matbaa makinesi.

**prior**[1] [ˈprâyə(r)]. Sabık, evvel; mukaddemki; kıdemli. **~ to ...,** ...den evvel. **~ity** [prâyˈoriti], evvellik; kıdem, tekaddüm. **~**[2]. Bir manastırın başkeşişi.

**priori** [prâyˈôrây]. **a ~,** kablî.

**prism** [prizm]. Menşur. **~atic** [–ˈmatik], menşurî.

**prison** [ˈprizn]. Hapishane. **to send to ~,** hapsetmek. **~er,** mahpus; esir: **to take ~,** esir almak: **~s' base,** esir kapmaca oyunu. **~-breaker,** hapishane kaçağı.
**pristine** [ˈpristâyn]. Evvelki.
**prithee** [ˈpriði]. (*esk.*) Lûtfen, rica ederim.
**privacy** [ˈprivəsi]. Mahremlik; yalnızlık; halvet. **a desire for ~,** kendi âleminde yaşama arzusu.
**private** [ˈprâyvit]. Hususî; şahsî; zata mahsus; alenî olmıyan; mahrem. Er, nefer. **~ car,** şahsî otomobil (*bir makama bağlı hususî otomobil* = **personal car**): **~ and confidential,** hususî ve mahremdir: **~ education,** hususî tahsil: **~ house,** mesken: **~ income,** şahsî gelir: **~ member,** hükûmete dahil olmıyan meb'us: **~ person,** resmî sıfati olmıyan kimse: **to talk to s.o. in ~,** birisile başbaşa konuşmak: **~ school,** hususî mekteb (*gen.* 'public school' lara talebe yetiştiren ilk mekteb).
**privateer** [ˌprâyvəˈtiə(r)]. Hususî şahıslara aid olup düşman gemilerine hücuma mezun ticaret gemisi; böyle bir geminin kaptan ve tayfası. Böyle bir gemi ile dolaşmak.
**privation** [prâyˈvêyşn]. Mahrumiyet; ihtiyac; yoksulluk.
**privilege** [ˈprivilic]. İmtiyaz; nasib; şeref. İmtiyaz vermek; muaf tutmak. **breach of ~,** imtiyazını suiistimal: **to be ~d to do stg.,** bir şeyi yapmak imtiyazına malik olmak: **it is my ~ to address you here tonight,** bu akşam size hitab etmekle şeref duyuyorum. **~d** *a.*, mümtaz; muaf.
**privy** [ˈprivi]. Abdesthane; *pl.* **~ies,** edeb yerleri. **to be ~ to stg.,** gizli bir şeyden haberdar olm.: **the P~ Council,** Kıraliçenin hususî meclisi: **the P~ Purse,** hazinei hassa: **Lord P~ Seal,** Kıraliçenin mühürdarı.
**prize**[1] [prâyz]. Mükâfat; ikramiye; ganimet; harb ganimeti olarak zabtedilen gemi. Takdir etm., ···e çok itibar etmek. **P~ Court,** deniz musadere mahkemesi: **he's a ~ fool,** (*coll.*) bulunmaz bir ahmaktır: **lawful ~,** meşru ganimet: **~ ox,** (bir ziraat sergisinde) madalya kazanan sığır. **~-fighter,** mükâfat için güreşen veya boks eden kimse; pek kuvvetli bir adam. **~-giving,** mükâfat dağıtma merasimi. **~-money,** zabtedilen bir geminin değerine göre zabteden gemi tayfasına nakden dağıtılan hisse. **~-winner,** bir musabaka veya imtihanda birinci gelen. **~**[2]. Manivelâ; manivelâ kuvveti. Bir şeyi manivelâ ile zorlamak.
**pro**[1] [proụ]. (*abb.*) **professional.** **~**[2] (*Lât.*) **~ forma,** şekle riyayeten, formalite icabı: **~ and con,** lehte ve aleyhte: **~ rata,** mütenasiben: **~ tempore (pro tem.),** muvakkat olarak. **~-** *pref.* ... yerinde; ... lehine. **~-Turkish,** Türk tarafdarı.
**probab·le** [ˈprobəbl]. Kuvvetle muhtemel. **~ility** [–ˈbiliti], ihtimal, muhtemel olma. **~ly,** ihtimal ki; galiba.
**probate** [ˈproụbêyt]. Vasiyetnamenin resmen isbat ve tasdik edilmesi.
**probation** [proˈbêyşn]. Tecrübe devresi; staj; (genc suçluyu) göz hapsine alınmak şartile salıverme: **~ officer,** böyle genclere nezaret eden memur. **~ary period,** tecrübe devresi; staj müddeti. **~er,** stajyer; **'probation'** altındaki çocuk.
**probe** [proụb]. Mil; sonda. Mil ile yoklamak; deşmek.
**probity** [ˈproụbiti]. Dürüstlük; namuskârlık.
**problem** [ˈproblem]. Mesele, dava. **he's a ~ to me,** onu hiç anlıyamam. **~atic** [–ˈmatik], şübheli, meşkûk.
**proce·dure** [proˈsîdyuə(r)]. Muamele, tarz, usul, kaide; gidiş. **~ed** [proˈsîd], ilerlemek; çıkmak; başlamak; davranmak; devam etm.; dava açmak. **to ~ against s.o.,** biri aleyhine dava açmak: **to ~ to blows,** yumruk yumruğa gelmek: **before we ~ any further,** daha fazla ilerlemeden evvel: **how do we ~ now?,** şimdi ne yapacağız [hangi tarafa gidecegiz]?: **to ~ from ...,** ···den ileri gelmek: **the negotiations now ~ing,** devam etmekte olan müzakereler: **to ~ with stg.,** bir şeye devam etmek. **~eding** [proˈsîdin(g)], hareket tarzı; muamele; gidiş: **~s,** harekât, müzakere; merasim; raporlar: **legal ~s,** takibat. **the ~s of a society,** bir cemiyetin müzakere zabıtları: **to take ~s against s.o.,** birine dava açmak. **~eds** [ˈproụsîdz], hasılat; kazanc.
**process** [ˈproụses]. Ameliye; muamele; tarz; usul; amel; metod. Muayyen bir muameleye tâbi tutmak. **during the ~ of construction,** inşaatın devamı esnasında: **it is a slow ~,** bu uzun bir iştir. **in ~ of**

time, zamanla. ~ion [proˈseşn], alay; mevkib; geçid resmi. **to walk in ~,** alay halinde geçmek.
**procla·im** [proˈklêym]. İlân etm., alenen bildirmek. **~mation** [ˌproklәˈmêyşn], ilân; beyanname.
**proclivity** [proˈkliviti]. Meyil; temayül.
**procrastinat·e** [proˈkrastinêyt]. Tehir etm.; işi geciktirmek. **~ion** [–ˈnêyşn], tehir.
**proctor** [ˈproktә(r)]. Üniversite talebelerinin inzibatından mesul olan kimse. **Queen's ~,** ingiliz adliyesinde miras ve boşanma davalarına müdahele hakkı olan memur.
**procure** [proˈkyuә(r)]. Elde et(tir)mek; istihsal etmek.
**prod** [prod]. Dürtüş(lemek), dürtme(k).
**prodigal** [ˈprodigl]. Müsrif; mirasyedi. **to be ~ of [with],** bol bol vermek. **~ son,** uzun bir sefahat hayatından sonra ailesine dönen genc. **~ity** [–ˈgaliti], müsriflik; bolluk.
**prodig·y** [ˈprodici]. Harika; ucube. **~ious** [–ˈdicәs], harikülâde; muazzam.
**produc·e** *n.* [ˈprodyûs]. Mahsul, istihsalât. *v.* [prәˈdyûs]. Hasıl etm., istihsal etm.; meydana getirmek; çıkarmak; ibraz etmek. **to ~ a play,** bir piyesi sahneye koymak. **~r,** müstahsıl; sahneye koyan. **~t** [ˈprodʌkt], mahsul, hasılat; muhassala. **~tion** [proˈdʌkşn], istihsal; imal; mahsul; meydana çıkarma; eser; sahneye koyma; ibraz. **~tive** [proˈdʌktiv], bereketli; müsmir; kazanclı. **~tivity** [–ˈtiviti], mahsuldarlık; mümbitlik.
**profan·e** [proˈfêyn]. Mukaddesata karşı hürmetsiz; dine aid olmıyan. Telvis etmek. **~ity** [–ˈfaniti], mukaddesata karşı hürmetsizlik; küfür.
**profess** [proˈfes]. Açıktan açığa itiraf etmek; iddiada bulunmak; öğretmek. **he ~es to be English,** İngiliz olduğunu iddia ediyor: **I do not ~ to be a scholar,** ben âlim olduğumu iddia etmiyorum. **~ed,** itiraf edilmiş; açıktan açığa (düşman vs.). **~edly,** açıkça, itiraf ettiği gibi. **~ion** [proˈfeşn], meslek; işgüç; beyan. **by ~,** meslek itibarile. **~ional,** meslekî. **~or** [proˈfesә(r)], profesör. **~orial** [ˌprofәˈsôriәl], profesörlüğe aid.
**proffer** [ˈprofә(r)]. Sunmak, arz ve teklif etmek.
**proficient** [proˈfişәnt]. Maharetli; vakif; ehliyetli.
**profile** [ˈproufâyl]. Yandan resim, görünüş. Yandan göstermek.
**profit** [ˈprofit]. Kâr, kazanc; istifade; temettü; menfaat. Kâr getirmek; fayda vermek. **at a ~,** kârla: **to ~ by ...,** ...den istifade etm., faydalanmak: **to make a ~ out of stg.,** bir şeyden kâr etm.: **to turn stg. to ~,** bir şeyden istifade etm. **~able,** kârlı, faydalı, kazanclı. **~eer** [–ˈtiә(r)], muhtekir, vurguncu; ihtikâr yapmak: **war ~,** harb zengini.
**profliga·cy** [ˈprofligәsi]. Sefahat; hovardalık. **~te,** sefih, hovarda.
**prof·ound** [prәˈfaund]. Pek derin. **~ly,** son derece. **~ secret,** çok gizli sır: **~ scholar,** mütebahhir. **~undity** [–ˈfʌnditi], derinlik.
**profus·e** [proˈfyûs]. Pek çok, bol; mebzul; müsrif. **to be ~ of [in],** bol vermek. **~ion** [–ˈfyûjn], bolluk; israf; mebzuliyet.
**progeny** [ˈprocәni]. Evlâd; nesil.
**prognos·is** [progˈnousis]. Hastalığın muhtemel neticesi hakkında mutalaa. **~tic** [progˈnostik], istikbale delâlet eden. **~ticate,** istikbal hakkında haber vermek.
**program(me)** [ˈprougram]. Program.
**progress**[1] *n.* [ˈprougres]. Terakki; ilerleme; gidiş; mühim bir zatın resmî seyahati. **the ~ of events,** hadiselerin cereyanı: **in ~ of time,** zamanla: **the work now in ~,** şimdi yapılmakta olan iş. **~**[2] *v.* [proˈgres]. İlerlemek; terakki etmek. **~ion** [–ˈgreşn], ilerleme. **~ive,** müterakki; adım adım ilerliyen: **in ~ stages,** tecriden.
**prohibit** [proˈhibit]. Yasak etm., mâni olmak. **~ion** [–ˈbişn], memnuiyet, yasak; (*Amer.*) alkollü içkilerin yasak olması. **~ive** [–ˈhibitiv], yasak edici,· menedici. **a ~ price,** yanaşılmaz fiat.
**project** *n.* [ˈprocekt]. Proje; tertib, tasavvur. *v.* [proˈcekt], Niyet etm., tasarlamak; plânını çizmek; fırla(t)mak; irtisam ettirmek. **~ile** [ˈprocektâyl], mermi; atılan şey. **~ing** [–ˈcektin(g)], çıkıntılı. **~ion, atma;** çıkıntı; (harita) irtisam. **~or,** projektör.
**proletar·ian** [ˌproliˈteәriәn]. İşçi sınıfından (kimse). **~iat** [–riat], işçi sınıfı.
**prolific** [proˈlifik]. Velûd, verimli,

mahsullü. **~acy** [–kəsi], velûdluk; bereket.

**prolix** [ˈprǫuliks]. Sözü uzatan.

**prologue** [ˈprǫulog]. Başlangıc.

**prolong** [proˈlon(g)]. Uzatmak, temdit etmek. **~ation** [–ˈgêyşn], uzatma, uzanma. **~ed** *a.*, uzatılmış.

**promenade** [proməˈnâd]. Gezinti; mesire; gezinti mahalli. Gezinmek.

**prominen·ce** [ˈprominəns]. Çıkıntı; tebarüz; ehemmiyet. **to come into ~**, şöhret kazanmak, sivrilmek. **~t**, çıkıntılı; bellibaşlı; mütebariz; mümtaz; şöhretli.

**promiscu·ous** [proˈmiskyuəs]. Karışık; rasgele. **~ity** [–ˈkyûiti], karmakarışıklık.

**promis·e** [ˈpromis]. Vaid; söz. Vadetmek; söz vermek; delâlet etmek. **breach of ~**, evlenme vadini tutmama: **to hold out a ~ to s.o.**, birine bir şeyi vadetmek: **Land of ~**, arzi mevud: **he shows great ~**, kendisinden çok şey ümid edilir: **the plan ~s well**, plân çok ümidli görünüyor: **you'll be sorry for it, I ~ you!**, pişman olacaksın, emin ol! **~ing** [ˈpromisin(g)], ümid verici.

**promontory** [ˈpromontəri]. (Denizde) dirsek, burun.

**promot·e** [proˈmǫut]. Terfi etm.; teşvik etm.; kolaylaştırmak. **to ~ a company**, bir şirket tesis etmek. **~er**, teşvik eden; tesis eden: **company ~**, anonim şirketler kurucusu. **~ion**, terfi.

**prompt** [prom(p)t]. İşini çabuk yapan; tez davranan. Akıl öğretmek; tahrik etm.; suflörlük etmek. **~ly**, tezelden, derhal: **~ delivery**, derhal teslim. **~er**, suflör. **~itude**, çabukluk.

**promulgate** [ˈproməlgêyt]. Neşretmek.

**prone** [prǫun]. Boyluboyuna uzanmış; mütemayil.

**prong** [pron(g)]. Yaba, çatal (dişi).

**pronoun** [ˈprǫunaun]. Zamir. **demonstrative ~**, işaret zamiri: **indefinite ~**, mübhem zamir: **relative ~**, nisbet zamiri.

**pron·ounce** [proˈnauns]. Telâffuz etm.; resmî eda ile söylemek. **to ~ for [in favour of] s.o.**, tarafını tutmak: **to ~ judgement**, (hâkim) hüküm vermek. **~ouncement**, beyan; kararın bildirilmesi. **~unciation** [proˈnʌnsiêyşn], telâffuz.

**proof**[1] [prûf] *n.* Delil; tecrübe; matbaa pruvası; bir içkinin alkol derecesinin ayarı. **to bring [put] stg. to the ~**, bir şeyi denemek: **to give [show] ~ of stg.**, bir şeye delâlet etm.: ⌜**the ~ of the pudding is in the eating**⌝, değeri tecrübe ile anlaşılır; 'Haleb oradaysa arşın burada'. **~**[2] *v.* Mukavim ve bilh. su geçmez hale koymak. *a.* **~ against stg.**, bir şeye dayanıklı. **-proof,** *suff.* ... korkmaz, dayanır, geçmez, tesir etmez, *mes.* **bullet-~**, kurşun işlemez; **water~**, su geçmez, gamsele.

**prop** [prop]. Destek; payanda; rükün; zahir. Yaslamak. **to ~ up**, desteklemek; zahir olmak.

**propagate** [ˈpropəgêyt]. Çoğaltmak; yaymak, neşretmek.

**propel** [proˈpel]. İleriye sürmek; sevketmek. **~lent**, sevkedici madde. **~lor**, pervane.

**propensity** [proˈpensiti]. Temayül.

**proper** [ˈpropə(r)]. Münasib, uygun, lâyık; usule göre: hususî, zatî; hakikî; (terbiye) pek fazla titiz; (*sl.*) yakışıklı. **~ly**, münasib bir şekilde; (*coll.*) enikonu, tamamile. **~noun**, has isim: **London ~**, asıl Londra: **to do the ~ thing by s.o.**, birine karşı vicdanen en doğru olan şeyi yapmak: **he very ~ly refused**, redetti, doğrusu da bu idi.

**propert·y** [ˈpropəti]. Mal, mülk, emlâk; hassa. **~ies**, emlâk; bir tiyatro kumpanyasının elbise, dekor vs.si. **a man of ~**, mal mülk sahibi: **that's my ~**, o benimdir: **that secret is public ~**, o sırrı sağır sultan bile duydu.

**prophe·cy** [ˈprofisi] *n.* Vahye istinaden gaibden haber verme, kehanet. **~sy** [ˈprofisây] *v.*, gaibden haber vermek; kehanette bulunmak: **to ~ rain**, yağmur yağacağını söylemek. **~t** [ˈprofit], peygamber; kâhin; gaibden haber veren kimse. **the P~**, Muhammed: ⌜**no man is a ~ in his own country**⌝, kimsenin kadri kendi muhitinde bilinmez. **~tic** [–ˈfetik], peygamberce; kehanet kabilinden.

**prophylactic** [ˌprofiˈlaktik]. Hastalıktan koruyan (ilâc, tedbir).

**propiti·ate** [proˈpişiêyt]. Gönül almak; yatıştırmak. **~ation** [–êyşn], yatıştırma; kefaret. **~ous** [proˈpişəs]. Müsaid; uğurlu. **the ~ moment**, en uygun zaman.

**proportion** [proˈpôşn]. Tenasüb;

nisbet; hisse, mikdar. **~s,** bir cismin genişlik, uzunluk ve derinliği. Münasib hisselere ayırmak; mikdar tayin etmek. in **~**, nisbetle: **in ~ as ..., ...** nisbeten: **out of ~**, nisbetsiz: **sense of ~**, nisbet mefhumu: **well ~ed,** mütenasib. **~al,** nisbî: **inversely ~**, makûsen mütenasib. **~ate,** mütenasib.

**propos·al** [proˈpǫuzl]. Teklif; evlenme teklifi. **~e,** teklif etm.; niyet etm.; evlenme teklifi yapmak: **to ~ a candidate,** birinin namzedliğini koymak: **to ~ s.o.'s health,** birinin sıhhatine içmeği teklif etm.: ⌜**man ~s, God disposes**⌝, ⌜tedbir bizden takdir Allahtan⌝. **~er,** teklifi yapan. **~ition** [ˌpropǫuˈzişn], teklif; mesele, teşebbüs. **a paying ~**, kârlı iş: **a tough ~**, halli güç mesele: **he's a tough ~**, Allahın belâsı adamdır.

**propound** [proˈpaund]. İleri sürmek.

**propriet·ary** [proˈprâyətəri]. Mal sahibliğine aid. **~ medicine,** reçetesi bir firmanın malı olan ilâc. **~or** [–ˈprâyətə(r)], mal veya mülk sahibi.

**propriet·y** [proˈprâyəti]. Uygunluk. **the ~ies,** terbiye icabları.

**propuls·ion** [proˈpʌlşn]. İleriye yürütme, fırlatma. **~ive,** fırlatıcı.

**prorog·ue** [proˈrǫug]. Muvakkaten tatil etmek. **~ation** [–ˈgêyşn], muvakkaten tatil.

**pros·aic** [prǫuˈzêy·ik]. Şairane mukabele; yavan, alelâde. **~e** [prǫuz], nesir. **~ody,** aruz. **~y,** yavan.

**proscri·be** [prǫuˈskrâyb]. Medenî haklardan iskat etm.; sürgüne yollamak; yasak etm. **~ption** [–ˈskripşn], nefyetme; yasak etme.

**prosecut·e** [ˈprosikyût]. Aleyhine dava açmak; takib etm.; bir işte devam etmek. **~ion** [–ˈkyûşn], adlî takibat; iş takibi. **~or,** davacı: **Public ~**, müddeiumumi.

**proselyte** [ˈprosəlâyt]. Mühtedi, dönme.

**prospect**[1] *n.* [ˈprospekt]. Manzara; görünüş; ümid; ihtimal. **his ~s are brilliant,** istikbali parlaktır. **~ive** [–ˈpektiv], muhtemel; istikbale aid. **~**[2] *v.* [prosˈpekt]. Toprağı ihtiva ettiği madenler bakımından tedkik etmek. **~or,** maden araştıran kimse. **~us** [prosˈpektəs], tarifname; rehber.

**prosper** [ˈprospə(r)]. Muvaffak olm.; refaha ermek; mamur olmak. Kolaylaştırmak, feyz vermek. **~ity** [–ˈperiti], refah mamurluk. **~ous** [ˈprospərəs], mamur, refah içinde; kârlı.

**prostitute** [ˈprostityût]. Fahişe.

**prostrat·e** *a.* [ˈprostrêyt]. Yüzükoyun uzanmış; mecalsiz. *v.* [prosˈtrêyt]. Yere atıp yatırmak; bitkin hale koymak. **to ~ oneseif,** yere kapanmak. **~ion** [–ˈtrêyşn], bitkinlik; secdeye varma.

**protagonist** [prǫuˈtagənist]. Mübariz; (bir piyes vs.de) kahraman.

**protect** [proˈtekt]. Korumak, muhafaza etmek. **~ion** [–ˈtekşn], koruma, himaye, muhafaza; siper. **~ionist,** himaye usulü tarafdarı. **~ive,** koruyucu. **~or,** hâmi. **~orate** [–ˈtektərit], büyük bir devletin himayesi altında bulunan küçük bir devlet.

**protégé** [ˈprotejêy]. Mahmi.

**pro tem** [ˈprǫuˈtem]. (*abb. Lât.*) **pro tempore,** muvakkaten.

**protest**[1] *n.* [ˈprǫutest], İtiraz; protesto(name), beyanname. **to do stg. under ~**, bir şeyi kerhen yapmak. **~**[2] *v.* [proˈtest]. İtirazda bulunmak; protesto etmek; iddia etm.

**protocol** [ˈprǫutokol]. Zabıtname; protokol.

**prototype** [ˈprǫutǫutâyp]. İlk örnek.

**protract** [prǫuˈtrakt]. Uzatmak. **~ed,** sürüncemeli. **~or,** minkale.

**protrud·e** [prǫuˈtrûd]. Dışarı fırlamak, çıkmak. **~ing,** fırlak, çıkıntılı.

**proud** [praud]. Mağrur, kibirli, azametli. **to be ~ of stg.,** ···ile gurur duymak: **to be ~ to do stg.,** bir şeyi yapmakla şeref duymak: **to do s.o. ~**, (*sl.*) birini fevkalâde ağırlamak: **to do oneself ~**, (*sl.*) kendisi için hiç bir şey esirgememek.

**prove** [prûv]. İsbat etm.; delâlet etm.; tahkik etm.; denemek; bulunmak, çıkmak. **what he said ~d to be correct,** söylediği doğru çıktı: **it remains to be ~d,** isbat edilsin bakalım: **to ~ a will,** bir vasiyetnameyi tasdik ettirmek. **~n,** (*yal.*) **not ~**, suçu sabit olmamış.

**provender** [ˈprovendə(r)]. Hayvanlara verilen ot vs.

**proverb** [ˈprovö(r)b]. Darbımesel, atalar sözü. **~ial** [–ˈvö(r)byəl], meşhur.

**provide** [proˈvâyd]. Tedarik etm., vermek; şart koymak. **to ~ against stg.,** bir şeye karşı tedbir almak; **to be ~d for,** ihtiyacları temin edilmek: **to ~ s.o. with stg.,** birine bir şeyi tedarik etm. **~nce** [ˈprovidəns], ihtiyat, takdir; Cenabı Hak. **~nt,** idareli, muktesid. **~ntial** [–ˈdenşl], Allahın hikmet ve takdirine aid; tam vaktinde yetişen. **~r** [proˈvâydə(r)], tedarik eden. **universal ~,** her şey satan mağaza.

**provinc·e** [ˈprovins]. Vilâyet; salâhiyet. **the ~s,** taşra: **that's outside my ~,** o benim salâhiyetim dahilinde değildir. **~ial** [–ˈvinşl], taşraya aid; taşralı; geri, darkafalı.

**provision** [proˈvijn]. Tedarik, tedbir; şart. **~s,** erzak, levazım. Erzak ve levazım vermek. **~ merchant,** erzak satıcı: **within the ~s of the law,** kanunun hükümleri altında: **there is no ~ to the contrary,** aksi hakkında hüküm yoktur: **to make ~ for stg.,** bir şeyi temini için lazımgelen tedbirleri almak: **to make ~ against stg.,** bir şeyin önüne geçmek için tedbir almak: **to make ~ for one's family,** ailenin ihtiyaclarını [istikbalini] temin etm.

**provisional** [proˈvijnl]. Muvakkat; iğreti.

**proviso** [proˈvâyzou]. Şart.

**provocat·ion** [ˌprovoˈkêyşn]. Kışkırtma; tahrik; meydan okuma. **~ive** [–ˈvokətiv], kışkırtıcı; tahrik edici.

**provok·e** [proˈvouk]. Kışkırtmak; tahrik etmek; meydan okumak; sebeb olm.; davet etmek. **to ~ an incident,** hadise çıkarmak: **to be ~d,** darılmak. **~ing,** cansıkıcı; darıltıcı.

**provost** [ˈprovəst]. Kolej müdürü; (İskoçya'da) belediye reisi. **~-marshal** [proˈvou], askerî inzibat âmiri.

**prow** [prau]. Geminin pruvası.

**prowess** [ˈpraues]. Yiğitlik; muvaffakiyet; maharet.

**prowl** [praul]. Sinsi sinsi (av peşinde) dolaşma(k). **to ~ about** [**be on the ~**], fena bir maksadla etrafta dolaşmak.

**prox.** [proks]. (*abb.*) **proximo.** Gelecek ay içinde.

**proxim·ate** [ˈproksimêyt]. Karib, en yakın. **~ity** [–ˈsimiti], yakınlık; civar.

**proxy** [ˈproksi]. Vekâlet; vekil. **by ~,** istinabe suretile.

**prud·e, ~ish** [prûd, -iş]. (Kadın hakkında) iffet hususunda ifrat derecede titiz olan. **~ery** [ˈprûdəri], ifrat derece utangaclık.

**pruden·ce** [ˈprûdəns]. İhtiyat; basiret. **~t,** ihtiyatlı, müdebbir.

**pry** [prây]. **to ~ about,** tecessüs etm., kolaçan etmek: **to ~ into s.o.'s affairs,** başkasının işlerine burnunu sokmak.

**P.S.** [ˈpîˈes]. (*abb. Lât.*) **post scriptum,** mektub haşiyesi.

**psalm** [sâm]. Mizmar; zebur. **~ist,** zebur müellifi, Davud Peygamber.

**pseudo-** [ˈpsyûdou] *pref.* Sahte; sözde, gûya. **~nym** [–nim], müstear ad; takma isim.

**pshaw** [(p)şô]. *Sabırsızlık veya istihfaf nidası.*

**psych·iatry** [(p)sâyˈkâyətri]. Akıl hastalıkları bilgisi. **~ic(al)** [ˈ(p)sâykik(l)], ruha aid; ispritizmeye aid. **~o-analyst** [ˈ(p)sâykouˈanalist], ruhî hal ve hastalıkları tahlil eden kimse. **~ological** [ˌ(p)sâykouˈlocikl], psikolojiye aid. **the ~ moment,** muayyen hal ve şartlar içinde harekete geçmek için en uygun zaman. **~ology** [–ˈkoləci], ruhiyat.

**pt.** (*abb.*) **pint.**

**pte.** (*abb.*) **private (soldier),** orduda erlerin adlarının önüne konulan unvan.

**p.t.o.** [ˈpîˈtîˈou]. (*abb.*) **please turn over!,** çeviriniz!

**pub** [pʌb]. (*abb.*) **public-house,** birahane.

**puberty** [ˈpyûbö(r)ti]. Bülûg, ergenlik çağı.

**public** [ˈpʌblik]. Umumî; herkese aid, hususî olmıyan; alenî. Halk, amme. **in ~,** alenen: **to go out in ~,** ortaya çıkmak: **the general ~** [**the ~ at large**], halkın ekserisi: **~ life,** memuriyet hayatı: **~ money,** milletin parası: **~ opinion,** umumî efkâr: **~ school,** (İngiltere'de) kibar tabakaya mahsus ve başlıca Oxford ve Cambridge üniversitelerine veya orduya hazırlayan lise derecesinde mekteb (*ismine rağmen bu mektebler tamamen hususîdir*): **~ service,** amme hizmeti; memurluk: **the ~ services,** amme hizmetleri: **his life was spent in ~ service,** hayatı halka hizmetle geçti: **his life was spent in the ~ service,** hayatı memurlukta geçti: **~ spirit,** hamiyet: **~ Works,** Nafia. **~-house,** birahane, meyhane. **~an** [ˈpʌblikən], birahaneci. **~ation** [ˌpʌbliˈkêyşn], neşretme; neşriyat. **~ist** [ˈpʌblisist], muharrir. **~ity** [pʌbˈlisiti], alenilik; ilâncılık; (*fig.*) tellallık.

**publish** [ˈpʌbliş]. Neşretmek, yaymak; ifşa etmek. **~er,** naşir.

**pucker** [ˈpʌkə(r)]. Buruşturmak; katlanmak; kıpkırışık olmak. Kırışık.
**puckish** [ˈpʌkiş]. Şakacı, şeytan.
**pudding** [ˈpudin(g)]. Puding. **rice ~,** bir nevi sütlac. **~-face,** ablak surat.
**puddle** [ˈpʌdl]. Su birikintisi; sıvacı çamuru. **to ~ about,** çamura bata çıka yürümek.
**pueril·e** [ˈpyuərâyl]. Çocukça; boş. **~ity** [–ˈriliti], çocukluk, ahmaklık.
**puff** [pʌf]. Nefha, üfleme; püf (böreği); mübalağalı ilân; pudra pomponu. Üflemek, püflemek; çok medhetmek. **to ~ and blow,** nefes nefese olm.: **to ~ oneself up,** kurum satmak: **to ~ out,** kabartmak: **to ~ up,** şişirmek; göklere çıkarmak. **~ed, ~ sleeves,** kabarmış yenler: **~ up,** kurumlu: **~ face,** şişirilmiş yüz. **~-adder,** zehirli bir cins yılan. **~-ball,** kurd mantarı. **~-box,** pudra kutusu. **~-pastry,** yufkalı hamurişi. **~iness** [ˈpʌfinis], şişkinlik; (*med.*) nefha. **~y,** şişkin; püfür püfür esen.
**pug** [pʌg]. Küçük buldog köpeği.
**pug·ilist** [ˈpyûcilist]. Boksör. **~nacious** [pʌgˈnêyşəs], kavgacı.
**pukka** [ˈpʌkə]. Hakikî, halis; sağlam.
**pull** [pul]. Çekmek; cezbetmek; asılmak. Çekiş, çekme; asılış; cezb; bir içim (bira vs.); kürek hamlesi; tutup çekecek şey. **to ~ at stg.,** bir şeyi çekmek: **to ~ a boat,** kürek çekmek: **to give a ~,** bir hamlede çekmek: **to have a ~,** nüfuz sahibi olm., arkalı olm.; (*sl.*) uzun bir yudum içmek: **to have a ~ over s.o.,** başkasına üstün olan bir tarafı olm.: **to ~ to pieces,** parça parça etm.: **to ~ s.o. to pieces,** birini şiddetle tenkid etm. **~ about,** oraya buraya sürüklemek, örselemek. **~ away,** çekip ayırmak, kopartmak. **~ down,** aşağı çekmek; indirmek; yıkmak; sıhhatini bozmak. **~ in,** içeriye çekmek; (atı) durdurmak. **~ off,** çıkarmak; kaldırmak; (*coll.*) muvaffak olmak. **~ on,** üstüne çekmek. **~ out,** çıkarmak; sökmek; (tren) istasyondan çıkmak: (*otom.*) **to ~ out from behind a vehicle,** bir araba vs. nin önüne geçmek için arkasından çıkıvermek. **~ over,** çekip devirmek: (*otom.*) **to ~ over to one side,** kenara çekmek. **~-over,** kazak. **~ round,** ayıl(t)mak; (hasta) iyileş(tir)mek. **~ through,** birini fena bir vaziyetten kurtarmak; (hasta) iyileşmek. **~ to, to ~ the door to,** kapıyı çekerek kapatmak. **~ together,** elbirliğiyle çalışmak: **to ~ oneself together,** kendini toplamak. **~ up,** yukarı çekmek, kaldırmak; sökmek; atı vs. durdurmak; (*fig.*) dizginlerini çekmek.
**pullet** [ˈpulit]. Pilic (dişi).
**pulley** [ˈpuli]. Makara; palanga. **~-block,** makara. **~-wheel,** makara dili.
**pulmonary** [ˈpʌlmonəri]. Akciğere aid.
**pulp** [pʌlp]. Sebzelerle meyvaların eti; kâğıd hamuru; her hangi sulu şekilsiz madde; lâpa; öz. Dövüp lâpa ve hamur gibi yapmak. **crushed to a ~,** ezilmis. **~y,** lâpa gibi; yumuşak.
**pulpit** [ˈpulpit[. Mimber; kürsü. **the influence of the ~,** kilisenin nüfuzu.
**puls·ate** [pʌlˈsêyt]. Nabız gibi atmak. **~e,** nabız; nabız atması. **to feel one's ~,** nabzını yoklamak.
**pulverize** [ˈpʌlvərâyz]. Toz haline getirmek; ezmek.
**puma** [ˈpyûma]. Amerika aslanı.
**pumice** [ˈpʌmis]. **~-stone,** süngertaşı.
**pump** [pʌmp]. Tulumba, pompa. Tulumba ile [gibi] çekmek. **to ~ s.o.,** (*sl.*) birinden malûmat edinmeğe çalışmak: **to ~ out,** tulumba ile çekip kurutmak: **to ~ up,** tulumba ile (su) çekmek; hava basıp (futbol) şişirmek.
**pumpkin** [ˈpʌmkin]. Helvacı kabağı.
**pumps** [pʌmps]. Rugan iskarpinler.
**pun** [pʌn]. Cinas (yapmak).
**punch**[1] [pʌnç]. Zımba; muşta; (*mech.*) nokta. Zımbalamak; yumrukla vurmak; nokta ile işaret etmek. **P~**[2], Polişinel; şişman ve kambur bir kukla: **~-and-Judy, ~ show,** Karagöz oyununa benziyen bir kukla oyunu.
**punctilious** [pʌnkˈtilyəs]. Merasime düşkün.
**punctual** [ˈpʌn(g)ktyuəl]. İşini vaktinde yapan; tam vaktinde; muntazam.
**punctuat·e** [ˈpʌn(g)ktyuêyt]. Noktalamak, işaretle teyid etm. **~ion** [–ˈêyşn], noktalama.
**puncture** [ˈpʌn(g)ktyuə(r)]. Sivri bir şeyle yapılan delik. Delmek. **to have a ~,** (*otom.*) lâstiği delinmek.
**pundit** [ˈpʌndit]. Allâme.
**pungent** [ˈpʌncənt]. Dokunaklı; müessir.
**punish** [ˈpʌniş]. Cezalandırmak, terbiyesini vermek. **~able,** cezaya

müstahak. **~ment,** ceza(landırma): **capital ~,** ölüm cezası.

**punitive** [ˈpyûnitiv]. Cezaya aid. **a ~ force,** tedib kuvveti.

**punkah** [ˈpʌn(g)ka]. Tavana asılı büyük bir yelpaze.

**punt**[1] [pʌnt]. Sırık ile yürütülen dibi düz bir cins kayık (kullanmak). **~**[2], top yere düşmeden ayak ile vurmak. **~er** [ˈpʌntə(r)], kumarcı.

**puny** [ˈpyûni]. Nahif, cılız; âciz.

**pup(py)** [pʌp(i)]. Köpek yavrusu; hoppa delikanlı. **to sell s.o. a ~,** birini kafese koymak.

**pupil**[1] [ˈpyûpl]. Talebe, mektebli. **~**[2]. Gözbebeği, hadeka.

**puppet** [ˈpʌpit]. Kukla.

**purblind** [ˈpö(r)blâynd]. Yarı kör; darkafalı.

**purchase**[1] [ˈpö(r)çis]. Satın alma(k); mubayaa; satın alınan şey. **your life would not have been worth an hour's ~,** bir saatten fazla yaşamazdınız. **~**[2]. Mihaniki kuvvet; cerrieskal; makara; istinad noktası. **to get [secure] a ~ on stg.,** bir şeye dayanmak, istinad et(tir)mek.

**purdah** [ˈpö(r)da]. (Hindistanda) kadınları erkek gözlerinden saklamak için perde.

**pure** [pyûə(r)]. Sâf; halis; afif, sade. **~ mathematics,** nazarî riyaziye. **~-blooded [~-bred],** sâf kan. **~-minded,** afif fikirli.

**purg·ative** [ˈpö(r)gətiv]. Müshil. **~atory,** âraf; ıstırab yeri. **~e** [pö(r)c], müshil (vermek); tasfiye etme(k).

**puri·fy** [ˈpyu̯ərifây]. Tasfiye etm., temizlemek. **~st** [ˈpyu̯ərist], (dil) fasahat meraklısı. **P~tan** [ˈpu̯əritən]. İngiltere'de 17 inci asırda zuhur eden ve İslamiyette Vehabiler gibi din ve ahlâk hususunda pek mutaassıb olan bir tarikate mensub kimse. **~tanical** [–ˈtanikl], din ve ahlâk meselelerinde mutaassıb. **~ty** [ˈpyu̯əriti], sâflık; temizlik; iffet; (dil) fasahat.

**purlieu** [ˈpö(r)lyû]. Hudud. **~s,** civar.

**purloin** [ˈpö(r)lôyn]. Aşırmak.

**purple** [ˈpö(r)pl]. Mor. **born in the ~,** yüksek (*esp.* hükümdar) ailesine mensub: **~ in the face,** pek hiddetlenmiş: **~ passages,** (bir yazıda) parlak sahifeler: **to be raised to the ~,** kardinal tayin olunmak.

**purport** *n.* [ˈpö(r)pôt]. Meal; mefhum; mufad. *v.* [pö(r)ˈpôt], manasında olm.; delâlet etmek. **it ~s to be a letter from X,** (mealine bakılırsa) bu mektubun X. tarafından yazıldığı iddia ediliyor.

**purpose** [ˈpö(r)pəs]. Maksad, niyet, emel, gaye, kasd. Niyet etm., kasdetmek. **for [with] the ~ of ...,** ... niyetiyle: **on ~ [of set ~],** kasden, mahsus: **to answer the ~,** işine uymak: **to answer [serve] several ~s,** muhtelif işlere yaramak: **to come to the ~,** sadede gelmek: **not to the ~,** sadedden haric: **to speak to the ~,** pek yerinde söylemek: **to no ~,** boş yere: **to work to good [some] ~,** iyi ve verimli olarak çalışmak: **to what ~?,** neye yarar?: **a novel with a ~,** tezli roman: **general ~s lorry,** her işe yarar kamyon. **~ly** [ˈpö(r)pəsli], kasden, mahsus.

**purr** [pö(r)]. Kedi mırıltısı. Mırıldamak.

**purse** [pö(r)s]. Para kesesi. **the public ~,** devlet hazinesi: **according to the length of one's ~,** servetine göre: **that is beyond my ~,** o benim harcım değil: ⌜**you can't make a silk ~ out of a sow's ear**⌝, ⌜arık etten yağlı tirit olmaz⌝. **~-proud,** zenginliği ile gururlanan. **~r,** gemi kâtibi ve levazım memuru. **~y,** şişman ve dar nefesli; buruşuk; zengin.

**pursu·ance** [pəˈsyûəns]. İfa, tatbik. **~ant,** mutabik, uygun. **~e** [pəˈsyû], takib etm.; kovalamak; devam etm.; aramak. **~it,** takib, peşinden gitme, kovalama; meşgale; araştırma: **to set off in ~,** takibe koyulmak.

**purvey** [pəˈvêy]. Erzak ve levazım tedarik etmek. **~or,** erzak müteahhidi.

**pus** [pʌs]. Cerahat.

**push**[1] [puş] *n.* İtme; kakma; dürtüş; teşebbüs; girginlik; ilerleme. **at a ~,** icabederse; mübrem vaziyette: **when it comes to the ~,** mesele ciddileşirse. **~**[2] *v.* İtmek, dürtmek, kakmak; sürmek; basmak, sokmak. **to ~ one's advantage,** elde edilen bir menfaatı son haddine kadar istismar etm.: **to ~ oneself (forward),** girginlik etm.: **he does not know how to ~ himself,** kendini satmasını bilmiyor. **~-bicycle,** bisiklet. **~-button,** elektrik düğmesi. **~-cart,** el arabası. **~ in,** itip içeri sokmak [girmek]. **~ off,** kayığı iterek iskeleden uzaklaştırmak: ~ off!, alarga!; (*sl.*) çek arabanı! ~ **on,** ileriye sürmek; ilerlemek: **to ~ on with the work,** işe sebatla devam etmek. ~ **out,** dışarıya itmek; sürmek; çık(ar)mak. ~ **to,** itip kapatmak. ~**ful** [**-ing**], girişken, pişkin; becerikli, girgin.

**pusillanimous** [ˌpyûsiˈlanimǝs]. Korkak.

**puss** [pus]. Kedi; haspa. ~ ~!, pisipisi! ~**sy**(**-cat**), kedi. ~**-in-the-corner,** köşe kapmaca oyunu.

**put** [put]. Koymak, vazetmek, yerleştirmek; ifade etm., arzetmek; tahmin etmek. **I ~ his age at 30,** yaşını 30 tahmin ediyorum: **to ~ a horse at** [**to**] **a fence,** bir atı atlatmak üzere bir maniaya sürmek: **to ~ (their) heads together,** baş başa vurmak: **to ~ into harbour,** limana girmek: **to ~ into English,** ingilizceye çevirmek: **I ~ it to you that ...,** müsaadenizle arzederim ki ...: **to ~ it mildly,** en hafif tabir ile: **to ~ a question,** sual sormak: **to ~ a resolution,** (bir meclis vs.de) bir teklif vermek: **to ~ to sea,** (gemi) alarga etm.: **to ~ s.o. to do stg.,** birine bir şeyi yaptırmak: **you can do anything if you are ~ to it,** insan mecbur olunca her şeyi yapar. ~ **about,** (gemiyi) geriye çevirmek; yaymak. ~ **across,** öbür tarafa geçirmek: **to ~ a deal across,** bir alışverişi muvaffakiyetle tamamlamak: **you can't ~ that across me,** ben bunu yutmam. ~ **away,** saklamak; (para) biriktirmek; (*sl.*) ziftlenmek. ~ **back,** yerine iade etm.; (saati) geri almak; (gemi) geri dönüp limana avdet etmek. ~ **by,** (parayı) saklamak; ayırmak. ~ **down,** aşağı koymak; yere koymak; bırakmak; yazmak; addetmek, isnad etm.; bağışlamak: **to ~ down passengers,** yolcuları indirmek: **to ~ down a rebellion,** bir isyanı bastırmak: ~ **that down to me,** onu benim hesabıma yaz: **I ~ it down to his youth,** gençliğine bağışladım: **I ~ him down as** [**for**] **an Englishman,** onu İngiliz sanıyorum. ~ **forth,** göstermek; sarfetmek; (ileri) sürmek. ~ **forward,** arzetmek; ilerletmek: **to ~ oneself forward,** girginlik etm.: **to ~ one's best foot forward,** adımlarını sıklaştırmak; elinden geleni yapmak. ~ **in,** içeri sokmak; dikmek; koymak; (*jur.*) ibraz etm.: **to ~ in a claim,** *etc.*, istida vs. resmen vermek: **to ~ in a good day's work,** tam bir günlük işi yapıp bitirmek: **to ~ in at a port,** bir limana uğramak: **to ~ in for a post,** bir mevkie namzedliğini koymak. ~ **off,** çıkartmak; tehir etm.; avutmak; tiksindirmek; çelmek; şaşırtmak; (gemi) iskeleden hareket etm.: **to ~ s.o. off doing stg.,** birini bir şey yapmaktan vazgeçirmek, cesaret veya iştah bırakmamak: **to ~ off one's guests,** misafirlere haber göndererek daveti tehir etm.: **to be ~ off stg.,** bir şeyden soğumak: **don't ~ me off!,** beni şaşırtma! ~ **on,** üzerine koymak; giymek; ilâve etm.; takınmak: *a.* yapma(cık). **to ~ the clock on,** saati ileri almak: **to ~ the radio on,** radyo işletmek: **to ~ the light on,** ışığı yakmak: **to ~ on a train,** bir treni servise koymak: **to ~ a play on,** bir piyesi sahneye koymak: **to ~ s.o. on to a job,** birine bir iş vermek: **who ~ you on to it?,** bunu size kim gösterdi?: ~ **me on to Oxford 120,** (telefon) Oxford 120'ı veriniz. ~ **out,** dışarı koymak; çıkarmak; uzatmak; söndürmek; yanıltmak; darıltmak; bozmak; **to ~ one's arm out,** (i) kolunu uzatmak; (ii) kolu çıkmak: **to ~ s.o.'s eyes out,** birinin gözlerine mil çekmek: **he was very ~ out,** fena bozuldu; pek dargındı: **he was not in the least ~ out,** hiç istifini bozmadı: **to ~ oneself out for s.o.,** birisi için zahmet çekmek: **don't let me ~ you out,** size zahmet vermeyim: **to ~ s.o. out in their reckoning,** birinin hesabını bozmak: **to ~ money out to interest,** parayı faize yatırmak: **all work is done on the premises, nothing is ~ out,** bütün iş dairede yapılır, dışarı verilmez. ~ **through,** iyi bir neticeye götürmek: **he ~ his foot through the ice,** ayağı buzu deldi: ~ **me through to the director,** (telefon) bana müdürü veriniz. ~ **together,** bitiştirmek, birleştirmek; çatmak; kurmak; birbirine katmak. ~ **up,** yukarı koymak; yüksekçe bir yere yapıştırmak; kurmak, inşa etm.; kaldırmak; ileri sürmek; namzedliğini koymak; namzed göstermek; istif etm.: **to ~ up at a place,** bir yerde konaklamak: **to ~ s.o. up (for the night),** birini evinde yatırmak: **to ~ up a bird,** *etc.*, avda bir kuş vs. kaldırmak: **to ~ up a fight,** karşı koymak: **to ~ up one's hands,** ellerini kaldırmak; teslim olm.: **to ~ up the**

**money for an undertaking,** bir teşebbüs için para yatırmak: **to ~ s.o. up to a thing,** aklına koymak: **to ~ stg. up for sale,** bir şeyi satışa çikarmak: **to ~ up an umbrella,** şemsiye açmak: **to ~ up with,** tahammül etm.; nazını çekmek. **~-up job,** danışıklı dövüş. **~ upon,** üstüne koymak: **to be ~ upon,** kendini ezdirmek: **I won't be ~ upon,** kimseyi enseme bindirmem; yağma yok!

**putative** [ˈpyûtətiv]. Meşru farzedilir.

**putr·efy** [ˈpyûtrifây]. Çürümek; taaffün etm. **~efaction** [–ˈfakşn], çürüme. **~id,** ufunetli, çürük; iğrenc.

**puttee** [ˈpʌti]. Dolak.

**putty** [ˈpʌti]. Camcı macunu.

**puzzle** [ˈpʌzl]. Muamma; şaşırtmaca; bilmece. Muamma gibi gelmek; düşündürmek. **to ~ stg. out,** düşünüp taşınarak bir şeyin içyüzünü keşfetmek: **to ~ over stg.,** bir şeyin üzerine zihnini yormak.

**pygmy** [ˈpigmi]. Cüce.

**pyjama** [piˈcâmə]. Pijama.

**pylon** [ˈpâylon]. Sütun, pilon.

**pyramid** [ˈpirəmid]. Ehram.

**pyre** [pâyə(r)]. **funeral ~,** ölünün cesedini yakmak için toplanan odun yığını.

**pyrotechnics** [ˌpâyroˈtekniks]. Hava fişekçiliği.

**Pyrrhic** [ˈpirik]. **a ~ victory,** (hakikatte bir felâket olan) zahirî muzafferiyet.

**python** [ˈpâyθon]. Kocaman yılan. **~ess** [ˈ–es], falcı kadın.

# Q

**Q** [kyû]. Q harfi.

**qua** [kwêy]. (*Lât.*) Mahiyetinde.

**quack**[1] [kwak]. (*ech.*) Ördek sesi. **~**[2]. Doktor taslağı; şarlatan. **~ery,** mutatabbiblik.

**quadr-** *pref.* Dört ..., dörtlü. **~angle** [ˈkwodran(g)gl], dört dılılı şekil; (*abb.* **quad**) bir mekteb veya üniversitenin büyük iç avlusu. **~ilateral** [ˌkwodriˈlatərəl], dört dılılı (şekil). **~uped** [ˈkwodruped], dört ayaklı. **~uple** [kwoˈdrûpl], dört kat; dört misli(ne çıkarmak). **~uplets** [kwoˈdrûplits], bir batında doğmuş dört çocuk.

**quaff** [kwaf]. Büyük yudumlarla içmek.

**quagmire** [ˈkwagmâyə(r)]. Batak(lık); çamur.

**quail**[1] [kwêyl]. Bıldırcın. **~**[2] *v.* Ürkmek, yılmak.

**quaint** [kwêynt]. Tuhaf; eski moda fakat hoş.

**quak·e** [kwêyk]. Titremek, deprenmek. **to ~ in one's shoes,** korkudan tiril tiril titremek. **~ing,** titreme; raşe. **Q~r** [kwêykə(r)]. İngiltere'de bir mezheb sâliki.

**quali·fication** [ˌkwolifiˈkêyşn]. Vasıf(landırma); ehliyet; şart, ihtiyat; tahdid. **without ~,** kayıdsız. **~fy** [ˈkwolifây], tavsif etm., vasıflandırmak; tadil etm.; bazı istisna iradıyla tahdid etm.; ehliyet vermek; hafifletmek. Ehliyet kesbetmek. **to ~ as a doctor,** (imtihan verip) doktor olm.: **to ~ s.o. for stg.,** birini bir iş veya vazife için ehliyetli kılmak: **to ~ oneself for a post,** bir iş için lâzım gelen şart ve vasıfları ihraz etmek. **~fied,** ehliyetli, ehliyetname sahibi; salâhiyetli; mahdud, şartlı: **to be ~ to do stg.,** lâzım gelen vasıfları haiz olm.: **~ approval,** şarta bağlı [mahdud] takdir.

**qualit·ative** [ˈkwolitətiv]. Keyfiyet ve mahiyete aid. **~y** [ˈkwoliti], keyfiyet; hassa; sıfat; mahiyet; vasıf. **a person of ~,** kibar bir adam: **a wine of ~,** iyi bir cins şarab.

**qualm** [kwâm]. Mide bulantısı; vicdan üzüntüsü; endişe. **to have no ~s about doing stg.,** bir şeyi yapmaktan hiç çekinmemek.

**quandary** [ˈkwondəri]. Müşkül vaziyet.

**quanti·tative** [ˈkwontitətiv]. Kemiyete aid. **~ty** [ˈkwontiti], mikdar; kemiyet; çokluk; bir hecenin uzunluğu. **a great ~ of ...,** pek çok: **in great ~ties,** çok mikdarda: **~-surveyor,** yapı işlerinde malzeme muhammini.

**quarantine** [ˈkworəntîn]. Karantina (ya koymak).

**quarrel** [ˈkworəl]. Kavga; bozuşma; ağız kavgası. Bozuşmak, kavga etmek. **to ~ with one's bread and butter,** kendi ekmeği ile oynamak: **I have no ~ with his behaviour,** hareketine diyeceğim yok: **to pick a ~ with s.o.,** birisile kavga aramak: **to take up s.o.'s ~,** bir kavgada birinin tarafını tutmak. **~some,** kavgacı.

**quarry**[1] [ˈkwori]. Avlanan hayvan.

~². Taş ocağı(ndan taş çıkarmak); (*fig.*) araştırmak. **~man,** taş ocağı amelesi.

**quart** [kwôt]. Bir istiab hacmı ölçüsü ki gallon'un dörtte biridir = 1·136 litre.

**quarter¹** [ˈkwôtə(r)]. Aman; af, merhamet. **to cry** [**ask for**] **~,** aman dilemek: **to give ~,** aman vermek. ~² *n.* Dörtte biri; çeyrek; üç aylık devir; kamerin terbii; (*abb.* **qr.**) 28 libre = 12·7 kilo; (*Amer.*) 25 sent; cihet; mahalle, semt; muhit: **~s,** ikametgâh, yatacak yer; (*naut.*) tayfaya tahsis olunan yer. *v.* Dört müsavi kısma taksim etm.; muhtelif evlere asker yerleştirmek. **to be ~ed with** [**on**] **s.o.,** (asker) birinin evine yerleştirilmek: **at close ~s,** yakından; **from all ~s,** her taraftan: **all hands to ~s!,** (harb gemisinde) herkes yerine!: **to beat** [**pipe**] **to ~s,** gemide harb hazırlığı emrini vermek: **to take up one's ~s with ...,** gidip ...in yanına yerleşmek: **we shall get no help from that ~,** o taraftan yardım görmiyeceğiz: **a ~'s rent,** üç aylık kira. **~-day,** İngiltere'de üç aylık kira veya maaşların tediye edildiği günler (25 mart. 24 haziran, 29 eylûl, 25 aralık). **~-deck,** geminin kıç güvertesi: **the ~,** deniz subayları. **~ly** [ˈkwôtəli], üç aylık, her üç ayda bir (neşredilen, tediye edilen). **~master** [–mâstə(r)], levazım subayı: **~ General,** bütün İngiliz ordusunun levazım reisi.

**quartet(te)** [kwôˈtet]. Dört kısımlık musiki parçası ve bunu çalan sanatkârlar.

**quartz** [kwôts]. Kuvars.

**quash** [kwoş]. Nakzetmek, ibtal etmek.

**quasi-** [ˈkwêysi] *pref.* Hemen hemen; yarı.

**quatrain** [kwoˈtrêyn]. Dört mısralık şiir parçası.

**quaver** [ˈkwêyvə(r)]. Sekizlik nota; ses titreme(k).

**quay** [kî]. Rıhtım.

**queasy** [ˈkwîzi]. Midesi çabuk bulanır.

**queen** [kwîn]. Kıraliçe; muhteşem kadın; (iskambilde) kız; (satrancda) ferz [vezir]. Ferz çıkmak. **~ly,** kıraliçe gibi, kıraliçeye yakışır. **~-bee,** arı beyi. **~-mother,** valide kıraliçe.

**queer** [kwiə(r)]. Tuhaf, acayib; keyifsiz; ne idüğü belirsiz ve biraz şübheli. Bozmak, altüst etmek. **I'm feeling rather ~,** bir hoşluğum var: **he's a ~ fish,** o bir âlem: **on the ~,** şübheli bir şekilde: **to be in ~ street,** parasız kalmak: **to ~ s.o.'s pitch,** el altından birinin tertibatını bozmak.

**quell** [kwel]. Bastırmak; teskin etmek.

**quench** [kwenç]. Söndürmek; gidermek.

**querulous** [ˈkweryuləs]. Sızlayıcı; mütemadiyen şikâyet eden.

**query** [kwiəri]. Sual, istifham (işareti). Sormak; şübhe etmek.

**quest** [kwest]. Araştırma.

**question** [ˈkwestyən]. Sual, sorgu; mesele, bahis mevzuu; şübhe. Sual sormak; şübhe etmek. **to be ~ed,** sorguya çekilmek: **to call in ~,** ...den şübhe etm.: **beyond (all) ~** [**past ~**], hiç şübhe yok: **in ~,** bahis mevzuu olan: **there is no ~ of his being dismissed,** işinden çıkarılması bahis mevzuu değildir: **out of the ~,** bahis mevzuu olamaz; mümkün değil: **to put a ~,** sual sormak: **to put the ~** (**to a meeting,** *etc.*), meseleyi reye koymak: **to raise a ~ about stg.,** bir şey hakkında itirazda bulunmak: **there was some ~ of ...,** ... mübhem bir şekilde bahis mevzuu idi: **without ~,** şübhesiz. **~able,** su götürür, şübheli. **~naire** [–ˈneə(r)], sualler listesi. **~-mark,** istifham işareti.

**queue** [kyû]. Bekliyen halk dizisi, kuyruk. **to ~ (up),** sıra olmak.

**quibble** [ˈkwibl]. Kaçamaklı söz. Kılı kırk yarmak.

**quick** [kwik]. *a.* Çabuk, tez, süratlı; çevik; kavrayışlı; canlı, diri. *n.* Canlı et. **the ~ and the dead,** canlılar ve ölüler: **to bite one's nails to the ~,** tırnaklarını (etine) kan oturuncaya kadar ısırmak: **to cut to the ~,** içine zehir gibi işlemek. **~en,** canlandırmak; çabuklaştırmak. **~lime,** sönmemiş kirec. **~ness,** çabukluk, sürat; çeviklik. **~sand,** deniz kıyılarındaki müteharrik kumlar. **~-silver,** civa. **~-firing,** seri ateşli. **~-tempered,** çabuk öfkelenen. **~-witted,** çabuk anlıyan.

**quid** [kwid]. (*coll.*) İngiliz lirası.

**quid pro quo** [ˈkwidprouˈkwou]. (*Lât.*) Mukabele.

**quiescent** [kwâyˈesənt]. Sakin; haretsiz.

**quiet** [ˈkwâyət]. Sakin; rahat; durgun; sessiz; hafif (ses); uslu; kendi halinde; gösterişsiz. Rahat, sükûnet. ~ *veya* **quieten,** teskin etm., yatıştırmak: ~ **down,** yatışmak. **be ~!,** sus!, rahat dur!: **to keep ~,** susmak; uslu oturmak; rahat dur(dur)mak: **to keep stg. ~,** bir şeyi örtbas etmek. **~ness,** [–nis], sessizlik, sükûnet, rahat.

**quietus** [kwây·îtəs]. **to give s.o. his ~,** öldürmek.

**quill** [kwil]. Kuş kanadının büyük tüyü; tüy kalem. **~-driver,** fena yazıcı.

**quilt** [kwilt]. Yorgan. Elbise veya yorgan gibi şeylerin içine pamuk koymak.

**quince** [kwins]. Ayva.

**quintessence** [kwinˈtesens]. Hulâsanın hulâsası.

**quintet** [kwinˈtet]. Beş çalgı ile çalınan parça veya bunu çalan sanatkârlar.

**quintuple** [kwinˈtyûpl]. Beş misli.

**quip** [kwip]. Dokunaklı cevab; nükteli söz.

**quirk** [kwö(r)k]. Kaçamaklı söz; yazıda süs.

**quit** [kwit]. Bırakmak, terketmek. Azade. **to ~ hold of,** salıvermek: **notice to ~,** kiracıya evin tahliyesi için ihbar: **~ you like men!,** (*esk.*) yiğitce davranınız! **~ter** [ˈkwitə(r)]. (*Amer. sl.*) hain; kaba soğan.

**quite** [kwâyt]. Bütün bütün; tamamen; pek çok. **~ good,** oldukça iyi. **~ right,** pek yerinde; tamam; doğru: **~ so!,** elbette!, şübhesiz!: **I don't ~ know,** pek iyi bilmiyorum.

**quits** [kwits]. **to be ~,** fit olm.: **I'll be ~ with him,** ben ondan acısını çıkarırım: **to cry ~,** 'artık yetişir' demek: **double or ~,** ya mars ya fit.

**quiver**[1] [ˈkwivə(r)]. Ok kuburu; tirkeş. **~ful,** kubur dolusu (ok): **a ~ of children,** bir ailenin kalabalık çocukları. ~[2]. Titreme(k); kıpırdamak.

**quixotic** [kwikˈsotik]. Don Kişot gibi idealist ve hayalperest.

**quiz** [kwiz]. Alaya almak; alaycı tecessüs ile bakmak; rasgele sualler sorarak bilgisini yoklama(k). Alaycı. **~zical,** tuhaf; alaycı.

**quoit** [kwôyt]. Kaydırmak. **~s,** yere çakılı bir kısa kazığa halkaları atıp geçirmekten ibaret bir oyun.

**quondam** [ˈkwondam]. Evvelce mevcud olan.

**quorum** [ˈkwôrəm]. Nısab.

**quota** [ˈkwoutə]. Hisse, pay.

**quot·ation** [kwouˈtêyşn]. İktibas; nakil, fiat tayini. **~e** [kwout], iktibas etm.; zikretmek; tırnak işareti koymak: **to ~ a price,** fiat tayin etmek.

**quoth** [kwouθ]. (*esk.*) Dedim, dedi vs.

**quotient** [ˈkwouşənt]. (Hesabda) harici kısmet.

**q.v.** [ˈkyûˈvî]. (*abb. Lât.* **quod vide.**) Buna bak!

# R

**R** [â(r)]. R harfi. **the three R's; reading, (w)riting, (a)rithmetic,** okuma, yazma, hesab, (ilk tahsilin esasları).

**R.A.** [âˈrêy]. (*abb.*) (i) **Royal Artillery,** Topçu sınıfı; (ii) **Royal Academy** veya **Academician,** Güzel Sanatler akademisi veya Akademi âzası.

**rabbi** [ˈrabây]. Haham.

**rabbit** [ˈrabit]. Ada tavşanı; korkak; acemi. **welsh ~,** üzerindeki peynirle birlikte kızartılmış ekmek.

**rabble** [ˈrabl]. Ayak takımından ibaret olan kalabalık. **the ~,** ayak takımı.

**rabi·d** [ˈrabid]. Kudurmuş; azgın. **~es** [ˈrêybi·îz], kuduz.

**rac·e**[1] [rêys]. Irk; nesil. **the human ~,** insan cinsi, beni Âdem. **~ial** [ˈrêyşyəl]. Irka aid, ırkî. ~[2]. Yarış, koşu; hızlı med akıntısı. Koşuya girmek, yarış etm.; alabildiğine koşmak, pek hızlı hareket etm.; bir makineyi pek hızlı işletmek. **~-horse,** yarış atı. **~-meeting,** at yarışı toplantısı. **~ing** [ˈrêysin(g)], yarışlar, *esp.* at yarışları. Pek hızlı koşan, akan. **~ing-stable,** yarış atları müessese.

**rack** [rak]. Parmaklık raf; yemlik; silâhlık; (*mech.*) eski bir işkence aleti. Bu alet ile insanın uzuvlarını germek. **to be on the ~,** işkence çekmek: **to ~ one's brains,** zihnini yormak: **to be ~ed with pain,** şiddetli acı duymak: **to go to ~ and ruin,** mahvolmak; iflâs etm.: **luggage ~,** file. **~-railway,** dişli demiryolu. **~-rent,** pek yüksek kira.

**racket**[1] [ˈrakit]. Raket. ~[2]. Samata; cümbüş; dolandırıcılık. **to kick up a ~,** gürültü yapmak: **to stand the ~,** acısını çekmek; dayanmak; masrafını

karşılamak. ~**eer** [–ˈtiə(r)], anaforcu, muhtekir.
**raconteur** [rakontö(r)]. Hikâyeci.
**racy** [ˈrêysi]. ~ **anecdote** [**style**], canlı ve orijinal hikâye [üslûb].
**raddle** [ˈradl]. ~**d face,** kabaca makiyajlanmış çehre.
**radia·l** [ˈrêydyəl]. Şua gibi; merkezden çıkan. ~**nce** [ˈrêydyəns]. Parlaklık, nur; şaşaa; şua halinde intişar eden şey. ~**nt,** parlak, şualar neşreden; güler yüzlü. Işık veya hararetin çıktığı nokta. ~**te** *a.* [ˈrêydiət], şua halinde; yıldız şeklinde. *v.* [–ˈêyt]. Şualar neşretmek; merkezî bir noktadan yay(ıl)mak; neşe saçmak. ~**tion** [–ˈêyşn], şualar neşretme; şua intişarı; radyasyon; merkezî bir noktadan yayılma. ~**tor,** radyatör.
**radical** [ˈradikl]. Köke aid; radikal; esasî. Radikal partisine mensub; kök.
**radio** [ˈrêydiou]. Radyo. ~**gram,** radyo telgraf.
**radish** [ˈradiş]. Turp.
**radium** [ˈrêydiəm]. Radyom.
**radius,** *pl.* **-dii** [ˈrêydiəs, -iây]. Nısıfkutur. ~ **of action,** gemi veya uçağın seyir siası: **within a** ~ **of three miles,** üç mil çevresinde.
**R.A.F.** [âˈrêyˈef]. (*abb.*) **Royal Air Force,** İngiliz Hava Kuvvetleri.
**raffia** [ˈrafyə]. Bir cins lif sicim.
**raffish** [ˈrafiş]. Rezil kabadayı, hovarda.
**raffle** [ˈrafl]. Eşya piyangosuna koymak.
**raft** [râft] Sal. ~**sman,** salcı.
**rafter** [ˈrâftə(r)]. Çatı kerestesi.
**rag¹** [rag]. Paçavra; değersiz şey. **in** ~**s and tatters,** lime lime: **to feel like a wet** ~, gevşek ve bitkin bir halde olm.: **meat cooked to a** ~, fazla pişirilerek parça parça olmuş et. ~**-and-bone man,** paçavracı. ~**-doll,** paçavradan yapılmış kukla. ~**-fair,** bit pazarı. ~**-picker** [**-man**], paçavracı. ~**-tag and bobtail,** esafil güruhu. ~**amuffin** [ˈragəmʌfin], baldırı çıplak; külhanbeyi. ~**ged** [ˈragid], partal; pejmürde; yırtık, lime lime; intizamsız. ~². Kaba şaka; gürültü. Gürültü ve muziplikle eğlenmek.
**rag·e** [rêyc]. Öfke, hiddet, şiddet; heves, ibtilâ. Köpürmek, kudurmak, tehevvür etm.; (rüzgâr) şiddetle esmek; (veba) salgın halinde hüküm sürmek. **to** ~ **at** [**against**] **stg.,** bir şeye köpürmek: **to be all the** ~, pek rağbette olm.: **to be in a** ~, çok öfkeli olm.: **to fly into a** ~, küplere binmek: **to have a** ~ **for stg.,** bir şeyin delisi olmak. ~**ing** [ˈrêycin(g)], kudurmuş; şiddetli, hiddetli; çılgın.
**ragout** [raˈgû]. Yahni.
**raid** [rêyd]. Akın (etm.); baskın (yapmak). ~**er,** akıncı; düşmanın ticaret gemilerini tahrib eden harb gemisi.
**rail¹** [rêyl]. Tırabzan üstü; küpeşte (üstü); bir bahçe etrafına konulan çitin demir (tahta) parçaları; demiryolu (tramvay) rayı; (*abb.*) **railway,** demiryolu. **to leave the** ~**s,** raydan çıkmak: **to get off the** ~**s,** raydan çıkmak; çığrından çıkmak: **live** ~, elektrikli ray: **price on** ~, trene teslim fiatı. ~**ing** [ˈrêylin(g)], (*gen.*) ~**s,** parmaklık(lı çit). ~**road** [ˈrêylroud], (*Amer.*) *v.* **railway.** ~**way** [ˈrêylwêy], demiryolu, şimendifer: ~**man,** demiryolu amelesi, memuru. ~². Şikâyet etm.; küfretmek; dil uzatmak. ~**lery** [ˈrêyləri], alaya alma; takılma.
**raiment** [ˈrêymənt]. Elbise.
**rain** [rêyn]. Yağmur (yağmak). Yağdırmak. **it's** ~**ing cats and dogs,** bardaktan boşanırcasına yağmur yağıyor: ⌜**it never** ~**s but it pours**⌝, felâket (*bazan* saadet) yalnız gelmez. ~**bow,** alaimisema. ~**coat,** yağmurluk. ~**drop,** yağmur damlası. ~**fall,** yağış; bir yerde muayyen bir zamanda yağan yağmur mikdarı. ~**proof,** yağmur geçmez. ~**y,** yağmurlu: **one must put by for a** ~ **day,** ⌜ak akçe kara gün içindir⌝.
**raise** [rêyz]. (Yukarı) kaldırmak; yükseltmek; dikmek; inşa etm.; yetiştirmek; çıkarmak; ileri sürmek; meydana getirmek; hasıl etm.; ayaklandırmak; (para, vergi) toplamak. **to** ~ **an army,** bir ordu toplamak: **to** ~ **a cry,** çığlık koparmak: **to raise s.o. from the dead,** bir ölüyü diriltmek: **to** ~ **hopes,** ümidlendirmek: **to** ~ **an objection,** itirazda bulunmak: **to** ~ **a ship,** batmış gemiyi yüzdürmek: **to** ~ **a smile,** dinliyenlerde tebessüm uyandırmak: **to** ~ **s.o.'s spirits,** birinin maneviyatını yükseltmek.
**raisin** [ˈrêyzn]. Kuru üzüm.
**raison d'être** [ˈrezonˈdetr]. (*Fr.*) Hikmeti vücud.

**raj** [râc]. Hakimiyet. **~ah** [ˈrâca], raca.

**rak·e**[1] [rêyk]. Hovarda; sefih adam. ~[2]. Tırmık; bahçevan tarağı. Tırmıklamak; taramak. **to ~ one's memory,** hafızasını yoklamak: **to ~ a ship,** bir gemiyi baştan başa topa tutmak. **~ in,** (gazinoda mizleri) tırmık ile toplamak: **to ~ in money,** çok para kazanmak. **~ off,** bir işte gayri meşru bir tarzda para almak. **~ out the fire,** ateşi söndürmek için kömürlerini çıkarmak. **~ over,** (toprağı) tırmalamak. **~ up,** eşelemek, kurcalamak. **~ish** [ˈrêykiş], çapkınca, serbest.

**rally**[1] [ˈrali]. Yeniden intizama girme(k); kuvvetlenme(k); canlanma(k); sıhhat kazanma(k); yeniden toplama(k); yeniden ihya etmek; (tenis) üstüste bir kaç vuruş. ~[2]. Takılmak, alaya almak.

**ram** [ram]. Koç; harb gemisini mahmuzu; şahmerdan. Vurarak pekiştirmek; zorla tıkıştırmak; sokmak; (gemi) pruvasile başka gemiye çarpmak.

**rambl·e** [ˈrambl]. Maksadsız dolaşma(k). Avare ve başıboş gezinme(k); ipsiz sapsız konuşmak (yazmak). **~er,** başıboş gezen adam; sarmaşık gülu. **~ing,** başıboş gezen; rabıtasız.

**ramif·y** [ˈramifây]. Dallanıp budaklanmak; şubelere ayrılmak. **~ication** [–fiˈkêyşn], dallanıp budaklanma; feri; şube.

**ramp**[1] [ramp]. Hafif meyil. ~[2]. Desise, pek yüksek fiat; anafor. ~[3]. Şahlanmak. **to ~ and rage,** kıyamet koparmak, küplere binmek. **~age** [ramˈpêyc], ~ *veya* **go on the ~,** cinleri tutmak; oraya buraya koşarak çılgınca gürültü yapmak. **~ageous,** gürültücü. **~ant** [ˈrampənt], şaha kalkmış; müfrit; dizginsiz.

**rampart** [ˈrampât]. Toprak tabya; sur; sed; müdafaa vasıtası.

**ramrod** [ˈramrod]. Harbi.

**ramshackle** [ˈramşakl]. Köhne.

**ran** *v.* run.

**ranch** [rânç]. Amerikada büyük çapta hayvan yetiştirmeğe mahsus çiftlik. Böyle bir çiftliği idare etm.

**rancid** [ˈransid]. Ekşimiş, kokmuş.

**ranco·ur** [ˈran(g)kö(r)]. Kin; kuyruk acısı. **~rous,** kinci.

**random** [ˈrandəm]. Rasgele; gelişi güzel. **to hit out at ~,** hem nalına hem mıhına vurmak.

**randy** [ˈrandi]. Şamatacı; şehvetli.

**rang** *v.* ring.

**range**[1] [rêync] *n.* Sıra, dizi; dağ silsilesi; saha; el, göz veya sesin gidebileceği yer; mıntaka; mesafe; menzil; atış mesafesi (meydanı); genişlik; mutfak ocağı. **out of ~,** menzil harici: **within [in] ~,** menzil dahilinde: **~ of speeds** (uçak) asgari ve azami sürat arasındaki fark: **~ of temperature,** hararet farkı: **the whole ~ of politics,** siyasetin sahası: **a wide ~ of patterns,** bir malın çok çeşidleri. **~-finder,** mesafe ölçüsü aleti. ~[2] *v.* Uzanmak; dolaşmak; gezmek; menzili ... olm.; sıra ile dizmek. **to ~ a gun,** topun menzilini tanzim etm.: **his activities ~ from music to shooting,** faaliyet sahası musikiden avcılığa kadar uzanır: **the temperature ~s from zero to eighty,** hararet sıfırla seksen arasında değişir: **to ~ oneself on s.o.'s side,** birine tarafdar olm.: **to ~ over the country,** memleketin her tarafına yayılıp dolaşmak. **~r** [ˈrêyncə(r)], devlet orman müfettişi.

**rank**[1] [ran(g)k]. Rütbe; saf. Saymak. **he ~s as England's greatest man,** İngiltere'nin en büyük adamı sayılır: **I don't ~ him very high,** ben onu o kadar mühim bulmuyorum: **~ and fashion,** en kibar sınıf: **~ and file,** efrat; aşağı tabaka: **other ~s,** efrat. ~[2] *a.* Bol; kaba; kokmuş. **~ poison,** safi zehir: **~ treachery,** halis hiyanet. **~ness,** mebzuliyet; kabalık kokmuşluk. **~er** [ˈran(g)kə(r)], alaylı zabit.

**rankle** [ˈran(g)kl]. İçine ukde olm.

**ransack** [ˈransak]. Arayarak altüst etm.; çapullamak.

**ransom** [ˈransəm]. Fidye (vermek). **to hold s.o. to ~,** bir esir için fidye istemek: ⌜**worth a King's ~**⌝, paha biçilmez.

**rant** [rant]. Atıp tutmak.

**rap**[1] [rap]. (*ech.*) Hafif darbe; kapıyı çalma. Hafifçe vurmak; çalmak. **to give s.o. a ~ on the knuckles,** birinin parmaklarına vurmak; birine haddini bildirmek: **to ~ out an oath,** bir küfür savurmak. ~[2]. Zerre. **not to care a ~,** metelik vermemek: **not worth a ~,** on para etmez.

**rapac·ious** [rəˈpêyşəs]. Haris; açgözlü; yırtıcı. **~ity** [–ˈpasiti], harislik, açgözlülük.

**rap·e** [rêyp]. Irzına geçme(k); (*esk.*) kız kaçırma. **~ine** [ˈrapâyn], yağma, çapulculuk.

**rapid** [ˈrapid]. Süratli, hızlı. **~s,** nehrin uçur gibi akıntılı yer. **~ slope,** dik yokuş. **~ity** [rəˈpiditi], sürat, hız.
**rapier** [ˈrêypiə(r)]. İnce uzun kılıc.
**rapprochement** [raˈproşmo(n)]. Barışma.
**rapscallion** [rapˈskalyən]. Haylaz; külhanbeyi.
**rapt** [rapt]. Vecde gelmiş. **~ attention,** can kulağı: **to be ~ in,** ···e dalmak. **~ure** [ˈraptyə(r)], vecid; sevinc deliliği: **to be in ~s,** etekleri zil çalmak: **to go into ~s over stg.,** bir şeye delice sevinmek. **~urous,** heyecanlı: **~ applause,** şiddetli alkış.
**rar·e** [reə(r)]. Nadir; seyrek; nefis; müstesna; kesif olmıyan; (*coll.*) fevkalâde. **~efy** [ˈreərifây], kesafetini azaltmak; hafifletmek. **~ity** [ˈreəriti], nadir şey; şey; hintkumaşı; nadirlik; kesafet azlığı.
**rascal** [ˈrâskl]. Çapkın; yaramaz; dolandırıcı. **~ity** [–ˈkaliti], çapkınlık. **~ly,** kurnaz.
**rash**[1] [raş] *n.* İsilik. **~**[2] *a.* Düşüncesiz; aceleci; ihtiyatsız. **~ness,** düşüncesizlik; atılganlık.
**rasher** [ˈraşə(r)]. **a ~ of bacon,** domuz pastırması dilimi.
**rasp** [râsp]. Kaba eğe; törpü (sesi). Törpülemek; eğe gibi ses vermek.
**raspberry** [ˈrâzbəri]. Ağac çileği.
**rat** (rat). Büyük fare, sıçan; hain. Fare avlamak; (*sl.*) düşman tarafina kaçmak. **~s!,** saçma! **like a drowned ~,** sırsıklam: **to smell a ~,** kuşkulanmak.
**ratchet** [ˈraçit]. Bir çarkın hep bir tarafa dönmesini temin eden alet.
**rate**[1] [rêyt] *n.* Bir şeyin başka şeye nispetle ölçüsü; nisbet; derece; sürat; faiz mikdarı, fiat; belediye resmi. **at any ~,** her halde: **at that ~,** o hesab ile; o halde; bu suretle: **at the ~ of fifty miles an hour (50 m.p.h.),** saatte elli mil süratle: **he was living at the ~ of ten pounds a week,** haftada on lira sarfederek yaşıyordu: **at the ~ of a shilling each,** tanesi bir şilinden: **the Bank Rate,** merkez Bankası (Bank of England) iskonto haddi: **to come upon the ~s,** belediye ianesi ile yaşamağa mecbur olm.: **the death ~ was 10 per thousand,** ölüm nisbeti binde ondu: **~ of discount,** iskonto fiatı: **railway ~s,** demiryolu nakliye ücretleri. **~**[2] *v.* Kıymet tahmin etm.; tasnif etm.; say(ıl)mak. **this house is ~d at £80 per annum (p.a.),** a (tahakkuk memurlarınca) bu evin senelik kirası 80 ingiliz lirası tahmin edilmiştir. **~**[3], Azarlamak.
**rather** [ˈrâðə(r)]. Bir az, oldukça; tercihen; daha doğrusu; hay hay!; **~ than ...,** ···den ziyade. **~ a lot,** oldukça: **~ fat,** şişmanca: **anything ~ than this,** bu olmasın da ne olursa olsun: **I'd ~ die than do that,** onu yapmaktansa ölürüm: **I'd ~ not,** müsaadenizle yapmayayım: **I ~ think we have met before,** evvelce görüştük gibime geliyor.
**rati·fy** [ˈratifây]. Tasdik etmek. **~fication** [–ˈkêyşn], tasdik.
**rating** [ˈrêytin(g)]. (Bahriyede) gemicinin rütbe ve sınıfı; belediye resimlerinin tahmini; (bir motor vs.nin) itibarî kuvveti. **the ~ authorities,** belediye resimlerini tahmin eden ve toplıyan idare.
**ratio** [ˈrêyşiou]. Nisbet.
**ration** [ˈraşn]. Tayın. Adam başına mikdar tayın etmek. **emergency [iron] ~,** ihtiyat tayını: **~ card,** karne.
**rational** [ˈraşənl]. Akıl sahibi; mantikî; insaflı; elverişli. **~ism,** akılcı felsefe. **~ize,** akla uydurmak; (bir endüstriyi) rasyonel bir şekilde teşkilatlandırmak.
**rat-(tat-)tat** [ˈratˈtatˈtat]. (*ech.*) Kapı halkasının çalınma sesi.
**ratter** [ˈratə(r)]. Fare tutan (kedi, köpek).
**rattl·e** [ˈratl]. Kaynana zırıltısı; takırtı; çıngıraklı yılanın çıngırağı. Takırda(t)mak; (*coll.*) şaşırtmak. **to ~ along,** (araba vs.) hızlı ve gürültülü gitmek: **to ~ a person,** birinin iki ayağını bir pabuca sokmak: **death ~,** ölüm hırıltısı: **to ~ off,** (dua vs.) çabukça okumak: **to ~ on,** cırcır ötmek. **~er,** (*Amer.*) *v.* **rattlesnake. ~esnake** [ˈratlsnêyk], çıngıraklı yılan. **~etrap** [ˈratltrap], köhne (araba); kırık dökük. **~ing** [ˈratlin(g)], takırtılı. **at a ~ pace,** dolu dizgin: **~ good,** (*coll.*) fevkalâde iyi.
**raucous** [ˈrôkəs]. Kısık, boğuk (ses).
**ravage** [ˈravic]. Zarar. **~s,** tahribat. Tahrib etm.; yağma etmek.
**rave** [rêyv]. Çıldırmak; sayıklamak; abuk sabuk söylemek. **to ~ about stg.,** bir şeye çıldırmak.
**ravel** [ravl]. İplik iplik karıştırmak.
**raven**[1] [ˈrêyvn]. Kuzgun. **~**[2] [ˈravn]. Şikâr peşinde dolaşmak; oburcasına

yemek. ~**ous** [~**ing**], çok acıkmış; haris.

**ravine** [rəˈvîn]. Dar ve derin bir dere.

**ravish** [ˈraviş]. Kapıp götürmek; ırzına tecavüz etm.; meftun etmek. ~**ing**, meftun edici, gönül kapıcı.

**raw** [rô]. Çiğ, ham; pişmemiş; (yara) pek hassas; soğuk ve rütubetli (hava); acemi. **a ~ hand**, acemi işçi: **~ material**, ham madde: **~ recruit**, acemi asker: **~ spirit**, saf ispirto: **to touch s.o. on the ~ (spot)**, bamteline basmak. **~-boned**, kemikleri fırlamış.

**ray** [rêy]. Şua; pertev; parıltı.

**rayon** [ˈrêyyon]. Sunî ipek.

**raze** [rêyz]. Temelinden yıkmak. **to ~ to the ground**, yerle yeksan etmek.

**razor** [ˈrêyzə(r)]. Ustura. **to set a ~**, usturayı bilemek. **~-strop**, berber kayışı.

**razzle-dazzle** [ˈrazlˈdazl]. **to go on the ~**, çümbüş yapmak.

**R.E.** [ˈâˈrî]. (*abb.*) **Royal Engineers**, istihkâm sınıfı.

**re** [rî]. **in ~ Jones v. Smith**, Jones'le Smith davası: **~ your letter of May 1st**, 1 mayıs tarihli mektubunuza cevab olarak (mektubunuz münasebetile).

**re-** *pref.* Geri; tekrar; yeniden; *mes.* **return** = geri dönmek; **re-write** = yeniden yazmak. *Bu önek hemen her fiilin başına konabileceği için* **re-** *ile başlıyan bütün fiiller lügate alınmamıştır. Bulamadığınız bir kelime için fiile bakınız ve manasını yukariki izahata göre değiştiriniz.*

**reach**[1] [rîç] *n.* Elin veya bir aletin yetişebileceği mesafe; topun menzili; saha; bir nehrin bükülmiyen düz kısmı. **to have a long ~**, (kol ile) çok ileriye uzanabilmek: **beyond [out of] ~**, yetişemiyecek yerde: **within ~**, elinin yetişebileceği yerde: **within easy ~ of the station**, istasyona yakın: **no help was within ~**, civarda yardım edecek kimse yoktu: **posts within the ~ of all**, herkesin elde edebileceği vazifeler: **within the ~ of small purses**, zengin olmıyanların erişebileceği fiatta. **~**[2] *v.* Yetişmek, ermek, varmak; uzanmak, uzatmak. **to ~ out**, uzatmak: **to ~ fifty**, ellisine basmak: **as far as the eye could ~**, göz alabildiğine: **it has ~ed my ears that ...**, ... kulağıma çalındı. **~-me-downs**, hazır elbise.

**react** [rîˈakt]. Aksülamel yapmak; mukabil tesir yapmak; tesir yapmak. **~ion** [–ˈakşn], aksülamel, mukabele; aksi tesir; irtica; tepki. **~ionary**, mürteci.

**read**[1] [rîd], *past* **read** [red]. Okumak; istihrac etm.; bakıp anlamak. **I can ~ him like a book**, onun içini dışını bilirim: **to ~ into a sentence stg. that is not there**, bir cümleden ifade etmediği bir mana çıkarmak: **to ~ medicine**, tıb tahsil etm.: **to ~ s.o. to sleep**, birini uyutmak için okumak: **to ~ up a subject**, bir mevzu hakkında okuyup malûmat edinmek: **this play ~s well but I doubt if it would act well**, bu piyesin okuması iyi fakat iyi temsil edileceği şübheli. **~ out**, yüksek sesle okumak. **~ through**, baştan başa okumak. **~**[2] [red]. **to take the minutes as ~**, bir toplantı zabıtlarını okunmuş sayarak kabul etm.: **well-~**, çok okumuş. **~able** [ˈrîdəbl], okumağa değer. **~ing** [ˈrîdin(g)], *n.* okuma; kıraat; tahsil; okuyuş; metin. *a.* Okuyan. **the ~ public**, okuyucular (sınıfı). **~ing-desk**, rahle.

**readjust** [ˌrîəˈcʌst]. Düzeltmek; tanzim etmek. **to ~ one's ideas**, fikirlerini yeni vaziyete uydurup değiştirmek.

**ready** [ˈredi]. Hazır; kolay, çabuk. (*mil.*) Hazırol vaziyeti. **to be ~ to do stg.**, bir şeyi yapmağa hazır olm.: **to make [get] ~**, hazırla(n)mak: **~ for action**, muharebeye hazır: **guns at the ~**, ateşe hazır toplar: **~ money**, peşin para; hazır para: **a ~ pen**, kolay yazar kimse: **these goods command a ~ sale**, bu mallar derhal satılabilir: **a ~ tongue**, kolay konuşur kimse: **rather too ~ to suspect people**, herkesten hemen şübhelenir: **he is very ~ with excuses**, mazeret bulmağa hazır: **a ~ wit**, hazırcevab (lık). **~-cooked**, önceden pişmiş. **~-made clothes**, hazır elbise. **~-reckoner**, hesab cedveli.

**real** [riə̯l]. Hakiki, asıl, gerçek; sahici; gayri menkul. **~istic** [–ˈlistik], tabiatı olduğu gibi gösteren; tabiate, hakikate uygun; amelî. **~ity** [riˈaliti], hakikat; gerçeklik; hakikî şey: **in ~**, hakikaten. **~ize** [ˈriə̯lâyz], tahakkuk ettirmek; icra etm.; hakikat olarak görmek; kavramak, anlamak; paraya tahvil etmek. **~-ization**, kuvveden fiile çıkma; tahakkuk ettirme; idrak. **~ly** [ˈriə̯li],

hakikaten; gerçekten; sahiden. **~?,** sahi mi?; yaa!: **not ~?,** olur mu?: **say what you ~ think,** olduğu gibi düşündüğünüzü söyleyiniz: **you ~ must have a talk with him,** onunla mutlaka görüşmelisiniz.
**realm** [relm]. Devlet; kırallık; saha.
**ream** [rîm]. 500 tabakalık kâğıd topu. **to write ~s,** sahifeler doldurmak.
**reap** [rîp]. Ekin biçmek; mahsul toplamak; elde etmek. ⌈**to ~ as one sows**⌉, ektiğini biçmek. **~er,** orakçı; orak makinesi: **the R~,** Ezrail.
**reappear** [ˈrîəˈpiə(r)]. Tekrar görünmek, tekrar ortaya çıkmak.
**rear**[1] [riə(r)] *n.* Arka, geri. *a.* Arka tarafa, geriye aid. **to bring up the ~,** en son gelmek. **~guard,** dümdar: **~ action,** ricat muharebesi. **~wards,** arkaya, geriye doğru. **~-admiral,** tuğ amiral. **~**[2] *v.* Dikmek, inşa etm.; yükseltmek. Şaha kalkmak.
**rearm** [rîˈâm]. Tekrar silâhlandırmak; yeni silâhlarla techiz etmek. **~ament,** yeni silâhlarla techiz etme.
**reason**[1] [ˈrîzn] *n.* Sebeb, mucib; akıl, idrak; insaf. **by ~ of,** ...den dolayı, sebebiyle: **there is ~ to believe that ...,** ... inanmak yerindedir: **for some ~ or other,** her nedense: **for no other ~ than that I forgot,** yegâne sebebi unutmuş olmamdır: **he complains with (good) ~ [not without ~],** haklı olarak şikâyet ediyor: **you have ~ to be proud,** iftihar etmekte haklısınız: **to hear [listen to] ~,** söz anlamak, makul olm.: **I cannot in ~ pay more,** bundan fazla para vermem makul değildir: **everything in ~,** makul olmak şartile her şey: **to lose one's ~,** aklını bozmak: **it's all the more ~,** ayrıca bir sebebdir: **it stands to ~ that ...,** ... aşikârdır: **the ~ why ...,** ... sebebi. **~**[2] *v.* Muhakeme etm.; netice çıkarmak; istidlâl etmek. **to ~ s.o. into [out of] doing stg.,** deliller ileri sürerek birini ikna edip bir şey yaptırmak [yapmaktan vazgeçirmek]: **to ~ with s.o.,** birini delillerle ikna etmeğe çalışmak. **~able,** insaflı, makul; haklı, münasib; kâfi mikdarda. **~ed,** sebebli, makul. **~ing,** muhakeme; mantıklı düşünme.
**reassure** [ˌrîəˈşô(r)]. Tatmin etm.; temin etmek.
**rebate** [ˈrîbêyt]. İskonto; tenzilat.
**rebel** *n.* & *a.* [ˈrebl]. Âsi; isyan eden. *v.* [riˈbel]. İsyan etm., ayaklanmak. **~lion** [riˈbelyən], isyan. **~lious,** âsi; itaatsiz.
**rebound**[1] [riˈbaund]. Geri sekme(k). **~**[2], *pp. of* **rebind.** Yeniden cildlenmiş.
**rebuff** [riˈbʌf]. Red; ters cevab. Şiddetle reddetmek.
**rebuke** [riˈbyûk]. Azar(lamak).
**recalcitrant** [riˈkalsitrənt]. İtaatsiz.
**recall** [riˈkôl]. Geri çağırmak; avdetini emretmek; (hükmü) feshetmek; hatırlamak.
**recant** [riˈkant]. Sözünü geri almak; mezhebinden dönmek.
**recapitulate** [ˌrîkaˈpityulêyt]. Tekrar hulâsa etmek.
**recast** [rîˈkâst]. Şeklini değiştirmek.
**reced·e** [riˈsîd]. Geri çekilmek; uzaklaşmak. **~ing chin,** kaçık çene.
**receipt** [riˈsît]. Makbuz; al(ın)ma. Ödeme makbuzu vermek. **~s,** varidat. **to acknowledge ~ of stg.,** bir şeyin alındığını haber vermek: **I am in ~ of your letter,** mektubunuzu aldım: **pay on ~,** alındığı zaman tediye etmek.
**receiv·e** [riˈsîv]. Almak; kabul etm.; hırsıza yataklık etmek. **~ed with thanks,** (makbuz üzerine) teşekkürle alınmıştır: **she is not ~ing today,** bugün misafir kabul etmiyor. **~er,** alıcı; hırsız yatağı. **~ing-order,** bir müflisin mallarını haczettiren karar. **~ ing-station,** telsiz alıcı merkezi.
**recent** [ˈrîsənt]. Yakın geçmişe aid; yeni; son zamanda vukuagelen: **~ly,** geçenlerde: **as ~ly as yesterday,** daha dün: **until quite ~ly,** çok yakın zamana kadar.
**receptacle** [riˈseptəkl]. Kab; zarf.
**reception** [riˈsepşn]. Kabul (tarzı); al(ın)ma; misafir kabulü; kabul resmi. **a cold ~,** istiskal: **to get a warm ~,** (i) hararetle karşılanmak; (ii) geldiğine geleceğine pişman olmak. **~ist,** (otelde vs.) misafiri kabul etmeğe memur kimse.
**receptiv·e** [riˈseptiv]. Kavrayıcı. **~ity** [–ˈtiviti], çabuk kavrayış; (radyo) alma kabiliyeti.
**recess** [riˈses]. Tatil; hücre, köşe.
**recipe** [ˈresipi]. Yemek yapma usulü; reçete; tedbir.
**recipient** [riˈsipyənt]. Alan kimse.
**reciproc·al** [riˈsiprəkl]. Karşılıklı; mütekabil; (*gram.*) müşarik. **~ate,** karşılıklı verip almak. **~ity** [ˌresiˈprositi], mütekabiliyet.

**recit·e** [ri'sâyt]. Ezber okumak; birer birer zikretmek. **~al,** hikâye etme; ezber okuma; resital. **~ation** [--'têyşn], ezber okuma.
**reckless** ['reklis]. Uluorta; pervasız.
**reckon** ['rekən]. Hesab etm.; tahmin etm.; zannetmek. **to ~ on,** ···e güvenmek: **to ~ in,** hesaba dahil etm.: **to ~ up,** yekûn etm.: **we shall have to ~ with ..., ...** hesaba katmalıyız. **~ing,** hesab: **day of ~,** hesab günü: **to be out in one's ~,** hesabı yanlış çıkmak.
**recla·im** [ri'klêym]. Islah etm.; ziraati elverişli bir hale getirmek. **past ~,** ıslah olmaz: **to ~ land from the sea,** denizi doldurmak.
**recline** [ri'klâyn]. Dayanmak; boylu boyuna uzanmak.
**recluse** [ri'klûs]. Münzevi; târiki dünya.
**recognition** [ˌrekog'nişn]. Tanı(n)ma; itiraf; tasdik. **in ~ of his past services,** geçmişteki hizmetlerine mükâfat olarak: **to alter past ~,** tanınmıyacak derecede değiş(tir)mek.
**recogniz·ance** [ri'kognizəns]. Kefaletname. **~e** ['rekəgnâyz], tanımak; itiraf etm., kabul ve tasdik etmek. **~d,** tanınmış; mukarrer; usulü dairesinde; mutcber.
**recoil** [ri'kôyl]. Geri tepme(k); geri çekilme(k); korkudan geri çekilme(k). **his evil deeds will ~ upon his head,** fenalıklarının cezasını çekecek.
**recollect** [ˌrekə'lekt]. Hatırlamak. **~ion** [-'lekşn], hatır(lam)a: **I have a dim ~ of it,** onu hayal meyal hatırlıyorum: **to the best of my ~,** hatırladığıma göre.
**recommend** [ˌrekə'mend]. Tavsiye etm.; emanet etm.; iltimas etm.; tenbih etm. **~ation** [-'dêyşn], iltimas; tavsiye (mektubu); nasihat.
**recompense** ['rekəmpens]. Mükâfat (vermek); telâfi (etm.); zarar tazmin etm.
**reconcil·e** ['rekənsâyl]. Barıştırmak; razı etm.; telif etmek. **to be ~ed,** barışmak: **to ~ oneself to ...,** ···e alışmak. **~iation** [-sili'êyşn], barışma.
**recondition** [ˌrîkon'dişn]. Yenilemek.
**reconnaissance** [ri'konisns]. İstikşaf; keşif.
**reconnoitre** [ˌrekə'nôytə(r)]. Keşif yapmak.
**reconstitute** [rî'konstityût]. Yeniden teşkil etm.; vakıaları birleştirip bir bütün teşkil etmek.
**reconstruct** [ˌrîkən'strʌkt]. Yeniden inşa etm., kurmak; bozup yapmak. **to ~ a crime,** bir cinayetin vuku buluş şeklini yeniden kurmak.
**record**[1] *n.* ['rekôd]. Sicil, kayıd; not; zabıtname; plâk; rekor. **~s,** evrak, dosya. **he has a bad ~,** sicili bozuktur: **to put on ~,** kaydetmek: **it is on ~ that ...,** ···diği vakidir: **service ~,** sicil. **~**[2] *v.* [ri'kôd]. Kaydetmek; yazmak; plâğa almak. **the ~ing angel,** Kiramen kâtibîn. **~er,** İngilterede bazı şehirlerin sulh hâkimi.
**recount** *v.* [ri'kaunt]. Anlatmak. **to ~ one's woes,** derdini dökmek.
**re-count** ['rîkaunt]. Tekrar saymak. Seçimlerde verilen reylerin tekrar sayılması.
**recoup** [ri'kûp]. Telâfi etmek. **to ~ oneself,** zararını çıkarmak.
**recourse** [ri'kôs]. Müracaat. **to have ~ to,** ···e başvurmak.
**re-cover** [rî'kʌvə(r)]. Yeniden kaplamak.
**recover** [ri'kʌvə(r)]. Geri almak; tekrar ele geçirmek; bulmak; telâfi etmek. İyileşmek; eski halini bulmak. **to ~ oneself,** kendini toplamak: **to ~ one's balance,** muvazenesini bulmak: **to ~ one's breath,** nefes almak: **to ~ consciousness,** ayılmak: **to ~ one's expenses,** masrafını çıkarmak: **to ~ lost ground,** kaybedilen nüfuzu telâfi etmek. **~y** [ri'kʌvəri], geri alma; elde etme; eski halini bulma; telâfii mafat; iyileşme; kalkınma. **to be past ~,** ümidsiz bir halde olm.: **the patient is making a good ~,** hasta çabuk iyileşiyor.
**re-create** [ˌrîkri'êyt]. Yeniden yaratmak.
**recreat·e** ['rekriêyt]. Dinlendirmek; eğlendirmek. **~ion** [-'êyşn], başını dinlendirme; eğlence; teneffüs: **~ ground,** oyun sahası. **~ive,** dinlendirici; yeni yaratıcı.
**recrimination** [riˌkrimi'nêyşn]. Karşılıklı şikâyet.
**recrudescence** [ˌrîkru'desəns]. Tekrar şiddetlenme.
**recruit** [ri'krût]. Acemi nefer; yeni âza. Askere almak; tarafdar toplamak. **~ one's health,** sıhhatini iyileştirmek.
**rectang·le** ['rektangl]. Mustatil. **~ular** [-'tangyulə(r)], mustatil; kaim zaviyeli.
**rectif·y** ['rektifây]. Düzeltmek; tas-

hih etm.; tasfiye ve taktir etm. **~ier,** düzeltici; (*elect.*) redresör.

**rectitude** [ˈrektityûd]. Dürüstlük.

**rector** [ˈrektô(r)]. Mahalle papazı; rektör. **~y,** mahalle papazının evi.

**recumbent** [riˈkʌmbənt]. Uzanıp yatan.

**recuperate** [riˈkyûpərêyt]. Sıhhatini vs. tekrar kazanmak.

**recur** [riˈkö(r)]. Tekrar vuku bulmak; tekrar hatıra gelmek. **~rence** [–ˈkʌrəns], tekrar vukua gelme. **~ring** [–ˈkö(r)rin(g)], arada sırada vukua gelen: **~ decimal,** devrî kesri aded.

**red** [red]. Kırmızı, kızıl; al. Kırmızı renk; müfrit solcu. **to see the ~ light,** tehlikeyi sezmek: **like a ~ rag to a bull,** kırmızı rengin boğayı kızdırması gibi öfkelendirici şey: **the R~ Sea,** Kızıldeniz: **to see ~,** gözünü kan bürümek: **~ tape,** kırtasiyecilik: **to turn [go] ~,** kızarmak, kızıllaşmak. **~-blooded,** dinc; cesur. **~-eyed,** ağlamaktan gözleri kızarmış. **~-handed,** suçüstü. **~-hot,** ateşte kıpkırmızı olmuş; kızgın: **~ communist,** azgın komünist. **~-letter day,** sayılı gün. **~breast** [ˈredbrest] *v.* **robin. ~coat** [ˈredkout], (*esk.*) İngiliz askeri. **~den** [ˈredn], kızarmak; kırmızılaştırmak. **~dish,** kırmızımsı. **~skin** [ˈredskin], kızıl derili; Şimalî Amerika yerlisi.

**redeem** [riˈdîm]. Rehinden çıkarmak; halâs etm.; (vad) ifa etm.; telâfi etm. **~ing feature,** kusurlarını unutturan iyi bir vasıf. **~able,** ödenmesi lâzım (sened). **~er, the R~,** Hazreti İsa.

**redemption** [riˈdemşn]. Tediye; rehinden çıkarma. **past ~,** ıslah edilmez.

**redolent** [ˈredolənt]. Kokulu.

**redouble** [rîˈdʌbl]. Bir kat daha artırmak.

**redoubtable** [riˈdautəbl]. Cesur; korkunc.

**redound** [riˈdaund]. **~ to,** artırmak. **this ~s to your credit,** bununla iftihar edebilirsiniz.

**redress** [riˈdres]. Tamir, telâfi, tazmin. Düzeltmek; tashih etm.

**reduc·e** [riˈdyûs]. Küçültmek; kısaltmak; azaltmak; bir hale sokmak; zayıflatmak; fethetmek. **to ~ stg. to ashes,** bir şeyi kül haline sokmak: **to ~ to poverty,** sefalete düşürmek: **to ~ to the ranks,** (*mil.*) rütbesini refetmek: **to ~ to silence,** susturmak: **to ~ to writing,** yaz(dır)mak. **~ed,** azalmış, indirilmiş: **in ~ circumstances,** darlık içinde. **~tion** [–ˈdʌkşn], azalma, küçültme; indirme; tenzil; ihtisar; fethetme; (*med.*) tecbir.

**redundant** [riˈdʌndənt]. Fazla; lüzumsuz.

**reed** [rîd]. Kamış, saz; düdük dili: **a broken ~,** güvenilmez. **~y,** kamışlık, sazlık.

**reef**[1] [rîf]. Camadan. **take in a ~,** yelkeni camadana vurmak; (*fig.*) ihtiyatla hareket etm. **~**[2]. Sığ kayalık; maden damarı.

**reek** [rîk]. Fena koku (yaymak). **he ~s of garlic,** sarmısak kokuyor: **the place ~s with drunkenness,** burada sarhoşluktan geçilmiyor.

**reel**[1] [rîl]. Makara(ya sarmak). **to ~ off,** sayıp dökmek. **~**[2]. Sendelemek. **~**[3]. Bir İskoç raksı.

**refectory** [riˈfektəri]. Yemekhane.

**refer** [riˈfö(r)]. Havale etm.; arzetmek; göndermek; müracaat etm.; reyine koymak; aid olm.; dokunmak; zikretmek. **~ to drawer,** (*abb.* **R.D.**), (banka) karşılığı bulunmıyan çeki sahibine havale ediniz!: **~ring to your letter of the 16th,** 16 tarihli mektubunuza cevab olarak: **in his letter he ~s to your book,** mektubunda kitabınızı zikrediyor: **this matter ~s to you,** bu mesele size aiddir: **he ~red to his watch,** saatine baktı: **we will not ~ to the matter again,** bir daha bu meseleye temas etmiyeceğiz: **never ~ to him in my presence!,** benim yanımda onun ismini ağzına alma! **~ee** [refəˈrî], hakem (olmak). **~ence** [ˈrefərens], müracaat; havale; münasebet; alâka; salâhiyet; zikir; hüsnühal kâğıdı, bonservis; (kitabda) fazla malûmat için hangi membalara müracaat edileceğine aid not. **with ~ to …,** münasebetiyle, …e dair: **to give s.o. as ~,** birisini referans göstermek: **to have ~ to …,** aid olm.: **to have good ~s,** bonservisleri olm.: **~ library,** tedkikat için müracaat olunan kütübhane: **to make ~ to,** zikretmek: **~ point,** nirengi noktası: **terms of ~ of a commission,** bir heyetin salâhiyeti: **without ~ to …,** sarfı nazar ederek; danışmaksızın: **work of ~,** müracaat kitabı. **~endum** [refəˈrendəm]. Tek bir mesele hakkında halkın reyine müracaat.

**refill** *v.* [rîˈfil] Yeniden doldurmak.

*n.* [ˈrîfil], yedek pil, kurşun, kâğıd. **~ing station,** benzin alma istasyonu.
**refine** [riˈfâyn]. Tasfiye etm.; kabalığını gidermek. **~d,** tasfiye edilmiş; ince; kibar; ince zevk sahibi. **~ment,** tasfiye; zariflik; kibarlik: **a ~ of cruelty,** akla gelmedik işkence. **~ry,** tasfiyehane.
**refit** *v.* [rîˈfit]. Yeniden techiz etmek; tamir etmek. *n.* [ˈrîfit] [**~ting**], techiz, tamir edilme.
**reflect** [riˈflekt]. Aksettirmek; düşünmek. **to ~ on** [**about**] **stg.**, hakkında düşünmek: **to ~ on s.o.,** namusuna dokunmak: **to ~ credit on s.o.,** ···e şeref kazandırmak: **to be ~ed,** aksetmek: **this speaker ~ed popular opinion,** bu hatib halkın fikirlerine mâkes oldu. **~ion,** aksetme, akis; düşünce; kusur bulma: **to cast ~s on s.o.,** birinin namusuna dokunmak: **on ~,** düşünüp taşındıkça. **~ive,** aksettirici; düşünceli. **~or,** mâkes; ayna.
**reflex** [ˈrîfleks]. Münakis (hareket). **~ive** [riˈfleksiv], **~ verb,** mutavaat fiili.
**re-form** [ˌrîˈfôm]. Yeniden teşkil etm.; tekrar sıraya girmek.
**reform** [riˈfôm]. Islah (etm.); tanzim (etm.). **~ation** [ˌrefôˈmêyşn]. Islah etme, olunma. **the R~,** 16 ıncı asırda Protestan kiliselerinin teessüsü ile neticelenen dinî inkilâb. **~atory** [–ˈfômətəri], ıslah edici; ıslahhane. **~er,** ıslahatçı.
**refract** [riˈfrakt]. Ziyayı inkisar ettirmek. **~ory,** âsi; erimez.
**refrain**[1] [riˈfrêyn] *n.* Nakarat. **~**[2] *v.* Kendini tutmak; ictinab etmek. **to be unable to ~ from,** ···den kendini alamamak.
**refresh** [riˈfreş]. Tazelemek; serinletmek; canlandırmak. **to ~ oneself,** açılmak; yorgunluğunu gidermek: **to ~ the inner man,** yiyerek, içerek canlanmak. **~er course,** kısmen unutulan bilgileri tazelemek için kurs. **~ment,** yemek, içki: **~ room,** istasyon büfesi.
**refrigerat·e** [riˈfricərêyt]. Soğutmak. **~or,** buz dolabı.
**reft** [reft]. Mahrum.
**refuel** [ˌrîˈfyûəl]. (Gemi, uçak) kömür, benzin almak.
**refuge** [ˈrefyûc]. Sığınak, melce. **to take ~,** sığınmak; iltica etm. **~e,** mülteci.
**refund** *n.* [ˈrîfʌnd]. Paranın iade edilmesi; geri verilen para. *v.* [riˈfʌnd]. (Para) geri vermek; telâfi etmek.
**refus·al** [riˈfyûzl]. Red. **to have the ~ of stg.,** kabul edip etmemeğe hakkı olm.: **to have the first ~ of stg.,** (ev vs.) satın almağa tâlib olanlar arasında rüchan hakkına malik olm.: **I will take no ~,** menfi cevab kabul etmem. **~e**[1] [riˈfyûz] *v.*, reddetmek; razı olmamak. **to ~ s.o. stg.,** birine bir şeyi vermemek. **~e**[2] [ˈrefyûs], kırpıntı; çöp, pislik. **~ dump** [**heap**], mezbele, çöplük.
**refute** [riˈfyût]. Cerhetmek.
**regain** [rîˈgêyn]. Tekrar ele geçirmek. **to ~ consciousness,** ayılmak: **to ~ one's house,** evine dönmek.
**regal** [ˈrîgl]. Kırala aid; şahane. **~ia** [–ˈgêylyə], hükümdarın tac ve sair resmî tezyinatı; farmasonların alâmetleri.
**regale** [riˈgêyl]. Yemekle ağırlamak; eğlendirmek.
**regard**[1] [riˈgâd] *n.* Münasebet; nazar; itibar; hürmet. **in** [**with**] **~ to,** hakkında; ···e nazaran: **in this ~,** bu hususta: **having ~ to,** nazarı dikkate alarak: **to have no ~ for,** hiçe saymak: **out of ~ for s.o.,** birinin hatırı için: **to pay no ~ to,** ···e hiç dikkat etmemek: **to send s.o. one's kind ~s,** birine selâm göndermek: **with kind ~s from,** selâmlarla. **~**[2] *v.* Bakmak; nazarı dikkate almak; saymak; aid olm. **as ~s ...,** ···e gelince. **~ful of,** ···i unutmıyarak; hatırını sayarak. **~ing,** hakkında; aid. **~less, ~ of,** bakmıyarak; (*coll.*) **to talk ~,** hem nalına hem mıhına vurmak: **to spend ~,** har vurup harman savurmak: **he was got up ~,** en pahalı tarzda giyinmiş kuşanmıştı.
**regatta** [riˈgata]. Kotra yarışı günü.
**regen·cy** [ˈrîcənsi]. Saltanat naibliği. **the R~,** Wales prensi George'un saltanat naibliği (1810–20). **~t,** saltanat naibi.
**regenerate** [rîˈcenərêyt]. Canlandırmak.
**regicide** [ˈrecisâyd]. Hükümdar kaatili veye katli.
**régime** [rêyˈjîm]. Hükümet şekli; perhiz.
**regiment** [ˈrecimənt]. (*mil.*) Alay; kalabalık. Sıkı ve yeknesak bir tarzda icbar etmek. **~al** [–ˈmentl], alaya mensub: **~s,** üniforma. **~ation,** hükûmetin ferdlerin işlerine müfrit şekilde müdahelesi.

**region** [ˈrîcən]. Ülke, mıntaka; nahiye. **the nether ~s,** cehennem. **~al,** muayyen mıntakaya aid.

**register**[1] [ˈrecistə(r)] *n.* Sicil; resmî defter; cedvel; fihrist; kayıd aleti; saat; sesin vüsati. **parish ~,** bir mahallede doğan ölen ve evlenenlerin defteri: **ship's ~,** geminin tabiiyet vesikası. **~**[2] *v.* Kaydetmek; taahhüdlü olarak göndermek; iyice tatbik etm.: (sinemada keder vs.yi) ifade etmek; intibak etm.; kaydolmak; (otelde) ismini deftere kaydetmek. **~ed,** taahhüdlü; kaydedilmiş.

**registr·ar** [recisˈtrâ(r)]. Evrak müdürü; sicil memuru; nüfus memuru: **~ general,** nüfus umum müdürü. **~ation,** kaydetme; tescil. **~y** [ˈrecistri], sicil dairesi; defterhane: **~ office,** evlenme memurluğu; hizmetçi idarehanesi: **certificate of ~,** geminin bayrak tasdiknamesi: **port of ~,** sicil limanı.

**regnant** [ˈregnənt]. Saltanat süren.

**regression** [riˈgreşn]. Gerileme; irtica.

**regret** [riˈgret]. Esef; pişmanlık. Teessüf etm.; pişman olm. **it is to be ~ted that ...,** yazık ki. **~ful,** nadim. **~table,** acınacak.

**regroup** [rîˈgrûp]. Tertiblerini değiştirmek.

**regular** [ˈregyulə(r)]. Muntazam; usule uygun; mutad; gedikli; meslekten; (*gram.*) kiyasî; (*coll.*) adamakıllı. Nizamiye askeri; mektebli. **it was a ~ battle,** adeta bir muharebe gibi idi: **he's a ~ nuisance,** tam başbelâsıdır. **~ity** [–ˈlariti], intizam; devam(lılık); usulü dairesinde olma. **~ize** [ˈregyulərâyz], usulüne uydurmak.

**regulat·e** [ˈregyulêyt]. Tanzim etm.; ayarlamak. **~ion** [–ˈlêyşn], nizam(name): **~s,** talimat(name). **~or,** regülatör; nâzım.

**rehabilitate** [ˌrîhaˈbilitêyt]. Namus ve itibarını iade etm.

**rehash** [rîˈhaş]. Eti ikinci defa pişirmek; (hikâye) bir az değiştirip tekrar ortaya koymak. Böyle bir yemek veya eser.

**rehear** [rîˈhiə(r)]. (Dava) yeniden dinlemek.

**rehears·e** [riˈhö(r)s]. Prova etm.; sayıp dökmek. **~al,** prova; sayıp dökme.

**reign** [rêyn]. Hüküm, saltanat sürmek. Saltanat devri.

**reimburse** [ˌrî·imˈbö(r)s]. Masraf telâfi etm.

**rein** [rêyn]. Dizgin. **to ~ in,** dizgin sıkmak: **to give a horse the ~,** dizginleri gevşetmek: **to give ~ to one's imagination,** *etc.*, kendini hayallere kaptırmak: **to keep a tight ~ over,** dizginini kısmak.

**reincarnation** [ˌrî·inkâˈnêyşn]. Bir ölünün ruhunun yeni bir vücude girmesi.

**reindeer** [ˈrêyndiə(r)]. Ren geyiği.

**reinforce** [ˌrî·inˈfôs]. Takviye etm. **~d concrete,** betonarme. **~ment,** takviye etme: **~s,** takviye kıtası.

**reinstate** [ˌrî·inˈstêyt]. Memuriyet iade etmek.

**reiterate** [rîˈitərêyt]. Tekrarlamak.

**reject** *v.* [riˈcekt]. Reddetmek; ıskartaya çıkarmak. *n* [ˈrîcekt]. Bir kusurdan dolayı ıskartaya çıkarılan. **~ion,** red.

**rejoic·e** [riˈcôys]. Sevinmek; neşelenmek; sevindirmek. **to ~ the heart of,** yüzünü güldürmek: **to ~ in stg.,** bir şeyden haz duymak. **~ing,** sevinc, neşe: **public ~s,** şenlik.

**rejoin** [riˈcôyn]. Tekrar birleştirmek; iltihak etm.; sert cevab vermek. **~der,** sert cevab.

**relapse** [riˈlaps]. Hastalığı nüksetme(k); yeniden dalâlete düşme(k).

**relat·e** [riˈlêyt]. (Hikâye) anlatmak; bağlamak. **~ed to,** ···e aid; bağlı; hısım. **~ion** [riˈlêyşn], münasebet; alâka, ilgi; hısım; hikâye etme. **~s,** akraba; münasebat. **to break off ~s, with s.o.,** birisile münasebatı kesmek: **in ~ to,** hususunda: **~ by marriage,** sıhrî akraba: **what ~ is he to you?,** sizin nenizdir? **~ship,** akrabalık. **~ive** [ˈrelətiv], akraba. Aid; ilgili; nisbî: **~ly,** nisbeten, oldukça: **~ to,** ···e dair: **~ pronoun,** izafî zamir: **he lives in ~ luxury,** başkalarına nisbetle lüks yaşıyor. **~ivity** [–ˈtiviti], izafiyet.

**relax** [riˈlaks]. Gevşe(t)mek; dinlen(dir)mek; yorgunluğunu gidermek. **~ed throat,** hafif gırtlak iltihabı. **~ation,** gevşetme, dinlenme, istirahat. **~ing,** liynet veren (ilac); rehavet veren (hava).

**re-lay** [rîˈlêy]. Yeniden döşemek.

**relay** [ˈrîlêy]. Menzil beygiri; nöbet(e çalışan işçiler); cereyanı takviye eden yardımcı batarya; (menzil) hayvan değiştirme(k); (radyo) bir merkezden alınan sesleri başka merkezlere yayma(k). **~ race,** bayrak koşusu.

**release** [ri'lîs] *n.* Salıverme; kurtuluş; serbest bırakma; gevşetme; (yenilik) piyasaya çıkarma; ayırma mekanizması. Serbest bırakmak; tahliye etm.; salıvermek; pazara çıkarmak.
**relegate** ['religêyt]. Madun bir mevkie düşürmek. **to ~ a matter to s.o.,** bir meseleyi birine havale etmek.
**relent** [ri'lent]. Yumuşamak; şiddetini kesmek. **~less,** amansız; fasılasız.
**relevant** ['reləvənt]. Alâkalı, ilgili.
**relia·ble** [ri'lâyəbl]. Güvenilebilir; emniyetli; inanılabilir. **~bility** [–'biliti], itimada şayanlık. **~nce** [ri'lâyəns], itimad; güven. **to place ~ on,** ···e güvenmek.
**relic** ['relik]. Bakiye; mukaddes emanet. **~s of the past,** eski eserler.
**relief**[1] [ri'lîf]. Yardım; kurtarma; hafifletme; ferahlık; nöbet değiştirme. **to go to the ~ of,** yardımına gitmek: **outdoor, ~,** evlere yapılan ictimaî yardım: **~ party,** kurtarma ekipi: **poor ~;** fakirlere yardım (teşkilatı). **~**[2]. Kabartma.
**relieve**[1] [ri'lîv]. Hafifletmek; ıstırabını dindirmek; ferahlatmak; (muhasaradan) kurtarmak; serbest bırakmak; nöbetini değiştirmek; yerine nöbete girmek. **to ~ s.o. of stg.,** birinden taşıdığı şeyi almak: **to ~ s.o. of his duties,** birini vazifesinden affetmek: **to ~ s.o. of his purse,** birinin kesesini aşırmak: **to ~ the watch,** nöbetçiyi değiştirmek: **to ~ congestion,** seyrüseferi kolaylaştırmak: **to ~ one's feelings,** içini boşaltmak: **to feel ~d,** ferahlamak. **~**[2]. Kabartma şekline koymak.
**relig·ion** [ri'licən]. Din; iman; dindarlık. **to get ~,** (*coll.*) birdenbire dindar olm.: **to make a ~ of doing stg.,** bir şeyi mukaddes bir vazife bilmek. **~ious** [–'licəs], dindar; sofu; dinî.
**relinquish** [ri'linkwiş]. Vazgeçmek; terketmek.
**reliquary** ['relikwəri]. Mukaddes eşyanın mahfazası.
**relish** ['reliş]. Tat, lezzet; çerez; cazibe; iştah. Lezzet almak; hoşlanmak. **I do not ~ the job,** bu iş hoşuma gitmiyor.
**eluctan·ce** [ri'lʌktəns]. İsteksizlik; çekingenlik. **with ~,** istemiyerek: **to affect ~,** nazlanmak, (istemem, cebime koy). **~t,** istemiyerek; gönülsüz.
**rely** [ri'lây]. **~ (up)on,** ···e güvenmek, emniyet etm.
**remain** [ri'mêyn]. Kalmak; baki kalmak; durmak; hâlâ mevcud olmak. **the fact ~s that ...,** bununla beraber şu var ki ...: **it ~s to be seen whether ...,** bakalım .... **~s,** bakiye; izler; harabe: **mortal ~,** cenaze. **~der,** artık; bakiye: **the ~,** ötekiler: **~s,** satılmamış nüshalar. **~ing,** geri kalan: **the ~,** öteki.
**remand** [ri'mând]. Muhakemesini ileriye bırakma(k). **he was ~ed for a week,** muhakemesi gelecek haftaya bırakıldı.
**remark** [ri'mâk]. Söz; ihtar; dikkat. Farketmek; dikkat etmek. Söylemek; şifahen veya tahriren bir mütalaa beyan etmek. **worthy of ~,** şayanı dikkat: **to make a ~,** bir ihtarda bulunmak: **it may be ~ed that ...,** şayanı dikkattir ki; şurasını kaydedelim ki: **to pass ~s upon s.o.,** birinin hakkında bir şeyler söylemek: **may I venture to ~ that ...,** musaadenizle şu noktaya işaret edebilir miyim ki. **~able,** şayanı dikkat; göze çarpan; acayib.
**remed·ial** [ri'mîdyəl]. Şifa verici; ilac gibi. **~y** ['remedi], çare; ilâc Tedavi etm.; çaresini bulmak; tamir etmek. **you have no ~ at law,** bu iş için dava açamazsın.
**rememb·er** [ri'membə(r)]. Hatırlamak; hatıra getirmek; anmak. **~ me (kindly) to them,** onlara benden selâm söyle: **to ~ oneself,** kendini toplamak; terbiyesini unutmamak: **it will be ~ed that ...,** hatırlardadır ki: **he ~ed me in his will,** vasiyetnamesinde beni unutmamış. **~rance** [ri'membrəns], hatırlama; hatıra; yadigâr. **~s,** selâmlar: **to the best of my ~,** hatırladığıma göre: **to call to ~,** hatıra getirmek: **to keep in ~,** hatırda tutmak: **I have no ~ of it,** onu hiç hatırlamıyorum: **in ~ of ...,** ···in hatırasına.
**remind** [ri'mâynd]. Hatırlatmak. **that ~s me!,** iyi ki aklıma geldi!: **he ~s one of his father,** babasını andırıyor. **~er,** hatırlatıcı söz, mektub vs.: **I'll send him a ~,** unutmasın diye bir daha yazarım.
**reminiscen·ce** [ˌremi'nisəns]. Hatırlama; hatırlanan şey. **to write one's ~s,** hatıralarını yazmak. **~t,** hatırlayan: **~ of,** ···i hatırlatan.
**remiss** [ri'mis]. İhmalci. **it was very**

~ of me not to have written, yazmamakla kabahat ettim.
**remi·ssion** [riˈmişn]. Affetme. **~t** [riˈmit], affetmek; havale etm.; göndermek; bir mahkemeden diğerine nakletmek. **~ttance,** gönderme; gönderilen para: **~-man,** ailesinden aldığı para ile geçinen kimse.
**remnant** [ˈremnənt]. Bakiye; kumaş parçası.
**remonstra·nce** [riˈmonstrəns]. Protesto, itiraz; tekdir. **~te, to ~ about stg.,** bir şeyi protesto etm.: **to ~ with s.o.,** birini tekdir etmek.
**remorse** [riˈmôs]. Vicdan azabı. **without ~,** vicdansız. **~ful,** vicdan azabı çeken. **~less,** amansız.
**remote** [riˈmo̜ut]. Uzak; hücra, dağbaşı olan. **~ ancestors,** eski atalar: **~ control,** (*elect.*) uzaktan idare: **~ prospect,** pek zayıf bir ihtimal: **a ~ resemblance,** cüzi bir benzeyiş.
**remount** *v.* [ˌrîˈma̜unt]. Tekrar binmek; tekrar yukarı çıkmak. *n.* [ˈrîma̜unt]. Yedek binek at.
**remov·e** [riˈmûv]. Kaldırmak; yerini değiştirmek; silmek; uzaklaştırmak; azletmek; nakletmek; taşınmak. Yer değiştirme. **it is but one ~ from ...,** ···e pek yakındir. **~able,** kaldırılabilir; azledilebilir. **~al,** nakil; kaldırma; yer değiştirme; taşınma; azil. **~ed,** uzak: **first cousins once ~,** kardeş torunları.
**remunerat·e** [riˈmyûnərêyt]. Hakkını ödemek; hizmetinin karşılığını vermek. **~ion,** ücret, hizmet mukabili. **~ive** [–ˈmyûnərətiv], kârlı.
**rena·issance** [riˈnêysəns]. Rönesans; uyanış. **~scent** [riˈnasənt], taze hayat bulan.
**rend** (**rent**) [rend, rent]. Yırtmak; çekip koparmak. **to ~ asunder,** yırtıp iki parçaya bölmek: **to ~ the garments,** döğünmek: **to ~ the heart,** yüreğini parçalamak: **to turn and ~ s.o.,** birdenbire birine sövüp saymak.
**render** [ˈrendə(r)]. (Karşılık olarak) vermek; ···haline koymak, ···laştırmak; çevirmek, tercüme etm.; (yağ) eritip tasfiye etmek. **to ~ beautiful,** güzelleştirmek: **to ~ dangerous,** tehlikeli bir hale koymak: **to ~ into Turkish,** türkçeye çevirmek: **to ~ safe,** temin etm.: **to ~ up,** teslim etm.: **to account ~ed,** *evvelce gönderilen bir faturadaki müfredatı tekrarlamak ikinci faturada kullanılan tabir.* **~ing** [ˈrendərin(g)], tercüme; tefsir; (piyes) temsil; (*mus.*) icra.
**rendezvous** [ˈrondivû]. Randevu.
**renegade** [ˈrenigêyd]. Mürted.
**renew** [riˈnyû]. Yenilemek; eskisinin yerine yenisini koymak. **to ~ a bill,** senedin vadesini yenilemek. **~al,** yenilenme; yeniden yapma; vade tecdidi.
**renounce** [riˈna̜uns]. Vazgeçmek; feragat etm.; reddetmek. **to ~ one's faith,** dininden dönmek: **to ~ a treaty,** bir muahedeyi feshetmek.
**renovate** [ˈrenəvêyt]. Yenilemek; tamir etmek.
**renown** [riˈna̜un]. Şöhret; nam; şan. **~ed,** meşhur.
**rent**[1] [rent]. *pp. of* **rend.** Yırtık; rahne. **~**[2]. Kira (ile tutmak); kiralamak. **~al,** alınan veya verilen kira. **~-day,** kira ödeme günü. **~-free,** kirasız.
**renunciation** [riˌnʌnsiˈêyşn]. Feragat; terk; vazgeçme.
**repaid** *v.* **repay.**
**repair** [riˈpe̜ə(r)]. Tamir (etm.); iyi bir hale koymak. Gitmek; müracaat etmek. **in good ~,** iyi halde: **in bad ~,** tamire muhtac: **to ~ one's fortunes,** servetini yeniden kurmak: **beyond ~,** tamir edilmez: **to be under ~,** tamirde olm.: **to ~ a wrong,** bir zarar telâfi etm.
**reparation** [ˌrepəˈrêyşn]. Tamir(at); tazmin.
**repartee** [ˌrepâˈtî]. Âni cevab. **to be quick at ~,** hazırcevab olmak.
**repast** [riˈpâst]. Yemek, taam.
**repatriate** [rîˈpatriêyt]. (Birini) kendi vatanına geri göndermek.
**repay** (**repaid**) [rîˈpêy, –pêyd]. Borc ödemek; karşılığını vermek; telâfi etmek. **to ~ an injury,** bir zararın acısını çıkarmak: **to ~ an obligation,** görülen bir iyiliğe mukabele etm.: **a book that ~s reading,** okumak zahmetine değer bir kitab. **~able,** ödenmesi lâzım. **~ment,** tediye; karşılık.
**repeal** [riˈpîl]. Fesih; ilga. Feshetmek.
**repeat** [riˈpît]. Tekrarlamak; tekrar söylemek; ezberden söylemek. Tekrarlama; (*comm.*) **~ order,** aynı şey [mikdarı] yeniden ısmarlama. **~ed,** mükerrer; tekrar tekrar yapılmış. **~er,** çalar ceb saati; mükerrer ateşli silâh.

**repel** [riˈpel]. Defetmek; reddetmek; iğrendirmek. **~ling** [**~lent**], dâfi; iğrenc.
**repent** [riˈpent]. Pişman olm.; tövbe etmek. **~ance,** pişmanlık; tövbe. **~ant,** tövbekâr.
**repercussion** [ˌrîpö(r)ˈkʌşn]. İnikâs; geri tepme.
**reperto·ire** [ˈrepətwâ(r)]. Bir tiyatro kumpanyasının vs. temsile hazır olduğu eserler. **~ry** [ˈrepətəri], fihrist; mecmua. **~ company,** her hafta yeni piyesi temsil eden bir tiyatro kumpanyası.
**repetition** [ˈrepitişn]. Tekrarlama.
**repine** [riˈpâyn]. Halinden şikâyet etm.
**replace** [riˈplêys]. Tekrar yerine koymak; başkasının yerine geçmek; yerine başkasını tayin etm.; telâfi etmek.
**replay** *v.* [rîˈplêy]. Berabere biten maçı tekrar oynamak. *n.* [ˈrîplêy]. Bu suretle oynanan maç.
**replet·e** [riˈplît]. Tıka basa doymuş. **~ion** [–ˈplîşn], doyma.
**replica** [ˈreplika]. Aynen taklid; suret.
**reply** [riˈplây]. Cevab (vermek); mukabele (etmek).
**report**[1] [riˈpôt] *n.* Takrir, rapor; tebliğ; şayia; şöhret; patlama sesi; talebenin not karnesi. **by general ~,** umumiyetle mütevatir olarak: **a man of good ~,** iyi şöhret sahibi adam: **to know of stg. by ~ only,** bir şeyi kulaktan bilmek: **there is a ~ that ...,** ... rivayet ediliyor: **law ~s,** mühim dava dosyalarından mürekkeb mecmua. **~**[2] *v.* Rapor etm.; haber vermek; anlatmak; birinin aleyhinde beyanatta bulunmak. **to ~ (oneself),** isbatı vücud etm.: **it is ~ed that ...,** ... rivayet ediliyor: **to ~ progress,** bir işin ilerleyişi hakkında malûmat vermek: **to ~ sick,** kendisinin hastalığını haber vermek: **to ~ (up)on stg.,** bir şey hakkında rapor vermek: **he is well ~ed on,** hakkında söylenenler iyidir. **~er** [riˈpôtə(r)], muhbir; gazeteci.
**repos·e** [riˈpǫuz]. İstirahat (etmek), dinle(ndir)me(k); sükûnet. **to ~ confidence in s.o.,** birine güvenmek. **~itory** [riˈpozitəri], mahzen, ambar.
**reprehen·d** [ˌrepriˈhend]. Azarlamak. **~sible,** azara lâyık.
**represent** [ˌrepriˈzent]. Temsil etm.; göstermek; tarif etm.; tasvir etm.; vekili olm. **he ~s himself as [to be]** ..., kendini ... gibi gösteriyor: **exactly as ~ed,** tarife tamamen uygun. **~ation,** tasvir; temsil; mümessillik; vekâlet; nezaketle yapılan ihtar: **to make false ~s to s.o.,** birine hakikati tahrif ederek söylemek: **the British Government has made ~s to Russia about the matter,** İngiliz hükûmeti bu hususta Rusya'ya protestoda bulunmuştur. **~ative,** mümessil; vekil; örnek; temsil eden; tipik.
**repress** [riˈpres]. Bastırmak. **~ion** [–ˈpreşn], bastırma; ihtibas. **~ive,** bastırıcı; zecrî.
**reprieve** [riˈprîv]. Ölüm cezasını affetme(k).
**reprimand** [ˌrepriˈmând]. Takbih (etmek); alenen ayıblama(k).
**reprint** *n.* [ˈrîprint]. Bir kitabın ikinci tab'ı; ayrı basım. *v.* [rîˈprint]. Yeniden tabetmek.
**reprisal** [riˈprâyzl]. Mukabelebilmisil.
**reproach** [riˈprǫuç]. Tekdir; itab; utanacak şey. İtab etm. **beyond ~,** kusursuz: **to ~ oneself,** kendini kabahatlı bulmak: **to ~ s.o. with stg.,** bir şey hakkında birine serzenişte bulunmak: **a term of ~,** takbih kabilinden söz. **~ful,** serzenişli.
**reprobate** [ˈreprobêyt]. Takdir etmemek. Habis; serseri.
**reproduc·e** [ˌrîprǫuˈdyûs]. Aynen taklid etm.; tekrar basmak; üretmek. **~tion** [–ˈdʌkşn], aynen taklid; yeniden basma; tenasül; üre(t)me.
**repro·of** [riˈprûf]. Hafifçe tekdir. **~ve** [riˈprûv], hafifçe tekdir etm.
**reptile** [ˈreptâyl]. Yılan ve kertenkele gibi zahifelerden hayvan; alçak adam.
**republic** [riˈpʌblik]. Cumhuriyet. **~an,** cumhuriyete aid; cumhuriyetçi.
**repudiate** [riˈpyûdiêyt]. Reddetmek; tanımamak; boşanmak.
**repugnan·ce** [riˈpʌgnəns]. Hoşlanmayış; istikrah; zıddiyet. **~t,** müstekreh.
**repuls·e** [riˈpʌls]. Muvaffakiyetsizlik; tard; defolunma. Defetmek; tardetmek; püskürtmek. **~ion** [–ˈpʌlşn], tiksinme; birbirini uzaklaştırma kuvveti. **~ive,** iğrenc.
**reput·able** [ˈrepyutəbl]. Namuslu. **~ation** [–ˈtêyşn], şöhret; tezkiye: **to get the ~ of ...,** adı ...e çıkmak: **with a bad ~,** tezkiyesi bozuk. **~e**

[ri'pyût], şöhret. **a doctor of ~,** tanınmış doktor: **to be held in ~,** sayılmak: **of ill ~,** dile düşmüş: **to know s.o. by ~,** birini ismen tanımak: **of no ~,** adı sanı belirsiz olan. **~ed,** gûya olan: **to be ~ wealthy,** zengin sanılmak.

**request** [ri'kwest]. Rica, taleb, dilek. Rica etm., taleb etmek. **at the ~ of s.o.,** birinin talebi üzerine: **to be in ~,** aranılmak; rağbet görmek: **bus-stop by ~,** ihtiyarî durak: **the public are ~ed to keep off the grass,** çimene basılmaması rica olunur.

**requiem** ['rekwiəm]. Ölünün ruhu için okunan dua; fatiha.

**requi·re** [ri'kwâyə(r)]. İstemek; taleb etm.; emretmek; muhtac olm.; icabetmek. **as ~red,** istenildiği gibi: **if ~red,** icabederse: **when ~red,** icabında. **~red,** lâzım; icabeden. **~rement,** ihtiyac; icab; şart. **~site** ['rekwizit], lâzımgelen; zarurî. Lâzımgelen şey: **~s,** levazımat; takım. **~sition** [ˌrekwi'zişn], el koymak; müsadere etm. Resmî taleb; elkoyma.

**requit·e** [ri'kwâyt]. Mukabele etm.; mislini ödemek; mükâfat [mücazat] vermek. **~al,** mükâfat; karşılık.

**rescind** [rî'sind]. Feshetmek.

**rescript** ['rîskript]. İrade.

**rescue** ['reskyû]. Kurtarmak; imdadına yetişmek. Kurtarma; imdad. **to the ~!,** imdad!: **~ party,** kurtarma ekipi.

**research** [ri'sö(r)ç]. Derin ve dikkatli tedkik; araştırma.

**reseat** ['rî'sît]. Bir sandalye, pantalon vs.nin oturacak yerini yenilemek. **to ~ oneself,** tekrar oturmak.

**resembl·e** [ri'zembl]. Benzemek; andırmak. **~ance,** benzeyiş.

**resent** [ri'zent]. (Bir şey) gücüne gitmek; ···den alınmak. **~ful,** küskün; kindar. **~ment,** küskünlük; gücenme; kin.

**reserv·ation** [ˌrezö(r)'vêyşn]. İhtiraz kaydi; istisna; tren vs.de tutulmuş yer; yer tutma. **to make a ~,** ihtirazî kayıd ileri sürmek: **game ~,** içinde yabani hayvanların avlanması yasak olan arazi; **Indian ~,** Amerika'da yerli hintliler için ayrılmış arazi: **mental ~,** içinden karar verme. **~e**[1] [ri'zö(r)v] *n.* İhtiyat; ihtiyat akçesi, kuvveti, kaydı; şart; muayyen bir maksad için ayrılmış arazi; çekingenlik. *a.* Yedek, ihtiyat. **the ~,** silâh altında olmıyan ihtiyatlar: **the ~s,** bir ordunun ihtiyata ayrılmış kuvvetleri: **to accept with ~,** kaydı ihtiyatla telâkki etm.: **to accept without ~,** olduğu gibi kabul etm.: **to break through s.o.'s ~,** birinin çekingenliğini yenmek: **to cast off ~,** açılmak: **in ~,** yedek [ihtiyat] olarak: **~ price,** son fiat: **to be sold without ~,** kayıdsız şartsız satılacak: **without ~,** uluorta; kayıdsız şartsız. **~e**[2] *v.* İhtiyat olarak saklamak; muhafaza etm.; tahsis etmek. **to ~ a seat for s.o.,** birine bir yer tutmak. **~ed** [ri'zö(r)vd], ketum; çekingen; münzevî tabiatli. **~ seat [place],** tren vs. de önceden tutulmuş yer: **all rights ~,** her hakkı mahfuzdur: **~ list,** yedek kadro. **~ist** ['rezö(r)vist], (*mil.*) yedek. **~oir** ['rezö(r)vwâ(r)], su haznesi, bend; (*fig.*) ihtiyat depo; hazne.

**reset** [rî'set]. Yeniden yerine koymak; (saati) doğrultmak.

**reside** [ri'zâyd]. İkamet etm., oturmak. **~nce** ['rezidəns], ikametgâh; ikamet etme. **~ncy,** müstemlekelerde umumî valinin resmî konağı. **~nt,** yerli; umumî vali; ikamet eden; sakin: **~ master,** bir mektebde yatıp kalkan muallim: **~ physician,** bir hastahanede yatıp kalkan ve tamamen orada çalışan doktor. **~ntial** [rezi'denşl] **district,** bir şehirde hususî evlerin bulunduğu kısım.

**residu·e** ['rezidyû]. Tortu; artık, bakiye. **~al** [rə'zidyuəl], artık nevinden; fazla ve baki kalan. **~ary** [–'zidyuəri], artık; baki.

**resign** [ri'zâyn]. İstifa etm.; vazgeçmek; feragat etm.; teslim etmek. **to ~ oneself (to),** tevekkül etm. **~ation** [ˌrezig'nêyşn], istifa; tevekkül. **~ed** [ri'zâynd], mütevekkil: **to become ~ to stg.,** istemiyerek razı olm.

**resilient** [ri'ziliənt]. Elâstikî, esneyen.

**resin** ['rezin]. Çam sakızı. **~ous,** sakızlı.

**resist** [ri'zist]. Dayanmak; mukavemet etm.; karşı koymak. **I could not ~ telling him,** ona söylemekten kendimi alamadım: **he could not ~ drink,** içkiye hiç yüzü yoktu. **~ance,** mukavemet; karşı koyma; dayanma; muhalefet: **to take the line of least ~,** nen kolay yolu tutmak. **~ant,** dayanıklı; mukavim.

**resolut·e** ['rezolyût]. Azimkâr; metin; cesur. **~ion** [–'yûşn], azim-

kârlık; karar; teklif: **good ~s,** iyi niyetler: **to put a ~ to the meeting,** bir teklifi reye koymak.
**resolve** [ri'zolv]. Karar (vermek), niyet(inde olm.); halletmek. **I am ~d not to go there again,** bir daha oraya gitmemeğe karar verdim.
**resonant** ['rezənənt]. Tannan.
**resort** [ri'zôt]. Merci; baş vurulacak yer; çare; çok gidilen yer. Müracaat etm.; kullanmak; gitmek. **in the last ~,** başka çare yoksa: **without ~ to force,** zora başvurmadan: **place of great ~,** çok ziyaret edilen yer: **seaside ~,** deniz kenarında sayfiye yeri.
**resound** [ri'zaund]. Tannan olm.; her taraftan duyulmak: (şöhret) yayılmak; aksettirmek. **~ing,** tannan; gürliyen: **a ~ success,** dillerde destan olan muvaffakiyet: **a ~ blow,** gürliyen darbe.
**resource** [ri'sôs]. Çare. **~s,** imkânlar. **his ~s are limited,** imkânları mahduddur. **~ful,** işin içinden çıkar, becerikli.
**respect** [ri'spekt]. Hürmet; riayet; itibar; münasebet. Hürmet etm., saygı göstermek; riayet etmek. **as ~s** ..., ···e gelince: **to have ~ for s.o.,** birinin hatırını saymak: **in every ~,** her hususta: **in many ~s,** bir çok bakımlardan: **in some ~s,** bazı hususlarda: **in this ~,** bu hususta: **out of ~ for,** ···in hatırı için: **to present one's ~s,** saygılarını sunmak: **with ~ to,** ···e gelince: **with all due ~ (to you),** hatırınız kalmasın: **without ~ of persons,** ⌜hem nalına hem mıhına⌝. **~able** [ri'spektəbl], hatırı sayılır; namuslu; kılığı kıyafeti düzgün; (*coll.*) oldukça iyi. **~ed,** hürmet gören; muhterem. **~er,** hürmet eden: **no ~ of persons,** hatır gönül bilmez. **~ful,** hürmetkâr: **to keep at a ~ distance,** hürmetkârane geri durmak: **I remain yours ~ly,** hürmetlerimi sunarım. **~ing,** ···e dair. **~ive,** kendi; sıraya göre; biri ... öteki: **the brothers are 10 and 12 ~ly,** iki kardeşten biri 12 öteki 10 yaşındadır.
**respir·e** [ri'spâyə(r)]. Nefes almak. **~ation** [ˌrespə'rêyşn], nefes alma. **~ator,** teneffüs cihazını koruyan maske.
**respite** ['respit, -pâyt]. Mühlet; muvakkat tehir; dinlenme. Mühlet vermek. **without ~,** fasılasız: **to give no ~,** nefes aldırmamak.
**resplendent** [ri'splendənt]. Parlak.
**respon·d** [ri'spond]. Cevab vermek; mukabele etm.; uymak; **to ~ to the controls,** (uçak) direksiyona itaat etmek. **~ent,** cevab veren; müddeaaleyh. **~se** [ri'spons], cevab; karşılık. **~sive,** hassas.
**responsib·le** [ri'sponsibl]. Mesul; ciddî; sadık. **~ility** [-'biliti], mesuliyet: **on one's own ~,** kendiliğinden.
**rest**[1] [rest] *n.* Dinlenme; rahat; durak; mola; sehpa. **to be at ~,** hareketsiz olm.; rahat olm.; ölmüş olm.: **to come to ~,** nihayet durmak: **to lay to ~,** defnetmek: **to set s.o.'s mind at ~,** birini ferahlatmak: **to take a [one's] ~,** bir az dinlenmek. **~**[2] *v.* Dinlen(dir)mek; istirahat et(tir)mek; sakin durmak; daya(n)mak: **~ assured that ...,** emin olunuz ki: **it ~s with you to ...,** ... sizin elinizdedir: **it does not ~ with me to ...,** ... benim elimde değil: **his glance ~ed upon,** bakışları ···in üzerinde durdu: **the matter cannot ~ here,** mesele burada bırakılamaz. **~-house,** misafirhane. **~**[3] *n.* Artık şey, bakiye; ötesi. **the ~,** diğerleri: **all the ~,** geri kalanlar: **for the ~,** ötesine gelince. **~ful** ['restfəl], rahat verici. **~ing-place** ['restin(g) 'plêys], istirahat yeri; mezar. **~ive** ['restiv], yürümek istemiyen; inadcı; aksi. **~less** ['restlis], rahatsız; tezcanlı; uykusuz.
**restaurant** ['restəro(n)]. Lokanta.
**restitution** [ˌresti'tyûşn]. Geri verme.
**restor·e** [ri'stô(r)]. Geri vermek; tamir etm.; yeniden kurmak. **~ation** [-'rêyşn], geri verme; yeniden kurma; tamir ederek eski haline koyma: **the R~,** 1660'ta Stuart hanedanının İngiliz tahtına yeniden geçmesi. **~ative,** kuvvet veren ilâc.
**restrain** [ri'strêyn]. Alıkoymak; menetmek; frenlemek. **a ~ing influence,** ayak bağı. **~ed,** zabtedilmiş; ölçülü. **~t,** zabt; tazyik; men; inzibat; itidal. **to fling aside all ~,** aklına geleni yapmak: **to keep s.o. under ~,** birini hapishane, tımarhanede tutmak: **lack of ~,** itidalsizlik; inzibatsızlık: **to speak without ~,** serbestçe söylemek: **to be under no ~,** serbest olmak.
**restrict** [ri'strikt]. Hasretmek; tahdid etm. **~ed,** mahdud; münhasır. **~ion** [-'strikşn], tahdid, inhisar. **~ive,** tahdid edici.

**result** [ri'zʌlt]. Netice; son; akibet. Neticelenmek; hâsıl olmak. **~ant,** hâsıl olan.
**resum·e** [ri'zyûm]. Yeniden başlamak; kesilmiş söze devam etmek. **to ~ one's seat,** tekrar yerine oturmak. **~ption** [ri'zʌmşn], yeniden başlama.
**résumé** ['rêyzyûmêy]. Hulâsa.
**resurrect** [ˌrezə'rekt]. Diriltmek; mezardan çıkarmak; yeniden rağbet kazandırmak. **~ion** [–'rekşn], diril(t)me; yeniden doğma; Hazreti İsa'nin mezardan çıkması.
**resuscitate** [ri'sʌsitêyt]. Canlandırmak.
**retail** ['rîtêyl]. Perakende satış. Perakende sat(ıl)mak; tafsilâtla anlatmak. **~er,** perakendeci: **~ of news,** havadis yayıcı.
**retain** [ri'têyn]. Alıkoymak; kendi elinde tutmak; muhafaza etm.; hatırında tutmak. **~er,** (i) uşak; (ii) hizmetini temin etmek için verilen pey akçesi: **~s,** maiyet efradı. **~ing wall,** destek olarak duvar.
**retaliat·e** [ri'taliêyt]. Karşılık yapmak. **~ion** [–'êyşn], mukabelebilmisil; öc: **in ~,** buna mukabil.
**retard** [ri'tâd]. Geciktirmek.
**retch** [reç]. Kusmağa çalışmak.
**retent·ion** [ri'tenşn]. Alıkoyma; kendi elinde tutma; muhafaza. **~ive,** bırakmaz: **~ memory,** kuvvetli hafıza.
**reticent** ['retisənt]. Bildiğini söylemez; ketum.
**retinue** ['retinyû]. Maiyet.
**retire** [ri'tâyə(r)]. Çekilmek; tekaüd olm.; yatmağa gitmek; geri çekmek; tekaüd etmek. **to ~ into oneself,** kabuğuna çekilmek. **~d,** emekli. **~ment,** ricat; tekaüd.
**retort** [ri'tôt]. Sert cevab (vermek).
**retouch** [rî'tʌç]. Rötuş (yapmak).
**retrace** [rî'trêys]. Kaynağına gitmek. **to ~ one's steps,** izini takiben geriye gitmek.
**retract** [ri'trakt]. Geriye çekmek; sözünü geri almak.
**retread** [rî'tred]. (Otomobilin lâstiğinin dış tabakasını) yenilemek.
**retreat** [ri'trît]. Ricat; melce. Ricat etmek. Geriye çekmek. **to beat a ~,** ricat etm.
**retrench** [ri'trenç]. Kısmak.
**retribution** [ˌretri'byûşn]. Ceza; intikam.
**retrieve** [ri'trîv]. Tekrar elde etm.; kurtarmak; (av köpeği) avı getirmek. **~r,** vurulan avı getiren köpek.
**retro-** ['rîtrou]. *pref.* Geriye doğru; geçmişe bakan.
**retrogr·ade** ['retrougrêyd]. Terakkiye karşı olan, mürteci. **~ess** [rîtrou'gres], geriye gitmek.
**retrospect** ['retrouspekt]. Geriye bakma. **~ive** [–'spektiv], makabline şamil.
**return**[1] [ri'tö(r)n] *n.* Dönüş; iade; mukabele, kazanc; resmî rapor; istatistik; bir mebusun intihab edilmesi. **~s,** kâr; irad; raporlar. **in ~,** karşılık olarak: **in ~ for,** ···e mukabil: **~ of income,** gelir beyannamesi: **many happy ~s of the day!,** nice senelere! (*birinin gününde söylenir*): **~ match,** intikam maçı: **by ~ of post,** (mektubu alır almaz) ilk posta ile: **~ ticket,** gidip gelme bilet. **~**[2] *v.* İade etm.; intihab etm.; karşılık yapmak; avdet etm.; geri gelmek. **to be ~ed for (such-and-such a district),** filan yerden mebus çıkmak: **the prisoner was ~ed guilty,** maznunun suçlu olduğuna karar verildi: **to ~ like for like,** mislile mukabele etm.: **~ing officer,** seçim memuru: **to ~ thanks,** (bir nutukla) teşekkür etmek.
**reuni·on** [rî'yûnyən]. Kavuşma; birleşme; toplantı. **~te** [ˌrîyû'nâyt]. Yeniden birleş(tir)mek; barış(tır)mak.
**rev** [rev]. (*abb.*) **Revolution,** devir. **to ~ up the engine,** motörü hızlatmak.
**reveal** [ri'vîl]. İfşa etm.
**reveille** [re'veli]. Kalk borusu.
**revel** ['revl]. Cümbüş, âlem (yapmak). **to ~ in stg.,** bir şeye pek düşkün olm. **~ler,** cümbüş, âlem yapan. **~ry,** cümbüş.
**revelation** [revə'lêyşn]. İfşa.
**revenge** [ri'venc]. İntikam (almak); hıncını çıkarmak. **out of ~,** intikam yüzünden. **~ful,** kin tutan.
**revenue** ['revənyu]. Gelir. **the public ~,** devlet varidatı: **~ officer,** rüsumat memuru.
**reverberate** [ri'vö(r)bərêyt]. Tannan olmak.
**revere** [ri'viə(r)]. Hürmet etm.; takdis etmek. **~nce** ['revərəns], hürmet, takdis (etm.): **to hold s.o. in ~,** birine tâzim göstermek: **to pay ~ to s.o.,** birine tâzim etm. **~nd,** muhterem (*rahibler için*). **~nt,** hürmetkâr. **~ntial** [–'renşl], hürmetten gelen.

**reverie** [ˈrevəri]. Dalgınlık.
**revers·al** [riˈvö(r)sl]. Taklib; fesih. **~e** [riˈvö(r)s], ters; aksi. Ters tarafı; aksi; mağlubiyet. Tersine çevirmek; ibtal etm.; muayyen bir şeyin aksini yapmak. **to ~ arms,** tüfekleri başaşağı etm.: **to ~ a car,** otomobili geri yürütmek: **to go into ~,** (*otom.*) geriye almak: **to be quite the ~ of stg.,** bir şeyin taban tabana zıddı olm.: **to take a position in ~,** bir mevkii arka tarafından zabtetmek: **to suffer a ~,** bir hezimete uğramak. **~ible** [riˈvö(r)sibl], devredebilir; tersine çevrilmesi mümkün.
**rever·sion** [riˈvö(r)şən]. Eski haline avdet; ilk sahibine dönme. **~t** [riˈvö(r)t], eski haline dönmek; eski sahibinin eline gelmek.
**revictual** [rîˈvitl]. İaşe etmek.
**review** [riˈvyû]. Geçid resmi; muayene; kitab tenkidi; mecmua. Tekrar gözden geçirmek; teftiş etm.; bir kitabın tenkidini yapmak. **~er,** kitab münekkidi.
**revile** [riˈvâyl]. Sövmek; küfretmek.
**revis·e** [riˈvâyz]. Tekrar gözden geçirmek; tashih, ıslah etm. **~ion** [–ˈvijn], yeniden gözden geçirme; tashih.
**reviv·e** [riˈvâyv]. Canlan(dır)mak; ayıl(t)mak; kurcalamak. **~al,** canlan(dır)ma; ayılma.
**revo·cable** [riˈvokəbl]. Feshedilebilir; geri alınması mümkün. **~cation** [–ˈkêyşn], fesih. **~ke** [riˈvo̱uk], feshetmek; geri almak.
**revol·t** [riˈvo̱ult]. İsyan (etm.), ayaklanma(k); iğrendirmek. **~ting,** iğrenc. **~ution** [ˌrevəˈlyûşn], devir, devrî hareket; inkilâb. **~utionary,** inkilâblara aid; tamamen değiştiren; inkilâbcı. **~utionize** [–âyz], tamamen değiştirmek. **~ve** [riˈvolv], dön(dür)mek, devret(tir)mek. **to ~ stg. in one's mind,** bir meseleyi düşünüp taşınmak. **~ver,** altıpatlar. **~ving,** dönen; devvar: **~ chair,** döner iskemle: **~ light,** yanar söner fener.
**revue** [riˈvyû]. Rövü.
**revulsion** [riˈvʌlşn]. Âni ve kuvvetli değişiklik; reaksiyon.
**reward** [riˈwôd]. Mükâfat(ını vermek).
**Re·x, ~gina** [reks, regînə]. Kıral (-içe). **~ v. Jones,** (İngiliz hukukunda) hükümetin vatandaş Jones'a karşı açtığı dava.
**Reynard** [ˈrenâd]. (Edebiyatta) tilki.
**rhapsod·y** [ˈrapsədi]. Heyecanlı yazı, güfte. **~ize, to ~ over stg.,** bir şeyi pek mübalağalı medhetmek.
**rhetoric** [ˈretərik]. Belâgat; tesirli hatiblik. **~al** [riˈtorikl], belâgate aid: **a ~ question,** yalnız tesir için ve cevabı beklenmiyen bir sual.
**rheum·atism** [ˈrûmətizm]. Romatizma. **~y** [ˈrûmi], **~ eyes,** kızarmış ve sulu gözler.
**rhinoceros** [râyˈnosərəs]. Gergedan.
**rhyme** [râym]. Kafiye; kısa şiir. Kafiyeli olm. **without ~ or reason,** ipsiz sapsız. **nursery ~,** çocuklar için tekerleme.
**rhythm** [ˈriðm]. Vezin; ahenk; mütenasib hareket. **~ic,** mevzun; ahenkli.
**rib** [rib]. Kaburga kemiği; şemsiye teli. **to poke s.o. in the ~s,** birini şaka ile dürtmek. **~bed** [ˈribd], yivli; muntazam girintili çıkıntılı.
**ribald** [ˈribôld]. Alay eden; küstah şakacı; açık saçık. **~ry,** soğuk şaka.
**ribbon, riband** [ˈribən(d)]. Kurdelâ. **blue ~,** dizbağı nişanı; her hangi sahada üstünlük işareti: **to cut to ~s,** lime lime kesmek.
**rice** [râys]. Pirinc. **~ pudding,** sütlâc. **~-field,** çeltik.
**rich** [riç]. Zengin; bereketli; (yemek) çok tatlı, yağlı vs.; ağır. **he ~ly deserved his fate,** başına geleni tamamen hak etti: **to grow ~,** zengin olm.: **the newly ~,** sonradan görmeler: **that's ~!,** ama lâf! **~es,** servet. **~ness,** zenginlik; bolluk; (yemek) ağırlık; (renk) parlaklık; (ses) gürlük.
**rick** [rik]. Ot, saman yığını.
**ricket·s** [ˈrikits]. Kırba illeti. **~y,** raşitik; sarsak, sağlam olmıyan.
**ricochet** [ˈrikəşêy]. (Taş vs.) atılan bir şeyin toprağa vurarak sekmesi. Böyle sekmek.
**rid** [rid]. Kurtarmak. Kurtulmuş. **to ~ the house of rats,** evi farelerden temizlemek: **to get ~ of [~ oneself of],** başından atmak. **~dance** [–əns], başından atma, kurtulma: **a good ~,** isabet oldu da ondan kurtulduk: **a good ~ of bad rubbish,** Allaha şükür musibetten kurtulduk.
**ridden** [ˈridn] *v.* **ride.** ···den mazlum; ···in el altında bulunan. **priest-~,** papaz istilâsına uğramış.
**riddle**[1] [ˈridl]. Muamma; bilmece. **~**[2]. Kalbur(dan geçirmek); delik deşik etmek.

**ride** (**rode, ridden**) [râyd, rQud, ridn]. ···e binmek; ata binmek. At, bisiklet vs. üzerinde gezinti; ormanda açılmış yol. **he ~s well,** iyi binicidir: **to ~ at anchor,** (gemi) demirli yatmak: **to ~ an idea, etc., to death,** bir fikir vs.yi ifrata vardırmak: **to ~ down,** atla giderek çiğnemek: **to ~ for a fall,** körükörüne bir felâkete doğru gitmek: **to go for a ~,** at vs.ye binerek gezinti yapmak: **to ~ hard,** (i) at vs.ile alabildiğine gitmek; (ii) gözü pek binici olm.: **it's a penny ~ on a bus,** otobüs ile bir penilik mesafedir: **to ~ a race,** at yarışına girmek: **to ~ (out) the storm,** (gemi) fırtınayı selâmetle atlatmak; (*fig.*) dayanıp fena bir vaziyetten kurtulmak: **to take s.o. for a ~,** (*Amer. sl.*) birisini öldürmek maksadile otomobile bindirip götürmek. **~r**[1] [ˈrâydə(r)], binici; süvarı. **~**[2], müzeyyel madde.

**ridge** [ric]. Sırt; dağ silsilesi; resif.

**ridicul·e** [ˈridikyûl]. Alay; hiciv. Alay etm.; gülünçleştirmek. **~ous** [–ˈdikyuləs], gülünç: **to make oneself ~,** âleme gülünç olmak.

**riding** [ˈrâydin(g)]. Binicilik; süvarilik. **~-lights,** demirli geminin fenerleri.

**rife** [râyf]. Çok bulunan; müstevli. **to be ~,** her tarafa yayılmak.

**riff-raff** [ˈrifraf]. Aburcubur kimseler.

**rifle**[1] [ˈrâyfl]. Ceblerini araştırıp soymak. **~**[2]. Yivli tüfek. **~man,** silâhşor; avcı eri. **~-range,** atış meydanı.

**rift** [rift]. Yarık; rahne. **a ~ in the clouds,** bulutlar arasında açık bir yer: **'a ~ in the lute',** iki dost arasında ufak bir nifak: **to create a ~ between ...,** aralarını açmak.

**rig**[1] [rig]. Hileli bir tarzda kurmak. **a ~ged court,** yalancıktan yapılan muhakeme: **to ~ the market,** piyasada fiatları hile ile yükseltmek veya alçaltmak. **~**[2]. Armanın tarz ve üslûbu; (*coll.*) kıyafet. Donatmak; giydirmek. **~ging,** gemi arması.

**right**[1] [râyt] *a. adv.* Doğru; haklı; elverişli; sağ (taraf). Doğru olarak; sağ tarafa; tam. **all ~,** pek iyi, hay hay: **he's all ~,** (i) bir şeyi yok, iyileşti; (ii) fena adam değildir: **it's all ~ for you to laugh,** senin için gülmek kolay: **~ angle, kaim** zaviye: **~ away!,** fayrap!; **to do stg. ~ away,** bir şeyi derhal yapmak: **to be ~,** haklı olm.: **things will come ~,** sonu iyi olacak: **that was him, ~ enough!,** hiç şübhe yok oydu: **am I ~ for London?,** bu Londra yolu mu (treni vs.mi)?: **he's not ~ in the head,** aklından zoru var: **~ ho!** hay hay! **~ and left,** sağda solda: **to hit out ~ and left,** hem nalına hem mıhına vurmak: **to be in one's ~ mind,** aklı başında olm.: **to go ~ on,** dosdoğru gitmek: **to put ~,** düzeltmek: **serves him ~!,** oh olsun!: **~ side up,** doğru (ters değil): **to get on the ~ side of s.o.,** birinin gözüne girmek: **he is on the ~ side of fifty,** yaşı elli yoktur: **that's ~!,** tamam!, doğru!: **to do the ~ thing by s.o.,** birine karşı insaflı davranmak: **he always says the ~ thing,** her zaman isabetli şey söyler. **~-about,** sağdan geri: **to send to the ~,** defetmek. **~-angled,** kaim zaviyeli. **~-down,** (*coll.*) tam; sapına kadar. **~-hand,** sağdaki: **s.o.'s ~ man,** birinin sağ eli (yerindeki adam). **~-handed,** sağ elile iş gören. **~-minded** [**-thinking**], doğru düşünceli. **~**[2] *n.* Hak; adalet; hakikat; sıhhat; salâhiyet; sağ el, taraf. **to be in the ~,** haklı olm.: **by ~s,** usûlen: **keep to the ~!,** sağdan gidiniz!: **on the ~,** sağ tarafta: **to possess stg. in one's own ~,** re'sen hak sahibi olm.: **to set to ~s,** yoluna koymak; düzeltmek: **~ of way,** mürur hakkı: **to be within one's ~s,** hak ve salâhiyeti dahilinde olm.: **I don't know the ~s and wrongs of it,** kimin haklı olduğunu bilmiyorum. **~**[3] *v.* Düzeltmek; hakkını ihkak etm.; tashih etmek. **~eous** [ˈrâytyəs], dürüst, müstakim; dindar. **~ indignation,** haklı hiddet. **~ful** [ˈrâytfəl], haklı; meşru; hak sahibi. **~ly** [ˈrâytli], doğru olarak.

**rigid** [ˈricid]. Eğilmez; sert. **~ity** [–diti], eğilmezlik; sertlik.

**rigmarole** [ˈrigmərQul]. Abuksabuk lâflar.

**rigo·rous** [ˈrigərəs]. Sert; şiddetli. **~ur,** sertlik, şiddet.

**rile** [râyl]. Sinirine dokunmak.

**rim** [rim]. Kenar; çerçeve.

**rind** [râynd]. Kabuk; deri.

**ring**[1] [rin(g)]. Halka; çerçeve; daire; yüzük; (ticarette) şebeke; (boks) ring. Etrafına halka çevirmek; etrafını kuşatmak; (hayvan) burnuna halka takmak. **the ~,** (i) boksörlük; (ii) (at yarışlarına) bahis defteri tutan adamlar: **to keep [hold] the ~,** bir mücadelede haricî müdahaleyi önlemek: **to make ~s round s.o.,** birisinden çok daha çabuk koşmak; birine taş çıkartmak: **split ~,** anahtar halkası: **wedding ~,** alyans. **~-finger,** yüzük parmağı. **~²**. Çan (çıngırak) sesi; çıngırtı. Çan çalmak; çalınmak. **there's a ~ at the door,** kapının zili çalınıyor: **there's a ~ on the telephone,** telefon çalıyor: **give me a ~ when you are back,** döndüğünüz zaman bana telefon ediniz: **to ~ true,** hakikiye benzemek. **~ down the curtain,** bir şeye nihayet vermek. **~ off,** telefonu kapatmak. **~ up,** telefon etmek. **~er** [ˈrin(g)ə(r)], kilise çancısı. **~leader** [ˈrin(g)lîdə(r)], elebaşı. **~let** [ˈrin(g)lit], kâkül; halkacık.

**rink** [rin(g)k]. Patinaj yeri.

**rinse** [rins]. Çalkama(k).

**riot** [ˈrâyət]. Kargaşalık; ayaklanma. Halk sokaklara üşüşüp gürültülü kargaşalık yapmak. **to read the R~ Act,** fesadcılara karşı ateş açmadan evvel ihtarda bulunmak; (*fig.*) şiddetle azarlamak: **to run ~,** başıboş hareket etm. **~ous,** gürültülü. **~ living,** hovardalık.

**R.I.P.** [ˈârâyˈpî]. (*abb. Lât.*) **Requiescat in pace,** Allah rahmet etsin!

**rip¹** [rip]. Yırtık. Yırtmak; dikişini sökmek. (*coll.*) **~ (along),** alabildiğine koş(tur)mak. **to ~ open,** yırtıp açmak; karnını delmek. **~²**. Uçarı çapkın.

**ripe** [râyp]. Kemale ermiş; olgun. **~ for execution,** tatbika hazır. **~n,** olgunlaş(tır)mak; kemale er(dir)mek.

**riposte** [ˈrîpost]. Hemen verilen cevab. Nükteli, sert cevab vermek.

**ripping** [ˈripin(g)]. (*sl.*) Mükemmel; âlâ.

**ripple** [ˈripl]. Ufacık dalga; çağlama(k); hafifçe dalgalanmak.

**ris·e¹** [râyz] *n.* Yükselme; çıkış; yokuş; çıkıntı; artma; doğuş. **to ask for a ~,** maaşının artmasını istemek: **to be on the ~,** artmakta olm.: **to get [take] a ~ out of s.o.,** birini hassas bir noktasına basıp tahrik ederek kızdırmak: **to give ~ to,** sebeb olm.: **to take its ~,** (nehir) membaından çıkmak. **~² (rose, risen)** [râyz, rouz, rizn] *v.* Kalkmak; çıkmak; yükselmek; artmak; erişmek, yetişmek; (güneş) doğmak; zuhur etm.; ayaklanmak; (meclis vs.) tatil edilmek: **to ~ to a remark,** mahsus tahrik maksadile söylenen bir söze kapılarak kızmak: **to ~ from the dead,** (ölü) dirilmek: **my hair rose,** saçlarım dimdik oldu: **my whole soul ~s against it,** bütün ruhum buna karşı isyan ediyor. **~ing** [ˈrâyzin(g)], yükselen; **artan;** doğan. Kalkış; doğuş; artma; yükseliş; isyan. **to be ~ four,** dört yaşına yaklaşmak: **the ~ generation,** yeni nesil: **a ~ man,** istikbali parlak adam.

**risk** [risk]. Tehlike; baht, şans. Tehlikeye atmak; göze almak. **to run a ~,** tehlikeye girmek: **to take a ~,** tehlikeyi göze almak: **to ~ one's own skin,** dayak yemeği göze almak. **~y,** tehlikeli; şanslı: **a ~ story,** açık saçık hikâye.

**rissole** [ˈrisoul]. Köfte.

**rit·e** [râyt]. Âyin. **~ual** [ˈrityuəl], dinî merasim; örf.

**rival** [ˈrâyvl]. Rakib. Rekabet etm.; müsavi olmak. **without ~,** emsalsiz: **nobody can ~ him in eloquence,** hitabette kimse ona çıkışamaz. **~ry,** rekabet.

**river** [ˈrivə(r)]. Nehir; ırmak; çay. **~side,** nehir kenarı.

**rivet** [ˈrivit]. Perçin çivisi. Perçinlemek. **to ~ one's attention,** dikkatini bir noktaya çivilemek.

**R.M.** (*abb.*) **Royal Marines.** Deniz silâhendazları.

**R.N.** (*abb.*) **Royal Navy.** Kıralî Bahriye.

**R.N.V.R.** (*abb.*) **Royal Naval Volunteer Reserve.** Deniz gönüllü yedekleri.

**road** [roud]. Yol, cadde. **the rule of the ~,** araba vs.nin yolun hangi tarafından gideceğini tayin eden nizam: **to take the ~,** yola çıkmak. **~man,** yol tamircisi. **~side,** yol kenarı: **~stead,** gemilerin demir yeri. **~way,** yolun ortası.

**roam** [roum]. Maksadsız gezmek.

**roar** [rô(r)]. Aslan sesi; gürleme. Aslan gibi bağırmak; gürlemek. **~s of laughter,** gürültülü kahkahalar: **to set the table in a ~,** sofrada herkesi gülmekten katıltmak. **~ing,** gürleyen; çayır çayır yanan (ateş): **the ~ forties,** 40°–50° cenub arz dairesinde bulunan deniz mıntakası: **in this heat the ice-cream sellers are doing a ~ trade,** bu sıcakta dondurma kapışılıyor.

**roast** [roust]. Kızartma(k); kavurmak.

**rob** [rob]. Çalmak; soymak. **to ~ s.o.**

of stg., birinin bir şeyini çalmak. ~ber, hırsız; haydud. ~bery, hırsızlık: highway ~, haydudluk.

**robe** [roub]. Rob; elbise; cübbe. Giydirmek. ~ of honour, kaftan.

**robin** [ˈrobin]. Muta. round ~, çok imzalı bir istida. R~ Hood [ˈhud], Zenginleri soyan fakat fıkaraya yardım eden meşhur bir İngiliz haydudu.

**robot** [ˈroubot]. Makine adam; otomatik.

**robust** [rouˈbʌst]. Gürbüz, dinc; kuvvetli. ~ious, gürültülü; kabadayı.

**rock**[1] [rok] *v.* Salla(n)mak; sars(ıl)mak. to ~ to sleep, beşiği sallayarak uykusunu getirmek. ~[2] *n.* Kaya(lık); taş. the R~, Cebelüttarık: to be on the ~s, kayalıkta oturmak; fena bir vaziyette bulunmak: to see ~s ahead, önünde tehlike görmek: to run upon the ~s, (gemi) kayalığa çarpmak. ~-bottom, en alçak nokta: ~ price, en aşağı fiat. ~-bound, kayalarla kuşatılmış. ~-garden [~ery] [ˈrokəri], kayalıklarda yetişen çiçekler için taş ve çakıldan mürekkeb küçük bahçe. ~er [ˈrokə(r)], beşik ayağı. off his ~, (*sl.*) bir tahtası eksik. ~y [ˈroki], kayalık, kaya gibi. the R~ Mountains [Rockies], Roki dağları.

**rocket** [ˈrokit]. Havaî fişek (gibi uçmak). ~-bomb, füze bombası.

**rod** [rod]. Çubuk; değnek; asâ; bir dayak aleti: ~ and line, sırıklı balık oltası: to make a ~ for one's own back, belâyı başına satın almak: to rule with a ~ of iron, çok sıkı bir disiplinle idare etm.: ⌜spare the ~ and spoil the child⌝, ⌜kızını dövmiyen dizini döver⌝.

**rode** *v.* ride.

**rodent** [ˈroudənt]. Kemirici (hayvan).

**roe**[1] [rou]. ~ (-deer), karaca. ~[2]. hard ~, balık yumurtası: soft ~, balık nefsi.

**Roger** [ˈrocə(r)]. the Jolly ~, korsan bayrağı.

**rogu·e** [roug]. Çapkın; yaramaz; dolandırıcı. a ~ elephant, sürüden ayrılmış ve tehlikeli olan fil: ~s' gallery, sabıkalıların resimlerini ihtiva eden ve polisçe maznunların teşhisinde kullanılan albüm. ~ery, dolandırıcılık; yaramazlık. ~ish, çapkınca; kurnaz.

**roisterer** [ˈrôystərə(r)]. Gürültücü.

**Roland** [ˈrouland]. to give s.o. a ~ for his Oliver, taşı gediğine koymak.

**role** [roul]. Rol.

**roll**[1] [roul] *n.* Tomar; (kumaş) top; üstüvane; top francala; sicil. a ~ of butter, yağ topağı: to call the ~, yoklama yapmak: to strike s.o. off the ~s, (bir avukatı) barodan çıkarmak. ~-call, yoklama. ~[2]. (Top; trampete vs.) gürleme(k). ~[3]. Yuvarla(n)mak; tomar yapmak; (sigara) sarmak; silindirle tesviye etm.; yalpa vurmak; salınmak. Yuvarlanma; yalpa. to ~ into a ball, top etm.: to ~ oneself into a ball, tortop olm.: to ~ one's eyes, gözlerini devirmek: he is ~ing in money, denizde kum onda para: to ~ one's r's, r harfini keskin telâffuz etm.: to walk with a ~, salınarak gezmek. ~ by, yuvarlanarak geçmek; (vakit) geçmek. ~ on, yuvarlanmağa devam etm.; geçmek. ~ out, üstüvane ile varak haline getirmek. ~ over, çevirmek; bir taraftan öbür tarafa yuvarlanmak: to ~ over and over, teker meker yuvarlanmak. ~ up, tomar yapmak, sarmak: to ~ up one's sleeves, kollarını sıvamak: to ~ oneself up in a blanket, bir yorgana sarılmak: the guests were beginning to ~ up, misafirler sökün etmeğe başladı. ~ed [rould], tomar halinde. ~ iron, haddeden geçirilmiş demir: ~ gold, altın kaplama. ~er [ˈroulə(r)]. Loğ; silindir; üstüvane; tekerleme (dalga). ~er-bearing, merdane şeklinde bilye. ~er-blind, istor. ~er-skate, tekerlekli ayak kızağı. ~er-towel, iki ucu birleşmiş tomar üzerine asılı havlu.

**rollick** [ˈrolik]. Cümbüş yapmak. a ~ing life, içkiyle eğlentiyle geçirilen ömür: a ~ing time, pek neşeli bir âlem.

**Rom·an** [ˈroumən]. Romalı. ~ Catholic, katolik: ~nosed, koç burunlu: ~ numerals, Roma rakamları. ~any [ˈromәni], Kıptı, çingene (lisanı). ~e [roum], Roma. ⌜when in ~ do as the Romans do⌝, muhite uymak lâzım: ⌜~ was not built in a day⌝, büyük işler zaman ister. ~ish [ˈroumiş], (*köt.*) katolik.

**roman·ce**[1] [rouˈmans]. Lâtinceden çıkan diller. ~[2]. Masal; hayal; roman gibi bir (aşk) macerası. Masal söylemek. ~tic, roman tarzında; hayalî; şairane; içli.

**romp** [romp]. Hoyratça oyun (oynamak). **to ~ home,** bir yarışı kolayca kazanmak.

**rood** [rûd]. Haç. **~-screen,** kilisede cemaat yeriyle koro mahfili arasındaki bölme.

**roof** [rûf]. Çatı, dam (koymak). **the ~ of the mouth,** damak eteği: **to have a ~ over one's head,** başını sokacak bir yeri olmak. **~ing,** çatı (levazımatı). **~less,** damsız; evsiz.

**rook**[1] [ruk]. Gök karga. **~ery** [ˈrukəri], gök kargaların yuvalarını bir arada yaptıkları yer. **~**[2]. Satranc oyununda ruh. **~**[3]. Hile ile parasını almak. **to be ~ed,** aldanmak.

**rookie** [ˈruki]. Acemi asker.

**room** [rûm]. Oda; yer. **~s,** apartman; pansiyon. (*Amer.*) apartman tutmak. **in s.o.'s ~,** birinin yerine: **to be cramped for ~,** yerin darlığından sıkışık olm.: **there is no ~ for doubt,** şübheye mahal yoktur: **there is ~ for improvement,** ıslaha muhtacdır: **to live in ~s,** bir evde oda tutup oturmak: **to make ~ for s.o.,** birine yer vermek: **his ~ is preferable to his company,** yokluğu hissedilmez: **to take up a great deal of ~,** çok yer tutmak. **... -~ed,** ... odalı. **~y,** ferah; bol. **~-mate,** oda arkadaşı.

**roost** [rûst]. Tünek. Geceleyin tünemek. **at ~,** tünemiş: **to go to ~,** (kuşlar) geceleyin tüneklerine konmak: **evil deeds come home to ~,** insan ettiğini bulur: **to rule the ~,** sözü geçmek. **~er,** horoz.

**root** [rût]. Kök; menşe; mastar; cezir. Köklemek; eşelemek. **to ~ up** [**~ out**], söküp çıkarmak; yok etmek. **~ and branch,** usul ve füru: **to destroy ~ and branch,** kökünü kurutmak: **to be ~ed to the spot,** olduğu yere mıhlanmak: **square** [**cube**] **~,** cezri murabba [mikâb]: **to take ~,** kök bağlamak. **~ed,** köklü; sabit; esaslı.

**rope** [roup]. İp; halat. İp ile bağlamak. **~ of pearls,** inci gerdanlık: **to know the ~s,** (bir işin) yolunu yordamını bilmek: **to put s.o. up to the ~s,** birine bir şeyin yolunu göstermek: **give him enough ~ and he'll hang himself,** sen hiç karışma o belâsını bulur. **~-ladder,** ip merdiven. **~'s-end,** dayak atmak için kullanılan ip parçası: **to get the ~,** dayak yemek. **~-walker,** ip cambazı. **~ in,** etrafını iple çevirmek: **to ~ s.o. in,** (*coll.*) birinin yardımını temin etmek. **~ off,** bir yerin bir kısmını iple ayırmak.

**rosary** [ˈrouzəri]. Tesbih; gül bahçesi.

**rose**[1] [rouz]. Gül (şeklinde şey); pembe renk. Pembe. **life is no bed of ~s,** bu dünya her zaman güllük gülistanlık değildir: ⌜**no ~ without its thorn**⌝, dikensiz gül olmaz: **under the ~,** sır olarak. **~ate,** gül renkli. **~bud,** gül konçası. **~-bed,** gül tarhı. **~-coloured,** pembe renkli: **to see things through ~ spectacles,** dünya güllük gülistanlık görmek. **~-window,** gül taklidi daire şeklinde pencere. **~**[2] *v.* **rise.**

**roster** [ˈrostə(r)]. Vazife nöbetlerini gösteren cedvel.

**rostrum** [ˈrostrəm]. Hitabet kürsüsü.

**rosy** [ˈrouzi]. Gül gibi; pembemsi. **the prospect is not ~,** ilerisi pek parlak görünmiyor.

**rot** [rot]. Çürü(t)mek. Çürüklük; maneviyatını kırma; boş lâkırdı. **to stop the ~,** maneviyatı düzeltmek: **what ~!,** saçma!

**rota** [ˈroutə] *v.* **roster.**

**rota·ry** [ˈroutəri]. Dönerek işliyen; devrî. **~te** [rouˈtêyt], bir mihver [merkez] etrafında dön(dür)mek; devret(tir)mek; (ziraatte) ekilen şeyi seneden seneye değiştirmek. **~tion** [–ˈtêyşn], dönme; deveran; nöbetleşe yapma: **by** [**in**] **~,** nöbetle; sırayla: **~ of crops,** ekin münavebesi: **four-course ~,** dörtlü münavebe. **~tory** [ˈroutətəri], devvar; çark gibi dönen.

**rote** [rout]. **to say** [**learn**] **stg. by ~,** bir şeyi makine gibi ezberlemek: **to do stg. by ~,** kör değneğini bellemiş gibi yapmak.

**rotte·n** [ˈrotn]. Çürük; kokmuş; sağlam olmıyan; değersiz. **I am feeling ~,** çok fenayım: **~ luck!,** aksilik. **~r** [ˈrotə(r)], soysuz; adam olmaz.

**rotund** [rouˈtʌnd]. Toparlak; şişman; tumturaklı. **~a,** kubbeli değirmi bina.

**rouge** [rûj]. Allık (sürmek), ruj.

**rough**[1] [rʌf]. *a.* & *n.* Pürüzlü; düzgün olmıyan; kaba; sert; dalgalı (deniz); rüzgârlı (hava); takribî; taslak. Taslak; külhanbeyi. **to cut up ~ about stg.,** (*coll.*) bir şeye çok kızmak: **at a ~ guess,** tahminen: **~ house,** gürültülü ve kavgalı kargaşalalık: ~ **justice,** mahallinde yerine getirilen adalet: **it's ~ (luck) on him!,** ona yazık oldu: **to sleep ~,** yataksız yatmak: **~ly speaking,** tahminen: **to take the ~ with the smooth,** ⌜her nimetin bir külfeti var⌝: **to have a ~ time of it,** eziyet çekmek: **to give s.o. the ~ side of one's tongue,** birini haşlamak. **~-and-ready,** işe yarar derecede; tahminî. **~-and-tumble,** itip kakma: **the ~ of life,** hayatın germü serdi. **~-cast,** kaba sıva(lı); taslağını yapmak. **~-coated,** uzun tüylü. **~-hew,** kabaca yontmak. **~-rider,** başıbozuk süvari. **~-spoken,** kaba, sert sözlü. ~² *v.* Pürüzlendirmek; cilâsını gidermek. **to ~ it,** (*coll.*) meşakkata katlanmak: **to ~ out,** taslağını yapmak. **~en** [ˈrʌfn], pürüzlendirmek; cilâsını gidermek; kabalaşmak; (deniz) dalgalı olmak. **~ness** [ˈrʌfnis], huşunet; hoyratlık; sertlik; kabalık; tüylülük; pürüzlülük; dalgalanma. **~shod** [ˈrʌfşod], buz çivileri ile nallanmış. **to ride ~ over s.o. or stg.,** birini hiçe sayarak fena muamele etm.

**roulette** [rûˈlet]. Rulet oyunu.

**round**¹ [raund] *a.* Yuvarlak; değirmi; daire şeklinde; toparlak; kesirsiz. **a ~ dozen,** en aşağı bir düzine: **eyes ~ with surprise,** faltaşı gibi gözler: **~ number,** kesirsiz aded: **~ oath,** okkalı küfür: **to go at a good ~ pace,** hızlıca gitmek: **with ~ shoulders,** omuzları kamburlaşmış: **~ style,** selis üslûb: **~ table conference,** kimseye kıdem verilmemek için yuvarlak bir masa etrafında toplanan konferans: **~ trip,** başladığı yerde biten seyahat. **~-shouldered,** bir az kamburca. **~-the-world,** devriâlem (seyahati). ~² *n.* Değirmi; yuvarlak şey; daire; bir adet fişek. **~(s),** kol, devriye. **there was a ~ of applause,** umumî bir alkış koptu: **a continual ~ of gaiety,** tükenmez eğlenceler: **the daily ~,** her günkü vazife: **to go the ~s,** kol gezmek: (**doctor**) **to go on** [**make**] **his ~,** (doktor) vizitelerine gitmek: **the story went the ~,** hikâye ağızdan ağza dolaştı: **out of the ~,** tamamen yuvarlak değil: **to stand a ~ of drinks,** grupta bulunan herkese içki ısmarlamak. ~³ *v.* Yuvarlak bir hale getirmek; (köşeyi vs.) dönmek. **~ off,** tamamlamak; düzeltmek. **~ on, to ~ on s.o.,** (i) birdenbire birine hücum etm.; (ii) birini koğulamak. **~ up,** (hayvanları) toplamak; (haydudları) sararak yakalamak. ~⁴ *adv.* & *prep.* Etrafında; her tarafında; devren; takriben; (*fiillerle olduğu zaman o fiile bakınız*). **~ about here,** bu civarda: **~ about thirty,** aşağı yukari otuz: **all the year ~,** bütün sene: **taken all ~,** umumiyet itibarile: **to argue ~ and ~ a subject,** karara varmadan bir mevzu üzerinde mütemadiyen münakaşa edip durmak: **to ask s.o. ~,** (civarda oturan) kimseyi davet etm.: **summer will soon come ~,** yaz yakında tekrar gelecek: **there is not enough to go ~,** bu hepsine yetişmez: **it's a long way ~,** o yol çok dolaşır. **~about** [ˈraundəbaut], *n.* atlıkarınca. ~ *a.* dolambaçlı. **~ers** [ˈraundö(r)z], bir top oyunu. **R~head** [ˈraundhed], 17 inci asırda İngiliz dahilî harbi esnasında Parlamento tarafındaki Cromwell tarafdarı. **~s** [raundz], kol, devriye. **~sman,** devriye: **milk ~,** muntazaman eve uğriyan sütçü.

**rouse** [rauz]. Yatağından çıkarmak; uyandırmak; canlandırmak; tahrik etm.; öfkelendirmek. **to ~ oneself,** silkinmek.

**rout**¹ [raut]. Hezimet; gürültülü kalabalık. **to ~ [put to ~],** bozguna uğratmak: **an utter ~,** kahkarî bir hezimet. ~². **~ about** [**~ up**], eşelemek: **~ out,** yatağından çıkarmak; çekildiği yerden çıkarmak.

**route** [rût]. Takib edilecek yol; yol. **en ~ for,** ···e gitmek üzere. **~-march,** (*mil.*) idman yürüyüşü.

**routine** [rûˈtîn]. Usul; âdet. **the daily ~,** her günkü işin gidişi: **~ work,** her günkü iş: **to do stg. as a matter of ~,** bir şeyi alışkanlık dolayısile yapmak.

**rov·e** [rouv]. ~ (**about**), ötedeberide dolaşmak; serserilik etmek. **his eyes ~d over the pictures,** gözlerini resimler üzerinde gezdirdi: **to ~ the seas,** korsanlık etmek. **~er,** başıboş gezinen; serseri; korsan; kıdemli izci. **~ing,** serserilik; başıboş gezinme. Serseri; göçebe.

**row**¹ [rou] *n.* Sıra, dizi, saf. ~² *v.* Kürek çekmek. **to go for a ~,** kürekli kayıkla gezmek. **~er** [ˈrouə(r)], kürekçi. ~³ [rau]. Gürültü, velvele; kavga. **to get into a ~,** başını belâya sokmak: **to have a ~ with s.o.,** birile kavga etm.: **hold your ~!,** sus!: **to make [kick up] a ~,** çok gürültü yapmak. **~dy** [ˈraudi], külhanbeyi; gürültücü. **~diness,** gürültücülük, yaramazlık.

**royal** [ˈrôyəl]. Kıral veya kıraliçeye aid; şahane; mükemmel. **His** [**Her**] **Royal Highness,** prenslerle prenseslere verilen unvan: **to have a** (**right**) **~ time,** son derece eğlenmek: **there is no ~ road to success,** muvaffakiyete kolay erişilmez. **~ist,** kıral tarafdarı. **~ty,** kırallık; hanedandan kimse; satılan nüsha başına müellifin aldığı ücret.

**R.S.V.P.** (*abb. Fr.*) **Répondez, s'il vous plaît,** lûtfen cevab veriniz.

**Rt. Hon.** (*abb.*) **the Right Honourable,** pek muhterem (*muşaviri haslar için*).

**rub** [rʌb]. Oğma, sürtünme, sürtüşme. Sürtmek; oğ(uştur)mak; aşın(dır)-mak; sürüşmek; temas etm. **to give stg. a ~ up,** bir şeyi parlatmak: **there's the ~!,** asıl müşkülât budur!: **to ~ shoulders with,** ···le haşır neşir olm.: **to ~ s.o. the wrong way** (**up**), birinin damarına basmak: **we ~ along somehow** [**we manage to ~ along**], geçinip gidiyoruz. **~ down,** silmek; tımar etm.; silip aşındırmak. **~ in,** (bir ilâc) sürerek yedirmek: **don't ~ it in!,** (*coll.*) haksız olduğumu biliyorum, tekrar edip durma! **~ off,** silip temizlemek. **~ out,** (yazı) silmek. **~ up,** silerek parlatmak; hafızasını tazelemek: **to ~ up against other people,** başka kimselerle temas etmek.

**rub-a-dub** [ˈrʌbəˈdʌb]. (*ech.*) Çabuk çalınan trampete sesi.

**rubber**[1] [ˈrʌbə(r)]. Silgi; tellâk. **~**[2]. Kauçuk, lâstik. Kauçuk veya lâstikten yapılmış. **~s,** galoş. **~ize,** lâstik ile kaplamak. **~neck** (*Amer. sl.*) etrafında her şeye merakla bakan seyyah. **~-stamp,** lâstik damga (ile damgalamak); (*fig.*) tedkik etmeden tasdik etmek. **~**[3]. (İskambil vs.), üç oyundan iki veya beşten üç kazanmak.

**rubbish** [ˈrʌbiş]. Süprüntü; çörçöp; değersiz şey; saçma. **~y,** değersiz. **~-bin,** çöp kutusu. **~-cart,** çöp arabası. **~-heap,** çöplük.

**rubble** [ˈrʌbl]. Moloz.

**Rubicon** [ˈrûbikon]. **to cross the ~,** (bir işte) dönülmez bir harekette bulunmak.

**rubicund** [ˈrûbikʌnd]. Kızıl çehreli.

**rubric** [ˈrûbrik]. Serlevha; fasıl başı.

**ruby** [ˈrûbi]. Yakut (rengi).

**ruck** [rʌk]. **the** (**common**) **~,** insanların ekserisi: **to get out of the ~,** alelâdenin üstüne çıkmak.

**ruck(le)** [ˈrʌk(l)]. Buruşturmak.

**rucksack** [ˈruksak]. Arka çantası.

**ruction** [ˈrʌkşn]. Gürültü, karışıklık. **there'll be ~s,** kıyamet kopar.

**rudder** [ˈrʌdə(r)]. Dümen.

**ruddy** [ˈrʌdi]. Kızıl.

**rude** [rûd]. Edebsiz, terbiyesiz, kaba; ibtidaî; sert. **would it be ~ to ask?,** (müsaadenizle) sorabilir miyim?: **in ~ health,** sapsağlam: **a ~ shock,** şiddetli bir darbe. **~ness,** terbiyesizlik; ibtidailik.

**rudiment** [ˈrûdimənt]. Kemale ermemiş uzuv. **~s,** bir şeyin elifbesi. **~ary** [–ˈmentəri], ibtidaî.

**rue** [rû]. (Bir şeye) pişman olmak. **you'll ~ the day you ever went there,** ne diye oraya gittim diye dövüneceksin. **~ful,** kederli.

**ruff**[1] [rʌf]. Kırmalı yakalık. **~**[2]. İskambilde koz kırıp almak.

**ruffian** [ˈrʌfyən]. Külhanbeyi; zorba. **you little ~!,** seni gidi çapkın seni! **~ly,** habis.

**ruffle** [ˈrʌfl]. Kırmalı yaka; suların üzerindeki kırışıklık. Düzgünlüğünü bozmak. **to ~ s.o.'s feelings,** birini kırmak: **nothing ever ~s him,** hiç bir şeyden kılı kıpırdamaz.

**rug** [rʌg]. Kilim, küçük halı; yol battaniyesi.

**Rugby** [ˈrʌgbi]. Hem el hem ayakla oynanan bir nevi futbol.

**rugged** [ˈrʌgid]. Yalçın; kayalık; sert.

**ruin** [ˈrûin]. Harabe; enkaz; yıkılma; iflâs. Harab etm.; yıkmak; berbad etm.; iflâs ettirmek. **to be ~ed,** mahvolmak, iflâs etm.: **to be** [**prove**] **the ~ of s.o.,** birinin mahvına sebeb olmak. **~ation** [–êyşn], tahrib etme; harabiyet sebebi: **it will be the ~ of him,** bu onu mahveder. **~ous** [ˈrûinəs], harab, iflâs edici; harabe halinde.

**rul·e**[1] [rûl] *n.* Kaide, usul, kanun, prensip; hüküm(et); karar; cedvel. **~s,** talimat. **that is against the ~s,** o yasaktır: **as a** (**general**) **~,** ekseriyetle: **to bear ~,** hâkim olm.: ⌜**the exception proves the ~**⌝, istisnalar kaideyi teyid eder: **large families were the ~ in Victorian days,** Victoria zamanında aileler umumiyetle kalabalıktı: **to make it a ~ to,** ···i kaide ittihaz etm. **~**[2] *v.* Saltanat sürmek;

idare etm.; hükmetmek. Hüküm sürmek. ~**d paper,** çizgili kâğıd. ~**r,** hükümdar; cedvel. ~ **out,** bertaraf etm.; çizgi ile ibtal etm. ~**ing** [ˈrûlin(g)], hüküm, karar. Hâkim. **the ~ classes,** devlet idare eden sınıflar: ~ **passion,** hâkim ihtiras.
**rum**[1] [rʌm]. (*sl.*) Tuhaf, acayib. **a ~ customer,** acayib bir adam. ~[2]. Rom.
**rumble** [ˈrʌmbl]. (*ech.*) Gümbürtü. Gurlamak.
**ruminate** [ˈrûminêyt]. Geviş getirmek; zihninde evirip çevirmek.
**rummage** [ˈrʌmic]. Altüst ederek aramak; kolaçan etmek. Yoklama. ~ **sale,** bir hayır müessesesi için muhtelif eski şeylerin satışı.
**rumour** [ˈrûmə(r)]. Rivayet, şayia. **it is ~ed that** [~ **has it that**] ..., rivayete göre ....
**rump** [rʌmp]. Sağrı; kıç; bakiye. ~-**steak,** sığırın bud etinden kesilen en iyi parçası.
**rumple** [ˈrʌmpl]. Buruşturmak; bozmak.
**rumpus** [ˈrʌmpəs]. Velvele, patırdı. **to kick up a ~,** kıyamet koparmak.
**run (ran, run)** [rʌn, ran, rʌn]. Koşmak; kaçmak; akmak; işlemek; varmak; (piyes) oynamak; geçmek. İşletmek; (atı) yarışta koşturmak. **a train ~ning at 50 miles an hour,** saatte 50 mil giden tren: **to ~ before the sea,** (gemi) dalgaların önünde gitmek: **to ~ before the wind,** (gemi) pupa yelken gitmek: **trains ~ning between London and Bristol,** L. ve B. arasında işliyen trenler: **to ~ s.o. as candidate,** birini namzed koymak: **to ~ s.o. close** [**hard**], birini pek yakından takib etm. (*fig.*): **I can't afford to ~ a car,** otomobil kullanacak kadar param yok: **there is nothing to do but to ~ for it,** kaçmaktan başka çare yok: **to ~ for office,** bir mevki için namzedliğini koymak: **the thought keeps ~ning through my head,** bu düşünce aklımdan çıkmıyor: **to ~ s.o. off his legs,** birini takatsız kalıncaya kadar koşturmak: **the letter ran like this,** mektub şöyle diyordu: **our stores are ~ning low,** istoklarımız azalıyor: **his nose was ~ning,** burnu akıyordu: **to ~ on the rocks,** (gemi) kayalığa oturmak: **prices ~ very high,** fiatlar umumiyetle çok yüksektir: **'he who ~s may read',** *kolayca anlaşılır şeyler hakkında söylenir*: **a wall ~s all round the garden,** bir duvar bahçenin etrafını çeviriyor: **a heavy sea was ~ning,** deniz pek dalgalı idi: **to ~ a ship ashore,** bir gemiyi karaya oturtmak: **so the story ~s,** hikâye edildiğine göre: **the talk ran on this subject,** konuşma bu mevzuda devam etti: **the time is ~ning short,** vakit daralıyor: **I can't ~ to more than £100,** yüz liradan fazla veremem: **the money won't ~ to a car,** para otomobil almağa yetişmez. ~ **about,** öteye beriye koşmak. ~ **across,** bir taraftan öbür tarafa koşmak; rasgelmek. ~ **after,** peşinden koşmak. ~ **against,** çarpmak; aksine gitmek. ~ **along,** boyunca gitmek: ~ **along now!,** haydi koş! ~ **at,** ...e saldırmak. ~ **away,** firar etm.; (at) gemi azıya almak: **to ~ away with ...,** ...le beraber kaçmak; alıp götürmek: **don't ~ away with the idea that ...,** fikrine kapılma: **that ~s away with a lot of money,** bu çok paraya patlar. ~ **down,** aşağıya koşmak; aşağıya akmak; (saat) kurulmadığı için durmak; (akümülatör) boşalmak; *a.* kurulmamış (saat); (adam) yorgun ve bitkin: **to ~ s.o. down,** (i) (otom. vs.ile) birini çiğnemek; (ii) birini kötülemek; (iii) yakalamak. ~ **in,** içeriye koşmak; (polis) karakola getirmek: **to get ~ in,** mahkemeye verilmek: **to ~ in an engine,** yeni makineyi sürtünme ile alıştırmak. ~ **into,** çarpmak; rasgelmek: **to ~ into debt,** borclanmak: **his income ~s into thousands,** geliri binlerce lirayı bulur. ~ **off,** sıvışmak: **to ~ off with stg.,** bir şeyi aşırmak: **to ~ off water from a tank,** bir sarnıcı boşaltmak: **to ~ off a letter on the typewriter,** bir mektubu makinede çabucak yazıvermek. ~ **on,** yoluna devam etmek: **he ran on and on,** (i) hiç durmıyarak koştu; (ii) uzun uzadıya konuştu: **the ship ran on the rocks,** gemi kayalara oturdu. ~ **out,** dışarı koşmak; akmak; uzanmak; sona ermek; tükenmek: dışarıya salıvermek; uzatmak: **the tide is ~ning out,** (cezir zamanı) denizin suyu çekiliyor: **we ran out of provisions,** erzakımız tükendi: **our lease has ~ out,** kira mukavelemiz bitti. ~ **over,** koşup karşıya geçmek; çiğnemek, gözden geçirmek; yoklamak; taşmak: **he has been ~ over,** (otomobil) altında kaldı. ~ **through,** koşarak

geçmek; göz gezdirmek: **to ~ through a fortune,** bir servetin altından girip üstünden çıkmak: **to ~ one's pen through a word,** kalemiyle bir kelimeyi çizmek: **to ~ s.o. through with a sword,** birine kılıc geçirmek. **~ up,** koşarak yukarı çıkmak; (fiat) yüksel(t)mek; (sancağı) çekmek: **to ~ up debts,** borclarını çoğaltmak: **to ~ up a house,** bir evi alacele inşa ettirmek: **to ~ up against s.o.,** tesadüfen yüz yüze gelmek, raslamak: **I shouldn't ~ up against him if I were you,** bana kalırsa sen ona zıd gitmesen iyi olur. **~about** [ˈrʌnəbaut], hafif otomobil. **~away** [ˈrʌnəwêy], kaçak (esir vs.); gemi azıya almış (at); (**truck**), kurtulmuş vagon: **a ~ match,** kızın âşıkıyle kaçarak evlenmesi: **a ~ victory,** kolayca kazanılmış zafer. **~ner** [ˈrʌnə(r)], koşucu; ulak; haberci; kızak (ayağı): **~-bean,** çalı fasulyası. **~-up,** bir musabakada ikinci kazanan. **~ning** [ˈrʌnin(g)], koşan, akan, işliyen; devamlı; cerahatli. Koşma; işleme; idare. **to be in the ~,** kazanması mümkün olm.: **to be out of the ~,** kazanması mümkün olmamak: **~ board,** (araba) basamak: **three days ~,** üç gün üstüste: **~ expenses,** umumî masraflar: **~ jump,** koşarak atlama: **in ~ order,** kullanılmağa elverişli. **~way** [ˈrʌnwêy], uçak meydanında vs.pist.

**rung**[1] [rʌn(g)]. El merdiveninin basamağı. **~**[2] *v.* **ring.**

**runnel** [ˈrʌnl]. Oluk; ark.

**runt** [rʌnt]. Kavruk adam veya hayvan.

**rupee** [rûˈpî]. Bir Hind parası.

**rupture** [ˈrʌptyə(r)]. Kırılma, kopma; münasebetlerin kesilmesi; fıtık. Kırmak, koparmak. **to be ~d,** fıtıklı olmak: **to ~ oneself,** fıtığı olmak. **~d,** fıtıklı.

**rural** [ˈrûrəl]. Kır ve köye aid, rustayi.

**ruse** [rûz]. Hile, dolab.

**rush**[1] [rʌş]. Saz. **~-bottomed chair,** oturacak yeri saz örgülü iskemle. **~light,** saz mumu. **~**[2] *n.* Hamle; hücum; saldırış; acele; furya. **the ~ hours,** (demiryol vs.) izdiham zamanları; (daire) işlerin baştan aşkın olduğu zamanlar: **we had a ~ to get the job done,** işi bitirmek için çok acele etmek lâzım geldi: **there was a ~ to read this paper,** bu gazete kapışa kapışa okunuyordu: **the ~ of modern life,** modern hayatın humması. **~**[3] *v.* Saldırmak; fırlamak; hamle yapmak; acele etm. Acele yaptırmak, koşturmak; iki ayağını bir pabuca sokmak. **to ~ a position,** (*mil.*) ansızın hücum etmek: **to ~ s.o. (into doing stg.),** birini dara getirmek: **I won't be ~ed,** dara gelemem.

**russet** [ˈrʌsit]. Kuru yaprak rengi.

**Russ·ia** [ˈrʌşə]. Rusya. **~ian,** rusyalı; rusça. **~ophil** [ˈrʌsǫufil], Rus dostu.

**rust** [rʌst]. Pas; (ekin) nebat pası. Paslan(dır)mak; maharet (bilgisini) kaybetmek. **~less,** passız; paslanmaz. **~y,** paslı.

**rustic** [ˈrʌstik]. Köye aid; köylü gibi. Köylü; çoban. **~ate,** köy hayatı geçirmek; bir talebeyi muvakkaten üniversiteden tardetmek.

**rustle** [ˈrʌsl]. (*ech.*) Hışırtı. Hışırdamak.

**rusty**[1] [ˈrʌsti]. **to cut up ~,** (*sl.*) darılmak. **~**[2]. *v.* **rust.**

**rut** [rʌt]. Tekerlek izi. **to get into a ~,** her günkü işin vs. itiyatlarına saplanmak: **to get out of the ~,** gündelik itiyatlardan kurtulmak.

**ruthless** [ˈrûθlis]. Merhametsiz; pek sert.

**rye** [rây]. Çavdar.

# S

**S** [es]. S harfi.

**sabbath** [ˈsabəθ]. Yahudilerin cumartesi, Hıristiyanların pazar günü; tatil günü. **to keep the ~,** dinî tatil gününün kaidelerine riayet etm.: **to break the ~,** bu kaidelere riayet etmemek.

**sable** [ˈsêybl]. Samur. Siyah.

**sabot** [ˈsabǫu]. Takunya.

**sabotage** [ˈsabətâj]. Baltalama(k); kundaklama(k).

**sabre** [ˈsêybə(r)]. Süvari kılıcı (ile vurmak). **~ rattling,** harb tehdidleri.

**sacerdotal** [sasö(r)ˈdǫutl]. Rahibliğe aid.

**sachet** [ˈsaşêy]. Küçük torba, kesecik.

**sack**[1] [sak]. Çuval, torba. **to ~ [give s.o. the ~],** (*coll.*) birini işinden koğmak: **to get the ~,** işinden çıkarılmak. **~**[2]. Yağma (etm.). **~-cloth** [ˈsakkloθ], çuval bezi; tövbe elbisesi. **in ~ and ashes,** keder ve

nedamet içinde. **~ing** [ˈsakin(g)], çuvallık bez.
**sacr·ament** [ˈsakrəmənt]. Hıristiyanların dinî (şarabla ekmek yeme) âyini; mukaddes ve mistik şey. **~ed** [ˈsêykrid], mukaddes; dinî. **~ duty,** vecibe: **~ to the memory of,** ...in hatırasına tahsis edilmiş: **nothing was ~ to him,** hiç bir şeye hürmet etmiyordu. **~ifice** [ˈsakrifâys], kurban; feda(kârlık). Kurban etm.; feda etmek. **he succeeded at the ~ of his health,** sıhhati pahasına muvaffak oldu: **to sell at a ~,** mecburen ziyanına satmak. **~ilege** [ˈsakrilic], dinî şeylere hürmetsizlik etme. **~istan** [saˈkristən], kilise kayyumu. **~isty** [ˈsakristi], kilisede âyin şeylerin saklandığı oda. **~osanct** [ˈsakrousan(g)kt], pek mukaddes.
**sad** [sad]. Kederli; acıklı; gam verici; hüzünlü; dönük (renk). **~ly,** hüzünlü bir tavırla; (*coll.*) çok. **a ~der and a wiser man,** sukutu hayale uğramış ve akıllanmış: **you are ~ly mistaken,** çok yanılıyorsunuz: **I am ~ly in need of a change,** bir tebdili havaya çok ihtiyacım var. **~den,** hüzün vermek. **~ness,** mahzunluk, gam.
**saddle** [ˈsadl]. Eyer, semer; dağ sırtı. Eyerlemek; üstüne atmak; yüklemek. **to be in the ~,** (*fig.*) dizginler elinde olm.: **to keep the ~,** at üzerinde durabilmek: **to put the ~ on the wrong horse,** bir şeyi yanlış yere birine atfetmek: **I am ~d with too big a house,** başımda çok büyük bir ev var: **~ gall,** yağır (yara). **~back,** balık sırtı. **~bag,** heybe. **~r,** sarac. **~ry,** saraclık; eyer ve koşum takımı.
**sadism** [ˈsêydizm]. Cebir ve eziyetle karışık şehvet.
**safari** [saˈfâri]. Av için yapılan sefer.
**safe**[1] [sêyf] *n.* Çelik kasa. **~**[2] *a.* Sâlim, sağlam; emniyette; tehlikesiz; emniyetli, emin, güvenilir; kurtulmuş. **it's a ~ bet that ..., ...** elde bir: **it is not ~ to go out alone,** sokağa yalnız çıkmak tehlikelidir: **it is ~ to say that ..., ...** demek yerindedir: **to be on the ~ side,** ne olur ne olmaz: **~ and sound,** sağ ve sâlim. **~guard,** himaye; muhafaza vasıtası; teminat; ihtiyat; korumak; temin etmek. **~-conduct,** mürur tezkeresi. **~-keeping,** emniyetle koru(n)ma: **it is in ~,** emniyettedir. **~ty** [ˈsêyfti], emniyet; selâmet; kurtuluş. **~ first,** evvelâ emniyet: **~-first policy,** ihtiyat politikası: **to play for ~,** (kumar) ihtiyatla oynamak. **~ty-catch,** emniyet kanadı. **~ty-lamp,** madenci lâmbası. **~ty-pin,** çengelli iğne. **~ty-razor,** jilet. **~ty-valve,** emniyet supabı.
**sag** [sag]. Bel verme(k); çökme(k); (fiatların) düşüklüğü; kıymetten düşmek; rüzgâr altına düşmek.
**saga** [ˈsâga]. (Ortaçağda) İskandinav destanı.
**sagaci·ous** [saˈgêyşəs]. Ferasetli; müdebbir; zeki. **~ty** [–ˈgasiti], feraset, anlayış.
**sage** [sêyc]. Hakîmane; akıllı. Hakîm.
**sahib** [ˈsâhib]. (Hindistanda) Avrupalı.
**sail** [sêyl]. Yelken; yeldeğirmeninin kanadı; bir aded yelkenli gemi; yelkenli gemide gezinti. Yelkenli ile gitmek; gemi ile gitmek; (gemi ile) sefere çıkmak; gemi gibi ağır ilerlemek. Yelkenle yürütmek; bir yelkenliyi idare etmek. **a fleet of fifty ~,** elli yelkenliden mürekkeb bir filo: **to go for [take a] ~,** yelkenli ile gezintiye çıkmak: **it is a month's ~ from America,** Amerika'dan yelkenli ile bir ayda gidilir: **to ~ before the wind,** pupa yelken gitmek: **to ~ close to the wind,** orsa gitmek; (hikâye) bir az yakası açık olm.; (hareket) sahtekârlığa yakın olm.: **to set ~,** sefere çıkmak: **to set the ~s,** yelkenleri düzeltmek: **to shorten ~,** yelkenleri azaltmak: **vessel under ~,** yelkenle yürüyen gemi. **~cloth,** yelken bezi. **~er, good [bad] ~,** (yelkenli) iyi [fena] giden. **~ing,** yelkenliyi kullanma; geminin yürümesi; geminin limandan hareketi: yelkenli: **port of ~,** geminin çıktığı liman: **it's all plain [smooth] ~,** bundan ötesi kolaydır. **~or,** gemici; bahriyeli: **he is a good ~,** onu deniz tutmaz: **I am a bad ~,** beni deniz tutar: **~ suit,** (çocuk) bahriyeli elbisesi.
**saint** [sêynt]. Aziz, veli; (*unvan olarak ismin başına gelirse* St. *yazılır*). **enough to try the patience of a ~,** 'Hazreti Eyyub'un sabrını taşırır'. **~ly,** evliya gibi.
**saith** [seθ]. (*esk.*) = says.
**sake** [sêyk]. *Yal.* for *ile kullanılır.* **for the ~ of,** hatırı için, ... için: **for my ~,** hatırım için: **for God's ~,** Allah aşkına: **for the ~ of one's country,** vatan uğrunda: **for old**

time's ~, mazinin hatırı için: **to talk for the ~ of talking,** konuşma zevki için konuşmak.
**salacious** [saˈlêyşəs]. Şehvanî.
**salad** [ˈsaləd]. Salata. **~ days,** genclik ve tecrübesizlik çağı. **~-dressing,** mayonez. **~-oil,** zeytinyağı.
**salar·y** [ˈsaləri]. Aylık. **~ied,** maaşlı.
**sale** [sêyl]. Satış; mezad; tenzilâtlı satış. **for ~,** satılık: **on ~,** satılıyor: **the ~s were enormous,** fevkalâde çok satıldı: **~ price,** tenzilâtlı fiat. **~able,** satışa elverişli. **~sman,** satış memuru. **~smanship,** satıcılık.
**salient** [ˈsêylyənt]. Çıkıntılı; haricî; göze çarpan; belli başlı. Çıkıntı; haricî zaviye.
**saline** [ˈsêylâyn]. Tuzlu. Müshil. **~ marshes,** sahillerde tuzlu toprak.
**saliva** [saˈlâyva]. Salya.
**sallow** [ˈsalou̯]. (Fıtraten) soluk benizli; renksiz.
**sally** [ˈsali]. Çıkış hareketi; alaylı nükte. **to ~ out [forth],** çıkmak.
**salmon** [ˈsamən]. Somon balığı. Pembemsi.
**saloon** [səˈlûn]. Büyük salon; (dans vs.) salonu; (*Amer.*) içki barı. **~ cabin,** birinci sınıf kamara: **~ car,** kapalı otomobil.
**salt** [solt]. Tuz(lu). Tuzlamak. **the ~ of the earth,** en mükemmel sınıf: **to take stg. with a grain of ~,** bir haberi ihtiyatla karşılamak: **in ~,** tuzlanmış: **to ~ a mine,** değersiz bir şeyi hile ile satmak: **an old ~,** ihtiyar gemici: **~ water,** tuzlu su, deniz suyu(na aid): **to weep ~ tears,** acı gözyaşları dökmek: **he is not worth his ~,** ekmeğini hak etmiyor. **~-cellar,** tuzluk. **~y,** tuzlu.
**salu·brious** [səˈlyûbri̯əs]. Sıhhate faydalı. **~tary** [ˈsalyutəri], sıhhat verici; faydalı.
**salut·e** [səˈlyût]. Selâmlamak; tazim göstermek; göze çarpmak. Selâm (verme). **to fire a ~,** topla selâmlamak: **to take the ~,** geçid resminde askerin selâmını almak. **~ation** [ˌsalyuˈtêyşn], selâmlama, selâm.
**salva·ble** [ˈsalvəbl]. Kurtarılabilir (eşya). **~ge** [ˈsalvic], (batmış bir gemiyi) yüzdürme(k); kazaya uğrıyan gemiyi kurtarma(k); tahlis ücreti; kurtarılmış eşya; (*jur.*) kurtarma ve yardım. **~tion** [salˈvêyşn], kurtuluş; selâmet; kurtar(ıl)ma. **S~ Army,** selâmet ordusu: **to work out one's own ~,** kurtuluşunu kendi kendine hazırlamak.
**salve**[1] [salv]. Merhem (sürmek). **to ~ one's conscience,** vicdanını müsterih kılmak. **~**[2] *v.* **salvage.**
**salvo** [ˈsalvou̯]. Selâm toplarının atılması.
**Sam** [sam]. (*abb.*) **Samuel. Uncle ~,** Birleşik Amerika Devletleri.
**Samaritan** [saˈmaritən]. **a good ~,** şefkatli ve mürüvvetli adam.
**Sambo** [ˈsambou̯]. Zencilere verilen lâkab.
**same** [sêym]. *Hemen daima* **the** *ile kullanılır.* Aynı, tıpkısı, farksız; mezkûr; gene o. **all the ~,** buna rağmen: **it's all the ~ to me,** bana göre hava hoş: **he said the ~ as you,** sizin söylediğinizin aynını söyledi: **he likes a holiday, the ~ as you,** sen nasıl tatil istersin o da ister: **he left the ~ day he came,** geldiği gün gitti: **~ here!,** (*coll.*) benden de al o kadar!: **he is just the ~ as ever,** tamamen eskisi gibi: **one and the ~,** tamamen aynı: **at the ~ time,** (i) ayni zamanda; (ii) bununla beraber: **'Happy New Year to you!' 'The ~ to you!',** 'Yeni yılınız kutlu olsun!' 'Sizin de!'. **~ness,** ayniyet, benzerlik; yeknesaklık.
**sample** [ˈsâmpl]. Mostra; nümune; örnek. Nümune almak; çeşnisine bakmak. **up to ~,** nümunesine uygun.
**sancti·fy** [ˈsan(g)ktifây]. Takdis etmek. **a custom ~fied by time,** zamanla kudsî bir hale gelmiş âdet. **~monious** [–ˈmou̯nyəs], sahte sofu. **~ty,** mukaddeslik.
**sanction** [ˈsan(g)kşən]. Tasvib; müeyyide. Münasib görmek, tasvib etm. **~ed by usage,** âdetle caiz sayılan.
**sanctu·ary** [ˈsan(g)ktyu̯əri]. Bir ibadethanenin en mukaddes yeri; melce; taarruzdan masuniyet temin eden yer. **to take ~,** böyle bir yere iltica etm.: **bird ~,** kuşların korunduğu yer. **~m** [ˈsan(g)ktəm], mukaddes yer; hususî oda.
**sand** [sand]. Kum (serpmek). **to build on ~,** çürük temel üzerine inşa etm. **~bag,** kum torbası (ile birinin başına vurmak); bir yeri kum torbalar ile muhafaza etmek. **~paper,** zımpara kâğıdı (ile cilâlamak). **~shoe,** lâstik tabanlı bez ayakkabı. **~storm,** kum fırtınası. **~y,** kumlu;

kumsal: ~-haired, sarımtrak kızıl saçlı. ~-bank, kumsal sığlık. ~-boy, as jolly as a ~, kanarya gibi neş'eli.

**sandal** [ˈsandl]. Çarık; sandal.

**sandwich** [ˈsan(d)wic]. Sandviç. **to ~ stg. between other things,** bir şeyi iki başka şey arasına sıkıştırmak. **~-man,** sırtında ve göğüsünde reklâm yaftaları dolaştıran adam.

**san·e** [sêyn]. Aklı başında; salim fikirli. **~ity** [ˈsaniti], akıl sıhhati; akliselim; muhakeme.

**sang** *v.* sing.

**sanguin·ary** [ˈsan(g)gwinəri]. Kanlı; zalim. **~e,** nikbin.

**sanit·ary** [ˈsanitəri]. Sıhhî. **~ation** [–ˈtêyşn], hıfzısıhha: **the ~ of the house is poor,** evin sıhhî tertibatı iyi değildir.

**sank** *v.* sink.

**Santa Claus** [ˌsantaˈklôz]. Noel baba.

**sap**[1] [sap]. Nebat usaresi. **~ling** [ˈsaplin(g)], fidan; delikanlı. **~**[2]. Duvar yıkmak için açılan hendek. Temelinden çürütmek. **~per** [ˈsapə(r)], istihkâm askeri; lağımcı.

**sapien·ce** [ˈsêypiəns]. Akıl. **~t,** ukalâ.

**sapphire** [ˈsafây(r)]. Gök yakut.

**Saracen** [ˈsarasen]. Haçlı seferleri zamanında Müslümanlara verilen ad.

**sarcas·m** [ˈsâkazm]. İstihza; dokunaklı alay. **~tic** [sâˈkastik], müstehzi, dokunaklı.

**sarcophagus** [sâˈkofəgəs]. Lâhid.

**sardine** [sâˈdîn]. **fresh ~,** ateşbalığı: **tinned ~s,** sardalya: **packed like ~s,** balık istifi.

**sardonic** [sâˈdonik]. Müstehzi, istihfafkâr; acı.

**sartorial** [sâˈtôriəl]. Terziliğe aid.

**sash**[1] [saş]. Kuşak. **~**[2]. Pencere kanadı. **~ window,** sürme pencere.

**sat** *v.* sit.

**Satan** [ˈsêytn]. Şeytan; iblis. **'~ reproving sin',** her kötülükten mesul olduğu halde başkasını ayıblayan. **~ic** [saˈtanik], şeytanî.

**satchel** [ˈsaçl]. Omuza asılan çantası.

**satellite** [ˈsatəlâyt]. Peyk.

**satiate** [ˈsêyşiêyt]. Doyurmak.

**satin** [ˈsatin]. Saten. **~y,** saten gibi.

**satir·e** [ˈsatây(r)]. Hiciv. **~ic(al)** [saˈtirik(l)], hicivli. **~ist,** hicivci. **~ize** [ˈsatirâyz], hicvetmek.

**satisfact·ion** [ˌsatisˈfakşn]. Hoşnudluk, memnuniyet; ikna; tarziye; tazmin; borc ödeme. **to give s.o. ~,** (i) birini ikna etm.; (ii) birine tarziye vermek: **to make full ~ to s.o.,** birinin zararını tamamen tazmin ve telâfi etmek. **~ory,** tatminkâr; ikna edici.

**satisf·ied** [ˈsatisfâyd]. Kani; razı, hoşnud; doymuş. **I am ~ that, ...** kanaatindeyim: **self-~,** kendini beğenmiş. **~y** [ˈsatisfây], ikna etm.; hoşnud etm.; tazmin etm.; (borcu) ödemek; kâfi gelmek; yerine getirmek. **to ~ a condition,** bir şartı yerine getirmek: **to ~ a longing,** bir hasreti gidermek. **~ying,** ikna edici; tatminkâr; doyuran.

**saturate** [ˈsatyurêyt]. İşba etmek.

**Satur·day** [ˈsatədêy]. Cumartesi. **~n** [ˈsatö(r)n], Zühal seyyaresi. **~nalia** [–ˈnêylyə], açıksaçık eğlenti. **~nine** [–ˈnâyn], gülmez, abus.

**satyr** [ˈsatö(r)]. (*mit.*) Kırların yarım tanrısı; şehvete düşkün adam.

**sauc·e** [sôs]. Salça; lezzet; (*sl.*) yüzsüzlük. **none of your ~!,** yüzsüzlüğün lüzûmu yok!: **what ~!,** ne yüzsüzlük! **~epan** [ˈsôspən], kulplu tencere: **double ~,** çift tencere. **~er** [ˈsôsə(r)], fincan tabağı. **~y** [ˈsôsi], utanmaz; zarif.

**Saul** [sôl]. **'Is ~ also among the prophets?',** *evvelce şiddetle muhalif olduğu bir fikir vs.ye birdenbire tarafdar olmak.*

**saunter** [ˈsôntə(r)]. Tembel tembel gezinme(k).

**sausage** [ˈsosic]. Sucuk; sosis. **~ roll,** sucuklu börek.

**savage** [ˈsavic]. Vahşi; yabani; merhametsiz. (*Yal.* at hakkında) ısırmak. **~ry,** vahşilik; gaddarlık.

**save**[1] [sêyv]. Kurtarmak; tahlis etm.; korumak; önüne geçmek; tasarruf etm. **God ~ the Queen!,** Allah Kıraliçe korusun (İngiliz millî marşı): **to ~ time,** vakit kazanmak: **to ~ s.o. the trouble of doing stg.,** birini bir zahmetten kurtarmak: **this has ~d me much work,** bu işimi çok hafifletti. **~ up,** para biriktirmek. **~**[2]. Ancak; ···den maada. **I am well ~ that I have a cold,** iyiyim, yalnız nezlem var. **~r** [ˈsêyvə(r)], kurtarıcı; hesabî; *mürekkeb kelimelerde* 'kurtaran'.

**savings** [ˈsêyvin(g)s] *n.pl.* Biriktirilmiş para.

**saviour** [ˈsêyvyə(r)]. Kurtarıcı; halâskâr. **Our S~,** İsa peygamber.

**savour** [ˈsêyvə(r)]. Tat, lezzet; çeşni;

(*fig.*) koku. Tadını alarak yavaş yavaş yemek. **to ~ of stg.,** tadı olm.; kokmak: **that ~s of treason,** bu ihaneti andırıyor. **~y,** lezzetli; iştah açıcı. Yemek sonunda yenen tuzlu şey.

**saw**[1] [sô]. Testere. *v.* (**sawed, sawn**) [sôd, sôn]. Testere ile kesmek; ileri geri hareket etmek. **to ~ off,** testere ile kesip ayırmak: **to set a ~,** testere dişlerini tashih etm.: **to ~ up wood,** odunu testere ile parça parça kesmek. **~bones,** (*şak.*) cerrah. **~dust,** testere tozu. **~-mill,** kereste fabrikası. **~**[2]. Atasözü. **~**[3] *v.* **see.**

**say** (**said**) [sêy, sed]. Söylemek; demek. Söz; söz sırası. **I ~ !,** *dikkati çekmek tabir*: **I cannot ~ when he will come,** (i) ne vakit geleceğini bilmiyorum; (ii) ... söylemeğe mezun değilim: **when all is said and done,** en nihayet: **you don't ~ so!,** acayib!: **it goes without ~ing that ...,** elbette ...: **give me a few, ~ five,** bir kaç tane, meselâ beş tane, ver: **to have one's ~,** bir meselede söyliyeceğini söylemek: **to have a ~ in a matter,** bir meselede söz sahibi olm.: **there is much to be said for this proposal,** bu teklifin lehinde çok şey söylenebilir: **there is no ~ing what will happen,** ne olacağını kimse bilmez: **to ~ nothing of ...,** ... de (üste) caba: **so to ~,** tabir caizse: **that is to ~,** yani: **'though I ~ it who shouldn't',** bunu söylemek bana düşmez amma ..., (kendini medhederken vs.); **what do you ~ to a drink?,** bir az içelim mi?: **well, ~ he does come, what then?,** pekiyi, diyelim geldi, ya sonra? **~ing** [ˈsêy·in(g)], *n.* söz; atasözü. **common ~,** meşhur tabir: **as the ~ goes,** dedikleri gibi.

**scab** [skab]. Yara kabuğu; uyuz; (*sl.*) greve iştirak etmiyen amele. **to ~ over,** (yara) kabuk bağlamak. **~by,** uyuz; (*sl.*) alçak.

**scabbard** [ˈskabö(r)d]. Kın.

**scaffold** [ˈskafo̤uld]. Yapı iskelesi; darağacı. Etrafında yapı iskelesi kurmak. **to go to [mount] the ~,** darağacına gitmek. **~ing,** yapı iskelesi.

**scald** [skôld]. Haşlamak; kaynar suda yıkamak; kaynar su ile yaralamak.

**scale**[1] [skêyl] *n.* Terazi gözü; mikyas; ıskala; **~s** *veya* **pair of ~s,** terazi. **to draw to ~,** ölçüye göre çizmek: **Fahrenheit ~,** F. derecesi: **on a large ~,** vâsi mikyasta: **~ of prices,** fiat cedveli: **~ of salaries,** barem: **to turn the ~,** ağır basarak vaziyete tesir etmek. **~**[2] *n.* Balık vs. pulu; kirec milhinden hasıl olan tortu. Pullarını ayıklamak; tortusunu gidermek. **~**[3] *v.* Tırmanarak kadem kadem çıkmak; terazi ile tartmak; ölçü ile resmetmek. **to ~ wages up,** bütün ücretleri ayni nisbet dahilinde artırmak.

**scallop** [ˈskalop]. Kab olarak kullanılan tarak kabuğu.

**scallywag** [ˈskaliwag]. Yaramaz, çapkın.

**scalp** [skalp]. Başın üst kısmı; başın saçlı olan derisi; kırmızı derililerin ve sair vahşilerin öldürdükleri düşmanlarının başlarından kesip zafer alâmeti olarak sakladıkları saçlı deri parçası. Başının derisini yüzmek.

**scamp**[1] [skamp] *n.* Yaramaz (*çocuklar hakkında ve kısmen muhabbet ifade eder*). **~**[2] *v.* (Bir işi) yarım yamalak yapmak. **~er** [ˈskampə(r)], çocuklar vs. neşe içinde koşmak.

**scan**[1] [skan]. Gözle iyice tedkik etm.; göz gezdirmek. **~**[2]. Şiirin hecelerini saymak. Vezne uygun olmak.

**scandal** [ˈskandl]. Rezalet; iftira; dedikodu. **~ize,** (uygunsuz bir söz vs. ile) utandırıp nefret ve infial uyandırmak. **~ous,** rezil; iftiralı (söz vs.).

**scant, ~y** [skant(i)]. Kıt, az; dar. **~ily clad,** yarı çıplak. **~iness,** kıtlık.

**scape·goat** [ˈskêypgo̤ut]. Herkesin kabahati kendisine yükletilen adam. **~grace** [ˈskêypgrêys], yaramaz.

**scar** [skâ(r)]. Yara izi (bırakmak). **to ~ over,** kabuk bağlamak.

**scarab** [ˈskarəb]. (Eski Mısır) mukaddes böceği.

**scarc·e** [ske̤əs]. Nadir; kıt; kâfi değil. **to make oneself ~,** sıvışmak. **~ely** [ˈske̤əsli], henüz; ancak; hemen hemen; hemen hiç. **~ any,** yok denecek kadar: **~ ever,** hemen hiç bir zaman: **he can ~ speak,** hemen hiç konuşamaz: **he is ~ ten years old,** on yaşında ya var ya yok: **I ~ know what to say,** ne söyliyeceğimi bilemiyorum: **I ~ know him,** onu hemen hiç tanımıyorum. **~ity** [ˈske̤əsiti], nadirlik; kıtlık.

**scare** [ske̤ə(r)]. Ansızın korku; esassız korku; endişe. Korkutmak. **to ~**

away, korkutup kaçırmak: **to be ~d to death [stiff, out of one's wits]**, ödü patlamak: **to give s.o. a ~**, birini ansızın ürkütmek: **to raise a ~**, ortalığı telaşa vermek. **~crow**, bostan korkuluğu. **~monger**, telaşçı.
**scarf** [skâf]. Boyun atkısı; kaşkol.
**scarify** [ˈskarifây]. Deriyi kazımak; (*fig.*) canını yakmak.
**scarlatina** [skâləˈtîna]. Kızıl hastalığı.
**scarlet** [ˈskâlit]. Al (renkli). **~ hat**, kardinal şapkası: **~ fever**, kızıl.
**scarred** [skâd]. Yara izi olan.
**scath·e** [skêyð]. Zarar; yara. **~less**, zarara uğramamış. **~ing** [ˈskêyðin(g)], pek dokunaklı, zehirli (söz vs).
**scatter** [ˈskatə(r)]. Saçmak, dağıtmak. Dağılmak; yayılmak. **~ed**, seyrek; aralıklı; dağınık. **~-brained**, sersem.
**scavenge** [ˈskavinc]. Süprüntüyü temizlemek. **~r**, çöpçü; süprüntü yiyen hayvan.
**scene** [sîn]. Sahne; (perde içinde) meclis; tiyatro sahnesi dekoru; vaka mahalli; manzara; rezalet. **behind the ~s**, perde arkasında; gizli kapaklı tarafı: **to make a ~**, rezalet çıkarmak. **~-shifter**, dekorları değiştiren adam. **~ry** [ˈsînəri], tabiî güzel manzara. sahne dekoru.
**scent** [sent]. Koku (hassası); güzel koku. Koklamak; koku yaymak. **to get [pick up] the ~**, kokuyu almak: **to be on the right ~**, iz [koku] üzerinde olm.: **to throw s.o. off the ~**, izini kaybettirmek. **~ed**, güzel kokulu: **keen-~ dog**, burnu keskin köpek.
**sceptic** [ˈskeptik]. Reybî, hiç bir şeye inanmaz kimse. **~al**, reybî; şübheli; hic bir şeye inanmaz. **~ism** [-ˈsizm], reybîlik; şübhe.
**sceptre** [ˈseptə(r)]. Saltanat asâsı.
**schedule** [ˈşedyûl]. Cedvel; program; zeyil. Listeye kaydetmek; program yapmak. **according to ~**, programa göre: **six hours behind ~**, miadından altı saat gecikmiş: **the train is ~d to arrive at 10 o'clock**, tarifeye göre tren saat onda gelecektir.
**scheme** [skîm]. Plân; tedbir, taslak; desise. Plânını kurmak; kumpas kurmak. **colour ~**, renklerin tertibi. **~r**, plâncı; entrikacı.
**schism** [ˈsizəm]. İtizal.
**schola·r** [ˈskolə(r)]. Talebe; âlim, edib; burslu talebe. **a fine ~**, çok âlim bir adam: **he is no ~**, tahsili az. **~rly**, âlimane. **~rship**, ilim; ilmî zihniyet; burs. **~stic** [skoˈlastik], mekteb ve üniversitelere aid; ukalâca.
**school** [skûl]. Mekteb; fakülte; balık sürüsü. Talim etm. **one of the old ~**, eski zaman adamı: **upper [middle, lower] ~**, bir mektebin büyük [orta, küçük] sınıfları: **what ~ were you at?**, hangi mektebde okudunuz? **~boy**, mektebli (erkek). **~fellow**, mekteb arkadaşı. **~girl**, mektebli (kız). **~master**, muallim. **~mistress**, kadın muallim. **~room**, dershane. **~treat**, yemekli müsamere, gezinti.
**schooner** [ˈskûnə(r)]. Uskuna.
**scien·ce** [ˈsâyəns]. İlim; bilgi; fen. **natural ~**, tabiî ilimler: **to study ~**, fen tahsil etm.: **~ student**, fen talebesi. **~tific** [ˌsâyənˈtifik], ilmî, fennî. **~tist** [ˈsâyəntist], fen adamı.
**scimitar** [ˈsimitə(r)]. Eğri kılıc.
**scintillate** [ˈsintilêyt]. Parıldamak.
**scion** [ˈsâyon]. **the ~ of a noble house**, asil bir ailenin ahfadından.
**sciss·ion** [ˈsijn]. Kesme. **~ors** [ˈsizö(r)z], *n.pl.* **~** *veya* **pair of ~**, makas.
**scoff** [skof]. Alay (etm.). İstihza. **to ~ at s.o.**, birini maskara etm.: **to ~ at stg.**, istihfaf etm.
**scold** [skould]. Azarlamak. Titiz hırçın kadın. **~ing** *n.*, azar.
**scoop** [skûp]. Kepçe; oyuk bir alet; tarak dubası kovası; vurgun; meraklı bir haberin rakib gazeteden evvel neşri. **to ~ out**, bir kepçe vs. ile boşaltmak: **to ~ a large profit**, büyük bir kâr vurmak: **at one ~**, bir hamlede.
**scoot** [skût]. Acele kaçış. **to ~ off [away, to do a ~]**, tabanları yağlamak. **~er**, çocuk için tekerlekli kızak; küçük motosiklet.
**scope** [skoup]. Saha; faaliyet sahası. **to give full ~ to one's imagination**, muhayyilesini dolu dizgin koşturmak.
**scorch** [skôç]. Kavurmak; hafifçe yakmak. Hafifçe yanmak, kavrulmak; pek sıcak olm.; (*sl.*) rüzgâr gibi gitmek. **~ing**, yakıcı: kavrulma; (*otom.*) çok hızlı gitme.
**score**[1] [skô(r)] *n.* Sıyrık; çizgi; hesab, oyunda kazanılan puvan; bir bestenin notası; sebeb; yirmi (tane). **a ~ of people**, yirmi kişi: **~s of people**, pek çok kimse: **a cheap ~**, zayıf bir nükte: **have no fear on that ~!**, o

cihetten korkma!: **on the ~ of ill-health,** sıhhatinin bozukluğu sebebile: **to pay off old ~s,** bir kuyruk acısını çıkarmak: **what's the ~?,** (oyunda) kim kazanıyor? ~[2] *v.* Sıyırmak, çizmek, yivlemek; (*mus.*) notaya geçirmek; hesab etm.; oyunda puvan kazanmak, puvanları yazmak. **to ~ a goal,** gol yapmak: **that's where he ~s,** işte üstünlüğü burada: **to ~ off s.o.,** bir münakaşada karşısındakini nükteli bir cevabla susturmak. **~r,** puvanları kaydeden kimse.

**scorn** [skôn]. Hor görme(k), istihfaf (etm.); istiğna (etm.); tenezzül etmemek. **to ~ s.o.'s advice,** birinin nasihatini tepmek: **to laugh s.o. to ~,** birile alay ede ede onu gülünç bir hale getirmek. **~ful,** istihfafkâr(ane).

**scorpion** [ˈskôpyən]. Akreb.

**Scot** [skot] *n.* İskoçyalı. **~ch** [skoç] *a.* İskoçyalı. **~land,** İskoçya: **~ Yard,** Londra Emniyet Müdürlüğü. **~s, ~tish,** İskoçyalı, İskoçya lehçesi.

**scotch** *v.* Hafifçe yaralamak; sakatlamak.

**scot-free** [ˈskotˈfrî]. Sağ ve salim; masrafsız.

**scoundrel** [ˈskaundrəl] *n.* Habis, hain. **~ly** *a.* habis; hainane.

**scour**[1] [skauə(r)]. Silerek temizleme(k); (nehir) aşındırmak. ~[2]. Her tarafa hızlı hızlı gezmek; araştırmak.

**scourge** [skö(r)c]. Kamçı; âfet. Kamçılamak; zulmetmek; (halka vs.) âfet olmak.

**scout**[1] [skaut]. Keşfe çıkan asker; gözcü; izci; Oxford universitesinde kolej hademesi. Keşfe çıkmak. **~-master,** izci oymak beyi. ~[2]. İstihfaf ile reddetmek.

**scowl** [skaul]. Kaş çatma(k); surat asmak. **to ~ at s.o.,** birine yan bakmak.

**scrag** [skrag]. Sıska adam, hayvan. **~gy,** zayıf, sıska.

**scramble** [ˈskrambl]. Manialı bir yerde güçlükle ilerleme; tırmanış. Tırmanarak ilerlemek. **to ~ for stg.,** kapışmak: **to ~ eggs,** yumurtayı çalkayarak pişirmek: **a general ~,** itişip kakışma.

**scrap**[1] [skrap]. Parça; kırıntı; hurda; artık. (Faydası yok diye) atmak; ıskartaya ayırmak. **a ~ of comfort,** en küçük bir teselli: **to catch ~s of a conversation,** bir konuşmanın bazı parçaları kulağına çalınmak: **~s of news,** kırık dökük haberler. **~-book,** öteden beriden toplanmış maktualar vs. yapıştırılan defter. **~-heap,** enkaz yığını: **to be thrown on the ~,** ıskartaya çıkarılmak. **~-iron,** hurda demir. ~[2]. (*coll.*) Kavga, dövüş(mek).

**scrape**[1] [skrêyp] *n.* Sıyırma, kaşıntı; çizgi; (*sl.*) başını belâya sokma. **to get into a ~,** başını belâya sokmak: **to get out of a ~,** işin içinden sıyrılmak: **we're in a nice ~,** ayıkla şimdi pirincinin taşını! ~[2] *v.* Kazımak, sıyırmak; hafifçe dokunmak; gıcırdamak. **to ~ acquaintance with s.o.,** tanışmak için birine yanaşmak: **to ~ along,** iyi kötü geçinip gitmek: **to bow and ~,** yerlere kadar eğilmek: **to ~ one's plate,** tabağını temizlemek: **to ~ through,** yakayı kurtarmak: **to ~ through an examination,** imtihanda güç belâ geçmek: **to ~ up some money,** dişinden tırnağından artırmak. **~r** [ˈskrêypə(r)], kaspa; demir çamurluk.

**scrappy** [ˈskrapi]. Yarım yamalak. **a ~ dinner,** artıklardan ibaret yemek.

**scratch**[1] [skraç] *n.* Tırnak vs. yarası; çizik; sıyrık; kaşınma sesi; yarışa başlama yeri. **to come up to ~,** beklenildiği gibi çıkmak: **to start from ~,** (yarış, işe vs.) avantajsız olarak en başından başlamak: **he came through the war without a ~,** burnu bile kanamadan harbden döndü. ~[2] *v.* Tırmalamak; kaşımak; çizmek; eşelemek; yarış vs.den vazgeçmek. '**~ my back and I'll ~ yours**', karşılıklı piyaz, birbirini öğmek: **tomorrow's match has been ~ed,** yarınki maç yapılmıyacak: **to ~ the surface,** üstünü kazımak; içine nüfuz etmemek. **~ out,** silmek: **to ~ s.o.'s eyes out,** birinin gözlerini çıkarmak. **~ up,** yeri kazıp çıkarmak. ~[3]. **a ~ meal,** [**team**], derme çatma yemek [takım]: **~ player,** birinci sınıf oyuncu. **~y** [ˈskraçi], gıcırtılı; kaşındıran; yarım yamalak.

**scrawl** [skrôl]. Kargacık burgacık. Okunmaz yazı yazmak.

**scream** [skrîm]. (*ech.*) Feryad, çığlık. Acı acı haykırmak; çığlık koparmak. **to ~ oneself hoarse,** sesi kısılıncaya kadar bağırmak: **~s of laughter,** kahkaha: (*coll.*) **it was a perfect ~,** aman ne komik şeydi! **~ing,** feryad eden; (*coll.*) pek komik.

**scree** [skrî]. Dağ yamacında basınca kayan küçük taşlar.

**screech** [skrîç]. (*ech.*) Keskin feryad (koparmak).
**screed** [skrîd]. Pek uzun ve usandırıcı mektub.
**screen** [skrîn]. Perde; ekran; kalbur. Gizlemek; siper etm.; kalburdan geçirmek; (kitab vs.yi) filme almak. **to ~ off,** paravana ile gizlemek.
**screw**[1] [skrû] *n.* Vida; uskur; (*sl.*) cimri adam; (*sl.*) maaş. **to have a ~ loose,** (*coll.*) bir tahtası eksik olm.: **there's a ~ loose somewhere,** bir yerde bir bozukluk var: **to put the ~ on,** sıkıştırmak: **to put a ~ on the ball,** hususî bir hareketle topun seyrini değiştirmek: **to give another turn to the ~,** bir daha sıkıştırmak. **~-coupling,** (borular vs.yi) vidalı başlık. **~-driver,** tornavida. **~ed,** vidalanmış, (*sl.*) çakır keyif. **~-driven,** uskurlu. **~**[2] *v.* Vidalamak. (Vidalı bir şey) dönmek; (*coll.*) dişinden tırnağından artırmak. **~ down,** vida ile sıkmak. **~ out, to ~ the truth out of s.o.,** birisinden hakikati güç belâ öğrenmek: **to ~ money out of s.o.,** birisinden domuzdan kıl çeker gibi para koparmak. **~ up, to ~ up the eyes,** gözlerini kısmak: **to ~ up one's courage,** cesaretini toplamak: **to ~ up one's face,** yüzünü buruşturmak: **to ~ up one's lips,** dudaklarını bükmek: **to ~ oneself up to do stg.,** kendini zorlamak: **to ~ stg. up in a piece of paper,** bir şeyi kâğıda sarıp bükmek.
**scribble** [ˈskribl]. Karışık ve okunmaz yazı; acele yazılmış mektub vs. Acele veya dikkatsiz yazmak. **~r,** fena yazıcı.
**scribe** [skrâyb]. Eski Yahudilerin din ilmi müfessiri; yazıcı.
**scrimmage** [ˈskrimic]. Göğüs göğüse kavga.
**scrimshank** [ˈskrimşan(g)k]. Yançizmek; vazifesinden kaçmak.
**scrip**[1] [skrip]. Dilenci torbası. **~**[2]. Muvakkat sened.
**script** [skript]. El yazısı, hat; ona benziyen matbaa harfleri; senaryo.
**scriptur·e** [ˈskriptyuə(r)]. **~** [**~s veya Holy ~s**], kitabı mukaddes. **~al,** ona aid.
**scroll** [skroul]. Tomar (şeklinde ziynet).
**scrounge** [skraunc]. (*mil. sl.*) Aşırmak. **to ~ around,** aşıracak şey var mı diye kolaçan etmek.
**scrub** [skrʌb]. Çalılık; aşınmış fırça; fırça gibi bıyık. Fırçalayarak yıkamak. **~bing-brush,** tahta fırçası. **~by** [ˈskrʌbi], çalılık; cılız; traşı uzamış; miskin.
**scruff** [skrʌf]. **by the ~ of the neck,** ensesinden.
**scrumptious** [ˈskrʌmşəs]. (*coll.*) Enfes.
**scrup·le** [ˈskrûpl]. Ufacık parça; kuruntu; vicdan üzüntüsü, endişe. **to ~ [have ~s],** tereddüd etm.; vicdanı üzülmek: **to have no ~s [make no ~s] about doing stg.,** vicdanen hiç tereddüd etmemek. **~ulous** [ˈskrûpyuləs], dikkatli, titiz; dürüst; vesveseli. **not over-~ in his dealings,** hareketlerinde pek dürüst değil.
**scrutin·y** [ˈskrûtini]. Dikkatli muayene; reylerin tasnifini tasdik. **to demand a ~,** reylerin yeniden tedkikini taleb etmek. **~eer** [–ˈniə(r)], rey tasnif memuru. **~ize,** tedkik etm.; dikkatle muayene etm.; gözden geçirmek.
**scud** [skʌd]. Hızlı gitme; hızlı uçan bulut. Hızla koşmak, uçmak. **to ~ before the wind,** rüzgârın önüne katılıp gitmek.
**scuffle** [ˈskʌfl]. Ehemmiyetsiz dövüş. Hafif tertip kavga etm.; ayaklarını yere sürmek.
**scull** [skʌl]. Çifte küreklerin biri. Sandalı çifte küreklerle yürütmek.
**scull·ery** [ˈskʌləri]. Bulaşıkhane: **~-maid,** bulaşıkçı kız. **~ion** [ˈskʌlyən], (*esk.*) aşçı yamağı.
**sculpt·or** [ˈskʌlptə(r)]. Heykeltraş. **~ure** [–tyə(r)], heykeltraşlık (etmek).
**scum** [skʌm]. Su yüzüne çıkan pislik. **the ~ of the people,** halkin en alçak tabakası.
**scupper** [ˈskʌpə(r)]. Frengi deliği. (*sl.*) Gemiyi delerek batırmak; baltalamak.
**scurf** [skö(r)f]. Baş kepeği. **~y,** kepekli.
**scurril·ity** [skʌˈriliti]. Kaba küfür. **~ous** [ˈskʌriləs], kaba küfürlü; pis iftiralı.
**scurry** [ˈskʌri]. Acele kaçış; şiddetli ve kısa süren kar fırtınası. Acele etm.
**scurvy**[1] [ˈskö(r)vi]. İskorpit. **~**[2]. Alçak.
**scuttle**[1] [ˈskʌtl]. Kömür kovası. **~**[2]. Lumbar ağzı. Gemiyi delerek, gemi altındaki muslukları açarak, batırmak. **~**[3]. Sıvışma(k); korkakça kaçmak.

**Scylla** [ˈsila]. **between ~ and Charybdis,** iki tehlike arasında kalmak.
**scythe** [sâyð]. Tırpan(lamak).
**sea** [sî]. Deniz(e aid). **to be at ~,** (i) deniz üzerinde olm., gemide bulunmak; (ii) şaşırmak: **by the ~,** deniz kenarında: **following [head] ~,** arkadan [önden] gelen dalgalar: **to go by ~,** gemi ile gitmek: **to go to ~ [take to the ~, follow the ~],** gemici olm.: **the high ~s [the open ~],** açık deniz: **to get one's ~ legs,** geminin hareketine alışıp ayakta durabilmek: **to put to ~,** denize açılmak: **the seven ~s,** bütün denizler: **to ship a (green) ~,** dalga gemiye girmek. **~-borne,** deniz yolu ile gönderilen. **~-chest,** gemici sandığı. **~-dog,** deniz kurdu. **~-front,** bir şehrin denize bakan kısmı. **~-going,** açık denizlere giden. **~-green,** açık mavimsi yeşil. **~-lawyer,** safsatacı ve daima kusur bulan gemici. **~-level,** deniz seviyesi. **~-lion,** ayı balığının büyük cinsi. **~-scout,** deniz izcisi. **~-serpent,** denizciler tarafından okyanusun dibinde yaşadığına inanılan iri yılan şeklinde bir canavar. **~-shell,** deniz kabuğu. **~-sick,** deniz tutmuş. **~board,** deniz kenarı. **~farer** [ˈsîfe̯ərə(r)], çok deniz yolculuğu eden. **~faring,** gemicilik. **~gull,** martı. **~man** [ˈsîmən], gemici. **able (-bodied) ~,** bahriye onbaşısı: **leading ~,** bahriye çavuşu: **merchant ~,** ticaret gemici: **ordinary ~,** bahriye neferi: **~like,** bir denizciye yakaşır surette: **~ship,** denizcilik. **~plane,** deniz uçağı. **~port,** liman. **~scape** [ˈsîskêyp]. Deniz manzarasını gösteren resim. **~side,** deniz sahili; yalı: **~ resort,** plaj. **~ward** [ˈsî-wö(r)d] *a.* **~s** *adv.* Denize doğru. **~weed** [–wîd], deniz yosunu. **~-worthy** [–wö(r)ði], denize çıkmağa elverişli.
**seal¹** [sîl]. Ayı balığı. **~er,** ayı balığı avcısı; onun gemisi. **~skin,** ayı balığı kürkü. **~².** Mühür; manalı işaret. Mühürlemek; kurşun mühür takmak; kapamak. **it is a ~ed book to me,** buna aklım ermez: **the book bears the ~ of genius,** kitabda dehanın damgası var: **his fate is ~ed,** akibeti taayyün etmistir: **the Great S~,** İngiliz hükümetin resmî mühürü: **my lips are ~ed,** bu sırrı kimseye söyleyemem: **to set one's ~,** mühürünü basmak: **to set the ~ on stg.,** bir meseleyi kökünden halletmek: **under the ~ of secrecy,** gizli kalmak şartile.
**seam** [sîm]. Dikiş yeri; maden damarı; ek yeri. **care had ~ed his face,** üzüntü yüzünü kırışıklarla kaplamıştı. **~less,** dikişsiz. **~stress** [ˈsemstris], dikişçi kadın. **~y** [ˈsîmi], **the ~ side of stg.,** bir şeyin fena tarafı.
**séance** [ˈsêyons]. (İspirtizme) toplantı.
**sear** [si̯ə(r)]. Kurutmak; soldurmak.
**search** [sö(r)ç]. Ara(ştır)ma(k); yoklama(k); tedkik (etm.). **in ~ of ...,** ...i bulmak için: **to ~ into stg.,** bir şeyi tedkik etm.: **to ~ for stg.,** bir şeyi araştırmak: **to ~ high and low,** fellek fellek aramak: **right of ~,** arama hakkı. **~ing, a ~ examination,** derin muayene; çok sıkı bir imtihan: **~ of the heart,** vicdanını yoklama: **a ~ regard [look],** nüfuz eden nazar: **~ questions,** inceden inceye sualler. **~light,** ışıldak.
**season¹** [ˈsîzn] *n.* Mevsim; vakit. **in due ~,** münasib bir zamanda, sırasına göre: **in and out of ~,** olur olmaz zaman: **the off ~,** mevsimi olmadığı zaman: **to last for a ~,** bir mevsimlik ömrü olm.: **out of ~,** mevsimsiz; yersiz: **the London ~,** Londra yüksek ailelerin şehirde kaldığı mevsim: **~ ticket,** abonman karnesi: **a word in ~,** yerinde bir söz. **~²** *v.* Çeşnilendirmek; olgunlaş(tır)mak; kurutmak. **to ~ justice with mercy,** adaleti merhametle telif etmek. **~able,** mevsime uygun; müsaid. **~al,** muayyen mevsime mahsus. **~ed,** olgun; kurutulmuş; çeşnili: **a ~ soldier,** harb görmüş asker **~ing,** çeşni.
**seat** [sît]. Oturulacak şey (yer); peyke; mevki; konak; (**valve**) yuva; kıç; pantalonun kıçı. Oturtmak. **to ask s.o. to be ~ed,** birine 'oturunuz' demek: **a car to ~ four,** dört kişilik otomobil: **a good ~ (on a horse),** (atın üzerinde) iyi oturma: **to keep one's ~,** (i) oturduğu yerde durmak; (ii) atın üzerinde durmak; (iii) tekrar mebus intihab edilmek: **to lose one's ~,** (i) attan düşmek; (ii) tekrar intihab edilmemek: **to ~ oneself [to take a ~],** oturmak: **to take a back ~,** bir kenara çekilmek; ehemmiyetini kaybetmek. **~er, single-~,** bir kişilik (uçak vs.): **two-~,** iki kişilik. **~ing,**

oturacak yerler; (makine) yatak: ~ **capacity,** oturacak yerlerin mikdarı.
**seccotine** [ˌsekəˈtîn]. Kuvvetli bir tutkal.
**sece·de** [siˈsîd]. Ayrılmak; itizal etmek. **~ssion** [siˈseşn]. Ayrılma; itizal.
**seclu·de** [siˈklûd]. İhtilattan menetmek. **to ~ oneself,** ihtilat etmemek; inzivaya çekilmek. **~ded,** münzevi; mahrem. **~sion** [–ˈklujn], inziva, üzlet.
**second**[1] [ˈsekənd] *a.* İkinci. **the ~ of January,** iki ocak: **every ~ day,** gün aşırı: **~ childhood,** bunaklık: **to be ~ to none,** hiç kimseden geri kalmamak: **~ sight,** kayıbdan haber verme. **~-best, it's a ~,** istediğimiz gibi değil, fakat olur: **to come off ~,** altta kalmak: **my ~ suit,** en iyi elbisemden sonra gelen elbisem. **~-class,** ikinci derecede; ikinci mevki. **~-hand** [ˌsekəndˈhand], kullanılmış; elden düşme: **~ dealer,** eskici: **to learn stg. ~,** bir şeyi başkasından öğrenmek. **~-rate,** ikinci derecede; silik. **~**[2] *n.* Saniye, an. **I'll come in a ~,** şimdi gelirim: **in a split ~,** bir anda. **~**[3] *v.* Yardım etmek. **to ~ a motion,** bir teklifi desteklemek: [seˈkond] **to ~ an officer for other service,** bir subayı başka bir vazife için ayırmak. **~er,** (bir teklifi) destekliyen. **~ary** [ˈsekəndəri], fer'î; ikinci derecede; ehemmiyetsiz: **~ school,** orta mekteb, lise.
**secre·cy** [ˈsîkrisi]. Ketumluk; sır olma; mahrem olma; gizlilik. **under pledge of ~,** mahrem olarak. **~t,** sır; gizli şey. Gizli; mektum; saklı, hafi. **in ~,** gizli olarak: **to let s.o. into the ~,** bir sırrı birine söylemek: **an open ~,** herkesin bildiği sır: **the S~ Service,** gizli haberalma teşkilatı: **to tell stg. as a ~,** bir şeyi mahrem olarak söylemek. **~te** [siˈkrît], saklamak; ifraz etmek. **~tion** [–ˈkrîşn], ifraz etme, ifrazat. **~tive,** ketum; fazla kapalı. **~tory,** ifraz edici.
**secretar·y** [ˈsekrətri]. Kâtib. **S~ of State,** Bakan. **~ial** [–ˈteəriəl], kâtib(liğ)e aid; yazı işlerine aid. **~iat** [–ˈteəriət], kâtibler heyeti.
**sect** [sekt]. Tarikat; cemaat. **~arian** [–ˈteəriən], tarikatçı; mutaassıb taraftar.
**section** [ˈsekşn]. Kesme, kesilmiş şey; dilim; parça; fasıl; bölge; şube; (*mil.*) manga. Kısımlara ayırmak. **all ~s of the population,** halkın bütün sınıfları: **made in ~s,** sökülüp takılır. **~al,** makta halinde; muayyen bir kısma aid; birbirlerine geçen ayrı kısımlardan yapılmış.
**sector** [ˈsektə(r)]. Daire dilimi; kıta, mıntaka.
**secular** [ˈsekyulə(r)]. Dünyevî; cismanî; lâik; asırlarca süren. **~ize,** lâikleştirmek.
**secur·e** [siˈkyuə(r)]. Emin; korkusu yok; sağlam. Temin etm.; sağla(mla)mak; elde etmek. **~ity,** emniyet; kefalet, rehin: **securities,** tahvilat: **to lend money on ~,** rehine karşı ödünc para vermek: **to stand ~ for s.o.,** birine kefil olmak.
**sedan-chair** [siˈdanˈçeə(r)]. Sedye.
**sedat·e** [siˈdêyt]. Temkinli; sakin. **~ive** [ˈsedətiv], yatıştırıcı (ilâc), müsekkin.
**sedentary** [ˈsedntəri]. Oturmuş; vaktini hep evde geçiren. **a ~ occupation,** oturduğu yerde yapılan iş.
**sedge** [sec]. Bataklıkta yetişen otlar.
**sediment** [ˈsedimənt]. Tortu, rüsub. **~ary** [–ˈmentəri], rüsubî.
**sedit·ion** [siˈdişn]. İsyan, ayaklanma. **~ious** [–şəs], âsi.
**seduc·e** [siˈdyûs]. Baştan çıkarmak, iğfal etm. **~tion** [–ˈdʌkşn], baştan çıkarma, iğfal. **~tive,** cazibeli; şuh.
**sedulous** [ˈsedyuləs]. Çalışkan; devamlı.
**see**[1] [sî] *n.* Bir piskoposun ruhani dairesi. **the Holy S~,** Papalık. **~**[2] *v.* **(saw, seen)** [sô, sîn]. Görmek, bakmak; görüşmek; anlamak. **you ~, ...,** *yerine göre* şimdi, efendim, anlatabildim mi? *vs. manaları var*: **as far as I can ~,** görebildiğim kadar: **to ~ s.o. to the door,** bir misafiri kapıya kadar teşyi etm.: **to ~ s.o. home,** birine evine kadar refakat etm.: **he can't ~ a joke,** şakadan anlamaz: **let me ~!,** dur bakayım!; efendime söyleyim: **we ~ a lot of each other,** birbirimizi sık sık görüyoruz: **he will never ~ fifty again,** elliyi çoktan aştı: **nothing could be ~n of him,** hiç görünürlerde yoktu: **one can't ~ to read,** çok karanlık, okunmuyor: **~ing that ...,** ···e göre. **~ after,** *v.* **~ to. ~ in,** just **~ him in, will you?,** onu içeri alır mısınız? **to ~ in the New Year,** yeni yılı merasimle kutlulamak. **~ into,** tedkik etmek. **~ off, to ~ s.o. off, [out],** (i) birini teşyi etm.; (ii) birini kapı dışarı etmek. **~ out, to ~ stg.**

out, bir şeyi sonuna kadar görmek. ~ **through**, **to** ~ **stg. through**, bir şeyi sonuna kadar götürmek; bir şeyin sonuna kadar dayanmak: **to** ~ **s.o. through**, birine müşkül bir zamanını atlatıncaya kadar yardım etm.: **to** ~ **through s.o.**, birinin içini okumak: **to** ~ **through s.o.'s plan**, birinin dolablarına kanmamak: **a ton of coal will** ~ **us through the winter**, bir ton kömür kışı çıkarır. ~ **to, I will** ~ **to it**, ben bu işe bakarım: **this stove must be** ~**n to**, bu sobaya baktırmak lâzım. ~**ing**, görme, bakma. Gören. ⌜~ **is believing**⌝, insan görünce inanır: ~ **that** ..., ...e göre: **within** ~ **distance**, göz görebildiği kadar.

**seed** [sîd]. Tohum (vermek; ekmek). **to go to** ~, tohuma kaçmak. ~**-corn**, tohumluk buğday vs. ~**-pearl**, ufak taneli inci. ~**ling**, fide. ~**y**, keyifsiz; köhne. **a** ~**-looking individual**, kılıksız bir herif.

**seek** (**sought**) [sîk, sôt]. Aramak; dilemek. **to** ~ **after stg.**, bir şeyin peşinde koşmak: **to** ~ **for**, aramak: **to** ~ **out**, arayıp bulmak: **the reason is not far to** ~, sebeb meydanda. ~**er**, arayan: **a** ~ **after knowledge**, bilgi arayan: **pleasure** ~, zevkine düşkün.

**seem** [sîm]. Görünmek; ... gibi gelmek. **it** ~**s as though** [**if**] ..., ... gibi görünüyor: **I** ~ **to have heard his name**, ismini duydum gibime geliyor: **it** ~**s not**, böyle olmadığı anlaşılıyor: **so it** ~**s**, öyle gibi: **there** ~**s to be some difficulty**, bu işin içinde bazı güçlükler var gibi görünüyor. ~**ing**, görünen; zahir; sureta: **in spite of his** ~ **indifference**, zahiren lâkayd görünmesine rağmen. ~**ly**, yakışır; münasib; terbiyeli. ~**liness**, edeb ve terbiye icabatı.

**seen** *v.* **see**².

**seep** [sîp]. Sızmak. ~**age** [–pic], sızıntı.

**seer** [siə(r)]. Kâhin.

**see-saw** [ˈsîsô]. Tahtarevalli; nöbetleşe hareket; inip çıkma. Kâh öyle kâh böyle olmak.

**seeth·e** [sîð]. Haşlamak; kaynaşmak. **to be** ~**ing with anger**, hiddetten köpürmek.

**segment** [ˈsegmənt]. Daire kıtası; parça.

**segregate** [ˈsegrigêyt]. Ayırmak, tecrid etmek. Büyük bir cisimden ayrılmak.

**seigneur** [ˈsêynyö(r)]. Derebeyi; büyük rütbeli asilzade. **grand** ~ [gra(n)–], yüksek bir aileden çok kibar tavırlı bir efendi.

**seism·ic** [ˈsâyzmik]. Zelzeleye aid. ~**ograph**, yer sarsıntılarını kaydeden alet.

**seiz·e** [sîz]. Kapmak; yakalamak; gasbetmek; kavramak; el koymak. ~ (**up**), (makine) yağsızlık vs.den dolayı yapışmak: **to be** ~**d with a desire to do stg.**, bir şey yapmak arzusuna kapılmak: **to be** ~**d with fear**, korkuya kapılmak: **to** ~ **hold of**, yakalamak: **to** ~ **the opportunity**, fırsatı ganimet bilmek. ~**ure** [ˈsîjə(r)]. Yakalama; gasbetme; elkoma; felc; (makine) yapışma. **to have a** ~, felce uğramak.

**seldom** [ˈseldəm]. Nadir olarak. ~, **if ever**, kırk yılda bir.

**select** [siˈlekt]. Seçmek. Seçme, güzide. ~**ion**, seçme; çeşid: **natural** ~, ıstıfa: ~**s from Shakespeare**, S.den seçme parçalar. ~**ive**, seçici.

**self**, *pl.* **selves** [self, selvz]. Kendi; kendi kendine. **all by one's very** (**own**) ~, tek başına: **all by himself**, (i) yapayalnız; (ii) tek başına: ~ **is his god**, kendine tapar: **he is quite his old** ~ **again**, tamamen eskisi gibidir: **one's second** ~, ⌜içtikleri su ayrı gitmez⌝: **ticket admitting** ~ **and friend**, kendiniz ve bir arkadaşınız için bilet. ~**ish** [ˈselfiş], hodbin, hodkâm. ~**less** [ˈselflis], kendini düşünmiyen.

**self-** *pref.* Self *ile yapılan mürekkeb kelimelerde* self *kendi kendine mânasını tazammum eder. Mürekkeb kelime başka bir mânaya geldiği zaman yerinde gösterilmiştir.* ~**-acting**, otomatik. ~**-apparent**, besbelli. ~**-assert·ion**, kendini beğenme: ~**ive**, yüzsüzce girişken. ~**-centred**, hodbin. ~**-colour**(**ed**), düz renkli; tabiî renkte. ~**-command**, kendini tutma. ~**-communion**, kendi kendine düşünme. ~**-conscious**, sıkılgan. ~**-contained**, kendi kendine yetişir: ~ **flat**, müstakil apartman dairesi. ~**-control**, kendine hâkim olma; soğukkanlılık: **to lose one's** ~, iradesini kaybetmek. ~**-defence**, kendini müdafaa: **the noble art of** ~, boks. ~**-denial**, nefsinden feragat. ~**-destruction**, intihar. ~**-determination**, bir milletin kendi mukadderatına kendisinin karar vermesi. ~**-esteem**, kendini beğenme. ~-

**government,** muhtariyet. **~-help,** başkasından yardım beklemeden şahsî gayret. **~-important,** kibirli. **~-indulgent,** zevkine düşkün. **~-made,** kendi kendini yetiştirmiş (adam). **~-possessed,** temkinli. **~-respect,** izzetinefis. **~-restraint,** kendini tutma. **~-righteous,** mürai. **~-sacrifice,** fedakârlık. **~-satisfied,** kendini beğenmiş. **~-seeking,** menfaatperest. **~-styled,** kendi verdiği adla. **~-sufficient,** müstağni. **~-sufficing,** kendi kendini idare eden. **~-supporting,** ekmeğini kendi kazanan. **~-willed,** inadcı.

**sell** (sold) [sel, sǫuld]. Sat(ıl)mak. (*sl.*) Dalavere. **to be sold,** (*sl.*) kafese konmak: **what a ~!,** ne dalavere!: **sold again!,** (i) yine yutturdular!; (ii) yağma yok!: **(house, etc,) 'to ~'** [**'to be sold'**], satılık (ev vs.): **this book ~s well,** bu kitab iyi satılıyor: **to ~ s.o. for a slave,** birini köle olarak satmak. **~ off, to ~ off one's belongings,** bütün eşyasını satıp savmak. **~ out,** bütün mevcudu satmak: **we are sold out of that book,** o kitab tamamen satıldı. **~ up,** müflisin malına elkoyup satmak. **~er** [ˈselə(r)], satıcı, bâyi. **this book is a good ~,** bu kitab iyi satılıyor: **best ~,** çok satılan kitab.

**selv·age, ~edge** [ˈselvic]. Kumaş kenarı.

**selves** *v.* **self.**

**semantic** [siˈmantik]. Mânaya aid.

**semaphore** [ˈseməfô(r)]. Semafor(la haberleşmek).

**semblance** [ˈsembləns]. Benzeyiş. **to bear the ~ of,** benzemek: **to put on a ~ of gaiety,** yalancıktan neşeli görünmek.

**semester** [siˈmestə(r)]. Sömestr.

**semi-** [ˈsemi] *pref.* Yarı .... **~-detached, ~ house,** yalnız bir taraftan bitişik müstakil ev. **~-final,** dömifinal. **~circ·le** [–sö(r)kl], yarım daire: **~ular** [–ˈsö(r)kyulə(r)], yarım daire şeklinde. **~colon** [–ˈkǫulən], noktalı virgül (;). **~tone** [ˈ–tǫun], yarım ses.

**seminary** [ˈseminəri]. (*esk.*) Mekteb; *şimdi yalnız katolik mekteblerine denir.*

**Semit·e** [ˈsîmâyt]. Samî; Yahudi. **~ic** [–ˈmitik], samî; yahudi.

**sempiternal** [ˌsempiˈtö(r)nl]. Ebedî.

**senat·e** [ˈsenət]. Senato; âyan meclisi; üniversite idare heyeti. **~or,** âyan meclisinin âzası. **~orial** [–ˈtôriəl], ona aid.

**send** (sent) [send, sent]. Göndermek. **to ~ for s.o.,** birini getirtmek, çağırmak: **to ~ s.o. for stg.,** birini bir şey için göndermek: **God ~ that ...,** Allah vere de ...: **(God) ~ him victorious,** Allah onu muzaffer eylesin!: **the blow sent him sprawling,** darbeyi yiyince yere yuvarlandı: **it sent a shiver down my spine,** bu bütün vücudümü ürpertti. **~ away,** uzaklaştırmak: **to ~ away for stg.,** bir şeyi başka yerden göndertmek. **~ down,** aşağıya göndermek; üniversiteden tardetmek. **~ forth,** dışarı göndermek; salmak. **~ in,** içeriye göndermek: **to ~ in a bill,** fatura göndermek: **to ~ in one's name,** ismini içeriye haber vermek: **to ~ in one's resignation,** istifasını vermek. **~ off,** yola vurmak; teşyi etm.: **to ~ off a letter,** mektubu postaya vermek. **~ on,** gelen bir şeyi başka bir yere göndermek; (bir emri) başkasına tebliğ etmek. **~ out,** dışarıya göndermek; fışkırtmak; neşretmek. **~ up,** yukarıya göndermek; artırmak; yükseltmek. **~er** [ˈsendə(r)], gönderen. **'return to ~',** gönderene iade.

**senil·e** [ˈsînâyl]. Bunak; ihtiyarlığa aid. **~ity** [–ˈniliti], bunaklık.

**senior** [ˈsînyə(r)]. Daha yaşlı; kıdemli. **I am three years ~ to you** [**I am three years your ~**], sizden üç sene büyüğüm: **Smith ~,** S. kardeşlerin en yaşlısı. **John Smith ~,** J. S. baba. **~ity** [ˌsîniˈorəti], daha yaşlılık; kıdemlilik.

**sensation** [senˈsêyşn]. His; duyma; ihtisas; heyecan. **the news caused a great ~,** haber büyük bir heyecan uyandırdı. **~al,** heyecan verici. **~alism,** halkı heyecanlandıracak şeylere düşkünlük.

**sense**[1] [sens] *n.* Beş hissin her biri; his, duygu; akıl, zekâ; mâna, meal. **the ~ of sight** [**hearing, etc.**], görme [işitme vs.] hissi: **common ~,** aklıselim: **to be in one's ~s,** aklı başında olm.: **to be out of one's ~s,** deli olm.: **to come to one's ~s,** kendine gelmek; aklı başına gelmek: **to talk ~,** (i) söylediğinde mâna olm.; (ii) saçmalamamak: **to take the ~ of the meeting,** bir toplanti vs.de halkın fikrini yoklamak: **to take a word in the wrong ~,** kelimeyi yanlış mânaya almak. ~² *v.* Farkında olm., hissetmek. **~less** [ˈsenslis], bayılmış; akılsız. **to fall ~**, kendinden geçerek düşmek: **to knock s.o. ~**, birine vurup bayıltmak.

**sensi·bility** [ˌsensiˈbiliti]. Hassaslık; içlilik. **~ble** [ˈsensibl], makul; akıllı; akla yakın; sezer; hissedilebilir. **be ~!**, makul ol!: **~ clothing**, uygun elbise: **to be ~ of one's danger**, tehlikeyi sezmek. **~tive** [ˈsensitiv], içli; alıngan; hassas. **very ~ to criticism**, tenkide gelmez. **~tiveness**, **~tivity** [–ˈtiviti], hassaslık.

**sensu·al** [ˈsensyuəl]. Şehvete düşkün. **the ~ pleasures**, nefsani zevkler. **~alist**, şehvet düşkünü. **~ality** [–ˈaliti], şehvet. **~ous** [ˈsensyuəs], hislere aid.

**sent** *v.* send.

**senten·ce** [ˈsentəns]. Cümle; mahkeme kararı; hüküm. (Cezaya) mahkûm etmek. **to pass ~ on s.o.**, birini mahkûm etm.: **to undergo one's ~**, mahkûmiyet müddetini geçirmek: **he is under ~ of death**, ölüme mahkûm olmuştur. **~tious** [senˈtenşəs], nasihat veren; fetva verir gibi söz söyliyen.

**sentient** [ˈsentiənt]. Hissedebilir.

**sentiment** [ˈsentimənt]. Fikir; his; hassaslık. **~al** [–ˈmentl], fazla hassas. **~alist** [–ˈmentəlist], hislerine fazla kapılır. **~ality** [–ˈtaliti], içlilik, fazla hassaslık.

**sent·inel, ~ry** [ˈsentinl, –tri]. Nöbetçi. **to stand ~ (over)**, nöbet beklemek; gözetlemek. **to relieve a ~**, nöbetçiyi değiştirmek. **~-box**, nöbetçi kulübesi.

**separa·ble** [ˈseprəbl]. Ayrılabilir. **~te** *a.* [ˈseprit], ayrı, ayrılmış; müstakil; müfrez . *v.* [ˈsepərêyt]. Ayırmak, ayrılmak; tefrik etmek. **to ~ milk**, sütün kaymağını almak. **~tion** [ˌsepəˈrêyşn], ayırma; ayrılma; ayrılık; hicran: **~ allowance**, asker ailelerine verilen tahsisat: **judicial ~**, mahkeme kararile ayrılık: **~ order**, ayrılık hükmü. **~tist** [ˈseprətist], muhtariyetçi. **~tor** [ˈsepərêytə(r)], sütün kaymağını (vs.) almağa mahsus makine.

**sepoy** [ˈsîpôy]. Eski hindli asker.

**September** [sepˈtembə(r)]. Eylûl.

**septennial** [sepˈteniəl]. Yedi senede bir (olan).

**septic** [ˈseptik]. Çürütücü; mikroblu.

**sepulchr·e** [ˈsepəlkə(r)]. Mezar; türbe. **a whited ~**, mürai; ⌈uzaktan gördüm bir yeşil türbe, içine girdim neuzübillah⌉. **~al** [–ˈpʌlkrəl], mezara aid; mevtaî: **a ~ voice**, mezardan gelir gibi bir ses.

**sequel** [ˈsîkwəl]. Bir şeyin mabadı, arkası; âkıbet, netice.

**sequence** [ˈsîkwəns]. Tevali; sıra; kâğıd sırası.

**sequest·er** [siˈkwestə(r)]. Haczetmek. **~ered**, münzevi; haczedilmiş. **~rate** [ˈsîkwestrêyt], haczetmek.

**sequin** [ˈsîkwin]. Eski Venedik altını; süs için elbiseye takılan madenî pul.

**seraglio** [siˈrâlyou]. Saray; harem.

**seraph**, pl. **-s**, **-im** [ˈseraf, –s, –im]. Yüksek sınıf melek. **~ic** [siˈrafik], melek gibi.

**Serb, ~ian** [sö(r)b,-ˈyən]. Sırp; sırpça. **~ia**, Sırbistan.

**sere** [siə(r)]. Kurumuş (yaprak); solmuş.

**serenade** [ˌserəˈnêyd]. Sevgilinin penceresi altında söylenen şarkı (söylemek).

**seren·e** [siˈrîn]. Sakin; huzur içinde; durgun (hava). **His S~ Highness**, *bazı prenslere verilen unvan*: **all ~!**, (*coll.*) işler tıkırında. **~ity** [–ˈreniti], sükûnet, durgunluk; huzur.

**serf** [sö(r)f]. Derebeylik devrinde demirbaş köle. **~dom**, kölelik.

**sergeant, serjeant** [ˈsâcənt]. Çavuş. **police ~**, polis komiseri muavini. **Common S~**, Londra Belediyesinin bir memuru: **~-at-arms**, Saray, Parlamento ve Londra memurlarına verilen unvan. **serjeant-at-law**, bir teşrifat memuru. **sergeant-major**, başçavuş.

**serial** [ˈsiəriəl]. Sıra ile devam eden; tefrika (halinde). **~ize**, tefrika etmek.

**series** [ˈsîrîz]. Sıra; seri. **in ~**, seri halinde.

**serious** [ˈsiəriəs]. Ciddî; vahim; temkinli. **to take stg. ~ly**, ciddiye almak.

**sermon** [ˈsö(r)mən]. Vaız. **~ize**, uzun uzadıya nasihat vermek.

**serpent** [ˈsö(r)pənt]. Yılan. **~ine** [–tâyn], yılankavi; hilekâr.

**serrated** [siˈrêytid]. Testere gibi dişli.

**serried** [ˈserid]. Sıkışık.

**serum** [ˈsiərəm]. Serom.

**servant** [ˈsö(r)vənt]. Hizmetçi, uşak; kul; memur. **general ~**, her işe bakan hizmetçi: **your humble ~**, hakîr kulunuz (*resmî mektubun sonunda*): **your obedient ~**, itaatli bendeniz (*resmî, ticarî bir mektubun sonunda*): **civil ~**, devlet memuru.

**serve** [sö(r)v]. Hizmet etm.; hizmetçilik etm.; yemeği sofraya koymak; misafire yemek vermek; işe yaramak; (tenis) servis yapmak. **to ~ one's apprenticeship**, çıraklık etm.: **to ~ in the army [navy]**, askerlik hizmetini orduda [bahriyede] yapmak: **~d with butter**, üzerine tereyağı gezdirilmiş: **dinner is ~d!**, yemeğe buyurun!: **when occasion ~s**, icabında; fırsad düşünce: **to ~ as a pretext**, bahane yerine geçmek: **to ~ out**, dağıtmak: **to ~ s.o. out**, birinden öc almak: **nothing but the best will ~**, en iyisi olmazsa olmaz: **it ~s the purpose**, işe yarar: **the new railway will ~ a large area**, yeni demiryolu büyük bir bölgenin ihtiyacını karşılayacak: **it ~s him right!**, belâsını buldu: **if my memory ~s me right**, hafızam beni aldatmıyorsa: **to ~ one's sentence**, mahkûmiyetini geçirmek: **he ~d me shamefully**, bana çok fena muamele etti: **to ~ one's time**, (i) çıraklık etm.; (ii) askerlik hizmetini yapmak; (iii) mahkûmiyetini geçirmek: **to ~ up food**, kotarmak.

**service**[1] [ˈsö(r)vis] *n.* Hizmet(çilik); vazife; yardım; idare; âyin; takım; (tenis) servis; (otel) servis. **active ~**, muharebe hizmeti: **on active ~**, cebhede: **I am at your ~**, emrinize âmadeyim: **to be of ~ to s.o.**, birine yardım etm.: **dinner ~**, sofra takımı: **will you do me a ~?**, size bir ricam var: **the fighting ~s**, kara, deniz, hava kuvvetleri: **to go into ~**, evlerde hizmetçilik etm.: **On Her Majesty's S~ (O.H.M.S.)**, devlet hizmetinde: **to see ~**, (asker) muharebe görmek: **this hat has seen much ~**, bu şapka çok görmüş geçirmiştir: **the Senior S~**, Bahriye: **to take ~ with s.o.**, birinin evine hizmetçi girmek. **~**[2] *v.* (Otomobil vs.ye) bakmak; tamir etmek. **~able**, işe yarar; faydalı.

**serviette** [sö(r)viˈet]. Peçete, peşkir.

**servil·e** [ˈsö(r)vâyl]. Köle gibi; körükörüne. **~ity** [–ˈviliti], zillet; aşağılık.

**serving** [ˈsö(r)vin(g)]. **a ~ soldier**, hizmette olan asker. **~-man**, hizmetçi, uşak.

**servitude** [ˈsö(r)vityûd]. Kulluk, kölelik. **penal ~**, kürek cezası.

**sesame** [ˈsesami]. Susam. **open ~**, açıl susam açıl.

**session** [ˈseşn]. Celse; ictima; oturuş. **the House is now in ~**, Parlamento ictima halindedir: **petty ~s**, sulh mahkemesi.

**set**[1] [set] *n.* Takım; seri; muhit, zümre; vaziyet; fidan; kaldırım taşı; (av köpeği) ferma. **~ of apartments**, apartman dairesi: **~ of a coat**, ceketin sırta oturuşu: **~ of the current**, akıntının istikameti: **to make a (dead) ~ at s.o.**, birine diş geçirmek; (kadın) bir erkeği avlamağa çalışmak: **we don't move in the same ~**, aynı muhite devam etmiyoruz: **~ of the sails**, yelkenlerin vaziyeti: **wireless ~**, radyo (alıcı makine). **~**[2] *a.* Sabit; muayyen. **to be all ~**, başlamağa hazır olm.: **~ fair**, (barometro) devamlı açık hava: **the fruit is ~**, meyvalar tuttu: **hard ~**, (çimento vs.) donmuş: **~ phrase**, klişe: **~ purpose**, katî maksad: **of ~ purpose**, taammüden: **a ~ smile**, daimî tebessüm: **a ~ speech**, klişe nutuk: **~ subject [book]**, imtihan için muayyen mevzu [kitab]. **~**[3] *v.* (Güneş, ay) batmak; (çimento) donmak, katılaşmak; meyletmek; (av köpeği) ferma etmek; koymak, yerleştirmek; başlatmak; tanzim etm.; vermek; (yazı) dizmek; (kıymetli taş) oturtmak. **to ~ a book for an exam.**, imtihan için bir kitab tesbit etm.: **(of a broken bone) to ~**, kırık kemik kaynamak: **to ~ a broken bone**, kırık kemiği yerine koyarak sarmak: **to ~ s.o. to do stg.**, birine bir iş vermek: **to ~ the dog barking**, köpeği havlatmak: **(of a dress) to ~ well [badly]**, elbise iyi (fena) oturmak: **to ~ a hen**, bir tavuğu kuluçkaya yatırmak: **to ~ one's hopes [mind, heart] on doing stg.**, bir şeyi candan istemek: **opinion is ~ting that way**, umumî efkâr o tarafa meylediyor: **to ~ right**, düzeltmek; yoluna koymak: **to ~ s.o. on his way**, birine yol göstermek: **to ~ words to music**, bir güfteyi bestelemek. **~ about**, **to ~ about doing stg.**, bir işe girişmek: **I don't know how to ~ about it**, bu işe nasıl girişeceğimi bilmiyorum: **to ~ a rumour about**, bir rivayeti yaymak: **to ~ about s.o.**, (*coll.*) birine hücum etmek. **~ against**, ···e dayamak: **to ~ one person against another**, birini başkası aleyhine çevirmek: **to ~ one thing against another,** bir şeyin değerini başka bir şeyinki ile ölçüye vurmak. **~ apart,** ayırmak; bir tarafa koymak; tahsis etm. **~ aside,** bir tarafa koymak; biriktirmek; ibtal etm.; **to ~ a will aside,** bir vasiyetnameyi ibtal etmek. **~ back, to ~ a house back from the road,** bir evi yoldan içeri almak: **(horse) to ~ back its ears,** (at) kulaklarını yatırmak: **to ~ back a clock,** saati geri almak. **~-back** *n.*, gerileme, muvakkat muvaffakiyetsizlik. **~ before,** önüne koymak: **to ~ Shakespeare before Dante,** Shakespeare'i Dante'ye tercih etmek. **~ down,** yere koymak; (yolcu vs.yi) çıkarmak; yazmak. **~ forth,** yola çıkmak; ileri sürmek. **~ in,** (kış vs.) gelip çatmak; (karanlık) basmak; meydana gelmek. **~ off,** yola çıkmak; boylamak; tebarüz ettirmek; güzelleştirmek; meydana çıkarmak; karşılık olarak koymak; takas yapmak: **this answer ~ them off laughing,** bu cevab onları güldürdü. **~-off,** karşılık. **~ on, to ~ a dog on to s.o.,** bir köpeği birine saldırtmak: **I was ~ on by a dog,** bir köpek bana saldırdı: **to be ~ on stg.,** bir şeyi canı çok istemek. **~ out,** yola çıkmak; izah etm.; teşhir etm.: **to ~ out to ...,** ...e koyulmak: **he ~ out to reform the world,** dünyayı ıslah etmeğe kalkıştı. **~ to,** (işe) koyulmak. **~-to,** dövüş. **~ up,** dikmek; kurmak; ileri sürmek: **to ~ up house,** ev kurmak: **to ~ s.o. up in business,** birini bir işe yerleştirmek: **to ~ up a shout,** feryad koparmak: **he ~s up to be a poet,** şairlik taslıyor: **to ~ up a manuscript,** (matbaa) bir yazıyı dizmek: **this medicine ~ me up,** bu ilac beni diriltti: **a well-~-up youth,** boylu boslu genc. **~ upon,** hücum etm., çullanmak.

**settee** [seˈtī]. Kanape.

**settle** [ˈsetl]. Yerleştirmek; iskân etm.; tesbit etm.; kararlaştırmak; teskin etm.; halletmek; düzeltmek; ödemek. Yerleşmek; sabit bir hale gelmek; konmak; bir işe koyulmak; durulmak; dibe çökmek; (bina) bir az yere çökmek. **to ~ an account,** bir hesabı tediye etm.: **to ~ an account with s.o.,** birile kozunu paylaşmak: **to ~ s.o.'s account,** birinin hesabını görmek: **to ~ an annuity on s.o.,** birine senelik gelir bağlamak: **to ~ definitely** [**once for all**], kesip atmak: **to ~ s.o.'s doubts,** birinin şübhesini gidermek: **it's as good as ~d,** oldu bitti sayılır: **(of a ship) to ~,** yavaş yavaş dibe batmak: **the snow is settling,** kar tutuyor: **that ~s it!,** mesele kendiliğinden halledildi. **~ down,** sükûnet bulmak; durulmak; bir yerde yerleşmek; oturmak; yeni bir muhite alışmak. **~ upon,** karar vermek; seçmek; (birine) bağlamak; üzerine konmak. **~d** [ˈsetld], sabit; devamlı; kararlaştırılmış; muayyen; değişmez; meskûn; ödenmiş. **~ in life,** ev bark ve işgüc sahibi; (*kız*) evlenmiş: **with a ~ job,** devamlı bir iş sahibi. **~ment** [ˈsetlmənt], yerleştirme; tanzim; ödeme; birine bağlanan irad; yeni imar ve iskân olunan yer: **marriage ~,** evlenme mukavelesi ile zevceye bağlanan irad vs. **~r** [ˈsetlə(r)], yeni bir memlekette yerleşen adam.

**seven** [ˈsevn]. Yedi. **~fold,** yedi misli. **~teen,** on yedi. **~th** [–θ], yedinci. **~tieth,** yetmişinci. **~ty,** yetmiş. **~-league boots,** (bir masalda) kim giyerse ona her adımda yedi fersah yol aldıran ayakkabı.

**sever** [ˈsevə(r)]. Ayırmak; kesmek; yarmak. **~ance,** ayırma; inkıta; kesilme.

**several** [ˈsevrəl]. Bir çok; bir kaç; müteaddid; muhtelif; ayrı ayrı.

**sever·e** [siˈviə(r)]. Şiddetli; vahim; sert. **~ity** [–ˈveriti], şiddet; sertlik; huşunet.

**sew** [sou]. Dikmek. **~ on,** dikerek takmak: **~ up,** dikip kapatmak. **~er** [ˈsouə(r)], dikişçi.

**sew·age** [ˈsyûic]. Lâğım pisliği. **~er** [ˈsyuə(r)], ana lâğım: **~erage** [–ric], lâğım pisliği.

**sex** [seks]. Cins(iyet). **the fair ~,** cinsi lâtif. **~-appeal,** cinsî cazibe. **~ed,** cinsiyeti olan: **over-~,** müfrit derecede cinsiyete düşkün. **~less,** cinsiyetsiz. **~ual** [ˈseksyuəl], cinsî; tenasülî. **~uality** [–ˈaliti], cinsiyet.

**sextet** [seksˈtet]. Altı seslik (çalgılık hava).

**sexton** [ˈsekstən]. Kilise kayyumu; mezarcı.

**shabby** [ˈşabi]. Kılıksız; pejmürde; süflî; alçak; cimri. **~-genteel,** zevahiri kurtarmağa çalışan düşkün kibar.

**shack** [şak]. Kulübe.

**shackle** [ˈşakl]. Ayak zinciri, **iki** zinciri birleştiren bakla. **~s,** köstek;

pranga; mani. Prangaya vurmak; zincirle bağlamak; menetmek.
**shad·e** [şêyd]. Gölge; renk derecesi; anat; lâmba fanusu. Gölge vermek; muhafaza etm.; örtmek. **~s of meaning,** ince mana farkları: **~s of difference,** incelikler: **green shading into blue,** maviye çalan yeşil: **the ~s of night,** karanlık: **to put s.o. in the ~,** birini gölgede bırakmak. **~iness,** gölgelik; şübheli olma. **~ow** [ˈşadou], gölge; saye; hayal. Gölgelendirmek; mübhem bir şekilde ima etm.; gizlice takib etmek. **to cast a ~,** gölge yapmak; (*fig.*) keder vermek: ⌜**coming events cast their ~s before them**⌝, olacak şey kendini belli eder: **there is not the ~ of doubt that ...,** zerre kadar şübhe yok ki: **a ~ of fear crossed his face,** yüzünde bir korku rüzgârı dolaştı: ⌜**may your ~ never grow less!**⌝, (*şak.*) Allah feyzini daim etsin! **he is a mere ~ of his former self,** nerede şimdi o eski hali?: **to quarrel with one's own ~,** beyhude yere üzülmek: **under the ~ of this disaster,** bu felâket havası içinde: **he is under a ~,** lekeli *veya* şübhelidir: **to wear oneself to a ~,** uğraşa uğraşa hayalifenere dönmek. **~owy** [ˈşadoui], hayal meyal; mübhem. **~y** [ˈşêydi], gölgeli; şübheli. **he is a ~ character,** sağlam ayakkabı değil: **to be on the ~ side of fifty,** ellisini aşmış olm.: **the ~ side of politics,** politikanın çirkin tarafı.
**shaft** [şâft]. Ok vs.nin sapı; ok; şua; sütun; şaft; maden kuyusu; hava cereyanı borusu. **~ing,** şaft donanımı.
**shaggy** [ˈşagi]. Kaba saçlı [kıllı]; pürüzlü.
**shah** [şâ]. Şah.
**shak·e** (**shook, shaken**) [şêyk, şuk, şêykn]. Silkmek; sarsmak; çalkamak. Sarsılmak; titremek. Sallanma; sarsıntı; çalkama; titreme. **to ~ all over,** tir tir titremek: **to ~ oneself free from stg.,** silkip kendini kurtarmak: **to ~ s.o.'s hand,** birinin elini sıkmak: **to ~ hands on it,** bir meselede uzlaşıp el sıkışmak: **he shook his head (to say no),** yok manasında başını salladı: **in a ~,** (*sl.*) hemencecik: **to ~ in one's shoes,** korkudan titremek. **~-down,** sarsa sarsa yere düşürmek; (*coll.*) muhitine alışmak. **~ off,** silkip atmak: **to ~ off a cold,** bir nezleyi savmak: **to ~ the dust off one's feet,** nefretle uzaklaşmak: **to ~ off a person,** sırnaşık birisinden yakasını kurtarmak. **~ out,** silkip tozunu vs. çıkarmak. **~ up,** çalkalamak; silkmek; (*coll.*) uyandırmak; gözünü açmak: **that's given him a bit of a ~ up,** bu onun aklını başına getirir. **~y,** sallanan; sarsılmış; sağlam değil; kuvvetsiz. **his English is ~,** ingilizcesi zayıfdır.
**shall** [şal]. *İstikbal sıygasını yapmak için kullanılan yardımci fiil; vurgulu olarak telaffuz edilirse vücubî sıygası mânasını ifade eder;* **shall not** *um.* **shan't** [şânt] *okunur ve bazan böyle yazılır.* **I ~ go,** gideceğim: **~ we go to the cinema?,** sinemaya gidelim mi?: **~ I tell him?,** bunu ona söyleyim mi?: **you ~ tell me tomorrow,** onu yarın bana söyleyeceksin: **you *shall* tell me,** mutlaka söylemelisiniz.
**shallow** [ˈşalou]. Sığ; sathî. Sığlık.
**sham** [şam]. Taklid; düzme; iğreti. Yalan; gözboyası; sahtekâr. Yalandan yapmak. **he is only ~ming,** mahsus yapıyor: **to ~ death [dead],** ölü gibi yapmak.
**shamble** [ˈşambl]. Şapşal yürüyüş; ayak sürtme. Böyle yürümek.
**shambles** [ˈşamblz]. Mezbaha. **the place was a ~,** kan gövdeyi götürüyordu.
**shame** [şêym]. Utanc; mahcubiyet; rezalet; günah; yazık. Utandırmak, mahcub etmek. **for ~!,** ne ayıb!: **it's a ~ to laugh at him,** onunla alay etmek doğru değil: **to ~ s.o. into doing stg.,** birini utandırarak bir şeyi yaptırmak: **to put to ~,** utandırmak: **~ upon you!,** ayıb sana!: **what a ~!,** ne yazık!; olur şey değil!: **without ~,** arsız. **~ful,** utandırıcı; utanacak; rezil; yüz kızartıcı. **~less,** arsız, utanmaz. **~faced,** utanmış; mahcub.
**shammy** (**-leather**) [ˈşamiˈleðə(r)]. Güderi.
**shampoo** [ˈşampû]. Şampuan (yapmak).
**shamrock** [ˈşamrok]. İrlanda'nın millî remzi olan bir nevi yonca.
**shandy**(**gaff**) [ˈşandigaf]. Zencefilli gazoz ile karıştırılmış bira.
**shanghai** [şan(g)ˈhây]. Afyonla sersemletip tayfa olarak gemiye alıp götürmek.
**shank** [şan(g)k]. İncik; gemi demiri bedeni. **to go on S~s's mare,** tabanvayla gitmek.
**shan't** = **shall not.** *v.* **shall.**

**shanty**[1] [ˈşanti]. Kulübe; derme çatma bina. ~[2]. Gemici şarkısı.
**shape** [şêyp]. Biçim (vermek); şekil (vermek); tertib etm.; yontmak; uydurmak; inkişaf etmek. **to ~ a course,** (gemi) filan cihete yol tutmak: **to ~ well,** ümid verici olmak. **~d,** ... şekilli, biçimli. **~less,** biçimsiz; çirkin. **~ly,** endamlı; güzel biçimli.
**share** [şeə(r)]. Pay, hisse (senedi). Paylaşmak; hissesini almak; iştirak etmek. **~ and ~ alike,** müsavi hisseler alarak: **you are not doing your ~,** hissene düşeni yapmıyorsun. **he came in for his full ~ of misfortune [good fortune],** lâyık olduğu felâkete [saadete] fazlasile uğradı: **to go ~s,** paylaşmak: **to ~ in [take a ~ in],** ···e iştirak etm.: **to ~ out,** taksim etmek. **~holder,** hissedar. **~-pusher,** değersiz hisse senedleri satan ruhsatsız simsar.
**shark** [şâk]. Köpek balığı; dolandırıcı.
**sharp** [şâp]. Keskin; sivri; hâd; köşeli; açıkgöz, zeki; acı, ekşi; şiddetli, haşin; açık, vazıh; pek meşru olmıyan; (*mus.*) yarım ton tiz. **at two o'clock ~,** tam saat ikide: **G ~,** sol diyez: **look ~!,** haydi çabuk!: **~ practice,** pek meşru olmıyan iş: **turn ~ left!,** tam sola dön!: **'~'s the word!',** haydi çabuk!: **that was ~ work!,** (i) maşallah ne çabuk bitti!; (ii) (*bazan*) bu iş bir az şübheli. **~-edged,** keskin. **~-eyed,** keskin gözlü. **~-featured,** yüzünün hatları keskin. **~-set,** iyi bilenmiş. **~en,** bilemek; şiddetlen(dir)mek. **to ~ one's wits,** zekâsını parlatmak. **~er** [ˈşâpə(r)], dolandırıcı, hilekâr. **~ness,** keskinlik; açıkgözlülük; şiddet; ekşilik. **~-shooter** [ˈ-şûtə(r)], keskin nişancı.
**shatter** [ˈşatə(r)]. Çatırdatarak kırmak; parça parça etm.; yok etmek. Gürültü ile kırılmak. **~ing,** ezici; çatırdıyan.
**shav·e** [şêyv]. Tıraş etme *veya* olma; dar kurtuluş. Tıraş etmek *veya* olmak; rendelemek; hafifçe dokunmak. **a close ~,** sinek kaydı traş; dar kurtulma: **to have a close [narrow] ~,** dar kurtulmak. **~en,** tıraş olmuş. **~ing,** tiraş olma [etme]; yonga: **~s,** rende talaşı.
**shawl** [şôl]. Şal.
**she** [şî]. O (müennes). **she-,** ... dişi: **~-bear,** dişi ayı.
**sheaf** [şîf]. Demet.
**shear** (**-ed** [*esk.* **shore**]; **-ed** [*veya* **shorn**]) [şiə(r), şiö(r)d, şô(r), şôn]. Kırkma(k); demirci makası ile kesmek. **~s [pair of ~s],** büyük makas. **to be shorn,** saçı kesilmiş olm.: **to be shorn of,** ···den mahrum olmak.
**shearwater** [ˈşiəwôtə(r)]. Yelkovan (kuş).
**sheath** [şîθ]. Kın; zarf; mahfaza. **~e** [–ð], kınına koymak; kaplamak: **to ~ the sword,** sulh yapmak: **copper ~d,** bakır kaplamalı. **~ing,** kaplama; zırh. **~-knife,** kınlı bıçak.
**sheaves** [şîvz]. **sheaf** *'in cemi.*
**she'd** [şîd] = **she had; she would.**
**shed**[1] [şed] *n.* Baraka; kulübe. ~[2] *v.* (**shed, shedding**). Dökmek; akıtmak; etrafa yaymak. **to ~ tears,** ağlamak: **to ~ light on a matter,** bir meseleyi aydınlatmak.
**sheen** [şîn]. Parlaklık; perdah.
**sheep,** *pl.* **sheep** [şîp]. Koyun. **the black ~ of the family,** bir ailenin işe yaramaz ve serseri ferdi: **to make ~'s eyes at s.o.,** birine mahcubane ve hasretle bakmak. **~fold,** koyun ağılı. **~ish,** süklüm püklüm, mahcub. **~skin,** koyun pöstekisi. **~-dog,** çoban köpeği. **~-run,** koyun otlağı.
**sheer**[1] [şiə(r)] *a.* Halis, hakikî; dimdik, amudî. **it is ~ robbery,** bu düpedüz soygunculuk: **a ~ waste of time,** bu vakit kaybetmekten başka bir şey değil. ~[2] *v.* Birdenbire yoldan sapmak. **to ~ off,** (*coll.*) savuşmak.
**sheet** [şît]. Yatak çarşafı; maden levhası; kâğıd yaprağı; gazete; geniş satıh. (Çarşaf vs.) ile örtmek. **~-anchor,** ocaklık demiri; en çok güvenilen şey (kimse).
**sheikh** [şêyk]. Şeyh.
**shekel** [ˈşekəl]. Miskal. **the ~s,** para.
**shelf** [şelf]. Raf; sığlık. **on the ~,** bir tarafa atılmış; ıskarta edilmiş; (*coll.*) evlenemiyecek (kadın).
**shell** [şel]. Kabuk; bina vs. kafesi, iskeleti; gülle; (*Amer.*) fişek. Kabuğunu kırmak, kabuğundan çıkarmak; bombardıman etmek. **to come out of one's ~,** açılmak: **to retire into one's ~,** kapanmak: **to ~ out,** (*sl.*) paraları sökülmek. **~-fire,** top ateşi: **to come under ~,** bombardımana tutulmak. **~-fish,** kabuklu deniz hayvanları. **~-hole,** merminin toprakta açtığı çukur. **~-proof,** top

işlemez (sığınak vs.). ~-**shock,** bombardımanın sinirler üzerindeki tesiri.
**shelter** [ˈşeltə(r)]. Sığınak; melce. Barın(dır)mak, sığınmak; himaye etmek. **under ~,** emniyetli, mahfuz: **to take ~,** sığınmak. ~**ed,** mahfuz; barınacak: ~ **industry,** ecnebi rekabetine karşı himaye edilen sanayi: **a ~ life,** mahfuz ve rahat hayat.
**shelve** [şelv]. Raflar yapmak; rafa koymak. Şevlenmek.
**shelves** [şelvz]. **shelf** *'in cemi.*
**shepherd** [ˈşepö(r)d]. Çoban. Koyunlara bakmak. **to ~ children across the street,** çocukları koruyarak caddenin karşı tarafına geçirmek: **the Good S~,** Hazreti İsa.
**sherbet** [ˈsö(r)bət]. Şerbet.
**sheriff** [ˈşerif]. İngiltere'de: kontluklarda kıralı temsil eden fahrî memur; Amerika'da: bir nevi polis müdürü.
**sherry** [ˈşeri]. Beyaz İspanyol şarabı.
**she's** [şîz] = **she is; she has.**
**shew** *v.* **show.**
**shibboleth** [ˈşiboleθ]. Rağbetten düşmüş fikir; bir fırkanın şiarı.
**shield** [şîld]. Kalkan; müdafi; siper. Korumak, siper olmak; himaye etmek. **the other side of the ~,** bir meselenin öbür (gizli) tarafı.
**shift** [şift]. Değiş(tir)me; taşınma; işçi takımı; iş nöbeti; çare; hile; kadın iç gömleği. Yerini değiştirmek, taşınmak; (rüzgâr) dönmek. ~ **of the cargo,** gemi ambarındaki yük istifinin bozulması: **to ~ one's ground,** yeni bir bahane ileri sürmek; yeni bir iddiada bulunmak: **an eight-hour ~,** sekiz saatte bir değişen iş nöbeti: **to be at one's last ~,** çaresiz kalmak: **to make a ~,** taşınmak: **to make (a) ~ to do stg.,** bir şeyi yapmanın yolunu bulmak: **to make (a) ~ with what one has,** mevcudla idare etm.: **to ~ for oneself,** kendi başının çaresine bakmak: **to ~ one's quarters,** taşınmak: **to ~ the responsibility of stg. on to s.o.,** bir şeyin mesuliyetini başkasının üzerine atmak: **to work in ~s,** bir işte nöbetle çalışmak. ~-**less,** haylaz; çul tutmaz. ~**y,** hilekâr; tilki gibi. ~ **look [eyes],** güvenilmez bakiş.
**shilling** [ˈşilin(g)]. Şilin. **to cut s.o. off with a ~,** mirastan mahrum etm.: **to take the Queen's ~,** eskiden İngiliz ordusuna gönüllü yazılmak.
**shilly-shally** [ˈşilişali]. Kararsızlık. Ne yapacağını bilmemek; tereddüd etm.
**shimmer** [ˈşimə(r)]. Parıltı. Pırıl pırıl olm.
**shin** [şin]. İncik. **to ~ up a tree,** (*sl.*) bir ağaca tırmanmak.
**shindy** [ˈşindi]. Patırdı, şamata. **to kick up a ~,** (*sl.*) gürültü etm.; kıyamet koparmak.
**shin·e** (**shone**) [şâyn, şon]. Parla(t)mak; ışıldamak. Parıltı; cilâ. **rain or ~,** hava nasıl olursa olsun. ~**ing** parlak. ~**y,** parlak, cilâlı; (elbise) havsız.
**shingl·e**[1] [ˈşingl]. Deniz kenarında çakıl. ~[2]. (Dam) padavra. ~[3]. Kadın saçını erkek çocuğunki gibi kesme(k). ~**y** [ˈşin(g)gli], çakıllı.
**ship** [şip]. Gemi (ile göndermek); gemiye yüklemek. ~**'s boy,** miço: **'when my ~ comes home',** farzı muhal zengin olursam: ~ **oars!,** fora kürek!: **to ~ a sea,** geminin içine dalga girmek: **to take ~ for ...,** ...e gitmek üzere gemiye binmek. ~-**breaker,** gemi enkazcısı. ~-**broker,** gemi simsarı. ~-**chandler,** gemi levazımı satan adam. ~-**load,** bir geminin yükliyebildiği eşya, yolcu vs. ~-**mate,** aynı gemide hizmet eden arkadaş. ~**build·er** [ˈ–bildə(r)], gemi inşaatçısı: ~**ing,** gemi inşaatı. ~**master** [–mâstə(r)], ticaret geminin kaptanı. ~**ment** [–mənt], gemiye yükletme; yüklenen eşya. ~**owner** [–ǫunə(r)], gemi sahibi. ~**per** [ˈşipə(r)], eşyayı gemi ile sevkeden tüccar. ~**ping** [ˈşipin(g)], gemiye yükletme. Gemiler; ticaret filosu. Gemilere, denizciliğe aid: ~ **charges,** navlun: **the harbour was full of ~,** liman gemi ile dolu idi. ~**shape** [–şêyp], muntazam. ~**wreck** [ˈşiprek], geminin batması, karaya oturmasi, kazaya uğraması. Gemiyi batırmak, karaya oturtmak. **to be ~ed,** kazaya uğramak: **to ~ a plan,** bir plânı suya düşürmek. ~**wright** [–râyt], gemi inşaatına çalışan marangoz vs. ~**yard** [–yâd], tezgâhi; tersane.
**shire** [şâyə(r)]. İngiltere'nin idare taksimatı, kontluk.
**shirk** [şö(r)k]. (...den) yançizmek. ~**er,** işten yan çizen; hạlaz; vazifesini ihmal eden.
**shirt** [şö(r)t]. Gömlek. **dress [starched, (*coll.*) boiled] ~,** kolalı suvare gömleği: **in ~ sleeves,** ceket-

siz: **to put one's ~ on a horse,** at yarışlarda varını yoğunu bir at üzerine koyarak bahse girmek. **~-front,** kolalı gömlek göğüslüğü.

**shiver**[1] [ˈşivə(r)]. Soğuk veya korkudan titreme(k); çok üşümek; yelken hafifçe sallanmak. **to send cold ~s down one's back,** tüylerini ürpertmek. ~[2]. Ufak parça. Parala(n)mak.

**shoal**[1] [şǫul]. Sığlık. Sığlaşmak. ~[2]. Balık sürüsü. **~s of,** pek çok.

**shock**[1] [şok]. Demet yığını. ~[2]. Bol ve karışık saç. ~[3]. Sadme; sarsıntı; çarpışma; şiddetli tesir; ağır bir yaranın vücuddeki tesiri; (*elect.*) şok. Hayret ve nefret uyandırmak, müteessir etm.; sarsmak; çarpıntıya uğratmak. **easily ~ed,** çabuk utanır: **I was ~ed to hear of his death,** ölümü haberi beni çok sarstı: **~ troops,** hücum kıtaları. **~er,** (**shilling**) **~,** (*coll.*) fevkalâde macera romanı. **~ing,** utandırıcı, nefret verici; müdhiş; berbad. **~-absorber,** amortisör.

**shod** *v.* **shoe.**

**shoddy** [ˈşodi]. Kaba ve âdi kumaş. Âdi; mezad malı.

**shoe** [şû]. Kundura; ayakkabı; at nalı. *v.* (**shod**) [şod]. Nallamak. **that's another pair of ~s,** o mesele başka: **to cast a ~,** (at) nalını düşürmek: **waiting for dead men's ~,** mirasına konmak (yerine geçmek) için birinin ölümünü bekleme: **to die in one's ~s,** gayritabiî şekilde ölmek, *esp.* asılmak: **you are not fit to black his ~s,** sen onun ayağının pabucu olamazsın: **I should not like to be in his ~s,** onun yerinde olmak istemem: ⌜**everyone knows where his own ~ pinches**⌝, herkes kendi derdini kendi bilir: **to put the ~ on the right foot,** kabahat kiminse onu itham etm.: **to step into another's ~s,** birinin yerini almak. **~black,** kundura boyacısı. **~horn,** ayakkabı çekeceği. **~maker,** kunduracı. **~-cream,** kundura cilâsı. **~-lace,** ayakkabı bağı.

**shone** *v.* **shine.**

**shoo** [şû]. **to ~ away,** kışt! diye koğmak.

**shook** *v.* **shake.**

**shoot**[1] [şût] *n.* Sürgün, filiz; av partisi; hususî av yeri; top atma. ~[2] *vb.* (**shot**) [şût, şot]. Fışkırmak; tüfek atmak; tüfek ile avcılık etm.; atılmak; (futbol) şut çekmek. Atmak, fırlatmak; kurşun vs. ile vurmak; akıntılı bir yerden kayıkla hızlı geçmek. **the car shot past,** otomobil uçar gibi geçti: **I'll be shot if I ...,** ···sem öleyim. **~ away, to ~ away all one's ammunition,** bütün mühimmatını sarfetmek: **he had one arm shot away,** bir kolunu gülle götürdü. **~ down,** top ateşile vurup düşürmek. **~ off,** ok gibi fırlamak; vurup ayırtmak: **to ~ off for a prize,** bir atış müsabakasında finale girmek. **~ out,** dışarıya fırlamak; birdenbire görünmek; (filiz) sürmek. **~ up,** yukarıya fırlamak; pek çabuk yükselmek; (çocuk) birdenbire boy atmak; ateş altına almak. **~ing,** atış; avcılık: **~ star,** haceri semavî: **~-box,** avcılık mevsimi için mahsus bir köşk: **~-gallery,** nişan atmağa mahsus kapalı yer: **~-stick,** açılır kapanır oturacak yeri olan baston.

**shop** [şop]. Dükkân, mağaza. **~ [go ~ping]**, çarşıya çıkmak, dükkânları gezmek. **all over the ~,** (*coll.*) karmakarışık: **to talk ~,** işten bahsetmek: **you have come to the wrong ~,** yanlış kapıyı çaldınız. **~keeper,** dükkâncı. **~keeping,** dükkâncılık. **~lifting,** dükkândan aşırma. **~-assistant,** tezgâhtar. **~-case,** dükkânların cam dolabı. **~-front,** dükkânın ön camekânı. **~-girl,** (mağazada) satıcı kız. **~-soiled,** uzun müddet dükkânda kalarak tazeliğini kaybetmiş. **~-walker,** büyük mağazalarda müşterilere yol gösteren memur.

**shore**[1] [ˈşô(r)]. Sahil, deniz kıyısı. **in ~,** karaya yakın: **off ~,** (gemi) açılmış; (rüzgâr) karadan esen: **on ~,** karada: **one's native ~,** vatan. ~[2]. Destek; payanda. **to ~ up,** desteklemek. ~[3] (**shorn**) *v.* **shear.**

**short** [şôt]. Kısa (boylu); az (zaman); eksik, noksan; gevrek. **to cut stg. ~,** kısa kesmek; birdenbire sona erdirmek: **to cut s.o. ~,** birinin sözünü birdenbire kesmek: **a ~ drink,** viski gibi az miktarda içilen içki: **to fall ~ of,** ···e erişmemek: **to fall ~ of the mark,** istenilen dereceye erişmemek: **to give ~ weight,** eksik tartmak: **to go ~ of stg.,** bir şeyden mahrum kalmak: **in ~,** hulâsa: **the official was £100 ~ in his accounts,** memurun yüz lira açığı çıktı: **to have a ~ memory,** çabuk unutmak: **a ~ ten miles,** pek on mil yok: **we are ~ of sugar,** şekerimiz azaldı: **to be ~ of**

breath, nefesi daralmak: **it is nothing [little] ~ of madness to do this,** bunu yapmak delilikten aşağı değil: **I would do anything ~ of murder to get some money,** para bulmak için her şeyi göze alırım (adam öldürmek müstesna): **time is running ~,** pek az vakit kaldı: **we are running ~ of coal [our coal is running ~],** kömürümüz azalıyor: **to sell ~,** açığa satış yapmak: **~ sight,** miyopluk: **to stop ~,** birdenbire durmak: **he stops ~ of nothing to achieve his ends,** maksadına erişmek isterse hiç bir şey ona mâni olamaz: **to be in ~ supply,** kıtlığına kıran girmek: **to be taken ~,** (abdesti) sıkışmak: **he has a ~ temper,** sabırsızdır: **he was very ~ with me,** bana ters muamele etti: **to make ~ work of stg.,** çabucak bitirmek. **~(-circuit),** kontakt (yapmak); (*fig.*) kestirme yol bulmak. **~-dated,** kısa vadeli. **~-handed,** işçisi veya yardımcısı az. **~-lived,** kısa ömürlü; geçici. **~-sighted,** miyop; basiretsiz. **~-tempered,** çabuk öfkelenir. **~-term,** kısa vadeli. **~-winded,** tıknefes. **~age** [–ic], kıtlık; eksiklik: **housing ~,** mesken buhranı. **~bread** [–bred], kurabiye. **~coming** [–kʌmin(g)], kusur; noksan. **~en** [ˈşôtn], kıs(alt)mak. **~-hand,** stenografi. **~ly,** hulâsa; yakında; soğuk bir şekilde: **~ afterwards,** biraz sonra. **~ness,** kısalık; eksiklik, kıtlık; sertlik.

**shot**[1] *v.* **shoot.** **~**[2] [şot]. *a.* Değişen renkli, yanar döner. **~**[3] *n.* Top vs.nin bir atışı; saçma; iyi [fena] nişancı; tecrübe; (futbol) şut. **that was a bad ~!,** hiç tutmadı: **at the first ~,** ilk hamlede: **he is a good ~,** iyi atıcıdır: **to have a ~ at stg.,** (*fig.*) bir şeyi bir kere tecrübe etm.: **several ~s were heard,** birkaç silâh sesi işitildi: **like a ~,** derhal: **to make a good ~ at stg.,** başarmak için iyi bir teşebbüs yapmak: **to be off like a ~,** ok gibi fırlayıp gitmek: **without firing a ~,** kurşun atmadan.

**should** [şud]. *Yardımcı fiildir. Ekseriya ya manevî bir mecburiyet yahud farazi bir mâna ifade eder, mes.* **you ~ go there,** oraya gitseniz iyi olur, gitmelisiniz; **~ you go there,** oraya giderseniz. **as it ~ be,** haklı olarak: **all is as it ~ be,** her şey yolundadır: **he ~ have arrived by this time,** şimdiye kadar gelmesi icab ederdi: **this ~ have been done yesterday,** bu dün yapılmalıydı: **'will he be at the party?' 'I ~ think so',** 'Toplantıya gelecek mi?' 'Zannederim, her halde': **'he is very sorry for what he did.' 'I ~ think so!',** 'Yaptığı şeye çok müteessirdir.' 'Elbette, bir de müteessir olmıyacak mıydı?': **why ~n't she ride a bicycle (if she wants to)?,** niçin bisiklete binmesin?: **whom ~ I meet but Ahmed?,** kime raslasam beğenirsin?, Ahmede.

**shoulder**[1] [ˈşo̤uldə(r)] *n.* Omuz; dağ kolu. **~ to ~,** omuz omuza: **his ~s are broad,** (*fig.*) o dayanıklıdır: **to cold ~ s.o. [to give s.o. the cold ~],** istiskal etm.: **to have a head upon one's ~s,** zeki olm.: **an old head on young ~s,** yaşına göre tecrübeli: **to stand head and ~s above the rest,** başkalarına kat kat üstün olmak: **I let him have it straight from the ~,** ona bütün kuvvetimle yumruğu yapıştırdım; (*fig.*) açtım ağzımı yumdum gözümü. **~-belt,** hamail. **~-blade,** kürek kemiği. **~**[2] *v.* Omuzlamak; sallasırt etm.; omuzla itmek; omuza vurmak. **to ~ arms,** tüfeği omuza almak: **to ~ (a responsibility,** *etc.*), (bir mesuliyeti vs.) üzerine almak.

**shout** [şa̤ut]. Bağırmak. Bağırma, nida. **~s of applause,** şiddetli alkışlar: **~s of laughter,** gürültülü kahkahalar: **to ~ out stg.,** ... diye bağırmak: **to ~ s.o. down,** birini yuhalamak.

**shove** [şʌv]. İtme(k); itip kakma(k); dürtme(k); omuz vurmak. **to ~ stg. into a drawer,** bir şeyi çekmeceye sokmak: **to ~ off,** (*naut.*) bir kayığı avara etm.: **to ~ one's way through,** ite kaka kendine yol açmak.

**shovel** [ˈşʌvl]. Kürek(le atmak).

**show**[1] *v.* **(-ed, shown)** [şo̤u, şo̤ud, şo̤un]. Göstermek; ibraz etm.; teşhir etm.; isbat etm.; anlatmak; öğretmek; belirtmek. Görülmek, görünmek; belirmek. **to ~ one's cards [hand],** kâğıt oyununda elini göstermek; maksadını belli etm.: **to ~ s.o. the door,** birini kapı dışarı etm.: **to ~ itself,** görünmek: **to ~ oneself,** kendini göstermek: **he has nothing to ~ for all his work,** bütün çalışmasına rağmen ortada bir şey yok: **on your own ~ing,** kendiniz itiraf ettiğiniz gibi: **to ~ s.o. to his room,** birini odasına götürmek: **what can I ~**

you, sir?, (dükkânda), ne istiyorsunuz, efendim? ~ **in,** (bir misafiri vs.) içeri almak. ~ **off,** güzel göstermek; teşhir etm.; gösteriş yapmak. ~ **out,** birini kapıya kadar uğurlamak; birini kapı dışarı etmek. ~ **through,** ... arkasından [arasından] görünmek; sırıtmak. ~ **up,** teşhir etm.; foyasını meydana çıkarmak; (renk vs.) belirmek; isbatı vücud etmek. **to be ~n up,** foyası meydana çıkmak; teşhir edilmek. ~[2] *n.* Gösteriş, alâyış; nümayiş; manzara; sergi, teşhir; (*coll.*) fırsat; (*coll.*) iş, mesele. **to give s.o. a fair ~,** (*coll.*) birine kendini göstermek için lâyik olduğu fırsatı vermek: **to give the ~ away,** (*coll.*) ağzından baklayı çıkarmak; ihanet ederek ifşa etm.; foyasını meydana çıkarmak: **to make a good ~,** (*coll.*) kendini göstermek; iyi bir tesir bırakmak: **a ~ place,** görmeğe değer muhteşem bir ev vs.: **the ~ pupil of the school,** mektebin örnek talebesi: **he claims, with some ~ of reason ...,** oldukça haklı olarak iddia ettiğine göre ...: **to run the ~,** (*coll.*) bir iş idare etm.; bir yerde hakikî patron olmak. **~-case,** dükkân içindeki camekân. **~-ground,** sergi sahası. **~iness** [ˈşouinis], gösteriş, debdebe. **~ing-off,** çalım; gösteriş. **~man,** sirk vs. müdürü: **~ship,** teşhir sanatı. **~room,** eşya teşhir salonu. **~y** [ˈşoui], gösterişli; göz alıcı.

**shower** [ˈşauə(r)]. Geçici hafif yağmur. Yağdırmak. **heavy ~,** sağanak. **~y,** ara sıra yağmur yağan. **~-bath,** duş.

**shrank** *v.* **shrink.**

**shred** [şred]. Dilim dilim kesmek; lime lime etmek. Dilim; paçavra. **not a ~ of evidence,** en küçük bir delil yok: **to tear to ~s,** doğramak.

**shrew** [şrû]. Kır faresi. (Kadın) cırlak, titiz. **~ish,** cırlak; hırçın.

**shrewd** [şrûd]. Zeki, cinfikirli; (hüküm vs.) isabetli.

**shriek** [şrîk]. (*ech.*) Feryad (etm.); çığlık. **to ~ with laughter,** gülmekten katılmak.

**shrift** [şrift]. **to give s.o. short ~,** derhal cezasını vermek.

**shrill** [şril]. Tiz, keskin sesli.

**shrimp** [şrimp]. Karides (avlamak); bodur boylu adam.

**shrine** [şrâyn]. Mukaddes yer; türbe.

**shrink** (**shrank, shrunk**) [şrin(g)k, şran(g)k, şrʌn(g)k]. Çekinmek; büzülmek; kısalmak, çekilmek. Daraltmak, kısaltmak. **~age,** daraltma; fire. **~ing,** çekingen; büzülür.

**shrivel** [ˈşrivl]. ~ (**up**), pörsü(t)mek; kuru(t)mak. **~led,** pörsük; kartalmış.

**shroud** [şraud]. Kefen(lemek); sarmak.

**shrove** [şrouv]. **~tide,** apukurya. **S~-Tuesday,** Hıristiyanların büyük perhizinin başlangıcı olan salı günü.

**shrub** [şrʌb]. Küçük ağac; çalı. **~bery** [ˈşrʌbəri], çalıdan mürekkeb ufak koru.

**shrug** [şrʌg]. Omuzlarını silkme(k).

**shrunk** *v.* **shrink.** **~en,** çekmiş; daralmış.

**shuck** [şʌk]. Kabuk (soymak). **~s!,** saçma!

**shudder** [ˈşʌdə(r)]. Hafifçe titreme(k): raşe.

**shuffle** [ˈşʌfl]. Ayak sürüme(k); iskambil kâğıdlarını karma(k); becayiş. **to ~ out of doing stg.,** estek etmek köstek etmek.

**shun** [şʌn]. ···den kaçınmak, sakınmak.

**'shun.** ((*abb.*) **Attention**) ~ !, hazırol!

**shunt** [şʌnt]. Trenin yolunu değiştirme. Treni yan yola geçirmek.

**shut** (**shut**) [şʌt]. *v.* Kapa(t)mak; kapanmak. *a.* Kapanmış. **to ~ one's eyes,** gözünü yummak: **to ~ one's eyes to stg.,** bir şeye göz yummak: **to ~ one's finger in the door,** parmağını kapıya kıstırmak. ~ **down,** kapağını indirip kapatmak. ~ **in,** kuşatmak. ~ **off,** kesmek. ~ **out,** içeri bırakmamak: **to ~ out a view,** bir manzarayı kapatmak. ~ **to,** (kapıyı vs.) kapatmak. ~ **up,** kapamak; hapsetmek; sus(tur)mak: **he ought to be ~ up,** Toptaşına [Bakırköyüne] göndermeli: **to ~ a house up,** bir evi kapayıp kullanmamak: **to ~ up shop,** işten vazgeçmek. **~ter** [ˈşʌtə(r)], kepenk; (*fot.*) kapak. **to put up one's ~s,** dükkânı kapatmak; bir işten vazgeçmek.

**shuttle** [ˈşʌtl]. Mekik. **~service,** gidiş geliş karşılıklı sefer. **~cock,** 'badminton' oyununda kullanılan ucu tüylü mantar bir oyuncak.

**shy**[1] [şây]. Çekingen, sıkılgan. (At) ürkmek. **to ~ at stg.,** bir şeyden ürküp atlamak: **to fight ~ of stg.,** bir şeyden kuşkulanmak: **to fight ~ of a**

job, bir işten çekinmek. ~², (*sl.*) Taş, top vs. atmak.

**Shylock** [ˈşâylok]. (*Shakespeare'in Venedik Taciri'nindeki bir Yahudi'nin adı.*) Hasis; tefeci; amansız alacaklı.

**Siamese** [ˈsâyəmîz]. **~ twins,** vücudleri yapışık olarak doğmuş ikizler.

**sibilant** [ˈsibilənt]. Safir, ıslıklı (harf).

**sibyl** [ˈsibil]. Kâhin kadın.

**sic** [sik]. Böyle (metinde aynen).

**sick** [sik]. Hasta; kusacak gibi. **to be ~,** hasta olm.; kusmak: **to be ~ of stg.,** bir şeyden bıkmak: **to be ~ of life,** dünyaya küsmek: **I'm ~ of you!,** senden illâllah!: **to feel ~,** midesi bulanmak: **~ at heart,** meyus. **~en,** hastalanmak: bıktırmak; mide bulandırmak: **~ing,** mide bulandırıcı, iğrendirici. **~ly,** hastalıklı, cılız; insanın içini bayıltıcı; (ışık, renk) sönük; (iklim) sıhhate muzır; (tebessüm) zoraki. **~ness,** hastalık; kusma. **~-allowance,** hastalık tahsisatı. **~-bay** [**-berth**], (gemi) hastalara mahsus kamara. **~-bed,** hasta yatağı. **~-headache,** yarım başağrısı.

**sickle** [ˈsikl]. Orak.

**side** [sâyd]. Yan; taraf; böğür; kenar; (*sl.*) kurum. Asıl olmıyan. **to ~ with,** tarafını tutmak: **he is on our ~,** o bizdendir: **the other ~ of the picture,** madalyonun ters tarafı: **you have the law on your ~,** kanun sizin tarafınızdadır: **this climate is on the cool ~,** bu iklim soğuğa kaçar: **on the heavy ~,** biraz ağırdır: **to be on the wrong** [**right**] **~ of forty,** kırk yaşından yukarı [aşağı] olm.: **to get on the soft ~ of s.o.,** birini zayıf tarafından yakalamak: **wrong ~ out,** (elbise vs.) ters. **~-car,** motosiklet yan arabası. **~-lamps,** otomobilin küçük lâmbaları. **~-line,** (demiryolu) tâli hat; bir fabrika vs. asıl istihsalı haricinde yaptığı şey; tâli iş. **~-saddle,** kadınların ata yan binmesine mahsus eyer. **~-show,** tâli bir mesele. **~-slip,** (bisiklet) yan savurma. **~-step,** yan basamak; bir yana adım atmak; bir mâniden kaçınmak. **~-stroke,** (yüzmede) yandan kulac atmak. **~-track,** (demiryolu) yan yolu; bir treni yan yola geçirmek; bir işi bir tarafa koymak. **~-walk,** (*Amer.*) kaldırım. **~-whiskers,** favori. **~board,** büfe. **-~d,** *suff.*... yanlı, ... taraflı. **~light,** (*naut.*) borda, yan feneri. **to throw a ~ on a subject,** bir meseleyi yeni bir tarafından aydınlatmak. **~long,** yan taraftan. **~ward** [ˈsâydwö(r)d], *a.* **~s,** *adv.* yan(a doğru). **~ways** [ˈsâydwêyz], yandan: **to walk ~,** yan yan yürümek.

**sidereal** [sâyˈdiəriəl]. Yıldızlara aid.

**siding** [ˈsâydin(g)]. Demiryolunda manevra için kullanılan yan hat.

**sidle** [ˈsâydl]. **to ~ along,** yan yan gitmek: **to ~ up to s.o.,** birine sokulmak.

**siege** [sîc]. Muhasara. **to lay ~ to,** muhasara etm.: **to raise a ~,** (i) muhasara edenleri çekilmeğe mecbur etm.; (ii) muhasarayı kaldırmak.

**siesta** [siˈesta]. Öğle uykusu.

**sieve** [siv]. Kalbur(dan geçirmek).

**sift** [sift]. Elekten geçirmek. **to ~ the evidence,** şehadetin delillerinden hakikisini yalanından ayırmak.

**sigh** [sây]. İç çekme(k), ah etme(k). **to ~ for stg.,** bir şeyin hasretini çekmek.

**sight** [sâyt]. Görme (kuvveti); nazar; müşahede; manzara; temaşa; rasad. Görmek; nişangâhını tanzim etmek. **~s,** tüfeğin gezle arpacığı; (şehir vs.) görülecek yerler. **at** [**on**] **~,** görür görmez: **to come into ~,** gözükmek; ortaya çıkıvermek: **his face was a ~!,** yüzünü görmeliydin!: **to find favour in s.o.'s ~,** birinin gözüne girmek: **at first ~,** ilk görüşte: **to get a ~ of,** bir kere görmek: **I can't bear the ~ of him,** onu görmeğe tahammül edemem: **to have good** [**bad**] **~,** gözleri iyi [fena] olmak: **in ~,** görünürde: **in the ~ of,** gözünde: **to be in ~ of,** görebilmek: **to keep in ~** [**not to let out of one's ~**], gözden kaçırmamak: **to know s.o. by ~,** birisile göz aşinalığı olm.: **long ~,** prezbitizm: **to lose one's ~,** kör olm.: **to lose ~ of,** gözden kaybetmek; **I've quite lost ~ of him,** ondan hiç haberim yok: **out of ~,** görünmez: ⌜**out of ~, out of mind**⌝, gözden uzak olan gönülden de uzak olur: **out of my ~!,** defol!: **to put out of ~,** gizlemek: **short ~,** miyopluk: ⌜**a ~ for sore eyes**⌝, (i) bir içim su; (ii) *uzun zaman görülmiyen bir dosta rastlayınca söylenir*: **to take a ~ at the sun,** güneşi rasad etm.: **what a ~ you are!,** bu ne hal! **~-seeing, to go ~,** seyredecek yerleri görmeğe gitmek. **~-seer** seyyah, turist. **~ing,** *n.* nişangâh. **-~ed, far-~,** uzağı gören, prezbit; basiretli: **near [short]-~,** yakını gören, miyop; basiretsiz: **weak-~,** gözü zayıf. **~less,** kör. **~ly,** yakışıklı.

**sign**[1] [sâyn] *n.* İşaret; iz; alâmet; levha; delil; belirti. **~-painter (-writer),** tabelâ ressamı, yazıcısı. **~**[2] *v.* İmzalamak; işaret etmek. **~ away,** bir mülkü vs. senedle başkasına terketmek. **~ off,** memur vs. işten çıkarken defteri imzalamak. **~ on,** memur vs. işe başlarken defteri imzalamak. **~post** [–poust], (yol gösteren) işaret direği.

**signal**[1] [ˈsignəl] *a.* Göze çarpan, parlak. **~**[2] *n.* İşaret; (*naut.*) haber (emir). *v.* İşaret etm.; (*naut.*) işaretle haber (emir) vermek. **~ize,** işaretle bildirmek. **~ler,** (*mil.*) işaretçi. **~man,** (demiryolu) işaret memuru; (bahriye) işaretçi. **~-box [-cabin],** demiryolu işaret kulesi. **~-cord,** (trende) imdad işareti.

**signat·ory** [ˈsignətəri]. İmza sahibi; muahid. **~ure** [ˈsignityuə(r)], imza.

**signet** [ˈsignit]. Mühür. **writer to the ~,** (İskoçya'da) avukat. **~-ring,** mühür yüzüğü.

**signif·y** [ˈsignifây]. Delâlet etm.; beyan etm.; tebliğ etm.; manası olmak. **it does not ~,** ehemmiyeti yok; zarar yok. **~icance** [–ˈnifikəns], mâna; ehemmiyet. **~icant,** mânalı; ehemmiyetli. **~ication** [–ˈkêyşn], mâna; ifade.

**silen·ce** [ˈsâyləns]. Sükût, sessizlik; susma. Susturmak; ateş kesmeğe mecbur etmek. ⌜**~ gives consent**⌝, sükût ikrardan gelir: **dead ~,** ölüm sükûtu: **to pass stg. over in ~,** bir şeyi sükûtla geçiştirmek. **~cer,** amortisör. **~t,** sessiz; sâkin: **to keep ~,** susmak.

**silhouette** [ˌsiluˈet]. Siluet(ini yapmak).

**silk** [silk]. İpek: **~ thread,** ibrişim. **to take ~,** avukatların en yüksek rütbesi olan **Queen's Counsel** tayın olunmak. **~-hat,** silindir şapka. **~en,** ipekli; ipek gibi. **~worm,** ipek böceği. **~y,** ipek gibi; (ses) fazla tatlı.

**sill** [sil]. Eşik.

**sill·iness** [ˈsilinis]. Ahmaklık, abdallık. **~y,** ahmak, abdal; gülünc; vahi. **to knock someone ~,** sersemletmek: **the ~ season,** gazetelerin, havadissizlikten, saçma sapan neşriyat yaptıkları devir.

**silt** [silt]. Suyun bıraktığı kum ve çamur. **to ~ up,** (liman vs.) bununla dol(dur)mak.

**silver** [ˈsilvə(r)]. Gümüş(ten yapılmış eşya); gümüş para. Gümüşten yapılmış; gümüş gibi. Gümüş kaplamak; gümüş gibi parlamak. **~ plate,** gümüş takımları: ⌜**to be born with a ~ spoon in one's mouth**⌝, (i) büyük ve zengin bir ailede doğmak; (ii) yıldızı parlak olmak. **~-gilt,** altın yaldızlı gümüş. **~ grey,** gümüşî. **~-haired,** beyaz saçlı. **~-headed,** beyaz saçlı; gümüş başlı (baston). **~-paper,** yaldız kâğıdı. **~-plate,** gümüş kaplama(k). **~-tongued,** talâkatlı. **~-wedding,** evlenmenin 25 inci yıldönümü. **~-smith** [–smiθ], gümüş eşya yapan kuyumcu. **~y,** gümüş gibi.

**simian** [ˈsimyən]. Maymun(a aid).

**simil·ar** [ˈsimilə(r)]. Misilli; benzer. **~arity** [–ˈlariti], benzerlik. **~e** [ˈsimili], teşbih. **~itude** [siˈmilityûd], benzerlik.

**simmer** [ˈsimə(r)]. Yavaş yavaş kayna(t)mak; (isyan vs.) patlamak üzere olmak.

**Simon** [ˈsâymən]. **Simple ~,** safdil.

**simony** [ˈsâymoni]. Mübarek eşya veya dinî memuriyetleri satma.

**simoon** [siˈmûn]. Sam yeli.

**simper** [ˈsimpə(r)]. Nazlı gülümseme(k).

**simpl·e**[1] [ˈsimpl] *a.* Basit, sade; kolay(ca anlaşılır); yalın; saf; safyürekli; tabiî. **~ folk,** kendi halinde kimseler: **it's a ~ matter,** işten bile değil: **you ~y must come,** muhakkak gelmelisiniz: **I was ~y delighted,** bilseniz ne kadar memnun oldum: **it's ~y ridiculous,** bu âdeta gülünc: **I ~y said that ...,** yalnız ... dedim. **~eton,** safdil adam. **~icity** [–ˈplisiti], basitlik; kolaylık; şafatasızlık; safderunluk. **~ify** [ˈsimplifây], basitleştirmek, kolaylaştırmak. **~e-minded,** sadedil.

**simula·crum** [ˌsimyuˈlêykrʌm]. Hayal; taklid. **~te** [ˈsimyulêyt]. Yalandan yapmak; taklidini yapmak; benzemek.

**simultaneous** [ˌsimʌlˈtêynyəs]. Aynı zamanda olan, yapılan, vukubulan. **~ly (with),** birlikte; hep beraber.

**sin** [sin]. Günah (işlemek). **for my ~s,** hangi günahım içinse: **like ~,** (*sl.*) şiddetle, alabildiğine: **more ~ned against than ~ning,** kabahat yalnız onun değil: **mortal ~,** Allahın affetmiyeceği günah: **original ~**, (Hıristiyanlarca) insanların yaratılışında olan günah işleme temayülü. **~ful**, günahkâr. **~ner** [ˈsinə(r)],günahkâr.

**since** [sins]. ···den beri; ···den sonra; madem ki; ... için. **ever ~ (then)**, o zamandan beri: **it is three years (ago) ~ I saw him**, onu üç seneden beri görmedim: **we have been here ~ March**, Marttan beri buradayız: **many years ~ [long ~]**, bundan çok sene evvel: **~ you say so, it must be true**, mademki siz söylüyorsunuz, doğrudur: **a more dangerous, ~ unknown, foe**, bilinmediği için daha tehlikeli bir düşman.

**sincer·e** [sinˈsiə(r)]. Samimî; riyasız; candan; halis. **yours ~ly**, hürmetlerimi sunarım. **~ity** [–ˈseriti], samimiyet; riyasızlık; hulûs: **in all ~**, tam bir hüsnüniyetle.

**sine** [ˈsine]. (*Lât.*) ···siz. **~ die** [dây·i], gün tayin etmeksizin: **~ qua non** [ˈsineˈkwêyˈnon], zarurî şey.

**sinecure** [ˈsâynikyuə(r)]. Hizmetsiz maaşlı memuriyet; arpalık.

**sinew** [ˈsinyu]. Veter; adale. **the ~s of war**, para. **~y**, adaleli, kuvvetli; (et) sert.

**sing** (**sang, sung**) [sin(g), san(g), sʌn(g)]. Şarkı söylemek; ötmek; (kurşun) vızıldamak; (kulak) çınlamak. **to ~ out**, bağırmak: **to ~ small**, kuyruğu kısmak. **~er**, muganni, hanende. **~-song**, cansıkıcı sesle söylenen; hep beraber şarkı söylemek için yapılan toplantı.

**singe** [sinc]. Azıcık yakmak; alevden geçirmek; saçların ucunu yakma(k). **to ~ one's wings**, maceralı bir işte zarar görmek; **to ~ the King of Spain's beard**, (16 ıncı asirda İngiliz gemicileri) İspanya sahillerini yağma etmek.

**single** [ˈsin(g)gl]. Tek; yeğane; bekâr. İki kişi arasında oyun. **to ~ out**, seçip bir tanesini almak: **to ~ out s.o.**, bir çok kimse arasından birini seçip ayırmak: **~ bed [bedroom]**, bir kişilik yatak [oda]: **~ combat**, iki adam arasında mücadele: **every ~ day**, tanrının günü: **not a ~ one**, bir tek bile yok: **the ~ state**, bekârlık. **~ness**, bekârlık: **~ of heart [-hearted]**, samimiyet: **~ of purpose**, garazsızlık; bir tek gayesi olma. **~-handed**, tek başına. **~-minded**, doğru fikirli. **~-seater**, tek kişilik (uçak).

**singlet** [ˈsin(g)glit]. İç gömleği.

**singular** [ˈsin(g)gyulə(r)]. (*gram.*) Müfred. Müfred; acayib; hususî; şayanı dikkat. **~ity** [–ˈlariti], garabet; hususiyet; tuhaflık.

**sinister** [ˈsinistə(r)]. Uğursuz, meş'um; netameli. **bend ~**, bir şahsın armasının sol tarafında, nesebinin gayrımeşru olduğunu gösteren işaret.

**sink**[1] [sin(g)k] *n.* Bulaşık çukuru; pisliğin biriktiği yer. **~**[2] *v.* (**sank, sunk**) [sin(g)k, san(g)k, sʌn(g)k]. Bat(ır)mak; dalmak; düşmek; alçalmak; çökmek. **to ~ by the bow [stern]**, (gemi) baş taraftan [kıçtan] batmak: **the building is ~ing**, bina çöküyor: **they sank their differences**, ihtilâflarını bertaraf ettiler: **that ~ing feeling**, insanın içine çöken o korku: **with ~ing heart**, gittikçe kasvete dalarak: **to ~ money in an enterprise**, bir işe parasını bağlamak: **the patient is ~ing**, hasta ölmek üzere dir: **my spirits sank**, içime kasvet çöktü: ⌜**here goes, ~ or swim!**⌝, haydi bakalım, ya batarız ya çıkarız!: **he was left to ~ or swim**, kendi mukadderatına terk edildi: **to ~ a well**, kuyu kazmak. **~ in(to)**, nüfuz etm.; tesir etm.; göm(ül)mek. **~ing-fund** [ˈsin(g)kin(g)ˌfʌnd], amortisman.

**sinuous** [ˈsinyuəs]. Dolambaçlı; yılankavi.

**sip** [sip]. Azar azar içmek. Azıcık içme.

**siphon** [ˈsâyfn]. Sifon; cam tüblü soda şişesi. Sifonla akıtmak.

**sir** [sö(r)]. Efendim!; **Baronet** *ve* **Knight** *'ların unvanı ki daima şahıs ismile kullanılır, mes.* **Sir George (Smith), Sir Peter (Jones)**: (**Sir Smith** *veya* **Sir Jones** *denmez*).

**sire** [sâyə(r)]. Baba; erkek hayvan, aygır. (Hayvan) babası olmak. (*Bir Kırala hitab ederken* **Sire, Sir** *yerinde kullanılır*).

**siren** [ˈsâyrən]. Cazibeli ve sehhar kadın; fettan; canavar düdüğü.

**sissy** [ˈsisi]. (*Amer. sl.*) Hanım evlâdı.

**sister** [ˈsistə(r)]. Kızkardeş; hemşire; **~ of mercy**, fukara ve hastalara bakan rahibe: **~ ships**, aynı tipte gemiler. **~hood**, kızkardeşlik; sörler birliği. **~ly**, kızkardeş gibi. **~-in-law**, görümce; baldız; yenge; elti.

**sit** (**sat**) [sit, sat]. Otur(t)mak; **to ~ for a borough** *etc.*, bir şehir vs.nin mebusu olm.: **to ~ on a committee,** *etc.*, bir komite vs.ye dahil olm.: **to ~ for an examination,** bir imtihana girmek: **to ~ a horse well [badly],** ata iyi [fena] binmek: **to ~ oneself down,** oturmak: **to ~ on s.o.,** (*coll.*) birini ezmek: **I won't be sat upon,** kendimi ezdirmem: **to ~ over a book,** bir kitaba kapanmak: **to ~ in Parliament,** Parlamento âzası olm.: **to ~ tight,** yerinden kımıldamamak; dediğinden vs. vazgeçmemek. **~ down,** oturmak: **to ~ down to table,** sofraya oturmak: **not to ~ down under an insult,** bir hakaretin altında kalmamak: **a ~ strike,** kolları kavuşturma grevi. **~ out,** (bir oyuna) iştirak etmemek: **to ~ out a dance with s.o.,** birisile dans etmeyip konuşmak: **to ~ a lecture out,** bir dersi, sabrederek, sonuna kadar dinlemek. **~ up,** doğru oturmak: **to ~ up in bed,** yatakta doğrulup oturmak: **to ~ up late,** geç vakte kadar (yatmayıp) oturmak: **to ~ up for s.o.,** birini bekliyerek yatmamak: **to make s.o. ~ up,** (*coll.*) birini şaşırtmak; şiddetle azarlamak: **to ~ up to the table,** iskemlesini masaya yaklaştırmak. **~ter** [ˈsitə(r)], (ressam için) poz alan kimse; kuluçka; (avcılıkta) kımıldanmıyan kuş; vurması kolay av. **I missed a ~,** vurması pek kolayken vuramadım. **~ting** [ˈsitin(g)], celse; ressam için poz alma; kuluçkalık. Oturan; kuluçka yatan; kımıldanmıyan (av). **a ~ shot,** vurması pek kolay bir av. **~ting-room,** oturma odası: **bed ~,** hem yatak hem oturma odası.

**site** [sâyt]. Mevki; (bir şeyin) bulunduğu veya vukubulduğu yer.

**situat·e(d)** [ˈsityuêyt(id)]. Yerleşmiş, kâin. **that is how I am ~,** işte vaziyetim budur: **a pleasantly ~ house,** yeri çok hoş bir ev. **~ion** [–ˈêyşn], mevki; bulunduğu yer; vaziyet; iş.

**six** [siks]. Altı. **coach and ~,** altı atlı araba: **two and ~,** iki buçuk şilin (2/6): **everything is at ~es and sevens,** her şey karmakarışık: ⌜**it's ~ of one and half a dozen of the other**⌝, ha o ha bu; al birini vur ötekine. **~fold,** altı misli. **~pence,** altı peni(lik); **~penny** (worth), altı penilik. **~teen,** on altı: **~th,** on altıncı. **~th,** altıncı. **~tieth,** altmışıncı. **~ty,** altmış. **~-footer,** altı kadem boyunda; çok uzun boylu.

**sizar** [ˈsâyzə(r)]. Üniversitede bir nevi burs talebesi.

**size¹** [sâyz]. Büyüklük; hacım; ölçü; numero; boy; çap. Büyüklüğüne göre tasnif etmek; **all of a ~,** hepsi aynı büyüklükte: **full ~,** tabiî büyüklük: **to take the ~ of stg.,** bir şeyi ölçmek: **to ~ up,** takdir etm., ölçmek: **to ~ s.o. up,** birini tartmak. **~able,** oldukça büyük. **~².** Çiriş(lemek); tutkal.

**sizzle** [ˈsizl]. (*ech.*) Cızırtı. Cızırdamak. **sizzling hot,** gayet sıcak.

**skate** [skêyt]. Paten (kaymak). **to ~ over thin ice,** pek nazik bir mevzua dokunmak.

**skedaddle** [skiˈdadl]. Acele kaçış. Tabanı kaldırmak.

**skein** [skêyn]. Çile. **tangled ~,** arabsaçı.

**skeleton** [ˈskeletən]. İskelet; çatı; kuru kemik. **~ crew,** çekirdek tayfa: ⌜**the ~ at the feast**⌝, bir toplantı vs.de neşe kaçıran şey: **family ~** [⌜**a ~ in the cupboard**⌝], bir ailenin utanılacak veya keder verici sırrı.

**sketch** [skeç]. Kabataslak resim; kroki; taslak; küçük piyes. Kroki yapmak; taslağını yapmak. **~y,** taslaklık; baştan savma: **~ knowledge,** derme çatma bilgi. **~-book,** kroki defteri.

**skew** [skyû]. Eğri(lik); mail.

**skewer** [ˈskyûə(r)]. Kebab şişi. Şişlemek.

**ski** [şî]. Kayak (yapmak).

**skid** [skid]. Takoz; tekerlek çariği; (otom.) kızak yapma(k); yana doğru kayma(k).

**skiff** [skif]. İskif.

**skil·ful** [ˈskilfəl]. Meharetli, becerikli. **~l,** meharet; ustalık; becerik. **~led,** meharetli, hünerli: **~ artisan [workman],** ehliyetli işçi.

**skilly** [ˈskili]. Sade suya yavan çorba.

**skim** [skim]. Köpüğünü almak; (sütten) kaymağını almak, sıyırmak. **to ~ along,** kayar gibi ilerlemek: **to ~ the cream off stg.,** (*fig.*) bir şeyin en iyi kısmını almak: **to ~ over [through] a book,** bir kitaba şöyle bir göz gezdirmek.

**skimp** [skimp]. Hasislik etm., kıt vermek. **to ~ one's work,** işini yarım yamalak yapmak. **~y,** dar; elverişli olmıyan.

**skin** [skin]. Deri; pösteki; kabuk;

tulum. Derisini yüzmek; (kabuğunu) soymak; sıyırmak. **mere ~ and bone,** bir deri bir kemik: **to keep one's eyes ~ned,** göz kulak olm.: **next to one's ~,** tenine: **to come off with a whole ~ [to save one's ~],** postu kurtarmak: **by the ~ of one's teeth,** dardarına. **-~ned,** ... derili: **thick-~** vurdumduymaz: **thin-~,** alıngan. **~flint,** cimri. **~ful, to have a ~,** (*sl.*) kafayı iyice çekmek. **~ny,** çok zayıf, çiroz gibi. **~-deep,** sathî.

**skip** [skip]. Zıplama(k); ip atlama(k). **~ (over),** atlamak. **~per,** ip atlayan.

**skipper** [ˈskipə(r)]. Kaptan, reis.

**skirl** [skö(r)l]. Gayda sesi.

**skirmish** [ˈskö(r)miş]. Hafif musademe. Karakol musademeleri yapmak. **~er,** (*mil.*) avcı.

**skirt** [skö(r)t]. Etek; (*coll.*) eksik etek. Kenarından geçmek. **~s,** civar. **~ing,** oda duvarının dip pervazı.

**skit** [skit]. Mizahî ve hicivli yazı. **~tish** [ˈskitiş], oynak; cilveli.

**skittles** [ˈskitlz] *v.* **ninepins.** ⌈**life is not all beer and ~**⌉, hayat eğlenceden ibaret değildir.

**skulk** [skʌlk]. Korkudan veya kötü niyetle gizlenmek; hırsızlama dolaşmak.

**skull** [skʌl]. Kafatası. **~ and cross-bones,** kafatası ve ince kemikler (korsan bayraklarında). **~-cap,** tepe takkesi.

**skunk** [skʌn(g)k]. Bir nevi kokarca; alçak ve pis herif.

**sky,** *pl.* **skies** [skây, –z]. Gök, sema. (Topu) havaya çelmek. **to laud to the skies,** göklere çıkarmak. **~lark,** tarla kuşu; (çocuk gibi) gürültü ile oynamak. **~light,** tepe penceresi; aydınlık. **~ward,** göğe doğru. **~-blue,** havai mavi.

**slab** [slab]. Büyük yassı parça; kalın dilim.

**slack** [slak]. Gevşek; tembel; mıymıntı; durgun. Laçka; tozlu kömür. Gevşetmek; tembel olm. **to ~ off,** laçka etm.: **to take up the ~,** boşunu almak. **~en,** yavaşla(t)mak; gevşe(t)mek; laçka etmek. **~er,** tembel; aylakçı.

**slag** [slag]. Cürüf.

**slain** *v.* **slay.**

**slake** [slêyk]. **to ~ one's thirst,** susuzluğunu gidermek: **to ~ lime,** kireci söndürmek.

**slam**[1] [slam]. (*ech.*) Şiddetle ve gürültü ile kapanma(k) veya kapatmak. **~**[2]. (İskambilde) mars, şelem.

**slander** [ˈslândə(r)]. İftira (etm.), zem(metmek). **~ous,** iftira nevinden.

**slang** [slan(g)]. Argo. (*coll.*) Küfretmek; azarlamak.

**slant** [slânt]. Meyil; eğrilik. Meyillenmek; meyilleştirmek. **on the ~,** verevlemesine. **~ing,** eğri; meyilli; **~ eyes,** çekik göz. **~ways,** eğri bir halde; çapraz.

**slap** [slap]. (*ech.*) Avucla vurma(k). **to ~ s.o. on the back,** şaka için birinin sırtına vurmak: **a ~ in the face,** şamar (gibi ters bir cevab); beklenmedik muvaffakiyetsizlik: **to run ~ into s.o.,** pat diye karşısına çıkmak: **~ on the spot,** şıp diye yapıştırma. **~stick,** şakşak. **~-bang,** şıp diye; hızla; beklenmedik şekilde. **~-dash,** acele ile; düşünmiyerek.

**slash** [slaş]. Uzun bir yara açmak; rasgele kesmek; kırbaç ile vurmak; şiddetle tenkid etmek. **to ~ about one,** sağa sola etrafa kılıc vs. vurmak. **~ing,** sert ve dokunaklı (tenkid); kamçılayan (yağmur).

**slat** [slat]. Tiriz; çıta; pancur tahtası. **~ted,** pancur gibi çıtalı.

**slate**[1] [slêyt]. Arduvaz; yazı taşı. Arduvaz kaplamak. **to clean the ~,** maziyi unutmak: **to start with a clean ~,** geçmişi unutarak yeni bir hayata başlamak. **~-coloured,** barudî. **~**[2]. (*coll.*) Azarlamak.

**slattern** [ˈslatö(r)n]. Hırpani kadın. **~ly,** hırpani, şapşal.

**slaughter** [ˈslôtə(r)]. Boğazlama(k); kesme(k); katliam. ⌈**like a sheep to the ~**⌉, kasablık koyun gibi. **~er,** mezbahacı. **~house,** mezbaha.

**slav·e** [slêyv]. Esir, köle; cariye. Köle gibi çalışmak; didinmek. **to ~ away at stg.,** bir işe dinlenmeden çalışmak. **~er,** esirci; esir gemisi. **~ery** [ˈslêyvəri], esirlik, kölelik: **to reduce to ~,** boyunduruk altına almak. **~ish,** köle gibi; aşağılık: **~ imitation,** körü körüne taklid.

**slay (slew, slain)** [slêy, slû, slêyn]. Öldürmek.

**sledge, sled** [slec, sled]. Kızak (ile gitmek). **~-hammer,** balyoz: **~ (blow,** *etc.*), pek şiddetli (vuruş vs.).

**sleek** [slîk]. (At vs.) tüyleri parlak; (saç) düz ve perdahlı; (söz) yüze gülücü.

**sleep** (**slept**) [slîp, slept]. Uyku. Uyumak. **to drop off to ~**, içi geçmek: **to go to ~**, uyumak; (ayak vs.) uyuşmak: **to put to ~**, yatırmak; (hayvanı) canını yakmadan öldürmek: **to ~ like a log [top]**, ölü gibi uyumak: **to send to ~**, uyutmak: **to talk in one's ~**, sayıklamak: **to walk in one's ~**, uykuda gezmek: **to ~ away the time**, vakti uykuda geçirmek: **to ~ in**, (hizmetçi) evde yatmak: **to ~ off the effects of stg.**, bir şeyi uyuyarak gidermek: **to ~ out**, açıkta yatmak; (hizmetçi) hizmet ettiği evde yatmamak: **to ~ over [upon] stg.**, bir mesele üzerinde bir gece düşünmek. **~er**, uyuyan adam; travers; (*coll.*) yataklı vagon: **a good [bad] ~**, uykusu iyi [fena] kimse. **~ing**, uyuma; uyuyan: ⌜**let ~ dogs lie**⌝, ⌜uyuyan yılanın kuyruğuna basma!⌝: **~ accommodation**, yatacak yer: **~ partner**, komanditer: **~-bag**, torba şeklinde yatak takımı: **~-car**, yataklı vagon: **~-draught**, uyku ilâcı: **~-sickness**, uyku hastalığı. **~y**, uykusu gelmiş; uyuşuk.

**sleet** [slît]. Sulu sepken (yağmak).

**sleeve** [slîv]. Yen; elbise kolu; manşon. **to laugh in [up] one's ~**, bıyık altından gülmek: **to roll up one's ~s**, paçaları sıvamak: **to have stg. up one's ~**, kozunu saklamak: **to wear one's heart on one's ~**, hislerini herkese göstermek.

**sleigh** [slêy]. Kızak. Kızakla gitmek.

**sleight** [slêyt]. **~ of hand**, elçabukluğu.

**slender** [slendə(r)]. İnce belli; narin; fidan gibi; az. **of ~ intelligence**, aklı kıt: **of ~ means**, dar gelirli.

**slept** *v.* **sleep.**

**sleuth** [slûθ]. **~ (hound)**, gayet keskin koku alan ve takibde kullanılan bir cins köpek; polis hafiyesi.

**slew** *v.* **slay.**

**slew** [slû]. Sap(tır)mak; dön(dür)mek.

**slice** [slâys]. Dilim; hisse. Dilimlemek; kesip biçme hareketi yapmak.

**slick** [slik]. (*coll.*) Düz; fazla cerbezeli; kurnaz. **be ~ about it!**, elini çabuk tut!: **~ in the eye**, tam gözüne.

**slid·e** (**slid**) [slâyd, slid]. Kay(dır)mak; kızak yapmak; sıvışmak; sokuşturmak. Kayma; çocukların kaydıkları buzlu yol; sürme; projeksiyon camı; bir aletin kayan parçası. **to let things ~**, ihmal etm.: **to ~ over stg.**, (*fig.*) üstünde durmamak; sükûtla geçiştirmek. **~ing**, kayıcı; sürme: **~ scale**, mütehavvil mikyas. **~e-rule**, hesab cedveli.

**slight**[1] [slâyt] *a.* İnce; pek az; hafif; ehemmiyetsiz. **there is not the ~est doubt**, zerre kadar şübhe yok. **~**[2]. Hatır kıracak söz (hareket); tahkir; saygısızlık. Hatırını kıracak harekette bulunmak; ···e saygısızlık göstermek. **to put a ~ on s.o.**, hor görmek. **~ing**, hatır kırıcı.

**slily** [ˈslâyli]. Kurnazca; el altından.

**slim** [slim]. İnce bel; fidan gibi; narin; (*sl.*) kurnaz. Kasden kendini zayıflatmak. **~ness**, ince bellilik, narinlik; (*coll.*) kurnazlık.

**slim·e** [slâym]. Balçık; sulu çamur; sümük. **~iness**, sulu çamur hali; sümüklülük; riyakârlık. **~y**, sümüklü; sulu çamurlu; yaltak ve riyakâr.

**sling** [slin(g)]. Sapan; askı; kol bağı. *v.* (**slung**) [slʌn(g)]. Sapan ile atmak; fırlatmak; asmak.

**slink** (**slunk**) [slin(g)k, slʌn(g)k]. Gizlice ve sinsi sinsi yürümek. **to ~ away [off]**, sıvışmak. **~ing**, sinsi.

**slip**[1] [slip] *n.* Kayma; ayak kayması; sürcme; ufak hata; (tekerlek) patinaj; yastık kılıfı; kombinezon; ufak deniz donu; gemi kızağı; fiş. **a ~ of a boy**, fidan gibi çocuk: **to give s.o. the ~**, birinin elinden sıvışarak kurtulmak: ⌜**there's many a ~ 'twixt the cup and the lip**⌝, ⌜çayı görmeden paçaları sıvama⌝. **~**[2] *v.* Kaymak; sürcmek; yanlışlık yapmak; salıvermek; gidivermek; (el vs.) sokuşturmak; sıkıştırmak; kaçırmak. **to ~ the anchor**, (gemi) demiri kaldıramıyıp zincirini salıvererek gitmek; ölmek: **to ~ home a bolt**, sürmeyi sürmelemek: **his name has ~ped my memory**, ismi hatırıma gelmiyor: **to ~ one's moorings**, (gemi) şamandıradan ayrılmak: **to ~ one's notice [attention]**, (bir şey) gözünden kaçmak: **to let ~ an opportunity**, fırsat kaçırmak: **I'll just ~ over to my mother's**, anneme şöyle bir uğrayacağım: **he ~ped the papers into his pocket**, kâğıtları cebine koyuverdi: **to let ~ a remark**, ağzından bir söz kaçırmak. **~-coach**, hareket halinde olan bir ekspres treninden bir istasyonda bırakılan vagon. **~-knot**, ilmek. **~-stream**, pervane suyu; pervane hareketinden hasıl olan hava cereyanı. **~-way**, gemi tezgâhı. **~**

**along,** süzülerek geçmek. ~ **away,** sıvışmak; (vakit) çabuk geçmek. ~ **by,** çabuk geçmek. ~ **down,** kayıp düşmek. ~ **off,** sıyrılmak; (elbise) sıyırmak. ~ **on,** giyivermek. ~ **out,** dışarı sıvışmak; ağzından kaçmak: **the secret ~ped out,** sır meydana çıkıverdi. ~ **up,** kayıp düşerek ayakları havaya kalkmak; yanılmak. **~per** [ˈslipə(r)], terlik. **~pered,** terlikli. **~pery,** kaypak; nazik tehlikeli (mevzu); kaçamaklı. **~shod** [ˈslipşod], yarım yamalak.

**slit** [slit]. Dar kesik; yarık; dar aralık. Yarmak; uzunluğuna kesmek, açılmak.

**slither** [ˈsliðə(r)]. Kayarak gitmek.

**sliver** [ˈslivə(r)]. Uzun ince parça.

**slobber** [ˈslobə(r)]. Salya(sı akmak). **to ~ over s.o.,** (*fig.*) ağlamalı surette muhabbet göstermek.

**slog** [slog]. (*sl.*) Şiddetli vuruş. Şiddetle vurmak. **to ~ along,** ağır ağır ve sebatla yürümek: **to ~ away at stg.,** bir şeye çok fazla çalışmak.

**slogan** [ˈslo̯ugən]. Düstur; şiar.

**sloop** [slûp]. (*esk.*) Küçük harb gemisi; (*şim.*) ronda flok yelkenli; (bahriye) gambot.

**slop** [slop]. Döküp saçmak. Taşmak. **to ~ about in the mud,** çamurda yürüyerek ıslanmak: **to ~ over stg.,** (gülünç bir şekilde) coşmak. **~s,** pis su; sulu yemek: **to live on ~,** sulu yiyeceklerle beslenmek. **~-pail,** bulaşık suyu vs. kovası. **~py,** ıslak ve kirli su dökülmüş, çamurlu; (insan) şapşal; gülünç şekilde hassas; yarım yamalak.

**slop·e** [slo̯up]. Bayır, yokuş; şev; meyil. Şevlen(dir)mek; meyilli olm. **~ arms!,** tüfek as!: **to ~ about,** (*coll.*) işsiz güçsüz gezmek: **to ~ down,** iniş teşkil etm.: **to ~ up,** yokuş teşkil etmek. **~ing,** meyilli; şevli.

**slot** [slot]. Müstatil delik; yiv (açmak).

**sloth** [slo̯uθ]. Tembellik. **~ful,** tembel.

**slouch** [sla̯uç]. Kamburunu çıkararak yürümek; hımbıl hımbıl yürümek. **~ing,** hımbıl, kamburu çıkmış.

**slough**[1] [sla̯u]. Bataklık. **~**[2] [slʌf]. Yılan gömleği; böceklerin soyulan derisi. (Yılan vs.) deri değiştirmek. **to ~ off [away],** düşmek, atılmak; atmak.

**sloven** [ˈslʌvn]. Şapşal kimse. **~ly,** şapşal, hırpani; (iş) fena yapılmış.

**slow** [slo̯u]. Ağır, yavaş hareket eden; uzun süren; geri kalmış; güç anlar; cansıkıcı. **to ~ up [down],** ağırlaş(tır)mak; hızını almak: **~ to anger,** kolayca hiddetlenmez: **to go ~,** acele etmemek; işi kasden yavaşlatmak: **to go ~ with one's provisions,** erzakını idare ile kullanmak; **he was not ~ to ...,** ...de gecikmedi: **to cook in a ~ oven,** ağır ateşte pişirmek: ⌜**~ but sure**⌝, yavaş fakat esaslı: **~ train,** her istasyona uğrıyan tren. **~ness,** yavaşlık, ağırlık; kalınkafalılık. **~-coach,** ağır yürüyen (çalışan) adam; mankafa. **~-match,** barutlu fitil. **~-motion,** yavaş çevrilen (filim). **~-worm,** kör yılan.

**sludge** [slʌc]. Yapışkan çamur; rüsub.

**slue** *v.* **slew**[2].

**slug** [slʌg]. Kabuksuz sümüklü böcek; tembel, yavaş yürüyen insan; ufak kurşun parçası. **~gard,** tembel, uykucu adam. **~gish,** tembel, uyuşuk; ağır; iyi işlemiyen (ciğer).

**sluice** [slûs]. Bend kapağı, savak. Su bendlerine kapak koymak; savağı açıp su ak(ıt)mak.

**slum** [slʌm]. Bir şehrin pis ve fakir mahallesi; teneke mahallesi. **to go ~ming,** hayır maksadile fakirleri ziyaret etmek.

**slumber** [ˈslʌmbə(r)]. Uyuklamak. Uyku.

**slump** [slʌmp]. (*ech.*) Çamura düşmek; birdenbire ve şiddetle düşme(k); (fiat vs.) ansızın düşme(k).

**slung** *v.* **sling.**

**slur** [slö(r)]. Leke, tahkir; heceleri karıştırarak fena telâffuz etme(k); iki notayı birleştirmek. **to cast a ~ on s.o.'s reputation,** birinin şerefini lekelemek: **to ~ over a word,** bir kelimenin hecelerini ayırd etmemek: **to ~ over a matter,** *etc.*, bir mesele üzerinden hafifçe geçivermek; gizlemek.

**slush** [slʌş]. Eriyen kar; sulu çamur; sahte teessür. **~y,** eriyen kar gibi.

**slut** [slʌt]. Pasaklı kadın. **~tish,** pasaklı.

**sly** [slây]. Sinsi, şeytan, kurnaz. **on the ~,** el altından: **a ~ dog,** cin gibi herif.

**smack** [smak]. (*ech.*) Şaplamak. Şamar, şaplak; az çeşni. Şap diye. **a ~ in the eye,** umulmıyan bir aksilik: **to ~ the lips,** dukaklarını şapırdatmak: **a ~ing noise,** şapırtı: **to ~ of stg.,** bir şey kokmak.

**small** [smôl]. Küçük; az; aşağılık.

the ~ of the back, boş böğür: ~ change, bozuk para: to cry [sing] ~, aşağıdan almak: to make s.o. cry ~, burnunu kırmak: he is a ~ eater, boğazlı değildir: to look [feel] ~, küçük düşmek: to make s.o. look ~, birini küçük düşürmek: to make oneself ~, göze görünmemek: ~ talk, havadan sudan konuşma: it is ~ wonder that ..., hiç şaşılacak şey değil. ~**holder**, küçük bir çiftlik sahibi. ~**holding**, küçük çiftlik. ~**ish**, oldukça küçük. ~**-arms**, hafif silâhlar. ~**-minded**, darkafalı. ~**pox**, çiçek hastalığı.

**smart**[1] [smât] *a.* Şık, zarif; açıkgöz; kurnaz; çabuk. **a ~ blow**, sert bir darbe: **a ~ fellow**, yaman adam: **look ~ about it!**, haydi, çabuk ol!: **to make oneself ~**, giyinip kuşanmak: **a ~ reply**, parlak, yerinde bir cevab: **~ society [the ~ set]**, yüksek sosyete: **he thinks it ~ to ...**, ···yi marifet zannediyor. ~[2]. Acı(mak), yanmak. **you shall ~ for this!**, sen bunun cezasını çekersin: **to ~ under an injustice**, bir haksızlık içinde ukde olmak. ~**en**, **to ~ up**, canlandırmak; düzeltmek: **to ~ oneself up**, süslenmek. ~**ness**, şıklık; uyanıklık; hazırcevablık.

**smash** [smaş]. (*ech.*) Çatır çatır parçalanma; çarp(ış)ma, kaza; iflâs. Çatır çatır parçala(n)mak; ezici bir darbe vurmak; tamamiyle bozguna uğratmak; iflâs etmek. **to ~ the door open**, kapıyı zorlayıp kırmak: **to [run] ~ into stg.**, bir şeye şiddetle çarpmak: **to ~ up**, parça parça etmek. ~**ing**, ezici, yıkıcı; (*coll.*) fevkalâde. ~**-and-grab raid**, camekânı kırıp (mücevher vs.yi) çalma. ~**-up**, büyük kaza; (otom. vs.) şiddetli çarpışma.

**smattering** [ˈsmatərin(g)]. Az buçuk bilme. **to have a ~ of French**, çatpat fransızca bilmek.

**smear** [smiə(r)]. Bula(ştır)ma(k); leke(lemek); hafifçe sürüş; sürmek.

**smell** (**smelt**) [smel, –t]. Koku; koklama hassası; fena koku. Kok(la)mak; kokusunu almak. **to ~ out**, (köpek) koklıyarak bulmak; (*fig.*) sır vs. keşfetmek. ~**ing-bottle**, amonyak şişesi. ~**ing-salts**, uçucu emlah. ~**y**, fena kokulu.

**smelt** *v.* İzabe etm.; filizi eritip maden çıkarmak. ~**ing furnace**, yüksek fırın.

**smil·e** [smâyl]. Gülümseme(k) tebessüm (etm.), yüzü gülmek. **he always comes up ~ing**, başına ne gelirse gelsin güler yüzle çıkar: **to keep ~ing**, ye'se kapılmamak.

**smirch** [smö(r)ç]. Leke(lemek).

**smirk** [smö(r)k]. Budalaca gülümseme(k).

**smite** (**smote, smitten**) [smâyt, smout, smitn]. Vurmak; şiddetli bir darbe indirmek. **to be smitten with the plague**, vebaya tutulmak: **to be smitten with s.o.**, birine abayı yakmak: **his conscience smote him**, vicdan azabı hissetti.

**smith** [smiθ]. Demirci, nalband. ~**y** [–ði], nalband dükkânı.

**smithereens** [ˈsmiðərînz]. **to knock [smash] to ~**, tuzla buz etm.

**smock** [smok]. Arkadan ilikli çocuk gögüslüğü. Elbiseyi kırmalı dikmek.

**smok·e** [smouk]. Duman; pipo, sigara içme(k); duman salıvermek; tütün içmek; tüt(süle)mek; islemek: **~ out**, duman ile öldürmek, kaçırmak. **to end in ~**, suya düşmek: **like ~**, (*sl.*) bal gibi, alabildiğine: ~**er**, tütün içen; (*sl.*) sigara içilen vagon: **he is a great ~**, çok tütün içer. ~**ing**, tüten; tütün içme; tütsüleme: **~ hot**, pek sıcak: **'no ~!'**, sigara içmek yasaktır! ~**y**, dumanlı, duman tüten. ~**e-screen**, sun'î sis. ~**ing-carriage**, sigara içilebilen vagon. ~**ing-jacket**, (*esk.*) smokin, (*şim.*) **dinner-jacket**, denir.

**smooth** [smûð]. Düz; durgun; cilâlı; tüysüz; tatlı; muntazam; sarsıntısız. Düzlemek; arızasını gidermek; teskin etm.; okşamak. **to ~ away (difficulty)**, (güçlük) ortadan kaldırmak: **to ~ over a matter [to ~ things out]**, meseleyi tatlıya bağlamak: **we are now in ~ water**, (*fig.*) güçlükleri atlattık. ~**-bore**, yivsiz (tüfek). ~**-shaven**, sakalı tıraşlı. ~**-spoken**, tatlı dilli; mürai.

**smote** *v.* **smite.**

**smother** [ˈsmʌðə(r)]. Boğmak; bastırmak.

**smoulder** [ˈsmouldə(r)]. Alevsiz yanmak; gizli olarak mevcud olmak. **a ~ing fire**, küllenmiş ateş.

**smudge** [smʌc]. Siyah leke; bulaşık leke. Bulaştırmak.

**smug** [smʌg]. Kendinden memnun.

**smuggle** [ˈsmʌgl]. Kaçakçılık yapmak, kaçırmak. ~**r**, kaçakçı; kaçakçı gemisi.

**smut** [smʌt]. Kurum tanesi; açık saçık konuşma. **~ty,** açık saçık (hikâye vs.).
**snack** [snak]. Meze, hafif yemek. **to have a ~,** safra bastırmak: **just a ~,** bir lokma. **~-bar,** hafif yemek ve meze veren birahane vs.
**snaffle** [ˈsnafl]. Çok hafif gem. (*sl.*) Kapmak, aşırmak.
**snag** [snag]. Bir şeyin çıkık pürüzlü ucu (*mes.* kırık dal); engel, mahzur. **to strike a ~,** bir engele raslamak: **there's a ~ somewhere,** altından çapanoğlu çıkabilir.
**snail** [snêyl]. Salyangoz. **at a ~'s pace,** kaplumbağa yürüyüşü ile.
**snak·e** [snêyk]. Yılan. **~y,** yılan gibi; yılankavi. **~e-bite,** yılan sokması. **~e-charmer,** yılan oynatıcı.
**snap** [snap]. (*ech.*) Koparma sesi; ısırmak istiyen köpeğin dişlerinin sesi; (çanta vs.) yaylı rabtiyesi; (*coll.*) gayret, enerji. Isırmağa çalışmak; birdenbire kopmak. **a cold ~,** kısa süren şiddetli soğuk: **to ~ one's fingers,** parmaklarını şıkırdatmak: **to ~ one's fingers at,** hiçe saymak: **to ~ s.o.'s head off,** birini şiddetle terslemek: **to ~ at** [**make a ~ at**], ısırmağa çalışmak: **put some ~ into it!,** haydi bir az gayret!: **to ~ out an order,** keskin ve şiddetli emir vermek: **the box shut with a ~,** kutu şırak diye kapandı: **to take a ~ of,** fotografını çekmek: **to ~ up,** kapışmak. **~pish,** ters huylu (köpek); öfkeli. **~py,** (*coll.*) canlı; çevik; yerinde (cevab): **make it ~!,** çabuk ol! **~shot,** enstantane fotograf. **~-fastener,** çıtçıt.
**snare** [sneə(r)]. Tuzak; dolab, hile. Tuzak ile tutmak. **to be caught in the ~,** tuzağa düşmek.
**snarl** [snâl]. (*ech.*) Hırlama(k).
**snatch** [snaç]. Kapmak; kavramak. Kapış; kısa müddet; parça. **to make a ~ at stg.,** bir şeyi kapmağa çalışmak: **to ~ a meal,** çabucak iki lokma bir şey yemek: **to get a ~ of sleep,** bir az kestirmek (uyumak): **to work in ~es,** intizamsız çalışmak.
**sneak** [snîk]. Gammaz; sinsi ve korkak kimse. Gammazlık etm.; sinsi sinsi dolaşmak. **to ~ away,** sıvışmak. **~ing,** hırsızlama, sinsi. Koğuculuk: **to have a ~ affection for s.o.,** birine karşı (kusurlarına rağmen) itiraf edilmez bir muhabbet beslemek.
**sneer** [sniə(r)]. Müstehzi bir tavırla gülme; ihtihza. İstihzala gülmek. **to ~ at,** istihfaf etm.: **to ~ at wealth,** zenginliğe dudak bükmek.
**sneeze** [snîz]. Aksırma(k). **an offer not to be ~ed at,** yabana atılmaz bir teklif.
**snick** [snik]. (*ech.*) Hafifçe kesik.
**sniff** [snif] (*ech.*) Koklamak için burnuna hava çekmek. **to have a ~ at stg.,** bir şeyi koklamak: **not to be ~ed at,** yabana atılmaz. **~le** *v.* **snuffle.**
**snip** [snip]. (*ech.*) Makasla kesmek. Makasla kesilmiş parça. **it's a ~!,** (*sl.*) o elde bir!
**snipe** [snâyp] *v.* Gizli bir yerden ateş etmek. **~r,** gizli ateş eden nişancı.
**snivel** [ˈsnivl]. Burnunu çekerek [yalancıktan] ağlamak. **~ling,** ağlamalı.
**snob** [snob]. Asalet ve servete fazla ehemmiyet veren züppe. **~bish,** snob gibi.
**snook** [snuk]. Nanik. **to cock a ~ at s.o.,** birine nanik yapmak.
**snooker** [ˈsnûkə(r)]. Bir nevi bilârdo oyunu. **to ~ s.o.,** (*sl.*) birini müşkül bir vaziyete sokmak.
**snooze** [snûz]. Kestirme(k).
**snore** [snô(r)]. (*ech.*) Horlama(k).
**snort** [snôt]. (*ech.*) Öfkeli bir at gibi kuvvetle burnundan nefes çıkarma(k).
**snot** [snot]. Sümük. **~ty,** sümüklü; (*sl.*) öfkeli. (*sl.*) Bahriye asteğmeni.
**snout** [snaut]. Hayvan burnu.
**snow** [snou]. Kar (yağmak). **to be ~ed in** [**up**], tamamen karla kapanmak; kardan dolayı dışarı çıkamamak: **to be ~ed under with work,** işten baş kaldıramamak. **~ball,** kar topu (atmak). **~drift,** kar yığıntısı. **~fall,** kar yağması. **~flake,** kuşbaşı kar. **~man,** kardan adam. **~plough,** kar temizleme makinesi. **~shoes,** karda yürüyebilmek için ayak raketi. **~storm,** tipi. **~y,** karlı. **~-blind,** kar tesiriyle gözleri kör olmuş. **~-bound,** kardan kapanmış. **~-line,** bir dağ vs. üzerindeki daimî karın hududu.
**snub**[1] [snʌb]. Terslemek; haddini bildirmek. İstihfaflı söz, hareket. **~**[2] *a.* Kısa ve ucu kalkık (burun).
**snuff** [snʌf]. Enfiye. *v. see* **sniff.**
**snuffle** [ˈsnʌfl]. (*ech.*) Burnunu çekmek. **to have the ~s,** nezleden burnu akmak.

**snug** [snʌg]. Rahat ve sıcak; mahfuz; konforlu. **to make all ~**, (*naut.*) fırtınaya karşı mahfuz kılmak.
**snuggle** [ˈsnʌgl]. **to ~ up to s.o.**, ısınmak için yanına sokulmak: **to ~ down in bed**, yatakta rahatça yerleşmek.
**so** [sǫu]. Öyle; böyle; şu kadar. **and ~**, nitekim; keza: **~ far**, şimdiye kadar; o kadar uzak: **~ long!**, (*coll.*) şimdilik Allaha ısmarladık: **he didn't ~ much as ask me to sit down**, bana otur bile demedi: **~ much ~ that ...**, o dereceye kadar ki: **~ much for that!**, bunun için bu kadar yeter: **~ much for his French!**, onun fransızcası da işte bu kadar!: **I regard it as ~ much lost time**, ben bunu kaybolmuş vakit sayıyorum: ⌜**~ many men, ~ many minds**⌝, ne kadar insan varsa o kadar fikir var: **and ~ on** [**and ~ forth**], ve daha bilmem ne: **I told you ~!**, ben sana demedim mi?: **you don't say ~!**, yok canım!: **quite ~!**, elbette!: **'you told me you knew French.' '~ I do!'**, 'Bana fransızca biliyorum demiştiniz.' 'Tabiî biliyorum': **~ help me God!**, (i) (yemin) Allah şahiddir; (ii) Allah muinim olsun!: **I know English and ~ does my brother**, ben ingilizce bilirim kardeşim de bilir: **it ~ happened that ...**, öyle oldu ki ...: **~ that ...**, ... için: **in a week or ~**, bir haftaya kadar filan: **~ ~**, şöyle böyle: **~ to speak** [**say**], tabir caizse: **~ you are not going to London?**, demek ki Londra'ya gitmiyorsun? **~-and-~**, filân. **~-called**, sözde; adlı.
**soak** [sǫuk]. Suya batırıp ıslatmak; sırsıklam etm. (olm.); çok içmek. **an old ~**, ayyaş.
**soap** [sǫup]. Sabun(lamak). **soft ~**, arabsabunu; dalkavukluk. **~y**, sabunlu; sabun gibi; göze girmeğe çalışan (kimse). **~-box**, sabun sandığı: **~ orator**, sokak hatibi. **~-bubble**, sabun kabarcığı.
**soar** [sô(r)]. Yükseklerde uçmak; kanadlarını açıp kımıldatmadan uçmak; (fiat) çok yükselmek. **~ing ambition**, hududsuz ihtiras.
**sob** [sob]. Hıçkırık(la ağlamak).
**sober** [ˈsǫubə(r)]. Ayık; çok içmemiş; az içki kullanan; mutedil; ciddî; gösterişsiz; (renk) donuk. Ayıltmak; aklını başına getirmek. **in ~ fact**, hakikatte: **in ~ earnest**, pek ciddî olarak: **to ~ down**, ayıl(t)mak. **~sides**, (*coll.*) pek ciddî.
**sobriety** [sǫuˈbrâyəti]. İmsak; itidal.
**soccer** [ˈsokə(r)]. = **association football**, futbol.
**sociab·le** [ˈsǫuşəbl]. Cemiyet halinde yaşıyan; munis; sokulgan. **~ility** [–ˈbiliti], muaşeret kabiliyeti.
**social** [ˈsouşl]. İctimaî, sosyal. **~ gathering**, eş dost toplantısı: **~ intercourse**, muaşeret. **~ism**, sosyalizm.
**soci·ety** [sǫuˈsâyəti]. Cemiyet; dernek; ictimaî heyet; şirket; sosyete. **he is fond of ~**, arkadaşlıktan ve sohbetten hoşlanır: **to go into** [**move in**] **~**, kibar âlemine girmek. **~ology** [ˌsǫusiˈoləci], ictimaiyat.
**sock** [sok]. Kısa çorab; taban. (*sl.*) (Darbe) indirmek. **to pull up one's ~s**, (*sl.*) kendini toplayıp çalışmağa başlamak.
**socket** [ˈsokit]. Sap deliği; içine sokulan oyuk; (*elect.*) dişi fiş. **eye-~**, göz evi.
**sod** [sod]. Çim parçası.
**sodden** [ˈsodn]. Sırsıklam. **~ with drink**, ayyaşlıktan abdallaşmış.
**sodomy** [ˈsodomi]. Livata.
**soever** [sǫuˈevə(r)]. **in any way ~**, nasıl olursa olsun: **how~ great it may be**, ne kadar büyük olursa olsun.
**sofa** [ˈsǫufə]. Kanape; sedir.
**soft** [soft]. Yumuşak; müşfik; mülâyim; tenperver. **~ drink**, alkolsuz içki: **~ fruits**, kiraz, çilek vs.: **~ goods**, mensucat: **~ job**, (*coll.*) kolay ve paralı iş: **to have a ~ place in one's heart for ...**, ···e karşı zâfı olm.: **~ water**, kirecsiz su. **~en**, yumuşa(t)mak: **to have ~ing of the brain**, beyni sulanmak. **~-boiled**, rafadan (yumurta). **~-headed** [**-witted**], abdal. **~-hearted**, merhametli. **~-spoken**, tatlı dilli; mürai.
**soggy** [ˈsogi]. Batak gibi; pek sulu (toprak).
**soil** [sôyl]. Toprak. Kirletmek; lekelemek. **one's native ~**, vatan: **son of the ~**, köylü: **night-~**, insan gübresi.
**sojourn** [ˈsocən]. Muvakkat(en) ikamet (etm.).
**solace** [ˈsoles]. Teselli (etm.).
**solar** [ˈsǫula(r)]. Güneşe aid; şemsî.
**sold** *v.* **sell**.
**solder** [ˈso(l)də(r)]. Lehim(lemek). **~ing iron**, havya.
**soldier** [ˈsǫulcə(r)]. Asker(lik etmek). **private ~**, nefer, er: **~ of fortune**,

rasgele her hangi bir memleketin hizmetinde askerlik eden kimse. **~ly,** askerce. **~y,** askerler.

**sole**[1] [soul]. Ayak tabanı; ayakkabı pençesi. Kunduraya pençe vurmak. **~**[2] *a.* Yegâne, biricik. **~ heir,** umumî mirascı. **~**[3]. Dil balığı.

**solecism** [ˈsolesizm]. Nahiv, şive hatası; adabı muaşereti ihlâl.

**solemn** [ˈsoləm]. Merasimli; muhteşem; vekarlı, ciddî. **this is the ~ truth [fact],** yemin ederim ki bu böyledir. **~ duty,** mukaddes vazife: **to keep a ~ face,** gülmemek için kendini tutmak. **~ity** [–ˈlemniti], temkinli merasim; temkinlilik, ciddiyet. **~ize** [-nâyz], merasimle kutlulamak.

**solicit** [soˈlisit]. Rica etm., yalvarmak; (fahişe, dilenci) taciz etmek. **~ation** [ˌsolisiˈtêyşn], ısrarla isteme; (fahişe vs.) taciz etme. **~or** [soˈlisitə(r)], mahkeme huzuruna çıkmıyan avukat; müşavir avukat. **~ous** [soˈlisitəs], ihtimamlı. **~ about stg.,** bir şey hakkında endişeli: **to be ~ for s.o.'s comfort,** birinin rahatına ihtimam göstermek. **~ude,** ihtimam.

**solid** [ˈsolid]. Sulb; katı; metin, sağlam. Sulb cisim; mücessem şekil. **~ tyre,** dolma lâstik: **~ vote,** müttefiken verilen rey: **to sleep for ten ~ hours,** tam on saat uyumak. **~arity** [–ˈdariti], tesanüd. **~ify** [–ˈlidifây], katılaş(tır)mak; tasallüb et(tir)mek. **~ity** [–ˈliditi], sulbiyet, katılık; metanet, sağlamlık.

**solilo·quize** [soˈlilokwâyz]. Kendi kendine konuşmak. **~quy,** kendi kendine konuşma.

**solit·aire** [soliˈteə(r)]. Yüzükte tek taş. **~ary** [ˈsolitəri]. Tenha; yalnız yaşıyan; tek. **~ude** [–tyûd], tenhalık; yalnızlık; üzlet.

**solo** [ˈsoulou]. Tek bir sanatkârın okuduğu, çaldığı hava. **~ist,** solo çalan sanatkâr.

**solstice** [ˈsolstis]. Gündönümü.

**solu·ble** [ˈsolyubl]. Eriyebilir; inhilâli kabil. **~tion** [–ˈlûşn], erime, inhilâl; çare.

**solv·e** [solv]. Halletmek, çözmek. **~able,** halledilebilir. **~ency,** ödeyebilme. **~ent,** inhilâl ettiren (mayi); borcunu ödeyebilir.

**sombre** [ˈsombə(r)]. Karanlık; muzlim; koyu (renk); endişeli.

**some** [sʌm]. Biraz, bir mikdar, bir kısım; bazı, bazısı; kimi; bir çok, bir kaç. **give me ~ bread,** bana ekmek ver: **~ came, ~ went,** kimi geldi kimi gitti: **in ~ degree [to ~ extent],** bir dereceye kadar: **~ sort of ...,** şöyle bir: **in ~ way or another,** her hangi bir şekilde; nasılsa: **ask ~ clever person!,** Akıllının birine sor!: **'Do you want ~ money?' 'No, I have ~',** 'Para ister misin?' 'Hayır, bende var': **I read it in ~ book or other,** bilmem hangi kitabda okudum: **I have been waiting ~ time,** bir hayli bekledim: **I will go there ~ time,** oraya münasib bir zamanda giderim: **~ sixty years ago,** altmış sene kadar evvel: **he's ~ doctor!,** (*coll.*) yaman doktor!: **'I hope we shall win.' '~ hope!',** 'inşallah yeneriz.' 'Bekle, yenersiniz! (*cont.*)': **he earns ten pounds a week and then ~,** (*coll.*) haftada su içinde on lira kazanıyor. **~body** [–bodi], bir kimse, birisi. **he is (a) ~,** o mühim bir şahsiyettir; **he thinks he's (a) ~,** kendini bir şey zannediyor. **~how** [–hau], **~ (or other),** her nasılsa; her nedense. **~one** [–wʌn] *v.* **somebody. ~thing** [–θin(g)], bir şey: **~ (or other),** bilmem ne: **the two ~ train,** ikiyi bilmem kaç geçe treni: **there is ~ in what you say,** söylediğinizde doğru bir taraf var: **he is ~ under fifty,** o elliden bir az aşağıdır: **that's ~ like a horse!,** işte at diye buna derler. **~time** [–tâym], bir zaman. **~ last year,** geçen sene içinde: **~ or other,** ileride bir gün: **~ soon,** yakında: **Mr. A., ~ mayor of B.,** eski B. belediye reisi Mr. A.: **~s,** bazan; arasıra. **~way** [–wêy], her halde. **~what** [–wot], biraz. **~where** [–weə(r)], bir yerde; her hangi bir yerde: **~ about 15 lira,** 15 lira filân: **~ else,** başka bir yerde: **I'll see him ~ first,** (*coll.* bir teklife karşı) (i) haddiyse yapsın bakalım!; (ii) avucunu yalasın!; (iii) cehennemin dibine!

**somersault** [ˈsʌməsôlt]. Taklak. **to turn a ~,** taklak atmak.

**somn·ambulist** [somˈnambyulist]. Uykuda gezer. **~olent** [–ələnt], uyku basmış.

**son** [sʌn]. Oğul. **~ny** [ˈsʌni], (*coll.*) Evlâdım.

**song** [son(g)]. Şarkı, türkü. **to make a ~ (and dance) about stg.,** mesele yapmak: **for a mere ~,** yok pahasına. **~ster,** hanende. **~-bird,** ötücü kuş.

**sonnet** [ˈsonit]. On dört mısralı şiir.
**sonor·ity** [soˈnoriti]. Tannanlık. **~ous** [ˈsonərəs], tannan.
**soon** [sûn]. Yakında; neredeyse; biraz sonra. **as ~ as he came,** gelir gelmez: **as ~ as possible,** mümkün olduğu kadar çabuk: **the ~er the better,** ne kadar çabuk olursa o kadar iyi: **~er or later,** eninde sonunda: **he no ~er came than he went,** gelmesi gitmesi ile bir oldu: **no ~er said than done,** demesile yapması bir oldu: **too ~,** pek fazla erken: **I would ~er not go,** gitmemeği tercih ederim: **I would ~er die,** ölürüm de bunu yapmam.
**soot** [sut]. İs. **to ~ up,** islenmek. **~y,** isli.
**sooth** [sûθ]. **in (very) ~,** hakikaten: **~ to say,** doğrusu.
**sooth·e** [sûð]. Teskin etm., dindirmek. **to ~ s.o.'s feelings,** birinin gönlünü almak. **~ing,** dindirici, teskin edici.
**sop** [sop]. Tirid; gönül almak için verilen şey. **to ~ up,** sünger gibi massetmek. **to throw [give] a ~ to,** önüne bir kemik atmak. **~ping (wet),** sırsıklam. **~py** [ˈsopi], ıslak; yağmurlu.
**sophis·m** [ˈsofizm]. Safsata. **~t,** safsatacı. **~ticated** [soˈfistikêytid], saflığını ve masumluğunu kaybetmiş; hayata alışmış. **~try,** safsata.
**soporific** [ˌsoupəˈrifik]. Uyutucu (ilâc).
**soprano** [souˈprânou]. Tiz kadın sesi.
**sorcer·er** [ˈsôsərə(r)]. Sihirbaz, büyücü. **~y,** sihirbazlık.
**sordid** [ˈsôdid]. Alçak, sefil; alçakça menfaatperest.
**sore** [sô(r)]. Dokunuldukça acıyan; yaralı; hassas; kırgın; şiddetli. Yalama, sıyrık, yara. **~ly wounded,** fena surette yaralı: **to be ~ all over,** vücudünün her tarafı ağrımak: **~ at heart,** mahzun: **a running ~,** irinli yara; devamlı ıstırab: **to touch s.o. on his ~ spot,** birinin bamteline basmak: **~ throat,** boğaz ağrısı.
**sorr·ow** [ˈsorou]. Keder, gam; ıstırab. Kederlenmek. **I saw to my ~,** teessürle gördüm ki. **~owful,** kederli, mustarib. **~y** [ˈsori]. Pişman; müteessir; mahzun; acınacak; miskin. **to be ~,** pişman olm.; üzülmek: **to be ~ about stg.,** bir şeye acınmak: **(I'm) ~!,** affedersiniz!: **you'll be ~ for this,** bunun acısını çekeceksin: **a ~ excuse,** saçma mazeret: **to cut a ~ figure,** rezil olm: **he's very ~ for himself,** süngüsü düşük.
**sort** [sôt]. Türlü; nevi, cins, çeşid; makule. **~(out),** tasnif etm.; ayıklamak. **an army of a ~ [after a ~, of ~s],** sözüm ona [iyi kötü] bir ordu: **he's a very good ~,** (*coll.*) çok iyi bir adamdır: **it is nothing of the ~,** hiç te öyle değil: **I ~ of expected it,** böyle bir şeyi âdeta bekledim (diyebilirim): **out of ~s,** keyifsiz: **this ~ of people,** bu gibi adamlar: **in some ~,** bir bakımdan: **that's the ~ of thing I mean,** böyle bir şey kasdediyorum. **~er** [ˈsôtə(r)], tasnif edici; ayıklayıcı.
**sortie** [ˈsôti]. (*mil.*) Çıkış hareketi.
**sot** [sot]. Ayyaş. **~tish,** içkiden abdallaşmış.
**sotto voce** [ˈsotouˈvouçêy]. Alçak sesle.
**sou** [sû]. **not worth a ~,** metelik etmez: **without a ~,** meteleksiz.
**sought** [sôt] *v.* **seek. much ~ after,** çok rağbette olan.
**soul** [soul]. Can, ruh; kişi. **he has a ~ above moneymaking,** para düşünecek adam değildir: **with all my ~,** candan: **the ship was lost with all ~s,** gemi içindekilerle beraber battı: **enough to keep body and ~ together,** bir lokma bir hırka: **I cannot call my ~ my own,** başımı kaşıyacak vaktim yok: **departed ~s,** ölüler: **God rest his ~,** nur içinde yatsın: **he's a good ~,** çok iyi adamdır: **he is the ~ of honour,** o mücessem namustur: **to be the life and ~ of a party,** toplantının ruhu olm.: **a lost ~,** dalâlete düşmüş: **not a ~,** kimsecikler yok: **there was not a ~ to be seen,** in cin yoktu: **poor ~!,** zavallı!, **upon my ~!,** vallahi, Allah bilir! **~ful,** içli. **~less,** ruhsuz. **~-destroying,** hayvanlaştırıcı. **~-stirring,** müheyyic.
**sound**[1] [saund]. Ses; sada; gürültü. Ses çıkartmak; sesi gelmek. **to ~ the alarm,** imdad düdügünü vs. çalmak: **it ~s bad to me,** bu bana fena görünüyor: **to ~ the charge,** hücum borusu çalmak: **I don't like the ~ of it,** pek aklım yatmıyor: **not a ~ was heard,** ses sada yok: **within ~ of,** sesi işitilecek mesafede. **~-detector,** uçak vs.yi dinleme aleti. **~**[2] *a.* Sağlam; sıhhatte; kusursuz; doğru; sadık: **~ly,** adamakıllı; selâmetle; doğruca. **as ~ as a bell,** sapsağlam: **~ly asleep,** mışıl mışıl uyuyan: **~**

sleep, deliksiz uyku: a ~ thrashing, temiz bir dayak. **~ness,** sağlamlık; sıhhat; doğruluk. ~[3] *v.* İskandil etm. to ~ s.o., birinin ağzını aramak. **~ing,** to take ~s, iskandil atmak. ~[4]. Deniz geçidi.

**soup** [sûp]. Çorba. **thick ~,** ezme çorbası: **clear ~,** süzme etsuyu. **we're in the ~!,** (*sl.*) hapı yuttuk. **~-kitchen,** imarethane.

**sour** [saụə(r)]. Ekşi; mayhoş; kekre; bozulmuş: hırçın. Ekşi(t)mek. ⌜**~ grapes!**⌝, ⌜kedi uzanamadığı ciğere pis der⌝: **poverty has ~ed him,** fakirlik onu ters ve huysuz yaptı. **~-faced,** suratsız. **~-tempered,** küskün.

**source** [sôs]. Kaynak; memba; esas.

**sous·e** [saụs]. Salamura(ya yatırmak); suya daldırmak; sırsıklam etm. **to get a good ~ing,** sırsıklam olmak.

**sous-entendu** [ˌsûzontondü]. Üstü kapalı; tahtında müstetir.

**south** [saụθ]. Cenub; (*naut.*) kıble. Cenubî. **~erly** [ˈsʌðəli], **a ~ course,** cenuba doğru rota: **~ wind,** cenubdan gelen rüzgâr. **~ern** [ˈsʌðən], cenuba aid; cenubda olan: **~er,** cenublu. **~ward(s)** [ˈsaụθwö(r)d], cenuba doğru. **~-east,** cenubuşark, keşişleme. **~-west,** cenubugarbî; lodos.

**sou'wester** [saụˈwestə(r)]. Lodos (rüzgâr); muşamba gemici şapkası.

**sovereign** [ˈsovrin]. Hükümdar; padişah; ingiliz lirası. Müstakil; hâkim. **a ~ remedy,** birebir ilâc. **~ty,** hakimiyet, istiklâl.

**sow**[1] [soụ] *v.* Ekmek; (*fig.*) yaymak. ⌜**as you ~, so will you reap**⌝, insan ektiğini biçer. **~er,** ekici. **~ing,** ekim. ~[2] [saụ] *n.* Dişi domuz. ⌜**to get the wrong ~ by the ear**⌝, bir kimse [şey] hakkında yanılmak.

**spa** [spâ]. Kaplıca (şehri).

**spac·e** [spêys]. Feza; yer; saha; meydan; fasıla; müddet; aralık; mesafe. **to ~ out [off],** aralıklı dizmek; fasıla vermek: **for a ~,** bir müddet zarfında. **~ious** [ˈspêyşəs], geniş; ferah.

**spade** [spêyd]. Bahçivan beli. ⌜**to call a ~ a ~**⌝, ⌜kör kadıya körsün demek⌝. **~-work,** bel işi; çok zahmetli hazırlık işi.

**spaghetti** [spaˈgeti]. Ince makarna.

**Spain** [spêyn]. İspanya.

**spake** [spêyk]. (*esk.*) = **spoke,** *v.* **speak.**

**span**[1] [span]. Karış; mesafe; fasıla; bir köprünün boyu. **to ~ a river with a bridge,** bir köprüyü bir nehrin bir tarafından öbür tarafına uzatmak. ~[2]. Çift koşulmuş öküz. ~[3] *v.* **spin.**

**spangle** [ˈspan(g)gl]. Pul(larla süslemek).

**Spani·ard** [ˈspanyəd]. İspanyol. **~sh,** ispanyol; ispanyolca. **~el** [ˈspanyəl]. Epanyöl.

**spank** [span(g)k]. (*ech.*) Kıçına şaplak vurmak. **to ~ along,** (*coll.*) çabuk gitmek.

**spanner** [ˈspanə(r)]. Somun anahtarı.

**spar**[1] [spâ(r)]. Seren; direk. ~[2]. Billur. ~[3]. Dostane boks maçı (etm.); münakaşa etmek.

**spare**[1] [speə(r)] *v.* Esirgemek; kıyamamak; arttırıp verebilmek. **can you ~ it?,** bunu size lâzım değil mi?: **there's enough and to ~,** yeter de artar: **to ~ no expense,** masrafı esirgememek: **to ~ s.o.'s feelings,** hislerine hürmet etm.: **to ~ the life of,** kıyamamak: **to have nothing to ~,** anca yetecek (parası vs.) olm.: ⌜**~ the rod and spoil the child**⌝, ⌜kızını dövmeyen dizini döver⌝: **I cannot ~ the time,** vaktim yok. ~[2] *a.* İnce yapılı; dar; fazla olarak; yedek (parça). **~ diet,** bol olmıyan yemek: **~ part,** yedek parça: **~ room,** misafir için yatak odası.

**sparing** [ˈspeərin(g)]. İdareli. **to be ~ with the butter,** yağı idareli kullanmak: **he is ~ of praise,** kolay kolay medhetmez.

**spark** [spâk]. Kıvılcım; zerre; canlı delikanlı. Kıvılcım saçmak. **not a ~ of life remained,** hayattan eser kalmadı. **~ing-plug,** buji.

**sparkl·e** [ˈspâkl]. Parıldamak; (şarab) köpüklenmek. Parlayış, revnak. **~ing,** parlayan; pırıl pırıl: **~ wine,** köpüklü şarab.

**sparrow** [ˈsparoụ]. Serçe. **~-hawk** atmaca.

**sparse** [spâs]. Kıt; seyrek.

**Spartan** [ˈspâtn]. Sparta'ya aid; meşakkate dayanıklı; her türlü lüksten mahrum.

**spasm** [ˈspazm]. Ispazmoz. **~odic** [–ˈmodik], teşennücî; devamsız, arada sırada.

**spat**[1] [spat]. Kısa tozluk, yarım getr. ~[2] *v.* **spit.**

**spate** [spêyt]. Şiddetli sel. **the river**

is in ~, nehir yükselmiş: **a ~ of words,** ağızkalabalığı: **a ~ of oaths,** ağız dolusu.

**spatial** [ˈspêyşl]. Mesafe, sahaya aid.

**spatter** [ˈspatə(r)]. Zifos (atmak). Birine çamur sıçratmak. **a ~ of rain,** serpinti.

**spawn** [spôn]. Balık, kurbağa yumurtası; (*istihfaf*) zürriyet. (Balık, kurbağa) yumurtlamak. **mushroom ~,** mantar filizi.

**S.P.C.A.** [ˈesˈpîˈsîˈêy] = **Society for the Prevention of Cruelty to Animals,** hayvanları koruma cemiyeti.

**speak** (**spoke, spoken**) [spîk, spouk, spoukn]. Söz söylemek, konuşmak; nutuk vermek. **do you ~ English?,** ingilizce bilir misiniz?: **English spoken,** burada ingilizce bilen var: **I know him to ~ to,** onunla aşinalığım var: **to ~ one's mind,** lâfını sakınmamak: **roughly ~ing,** aşağı yukarı: **so to ~,** tabir caizse. **~ for, to ~ for s.o.,** birinin namına konuşmak; birinin lehinde konuşmak: **~ing for myself,** bence: **the facts ~ for themselves,** vaziyet besbellidir: **that ~s well for his perseverance,** bu onun sebatını isbat eder: **that ~s ill for his education,** bu onun tahsilinin ne derece olduğunu gösterir. **~ of, ~ing of ...,** ...e gelince: **we were ~ing of you,** sizden bahsediyorduk: **to ~ well [highly] of s.o.,** birini medhetmek: **he has no money to ~ of,** parası var denmez: **his sunken cheeks spoke of his sufferings,** çökük yanakları çektiklerinin deliliydi. **~ out,** yüksek sesle konuşmak; âleme söylemek. **~ to, I can ~ to his having been there,** kendisinin orada bulunduğuna şahidim. **~ up,** yüksek sesle konuşmak, sesini yükseltmek: **to ~ up for s.o.,** birinin lehinde konuşmak. **~-easy** [ˈspîkˈîzi], gizli meyhane. **~er** [ˈspîkə(r)], hatib; spiker: **the S~,** Avam Kamarasının reisi; **to catch the ~'s eye,** İngiliz Parlamentosunda reisten söz almak. **~ing** [ˈspîkin(g)], konuşma; konuşan: **a ~ likeness,** yaşıyan bir resim: **not to be on ~ terms with ...,** ...le dargın olmak. **~ing-trumpet,** megafon.

**spear** [spiə(r)]. Mızrak(la vurmak).

**special** [ˈspeşl]. Mahsus; hususî; has; fevkalâde. Hususî tren; gazetenin hususî nüshası. **~ friend,** en yakın dost. **~ist,** mütehassıs. **~ity** [–ˈaliti], hususiyet: **to make a ~ of stg.,** bir şeyi kendine ihtisas yapmak. **~ize,** ihtisas yapmak.

**specie** [ˈspîşi·î]. Madenî para.

**species** [ˈspîşîz]. Nevi, cins.

**specif·ic** [speˈsifik]. Has; bir cinse aid; izafî; sarih, kat'î. Bir hastalığa mahsus ilâc. **~y** [ˈspesifây], tasrih etm.; açıkça izah etm.; tayin etmek. **unless otherwise ~ied,** hilâfı bildirilmedikçe. **~ication** [–ˈkêyşn], tasrih; belirtme; tafsilât veren takrir; mukavele şartnamesi.

**specimen** [ˈspesimən]. Nümune; örnek. **a queer ~,** (*coll.*) antika.

**specious** [ˈspîşəs]. Zahiren doğru fakat hakikatte yanlış; makul görünür.

**speck** [spek]. Nokta; ufacık leke; zerre; ben; azıcık şey. Beneklendirmek; (*coll.*) belli belirsiz yağmur yağmak.

**speckle** [ˈspekl]. Benek; ufacık nokta. **~d** benekli; abraş.

**specta·cle** [ˈspektəkl]. Manzara; temaşa. **~cles,** gözlük. **~cled,** gözlüklü. **~cular** [specˈtakyulə(r)], pek gösterişli; hayret verici. **~tor** [spekˈtêytə(r)], seyirci.

**spectr·al** [ˈspektrəl]. Heyulâ gibi; tayfî. **~e,** [–tə(r)], heyulâ, hayalet. **~um,** *pl.* **-tra** [ˈspektrʌm, –tra], tayf.

**speculat·e** [ˈspekyulêyt]. Borsada hava oyunları yapmak. **~ on [about],** nazarî olarak düşünmek; tahmin etmek. **~ion** [–ˈlêyşn], nazariye(cilik); tahmin etme; hava oyunu. **~ive,** nazariye şeklinde; kumar nevinden. **~or,** acyocu.

**sped** *v.* **speed.**

**speech** [spîç]. Konuşma kabiliyeti; nutuk; dil, telâffuz. **direct [indirect] ~,** vasıtasız [vasıtalı] ifade: **figure of ~,** timsal: **free ~,** söz hürriyeti; **parts of ~,** kelimenin kısımları: **to be slow of ~,** ağır konuşmak. **~ify,** nutuk paralamak. **~less,** dili tutulmuş. **~-day,** mekteblerde mükâfat dağıtma günü.

**speed** [spîd]. Sürat, hız. *v.* (**sped**) [sped]. Çabuk gitmek. Hızlandırmak. **at ~,** hızlı giderek: **at full ~,** alabildiğine koşarak vs.: **to ~ the parting guest,** giden misafiri uğurlamak: **to put on ~,** hızını arttırmak: **to ~ up the work,** işe hız vermek. **~ometer** [–ˈdomitə(r)], sürat ölçen alet. **~y,** çabuk; tez.

**spell**[1] (**spelt**) [spel, –t] *v.* Kelimelerin imlâlarını doğru yazmak. **to ~ out,** hecelemek; **how is it spelt?,** nasıl yazılır?: **this move ~s disaster,** bu hareketin sonu felâkettir. **~ing,** imlâ. **~²** *n*. Tılsım; büyü; afsun. **the ~ is broken,** tılsım bozuldu: **to cast a ~ over s.o.,** birini büyülemek. **~binder** [ˈspelbâyndə(r)], sihirli hatib. **~bound,** sihirlenmiş. **~³** *n*. Nöbet vakti; müddet. **four hours at a ~,** fasılasız dört saat: **to take ~s at a job,** bir işte nöbetleşe çalışmak; münavebe ile yapmak.

**spend** (**spent**) [spend, spent]. Sarfetmek; harcamak; (vakit) geçirmek. **to ~ money on s.o.,** birisi için para sarfetmek: **to ~ money on stg.,** bir şeye para sarfetmek: **well spent,** mahalline masruf: **to ~ oneself in a vain endeavour,** boş yere kendini yormak: **the bullet has spent its force,** kurşun hızını kaybetmişti. **~ing,** sarfetme: **~ power,** satınalma kabiliyeti. **~thrift** [–θrift], mirasyedi; müsrif.

**spent** *v*. **spend.** Tükenmiş sönmüş. **~ cartridge,** boş fişek: **the day was far ~,** akşam yaklaşıyordu.

**spew** [spyû]. Kus(tur)mak.

**spher·e** [sfiə(r)]. Küre; saha. **~ical** [ˈsferikl], kürevi.

**sphinx** [sfinks]. Ebülhevl; isfenks; esrarengiz adam.

**spic·e** [spâys]. Bahar (katmak). **a ~ of irony,** bir istihza kokusu. **~y,** baharlı; nükteli; biraz açık saçık.

**spick** [spik]. **~ and span,** taptaze; yepyeni.

**spider** [ˈspâydə(r)]. Örümcek. **~y,** pek ince ve uzun: **~ handwriting,** iğri büğrü yazı.

**spik·e** [spâyk]. Sivri uclu demir (ile delmek). **to ~ a gun,** topun falya deliğini tıkamak. **~ed,** sivri uclu. **~y,** diken diken.

**spill** [spil]. (İstemiyerek) dökmek; dökülmek. **to ~ blood,** kan dökmek: **to have a ~,** düşmek.

**spin** (**span, spun**) [spin, span, spʌn]. Eğirmek; bükmek; dön(dür)mek; fırıldanmak. Dönme; (kriket vs.de) topun seyrini değiştirme. **to go for a ~,** bisiklet ile bir gezinti yapmak: **to get into a ~,** (uçak) döne döne inmek: **to send s.o. ~ning,** birini yere yuvarlamak. **~ along,** hızlı gitmek. **~ out,** uzunuzadıya anlatmak: **to ~ out one's money,** parayı yetiştirmek. **~ round,** mihver etrafında dönmek. **~ner** [ˈspinə(r)], eğirici. **master ~,** iplik fabrikatörü. **~ning,** fırıldak gibi dönen; dönme; iplik imali: **~-mill,** iplikhane: **~-wheel,** çıkrık: **~-top,** topac. **~ster** [ˈspinstə(r)], evlenmemiş kadın; ihtiyar kız.

**spinach** [ˈspinic]. Ispanak.

**spin·al** [ˈspâynl]. Belkemiğine aid, şevkî. **~e,** belkemiği; şevk; diken. **~eless,** gevşek, karaktersiz; dikensiz. **~y,** dikenli.

**spindle** [ˈspindl]. İğ; mil; mihver. **~-shanks,** leylek bacaklı.

**spinney** [ˈspini]. Koru; çalılık.

**spiral** [ˈspâyrəl]. Helezon; helis; burmalı.

**spire** [spâyə(r)]. Çankulesi tepesi; kule veya minare külâhı.

**spirit¹** [ˈspirit] *n*. Ruh, can; maneviyat; şevk; cesaret; melek, peri; ispirto. **~s,** ispirtolu içkiler; ervah. **high ~s,** keyif: **to be in high ~s,** keyfi yerinde olm.: **low ~s,** keder: **to be in low ~s,** süngüsü düşük olm.: **to keep up one's ~s,** cesaretini kaybetmemek: **to enter into the ~ of stg.,** bir şeyin ruhuna nüfuz etm.: **in a ~ of mischief,** muziblikle: **to take stg. in a wrong ~,** bir şeyi fenaya çekmek. **~²** *v*. **to ~ s.o. [stg.] away,** kayıplara karıştırmak. **~ed,** canlı, heyecanlı. **~less,** cansız; korkak. **~ual** [–yuəl], ruhanî; manevî; dinî.

**spirt** [spö(r)t]. Fışkır(t)ma(k).

**spit¹** [spit]. Kebab şişi; şişe saplamak. **a ~ of land,** dil, sığlık. **~²** (**spat**) [spit, spat], Tükrük. Tükürmek; (kedi) tıslamak; yağmur çiselemek. **to ~ stg. out,** bir şeyi tükürmek: **he is the very ~ of him,** hık demiş burnundan düşmüş. **~fire** [–fâyə], (öfkeli kedi gibi) ateş püsküren. **~tle** [ˈspitl], salya. **~toon** [spiˈtûn], tükrük hokkası.

**spite** [spâyt]. Kin, garaz; nisbet. **to ~ s.o.,** ona nisbet olsun diye bir şeyi yapmak: **in ~ of,** ···e rağmen: **out of ~,** nisbet için. **~ful,** nisbetçi, garazkâr.

**splash** [splaş]. (*ech.*) Suya çarpma sesi; su zifoz. Su vs. sıçra(t)mak. **to make a ~,** fiyaka yapmak.

**spleen** [splîn]. Dalak; kara sevda. **to vent one's ~ on s.o.,** hıncını birisinden çıkarmak.

**splend·id** [ˈsplendid]. Debdebeli; mükemmel. **that's ~!,** ha şöyle! **~our** [–də(r)], parlaklık, debdebe, ihtişam.

**splice** [splâys]. İki ip ucunu birbirine örerek bağlamak; iki tahtayı birbirine eklemek. **to get ~d,** (*coll.*) evlenmek: **to ~ the main brace,** (*naut.*) bir donanmada bir şeyi tes'id etmek için rom dağıtmak.

**splint** [splint]. Cebire, kırık tahtası. **~er** [–tə(r)], kıymık; kırık. Uzun parçalarını ayırmak; uzun sivri parçalara ayrılmak.

**split**[1] **(split)** [split]. Uzunluğuna yar(ıl)mak; kır(ıl)mak; yırt(ıl)mak; taksim etmek. Yar(ıl)ma; çatlak; ihtilâf. **to ~ the atom,** atomu parçalamak: **to ~ hairs,** kılı kırk yarmak: **to ~ on s.o.,** (*coll.*) birini ele vermek: **to ~ one's sides with laughter,** katılırcasına gülmek: **to ~ up,** bölunmek; taksim etm. **~**[2] *a.* Yarılmış; yarık, çatlak. **~ personality,** (psikoloji) ikiz şahsiyet: **~ second,** saniyenin kesri.

**splotch** [sploç]. Büyük ve şekilsiz boya lekesi (yapmak).

**splutter** [ˈsplʌtə(r)] *v.* **sputter.**

**spoil (-ed, -t)** [spôyl, –d, –t]. Bozmak; bozulmak; zarar vermek; şımartmak; yağma etmek. **~(s),** ganimet. **⌜to ~ the Egyptians⌝,** düşmanından mümkün olan her menfaati çıkarmak: **to be ~ing for a fight,** kavgaya susamak. **~t,** bozulmuş; şımarık, nazlı. **~-sport,** oyun bozan.

**spoke**[1] [spo͡uk] *v.* **~n,** *v.* **speak. pleasant ~,** tatlı dilli: **plain-~,** dobra dobra söyliyen: **he is very well ~ of,** onu çok medhediyorlar. **~sman,** sözcü. **~**[2] *n.* Tekerlek parmaklığı. **to put a ~ in s.o.'s wheel,** birinin işine engel olmak.

**spoliation** [ˈspoliêyşn]. Yağma etme.

**spong·e** [ˈspʌnc]. Sünger (ile silmek). **to ~ a meal,** anafordan yemek yemek: **to ~ on another,** otlakçılık etm.: **to ~ down,** temizlemek: **to ~ off [out],** üzerinden sünger geçirerek silmek: **to throw up the ~,** yenildiğini itiraf etm. **~er,** otlakçı; dalkavuk. **~ing,** otlakçı(lık). **~y,** sünger gibi.

**sponsor** [ˈsponsə(r)]. Kefil; vaftiz babası. Kefil olm.

**spontane·ous** [sponˈtêynyəs]. Kendiliğinden olan; ihtiyarî; içten doğan. **~ity** [–ˈni͡əti], kendiliğinden yapma vs., içinden gelme.

**spoof** [spûf]. (*sl.*) Hile. Kafese koymak.

**spook** [spûk]. Hayal, hortlak.

**spool** [spûl]. Makara(ya sarmak).

**spoon** [spûn]. Kaşık(la almak); (*sl.*) flört yapmak. **~-fed,** kaşıkla yedirilen (çocuk); devlet tarafından sunî bir şekilde teşvik edilen (sanayi).

**spoonerism** [ˈspûnərizm]. Bir kelimenin harfleri veya heceleri arasında yanlışlıkla veya şaka için yapılan yer değiştirme (*mes. 'sesini kes' yerinde 'kesini ses' demek*).

**spoor** [spô(r)]. Vahşi hayvan izi(ni takib etmek).

**sporadic** [spoˈradik]. Münferid; arada sırada vukubulan.

**spore** [spô(r)]. Büzeyr.

**sporran** [ˈsporən]. İskoçyaların eteklik önüne astığı kürk kaplı torba.

**sport** [spôt]. Avcılık; spor; eğlence; maskara; **~s** *gen.* atletik sporlar (avcılık vs.ye **field ~s,** denir, ve futbol vs.ye **games** denir). Oynamak; (*istihfaf*) giymek. **in ~,** şaka yollu: **to make ~ of,** maskara etm.: **to be the ~ of fortune,** kaderin cilvesine tâbi olm.: **to have good ~,** (avda) işi iyi gitmek: **to be a ~,** hatırı için iştirak etmek. **~ing,** avcılığa aid; sportmene yakışacak. **~ive,** oyunbaz, şen. **~sman,** avcılığa düşkün; sportmen. **~smanlike,** sportmenliğe yaraşan.

**spot** [spot] *n.* Benek; nokta; ufacık leke; damla; yer. *v.* Beneklemek; lekelemek; (*coll.*) farketmek. **on the ~,** tam yerinde; derhal: **s.o.'s weak ~,** birinin zayıf damarı: **the weak ~ of stg.,** bir işin püf noktası: **to knock ~s off s.o.,** (*sl.*) birini adamakıllı yenmek. **~ted,** benekli, lekeli: **~ fever,** lekeli humma. **~less,** lekesiz, tertemiz. **~light,** tiyatro projektörü: **in the ~,** herkesin ağzında. **~ty,** benekli, lekeli.

**spouse** [spa͡uz]. Koca *veya* karı.

**spout** [spa͡ut]. Oluk ağzı; fışkırma. Fışkır(t)mak; (*sl.*) yüksek sesle aktör gibi söylemek.

**sprain** [sprêyn]. Burkulma. Burkmak.

**sprang** *v.* **spring**[4].

**sprat** [sprat]. **⌜to throw a ~ to catch a mackerel⌝,** büyük bir istifade için küçük bir şey vermek (⌜kaz gelen yerden tavuk esirgenmez⌝).

**sprawl** [sprôl]. Yere serpilme(k). **to send s.o. ~ing,** birini yere yuvarlamak: **~ing handwriting,** iri ve biçimsiz yazı.

**spray**[1] [sprêy]. Çiçekli dal. **~**[2]. Serpinti; toz halinde serpilen ilâc; dalga serpintisi; vaporizatör. Su serpmek; püskürtmek. **to ~ a tree,** ağaca ilâc püskürtmek. **~er,** püskürtme.

**spread (spread)** [spred]. Yay(ıl)mak; ser(il)mek; sürmek; genişle(t)mek. Yayılma, intişar; (kuş vs.) kanadların açıklık derecesi; (*sl.*) ziyafet. **to ~ out,** germek; teşhir etm.: **to ~ the table,** sofra kurmak. **~-eagle,** kollarını ayaklarını gerip bağlamak.

**spree** [sprî]. (*coll.*) Cümbüş. **to go on the ~,** âlem yapmak.

**sprig** [sprig]. İnce dal.

**sprightly** [ˈsprâytli]. Şetaretli, şen, canlı.

**spring¹** [sprin(g)]. İlk bahar. **~-clean,** evde büyük temizlik yapmak. **~-time,** ilk bahar mevsimi. **~².** Pınar. **~-water,** pınar suyu. **~³** *n.* Sıçrama; yay (kuvveti). **~-,** yaylı. **~-balance,** yaylı kantar. **~-board,** sıçrama tahtası. **~y,** elâstikî, esnek. **~⁴** *v.* **(sprang, sprung)** [sprin(g), spran(g), sprʌn(g)]. Yay gibi fırlamak; sıçramak; çıkmak, doğmak. (Tuzak) kapatmak. **to ~ at s.o.,** birine saldırmak: **to ~ to one's feet,** yerinden fırlayıp kalkmak: **to ~ a leak,** (gemi) su etmeğe başlamak: **to ~ a surprise on s.o.,** birine bir sürpriz yapmak: **to ~ up,** birdenbire kalkmak; baş göstermek.

**sprinkl·e** [ˈsprin(g)kl]. Serpmek. **~er,** pülverizatör. **~ing,** azıcık bir serpinti: **a ~ of knowledge,** bir az malûmat: **there was a ~ of Englishmen at the meeting,** toplantıda tek tük İngilizler vardı.

**sprint** [sprint]. Sürat koşusu. Tabana kuvvet koşmak.

**sprite** [sprâyt]. Peri; şakacı cin.

**sprout** [spraut]. Filiz (sürmek). **~s,** frenk lâhanası.

**spruce¹** [sprûs]. Lâdin. **~².** Şık. **to ~ oneself up,** kendine çekidüzen vermek.

**sprung** *v.* spring⁴.

**spry** [sprây]. Faal; açıkgöz.

**spud** [spʌd]. (*sl.*) Patates.

**spun** [spʌn] *v.* **spin.** **~ glass,** cam ipliği: **~ silk,** kaba bir ipekli kumaş: **~ yarn,** sicim.

**spunk** [spʌnk]. Cesaret. **~y,** (*coll.*) atak.

**spur** [spö(r)]. Mahmuz(lamak); teşvik (etmek); dağ kolu. **to ~ s.o. on,** tahrik etm.: **on the ~ of the moment,** boş bulunarak: **one can't find it on the ~ of the moment,** ha deyince bulunmaz: **to win one's ~s,** liyakatını isbat etmek.

**spurious** [ˈspyuriəs]. Kalp, sahte.

**spurn** [spö(r)n]. Tepmek; istihfaf ile reddetmek.

**spurt** [spö(r)t]. Yarışta âni hamle; kısa müddet için fevkalâde gayret etmek.

**sputter** [ˈspʌtə(r)]. (*ech.*) Tükürür gibi konuşma(k); öfkeli söyleme(k); (kalem) mürekkeb saçmak.

**spy** [spây]. Hafiye, casus(luk etmek). Farketmek. **to ~ on s.o.,** birini gizlice gözetlemek: **to ~ out the ground,** etrafını iyice keşfetmek. **~-glass,** eski usul dürbün. **~-hole,** gözetleme deliği.

**squabble** [ˈskwobl]. Ağız kavgası(na tutuşmak).

**squad** [skwod]. Takım, müfreze. **the awkward ~,** acemi takımı: **the Flying ~,** İngiliz polisinin müteharrik kolu: **firing ~,** idam mahkûmunu kurşuna dizen müfreze.

**squadron** [ˈskwodrən]. Süvari taburu; küçük filo; uçak filosu. **~-leader,** hava binbâşı.

**squalid** [ˈskwolid]. Pis, sefil, miskin.

**squall** [skwôl]. Bora; sağanak; yaygar (koparmak). **~ing,** yaygaracı. **~y,** boralı.

**squalor** [ˈskwolə(r)]. Fakirlikten gelen sefalet ve pislik.

**squander** [ˈskwondə(r)]. İsraf etm.; heba etm. **~er,** mirasyedi.

**square¹** [skweə(r)]. Murabba; kare; meydan; gönye. Murabbaî; amud; dürüst. **to be all ~,** ödeşmek: **a ~ deal,** dürüst bir muamele: **to beat s.o. fairly and ~ly,** birini adamakıllı ve haklı olarak yenmek: **to get things ~,** işleri yoluna koymak: **on the ~,** kaim zaviyeli: **to act on the ~,** dürüst hareket etm.: **out of ~,** kaim zaviyeli olmıyan: **~ root,** cezir. **~-shouldered,** geniş omuzlu. **~²** *v.* Terbi etm.; dört köşe yapmak; tanzim etm.; (hesabı) ödemek; rüşvet yedirmek. **to ~ the circle,** imkânsız bir şeyi yapmağa çalışmak: **to ~ up to s.o.,** birine karşı kavga vaziyeti almak: **to ~ up with s.o.,** hesablaşmak.

**squash** [skwoş]. Ezmek; pelte haline getirmek. Kalabalık; bal kabağı. **to ~ s.o.,** (*coll.*) birine haddini bildirmek, ezmek. **~y,** yumuşak ve sulu.

**squat** [skwot]. Yerden yapma, bastıbacak. Çömelmek; (*sl.*) oturmak. **to ~ upon a piece of land,** bir arsada

oturup ona tasarruf iddia etmek. **~ter,** (*Amer.*) boş topraklara yerleşen muhacir; (Avustralyada) koyun sürüsü sahibi; (*Eng.*) haksız olarak bir eve yerleşen kimse.
**squaw** [skwô]. Kızılderili kadın.
**squawk** [skwôk]. (*ech.*) Cıyak(lamak).
**squeak** [skwîk]. (*ech.*) (Fare) ince ses (çıkarmak). **a narrow ~,** (*coll.*) dar kurtuluş.
**squeal** [skwîl]. (*ech.*) Uzun ve ince bir ses ile bağırma(k); (*coll.*) şikâyet etm.; (*sl.*) suç ortaklarını ele vermek.
**squeamish** [ˈskwîmiş]. Hemen midesi bulanır; fazla titiz.
**squeeze** [skwîz]. Sık(ıştır)mak; sıkıp suyunu çıkarmak. İte kaka araya sokulmak. Sıkma; izdiham. **to give s.o. a ~,** birini kolları arasında sıkmak: **it was a tight ~,** çok sıkışıktı: **a ~ of lemon,** bir kaç damla limon suyu: **to ~ money out of s.o.,** birisinden para sızdırmak.
**squelch** [skwelç]. (*ech.*) Sulu çamurda yürürken çıkan ses (vermek); çiğnemek.
**squib** [skwib]. Kestane fişeği. **a damp ~,** muvaffakiyetsiz teşebbüs.
**squint** [skwint]. Şaşılık. Şaşı olm.; şaşı gibi bakmak. **to have a ~ at stg.,** (*sl.*) bir şeye bakıvermek. **~ing,** şaşı.
**squire** [skwâyə(r)]. Bir köyün bellibaşlı emlâk sahibi; kadına kavalyelik eden erkek; (*esk.*) şövalyeye refakat eden genc asılzade.
**squirm** [skwö(r)m]. Kıvranmak.
**squirrel** [ˈskwirəl]. Sincab.
**squirt** [skwö(r)t]. Çocuk fıskıyesi, nane molla. Dar bir delikten fışkır(t)mak.
**s.s.** [ˈesˈes] = **steamship,** vapur.
**St.** = (i) **street,** cadde; (ii) **saint,** aziz.
**stab** [stab]. Hançerlemek; bıçak vs. saplamak. Bıçak yarası, sivri bir silâhla vurma. **to ~ at s.o.,** birine bıçak savurmak: **a ~ in the back,** arkadan vurma.
**stabil·ity** [staˈbiliti]. Muvazene; istikrar; tevazün. **~ize** [ˈstêybilâyz], tesbit etm.; tevzin etm., denkleştirmek. **~izer,** stabilizatör.
**stable**[1] [ˈstêybl] *a.* Sabit; yıkılmaz; metin. **~**[2]. Ahır; bir ahırdaki atlar. Ahıra koymak. **he has a fine ~,** pek güzel atları vardır: **~ companion,** aynı ahırdan gelen at. **~man,** at uşağı. **~-boy,** seyis yamağı.
**staccato** [stəˈkâtou]. (*mus.*) Kesik kesik.
**stack** [stak]. Tınaz; yığın; baca. Tınaz haline koymak; yığmak.
**stadium** [ˈstêydyəm]. Spor meydanı.
**staff** [stâf]. Erkânı harbiye; kurmay; kadro; maiyet; bir evin hizmetçileri. Kadrosu tedarik etmek. **to be over [under]-~ed,** kadrosu fazla dolu [eksik] olmak.
**stag** [stag]. Erkek geyik. **~hound** [–haund], zağar.
**stage** [stêyc]. Menzil, durak yeri; merhale, derece; devre; iş sahası; sahne, tiyatro; yapı iskelesi; iskele. Sahneye koymak; meydana getirmek. **the ~,** tiyatro: **to go on the ~,** aktör olm.: **~ directions,** sahne izahatı: **at this ~ an interruption occurred,** tam o safhada kesildi: **he is still in the schoolboy ~,** o henüz çocukluk çağındadır: **to travel by easy ~s,** sık sık mola vererek seyahat etm.: **~ fever,** aktör olmak arzusile yanma: **~ whisper,** başkaları tarafından işitilen fısıltı. **~-coach,** menzil arabası. **~-craft,** sahne tekniği. **~-fright,** bir artistin sahneye (ilk defa) çıkarken hissettiği korku ve heyecan. **~-manager,** rejisör. **~r** [ˈstêycə(r)], **old ~,** eski kurt.
**stagger** [ˈstagə(r)]. Sendelemek. Sersem(let)mek; afallaştırmak; zikzakvari koymak; mütenaviben tanzim etmek. **to ~ to one's feet,** sendeleye ayağa kalkmak. **~ed,** zikzakvari; münavebeli. **~ing,** sendeliyen; sersemletici.
**stagna·nt** [ˈstagnənt]. Durgun (su); rakit. **~te** [stagˈnêyt], durgunlaşmak; rakit hale gelmek.
**staid** [stêyd]. Ağırbaşlı, ciddî.
**stain** [stêyn]. Leke(lemek); boya(mak). **without a ~ on his character,** alnının akıyle. **~less,** lekesiz, tertemiz; afif: **~ steel,** paslanmaz çelik.
**stair** [steə(r)]. Merdiven basamağı. **~s [~case],** merdiven: **below ~s,** alt katta; hizmetçilerin arasında.
**stake** [stêyk]. Kazık; nebat desteği; dayak; bahsolunan şey. Desteklemek; bahse koymak; rest çekmek. **to ~ out [off],** kazıklar ile taksim etmek: **to ~ one's all,** (bir maksad için) her şeyini tehlikeye koymak: **his life is at ~,** hayatı mevzuu bahistir: **to have a ~ in stg.,** bir işte menfaati olm.: **to hold the ~s,** kumarda ortaya konan parayı muhafaza etm.:

to ~ one's hopes on, ···e ümidini bağlamak: to lay the ~s, kumarda para koymak: to perish at the ~, diri diri yakılmak.

**stale** [stêyl]. Bayat; kurumuş; ekşimiş; basmakalıb; (atlet) fazla idmandan yorulmuş. Ekşimek; bayatlamak. **a pleasure that never ~s,** doyulmaz bir zevk. **~mate** [-ˈmêyt], (satranc) pat; çıkmaz. Pat etmek; (*fig.*) birini çıkmaza sokmak.

**stalk**[1] [stôlk]. Sap; sâk. **~ed,** saplı. **~**[2]. Avı gizlice takib etme(k). **to ~ along,** azametli adımlarla yürümek.

**stall**[1] [stôl]. Satış sergisi; ahırın bölmesi; (tiyatroda) koltuk; (kilisede) hususî koltuk. **newspaper ~,** gazete kulübesi: **finger-~,** parmak kılıfı. **~-fed,** ahırda besili (sığır). **~**[2]. (Otomobil) istemiyerek dur(dur)mak; (uçak) düşecek hale getirmek.

**stallion** [ˈstalyən]. Damızlık; aygır.

**stalwart** [ˈstôlwö(r)t]. Gürbüz, kuvvetli (adam).

**stamina** [ˈstaminə]. Dayanıklılık.

**stammer** [ˈstamə(r)]. Pepeleme(k).

**stamp** [stamp]. Posta pulu; damga; zımba; ıstampa; ayağını yere vurma(k); damgalamak; pul yapıştırmak; darbetmek. **to ~ about,** tepinmek: **to ~ on,** çiğnemek: **to ~ stg. on the mind,** bir şeyi dimağına hâkketmek: **that ~s him as a fool,** deli olduğunu bu gösteriyor: **to ~ out,** ayaklarla söndürmek: ezmek; bastırmak: **men of that ~,** bu karakterde adamlar. **~-duty,** pul vergisi.

**stampede** [stamˈpîd]. Panik; hezimet. Paniğe uğratmak. Karmakarışık bir halde kaçmak; paniğe kapılmak.

**stance** [stans]. Vücudün vaziyeti.

**stanch** [stânç]. Akan kanı durdurmak.

**stanchion** [ˈstânçən]. Puntal; destek.

**stand**[1] [stand] *n.* Duruş, vaziyet; durma; mukavemet; destek, ayak(lık); sergi; tribün; şapka vs. askısı. **to come to a ~,** duraklamak: **to make a ~ (against),** (···e karşı) mukavemet etm.: **to take a firm ~,** ayakta sımsıkı durmak: **I take my ~ on the principle that ...,** ben şu prensipe dayanırım ...: **to take up one's ~ near the door,** kapının yanına gidip durmak. **~**[2] *v.* (**stood**) [stand, stud]. (Ayakta) durmak; bulunmak. Dayamak; koymak; tahammül etm. **to buy stg. as it ~s,** bir şeyi olduğu gibi satın almak: **~ and deliver!,** ya keseni, ya canını!: **to ~ s.o. a drink,** *etc.*, birine içki vs. ikram etm.: **to ~ six feet high,** boyunun uzunluğu altı kadem olm.: **how do we ~ in the matter of horses?,** at vaziyetimiz nasıl?: **to let ~,** bırakmak: **to let the tea ~,** çayı demlenmek için bırakmak: **I ~ to lose £100 in this matter,** bu iş bana yüz liraya patlıyabilir: **he ~s to lose nothing,** bu işte onun kaybedecek bir şeyi yoktur: **as matters ~,** şimdiki halde: **the house ~s in my name,** ev benim üstümedir: **the thermometer stood at 80,** termometro 80 (F.) dereceyi gösteriyordu: **to ~ to the south,** (gemi) cenuba teveccüh etm.: **to ~ well with s.o.,** birinin nazarında itibarı olmak. **~-offish,** (*coll.*) burnundan kıl aldırmaz. **~ aside,** bir tarafa çekilip durmak: **to ~ aside in favour of s.o.,** birinin lehine çekilmek. **~ back,** geriye çekilmek: **the house ~s back from the road,** ev yol üzerinde değildir. **~ by,** hazır olm.; yanında durmak; yardım etm.: **to ~ by one's word,** sözünden dönmemek: **I ~ by what I said,** söylediğimden şaşmam. **~-by,** baş vurulacak şey; çare; güvenilir şey; yedek. **~ down,** bitirip çekilmek: **to ~ down in favour of s.o.,** başkasının lehine namzedliğini geri almak. **~ for,** temsil etm.; demek; bir yere namzed olm; tarafını tutmak: **MS. ~s for manuscript,** MS. kısaltması manuscript kelimesini gösterir. **~ in,** içinde durmak; bulunmak: **to ~ in with others,** müşterek masrafa iştirak etm.: **to ~ in to land,** (*naut.*) gemiyi karaya doğru yürütmek. **~ off,** uzak durmak; (*naut.*) açılmak; (bir işçiye) iş olmadığı için yol vermek. **~ out,** çıkıntı teşkil etm.; göze çarpmak: **to ~ out to sea,** engine çıkmak: **to ~ out against,** ···e kafa tutmak; tebarüz etmek. **~ over,** yanında durmak; tehir edilmek: **he does no work unless one ~s over him,** başında durmadıkça iş yapmaz. **~ up,** ayağa kalkmak: **to ~ up for,** tarafını tutmak: **~ up for your rights!,** hakkını ara!: **to ~ up to [against],** ···e kafa tutmak.

**standard** [ˈstandö(r)d]. Bayrak; mikyas, derece; model; miyar olarak; umumî. **the ~ answer,** basmakalıb cevab: **~ of living,** hayat seviyesi: **not up to ~,** matluba muvafık olmıyan. ~**ize,** (sanayide) bütün iş aletlerini bir modele uydurmak.

**standing**[1] [ˈstandin(g)]. *a.* Ayakta; sabit. ~ **army,** hazarî ordu: ~ **back,** (ev) içerlek: ~ **crops,** biçilmemiş mahsul: ~ **joke,** umumî alay mevzuu: ~ **rule,** daimî nizamname: ~ **water,** durgun su: **to be brought up all ~,** birdenbire ve tamamen durdurulmak: **to leave s.o. ~,** birini fersah fersah geçmek: **to be left ~,** kala kalmak; (*fig.*) bir işte yaya kalmak: ~ **room only!,** yalnız ayakta duracak yer. ~[2] *n.* Kıdem, rütbe; mevki; şöhret. **an officer of six months' ~,** altı ay hizmet görmüş subay: **of long ~,** eski: **men of high ~,** yüksek zevat: **man of no ~,** chemmiyetsiz bir adam: **a firm of recognized ~,** tanınmış bir firma.

**stand·point** [ˈstandpôynt]. Noktai nazar. ~**still** [ˈstanstil], duraklama, sekte. **at a ~,** durgun; işlemez: **to bring to a ~,** durdurmak; sekteye uğratmak: **to come to a ~,** sekteye uğramak.

**stank** *v.* **stink.**

**stanza** [ˈstanzə]. Şiir kıtası.

**staple**[1] [ˈstêypl]. Bir memleketin başlıca mahsulü veya eşyası; ham madde. Başlıca, esaslı. ~[2]. (Pamuk ve yün) lif, tel.

**star** [stâ(r)]. Yıldız; sakar; baht. Yıldızlarla süslemek; (aktör vs.) birinci rolu oynamak. **his ~ is in the ascendant,** yıldızı parlak: **to be born under a lucky ~,** talihi yaver olm.: **to see ~s,** şeşi beş görmek: **shooting ~,** akan yıldız : **the S~s and Stripes** [**the S~-spangled Banner**], Amerika Birleşik Devletlerinin bayrağı: **I thank my ~s that ...,** çok şükür ki .... ~**-gazing,** müneccimlik; dalgınlık. ~**-lit,** yıldızlarla aydınlanmış. ~**-board** [ˈstâbəd], sancak. **hard a ~,** alabanda sancak. ~**light,** yıldızların ışığı. ~**ry,** yıldızlarla aydınlanmış; yıldızlı.

**starch** [stâç]. Nişasta, kola(lamak). ~**ed,** kolalı. ~**y,** nişastalı.

**star·e** [steə(r)]. Israrla bakma(k); dik dik bakma(k). **to ~ at,** ···e pek dikkatle bakmak: **to ~ s.o. out of countenance,** dik dik bakarak birini utandırmak: **it's ~ing you in the face,** kör kör parmağım gözüne. ~**ing,** ısrarla bakan: ~ **colour,** çiğ renk: ~ **eyes,** faltaşı gibi gözler.

**stark** [stâk]. Katı; tamam. ~ **(naked)** çırçıplak: ~ **(staring) mad,** zırdeli: ~ **nonsense,** saçma sapan: **the ~ truth,** olduğu gibi hakikat.

**start**[1] [stât] *n.* Ürkme; âni hareket; hareket etme, seyahate çıkış; başlangıc; (yarış) çıkış. **at the ~,** başlangıcda: **false ~,** çıkış hatası: **flying ~,** yarışa çıkış noktasından evvel başlama: **from ~ to finish,** çıkıştan bitişe kadar: **to get the ~ of s.o.,** birinden daha evvel başlamak: **to give a ~,** ürküp sıçramak: **to give s.o. a ~,** (i) birdenbire çıkıverip birini ürkütmek: (ii) bir işte birini ortaya çıkarmak, ona yardım etm.; (iii) bir yarışta birine avans vermek: **to make a ~,** başlamak. ~[2] *v.* Ürkmek; ürküp sıçramak; silkinmek; âni bir hareket yapmak; fırlamak; yola çıkmak; başlamak; kalkmak; (perçin) gevşemek. Başlatmak; yürütmek, harekete getirmek. **to ~ away** [~ **off,** ~ **out,** ~ **on one's way**], yola çıkmak: **his eyes were ~ing from his head,** gözleri fırlamıştı: **to ~ a fire,** yangın çıkarmak: **once you ~ him talking he never stops,** bu adamı konuşmağa başlatırsan susmak bilmez: **to ~ out to do stg.,** önce ... yapmak niyetile işe başlamak: **tears ~ed from his eyes,** gözleri yaşardı: **new factories ~ed up everywhere,** her yerde yeni fabrikalar baş gösterdi: ~ **up the engine!,** makineyi harekete geçir!: **to ~ with,** ilkönce başlangıcda. ~**er,** hareket işaretini veren kimse; hareket edici cihaz. ~**ing-,** ~**handle,** ilk hareket kolu: ~**line,** çıkış hattı: ~**post,** çıkış direği: ~**price,** (at) yarış başlamadan evvelki son bahis.

**startl·e** [ˈstâtl]. Ürkütmek; şaşırtmak. ~**ing,** hayret verici, heyecanlı; ürkütücü.

**starv·ation** [stâˈvêyşn]. Şiddetli açlık; ölüm derecesi açlık. ~ **wages,** açlıktan ölecek derecede ücret. ~**e,** açlıktan ölecek hale gelmek; çok acıkmak. Yiyeceksiz bırakmak; mahrum etmek. **to ~ to death,** açlıktan öl(dür)mek: **to ~ out a town,** bir şehri aç bırakarak zabtetmek.

**state**[1] [stêyt] *n.* Hal, vaziyet; mevki, mertebe; debdebe, merasim; devlet, hükümet. *a.* Devlete aid, resmî. **affairs of ~,** devlet işleri: ~ **apartments,** bir sarayda mükellef daire: ~ **ball,** sarayda verilen balo: ~ **carriage** [**coach**], büyük merasim arabası: **to**

dine in ~, mükellef bir ziyafette bulunmak: **to keep great ~ [to live in ~]**, ihtişam içinde yaşamak, debdebeli bir hayat sürmek: **to lie in ~**, (büyük bir adamın cenaze merasimde) tabut teşhir edilmek: **to sit in ~**, kurulup oturmak: **~ of life**, ictimaî mevki: **~ of mind**, ruh haleti: **to be in a great ~ (of mind)**, etekleri tutuşmak: **what a ~ you are in!**, bu halin nedir!: **here's a pretty ~ (of affairs)!**, gel ayıkla pirincin taşını!: **~ robes**, merasim elbisesi: **Secretary of S~**, (*Eng.*) bazı bakanlara verilen unvan; (*Amer.*) Hariciye Vekili: **~ service**, devlet hizmeti: **the United S~s (U.S.A.)**, Amerika Birleşik Devletleri. **~-room**, merasim odası; (*naut.*) lüks hususî kamara. **~**[2] *v.* beyan etm., ifade etm., bildirmek, ilân etm.; tayin etmek. **~craft** [-krâft], siyasî maharet. **~less**, hiç bir devletin tabiiyetinde olmıyan. **~ly**, muhteşem, heybetli. **~ment**, ifade, beyan; tebliğ; hesab. **~sman**, devlet adamı: **~like**, mahir bir devlet adamına lâyık; tedbirli: **~ship**, siyaset; devlet idaresi sanati.

**static** [ˈstatik]. Değişmiyen; sükûnette. **~s**, müvazene bahsi; (radyo) parazit.

**station** [ˈstêyşn]. Durak yeri; konak; istasyon; mevki; merkez; karakol. Yerleştirmek; bir yere tayin etmek. **~ in life**, ictimaî mevki: **to marry below one's ~**, küfvü olmıyanla evlenmek. **~-master**, istasyon müdürü. **~ary** [ˈstêyşənəri], hareket etmiyen, sabit.

**stationer** [ˈstêyşənə(r)]. Kâğıdcı; kırtasiyeci. **~'s Hall**, İngilitere'de bir kitabın telif hakkını temin etmek üzere kaydedilen daire. **~y**, kırtasiye, yazı eşyası; kâğıdcı eşyası: **H. M. S~ Office**, Devlet Neşriyat Müdürlüğü.

**statistics** [stəˈtistiks]. İstatistik.

**statue** [ˈstatyu]. Heykel. **~sque** [-ˈesk], heykel gibi. **~tte**, küçük heykel.

**stature** [ˈstatyuə(r)]. Boy; kamet.

**status** [ˈstêytəs]. Hal, vaziyet; ictimaî *veya* hukukî vaziyet; sıfat; salâhiyet. **~ quo** [kwou], şimdiki vaziyet.

**statut·e** [ˈstatyût]. Kanun. **~ary**, kanunî. **~e-book**, kanunname. **~e-law**, yazılı kanun.

**staunch** [stônç]. Samimî, emin; muhkem.

**stave**[1] [stêyv] *n.* Fıçı tahtası; değnek. **~**[2] *v.* (**stove**) [stouv]. **to ~ in**, üzerine vurarak kırmak, delmek: **to ~ off**, savmak, önüne gelmek; ···den muvakkaten kurtulmak.

**stay**[1] [stêy]. Istralya; destek. **~s**, korse. Istralya ile takviye etm.; desteklemek. **~**[2]. Kalmak; ikamet etm.; durmak; misafir olm.; dayanmak. Durdurmak. İkamet; bir yerde muvakkaten oturma. **to ~ to dinner**, akşam yemeğine kalmak: **to ~ the night**, gecelemek: **to ~ one's hand**, harekete geçmemek: **this horse cannot ~ three miles**, bu at üç mil dayanamaz. **~-at-home**, hep evde oturan. **~ away**, gelmemek. **~ in**, sokağa çıkmamak; izinsiz olmak. **~ on**, bir az daha kalmak. **~ out**, içeri girmemek: **to ~ out all night**, geceleyin kendi evinde bulunmamak. **~ up**, yatmamak: **to ~ up late**, geç vakte kadar yatmamak. **~er** [ˈstêyə(r)], **a good ~**, dayanıklı. **~ing** [ˈstêy·in(g)], **~ power**, dayanıklılık.

**stead** [sted]. **in s.o.'s ~**, birinin yerine: **to stand s.o. in good ~**, birine pek faydalı olm.

**stead·fast** [ˈstedfəst]. Metin, sarsılmaz. **~iness**, sebat, metanet. **~y**[1] *a.*, devamlı; sebatkâr, metin, sallanmaz; muntazam; ağırbaşlı. **~ (on)!**, yavaş!: **to keep ~**, (i) kımıldamamak; (ii) *veya* **lead a ~ life**, namuslu ve muntazam bir tarzda yaşamak: **~ weather**, sabit hava: **his health gets steadily worse**, sıhhati gittikçe fenalaşıyor. **~**[2] *v.* Metanet vermek; kımıldamasını menetmek; yatıştırmak. **~ (down)**, sakin olm.: **to ~ oneself against stg.**, bir şeye dayanmak: **marriage will ~ him down**, evlenince yatışır.

**steak** [stêyk]. Kalın bir dilim et; biftek.

**steal** (**stole, stolen**) [stîl, stoul, stouln]. Çalmak; hırsız olmak. Hırsızlama yürümek. **to ~ away**, sıvışmak: **to ~ a glance at s.o.**, gizlice bakmak: **to ~ a march on s.o.**, başkasından evvel davranmak. **~th** [stelθ], gizli teşebbüs: **by ~**, gizlice; el altından: **~thy**, sinsi; hırsızlama.

**steam** [stîm]. İstim; buhar; buğu. Buhar salıvermek; (vapur, tren) gitmek. Buğuda pişirmek. **at full ~**, bütün süratle: **full ~ ahead**, tam süratle ileri: **to get up [raise] ~**, istim getirmek: **~ing hot**, çok sıcak: **to let**

off [blow off] ~, (i) kazandan istimi salıvermek; (ii) ağzını açıp gözünü yumarak hiddetini hafifletmek: **to ~ open an envelope,** zarfın zamkını buğu ile eriterek açmak: **to proceed under its own ~,** (gemi) kendi makinesiyle yürümek. **~boat,** vapur. **~er,** vapur; buhar tenceresi. **~ship,** vapur. **~y,** buharlı; buğulu. **~-engine,** buhar makinesi; lokomotif. **~-roller,** istim ile işliyen silindir.

**steed** [stîd]. At.

**steel** [stîl]. Çelik; çakmak; masad. Çelik gibi yapmak. **to ~ one's heart,** kalbini katılaştırmak. **~y,** çelik kadar katı.

**steep**[1] [stîp] *a.* Sarp, dik, yalçın. **a ~ climb,** pek güç bir çıkış: **~ story,** inanılmaz hikâye: **~ price,** fahiş fiat: **that's a bit ~!,** (*coll.*) bu kadarı da bir az fazla! **~en,** daha sarp olm. **~ness,** sarplık, diklik. **~**[2] *v.* Suya batırıp bırakmak Suda çok durmak; meşbu olm.

**steeple** [ˈstîpl]. Çan kulesi. **~chase,** manialı yarış. **~jack,** kule veya baca tamircisi.

**steer** [stiə(r)]. Dümenle idare etm.; dümen dinlemek. **to ~ clear of,** ···den sakınmak. **~age,** dümen tesiri; üçüncü mevki güvertesi: **to travel ~,** üçüncü mevkide seyahat etm. **~ing,** dümen cihazı: **~-column,** direksyon mili: **~-gear,** direksyon cihazı: **~-wheel,** dümen dolabı; direksyon çarkı. **~sman,** serdümen.

**stellar** [ˈstela(r)]. Yıldızlara aid.

**stem**[1] [stem]. Sap, sâk; ağac gövdesi; (kadeh) sap; (pipo) boru; (kelime) kök; (aile) silsile; (gemi) prova. Saplarını koparmak. **from ~ to stern,** (*naut.*) baştan kıça kadar. **~med,** saplı. **~**[2] *v.* Sed çekmek; önlemek. **to ~ the current,** akıntıya karşı ilerlemek.

**stench** [stenç]. Pis koku; taaffün.

**stencil** [ˈstensl]. Delikli karton vs. den resim vs. kalıb. Delikli kalıb vasıtasile harf vs. çizmek.

**stenographer** [stəˈnogrəfə(r)]. İstenograf.

**stentorian** [stənˈtôriən]. (İnsan sesi) gök gürültüsü gibi.

**step**[1] [step] *n.* Adım; kısa mesafe; basamak; kademe, derece; ayak sesi; tedbir, teşebbüs. **~s,** evin taş merdiveni: **~s [pair of ~s],** ayaklı merdiven: **~ by ~,** adım adım: **to be in ~ [keep ~],** adım uydurmak: **to be out of ~,** bozuk adım atmak: **the first ~ will be to ...,** yapılacak ilk şey ...: **flight of ~s,** merdiven: **to follow in s.o.'s ~s,** birinin eserini takib etm., birinin yolundan gitmek: **it's a good ~ to ...,** ... epeyce uzaktır: **a short ~,** kısa adım; yakın mesafe: **to take ~s to,** ... için tedbir almak, teşebbüse girişmek: **watch your ~!,** ayağını denk al! **~**[2] *v.* Adım atmak; yürümek. Adımlamak; adımlar ile ölçmek. **to ~ a mast,** geminin direğini ıskaçaya oturtmak. **~ this way!,** bu tarafa gel! **~ping-stone,** atlama taşı; vasıta. **~ down,** inmek: (voltaj vs.) azaltmak. **~ in,** içeriye girmek; müdahele etmek. **~ off,** (araba vs.)-den inmek. **~ on,** ayak bastırmak; çiğnemek: **to ~ on the gas,** (*Amer. sl.*) otomobili hızlandırmak. **~ out,** adımlarını açmak; çıkmak. **~ over,** üzerinden geçmek: **to ~ over to the opposite house,** karşıki eve geçivermek. **~ up,** yukarıya çıkmak: arttırmak.

**step-,** Üvey ...: **~brother,** üvey birader, vs.

**steppe** [step]. Bozkır.

**stereotype** [ˈstiəriotâyp]. Kalıbla basılmış (eser); stereotipi basmak; tesbit etmek. **~d,** beylik: **~ remark,** basmakalıb.

**steril·e** [ˈsterâyl]. Kısır; akım; aseptik. **~ity** [stəˈriliti], kısırlık; akamet. **~ize** [ˈsterilâyz], tâkim etm.; isterilize etm., kısırlaştırmak.

**sterling** [ˈstö(r)lin(g)]. İngiliz lirası; hakikî; kıymetli; tam ayar.

**stern**[1] [stö(r)n] *a.* Sert; merhametsiz. **~ness,** sertlik. **~**[2] *n.* Geminin kıç tarafı; (*coll.*) kıç. **to anchor by the ~,** kıçtan demirlemek: **to sink ~ foremost,** (gemi) kıçtan batmak.

**stet** [stet]. Kalsın! (*Eski tashihi ibtal etmek için kullanılan tabir.*)

**stevedore** [ˈstîvədô(r)]. İstifçi, tahliyeci.

**stew** [styû]. Yahni; bastı; (*esk.*) umumhane. Yavaş yavaş kaynatmak; (*sl.*) sıcaklıktan boğulmak. **to be in a ~,** (*sl.*) etekleri tutuşmak: **~ed apples,** *etc.*, elma vs. kompostosu: ⌜**let him ~ in his own juice**⌝, bırak, ne hali varsa görsün. **~pan,** tencere.

**steward** [ˈstyûəd]. Vekilharc; kâhya; idare memuru; (*naut.*) kamarot. **shop ~,** bir fabrikada sendika memuru. **~ess,** kadın kamarot. **~ship,** kâhyalık: to give an account of one's ~, idaresinin hesabını vermek.

**stick**[1] [stik] *n.* Değnek, baston; sırık; dal. *v.* Sırık ile desteklemek. **the big ~,** kuvvet kullanma: **⌜any ~ to beat a dog⌝,** sevmediği bir adamı küçük düşürmek her şeyine kulp takmak caizdir: **to get the ~,** dayak yemek: **to gather ~s,** kuru dal toplamak: **not a ~ was saved,** bir çöp bile kurtulmadı: **walking-~,** baston: **⌜to have hold of the wrong end of the ~⌝,** bir şeyi ters anlamak: **without a ~ of furniture,** döşeme namına hiç bir şey yok. **~**[2] *v.* (**stuck**) [stʌk]. Yapıştırmak; sokmak; hançerlemek; koymak; atıvermek; tahammül etmek. Yapışmak, yapışık kalmak; kalmak; vazgeçmemek; saplanmak. **to ~ it,** dayanmak: **I can't ~ him,** ona tahammül edemem: **here I am and here I ~!,** buradayım ve buradan kımıldanmam: **to be stuck,** saplanmak; işin içinden çıkamamak; anlıyamamak: **I lent Ali my dictionary, but he's stuck to it,** Aliye lûgatimi ödünc verdim, üstüne oturdu: **some of the money stuck to his fingers,** paranın bir kısmını ziftlendi [deve yaptı]: **to ~ by [to] a friend,** bir dosta sadık kalmak: **to ~ to one's guns,** sebat etm.: **the lift has stuck,** asansör işlemiyor: **to ~ to one's post,** vazifesinden ayrılmamak: **it ~s in my throat,** ben bunu hazmedemiyorum: **to ~ together,** (i) 'anca beraber kanca beraber' olm.: (ii) (iki şeyi) birbirine yapıştırmak: **to ~ to one's word,** sözünden şaşmamak. **~-in-the-mud,** gayri müteşebbis, eski kafalı. **~ at, to ~ at a job,** bir işe durmadan devam etm.: **he ~s at nothing,** hiç bir şeyden çekinmez: **I rather ~ at doing that,** doğrusu, bunu yapmağa düşünürüm. **~ down, ~ it down anywhere!,** nereye olursa olsun koyuver!: **to ~ down an envelope,** zarfı yapıştırmak: **to ~ down in a notebook,** not defterine yazıvermek. **~ on,** üzerine yapıştırmak; üzerinde yapışık kalmak: **to ~ it on,** (*sl.*) çok pahalıya satmak; hesaba ilâveler yapmak. **~ out,** çıkıntılı olm.; kaba durmak: **to ~ it out,** dayanmak: **to ~ out one's chest,** göğsünü şişirmek: **to ~ out one's hand before stopping,** (otomobilde) duracağını göstermek için elini uzatmak: **to ~ out for higher wages,** ısrarla fazla ücret istemek. **~ up,** (ilânı vs.) duvara vs.ye yapıştırmak; dik durmak: **to ~ up for s.o.,** birini müdafaa etm. **~-iness,** yapışkanlık. **~ing-plaster,** ingiliz yakısı. **~y,** yapışkan; aksi: **he will come to a ~ end,** (*sl.*) akibeti fena olacak.

**stickler** [ˈstiklə(r)]. **a ~ for stg.,** (bir hususta) titiz: **a ~ for discipline,** disiplin meraklısı.

**stiff** [stif]. Eğilmez, bükülmez; katı; nezaketsiz; güç: (içki) kuvvetli: (fiat) fahiş. (*sl.*) Ceset. **to be ~ (from sitting still),** her tarafı tutulmak; **(from great exertion)** her tarafı ağrımak: **to have a ~ neck,** boyunu tutulmak: **to offer ~ resistance,** şiddetli mukavemet göstermek: **that's a bit ~!,** (*coll.*) bu kadarı da bir az fazla! **~en** [ˈstifn], katılaş(tır)mak; sertleş(tir)mek; takviye etmek.

**stifl·e** [ˈstâyfl]. Boğmak; nefesini tıkamak; bastırmak. **~ing,** boğucu.

**stigma** [ˈstigma]. Namus lekesi; rezalet. **~tize,** damgalamak; terzil etm.; fena bir isnadda bulunmak.

**stile** [stâyl]. Çitten asmak için basamak. **⌜to help a lame dog over a ~⌝,** yardıma muhtac olana yardım etmek.

**stiletto** [stiˈletou]. Sivri kama.

**still**[1] [stil]. Hareketsiz; sakin, durgun; sessiz. Sükûnet. Teskin etm.; gidermek. **to stand [lie, keep] ~,** kımıldanmamak: **~ life,** natürmort: **~ wines,** köpüklü olmıyan şarablar: **the ~ small voice,** vicdan: **⌜~ waters run deep⌝,** (i) ⌜yumuşak huylu atın çiftesi pektir⌝; (ii) ⌜yere bakar yürek yakar⌝. **~ness,** sükûn, durgunluk. **~born,** ölü doğmuş. **~**[2]. Hâlâ; bununla beraber; maamafih; daha. **~**[3]. İmbik.

**stilt** [stilt]. Yerden yüksekte yürümek için ayaklıkları bulunan uzun koltuk değnekleri. **~ed,** (üslûb) sun'î.

**stimul·ant** [ˈstimyulənt]. Münebbih; alkollu içki. **~ate,** tahrik ve teşvik etm., gayrete getirmek. **~us,** münebbih: **to give a ~ to,** teşvik etm.

**sting** (**stung**) [stin(g), stʌn(g)]. (Arı, akreb vs.) iğne; sokma, ıstırab. (Arı vs.) sokmak; haşlamak; yakmak; yanmak; ıstırab vermek. **that cane ~s,** o değnek insanın canını yakar: **his conscience stung him,** vicdan azabı duydu: **the remark stung him (to the quick),** bu söz (zehir gibi) içine işledi: **his speech had no ~ in it,** nutku çok cansızdı. **~ing,** sokucu; yakıcı.

**sting·iness** [ˈstincinis]. Cimrilik. **~y** [–i], cimri.

**stink** (**stank** veya **stunk, stunk**) [stin(g)k, stan(g)k, stʌn(g)k]. Pis kokmak. Pis koku; taaffün. **to ~ out,** fena koku ile kaçırmak: ⌜**to cry ~ing fish**⌝, kendi malını kötülemek (⌜yoğurtçu yoğurdum karadır⌝ demesi gibi): **it ~s in the nostrils,** koklayanın burnu düşer. **~er,** (*sl.*) pis herif: **to write s.o. a ~,** birine pek şiddetli bir mektub yazmak: **the chemistry paper was a ~,** kimya sualleri berbattı.

**stint** [stint]. Kısmak. **to ~ oneself,** kendini bir şeyden mahrum etm.: **without ~,** bol bol: **to give without ~,** ibzal etmek.

**stipend** [ˈstâypend]. (Papaz) maaş. **~iary** [–ˈpendiəri], muayyen maaş alan: **~ magistrate,** büyük şehirlerde maaşlı sulh hâkimi.

**stipulat·e** [ˈstipyulêyt]. Şart koymak; şartları tayin etmek. **~ion** [–ˈlêyşn], şart.

**stir** [stö(r)]. Kımılda(t)ma; hareket; karıştırma; heyecan, telâş. Kımıldatmak; karıştırmak; faaliyete geçirmek; heyecanlandırmak; tahrik etmek. Kımıldanmak. **there is not a ~,** ortalıkta çıt yok: **there is not a breath of air ~ring,** yaprak kımıldamıyor: **to ~ one's blood,** kanını tutuşturmak: **he is not ~ring yet,** leş gibi yatıyor: **to create a ~,** heyecan uyandırmak: **to ~ s.o. to pity,** birinin merhametini tahrik etm.: **if you ~, I shoot!,** davranma, yakarım!: **to ~ s.o.'s wrath,** birini gazaba getirmek. **~ring** *a.* heyecanlı. **~ up,** karıştırmak; teşvik etm.

**stirrup** [ˈstirəp]. Üzengi.

**stitch** [stiç]. Dikiş (dikmek); örgülerde ilmik; geğrek ağrısı. **to ~ up,** yırtığı, yarayı dikmek: ⌜**a ~ in time saves nine**⌝, zamanında tamir edilen küçük bir hata büyük fenalıkların önüne geçer: **without a ~ of clothing,** çırıl çıplak: **with every ~ of canvas set,** (*naut.*) bütün yelkenleri fora.

**stock**[1] [stok]. *n.* Mevcud; stok mal; soy; kundak; et suyu; çiftlik hayvanları; devlet eshamı. *a.* Beylik. **~s,** gemi tezgâhı; esham ve tahvilat; (*esk.*) suçlunun ayaklarının geçirilerek teşhir edildiği tahta kanape. **~-breeder,** hayvan yetiştiren kimse. **~-broker,** borsa simsarı. **~holder,** hissedar. **~-book,** eşya defteri. **~-in-trade,** mağaza mevcudu; klişe. **~**[2] *v.* Bir dükkân, çiftlik vs.ye lâzım olan şeyleri, hayvanları, alıp koymak; stok yapmak; mağazada satılacak şey tutmak.

**stockade** [stoˈkêyd]. Şarampol; kazıklarla yapılmış sed.

**stockinette** [ˌstokiˈnet]. İç çamaşırı için kullanılan ince örgülü bir kumaş.

**stocking** [ˈstokin(g)]. Çorab.

**stocky** [ˈstoki]. Kısa boylu fakat sağlam yapılı.

**stodgy** [ˈstoci]. Ağır, sıkıcı (kitab vs.); doyurucu (yemek).

**stoic** [ˈstouik]. Revaki; metın. **~al,** revakilere mensub; metin.

**stoke** [ˈstouk]. Ocak veya kazanına kömür atmak, karıştırmak vs. **to ~ up,** ocak vs.nin ateşini arttırmak; (*sl.*) karnını doyurmak. **~r,** vapur ateşçisi. **~-hold,** vapurun ocak dairesi. **~-hole,** kazan dairesi; külhan.

**stole** [stoul]. (Kürk) kadın atkısı.

**stole(n)** *v.* **steal.** Çalınmış: **~ glance,** kaçamak bakış.

**stolid** [ˈstolid]. Teessürsüz; duygusuz; zahiren abdal.

**stomach** [ˈstʌmək]. Mide, karın. (*Yalnız menfi ve mecazi mânada*) hazmetmek; tahammül etmek. **on an empty ~,** açkarnına: **on a full ~,** yemek üstüne: **he has no ~ for adventure,** sergüzeştlere hevesi yok: **it makes my ~ rise,** midemi bulandırıyor: **to turn one's ~,** mide bulandırmak. **~-ache,** karnağrısı.

**stone** [stoun]. Taş; kıymetli taş; çekirdek; 14 librelik (*6·35 kg.*) ingiliz ölçüsü; kum hastalığı. Taşlamak. ⌜**a rolling ~ gathers no moss**⌝, yuvarlanan taş yosun tutmaz: **to leave no ~ unturned,** başvurmadığı çare kalmamak: **not to leave a ~ standing,** yerle yeksan etm.: **a ~'s throw,** yirmi otuz adım. **~mason,** taşçı. **~wall,** (siyasette) obstrüksyonculuk yapmak. **~ware,** kumlu taştan yapılan saksı vs. **~work,** taşçı işi. **~-cold,** çok soğuk. **~-dead,** taş gibi ölü. **~-deaf,** hiç işitmez sağır. **~-fruit,** çekirdekli meyva.

**stony** [ˈstouni]. Taşlık; taş gibi. **a ~ look,** soğuk bakış: **~ politeness,** buz gibi nezaket. **~-broke,** (*coll.*) dımdızlak. **~-hearted,** taş yürekli.

**stood** *v*. **stand.**
**stooge** [stûc]. Başkasının aleti olan insan.
**stook** [stûk]. Ekin demetleri yığını.
**stool** [stûl]. Tabure; büyük abdest. ⌈**to fall between two ~s**⌉, ⌈iki cami arasında beynamaz⌉: **~ pigeon,** polis hafiyesi; birinin suçortağı.
**stoop** [stûp]. Öne doğru eğilmek; azıcık kambur olm.; alçalmak. **I wouldn't ~ to such a thing,** böyle şeye tenezzül etmem. **~ing,** hafifçe kamburlaşmış.
**stop¹** [stop]. Durma; durdurma; durak(lama); sekte; durdurma vasıtası; org düğmesi; adese perdesi; nokta. **to bring to a ~,** durdurmak: **to come to a ~,** bir durağa gelmek; durmak; sekteye uğramak: **to make a ~,** durmak; bir yerde muvakkaten kalmak: **to put a ~ to,** ···e nihayet vermek. **~²** *v*. Dur(akla)mak; kesilmek; misafir kalmak Durdurmak, alıkoymak; tıkamak; (diş) doldurmak; tatil etm.; menetmek. **~ it!,** artık yeter!: **to ~ at nothing,** hiç bir şeyden çekinmemek: **he did not ~ at that,** bununla kalmadı: **to ~ away,** gelmemek: **to ~ (payment of) a cheque,** bir çekin tediyesini durdurmak: **to ~ dead [short],** birdenbire durmak: **to ~ down a lens,** adese perdesini küçültmek: **to ~ for s.o.,** birini beklemek; birini almak için araba vs.yi durdurmak: **to ~ a gap,** delik tıkamak; bir eksikliği tamamlamak: **he never ~s talking,** o susmak bilmez: **to ~ up,** tıkamak. **~page,** durdurma; alıkoyma; maaştan kesilen mikdar; tıkanma. **~per,** tapa. **~ping,** (diş) dolgu. **~-gap,** muvakkat çare; yasak savma. **~-press,** makineye verilirken (havadis). **~-watch,** zamanı saniyeye kadar ölçen ceb saati.
**stor·age** [ˈstôric]. Biriktirme; depoya koyma; ambar; ardiye (ücreti). **cold ~,** buzdolabında muhafaza. **~e¹** [stô(r)] *n*. Depo, ambar, ardiye; mağaza; (*Amer.*) dükkân; biriktirilmiş şeyler; stok; bolluk. **~s,** erzak; levazımat; muhtelif eşya satan büyük mağaza. **this book is a ~ of information,** bu kitab malûmat hazinesidir: **departmental ~s,** her türlü eşya satan büyük mağazalar: **to hold [keep] in ~,** ileride kullanmak için saklamak: **we do not know what the future has in ~ for us,** istikbalın bize ne hazırladığını bilmiyoruz: **I have a great surprise in ~ for you,** size büyük bir sürprizim var: **to lay in a ~ of stg.,** bir şeyi depo edip iddihar etm.: **to set great ~ by stg.,** bir şeye çok ehemmiyet vermek: **to set little ~ by stg.,** bir şeyi hiçe saymak: **war ~s,** harb mühimmat ve levazımı. **~house,** ardiye: **he is a ~ of information,** malûmat kumkumasıdır. **~-keeper,** ambar memuru; kilerci; (*Amer.*) dükkâncı. **~-room,** kiler. **~²** *v*. İleride kullanmak için saklamak; ardiyeye koymak; erzak ile doldurmak. **to ~ up,** biriktirip saklamak.
**storied** [ˈstôrid]. Tarihî. **-~** *suff*. ... katlı (ev).
**stork** [stôk]. Leylek.
**storm** [stôm]. Fırtına; (alkış vs.) tufan; hücum; telâş. Kıyamet koparmak; çok hiddetlenmek. (Bir şehri) hücumla zabtetmek. **to bring a ~ about one's ears,** başına belâ çıkarmak: **~ centre,** kasırga merkezi: **to stir up a ~,** kıyamet kopartmak: **to take by ~,** şiddetle hücum ederek zabtetmek: **to take the audience by ~,** dinleyicileri teshir etm.: ⌈**a ~ in a tea-cup**⌉, ⌈bir bardak suda fırtına⌉. **~y,** fırtınalı; gürültülü. **~-bound,** (gemi) fırtına sebebiyle bir yerde durmuş. **~-troops,** hücum kıtalari.
**story¹** [stôri]. Hikâye, masal; rivayet; martaval. **but that's another ~,** onu başka bir zaman anlatırım: **that's quite another ~,** o büsbütün başka: **it's quite another ~ now,** ⌈eski çamlar bardak oldu⌉: **to make a long ~ short,** sözü uzatmıyayım: **the same old ~,** ⌈eski hamam eski tas⌉: **to tell stories,** gammazlık etm.: yalan söylemek: **these empty bottles tell their own ~,** bu boş şişeler kâfi derecede izah ediyor. **~-teller,** hikâyeci; (*coll.*) martavalcı. **~² [storey].** Kat (ev).
**stout¹** [staut] *a*. Şişman; sağlam; cesur. **a ~ fellow,** yaman bir adam. **~²** *n*. Siyah bira.
**stove¹** [stouv]. Soba; fırın. **~²** *v*. **stave.**
**stow** [stou]. İstif etm.; yerine koymak. **~ it!,** (*sl.*) sus!; yapma! **~age** [ˈstouic], istif (yeri, ücreti). **~away,** parasız seyahat için gemide saklanan kimse. **~ away,** bir yerde saklamak; parasız seyahat için bir gemide saklanmak.

**straddle** [ˈstradl]. Apış(tır)mak; apışarak bir şeyin üzerinde oturmak.

**straggl·e** [ˈstragl]. Sürüden ayrılmak; dağılarak gitmek. **~er,** geri kalan; sürüden ayrılmış. **~ing,** dağınık; seyrek.

**straight** [strêyt]. Doğru; müstakim; dümdüz; dürüst; açık. Düz hat şeklinde olan kısım. Açıkça; hemen; sapmadan; tam; doğrudan doğruya. **to act on the ~,** dürüst hareket etm.: **to be ~ with s.o.,** birine hakikati söylemek; birine karşı dürüst hareket etm.: **to drink ~ from the bottle,** şişeden içmek: **to look s.o. ~ in the face,** birinin gözüne bakmak: **he hit me ~ in the face,** tam yüzüme vurdu: **a ~ fight,** iki kişi arasında kavga; (seçimde) yalnız iki namzed arasında mücadele: **~ off,** derhal: **~ out,** açıkça: **out of the ~,** eğri: **to put ~,** düzeltmek: **I tell you ~,** size açıkça söyliyorum: **to read a book ~ through,** bir kitabı baştan başa okumak: **this train goes ~ through to London,** bu tren aktarmasız Londraya kadar gider. **~en,** doğrul(t)mak; düzeltmek: **to ~ up,** kalkmak. **~forward** [–ˈfôwəd], dürüst, hilesiz. **~ness** [ˈstrêytnis], doğruluk; dürüstlük. **~way** [–wêy], derhal; tezelden.

**strain**[1] [strêyn] *v.* Germek; kasmak; zarar vermek; zorlamak; burkulmak; süzmek; elemek. Kendini zorlamak; olanca kuvvetini sarfetmek; sıkınmak. **to ~ at** [**after**] **stg.,** bir şeyi elde etmek için kendini zorlıyarak çabalamak: **to ~ one's back,** belini incitmek: **to ~ the law,** bir kanunun müsaade ettiği hududu zorlamak: **to ~ stg. out of a liquid,** bir mayii süzerek içindeki şeyi elde etm.: **to ~ every nerve,** bütün gayretini sarfetmek: **to ~ oneself,** kendini fazla yormak; vücudün bir kısmı burkulmak: **to ~ a point,** bir fikri vs.yi ifrata götürmek [zorlamak]: **to ~ off the vegetables,** zerzevatın suyunu süzmek. **~**[2] *n.* Ger(il)me; gerginlik; kuvvet, zor(luk); tazyik; burkulma; tavır; nağme. **martial ~s,** askerî havaları: **the ~ of modern life,** modern hayatın sinirleri geren faaliyeti: **mental ~,** zihnî yorgunluk: **the ~ on the rope was tremendous,** ip çok fazla gerilmişti: **the education of my boy puts a great ~ on my resources,** çocuğumun tahsil masrafı omuzumda ağır bir yüktur: **he said a lot more in the same ~,** bu mealde daha çok söyledi: **to stand the ~,** zora dayanmak: **their friendship stood the ~,** her şeye rağmen dost kaldılar: **parts under ~,** tazyik altında olan kısımlar. **~**[3] *n.* İrsî bir hususiyet; damar. **he has a ~ of German blood,** onda bir az Alman kanı var. **~ed,** gerilmiş; gergin; burkulmuş; zoraki. **~er,** süzgeç; gergi.

**strait** [strêyt]. Boğaz. Sıkı, dar. **~s,** sıkıntı, müzayaka. **to be in great ~s,** sıkıntı içinde olm.: **~ jacket,** deli gömleği. **~en,** daraltmak: **to be in ~ed circumstances,** darlık içinde olmak. **~-laced,** (ahlâk) pek titiz.

**strand**[1] [strand]. Sahil. Karaya otur(t)mak. **to be ~ed,** fena vaziyette bulunmak; yaya kalmak: **to leave s.o. ~ed,** birini yüzüstü bırakmak. **~**[2]. Halat kolu; ipin elyafı.

**strange** [strêync]. Garib, tuhaf; yabancı; yeni. **I am ~ to the work,** bu işe alışık değilim: **to find** [**feel**] **~,** yadırgamak. **~r,** yabancı; tanınmıyan kimse. Daha garib vs. **I am a ~ to these parts,** buranın yabancısıyım: **you're quite a ~ !,** seni gören hacı olur.

**strang·le** [ˈstran(g)gl]. Boğmak. **~lehold, to have a ~ on s.o.,** birini boğazından yakalamak. **~ulation** [–yûˈlêyşn], boğazın sıkılması, boğulma, ihtinak.

**strap** [strap]. Kayış(la bağlamak). **to give s.o. the ~,** (*coll.*) kayışla dayak atmak: **to ~ up,** kayışla bağlamak. **~hanger,** otobüste ayakta kalan yolcu. **~ping,** iri yarı ve dinc (insan).

**strat·agem** [ˈstratəcem]. Harb hilesi. **~egic(al)** [strəˈtîcik(l)]. Sevkülceyşe aid, stratejik.

**strat·osphere** [ˈstratousfiə(r)]. Hava tabakasının üstü. **~um,** *pl.* **-ta** [ˈstrêytəm, –ta]. Kat; tabaka.

**straw** [strô]. Saman(dan yapılmış); hasırdan yapılmış. ⌜**one can't make bricks without ~**⌝, icab eden malzeme olmadıkça yapılamaz: **I don't care a ~,** bu bana vızgelir: ⌜**the ~ that breaks the camel's back**⌝, bardağı taşıran damla: **to draw ~s,** çöple kura çekmek: ⌜**a drowning man will clutch at a ~**⌝, ⌜denize düşen yılana sarılır⌝ *ve bundan* **to clutch at a ~,** müşkül bir vaziyette her çareye başvurmak: **that's the last ~ !,** bir bu eksikti!: **a man of ~,** uydurma adam, varlıksız adam: ~ **mattress,** samanla doldurulmuş minder: ⌜**a ~ shows which way the wind blows**⌝, küçük bir işaret vaziyetin değiştiğini gösterir: **not worth a ~,** metelik etmez. **~-bottomed,** oturacak yeri hasırlı (iskemle). **~-hat,** kanotye. **~berry** [ˈstrôbri], çilek.

**stray** [strêy]. Başıboş; avare; serseri; rasgele; tek tük. Başıboş gezen hayvan. Yoldan sapmak; başıboş avare gezmek. **waifs and ~s,** kimsesiz çocuklar: **to let one's thoughts ~,** dalmak.

**streak** [strîk]. İntizamsız boyalı çizgi; şua. Şekilsiz çizgilerle boyamak; pek hızlı gitmek. **the first ~ of dawn,** sabahın ilk ışığı: **like a ~ of lightning,** şimşek gibi: **there is a ~ of humour in him,** onun mizahtan anlıyan bir tarafı var. **~y,** intizamsız çizgilerle boyanmış; yol yol: **~ bacon,** yağlı ve yağsız karışık domuz pastırması.

**stream** [strîm]. Çay; su; akıntı; sel; akan su gibi hareket. Akmak; tevali etm. Suya atmak. **a ~ of abuse,** küfür yağmuru: **a ~ of cars,** sel halinde otomobiller: **down ~,** akıntı ile: **to ~ forth** [**out**], sel gibi çıkmak: **up ~,** akıntıya karşı: **with the ~,** akıntı ile. **~er,** bandırol; kâğıd şeridi. **~ing,** akan; ağlayan: **~ with sweat,** ter içinde.

**street** [strît]. Sokak, cadde. **you are ~s better than him,** onu fersah fersah geçersiniz: **~ level,** sokak hizası: **the man in the ~,** alelâde insan: **~ musician,** sokak çalgıcısı: **not in the same ~ with ...,** (*coll.*) ···le aynı seviyede değil: **to turn s.o. into the ~,** birini sokağa atmak. **~-arab,** afacan.

**strength** [stren(g)gθ]. Kuvvet; güc; mukavemet; kadro, mevcud. **to bring a battalion up to ~,** bir taburun mevcudunu tamamlamak: **on the ~,** kadroda dahil: **on the ~ of,** ···e binaen, mucibince, ···e güvenerek: **to strike off the ~,** birini kadrodan çıkarmak. **~en,** kuvvetlendirmek; takviye etmek.

**strenuous** [ˈstrenyuəs]. Gayretli; şiddetli; güç, yorucu.

**stress** [stres]. Bir şeye tatbik olunan kuvvet; tazyik; ıstırab; vurgu. Tazyik etm.; ehemmiyet vermek; üzerine aksan koymak. **to ~** [**lay ~ on**] **the importance of a matter,** bir meseleyi çok ehemmiyet vermek.

**stretch**[1] [streç] *n.* Ger(il)me; uzatma; saha; müddet. **at a ~,** fasılasız: **at full ~,** tamamen uzanmış bir halde: **to give a ~,** gerinmek: **to go for a ~,** biraz gezinmek: **by a ~ of the imagination,** muhayyeleyi zorlıyarak: **~ of wing,** açık kanadlar arasındaki mesafe. **~**[2] *v.* Ger(il)mek; uza(t)mak; çekmek; sermek; esnemek; yayılmak; gevşemek. **to ~ oneself,** gerinmek: **to ~ one's legs,** uyuşukluğunu gidermek: **to ~ a privilege,** bir imtiyazı biraz suiistimal etm.: **to ~ a point,** bir noktayı zorlamak. **~ out,** uzatmak; elini uzatmak; serilmek. **~er,** gergi; teskere. **~er-bearer,** teskereci.

**strew** [strû]. Serpmek; dağıtmak.

**stricken** [ˈstrikn] *v.* **strike.** *a.* Tutulmuş; felâkete uğramış. **~ in years,** pek yaşlı.

**strict** [strikt]. Sert; şiddetli; müsamahasız; tam. **he is a ~ Moslem,** koyu bir müslümandır: **in the ~est sense of the word,** kelimenin tam mânasile: **~ly speaking,** doğrusunu söylemek lâzım gelirse: **smoking is ~ly prohibited,** sigara içmek katiyen yasaktir. **~ure** [–yuə(r)], takbih; tenkid: **to pass ~s on s.o.,** birini zemmetmek.

**stride** (**strode**) [strâyd, stroud]. Uzun adım; bir adımlık mesafe. Uzun adımlarla yürümek. **to get into one's ~,** tam yoluna girmek: **to make great ~s in stg.,** bir şeyde çok ilerlemek: **to take stg. in one's ~,** bir şeyi kolayca yapıvermek: **to ~ over,** uzun adımlarla geçmek.

**strident** [ˈstrâydənt]. Keskin, acı (ses).

**strife** [strâyf]. Kavga, bozuşma.

**strike**[1] [strâyk]. Grev (yapmak). **to be on ~,** grev halinde olm.: **to come out on ~,** grev yapmak. **~**[2] *n.* Maden filizini bulma. **a lucky ~,** turnayı gözünden vurma. **~**[3] *v.* (**struck, struck,** (*esk.*) **stricken**) [strâyk, strʌk, strikn]. Vurmak; çalmak; çarpmak. **to ~ against stg.,** bir şeye çarpmak: **to be struck on s.o.,** (*coll.*) birine abayı yakmak: **it ~s me that ...,** bana öyle geliyor ki ...: **how did he ~ you?,** onu nasıl buldun?: **he struck me as (being) rather conceited,** o bana biraz kibirli gibi geldi: **to ~ the eye,** göze çarpmak: **to ~ his flag,** (amiral) forsunu indirmek (kumandasını terk etm.): **to ~ its flag** [**colours**], (gemi) bayrağını indirmek, teslim olm.: to ~ **the hour,** saati çalmak: **it has just struck ten,** saat şimdi onu çaldı: **the hour has struck,** mühim an geldi çattı: **his hour has struck,** sonu yaklaştı; *bazan* hayatının en mühim zamanı geldi: **to ~ a match,** kibrit çakmak: **to ~ oil,** *etc.*, (maden arayıcı) petrol vs.ye raslamak: **the plant has struck (root),** dikilen nebat tuttu: **the road now ~s north,** yol burada şimale sapiyor: **the thought struck him that ...,** birdenbire aklına geldi ki ...: **I've struck upon an idea,** aklıma bir fikir geldi: **we've struck (upon) just the right man,** tam adamına çattık: **to ~ s.o. with wonder,** birini hayrete düşürmek: **to ~ terror into s.o.,** birinin içine dehşet salmak. **~ down,** vurup yere düşürmek. **~ in,** vurup saplamak: **he struck in with a new proposal,** yeni bir teklifle lâfa karıştı. **~ off,** vurup koparmak: **to ~ a name off a list,** bir ismi listeden çıkarmak; **to ~ off 100 copies,** 100 nüsha basmak. **~ out,** çizmek: **to ~ out right and left,** sağa sola vurmak: **to ~ out for oneself,** kendi açtığı çığırda ilerlemek: **to ~ out for the shore,** sahile doğru yüzmek. **~ up,** **to ~ up a tune,** bir makam tutturmak: **the band struck up,** mızıka çalmağa başladı.

**strik·er.** Vurucu; grevci; (saat) tokmak; (tüfek) horoz. **~ing** *a.* Göze çarpan; göz alıcı; dikkate şayan. **a ~ clock,** çalar saat: **~ly beautiful,** göz alacak derecede güzel: **a ~ likeness,** şaşılacak benzerlik: **within ~ distance,** vuracak mesafede.

**string**[1] [strin(g)] *n.* Sicim, ip; kiriş; tel; dizi; sıra. **the ~s,** (*mus.*) telli çalgılar: ⌜**to have two ~s to one's bow**⌝, ümidini yalnız bir yere bağlamamak: **to harp on one ~,** aynı mevzuu diline vird etm.: **Oxford's first ~,** (yarışta) Oxford'un birinci mümessili: **to pull ~s,** iltimas yaptırmak: **he can pull ~s,** dümende dayısı var: **to play second ~ to s.o.,** aynı işte ikinci derecede bırakılmak. **~y,** sicim gibi; sert (et). **~**[2] *v.* (strung) [strin(g), strʌn(g)]. Dizmek; kirişlemek; tel takmak. **to ~ s.o. up,** (*sl.*) birini asmak: **to be all strung up,** asabileşmek: **highly strung,** sinirli: **~ed instrument,** telli çalgı.

**stringen·cy** [ˈstrincənsi]. Sertlik. **~t,** sert; katî.

**strip**[1] [strip] *n.* Uzun ve dar parça; şerid. **~**[2] *v.* Soy(un)mak; soyup soğana çevirmek. **to ~ to the skin,** tamamen soyunmak.

**stripe** [strâyp]. Kumaş yolu; darbe; (çavuş) kol işareti. **to lose one's ~s,** (çavuş) rütbesi elinden alınmak. **~d,** çubuklu.

**stripling** [ˈstriplin(g)]. Genc, delikanlı.

**strive** (**strove, striven**) [strâyv, stroʊv, strivn]. Çabalamak; uğraşmak. **to ~ for [after]** stg., bir şeyi elde etmeğe çabalamak: **to ~ with [against] ...,** ···le uğraşmak.

**strode** *v.* **stride.**

**stroke**[1] [stroʊk] *n.* Vuruş, darbe; hareket; çizgi; felç; hamlacı; yüzme tarzı. **a good ~ of business,** kârlı bir iş: **finishing ~,** işini bitiren darbe: **a ~ of genius,** dahiyane bir hareket vs.: **he had a ~,** ona inme indi: **to be killed by a ~ of lightning,** yıldırım çarparak ölmek: **on the ~ of nine,** saat tam dokuzda: **to row a fast [slow] ~,** hızlı [ağır] darbelerile kürek çekmek: **he has not done a ~ of work,** elini hiç bir işe sürmedi. **~**[2] *v.* Okşamak. **to ~ a boat,** kürek sandalının hamlacısı olm.: **to ~ s.o. the wrong way,** birinin damarına basmak.

**stroll** [stroʊl]. Gezinmek. Gezinti. **to take a ~,** kısa bir mesafe içinde gezmek.

**strong** [stron(g)]. Kuvvetli; sağlam; dokunaklı. **~ language,** sert sözler; küfürbazlık: **patience is not his ~ point,** hiç sabırlı değildir: **~ room,** değerli eşyanın muhafaza edildiği muhkem oda. **~hold,** kale. **~-box,** kasa. **~-minded,** azimkâr.

**strop** [strop]. Ustura kayışı.

**strove** *v.* **strive.**

**struck** *v.* **strike.**

**structur·e** [ˈstrʌktyə(r)]. Yapı; bünye; yapılış tarzı. **~al,** bünyevî; yapıya aid.

**struggl·e** [ˈstrʌgl]. Savaş; çabalama; mücadele. Çabala(n)mak, uğraşmak; mücadele etm. **to ~ to one's feet,** zahmetle ayağa kalkmak: **we ~d through,** düşe kalka çıktık; uğraşa uğraşa bütün müşkülleri yendik. **~ing,** çırpınan: **a ~ artist,** geçinebilmek için çabalıyan sanatkâr.

**strum** [strʌm]. Piyano gelişi güzel çalmak.

**strung** *v.* **string.**

**strut**[1] [strʌt]. Demir kuşak; istinad

kemeri. Desteklemek. ~². Kurumlu yürüyüş. Babahindi gibi gezmek.
**stub** [stʌb]. Kütük; kurşun kalem parçası; izmarit. **to ~ one's toe against stg.**, ayağı bir şeye çarpmak.
**stubbl·e** [ˈstʌbl]. Yerde kalan saman kökleri; pek kısa kesilmiş saç; uzun tıraş. **~y,** anızlı (tarla); tıraşlı; kıllı.
**stubborn** [ˈstʌbən]. İnadcı, anud.
**stuck** *v.* **stick. ~-up,** kibirli; şımarık.
**stud¹** [stʌd]. Bir şahsın beslediği atlar. **~-book,** cins atların şecere defteri. **~-farm,** hara. **~-horse,** damızlık. ~². Yaka düğmesi; büyük başlı çivi. İri başlı çivilerle donatmak. **~ded,** iri başlı çivilerle vs. süslenmiş: **the sky was ~ with stars,** gökyüzü yıldızlarla serpili idi.
**stud·ent** [ˈstyûdənt]. Talebe; araştırıcı. **~entship,** talebeye verilen burs. **~ied** [ˈstʌdid], *v.* **study.** *a.* Kasdî; pek dikkatli; zoraki. **~io** [ˈstyûdyou], atölye, studyo; salon. **~ious** [ˈstyûdyəs], çalışkan; dikkatli; istekli. **he ~ly avoided me,** kasden bana görünmek istemedi. **~y** [ˈstʌdi]. Mütalaa, okuma, tahsil; çalışma odası; küçük kütübhane; resim taslağı; dalgınlık. Mütalaa etm., okumak; tahsil etm.; çalışmak; tedkik etm., dikkatle muayene etm.; ihtimam etmek. **to ~ for the bar,** hukuk tahsil etm.: **brown ~,** dalgınlık: **to finish one's studies,** tahsilini bitirmek: **to ~ one's health,** sıhhatine ihtimam etm.: **to make a ~ of stg.**, bir şeyi bilhassa tedkik etmek.
**stuff¹** [stʌf] *n.* Madde; şey; kumaş; yave. **garden ~,** sebze: **there's good ~ in him,** bu adamda cevher var: **this book is sorry ~,** bu kitab pek yavan: **how are we to get the ~ home?,** bu eşyayı eve nasıl taşıyacağız?: **he is the ~ heroes are made of,** bu adam kahramanların yoğrulduğu hamurdan: **~ and nonsense!,** saçma sapan!: **he's hot ~,** (*sl.*) yamandır: **that's the ~ to give them!,** (*sl.*) ha şöyle! ~² *v.* Doldurmak; tıkamak; istif etm.; çok yedirmek; dolma doldurmak; ölü hayvanın derisini saman ile doldurmak, tahnit etmek. Çok yemek, tıkınmak. **to ~ up,** tıkamak: **my nose is ~ed up,** burnum tıkandı. **~iness,** havasızlık; burnun tıkanıklığı. **~ing,** dolma; yastık vs. içine doldurulan şey: **to knock the ~ out of s.o.,** (i) birine dayak atmak; (ii) birinin burnunu kırmak (*fig.*). **~y,** havasız; (*coll.*) eski kafalı; (*Amer.*) dargın.
**stultify** [ˈstʌltifây]. Tesirini azaltmak.
**stumbl·e** [ˈstʌmbl]. Sürçme(k); yanılmak. **to ~ across [upon],** ...e rasgelmek: **to ~ over stg.**, ayağı bir şeye çarparak sekmek: **to ~ in one's speech,** söz söylerken duralamak. **~ing-block,** mania; zorluk.
**stump** [stʌmp]. Kesilmiş [kırılmış] şeyin geri kalan kısmı; kütük; izmarit; gölge kalemi; tahta ayak; (kriket) kazık. **~s,** (*sl.*) bacaklar. Şaşırtmak; cevab veremez bir hale getirmek: **to ~ along [about],** tahta ayaklı gibi gürültü yaparak gezmek: **to ~ up,** (*sl.*) ödemek: **to be on the ~,** seçimlerde nutuk söylemek gezmek: **to draw ~s,** kriket oyununu bitirmek: **~ orator,** sokak hatibi: **to stir one's ~s,** (*sl.*) kımıldanmak. **~y,** bodur; küt.
**stun** [stʌn]. Sersemletmek. **~ning,** sersemletici; (*sl.*) mükemmel.
**stung** *v.* **sting.**
**stunt¹** [stʌnt]. Büyümesine mâni olm.; bodur bırakmak. **~ed,** bodur; büyümemiş. ~². (*coll.*) Göze çarpmak [reklam yapmak] için gösterilen marifet. (Uçak) akrobatik hareketler yapmak. **~ flying,** maharetli uçuş.
**stupef·y** [ˈstyupifây]. Uyuşturmak; şaşkına çevirmek, sersemletmek. **~action** [–ˈfakşn], şaşkınlık; uyuşukluk. **~ied,** şaşkın; abdallaşmış.
**stupendous** [styuˈpendəs]. Hayret verici; harikulâde.
**stupid** [ˈstyupid]. Akılsız; ahmak. **to drink oneself ~,** abdallaşmağa kadar içmek. **~ity** [–ˈpiditi], akılsızlık; hamakat.
**stupor** [ˈstyûpə(r)]. Uyuşukluk; sersemlik.
**sturd·y** [ˈstö(r)di]. Gürbüz; güclü kuvvetli. **~iness,** gürbüzlük; celâdet.
**stutter** [ˈstʌtə(r)]. (*ech.*) Keke(lemek), pepe(lemek).
**sty¹** [stây]. Domuz ahırı; pis yer. ~² [**stye**]. Arpacık.
**stygian** [ˈsticiən]. Karanlık, muzlim.
**styl·e** [stâyl]. Üslûb; tarz; nevi; zevk; unvan; mil. Unvan vermek; tarif etmek. **to do stg. in ~,** bir şeyi mükellef bir şekilde yapmak: **in fine ~,** mükemmel bir şekilde: **in the latest ~,** son modaya göre: **to live in great ~,** saltanatla yaşamak: **she has**

no ~, onda bir kibarlık yok. **~ish,** zarif; gösterişli. **~ist,** üslûbcu; iyi üslûb sahibi muharrir.
**suasion** [ˈswêyjn]. İkna; tatlılıkla kandırma.
**suav·e** [swêyv]. Tatlı, nazik; (*bazan*) kuşkulandıracak derecede tatlı dilli. **~ity** [ˈswaviti], (söz) tatlılık; fazla nezaket.
**sub-** *pref.* Altında; altı; aşağıda; ikinci derecede; tahtel···; *mes.* **acid** = acı, **sub-acid** = biraz acı; **agent** = acenta, **sub-agent** = acenta muavin: **marine** = denize aid, **submarine** = denizaltı.
**subaltern** [ˈsʌbəltən]. Teğmen.
**subconscious** [sʌbˈkonşəs]. Şuuraltı; gayri meş'ur: **~ly,** gayrışuurî olarak.
**subdivide** [ˌsʌbdiˈvâyd]. Taksim edilmiş parçaları tekrar taksim etmek.
**subdue** [sʌbˈdyû]. Râmetmek; inkıyad altına almak: zabtetmek; hafifletmek. **~d,** mağlub; sakin; donuk; uslu: ~ **conversation,** pesten konuşma: **he seems rather ~ today,** bugün biraz keyifsiz.
**sub-edit** [sʌbˈedit]. (Bir yazıyı) gazeteye münasib bir şekilde koymak. **~or,** yazı işleri müdürü.
**sub-human** [sʌbˈhyûmən]. Yarı insan; tam beşerî olmıyan.
**subject**[1] [ˈsʌbcekt] *n.* Mevzu; bahis mevzuu; (*gram.*) fail; tebaa; sebeb; saded. **to change the ~,** lâkırdıyı değiştirmek: **the pupil must pass in five ~s,** talebenin beş dersten muvaffak olması lâzımdır: **the ~ of much ridicule,** bir haylı alay mevzuu. **~**[2] *a.* Tâbi; maruz; arasıra mübtelâ olan; bir şarta bağlı. **~ to correction,** yanlış olabilir. **~**[3] *v.* [sʌbˈcekt]. İnkıyad ettirmek; itaat altına almak; maruz bırakmak; uğratmak; duçar etmek. **~ive,** enfüsi. **~ case,** mücerred hali.
**subjugate** [ˈsʌbcugêyt]. İnkıyad ettirmek; zabtetmek; boyun eğdirmek.
**subjunctive** [sʌbˈcʌnktiv]. ~ **(mood),** iltizamî sıyga.
**sublet** [sʌbˈlet]. Devren kiraya vermek.
**sublim·ate** [sʌblimêyt] *v.* Tasfiye etm.; ulvileştirmek. **~e** [sʌbˈlâym], ulvî; âlî; son derece. **the ~,** ulviyet: **~ impudence,** inanılmaz hayasızlık: **the S~ Porte,** Babiâlî: **~ly unconscious of,** ···den serapa bihaber. **~ity,** ulviyet.
**submarine** [ˈsʌbmərîn]. Denizaltı; tahtelbahir.
**submerge** [sʌbˈmö(r)c]. Suya bat(ır)mak. **~d,** su altında: **the ~ (tenth),** düşkünler.
**submiss·ion** [sʌbˈmişn]. Boyun eğme, inkıyad, teslim; tevazu; arz. **my ~ is that ...,** ... arz ediyorum: **to starve into ~,** aç bırakarak inkıyad altına almak. **~ive,** itaatlı.
**submit** [sʌbˈmit]. Arzetmek; teslim etm. Râmolmak, inkıyad etm.
**subordinate** *a.* & *n.* [sʌbˈôdinit]. Madun; tâbi; tâli. *v.* [–nêyt]. Tâbi etm.; tâli bir hale koymak.
**suborn** [sʌˈbôn]. Rüşvetle teşvik etm.
**subpoena** [sʌbˈpînə]. (*jur.*) Mahkemeye resmen celbetmek.
**subscri·be** [sʌbˈskrâyb]. Abone olm.; iane vermek. İmza etmek. **I cannot ~ to that,** bunu kabul edemem. **~ber,** abone. **~ption** [sʌbˈskripşn]. Abone; iane; imza. **to open a ~ list,** defter açmak: **to raise ~s,** iane toplamak.
**subsequent** [ˈsʌbsikwent]. Muahhar; müteakıb. **~ly,** müteakıben: **~ to,** ···den sonra.
**subservien·ce** [sʌbˈsö(r)viəns]. Uşak ruhluluk; yaranma; mizacgirlik. **~t,** tâli; mizacgir, uşak ruhlu: **to be ~,** yaranmak.
**subside** [sʌbˈsâyd]. Çökmek; inmek; yatışmak; durmak. **~nce** [ˈsʌbsidəns], çökme; azalma.
**subsidiary** [sʌbˈsidyəri]. Tâli. Şube.
**subsid·ize** [ˈsʌbsidâyz]. ···e tahsisat bağlamak; iane vermek. **~y,** tahsis olunan meblağ; iane.
**subsist** [sʌbˈsist]. Geçinmek; devam etmek. **~ence,** geçim; nafaka: **means of ~,** maişet.
**substan·ce** [ˈsʌbstəns]. Madde; cisim; öz; hulâsa; servet. **a man of ~,** varlıklı adam. **~tial** [sʌbˈstanşəl], cismanî; mevcud; sağlam; mühim; varlıklı; gıdalı. **~ly the same,** esas itibarile aynı. **~tiate** [–şiêyt], doğruluğunu isbat etm. **~tive** [ˈsʌbstantiv], isim. **~ rank,** aslî rütbe.
**substitute** [ˈsʌbstityût]. Vekil; bedel; ikame madde. Başkasının yerine koymak. **to ~ for s.o.,** birinin yerini tutmak.
**subterfuge** [ˈsʌbtəfyûc]. Kaçamak; hile.
**subterranean** [ˌsʌbtəˈrêynyən]. Yeraltı.

**subtle** [ˈsʌtl]. İnce(lmiş); kolayca farkedilmez; kurnaz. **~ty,** incelik; ince fikirlilik; kurnazlık.
**subtract** [sʌbˈtrakt]. Tarhetmek; çıkarmak. **~ion** [–ˈtrakşn], tarh; çıkarma.
**suburb** [ˈsʌbö(r)b]. Varoş. **~an** [–ˈbö(r)bən], mahalleli; şehir civarına giden (tren).
**subver·sive** [sʌbˈvö(r)siv]. Müfsid; yıkıcı. **~t,** altüst etm.; yıkmak.
**subway** [ˈsʌbwêy]. Yeraltı geçidi; tünel.
**succeed** [sʌkˈsîd]. Yerine geçmek; halef olm.; takib etm.; vâris olmak; muvaffak olmak. **to ~ in,** başarmak: **to ~ to the throne,** tahta geçmek: **to ~ to an estate,** bir mülke vâris olmak. **~ing,** takib eden; sonra gelen.
**success** [sʌkˈses]. Muvaffakiyet; iyi netice. **to make a ~ of stg.,** bir işi başarmak; bir işte kazanc elde etm.: **to meet with ~,** muvaffak olmak. **~ful,** muvaffakiyetli. **~ion** [sʌkˈseşn], tevali; istihlâf; tevarüs; silsile. **after a ~ of defeats,** üstüste mağlubiyetlerden sonra: **in ~,** biribiri arkasına; sıra ile: **for three years in ~,** üstüste üç sene: **in rapid ~,** süratle birbiri arkasında. **~ive,** mütevali; birbirini takib eden. **~or** [sʌkˈsesə(r)], halef; vâris.
**succinct** [sʌkˈsin(g)t]. Veciz, kısa.
**succour** [ˈsʌkə(r)]. Yardım (etm.); imdad.
**succulent** [ˈsʌkyulənt]. Sulu, lezzetli.
**succumb** [sʌˈkʌm]. Yenilmek; çökmek; kapılmak. **to ~ to one's injuries,** aldığı yaralardan ölmek.
**such** [sʌç]. Öyle, şöyle; bu(nun) gibi; bu kadar; gibi. **~ and ~,** filân ve filân: **the food, ~ as it is, is abundant,** yiyecek pek iyi değilse de boldur: **Latin, as ~, is not very useful,** Lâtince haddi zatında o kadar faydalı değildir: **I am a doctor and, as ~, must refuse to do this,** ben doktorum ve bu sıfatla bunu yapamam: **until ~ time as ...,** ...ince kadar: **all ~ as are of my opinion,** benim fikrimde olanlar: **we know of no ~,** böyle bir şey bilmiyoruz: **there are no ~ things as fairies,** peri diye bir şey yoktur: **~ people [people ~ as these],** bu gibiler: **in Bristol or some ~ place,** Bristol'da veya bunun gibi bir yerde. **~like,** bu gibi.
**suck** [sʌk]. Emme(k). **to ~ at stg.,** emmek; (pipo vs.) çekmek: **to ~ down,** yutmak: **to ~ dry,** emerek suyunu kurutmak; (birini) sızdırmak: **to ~ in,** yutmak: **to ~ up,** massetmek: **to ~ up to s.o.,** (*sl.*) birine çanak yalayıcılık etm. **~er,** emici uzuv; fışkın; (*sl.*) safdil.
**suckle** [ˈsʌkl]. Emzirmek.
**suction** [ˈsʌkşn]. Emme. **to adhere by ~,** emerek (havayı çekerek) yapışmak.
**sudden** [ˈsʌdn]. Âni; birden; umulmadık. **all of a ~,** ansızın, birdenbire.
**suds** [sʌdz]. **(soap-)~,** sabun köpüğü.
**sue** [syû]. Dava etmek. **to ~ s.o. for damages,** birinin aleyhine zarar ve ziyan davası açmak: **to ~ for peace,** sulh taleb etmek.
**suède** [swêyd]. Podösüet.
**suet** [ˈsyuit]. İç yağı.
**suffer** [ˈsʌfə(r)]. Tahammül etm.; çekmek; müsaade etmek. Cefa çekmek; acı duymak; zarar görmek. **the battalion ~ed heavy losses,** tabur ağır zayiat verdi: **to ~ for one's misdeeds,** yaptığı fenalıkların acısını çekmek: **to ~ from a weak heart,** kalbi zayıf olmak. **~ance,** müsamaha: **on ~,** müsamaha yüzünden. **~er,** acı çeken kimse; kazazede.
**suffic·e** [sʌˈfâys]. Kâfi gelmek; yetişmek. **~ it to say that ...,** yalnız şu kadarını söyliyeyim ki. **~iency** [–ˈfişənsi], kifayet; kâfi mikdar; geçinecek kadar gelir. **~ient** [–ˈfişənt], kâfi (miktarda): **⌜~ unto the day is the evil thereof⌝,** bugünün derdi yeter (yarın Allah kerim)
**suffix** [ˈsʌfiks]. Son ek.
**suffocat·e** [ˈsʌfəkêyt]. Boğ(ul)mak. **~ing,** boğucu.
**suffrage** [ˈsʌfric]. Rey. **universal ~,** umumî seçim hakkı.
**suffuse** [sʌˈfyûz]. Üzerinde yayılmak.
**sugar** [ˈşugə(r)]. Şeker(lemek). **burnt ~,** karamelâ: **brown ~,** ham şeker: **lump ~,** kesme şeker: **castor ~,** toz şeker: **to ~ the pill,** hapı yaldızlamak. **~y,** şekerli; pek tatlı. **~-loaf,** kelle şeker.
**suggest** [sʌˈcest]. Teklif etm.; ilham etm.; fikir vermek; hatıra getirmek. **I ~ that ...,** (avukat) ... ileri sürüyorum. **~ion** [sʌˈcesçən], teklif; telkin; fikir: **with just the ~ of a foreign accent,** pek az hissedilir bir ecnebi şivesile. **~ive,** telkin edici; fikir verici; yakası açık.
**suicid·e** [ˈsyûisâyd]. İntihar (eden kimse). **to commit ~,** intihar etm.:

~ **squad,** fedai müfreze. ~**al** [–ˈsâydl], intihar sayılacak: ~ **tendencies,** intihara temayül: **it would be ~ to, ...** yapmak intihardır.

**sui generis** [ˈsyûâyˈceneris]. (*Lât.*) Nevi şahsına münhasır.

**suit**[1] [syût] *n.* (Elbise vs.) takım; (iskambil) takım; dava; kur; izdivac talebi. **to follow ~,** (iskambil) aynı renkten oynamak; (*fig.*) aynı şeyi yapmak: **generosity is not his strong ~,** onda pek cömerdlik arama. ~**or** [ˈsyûtə(r)], bir kıza talib; davacı. ~[2] *v.* Uy(dur)mak; uygun gelmek; yakışmak; mutabık gelmek. **it ~s my book to put up with him,** ona tahammül etmek işime geliyor: **that hat does not ~ you,** o şapka size yakışmıyor. ~**ability** [–ˈbiliti], uygunluk, yakışık. ~**able** [–əbl], uygun; yakışır; münasib.

**suite** [swît]. Maiyet. ~ **of rooms,** daire: ~ **of furniture,** aynı desende mobilya.

**sulk** [sʌlk]. Somurtma(k); küsmek; küskünlük. **to be in the ~s [to have (a fit of) the ~s],** somurtmak. ~**y,** somurtkan.

**sullen** [ˈsʌlən]. Asık suratlı; küskün; kapanık. **to do stg.** ~**ly,** bir şeyi istemiyerek yapmak.

**sully** [ˈsʌli]. Lekelemek, kirletmek.

**sulphur** [ˈsʌlfə(r)]. Kükürt.

**sultan** [ˈsʌltən]. Padişah; sultan. ~**a** [–ˈtânə], Padişahın karısı veya kızı; kuru İzmir üzümü.

**sultry** [ˈsʌltri]. Sıcak ve sıkıntılı.

**sum** [sʌm]. Yekûn; meblağ; (riyaziyede) mesele; hulâsa. **to ~ up,** icmal etm.; yekûn yapmak; hulâsa etmek. **a ~ of money,** bir mikdar para: **I can't do this ~,** bu hesabı yapamıyorum: **he is very bad at his ~s,** hesabı çok fenadır: **in ~,** hulâsa: **to ~ s.o. up,** birisi hakkında hüküm vermek: **to ~ up the situation at a glance,** vaziyeti bir bakışla takdir etmek. ~**ming-up,** (*jur.*) delillerin ikamesinden sonra hâkimin jüri âzalarına yaptığı hulâsa. ~**marize** [ˈsʌmərâyz], hulâsa etmek. ~**mary,** hulâsa; icmal; kestirme.

**summer** [ˈsʌmə(r)]. Yaz(ı geçirmek). ~ **time,** yaz saati. ~**time,** yaz mevsimi. ~**-house,** çardak.

**summit** [ˈsʌmit]. Zirve; şahika.

**summon** [ˈsʌmən]. Çağırmak, celbetmek. **to ~ a town to surrender,** bir şehri teslim olmağa davet etm: **to ~ up one's courage,** cesaretini toplamak. ~**s,** celbname; resmî davet; mahkemeye celbetmek: **to take out a ~ against s.o.,** birini mahkemeye vermek: **to serve a ~ on s.o.,** birine celbname tebliğ etm.

**sumptuous** [ˈsʌmtyuəs]. Mükellef, tantanalı.

**sun** [sʌn]. Güneş. **to ~ oneself,** güneşlenmek: **against the ~,** (i) güneş karşısında olarak; (ii) sağdan sola: **with the ~,** (i) güneş arkasında olarak; (ii) soldan sağa: **to demand a place in the ~,** (bir millet) yeryüzünde muhtac olduğu sahayı istemek: **his ~ is set,** yıldızı söndü: **he had a touch of the ~,** onu güneş çarptı. ~**bathe,** güneşlenmek. ~**beam,** güneş şuaı. ~**burn,** güneşten yanma: ~**burnt,** güneşten yanmış, esmerleşmiş. ~**-down,** güneşin batması. ~**flower,** ay çiceği. ~**light,** güneş ışığı. ~**lit,** güneşle aydınlanmış. ~**ny,** güneşli: **the ~ side of the picture,** meselenin hoş tarafı. ~**rise,** güneş doğması. ~**set,** gurub. ~**shade,** yazlık şemsiye. ~**shine,** güneş ışığı; neşe. ~**stroke,** güneş çarpması. ~**-glasses,** renkli gözlük. ~**-helmet,** koloniyal şapka.

**Sunday** [ˈsʌndi]. Pazar günü. **a month of ~s,** uzun bir müddet: **one's ~ best,** bayramlık elbise.

**sunder** [ˈsʌndə(r)]. *va.* Ayırmak; yarmak.

**sundry** [ˈsʌndri]. Öteberi; muhtelif. **all and ~,** cümle âlem. *pl.* **sundries,** müteferrika.

**sung** *v.* **sing.**

**sunk** *v.* **sink.** ~**en** [ˈsʌn(g)kn], batırılmış; (yer) münhat: ~ **cheeks,** çökük yanaklar: ~ **eyes,** çukura kaçmış gözler.

**sup** [sʌp]. Yudum. Yudum yudum içmek; akşam yemeği yemek.

**super-** [ˈsyûpə(r)] *pref.* Üzerinde, fevkinde; fazla, ziyade.

**superannuat·e** [ˌsyûpərˈanyuêyt]. Emekliye ayırmak. ~**ion** [–anyuˈêyşn], yaşlılık sebebiyle tekaüd: ~ **fund,** tekaüd sandığı.

**superb** [syuˈpö(r)b]. Muhteşem; enfes.

**supercilious** [ˌsyûpəˈsilyəs]. Müstağni; tepeden bakan.

**superficial** [ˌsyûpəˈfişəl]. Sathî; yarım yamalak.

**superflu·ity** [ˌsyûpəˈflûiti]. Lüzumundan fazla mikdar; lüzumsuz şey. ~**ous** [–ˈpö(r)fluəs], lüzu msuz; fazla

**superhuman** [ˌsyûpəˈhyûmən]. Fevkalbeşer.
**superintend** [ˌsyûprinˈtend]. Nezaret etm., kontrol etmek. **~ent,** müfettiş; kontrol memuru.
**superior** [syuˈpiərię(r)]. Üstün, faik; müstağni. Âmir; reis. **~ity** [–ˈoriti], üstünlük, faikiyet.
**superlative** [syuˈpö(r)lətiv]. Eşsiz. Tafdil sıygası.
**superman** [ˈsyûpəman]. Fevkalbeşer; üstün insan.
**supernatural** [ˌsyûpəˈnatyərəl]. Fevkattabia; mucizevî.
**superscription** [ˌsyûpəˈskripşn]. Kitabe.
**superse·de** [ˌsyûpəˈsîd]. Başkasının yerine geç(ir)mek. **this method has now been ~d,** bu usulün yerini şimdi baskası tutmuştur. **~ssion** [–ˈseşn], yerine başkasını koyma.
**supersonic** [ˌsyûpəˈsonik]. Sesten daha süratli.
**superstit·ion** [ˌsyûpəˈstişn]. Batıl itikad; hurafe(lere inanma). **~ious,** [–ˈstişəs], batıl itikadlara inanan; hurafatçı.
**superstructure** [ˈsyûpəstrʌktyə(r)]. Bir yapının üst kısmı; gemi teknesinin üzerindeki kısım.
**supervene** [ˌsyûpəˈvîn]. Zuhur edivermek.
**supervis·e** [ˈsyûpəvâyz]. Nezaret etm.; bakmak; idare etmek. **~ion** [–ˈvijn], nezaret, teftiş. **~or,** müfettiş; âmir; nezaretçi.
**supper** [ˈsʌpə(r)]. Akşam yemeği. **the Last S~,** Hazreti İsa'nın havarileriyle son yemeği: **the Lord's S~,** (Hıristiyanlıkta) şarablı ekmek yeme âyini.
**supplant** [sʌˈplânt]. Bir kimsenin yerini almak.
**supple** [ˈsʌpl]. Kolayca eğilir; uysal.
**supplement** *n.* [ˈsʌplimənt]. Zeyil, ilâve, zam. *v.* [–ˈment]. ···e ilâve etm.; zam ederek tamamlamak. **~ary** [–ˈmentəri], zeyil veya ilâve şeklinde.
**suppli·ant** [ˈsʌpliənt]. Yalvaran; niyazkâr. **~cate** [ˈsʌplikêyt], yalvarmak; niyaz etmek.
**suppl·ier** [sʌˈplâyə(r)]. Tedarik eden; lâzım olan şeyleri veren. **~ies** [–ˈplâyz], levazım, erzak; tahsisat. **~y¹** [sʌˈplây], *n.* bir maddenin mevcudu; mikdar; levazımı tedarik etme. **~ and demand,** arz ve taleb: **England has large ~ies of coal,** İngiltere'nin büyük mikdarda kömürü var: **to be in short ~,** kıt olm.: **to vote ~ies,** (Parlamento) tahsisat kabul etmek. **~y²** *v.* Tedarik etm.; temin etm.; arzetmek; bir eksiğini doldurmak.
**support¹** [sʌˈpôt] *n.* Destek; istinad; yardım. **documents in ~ of a claim,** bir iddiayı teyid eden vesikalar: **to speak in ~ of s.o.,** birinin lehinde söz söylemek: **the sole ~ of his old age,** ihtiyarlığında yegâne dayandığı kimse. **~²** *v.* Desteklemek; istinad etm.; teyid etm.; yardım etm.; beslemek; kaldırmak; tahammül etm. **to ~ oneself,** (i) geçinmek; (ii) dayanmak. **~er,** tarafdar; yardımcı.
**suppos·e** [sʌˈpǫuz]. Farzetmek; zannetmek, inanmak. **creation ~s a creator,** hilkatın varlığı hâlikın varlığını gösterir: **~ing that you are right,** farzedelim ki haklısınız: **I don't ~ he will come,** geleceğini zannetmiyorum: **'Will you go there? 'I ~ so',** 'Oraya gidecek misin?' 'Her halde': **~ we change the subject,** mevzuu değiştirsek nasıl olur?: **he is ~ed to be very rich,** pek zengin olduğunu söyliyorlar: **I am not ~ed to know,** (i) onu benim bilmemem mefruz; (ii) onu benim bilmediğimi zannediyorlar. **~ed,** mefruz; farazî; zannedilmiş: **~ly,** gûya. **~ing, ~ (that)** ..., ... takdirde. **~ition** [ˌsʌpǫuˈzişn], farz; zan.
**suppress** [sʌˈpres]. Bastırmak; kaldırmak; tenkil etm.; zabtetmek. **~ion** [–ˈpreşn], tenkil, bastırma.
**suprem·acy** [syûˈpreməsi]. Üstünlük hakimiyet. **~ over,** ···e üstünlük. **~e** [–ˈprîm], en yüksek; en mühim; üstün. **~ happiness,** tam bahtiyarlık: **to hold s.o. in ~ contempt,** birini şiddetle istihfaf etm.: **to reign ~,** mutlak bir şekilde hâkim olmak.
**surcharge** [ˈsö(r)çâc]. Zammedilmiş vergi, resim. Fazla yükletmek; fazla resim tarhetmek.
**sure** [şûə(r)]. Emin; muhakkak; şübhesiz; müsbet, sağlam; muhkem; (*Amer.*) hay hay! **~ly,** elbette: **to be ~ of stg.,** bir şeyden emin olm.: **I'm ~ I don't know!,** vallahi bilmem: **be ~ not to forget!,** unutma!: **well I'm ~! [well to be ~!],** Allah! Allah!: **to be ~!,** hakkınız var: **she's not very pretty to be ~, but ...,** güzel olmadığı muhakkak, fakat ...: **he is ~ to come,** muhakkak gelir: **I said it would rain and ~ enough it did,** yağmur yağacak

dedim, bak işte yağdı: **he will come, ~ enough,** korkma, gelir: **and ~ enough he died next week,** hakikaten de ertesi hafta öldü: **I don't know for ~,** katî olarak bilmiyorum: **to make ~ of,** ···i temin etm.: **to make ~ of a fact,** bir vakayı tahkik etm.: ⌜**slow but ~**⌝, ağır fakat esaslı: **a ~ thing,** elde bir: **~ly you don't believe that!,** buna nasıl inanırsın? **~footed** [şûəˈfutid], düşmez; ayağı hiç kaymaz. **~ty** [ˈşûə(r)ti], kefil; şübhesizlik. **to stand [go] ~ for s.o.,** birine kefalet etmek.

**surf** [sö(r)f]. Deniz sahilindeki köpüklü dalgalar. **~-riding,** dalgalar üzerinde bir nevi kayakla kaymaktan ibaret deniz sporu.

**surface** [ˈsö(r)fis]. Satıh; yüz; bir şeyin üst kısmı; görünüş. Üzerine yüz kaplamak. (Denizaltı) suyun üstüne çıkmak. **one never gets below the ~ with him,** bu adamın içine nüfuz etmek imkânsızdır.

**surfeit** [ˈsö(r)fit]. Yemekte ifrat; tokluk; fazla bolluk. Fazla ye(dir)mek. **to ~ oneself with stg.,** bir şeyden midesi bulanacak kadar yemek (& *fig.*).

**surge** [sö(r)c]. Büyük dalga (gibi) kabarma(k).

**surg·eon** [ˈsö(r)cən]. Cerrah. **~ery,** cerrahî ilmi; doktor muayenehanesi. **~ical** [–cikl], ameliyata aid.

**surly** [ˈsö(r)li]. Abus, hırçın.

**surmise** [səˈmâyz]. Tahmin; şübhe. Tahmin etm.; delil olmadığı halde inanmak.

**surmount** [səˈmaunt]. Üstünden gelmek.

**surname** [ˈsö(r)nêym]. Soyadı (takmak).

**surpass** [səˈpâs]. Üstün çıkmak. **~ing,** eşsiz: **~ly,** son derece.

**surplice** [ˈsö(r)plis]. (Papaz) beyaz cübbe.

**surplus** [ˈsö(r)plʌs]. Artık; fazla; arka kalan.

**surpris·e** [səˈprâyz]. Umulmadık şey; hayret. Hayret vermek; şaşırtmak. **to be ~ed at stg.,** bir şeye şaşmak: **to take s.o. by ~,** birini gafil avlamak. **~ing,** hayret verici; acayib.

**surrender** [səˈrendə(r)]. Teslim (etm.), terk (etmek). Teslim olmak.

**surreptitious** [ˌsʌrepˈtişəs]. El altından; gizli.

**surround** [sʌˈraund]. Çerçeve. Kuşatmak; sarmak. **~ing,** etrafında olan; muhit. **~ings,** muhit; çevre.

**surtax** [ˈsö(r)taks]. Artırılmış vergi.

**surveillance** [səˈvêyəns]. Gözetme; nezaret.

**survey** *n.* [ˈsö(r)vêy]. Muayene; mütalaa; göz gezdirme; mesaha; hulâsa. *v.* [sö(r)ˈvêy]. Göz gezdirmek; muayene etm.; mesaha etm., haritasını çıkarmak. **~ing,** mesaha ilmi **~or,** mesaha memuru; müfettiş.

**surviv·al** [səˈvâyvl]. Beka; artakalma; hayatta kalma. **~e,** artakalmak; başkasından daha çok yaşamak; hâlâ yaşamak: **to ~ an accident,** kazadan canını kurtarmak.

**suscepti·ble** [sʌˈseptibl]. Hassas; alıngan; çabuk müteessir olan, şıpsevdi. **~ of proof,** isbat edilebilir: **~ to a disease,** bir hastalığa istidadı olan. **~bility** [–ˈbiliti], hassaslık; alınganlık; istidad.

**suspect** *a.* [ˈsʌspekt]. Şübheli; maznun; bulaşık. *v.* [sʌsˈpekt]. Şübhelenmek; kuşkulanmak; hakkında suizanda bulunmak. **to ~ s.o. of a crime,** birinin bir cinayeti işlediğinden şübhe etm.: **I ~ed as much,** ben de bundan şübhe ediyordum.

**suspen·d** [sʌsˈpend]. Asmak; tehir etm.; muvakkaten tatil etm.; muvakkaten işten el çektirmek. **~ded,** muallak; asılı; muvakkaten tatil edilmiş. **~ders,** çorab askısı. **~se** [sʌsˈpens], muallaklık; merak. **to be in ~,** askıda kalmak; (insan) merakta kalmak: **to keep s.o. in ~,** birini merakta bırakmak: **matters in ~,** muallakta kalan meseleler. **~sion** [–ˈpenşn], asma; talik; muvakkaten tatil olunma.

**suspic·ion** [sʌsˈpişn]. Suizan; şübhe; kuşkulanma. **above ~,** şübhe edilemez. **~ious** [–ˈpişəs], vesveseli; suizan sahibi; şübheli: **to be ~ about,** ···den kuşkulanmak: **it looks to me ~ly like measles,** bence kızamık olması pek muhtemeldir.

**sustain** [sʌsˈtêyn]. Ağırlığını taşımak; kuvvet ve ümid vermek; beslemek; teyid etmek. **to ~ an injury,** yaralanmak, zarar görmek: **to ~ an objection,** (*jur.*) bir itirazı kabul etmek. **~ed,** devamlı.

**sustenance** [ˈsʌstənəns]. Gıda; besleme.

**suzerain** [ˈsyûzərən]. Metbu. **~ty,** metbuiyet.

**swab** [swob]. İp süpürge; bulaşık bezi; (*sl.*) alçak herif. **~ down,** ip süpürge ve su ile temizlemek.

**swaddle** [ˈswodl]. Kundaklamak.
**swag** [swag]. (*sl.*) Hırsızlık elde edilen şey.
**swagger** [ˈswagə(r)]. (Caka) satmak; kabadayılık etmek. (*coll.*) Gösterişli.
**swain** [swêyn]. (*şair.*) Çoban; âşık.
**swallow**[1] [ˈswolou] *n.* Kırlangıc. ⌜**one ~ does not make a summer**⌝, bir çiçekle bahar olmaz. **~-tail coat,** jaketatay. **~**[2] *v.* Yutmak. **to drink stg. at one ~,** (bir içki vs.yi) dikmek: **to ~ an insult,** hakareti sineye çekmek: **to ~ one's pride,** gururu bir tarafa bırakmak: **to ~ one's words,** tükürdüğünü yalamak: **to ~ up,** yutmak.
**swam** *v.* **swim.**
**swamp** [swomp]. Bataklık. Dalga içine girip batırmak. **~y,** bataklık.
**swan** [swon]. Kuğu. **the ~ of Avon,** Shakespeare. **~-necked,** kuğu boynu şeklinde. **~-song,** bir şairin vs. son eseri.
**swank** [swan(g)k]. (*coll.*) Çalım (satmak); gösteriş.
**swap** [swop]. Trampa (etm.). **to ~ stg. for stg.,** bir şeyi bir şeyle trampa etm.: **to ~ places with s.o.,** birile yer değiştirmek: ⌜**don't ~ horses in midstream**⌝, sıkı bir zamanda idare edenleri değiştirme.
**sward** [swôd]. Çimenli yer.
**swarm**[1] [swôm]. Arı kümesi; kalabalık. Kaynamak; küme teşkil etmek. **~s of,** pek çok: **this place ~s with foreigners,** bu yerde ecnebiler kum gibi kaynıyor. **~**[2]**, to ~ up a tree,** bir ağaca tırmanmak.
**swarthy** [ˈswôði]. Koyu esmer.
**swashbuckler** [ˈswoşbʌklə(r)]. Kabadayı.
**swastika** [ˈswastikə]. Gamalı haç.
**swat** [swot]. (*coll.*) Vurmak; ezmek.
**swathe** [swêyð]. Kundaklamak; sarmak.
**sway**[1] [swêy]. Salla(n)mak. **~**[2]. Hüküm; nüfuz; hakimiyet. Bir tarafa meylettirmek; fikrini çelmek. **to hold ~ over,** ···e hâkim olm.: **to bring a people under one's ~,** bir milleti hakimiyeti altına almak.
**swear** (**swore, sworn**) [sweə(r), swô(r), swôn]. Yemin etm.; küfür etmek. **to ~ at s.o.,** birine küfür etm.: **to ~ away s.o.'s life,** yalan yere yemin ederek birinin idamına sebeb olm.: **to ~ by stg.,** bir şey üzerine yemin etm.; (*coll.*) bir şeye çok itimad etm.: **to ~ stg. on the Bible,** İncil üzerine yemin etm.: **to ~ s.o. in,** bir vazifeye tayin edilen birine yemin ettirmek: **to ~ off stg.,** bir şeye tövbe etm.: **to ~ s.o. to secrecy,** kimseye söylemiyeceğini yemin ettirmek. **~ing,** küfür.
**sweat** [swet]. Ter; (*sl.*) angarya. Terlemek; (*sl.*) çok çalışmak. Terletmek; az ücret ile çalıştırmak. **to be in a ~,** terlemek; (*sl.*) etekleri tutuşmak: **~ed labour,** az ücretli iş. **~er,** sveter, kazak.
**Swed·e** [swîd]. İsveçlı. **~en,** İsveç. **~ish,** isveçli; isveççe.
**sweep**[1] [swîp] *n.* Süpürme; baca süpürücü; akış; kavis; saha; bir şeyin faaliyet ve tesir sahası; (*sl.*) kirli çocuk. **to make a clean ~ of,** silip süpürmek: **with a ~ of the arm,** bir kol hareketi ile. **~**[2] (**swept**) [swept] *v.* Süpürmek; taramak. Hızla ve mağrurane ilerlemek; kavis yaparak dönmek. **to ~ all before one,** devamlı bir şekilde muvaffak olm.: **to ~ away,** silip süpürmek: **to ~ the board,** mümkün olan her şeyi kazanmak: **pirates swept down on the town,** korsanlar şehre çullandılar: **to ~ the horizon with a telescope,** ufku bir dürbünle bir baştan bir başa tedkik etm.: **to be swept off one's feet,** dalga vs. ile sürüklenmek; (*fig.*) heyecana kapılmak: **the maid swept out the room,** hizmetçı odayı baştan başa süpürdü: **the lady swept out of the room,** hanım vardakosta bir eda ile odadan çıktı: **to ~ the seas of one's enemies,** denizleri düşmandan temizlemek: **the shore ~s to the south in a wide curve,** sahil büyük bir kavisle cenuba doğru kıvrılıyor. **~er,** sokak süpürücü; süpürücü makine. **~ing** *a.* geniş, umumî. **~ings,** süprüntü. **~stake** [–stêyk] (*abb.* **sweep**). (Bir at yarışlarina mahsus) piyango.
**sweet** [swît]. Tatlı; şekerli; lezzetli; hoş; taze. Şekerleme; tatlı (yemek). **to be ~ on s.o.,** (*sl.*) birine düşkün olm.: **to say ~ nothings to s.o.,** havadan sudan güzel sözler söylemek: **to have a ~ tooth,** tatlı şeylere düşkün olm.: **at one's own ~ will,** keyfine göre. **~en,** şekerlen(dir)mek; tatlılaş(tır)mak. **~ening,** şekerlendirici şey. **~heart,** sevgili. **~meat,** şekerleme. **~ness,** tatlılık.
**swell** (**swelled, swollen**) [swel, sweld, swouln]. Şiş(ir)mek; kabar(t)mak; art(tır)mak. Kabarma; artma; ölü

dalga (deniz); (*coll.*) (i) züppe; (ii) mühim bir şahıs. Gösterişli; parlak. **~ing,** şiş; kabarma.
**swelter** [ˈsweltə(r)]. Sıcaktan bayılma(k). **~ing heat,** bayıltıcı sıcaklık.
**swept** *v.* **sweep.** **~-back wings,** (uçak) geriye dönük kanadlar.
**swerve** [swö(r)v]. Sapma(k); doğru yoldan çıkma(k).
**swift** [swift]. Süratli, hızlı, tez.
**swig** [swig]. Bir yudumda içmek.
**swill** [swil]. Bol su ile yıkama(k); (*sl.*) iştahla içmek. Yıkama; içme; sulu domuz yemi.
**swim** **(swam, swum)** [swim, swam, swʌm]. Yüzme(k). **eyes ~ming with tears,** gözünden yaşlar boşanarak: **my head is ~ming,** başım dönüyor. **~mingly,** (*sl.*) gül gibi. **~ming-bath** [**-pool**], yüzme havuzu.
**swindle** [ˈswindl]. Dolandırmak; kafese koymak. Hile, dalavere. **~r,** dolandırıcı, dalavereci.
**swin·e** [swâyn]. (*sing. & pl.*) Domuz(lar); hınzır. **~ish,** hınzırcasına; pek alçak.
**swing**[1] **(swung)** [swin(g), swʌn(g)] *v.* Salla(n)mak; salınmak; as(ıl)mak; rakkas gibi hareket etmek. **the car swung round,** otomobil tekerlekleri kayıp geri döndü: **the car swung round the corner,** otomobil köşeyi dönüverdi: **to ~ oneself into the saddle,** eyerin üzerine sıçramak. **~(e)ing,** sallanan: **a ~ blow,** (*coll.*) şiddetli bir yumruk. **~**[2] *n.* Salla(n)ma; sallantı; salıncak. **the ~ of the pendulum,** rakkasın hareketi; (*fig.*) münavebeye meyletme, (*esp.* seçimlerde) partilerin münavebe ile seçilmesi temayyülü: **to be in full ~,** tam faaliyette olm.: **everything went with a ~,** her şey tam yolunda gitti: **to walk with a ~,** hızla ve intizamla yürümek.
**swipe** [swâyp]. Hızlı vuruş. Var kuvvetile vurmak; (*sl.*) çalmak.
**swirl** [swö(r)l]. Girdab gibi hareket (etm.); anafor.
**swish** [swiş]. (*ech.*) Hışırtı. Hışırdamak; (*coll.*) huş dalı ile dövmek.
**Swiss** [swis]. İsviçreli.
**switch** [swiç]. İnce dal, değnek; priz; elektrik düğmesi; demiryolu makası. İnce değnek ile vurmak; (treni) bir yoldan diğerine geçirmek. **to ~ the tail,** (hayvan) kuyruğunu savurmak: **to ~ off,** elektriği söndürmek: **to ~ on,** elektriki yakmak: **to ~ over,** başka cihete çevirmek. **~back,** inişli yokuşlu (yol). **~board,** (*elect.*) dağıtma tablosu.
**Switzerland** [ˈswitzəland]. İsviçre.
**swivel** [ˈswivl]. Fırdöndü. Mil etrafında dönmek.
**swollen** [ˈswoùln] *v.* **swell.** Şişmiş, kabarmış. **to suffer from a ~ head,** küçük dağları ben yarattım demek.
**swoon** [swûn]. Bayılma(k); kendinden geçme(k).
**swoop** [swûp]. Doğan gibi şikâr vs.-nin üzerine saldırma(k); üzerine atılma(k). **to ~ down upon stg.,** şiddetli hücum etm.: **at one fell ~,** müdhiş anî bir darbe ile.
**swop** *v.* **swap.**
**sword** [sôd]. Kılıc. **to cross ~s with s.o.,** kavga etm.; birile boy ölçüşmek: **to draw the ~,** kılıc çekmek: **to put to the ~,** kılıcdan geçirmek. **~sman,** iyi kılıc kullanan kimse. **~-play, verbal ~,** söz düellosu.
**swore, sworn** *v.* **swear.**
**swot** [swot]. (*sl.*) Ağır iş; çok çalışan talebe. Durmadan çalışmak.
**swum** *v.* **swim.**
**swung** *v.* **swing.**
**sybarite** [ˈsibərâyt]. Tenperver kimse.
**sycophant** [ˈsikoùfant]. Dalkavuk.
**syllab·ic** [siˈlabik]. Heceye aid. **~le** [ˈsiləbl], hece.
**syllabus** [ˈsiləbəs]. Hulâsa; cedvel; müfredat programı.
**sylph** [silf]. Hava perisi; güzel kız.
**sylvan** [ˈsilvən]. Ormana aid.
**symbol** [ˈsimbəl]. Remiz, temsil. **~ic(al)** [–ˈbolik(l)], temsilî; remzî. **~ize** [ˈsimbəlâyz], temsil etm.; remiz teşkil etmek.
**symmet·ry** [ˈsimitri]. Tenasüb. **~rical** [–ˈmetrikl], mütenasib.
**sympath·y** [ˈsimpəθi]. Başkasının hislerine iştirak etme; müşterek his; şefkat; taziye. **to be in ~ with s.o.'s ideas,** birinin fikirlerine iştirak etm.: **I have no ~ for him,** ona hiç acımam: **popular ~ies are on his side,** umumî efkâr onun lehine mütemayildir. **~etic** [–ˈθetik], başkasının hislerine iştirak eden; acıyan; sevimli. **~ize** [ˈsimpəθâyz], başkasının hislerine iştirak etm.; taziye etmek. **~izer,** başkasının hislerine iştirak eden kimse; tarafdar.
**symphony** [ˈsimfəni]. Senfoni.
**symptom** [ˈsimptəm]. Hastalık alâmeti; emare. **~atic** [–ˈmatik], emaresi olan; delâlet eden.

**synagogue** [ˈsinəgog]. Havra.
**synchronize** [ˈsinkrǫnâyz]. Aynı zamana tesadüf et(tir)mek.
**syndicate** [ˈsindikit]. Ticarî firmalar birliği.
**synod** [ˈsinod]. Ruhanî meclis.
**synonym** [ˈsinənim]. Müteradif. **~ous** [–ˈnomiməs], **~ (with)**, (ile) müteradif.
**synopsis** [siˈnopsis]. Hulâsa.
**syntax** [ˈsintaks]. Nahiv.
**synthe·sis** [ˈsinθesis]. Terkib. **~tize**, terkib etmek. **~tic** [–ˈθetik], terkibî; sunî.
**Syria** [ˈsirį̈ə]. Suriye. **~n**, suriyeli. **~c**, Süryani dili.
**syrup** [ˈsirəp]. Şurub: **golden ~**, şeker pekmezi. **~y**, şurublu.
**system** [ˈsistim]. Usul, kaide; metod; manzume; (jeoloji) devir. **the ~**, vücud: **the Solar ~**, güneş manzumesi: **to lack ~**, sistemsiz olmak. **~atic** [–ˈmatik], sistemli. **~atically**, muntazaman. **~atize** [–tâyz], sistemleştirmek.

# T

**T** [tî]. T harfi; T şeklinde: **to a T**, tıpkısı; tamamiyle.
**ta** [tâ]. (*çocuk dilinde*) Teşekkür ederim.
**taal** [tâl]. Cenubî Afrika felemenkçesi.
**tab** [tab]. Flâpa, uc, dil: **a red ~**, erkânı harb subayı.
**tabby** [ˈtabi]. **~ (-cat)**, siyah çizgili tekir kedi; acuze.
**tabernacle** [ˈtabənakl]. Kilise; mukaddes ve mubarek bir şeyin muhafaza edildiği yer.
**table** [ˈtêybl]. Masa; sofra; cedvel. Masaya koymak; listeye geçirmek. **at ~**, sofra başında: **to ~ a bill**, bir lâyihayı Parlamentoya takdim etm.; (*Amer.*) bir lâyihayı tehir etm.: **to clear the ~**, sofrayı kaldırmak: **to keep the ~ amused**, sofrada herkesi eğlendirmek: **he keeps a good ~**, sofrası zengindir: **to lay [set] the ~**, sofra kurmak: **to sit down to ~**, sofraya oturmak: **to turn the ~s on s.o.**, vaziyeti tamamen birinin aleyhine çevirmek. **~d'hote** [tâblˈdǫut], tabldot. **~land**, yayla; yüksek ova. **~-linen**, sofra örtüsü, vs. **~-spoon**, büyük kaşık. **~-talk**, sofra sohbeti. **~-turning**, ispiritizmede masayı oynatma. **~-water**, maden suyu.
**tabl·et** [ˈtablit]. Levha; kitabe; komprime; küçük kalıb (sabun). **~oid** [ˈtablôyd], komprime.
**taboo** [taˈbû]. Tabu (yapmak); yasak (etm.); kullanılmasını menetmek. **the subject is ~ here**, bu mevzu burada konuşulmaz.
**tabul·ar** [ˈtabyulə(r)]. Masa gibi düz; cedvel halinde tertib edilmiş. **~ate** [–êyt], cedvele geçirmek; tasnif etmek.
**tacit** [ˈtasit]. Sâkit; zımnî.
**taciturn** [ˈtasitö(r)n]. Az konuşur; sükûtî.
**tack**[1] [tak]. İri başlı küçük çivi; teyel. Teyellemek. **to ~ down**, bu çivilerle çivilemek: **to get down to brass ~s**, asıl mevzua gelmek: **to ~ oneself on to s.o.**, birine takılmak. **~**[2]. Tiramola etme(k). **to be on the right ~**, doğru yolda olm.: **to try another ~**, başka bir tedbire başvurmak: **to be [sail] on the starboard [port] ~**, kontralar sancaktan [iskeleden] seyretmek. **~**[3]. Gıda. **hard ~**, katı bisküvi.
**tackle**[1] [ˈtakl]. Takım, cihaz; palanga. **~**[2]. (Futbol) topu karşısındakinin ayağından alma(k). Uğraşmak; yakalamak; girişmek.
**tact** [takt]. (Muaşerette) muamele ve usul. **without ~**, patavatsız. **~ful**, muamele bilir: **to handle s.o. ~ly**, birini idare etm.; ⌜nabzına göre şerbet vermek⌝. **~less**, zarafetsiz, patavatsız.
**tactic·al** [ˈtaktikl]. Tâbiyeye aid. **~s** [ˈtaktiks], tâbiye; hattı hareket.
**tadpole** [ˈtadpǫul]. Ayaksız kurbağa yavrusu.
**taffeta** [ˈtafitə]. Tafta; canfes.
**tag** [tag]. Küçük etiket; kundura kulağı; bir şeyin sarkan ucu; meşhur bir mesel.
**tail** [têyl]. Kuyruk; arka; son. Kuyruğu ile tutmak; (yemiş) sapını ayıklamak. **to ~ behind s.o.**, birinin peşinden gitmek: **to ~ off**, gittikçe azalmak: **to turn ~**, gerisin geriye kaçmak: **with his ~ between his legs**, kuyruğunu kısmış (köpek); (insan) süklüm püklüm: ⌜**the sting is in the ~**⌝, (eşek arısının) iğnesi kuyruğundadır; *bir mektub vs.nin dokunakli kısmı en sonunda olduğu zaman söylenir*: **to wear ~s**, frak giymek. **~-coat**, frak. **~-end**, en son kısım: **to come in at the ~**, sonuna doğru

gelmek. ~**-lamp,** arka lâmbası. ~**-plane,** uçağın kuyruk sathı.

**tailor** [ˈtêylə(r)]. Terzi. Elbiseyi dikmek.

**taint** [têynt]. Leke(lemek); fena koku (vermek). **free from ~,** kusursuz; taptaze. **~ed,** kokmuş; lekelenmiş.

**take** (**took, taken**) [têyk, tuk, têykn]. Almak; kapmak; kabul etm.; zabtetmek; götürmek; saymak; içine almak; kazanmak; icab etmek. Tutulan (balık vs.). **you must ~ us as you find us,** bizi olduğumuz gibi kabul etmelisiniz: **it ~s a strong man to do that,** bunu ancak kuvvetli bir adam yapabilir: **to ~ s.o. for another,** birini başka birine benzetmek: **to ~ s.o. for a fool,** birini abdal yerine koymak: **what do you ~ me for?,** beni ne zannettiniz?: **you can ~ it from me,** inan olsun: **to ~ it into one's head to do stg.,** bir şey aklına esmek: **this journey ~s two hours,** bu yol iki saat sürer: **it won't ~ long,** uzun sürmez: **to be ~n ill,** hastalanmak: **~ it or leave it!,** ister beğen ister beğenme!: **to ~ a matter seriously,** bir işi ciddiye almak: **to ~ s.o. over a house,** birine bir evi gezdirmek: **I ~ it that ...,** ... farzediyorum: **the vaccination did not ~,** aşı tutmadı: **what took him there?,** ne diye oraya gitti?: **he does not ~ well (photo),** (*fot.*) onun resmi hiç iyi çıkmaz: **to ~ things as they come,** vaziyeti olduğu gibi kabul etm.: **to be ~n with an idea,** bir fikirden hoşlanmak: **I was not ~n with him,** o beni sarmadı. **~ about,** gezdirmek. **~ after,** benzemek: **he ~s after his father,** babasına çekmiş. **~ away,** götürmek; kaldırmak; çıkarmak. **~ back,** geri almak; geldiği yere geri götürmek. **~ down,** indirmek; aşağıya götürmek; (makineyi) sökmek: **to ~ down in writing,** yazmak. **~ in,** içeriye götürmek; içine almak; kavramak; aldatmak; yutmak; dahil etm.; barındırmak; (gazeteye) abone olm.: **to ~ in at a glance,** bir bakışta görüp kavramak: **to ~ in lodgers,** kiracı almak: **to be ~n in,** aldanmak. **~-in,** faka basma. **~ off,** kaldırmak; götürmek; taklid etm.; hareked etm.: **to ~ s.o.'s attention off stg.,** birinin dikkatini bir şeyden çelmek: **to ~ off one's clothes,** soyunmak: **to ~ oneself off,** çekilmek: **to ~ so much off the price,** fiattan şu kadar indirmek: **I'll ~ a morning off today,** bugün öğleden evvel çalışmıyacağım: **to ~ s.o. off,** birinin taklidini yaparak alay etmek. **~-off,** ilk sıçrayış; havalanma. **~ on,** deruhde etm.; hizmetini almak; (*coll.*) müteessir olm.: **(train) to ~ on passengers,** (tren) yolcu almak: **to ~ s.o. on at tennis,** *etc.*, bir oyunda birinin meydan okumasını kabul etm.: **don't ~ on so!,** kızma! **~ out,** çıkarmak: **to ~ s.o. out to dinner,** birini yemeğe lokantaya götürmek: **to ~ out an insurance policy,** bir şeyi sigorta ettirmek: **this work ~s it out of one,** bu iş pek yorucudur: **I'll ~ it out of him!,** ben ona gösteririm! **~ over,** devralmak; tesellüm etm.: **to ~ over the liabilities,** borcları kendi üzerine almak: **to ~ s.o. over a house,** *etc.*, birine bir ev vs.yi gezdirmek, göstermek. **~ round,** gezdirmek. **~ to,** **to ~ to s.o.,** birini gözü tutmak: **to ~ to one's bed,** hasta yatmak: **to ~ to drink,** içki ibtilâsına düşmek: **to ~ to writing poetry,** şiir yazmağa başlamak. **~ up,** kaldırmak; tevkif etm.; başlamak; kısaltmak; massetmek: **to ~ up all one's attention,** tamamen meşgul etm.: **to be ~n up with stg.,** bir şeyle bozmak: **to ~ up one's duties again,** vazifesine tekrar başlamak: **to ~ up an idea,** bir fikri kabul edip tatbik etm.: **to ~ a matter up,** bir işi ele almak: **to ~ up one's pen,** kalemini almak: **to ~ up a profession,** bir mesleğe intisab etm.: **to ~ up a lot of room,** çok yer kaplamak: **to ~ s.o. up shortly,** birini terslemek; **chess ~s up a lot of time,** satranc çok vakit alır: **to ~ up with s.o.,** birisile düşüp kalkmak. **~ upon, to ~ upon oneself,** deruhde etm. **~r,** alıcı; bahse girişen kimse; kabul eden kimse.

**taking** [ˈtêykin(g)]. Can alıcı. **~s,** kazanc.

**tale** [têyl]. Masal, hikâye; aded. **to tell ~s (out of school),** koğuculuk etm.: **let me tell my own ~,** bir de ben anlatayım: **that tells its own ~,** bu kâfidir: **don't tell such ~s to me!,** (i) atma Receb!; (ii) yalan istemem. **~bearer,** gammaz.

**talent** [ˈtalənt]. İstidad; hüner; kabiliyet; dâdihak. **he has a ~ for languages,** lisana kabiliyeti var. **~ed,** hünerli.

**talisman** [ˈtalizman]. Tılsım.
**talk** [tôk]. Konuşma(k), sohbet; söz; dedikodu. **to ~ big,** (yüksekten) atıp tutmak; dem vurmak: **to ~ of doing stg.,** bir şey yapacağından bahsetmek: **to get oneself ~ed about,** kendini dile düşürmek: **to ~ oneself hoarse,** sesi kısılıncaya kadar konuşmak: **to ~ s.o. into doing stg.,** dil dökerek birini kandırmak: **I know what I am ~ing about,** bu iş hakkında bilerek konuşuyorum: **he knows what he is ~ing about,** o bu işin ehlidir: **now you're ~ing!,** (*sl.*), ha şöyle!: **to ~ (severely) to s.o. [to give s.o. a good ~ing to],** birini azarlamak: **~ing of that ...,** bu münasebetle ...: **who do you think you are ~ing to?,** karşındakinin kim olduğunu zannediyorsun? **~ative** [ˈtôkətiv], geveze; çeneli. **~er,** konuşkan: **a great ~,** çenesi düşük: **a good ~,** hoşsohbet. **~ing-to, to give s.o. a good ~,** birini tekdir etm. **~ at,** birine taş atmak (ʳkızım sana söyliyorum gelinim sen anla!ʾ). **~ away,** durmadan söylemek: **to ~ away the time,** konuşarak vakit geçirmek. **~ down, to ~ down to one's audience,** dinleyicilerin seviyesine hitab etm.: **to ~ s.o. down,** birini susturuncaya kadar konuşmak. **~ over,** (mesele) görüşmek; (kimse) dil dökerek ikna etmek. **~ round,** (mesele) etrafında dönüp dolaşmak.
**tall** [tôl]. Uzun boylu; yüksek. **that's a ~ order!,** (*sl.*) artık bu kadarı da fazla!; **that's a ~ story!,** buna kim inanır?
**tallow** [ˈtalou]. Donyağı.
**tally** [ˈtali]. Çetele (hesabı); etiket. Çetele tutmak. Mutabık olm. **these accounts do not ~ (with each other),** bu hesablar birbirini tutmuyor.
**tally-ho** [ˈtaliˈhou]. *Avcıların avı görünce bağırdıkları kelime.*
**talon** [ˈtalən]. Yırtıcı kuş pençesi.
**tambour** [ˈtambur]. Davul. **~ine** [–ˈrîn], tef.
**tame** [têym]. Ehlî; uysal; yavan. Ehlileştirmek; uslandırmak. **to submit ~ly,** mukavemet etmeden boyun eğmek.
**tam-o'shanter** [ˈtamiˈşantə(r)], **tammy** [ˈtami]. İskoçya beresi.
**tamper** [ˈtampə(r)]. **to ~ with,** (makine vs.yi) kurcalamak; (defter vs.yi) tahrif etm.; (şahidi) rüşvetle kandırmak.
**tan** [tan]. Güneş yanığı. Koyu sarı renkli. (Rüzgâr, güneş) yakmak; sepilemek; (*sl.*) dayak atmak. **~ned,** güneş, rüzgârdan yanmış (deri).
**tandem** [ˈtandəm]. İki kişilik bisiklet.
**tang** [tan(g)]. Kuvvetli çeşni veya koku.
**tangent** [ˈtancənt]. Mümas. **to go [fly] off at a ~,** bir fikir silsilesinden birden bire başkasına geçmek.
**tangerine** [ˌtancəˈrîn]. Mandalina.
**tangible** [ˈtancibl]. Elle tutulur; mevcud; hakikî.
**Tangier(s)** [tanˈciəz]. Tanca.
**tangle** [ˈtan(g)gl]. Karışıklık; arapsaçı. Karıştırmak.
**tank** [tan(g)k]. Sarnıc; su haznesi; benzin deposu; tank. **~er,** sarnıc gemisi.
**tankard** [ˈtan(g)kö(r)d]. Büyük bira bardağı.
**tanner**[1] [ˈtanə(r)]. Debbağ, tabak. **~y,** tabakhane. **~**[2]. (*sl.*) Altı peni.
**tantalize** [ˈtantəlâyz]. Bir şeyi verir gibi uzatıp geri çekmek suretiyle kızdırmak.
**tantamount** [ˈtantəmaunt]. **to be ~ to,** ···e müsavi olm.; sayılmak.
**tantrum** [ˈtantrəm]. Öfke nöbeti. **to get into a ~,** hiddetten ter ter tepinmek.
**tap**[1] [tap]. Musluk; fıçı tapası. Fıçı delip içindekini akıtmak. **to ~ a new country,** yeni bir memleketi ticarete açmak: **to ~ a lung,** akciğeri delmek: **to ~ a telegraph wire,** telgraf telinden muhabereyi kapmak: **to ~ a tree,** bir ağacdan recine vs.yi çıkarmak: **beer,** *etc.*, **on ~,** (bira vs.) fıçıdan: **to ~ s.o. for a fiver,** (*sl.*) birisinden beş lira sızdırmak. **~-room,** otel ve handa içki odası. **~-water,** musluk suyu. **~**[2]. (*ech.*) Hafif darbe, hafifçe vurmak. **to ~ at the door,** kapıyı hafifçe çalmak.
**tape** [têyp]. (Keten) şerid; kurdelâ. Şerid ile bağlamak; mesaha şeridi ile ölçmek. **insulating ~,** tecrid şeridi: **red ~,** kırmızı kurdelâ; kırtasiyecilik.
**taper**[1] [ˈtêypə(r)]. Mahrutilik. Bir nihayetine doğru incel(t)mak. **to ~ away,** sivrileşmek. **~ed, ~ing,** gittikçe incelip sivrileşen. **~.**[2] Şamalı fitil.
**tapestry** [ˈtapestri]. Duvara örtülü kaneviçe ve gergef işi.
**tar** [tâ(r)]. Katran; (*coll.*) bahriye neferi. Katranlamak. ʳ**they are both**

~red with the same brush⌝, ikisinin de kusuru aynı; 'al birini vur ötekini': to ~ and feather s.o., birini, ceza olarak, önce katrana sonra tüye bulamak: ⌜to spoil the ship for a ha'porth of ~⌝, az bir masraftan kaçınıp büyük bir zarara girmek. ~-brush, to have a touch of the ~, damarlarında bir az zenci kanı bulunmak. ~ry, katranlı.

**tarboosh** [ˈtâbûş]. Fes.

**tardy** [ˈtâdi]. Ağır, yavaş; gecikmiş.

**tare** [tẹə(r)]. Dara. to allow for ~, darasını çıkarmak.

**target** [ˈtâgit]. Hedef; atış nişanı; amac.

**tariff** [ˈtarif]. Tarife (fiat cedveli).

**tarmac** [ˈtâmak]. Asfalt(lamak).

**tarn** [tân]. Küçük dağ gölü.

**tarnish** [ˈtâniş]. Donukluk; leke. Donuklaş(tır)mak; lekele(n)mek.

**tarpaulin** [tâˈpôlin]. Katranlı muşamba.

**tarry** [ˈtari]. Geride kalmak; beklemek.

**tart**[1] [tât] *a.* Ekşi; keskin. ~ness, ekşilik. ~[2] *n.* Meyvalı börek; reçelli pasta; (*sl.*) sokak kadını.

**tartan** [ˈtâtən]. Kareli İskoçya yünlüsü.

**Tartar** [ˈtâtâ(r)]. Tatar; şırret. to catch a ~, zorlu adama çatmak. ~y, orta Asya'nın eski ismi.

**task** [tâsk]. Götürü iş; vazife. Çalıştırmak; yormak. to take s.o. to ~ for stg., birini bir şeyden dolayı tekdir etmek. ~master, iş veren kimse: a hard ~, çok çalıştıran patron.

**tassel** [ˈtasl]. Püskül. ~led, püsküllü.

**tast·e** [têyst]. Tat; lezzet; çeşni; azıcık parça yemek; zevk; meyil. Tatmak. Tadı olmak. to ~ of stg., tadı bir şeye benzemek: bad ~, zevksizlik: to find stg. to one's ~, bir şeyi zevkine uygun bulmak: to have a ~ for stg., bir şeyden zevk almak: to have no ~ for music, *etc.*, musiki vs.den zevk almamak; anlamamak: ⌜everyone to his ~⌝, bu zevk meselesidir: I've ~d nothing for two days, iki gündenberi ağzıma bir lokma koymadım: to take just a ~ of stg., bir lokmacık yemek. ~eful, lezzetli; zarif; zevkli. ~eless, lezzetsiz; yavan; zevksiz. ~er, çeşnici. ~y, lezzetli; tatlı. ~iness, lezzet.

**ta-ta** [ˈtaˈtâ]. (*çocuk dilinde*) Allaha ısmarladık; 'attâ'.

**tatter** [ˈtatə(r)]. Paçavra. in ~s, lime lime: to tear to ~s, parça parça etmek. ~demalion [-dəˈmalyʌn], çapaçul. ~ed, partal; lime lime.

**tattle** [ˈtatl]. Dedikodu (yapmak); çene çalmak.

**tattoo**[1] [taˈtû]. Askerin koğuş trampetesi. ~[2]. Ten üzerine dövme (yapmak).

**taught** *v.* teach.

**taunt** [tônt]. Yüzüne vurma; hakaret; istihza. İğnelemek. to ~ with, ···i yüzüne vurmak; ···i başına kakmak.

**taut** [tôt]. Gergin, gerili, sıkı. to pull ~, kasa etmek. ~en, germek için çekmek.

**tavern** [ˈtavö(r)n]. Meyhane.

**tawdr·y** [ˈtôdri]. Bayağı; mezad malı; zevksiz. ~iness, bayağılık.

**tawny** [ˈtôni]. Sarımtrak kahve renginde.

**tax** [taks]. Vergi, resim. Vergi tarhetmek. direct ~es, vasıtasız vergiler: indirect ~es, vasıtalı vergiler: to be a ~ on s.o., birine yük olm.: to ~ the patience [courage] of s.o., birinden çok sabır [cesaret] istemek: to ~ s.o. with doing stg., birini bir şeyle ittiham etmek. ~able, vergiye tâbi tutulabilir; mükellef. ~ation [takˈsêyşn], vergi tarhı.

**taxi(cab)** [ˈtaksi(kab)]. Kira otomobili, taksi (ile gitmek); (*otom.*) makineyi durdurarak ilerlemek; (uçak) yerde yürümek. ~-driver, taksi şoförü.

**tea** [tî]. Çay. ~-caddy, çay kutusu. ~-cake, bir nevi çörek. ~cup, çay fincanı. ~-garden, (i) çay çiftliği; (ii) çay kahve ve hafif yemekler satılan bahçe. ~pot, çaydanlık. ~-service [~-things], çay takımı.

**teach** (taught) [tîç, tôt]. Öğretmek; tahsil ettirmek; okutmak. Muallimlik etm., ders vermek. that will ~ him (a lesson)!, o ona Hanya'yı Konya'yı gösterir: to ~ s.o. a thing or two, birinin gözünü açmak: I'll ~ you to speak to me like that!, bana böyle konuşmayı sana gösteririm. ~er, muallim, hoca. ~ing, muallimlik; ders verme; ders.

**teak** [tîk]. Hind meşesi, tik ağacı.

**team** [tîm]. Birlikte koşulmuş iki veya daha çok at vs.; takım; ekip. to ~ up with s.o., birisile birleşmek; beraber çalışmak: ~ games, futbol gibi takım ile oynanan oyunlar: the ~ spirit, müşterek ve ekip halinde çalışma ruhu: ~ **work,** takım halinde çalışma.

**tear**[1] [tiə(r)]. Göz yaşı. **to burst into ~s,** gözlerinden yaş boşanmak: **to move s.o. to ~s,** gözünü yaşartmak: **to shed ~s,** gözyaşı dökmek. **~ful** [ˈtiəfl], gözleri yaşlı; ağlamalı (ses). ~[2] (tore, torn) [teə(r), tô(r), tôn]. Yırt(ıl)mak, yar(ıl)mak; pek hızlı gitmek. Yırtık, yarık, rahne. **to be torn by conflicting emotions,** birbirine zıd hislerle kıvranmak: **the country was torn by faction,** memleket baştan başa tefrika içinde idi: **to ~ one's hair,** saçını başını yolmak: **to ~ stg. open,** bir şeyi yırtıp açmak: **to ~ stg. from s.o.,** birisinden bir şeyi zorla kapmak. ~ **away,** koparmak; çabuk ayrılmak: **I could not ~ myself away from this lovely spot,** bu güzel yerden bir türlü ayrılmak istemedim. ~ **down,** yırtıp koparmak; (*coll.*) alabildiğine aşağıya doğru koşmak. ~ **off,** yırtıp koparmak; koşarak gitmek. ~ **out,** yırtıp çıkarmak, sökmek, dışarıya fırlamak. ~ **up,** parça parça etm.; kökünden koparmak; yukarıya fırlamak. **~ing, to be in a ~ hurry,** (telâştan) aceleden etrafını görememek: ~ **rage,** kudurmuş bir hiddet: ~ **wind,** insanı uçuracak gibi rüzgâr.

**tease** [tîz]. Taciz etm.; takılmak; ditmek. **~r,** taciz eden kimse; (*coll.*) çetin bir sual; zor bir mesele.

**teat** [tît]. Meme başı; emcik.

**techn·ical** [ˈteknikl]. Fennî, ilmî; teknik; usule göre. **a ~ offence,** (hakikatte mühim değil) yalnız kanun nazarında suç: ~ **school,** sanat mektebi: **judgement quashed on a ~ point,** şekle aid bir hata yüzünden nakzedilen karar. **~icality** [–ˈkaliti], teknik teferrüat. **~ician** [–ˈnişn], teknisyen. **~ique** [–ˈnîk], bir şeyi yapma usulü. **~ology** [–ˈnoləci], sanayi bilgisi.

**Teddy** [ˈtedi]. (*abb.*) Edward. ~ **bear,** pelüşten yapılmış oyuncak ayı. ~ **boy,** külhanbeyi.

**tedi·ous** [ˈtîdyəs]. Cansıkıcı; bıktırıcı. **~um,** [–əm], can sıkıntısı.

**tee** [tî]. T harfi (şeklinde olan şey).

**teem** [ˈtîm]. Kaynamak; bol olmak. **the river is ~ing with fish,** nehirde balık kum gibi kaynıyor.

**teens** [tînz]. (*coll.*) On üçten on dokuza kadar yaş. **to be out of one's ~,** yaşı on dokuzdan fazla olmak.

**teeny** [ˈtîni]. ~ **(-weeny),** mini mini.

**teeth** *v.* tooth. **~e** [ˈtîð], (çocuk) diş çıkarmak.

**teetotal** [tîˈtoutl]. İçki içmiyen; alkolsuz (içki). **~ism,** alkollu içkilere tövbe etme. **~ler,** alkollu içkilere tövbe eden kimse.

**teetotum** [tîˈtoutəm]. Parmaklarla çevirilen topaç.

**tele-** [ˈteli] *pref.* Uzaktan veya uzağa işliyen.

**telegr·am** [ˈteligram]. Telgraf (name). **~aph,** telgraf (çelmek); telgrafla bildirmek. **~aphic** [–ˈgrafik], telgrafa aid. **~aphist** [–ˈlegrəfist], telgrafçı. **~aphy** [–ˈlegrəfi], telgrafçılık.

**telepathy** [teˈlepəθi]. Telepati.

**telephone** [ˈteləfoun]. Telefon(la görüşmek, bildirmek). **~-box,** telefon kulübesi.

**teleprinter** [ˈteliprintə(r)]. Telgrafla gönderilen haberi tabeden makine.

**telescop·e** [ˈteliskoup]. Teleskop; dürbün; teleskop kısımları gibi iç içe geçmek. **~ic** [ˈskopik], teleskopa aid; teleskop gibi iç içe geçen.

**televis·ion** [ˌteliˈvijn]. Televizyon. **~e,** [–vâyz], televizyonla göstermek.

**tell** (told) [tel, tould]. Anlatmak, nakletmek, söylemek; bildirmek, haber vermek; ifşa etm.; farketmek; saymak. Tesir etmek. ~ **me another!,** (*sl.*) külâhıma anlat!: **'blood will ~',** kan (asalet) belli olur: **his modesty ~s in his favour,** tevazuu onun lehinedir: **I have heard ~ that ...,** kulağıma çalındığına göre ...; **how do you ~ which button to press?,** hangi düğmeye basılacağını nasıl biliyorsunuz?: **I never can ~ those two apart,** bunların ikisini hiç birbirinden ayırd edemem: **he looks honest but you never can ~,** namuslu görünüyor fakat belli olmaz: **his age is beginning to ~ on him,** yaş onun üzerinde tesirini gösteriyor: **I told you so!,** ben sana demedim mi? **~-tale,** koğucu: ~ **signs,** haber veren işaretler: **a ~ blush,** kabahat vs.ye delâlet eden yüz kızarması. ~ **off,** (bir adama) muayyen bir iş vermek; (*coll.*) azarlamak. **~er** [ˈtelə(r)], anlatıcı vs.; Avam Kamarasında reyleri saymağa memur dört kişiden biri; banka vs. veznedar. **~ing** [ˈtelin(g)], anlatış, nakil; ifşa etme. Müessir. **there's no ~ what he will do,** onun ne yapacağı bilinmez.

**temerity** [tə'meriti]. Cüret; yüzsüzlük. **he had the ~ to say ...,** ···i söylemek küstahlığını gösterdi.

**temper**[1] ['tempə(r)]. Huy, tabiat; huysuzluk, öfke. **to be in a ~,** hiddetli olm.: **to be out of ~,** öfkesi üstünde olm.: **to have a bad ~,** huysuz olm.: **to have a good ~,** iyi huylu olm.: **to keep one's ~,** öfkesini tutmak: **to lose one's ~,** hiddetlenmek: **to show ~,** hiddetini belli etm. **~**[2]. Çeliğe verilen su; kıvam. (Çeliğe) su vermek; tadil etm.; hafifletmek. **~ament** ['temprəmənt], mizac; tabiat; hilkat; huy. **of even [equable] ~,** temkinli. **~amental** [-'mentl], mizaca aid; sebatsız; çabuk kızan: **~ly,** mizac itibarile.

**temper·ance** ['tempərəns]. İtidal; imsak; az içki kullanma. **~ hotel,** içkisiz otel: **~ society,** içki aleyhdarları cemiyeti. **~ate,** mutedil; mümsik; az içki kullanan. **~ature** ['temprətyə(r)], hararet derecesi; sühunet; ateş. **to run a ~,** ateşi olm.: **to take s.o.'s ~,** derecesini almak.

**tempest** ['tempist]. Şiddetli fırtına. **~uous** [-'pestyuəs], pek fırtınalı; gürültülü.

**Templar** ['templa(r)]. Haçlılardan 'Temple' tarikatine mensub şövalye.

**temple**[1] ['templ]. Mabed; ibadethane; *esp.* mabedi Süleyman. **~**[2]. Şakak.

**temporal** ['temporəl]. Fani, geçici; dünyevî (ruhanî mukabili). **the ~ power,** cismanî kuvvet.

**tempor·ary** ['tempərəri]. Muvakkat; geçici; iğreti. **~ize** [-âyz], Vakit kazanmağa çalışmak.

**tempt** [tempt]. İğva etm.; günaha teşvik etm.; imrendirmek. **I am ~ed to ...,** şeytan diyor ki ...: **I am strongly ~ed to accept,** kabul etmeğe kuvvetle mütemayilim: **to ~ God [Providence],** kudretinin fevkinde bir teşebbüse girişmek: **it is ~ing Providence to go out in that old boat,** o eski kayıkla çıkmak tehlikeye [ölüme] susamaktır. **~ation** [tem'têyşn], iğva; günaha teşvik; şeytanın iğvası. **it's a great ~ to ...,** şeytan diyor ki ...: **to yield to ~,** iğvaya kapılmak. **~er,** baştan çıkarıcı; şeytan.

**ten** [ten]. On. **to count in ~s,** onar onar saymak: **~ to one he'll forget,** yüzde yüz unutur. **~th,** onuncu.

**tenable** ['tenəbl]. Tutulabilir; muhafazası mümkün; kabul edilebilir.

**tenac·ious** [te'nêysəs]. Bırakmaz; inadçı; yapışkan. **~ity** [-'nasiti], azimlilik; inad; yapışkanlık.

**tenan·cy** ['tenənsi]. Kiracılık; kira müddeti. **a life ~,** kaydı hayat şartiyle. **~t,** kiracı; müstecir: **life ~,** kaydı hayat şartiyle kiracı. **~try,** bir malikânenin bütün ev ve çiftliklerinin kiracıları.

**tend**[1] [tend]. Bakmak; gözetmek. **~**[2]. Meyletmek; yüz tutmak; kaçmak. **blue ~ing to green,** yeşile çalan mavi. **~ency** ['tendənsi], meyil; yüz tutma; istidad. **~entious** [ten'denşəs], bir maksada müstenid.

**tender**[1] ['tendə(r)]. *a.* Nahif; zayıf; hassas; müşfik; taze. **of ~ years,** pek genc: **to touch s.o. on his ~ spot,** yarasına dokunmak. **~-hearted,** müşfik. **~foot,** acemi. **~ness,** şefkat; hassaslık; narinlik. **~**[2] *n.* Bakıcı; lokomotifin arkasına bağlı su ve kömür vagonu. **~**[3] *v.* Arzetmek; teklif mektubu vermek; sunmak. Teklif (mektubu). **by ~,** eksiltme usuliyle: **to ~ for stg.,** eksiltmede teklif mektubu vermek: **to put to ~,** münakaşaya koymak: **sealed ~,** kapalı zarf usuliyle eksiltme: **legal ~,** bir alacaklının kabul etmeğe mecbur olduğu memleket parasının nevileri: **to ~ thanks,** teşekkürlerini sunmak.

**tendon** ['tendon]. Veter; sinir.

**tenement** ['tenəmənt]. Kira evi; bir aile tarafından işgal edilen apartman (*um. sade işçi evlerine denir*); (*jur.*) mülk.

**tenet** ['tenit]. Akide; katî fikir.

**tennis** ['tenis]. Tenis. **~-court,** tenis sahası. **~-elbow,** fazla tenis oynamaktan dirsek sinirlerinin burkulması.

**tenor**[1] ['tenə(r)]. Meal; mâna; temayül. **~**[2]. Tenor sesi.

**tense**[1] [tens] *n.* Fiil sıygası. **~**[2] *a.* Gergin; meraklı, heyecan verici.

**tens·ile** ['tensâyl]. Gerilip uzamağa kabiliyetli. **~ strength,** gerilme kuvveti. **~ion** ['tenşən], gerginlik; gerilme kuvveti.

**tent** [tent]. Çadır. **~-peg,** çadır kazığı.

**tentacle** ['tentəkl]. İnce uzun bir uzuv.

**tentative** ['tentətiv]. Deneme; tecrübe.

**tenterhook** ['tentəhuk]. **to be on ~s,** son derece şübhe ve merak içinde olmak.

**tenu·ity** [te'nyûiti]. İncelik; seyreklik. **~ous** ['tenyuəs], ince; seyrek; zayıf.

**tenure** [ˈtenyuə(r)]. Tasarruf şartı; gedik; memuriyet müddeti.
**tepid** [ˈtepid]. Ilık; hararetsiz.
**term**[1] [tö(r)m] *v.* İsim vermek; demek; tabir etmek. ~[2] *n.* Müddet, vade, devre; trimestr; mahkemelerin ictima devresi; had, son; tabir, terim. **~s,** şartlar; münasebetler; meal. **to put [set] a ~ to stg.,** bir şeye had tayin etm.: **long [short] -~,** uzun [kısa] vadeli: **they are on bad [good] ~s,** araları bozuk [iyi]: **to be on the best of ~s with s.o.,** birisile arası fevkalâde olm.: **not to be on speaking ~s,** araları bozuk olduğu için birbirile konuşmamak: **not on any ~s,** hiç bir şekilde: **to come to ~s [to make ~s],** şartlar üzerinde uyuşmak: **make [name] your own ~s,** şartlarınızı kendiniz tayin ediniz: **his ~s are two pounds a lesson,** ders başına iki lira ücret istiyor: **the ~s of his letter,** mektubunun meali: **by the ~s of the treaty,** muahede mucibince: **one cannot reckon happiness in ~s of worldly success,** saadet maddî muvaffakiyetle ölçülemez.
**termagant** [ˈtö(r)məgənt]. Cadoloz, acuze.
**termin·able** [ˈtö(r)minəbl]. Bitirilebilir; muayyen müddeti olan. **~al,** ucda bulunan; ucunu teşkil eden; devreye aid; son durak, istasyon; (*elect.*) bağlama vidası. **~ate** [–êyt], son vermek, bitirmek. Sona ermek, bitmek. **~ation** [–ˈnêyşn], son, netice; inkıza; son ek. **~ology** [ˌtö(r)miˈnoləci], bir ilim vs.'nin ıstılahları. **~us** [ˈtö(r)minʌs], (tramvay vs.) son durak veya istasyon.
**terrace** [ˈteris]. Tabiî veya sunî yüksek düz yer; taraça; bir sed üzerinde bir sıra evler. Bir yokuşta bir sıra sahanlık yapmak.
**terra-cotta** [ˈteraˈkota]. Koyu turunc renk.
**terra firma** [ˈteraˈfö(r)ma]. Arzın kara kısmı.
**terrestrial** [təˈrestriəl]. Yeryüzüne aid; arzî.
**terribl·e** [ˈteribl]. Müdhiş, korkunc, dehşetli. **~y,** (*coll.*) ziyadesile.
**terrier** [ˈteriə(r)]. Ufak av köpeği.
**terrif·ic** [təˈrifik]. Korkunc, müdhiş; son derece. **~ pace,** başdöndürücü sürat. **~y** [ˈterifây], tedhiş etm.: **to be ~ied,** dehşet duymak.
**territor·y** [ˈteritori]. Ülke; arazi; mıntaka. **~ial** [–ˈtôriəl], araziye aid; muayyen bir mıntakaya aid; İngiltere'de bir nevi gönüllü askeri teşkilatına mensub kimse: **~ integrity,** toprak bütünlüğü: **~ waters,** kara suları.
**terror** [ˈterə(r)]. Dehşet; gözyılgınlığı; dehşet verici şey. **~ism,** tedhiş siyaseti. **~ist,** tedhişçi.
**terse** [tö(r)s]. Veciz; keskin ve kısa.
**test** [test]. Tecrübe; muayene. Denemek, tecrübe etm., prova etm.; muayeneye tâbi tutmak; tahlil etm.; tasdik etmek. **~ case,** (i) emsal teşkil eden dava; (ii) tecrübe için yapılan şey: **intelligence ~,** zekâ testi: **driving ~,** şoförlük imtihanı: **~ run,** (*otom.*) tecrübe için kullanma. **~-tube,** kimyager tübü. **~er** [ˈtestə(r)], muayene memuru; prova eden kimse.
**testa·ment** [ˈtestəmənt]. Vasiyetname; ahid. **New ~,** ahdicedid: **Old ~,** ahdiatîk. **~te** [ˈtestêyt], muteber vasiyetname bırakarak ölen. **~tor** [–ˈtêytə(r)], vasiyet eden erkek. **~trix,** vasiyet eden kadın.
**testicle** [ˈtestikl]. Husye, taşak.
**testify** [ˈtestifây]. Şahadet etm.; teyid etm. **to ~ to a fact,** şahadetiyle bir fiili tasdık etm., teyidetmek.
**testimonial** [testiˈmouniəl]. Takdirname; teşekkür için verilen hediye.
**testimony** [ˈtestiməni]. Şahadet; delil.
**testy** [ˈtesti]. Çabuk öfkelenen; haşin.
**tête-à-tête** [ˈtêytaˈtêyt]. Başbaşa görüşme; iki kişilik kanape.
**tether** [ˈteðə(r)]. Otlatmakta olan hayvanı bağlamak için kullanılan ip. (Hayvanı) iple bir kazığa bağlamak. ⌜**to be at the end of one's ~**⌝, sabrı (tahammülü, imkânları) tükenmek.
**tetra-** [ˈtetra] *pref.* Dörtlü, dörtten mürekkeb ....
**Teuton, -ic** [ˈtyûtən, ˈtyûˈtonik]. Alman.
**text** [tekst]. Metin; İncilden kısa bir parça; mevzu. **to stick to one's ~,** sadedden ayrılmamak. **~-book,** (i) elkitabı; (ii) bir imtihan için tayin edilen kitab. **~ual** [ˈtekstyuəl], Metne aid.
**textile** [ˈtekstâyl]. Dokuma işi. **~s,** mensucat.
**texture** [ˈtekstyuə(r)]. Bir nescin yapısı; örgü.
**Thames** [temz]. Taymis nehri. ⌜**to set the ~ on fire**⌝, meşhur olmak.
**than** [ðan]. *Mukayese için kullanılan*

*edat.* ···den. **he is stronger ~ you,** o sizden daha kuvvetlidir: **more ~ once,** bir çok defa: **I would rather go by ship ~ fly,** tayyare ile gitmektense vapurla gitmeği tercih ederim: **he is no more an American ~ I am,** ne münasebet! o Amerikalı değildir.

**thank** [θan(g)k]. Şükretmek; teşekkür etmek. **~ you!,** teşekkür ederim: **~ God [heaven, goodness],** çok şükür, elhamdülillah!: **you have your friends to ~ for this,** bunu dostlarınıza borclusunuz: **you have only yourself to ~ for this,** kabahat sende!: **I'll ~ you to mind your own business,** siz kendi işinize baksanız daha iyi olur: **~ your lucky stars!,** talihine şükret! **~ful,** müteşekkir. **~less,** nankör (iş vs.). **~s,** teşekkür, şükür: **~!,** teşekkür ederim: **~ to ...,** ···in sayesinde: **that's all the ~ I get!,** işte bana böyle teşekkür ediyorlar!: **to give ~ to s.o. for stg.,** birine bir şey için teşekkür etmek. **~sgiving,** teşekkür; şükran duası: **~ Day,** (*Amer.*) Kasımın son perşembesinde kutlanan şükran yortusu. **~-offering,** bir nimete şükretmek için verilen şey.

**that**[1] [ðat], *pl.* **those** [ðouz]. O, şu; o şey. **~ book is mine,** o kitab benimdir: **English soil is more fertile than ~ of Germany,** Ingiltere'nin topragı Almanya'nınkinden daha mümbittir. **~**[2]. Ki. **he said ~ ...,** dedi ki ...; **he said ~ he would come,** geleceğini söyledi: **he was so ill ~ he could not speak,** konuşamıyacak kadar hasta idi: **oh ~ I could be in England now!,** şimdi İngiltere'de olabilsem! **~**[3] *rel. pron.* **who, when, which** *yerine kullanılır.* **the child ~ I saw,** gördüğüm çocuk: **the letter ~ I sent you,** size gönderdiğim mektub.

**thatch** [θaç]. Saman veya sazdan dam örtüsü (ile kaplamak).

**thaw** [θô]. Kar ve buzu eritmek; (don) erimek, çözülmek; (insan) açılmak. Karın erimesi. **the conversation began to ~ a little,** buzlar çözülmeğe başladı.

**the** [ðî, ðə]. *Muayyen harfi tarif.* **I saw a man,** bir adam gördüm, *fakat* **I saw ~ man,** adamı gördüm: **I went to a house,** bir eve gittim, *fakat* **I went to ~ house,** o eve gittim. *Yalnız bir tane olan şeylerin başına konur,* *mes.* **the sun,** güneş; **the moon,** ay; **the year 1950,** 1950 senesi.

**theatr·e** [ˈθiətə(r)]. Tiyatro; meydan, sahne; ameliyathane. **the ~ of war,** darülharb. **~ical** [θiˈatrikl], tiyatroya aid; dramatik; gösterişli. **~icals,** amatörler tarafından oynanan piyesler.

**thee** [ðî]. *acc. of* thou (*esk.*). Seni.

**theft** [θeft]. Hırsızlık, çalma. **petty ~,** aşırma.

**their** [ðeə(r)]. Onların: **~s,** onlarınki. **this house is ~s,** bu ev onlara aid: **this is a house of ~s,** bu onların evlerinden biridir.

**them** [ðem]. Onları: **of ~,** onların; **to ~,** onlara: **from ~,** onlardan. **~selves,** kendiler(i).

**theme** [θîm]. Mevzu; talebe için vazife.

**then** [ðen]. O zaman; ondan sonra; şu halde; hem de. **now and ~,** arasıra, vakit vakit: **~ and there,** hemen: **I haven't the time and ~ it is not my business,** vaktim yok, zaten vazifem de değil: **by ~,** o zaman(a kadar): **now one boy does best, ~ another,** kâh bir çocuk iyi yapar kâh başkası: **the ~ Vali,** o zamanki Vali: **suppose he refuses, what ~?,** ya reddederse ne olacak?: **you knew all the while ~,** demek şimdiye kadar bunu sen hep biliyordun.

**thence** [ðens]. Oradan; o sebebden dolayı. **~forth [~forward],** o zamandan beri.

**theocracy** [θîˈokrəsi]. Ruhanî idare veya hükümet.

**theolog·y** [θîˈoləci]. İlâhiyat. **~ian** [–ˈloucicən], ilâhiyat mütehassısı.

**theorem** [ˈθîorem]. Riyaziyede mesele.

**theor·y** [ˈθîəri]. Nazariye. **in ~,** nazariyat itibariyle. **~etical** [–ˈretikl], nazarî. **~ist,** nazariyatçı. **~ize,** nazariye kurmak.

**therapeutic** [θeraˈpyûtik]. Tedaviye aid.

**there** [ðeə(r)]. Ora(sı); orada; oraya; burada; haydi haydi; işte! **~ is,** var: **~ is not,** yok: **~ was,** vardı: **~ will be,** olacak: **he's all ~,** çok açıkgözdür: **he is not all ~,** bir tahtası eksik: **it is ten miles ~ and back,** oraya gidip gelme on mildir: **hurry up, ~!,** çabuk olalım!: **'where's your father?' '~ he is!'** 'Babanız nerede?' '(*göstererek*) İşte şurada!': **~ you have me!** (*sl.*) vallahi bilmem, buna cevab veremem: ~, ~!, never mind!, (*çocuğa*) haydi haydi, zarar yok, üzülme!: ~ you are!, (i) ben demedim mi?; (ii) demek geldin ha!; (iii) buyurun! ~**abouts** [ˈðeə(r)əbauts], oraya yakın; oralarda: **at two o'clock or ~**, saat iki raddelerinde. ~**after** [–ˈâftə(r)], ondan sonra. ~**at**, o zamanda; orada; ondan dolayı. ~**by** [–ˈbây], o münasebetle; o vechile. ~**fore** [ˈðeəfô(r)], binaenaleyh; bu sebebden; o halde. ~**from** [–ˈfrom], oradan. ~**in** [–ˈin], onun içinde; orada: **~ you are mistaken**, burada yanılıyorsunuz. ~**of** [–ˈov], ondan; onunki. ~**on** [–ˈon], bunun üzerine. ~**to** [–ˈtû], oraya; buna, ona. ~**-upon** [ˌðeərəˈpon], onun üzerinde; bunun üzerine. ~**with** [–ˈwið], onunla; hemen, derhal.

**therm** [θö(r)m]. İngiliz hararet vahidi. ~**al** [ˈθö(r)məl], hararete aid. **~ spring**, kaplıca. ~**ometer** [–ˈmomitə(r)], termometre; derece. ~**os** [ˈθö(r)mos], ~**(-flask)**, termos (şişesi). ~**ostat**, harareti kendiliğinden tanzim eden alet.

**thesaurus** [θîˈsôrʌs]. Büyük lûgat; hazine.

**these** [ðîz]. *pl. of* **this**. Bu, bunlar.

**thesis**, *pl.* **-ses** [ˈθîsis, -sîz]. Dava; tez.

**they** [ðêy]. Onlar. **~ say that ...**, diyorlar ki ..., denildiğine göre ...: **they'd = they had, they would; they'll = they will; they're = they are.**

**thick** [θik]. Kalın, kesif; sık; koyu; kalın kafalı; (hava) sisli; (ses) boğuk. **to be very ~ with s.o.**, birisile senli benli olm.: **they are as ~ as thieves**, aralarından su sızmaz: **that's a bit ~!**, (*sl.*) bu ne pişkinlik, bu kadarı da fazla!: **in the ~ of the fight**, muharebenin en civcivli zamanında: **to stick to s.o. through ~ and thin**, birine hem iyi hem fena günlerinde sadık kalmak. ~**-set**, kısa boylu ve tıknaz. ~**en** [ˈθikn], kalınlaş(tır)mak; koyulaş(tır)mak; kesif olm. (yapmak); koyultmak. **the plot ~s**, (roman vs.) vakalar karışıyor [çatallaşıyor]. ~**et** [ˈθikit], çalılık; koru. ~**ness** [ˈθiknis], kalınlık; kesafet; tabaka.

**thief** [θîf]. Hırsız. ⌜**honour among thieves**⌝, hırsızlar arasında bile bir namus telâkkisi vardir; merdlik merdliktir: ⌜**set a ~ to catch a ~**⌝, ⌜çivi çiviyi söker⌝.

**thiev·e** [θîv]. Hırsızlık yapmak; çalmak. ~**ish**, [~**ing**], hırsız gibi; hırsızlık (yapan). ~**es**, *pl. of* **thief.**

**thigh** [θây]. But; oyluk.

**thimble** [ˈθimbl]. Yüksük.

**thin** [θin]. İnce; seyrek; zayıf, eti yağı az; sulu. Seyrekleş(tir)mek; inceltmek; zayıfla(t)mak. **as ~ as a lath**, bir deri bir kemik: **that's a bit ~**, (mazeret) ikna etmez: ~**ly clad**, ince giyinmiş; fakir kıyafetli: **to have a ~ time of it** (*coll.*), çok eziyet çekmek: **a ~ly veiled threat**, pek kapalı olmıyan tehdid. **~ down**, (tahta vs.) inceltmek; (boya vs.) sulandırmak. **~ out**, seyrekleş(tir)mek.

**thine** [ðâyn]. (*esk.*) Seninki; senin.

**thing** [θin(g)]. Şey, nesne; mesele; mahluk. ~**s**, eşya; ahval, ortalık, iş. **~s are looking pretty bad**, ortalık çok fena, ahval kötü: **to be ⌜all ~s to all men⌝**, herkesin ⌜nabzına göre şerbet vermek⌝; **to talk of one ~ and another**, şundan bundan bahsetmek: **to clear away the ~s**, sofrayı toplamak: **a dear old ~**, sevimli bir ihtiyar kadıncağız [adamcağız]: **dumb ~s**, hayvanlar: **I'm not feeling at all the ~**, bir az keyifsizim: **that's the ~ for me!**, işte tam istediğim!: **how are ~s?**, ne var ne yok?, işler nasıl?: **he knows a ~ or two**, çok bilmiştir, şeytandır: **the latest ~ in hats**, en son şapka modası: **to make a good ~ [out] of stg.**, bir şeyden kâr çıkarmak: **to make a mess of ~s**, işi berbad etm.: **it's not the ~ (to do)**, bu yapılmaz: **for one ~ ...**, bir kere ..., evvelâ ...: **to pack up one's ~s**, pılısını pırtısını toplamak: **she is very ill, poor ~!**, zavallı, çok hastadır: **to take off one's ~s**, şapka ve paltosunu çıkarmak: **to take ~s too seriously**, hadiseleri fazla ciddiye almak: **to go the way of all ~s**, her şey gibi sona ermek (*gen.* ölmek).

**thing-amy, -ummy, -umajig, -mibob** [ˈθin(g)emi, ˈθin(g)aməcig, ˈθin(g)mibob]. Şey; hani, ne derler.

**think** (**thought**) [θin(g)k, θôt]. Düşünmek, zannetmek; addetmek, saymak; hükmetmek; farzetmek; tasavvur etm.; ummak. **I don't ~!**, (*sl.*) ne münasebet!: **I should hardly ~ so**, pek zannetmem: **let me ~!**, dur bakayım: **I thought as much**, zaten bunu bekliyordum: **I couldn't ~ of it!** dünyada böyle bir şey yapamam!:

I ~ **very highly of him,** benim nazarımda onun kıymeti büyüktür: **I would never have thought that of you,** senin böyle bir şey yapacağın hiç aklıma gelmezdi: **to ~ too much of oneself** (i) kendini beğenmek; (ii) hep kendini düşünmek: **he is well thought of,** itibarı yüksektir: **I told him what I thought of him,** açtım ağzımı yumdum gözümü: **~ing to please me, he said ...,** gözüme girmek maksadile ... dedi: **to ~ that he was once rich!,** şimdiki halini görüp de onun vaktile zengin olduğuna kim inanır?: **while I ~ of it,** hatıramdayken: **who'd have thought it! well, ~ of that!,** acayib!, kimin aklına gelirdi?: **what am I ~ing about!,** na kafa! **~ out,** tasarlamak. **~ over,** düşünüp taşınmak. **~able** [ˈθin(g)kəbl], tasavvur edilebilir; kabul edilebilir. **~er,** mütefekkir; filozof. **~ing,** düşünceli; aklı başında; düşünme: **that's my way of ~,** ben böyle zannediyorum: **to my way of ~,** bana göre: **to put on one's ~ cap,** düşünüp taşınmak.

**third** [θö(r)d]. Üçüncü. Üçte bir. **~ degree,** Amerika'da bir maznunu şiddet kullanarak istintak etme: **~ party,** bir sigorta mukavelesi vs.de taraflardan olmıyan uçüncü şahıs. **~-class,** (demiryol vs.de) üçüncü mevki; âdi. **~-hand, information at ~,** üçüncü elden alınan haber. **~-rate,** âdi, değersiz.

**thirst** [θö(r)st]. Susama(k); susuzluk. **to ~ for [after],** ···e susamak. **~y,** susamış; kurak (toprak).

**thirt·een** [θö(r)ˈtîn]. On üç: **~th,** on üçüncü; on üçte bir. **~ieth,** otuzuncu; otuzda bir. **~y,** [ˈθö(r)ti], otuz: **~-first,** otuz birinci.

**this** [ðis], *pl.* **these** [ðîz]. Bu; şu. **~ day fortnight,** iki hafta sonra bu gün: **~ far,** şuraya kadar: **~ high,** şu boyda: **it's Ahmed ~ and Ahmed that,** Ahmed aşağı Ahmed yukarı: **to put ~ and that together,** vakaları yanyana koymak: **it was like ~,** (i) buna benzerdi; (ii) şöyle oldu ...: **~ is where he lives,** burada oturuyor.

**thistle** [ˈθisl]. Devedikeni. **~down,** şeytan arabası.

**thither** [ˈðiðə(r)]. Oraya.

**tho'** *v.* **though.**

**thorn** [θôn]. Diken(li çalı). **to be a ~ in one's side,** başına derd olmak. **~y,** dikenli: **a ~ question,** çatallı bir mesele.

**thorough** [ˈθʌrə]. Tam; mükemmel; bütün bütün; baştan başa; yaman; **~ly,** adamakıllı, enikonu, tam. **~-bred,** cins (at), safkan. **~fare,** umumî geçid, büyük cadde: **no ~,** çıkmaz sokak. **~going,** tam, son derece; müdhiş; vicdanlı. **~ness,** tamlık; mükemmellik. **~-paced,** yaman, müdhiş.

**those** [ðo͜uz]. *pl. of* **that.** O ···lar; onlar.

**thou** [ða͜u]. (*esk. ve şair.*) Sen.

**though** [ðo͜u]. ···diği halde; her ne kadar; gerçi; öyle de olsa; bununla beraber. **~ I am poor I am honest,** fakır de olsam namusluyum: **he acts as ~ he were mad,** deli imiş gibi hareket ediyor: **it isn't as ~ you were a young man,** genc değilsin ki!; **strange ~ it may seem,** garib görünüyor ama: **what ~ the way be long,** yol uzun olsa da: **I wish you had told me ~,** öyle amma keşke bana söyleseydiniz: **'he said you were mad'. 'Did he, ~?',** 'Senin için "delidir" dedi.' 'Bak yediği naneye!'

**thought**[1] [θôt] *v.* **think.** **~**[2] *n.* Düşünce; düşünme; fikir. **he is a ~ too self-confident,** kendine bir parçacık fazla güveniyor: **the mere ~ of it infuriates me,** bunun tasavvuru bile beni deli ediyor: **he has no ~ for others,** başkalarını hiç düşünmez: **I had no ~ of offending him,** hiç onu kırmak istemedim: **second ~s are best,** düşüne taşına verilen kararlar en iyisidir: **on second ~s I decided not to go,** sonradan düşününce gitmemeğe kararverdim. **~ful,** düşünceli; dalgın; nazik: **a ~ book,** derin bir kitab: **to be ~ of others,** başkalarını düşünmek. **~less,** düşüncesiz; ihtimamsız.

**thousand** [ˈθa͜uzənd]. Bin. **~th,** bininci; binde bir.

**thral·l** [θrôl]. Köle. **~dom,** kölelik.

**thrash** [θraş]. Dövmek; şiddetli dayak atmak; mağlub etm.; *v.* **thresh. to ~ out,** tamik etm.; inceden inceye tedkik etmek. **~ing,** dayak, dövme; mağlubiyet.

**thread** [θred]. İplik; tel; vida dişi; file. **to ~ a needle,** ipliği iğnenin gözünü geçirmek: **to ~ beads,** *etc.*, boncuk vs.yi ipliğe dizmek: **to ~ one's way,** kalabalık vs. içinden sıyrılarak ilerlemek: **to hang by a ~,** kıl üstünde olm.: **to lose the ~ of one's**

argument, fikir silsilesini kaybetmek. ~**bare**, havı dökülmüş; beylik.

**threat** [θret]. Tehdid; tehlike. ~**en** [ˈθretn], tehdid etm.; ... tehlikesi olm.: **to ~ s.o. with stg.**, birini bir şeyle tehdid etm.: **the sky ~s rain**, gökte yağmur tehlikesi var: **a storm is ~ing**, bir fırtına tehlikesi var. ~**ening**, tehdidkâr; ... tehlikesine işaret olan.

**three** [θrî]. Üç ~**fold**, üç misli. ~**pence** [ˈθrepəns], üç peni. ~**penny** [ˈθrepəni], üç penilik. ~**quarter** [ˈθrîˈkwôtə(r)], dörtte üç büyüklükte. ~**score** [ˈθrîskô(r)], altmış.

**thresh** [θraş]. Harman dövmek; suyu dövmek. ~**er**, harmancı; döven makinesi. ~**ing**, harman etme: ~-**floor**, harman yeri: ~-**machine**, döven makinesi.

**threshold** [ˈθreşhǫuld]. Eşik; başlangıc.

**threw** *v.* throw.

**thrice** [θrâys]. Üç defa.

**thrift** [θrift]. İdare, iktisad, tasarruf. ~**less**, idaresiz, müsrif. ~**y**, muktesid, idareli.

**thrill** [θril]. Teessür veya heyecandan titreme; râşe. Titretici bir tesir yapmak; heyecanlandırmak. Heyecanla titremek. ~**er**, (*coll.*) pek heyecanlı roman vs. ~**ing**, heyecanlı.

**thriv·e** (**throve, thriven**) [θrâyv, θrǫuv, θrivn]. (İş) iyi gitmek; (çocuk vs.) iyi gelişmek, uygun şartlar içinde büyümek; muvaffak olmak. ~**ing**, mamur; sağlam; muvaffak.

**throat** [θrǫut]. Boğaz; gırtlak. **to clear one's ~**, boğazını temizlemek: **to cut one's (own) ~**, (i) boğazını keserek intihar etm.; (ii) bindiği dalı kesmek: **to cut one another's ~s**, birbirinin iflâsına sebeb olacak derecede, rekabete girişmek: **to have a sore ~**, boğaz olmak: **to thrust stg. down s.o.'s ~**, bir fikir vs.yi ileri sürmekte ısrar etmek. ~**y**, gırtlakta hasıl olan (ses).

**throb** [θrob]. (Kalb) çarpmak; zonklamak. ~ *veya* ~**bing**, çarpıntı, zonklama.

**throes** [θrǫuz] Istırab, ağrı. **to be in the ~ of death**, can çekişmek.

**throne** [θrǫun]. Taht. **to come to the ~**, tahta çıkmak, cülûs etmek.

**throng** [θron(g)]. Kalabalık (etm.), üşüşmek. **to be ~ed with ...**, ... dolup dolup taşmak: **to join the ~**, kervana katılmak.

**throttle** [θrotl]. Boğazını sıkmak; boğmak; istim vs.yi kesmek veya kısmak. Gırtlak; kısma azaltma cihazı. **to ~ down**, istim kısmak: **at full ~**, tam istimle: **to open the ~**, fazla istim vermek.

**through** [θrû]. Bir yandan öbür yana; içinden; arasından; ···den dolayı; delâletiyle. Ucdan uca giden; doğrudan doğruya giden. **all ~ my life**, hayatım boyunca: **it was all ~ you that we got in such a mess**, başımıza bu işler sizin yüzünüzden geldi: **it was ~ you that I got this post**, bu memuriyeti sizin sayenizde elde ettim: **you are ~**, (telefonda) irtibat tesis edilmiştir: **to be ~**, bitirmek: **to be ~ with stg.**, bir işi bitirmek; bir şeyden vazgeçmek: **to have been ~ stg.**, bir şeyi çekmek: **nobody knows what I've been ~**, başımdan neler geçtiğini kimse bilmez: **he only did it ~ ignorance**, bunu cahillikten yaptı: **to let s.o. ~**, birini geçirmek: **he looked me ~ and ~**, içimi okur gibi baktı: **to look ~ a telescope**, teleskopla bakmak: **to send stg. ~ the post**, bir şeyi posta ile göndermek: **he put his hand ~ the window**, kazara elile pencereyi kırdı: **I have read the book right ~**, kitabı baştan başa okudum: **I've read it ~ and ~**, başından sonuna kadar okudum: **I'm ~ with you!**, aramızda her şey bitti. ~**out** [θrûˈaut], baştan başa; her kısmında; tamamiyle. ~ **the year**, bütün sene.

**throve** *v.* thrive.

**throw**[1] [θrǫu] *n.* Atma, fırlatma; (güreşte) yere serme. **a stone's ~**, pek yakın. ~[2] *v.* (**threw, thrown**) [θrǫu, θrû, θrǫun]. Atmak; fırlatmak; yere atmak. **to be ~n upon one's own resources**, kendi yağı ile kavrulmağa mecbur olm.: **to ~ open the door**, kapıyı birdenbire ardına kadar açmak: **to ~ two rooms into one**, iki odayı birleştirip bir oda yapmak: **to ~ s.o. out of work**, birini işinden etmek. ~ **about**, etrafa saçmak; öteye beriye atmak: **to ~ oneself about**, kendini yerden yere atmak. ~ **away**, istenilmiyen şeyi atmak; ıskarta etm.; israf etm.; heba etm.: **(of a girl) to ~ herself away**, (*kız*) kendini ziyan etm.: **to ~ away an opportunity**, bir fırsatı elinden kaçırmak. ~ **back**, geri atmak; aksettirmek; geciktirmek; ataya çekmek: **to be ~n back upon ...**, ···e baş vurmağa mecbur olmak. **~-back,** muvafakkiyetsizlik; gerileme; atalardan birine çekme. **~ in,** içeri atmak; caba vermek: **to ~ in a word,** söze karışıp bir şey söylemek. **~ off,** çıkarıp atmak; üstünden atmak. **~ out,** dışarı atmak; kovmak; neşretmek; reddetmek; ıskarta etm.; ileri sürmek; ortaya atmak; şaşırtmak; (fidan vs.) sürmek: **to ~ out one's chest,** göğsünü şişirmek. **~-out,** ıskarta edilen şey. **~ over,** öbür tarafa atmak; terketmek. **~ together,** derme çatma kurmak; rasgele birleştirmek. **~ up,** yukarı atmak; kusmak; …den vazgeçmek; acele kurmak: **to ~ up one's job,** istifa etm.

**thrum** [θrʌm]. (Piyano vs.) acemice çalmak.

**thrush** [θrʌş]. Ardıc kuşu.

**thrust** [θrʌst]. Şiddetle ve ansızın itmek, sürmek, sokmak. Ansızın itiş, sokuş; hamle. **to ~ at s.o.,** süngü vs. ile birine hamle etm.: **to ~ oneself upon s.o.,** kendini zorla birine kabul ettirmek: **to ~ one's way through,** ite kaka yol açmak. **~er,** kendini zorla ileri atan kimse.

**thud** [θʌd]. (*ech.*) Mat ses(le düşmek).

**thug** [θʌg]. Kaatil, amansız haydud.

**thumb** [θʌm]. Baş parmak. Bir kitabı çok kullanarak parmaklarla kirletmek; acemice kullanmak. **to bite one's ~s,** hiddetten dudaklarını ısırmak: **his fingers are all ~s,** (el işlerinde) pek beceriksizdir: **under the ~ of,** …in tahakkümü altında: **rule of ~,** kararlama: **~s up!,** yaşasın! **~-index,** harf yerleri oyuk indeks. **~-nail,** baş parmağın tırnağı: **~ sketch,** pek küçük kroki. **~-screw,** (i) baş parmakları sıkmağa mahsus eski bir işkence aleti; (ii) kanadlı vida.

**thump** [θʌmp]. (*ech.*) Yumruk darbesi ve onun sesi. Yumruklamak. **~ing,** (*sl.*) iri: **a ~ lie,** kuyruklu yalan.

**thunder** [θʌndə(r)]. Gök gürlemesi. (Gök) gürlemek. **~ of applause,** alkış tufanı: **to steal s.o.'s ~,** başkasından evvel davranarak onun usulünü vs. kullanmak. **~bolt,** yıldırım; ezici ve şaşırtıcı haber. **~ous,** gürliyen. **~struck,** son derece şaşkın. **~-clap,** göğün bir kere gürlemesi.

**Thursday** [ˈθö(r)zdêy]. Perşembe.

**thus** [ðʌs]. Böyle(ce); şöyle ki, nitekim. **~ far,** buraya kadar.

**thwart** [θwôt]. İstediğini yaptırmamak; önüne geçmek.

**thy** [ðây]. (*esk.*) Senin. **~self,** sen kendin.

**thyme** [tâym]. Kekik.

**tiara** [tâyˈârə]. Papanın tacı; mücevherli ziynet tacı.

**tic** [tik]. Yüz sinirlerinin kendiliğinden ihtilacı.

**tick¹** [tik]. Sakırga; küçük işaret. **he's a ~!,** mikrob gibidir: **to ~ off,** işaret etm.; (*sl.*) azarlamak. **~².** (*ech.*) Tıkırtı; (*sl.*) an. Tıkırdamak. **half a ~!,** bekle bir az! **in a ~** [**in two ~s**], kaşla göz arasında; hemen: **on ~,** dakika dakikasına. **~³.** (*sl.*) Veresi(ye). **to buy on ~,** veresiye almak. **~er** [ˈtikə], teleemprimör; (*sl.*) ceb saati.

**ticket** [ˈtikit]. Bilet; yafta; (*Amer.*) siyasî fırkanın programı. Üzerine yafta koymak. **single ~,** yalnız gitme bilet: **return ~,** gidip gelme bilet: **to let out on ~ of leave,** bir mahpusu bazı şartlarla serbest bırakmak. **~-collector,** biletçi. **~-holder,** bileti olan kimse; abone.

**tickl·e** [ˈtikl]. Gıdıkla(n)ma(k). **to be ~ed to death,** (*coll.*) (bir şeyi işitince vs.) son derece hoşlanmak: **to ~ one's fancy,** garib bir şekilde hoşuna gitmek: **to ~ the palate,** (yemek) tecessüsünü tahrik etm. **~ish,** gıdıklanır; nazik (mesele).

**tid·al** [ˈtâydl]. Med ve cezre aid; (liman vs.) med ve cezre maruz. **~ wave,** med dalgası. **~e,** med ve cezir; (*esk.*) zaman; (*fig.*) cereyan. **high ~,** med: **low ~,** cezir: **the ~ is coming in,** deniz doluyor: **the ~ is going out,** deniz çekiliyor: **to go against the ~,** akıntıya karşı gitmek: **to ~ over stg.,** bir işin içinden çıkmak: **£100 will ~ us over the winter,** yüz lira bize kışı çıkarır: **the ~ has turned,** akıntı değişti; (*fig.*) talih döndü: ⌜**time and ~ wait for no man**⌝, zaman kimseyi beklemez. **~less,** med ve cezre tâbi olmiyan. **~-race,** med ve cezirin çok sür'atli olduğu yer.

**tiddler** [ˈtidlə(r)]. Pek küçük balık

**tid·iness** [ˈtâydinis]. İntizam, temizlik. **~y,** muntazam; üstü başı temiz; toplu; (*sl.*) büyükçe. İntizama koymak. **to ~ oneself,** kendine çekidüzen vermek: to ~ up, derlemek toplamak.

**tidings** [ˈtâydin(g)z]. Haber; havadis. **glad ~**, müjde.

**tie**[1] [tây] *n.* Bağ; düğüm; boyunbağı; mâni; ağac kuşak; berabere kalma. **the election ended in a ~**, seçimde iki namzed müsavi rey aldı: **children are a great ~**, çocuklar insanın ayağını bağlar. **~-pin**, boyunbağı iğnesi. **~**[2] *v.* (**tying, tied**) [ˈtây·in(g), tâyd]. Bağlamak; rabtetmek; birleştirmek; düğümlemek. Berabere kalmak. **to be ~d**, bağlanmak; serbest olmamak: **to ~ with s.o.**, birile berabere kalmak. **~ down**, bir yere bağlamak. **~ on**, sicim ile bir yere bağlamak. **~ up**, sicim ile bağlamak: (yara) sarmak; (at vs.yi) bir direğe vs.ye bağlamak: **to ~ up an estate**, bir mülkün satılmasını şartlara bağlamak. **~d** [tâyd], bağlı; serbest olmıyan. **~ cottage**, bir çiftlikte çalışan işçiye mahsus ev: **~ (public-)house**, yalnız muayyen bir fabrikanın içkilerini satan birahane.

**tier** [tiə(r)]. Sıra; dizi; üstüste konmuş dizi. **~ed**, dizili, sıralı.

**tiff** [tif]. Atışma; güceniklik.

**tiffin** [ˈtifin]. Hafif öğle yemeği.

**tiger** [ˈtâygə(r)]. Kaplan. **he's a ~ for work**, müdhiş çalışkandır. **~ish**, vahşi.

**tight** [tâyt]. Sıkı; dar; sızmaz, su geçirmez; eli sıkı; (*coll.*) sarhoş. **to be in a ~ corner**, müşkül bir vaziyette bulunmak: **to keep a ~ hold [hand] over s.o.**, birini sıkı altında tutmak: **money is ~ just now**, bu günlerde para kıt. **~ness**, gerginlik, sıkılık; darlık; hasislik. **~-fisted**, eli sıkı. **~-fitting**, dar, sıkı. **~-laced**, (*fig.*) sofu. **~-rope walker**, ip cambazı. **~en** [ˈtâytn], pekiş(tir)mek; sık(ıştır)mak; ger(in)mek; kısmak: **to ~ one's belt**, kuşağını sıkmak; yiyeceğinden kısmak: **to ~ up**, sıkıştırmak; şiddetlendirmek. **~s**, cambaz elbisesi.

**tigress** [ˈtâygris]. Dişi kaplan.

**Tigris** [ˈtâygris]. Dicle nehri.

**tile** [tâyl]. Kiremit, çini (ile kaplamak); tuğla, mermer ile döşemek. **to have a ~ loose**, (*coll.*) bir tahtası eksik olmak. **~d**, kiremit, çini ile kaplanmış; tuğla, mermer ile döşenmiş.

**till**[1] [til] *n.* Para çekmecesi. **~**[2] *v.* Çift sürmek; (toprağı) işlemek. **~**[3], ···e kadar. **he will not come ~ eight**, saat sekizden evvel gelmiyecek: **I shan't go ~ I'm invited**, davet edilmedikçe gitmem: **to laugh ~ one cries**, gözlerinden yaş gelinceye kadar gülmek.

**tiller**[1] [ˈtilə(r)]. Dümen yekesi. **~**[2]. Çiftçi.

**tilt**[1] [tilt]. Meyil, eğilme. Hafifçe meylet(tir)mek; eğerek boşaltmak. **to be on the ~**, bir az eğri olm: **to ~ over**, eğilmek, devrilmek: **to ~ up**, (eğilebilir bir şey) yukarı kalkmak veya kaldırmak. **~**[2]. At üzerinde mızrak oyunu oynatmak. **to ~ at**, üzerine hamle etm.: **to ~ at windmills**, (Don Kişot gibi) yel değirmenlerine saldırmak: **at full ~**, alabildiğine koşarak vs.: **to run full ~ into s.o.**, rap diye karşısına çarpmak.

**timber** [ˈtimbə(r)]. Kereste(lik ağaclar); gemi kaburgası. **standing ~**, daha kesilmemiş kerestelik agaclar. **~ed**, eski usul yarı kereste yarı kârgir (ev). **~-yard**, kereste mağazası.

**time**[1] [tâym] *v.* (Yarış) zamanını hesablamak. **well-~d**, tam zamanında. **~**[2] *n.* Vakit, zaman; müddet; mühlet; devre, saat; kere, defa, sefer; tempo, usul. **the ~s**, devir, zaman: **three ~s four is twelve**, üç kere dört on iki: **ten ~s as big as ...**, ···den on defa daha büyük: **'and about ~ too!'**, (*coll.*) 'zamanı geldi de geçiyor bile': **~ and again [~ after ~]**, tekrar tekrar: **it's a race against ~**, vakit pek dardır: **to work against ~**, bir iş için vakti dar olm.: **all the ~**, (i) mütemadiyen: (ii) bütün bu zaman zarfında: **any ~ you like**, ne zaman isterseniz: **he may turn up any ~**, (i) şimdi neredeyse gelir; (ii) her hangi bir zamanda gelebilir: **at ~s**, bazan: **three at a ~**, üç tane birden; üçer üçer: **to do two things at a ~**, iki işi birden yapmak: **for weeks at a ~**, üstüste haftalarca: **at no ~**, hiç bir zaman: **one-~ Governor**, sabık Vali: **to beat ~**, tempo tutmak: **to be behind the ~s**, eski kafalı olm.: **~ bomb**, ayarlı bomba: **by the ~ we get there**, biz oraya vardığımız zaman: **to do one's ~**, mahkûmiyetini geçirmek: **Father Time**, *vaktin müşahhas timsali*: **for the ~ being**, şimdilik muvakkaten: **to keep ~**, tempoya uymak: **I've no ~ for him.**

(*sl.*) bu adama tahammül edemem: **from ~ to ~**, arada sırada: **give me ~**, bana mühlet ver: **to have a good ~ (of it)**, eğlenceli vakit geçirmek: **to have a (rough) ~ of it**, eziyet çekmek: **what a ~ I had getting here!**, buraya gelinceye kadar neler çektim!: **I had the ~ of my life**, ömrümde o kadar eğlenmedim: **in good ~**, erken: **all in good ~!**, sırası gelecek: **in his own good ~**, ne zaman canı isterse: **you will learn in good ~**, sırası gelince öğrenirsiniz: **I hope we shall arrive in ~**, inşallah geç kalmayız: **in no ~**, çabucak: **this watch keeps good ~**, bu saat doğru gidiyor: **this house will last our ~**, bu ev bizim ömrümüzün sonuna kadar dayanır: **to look at the ~**, saate bakmak: **next ~**, gelecek sefer: **to arrive on [up to] ~**, tam vaktinde gelmek: **out of ~**, (i) temposu bozuk; (ii) vakitsiz: **my ~ is my own**, serbestim (vaktimi istediğim gibi kullanırım): **to pass the ~ of day**, hosbeş etm.: **to serve one's ~**, çıraklık etm.; askerlik etm.: **for some ~ past**, epeyi bir zamandan beri: **for some ~ to come**, daha uzun bir müddet: **to take a long ~ over stg.**, bir işi fazla uzatmak: **to take one's ~ over stg.**, bir işi yavaş yavaş ve itina ile yapmak (*bazan istihza ile soylenir*): **to tell the ~**, saatin kaç olduğunu söylemek: **have you the ~ on you?**, saate bakar mısınız?: **this ~ tomorrow**, yarın bu saatte: **~'s up!**, vakit geldi; bitti!: **in a week's ~**, bir hafta sonra: **what ~ is it?**, saat kaç? **~keeper**, (yarış vs.de) kontrol memuru: **my watch is a good ~**, saatim iyi işliyor. **~less**, ebedî, sonsuz. **~ly**, vaktinde, mevsiminde, yerinde. **~-piece**, saat (alet). **~-expired**, askerlik hizmetini bitirmiş. **~-exposure**, (*fot.*) poz. **~-honoured**, eski ve muteber. **~-lag**, birbirile alâkalı iki hadise arasındaki müddet. **~-server**, zamane adamı. **~-sheet**, amelenin iş saatinin kaydolunduğu kâğıd. **~-switch**, kontakt saati. **~-table**, (tren vs.) tarife; (mekteb vs.) ders programı. **~-work**, saatine göre verilen iş. **~-worn**, zamanla aşınmış.

**timid** [ˈtimid]. Çekingen; sıkılgan; ürkek. **~ity** [-ˈmiditi], çekingenlik, ürkeklik.

**timorous** [ˈtimərəs]. Ürkek.

**tin** [tin]. Kalay; teneke (kutu). Kalaydan yapılmış, teneke. Kalaylamak; teneke kablara koymak. **~-hat**, madenî miğfer. **~-opener**, konserve anahtarı. **~-plate**, saç; kalaylamak. **~-pot**, (*coll.*) değersiz. **~-whistle**, çığırtma. **~foil** [ˈtinfôyl], kurşun kâğıdı. **~ker** [ˈtin(g)kə(r)], gezici tenekeci; fena ve kaba tamirci. **to ~ with**, kaba tamir yapmak; ufak tefek kusurları düzeltmek. **~ned**, kalaylı, kalaylanmış; teneke kutuya konmuş (yiyecek). **~ny**, teneke gibi ses çıkaran. **~sel** [ˈtinsl], kılaptan; lâme; sahte parlaklık. **~smith** [ˈtinsmiθ], tenekeci.

**tincture** [ˈtin(g)ktyə(r)]. Tentür; boya hafif tesir. Hafifçe boyamak.

**tinder** [ˈtində(r)]. Kav; kuru ve yanıcı şey.

**ting** [tin(g)]. (*ech.*) Çın(lamak). **~-a-ling**, ufak çıngırak sesi.

**tinge** [tinc]. Hafif renk. Hafifçe boyamak. **admiration ~d with envy**, bir az hased karışık bir hayranlık.

**tingle** [ˈtin(g)gl]. (Tokat vs.den sonra) sızlamak; karıncalanmak; yanmak. **his cheeks ~d with shame**, utancından başından aşağı kaynar sular döküldü.

**tinkle** [ˈtin(g)kl]. (*ech.*) Çınlamak.

**tint** [tint]. Renk çeşidi; hafif renk (vermek).

**tiny** [ˈtâyni]. Minimini, ufacık.

**tip**[1] [tip]. Uc(una takılan başlık). Uc geçirmek; ucunu teşkil etmek. **~-toe** [ˈtiptou], ayaklarının ucuna basmak: **on ~**, ayaklarının ucuna basarak: **to be on the ~ of expectation**, büyük bir merakla beklemek. **~top** [ˈtiptop], en yüksek nokta. Son derece iyi. **~**[2]. Meyil; hafif itme; **(rubbish-)~**, çöp dökülen yer. Eğ(il)mek; meylet(tir)mek; boşaltmak. **~ out**, eğip boşaltmak. **~ over**, devrilmek. **~-up**, (devirerek) boşaltılabilir (araba vs.). **~-and-run**, bir nevi kriket oyunu. **~**[3]. Bahşiş; yarış atları hakkında malûmat verme; bir hususta faydalı bir tavsiye. Bahşış vermek. **if you take my ~ ...**, beni dinlerseniz .... **~ping**, bahşiş verme usulü.

**tipple** [ˈtipl]. İçki(ye düşkün olmak).

**tipsy** [ˈtipsi]. Çakır keyif; sendeliyerek.

**tirade** [tâyˈrêyd]. Uzun ve şiddetli tenkid; hücum.

**tir·e** [ˈtâyə(r)]. Yor(ul)mak; usan(dır)mak. **to be ~d**, yorulmak;

uykusu gelmek: to be ~d of stg., bir şeyden bıkmak: to ~ out, takatini tüketmek: to be ~d out, yorgunluktan bitkin olmak. ~ed, yorgun. ~eless, yorulmak bilmez. ~esome, yorucu; usandırıcı: how ~!, Allah belâsını versin! ~ing, yorucu.

**'tis** [tiz] = it is.

**tissue** [ˈtisyu]. Nesic; dokuma. a ~ of lies, yalan dolan: ~ paper, ipek kâğıdı.

**tit**[1] [tit]. Baştankara. ~[2]. ~ for tat, mukabelebilmisil.

**Titan** [ˈtâytən]. Dev, titan. ~ic [tâyˈtanik], dev gibi, muazzam.

**tit-bit** [ˈtitbit]. Küçük ve lezzetli lokma.

**tithe** [tâyð]. Aşar vergisi; ondalık.

**titillate** [ˈtitilêyt]. Gıdıklamak.

**titivate** [ˈtitivêyt]. Süslemek.

**title** [ˈtâytl]. Unvan; kitab ismi; sened, hüccet; hak. persons of ~, asalet unvanı sahibleri. ~d, asalet unvanı olan. ~-deed, tapu senedi. ~-page, baş sahife.

**titter** [ˈtitə(r)]. (*ech.*) Kıskıs gülme(k).

**tittle** [ˈtitl]. Zerre. ~-tattle, dedikodu.

**titular** [ˈtityulə(r)]. Unvan sahibi olan; asîl (memur); itibarî.

**to** [tû, tu]. *Fiillerin başına gelerek mastar teşkil eden edat, mes.* to go, gitmek. ···e; için. to London, Londra'ya: to turn to the right, sağa dönmek: give it to him, bunu ona ver: three to ten, (i) saat ona üç var; (ii) onda üç ihtimal; (iii) three to ten cement, üç ölçü cimento on ölçü kum vs.: there is nothing to see, görecek bir şey yok: easy to understand, anlaması kolay.

**toad** [toud]. Kara kurbağa; iğrenc kimse. ~stool, zehirli mantar. ~y, dalkavuk(luk etmek).

**toast** [toust]. Kızarmış ekmek; kadeh kaldırma. Kızartmak; sıhhatine veya şerefine içmek. to give a ~, kadeh kaldırmak: to have s.o. on ~, (*sl.*) birinin yakası elinde olmak. ~-rack, kızarmış ekmek kabı.

**tobacco** [təˈbakou]. Tütün. ~nist [–ˈbakənist], tütüncü. ~-pouch, tütün kesesi.

**toboggan** [təˈbogən]. Alçak bir nevi kızak (üzerinde kaymak).

**tocsin** [ˈtoksin]. Tehlike çanı.

**today** [təˈdêy]. Bugün.

**toddle** [ˈtodl]. Tıpış tıpış yürümek. ~r, yürümeğe başlıyan çocuk.

**toddy** [ˈtodi]. Viski ile şeker ve sıcak sudan yapılmış içki.

**to-do** [təˈdû]. Karışıklık. to make a ~, iş çıkarmak: to make a ~ about stg., bir şeyi mesele yapmak.

**toe** [tou]. Ayak parmağı; kundura veya çorabın burnu. to ~ the line, hizaya gelmek; yola gelmek: to stand on the tip of one's ~s, ayaklarının ucuna basıp yükselmek: to turn up one's ~s, nalları dikmek.

**toff** [tof]. (*coll.*) Pek şık adam; (*cont.*) kibar adam.

**toffee** [ˈtofi]. Kavrulmuş şeker ile tereyağından yapılmış şekerleme.

**together** [təˈgeðə(r)]. Beraber; birlikte. to come [meet] ~, bir araya gelmek: for months ~, aylarca hep ...: to stand or fall ~, 'anca beraber kanca beraber' olm.: to strike two things ~, iki şeyi birbirine çatmak.

**toil** [tôyl]. Zahmet; pek zor iş. Çok çalışmak, zahmet çekmek. to ~ up a hill, ıkına sıkına bir tepeye çıkmak. ~-worn, bitkin; didinmiş. ~s [tôylz], ağ; tuzak. ~some [ˈtôylsəm], yorucu; zahmetli.

**toilet** [ˈtôylit]. Tuvalet; çekidüzen. ~-paper, taharet kâğıdı.

**token** [ˈtoukn]. Alâmet, işaret; remiz; yadigâr. as a ~ of [in ~ of], işareti veya alâmeti olarak: book ~, kitab bonosu: by the same ~, bundan başka: ~ payment, bir borcun veya hakkın tanındığına işaret olarak ödenen para. ~-money, itibarî bir kıymet taşıyan para.

**told** *v.* tell.

**toler·able** [ˈtolərəbl]. Tahammül olunur; dayanılabilir; iyice. it is tolerably certain that ~, -diği aşağı yukarı katîdir. ~ance [–rəns], müsamaha; mülâyemet; tecviz; (*med.*) tahammül; (*mech.*) ihtiyat paye. ~ant, müsamahakâr. ~ate, müsamaha etm.; tahammül etm.; dayanmak; müsaade etmek. ~ation [–ˈlêyşn], müsamaha; tahammül etme.

**toll**[1] [toul] *n.* Müruriye; (yol, köprü) parası. the ~ of war, harbin ceremesi: the ~ of the roads, vesaiti nakliye kazaları. ~-bar [~-gate], müruriyeli geçid yeri. ~-call, yakın şehirler arasında telefonla konuşma. ~[2] *v.* (Birinin ölümünü veya cenaze alayını haber vermek için) bir çan ağır ağır ve muntazaman çalmak.

**Tom** [tom]. (*abb.*) Thomas. ~, **Dick, or Harry,** Ali Veli: ~ **cat,** erkek kedi.

**tomahawk** [ˈtomahôk]. Kızıl derililerin harb baltası.

**tomato** [to̤ṳˈmâto̤ṳ]. Domates.

**tomb** [tûm]. Mezar; türbe. ~**stone,** mezartaşı.

**tom·boy** [ˈtombôy]. Erkek tabiatli kız. ~**fool** [–ˈfûl], ahmak. Abdalca davranmak. ~**foolery,** maskaralık.

**tome** [to̤ṳm]. Büyük kitab.

**Tommy** [ˈtomi]. (*abb.*) Thomas. Bir nefer, er. ~ **Atkins,** 'Mehmedcik'. ~**gun,** kısa makineli tüfek. ~**-rot,** (*coll.*) saçma.

**tomorrow** [təˈmoro̤ṳ]. Yarın; ~'s, yarınki: ~ **week,** gelecek çarşamba vs.

**tomtit** [tomˈtit]. Baştankara.

**tomtom** [ˈtomtom]. Darnuka.

**ton** [tʌn]. Ton. **English** [**long**] ~ = 1,016 kg.; **American** [**short**] ~ = 907 kg.; **metric** ~ = 1,000 kg. **there's** ~**s of it,** ondan yığınlarla var. ~**nage** [ˈtʌnic], tonilâto; geminin istiab derecesi. **gross** ~, gayri sâfi tona (brüt tona): **registered** ~, rüsum tonilâtosu, sâfi tona.

**tone**[1] [to̤ṳn] *n.* (*mus.*) Perde; ton; ses ahengi; renklerde açıklık ve koyuluk derecelerinden her biri; vücudun hali. **to change one's** ~, ağız değiştirmek: **I don't like his** ~, ağzını beğenmiyorum: ~ **of voice,** sesin tonu: **the** ~ **of the school is excellent,** mektebin maneviyatı yüksektir. ~[2] *v.* Bir şeye tarz ve hususiyet vermek; renk vermek; ahenk vermek. **to** ~ **with,** renk, çeşid vs. cihetiyle uygun düşmek: **to** ~ **down,** hafifletmek: **to** ~ **up,** kuvvetlendirmek. ~**d, low** [**high**] ~, alçak [yüksek] perdeli: **low-**~ **conversation,** alçak sesle konuşma. ~**less,** donuk; renksiz.

**tongs** [ton(g)z]. Maşa.

**tongue** [tʌn(g)g]. Dil; lisan; çıngırak topuzu; ayakkabının dili. **to find one's** ~, dile gelmek: **to give** ~, (köpek) havlamak, *esp.* avını görünce havlamak: **to keep a civil** ~ **in one's head,** terbiye dairesinde konuşmak: **with one's** ~ **in one's cheek,** ciddî olmıyarak; yarım ağızla: **the gift of** ~**s,** lisan öğrenmek kabiliyeti: **furred** ~, paslı dil: **to put out the** ~, dilini cıkarmak. ~**-tied,** dili tutuk. ~**-twister,** şaşırtmaca.

**tonic** [ˈtonik]. Mukavvi; vurgulu; (*mus.*) baş notaya aid. **to act as a** ~, canlandırmak.

**tonight** [təˈnâyt]. Bu gece.

**tonsil** [ˈtonsil, -əl]. Bademcik. ~**itis** [–ˈlâytis], bademcik iltihabı.

**tonsure** [ˈtonşu̯ə(r)]. Katolik papazlarının tepelerinin tıraş edilmesi veya tıraş edilen yer.

**too** [tû]. De, dahi; çok fazla, lüzumundan fazla. **will you come** ~?, sen de gelecek misin?: **and very nice** ~!, hem de ne güzel! **it is** ~ **hot to work,** çalışamıyacak kadar sıcak: **this hat is** ~ **big for me,** bu şapka bana çok büyük geliyor: **he is just** ~ **lazy,** tahammül edilemez derecede tembel: **I shall be only** ~ **glad to help you,** size yardım etmek benim için bir zevktir.

**took** *v.* **take.**

**tool** [tûl]. Alet; (torna) kalem. Meşin, *esp.* kitab cildini süslemek; bir maddeyi aletle işlemek. **to be the (mere)** ~ **of,** …in elinde oyuncak olm.: **to down** ~**s,** (işçi hiddetten vs.) işi bırakmak; grev yapmak: ⌜**a bad workman blames his** ~**s**⌝, kötü işçi aletlerine kabahat bulur.

**tooth,** *pl.* **teeth** [tûθ, tîθ]. Diş. **armed to the teeth,** tepeden tırnağa kadar silâhlı: **to cast stg. in s.o.'s teeth,** bir şeyi birinin yüzüne vurmak: **to cut a** ~, (çocuk) diş çıkarmak: **to escape by the skin of one's teeth,** daradar kurtulmak: **to fight** ~ **and nail,** canını dişine takarak mücadele etm.: **in the teeth of …,** …e rağmen; …e mukabil: **to be long in the** ~, (*şak.*) yaşlı olm.: **to have a** ~ **out,** bir dişini çektirmek: **to set one's teeth,** dişini sıkmak. ~**ache,** dişağrısı. ~**brush,** diş fırçası. ~**pick,** kürdan. ~**some,** lezzetli. ~**-paste,** diş macunu. ~**-powder,** diş tozu.

**tootle** [ˈtûtl]. (Flâvtayı) hafifçe çalmak.

**top**[1] [top] *n.* Topaç. **to sleep like a** ~, deliksiz bir uyku uyumak. ~[2] *v.* **to** ~ **a tree,** bir ağacın tepesini kesmek: **to** ~ **a class,** sınıfta birinci olm.: **to** ~ **a hill,** tepenin üstüne çıkmak: **to** ~ **all,** üstelik, en fenası [en iyisi]: **to** ~ **s.o. by a head,** birinden bir baş boyu daha uzun olm.: **a statue** ~**s the column,** sütunun üstünde bir heykel var. ~ **up,** doldurmak. ~[3] *n.* Bir şeyin en yüksek yeri; tepe; üst; zirve; baş. Üst, en yukarı, başta olan. **to be at the** ~ **of one's form,** (i) sınıfın başında olm.: (ii) (oyuncu) tam kıvamında olm.: **at ~ speed,** âzamî süratle: **at the ~ of the tree,** mesleğinde en yüksek derecede: **at the ~ of one's voice,** avazı çıktığı kadar: **to come out on ~,** üst gelmek: **from ~ to bottom,** baştan aşağı: **from ~ to toe,** tepeden tırnağa kadar: **the ~ of the morning to you!,** sabahı şerefler hayırlar olsun: **on ~,** üstte: **on ~ of that,** bunun üstüne: **one thing happens on ~ of another,** dokuz ayın çarşambası bir araya geldi: **to go to bed on ~ of one's supper,** yemek üstüne yatmak: **to feel on ~ of the world,** kendini çok mes'ud hissetmek: **to go over the ~,** siperden çıkıp hücum etmek. **~most,** en yüksek. **-~ped,** üstü ... olan: **cloud-~ mountains,** tepeleri bulutla kaplı dağlar. **~per,** (*sl.*) silindir şapka; pek mükemmel şey. **~ping,** (*sl.*) en âlâ, mükemmel. **~sail,** gabya. **~-boots,** çizme. **~-coat,** palto **~-hat,** silindir şapka. **~-heavy,** alt kısmına nisbeten üst kısmı ağır olan; havaleli. **~-hole,** (*sl.*) en âlâ, mükemmel.

**topaz** [ˈtoupaz]. Sarı yakut.

**tope** [toup]. Ayyaşlık etmek. **~r,** ayyaş.

**topee** [ˈtoupi]. Kolonyal şapka.

**topic** [ˈtopik]. Bahis mevzuu; mesele. **~al,** muayyen yere aid; günün meselelerine aid.

**topple** [ˈtopl]. Düşecek gibi olmak. **to ~ over,** yuvarlanmak. Devirmek.

**topsy-turvy** [ˈtopsiˈtö(r)vi]. Baş aşağı; altüst. **to turn ~,** altüst etmek.

**torch** [tôç]. Meşale. **electric ~,** ceb feneri. **~-bearer,** meşale taşıyan. **~light,** meşale ışığı: **~ procession,** fener alayı.

**tore** *v.* **tear.**

**toreador** [ˌtoriəˈdô(r)]. Boğa güreşçisi.

**torment** [ˈtômənt]. *n.* Azab; eziyet; baş belâsı. *v.* [tôˈment]. Cefa etm., eziyet etm.; başının etini yemek. **~or,** eziyet veren kimse.

**torn** *v.* **tear.**

**tornado** [tôˈnêydou]. Kasırga; şiddetli fırtına.

**torpedo** [tôˈpîdou]. Torpido (ile vurmak). **~-boat,** muhrib. **~-net,** şıpka. **~-tube,** torpido kovanı.

**torp·id** [ˈtôpid]. Uyuşuk; tembel. **~or** [ˈtôpə(r)], uyuşukluk, cansızlık.

**torque** [tôk]. Devir hasıl eden kuvvet.

**torrent** [ˈtorənt]. Sel; şiddetli akış. **~ial** [təˈrenşiəl], sel gibi.

**torrid** [ˈtorid]. Pek sıcak.

**torsion** [ˈtôşən]. Bük(ül)me; burma.

**torso** [ˈtôsou]. İnsanın gövdesi.

**tort** [tôt]. (*jur.*) Haksızlık, zarar.

**tortoise** [ˈtôtəs]. Kaplumbağa. **~-shell,** bağa.

**tortuous** [ˈtôtyuəs]. Dolambaçlı.

**torture** [ˈtôtyə(r)]. İşkence (etmek); azab; işkence cezası. **to put s.o. to the ~,** birini söyletmek için işkence etmek.

**Tory** [ˈtôri]. İngiliz muhafazakâr partisine aid.

**tosh** [toş]. (*sl.*) Saçma.

**toss** [tos]. Havaya fırlatmak; (boğa vs.) boynuzla adamı havaya atmak; **~ (up),** yazı tuğra atma(k); havaya atma. **to ~ about,** (gemi) dalgalar üzerinde sallanmak; (insan) yatakta dönüp durmak: **to ~ off,** (bir kadehi) yuvarlamak: **to ~ one's head,** başını arkaya doğru silkmek; **pitch and ~,** *n.* yazı tuğra oyunu; *v.* (gemi) hem yalpa etmek hem baş vurmak: **to take a ~,** (attan) düşmek. **~-up,** **it's a ~,** baht işidir, belli olmaz.

**tot**[1] [tot]. Minimini çocuk; bir yudum. **~**[2], **to ~ up figures,** rakamları toplamak: **(of expenses) to ~ up,** (masraf) artmak.

**total** [ˈtoutl]. Yekûn, mecmuu. Tamam, tam; ... yekûnu. ···e baliğ olmak. **~itarian** [ˌtoutaliˈteəriən], tek partili hükümet. **~ity** [–ˈtaliti], yekûn. **~izator** [–lâyzêytə(r)], at yarışında müşterek bahislerin hesabını yapan makine. **~ly,** [ˈtoutəli], tamamen, tamamiyle.

**tote** [tout] = **totalizator.**

**totem** [ˈtoutem]. Amerika yerlileri arasında bir klanın remzi olan hayvan.

**totter** [ˈtotə(r)]. Sendeleme(k). **~ing,** sendeliyen; sarsılmış. **~y,** sarsak.

**touch**[1] [tʌç] *n.* Temas; dokunuş; el yordamı; boya vurma tarzı; iz. **sense of ~,** dokunma hissi: **a ~ of fever,** hafif bir sıtma nöbeti: **a ~ of salt,** *etc.*, bir parçacık tuz vs.: **a ~ of the sun,** biraz güneş çarpması: **to get into ~ with s.o.,** birisile temasa girmek: **to lose ~ with s.o.,** birinin izini kaybetmek: **it was ~ and go whether he would die of his illness,** hastalıktan ölmesine bıçak sırtı kalmıştı. **~**[2] *v.* Dokunmak; ellemek; ···e el sürmek;

temas etm.; müteessir etmek. **to be ~ed,** mütessir olm.; **he seems a little ~ed,** (i) biraz müteessir görünüyor; (ii) biraz oynatmışa benziyor: **he ~ed the bell,** zili hafifçe çaldı: **I can ~ the ceiling,** tavana yetişebilirim: **he ~ed me for a fiver,** (*coll.*) benden beş lira sızdırdı: **no one can ~ him in teaching English,** İngilizce hocalığında hiç kimse ona yaklaşamaz: **(ship) to ~ at a port,** (gemi) bir limana uğramak: **to ~ off a mine,** lağım atmak: **to ~ on a subject,** bir mevzua temas etm.: **I couldn't ~ the algebra paper,** (*coll.*) cebir imtihan suallerinin içinden çıkamadım: **to ~ up a picture,** bir resmi rötuş yapmak: **~ wood!,** nazar değmesin! **~ed,** müteessir; (*sl.*) bir tahtası eksik. **~ing,** dokunaklı, acıklı; bitişik; dair. **~iness,** alınganlık. **~stone,** mihenk (taşı). **~-wood,** kav. **~y,** alıngan: **he is rather ~ on that point,** bu mevzua karşı hassastır.

**tough** [tʌf]. Meşin gibi sağlam ve kırılmaz; metin; çiğnenmez; güç. Külhanbeyi, şirret adam. **a ~ customer,** zorlu ve netameli adam; aksi ve inadcı adam: **a ~ proposition,** güç bir iş. **~en,** sağlam ve kırılmaz bir hale sokmak; dayanıklı bir hale getirmek.

**tour** [tuə(r)]. Gezinti; turne; cevelân. Seyahat, tur yapmak. **~ist,** gezmek maksadı ile seyahat eden.

**tournament** [ˈtuənəmənt]. Turnuva; mızrak oyunu.

**tousle** [ˈtauzl]. (Saçları) karmakarışık etmek.

**tout** [taut]. Çığırtkan; müşteri toplayıcı. **to ~ for customers,** müşteri aramak.

**tow** [tou]. Yedek çekmek; cer etmek. Cer; yedek çekme. **to take in ~,** yedeğe almak.

**toward(s)** [təˈwôd(z)]. Cihetinde; ···e doğru; için. **~ morning,** sabaha karşı.

**towel** [ˈtauəl]. Havlu; silecek.

**tower** [ˈtauə(r)]. Kule; burc. Yükselmek. **to ~ above stg.,** bir şeyden çok daha yüksek olm.: **a ~ of strength,** güvenilen ve dayanılan kimse. **~ing,** çok yüksek: **~ ambition,** hududsuz ihtiras: **to be in a ~ rage,** son derece öfkelenmek.

**town** [taun]. Şehir; kasaba. **a man about ~,** sosyete adamı: **country ~,** taşra kasabası: **to live in ~,** Londra'da ikamet etm.: **he is out of ~,** taşraya gitti: **it is the talk of the ~,** bütün şehrin ağzındadır. **~ship,** nahiye. **~sman,** şehirli. **~speople,** şehir ahalisi. **~-clerk,** belediye evrak müdürü. **~-council,** belediye meclisi. **~-crier,** şehir tellâlı. **~-hall,** belediye dairesi. **~-house,** konak. **~-planning,** şehircilik.

**toxic** [ˈtoksik]. Zehirli.

**toy** [tôy]. Oyuncak. **to ~ with stg.,** bir şeyle oynamak; gayri ciddî bir tarzda bir şeyle meşgul olm.: **to ~ with one's food,** yemeğini isteksizce yemek: **to ~ with an idea,** zevk aldığı bir fikri zihninde everip çevirmek: **to ~ with s.o.,** birini okşamak.

**trace**[1] [trêys]. İz; alâmet; şemme. İzini takib etm.; şeklini çizmek; taslağını resmetmek; saman kâğıdıyle kopya etm.; sözle tasvir etmek. **to ~ out a scheme,** bir projeyi tasarlamak: **I cannot ~ any letter of that date,** bu tarihli bir mektub bulamadım: **to ~ stg. back to its source,** bir şeyi menşelerine irca etmek. **~r** [ˈtrêysə(r)], **~ bullet,** *etc.*, (havada duman vs.den) iz bırakan mermi. **~ry** [ˈtrêysəri], ağ şeklinde taş süsü; yaprak damarlarının şebekesi. **~**[2]. Koşum kayışı. **to kick over the ~s,** serkeşlık etmek.

**tracing** [ˈtrêysin(g)]. İnce resim kopyesi. **~-paper,** saman kâğıdı.

**track** [trak]. Gelip geçmekle kendiliğinden hasıl olan yol; iz; demiryol hattı; dümen suyu. İzini takib etmek. **to ~ down,** izini takib ederek keşfetmek: **to be on the ~ of ...,** ···in izi üzerine olm.: **to be off the ~,** yoldan sapmış olm.: **the beaten ~,** herkesin yürüdüğü yol: **to cover up one's ~s,** izini gizlemek: **to keep ~ of,** izini kaybetmemek: **to make ~s,** (*coll.*) sıvışmak: **to throw s.o. off the ~,** birine izini kaybettirmek. **~er,** (avda) izci. **~less,** 'kuş uçmaz kervan geçmez'.

**tract**[1] [trakt]. Mıntaka; saha. **~**[2]. Risale, *esp.* dinî risale.

**tract·able** [ˈtraktəbl]. Uslu, mazlum; kolayca işlenir. **~ion** [ˈtrakşn], çekme, cer. **~ engine,** âdi yollarda ağır yükleri çekmeğe mahsus lokomotif. **~or,** traktör.

**trad·e** [trêyd]. Ticaret; tüccarlık; iş, meslek. Alıp satmak; ticaret yapmak; mubadele etm. **to be in ~,**

tüccarlık etm.: **Board of T~**, İngiliz Ticaret Bakanlığı: **'everyone to his ~'**, herkes kendi işine bakmalı: **to ~ on stg.**, bir şeyi istismar etm.: **~ price**, tüccar arasındaki satış fiatı: **~ union**, işçi sendikası. **~er**, tüccar; ticaret gemisi. **~esman**, *pl.* **men** [**~es-people**], dükkâncı, esnaf sınıfı. **~ing**, alışveriş; ticarete aid. **~e-mark**, alâmeti farika. **~e-wind**, alize.

**tradition** [trəˈdişn]. Anane; ecdaddan naklolunan rivayet; hadiz. **~al** [–ˈdişnl], ananevi; menkul.

**traffic** [ˈtrafik]. Alışveris, ticaret; gidip gelme, seyrüsefer. Alışveriş etm. **~ indicator**, (*otom.*) işaret kolu: **~ policeman**, işaret memuru: **~ regulations**, seyrüsefer nizamnamesi. **~er**, alışveriş eden kimse (*um. fena mânada*).

**trag·edy** [ˈtracidi] Facia; trajedi; haile; feci vaka. **~edian** [–ˈcîdyən], haile muharriri, aktörü. **~ic(al)** [–cikl], facia nevinden, feci. **~i-comedy**, gülünc vakalarla karışık facia.

**trail** [trêyl]. Hareket eden kimsenin arkasında bırakılan şey; iz; kuyruk; orman vs.de çiğnenerek açılan yol. Peşinden sürükle(n)mek; izini takib etmek. **to ~ stg. along**, bir şeyi sürüklemek: **to pick up the ~**, izi bulmak: **the car left a ~ of dust behind it**, otomobil arkasında bir toz bulutu bıraktı. **~er**, otomobilin arkasına takılan araba. **~ing**, sürünen: **~ edge**, (uçak) kanadın arka kenarı.

**train**[1] [trêyn] *n.* Tren; katar; kafile; maiyet; yere sürünen uzun etek. **~ of events**, hadiselerin zinciri: **a ~ of powder**, barut serpintisi: **~ of thought**, düşünce silsilesi: **war brings famine in its ~**, harb arkasından açlık getirir. **~load**, tren dolusu. **~**[2] *v.* Alıştırmak, terbiye etm.; idman ettirmek; (topu) dirisa etmek. İdman etm.; alışmak. **~ed**, terbiye edilmiş: **~ nurse**, diplomalı hastabakıcı. **~er**, mürebbi; antrenör. **~ing**, talim; idman; dirisa: **to be in good ~**, idmanlı olm.: **to go into ~**, idman yapmak: **to be out of ~**, idmansız olmak: **~-ship**, mekteb gemisi.

**trait** [trêy(t)]. Hususiyet.

**traitor** [ˈtrêytə(r)]. Vatan haini. **to turn ~**, (vatana) ihanet etmek. **~ous**, hain.

**tram** [tram]. Tramvay.

**trammel** [ˈtraml]. Mâni. Serbest hareket etmesine mâni olmak.

**tramp** [tramp]. Ağır ayak sesi; yayan yolculuk; serseri. Ağır adımlarla yürümek; yayan gitmek. **~ (steamer)**, muntazam sefer yapmıyan yük gemisi. **to ~ the country**, kırlarda gezip dolaşmak: **to go for a long ~**, uzun bir yürüyüşe çıkmak.

**trample** [ˈtrampl]. **to ~ stg. under foot** [**to ~ on stg.**] bir şeyi ayak altında çiğnemek: **to ~ on s.o.'s feelings**, birinin hislerini çiğnemek.

**trance** [trâns]. Vecid, kendinden geçme; hipnotizma haleti.

**tranquil** [ˈtrankwil]. Sâkin; âsude. **~lity** [–ˈkwiliti], âsudelik, sükûn. **~lize** [–lâyz], teskin etm.

**trans-** [trânz] *pref. Yer, hal ve şart değişmesini ifade eden ön ek*; öte; mavera. *Mes.* **~atlantic**, Atlantiğin ötesine aid: **~form**, şeklini değıştirmek.

**transact** [trânˈzakt]. **to ~ business with s.o.**, birisile muamele yapmak. **~ion** [–ˈakşn], (işe aid) muamele: **the ~s of a society**, bır cemıyetin zabıtnameleri veya raporları.

**transatlantic** [ˌtrânsatˈlantik]. Atlas Okyanusunun ötesinde bulunan. Amerika ile Avrupa arasında işleyen yolcu vapuru.

**transcend** [tranˈsend]. Hududunu geçmek; üstün olmak. **~ent**, üstün; âlâ; mücerred.

**transcri·be** [trânˈskrâyb]. Aynını kopye etm. **~pt** [ˈtranskript], bir metni hususî harflerle istinsah; kopya.

**transept** [ˈtrânsept]. Bazı kilise binalarının yan kolları.

**transfer** *n.* [ˈtrânsfə(r)]. Nakil; havale; yer değiştirme; (resim vs.) çıkartma. *v.* [transˈfö(r)]. Nakletmek; başka bir yere geçirmek; devretmek; havale etm.; intikal etmek. **~able** [–ˈfö(r)rəbl], devredilebilir: **'not ~'**, (bilet vs.) başkasına devredilemez. **~ence** [ˈtrânsfərəns], nakil, havale etme veya edilme.

**transfigure** [trânsˈfigə(r)]. Şekil ve suretini değıştirmek.

**transfix** [trânsˈfiks]. Mıhlamak.

**transform** [trânsˈfôm]. Şeklini değiştirmek; tahvil etm. **~ation** [–ˈmêyşn], tahavvül. **~er**, transformatör; muhavvile.

**transfus·e** [transˈfyûz]. Kabdan kaba nakletmek. **~ion** [–ˈfyûjn], **blood ~**, insandan insana kan nakletme.

**transgress** [trâns'gres]. Günah işlemek; tecavüz etmek. **~ion** [-'greşn], günah; tecavüz.
**tranship** [trân'şip]. Bir gemi vs.den diğerine aktarma etmek. **~ment,** aktarma.
**transient** ['trânsiənt]. Geçici; fâni.
**transit** ['trânsit]. Yerden yere geçme; mürur. **~ion** [-'sişn], istihale; halden hale geçiş; intikal. **~ive,** müteaddi. **~ory,** geçici, fâni; süreksiz.
**translat·e** [trâns'lêyt]. Tercüme etm.; başka dile çevirmek; tefsir etm.; (piskoposu) başka bir piskoposluğa nakletmek. **Shakespeare does not ~ well,** Shakespeare'in eserleri tercumede çok kaybeder. **~ion** [-'lêyşn], tercüme (etme).
**translucent** [trâns'lyûsənt]. Yalnız ziyayı geçiren; nim şeffaf.
**transmission** [trâns'mişn]. Nakil; hareket nakili; (*otom.*) nakil cihazı.
**transmit** [trâns'mit]. Göndermek; nakletmek; geçirmek; sirayet ettirmek. **~ter,** gönderen; nakledici; (radyo) verici cihaz.
**transmute** [trâns'myût]. Şeklini değiştirmek.
**transparen·t** [trâns'peərənt]. Şeffaf; berrak; vazih. **~cy,** şeffaflık; şeffaf bir madde üzerinde fotograf.
**transpire** [trâns'pâyə(r)]. Duman halinde salıvermek; ifşa edilmek; (*coll.*) vukubulmak.
**transport** *n.* ['trânspôt]. Nakil; nakliye gemisi; nakliyat; vecid, taşkınlık. *v.* [-'pôt]. Nakletmek: **to be ~ed with joy [to be in ~s of joy],** etekleri zil çalmak. **~ation** [-'têyşn], nakletme; sürgün.
**transpose** [trâns'pouz]. Yerini değiştirmek.
**transverse** [tranz'vö(r)s]. Eğri; münharif.
**trap** [trap]. Tuzak; dolab; atlı hafif araba. Tuzağa düşürmek; tuzak ile tutmak. **~s,** (*coll.*) pılıpırtı: **~ped by the flames,** alevler tarafından hapsedilmiş: **to set a ~,** tuzak kurmak. **~-door,** ayak altında veya tavanda kapan gibi kapı. **~per** ['trapə(r)], kürk hayvanları tuzakçısı. **~pings** ['trapin(g)z], süslü koşum takımı.
**trash** [traş]. Değersiz şeyler; kötü eser; saçma. **~y,** değersiz.
**travail** ['travêyl]. Meşakkatli iş; doğum ağrıları. **woman in ~,** doğurmakta olan kadın.
**travel** ['travl]. Seyahat etme; yolculuk. Seyahat etm.; hareket etmek. **to ~ in textiles,** *etc.*, bir mensucat vs. ticarethanesinin seyyar memuru olmak. **~led, much- [well-] ~,** çok seyahat etmiş olan. **~ler,** seyyah, yolcu. **~ling,** seyahat etme, yolculuk; seyyar.
**traverse** ['travö(r)s]. Her şeyin çapraz kısmı. [-'vö(r)s]. Karşıdan karşıya geçmek; yandan yana hareket et(tir)mek.
**travesty** ['travesti]. Gülünc bir taklid; ciddî bir şeyi gülünc yapma(k).
**trawl** [trôl]. Sürütme usulile balık avlamak. **~er,** taraklı ağ çeken balıkçı gemisi.
**tray** [trëy]. Tepsi; mektub sepeti. **ash ~,** sigara tablası.
**treacher·y** ['treçəri]. Hainlik, ihanet. **~ous,** hain; hilekâr; güvenilmez.
**treacle** ['trîkl]. Şeker tortusu.
**tread**[1] [tred] *n.* Ayak sesi; yürüyüş; merdiven basamağı; otomobil lâstiğinin yere basan kısmı. **~**[2] **(trod, trodden)** [tred, trod, trodn]. Ayağını yere basmak; yürümek; ayakla basmak. **to ~ under foot,** ayak altında çiğnemek: **to ~ water,** ayaklarını vurarak su içinde dik durmak. **~ down,** ayakla ezmek. **~ on,** üzerine yürümek. **~ out,** (şarablık üzüm) çiğnemek; (ateş) üzerine ayakla basıp söndürmek. **~le** ['tredl], pedal. **~mill** ['tredmil], eskiden ceza olarak mücrimlere bastırıp işlettirilen ayak değirmeni; (*fig.*) her gün yapılması lâzım olan meşakkatli iş.
**treason** ['trîzən]. İhanet. **high ~,** vatana hiyanet. **~able,** vatana hiyanet nevinden.
**treasur·e** ['trejə(r)]. Hazine; define; pek kıymetli şey [şahış]. Pek kıymetli saymak. **buried ~,** define: **a real ~ of a (servant,** *etc.*), bulunmaz (hizmetçi vs.): **to ~ up,** kıymetli sayarak saklamak: **to ~ up wealth,** para biriktirmek. **~er,** hazinedar; veznedar. **~e-house,** hazine. **~y** ['trejəri], hazine; devlet hazinesi; maliye dairesi: **the T~ Bench,** Avam Kamarasında bakanlara mahsus mevki: **~ note,** on şilin veya bir İngiliz liralık kâğıd para: **~ of verse,** seçme şiirler kitabı.
**treat**[1] [trît] *n.* Her zamankinden fazla bir zevk; çocuklara vs. bilhassa verilen yemek. **it is a ~ to listen to him,** onu dinlemek bir zevkdir: **to give**

**oneself a ~,** her zamankinden farklı bir şeyin zevkini sürmek: **to stand ~,** başkalarına verilen içki vs.nin masrafını ödemek. ~[2] *v.* Muamele etm.; idare etm.; hakkında davranmak; tedavi etm.; başkasının içki vs. ödemek. **to ~ s.o. like a child,** birini çocuk gibi idare etm.: **I will ~ you to a drink,** size bir içki ikram edeceğim: **I shall ~ myself to a new hat,** paraya kıyıp kendime bir şapka alacağım: **to ~ stg. as a joke,** bir şeyi şaka saymak: **to ~ for peace,** sulh müzakere etm.: **to ~ s.o. for rheumatism,** birinin romatizmasını tedavi etm.: **to ~ of a subject,** bir mevzudan bahsetmek: **to ~ with s.o.,** birisile müzakere etmek. **~ise** [ˈtrîtiz], ilmî eser; risale. **~ment,** muamele; tedavi; ameliye. **his ~ of the subject is superficial,** mevzuu sathî bir şekilde ele almıştır. **~y** [ˈtrîti], muahede; mukavele.

**treble**[1] [trebl]. Üç misli (olm.). ~[2]. Tiz sesli (*esp.* erkek çocuğun sesi).

**tree** [trî]. Ağac; şecere. Bir ağaca sığınmağa mecbur etmek. **to be at the top of the ~,** mesleğinin en yüksek mevkiinde olm.: **up a ~,** müşkül bir vaziyette.

**trek** [trek]. Öküz arabasında seyahat (etm.); (*coll.*) yol (etm.); göçmek.

**trellis** [ˈtrelis]. Kafes şeklinde bölme.

**trembl·e** [ˈtrembl]. Titreme(k). **to be all of a ~,** tir tir titremek. **~ing,** titreme: **in fear and ~,** korkudan titreyerek.

**tremendous** [triˈmendəs]. Kocaman; hadsiz hesabsız; korkunc. **~ly** (*coll.*) pek çok.

**trem·or** [ˈtremə(r)]. Titreme; raşe. **earth ~,** yer sarsıntısı: **without a ~,** kılı bile kıpırdamadan. **~ulous** [ˈtremyuləs], titrek.

**trench** [trenç]. Kirizma (yapmak); siper(ler ile muhafaza etmek). **to ~ on [upon],** tecavüz etm. **~-coat,** askerî muşamba.

**trend** [trend]. Yön(elmek); meyil.

**trepidation** [ˌtrepiˈdêyşn]. Korku; telâş.

**trespass** [ˈtrespəs]. Kanuna muhalefet; günah; başkasının arazisine haksız yere ayak basma(k). **to ~ against the law,** kanunu ihlâl etm.: **to ~ on s.o.'s preserves,** (*fig.*) başkasının faaliyet sahasına tecavüz etm.: **I fear I am ~ing on your time,** korkarım vaktinizi alıyorum. **~er,** müsaade almaksızın başkasının arazisine giren kimse: '**~s will be prosecuted**', 'girenler hakkında kanunî takibat yapılacaktır'.

**tri-** [trây]. *pref.* Üçlü; üçte bir.

**trial** [ˈtrâyəl]. Muhakeme; tecrübe; gaile. Tecrübe için yapılan. **~ balance,** zimmet ve matlubun muvakkat mukayesesi: **~ and error,** deneyerek ve yanlış yapa yapa: **we will give it a ~,** onu bir deneyelim: **he is rather a ~ to me,** ona tahammül etmek kolay değildir: **on ~,** tecrübe için; muhakeme edilmekte: **~ order,** tecrübe için sipariş.

**triang·le** [ˈtrâyan(g)gl]. Müselles; üçgen: **the eternal ~,** iki erkekle bir kadın veya aksi. **~ular** [–ˈan(g)gyulə(r)], üç köşeli; müselles şeklinde.

**trib·al** [ˈtrâybl]. Bir kabileye aid. **~e,** kabile; güruh. **~sman,** bir kabileye mensub kimse.

**tribulation** [tribyuˈlêyşn]. Keder.

**tribunal** [trâyˈbyûnl]. Mahkeme; hâkim makamı.

**tribune** [ˈtribyûn]. Hitabet kürsüsü.

**tribut·e** [ˈtribyût]. Harac; hürmet veya takdir nişanesi. **to lay under ~ [to levy ~ on],** haraca kesmek: **to pay a ~ to s.o.,** birine karşı hürmet veya takdir nişanesi göstermek. **~ary,** haraca bağlı; tâbi; büyük ırmağa dökülen küçük bir nehir.

**trice** [trâys]. **in a ~,** bir çırpıda.

**trick** [trik]. Hile; kurnazca bir oyun veya çare; işin sırrı; el çabukluğu; garib veya nahoş âdet; (iskambil) leve. Aldatmak; kafese koymak. **the whole bag of ~s,** takım taklavat: **that'll do the ~,** (*coll.*) bu işimizi görür: **he knows a ~ or two [he's up to every ~],** o ne kurnazdır: **I know a ~ worth two of that,** ben bundan âlâsını bilirim: **he has been up to his old ~s,** yine her zamanki marifetlerini yaptı: **to ~ out,** süslemek: **to play a ~ on s.o.,** birine azizlik etm. **~ery,** hilekârlık. **~iness,** hilekârlık, kurnazlık. **~ster,** dolandırıcı. **~y,** hilekâr; kullanması nazik ve çok dikkat icabeden.

**trickle** [ˈtrikl]. Damla damla akma(k); ağır ağır geçmek.

**tricolour** [ˈtrâykʌlə(r)]. Üç renkli; [ˈtrikələ(r)], üç renkli bayrak; fransız bayrağı.

**tricycle** [ˈtrâysikl]. Üç tekerlekli bisiklet.

**trident** [ˈtrâydənt]. Üç çatallı zıpkın.
**tried** [trâyd] *v.* **try**: tecrübeli; güvenilir: **much ~,** çok zahmet görmüş: **well-~,** denenmiş.
**triennial** [trâyˈenyəl]. Üç senede bir olan.
**trier** [ˈtrâyə(r)]. **he is a ~,** elinden geleni yapar.
**trifl·e** [ˈtrâyfl]. Kıymetsiz ehemmiyetsiz şey. Gayri ciddî davranmak. **to ~ away one's time,** vaktini ehemmiyetsiz şeylerle geçirmek: **a mere ~,** devede kulak: **to ~ with s.o.,** birini oynatmak. **~er,** işini ciddiye almıyan kimse. **~ing,** cüzi; ehemmiyetsiz.
**trigger** [ˈtrigə(r)]. Tetik. **~-finger,** şahadet parmağı.
**trigonometry** [ˌtrigəˈnomətri]. Müsellesat.
**trill** [tril]. Ses titremesi; kuş ötmesi.
**trillion** [ˈtrilyən]. (*Eng.*) $10^{18}$; (*Amer.*) $10^{12}$.
**trilogy** [ˈtriləci]. Üç müstakil parçadan mürekkeb fakat bir bütün teşkil eden roman.
**trim** [trim]. Muntazam; zarif. Nizam. İntizama koymak; süslemek; (kesip) düzeltmek. **to ~ a boat,** bir kayığının muvazenesini tanzim etm.: **in fighting ~,** her şey muharebe için hazır: **to be in good ~,** iyi bir halde olm. **~mer,** zaman adamı.
**Trinity** [ˈtriniti]. Üçlü bir. **the (Holy) ~,** teslis. **~ House,** Büyük Britanya sularında kılavuzluk, fener kulelerini vs.nin idaresi.
**trinket** [ˈtrin(g)kit]. Cicibici; biblo.
**trio** [ˈtrîou]. Üçlük takım.
**trip**[1] [trip]. Kısa bir seyahat. **round ~,** bir yere gidip gelme. **~per,** seyahate çıkan kimse. **~**[2]. Sürç(me); birinin ayağını çelme. Sürçmek; çelme takmak; yanlışlık etmek. **to ~ (up),** ayağını çelmek: **to ~ along,** hafifçe yürümek: **to catch s.o. ~ping,** birinin hatasını yakalamak. **~-wire,** birinin geçtiğini haber veren gerilmiş tel.
**tripartite** [trâyˈpâtâyt]. Üç parti tarafından yapılmış (muahede vs.).
**tripe** [trâyp]. İşkembe; (*sl.*) saçma.
**triple** [ˈtripl]. Üç misli (çıkarmak). **~t,** üçüz.
**triplex** [ˈtripleks]. Üç tabakadan mürekkeb.
**triplicate** [ˈtriplikêyt]. Üç nüsha yapmak. [–kit], üçlü.
**tripod** [ˈtrâypod]. Sehpa.
**Tripoli** [ˈtripəli]. Trablus.
**trisyllabic** [ˌtrâysiˈlabik]. Üç heceli.
**trite** [trâyt]. Eskimiş, basmakalıb.
**triumph** [ˈtrâyəmf]. Eski Romada zafer alayı; parlak muvaffakiyet; zafer (şenliği). Muzaffer olm.; galib gelmek. **to ~ over s.o.,** birine galebe çalmak. **~al** [–ˈʌmfəl], zafere **aid: ~ arch,** takı zafer. **~ant** [–ˈʌmfənt], muzaffer(ane); gururlu.
**trivet** [ˈtrivit]. Sacayak. **right as a ~,** mükemmel bir halde.
**trivial** [ˈtrivyəl]. Ehemmiyetsiz; cüzi. **~ity** [–ˈaliti], ehemmiyetsizlik; bayağılık.
**trod** *v.* **tread.**
**Trojan** [ˈtroucən]. Eski Trova'ya aid; Trovalı. **to work like a ~,** domuz gibi çalışmak.
**trolley** [ˈtroli]. El arabası.
**trollop** [ˈtroləp]. Pasaklı kadın; sürtük.
**troop** [trûp]. Takım; sürü; süvari bölüğü. **~s,** askerler. **to ~ in,** sürü halinde girmek: **to ~ together,** bir araya toplamak. **~er,** süvari asker. **~ing, ~ the colours,** geçid resmi yapan bir alayın ortasında sancak taşıma merasimi. **~ship,** askerî nakliye gemisi. **~-train,** asker nakleden tren.
**trophy** [ˈtroufi]. Ganimet; zafer hatırası.
**tropic** [ˈtropik]. Medar. **the ~s,** sıcak memleketler. **~al,** sıcak memleketlere aid.
**trot** [trot]. Tırış (gitmek); (*coll.*) gitmek. **to go at a ~,** link gitmek: **to break into a ~,** link gitmeğe başlamak. **to ~ out one's knowledge,** (*coll.*) bilgi satmak: **he ~ted out the usual excuse,** her zamanki bahaneyi ileri sürdü. **~ter** [ˈtrotə(r)], link atı. **pig's ~s,** domuz paçası.
**troth** [trouθ]. Sadakat. **to plight one's ~,** sadakatini tasdık etm.; evlenme vadetmek.
**trouble**[1] [ˈtrʌbl] *n.* Zahmet; derd; sıkıntı; belâ; tasa; rahatsızlık; bozukluk. **~s,** kargaşalık. **to be in ~,** başına iş açılmak: **to bring ~ upon oneself,** başına iş açmak: **engine ~,** makinede bozukluk: **to get oneself into ~,** başını belâya sokmak: **to get s.o. into ~,** birinin başına iş açmak: **to have heart ~,** kalbinden rahatsız olm.: **the ~ is that ...,** işin fenası ...: **he is looking for ~,** başının belâsını

arıyor: **it's no ~ at all to ..., ...** işten bile değil: **nothing is too much ~ for him,** hiç bir şeyden yüksünmez: **to put s.o. to ~,** birini zahmete sokmak: **to take ~ over stg.,** bir şeyi itina ile yapmak: **what's the ~?,** ne var?, ne oldu? **~-maker,** fitneci. **~²** *v.* Zahmet vermek; rahatsız etm.; eziyet vermek. **to ~ oneself about stg.,** bir işte kendine zahmet vermek; bir şey hakkında merak etm.: **to be ~d with rheumatism,** romatizmaya mübtelâ olm.: **may I ~ you to pass the water?,** suyu zahmet eder misiniz? **~d,** rahatsız, kederli; bulanık: ⌜**to fish in ~ waters**⌝, bulanık suda balık avlamak. **~r,** fitneci; rahat bozucu. **~some,** rahatsız edici; belâlı; zahmetli; yorucu.

**trough** [trof]. Tekne; yalak; derin yer. **~ of the sea,** iki dalga arasındaki çukur.

**trounce** [trauns]. Dövmek; yenmek.

**troupe** [trûp]. Takım, trup.

**trousers** [ˈtrauzö(r)z]. **(pair of) ~,** pantalon. **she wears the ~,** o kadın evde hâkimdir.

**trousseau** [ˈtrûsou]. Çeyiz.

**trout** [traut]. Alabalık.

**trowel** [ˈtrauəl]. Mala. ⌜**to lay it on with a ~**⌝, ballandıra ballandıra medhetmek.

**troy** [trôy]. **~ weight,** kuyumculukta kullanılan ağırlık ölçüsü sistemi.

**truan·cy** [ˈtrûənsi]. Dersi asma. **~t,** mektebine mazeretsiz gitmiyen kimse. **to play ~,** dersi asmak.

**truce** [trûs]. Mütareke; muvakkat kurtuluş.

**truck¹** [trʌk]. Trampa (etm.); ufak tefek, âdi şeyler. **to have no ~ with s.o.,** birisinden elini ayağını kesmek. **~².** İki tekerlekli el arabası; demiryol yük vagonu.

**truckle** [ˈtrʌkl]. **to ~ to s.o.,** birine yaranmak.

**truculen·ce** [ˈtrʌkyuləns]. Kabadayılık; huşunet. **~t,** kabadayı, haşin.

**trudge** [trʌc]. Yorgun argın yürüme(k).

**tru·e** [trû]. Doğru, hakikî; sahih; halis; sadık. Doğrultmak; düzeltmek; tesviye etmek. **to breed ~,** aslına uygun olarak çoğalmak: **to come ~,** doğru çıkmak: **this is also ~ for ...,** bu ... için de variddir: **out of ~,** amudî [ufkî] değil; eğri. **~-blue [~-hearted],** sadık, samimî. **~-born Turk,** su katılmamış Türk. **~ism** [ˈtrûizm], malûmu ilâm. **~ly** [ˈtrûli], hakikaten; sadakatle. **I am, yours ~ ...,** (mektubun sonunda) hürmetlerimi sunarım: **yours ~,** (*şak.*) bendeniz.

**trump¹** [trʌmp]. Boru. **the last ~ [the ~ of doom],** İsrafilin sûru. **~².** İskambil kozu; (*sl.*) merd adam. Koz oynamak. **he always turns up ~s,** onun horozu bile yumurtlar: **to ~ up a charge,** birini uydurma suçla ittiham etmek. **~ed-up,** uydurma. **~ery** [ˈtrʌmpəri], iğreti; değersiz: **a ~ excuse,** sudan bir bahane.

**trumpet** [ˈtrʌmpit]. Boru (şeklinde şey). Boru gibi ses çıkarmak. **to blow one's own ~,** kendini medhetmek. **~er,** borazan(cı).

**truncate** [ˈtrʌnkêyt]. Ucunu kesmek.

**truncheon** [ˈtrʌnçən]. Kısa (polis) sopası.

**trundle** [ˈtrʌndl]. Yuvarlamak.

**trunk** [trʌn(g)k]. Gövde; bavul; fil hortumu; ana yol. **~s,** kısa don. **~-call,** muayyen bir bölge haricinine telefon etme. **~-line,** (demiryol) ana hat; (telefon) uzak yerler arasında telefon.

**truss** [trʌs]. Saman demeti; köprü makası; kasık bağı. Demet yapmak; bağlamak.

**trust** [trʌst]. Güven, itimad; emniyet; emanet; tevliyet. Güvenmek, itimad etm.; inanmak; emniyet etm.; ümid etmek. **to ~ s.o. to do stg.,** birinin bir şeyi yapacağına güvenmek: **to ~ s.o. with stg.,** bir şeyi birine tevdi etm.: **to betray one's ~,** birinin itimadını suiistimal etm.: **I ~ you will soon be better,** inşallah yakında iyileşirsiniz: **~ him!,** (*istihza ile*) hiç korkma, yapar! **~ed,** mutemed. **~ee** [–ˈtî], mütevelli; emanetçi; mutemed. **~ful [~ing],** emniyet ve itimad eden; çabuk inanır. **~worthy,** güvenilebilir, itimada şayan; sadık. **~y,** sadık. **~-deed,** tevliyet senedi.

**truth** [ˈtrûθ]. Hakikat, doğruluk. **in ~ [of a ~, ~ to say, ~ to tell],** hakikaten, doğrusu(nu isterseniz): ⌜**~ will out**⌝, ⌜yanlış hesab Bağdaddan döner⌝. **~ful,** doğru sözlü; hakikate uygun.

**try¹** [trây] *n.* Deneme; tecrübe; teşebbüs. **to have a ~ at doing stg.,** bir şeyi şöyle bir denemek. **~² (tried)** [trây, –d]. *v.* Denemek, tecrübe etm.; muhakeme etm.; acı ve ıstırab vermek. Çalışmak; gayret etmek. **to ~ one's best,** elinden geleni yapmak: **his courage was severely tried,** onun cesareti için bu bir imtihan oldu: **to ~ the door,** kapıyı yoklamak: **to ~ one's eyes,** gözlerini yormak: **to ~ for stg.**, bir şeyi elde etmeğe çalışmak: **don't ~ my patience too far,** sabrımı tüketme. **~ on,** elbiseyi prova etm.: **don't ~ that on with me!**, bu oyunu bana oynıyamazsın. **~-on,** blöf. **~ out,** denemek. **~-out,** deneme, prova. **~ over,** bir besteyi tecrübe etmek. **~ing** [ˈtrây·in(g)], çetin; üzücü; zahmetli.

**tryst** [trist]. Buluşma vadi.

**tsar** [zâ(r)]. Çar. **~evitch,** çareviç. **~ina,** çariçe.

**tub** [tʌb]. Badya; banyo; çamasır leğeni; biçimsiz gemi. Banyo vermek. **~by,** fıçı gibi. **~-thumper,** (*coll.*) palavracı hatib.

**tub·e** [tyûb]. Embube; tüb; boru; yeraltı demiryolu. **inner ~,** otomobil iç lâstiği. **~ular** [–yulə(r)], boru şeklinde; embubî.

**tuberc·le** [ˈtyûbö(r)kl]. Küçük yumru kök; verem şişi. **~ular** [–ˈbö(r)-kyulə(r)], veremli. **~ulosis** [–ˈlousis], verem.

**tuck** [tʌk]. Kırma; (*sl.*) yemek. (Elbise) kasmak; kırma yapmak; sokmak. **to ~ a rug round s.o.**, birini bir battaniyeye sarmak: **to ~ in,** içeri sokmak; (*sl.*) iştahla yemek: **to ~ up one's sleeves,** kolları çemrenmek: **to ~ a child up in bed,** çocuğu yatakta sarıp sarmalamak. **~-shop,** mektebliler için pastacı dükkânı.

**Tuesday** [ˈtyûzdêy]. Salı günü.

**tuft** [tʌft]. Perçem; püskül. **~ed,** püsküllü, sorguçlu.

**tug** [tʌg]. Anî ve kuvvetli çekiş; römorkör. Şiddetle çekmek. **to ~ at stg.**, tekrar tekrar ve şiddetle çekmek: **to ~ stg. along,** bir şeyi sürüklemek. **~-boat,** römorkör. **~-of-war,** halat çekme; mücadele.

**tuition** [tyuˈişn]. Tedris; öğretme. **private ~,** hususî dersler.

**tulip** [ˈtyûlip]. Lâle.

**tumble** [ˈtʌmbl]. Düşme; takla; karışıklık. Birdenbire düşmek; yıkılmak; karıştırmak. **to ~ into bed** [(*coll.*) **~ in**], kendini yatağa atıvermek: **to ~ into one's clothes,** acele giyinmek: **to ~ to stg.**, (*coll.*) kavramak: **to ~ down** [**over**], yere düşmek; yıkılmak; çökmek; düşürmek: **to ~ on (to) stg.**, bir şeye rasgelmek, tesadüfen bulmak: **to ~ out,** dışarı düşmek; (*coll.*) yataktan kalkmak: **people ~d over each other to buy the papers,** gazeteler kapışıldı. **~-down,** yıkık, harab, köhne. **~r,** bardak; perendebaz; taklakçı güvercin.

**tumbril** [ˈtʌmbril]. Müteharrik sandıklı boşaltma arabası; fransız ihtilâlinde mahkümları idam yerine götüren araba.

**tummy** [ˈtʌmi]. (*sl.*) = **stomach,** karın.

**tumour** [ˈtyûmə(r)]. Şiş, kabarcık.

**tumult** [ˈtyûmʌlt]. Gürültülü kargaşalık. **~uous** [tyuˈmʌltyuəs], gürültülü; dağdağalı; telaşlı.

**tumulus** [ˈtyûmyuləs]. Höyük.

**tundra** [ˈtʌndra]. Şimalî Asya'da vâsi ağacsız ova.

**tun·e** [tyûn]. Nağme, hava; akord (etm.); (makine vs.yi) iyi bir hale getirmek. **in ~,** akordlu; uygun: **out of ~,** akordsuz; uymıyan: **to change one's ~,** nağmeyi değiştirmek; alçaktan almak: **to the ~ of £100,** yüz liraya kadar: **to ~ in to a station,** radyoda bir istasyonu bulmak: **to ~ up,** (*otom.* vs.) ayarlıyarak mükemmel bir hale getirmek; (orkestra) akord etmek. **~eful,** akordlu. **~er,** akordcu. **~ing,** akord etme: **~-fork,** diyapazon.

**tunic** [ˈtyûnik]. Kadınların ceketi; asker veya polis ceketi.

**tunnel** [ˈtʌnl]. Tünel (açmak).

**turban** [ˈtö(r)bən]. Sarık (şeklinde kadın başlığı).

**turbid** [ˈtö(r)bid]. Bulanık; çamurlu.

**turbot** [ˈtö(r)bʌt]. Kalkan balığı.

**turbulen·ce** [ˈtö(r)byuləns]. Kargaşalık; gürültü. **~t,** şamatacı; anaforlu (su).

**turco-** [ˈtö(r)kou] *pref.* Türk ... **T~phil,** Türk dostu. **T~man,** Türkmen.

**turf** [tö(r)f]. Çimen(lik); kesek. Çimen döşemek. **the ~,** at yarışlarına aid her şey: **to ~ s.o. out,** (*sl.*) birini kovmak.

**turgid** [ˈtö(r)cid]. Şiş kabarmış; tumturaklı.

**Turk** [tö(r)k]. Türk. **he's a young ~,** haşarı çocuktur. **~ey,** Türkiye: **~ red,** parlak kırmızı (kumaş). **~ish,** türk; türkçe: **~ delight,** lokum.

**turkey** [ˈtö(r)ki]. Hindi. **~ cock,** babahindi.

**turmoil** [ˈtö(r)môyl]. Hengâme; gürültü.

**turn**[1] [tö(r)n] *n.* Nöbet, sıra; dönme, devir; köşe; hal; biçim; istidad, meyil; kısa bir gezinti; numara. **~ and ~ about,** sıra ile: **by ~s,** nöbetle: **done to a ~,** (yemek) tam kara pişmiş: **at every ~,** her anda, her yerde: **the sight gave me quite a ~,** (*coll.*) bu manzara beni adamakıllı sarstı: **to do s.o. a good [bad] ~,** birine iyilik [fenalık] yapmak: **⌜one good ~ deserves another⌝,** iyiliğin karşılığı iyiliktir: **a car with a good ~ of speed,** çok süratlı bir otomobil: **in ~,** nöbetle: **he is of a mechanical ~,** makineye istidadı var: **the boy has a serious ~ of mind,** bu çocuk ciddî hallidir: **the milk is on the ~,** süt bozulmağa başladı: **this will serve my ~,** bu benim işimi görür: **to take ~s at doing stg.,** bir işi nöbetleşe yapmak: **the matter has taken a political ~,** mesele siyasî bir mahiyet aldı: **the matter has taken a serious ~,** iş sarpa sardı: **things are taking a ~ for the better,** işler düzelmeğe başladı: **the ~ of the tide,** (i) medle cezir arasında; (ii) işin dönüm noktası. **~**[2] *v.* Dönmek; devretmek; bozulmak; sapmak; müracaat etm.; değişmek; olmak. Çevirmek; döndürmek; bozmak; (mideyi) bulandırmak; aklı çelmek; torna ile şekil vermek; (elbiseyi) tornistan yapmak. **about ~!,** (*mil.*) geriye dön!: **he can ~ his hand to anything,** eli her işe yatar: **to ~ s.o.'s head,** başını döndürmek: **he has ~ed fifty,** ellisini aştı: **the leaves are beginning to ~,** yapraklar sararmağa başladı: **it's ~ed six,** saat altıyı geçti: **not to know where to ~,** nereye başvuracağını bilmemek. **~-table,** (lokomotif vs.ye mahsus) döner plâtform. **~ aside,** bir tarafa çevirmek; sapıtmak. **~ away,** başka tarafa dönmek; defetmek; bir yana çevirmek. **~ back,** geri dönmek; geri çevirmek. **~ down,** indirmek; (lâmbayı) kısmak; reddetmek; aşağıya sapmak. **~-down,** devrik. **~ in,** kıvırmak; (*coll.*) yatmak: **his toes ~ in,** ayakları içeri dönük. **~ into,** *veya* **be ~ed into,** ···e tahavvül etm.; kesilmek: **~ into,** tahvil etm.: **don't ~ the matter into a joke!,** işi alaya dökmeyiniz! **~ off,** sap(ıt)mak: (havagazi vs.yi) kesmek; (musluğu) kapamak. **~ on,** (havagazi vs.yi) açmak; **to ~ on s.o.,** birine saldırmak: **everything ~s on his answer,** her şey onun cevabına bağlıdır; **to ~ s.o. on to do stg.,** birini bir işe koymak. **~ out,** çıkmak; olmak; neticelenmek; yataktan kalkmak; kapı dışarı etm.; kovmak; (havagazı vs.yi) kesmek; imal etm.; (çekmece vs.yi) boşaltmak; yataktan kaldırmak: **as it ~ed out,** halbuki neticede ...: **it has ~ed out as you said,** dediğin çıktı; **it ~s out that ...,** anlaşıldı ki ...: **his son ~ed out badly,** oğlu fena çıktı: **the dog ~ed out to be mine,** meğer o köpek benim köpekmiş: **everyone ~ed out to see the King,** herkes çıkıp kıralı görmeğe geldi: **his feet ~ out,** ayakları dışarıya doğru çevrik: **this factory ~s out all sorts of goods,** bu fabrika her türlü şey imal ediyor: **to ~ out the government,** hükümeti düşürmek: **to ~ out the guard,** nöbetçileri çağırmak: **to ~ a horse out (to grass),** atı otlağa çıkarmak. **~-out,** ictima; kılık kıyafet. **~ over,** yattığı yerde bir taraftan bir tarafa dönmek; (araba vs.) yana devrilmek; (kayık) albura olm.: devirmek; altüst etm.; çevirmek; havale etm.: **the shop ~s over £500 a week,** bu dükkânda haftada 500 lira döner. **~-over,** bir işte muayyen bir müddette dönen para. **~ round,** dönmek; devretmek; fikrini değiştirmek: tersine çevirmek; döndürmek; çevirmek; devrettirmek. **~ to,** işe girişmek; gayret etm.: **the rain ~ed to snow,** yağmur kara çevrildi. **~ up,** çıkıvermek; gelmek; yukarı çevirmek: **to ~ up a lamp,** lâmbayı açmak: **to ~ up the nose,** burun kıvırmak: **to ~ up the nose at stg.,** bir şeye burun kıvırmak: **to ~ up one's sleeves,** yenlerini kıvırmak: **to ~ up a word in the dictionary,** bir kelimeyi lûgatte aramak: **the stink ~ed me up,** (*coll.*) fena koku beni kusturdu. **~coat** [–kout], partiden dönen adam. **~er,** tornacı. **~ing,** devvar; dönen: köşe, dirsek: **~-point,** dönüm noktası. **~key** [–kî], zındacı. **~pike** [–pâyk], müruriye ile geçilen yol. **~screw** [–skrû], tornavida. **~stile** [–stâyl], turnike.

**turnip** [ˈtö(r)nip]. Şalgam.

**turpentine** [ˈtö(r)pəntâyn]. Terementi.

**turpitude** [ˈtö(r)pityûd]. Denaet, alçaklık.

**turquoise** [ˈtö(r)kwâz]. Firuze.
**turret** [ˈtʌrit]. Küçük kule. **~ed,** onlarla süslenmiş (bina).
**turtle** [ˈtö(r)tl]. Su kaplumbağası. **to turn ~,** (gemi vs.) kâmilen devrilmek. **~-dove,** kumru.
**tusk** [tʌsk]. Fil vs.nin büyük dişlerinden her biri. **~er,** büyük dişleri iyi gelişmiş fil.
**tussle** [ˈtʌsl]. Güreşme(k); uğraşma(k).
**tussock** [ˈtʌsək]. Topak şeklinde yetişen uzun çimen.
**tut** [tʌt]. (*ech.*) *Sabırsızlık nidası*; çıkçık.
**tutelage** [ˈtyûtəlic]. Vasilik; vesayet altında bulunma.
**tutor** [ˈtyûtə(r)]. Hususî muallim. **~ial** [–ˈtôriəl], mürebbiye aid; hususî ders.
**tuxedo** [tʌkˈsîdou]. (*Amer.*) Smokin.
**twaddle** [ˈtwodl]. Lâklâkiyet (söylemek).
**twain** [twêyn]. (*şair.*) İki. **to cleave in ~,** (kılıcla) ikiye bölmek.
**twang** [ˈtwan(g)]. (*ech.*) Hımhım. (Gerilmiş kiriş) ses vermek; gitara çalmak; genizden konuşmak.
**'twas** [twoz] = **it was.**
**tweed** [twîd]. İskoç kumaşı. **~s,** bu kumaştan yapılan elbise.
**'tween** [twîn] = **between.**
**tweet** [twît]. (*ech.*) Küçük kuşların cıvıltısı.
**tweezers** [ˈtwîzö(r)z]. Cımbız.
**twelfth** [twelfθ]. On ikinci; on ikide bir. **the ~, grouse** avının başlangıcı (12 Ağustos): **~ Night,** noelden sonra on ikinci gece.
**twelve** [twelv]. On iki. **~month,** bir sene.
**twent·y** [ˈtwenti]. Yirmi. **~ieth** [–əθ], yirminci; yirmide bir.
**twice** [twâys]. İki kere. **to think ~ before doing stg.,** bir şeyi yapmağa çekinmek: **he did not have to be asked ~,** o bu işe dünden hazırdı.
**twiddle** [ˈtwidl]. Çevirip oynamak.
**twig¹** [twig]. *n.* İnce dal. **~²** *v.* (*sl.*) Kavramak, 'çakmak'.
**twilight** [ˈtwâylâyt]. Alaca karanlık.
**'twill** = **it will.**
**twin** [twin]. İkiz; eşit; çift (olan).
**twine** [twâyn]. Kınnap, kalın sicim; büküm. Bük(ül)mek; sar(ıl)mak. **to ~ itself,** kıvrılmak.
**twinge** [twinc]. Süreksiz ince acı. Sancımak. **a ~ of conscience,** anî vicdan azabı.
**twinkl·e** [ˈtwinkl]. Parıltı. Yıldız gibi titreye titreye parıldama(k). **in the ~ing of an eye,** kaşla göz arasında.
**twirl** [twö(r)l]. Kıvrım; fırıldak gibi dönüş. Fırıldatmak, fırıldanmak. **to ~ one's moustache,** bıyığını burmak.
**twist¹** [twist] *n.* Bur(ul)ma; büküm; yılankavi şekil; (top) *v.* **spin. a mental ~,** (zihniyet) gariblik: **the road takes a ~,** yol sapıyor: **with a ~ of the wrist,** bileğini hafifçe bükerek. **~²** *v.* Burmak; kıvırmak; sarmak; ters mana vermek; dolambaçlı olmak. **to '~ s.o. round one's little finger',** birini parmağında oynatmak: **the road ~s and turns,** yol sağa sola kıvrılıyor: **to ~ a rope round a post,** bir ipi bir direğe sarmak. **~ed,** bükülmüş; burmalı: **to give a ~ meaning to stg.,** bir şeye ters mâna vermek. **~er,** cevabı zor bir sual; (*sl.*) hilekâr.
**twit** [twit]. İğnelemek, takılmak. **to ~ s.o. with stg.,** bir şeyi birinin başına kakmak.
**twitch** [twiç]. Seğirme; anî ve asabî hareket; tik. Seğirmek; oynatmak.
**twitter** [ˈtwitə(r)]. Cıvıltı. Cıvıldamak.
**'twixt** [twikst] = **betwixt.**
**two** [tû]. İki. **~ by ~** [**in ~s**], ikişer ikişer: **to put ~ and ~ together,** olanı biteni (sözleri) birbirine eklemek. **~fold,** iki katlı. **~pence** [ˈtʌpens], iki peni. **~penny** [ˈtʌpəni], iki penilik: **~-halfpenny,** iki buçuk penilik; değersiz. **~-edged,** iki taraflı (kılıc vs.): **a ~ sword,** acem kılıcı. **~-seater,** iki kişilik (otomobil). **~-stroke,** iki zamanlı.
**tying** *v.* tie.
**tyke** [tâyk]. Âdi köpek.
**type** [tâyp]. Örnek, nümune; tip; matbaa harfi. Yazı makinesiyle yazmak. **to set ~,** yazı dizmek: **true to ~,** aslına çekmiş. **~write,** yazı makinesiyle yazmak. **~writer,** yazı makinesi. **~-script,** makine ile yazılmış yazı.
**typh·oid** [ˈtâyfôyd]. Tifo. **~us** [ˈtâyfʌs], tifüs.
**typical** [ˈtipikl]. Tipe uygun; timsalî. **that is ~ of him,** tam ondan beklenecek bir şey.
**typ·ist** [ˈtâypist]. Daktilo. **~ography** [–ˈpogrəfi], matbaacılık.
**tyran·ny** [ˈtirəni]. Müstebid idare; gaddarlık; şiddetli nüfuz. **~nical**

[–ˈranikl], müstebidce; zalimane. **~-nize** [ˈtirənâyz], müstebidce davranmak; zulmetmek: **to ~ over,** kasıp kavurmak. **~t** [ˈtâyrənt], müstebid, zalim hükümdar.
**tyre** [tâyə(r)]. Tekerlek çemberi; otomobil lâstiği. **pneumatic-~d,** şişirme lâstikli. **rubber-~d,** lâstikli: **solid-~d,** som lâstikli.
**tyro** [ˈtâyrou]. Acemi.

# U

**U** [yû]. U harfi. **U-boat,** Alman denizaltısı.
**ubiquitous** [yûˈbikwitəs]. Her yerde bulunan.
**udder** [ˈʌdə(r)]. Hayvan memesi.
**ugh** [uf]. Öf!
**ugl·y** [ˈʌgli]. Çirkin; biçimsiz; sakil. **an ~ customer,** zorlu ve tehlikeli adam; netameli kimse: **things are looking ~,** ortalık tehlikeli görünüyor. **~iness,** çirkinlik; biçimsizlik.
**U.K.** [ˈyûˈkêy] = **United Kingdom,** Büyük Britanya ve Şimalî İrlanda.
**ulcer** [ˈʌlsə(r)]. Kendi kendine hasıl olan cerahatli yara; çıban.
**ulster** [ˈʌlstə(r)]. Uzun be bol kemerli palto.
**ult.** = **ultimo.**
**ulterior** [ʌlˈtiəriə(r)]. Ötede olan; daha uzak; sonraki; gizli. **~ motive,** gizli maksad: **without ~ motive,** ivazsız.
**Ultima** [ˈʌltima]. **~ Thule** [ˈθûlî], Şimalin en son yeri; en son varılacak nokta.
**ultim·ate** [ˈʌltimit]. Son; en sonraki; nihaî; asıl. **~atum** [–ˈmêytʌm], ültimatom. **~o,** geçen ay.
**ultra** [ˈʌltra]. Aşırı; son derece; müfrit. **~ vires** [ˈvâyrîz], salâhiyet haricinde. **ultra-,** *pref.* öbür tarafta; ifrat derecede, *mes.* **~-microscopic,** mikroskopla görülmiyecek kadar ufak: **~-conservative,** son derecede muhafazakâr.
**ultramarine** [ˌʌltramaˈrîn]. Lâciverd; deniz aşırı.
**umber** [ˈʌmbə(r)]. Ombra boyası.
**umbilical** [ʌmˈbilikl]. **~ cord,** göbek bağı.
**umbrage** [ˈʌmbric]. Küskünlük. **to take ~ at stg.,** bir şeyden alınmak.
**umbrella** [ʌmˈbrelə]. Kış şemsiyesi. **~-stand,** şemsiyelik.
**umpire** [ˈʌmpâyə(r)]. Hakem.
**umpteen** [ʌmpˈtîn]. (*sl.*) Bir çok.
**un-** [ʌn] *pref. Menfilik ifade eden ön ek; mes.* **true,** hakikî; **untrue,** yalan. *Bu ek hemen her sıfatın başına ilâve edilerek kelimenin aksi mânasını ifade ettiğinden lûgatte yalnız çok kullanılan kelimelerle* **un** *ekinin menfiden başka bir mâna tazammun ettiği kelimeler alınmıştır.* **Un** *ile başlayıp lûgatte bulunmıyan kelimeler için asıl sıfata bak.*
**U.N.** [ˌyûˈen] = **United Nations,** Birleşmiş Milletler.
**unable** [ʌnˈêybl]. Elinden gelmez; gayri muktedir. **to be ~ to do,** yapamamak: **we were ~ to go,** gidemedik.
**unaccompanied** [ʌnaˈkʌmpənid]. Refakatsiz; yalnız.
**unaccountable** [ʌnaˈkauntəbl]. Anlaşılmaz; garib.
**unacquainted** [ʌnaˈkwêyntid]. **to be ~ with,** bilmemek; tanımamak.
**unadorned** [ʌnaˈdônd]. Süssüz; sade.
**unadulterated** [ʌnaˈdʌltərêytid]. Halis muhlis; saf. **~ nonsense,** deli saçması.
**unaffected** [ʌnaˈfektid]. Müteessir olmıyan; tabiî, sadık.
**unaided** [ʌnˈêydid]. Yardımsız. **with the ~ eye,** dürbün vs. kullanmadan.
**unalloyed** [ʌnaˈlôyd]. Mağşuş olmıyan. **~ happiness,** tam saadet.
**unamended** [ʌnaˈmendid]. Olduğu gibi.
**unanim·ity** [ˌyûnaˈnimiti]. Reylerin ittifakı; fikir ittifakı. **~ous** [–ˈnaniməs], aynı fikirde; muttehid.
**unanswerable** [ʌnˈânsərəbl]. Cevab verilemez; cerhedilemez.
**unarmed** [ʌnˈâmd]. Silâhsız.
**unasked** [ʌnˈâskd]. İstenmemiş; sorulmamış; davetsiz. **to do stg. ~,** kendiliğinden yapmak.
**unassailable** [ʌnaˈsêyləbl]. Hücum edilemez; münakaşa götürmez.
**unassuming** [ʌnaˈsyûmin(g)]. Mütevazı, alçakgönüllü.
**unattached** [ʌnaˈtaçt]. Bağlı olmıyan; başlı başına; evli olmıyan.
**unavailing** [ʌnaˈvêylin(g)]. Nafile; semeresiz.
**unavoidabl·e** [ʌnaˈvôydəbl]. Kaçınılmaz; çaresiz; önüne geçilmez. **~y absent,** elinde olmıyan sebeblerden dolayı hazır bulunmıyan.
**unaware** [ʌnaˈweə(r)]. Habersiz. **to be ~ of stg.,** bir şeyden haberi olmamak: **I am not ~ that ...,** ...

bilmez değilim. **~s,** bilmiyerek: **to catch ~,** gafil avlamak.
**unbalance** [ʌnˈbaləns]. Muvazenesini bozmak. **~d,** muvazenesiz.
**unbearable** [ʌnˈbeərəbl]. Tahammül edilmez, dayanılmaz.
**unbeaten** [ʌnˈbîtn]. Yenilmemiş. **~ path,** çiğnememiş patika.
**unbeknown** [ʌnbiˈnoun]. Tanınmıyan; mechul. **to do stg. ~ to anyone,** bir şeyi kimsenin haberi olmadan yapmak.
**unbelie·f** [ʌnbiˈlîf]. İmansızlık; inanmazlık. **~vable** [-ˈlîvəbl], inanılmaz, akla sığmaz. **~ver,** dinsiz, imansız.
**unbend** [ʌnˈbend]. Doğrultmak; çözmek. Açılmak; ciddiyetini biraz bırakmak. **~ing,** eğilmez; sabıt; ciddî.
**unbidden** [ʌnˈbidn]. Davetsiz; emir almadan; kendi başına.
**unblushing** [ʌnˈblʌşin(g)]. Yüzsüz, utanmaz.
**unborn** [ʌnˈbôn]. Henüz doğmamış. **generations yet ~,** gelecek nesiller.
**unbosom** [ʌnˈbuzʌm]. **to ~ oneself,** kalbini açmak; derdini dökmek.
**unbound** [ʌnˈbaund]. Çözülmüş; cildlenmemiş. **~ed,** hadsiz; ölçüsüz.
**unbridled** [ʌnˈbrâydld]. Dizginsiz; müfrit.
**unbroken** [ʌnˈbroukn]. Daha kırılmamış; fasılasız; terbiye görmemiş (at); zabtedilmemiş; işlenmemiş (arazi). **an ~ custom,** öteden beri devam eden âdet: **his record for 100 metres is still ~,** onun 100 metredeki rekoru henüz kırılmamıştır: **he has an ~ record of service,** o fasılasız hizmet görmüştür.
**unburden** [ʌnˈbö(r)dn]. Yükten kurtarmak. **to ~ oneself [one's heart],** derd yanmak, içini dökmek: **to ~ oneself of a secret,** bir sır ifşa ederek ferahlamak.
**uncalled** [ʌnˈkôld]. **~ capital,** tediyesi henüz taleb edilmemiş sermaye. **~ for,** yersiz; haksız.
**uncanny** [ʌnˈkani]. Esrarengiz; acayib.
**uncared-for** [ʌnˈkeədfô(r)]. Bakımsız.
**unceasing** [ʌnˈsîsin(g)]. Durmıyan, fasılasız; mütemadi.
**unceremonious** [ˌʌnseriˈmouniəs]. Teklifsiz; lâübali.
**uncertain** [ʌnˈsö(r)tən]. Gayri muayyen; katî olmıyan; kararsız; şübheli; belli olmıyan. **to be ~ whether ..., ...** katî olarak bilmemek.
**unchallenged** [ʌnˈçalincd]. **his superiority is ~,** üstünlügü su götürmez: **to let s.o. pass ~,** (nöbetçi vs.) hüviyet sormadan geçmesine müsaade etm.; **to let stg. pass ~,** bir şeyi sükûtla karşılamak.
**unchang·eable** [ʌnˈçêyncəbl]. Değişemez. **~ed,** eskisi gibi. **~ing,** değişmez.
**uncharted** [ʌnˈçâtid]. Haritası yapılmamış; mechul.
**unchecked** [ʌnˈçekt]. Serbest.
**uncivil** [ʌnˈsivil]. Nezaketsiz.
**uncle** [ˈʌnkl]. Amca; dayı; (*sl.*) rehinci.
**unclean** [ʌnˈklîn]. Pis, murdar, kirli.
**unclothed** [ʌnˈklouðd]. Çıplak.
**unclouded** [ʌnˈklaudid]. Bulutsuz; berrak.
**uncoil** [ʌnˈkôyl]. Kangalı açmak.
**uncomfortable** [ʌnˈkʌmfətəbl]. Rahatsız. **to be ~ about stg.,** biraz vicdan azabı duymak.
**uncommon** [ʌnˈkomən]. Nadir; az bulunan. **~ly,** fevkalâde: **not ~,** çok defa.
**uncommunicative** [ˌʌnkəˈmyûnikətiv]. Az konuşur; ketum; çekingen.
**uncomplaining** [ˌʌnkəmˈplêynin(g)]. Şikâyet etmez; sabırlı.
**uncomplimentary** [ˌʌnkompliˈmentəri]. Pek takdir edici olmıyan; zemmedici.
**uncompromising** [ʌnˈkomprəmâyzin(g)]. Uzlaşmaz; eğilmez; katî.
**unconcern** [ˌʌnkonˈsə(r)n]. Fütursuzluk. **~ed,** fütursuz; vazifesiz.
**unconditional** [ˌʌnkonˈdişənl]. Şartsız; mutlak.
**unconfirmed** [ˌʌnkonˈfö(r)md]. Tasdik [teyid] edilmemiş.
**unconscionable** [ʌnˈkonşənəbl]. Vicdansız: **an ~ rogue,** yaman bir madrabaz.
**unconscious** [ʌnˈkonşəs]. Kendinden geçmiş; baygın bir halde; bihaber.
**unconsidered** [ˌʌnkonˈsidö(r)d]. Değersiz; düşüncesizce söylenmiş.
**unconstrained** [ˌʌnkənˈstrêynd]. Serbest.
**unconventional** [ˌʌnkənˈvenşənl]. Mûtad hilâfına; kalender; teklifsiz.
**uncouple** [ʌnˈkʌpl]. (Birbirine bağlanmış iki şeyi) çözmek.
**uncouth** [ʌnˈkûθ]. Hoyrat; çirkin; görgüsüz.
**uncover** [ʌnˈkʌvə(r)]. Örtüsünü kal-

dırmak; meydana çıkarmak. **~ed,** örtüsüz; sigortasız: **to remain ~,** şapkasını elinde tutmak.
**uncrowned** [ʌnˈkraund]. Henüz tetvic edilmemiş; kıral olmadığı halde kıral kadar kudretli.
**unct·ion** [ˈʌnkşn]. Tedavi veya takdis için yağ sürme. **extreme ~,** ölmekte olan birisine mukaddes yağ sürmek. **~uous** [ˈʌnktyuəs], nahoş bir şekilde fazla nazik.
**undeceive** [ˌʌndiˈsîv]. Aldanmış bir kimsenin gözünü açmak. **to ~ oneself,** gafletten uyanmak. **~d,** (i) aldanmamış; (ii) aldanmış olduğunu anlıyan.
**undecided** [ˌʌndiˈsâydid]. Kararsız; karar verilmemiş; askıda olan.
**undefin·able** [ˌʌndiˈfâynəbl]. Tarif edilemez. **~ed,** gayri muayyen; mübhem.
**undelivered** [ˌʌndiˈlivö(r)d]. Kurtarılmamış. **~ letter,** teslim edilmemiş mektub.
**undemonstrative** [ˌʌndiˈmonstrətiv]. Hislerini saklıyan; sakin.
**undeniable** [ˌʌndiˈnâyəbl]. İnkâr olunamaz; su götürmez.
**under** [ˈʌndə(r)]. Altında; ···da; aşağı; mucibince. **~ the table,** masanın altında: **to be ~ discussion,** müzakere edilmekte olm.: **~ one's eyes,** gözünün önünde: **~ Queen Elizabeth,** Kıraliçe Elizabeth'in saltanatı zamanında: **~ repair,** tamirde: **to study ~ s.o.,** ···den tahsil etm.: **~ the terms of the treaty,** muahede şartları mucibince.
**under-** *pref. Bu ek ekseriya bitişik yazılır. Kelime başına gelerek*: altında, daha aşağıda, daha küçük, daha az, lüzumundan daha az *mânalarını tazammun eder,* **underground,** yeraltı; **under-nourished,** iyi gıda almıyan; **undersized,** yaşına göre küçük.
**underbid** [ˌʌndəˈbid]. Eksiltmek.
**undercoat** [ˌʌndəˈkout]. Astar boya.
**undercharge** [ˌʌndəˈçâc]. Bir şey için hakikî değerden az para istemek.
**undercloth·es** [ˌʌndəˈklouðz], **~ing** [–ˈklouðin(g)]. İç çamaşırı.
**undercurrent** [ˌʌndəˈkʌrənt]. Su sathının altından akan akıntı; zahirde görünmiyen fakat hakikatte mevcud olan his vs.
**undercut** [ˌʌndəˈkʌt]. Alt kısmını kesmek; birisinden daha ucuza satmak.
**underdo** (**underdid,** **underdone**) [ˌʌndəˈdû, –ˈdid, –ˈdʌn]. Az pişirmek. **~ne,** az pişirilmiş; kanlı (et).
**underdog** [ˌʌndəˈdog]. Tazyik gören kimse veya millet; mazlum.
**under·feed** (**underfed**) [ˌʌndəˈfîd, –ˈfed]. Kâfi yemek vermemek. **~fed,** kâfi gıda almıyan; bakımsız.
**underfoot** [ˌʌndəˈfut]. Ayak altında.
**undergo** (**underwent,** **undergone**) [ˌʌndəˈgou, –ˈwent, –ˈgon]. Çekmek; geçmek; duçar olm.; uğramak. **to ~ an operation,** ameliyat olmak.
**undergraduate** [ˌʌndəˈgradyuit]. Üniversite talebesi.
**underground** [ˌʌndəˈgraund]. Yeraltı; gizli. **the ~,** yeraltı demiryolu: **to go ~,** (*fig.*) polis vs.den gizlenmek.
**undergrow·n** [ˌʌndəˈgroun]. Cılız; yaşına göre küçük. **~th,** ormanda büyük ağaclar altında yetişen çalılar.
**underhand** [ˌʌndəˈhand]. Gizli; alçak; (oyunlarda) topa aşağıdan vurarak.
**under·lie** (**-lying, -lay, -laid**) [ˌʌndəˈlây, –ˈlây·in(g), –ˈlêy, –ˈlêyd]. Altında bulunmak; temeli olmak. **~lying** alttaki; esaslı.
**underline** [ˌʌndəˈlâyn]. Satırın altını çizmek; bir şeyin ehemmiyetini tebarüz ettirmek.
**underling** [ˈʌndəlin(g)]. Madun; ehemmiyetsiz memur.
**underman** [ˌʌndəˈman]. Lâzım olan işçi vs. vermemek. **~ned,** lâzım olan işçi vs.si eksik.
**undermentioned** [ˌʌndəˈmenşənd]. Aşağıda zikredilen.
**undermine** [ˌʌndəˈmâyn]. Temelini çürütmek; (su) oymak; baltalamak; gizli entrikalarla zarar vermek. **to ~ one's health,** sıhhatini tedricen bozmak.
**under·most** [ˌʌndəˈmoust]. En aşağıdaki. **~neath** [ˌʌndəˈnîθ], aşağısında; aşağıdaki.
**underpa·y** (**-paid**) [ˌʌndəˈpêy, –pêyd]. Hak ettiği ücretten daha az para vermek. **~id,** noksan ücret alan.
**underpin** [ˌʌndəˈpin]. Desteklemek.
**underrate** [ˌʌndəˈrêyt]. Kıymetinden az değer vermek; küçümsemek.
**undersell** (**-sold**) [ˌʌndəˈsel, –ˈsould]. (Birisinden) daha ucuz satmak.
**undersigned** [ˌʌndəˈsâynd]. **the ~,** imzası aşağıda yazılı olan.
**undersized** [ˌʌndəˈsâyzd]. Normal hacımdan aşağı; cılız.
**understand** (**-stood**) [ˌʌndəˈstand,

–ˈstud]. Anlamak, kavramak; takdir etm.; bilmek. **to give s.o. to ~,** birine münasib bir şekilde anlatmak: **I am given to ~ that ...,** malûmatıma göre ...: **to ~ horses,** attan anlamak: **'yes, so I ~',** evet, ben de böyle işittim: **we ~ that ...,** öğrendiğimize göre .... **~ing,** anlayışlı, zeki; kavrayış, intikal; anlaşma, ittifak; şart: **to come to an ~ with s.o.,** birisile anlaşmak: **on the ~ that ...,** ... şartile. **~able,** anlaşılabilir; tabiî.

**understate** [ˌʌndəˈstêyt]. Tefrit etmek. **you ~ the case,** vaziyetin ehemmiyetini lâyıkı ile göstermiyorsunuz. **~ment,** tefrit; bir şeyi olduğundan az göstermek.

**understood** [ˌʌndəˈstud] *v.* **understand.** Anlaşılmış; müsellem; (*gram.*) tahtında müstetir. **it is an ~ thing that ...,** malûm bir şeydir ki; ... âdettir.

**understudy** [ˌʌndəˈstʌdi]. İcabında bir aktörün yerine oynamak için rolünü ezberlemek. Yedek aktör.

**undertak·e (-took, taken)** [ˌʌndəˈtêyk, –ˈtuk, -ˈtêykn]. Üzerine almak; deruhde etm.; söz vermek; girişmek. **~er,** cenaze müteahhidi. **~ing,** teşebbüs; taahhüd: **to give an ~,** bir taahhüde girmek.

**undertone** [ˈʌndətou̯n]. Pes ses; talî renk. **there is an ~ of bitterness in all he writes,** bütün yazılarında bir acılık şemmesi var.

**undertow** [ˈʌndətou̯]. Bir nehir veya deniz sathının altında akıntı; sahile çarpan dalgaların geri gitmesi.

**undervalue** [ˌʌndəˈvalyu]. Kıymetinden az değer vermek; hakikî kıymetini takdir etmemek; küçümsemek.

**underwear** *v.* **underclothes.**

**underwent** *v.* **undergo.**

**underworld** [ˈʌndəwö(r)ld]. Ahret; ruhlar diyarı; cehennem. Cemiyetin en aşağı tabakası; caniler âlemi.

**underwrite** [ˌʌndəˈrâyt]. Gemiyi sigorta etmek.

**undesirable** [ʌndiˈzâyrəbl]. İstenmiyen. Bir memlekete girmesi istenilmiyen kimse.

**undeterred** [ʌndiˈtö(r)d]. Azminden dönmiyen. **~ by his failure,** muvaffak olmadığı için yılmadı.

**undeveloped** [ʌndiˈveləpt]. Gelişmemiş; işlenmemiş; kemale ermemiş.

**undeviating** [ʌnˈdîviêytin(g)]. Hiç şaşmıyan; doğru gider.

**undid** *v.* **undo.**

**undiluted** [ˌʌndâyˈlyûtid]. Su katılmadık; saf.

**undischarged** [ˌʌndisˈçâcd]. Dolu (top, tüfek); ödenmemiş. **~ bankrupt,** iflâsı kaldırılmamış olan kimse.

**undisguised** [ˌʌndisˈgâyzd]. Kılığını değiştirmemiş; açık.

**undismayed** [ˌʌndisˈmêyd]. Müşkülat veya tehlikeden yılmıyan.

**undisputed** [ˌʌndisˈpyûtid]. Hakkında münazaa edilmez; sarih.

**undistinguished** [ˌʌndisˈtin(g)gwişt]. Mümtaz olmıyan; alelâde.

**undisturbed** [ˌʌndisˈtö(r)bd]. Karıştırılmamış; rahat; istifi bozulmamış.

**undivided** [ˌʌndiˈvâydid]. Taksim edilmemiş; bütün; ayrılmamış.

**undo (undid, undone)** [ʌnˈdû, ʌnˈdid, ʌnˈdʌn]. Çözmek; bozmak; ihlâl etmek. **to ~ the harm that has been done,** yapılan zararı telâfi etm.; bir şeyi keenlemyekün saymak: **what's done can't be undone,** olan oldu; ok yaydan çıktı. **~ing,** inkıraz sebebi: **drink was his ~,** içki onu mahvetti. **~ne,** yapılmamış; çözülmüş: **to be ~,** çözülmek; yanmak: **we are ~,** hapı yuttuk, yandık: **to leave nothing ~,** etmediğini bırakmamak.

**undoubted** [ʌnˈdau̯tid]. Şübheli olmıyan; muhakkak; şübhesiz.

**undreamt** [ʌnˈdremt]. **~ of,** hıç düşünülmemiş; tasavvur edilemez; akla hayale gelmez.

**undress** [ʌnˈdres]. Elbisesini çıkarmak; soyunmak. Gündelik elbise. **~ uniform,** gündelik üniforma. **~ed,** elbisesini çıkarmiş: **to get ~,** soyunmak: **~ leather,** terbiye edilmemiş deri.

**undue** [ʌnˈdyû]. Fazla; yersiz; ifrat derecede; usule muvafık olmıyan.

**undulate** [ˈʌndyulêyt]. Dalgalanmak.

**unduly** [ʌnˈdyûli]. Lâyık olmıyan bir tarzda; lüzumundan fazla.

**undying** [ʌnˈdây·in(g)]. Lâyemut, ölmez; zeval bulmaz, ebedî.

**unearned** [ʌnˈö(r)nd]. Çalışarak kazanılmamış; müstahak olunmamış. **~ income,** çalışılmadan kazanılan irad.

**unearth** [ʌnˈö(r)θ]. Toprağı kazarak keşfetmek; bulup meydana çıkarmak. **~ly,** dünyaya aid olmıyan; fevkettabia; müdhiş.

**uneasy** [ʌnˈîzi]. Rahatsız, huzursuz; kurtlu. **to feel ~,** endişe etm.;

pirelenmek: **to make s.o. ~,** birinin huzurunu kaçırmak.

**unemploy·ed** [ˌʌnemˈplôyd]. İşsiz, boş, açıkta. **the ~,** işsizler. **~ment,** işsizlik.

**unending** [ʌnˈendin(g)]. Sonsuz; tükenmez.

**unenviable** [ʌnˈenviəbl]. Gıbta edilmiyecek; üzüntülü.

**unequal** [ʌnˈîkwəl]. Müsavi olmıyan. **to be ~ to the task,** işe gücü yetmemek: **I feel ~ to going there today,** bugün oraya gidecek halim yok. **~led,** eşsiz.

**unequivocal** [ʌniˈkwivəkl]. Mânası sarih.

**unerring** [ʌnˈö(r)rin(g)]. Hedefi şaşmıyan.

**unessential** [ʌneˈsenşl]. Tâli; esaslı olmıyan; ehemmiyetsiz.

**uneven** [ʌnˈîvn]. Ârızalı; muntazam [düz] olmıyan. **~ temper,** mizacı belli olmaz.

**unexcelled** [ˌʌnekˈseld]. Emsalsiz; kimse tarafından geçilmemiş.

**unexceptionable** [ˌʌnekˈsepşənəbl]. Bir diyecek yok; mahzursuz.

**unexpected** [ˌʌnekˈspektid]. Umulmadık.

**unexplored** [ˌʌnekˈsplôd]. Henüz keşf [tedkik] edilmemiş; ayak basılmamış.

**unfading** [ʌnˈfêydin(g)]. Solmaz; ebedî.

**unfailing** [ʌnˈfêylin(g)]. Şaşmaz; bitmez tükenmez; hiç eksik olmaz.

**unfair** [ʌnˈfeə(r)]. İnsafsız, haksız; tarafsız olmıyan; hileli. **~ play,** mızıkçılık.

**unfaltering** [ʌnˈfoltərin(g)]. Tereddüdsüz.

**unfamiliar** [ˌʌnfəˈmilyə(r)]. Garib; yabancı; alışılmamış; alışkın olmıyan.

**unfashionable** [ʌnˈfaşənəbl]. Modası geçmiş; modaya uygun olmıyan.

**unfathom·able** [ʌnˈfaðəməbl]. Dibi bulunmaz; sırrına erişilemez.

**unfed** [ʌnˈfed]. Gıda almamış, aç.

**unfeeling** [ʌnˈfîlin(g)]. Hissiz; merhametsiz.

**unfeigned** [ʌnˈfêynd]. Samimî; yapma olmıyan.

**unfilial** [ʌnˈfilyəl]. Oğula yakışmaz.

**unfit** [ʌnˈfit]. Uymaz ehliyetsiz, elverişsiz; çürük, sıhhati bozuk. Elverişsiz hale koymak. **to be discharged as ~,** çürüğe çıkmak. **~ted, ~ for [to do] stg.,** bir şeye yaramıyan. **~ting,** yakışmaz; uygun olmıyan.

**unflagging** [ʌnˈflagin(g)]. Yorulmak bilmez; durmaz, devamlı.

**unflattering** [ʌnˈflatərin(g)]. Zemmedici; pek takdirkâr olmıyan.

**unfledged** [ʌnˈflecd]. Henüz tüylenmemiş; dünyayı bilmiyen.

**unflinching** [ʌnˈflinçin(g)]. Çekinmez; yılmaz.

**unfold** [ʌnˈfould]. Katlanmış birşeyi açılmak; yayılmak; meydana koymak; anlatmak.

**unforeseen** [ˌʌnfôˈsîn]. Gayri melhuz.

**unforgettable** [ˌʌnfəˈgetəbl]. Unutulmaz.

**unforgiv·able** [ˌʌnfəˈgivəbl]. Affedilmez. **~ing,** affetmez, kindar.

**unformed** [ʌnˈfômd]. Çelimsiz, şekilsiz.

**unfortunate** [ʌnˈfôtyunit]. Talihsiz; aksi. **~ly,** ne yazık ki; maalesef.

**unfounded** [ʌnˈfaundid]. Esassız; sebebsiz.

**unfrequented** [ˌʌnfriˈkwentid]. Issız; tenha.

**unfriendly** [ʌnˈfrendli]. Dostane olmıyan; soğuk. **to meet with an ~ reception,** soğuk karşılanmak.

**unfrock** [ʌnˈfrok]. Papazlıktan çıkarmak.

**unfulfilled** [ˌʌnfulˈfild]. İcra edilmemiş; tatmin edilmemiş (arzu).

**unfurl** [ʌnˈfö(r)l]. (Bayrağı) açmak.

**ungainly** [ʌnˈgêynli]. Hantal, biçimsiz.

**ungallant** [ʌnˈgalənt]. Kadına karşı nezaketsiz.

**un-get-at-able** [ˌʌngetˈatəbl]. Erişilemez.

**ungodly** [ʌnˈgodli]. Dinsiz; (*coll.*) Allahın belâsı.

**ungovernable** [ʌnˈgʌvənəbl]. Zabtolunmaz; taşkın.

**ungracious** [ʌnˈgrêyşəs]. Nezaketsiz; ters.

**ungrat·eful** [ʌnˈgrêytfəl]. Nankör. **~ified** [–ifâyd], tatmin edilmemiş.

**ungrounded** [ʌnˈgraundid]. Esassız.

**ungrudging** [ʌnˈgrʌcin(g)]. Esirgemiyen; memnuniyetle verilen; cömerd.

**unguarded** [ʌnˈgâdid]. Muhafaza edilmemiş; dikkatsiz(ce yapılmış).

**unhallowed** [ʌnˈhaloud]. Takdis edilmemiş; dinsiz. **~ joy,** şeytanî neşe.

**unhampered** [ʌnˈhampö(r)d]. Serbest.

**unhandy** [ʌnˈhandi]. Beceriksiz; kullanışsız.
**unhanged** [ʌnˈhan(g)d]. İpten kazıktan kurtulmuş.
**unhapp·y** [ʌnˈhapi]. Mahzun; bedbaht; mesud olmıyan. **an ~ expression,** *etc.*, yersiz bir tabir vs.: **in an ~ moment** uğursuz bir dakikada: **to make oneself ~ about stg.,** bir şeyi tasa etmek. **~ily,** mesud olmıyarak; maalesef; aksi gibi: **that was rather ~ put,** ifade şekli uygun düşmedi.
**unharmed** [ʌnˈhâmd]. Sağ ve salim; zarara uğramamış.
**unhealthy** [ʌnˈhelθi]. Hastalıklı; sıhhate muzır. **an ~ curiosity,** marazî bir tecessüs.
**unheard** [ʌnˈhö(r)d]. **to condemn s.o. ~,** birini dinlemeden mahkûm etm.: **~ of,** hiç işidilmemiş; şöhretsiz.
**unhinge** [ʌnˈhinc]. Aklını oynatmak.
**unholy** [ʌnˈhǫuli]. Dine mugayir; habis. **an ~ muddle [mess],** (*coll.*) müdhiş karışıklık.
**unhoped** [ˈʌnˈhǫupt]. **~ for,** umulduğundan iyi; hiç beklenmiyen (iyilik).
**unhorse** [ʌnˈhôs]. Attan düşürmek.
**unhurt** [ʌnˈhö(r)t]. Zarar görmemiş.
**uni-** [ˈyûni] *pref.* Bir olan; bir ···den yapılmış: tek ....
**unicorn** [ˈyûnikôn]. Tek boynuzlu at cinsinden efsanevi bir hayvan.
**unidentified** [ˌʌnâyˈdentifâyd]. Hüviyeti tesbit edilmemiş.
**unif·ication** [ˌyûnifiˈkêyşn]. Birleştirme. **~y** [–fây], bir yapmak; birleştirmek.
**uniform** [ˈyûnifôm]. Üniforma; resmi elbise. Yeknesak; birbirlerine benzer; mütecanis; değişmiyen; dümdüz. **full ~,** büyük üniforma. **~ity** [–ˈfômiti], yeknesaklık, tecanüs.
**unilateral** [ˌyûniˈlatərəl]. Tek taraflı.
**unimagin·able** [ˌʌniˈmacinəbl]. Tasavvur edilemez; akla sığmaz. **~ative,** muhayyilesi dar.
**unimpeachable** [ˌʌnimˈpîçəbl]. Şübhe edilemez; cerhedilemez.
**uninformed** [ˌʌninˈfômd]. Bihaber; cahil.
**uninspired** [ˌʌninˈspâyö(r)d]. İlhamsız.
**uninterest·ed** [ʌnˈintrestid]. Alâkasız; bigâne. **~ing,** enteresan olmıyan.
**uninterrupted** [ʌnintəˈrʌptid]. Fasılasız: **~ly,** ara vermeden.
**uninviting** [ʌninˈvâytin(g)]. Cazibesiz; iştiha celbetmiyen.
**union** [ˈyûnyən]. Birlik; cemiyet; sendika; ittihad; iki şeyin birleştiği yer. **the U~ Jack,** İngiliz bayrağı. **~ist,** ittihadcı; sendikacı.
**unique** [yûˈnîk]. Yegâne, biricik; eşsiz.
**unison** [ˈyûnizn]. **all in ~,** hepsi aynı zamanda; hep beraber.
**unit** [ˈyûnit]. Vahdet; vahidi kıyasî; birlik; bir bütün teşkil eden şey. **~e** [yûˈnâyt], birleş(tir)mek. **~ed,** birleşik, birleşmiş; müttehid: **the U~ Kingdom (U.K.),** Büyük Britanya: **the U~ States (U.S.A.),** Amerika Birleşik Devletleri. **~y** [ˈyûniti], vahdet; teklik; ittihad; birlik.
**univers·al** [yûniˈvö(r)sl]. Küllî; âlemşümul; umumî. Külliyet. **~e** [ˈyûnivö(r)s], kâinat; mevcud olan şeyler; âlem; dünya. **~ity** [–ˈvö(r)siti]. Üniversite.
**unjust** [ʌnˈcʌst]. Haksız; adaletsiz.
**unkempt** [ˌʌnˈkempt]. Taranmamış; hırpani.
**unkind(ly)** [ʌnˈkâynd(li)]. Dostane olmıyan; sert; insafsız. **an ~ fate,** zalim bir talih: **don't take it ~ly,** hatırınız kırılmasın!
**unknown** [ʌnˈnǫun]. Mechul; tanınmıyan. **~ to me,** benim haberim olmadan.
**unlace** [ʌnˈlêys]. Bağları çözmek.
**unladylike** [ʌnˈlêydilâyk]. Bir hanıma yakışmaz.
**unlawful** [ʌnˈlôfəl]. Gayrikanunî; haram.
**unlearn** [ʌnˈlö(r)n]. Alıştığından vazgeçmek. **~ed,** [–ˈlö(r)nd], öğrenilmemiş (ders); [–ˈlö(r)nid], cahil.
**unless** [ʌnˈles]. ···mezse; ···medikçe. **I shan't go ~ I hear from you,** sizden haber almazsam gitmeyeceğim: **I won't go with anyone ~ it be you,** sizden maada kimse ile gitmem.
**unlettered** [ʌnˈletö(r)d]. Okumamış; ümmî.
**unlicensed** [ʌnˈlâysənst]. Vesikasız; ehliyetnamesiz; ruhsatsız.
**unlicked** [ʌnˈlikt]. **an ~ cub,** yontulmamış delikanlı.
**unlike** [ʌnˈlâyk]. Benzemez; ... gibi olmıyan. **that was most ~ you!,** bunu size hiç yakıştıramadım: **~ his father he dislikes music,** babasının aksine musikiyi sevmez. **~ly,** muhtemel değil: **in the ~ event of ...,** farzı muhal. **~lihood,** muhtemel olmayış.
**unlimited** [ʌnˈlimitid]. Hadsız hesabsız.

**unload** [ʌnˈloud]. (Yükünü) boşaltmak [indirmek]. **~ed,** boş; tahliye edilmiş.
**unlock** [ʌnˈlok]. Kilidini açmak. **~ed,** kilidlenmemiş.
**unlooked** [ʌnˈlukt]. **~ for,** beklenmedik.
**unloose** [ʌnˈlûs]. Çözmek. **to ~ s.o.'s tongue,** birini söyletmek.
**unlovely** [ʌnˈlʌvli]. Cazibesiz; çirkin.
**unluck·y** [ʌnˈlʌki]. Bedbaht; talihsiz. **it's ~ that he saw you,** seni görmesi aksilik oldu. **~ily,** maalesef.
**unmake** [ʌnˈmêyk]. Yaptığını bozmak.
**unman** [ʌnˈman]. Cesaretini kırmak; erkeği kadın gibi ağlatmak. **~ly,** erkeğe yakışmaz.
**unmanageable** [ʌnˈmanicəbl]. İdare edilemez.
**unmannerly** [ʌnˈmanö(r)li]. Terbiyesiz.
**unmask** [ʌnˈmâsk]. Maskesini açmak; meydana çıkarmak.
**unmatched** [ʌnˈmaçt]. Eşsiz, emsalsiz.
**unmean·ing** [ʌnˈmînin(g)]. Mânasız. **~t** [ʌnˈment], kasdedilmemiş.
**unmeasured** [ʌnˈmejö(r)d]. Ölçülmemiş; ölçüsüz; hadsız.
**unmentionable** [ʌnˈmenşənəbl]. Ağza alınmaz; iğrenc.
**unmerciful** [ʌnˈmö(r)sifəl]. Amansız.
**unmindful** [ʌnˈmâyndfəl]. **~ of one's duty,** vazifesini unutarak.
**unmistakable** [ˌʌnmisˈtêykəbl]. Aşikâr.
**unmitigated** [ʌnˈmitigêytid]. Son derece.
**unmixed** [ʌnˈmikst]. Mahlût olmiyan; halis. **it is not an ~ blessing,** mahzuru yok değil.
**unmounted** [ʌnˈmauntid]. Atlı olmıyan; yaya; takılmamış (mücevher); kartonsuz (fotograf).
**unmoved** [ʌnˈmûvd]. Kımıldanmamış; müteessir olmamış; niyetinden dönmemiş.
**unnecessary** [ʌnˈnesəsəri]. Lüzumsuz; fuzuli.
**unnerve** [ʌnˈnö(r)v]. Cesaretini kırmak.
**unnoticed** [ʌnˈnoutist]. Gözden kaçmış. **to leave ~,** (i) bir şeye göz yummak; (ii) göze çarpmadan gitmek.
**unnumbered** [ʌnˈnʌmbö(r)d]. Sayısız; numarası konmamış.
**unobjectionable** [ˌʌnobˈcekşənəbl]. Aleyhine diyecek bir şey yok.
**unobserved** [ˌʌnobˈzö(r)vd]. Görülmemiş.
**unobtrusive** [ˌʌnobˈtrûsiv]. Göze çarpmaz.
**unopposed** [ˌʌnoˈpouzd]. Muhalefetsiz.
**unpack** [ʌnˈpak]. Eşyasını bavulundan çıkarmak; sandık açmak.
**unpaid** [ʌnˈpêyd]. Maaşını almamış; ödenmemiş; maaşsız çalışan.
**unpalatable** [ʌnˈpalətəbl]. Tatsız; nahoş.
**unparalleled** [ʌnˈparəleld]. Emsalsiz; naziri yok.
**unparliamentary** [ˌʌnpâləˈmentəri]. Parlamento usullerine mugayir: **~ language,** terbiyesiz sözler.
**unperceived** [ˌʌnpəˈsîvd]. Farkına varılmamış; görülmemiş olarak.
**unperturbed** [ˌʌnpəˈtö(r)bd]. Fütursuz.
**unpin** [ʌnˈpin]. Toplu iğnelerini çıkarmak.
**unplaced** [ʌnˈplêyst]. Yarışta yer kazanmıyan; imtihanda derece almıyan.
**unpolished** [ʌnˈpolişt]. Cilâsız; (elmas vs.) ham; kaba.
**unpopula·r** [ʌnˈpopyulə(r)]. Halkın hoşuna gitmiyen. **~ted,** iskân edilmemiş.
**unpractical** [ʌnˈpraktikl]. Kullanışlı olmıyan; iş adamı değil.
**unprecedented** [ʌnˈprîsidentid]. Emsali görülmemiş.
**unpredictable** [ʌnprîˈdiktəbl]. Önceden bilinmez.
**unprepared** [ˈʌnpriˈpeərd]. Hazırlanmamış; irticalen söylenmiş. **to be caught ~,** gafil avlanmak.
**unprincipled** [ʌnˈprinsipld]. Ahlâksız.
**unprintable** [ʌnˈprintəbl]. Tabedilemiyecek kadar (kaba vs.).
**unprofessional** [ˌʌnprouˈfeşənl]. Meslek usullerine aykırı.
**unprompted** [ʌnˈpromptid]. Kendiliğinden.
**unprovided** [ˌʌnprouˈvâydid]. **~ with stg.,** bir şeyden mahrum: **his family was left ~ for,** ailesini geçindirecek bir şey bırakmadı.
**unprovoked** [ˌʌnprouˈvoukt]. Sebebsiz.
**unqualified** [ʌnˈkwolifâyd]. İhtisassız; diplomasız; tam, şartsız.
**unquestion·able** [ʌnˈkwesçənəbl]. Su götürmez; şübhesiz; muhakkak. **~ing, ~ obedience,** tam itaat.

**unquiet** [ʌnˈkwâyət]. Rahatsız; meraklı.
**unread** [ʌnˈred]. Okunmamış (kitab vs.); tahsil görmemiş.
**unreal** [ʌnˈriəl]. Hakikî olmıyan; hayalî.
**unreason·able** [ˈʌnˈrîznəbl]. İnsafsız; müfrit. **~ing**, mantıksız.
**unredeemed** [ˌʌnriˈdîmd]. Rehinden çıkarılmamış; telâfi edilmemiş.
**unregenerate** [ˌʌnriˈcenərit]. Islah olunmamış; tövbekâr olmamış.
**unrelated** [ˌʌnriˈlêytid]. Başka şeyle alâkası olmıyan; akraba olmıyan.
**unrelenting** [ˌʌnriˈlentin(g)]. Katı yürekli.
**unrelieved** [ʌnriˈlîvd]. Derdi hafiflememiş; yeknesak.
**unremitting** [ˌʌnriˈmitin(g)]. Fasılasız.
**unrequited** [ˌʌnriˈkwâytid]. Mükâfatı verilmemiş. **~ love,** mukabele görmiyen aşk.
**unreservedly** [ˌʌnriˈzö(r)vidli]. Şartsız; büsbütün; açıkça.
**unrest** [ʌnˈrest]. Rahatsızlık; kargaşalık.
**unrestricted** [ˌʌnriˈstriktid]. Serbest.
**unrivalled** [ʌnˈrâyvld]. Rakibsiz; eşsiz.
**unroll** [ʌnˈroul]. Aç(ıl)mak; gözlerinin önüne ser(il)mek.
**unruffled** [ʌnˈrʌfld]. İstifini bozmıyan.
**unruly** [ʌnˈrûli]. Ele avuca sığmaz.
**unsafe** [ʌnˈsêyf]. Emin [sağlam] olmıyan; tehlikeli.
**unsaid** [ʌnˈsed] *v.* **unsay. to leave ~,** meskût geçmek.
**unsatis·factory** [ˌʌnsatisˈfaktəri]. Kusurlu; memnun etmiyen. **~fied** [–ˈsatisfâyd], doymamış; gayrimemnun.
**unsavoury** [ʌnˈsêyvəri]. Fena kokulu; tadı fena; rezil.
**unsay** (unsaid) [ʌnˈsêy, –ˈsed]. Söylediğini geri almak.
**unscrupulous** [ʌnˈskrûpyuləs]. Vicdansız.
**unseal** [ʌnˈsîl]. Mühürünü kırmak.
**unseasonable** [ʌnˈsîzənəbl]. Mevsimsiz.
**unseat** [ʌnˈsît]. Attan düşürmek; mebusluktan çıkarmak.
**unsecured** [ˌʌnsiˈkyuəd]. İyi bağlamamış; kefaleti olmıyan.
**unseemly** [ʌnˈsîmli]. Yakışmaz. **~ words,** ileri geri sözler.
**unseen** [ʌnˈsîn]. Görülmemiş; gizli. **the ~**, öbür dünya; iyi saatte olsunlar: **~ translation,** hazırlamadan yapılan tercüme.
**unselfish** [ʌnˈselfiş]. Hodbin olmıyan.
**unsettle** [ʌnˈsetl]. Bulandırmak; karıştırmak. **~d,** tesbit edilmemiş; (hava) değişik; (memleket) gayri meskûn; (insan) mütereddid; (hesab) ödenmemiş.
**unshaven** [ʌnˈşêyvn]. Tıraşı uzamış.
**unsheathe** [ʌnˈşîð]. Kınından çıkarmak.
**unshod** [ʌnˈşod]. Nalsız; yalın ayak.
**unshrinkable** [ʌnˈşrin(g)kəbl]. Çekmez.
**unsightly** [ʌnˈsâytli]. Göze batan; çirkin.
**unsinkable** [ʌnˈsin(g)kəbl]. (Suya) batmaz.
**unskil·ful** [ʌnˈskilfl]. Naharetsiz. **~led,** tecrübesiz: **~ labour,** ihtısassız işçiler.
**unsolicited** [ˌʌnsouˈlisitid]. İstenilmemiş.
**unsophisticated** [ˌʌnsouˈfistikêytid]. Tabiî, sadedil.
**unsound** [ʌnˈsaund]. Sağlam olmıyan; çürük; bozuk. **of ~ mind,** şuuru muhtel.
**unsparing** [ʌnˈspeərin(g)]. Esirgemiyen; bol. **~ of others,** başkalarına karşı insafsız.
**unspeakable** [ʌnˈspîkəbl]. Ağza alınmaz; iğrenc; tarif edilemez.
**unsplinterable** [ʌnˈsplintərəbl]. Parçalanmadan kırılan (cam).
**unspoken** [ʌnˈspoukn]. Söylenmemiş; zımnî.
**unsteady** [ʌnˈstedi]. Kararsız; sendeliyen.
**unstinted** [ʌnˈstintid]. Bol.
**unstitch** [ʌnˈstiç]. Dikişini sökmek.
**unstop** [ʌnˈstop]. Tıpasını çıkarmak. **~ped,** tıpası çıkmış: **(tooth) to come ~,** (dişin) dolgusu çıkmak.
**unstuck** [ʌnˈstʌk]. **to come ~,** (yapışık bir şey) ayrılmak; (*sl.*) hata etm.
**unsuited** [ʌnˈsyûtid]. **~ to [for],** ···e uymaz, yakışmaz, münasib olmıyan.
**unsurpassed** [ˌʌnsəˈpâst]. Emsalsiz.
**unsuspected** [ˌʌnsʌsˈpektid]. Umulmadık; hakkında şübhe olmıyan.
**unswerving** [ʌnˈswö(r)vin(g)]. Sapmaz; metin.
**unsympathetic** [ˌʌnsimpaˈθetik]. Başkasının hislerine iştirak etmiyen; soğuk.
**untameable** [ʌnˈtêyməbl]. Ehlileştirilemiyen; zabtolunmaz.

**untapped** [ʌnˈtapt]. Delinmemiş (fıçı). ~ **resources,** henüz kullanılmamış kaynaklar vs.
**untaught** [ʌnˈtôt]. Cahil.
**unthink·able** [ʌnˈθin(g)kəbl]. Düşünülmez, akla gelmez. **~ing,** düşüncesiz; dalgın; dikkatsizce yapılmış.
**untimely** [ʌnˈtâymli]. Vaktinde olmıyan. **to come to an ~ end,** vaktinden evvel ölmek; muvaffak olmadan çabucak sona ermek.
**untiring** [ʌnˈtâyərin(g)]. Yorulmak bilmez.
**unto** [ˈʌntû]. **to**'*in eski şekli.*
**untold** [ʌnˈtould]. Nakledilmemiş; sayılmaz. **worth ~ gold,** dünya kadar altın değerinde.
**untouch·able** [ʌnˈtʌçəbl]. El sürülemez; (Hindistanda) en aşağı sınıfa mensub. **~ed,** dokunulmamış; zarar görmemiş; tamam; müteessir olmıyan: **he left his food ~,** yemeğine el sürmedi.
**untoward** [ʌnˈtouö(r)d]. Talihsiz; aksi.
**untrammelled** [ʌnˈtraməld]. Serbest. **~ by convention,** her türlü tekellüften âri.
**untried** [ʌnˈtrâyd]. Tecrübe edilmemiş.
**untrodden** [ʌnˈtrodn]. Ayak basılmamış; bakir (orman); kuş uçmaz kervan geçmez. **~ path,** (*fig.*) yeni bir çığır.
**untroubled** [ʌnˈtrʌbld]. Rahat; sakin.
**untru·e** [ʌnˈtrû]. Yalan; vefasız. **~th,** yalan.
**untutored** [ʌnˈtyûtö(r)d]. Cahil; fıtrî.
**unused** [ʌnˈyûzd]. Kullanılmamış; kullanılmıyan. [ʌnˈyûst], alışmamış.
**unusual** [ʌnˈyûjuəl]. Mûtad hilâfına; nadir; müstesna.
**unutterable** [ʌnˈʌtrəbl]. Ağza alınmaz.
**unvaried** [ʌnˈveərid]. Değişmez; yeknesak.
**unvarnished** [ʌnˈvânişt]. Verniklenmemiş. **a plain ~ tale,** süssüz ilâvesiz hikâye.
**unveil** [ʌnˈvêyl]. Peçesini kaldırmak. **to ~ a statue,** *etc.*, bir heykel vs.yi merasimle açmak.
**unversed** [ʌnˈvö(r)st]. **~ in,** ···in cahili.
**unvoiced** [ʌnˈvôyst]. Sözle söylenmemiş; ifade edilmemiş.
**unwarrant·able** [ʌnˈworəntəbl]. Haklı telâkki edilemez. **~ed,** mazur görülmez; haksız.
**unwary** [ʌnˈweəri]. Gafil; ihtiyatsız.
**unwashed** [ʌnˈwoşt]. Yıkanmamış. **the Great U~,** ayak takımı.
**unwavering** [ʌnˈwêyvərin(g)]. Sabit; tereddüd etmiyen.
**unwear·ied** [ʌnˈwiərid]. Yorulmamış, yorulmaz. **~ying,** yorulmak bilmez.
**unwelcome** [ʌnˈwelkəm]. Nahoş.
**unwell** [ʌnˈwel]. Hafifçe hasta, keyifsiz.
**unwept** [ʌnˈwept]. Ağlanmamış.
**unwholesome** [ʌnˈhoulsəm]. Sıhhate muzır; ağır (yemek); ahlâk bozucu.
**unwieldy** [ʌnˈwîldi]. Havaleli; hantal.
**unwilling** [ʌnˈwilin(g)]. İsteksiz, istemiyerek.
**unwind** (**unwound**) [ʌnˈwâynd, -ˈwaund]. Dolanmış [dürülmüş] şeyi açmak.
**unwise** [ʌnˈwâyz]. Akıllı olmıyan; akıllı iş değil.
**unwittingly** [ʌnˈwitin(g)li]. Kasden olmamış; bilmiyerek.
**unwonted** [ʌnˈwountid]. Alışılmamış.
**unworkable** [ʌnˈwö(r)kəbl]. İşlenmez; tatbik edilemez.
**unworldly** [ʌnˈwö(r)ldli]. Dünyevî olmıyan.
**unworthy** [ʌnˈwö(r)ði]. Lâyık olmıyan; müstahak olmıyan.
**unwrap** [ʌnˈrap]. Açmak; sargı çıkarmak.
**unwritten** [ʌnˈritn]. Yazılmamış; şifahî. **the ~ law,** teamüle müstenid kanun.
**unyielding** [ʌnˈyîldin(g)]. Boyun eğmez; teslim olmıyan; sert.
**up** [ʌp]. Yukarı, yukarıda, yukarıya; (fiat) yükselmiş. [*Fiille birlikte:* (i) *yukarıya doğru hareket ifade eder, mes.* **to go up,** yukarı çıkmak; **to raise up,** yükseltmek; (ii) *sonuna kadar götürülen bir hareket ifade eder, mes.* **to use up,** kullanıp bitirmek.] **to be ~ and about,** (hasta) iyileşip gezip yürümek: **we're ~ against it!,** işte şimdi çattık!: **it's all ~ with us,** hapı yuttuk; yandık: **his blood was ~,** öfkelendi: **we must be ~ and doing,** haydi iş başına: **~s and downs,** değişiklik; iniş yokuş; kâh düşme kâh kalkma: **the ~s and downs of life,** feleğin germü serdi: **to curse s.o. ~ and down,** birini tepeden tırnağa donatmak: **to walk ~ and down,** bir aşağı bir yukarı gezmek: **is he ~ (yet)?,** kalktı mı?: **it's not ~ to much,**

pek bir şeye benzemiyor: **after his illness he is not ~ to much,** hastalığından beri pek işe yaramıyor: **Parliament is ~,** Parlamento kapalı: **there's something ~,** bir şeyler oluyor: **~ to ...,** ···e kadar: **what are you ~ to?,** ne halt ediyorsunuz?: **he's ~ to something or other,** her halde bir dolab çeviriyor: **the next step is ~ to you [it is ~ to you to take the next step],** bundan sonraki teşebbüsü yapmak size aiddir: **what's ~?,** ne var?: **what's ~ with you?,** ne oluyorsunuz? **~-and-down motion,** yukarı aşağı hareket. **~-end,** dikine oturtmak. **~-grade,** yokuş: **to be on the ~,** ilerlemekte olmak. **~-hill,** yukarıya doğru giden; güç, çetin. **~-river,** nehrin yukarı tarafında [tarafına]. **~-stream,** akıntıya karşı. **~-to-date,** asrî, son moda.

**upbraid** [ʌpˈbrêyd]. Azarlamak.

**upbringing** [ʌpˈbrin(g)in(g)]. Çocuk terbiyesi. **what was his ~?,** nerede yetişmiş?

**upheaval** [ʌpˈhîvəl]. Arzın kabuğunun kabarması; âni ve büyük değişiklik.

**uphold** (**upheld**) [ʌpˈhould, –held]. Tutmak; muhafaza etm.; tasdik etmek.

**upholster** [ʌpˈhoulstə(r)]. Koltuk vs.yi doldurup kumaşla kaplamak. **~er,** döşemeci. **~y,** döşeme(cilik).

**upkeep** [ˈʌpkîp]. Bakım; idame (masrafı).

**upland** [ˈʌplənd]. Yaylada bulunan. **the ~s,** yayla.

**uplift** *n.* [ˈʌplift]. Yükseltme; kalkınma. *v.* [ʌpˈlift]. Yükseltmek.

**upon** [ʌˈpon]. Üzerine, üzerinde; *v.* **on. winter is ~ us,** kış gelip çattı.

**upper** [ˈʌpə(r)]. Üst; üstteki; yukarıdaki. **~s,** kunduranın yüsü. **to be (down) on one's ~s,** zarure içinde olm. **the ~ classes [the ~ ten (thousand)],** kibar tabaka: **to get the ~ hand,** üstün gelmek: **the ~ House,** Lordlar Kamarası. **~most,** en üsttedeki; birinci.

**uppish** [ˈʌpiş]. (*coll.*) Şımarık, yüzsüz.

**upright** [ˈʌprâyt]. Dik; amudî; namuslu. Amudî kısım, direk. **bolt ~,** dimdik: **out of the ~,** amudî olmıyan: **~ piano,** kuyruksuz piyano. **~ness,** doğruluk.

**uprising** [ʌpˈrâyzin(g)]. Ayaklanma; kalkış.

**uproar** [ˈʌprô(r)]. Velvele; kargaşalık. **~ious** [–ˈrôriəs], şamatalı: **to laugh ~ly,** kahkahalarla gülmek.

**uproot** [ʌpˈrût]. Kökünden söküp çıkarmak.

**upset** (**upset**) [ʌpˈset]. Devirmek; bozmak; üzmek, canını sıkmak; musallat olm.; dokunmak. Devrilmek; alabura olmak. Devrilmiş; altüst; bozulmuş; müteessir; keyfi kaçmış. Devrilme; karışıklık. **beer ~s me,** bira bana dokunur: **he is easily ~,** en kücük şeye üzülür: **don't ~ yourself!,** üzülme!

**upshot** [ˈʌpşot]. Netice, akibet.

**upside-down** [ˌʌpsâydˈdaun]. Altüst; ters. **to hold stg. ~,** bir şey baş aşağı tutmak: **to turn ~,** altüst etmek.

**upstairs** [ʌpˈsteəz]. Üst katta, üst kata.

**upstanding** [ʌpˈstandin(g)]. Boyu bosu yerinde; dik.

**upstart** [ˈʌpstât]. Türedi; zıpçıktı.

**uptake** [ˈʌptêyk]. Kavrayış, intikal; yukarıya çekme. **quick in the ~,** çabuk kavrar: **slow in the ~,** kalınkafalı.

**upturn** [ʌpˈtö(r)n]. (Toprağı) çevirmek. Yükselme: **prices are on the ~,** fiatlar yükseliyor. **~ed,** yukarıya çevrilmiş [bakan]; kalkık (burun).

**upward** [ˈʌpwö(r)d]. Yukarıya doğru giden. **~s,** yukarıya doğru; ziyade. **five pounds and ~,** beş liradan itibaren: **~ of 100 planes,** yüzden fazla uçak: **from ten years of age ~,** on yaşından itibaren.

**urban** [ˈö(r)bən]. Şehre mensub veya aid. **~ization** [–âyˈzêyşn], şehrileştirme. **~e** [–ˈbêyn], nazik, çelebi. **~ity,** [–ˈbaniti], nezaket, çelebilik.

**urchin** [ˈö(r)çin]. Küçük erkek çocuk, afacan.

**Urdu** [ˈûədû]. Orduca.

**urge** [ö(r)c]. Sevk, saik. Sevketmek; sürmek; ısrar etm. **to feel an ~ to do stg.,** içinden dürtüyorlar gibi bir şeyi yapmak istemek.

**urgen·t** [ˈö(r)cənt]. Mübrem; mühim. **~cy,** mübremiyet, ehemmiyet.

**urin·al** [ˈyuərinl]. (Küçük abdest için) helâ. **~e.** İdrar.

**urn** [ö(r)n]. Eskiden ölünün küllerinin saklandığı kab; ayaklı kab; semaver.

**us** [ʌs]. Bizi. **to ~,** bize: **from ~,** bizden.

**U.S.** = **United States. U.S.A.** = **United States of America,** Amerika Birleşik Devletleri.

**us·able** [ˈyûzəbl]. Kullanılabilir. **~age** [ˈyûzic], kullanış; muamele; âdet.

**use**[1] *n.* [yûs]. Fayda; kullan(ıl)ma; istimal; kullanış; âdet. **to come into ~,** kullanılmağa başlamak: **in everyday ~,** her gün kullanılan: **for the ~ of,** ···in için, ···in istimaline mahsus olarak: **we will find a ~ for it,** belki bir gün bir şey için kullanırız: **to have the ~ of,** kullanabilmek: **he has lost the ~ of his right hand,** sağ elini kullanamıyor: **to make (good) ~ of stg. [to put stg. to (good) ~],** bir şeyden istifade etm., çok kullanmak: **to make a good [bad] ~ of stg.,** bir şeyi iyi [fena] bir maksad için kullanmak: **it's no ~,** faydası yok; beyhude: **of no ~,** faydasız: **of ~,** faydalı: **to be of ~ for stg.,** bir şeye yaramak: **can I be of any ~ to you?,** size yardım edebilir miyim?: **out of ~,** kullanılmıyor; şimdi kullanılmaz: **to go out of ~,** artık kullanılmamak: **what's the ~?,** ne fayda? **~**[2] *v.* [yûz]. Kullanmak, istimal etm.; istifade etm.; işletmek; muamele etmek. **to ~ up,** tüketmek: **to be ~d up,** tükenmek; suyunu çekmek. **~**[3] *v. Yalnız mazi sıygasında* (**used** [yûst]) *kullanılır.* Âdeti olm., âdet edinmek. **to be ~d to,** ···e alışık olmak. **I ~d to go there every day,** ben oraya her gün giderdim: **I am not ~d to such treatment,** böyle muamele edilmeğe alışık değilim: **I ~d not to like whisky,** eskiden viskiden hoşlanmazdım: **things aren't what they ~d to be,** dünya eskisi gibi değil.

**used**[1] [yûzd] *v.* **use**[2]. Kullanılmış; yeni değil. **~**[2] [yûst] *v.* **use**[3].

**use·ful** [ˈyûsfəl]. Faydalı; işe yarar. **to make oneself ~,** faydalı bir iş görmek; yardım etmek: **~ness,** faydalı olma. **~less** [ˈyûslis], faydasız; bir işe yaramaz; beyhude. **~r** [ˈyûzə(r)], kullanan.

**usher** [ˈʌşə(r)]. Mubassır; teşrifatçı; (tiyatro vs.de) yer gösteren memur. **to ~ in,** salon vs.ye almak: **to ~ in a new epoch,** yeni bir devir açmak.

**U.S.S.R.** [ˈyûˈesˈesˈâ]. = **Union of Soviet Socialist Republics,** Sovyet Rusya.

**usual** [ˈyûjuəl, -jul, -jəl]. Mûtad; alelâde; olağan. **as ~,** her zamanki gibi.

**usur·er** [ˈyûzyuərə(r)]. Mürabahacı; faizci. **~ious** [yûˈzyûriəs], mürabahalı; fahiş faiz alan. **~y,** mürabahacılık; fahiş faiz.

**usurp** [yûˈzö(r)p]. Gasbetmek; kabullenmek. **~ation** [–pêyşn], gasbetme. **~er,** gasbedici; gayrimeşru vasıtalarla saltanatı elde eden kimse.

**utensil** [yûˈtensl]. Kab; mutfak takımı; alet.

**utilit·arian** [ˌyûtiliˈteəriən]. Faydacı. **~y,** faydalı olma; yarama. **~ clothes, boots,** *etc.*, harb zamanında ham malzemeden tasarruf için hükûmetçe tesbit edilen kalite ve fiatta elbise vs.: **general ~ wagon,** her işe yarıyan araba: **public ~ies,** âmme müesseseleri.

**utilize** [ˈyûtilâyz]. Kullanmak; istifade etm.

**utmost** [ˈʌtmoust]. Son derece; en uzak; en ziyade. **to do one's ~,** elinden gelen her şeyi yapmak: **to the ~,** alabildiğine.

**Utopia** [yuˈtoupiə]. Muhayyel yeryüzü cenneti; hayali ham. **~n,** hayalî; çok istenen fakat yapılamıyan.

**utter**[1] [ˈʌtə(r)] *a.* Bütün bütün; tam; sapına kadar. **~ ass,** şeddeli eşek: **~ rot,** deli saçması. **~most** *v.* **utmost.** **~**[2] *v.* Ağza almak; seslenmek; (kalp para vs.yi) sürmek. **~ance,** ifade; ses çıkarma; söz: **to give ~ to one's feelings,** hislerini sözle ifade etmek.

**uxorious** [ʌkˈsôriəs]. Karısına son derece düşkün olan.

# V

**V** [vî]. V harfi. (i) = *vide,* buna bak; (ii) = **versus,** karşı. **V-shaped,** V şeklinde.

**vacan·t** [ˈvêykənt]. Boş; açık; bön. **house to be sold with ~ possession,** derhal taşınılabilen satılık ev. **~cy,** boşluk; münhal yer; bön bön bakma.

**vacat·e** [vəˈkêyt]. Terketmek; boş bırakmak. **to ~ office,** istifa etmek. **~ion,** tatil (vakti): **the long ~,** üniversitenin uzun yaz tatili.

**vaccin·ate** [ˈvaksinêyt]. Aşılamak. **~ation** [–ˈnêyşn], aşılama (*um. yalnız çiçek aşısı*). **~e** [ˈvaksîn], aşı (madde).

**vacillate** [ˈvasilêyt]. İki şık arasında tereddüd etm.

**vacu·ous** [ˈvakyuəs]. Mânasız, bön. **~um** [ˈvakyuəm], boşluk: **~-clean-**

er, elektrik süpürgesi: ~-flask, termos.

**vade mecum** [ˈvêydiˈmîkəm]. Her zaman üstte taşınılan faydalı bir şey ve bilhassa küçük rehber.

**vae victis** [ˈvîˈviktis]. Veyl mağluba!

**vagabond** [ˈvagabond]. Serseri; derbeder.

**vagary** [vəˈgeəri]. Kapris; akla esen şey.

**vagran·t** [ˈvêygrənt]. Serseri; dilenci; avare. **~cy**, serserilik, dilencilik.

**vague** [vêyg]. Mübhem; gayri muayyen; hayal meyal. **I haven't the ~st idea,** zerre kadar malûmatım yok. **~ness**, mübhemlik.

**vain** [vêyn]. Beyhude, nafile; semeresiz; kendini beğenmiş. **in ~**, nafile: **to take God's name in ~**, Allahın ismini hürmetsizce kullanmak. **~-glorious** [–ˈglôriəs], mağrur; kendini beğenen. **~glory**, gurur.

**valance** [ˈvaləns]. Kısa perde.

**vale**[1] [vêyl]. Dere. **this ~ of tears,** bu mihnet diyarı (bu dünya). **~**[2] [ˈvali]. Elveda. **~diction**, veda. **~dictory** [–ˈdiktəri], veda nevinden.

**Valentine** [ˈvaləntâyn]. **St. ~'s Day** = 14 şubatta bir kimsenin sevgilisine gönderdiği resimli kart.

**valet** [ˈvalit, ˈvalêy]. Şahsî uşak. Birine oda hizmetçisi olmak.

**valiant** [ˈvalyənt]. Cesur; yiğit.

**valid** [ˈvalid]. Mer'i; müteber; hükmü cari olan. **~ity** [–ˈliditi], mer'iyet; müteber olma.

**valise** [vaˈlîz]. Bavul; yol çantası.

**valley** [ˈvali]. Vâdi; dere.

**valo·rous** [ˈvalərəs]. Yiğit, cesur. **~ur** [ˈvalə(r)], cesaret.

**valu·able** [ˈvalyuəbl]. Kıymetli (şey). **~s**, küçük fakat kıymetli şeyler. **~ation** [ˌvalyuˈêyşn], kıymet takdiri; tahmin: **to make a ~ of the goods,** eşyayı ehlihibreye keşfettirmek: **at his own ~**, kendi anlattıklarına göre. **~e** [ˈvalyu], kıymet; ehemmiyet. Kıymetini takdir etm.; kıymet biçmek: **to be of ~**, kıymetli olm.: **to be of no ~**, değersiz olm.: **if you ~ your life,** canının kıymetini biliyorsan: **to get good ~ for one's money,** sarfettiği paranın tam mukabilini almak: **these things are very good ~**, bu eşya fiatına göre ucuz sayılır: **to set a high ~ on stg.**, bir şeyi pek fazla takdir etmek. **~eless**, değersiz. **~er**, muhammin.

**valve** [valv]. Supap, ventil; (radyo) lâmba; dessame.

**vamoose** [vaˈmûs]. (*Amer. sl.*) Sıvışmak.

**vamp**[1] [vamp]. Yenileştirmek. **~**[2]. Piyanoda tulûat yapmak. **~**[3], (*sl.*) Fındıkçı kadın.

**vampire** [ˈvampây(r)]. Geceleri mezarından çıkıp yaşıyanların kanını emdiğine inanılan hortlak; insan ve hayvan kanı emen büyük bir yarasa.

**van**[1] [van]. *v.* **vanguard.** **~**[2]. Furgon.

**vandal** [ˈvandl]. (Sanat ve) ilim eserlerini tahrib eden kimse. **~ism**, güzel şeyleri yıkıp bozma meyli.

**vane** [vêyn]. Yelkovan; yeldeğirmeni vs.nin kanadı.

**van(guard)** [ˈvangâd]. Pişdar; ordunun ön safı. **to be in the ~ of progress,** terakkiye önayak olmak.

**vanish** [ˈvaniş]. Gözden kaybolmak; yok olmak. **~ing-point**, intiha noktası.

**vanit·y** [ˈvaniti]. Benlik davası; geçicilik; beyhudelik. **~ies**, mâsiva. **~ bag [case]**, kadınların pudra kutusu vs.leri bulunan küçük el çantası.

**vanquish** [ˈvankwiş]. Yenmek.

**vantage** [ˈvântic]. Daha iyi vaziyet. **point of ~**, bir şeyi seyretmek için müsaid olan bir yer.

**vapid** [ˈvapid]. Yavan, tatsız.

**vapo·ur** [ˈvêypə(r)]. Buğu, buhar; ruh. Buhar çıkarmak. **~ings**, boş lâf. **~rize**, tebhir etm.; buhar haline gelmek, getirmek. **~rizer**, tebhir makinesi. **~rous**, buharlı.

**var·iable** [ˈveəriəbl]. Değişir; kararsız. **~iance** [ˈveəriəns], ihtilâf; uyuşmazlık: **to be at ~ with s.o.**, birisile uyuşamamak: **to set two people at ~**, iki kişinin arasını bozmak: **this theory is at ~ with the facts,** bu nazariye vakiaya uymaz. **~iation** [veəriˈêyşn], değişme; tahavvül; fark. **~i-coloured** [ˈveəriˈkʌlö(r)d], rengârenk. **~ied**, çeşidli. **~iegated** [ˈveəriəgêytid], rengârenk; alacalı. **~iety** [vəˈrâyəti], çeşidlilik, başka başka olma; nevi, cins: **~ entertainment,** varyete. **~ious** [ˈveəriəs], muhtelif, çeşidli; türlü türlü; bir kaç: **for ~ reasons,** bir çok sebeblerden dolayı: **at ~ times,** muhtelif zamanlarda. **~y** [ˈveəri], değiş(tir)mek; çeşitle(n)mek.

**varicose** [ˈvarikous]. **~ veins,** damar şişmesi.

**varlet** [ˈvâlit]. (*esk.*) Uşak; çapkın; herif.
**varnish**[ˈvâniş]. Vernik (lemek). **~ed,** vernikli.
**varsity** [ˈvâsiti]. (*coll.*) Üniversite.
**vase** [vâz]. Saksı; vazo.
**vaseline** [ˈvasəlîn, ˈvazilîn]. Vazelin.
**vassal** [ˈvasl]. (Metbua nisbetle) tâbi; kül. **~age** [ˈvasəlic], tâbilik.
**vast** [vâst]. Vâsi; çok büyük.
**vat** [vat]. Büyük fıçı.
**Vatican** [ˈvatikən]. Papalık makamı; Papalık idaresi.
**vaudeville** [ˈvo̤udəvil]. Vodvil.
**vault**[1] [volt, vôlt]. Sıçrama; atlayış. Atla(t)mak. **~**[2]. Kubbe; lâhit. **the ~ of heaven,** gök kubbe.
**vaunt** [vônt]. Öv(ün)me(k). **~ing,** övüngen.
**V.C.** [vîˈsî] = **Victoria Cross.**
**V.D.** [ˈvîˈdî] = **venereal disease,** zührevi hastalık.
**veal** [vîl]. Dana eti.
**vee** [vî]. V şeklinde olan.
**veer** [ˈvi̯ə(r)]. Dönmek; ciheti değiş(tir)mek. **to ~ away,** dışarı vermek.
**veget·able** [ˈvecitəbl]. Sebze; nebat. Nebatî. **~s,** zerzevat. **~ garden,** bostan: **~ marrow,** kabak. **~arian** [–ˈte̤əri̯ən], etyemez. **~ate** [ˈveciteyt], tenebbüt etm. **~ation** [–ˈtêyşn], tenebbüt; otlar vs.
**vehemen·ce** [ˈvîəməns]. Hiddet, şiddet. **~t,** hiddetli, ateşli, şiddetli.
**vehic·le** [ˈvîəkl]. Nakil vasıtası; araba. **~ular** [viˈhikyulə(r)], nakil vasıtalarına aid.
**veil** [vêyl]. Peçe; yüz örtüsü; bahane. Peçe ile örtmek; saklamak. **beyond the ~,** öbür dünyada: **to draw a ~ over,** bir şeyi örtmek; ötesini saklamak: **to take the ~,** rahibe olmak. **~ed,** peçeli; üstü örtülü. **~ing,** peçelik.
**vein** [vêyn]. Damar; kara kan damarı; verid; hususî hal. Renkli çizgilerle boyamak. **to be in the ~ for doing stg.,** bir şey yapmağı canı istemek: **there is a ~ of humour in all he says,** her sözünde bir mizah kokusu var. **~ed,** damarlı.
**veldt** [velt]. Cenub Afrikasında bozkır.
**vellum** [ˈveləm]. Tirşe.
**velocity** [vəˈlositi]. Hız; sürat.
**velour(s)** [vəˈlu̯ə(r)]. Yünlü kadife.
**velvet** [ˈvelvit]. Kadife(den yapılmış); kadife gibi; pek yumuşak. **to be on ~,** iyi bir vaziyette olm.: ⌜**an iron hand in a ~ glove**⌝, ⌜aba altında değnek⌝. **~een** [–ˈtîn], pamuk kadife. **~y,** kadife gibi, yumuşak.
**venal** [ˈvînl]. Para için her şey yapan; mürteşi. **~ity** [–ˈnaliti], irtişa.
**vend** [vend]. Satmak. **~or,** satıcı.
**vendetta** [venˈdeta]. Kan davası.
**veneer** [vəˈni̯ə(r)]. İyi cins tahta ile kaplamak. Kaplamalık tahta; gösteriş.
**venera·ble** [ˈvenrəbl]. İhtiyar ve muhterem. **~te,** hürmet etm.; takdis etmek. **~tion** [–ˈrêyşn], ihtiram; derin hürmet.
**venereal** [vəˈnîri̯əl]. **~ diseases,** zührevi hastalıklar.
**Ven·etian** [vəˈnîşən]. Venedikli. **~ blind,** jaluzi. **~ice** [ˈvenis], Venedik.
**venge·ance** [ˈvencəns]. İntikam. **with a ~,** şiddetli bir halde; alabildiğine. **~ful,** hıncli.
**veni·al** [ˈvînyəl]. Affedilebilir. **~ offence,** küçük günah. **~ality** [–ˈaliti], (bir suç) affedilebilme.
**venison** [ˈvenzən]. Geyik ve karaca eti.
**venom** [ˈvenəm]. Yılan vs.nin zehiri; zehirli kin. **~ous,** zehirli.
**vent** [vent]. Delik; hava borusu. İzhar etm.; çıkarmak. **to ~ one's anger on s.o.,** hiddetini birisinden çıkarmak: **to give ~ to one's feelings,** ağzını açıp gözünü yummak. **~-hole,** menfez; hava deliği.
**ventil·ate** [ˈventilêyt]. Taze hava vermek; serbestçe ifade etmek. **~-ation** [–ˈlêyşn], havalandırma. **~-ator,** hava değiştirme cihazı.
**ventriloquis·m** [venˈtrilokwizm]. Karnından konuşma. **~t,** vantrilok.
**venture** [ˈventyə(r)]. Baht işi; tehlikeli iş. Baht işine atılmak; cüret etmek. **at a ~,** rasgele: ⌜**to draw a bow at a ~**⌝, ⌜boş atıp dolu tutmağa⌝ çalışmak: '**At a ~, £100,**', vallahi bilmem, belki yüz lira: ⌜**nothing ~ nothing win**⌝, hiç bir riske girmezsen hiç bir şey kazanamazsın. **~some,** gözü pek; cesur; maceraperest; tehlikeli.
**venue** [ˈvenyu]. Randevu.
**Venus** [ˈvînʌs]. Aşk ilâhesi; Zühre.
**veraci·ous** [vəˈrêyşəs]. Doğru sözlü. **~ty** [–ˈrasiti], doğru sözlülük.
**veranda(h)** [vəˈrandə]. Üstü örtülü taraça.
**verb** [vö(r)b]. Fiil. **~al,** şifahî; fiile aid: **~ translation,** harfi harfine tercüme. **~atim** [vö(r)ˈbêytim]. Harf beharf; aynen. **~iage** [ˈvö(r)bi̯əc],

haşviyat; yave. **~ose** [vö(r)ˈbo̤us], çok sözlü. **~osity** [–ˈbositi], ıtnab.
**verd·ant** [ˈvö(r)dənt]. Yeşil; toy. **~ure** [ˈvö(r)dyu̥ə(r)], yeşillik.
**verdict** [ˈvö(r)dikt]. Jüri heyeti hükmü; hüküm; ilâm; karar. **open ~,** mücrimin mechul bulunduğunu ifade eden hüküm.
**verdigris** [ˈvö(r)digrîs]. Bakır pası.
**verge** [vö(r)c]. Kenar; hudud. Meyletmek, yaklaşmak. **blue ~ing on green,** yeşile çalan mavi: **to be on the ~ of fifty,** ellisine yaklaşmak: **to be on the ~ of war,** harbe girmeğe ramak kalmak.
**verger** [ˈvö(r)cə(r)]. Kayyum; zangoç.
**verif·y** [ˈverifây]. Tahkik etm.; tasdik etmek. **~ication** [–fıˈkëyşn], tahkik.
**verily** [ˈverili]. Filhakika.
**vermilion** [vö(r)ˈmilyən]. Zincifre boyası.
**vermin** [ˈvö(r)min]. Bit gibi pis haşarat; sıçan güvercin gibi zararlı hayvan ve kuşlar; alçak ve muzir kimseler. **~ous,** bitli; muzır hayvanlarla dolu; alçak.
**vermouth** [ˈvö(r)mût]. Vermut.
**vernacular** [vö(r)ˈnakyulə(r)]. Yerli (dil). **the ~,** yerli dil; halk dili.
**vernal** [ˈvö(r)nəl]. İlkbahara aid.
**versatil·e** [ˈvö(r)sətâyl]. Hezarfen; bir dalda durmaz. **~ity** [–ˈtiliti], hezarfenlik.
**verse** [vö(r)s]. Nazım; şiir; ayet. **in ~,** manzum. **~d** [vö(r)st], **~ in ...,** ···de üstad olan.
**versify** [ˈvö(r)sifây]. Nazım yapmak.
**version** [ˈvö(r)şn]. Tercüme; muhtelif tercümelerin [rivayetlerin veya ifadelerin] her biri. **according to his ~,** onun ifadesine göre.
**verso** [ˈvö(r)so̤u]. Bir kitabın sol sahifesi; bir sikkenin tersi.
**versus** [ˈvö(r)səs]. ···karşı.
**vertebra,** *pl.* **-ae** [ˈvö(r)tibra, –rî]. Omurganın her faslı. **~te,** [–bret], fıkralı (hayvan).
**vert·ex,** *pl.* **-tices** [ˈvö(r)teks, –tisîz]. Zirve; bir zaviyenin ucu. **~ical** [ˈvö(r)tikl], şakulî, amudî. **~iginous** [–ˈticinəs], baş döndürücü. **~igo** [ˈvö(r)tigo̤u], baş dönmesi.
**verve** [vö(r)v]. Gayret, şevk.
**very**[1] [ˈveri] *a.* Hakikî; tam. **this ~ day,** bugünkü gün: **to the ~ day,** günü gününe: **you're the ~ man I was looking for,** tam aradığım adamsın: **the ~ beggars despise it,** onu dilenciler bile hakîr görür: **the ~ idea!,** daha neler!: **she wept for ~ joy,** sadece sevincinden ağladı: **soldiering is the ~ thing for you,** askerlik senin için biçilmiş kaftandır: **this is the ~ thing for a headache,** bu baş ağrısı için birebirdir: **the veriest fool knows that,** bunu bilmiyecek abdal yoktur. **~**[2] *adv.* Pek, çok, ziyadesile. **the ~ best,** en iyisi: **the ~ first,** tam ilki: **at the ~ most,** en fazla: **at the ~ latest,** en geç: **not so ~ small,** pek de o kadar küçük değil: **my ~ own,** kendi öz malım. **V~**[3], **~ light,** işaret fişeği.
**vesper** [ˈvespə(r)]. Akşam(a aid); akşam yıldızı. **~s,** akşam duası.
**vessel** [ˈvesl]. Gemi; tekne; damar.
**vest**[1] [vest] *n.* İç gömleği; (terzi tabiri) yelek. **~**[2] *v.* (Salâhiyet vs.yi) vermek; temlik etmek. **~ed interests,** mükteseb haklar.
**vesta** [ˈvestə]. Şamalı kibrit.
**vestal** [ˈvestəl]. **~ virgin,** ateş ve ocak ilâhesi Vesta'nın hizmetine mahsus bakire.
**vestibule** [ˈvestibyûl]. Dehliz, hol.
**vestig·e** [ˈvestic]. İz; zerre. **~ial** [–ˈticiəl], artakalan izi olan; bakiye olan.
**vest·ment** [ˈvestmənt]. Elbise *esp.* papaz elbisesi. **~ry** [ˈvestri], kiliselerde papazın giyinme odası.
**vet** [vet]. (*coll.*) = **veterinary surgeon,** baytar. Muayene ve teftiş etmek.
**veteran** [ˈvetərən]. Eski asker; bir meslekte ihtiyarlamış adam. Emekdar; kıdemli ve tecrübeli.
**veto** [ˈvîto̤u]. Reddetmek (hakkı).
**vex** [veks]. İncitmek, canını sıkmak; küstürmek. **to be ~ed at stg.,** bir şeye küsmek: **to be ~ed with s.o.,** birine gücenmek. **~ed,** gücenmiş; küskün: **a ~ question,** münakaşalı bir mesele. **~ation** [vekˈsêyşn], küskünlük; aksilik. **~atious,** cansıkıcı; aksi.
**via** [ˈvâyə]. Yolu ile; tarikiyle.
**viaduct** [ˈvâyədʌkt]. Bir (demir-) yolunu bir dere üzerinden aşırmak üzere inşa edilen bir sıra kemer.
**vial** [ˈvâyəl]. Küçük şişe. **to pour out the ~s of one's wrath,** öfkesini meydana vurmak.
**viands** [ˈvâyəndz]. Yiyecekler.
**vibra·te** [vâyˈbrêyt]. Titremek; sallanmak. **~nt** [ˈvâybrənt], tannan. **~tion** [–ˈbrêyşn], titreme.
**vicar** [ˈvikə(r)]. İngiltere'de bir

mahalleye kaydi hayatla tayin edilen papaz. ~ **of Bray,** menfaat yüzünden sık sık mezheb değiştiren yarı tarihî bir rahib; zamane adamı: **the ~ of Christ,** Papa. **~age** [ˈvikəric], bir vicar'ın resmî evi. **~ious** [vây-ˈkeəriəs], başkası yerine.

**vice**[1] [vâys]. Günah; kötü âdet; çapkınlık. ~[2]. Mengene. ~[3] [ˈvâysi]. Yerinde. **vice-** [vâys] *pref.* ikinci ..., ... muavini. **~-admiral,** tümamiral. **~-chancellor,** üniversite rektörü. **~-consul,** konsolos muavini. **~-marshal, Air ~,** hava tümgeneral. **~-president,** ikinci reis. **~regal** [–ˈrîgl], kıral naibine aid; umumî valiliğe aid. **~roy** [ˈvâysrôy], kıralın naibi; umumî vali. ~ **versa** [ˌvâysi-ˈvö(r)sa], karşılıklı olarak; ve aksi.

**vicinity** [viˈsiniti]. Civar yerler; etraf. **in the ~ of,** civarında.

**vicious** [ˈvişəs]. Fasid; kötü; hırçın. ~ **circle,** fasid daire.

**vicissitude** [vâyˈsisityûd]. Değişiklik; inkilâb. **the ~s of life,** hayatın germü serdi.

**victim** [ˈviktim]. Kurban; mağdur. **the ~ of an accident,** bir kazaya uğrayan: **to make a ~ of oneself,** mağdur taslamak. **~ization** [–mây-ˈzêyşn], mağduriyet; zulüm. **~ize,** iğfal etm.; aldatmak; bir grev vs.-nin sonunda elebaşıları cezalandırmak.

**victor** [ˈviktə(r)]. Galib; fatih. **~ious** [–ˈtôriəs], muzaffer. **~y** [ˈviktəri], zafer, galebe.

**Victoria** [vikˈtôriə]. Kadın ismi, *esp.* meşhur İngiliz Kıraliçesinin ismi. ~ **Cross,** İngiltere'nin en yüksek şecaat nişanı. **~n,** Kıraliçe Victoria devrine aid.

**victual** [ˈvitəl]. Erzak vermek. **~ler,** erzak müteahhidi: **licensed ~,** ruhsatlı içki satıcısı. **~s,** erzak; yemek.

**vide** [ˈvâydi]. (*Lât.*) Bak.

**videlicet** [vâyˈdîliset]. (*Um. kısası* **viz.** *yazılır ve söylenir*). Yani; demek ki.

**vie** [vây]. Rekabet etmek.

**view** [vyû]. Manzara; görünme; nazar; fikir, rey; niyet. Bakmak; muayene etm.; telâkki etm. **front ~ of the house,** evin önden görünüşü: ~ **from the front of the house,** evin önündeki manzara: **to hold extreme ~s,** fikirleri aşırı olm.: **in ~,** görünürde: **in ~ of ...,** ···den dolayı: **in ~ of everyone,** ele güne karşı: **we were in ~ of land,** kara görünüyordu: **to have stg. in ~,** bir şey hakkında bir niyeti olm.: **the town came into ~,** şehir göründü: **to keep stg. in ~,** bir şeyi gözden kaybetmemek: **on ~,** teşhir edilen: **out of ~,** görünmiyen: **point of ~,** noktai nazar: **with a ~ to,** ... maksadiyle, ... niyetinde. **~-finder,** (*fot.*) vizör. **~-halloo,** avcının tilki görünce bağırışı.

**vigil** [ˈvicil]. Uyanık durma; gece ibadeti; bir yortunun arifesi. **to keep ~ over s.o.,** gece uyumayıp birine bakmak. **~ance,** uyanıklık; gözaçıklığı; dikkat. **~ant,** uyanık.

**vigo·ur** [ˈvigə(r)]. Kuvvet; dinclik; faaliyet. **~rous,** dinc; kuvvetli; gayretli.

**Viking** [ˈvâykin(g)]. 8–10 uncu asırlarda Norveçten çıkan korsan.

**vil·e** [vâyl]. Kötü, iğrenc; şeni; pis. **~ify** [ˈvilifây], zemmetmek; kötülemek.

**villa** [ˈvila]. Köşk, sayfiye.

**village** [ˈvilic]. Köy. **~r,** köylü.

**villain** [ˈvilən]. Cani, habis; külhanbeyi. **the ~ (of the piece),** (piyeste) fena adam: **you little ~!,** seni gidi!. **~ous,** habis; pek fena. **~y,** habislik; cürüm.

**villein** [ˈvilən]. Ortaçağda yarı hür yarı köle olan köylü.

**vim** [vim]. (*coll.*) Kuvvet, faaliyet.

**vindicat·e** [ˈvindikêyt]. Doğruluğunu isbat etm.; müdafaa etm. **~ion** [–ˈkêyşn], müdafaa; haklı çıkartma: **in ~ of his conduct,** hareketini mazur göstermek üzere.

**vindictive** [vinˈdiktiv]. Öc alıcı; kinli.

**vine** [vâyn]. Asma; üzüm kütüğü. **~yard** [ˈvinyəd], üzüm bağı. **~gar** [ˈvinigə(r)], sirke: **~ish** [–gəriş], ekşi; hırçın.

**vingt-et-un** [ˈvantêyˈʌn]. İskambilde yirmi bir oyunu.

**vint·age** [ˈvintic]. Bağbozumu [mahsulü]; filân bağ [sene]nin şarab mahsulü. ~ **wine,** iyi cins şarab: **a ~ year,** şarab mahsulü mükemmel bir sene. **~ner,** şarabcı.

**viola** [viˈoula]. Keman ile viyonsel arasında orta bir keman.

**violat·e** [ˈvâyəlêyt]. İhlâl etm.; bozmak; tecavüz etm.; ırzına geçmek. **~ion** [–ˈlêyşn], ihlâl, bozma; ırza tecavüz.

**violen·ce** [vâyələns]. Zor, şiddet; zorbalık. **to do ~ to,** zorlamak: **to resort to ~,** cebre müracaat etmek.

~t, şiddetli; azılı; zorlu: **to become ~**, hiddete kapılmak; cebir kullanmak: **a ~ death**, cebir [kaza] ile ölüm.
**violet** [ˈvâyəlit]. Menekşe. Mor.
**violin** [ˌvâyəˈlin]. Keman. **~ist**, kemancı.
**viper** [ˈvâypə(r)]. Engerek; yılan. **to ⌜nourish a ~ in one's bosom⌝**, ⌜besle kargayı oysun gözünü⌝. **~ish**, yılan gibi; zehirli.
**virago** [viˈrâgou, -êyg-]. Şirret kadın; cadaloz.
**virgin** [ˈvö(r)cin]. Kız, bakire. Kız, bakir, el sürülmemiş. **~ forest**, balta girmemiş orman: **the V~ Mary**, Meryemana: **the V~ Queen**, İngiliz kıraliçesi Ici Elizabeth. **~al**, bakireye aid. **~ity** [-ˈciniti], bakirelik, iffet.
**viril·e** [ˈvirâyl]. Erkeğe aid; kuvvetli; yiğit. **~ity** [viˈriliti], erkeklik; kuvvet.
**virtu·al** [ˈvö(r)tyuəl]. Bilkuvve mevcud olan; hakikî; zımnî. **~e** [-tyû], meziyet; fazilet; doğruluk; hassa; iffet. **in [by] ~ of**, …e binaen: **to make a ~ of necessity**, mecburî bir vaziyetten fazilet hissesi çıkarmak. **~oso** [-tyuˈousou], üstad. **~osity**, [-ˈositi], ustalık. **~ous** [ˈvö(r)tyuəs], faziletli; afif.
**virulenc·e** [ˈviryuləns]. Şiddet; zehirlilik; keskinlik. **~t**, şiddetli, keskin; zehirli.
**virus** [ˈvâyrʌs]. Virüs; hastalıktan ileri gelen zehir; manevî zehir.
**visa** [ˈvîza]. Vize.
**visage** [ˈvizic]. Çehre, yüz.
**vis-à-vis** [ˌvîzaˈvî]. Yüz yüze; karşı karşıya. **my ~**, karşımdaki kimse.
**visc·id** [ˈvisid]. Lüzucetli; cıvık. **~ous** [ˈviskʌs]. Lüzucetli.
**viscount** [ˈvâykaunt]. Vikont. **~cy**, vikontluk. **~ess**, vikontes.
**visib·le** [ˈvizibl]. Gözle görülür; mer'î. **~ility** [-ˈbiliti], rüyet imkânı.
**vision** [ˈvijn]. Görme hissi; görüş; rüyet; rüya; muhayyile kuvveti. **beyond our ~**, rüyet dairemizin dışında: **field of ~**, rüyet sahası: **a man of ~**, ileriyi gören adam. **~ary**, hayalperest; hayalî.
**visit** [ˈvizit]. Ziyaret; (doktor) vizita. Ziyaret etm.; görmeğe gitmek; uğramak. **to ⌜~ the sins of the fathers upon the children⌝**, babaların günahını çocuklara çektirmek. **~ation** [-ˈtêyşn], resmî teftiş ve muayene; Allahtan gelen ceza; felâket. **~or**, misafir; müfettiş.
**visor** [ˈvâyzə(r)]. Tulga siperliği.
**vista** [ˈvistə]. Açık saha; manzara.
**visual** [ˈvijuəl]. Görmeğe aid; basarî. **the ~ nerve**, göz siniri. **~ize**, gözünün önüne getirmek.
**vital** [ˈvâytl]. Hayatî; hayata lâzım; esaslı. **~s**, vücudün kalb vs. gibi hayat için en lâzım uzuvları. **a ~ blow**, öldürücü bir darbe: **~ point**, can alıcı yer: **~ statistics**, doğum, ölüm vs.ye aid istatistik. **~ity** [-ˈtaliti], dirilik; hayatiyet; canlılık. **~ize** [ˈvâytəlâyz], can vermek; diriltmek.
**vitiate** [ˈvişiêyt]. Bozmak; ifsad etmek.
**vitreous** [ˈvitriəs]. Cam gibi.
**vitriol** [ˈvitriol]. Zaç. **~ic** [vitriˈolik], zaçyağlı; gayet acı, zehirli (söz vs.).
**vituperat·e** [vâyˈtyûpərêyt]. Hakaret etm.; küfretmek. **~ive**, tahkir edici.
**vitus** [ˈvâytʌs]. **St. ~' dance**, daürrakıs.
**viva** [ˈvâyvə]. **~ (voce)** [ˈvousi], şifahi (imtihan).
**vivaci·ous** [vâyˈvêyşəs]. Canlı, şetaretli. **~ty** [ˈvasiti], canlılık, neşelilik.
**vivid** [ˈvivid]. Canlı; keskin; parlak.
**vivisect** [ˈvivisekt]. Canlı mahluk üzerine teşrih yapmak.
**vixen** [ˈviksən]. Dişi tilki; şirret kadın.
**viz.** [viz]. Yani, demek ki.
**vizier** [viˈziə(r)]. Vezir. **Grand ~**, sadrazam.
**vocabulary** [voˈkabyuləri]. Küçük lûgat kitabı; bir dilde bulunan kelimeler; bir kimsenin kullandığı kelimeler.
**vocal** [voukl]. Sese aid; sesli; şifahi; his ve fikirlerini ifadeye hazır. **~ cords**, ses telleri: **~ music**, hanendelik: **he becomes very ~ on this subject**, bu mevzu açılınca bülbül kesilir. **~ist**, hanende. **~ize**, ses ile ifade etm.; sessiz harfi sesli yapmak.
**vocation** [vouˈkêyşn]. Meslek; istidad; takdiri ilâhî.
**vocative** [ˈvokətiv]. **~ (case)**, bir ismin hitab hali.
**vocifer·ate** [vouˈsifərêyt] Bar bar bağırmak. **~ous**, bağırıp çağıran.
**vogue** [voug]. Moda; rağbet. to

become the ~, alıp yürümek; rağbet kazanmak.
**voice** [vôys]. Ses; sada; söz; sıyga. İfade etm., söylemek. **active** [**passive**] ~, malûm [mechul] sıyga: **he has no ~ in the matter,** bu meselede fikrini söylemeğe hakkı yok: **with one ~,** hep bir ağızdan. **~d,** sesli.
**void** [vôyd]. Boş; açık; batıl; beyhude. Boşluk; boş yer. Boşaltmak; çıkarmak. **~ of,** ···den âri, mahrum: **an aching ~ in his heart,** kalbinde sızlayan bir boşluk. **~able,** ibtali mümkün.
**vol.** (*abb.*) = **volume.**
**volatil·e** [ˈvolətâyl]. Kolayca tebahhur eden; uçar; gelgeç.
**volcan·o** [volˈkêynou]. Yanardağ, volkan. **~ic** [–ˈkanik], yanardağa aid.
**vole** [voul]. Tarla faresi.
**volition** [voˈlişn]. İrade; ihtiyar.
**volley** [ˈvoli]. Yaylım ateş (etm.); bir topa daha yere değmeden vurmak.
**volte-face** [ˈvoltˈfas]. Yüzgeri; faça.
**volub·le** [ˈvolyûbl]. Çabuk ve çok konuşan; çenebaz. **~ility** [–ˈbiliti], çabuk ve kolayca konuşma; cerbeze.
**volum·e** [ˈvolyûm]. Cild; hacim. **~s of smoke,** yığın yığın çıkan duman: **it speaks ~s for him,** (*coll.*) bu onun çok lehine kaydedilecek bir şeydir. **~inous** [voˈlyûminəs], bir çok cildlerden mürekkeb; büyük hacimli; bol.
**volunt·ary** [ˈvoləntəri]. İstiyerek; ihtiyarî; iradî; gönüllü. **~eer** [volənˈtiə(r)], gönüllü asker; bir işe kendi isteği ile giren kimse. Gönüllü asker olm.; her hangi bir hizmete kendi şahsını teklif etm.: **to ~ information,** kendiliğinden malûmat vermek: **to ~ one's services,** hizmetini arzetmek.
**voluptu·ary** [volˈlʌptuəri]. Zevkine meclûb; şehvete düşkün. **~ous,** zevka düşkün; şehvetli.
**vomit** [ˈvomit]. Kusmak. Kusuntu.
**voodoo** [ˈvûdû]. Amerikalı zencilerin büyüsü. **~ doctor,** zenci büyücü.
**voraci·ous** [vôˈrêyşəs]. Doymak bilmez; oburcasına yiyen. **~ty** [–ˈrasiti], oburluk.
**vortex,** *pl.* **-ices** [ˈvôteks, –tisîz]. Girdab.
**votary** [ˈvoutəri]. Perestişkâr; âbid.
**vote** [vout]. Rey; oy; rey vermek (hakkı). Lehine rey vermek; reyle kabul etmek. **I ~ we go back,** (*coll.*) geri gitmemizi teklif ediyorum. **~r,** rey veren. **~ve** [ˈvoutiv], nezir olarak verilmiş.
**vouch** [vauç]. **to ~ for,** tasdik etm.; tekeffül etm.; temin etmek. **~er,** müsbit varaka; fiş. **~safe** [–ˈsêyf], ihsan etm.; nasib etmek. **to be ~d,** nasib olmak.
**vow** [vau]. Adak, nezir; tövbe; yemin. Adamak; yemin etmek. **to ~ not to,** tövbe etm.; **to take ~s,** rahib(e) olmak.
**vowel** [ˈvauəl]. Sesli harf.
**voyage** [ˈvôyec]. Deniz seyahati; sefer. Denizde seyahat etmek. **on the ~ home,** memlekete dönerken: **on the ~ out,** harice sefer esnasında.
**vs.** = **versus,** karşı.
**Vulcan** [ˈvʌlkan]. Ateş ve madencilik ilâhı. **~ite** [ˈvʌlkənâyt], sert kauçuk. **~ize,** kauçuğu hararet ile daha elâstiki ve dayanıklı yapmak.
**vulgar** [ˈvʌlgə(r)]. Pespaye; âdi, bayağı; galiz; zevksiz. **~ fraction,** âdi kesir: **the ~ herd,** ayak takımı: **the ~ tongue,** halk dili. **~ism,** kaba söz; kabalık. **~ity** [–ˈgariti], âdilik, bayağılık; kabalık. **~ize** [ˈvʌlgərâyz], âdileştirmek.
**Vulgate** [ˈvʌlgêyt]. İncilin Lâtince tercümesi.
**vulnerable** [ˈvʌlnərəbl]. Kolayca yaralanır; hücuma maruz; müdafaası zor.
**vulture** [ˈvʌltyə(r)]. Akbaba; haris.
**vying** *v.* **vie.**

# W

**W** [ˈdʌblyu]. W harfi.
**W.A.A.F.** [waf]=**Women's Auxiliary Air Force,** kadınlardan mürekkeb hava ordusu yardımcı kuvveti.
**wad** [wod]. Tıkaç; tapa; kâğıd para destesi. (Yorgan vs.yi) pamuklamak. **~ded,** içi pamuklu. **~ding,** pamuk gibi yumuşak şey.
**waddle** [ˈwodl]. Badi badi yürümek (yürüyüş).
**wade** [wêyd]. Su içinde veya kar vs. gibi yarı sulu şey içinde yürümek. **to ~ through a book,** bir kitabı güç belâ sonuna kadar okumak. **~r,** uzun bacaklı bataklık kuş.
**wadi** [ˈwodi]. Arabistan ve Şimalî Afrikada yağmursuz mevsimlerde kuruyan dere.
**wafer** [ˈwêyfə(r)]. Pek ince bir bisküi; kâğıd helvası.

**waffle** [ˈwofl]. İzgara pidesi.
**waft** [wâft]. Havada ağır ağır taşımak. Hafif esinti.
**wag**[1] [wag]. Nükteci; şakacı. **~gish** [ˈwagiş], nükteli. ~[2]. Salla(n)ma(k). **to ~ one's finger at s.o.,** parmağını sallayarak tehdid etm.: ⌈**the tail ~s the dog**⌉, ekseriyetin değil ekalliyetin sözü geçiyor: **to set people's tongues ~ging,** dillere destan olm.
**wage**[1] [wêyc] *n.* (*gen.* ~s). Ücret (*haftalık ücrete* **wages** *denir, aylık ve senelik ücrete* **salary** *denir*); mükâfat. ⌈**the ~s of sin is death**⌉, günahın kefareti ölümdür. **~-earner,** haftalık ücretli kimse; bir aileyi geçindiren kimse. ~[2] *v.* **to ~ war (on s.o.),** (birine karşı) harb etmek. **~r** [ˈwêycə(r)]. Bahis; bahse konulan şey. Bahis tutuşmak, bahse girmek.
**waggle** [ˈwagl] *v.* **wag.**
**wag(g)on** [ˈwagən]. Dört tekerlekli yük arabası; demiryolunda eşya vagonu. **to be on the water ~,** (*Amer.*) alkollü içkiler içmemek. **~er,** yük arabacısı.
**waif** [wêyf]. Kimsesiz çocuk, hayvan. **~s and strays,** kimsesiz başıboş gezen çocuklar.
**wail** [ˈwêyl]. Figan (etm.); yüksek sesle ağlama(k). **to ~ over stg.,** bir şeye hayıflanmak.
**wain** [wêyn]. (*esk.*) Büyük yük arabası. **Charles's ~,** Büyük Ayı.
**wainscot** [ˈwêynskot]. Oda duvarlarını tahta ile kaplamak. Tahta kaplaması.
**waist** [wêyst]. İnsanın beli; bir şeyin orta ve dar kısmı. **~coat** [ˈwêys(t)-kout, ˈweskət], yelek. **~line,** bel kısmı. **~-deep** [**~-high**], bele kadar derinlikte.
**wait**[1] [wêyt] (*gen.* ~s). Noel yortusunda geceleri sokakta ilâhi okuyanlar. ~[2]. Bekleme(k). **~ for,** ···i beklemek. ⌈**everything comes to him who ~s**⌉, ⌈bekliyen derviş muradına ermiş⌉: **to keep s.o. ~ing,** birini bekletmek: **to lie in ~,** pusuya yatmak: **to ~ a meal for s.o.,** birini beklemek için yemeği geciktirmek: **~ and see!,** beklersen görürsün: **to ~ at table,** sofrada hizmet etm.: **he did not ~ to be told twice,** iki defa söyletmedi: **to wait on ...,** ···e hizmet etm.; maiyetinde bulunmak: **to ~ on s.o. hand and foot,** birine canla başla hizmet etm.: **to ~ up for s.o.,** yatmayıp birini beklemek. **~er,** garson. **~ress,** kadın garson. **~ing-room,** bekleme salonu.
**waive** [wêyv]. ···den vazgeçmek; ısrar etmemek.
**wake**[1] [wêyk] *n.* Dümensuyu. **in the ~ of,** peşinde; izini takip ederek. ~[2] *v.* (**woke, woken**) [wêyk, wouk, woukn]. **~ (up),** uyan(dır)mak; canlan(dır)mak. **he badly wants waking up,** onu kendine getirmek lâzım: **he woke to find himself famous,** bir günde meşhur oldu: **to ~ up to what is happening** [**to the truth**], ayağı suya ermek. **~ful,** uyanık; uykusuz. **~n** *v.* **wake.**
**Wales** [wêylz]. Gal memleketi. **the Prince of ~,** Gal Prensi (İngiliz veliahdinin unvanı).
**walk** [wôk]. Yürüme gezinti (yeri); bahçede yol; kaldırım. Yürümek, yayan gitmek. Yürütmek; üzerinden gezmek. **to fall into a ~,** (koşan at) yürümeğe başlamak: **to go for a ~,** gez(in)mek: **he ~ed me off my legs,** benim dayanamıyacağım kadar yürüdü: **~ in life,** meslek; ictimai seviye: **to ~ in one's sleep,** uykuda gezmek. **~ away,** ayrılmak: **to ~ away from a competitor,** bir rakibden fersah fersah üstün olmak. **~ in,** girmek. **~ into,** ···e girmek; çatmak; burun buruna gelmek; (*coll.*) saldırmak. **~ off,** ayrılmak: **to ~ off a big meal or too much drink,** fazla yemek veya içkinin tesirini gidermek için dolaşmak: **to ~ off with,** aşırmak. **~ out,** çıkıp gitmek: **to ~ out of,** ···den çıkmak: **to ~ out with a young man,** (kız) bir gencle dolaşmak. **~ over,** (bir yeri) yaya olarak dolaşmak; (bir müsabakada) kolayca [rakib bulunmadığı için] kazanmak. **~ing,** yayan (gitme): **a ~ library,** ayaklı kütübhane: **~-stick,** baston.
**wall** [wôl]. Duvar; sur. **to ~ in,** duvar ile kuşatmak: **to ~ up a door,** *etc.*, kapı vs.yi örerek kapatmak. ⌈**~s have ears**⌉, yerin kulağı var: **with one's back to the ~,** mezbuhane: **to go to the ~,** ezilmek: **to run one's head against a brick ~,** olmıyacak şeyi zorlamak. **~ed,** duvarla çevrilmiş. **~paper,** duvar kâğıdı.
**wallah** [ˈwola]. (Hindli) memur, uşak.
**wallet** [ˈwolit]. Cüzdan, portföy.
**wallop** [ˈwolop]. Şiddetli dayak atmak. Şiddetli vuruş. **~ing,** şiddetli dayak atma; iri yarı: **a ~ lie,** kuyruklu yalan.

**wallow** [ˈwalǫu]. Çamura vs.ye yatıp yuvarlanmak. Çamurda yuvarlanma. **to ~ in blood,** kan içinde yüzmek.
**walnut** [ˈwolnʌt]. Ceviz (tahtası).
**walrus** [ˈwolrʌs]. Deniz ayısı; mors.
**waltz** [wols]. Vals (oynamak).
**wan** [won]. Solgun; donuk.
**wand** [wond]. İnce asâ, sihirbaz değneği.
**wander** [ˈwondə(r)]. Başıboş gezme(k); yoldan sapmak. **to let one's thoughts ~,** hayalata dalmak: **my thoughts were ~ing,** dalgındım: **to ~ in one's mind,** sayıklamak. **~ing,** başıboş gezinti; dalgınlık seyyar; göçebe; dalgın; (hasta) sayıklayan, hezeyan içinde.
**wane** [wêyn]. (Ay) küçülmek; zeval bulmak. **on the ~,** zevale yüz tutan: **his star is on the ~,** yıldızı sönüyor.
**wangle** [ˈwan(g)gl]. (*sl.*) Hile ile elde etmek. **to ~ leave,** izin koparmak.
**want¹** [wont] *n.* Yokluk; noksan; sefalet; ihtiyac; istek. **~ of ...,** ···sizlik: **to be in ~ of stg.,** bir şeye muhtac olm.: **~ of courage,** cesaretsizlik: **for ~ of, ...** bulunmadığı için; ···sizlikten dolayı: **for ~ of stg. better,** daha iyisi olmadığı için. **~²** *v.* İstemek; muhtac olm.; ···den mahrum olm.; icabetmek. Eksik olm.; zaruret içinde olmak. **to be ~ed,** istenilmek; ···e ihtiyac olm.: **you are ~ed,** sizi istiyorlar: **he is a little ~ing,** biraz kaçık: **I don't ~ it known,** onun bilinmesini istemiyorum: **to ~ for nothing,** hiç bir eksiği olmamak; dört başı mamur olm.: **he ~s patience,** bir kusuru varsa o da pek sabırsızdır: **it ~s a lot of patience,** çok sabır icabeder: **it ~ed but a few days to Christmas,** Noele ancak bir kaç gün kalmıştı: **what does he ~ with me?** beni ne diye istiyor?: **you ~ to see a doctor at once,** hemen bir doktora gitmelisiniz: **you don't ~ to go there,** oraya gitmek istemiyorsunuz; *bazan,* oraya gitmenize lüzum yok. **~ed,** lâzım; matlub; polis tarafından aranan (mücrim). **~ing,** (bir tahtası) eksik.
**wanton** [ˈwontən]. Sebebsiz; oyunbaz; iffetsiz. Aşifte.
**war** [wô(r)]. Harb; muharebe. **to ~ against ...,** ···e karşı savaşmak: **to be at ~,** harb halinde olm.: **to go to ~,** harbe girişmek: **you look as though you had been in the ~s,** muharebeden mi çıktın?: **on a ~ footing,** seferî vaziyette: **W~ Office,** Harbiye Vekâleti. **~-cry,** harb narası. **~-dance,** vahşiler tarafından harbe hazırlık olarak veya zafer için yapılan dans. **~-fever,** harb humması. **~-head,** torpitonun cebhaneli harb başlığı. **~-horse,** muharebe atı: **an old ~,** tecrübeli asker. **~-lord,** bilhassa Çin dahili harbindeki kumandanlar gibi mahallî diktatör. **~-monger,** harbe kışkırtan. **~-path, to be on the ~,** kavgaya hazır olmak. **~-whoop,** vahşilerin harb narası. **~-worn,** harbde yıpranmış. **~fare** [–fęə(r)], harb hali veya icrası. **~like** [–lâyk], cengâver; harbe mahsus; harb olacak gibi. **~paint** [–pêynt], vahşilerin harbe girerken yüz ve vücudlerine sürdükleri boya: **in full ~,** giyinmiş kuşanmış. **~ship** [–şip], harb gemisi.
**warble** [ˈwôbl]. Ötmek. **~r,** küçük öten kuş.
**ward** [wôd]. Gözetme; muhafaza; vesayet altında olan çocuk; şehrin adası; (hastahane vs.de) koğuş. Korumak. **~s of a lock,** kilidin dilleri. **to ~ off,** defetmek; önüne geçmek. **~ in chancery,** mahkeme vesayetinde olan çocuk. **~en,** muhafız. **~er** [ˈwôdə(r)], zindancı; muhafız. **~ress,** kadın zindancı.
**wardrobe** [ˈwôdrǫub]. Portmanto; elbise dolabı.
**wardroom** [ˈwôdrûm]. Harb gemisinde subaylar salonu.
**ware** [węə(r)]. Mamul eşya; mal; çini. **~s,** satılık eşya. **~house** [ˈwęəhaus], ambar(a koymak): **bonded ~,** gümrük antreposu.
**'ware.** (*abb.*) = **Beware.** Dikkat!; sokulma!
**wariness** [ˈwęərinis]. İhtiyat, açıkgözlülük.
**warm¹** [wôm] *a.* Bir az sıcak (*mes. su için 38–44 sant.*); (insan, hava) rahat derecede sıcak; hararetli (münakaşa vs.); (renkler) kırmızı veya sarıya çalan. **you are getting ~,** *bazı oyunlarında aranılan şeye veya doğru cevabına yaklaşıldığı zaman söylenir:* **he has a ~ heart,** iyi kalblidir: **to make it ~ for s.o.,** anasından emdiğini burnundan getirmek: **a ~ scent,** avın taze kokusu: **it's ~ work!,** bu terletici iştir! **~-blooded,** sıcak kanlı (hayvanlar). **~-hearted,** iyi

kalbli. ~² *v*. Isıtmak; ısınmak. **to ~ up,** yeniden ısıtmak; hararetlenmek. **~th** [wômθ], hararet.

**warn** [wôn]. Tenbih etm.; ihtar etm.; önceden haber vermek; ibret olmak. **to ~ s.o. off,** iştirak etmekten menetmek. **~ing,** tenbih, ihtar; ibret; tenbih edici: **as a ~ to others,** ibret için: **to take ~,** ibret almak.

**warp¹** [wôp]. Arış; palamar. **~².** Biçimini değiştirmek. Eğrilmek. **~ed,** çarpık.

**warrant** [ˈworənt]. Ruhsat [salâhiyet] veren sened; tevkif müzekkeresi; berat; kefalet; haklı sebeb. Tekeffül etm.; ruhsat [salâhiyet] vermek. **I'll ~ you!,** emin olunuz!: **he had no ~ for his hopes,** ümidini destekleyecek hiç bir sebeb yoktu: **nothing can ~ such rudeness,** hiç bir şey bu kabalığı mazur gösteremez. **~y,** kefalet(name); salâhiyet. **~-officer,** gedikli.

**warren** [ˈworən]. Aynı yerde bulunan bir çok ada tavşanı yuvaları.

**warrior** [ˈworiə(r)]. Cengâver; muharib.

**Warsaw** [ˈwôsô]. Varşova.

**wart** [wôt]. Siğil. **~y,** siğilli.

**wary** [ˈweəri]. Açıkgöz; uyanık; ihtiyatlı. **to be ~ of,** ···den sakınmak.

**was** [woz] *v*. **be.** İdi; oldu.

**wash¹** [woş] *n*. Yıka(n)ma; çamaşır; geminin hareketinden hasıl olan dalgalar; badana. **to give stg. a ~,** bir şeyi yıkamak: **pig ~,** domuza yedirilen sulu yem: **the ~ of the waves,** dalgaların çağılışı. **~²** *v*. Yıka(n)mak; su ile silmek; hafif boya ile kaplamak; badana etmek; dalga gibi çarpmak. **to ~ stg. ashore,** (deniz) bir şeyi sahile atmak: **to ~ one's hands of,** ···le bütün alâkasını kesmek: **a sailor was ~ed overboard,** bir gemiciyi dalga alıp götürdü: **that story won't ~,** (*sl.*) bu masalı kimse yutmaz. **~-basin,** leğen; lavabo. **~-tub,** çamaşırlık. **~ away,** (leke vs.yi) su ile temizlemek; (dalga, su vs.) alıp götürmek; aşındırmak. **~ down,** çok su ile temizlemek; su bir şeyi alıp aşağıya götürmek: **to ~ down one's dinner with a glass of beer,** yemeğin sonunda bir bardak bira içmek. **~ off,** yıkayarak çıkarmak. **~ out,** (leke vs.yi) su ile kaldırmak; (*coll.*) silmek: **you can ~ that out,** bunu sil. **~ed out,** soluk; halsiz. **~-out,** fiyasko; selin yaptığı tahribat. **~ up,** bulaşık yıkamak; (deniz) sahile atmak. **~able** [ˈwoşəbl], yıkanabilir. **~er¹** [ˈwoşə(r)], yıkayan kimse: **~-up,** bulaşıkçı. **~er²,** rondelâ, pul. **~erwoman** [ˈwoşəwumən], çamaşırcı kadın. **~ing** [ˈwoşin(g)], yıka(n)ma. **the ~,** çamaşır. **~y** [ˈwoşi], sulu; yavan.

**wasp** [wosp]. Eşek arısı. **~ish,** eşek arısı gibi; kinci. **~-waisted,** incecik belli.

**wastage** [ˈwêystic]. Heder; israf; fire.

**waste¹** [wêyst] *n*. İsraf; heder; lüzumsuz ziyan; döküntü; çöl, ıssız yer. Metrûk; kullanılmış; ıskartaya çıkarılmış. **to go to ~,** heder olm.: **to lay ~,** tahrib etmek. **~-paper basket,** kâğıd sepeti. **~-pipe,** fazla su borusu. **~²** *v*. Heder etm.; telef etm.; israf etm.; faydasız ve nafile yere harcamak; (vücudü) harab etmek. **to ~ away,** zayıflıya zayıflıya eriyip gitmek: **don't ~ your breath!,** beyhude çeneni yorma!: **to ~ one's time,** vakti boşuna geçirmek, abesle uğraşmak. **~d,** harab; zayıflamış; heder edilmiş; beyhude yere sarfedilmiş. **~ful,** müsrif; idareli olmıyan. **~r** [ˈwêystə(r)], **wastrel** [ˈwêystrəl], haylaz kimse; adam olmaz.

**watch¹** [ˈwoç] *n*. Ceb saati; nöbet (çilik); vardıya; gözetleme; tarassud; nöbetçi, bekçi. **to keep ~,** gözetlemek: **in the ~es of the night,** geceleyin: **officer of the ~,** vardıya zabiti: **to be on the ~,** gözetlemek: **to be on the ~ for s.o.,** birinin yolunu beklemek. **~-chain,** saat köstеği. **~-dog,** bekçi köpeği. **~-keeper,** (gemide) nöbetçi. **~²** *v*. Seyretmek; gözlemek; dikkat etm.; göz hapsine almak; tarassud etmek. **to ~ for s.o.,** birini(n yolunu) beklemek: **to ~ one's opportunity [time],** fırsat gözlemek: **to ~ out,** göz kulak olm.: **~ out!,** dikkat!: ⌜**a ~ed pot never boils**⌝, hiç bir şeyin üzerinde düşmemeli: **to ~ by a sick person,** bir hastanın yanında beklemek: **to ~ over a flock,** bir sürüye bakmak. **~er** [ˈwoçə(r)], bekçi; bakıcı: **bird ~,** kuşları tedkike meraklı kimse. **~ful** [ˈwoçfəl]. uyanık; müdebbir. **~maker** [–mêykə(r)], saatçi. **~man** [ˈwoçmən], bekçi. **~word** [–wö(r)d], parola; şiar.

**water** [ˈwôtə(r)]. Su. Sula(ndır)mak.

(Ağzı) sulanmak; (göz) yaşarmak. ~ **on the brain,** istiskaı re's: **by** ~, deniz yoluyle: **to throw cold** ~ **on a scheme,** bir plânı istihfaf etm.: **a diamond of the first** ~, en iyi cinsten elmas: **to** ~ **down,** ···e su katmak; hafifletmek: **head of** ~, suyun enerji istihsal kuvveti: **not to hold** ~, sızmak; su götürmek: ~ **on the knee,** diz mafsalı boşluğunda su dolması: **on land and** ~, hem karada hem denizde: **to be in low** ~, kederli; parasızlıktan sıkıntıda olm.: **to make** ~, (insan) işemek; (gemi) su almak: **to take the** ~, (yüzücü) suya girmek; (gemi) suya indirilmek: **to take the** ~**s,** içmelere gitmek: **to be under** ~, su altında olm.; (araziyi) su basmış olmak. ~**-borne,** gemide taşınan (eşya). ~**-buffalo,** manda. ~**-butt,** yağmur suyunu toplamağa mahsus fıçı. ~**-can,** ibrik; bahçe kovası. ~**-cart,** su getirmeğe mahsus araba. ~**-closet,** abdeshane. ~**-diviner** *v.* **diviner.** ~**-ice,** su ile yapılan dondurma. ~**-jump,** su ile dolu manialı hendek. ~**-lily,** nilüfer. ~**-main,** ana su borusu. ~**-melon,** karpuz. ~**-power,** su kuvveti. ~**-rate,** su tedariki için tarhedilen vergi. ~**-softener,** suyun kirecini gideren ilâc vs. ~**-tower,** kule üzerinde su hazinesi. ~**-wheel,** su değirmeninin çarkı. ~**-wings,** (yüzme öğrenci için) küçük balonlar. ~**bottle** [–botl], su sürahisi. ~**colour** [–kʌlə(r)], sulu boya (resim). ~**course** [–kôs], su kanalı; dere. ~**ed,** su katılmış; sulanmış; menevişli. ~**fall** [–fôl], şelâle; çağlıyan. ~**fowl** [–faul], su kuşlarının umumî adı. ~**hole** [–houl], su birikintisi. ~**ing,** sulama; (ipek vs.) dalga. ~**less** [–lis], susuz; çorak. ~**line** [–lâyn], geminin su kesimi. ~**logged** [–logd], (gemi) içinde su dolmuş; fazla yağmurdan dolayı su ile meşbu (arazi). **W**~**loo** [ˌwôtəˈlû], Waterloo muharebesi; Londra'nın ana istasyonlarından biri: **to meet one's** ~, nihaî hezimete uğramak. ~**man** [ˈwôtəmən], nehir sandalcısı; iyi kürekçi. ~**mark** [–mâk], sudan hasıl olan nişan. ~**proof** [–prûf], yağmurluk, muşamba; su geçmez (hale getirmek). ~**shed** [–şed], havza. ~**side** [–sâyd], su kenarı(nda bulunan). ~**spout** [–spaut], hortum (kasırga). ~**tight** [–tâyt], su sızmaz. ~**way** [–wêy], seyre elverişli kanal vs. ~**works** [–wö(r)ks], bir şehrin su tesisatı: **to turn on the** ~, (*coll.*) ağlamak. ~**y,** sulu; yağmurlu; (renk) soluk.

**watt** [wot]. Vat. ~**age,** vatlık.

**wattle** [ˈwotl]. İnce çubuklardan sepet örgüsü.

**wave** [wêyv]. Dalga; el vs. sallama. Dalgalan(dır)mak; el, mendil vs.yi sallıyarak işaret etmek. **to** ~ **s.o. aside,** birine 'istemez' işareti yapmak: **to** ~ **aside an objection,** bir itirazı kabul etmemek: **the enemy attacked in** ~**s,** düşman dalgalar halinde hücum ediyordu: **to** ~ **the hair,** saçlara ondülâsyon yapmak. ~**d,** ondüle; dalgalı. ~**-length,** dalga uzunluğu.

**waver** [ˈwêyvə(r)]. Tereddüd etm.; gevşemek; (ışık) titremek. ~**ing,** kararsız; titriyen (alev). Tereddüd.

**wavy** [ˈwêyvi]. Dalgalı; ondüleli.

**wax**[1] [waks]. Balmumu; kulak kiri. Mumlu. Balmumu sürmek. **sealing-**~, mühür mumu. ~**en,** balmumundan yapılmış. ~**work,** balmumundan şekil: ~**s,** balmumu şekiller sergisi. ~[2] *v.* (Ay) hacmı büyümek; artmak.

**way** [wêy]. Yol; tarik; cihet; usul; âdet; çare. **in a bad** ~, fena suretle; fena halde: **by the** ~, yolda; hatırımda iken: **all this is by the** ~, şimdi esas meseleye gelelim: **by** ~ **of confirmation,** tasdık makamına: **to give** ~ **to** ..., ···e yol vermek; ···e teslim olm.; ···in fikrini vs. kabul etm.: **get out of the** ~**!,** önümden çekil!: **to get out of the** ~ **of doing stg.,** bir itiyadı kaybetmek: **to get into the** ~ **of doing stg.,** itiyadını kazanmak; alışmak: **to get in s.o.'s** ~, mâni olm.: **to get [have] one's** ~, istediğini yap(tır)mak: **to go the** ~ **of all things,** eskiyip ortadan kalkmak; ölmek: ~ **in,** giriş: **to be in the** ~, mâni olm.: **he has nothing in the** ~ **of relations,** akraba namına kimsesi yok: **to know one's** ~ **about a place,** bir yerin girdisini çıktısını bilmek: **a long** ~, uzak: **he is a long** ~ **the best,** bu çok büyük bir farkla en iyisidir: **to make a penny go a long** ~, idareli olm.: **to make** ~, ilerlemek; yol vermek: **there is no** ~ **out,** çıkar yol yok: **in the ordinary** ~, alelâde: **on the** ~, yolda: **he is on the** ~ **to ruin,** mahve gidiyor: ~ **out,** çıkış: **out of the** ~, hücradır: **to go out of one's** ~ **to do stg.,** bir şeyi yapmak için bilhassa zahmete girmek: **he is nothing out of the ~**, hiç bir fevkalâdeliği yok: **to go one's own ~**, bildiğini okumak: **all right, have it your own ~!**, siz bilirsiniz!; haydi, sizin dediğiniz olsun!: **he lives in a small ~**, mütevazı bir şekilde yaşıyor: **in some ~**, bazı cihetlerden: **come this ~!**, buradan!: **to get under ~**, (gemi) hareket etm.; (*fig.*) işe başlamak. **'~ = away. ~ back in 1900**, ta 1900 de. **~far·er** [ˈwêyfe͜ərə(r)], yolcu: **~ing**, yolculuk; seyyah. **~lay (-laid)** [wêyˈlêy, -ˈlêyd], yolunu kesmek; pusuya yatıp beklemek. **~side** [–ˈsâyd], yol kenarı(nda bulunan). **~ward** [–wö(r)d], nazlı; ters; keyfine tâbi.

**w.c.** [ˌdʌblyûˈsî] = **water-closet**, abdesane.

**weak** [wîk]. Kuvvetsiz, zayıf; iradesiz; âciz; hafif (çay vs.). **~ spot**, püf noktası. **~en**, kuvvetten düş(ür)mek; gevşe(t)mek. **~ling**, cılız kuvvetsiz kimse; iradesiz karaktersiz kimse. **~ly**, cılız. **~ness**, kuvvetsizlik; zaaf: **to have a ~ for ...**, ···e inhimak etm., ···e karşı zâfı olmak. **~-kneed** [**~-minded**], iradesiz.

**weal**¹ [wîl]. Et üzerinde kırbaçla yapılan bere. **~²**. Hayır. **the public ~**, umumî menfaat: **~ and woe**, iyi ve kötü zamanlar. **~th** [welθ], zenginlik; bolluk: **~y**, zengin.

**wean** [wîn]. Sütten kesmek.

**weapon** [ˈwepən]. Silâh.

**wear**¹ [we͜ə(r)] *n.* Giyme; eski(t)me; kullanma; yıpranma; dayanma. **children's ~**, çocuk elbiseleri: **stuff that will stand hard ~**, dayanıklı kumaş: **these shoes have still a lot of ~ in them**, bu ayakkabılar daha çok giyilir: **~ and tear**, kullanılmakla tabiî olan eskime: **the worse for ~**, eskimiş. **~² (wore, worn)** [we͜ə(r), wô(r), wôn]. Giymek; giyip kullanmak; taşımak; takmak; eski(t)mek; dayanmak. **to ~ oneself to death**, kendini fazla yormak: **to ~ the hair long**, saçlarını uzun bırakmak: **the day was ~ing to its close**, günün sonu yaklaşıyordu: **to ~ stg. into holes**, delik deşik oluncaya kadar giymek: **to be worn to a shadow (with care)**, üzüntüden iğne ipliğe dönmek: **to ~ well**, (eşya) iyi dayanmak: **to ~ (one's years) well**, (ihtiyar) dinc kalmak: **a well-worn joke**, bayat nükte. **~ away**, aşın(dır)mak. **~ down**, aşındırmak; yormak. **~ off**, vakit geçtikçe zail olm.; yavaş yavaş azalmak. **~ on, as the evening wore on**, gece ilerledikçe. **~ out**, bütün bütün eskitmek; yıpratmak: **to ~ oneself out**, bitkin bir hale gelmek. **~able** [ˈwe͜ərəbl], giyinebilir. **~er** [ˈwe͜ərə(r)]. Giyen; takan.

**wear·y** [ˈwi͜əri]. Yorgun; bıkkın; usandırıcı. Yor(ul)mak; usan(dır)mak. **to ~ of s.o.**, birisinden bıkmak. **~iness**, yorgunluk; bıkkınlık. **~-isome**, bıktırıcı; yorucu.

**weasel** [ˈwîzl]. Gelincik (*hayvan*).

**weather**¹ [ˈweðə(r)] *n.* Hava. *a.* Rüzgâra duçar olan (taraf). **to make heavy ~**, (gemi) fırtınada çalkalanmak: **to make heavy ~ of stg.**, bir şeyi lüzumundan fazla güç bulmak: **to keep one's ~ eye open**, muhtemel tehlikeye göz kulak olm.: **(wind and) ~ permitting**, hava müsaid olduğu takdirde: **to be under the ~**, keyifsiz olmak. **~bound**, fena hava sebebiyle limandan çıkarmıyan. **~-beaten**, rüzgar vs.den hırpalanmış; yanık (çehre). **~-cock**, fırıldak. **~-wise**, havadan anlar. **~²** *v.* (Rüzgâr vs.) aşındırmak, rengini değiştirmek, çatlatmak; (fırtına karşı) mukavemet göstermek; rüzgâr ters iken gemi bir burun vs.den geçmek. Rüzgar vs.den aşınmak, solmak. **~ed**, rüzgâr vs.den aşınmış, solmuş vs.

**weav·e (wove, woven)** [wîv, wo͜uv, wo͜uvn]. Dokumak; (*fig.*) terkib etmek. Dokuma tarzı. **~er**, dokumacı. **~ing**, dokuma(cılık).

**web** [web]. Nesic; ağ; yüzücü kuşların parmakları arasında perde. **a ~ of lies**, yalan dolan. **~bed**, perdeli, zarlı (kuş ayağı). **~bing**, kuvvetli dokunmuş kumaş. **~-footed**, perdeayaklı.

**wed** [wed]. Nikâhla almak; ··· evlen(dir)mek; birleştirmek. **to be ~ded to an opinion**, *etc.*, bir fikir vs.ye iyice bağlanmak. **~ded**, evlenmiş; evli: **his ~ wife**, nikâhlı karısı: **~ life**, evlilik hayatı. **~ding** [ˈwedin(g)], evlenme merasimi; düğün: **silver ~**, bir evlenmenin 25 inci yıldönümü: **golden ~**, 50 inci yıldönümü: **diamond ~**, 60 ıncı yıldönümü: **~ day**, evlenme günü [yıldönümü]: **~-present**, düğün hediyesi: **~-ring**, nikâh yüzüğü. **~lock** [ˈwedlok], evlenme; evlilik: **child born in [out of], ~**, meşru [gayri meşru] çocuk

**wedge** [wec]. Kama; takoz. Kama ile tesbit etmek; araya sokmak. **the thin end of the ~**, gittikçe büyük olan bir hareketin ilk adımı.

**Wednesday** [ˈwenzdi]. Çarşamba. **Ash ~**, büyük perhizin ilk günü.

**wee** [wî]. Minimini, ufacık.

**weed** [wîd]. Yaramaz ot; yaprak cıgara. Bahçedeki yaramaz otları ayıklamak. **the ~**, tütün: **to ~ out**, faydasız kimseleri [şeyleri] söküp atmak. **~y**, yaramaz otlu; çelimsiz insan. **~s, widow's ~**, dul kadınların matem elbisesi.

**week** [wîk]. Hafta; yedi gün. **this day ~ [a ~ today, a ~ from now]**, gelecek hafta bugün: **what day of the ~ is it?**, bugün ne? **twice a ~**, haftada iki kere: **~ in ~ out**, fasılasız: **to knock s.o. into the middle of next ~**, (*coll.*) birine dehşetli bir yumruk indirmek: **to be paid by the ~**, haftalık almak: **a ~ of Sundays**, çok uzun zaman: **tomorrow ~**, gelecek hafta yarınki gün: **it will be a ~ tomorrow that he died**, yarın öleli bir hafta olacak: **yesterday ~**, geçen hafta dünkü gün. **~day**, hafta günü (pazar günü değil). **~ly**, haftalık; haftada bir olan: **twice ~**, haftada iki kere. **~-end**, hafta sonu (cumartesi — pazartesi).

**ween** [wîn]. (*esk. ve şair.*) Zannetmek.

**weep** [wîp]. Ağlamak. **to have a good ~**, iyice ağlayıp içini boşaltmak: **that's nothing to ~ about**, ağlanacak bir şey değil, (*bazan*) daha iyi ya! **~ing**, ağlayan: **~ willow**, salkım söğüt.

**weft** [weft]. Atkı, argac.

**weigh** [wêy]. Tartmak; düşünüp taşınmak; hesab etmek. Ağır gelmek; sıkleti olmak. **to ~ anchor**, demir almak: **to ~ down**, …den daha ağır gelmek; (keder) bastırmak: **branch ~ed down with fruit**, meyvalarla yüklenmiş dal: **to ~ in**, (cokey) yarıştan evvel tartılmak: **~ out**, (şeker vs.yi) lâzımgelen miktarda tartmak. **~bridge**, araba vs. tartmağa mahsus bir nevi baskül. **~ing-machine**, kantar, terazi. **~t** [wêyt], ağırlık; sıklet; tartı; kantar taşı; ağır yük; ehemmiyet, tesir. …e ağır bir şey takmak. **his word carries ~**, sözünün tesiri var: **worth its ~ in gold**, ağırlığınca altın eder: **to lose ~**, zayıflamak: **that's a ~ off my mind**, bunun üzerine ferahladım: **of no ~**, ehemmiyetsiz; nüfuzu az: **people of ~**, nüfuzlu insanlar: **to pull one's ~**, hissesine düşen işi yapmak: **to put the ~**, gülle atmak: **to sell by ~**, tartı ile satmak: **to throw one's ~ about**, (*coll.*) yüksekten atmak. **~ty**, ağır; mühim; nüfuzlu.

**weir** [wiə(r)]. Nehrin suyunu kesen sed.

**weird** [wiə(r)d]. Esrarengiz; tekin olmıyan; garib ve biraz korkutucu.

**welcome** [ˈwelkəm]. Hoş geldin!, safa geldiniz! Makbul; hoş. Karşılama; 'hoş geldin' deme; iyi kabul. Misafire 'hoş geldin' demek; iyi karşılamak; memnuniyetle kabul etmek. **you're ~!**, bir şey değil: **you are ~ to it**, (i) buyurun; (ii) gözüm yok: **you are ~ to try**, tecrübe edebilirsiniz: **to give s.o. a cold ~**, birini soğuk karşılamak: **to give s.o. a warm ~**, (i) birini hararetle karşılamak; (ii) birini geldiğine pişman etm.: **to overstay one's ~**, iyi karşılanan bir yerde fazla kalmak.

**weld** [weld]. Kaynak (yapmak).

**welfare** [ˈwelfeə(r)]. Saadet; refah; sıhhat. **child ~**, çocuk bakımı ve terbiyesi: **the public ~**, âmme menfaati: **~ worker**, hayır işlerile uğraşan kimse.

**welkin** [ˈwelkin]. (*şair.*) Gök kubbe. **to make the ~ ring**, kubbeleri çınlatmak.

**well**[1] [wel]. Kuyu; (ufak gemide) kaporta ağzı; (evde) merdiven, asansör yeri; menba. **to ~ up**, fışkırmak. **~**[2]. İşte; şu halde. **~!**, *hayret ifade eder*: **~ ~! [~ I never!, ~ really!]**, Allah Allah!: **~ then**, şu halde. **~**[3]. İyi, alâ; tamamiyle; münasib. İyilik. **as ~**, dahi: **as ~ as**, kadar (iyi): **he is learned as ~ as rich**, hem malûmatlı hem de zengindir: **one might as ~ say …**, aynı şekilde … denebilir de: **you may (just) as ~ go**, gitseniz de olur: **~ done**, iyi pişmiş: **~ done!**, aferin!: **~ and good**, ne alâ: **it serves him jolly ~ right!**, oh olsun!: **to let ~ alone**, işi tadına bırakmak: **~ off**, hali vakti yerinde: **you don't know when you are ~ off**, elindeki nimetin farkında değilsin: **we are very ~ off for potatoes this year**, bu sene patatesimiz bol: **~ on in years**, yaşı ilerlemiş: **it is ~ on midnight**, gece yarısı yaklaşıyor: **you are ~ out of it**, bundan kurtulduğuna şükret: **pretty ~ all**, hemen hemen hepsi: **to be ~ up**

**in a subject,** bir mevzu iyi bilmek: **that's all very ~ (and good) but ...,** hepsi iyi hoş, amma ...: **to wish s.o. ~,** birinin iyiliğini istemek. **~-behaved,** terbiyeli. **~-being,** saadet, refah. **~-bred,** terbiyeli. **~-built,** (insan) boyu bosu yerinde; (ev) sağlam. **~-chosen,** iyi seçilmiş; münasib. **~-conducted,** terbiyeli; iyi idare edilen. **~-connected,** iyi aileden. **~-doing,** hayırseverlik. **~-earned,** müstahak. **~-grown,** yaşına göre büyük. **~-informed,** malûmatlı; salâhiyetli. **~-knit,** sıkı yapılı. **~-known,** tanınmış, meşhur. **~-made,** iyi yapılmış. **~-marked,** aşikâr; belli. **~-meaning,** iyi kalbli; (aslında) iyi niyetli; (yanlış bir hareket yapsa bile) niyeti iyi. **~-meant,** iyi niyetle yapılmış. **~-nigh,** hemen hemen. **~-off,** hali vakti yerinde olan. **~-to-do,** refah içinde yaşıyan. **~-worn,** çok kullanılmış; aşınmış.

**Wellingtons** [ˈwelin(g)tənz]. (Lâstik) çizme.

**Welsh**[1] [welş]. Gal'li; gal dili. **~man,** Gal'li. **~**[2] *v.* (Bahis tutan adam) kaybettiği parayı ödemeden sıvışmak.

**welt**[1] [welt] *v.* **weal**[1]. Dövmek. **~**[2], Zıh (vs.yi takmak).

**welter** [ˈweltə(r)]. Çamur, kan vs. içinde yatıp yuvarlamak. Karmakarışık. **in a ~ of blood,** kan revan içinde.

**wench** [wenç]. Kız veya genc kadın; haspa. Zamparalık etmek.

**wend** [wend]. **to ~ one's way,** yürümek.

**went** *v.* **go.**

**wept** *v.* **weep.**

**were** *v.* **be.**

**we're** [wiə(r)] = **we are.** *v.* **be.**

**werewolf** [ˈweəwulf]. Kurt şeklinde dolaştığına inanılan insan.

**wert** [wö(r)t]. (*esk.*) *2nd pers. sing. past of* **be,** oldun, idin.

**west** [west]. Garb, batı. **the W~,** birleşik Amerika'nın garb devletleri: **the ~ End,** Londra'nın kibar mahallelerinin ve büyük mağaza ve tiyatrolarının bulunduğu kısım: **to go ~,** ölmek; mahvolmak. **~erly,** (rüzgâr) garbdan esen. **~ern,** garbda. **~erner,** garblı. **~ward,** garb cihetine giden: **~s,** garba doğru.

**wet** [wet]. Islak, rutubetli; yağmurlu; (*Amer.*) içki yasak olmıyan. Islaklık; yağmur. Islatmak. **~ blanket,** oyun bozan: **~ nurse,** süt nine: **~ to the skin [~ through, soaking ~],** sırsıklam: **to ~ one's whistle,** içki içmek. **~ting, to get a ~,** yağmurdan veya suya düşerek ıslanmak.

**whack** [wak]. (*ech.*) Dayak atmak. **to have a ~ at stg.,** (*sl.*) bir şeyi şöyle bir denemek. **~ing,** dayak; (*sl.*) kocaman; **a ~ lie,** kuyruklu yalan.

**whale** [wêyl]. Balina; (*sl.*) dev gibi. Balina avlamak. **~bone,** korse balinası. **~r,** balina gemisi.

**wharf** [wôf]. Rıhtım, iskele. **~age,** rıhtım ücreti.

**what** [wot]. Ne; ne?; hangi. **tell me ~ you saw,** gördüğünüzü söyleyiniz: **~ about a game of tennis?,** tenis oynıyalım mı?: **~ about the others?,** ya ötekiler?: **there wasn't a day but ~ it rained,** yağmur yağmadık gün olmadı: **~ for?,** ne için, niçin?: **~ did he do that for?,** ne diye bunu yaptı?: **~ if he does not come?,** gelmezse ne olacak? **~ if it *is* true?,** doğru olsa bile ne çıkar?: **I know ~! [I'll tell you ~!],** buldum!: **and ~'s more ...,** hem de: **~ next?,** bundan sonra ne var?: **~ next!,** daha neler!: **... and ~ not,** ... ve saire: **well, ~ of it?,** olsun, ne çıkar?: **I'll show you ~'s ~!,** ben sana dünyanın kaç bucak olduğunu gösteririm: **~ though we are poor,** fakirsek ne çıkar? **~-d'ye-call-'em, -him, -her, -it,** *adı hatıra gelmiyen bir şeyi yahud bir kimseyi anlatmakta kullanılır;* **I saw Mr. ~-him,** Şeyi gördüm. **~-ho!,** yahu! **~-not,** etajer. **~(so)-ever** [wot(sou)ˈevə(r)], herhangi; her ne: **do ~ you like,** ne istersen yap: **I cannot see anyone ~,** (i) kim olursa olsun hiç kimseyi göremem; (ii) görünürde in cin yok: **~ I have is yours,** nem varsa senindir: **~ happens I will remain your friend,** ne olursa olsun size dost kalacağım: **he has no luck ~,** hiç mi hiç talihi yoktur.

**wheat** [wît]. Buğday. **~en,** buğdaydan yapılmış.

**wheedle** [ˈwîdl]. Dil dökerek kandırmak. **to ~ stg. out of s.o.,** tatlı sözle birisinden bir şey sızdırmak.

**wheel** [wîl]. Çark; tekerlek; direksyon volanı; (gemi) dümen dolabı. (*mil.*) Çark etmek. Tekerlekli bir şeyi el ile yürütmek. **left [right] ~!,** (*mil.*) sola [sağa] çark!: **the ~s of government,** idare makinesi: **he ~ed round,**

birdenbire olduğu yerde geri döndü: **to run on ~s**, tekerlekle hareket etm.: **to take the ~**, otomobili idare etm.; geminin dümenini kullanmak: ⌜**there are ~s within ~s**⌝, ⌜işin içinde iş var.⌝. **~barrow,** tek tekerlekli el arabası. **~ed,** tekerlekli. **~wright,** araba yapan, tamir eden usta. **~-chair,** tekerlekli hasta sedyesi.

**wheez·e** [wîz]. (*ech.*) Hışıltı; (*sl.*) şaka. Hışıldamak. **~y,** hışıltılı.

**whelp** [welp]. Enik; edebsiz çocuk.

**when** [wen]. Ne vakit? ne zaman?; ne zaman ki ..., ···diği zaman; iken. **~ will you come?,** ne vakit geleceksiniz?: **~ you come,** geldiğiniz zaman: **~ a child I used to be afraid of the dark,** çocukken karanlıktan korkardım: **~ ever** [**~ on earth**] **will he come?,** ne halt etmeğe bu kadar gecikti?: **why walk ~ you can ride?,** atla gitmek mümkün iken niçin yürüyeceksiniz?: **tell me the ~ and the how of it,** bunun ne zaman ve nasıl olduğunu söyle. **~ever** [wenˈevə(r)], her ne zaman. **I go ~ I can,** imkân olduğu zaman giderim: **you can come ~ you wish,** ne zaman isterseniz gelebilirsiniz.

**whence** [wens]. Nereden?; nasıl?; geldiği yerden. **let him return to the land ~ he came,** geldiği yere gitsin: **~ we can understand that ...,** işte bundan anlıyoruz ki ....

**where** [we͜ə(r)]. Nerede?; nereye?; ···diği yerde. **~ from?,** nereden?: **~ do you come from?,** nereden geliyorsunuz?: **I shall stay ~ I am,** bulunduğum yerde kalacağım: **I don't know ~ I am,** nerede bulunduğumu bilmiyorum; vaziyeti (ne yapacağımı) bilmiyorum: **~ was I?,** nerede kalmıştım?: **that is ~ you are mistaken,** işte burada yanılıyorsunuz. **~abouts**[1] [we͜ərəˈba͜uts], nerelerde?; takriben nerede? **~**[2] [ˈwe͜ərəba͜uts]. Bir kimse veya şeyin tahmîni olarak bulunduğu yer; neresi. **nobody knows his ~,** nerede bulunduğunu kimse bilmiyor. **~as** [we͜ərˈaz], halbuki; mademki. **~at** [we͜ərˈat], ki ondan sonra. **I laughed; ~ he became very angry,** ben gülünce kızdı. **~by** [we͜əˈbây], ki ondan; onun ile. **is there no way ~ we can learn the truth?,** hakikati öğrenmenin bir yolu yok mu? **~fore** [ˈwe͜əfô(r)], niçin; onun için. **~in** [we͜ərˈin], ne hususta?; ki içinde. **~of** [we͜ərˈov], neden?; ki ondan. **~on** [we͜ərˈon], ki üzerinde. **the stone ~ he sat,** üzerinde oturduğu taş. **~soever** [we͜əso͜uˈevə(r)], nerede olursa. **~upon** [we͜ərʌˈpon], bunun üzerine. **~ver** [we͜ərˈevə(r)] her nerede: nerede? **~with** [we͜əˈwiθ], ki bunun ile: **~al,** icab eden şey; para.

**whet** [wet]. Bilemek; (iştahını) açmak.

**whether** [ˈweðə(r)]. ... mi?; hangisi. **~... or ...,** ... mi, ... mi. **I don't know ~ we shall find him at home,** onu evde bulup bulamıyacagımızı bilmiyorum: **~ it rains or not,** yağmur yağsa da yağmasa da: **I wonder ~ he will come,** acaba gelir mı?: **~ you like it or not,** isteseniz de istemeseniz de: **~ or no,** her halde: **he asked me ~ I liked his book,** kitabını sevip sevmediğimi sordu.

**whew** [ˈhwu, hi͜u]. (*ech.*) *Yorgunluk veya hayret nidasi.*

**which** [wiç]. Hangi?, hangisi: ... ki. **the book ~ is on the table,** masanın üstündeki kitab: **the house in ~ we live,** (içinde) oturduğumuz ev: **this is the house of ~ I was speaking,** kendisinden bahsettiğim ev budur: **~ way?,** ne tarafa?; nasıl? **~ever** [–ˈevə(r)], her hangisi: **take ~ you like best,** hangisinden en çok hoşlanırsanız onu alınız: **~ way he looked,** hangi tarafa baktı ise.

**whiff** [wif]. Püf; duman vs. dalgası; hafif koku. Hafifçe püflemek. **I must get a ~ of fresh air,** bir parçacık hava almalıyım: **there wasn't a ~ of wind,** hiç rüzgâr yoktu.

**Whig** [wig]. İngiliz liberal partisine aid.

**while**[1] [wâyl]. İken; esnasında. **~ I live,** ömrüm oldukça: **~ admitting the thing is difficult,** gerçi bunun güç olduğunu kabul ederim. **~**[2] *n.* Müddet. **the ~,** bu esnada: **between ~s,** arada sırada: **after a ~,** bir az sonra: **once in a ~,** kırk yılda bir: **worth ~,** değer: **worth one's ~,** zahmete değer: **I'll make it worth your ~,** sizi memnun ederim. **~** *v.* **to ~ away the time,** vakti hoş geçirmek.

**whilst** [wâylst]. İken.

**whim** [wim]. Geçici arzu, heves; kapris. **~sical** [ˈwimzikl], gelgeç, maymun iştahlı; tuhaf.

**whimper** [ˈwimpə(r)]. Ağlıyacak gibi ses çıkarma. Ağlayıp sızlamak.

**whine** [wâyn]. (*ech.*) Sızlanma(k).
**whinny** [ˈwini]. Kişneme(k).
**whip**¹ [wip] *n.* Kamçı, kırbaç; (parlamento) parti disiplinini koruyan âza; mühim ve acele bir mesele hakkında parti âzasına dağıtılan kısa tamim. **four-line ~**, altı dört kere çizilmiş tamim ('whip'): **to have the ~ hand of s.o.**, birine karşı üstün bir vaziyette olmak. **~**² *v.* Kamçılamak, kırbaçla dövmek; yumurta vs.yi çalkayıp köpürtmek; ansızın bir hareket ile çıkarmak. **to ~ round the corner,** köşeyi hızla dönmek: **to ~ a revolver out of one's pocket,** cebinden bir tabancayı süratle çekmek. **~ back,** (fazla gerilen ip vs.) kopup geri sıçramak. **~ in,** (köpekleri) kamçı ile toplamak. **~ off,** ansızın kaldırmak. **~ round,** birdenbire dönmek: **to have a ~ round for subscriptions,** bir çok kimseye müracaat edip iane toplamak. **~ up,** (*fig.*) kamçılamak; tahrik etmek. **~per-snapper** [ˈwipə(r)ˈsnapə(r)], kendini beğenmiş genc. **~ping** [ˈwipin(g)], kamçılama; façuna: **~-boy,** başkasının yaptığı kabahatin cezasını çeken kimse.
**whirl** [wö(r)l]. Fırıldanma(k); fırıldatmak; kasırga gibi dönmek; pek hızlı hareket et(tir)mek. **my head is in a ~,** başım dönüyor: **a ~ of pleasures,** zevk ve eğlence fırtınası. **~igig,** fırıldak. **~pool,** girdab. **~wind,** kasırga: **to sow the wind and reap the ~,** rüzgâr ekip fırtına biçmek.
**whirr** [(h)wö(r)]. (*ech.*) Kanad sesi. Uğuldamak.
**whisk** [wisk]. (*ech.*) Hafif ve hızlı hareket; yumurta çalkama aleti. Hızla ve az fırla(t)mak; yumurta çalkamak. **~er(s)** [ˈwiskö(r)·z], favori; hayvan bıyığı.
**whisper** [ˈwispə(r)]. (*ech.*) Fısıltı. Fısıldamak. **it is ~ed that ...,** kulaktan kulağa fısıldandığına göre. **~ing,** fısıltı.
**whistle** [ˈwisl]. Islık (çalmak); düdük (çalmak). **to ~ for s.o.,** ıslık çalarak çağırmak: ⌜**he can ~ for his money**⌝, ⌜o paranın üstüne bir bardak su içsin⌝.
**whit** [wit]. Zerre. **not a ~,** hiç: **every ~ as good as ...,** tamamen aynı derecede iyi.
**white** [wâyt]. Beyaz, ak, ağarmış. Beyaz kısmı [renk, adam]. **to go [turn] ~,** (benzi) sararmak: **~ heat,** narı beyza: **the W~ House,** Amerika Cumhurreisinin resmî dairesi: **a ~ man,** beyaz adam; temiz yürekli adam: **~ metal,** gümüş taklidi maden: **in a ~ rage,** hiddetten kudurmuş bir halde: **~ sale,** çamaşır satışı. **~-headed,** beyaz saçlı. **~-lipped,** korkudan dudakları beyazlaşmış. **~-livered,** korkak. **W~hall,** Londra'da hükûmet dairelerinin bulunduğu mahalle; (*fig.*) hükûmet. **~n** [ˈwâytn], ağar(t)mak; beyazlatmak. **~ness,** beyazlık. **~wash** [–woş], badana; kirecsuyu. Badanalamak; (birini) temize çıkarmak.
**whither** [ˈwiðə(r)]. Nereye. **~soever,** her hangi bir yere.
**Whit Monday** [ˈwitˈmʌndi]. **Whit Sunday** 'den sonra gelen pazartesi (İngiltere'de umumî tatil günlerinden biri). **~sun(tide),** Paskalyadan yedi hafta sonra olan yortu.
**whittle** [ˈwitl]. **to ~ down [away],** kese kese mikdarını azaltmak.
**whizz** [wiz]. (*ech.*) Vızıltı. Vızıldamak.
**who** [hû]. Kim; o ki; onlar ki. **the man ~ came,** gelen adam: **~ goes there?,** kimdir o?: **no matter ~,** kim olursa olsun: **Who's ~,** meşhur adamların tercümei hal kitabı. **~ever** [hûˈevə(r)], her kim; kim olursa olsun. **~m** [hûm], **who**'*nun mefulünbihi.* Kimi, ki onu. **of ~,** kimden, ki ondan. **the man ~ we met,** rastladığımız adam: **the person to ~ you sent the letter is dead,** kendisine mektub gönderdiğiniz şahıs ölmüştür. **~(so)ever** [–(sou̯)ˈ-evə(r)], kim olursa olsun; her kim ise. **~se** [hûz], **who** *'nun muzafünileyh hali.* Kimin; ki onun. **the man ~ house you saw,** evini gördüğünüz adam. **~soever** [hûsou̯ˈevə(r)]. Her kim olursa olsun.
**whoa** [wou̯]. Çüş!; dur!
**whole** [hou̯l]. Bütün, tam; kusursuz; sağlam. Kül; tam şey; yekûn. **~ brother,** ana baba bir kardeş: **on the ~,** umumiyetle: **he swallowed it ~,** (i) çiğnemeden yuttu; (ii) hepsini yuttu (inandı): **taken as a ~,** bir bütün olarak: **the ~ (of the) world,** tam dünya; herkes. **~sale,** toptan; küme halinde; büyük mikyasta: **~r,** toptancı. **~-hearted,** candan; samimî. **~-time job [work],** insanın bütün vaktini alan iş. **~some**

[ˈhoulsəm], sıhhate nafı; sağlam; kolayca hazmedilir: **not a ~ book for the young,** çocuklar için uygun olmıyan kitab.

**wholly** [ˈhouli, houl·li]. Tamamiyle, büsbütün.

**whoop** [(h)wûp]. (*ech.*) Nida; bağırma; haykırma. Bağırmak; boğmaca öksürüğüne tutulmuş gibi öksürmek. **he ~ed with joy,** sevincinden bağırdı.

**whop** [(h)wop]. (*ech.*) Darbe, dayak atmak. **~ping,** dayak; pek büyük: **a ~ lie,** [**a whopper**], kuyruklu yalan.

**whore** [hô(r)]. Orospu. Zamparalık etmek.

**why** [(h)wây]. Niçin?, neden?, ne diye?; tuhaf şey! **that is the reason ~ ...,** işte bundan dolayıdır ki ....

**wick** [wik]. Fitil.

**wicked** [ˈwikid]. Habis, pek kötü, hain.

**wicker** [ˈwikə(r)]. Sepet işi; hasır.

**wicket** [ˈwikit]. Bahçe vs.nin küçük kapısı; büyük kapının içinde [yanında] ufak kapı. Kriket oyununda kullanılan ve kaleyi teşkil eden üç kazık; kalelerin arasındaki yer.

**wide** [wâyd]. Geniş; vâsi; enli; açık; bol. **~ awake,** tamamen uyanmış: **~ of the mark,** hedeften uzak; çok yanlış: **~ open,** ardına kadar açık: **~ly read,** çok okunan; malûmatlı: **the ~ world,** şu koca dünya. **~awake,** açıkgöz; tetik. **~n,** genişle(t)mek; büyü(t)mek. **~-spread,** yayılmış; umumî.

**widow** [ˈwidou]. Dul kadın. Dul bırakmak. **grass ~,** kocası muvakkaten başka yerde bulunan kadın: ⌜**the ~'s mite**⌝, ⌜çok veren maldan az veren candan⌝, bir fakir tarafından verilen küçük iane. **~ed,** dul. **~er,** dul erkek. **~hood,** dulluk.

**width** [widθ]. En; genişlik; vüsat.

**wield** [wîld]. Elle tutup kullanmak; savurmak. **to ~ power,** tahakküm etm.: **to ~ the sceptre,** saltanat etmek.

**wife,** *pl.* **wives** [wâyf, wâyvz]. Karı, zevce, refika. **the ~,** refikam: **to take a ~,** evlenmek: **to take s.o. to ~,** birisile evlenmek: **an old wives' tale,** tandırname. **~ly,** zevceliğe yakışır.

**wig** [wig]. Takma saç; peruka. **~ging,** (*sl.*) azarlama.

**wigwam** [ˈwigwam]. Şimalî Amerika kızılların kulübesi.

**wild** [wâyld]. Vahşi; yabani; delişmen; fırtınalı; haşarı; hedeften uzak. **the ~s,** beyaban; çöl. **to be ~ to do stg.,** bir şeyi yapmak için yanıp tutuşmak: **to be ~ with joy,** sevincinden çıldırmak: **~ exaggeration,** pek fazla mübalâğa: **to lead a ~ life,** hovardalık etm.: **to make s.o. ~,** birini çıldırtmak: **to run ~,** (nebat) azmak; (çocuk) başıboş dolaşmak: **~ talk,** palavra. **~-boar,** yaban domuzu. **~-cat,** yaban kedisi: **~ scheme,** (ticaret) çılgınca bir plân. **~-goose,** yabani kaz: **~ chase,** ahmakça ve sonsuz bir teşebbüs. **~erness** [ˈwildənis], beyaban; çöl: **a voice in the~,** çölde bir vaız. **~fire** [ˈwâyldfâyə(r)], **to spread like ~,** etrafı alev gibi sarmak. **~fowl** [-faul], yabani (av) kuşları. **~ness,** vahşilik, yabanilik; şiddet; fırtınalılık; delilik.

**wil·e** [wâyl] (*gen.* ~s). Hileler; kurnazlık. Hile ile cezbetmek. **~iness** [ˈwâylinis], kurnazlık. şeytanlık. **~y** [ˈwâyli], cin fikirli; kurnaz.

**wilful** [ˈwilfəl]. Söz anlamaz; inadcı; kasdî.

**will**[1] [wil] *n.* İrade; istek; keyif; vasiyet(name). **at ~,** istediği zaman; keyfine göre: **with the best ~ in the world I can't do it,** bütün arzuma rağmen yapamam: **free ~,** iradei cüz'iye: **of one's own free ~,** kendi isteğile: **good ~,** hüsnüniyet: **ill ~,** suiniyet: **the last ~ and testament of ~,** ···in son vasiyetnamesi *ki vasiyetnameye başlarken kullanılan mûtad cümledir*: **to make one's ~,** vasiyetnamesini yazmak: **to have a ~ of one's own,** inadcı olmak; aklına geleni yapmak istemek: **to take the ~ for the deed,** bir iyilik yapma arzusunu iyilik saymak: ⌜**where there's a ~ there's a way**⌝, meramın elinden bir şey kurtulmaz: **to work one's ~ upon s.o ,** bir adama istediğini yapmak. **~**[2] *v.* İstemek; razı olm.; azmetmek; vasiyetname ile birakmak. *Yardımcı fiil olarak istikbal sıygasının teşkiline yarar, mes.*: **I ~ write,** yazacağım; *fakat bazan sadece bir rica ifade eder, mes.*: **~ you close the window!,** pencereyi kapatır mısınız?: '**~ you be there?**' '**I ~**', 'Siz orada bulunacak mısınız?' 'Bulunacağım': **as you ~!,** siz bilirsiniz!: **accidents** *will* **happen,** kazanın önüne geçilmez: **this car ~ do 40 miles to the gallon,** bu otomobil bir galon benzinle 40 mil gider: **this cár ~ take five people easily,** bu otomobil ferah ferah beş kişi alır: **he *will* have it that ...,** ... diye inad ediyor: **he ~ have none of it,** (i) onun hissesine bundan hiç bir şey düşmiyecek; (ii) bunu hiç kabul etmiyor: **to ~ s.o. into doing stg.,** birine bir şeyi irade kuvvetile yaptırtmak: **say what you ~, no one ~ believe you,** ne istersen söyle, kimse sana inanmaz: **~ you sit down?** [**won't you sit down?**], oturmaz mısınız?: ***will* you sit down!** (*hiddetle*) oturacak mısın? **~ed** [wild], ... iradeli. **~ing,** razı; istekli; candan: **~ helpers,** candan yardım edenler. **~ingly,** istiyerek; memnuniyetle. **~ingness,** isteklilik: **with the utmost ~,** canü gönülden. **~y-nilly** [ˈwiliˈnili], çarnaçar, ister istemez.

**will-o'-the-wisp** [ˌwiloðiˈwisp]. Boş gaye.

**willow** [ˈwilou]. Söğüd; (*sl.*) kriket çomağı. **~y,** fidan gibi; eğilir bükülür. **~-pattern,** meşhur Çin porselenlerinde bulunan söğüt resmi.

**wilt**[1] [wilt]. (*esk.*) *2nd pers. sing. pres. of* **will.** **~**[2]. Tazeliğini kaybet(tir)mek.

**win** [win]. Kazanmak; kesbetmek; galebe çalmak, yenmek. Galebe, zafer. **to ~ over** [**round**], kendi tarafına celbetmek: **to ~ through,** muvaffak olm.; müşkülatın hakkından gelmek: **to ~ the shore,** *etc.*, sahil vs.ye varmak. **~ner** [ˈwinə(r)], kazanan kimse veya at; muvaffakiyetli bir iş.

**winc·e** [wins]. Acıdan birdenbire ürkme(k). **without ~ing** [**without a ~**] göz kırpmadan.

**winch** [winç]. Vinç; bocurgat.

**wind**[1] [wind] *n.* Rüzgâr; hava; nefes; osuruk. **the ~,** (*mus.*) bir orkestranın nefes çalgıları: **the wood ~,** flâvta cinsinden nefes çalgıları: **to sail before the ~,** pupa yelken gitmek: **between ~ and water,** su kesiminde: **to fling prudence,** *etc.*, **to the ~s,** ihtiyat vs.ye hiç aldırmamak: **to get ~ of,** kokusunu almak: **there's something in the ~,** ortada [havada] bir şeyler var: **to lose one's ~,** nefesi kesilmek: **to recover one's ~,** nefes almak için dinlemek: **to get one's second ~,** bir koşu vs.de bir kere nefesi kesildikten sonra tekrar nefes almak: **to raise the ~,** ne yapıp yapıp para elde etm.: **sound in ~ and limb,** sapasağlam: **to get the ~ up,** (*sl.*) telaşa düşmek: **to put the ~ up s.o.,** birini telaşa düşürmek: **to see which way the ~ blows,** havayı koklamak. **~-sleeve,** rüzgârın cihetini gösteren bez manika. **~-tunnel,** aerodinamik tünel. **~bag,** lâkırdı kumkuması. **~fall** [–fôl], rüzgârın düşürdüğü şey, *esp.* elma: yağlı lokma; umulmadık miras. **~iness** [–inis], rüzgârlı olma. **~jammer** [–camə(r)], geçen asırda büyük yelkenli ticaret gemisi. **~mill,** yel değirmeni. **~pipe** [–pâyp], nefes borusu; gırtlak. **~screen** [–skrîn], rüzgâr siperi; otomobilin ön camı: **~-wiper,** cam silicisi. **~swept,** rüzgârlı. **~ward** [–wö(r)d], rüzgâr üstü. **~y** [ˈwindi], rüzgârlı; uzatıcı (nutuk vs.); rihî; (*sl.*) endişeli. **~**[2] [wind] *v.* Kokusunu almak; soluğunu kesmek. **to ~** [wâynd] **the horn,** av borusunu çalmak. **~ed,** soluğu kesilmiş. **~**[3] (**wound**) [wâynd, waund] *v.* Dolaş(tır)mak; çevirmek; sarmak; (saat vs.) kurmak. **to ~ about,** dolaşmak; yılankavi olm.: **to ~ wool into a ball,** ipliği yumak yapmak: **the plant ~s round the pole,** nebat sırığa sarılıyor. **~ up,** çark ile kaldırmak; kurmak; bitirmek. Sona ermek: **to ~ up a company,** bir şirketi tasfiye etm.: **how does the play ~ up?,** piyes nasıl bitiyor? **~ing** [ˈwâyndin(g)], dolambaçlı; yılankavi: dönemec; dolambaç; sarma: **~-gear,** [**-plant**], maden ocağında asansör işleten makine: **~-sheet,** kefen. **~lass** [ˈwindləs], ırgat; çıkrık.

**window** [ˈwindou]. Pencere; camekân; gişe. **~-box,** pencere önünde çiçek kutusu. **~-dressing,** satılacak eşyanın açılıp serilmesi; (*fig.*) gösteriş. **~-pane,** pencere camı.

**wine** [wâyn]. Şarab. **~-bibber,** şarab ayyaşı. **~-coloured,** şarabî renk. **~-glass,** şarab kadehi. **~-grower,** şarab üzümü yetiştiren bağcı. **~-press,** şarab baskısı. **~-vault,** yeraltı şarab deposu; yeraltı meyhanesi.

**wing** [win(g)]. Kanad; kol; kulis; (*otom.*) çamurluk. Uç(ur)mak; kanad takmak. **to be on the ~,** uçmakta olm.: **to take ~,** uçmak: **to take s.o. under one's ~,** birini himaye altına almak. **~ed,** kanadlı; kanadından yaralı.

**wink** [win(g)k]. Göz kırpma. Göz kırpmak; gözle işaret etm.: (ışık) yanıp sönmek. **to ~ at stg.**, bir şeye göz yummak: **forty ~s**, kısa uyku: **I did not sleep a ~**, kirpiğimi kırpmadım.

**winnow** [ˈwinou]. Hububat savurmak. **to ~ out**, savurarak ayıklamak; (*fig.*) fenadan iyiyi ayırmak.

**winsome** [ˈwinsəm]. Alımlı, cazibeli.

**wint·er** [ˈwintə(r)]. Kış(lık). Kışla(t)mak. **in ~**, kışın: **the depth of ~**, kara kış: **~ quarters**, kişlak. **~ry**, kış gibi; pek soğuk.

**wipe** [wâyp]. Silme(k); bez ile temizleme(k). **to ~ away [off]**, silip çıkartmak: **to ~ s.o.'s eyes**, (*coll.*) başkasından evvel davranmak: **to ~ out**, silip temizlemek: **to ~ out an insult**, bir hakareti temizlemek.

**wire** [wâyə]. Tel; telgrafname. Tel takmak; tel ile bağlamak; (ev vs.de) elektrik tesisatı kurmak; telgraf çekmek. **~ entanglement**, tel mania. **~d**, telli; tel ile çevrilmiş; elektrik telleri konmuş. **~less**, telsiz; radyo; radyo ile telgraf çekmek: **~ set**, radyo: **~ receiver**, âhize. **~-haired terrier**, dik kıllı fino.

**wiry** [ˈwâyri]. Tel gibi; ince fakat adaleli.

**wisdom** [ˈwizdəm]. İrfan; akıl; hikmet. **~ tooth**, akıl dişi.

**wise**[1] [wâyz] *n.* Tarz, suret. **in no ~**, hiç bir surette: **in some ~**, bir surette. **-wise**, *suff.* ···vari, ... gibi. **~**[2] *a.* Ârif; tecrübeli ve akıllı; hakîm. **to be ~ after the event**, iş işten geçtikten sonra akıllanmak: **to look ~**, işten anlar gibi bakmak: **I am none the ~r**, daha iyi öğrenmiş değilim: **no one will be any the ~r**, ⌜kim kime dum duma⌝: **put me ~ about it**, (*Amer.*) bana bu işi anlat. **~acre**, ukalâ dümbeleği. **~-crack**, nükte; (*bazan*) ukalâlık.

**wish** [wiş]. İstemek, arzu etm.; temenni etmek. Arzu, istek; temenni. **I ~ I were there**, keşki orada olsam: **what more can you ~?**, daha ne istersen?: **one would ~ that ...**, gönül ister ki. **~bone** [ˈwişboun], lâdes kemiği. **~ful** [–fəl], istekli; hasretli: **~ thinking**, hüsnükuruntu. **~ing-well** [ˈwişin(g)wel], dilek kuyusu.

**wishy-washy** [ˈwişiwoşi]. Yavan, değersiz.

**wisp** [wisp]. Demetçik; ince saç lülesi.

**wistful** [ˈwistfəl]. Hasretli; arzulu.

**wit**[1] [wit]. **to ~**, yani, demek ki: **God wot**, Allah bilir. **~tingly** [ˈwitin(g)li], kasden; bile bile. **~**[2] *n.* Akıl; nüktecilik; nükteci. **to be at one's ~'s end**, ne yapacağını bilmemek: **to collect one's ~s**, aklını başına toplamak: **to keep one's ~s about one**, uyanık olmak: **to be out of one's ~s**, aklını oynatmak: **one who lives by his ~s**, kaparozcu. **~less** [ˈwitlis], budala; akılsız. **-~ted** [ˈwitid] *suff.* ···kafalı, *mes.* **slow-~**, kalın kafalı. **~ticism** [ˈwitisizm], nükteli söz. **~ty** [ˈwiti], nükteli, nekre.

**witch** [wiç]. Büyücü kadın. Teshir etmek. **~craft**, büyücülük. **~ery**, büyücülük; sihir. **~-doctor**, vahşi kabilelerde büyücü hekim.

**with** [wið]. İle; ···le; ···ce; ···den; yanında, beraber, birlikte; arasında, nezdinde. **~ all his wealth he is not happy**, bütün servetine rağmen mesud değildir: **the difficulty ~ him is that ...**, bu adamın müşkül tarafı...: **~ that he left the room**, bunun üzerine odadan çıktı. **~al** [wiðˈôl], dahi; bununla beraber.

**withdraw** (**-drew, -drawn**) [wiðˈdrô, -drû, -drôn]. Geri çekmek; feragat etm.; (bankadan para) çekmek. Rücu etm.; çekilmek. **~al**, geri çek(il)me; rücu; geri çağır(ıl)ma: **~ of an order**, bir emri geri alma: **~ of a coin from circulation**, bir parayı tedavülden kaldırma.

**wither** [ˈwiðə(r)]. Kuru(t)mak: solmak. **to ~ s.o. with a look**, birini bir bakışla yerin dibine geçirmek.

**withhold** (**-held**) [wiðˈhould, -held]. Vermemek; esirgemek; tutmak. **to ~ the truth from s.o.**, birinden hakikati gizlemek.

**within** [wiðˈin]. İçinde, içerisinde; dairesinde. **to live ~ one's income**, geliri ile mütenasib yaşamak: **they are ~ a few months of the same age**, bir kaç ay farkla aynı yaştadırlar: **~ reason**, makulât dairesinde: **well ~ the truth**, en aşağı tabirle.

**without** [wiðˈaut]. ···siz; ···sizin; bilâ···; ... haricinde. Dışında. **~ walls**, duvarsız: **~ the walls**, surların dışında: **to go ~ stg.**, kendini bir şeyden mahrum etm.: **a child can't do ~ games**, çocuk oynamadan edemez: **that goes ~ saying**, bu bedihidir: **not ~ difficulty,** oldukça müşkülâtla: **he passed ~ seeing us,** bizi görmeden geçti.

**withstand** (-stood) [wiðˈstand, -stud]. Karşı koymak; dayanmak.

**witness** [ˈwitnis]. Şahid; delil. Müşahede etm., görmek; şehadet etm. **to ~ a signature,** kendi imzasile başka bir imzayı tasdik etmek. **~-box,** mahkemelerde şahid yeri.

**wives** *v.* **wife.**

**wizard** [ˈwizö(r)d]. Sihirbaz; büyücü. **~ry,** sihirbazlık.

**wizened** [ˈwizənd]. Kart, kurumuş.

**wobbl·e** [ˈwobl]. Sallanma(k); sendeleme(k); zikzak yapma(k); (ses) titreme(k). **~y,** sallanan, sendeliyen; titrek.

**woe** [wou]. Keder, elem; dert. Vay!, hayıf! **~ is me!,** vay bana!: **~ be to ...,** (*lânet*) ···in Allah belâsını versin!: **~ to the vanquished!,** veyl mağluba! **~begone,** hazin, kederli. **~ful,** hüzün verici: **he is ~ly weak in French,** fransızcası fecidir.

**woke** *v.* **wake.**

**wold** [would]. Yüksek ârizalı çıplak kır. **the Wolds,** Yorkshire'da çıplak ve yalçın tepeler.

**wolf,** *pl.* **wolves** [wulf, -vz]. Kurt. Yemeği aç kurt gibi yutmak. **to cry ~,** yalan yere tehlike ilân etm.: **to keep the ~ from the door,** (ailesini vs.) açlıktan korumak: ⌜**a ~ in sheep's clothing**⌝, koyun postuna girmiş kurt. **~ish,** kurt gibi; vahşi; açgözlü.

**woman,** *pl.* **-men** [ˈwumən, ˈwimin]. Kadın. **~hood,** kadınlık. **~ish,** kadın gibi. **~ize,** zamparalık etmek. **~kind,** kadın kısmı. **~ly,** kadına yakışır.

**womb** [wûm]. Rahim.

**women** *v.* **woman. ~folk,** kadınlar; kadın kısmı.

**won** *v.* **win.**

**wonder** [ˈwʌndə(r)]. Hayran olm.; şübhe veya meraka düşmek. Hayret; acibe. **I ~ whether ...,** acaba ...: **no ~ ...,** hayret edilmez; pek tabiî: **I ~ he was there,** orada olmasına şaştım: **I ~ if he was there,** acaba orada mıydı: **a nine days' ~,** kısa bir zaman için herkese hayret veren bir şey: **it's a ~ he wasn't killed,** ölmemiş olması mucizedir: **I don't ~ you are annoyed,** ben de olsam kızarım: **I shouldn't ~ if he doesn't come,** gelmezse şaşmam: **'She'll soon find a husband.' 'I ~ !',** 'Yakında bir koca bulur.' 'Pek zannetmem': **one of the Seven Wonders of the world,** acaibi seb'ai âlemden biri: **a ~ of delicate workmanship,** ince işçilik harikası: **he wasn't late today, for a ~,** bugün mucize kabilinden vaktinde geldi; **a little praise works ~s,** azıcık öğme mucize gibi tesir ediyor. **~-worker,** mucize yapan kimse. **~ful** [ˈwʌndəfəl], şayanı hayret; fevkalâde: **~ to relate,** şaşılacak şey şudur ki. **~land** [-land], harikalar diyarı. **~struck** [-strʌk], şaşmiş.

**wondrous** [ˈwʌndrəs]. Hayret verici; fevkalâde.

**won't** = **will not.**

**wont** [wount]. Alışmış, âdet edinmiş. Âdet; mûtad. **according to his ~,** mûtadına gore: **to be ~ to do stg.,** bir şeyi yapmak âdetinde olmak. **~ed,** mûtad.

**woo** [wû]. Kur yapmak; celbetmeğe çalışmak; elde etmeğe çalışmak.

**wood** [wud]. Odun, tahta, ağac, ahşab; küçük orman. **beer drawn from the ~,** fıçıdan çekilmiş bira: **wine in the ~,** fıçı şarabı: **touch ~!,** şeytan kulağına kurşun!: **to be unable to see the ~ for the trees,** esas mesele teferrüat içinde boğulmak: **we are not out of the ~ yet,** daha işin içinden çıkmadık. **~-carving,** tahta oyması. **~-pulp,** tahta hamuru. **~-shed,** odunluk. **~-wind,** tahtalı nefes çalgıları. **~craft** [ˈwudkrâft], orman hayatı *ve bilhassa* avcılık bilgisi. **~cut** [ˈwudkʌt], tahta kalıbdan basılmış resim. **~cutter,** oduncu. **~ed** [ˈwudid], ağaclı. **~en** [wudn], tahtadan yapılmış; ahşab; alık (yüz); kazık gibi (duruş); resmî: **the ~ spoon,** bir yarışta en son geleni verilen tahta kaşık. **~land** [ˈwudlənd], ağaclık; ormanlık. **~man** [ˈwudmən], oduncu, ormancı. **~pecker** [ˈwudpekə(r)], ağackakan. **~work** [ˈwudwö(r)k], bir binanın ahşab kısmı; marangozluk. **~worker,** marangoz. **~y** [ˈwudi], haşebî; odun gibi.

**wool** [wul]. Yün. **dyed in the ~,** dokunmadan evvel boyanmış; **dyed-in-the-~ communist,** koyu bir komünist: **steel ~,** çelik talaşı. **~len,** yünden yapılmış; yünlü: **~s,** yünlü elbiseler. **~liness,** yünlülük; vuzuhsuzluk. **~ly,** yünlü; vuzuhsuz; keskin olmıyan. **~pack,** yün balyası. **~sack,** yün çuvalı: **the W~,** Lordlar Kamarasında Lord Chancellor'un oturduğu yer: **to be raised to the ~,** Lord Chancellor olmak. **~-bearing,** yün hasıl eden (hayvan). **~-gathering,** dalgın(lık): **to go [be] ~,** dalıp gitmek.

**word** [wö(r)d]. Kelime, söz. Kelime ile ifade etmek. **~s** (*mus.*) güfte. **to bring ~,** haber vermek: **you say the ~,** siz emredin!: **the W~ of God,** kitabı mukaddes: **to be as good as one's ~,** dediğini yapmak: **~ for ~,** kelimesi kelimesine: **he never has a good ~ for anyone,** herkesi kötüler: **to have a ~ with s.o.,** (i) birisile görüşmek; (ii) birine bir çift söz söylemek: **to have ~s with s.o.,** birisile atışmak: **high ~s,** hiddetli sözler: **in a ~,** hulâsa: **to have the last ~,** (i) bir münakaşada son sözü söylemek; (ii) son söz birinin olm.: **a man of his ~,** sözünün eri: **I am a man of my ~,** söz bir, Allah bir: **you have taken the ~s out of my mouth,** ben de tam bunu söyliyecektim: **my ~!,** maşallah!: **bad is not the ~ for it!,** bunu fena demek azdır: **in other ~s,** diğer tabirle: **~s have passed between them,** atıştılar: **to put in a good ~ for s.o.,** birinin lehinde bir şey söylemek: **may I put a ~ in?,** ben de bir şey söyliyebilir miyim?: **to send ~,** haber yollamak: **he told me in so many ~s, to go to Hell,** bana aynen 'cehennem ol!' dedi: **he did not say it in so many ~s,** aynen böyle demedi (fakat böyle demeğe getirdi): **take my ~ for it!,** sözüme inan: **I took you at your ~,** ben sizin sözünüze güvendim de: **he is too stupid for ~s,** tarif edilmez derecede abdal: **upon my ~!,** vallahi!: ⌜**a ~ to the wise (is sufficient)**⌝, ⌜anlayana sivrisinek saz ...⌝. **~-book,** sözlük. **~-perfect,** rolünü tamamen ezberlemiş. **~ing,** bir mektub vs.de kullanılan sözler: **we must be very careful about the ~ of the agreement,** anlaşmanın yazılış şekline çok dikkat etmeliyiz. **~y,** itnablı; çok uzun (söz): **~ warfare,** söz kavgası.

**wore** *v.* **wear.**

**work**[1] [wö(r)k] *n.* İş; vazife; iş güç; çalışma; el işi; eser, kitab; istihkâm; **~s,** atölye; (saat vs.nin) makinesi. **to be at ~,** iş başında olm.: **to have one's ~ cut out for one,** başında çok güç bir iş olmak: **it's all in the day's ~,** iş böyledir: **the poison had done its ~,** zehir tesir etti: **to go the right way to ~,** bir işe uygun usulle girişmek: **the ~s of God,** kâinat: **to make ~ (unnecessarily),** (hiç yoktan) iş çıkarmak: **Public Works,** nafia işleri: **to set to ~,** işe girişmek: **to set s.o. to ~,** birini bir işe oturtmak. **~**[2] *v.* **(worked** [wö(r)kt], *bazı tabirlerde* **wrought** [rôt]). Çalış(tır)mak; iş görmek; işle(t)mek; tesir yapmak; meydana getirmek; şekil vermek. **to ~ at history,** tarihe çalışmak: **to ~ a district,** (bir şirket memuru vs.) bir bölgeden mesul olm.: **he ~ed himself into a rage,** üzerinde dura dura gittikçe hiddetlendi: **to ~ loose,** laçka olm.: **we need more facts to ~ on,** üzerinde işlemek için daha fazla vakialara ihtiyacımız var: **to ~ one's passage,** yol ücretini gemide çalişarak ödemek: **my plan did not ~,** plânım muvaffak olmadı. **~ in,** içine işlemek. **~ off,** gidermek; (somun vs.) oynıyarak yerinden çıkmak: **to ~ off one's anger on s.o.,** öfkesini birinden almak: **to ~ off one's fat,** çalışıp zayıflamak. **~ on,** tesir etmek. **~ out,** yapıp bitirmek; halletmek; hesab etm.; yerinden oynamak: **(apprentice) to ~ out his time,** (çırak) muayyen müddeti doldurmak: **it ~ed out very well for me,** sonunda benim menfaatime uygun geldi: **how much does it ~ out at?,** kaça çıkar? **~ up,** hazırlamak: **what are you ~ing up to?,** sözü nereye getirmek istiyorsun?: **to get ~ed up,** heyecanlanmak: **he ~ed himself up into a temper,** konuştukça vs. hiddetlendi: **the symphony ~s up to a magnificent finale,** senfoni tedricen muhteşem bir finaleye doğru inkişaf ediyor. **~able** [-əbl], işlenebilir. **~aday** [-ədêy], âdi günlerde kullanılan: **~ clothes,** gündelik elbise: **this ~ world,** bu yeknesak cansıkıcı dünya. **~er** [-ə(r)], işçi, amele: **he is a hard ~,** çok çalışkandır: **a ~ of miracles,** mucize yapan kimse: **~-bee,** işçi arı. **~house** [-haus], darülâceze; düşkünler evi: **to bring s.o. to the ~,** birini yoksul etm. **~ing,** iş gören; işe aid; işliyen; işleme: **~ agreement,** işbirliği anlaşması; muvakkat anlaşma şekli: **~ capital,** mütedavil sermaye: **the ~ classes,** işçi sınıfı: **~ clothes,** iş elbisesi: **~ day,** iş günü: **~ expenses,** umumî masraflar: **~ majority,** kâfi ekseriyet: **a ~ man,** işçi sınıfına mensub adam: **~ party,**

ekip: **not ~**, çalışmıyan; işlemiyen, bozuk. **~man,** işçi, amele: **~like,** başarılı; üstad elinden çıkmış: **~ship,** sanatkârlık. **~people** [-pîpl], işçiler. **~shop** [-şop], atölye.

**world** [wö(r)ld]. Dünya; cihan; âlem; herkes. **a ~ of ...,** pek çok: **a ~ of money,** dünya kadar para: **all the ~,** herkes: **all the ~ and his wife were there,** bütün sosyete herkes orada idi: **for all the ~ like ...,** tıpkı ... gibi: **I would not do it for all the ~,** dünyayı verseler yapmam: **you have the ~ before you,** önünde koca bir hayat var: **to go to a better ~,** bu dünyaya gözünü yummak: **all the difference in the ~,** dağlar kadar fark: **to come down in the ~,** ictimaî mevkice vs. düşmek: **~ without end,** ebediyen: **this ~'s goods,** dünyalık: **he is not long for this ~,** çok yaşamaz: **a man of the ~,** görmüş geçirmiş adam: **the next** [**other**] **~** [**the ~ to come**], ahret: **one who has seen the ~,** dünya görmüş adam; feleğin çemberinden geçmiş: **to think the ~ of s.o.,** birini son derece sevmek veya fevalâde takdir etm.: **what in the ~ are you doing?,** ne yapıyorsun, yahu? **~liness,** dünyaperestlik. **~ling,** dünyaperest. **~ly,** dünyevî; maddî. **~-wide,** âlemşumul.

**worm**[1] [wö(r)m] *n.* Solucan; kurt. **to have ~s,** karnında şerid bulunmak: ⌜**even a ~ will turn**⌝, en pısırık adam bile ancak bir hadde kadar sabreder: **he's rather a ~,** miskinin biridir. **~-gear,** sonsuz vida dişlisi. **~**[2] *v.* **to ~ (oneself, one's way) through undergrowth,** *etc.*, çalılar vs. arasından kaymak: **to ~ oneself into s.o.'s favour,** hile ile birine sokulmak: **to ~ a secret out of s.o.,** hile ile sırrını ağzından kapmak. **~eaten** [-îtn], kurt yenikli.

**worn** [wôn] *v.* **wear.** Yıpranmış, aşınmış. **~ out,** bitkin, pek yorgun; eskimiş.

**worried** [ˈwʌrid] *v.* **worry.** Tasalı; efkârlı; kaygılı.

**worry** [ˈwʌri]. Tasa, üzüntü, sıkıntı. Musallat olm.; üzmek, tasa etmek. Üzülmek; endişeli olm.; tasa çekmek; (köpek koyunları) hırpalamak. **to ~ along (somehow),** yuvarlanıp gitmek (*fig.*): **please don't ~!,** zahmet etmeyin!: **to ~ the life out of s.o.,** birinin başının etini yemek: **to ~ oneself,** üzülmek; beyhude yere üzüntüye girmek: **don't ~ yourself!,** kalbini bozma!: *he* **has nothing to ~ about,** onun tuzu kuru.

**worse** [wö(r)s]. Daha fena. **to go from bad to ~,** gittikçe fenalaşmak: **it is getting ~ and ~,** gittikçe fenalaşıyor: **to make matters ~ ... [and what's ~ ...],** üstelik: **he was none the ~ for his long journey,** memûl hilâfına bu uzun yolculuk ona hiç tesir etmedi: **I think none the ~ of you for refusing,** bunu kabul etmediniz diye size karşı hislerim değişmedi: **he got off with nothing ~ than a wetting,** bir ıslanmakla kurtuldu: **he is ~ off now than ten years ago,** on sene evveline nisbetle maddî vaziyeti daha fenadır: **so much the ~ for him!,** yazıklar olsun ona!: **he is in a ~ way than you,** sizden daha fena bir vaziyettedir. **~n,** daha fena hale koymak. Fenalaşmak.

**worship** [ˈwö(r)şip]. Tapmak; perestiş olmak. Tapınma(k); ibadet (etmek). **His W~,** *belediye reislerine verilen ünvan.* **~ful,** pek muhterem; tapıcı. **~per,** tap(ın)an; ibadet eden.

**worst**[1] [wö(r)st]. En fena (şey, hal vs.). **the ~ of the winter is over,** kışın en şiddetli kısmı geçti: **if the ~ comes to the ~,** pek sıkışırsan: **to get the ~ of it,** altta kalmak: **do your ~!,** elinden geleni arkana koyma! **~**[2] *v.* Yenmek, mağlub etmek. **to be ~ed,** altta kalmak. **~ed** *a.* [ˈwustid], yün iplik(ten örülmüş).

**worth** [wö(r)θ]. Kıymet; değer; kadir. Değer, eder. **to be ~ ..., ...** değmek: **he is ~ £10,000,** on bin liralık adamdır: **to die ~ a million,** bir milyon bırakarak ölmek: **to do stg. for all one is ~,** bir şeyi bütün kuvvetile yapmak: **it would be as much as my life is ~ to do this,** bunu ancak hayatım pahasına yapabilirim; **it is ~ the money,** bu fiata değer: **to get one's money's ~,** sarfettiği paranın değerini çıkarmak: **give me a shilling's ~ of cheese,** bana bir şilinlik peynir veriniz: **I tell you this for what it is ~,** pek mühim değil (*bazan,* doğru olup olmadığını bilmiyorum) fakat size söyliyeyim: **~ while,** değer, dişe dokunur: **it isn't ~ while,** değmez. **~less,** değersiz; beş **para** etmez.

**worth·y** [ˈwö(r)ði]. Müstahak; yakışır; şayan; lâyik; değerli. **to be ~ of stg.**, bir şeye müstahak olm.: **a ~ man**, değerli bir adam; kendi halinde bir adam: **that is not ~ of you**, o sana yakışmaz; bunu sana hiç yakıştırmam: **~ of respect**, şayanı hürmet: **a ~ successor**, hayrühalef: **the village worthies**, köyün ileri gelenleri. **~iness**, liyakat; değer.

**would** [wud]. **will** *fiilin mazi sıygası. Mazi istikbal ifade etmek için* **will** *yerine kullanılır, mes.* **they said they ~ do it at once**, derhal yapacaklarını söylediler. *Şart cümlelerinde de kullanılır, mes.* **if you ate this you ~ die**, bunu yerseniz ölürsünüz: **if you had invited me I ~ have come**, beni davet etseydiniz gelirdim. *Bazan itiyad ifade eder, mes.* **he ~ come and see me every week**, her hafta beni görmeğe gelirdi: **~ you kindly shut the door?**, lûtfen kapıyı kapar mısınız?: **~ (to heaven) it were not true**, keşki doğru olmasaydı: **it *would* rain today!**, aksi gibi bugün yağmur yağıyor: **the wound ~ not heal**, yara bir türlü iyi olmuyordu: **I ~ I were rich**, keşki zengin olsam. **~-be, ...** taslağı; sözüm ona.

**wound**[1] [wûnd]. Yara; ceriha; acı. Yaralamak. **to ~ s.o.'s feelings**, birinin kalbini kırmak. **~**[2] *v.* **wind**[3].

**wove** *v.* **weave. ~n** *v.* **weave.** Dokuma.

**wrack** [rak]. (i) = **rack**, harabiyet; (ii) dalga tarafından sahile atılan şeyler.

**wraith** [rêyθ]. Birisinin ölmünden evvel veya biraz sonra görülen hayali.

**wrangle** [ˈran(g)gl]. Kavga; ağız dalaşı, münazaa (etmek). **~r**, nizacı; Cambridge üniversitesinde riyaziye imtihanında birinci derece kazananlardan biri.

**wrap** [rap]. Şal, atkı vs. gibi örtü. Sarmak; örtmek; ambalaj yapmak. **to ~ stg. up in paper**, bir şeyi kâğıda sarmak: **to ~ oneself up**, sarınmak; (soğuk havada) iyi giyinmek. **~ped, affair ~ in mystery**, esrara bürünmüş şey: **~ in meditation**, tefekküre dalmış: **~ up**, kâğıd vs.de sarılmış; şal vs. ile örtülmüş. **~per**, saran şey; yeni bir kitabın dış kabı. **~ping**, sargı, ambalaj. **~t** *v.* **wrapped.**

**wrath** [rôθ]. Öfke, hiddet. **the Day of W~**, kıyamet günü. **~ful**, öfkeli.

**wreak** [rîk]. **to ~ one's wrath on s.o.**, hiddetini birisinden çıkarmak: **to ~ vengeance on s.o.**, birisinden öc almak.

**wreath** [rîθ] *n.* Çelenk. **~e** [rîð] *v.* Çelenklerle bezemek; sarıp örtmek. (Duman vs.) çelenk şeklinde olarak savrulmak. **~d in smiles**, tebessümlerle kaplı.

**wreck**[1] [rek] *v.* (Gemiyi) karaya oturtup kazaya uğratmak; (plân vs.yi) bozmak, baltalamak. **to be ~ed**, kazazede olm.; karaya oturup tahrib olunmak; muvaffakiyetsizliğe uğramak. **~**[2] *n.* Geminin kazası; kazazede gemi; gemi leşi; enkaz; harab olmuş kimse. **he is a perfect ~**, sıhhati mahvolmuştur: **to be a nervous ~**, sinirleri harab olmak. **~age**, (denizin sahile attığı) enkaz. **~er**, soygunculuk için gemi vs.yi kazaya uğratan kimse; bir teşebbüsü baltalayan kimse.

**wren**[1] [ren]. Çalı kuşu. **W~**[2], = **Women's Royal Naval Service**, İngiliz bahriyesinde hizmet gören kadın.

**wrench** [renç]. Burkulma; ingiliz anahtarı; (*fig.*) acıklı bir ayrılma. Burkmak; zorla yerinden koparmak.

**wrest** [rest]. Bükmek; zorla çevirip sökmek; zabtetmek. **to ~ from its meaning**, ···e ters mana vermek.

**wrestl·e** [ˈresl]. Güreşmek; uğraşmak. **to ~ with**, (bir müşkülü vs.) yenmeğe çalışmak. **~er**, güreşçi. **~ing**, güreş.

**wretch** [reç]. Biçare adam; habis herif. **~ed** [ˈreçid], biçare; sefil; miskin; alçak: **I'm feeling pretty ~**, hiç keyfim yok: **I can't find the ~ thing**, bu Allahın belâsını bulamıyorum: **~ weather**, berbad hava.

**wriggle** [ˈrigl]. Solucan gibi kıvrılma(k). **to ~ into s.o.'s favour**, ustalıkla birinin gözüne girmek: **to ~ one's way out [through, into]**, kıvrıla kıvrıla çıkmak [geçmek, sokulmak]: **to ~ out of a difficulty**, bir müşkülden meharetle sıyrılıp çıkmak: **to try to ~ out of it**, bir kaçamak yolu aramak.

**wring** (**wrung**) [rin(g), rʌn(g)]. Bükerek sıkma(k) [kırma(k)]. **to ~ a bird's neck**, bir kuşun boynunu büküp öldürmek: **to ~ one's hands**, heyecanla ellerini sıkmak: **to ~ stg. out of s.o.**, tazyik [hile] ile bir sır vs.yi söyletmek. **~ing wet**, sırsıklam.

**wrinkle** [ˈrin(g)kl]. Buruşuk. Buruşturmak; porsutmak. **to give s.o. a ~**, (bir iş hakkında) birine faydalı bir tavsiye vermek. **~d,** buruşuk; porsuk.

**wrist** [rist]. Bilek. **~let,** bilezik: **~ watch,** kol saati. **~-watch,** kol saati.

**writ** [rit]. Yazı; ferman; hukukî emir. Yazılmış. **Holy W~,** kitabı mukaddes: **~ large,** büyük harflerle yazılmış; hakkında yanılınmaz: **his ~ does not run here,** onun emirleri burada geçmez: **to serve a ~ on s.o.,** birine bir mahkeme emrini tebliğ etmek.

**write** (**wrote, written**) [râyt, rout, ritn]. Yazmak; muharrirlik etmek. **to ~ for stg.,** bir şeyi mektubla ısmarlamak: **guilt is written all over him,** suçluluk üstünden akıyor: **it's nothing to ~ home about,** (*coll.*) hiç bir fevkalâdeliği yok. **~ back,** mektubla cevab vermek. **~ down,** yazmak, kaydetmek; bir şirketin itibarî sermayesini azaltmak: **I wrote him down as a fool,** onun abdallığını anladım. **~ in,** (bir yazıya bir kelimeyi) dercetmek. **~ off,** yazıvermek; çizmek; ibtal etm.: **to ~ off a bad debt,** bir alacak hesabını silmek: **to ~ off capital,** defterde gösterilen sermayeyi indirmek: **to ~ off so much for wear and tear,** aşınma için şu kadar tenzil etm.: **you can ~ that off!,** bunu unut! **~ out,** suretini yazmak: **to ~ stg. out in full,** tamamını yazmak: **to ~ out a prescription** [**a cheque**], reçete [çek] yazmak: **to ~ oneself out,** yazma kudretini tüketmek. **~ up,** (bir mevzuu) kaleme almak; yazı ile medhedmek; (bir defter vs.yi) ikmal etmek. **~r** [ˈrâytə(r)], muharrir; müellif. **to be a good ~,** iyi bir muharrir [hattat] olmak.

**writhe** [râyð]. Ağrıdan kıvranmak.

**writing** [ˈrâytin(g)]. El yazısı; yazma; eser. **to answer in ~,** yazı ile cevab vermek. **~-pad,** sumen.

**written** [ˈritn] *v.* **write.**

**wrong** [ron(g)]. Yanlış, hatalı; ters; yanılmış; haksız; insafsız. Haksızlık; gadir; günah. Günahına girmek; hakkını yemek; ···e haksız muamele etmek. **to be ~,** yanılmak; haksız olm.: **to be in the ~,** haksız tarafta olm.; kabahatli olm.: **to do s.o. a ~** [**to do a ~ to s.o.**], birine bir haksızlık yapmak: **to go ~,** (insan) yanılmak; baştan çıkmak; (şey) bozulmak; berbad olm.: **I hope there's nothing ~,** inşallah fena bir havadis yok!: **to be in the ~ place,** yanlış yerde olm.: **to put s.o. in the ~,** birini haksız çıkarmak: **on the ~ side of forty,** kırkını aşmış: **the ~ side of a material,** bir kumaşın ters yüzü: **to be ~ side up,** ters çevrilmek: **to say the ~ thing,** pot kırmak: **to take a ~ turning,** yanlış yola sapmak: ⌜**two ~s don't make a right**⌝, haksızlığa haksızlıkla mukabele etme: **the ~ way round,** ters: **(food) to go down the ~ way,** (yemek) genzine kaçmak: **what's ~ with you?,** size ne oldu?; neniz var?: **what's ~ with the bicycle?,** bu bisikletin neresi bozuk?: **what's ~ with this place?,** bu yerin ne kusur var?; bu yerin suyu mu çıktı?: **there's something ~ with him,** ona bir hal oldu; bu adamın şübheli bir tarafı var. **~-headed,** yanlış fikirli; inadcı. **~doer** [–ˈdûə(r)], günahkâr; mücrim; kabahat yapan kimse. **~doing,** haksızlık; kabahat; kanunu ihlâl etme. **~ful,** haksız, insafsız; hatalı.

**wrote** *v.* **write.**

**wrought** [rôt] *v.* **work.** İşlenmiş. **to be ~ up,** heyecanlanmış olm.: **~ iron,** dövme demir.

**wrung** *v.* **wring.**

**wry** [rây]. Çarpık, iğri. **to make a ~ face,** yüzünü ekşitmek.

# X

**X** [eks]. X harfi.

**xenophob·e** [ˈzenofoub]. Ecnebi düşmanı. **~ia** [–ˈfoubiə], ecnebi düşmanlığı.

**Xmas** = **Christmas,** Noel.

**X-ray** [eksˈrêy]. Röntgen şuaına aid. Röntgenini almak. **~s,** Röntgen şuaı.

# Y

**Y** [wây]. Y harfi.

**yacht** [yot]. Yat(ile seyahat yapmak). **~ing,** kotracılık. **~sman,** yat sahibi; yelkenli kayıkla spor yapan kimse.

**yah** [yâ]. *İstihza nidası.*

**yank** [yan(g)k]. Hızlı çekmek.

**Yank(ee)** [ˈyan(g)ki]. Amerikalı.

**yap** [yap]. (*ech.*) Küçük bir köpek gibi havlama(k).

**yard**[1] [yâd]. Yarda; 3 kadem, 36 pusluk bir İngiliz ölçüsü (0·914 metre); (*naut.*) seren. ~[2]. Avlu; tersane; saha; açık havadaki depo. **coal ~**, kömür deposu: **goods ~**, demiryolu eşya deposu.

**yarn**[1] [yân]. Dokuma iplik. **spun ~**, ısparçına. ~[2]. Hikâye; maval. Masal söylemek. **to have a ~**, hoşbeş etm.: **to spin a ~**, hikâye söylemek; maval okumak: **it's no good spinning those ~s to me!**, bana maval okuma!

**yawn** [yôn]. Esneme(k). **to ~ one's head off**, çenesi düşecekmiş gibi esnemek: **a precipice ~ed at his feet**, önünde bir uçurum açıldı. **~ing**, ağzı açık ve geniş.

**ye** [yî]. (*esk.*) Siz.

**yea** [yêy]. Evet; hem de.

**year** [yiə(r)]. Sene, yıl; yaş. **~ after ~**, her sene üstüste: **to be getting on (in ~s)**, yaşlanmak: **~ in ~ out**, bütün sene zarfında: **he was in my ~**, (üniversite vs.ye) aynı sene girdik: **to be 20 ~s old**, yirmi yaşında olm.: **he is old for his ~s**, (i) yaşlı gösteriyor; (ii) (çocuk) yaşına göre büyük. **~ling**, bir yaşında (hayvan). **~ly**, senede bir kere; her sene vukubulan. **~-book**, salname, yıllık. **~-long**, bir sene süren.

**yearn** [yö(r)n]. **to ~ for [after]**, hasret çekmek, çok arzu etmek.

**yeast** [yîst]. Bira [ekmek] mayası. **~y**, köpüklü, maya gibi (işliyen).

**yell** [yel]. Avazı çıktığı kadar bağırmak; yırtınmak. Feryad, bağırma.

**yellow** [ˈyelou]. Sarı (renk); (*Amer. sl.*) korkak. **to turn ~**, sararmak: **~ fever**, sarı humma: **~ jack**, karantina flâması. **~ish**, sarımsı.

**yelp** [yelp]. Vurulmuş köpeğin kısa ve keskin bağırması. Böyle bağırmak.

**yeoman** [ˈyoumən]. Küçük mülk sahibi yahud çiftçi: **yeomanry**'ye mensub asker. **~ service**, dürüst ve gayretli hizmet: **Yeoman of the Guard**, *v.* **beefeater**. **~ry**, çiftçi sınıfı; İngiliz ordusunda gönüllü süvari alayı.

**yes** [yes]. Evet. **~-man**, evetefendimci.

**yesterday** [ˈyestədêy]. Dün. **the day before ~**, evvelki gün.

**yet** [yet]. Daha; henüz; hâlâ; lâkin. **as ~**, şimdiye kadar: **it is strange ~ true**, garibdir fakat doğrudur.

**yew** [yû]. Porsuk ağacı.

**Yiddish** [ˈyidiş]. İbranice ile karışık bir alman lehçesi.

**yield**[1] [yîld]. Mahsul(ün mikdarı); hasılat; kazanc. Hasıl etm.; vermek. ~[2]. Teslim olm.; ramolmak; gevşemek, kapılmak. Teslim etm.; terketmek. **to ~ to temptation**, iğvaya kapılmak: **I ~ to none in admiration of his work**, onun eserini takdirde kimseden geri kalmam. **~ing**, yumuşak; gevşek; uysal.

**yoke** [youk]. Boyunduruk; sakaların omuz sırığı; bir gömleğin üst tarafına eklenen parçası. Boyunduruğa koşmak. **to cast off the ~**, boyunduruktan kurtulmak: **a ~ of oxen**, bir çift öküz.

**yokel** [youkl]. Köylü; hödük.

**yolk** [youk]. Yumurta sarısı.

**yon(der)** [yon(də(r))]. Şurada(ki); ötede(ki); orada(ki).

**yore** [yô(r)]. Eski zaman. **of ~**, eskiden.

**Yorkshire** [ˈyôkşə(r)]. En büyük İngiliz kontluğun ismi. **~ pudding**, bir nevi hamur işi.

**you** [yû]. Sen, siz; seni, sizi; sana, size. **~ (there)!**, bana bak!: **if I were ~**, sizin yerinizde olsam.

**young** [yʌn(g)]. Genc. Yavru. **with ~**, gebe: **Y~ England**, bugünkü İngiliz gencleri: **~ Mr. Jones**, (i) genc Mr. J.; (ii) Mr. Jones'in oğlu; J. kardeşlerden küçüğü: **~ man**, delikanlı: **my ~ man [woman]**, sevgilim: **the day [night, year] is ~ yet**, daha günün vs. başındayız: **I am not as ~ as I was**, eskisi gibi genc değilim. **~er**, daha genc: **~ son**, küçük oğul: **you are looking years ~**, maşallah çok gencleşmişsiniz!: **when I was forty years ~**, kırk sene evvel (gençliğimde). **~ster** [ˈyʌn(g)stə(r)], çocuk; delikanlı.

**your** [yô(r)]. Senin; sizin. **you cannot alter ~ age**, kimse yaşını değiştiremez: **~ true patriot will die for his country**, hakikî vatanperver vatanın uğrunda canını verir; (*bu gibi cümlelerde* **you** 'umumiyetle' *mânasına gelir*). **~s**, seninki, sizinki: **is he a friend of ~?**, o sizin dostlarınızdan mı?: **that is a bad habit of ~**, bu sizin fena bir âdetiniz. **~self** [yôˈself], kendin, kendiniz (*bir kimse hakkında*): **~selves**, kendiniz (*iki veya fazla kimse hakkında*).

**youth** [yûθ]. Genclik; genc bir adam; gencler. **~ful**, genc (gibi).

**yowl** [yaul]. (*ech.*) Miyavlamak; ürümek.
**Yule** [yûl]. Noel. ~**-log,** Noel yortusunda yakılan kütük. ~**-tide,** Noel yortusu.

# Z

**Z** [zed]. Z harfi.
**zany** [ˈzêyni]. Soytarı; budala.
**Zarathustra** [ˌzaraˈθûstra]. Zerdüşt.
**zeal** [zîl]. Gayret(keşlik), himmet, hamiyet. ~**ot** [ˈzelot], mutaassıb; fazla derecede gayretkeş. ~**ous** [ˈzeləs], gayretkeş, hamiyetli.
**zebra** [ˈzîbra]. Afrikaya mahsus derisi yollu yaban eşeği.
**zenana** [zeˈnâna]. Hindistanda harem dairesi.
**zenith** [ˈzeniθ]. Semtürres; evc. **at the** ~ **of his career,** meslek hayatının zirvesinde.
**zephyr** [ˈzefə(r)]. Sabah [pek hafif] rüzgâr; pek hafif kumaş.
**zero** [ˈziərou]. Sıfır. ~ **hour,** (*mil.*) bir taarruz vs.nin başlayacağı saat.
**zest** [zest]. Tat; hoşnudluk; şevk.
**zigzag** [ˈzigzag]. Zikzak (yapmak).
**zinc** [zin(g)k]. Çinko (ile kaplamak).
**Zion** [ˈzâyən]. Kudüs; cennet. ~**ist,** Siyonist.
**zip** [zip]. (*ech.*) Havada giden bir kurşunun sesi. ~**-fastener,** 'zipli' kapak vs.
**zodiac** [ˈzoudiak]. Burclar mıntakası.
**zon·al** [ˈzounəl]. Bir mıntakaya aid. ~**e,** mıntaka.
**zoo** [zû] = **zoological garden,** vahşi hayvanlar bahçesi. ~**logy** [zûˈoləci], hayvanat ilmi.
**zoom** [zûm]. (*ech.*) (Uçak) birdenbire ve pek hızlı yukarıya çıkmak; bu hareketin gümbürtüsü.

# ERRATA

**letter** [ˈletə (r)]. *n.* Harf; mektub. *v.* Harflerle basmak *veya* süslemek. ~**s,** edebiyat: ~**-box,** posta kutusu.

**lettuce** [ˈletis]. Salata. **Cos** ~, marul.